WINE GRAPES

WINE GRAPES

A complete guide to 1,368 vine varieties, including their origins and flavours

JANCIS ROBINSON
JULIA HARDING
JOSÉ VOUILLAMOZ

ALLEN LANE
an imprint of
PENGUIN BOOKS

WINE GRAPES
A complete guide to 1,368 vine varieties, including their origins and flavours
Jancis Robinson, Julia Harding and José Vouillamoz

UK | USA | Canada | Ireland | Australia
India | New Zealand | South Africa

Allen Lane is part of the Penguin Random House group of companies
whose addresses can be found at global.penguinrandomhouse.com.

Penguin
Random House
UK

First published 2012
008

The authorized representative in the EEA is Penguin Random House Ireland, Morrison Chambers, 32 Nassau Street, Dublin D02 YH68

Designed by Here Design
Printed in China

ISBN: 978–1–846–14446–2

www.greenpenguin.co.uk

FSC www.fsc.org MIX Paper | Supporting responsible forestry FSC® C018179

Penguin Random House is committed to a sustainable future for our business, our readers and our planet. This book is made from Forest Stewardship Council® certified paper.

CONTENTS

PEDIGREE DIAGRAMS

PREFACE

I find it difficult to express how thrilled I am to have created this entirely new book with co-authors as knowledgeable and gifted as Julia Harding MW and Dr José Vouillamoz.

In the mid 1980s I wrote the first consumers' guide to wine grape varieties, *Vines, Grapes & Wines*, published in 1986 and still in print, an amalgamation of all that I could find out about the vines then planted for wine production around the world. At that stage all that was available were the highly academic works of French vine expert Pierre Galet of Montpellier and an American adaptation of part of them by his pupil Lucie T Morton. The French in particular regarded varietal wines, which they called *vins de cépage*, wines named after the grape variety from which they were made rather than after the place they were grown, as inherently inferior. It was all the more surprising then to find that the book was so sought after, not just in English and German but also in French as *Le Livre des Cépages*. It seemed as though not just wine consumers but also French vine growers had a distinct thirst for information about vine varieties and how they were related.

I have always been fascinated by grape varieties, especially by their extremely complex synonymy and how one vine can turn out to be identical to another with a quite different name and reputation in another country. I have paid particular attention to the grape variety entries in all three editions of *The Oxford Companion to Wine* and Oxford University Press even published a collection of them in pocket-book form in 1996 as *Jancis Robinson's Guide to Wine Grapes*, a book I know many people still consult. But they should all throw it out immediately because this book, while hardly being pocket size, is so much more up to date and comprehensive than either of my previous works on the subject.

While the range of varieties now knowingly and deliberately used for commercial wine production has expanded enormously in this new century, our knowledge of the relationships between these varieties is infinitely greater than it has ever been. Thanks to the DNA profiling that has been applied to grapevines since 1993, specialists such as Dr José Vouillamoz (who identified the parents of Sangiovese in 2004 and two years later discovered a fascinating family relationship between Pinot and Syrah) now understand so much more about the pedigrees of the world's plants, including grapevines.

As those who read this book will learn, a mere handful of 'founder varieties' (similar to 'founder crops', see p XXIV) such as Pinot, Savagnin, Nebbiolo and Tribidrag are ancestors of dozens of often much more familiar grapes, and it looks increasingly as though an enormous proportion of the grapes we know today are members of a few key families, all clearly illustrated in our unique pedigree diagrams (see p VI). As ever, the study of wine grapes tells us much about social history and the movement of people and conquerors.

With its detailed information on the viticultural characteristics of each variety, this book should prove a boon for growers trying to decide which vines might thrive in their particular circumstances, while its unrivalled wealth of topical detail about where each variety is grown, to what extent, and how the wines made from that variety taste, will, we feel confident, fill a significant gap in the current literature available to wine lovers.

The total number of different vine varieties is about 10,000, members of half a dozen species. In order to make this book manageable, we have decided to limit ourselves to those varieties which we know at the time of writing produce commercial quantities of wine, all 1,368 of them. But such is the current explosion of interest in lesser-known grapes that we fully acknowledge, and indeed welcome, the fact that some wines produced in 2012, and possibly some 2011s, will have been made from grape varieties not mentioned in this book. We welcome suggestions for varieties to be included in any subsequent edition, and ask that they be sent to contact@winegrapes.org.

It is worth pointing out, however, that most grape varieties travel under a wide variety of aliases. We have tried to include all correct, and the most prominent incorrect, synonyms. How to use this book (see below) outlines our procedure and conventions in detail.

HOW TO USE THIS BOOK

The 1,368 grape varieties described in this book are presented in alphabetical order. Where a variety is known under several different names, we have selected a 'prime name', which is generally the name most commonly used in the variety's country or region of origin. For example, Garnacha is the prime name rather than Grenache because the variety originates from Spain, where Garnacha is the most common as well as the official name. Well-known synonyms such as Grenache do appear in this alphabetically ordered book but only to point to the entry under the variety's prime name (eg 'Grenache, see GARNACHA').

Every prime name and every synonym is listed in the index so that you can easily find a variety even if you don't know its prime name.

As explained on pp XIV–XVI, we define a variety according to its genetic identity and not by its physical appearance or the wines it produces. Therefore, even though Pinot Noir and Pinot Blanc, for example, look different in the vineyard and make very different wines, they are both found in the single entry headed PINOT. Similarly, both Garnacha Tinta and Garnacha Blanca are in the GARNACHA entry.

However, when there are several varieties that have similar names but are genetically distinct (eg Malvasia di Candia Aromatica, Malvasia di Casorzo, Malvasia di Lipari, etc), they are given separate entries. In such instances, there is usually another short introductory entry that explains the possible connections or distinctions among them (in this instance, the entry headed MALVASIA).

Within each entry, the first instance of a prime name in any paragraph is highlighted in SMALL CAPITAL LETTERS so that you can easily turn to those entries for more information.

As there is so much confusion over synonyms, and many instances where one variety is wrongly considered to be the same as another, most entries begin with a list of synonyms for that variety and a list of varieties that are commonly mistaken for it. Where such synonyms and mistaken identities have been proven by DNA analysis, the variety name or synonym is marked ⧖.

The berry colour of each variety and for the most important colour mutations (see p XV) is shown using these symbols:

BERRY COLOUR: ● ● ● ● ●

The colours used are idealized and the colour of the berry underlined is the closest approximation to the actual colour of the fruit on the vine, which is rarely uniform.

For each variety we give the total vineyard area planted with that variety for each country and/or region where relevant, based on the most recent statistics we could find. These are usually based on a vineyard census undertaken within the past few years but the Italians carry out a survey of their vines only once a decade, so figures for plantings in Italy are necessarily dated. The anticipated 2010 Italian viticultural census had yet to be published at the end of 2011, unfortunately.

In every entry, in the section entitled 'Where it's grown and what its wine tastes like', we have discussed each relevant country or region in the same order (France, Italy, Spain, Portugal, Northern Europe, Germany, Austria, Switzerland, Eastern Europe, the eastern Mediterranean, California, Washington, Oregon, the rest of the US, Canada, Mexico, South America, Australia, New Zealand, South Africa and finally Asia) no matter where that variety is most important. Thus, for example, the Tempranillo section begins by considering the limited extent to which this Spanish grape is found in France.

Complex and unexpected relationships between varieties are revealed by the fourteen pedigree diagrams, listed on p VI and referred to in the relevant entries. We have chosen to illustrate these fourteen pedigrees because they show the remarkable network of parent–offspring relationships that connect so many of the varieties discussed in this book. The diagrams can be read in the same way as a human family tree and, where possible, colour is used to show the variety's region of origin. When one of the parents of a variety is unknown – represented by a question mark – it is usually impossible to know which is the parent and which is the offspring and in such instances the pedigree we have illustrated is only one possible interpretation of the currently known genetic relationships. In some pedigrees, the same variety appears several times, simply to make the diagram clearer.

A list of all the varieties grouped by country of origin can be found on pp XXVIII–XXXV.

The full details of articles and books cited in the text using the author's name and the date of publication (eg Smith 1999) are given in the bibliography on pp 1183–1202. Where an argument is supported by relationships revealed by José Vouillamoz's extensive database of DNA profiles but has not been published in a scientific paper, it is followed by '(Vouillamoz)' with no date.

Selected viticultural and winemaking terms are explained in the glossary on pp 1179–82.

INTRODUCTION

THE IMPORTANCE OF GRAPE VARIETIES

Until the mid twentieth century, wine consumers knew hardly anything about grape varieties. They may have encountered words such as Riesling on Germanic bottles and Verdelho on bottles of Madeira but few wine drinkers understood that these were grape names. Virtually all other wine was named after the region, village, vineyard or property where it was grown – thus Burgundy, Pommard, Chambertin or Château Haut-Brion. It was the energetic American wine writer and merchant Frank Schoonmaker who, partly out of respect for French wine, encouraged California producers to abandon borrowed generic terms such as Burgundy for their wines and to call wines after the grape varieties responsible for them.

Thus from the mid to late twentieth century, wine drinkers, particularly those outside Europe, were introduced to bottles labelled Cabernet Sauvignon and Chardonnay, the varieties, their makers were swift to point out, that were responsible respectively for the most famous red bordeaux and white burgundies. This so-called 'varietal labelling'* is in many ways much easier to understand than the complexities of traditional geographical labelling. This was especially true in the early days of varietal labelling, when there were so relatively few varieties in general commercial circulation. In the 1980s and early 1990s, for example, it sometimes seemed as though learning the names of a handful of the most popular grape varieties – Cabernet Sauvignon, Merlot, Syrah/Shiraz, Chardonnay and Sauvignon Blanc – was enough to unlock the entire world of wine, so much did they dominate vineyards and therefore labels.

So popular had varietal labelling become by the 1990s that, indicating a significant swing in the balance of power, some producers in even the most classic European wine regions, Bordeaux and Burgundy, pleaded with the French authorities to be allowed to add grape names to their labels to make their wines as easy to identify and sell as the popular varietals from the United States and the southern hemisphere. Eventually Chardonnay and Pinot Noir were allowed on the labels of some of the less expensive burgundies, while the odd branded bordeaux was sold as a Merlot-Cabernet.

One benign effect of varietal labelling meanwhile has been dramatically to increase interest on the part of both producers and consumers in grape varieties themselves, and this has resulted in a huge increase in the number of them that are deliberately being cultivated (as opposed to growing more randomly, sometimes even unacknowledged, as part of a field blend). Perhaps partly as a reaction to boredom with such a restricted diet of grape varieties as we were threatened with in the early 1990s, we are now benefiting from a much wider array of them, and a general interest in regions' local, indigenous varieties that is unparalleled in the history of wine. In Italy alone, where we reckon there are currently about 380 different varieties responsible for wine in commercial circulation, there are initiatives all over the country to save from extinction and recuperate historic vine varieties that until recently have been known

* **Varietal**, incidentally, is an adjective and is most logically applied to a wine or the way it is labelled. The associated noun is **variety**, which applies to the particular sort of vine used to produce a wine. There has recently been a tendency, regretted by those of us who treasure precision, to describe plants as varietals. It may become so widespread as to be unstoppable, but not for want of trying on our part.

to perhaps only one or two farmers. Earlier, similar activity in Switzerland virtually saved such autochthonous varieties as Rouge du Pays, Humagne and Himbertscha, while the Plaimont group of co-operatives in the Gers has achieved similar results in south-west France. The same sort of initiative, a sort of World Wine Life Fund, can now be seen throughout Europe, and the result is that, even in mass-market circulation, the range of varieties on offer is immeasurably wider than it was fifteen or even ten years ago.

Climate–maturity groupings based on relationships between phenological requirements and growing season average temperatures for high- to premium-quality wine production in the world's benchmark regions for many of the world's most common cultivars. The dashed line at the end of the bars indicates that some adjustments may occur as more data become available, but changes of more than +/- 0.2-0.5 °C are highly unlikely (Jones 2006). Also note that table grapes* and raisins* are grown up to 24 °C or more. The figure and the research behind it are a work in progress and are used with the permission of the author, Dr Gregory V Jones, Southern Oregon University.

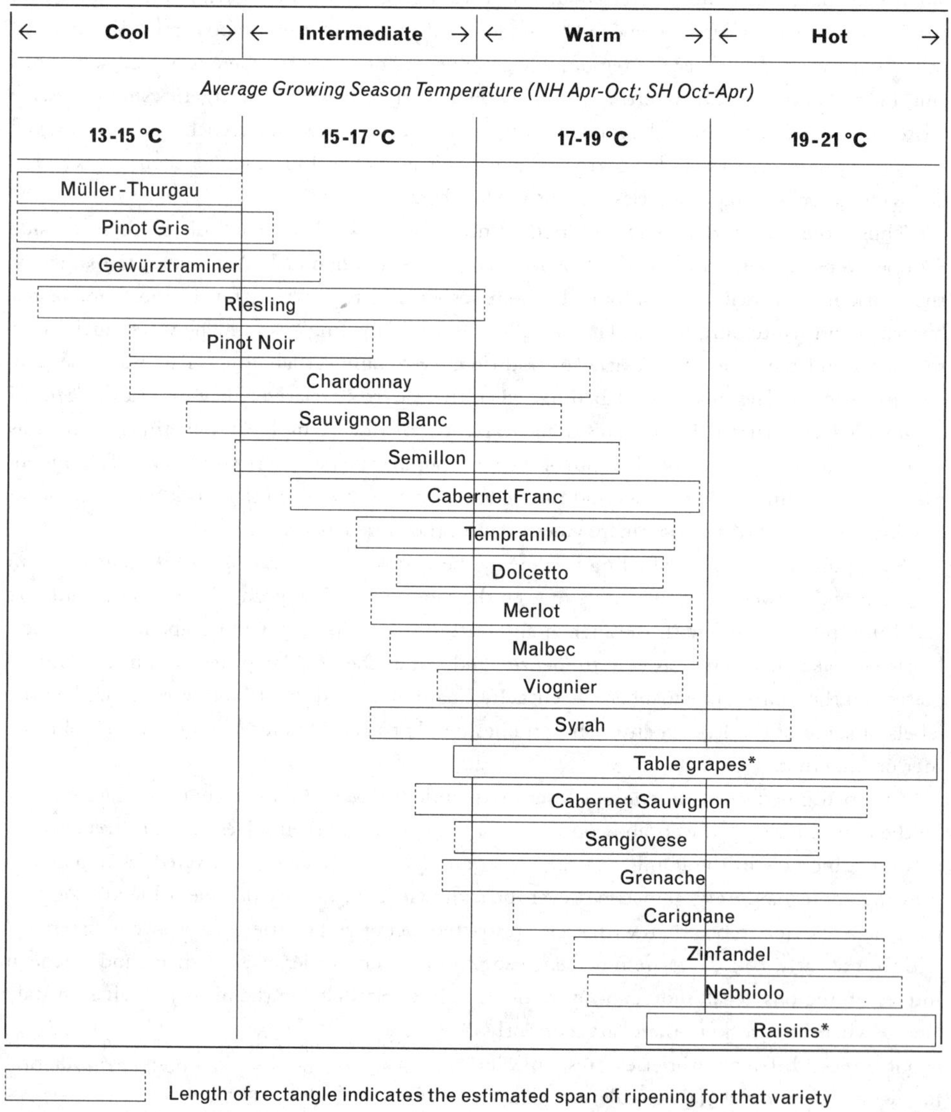

But the world of wine is extremely small and well connected nowadays. Once growers in, say, the Adelaide Hills or Santa Ynez Valley read about or taste wine made from an exciting and novel Iberian grape variety, chances are that one or two of them will want to try planting it and vinifying its grapes themselves. Within the strictures of plant quarantines, there is an international traffic in cuttings of grapevines that would have been unthinkable in the last century. This plant material typically makes its way from one of the major vine nurseries in Italy or France to a grower outside Europe but there is an increasing number of serious vine nurseries all over the world today – all of them keen to have as comprehensive a collection of different varieties as possible.

There is a real curiosity about and passion for these 'heritage varieties' on the part of consumers and, particularly, producers, as well as for older vines (see p XX). We hope that this book can not only help answer the sort of questions that wine drinkers may have about what they are tasting but also help growers identify which variety might best be suited to their region and vineyard land.

We have tried to give as much information as possible on when each variety ripens, which pests and diseases it is particularly prey to and what sort of wine the variety produces. Although some varieties are quite adaptable to site and climate, others are much more fussy and climatically sensitive. Some need longer to ripen than others and produce their best results only when the growing season and the temperatures and available sunlight during that season closely match the needs of the variety. There is little point in planting a very early ripening variety in a hot climate as the grapes will be ripe in certain respects (enough sugar in the berries to be fermented into a respectable level of alcohol) but will not have accumulated much flavour, while a late-ripening variety planted in a cool climate will probably fail to notch up sufficient sugar levels most years. Varieties that bud early are at particular risk in regions notorious for the incidence of spring frosts. The diagram left, developed by Professor Greg Jones, illustrates the relationship between climate and grape ripening for some of the world's most well-known varieties.

Another observable development in the wines being made today is an increase in the number of wines labelled with more than one variety. Within Europe and most wine-producing countries, a wine that is labelled with the name of a single grape variety must contain at least 85% of that variety (75% within the US). This means in practice that some of the wines in commercial circulation labelled with the name of a highly prized and expensive variety may in fact contain 15% of a cheaper or more plentiful variety or varieties.

However, there is a distinct trend towards deliberate marketing of blends of different varieties (a practice that is de rigueur in Bordeaux and the southern Rhône) with the aim of creating more interesting and complex flavours or to emphasize the region of origin rather than the varietal make-up. In our book we have tried to indicate which are the most common and successful blends: the so-called Bordeaux blend of Cabernet Sauvignon, Merlot and some of Cabernet Franc, Petit Verdot, Malbec (see the COT entry) and Carmenère for reds and Sémillon and Sauvignon Blanc for whites, for example; Grenache (see GARNACHA), Syrah and Mourvèdre (see MONASTRELL) in southern France and, increasingly, elsewhere. Cocktails of the various white wine grapes commonly grown in southern France such as Viognier, Marsanne, Roussanne and Grenache Blanc have also become increasingly common around the world of wine. And the standard Champagne recipe of blending Pinot Noir, Pinot Meunier and Chardonnay is much emulated by sparkling-wine producers everywhere. By law, a wine labelled with the names of more than one grape variety must have those names listed in descending order of their importance in the blend.

THE VINE FAMILY

Each variety described in this book is either a member of single vine species or, less commonly, a hybrid of more than one species. The wine grapevine belongs to the botanical family of woody climbing plants called *Vitaceae* and to the botanical genus *Vitis* comprising 65–70 species. Most *Vitis* species are indigenous to North America (eg *Vitis labrusca*, *Vitis berlandieri* and *Vitis riparia*) and East Asia (eg *Vitis amurensis*), where the grapes are sometimes vinified, but the vast majority of wines around the world are made from the only indigenous species in Eurasia: *Vitis vinifera*, meaning 'wine-bearing grape', which has consistently proved its superiority in terms of wine quality and flavour. (The full technical name is *Vitis vinifera* L. because the naturalist Carl von Linné was the first to name this species.) Wine made from American species, particularly *Vitis labrusca*, can have a very distinctive flavour – definitely an acquired taste – combining animal fur and candied fruits, often described as 'foxy'. A similar aroma and taste are often found in the juice and jelly made from Concord grapes that are widely available in the United States.

Vitis vinifera is sometimes divided into two subspecies: *vinifera* (full name *Vitis vinifera* L. subsp. *vinifera*, also called subsp. *sativa*), the cultivated Eurasian vine; and *silvestris* (full botanical name *Vitis vinifera* L. subsp. *silvestris* (C. C. Gmel.) Hegi), the wild vine that is found growing naturally, often climbing up trees, from Portugal to Tajikistan as well as along the main continental rivers of Western Europe and in North Africa (Arnold *et al.* 1998).

However, this division is questionable since the physical characteristics of the two subspecies overlap and differences are probably the result of domestication by humans (see p XXIII) rather than evolution (Zohary and Hopf 2000). The most clear-cut distinction between wild and cultivated vines – and the difference that is extemely significant for successful cultivation – is the sex of the flowers. *Silvestris* vines are dioecious, which means that all the flowers on the same vine are either all male or all female, whereas *vinifera* vines are hermaphroditic, which means the flowers are both male and female and are therefore much easier to cultivate and more likely to be fruitful since they don't need any other variety planted nearby to achieve pollination, as explained below. Significantly, a tiny percentage of *silvestris* vines are hermaphroditic and were undoubtedly the starting point of vine domestication.

GRAPE VARIETY, MUTATION AND CLONE

What is a grape variety and how does a new one spontaneously appear? These apparently straightforward questions have complex answers that allow us to distinguish between the terms variety, mutation and clone, which are often confused or misused but which enable us to make important distinctions in this book. They also help to explain the enormous genetic diversity among grape varieties in the world today that gives wine drinkers a bewildering choice of wines.

Every single **grape variety** is the result of a single vine grown from a single seed, and this seed is the result of sexual reproduction between two parent varieties. The following set of conditions has to be fulfilled for a new variety to be born:

- a flower in a grape bunch is fertilized, resulting in the formation of a berry
- the berry falls to the ground or is eaten and carried away by an animal, very often a bird
- the grape seed from that berry falls onto the soil and germinates, producing a seedling
- the plantlet has to survive at least three years in order to bear fruit and to be identified as an interesting grapevine.

When all these criteria are met, a new variety is born, with one father and one mother. This sequence represents the natural birth process of all varieties in the world.

Like most plants, flowers of almost all cultivated grape varieties are hermaphroditic and have both male organs (stamens that produce pollen) and a female organ (a pistil that contains the ovary to be fertilized). There are a few exceptions: functionally female varieties that require the close proximity of male plants in order to be fertilized (see, for example, LAMBRUSCO DI SORBARA).

The pollen that reaches the pistil of a grape flower, thanks to the wind or to insects such as flies, may come from the same flower, from another flower on the same plant, or from a flower of another plant of the same grape variety. In each case, this is known as self-pollination or selfing. Cross-pollination occurs when the pollen comes from a flower of another plant of another grape variety. In modern, regimented, monovarietal vineyards, self-pollination is far more common, ensuring fruitful harvests every year, and thus the economic well-being of the grape grower and wine producer.

The incredible genetic diversity among grape varieties seems to indicate that cross-fertilization is the origin of all the varieties in existence today, with the combination of distinct genomes leading to more distinctive and interesting new varieties and an ever-expanding palette for the wine producer. This cross-fertilization was common until the early twentieth century because several grape varieties were often intermixed in the vineyards, facilitating natural crossings between distinct grape varieties. A vine breeder may replicate this process by deliberately fertilizing one variety with a pollen grain from another.

However, not all new varieties are likely to increase the grape grower's options or the wine drinker's pleasure. If one of the countless grape seeds produced on the planet every year by self-pollination is planted, accidentally or deliberately, the seedling will represent a new and distinct variety that is highly consanguineous with the mother plant (inbreeding), resulting in plants that are usually degenerative and more susceptible to diseases.

To maintain the desirable characteristics of a particular plant of a particular variety – in terms of its behaviour in the vineyard as well as the flavour of the wine it produces – it must be vegetatively propagated by cuttings or by layering (see the glossary on pp 1179–82). However, changes cannot necessarily be avoided. At each cell division during plant growth, when the DNA is replicated, errors are likely to occur, resulting in spontaneous **mutations**. (Note that the word mutation may be used to refer to the natural process described below as well as the result of that process.) The vast majority of these mutations have no effect on the plant or on the wine it produces because they affect non-coding portions of DNA (see DNA profiling below).

After decades or centuries of vegetative propagation, each grapevine plant harbours hundreds of thousands of new mutations. Most of them will go unnoticed, and only those having a desired effect that is observable to the human eye – smaller berries, bigger crops, for example – will be noticed and propagated separately: these are **clones**. Any vine grower will tell you that there can be huge differences between different clones of the same variety. They may be genetically almost identical but some clones will ripen much more easily than others, some will be more vigorous, and the thickness and colour of the skins can vary and therefore affect the style of the wines produced. Clones are specially selected in a nursery or by growers for particular characteristics, and for some varieties there can be as many as hundreds of different clones in circulation. (See the glossary for the difference between clonal and mass selection.) We feel this book is quite long enough as it is, so have mentioned individual clones only for varieties where we feel they are a particularly integral part of their, often fascinating, story.

However spectacular the physical changes in a particular variety, the accumulation of mutations over time does not result in new grape varieties, only in clones of the original one. Therefore, a **grape variety** (known as a cultivar in the International Code of Nomenclature for Cultivated Plants) can be defined as a group of morphologically distinct clones originating from a single initial seed which was sexually produced by a father variety fertilizing a mother variety (Boursiquot and This 1999).

This is why we consider Pinot Noir (dark-skinned), Pinot Gris (grey- or pink-skinned) and Pinot Blanc (light-skinned) to be mutations of the single variety Pinot and not distinct varieties. They are simply skin-colour variations of the same variety produced by a particular type of mutation: the insertion of a transposon (or 'jumping gene') in the DNA region that regulates a gene responsible for the synthesis of anthocyans causes the loss of black berry-skin colour (This *et al.* 2007). Pinot Noir, Pinot Gris and Pinot Blanc are indistinguishable by standard DNA profiling (described below) and are therefore considered to be genetically one and the same variety.

The older a variety is, the more mutations it is likely to have. Since such mutations may be selected for propagation, the more mutations a vine has undergone, the more clones it is likely to have. (See Changes in vineyard composition below.) For example, Pinot has hundreds of clones simply because it is very old, not because it is naturally more prone to mutation than other varieties, as has been mistakenly suggested (see the PINOT entry).

Some ampelographers and writers have used the term 'polyclonal' to describe certain varieties such as Sangiovese in order to account for the significant diversity seen in the vineyard (see the SANGIOVESE entry for more details). However, this is not compatible with the definition of 'variety' used in this book; a vineyard population identified as 'polyclonal' is in fact composed of the true variety and its progenies, most of which are the result of self-pollination (see above).

VINE BREEDING

Many of the varieties described in this book have been intentionally created by breeders, usually to enhance desirable characteristics such as productivity or flavour or to overcome difficulties caused by the climate, diseases or pests. The deliberate crossing of different varieties, mimicking some of the spontaneous crosses that have occurred in the vineyard in past centuries (as described above), have given rise to many of the varieties we are familiar with today. Here and throughout the book we use the term **cross** to refer to a variety that is the result of crossing two varieties from the same species (also known as an intraspecific cross) and **hybrid** to refer to the product of crossing varieties from two different species (an interspecific cross). We use the term **natural cross** to describe a cross that has occurred spontaneously in the vineyard.

Grape breeding was a particularly popular activity in the early to mid twentieth century in Europe in the wake of the arrival of the mildews and phylloxera (see below), when French breeders such as Seyve, Villard, Seibel and the Bouschets presented growers with a wide array of new varieties promising greater resistance to pests and diseases and/or increased productivity. The best-known and most widely planted today is perhaps the *vinifera* variety Alicante Bouschet (see the ALICANTE HENRI BOUSCHET entry), but hybrids such as Villards Blanc and Noir and various others carrying the name of the hybridizer Seibel were very common in France in particular in the 1950s and 1960s. In the impoverished post-war era, high yields were much more important to most farmers than nuances of quality and a lack of non-*vinifera* characteristics. By the late twentieth century, however, most hybrids had been eliminated from European vineyards and they are generally barred from producing quality wine within the European Union.

The much-vaunted German Wine Law of 1971, which enshrined high sugar levels as the single most desirable element of a grape, encouraged German vine breeders to outdo each other for the sugar levels they could achieve in the varieties they created – the likes of Morio-Muskat, Optima and Siegerrebe – although it is notable that it took well over a century to elucidate the parentage of the most popular crossing of all, Müller-Thurgau, created by the Swiss-born Hermann Müller in 1882. Fortunately, these body-builder creations are well in retreat in favour of varieties that produce nobler wine in modern Germany.

Austrian grape breeders on the other hand have left a decent legacy in the likes of Blauburger, Neuburger and Zweigelt. Breeders in particularly cold climates such as those of upstate New York, Russia and Ukraine have tended to concentrate on developing varieties specifically to withstand the cold, often hybrids rather than crosses of two *vinifera* varieties, sometimes incorporating genes from the cold-hardy Asian vine species *Vitis amurensis* (see those varieties with the suffix Severny, for example).

The mid-twentieth-century grape breeders of California are remembered today via the still-substantial plantings of Ruby Cabernet around the world and the rather less noble Rubired that is today basically used to deepen the colour of red wines.

Today's new varieties are much more likely to be bred for the quality of the resulting wine than earlier versions but there currently seems to be more interest in rediscovering old varieties than in creating new ones – despite the best efforts of breeders in the south of France who have given us, for example, Caladoc, Chasan and Portan, and those of Australia who have had reasonable success with the likes of Tarrango and Cienna. Another hotspot for modern vine breeding is the American Midwest, where cold-hardy hybrids such as Traminette and Frontenac have been developed, although many breeders here seem to be motivated as much by enthusiasm for, and the local popularity of, something home-grown as by the need to create cold-hardy and disease-resistant varieties.

Another subgroup of new varieties are those hybrids that have been bred specifically to be resistant to common vine diseases and therefore to need far fewer agrochemical inputs than most conventional varieties. Hybrids such as Regent, Rondo and Solaris can be particularly useful at the coolest, wettest limits of the wine-producing world. It is not just global warming but the emergence of vine varieties such as these that has helped to push the limits of viticulture towards the North Pole in recent years. An alternative to the complex vine-breeding programmes, often involving varieties from three or more species, that have yielded these exceptionally hardy varieties (see, for example, the entry for L'ACADIE BLANC) would be genetic modification of existing varieties, but wine consumers have so far proved as resistant to GM as are the guardians of food purity.

PESTS AND DISEASES

Pests and diseases can affect both yield and wine quality, posing a significant economic threat to grape growers and motivating vine breeders to develop resistant clones or varieties.

The European grapevine, responsible for the vast majority of the world's wines, is prey to a wide range of pests and diseases but is particularly susceptible to fungal diseases. The introduction from North America in the nineteenth century of two of the major ones, and a fatal pest, was to have far-reaching consequences for the history of viticulture. The Victorians loved collecting botanical specimens. Once steamships could travel fast enough across the Atlantic, not only could North American plants reach Europe alive, but the organisms responsible for North American plant diseases were also accidentally imported with them. Powdery mildew, or *oïdium*, was first identified in England in 1845 and was to spread through mainland Europe, leaving vines weakened and diseased. Applications of sulphur were observed to help combat the resulting fungal disease and they continue to this day.

In the 1860s, southern French vines began to display another, puzzling, set of symptoms whereby vines appeared weakened in summer and failed to grow the following spring. Not before the spread of this phenomenon to most of Europe was it established that thousands of hectares of vines had succumbed to a North American aphid known popularly today as phylloxera that would munch its way through the roots of *Vitis vinifera* vines and kill them. So devastating were phylloxera's effects that total French wine production fell by three quarters between 1875 and 1889, and the pest was to spread gradually to virtually all major wine regions.

This pest was finally identified as being identical to one known in North America which seemed to have little effect on native vines of North American species. Eventually, after panicked experimentation with all manner of 'cures', it was realized that grafting European varieties onto phylloxera-resistant North American rootstocks (see below) was the solution. Only a small proportion of the world's wine drinkers are fans of wines made from vine varieties belonging to North American species, but all of them have reason to be grateful to these phylloxera-resistant American vines (which in fact vary in their individual resistances).

In the late 1870s, the world's European grapevine species fell prey to another fungal disease from North America, downy mildew, or *peronospora*. Spraying with copper sulphate, which came to be known as Bordeaux mixture (*bouillie bordelaise* in French), had been identified as the most effective treatment by the mid 1880s and is still widely used, leading to considerable copper toxicity in the soil – causing harm to soil microorganisms and damage to vine health – of many of the world's damper wine regions today.

Grafting is an expensive and time-consuming business, so many growers in wine regions such as South Australia and Oregon in the United States in which phylloxera has not been rife have elected to plant European vines directly into the soil, but extreme caution is needed to keep phylloxera at bay. It is still unknown in the state of South Australia, but thanks only to extremely stringent quarantines imposed on its borders with phylloxera-infested Victoria. Phylloxera surfaced in Oregon in the 1990s and resurfaced in the Yarra Valley in the Australian state of Victoria quite recently. Phylloxera is yet to make a major impact on Chinese viticulture but it is likely that it will, not least because of the Chinese producers' enthusiastic importation of vine cuttings from abroad. Chile is the only significant wine-producing country where phylloxera is, so far, unknown – although some producers choose to graft European vines onto American rootstocks for other reasons, eg to reduce vine vigour and divert the vine's energy to grape ripening.

Some more recent widespread threats from vine pests and diseases have come from fungal diseases that affect vine wood specifically, such as esca and eutypa dieback, which have had serious effects in some French wine regions, and from two vine diseases for which there is so far no known cure: grapevine yellows (*flavescence dorée*), which has been observed in many corners of the wine world, and Pierce's disease, which has been killing vines in the southern states of the United States and in parts of California.

Different varieties have different susceptibilities to individual pests and diseases and we have given details of these in the Viticultural characteristics section of each variety's description. These susceptibilities can affect which variety is planted where and whether a variety is likely to produce high-quality fruit and at the same time be economically viable.

ROOTSTOCKS, GRAFTING AND FASHION

It would be quite possible to write a similar book to this exclusively about the scores of different varieties of rootstock that are currently planted by wine producers. Rootstocks may be chosen because of their tolerance of local conditions such as certain soil types or weather patterns, or because of their resistance to particularly prevalent pests, or their ability to make the vines less vigorous. But since rootstocks are of much less direct interest to wine lovers because they have a less dramatic influence on the way a wine tastes, we have concentrated solely on the hundreds of different varieties of vine that produce grapes used for wine, including those from species other than *Vitis vinifera*.

Because different varieties of grapevine have become so much more important in recent years, not least because of varietal labelling, and because there are fashions in the fortunes of different varieties, it is by no means uncommon for a grower to graft a different variety, one that is perceived to have more value than its predecessor, onto an established rootstock. This practice is known as top grafting. In the late 1970s and 1980s, for example, when temperature control in the winery had become de rigueur and demand for newly fresh and fruity white wines was at its peak, Chardonnay was regarded as the most desirable variety of all. Thanks to top grafting, many a workhorse variety such as Colombard or Chenin Blanc was replaced by the more glamorous Chardonnay. In the United States, the spotlight of fashion shone briefly on Merlot next, and then Pinot Noir. Top grafting enables rapid switches to be made from one variety to another (and can have a disruptive effect on statistics about the area planted to individual varieties).

VINE AGE

Most vines produce commercial quantities of wine from their third year in the ground, with the amount of wine produced reaching a plateau at approximately 6–20 years old and then decreasing steadily thereafter. The quality of wine produced is conventionally thought to increase with age, however (although the very first, necessarily small, crop can sometimes produce good-quality wine) – perhaps partly because an old vine's root system is so extensive and the vine is by then so well acclimated to the site. For this reason, old vines carry such cachet that they are often cited on wine labels variously as Old Vines, Vieilles Vignes, Vinhas Velhas, Viñas Viejas, Alte Reben, and so on.

Because their yields are relatively low, and even though they can produce high-quality wine with apparently effortless concentration, old vines can be prime targets for uprooting, especially when, as in the European Union recently, there have been financial incentives for growers to reduce their total vineyard area. Considering how important viticulture is to Europe and that it has been so for many centuries, some observers may be surprised by how relatively few really old vines are still in production there. Burgundy has some old vines, as have the Languedoc and Roussillon, and there are pockets elsewhere throughout Europe, from the Greek islands and the Mosel to Priorat and the Douro, but, in general, European wine producers have been too pragmatic about the need to maintain yields to preserve ancient vines and most have systematically replaced their plants every 25–30 years, uprooting the old vines and replacing them with grafted plants. Less commonly, growers replace individual vines but this is generally worthwhile only when the produce of their old vines can command high enough prices to compensate for tiny yields.

The world's most extensive plantings of really old vines are, perhaps contrary to expectations, in California. Prohibition helped preserve many a vine planted in the late nineteenth and early twentieth centuries. Because the wine industry collapsed in the 1920s, it was simply not worth the cost of replacing vines, and many were virtually abandoned. Barossa in South Australia, with its particularly dry climate and many arch-traditionalist vine growers of Silesian origin who have been reluctant to succumb to the whims of varietal fashion, is another hotspot for centenarian vines, and there are some dotted around South America and South Africa.

In this era when quality is so much more important than quantity for wine production, there is at long last a realization that many, if not most, old vines are worth preserving on the grounds of the intensity of the wine they produce. In California, the Historic Vineyards Society has been formed by a group of high-profile wine producers to preserve and record in minute detail the state's stocks of gnarled, ancient vines, particularly from the threat of being pulled out to make way for whichever variety happens to be most fashionable. There is a similar initiative in the Barossa Valley and an increasing number of producers are proudly citing the age of the vines responsible on the back labels of their old-vine bottlings, although the term 'old vines' is unhelpfully relative and has as yet no internationally agreed definition.

CHANGES IN VINEYARD COMPOSITION

The phylloxera plague that hit Western European vineyards from the 1860s on, described in more detail in Pests and diseases above, drastically reduced the diversity of grape varieties. When growers had to replant their vineyards by grafting their *vinifera* varieties onto phylloxera-resistant American rootstocks, they tended to discard many ancient native local varieties in favour of a small number of more fashionable or easily grown ones. Today there is a growing interest in long-forgotten varieties and in the many growers who are committed to rescuing varieties on the verge of extinction (see the LIMNIONA entry, for example).

The number of clones for each variety was also severely reduced because vineyards were replanted with a small number of clones selected for their resistance to diseases or their productivity (clonal selection). However, some growers continued – and continue – to select the best vines in their old vineyards in order to maintain clonal diversity (a practice known as mass selection or *séléction massale*).

Cultivation practices also changed radically in the late nineteenth century. The old tradition of propagation by layering (see the glossary on pp 1168–71), a simple way to fill in gaps in the vineyard left by dead vines, was replaced by the direct planting of grafted vines, enabling vine nurseries to flourish. The corollary was the abandonment of vineyard field blends, in which white and red, late or early varieties were interplanted and often harvested together. With the exception of a minute number of pre-phylloxera vineyards which have miraculously survived and some recent experiments, modern vineyards are now generally planted to a single variety, often a single clone, which means that varieties can no longer spontaneously cross in the vineyard and generate new ones.

LABELLING AND NAMING

Single-variety vineyards are much more convenient when it comes to labelling and marketing. Grape variety names now matter. Not just producers but also consumers demand to know which varieties go into the wines they drink, whether they are stated on the front label, the back label or just in the sales pitch. Towards the end of the twentieth century, when the wine world was apparently ruled by a handful of internationally famous grape varieties, one anti-GM argument advanced even by the scientists working on genetic modification was that genetically modified varieties would need new, unfamiliar names, and that they would therefore be impossible to sell. Today's panoply of grape names (see the index for all those mentioned in this book) shows just how much consumer taste has evolved since then.

But the naming of varieties can be highly confusing – not least because the same variety can be called so many different names, certainly in different places but sometimes even in the same place. DNA profiling (see below) has shown, for example, that the Zinfandel of California is genetically identical to the Primitivo of Puglia and also, much more recently, to an obscure Croatian variety memorably called Tribidrag. But it has taken many years for some Californians to admit that the Zinfandel so familiar to them could be one and the same as an Italian import – and, as we write, few Americans have even heard of Tribidrag. In the vineyards of Liguria around Genoa, growers and drinkers were particularly proud of two of their local white wine varieties, so good with their fine seafood, Pigato and Vermentino. It came as something of a shock therefore to be told only quite recently by DNA profilers that the two varieties are genetically identical, and also identical not just to the Vermentino of Sardegna (which they had doubted) but also to the Favorita of Piemonte and the Rolle of southern France (but not the Rollo of Liguria, which might just be Tuscan in origin).

We fervently hope that this book, with its uniquely comprehensive survey of DNA results, can help to unravel some of the mysteries of grape identification.

DNA PROFILING

The identification of grape varieties has traditionally been based on the analysis and comparison of the physical characteristics of a vine's leaves, shoot tips, fruit bunches and berries, the area of study known as ampelography (from the Greek *ampelos*, meaning 'grapevine', and *graphos*, meaning 'description'). However, classical ampelography has its limitations: morphological characteristics (see the glossary on pp 1179–82) can be modified by environmental factors and diseases; seedlings are often impossible to identify before the vine reaches the age of four or five; some closely related varieties have very similar morphology; and no ampelographer on earth, however skilful, would be able to identify all the 10,000 or so varieties that exist today.

In recent years, ampelography has been significantly complemented by DNA profiling, a method of analysis routinely used in forensics today and described in more detail below. In grape genetics, the first DNA profile was established in 1993 by M R Thomas, S Matsumoto, P Cain and N S Scott in Australia. The first successful paternity test was done in 1997, when Carole Meredith and her doctoral student John Bowers at the University of California at Davis showed that Cabernet Sauvignon, the noble Bordeaux grape variety, was probably derived from a natural cross between Cabernet Franc and Sauvignon Blanc, explaining the similarities in aromas and flavours found in the grapes and wines of parents and offspring, and showing that it was perfectly possible for a dark-skinned variety to have one light-skinned parent.

This discovery came as quite a surprise to the wine world and since then the study of the relationships between grape varieties has entered a new era, with the unveiling of several other unexpected parentages, as described in this book. When one of the parents is missing (extinct or not yet analysed), it is still possible to detect parent–offspring relationships between two varieties. However, it is not possible to determine which is the parent and which the progeny because the age of a variety is unknown. When both parents are missing, the use of a recently developed probabilistic approach can detect siblings, half-siblings, grandparents or grandchildren – see the complex pedigree of Syrah for an example.

This book provides readers with virtually every DNA result published before 31 August 2011 as well as with dozens of unpublished results and fourteen pedigree diagrams exclusive to us that reveal many new, sometimes unexpected, familial relationships.

Since some grape-variety entries in this book refer to specific DNA test results and in some instances contradict earlier conclusions, it is important to explain not only how DNA profiling works in this context but also the importance of accurately labelled examples of vine varieties in the various vine collections around the world, so-called reference samples.

DNA profiling is based on the analysis of specific regions of DNA that scientists refer to as SSRs (Simple Sequence Repeats) or 'microsatellites'. Because microsatellite length is stable within an individual grape variety but differs greatly from one individual to another, a DNA profile based on 6–8 distinct regions is enough to discriminate almost all the varieties in the world, although more than eight regions may be required to distinguish closely related varieties.

Simple as this may sound, without accurate reference samples, vine plants can easily be misidentified, leading to the sort of confusion that recently occurred in Australia, where wines labelled Albariño, a Spanish variety with an increasingly high reputation and market value, turned out to be made from the ancient but less sought-after French variety Savagnin Blanc (see the ALVARINHO entry for more detail). Producers were forced to relabel their wines and change their marketing strategies – or replant with the 'true' Albariño. Mislabelling was also the cause of the long-standing error over the parentage of Müller-Thurgau (see that entry for more details).

HISTORICAL PERSPECTIVE

GRAPEVINE DOMESTICATION: WHY, WHERE AND WHEN?

Why would a Paleolithic man or woman climb a tall forest tree to reach black or dark-red berries that may be attractive to birds but offer so little food value?

Serendipitous inebriation might be the answer. This idea, christened the 'Paleolithic hypothesis' by oeno-archaeologist Patrick McGovern (2003), suggests that a group of early humans might have been attracted by the berries only to find they rather liked their sugary, tart taste. They would then have collected several bunches in an animal hide, a wooden container or a rock crevice and, after a few days, the juice from the berries at the bottom produced a low-alcohol wine thanks to the fermentation yeasts that occur naturally in the atmosphere. Having eaten all the grapes, they might have tasted this beverage, enjoying a pleasant euphoria and hurrying off to pick more berries. Without a proper, sealed container, this 'Stone Age Beaujolais Nouveau', as McGovern calls it, had to be drunk quickly before it turned to vinegar.

According to this hypothesis, things changed about 10,000–8000 BC with the transition from a nomadic existence to sedentism, the process of permanent settlement that led to the beginnings of agriculture, the development of dense populations and the long-term storage of food. The 'Hermaphroditic hypothesis' might be used to explain the first example of grapevine domestication at this time: to secure a sufficient grape supply and avoid the dangers associated with climbing trees, some early Mesolithic or Neolithic men or women must have started to cultivate wild grapevines, either by sowing seeds or by burying shoots (still attached to the parent vine) to produce a new plant, a method known as layering which mimics what *Vitis vinifera* does naturally.

If a male plant were chosen, the vine would never produce berries and would be quickly abandoned. If a female plant were chosen (see p XVI), the vine would produce berries only if a male plant was near enough for pollination, otherwise the plant would appear sterile and would also be abandoned. If a hermaphroditic plant were chosen, a grape harvest would be produced each year and the vine would be preserved.

It is therefore very likely that grapevine domestication was made from the 2–3% of hermaphroditic wild grapevines generally found in natural vine populations. The domesticated vines would then have been propagated by sowing seed or planting cuttings. Vines with larger berries, bigger bunches, higher sugar content or better aromas would then have been selected and further propagated over the centuries, leading to the 10,000 or so grape varieties in existence today.

Where was the first wine grapevine domesticated? Most ampelographers (Negrul 1938; Levadoux 1956), archaeologists (McGovern *et al.* 1997; Zohary and Hopf 2000), botanists (Candolle 1883; Hegi 1925; Vavilov 1926) and grape geneticists (Olmo 1995; Alleweldt 1997) agree that the origins of viniculture (ie viticulture and winemaking) are located in what might be described as the grapevine's 'Fertile Triangle', a vast upland region between the Taurus mountains (eastern Turkey), the northern Zagros mountains (western Iran) and the Caucasus mountains (Georgia, Armenia and Azerbaijan). Prominent Russian botanist and geneticist Nikolai Ivanovich

Vavilov (1926) suggested that the area of greatest morphological diversity usually corresponds to the centre of origin of a crop, which led his disciple Aleksandr Mikhailovich Negrul (1938; 1968) to the conclusion that Transcaucasia (the area between the Black and Caspian seas extending from the Greater Caucasus to the Turkish and Iranian borders, and comprising the Republics of Georgia, Armenia and Azerbaijan) or southern Anatolia (the Asian part of Turkey) can be considered the birthplace of the cultivated grapevine, since they are the areas showing the greatest diversity within *Vitis vinifera*.

A recent DNA quest to discover where and when the wild Eurasian grapevine was first domesticated was baptized the 'Noah hypothesis' (McGovern 2003), a parallel to the Eve hypothesis that traces humankind back to a single ancestor through mitochondrial DNA analysis, as well as a reference to the story of Noah planting the first vineyard on Mount Ararat (a volcanic cone in today's Turkey on which Noah's ark is said to have run aground but where neither wild nor cultivated grapevines have ever been observed). Analysis of the relatedness of the *silvestris* and *vinifera* subspecies throughout Eurasia supported the thesis that the grapevine was first domesticated in the Near East (Myles *et al.* 2011). A focus on the grapevine's Fertile Triangle showed that the closest genetic relationship between local wild grapevines and traditional cultivated grape varieties from southern Anatolia, Armenia and Georgia was observed in southern Anatolia, favouring the headwaters of the Tigris and Euphrates rivers in the Taurus mountains as the possible location of the first grapevine domestication (Vouillamoz, Grando *et al.* 2004).

This is consistent with the fact that this area of former Mesopotamia is still home to natural populations of wild grapevines, and with McGovern's (2003) report of abundant evidence for the presence of both wild and cultivated *Vitis vinifera* in archaeological sites from the Neolithic (8500–4000 BC) to the Early Bronze Age (3300–2100 BC) along the upper reaches of the Tigris and Euphrates rivers, at the foot of the Taurus mountains (in Çayönü north-west of Diyarbakır, Hacinebi, Hassek Höyük, Korucutepe, Kurban Höyük, Tepecik, etc.). This area also includes the Karacadağ region in the northern part of the Fertile Crescent that has recently been acknowledged by archaeological and genetic studies as the centre of domestication of several agricultural 'founder crops' (those on which systematic agriculture throughout the world is based): einkorn wheat, peas, chickpeas, lentils, emmer and rye (Salamini *et al.* 2002). The northern part of the Fertile Crescent, incidentally, has also been shown by glottochronology, the part of linguistics dealing with the chronological relationships between languages, to be the place of origin of all the Indo-European languages (Gray and Atkinson 2003). This suggests that the Fertile Crescent was the cradle of modern civilization and it seems that all the conditions for successful vine domestication and early winemaking were met in south-eastern Anatolia.

Although genetics, archaeology and linguistics tend to point to southern Anatolia as the possible cradle of grapevine domestication, Transcaucasia still remains a serious candidate: ancient grape remains were excavated from Neolithic sites in Chokh in Dagestan, in Shomu-Tepe in Azerbaijan as well as in Shulaveris-Gora near Tbilisi in Georgia (McGovern 2003), and chemical evidence was recently found for wine production in the Areni wine region south east of Yerevan in Armenia around 4000 BC (Barnard *et al.* 2011).

When was the first grapevine domesticated? We will probably never know but analysis of archaeological grape remains and jars might provide some clues. However, remains found in archaeological digs usually consist of carbonized seeds and charred wood and the carbonization process, together with the enormous variability among both wild (*silvestris*) and cultivated (*vinifera*) vines, makes it difficult to identify them as one or the other.

Carbonized pips have been excavated from numerous prehistoric archaeological digs in Europe (Greece, Bosnia and Herzegovina, Italy, Switzerland, Germany, France and Spain; Rivera Nunez and Walker 1989) and in south-west Asia (Zohary and Hopf 2000) but these ancient remains most likely came from wild grape berries that were collected long before domestication. According to Georgian ampelographer Revaz Ramishvili, six 8,000-year-old pips found in the Neolithic site in Shulaveris-Gora in the hills south of Tbilisi, one of the earliest known permanent settlements in Georgia, are pear-shaped like those more typical of *vinifera* and allegedly constitute evidence of seeds from the first domesticated vine (McGovern 2003) but their identification is still in doubt. For archaeobotanists Daniel Zohary and Maria Hopf (2000), grape seeds found in the Copper Age (3300–3000 BC) site in Tell esh-Shuna (northern Jordan) constitute the earliest convincing evidence of grapevine cultivation because *silvestris* is not found in Jordan today. However, even if it is fairly unlikely that these very dry regions were home to wild grapevines some five or six millennia ago, one could argue that they have now simply disappeared from this area.

Less arguable is the evidence found in Early Bronze Age (3400–3000 BC) sites in Jericho (Palestine), Lachish (Israel), Numeira (Dead Sea, possibly Gomorrah), Arad (Israel) and Kurban Höyük (near Urfa, southern Turkey) because archaeologists have uncovered not only carbonized seeds but also charred grapevine wood as well as whole dried berries, providing solid evidence of cultivation (Zohary and Hopf 2000). However, early grapevine cultivation must be much older than this because the earliest chemical evidence of winemaking dates back to the sixth millennium BC. Using infrared spectrometry to identify tartaric acid – a sure sign of the presence of grapes – in Neolithic jar deposits, oeno-archaeologists Patrick McGovern *et al.* (1996) found that wine was already produced in 5400–5000 BC in Hajji Firuz Tepe (in the Zagros mountains in northern Iran), on the periphery of the area in which wild vines are found today. These clay-stoppered jars were laid on their side, embedded in the earthen floor of the 'kitchen', indicating that they contained wine rather than grape juice. Terebinth tree resin was added to this retsina-like wine as a preservative. Possibly older evidence of winemaking has been found, in Shulaveris-Gora in Georgia, for example, but has not yet been verified by chemical analysis.

WESTWARD EXPANSION OF VINICULTURE

From the primary domestication site in the Near East, be it eastern Anatolia or Transcaucasia, grapevine cultivation gradually spread to Lower Mesopotamia (Zagros mountains in Iran, and in Syria and Iraq) and to the Levant (Lebanon, Palestine and Jordan) between 3500 and 3000 BC. From Palestine, wine was exported to Egypt, where wild vines have never grown, as early as 3200–2700 BC: 700 jars of resinated wine were found in Abydos in the tomb of Scorpion I, king of Egypt, around 3150 BC (McGovern 2003). Soon after, grapevine cultivation was introduced to Egypt, as evidenced by hieroglyph signs showing that viniculture was already far advanced in Egypt some five millennia ago: under the reign of Den, fourth king of the First Dynasty (*c*.3100–2900 BC), jar stoppers specified the names of the pharaoh, the vineyard and the winery where the wine was made – the earliest 'wine label' in history; in Pharaoh Khasekhemwy's tomb (Second Dynasty, *c*.2900–2696 BC), a jar stopper showed a grapevine growing on a trellis; on tomb walls of the Third Dynasty (*c*.2660–2585 BC), grape-harvest as well as winemaking scenes appeared (McGovern *et al.* 1997).

A few centuries later, grapevine cultivation spread around the Mediterranean and reached Crete, southern Greece, Cyprus and the southern Balkans by about 2200–2000 BC (Kroll 1991; Valamoti *et al.* 2007). In Italy, the earliest testimonies date back to around 900 BC, while viniculture is said to have been introduced around 600 BC to Spain and the Maghreb by the Phoenicians and to France by the Greek Phocaeans, via the foundation of Marseille (Massalia) (Terral *et al.* 2010). The general spread of viniculture throughout Western Europe is closely linked with the expansion of the Roman Empire (44 BC to AD 1453) and Christianity (AD 380 to the present), following the main trade routes, notably along rivers such as the Rhine, Rhône, Danube and Garonne. At the same time, the spread of Islam to North Africa, Spain and the Middle East had a powerful impact on grapevine cultivation by favouring table grapes over wine grapes. The introduction of viniculture to the New World took place much later, from the sixteenth century on, mainly by missionaries who introduced varieties as seeds (see, for example, the CRIOLLA entry) or by early settlers to America, South Africa, Australia, and so on, who took cuttings with them.

The general assumption that the grapevine spread westward then northward is based on archaeological and historical data and presupposes a single centre of grapevine domestication. However, recent studies of chloroplast DNA (ie non-nuclear DNA found in chloroplasts, the photosynthesis subunits of plant cells) suggest the possible existence of secondary domestication centres from *silvestris* populations in the Iberian Peninsula (Arroyo-García *et al.* 2006) and on Sardinia (Grassi *et al.* 2003), which would account for the enormous genetic diversity among grape varieties today (This *et al.* 2006). However, the existence of secondary domestication centres is still disputed because (a) some of the so-called dioecious wild grapevines might be crossings between real wild grapevines and feral cultivated varieties, or hybrids between *Vitis vinifera* and naturalized non-European *Vitis* species (Arrigo and Arnold 2007) and (b) the majority of recent nuclear-DNA-based studies of western germplasms have rejected any genetic relationship between wild and cultivated grapevines in Western Europe (This *et al.* 2006).

AMPELOGRAPHIC GROUPS

During the westward expansion of viniculture, grapevines were transplanted by cuttings or by seeds, leading to the 10,000 or so grape varieties in existence today. As explained above (see p XVI), vegetative propagation using cuttings ensures the preservation of a vine's desired characteristics (with the exception of mutations acquired later), while sexual propagation by seeds generates a new variety each time. Given the huge genetic diversity among modern varieties, it seems that sexual reproduction by seeds has played the major role in human-driven grapevine evolution, as shown by the parentages recently discovered for varieties such as Cabernet Sauvignon (Bowers and Meredith 1997), Chardonnay (Bowers *et al.* 1999), Syrah (Bowers *et al.* 2000), Sangiovese (Vouillamoz *et al.* 2007) and Merlot (Boursiquot *et al.* 2009).

A more recent global study has shown that 384 varieties in the US Department of Agriculture's grape germplasm collection are interconnected by a series of first-degree relationships (parent–offspring or siblings; Myles *et al.* 2011). Such interconnections are also found in our reconstruction of a ground-breaking pedigree of 156 Western European varieties that are related by natural crossings (see PINOT entry). This suggests that the diversity of Western European grape varieties derives from a limited number of founder varieties, possibly ancestors of Pinot, Gouais Blanc, Savagnin, Cabernet Franc, Mondeuse Noire, etc.

Long before modern molecular studies, Spanish ampelographer Simón de Rojas Clemente y Rubio (1807) first classified grape varieties from Andalucía into natural groups. Comte Alexandre-Pierre Odart (1845) attempted to do the same for French varieties, as did Russian ampelographer Andrasovsky (1933) for Hungarian varieties. Russian botanist and ampelographer Aleksandr Mikhailovich Negrul (1938; 1946) was the first to propose a large-scale classification of *Vitis vinifera* grape varieties, dividing them into three main groups according to their morphology, ecology and geographic origins: *Proles occidentalis*, *Proles pontica* and *Proles orientalis*.

Large-scale classification of major grape varieties into eco-geographic groups (*proles*) as proposed by Negrul (1938; 1946), Levadoux (1956) and Bisson (2009). Note that recent DNA studies challenge the classification of those marked with this symbol⧖.

Proles occidentalis Negr.	*Proles pontica* Negr.	*Proles orientalis* Negr.
Western Europe	**Balkans, Bulgaria, Georgia, Greece, Hungary, Asia Minor, Moldova, Romania**	**Afghanistan, Armenia, Azerbaijan, Caspian Basin, Iran, Middle Asia**
Aleatico, Aligoté, Cabernet Franc, Cabernet Sauvignon, Chardonnay, Folle Blanche, Gamay Noir, Graciano, Monastrell, Muscadelle, Petit Verdot, Pinot, Riesling, Sauvignon Blanc, Sémillon, Sercial, Touriga Nacional, Traminer, Verdelho	Alvarna, Çavuş, Clairette, Dodrelaby, Furmint, Hárslevelű, Kokur Bely, Korinthiaki, Mtsvane Kakhuri, Plavaï, Rkatsiteli, Saperavi, Vermentino⧖	Charas, Chasselas⧖, Cinsaut⧖, Cornichon Blanc, Katta Kurgan, Khalili, Muscat Blanc à Petits Grains⧖, Muscat of Alexandria⧖, Ohanes, Rish Baba, Sultaniye

Proles pontica are grape varieties from Transcaucasia and Anatolia considered to comprise the oldest group. Western European *Proles occidentalis*, exclusively wine-grape varieties, and *Proles orientalis* in the Near East, are derived from *Proles pontica*. The outline of this classification has recently been confirmed by DNA analysis (Aradhya *et al.* 2003; Arroyo-García *et al.* 2006; Myles *et al.* 2011) and it is pretty safe to assume that all the 156 grape varieties included in the Pinot pedigree (see that entry) belong to the *Proles occidentalis*. However, Vermentino, Chasselas, Cinsaut, Muscat Blanc à Petits Grains and Muscat of Alexandria (marked with ☿ in the table above) should probably be placed in the *Proles occidentalis* group.

Ampelographer Louis Levadoux (1948; 1956) and later his disciple Jean Bisson (1999; 2009) proposed a detailed classification of French *Proles occidentalis* into groups called eco-geogroups or sorto-types, as shown on the map below and in the table opposite. Some of the varieties listed in the table do not appear in our book because their wines are no longer commercially available.

Geographic distribution of the eco-geogroups (or sorto-types) of French grape varieties as proposed by Levadoux (1956) and Bisson (2009), adapted from Bisson (2009).

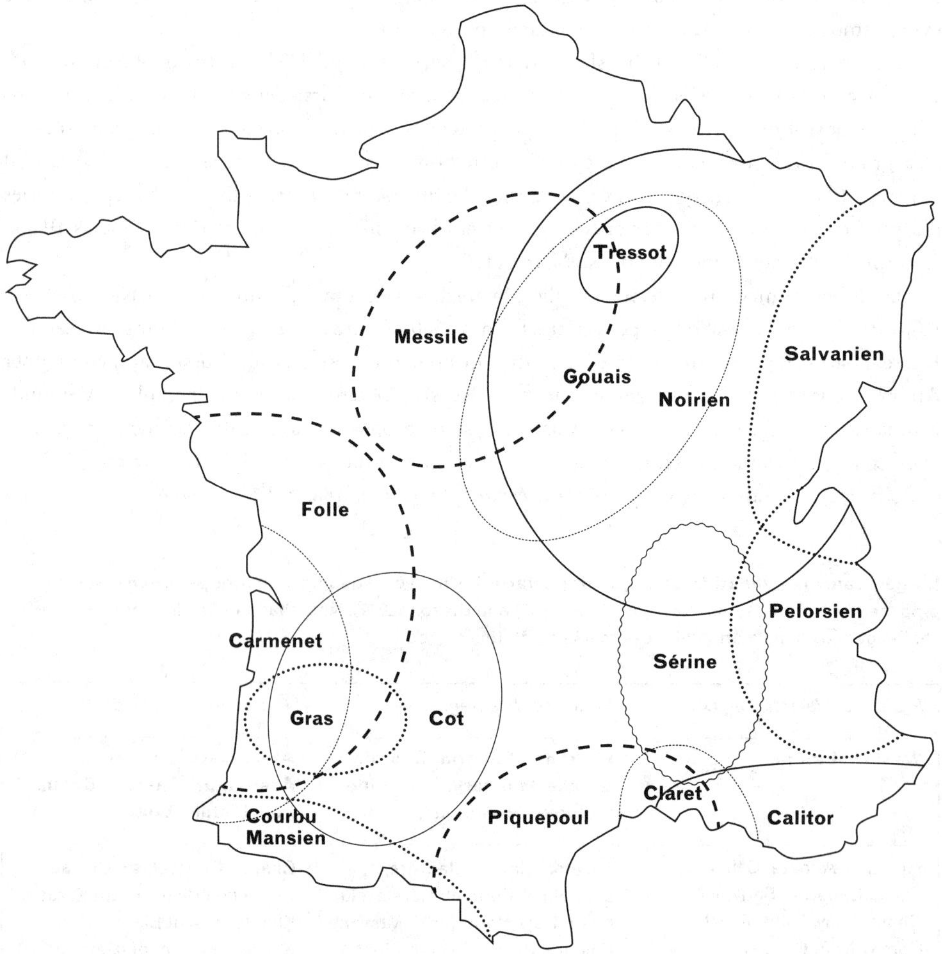

Composition of the eco-geogroups (or sorto-types) of French grape varieties proposed by Levadoux (1948; 1956) and Bisson (1999; 2009). Grape variety names have been amended where necessary to match the prime names used in the rest of this book (see p IX) and colour variations have been omitted.

Eco-geogroup	**Representative variety**	**Other varieties in the group**
Calitor	Calitor Noir	Braquet Noir, Ribier Noir
Carmenet (old synonym for Cabernet Franc)	Cabernet Franc	Arrouya, Arrufiac, Cabernet Sauvignon, Carmenère, Castets, Fer, Gros Cabernet, Gros Verdot, Lercat, Merlot, Merlot Blanc, Hondarribi Beltza, Pardotte, Petit Verdot
Claret	Clairette	Barbaroux, Bourboulenc, Bouteillan Noir, Brun Fourca, Calitor Blanc, Colombaud, Grec Rouge, Mauzac Noir(?), Pascal Blanc, Plant Droit, Téoulier Noir
Cot	Cot	Abondance, Duras, Gascon, Gibert, Grapput, Lignage, Malpé, Mérille, Négrette, Noual, Prunelard, Romorantin*, Saint-Macaire, Valdiguié
Courbu	Courbu Blanc	Baroque, Crouchen, Lauzet, Manseng Noir, Petit Courbu, Tannat
Folle	Folle Blanche	Balzac Blanc, Bouillet, Fuella Nera, Jurançon Blanc, Jurançon Noir, Montils, Ondenc, Roublot*, Sacy*, Sémillon
Gras	Graisse	Ardounet, Blanc Dame, Blancart, Camaraou Blanc, Claverie, Guillemot, Mouyssaguès, Raffiat de Moncade
Gouais	Gouais Blanc	Aligoté, Enfariné Noir, Gueuche Noir, Muscadelle, Saint-Pierre Doré
Mansien	Petit Manseng	Ahumat, Camaralet de Lasseube, Camaraou Noir, Courbu Noir, Gros Manseng, Mancin
Messile	Gros Meslier	Béquignol Noir, Chany Gris, Chenin Blanc, Colombard, Menu Pineau, Meslier Saint-François, Mézy, Petit Meslier, Pineau d'Aunis, Plant Vert, Sauvignon Blanc, Sauvignonasse
Noirien (old synonym for Pinot)	Pinot Noir	Auxerrois, Chardonnay, Franc Noir de la Haute Saône, Gamay Noir, Gamay Teinturier de Bouze, Genouillet, Gouget Noir, Melon, Troyen
Pelorsien (named after Peloursin)	Durif	Béclan, Bia Blanc, Dureza, Exbrayat, Jacquère, Joubertin, Mondeuse Blanche, Peloursin, Servanin, Verdesse
Piquepoul	Piquepoul Noir	Aubun, Bourrisquou, Cinsaut, Counoise, Picardan, Rivairenc, Terret, Tibouren
Salvanien	Savagnin	Aubin Vert*, Bargine, Gringet**, Knipperlé, Persagne, Trousseau
Sérine (old synonym for Syrah)	Syrah	Brun Argenté, Chatus, Chichaud, Douce Noire, Etraire de l'Aduï, Marsanne, Mondeuse Noire, Muscardin, Persan, Pougnet, Roussanne, Roussette d'Ayze, Viognier
Tressot	Tressot	Arbane, Bachet Noir*, Peurion

* These are natural Pinot × Gouais Blanc crosses and their assignment to these ampelographic groups might therefore need to be revised.

** Recent DNA profiling has shown that Gringet (see the entry) is not genetically related to Savagnin, which might challenge its membership of this ampelographic group.

In the text we refer where useful to the relevant groups, and would expect further work to group non-French varieties.

VARIETIES BY COUNTRY OF ORIGIN

Countries are listed in descending order of the number of varieties included in this book. Where the origin is particularly uncertain, alternatives are given in brackets.

ITALY
377 varieties

Abbuoto
Abrusco
Acitana
Addoraca
Aglianico
Aglianicone
Albana
Albanella
Albanello
Albaranzeuli Bianco
Albaranzeuli Nero
Albarola
Albarossa
Aleatico
Alionza
Ancellotta
Arilla
Arneis
Arvesiniadu
Avanà
Avarengo
Baratuciat
Barbarossa
Barbera
Barbera Bianca
Barbera del Sannio
Barbera Sarda
Bariadorgia
Barsaglina
Bellone
Besgano Bianco
Biancame
Bianchetta Trevigiana
Bianco d'Alessano
Biancolella
Biancone di Portoferraio
Bigolona
Blatterle
Bombino Bianco
Bombino Nero
Bonamico
Bonarda Piemontese
Bonda
Bosco
Bracciola Nera
Brachetto del Piemonte
Brugnola
Brustiano Bianco
Bubbierasco
Bussanello
Cacamosca
Caddiu
Calabrese di Montenuovo
Caloria
Canaiolo Nero
Cannamela
Caprettone
Cargarello
Carica l'Asino
Carricante
Casavecchia
Cascarolo Bianco
Casetta
Castagnara
Castiglione
Catalanesca
Catanese Nero
Catarratto Bianco
Cavrara
Centesimino
Cerreto
Cesanese
Cianorie
Ciliegiolo
Ciurlese
Cividin
Cococciola
Coda di Cavallo Bianca
Coda di Pecora
Coda di Volpe Bianca
Colombana Nera
Colorino del Valdarno
Cordenossa
Cornalin
Cornarea
Cortese
Corva
Corvina Veronese
Corvinone
Croatina
Crovassa
Damaschino (or Mediterranean coast?)
Dindarella
Dolcetto
Dolciame
Dorona di Venezia
Doux d'Henry
Drupeggio
Durella
Enantio
Erbaluce
Erbamat
Ervi
Falanghina Beneventana
Falanghina Flegrea
Fenile
Fertilia
Fiano
Flavis
Fogarina
Foglia Tonda
Forastera
Forgiarin
Forsellina
Fortana
Francavidda
Frappato
Fraueler
Freisa
Fubiano
Fumin
Gaglioppo
Galatena
Gallioppo delle Marche
Gallizzone
Gamba di Pernice
Garganega

Ginestra
Girò
Grapariol
Grechetto di Orvieto
Greco
Greco Bianco
Greco Nero
Greco Nero di Sibari
Greco Nero di Verbicaro
Grignolino
Grillo
Grisa Nera
Groppello di Mocasina
Groppello di Revò
Groppello Gentile
Gruaja
Guardavalle
Impigno
Incrocio Bianco Fedit 51
Incrocio Bruni 54
Incrocio Manzoni 2.15
Incrocio Terzi 1
Invernenga
Inzolia
Italia
Italica
Lacrima di Morro d'Alba
Lagarino Bianco
Lagrein
Lambrusca di Alessandria
Lambrusca Vittona
Lambruschetto
Lambrusco Barghi
Lambrusco di Fiorano
Lambrusco di Sorbara
Lambrusco Grasparossa
Lambrusco Maestri
Lambrusco Marani
Lambrusco Montericco
Lambrusco Oliva
Lambrusco Salamino
Lambrusco Viadanese
Lanzesa
Luglienga
Lumassina
Maceratino
Magliocco Canino
Magliocco Dolce
Maiolica
Maiolina
Malbo Gentile
Maligia
Malvasia Bianca di Basilicata
Malvasia Bianca di Candia
Malvasia Bianca di Piemonte
Malvasia Bianca Lunga
Malvasia del Lazio
Malvasia di Candia Aromatica
Malvasia di Casorzo
Malvasia di Lipari
Malvasia di Schierano
Malvasia Nera di Basilicata
Malvasia Nera di Brindisi
Malvasia Nera Lunga
Mammolo
Mantonico Bianco
Manzoni Bianco
Manzoni Moscato
Manzoni Rosa
Marchione
Maruggio
Marzemina Bianca
Marzemino
Mayolet
Mazzese
Melara
Minella Bianca
Minutolo
Molinara
Monica Nera
Montepulciano
Montonico Bianco
Montù
Moradella
Morone
Moscatello Selvatico
Moscato di Scanzo
Moscato di Terracina
Moscato Giallo
Moscato Rosa del Trentino
Mostosa
Muscat Blanc à Petits Grains (or Greece?)
Nascetta
Nasco
Nebbiera
Nebbiolo
Nebbiolo Rosé
Negrara Trentina
Negrara Veronese
Negretto
Negroamaro
Ner d'Ala
Nerello Cappuccio
Nerello Mascalese
Neret di Saint-Vincent
Neretta Cuneese
Neretto di Bairo
Neretto Duro
Neretto Gentile
Neretto Nostrano
Nero Buono di Cori
Nero d'Avola
Nero di Troia
Nieddera
Nigra
Nocera
Nosiola
Notardomenico
Nuragus
Ortrugo
Oseleta
Pallagrello Bianco
Pallagrello Nero
Pampanuto
Paolina
Pascale
Passau
Passerina
Pavana
Pecorino
Pelaverga
Pelaverga Piccolo
Pepella
Perera
Perricone
Petit Rouge
Piccola Nera
Picolit
Piculit Neri
Piedirosso
Pignola Valtellinese
Pignoletto
Pignolo
Pinella
Plassa
Pollera Nera
Prié
Primetta
Prodest
Prosecco
Prosecco Lungo
Prunesta
Pugnitello
Quagliano
Raboso Piave
Raboso Veronese
Raspirosso
Rastajola
Rebo
Recantina
Refosco dal Peduncolo Rosso
Refosco di Faedis
Retagliado Bianco
Ribolla Gialla
Ripolo
Rollo
Rondinella
Rossara Trentina
Rossese Bianco
Rossese Bianco di Monforte
Rossese Bianco di San Biagio
Rossese di Campochiesa
Rossetto
Rossignola
Rossola Nera
Rossolino Nero
Roussin
Rovello Bianco
Ruchè
Ruggine
Ruzzese
Sabato
Sagrantino
San Giuseppe Nero
San Lunardo
San Martino
San Michele
San Pietro
Sangiovese

Sant'Antonio
Santa Maria
Santa Sofia
Schiava Gentile
Schiava Grigia
Schiava Grossa
Schiava Lombarda
Schioppettino
Sciaglìn
Sciascinoso
Scimiscià
Semidano
Sgavetta
Soperga
Sorbigno
Spergola
Suppezza
Susumaniello
Tamurro
Tazzelenghe
Termarina Rossa
Teroldego
Terrano
Tignolino
Timorasso
Tintilia del Molise
Tintore di Tramonti
Torbato
Trebbiano d'Abruzzo
Trebbiano Giallo
Trebbiano Modenese
Trebbiano Romagnolo
Trebbiano Spoletino
Trebbiano Toscano
Trevisana Nera
Tronto
Ucelùt
Uva della Cascina
Uva Longanesi
Uva Rara
Uva Tosca
Uvalino
Valentino Nero
Vega
Verdea
Verdeca
Verdello
Verdicchio Bianco
Verdiso
Verduzzo Friulano
Verduzzo Trevigiano
Vermentino
Vermentino Nero
Vernaccia di Oristano
Vernaccia di San Gimignano
Versoaln
Vespaiola
Vespolina
Vien de Nus
Vitovska (or Slovenia?)
Vuillermin

FRANCE
204 varieties

Abouriou
Ahumat
Alicante Henri Bouschet
Aligoté
Altesse
Aramon Noir
Aranel
Arbane
Arinarnoa
Arriloba
Arrouya
Arrufiac
Aspiran Bouschet
Aubin Blanc
Aubin Vert
Aubun
Aurore
Auxerrois
Bachet Noir
Baco Blanc
Baco Noir
Balzac Blanc
Barbaroux
Baroque
Beaunoir
Béclan
Béquignol Noir
Berdomenel
Biancu Gentile
Blanc Dame
Blanqueiro
Bouchalès
Bourboulenc
Bourrisquou
Bouteillan Noir
Braquet Noir
Brun Argenté
Brun Fourca
Cabernet Sauvignon
Cacaboué
Caladoc
Calitor Noir
Camaralet de Lasseube
Camaraou Noir
Canari Noir
Carmenère
Cascade
Castets
César
Chambourcin
Chancellor
Chardonnay
Chasan
Chatus
Chelois
Chenanson
Chenin Blanc
Chichaud
Chouchillon
Cinsaut
Clairette
Claverie
Codivarta
Colobel
Colombard
Colombaud
Cot
Couderc Noir
Counoise
Courbu Blanc
Courbu Noir
Crouchen
De Chaunac
Douce Noire
Duras
Dureza
Durif
Ederena
Egiodola
Ekigaïna
Enfariné Noir
Etraire de l'Aduï
Fer
Florental
Folignan
Folle Blanche
Fuella Nera
Gamay Noir
Gamay Teinturier de Bouze
Ganson
Gascon
Genouillet
Goldriesling
Gouais Blanc (or Germany?)
Gouget Noir
Graisse
Gramon
Grand Noir
Gringet
Grolleau Noir
Gros Manseng
Gros Verdot
Gueuche Noir
Jacquère
Jurançon Blanc
Jurançon Noir
Knipperlé
Landal
Landot Noir
Lauzet
Len de l'El
Léon Millot
Liliorila
Lucie Kuhlmann
Madeleine Angevine
Mandrègue
Manseng Noir
Maréchal Foch
Marsanne
Marselan
Mauzac Blanc
Mauzac Noir
Mècle de Bourgoin
Melon
Menu Pineau
Mérille
Merlot
Meslier Saint-François
Mézy
Milgranet
Molette
Mollard
Monbadon
Mondeuse Blanche
Mondeuse Noire
Monerac
Montils
Mornen Noir

Morrastel Bouschet
Mouyssaguès
Muscadelle
Muscardin
Muscat Fleur d'Oranger (?)
Muscat Ottonel
Négret de Banhars
Négrette
Noir Fleurien
Oeillade Noire
Onchette
Ondenc
Pascal Blanc
Peloursin
Persan
Petit Bouschet
Petit Courbu
Petit Manseng
Petit Meslier
Petit Verdot
Picardan
Pineau d'Aunis
Pinot
Piquepoul
Plant Droit
Plantet
Portan
Poulsard
Précoce de Malingre
Prunelard
Raffiat de Moncade
Ravat Blanc
Rayon d'Or
Rivairenc
Romorantin
Rosé du Var
Rosette
Roussanne
Roussette d'Ayze
Sacy
Saint-Macaire
Salvador
Sauvignon Blanc
Sauvignonasse
Savagnin
Segalin
Seibel 10868
Sémillon
Servant
Seyval Blanc
Seyval Noir
Syrah
Tannat
Teinturier
Téoulier Noir
Terret
Tibouren
Tressot
Triomphe
Trousseau
Valdiguié
Verdesse
Vidal
Vignoles
Villard Blanc
Viognier

SPAIN
84 varieties

Airén
Alarije
Albarín Blanco
Albillo de Albacete
Albillo Mayor
Albillo Real
Alcañón
Beba
Belat
Bobal
Cabernet Franc
Caíño Blanco (or Portugal?)
Calagraño
Callet
Cañocazo
Carrasquín
Cayetana Blanca (or Portugal?)
Chelva
Coloraillo
Corredera
Doradilla
Escursac
Fogoneu
Forcallat Tinta
Garnacha
Garrido Fino
Gibi
Giró Blanc
Godello
Gorgollasa
Graciano
Hondarribi Beltza
Juan García
Lado
Lairén
Listán de Huelva
Listán Negro
Listán Prieto
Macabeo
Malvasía de Lanzarote
Mandón
Manto Negro
Marmajuelo
Maturana Blanca
Mazuelo
Mencía
Merseguera
Monastrell
Monstruosa
Moravia Agria
Moristel
Negramoll
Palomino Fino
Pardillo
Parellada
Parraleta
Pedro Ximénez
Perruno
Picapoll Blanco
Planta Nova
Prensal
Prieto Picudo
Rión
Romé
Romero de Híjar
Royal de Alloza
Serodio
Sumoll
Sumoll Blanc
Teca
Tempranillo
Tinta Castañal
Tinto Velasco
Trepat
Trepat Blanc
Verdejo
Verdejo Serrano
Verdil
Verdoncho
Vidadillo de Almonacid
Vijariego
Xarello
Zalema
Zamarrica

GREECE
77 varieties

Agiomavritiko
Agiorgitiko
Aïdani
Araklinos
Areti
Assyrtiko
Athiri
Avgoustiatis
Batiki
Bekari
Chidiriotiko
Chondromavro
Dafni
Debina
Fokiano
Gaidouria
Goustolidi
Kakotrygis
Karnachalades
Katsakoulias
Katsano
Kolindrino
Korinthiaki
Koriostafylo
Kotsifali
Koutsoumpeli
Krassato
Kydonitsa
Liatiko
Limnio
Limniona
Malagousia
Mandilaria
Mavro Kalavritino
Mavro Messenikola
Mavrodafni
Mavrotragano
Mavroudi Arachovis
Monemvassia
Moschofilero
Moschomavro
Muscat of Alexandria (or Italy?)
Negoska
Nigrikiotiko
Opsimo Edessis
Petrokoritho
Petroulianos
Platani
Plyto

Potamissi
Priknadi
Ritino
Robola
Roditis
Rokaniaris
Romeiko
Savatiano
Sefka
Sideritis
Sklava
Skopelitiko
Skylopnichtis
Stavroto
Sykiotis
Syriki
Theiako Mavro
Thrapsathiri
Tourkopoula
Tsaoussi
Vertzami
Vidiano
Vilana
Violento
Vlachiko
Volitsa Mavri
Xinomavro
Zakynthino

PORTUGAL
77 varieties

Agronómica
Água Santa
Alfrocheiro
Alvarelhão
Alvarinho
Amaral
Antão Vaz
Arinto de Bucelas
Avesso
Azal
Baga
Barcelo
Batoca
Bical
Borraçal
Camarate
Caracol
Carrega Branco
Casculho
Castelão
Cerceal Branco
Côdega de Larinho
Complexa
Cornifesto
Diagalves
Dona Branca
Donzelinho Branco
Donzelinho Tinto
Dorinto
Encruzado
Espadeiro
Fernão Pires
Folgasão
Fonte Cal
Galego Dourado
Generosa (Portugal)
Gouveio Real
Grossa
Jampal
Loureiro
Malvasia Branca de São Jorge
Malvasia de Colares
Malvasia Fina
Malvasia Preta
Marufo
Monvedro
Moreto do Alentejo
Padeiro
Pedral
Preto Martinho
Rabigato
Rabo de Ovelha
Ramisco
Rio Grande
Rufete
Seara Nova
Sercial
Sercialinho
Síria
Tamarez
Terrantez
Terrantez da Terceira
Terrantez do Pico
Tinta Barroca
Tinta Carvalha
Tinta Francisca
Tinto Cão
Touriga Fêmea
Touriga Franca
Touriga Nacional
Trajadura
Trincadeira
Trincadeira das Pratas
Verdelho
Vinhão
Viosinho
Vital

GERMANY
76 varieties

Acolon
Affenthaler
Albalonga
Allegro
Arnsburger
Bacchus
Blauer Urban
Breidecker
Bronner
Bukettraube
Cabernet Carbon
Cabernet Carol
Cabernet Cortis
Cabernet Cubin
Cabernet Dorio
Cabernet Dorsa
Cabernet Mitos
Calandro
Dakapo
Dalkauer
Deckrot
Domina
Dornfelder
Dunkelfelder
Ehrenfelser
Elbling
Faberrebe
Freisamer
Gänsfüsser
Geisenheim 318-57
Gf-Ga 48-12
Gutenborner
Helfensteiner
Helios
Heroldrebe
Hibernal
Hölder
Huxelrebe
Johanniter
Juwel
Kanzler
Kerner
Madeleine × Angevine 7672
Merzling
Monarch
Morio-Muskat
Müller-Thurgau
Nobling
Optima
Orangetraube
Oraniensteiner
Orion
Orleans Gelb
Ortega
Osteiner
Perle
Phoenix
Prinzipal
Prior
Räuschling
Reberger
Regent
Regner
Reichensteiner
Rieslaner
Riesling
Rondo
Rosetta
Rotberger
Saphira
Scheurebe
Schönburger
Siegerrebe
Solaris
Tauberschwarz
Würzer

US
76 varieties

Adalmiina
Alexander
Blanc du Bois
Bordô
Brianna
Cabernet Pfeffer
Campbell Early
Canada Muscat
Cardinal
Carlos
Carmine
Carnelian
Catawba
Cayuga White

Centurian
Chardonel
Chisago
Clinton
Concord
Corot Noir
Delaware
Delisle
Dutchess
Early Muscat
Edelweiss
Elvira
Emerald Riesling
Eona
Espirit
Flora
Fredonia
Frontenac
Golden Muscat
GR 7
Herbemont
Isabella
Ives
Jacquez
Kay Gray
La Crescent
La Crosse
Louise Swenson
Lydia
Magnolia
Marquette
Melody
Moore's Diamond
Muscat Swenson
New York Muscat
Niagara
Noah
Noble
Noiret
Norton
Petite Amie
Petite Pearl
Pionnier
Prairie Star
Radisson
Royalty
Rubired
Ruby Cabernet
Sabrevois
Scuppernong
St Croix
St Pepin
St Vincent
Steuben
Swenson Red
Swenson White
Symphony
Temparia
Traminette
Valiant
Valvin Muscat
Van Buren

CROATIA
39 varieties

Babić
Babica
Bogdanuša
Bratkovina Bijela
Cetinka
Debit
Dišeća Ranina
Dobričić
Drnekuša
Duranija
Gegić
Glavinuša
Graševina
Grk
Hrvatica
Kraljevina
Kujundžuša
Kupusar
Lasina
Ljutun
Malvazija Istarska
Medna
Mladinka
Ninčuša
Plavac Mali
Plavec Žuti
Plavina
Pošip Bijeli
Prč
Ranac Bijeli
Škrlet
Sušćan
Trbljan
Tribidrag
Trnjak
Vlaška
Vugava
Žlahtina
Zlatarica Vrgorska

SWITZERLAND
39 varieties

Amigne
Arvine
Birstaler Muskat
Bondola
Bondoletta
Cabernet Blanc
Cabernet Colonjes
Cabernet Jura
Cabertin
Carminoir
Charmont
Chasselas
Completer
Diolinoir
Doral
Eyholzer Rote
Galotta
Gamaret
Garanoir
Himbertscha
Hitzkircher
Humagne
Kalina
Lafnetscha
Mara
Mennas
Millot-Foch
Muscat Bleu
Pinotin
Plantscher
RAC 3209
Réselle
Rèze
Riesel
Rouge de Fully
Rouge du Pays
Siramé
VB 32-7
VB 91-26-4

HUNGARY
34 varieties

Arany Sárfehér
Bianca
Bíborkadarka
Budai Zöld
Csaba Gyöngye
Cserszegi Fűszeres
Csókaszőlő
Ezerfürtű
Ezerjó
Furmint
Generosa (Hungary)
Hárslevelű
Irsai Olivér
Juhfark
Kabar
Kadarka (or Serbia? or Croatia?)
Karát
Kéknyelű
Királyleányka
Kövidinka
Kunleány
Leányka
Magyarfrankos
Mátrai Muskotály
Menoir
Mézes Fehér
Rubintos
Turán
Zalagyöngye
Zefír
Zengő
Zenit
Zéta
Zeusz

UKRAINE
29 varieties

Antey Magarachsky
Aromatny
Bastardo Magarachsky
Cevat Kara
Citronny Magaracha
Dnestrovsky Rozovy
Ekim Kara
Golubok
Kapitan Jani Kara
Kapselsky
Kefessiya
Kok Pandas
Kokur Bely
Krona
Lapa Kara
Muscat Odessky
Odessky Cherny
Ovidiopolsky
Pervenets Magaracha
Podarok Magaracha

Riesus
Rubin Golodrigi
Rubin Tairovsky
Rubinovy Magaracha
Sary Pandas
Soldaia
Solnechnodolinsky
Sukholimansky Bely
Zagrei

GEORGIA
27 varieties

Aladasturi
Aleksandrouli
Ashughaji
Asuretuli Shavi
Avasirkhva
Budeshuri Tsiteli
Chinuri
Chkhaveri
Dzvelshavi Obchuri
Goruli Mtsvane
Kachichi
Kapistoni Tetri
Khikhvi
Kisi
Krakhuna
Mtsvane Kakhuri
Mujuretuli
Ojaleshi
Otskhanuri Sapere
Rkatsiteli
Saperavi
Shavkapito
Tavkveri
Tsitska
Tsolikouri
Tsulukidzis Tetra
Usakhelouri

TURKEY
26 varieties

Adakarası
Boğazkere
Çalkarası
Çavuş
Dimrit
Emir
Foça Karası
Hasandede
Horozkarası
İri Kara
Kalecik Karası
Karalahna
Karasakız
Köhnü
Kolorko
Kösetevek
Merzifon Karası
Narince
Öküzgözü
Papazkarası
Sidalan
Sultaniye
Urla Karası
Vasilaki
Yapıncak
Yediveren

AUSTRIA
19 varieties

Blauburger
Blauer Portugieser
Blauer Wildbacher
Blaufränkisch
Brauner Veltliner
Frühroter Veltliner
Goldburger
Grüner Veltliner
Jubiläumsrebe
Neuburger
Österreichisch Weiss
Ráthay
Roesler
Roter Veltliner
Rotgipfler
Sankt Laurent
Silvaner
Zierfandler
Zweigelt

AZERBAIJAN
16 varieties

Ag Aldara
Ag Kalambir
Ag Malayi
Agdam Gyzyl Uzumu
Arna-Grna
Bayanshira
Gara Aldara
Gara Ikeni
Gara Sarma
Hamashara
Khindogni
Madrasa
Shirvanshahy
Sibi Abbas
Sysak
Tatly

BULGARIA
16 varieties

Buket
Dimyat
Evmolpia
Keratsuda
Mavrud
Melnik 82
Misket Cherven
Misket Varnenski
Misket Vrachanski
Pamid
Ranna Melnishka Loza
Rubin
Ruen
Shevka
Shiroka Melnishka
Storgozia

SERBIA
15 varieties

Bačka
Bagrina (or Romania? or Hungary?)
Kreaca (or Romania? or Hungary?)
Morava
Neoplanta
Panonia
Petra
Probus
Prokupac
Rubinka
Sila
Sirmium
Slankamenka (or Romania? or Hungary?)
Začinak
Župljanka

ARMENIA
14 varieties

Akhtanak
Areni
Charentsi
Garandmak
Kakhet
Kangun
Karmrahyut
Lalvari
Megrabuir
Mskhali
Nerkarat
Nerkeni
Tigrani
Voskeat

RUSSIA
13 varieties

Bessarabsky Muskatny
Cabernet Severny
Fioletovy Ranny
Krasnostop Zolotovsky
Kumshatsky Cherny
Plechistik
Pukhliakovsky
Saperavi Severny
Severny
Sibirkovy
Tsimladar
Tsimlyansky Cherny
Zhemchuzhina Oskhi

SLOVAKIA
12 varieties

Breslava
Devín
Dunaj
Hetera
Hron
Mília
Nitranka
Noria
Rimava
Rudava
Torysa
Váh

ROMANIA
12 varieties

Băbească Neagră
Busuioacă de Bohotin
Crâmpoşie Selecţionată
Fetească Regală
Frâncuşă
Galbenă de Odobeşti
Grasă de Cotnari
Mustoasă de Măderat
Negru de Drăgăşani
Novac
Şarbă
Zghihară de Huşi

CZECH REPUBLIC
11 varieties

Agni
André
Aurelius
Cabernet Moravia
Laurot
Malverina
Muškát Moravský
Neronet
Pálava
Rubinet
Veritas

ARGENTINA
7 varieties

Caberinta
Cereza
Criolla Grande
Pedro Giménez
Torrontés Mendocino
Torrontés Riojano
Torrontés Sanjuanino

ALBANIA
7 varieties

Debine e Bardhë
Debine e Zezë
Pules
Serina e Zezë
Shesh i Bardhë
Shesh i Zi
Vlosh

CHINA
6 varieties

Beichun
Crystal
French Wild
Longyan
Rose Honey
Tuo Xian

SLOVENIA
6 varieties

Bouvier
Klarnica
Ranfol
Vitovska Grganja
Žametovka
Zelen

AUSTRALIA
5 varieties

Cienna
Cygne Blanc
Taminga
Tarrango
Tyrian

CYPRUS
5 varieties

Maratheftiko
Mavro
Ofthalmo
Spourtiko
Xynisteri

JAPAN
5 varieties

Black Queen
Koshu
Muscat Bailey A
Ryugan
Yamabudo

UZBEKISTAN
5 varieties

Bakhtiori
Bishty
Parkent
Pervomaisky
Soyaki

BOSNIA & HERZEGOVINA
4 varieties

Bena
Blatina
Krkošija
Žilavka

CANADA
4 varieties

Dragon Blue
L'Acadie Blanc
Vandal-Cliche
Vincent

SOUTH AFRICA
4 varieties

Chenel
Colomino
Nouvelle
Pinotage

MOLDOVA, REPUBLIC OF
3 varieties

Doina
Fetească Albă (or Romania?)
Fetească Neagră (or Romania?)

MONTENEGRO
3 varieties

Krstač
Vranac
Zizak

ISRAEL
2 varieties

Argaman
Neheleschol

MACEDONIA, REPUBLIC OF
2 varieties

Ohridsko Crno
Stanušina Crna

MALTA
2 varieties

Ġellewża
Girgentina

TURKMENISTAN
2 varieties

Kara Izyum Ashkhabadsky
Terbash

PERU
1 variety

Quebranta

THAILAND
1 variety

Malaga Blanc

UNITED KINGDOM
1 variety

Muscat of Hamburg

COLOUR ILLUSTRATIONS

TAKEN FROM *AMPÉLOGRAPHIE* BY VIALA AND VERMOREL (1901–10)

Published in French between 1901 and 1910, the magnum opus *Ampélographie* by Pierre Viala and Victor Vermorel has long been the absolute reference for grape researchers worldwide. To achieve this, author and editor Pierre Viala (1859–1936), a professor of viticulture in Montpellier and Paris, received substantial financial support from Victor Vermorel (1848–1927), an industrialist but also director of the viticulture research station in Villefranche-sur-Saône (Rhône-Alpes, France), and surrounded himself with no fewer than eighty-five grape experts from several countries. The seven volumes (3,200 pages) cover 5,200 wine- and table-grape varieties or clones, of which the 627 most important ones are described in detail and illustrated with 500 magnificent paintings by Jules Troncy and Alexis Kreÿder.

In homage to this massive work containing many visionary hypotheses that have recently been confirmed by DNA profiling, we have chosen eighty of these paintings to show varieties of different origins, colours and degrees of familiarity to illustrate *Wine Grapes*.

An ampelographic tragedy occurred on 6 December 2006: most of the original paintings kept at the Maison du Patrimoine in Villefranche-sur-Saône were destroyed in a fire. Thanks to the generosity of Neil Tully MW and Josef-Marie and Marlis Chanton, fortunate owners of the original 1901–10 edition, the illustrations in *Wine Grapes* are as true as possible to the original paintings. While the original edition has become very hard to find, readers can still make do with the full seven volumes of *Ampélographie* that were republished in 1991 by Jeanne Laffitte in Montpellier.

ACKNOWLEDGEMENTS

A book as comprehensive as this necessarily leans heavily on a great number of informants and collaborators. While apologizing deeply to anyone mistakenly omitted, we would like to thank all those named below, and especially those whose names are highlighted in bold, whose contributions were even greater than we dared hope.

Eduardo Abade, Francesco Alia, Max Allen, César Almeida, Lucy Anderson, Jane Anson, Levon Bağış, Anna Baddeley, Irina Ban, Ali Başman, Remo Becci, Klime Beleski, Susan Bell, Amanda Blasko, Bruce Bordelon, Olivier Bourdet-Pees, Emmanuel Bourguignon, **Yılmaz Boz**, Dumitru Bratco, Jorge Brites, Gilbert Brochot, Franco Bronzat, Ernst Büscher, Christèle del Campo-Auberger, Marco Carassi, **Umay Çeviker**, Josef-Marie Chanton, Marlis Chanton, Juan Chavarri, Mark Chien, Peter Christensen, Lis Clément, Frank Cornelissen, Laura Costantini, Michael Cox, **Manna Crespan**, Jakob Cripp, Luis Cruz Carneiro, Oksana Deineko, Dénes Dienes, Adam Dijkstra, Ricardo Diogo V Freitas, Paul A Domoto, Janet Dorozynski, Pero Drljevic, J E Eiras Dias, **Rolando Faustino**, Carolina Ferreira, Matthew Fidelibus, Helen Fisher, Boris Gasparyan, Sami Ghosn, Caroline Gilby, Anna Giorgieva, Gianpoalo Girardi, Vazha Gochiridze, Robert Gorjak, Nami Goto-Yamamoto, Maria Stella Grando, Michel Grisard, Joaquim Guerra, Samuel Guibert, Natalie Guinovart, Luis Gutiérrez, Raymond Haak, Lars Hagerman, Stefano Haldemann, Philip Harmandjiev, Mark Hart, David Harvey, Jennifer Hashim, Paula Hegele, **Shigeyuki Hirayama**, Amy Hopkinson, Thomas Horan, Terry Hughes, Karim Hwaidak, Angelos Iatrides, **Serena Imazio**, Angel Ivanov, Gizella Jahnke, **Tim James**, Jaroslav Ježek, Michael Karam, Đurđa Katić, András Kató, Richard Kelley, Tina Kezeli, Malkaz Kharbedia, Erzsébet Kiss, Vesna Klikovac-Katanić, Yunus Emre Kocabaşoğlu, Katerina Kostovska, Stefanos Koundouras, Pál Kozma, Jakob Kripp, Pavel Krška, Thierry Lacombe, James Lawther, Konstantinos Lazarakis MW, Margarita Levieva, Paris Livadiotis, **Wink Lorch**, Virgilio Loureiro, **Mário Louro**, Bart Lyrarakis, Vasco Magalhaes, David Maghradze, Miroslav Majer, Aleksandar Makedonski, **Edi Maletić**, **Nico Manessis**, Vesna Maras, Neil Martin, María-Carmen Martínez, Liliana Martínez, Maria del Carmen Martínez Rodríguez, Antero Martins, Tim Martinson, Jose Luis Mateo García, Ivica Matošević, Richard Mayson, Georges Meekers, Levan Mekhuzla, Carole Meredith, Emilie Merienne, Susana Mestre, **Gabriella Mészáros**, Goran Milanov, Julio Molino, Antonella Monaco, Adam Montefiore, Teresa Mota, Linda Murphy, Rebecca Murphy, Ciprian Neascu, **Richard Nemes**, Igor Nykolyn, Daniel O'Donnell, Michael Oates-Warmer, Asli Odman, Taner Öğütoğlu, **Ken Ohashi**, Tariel Panahov, Jaroslava Kaňuchová Pátková, Luis Pato, Marko Pavlac, Vanda Pedroso, **Ivan Pejić**, Carolina Peralta, Ronny Persson, Mikhail Petkov, Tom Plocher, Dorota Pospíšilová, **Volodymyr Pukish**, **Carlos Quintas**, Jorge Quintas, Barbara Raifer, Zurab Ramazashvili, Doron Rav Hon, Bruce Reisch, Francisco Roig, Louisa Rose, René Rougier, Michel Rougier, Saario Meeri, Vugar Salimov, Dimitris Saltabassis, Vicky Scharlau, Edel Schaub Moss, David Schildknecht, Steffen Schindler, Michael Schmidt, Anna Schneider, Attilio Scienza, **Victor de la Serna**, Araceli Servera Ribas, Maria José Sevilla, William Shoemaker, Pierre de Sigoyer, Reva Singh, Derek Smedley MW, **Lisa Smiley**, Danilo Šnajder, Barbara Snider, Matos Soares, Gökhan Söylemezoğlu, **Walter Speller**, Susan Spence, Barbara Spinola, **Haroula Spinthiropoulou**, Saša Špiranec, Vlaho Srdjan Zitkovich, Susanne Staggl, Nataša Štajner, **Richard Stavek**, Paolo Storchi, Gary Strachan, Constantin Stratan, Ed Swanson, Charles Sydney, Paul Symington, József Szentesi, Michael Tabone, Shaoyun Tan, Sabrina Tedeschi, Giorgi Tevzadze, Étienne Thiebaud, Geoff Thorpe, Carolina Tonnelier, Marta Trabal, Jim Trezise, Max Troychuk, Vladimir Tsapelik, **Neil Tully MW**, Ognyan Tzvetanov, Levan Udjmadjuridze, Lorenzo Valenzuela, Luis Vaz, Rita Vignani, Mario Vingerhoets, Viacheslav Vlasov, Sophocles Vlassides, Ken Volk, Randall Vos, Roland Wahlquist, Rob Walker, **Fongyee Walker and the Dragon Pheonix team**, Hans Walter-Peterson, Gary Werner, Jeremy Wilkinson, Jim Wolpert, John Worontschak, Doug Wregg, Olivier Yobregat, **Jonathan Wurdeman**, **Jesús Yuste Bombín**, Goran Zdunić, Carlos Željko Bročilović, Jorge Zerolo, Anne-Dominique Zufferey, Eva Zwahlen.

A

ABBUOTO
ABOURIOU
ABRUSCO
ACITANA
ACOLON
ADAKARASI
ADALMIINA
ADDORACA
AFFENTHALER
AG ALDARA
AG KALAMBIR
AG MALAYI
AGDAM GYZYL UZUMU
AGIOMAVRITIKO
AGIORGITIKO
AGLIANICO
AGLIANICONE
AGNI
AGRONÓMICA
ÁGUA SANTA
AHUMAT
AÏDANI
AIRÉN
AKHTANAK
ALADASTURI
ALARIJE
ALBALONGA
ALBANA
ALBANELLA
ALBANELLO
ALBARANZEULI BIANCO
ALBARANZEULI NERO
ALBARÍN BLANCO
ALBAROLA
ALBAROSSA
ALBILLO DE ALBACETE
ALBILLO MAYOR
ALBILLO REAL
ALCAÑÓN
ALEATICO
ALEKSANDROULI
ALEXANDER
ALFROCHEIRO
ALICANTE HENRI BOUSCHET
ALIGOTÉ
ALIONZA
ALLEGRO
ALTESSE
ALVARELHÃO
ALVARINHO

⚯ this symbol on the following pages indicates that the synonyms, and mistaken identities, are confirmed by DNA evidence

AMARAL
AMIGNE
ANCELLOTTA
ANDRÉ
ANTÃO VAZ
ANTEY MAGARACHSKY
ARAKLINOS
ARAMON NOIR
ARANEL
ARANY SÁRFEHÉR
ARBANE
ARENI
ARETI
ARGAMAN
ARILLA
ARINARNOA
ARINTO DE BUCELAS
ARNA-GRNA
ARNEIS
ARNSBURGER
AROMATNY
ARRILOBA
ARROUYA
ARRUFIAC
ARVESINIADU
ARVINE

ASHUGHAJI
ASPIRAN BOUSCHET
ASSYRTIKO
ASURETULI SHAVI
ATHIRI
AUBIN BLANC
AUBIN VERT
AUBUN
AURELIUS
AURORE
AUXERROIS
AVANÀ
AVARENGO
AVASIRKHVA
AVESSO
AVGOUSTIATIS
AZAL

ABBUOTO

Rare, central Italian variety producing plenty of alcohol, generally blended.

BERRY COLOUR:

PRINCIPAL SYNONYMS: Aboto, Cecubo

ORIGINS AND PARENTAGE

Abbuoto comes from the Frosinone area of Lazio, central Italy. The synonym Cecubo derives from the erroneous belief that this was the variety that produced the famous Caecubum, one of the best wines in Italy in Roman times, mentioned by Horace, Pliny the Elder and so on. However, Caecubum was a white wine, the name of a cru south east of Roma, and it had disappeared by the second century AD.

Recent DNA parentage analysis has suggested that Abbuoto may be a natural PIEDIROSSO × CASAVECCHIA cross (Cipriani *et al.* 2010) but this is contradicted by the inclusion of Abbuoto in Casavecchia's parentage.

VITICULTURAL CHARACTERISTICS

Medium to large berries, thick skins, sometimes irregular yields, mid ripening.

WHERE IT'S GROWN AND WHAT ITS WINE TASTES LIKE

Abbuoto, grown mainly in north-west and central **Italy**, particularly in Fiuggi, Sessa Aurunca, Fondi, Formia, Monte San Biagio and Terracina in Lazio, produces generous wines generally high in alcohol and with good ageing potential but which tend quickly to lose their colour. The Campanian producer Villa Matilde has about 2 ha (5 acres) of Abbuoto, which is the main component in their Cecubo blend, with PIEDIROSSO and TRIBIDRAG, known as Primitivo in Italy. There were 717 ha (1,772 acres) in Italy in 2000.

ABOURIOU

Declining, early-ripening south-west French variety making tannic red.

BERRY COLOUR:

PRINCIPAL SYNONYMS: Beaujolais (Lot-et-Garonne), Early Burgundy (Napa Valley and Australia), Gamay Beaujolais (Puy-l'Évêque, Entre-Deux-Mers), Gamay-Saint-Laurent (Aveyron), Loubéjac (Dordogne), Malbec Argenté (Médoc), Plant Abouriou, Précoce Naugé (Lot-et-Garonne), Précoce Noir (Dordogne), Pressac de Bourgogne (around Libourne)

VARIETIES COMMONLY MISTAKEN FOR ABOURIOU: COT, GAMAY NOIR

ORIGINS AND PARENTAGE

Abouriou, probably from around Villeréal, was widespread in the Lot-et-Garonne until the mid nineteenth century. It had almost disappeared when it was rescued by Numa Naugé, a vine breeder in Sénézelles near Casseneuil (between Bordeaux and Toulouse), who presented it in 1882 as a chance seedling found by a local farmer some forty years earlier, growing up the wall of a ruined castle in the area of Villeréal (Lot-et-Garonne). That is why Abouriou, meaning 'early' in the local dialect, is also called Précoce Naugé.

Abouriou has no relationship with GAMAY NOIR but DNA profiling has recently discovered a possible parent–offspring relationship between Abouriou and Magdeleine Noire des Charentes, the mother of MERLOT and COT (Boursiquot *et al.* 2009). As a consequence, Abouriou is either a grandparent or a half-sibling of both Merlot and Cot (see PINOT pedigree diagram).

Abouriou was used to breed EDERENA and EGIODOLA and was crossed with CASTETS in the 1970s in Slovakia to breed HRON, NITRANKA, RIMAVA and VÁH.

VITICULTURAL CHARACTERISTICS

Early ripening and quite good resistance to botrytis bunch rot and other fungal diseases.

WHERE IT'S GROWN AND WHAT ITS WINE TASTES LIKE

Total plantings in **France** in 2008 were 338 ha (835 acres), about half what they were fifty years ago, mainly in the Lot-et-Garonne (200 ha/494 acres) and the Loire-Atlantique (100 ha/247 acres) *départements* and especially around Cocumont south west of Marmande. The wines are deeply coloured, spicy and tannic but lack acidity and the vine was valued mostly for its precocity and resistance to disease. The variety is authorized in the Côtes du Marmandais, where Elian Da Ros makes a varietal wine as well as including it in his Clos Baquey blend. Château Lassolle's Le Rouge qui Tache (meaning 'the red that stains') is another example and Abouriou is found in several south-western vins de pays.

The variety once grown in **California** under the name Early Burgundy was identified by Paul Truel as identical to Abouriou, and Luddite make a varietal version from the Gibson-Martinelli vineyard in Russian River Valley. However, recent DNA analysis has shown that some Early Burgundy vines in California are actually BLAUER PORTUGIESER (Mike Officer, personal communication).

ABROSTINE

See ABRUSCO.

ABRUSCO

Obscure and endangered late-ripening Italian red adding colour to blends.

BERRY COLOUR: ● ● ● ● ●

PRINCIPAL SYNONYMS: Abrostalo, Abrostine⌘, Abrostino, Abrusco Nero, Abrusio, Colore or Colorino, ?Lambrusco, ?Raverusto

VARIETIES COMMONLY MISTAKEN FOR ABRUSCO: COLORINO DEL VALDARNO⌘, Muscat Rouge de Madère⌘

ORIGINS AND PARENTAGE

Abrusco is an ancient Tuscan variety used in Chianti blends. An ampelographic and DNA study carried out in 2009 at the University of Siena has shown it to be identical to Abrostine (Rita Vignani, personal communication), contrary to what had previously been published by Vignani *et al.* (2008). Soderini's (1600) treatise on grapevine cultivation mentions Abrostino or Colore, used even then to add colour to pale reds. Abrusco is distinct from COLORINO DEL VALDARNO, another Tuscan grape valued for its deep colour that is often cultivated with, and mistaken for, Abrusco/Abrostine. In addition, some sources have suggested that Muscat Rouge de Madère (not commercially cultivated) could also be a synonym for Abrusco but this has been disproved by DNA analysis (Vouillamoz). The names Abrusco and Abrostine supposedly derive from the Latin *lambrusco*, meaning 'wild vine'.

VITICULTURAL CHARACTERISTICS

Mid ripening.

WHERE IT'S GROWN AND WHAT ITS WINE TASTES LIKE

Le Tre Stelle's Agino, an IGT Toscana, is one of the very few varietal Abrusco wines in production. Twenty Abrusco plants were discovered in their old vineyards and the producer has established a vine adoption scheme to help save the variety. The 2000 census recorded just 6 ha (15 acres) in **Italy** under the name Abrusco, and Abrostino is mentioned only (and incorrectly) as a synonym for COLORINO DEL VALDARNO. Federico Staderini's varietal wine is labelled Abrostine.

Although the berries have pale flesh, they produce deeply coloured, well-structured, slightly spicy wines that are usually blended.

ACITANA

Rare dark-skinned variety from north-east Sicilia used in blends.

BERRY COLOUR: ● ● ● ● ●

PRINCIPAL SYNONYMS: Citana Nera

ORIGINS AND PARENTAGE

The variety and the name probably come from a group of communes north-east of Catania in Sicilia with names beginning with 'Aci': Aci Bonaccorsi, Acireale, Aci Castello, Aci Catena, Aci Platani, Aci San Filippo, Aci Sant'Antonio, Aci Santa Lucia and Aci Trezza.

WHERE IT'S GROWN

Very rare variety formerly cultivated in the provinces of Messina and Catania in north-east Sicilia, **Italy**. Today it is grown around Messina, where it finds its way into Palari's Faro blend (though the Faro regulations do not specifically authorize it). It is generally blended with NERELLO MASCALESE, NERELLO CAPPUCCIO and other local varieties.

ACOLON

Recently authorized German cross favoured for its colour and early ripening.

BERRY COLOUR: ● ● ● ● ●

PRINCIPAL SYNONYMS: Weinsberg 71-816-102

ORIGINS AND PARENTAGE

A BLAUFRÄNKISCH × DORNFELDER cross created at the Weinsberg research centre in Baden-Württemberg in Germany by Bernd Hill in 1971 and included on the official register in 2002.

VITICULTURAL CHARACTERISTICS

Vigorous, yields well, ripens early, and the grapes are deeply coloured with good sugar levels.

WHERE IT'S GROWN AND WHAT ITS WINE TASTES LIKE

Despite its short life on the official list, there are already 478 ha (1,181 acres) in **Germany**, just under half in Württemberg and about a quarter in the Pfalz. The wines are fairly similar to those of its parent BLAUFRÄNKISCH (or Blauer Limberger): quite full-bodied with good fruit and moderate tannins. The more concentrated versions can benefit from restrained use of oak. Varietal wines are made by Medinger and Sturmfeder in Württemberg, Anselmann and Karl Dennhardt in the Pfalz, and Kochertalkellerei in Baden.

There are also nearly 3 ha (7 acres) in **Switzerland** and the variety has been planted in **England**, eg at Biddenden Vineyards in Kent.

ADAKARASI

Vine grown mainly on the Turkish island of Avşa, making soft, dark reds.

BERRY COLOUR: ● ● ● ● ●

PRINCIPAL SYNONYMS: Ada Karası

VARIETIES COMMONLY MISTAKEN FOR ADAKARASI: ÇALKARASI (Denizli), HOROZKARASI ⌘ (Kilis), KALECIK KARASI (Ankara), PAPAZKARASI (Avşa)

ORIGINS AND PARENTAGE

Adakarası, literally 'black of the islands', most likely originates from the Turkish island of Avşa in the Sea of Marmara.

OTHER HYPOTHESES

Although they have quite distinct cultivation areas, Adakarası, ÇALKARASI, HOROZKARASI, KALECIK KARASI and PAPAZKARASI are often thought to be one and the same variety. However, there is no convincing ampelographic evidence to date, and DNA analysis has shown that Adakarası and Horozkarası, at least, are distinct (Vouillamoz).

VITICULTURAL CHARACTERISTICS

Large berries with thick skins. A late-ripening variety that needs lots of sunshine. Productive but susceptible to powdery mildew. Produces its most aromatic wines in sandy-loam soils.

WHERE IT'S GROWN AND WHAT ITS WINE TASTES LIKE

Adakarası is found mainly on the island of Avşa in the Sea of Marmara, **Turkey**, where there were 92 ha (227 acres) in 2010, and also to a very limited extent around Balıkesir in the province of the same name, in that part of the Marmara region south of the Black Sea, and in the Şarköy district of the Tekirdağ province on the Sea of Marmara. Wines are generally soft and dark but still have good ageing potential. Producers of varietal wines include Bortaçına and Büyülübağ.

ADALMIINA

Minor but complex, cold-hardy American hybrid grown mostly in Québec.

BERRY COLOUR: ● ● ● ● ●

PRINCIPAL SYNONYMS: Aldemina, ES 6-16-30

ORIGINS AND PARENTAGE

An Elmer Swenson 2-3-17 × Elmer Swenson 35 hybrid created in the mid to late 1980s by grape breeder Elmer Swenson in Osceola, Wisconsin, US, in which:

- Elmer Swenson 2-3-17 is an Elmer Swenson 283 × Elmer Swenson 193 hybrid
- Elmer Swenson 283 is an Elmer Swenson 114 × SEYVAL BLANC hybrid (see ST PEPIN for the pedigree of Elmer Swenson 114)
- Elmer Swenson 193 is a Minnesota 78 × Seneca hybrid (see BRIANNA for the complete pedigree of Minnesota 78)
- Seneca is a LUGLIENGA × Ontario hybrid (see CAYUGA WHITE for the pedigree of Ontario)
- Elmer Swenson 35 is a Minnesota 78 × Dunkirk hybrid
- Dunkirk is a Brighton × Jefferson hybrid (see BRIANNA for the pedigree of Brighton)
- Jefferson is a CONCORD × Iona hybrid (see MOORE'S DIAMOND for the pedigree of Iona).

Adalmiina is therefore a complex *Vitis riparia*, *Vitis labrusca*, *Vitis vinifera*, *Vitis lincecumii* and *Vitis rupestris* hybrid. In 1997 the variety was sent to Meeri Saario, a horticulturist then at the University of Helsinki, who named it Adalmiina and began to propagate it along with several other Swenson varieties.

VITICULTURAL CHARACTERISTICS

Vigorous and productive, winter-hardy but not reliably so, early ripening. Good all-round disease resistance but prone to berry splitting at maturity.

WHERE IT'S GROWN AND WHAT ITS WINE TASTES LIKE

Adalmiina is grown to a limited extent in Québec, **Canada**, where it is generally used in blends to compensate for its low acidity. Producers of such blended wines include Le Chat Botté, Coteau St-Paul, Domaine Mont Vézeau and Vignoble Le Nordet. Varietal wines have no foxy flavour and are not unlike Muscadet except in their slight lack of acidity, sometimes also with mineral and tropical-fruit notes.

Meeri Saario and a few others in **Finland** and the Baltic States grow Adalmiina under plastic but for the table only.

ADDORACA

Extremely rare Calabrian white that contributes to an equally rare dessert wine.

BERRY COLOUR: ● ● ● ● ●

PRINCIPAL SYNONYMS: Odoacra

ORIGINS AND PARENTAGE

Addoraca comes from the province of Cosenza in Calabria, southern Italy. Its name means 'perfumed' in Calabrian dialect.

WHERE IT'S GROWN AND WHAT ITS WINE TASTES LIKE

Extremely rare variety cultivated exclusively in the commune of Saracena in the province of Cosenza in Calabria, **Italy**. It is blended with CODA DI VOLPE BIANCA (under its Calabrian synonym Guarnaccia Bianca), MALVASIA BIANCA DI CANDIA and MUSCAT BLANC À PETITS GRAINS (known locally as Moscatello di Saracena) in the historical Moscato di Saracena passito-style dessert wine made by a small number of producers such as Cantine Viola (supported by Slow Food) and Feudo dei Sanseverino.

AFFENTHALER

Very rare old variety from southern Germany.

BERRY COLOUR: ● ● ● ● ●

PRINCIPAL SYNONYMS: Blauer Affenthaler, Kleiner Trollinger, Säuerlicher Burgunder

ORIGINS AND PARENTAGE

Affenthaler is an old variety from Affental near Baden-Baden in Baden-Württemberg, southern Germany, recently resurrected from two old vines discovered in Kaisersbach and Steinheim in 2004. Affenthaler was first mentioned by Johann Michael Sommer in 1791 (Krämer 2006). It used to be cultivated along the Neckar, in Enztal, Remstal and Hohenlohe, together with BLAUFRÄNKISCH (aka Blauer Limberger) and SCHIAVA GROSSA (aka Trollinger). That's probably why Affenthaler is sometimes called Kleiner Trollinger.

Although it is also called Säuerlicher Burgunder (meaning 'acidic Pinot'), DNA parentage analysis has shown that Affenthaler is not related to PINOT, but has a parent–offspring relationship with GOUAIS BLANC (Boursiquot *et al.* 2004). Given Gouais Blanc's ancestry, Affenthaler is most likely a natural cross between Gouais Blanc and another, unknown (extinct?) variety.

VITICULTURAL CHARACTERISTICS

Late ripening, frost-resistant.

WHERE IT'S GROWN AND WHAT ITS WINE TASTES LIKE

Affenthaler's cultivation was almost abandoned after the Second World War and it has now become very rare even in its homeland of Baden-Württemberg in southern **Germany**. It produces dark, tannic wines with significant acidity. The Affenthaler Winzergenossenschaft Bühl in Baden make one example.

The variety should not be confused with a wine known as Affentaler or Affenthaler, a regional speciality of the area around Baden-Baden, made from Spätburgunder (aka PINOT NOIR).

AG ALDARA

Ag Aldara, meaning 'white of Aldara' and also known as Agaldere, Alani Khagokh or Aldara Spitak, is a mid- to late-ripening variety from **Armenia** or **Azerbaijan**, sometimes blended with BAYANSHIRA but also grown for the table. It is used for both dry and fortified port-style wines.

AG KALAMBIR

A white-berried variety from **Azerbaijan** used for grape juice and to make dry and dessert wines as well as a cognac-like spirit.

AG MALAYI

Light-skinned Azeri variety producing light, crisp wines in Nakhchivan.

BERRY COLOUR:

ORIGINS AND PARENTAGE

Ag Malayi is local to the Nakhchivan Autonomous Republic, an exclave of Azerbaijan bordering on Armenia, Iran and Turkey. It is not a colour mutation of ARENI, called Malayi in this region (Tariel Panahov, personal communication).

VITICULTURAL CHARACTERISTICS

Late ripening.

WHERE IT'S GROWN AND WHAT ITS WINE TASTES LIKE

Ag Malayi is used to produce dry white wines in **Azerbaijan** with relatively low alcohol levels.

AGDAM GYZYL UZUMU

A pink-berried variety from **Azerbaijan** used for grape juice and to make both dry and dessert whites.

AGIOMAVRITIKO

Almost extinct Greek island variety.

BERRY COLOUR: ● ● ● ● ●

ORIGINS AND PARENTAGE

Agiomavritiko may originate on the Greek island of Zákynthos, where it has traditionally been planted, but since DNA profiling has not yet been done, it is not known whether it is a unique variety or whether it matches any other Greek variety.

WHERE IT'S GROWN AND WHAT ITS WINE TASTES LIKE

Agiomavritiko is found on the island of Zákynthos in the Ionian Sea and in Larísa in Thessalía in the centre of mainland **Greece**. Comoutos on Zákynthos include it in one of their red blends.

AGIORGITIKO

Greece's most widely planted dark-skinned variety, capable of both high volumes and high quality as well as a wide range of styles.

BERRY COLOUR: ● ● ● ● ●

PRINCIPAL SYNONYMS: Aghiorghitiko, Mavro Nemeas (Neméa), Mavrostaphylo Mavraki, Mavroudi Nemeas, Nemeas Mavro

ORIGINS AND PARENTAGE

This variety probably comes from Argolída and Korinthía in the eastern part of the Pelopónnisos (Peloponnese) in Greece. Agiorgitiko, literally meaning 'St George's grape', was apparently named after the tiny chapel of St George in Neméa in the north-eastern part of the Pelopónnisos (Manessis 2000) or after a village of that name.

OTHER HYPOTHESES

Like many old Greek varieties, Agiorgitiko is said to have been cultivated since classical Greece but this highly doubtful hypothesis has not so far been confirmed by any historical or genetic evidence.

The name Agiorgitiko is often thought to refer to St George's Day, the date when the grapes are said to be ripe, but the saint's day falls far too early for that to be true – in April or May, depending on which calendar you use.

VITICULTURAL CHARACTERISTICS

Potentially high yielding, highly susceptible to virus diseases, and also to downy and powdery mildews and botrytis bunch rot; prone to potassium deficiency and water stress. Best suited to poor soils and relatively dense planting. Late budding and ripening. The better vines have small bunches of small, thick-skinned berries.

WHERE IT'S GROWN AND WHAT ITS WINE TASTES LIKE

Agiorgitiko is capable of making a wide range of styles in **Greece**, depending mainly on the vineyard location and the extent to which the yields are controlled – from fresh, fruity rosé to concentrated, tannic and ageworthy reds, as well as just about everything in between, including light-bodied versions made by carbonic maceration and large volumes of easy-drinking soft reds. The most serious examples are deeply coloured, have aromas of spicy red fruits, marry well with oak and are rich and full-bodied without excessive alcohol. Vines planted on the valley floor may sometimes produce wines that are very soft, lacking in acidity and high in alcohol, which is why the most highly regarded sites are at higher altitudes, eg on the hillsides in Koutsi at around 500–600 m (1,640–1,970 ft) or up to 900 m (2,950 ft) on the Asprokampos plateau. There is much current debate about how to incorporate some form of subregional classification into the existing legislation.

In recent years growers have been held back by the fact that virtually all Agiorgitiko vines planted in Greece are virused, causing problems with yield, ripening and quality. Current research should result in new clones being available in 2012, clones that not only are virus-free and therefore able to attain higher levels of sugar and phenolics in the berries but also show preferred characteristics such as smaller berries with thicker skins that will have greater resistance to botrytis bunch rot.

Agiorgitiko is the most widely planted red wine grape in Greece, found mainly in the Pelopónnisos (3,204 ha/7,917 acres) and in Attikí on the southern mainland (5,202 ha/12,854 acres). It is the sole variety authorized for the Neméa appellation in the north west of the Pelopónnisos – for reds (both dry and sweet) but not for rosé – and it has shown itself to be a good blending partner for CABERNET SAUVIGNON grown much further north in the Metsovo region in Ípeiros (Epirus) on the mainland. Recommended producers in Neméa include Driopi, Gaia, Parparoussis (based in Pátra but fruit from Gymno in Neméa) and Skouras. While Agiorgitiko and Neméa are virtually synonymous, the variety is also planted in much of the central and northern Pelopónnisos, in Attikí to the east and Makedonía in the north of the country. Biblia Chora make a very good example in this last region.

AGLIANICO

High-quality, late-ripening, tannic and ageworthy southern Italian red.

BERRY COLOUR: ● ● ● ● ●

PRINCIPAL SYNONYMS: Aglianco di Puglia, Aglianica, Aglianichella, Aglianichello (Napoli in Campania), Aglianico Amaro (Benevento and Caserta in Campania), Aglianico del Vulture (Basilicata), Aglianico di Castellaneta, Aglianico di Taurasi (Avellino in Campania), Aglianico Nero, Aglianico Pannarano, Aglianicuccia, Agliano, Agliatica, Agliatico, Agnanico, Agnanico di Castellaneta, Cascavoglia, Cerasole, Ellanico, Ellenico, Fiano Rosso, Fresella, Gagliano, Ghiandara, Ghianna, Ghiannara, Glianica, Gnanica, Gnanico, Granica, Olivella di S. Cosmo, Ruopolo, Spriema, Tringarulo, Uva Catellaneta, Uva dei Cani, Uva di Castellaneta, Uva Nera

VARIETIES COMMONLY MISTAKEN FOR AGLIANICO: Aglianico di Galluccio, AGLIANICONE, ALEATICO, CILIEGIOLO, PIGNOLO, TRONTO (under the name Aglianico di Napoli)

ORIGINS AND PARENTAGE

On the basis of arguable Greek etymology, many authors have suggested that Aglianico was introduced by the Greeks when they colonized the Tirrenian coast (seventh and sixth centuries BC). However, unlike the vast majority of sources, we do not support the hypothesis of a Greek origin for Aglianico, neither the name nor the grape.

The earliest written mention of the name Aglianico appeared in 1520: an archive document testifies that Giulio Antonio Acquaviva d'Aragona, Count of Conversano, owned fields planted with grapevines called Aglianiche (plural of Aglianico). The supposed Greek origin of Aglianico dates back to 1581, when the Neapolitan scholar Giambattista della Porta (1535–1615) tentatively identified the Helvola grape mentioned by Pliny the Elder with a hypothetical *hellanico* or hellenic grape. Subsequently, the shortcut Aglianico = *hellanico* = grape from Greece (*Hellas*) has been blindly assumed by most authors since the sixteenth century, which is revealed in the use of its modern synonyms Ellanico/Hellanico or Ellenico/Hellenico. However, this etymology is highly uncertain because:

- Helvola was a yellowish wine, while Aglianico has black berries.
- Linguists have established that it is impossible to obtain the word *aglianico* from *ellenico* or *hellenico*.
- In Pliny the Elder's and other Roman authors' works, the adjective used to indicate a Greek origin was *graecus* and never *hellenicus*, not least because this adjective simply didn't exist at this time.
- No Roman author ever used the name 'hellenic grapes', but always 'Greek grapes', because the concept of Hellenism came into use only at the Renaissance period to designate the Grecian realm or culture.

The name Aglianico most likely derives from the Spanish *llano*, meaning 'plain', since the Spanish occupied central Italy during the fifteenth and sixteenth centuries, a period consistent with the first appearance of the name Aglianico. In addition, the Spanish *lla* is pronounced 'glia' in Italian. Thus, Aglianico would be 'the grape of the plain', which brings to mind its old dialect name A Glianica or La Glianica.

Another less likely hypothesis suggests that the name Aglianico could come from the Greek *aglaos*, meaning 'beautiful' or 'bright', supposedly in contrast to other dark wines. Another even less likely hypothesis claims that the name Aglianico could come from the month of July (*Luglio* in Italian), when the grape ripens. However, this makes no sense given that Aglianico is a very late-ripening variety. Finally, little credibility is given to some attempts to derive Aglianico from Gauranum or Gauranicum, a type of Falerno wine described by Pliny the Elder and other Roman writers.

DNA profiling has shown that Aglianico is not closely related to any modern Greek grape variety (Scienza and Boselli 2003). It has also shown that Aglianico is closely related to many varieties cultivated in Campania or Basilicata (Costantini *et al.* 2005), in particular to AGLIANICONE, of which Aglianico could be a parent (Vouillamoz).

Neither linguistic nor genetic evidence supports a Greek etymology and origin for Aglianico. Therefore, it seems most likely that Aglianico is an ancient grape from southern Italy.

OTHER HYPOTHESES

Some authors speculated that Aglianico was the grape used for the famous Falerno wine of Roman times, and that it was domesticated from wild local grapes.

VITICULTURAL CHARACTERISTICS

Early budding and very late ripening (sometimes as late as November), grows vigorously and yields need to be controlled. Good resistance to powdery mildew but susceptible to botrytis bunch rot.

WHERE IT'S GROWN AND WHAT ITS WINE TASTES LIKE

Aglianico is grown almost exclusively in southern **Italy**, where it is the most widespread black variety, in particular in the provinces of Avellino and Benevento in Campania, and in the provinces of Potenza and Matera in Basilicata, where it thrives in cool but dry and sunny mountainous conditions between 200 and 600 m above sea level. There's also a scattering of vines in Calabria, Puglia, Molise, and on the island of Procida near Napoli. Total vineyard plantings in 2000 were around 9,890 ha (24,400 acres).

In the volcanic soils of Taurasi DOCG, north east of Avellino in Campania, and Aglianico del Vulture DOC, just over the border in Basilicata, the vine gives its best results: dark, concentrated and tannic wines that improve with age. The long ripening season is essential to tame these tannins. The best wines are deep in colour with aromas suggestive of the volcanic origin of the vineyards or of chocolate and plum. They have fine-grained tannins and marked acidity on the palate, becoming more subtle and tarry as they mature. The high levels of tannin and acidity have earned this variety the sobriquet 'Barolo of the south'. Taurasi may include up to 15% of other permitted local red varieties whereas Aglianico del Vulture and the more recent Aglianico del Taburno DOC must be 100% Aglianico. Best producers include d'Angelo, Galardi, Mastroberardino, Paternoster, Odoardi, Pietracupa and Villa Matilde.

Aglicanico's love of the southern Italian climate had encouraged a handful of growers in **Australia** concerned by climate change to begin experimental plantings. Westend Estate in Griffith, New South Wales (owned by the Calabria family, descended from Italian immigrants who arrived in the late 1920s), have been particularly successful, winning the trophy for best red wine in the 2010 Australian Alternative Varieties Wine Show. It is also found in **California**, where Kenneth Volk in Paso Robles, for example, makes an Italianate but more rounded version, and it has been planted in **China** by Grace Vineyard.

AGLIANICONE

This high-yielding but generally low-quality black-berried Campanian variety is cultivated both in Campania and in Matera and Potenza in Basilicata in **Italy**, where there were 140 ha (346 acres) in 2000. It has been erroneously identified with the old Tuscan variety CILIEGIOLO because the DNA comparison was based on the wrong plant (Crespan *et al.* 2002). Aglianicone is not a mutation of AGLIANICO (Costantini *et al.* 2005) but DNA paternity testing suggests it could be a close relative of it (Vouillamoz).

AGNI

Recent Czech cross producing unusually aromatic reds.

BERRY COLOUR: ● ● ● ● ●

PRINCIPAL SYNONYMS: (AN × IO) PE-11/47

ORIGINS AND PARENTAGE

Agni is an ANDRÉ × IRSAI OLIVÉR cross obtained recently by Jan Havlík, František Zatloukal and Ludvík Michlovský, grape breeders at the research centres of ŠSV in Velké Pavlovice and of Perná, both in southern Morava in the Czech Republic, and included in the official Czech register of varieties in 2001.

VITICULTURAL CHARACTERISTICS

Vigorous and early ripening. Small to medium-sized bunches of small berries.

WHERE IT'S GROWN AND WHAT ITS WINE TASTES LIKE

Agni produces full-bodied, fresh, aromatic wines with Muscat and rose-like aromas and red fruit flavours. It may have potential for dessert as well as dry red wines. Producers of varietal wines include František Foretník, Michlovský and Jakub Šamšula. There were 6 ha (15 acres) in Morava in 2009, the majority in Mikulovská in the south east of the **Czech Republic**.

AGRONÓMICA

Very minor, smoky-flavoured Portuguese cross found on the Açores but rarely grown on the Portuguese mainland.

BERRY COLOUR: ● ● ● ● ●

ORIGINS AND PARENTAGE

According to the Vitis International Variety Catalogue, a CASTELÃO × MUSCAT OF HAMBURG cross made in Portugal but this parentage is not consistent with the DNA profiles of Agronómica and Castelão in Veloso *et al.* (2010) and Muscat of Hamburg in several sources (Crespan *et al.* 1999; Gianetto *et al.* 2010; Santana *et al.* 2010, etc.).

VITICULTURAL CHARACTERISTICS

Early to mid ripening.

WHERE IT'S GROWN AND WHAT ITS WINE TASTES LIKE

Wines tend to be soft and have a strange, heavy, smoky aroma and taste. There were 320 ha (791 acres) of Agronómica in **Portugal** in 2010. It's authorized for IGP Açores and the Pico co-operative use it in a vinho de mesa red blend.

ÁGUA SANTA

Unexciting cross used mainly in blends in central Portugal.

BERRY COLOUR: ● ● ● ● ●

ORIGINS AND PARENTAGE

Água Santa, meaning 'holy water', is a CAMARATE (under the name Mortágua) × CASTELÃO (under the name João Santarém) cross obtained in 1948 by José Leão Ferreira de Almeida at the national agronomy research centre in Oeiras, east of Lisboa in Portugal, as confirmed by DNA profiling (Zinelabidine *et al.* 2012). Água Santa is therefore a 'double grandchild' of CAYETANA BLANCA and ALFROCHEIRO (see the pedigree diagram in Cayetana Blanca).

OTHER HYPOTHESES

Água Santa is sometimes said to be a TRINCADEIRA × CASTELÃO cross but the DNA profile of Trincadeira (Almadanim *et al.* 2004; Veloso *et al.* 2010) is not consistent with this parentage.

VITICULTURAL CHARACTERISTICS

Vigorous, productive, mid budding and ripening. Large, loose bunches of big berries. Susceptible to powdery mildew.

WHERE IT'S GROWN AND WHAT ITS WINE TASTES LIKE

Água Santa is found mainly in the Bairrada and Lisboa regions of **Portugal**, producing low-acid, medium-bodied, slightly astringent wines with quite high alcohol that generally ends up in blends. There were said to be 133 ha (329 acres) in Portugal in 2010 but the variety is in decline.

AHMET BEY

See HASANDEDE.

AHUMAT

Virtually extinct Sauvignon Blanc-like vine from south-west France.

BERRY COLOUR: ● ● ● ● ●

VARIETIES COMMONLY MISTAKEN FOR AHUMAT: SAUVIGNON BLANC

ORIGINS AND PARENTAGE

Ahumat is an old variety from the Jurançon area in south-west France, where isolated vines are still scattered in the vineyards. The name Ahumat means 'smoked out' or 'smoky' in Béarnese (a Gascon dialect spoken in Béarn), in reference to the white bloom on the berries. Ahumat is often considered to be identical to the grey-berried colour mutation of SAUVIGNON BLANC but this has been disproved by DNA profiling (Bordenave *et al.* 2007). However, it is morphologically close to CAMARALET DE LASSEUBE (Bisson 2009).

VITICULTURAL CHARACTERISTICS

Buds and ripens early and is therefore at risk from spring frosts. Small, compact bunches.

WHERE IT'S GROWN AND WHAT ITS WINE TASTES LIKE

This variety, once trained up fruit trees rather than on a trellis, is local to Jurançon and Madiran, **France**, and has virtually disappeared, not helped by the fact that it has never been included in the official French register. Some vines have been found in pre-phylloxera vineyards in the village of Viella in the Gers. Producteurs Plaimont's study of this variety at their Saint-Mont nursery suggests it has some similarities to SAUVIGNON BLANC, though it is more robust.

AÏDANI

Minor, aromatic Greek variety that plays a small but significant role in the wines of Santoríni.

BERRY COLOUR: ● ● ● ● ●

PRINCIPAL SYNONYMS: Adani, Aedano Leyko, Aïdani Aspro, Aïdani Blanc, Aïdani Lefko, Aspaedano, Aspraïdano, Moschaïdano

ORIGINS AND PARENTAGE

Aïdani or Aïdani Aspro (*aspro* means 'white') comes from the Kykládes (Cyclades) islands in Greece, possibly from Santoríni, where it has traditionally been cultivated (Galet 2000).

Contrary to local belief, recent genetic investigations have suggested that the black-berried Aïdani Mavro (*mavro* means 'black') is not a distinct variety but simply a colour mutation of Aïdani Aspro (or vice versa; Biniari and Stavrakakis 2007). This dark-skinned mutation is referred to by various synonyms such as Aedano Mavro, Aïdani Noir, Aïdano Mavro, Ithani Mavro, Mavraïdano, Mayraedano and Santorin.

OTHER HYPOTHESES

Aïdani is said by some to originate in the city of Adana in south-eastern Turkey (Manessis 2000; Nikolau and Michos 2004) but there is no genetic evidence to confirm this hypothesis. It is also said to have been an ingredient in the blended historical wines Aperanthitis, Malvasia and Naxos (Nikolau and Michos 2004).

VITICULTURAL CHARACTERISTICS

Drought-resistant but susceptible to powdery and downy mildews. High yielding, late budding and ripening. Large bunches of thick-skinned berries.

WHERE IT'S GROWN AND WHAT ITS WINE TASTES LIKE

Aïdani of **Greece** produces wines marked by floral and sometimes tropical-fruit aromas but it is rarely found in a starring role. Lower in alcohol and acidity than ASSYRTIKO, its most common use is as a minor player, along with ATHIRI, to Assyrtiko's lead in the dry, steely, volcanic whites of Santoríni, although it is also found on the smaller neighbouring islands of Páros, Náxos and Amorgós. It is also blended with Assyrtiko to produce Vinsanto, at its best – in the hands of producers such as Sigalas and Hatzidakis – one of the most concentrated and exotically complex styles of sweet wine made from grapes harvested late once they have started to shrivel and then further dried off the vine. Argyros Estate on Santoríni make a rare dry varietal Aïdani using cryoextraction to concentrate the juice.

There are very limited plantings of Aïdani Mavro on the island of Páros, where Moraitis blend it with CABERNET SAUVIGNON or MANDILARIA. It is also found on neighbouring Náxos and Santoríni and was traditionally used to make a sweet wine known as Liasto (Manessis 2000), meaning 'sun dried'.

AÏDANI MAVRO

See AÏDANI.

AIRÉN

Spain's most widely planted light-skinned variety making brandy and rather neutral, fresh white wines.

BERRY COLOUR: ● ● ● ● ●

PRINCIPAL SYNONYMS: Aidén (Albacete), Blancón (Zamora), Burra Blanca (Islas Canarias), Colgadera (Toro), Forcallada, Forcallat or Forcallat Blanca (Catalunya), Manchega (Albacete), Valdepeñera Blanca or Valdepeñas (Ciudad Real)

VARIETIES COMMONLY MISTAKEN FOR AIRÉN: LAIRÉN (Córdoba in Andalucía)

ORIGINS AND PARENTAGE

Airén is a very old variety from the region of Cuenca in Castilla-La Mancha in central Spain, as suggested by the synonym Manchega.

The name Forcallat Blanca is often used as a synonym for Airén, especially in Catalunya, but it is also the name of a distinct variety from Valencia that is no longer cultivated commercially but has a unique DNA profile (Ibañez *et al.* 2003), which means that it is not a white mutation of FORCALLAT TINTA from Castilla-La Mancha. It is nevertheless authorized in Vino de la Tierra Murcia in south-east Spain.

Airén has often been confused with the variety called LAIRÉN in Andalucía that was already mentioned by the Spanish agronomist Gabriel Alonso de Herrera (1470–1539) in 1513 in his description of varieties of Castilla, Extremadura and Andalucía (Alonso de Herrera 1790), but DNA profiling has shown that Lairén and Airén have distinct DNA profiles (Ibañez *et al.* 2003; Santana *et al.* 2010).

OTHER HYPOTHESES

In Italy, DNA studies suggested a close genetic relationship between CLAIRETTE, Airén from Spain and RODITIS from Greece (Labra *et al.* 1999) but this is highly questionable (see INZOLIA).

VITICULTURAL CHARACTERISTICS

Big bunches of medium-sized berries. Very late budding and ripening. Very high fertility and yields. Good drought resistance and well suited to poor soils. Good general resistance to pests and diseases.

WHERE IT'S GROWN AND WHAT ITS WINE TASTES LIKE

A staggering total of 284,623 ha (703,319 acres) of **Spain** was planted with Airén in 2008, almost 26% of the national vineyard. This means that the variety is grown on more land than any vine variety in the world thanks to the low-density bushvine plantings in arid La Mancha, where the variety's drought resistance is a huge advantage and probably its saving grace. Nevertheless, recent vine-pull schemes aimed at reducing the European wine surplus have resulted in a reduction in the overall area – down from 305,000 ha (753,671 acres) in 2004. The vast majority of the vines (276,796 ha/683,978 acres) are in Castilla-La Mancha but there are also plantings in the autonomous communities of Madrid, Murcia and Andalucía.

Airén plays a major role in the Spanish brandy industry and was traditionally blended with the red variety Cencibel (TEMPRANILLO) to produce light red wines in Valdepeñas and La Mancha, for example. Today, however, it is increasingly vinified as a rather neutral, crisp dry white, playing a similar role to France's Ugni Blanc (TREBBIANO TOSCANO). It also blends well with MACABEO and Malvar (LAIRÉN). Producers who take the variety reasonably seriously include Félix Solís and José María Galán León in Valdepeñas; Bodegas 1890, Huertas and Pedro Luis Martínez in Jumilla; and Evaristo Mateos, Naranjo and Santa Rita in La Mancha. The first white wine to carry the name Pesquera, Alejandro Fernández' famous Ribera del Duero bodega, was Alejairén, an oak-aged 100% Airén from La Mancha.

It is difficult to see why anyone would want to plant this variety outside Spain.

AKHTANAK

Versatile, late-ripening Armenian complex cross.

BERRY COLOUR: ● ● ● ● ●

PRINCIPAL SYNONYMS: 2-18-23, Akchtanak, Haghtanak, Hakhtanak

ORIGINS AND PARENTAGE

Akhtanak is a complex cross obtained in 1977 by P K Aivazyan at the Armenian viticultural research centre in Merdzavan, just west of Armenia's capital Yerevan, by crossing Sorok Lyet Oktyabrya with SAPERAVI, in which Sorok Lyet Oktyabrya is a Kopchak × ALICANTE HENRI

BOUSCHET cross, and Kopchak is a Moldovan variety that is no longer cultivated. The synonym Haghtanak means 'victory'.

VITICULTURAL CHARACTERISTICS

Very productive and very late ripening.

WHERE IT'S GROWN AND WHAT ITS WINE TASTES LIKE

Akhtanak, one of **Armenia**'s most promising new varieties, is grown in the Ararat region in west-central Armenia and is used to produce a wide range of wine styles – dry, sweet, fortified and sparkling – as well as being grown for the table. Producers of semi-sweet, sweet and fortified wines include the Ijevan Wine Factory in Tavush and VAN 777 in Ararat.

ALADASTURI

Minor Georgian variety responsible for alcoholic reds.

BERRY COLOUR:

PRINCIPAL SYNONYMS: Aladastouri, Anadassaouli, Anadastouri

ORIGINS AND PARENTAGE

Aladasturi originates from the Guria region on the Black Sea in western Georgia.

VITICULTURAL CHARACTERISTICS

Large, relatively compact bunches of large to medium-sized berries. Vigorous, early budding and late ripening. Susceptible to powdery mildew.

WHERE IT'S GROWN AND WHAT ITS WINE TASTES LIKE

Aladasturi is planted mainly in Imereti in central western **Georgia**, producing wines that are typically high in alcohol and moderate in colour, often made as a rosé. Khareba produce a varietal example although the variety is often blended. There were 44 ha (109 acres) in Georgia in 2004.

ALARIJE

Light-skinned variety from south-west Spain travelling under many aliases.

BERRY COLOUR:

PRINCIPAL SYNONYMS: Acería (Badajoz), Alarije Dorada or Alarije Dorado or Alarije Verdoso (Cáceres), Arín (Córdoba), Arís (Guadalajara), Barcelonés (Tarragona), Cagazal (Navarra), Coloraillo Gordo or Coloraillo Pequeno (Campillo de Altobuey in Cuenca), Esclafacherri (Albacete), Malfar, Malvasía (Calatayud, Navarra and Tarragona), Malvasía Riojana (Álava, La Rioja, Navarra), Pirulés Dorada and Pirulés Verde (Ribera del Duero), Rojal (La Rioja), Subirat (Tarragona), Subirat Parent (Catalunya and Tarragona)

ORIGINS AND PARENTAGE

Alarije is a very old variety from Extremadura in south-west Spain, where it was first mentioned in 1448 in the manuscript *Libro de Oficios* of the Real Monasterio de Santa María de Guadalupe (Asensio Sánchez 2000) as the main white variety in the area. In 1513, Gabriel Alonso de Herrera described this variety under the synonym Alarize, using the plural Alarixes (Alonso de Herrera 1790): 'Las Alarixes son uvas que hacen las cepas altas, a manera de las Albillas ... son uvas muy bermejas, y que las abejas las comen mucho' ('The Alarixes are grapes that make tall vines, like the Albillos ... the grapes are reddish and the bees eat them a lot').

Although Alarije has several synonyms based on the name Malvasia, it is genetically quite distinct from the true Malvasias grown in other Spanish regions: MALVASIA DI LIPARI (under the names Malvasía de Sitges and Malvasía Rosada) and MALVASÍA DE LANZAROTE (Rodríguez-Torres *et al.* 2009).

In Madrid, Alarije is sometimes called Torrontés or Turrontés, which is very confusing because DNA profiling has shown that, in Portugal, Torrontés is also the name for Arinto (ARINTO DE BUCELAS), DONA BRANCA, FERNÃO PIRES and MALVASIA FINA (under the names Assario and Boal Branco; Lopes *et al.* 1999; Pinto-Carnide *et al.* 2003; Gago *et al.* 2009; Vouillamoz). The risk of confusion is further increased because TORRONTÉS MENDOCINO, TORRONTÉS RIOJANO and TORRONTÉS SANJUANINO are distinct varieties in South America.

DNA parentage analysis has recently shown that Alarije (under the synonym Subirat Parent) is the result of a cross between two Spanish table grapes: GIBI, once cultivated in Spain and the south of France, and PLANTA NOVA (under the synonym Tortozón) from the Valencia region (Lacombe *et al.* 2007). Since Gibi is also the likely mother of PEDRO XIMÉNEZ (Vargas *et al.* 2007), Alarije and Pedro Ximénez turn out to be half-siblings.

OTHER HYPOTHESES

Some authors as well as growers from Cañamero (in the province of Cáceres, Extremadura) consider Alarije Dorado (*golden*) and Alarije Verdoso (*greenish*) to be two distinct varieties but isoenzymatic profiling and DNA studies clearly show that they are clonal variations of Alarije (Asensio Sánchez 2000).

VITICULTURAL CHARACTERISTICS

Big compact bunches of small to medium-sized, thick-skinned berries. Mid to late budding and late ripening. Very productive and susceptible to fungal diseases.

WHERE IT'S GROWN AND WHAT ITS WINE TASTES LIKE

Alarije is recommended and grown in many parts of Extremadura in south-west **Spain** but especially around Cañamero in the north east of the region. It is authorized in the Ribera del Guadiana appellation but is less widely planted than Pardina (CAYETANA BLANCA). Wines are yellowish green and have a propensity to oxidize so that they become golden as they age. They seem well suited to oak but are also used to make sweet and fortified wines. Official Spanish statistics do not seem to have caught up with DNA findings and record 30 ha (74 acres) of Alarije in Castilla-La Mancha (none in Extremadura) and 90 ha (222 acres) of Subirat Parent in Catalunya. According to Rioja's Abel Mendoza, possibly the only producer of a varietal Malvasía Riojana in the region, there are about 70 ha (173 acres) of Malvasía Riojana in Rioja but few producers pay it any attention. Abel Mendoza's is a golden, rich yet fresh, barrel-fermented version. In Catalunya, Jané Ventura includes Subirat Parent in his white Penedès blend, along with XARELLO and other varieties, and it may be used in the production of Cava.

ALBALONGA

Fresh and aromatically complex German cross planted and valued mainly in Rheinhessen.

BERRY COLOUR:

PRINCIPAL SYNONYMS: Würzburg B 51-2-1

ORIGINS AND PARENTAGE

A RIESLANER × SILVANER cross, initially called Würzburg B 51-2-1, obtained in 1951 by Hans Breider at the Würzburg horticulture institute and included on the official register in 1971. Its name means 'long white', in reference to the shape of the bunches. Albalonga is genetically very similar to Silvaner since the latter appears twice in its pedigree.

VITICULTURAL CHARACTERISTICS

Good level of acidity. Small to medium-sized berries. Yields must be restricted to achieve quality. Generally late ripening and suited to late-harvest or botrytized wines.

WHERE IT'S GROWN AND WHAT ITS WINE TASTES LIKE

Germany has a total of 13 ha (32 acres), mostly in Rheinhessen and the Pfalz. The wines range from medium to very sweet, retaining good acidity and tasting of exotic fruits. Sometimes the aromas are more floral and black-fruited. Wittmann in Westhofen, Rheinhessen, have championed the variety and regularly produce an impressive TBA version from grapes that ripen early. Krebs-Grode, also in Rheinhessen, produce another good varietal example.

It is also planted to a very limited extent in **England**, at Bothy Vineyard in Oxfordshire.

ALBANA

Ancient but inherently neutral Italian variety that can produce very good sweet wines.

BERRY COLOUR:

PRINCIPAL SYNONYMS: Albana a Grappolo Fitto, Albana a Grappolo Lungo, Albana della Forcella, Albana di Romagna, Albana Gentile, Albana Grossa, Albanone, Ribona, Riminèse🧬 (Corse), Sforcella

ORIGINS AND PARENTAGE

From Emilia-Romagna in central Italy, including the provinces of Bologna, Ravenna and Forlì-Cesena, this variety has long been widely cultivated in different places under various names, as suggested by its long list of synonyms. The earliest mention of Albana is the Italian jurist and oenophile Pietro de Crescenzi's reference to it in *c.*1305 as a white-berried variety producing powerful Romagna wine, the best being made in Forlì. The name Albana might derive from the Alban Hills (Colli Albani), a volcanic hill complex in Lazio, south of Roma. This etymology seems more likely than the simplistic derivation from the Latin word *alba*, meaning 'white'. Recent DNA profiling has confirmed that the variety cultivated on the French island of Corse under the name Riminèse is Albana and parentage analysis has shown that Albana has a parent–offspring relationship with GARGANEGA, one of the most widespread and ancient white varieties of Italy (Di Vecchi Staraz, This *et al.* 2007; Crespan, Calò *et al.* 2008). Since Garganega also has parent–offspring relationships with eight other varieties (CATARRATTO BIANCO, DORONA DI VENEZIA, MALVASIA BIANCA DI CANDIA, MARZEMINA BIANCA, MONTONICO BIANCO, MOSTOSA, SUSUMANIELLO and TREBBIANO TOSCANO), Albana is either their half-sibling or their grandparent (see Garganega for an illustrated pedigree).

Albana has several clones, the most widespread ones being Albana Gentile di Bertinoro, followed by Albana della Serra or Albana della Forcella, Albana della Compadrana, Albana della Bagarona and Albana della Gaiana. However, some of these may be the result of seed propagation, which means they are not true Albana vines but offspring of Albana. DNA profiling has not yet been applied to these clones.

OTHER HYPOTHESES

Some authors have suggested that Albana already existed in Roman times and was introduced to Emilia-Romagna by the Romans. There is no evidence to support this.

VITICULTURAL CHARACTERISTICS

Generally vigorous, susceptible to botrytis bunch rot, and needing a good supply of water to ripen fully. Ripens late.

WHERE IT'S GROWN AND WHAT ITS WINE TASTES LIKE

According to the agricultural census in **Italy**, in 2000 total plantings were 2,800 ha (6,900 acres), mainly in Emilia-Romagna but also in La Spezia (Liguria) and Mantova (Lombardia), though the area has declined considerably since then. This has led the town of Imola in the province of Bologna to initiate a conservation programme involving the evaluation and promotion of old clones. The best Albana comes mainly from the red clay soils of the hills between Faenza and the River Ronco to the east of Forlì, together with the calcareous subzone of Bertinoro.

Albana doesn't have that much flavour but its thin yet resistant skins make it a ready candidate for drying and it is used to produce wines of varying degrees of sweetness, from dry to a sweet passito version made from grapes dried on or off the vine. The sweet wines are the most promising and some producers have experimented with botrytized and barrel-aged versions. Fattoria Zerbina's Scaccomatto is a good example of the passito style. There's also a sparkling Albana Spumante. Albana di Romagna was the first Italian white wine to be promoted to DOCG status, in 1986 – a surprise to all but close students of vini-politics.

Riminèse has virtually disappeared from the French island of Corse but it is still included in Comte Abbatucci's Cuvée du Général blend.

ALBANELLA

Dry, ageworthy whites.

BERRY COLOUR: ● ● ● ● ●

VARIETIES COMMONLY MISTAKEN FOR ALBANELLA: ALBANELLO (Sicilia), TREBBIANO TOSCANO

ORIGINS AND PARENTAGE

From the Marche in central Italy, Albanella is similar to but distinct from ALBANELLO from Siracusa (Sicilia). Albanella has long been confused with TREBBIANO TOSCANO but recent ampelographic and genetic analyses have shown that they are distinct (Attilio Scienza and Serena Imazio, personal communication). It is also morphologically similar to GRECO BIANCO and possible relationships with ALBARANZEULI BIANCO from Sardegna and, more questionably, with ELBLING from Germany have been suggested.

WHERE IT'S GROWN AND WHAT ITS WINE TASTES LIKE

Once much more widespread in the Marche, Albanella is now confined to the Pesaro region of **Italy**, where it is a principal component in the Colli Pesaresi Bianco DOC. Albanella produces dry whites with good ageing potential, especially at Fattoria Mancini.

ALBANELLO

Disappearing Sicilian variety once used for sweet wines.

BERRY COLOUR: ● ● ● ● ●

PRINCIPAL SYNONYMS: Albanello di Siracusa

VARIETIES COMMONLY MISTAKEN FOR ALBANELLO: ALBANELLA (Marche)

ORIGINS AND PARENTAGE

Albanello is one the most ancient white wine varieties in Sicilia, historically cultivated in the provinces of Siracusa, Ragusa, Caltanissetta and Catania. Its sweet wine, obtained from grapes sun-dried on mats for several days, was already rare and expensive in the eighteenth century. Albanello is similar to but distinct from ALBANELLA from Pesaro (in the Marche, Italy) and is also morphologically similar to ALBARANZEULI BIANCO from Sardegna.

WHERE IT'S GROWN AND WHAT ITS WINE TASTES LIKE

The late-ripening Albanello is slowly disappearing but in 2000 was still cultivated in **Italy** on 125 ha (309 acres) in the Sicilian provinces of Siracusa, Caltanissetta and Enna, where it was once used to produce Ambrato di Comiso, a Marsala-like wine abandoned many years ago (Alla 2003). Albanello is part of the light dry blend that makes up Eloro Bianco DOC, together with CATARRATTO BIANCO, GRILLO, etc.

ALBARANZEULI BIANCO

Virtually extinct white grape from Sardegna.

BERRY COLOUR: ● ● ● ● ●

PRINCIPAL SYNONYMS: Laconari, Lacconargiu or Lacconarzu

VARIETIES COMMONLY MISTAKEN FOR ALBARANZEULI BIANCO: ALBILLO REAL (Spain)

ORIGINS AND PARENTAGE

Like many Sardinian varieties, Albaranzeuli Bianco may have been introduced from Spain during Spanish rule of the island (1323–1720), as suggested by its Spanish-sounding name and the confusion with ALBILLO REAL. However, its Spanish origin has recently been challenged: Cipriani *et al.* (2010) have suggested through DNA parentage analysis that Albaranzeuli Bianco is a possible natural cross between GIRÒ and the table grape Molinera (under the name Panse Rosa di Málaga). This parentage showed some DNA discrepancies and needs further investigation. A possible genetic link between Albaranzeuli Bianco and ALBARANZEULI NERO as well as ALBANELLA has been suggested but not yet confirmed.

VITICULTURAL CHARACTERISTICS

High yields, resistant to fungal diseases, late ripening.

WHERE IT'S GROWN AND WHAT ITS WINE TASTES LIKE

Albaranzeuli Bianco is on the brink of extinction in **Italy** and is cultivated only with other local varieties in old vineyards in the Nuoro and Oristano provinces of Sardegna, where it is called Laconari, Lacconargiu or Lacconarzu. Usually blended with other varieties, Albaranzeuli Bianco is authorized in several IGT wines (dry and sparkling). There were just 75 ha (185 acres) remaining in 2000.

ALBARANZEULI NERO

A pinky-red-skinned Sardinian variety still very occasionally used in blends.

BERRY COLOUR: ● ● ● ● ●

PRINCIPAL SYNONYMS: Albarenzelin Nero, Uva Melone

ORIGINS AND PARENTAGE

Like many Sardinian varieties, Albaranzeuli Nero may have been introduced from Spain during Spanish rule (1323–1720), as suggested by its Spanish-sounding name. A possible genetic link with the white ALBARANZEULI BIANCO has not yet been studied.

OTHER HYPOTHESES

For some authors, Albaranzeuli Nero is similar to GIRÒ.

VITICULTURAL CHARACTERISTICS

Mid budding, late ripening. Medium-sized, thin-skinned berries.

WHERE IT'S GROWN AND WHAT ITS WINE TASTES LIKE

Recommended in the province of Nuoro but it is little planted and always blended with other varieties. It is mostly used to produce rosé wines and has typically moderate levels of alcohol and acidity. In 2000 there were just 42 ha (104 acres) left in **Italy**.

ALBARÍN BLANCO

Very minor variety from the far north of Spain.

BERRY COLOUR: ● ● ● ● ●

PRINCIPAL SYNONYMS: Blanco Legítimo, Blanco País (Galicia), Blanco Verdín (Galicia), Branco Lexítimo (Betanzos in north-west Spain), Raposo (Galicia), Tinta Fina (Valles de Benavente in Zamora)

VARIETIES COMMONLY MISTAKEN FOR ALBARÍN BLANCO: ALBILLO REAL, ALVARINHO, SAVAGNIN BLANC

ORIGINS AND PARENTAGE

Albarín Blanco is a traditional variety from Asturias in the far north west of Spain, also known in Galicia in the north west under various synonyms. Comparison of the DNA profiles of Albarín Blanco (Martín *et al.* 2006; Gago *et al.* 2009; Santana *et al.* 2010) and Blanco Lexítimo (Vilanova *et al.* 2009) from Betanzos in north-western Spain shows that they are identical varieties but the name Blanco Lexítimo, or Branco Lexítimo in Galician, is given to more than one variety in Galicia (Jesús Yuste Bombín, personal communication).

DNA parentage analysis has suggested that Albarín Blanco and Budelho (also known as Pedro Ximenes Canario but no longer cultivated) are siblings (Santana *et al.* 2010).

Albarín Blanco has been confused with Albariño (ALVARINHO; Galet 2000) and with ALBILLO REAL but DNA profiling has shown them to be distinct varieties (Ibañez *et al.* 2003). In addition, Albarín Blanco should not be confused with ELBLING or GOUAIS BLANC, even though all three have the synonym Blanco Verdin in the Vitis International Variety Catalogue. Like Albariño, Albarín Blanco has been and still is mistaken for SAVAGNIN BLANC (introduced to Galicia after the phylloxera epidemic in the late nineteenth century) due to a labelling or identification error in Misión Biológica de Galicia's collection (Santiago *et al.* 2005).

Albarín Negro from the Asturias is not a colour mutation of Albarín Blanco but rather a synonym for ALFROCHEIRO from Portugal (González-Andrés *et al.* 2007).

VITICULTURAL CHARACTERISTICS

Early ripening, good resistance to downy mildew. Medium to small bunches of thick-skinned, medium-sized berries.

WHERE IT'S GROWN AND WHAT ITS WINE TASTES LIKE

Albarín Blanco is still planted to a limited extent in Asturias in north-west **Spain**. It is provisionally included in the official Spanish register of commercial varieties and recommended in Vino de la Tierra de Cangas. There were about 40 ha (99 acres) around Cangas del Narcea in 2008. Wines can be floral and spicy like Moscatel but the acidity is higher and there are fewer phenolics in the skins. Pardevalles make a varietal version in Tierra de León, where Albarín Blanco is recommended, and Gordonzello, in the same region, made their first varietal wine in 2011.

Under the Galician name Branco Lexítimo, the variety is authorized in Betanzos but it tends to be used in blended bulk wines for strictly local consumption.

ALBARIÑO

See ALVARINHO.

ALBAROLA

Neutral white grape of Liguria.

BERRY COLOUR: ● ● ● ● ●

PRINCIPAL SYNONYMS: Bianchetta Genovese, Calcatella

VARIETIES COMMONLY MISTAKEN FOR ALBAROLA: BIANCHETTA TREVIGIANA, SCIMISCIÀ (Genova in Liguria)

ORIGINS AND PARENTAGE

Several authors locate its origin in Val Polcevera on the Ligurian coast west of Genova in north-west Italy. Morphological and DNA studies have shown that Bianchetta Genovese is identical to Albarola, and that in the region of Lavagna a mysterious and local Albarola has a unique DNA profile (Botta, Scott *et al.* 1995; Torello Marinoni, Raimondi, Ruffa *et al.* 2009).

WHERE IT'S GROWN AND WHAT ITS WINE TASTES LIKE

This late-ripening variety is called Bianchetta Genovese in the neighbourhood of Genova and Albarola along the rest of **Italy**'s Ligurian coast. Albarola used to be widely cultivated from Liguria all the way down to Toscana, for table grapes as well as for wine, but is now found only on the Ligurian coast, where it is blended with BOSCO and VERMENTINO in the highly sought-after Cinque Terre Sciacchetrà dessert wine, such as that made by Possa. It is the major grape in Bianchetta del Golfo di Tigullio DOC (Vini Bisson produce a varietal version) and it is authorized in DOCs such as Cinque Terre, Candia dei Colli Apuani and Colline di Levanto. Total vineyard area was 470 ha (1,160 acres) in 2000.

ALBAROSSA

Minor Piemontese dark-skinned cross recently replanted by Michele Chiarlo.

BERRY COLOUR: ● ● ● ● ●

PRINCIPAL SYNONYMS: Incrocio Dalmasso XV/31

ORIGINS AND PARENTAGE

From the Veneto in northern Italy. Italian vine breeder Giovanni Dalmasso claimed that Albarossa was a NEBBIOLO × BARBERA cross made in 1938 at the Conegliano research centre with the aim of creating a variety that combined the quality of Nebbiolo with the resistance and high yields of Barbera. However, DNA profiling has recently shown that the Nebbiolo used by Dalmasso was in fact Nebbiolo di Dronero, which is not Nebbiolo at all but a synonym for CHATUS (Torello Marinoni, Raimondi, Ruffa *et al.* 2009). Albarossa is therefore a Chatus × Barbera cross, just like its siblings CORNAREA, NEBBIERA, SAN MICHELE and SOPERGA. Albarossa entered the Italian national register of varieties in 1977.

VITICULTURAL CHARACTERISTICS

Likes poor, dry chalky soils on steep sunny slopes and ripens late. Compact bunches of small berries with very thick skins, high levels of sugar, anthocyanins, polyphenols and good acidity.

WHERE IT'S GROWN AND WHAT ITS WINE TASTES LIKE

Total plantings were fewer than 10 ha (25 acres) in **Italy** in 2000, although the Piemonte producer Michele Chiarlo recently planted 1 ha and made his first varietal Albarossa in 2006.

ALBILLO DE ALBACETE

Also known as Albilla or Albilla de Manchuela, this rare white-berried variety from the province of Albacete in Castilla-La Mancha, south-east **Spain**, is often confused with ALBILLO REAL and PALOMINO FINO, but comparison of their respective DNA profiles (Martín *et al.* 2003; Ibañez *et al.* 2003; Santana *et al.* 2010) shows that they are distinct varieties. Under the name Albilla de Manchuela, it has recently been revived in the Manchuela DO by Juan Antonio Ponce, who makes a dry varietal version called El Reto, a full-bodied but nervy wine with aromas of honey and apple-skin and great tension.

ALBILLO MAYOR

Ribera del Duero variety used mostly in blends and now partly disentangled from the Albillo confusion.

BERRY COLOUR: ● ● ● ● ●

PRINCIPAL SYNONYMS: Pardina (Ribera del Duero), Turruntes or Turruntés

VARIETIES COMMONLY MISTAKEN FOR ALBILLO MAYOR: ALBILLO REAL (Valladolid), CALAGRAÑO, CAYETANA BLANCA (under the name Pardina)

ORIGINS AND PARENTAGE

The name Albillo has been applied to a wide range of distinct and unrelated varieties (see ALBILLO REAL). Among those with a unique DNA profile, only Albillo Real and Albillo Mayor are cultivated. The latter comes from Ribera del Duero in north-central Spain.

According to Santana *et al.* (2010), Albillo Mayor most likely has a parent–offspring relationship with TEMPRANILLO from Rioja and DORADILLA from Málaga.

VITICULTURAL CHARACTERISTICS

Vigorous. Early budding and ripening. Small to medium-sized compact bunches of small, thin-skinned berries. Suits long pruning and prefers well-drained soils and a not too sunny exposition. Little susceptibility to botrytis bunch rot or downy mildew but at risk from spring frosts. Low yields.

WHERE IT'S GROWN AND WHAT ITS WINE TASTES LIKE

In 2008 there were 1,443 ha (3,566 acres) in **Spain**, all in Castilla y León. This is one of the few Albillo Somethings that have been identified but it is still likely to be confused with ALBILLO REAL since this (and other varieties called Albillo Something) is often referred to simply as Albillo. At the moment, neither DO regulations nor wine labels distinguish between Albillo Mayor and Albillo Real (and it is therefore impossible to recommend specific wines or producers). Albillo Mayor produces wines that are typically golden yellow, low in acidity, aromatic, fairly alcoholic, full-bodied and used mainly in blends, eg with the more acidic MACABEO.

Both Albillo Mayor and ALBILLO REAL have been planted in **California**'s San Joaquin Valley as part of a University of California at Davis field trial to identify new varieties that could thrive in the state.

ALBILLO REAL

Aromatic Spanish variety producing smooth wines that are used mostly in blends, both red and white.

BERRY COLOUR: ● ● ● ● ●

PRINCIPAL SYNONYMS: Albillo, Albillo de Cebreros (province of Ávila), Albillo de Madrid

VARIETIES COMMONLY MISTAKEN FOR ALBILLO REAL: ALBARANZEULI BIANCO (Sardegna in Italy), ALBARÍN BLANCO (Asturias), ALBILLO DE ALBACETE (Castilla-La Mancha), ALBILLO MAYOR (Castilla y León), Albillo Real Extremadur (Valle del Tiétar and Zamora), ALVARINHO (Asturias and Galicia), CHENIN BLANC (Australia), CAYETANA BLANCA (under the name Pardina), PARDILLO (La Mancha)

ORIGINS AND PARENTAGE

The name Albillo was mentioned by the Spanish agronomist Gabriel Alonso de Herrera (1470–1539) as early as 1513 in his description of the varieties of Castilla, Extremadura and Andalucía (Alonso de Herrera 1790), under the plural form Albillas (see ALARIJE for the full citation). Indeed, several distinct varieties have been named Albillo (Cervera *et al.* 2001), derived from *alba*, meaning 'white', and significant confusion reigns over Albillo's identity, as demonstrated by recent DNA profiling studies:

- Albilla from Guadalajara in Ibañez *et al.* (2003) is identical to Malvasia Rei (PALOMINO FINO) in Lacombe *et al.* (2007)
- an accession called Albillo in Ibañez *et al.* (2003) is identical to ALBILLO DE ALBACETE in Martín *et al.* (2003), a very rare variety from the province of Albacete in Castilla-La Mancha
- accessions called Albillo from Salamanca in Ibañez *et al.* (2003) and Albillo Negro from Cigales in Santana *et al.* (2007) turned out to be identical to CHASSELAS
- accessions called Albillo in Ibañez *et al.* (2003) and in Santana *et al.* (2010) correspond to Albillo Real in Santana *et al.* (2007)
- Albillo Blanco from Vizcaya in Ibañez *et al.* (2003) has a unique DNA profile
- Albillo Criollo from the Islas Canarias has a unique DNA profile (Jorge Zerolo Hernández, personal communication; Vouillamoz)
- accessions called Albillo Real in Ibañez *et al.* (2003) and in Martín *et al.* (2003) correspond to Albillo Real Extremadur, a variety from the Valle del Tiétar and Zamora but no longer cultivated, in Santana *et al.* (2010)
- ALBILLO MAYOR from several sources (Ibañez *et al.* 2003; Martín *et al.* 2003; Fernández-González, Mena *et al.* 2007; Santana *et al.* 2010) has a unique DNA profile
- Albillo de Nava from Rueda in Yuste *et al.* (2006) and in Santana *et al.* (2007) is in fact VERDEJO
- Albillo di Toro found in Tunisia was identified as MUSCAT OF ALEXANDRIA (Di Vecchi Staraz, This *et al.* 2007).

Among the varieties listed above that were not identified with another well-known variety, only Albillo Real, from the Valladolid area in north-central Spain, and ALBILLO MAYOR, from Ribera del Duero, from Madrid and from Toledo in Castilla y León, are still cultivated. Since only Albillo Real is referred to as simply Albillo, it seems likely that most references to Albillo since Alonso de Herrera in 1513 designate Albillo Real (Cervera *et al.* 2001) from Valladolid.

According to Schneider *et al.* (2010), Albillo Real could have a parent–offspring relationship with Legiruela, a variety from the Provincia de Ávila in Castilla y León that unexpectedly turned out to be identical to PRIÉ, cultivated over centuries in the Valle d'Aosta, which makes it a half-sibling (or grandchild or grandparent) of LAIRÉN, LUGLIENGA, MAYOLET and PRIMETTA (see the complete pedigree of PRIÉ). Albillo Real may also be genetically close to PARELLADA (Ibañez *et al.* 2003).

VITICULTURAL CHARACTERISTICS

Vigorous but not particularly fertile and low yields. Very early budding and ripening, very small (short) bunches of medium-sized, thin-skinned berries. Good drought resistance and well suited to sandy soils. Good general resistance to pests and diseases but highly prone to spring frosts.

WHERE IT'S GROWN AND WHAT ITS WINE TASTES LIKE

Until the middle of the last century, Albillo was used mainly for the table but now it is also used to make soft, aromatic golden wines with a slightly bitter aftertaste, sometimes fermented in barrique but not generally for long ageing. It may also be used to add perfume to blended wines. Official wine-grape statistics in **Spain** perpetuate the confusion described above by listing not only Albillo Real (415 ha/1,025 acres, all in Castilla y León) and ALBILLO MAYOR (1,443 ha/3,566 acres, also in Castilla y León) but also plain old Albillo (893 ha/2,207 acres, mostly in Madrid but also in Castilla-La Mancha, Extremadura, Islas Canarias and Galicia). As suggested above, we can only assume that the figure for Albillo refers to Albillo Real and/or any of the others described above. Similarly, the Wines from Spain website lists only Albillo, describing it as rather neutral but with a smooth texture from high levels of glycerol, and the regulations for the various DOs in which Albillo is a principal (Viños de Madrid) or permitted (Ribeiro and Ribera del Duero) ingredient specify nothing more precise than 'Albillo'.

In Ribera del Duero, a very small amount of Albillo is sometimes blended with TEMPRANILLO, eg in Reyes' Téofilo Reyes Reserva (2%) and in Bodegas del Campo's Pagos de Quintana Reserva (5%). Lezcano-Lacalle do the same thing in Cigales. Recommended producers of white blends containing Albillo (most likely Albillo Real) include Quinta de la Quietud in Toro, as well as Viña Mein and Manuel Formigo de la Fuente further north in Galicia's Ribeiro. Albillo wines are soft and aromatic.

Both ALBILLO MAYOR and Albillo Real have been planted in **California**'s San Joaquin Valley as part of a University of California at Davis trial to identify new varieties that could thrive in the state.

ALCAÑÓN

Rare variety producing light but characterful whites in Somontano.

BERRY COLOUR: ● ● ● ● ●

PRINCIPAL SYNONYMS: Blanco Castellano☒, Bobal Blanca☒

VARIETIES COMMONLY MISTAKEN FOR ALCAÑÓN: MACABEO☒

ORIGINS AND PARENTAGE

Originating from the Somontano area in north-east Spain, Alcañón is widely considered to be identical to MACABEO, but a comparison of the DNA profiles of Alcañón (Martín *et al.* 2003; Fernández-González, Mena *et al.* 2007) and Macabeo (Ibañez *et al.* 2003; Martín *et al.* 2003) clearly shows that they are distinct varieties (Vouillamoz). In addition, the DNA profile of Blanco Castellano in Ibañez *et al.* (2003) matches Alcañón's (Vouillamoz), which means that Blanco Castellano cultivated in Castelló in Valencia and Alcañón are probably one and the same. A possible genetic link with ZALEMA from Andalucía, south-west Spain, has been suggested (Ibañez *et al.* 2003).

WHERE IT'S GROWN AND WHAT ITS WINE TASTES LIKE

According to official statistics, there were just 11 ha (27 acres) of Alcañón in Aragón in north-eastern **Spain** in 2008, but since producers in Somontano (and further afield) seem frequently to confuse MACABEO with Alcañón, it is hard to tell who has the real thing. Wines are light and characterful but increasingly rare.

ALEATICO

Unusually perfumed and Muscat-like Italian red.

BERRY COLOUR:

PRINCIPAL SYNONYMS: Aleaticu (Corse), Halápi (Hungary), Moscatello Nero, Moscato Nero, Vernaccia di Pergola (Marche), Vernaccia Moscatella

VARIETIES COMMONLY MISTAKEN FOR ALEATICO: AGLIANICO, CILIEGIOLO, LIATIKO, MOSCATO DI SCANZO, MUSCAT OF HAMBURG (under the name Moscato Nero)

ORIGINS AND PARENTAGE

Native to Toscana in Italy, whence it has spread to southern Italy. The name Aleatico might derive from *Luglio* (Italian for 'July'), the month in which the berry matures.

DNA profiling has shown that it is not the same as the modern Greek variety LIATIKO (Vouillamoz) and does not have Greek origins. In fact, DNA paternity testing recently demonstrated that Aleatico is not a dark-berried mutant of MUSCAT BLANC À PETITS GRAINS but instead is either a parent or a progeny of it (Crespan and Milani 2001). It also showed that Aleatico is genetically close to GRECO (aka Greco di Tufo of Campania; Costantini *et al.* 2005). DNA tests also proved that Aleatico is the same as Moscatello Nero (Filippetti *et al.* 2002) and the same as Vernaccia di Pergola (Crespan *et al.* 2003), named after the village of Pergola in the province of Pesaro in the Marche. More surprisingly, comparison of several DNA profiling sources shows that the obscure Hungarian variety Halápi is identical to Aleatico (Vouillamoz). In addition, DNA comparative studies showed that Aleatico might be genetically close to LACRIMA DI MORRO D'ALBA, SANGIOVESE and GAGLIOPPO (Filippetti, Silvestroni, Thomas and Intrieri 2001).

Aleatico was used to breed BESSARABSKY MUSKATNY and PERVOMAISKY.

OTHER HYPOTHESES

Because of the existence of a variety named Leaticos in ancient Greece, Aleatico is often said to have been introduced to Toscana by the Greeks and then to have spread to Lazio and Puglia. Some authors erroneously suggested it was identical to AGLIANICO, probably based on the similarity of the names.

VITICULTURAL CHARACTERISTICS

Buds early but ripens midseason. Sensitive to powdery mildew and botrytis bunch rot.

WHERE IT'S GROWN AND WHAT ITS WINE TASTES LIKE

Aleatico has a strong Muscat flavour and produces a generally pale but fine, unusually fragrant red wine.

In **Italy** it's grown mainly in Lazio, in the Tuscia area, and also in Puglia. It also has a long tradition of making sweet, grapey red on the island of Elba. Under the name Vernaccia di Pergola, the grape almost disappeared from the province of Pesaro in the Marche but is now undergoing a revival thanks to the efforts of local viticulturists and scholars at Ascoli Piceno and is producing the locally acclaimed Pergola DOC. The 2000 Italian agricultural census shows total plantings of just over 500 ha (1,235 acres).

Formerly widespread on the island of Corse, **France**, today it is grown mainly near Aleria (10 ha/25 acres), and is authorized in Vin de Pays de l'Île de Beauté. Aleatico is also planted in **California** (mainly in Sonoma, for late-harvest wines) and, less predictably, in **Kazakhstan** and **Uzbekistan**.

ALEKSANDROULI

Rare but highly promising Georgian variety often late harvested for violet-scented semi-sweet wines.

BERRY COLOUR: ● ● ● ● ●

PRINCIPAL SYNONYMS: Aleksandroouly, Aleksandrouli Shavi, Alexandrouli

ORIGINS AND PARENTAGE

Aleksandrouli comes from the Racha-Lechkhumi region in north-western Georgia. Although it is also known as Kabistoni, Aleksandrouli is not the black-skinned version of KAPISTONI TETRI, as shown by morphological and genetic studies (Chkhartishvili and Betsiashvili 2004; Maghradze *et al.* 2009).

VITICULTURAL CHARACTERISTICS

Mid budding, late ripening and often harvested late. Berries reach high sugar levels while retaining good levels of acidity.

WHERE IT'S GROWN AND WHAT ITS WINE TASTES LIKE

Aleksandrouli is the main variety used to produce the distinctive, late-harvest, violet-scented, naturally semi-sweet wines entitled to the Khvanchkara appellation in **Georgia**, as well as for dry red wines that typically have flavours of cherry and mulberry as well as that marked floral aroma. For both, Aleksandrouli is often blended with MUJURETULI. There were 163 ha (403 acres) in Georgia in 2004, almost all in Racha-Lechkhumi and especially on the right bank of the Rioni River in villages such as Khvanchkara, Chorjo, Bostana, Chrebalo, Joshkhi, Sadmeli and Tola. Recommended producers include Bugeuli and Rachuli Wine.

ALEXANDER

American hybrid of purely historical significance in the US.

BERRY COLOUR: ● ● ● ● ●

PRINCIPAL SYNONYMS: Alexandria, Black Cape, Buck Grape, Cape, Clifton's Constantia, Clifton's Lombardia, Columbian, Constantia, Farkers Grape, Madeira of York, Rothrock, Rothrock of Prince, Schuylkill, Schuylkill Muscadel, Schuylkill Muscadine, Springmill Constantia, Tasker's Grape, Vevay, Vevay Winne, Winne, York Lisbon

ORIGINS AND PARENTAGE

This variety was the earliest named American hybrid and was used to make the first commercial wines in the United States. It is a spontaneous hybrid between *Vitis labrusca* and *Vitis vinifera* that was found growing in the woods along the Schuylkill River in the vicinity of Springettsbury, just north west of Philadelphia, Pennsylvania, around 1740 by James Alexander, then gardener to Governor Thomas Penn (although some accounts suggest it was named after his son, John Alexander, gardener to Governor John Penn). It was found near a vineyard planted in the early 1680s with vines from Europe by Andrew Doz for William Penn, hence the *vinifera* parent. About fifteen years later, probably in 1756, Colonel Benjamin Tasker Jr planted it on his sister's estate in Prince George's County, Maryland, whence it quickly spread to other regions, thanks to Tasker's successful wine (Pinney 1989). According to Galet (2000), it was planted in Indiana in the early nineteenth century but was gradually replaced by the more disease-resistant CATAWBA.

VITICULTURAL CHARACTERISTICS

Very late ripening.

ALFROCHEIRO

Portuguese variety with intense fruit aromas and high quality potential.

BERRY COLOUR:

PRINCIPAL SYNONYMS: Albarín Negro (Asturias in Spain), Albarín Tinto (Asturias), Alfrocheiro Preto, Alfrucheiro, Baboso Negro (Islas Canarias), Bastardo Negro, Bruñal (Arribes del Duero), Caíño Gordo (Galicia in north-west Spain), Tinta Bastardinha, Tinta Francesa de Viseu (Viseu in Dão-Lafões in central Portugal)

VARIETIES COMMONLY MISTAKEN FOR ALFROCHEIRO: TROUSSEAU

ORIGINS AND PARENTAGE

Alfrocheiro, also known as Alfrocheiro Preto (*preto* means 'black' in Portuguese), originates from the Dão region of central Portugal or from the Alentejo further south. Because it shows very little morphological diversity, Alfrocheiro is thought to be a relatively young variety, perhaps introduced to Portugal during the post-phylloxera replanting at the beginning of the twentieth century (Rolando Faustino, personal communication). However, recent DNA parentage studies have shown that Alfrocheiro has its roots deep in Iberian soil:

- CORNIFESTO, MALVASIA PRETA, CAMARATE and CASTELÃO from Portugal as well as JUAN GARCÍA from Spain (also found in Portugal under the name Gorda or Tinta Gorda) are all sibling varieties born from natural crosses between Alfrocheiro and CAYETANA BLANCA (Zinelabidine *et al.* 2012), the latter a white-berried variety that is widespread in the southern central Iberian peninsula (see Cayetana Blanca for the pedigree diagram).
- It is genetically close to TROUSSEAU (under the name Bastardo), a French variety that was introduced to the Iberian peninsula a long time ago (Almadanim *et al.* 2007). This relationship explains why the two varieties are sometimes confused, Alfrocheiro Preto often being called Bastardo Negro.
- It is a possible sibling of PRIETO PICUDO from Tierra de León in Spain (Santana *et al.* 2010).
- It is also genetically close to CARRASQUÍN from Asturias in the far north west of Spain (Martín *et al.* 2006).

The name Tinta Francesa, still in use in the Viseu region, might indicate that this variety was introduced from France, which would be consistent with the genetic relationship with TROUSSEAU, but there is no other historical background to confirm this origin.

OTHER HYPOTHESES

Alfrocheiro, traditionally called Albarín Negro in the Asturias region on the north coast of Spain, is neither a colour mutation of ALBARÍN BLANCO (González-Andrés *et al.* 2007) nor a colour mutation of Alfrocheiro Branco in Portugal, the latter being a synonym for the Italian TREBBIANO TOSCANO.

VITICULTURAL CHARACTERISTICS

Fertile, productive, early budding and early ripening. Small, compact bunches of small berries. Susceptible to powdery mildew and botrytis bunch rot.

WHERE IT'S GROWN AND WHAT ITS WINE TASTES LIKE

Alfrocheiro was until recently grown mainly in Dão, central **Portgual**, and to a lesser extent in the Alentejo. However, appreciation of its quality and potential has encouraged growers elsewhere to try their luck with this variety – in the Bairrada and Tejo regions, for example. There were 1,492 ha (3,687 acres) in Portugal in 2010. Wines are richly coloured, show good balance between body and freshness and have fine-grained tannins, but they are distinguished by their aroma: intense notes of blackberries and ripe strawberries that are still evident when the variety is occasionally made in a rosé style or, even more occasionally, as a white wine. Red wines produced in Dão are generally smooth, for easy drinking and not designed for long ageing since those intense fresh-fruit aromas soon turn into something more complex but less typical. In the heat and sunshine of the Alentejo, where water stress is common, the wines tend to be paler in colour, richer and more alcoholic, a little rustic and less redolent of strawberries. Alfrocheiro also blends well with varieties such as Aragonez (aka TEMPRANILLO) and TOURIGA NACIONAL. Recommended producers include Adega Cooperativa de Borba and Ervideira in the Alentejo, Quinta da Lagoalva de Cima in the Tejo and Quinta dos Roques in Dão.

On **Spain**'s Islas Canarias, Baboso Negro is usually blended with other local varieties, although producers such as Frontos, Tananjara and Viñátigo make varietal wines.

ALIBERNET

See ODESSKY CHERNY.

ALICANTE BRANCO

See DAMASCHINO.

ALICANTE HENRI BOUSCHET

Productive, widely dispersed, red-fleshed southern French cross on the wane in France but waxing in southern Portugal.

BERRY COLOUR: ● ● ● ● ●

PRINCIPAL SYNONYMS: Alicante, Alicante Bouschet, Alicante Bouschet no. 2, Dalmatinka (Croatia), Garnacha Tintorera (Spain), Kambuša (Bosnia and Herzegovina), Sumo Tinto (Pinhel in Beiras in Portugal), Tintorera (Spain)

VARIETIES COMMONLY MISTAKEN FOR ALICANTE HENRI BOUSCHET: GARNACHA, GRAND NOIR (Portugal), TINTO VELASCO (La Mancha in Spain)

ORIGINS AND PARENTAGE

At the Domaine de la Calmette in Mauguio (Hérault), Henri Bouschet created several varieties by crossing Grenache (GARNACHA) with PETIT BOUSCHET (an ARAMON NOIR ×

TEINTURIER cross obtained by Henri's father, Louis). In 1855, he obtained Alicante Henri Bouschet and Alicante Bouschet no. 1, and in 1865 he obtained Alicante Bouschet nos. 2, 5, 6, 7, 12 and 13. All these sibling varieties were morphologically very similar but produced wines of varying quality. While some of them were propagated in the vineyards under the misleading generic name Alicante Bouschet, others remained solely in grape collections. By the end of the nineteenth century, all but the most productive one – Alicante Henri Bouschet – had disappeared. However, since it was impossible to distinguish the Alicante Henri Bouschet obtained in 1855 from the Alicante Bouschet no. 2 obtained in 1865, Alicante Henri Bouschet vineyards today are a mix of two similar but distinct sibling varieties. DNA profiling could be used to distinguish them. In practice, it is always referred to simply as Alicante Bouschet.

Alicante Henri Bouschet was used to breed a host of other red-fleshed varieties, of which only ODESSKY CHERNY is still cultivated today.

VITICULTURAL CHARACTERISTICS

Red-fleshed berries. Early budding and early to mid ripening and therefore at risk of spring frosts. Quite vigorous and needs to be pruned short. Moderate but consistent yields in dry climates. Sensitive to wind and drought. Resistant to powdery mildew but susceptible to downy mildew, phomopsis, flavescence dorée and very susceptible to bacterial diseases. There are two mutations: one with colourless flesh and the other with hairy leaves.

WHERE IT'S GROWN AND WHAT ITS WINE TASTES LIKE

There was a surge in Alicante Bouschet's popularity in France after phylloxera, in the late 1880s, due especially to its deep colour and impressive yields (up to 200 hl/ha or 11 tons/acre) but it has been declining since the 1980s and now ranks thirteenth in **France**, where there was a total of 5,680 ha (14,036 acres) in 2008, down from 15,769 ha (38,966 acres) in 1988. The majority of plantings are in the south, from the Lot-et-Garonne in the west to the Drôme in the east, but also further north in the Jura and in the Val de Loire. The greatest concentration remains in its Hérault homeland (2,455 ha/6,066 acres). It has a major advantage in being one of the few red-fleshed *Vitis vinifera* varieties, which means it is sanctioned by the French authorities for the production of quality wine (unlike hybrids).

Wines are deeply coloured and, if the yields are controlled, it can produce soft, fruity wines, but even today it tends to be used to darken the colour in blends. Rare examples of French varietal wines include Domaine La Sorga's Allé Canto and Domaine Zelige-Caravent's Nuit d'Encre from Pic-Saint-Loup (though the latter includes 5% Grenache).

Spain has an even greater area given over to this variety than France: in 2008 there were 22,251 ha (54,983 acres) of Garnacha Tintorera, more than half of that total in Castilla-La Mancha but also significant plantings in Galicia in the north west (6,319 ha/15,615 acres) and Valenciana (1,168 ha/2,886 acres). In 1990 there were only 16,000 ha (39,537 acres).

Alicante Bouschet's popularity is also increasing in **Portugal** (total plantings in 2010 were 2,203 ha/5,444 acres), especially in the arid Alentejo in the south, where an increasing number of producers are taking the variety more seriously and producing impressive, deeply coloured and deep-fruited varietal wines. The variety is thought to have been introduced to the region before the 1950s by John Reynolds at his family property, Herdade do Mouchão. The success of the property's top wine, Mouchão, a blend of Alicante Bouschet with some TRINCADEIRA, encouraged others to plant this variety in the Alentejo. Recommended producers include Caves Bonifácio and Félix Rocha in the Lisboa region, Esporão, Francisco Nunes Garcia, Herdade dos Grous and Terras de Alter in the Alentejo.

In **Italy**, the official census of 2000 distinguishes between Alicante (488 ha/1,206 acres) and the less widely planted Alicante Bouschet (73 ha/180 acres, mainly in Sardegna, Toscana and Calabria) but Alicante is likely to refer to GARNACHA, not to Alicante Henri Bouschet.

Turkey had 550 ha (1,359 acres) in 2010 and the variety is also found to a limited extent in **Hungary** (25 ha/62 acres in 2008), the **former Yugoslavia**, **Cyprus** (81 ha/200 acres in 2010) and **Switzerland**. Some has been grown in **Croatia**, **Bosnia and Herzegovina**, **Israel** and **North Africa**.

Outside Europe, in North America, **California** still had 1,175 acres (476 ha) in 2008, mostly in the hot Central Valley counties of Fresno, San Joaquin and Tulare, where this variety was planted during Prohibition; and in South America in 2008, **Chile** had 438 ha (1,082 acres), **Argentina** 215 ha (531 acres) and **Uruguay** 11 ha (27 acres).

ALIGOTÉ

Burgundy's tart 'other' white wine grape, particularly popular in Eastern Europe.

BERRY COLOUR: ● ● ● ● ●

PRINCIPAL SYNONYMS: Aligotte, Alligotay, Alligoté or Alligotet (Côte de Nuits), Beaunié or Beaunois, ?Carcairone (Val di Susa in Italy), Chaudenet Gras (Côte Châlonnaise), Giboulot or Giboudot (Rully and Mercurey), Griset Blanc (Beaune), Mahranauli (Republic of Moldova), Plant de Trois (Gevrey, Dijon, Selongey), Plant Gris (Meursault), Troyen Blanc (Gevrey), Vert Blanc (Salins in the Jura)

VARIETIES COMMONLY MISTAKEN FOR ALIGOTÉ: CHARDONNAY, GRAŠEVINA (aka Riesling Italico), MELON, SACY (known as Aligoté Vert in the Côte d'Or in Burgundy)

ORIGINS AND PARENTAGE

This variety was probably first cited by Dupré de Saint-Maur in 1780 under the name Plant de Trois, an old synonym referring to the fact that its branches usually bear three bunches each. Under its modern name, Aligoté, a document from the Côte d'Or dated 1807 proposed pulling it up rather than planting it. Various subsequent mentions showed different spellings such as Alligotay, Aligotté, Alligoté or Alligotet in the Vallée de la Saône where Aligoté most likely originated, according to Durand and Pacottet (1901b).

DNA parentage analysis has established that Aligoté is a natural progeny of PINOT and GOUAIS BLANC, just like its siblings CHARDONNAY, GAMAY NOIR, MELON and other minor varieties of north-eastern France (Bowers *et al.* 1999; see Pinot pedigree diagram). This is consistent with the birthplace of Aligoté being in north-eastern France, most likely in Burgundy.

The name Aligoté probably derives from Gôt, an old synonym for GOUAIS BLANC in the Arrières-Côtes de Beaune, where the mother was replaced by its child, Aligoté. Aligoté is a member of the same ampelographic group as Gouais Blanc (see p XXVII; Bisson 2009).

OTHER HYPOTHESES

According to Bazin (2002), Aligoté was cited in 1667 in Burgundy under the name Beaunié, which is very unlikely since Beaunié is an old synonym for SAVAGNIN BLANC. The name Aligoté was also said to derive from the Latin *alligatum*, meaning 'vigne palissée' or 'trellised vine'. This etymology is questionable because this vine-training system was not traditional in north-east France. Another unlikely hypothesis claims the name derives from the Germanic *hariôn*, which gave *harigoter* in ancient French, meaning 'to cut in pieces'. However, it seems rather strange to name a plant after the desire to pull it up. Aligoté is also said to derive from Aligot, the traditional melted-cheese dish from Aubrac (Aveyron, Cantal and Lozère), with which its wine was usually drunk. However, this hypothesis is also unlikely because Aligoté has never been cultivated in this region.

VITICULTURAL CHARACTERISTICS

Vigorous, early budding and thus at risk of spring frosts. Early ripening and susceptible to downy mildew and botrytis bunch rot. Yields vary considerably according to vineyard site.

WHERE IT'S GROWN AND WHAT ITS WINE TASTES LIKE

The total area of Aligoté vineyards in **France** is almost exactly the same as it was fifty years ago, despite a decline in the 1980s. In 2008 there were 1,946 ha (4,809 acres), almost entirely in Burgundy, both in the Côte d'Or and further south, and to a lesser extent in the Chablis region, which explains why it is often thought of as Burgundy's 'other white variety'.

Top-quality wines, in which the variety's naturally high acidity is tempered by all-important ripeness, are fine and full of nerve but demand the best sites, and it is more profitable to use such precious land for CHARDONNAY or PINOT NOIR. It is therefore generally planted on the highest and lowest slopes, producing relatively neutral, sometimes eye-wateringly tart, everyday wines. These are greatly improved by a very small addition of crème de cassis (blackcurrant liqueur) to make the apéritif known as kir. The two main appellations are Bourgogne Aligoté and the finer Bouzeron, where yields are restricted to 65 hl/ha (2.5 tons/acre) rather than the usual 72 hl/ha (3.3 tons/acre). Top producers include Michel Bouzereau, Arnaud Ente, Michel Lafarge, Ponsot (in Monts Luisants, Morey-Saint-Denis) and A & P de Villaine. Aligoté used to be planted in the cru of Meursault in the nineteenth century, and in 1930 a legal judgement authorized the production of Corton-Charlemagne with Aligoté

Switzerland has 21 ha (52 acres), mainly in the canton of Genève. Recommended producers include Domaine des Balisiers and Domaine Les Perrières.

The variety is widely planted in Eastern Europe. **Ukraine** had 9,625 ha (23,784 acres) in 2009 (11% of the total vineyard area); the **Republic of Moldova** had 15,790 ha (39,018 acres) in 2009 in the south and centre of the country; **Romania** had 7,203 ha (17,799 acres) in the Moldova region in the east, in Vrancea, Tulcea on the Black Sea coast, and Iași and Vaslui in the north east; **Bulgaria** had 1,095 ha (2,706 acres) in 2009; **Russia**'s Krasnodar Krai region had 414 ha (1,023 acres) capable of making some of the most confident dry whites made on the Black Sea coast and it is also widely planted in other Russian wine regions such as Rostov (620 ha/1,532 acres in 2010). Aligoté seems to thrive in much of Eastern Europe, with an excess of acidity rarer than in Burgundy. In Russia it has also been used, often in blends, for sweet wines but in general Aligoté is best drunk in the first few years.

A little is also planted for devotees of burgundy in **California** (Au Bon Climat, Calera), **Washington** (Steele Wines), **Canada** (Château des Charmes in Ontario) and **Australia** (Hickinbotham in South Australia).

ALIONZA

Rare but ancient white on the verge of extinction.

BERRY COLOUR: ● ● ● ● ●

PRINCIPAL SYNONYMS: Aleonza, Alionga Bianca del Bolognese, Allionza, Glionza, Leonza, Uva Lonza, Uva Schiava

VARIETIES COMMONLY MISTAKEN FOR ALIONZA: SKLAVA (Greece)

ORIGINS AND PARENTAGE

From Bologna and Modena provinces in Emilia-Romagna in north-central Italy; mentioned as early as the fourteenth century by Pietro de Crescenzi. DNA profiling suggests that Alionza could be closely related to TREBBIANO TOSCANO (Filippetti, Silvestroni, Thomas and Intrieri 2001).

The synonym Uva Schiava was probably given because both Alionza and the original SCHIAVA GROSSA were traditionally grown on horizontal wires.

VITICULTURAL CHARACTERISTICS

Grows in open, warm, hilly soils. Yields are regular and consistent and it shows good resistance to powdery mildew, botrytis and light frosts. Late ripening.

WHERE IT'S GROWN

In **Italy**, Alionza is used for blending in the provinces of Bologna and Modena, but, with a mere 43 ha (106 acres) for the whole of Italy, Alionza is unfortunately slowly disappearing. Some French ampelographers suggested in the late nineteenth century that Alionza was also present in the vineyards of southern France, which seems unlikely. Formerly cultivated in the region of Brescia and Mantova (Lombardia). In the past, Alionza was also considered to be a good table grape.

ALLEGRO

German hybrid with good disease resistance.

BERRY COLOUR: ● ● ● ● ●

PRINCIPAL SYNONYMS: Geisenheim 8331-1

ORIGINS AND PARENTAGE

A complex hybrid obtained by crossing CHANCELLOR × RONDO at Geisenheim in 1983. An application for its inclusion in the official list of varieties was filed in 2002.

VITICULTURAL CHARACTERISTICS

Good resistance to rot, thanks to open bunches, and to both downy and powdery mildews. Good yields and moderate acidity.

WHERE IT'S GROWN AND WHAT ITS WINE TASTES LIKE

Wines have aromas of spice and ripe red fruits and rounded tannins. The Bergsträsser Winzergenossenschaft (co-op) in Hessische-Bergstrasse, **Germany**, produce a varietal wine.

ALTESSE

Savoie's finest grape making delicate whites.

BERRY COLOUR: ● ● ● ● ●

PRINCIPAL SYNONYMS: Anet (Isère), Fusette d'Ambérieu, Marestel (Bourget), Plant d'Altesse, Prin Blanc, Roussette (Savoie), Roussette de Montagnieu (Ain), Roussette de Seyssel, Roussette Haute (Seyssel)

VARIETIES COMMONLY MISTAKEN FOR ALTESSE: FURMINT, JACQUÈRE, ROUSSANNE, ROUSSETTE D'AYZE, VIOGNIER

ORIGINS AND PARENTAGE

The earliest mention of Altesse was provided by the Marquis Costa de Beauregard in 1774: 'le vin d'Altesse, qu'on prend pour un vin étranger excellent [. . .] est un plant de Chypre, qu'un de nos Princes a fait venir en Savoye' ('Altesse wine, which is considered to be an excellent foreign wine [. . .] is a vine from Cyprus that was brought to Savoie by one of our princes'). The Marquis reported an ancient local tradition saying that Altesse took its name from the title of the Dukes of Savoie (the French *altesse* means 'highness'), who allegedly brought it back from the Orient (see below).

Many other authors support an eastern origin for Altesse, including Galet (1990; 2000), who claimed that it had no link with any other French varieties and suggested it could be identical to the Hungarian variety FURMINT. However, based on morphological observations, Foëx (1901) had reservations about an eastern origin for Altesse, believing it to come from the Coteau des Altesses near the Lac du Bourget, Altesses being the local name of the 'terraces' where the variety was cultivated. Although this toponym could not be located (Rézeau 1997), the synonym Plant d'Altesse that was widely used in the nineteenth century largely supports this etymology.

Recent genetic studies back up Foëx in rejecting an eastern origin for Altesse because it is genetically close to CHASSELAS, whose origin was shown to be around Lac Léman (Lake Geneva; Vouillamoz and Arnold 2009). In addition, there is no genetic link between Altesse and FURMINT (Vouillamoz), and Altesse can therefore be considered to be an old autochthonous variety of Savoie.

OTHER HYPOTHESES

The most widespread theory about the eastern origin of Altesse was reported by Tochon (1887): 'On sait que Louis II, Duc de Savoie, succéda à Amédée IV et que Louis, son second fils, épousa vers le milieu du XVe siècle, Charlotte de Lusignan, unique héritière d'Arménie, de Chypre et de Jérusalem; c'est à cette époque qu'il faudrait faire remonter l'introduction en Savoie du plant d'Altesse, car peu après, Louis et Charlotte furent chassés de leur royaume par Jacques, frère naturel de la reine' ('It is known that Louis II, Duke of Savoy, succeeded Amédée IV and that Louis, his second son, married Charlotte de Lusignan, the unique heir of Armenia, Cyprus and Jerusalem in the middle of the fifteenth century; the introduction of Altesse to Savoy certainly dates back to this period because some time later, Louis and Charlotte were expelled from their realm by Jacques, the Queen's brother'). Other legends say that Altesse was brought back from Byzantium (today Istanbul) in 1367 by Amédée VI de Savoie (or Comte Vert), or that it was brought during the Middle Ages from Cyprus to Lucey, where Claude de Mareste planted Altesse in 1530 in the Aimavigne vineyard (Jongieux), renamed Marestel in 1563. Finally, based on an erroneous identification with FURMINT, Galet (1990) suggested that Altesse could originate in Hungary and that it was so named because the famous Furmint-based Tokaji wines were praised by royal courts.

VITICULTURAL CHARACTERISTICS

Mid to late ripening, its berries become reddish (*roux/rousse* is French for 'red' or 'russet') at full maturity, hence its synonym Roussette. A delicate variety susceptible to downy and powdery mildews and to botrytis bunch rot. Generally trained as a bushvine or on a single cordon or guyot.

WHERE IT'S GROWN AND WHAT ITS WINE TASTES LIKE

Varietal examples are produced within the Roussette de Savoie appellation in **France** and its more specific crus or sub-appellations Frangy, Marestel, Monterminod and Monthoux. Recommended producers include Dupasquier, Prieuré Saint-Christophe, Chevalier-Bernard and Jean Perrier. Less well known and slightly further north is the Altesse de Bugey appellation, well represented by Franck Peillot. Such wines may be nutty or more exotically perfumed – even violet-scented – with flavours of honey and almonds, good acidity and the potential to age in bottle. Altesse is often blended with varieties such as JACQUÈRE, ALIGOTÉ, GRINGET or CHARDONNAY within the Vin de Savoie appellation and its more specific crus, including Abymes, Apremont, Ayze, Crépy and so on. The Seyssel appellation is best known for sparkling wines made from Altesse, possibly with CHASSELAS and MOLETTE, though its still wines must be 100% Altesse.

There were 357 ha (882 acres) in eastern France in 2008, mainly in Savoie, Ain, Isère and Haute-Savoie, where it is officially recommended, although it is also authorized further south in the Var and the Vaucluse. Henri Cruchon makes a rare varietal example in **Switzerland**.

ALTINTAŞ

See VASILAKI.

ALVA

See SÍRIA.

ALVADURÃO

See SÍRIA.

ALVARELHÃO

Declining but potentially high-quality variety from northern Portugal.

BERRY COLOUR: ● ● ● ● ●

PRINCIPAL SYNONYMS: Albarello (Galicia in Spain), Alvarelhao, Alvarelho, Alvarellao, Brancelho⚮ (Vinho Verde), Brancellao⚮ (Galicia), Broncellao, Locaia (Minho), Pilongo (Dão), Pirruivo (Arouca), Serradelo⚮, Serradillo, Uva Gallega, Varancelha, Verancelha

ORIGINS AND PARENTAGE

Alvarelhão probably comes from the Douro or Dão in Portugal. The level of genetic diversity suggests that it is a moderately old variety (Rolando Faustino, personal communication). According to DNA studies, Alvarelhão could be genetically very close to SERCIAL (under the name Esgana Cão; Lopes *et al.* 1999).

Although the white-berried PLANTA NOVA from Valencia in eastern Spain is also known as Alvarelhão Branco, it is not a colour mutation of Alvarelhão (Vouillamoz).

VITICULTURAL CHARACTERISTICS

Vigorous, susceptible to powdery mildew. Medium-sized and fairly loose bunches of medium-sized, thinnish-skinned berries (although Brancellao in Galicia is said to have small, thick-skinned berries). Mid budding, early ripening.

WHERE IT'S GROWN AND WHAT ITS WINE TASTES LIKE

Until the 1970s, Alvarelhão was an important and highly regarded variety in **Portugal**, for port in the Douro and for unfortified wine in Dão and the surrounding area of Beiras and, to a lesser extent, in the Vinho Verde region, where it is particularly recommended in the Monção subregion close to the Spanish border. However, it has been a victim of changes in consumer preferences and total Portuguese plantings were just 67 ha (166 acres) by 2010, although there has been recent renewed interest for the production of rosé (Rolando Faustino, personal communication). Wines are typically fresh, with low extract but moderate to high alcohol, delicately aromatic and best drunk young. In Minho, the Adega Cooperativa de Monção make a good rosé and Alvaianas a good red. Campolargo also make a varietal red in Bairrada.

Alvarelhão is known as Brancellao in Galicia in north-west **Spain**, particularly in Rías Baixas, where there were 17 ha (42 acres) in 2008 and where Coto de Gomariz and Sameirás use it in blends. It is highly likely that Quinta da Muradella's Albarello is also Alvarelhão.

In **California**'s Central Valley, a few producers of port-style wines use Alvarelhão in their blend and Yarra Yering in **Australia**'s Yarra Valley include it in their Dry Red No. 3.

ALVARINHO

Fashionable, high-quality, fresh, aromatic variety straddling the Spanish–Portuguese border.

BERRY COLOUR:

PRINCIPAL SYNONYMS: Albariño (Spain), Albelleiro (Rosal in Galicia in Spain), Alvarin Blanco, Azal Blanco, Galego, Galeguinho, Padernã

VARIETIES COMMONLY MISTAKEN FOR ALVARINHO: ALBARÍN BLANCO (Asturias in northern Spain), ALBILLO REAL, AZAL, CAIÑO BLANCO (Rosal in Galicia in Spain, and also in Portugal), GALEGO DOURADO, SAVAGNIN BLANC (Galicia, France, Australia), VERDECA (Puglia in Italy)

ORIGINS AND PARENTAGE

This variety originates from north-eastern Portugal or over the border in Galicia in north-west Spain, where it was once exclusively cultivated – as Alvarinho and Albariño respectively. In Portugal, Alvarinho shows great morphological diversity, which usually indicates that the variety is quite old, and in Spain it is considered to be among the oldest varieties of the north west (Santiago *et al.* 2005). Even though the variety was not mentioned in Galicia until 1843, a parcel of forty 200- to 300-year-old Albariño vines has been observed there (Boso *et al.* 2005; Santiago *et al.* 2007).

DNA parentage analysis has suggested that AMARAL (under the name Caíño Bravo) and Alvarinho (under the name Albariño) gave birth to CAÍÑO BLANCO (Díaz-Losada *et al.* 2011), all three traditionally grown in the north-western Iberian peninsula, which explains why Alvarinho and Caíño Blanco were often confused.

DNA profiling has established that Alvarinho/Albariño is not identical to ALBARÍN BLANCO, also known as Blanco Lexítimo, and is not a colour mutation of ALFROCHEIRO, traditionally called Albarín Negro in the Asturias in Spain (Ibañez *et al.* 2003; Vouillamoz).

In Spain, both Albariño and ALBARÍN BLANCO have long been mistaken for SAVAGNIN BLANC due to a labelling or identification error in the collection of the Misión Biológica de Galicia but this can be rejected by DNA analysis (Santiago *et al.* 2005). How much Savagnin Blanc is still grown in the Spanish vineyards under the names Albarín Blanco and Albariño is not yet known. Comparison of DNA profiles indicates a possible close relationship with LOUREIRO (Ibañez *et al.* 2003).

OTHER HYPOTHESES

Some authors have suggested that the presence of 200- to 300-year-old Alvarinho vines and wild grapevines (*Vitis vinifera* subsp. *silvestris*) in Galicia proves that Alvarinho is native to the area (Santiago *et al.* 2007) but there is no genetic evidence so far.

VITICULTURAL CHARACTERISTICS

Moderately vigorous, robust and fertile. Mid budding, early to mid ripening. Small bunches of medium-sized, relatively thick-skinned berries. Susceptible to downy and powdery mildews and especially to mites. Best suited to dry soils.

WHERE IT'S GROWN AND WHAT ITS WINE TASTES LIKE

Alvarinho is widely planted in north-west Portugal, having spread from its original heartland in the municipality of Monção, right on the country's northern border with Spain, where it is known as Albariño. However, the quality of the variety and the wind of fashion have taken the variety beyond north-west Iberia as far as North America and the Antipodes. The best varietal wines combine aromas and flavours that are both fruity and floral – from linden, orange and acacia blossom through lemongrass and honeysuckle to orange, dried orange peel, grapefruit, bergamot, peach and, in some cases, green apple. Fresh acidity balances the full body and firm structure and there can often be a marine note, reminding the taster how well these wines go with seafood. There are nevertheless still plenty of rather weedier examples in the region, especially where yields are not controlled on traditional pergolas.

There were 2,340 ha (5,782 acres) in **Portugal** in 2010, mostly in the Minho region in the north west but now also further south in Dão, the Península de Setúbal peninsula and Lisboa. Within the Vinho Verde DOC in the north west, there is specific recognition for the quality of Alvarinho in that varietal wines may be produced in the subregions of Monção and Melgaço even though most Vinho Verde wines are blends, generally led by LOUREIRO and TRAJADURA. It is also used to make sparkling wine. Recommended producers include Anselmo Mendes, Quinta do Feital and Quinta da Soalheiro.

In **Spain**, there were 5,320 ha (13,146 acres) of Albariño in 2008, virtually all in Galicia, where it is was one of the first Spanish wines to be labelled by variety. It dominates the Rías Baixas DO and is becoming increasingly popular and fashionable at home and abroad. Recommended producers include Castro Martín, Do Ferreiro (notably the Cepas Vellas from a 200-year-old vineyard), Fillaboa, Pazo de Señorans (especially the late-released Selección Anãda), Pazos de Lusco, Raúl Pérez, Tricó and La Val; Palacio de Fefiñanes make a successful unoaked version as well as 1583 Albariño de Fefiñanes, one of the very few notable barrel-fermented examples.

Under the name Albariño, the variety is also grown in **California**, **Oregon** and **Washington** State and until 2009 (see below) was thought to be the source of a growing number of fresh, peachy wines in Australia.

California's acreage rose from 28 acres (11 ha) in 2008 to 128 acres (52 ha) in 2009, confirming that this variety is on a roll. Recommended producers include Bokisch (Lodi), Bonny Doon (Monterey), Mahoney (Carneros), Tangent (Edna Valley). **Oregon**'s pioneer of Albariño is Abacela in the south of the state but others are starting to plant in their wake, and in **Washington** State a few producers are following suit, eg Coyote Canyon.

In **Uruguay**, Bouza have planted Albariño in their vineyards not far from Montevideo and make a pure, fragrant wine.

The unwelcome revelation that the majority of **Australia**'s Albariño vines were in fact SAVAGNIN BLANC, owing to a labelling error in a Spanish vine collection, perpetuated in France and again once the vines arrived and were propagated in Australia, means that no one really knows how much true Albariño is planted there. However, **New Zealand**'s leading vine nursery has assured growers across the Tasman Sea that their Albariño is true to type and Cooper's Creek, a pioneer of alternative varieties in New Zealand, produced their first vintage of Albariño in 2011.

Despite all the confusion over Albariño in Australia, the qualities and – possibly more importantly – market potential of this Portuguese variety have persuaded the CTPS, the official French committee in charge of plant selection and authorization, to agree to its inclusion in the French national register, which currently includes 288 different vine varieties. Unusually, it will not be subject to an experimental period although it is unlikely that French growers will be granted a subsidy to plant it.

AMABILE DI GENOVA

See MALBO GENTILE.

AMARAL

High-acid Portuguese red found mainly in Vinho Verde.

BERRY COLOUR: ● ● ● ● ●

PRINCIPAL SYNONYMS: Azal Tinto, Azar, Cainho Bravo, Cainho Miúdo, Caíño Bravo⚲ (Galicia in Spain), Cainzinho

VARIETIES COMMONLY MISTAKEN FOR AMARAL: BORRAÇAL⚲, Melhorio⚲ (Vinho Verde), Sousão Galego⚲ (Minho), TOURIGA NACIONAL⚲, VINHÃO⚲

ORIGINS AND PARENTAGE

Officially named Amaral but frequently called Azal Tinto, even though it is not related to the light-skinned AZAL, this variety most likely originates in western Portugal, where it has had a significant genetic influence: recent DNA parentage analyses have suggested possible parent–offspring relationships with LOUREIRO, TINTA CASTAÑAL and VINHÃO as well as with Melhorio and Sousão Galego, two local and no longer cultivated varieties, and that Amaral (under the name Caíño Bravo) and ALVARINHO (under the name Albariño) gave birth to CAIÑO BLANCO, indicating that Amaral is a fairly old variety (Castro *et al.* 2011; Díaz-Losada *et al.* 2011).

VITICULTURAL CHARACTERISTICS

Low yields, mid budding, late ripening. Small, compact bunches of small berries. Often pergola trained.

WHERE IT'S GROWN AND WHAT ITS WINE TASTES LIKE

There were 123 ha (304 acres) of Amaral in **Portugal** in 2010 making acidic, well-coloured, relatively full-bodied but generally dull wines. It is authorized in the Vinho Verde region and particularly recommended in southern subregions such as Amarante but it is usually part of a blend; producers include Casa de Vilacetinho and Casa do Valle.

AMIGNE

Ancient but minor Swiss variety making mainly sweeter styles of wine.

BERRY COLOUR: ● ● ● ● ●

PRINCIPAL SYNONYMS: Amigne Blanche, Grande Amigne, Grosse Amigne, Petite Amigne

ORIGINS AND PARENTAGE

Amigne is indigenous to the Valais in Switzerland, where it was first mentioned in 1686 as growing between Sion and Sierre, a dozen kilometres away from Vétroz, its current preferred location. DNA parentage analysis has suggested a genetic relationship (grandchild, nephew/niece or half-sibling) with AVANÀ from Piemonte in Italy and has established that Amigne is a likely grandchild of PETIT MESLIER from Champagne and the Franche-Comté in France (Vouillamoz and Moriondo 2011). In this case, Amigne would be a great-grandchild of Petit Meslier's parents, SAVAGNIN and GOUAIS BLANC, both also cultivated in the Valais since at least the sixteenth century. It is therefore quite possible that one or both of Amigne's extinct parents were introduced from north-eastern France together with Savagnin and Gouais Blanc.

Labra *et al.* (2002) have suggested a close genetic relationship between Amigne and ROUSSIN from the Valle d'Aosta in Italy but this has been disproved by more detailed DNA analyses (Vouillamoz and Moriondo 2011).

OTHER HYPOTHESES

Amigne is often said to have been brought to the Valais by the Romans, a hypothesis based on the various 'Aminea' cited as wine-grape varieties of southern Italy by Roman authors such as Cato the Censor, Pliny the Elder or Columella. However, there is no botanical evidence that these are the modern Amigne, and it is much more likely that they simply share the same etymology, derived from the Latin *amoenus*, meaning 'delicious' or 'enjoyable'.

VITICULTURAL CHARACTERISTICS

High but irregular yields due to susceptibility to coulure and millerandage. Mid to late budding and ripening. Long, very loose bunches of small berries make it ideal for late harvesting. Best suited to well-exposed, moderately fertile slopes.

WHERE IT'S GROWN AND WHAT ITS WINE TASTES LIKE

Almost all the 43 ha (106 acres) of Amigne in 2009 were planted around the town of Vétroz in the Valais, south-west **Switzerland**. The wines are complex with citrus notes of orange zest and tangerine and crisp acidity. They used to be dry but most today are made sweet and occasionally too heavy. Amigne flétrie, made from grapes left to shrivel and concentrate on the vine, such as that from Cave des Tilleuls, can be voluptuous but well balanced. Recommended producers of dry or medium-dry wines include La Madeleine, Les Ruinettes and Vieux-Moulin, and some of the best sweet wines are made by Jean-René Germanier and La Tine.

AMOR-NÃO-ME-DEIXES

See ARAMON NOIR.

AMORGIANO

See MANDILARIA.

ANCELLOTTA

Useful, deeply coloured blending ingredient valued more for its colour, tannins and acidity than for its flavour.

BERRY COLOUR: ● ● ● ● ●

PRINCIPAL SYNONYMS: Ancellotta di Massenzatico or Lancellotta

ORIGINS AND PARENTAGE

From Emilia-Romagna in north-central Italy. Its name might derive from the Lancellotti family of Modena (fourteenth-fifteenth centuries), who were responsible for the spread of this variety.

Ancellotta was used to breed GALOTTA.

VITICULTURAL CHARACTERISTICS

Vigorous with small dark berries, relatively late ripening.

WHERE IT'S GROWN AND WHAT ITS WINE TASTES LIKE

In **Italy**, Ancellotta is cultivated mainly in the province of Reggio Emilia (eg for Reggiano Rosso DOC) and used mostly for the production of concentrated musts to reinforce the colour of lighter wines such as those made from LAMBRUSCO SALAMINO and LAMBRUSCO MARANI. After the First World War, Ancellotta was introduced in the province of Trentino (in Mori and Avio), as well as in small areas in Friuli, Lombardia, Veneto, Toscana, Apulia and Sardegna. There were nearly 4,500 ha (11,120 acres) in Italy in 2000.

Ancellotta is probably more highly valued in **Brazil** (180 ha/445 acres in 2007) than it is in Italy, although even there it is still used mainly in blends. Don Guerino's very successful, rich, deeply flavoured but fresh varietal example is an exception. Also found in the Valais in **Switzerland** (19 ha/47 acres), where it is used to bring additional colour to some PINOT-based reds.

ANDRÉ

Late-ripening Czech cross producing well-structured reds if harvested fully ripe.

BERRY COLOUR: ● ● ● ● ●

PRINCIPAL SYNONYMS: Andrea, Semenac A 16-76

ORIGINS AND PARENTAGE

André is a BLAUFRÄNKISCH × SANKT LAURENT cross bred in 1961 by Jaroslav Horak at the ŠSV research centre in Velké Pavlovice, southern Morava, in Czechoslovakia (now the Czech Republic), which makes it a sibling of ZWEIGELT. André was named after the Czech naturalist Christian Carl André (1763–1831), who introduced grape-breeding techniques to the research centre. André was included in the official Czechoslovak register of varieties in 1980.

André was used to breed AGNI and RUBINET.

VITICULTURAL CHARACTERISTICS

Medium-sized bunches of small, thick-skinned berries. Resistant to botrytis bunch rot as well as to winter frosts. Productive, ripens late, and needs deep, fertile soils and warm sites to reduce the high level of malic acid in the berries.

WHERE IT'S GROWN AND WHAT ITS WINE TASTES LIKE

If the grapes are fully ripe, wines are full-bodied, well structured, with aromas of ripe blackberries, and suited to barrel ageing and a few years in bottle, but growers looking for high yields produce harsh, astringent wines. Total plantings in the **Czech Republic** in 2009 amounted to 265 ha (655 acres), almost all in Morava in the south east. Recommended producers include Patria Kobylí.

Slightly more hectares of André are planted in **Slovakia** (283 ha/699 acres in 2009), where Jozef Kotuček and Karpatská Perla produce varietal wines, the latter a late-harvest version.

ANSONICA

See INZOLIA.

ANTÃO VAZ

High-quality Portuguese variety at home in the hot, dry south.

BERRY COLOUR: ● ● ● ● ●

PRINCIPAL SYNONYMS: Antonio Vaz

ORIGINS AND PARENTAGE

Antão Vaz is an old variety that probably comes from the south of the Alentejo in Portugal, where it has long been cultivated, particularly in the region of Vidigueira, south east of Lisboa.

DNA parentage analysis has revealed a parent–offspring relationship with the Iberian variety CAYETANA BLANCA (Zinelabidine *et al.* 2012; see pedigree diagram in Cayetana Blanca).

VITICULTURAL CHARACTERISTICS

Sturdy, vigorous and well adapted to the hot, dry Alentejo conditions. Productive, ripens evenly, good general disease resistance. Large bunches of large, tough-skinned berries – rather like a table grape. Mid budding and ripening.

WHERE IT'S GROWN AND WHAT ITS WINE TASTES LIKE

Until recently, Antão Vaz was grown almost exclusively in Vidigueira in the southern Alentejo, south east of Lisboa in **Portugal**, but it has now spread to subregions such as Reguengos, Évora and Moura to become one of the region's main varieties, very popular with growers because it is productive, resilient and capable of producing good-quality white wine in the heat of the Alentejo. It is also grown to a lesser extent closer to Lisboa and in the Península de Setúbal. There were 1,209 ha (2,987 acres) in 2010. Wines typically have tropical-fruit flavours and a good balance between alcohol and acidity. Picked early, the wines are fresh and crisp and low in alcohol, but riper styles have broader flavours and the structure to benefit from barrel fermentation and/or blending with more acidic varieties such as ARINTO DE BUCELAS or SÍRIA (Mayson 2003). Recommended producers of varietal wines include Exquisite Wine, Herdade da Calada, Herdade Fonte Paredes and Quinta do Quetzal in the Alentejo.

ANTEY MAGARACHSKY

Ukrainian hybrid used to make sweet red wines.

BERRY COLOUR: ● ● ● ● ●

PRINCIPAL SYNONYMS: Antei, Antei Magarachskii, Magarach 70-71-52

ORIGINS AND PARENTAGE

Antey Magarachsky is a hybrid obtained by P Y Golodriga, V T Usatov, L P Troshin,

Y A Malchikov and N Dubovenco in 1971 at the Magarach research centre in Yalta, southern Ukraine. It was initially thought to be a cross between RUBINOVY MAGARACHA and Magarach 85-64-16, the latter being a progeny of Perle Noire (or Seyve-Villard 20-347) fertilized with a pollen mixture of unknown *Vitis vinifera* varieties. However, recent DNA parentage analysis has shown that Perle Noire is the parent not the grandparent (Goryslavets *et al.* 2010), so that the parentage of Antey Magarachsky was corrected to Rubinovy Magaracha × Perle Noire (see MUSCAT BLEU for the pedigree of Perle Noire). Antey Magarachsky was included in Ukraine's official register of varieties in 1988.

VITICULTURAL CHARACTERISTICS

Thin-skinned berries. Resistant to phylloxera, downy and powdery mildews. Cold-hardy to –24 °C (–11 °F).

WHERE IT'S GROWN AND WHAT ITS WINE TASTES LIKE

Antey Magarachsky is grown for the table as well as for the vat in **Ukraine**. The Magarach Institute produce a strong (16%) sweet red wine by blending it with other Magarach-bred hybrids such as BASTARDO MAGARACHSKY.

ARAGONEZ

See TEMPRANILLO.

ARAIGNAN BLANC

See PICARDAN.

ARAKLINOS

Rare Greek variety used mainly in red blends.

BERRY COLOUR: ● ● ● ● ●

PRINCIPAL SYNONYMS: Araclinos, Araklino, Raklino

ORIGINS AND PARENTAGE

Araklinos may be from Greece's western Pelopónnisos (Peloponnese) or from the island of Kefaloniá. According to Myles *et al.* (2011), the reference sample of Araklinos (under the name Araclinos) in the United States Department of Agriculture's National Clonal Germplasm Repository could be identical to VRANAC from Montenegro, but this needs to be verified with ampelographic observations and with DNA profiling of local reference samples in Greece and Montenegro.

VITICULTURAL CHARACTERISTICS

High yielding, early to mid budding, mid ripening. Very sensitive to drought. Big, compact bunches of berries that taste sweet and sour.

WHERE IT'S GROWN AND WHAT ITS WINE TASTES LIKE

Araklinos is cultivated in **Greece** to a very limited extent in the western Pelopónnisos and also on Kefaloniá, where Domaine Foivos include a small percentage of both Araklinos and THEIAKO MAVRO in a blend with MAVRODAFNI. Varietal wines have high alcohol and medium acidity.

ARAMON BLANC

See ARAMON NOIR.

ARAMON GRIS

See ARAMON NOIR.

ARAMON NOIR

Very productive vine once responsible for much of France's least noble wine.

BERRY COLOUR:

PRINCIPAL SYNONYMS: Amor-Não-Me-Deixes (Alentejo in Portugal), Aramonen, Gros Bouteillan (Draguignan), Pisse-Vin (Hyères), Plant Riche, Rabalaïré (Var), Ramonen, Réballaïré (Var), Ugni Noir (Var and Bouches-du-Rhône)

VARIETIES COMMONLY MISTAKEN FOR ARAMON NOIR: BOUTEILLAN NOIR, JUAN GARCÍA

ORIGINS AND PARENTAGE

Like many old varieties, Aramon appears in three different colours – Noir, Gris and Blanc – all found in southern France, where this variety most likely originates. Surprisingly, the white version (which actually has rather pink berries) was the first one to be mentioned: the eminent Swiss botanist Augustin-Pyramus listed 'Aramon Blanc, Hérault' in the catalogue of the numerous varieties he had gathered in his collection at the Botanic Garden in Genève. Even more surprisingly, the black version was first mentioned in the UK in the Royal Horticultural Society's *Catalogue of fruits* (1826), before being described by the famous grape breeder Louis Bouschet (1829).

Aramon is said to take its name from the village of Aramon in the Gard near Nîmes, though it has never been widely cultivated there. A more convincing hypothesis is that the name Aramon derives from *rameux* (meaning 'branched'), which would explain its synonyms Aramonen and Ramonen.

DNA profiling has established that the Malvoisie Espagnole in the INRA (Institut National de la Recherche Agronomique) grape collection at Vassal in Montpellier, in southern France, is identical to Aramon Gris (Lacombe *et al.* 2007), which has the same profile as Aramon Noir. In addition, the DNA profile of the strangely named Portuguese (Alentejo) variety Amor-Não-Me-Deixes (meaning 'my love, don't leave me') published in Almadanim *et al.* (2004) turned out to be identical to that of Aramon (Vouillamoz). DNA analysis has also discovered a natural parent–offspring relationship with GOUAIS BLANC (Boursiquot *et al.* 2004), which means that Aramon is a half-sibling of at least eighty Western European varieties.

Although Aramon Noir is also called Gros Bouteillan and Ugni Noir, it is not related to either BOUTEILLAN NOIR or TREBBIANO TOSCANO, known as Ugni Blanc in France.

Aramon Noir was used to breed GRAMON, GRAND NOIR, MONERAC and PETIT BOUSCHET.

OTHER HYPOTHESES

It has been suggested that the Marquis d'Aramon or his ancestors could have introduced the variety from Spain to the village of Aramon in the Gard, but this is doubtful since Aramon has nothing in common with any Spanish variety. Alternatively, Aramon is said to have been introduced to Provence by the Phocaeans from central Greece.

VITICULTURAL CHARACTERISTICS

Early budding and therefore at risk of spring frosts. Late ripening and highly productive with big bunches and berries so needs to be pruned short (bushvine or cordon). Sensitive to downy mildew, botrytis bunch rot, phomopsis and mites.

WHERE IT'S GROWN AND WHAT ITS WINE TASTES LIKE

In **France** all three colours are listed separately in the official register of varieties and all have declined considerably in terms of planting area in the last fifty years, with the Gris and Blanc versions virtually extinct. Aramon was widely embraced in the Languedoc in the late nineteenth century for its resistance to powdery mildew and huge yields of extremely neutral wines often tinted with ALICANTE HENRI BOUSCHET. Carignan (MAZUELO) overtook it only after the 1958 census, which found 150,230 ha (371,226 acres) of Aramon, but by 2008 there were only 3,304 ha (8,164 acres), mainly in the Hérault. In its heyday, Aramon Noir was also planted in Algeria. In 2006 there were 22 ha (54 acres) of Aramon Blanc and 41 ha (101 acres) of Aramon Gris in the Languedoc, where they tend to disappear in very ordinary white-wine blends, although since 2006 Paul Reder of Domaine Hautes Terres de Comberousse has been producing a varietal Aramon Gris called GriGri from vines planted in 2000. Domaine La Sorga's Ah Ramon! is a rare example of varietal Aramon Noir.

ARANEL

Recent, promising French cross.

BERRY COLOUR:

ORIGINS AND PARENTAGE

A cross between GARNACHA ROJA (pinkish grey-berried colour mutation of GARNACHA) and Saint-Pierre Doré (not commercially cultivated) made at the Domaine de Vassal in 1961 by Paul Truel. Since Saint-Pierre Doré has a parent–offspring relationship with GOUAIS BLANC, Aranel is a grandchild (or a half-sibling) of Gouais Blanc. It was named after Arnel, a lake near the Domaine du Chapitre, the commercial winemaking arm of SupAgro, the agronomy college in Montpellier.

VITICULTURAL CHARACTERISTICS

Mid ripening. Good resistance to botrytis bunch rot.

WHERE IT'S GROWN AND WHAT ITS WINE TASTES LIKE

Aranel was officially authorized in 1967 and has been recommended in much of the south of **France** since 1992, though plantings are still extremely limited (5 ha/12 acres in 2006). It is said to have the potential to accumulate good levels of both sugar and acid and show both fruity and floral aromas. Domaine de Lunard make a rare varietal wine, as do the Domaine du Chapitre. It is also grown in New South Wales, **Australia**.

ARANY SÁRFEHÉR

Hungarian variety making simple but useful, high-acid wines and fizz on the Alföld (Great Plain).

BERRY COLOUR: ● ● ● ● ●

PRINCIPAL SYNONYMS: Fehér Kadarka, Fehér Muskotály, Huszár Szőlő, Izsáki, Izsáki Fehér, Izsáki Sárfehér, Német Dinka, Vékonyhéjú

VARIETIES COMMONLY MISTAKEN FOR ARANY SÁRFEHÉR: Sárfehér

ORIGINS AND PARENTAGE

Arany Sárfehér, literally 'gold white mud', is a Hungarian variety that until 2000 was known as Izsáki Sárfehér, after the town of Izsák in the county of Bács-Kiskun, its possible place of origin.

Sárfehér is a distinct but now almost extinct variety, no longer grown commercially, but at one time more widely planted in Sopron, Neszmély and Somló in the north of the country.

Arany Sárfehér has sometimes been considered to be a white mutation of KADARKA because it was selected from a Kadarka vineyard but this has been disproved by DNA profiling (Galbács *et al.* 2009). Arany Sárfehér was used to breed MÁTRAI MUSKOTÁLY.

VITICULTURAL CHARACTERISTICS

Late budding, very late ripening. Large bunches of medium-sized berries with a waxy skin. Resistant to botrytis bunch rot but poor drought tolerance. At risk from frost. Well suited to sandy and loess soils.

WHERE IT'S GROWN AND WHAT ITS WINE TASTES LIKE

Arany Sárfehér is grown mainly in Kunság and to a far lesser extent in Hajós-Baja in central southern **Hungary**, the area traditionally known as the Alföld, and was once more popular than it is today, particularly during the first half of the twentieth century, when it was exported to Germany as a table grape, although it is starting to experience a return to fashion. Producers of varietal wines in that region include Gedeon and Birkás. Izsáki Arany Sárfehér is a specific appellation for varietal wines produced from a group of villages around the town of Izsák in the Kunság region. Wines are typically low in alcohol and very high in acidity and are therefore used mostly to add freshness to blends or in the production of sparkling wines. There were 1,634 ha (4,038 acres) in Hungary in 2008.

ARBANE

Old, aromatic but virtually extinct variety from southern Champagne.

BERRY COLOUR: ● ● ● ● ●

PRINCIPAL SYNONYMS: Albane (Aube), Arbanne, Arbenne (Les Riceys and Tonnerois), Arbone, Crène (Balnot and Polisot in the Aube), Crénillat (Gier in the Loire), Darbanne (Aube)

ORIGINS AND PARENTAGE

Arbane is an old variety of the Aube *département*, where it has long been cultivated, mainly in

Bar-sur-Aube. Its earliest mention allegedly appeared in 1388 under the name Alban in Les Riceys but this is questionable. The first reliable citation was found in Chaptal *et al.* (1801), who mentioned this variety in Morveaux near Baroville (Bar-sur-Aube): 'Albane et Fromenteau, espèces dont le vin est considéré dans le pays' ('Albane and Fromenteau, species whose wine is highly regarded in the region'). The name Arbane most likely comes from the medieval Latin *albana*, meaning 'white grape', which is corroborated by its well-known synonym Albane. Arbane is a member of the Tressot ampelographic group (see p XXVII; Bisson 2009).

VITICULTURAL CHARACTERISTICS

Vigorous, early budding but mid to late ripening. Low yields. Needs a protected site. Susceptible to downy mildew. Small bunches and berries.

WHERE IT'S GROWN AND WHAT ITS WINE TASTES LIKE

Despite its potential, less than 1 ha (2.5 acres) of **France** was planted with it in 2006. The champagne house Moutard-Diligent in Buxeuil continue to make an extremely rare Vieilles Vignes bottling from 100% Arbane, as well as a cuvée that includes Arbane, PETIT MESLIER, PINOT BLANC, CHARDONNAY, PINOT NOIR and PINOT MEUNIER.

ARBOIS BLANC

See MENU PINEAU.

ARENI

Armenia's signature variety making top-quality dry and sweet reds.

BERRY COLOUR: ● ● ● ● ●

PRINCIPAL SYNONYMS: Areni Chernyi, Areni Noir, Areny, Areny Tcherny, Malai Sev, Malayi (Nakhchivan Autonomous Republic in Azerbaijan), Sev Areni, Urza Sev

ORIGINS AND PARENTAGE

Areni is a very old variety that most likely originates in the village of the same name in the west of the province of Vayots Dzor in southern Armenia, on the border with Azerbaijan, where it has traditionally been cultivated and where the oldest winery remains (4000 BC) have recently been excavated (Barnard *et al.* 2011). Areni was used to breed TIGRANI.

DNA profiling could easily determine whether the white-berried Areni Spitak, considered by Galet (2000) to be a distinct variety, is a colour mutation of Areni.

VITICULTURAL CHARACTERISTICS

Mid budding, late ripening. Medium-sized to large, compact bunches of largish, thick-skinned berries. Moderate resistance to fungal diseases and good resistance to winter frosts.

WHERE IT'S GROWN AND WHAT ITS WINE TASTES LIKE

Areni, **Armenia**'s best variety, is widely used to make high-quality, medium- to full-bodied, ageworthy, dry and semi-sweet table wines as well as brandy and the occasional rosé in the east and south of Armenia and in **Azerbaijan**. Armenian producers include 365 Wines, Areni Winery, Brest, Getnatun, Ginekar, Ginetas, Kimley, Maran and Vedi Alco, as well as at the Shahbuz Winery in Azerbaijan (using the synonym Malayi). The Italian consultant Alberto

Antonini, who likens the variety to a cross between SANGIOVESE and PINOT NOIR, is working with Zorik Gharibian at Zorah to make a fresh, sinewy, minerally wine from fruit planted at 1,300 m (4,265 ft) above sea level.

ARETI

Very rare Greek from the island of Zákynthos.

BERRY COLOUR: ● ● ● ● ●

ORIGINS AND PARENTAGE

Areti, literally meaning 'virtue', is a rare variety indigenous to the Greek island of Zákynthos in the Ionian Sea.

WHERE IT'S GROWN AND WHAT ITS WINE TASTES LIKE

Comoutos of **Greece** on the island of Zákynthos include Areti in their traditional white Verdea blend (a mix of very ripe and less ripe grapes that used to be made in an oxidative style) together with Skiadopoulo (see FOKIANO), GOUSTOLIDI, Pavlos (ie MALVASIA BIANCA LUNGA) and ROBOLA.

ARGAMAN

Recent Israeli cross bred to replace Carignan and used mainly in blends.

BERRY COLOUR: ● ● ● ● ●

ORIGINS AND PARENTAGE

Argaman, meaning 'deep purple' in Hebrew, is a VINHÃO (under the name Souzão) × MAZUELO (under the name Carignan) cross obtained in 1972 by Pinchas Spiegel-Roy at the Volcani research centre, Bet-Dagan, Israel, with the aim of creating a highly coloured and productive variety to replace Carignan (Spiegel-Roy *et al.* 1996). It was first planted in 1984 and patented to the breeding institute in 1992.

VITICULTURAL CHARACTERISTICS

Vigorous, productive, early to mid ripening. Susceptible to powdery mildew (but less so than Carignan) and to botrytis bunch rot when ripe.

WHERE IT'S GROWN AND WHAT ITS WINE TASTES LIKE

Wines are deeply coloured and light-bodied, with average acidity and tannins, and are generally found in blends, particularly where the vine yields prolifically, as in the Shimshon region of the Judean Plain of **Israel**. Better results are achieved in Galil, where yields are lower. Carmel launched the first Argaman varietal in 1996 and Segal, who have made a speciality of this variety, produce a single-vineyard, oak-aged, medium-bodied, red-fruited varietal wine from their Dovev vineyard. There were around 150 ha (371 acres) in Israel in 2010, mainly in the central coastal plain. Although Argaman was bred to provide a deeper-coloured blending grape to replace Carignan, that role has since been taken by CABERNET SAUVIGNON and MERLOT and Carignan has made something of a comeback.

ARGANT

See GÄNSFÜSSER.

ARILLA

Rare but productive island white from southern Italy.

BERRY COLOUR: ● ● ● ● ●

PRINCIPAL SYNONYMS: Agrilla, Arillo, Rille, Uva Rilla

ORIGINS AND PARENTAGE

Arilla has long been cultivated on the Italian island of Ischia, near Napoli, in particular on south-facing slopes. It is thought to originate on Sicila.

VITICULTURAL CHARACTERISTICS

Very productive.

WHERE IT'S GROWN

Grown exclusively on the island of Ischia, **Italy**, Arilla can produce up to 10 kg per plant – a level of production unlikely to produce much flavour, which is probably why it is usually blended with other local varieties such as FORASTERA, BIANCOLELLA and Don Lunardo (SAN LUNARDO).

ARINARNOA

Successful, widely dispersed French cross.

BERRY COLOUR: ● ● ● ● ●

ORIGINS AND PARENTAGE

Arinarnoa was presented as a MERLOT × PETIT VERDOT cross obtained in 1956 by Pierre Marcel Durquéty at INRA (Institut National de la Recherche Agronomique) in Bordeaux, south-west France, but recent DNA parentage analysis at INRA in Montpellier corrected it to a TANNAT × CABERNET SAUVIGNON cross, which makes it a grandchild of CABERNET FRANC and SAUVIGNON BLANC. It was named after the Basque dialect words *arin*, meaning 'light', and *arno*, meaning 'wine'.

VITICULTURAL CHARACTERISTICS

Late ripening, good resistance to botrytis bunch rot. Open bunches and thick-skinned berries.

WHERE IT'S GROWN AND WHAT ITS WINE TASTES LIKE

Officially approved in 1980, and recommended in many southern and south-western *départements* of **France** since the mid 1980s. There were 164 ha (405 acres) in 2006, mainly in Languedoc-Roussillon, although Laurent and Annie André make a varietal wine north of Cognac in the

Charente-Maritime. Wines tend to be deeply coloured with firm but fine tannins, sometimes with a herbaceousness similar to that of CABERNET FRANC but a little sweeter.

The variety has spread far if not wide. Carlos Esteva at Can Rafols dels Caus in Penedès, **Spain**, has about 1.5 ha (4 acres), planted in the early 2000s, and is very enthusiastic about the variety, which he uses in the Petit Caus and Gran Caus red blends and also in the Gran Caus rosé. In **Lebanon**, Chateau Ksara's Le Souverain is a 50/50 blend of Arinarnoa with CABERNET SAUVIGNON; Giménez Méndez in **Uruguay** produces both a varietal Arinarnoa and a blend with TANNAT; in **Argentina**, where Familia Zuccardi produce a varietal version, there were just 2 ha (5 acres) planted in 2008; and **Brazil**'s Rio Grande do Sul had 7 ha (17 acres) in 2007.

ARINTO DE BUCELAS

High-quality, high-acid, adaptable Portuguese variety from just north of Lisboa.

BERRY COLOUR:

PRINCIPAL SYNONYMS: Arinto, Arinto Cercial, Arinto d´Anadia, Arinto Galego, Arintho, Azal Espanhol, Chapeludo, Pedernã, Terrantez da Terceira (Açores)

VARIETIES COMMONLY MISTAKEN FOR ARINTO DE BUCELAS: Arinto de Colares, BICAL, DORINTO (Douro), FOLGASÃO (Lisboa), LOUREIRO, MALVASIA FINA, SERCIAL, TAMAREZ

ORIGINS AND PARENTAGE

As its name suggests, Arinto de Bucelas most likely comes from the Bucelas area north of Lisboa, Portugal. It was first mentioned under the name Arinto in 1712 by Vicêncio Alarte in his *Agricultura das vinhas e tudo o que pertence ao lias até o perfeito recolhimento do vinho*, the first viticultural treatise ever written in Portugal. This means it is among the oldest varieties in Portugal, confirmed by the morphological diversity observed in this variety (Rolando Faustino, personal communication). From the area around Lisboa, Arinto de Bucelas first expanded north to Bairrada and then to the Vinho Verde region, where it is today considered to be a traditional variety. Lopes *et al.* (1999) have found that Arinto de Bucelas is genetically very close to TRAJADURA, an old variety cultivated in Galicia in north-west Spain and in the Douro region in north-east Portugal.

Arinto Roxo is a colour mutation of Arinto de Bucelas (*roxo* meaning 'purple' in Portuguese).

Arinto de Bucelas should not be confused with the following white-berried varieties: Arinto Branco (synonym for LOUREIRO), Arinto de Alcobaça (synonym for BICAL), Arinto de Colares (no longer cultivated), Arinto do Dão (synonym for MALVASIA FINA), Arinto dos Açores, Arinto Gordo (synonym for TAMAREZ), or Arinto no Douro (now officially known as DORINTO but also known as Arinto de Trás-os-Montes, Arinto Cachudo, Arinto do Douro and Arinto do Interior).

Although TEMPRANILLO is sometimes called Arinto Tinto, it is not a red mutation of Arinto de Bucelas.

VITICULTURAL CHARACTERISTICS

Late budding and ripening. Vigorous. Medium-sized, very compact bunches of small berries rich in acidity. Copes well with water shortages in summer. Very adaptable to different climatic conditions. Attractive to leafhoppers, which damage the leaves, and sometimes susceptible to botrytis bunch rot.

WHERE IT'S GROWN AND WHAT ITS WINE TASTES LIKE

Arinto de Bucelas is more often referred to as Arinto but this causes a good deal of confusion, as explained above. Thanks to Arinto's ability to retain acidity in a wide range of sites and its attractively fresh lemony character, it is widely planted throughout **Portugal**'s wine-growing regions: from the temperate and humid regions of Vinho Verde in the north west (under the name Pedernã and often partnered by LOUREIRO and TRAJADURA) to the Alentejo in the hot, dry south, where it generally adds freshness to a blend. It is also planted in Bairrada, for both still and sparkling wines, as well as in the Tejo and in Lisboa.

However, the best Arinto wines are still mostly found in Bucelas, where the maritime influence helps to produce the most elegant wines, which have not only lemon, lime or green-apple freshness but also a steely, mineral backbone, thanks also to the calcareous soils of the region, resulting in wines that age well and gain in complexity but may also be drunk young and fresh. Arinto has particularly benefited from the introduction of longer, cooler fermentation styles. It must constitute at least 75% of wines labeled Bucelas DOC, with SERCIAL and RABO DE OVELHA as optional blending partners. Arinto plus SAUVIGNON BLANC is a more recent blend to appear in Portugal.

Despite this popularity and usefulness in blends, some fine varietal wines are being made, eg by Casa de Laraias in the Vinho Verde region, Quinta da Romeira (part of the Companhia das Quintas) in Bucelas, Quinta de Chocapalha in the Tejo and Julian Reynolds in the Alentejo. There were 3,175 ha (7,846 acres) in Portugal in 2010. How long before it crops up in quality-conscious vineyards elsewhere?

ARINTO DO DÃO

See MALVASIA FINA.

ARNA-GRNA

Light-skinned Azeri variety found mainly in the Nakhchivan Autonomous Republic.

BERRY COLOUR: ● ● ● ● ●

PRINCIPAL SYNONYMS: Alagura, Arha-Grna, Khana Crna

ORIGINS AND PARENTAGE

Arna-Grna is indigenous to the Nakhchivan Autonomous Republic, an exclave of Azerbaijan bordering on Armenia, Iran and Turkey. Galet (2000) suggests that it could be identical to Arevik from Armenia but this has not yet been tested.

VITICULTURAL CHARACTERISTICS

Regularly productive, late ripening. Compact bunches. Functionally female flowers.

WHERE IT'S GROWN AND WHAT ITS WINE TASTES LIKE

Producers such as Tumbul Sharab and Babek Wine-2 Yeddiler in the Babek region of Nakhchivan, **Azerbaijan**, make dry white wines with this variety, although it is also used for juice.

ARNEIS

Piemonte's scented and full-bodied signature dry white.

BERRY COLOUR: ● ● ● ● ●

PRINCIPAL SYNONYMS: Bianchetta di Alba, Bianchetto, Nebbiolo Bianco

ORIGINS AND PARENTAGE

From Roero in Piemonte, north-west Italy, Arneis was possibly first mentioned in 1432 in Chieri in the province of Torino under the Latin name Ranaysii, which could derive from the toponym Bric Renesio near Canale, already known in 1478 as Reneysium. On the other hand, the place may have taken its name from the presence of Arneis in the vineyards (Comba and Dal Verme 1990). During the seventeenth and eighteenth centuries, the contraction of *renesi* supposedly produced Arneis. Alternatively, the local dialect *arnèis*, used to describe a wily and temperamental person, could have given its name to the grape, reflecting how difficult it is to grow and vinify. However, the earliest written mention of Arneis was in Di Rovasenda (1877). Despite its synonym Nebbiolo Bianco, Arneis is not related to the black-berried NEBBIOLO.

VITICULTURAL CHARACTERISTICS

Vigorous, tolerant of downy mildew but susceptible to powdery mildew; ripens in the second half of September.

WHERE IT'S GROWN AND WHAT ITS WINE TASTES LIKE

Mainly cultivated in the Roero and Langhe areas of Piemonte in north-west **Italy** but also in small amounts in Liguria and on Sardegna. According to local tradition, Arneis used to be planted together with NEBBIOLO to attract birds with its strong flavour, thus protecting Nebbiolo, which had (and still has) a better market value. It was also used to soften Nebbiolo, which may explain the synonym Nebbiolo Bianco.

Arneis virtually disappeared in the early 1970s, when only Vietti and Bruno Giacosa were bottling it. However, it was saved from extinction by a welcome revival of white Piemonte wines in the 1980s. The wines are generally unoaked, subtly fruit-scented, full-bodied and tasting of ripe pears but can lack acidity, especially when later picked; best drunk young.

The Italian census of 2000 recorded 745 ha (1,840 acres) and the best producers in Piemonte are Malvirà, Deltetto, Cascina Chicco and Bruno Giacosa.

Beyond Europe, it is provoking growing interest in **Australia**, particularly in Victoria, where it has been planted by producers such as Yandoit Hill, Chrismont, Pizzini, Dal Zotto, Gary Crittenden, Yarra Loch and Rutherglen Estates. Other producers include Tertini in the Southern Highlands (NSW). Coopers Creek in Gisborne, **New Zealand**, have produced a fresh, off-dry version of this wine with fine, pure floral and citrus flavours. There are minor plantings in the US, notably by Seghesio in Sonoma County, **California**, and Ponzi in the Willamette Valley, **Oregon**.

ARNSBURGER

Scarce German cross used to make base wines for inexpensive fizz.

BERRY COLOUR: ● ● ● ● ●

PRINCIPAL SYNONYMS: Geisenheim 22-74

ORIGINS AND PARENTAGE

Arnsburger was described at its introduction as a selfling (see p 1182) of two clones of RIESLING (Riesling 88 GM × Riesling 64 GM) developed in 1939 by Heinrich Birk at the Geisenheim research centre in the Rheingau, Germany, and put on the official list of varieties in 1984. If it is a selfling of RIESLING, then Arnsburger would reach the maximum level of consanguinity. However, recent genetic analyses based on newly developed DNA markers have suggested that Arnsburger is more likely to be a MÜLLER-THURGAU × CHASSELAS (under the name Chasselas Tokay) cross (Myles *et al.* 2011). This parentage still needs to be confirmed by standard DNA markers (microsatellites) but it would be consistent with the purpose of the cross, bred to provide a better alternative to Müller-Thurgau for the production of sparkling white wines.

The name Arnsburger is intended to reflect the wine-growing tradition of the Cistercians, who founded the Abbey of Arnsburg in the Wetterau, north of Frankfurt, and played a major part in the vinous history of the Rheingau.

Arnsburger was crossed with the hybrid Seyve-Villard 1-72 to create SAPHIRA.

VITICULTURAL CHARACTERISTICS

Productive and disease-resistant, with loose bunches that make it less susceptible to botrytis bunch rot. Rather low sugar levels and high acidity (in Germany). Early to mid ripening. Large bunches of firm-skinned berries.

WHERE IT'S GROWN AND WHAT ITS WINE TASTES LIKE

Developed in **Germany** for its high yields and high acidity, suitable for making the base wines for sparkling wine, although German plantings are only around 3 ha (7 acres). Schönhals in Rheinhessen uses it but only in a blend.

Introduced to the **Portuguese** island of Madeira in the mid 1980s, Arnsburger is now more common on the north side of that Atlantic island than in Germany, thanks to an EU-funded project (with a German consultant). There are currently 14 ha (35 acres) used to make rather light but dull unfortified dry white wines for the local market and classified as Terras Madeirenses. When the fruit is picked fully ripe in Madeira's much hotter climate, the wines can lack freshness. It is not authorized for the production of Madeira. Recommended producers include Adega de São Vicente, Justino Henriques and João Mendes.

In **New Zealand**, 5 ha (12 acres) were planted in 2007, of which only 2 ha (5 acres) remain, mainly in the Gisborne region.

AROMATNY

A pinky-white-berried hybrid obtained by L F Meleshko, Y P Chebanenko, M I Tulaeva, M G Bankovskaya, L M Pismennaya and I V Gerus at the Tairov research centre in Odessa, **Ukraine**, by crossing Vertes Csillaga (a Hungarian Eger 1 × MENOIR hybrid) with Romulus (an American Ontario × SULTANIYE hybrid), in which Eger 1 is an open pollination of Seyve-Villard 12–286, itself a Seibel 6468 × Subéreux hybrid (see VILLARD BLANC for the complete pedigrees of the last two, and CAYUGA WHITE for Ontario). It was registered in 2009. This vigorous and productive variety is winter-hardy and ripens early to produce dry, fruity and, as its name suggests, aromatic table wines.

ARRILOBA

Recent light-skinned cross that is widely recommended but little planted.

BERRY COLOUR: ● ● ● ● ●

ORIGINS AND PARENTAGE

A RAFFIAT DE MONCADE × SAUVIGNON BLANC cross obtained in 1954 by Pierre Marcel Durquéty at INRA (Institut National de la Recherche Agronomique) in Bordeaux, south-west France. Arriloba is therefore a half-sibling of CABERNET SAUVIGNON. Its made-up name may possibly be a combination of the Basque dialect words *arr-*, an intensifier, and *iloba*, meaning 'nephew', ie 'great-nephew'; alternatively from *(h)arri*, meaning 'stone', plus *iloba*. But neither explanation makes a great deal of sense.

VITICULTURAL CHARACTERISTICS

Early ripening, vigorous, fertile and productive. Occasionally subject to drying out of the stems but the thick skins of the very small berries protect them from botrytis bunch rot.

WHERE IT'S GROWN AND WHAT ITS WINE TASTES LIKE

Arriloba is recommended in just about every French *département* but there were just 59 ha (146 acres) planted in **France** in 2006, almost the same as in 2000. The Cave des Vignerons Landais' Douceur de Chalosse is one of the few varietal wines produced, from vines to the north west of Bayonne, and Brard Blanchard include Arriloba in their Vin de Pays Charentais white blend.

ARROTELAS

See RETAGLIADO BIANCO.

ARROUYA

Virtually extinct old Jurançon vine.

BERRY COLOUR: ● ● ● ● ●

PRINCIPAL SYNONYMS: Dourec Noir (Béarn), Eremachaoua (Béarn), Erematxahua (Pays Basque)

VARIETIES COMMONLY MISTAKEN FOR ARROUYA: CABERNET FRANC⌛ (Pyrénées), FER⌛ (Jurançon), MANSENG NOIR⌛ (Jurançon)

ORIGINS AND PARENTAGE

Arrouya is an old variety from the Jurançon region in the Pyrénées-Atlantiques, south-west France, where it was mentioned as early as 1783 (Rézeau 1997): 'Arrouya rouge … il mûrit tard' ('red Arrouya … it ripens late'), and again in many subsequent citations in the nineteenth century. The name Arrouya derives from the Béarnese dialect word *arro(u)y*, meaning 'red'. Ampelographers consider that Arrouya has some morphological similarities with CAMARAOU NOIR, FER and PETIT COURBU (Bordenave *et al.* 2007) and belongs to the Carmenet ampelographic group (Bisson 2009; see p XXVII), together with CABERNET FRANC, which is why Arrouya was often confused with Cabernet Franc in the Hautes- and Basses-Pyrénées up to the end of the nineteenth century.

VITICULTURAL CHARACTERISTICS

Vigorous, late ripening and difficult to ripen in some years. Susceptible to downy mildew and black rot, and its compact, small-berried bunches make it rather susceptible to botrytis bunch rot.

WHERE IT'S GROWN AND WHAT ITS WINE TASTES LIKE

Isolated vines are still occasionally found in mixed plantings with Camaraou Blanc and CAMARAOU NOIR in old vineyards in the Jurançon region, **France**. Wines tend to be rather pale, acidic and sometimes astringent or bitter, and it is not hard to see why the variety is on the way out, even though it is still authorized in the Pyrénées-Atlantiques.

ARRUFIAC

South-west France relic starting to regain a little local respect.

BERRY COLOUR: ● ● ● ● ●

PRINCIPAL SYNONYMS: Arrefiat, Arruffiac, Arruffiat, Arrufiat, Bouisselet (Pacherenc), Raffiac, Raffiat, Réfiat, Rufiat, Zurizerratia (Pays Basque)

VARIETIES COMMONLY MISTAKEN FOR ARRUFIAC: COURBU BLANC (Jurançon region), RAFFIAT DE MONCADE (Béarn and Landes)

ORIGINS AND PARENTAGE

Arrufiac is a traditional variety of the Jurançon region and the Vallée de l'Adour in the Pyrénées-Occidentales, south-west France, where it was mentioned for the first time in 1802, in Flourens just east of Toulouse in the Haute-Garonne, under its old synonym Bouisselet in the proceedings of an agricultural meeting. A few years later, Baron Picot de Lapeyrouse (1814) wrote: 'J'ai choisi dans nos cépages les plus fins, les plus hâtifs: le bouisselet, le négret, le redondal, l'escessalaire blanc' ('I have chosen from our finest vines the earliest ripening: bouisselet, négret, redondal, escessalaire blanc'). The name Arrufiac could derive from the Languedocien word *rufe* or the Gascon word *arrufe*, meaning 'rough' or 'proud' but also 'gathered' or 'wrinkled', perhaps hinting at the character of the resultant wine (Lavignac 2001).

Recent genetic and ampelographic studies have suggested that Arrufiac could be genetically close to PETIT VERDOT (Bordenave *et al.* 2007). Indeed, Bisson (2009) places Arrufiac within the Carmenet ampelographic group (see p XXVII), and also indicates similarities with the Courbu group.

Although they share some synonyms, Arrufiac and RAFFIAT DE MONCADE are not related.

VITICULTURAL CHARACTERISTICS

Big bunches of small, thick-skinned berries. Vigorous and rather susceptible to downy mildew and black rot. Late ripening.

WHERE IT'S GROWN AND WHAT ITS WINE TASTES LIKE

Arrufiac is today cultivated mainly in the Pacherenc du Vic-Bilh appellation of south-west **France**, in which it may be blended with the principal varieties PETIT MANSENG, GROS MANSENG, COURBU BLANC, PETIT COURBU, as well as further north and west in the Côtes de Saint-Mont. Rescued from obscurity by André Dubosc of the Plaimont co-operative in the 1980s, Arrufiac is now being treated with greater respect and used in greater measure by certain producers, for example: Cru du Paradis (30%), Château Viella (20%) and Producteurs Plaimont's Saint-Mont Les Vignes Retrouvées (20%). Wines are typically slightly tannic with fresh acidity and a hint of almonds. They can be bitter on the finish and are best in a blend. There were 93 ha (230 acres) in France in 2006.

ARVESINIADU

Rare Sardinian white that disappears into blends.

BERRY COLOUR: ● ● ● ● ●

PRINCIPAL SYNONYMS: Argu Ingianau, Arvesimiadu Bianco, Arvu Siniadu, Uva Oschirese

ORIGINS AND PARENTAGE

Obscure and almost extinct variety from the Italian island of Sardegna. The earliest and only historical reference to this vigorous and late-ripening variety is found in Manca dell'Arca (1780).

WHERE IT'S GROWN AND WHAT ITS WINE TASTES LIKE

There are very limited plantings in the provinces of Sassari and Campidano in Sardegna, **Italy**, and the 2000 census recorded just 155 ha (380 acres). Arvesiniadu is generally blended with other Sardinian varieties to produce local whites but, within the IGT Isola dei Nuraghi, Mulas produce both dry and sweet varietal wines.

ARVINE

The finest indigenous Valais variety making lively, sometimes intense, whites both dry and sweet.

BERRY COLOUR: ● ● ● ● ●

PRINCIPAL SYNONYMS: Arvena, Arvina, Petite Arvine

ORIGINS AND PARENTAGE

Arvine is indigenous to the Valais in Switzerland, where it was first mentioned in Sion in 1602 under the dialect name Arvena, which could derive from the old Latin *arvena*, meaning 'just arrived', a name that was possibly given to this variety at its introduction or birth (Vouillamoz and Moriondo 2011).

Since the late nineteenth century, it has often been called Petite Arvine ('small Arvine') to distinguish it from Grosse Arvine ('big Arvine'), a variety that has been shown to be a progeny of RÈZE and a possible grandchild or nephew/niece of Arvine but is no longer cultivated (Vouillamoz and Moriondo 2011). No other close genetic relationships could be uncovered by DNA parentage analysis but it is possible to speculate about some distant ancestors in the Valle d'Aosta, Savoie and north-eastern France.

Arvine has no genetic link with Arvino Nero, a synonym of GAGLIOPPO in Calabria, or with Alvino Verde, a synonym of VERDECA in Campania, but they might all share a common etymology (Vouillamoz and Moriondo 2011).

OTHER HYPOTHESES

Like many other varieties from the Valais, Arvine is often said to have been introduced by the Romans, and, according to the philologist Jacques André (1953), the name Arvine could derive from Helvola, a variety mentioned by Cato the Censor. This is highly unlikely because there is no botanical evidence that ancient and modern varieties are identical, and also because the historian André Tchernia (1986) has established that Helvola had black berries.

VITICULTURAL CHARACTERISTICS

Productive, very early budding, late ripening. Prone to wind damage, susceptible to downy mildew, to botrytis bunch rot towards harvest, also to mites. Prefers well-drained sites not subject to water stress. Compact bunches of small berries.

WHERE IT'S GROWN AND WHAT ITS WINE TASTES LIKE

Arvine is the finest of the Valais varieties and produces wines that are nervy but also have substance, typically with flavours of grapefruit and a mineral saltiness. There were 154 ha (378 acres) in **Switzerland** in 2009, almost all in the Valais. Recommended producers of dry varietal wines include John and Mike Favre, Mandolé, Domaine des Muses and La Rayettaz; good sweet versions are made by producers such as Gérald Besse and La Liaudisaz (Marie-Thérèse Chappaz).

In 2010, the French CTPS (Comité Technique Permanent de la Sélection des Plantes Cultivées) authorized Arvine in **France**, some years after Michel Chapoutier abandoned his experimental vineyard in the northern Rhône, and it was included in the official register of varieties in 2011. Before that Angelo Gaja had tried his hand at Arvine in the Langhe in north-west **Italy** but had severe problems with coulure and broken shoots. Arvine seems happier at home. Nevertheless, there are just a few hectares in the Valle d'Aosta in the far north west of Italy, where Grosjean Frères and Les Crêtes are making good examples of varietal wines.

ASHUGHAJI

Minor Georgian variety producing fresh, aromatic reds.

BERRY COLOUR: ● ● ● ● ●

PRINCIPAL SYNONYMS: Achougage, Ashugazh

ORIGINS AND PARENTAGE

Ashughaji is indigenous to Apkhazeti, north-western Georgia.

VITICULTURAL CHARACTERISTICS

Compact bunches of medium-sized berries producing slightly pink juice. Mid budding, late ripening, low yielding.

WHERE IT'S GROWN AND WHAT ITS WINE TASTES LIKE

Ashughaji produces deeply coloured, aromatic, dry red wines with high acidity and is grown mainly in Apkhazeti, north-western **Georgia**.

ASPIRAN BOUSCHET

Deeply coloured but almost extinct.

BERRY COLOUR: ● ● ● ● ●

PRINCIPAL SYNONYMS: Grand Noir de Laques (Puy-l'Évêque in the Lot), Spigamonte (Valpo)

ORIGINS AND PARENTAGE

A Gros Bouschet × RIVAIRENC (under the name Aspiran Noir) cross obtained in 1865 by Henri Bouschet at the Domaine de la Calmette in Mauguio near Montpellier. Aspiran Bouschet is therefore a grandchild of Gros Bouschet's parents, ARAMON NOIR and TEINTURIER.

WHERE IT'S GROWN AND WHAT ITS WINE TASTES LIKE

At the turn of the century there were just a few hectares left in **France** according to Galet (2000), but in 2008 there were a surprising 1,912 ha (4,725 acres) in **Argentina** (where it is known as Aspirant Bouchet), mainly in Mendoza. Here and in **Chile** very small amounts may be added to deepen the colour of some blends.

ASPIRAN NOIR

See RIVAIRENC

ASPRINIO

See GRECO.

ASPRO

See XYNISTERI.

ASPROFILERO

See MOSCHOFILERO.

ASPROUDA

Asprouda (and variant spellings such as Asproudes, Asproudi, Asprudi) is a generic name meaning 'whitish' that has been given to several distinct white varieties all over **Greece**, especially in the south. According to the viticulturist Haroula Spinthiropoulou, growers use the name for just about any unidentified light-berried variety.

The official list of Greek varieties authorized for winemaking includes Asprouda Patron, Asprouda Santorinis and Asprouda Serron, but the grape collection at the United States Department of Agriculture's National Clonal Germplasm Repository in Davis, California, for example, includes the following seven samples all with distinct and unique DNA profiles: Asprouda Mykinon, Asprouda Patron, Aspruda Ariloghi, Aspruda Halkidos, Aspruda Mikynon, Aspruda Santorini and Aspruda Zakintho (each one named after the place in which it was collected). It is therefore often impossible to know which variety is being referred to when the name Asprouda or Aspruda is used.

To complicate matters further, producers such as Antonopoulos in Pátra refer to a field blend which is a component in their Adoli Ghis white as 'Asproudes' – it includes Santameriana, White Volitsa, Mygdali and Brena (the latter may be SAVATIANO). Down in Monemvasia, both Vatistas and Monemvasia Winery have a single variety they refer to as Asprouda (made as a varietal as well as contributing to their sweet wines) which is also known locally as Asprovaria or Varia. Four different Asproudas have been identified in this area alone and DNA analysis is long overdue.

ASSARIO

See MALVASIA FINA.

ASSYRTIKO

Top-quality, intense, highly structured Greek island white that retains acidity even in the heat.

BERRY COLOUR: ● ● ● ● ●

PRINCIPAL SYNONYMS: Assirtico, Assyrtico, Asyrtico, Asyrtiko

ORIGINS AND PARENTAGE

Assyrtiko is a variety of the Greek Aegean Islands that most likely comes from Santoríni. Recent DNA profiling has suggested a possible parent–offspring relationship with two other Aegean varieties, GAIDOURIA (under the name Guydouria) and PLATANI (Myles *et al.* 2011).

Although some growers refer to FlaskAssyrtiko and Arseniko simply as different clones, the former is a distinct and currently unidentified variety and the latter is virus-affected Assyrtiko (Lazarakis 2005).

OTHER HYPOTHESES

It has been suggested that the name Assyrtiko could come from *assyrico*, meaning 'Assyrian' (Boutaris 2000), suggesting an origin in Mesopotamia (present-day Iraq), or alternatively from *seri*, the local pronunciation of 'sherry', in keeping with the belief that Assyrtiko could have been introduced to Santoríni from Jerez in Spain (Boutaris 2000). Both these etymologies are very unlikely because no genetic link with either Spanish or Near Eastern varieties has been detected (Vouillamoz).

VITICULTURAL CHARACTERISTICS

Vigorous and relatively productive. Resistant to downy and powdery mildews and to drought, and its hard wood protects it from the wind (Lazarakis 2005). Compact clusters of large berries that retain high levels of tartaric acid. Very late budding and late ripening.

WHERE IT'S GROWN AND WHAT ITS WINE TASTES LIKE

Assyrtiko is most closely associated with the powerful, intense yet fresh and mineral-scented whites of the volcanic island of Santoríni, although its excellent quality potential and ability to retain acidity in hot climates have encouraged growers in most parts of **Greece** to try their hand at this variety, despite its tendency to oxidize. There were 1,704 ha (4,211 acres) in Greece in 2008, making it the third most planted white wine variety in the country after SAVATIANO and RODITIS.

On Santoríni it represents around 70% of the vineyard area (Lazarakis 2005) and is used to make not only taut, mineral dry whites, sometimes softened by a small percentage of ATHIRI and/or AÏDANI, but also Vinsanto (for details see Aïdani). Argyros, Hatzidakis and Sigalas make particularly fine examples of dry whites, sometimes oaked, along with Boutaris, Gaia and the Santoríni Co-op. Assyrtiko vines on Santoríni can reach a tremendous age because they are ungrafted and because of the way they are trained, the canes woven nest-like around the fruit to protect it from the wind. It is said that some vines may be as old as 500 years, although this refers to the roots since the canes are annually pruned back.

When Assyrtiko is planted in Attikí or further north in Dráma and Halkidikí, the wines tend to be broader and more fruity but they are still both full-bodied and fresh. Here they are more often blended than is the case on Santoríni. Biblia Chora, for example, blend Assyrtiko with SAUVIGNON BLANC and, in Ovilos, with SÉMILLON but their Areti white is 100% Assyrtiko.

It is surely only a matter of time before this exciting variety goes travelling – it has already arrived in northern **Italy** thanks to Alois Lageder, and in **Australia** thanks to Jim Barry Wines in the Clare Valley.

ASURETULI SHAVI

Minor Georgian variety once popular with German colonists.

BERRY COLOUR: ● ● ● ● ●

PRINCIPAL SYNONYMS: Asuretuli, Schalltraube, Shadi Traube, Shala

ORIGINS AND PARENTAGE

Asuretuli Shavi, meaning 'black from Asureti', is indigenous to the the Kartli region of southern central Georgia, and was discovered growing wild in woods near the German settlement of Asureti by the German colonist Schall some time between 1825 and 1845, hence the synonym Schalltraube ('Schall's grape'). The variety was propagated mainly in German colonies prior to the colonists' expulsion in 1941 and decreased significantly thereafter (Ortoidze *et al.* 2010).

VITICULTURAL CHARACTERISTICS

Medium-sized, compact clusters of small berries. Mid budding, late ripening, yields sometimes low, owing mainly to functionally female flowers in some clones. Good resistance to spring frosts.

WHERE IT'S GROWN AND WHAT ITS WINE TASTES LIKE

Asuretuli Shavi wines typically have moderate alcohol and acidity with an interesting varietal aroma, and are best drunk young, not unlike lighter styles of Beaujolais. The variety is grown mainly in the village of Asureti in Kartli, southern central **Georgia** (hence the name), where the Asuretian Wine Cellar produce a dry varietal version. It may also be made in a medium-dry rosé style. There were just 5 ha (12 acres) planted in Georgia in 2004.

ATHIRI

An old, easy-to-grow Greek variety producing soft, dry, fruity whites.

BERRY COLOUR: ● ● ● ● ●

PRINCIPAL SYNONYMS: Asprathiri (Ródos/Rhodes, Santoríni), Asprathiro, Athiri Aspro, Athiri Lefko, Athiri Leyko

VARIETIES COMMONLY MISTAKEN FOR ATHIRI: THRAPSATHIRI

ORIGINS AND PARENTAGE

Athiri or Athiri Aspro (*aspro* means 'white') probably comes from the Aegean region of Greece. It has long been thought to be close or identical to THRAPSATHIRI but this has been contradicted by DNA profiling (Biniari and Stavrakakis 2007).

Athiri Mavro, which is said to be a black mutation of Athiri Aspro (or vice versa), is found to a very limited extent in the Kykládes (Cyclades; Manessis 2000).

OTHER HYPOTHESES

Legend has it that this variety was already known in classical Greece under the name Thiriaki or Theriaki, deriving from Thira, the old name for Santoríni, which became Thirea, Thiri and then Athiri, and that it was used in the Middle Ages in the blend of the famous Malvasia wine (Manessis 2000).

VITICULTURAL CHARACTERISTICS

Vigorous, drought resistant but susceptible to powdery mildew. Mid budding and mid ripening (a few days after ASSYRTIKO). Thin-skinned berries. Adaptable but best suited to light, calcareous or argilo-calcareous soils (Lazarakis 2005).

WHERE IT'S GROWN AND WHAT ITS WINE TASTES LIKE

Athiri's popularity is not surprising: it is easy to grow, is adaptable and produces good-quality whites even at high yields. Wines are lemony, fresh and fruity even though the acidity is quite soft and the alcohol moderate, making it a good choice to slightly soften ASSYRTIKO's steeliness on Santoríni. It is authorized for quality wines in the Santoríni, Ródos and Côtes de Meliton appellations as well as for various regional wines. Although it is not officially allowed in Vinsanto, many of the vineyards are interplanted so it may well be included anyway.

In 2008 there were 671 ha (1,658 acres) in **Greece**, mainly on the southern Aegean Islands, making Athiri the second most planted white variety there after ASSYRTIKO, notably on Santoríni, Ródos and Kríti (Crete). According to Lazarakis (2005), the total plantings in Greece in the early part of this century were almost 1,000 ha (2,471 acres), an indication of its spread to other parts of the country such as Lakonía (southern Pelopónnisos/Peloponnese) and Halkidikí and Makedonía (northern mainland Greece). Tsantali's easy-drinking Athiri epitomizes the variety's fresh, fruity, crowd-pleasing style. More serious varietal wines are fairly rare but the best come from the higher vineyards on the island of Ródos, eg Emery's Mountain Slopes Athiri.

AUBIN BLANC

Virtually extinct Lorraine speciality.

BERRY COLOUR: ● ● ● ● ●

PRINCIPAL SYNONYMS: Albin Blanc, Aubain (Moselle), Aubier (Joinville and Poissons in the Vallée de la Marne), Aubin, Blanc de Creuë (Meuse), Blanc de Magny, Gros Vert de Crenay

VARIETIES COMMONLY MISTAKEN FOR AUBIN BLANC: AUBIN VERT

ORIGINS AND PARENTAGE

Aubin Blanc is an old Lorraine variety, formerly cultivated in the Moselle region, where it was first mentioned in a ruling of the parliament at Metz in August 1722 (Rézeau 1997): 'toutes les Vignes ... seront arrachées incessamment; sçavoir ... toutes les Plantes de Raisins blancs, soit qu'elles soient séparées ou entremêlées, dans les Vignes, de Raisins noirs ou Auxerrois, à la reserve des Fromentaux de Sainte Ruffine, & des Aubins des Villages de Peltre, Crepy & Magny' ('all vineyards ... will be incessantly pulled up; that is ... all the white-berried varieties, either in separate vineyards or interplanted in the vineyards with black-berried varieties or Auxerrois, with the exception of the Fromentaux of Sainte Ruffine, & of the Aubins in the villages of Peltre, Crepy & Magny'). The plural form Aubins most likely refers not only to Aubin Blanc but also to AUBIN VERT, mentioned in 1829. The name Aubin derives from the Latin *albanus*, meaning 'whitish'.

DNA parentage analysis has indicated that Aubin Blanc is a natural GOUAIS BLANC × SAVAGNIN cross (Bowers *et al.* 2000), which makes it a sibling of PETIT MESLIER and RAÜSCHLING, as well as a half-sibling of AUBIN VERT and of the numerous other progenies of either Gouais Blanc or Savagnin (see PINOT pedigree). This explains why Aubin Blanc was considered by Mouillefert (1902a) to be close to Petit Meslier and MESLIER SAINT-FRANÇOIS. Aubin Blanc belongs to the Salvanien ampelographic group (see p XXVII), named after Savagnin (Bisson 2009). Aubin Blanc has no relationship with the AUBUN variety from the Vaucluse.

VITICULTURAL CHARACTERISTICS

Very early budding, therefore at risk from spring frosts. Early ripening. Small berries.

WHERE IT'S GROWN AND WHAT ITS WINE TASTES LIKE

This is a very rare variety with less than a hectare (2.5 acres) in **France** in 2006 (down from 3 ha/7 acres in 1958). It is recommended in the Meurthe-Moselle, Meuse and Moselle *départements*, and the Côtes de Toul appellation west of Nancy in Lorraine is a blend of Aubin Blanc and AUXERROIS. Wines are not particularly aromatic with only moderate acidity.

AUBIN VERT

Virtually extinct Lorraine variety.

BERRY COLOUR: ● ● ● ● ●

PRINCIPAL SYNONYMS: Blanc d'Euvézin, Vert Blanc

VARIETIES COMMONLY MISTAKEN FOR AUBIN VERT: AUBIN BLANC

ORIGINS AND PARENTAGE

Like AUBIN BLANC, Aubin Vert is an old variety of Lorraine, formerly cultivated in the Moselle area (near Metz), where it was first mentioned in 1722 in the plural form Aubins (see Aubin Blanc), and later in 1829 as Aubin Vert.

DNA parentage analysis has established that Aubin Vert is a natural PINOT × GOUAIS BLANC cross (Bowers *et al.* 1999), which makes it a sibling of CHARDONNAY, GAMAY NOIR, ALIGOTÉ, MELON, etc., as well as a half-sibling of AUBIN BLANC and of the numerous other progenies of Gouais Blanc and Pinot (see PINOT pedigree). Aubin Vert was included with Aubin Blanc in the Salvanien ampelographic group (see p XXVII), named after SAVAGNIN (Bisson 2009) but this is questionable now that the parentage of Aubin Vert is known.

VITICULTURAL CHARACTERISTICS

Early-ripening, high-acid variety.

WHERE IT'S GROWN AND WHAT ITS WINE TASTES LIKE

Although this variety is authorized in the Meurthe-Moselle, Meuse and Moselle *départements* of **France**, plantings are negligible.

AUBUN

Minor blending component from the southern Rhône.

BERRY COLOUR: ● ● ● ● ●

PRINCIPAL SYNONYMS: Carignan de Bédoin, Carignan de Gigondas (Vaucluse), Grosse Rogettaz (Savoie), Morescola or Murescola (Corse), Moustardier, Moutardier

VARIETIES COMMONLY MISTAKEN FOR AUBUN: COUNOISE☒ (Vaucluse)

ORIGINS AND PARENTAGE

The first mention of Aubun appeared in 1885 in the region of Bédoin near Mont Ventoux in the Vaucluse, where it was supposedly growing as a wild grape. Like AUBIN BLANC and AUBIN VERT from Lorraine, with which it has no relationship, Aubun was named after the Latin *albanus*, meaning 'whitish', supposedly referring to a now extinct white mutation. Contrary to local belief, Aubun is not identical to COUNOISE, and both belong to the Piquepoul ampelographic group (see p XXVII).

VITICULTURAL CHARACTERISTICS

Late budding and mid ripening. A rather rustic vine with large bunches and thick-skinned berries. Susceptible to downy mildew and leafhoppers.

WHERE IT'S GROWN AND WHAT ITS WINE TASTES LIKE

Aubun was widely planted in **France** during the phylloxera infestations in the late nineteenth century because it was thought to show some resistance to this pest but it turned out that this was due simply to the fact that the local soils in which it was planted were relatively resistant to it (Galet 2000).

Wines tend to be rather ordinary, sometimes like a softer version of Carignan (MAZUELO), hence the synonyms Carignan de Bédoin and Carignan de Gigondas, and the variety may be better suited to rosé, such as the varietal wines produced by Domaine de Lansac north of Arles and Domaine des Moulines between Lunel and Montpellier. Antony Tortul's La Sorga L'Aubunite (a clever pun on *l'eau bénite*, meaning 'holy water'), however, is a Languedoc red made from 85%

Aubun (biodynamically grown near Carcassonne). There were 720 ha (1,779 acres) in France in 2008, mainly in the Gard and the Aude, although it is authorized throughout the Languedoc and in Provence. This is considerably less than the 5,882 ha (14,535 acres) of thirty years ago.

Pockets of Aubun are found in **California**. Copain and Donkey & Goat, for example, both produce a McDowell Valley Syrah from Mendocino which is made from a field blend that includes a small amount of Aubun. The variety was part of the original collection taken by James Busby to **Australia** in 1833.

AURELIUS

Czech cross producing aromatic vaguely Riesling-like wines that can reach Spätlese-like richness.

BERRY COLOUR: ● ● ● ● ●

ORIGINS AND PARENTAGE

Aurelius is a NEUBURGER × RIESLING cross bred in 1953 by Josef Veverka at the ŠSV research centre in Velké Pavlovice, southern Morava, in the Czech Republic, which Veverka had to leave in 1959 for political reasons. The seedlings were then moved to the research centre in Perná, also in southern Morava, where Aurelius was initially selected as NE × RR 45/18 by Veverka and his colleague František Zatloukal. It was officially named and included in the Czech register of varieties in 1983.

VITICULTURAL CHARACTERISTICS

Productive and able to attain high sugar levels. Tender-skinned berries susceptible to botrytis bunch rot unless planted in the driest sites, where it is suited to late harvesting.

WHERE IT'S GROWN AND WHAT ITS WINE TASTES LIKE

Aurelius shows its RIESLING parentage in wines that are aromatic and occasionally spicy, often achieving high sugar levels when left on the vine, with flavours of quince and lime flowers, but they tend to be less crisp and elegant than Riesling wines. Recommended producers include Josef Dufek and Château Valtice. There were just 49 ha (121 acres) in 2009, almost all in Morava in the south east of the **Czech Republic**.

AURORE

Complex hybrid more successful in North America than in its homeland France.

BERRY COLOUR: ● ● ● ● ●

PRINCIPAL SYNONYMS: Aurora, Feri Szölö, Financ Szölö, Redei, Seibel 5279

VARIETIES COMMONLY MISTAKEN FOR AURORE: Aurora (table grape)

ORIGINS AND PARENTAGE

Albert Seibel in Aubenas south west of Valence in the Ardèche obtained this by crossing Seibel 788 × Seibel 29, where:

- Seibel 788 is a Sicilien × Clairette Dorée Ganzin hybrid (see HELIOS for the complete pedigree of Clairette Dorée Ganzin)
- Sicilien is a Bicane × PASCAL BLANC cross, the former a French wine and table grape that is no longer cultivated
- Seibel 29 is a Munson × 'unknown' *Vitis vinifera* subsp. *vinifera* variety hybrid
- Munson is a Jaeger 43 × *Vitis rupestris* hybrid, with Jaeger 43 being a variety of *Vitis lincecumii* Buckley.

It was originally bred for the table as well as for wine but it turned out to be a poor table grape because it did not travel well and the grapes tended to fall off the stems. Aurore is named after Aurora, the Roman goddess of the dawn.

VITICULTURAL CHARACTERISTICS

Vigorous, productive, very early ripening (before CHASSELAS), good resistance to downy mildew but susceptible to botrytis bunch rot and black rot. Berries sometimes have a pink tinge.

WHERE IT'S GROWN AND WHAT ITS WINE TASTES LIKE

According to Galet (2000), there were once 288 ha (712 acres) in **France**, in both the north east and the south west. It was once widely planted across the north-eastern US and **Minnesota** and is still found in some colder American states with short growing seasons such as **New York** (3,530 tons crushed in 2009, mostly for wine). In **Canada** most of the vines in British Columbia were pulled up at the end of the 1980s and Ontario growers have harvested just 198 tons over the last five years. Producers of varietal wines include Jewell Towne Vineyards (**New Hampshire**), Schilling Bridge (**Nebraska**) and Domaine des Côtes d'Ardoise (**Québec**). Wines are light and relatively neutral but may have a slight foxy character.

AUXERROIS

Low-acid variety useful in cool climates.

BERRY COLOUR:

PRINCIPAL SYNONYMS: Aucerot (Australia), Auxera, Auxerrois de Laquenexy (Moselle), Auxois, Auzerrois Blanc, Blanc de Kienzheim, Ericey de la Montée, Kleiner Heunisch (Baden-Württemberg in Germany), Okseroa, Pinot Auxerrois (Alsace and Canada), Riesling Jaune de la Moselle

VARIETIES COMMONLY MISTAKEN FOR AUXERROIS: CHARDONNAY⧖, PINOT⧖ (BLANC or GRIS)

ORIGINS AND PARENTAGE

Auxerrois is a variety from Alsace-Lorraine that was allegedly mentioned as a wine name in the thirteenth-century fable 'La Borgoise d'Orliens': 'Et puis burent a grant plantez / Et des blans et des auchorrois' ('And they drank large amounts / Of white wines and auchorrois'). However, this reference should be treated with caution since Auxerrois is a white variety, and since the name Auxerrois has been given to several distinct varieties of north-eastern France, such as PINOT (BLANC or GRIS) and CHARDONNAY. The earliest reliable reference to this variety appeared in 1816 in the Moselle: 'les espèces de raisins de grosse race sont beaucoup moins bonnes que nos maurillons, nos francs-pineaux, nos meûniers et nos auxerrois' ('the very vigorous grape varieties are far less good than our maurillons, our francs-pineaux, our meûniers and our auxerrois').

DNA parentage analysis has shown that Auxerrois is one of the numerous natural progenies of PINOT and GOUAIS BLANC (Bowers *et al.* 1999). It is therefore a sibling of CHARDONNAY, GAMAY NOIR, ALIGOTÉ, etc., which explains the confusion with PINOT BLANC and CHARDONNAY as well as its synonyms Pinot Auxerrois in Alsace and Kleiner Heunisch (little Gouais Blanc) in Baden-Württemberg. Auxerrois is a member of the Noirien ampelographic group (see p XXVII).

The name Auxerrois does not refer to the city of Auxerre in the Yonne but most likely derives from Auxois, the old name for Alsace as well as an old synonym for Auxerrois, giving further evidence for an Alsace-Lorraine origin for this variety. Curiously, Auxerrois has also been a synonym for the red COT in Cahors since at least 1589, because it was said to have been brought from Auxerre.

VITICULTURAL CHARACTERISTICS

Early ripening. Well suited to northern France and limestone soils. Susceptible to downy mildew, botrytis bunch rot, grape moth and especially powdery mildew. Small bunches and berries.

WHERE IT'S GROWN AND WHAT ITS WINE TASTES LIKE

In 2008 there were 2,330 ha (5,758 acres) in **France**, mostly in Alsace, where it is the second most planted white variety after RIESLING, and the French Moselle, although there are still a few vines in the Loire. It is also significant in **Luxembourg** (184 ha/455 acres in 2008), can be useful in the **Netherlands** and a few hectares were recorded in **England** in 1990. Wines tend to be relatively neutral and low in acidity, a valued attribute in cool climes, but if yields are curbed, wines can be positively rich and honeyed.

In Alsace, Auxerrois is rarely seen on the label, though that would be perfectly legal; instead it is generally blended into wine sold as Pinot Blanc (occasionally Clevner/Klevner), or goes into either Crémant or the multi-varietal Edelzwicker, to which it can add weight and substance. Bizarrely in a region which prides itself on varietal labelling, a wine labelled Pinot Blanc could comprise 100% Auxerrois. Since Auxerrois is not permitted in Alsace Grand Cru wines, some of the best sites have been grubbed up in favour of the more noble varieties. Similarly, this often hidden variety is not allowed in the sweet Alsace styles Vendange Tardive or Sélection de Grains Nobles.

Alsace producers who take Auxerrois seriously include Philippe Blanck, Josmeyer, Kientzler and Louis Sipp. Armand Hurst has produced an SGN-style sweet Auxerrois.

The Côtes de Toul appellation west of Nancy specifies a blend of AUBUN and Auxerrois for white wines and Auxerrois may play a supporting role to GAMAY NOIR and PINOT NOIR in the region's vin gris.

Germany's 285 ha (704 acres) are mostly in Baden and the Pfalz, although there have been some recent plantings by top growers in the middle Nahe and Rheinhessen.

In **Canada**, Ontario's plantings are in decline and in 2006 were about half what they were in 2002 (roughly 45 acres/18 ha compared with 85 acres/34 ha). There were 43 acres (17 ha) in BC in 2008.

In **South Africa** in the 1980s it was discovered that vines purported to be early CHARDONNAY plantings were in fact Auxerrois.

AVANÀ

Light, early-drinking Piemontese red, rarely made as a varietal.

BERRY COLOUR: ● ● ● ● ●

PRINCIPAL SYNONYMS: Avanale, Avanas, Avanato, Avanè, Avenà, Avenai, Hibou Noir☒ (Isère and Hautes-Alpes in France), Vermaglio (near Saluzzo)

ORIGINS AND PARENTAGE

From Val di Susa in Piemonte, north-west Italy, Avanà was most probably the variety referred to as Avanato by the agronomist Croce in 1606. Some early authors found morphological similarities with the French varieties GAMAY NOIR (Gamay d'Orléans) and Troyen (Troyan), the latter no longer cultivated. Although DNA profiling has not confirmed this precise relationship, it has proved the link with France since Avanà is identical to Hibou Noir, an old variety from Savoie and Dauphiné that has almost disappeared (Schneider *et al.* 2001). In addition, DNA parentage analysis strongly suggests that Avanà is either a parent or an offspring of CACABOUÉ, another old, and almost extinct, Savoie variety, and a grandchild, nephew/niece or half-sibling of AMIGNE, an old variety from the Valais in Switzerland (Vouillamoz and Moriondo 2011).

These genetic links between Savoie, the Valais and Piemonte are not really surprising given that Piemonte was part of the Duchy of Savoy, also including the Valais, between 1418 and 1713, when Hibou Noir could have been brought to Piemonte and have taken the name Avanà, or vice versa. Moreover, Avanà is mostly located in the Val di Susa, the valley linking Torino to the French Savoie and Hautes-Alpes.

VITICULTURAL CHARACTERISTICS

Irregular yields, susceptible to powdery mildew, early to mid ripening.

WHERE IT'S GROWN AND WHAT ITS WINE TASTES LIKE

Still cultivated today in north-west **Italy**, often on pergolas, in the Val di Susa and Val Chisone in the Pinerolese in Piemonte, Avanà is usually blended with other local varieties (BARBERA, PERSAN under the name Becuét, FREISA, NERETTA CUNEESE). Since 1996, Avanà has been authorized in the Pinerolese DOC, in particular in the Pinerolese's traditional blend called Ramìe, which is light, fruity, crisply fresh and slightly astringent. Since 1997, it has been authorized on its own or in a blend (with Barbera, DOLCETTO, Neretta Cuneese, etc.) in the Valsusa DOC. A few wineries such as Sibille, Martina and Clarea make a 100% Avanà. It produces fresh, light-bodied wines with light tannins and moderate acidity for early drinking.

Just over 30 ha (75 acres) were recorded in the 2000 census.

AVARENGO

Low-yielding, perfumed Piemontese red.

BERRY COLOUR: ● ● ● ● ●

PRINCIPAL SYNONYMS: Amarene, Avarena, Avarenc, Mustèr (in the Canavese region in the province of Torino), Riondosca or Riundasca (province of Biella)

ORIGINS AND PARENTAGE

Avarengo is an old variety from Val Chisone to Pomaretto in the Pinerolese area of Piemonte, north-west Italy. DNA parentage analysis has clearly established a parent–offspring relationship between Avarengo and GRISA NERA, another old Piemontese variety, thus anchoring the origin of both grapes in this area (Anna Schneider and José Vouillamoz, unpublished data). In addition, DNA parentage analysis suggests a close genetic link between Avarengo and NER D'ALA from northern Piemonte (Vouillamoz and Moriondo 2011).

VITICULTURAL CHARACTERISTICS

Mid ripening.

WHERE IT'S GROWN AND WHAT ITS WINE TASTES LIKE

Blended with other local varieties in Pinerolese Ramìe DOC (minimum 15%), north-west **Italy**. A few vineyards remain in the Canavese but it is being slowly abandoned because of its low productivity, as suggested by the derivation of its name from *avaro*, Latin for 'stingy'. There were 1,680 ha (4,151 acres) recorded in the Italian census of 2000. The wines are fresh, pale and perfumed.

AVASIRKHVA

Rare Georgian variety grown in Apkhazeti.

BERRY COLOUR: ● ● ● ● ●

PRINCIPAL SYNONYMS: Ajiche, Ajishi, Auasirkhva, Avasarkhva, Avassirkhva

ORIGINS AND PARENTAGE

Avasirkhva is indigenous to Apkhazeti, north-western Georgia.

VITICULTURAL CHARACTERISTICS

Mid budding, late ripening, moderate yields.

WHERE IT'S GROWN

Although Avasirkhva shows very good potential for high-quality dry whites in a subtropical climate, there were just 7 ha (17 acres) in the Apkhazeti region of **Georgia** in 2004. The wines are mostly produced for home consumption.

AVESSO

Portuguese variety with more body than others making Vinho Verde.

BERRY COLOUR: ● ● ● ● ●

PRINCIPAL SYNONYMS: Bornal, Bornão, Borral

VARIETIES COMMONLY MISTAKEN FOR AVESSO: CAYETANA BLANCA (under the name Mourisco Branco in Vinho Verde)

ORIGINS AND PARENTAGE

Avesso, meaning 'contrary' in Portuguese (Mayson 2003), perhaps because of its ability to achieve higher alcohol levels than other varieties in its Vinho Verde homeland in northern Portugal, was first mentioned by Menezes in 1896 (Cunha *et al.* 2009). This reference and the low level of genetic diversity (Rolando Faustino, personal communication) suggest it is not a particularly old variety.

VITICULTURAL CHARACTERISTICS

Vigorous, relatively productive, early budding and ripening. Medium-sized to large berries. Best on dry, low-fertility granite-based soils. Susceptible to downy mildew and botrytis bunch rot but not to powdery mildew.

WHERE IT'S GROWN AND WHAT ITS WINE TASTES LIKE

Avesso is found in the Minho region of northern **Portugal**, predominantly in Baião, Cinfães and Resende near the lower reaches of the Douro, where it is an important ingredient in the Vinho Verde appellation and can often achieve higher alcohol levels (12–13%) than the 11.5% maximum permitted in the regulations (Mayson 2003). This fuller body is accompanied by fresh acidity and fruit, leaning towards stone fruit and almost tropical-fruit flavours. Total plantings in 2010 were 730 ha (1,804 acres). Recommended producers of varietal wines include Anselmo Mendes, Quinta do Ferro and Casa Santa Eulália, all in the Vinho Verde region.

AVGOUSTIATIS

Minor Greek variety found mainly on the island of Zákynthos and in the western Pelopónnisos (Peloponnese).

BERRY COLOUR: ● ● ● ● ●

PRINCIPAL SYNONYMS: Avgoustella, Aygoustiates

ORIGINS AND PARENTAGE

Avgoustiatis probably comes from the Greek Ionian Islands or the Kykládes (Cyclades). It was named after the month of August, when the berries ripen. Although some of its synonyms in the Vitis International Variety Catalogue overlap with those given for the white-skinned variety GOUSTOLIDI, they do not appear to be related. DNA profiling with only eight markers has suggested a possible genetic relationship with MAVRODAFNI (Boutaris 2000) but this would need to be confirmed with many more markers.

VITICULTURAL CHARACTERISTICS

Mid ripening, susceptible to downy mildew but less so to powdery. Big, compact bunches.

WHERE IT'S GROWN AND WHAT ITS WINE TASTES LIKE

Avgoustiatis produces deeply coloured wines with soft tannins and highish acidity. It is grown in **Greece** on the island of Zákynthos and slightly further east on the western side of the Pelopónnisos, near Pýrgos, where Mercouri have in the past made a varietal wine and currently make a blend with MONASTRELL (Mourvèdre). The nearby estate of Brintziki also make a varietal wine and Comoutos (on Zákynthos) include Avgoustiatis in their red blends. In 2008 it was the second most planted red grape on the Ionian Islands (108 ha/267 acres) after KORINTHIAKI. Also found on the island of Kérkyra (Corfu).

AZAL

High-acid variety losing ground but still widespread in northern Portugal.

BERRY COLOUR: ● ● ● ● ●

PRINCIPAL SYNONYMS: Asal, Azal Branco, Asal da Lixa or Azal da Lixa (Amarante in Minho), Carvalha or Carvalhal, Gadelhudo (Felgueiras in Minho), Pinheira (Lousada in Minho)

VARIETIES COMMONLY MISTAKEN FOR AZAL: ALVARINHO

ORIGINS AND PARENTAGE

Azal, probably from the Vinho Verde region in northern Portugal, is genetically closely related to TRAJADURA, ARINTO DE BUCELAS and LOUREIRO, all from northern Portugal or north-western Spain (Lopes *et al.* 1999; Almadanim *et al.* 2007; Castro *et al.* 2011).

Azal is not a white mutation of Azal Tinto, which is a synonym for AMARAL (Castro *et al.* 2011).

VITICULTURAL CHARACTERISTICS

Vigorous, consistently productive, early budding, mid to late ripening (the last to be harvested for Vinho Verde). Best suited to dry, well-exposed sites. Susceptible to downy mildew, botrytis bunch rot and especially powdery mildew. Compact bunches of short but medium-sized, thick-skinned berries.

WHERE IT'S GROWN AND WHAT ITS WINE TASTES LIKE

Azal produces delicate, fresh wines with flavours of lemon and green apple but fear of rot sometimes spurs growers to pick too early, resulting in screeching acidity. It is found mainly in the Penafiel, Amarante and Basto subregions of Vinho Verde in the Minho region of northern **Portugal**. In the early twentieth century it was the most important light-skinned variety here but the area had declined to 1,480 ha (3,657 acres) by 2010, probably because it is one of the least interesting in the Vinho Verde appellation. Recommended producers include Quinta de Linhares, Casa de Vilacetinho and Manuel Nunes da Costa Camizão.

AZAL TINTO

See AMARAL.

B

BĂBEASCĂ NEAGRĂ
BABIĆ
BABICA
BACCHUS
BACHET NOIR
BAČKA
BACO BLANC
BACO NOIR
BAGA
BAGRINA
BAKHTIORI
BALZAC BLANC
BARATUCIAT
BARBAROSSA
BARBAROUX
BARBERA
BARBERA BIANCA
BARBERA DEL SANNIO
BARBERA SARDA
BARCELO
BARIADORGIA
BAROQUE
BARSAGLINA
BASTARDO MAGARACHSKY
BATIKI
BATOCA
BAYANSHIRA
BEAUNOIR
BEBA
BÉCLAN
BEICHUN
BEKARI
BELAT
BELLONE
BENA
BÉQUIGNOL NOIR
BERDOMENEL
BESGANO BIANCO
BESSARABSKY MUSKATNY
BIANCA
BIANCAME
BIANCHETTA TREVIGIANA
BIANCO D'ALESSANO
BIANCOLELLA
BIANCONE DI PORTOFERRAIO
BIANCU GENTILE
BÍBORKADARKA
BICAL
BIGOLONA
BIRSTALER MUSKAT
BISHTY
BLACK QUEEN
BLANC DAME
BLANC DU BOIS

this symbol on the following pages indicates that the synonyms, and mistaken identities, are confirmed by DNA evidence

BLANQUEIRO
BLATINA
BLATTERLE
BLAUBURGER
BLAUER PORTUGIESER
BLAUER URBAN
BLAUER WILDBACHER
BLAUFRÄNKISCH
BOBAL
BOĞAZKERE
BOGDANUŠA
BOMBINO BIANCO
BOMBINO NERO
BONAMICO
BONARDA PIEMONTESE
BONDA
BONDOLA
BONDOLETTA
BORDÔ
BORRAÇAL
BOSCO
BOUCHALÈS
BOURBOULENC
BOURRISQUOU
BOUTEILLAN NOIR
BOUVIER
BRACCIOLA NERA
BRACHETTO DEL PIEMONTE
BRAQUET NOIR
BRATKOVINA BIJELA
BRAUNER VELTLINER
BREIDECKER
BRESLAVA
BRIANNA
BRONNER
BRUGNOLA
BRUN ARGENTÉ
BRUN FOURCA
BRUSTIANO BIANCO
BUBBIERASCO
BUDAI ZÖLD
BUDESHURI TSITELI
BUKET
BUKETTRAUBE
BUSSANELLO
BUSUIOACĂ DE BOHOTIN

BĂBEASCĂ NEAGRĂ

High-acid Romanian variety producing fairy light reds.

BERRY COLOUR: ● ● ● ● ●

PRINCIPAL SYNONYMS: Asîl Kara (Daghestan), Băbească, Căldăruşă, Chernyi Redkii (Ukraine), Crăcană, Crăcănată, Poama Rară Neagră (Republic of Moldova), Rarăneagră or Rară Neagră (Republic of Moldova), Răşchirată, Rastriopa (Republic of Moldova), Serecsia Ciornaia (Republic of Moldova), Sereksia (Ukraine), Sereksiya (US)

ORIGINS AND PARENTAGE

Băbească Neagră, literally 'grandmother's black', is a very old Romanian variety that most likely originated in the Nicoreşti vineyards in the county of Galaţi in Romanian Moldova, bordering Ukraine, where it was allegedly mentioned in documents as early as the fourteenth century and where it still shows significant intravarietal diversity (an indication of its age), including colour mutations with white (Băbească Albă) and green-grey or pink berries (Băbească Gri) (Dejeu 2004).

VITICULTURAL CHARACTERISTICS

Vigorous, mid budding, late ripening. Prone to millerandage, especially if yields are high. Medium-sized to large loose bunches of thin-skinned berries. Susceptible to botrytis bunch rot, powdery and downy mildews and drought but resistant to winter frosts (to –18 °C/–0.4 °F).

WHERE IT'S GROWN AND WHAT ITS WINE TASTES LIKE

Wines are typically high in acidity, not very deeply coloured, with red-fruit flavours, mostly sour cherry. In **Romania**, there were 4,516 ha (11,159 acres) in 2008, meaning it was quite widely planted but with the majority in Galaţi in the east. Most wines are ordinary and consumed locally but some higher-quality examples come from Nicoreşti and Odobeşti, the former giving its name to an appellation for this variety. Senator are unusual in producing a more serious varietal wine from the sandy terraces north of the Dunărea (Danube) in the far south of the country.

Senator have just a few hectares of the pink-berried Băbească Gri in Huşi in the far east of Romania and make a clean, very fresh white wine with more mineral than fruit flavours and a note of lime on the finish.

The **Republic of Moldova** had 80 ha (198 acres) of Rară Neagră in 2009. Recommended producers include Dionysos Mereni. According to Galet (2000), also planted in **Ukraine**.

BABIĆ

Lively and potentially high-quality Croatian variety well suited to barrel ageing.

BERRY COLOUR: ● ● ● ● ●

PRINCIPAL SYNONYMS: Babić Crni, Rogoznička

VARIETIES COMMONLY MISTAKEN FOR BABIĆ: BABICA, PLAVAC MALI

ORIGINS AND PARENTAGE

Babić comes from the southern part of the Dalmatian coast, where it has been cultivated mainly near Šibenik and Primošten, north west of Split. Recent DNA profiling has shown that Rogoznička is a local synonym for Babić but that, contrary to common belief, BABICA, a local variety grown in the Kaštela area, is not a synonym for Babić (Zdunić, Pejić *et al.* 2008). In addition, parentage analysis has found a possible parent–offspring relationship between Babić and DOBRIČIĆ and has shown that Babica is a progeny of PLAVAC MALI (Zdunić, Pejić *et al.* 2008). Since Dobričić is a parent of Plavac Mali, Babić is either a grandparent or half-sibling of Plavac Mali, and either a great-grandparent or an uncle/aunt of Babica (see TRIBIDRAG for the full pedigree).

VITICULTURAL CHARACTERISTICS

High and consistent yields, best pruned short. Mid to late ripening. Susceptible to botrytis bunch rot.

WHERE IT'S GROWN AND WHAT ITS WINE TASTES LIKE

The quality of the wines varies enormously from site to site. If the vines are planted on deep, fertile soils and yields are high, the wine is rather ordinary. But with lower yields and in the best, old-vine, windy sites in Primošten, for example, wines are deeply coloured and full-bodied with dense, peppery berry fruit and soft tannins, making them approachable in youth, although they tend to be much more interesting after some time in oak. There were around 600 ha (1,483 acres) in **Croatia** in 2009, 1.85% of the vineyard area, mostly in the Šibenik area north west of Split, although it is also found in old interplanted vineyards in the rest of Dalmacija (Dalmatia), including the island of Korčula. The best wines come from the rocky terraces of Primošten, between Split and Šibenik. Recommended producers include Leo Gracin, Piližota, Suha Punta and Tomić.

BABICA

Rare variety from Croatia's Adriatic coast.

BERRY COLOUR:

PRINCIPAL SYNONYMS: Kaštelanka

VARIETIES COMMONLY MISTAKEN FOR BABICA: BABIĆ

ORIGINS AND PARENTAGE

Babica is an indigenous variety from the Kaštela region north of Split on the Adriatic coast of Croatia. Recent DNA profiling has shown that Babica is not a synonym for BABIĆ, and DNA parentage analysis has found that Babica is a progeny of PLAVAC MALI (Zdunić, Pejić *et al.* 2008), which makes it a grandchild of TRIBIDRAG (see TRIBIDRAG for the pedigree diagram).

VITICULTURAL CHARACTERISTICS

Late budding, mid to late ripening. Good, regular yields. Thin-skinned berries susceptible to botrytis bunch rot.

WHERE IT'S GROWN AND WHAT ITS WINE TASTES LIKE

In the last fifty years, Babica has become the main variety in Kaštela, around the town of Split, **Croatia**. It is also found to a lesser extent further north along the coast in Trogir, Benkovac and Zadar (Zdunić 2005). It is one of the varieties used in the Kaštela region to produce a light red wine known as *opolo*, often made from PLAVAC MALI or a blend of dark-skinned varieties, *opolo* being a wine description rather than an appellation. Babica, along with several other autochthonous varieties, is part of a revitalization project involving clonal selection and replanting.

BACCHUS

One of the most important German crosses producing highly aromatic whites, not least in England.

BERRY COLOUR: ● ● ● ● ●

PRINCIPAL SYNONYMS: Bacchus Weiss, Geilweilerhof 33-29-133

ORIGINS AND PARENTAGE

This (SILVANER × RIESLING) × MÜLLER-THURGAU cross was bred in 1933 by Peter Morio of the Geilweilerhof research centre at Siebeldingen in the Pfalz in Germany (now part of the Julius Kühn-Institut). It was authorized in Germany in 1972. Its parentage was confirmed by DNA profiling (Grando and Frisinghelli 1998). Bacchus and OPTIMA have the same parents, including Riesling twice, and are therefore siblings with a significant level of consanguinity. The name obviously refers to the Greek God of wine. Bacchus was subsequently used to breed BIRSTALER MUSKAT, PHOENIX and RÉSELLE.

VITICULTURAL CHARACTERISTICS

A productive variety that ripens relatively early to produce grapes that have good sugar levels, all-important in cooler climates such as Germany and England, but only moderate levels of acidity unless it fails to ripen fully. Susceptible to botrytis bunch rot.

WHERE IT'S GROWN AND WHAT ITS WINE TASTES LIKE

Germany has 2,016 ha (4,982 acres), mainly in Rheinhessen and Franken but with smaller areas in the Nahe and the Pfalz, even smaller in Baden. The area devoted to Bacchus has declined quite significantly since 1990 when plantings reached their peak at around 3,500 ha (8,649 acres). When fully ripe, the grapes produce highly aromatic wines which bear some grassy similarity to SAUVIGNON BLANC, though they tend to be more floral and exuberant, sometimes with strong notes of elderflower. In Germany it is often blended with the more grapey MÜLLER-THURGAU in basic off-dry whites but recommended varietal wines are made in Franken by Schmitt's Kinder and Waldemar Braun.

It was first planted in **England** in 1973 and became a recommended variety in 1998. There were around 130 ha (321 acres) in 2009, which is just under 10% of the total UK vineyard, making it the second most planted white variety in England and Wales after CHARDONNAY. It is most often made as a varietal wine and is valued for its ability to produce good yields of fully ripe grapes and make distinctively aromatic fresh wines, the UK's aromatic alternative to SAUVIGNON BLANC. The cooler UK climate means the wines tend to have better freshness than in Germany. Camel Valley, Chapel Down and Three Choirs all make good examples.

It is also found to a very limited extent in **Switzerland** (fewer than 2 ha/5 acres), where Jauslin make a varietal wine. There are limited plantings in **Japan** (25 ha/62 acres in 2009), eg by Furano Winery in Hokkaido.

BACHET NOIR

Ancient and almost extinct.

BERRY COLOUR:

PRINCIPAL SYNONYMS: Bachey (Aube and Aujon valleys), François Noir (Aube and Haute-Marne), Gris Bachet (Haute-Marne)

ORIGINS AND PARENTAGE

Bachet Noir is an old and almost extinct variety from the Aube *département* in the south of the Champagne region where it was first mentioned in a less than complimentary way in 1821: 'Le français, ou bachet rouge, passe pour donner du vin mediocre' ('The français, or bachet rouge, is said to give a mediocre wine'). It has been suggested that the name Bachet could derive from Buchey, a village in the Haute-Marne. However, it is more likely that this variety was named after the family name Bachet that is widespread in north-eastern France.

Ampelographers have placed Bachet Noir in the Tressot group (see p XXVII; Bisson 2009). However, DNA parentage analysis contradicts this in establishing that Bachet Noir is a natural PINOT × GOUAIS BLANC cross, just like its siblings CHARDONNAY, GAMAY NOIR, ALIGOTÉ, etc. (Bowers *et al.* 1999). These findings also explain its morphological similarity with its brother BEAUNOIR.

VITICULTURAL CHARACTERISTICS

Susceptible to chlorosis, downy and powdery mildews, and botrytis bunch rot, and vulnerable to drought, though it shows good resistance to winter frosts. Small berries. Early ripening.

WHERE IT'S GROWN AND WHAT ITS WINE TASTES LIKE

Wines are reputed to have quite high alcohol and acidity levels and a certain herbaceous character. Even though the variety is officially authorized in the Aube, it has more or less disappeared, with less than a hectare (2.5 acres) in **France** in 2006.

BAČKA

A pink-berried PETRA × BIANCA hybrid obtained in **Serbia** by P Cindrić, Nada Korać and V Kovač and named after that part of the Pannonian Plain that straddles Serbia and Hungary, though it is found mostly in northern Serbia. Authorized in 2002, it is valued mostly for its good yields and resistance to frost and fungal diseases. Varietal wines are produced by, for example, Daniel Celovski and Jaroslav Žila.

BACO BLANC

Light-skinned French hybrid still allowed in Armagnac but declining.

BERRY COLOUR: ● ● ● ● ●

PRINCIPAL SYNONYMS: Baco 22 A, Maurice Baco, Piquepoul du Gers (Armagnac)

VARIETIES COMMONLY MISTAKEN FOR BACO BLANC: PIQUEPOUL BLANC

ORIGINS AND PARENTAGE

A hybrid obtained in 1898 by François Baco in Bélus in the Landes, south-west France, by crossing FOLLE BLANCHE with NOAH. It is therefore a half-sibling of BACO NOIR. Both crosses were made with the intention of creating a hybrid with resistance to black rot. Its synonym Maurice Baco was given by François Baco in memory of his son, who died at seventeen, whereas Piquepoul du Gers is a name used in the Charente misleadingly suggesting all-*vinifera* origins.

VITICULTURAL CHARACTERISTICS

Susceptible to mites, chlorosis, flavescence dorée, and powdery and downy mildews but resistant to botrytis bunch rot. Late ripening, the berries have a slight foxy flavour.

WHERE IT'S GROWN AND WHAT ITS WINE TASTES LIKE

Baco Blanc is the only hybrid used in an appellation d'origine contrôlée: it is one of the ten varieties authorized to make the base wine for armagnac, the dark spirit from south-west **France**. There were 827 ha (2,044 acres) in France in 2008 but it is gradually declining in favour of Ugni Blanc (TREBBIANO TOSCANO). However, the threat to ban it altogether was reversed in 2005. It was once grown in **New Zealand** to a limited extent.

BACO NOIR

Dark-skinned French hybrid faring better across the Atlantic than at home.

BERRY COLOUR: ● ● ● ● ●

PRINCIPAL SYNONYMS: Baco 1, Baco 24-23, Bacoi, Bago, Bako Speiskii, Bakon

ORIGINS AND PARENTAGE

Hybrid obtained in 1902 by François Baco in Bélus in the Landes, south-west France, by crossing FOLLE BLANCHE (under the name Piquepoul du Gers) with *Vitis riparia* Grand Glabre. It is therefore a half-sibling of BACO BLANC. However, as noted by Galet (1988), since *Vitis riparia* Grand Glabre has female-only flowers with sterile pollen, Darrigan considered this hybrid to be Folle Blanche × *Vitis riparia* Grand Glabre + *Vitis riparia* ordinaire, which suggests that Baco used a mix of *Vitis riparia* pollen taken from two distinct varieties.

It was first called Baco 24-23 and then renamed Baco 1 by Prosper Gervais after his visit to Baco in 1910. It has been known as Baco Noir since 1964 (Galet 1988). It was used to breed GR7.

VITICULTURAL CHARACTERISTICS

Early budding and therefore at risk from spring frosts. Best suited to heavy soils. Vigorous and early ripening. Good resistance to downy and powdery mildews but highly susceptible to black rot and crown gall. Small to medium-sized bunches of small berries that are high in acids but low in tannins.

WHERE IT'S GROWN AND WHAT ITS WINE TASTES LIKE

Baco Noir was at one time quite widely planted in **France**, in regions as diverse as Burgundy, Anjou and its home in the Landes, but the area has dwindled considerably, down to around 11 ha (27 acres) in 2008 because it is not included in the French register of authorized varieties. There's just 1 ha (2.5 acres) over the border in **Switzerland**. Varietal wines are deeply coloured and, more importantly, they are not marked by the foxy flavours generally associated with North American vine species. Rare examples are made by Bosshardt Weine in Switzerland and by Domaine Mondon-Demeure in France.

The variety has fared better in cooler parts of North America since its introduction in the early 1950s. **New York** State crushed 820 tons in 2009 (almost double the figure for 2007) in both the Hudson River and Finger Lakes regions and in southern Oregon, but the variety is most popular in Ontario, **Canada**, where there were around 565 acres (229 ha) in 2006. Notable producers of these fresh, berry-flavoured reds include Henry of Pelham and Peller Estates. MARÉCHAL FOCH is a more popular hybrid in British Columbia but Summerhill are still making a varietal Baco Noir there. It is also found in both Novia Scotia and Québec.

BAGA

Controversial, demanding northern Portuguese variety capable of making great, ageworthy wines.

BERRY COLOUR: ● ● ● ● ●

PRINCIPAL SYNONYMS: Baga de Louro (Dão and Bairrada), Carrasquenho (Bairrada), Carrega Burros, Poeirinho (Dão and Bairrada), Tinta Bairrada or Tinta da Bairrada (Douro), Tinta de Baga (Dão and Barraida)

VARIETIES COMMONLY MISTAKEN FOR BAGA: MORETO DO ALENTEJO, PEDRAL (north-west Portugal)

ORIGINS AND PARENTAGE

Baga probably originates from Dão in the Beiras region of Portugal, where it shows the greatest morphological diversity, and from there it spread west towards the coast and the neighbouring Barraida region, where the vines show much less variability (Rolando Faustino, personal communication). The name Baga simply means 'berry' in Portuguese.

VITICULTURAL CHARACTERISTICS

Long, compact bunches of relatively large, thin-skinned berries, susceptible to botrytis bunch rot but resistant to powdery mildew. Relatively vigorous, productive, late ripening.

WHERE IT'S GROWN AND WHAT ITS WINE TASTES LIKE

Baga tends to be either loved or loathed – by growers as well as consumers. Growers face the dilemma of leaving the grapes on the vine long enough to achieve full ripeness, particularly of the tannins, versus the high risk of rot if it rains in September, as it often does in this Atlantic-influenced part of **Portugal**. Baga can therefore make the best of wines and the worst of wines: the best, generally those allowed to reach around 13% alcohol, have aromas of forest fruits when young but develop depth and complexity of black plums, herbs, olives, smoke and tobacco as the tannins soften in the bottle; the worst are thin, pale, green and astringent, especially if the wines have been fermented on the stalks, as used to be the norm. Early-picked fruit may nevertheless be useful for rosé or as the base wine for fizz.

Baga dominates the Bairrada region and DOC, the part of Beiras closest to the coast where Baga represents as much as 90% of the region's red grapes, producing its best wines in calcareous-clay soils. In Dão, further inland and higher, it is still the predominant red grape variety although the area is now declining and Baga is not authorized in Dão DOC wines, only in IGP Beiras. It is also planted elsewhere in Beiras as well as in the northern part of the Lisboa region and the Tejo, and is authorized for table wines in the Douro. There were 9,885 ha (24,426 acres) in total in Portugal in 2010.

One of the most passionate and successful proponents of Baga is Luís Pato, who makes not only fine ageworthy reds, one – Quinta do Ribeirinho Pé Franco – from ungrafted vines planted in 1988, but also an interesting sparkling red Baga. Other recommended producers in Bairrada include Campolargo, Casa de Saima, Caves São João, Adega Cooperativa de Cantanhede, Dulcinea dos Santos Ferreira, Filipa Pato and Kompassus.

BAGRINA

Pink-skinned Balkan variety making fresh and ageworthy whites.

BERRY COLOUR: ● ● ● ● ●

PRINCIPAL SYNONYMS: Bagrina Crvena (Serbia and Montenegro), Bagrina Krajinska (Serbia and Montenegro), Bagrina Roșie (Serbia and Montenegro), Braghina, Braghină (Romania and the Republic of Moldova), Braghină de Drăgășani (Romania and the Republic of Moldova), Bragina, Bragina Rara (Romania)

ORIGINS AND PARENTAGE

Bagrina is a very old Balkan variety, possibly originating from the Banat, the historical region that is today part of Romania, Serbia and Hungary although it was once all part of Hungary.

VITICULTURAL CHARACTERISTICS

Mid to late ripening. Functionally female flowers. Largish and often loose bunches of thin-skinned berries.

WHERE IT'S GROWN AND WHAT ITS WINE TASTES LIKE

Bagrina is found mostly in Drăgășani in south-west **Romania**, in Timok in eastern **Serbia** and, to a lesser extent, in **Bulgaria**. It needs to be planted near another variety that will act as a pollinator. Wines are said to be fresh and harmonious and to age well.

BAKHTIORI

Vigorous and early-ripening Central Asian variety probably originating in **Uzbekistan**. Used to produce both wine and spirits.

BALAU

See NERETTO DURO.

BALSAMINA

See MARZEMINO.

BALZAC BLANC

Almost extinct variety from Cognac country.

BERRY COLOUR: ● ● ● ● ●

PRINCIPAL SYNONYMS: Balzard Blanc, Balzat, Blanc Limousin, Margnac Blanc, Ressière

ORIGINS AND PARENTAGE

Balzac Blanc was first mentioned in the Charente by Boutard (1842), who considered it to be a variant of Balzac Noir (a synonym of MONASTRELL). However, this has been disproved by DNA profiling because Balzac Blanc has been shown to be a natural GOUAIS BLANC × CHENIN BLANC cross, just like its siblings MESLIER SAINT-FRANÇOIS and COLOMBARD (Bowers *et al.* 2000). Balzac Blanc is morphologically similar to Colombard and is a member of the Folle ampelographic group (see p XXVII; Bisson 2009).

VITICULTURAL CHARACTERISTICS

Mid ripening, susceptible to powdery and downy mildews.

WHERE IT'S GROWN

Balzac Blanc is virtually extinct and has never been included in the official register of varieties in **France**.

BÁNÁTI RIZLING

See KREACA.

BARATUCIAT

Rare but recovered Piemontese white that tastes a bit like Sauvignon Blanc.

BERRY COLOUR: ● ● ● ● ●

PRINCIPAL SYNONYMS: Bertacuciàt

ORIGINS AND PARENTAGE

From the Val di Susa in Piemonte, north-west Italy, Baratuciat does not appear in local historical or ampelographic documents but was recovered here a few years ago. Its strange name supposedly comes from the local dialect meaning 'cat's testicles'.

VITICULTURAL CHARACTERISTICS

Vigorous, productive, mid ripening. Resistant to botrytis bunch rot thanks to its thick skins. Significant level of acidity in the berries.

WHERE IT'S GROWN AND WHAT ITS WINE TASTES LIKE

Baratuciat, used for both table and wine grapes, is traditionally and exclusively cultivated in the Val di Susa, mainly in Almese, Villardora, Rubiana, Rosta and Buttigliera in north-west **Italy**. Until recently, it was grown and vinified on a very small scale, for home consumption, sometimes resulting in good late-harvest wines. However, since the early 2000s, Giorgio Falca in Almese has been producing a commercial bottling, albeit in tiny quantities, and the variety entered the official register in 2008, so other examples may follow. The wine's aromas can be quite similar to those of SAUVIGNON BLANC (box and elderflower), with a slightly bitter finish.

BARBAROSSA

Confusing name for several, possibly unrelated, varieties.

BERRY COLOUR: ● ● ● ● ●

ORIGINS AND PARENTAGE

Italian for 'red beard', the name Barbarossa has been given to many different varieties but knowledge about their true identity is currently so limited that it makes sense to treat them together.

- A grape in Toscana named Barbarossa was mentioned by Soderini (1600) but it is now said to have disappeared.
- An extremely rare Barbarossa grape is cultivated in Emilia-Romagna, where it was found in 1955 in a centenarian vineyard. It was named Barbarossa after the nickname of the Holy Roman Emperor Frederick I, who lived in that region.
- Another rare Barbarossa is cultivated in Liguria in the provinces of Savona and Genova, where it is also called Barbarossa di Finalborgo, Verduna or Verdona. It is said to be of Tuscan origin and Galet (2000) considers it to be identical to the Barbarossa from the French island of Corse. However, there is so far no DNA evidence for this.
- Two additional Barbarossa vines were mentioned in Puglia and Campania, both having distinct and unique DNA profiles (Zulini *et al.* 2002; Costantini *et al.* 2005).
- According to Galet (2000), the wine and table grape BARBAROUX cultivated in Provence (and authorized in the Cassis appellation) could be identical to Barbarossa du Piémont, which is also a table grape.

Morphological comparisons and DNA profiling of all existing Barbarossas could resolve this confusion.

WHERE IT'S GROWN AND WHAT ITS WINE TASTES LIKE

Fattoria Paradiso in Emilia-Romagna in eastern **Italy** claim to produce the only 100% Barbarossa, from vines that are the result of cuttings taken from a 100-year-old vineyard. In Liguria in the north west, Barbarossa di Finalborgo is blended in the local table wine Nostralino. The vine known as Barbarossa on the island of Corse (fewer than 150 ha/370 acres planted according to Galet 2000) is a principal component in the Ajaccio appellation and a lesser one in Vin de Corse.

BARBAROUX

Almost extinct, relatively undistinguished pink-skinned Provençal variety.

BERRY COLOUR: ● ● ● ● ●

PRINCIPAL SYNONYMS: Barbarons (Toulon, Provence), Grec Rose, Roussée

VARIETIES COMMONLY MISTAKEN FOR BARBAROUX: BARBAROSSA (Liguria and Piemonte), GÄNSFÜSSER

ORIGINS AND PARENTAGE

Barbaroux is an old table and wine grape already cited in 1667 in the area of Toulon in Provence, southern France, under its synonym Barbarons: 'le Maroquin ou barbarons, est un gros Raisin violet, qui a des grappes d'une grosseur extraordinaire' ('Maroquin or barbarons is a big violet-berried grape that has extraordinarily large bunches'). Several other references were recorded between the seventeenth and nineteenth centuries in Provence, where its modern spelling first appeared in 1715. The name Barbaroux supposedly derives from the Provence dialect *barbarous*, meaning 'variety with reddish berries'.

DNA profiling recently showed that Barbaroux is different from all the various grape varieties collectively called BARBAROSSA in Liguria and in Piemonte (Torello Marinoni, Raimondi, Ruffa *et al.* 2009). However, it is not yet known whether Barbaroux is identical to the variety cultivated on the French island of Corse under the name Barbarossa. It should also be noted that Barbaroux is sometimes called Gänsfüssler, although it is not genetically related to true GÄNSFÜSSER from Rheinland-Pfalz in Germany (Vouillamoz).

VITICULTURAL CHARACTERISTICS

Fertile but not especially vigorous. Early budding, mid ripening. Susceptible to powdery mildew and botrytis bunch rot. Big berries and bunches that require a long ripening season.

WHERE IT'S GROWN AND WHAT ITS WINE TASTES LIKE

Barbaroux vineyards are becoming increasingly rare in Provence, **France**, and there remained just 38 ha (94 acres) in 2008. Wines are generally deep gold in colour, aromatic but light and soft. The variety is authorized as a minor component in the red and rosé versions of the Cassis appellation. It may also play a minor role in Côtes de Provence red and rosé but only if the vines were planted before 1994, suggesting the variety's demise is pretty certain.

BARBERA

With origins still unknown, this versatile variety makes crisp, sweet Italian reds in a wide range of styles and locations.

BERRY COLOUR:

PRINCIPAL SYNONYMS: Barbera a Peduncolo Rosso, Barbera a Raspo Verde, Barbera Amaro, Barbera d'Asti, Barbera Dolce, Barbera Fina, Barbera Grossa, Barbera Nera, Barbera Nostrana, Barbera Vera, Barberone, Gaietto, Lombardesca, Sciaa

VARIETIES COMMONLY MISTAKEN FOR BARBERA: BARBERA DEL SANNIO (Campania), BARBERA SARDA (Sardegna), Barberùn (Piemonte), MAMMOLO (under the name Sciaccarello on Corse), NERETTO DURO (under the name Barbera Rotonda in Piemonte), PERRICONE (Sicilia)

ORIGINS AND PARENTAGE

Some authors have suggested that the name Barbera derives from Vinum Berberis, a medieval fermented drink made from barberry with an intense red colour and an acidic and astringent taste, a bit like Barbera wines.

Barbera is usually believed to have originated in the hills of Monferrato in central Piemonte (see below). However, to the best of our knowledge, the first reliable reference to the name Barbera came at the end of the eighteenth century, when Conte Nuvolone Pergamo (1787–98) mentioned a variety known as Barbera near Asti in north-west Italy, as Ughetta near Vercelli and in the Canavese, and as VESPOLINA near Novara. (Ughetta is used as a synonym for both Barbera and Vespolina, which are completely distinct varieties.) The Count also mentioned a *Vitis vinifera Montisferratensis* that was often identified with Barbera, which is doubtful because this geographical name could be applied to any other historical grape from the same region (eg FREISA).

Since the nineteenth century, Barbera has often been described as typical of the Asti region. However, it didn't become important in Piemonte as a whole until after phylloxera. The lack of trustworthy records for Barbera in Piemonte before 1798 and its relatively recent expansion suggest that its origin might be elsewhere. Indeed, a recent DNA study unexpectedly showed that Barbera bears little relation to the other grapes of Piemonte, which suggests that Barbera was probably introduced to the region relatively recently (Schneider *et al.* 2003). In addition, the lack of any other record of Barbera in the rest of Italy before the eighteenth century suggests that it began life relatively recently through a natural cross in a vineyard somewhere, though probably not in Piemonte, for the reasons given above. Additional genetic research is needed to uncover Barbera's birthplace.

DNA profiling has also shown that BARBERA BIANCA and BARBERA DEL SANNIO are not related to Barbera (Sefc *et al.* 2000; Costantini *et al.* 2005). In Piemonte, Barbera is also distinct from the Barberùn scattered in old vineyards (Schneider and Mannini 2006).

Barbera was used to breed ALBAROSSA, CORNAREA, ERVI, INCROCIO TERZI 1, NEBBIERA, NIGRA, PRODEST, SAN MICHELE and SOPERGA.

OTHER HYPOTHESES

Legend has it that Barbera was introduced to Piemonte by the Lombardians in the seventh century AD. Later, a document found in the capitular archives of Casale Monferrato dated 7 November 1249 reports the church of Sant'Evasio lending to a local farmer a parcel to be planted 'bonis vitibus berbexinis'. Several historians and wine writers erroneously transcribed *berbexinis* as *barbexinis* and hypothesized that this grapevine was Barbera. Yet it is much more likely that *berbexinis* refers to Berbesino, mentioned by several ampelographers as a synonym for GRIGNOLINO, and sometimes for ORTRUGO or NEBBIOLO, but never for Barbera.

Others have speculated that the Grisa mentioned by Pietro de Crescenzi in his *Trattato dell'agricoltura* (1304), the Grissa found in 1514 in the archives of Chieri in Piemonte and the Grisa Maggiore mentioned by Croce (1606) were all Barbera. However, since these names refer to the grey colour of the berries, it is doubtful that they would correspond to Barbera, and those names could have been given to several other grapes.

One recent source suggested that Barbera may be related to MONASTRELL (Mourvèdre) but DNA data strongly disprove this hypothesis (Vouillamoz).

Other hypotheses suggested that Barbera was identical to Sicilia's PERRICONE or to Corse's Sciaccarello – both disproved by DNA profiling, Sciaccarello being identical to MAMMOLO (Vouillamoz).

VITICULTURAL CHARACTERISTICS

Productive, ripens relatively late (after DOLCETTO but before NEBBIOLO) but retains a high level of acidity even when fully ripe and a deep ruby colour.

WHERE IT'S GROWN AND WHAT ITS WINE TASTES LIKE

Barbera is the most typical and widespread variety in Piemonte. It is also cultivated in Lombardia, Emilia-Romagna as well as in some spots in southern **Italy**. The Italian census of 2000 shows total plantings of 28,365 ha (70,090 acres), a dramatic decrease since the mid-1980s when more than thirty people died after drinking methanol-enhanced Barbera, but it is still Italy's third most planted black variety after SANGIOVESE and MONTEPULCIANO. More than 60% of the total is in Piemonte.

At their best, Barbera wines are deeply coloured, fresh and relatively soft wines with bright cherry fruit flavours.

In its Piemonte heartland, Barbera gives its name to Barbera d'Asti DOCG (elevated from the 2008 vintage and allowing up to 15% FREISA, GRIGNOLINO and/or DOLCETTO), Barbera d'Alba DOC and Barbera di Monferrato DOC (which also allows up to 15% Freisa, Grignolino and/or Dolcetto). The best areas are thought to be the hills immediately to the north and south of Alba and Monforte d'Alba in the Alba DOC and, in the province of Asti, the area from Nizza Monferrato north west towards Vinchio, Castelnuovo Calcea, Agliano, Belveglio and Rocchetta Tanaro. In recognition of this, Nizza, Tinella and Colli Astiano (or simply Astiano) have been recognized as specific subzones of Barbera d'Alba, and Barbera Monferrato Superiore had been granted DOCG status from the 2008 vintage. Growers in Asti argue that Barbera d'Asti can be superior to the generally higher-priced Barbera d'Alba because the best vineyards around the town of Alba are given over to NEBBIOLO (for the likes of Barolo and Barbaresco), whereas those around Asti are more likely to be planted with the local speciality Barbera.

Where the vine is overcropped (the yield allowances are generous – up to 70 hl/ha in Alba), the wines can be thin and mean. However, Barbera has been taken more seriously in the last twenty or so years and is now made in a wide range of styles, from cheap, young, simple, tart and fruity to far more expensive, dense oaked versions that benefit from ageing. This is partly due to the size and heterogeneity of the area but also to the divergent ambitions of the producers. The French wine consultant Émile Peynaud had suggested the use of oak in the 1970s but the first to experiment with this style was Giacomo Bologna with his pioneering Bricco dell'Uccellone. The use of new barriques – not to mention the more concentrated wines that tend to accompany this approach – certainly firms up the structure of such wines and adds extra body and spicy flavours but it is not universally admired and tends to significantly change the more traditional bright cherry flavours of the variety. The pure sweet fruit of the Barbera grape is arguably more apparent in less heavily oaked examples.

Most producers of great Barolo and Barbaresco in Piemonte are also able to produce fine Barbera so there is a host of possibilities (see NEBBIOLO), but some of those whose Barberas are obviously superior are Elio Altare, Domenico Clerico, Conterno-Fantino, Corregia and Scavino.

In Lombardia, and more specifically in Oltrepò Pavese, Barbera is widely planted and produces easy varietal wines, some of them fizzy, or is often blended with CROATINA or BONARDA PIEMONTESE. It is also permitted though used less and less in the Terre di Franciacorta DOC and appears in many a vino da tavola. In Emilia-Romagna, it constitutes 55–70% of the Guttornio version of Colli Piacentini, the hills above Piacenza, up to 60% in the Novello version and 85–100% in the varietal wines. Lighter versions, frequently fizzy, are produced in the Colli Bolognesi and Colli di Parma. In central and southern Italy its role is generally to add acidity to blends.

Further from home, Italian immigrants made sure Barbera had a home in **California** (more than 17,300 acres/7,000 ha in 2008). It has proved more popular than the noble NEBBIOLO in California with older vines in the Sierra Foothills proving particularly successful, as witness the wines of Yorba and Jeff Rundquist. The variety benefited from the Cal-Ital vogue.

Argentina had nearly 800 ha (1,977 acres) in 2007, with Norton making one of the better versions.

Along with several other Italian varieties, Barbera's reputation has recently begun to climb in **Australia**, despite slow progress since it was introduced in the 1960s. Recommended producers include Brown Bros, Gary Crittenden, Dal Zotto, Gapsted, Mount Langi Ghiran and Michelini. Centennial Vineyards in New South Wales make an Amarone-style wine.

Limited plantings are also found in **South Africa** (50 ha/124 acres in 2008), **Slovenia** (139 ha/343 acres), **Greece** (0.5 ha/1.24 acres on Alpha Estate in Makedonía) and **Israel**.

BARBERA BIANCA

Rare Piemontese variety unrelated to dark-skinned Barbera.

BERRY COLOUR: ● ● ● ● ●

PRINCIPAL SYNONYMS: Bertolino, Peisìn

VARIETIES COMMONLY MISTAKEN FOR BARBERA BIANCA: CARICA L'ASINO

ORIGINS AND PARENTAGE

Barbera Bianca supposedly has clusters and berry shape similar to those of the red BARBERA, as well as similarly high acidity. However, it has a distinct DNA profile and therefore cannot be considered a colour mutation of it (Sefc *et al.* 2000; Schneider and Mannini 2006). Barbera Bianca has been known in Piemonte in north-west Italy since only 1825.

VITICULTURAL CHARACTERISTICS

High yields, mid ripening, high tartaric-acid content.

WHERE IT'S GROWN AND WHAT ITS WINE TASTES LIKE

Barbera Bianca is now cultivated mainly in old vineyards in the Acqui and Alessandria regions in Piemonte, north-west **Italy**, where it is allowed in the Colli Tortonesi Bianco DOC. Barbera Bianca is usually blended with MUSCAT BLANC À PETITS GRAINS (Moscato) and other local varieties such as CORTESE, VERMENTINO (Favorita) and TIMORASSO. Some producers use it for a light sparkling wine because of its high acidity. The 2000 census records a total of 280 ha (690 acres) in Italy.

BARBERA DEL SANNIO

Increasingly well-made central Italian variety unrelated to more northerly Barberas.

BERRY COLOUR: ● ● ● ● ●

PRINCIPAL SYNONYMS: Barbetta

VARIETIES COMMONLY MISTAKEN FOR BARBERA DEL SANNIO: BARBERA (Piemonte)

ORIGINS AND PARENTAGE

Barbera del Sannio is local to Castelvenere in the Valle Telesina, in the province of Benevento, Campania, central Italy. It was possibly first described in 1844 by Gasparrini under the vague name Lugliese or Lugliatica (*Luglio* means 'July' in Italian) and later renamed Barbera by Froio (1875) because of its similarity to the variety that is today famous in Piemonte and with which it is still confused in the wine regulations (Manzo and Monaco 2001). However, DNA profiling has shown that Barbera del Sannio is distinct from BARBERA and is genetically related to CASAVECCHIA, CATALANESCA and Summariello (NERO DI TROIA), although they are all quite different from other Campanian varieties and might well represent recent introductions (Costantini *et al.* 2005).

VITICULTURAL CHARACTERISTICS

Vigorous. Relatively resistant to botrytis bunch rot thanks to thick skins.

WHERE IT'S GROWN AND WHAT ITS WINE TASTES LIKE

Barbera del Sannio used be a rather anonymous colouring ingredient in blends. Today, however, an increasing number of producers are making deeply coloured, medium-bodied, perfumed varietal wines which have been authorized within the Sannio appellation in Campania, **Italy**, since 1997. Recommended producers include 'A Canc'llera and Antica Masseria Venditti (using the synonym Barbetta) and Anna Bosco.

BARBERA SARDA

Renascent Sardinian red valued for its fresh acidity.

BERRY COLOUR: ● ● ● ● ●

VARIETIES COMMONLY MISTAKEN FOR BARBERA SARDA: BARBERA, MAZUELO

ORIGINS AND PARENTAGE

From Sardegna, Barbera Sarda is identical to neither BARBERA nor MAZUELO (Carignan), as has been suggested. Some authors consider it to be an offspring of Barbera from Piemonte but this has not yet been proven by DNA profiling.

VITICULTURAL CHARACTERISTICS

Late ripening.

WHERE IT'S GROWN AND WHAT ITS WINE TASTES LIKE

Barbera Sarda is cultivated only on Sardegna, **Italy**, where it is recommended mainly in the provinces of Nuoro, Cagliari, Oristano and Sassari. It is generally used to bolster acidity in blends. The last years of the 2000s saw the start of a revival, eg in Cantina di Dolianova's Terresicci regional wine (85% Barbera Sarda, 10% SYRAH, 5% MONTEPULCIANO).

BARCELO

Minor Portuguese found in white Dão blends.

BERRY COLOUR:

PRINCIPAL SYNONYMS: Barcello, Barcelos

ORIGINS AND PARENTAGE

Barcelo is a Portuguese variety that may well come from the Dão region in Portugal. Its DNA profile (Veloso *et al.* 2010) is almost identical to that of AMARAL (Pinto-Carnide *et al.* 2003) but further genetic and ampelographic analysis is needed to confirm their identity.

VITICULTURAL CHARACTERISTICS

Early to mid budding, mid ripening. Susceptible to coulure.

WHERE IT'S GROWN AND WHAT ITS WINE TASTES LIKE

Barcelo is found mainly in the Beiras region of central **Portugal** and is authorized in Dão DOC, usually as part of a blend, and in IGP Beiras. Wines are typically fruity, fresh, perfumed and well balanced – not unlike ENCRUZADO but with lower alcohol. There were 35 ha (86 acres) in Portugal in 2010.

BARIADORGIA

Virtually extinct Sardinian white, surviving on the French island of Corse as Carcajolo Blanc.

BERRY COLOUR:

PRINCIPAL SYNONYMS: Bariadorgia Bianca, Bariadorza, Barria Dorgia or Barriadorgia (Sardegna in Italy), Carcaghjolu Biancu, Carcajola, Carcajola Bianco (Sardegna), Carcajolo Blanc (Corse in France), Gregu Bianco (Sardegna)

VARIETIES COMMONLY MISTAKEN FOR BARIADORGIA: BIANCU GENTILE (Corse)

ORIGINS AND PARENTAGE

Bariadorgia is from the area of Alghero on the Italian island of Sardegna and identical to Carcajolo Blanc, specific to the area of Sartène in the south west of the French island of Corse, where it was first mentioned in 1822 under the name Carcajola. DNA profiling suggests that Bariadorgia is also identical to Gregu Bianco, specific to the Campidano area on Sardegna, because the DNA profile of Gregu Bianco published by De Mattia *et al.* (2007) turned out to be identical to that of Carcajolo Blanc (Vouillamoz). This is consistent with the theory that Carcajolo was introduced from Sardegna (Foëx 1904). The name Carcajolo comes from the Italian *caricagiola*, meaning 'heavily laden'. On Corse, Carcajolo Blanc is often confused with another Corsican variety, BIANCU GENTILE.

Carcajolo Blanc is not a colour mutation of Carcajolo Noir, the latter being identical to PARRALETA (called Caricagiola Nera on Sardegna).

VITICULTURAL CHARACTERISTICS

Mid ripening, needing a warm, well-exposed site to ripen fully. Susceptible to powdery mildew and botrytis bunch rot. Large bunches, biggish berries in which sugar levels tend to be on the low side.

WHERE IT'S GROWN AND WHAT ITS WINE TASTES LIKE

In Sardegna, **Italy**, Bariadorgia is almost extinct. Within **France**, Bariadorgia is authorized only on the island of Corse, where it is known as Carcajolo Blanc or Carcajola Bianco but where only a few vines survive in Figari and Sartène in the south. Wines tend to be high in acidity with only moderate alcohol levels. Plantings are so limited that the variety does not appear in the French official statistics for 2008.

BAROQUE

South-west French variety given an injection of life from a renowned local chef.

BERRY COLOUR:

PRINCIPAL SYNONYMS: Baroca (Tursan), Barroque (Tursan), Bordalès (Landes), Bordelais Blanc (Landes), Bordeleza Zuria (Pays Basque)

ORIGINS AND PARENTAGE

Baroque is a variety of the Landes *département*, south-west France, most likely originating from the area of Tursan and Chalosse in the Adour basin. Its earliest mention was in 1894 in a local vine nursery. The name Baroque could derive from the name of a hamlet near Eyres-Moncube, where there were significant plantings. Lavignac (2001) offers an alternative etymology: Baroque is a French adaptation of the Gascon *varòc* or *varòca*, from *vara*, meaning 'pole' or 'stick'. In Castilian, *vara* means a long cane left during pruning and the suffix *–òc* points to the variety's tendency to produce stick-like buds. The variety is usually pruned to long canes.

It belongs to the Courbu ampelographic group (see p XXVII), and is morphologically close to LAUZET and to TANNAT (Bordenave *et al.* 2007; Bisson 2009).

Baroque was used to breed LILIORILA.

OTHER HYPOTHESES

Baroque is sometimes said to be a natural FOLLE BLANCHE × SAUVIGNON BLANC cross but this has been disproved by DNA parentage analysis (Vouillamoz).

VITICULTURAL CHARACTERISTICS

Vigorous, late ripening, sensitive to drought. Bunches are very compact, making them highly susceptible to botrytis bunch rot. Good resistance to powdery mildew and black rot.

AGLIANICO

pp 11 – 13

ALIGOTÉ

pp 34–5

ALTESSE

pp 36–8

AMIGNE

pp 41 – 2

ARVINE

pp 58–9

BAGRINA

p 81

BIANCOLELLA

p 104

PORTUGAIS BLEU

see BLAUFRÄNKISCH

pp 116–18

CABERNET FRANC

pp 153–7

CABERNET SAUVIGNON

pp 160–68

CANAIOLO NERO

pp 182–3

CARMENÈRE

pp 189–92

CATARRATTO BIANCO

pp 204–5

CHAOUCH

see CAVUŞ

JAÉN BLANCO

see CAYETANA BLANCA

pp 208–11

PINOT BLANC CHARDONNAY

see CHARDONNAY

pp 221 – 7

WHERE IT'S GROWN AND WHAT ITS WINE TASTES LIKE

Plantings of Baroque, which became popular in the late nineteenth and early twentieth centuries because of its resistance to powdery mildew, have declined significantly in the last fifty years. There were 112 ha (277 acres) in **France** in 2008 compared with 5,656 ha (13,976 acres) in 1958. The Tursan appellation requires between 30% and 80% of this variety, blended with local varieties PETIT MANSENG or GROS MANSENG or international varieties SAUVIGNON BLANC and CHENIN BLANC. Recent investment in the appellation by the local chef Michel Guérard of Eugénie-les-Bains has saved this characterful variety from extinction and it now represents about 60% of the white wine vines in Tursan. The variety may also be a minor component in the vin de liqueur Floc de Gascogne. Varietal wines tend to be slightly floral or nutty and fresh although Guérard's Baron de Bachen has high alcohol and a fine aroma of ripe pears.

BARSAGLINA

Deeply coloured Tuscan red that has all but disappeared.

BERRY COLOUR: ● ● ● ● ●

PRINCIPAL SYNONYMS: Barsullina, Bersaglina, Massareta (Liguria), Massaretta☒ (Liguria)

ORIGINS AND PARENTAGE

Some have suggested that this Tuscan variety from Massa-Carrara, hence the synonym Massaretta in Liguria (Torello Marinoni, Raimondi, Ruffa *et al.* 2009), shares certain characteristics with SANGIOVESE.

VITICULTURAL CHARACTERISTICS

Vigorous, sensitive to powdery mildew and ripens in the second half of September. Rich in polyphenols, mainly anthocyanins, hence its deep colour.

WHERE IT'S GROWN AND WHAT ITS WINE TASTES LIKE

The main area of cultivation used to be in the province of Massa-Carrara in Toscana, **Italy**, where it had almost disappeared but was rescued by Paolo Storchi from a few vines at Podere Scurtarola, where it makes up around 15% of the Scurtarola Rosso blend. Mannucci Droandi produce a varietal Barsaglina labelled IGT Toscana. Wines have an intense ruby colour, fruity aromas and hints of violets and are quite tannic. Barsaglina is authorized as part of a blend in Toscana's Colli di Luni DOC and in Val di Magra IGT. Fewer than 30 ha (75 acres) were recorded in the Italian census of 2000. It is also said to be grown in Liguria under the name Massaretta.

BASTARDO

See TROUSSEAU.

BASTARDO MAGARACHSKY

Versatile Ukrainian cross.

BERRY COLOUR: ● ● ● ● ●

PRINCIPAL SYNONYMS: Bastard de Magaraci, Bastardo Magarach, Magaratch 217, Magarach Bastardo

ORIGINS AND PARENTAGE

Bastardo Magarachsky is a TROUSSEAU × SAPERAVI cross obtained in 1949 by N Paponov and V Zotov at the Magarach research centre in Yalta, southern Ukraine, and named after the breeding institute and Bastardo, a synonym for Trousseau. Included in the official register of varieties in 1969.

VITICULTURAL CHARACTERISTICS

Compact bunches of medium-sized to large berries. Budburst to maturity: 146 days. Relatively resistant to downy and powdery mildews and to drought.

WHERE IT'S GROWN AND WHAT ITS WINE TASTES LIKE

Bastardo Magarachsky is widely planted in **Ukraine**, with a total of 1,330 ha (3,287 acres) in 2009. It is used to produce wines at various sweetness levels and for the table. Magarach Winery make a sweet varietal version and use it in dry, sweet and fortified blends with other local varieties or CABERNET SAUVIGNON. Inkerman make a dry varietal wine and Massandra also blend it with Cabernet Sauvignon to make an oak-aged dessert wine. It is said to have chocolate-like flavours. Sun Valley (Solnechnaya Dolina) make a dry blend with PINOT NOIR called Meganom.

The variety is also found in the **Republic of Moldova, Romania, Russia** and **Central Asia**, apparently.

BATIKI

Variety grown mainly in central Greece making soft whites for blending, retsina and the table.

BERRY COLOUR: ● ● ● ● ●

PRINCIPAL SYNONYMS: Bantiki, Dembatiki, Deve Baliki, Dimbatiki, Dimi Batiki, Timbi Batiki

ORIGINS AND PARENTAGE

Batiki is said to originate in the region of İzmir, historically known as Smyrna, on Turkey's Aegean coast (Nikolau and Michos 2004).

VITICULTURAL CHARACTERISTICS

Vigorous, productive, late budding, early to mid ripening. Large compact bunches of large berries, prone to attack by bees, birds and wasps but resistant to drought. Susceptible to downy and powdery mildews and botrytis bunch rot.

WHERE IT'S GROWN AND WHAT ITS WINE TASTES LIKE

Batiki, found mainly in Tirnavos in Thessalía, central **Greece**, but also further north in Makedonía, Thráki (Thrace), Dráma and Thessaloníki, produces lightly aromatic wines low in alcohol and acidity, used mainly in blends and in resinated wines such as retsina. Dougos in Larísa (Thessalía) blend Batiki with RODITIS and Yannis Voyatzis has recently planted Batiki at his family vineyard in Velvendos. Batiki has traditionally been used for the table as well as for wine.

BATOCA

Minor, unexceptional white from Vinho Verde country.

BERRY COLOUR: ● ● ● ● ●

PRINCIPAL SYNONYMS: Alvadurão Portalegre, Alvaraça (Douro), Alvarça, Alvaroça, Alvaroco, Asal Espanhol (Amarante), Batoco, Blanca Mar✕ (Galicia in Spain), Espadeiro Branco (Lousada), Sa Douro or Sedouro (south of Minho).

ORIGINS AND PARENTAGE

Batoca is a variety from the Vinho Verde region in north-western Portugal. Comparison of DNA profiles reveals that Batoca is identical to the Spanish variety Blanca Mar in Vilanova *et al.* (2009). Although it is called Espadeiro Branco in Lousada, it is ampelographically distinct from, and not a colour mutation of, ESPADEIRO (Veloso *et al.* 2010).

VITICULTURAL CHARACTERISTICS

High but irregular yields due to its susceptibility to fungal diseases.

WHERE IT'S GROWN AND WHAT ITS WINE TASTES LIKE

Batoca produces soft, moderate-quality wines and although it is authorized in **Portugal** in the Vinho Verde DOC (except in the Monção, Lima and Amarante subregions), there were just 15 ha (37 acres) in Portugal in 2010. Quinta Santa Cristina make a rare varietal example.

BAYANSHIRA

Lacklustre Azeri variety with high acidity and generally low alcohol.

BERRY COLOUR: ● ● ● ● ●

PRINCIPAL SYNONYMS: Ag Shirei, Ag Üzüm (Armenia), Bahïan Chireï, Bajac Shirei, Banants (Armenia), Bayan Shirei, Bayan Shirey or Bayanshire, Shirei, Spitak Khagog (Armenia)

ORIGINS AND PARENTAGE

Bayanshira most likely originates in the village of Bayan (Banants in Armenian, hence the synonym) in the Dashkasan region of Azerbaijan (Chkhartishvili and Betsiashvili 2004).

OTHER HYPOTHESES

According to Kolenati in Chkhartishvili and Betsiashvili (2004), Bayanshira could originate in the Mount Ararat area in Armenia, where numerous wild grapevines (*Vitis vinifera* subsp.

silvestris) supposedly thrive, but this is unlikely because no wild vines have ever been recorded in this area.

VITICULTURAL CHARACTERISTICS

Vigorous, productive, early budding, late ripening. Highly susceptible to pests and fungal diseases and at risk from winter frosts but it tolerates summer drought.

WHERE IT'S GROWN AND WHAT ITS WINE TASTES LIKE

Bayanshira, which makes wines in **Azerbaijan** which are typically crisp, moderate in alcohol but overall pretty inferior, is grown for the table and for juice as well as for a range of wine styles. Zaman in the Samukh region, for example, use it to make sparkling, dry, dessert and fortified wines. It is sometimes distilled into brandy and is also found in other **former Soviet republics**.

BEAUNOIR

Virtually extinct black-skinned variety with one illustrious and one promiscuous parent.

BERRY COLOUR: ● ● ● ● ●

PRINCIPAL SYNONYMS: Beaunoire (Vallée de l'Ource), Cep Gris, Mourillon (Gyé-sur-Seine), Pinot d'Aï, Pinot d'Ailly, Pinot d'Orléans (Riceys), Sogris (Vallée de la Seine, Vallée de la Laignes)

ORIGINS AND PARENTAGE

Beaunoir is an old and almost extinct variety of the Aube *département*, in the south of the Champagne region in north-east France. The literal meaning of the name is 'beautiful black' but it may also derive from the city of Beaune in Burgundy, where the variety could originate. Indeed, DNA parentage analysis has shown that Beaunoir is a natural PINOT × GOUAIS BLANC cross, just like its siblings CHARDONNAY, GAMAY NOIR, ALIGOTÉ, BACHET NOIR, etc. (Bowers *et al.* 1999). This probably explains why Beaunoir is also called Mourillon, Pinot d'Aï, Pinot d'Ailly and Pinot d'Orléans, and also why Bisson (2009) described Beaunoir as morphologically close to its sibling Bachet Noir. Misleadingly, the name Beaunoir is sometimes given to PINOT BLANC.

VITICULTURAL CHARACTERISTICS

Compact bunches, vigorous and prolific. Mid ripening.

WHERE IT'S GROWN AND WHAT ITS WINE TASTES LIKE

Beaunoir is not included in France's official register of varieties and is not therefore authorized for quality wines. It has more or less disappeared from the Aube and from its other erstwhile home further to the south and east in Châtillon-sur-Seine in the far north of Burgundy. According to Galet (2000), the wines tend to be light in colour and alcohol and rather ordinary.

BEBA

Andalusian variety grown mainly in Extremadura today.

BERRY COLOUR:

PRINCIPAL SYNONYMS: Beba de los Santos, Beba Dorada de Jerez, Blanca de Mesa (Valladolid), Blanca Superior Parral, Boal de Praça (Portugal), Breval (Islas Canarias), Calop Blanco, Eva, Malvoisie de la Chartreuse or Malvoisie des Chartreux (France), Teta de Vaca, Uva de Planta, Valencí Blanco

VARIETIES COMMONLY MISTAKEN FOR BEBA: CHELVA, DORADILLA

ORIGINS AND PARENTAGE

Beba was first mentioned by Clemente y Rubio (1807) as a table and wine grape cultivated in Sanlúcar de Barrameda, Jerez, Trebujena, Chipionia and Conil in Andalucía, southern Spain. It is often called Beba Dorada de Jerez because the berries take on a golden (*dorada*) tint under the sun. Although Beba is sometimes called Chelva, it should not be confused with the true CHELVA from Extremadura that has a distinct DNA profile (Fernández-González, Mena *et al.* 2007). Recent DNA analyses have mistakenly identified the reference sample of Beba in the United States Department of Agriculture's National Clonal Germplasm Repository as TROUSSEAU (Myles *et al.* 2011) but other published DNA profiles of Beba contradict this (Vouillamoz).

The red-berried **Calop Rojo** is a colour mutation of Calop Blanco, and both have the same DNA profile as Beba (Martín *et al.* 2003; Laiadi *et al.* 2009). Neither is grown commercially today.

VITICULTURAL CHARACTERISTICS

Loose bunches of medium-sized to large berries. Late ripening. Vigorous but with irregular yields because it is very sensitive to poor weather at flowering.

WHERE IT'S GROWN AND WHAT ITS WINE TASTES LIKE

Beba is grown mainly in the province of Badajoz in Extremadura, south-west **Spain**, in particular in Los Santos de Maimona, and to the north in the province of Cáceres and south around Huelva and Sevilla (Asensio Sánchez 2000). There was a total of 3,547 ha (8,765 acres) in Spain in 2008 but it is difficult to know how much is grown for wine and how much for the table – and it is authorized in several DOs in the Islas Canarias and in Andalucía. Wines are fresh and fruity, often produced within the Ribera del Guadiana DO.

It is also found to a very limited extent over the border in **Portugal**'s Algarve – less than a hectare (2.5 acres), according to official 2010 statistics.

BÉCLAN

Virtually extinct Jura vine.

BERRY COLOUR:

PRINCIPAL SYNONYMS: Baclan, Baclans, Becclan, Petit Béclan, Petit Dureau, Petit Margillin, Roussette Noire (Lons-le-Saunier), Saunoir (Saint-Amour)

VARIETIES COMMONLY MISTAKEN FOR BÉCLAN: DURIF (under the name Petite Sirah), PELOURSIN, PINOT NOIR

ORIGINS AND PARENTAGE

Béclan is a variety from the Jura in eastern central France, where it was first mentioned in 1732 under its old synonym Baclans in a decree of the Besançon parliament: 'ne laisser subsister en plants noirs ou blancs que le pulsard noir, le gros & le petit noirin dit pinot, le baclans ... qui sont les seuls bons plants' ('for black or white grape varieties, keep only pulsard noir, gros & petit noirin known as pinot, baclans ... which are the only good varieties'). The name Béclan could derive from the local dialect *bècle*, meaning 'climbing vine' (French *treille*). In the Jura, this variety is called Petit Béclan to distinguish it from Gros Béclan, a synonym for PELOURSIN. The similarity between the two varieties is confirmed by ampelographers who place Béclan in the Pelorsien ampelographic group (see p XXVII), named after Peloursin (Bisson 2009).

Béclan has been confused with PELOURSIN (called Gros Béclan), DURIF (under the name Petite Sirah in California) and PINOT NOIR (Rouget 1897) but all these hypotheses have been shown by DNA profiling to be incorrect (Meredith *et al.* 1999).

VITICULTURAL CHARACTERISTICS

Mid ripening, susceptible to downy mildew. Small, compact bunches and small berries.

WHERE IT'S GROWN AND WHAT ITS WINE TASTES LIKE

Officially less than 1 ha (2.5 acres) is left in **France**. Domaine Jean Bourdy planted some Béclan vines in 2006 as part of a project to conserve local varieties and Domaine Ganevat in Rotalier have replanted a small area and include the grapes in an early-drinking red blend. It can produce lively wines with moderate alcohol.

BÉCUET

See PERSAN.

BEGLERI

See THRAPSATHIRI.

BEICHUN

Winter-hardy Chinese hybrid.

BERRY COLOUR: ● ● ● ● ●

PRINCIPAL SYNONYMS: Bei Chun

ORIGINS AND PARENTAGE

Beichun is a hybrid obtained in 1954 at the Institute of Botany of the Chinese Academy of Sciences by crossing MUSCAT OF HAMBURG with a *Vitis amurensis* Ruprecht vine, a Chinese wild species. Its name is a contraction of Beijing and *chun,* meaning 'alcohol'. It is one in a series of hybrids in the so-called 'Bei' series of crossings of *vinifera* with *amurensis* varieties, along with Beimei, Beihong and Beiquan, for example.

VITICULTURAL CHARACTERISTICS

Vigorous, productive, late ripening. Good resistance to winter cold, drought and humidity but susceptible to downy and powdery mildews.

WHERE IT'S GROWN AND WHAT ITS WINE TASTES LIKE

In the 1970s and 1980s Beichun's popularity in **China** increased considerably. With national total plantings of around 7,000 ha (17,297 acres; Wang *et al.* 2010), it is widely planted around Beijing in the province of Hebei as well as in Shandong, Henan and Liaoning. It can survive without being buried in winter in northern China and fares better in the humid conditions of southern China than any pure *Vitis vinifera* variety (Li *et al.* 2007). Wines are ruby-coloured, soft and refreshing but the flavour is dominated by the *Vitis amurensis* genes.

BEKARI

Minor variety local to Ípeiros (Epirus) in western Greece.

BERRY COLOUR: ● ● ● ● ●

PRINCIPAL SYNONYMS: Bekaro, Mbekari Mavro or Mpekari Mavro, Mbekaro or Mpekaro

ORIGINS AND PARENTAGE

Bekari probably originates from Ioánnina or Thesprotía in the Ípeiros region in the west of mainland Greece.

VITICULTURAL CHARACTERISTICS

Productive, disease-resistant, sensitive to drought. Large berries. Mid ripening.

WHERE IT'S GROWN AND WHAT ITS WINE TASTES LIKE

Grown to a limited extent in Ípeiros in western **Greece**, where it was traditionally blended with the white variety DEBINA and the red VLACHIKO to make a sparkling rosé, still produced by, for example, the Ioánnina co-operative and Glinavos, both in Zítsa, although such wines do not qualify for the Zítsa appellation. Glinavos also use it in a red blend with AGIORGITIKO and Vlachiko and in another that adds CABERNET SAUVIGNON to the mix. Wines have moderate acidity and colour and are generally fairly alcoholic.

BELAT

Recently rescued from old vineyards near Barcelona.

BERRY COLOUR: ● ● ● ● ●

ORIGINS AND PARENTAGE

Indigenous to Penedès in Catalunya, north-eastern Spain, this as yet unknown variety is part of a conservation project undertaken by the producer Albet i Noya, north east of Vilafranca del Penedès. It has a unique DNA profile and was initially called Número 2 (tinta), then Velat, meaning 'veiled' in Catalan, because it had gone unnoticed in the vineyard, and then finally Belat, an anagram of Albet.

VITICULTURAL CHARACTERISTICS

Mid budding and ripening. Small to medium-sized thin-skinned berries.

WHERE IT'S GROWN AND WHAT ITS WINE TASTES LIKE

Belat was part of an experimental project begun in **Spain** in 1999 by Josep Maria Albet of Albet i Noya, south west of Barcelona. An old vine was discovered in an abandoned vineyard and the variety is now being commercially cultivated on a small scale. The wine is relatively light in colour, and has soft tannins plus aromas and flavours of red fruits, tobacco and a touch of spice.

BELLONE

Vigorous Roman white that rarely appears solo.

BERRY COLOUR: ● ● ● ● ●

PRINCIPAL SYNONYMS: Arciprete Bianco, Cacchione, Fagotto and many more

ORIGINS AND PARENTAGE

Possibly originating from Lazio in central Italy, it is considered by some to be the Uva Pantastica mentioned by Pliny the Elder but no convincing evidence can be found to support this hypothesis.

VITICULTURAL CHARACTERISTICS

Vigorous, high yielding, late ripening.

WHERE IT'S GROWN AND WHAT ITS WINE TASTES LIKE

Cultivated mainly in Lazio, **Italy**, particularly south of Roma in Anzio, Marino, Nettuno and Velletri, it produces light, fresh wines that are quite delicate but finish slightly bitter and are traditionally softened with a little residual sugar. It frequently disappears into blends such as Frascati or Castelli Romani, although a 100% Bellone is authorized in the Marino and Nettuno DOCs. Terre delle Ginestre's Lentisco and the Cantina Cincinnato's Bellone are examples of the few varietal examples available.

One particular clone of Bellone, oddly nicknamed Arciprete Bianco (meaning 'white archpriest'), is used by Azienda Agricola Marco Carpineti to produce Ludum, a singular dessert wine made from grapes dried on the vine, as well as their dry Collesanti, which has aromas of apples and apricots. Bellone was also used in the past to make Cannellino, a delicate late-harvest Frascati. Total plantings in Italy according to the 2000 census were 1,480 ha (3,660 acres).

BENA

Minor Bosnian white generally grown around Mostar and blended with Žilavka.

BERRY COLOUR: ● ● ● ● ●

ORIGINS AND PARENTAGE

Bena is indigenous to Bosnia and Herzegovina.

WHERE IT'S GROWN AND WHAT ITS WINE TASTES LIKE

Bena has traditionally been planted together with ŽILAVKA and KRKOŠIJA in the Mostar region in the south of **Bosnia and Herzegovina** and is generally used as a minor blending component with the dominant Žilavka, as in Čitluk Winery's Kameno bottling.

BÉQUIGNOL NOIR

Rare dark-skinned French variety much more common in Argentina.

BERRY COLOUR:

PRINCIPAL SYNONYMS: Béquin Rouge (Saint-Macaire and La Réole), Breton (Dissay in the Vienne), Cabernet (Dissay), Chalosse Noire (Bas-Médoc), Embalouzat (Entre-Deux-Mers), Mançais Noir (Indre-et-Loire)

VARIETIES COMMONLY MISTAKEN FOR BÉQUIGNOL NOIR: CABERNET FRANC (Dissay), CASTETS, DURIF (Lot-et-Garonne), FER (Lot), PRUNELARD (Dordogne and Charente)

ORIGINS AND PARENTAGE

The first mention of Béquignol Noir, from the Gironde, south-west France, appeared in the 'Livre de raison d'Antoine Feuilhade', an account book written between 1763 and 1777 by the mayor of Libourne (Garde 1946). Its name comes from the Gascon *bèc* with the suffix *inhol*, literally meaning 'little beak', referring to the shape of the berries (Rézeau 1997).

Béquignol Noir belongs to the Messile ampelographic group (see p XXVII). Grey and white mutations of Béquignol Noir have been observed. Recent DNA profiling has suggested a possible parent–offspring relationship with SAVAGNIN (Myles *et al.* 2011).

Béquignol Noir has often been confused with FER (eg in Viala and Vermorel 1901–10), CABERNET FRANC, DURIF and PRUNELARD but DNA profiling has shown them to be distinct (Bowers *et al.* 1999).

VITICULTURAL CHARACTERISTICS

Vigorous, fertile and very productive. Copes well under dry conditions. Mid to late ripening. Small berries. Susceptible to mites.

WHERE IT'S GROWN AND WHAT ITS WINE TASTES LIKE

There was less than 1 ha (2.5 acres) in **France** in 2008 (down from 223 ha/551 acres in 1958), though it is still a recommended variety in the Gironde and the Vendée. Wines are light and destined for early drinking.

There were a surprising 919 ha (2,271 acres) in **Argentina** in 2008, mostly in Mendoza. It is known locally and colloquially as 'red Chenin' because of its ampelographic similarity – eg the leaf and berry shapes – with that variety. Like Argentina's Bonarda (ie DOUCE NOIRE), it is not very highly regarded and tends to be used to add colour to blends.

BERDOMENEL

Virtually extinct Ariège variety recently replanted.

BERRY COLOUR: ● ● ● ● ●

PRINCIPAL SYNONYMS: Berdanel (Pamiers in the Ariège)

ORIGINS AND PARENTAGE

Berdomenel is a variety exclusive to Pamiers in the Ariège *département* in the French Pyrénées, between Toulouse and the Spanish border, where it is today threatened with extinction. Its name could derive from the Occitan *verdau menèl*, meaning 'small greenish-berried', but there is no link with PETIT VERDOT (Lavignac 2001), which has roughly the same meaning.

VITICULTURAL CHARACTERISTICS

Vigorous but low yielding. Mid ripening. Susceptible to powdery mildew and botrytis bunch rot.

WHERE IT'S GROWN

Berdomenel was almost extinct in **France** when Les Vignerons Ariégeois in Montégut-Plantaurel (just west of Pamiers) came to the rescue by planting 30–40 vines of this and other endangered local varieties, including CAMARALET DE LASSEUBE, CANARI NOIR, MANDRÈGUE and Tortozon (PLANTA NOVA; Carbonneau 2005). They aim to multiply these and to produce a commercial wine if their experiments yield encouraging results. Berdomenel is not included in the French official register of varieties.

BERGERON

See ROUSSANNE.

BERMEJUELA

See MARMAJUELO.

BESGANO BIANCO

Very rare northern Italian white once used to produce Vin Santo.

BERRY COLOUR: ● ● ● ● ●

PRINCIPAL SYNONYMS: Colombana Bianca

ORIGINS AND PARENTAGE

Besgano Bianco originated in either the Bobbio region in the province of Piacenza, Emilia-Romagna, or around Voghera in the province of Pavia, Lombardia. It is also known as Colombana Bianca but is not a colour mutation of COLOMBANA NERA.

WHERE IT'S GROWN AND WHAT ITS WINE TASTES LIKE

Until the 1940s, Besgano Bianco was quite widely planted in Val Trebbia and Val Nure in the provinces of Genova and Piacenza in **Italy** to produce a Vino Santo, together with MALVASIA BIANCA LUNGA and VERDEA. Today it is such a rarity that it is not even mentioned in the 2000 Italian census. Used for both table and wine grapes.

BESSARABSKY MUSKATNY

A Seibel 13666 × ALEATICO white-berried hybrid bred in **Russia** and found, for example, in Vityazevo Winery's dessert wines. Seibel 13666 is a PLANTET × Seibel 6468 hybrid (see HELIOS for the pedigree of Seibel 6468).

BIANCA

Hungarian hybrid producing generally unexciting wines, also in Russia.

BERRY COLOUR: ● ● ● ● ●

PRINCIPAL SYNONYMS: Bianka, Ec 40, Ecs 40, Egri Csillagok 40, May Rot

ORIGINS AND PARENTAGE

Bianca is an Eger 2 × BOUVIER hybrid obtained in 1963 by József Csizmazia and László Bereznai at the Kölyuktetö research centre in the Eger region of northern Hungary, in which Eger 2 is a selfling of VILLARD BLANC called Seyve-Villard 12-375 SP. Initially named Egri Csillagok 40 (meaning 'star of Eger 40'), Bianca was registered in 1982. It was used to breed BAČKA and RUBINKA.

VITICULTURAL CHARACTERISTICS

Early to mid budding, early ripening. Fertile, vigorous and productive but prone to coulure. Medium-sized to large bunches of small, waxy-skinned berries. Winter-hardy and good resistance to fungal diseases.

WHERE IT'S GROWN AND WHAT ITS WINE TASTES LIKE

Hungary's 1,280 ha (3,163 acres) of Bianca are quite widely dispersed but the vast majority (1,137 ha/2,810 acres) are in Kunság in the centre of the country. Wine styles vary considerably, depending on harvest date and winemaking, but varietal wines are typically rather neutral, occasionally slightly flowery, with moderate alcohol, becoming a little more exotic but losing freshness if harvested later. The variety is prone to oxidation in the winery but its disease resistance endears it to organic growers. Recommended producers of varietal wines include Hárs and Hilltop Neszmély but Bianca is mostly used for inexpensive everyday wines.

Bianca is widely planted in **Russia**'s Krasnodar Krai wine region (2,731 ha/6,748 acres in 2009) and there are 15 ha (37 acres) in the **Republic of Moldova**.

BIANCAME

This vigorous, mid- to late-ripening white variety from the Marche in east-central **Italy** is grown mainly in Ancona, in Ascoli Piceno, and in the provinces of Forlì and Pesaro e Urbino. It is responsible for Bianchello del Metauro DOC wines (with the addition of up to 5% MALVASIA BIANCA LUNGA, under the synonym Malvasia Toscana), which is why the grape itself is sometimes called Bianchello. It is also grown further north in Emilia-Romagna, where it can be part of a blend or solo in the Colli di Rimini DOC. It is sometimes confused with PASSERINA or TREBBIANO TOSCANO and may possibly be known as Morbidella. Wines tend be fresh with delicate floral and fruit aromas. There were around 2,080 ha (5,140 acres) in 2000, according to the Italian agricultural census.

BIANCHETTA GENOVESE

See ALBAROLA.

BIANCHETTA TREVIGIANA

Vigorous, thick-skinned northern Italian white producing rather ordinary, slightly astringent wines that are generally blended.

BERRY COLOUR: ● ● ● ● ●

PRINCIPAL SYNONYMS: Bianca Gentile di Fonzaso, Bianchetta Gentile, Bianchetta Semplice, Senese (Breganze), Vernanzina (Colli Berici near Vicenza), Vernassina (Colli Euganei)

VARIETIES COMMONLY MISTAKEN FOR BIANCHETTA TREVIGIANA: ALBAROLA

ORIGINS AND PARENTAGE

Numerous and distinctively different varieties have been cultivated under the name Bianchetta, meaning 'little white'. Bianchetta Trevigiana takes its name from Treviso in the Veneto in northern Italy, where it is considered an indigenous variety and where it is said to have first been mentioned under the name Bianchetta Gentile by Agostinetto di Cimadolmo in Treviso in 1679.

Isoenzymatic and DNA profiling provided evidence that Bianchetta Trevigiana, Senese, Vernanzina and Vernassina are synonyms (Crespan *et al.* 2003). On the other hand, comparison of the DNA profiles showed that Bianchetta Trevigiana is distinct from the supposedly synonymous ALBAROLA and the non-cultivated Pavana Bianca and Vernaccia Trentina (Vouillamoz). In addition, Cipriani *et al.* (2010) have recently suggested through DNA parentage analysis that Bianchetta Trevigiana is a possible Brambana × DURELLA natural cross, the former a black-berried variety that is not commercially cultivated. This parentage showed some DNA discrepancies and needs further investigation.

VITICULTURAL CHARACTERISTICS

Thick-skinned and susceptible to powdery mildew, it ripens late September to early October.

WHERE IT'S GROWN AND WHAT ITS WINE TASTES LIKE

In the provinces of Belluno and Verona, Bianchetta Trevigiana has been cultivated in **Italy** since the seventeenth century but it has declined since then because of its susceptibility to powdery mildew and its lack of frost resistance. Bianchetta Trevigiana was supposedly also cultivated in Trentino-Alto Adige under the name Vernaccio or Vernazza, as well as in Val Sacra, Valsugana and in the Val d'Adige up to Bolzano. It is never made as a varietal wine but it is part of the blends of a wide variety of DOC wines, such as Colli di Conegliano (including Torchiato di Fregona dry and sweet aged whites), Lugana, Montello–Colli Asolani, Prosecco and Valdadige. It was also formerly used in the preparation of vermouth. It makes relatively simple, sometimes slightly astringent, wines. Total plantings in Italy in 2000 were 65 ha (160 acres).

BIANCO D'ALESSANO

Southern Italian white frequently blended with Verdeca, enjoying success in Australia.

BERRY COLOUR: ● ● ● ● ●

PRINCIPAL SYNONYMS: Acchiappapalmento, Bianco di Lessame, Iuvarello⊗ (Calabria), Verdurino⊗, Vuiono⊗ (Calabria)

ORIGINS AND PARENTAGE

A recent ampelographic study has shown that Iuvarello from Calabria is identical to Bianco d'Alessano from Puglia (Schneider, Raimondi, Grando *et al.* 2008).

VITICULTURAL CHARACTERISTICS

Mid to late ripening.

WHERE IT'S GROWN AND WHAT ITS WINE TASTES LIKE

Grown throughout Puglia in southern **Italy** but especially in Murgia (near Taranto) and in the Valle d'Itria. It contributes to DOC wines such as Gravina, Martinafranca, Locorotondo, Ostuni and Lizzano and is often blended with the more widespread and fertile VERDECA. Its recent revival gave rise to interest in producing monovarietal Bianco d'Alessano wines. Sometimes used for vermouth production. 960 ha (2,370 acres) were recorded in the Italian census of 2000.

Salena Estate in the Riverland, **Australia**, has had great success with this variety. After blending it away for several years, it has now been bottled as a varietal wine and scooped three trophies at the Australian Alternative Varieties Show in 2010. It ripens late here but retains high acidity levels, making it ideal for this warmer inland region. There are currently just 4 ha (10 acres) planted.

BIANCOLELLA

Light white grape from Napoli and the neighbouring islands inspires some heroic production.

BERRY COLOUR: ● ● ● ● ●

PRINCIPAL SYNONYMS: Bianculillo, Jancolella, San Nicola☒

ORIGINS AND PARENTAGE

Biancolella is indigenous to the island of Ischia in the bay of Napoli in Campania, southern Italy, where it was not mentioned until the nineteenth century. DNA profiling has established that the San Nicola grape variety traditionally cultivated on the nearby island of Capri is identical to Biancolella (Costantini *et al.* 2005). Recent DNA parentage analysis has suggested that Biancolella is a possible parent of SAN LUNARDO, also from the island of Ischia (Cipriani *et al.* 2010). It is also closely related to FALANGHINA BENEVENTANA (Costantini *et al.* 2005).

OTHER HYPOTHESES

Like many other Italian grapes, Biancolella is said to be of Greek origin, mainly because Ischia is a former Greek colony. However, there is no convincing evidence for this hypothesis, and Biancolella's DNA has no relationship with those of modern Greek varieties (Vouillamoz).

VITICULTURAL CHARACTERISTICS

Late ripening.

WHERE IT'S GROWN AND WHAT ITS WINE TASTES LIKE

Cultivated mainly on the island of Ischia in the bay of Napoli, southern **Italy**, but also grown on other volcanic soils in the provinces of Napoli and Caserta, and also possibly near Bastia and San Martino di Lota on Corse, **France**, under the name Petite Blanche (identity yet to be verified). It is authorized in DOCs such as Ischia, Capri, Campi Flegrei, Penisola Sorrentina and Costa d'Amalfi. Only 293 ha (725 acres) were recorded in the 2000 census of Italian vineyards.

On Ischia, Biancolella is usually blended with other local varieties but a few growers produce 100% Biancolella wines. Some of them represent perfect examples of 'heroic viticulture'. Casa d'Ambra's Biancolella Frassitelli, for example, is harvested with the help of a monorail train on the steep hillside vineyards of Monte Epomeo at an elevation of 600 m above sea level. The most extreme example must be Cantine Antonio Mazzella's Vigna del Lume, a 100% Biancolella from a vineyard 150 m above the sea that is reachable only from an old footpath. Late-picked grapes are pressed and vinified in old cellars dug in soft tuff rock nearby. The wine is then transported by sea on wooden boats and bottled on the mainland. This 'impossible', straw-coloured wine has aromas of pineapple, almonds and white flowers, good balance, nice minerality (volcanic soil) and good structure with medium to high alcohol content.

BIANCONE DI PORTOFERRAIO

Ordinary white exclusive to Elba.

BERRY COLOUR: ● ● ● ● ●

PRINCIPAL SYNONYMS: ?Biancone (Corse in France), Pagadebiti di Porto S. Stefano

VARIETIES COMMONLY MISTAKEN FOR BIANCONE DI PORTOFERRAIO: BIANCU GENTILE (Corse), ?ROLLO (under the name Livornese Bianca)

ORIGINS AND PARENTAGE

Indigenous to the island of Elba in Toscana, or possibly originating from the French island of Corse under the name Uva Bianca or Biancone. Parentage analysis has revealed a possible parent–offspring relationship between MAMMOLO and the Corsican Biancone as well as with several other Tuscan varieties, CALORIA, COLOMBANA NERA and POLLERA NERA, which turn out to be half-siblings of Biancone (Di Vecchi Staraz, This *et al.* 2007), although it is not sure that Biancone from Corse in their study is truly identical to Biancone di Portoferraio.

WHERE IT'S GROWN AND WHAT ITS WINE TASTES LIKE

This late-ripening variety is grown exclusively on Elba, **Italy**, to produce a simple table wine.

BIANCU GENTILE

Rare, aromatic Corsican.

BERRY COLOUR: ● ● ● ● ●

PRINCIPAL SYNONYMS: Biancone Gentile (Corse)

VARIETIES COMMONLY MISTAKEN FOR BIANCU GENTILE: BARIADORGIA⌘, BIANCONE DI PORTOFERRAIO (Elba in Italy), ROLLO⌘ (aka Livornese Bianca, also called Biancone in Toscana)

ORIGINS AND PARENTAGE

Biancu Gentile is an old variety from the French island of Corse, mentioned as far back as 1822 as growing in Pila Canale in the southern part of the island, south east of Ajaccio. It is often confused with another Corsican variety, Carcajolo Blanc (BARIADORGIA), as well as with BIANCONE DI PORTOFERRAIO and ROLLO (under the name Livornese Bianca), both having Biancone among their synonyms, with which Biancu Gentile is supposedly closely related (Galet 2000). Biancu Gentile literally means 'nice white' or 'noble white' but *gentile* also has the sense 'grafted' or the opposite of wild.

VITICULTURAL CHARACTERISTICS

Vigorous, very early budding and early ripening and able to accumulate high levels of sugar in a short time. Very susceptible to powdery mildew and botrytis bunch rot.

WHERE IT'S GROWN AND WHAT ITS WINE TASTES LIKE

Biancu Gentile is exclusive to the island of Corse, in **France**, where it is a recommended variety. There are very limited plantings in Patrimonio and a little more in the south, in Sartène and Figari, but the total vineyard area was just 6 ha (15 acres) in 2008. Varietal wines may be produced within the Île de Beauté classification but it is often successfully blended with VERMENTINO. It was rediscovered during a viticultural census of the island and interest in the variety revived in the mid 1990s thanks to Antoine Arena in Patrimonio and Yves Canarelli in Figari, both of whom make varietal examples.

Biancu Gentile produces wines that have intense aromas of exotic fruits and citrus and tend to be full-bodied and powerful, but the grapes need to be harvested before the acidity starts to drop to avoid heaviness.

There have been some experimental plantings of Biancu Gentile, along with a wide range of other French, Italian, Portuguese and Spanish varieties, in **California**'s San Joaquin Valley, in trials by the University of California at Davis viticulture extension programme.

BÍBORKADARKA

Minor red-fleshed Hungarian cross used mainly for colour.

BERRY COLOUR: ● ● ● ● ●

PRINCIPAL SYNONYMS: Bíbor Kadarka, Biborkadarsa, Cs 4

ORIGINS AND PARENTAGE

Bíborkadarka, meaning 'purple Kadarka', registered in 1974, is a KADARKA × Muscat Bouschet cross obtained in 1948 by Pál Kozma at the Szent István University in Hungary, in which Muscat Bouschet is a black-berried Muscat Rouge de Madère × PETIT BOUSCHET cross that is no longer commercially cultivated.

VITICULTURAL CHARACTERISTICS

High yielding, late ripening. Big bunches of small, waxy-skinned, red-fleshed berries. Susceptible to botrytis bunch rot and sensitive to winter cold.

WHERE IT'S GROWN AND WHAT ITS WINE TASTES LIKE

Wines are typically deeply coloured and high in acidity, produced in the regions of Eger and Mátra in north-east **Hungary** and Szekszárd and Kunság in the centre and south. They are generally used to add colour to blends. There was a total of 154 ha (381 acres) across the country in 2008. Bálint Losonci in Mátra is one of the rare producers of a successful varietal wine.

BICAL

Aromatic, early-ripening variety widely planted in western central Portugal.

BERRY COLOUR: ● ● ● ● ●

PRINCIPAL SYNONYMS: Arinto de Alcobaça (Alentejo and Tejo), Bical de Bairrada, Borrado das

Moscas (Dão), Fernão Pires Galego (Ançã-Cantanhede), Pintado das Moscas (Dão), Pintado dos Pardais (Lisboa)

VARIETIES COMMONLY MISTAKEN FOR BICAL: ARINTO DE BUCELAS, FERNÃO PIRES

ORIGINS AND PARENTAGE

The origins of Bical lie somewhere within the Bairrada and Dão regions in central Portugal, its traditional area of cultivation (Rolando Faustino, personal communication). Its synonym Borrado das Moscas, literally 'fly droppings', derives from the little brown dots that appear on fully ripe berries.

Bical is also called Arinto de Alcobaça in the Alentejo and Tejo regions, but in Alcobaça in northern Estremadura the variety that is known simply as Arinto is in fact ARINTO DE BUCELAS, which is also grown further south, closer to Lisboa. Similarly, Bical should not be confused with FERNÃO PIRES in the Ançã-Cantanhede region near Coimbra, where it is known as Fernão Pires Galego. Just to add to the confusion, Bical is sometimes used as a synonym for TOURIGA NACIONAL, even though the latter has black berries. The alternative Bical Tinto is marginally less confusing.

VITICULTURAL CHARACTERISTICS

Vigorous and moderately productive. Medium-sized bunches of small berries. Very susceptible to downy and powdery mildews and to coulure, less so to botrytis bunch rot. Early budding and very early ripening.

WHERE IT'S GROWN AND WHAT ITS WINE TASTES LIKE

Bical is grown mostly in its traditional homelands of Bairrada and Dão in **Portugal** but is also found further north. There were 1,465 ha (3,620 acres) in Portugal in 2010. Ripening early allows Bical to escape the September rains that are common in the region and wines therefore have respectable alcohol levels, fresh acidity as well as attractive peachy aromas, sometimes verging on tropical fruits. With time in bottle, the best examples can develop aromas akin to those of aged Riesling. Wines with lower acidity may benefit from blending with ARINTO DE BUCELAS or CERCEAL BRANCO. The variety also seems well suited to barrel fermentation and ageing on the lees. The best wines are grown on the east-facing calcareous soils of Óis do Bairro north of Coimbra. On its own or blended with Arinto, it is often used to produce sparkling wines.

Bical is authorized in the Bairrada, Beira Interior, Dão, Távora-Varosa and Douro DOCs. Recommended producers in Bairrada include Aliança, Campolargo, Filipa Pato, Luís Pato and Quinta do Encontro.

BIGOLONA

Rare white from the Valpolicella region of northern Italy used in sweet-wine blends.

BERRY COLOUR: ● ● ● ● ●

PRINCIPAL SYNONYMS: Bigolara, Bigolona Bianca, Bigolona Veronese, Sampagna, Smarzirola

ORIGINS AND PARENTAGE

Bigolona was mentioned among the varieties of the Valpolicella region in 1818–23 (Costacurta *et al.* 1980).

VITICULTURAL CHARACTERISTICS

Vigorous and productive, mid ripening. Well suited to both botrytized and late-harvest sweet wines and traditionally used to produce the local Vino Santo. Its synonym Smarzirola comes from the Italian *marcire*, meaning 'to rot', referring to the fact that the variety responds well to the noble form of botrytis bunch rot.

WHERE IT'S GROWN AND WHAT ITS WINE TASTES LIKE

Since the 1970s, Bigolona and other rare local varieties have been rescued from extinction and studied at the Conegliano and Verona research centres in the Veneto, northern **Italy**. Bigolona is not included in the national register of varieties and is grown to only a very limited extent in Valpolicella and Illasi for the production of late-harvest and passito wines. For example, Tommaso Bussola includes about 3% Bigolona in his Peagnà Passito Bianco, IGT Veneto, with GARGANEGA and other white varieties.

BIRSTALER MUSKAT

Light-skinned Swiss hybrid used mainly for the table.

BERRY COLOUR: ● ● ● ● ●

PRINCIPAL SYNONYMS: Birchstaler Muscat, Muscat de la Birse, VB 86-6

ORIGINS AND PARENTAGE

Birstaler Muskat is a SEYVAL BLANC × BACCHUS hybrid obtained in 1986 by Valentin Blattner in Soyhières in the Jura, north-west Switzerland, and therefore a sibling of RÉSELLE. It is named after the River Birs (*Tal* is German for 'valley'), a tributary of the Rhein.

VITICULTURAL CHARACTERISTICS

Early budding, early to mid ripening. Large, open bunches. Relatively good resistance to downy and powdery mildews and frost but susceptible to coulure.

WHERE IT'S GROWN AND WHAT ITS WINE TASTES LIKE

Birstaler Muskat is typically a table grape but it is occasionally made into wine (officially less than half a hectare was recorded for winemaking in **Switzerland** in 2009): eg the sweet Muscat de la Birse bottling by the variety's breeder, Valentin Blattner. Wines have a light Muscat taste and low acidity.

It is also planted in **Belgium** for the table.

BISHTY

Also known as Bichty, this white-berried variety probably comes from **Uzbekistan**, where it is found mainly in the Boukhara region (Galet 2000).

BLACK QUEEN

Japanese complex hybrid widely used for winemaking in Asia.

BERRY COLOUR: ● ● ● ● ●

PRINCIPAL SYNONYMS: Pokdum or Pok Dum (Thailand)

ORIGINS AND PARENTAGE

Complex hybrid obtained by Zenbei Kawakami (1868–1944) at his Iwanohara Winery in Niigata Prefecture in Japan by crossing Bailey and Golden Queen (see MUSCAT BAILEY A for the pedigree of Bailey), in which:

- Golden Queen is a Black Alicante × Ferdinand de Lesseps hybrid
- Black Alicante is a Spanish table grape variety that must not be confused with TRINCADEIRA, which is called Black Alicante in Australia
- Ferdinand de Lesseps is a CHASSELAS × ISABELLA hybrid.

Black Queen is therefore a *Vitis vinifera*, *Vitis labrusca* and *Vitis lincecumii* hybrid.

In Thailand, Pokdum is a mutation of Black Queen (*pok dum* in Thai) found in the 1980s by wine grower Nong Pok in Pak Chong in the Nakhon Ratchasima province. It should be noted that this origin is not consistent with the theory that Pokdum is identical to Pione, a Kyoho × MUSCAT OF ALEXANDRIA table grape hybrid from Japan.

VITICULTURAL CHARACTERISTICS

Vigorous, late budding, very late ripening. Large bunches of medium-sized berries. Susceptible to botrytis bunch rot when ripe.

WHERE IT'S GROWN AND WHAT ITS WINE TASTES LIKE

As one might expect, Black Queen is one of the dozens of vine varieties used for winemaking in **Japan** (469 ha/1,159 acres in 2009). Black Queen is the dominant grape used in **Taiwan**'s fledgling wine industry and is also important in those of **Vietnam** and **Cambodia.** As Pokdum, it accounts for about 10% of **Thailand**'s vineyards and is grown, for ex-ample, by the Siam Winery in 'floating vineyards' in the Chao Praya Delta.

BLANC DAME

Almost abandoned by south-west French vignerons but not quite by armagnac distillers.

BERRY COLOUR: ● ● ● ● ●

PRINCIPAL SYNONYMS: Blanc Madame (around Jurançon), Blanquette Grise (Gers), Clairette de Gascogne, Claret, Claret de Gascogne, Plan de Dame (Hautes-Pyrénées)

ORIGINS AND PARENTAGE

Blanc Dame is a variety of south-west France that has virtually disappeared. It was possibly mentioned in 1840 under the name Plan de Dame in Rabastens in the Hautes-Pyrénées, and definitely in 1887 in the Béarn under its modern name *blandame* or *Blanc-madame*. The suffix *-dame* or *-madame* has often been used in French to designate various fruits, eg the pear called Cuisse-madame or Cuisse de Dame.

Like more than eighty other varieties of Western Europe, Blanc Dame is likely to be a natural progeny of the prolific GOUAIS BLANC (Boursiquot *et al.* 2004). Blanc Dame belongs to the Gras ampelographic group (see p XXVII; Bisson 2009).

VITICULTURAL CHARACTERISTICS

Early budding, mid ripening and fertile. Susceptible to fungal diseases.

WHERE IT'S GROWN

Blanc Dame is authorized in **France** in the Gers and in the Pyrénées-Atlantiques but there were only 20 ares (half an acre) remaining in 2008, no doubt in part thanks to the subsidies offered to growers by the EU to grub up varieties such as this. What remains plays a very minor part in the base wines for armagnac, although it is also suited to the table.

BLANC DU BOIS

Potentially high-quality, aromatic hybrid from Florida with useful resistance to Pierce's disease.

BERRY COLOUR: ● ● ● ● ●

PRINCIPAL SYNONYMS: Blanc DuBois, Florida H18-37

ORIGINS AND PARENTAGE

A complex *Vitis aestivalis*, *Vitis labrusca*, *Vitis simpsonii* and *Vitis vinifera* hybrid obtained in 1968 by John A Mortensen at the Central Florida Research and Education Center of the University of Florida in Leesburg by crossing Florida D6-148 with CARDINAL (Mortensen 1987), where:

- Florida D6-148 is a self-pollination of Florida A4-23 that is resistant to Pierce's disease
- Florida A4-23 is a Florida 449 × Florida W907 hybrid obtained in 1956
- Florida 449 is a seedling of *Vitis aestivalis* subsp. *smalliana* with an unknown parent, found in 1946
- Florida W907 is a Florida W381 × CARDINAL hybrid obtained in 1949
- Florida W381 is a Pixiola × GOLDEN MUSCAT hybrid obtained in 1945
- Pixiola is a seedling of *Vitis simpsonii* Munson with an unknown parent.

At the suggestion of Lafayette Vineyards & Winery, the variety was named after the French winemaker Émile Dubois, a prominent grape grower and wine producer in Tallahassee, Florida, in the late nineteenth century. It was released for commercial planting in 1987.

VITICULTURAL CHARACTERISTICS

Vigorous, resistant to Pierce's disease, even in the humid conditions of states such as Florida and Texas, and also to downy mildew but susceptible to black rot and anthracnose. Good resistance to nematodes but not suited to poorly drained soils with a high pH. Does not generally need to be grafted. Very early ripening (4–14 July in Texas).

WHERE IT'S GROWN AND WHAT ITS WINE TASTES LIKE

Lakeridge in **Florida** were the first to plant Blanc du Bois and Haak Winery near Galveston in **Texas** were one of the first in Texas in the late 1980s. The latter make both dry and off-dry varietal wines, including a barrel-fermented version, as well as wines in the style of Madeira and white port. As an unadorned dry varietal wine, Blanc du Bois has light citrus aromas and no foxiness. Other Texan producers such as Chisholm Trail, La Cruz de Comal, Enoch's Stomp, Loan Oak, Los Pinos Ranch, Pleasant Hill, Tara and Tehuacana Creek have followed suit, as well as Feliciana, Landry and Ponchartrain in Louisana, and San Sebastian in Florida. Wines are typically aromatic, ranging from floral and occasionally Muscat-like to citrus-flavoured, and the drier wines show the freshness of SAUVIGNON BLANC, ALVARINHO or TORRONTÉS. According to Raymond Haak, there are currently around 100 acres (40 ha) planted in the US, mainly in Texas and Florida, but there are also minor plantings in **Louisiana**, **Georgia** and **South Carolina**.

BLANCO LEXÍTIMO

See ALBARÍN BLANCO.

BLANQUEIRO

Very minor Provençal variety.

BERRY COLOUR: white (first of five swatches marked)

PRINCIPAL SYNONYMS: Blanqueirol, Blanqueiron, Blanquerel, Pignairon

ORIGINS AND PARENTAGE

Blanqueiro probably comes from the vineyards of Saint-Jeannet and La Gaude in the Alpes-Maritimes.

VITICULTURAL CHARACTERISTICS

Compact bunches of small berries. Mid to late ripening.

WHERE IT'S GROWN

Blanqueiro is authorized as a secondary variety in the Bellet appellation in Provence, south-east **France**, but is today rarely found and does not feature in the official French vineyard statistics.

BLATINA

Bosnia and Herzegovina's most important red variety.

BERRY COLOUR: black (last of five swatches marked)

PRINCIPAL SYNONYMS: Blathina, Blatina Crna, Blatina Hercegovacka, Blatina Mala, Blatina Velika, Praznobačva, Zlorod

VARIETIES COMMONLY MISTAKEN FOR BLATINA: CETINKA (Croatia), TRIBIDRAG (Metković in Croatia)

ORIGINS AND PARENTAGE

Blatina is the main indigenous red wine variety in Bosnia and Herzegovina, possibly originating from the region of Mostar in the south of the country. Croatia's CETINKA is also known as Blatinka but they are distinct varieties (Goran Zdunić, personal communication). DNA profiling has shown that Blatina is distinct from TRIBIDRAG (Nataša Štajner, personal communication), as it is sometimes called in the Metković region of Croatia, just over the border.

VITICULTURAL CHARACTERISTICS

Late ripening. Erratic fertility due to functionally female-only flowers and therefore needs another variety nearby for pollination. Good general disease resistance but sensitive to winter frosts.

WHERE IT'S GROWN AND WHAT ITS WINE TASTES LIKE

In **Bosnia and Herzegovina**, Blatina is often produced as a varietal wine, or with a small percentage of other varieties such as ALICANTE HENRI BOUSCHET (under the name Kambuša) or TRNJAK that are interplanted in the vineyard to ensure pollination, as in Čitluk Winery's oaked and unoaked versions. Wines are typically medium-bodied and full-flavoured but retain good acidity even this far south. Other producers include Hepok in Ljubuški, south west of Mostar, who age their deeply coloured, berry-fruited, firmly structured example in oak, and Škegro.

BLATINKA

See CETINKA.

BLATTERLE

Old and almost extinct white-berried variety from the Alto Adige in northern **Italy**, with a unique DNA profile (Vouillamoz and Barbara Raifer, personal communication). Its name, also spelt Platterle, means 'little leaf' in the local dialect. Formerly grown around Bolzano and in the Valle Isarco (Eisacktal in German), this early-ripening variety is today cultivated by a handful of producers, including Weingut Nusserhof, who produce a varietal vino da tavola labelled Blaterle (changing the spelling to avoid regulatory issues). Blatterle is also a component in Unterortl's Juval Glimmer blend from the Val Vinosta (Vinschgau), together with FRAUELER, another rare local variety, plus MÜLLER-THURGAU.

BLAUBURGER

Light but deeply coloured Austrian cross contributing colour to blends.

BERRY COLOUR: ● ● ● ● ●

VARIETIES COMMONLY MISTAKEN FOR BLAUBURGER: Blauburgunder (German name for PINOT NOIR)

ORIGINS AND PARENTAGE

A BLAUER PORTUGIESER × BLAUFRÄNKISCH cross bred in 1923 by Fritz Zweigelt at the Klosterneuburg research centre west of Wien (Vienna), making Blauburger a grandson of GOUAIS BLANC. It was used to breed RÁTHAY.

VITICULTURAL CHARACTERISTICS

Early to mid ripening, susceptible to powdery mildew and various fungal diseases such as esca and phomopsis.

WHERE IT'S GROWN AND WHAT ITS WINE TASTES LIKE

There were just over 1,000 ha (2,471 acres) planted in **Austria** in 2007, mainly in Niederösterreich (Lower Austria), up from 884 ha (2,184 acres) at the end of the 1990s. The wines are deeply coloured but they are relatively neutral and lightweight, without much tannin or acidity, and are therefore generally used to add colour to blends, although a few producers do make a varietal Blauburger.

There were a reported 10 ha (25 acres) in **Germany** in 2006, a quarter of which were in Franken. **Hungary** recorded nearly 400 ha (1,000 acres) in 2008, with more than half in the Eger region, where the variety is authorized in Egri Bikavér (Bull's Blood of Eger).

BLAUER ARBST

See PINOT NOIR in the PINOT entry.

BLAUER PORTUGIESER

Prolific Austrian red wine variety producing rather dull, low-acid wines.

BERRY COLOUR: ● ● ● ● ●

PRINCIPAL SYNONYMS: Autrichien, Badner, Early Burgundy⚮ (Napa Valley in Californa), Modrý Portugal (Czech Republic), Oporto (Slovakia), Portugais Bleu (Jura), Portugalkja, Portugieser⚮, Portugieser Blau⚮, Portugizac Crni (Croatia), Português Azul⚮ (Portugal), Vöslauer (Austria); Kékoportó is no longer officially used in Hungary

VARIETIES COMMONLY MISTAKEN FOR BLAUER PORTUGIESER: KRALJEVINA⚮ (north-western Croatia), MORETO DO ALENTEJO⚮ (Alentejo in Portugal)

ORIGINS AND PARENTAGE

Despite its name, Blauer Portugieser is not likely to be of Portuguese origin. **Grauer Portugieser** (grey berries) and **Grüner Portugieser** (green berries) could not be distinguished through DNA profiling and simply represent colour mutations of Blauer Portugieser (Sefc, Steinkellner *et al.* 1998). The existence in Austria of colour mutations and numerous synonyms tends to favour this country as the geographical origin.

Breeding reports and parentage analysis showed that Grauer Portugieser and FRÜHROTER VELTLINER are the parents of JUBILÄUMSREBE (Sefc, Steinkellner *et al.* 1998). Blauer Portugieser was a parent in several other crosses that are relatively important today: BLAUBURGER, DAKAPO, DOMINA, HEROLDREBE, PORTAN and SEGALIN.

OTHER HYPOTHESES

It is sometimes argued that Portugieser was brought to Austria from Portugal, supposedly imported from Porto in 1772 by von Fries, an Austrian ambassador who planted the variety in the region of Vöslau, and the nineteenth-century ampelographer Goethe (1887) suggested that Blauer Portugieser was identical to MORETO DO ALENTEJO. Although recent genetic studies could not rule out the hypothesis that Blauer Portugieser and Moreto might have had a common ancestor (Regner *et al.* 1999), comparison of DNA profiles from all over Europe strongly supported an Austrian/German origin for Blauer Portugieser (Sefc *et al.* 2000). In addition, the Portuguese synonym Português Azul seems to be a literal translation of Blauer Portugieser rather than vice versa.

VITICULTURAL CHARACTERISTICS

Early to mid ripening, at risk from spring and autumn frosts, botrytis bunch rot and esca. Consistently productive.

WHERE IT'S GROWN AND WHAT ITS WINE TASTES LIKE

Thanks to Portugieser's generous yields, it has spread to much of Central Europe and was even at one time quite widespread in south-west **France**. Domaine Ganevat in Rotalier in the Jura have replanted a small area of Portugais Bleu and include it in an early-drinking red blend.

The variety was taken to **Germany** in the nineteenth century and plantings are more extensive there than in Austria, totalling 4,354 ha (10,759 acres), mainly in the Pfalz (2,178 ha/5,382 acres) and Rheinhessen (1,661 ha/4,104 acres). A good deal of the crop goes into pink Weissherbst and most of the red wines, wherever they are produced, are soft, pale, sometimes thin; ordinary at best. They tend to need significant chaptalization to give them body and are best drunk young.

Blauer Portugieser is the third most planted dark-berried variety in **Austria**, with 2,223 ha (5,490 acres) in 2007, down very slightly on 1999. The vast majority of vines are in Niederösterreich (Lower Austria), particularly in the Thermenregion south of Wien (Vienna) and north west of the capital in Kamptal.

In **Hungary**, where it was once more commonly known as Kékoportó, there were 1,250 ha (3,090 acres) in 2008, with the majority of vines in Villány and Kunság in the south and Eger in the north east. Producers in Villány seem to be able to coax the most concentration from the variety, presumably by restricting its yields, and some examples have enough body to benefit from oak ageing. Top producers include Gábor Kiss, József Szemes and Béla Jekl.

It is also grown in northern **Croatia** as Portugizac Crni, or Portugaljka, and in the **Czech Republic**, which recorded 670 ha (1,655 acres) in 2008, mostly in Velkopavlovická and Slovácká. In **California**, recent DNA tests have shown that some Early Burgundy vines are in fact Blauer Portugieser (Mike Officer, personal communication), although it seems likely that others are ABOURIOU, as suggested by the ampelographer Paul Truel in the 1970s.

BLAUER URBAN

Rarity just about surviving in southern Germany.

BERRY COLOUR: ● ● ● ● ●

PRINCIPAL SYNONYMS: Schwarzer Urban, Urban, Urban Blau, Urban Blauer

ORIGINS AND PARENTAGE

This variety, probably distinct from Roter Urban (no longer cultivated), may originate in the Alto Adige in northern Italy.

VITICULTURAL CHARACTERISTICS

Productive.

WHERE IT'S GROWN

At one time Blauer Urban was widely planted in Central Europe, Italy and France but now it is scantily planted in southern **Germany**, where Graf Adelmann in Württemberg produces a good example. It was traditionally used as table or wine grape, planted together with CHASSELAS (Gutedel) and GRÜNER VELTLINER.

BLAUER WILDBACHER

Distinctively perfumed Austrian red wine grape producing eye-watering pinks.

BERRY COLOUR: ● ● ● ● ●

PRINCIPAL SYNONYMS: Schilcher (Steiermark/Styria), Wildbacher (Italy), Wildbacher Blau

VARIETIES COMMONLY MISTAKEN FOR BLAUER WILDBACHER: Wildbacher Spätblau⊗

ORIGINS AND PARENTAGE

Blauer Wildbacher is an ancient traditional variety named after the small village of Wildbach south of Graz in south-eastern Austria. The name Wildbacher should be handled with care: in a recent DNA study, the more widespread and early-ripening Blauer Wildbacher and the almost-extinct and late-ripening Wildbacher Spätblau turned out to have distinct DNA profiles, as well as possibly having a parent–offspring relationship (Meneghetti *et al.* 2009). Further parentage analysis has also suggested that Blauer Wildbacher has a parent–offspring relationship with GOUAIS BLANC, the second parent supposedly being an extinct wild vine (Meneghetti *et al.* 2009).

OTHER HYPOTHESES

Several nineteenth-century ampelographers were convinced that Blauer Wildbacher was domesticated from wild vines (Meneghetti *et al.* 2009).

VITICULTURAL CHARACTERISTICS

Very late ripening with small berries and robust enough to cope with the climate of the Weststeiermark (Western Styria).

WHERE IT'S GROWN AND WHAT ITS WINE TASTES LIKE

There are about 10 ha (25 acres) in **Italy**, where it is known as Wildbacher. Producers of varietal Wildbacher include Cantine Collato (Colli Trevigiani IGT) and Tenuta Col Sandago.

This high-acid, spicily perfumed variety with aromas of nettles is ideally suited to be the basis for Schilcher, the eye-wateringly tart pink speciality of the Weststeiermark in central south **Austria**, where the variety is also known as Schilcher. About 3% is sparkling and even less is made into a still red. Virtually all the Austrian plantings are in the Steiermark and largely in its western enclave (357 ha/880 acres out of a total 454 ha/1,120 acres in 2008). Domäne Müller even make a TBA version. The best producers include Langmann vulgo Lex, Franz Strohmeier, Christian Reiterer and Friedrich Christian.

BLAUFRÄNKISCH

Dark-skinned, dark-fruited Austro-Hungarian variety increasingly well grown and vinified, gaining ground.

BERRY COLOUR: ● ● ● ● ●

PRINCIPAL SYNONYMS: Blauer Limberger (Germany), Borgonja🍇 (Croatia), ?Burgund Mare (Romania), Franconia🍇 or Franconia Nera (Italy), Frankovka🍇 (Croatia, Serbia and Czech Republic), Frankovka Modrá (Slovakia), Frankovna Crna (Slovenia), Gamé (Bulgaria), Kékfrankos🍇 (Hungary), Limberger🍇 or Lemberger (Germany), Modra Frankinja (Slovenia), Nagyburgundi (Hungary), Sura Lisicina (Serbia)

VARIETIES COMMONLY MISTAKEN FOR BLAUFRÄNKISCH: GAMAY NOIR🍇, PINOT NOIR🍇

ORIGINS AND PARENTAGE

From the Middle Ages onwards, the name Fränkisch was given to several high-quality varieties, indicating their superiority over the inferior Heunisch varieties. The word Fränkisch derives from Franconia, a historic region of Germany that today straddles parts of northern Bavaria, southern Thüringen and Heilbronn-Franken in Baden-Württemberg. One of these high-quality varieties might have been Blaufränkisch, whose name did not appear until 1862 at a grape-variety exhibition in Wien (Vienna; Aeberhard 2005) and was officially adopted by the international ampelographic commission in Colmar, France, in 1875. It was later mentioned in Germany under the synonyms Lemberger (1877) and Limberger (end of the nineteenth century), both deriving from the Austrian locations from which the variety was exported to Germany: Lemberg in the Steiermark (Styria) and Limburg in Maissau (Niederösterreich/Lower Austria). In Hungary, the variety was first mentioned in 1890 under the name Kékfrankos, the literal translation of Blaufränkisch.

DNA profiling has recently confirmed that the Hungarian Kékfrankos is identical to Blaufränkisch (Jahnke *et al.* 2009) and has shown that Blaufränkisch is also cultivated in Croatia under the name Borgonja (Edi Maletić and Ivan Pejić, personal communication; Vouillamoz), long thought to be identical to GAMAY NOIR or PINOT (Borgonja means 'Bourgogne').This finding makes more sense in the light of DNA parentage analysis: Blaufränkisch has a parent–offspring relationship with GOUAIS BLANC (Regner *et al.* 1998; see Pinot pedigree diagram), which is a parent of Gamay Noir (Bowers *et al.* 1999). As a consequence, the birthplace of Blaufränkisch is most likely to be within the area now demarcated as Dalmacija, Austria and Hungary, although the exact location is still unknown.

The use of the synonym Gamé in Bulgaria has often led to – and may well be due to – the mistaken belief that Blaufränkisch was a clone of GAMAY NOIR.

Two crosses between Blaufränkisch and SANKT LAURENT produced ANDRÉ and ZWEIGELT, while another one with BLAUER PORTUGIESER produced BLAUBURGER. Blaufränkisch was a parent in the crosses that created ACOLON, CABERNET CUBIN, CABERNET MITOS, HEROLDREBE, MAGYARFRANKOS, RUBINTOS and REBERGER.

VITICULTURAL CHARACTERISTICS

Vigorous, early budding and late ripening, thus needing a relatively warm climate. It is susceptible to downy and powdery mildews.

WHERE IT'S GROWN AND WHAT ITS WINE TASTES LIKE

Italy has strictly limited plantings of this variety (127 ha/310 acres), called Franconia Nera, in Friuli in the north east, where it can be turned into fruity, zesty, varietal wines in the Friuli Isonzo and Friuli Latisana DOCs. The German-descended Friedrich Schatz of Málaga, in southern

Spain, makes a varietal Lemberger called Acinipo, which is sweet and lively with good acidity and benefits from a few years in bottle.

In **Germany**, as (Blauer) Lemberger or Limberger, the variety is grown on 1,729 ha (4,272 acres) of vineyard, mainly in Württemberg in the south around Stuttgart. The wines tend to be lighter and a little softer than their Austrian counterparts.

Blaufränkisch is **Austria**'s second most planted dark-berried variety after ZWEIGELT, with total plantings in 2008 of 3,340 ha (8,250 acres), up from 2,640 ha (6,524 acres) in 1999, though this is still little more than 6% of the total vineyard area. The vast majority of the vines are in Burgenland, particularly in Mittelburgenland (1,194 ha/2,950 acres), where the warmth and shelter provided by the hills to the west, north and south of the region, in addition to the influence of the warm dry winds from the Pannonian Plain to the east, create ideal conditions for fine, well-structured reds. There are also significant plantings in Neusiedlersee (807 ha/1,994 acres) and Neusiedlersee-Hügelland (962 ha/2,377 acres).

Blaufränkisch can produce deeply coloured, racy and characterful wines, encouraging more ambitious producers to make oak-aged versions, which are more likely to qualify for the Burgenland DAC Reserve classification, with lighter, unoaked versions designated Mittleburgenland DAC Classic. Indeed, the variety is capable of a wide range of styles that express their very specific origins, from light, fresh and fruity to deeply flavoured, dark-fruited, peppery and firmly structured but still fresh and ageworthy wines. Producers in the foothills of the Leithaberg mountains are able to produce particularly exciting, nervy and elegant examples within the Leithaberg DAC (minimum 50% Blaufränkisch) thanks to the cooler sites further away from the Neusiedlersee and the complex slate and limestone soils. Closer to the lake, wines tend to be richer and more full-bodied. South of the lake, in the northern part of Südburgenland, clay-rich soils produce spicy, full-bodied wines that in the past were often overly oaky but this region has some exceptional sites that benefit from Pannonian warmth as well as cooler, higher slopes, giving a long ripening season and a perfect combination for finesse and complexity, exemplified by the wines of Roland Velich at Moric. More recently, fine wines with distinctive regional characteristics are being produced in the iron-rich Eisenberg DAC area in southern Südburgenland and in the slaty soils of Spitzerberg in Carnuntum. Particularly fine examples of Austrian Blaufränkisch are made by K+K Kirnbauer, Nittnaus, Pichler-Kruzler, Pittnauer, Prieler, Triebaumer and Wohlmuth.

Hungary has just over 8,000 ha (19,770 acres) of Kékfrankos, a direct translation of Blaufränkisch (though it is occasionally known as Nagyburgundi), spread across most wine regions but with the greatest presence in Kunság in the south and centre, Eger to the north east and Sopron on the border with Austria. It is considered to be one of Hungary's best red wine varieties and the Austrian Franz Weninger makes fine examples in a range of styles on both sides of the border, in Burgenland and in Sopron.

Blaufränkisch, known locally as Frankovka Modrá, is the most widely planted dark-skinned variety in **Slovakia**, where it is grown on 1,742 ha (4,305 acres) and 9% of the total vineyard area. Recommended producers include Villa Vino Rača, who organize an annual international Frankovka Modrá festival, and Château Topol'čianky. Mrva & Stanko is a smaller producer with a fine reputation for this variety.

In **Bulgaria** it is known as Gamé, probably a reflection of the fact that for many years the variety was thought to be the same as the GAMAY NOIR of Beaujolais. In **Romania**, where it goes by the name of Burgund Mare, there are 891 ha (2,202 acres), mainly in the south. It is authorized in south-central regions such as Dealu Mare and Ştefăneşti but also further east towards the Black Sea.

The variety is called Frankovka in the **Czech Republic** and in Vojvodina, **Serbia**, where it can produce lively, fruity, vigorous wines for early consumption. There were 693 ha (1,712 acres) of Modra Frankinja in Slovenia in 2009.

In **Croatia**, Frankovka represented 2.7% of the total vineyard area in 2009, ie around 880 ha (2,175 acres) but this does not include the Borgonja plantings as these are limited and have only recently been shown to be Blaufränkisch (see above). Frankovka is found mainly in Kontinentalna

Hrvatska (Continental Croatia) in the north east and Borgonja in the north-western Istra peninsula, where it produces relatively light wines.

As Lemberger, Blaufränkisch is also planted in British Columbia, **Canada**, and was at one time quite popular in **Washington** State's Yakima Valley, although by 2011 Washington's total plantings had fallen back to just 73 acres (30 ha). There are limited plantings in **New York** State's Finger Lakes region, where it may be blended with CABERNET FRANC to add richer, darker fruit flavours or made as a varietal. Peter Bell at Fox Run notes that in some vintages the wines need to be deacidified a little. Other producers of varietal wines include Anthony Road, Goose Watch and Swedish Hill in New York State. On Long Island, Channing Daughters label their wine Blaufränkisch.

In **Australia**, one lone producer in the Adelaide Hills, Hahndorf Hill, have made a very successful Blaufränkisch, having for some years blended it with Trollinger (SCHIAVA GROSSA) to make a rosé. Mac Forbes also has plans to make an Australian Blaufränkisch.

Lemberger is also found in **Japan**.

BOAL BRANCO

See MALVASIA FINA.

BOBAL

Important and increasingly respected variety producing velvety wines in Spain's Levante.

BERRY COLOUR: ● ● ● ● ●

PRINCIPAL SYNONYMS: Balau, Benicarló, Bobos, Carignan Espagnol or Carignan d'Espagne (Aude in France), Coreana, Espagnol (Var in France), Moravio, Pobretón, Provechón, Rajeno, Requena, Terret d'Espagne (France), Tinta Madrid (Cigales in Spain), Tonto de Zurra, Valenciana

VARIETIES COMMONLY MISTAKEN FOR BOBAL: GRACIANO, MAZUELO, MONASTRELL, MORAVIA AGRIA (Albacete), PRUNELARD (Bobal is called Prunelar in the Ariège in France)

ORIGINS AND PARENTAGE

Bobal comes from the region of Utiel-Requena in central eastern Spain, where it is the predominant variety and where it is said to have been mentioned as early as the fifteenth century in Jaume Roig's misogynist diatribe *Espill o llibre de les dones*. The old Spanish synonym Benicarló, named after a port on the Mediterranean coast, is no longer in use and is best avoided because the name Beni Carlo is mistakenly used for MONASTRELL in the Dordogne.

Bobal Blanca is not a colour mutation of Bobal because it is identical to ALCAÑÓN (Fernández-González, Mena *et al.* 2007). Although Bobal is called Prunelar in the Ariège, it should not be confused with the true French variety PRUNELARD.

VITICULTURAL CHARACTERISTICS

Big, compact bunches of medium-sized berries. Mid to late budding and ripening. Vigorous and fertile with quite high yields. Well suited to short pruning. Extremely drought-resistant but susceptible to powdery mildew and botrytis bunch rot.

WHERE IT'S GROWN AND WHAT ITS WINE TASTES LIKE

Bobal is grown mainly in the provinces of Valencia, Cuenca and Albacete in eastern central **Spain**. It is particularly important in the Utiel-Requena and Manchuela DOs, where it is the dominant variety. Total plantings in Spain in 2007 were officially 85,124 ha (210,346 acres), mostly in Castilla-La Mancha (51,277 ha/126,708 acres) and Valencia (33,389 ha/82,506 acres), but the real total is probably a good deal higher. It tends to ripen unevenly so the wines can be slightly rustic, but the best of them are deeply coloured and smooth. Bobal has traditionally been relegated to bulk-wine production but producers working at higher altitudes (700–770 m above sea level), such as Mustiguillo in Utiel-Requena and Finca Sandoval in Manchuela, have been taking this variety more seriously and giving it the attention it deserves, particularly since it retains its acidity better than MONASTRELL and tends to be lower in alcohol, producing well-structured but velvety and fresh reds. At Finca Sandoval, Victor de la Serna includes it in his SYRAH-dominated blends. Bobal is authorized in the Utiel-Requena, Valencia, Manchuela and Ribera del Júcar DOs. Quaffable, fresh Bobal rosés are also on the increase.

BOĞAZKERE

Tannic but potentially high-quality, quite common Turkish variety.

BERRY COLOUR: ● ● ● ● ●

PRINCIPAL SYNONYMS: Şaraplık Siyah🜨 (formerly in Çermik near Diyarbakır)

ORIGINS AND PARENTAGE

Boğazkere probably comes from the Elâzığ or Diyarbakır provinces in eastern Anatolia. DNA profiling has shown that Şaraplık Siyah, once cultivated in the Çermik region near Diyarbakır in south-eastern Anatolia, is in fact identical to Boğazkere, and DNA parentage analysis has suggested a parent–offspring relationship between Boğazkere and Morek, an almost extinct variety from the Ergani region near Diyarbakır (Vouillamoz *et al.* 2006). A recent genetic study based on forty-four varieties from Turkey and elsewhere suggested that Boğazkere could be genetically close to Kabarcık Beyaz (no longer cultivated) and ÖKÜZGÖZÜ (Sabir *et al.* 2009), but this is not consistent with the analysis of sixty-two varieties from Turkey compared with fifty-four additional ones from neighbouring Armenia and Georgia that clustered Boğazkere with three extinct varieties: Morek and Belelük from the Çüngüş region near Diyarbakır and Khupishizh from Apkhazeti in Georgia (Vouillamoz *et al.* 2006). Boğazkere means 'burns the throat', hinting at the variety's marked tannins.

VITICULTURAL CHARACTERISTICS

Mid to late ripening. Best suited to hot, dry climates but shows good cold resistance. Does well in poor, permeable, calcareous-clay soils and at higher altitudes (cooler nights). Small berries with thick skins and good general disease resistance. Grapes generally have high levels of acidity but can also reach high sugar levels.

WHERE IT'S GROWN AND WHAT ITS WINE TASTES LIKE

Boğazkere is found mainly in the Çermik and Çüngüş districts of Diyarbakır, south-eastern Anatolia, but also in the neighbouring provinces of Elâzığ and Malatya as well as further to the south west in the province of Gaziantep. Total plantings in **Turkey** in 2010 were 1,486 ha (3,672 acres). The variety's increasing popularity has encouraged growers in western Anatolia to plant it, both on the Denizli plateau in south-western Turkey and further north and west in Manisa, with varying degrees of success. Varietal wines are not very deeply coloured, and have marked tannins, flavours of dark or dried fruit, and moderate acidity. They can be rather rustic or tough but the best develop complex, earthy flavours such as leather and tobacco as the tannins calm down with time in bottle. The variety marries well with oak and is often blended with Turkey's other widely planted indigenous variety, ÖKÜZGÖZÜ, which has softer tannins and fresher acidity. Producers include Corvus, Doluca, Kavaklidere, Kayra and Sevilen.

BOGDANUŠA

Makes light, fresh white on the Croatian island of Hvar.

BERRY COLOUR: ● ● ● ● ●

PRINCIPAL SYNONYMS: Bogdanjuša, Bogdanuša Bijela, Bojdanuša (Hvar and Makarska), Hvarka, Vrbanjka (Supetar on the island of Brač)

VARIETIES COMMONLY MISTAKEN FOR BOGDANUŠA: MLADINKA

ORIGINS AND PARENTAGE

Bogdanuša comes from the Croatian island of Hvar, south of Split. The name Bogdanuša, meaning 'given by God' in Croatian, was supposedly chosen because this wine was drunk during religious celebrations.

OTHER HYPOTHESES

Ampelographic studies and DNA profiling have shown that, contrary to common local belief, MLADINKA is not identical to Bogdanuša (Goran Zdunić, personal communication).

VITICULTURAL CHARACTERISTICS

Late ripening. Very susceptible to botrytis bunch rot.

WHERE IT'S GROWN AND WHAT ITS WINE TASTES LIKE

Bogdanuša is officially authorized only in the Srednja i Južna Dalmacija (Central and Southern Dalmatia) wine region in **Croatia** and most of the plantings are on the island of Hvar, mainly in the Stari Grad and Jelsa regions and more often than not for sweet wines or for use in blends. Varietal wines tend to have fresh acidity, moderate alcohol and sometimes a slight bitterness on the finish, and are produced by, for example, Plančić (Hvar) and Ivo Carič (Hvar). Plančić's Ager is an oaked blend of Bogdanuša and PRČ, both indigenous to the island, while Zlatan Otok's Zavala blends POŠIP BIJELI, Maraština (MALVASIA BIANCA LUNGA) and Bogdanuša. Plenkovič and Tomić use the variety in blends.

BOMBINO BIANCO

Productive Italian white found mainly in the south and only rarely distinguished.

BERRY COLOUR: ● ● ● ● ●

PRINCIPAL SYNONYMS: ?Bambino, Buonvino, Ottonese (Lazio in central Italy), Pagadebit, Straccia Cambiale

VARIETIES COMMONLY MISTAKEN FOR BOMBINO BIANCO: MOSTOSA, PASSERINA, TREBBIANO D'ABRUZZO

ORIGINS AND PARENTAGE

While some authors argue that Bombino Bianco came from Spain, most consider it to be indigenous to Italy, possibly Puglia. Bombino means 'small bomb', probably related to the shape of the bunches. DNA profiling has confirmed that Ottonese cultivated in Lazio is identical to Bombino Bianco (Manna Crespan, personal communication). Although Bombino Bianco is not a white mutation of BOMBINO NERO, DNA analysis suggests they could be closely related (Vouillamoz). It is not yet known whether Bombino Bianco and TREBBIANO D'ABRUZZO are distinct varieties (see Trebbiano d'Abruzzo) but both are unrelated to TREBBIANO TOSCANO and Trebbiano di Soave, the latter being identical to VERDICCHIO BIANCO. Recent DNA parentage analysis has suggested that Bombino Bianco is a possible parent of both IMPIGNO and MOSCATELLO SELVATICO from Puglia (Cipriani *et al.* 2010).

VITICULTURAL CHARACTERISTICS

High yields and late ripening. Good resistance to most fungal diseases.

WHERE IT'S GROWN AND WHAT ITS WINE TASTES LIKE

Widely planted in southern **Italy**, mainly in Puglia and particularly in the provinces of Foggia (where it constitutes 40–60% of the San Severo DOC), Bari and Lecce, but also in Emilia-Romagna (at least 85% required for Pagadebit di Romagna DOC), Lazio (where it is called Ottonese in the province of Frosinone, is part of a traditional white blend such as Frascati DOC, and may be a varietal in the Marino DOC), the Marche and Abruzzo. Trere's sparkling Pagadebit di Romagna DOC is a blend of 85% Bombino Bianco with 15% CHARDONNAY, whereas Guarini's version of this DOC is still and, most unusually, made from 100% Bombino Bianco.

The variety produces relatively neutral whites but is a useful ingredient in EU table wine and in cheap German Sekt, where it is generally blended with a more aromatic partner such as MORIO-MUSKAT. Better producers coax aromas of exotic fruits, occasional herbaceous or citrus notes, and a light minerality from this variety. Producers who make a varietal wine labelled Ottonese include Proietti and the Cantina Sociale Olevano Romano. The variety is also used in the production of vermouth. There were almost 3,000 ha (7,410 acres) in Italy in 2000.

BOMBINO NERO

Southern Italian dark-skinned variety much less common than its pale namesake.

BERRY COLOUR:

PRINCIPAL SYNONYMS: ?Bambino, Buonvino

ORIGINS AND PARENTAGE

Bombino Nero is most likely from Puglia. Although Bombino Nero is not the black version of BOMBINO BIANCO, DNA profiling suggests they could be closely related (Vouillamoz). Bombino means 'small bomb', probably a reference to the shape of the bunches.

VITICULTURAL CHARACTERISTICS

Vigorous, very late ripening and high in anthocyanins, thus giving good colour, even though it is often used for rosé.

WHERE IT'S GROWN AND WHAT ITS WINE TASTES LIKE

Cultivated mainly in Puglia, **Italy** (alone or as part of a blend in Castel del Monte DOC and optional in Lizzano DOC), for red and especially rosé wines, but also in Basilicata, Lazio and on Sardegna. The 2000 census records plantings of 1,170 ha (2,890 acres). Wines tend to be light in colour with simple, soft, red-fruit flavours.

BONAMICO

High-yielding Italian black variety hanging on in the Pisa region.

BERRY COLOUR:

PRINCIPAL SYNONYMS: Giacomino⧖, Uva di Palaia (after the name of a town in Toscana)

WHERE IT'S GROWN AND WHAT ITS WINE TASTES LIKE

Bonamico, literally 'good friend', was planted to a limited extent all over Toscana, **Italy**, until the 1960s but is now found only in old vineyards near Pisa, Pistoia and Lucca in Toscana. The 2000 census registered just over 100 ha (250 acres). High yields and late maturation (towards the end of September) mean that it is sometimes used as a table grape. In the winery, it is usually blended with other local grapes. Its naturally high yields result in musts low in alcohol and high in acid, and the risk of unripe tannins. However, if yields are restricted, Bonamico can produce medium-bodied, balanced wines with good acidity and floral aromas.

BONARDA PIEMONTESE

Aromatic, well-coloured Italian red with small bunches and soft tannins.

BERRY COLOUR: ● ● ● ● ●

PRINCIPAL SYNONYMS: Bonarda del Monferrato, Bonarda dell'Astigiano, Bonarda di Chieri

VARIETIES COMMONLY MISTAKEN FOR BONARDA PIEMONTESE: CROATINA (Oltrepò Pavese), DOUCE NOIRE (known as Bonarda in Argentina), Durasa (Novara), NERETTA CUNEESE (Cuneo), NERETTO DURO (Torino), REFOSCO DAL PEDUNCOLO ROSSO (Roero), UVA RARA (Novara)

ORIGINS AND PARENTAGE

Bonarda is a misleading name applied to at least six distinct varieties in northern Italy (see above). In Piemonte in north-west Italy, Bonarda Piemontese was first mentioned by Nuvolone Pergamo (1787–98), Conte di Scandaluzza, who refers to its presence in the Torino hills. For details of the commonly planted Bonarda of Argentina, see DOUCE NOIRE.

VITICULTURAL CHARACTERISTICS

Bonarda Piemontese is generally trained on pergolas and has good resistance to diseases.

WHERE IT'S GROWN AND WHAT ITS WINE TASTES LIKE

Bonarda Piemontese is cultivated mainly in the Pinerolese, in Astigiano, and in the Collina Torinese in northern **Italy**, and used in blends to provide colour, or sugar to increase alcohol levels, and the variety is often used to soften the tannins of NEBBIOLO. A varietal Bonarda is authorized in DOCs such as Collina Torinese, Colline Novaresi, Pinerolese and Piemonte, although the regulations do not always make it clear which Bonarda is being referred to. Such wines are characterized by fresh floral aromas, low acidity, powdery tannins and a balanced structure. Paolo de Marchi at Villa Sperino has been instrumental in reviving Coste della Sesia DOC, which allows for a maximum of 50% Bonarda Piemontese (though a varietal Bonarda delle Coste della Sesia is also possible). Benotto, Carlo Daniele Ricci and Cascina Gilli all produce varietal Bonarda Piemontese, though it is rarely exported.

Piemonte regional government statistics recorded 218 ha (540 acres) in 2001.

BONDA

Virtually extinct red from Italy's Valle d'Aosta.

BERRY COLOUR: ● ● ● ● ●

VARIETIES COMMONLY MISTAKEN FOR BONDA: PRIMETTA

ORIGINS AND PARENTAGE

Bonda probably has its roots in northern Italy but it is very distinct from all other Valle d'Aosta varieties. DNA profiling has shown that Bonda and PRIMETTA are distinct varieties, even though they share the misleading synonym Prié Rouge, and that they are not colour mutations of the white PRIÉ (Vouillamoz and Moriondo 2011). A close relationship with Mossana from

Piemonte (no longer cultivated) was suspected as far back as the nineteenth century and has been confirmed by DNA study, as has Bonda's close relationship with NOSIOLA from Trentino (Vouillamoz and Moriondo 2011).

WHERE IT'S GROWN AND WHAT ITS WINE TASTES LIKE

The late-ripening Bonda appears only in blends and is close to extinction, with just a few plants surviving in old vineyards between Châtillon and Quart in the Valle d'Aosta, northern **Italy** (fewer than 3 ha/7 acres were recorded in the 2000 census).

BONDOLA

Old, rare Ticino variety that has been overwhelmed by Merlot.

BERRY COLOUR:

PRINCIPAL SYNONYMS: Bondola Nera, Briegler or Brieger (Zürich), Bundula (Ticino), Longobardo (Aargau)

VARIETIES COMMONLY MISTAKEN FOR BONDOLA: BONDOLETTA, HITZKIRCHER

ORIGINS AND PARENTAGE

Bondola was once the most important grape variety in Ticino, the southern, Italian-speaking canton of Switzerland, where it was first mentioned in 1785 by H R Schinz, who described it as a very fine variety (Marcel Aeberhard, personal communication). DNA profiling has revealed that Briegler, an almost extinct variety once cultivated in German-speaking Swiss cantons (mainly Zürich, Aargau, Luzern and Bern), is identical to Bondola (Aeberhard 2005; Frei *et al.* 2006). DNA parentage analysis shows that Bondola and COMPLETER from Graubünden in the south east are the parents of BONDOLETTA and HITZKIRCHER, two minor Swiss varieties that have sometimes been confused with Bondola (Vouillamoz and Moriondo 2011).

A white mutation called Bondola Bianca has been mentioned (Marcel Aeberhard, personal communication) but what the Ticino grape grower Stefano Haldemann has recently called Bondola Bianca is a completely distinct variety (Vouillamoz).

VITICULTURAL CHARACTERISTICS

Vigorous, consistently productive, early to mid budding, mid ripening. Large, thin-skinned berries, susceptible to botrytis bunch rot.

WHERE IT'S GROWN AND WHAT ITS WINE TASTES LIKE

Bondola used to be widely cultivated in the Sopraceneri, the northern part of Ticino, Italian **Switzerland**, especially in the areas of Bellinzona, Biasca and Giornico (Marcel Aeberhard personal communication), but MERLOT has bulldozed its way across the region in the last century and plantings of Bondola and other old local varieties have been drastically reduced. Bondola gives typically deeply coloured, fruity wines with hints of red cherries and blueberries, showing rustic yet pleasant tannins and a vinous, slightly acidic, bitter and refreshing finish. In some ways, Bondola, lacking structure and complexity, is to Merlot in Ticino what DOLCETTO is to NEBBIOLO in Piemonte.

There were just 13 ha (32 acres) in Ticino in 2009 and only a handful of producers of varietal wines, including Azienda Mondò and Stefano Haldemann at La Segrisola. Nostrano, meaning 'ours' in Italian, is a locally popular Ticino wine that blends Bondola with up to thirty other varieties, including BARBERA, FREISA and American hybrids such as ISABELLA and CLINTON.

The variety is even more elusive under the name Briegler but the Rebbaugesellschaft

Hitzkirch in the canton of Luzern have some in a mixed vineyard that is used to make their Schillerwein, a rosé made from a mix of red and white grapes.

BONDOLETTA

Extremely rare but recently rescued Swiss variety.

BERRY COLOUR:

VARIETIES COMMONLY MISTAKEN FOR BONDOLETTA: BONDOLA ☒

ORIGINS AND PARENTAGE

Bondoletta is a local variety from Ticino, Italian-speaking Switzerland, long considered a type of the more widespread but also local BONDOLA. DNA profiling has recently shown that they are distinct (Vouillamoz, Frei and Arnold 2008) and that Bondoletta is a cross between Bondola and COMPLETER from Graubünden (Vouillamoz and Moriondo 2011).

VITICULTURAL CHARACTERISTICS

Early to mid budding and ripening. Berries smaller than BONDOLA and prone to shrivel.

WHERE IT'S GROWN AND WHAT ITS WINE TASTES LIKE

The only producer of this rare variety is Stefano Haldemann in Tenero in Ticino, southern **Switzerland**, who discovered a sole surviving vine in the 1990s and used it to replant a 50-ares (1.24 acres) plot. His La Segrisola is a blend of 65% BONDOLA and 35% Bondoletta. Bondoletta wines are an intense violet colour, have lower acidity than Bondola, smooth tannins only if fully ripe, fruity aromas but a certain rusticity.

BONICAIRE

See TREPAT.

BONVEDRO

See PARRALETA.

BORBA

See GRAŠEVINA.

BORDÔ

American hybrid used to make cheap, sweetish reds and juices in Brazil.

BERRY COLOUR:

PRINCIPAL SYNONYMS: Grana d'Oro

VARIETIES COMMONLY MISTAKEN FOR BORDÔ: ?IVES (US), ?York Madeira (US)

ORIGINS AND PARENTAGE

Bordô is a variety grown in Brazil that is genetically very close to the American hybrids ISABELLA, CONCORD and NIAGARA (Schuck *et al.* 2009), suggesting it is hybrid of American origin too. It would be very useful to compare its DNA profile with those of IVES from Ohio and York Madeira from Pennsylvania, two American hybrids with the synonym Bordo for which DNA information is not available. Bordo is also a synonym for both CABERNET SAUVIGNON and CABERNET FRANC in Romania and for CARMENÈRE in Italy.

VITICULTURAL CHARACTERISTICS

Very productive, resistant to anthracnose and, to a lesser extent, to various types of rot.

WHERE IT'S GROWN AND WHAT ITS WINE TASTES LIKE

Bordô is the second most planted variety in **Brazil**, where it is valued for its deep colour and used mainly to produce inexpensive, slightly sweet table wines as well as juice. There were 6,726 ha (16,620 acres) in the state of Rio Grande do Sul alone in 2007.

BORGONJA

See BLAUFRÄNKISCH.

BORNOVA MISKETI

See MUSCAT BLANC À PETITS GRAINS.

BORRAÇAL

Makes potent, tart, sometimes tough wines in north-west Spain as well as in its native Vinho Verde.

BERRY COLOUR:

PRINCIPAL SYNONYMS: Azedo, Bogalhal (Basto), Caíño Tinto (Galicia in Spain), Espadeiro Redondo (Monção), Olho de Sapo (Basto), Tinta Femia

VARIETIES COMMONLY MISTAKEN FOR BORRAÇAL: AMARAL (Portugal)

ORIGINS AND PARENTAGE

Borraçal is a variety from the Vinho Verde region in north-west Portugal, where it was mentioned in the Minho region as early as 1790 by Lacerda Lobo (Galet 2000).

VITICULTURAL CHARACTERISTICS

Mid budding, late ripening. Medium to low, irregular yields. Susceptible to coulure, millerandage, powdery mildew and sunburn; very susceptible to botrytis bunch rot. Well suited to poor, dry soils but hard to ripen on cool sites.

WHERE IT'S GROWN AND WHAT ITS WINE TASTES LIKE

Borraçal wines are typically ruby coloured, high in both alcohol and acidity, and in the Minho region of northern **Portugal**, where it is generally recommended (but not in the Monção subregion), it is often blended with VINHÃO. Recommended producers include the Cooperativa de Ponte da Barca and the Quinta da Lixa. There were 187 ha (462 acres) in Portugal in 2010.

As Caíño Tinto in **Spain**, the variety is found mainly in Galicia in the north west, where it is said to be productive and have small compact bunches of thick-skinned berries. Wines are generally high in alcohol and are often made using carbonic maceration to overcome the variety's characteristically astringent, often bitter, tannins. Older vines and limited yields can improve wine quality.

BOSCO

Ligurian white speciality making excellent sweet wines.

BERRY COLOUR: ● ● ● ● ●

PRINCIPAL SYNONYMS: Bosco Bianco del Genovese, Bosco Bianco di Savona

VARIETIES COMMONLY MISTAKEN FOR BOSCO: SCIMISCIÀ ✗ (Liguria)

ORIGINS AND PARENTAGE

Bosco, one of the most traditional grapes of Liguria in north-west Italy, where it probably comes from, means 'woodland', supposedly because it was introduced to Cinque Terre in the south-east of Liguria from cuttings taken in the woods around Villa dei Marchesi Durazzo in nearby Genova.

VITICULTURAL CHARACTERISTICS

Very vigorous, early to mid budding, mid ripening. Moderate resistance to downy mildew and sour rot, more susceptible to powdery mildew. Loose bunches of medium-sized, thick-skinned berries.

WHERE IT'S GROWN AND WHAT ITS WINE TASTES LIKE

Cultivated now in the provinces of Genova and La Spezia, north-west **Italy**, Bosco is the main variety in the locally famous and expensive Cinque Terre Sciacchetrà dessert wine, often blended with ALBAROLA and VERMENTINO. Amber-coloured, with aromas of nuts, almonds, honey and dried figs and tasting of apricot jam, this rich wine shows excellent balance between sweetness and acidity. Walter de Battè and Possa make very good examples. It is also authorized in Colline di Levanto and one or two other local DOC white wines. Just over 100 ha (250 acres) were recorded in the 2000 census.

BOUCHALÈS

Declining dark-skinned variety surviving around Bordeaux.

BERRY COLOUR: ● ● ● ● ●

PRINCIPAL SYNONYMS: Bouchalets (Estillac in Lot-et-Garonne), Bouissalet (Dordogne), Gros Bouchalès (Gers and Lot-et-Garonne), Gros Boucharès (Dordogne), Grappu or Grapput (Gironde), Picardan Noir, Prolongeau

VARIETIES COMMONLY MISTAKEN FOR BOUCHALÈS: COT (Haute-Garonne and Ariège), MÉRILLE (Ariège)

ORIGINS AND PARENTAGE

Bouchalès was mentioned in 1783–4 in several places in south-west France:

- Estillac (Lot-et-Garonne): 'Bouchalets ... il est extraordinairement noir, et très bon' ('Bouchalets ... is extraordinarily black, and very good')
- Auch (Gers): 'Gros bouchalès, raisin noir ... de très bon gout, excellent pour le vin' ('Gros bouchalès, black variety ... tastes good, excellent for wine')
- Agen (Lot-et-Garonne): 'Bouchalès grand ... Bouchalès petit ... Le vin en seroit excellent' ('Bouchalès grand ... Bouchalès petit ... The wine is said to be excellent')
- Saint-Barthélémy (Dordogne): 'Gros Boucharès. Petit Boucharès'
- Tonneins (Lot-et-Garonne): 'Gros Bouchalès, très bon, son vin est bien noir' ('Gros Bouchalès, very good, its wine is really black')

Several subsequent references in the south west support the Vallée de la Garonne as Bouchalès' place of origin. Although it is also called Bouissalet in the Dordogne, it is not a colour mutation of ARRUFIAC, which is called Bouisselet in Pacherenc.

According to some scholars, the name Bouchalès could come from the Occitan *boish* meaning 'box tree' (Lavignac 2001). However, many varieties may resemble box and a simple alternative etymology is that since Bouchalès is a family name that still exists in the Lot-et-Garonne, the likely birthplace of this variety, it is quite possible that the variety inherited its name from its propagator.

Bouchalès was often confused with COT, perhaps because both belong to the Cot ampelographic group (see p XXVII; Bisson 2009).

Picardan Noir is an old and widespread synonym for Bouchalès but it is a source of confusion because it is also a synonym for the dark-skinned CINSAUT.

VITICULTURAL CHARACTERISTICS

Early budding, mid ripening, vigorous and productive. Susceptible to black rot.

WHERE IT'S GROWN

Bouchalès is still cultivated to a limited extent in the Vallée de la Garonne, **France**, a total of 110 ha (272 acres) in 2008 compared with nearly 5,000 ha (12,355 acres) fifty years ago. It is not easy to graft or grow and therefore continues to decline. At the Château de la Vieille Chapelle, north west of Libourne in Bordeaux, a few very old vines from their oldest vineyard were identified as Bouchalès by DNA profiling at the Institut Français de la Vigne et du Vin Pôle Sud-Ouest in 2009. These centenarian vines might have survived phylloxera thanks to the ancient tradition of flooding the vineyards every winter with the waters of the neighbouring Dordogne, disrupting the pest's life cycle. The grapes from these vines are blended with MERLOT from the same vineyard in the special Cuvée A&A (*Amis & Associés*, 'Friends and Associates').

BOURBOULENC

Useful vine making interesting white throughout the south of France.

BERRY COLOUR: ● ● ● ● ●

PRINCIPAL SYNONYMS: Blanquette (Pyrénées-Orientales and south-west France), Bourboulenco, Bourbouleng, Bourboulenque, Bourbounenco, Clairette à Grains Ronds (Gard), Clairette Dorée (Var, Hérault), Doucillon, Malvoisie (La Clape in the Aude), Picardan

VARIETIES COMMONLY MISTAKEN FOR BOURBOULENC: Roussaou or Roussette (Vaucluse)

ORIGINS AND PARENTAGE

Bourboulenc is an old variety from the Vaucluse in Provence, southern France, probably first mentioned in 1515 in Cavaillon as Borbolenques: 'ex bona et fertile vinea videlicet espace, buguera nigra, espagnens, ribiers, brossonaulx, rodeilhac, borbolenques' ('from a good and fertile vineyard, that is to say . . .'), and in 1538 near Avignon, also under the name Borbollenque. Numerous additional references have been recorded in the Vaucluse and in the southern Rhône. The name Bourboulenc is likely to derive from Barbolenquiera, the old name of a small vineyard in Aubignan near Avignon. Bourboulenc is a member of the Claret ampelographic group (see p XXVII; Bisson 2009).

OTHER HYPOTHESES

The name Bourboulenc was said to come from Bourboun, the local nickname of the inhabitants of Aix-en-Provence who helped the Bourbons to invade the city, but this hypothesis is not supported by any historical mention of Bourboulenc in Aix.

Galet (1990) says that this variety is described twice in Viala and Vermorel's (1901–10) *Ampélographie*: once under Bourboulenc, and once under Roussette. Under Bourboulenc, it is considered to be an old variety from the Vaucluse, of unknown origin. Under Roussette, also called Clairette Rousse or Roussaou (not commercially cultivated today and completely distinct from Savoie's ALTESSE), it is considered to be a variety of Greek origin, simply because a vine grower from La Cadière d'Azur in the Var had supposedly received this variety from Kumi under the name Asprokondura. This hypothesis has little credibility, not only because there are no Greek references to Asprokondura but also because Bourboulenc clearly belongs to the Claret ampelographic group (see p XXVII), which has no link with Greek varieties. In addition, the name Roussaou is also given to GOUAIS BLANC, and the true Roussaou is a possible offspring of Gouais Blanc (Boursiquot *et al.* 2004). Since Bourboulenc and Roussaou are often confused, this might explain why Bourboulenc is sometimes said to be an offspring of Gouais Blanc, although this has not so far been supported by DNA evidence. This ampelographic conundrum should be clarified by DNA profiling.

VITICULTURAL CHARACTERISTICS

Slow, late ripening, suited to warm, dry conditions. Susceptible to powdery mildew but resistant to botrytis bunch rot thanks to its long loose bunches and thick skins.

WHERE IT'S GROWN AND WHAT ITS WINE TASTES LIKE

There were 622 ha (1,537 acres) in **France** in 2008, where it is authorized or recommended in many *départements* of southern France. This is about half the area compared with fifty years ago but plantings see-sawed in the 1970s and 1980s, during a period of re-evaluation and greater appreciation of its ability to retain acidity in the heat. It may be included in many an appellation, eg Châteauneuf-du-Pape, Costières de Nîmes, Tavel in the southern Rhône or Cassis and Bandol in Provence, though it rarely contributes more than 20–30% to the blend. However, it is listed as a principal variety for whites such as Languedoc, Corbières and Minervois, where it is usually blended with other regional varieties such as Grenache Blanc (GARNACHA BLANCA), Macabeu (MACABEO), MARSANNE or CLAIRETTE. It is particularly successful on the La Clape massif south of Narbonne, where it takes on a marine note.

Wines are generally moderate in alcohol, and semi-aromatic with citrus aromas and a hint of smoke as long as the grapes are picked late enough, though they are sometimes a little rustic. Recommended producers include Château d'Anglès and Château des Karantes, both using 50% Bourboulenc in their white La Clape blends. Christophe Barbier of Château Bouisset goes one step further and makes two varietal wines: Les Terres Salées, made from ungrafted vines and aged in new oak, and the unoaked Bourboulenc Classique.

BOURRISQUOU

Very rarely makes wine; better known as the parent of various rootstocks.

BERRY COLOUR: ● ● ● ● ●

PRINCIPAL SYNONYMS: Bourriscou (Ardèche), Mourrisquou de Romani, Romanet (Ardèche)

ORIGINS AND PARENTAGE

There are several stories about the origin of Bourrisquou. According to Rougier (1904), a Dr Silhol wrote in 1887 that Bourrisquou was found in a wood in Lanas near the village of Vogüé (west of Montélimar in the Ardèche) by a vine grower named Romanet, hence the local synonym. However, Silhol did not believe that Bourrisquou originated from the region and suggested instead that it was introduced from Spain to Aubenas, north of Vogüé, by a soldier coming back from the war. Then birds ate the berries in Aubenas and ejected the seeds in Lanas, where Romanet found the grapevine. It should be noted that this hypothesis does not make sense because if the vine had grown from the grape seed, the Lanas vine would have been a progeny of the Aubenas vine, and thus a distinct variety.

After a thorough *in situ* investigation, Rougier confirmed that it was Bourrisquou that Romanet found, and suggested a convincing story for its etymology: the vine was propagated in the Aubenas area by a hawker named Bonnand, who was selling various objects carried on the back of his donkey. Bonnand heard many vine growers complaining about their vines being attacked by powdery mildew and persuaded them to buy and plant cuttings of Romanet's disease-free vines, thus propagating Bourrisquou in the whole region. The link is that *bourrisquou* means 'donkey' in the local dialect (corresponding to the French *bourrique*) and the vine that Bonnand used to sell with his donkey was simply called Bourrisquou, which was also Bonnand's nickname . . .

The Spanish origin of Bourrisquou is highly questionable, and Rougier wisely considers that Bourrisquou was born in the Lanas wood. The true origin of Bourrisquou still has to be established by DNA analysis.

Bourrisquou was used by Couderc to breed several hybrids, often used as rootstocks.

VITICULTURAL CHARACTERISTICS

Small berries, late ripening.

WHERE IT'S GROWN

Bourrisquou is not included in the official register of varieties in **France** and has almost disappeared from the Ardèche. Only a few producers still have it in their vineyards but Mas de La

Salle in the Cévennes have recently planted a few experimental ares of Bourrisquou, CHICHAUD and Vituaigne and will increase the area if the results of vinification are encouraging.

BOUTEILLAN NOIR

Virtually extinct Provençal variety.

BERRY COLOUR: ● ● ● ● ●

PRINCIPAL SYNONYMS: Fouiral (Hérault), Moulas (Vaucluse), Psalmodi Noir, Sigoyer (Alpes-de-Haute-Provence)

VARIETIES COMMONLY MISTAKEN FOR BOUTEILLAN NOIR: ARAMON NOIR (under the name Gros Bouteillan), CALITOR NOIR

ORIGINS AND PARENTAGE

The earliest mention of this variety appeared in 1715 in the Vaucluse, in southern France: 'Bouteillan. Quoique cette espece soit assez commune dans nôtre [sic] terroir, elle l'est encore plus dans ceux de Rians, de Pertuis, de Cadenet, de Cucuron, etc.' ('Bouteillan. Although this species is rather common in our region, it is even more common in Rians, Pertuis, Cadenet, Cucuron, and so on'). Several other mentions of Bouteillan Noir were later recorded, mainly in the Vaucluse. Its name could derive from the shape of its cluster, which resembles a small bottle (*bouteille*).

Although ARAMON NOIR is also called Gros Bouteillan, it is not related to Bouteillan Noir. According to Bisson (2009), Bouteillan Noir is ampelographically close to COLOMBAUD, which might explain why this latter is sometimes erroneously considered a white mutation of the former.

VITICULTURAL CHARACTERISTICS

Fertile, late budding, late ripening, erratic yields because subject to coulure. Very susceptible to powdery mildew and botrytis bunch rot.

WHERE IT'S GROWN AND WHAT ITS WINE TASTES LIKE

Bouteillan Noir has more or less disappeared from the vineyard in **France** but, according to Galet (2000), the wines are pale and low in alcohol.

BOUVIER

Low-acid Central European variety used for many styles of wine.

BERRY COLOUR: ● ● ● ● ●

PRINCIPAL SYNONYMS: Bela Ranina, Bouvier Blanc (France), Bouvierovo Hrozno (Slovakia), Bouviertraube (Austria), Radgonska Ranina (Slovenia), Ranina or Ranina Bela (Slovenia)

ORIGINS AND PARENTAGE

The Austrian-born banker and winery owner Clotar Bouvier (1853–1930) discovered this and other varieties in 1900 in his vineyards in Hercegovščak (Herzogenberg) near Gornja Radgona (Oberradkersburg), in north-east Slovenia. Bouvier first used it to breed other varieties in his own nursery and later started to sell it in Austria and Slovenia. DNA parentage analysis has suggested that Bouvier could be a natural cross between PINOT and an unknown variety (Regner 2000a).

Bouvier was used by its discoverer and others to breed BIANCA, KABAR, VERITAS, ZENGŐ, ZENIT, ZÉTA and ZEUSZ, all grown in Hungary.

VITICULTURAL CHARACTERISTICS

Mid budding, very early ripening. Low and unreliable yields. Small berries, with a slight pink cast at maturity. Good resistance to winter frost but susceptible to downy and powdery mildews and botrytis bunch rot, as well as to chlorosis.

WHERE IT'S GROWN AND WHAT ITS WINE TASTES LIKE

Bouvier, grown in Central Europe for the table and for wine, is used for several very different wines in **Austria**: the frothy, part-fermented Sturm wine sold soon after harvest; lightly Muscat-scented dessert wines up to TBA sweetness levels, which are aromatic but tend to lack acidity and are therefore often blended; as well as for simple early-drinking whites. There were 234 ha (578 acres) in Austria in 2009 – a decline of nearly 40% in the last ten years – the vast majority (204 ha/504 acres) being in Burgenland and more especially around the Neusiedlersee but also further north in the Weinviertel. Recommended producers include Heribert and Pamela Brandl, Laurer, Willi Opitz, Franz Schindler and Wenzl-Kast.

Recommended producers in **Slovenia** (40 ha/99 acres in 2009), where it is planted mostly in the Podravje region in the north east, include Radgonske Gorice and Steyer. Bouvier is also planted in **Slovakia**, under the name Bouvierovo Hrozno, and to a limited extent in Eger in north-western **Hungary**.

BOVALE GRANDE

See MAZUELO.

BRACCIOLA NERA

Obscure, high-yielding, late-ripening black Italian variety found in Toscana and Liguria.

BERRY COLOUR: ● ● ● ● ●

PRINCIPAL SYNONYMS: Bracciuola, Braciola, Braciuola, Brassola

ORIGINS AND PARENTAGE

It is often assumed that Bracciola Nera was first mentioned by Soderini (1600) under the name Bracciuola. However, Soderini's Bracciuola was cited among the best white-berried varieties, together with Trebbiano Perugino (no longer commercially cultivated) and COLOMBANA NERA, and there's no evidence that Bracciuola was a white version of Bracciola Nera. This means that the earliest mention of the latter appears in Acerbi (1825), under the name Braciola in Cinque Terre in Liguria, in north-west Italy, suggesting its origins may lie somewhere on the Levant coast.

VITICULTURAL CHARACTERISTICS

Productive, late ripening.

WHERE IT'S GROWN AND WHAT ITS WINE TASTES LIKE

Cultivation of this variety with good acidity is today limited to the provinces of Massa-Carrara in Toscana and La Spezia in Liguria, **Italy**, where it is found only in the Colli di Luni DOC, traditionally blended with VERMENTINO NERO and with the white varieties TREBBIANO TOSCANO and VERMENTINO. 110 ha (270 acres) were recorded in the 2000 census.

BRACHETTO DEL PIEMONTE

Rose-scented Piemontese variety making very light, sweet, frothy reds.

BERRY COLOUR: ● ● ● ● ●

PRINCIPAL SYNONYMS: Bracchetto, Brachetto d'Acqui

VARIETIES COMMONLY MISTAKEN FOR BRACHETTO DEL PIEMONTE: BRAQUET NOIR (Nice in France)

ORIGINS AND PARENTAGE

The name Brachetto has been given to many distinct varieties, which is why we opted for the more specific Brachetto del Piemonte for this variety of uncertain origins, possibly originating from the hills of Monferrato spanning the provinces of Asti and Alessandria, where it was first mentioned by Di Rovasenda (1877).

VITICULTURAL CHARACTERISTICS

Early ripening with average to low yields. Highly susceptible to viruses.

WHERE IT'S GROWN AND WHAT ITS WINE TASTES LIKE

Cultivated especially around Acqui Terme in the provinces of Asti and Alessandria in the south east of Piemonte in north-west **Italy**, supposedly since the Roman period under the name Vinum Acquense. It is also recommended in the Piemonte provinces of Cuneo and Torino. In the 1970s, the producer Arturo Bersano invented Brachetto Spumante, a dessert sparkling wine fermented in pressurized tanks that has now become popular as the sweet, low-alcohol Brachetto d'Aqui DOCG. Bersano's Brachetto d'Acqui Castelgaro has aromas of wild roses and fresh red fruit, with a sweet yet fresh palate – a light red equivalent of Moscato d'Asti. Dry, aromatic varietal Brachetto – both still and sparkling – is also produced within the Piemonte DOC but is much less common. There were 1,280 ha (3,160 acres) in Piemonte in 2001.

Pizzini produce some in the King Valley, Victoria, **Australia**.

BRAJDENICA

See DURANIJA.

BRANCELLAO

See ALVARELHÃO.

BRAQUET NOIR

Very minor but historic vine making aromatic light reds around Nice.

BERRY COLOUR: ● ● ● ● ●

PRINCIPAL SYNONYMS: Brachet (Nice)

VARIETIES COMMONLY MISTAKEN FOR BRAQUET NOIR: BRACHETTO DEL PIEMONTE (Piemonte), CALITOR NOIR

ORIGINS AND PARENTAGE

Braquet Noir is an old variety from the regions of Bellet, Nice, Grasse and Cannes, where it was already mentioned in 1783–4: 'Braquet, grain de moyenne grosseur, rond, et noir clair, bon à manger et à faire du bon vin par son mélange avec d'autres raisins' ('Braquet, medium-sized, round and light black berries, good to eat and to make good wine in a blend with other varieties'). This variety is often confused with CALITOR NOIR, and also with BRACHETTO DEL PIEMONTE aka Brachetto d'Acqui from Piemonte (Galet 1990). Braquet Noir belongs to the Calitor ampelographic group (see p XXVII; Bisson 2009). Its name could derive from an old local family name.

Braquet Blanc de Nice is a white mutation of Braquet Noir.

VITICULTURAL CHARACTERISTICS

Thrives in hot, dry conditions and poor soils. Very susceptible to botrytis bunch rot. Mid ripening. Low yields.

WHERE IT'S GROWN AND WHAT ITS WINE TASTES LIKE

There were just 12 ha (30 acres) in **France** in 2008, mainly in the area around Nice. Under the names Braquet and Brachet, the latter used in the French official catalogue of varieties, it is today cultivated mainly in the Bellet appellation and blended with FUELLA NERA. These two principal varieties must contribute at least 60% but Grenache (GARNACHA) and CINSAUT may also be added to produce red and rosé wines. The variety is aromatic, often floral, but low in colour, which is probably why the rare varietal wines, such as those made by Château de Bellet and Le Clos Saint-Vincent, are rosés.

BRATKOVINA BIJELA

Rare Croatian island white adding freshness to Pošip blends.

BERRY COLOUR: ● ● ● ● ●

PRINCIPAL SYNONYMS: Brabkovica, Mesnac (Zadar, midway between Split and Rijeka)

ORIGINS AND PARENTAGE

Bratkovina Bijela most likely originates from the island of Korčula in Dalmacija (Dalmatia), roughly halfway between Split and Dubrovnik in Croatia. DNA parentage analysis has established that it is a parent of POŠIP BIJELI (Piljac *et al.* 2002).

The black-berried Bratkovina Crna from the same island is a distinct variety but is not commercially cultivated, not least because its female flowers make pollination difficult and yields therefore unpredictable.

VITICULTURAL CHARACTERISTICS

Late budding, mid to late ripening, very productive. Medium to large, loose clusters of thick-skinned berries. Relatively low sugar levels in the grapes.

WHERE IT'S GROWN AND WHAT ITS WINE TASTES LIKE

Bratkovina Bijela is officially authorized only in the Srednja i Južna Dalmacija (Central and Southern Dalmatia) wine region of **Croatia**, which includes the island of Korčula. Here it is often blended with its progeny POŠIP BIJELI to add freshness. The Saint-Marelić Winery in Cara, for example, include 15% Bratkovina Bijela in their top Pošip Sveti Ivan.

BRAUNER VELTLINER

Almost extinct Austrian, unrelated to Roter Veltliner.

BERRY COLOUR: ● ● ● ● ●

PRINCIPAL SYNONYMS: Todträger, Veltliner Braun

VARIETIES COMMONLY MISTAKEN FOR BRAUNER VELTLINER: ÖSTERREICHISCH WEISS ☓

ORIGINS AND PARENTAGE

Brauner Veltliner is a variety from Niederösterreich (Lower Austria) with brownish-greyish berries that is used to make white wines. It is often erroneously considered to be a colour mutation of ROTER VELTLINER or a synonym for ÖSTERREICHISCH WEISS but this is rejected by DNA profiling (Regner *et al.* 1996).

WHERE IT'S GROWN AND WHAT ITS WINE TASTES LIKE

Formerly widespread in Niederösterreich (Lower **Austria**) and in **Hungary**, Brauner Veltliner has today almost disappeared. A rare varietal Brauner Veltliner is crafted by Alois Raubal (Bioweinbau-Raubal) in Austria's Thermenregion.

BREIDECKER

Relatively neutral German hybrid that has found limited favour in New Zealand.

BERRY COLOUR:

PRINCIPAL SYNONYMS: Geisenheim 49-84

ORIGINS AND PARENTAGE

Complex hybrid (RIESLING × SILVANER) × CHANCELLOR bred in 1949 by Heinrich Birk at Geisenheim and released in 1962. The complete pedigree of Chancellor can be found under PRIOR. It should be noted that the female parent (Riesling × Silvaner) is not identical to MÜLLER-THURGAU (formerly and erroneously called Riesling × Silvaner).

VITICULTURAL CHARACTERISTICS

A hybrid with good resistance to botrytis bunch rot and downy mildew.

WHERE IT'S GROWN AND WHAT ITS WINE TASTES LIKE

Named after Heinrich Breidecker, one of New Zealand's pioneer grape growers, this variety is planted to a very limited extent (7 ha/17 acres) on the South Island of **New Zealand**, as far apart as Marlborough in the north (Hunter's, for example, produce a varietal with strong pear and apple aromas) and Central Otago in the south. It was imported from Geisenheim in 1979, presumably on the recommendation of Helmut Becker, then advising the fledgling New Zealand wine industry.

BRESLAVA

One of the earliest Slovakian crosses but little dispersed.

BERRY COLOUR:

PRINCIPAL SYNONYMS: CHRTČ × St Dc ALC 10/28 or Chrtc × St M D#Alc 10/28

ORIGINS AND PARENTAGE

Breslava, originally known snappily as Chrtc × St M D#Alc 10/28, is a (Chasselas Rose × GEWÜRZTRAMINER) × Santa Maria d'Alcantara cross (with Slovakian names: (Chrupka Červená × Tramín Červený) × Santa Maria d'Alcantara) obtained in 1960 by Dorota Pospíšilová at the VUVV research centre for oenology and viticulture in Bratislava, Slovakia, in which Chasselas Rose (or Chrupka Červená) is a colour mutation of CHASSELAS, and Santa Maria d'Alcantara is a white-berried variety of unknown origin. The variety was later christened Breslava, the historical name of Bratislava, capital city of Slovakia, a name that was authorized in 2011.

VITICULTURAL CHARACTERISTICS

Vigorous, regularly productive, mid budding and ripening. At risk from spring and heavy winter frosts and from mites but resistant to botrytis bunch rot. Largish, thick-skinned berries

BRIANNA PEDIGREE DIAGRAM

The complex pedigree of BRIANNA, a hybrid created in Wisconsin (US), involves deliberate and natural crosses between 93 distinct parents encompassing the species *Vitis aestivalis*, *Vitis berlandieri*, *Vitis cinerea*, *Vitis labrusca*, *Vitis lincecumii*, *Vitis riparia*, *Vitis rupestris* and *Vitis vinifera.* In this pedigree, some varieties appear more than once simply to make the illustration easier to read, and *Vitis* is abbreviated to *V.* in some instances to save space.

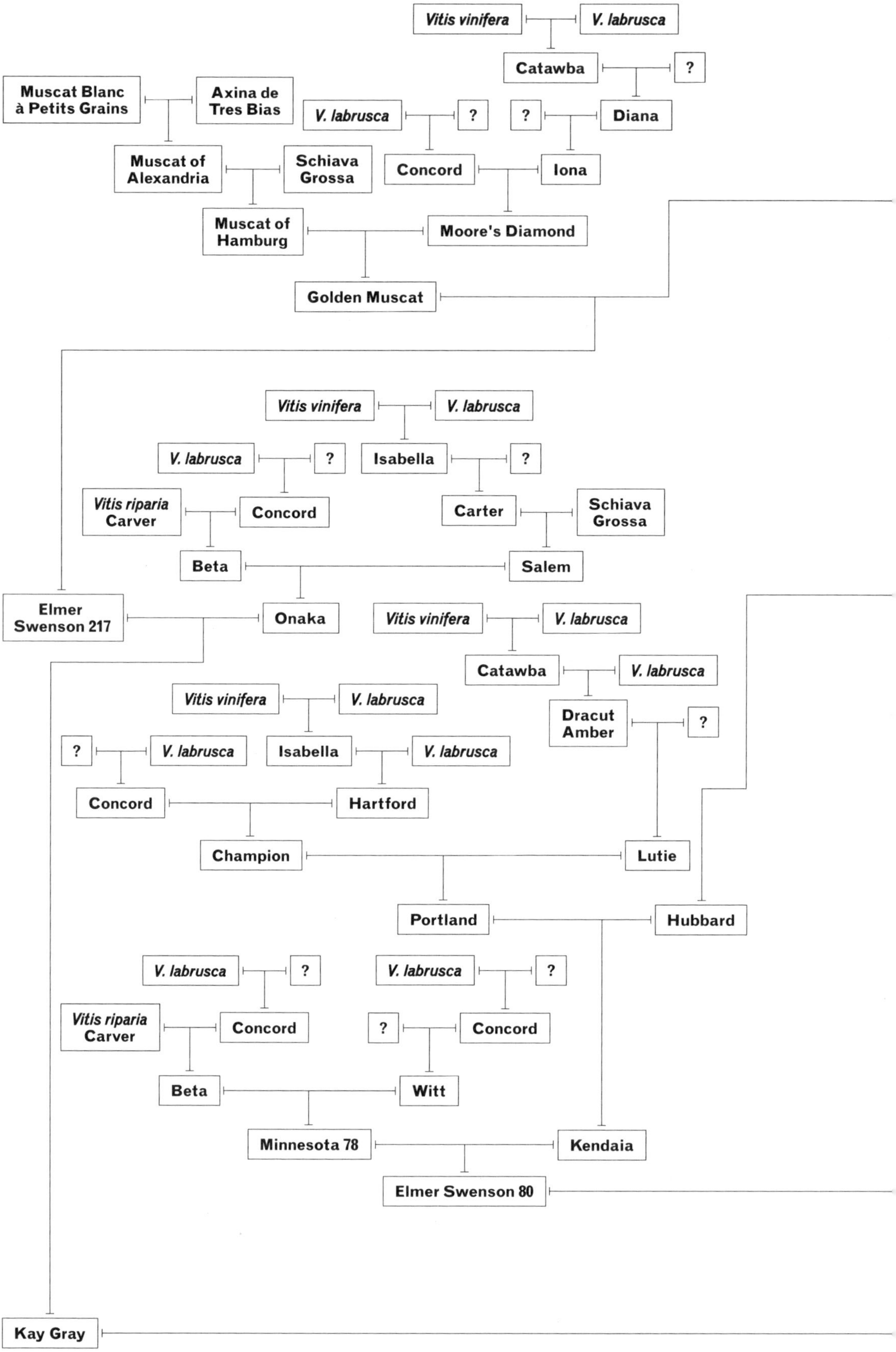
Vitis vinifera
V. labrusca
Catawba
?
Muscat Blanc à Petits Grains
Axina de Tres Bias
V. labrusca
?
?
Diana
Muscat of Alexandria
Schiava Grossa
Concord
Iona
Muscat of Hamburg
Moore's Diamond
Golden Muscat
Vitis vinifera
V. labrusca
V. labrusca
?
Isabella
?
Vitis riparia Carver
Concord
Carter
Schiava Grossa
Beta
Salem
Elmer Swenson 217
Onaka
Vitis vinifera
V. labrusca
Catawba
V. labrusca
Vitis vinifera
V. labrusca
Dracut Amber
?
?
V. labrusca
Isabella
V. labrusca
Concord
Hartford
Champion
Lutie
Portland
Hubbard
V. labrusca
?
V. labrusca
?
Vitis riparia Carver
Concord
?
Concord
Beta
Witt
Minnesota 78
Kendaia
Elmer Swenson 80
Kay Gray

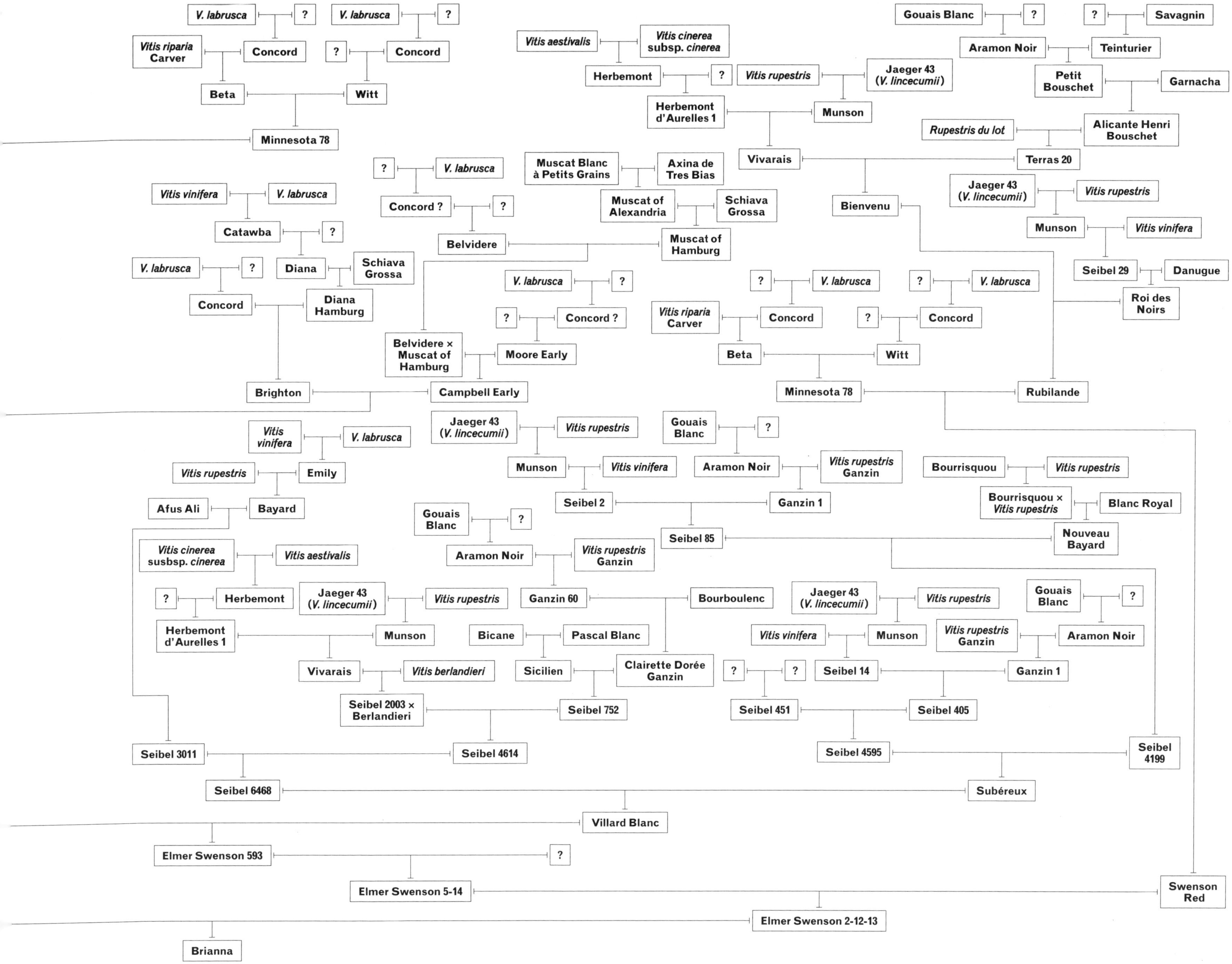
V. labrusca
?
V. labrusca
?
Vitis riparia Carver
Concord
?
Concord
Beta
Witt
Minnesota 78
Vitis aestivalis
Vitis cinerea subsp. cinerea
Herbemont
?
Vitis rupestris
Jaeger 43 (V. lincecumii)
Herbemont d'Aurelles 1
Munson
Vivarais
Gouais Blanc
?
?
Savagnin
Aramon Noir
Teinturier
Petit Bouschet
Garnacha
Rupestris du lot
Alicante Henri Bouschet
Terras 20
Bienvenu
?
V. labrusca
Concord ?
?
Belvidere
Vitis vinifera
V. labrusca
Catawba
?
Diana
Schiava Grossa
V. labrusca
?
Concord
Diana Hamburg
Muscat Blanc à Petits Grains
Axina de Tres Bias
Muscat of Alexandria
Schiava Grossa
Muscat of Hamburg
Jaeger 43 (V. lincecumii)
Vitis rupestris
Munson
Vitis vinifera
Seibel 29
Danugue
Roi des Noirs
V. labrusca
?
?
Concord ?
Moore Early
Belvidere × Muscat of Hamburg
Campbell Early
Brighton
?
V. labrusca
Vitis riparia Carver
Concord
Beta
?
V. labrusca
?
Concord
Witt
Minnesota 78
Rubilande
Vitis vinifera
V. labrusca
Emily
Vitis rupestris
Afus Ali
Bayard
Jaeger 43 (V. lincecumii)
Vitis rupestris
Munson
Vitis vinifera
Seibel 2
Gouais Blanc
?
Aramon Noir
Vitis rupestris Ganzin
Ganzin 1
Seibel 85
Bourrisquou
Vitis rupestris
Bourrisquou × Vitis rupestris
Blanc Royal
Nouveau Bayard
Gouais Blanc
?
Aramon Noir
Vitis rupestris Ganzin
Ganzin 60
Bourboulenc
Vitis cinerea susbsp. cinerea
Vitis aestivalis
?
Herbemont
Herbemont d'Aurelles 1
Jaeger 43 (V. lincecumii)
Vitis rupestris
Munson
Vivarais
Vitis berlandieri
Seibel 2003 × Berlandieri
Bicane
Pascal Blanc
Sicilien
Clairette Dorée Ganzin
Seibel 752
Seibel 4614
Jaeger 43 (V. lincecumii)
Vitis rupestris
Vitis vinifera
Munson
Gouais Blanc
?
Vitis rupestris Ganzin
Aramon Noir
?
?
Seibel 14
Ganzin 1
Seibel 451
Seibel 405
Seibel 4595
Seibel 4199
Seibel 3011
Seibel 6468
Subéreux
Villard Blanc
Elmer Swenson 593
?
Elmer Swenson 5-14
Swenson Red
Elmer Swenson 2-12-13
Brianna

WHERE IT'S GROWN AND WHAT ITS WINE TASTES LIKE

Wines are aromatic, with fresh flavours of grapefruit, and better quality than, say, wines produced from MÜLLER-THURGAU. Breslava is not yet widely planted (10 ha/25 acres in 2011) but can be found in Malokarpatská in south-western **Slovakia**. Producers include Karol Braniš and Vino Matyšák.

BRIANNA

Minor but usefully winter-hardy and increasingly popular American hybrid.

BERRY COLOUR: ● ● ● ● ●

PRINCIPAL SYNONYMS: ES 7-4-76

ORIGINS AND PARENTAGE

A KAY GRAY × Elmer Swenson 2-12-13 hybrid obtained in 1983 by Elmer Swenson (1913–2004), a grape breeder from Osceola, Wisconsin, US. The reconstruction of the complete pedigree of Brianna (see diagram), shows it to be an extremely complex *Vitis riparia*, *Vitis labrusca*, *Vitis vinifera*, *Vitis aestivalis*, *Vitis lincecumii*, *Vitis rupestris*, *Vitis cinerea* and *Vitis berlandieri* hybrid. Brianna was selected in 1989 as a table grape and in 2001 as a wine grape. It was named in 2002 by Ed Swanson of Cuthills Vineyards in Pierce, Nebraska.

VITICULTURAL CHARACTERISTICS

Cold-hardy but susceptible to black rot and botrytis bunch rot and highly susceptible to crown gall. Small to medium-sized compact bunches of thick-skinned, large to medium-sized berries.

WHERE IT'S GROWN AND WHAT ITS WINE TASTES LIKE

Brianna is planted to a limited extent in the upper Midwest, particularly in **Iowa** and **Nebraska** but also in **Minnesota** and **South Dakota**, with recent plantings in **New Hampshire** and in **New York** State's Hudson Valley; it is one of the most popular and valued of the recently released American cold-hardy hybrids. Producers of varietal wines include Two Saints in Iowa, Cuthills and Mac's Creek in Nebraska, and Parley Lake and Indian Island in Minnesota. Brianna wines typically have tropical fruit flavours but the grapes need to be picked early enough to avoid increased foxy flavours from the variety's *labrusca* parentage (Smiley 2008).

BRONNER

German hybrid bred for its disease resistance.

BERRY COLOUR: ● ● ● ● ●

PRINCIPAL SYNONYMS: Freiburg 250-75

VARIETIES COMMONLY MISTAKEN FOR BRONNER: Bronner (seedling of PINOT BLANC)

ORIGINS AND PARENTAGE

Complex hybrid obtained in 1975 by Norbert Becker at the Freiburg research centre, in southern

Germany, by crossing MERZLING × Geisenheim 6494 (see PRIOR for the complete pedigree of Geisenheim 6494). It was named in honour of Johann Philipp Bronner (1792–1864), a pharmacist as well as a creative winemaker and a wine boffin, who is credited with having introduced SANKT LAURENT and BLAUER PORTUGIESER to Germany from Austria.

The variety Bronner should not be confused with the PINOT BLANC seedling also named Bronner, in this case after its breeder in Wissloch, just south of Heidelberg, Germany, mainly used as a table grape.

VITICULTURAL CHARACTERISTICS

Vigorous, mid to late ripening, with very good resistance to downy mildew and generally quite resistant to powdery mildew and botrytis bunch rot. Bunch thinning essential to achieve good quality.

WHERE IT'S GROWN AND WHAT ITS WINE TASTES LIKE

This disease-resistant variety is grown to a limited extent in **Italy**, **Belgium**, **Germany** and **Switzerland**. The wines are relatively neutral on the nose, with an occasional note of ripe apples. Well ripened, it is a little like PINOT BLANC. If the yields are too high, it can taste bitter on the finish. Lieselehof in Trentino-Alto Adige in northern Italy make a varietal wine. German producers include Zang in Franken and Goldene Gans and Stadt Lahr in Baden.

BRUGNOLA

Recently proven distinct Valtellina relative of Nebbiolo.

BERRY COLOUR: ● ● ● ● ●

VARIETIES COMMONLY MISTAKEN FOR BRUGNOLA: FORTANA

ORIGINS AND PARENTAGE

Long cultivated in Valtellina, Brugnola has nearly always been considered simply a synonym of FORTANA. However, recent DNA profiling has shown that they are distinct varieties and that Brugnola has a parent–offspring relationship with NEBBIOLO, which has been cultivated for centuries in Valtellina under the name Chiavennasca (Schneider *et al.* 2005–6).

VITICULTURAL CHARACTERISTICS

Good yields. Resistant to botrytis bunch rot and downy mildew, sensitive to late attacks of powdery mildew.

WHERE IT'S GROWN AND WHAT ITS WINE TASTES LIKE

Brugnola is cultivated both for the table and for wine in Valtellina in the far north of **Italy**. A small percentage of Brugnola can be found blended with other local grapes such as NEBBIOLO and ROSSOLA NERA in several Valtellina wines, such as Nino Negri's Valtellina Superiore Grumello DOCG. It is still considered a synonym of FORTANA in the 2000 agricultural census so no vineyard statistics are available for this separate variety.

BRUJIDERA

See MARUFO.

BRUN ARGENTÉ

Very minor southern Rhône variety.

BERRY COLOUR: ● ● ● ● ●

PRINCIPAL SYNONYMS: Camarèse or Camarèze (Gard), Vaccarèse, Vacarèse or Vaccarèze (southern Rhône)

VARIETIES COMMONLY MISTAKEN FOR BRUN ARGENTÉ: CINSAUT

ORIGINS AND PARENTAGE

The earliest mention of this variety was found under its synonym Vaccarèse in 1538 in Saint-Saturnin-lès-Avignon: 'vaqueyresse, unharum, borbollenque' (Vaccarèse, ???, BOURBOULENC). Additional references to Vaccarèse near Avignon strongly support this region as the birthplace of Brun Argenté. The synonym Camarèse was first mentioned in 1806, whereas the name Brun Argenté appeared only in 1909 in Viala and Vermorel (1901–10), without any clear ampelographic description. However, the name Vaccarèse is specific to the area of Châteauneuf-du-Pape and Camarèse to the area of Chusclan in the Gard, while Brun Argenté is used in a larger area.

The name Brun Argenté clearly refers to the dark berries (*brun* means 'brown') and the silvery look of the down on the leaves (*argent* means 'silver'). The name Vaccarèse could derive from Vaccarès, a region in the Camargue, and the name Camarèse could come from Camarès, a locality in the Aveyron (Midi-Pyrénées), whence it might have been transported to the Gard.

Brun Argenté is a member of the Sérine ampelographic group (see p XXVII; Bisson 2009).

VITICULTURAL CHARACTERISTICS

Fertile, late ripening, needs to be pruned short. Prefers hot, low-vigour, well-exposed sites. Very sensitive to botrytis bunch rot. Big bunches and berries.

WHERE IT'S GROWN AND WHAT ITS WINE TASTES LIKE

There were just 12 ha (30 acres) planted in **France** in 2008, cultivated mainly under the name Camarèse in Chusclan in the Gard (one of the named Côtes du Rhône villages), where it is generally blended with Grenache (GARNACHA) to produce rosé wines (up to a maximum of 20% but usually less). As Vaccarèse, it is authorized in the Châteauneuf-du-Pape appellation but only a few producers such as Château de Beaucastel and Domaine du Pegau actually cultivate (and then blend) it. Although the appellation regulations and the French official register of varieties use the name Brun Argenté, neither it nor Vaccarèse is seen to any extent on labels.

BRUN FOURCA

Ancient but now virtually extinct western Provençal vine.

BERRY COLOUR: ● ● ● ● ●

PRINCIPAL SYNONYMS: Brun d'Auriol, Brun Fourcat, Farnous (Var), Flouron (Drôme), Mançonnet (Ardèche), Moulan (Hérault), Moureau or Mouzeau (Gard), Mourrastel-Flourat (Hérault)

VARIETIES COMMONLY MISTAKEN FOR BRUN FOURCA: CHICHAUD🧬, PINOT NOIR🧬

ORIGINS AND PARENTAGE

Brun Fourca used to be cultivated mainly in Provence, southern France, its most likely place of origin, as well as in the Languedoc. The first mention of a grape called Brun Fourca appeared in 1772 but it was used as a synonym for PINOT. The earliest mention of the true Brun Fourca was found in 1783–4 in Aix-en-Provence and in Aubagne (Bouches-du-Rhône). Its name comes from the characteristic shape of its bunches: divided in two parts like a fork (French *fourche*).

VITICULTURAL CHARACTERISTICS

Mid ripening, with small bunches but with big berries that have a tendency to fall off when ripe. Susceptible to botrytis bunch rot and even more so to powdery mildew.

WHERE IT'S GROWN AND WHAT ITS WINE TASTES LIKE

Brun Fourca may be included as a supplementary variety, alongside the principal players Cinsault (CINSAUT), Grenache (GARNACHA) and Mourvèdre (MONASTRELL), in the Palette appellation around Aix-en-Provence, **France**, but there was less than 1 ha (2.5 acres) remaining in 2006. Nevertheless, Château Crémade claim to include it in their red Palette and Château Simone still have a few old vines.

According to the French official register of varieties, Brun Fourca produces wines that sound barely memorable – 'moderately coloured with a certain acidity'.

BRUÑAL

See ALFROCHEIRO.

BRUNELLO

See SANGIOVESE.

BRUSTIANO BIANCO

Scarce white variety found on the Italian island of Sardegna and reintroduced on the French island of Corse.

BERRY COLOUR: ● ● ● ● ●

PRINCIPAL SYNONYMS: Calitrano or Colitrano (Sartène on Corse), Licronaxu Bianco⚥ (Sardegna)

VARIETIES COMMONLY MISTAKEN FOR BRUSTIANO BIANCO: RETAGLIADO BIANCO⚥, VERMENTINO⚥

ORIGINS AND PARENTAGE

Comparison of DNA profiles from independent sources provides evidence that Brustiano Bianco from the French island of Corse is identical to Licronaxu Bianco and Licronaxu Nero on the Italian island of Sardegna (Vouillamoz), the latter being a simple colour mutation (De Mattia *et al.* 2007).

VITICULTURAL CHARACTERISTICS

Productive and very sensitive to powdery mildew.

WHERE IT'S GROWN AND WHAT ITS WINE TASTES LIKE

It is cultivated on Sardegna and on Corse, where it has recently been tested for reintroduction and can already be found in wines such as Domaine Abbatucci's Cuvée Collection Il Cavaliere Diplomate d'Empire, a blend of BIANCU GENTILE, Brustiano Bianco, SCIMISCIÀ (using the synonyms Genovese and Rossala Bianca) and VERMENTINO.

BUBBIERASCO

BERRY COLOUR: ● ● ● ● ●

ORIGINS AND PARENTAGE

An old variety from the Saluzzo region in the province of Cuneo in Piemonte, north-west Italy. A recent DNA parentage study demonstrated that it is a natural cross between NEBBIOLO, formerly widely cultivated in the Saluzzo region, and Bianchetta di Saluzzo, another old and non-cultivated white variety from the same area (Schneider *et al.* 2005–6).

WHERE IT'S GROWN

Scattered in old vineyards in the province of Cuneo, especially in the Val Bronda.

BUDAI ZÖLD

Very minor Hungarian variety making fresh, simple whites.

BERRY COLOUR: ● ● ● ● ●

PRINCIPAL SYNONYMS: Budai, Zöld Budai, Zöld Szőlő, Zöldfehér, Zöldszőlő

ORIGINS AND PARENTAGE

Budai Zöld, literally the 'green of Buda', probably originates in the region around Buda, the old capital city of Hungary on the west bank of the Duna (Danube) that now constitutes part of Budapest, where Budai Zöld used to be one of the most important varieties.

VITICULTURAL CHARACTERISTICS

Early budding, mid ripening, high yielding. Big bunches of slightly waxy, medium-sized berries. Susceptible to botrytis bunch rot, downy mildew and low winter temperatures.

WHERE IT'S GROWN AND WHAT ITS WINE TASTES LIKE

Wines are deeply coloured, full-bodied but relatively simple with fresh green-apple acidity and a vegetal character. There were just 6 ha (15 acres) in **Hungary** in 2008, partly on the southern slopes of Mount Badacsony, on the northern side of Lake Balaton in western Hungary, and to a lesser extent further north in Somló, although it was quite widely cultivated in the Buda region in northern central Hungary before phylloxera struck in the 1870s. The variety is usually planted adjacent to KÉKNYELŰ to improve the pollination, and therefore yields, of the latter, although Szeremley make a respectable everyday-drinking varietal wine in Badacsony.

BUDESHURI TSITELI

Very minor Georgian variety usually blended with Saperavi.

BERRY COLOUR: ● ● ● ● ●

PRINCIPAL SYNONYMS: Budeshuri Saperavi, Budeshuri Shavi, Tsiteli Budeshuri

ORIGINS AND PARENTAGE

Budeshuri Tsiteli probably comes from eastern Georgia. The word *budeshuri* refers in Georgian to the oval-shaped berries. Budeshuri Tsiteli should not be confused with the white-berried Budeshuri Tetri (Chkhartishvili and Betsiashvili 2004).

VITICULTURAL CHARACTERISTICS

High yields, early to mid ripening and resistant to fungal diseases.

WHERE IT'S GROWN

Kindzmarauli Marani in Kakheti, south-eastern **Georgia**, blend Budeshuri Tsiteli with SAPERAVI but there were just 2 ha (5 acres) in Georgia in 2004.

BUJARIEGO

See VIJARIEGO.

BUKET

Relatively unknown Bulgarian cross with potential for well-structured, full-flavoured reds.

BERRY COLOUR: ● ● ● ● ●

PRINCIPAL SYNONYMS: Bouquet

ORIGINS AND PARENTAGE

A PINOT NOIR × MAVRUD cross obtained in 1951 at the Pleven viticulture and oenology institute in northern Bulgaria.

Buket was used to breed STORGOZIA.

The German variety BUKETTRAUBE is also sometimes called Buket but it has white berries and a distinct DNA profile (Vouillamoz).

VITICULTURAL CHARACTERISTICS

Late budding and mid to early ripening. Sensitive to drought though quite resistant to botrytis bunch rot thanks to its small, thick-skinned berries.

WHERE IT'S GROWN AND WHAT ITS WINE TASTES LIKE

Grown mainly in the northern half of **Bulgaria**, particularly in the province of Pleven, Buket was for long planted and harvested with Gamza (KADARKA) and not separated in the winery.

However, it makes wines with good acid, tannin and alcohol levels and seems to have the potential to make serious reds that take well to oak maturation. Borovitza produce a dense, well-structured, cherry-fruited and spicy varietal example. There were 259 ha (640 acres) in Bulgaria in 2008.

BUKETTRAUBE

Aromatic cross bred in Germany but most widely planted in South Africa for sweet wines.

BERRY COLOUR:

PRINCIPAL SYNONYMS: Bocksbeutel, Bouquettraube, Buket, Bukettrebe, Würzburger

ORIGINS AND PARENTAGE

This originated from a seedling selected by Sebastian Englert (1804–80) of Randersacker (close to Würzburg/Franken) in Germany, who founded the first viticultural school in Franken. DNA parentage analysis at INRA (Institut National de la Recherche Agronomique) in Montpellier has shown that Bukettraube is a cross between SILVANER and SCHIAVA GROSSA (aka Trollinger). A variety known as BUKET is planted in Bulgaria but it has black berries and a distinct DNA profile (Vouillamoz).

VITICULTURAL CHARACTERISTICS

Mid ripening, susceptible to powdery mildew.

WHERE IT'S GROWN AND WHAT ITS WINE TASTES LIKE

Although this variety was bred in Germany, it is no longer grown there because of its susceptibility to powdery mildew. There may still be a few hectares in Alsace, **France**, where it is known as Bouquettraube.

South Africa has more of this variety than any other country (88 ha/217 acres), the majority in Malmesbury, and only in South Africa is it bottled as a varietal wine. These are generally sweet or semi-sweet, occasionally botrytized, and have a distinctively floral, grapey, Muscat-like aroma. Cederberg produce a well-made, exotically fruited version with around 25 g/l residual sugar.

BUSSANELLO

Aromatic white Italian cross recommended in parts of Piemonte.

BERRY COLOUR:

PRINCIPAL SYNONYMS: Incrocio Dalmasso 12/37

ORIGINS AND PARENTAGE

A GRAŠEVINA (Riesling Italico) × FURMINT cross obtained by the Italian vine breeder Giovanni Dalmasso at the Conegliano research centre in the 1930s which did not enter the national variety catalogue until 1977.

VITICULTURAL CHARACTERISTICS

Ripens late September to early October.

WHERE IT'S GROWN AND WHAT ITS WINE TASTES LIKE

Bussanello was recommended in 2000 for the Piemontese provinces of Alessandria, Asti and Cuneo in **Italy**, partly thanks to the promising research results at the National Council for Research in Torino in the 1990s. It is also found in the Langhe Bianco DOC of Piemonte and in Friuli-Venezia Giulia. It has inherited intense fruity and flowery aromas from its parents, with hints of jasmine, pear and red apples.

BUSUIOACĂ DE BOHOTIN

Aromatic pink-skinned Romanian variety making sweet, well-balanced Muscat-like wines.

BERRY COLOUR: ● ● ● ● ●

PRINCIPAL SYNONYMS: Busuioacă Neagră, Busuioacă Vânată de Bohotin, Tămâioasă de Bohotin, Tămâioasă Violetă

VARIETIES COMMONLY MISTAKEN FOR BUSUIOACĂ DE BOHOTIN: Muscat Rouge de Madère or Moscato Violetto (France and Italy)

ORIGINS AND PARENTAGE

In 1926–8, a pink-berried Muscat was introduced by Professor Hogaş in Bohotin in Iaşi, eastern Romania, and the variety was then propagated under the name Busuioacă de Bohotin (Dejeu 2004). While Busuioacă de Bohotin is often considered to be a synonym for Muscat Rouge de Madère, aka Moscato Violetto (Dejeu 2004), its DNA profile in Harta *et al.* (2010) does not correspond to that of Muscat Rouge de Madère in Di Vecchi Staraz, This *et al.* (2007) although it does correspond to MUSCAT BLANC À PETITS GRAINS, making this reference sample of Busuioacă de Bohotin a colour mutation of the latter (Vouillamoz). However, the DNA profile of the Busuioacă de Bohotin in Motoc (2009) is unique, so that the true-to-type reference variety still needs to be identified. The name Busuioacă possibly derives from *busuioc*, meaning 'basil', referring to the aroma of the wine.

VITICULTURAL CHARACTERISTICS

Mid to late budding, mid ripening. Compact bunches of small to medium-sized, thick-skinned berries. Susceptible to downy and powdery mildews, resistant to botrytis bunch rot. Some resistance to drought and cold-hardy to −18 °C (−0.4 °F).

WHERE IT'S GROWN AND WHAT ITS WINE TASTES LIKE

This Muscat-flavoured variety makes pink wines in semi-sweet and sweet styles, with flavours of honeysuckle, roses and peaches, the sweetness well balanced by the acidity. The variety is grown mainly in Huşi in Romanian Moldova in the east and further south and west in Dealu Mare, with a total of 90 ha (222 acres) in **Romania** in 2008. Recommended producers include Basilescu and Senator.

C

CABERINTA
CABERNET BLANC
CABERNET CARBON
CABERNET CAROL
CABERNET COLONJES
CABERNET CORTIS
CABERNET CUBIN
CABERNET DORIO
CABERNET DORSA
CABERNET FRANC
CABERNET JURA
CABERNET MITOS
CABERNET MORAVIA
CABERNET PFEFFER
CABERNET SAUVIGNON
CABERNET SEVERNY
CABERTIN
CACABOUÉ
CACAMOSCA
CADDIU
CAÍÑO BLANCO
CALABRESE DI MONTENUOVO
CALADOC
CALAGRAÑO
CALANDRO
CALITOR NOIR
ÇALKARASI
CALLET
CALORIA
CAMARALET DE LASSEUBE
CAMARAOU NOIR
CAMARATE
CAMPBELL EARLY
CANADA MUSCAT
CANAIOLO NERO
CANARI NOIR
CANNAMELA
CAÑOCAZO
CAPRETTONE
CARACOL
CARDINAL
CARGARELLO
CARICA L'ASINO
CARLOS
CARMENÈRE
CARMINE
CARMINOIR
CARNELIAN
CARRASQUÍN
CARREGA BRANCO
CARRICANTE
CASAVECCHIA
CASCADE
CASCAROLO BIANCO
CASCULHO
CASETTA
CASTAGNARA
CASTELÃO
CASTETS
CASTIGLIONE
CATALANESCA
CATANESE NERO
CATARRATTO BIANCO
CATAWBA
CAVRARA
ÇAVUŞ
CAYETANA BLANCA
CAYUGA WHITE
CENTESIMINO
CENTURIAN
CERCEAL BRANCO
CEREZA
CERRETO
CESANESE
CÉSAR
CETINKA

ᛝ this symbol on the following pages indicates that the synonyms, and mistaken identities, are confirmed by DNA evidence

CEVAT KARA
CHAMBOURCIN
CHANCELLOR
CHARDONEL
CHARDONNAY
CHARENTSI
CHARMONT
CHASAN
CHASSELAS
CHATUS
CHELOIS
CHELVA
CHENANSON
CHENEL
CHENIN BLANC
CHICHAUD
CHIDIRIOTIKO
CHINURI
CHISAGO
CHKHAVERI
CHONDROMAVRO
CHOUCHILLON
CIANORIE
CIENNA
CILIEGIOLO
CINSAUT
CITRONNY MAGARACHA
CIURLESE
CIVIDIN
CLAIRETTE
CLAVERIE
CLINTON
COCOCCIOLA
CODA DI CAVALLO BIANCA
CODA DI PECORA
CODA DI VOLPE BIANCA
CÔDEGA DE LARINHO
CODIVARTA

COLOBEL
COLOMBANA NERA
COLOMBARD
COLOMBAUD
COLOMINO
COLORAILLO
COLORINO DEL VALDARNO
COMPLETER
COMPLEXA
CONCORD
CORDENOSSA
CORNALIN
CORNAREA
CORNIFESTO
COROT NOIR
CORREDERA
CORTESE
CORVA
CORVINA VERONESE
CORVINONE
COT
COUDERC NOIR
COUNOISE
COURBU BLANC
COURBU NOIR
CRÂMPOŞIE SELECŢIONATĂ
CRIOLLA GRANDE
CROATINA
CROUCHEN
CROVASSA
CRYSTAL
CSABA GYÖNGYE
CSERSZEGI FŰSZERES
CSÓKASZŐLŐ
CYGNE BLANC

CABERINTA

Productive but not very popular Argentine hybrid.

BERRY COLOUR: ● ● ● ● ●

PRINCIPAL SYNONYMS: CG 14892, Gargiulo 14892

ORIGINS AND PARENTAGE

Caberinta is a Ruby × RUBY CABERNET hybrid obtained by Angel Gargiulo at INTA (the National Institute of Agricultural Technology), in Buenos Aires, Argentina, in which Ruby is a Keuka × Ontario hybrid (see CAYUGA WHITE for the pedigree of Ontario) and Keuka is a Chasselas Rose × Mills hybrid (see GR 7 for the pedigree of Mills); Chasselas Rose is a colour mutation of CHASSELAS. The name is a contraction of Cabernet plus INTA.

VITICULTURAL CHARACTERISTICS

Highly productive.

WHERE IT'S GROWN AND WHAT ITS WINE TASTES LIKE

This home-grown hybrid has not really caught on despite its high yields and ability to retain fresh acidity. In 2009 there were 69 ha (171 acres) in **Argentina**, mostly in Mendoza, down from 90 ha (222 acres) in 2008. It has also been found in **Morocco**.

CABERNET BLANC

Recent Swiss hybrid that combines fresh, clean flavours and good disease resistance.

BERRY COLOUR: ● ● ● ● ●

PRINCIPAL SYNONYMS: VB 91-26-1

ORIGINS AND PARENTAGE

Cabernet Blanc is a hybrid between CABERNET SAUVIGNON and an undisclosed disease-resistant variety (known in German as a *Resistenzpartner*) obtained in 1991 by the private breeder Valentin Blattner in Soyhières, in the Swiss Jura, possibly a Cabernet Sauvignon × (SILVANER × (RIESLING × *Vitis vinifera*) × (JS 12417 × CHANCELLOR)) complex hybrid, just like PINOTIN and CABERTIN.

VITICULTURAL CHARACTERISTICS

Vigorous. Loose bunches of small, thick-skinned berries. Resistant to downy and powdery mildews, botrytis bunch rot and winter frosts. Prone to millerandage so that bunches are often a mix of tiny seedless berries (high sugar and extract) and more normal seeded ones.

WHERE IT'S GROWN AND WHAT ITS WINE TASTES LIKE

Wines are said to have aromas similar to those of SAUVIGNON BLANC and a flavour profile that is somewhere between those of RIESLING and Sauvignon Blanc. As well as providing good resistance to disease, the berries' thick skins mean the grapes can be picked quite late, allowing producers such as Anselmann in the Pfalz in **Germany** to produce sweet as well as dry varietal wines. The variety seems to have caught on mainly in the Pfalz, where there were close to 7 ha (17 acres) in 2007, and a request for inclusion in the official list of varieties was made in 2004. However, it is also found in **Austria**, and the **Netherlands**, where Colonjes in Gelderland and Gelders Laren in Achterhoek make very crisp varietal wines.

(The name Cabernet Blanc is sometimes seen on labels of blush (rosé) or white wines made from CABERNET SAUVIGNON, particularly in the US.)

CABERNET CARBON

Recent tannic German, very complex, hybrid bred for superior disease resistance.

BERRY COLOUR: ● ● ● ● ●

PRINCIPAL SYNONYMS: Freiburg 377-83

ORIGINS AND PARENTAGE

Complex hybrid obtained in 1983 by Norbert Becker at the Freiburg research centre, southern Germany, by crossing CABERNET SAUVIGNON with Freiburg 236-75, where Freiburg 236-75 is a MERZLING × (Zarya Severa × SANKT LAURENT) hybrid, Zarya Severa is a Seyanets Malengra × *Vitis amurensis* hybrid, and Seyanets Malengra is a seedling of PRÉCOCE DE MALINGRE.

VITICULTURAL CHARACTERISTICS

Good resistance to downy mildew. Large berries. Mid to late ripening.

WHERE IT'S GROWN AND WHAT ITS WINE TASTES LIKE

Like several other Cabernet Something varieties, this was bred for its enhanced disease resistance. In **Germany**, it entered the official register of varieties in 2003 but the plantings are too few to make it into the official German vineyard statistics. Wines are deeply coloured, have plenty of tannins and taste spicy with intense Cabernet-like cassis notes. Producers of varietal wines include Helde and Isele in Baden.

After several years of trials in the experimental vineyard of Agroscope Changins-Wädenswil at Pully (south east of Lausanne, on the shore of Lac Léman/Lake Geneva), it is now cultivated on less than half a hectare (about an acre) in **Switzerland**.

CABERNET CAROL

Barely planted disease-resistant German hybrid.

BERRY COLOUR: ● ● ● ● ●

PRINCIPAL SYNONYMS: Freiburg 428-82 R

ORIGINS AND PARENTAGE

Complex hybrid obtained in 1982 by Norbert Becker at the Freiburg research centre, southern

Germany, by crossing CABERNET SAUVIGNON with SOLARIS, which makes it a sibling of CABERNET CORTIS. (See Solaris for full pedigree.)

VITICULTURAL CHARACTERISTICS

Vigorous with generally good disease resistance, particularly to downy mildew.

WHERE IT'S GROWN

There are minuscule plantings in **Switzerland** (less than 1 ha/2.5 acres), including one tenth of a hectare (0.25 acres) planted in 2003 by a grower in the Valais, but it barely shows up on the radar anywhere else, except in Baden in southern **Germany**, where Kaufmann, Staatsweingut Freiburg & Blankenhornsberg and Rabenhof include it in blends.

CABERNET COLONJES

Minor dark-skinned Swiss hybrid found mainly in the Netherlands.

BERRY COLOUR: ● ● ● ● ●

PRINCIPAL SYNONYMS: VB 91-26-5, VB 91-26-05

ORIGINS AND PARENTAGE

A hybrid obtained in 1991 by Valentin Blattner in Soyhières, in the Swiss Jura, by crossing CABERNET SAUVIGNON with an undisclosed disease-resistant variety (known in German as a *Resistenzpartner*). It was initially named VB 91-26-5 by its breeder and renamed Cabernet Colonjes by Wijnhoeve Colonjes, the first producer in the Netherlands to plant this variety.

VITICULTURAL CHARACTERISTICS

Vigorous, early budding, very early ripening although ripening is often uneven. Susceptible to powdery mildew near harvest. Compact bunches of small to medium-sized, relatively thin-skinned berries are susceptible to botrytis bunch rot.

WHERE IT'S GROWN AND WHAT ITS WINE TASTES LIKE

Cabernet Colonjes is grown almost exclusively in the **Netherlands**, eg by Colonjes in Groesbeek, who use it to add fruit to their red blend with CABERNET CORTIS, PINOTIN and REGENT, blended with Regent in their rosé, and as a blanc de noirs. Cabernet Colonjes' greatest attribute is its fruitiness and the wines have been likened to MERLOT. Also found in **Sweden**.

CABERNET CORTIS

Recent tannic German hybrid bred for superior disease resistance.

BERRY COLOUR: ● ● ● ● ●

PRINCIPAL SYNONYMS: Freiburg 437-82 R

ORIGINS AND PARENTAGE

Complex hybrid obtained in 1982 by Norbert Becker at the Freiburg research centre in Baden-

Württemberg, southern Germany, by crossing CABERNET SAUVIGNON with SOLARIS, which makes it a sibling of CABERNET CAROL. (See Solaris for a pedigree.)

VITICULTURAL CHARACTERISTICS

Good resistance to downy mildew and botrytis bunch rot but highly susceptible to powdery mildew. Large berries. Early ripening (before PINOT NOIR).

WHERE IT'S GROWN AND WHAT ITS WINE TASTES LIKE

Like several other Cabernet Something varieties, this was bred for its enhanced disease resistance. Wines are deeply coloured and tannic, with herbal notes. It was authorized in **Italy** in 2011. **Denmark** had around 3 ha (7 acres) in 2010.

It entered **Germany**'s official register of varieties in 2003 but the plantings are too few to make it into the official German vineyard statistics. Wines taste generally Cabernet-like. Becker (in Malsch) and Rabenhof, both in Baden, make varietal examples.

After several years of trials in the experimental vineyard of Agroscope Changins-Wädenswil at Pully (south east of Lausanne, on the shore of Lac Léman/Lake Geneva), it is now cultivated on 2 ha (5 acres) in **Switzerland**.

CABERNET CUBIN

Deeply coloured, tannin-rich German cross that needs the best sites to ripen fully.

BERRY COLOUR: ● ● ● ● ●

PRINCIPAL SYNONYMS: Weinsberg 70-281-35

ORIGINS AND PARENTAGE

A BLAUFRÄNKISCH × CABERNET SAUVIGNON (Levadoux clone) cross created at the Weinsberg research centre in Baden-Württemberg, southern Germany, in 1970. It is therefore a sibling of CABERNET MITOS. Included on the official register in 2005.

VITICULTURAL CHARACTERISTICS

Late ripening with reliable yields.

WHERE IT'S GROWN AND WHAT ITS WINE TASTES LIKE

At the end of 2006 there were nearly 70 ha (173 acres) planted in **Germany**, some still experimental. Wines are deeply coloured with firm tannins that need time to soften, reflecting their Cabernet inheritance, and well disposed to barrel ageing. Gemmrich in Württemberg produced a good varietal example.

It is also found to a very limited extent in **Switzerland** (3 ha/7 acres). Schmidheiny's Zeus, for example, is a varietal Cabernet Cubin aged eighteen months in barrique.

CABERNET DORIO

Recent German cross with a certain Cabernet likeness.

BERRY COLOUR: ● ● ● ● ●

PRINCIPAL SYNONYMS: Weinsberg 71-817-89

ORIGINS AND PARENTAGE

A DORNFELDER × BLAUFRÄNKISCH cross created at the Weinsberg research centre in Baden-Württemberg, southern Germany, in 1971. It is therefore a sibling of CABERNET DORSA.

VITICULTURAL CHARACTERISTICS

Mid to late ripening and producing grapes with high sugar content.

WHERE IT'S GROWN AND WHAT ITS WINE TASTES LIKE

There were approximately 40 ha (99 acres) planted in **Germany** at the end of 2006, some still experimental – good progress given that it was not included in the official register of varieties until 2004. Recommended producers of varietal wines include Ökonomierat Geil Erben and Weingut der Stadt Mainz, both in Rheinhessen. When the grapes are fully ripe, the wine reflects its parentage.

It is also found to a very limited extent in **Switzerland** (less than 1 ha/2.5 acres).

CABERNET DORSA

The most successful of the new German Cabernets.

BERRY COLOUR: ● ● ● ● ●

PRINCIPAL SYNONYMS: Weinsberg 71-817-92

ORIGINS AND PARENTAGE

A DORNFELDER × BLAUFRÄNKISCH cross created at the Weinsberg research centre in Baden-Württemberg, southern Germany, in 1971. It is therefore a sibling of CABERNET DORIO. Included on the official register in 2004.

VITICULTURAL CHARACTERISTICS

Early budding, mid ripening. Smallish berries with generally high sugar levels. Consistently productive. Highly susceptible to downy mildew.

WHERE IT'S GROWN AND WHAT ITS WINE TASTES LIKE

At the end of 2006 there were more than 230 ha (568 acres) planted in **Germany**, some still on an experimental basis. Given that it was not included on the official register of varieties until 2004, it is clearly making good progress. The wines are full-bodied with potentially firm tannins and show the fruit flavours and structural characteristics of both parents – with notable dark-fruit aromas. Producers of varietal wines include Fuchs and Gruber in Rheinhessen, Amlinger in the Mosel, Marx in the Nahe, Anselmann and Fitz-Ritter in the Pfalz, Karl H Johner in Baden, and

Angelika Knauer and Heuchelberg Weingärtner in Württemberg. This variety is expected to gain ground in Germany.

It is also planted to a limited extent in **Switzerland** (20 ha/49 acres) and in **Sweden**.

CABERNET FRANC

Fragrant, well-structured parent of Cabernet Sauvignon that shines in the Loire and in Bordeaux blends.

BERRY COLOUR:

PRINCIPAL SYNONYMS: Achéria (Basque Country, mainly in Irouléguy), Ardounet (Béarn), Bidure (Graves), Bordeaux (Switzerland), Bordo (Romania), Boubet (Pyrénées-Atlantiques), Bouchet Franc or Gros Bouchet (Saint-Émilion and Pomerol), Bouchy (Madiran and Béarn), Breton (Val de Loire), Cabernet Gris, Cabrunet (Pomerol), Capbreton Rouge (Landes), Carmenet (Médoc), Couahort (Béarn), Plant Breton or Plant de l'Abbé Breton (Chinon in Indre-et-Loire), Sable Rouge (Tursan), Trouchet (Béarn), Tsapournako (Greece), Verdejilla Tinto (Aragón in Spain), Véron (Nièvre and Deux-Sèvres), Vidure, Vuidure or Grosse Vidure (Graves)

VARIETIES COMMONLY MISTAKEN FOR CABERNET FRANC: Ardonnet (Béarn, no longer cultivated), ARROUYA (Pyrénées), BÉQUIGNOL NOIR (Dissay in the Vienne), CABERNET SAUVIGNON, CARMENÈRE, Gros Cabernet (Médoc, no longer cultivated), HONDARRIBI BELTZA (País Vasco in Spain)

ORIGINS AND PARENTAGE

Cabernet Franc is undoubtedly one of the most important and ancient varieties in the Bordeaux region. However, recent genetic and historical studies point to the Spanish País Vasco (Basque Country) for its place of origin:

- DNA parentage analysis has shown that Cabernet Franc has parent–offspring relationships with two very old cultivars from the Basque Country, Morenoa and HONDARRIBI BELTZA (often confused with Cabernet Franc in the Txakoli vineyards), thus supporting the hypothesis that Cabernet Franc is a native of the Spanish País Vasco (Basque Country) (Boursiquot *et al.* 2009).
- Founded in the twelfth century, the collegiate church of Roncesvalles (Roncevaux in French and Orreaga in Basque) in the País Vasco, situated close to the French border, was an important stop-over for pilgrims on their way to Santiago de Compostela. Since their early days, Roncesvalles' priests established vineyards, from the French town of Irouléguy to the Basque city of Hondarribia (Fontarrabie in French), using local varieties such as Achéria (meaning 'fox'), which is the Basque name for Cabernet Franc and also morphologically the most archaic or primitive of the clones of Cabernet Franc (Levadoux 1956; Lavignac 2001; Bordenave *et al.* 2007).

Yet there is no historical document mentioning Achéria before 1905 (Rézeau 1997), and the earliest mention of Cabernet Franc was supposedly recorded in the Val de Loire under the old local synonym Breton, referred to by the illustrious French writer François Rabelais in his *Gargantua*, published in 1534 (Rézeau 1997): 'ce bon vin breton poinct ne croist en Bretaigne, mais en ce bon pays de Verron' ('this good breton wine does not grow in Brittany, but in this good country of Verron' (today's Beaumont-en-Véron near Chinon, at the confluence of the Loire and Vienne rivers)).

Although it is far from proven that Rabelais' *vin breton* and Cabernet Franc are one and the same, the presence of a Bordeaux variety in this area is quite possible: around 1050, the Comte d'Anjou Geoffroi Martel made a request to the Abbey of Ronceray in Angers for a piece of land near the confluence of the Loire and the Maine to plant a vineyard with grapevines from Bordeaux (*vineam burdegalensem*; Dion 1959).

The name Breton would suggest that this Bordeaux variety came to the Val de Loire via Brittany, where vineyards existed in the Middle Ages, when the climate was milder. This is supported by the fact that Magdeleine Noire des Charentes, which gave birth to MERLOT through a natural cross with Cabernet Franc, was also found as an ancient, remnant vine near Suliac in Brittany (Boursiquot *et al.* 2009). However, Comte Alexandre-Pierre Odart (1845) provides another explanation for the use of the name Breton in the Val de Loire: in 1631, Cardinal de Richelieu sent out thousands of vines of the best Bordeaux variety to his steward, Abbé Breton, who planted them in Chinon and Bourgueil. The vine was later named Plant de l'Abbé Breton, and then simply Breton. For Lavignac (2001), the name Breton was possibly given to this variety because Breton sailors used to transport wine from the Gironde, from port to port northwards along the French coast, arriving in Brittany via the Loire estuary. Alternatively, it could be a simplified version of Capbreton, a commune in the Landes.

The old synonym Vidure was first mentioned in 1675 by Adrien de Valois in his *Notitia Galliarum*, in which he was the first to consider Vidure as a derivation of Biturica (see Other hypotheses below). For other authors, Vidure might derive from the Gascon dialect *bit duro*, meaning 'hard vine' (French *vigne dure*), most likely because this variety was hard to prune (Rézeau 1997). Under the old synonym Cabrunet, the variety was mentioned in 1716 in Pomerol: 'il seroit necessaire d'arracher cette partie de vigne pour replanter le terrain en bouchet, noir de pressac, et cabrunet' ('it would be necessary to grub up this part of the vineyard to replant the land with bouchet, noir de pressac, and cabrunet'; Rézeau 1997).

Finally, the first mention of the modern spelling Cabernet did not appear until Rozier (1823).

The etymology of Cabernet has been the subject of much speculation. The most convincing hypothesis is based on the Latin *carbon*, meaning 'black', in reference to the dark colour of the berries; *carbonet* became *carbenet* and metathesis (an inversion of consonants) gave *cabernet* (Rézeau 1997). However, the name Cabernet and its old synonyms (Bidure, Breton, Cabrunet, Carmenet, Véron or Vidure) were often given to several distinct varieties, as evidenced by the following entries in a list of grape varieties established in 1859 by Henry-Xavier Bouchereau (1863), then proprietor of Château Carbonnieux, where he had built up a collection of 1,242 grape varieties from France and beyond:

- Carmenet-Sauvignon or Cabernet-Sauvignon, the modern CABERNET SAUVIGNON
- Carmenère, the modern CARMENÈRE
- Gros Cabernet, possibly corresponding to Cabernet Franc but it is also the name of a distinct but no longer cultivated variety (see the CABERNET SAUVIGNON pedigree diagram)
- Petit Cabernet, corresponding to CABERNET SAUVIGNON.

As a consequence, Cabernet Franc, CABERNET SAUVIGNON and CARMENÈRE were frequently confused until the end of the nineteenth century. However, the oldest references most likely referred to Cabernet Franc since it has been established by DNA parentage analysis that Cabernet Franc is a parent of Cabernet Sauvignon (Bowers and Meredith 1997), Carmenère and MERLOT (Boursiquot *et al.* 2009). It is also likely that the parents of a variety as old as Cabernet Franc have long disappeared, making it an orphan. The frequent confusion among these varieties and the parentages discovered are consistent with the Carmenet ampelographic group (see p XXVII).

Cabernet Franc was used to breed CABERNET MORAVIA in the Czech Republic and INCROCIO TERZI I in Italy.

OTHER HYPOTHESES

Several authors have suggested that the old synonym Vidure comes from Bidure, itself deriving from Biturica or Biturigiaca mentioned by Pliny the Elder and Columella as the grape variety of the Bituriges Vibisci, the celtic inhabitants of what is today the Gironde, Burdigala (Bordeaux) being the main city even then (Roudié 1994). Columella grouped Biturica within the same family as Balisca, another ancient variety also called Cocolubis and cultivated in Cantabria (Spain). That's why Roger Dion (1959) suggested a Spanish origin for Cabernet Franc, considering it to be a descendant of Biturica. Some other scholars suggested that Biturica was introduced from Durazzo (now Dürres) in Epirus (Albania), thus suggesting an Albanian origin for Cabernet Franc. Needless to say, there is no scientific evidence for any of these theories.

Some authors have suggested that the name Cabernet might derive from the Latin *carminium*, designating a red dye, also the origin of the colour carmine, but this does not really correspond to the colour of Cabernet Franc's berries (Lavignac 2001). According to another hypothesis, the name derives from the *carmenes*, as the courtyards with fruit trees and vines are known in the region of Granada in Andalucía (Lavignac 2001).

Like many old varieties, Cabernet Franc is often described as a 'population-variety' in south-western France, which implies that various morphological types as well as several distinct varieties build up the whole variety, a concept not embraced in this book.

VITICULTURAL CHARACTERISTICS

Mid ripening, rather vigorous, suited to clay-limestone soils but also performs well on sandy soils if there is no water stress. Small berries. Very similar to CABERNET SAUVIGNON but with less dramatically indented leaves. Buds and matures earlier than Cabernet Sauvignon and thus more prone to coulure but easier to ripen fully. Very hard wood and small berries.

WHERE IT'S GROWN AND WHAT ITS WINE TASTES LIKE

Wine made from Cabernet Franc is generally paler, lighter, crisper, softer and more obviously aromatic than that of its progeny CABERNET SAUVIGNON. Its aroma spans a fine line between refreshingly leafy and aggressively herbaceous, depending on ripeness level. Getting the yield right can be crucial; lack of ripeness can result in high levels of 'green'-smelling methoxypyrazines.

With 36,948 ha (91,300 acres) in 2009, it is **France**'s sixth most planted red wine variety, a long way behind her fifth, Carignan (MAZUELO), with 53,155 ha (131,349 acres). Cabernet Franc's French heartland is the Val de Loire, where, thanks to improved viticultural techniques, not least lower yields, it is generally producing much riper wines. Although elsewhere Cabernet Franc is chiefly used as a blending grape, it is in the Val de Loire that it shines as the dominant and often only variety. There are 5,953 ha (14,710 acres) in the mid Val de Loire, mostly in the Indre-et-Loire (5,356 ha/13,235 acres) *département* in such appellations as Chinon and Bourgueil and to a lesser extent in the Loir-et-Cher (540 ha/1,334 acres). In the Pays de Loire region even more is planted, 9,176 ha (22,674 acres), mainly in Maine-et-Loire (8,671 ha/21,427 acres) for Saumur-Champigny and Anjou-Villages, with 409 ha (1,011 acres) in Loire-Atlantique. Good-quality Cabernet Franc grown here results in fragrant, medium-bodied, well-balanced, silky, fairly early-maturing wines that can smell of pencil shavings, sometimes raspberries. Dependable examples include Chinons from Philippe Alliet, Bernard Baudry, Baudry-Dutour, Château de Coulaine, Charles Joguet (whose Les Varennes du Grand Clos, Cuvée Cabernet Franc de Pied is made from ungrafted vines) and Olga Raffault; Bourgueils and Saint-Nicolas de Bourgueils from Yannick Amirault, Dom de la Butte, Catherine & Pierre Breton, Dom de la Cotellaraie and Pierre-Jacques Druet; and Saumurs and Saumur-Champignys from Château Fouquet, Château de Hureau and Clos Rougeard.

Bordeaux is the world's other Cabernet Franc stronghold and here it has long been planted as an insurance policy against poor viticultural performance by CABERNET SAUVIGNON and MERLOT because it tends to bud and flower at different times from them. Cabernet Franc has been giving way to its routine blending partners, however: first Merlot and then Cabernet Sauvignon now that climate change seems to be making it easier to ripen there. In 2009 the total plantings had dropped to 12,396 ha (30,631 acres; less than half the area planted with Cabernet Sauvignon and less than a fifth Merlot's total area). It is much more popular in the cooler vineyards on the right bank of the Gironde than in the Médoc because it is so much easier to ripen there than Cabernet Sauvignon. Often called Bouchet, with Merlot it plays an important part in virtually all Saint-Émilions. It constitutes the majority of the blend in some vintages of top Saint-Émilion Château Cheval Blanc, which probably constitute the greatest, and certainly longest-living, examples.

Like its systematic Bordeaux blending partners, Cabernet Franc is also widely planted in south-west France and is particularly popular in the relatively cool, inland vineyards of Bergerac. Throughout the south west it is used to soften more tannic varieties such as TANNAT, CABERNET SAUVIGNON and FER. It is allowed in a wide range of appellations such as (with Tannat) Irouléguy, Madiran and Tursan and (with NÉGRETTE) in Fronton. There were 385 ha (951 acres) in the Gers *département* (where Madiran is produced) and 403 ha (996 acres) in Tarn-et-Garonne and 153 ha (378 acres) in the Lot (Cahors country). To the north of Bordeaux, 1,054 ha (2,604 acres) were planted in Poitou-Charentes.

CABERNET SAUVIGNON has rarely dazzled in the Languedoc but there has been a stirring of interest in aromatic varietal Cabernet Franc, with all of 2,265 ha (5,597 acres) planted in the Aude *département*.

Italy's 2000 survey found 7,085 ha (17,507 acres) of Cabernet Franc, although it is mainly concentrated in the far north east, where there has been considerable confusion between this variety and CARMENÈRE, so the total found in the 2010 vineyard survey may well be very different – not least because the market has seemed much more enthusiastic about MERLOT. As in the Loire, vines were for long allowed to yield too heavily, producing decidedly 'green' wines, but this is slowly improving. An exceptional champion of true and subtle Cabernet Franc in central Italy is Tenuta di Trinoro just south of Montepulciano.

In **Spain** in 2008 there were 736 ha (1,819 acres), mostly in Castilla-La Mancha and Catalunya, used mainly for Bordeaux blends although there are some Catalan varietal examples. Just 21 ha (52 acres) were recorded in **Portugal** in 2010.

Cabernet Franc is planted to a limited extent in much of Eastern Europe but is generally much, much less popular than CABERNET SAUVIGNON. Over the border from Friuli in western **Slovenia**, Cabernet Franc tends to be blended with Cabernet Sauvignon, MERLOT and, often, the local REFOSCO, although total plantings in 2009 were only 27 ha (67 acres) in Goriška Brda and 17 ha (42 acres) in Slovenska Istra.

Romania – Cabernet-land – had just 8 ha (20 acres) in 2008, Krasnodar Krai in southern **Russia** just 20 ha (49 acres) in 2009. There are tiny plantings in **Serbia**, **Kosovo**, possibly **Kazakhstan**. The exception to this is **Hungary**, which had a total of 1,243 ha (3,072 acres) in 2008, most successfully in Villány in the south, where some of the country's finest reds, both Bordeaux blends and varietals, depend on Cabernet Franc from the likes of Malatinszky, Attila Gere with Weninger, and Vylyan, but it is also planted further north and east in Kunság and Szekszárd and in Eger even further north. Often in Bordeaux blends but also increasingly successful as a varietal.

Cabernet Franc is planted to a limited extent throughout northern **Greece**, nearly always blended in regional wines. However, the Côtes de Meliton appellation in Halkidikí requires 70% LIMNIO and 30% CABERNET SAUVIGNON and/or Cabernet Franc. In Velventós, western Makedonía, Yannis Voyatzis makes a varietal Tsapournako, shown by DNA profiling to be a clone of Cabernet Franc.

In the eastern Mediterranean, **Cyprus** has 267 ha (660 acres), **Israel** (where Margalit makes a particularly fine example) 100 ha (247 acres) and **Turkey** 39 ha (96 acres). It is also known on **Malta**.

In most of the US, Cabernet Franc is grown chiefly to provide a minor ingredient in Bordeaux blends, but in Virginia, New York State and to a certain extent Washington State the variety is appreciated in its own right as being particularly suitable for local conditions.

California's total area of Cabernet Franc, 3,480 acres (1,408 ha) in the ground in 2009, has remained virtually unchanged for years, although there are signs of a slight increase in interest. Most of it is grown in the Napa Valley and Sonoma, although Rusack grow the fine Anacapa in Santa Ynez Valley far to the south. Exceptional wines dominated by Cabernet Franc come from

Arietta, Crocker & Starr, Detert Family Vineyards, Favia, Lang & Reed and Viader in the Napa Valley and Pride Mountain and Raymond Burr in Sonoma County.

In the much more continental climate of eastern **Washington** State, Cabernet Franc has the advantage over MERLOT of harder wood and consequent ability to withstand the extremely cold winters, but it has been losing ground, not least to the much more fashionable SYRAH, and there were only 972 acres (393 ha) in 2011. Recommended producers of varietal versions include Barrister, Spring Valley Vineyard and Tamarack Cellars. **Oregon** had just 98 acres (40 ha) in 2008.

Bowers Harbor Vineyards' 2007 varietal Cabernet Franc was voted best red in the 2010 **Michigan** wine competition and the state had 86 acres (35 ha) in 2006. It is also grown in **Indiana**. **New York** State's total plantings were 498 acres (202 ha) in 2008 and the maritime climes of Long Island have long been regarded as propitious for the variety, with Schneider Vineyards' Le Breton showing why. Superior varietal examples in the more continental Finger Lakes region have been made by King Family Estate, Shalestone Vineyards and Swedish Hill.

In **Virginia**, much faith has been invested in Cabernet Franc ever since Horton showed the way with a particularly aromatic, varietally true example in 1991. Perhaps partly because of its cold hardiness, it is seen by most as the state's most promising red wine grape, although the total planted, just over 300 acres (121 ha), represents only about 10% of the state's total vineyard area. Barboursville Vineyards Reserve is one of the most successful.

Canada's relatively short growing season means that Cabernet Franc can be ripened more easily than CABERNET SAUVIGNON in many areas. It plays a relatively minor role in the sun-baked Okanagan Valley in British Columbia, where there are just 391 acres (158 ha), but it is becoming increasingly popular in Ontario, where there are already 1,565 acres (633 ha).

In South America, there are 301 ha (744 acres – 3.6% of vines planted for quality wine) in **Uruguay**. **Brazil** has 352 ha (870 acres) in Rio Grande do Sul and, as elsewhere in the world, most of these vines' produce disappears into Bordeaux blends. **Argentina** had 622 ha (1,537 acres) in 2008, mainly in Mendoza (352 ha/870 acres) and to a lesser extent in San Juan (153 ha/378 acres). Doña Paula makes an outstanding varietal version from Gualtallary in Mendoza. **Chile** is the prime grower of Cabernet Franc in South America with 1,143 ha (2,824 acres). O Fournier have made an exceptional varietal from old vines in Maule.

In **Australia**, the variety's 693 ha (1,712 acres) are used mainly as minor ingredients in Bordeaux blends, although the number of varietal versions such as Redgate's gold-medal winner is increasing. Across the Tasman, **New Zealand** has a steady 166 ha (410 acres).

Cabernet Franc is also found in **China** (456 ha/1,127 acres in 2009), where it plays a very subsidiary role to the dominant variety, CABERNET SAUVIGNON. There's a tiny amount in **Japan**, used as usual for blending purposes.

South Africa's area of Cabernet Franc tripled in the first decade of this century but still represents less than 1% of the South African vineyard area: 327 ha (808 acres) in 1998 up to 979 ha (2,419 acres) in 2008. Notable producers of fine varietal versions are Raats and Warwick Estate of Stellenbosch.

CABERNET GERNISCHT

See CARMENÈRE.

CABERNET JURA

A recent dark-skinned, disease-resistant Swiss hybrid that has so far remained in its homeland.

BERRY COLOUR: ● ● ● ● ●

PRINCIPAL SYNONYMS: VB 5-02

ORIGINS AND PARENTAGE

Cabernet Jura is a hybrid between CABERNET SAUVIGNON and an undisclosed disease-resistant variety (known in German as a *Resistenzpartner*) obtained by the private breeder Valentin Blattner in Soyhières, in the Swiss Jura.

VITICULTURAL CHARACTERISTICS

Vigorous, mid ripening. A waxy layer on the berry skin gives resistance to powdery and downy mildews and botrytis bunch rot. Good resistance to winter frosts.

WHERE IT'S GROWN AND WHAT ITS WINE TASTES LIKE

Wines are deeply coloured and aromatic. Even though there were just 19 ha (47 acres) planted in **Switzerland** in 2009, scattered across many different cantons, varietal wines are made by several producers, including Valentin Blattner, Simmendinger and Stegeler.

CABERNET MITOS

Recent, deeply coloured German cross making wines useful in blends.

BERRY COLOUR: ● ● ● ● ●

PRINCIPAL SYNONYMS: Weinsberg 70-77-4F

ORIGINS AND PARENTAGE

A BLAUFRÄNKISCH × CABERNET SAUVIGNON (Levadoux clone) cross created at the Weinsberg research centre in Baden-Württemberg, southern Germany, in 1970. It is therefore a sibling of CABERNET CUBIN.

VITICULTURAL CHARACTERISTICS

Vigorous and ripens late to very late with high sugar levels and moderate acidity.

WHERE IT'S GROWN AND WHAT ITS WINE TASTES LIKE

There are currently 320 ha (791 acres) of this variety planted in **Germany**, mainly in the Pfalz, Baden and Reinhessen. It entered the official variety register in 2000. Wines are deeply coloured, high in extract, reasonably tannic and therefore ideal for beefing up blends of otherwise vapid reds. Producers of varietal examples include Fuchs and Dr Lawall in Rheinhessen, Anselmann and Winzergenossenschaft Edenkoben in the Pfalz, Karl H Johner in Baden, and Willy in Württemberg.

It is also found to a very limited extent in **Switzerland** (less than 2 ha/5 acres).

CABERNET MORAVIA

High-quality, late-ripening Czech cross producing Cabernet-like wines.

BERRY COLOUR: ● ● ● ● ●

PRINCIPAL SYNONYMS: M-43

ORIGINS AND PARENTAGE

Cabernet Moravia is a CABERNET FRANC × ZWEIGELT cross obtained in 1975 by the private breeder Lubomír Glos in Moravská Nová Ves in southern Morava, in the Czech Republic. It was included in the official Czech register of varieties in 2001.

VITICULTURAL CHARACTERISTICS

Vigorous and regularly high yielding. Medium-sized to large bunches of largish, thick-skinned berries. Very late ripening, requiring the warmest regions and sites. Resistant to botrytis bunch rot.

WHERE IT'S GROWN AND WHAT ITS WINE TASTES LIKE

Although Cabernet Moravia's yields must be controlled if high-quality wine is to be produced, the best examples are deeply coloured with the cassis aromas and structure typical of the Cabernets. Recommended producers include Radomil Baloun, Vajbar Bronislav, Kubík and Patria Kobylí. There were 199 ha (492 acres) planted in the **Czech Republic** in 2009, almost all in Morava.

CABERNET PFEFFER

Rare, peppery California cross.

BERRY COLOUR: ● ● ● ● ●

PRINCIPAL SYNONYMS: Pfeffer Cabernet

VARIETIES COMMONLY MISTAKEN FOR CABERNET PFEFFER: GROS VERDOT ☒, TROUSSEAU

ORIGINS AND PARENTAGE

Does Cabernet Pfeffer really exist? It is said to be a cross between CABERNET SAUVIGNON and an unknown parent obtained by William Pfeffer in the 1880s in the Santa Clara Valley, California. It was allegedly destroyed by phylloxera in the 1890s and Pfeffer is reported to have bred a better version as a seedling and grafted it onto resistant rootstocks (Sullivan 1998), so that this 'new' Cabernet Pfeffer is in reality a sibling of the original one. However, Cabernet Pfeffer is also said to be identical to the old Bordeaux variety GROS VERDOT, and recent DNA profiling by Kenneth Volk in Santa Barbara has shown this to be true in the Volk vineyards (Ken Volk, personal communication). But if Cabernet Pfeffer really is a seedling bred in California, it could not be identical to a variety that was known in the Gironde much earlier than the date that the cross was made. More research is now needed to find out if all the Cabernet Pfeffer plantings in California are identical to Gros Verdot or if the true Cabernet Pfeffer is indeed still cultivated.

OTHER HYPOTHESES

It is also confused with TROUSSEAU and it is sometimes said that Cabernet Pfeffer is a cross between CABERNET SAUVIGNON and Trousseau but this has not yet been tested by DNA profiling.

WHERE IT'S GROWN AND WHAT ITS WINE TASTES LIKE

Plantings of vines called Cabernet Pfeffer are so limited that the variety is not even mentioned in the official **California** grape acreage report for 2010 although there were apparently 14.8 tons crushed in the state in 2007. DeRose apparently have a few hectares in the Cienega Valley AVA in San Benito County, as do Casa Nuestra in the Napa Valley (as part of a field blend), and, further afield, Page Springs in **Arizona** also grow it. The variety seems to live up to its name – even though it was named after the breeder and not because of its taste – and produces fresh, peppery or spicy wines that are relatively light but can still be pretty tannic. Most producers use it in a blend but DeRose make a varietal wine they label Cabernet Pfeffer.

CABERNET SAUVIGNON

The world's best-travelled red wine variety making concentrated, tannic wines for particularly long ageing.

BERRY COLOUR: ● ● ● ● ●

PRINCIPAL SYNONYMS: Bidure (Graves), Bordeaux (Switzerland), Bordo (Romania), Bouchet or Bouchet Sauvignon or Petit Bouchet (Saint-Émilion and Pomerol), Burdeos Tinto (Spain), Cabernet Petit, Carbonet or Carbouet (Bazadais and Petites Graves), Carmenet, Lafit or Lafite (Bulgaria, Russia and the Republic of Moldova), Marchoupet (Castillon), Navarre (Dordogne), Petit Cabernet⚭, Petit Cavernet Sauvignon (Pauillac), Sauvignon (Médoc), Sauvignonne (Graves), Vidure or Vidure Sauvignonne or Petite Vidure (Graves)

VARIETIES COMMONLY MISTAKEN FOR CABERNET SAUVIGNON: CABERNET FRANC⚭, CARMENÈRE⚭

ORIGINS AND PARENTAGE

Cabernet Sauvignon comes from the Gironde in south-west France. The earliest mention of the variety appeared under the name Petit Cabernet in the 'Livre de raison d'Antoine Feuilhade', an account book written between 1763 and 1777 by the mayor of Libourne (Garde 1946). It was then listed in a catalogue of grape varieties established in Pauillac by Dupré de Saint-Maur in 1784: 'Gros cavernet sauvignon, noir, première qualité pour faire du bon vin, assé productif & coloré, mais sujet au coulage. Petit cavernet sauvignon, noir, seconde qualité pour faire du bon vin, agreable & delicat, moins productif que le precedent' ('Gros cavernet sauvignon, black, top quality for making good wine, rather productive and deeply coloured, but subject to coulure. Petit cavernet sauvignon, black, secondary quality for making good wine, agreeable and delicate, less productive than the former'), and the first mention of the modern Cabernet Sauvignon spelling did not appear until 1840 (Rézeau 1997).

Until the end of the nineteenth century, Cabernet Sauvignon was possibly confused with CABERNET FRANC under old synonyms such as Cabrunet, Carmenet, Vidure, or simply the French plural Cabernets (see Cabernet Franc). It is also possible that Cabernet Sauvignon simply did not exist before the eighteenth century.

The parentage of Cabernet Sauvignon was fortuitously discovered by Carole Meredith and her PhD student John Bowers at the University of California at Davis as recently as 1996. While generating a database for the DNA profiles of the most important grape varieties in their collection, Napa Valley-born Bowers noticed that the DNA profile of Cabernet Sauvignon was perfectly consistent with its being an offspring of CABERNET FRANC and SAUVIGNON BLANC

(Bowers and Meredith 1997). This was the first time anyone had identified the parents of a classic wine grape variety, and it came as a big surprise to the wine world. Although ampelographers had already noticed many morphological similarities between both Cabernets, and although Cabernet Sauvignon apparently got its name because its wood and leaves resemble those of Sauvignon Blanc (Rézeau 1997), this parentage was totally unexpected, not least because no one had ever thought that a black-berried variety could have one white-berried parent. Gagnaire (1872) wrote at length in defence of this impossibility.

The cross that gave birth to Cabernet Sauvignon was most likely spontaneous and occurred somewhere in the Gironde before the mid eighteenth century. Since MERLOT is also a progeny of CABERNET FRANC, Cabernet Sauvignon and Merlot turn out to be half-siblings (see pedigree diagram overleaf). Most of these varieties belong to the Carmenet ampelographic group (see p XXVII). In addition, a probabilistic analysis of DNA data has shown that SAUVIGNON BLANC and CHENIN BLANC are siblings – the offspring of SAVAGNIN and an unknown parent (Vouillamoz). This makes Cabernet Sauvignon a nephew/niece of Chenin Blanc from the Loire and a grandchild of Savagnin from north-east France.

A pinky bronze-berried mutation of Cabernet Sauvignon was observed in 1977 in the Clegett Wines vineyard in Langhorne Creek in South Australia and it was registered under the name Malian. In 1991, one of those vines produced white berries, now registered as Shalistin. Clegett make a sweet and a dry version of Shalistin.

Cabernet Sauvignon was used to breed numerous varieties around the world, some of them cultivated for commercial winemaking. The following are therefore all half-siblings:

- Australia: CIENNA and TYRIAN
- Bulgaria: RUEN
- France: ARINARNOA, EKIGAÏNA and MARSELAN
- Germany: CABERNET CARBON, CABERNET CAROL, CABERNET CORTIS, CABERNET CUBIN, CABERNET DORIO, CABERNET DORSA, CABERNET MITOS
- Italy: INCROCIO MANZONI 2.15
- Serbia: PROBUS
- Romania: Arcas
- Switzerland: ?CABERNET BLANC, CABERNET COLONJES, CABERNET JURA, ?CABERTIN, CARMINOIR, ?PINOTIN, RIESEL and VB 91-26-4
- Ukraine: ODESSKY CHERNY and RUBINOVY MAGARACHA
- US: RUBY CABERNET.

OTHER HYPOTHESES

Since the word *sauvignon* derives from the French *sauvage* ('wild'), it has been argued that Cabernet Sauvignon was domesticated from wild grapevines in the Gironde, notwithstanding the fact that *Vitis vinifera* subsp. *silvestris* has never been recorded in this area.

Like CABERNET FRANC, Cabernet Sauvignon, one of whose early synonyms was Bidure, has been said to derive from the Biturica of Roman times but this theory can be rejected now that its parentage is known. Assuming that Biturica was identical or similar to Balisca, another variety mentioned by Pliny the Elder and Columella and supposedly cultivated in Spain under the name Cocolubis, some authors suggested that Cabernet Sauvignon was introduced from Spain.

A recent essay created quite a buzz by claiming that Cabernet Sauvignon was of Greek origin, its ancestor being the obscure Greek grape variety Volitsa (VOLITSA MAVRI; Lambert-Gocs 2007). Based on the vague descriptions given by Pliny the Elder and Columella, the author concluded that the modern Volitsa of the Pelopónnisos (Peloponnese), supposedly identical to a variety called VLOSH in Albania, is also identical to the antique Balisca and is therefore a direct ancestor of Cabernet Sauvignon, hence the Greek-origin hypothesis for the noble Bordeaux variety. However, DNA profiling is categorical: there is no direct parentage, nor any close genetic similarity, between Volitsa and Cabernet Sauvignon (Vouillamoz).

CABERNET SAUVIGNON PEDIGREE DIAGRAM

The discovery of the natural parents of CABERNET SAUVIGNON in 1997 opened the new era of 'historical genetics'. The reconstruction of this complete pedigree of natural crosses traverses France and unexpectedly includes such world-famous varieties as SAVAGNIN (Traminer), CHENIN BLANC, MERLOT and COT (Malbec). Since question marks represent unknown (and probably extinct) varieties, the inverse relationships are also possible, so that in theory ABOURIOU, Morenoa and HONDARRIBI BELTZA could be the parents rather than – as illlustrated here – the offspring (this is not true for SAUVIGNON BLANC and Chenin Blanc because they have been shown to be siblings).

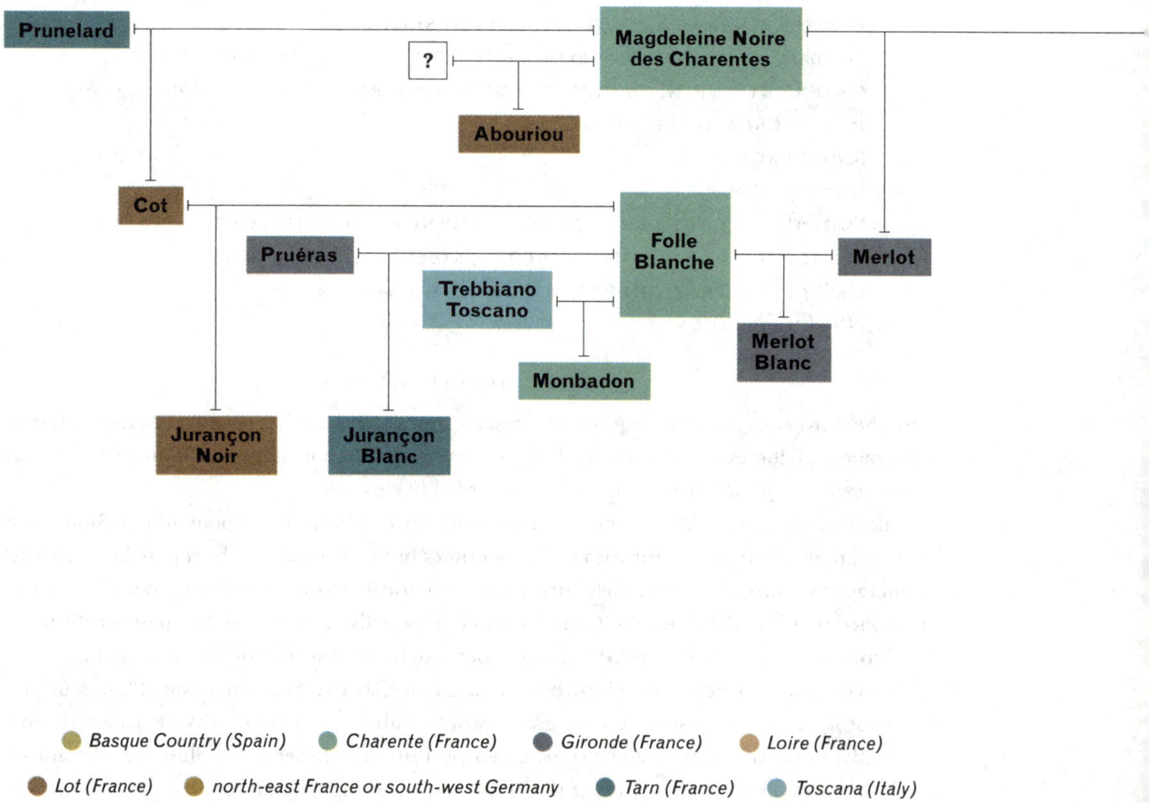

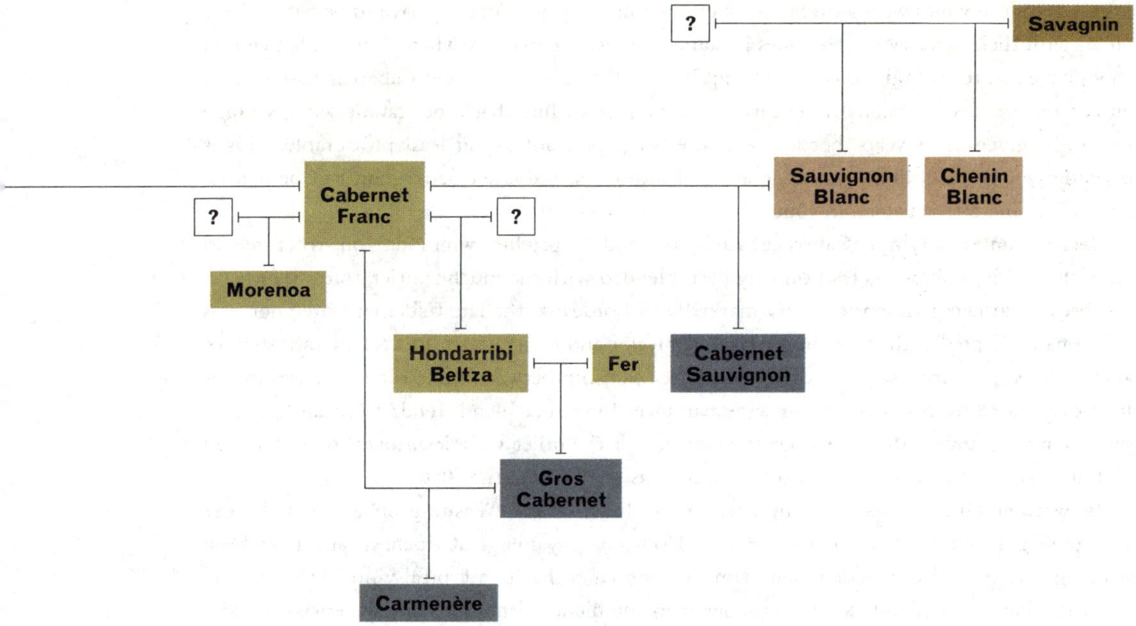
?
Savagnin
Sauvignon Blanc
Chenin Blanc
Cabernet Franc
?
?
Morenoa
Hondarribi Beltza
Fer
Cabernet Sauvignon
Gros Cabernet
Carmenère

VITICULTURAL CHARACTERISTICS

Vigorous, late budding, mid to late ripening. Thrives on well-drained gravel soils, preferably acid and well exposed. Very susceptible to fungal diseases that affect the wood such as eutypa dieback and esca and also to powdery mildew. Occasionally susceptible to drying out of the stems (especially when grafted onto SO4 rootstock). Bunches and berries are small, the skins thick and distinctively blueish. Its hard wood makes it suitable for mechanical harvesting and generally safe from the risk of winter freeze.

WHERE IT'S GROWN AND WHAT ITS WINE TASTES LIKE

Because Cabernet Sauvignon is the dominant grape in the assemblages responsible for all of the most famous wines made (in quantity) in Bordeaux's Médoc and Graves regions, including all five 'first growths', Cabernet Sauvignon has been admired and planted almost wherever vines are grown for wine production – except in regions too cool for this relatively late ripener. This global spread was particularly marked during the late-twentieth-century's love affair with international varieties.

It has a particularly powerful identity. Thanks to the small, thick-skinned grapes, the wines tend to be deep in colour, high in tannin and relatively high in acidity. There's a recognizable nobility and stability to its aroma which can vary from blackcurrant to cedar, although such is its affinity with oak that in young wines it can be difficult to dissociate the fruit from the wood. Few varieties produce wines with a higher level of phenolics and winemakers have to be particularly careful with their extraction techniques in some vintages – especially when Cabernet Sauvignon grapes have had to struggle to ripen. The upshot of all this tannin is that Cabernet-based wines can last for decades, occasionally centuries, and may need, like that other tannic wine, vintage port, to be given many years' bottle age before being palatable, or at least pleasurable. This is particularly true in climates as moderate as Bordeaux's. Cabernet Sauvignon can be a mite monolithic, and often austere, in youth.

Because, unless very ripe, Cabernet Sauvignon tends to produce wines long on structure and rather shorter on flesh, it has traditionally been blended with the much fruitier, softer MERLOT. And because growing seasons can vary markedly in Bordeaux, the late Cabernet Sauvignon was traditionally planted with the more precocious Merlot and CABERNET FRANC as an insurance policy against poor fruit set. The close genetic relationship between these three varieties makes them easy blending partners and in wine parlance 'Bordeaux blend' tends to mean Cabernet Sauvignon with any of Merlot, Cabernet Franc and/or the other varieties grown to produce red bordeaux: PETIT VERDOT, Malbec (COT) and occasionally CARMENÈRE.

In warmer climes, especially in northern California and Washington State, Cabernet Sauvignon grapes can reach such an advanced state of physiological ripeness, and have been encouraged to do so by extended 'hang time' on the vine, that in a typical vintage they hardly need the added flesh of a MERLOT component in the blend, although some winemakers add a seasoning of other grape varieties for aromatic complexity.

What is remarkable about Cabernet Sauvignon is that it imprints its identity so firmly on any wine that includes it, even in quite small amounts, although it manages to produce great wine in only a small proportion of the hundreds of regions in which it is planted. The most obvious established candidates are the Médoc, Pessac-Léognan, Penedès, Napa, Bolgheri, Sonoma, Santa Cruz Mountains, parts of Washington State, Puente Alto in Chile, and Coonawarra and Margaret River in Australia.

In **France**, Cabernet Sauvignon was the fourth most planted red wine variety after MERLOT, Grenache (GARNACHA) and SYRAH in 2009, with a total of 56,386 ha (139,333 acres). In the greater Bordeaux region, it was a poor second to the much easier to ripen Merlot: 26,790 ha (66,200 acres) to Merlot's 69,053 (170,634 acres).

Throughout south-west France, it is one of the principal varieties in appellations on the fringes of Bordeaux such as Buzet, Bergerac and Côtes de Duras and also in Irouléguy on the western Spanish border. It is a supplementary variety in the likes of Béarn, Madiran, Côtes de Saint-Mont, Tursan, Marcillac, Gaillac and Fronton (adding structure to NÉGRETTE in the last two).

It is the fifth most planted variety in Languedoc-Roussillon, with a total of 18,722 ha (46,263 acres) in 2009, mainly in the Languedoc. While it's a permitted ingredient in appellations such as Cabardès and Malepère, it is used mainly for varietal wines. Perhaps the most famous is Mas de Daumas Gassac, where Cabernet Sauvignon was planted on the advice of the late Bordeaux oenologist Émile Peynaud, no less. With Aniane neighbour Grange des Pères, these are two of the most successful Cabernet Sauvignon-based wines from the Languedoc, but too often less ambitious examples lack conviction and concentration, perhaps because of excessive yields or poor-quality clones.

Cabernet Sauvignon grown to the east in Provence – 3,483 ha (8,607 acres), mainly in the southern Rhône and the Var – seems much more convincing and generally has no shortage of concentration. Indeed it sometimes has a sort of dry-climate dustiness. It has traditionally been blended with SYRAH in reds and rosés, although there is more Syrah planted here and much more Grenache (GARNACHA). Domaine de Trévallon is the standard-bearer.

In the Loire and central France, Cabernet Sauvignon's total of 1,283 ha (3,170 acres) makes it very minor indeed compared with CABERNET FRANC (less than a tenth as much), partly because it is much harder to ripen this far from the equator, but Cabernet Sauvignon is the principal variety in Cabernet d'Anjou rosé, can be significant in appellations such as Anjou-Villages and is permitted (usually playing a bit part to Cabernet Franc's lead) in Bourgueil and Chinon.

Cabernet Sauvignon is important in **Italy**, where it has been grown since the 1820s, when it was introduced to Piemonte. As in Spain, Italy has seen a period of Cabernet worship, particularly in the revitalized Maremma, where the variety really does seem at home, as evinced by Sassicaia, based on cuttings from Bordeaux, as long ago as the late 1960s. The variety was enthusiastically planted all over Italy in the late twentieth and early twenty-first centuries – particularly in Toscana, where it, along with MERLOT and occasionally SYRAH, was thought to add body and international appeal to what were seen as the shortcomings of the local SANGIOVESE. This was perfectly legal in Chianti Classico, not so for Brunello di Montalcino, which in 2008 was riven by ideological dispute over this issue. Cabernet Sauvignon, long a required ingredient in the Tuscan Carmignano DOC, was a vital element in many of the so-called Supertuscans that followed in Sassicaia's footsteps and enjoyed fame in the late twentieth century, such as Solaia, Sammarco, Saffredi, Paleo Rosso, La Vigna d'Alceo and hundreds more. To a much more limited extent it was also planted throughout northern Italy, where its noblest examples include Gaja's Darmagi in Piemonte, Maurizio Zanella in Lombardia, Lageder and San Leonardo in Alto Adige. Here and in Friuli it can be quite light and herbaceous.

The most recent official Italian viticultural census, in 2000, found 8,042 ha (19,872 acres) but the current total is likely to be a good deal higher, especially considering the fact that Sicilia alone had 3,992 ha (9,864 acres) planted in 2008. There are many DOCs with Cabernet in their name throughout northern and central Italy as well as further south in Puglia and Sicilia, some specifying Cabernet Sauvignon or CABERNET FRANC, some leaving it to the discretion of the producer. Cabernet Sauvignon-dominated wines of great quality are relatively rare this far south but Sicilia can boast examples from Planeta and Tasca d'Almerita, and the Montevetrano blend from Campania has also attracted international attention.

Cabernet Sauvignon was also introduced into **Spain** to a limited extent in the mid nineteenth century by the Marqués de Riscal in Rioja and Eloy Lecanda of Vega Santa Cecilia (later Vega Sicilia) in Ribera del Duero. But it was its espousal by Miguel Torres, Jr, and Jean León in the 1960s in Penedès that encouraged Spanish wine producers to value this French intruder so highly. A Torres, Mas La Plana Cabernet Sauvignon 1970, did remarkably well in a 1979 rerun of the famous 1976 Judgement of Paris blind-tasting comparison between top French wines and those emulating them elsewhere.

In 2008 Cabernet Sauvignon was the fourth most planted red wine variety in Spain, its total of 19,430 ha (48,013 acres) being a considerable increase on the 10,000 ha (24,710 acres) in the ground in 2004. It is particularly popular in such northern DOs as Navarra and Somontano and in La Mancha but it has become a significant minor ingredient with TEMPRANILLO in Ribera del Duero, and varietal Cabernet Sauvignon can be found all over the country, including in the newer high-altitude vineyards of Andalucía. Many of those who have planted new vineyards in Spain seem to view Cabernet Sauvignon as an essential accoutrement of sophistication, although only in Penedès does it have much of a track record. Also in Catalunya, it can be found in some Priorat blends too.

The Portuguese are more inured to Cabernet-mania and there were just 2,271 ha (5,612 acres) planted in 2010 in **Portugal.** Palmela on the Península de Setúbal is the only denomination that embraces it but it pops up in various Vinho Regionals all over the country, most successfully in the Tejo.

In Eastern Europe, Cabernet Sauvignon is very important in **Bulgaria**, where with a 2009 total of 15,827 ha (39,109 acres) it accounts for 15% of the country's vineyard area and is almost as widely planted as the common PAMID. It is planted mainly in the south, although the wines tend to show more finesse in the north. Single-estate Bordeaux blends are the most popular wines produced by the new wave of ambitious Bulgarian producers.

Cabernet Sauvignon is widely planted in **Ukraine** – 8,468 ha (20,925 acres) in 2009 – and it is also a hugely popular variety over the Black Sea to the east in **Russia**, where there were 3,525 ha (8,710 acres) in Krasnodar Krai alone in 2010, as well as a more modest 68 ha (168 acres) in Rostov, and over the border to the west in the **Republic of Moldova**, where 7,590 ha (18,755 acres) were planted. **Romania** had 3,128 ha (7,729 acres) planted in 2008 but MERLOT is much more popular. All of these countries are capable of making very respectable wines from the variety but have until recently been hampered by poor viticulture and sometimes excessive yields.

Cabernet grown in **Slovenia** is generally a much more serious proposition, as demonstrated by the likes of Rojac in Istra (Istria) and Scurek and Marjan Simčič in Brda. In 2009 there were 470 ha (1,161 acres) planted, mainly in Vipavska Dolina and Goriška Brda near the Italian border but also further south near the coast in Slovenska Istra.

In **Croatia**, there were 894 ha (2,209 acres) in 2009, less than 3% of the country's vines, while Cabernet's 228 ha (563 acres) in the much, much cooler **Czech Republic** are predominantly in Morava in the south east, where it is most likely to stand a chance of ripening. There is a tiny amount planted in **Slovakia** – and indeed an increasing amount to the north in **Germany**, where global warming has allowed some growers such as Hammel and Martin Wassmer in the southern regions of the Pfalz and Baden respectively to ripen it successfully.

The warmer Pannonian climate of **Hungary** is much more suitable for this late-ripener even though CABERNET FRANC is the favoured variety here. In 2008 Hungary had a total of 2,752 ha (6,800 acres), with the largest areas in Villány (where star performers include Malatinszky and Atilla Gere) and Eger, where it is a permitted ingredient in Egri Bikavér (Bull's Blood of Eger).

Across the border in **Austria**, Cabernet Sauvignon's 519 ha (1,282 acres) made it the third most planted red wine grape in 2007 but it is far less important than BLAUFRÄNKISCH and ZWEIGELT and is generally used as an ingredient in a blend, or cuvée. Plantings are mainly in Burgenland, particularly in Neusiedlersee (151 ha/373 acres), where the area has almost doubled since 1999. In **Switzerland**, too, fully ripe Bordeaux blends, and blends with indigenous Swiss varieties, have become more prevalent thanks to global warming, not just in Ticino but in the French-speaking Swiss cantons, eg in Genève, where Domaine des Balisiers make their Comte de Penay cuvée in amphorae.

Cabernet Sauvignon has played a part in the modern wine revolution in **Greece**, where in 2008 there were 1,743 ha (4,307 acres), particularly in Dráma and Kavala in north-eastern Makedonía, Messinía on the west of the Pelopónnisos and Attikí south of Athens. Antonopoulos, Biblia Chora, Costa Lazaridi, Pavlidis, Katogi Strofilia, Tsantali and Tselepos have all made very creditable examples. It plays a more minor role on the island of **Cyprus**, where there are only 279 ha (689 acres). The tiny Mediterranean island of **Malta** boasts about 100 ha (247 acres).

As **Turkey**'s wine industry has joined the mainstream, Cabernet Sauvignon has increasingly been planted, so that by 2010 there were 493 ha (1,218 acres) and some pretty serious blends, such as those of Corvus, Büyülübağ and Kayra.

Cabernet Sauvignon has played a crucial role in the modern wine industry in **Lebanon**, not least for providing the backbone of its best-known export, Château Musar's spicy red. The variety is also, not surprisingly, the most popular variety in **Israel**'s cosseted vineyards, where, with a 2009 total of 1,250 ha (3,089 acres), it constitutes a quarter of all plantings and is particularly successful in the hands of Ella Valley, Galil Mountain, Golan Heights and Margalit.

Cabernet's popularity in Israel is a reflection of its huge importance in the United States. With the exception of a short-term fling with MERLOT and a *Sideways*-inspired flirtation with PINOT NOIR, Cabernet Sauvignon *is* red wine for Americans. It is planted in every state that can ripen it and in **California**, the nation's dominant wine producer, is the most planted red wine variety by far, its 2010 total of 77,602 acres (31,404 ha) second only to CHARDONNAY. It grows like a sprawling weed in the Central Valley and is planted the length of the state but reaches its apogee in Napa Valley's sun-kissed, fog-cooled climes as well as in Alexander Valley, Sonoma Valley and various other spots protected from exaggerated Pacific influence, such as on well-exposed sites above the fog line of the Santa Cruz Mountains. If price is any measure, then California has definitively proved to be one of Cabernet Sauvignon's sweetest spots. The dotcom boom fuelled demand for limited-production, so-called cult Cabernets, so that for a time they sold like hot cakes via over-subscribed mailing lists. So ripe are many of these wines that they can be drunk on release and some taste so sweet and alcoholic that they can sometimes seem, to a palate raised on the more angular charms of red bordeaux, closer to hot chocolate than wine.

The variety is also grown in **Mexico**, particularly in Baja California, Coahulia de Zaragoza and Zacatecas.

Much of **Oregon** is too cool to ripen Cabernet of either sort, except for the Columbia River region in the north east, on the border with Washington State and, to a lesser extent, in the Rogue Valley in the south west, near the California border. In 2008 a total of 523 acres (212 ha) was planted. In semi-desert eastern **Washington** State, on the other hand, Cabernet Sauvignon has become by far the most popular variety, with plantings totalling 10,293 acres (4,167 ha) by 2011. Here Cabernet Sauvignon can really shine, the fruit often showing appetizing brightness. Leading vineyards include Champoux in Horse Heaven Hills, Ciel du Cheval and Klipsun vineyards in Red Mountain, and Wahluke Slope. Whereas in California there is an army of admired producers of Cabernet Sauvignon, Washington's constitute more of a cohort, with Leonetti, Powers, Quilceda Creek, Andrew Will and Woodward Canyon in the front line but dozens more only a step behind.

Virtually all American states make Cabernet Sauvignon and the majority of them grow it, with **New York** State's 340 acres (138 ha) mainly on Long Island, and the **Texas** wine industry is dominated by it. Stylistically most states' winemakers are heavily influenced by the California style of ripe fruit given a bit of discipline in expensive French, and occasionally American, oak. **Virginia**'s 260 acres (105 ha) of Cabernet Sauvignon are a little more difficult to ripen than the CABERNET FRANC favoured by many growers, although RdV have proved that, with care, Virginia Cabernet Sauvignon can be magnificent.

In **Canada**, the balmy temperatures of the southern Okanagan Valley in British Columbia are much better suited to Cabernet Sauvignon than anywhere else and the BC total in 2008 was 1,683 acres (681 ha), representing nearly 15% of red wine varieties. Mission Hill's top bottling and Osoyoos Larose are outstanding Cabernet-based wines by any measure. Ontario had only about a third as much in 2006 and very careful site selection is needed here for this late-ripening variety.

Cabernet Sauvignon is also hugely important to South American wine – especially in **Chile**, where cuttings were brought directly from Bordeaux before phylloxera. The 2008 total of 40,728 ha (100,641 acres and far more even than in California) represents about a third of all vines planted and the variety can be found in all regions, especially in Colchagua, Maule, and to a lesser extent in Curicó, Maipo and Cachapoal, where it is the most planted variety. It was notable that some of the world's most garlanded producers of Cabernet Sauvignon-dominated wines elsewhere – the likes of Robert Mondavi of California and Baroness Philippine de Rothschild, Bruno Prats and Paul Pontallier of Bordeaux – chose to invest in Chilean Cabernet Sauvignon many years ago. Chilean Cabernet, much of it grown on ungrafted vines in a disease-free environment, tastes particularly fruity and easy to appreciate young, even if there can be an edge of greenness to wines from over-burdened vines. But there is also a deepening top layer of more complex, ageworthy examples such as Almaviva, Altaïr, Aurea Domus, Casa Lapostolle, Concha y Toro's Don Melchior, De Martino, Errazuriz/Viñedo Chadwick/Seña, and top bottlings from the likes of Gilmore, Haras de Pirque, Montes, O Fournier, Perez Cruz and others.

In **Argentina**, Malbec (COT) and Bonarda (DOUCE NOIRE) are much more important than Cabernet Sauvignon, which is a shame since Argentine Cabernet can present a magnificent combination of ripeness and minerality (and occasionally tart greenness if yields are too high). In 2008 there was a total of 17,746 ha (43,851 acres), the majority (13,342 ha/32,968 acres) in Mendoza, but it is also planted in several other regions, notably San Juan (2,182 ha/5,392 acres), La Rioja (1,018 ha/2,515 acres) and Salta (444 ha/1,097 acres). It is also often used as a stiffener for Malbec. Some of those who produce superior Cabernet Sauvignons include Catena Zapata, Viña Cobos, Cuvelier Los Andes, Mendel, Terrazas de los Andes, Fabre Montmayou and Vistalba, most of them having close links to either Bordeaux or the Napa Valley.

Brazil had 1,868 ha (4,616 acres) of Cabernet Sauvignon in 2007 but it can be difficult to ripen it fully in traditional damp regions, although Lidio Carraro, Miolo and Salton have all made creditable examples.

TANNAT and even MERLOT are more important in **Uruguay**, which had 731 ha (1,806 acres) of Cabernet Sauvignon in 2008.

In the 1970s and 1980s, Cabernet Sauvignon was worshipped in **Australia** but it is now viewed as a poor second to Shiraz (SYRAH), both qualitatively and quantitatively. In 2008 the total area planted with Cabernet Sauvignon was 27,553 ha (68,085 acres), considerably less than Shiraz. By far the majority of plantings are in the Riverland (3,496 ha/8,639 acres), for inexpensive, fruity varietals, and in Coonawarra (3,444 ha/8,510 acres), which was the first region in Australia ever to establish a varietal speciality, Cabernet Sauvignon. In recent years a lack of fine-tuned viticulture has tarnished Coonawarra Cabernet's reputation but the potential for top-quality wine from its terra rossa soils is indubitably there. Cabernet Sauvignon is also important in Langhorne Creek (1,921 ha/4,747 acres), which produces a rather soft style for blending, Barossa Valley (1,336 ha/3,301 acres), whose Cabernet is particularly dense and chocolatey, and Western Australia's Margaret River (1,291 ha/3,190 acres), which makes sophisticated, well-balanced examples – on a west coast just like the variety's other homes in Bordeaux, Napa Valley and Bolgheri. (There is a theory that rays from the setting sun are particularly propitious for Cabernet Sauvignon vines.) The Margaret River tradition is to blend Cabernet Sauvignon with MERLOT for less serious bottlings and to make varietal Cabernet Sauvignon for the long term. Australia's most accomplished Cabernet Sauvignon producers include Cape Mentelle, Cullen, Howard Park, Moss Wood, Vasse Felix and Woodlands in Margaret River; Balnaves, Bowen, Majella, Parker and Wynns in Coonawarra; Henschke, Penfolds and Wendouree from further north in South Australia; Mount Mary and Yarra Yering in Victoria; Lake's Folly in Hunter Valley. Cabernet Sauvignon/Shiraz blends are part of Australian wine's long and glorious history.

New Zealand is PINOT NOIR not Cabernet country. There were only 517 ha (1,278 acres) of Cabernet Sauvignon vines in the ground in 2009 – much less than Pinot Noir's 4,240 ha (10,477 acres) and way behind MERLOT's 1,363 ha (3,368 acres). Most regions are too cool or wet for the variety but it clearly thrives in the appropriately named Gimblett Gravel soils and elsewhere in Hawke's Bay on the North Island. Reliable producers of varietal Cabernet Sauvignon and, generally more successful, Bordeaux blends include Cape Kidnappers, Craggy Range, Esk Valley, C J Pask, Stonecroft, Stonyridge, Te Awa Farm, Te Mata, Trinity Hill, Unison and Villa Maria.

As for **China**, the world's fastest-growing wine producer, Cabernet Sauvignon has played the crucial role, being by far the most favoured variety since red bordeaux has been seen as the archetypal wine. By 2009 plantings were reckoned to total 20,352 ha (50,288 acres), in all of China's wine regions, and they must already be considerably more than this – very possibly more than in Bordeaux by now. Inland Ningxia is looking like the province with the greatest potential, although the Rothschilds of Château Lafite have bet on Shangdong on the coast. Much of the wine produced is extremely lean but producers such as Grace, Silver Heights and Jade Mountain have broken the mould.

The variety has been planted in virtually all new territories for the vine, including **Japan** (469 ha/1,159 acres in 2009, mostly in Yamanashi and Hyogo), which, with extreme vine-by-vine care, can produce quite creditable copies of ripe red bordeaux from its limited plantings. In 1988 Grover pioneered Cabernet Sauvignon in **India**, with several harvests a year on high ground north of Bangalore in Karnataka State.

Cabernet Sauvignon is taken very seriously in **South Africa**, even if work has been needed to eradicate the ripening difficulties associated with the leafroll virus endemic in some vineyards. Cabernet Sauvignon is traditionally blended with MERLOT here and in 2008 total plantings were 12,697 ha (31,375 acres), making Cabernet Sauvignon the Cape's most widely planted red wine variety, accounting for 28% of red plantings and second only to CHENIN BLANC overall. The main areas are Stellenbosch, Paarl and Malmesbury, and those making particularly distinctive Cabernets include Glenelly Hill, Jordan, Morgenster, Reyneke, Rustenberg, Simonsig, Thelema, Vergelegen, Vilafonté, Warwick and Waterford.

CABERNET SEVERNY

Winter-hardy hybrid with Asian not Cabernet genes but marked Cabernet colour and flavours.

BERRY COLOUR: ● ● ● ● ●

PRINCIPAL SYNONYMS: Cabernet Szevernuej, Kaberne Severnyi

ORIGINS AND PARENTAGE

Cabernet Severny, literally 'northern Cabernet', is a complex hybrid bred specifically for its winter hardiness by Ya I Potapenko, L I Proskurnya and A S Skripnikova at the All-Russia research and development centre for viticulture and winemaking in Novocherkassk, in the province of Rostov, Russia, by crossing a DIMYAT (under the name Galan) × *Vitis amurensis* hybrid with a pollen mixture from other *Vitis vinifera* × *Vitis amurensis* hybrids. Included in the official register of varieties in 1990. Although no Cabernet was involved in the pedigree, it was christened Cabernet Severny because its aroma and flavour were thought to be similar to those of CABERNET SAUVIGNON (Taras Hiabahov, personal communication).

VITICULTURAL CHARACTERISTICS

Productive, mid ripening. Small, dense bunches of small, thick-skinned berries. Cold-hardy to –25 to –26 °C (–13 to –15 °F) but susceptible to powdery mildew, less so to downy mildew.

WHERE IT'S GROWN AND WHAT ITS WINE TASTES LIKE

Cabernet Severny's greatest attribute is its winter hardiness, which is why it is grown not only in **Russia** – in the Krasnodar and Rostov regions – but also in **Canada**, notably in Québec, where producers such as Carone have shown that this deeply coloured Cabernet-like variety is well suited to oak ageing. In Russia it is used both for dry and for dessert wines. It is also being considered by Pernod Ricard as suitable for planting in their Helan Mountain project in Ningxia province, **China**.

CABERTIN

Recent disease-resistant Swiss hybrid with potential.

BERRY COLOUR: ● ● ● ● ●

PRINCIPAL SYNONYMS: VB 91-26-17

ORIGINS AND PARENTAGE

Cabertin is a hybrid between CABERNET SAUVIGNON and an undisclosed disease-resistant variety (known in German as a *Resistenzpartner*) obtained in 1991 by the private breeder Valentin Blattner in Soyhières, in the Swiss Jura, possibly a Cabernet Sauvignon × (SILVANER × (RIESLING × *Vitis vinifera*) × (JS 12417 × CHANCELLOR)) complex hybrid, just like PINOTIN and CABERNET BLANC.

VITICULTURAL CHARACTERISTICS

Small, loose bunches of small, thick-skinned berries. Resistant to botrytis bunch rot, downy and powdery mildews and winter frosts. Mid to late ripening (two weeks earlier than CABERNET SAUVIGNON) but early budding and therefore at risk of spring frosts.

WHERE IT'S GROWN AND WHAT ITS WINE TASTES LIKE

Wines are deeply coloured, with aromas of fruits of the forest and cassis. They have sufficient structure to withstand barrel ageing. Plantings are very limited but varietal wines are produced by, for example, Kirsch in Jena, eastern **Germany**, and Domaine Viticole du Chenoy in Namur, **Belgium**, and there are further small vineyards in the Pfalz in Germany.

CACABOUÉ

Virtually extinct Savoie vine.

BERRY COLOUR: ● ● ● ● ●

PRINCIPAL SYNONYMS: Caca d'Oie, Cacabois, Cacabouet, Persan Blanc, Saint-Péray

VARIETIES COMMONLY MISTAKEN FOR CACABOUÉ: Grosse Jacquère, MOLETTE (Lac du Bourget and Chambéry)

ORIGINS AND PARENTAGE

Cacaboué is a local and almost extinct variety from Savoie but its age is hard to determine. The name Cacaboué is a dialect form of 'Caca d'Oie', literally meaning 'goose poo', in reference to the shape of the pulp remaining on the stem after pulling off the berry. DNA parentage analysis strongly suggests that Cacaboué is either a parent or an offspring of AVANÀ from Piemonte, in north-west Italy (Vouillamoz), a variety that used to be cultivated in Savoie under the name Hibou Noir.

OTHER HYPOTHESES

DNA profiling reveals several cases of mistaken identity (Vouillamoz):

- Cacaboué is not identical to Grosse Jacquère (not commercially cultivated), as has been suggested by many authors (Galet 2000; Schneider *et al.* 2001), and is therefore not an offspring of GOUAIS BLANC, as the Vitis International Variety Catalogue wrongly indicates (but Grosse Jacquère is indeed an offspring of Gouais Blanc)

- Cacaboué is not identical to MOLETTE
- Cacaboué is not a colour mutation of PERSAN, although it is sometimes called Persan Blanc.

VITICULTURAL CHARACTERISTICS

Vigorous, fairly productive, mid ripening.

WHERE IT'S GROWN

Cacaboué used to be cultivated in particular in Charpignat near the southern end of the Lac du Bourget in Savoie, eastern **France**. Charpignat was once an officially designated cru (like Apremont or Arbin today) but the vineyards have lost the battle with the builders and only a few vines remain. However, Domaine Langain produce tiny amounts (10 hl) of a varietal wine from a few ares planted above the Lac du Bourget, classified as Vin de Pays des Allobroges and sold in a local Michelin-starred restaurant. This and other obscure or endangered varieties are currently being studied by the Centre d'Ampélographie Alpine at Montmélian (Savoie) and INRA (Institut National de Recherche Agronomique) in Montpellier, southern France.

CACAMOSCA

Rare, low-yielding Campanian white.

BERRY COLOUR: ● ● ● ● ●

PRINCIPAL SYNONYMS: ?Riciniello Bianco

ORIGINS AND PARENTAGE

Old variety from Campania in southern Italy. The odd name of this white grape variety refers to the characteristic presence of numerous brown spots scattered on the surface of the berries that look like 'fly's dung'. It is genetically close to FENILE, also from Campania (Costantini *et al.* 2005).

VITICULTURAL CHARACTERISTICS

Low vigour, very low yields, small clusters. Mid ripening.

WHERE IT'S GROWN AND WHAT ITS WINE TASTES LIKE

Cacamosca used to be cultivated on the Amalfi and Sorrento coasts as well as on the Collina di Posillipo near the bay of Napoli, in **Italy**, where it produced a reputedly excellent wine which was even exported in the eighteenth and nineteenth centuries. Today it survives only in old and ungrafted vineyards in Gragnano and Lettere near Amalfi, in the province of Napoli, and generally disappears into blends. The Cacamosca grapes tend to have high sugar and low acidity.

CADDIU

Sardinian blending and sometimes table grape.

BERRY COLOUR: ● ● ● ● ●

PRINCIPAL SYNONYMS: Caddeo, Caddiu Nieddu (Oristano), Caddu (Bosa), ?Pàmpinu

ORIGINS AND PARENTAGE

Not much is known about Caddiu's origin but it probably comes from the Italian island of Sardegna, where it has allegedly been known since 1780 under the name Pàmpinu. However, the true identity of Caddiu has yet to be established because the DNA profile of its supposed synonym Nieddu Pedra Serra in Zecca *et al.* (2010) turns out to be identical to that of PASCALE in De Mattia *et al.* (2007), another Sardinian variety (Vouillamoz).

VITICULTURAL CHARACTERISTICS

Vigorous, though yields are no more than average, good resistance to frost and fungal diseases. Ripens late.

WHERE IT'S GROWN AND WHAT ITS WINE TASTES LIKE

Used both for the table and for wine, Caddiu is cultivated only in the Bassa Valle del Tirso in Cagliari, Sardegna, **Italy**, and is always blended with other varieties, such as NIEDDERA or MONICA NERA. Contini's Barrile IGT, for example, based on Nieddera, includes 10–15% Caddiu. 1,100 ha (2,720 acres) were recorded in the 2000 census.

CAGNULARI

See GRACIANO.

CAÍÑO BLANCO

Rare Iberian white that ripens late but keeps its acidity.

BERRY COLOUR: ● ● ● ● ●

PRINCIPAL SYNONYMS: Cainho de Moreira☒ (Vinho Verde), Caíño Branco (Galicia and Portugal)

VARIETIES COMMONLY MISTAKEN FOR CAÍÑO BLANCO: Albariño☒ (ALVARINHO)

ORIGINS AND PARENTAGE

Caíño Blanco is indigenous to the area of the Iberian peninsula between Vinho Verde in northern Portugal and Galica in north-west Spain, where it was allegedly first mentioned as early as 1722 (Huetz de Lemps 2009). It has often been confused with Albariño (ALVARINHO) but ampelographic and DNA studies have rejected this identity (Santiago *et al.* 2007; Gago *et al.* 2009). A more recent DNA parentage analysis has suggested that Caíño Blanco is a likely natural cross between Albariño and Caíño Bravo (AMARAL) that occurred before the eighteenth century

(Díaz-Losada *et al.* 2011), but this parentage was based on twenty-seven DNA markers only and would require confirmation with additional markers, especially given the complexity of the Caíño group. There is some confusion in Vinho Verde, where the name Alvarinhão is used as a synonym for both Caíño Blanco and FERNÃO PIRES.

VITICULTURAL CHARACTERISTICS

Early to mid budding, mid to late ripening. Susceptible to downy and powdery mildews.

WHERE IT'S GROWN AND WHAT ITS WINE TASTES LIKE

Caíño Blanco is grown to a very limited extent in the Rosal area within the Rías Baixas DO, **Spain**, where Terras Gauda, for example, blend it with Albariño (ALVARINHO) to add freshness and zest. Although wines have relatively high alcohol thanks to late ripening, they retain fresh acidity and noticeable minerality. Caíño Blanco is authorized for both the Rías Baixas and Monterrei DOs. There were 58 ha (143 acres) in Spain in 2008, exclusively in Galicia, and plantings have been increasing since the 1990s (Böhm 2011). **Portugal** had just 7 ha (17 acres) in 2010.

CAÍÑO BRAVO

See AMARAL.

CAÍÑO TINTO

See BORRAÇAL.

CALABRESE DI MONTENUOVO

Recently identified Calabrian parent of Sangiovese.

BERRY COLOUR: ● ● ● ● ●

VARIETIES COMMONLY MISTAKEN FOR CALABRESE DI MONTENUOVO: NERO D'AVOLA (also called Calabrese), SANGIOVESE

ORIGINS AND PARENTAGE

During a survey of the old local varieties of Campania, Italy (Costantini *et al.* 2005), half a dozen vines of this odd grape variety were fortuitously found in a vineyard near a winery around Lago d'Averno in Campi Flegrei in the province of Napoli, and subsequently in two other nearby vineyards. Known locally as Calabrese di Montenuovo, simply because the previous owners of the winery came from Calabria and because the hill on which the vines grow is called Montenuovo, the variety's DNA turned out to be unique (Vouillamoz). However, this obscure variety remained unnoticed until it was found to be the 'missing parent' of SANGIOVESE. Crespan *et al.* (2002) had established a possible parent–offspring relationship between Sangiovese and the Tuscan variety CILIEGIOLO but DNA research proved in 2004 that Ciliegiolo was in fact one parent of Sangiovese and Calabrese di Montenuovo the other, thanks to a natural crossing in the vineyard (Vouillamoz, Monaco *et al.* 2007). Further investigation found no DNA matches between

Calabrese di Montenuovo and 194 additional old varieties, mainly from the collection at Librandi in Calabria, but a parent–offspring relationship could be established with MAGLIOCCO CANINO from Calabria (Vouillamoz, Monaco *et al.* 2007; Vouillamoz, Monaco *et al.* 2008). This parentage strongly supports a Calabrian origin for Calabrese di Montenuovo, and provides evidence that Sangiovese is half-Tuscan, half-Calabrian.

WHERE IT'S GROWN

To a very limited extent in Campi Flegrei in the province of Napoli in southern **Italy**.

CALABRESE

See NERO D'AVOLA.

CALADOC

Usefully reliable cross created in southern France but now well dispersed.

BERRY COLOUR: ● ● ● ● ●

PRINCIPAL SYNONYMS: Kaladok

ORIGINS AND PARENTAGE

A GARNACHA × COT cross obtained in 1958 by Paul Truel at INRA (Institut National de la Recherche Agronomique) in Montpellier, southern France, in order to produce a variety that was resistant to coulure. The cross was so successful that Rhône nurserymen described it as 'un Grenache qui ne coule pas' ('a Grenache which does not suffer from coulure'). According to Galet (2000), the name is a contraction of Galabert, with a substitution of C for G, and Languedoc, Galabert being a lake in the Bouches-du-Rhône, where this variety was planted.

VITICULTURAL CHARACTERISTICS

Mid to late ripening, average vigour. Can be pruned short and trained as a gobelet (bush) vine. Consistent production thanks to its resistance to coulure. Also resistant to botrytis bunch rot and not generally susceptible to powdery mildew.

WHERE IT'S GROWN AND WHAT ITS WINE TASTES LIKE

Caladoc is recommended in a dozen *départements* across southern **France**, where there were total plantings of 2,449 ha (6,052 acres) in 2008, but is not permitted in any appellation wines. After a slow start (just 64 ha/158 acres in 1998), the variety has spread quite rapidly and still seems to be on the increase. Wines tend to be full-bodied, deeply coloured and tannic but the variety is also useful for fruity rosés. Mas de Rey in Provence make a varietal version.

In **Spain**, Carlos Esteva of Can Rafols dels Caus in Penedès planted one third of a hectare of Caladoc vines at the beginning of the 1980s, along with the CHENANSON and PORTAN varieties, which together constitute the blend of their Petit Caus red. More recently in the same region, Albet i Noya have also planted Caladoc. **Portugal** had 1,115 ha (2,755 acres) in 2009, mainly in Lisboa. DFJ Vinhos' Grand'Arte is 100% Caladoc but they also produce several blends with ALICANTE HENRI BOUSCHET and TEMPRANILLO (Tinta Roriz) in the Lisboa region.

It is also found in **Lebanon**, and in **Morocco** Dom de la Zouina make a varietal vin gris.

There are minor plantings in **Argentina** (17 ha/42 acres in 2008), where Zuccardi make a Mendoza varietal, and a few rows in Rio Grande do Sul, **Brazil**.

CALAGRAÑO

Virtually extinct Rioja vine making coarse, uninteresting wine.

BERRY COLOUR: ● ● ● ● ●

PRINCIPAL SYNONYMS: Calagraña (Rueda), Navès

VARIETIES COMMONLY MISTAKEN FOR CALAGRAÑO: ALBILLO MAYOR (Ribera del Duero), CAYETANA BLANCA

ORIGINS AND PARENTAGE

Calagraño is an old and almost extinct variety from Rioja in northern Spain, where it still survives near San Asensio. It is often considered to be a synonym for CAYETANA BLANCA (referred to by Galet (2000) as Jaén Blanco but also known as Pardina) but their DNA profiles are distinct and Calagraño appears to be genetically similar to GARNACHA. The name Pardina is particularly confusing because it is also used as a synonym for ALBILLO MAYOR in Ribera del Duero.

WHERE IT'S GROWN AND WHAT ITS WINE TASTES LIKE

A few vines can still be found around the town of San Asensio, north west of Logroño, and elsewhere in Rioja (Carreño *et al.* 2004), although the variety is not authorized for the Rioja DO unless the vines were planted before 1970. In 2008, there were 22 ha (54 acres) in **Spain**: 10 ha (25 acres) in La Rioja and 6 ha (15 acres) in both the País Vasco and Valencia. Wines are apparently rather hard and coarse but can respond to oak ageing.

CALANDRO

Recent German hybrid awaiting official recognition.

BERRY COLOUR: ● ● ● ● ●

PRINCIPAL SYNONYMS: Geilweilerhof 84-58-1233

ORIGINS AND PARENTAGE

A DOMINA × REGENT hybrid obtained in 1984 by Rudolf Eibach and Reinhard Töpfer at the Geilweilerhof research centre in Germany.

VITICULTURAL CHARACTERISTICS

Good resistance to powdery mildew and botrytis bunch rot. Slightly lower yields than REGENT but slightly higher sugar levels. Ripens at the same time as Regent.

WHERE IT'S GROWN AND WHAT ITS WINE TASTES LIKE

Not yet on the official variety register in **Germany** although the licence was applied for in 2004. Full-bodied, tannic wines with smoky, berry-fruit aromas and good ageing potential.

CALITOR NOIR

Declining, low-quality Provençal variety with more synonyms than hectares under vine.

BERRY COLOUR:

PRINCIPAL SYNONYMS: ColTor or Colitor or Coytor, Garriga (Roussillon), Pécoui-Touar (Bandol), Picpoul de Fronton

VARIETIES COMMONLY MISTAKEN FOR CALITOR NOIR: BOUTEILLAN NOIR, BRAQUET NOIR (Nice), JURANÇON NOIR, PIQUEPOUL NOIR

ORIGINS AND PARENTAGE

The earliest citation of this variety was found under its old spelling of Colitor in a list of varieties compiled by Olivier de Serres (1600): 'non plus aujour-d'hui ne sont indifféremment recogneus par toutes les Prouinces, les noms des raisins, dót l'on use le plus en diuers endroits de ce Roiaume, qui sont, *Nigrier, Pinot, Pique-poule, Meurlon, Foirard, Brumestres, Piquardant, Ugnes, Caunès, Samoyran, Ribier, Beccane, Pounhete, Rochelois, Bourdelois, Beaunois, Maluoisie, Meslier, Marroquin, Bourboulenc, Colitor, Veltoline, Corinthien ou Marine-noire, Grecs, Salers, Espaignols, Augibi, Clerete, Prunelat, Gouest, Abeillane, Pulceau, Tresseau, Lombard, Morillon, Sarminien, Chatus, la Bernelle,* & autres infinis' ('consistently distinguished today in all provinces are the grape names that are the most commonly used in this kingdom, which are . . .' etc.). The name Colitor was then also mentioned in 1656 in Lapalud, north of Orange in the Vaucluse.

The name Calitor most likely derives from a Provençal dialect word *col*, meaning 'peduncle' or 'stalk', and *tor*, meaning 'twisted', in reference to the right-angle twist of the bunch stalk, which would explain the local dialect synonyms Col Tor and Pécoui Touar, literally 'twisted stalk'.

Two colour mutations are known, **Calitor Blanc** and **Calitor Gris**, the white-berried version being cited as early as 1782 as a variety giving excellent wines in Nîmes (Rézeau 1997). The Calitor ampelographic group (see p XXVII) also contains BRAQUET NOIR, with which Calitor Noir is sometimes confused, and Ribier Noir (Bisson 2009).

VITICULTURAL CHARACTERISTICS

Late ripening, fertile and productive. May be pruned short. Susceptible to downy mildew and botrytis bunch rot but resistant to powdery mildew. Big bunches and berries.

WHERE IT'S GROWN AND WHAT ITS WINE TASTES LIKE

According to the official French register of varieties, it produces wines that are light, dull, low in acidity and pale – which is probably why there were only 31 ha (77 acres) in **France** in 2008, down from 319 ha (788 acres) in 1968 and still declining, although it was already losing ground earlier in the century in favour of the more productive ARAMON NOIR. It is authorized in the southern Rhône (eg in the Tavel appellation, maximum 10%) but also as far south and west as the Pyrénées-Orientales and in Côtes de Provence wines. However, the regulations for this last stipulate that it may be included only for vines planted before 1994 – it is clearly no longer encouraged. Similarly, it was once a permitted ingredient in Bandol but has been banned since 1992. Dom de la Mordorée in Tavel describes it as out of fashion and had only a few vines remaining.

ÇALKARASI

Produces light, fresh reds and rosés in south-west Turkey.

BERRY COLOUR: ● ● ● ● ●

PRINCIPAL SYNONYMS: Çal Karası

VARIETIES COMMONLY MISTAKEN FOR ÇALKARASI: ADAKARASI (Avşa), HOROZKARASI (Kilis), KALECIK KARASI (Ankara), PAPAZKARASI (Avşa)

ORIGINS AND PARENTAGE

Çalkarası (meaning 'black of Çal') comes from the Çal district of the province of Denizli, in western Turkey.

OTHER HYPOTHESES

See ADAKARASI.

VITICULTURAL CHARACTERISTICS

Productive and mid to late ripening, with good general disease resistance. Berries can reach high sugar levels while retaining relatively high acidity.

WHERE IT'S GROWN AND WHAT ITS WINE TASTES LIKE

Çalkarası is grown on the Çal peninsula in the Pamukkale region, at a height of 600–850 m above sea level in the Denizli province of south-western **Turkey**, where the climate is moderated by the proximity of the Mediterranean and soils are generally calcareous clay. There were 833 ha (2,058 acres) planted in 2010. Although it can be pressed into service to make respectable but light reds or rosés, it is sometimes used as a filler in inexpensive blends or distilled for the production of raki. Kavaklidere produce several rosé versions (still, dry, sparkling and semi-sweet), Pamukkale, Erdoğuz and Küp, among others, produce dry reds. Wines are typically fresh and light, red-fruit-flavoured with soft tannins.

CALLET

Rare Mallorcan making increasingly interesting wines, usually blended.

BERRY COLOUR: ● ● ● ● ●

ORIGINS AND PARENTAGE

Callet is a Mallorcan variety that was declared to be a PALOMINO FINO × Villardiel cross in the Vitis International Variety Catalogue but recent DNA parentage analysis has suggested instead that Callet is a natural cross between two other varieties from Mallorca, FOGONEU and Callet Cas Concos (no longer commercially cultivated), which may be a natural cross between BEBA from Andalucía and GIRÒ from Sardegna (García-Muñoz *et al.* 2012). This would make Callet a grandchild of Beba, and Giró and a half-sibling of MANTO NEGRO from Mallorca. However, these parentages were based on too few DNA markers (twenty) and need to be confirmed. *Callet* means 'black' in the Mallorcan dialect.

VITICULTURAL CHARACTERISTICS

Productive in deep soils, late ripening. Very good general disease resistance but susceptible to sour rot in sites where the bunches are more compact and the berries smaller, and sometimes affected by downy mildew. Copes well with drought.

WHERE IT'S GROWN AND WHAT ITS WINE TASTES LIKE

There were 134 ha (331 acres) in **Spain** in 2008, all on the island of Mallorca, though the variety is often interplanted in the vineyard with its suggested parent FOGONEU, with which it is then blended. Its suggested half-sibling MANTO NEGRO may well be part of the blend, adding fruit but not tannins. Callet is authorized in both the island's DOs, Binissalem and Plà i Llevant, but thrives in the clay and terra rossa or sandy soils in the east around Felanitx, especially where yields are naturally lower. Plantings had increased in the mid 2000s but seem to be declining once more, although indigenous varieties are still gaining in popularity.

Wines have traditionally been rather rustic, light red or rosé in style and typically low in alcohol, but there is an increasing trend towards higher quality, finer tannins, still at moderate alcohol levels, sometimes with a dash of non-native CABERNET SAUVIGNON, SYRAH or TEMPRANILLO in the blend, although the first tends to overwhelm the Callet. Acid levels are moderate but generally high enough to achieve balance and freshness in these minerally, food-friendly wines that are typically red-fruited and sometimes violet-scented. Outside Mallorca, the best-known Mallorcan producer is Àn Negra, who use Callet to make both varietal and blended wines. Other recommended producers of Callet blends include 4 Kilos, Can Majoral, Macía Batle, Oliver Moragues (OM) and Ribas. Toni Gelabert's mineral, iodine-inflected dense but elegant Negra de sa Colònia is a varietal Callet made from vines so close to the sea at Colònia Sant Pere that they have to be protected by a wall and are sometimes burnt by the salt.

CALORIA

Old variety in the far north of Toscana, where it still occasionally finds its way into blends.

BERRY COLOUR: ● ● ● ● ●

ORIGINS AND PARENTAGE

Old local grape that is probably indigenous to the province of Massa-Carrara in Toscana, central Italy. Parentage analysis has revealed possible parent–offspring relationships between MAMMOLO and Caloria as well as several other Tuscan varieties, BIANCONE DI PORTOFERRAIO, COLOMBANA NERA and POLLERA NERA, which turn out to be half-siblings of Caloria (Di Vecchi Staraz, This *et al.* 2007).

VITICULTURAL CHARACTERISTICS

Late ripening. Low in anthocyanins and polyphenols.

WHERE IT'S GROWN

In the province of Massa-Carrara in Toscana, **Italy**; generally blended. There were just 160 ha (395 acres) in Italy in 2000, according to the census.

CAMARALET DE LASSEUBE

Low fertility threatens this south-west France vine that can produce full-bodied, aromatic whites.

BERRY COLOUR:

PRINCIPAL SYNONYMS: Camaralet, Camaralet Blanc, Camaralet de la Seube, Camarau Blanc or Kamarau (País Vasco in Spain), Gentil Aromatique, Moustardet, Petit Camarau

VARIETIES COMMONLY MISTAKEN FOR CAMARALET DE LASSEUBE: Camaraou Blanc (also called Camarau in País Vasco in Spain)

ORIGINS AND PARENTAGE

A grape variety called Camarau Blanc, an old synonym for Camaralet de Lasseube in the País Vasco (Basque Country), northern Spain, was mentioned in 1783–4 in the Pyrénées-Atlantiques (Rézeau 1997): 'Camarau blanc ... les grains sont ronds et dorés, le gout en est piquant mais excellent' ('Camarau blanc ... the berries are round and golden, the taste is spicy but excellent'). However, it is not known whether Camarau Blanc corresponded to Camaralet de Lasseube or to Camaraou Blanc (barely cultivated) since both were called Camarau in the País Vasco. The modern name Camaralet did not appear until 1920, in Saint-Faust (Rézeau 1997).

The name Camaralet could come from the Béarnese (a Gascon dialect spoken in Béarn) *camarau*, possibly referring to the commune of Camalès north of Tarbes in the Hautes-Pyrénées, and Lasseube is a commune in the Pyrénées-Atlantiques about 25 km south east of Pau. However, Lavignac (2001) suggests that the Béarnese *camarau* derives from the Gascon *cama*, meaning 'leg, trunk' and would designate a vine with a long trunk, possibly referring to its cultivation up trees.

According to ampelographers, Camaralet de Lasseube is morphologically close to AHUMAT, ARROUYA and Penouille (Bordenave *et al.* 2007), and it is a member of the Mansien ampelographic group (see p XXVII; Bisson 2009).

VITICULTURAL CHARACTERISTICS

Female-only flowers, therefore highly susceptible to coulure and low yields. Mid ripening. Small berries in small loose bunches, therefore not particularly susceptible to botrytis bunch rot. The juice is sensitive to oxidation.

WHERE IT'S GROWN AND WHAT ITS WINE TASTES LIKE

Camaralet de Lasseube is recommended in the Haute-Garonne, the Gers and the Pyrénées-Atlantiques and may be included in the blend in the Béarn and Jurançon appellations in south-west **France** but was almost abandoned because of its female flowers and consequent poor fertility and low yields. However, the wine is well regarded for its finesse and full, honeyed, peachy flavours with aromas of cinnamon, pepper and fennel. It also has the advantage of lower final alcohol levels than either Manseng, which makes it a useful blend component in dry Jurançon wines. In this appellation, Domaine Cauhapé have 3–4 ha (7–10 acres) of Camaralet, which is blended with GROS MANSENG and COURBU BLANC in Geyser, a dry Jurançon. Domaine Nigri include 10% Camaralet de Lasseube in their version and Domaine Lapeyre use a small amount in their top dry wine, Mantoulan. There was a grand total of 4 ha (10 acres) in France in 2008.

Recently, Les Vignerons Ariégeois in Montégut-Plantaurel (just west of Pamiers) planted

30–40 vines of this and other endangered local varieties (Carbonneau 2005), of which Camaralet de Lasseube shows the greatest promise so far.

CAMARAOU NOIR

Vigorous south-west French vine makes basic wine, mainly in Galicia.

BERRY COLOUR: ● ● ● ● ●

PRINCIPAL SYNONYMS: Caíño Redondo☓ (Galicia in north-west Spain), Camarau or Kamarau (Pays Basque), Moustardet

VARIETIES COMMONLY MISTAKEN FOR CAMARAOU NOIR: ESPADEIRO☓

ORIGINS AND PARENTAGE

Camaraou Noir was not mentioned until Viala and Vermorel (1901–10): 'Camaraou ... cépage rouge spécial aux Basses Pyrénées, nommé aussi camarau' ('Camaraou ... red grape variety specific to the Basses Pyrénées, also named camarau') (see CAMARALET DE LASSEUBE for the etymology).

Camaraou Noir is not a black mutation of Camaraou Blanc (which has now disappeared from the vineyards of Béarn and Jurançon), and both are members of the Mansien ampelographic group (see p XXVII; Bisson 2009). In addition, Camaraou Noir is ampelographically close to ARROUYA and Penouille (Bordenave *et al.* 2007).

Quite unexpectedly, comparison of the DNA profiles of Caíño Redondo (Santiago *et al.* 2005) and Espadeiro (Martín *et al.* 2003) in Galicia, Spain (distinct from the true ESPADEIRO in Portugal), shows that they are identical to Camaraou Noir (Vouillamoz).

VITICULTURAL CHARACTERISTICS

Vigorous, late ripening. Large, compact bunches of small berries.

WHERE IT'S GROWN

Camaraou Noir has now almost disappeared from the vineyards of south-west **France** and can be found only as isolated vines in old Béarn and Jurançon vineyards (Bordenave *et al.* 2007). Traditionally grown up trees, it produced a very ordinary wine.

The variety also survives in Galicia, north-west **Spain**, under the confusing synonym Espadeiro, authorized in the Rías Baixas DO and long cultivated in the provinces of Orense and Pontevedra. Producers of varietal wines include Forjas del Salnés; others such as Coto Redondo and Francisco Alfonso Reboreda blend it with other local varieties such as MENCÍA or Caíño Tinto (BORRAÇAL). In 2008 there were 137 ha (339 acres) in Galicia.

CAMARATE

Sun-loving Portuguese variety making soft, easy-drinking reds.

BERRY COLOUR: ● ● ● ● ●

PRINCIPAL SYNONYMS: Camarate Tinto, Castelão da Bairrada (Tejo), Castelão do Nosso (Tejo), Castelão Nacional (Tejo), Moreto do Douro (Tejo), Moreto de Soure (Tejo), Mortágua (Arruda), Mortagua de Vide Preta (Tejo), Negro Mouro (Lisboa, Dão, Pinhel and Tejo), Vide Preta (Tejo)

VARIETIES COMMONLY MISTAKEN FOR CAMARATE: CASCULHO, CASTELÃO (Tejo and Oeste), MORETO DO ALENTEJO (Barraida, Douro and Tejo)

ORIGINS AND PARENTAGE

Camarate's low level of morphological diversity observed in the vineyard suggests that it is a relatively young variety (Rolando Faustino, personal communication). Indeed, recent DNA parentage analysis has shown that Camarate is a natural cross between the Iberian variety CAYETANA BLANCA and the Portuguese ALFROCHEIRO (Zinelabidine *et al.* 2012), which possibly occurred in the Bairrada region of western Portugal, its main area of cultivation. Camarate is therefore a sibling of CORNIFESTO, MALVASIA PRETA, JUAN GARCÍA and CASTELÃO (see Cayetana Blanca for the pedigree diagram). Camarate was used to breed ÁGUA SANTA.

VITICULTURAL CHARACTERISTICS

Vigorous, early to mid budding and ripening. Small to medium-sized compact bunches of medium-sized berries. Susceptible to downy and powdery mildews, esca, eutypa dieback and coulure and highly susceptible to sour rot. Best suited to deep, low-fertility, well-drained soils and hot, well-exposed sites.

WHERE IT'S GROWN AND WHAT ITS WINES TASTE LIKE

Camarate makes relatively full-bodied but soft wines with raspberry fruit flavours and smooth tannins, good for blending or for early drinking. The variety's heartland, Bairrada in **Portugal**, where it is the second most planted red variety (796 ha/1,967 acres in 2010), produces the best examples but it has also adapted well to the Lisboa, Tejo and Dão regions, where it is authorized in a number of appellations such as Alenquer and Torres Vedras. Recommended producers in Bairrada include Campolargo and Caves de São João.

CAMARÈSE

See BRUN ARGENTÉ.

CAMPBELL EARLY

North American hybrid now more widely grown in Asia, often for the table.

BERRY COLOUR:

PRINCIPAL SYNONYMS: Campbell, Campbell's Early, Island Belle

ORIGINS AND PARENTAGE

A complex *Vitis labrusca* and *Vitis vinifera* hybrid obtained in the early 1890s by G W Campbell in Delaware, Ohio, by crossing Moore Early with (Belvidere × MUSCAT OF HAMBURG), where Moore Early and Belvidere are both CONCORD seedlings used as table grapes (Munson 1909).

VITICULTURAL CHARACTERISTICS

Early to mid ripening. Big berries and big clusters, although there are typically some undersized bunches among them. Likes fertile soil. Hardiness (to –15 °F/–26 °C) and disease resistance similar to CONCORD's. Develops colour ahead of sugar ripeness.

WHERE IT'S GROWN AND WHAT ITS WINE TASTES LIKE

The fruit tastes rather like CONCORD but is sweeter and has much less of the foxy flavour associated with *Vitis labrusca* parentage. Although the variety is now rarely found in the US, it is popular in Asia. In **Korea**, for example, there were 15,419 ha (38,101 acres) in 1997 and in 2008 the variety represented 75% of the the total vineyard area, although almost all of this fruit is used for purposes other than winemaking. **Japan** had around 60 ha (148 acres) for wine in 2009, mainly in Hokkaido, but the variety is also planted more extensively for the table than for winemaking.

CANADA MUSCAT

Grapey hybrid just about surviving in Australia.

BERRY COLOUR:

PRINCIPAL SYNONYMS: NY 17806

ORIGINS AND PARENTAGE

A complex *Vitis vinifera* and *Vitis labrusca* hybrid made in 1928 by Richard Wellington at New York State's Cornell University by crossing MUSCAT OF HAMBURG with Hubbard (see BRIANNA for the complete pedigree of Hubbard). Canada Muscat was selected from among twelve seedlings in 1936 but not officially introduced until 1961.

VITICULTURAL CHARACTERISTICS

Vigorous and relatively hardy. Small to medium-sized bunches of thick-skinned, Muscat-flavoured berries. Ripens a few days after CONCORD.

WHERE IT'S GROWN AND WHAT ITS WINE TASTES LIKE

Canada Muscat was once planted to a very limited extent in Ontario, **Canada** (hence part of the name), but it now appears to survive – by the skin of its teeth – in **Australia**'s McLaren Vale, in the state of South Australia, where Patritti makes a blend of Canada Muscat and MUSCAT BLANC À PETITS GRAINS.

CANAIOLO BIANCO

See DRUPEGGIO.

CANAIOLO NERO

Soft, full-bodied, central Italian red used mainly to complement Sangiovese.

BERRY COLOUR:

PRINCIPAL SYNONYMS: Cannaiola di Marta, Cannaiola Macchie di Marta

VARIETIES COMMONLY MISTAKEN FOR CANAIOLO NERO: COLORINO DEL VALDARNO

ORIGINS AND PARENTAGE

Canaiolo Nero was already mentioned in 1303 as Canajuola in Pietro de Crescenzi's *Opus Commodorum Ruralium*. It was then praised as one of the best red grapes in Toscana by the Florentine agronomist Soderini (1600), under the name Canaiuola. Canaiolo Bianco (DRUPEGGIO) is not a colour mutation of Canaiolo Nero (Storchi *et al.* 2011). **Canaiolo Rosa**, on the other hand, is a colour mutation of Canaiolo Nero, isolated during a clonal selection programme in the province of Lucca.

OTHER HYPOTHESES

Some authors consider that Canaiolo was already known by the Etruscans and call it *Vitis vinifera etrusca*. This is not proven, not least because many other varieties could have had this name.

VITICULTURAL CHARACTERISTICS

Uneven production and difficult to graft; few good clones available. Late ripening.

WHERE IT'S GROWN AND WHAT ITS WINE TASTES LIKE

Canaiolo was more popular in central **Italy** than SANGIOVESE until the eighteenth century. Today, it is one of the major components of the 20% of other local red varieties that may be blended with Sangiovese in Chianti and Chianti Classico DOCG, the velvety texture and full body of the best examples rounding out Sangiovese's firm tannins. Small percentages are allowed in various other Tuscan and central Italian DOCs and a varietal wine is possible in the Tuscan DOC Pietraviva and in Umbria's Rosso Orvietano, for example. Canaiolo is also grown to a limited extent in Lazio, Liguria, Sardegna and the Marche but in 2000 total plantings in Italy were 2,760 ha (6,820 acres) and declining.

Good-quality Canaiolo is produced at the two Ricasoli properties of Castello di Brolio and Castello di Cacchiano in Gaiole in Chianti, in Barbarino Val d'Elsa (where the Castello della Paneretta and neighbour Isole e Olena both use Canaiolo in their Chianti Classico), and in the Vino Nobile di Montepulciano production zone.

Recently, a few producers have begun making wines from 100% Canaiolo, such as Le Poggette's Canaiolo Umbria IGT, Castelvecchio's Numero Otto Toscano Rosso IGT, Possa's Ü Neigru Rosso di Possaitara from Cinque Terre in Liguria, and Bibi Graetz's rare Canaiolo di Testamatta.

CANARI NOIR

Hanging on by a thread in the Ariège.

BERRY COLOUR: ● ● ● ● ●

PRINCIPAL SYNONYMS: Batista (Spain), Canaril, Canarill (Ariège and Haute-Garonne), Carcassès (Ariège), Luverdon (Val di Susa and Chisone in Piemonte, Italy)

VARIETIES COMMONLY MISTAKEN FOR CANARI NOIR: COT (Anjou)

ORIGINS AND PARENTAGE

Canari has today almost disappeared from its original home in the Ariège and the Haute-Garonne in the French Pyrénées. There are two reasons to think that it was more widespread in the past. First because Luverdon, a variety found in Italy in the high-altitude vineyards of the Val di Susa and the Val Chisone in Piemonte, where it is also erroneously called Gamay Luverdon, was unexpectedly shown by DNA profiling to be identical to Canari Noir (Schneider *et al.* 2001).

Secondly because the DNA profiles of Canari Noir and the Spanish variety Batista (Martín *et al.* 2003) turned out to be identical (Vouillamoz).

The colour mutations **Canari Blanc** and **Canari Gris** have been observed in the Haute-Garonne (Galet 1990).

VITICULTURAL CHARACTERISTICS

Vigorous, productive, mid to late ripening. Highly susceptible to botrytis bunch rot.

WHERE IT'S GROWN

This is one of the nearly extinct varieties planted by Les Vignerons Ariégeois in Montégut-Plantaurel, **France** (Carbonneau 2005). They aim to multiply these and to produce a commercial wine if their experiments yield encouraging results.

CANINA NERA

See FORTANA.

CANNAMELA

Vigorous black variety from the island of Ischia in the bay of Napoli, **Italy**, usually blended in the Ischia Rosso DOC. Cannamela means 'small apple', almost certainly in reference to the shape and colour of the berries. Galet (2000) describes the flesh as sweet and perfumed.

CANNONAU

See GARNACHA.

CAÑOCAZO

Virtually extinct variety once popular in sherry country.

BERRY COLOUR: ● ● ● ● ●

PRINCIPAL SYNONYMS: False Pedro (Australia), Hardskin Pedro (Australia), Mollar Blanco, Pedro

ORIGINS AND PARENTAGE

Cañocazo is an old variety from Jerez de la Frontera in Andalucía, south-west Spain, where it was mentioned by the Spanish ampelographer Simón de Rojas Clemente y Rubio (1807) as growing in Jerez, Sanlúcar de Barrameda and Trebujena. Although it is also called Mollar Blanco in Sanlúcar de Barrameda, Cañocazo is not related to the dark-skinned NEGRAMOLL, for which Mollar is a widespread synonym. Further confusion is caused by the use of the name False Pedro in South Africa for what was imported as PEDRO XIMÉNEZ but turned out to be the Portuguese variety GALEGO DOURADO, which is now virtually extinct (Kerridge and Antcliff 1999).

VITICULTURAL CHARACTERISTICS

Mid ripening. Susceptible to downy and powdery mildews and to botrytis bunch rot.

WHERE IT'S GROWN AND WHAT ITS WINE TASTES LIKE

In the nineteenth century, before phylloxera devastated the Jerez region, Cañocazo was grown there alongside PALOMINO FINO and PEDRO XIMÉNEZ, notably in Trebujena but also in Jerez and Sanlúcar de Barrameda. The grapes were used mainly for drying for use in sweet wines, although the sugar level in the berries was generally quite moderate. It has virtually disappeared from **Spain** apart from a few vines in Andalucía but is not authorized for quality wines. Its susceptibility to disease has been a major cause of its fall from favour. However, the Rancho del Merced, the experimental centre for ampelography in Jerez, has some experimental plantings of this and other obscure varieties to assess their suitability for distillation.

The name False Pedro was adopted for this variety in **Australia** to avoid confusion with PEDRO XIMÉNEZ, with Hardskin Pedro as a colloquial alternative (Kerridge and Antcliff 1999). Wines are generally low in acidity.

CAPRETTONE

White, southern-Italian variety long thought to be the same as Coda di Volpe Bianca.

BERRY COLOUR: ● ● ● ● ●

VARIETIES COMMONLY MISTAKEN FOR CAPRETTONE: CODA DI VOLPE BIANCA ☒

ORIGINS AND PARENTAGE

Caprettone comes from the province of Napoli in Campania, southern Italy. Its name might refer to the fact that the cluster looks like a sheep's (*capra* in Italian) beard, or to the fact that it was primarily cultivated by shepherds. Caprettone has long been considered a clone of CODA DI VOLPE BIANCA but recent ampelographic and DNA studies have shown they are not related, and that Caprettone is closely related to PIEDIROSSO and GINESTRA (Costantini *et al.* 2005). Some sources suggest a relationship with CATALANESCA and Uva Rosa but this has been disproved by DNA data (Costantini *et al.* 2005).

VITICULTURAL CHARACTERISTICS

Ripens mid to late September, moderate acidity.

WHERE IT'S GROWN

Cultivated in only fifteen villages in the Vesuvio area in the province of Napoli in Campania, southern **Italy**, Caprettone constitutes a tiny percentage of Lacryma Christi del Vesuvio Bianco DOC and is not produced as a varietal. However, since the Lacryma Christi regulations say that Caprettone is a local name for CODA DI VOLPE BIANCA, it is impossible to know whether what producers include in the blend is true Caprettone or in fact Coda di Volpe Bianca.

CARACOL

Very minor variety found on the Portuguese island of Porto Santo.

BERRY COLOUR: ● ● ● ● ●

PRINCIPAL SYNONYMS: Olho de Pargo, Uva das Eiras

ORIGINS AND PARENTAGE

This obscure variety was supposedly introduced to Porto Santo, an island in Portugal's Madeira archipelago, under the name Olho de Pargo (meaning 'eye of snapper'), in the 1930s by a Portuguese emigrant to South Africa. He gave some vines to his friend João da Silva, who planted them in Eiras, hence the first local name, Uva das Eiras. It was later renamed Caracol (meaning 'snail'), da Silva's nickname.

DNA profiling of Caracol at the University of California at Davis did not show a match with any other known variety.

VITICULTURAL CHARACTERISTICS

Fairly productive and susceptible to botrytis bunch rot.

WHERE IT'S GROWN

Caracol is used both for the table and for wine on the island of Porto Santo, **Portugal**, where there were 9 ha (22 acres) in 2010. The variety is also authorized for Madeira wines.

CARCAJOLO BLANC

See BARIADORGIA.

CARCAJOLO NERO

See PARRALETA.

CARDINAL

Red-skinned North American table grape occasionally used to make wine in parts of Asia.

BERRY COLOUR: ● ● ● ● ●

PRINCIPAL SYNONYMS: Apostoliatiko, Francesa, G 10-30, Karaburnu Rannii, Kardinal, Rannii Carabournu

ORIGINS AND PARENTAGE

A cross obtained in 1939 by E Snyder and F Harmon at the Fresno Horticultural Field Station in California, initially said to be a crossing of Ahmeur Bou Ahmeur (or Flame Tokay) × Alphonse Lavallée (or Ribier, although unrelated to the French Ribier Noir) but recently corrected after DNA parentage analysis to Szőlőskertek Királynője (also known as Königin der Weingärten) ×

Alphonse Lavallée (Akkak *et al.* 2007; Ibañez, Vargas *et al.* 2009), where Szőlőskertek Királynője is a table grape obtained in Hungary in 1916 by crossing Afus Ali (a Lebanese table grape also known as Dattier de Beyrouth) × CSABA GYÖNGYE (aka Perle von Csaba), and Alphonse Lavallée is another table grape obtained in France by crossing Kharistvala Kolkhuri × MUSCAT OF HAMBURG.

Cardinal was used to breed BLANC DU BOIS, CORREDERA and ZHEMCHUZHINA OSKHI.

VITICULTURAL CHARACTERISTICS

Late budding and early ripening. Vigorous and productive. Large bunches of large, thick-skinned berries. Susceptible to downy and powdery mildews.

WHERE IT'S GROWN

Cardinal is widely planted around the world as a table grape and was harnessed for winemaking by Allied Domecq when they tried to establish a winery in northern **Vietnam**.

CARGARELLO

White variety on the verge of extinction in the Rimini area of northern Italy.

BERRY COLOUR: ● ● ● ● ●

PRINCIPAL SYNONYMS: Cargarèl

VITICULTURAL CHARACTERISTICS

Vigorous, productive, mid ripening. Susceptible to botrytis bunch rot.

WHERE IT'S GROWN

Cargarello used to be cultivated mainly in the Valle del Conca near Rimini on the east coast of **Italy**. Today only a few vines survive in the area of San Giovanni in Marignano (near Rimini) and San Teodoro di Mondaino (near Montefeltro), eg in the vineyards of the Rocche Malatestiane co-operative in Rimini. Some Cargarello is also found in the Republic of San Marino thanks to the Consorzio Vini Tipici di San Marino.

CARICA L'ASINO

Rare Italian white that disappears into blends in Piemonte.

BERRY COLOUR: ● ● ● ● ●

VARIETIES COMMONLY MISTAKEN FOR CARICA L'ASINO: BARBERA BIANCA, VERMENTINO

ORIGINS AND PARENTAGE

Possibly introduced from Liguria to the provinces of Alessandria and Novara in Piemonte after the post-phylloxera reconstruction of vineyards. Its name means 'charges the donkey' and most likely comes from the variety's high productivity. It has also been suggested that the grapes were carried out of the vineyards with the help of donkeys, the only possible means of transport there and then.

VITICULTURAL CHARACTERISTICS

Ripens at the end of September.

WHERE IT'S GROWN AND WHAT ITS WINE TASTES LIKE

Exclusive to Piemonte, north-west **Italy**, it is generally blended with other local varieties such as CORTESE, TIMORASSO, BARBERA BIANCA. However, Patrizia Marenco of Marenco di Strevi rescued the variety from a local vineyard and replanted 3,000 vines to produce Carialoso, a 100% Carica l'Asino, a dry, aromatic wine tasting of peaches and apricots (according to the producer).

CARICAGIOLA

See PARRALETA.

CARIGNAN

See MAZUELO.

CARIÑENA

See MAZUELO.

CARIÑENA BLANCA

See MAZUELO.

CARLOS

Light-skinned Muscadine hybrid very popular in North Carolina.

BERRY COLOUR: ● ● ● ● ●

ORIGINS AND PARENTAGE

Carlos is a Muscadine hybrid (mainly *Vitis rotundifolia* with a hint of both *Vitis labrusca* and *Vitis vinifera*) obtained in 1951 by W B Nesbitt, V H Underwood and D E Carroll at the Department of Horticultural Science of North Carolina State University by crossing Howard × North Carolina 11-173, in which:

- Howard is a SCUPPERNONG × Rotundifolia Male Black hybrid, the latter being a selected *Vitis rotundifolia* vine
- North Carolina 11-173 is a Topsail × Tarheel hybrid
- Topsail is a Latham × Burgaw hybrid
- Latham is a selected vine of *Vitis rotundifolia*
- Burgaw is a Thomas × North Carolina V19 R7 B2 hybrid
- Thomas is a selected vine of *Vitis rotundifolia*
- North Carolina V19 R7 B2 is a SCUPPERNONG × New Smyrna hybrid
- New Smyrna is a selected vine of *Vitis rotundifolia*

- Tarheel is a Luola × North Carolina V36 R15 B4 hybrid
- Luola is a selected vine of *Vitis rotundifolia*
- North Carolina V36 R15 B4 is an Eden × North Carolina V23 R4 B2 hybrid
- Eden is an Ontario × New York 10085 hybrid (see CAYUGA WHITE for the pedigree of Ontario)
- New York 10085 is a Triumph × Mills hybrid (see CASCADE for the pedigree of Triumph and GR 7 for the pedigree of Mills)\
- North Carolina V23 R4 B2 is an Eden × *Vitis rotundifolia* var. *munsoniana* hybrid.

VITICULTURAL CHARACTERISTICS

Relatively cold-hardy but early budbreak makes it prone to damage from spring frosts. Early to mid ripening. Vigorous, productive. Small berries with bitter, tough skins.

WHERE IT'S GROWN

Carlos is the leading Muscadine variety in **North Carolina**, with 410 acres (166 ha) in 2006, representing 95% of Muscadine production in the state. It is used mainly for wine and juice but also for the table. Producers of varietal wines include Colony in **Texas** and Bannerman in North Carolina.

CARMENÈRE

Produces deeply coloured, sometimes herbaceous wines, in Chile far more than in its native Bordeaux.

BERRY COLOUR:

PRINCIPAL SYNONYMS: Bordo (Reggio Emilia in Italy), Cabernelle (Médoc), Cabernet Gernicht (China), Cabernet Gernischet (China), Cabernet Gernischt (China), Cabernet Shelongzhu (China), Carbonet (Médoc), Carbouet (Graves), Caremenelle (Médoc), Carménègre, Carménère, Carmeneyre (Bergerac), Grosse Vidure (Médoc)

VARIETIES COMMONLY MISTAKEN FOR CARMENÈRE: CABERNET FRANC (Gironde and north-eastern Italy), CABERNET SAUVIGNON (Gironde), MERLOT (Chile)

ORIGINS AND PARENTAGE

Carmenère is an old variety from the Gironde, where it was first mentioned in 1783–4 in Bergerac (Dordogne) under the name Carmeynere, together with CABERNET FRANC (Carmenet): 'le Carmenet, que l'on nomme de même en Médoc. La Carmeneyre. L'un et l'autre de ces deux cépages produit peu et principalement la Carmeneyre, mais le vin est très bon' ('Carmenet, with the same name in the Médoc. Carmeneyre. Both these varieties have low yields, in particular Carmeneyre, but the wine is very good'; Rézeau 1997). Its modern spelling first appeared in Jullien (1816): 'le carmenet, la carmenère, le malbek et le verdot peuplent seuls les vignobles du Médoc, qui fournissent les vins rouges de première qualité' ('carmenet, carmenère, malbek and verdot alone constitute the Médoc vineyards that give top-quality red wines'). Although some have suggested a link with the dark-red colour *carmin* (carmine), the etymology of Carmenère, like that of Carmenet, remains unclear.

DNA parentage analysis has recently established that Carmenère is a natural cross between CABERNET FRANC and Gros Cabernet (see the pedigree diagram in the CABERNET SAUVIGNON entry), a very old variety from the Gironde and the Tarn that was often confused with Cabernet Franc (Boursiquot *et al.* 2009). Gros Cabernet is no longer cultivated but is likely to be a FER × HONDARRIBI BELTZA cross, both varieties being of Spanish Basque origin

(Boursiquot *et al.* 2009). Hondarribi Beltza, also called Txakoli or Chacolí, is in turn a likely progeny of Cabernet Franc (Boursiquot *et al.* 2009). This means that Carmenère and Cabernet Franc have a high level of consanguinity, because Carmenère is at the same time a progeny and a great-grandchild of Cabernet Franc. Carmenère is also a half-sibling of both Cabernet Sauvignon and MERLOT. All these close relationships explain the frequent confusion between Carmenère, Cabernet Franc, Cabernet Sauvignon and Merlot, all belonging to the Carmenet ampelographic group (see p XXVII; Bordenave *et al.* 2007; Bisson 2009).

In Italy, ampelographic and molecular analyses have established that most of the vines planted under the CABERNET FRANC type called 'Italian' in the north-eastern part of the country were in fact Carmenère, while the Cabernet Franc type called 'French' was the true Cabernet Franc (Calò, Di Stefano *et al.* 1991). The same confusion between Cabernet Franc and Carmenère occurred in California. In Chile, a few vines thought to be MERLOT were discovered in 1994 to be Carmenère, and a subsequent DNA profiling survey revealed that the majority of the 'Merlot Noir' then grown in Chile was actually Carmenère (Hinrichsen *et al.* 2001).

In China, the true identity of Cabernet Gernischt has long been debated, and it has been described as:

- a cross between CABERNET SAUVIGNON and CABERNET FRANC brought to Changyu (or Zhangyu) Winery in China at the time of, or shortly after, the founding of the winery in 1892
- CABERNET FRANC introduced to China by Changyu Winery in 1892, which would be consistent with the name of this variety being a misspelling of *Cabernet gemischt*, German for 'mixed Cabernet', referring to the initial mixture of imported cuttings (Luo 1999)
- a cross developed in 1931 by Changyu Winery from newly introduced and undisclosed varieties from Europe (Zhengping 2011)
- the French variety Carmenère, introduced as 'mixed Cabernet' for the same reasons as above (Freeman 2000; Pszczólkowski 2004).

Some ampelographic and molecular studies in China suggested that Cabernet Gernischt is identical to CABERNET FRANC (Yin *et al.* 1998; Song *et al.* 2005), while another recently suggested that it is identical to Carmenère (Li *et al.* 2008). DNA profiling of genuine samples of Cabernet Gernischt from the Changyu Winery definitely shows that it is identical to Carmenère (Vouillamoz).

The name Cabernet Gemischt, of which Cabernet Gernischt is said to be a misspelling, is often believed to correspond to an old, extinct European grape variety, possibly an ancestor of CABERNET FRANC. This is not verified in any ampelographic or viticultural text in Europe, where the words 'Cabernet, gemischt' were used only in relation to oenological analyses of wines that were a blend of CABERNET SAUVIGNON and Cabernet Franc. In China, this variety is also widely called Cabernet Shelongzhu, literally 'Cabernet snake pearl', an alternative name chosen by Changyu Winery because it was easier to pronounce, at least for Chinese speakers.

VITICULTURAL CHARACTERISTICS

Vigorous and not very fertile (low basal-bud fertility requires cane pruning), producing small bunches of medium-sized deep blue berries. Susceptible to coulure and root infestation. Ripens late (4–5 weeks after MERLOT in Chile). In China, relatively hardy and resistant to anthracnose, well suited to sandy soils such as those in the east of Helan Mountain and the Hexi Corridor in Gansu province.

WHERE IT'S GROWN AND WHAT ITS WINE TASTES LIKE

Carmenère was widely cultivated in the Médoc, south-west **France**, in the early eighteenth century but largely abandoned in Bordeaux after the phylloxera invasion in the 1870s because of its poor fruit set and consequently unreliable yields: there were just 21 ha (52 acres) in France in

2008. Examples of what remains can be found at classed growth Château Clerc Milon in Pauillac, which still has about 0.4 ha (1 acre or around 1% of its vineyard area, like Château Belle-Vue of the Haut-Médoc) and Clos Louie in the Côtes de Castillon, where there are some pre-phylloxera Carmenère vines interplanted with ancient examples of other Bordeaux varieties, in its 125-year-old, ungrafted vineyard. Château Brane-Cantenac planted 0.5 ha (1.2 acres) in 2007 and have included some of the fruit in their 2011 *grand vin*.

In **Italy**, the census recorded 45 ha (111 acres) in 2000, mostly in the north east, but there is probably a good deal more that goes unrecorded because in many cases growers discovered that what they thought was CABERNET FRANC was in fact Carmenère. When the confusion became clear, Ca'del Bosco in Lombardia began to produce a successful varietal wine called Carmenero, classified as a vino da tavola because the variety is not authorized for DOC or IGT varietal wines, though it is a permitted ingredient in many blended appellation wines such as Garda Cabernet and Monti Lessini. Inama's Carmenere Puì is a blend of Carmenère, MERLOT and other local varieties from the Colli Berici but classified as IGT Veneto, and Ca Orologico's Relógia is mainly Carmenère with 20% CABERNET SAUVIGNON.

A few **California** producers have shown interest in Carmenère: in 2009 plantings were 57 acres (23 ha), just slightly higher than in Bordeaux, perhaps partly encouraged by the Meritage Alliance, whose Bordeaux-blend recipe includes Carmenère. Guenoc, for example, blend it with the usual suspects CABERNET SAUVIGNON, CABERNET FRANC, PETIT VERDOT and MERLOT. Dover Canyon make a rare varietal wine using grapes from Colbert Vineyard in East Paso Robles. Varietal Carmenère is by no means uncommon in the warmth of **Washington** State, where ripening conditions are not unlike those of Chile.

Black Hills, British Columbia, **Canada**, produce a single-vineyard Carmenère and use it in blends, though official 2008 statistics for the province show less than an acre (0.4 ha) of this variety.

The first Carmenère vines were brought to **Chile**, Carmenère's current power base, in the mid nineteenth century, where it was planted as a field blend, mainly with MERLOT, though growers probably thought they had Merlot exclusively, and thrived thanks to a warmer and drier climate than south-west France, allowing a longer ripening period. The French ampelographer Claude Valat was the first to notice that not all Chilean Merlot was the real McCoy, and in 1994 J-M Boursiquot identified it as Carmenère, which was confirmed by DNA analysis in 1997. The Chilean authorities recognized Carmenère as an official variety for the country in 1998. Thanks to this identification process, the reported area planted with Carmenère has shot up: from zero in 1996 to 7,183 ha (17,750 acres) in 2006, apparently declining a little thereafter to 7,054 ha (17,431 acres) in 2008, although plantings of Merlot also continued to increase after the identification of Carmenère and it is hard to be sure at what point the varieties were accurately distinguished in the official statistics.

Much has been written about this confusion between MERLOT and Carmenère but Richards (2006) suggests that most growers recognized the difference, referring to the former as Merlot and the latter as Merlot Chileno.

If the grapes are harvested before they are fully ripe, wines have strong capsicum and herbaceous flavours. These turn into red berry and sometimes black pepper or tomato when the berries are just ripe, and then at full maturity blackberry and blueberry with overtones of chocolate and coffee and soy sauce, though the variety also tends to lose acidity at this stage. There are many fine varietal examples in Chile from producers such as Antiyal, Concha y Toro, Errázuriz, Montes and Ventisquero. The variety may also make a significantly refreshing if quantitatively minor contribution to many of Chile's top blends such as Almaviva. De Martino and others are working to find the best possible sites for the variety and their range now includes a single-vineyard wine from the Alto de Piedras vineyard in the Isla de Maipo in the Maipo Valley.

Argentina had 32 ha (79 acres) in 2008, almost all in the Mendoza region.

Brown Brothers produce limited quantities of varietal Carmenère in Victoria, **Australia.**

In **China**, as Cabernet Gernischt planted there in the late nineteenth century, Carmenère wines seem to have the herbaceous flavours often associated with CABERNET FRANC and a texture similar to that of CABERNET SAUVIGNON. It is widely planted in Jiaodong in the province of Shandong and also grown in the southern part of north-east China, northern China and north-west China. Recommended producers of varietal wines include Changyu Wine Group and China Great Wall Wine. There were 1,218 ha (3,010 acres) of Cabernet Gernischt in China in 2009.

CARMINE

Minor North American cross with marked Cabernet characteristics.

BERRY COLOUR: ● ● ● ● ●

ORIGINS AND PARENTAGE

Carmine is an F2-7 × MERLOT cross, where F2-7 is a MAZUELO × CABERNET SAUVIGNON cross, bred by Harold P Olmo in 1946 at the University of California at Davis and introduced onto the market in 1976 (Walker 2000). At that time, the recently discovered parentages of Cabernet Sauvignon and Merlot were unknown and Olmo was obviously not aware that he was crossing two half-siblings.

VITICULTURAL CHARACTERISTICS

Late budding and late ripening. Productive, with large bunches.

WHERE IT'S GROWN AND WHAT ITS WINE TASTES LIKE

Although Olmo's intention was to breed a variety that would be suited to the cooler coastal sites of California, it never really caught on in the state. It is now planted to a limited extent in states such as **Pennsylvania**, **Michigan**, **Oregon**, **Colorado** and **Kentucky**. Wines are typically deeply coloured and tannic with capsicum aromas and flavour and tend to be used in blends.

CARMINOIR

Recent Swiss cross that seems to have inherited more from Cabernet Sauvignon than from Pinot.

BERRY COLOUR: ● ● ● ● ●

ORIGINS AND PARENTAGE

Carminoir is a PINOT NOIR × CABERNET SAUVIGNON cross obtained in 1982 at the Caudoz viticultural research centre in Pully, just outside Lausanne, Switzerland, today part of Agroscope Changins-Wädenswil (Vouillamoz 2009c).

VITICULTURAL CHARACTERISTICS

Fertile, medium to high vigour, moderate but regular production. Compact, medium-sized to small bunches of small berries. Mid budding, mid to late ripening. Good resistance to botrytis bunch rot. Needs low-vigour sites and rootstocks.

WHERE IT'S GROWN AND WHAT ITS WINE TASTES LIKE

Carminoir ripens fully only in the best sites in **Switzerland**, ie those sites where CABERNET SAUVIGNON reaches full maturity, which is limiting its spread. There were 10 ha (25 acres) in 2009, mainly in the Valais (6 ha/15 acres), where good varietal examples are made by Vieux Moulin and Jean-Camille Juilland, and in Ticino (2.5 ha/6 acres), where it is usually blended with MERLOT, eg by Daniel Huber and by Cantina dell'Orso. Carminoir wines, usually aged in oak, are deeply coloured and full-bodied with powerful tannins and aromas of dried plums, spices and elderberry.

CARNELIAN

North American cross bred to retain Cabernet Sauvignon's characteristics in hot climates. A surprising success in Western Australia.

BERRY COLOUR: ● ● ● ● ●

ORIGINS AND PARENTAGE

An F2-7 × GARNACHA cross, where F2-7 is a MAZUELO × CABERNET SAUVIGNON cross, bred by Harold P Olmo in 1949 at the University of California at Davis and introduced onto the market in 1974 (Walker 2000).

VITICULTURAL CHARACTERISTICS

Medium to large compact clusters of small to medium-sized berries, vigorous and suited to warm to hot climates but with a tendency to overcrop. Early to mid ripening. Berries can reach high sugar levels but are quite difficult to harvest.

WHERE IT'S GROWN AND WHAT ITS WINE TASTES LIKE

Harold Olmo developed Carnelian specifically for warmer sites in **California**, as a sort of Cabernet substitute for the Central Valley, although the GARNACHA characteristics of high sugar and being difficult to pick have limited its success and spread. Although several thousand acres were planted soon after the variety was released, the area has declined since then and stood at 893 acres (340 ha) in 2008, mostly in Fresno County but also in Kern and Tulare. When yields are not restricted, wines tend to lack character and are usually hidden in blended jug wines. It was also planted to a limited extent in **Texas** and **Hawaii**, although in Hawaii most of the Carnelian vines at the Tedeschi Vineyard on Maui have been replaced by better-quality varieties such as SYRAH, CHENIN BLANC, VIOGNIER and Malbec (COT).

In **Australia**, Carnelian has been planted in several vineyards in the state of Western Australia, thanks to a vine nursery mix-up (they thought it was SANGIOVESE), but the results have been surprisingly good, showing CABERNET SAUVIGNON-like characteristics while retaining good acidity in the cooler parts of this region. Producers of varietal wines include Mad Fish (a Howard Park brand), Oakover, Western Range and Windshaker Ridge. Peos Estate, in Manjimup, describe the wines as having high tannin and acidity and good ageing potential but being somewhat unapproachable in youth and therefore generally best in blends, although Carnelian apparently also shows good potential for rosé.

CARRASQUÍN

Very minor Asturian variety used in blends or for rosé.

BERRY COLOUR: ● ● ● ● ●

VARIETIES COMMONLY MISTAKEN FOR CARRASQUÍN: Carrasco✕ or Carrasco Negro (Asturias)

ORIGINS AND PARENTAGE

Carrasquín, from Asturias in the far north west of Spain, is genetically very close to Albarín Negro (Martín *et al.* 2006), the name given to ALFROCHEIRO in Asturias, and to PRIETO PICUDO (Martín *et al.* 2003) from the province of León, just south of Asturias.

OTHER HYPOTHESES

Carrasquín is often considered to be identical to Carrasco (no longer cultivated) from Asturias but comparison of their respective DNA profiles (Martín *et al.* 2006; Gago *et al.* 2009) shows that they are distinct varieties (Vouillamoz).

WHERE IT'S GROWN AND WHAT ITS WINE TASTES LIKE

Plantings in Asturias in north-west **Spain** are too minor to feature in official statistics but the wines are apparently pale, low in acidity, rather neutral and used mainly for rosé wines. Monasteria de Corias blend Carrasquín with MENCÍA, Verdejo Negro (TROUSSEAU) and Albarín Negro (ALFROCHEIRO) and the variety is authorized in the Vino de la Tierra de Cangas, the regional wine from south-western Asturias.

CARREGA BRANCO

Portuguese variety making light, unexceptional wines.

BERRY COLOUR: ● ● ● ● ●

ORIGINS AND PARENTAGE

This little-known variety probably comes from the Douro or the Trás-os-Montes region in north-eastern Portugal.

VITICULTURAL CHARACTERISTICS

Susceptible to botrytis bunch rot. Medium-sized bunches of thin-skinned berries. Mid to late ripening.

WHERE IT'S GROWN AND WHAT ITS WINE TASTES LIKE

Grown mainly in the Douro in northern **Portugal** and to a lesser extent in Beiras, there were 460 ha (1,137 acres) planted in 2010. Wines are typically light and simple and best blended with higher-acid varieties such as RABIGATO.

CARRICANTE

Potentially very fine, crisp and distinctive Sicilian white variety.

BERRY COLOUR: ● ● ● ● ●

PRINCIPAL SYNONYMS: Catanese Bianco

VARIETIES COMMONLY MISTAKEN FOR CARRICANTE: CATARRATTO BIANCO

ORIGINS AND PARENTAGE

Local tradition refers to Viagrande in the province of Catania, on the Italian island of Sicilia, as the cradle of Carricante. Its name probably refers to the variety's abundant yields (*carica* means 'load'). In 1774, Sestini tells that Etna vintners used to leave Carricante wines on their lees in barrel to induce malolactic fermentation in the spring and thus reduce their naturally high acidity (Giavedoni and Gily 2005). Cipriani *et al.* (2010) have recently suggested through DNA parentage analysis that Carricante is a possible Montonico Pinto × Scacco natural cross, the former a black-berried variety that is not commercially cultivated and the latter a white-berried Forcelese d'Ascoli × DINDARELLA cross (Forcelese d'Ascoli is a light-skinned but non-cultivated variety). This parentage showed some DNA discrepancies and needs further investigation.

VITICULTURAL CHARACTERISTICS

Late ripening, usually harvested at the end of September or in early October, sometimes even later, in order to allow the acidity to drop. Bushvines are common.

WHERE IT'S GROWN AND WHAT ITS WINE TASTES LIKE

Carricante is cultivated on the eastern slopes of Etna on the eastern side of Sicilia, **Italy**, mainly in the province of Catania (Viagrande, Trecastagni, Zafferana Etnea and Milo). In other parts of Sicilia, Carricante is sometimes erroneously called Catarratto. It is often blended with other Sicilian white varieties such as CATARRATTO BIANCO, INZOLIA and MINELLA BIANCA, though there must be at least 60% Carricante in the blend for Etna Bianco and 80% for Etna Bianco Superiore (for which the grapes can come only from Milo). It may occasionally be used to soften red varieties such as NERELLO MASCALESE. However, 100% Carricante wines can be impressive and delicious, showing aromas of orange and grapefruit, orange flowers, aniseed and white fruits, with fresh acidity and notable minerality within a sinuous framework. Even in young wines there may be a lightly honeyed note. Total Italian plantings in 2000 were 264 ha (652 acres) and Sicilian plantings totalled 101 ha (250 acres) by 2008 but the variety is increasingly appreciated.

Barone di Villagrande's organically grown Etna Bianco Superiore is one example of a varietal Carricante. Benanti make two fine Etna Biancos: the less expensive Bianco di Casele and the more concentrated, minerally and ageworthy old-vine Pietramarina Superiore. Both display those characteristic orange and aniseed notes but with a herbal, even cedary, edge and a modest 12–12.5% alcohol.

CASAVECCHIA

Ancient, resurrected dark-skinned Italian variety exclusive to parts of Campania.

BERRY COLOUR: ● ● ● ● ●

ORIGINS AND PARENTAGE

According to oral tradition, cuttings of Casavecchia were taken from a centenarian vine in the mid nineteenth century by a local grower who then handed them down to other growers and family members (Masi *et al.* 2001). The original vine was supposedly found near an old stone farmhouse in Funari di Pontelatone, north of Caserta in Italy's Campania, which gave its name to this mysterious variety (*casa* means 'house', *vecchia* means 'old').

According to recent DNA studies, Casavecchia's profile does not match that of any other variety in Campania or anywhere else, and possible genetic links with CATALANESCA from Napoli and PALLAGRELLO NERO were suspected (Costantini *et al.* 2005), though parent–offspring relationships can be ruled out (Vouillamoz).

OTHER HYPOTHESES

Some authors have suggested that Casavecchia could correspond to the Trebulanum mentioned by Pliny the Elder. Cipriani *et al.* (2010) have recently suggested through DNA parentage analysis that Casavecchia is possibly a natural cross between MALVASIA BIANCA DI CANDIA and ABBUOTO (under the name Cecubo) but this is unlikely because the same study suggests that Abbuoto itself may be a natural cross between PIEDIROSSO and Casavecchia. These parentages therefore need further investigation.

VITICULTURAL CHARACTERISTICS

Ripens end of September to mid October.

WHERE IT'S GROWN AND WHAT ITS WINE TASTES LIKE

Casavecchia is highly localized, planted mainly in old vineyards in nine communes in the north of Caserta in Campania (mainly in Pontelatone, Caiazzo, Formicola, Liberi and Castel di Sasso), **Italy**. Fervent producers and conscientious scholars have inspired the recent efforts to safeguard and promote this variety, which is used in blends and increasingly to make varietal wines. These offer aromas of herbs, dried leaves, humus, dried mushrooms, green pepper and liquorice.

Terre del Principe's Cantomoggia Casavecchia and Viticoltori del Casavecchia's Corterosa are good examples. Fattoria Alois make reference to Pliny the Elder (see above) in the name of their varietal Casavecchia, Trebulanum (IGT Campania), which is a powerful, oak-aged wine with aromas of dark fruit and leather, firm, dense tannins and savoury, mocha flavours.

CASCADE

Early ripening but dangerously virus-prone dark-skinned hybrid.

BERRY COLOUR: ● ● ● ● ●

ORIGINS AND PARENTAGE

A complex hybrid obtained at the beginning of the twentieth century by Albert Seibel in Aubenas in the Ardèche, southern France, by crossing Seibel 7042 with Gloire de Seibel, where:

- Seibel 7042 is a Seibel 5351 × Seibel 6268 cross
- Seibel 5351 is a Seibel 880 × Seibel 2679 cross (see PRIOR for the complete pedigree of Seibel 880)
- Seibel 2679 is a Triumph × Alicante Terras 20 cross
- Triumph is a CONCORD × Chasselas Musqué cross (the latter a Muscat-flavoured mutation of CHASSELAS)
- Alicante Terras 20 is an ALICANTE HENRI BOUSCHET × Rupestris du Lot (a variety of *Vitis rupestris* used as a rootstock) hybrid
- Seibel 6268 is a Seibel 4614 × Seibel 3011 cross (see HELIOS for the complete pedigrees of Seibel 4614 and Seibel 3011)
- Gloire de Seibel is a Seibel 867 × Seibel 452 hybrid (see NOIRET for the pedigree of Seibel 867)
- Seibel 452 is an Alicante Ganzin × Seibel 1 hybrid (see ROYALTY for the pedigree of Alicante Ganzin)
- Seibel 1 is a Munson × *Vitis vinifera* subsp. *vinifera* hybrid, the latter an unknown *vinifera* variety (see PRIOR for the complete pedigree of Munson).

Cascade has been commercially available since 1938. It was later used to breed L'ACADIE BLANC in Canada.

VITICULTURAL CHARACTERISTICS

Productive, hardy and very early ripening. Its loose clusters give it good resistance to fungal diseases but it is highly susceptible to viruses.

WHERE IT'S GROWN AND WHAT ITS WINE TASTES LIKE

This hybrid was once planted to a limited extent in **New York** State (183 acres/74 ha in 1975) thanks to its ability to ripen very early, useful in the cool Northeast. However, its extreme susceptibility to virus diseases has limited its distribution and popularity and the last blocks have recently been pulled up. Wines tended to be light and low in acidity.

CASCAROLO BIANCO

Ancient and barely surviving white variety from Piemonte.

BERRY COLOUR: ● ● ● ● ●

PRINCIPAL SYNONYMS: Cascarala, Cascarecul, Cascarelbo

VARIETIES COMMONLY MISTAKEN FOR CASCAROLO BIANCO: Augster Weisser ✕

ORIGINS AND PARENTAGE

Cascarolo Bianco is a very old Piemontese variety cited by Croce in the Torino hills as early as 1606. Its name probably comes from its susceptibility to coulure, the Italian word *cascolare* meaning 'to fall'. It has been suggested that the non-cultivated Fehér Gohér (or Augster Weisser) from Hungary could be a synonym for Cascarolo Bianco (Goethe 1887) but DNA profiling strongly rejects this hypothesis (Vouillamoz).

Recent DNA studies provided evidence that Cascarolo Bianco is a likely progeny of RÈZE, a very old variety from the Swiss Alps (Vouillamoz, Schneider *et al.* 2007), and a close relationship with ERBALUCE is suspected (Vouillamoz).

WHERE IT'S GROWN AND WHAT ITS WINE TASTES LIKE

Cascarolo Bianco used to be widespread in Piemonte (around Pinerolo and Chieri, in Casale Monferrato and in Moncalvo), north-west **Italy**, where its wine was highly esteemed. Today it has almost disappeared, scattered throughout old vineyards, especially at the foot of the Alps.

CASCULHO

Minor northern Portuguese variety generally found in blends.

BERRY COLOUR:

PRINCIPAL SYNONYMS: Cascudo

VARIETIES COMMONLY MISTAKEN FOR CASCULHO: CAMARATE

ORIGINS AND PARENTAGE

Casculho probably originates in the Trás-os-Montes region in northern Portugal. Although Casculho is also a synonym for CAMARATE from the Bairrada region, they have distinct DNA profiles (Veloso *et al.* 2010) and are therefore not identical.

VITICULTURAL CHARACTERISTICS

Large bunches of small, thin-skinned berries.

WHERE IT'S GROWN AND WHAT ITS WINE TASTES LIKE

Casculho is authorized for Douro table wines as well as for regional wines from the Península de Setúbal but is rarely found as a varietal. Quinta do Carrenho's Dona Berta Vinha Centenária, made from an old-vine field blend in the Douro, is an exception. There were 730 ha (1,804 acres) in **Portugal** in 2010.

CASETTA

Northern Italian dark-skinned variety snatched from the jaws of extinction and making highly structured and distinctive wines.

BERRY COLOUR:

PRINCIPAL SYNONYMS: Lambrusco a Foglia Tonda, Lambrusco Casetta, Maranela

VARIETIES COMMONLY MISTAKEN FOR CASETTA: ENANTIO, FOGLIA TONDA (Toscana)

ORIGINS AND PARENTAGE

Casetta is local to the Terra dei Forti, a valley in the Vallagarina region south of Trento in the far north of Italy. The name Casetta is most likely to come not from its simple Italian meaning ('small house') but from the name of a family in Marani di Ala, which also explains the variety's synonym Maranela.

Casetta is often confused with ENANTIO, another autochthonous variety from Vallagarina. Indeed, Enantio is a recent name given to the traditional Lambrusco a Foglia Frastagliata (meaning 'toothed leaf') and, in the local dialect, Casetta is often called Lambrusco a Foglia Tonda, while Enantio is called Foja Tonda (meaning 'round leaf'). However, Enantio is morphologically quite different from Casetta (Calò *et al.* 2006) and comparison of their DNA profiles therefore rules out any parent–offspring relationship (Vouillamoz).

Although it is sometimes called Lambrusco Casetta or Lambrusco a Foglia Tonda, Casetta is genetically distinct from all other Lambruscos in Italy.

OTHER HYPOTHESES

Some authors suggest that Casetta could be the result of the domestication of local wild grapes (Calò *et al.* 2006) but this hypothesis has not been confirmed by genetic studies.

VITICULTURAL CHARACTERISTICS

Susceptible to fungal diseases, especially botrytis bunch rot. Mid ripening.

WHERE IT'S GROWN AND WHAT ITS WINE TASTES LIKE

Casetta was once widespread in Trentino and the Veneto in **Italy**, mainly in Val Cipriana and Val San Valentino (in particular in Marani and in San Margherita near Ala). Due to its susceptibility to fungal diseases, Casetta was virtually abandoned in the 1960s but was rescued from near extinction by the northern Italian producer Albino Armani, who discovered ungrafted vines that were more than 100 years old and began new plantings in the 1990s with the help of the Istituto Agrario di San Michele all'Adige. Today, Casetta's vineyards cover a mere 12 ha (30 acres), a mix of new and old plantings, often ungrafted, in the Vallagarina region.

Since 2006, Casetta has been authorized in the Terra dei Forti DOC, often blended with other local varieties and occasionally made as a varietal wine. Casetta wine is rich in anthocyanins, tannins, acidity and alcohol, and needs time to gain balance. Typical aromas include dried plums, maraschino cherry, cinnamon, tobacco and musk.

Albino Armani was the first to produce a 100% Casetta wine, called Foja Tonda, in 1994, and is still the most important producer of this variety.

CASTAGNARA

Low-acid Italian red wine variety improved by blending.

BERRY COLOUR:

PRINCIPAL SYNONYMS: Santa Maria Nera, Sarnese

ORIGINS AND PARENTAGE

DNA analysis at the University of Portici did not find any genetic relationship with other Campanian varieties (Giavedoni and Gily 2005).

VITICULTURAL CHARACTERISTICS

High yields, usually pergola trained, often ungrafted in old vineyards and susceptible to coulure. Harvested in the first ten days of October.

WHERE IT'S GROWN AND WHAT ITS WINE TASTES LIKE

Castagnara had a good reputation in the nineteenth century and was widespread in **Italy**'s Campanian provinces of Napoli and Avellino. Today it is scantily cultivated in the province of Napoli, especially in Lettere, Casola, Gragnano and Sant'Antonio Abate. It is authorized in the Penisola Sorrentina Rosso DOC, where it is usually blended with PIEDIROSSO and/or SCIASCINOSO and/or AGLIANICO. Wines are notably low in acidity.

CASTELÃO

Portugal's most common variety: hardy, adaptable and widely grown under many different aliases.

BERRY COLOUR: ● ● ● ● ●

PRINCIPAL SYNONYMS: Bastardo Castico, Bastardo Espanhol, Castelão Francês (Alentejo, Douro, Lisboa, Tejo and Setúbal), Castellao Portugues, Castico, João Santarém or João de Santarém (Oeste), Periquita (Península de Setúbal), Piriquita or Piriquito (Algarve, Alentejo, Bairrada, Tejo)

VARIETIES COMMONLY MISTAKEN FOR CASTELÃO: CAMARATE, MORETO DO ALENTEJO, PEDRAL, TRINCADEIRA, TROUSSEAU

ORIGINS AND PARENTAGE

Castelão is a very old Portuguese variety first mentioned in the Douro in 1531 by Rui Fernandes (1531–2), in the 'Descripção do terreno em roda da cidade de Lamego duas leguas' under the name Catelão, which is undoubtedly a misspelling of Castelão. Recent DNA parentage analyses have revealed that Castelão is a natural cross between CAYETANA BLANCA, widespread throughout the south and centre of the Iberian peninsula, and ALFROCHEIRO from the south or centre of Portugal (Zinelabidine *et al.* 2012). Castelão is therefore a sibling of CORNIFESTO, MALVASIA PRETA, JUAN GARCÍA and CAMARATE (see Cayetana Blanca for the pedigree diagram), and also possibly of JAMPAL (Myles *et al.* 2011).

In some regions, Castelão used to be called Trincadeira, which is a misleading synonym because it is the official name of a distinct black variety (see TRINCADEIRA) and is also used for several other black- or white-berried varieties (eg TAMAREZ, Trincadeiro Branco). Similarly, the synonyms Bastardo Castico and Bastardo Espanhol could create confusion with TROUSSEAU (known as Bastardo in Dão and the Douro) and should be avoided. The origin of the synonym Periquita (literally, 'parakeet') is unknown.

Castelão was used to breed numerous varieties, including ÁGUA SANTA and possibly AGRONÓMICA (see that entry).

VITICULTURAL CHARACTERISTICS

Hardy and adaptable. Early budding, early to mid ripening. Small, relatively compact bunches of medium-sized, thick-skinned but not very juicy berries. Susceptible to coulure and millerandage, and therefore sometimes patchy ripening, but good all-round disease resistance.

WHERE IT'S GROWN AND WHAT ITS WINE TASTES LIKE

Castelão is the most widely planted variety in **Portugal**: 18,550 ha (45,838 acres) in 2010, down from around 20,000 ha (49,420 acres) in 2004. Its popularity is due in large measure to its adaptability to a wide range of environments, from the arid south to the damper, Atlantic-influenced west, and to its general hardiness. Particularly dominant in the south and centre of the country – Tejo, Lisboa, Setúbal and Alentejo – it is also grown further north in the Douro. This versatility extends to winemaking since it is used not only for reds, both fortified and not, but also for rosé and as a base for fizz. Wines vary in style, even within the same region. In the Península de Setúbal, south of Lisboa, for example, in the sandy soils of Poceirão and Fernando Pó, in good years Castelão produces respectable yields of around 50 hl/ha (*c*.3 tonnes/acre) and wines that are deeply coloured, aromatic, full-bodied, fleshy, ageworthy, well suited to oak maturation, and tasting of red fruits and fruits of the forest, whereas vines planted in the calcareous-clay soils of the Arrábida mountains have lower yields and produce much lighter, paler wines with overly high acidity and little capacity to age in bottle. Recommended producers include Casa Ermelinda Freitas and Bacalhôa Vinhos de Portugal in the Península de Setúbal plus Adega Cooperativa de Pegões and Casa Agrícola Horácio Simões in Palmela. Jose Maria da Fonseca's brand Periquita, made in Terras do Sado, combines a majority of Castelão with TOURIGA NACIONAL and TOURIGA FRANCA to impressive and ageworthy effect.

CASTETS

Very minor western Pyreneen variety clinging on in France; a parent in Slovakia.

BERRY COLOUR: ● ● ● ● ●

PRINCIPAL SYNONYMS: Engrunat (Saint-Loubès), Machouquet (Saint-Loubès), Nicouleau (Saint-Macaire)

VARIETIES COMMONLY MISTAKEN FOR CASTETS: BÉQUIGNOL NOIR (Lot-et-Garonne), FER (Tarn)

ORIGINS AND PARENTAGE

According to Cazeaux-Cazalet (1901), this variety was found in a wood in the region of Saint-Macaire in the Gironde, south-west France, by a certain M Nicouleau around 1870. Later on, it was propagated by a M Castets from Saint-André-du-Bois (Gironde), most likely between 1870 and 1874 because the earliest mention of the name Castets for this variety appeared in 1874 in Saint-Pierre d'Aurillac (Gironde). However, Castets-en-Dorthe is also the name of a village very close to Saint-Macaire, so it is quite possible that the variety was named after the village where it was found.

From 1882, Castets was recognized for its resistance to downy mildew and was widely propagated by Dr Guilbert across all of Aquitaine and to the centre and south of France. According to Lavignac (2001), Castets belongs to the Carmenet ampelographic group (see p XXVII), which also includes FER, and explains why they have often been confused, and locates its birthplace in the Pyrénées-Atlantiques rather than the Gironde, where it has virtually disappeared.

Castets was crossed with ABOURIOU and other varieties in the 1970s to breed several new varieties in Slovakia, of which HRON, NITRANKA, RIMAVA, RUDAVA, TORYSA and VÁH are currently commercially cultivated.

OTHER HYPOTHESES

It has been suggested that Nicouleau did not find this variety in Saint-Macaire but introduced it from the Pyrénées.

VITICULTURAL CHARACTERISTICS

Vigorous but not particularly fertile and needs to be pruned long. Late budding, mid ripening. Susceptible to powdery mildew. Small berries

WHERE IT'S GROWN AND WHAT ITS WINE TASTES LIKE

Wines are deeply coloured, relatively high in alcohol and able to age in bottle, though the acidity may be only moderate. The variety is recommended in the Aude and Cantal *départements* of **France** and specifically authorized for inclusion as a minor optional ingredient in the red and rosé wines of Vins d'Estaing north of Rodez, blended with varieties such as FER, GAMAY NOIR, etc., and in the Palette appellation around Aix-en-Provence (with CINSAUT, Grenache/GARNACHA, Mourvèdre/MONASTRELL, etc.) but in 2008 there was less than a hectare remaining in France. Varietal wines appear not to exist but Château Simone include Castets in their red Palette.

Castets has recently been identified in an old vineyard in the Russian River Valley, **California**.

CASTIGLIONE

Dark-skinned Italian variety almost exclusive to Calabria and usually blended.

BERRY COLOUR:

PRINCIPAL SYNONYMS: Castigliono, Mantonico Nero, Marchesana, Zagarese, Zagarolese

VARIETIES COMMONLY MISTAKEN FOR CASTIGLIONE: MAGLIOCCO CANINO, MAGLIOCCO DOLCE

ORIGINS AND PARENTAGE

Said to be a recent introduction to Calabria (Calò *et al.* 2006). A DNA study showed that the Castiglione of Calabria also exists in Campania, near Lago d'Averno, where it is simply called Calabrese (Costantini *et al.* 2005), a name confusingly used as a synonym for several different varieties. It is sometimes locally confused with MAGLIOCCO CANINO (Schneider, Raimondi *et al.* 2009).

VITICULTURAL CHARACTERISTICS

Late ripening.

WHERE IT'S GROWN

Cultivated mainly in the provinces of Reggio Calabria and Cosenza in Calabria, southern **Italy**. Rarely made as varietal and usually blended with other varieties, such as PRUNESTA, Castiglione is authorized in the red and rosé versions of Bivongi DOC, in which it is blended with varieties such as GAGLIOPPO, the obscure GRECO NERO, NOCERA, NERO D'AVOLA and Prunesta. The Italian census of 2000 records a total of 93 ha (230 acres).

CATALANESCA

Very old, high-acid white Italian variety suited to both table and glass and exclusive to Campania.

BERRY COLOUR: ● ● ● ● ●

PRINCIPAL SYNONYMS: Catalana, Uva Catalana

ORIGINS AND PARENTAGE

Although Catalanesca is listed as a table grape in the Italian agricultural census, its ability to make wine in the Vesuvio area of Campania in southern Italy has been mentioned by ampelographers since the sixteenth century. The name of this variety suggests a Spanish origin, and local tradition says it was introduced in 1450 on the slopes of Monte Somma, between Somma Vesuviana and Terzigno, by Alfonso, king of Aragón. However, the DNA profile of Catalanesca does not match any known Spanish variety but suggests that it could be genetically close to CASAVECCHIA, Summariello (NERO DI TROIA) and BARBERA DEL SANNIO (Costantini *et al.* 2005). That being said, all those varieties appear to be genetically completely separate from other Campanian cultivars, suggesting that they might represent recent introductions.

VITICULTURAL CHARACTERISTICS

Vigorous growth, usually trained on pergola (as you might expect for table grapes). Late ripening, harvest between October and November, sometimes even later. Good resistance to fungal diseases thanks to its thick skins.

WHERE IT'S GROWN AND WHAT ITS WINE TASTES LIKE

Catalanesca is recommended in Campania and authorized on Sardegna, **Italy**. Since October 2006, Catalanesca has been officially included among the wine grapes of Campania but is not yet allowed for IGT wines.

The best wines usually come from the north-western slopes of Vesuvio (in communes such as Somma Vesuviana, Sant'Anastasia, Ottaviano, San Sebastiano, Massa di Somma and Pollena Trocchia). Dry-tasting Catalanesca wines usually have some residual sugar to balance the marked acidity, and offer aromas of acacia, apricot and honey. Varietal wines are produced by Casa Barone and Sorrentino, for example. The 2000 census notes total plantings of 99 ha (245 acres).

CATANESE NERO

Minor Sicilian red variety.

BERRY COLOUR: ● ● ● ● ●

PRINCIPAL SYNONYMS: Vesparola

ORIGINS AND PARENTAGE

The name suggests that this variety comes from the province of Catania, around Etna in north-east Sicilia.

VITICULTURAL CHARACTERISTICS

Not very disease-resistant, which may explain its virtual abandonment. Mid to late ripening.

WHERE IT'S GROWN AND WHAT ITS WINE TASTES LIKE

Although it may have originated from the north-eastern part of Sicilia, **Italy**, there is no Catanese Nero in this area today and the variety is mainly cultivated in the north-western part of Sicilia, in the provinces of Palermo, Trapani and Agrigento. It is always blended with other varieties, mainly for rosé, and is not authorized for any DOC wines. There were 80 ha (198 acres) in 2000.

CATARRATTO BIANCO

Widely planted and variously named Sicilian white with potential for quality in the right hands.

BERRY COLOUR: ● ● ● ● ●

PRINCIPAL SYNONYMS: Catarratteddu, Catarratto Bertolaro, Catarratto Bianco Comune⸸, Catarratto Bianco Lucido⸸, Catarratto Bianco Lucido Serrato, Catarratto Corteddaro, Catarratto Latino, Catarrattu Lu Nostrum

VARIETIES COMMONLY MISTAKEN FOR CATARRATTO BIANCO: CARRICANTE⸸, FRANCAVIDDA⸸ (Puglia)

ORIGINS AND PARENTAGE

As early as 1696, Catarratto Bianco was described as a Sicilian wine grape in Cupani's *Hortus catholicus*. A distinction is usually made between two types: Catarratto Bianco Comune and Catarratto Bianco Lucido, the former thought to derive from the latter. These two types are often considered to be distinct varieties (as in the vineyard statistics below) but a recent ampelographic study provided evidence that they are both simply clones of the same variety (Di Vecchi Staraz, This *et al.* 2007), which is why it is best to use the name Catarratto Bianco for this variety. In addition, outside the typical CARRICANTE cultivation area, Carricante is the name wine growers in Etna often give their Catarratto Bianco.

Like most ancient varieties, several different types of Catarratto Bianco have been described, often with local names such as Bagascedda, Catarratto Ammantiddatu, Fimminedda, Mattu, etc. Because of this biodiversity, some say that Catarratto Bianco has a polyclonal origin, ie originates from several grape seeds. However, since every grape seed gives birth to a distinct variety, it seems most likely that Catarratto Bianco vineyards are interplanted with other, unidentified varieties. DNA studies could easily verify whether they are clonal variations or distinct varieties (most probably progenies of self-fertilized Catarratto Bianco).

Recent parentage analysis (Di Vecchi Staraz, This *et al.* 2007; Crespan, Calò *et al.* 2008) unexpectedly revealed that GRILLO, another important Sicilian white grape, is most likely the result of a natural cross between Catarratto Bianco and MUSCAT OF ALEXANDRIA, and that Catarratto Bianco most likely has a parent–offspring relationship with GARGANEGA, one of the most ancient and widespread white grapes in Italy. Since Garganega also has parent–offspring relationships with eight other varieties (ALBANA, DORONA DI VENEZIA, MALVASIA BIANCA DI CANDIA, MARZEMINA BIANCA, MONTONICO BIANCO, MOSTOSA, SUSUMANIELLO and TREBBIANO TOSCANO), Catarratto Bianco is either their half-sibling or their grandparent (see Garganega for an illustrated pedigree).

VITICULTURAL CHARACTERISTICS

Vigorous, mid to late ripening. Catarratto Bianco Comune has good resistance to cryptogamic diseases but Catarratto Bianco Lucido's compact bunches make it susceptible to powdery mildew and botrytis bunch rot.

WHERE IT'S GROWN AND WHAT ITS WINE TASTES LIKE

Catarratto Bianco is cultivated all over Sicilia, **Italy**, but mainly in the provinces of Trapani, Palermo and Agrigento. It is the island's most widely planted variety and the second most planted variety in all of Italy. The Italian agricultural census distinguishes between Comune and Lucido (see above), recording total plantings in 2000 of 43,247 ha (106,866 acres) for the former and 7,548 ha (18,652 acres) for the latter, even though Lucido seems to have the potential to make higher-quality wine. In Sicilia in 2008 there were 38,079 ha (94,095 acres) of Comune and 3,543 ha (8,755 acres) of Lucido.

Much of the wine made from Catarratto Bianco is very ordinary and surplus to requirements: in the past it went into the production of sweet, fortified marsala but today it is more likely to be compulsorily distilled or turned into grape concentrate to boost the sugar levels of grape juice in cooler parts of Europe.

But Catarratto Bianco can make more interesting wine. It is authorized in quite a few DOCs, as a varietal or as part of a blend in Alcamo, Contea di Sclafani, Erice, Monreale, Salaparuta and Santa Margherita di Belice and in yet more as part of a blend. And some of the best examples are IGT wines made by those who have shown the variety greater respect than it is locally accustomed to thanks to its profusion. At its best, it has bright aromas of citrus and herbs with both body and freshness and a minerally aftertaste. It sometimes has a nutty flavour and shows some resemblance to VIOGNIER when fully ripe. Recommended producers include Calatrasi, who make a barrel-aged version, Feudo Montoni and Porta del Vento.

California had 197 acres (80 ha) in 2008, down from 388 acres (157 ha) in 2007.

CATAWBA

Historically important American or part-American variety that was very popular in the nineteenth century but ripens inconveniently late and is now in decline.

BERRY COLOUR: ● ● ● ● ●

PRINCIPAL SYNONYMS: Arkansas, Catawba Rosa, Cherokee, Francher, Lincoln, Mammoth Catawba, Meads Seedling, Mecleron, Merceron, Michigan, Munipale Red, Omega, Rose of Tennessee, Saratoga, Singleton, Tekomah, Tokay

ORIGINS AND PARENTAGE

The much-debated history of Catawba was cleared up by Mosher (1853), Bailey (1906) and Sondley (1918). In 1802, the recently settled Samuel Murray (1739–1817), who was the owner of an inn fronting the Howard Gap road on Cane Creek, about half a mile away from what later became Fletcher near Ashville, Henderson County, North Carolina, had found several grapevines growing wild in the woods near the summit of Black Ridge, just above his inn. One of those vines developed particularly well after the trees were cut down and Murray planted it next to his farm. In 1807, General Davy, a Senator, transplanted some of these vines to his residence on the Catawba River in Rocky Mount in North Carolina, not far from Murray's home. Some time between 1807 and 1816, Davy propagated the variety in Washington DC and Maryland and christened it Catawba after the river of that name. The first ever wine was made in 1822 in Georgetown, Washington DC, by Major John Adlum. After a few years, Adlum sent cuttings to Nicholas Longworth of Cincinnati, who developed and popularized Catawba's cultivation in the west, using it to make dry, sweet or sparkling wines.

While the historical origin of Catawba is more or less clear, its genetic origin is not. Three hypotheses have been proposed (Galet 2000):

- Catawba is a pure, wild *Vitis labrusca* native variety
- Catawba is a *Vitis labrusca* × *Vitis aestivalis* natural hybrid
- Catawba is a natural seedling produced by a cross between a native *Vitis labrusca* vine and an undetermined *Vitis vinifera* subsp. *vinifera* variety that had previously been imported to the United States. In this case, it is assumed that *vinifera* pollen from neighbouring vineyards – carried by the wind or insects – pollinated a wild *labrusca* somewhere in the area, and that the berries of this vine were eaten by birds who deposited the seeds on the Black Ridge hill.

The results of ampelographic studies on the question of Catawba's pedigree are contradictory, and these hypotheses have unfortunately not yet been tested by genetic analysis.

VITICULTURAL CHARACTERISTICS

Vigorous, hardy and productive but very late ripening, thus best in sites with long growing seasons. Medium-sized bunches of large berries. Susceptible to black rot, downy mildew and, to a lesser extent, to powdery mildew.

WHERE IT'S GROWN AND WHAT ITS WINE TASTES LIKE

Catawba was one of the first significant varieties to be planted in North America, even before CONCORD, and was very popular in the nineteenth century – in 1865, for example, 20,000 bottles of sparkling Catawba were made at the Pleasant Valley wine company, the oldest winery in the Finger Lakes region in New York State. Although it is now in decline, due in large part to its unsuitability for high-quality dry wines, with limited plantings in states such as **Pennsylvania**, **Missouri**, **Indiana**, **Iowa**, **Michigan** and **Illinois**, as well as in Ontario, **Canada**, it is still widely planted in **New York** State, for juice as well as for wine, with a total of 1,291 acres (522 ha) in 2006. Wine styles range from dry to very sweet (including icewine), still to sparkling, and pink to white, typically with some foxy *labrusca* flavours, a certain spiciness and high acidity. Lakewood in New York State produce a good example.

CAVRARA

Virtually extinct red of the Veneto.

BERRY COLOUR: ● ● ● ● ●

PRINCIPAL SYNONYMS: Bassanese dal Peduncolo Rosso, Caprara, Cavarada, Cavarara

ORIGINS AND PARENTAGE

Cavrara was mentioned in the Veneto as early as 1754 (Giavedoni and Gily 2005) but was more or less replaced by other varieties during the nineteenth and twentieth centuries.

VITICULTURAL CHARACTERISTICS

Vigorous, high yields and very disease-resistant. Leaves turn red in the autumn. Mid to late ripening.

WHERE IT'S GROWN

Used to be widespread in the province of Vicenza, mainly in Colli Berici in north-eastern **Italy**. Cavrara was considered to be extinct by the mid twentieth century but since then it has been rediscovered in several old vineyards in the Veneto. Always blended.

ÇAVUŞ

Eastern table grape showing potential for citrus-flavoured wines in Turkey and Greece.

BERRY COLOUR: ● ● ● ● ●

PRINCIPAL SYNONYMS: Čauš Beli (Serbia), Čauš Bijeli (Dalmacija/Dalmatia in Croatia), Ceaus Alb (Romania), Chaouch (France), Damascenka (Czech Republic and Slovakia), Feher Tökszölö (Hungary), Panse de Constantinople (France), Parc de Versailles (Paris in France), Tchaouch (Russia), Tsaousi☒ or Tsaoussi (Greece, apart from Kefaloniá, Makedonía and Thráki/Thrace), Turceasca (Odobeşti in Romania)

VARIETIES COMMONLY MISTAKEN FOR ÇAVUŞ: TSAOUSSI

ORIGINS AND PARENTAGE

Çavuş is a widespread variety known under many names in the Mediterranean and the Near East, where it is used mostly for table grapes and only sometimes to make wine. According to Galet (2000), the origins of this variety are controversial, but the most plausible hypothesis is that a Turkish sergeant (*çavuş* in Turkish) brought it back from Taïf near Mecca, in the seventeenth century, to present it to the Sultan, who had the vines multiplied and propagated. This hypothesis is backed up by the thirty-seven vines of Çavuş (under the spelling Tchaouch) that were sent by Sultan Ahmet III in 1720 to the first Ottoman embassy in France, vines that were transplanted to Versailles on the order of Louis XV, hence the French synonym of Parc de Versailles.

The Eastern origin of Çavuş is supported by DNA profiling, which has shown that the variety Mor Üzüm, previously considered to be indigenous to Uçhisar and Nevşehir in Central Anatolia in Turkey, is in fact identical to Çavuş and is closely related to other varieties from Turkey or neighbouring Armenia and Georgia (Vouillamoz *et al.* 2006). Indeed, although Lefort and Roubelakis-Angelakis (2001) have shown close genetic relationships between Çavuş (under the name Tsaousi) and some Greek varieties, they consider this variety to be of Eastern origin.

A pink-skinned mutation has also been observed.

VITICULTURAL CHARACTERISTICS

Productive, although it has female-only flowers and needs a pollinator. Mid ripening, susceptible to botrytis bunch rot. Well suited to calcareous-clay soils and dry conditions. At risk from winter frost. Compact bunches of large, thin-skinned berries.

WHERE IT'S GROWN AND WHAT ITS WINE TASTES LIKE

Çavuş is grown mainly as a table grape in **Turkey**. However, Resit Soley at Corvus on the windy and sandy Turkish island of Bozcaada, off the west coast, was one of the first to vinify it. His wine, aged on the lees, is rich in melon, citrus and green-apple fruit flavour. It is not generally considered to be a variety that benefits from bottle ageing but Soley suggests that the best examples maintain their structure as they age. There were just under 30 ha (74 acres) planted for wine grapes in Turkey in 2010.

This variety is also grown for the table in **Algeria**, **Bulgaria**, **Russia**, **Ukraine**, **Serbia**, **Romania** and **Greece**. The variety known as TSAOUSSI on the Greek island of Kefaloniá, and to a lesser extent in Makedonía and Thráki (Thrace), is a distinct variety but there is frequent confusion because of the use of the synonym Tsaoussi to refer to the table grape Çavuş.

CAYETANA BLANCA

Lesser-quality Iberian variety known in Portugal as Mourisco Branco but more common in Spain under various synonyms.

BERRY COLOUR:

PRINCIPAL SYNONYMS: Amor Blanco (Islas Canarias), Aujubi, Baladi, Baladi-Verdejo (Cordoba in Spain), Balay (Cordoba), Belledy, Blanca Cayetana (Badajoz in Spain), Blanco Jaén, Cagazal (La Rioja), Cayetana (Badajoz and Cáceres in Extremadura), Cazagal (La Rioja), Charello, Charelo, Chaselo, Cheres, Cirial (Jaén in Spain), Dedo, Dedro, Djiniani (Maghreb in Morocco), Doradillo (Australia), Farta Gosos (Sagunto in the province of Valencia), Garillo (Sevilla), Garrida (Sevilla), Garrido (Sevilla), Garriga, Garrilla, Hoja Vuelta, Jaén Blanco, Jaén Doradillo, Jaén Empinadillo, Jaén Prieto Blanco, Jaenes, Jainas (La Rioja), Jarime, Jean de Castilla, Jean de Letur, Jean de Letur de Maratella, Jean Doradillo, Jean Dore, Jean Prieto, Machuenco, Maizancho (Ciudad Real and Valdepeñas in Spain), Malvasia or Malvoisie Espagnole (France), Mariouti, Morisco, Mourisco Arsello (Portugal), Mourisco Branco (Portugal), Mourisco Portalegre (Portugal), Naves, Naves Cazagal, Neruca, Padero, Parda, Pardina (Badajoz), Pirulet, Plateadillo, Robal (Zaragoza and Calatayud in Spain), Sarigo (Trás-os-Montes and Douro in Portugal), Tierra de Barros

VARIETIES COMMONLY MISTAKEN FOR CAYETANA BLANCA: ALBILLO MAYOR (Ribera del Duero), AVESSO (Vinho Verde), CALAGRAÑO or Calegraño (La Rioja), DORADILLA, Jaén Colorado (Léon), XARELLO (Catalunya)

ORIGINS AND PARENTAGE

Cayetana Blanca is an old variety from the Iberian peninsula that is widespread in Spain, often under the names Pardina and Jaén Blanco, and also found in Portugal as Mourisco Branco. The name Jaén, which comes from the city of that name in north-eastern Andalucía, is a source of confusion. For example, in 1513 in *Agricultura general* Spanish agronomist Alonso de Herrera (1790) described both Cayetana Blanca in Extremadura and Jaén Blanco in Castilla along with a dozen other varieties; and Spanish ampelographer Simón de Rojas Clemente y Rubio (1807) said: 'El primero que dé á conocer con exâctitud todos los vidueños que llaman Jaenes en España hará á su patria un servicio real' ('The first person to know precisely all the varieties called Jaén in Spain will do his country a real service').

A first DNA parentage analysis had suggested that Cayetana Blanca (under the name Cayetana *tout court*) was an offspring of ANTÃO VAZ and RABO DE OVELHA, both originating from the Alentejo in southern Portugal (Lopes *et al.* 2006), which would have suggested a possible Alentejo birthplace for Cayetana Blanca, even though it is now quite rare in that region under the name Mourisco Branco. However, a more recent DNA parentage analysis based on a new kind of marker (Single Nucleotide Polymorphism or SNP) has rejected this parentage and has revealed, in addition to other previously discovered genetic relationships (Lopes *et al.* 2006; Santana *et al.* 2010), a complex network of relationships for Cayetana Blanca throughout the Iberian peninsula (Zinelabidine *et al.* 2012; see pedigree diagram overleaf):

– Five varieties that are currently cultivated in the Iberian peninsula (mainly in Portugal) are the results of natural crosses that occurred at different times and places between Cayetana Blanca and the Portuguese variety ALFROCHEIRO: CORNIFESTO, MALVASIA PRETA, CAMARATE, Mouratón (JUAN GARCÍA) and Periquita (CASTELÃO), making these varieties siblings.
– Jaén Tinto, an old and almost extinct variety from southern Spain that should not be confused with Mencía (also called Jaén Tinto in Bierzo in Castilla y León), is a natural cross between Cayetana Blanca (father) and LISTÁN PRIETO (mother) from Castilla-La Mancha. This parentage might explain why Cayetana Blanca is also called Jaén Blanco all over Spain.

- Cayetana Blanca has parent–offspring relationships with at least ten additional grape varieties from the Iberian peninsula, many of them no longer cultivated: in Spain, Castillo de Arcos, Plateado (DORADILLA), Garrido Macho, Rocia and Puerto Alto from Andalucía, as well as Castellana Blanca from Castilla y León; in Portugal, Gouveio Estimado from the Douro, Cigüente (SÍRIA) from Dão as well as ANTÃO VAZ and RABO DE OVELHA from the Alentejo. These varieties are therefore all half-siblings, although, in theory, one of them could be a parent of Cayetana Blanca and the grandparent of the other nine.

This complex pedigree illustrates the central genetic role played by Cayetana Blanca in both Portugal and Spain and suggests that it originated many centuries ago from unknown (and probably extinct) parents somewhere in the south-west Iberian peninsula, where it has had so many progenies (Zinelabidine *et al.* 2012).

Under the name Jaén Blanco, this variety has a pink-berried mutation called **Jaén Rosado** with the same DNA profile as Cayetana Blanca, whereas the white-berried Jaén Colorado and the black-berried Jaén Tinto are distinct varieties (Ibañez *et al.* 2003; Martín *et al.* 2003). Jaén Blanco was generally thought to be a synonym for CALAGRAÑO from La Rioja until DNA profiling proved them to be distinct (Carreño *et al.* 2004).

Under the name Pardina, this variety should not be confused with ALBILLO MAYOR, which is also known as Pardina in Ribera del Duero in north-west Spain.

Mourisco Branco is not related to the Minho variety AVESSO, even though the Vitis International Variety Catalogue lists Avesso do Minho as a synonym for Cayetana Blanca and Galet gives it as a synonym for Jaén Blanco.

In Australia, Cayetana Blanca was introduced from Spain in the nineteenth century under the name Doradillo (Galet 2000) but it should not be confused with the Andalusian variety DORADILLA, which is quite distinct.

VITICULTURAL CHARACTERISTICS

Very vigorous, fertile and productive. Large compact bunches of medium-sized, thin-skinned berries. Late budding and very late ripening. Resistant to drought and best suited to poor, sandy soils. Highly susceptible to powdery mildew and botrytis bunch rot and also prone to mites.

WHERE IT'S GROWN AND WHAT ITS WINE TASTES LIKE

Cayetana Blanca is now mainly cultivated in **Spain** under the names Jaén Blanco, Pardina and Blanca Cayetana. Official Spanish statistics have not yet caught up with DNA discoveries and still list Pardina, Jaén Blanco and Blanca Cayetana separately: there were 4,973 ha (12,289 acres) of Jaén Blanco in 2008, mostly in Castilla-La Mancha, Andalucía and around Madrid; Pardina notched up a massive 31,440 ha (77,690 acres) in 2008, almost entirely in Extremadura in south-west Spain, especially in the province of Badajoz, where it is authorized in the Ribera del Guadiana DO but used more widely for Extremadura regional wines or for distillation into brandy de Jerez; the same is true for the 11,850 ha (29,282 acres) of Blanca Cayetana in Extremadura, also predominantly in Badajoz. Wines are typically neutral and moderate in alcohol with medium to high acidity and they are prone to oxidation.

Varietal wines labelled Pardina are produced by the Nuestra Señora de la Soledad and Montevirgen co-operatives in Badajoz and by Marcelino Díaz and San Antonio (Ribera del Guadiana). Varietal wines labelled Cayetana are produced by, for example, the Bodegas San Marcos (though the label states 50% Pardina and 50% Cayetana), Vinificaciones Extremeñas and Viticultores de Barros.

Mourisco Branco vines are now relatively few in the Alentejo, **Portugal** – just 141 ha (348 acres) in 2010.

There were still 74 ha (183 acres) of Doradillo in **Australia** in 2008, mostly in the hot, irrigated Riverland, where it produces fruit better suited to distillation or to the production of fortified wines.

CAYETANA BLANCA PEDIGREE DIAGRAM

This very old variety has played a central role in the Iberian peninsula by giving birth to no fewer than sixteen natural offspring, arranged here according to their geographical origins, from left (Portugal) to right (Spain) and from top (northern Iberian peninsula) to bottom (southern Iberian peninsula). The presence of unknown (and probably extinct) varieties (?) means that the inverse relationships are also theoretically possible (see p x).

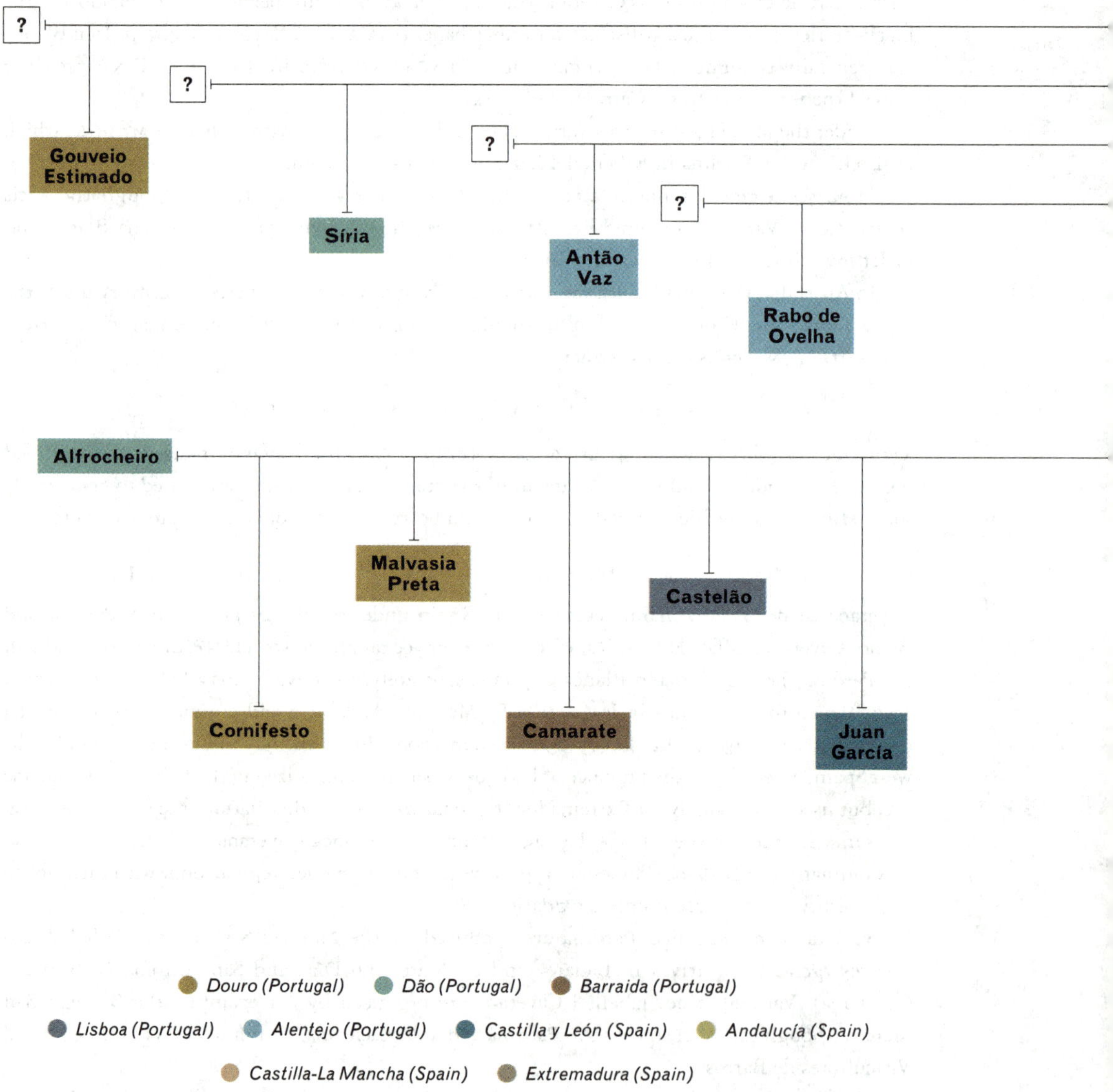

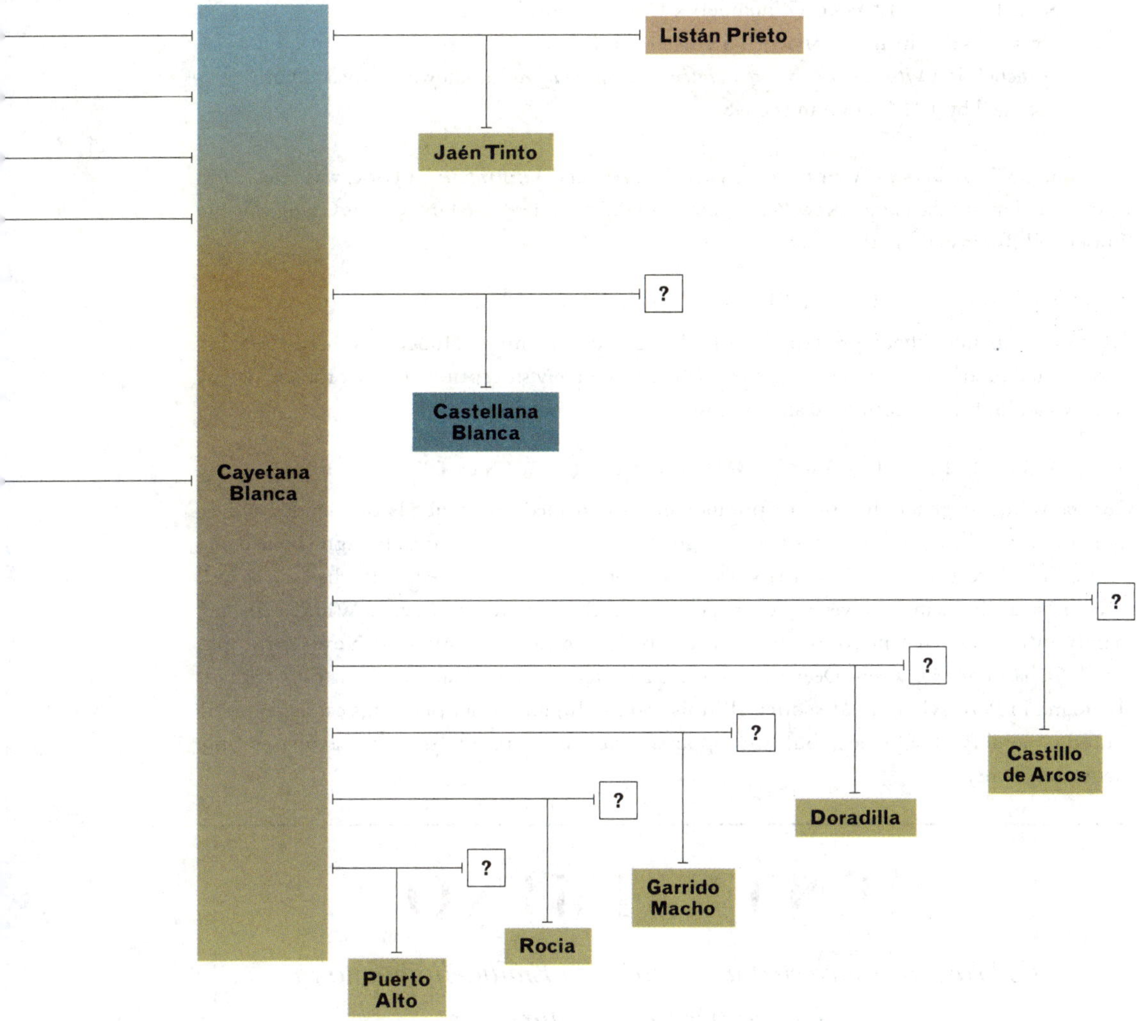

Listán Prieto
Jaén Tinto
?
Castellana Blanca
Cayetana Blanca
?
?
?
?
?
Castillo de Arcos
Doradilla
Garrido Macho
Rocia
Puerto Alto

CAYUGA WHITE

Versatile New York hybrid.

BERRY COLOUR: ● ● ● ● ●

PRINCIPAL SYNONYMS: Geneva White 3, GW 3, New York 33403, NY 33403

ORIGINS AND PARENTAGE

A SEYVAL BLANC × Schuyler hybrid obtained by John Einset and Willard B Robinson in 1945 at NYSAES, the New York State Agricultural Experiment Station in Geneva, where:

- Schuyler is a TRIBIDRAG (Zinfandel) × Ontario hybrid
- Ontario is a Winchell × MOORE'S DIAMOND hybrid
- Winchell is a *Vitis labrusca* × *Vitis vinifera* subsp. *vinifera* (unknown cultivar) hybrid obtained by J M Clough in the US.

This complex *Vitis labrusca*, *Vitis vinifera*, *Vitis lincecumii* and *Vitis rupestris* hybrid was selected in 1952, tested under the names New York 33403 or GW 3, and released in 1972 as Cayuga White (Einset and Robinson 1972).

VITICULTURAL CHARACTERISTICS

Vigorous with long shoot growth. Late budding, mid ripening. Moderately hardy. Highly vigorous and productive with large, open bunches. Moderately susceptible to black rot and downy mildew and highly susceptible to anthracnose.

WHERE IT'S GROWN AND WHAT ITS WINE TASTES LIKE

Cayuga White is generally used to produce off-dry to medium varietals or blended wines, sometimes even icewine, but also a few very good dry whites. They tend to be light-bodied and a little like Muscat; the best have RIESLING-like citrus notes. Harvested early, the fruit is well suited to sparkling wines; harvesting overripe increases the foxy aromas. Cayuga White is planted mainly in the eastern United States, with the biggest area in the State of **New York** (404 acres/163 ha in 2006), where Deer Run produce a better example than most, but the variety is also found in **Pennsylvania**, **Missouri**, **Illinois** and **Michigan**. Minor plantings can be found in **Canada**, notably in Québec, but the opportunities are limited because Cayuga is only moderately hardy.

CENTESIMINO

Red variety resurrected in a garden in Emilia-Romagna in the mid twentieth century.

BERRY COLOUR: ● ● ● ● ●

PRINCIPAL SYNONYMS: Alicante del Faentino, Sa(u)vignon Rosso, Savignôn Rosso

ORIGINS AND PARENTAGE

Until the nineteenth century, this grape from the province of Ravenna in Emilia-Romagna, north-east Italy, was called Savignôn Rosso, although it is not a red mutation of SAUVIGNON BLANC. In the mid twentieth century, during the reconstruction of vineyards after phylloxera, Pietro Pianori propagated this obscure variety from old vines growing in his walled garden at Podere Terbato in Faenza. Those vines had fortunately escaped phylloxera, and all modern plantings of this variety derive from cuttings taken from Pianori's garden. Pianori was nicknamed 'Centesimino' (meaning 'little cent'), which gave the grape its modern name. Both names were given on the Italian official register of varieties when it was included in 2004.

OTHER HYPOTHESES

Tradition has it that this grape came to Emilia-Romagna from Spain, which explains why Centesimino is sometimes erroneously called Alicante del Faentino.

VITICULTURAL CHARACTERISTICS

Mid to late budding, mid ripening. Smallish, compact bunches of smallish berries. Susceptible to powdery mildew.

WHERE IT'S GROWN AND WHAT ITS WINE TASTES LIKE

This grape variety was officially registered under the nickname of its discoverer. Today it is cultivated by a consortium of eight wineries in Oriolo dei Fichi near Faenza in the province of Ravenna, eastern **Italy**, where wines are marketed under the misleading synonym of Savignôn (or Savignon) Rosso. It is also cultivated in Forlì and in the Brisighella Valley. Wines may be classified as Ravenna Rosso IGT and appear in both oaked and unoaked guises. Wines can have both floral and spicy notes. Some producers such as Poderi Morini and Il Pratello make a passito version. They tend to have good acidity.

CENTURIAN

California cross as unsuccessful as its sibling Carnelian.

BERRY COLOUR: ● ● ● ● ●

PRINCIPAL SYNONYMS: Centurion

ORIGINS AND PARENTAGE

A cross between (MAZUELO (under the name Carignan) × CABERNET SAUVIGNON) and GARNACHA – the same as CARNELIAN – obtained by Harold P Olmo in 1974 at the University of California at Davis. It is sometimes mistakenly spelt Centurion.

VITICULTURAL CHARACTERISTICS

Productive.

WHERE IT'S GROWN

As with its sibling CARNELIAN, this cross was intended to produce a variety that tolerated the climate of California's hot Central Valley but retained the characteristics of CABERNET SAUVIGNON. In both instances, the aim was not achieved and the lack of success of Carnelian did nothing to encourage growers to plant Centurian. The area planted to Centurian in **California** peaked at about 1,000 acres (405 ha) but continues to decline, down to around 300 acres (121 ha) by the early 2000s and down further to 87 acres (35 ha) in 2008, mostly in the San Joaquin Valley. No varietal versions encountered.

CERCEAL BRANCO

Fresh Douro variety not to be confused with Madeira's Sercial.

BERRY COLOUR: ● ● ● ● ●

PRINCIPAL SYNONYMS: Cercial (Bairrada), Cercial do Douro

VARIETIES COMMONLY MISTAKEN FOR CERCEAL BRANCO: JAMPAL, SERCIAL (Madeira; known as Esgana Cão on the Portuguese mainland)

ORIGINS AND PARENTAGE

Cerceal Branco grown in the Douro and Dão is identical to the variety known as Cercial in Bairrada but distinct from both SERCIAL on Madeira and Cercial in Pinhel, although the local spelling variants cause considerable confusion, as evidenced by various DNA studies:

- Cerceal Branco grown in Dão (Almadanim *et al.* 2007) is identical to Cercial in Bairrada (Martín *et al.* 2006) and to Cercial do Douro (Lopes *et al.* 2006); Cercial in Bairrada is probably a progeny of MALVASIA FINA and SERCIAL (Lopes *et al.* 2006)
- Cercial from Madeira is more commonly spelt SERCIAL and is identical to Esgana Cão (Lopes *et al.* 2006)
- Cercial from Pinhel in Beiras is identical to JAMPAL (Lopes *et al.* 2006).

VITICULTURAL CHARACTERISTICS

Compact, medium-sized bunches of thick-skinned, medium-sized berries. Late ripening. Highly susceptible to downy mildew, and also, but to a lesser extent, to powdery mildew and botrytis bunch rot.

WHERE IT'S GROWN AND WHAT ITS WINE TASTES LIKE

Cerceal Branco, grown mostly in the Douro but also found in Dão and Bairrada, produces wines that are light-bodied and fresh, generally adding acidity to white blends. It is also authorized further south in the Tejo and Alentejo DOCs. There were 113 ha (279 acres) in **Portugal** in 2010. Recommended producers in Bairrada include Campolargo, who include it in the blend of one of their sparkling wines, and Vadio.

CEREZA

Pink-skinned native of Argentina. The country's most planted vine, used for very basic rosés and whites.

BERRY COLOUR: ● ● ● ● ●

ORIGINS AND PARENTAGE

Cereza, meaning 'cherry', is local to Argentina, a natural cross between LISTÁN PRIETO (Criolla Chica) and MUSCAT OF ALEXANDRIA, and therefore a member of the group of varieties known as the Criollas (see CRIOLLA entry and Muscat pedigree diagram). According to Martínez, Cavagnaro *et al.* (2006), Cereza could also be related to CRIOLLA GRANDE, while Tapia *et al.* (2007) suggest a close relationship with Criolla Mediana, which is no longer commercially cultivated.

VITICULTURAL CHARACTERISTICS

Very vigorous and productive with a long vegetative cycle. Late ripening. Big, loose bunches of large, thin-skinned berries, irregular coloration (Galet 2000).

WHERE IT'S GROWN AND WHAT ITS WINE TASTES LIKE

Cereza is the most widely planted variety in **Argentina**, just ahead of Malbec (COT) in 2009, but destined mainly for very ordinary pale rosé and white wine for early consumption within Argentina. According to Galet (2000), Cereza is Argentina's best table grape although it does not keep very well. It's also used for grape concentrate. There were 30,054 ha (74,265 acres) in Argentina in 2009, mostly in eastern Mendoza (16,578 ha/40,965 acres) and in San Juan (11,641 ha/28,766 acres) and to a lesser extent in Catamarca.

CERRETO

White-berried variety from Campania, central **Italy**, grown exclusively in the Costa d'Amalfi south of Napoli and probably named after the city of Cerreto Sannita in the province of Benevento. Used occasionally in blends in DOCs such as Sannio and Solopaca, eg by Antica Masseria Venditti. It should not be confused with IGT Cerreto, which is for SANGIOVESE-based reds only.

CESANESE

Lazio's prime red wine grape, aromatic and difficult to ripen.

BERRY COLOUR: ● ● ● ● ●

PRINCIPAL SYNONYMS: Bonvino Nero, Cesanese Comune, Cesanese di Affile (province of Roma), ?Cesanese Nostrano (Ciociaria in the province of Frosinone), Nero Ferrigno, Sanguinella, Uva di Affile

ORIGINS AND PARENTAGE

The variety may have been noted in 1600 under the name Cesenese around Firenze in Toscana by Soderini in his treatise on grapevines and wines, and more certainly in Lazio by Acerbi (1825) and subsequent authors. Its name comes from Cesano, a village in the Castelli Romani, south of Roma.

VITICULTURAL CHARACTERISTICS

Late ripening, with consistent and high yields. Susceptible to powdery mildew. Cesanese Comune has larger berries than Cesanese di Affile.

WHERE IT'S GROWN AND WHAT ITS WINE TASTES LIKE

It is still grown almost exclusively in Lazio, **Italy**. Two main clones, morphologically slightly different, are generally distinguished: Cesanese Comune, the most widespread, and the potentially better quality Cesanese di Affile, a version with the aroma of lychees supposedly brought in 1820 by a Sicilian monk and planted in Affile (south of Roma), where it is still mainly cultivated now. The Italian agricultural census of 2000 recorded 477 ha (1,179 acres) of the former and 615 ha

(1,520 acres) of the latter. Before the growth in popularity of local white wines, Cesanese was far more widely planted. Cesanese is not easy to grow or vinify and high expectations are only occasionally realized, both historically and today. The best wines come from the areas around the towns of Anagni and Piglio in the province of Frosinone and Olevano Romano in the province of Roma.

Three DOCs allow 100% Cesanese wines in the Castelli Romani region: Cesanese del Piglio, Cesanese di Affile and Cesanese di Olevano Romano. Cesanese is also blended in other Lazio DOCs, eg Lago di Corsara and Rosso Orvietano. Traditionally sweetish frothy Cesanese was common but today they are more usually dry and fruity, with velvety tannins and a distinctive perfume reminiscent of mulberries and pimento. Varietal versions are produced by Terre Rubre, Marcella Giuliani, Coletti Conti and Cantine Ciolli, among others.

Andrea Franchetti, who experimented with the variety at Tenuta di Trinoro in southern Toscana but abandoned it there because it either failed to ripen or was too emphatically aromatic, grows some on Etna, Sicilia.

CÉSAR

Tannic variety playing a very minor and local role in northern Burgundy.

BERRY COLOUR: ● ● ● ● ●

PRINCIPAL SYNONYMS: Céear, Céelar (Vermenton and Vallée de la Cure in the Yonne), Célar, Picarniau (Cravant, Irancy and Tonnerre), Romain (Auxerre), Römer (Rheinland-Pfalz in Germany)

VARIETIES COMMONLY MISTAKEN FOR CÉSAR: LAMBRUSCA DI ALESSANDRIA ☒ (Italy)

ORIGINS AND PARENTAGE

César is local to the Yonne *département*, south and west of Paris in north-central France (Chablis territory), where it was first mentioned in 1783 under its synonym Romain, in Auxonne, south west of Dijon in the Côte d'Or, and in Irancy, south east of Auxerre, in 1783–4 as 'César ou Romain' (Rézeau 1997). Its cradle is said to be in nearby Coulanges-la-Vineuse, a village which may take its name from the Latin *colongiae vinosae* ('vinous colonies').

DNA parentage analysis has revealed that César is a natural PINOT × GÄNSFÜSSER (aka Argant) cross (Bowers *et al.* 2000), which speaks in favour of a north-eastern France origin for this variety. However, César is also known in Rheinland-Pfalz under the name Römer, and the French ampelographer Adrien Berget found that Argant/Gänsfüsser was morphologically distinct from any French variety and considered it to be of German origin. The birthplace of César is therefore likely to be somewhere between the Yonne and Rheinland-Pfalz.

OTHER HYPOTHESES

César is often said to be the oldest French variety because it was supposedly introduced to the Yonne by the Roman legions of Julius Caesar (*César* in French) but this is disproved by its parentage. In addition, it has sometimes been suggested that César is identical to the Italian variety LAMBRUSCA DI ALESSANDRIA, which has also been disproved by DNA analysis (Vouillamoz).

VITICULTURAL CHARACTERISTICS

Vigorous and mid ripening. Very sensitive to wind in the spring. Rather susceptible to downy and powdery mildews.

WHERE IT'S GROWN AND WHAT ITS WINE TASTES LIKE

There were just 10 ha (25 acres) remaining in **France** in 2008. It is officially recommended in northern Burgundy around the towns of Chablis and Auxerre for generic burgundy (eg Bourgogne-Chitry, Bourgogne-Épineuil) and for Irancy. It is not authorized for burgundy produced in *départements* other than the Yonne. It generally produces wines that have red-fruit flavours and sufficient tannins to allow ageing in bottle, and therefore can add power and longevity to PINOT NOIR in basic red burgundy. In practice this happens mainly, but still only occasionally, in Irancy. Simmonet Febvre used to make a varietal César but now just use the grape in their Irancy blend (maximum 10%).

Domaine Sorin-Coquard's Cuvée Antique is a rare example of a varietal César, as is Cave de la Tourelle's Bourgogne César at Saint-Bris le Vineux.

There are a few hectares of 'César' in Isla de Maipo in **Chile**, known locally as Romano, but it is uncertain if this is the same variety. It used to be more widely planted but was mostly grubbed up because of its excessive yields (although that was why it was introduced) and consequent poor ripening.

CETINKA

Extremely rare Croatian island white.

BERRY COLOUR: ● ● ● ● ●

PRINCIPAL SYNONYMS: Blajka (Hvar), Blatinka (Korčula), Blatka (Hvar), Blatska (Hvar), Cetinjka, Cetinjka Bijela, Cetinka Bijela, Poserača, Potomkinja

VARIETIES COMMONLY MISTAKEN FOR CETINKA: BLATINA

ORIGINS AND PARENTAGE

Cetinka most likely comes from the islands of Korčula and Hvar, off the Adriatic coast of Croatia, just south of Split.

VITICULTURAL CHARACTERISTICS

Late budding, mid to late ripening. Medium-sized to large, loose clusters so not particularly susceptible to botrytis bunch rot. Female-only flowers but nevertheless productive and with consistent yields. Relatively low sugar levels attained in the berries.

WHERE IT'S GROWN AND WHAT ITS WINE TASTES LIKE

Cetinka is officially authorized in the Srednja i Južna Dalmacija (Central and Southern Dalmatia) wine region of **Croatia** but it is found mainly on the islands of Korčula and Hvar, and to a lesser extent on the Peljesac peninsula and the islands of Mljet, Lastovo and Vis. Cetinka is often blended with Maraština (MALVASIA BIANCA LUNGA) or GRK. The Blato co-operative on Korčula produces a varietal wine. Wines tend to be light, fresh and slightly acidic.

CEVAT KARA

Southern Ukrainian variety used in fortified blends.

BERRY COLOUR: ● ● ● ● ●

PRINCIPAL SYNONYMS: Djevat Kara, Dshevat-Kara, Dzhevat Kara, Jewath, Polkovnik Kara

ORIGINS AND PARENTAGE

Cevat Kara, literally 'colonel's black', comes from Solnechnaya Dolina in eastern Krym (Crimea), in the far south of Ukraine.

VITICULTURAL CHARACTERISTICS

Productive, mid ripening. Medium-sized berries with thick but fragile skins.

WHERE IT'S GROWN AND WHAT ITS WINE TASTES LIKE

Used mostly in blended fortified wines such as Massandra's eponymous Black Colonel and their Black Doctor (see EKIM KARA) in **Ukraine**.

CHAMBOURCIN

French humidity-tolerant hybrid popular in the US and Australia but on the decline in its homeland.

BERRY COLOUR: ● ● ● ● ●

PRINCIPAL SYNONYMS: Joannes Seyve 26-205

ORIGINS AND PARENTAGE

Chambourcin is a complex hybrid obtained after 1945 by Joannes Seyve by crossing Seyve-Villard 12-417 with CHANCELLOR, where Seyve-Villard 12-417 is a Seibel 6468 × Subéreux cross (for the complete pedigrees, see BRIANNA for Seibel 6468, and PRIOR for Subéreux). It was named after the lieu-dit Chambourcin in the village of Bougé-Chambalud in Isère, south-eastern France, where Seyve owned an experimental vineyard. It was first commercialized in 1963 and was used to breed REGENT.

VITICULTURAL CHARACTERISTICS

Very vigorous and may be pruned short but sometimes prone to millerandage. Resistant to winter cold but susceptible to chlorosis and drought. Mid to late ripening. Good resistance to fungal diseases, especially downy mildew, and therefore tolerant of humid conditions.

WHERE IT'S GROWN AND WHAT ITS WINE TASTES LIKE

There were 800 ha (1,977 acres) in **France** in 2008 (down from 948 ha (2,343 acres) in 2006), predominantly in the Pays de Loire regions of central France, although it is authorized in many French *départements*. Although the overall trend is down, from a peak of 3,363 ha (8,310 acres) in the late 1970s, when it was particularly popular in the Muscadet region, it is still one of the most widely planted hybrids in France. Wines are full-flavoured and aromatic and do not have the intrusive flavours sometimes associated with hybrids.

The variety's cold hardiness endears it to growers in Canada and in America's Northeast and Midwest. There are around 120 acres (49 ha) in **Canada**, mostly in Ontario (85 acres/34 ha in 2006, down from 128 acres/52 ha in 2002) but also in Québec. There were 154 acres (62 ha) in **Missouri** in 2009, 146 acres (59 ha) in **Pennsylvania** in 2008, 31 acres (13 ha) in **New York** State in 2006 and 42 acres (17 ha) in **Michigan** in the same year. Minor plantings are found elsewhere in the US, including **North Carolina**, where Davesté Vineyards make a varietal Chambourcin, **Virginia** (96 acres/39 ha in 2008), **Illinois** (105 acres/42 ha in 2009), **Indiana** (65 acres/26 ha in 2009), **Ohio** (15 acres/6 ha) and **Nebraska** (5 acres/2 ha in 2007). It is also recommended in the southern half of **Iowa**. Butler Winery in Indiana produce a good example.

Chambourcin is apparently the most successful hybrid in **Australia**. According to the *Australian Wine Industry Directory 2010*, there are eighty-six producers in Australia, many in warm and humid regions such as the Hunter Valley in New South Wales because of its disease resistance but also up the east coast to Queensland. Cassegrain at Port Macquarie in the Hastings River region of NSW, credited with pioneering Chambourcin in Australia, produce two separate varietal wines. By contrast, there are just a few hectares in **New Zealand**. It also shows promise in **Vietnam**, thanks not only to its ability to cope with humidity but also to the influence of Australian consultants.

CHANCELLOR

Minor French-American hybrid now more widely planted in North America than in France.

BERRY COLOUR: ● ● ● ● ●

PRINCIPAL SYNONYMS: Seibel 7053

ORIGINS AND PARENTAGE

Complex hybrid obtained by Albert Seibel, most likely in the early twentieth century, in Aubenas in the Ardèche between Nîmes and Lyon, France, by crossing Seibel 5163 × Seibel 880 (see PRIOR for the complete pedigree of Chancellor). It was given the name Chancellor in New York State in 1970.

Chancellor was used to breed ALLEGRO, BREIDECKER, CHAMBOURCIN and GEISENHEIM 318-57.

VITICULTURAL CHARACTERISTICS

Early budding and therefore at risk from spring frosts. Early ripening. Susceptible to downy and powdery mildews.

WHERE IT'S GROWN AND WHAT ITS WINE TASTES LIKE

There were 46 acres (19 ha) in **New York** State and 35 acres (14 ha) in **Michigan** in 2006, and more limited plantings in US states such as **Indiana** and **Nebraska**. In **Canada**, there were 10 acres (4 ha) in British Columbia in 2008 and minor plantings in Québec and Ontario. It produces deeply coloured wines of better quality than many French-American hybrids but they are sometimes slightly bitter. Joseph's Estate in Niagara-on-the-Lake, Ontario, make a Chancellor/MERLOT blend.

CHARBONO

See DOUCE NOIRE.

CHARDONEL

New York hybrid that is more hardy and productive than its Chardonnay parent, adopted in some northern US states.

BERRY COLOUR: ● ● ● ● ●

PRINCIPAL SYNONYMS: GW 9, NY 45010

ORIGINS AND PARENTAGE

Chardonel is a complex *Vitis rupestris*, *Vitis vinifera* and *Vitis lincecumii* hybrid obtained by Bruce Reisch, Robert Pool and John Einset in 1953 at the Geneva research centre, New York State, by crossing SEYVAL BLANC with CHARDONNAY. The first fruits were observed in 1958 and the original vine was propagated in 1960 under the number NY 45010 (New York 45010). In later trials it was renamed GW 9 (Geneva White 9) for ease of identification (Reisch *et al.* 1990). The first wines were made in 1966 and the variety was released in 1990.

VITICULTURAL CHARACTERISTICS

Vigorous and productive, late budding and late ripening. More cold-hardy than CHARDONNAY (–10 to –15 °F/–23 to –26°C) and almost as hardy as SEYVAL BLANC. Bunches less compact than either of its parents and thus only moderately susceptible to downy and powdery mildews and botrytis bunch rot.

WHERE IT'S GROWN AND WHAT ITS WINE TASTES LIKE

Since its introduction in 1990, Chardonel has proved popular and is now planted in US states such as **Missouri** (190 acres/77 ha in 2009), **Arkansas**, **Michigan** (24 acres/10 ha in 2006), **Pennsylvania** and **Ohio** (6 acres/2.4 ha in 2008). In **New York** State, it is recommended only for warmer sites where the growing season is longer than average. Wines are generally light and pleasant, with the better examples being fuller-bodied and resembling CHARDONNAY while the lesser ones are more like SEYVAL BLANC. In some years or sites wines can be herbaceous but the high level of acidity also makes the variety suitable for the production of sparkling wine. The best examples benefit from barrel fermentation. Chardonel is produced in a variety of syles, ranging from dry and oaked through medium dry to sweet.

Superior producers of varietal wines include Chaumette in Ozark Mountain AVA, Missouri; St James Winery in Ozark Highlands AVA, Missouri; Tabor Hill in Lake Michigan Shore AVA, Michigan; and Butler Winery in **Indiana**.

CHARDONNAY

Hugely popular, versatile and widely planted international white variety equally capable of extreme mediocrity and regal splendour.

BERRY COLOUR: ● ● ● ● ●

PRINCIPAL SYNONYMS: Aubaine (Saône-et-Loire), Auvernat (Auvergne, Orléans), Auxerrois, Auxois (Moselle), Beaunois (Yonne), Chaudenay (Touraine), Clevner or Clävner, Gamay Blanc (Jura), Luisant (Franche-Comté), Melon à Queue Rouge (Jura, Franche-Comté), Melon d'Arbois (Jura), ?Obaideh (Lebanon), Pinot Blanc Chardonnay, Wais Edler or Waiser Clevner (Republic of Moldova)

VARIETIES COMMONLY MISTAKEN FOR CHARDONNAY: ALIGOTÉ, AUXERROIS, MELON, PINOT BLANC, SACY, SAVAGNIN BLANC

ORIGINS AND PARENTAGE

The historical cultivation area of Chardonnay, which comes from the Saône-et-Loire, between Lyon and Dijon in eastern central France, ranged from Burgundy to Champagne, mainly in the Côte d'Or, Saône-et-Loire and Marne. Under its synonym Beaunois, 'from Beaune', this variety was supposedly first mentioned in 1583 but the same name was also given to other varieties such as ALIGOTÉ.

The earliest reliable mention of Chardonnay appeared in 1685–90 in the village of Saint-Sorlin, today called La Roche-Vineuse in the Saône-et-Loire, where 'chardonnet' was said to make the best wine. Indeed, the variety takes its name from the village of Chardonnay near Uchizy in the Mâconnais, southern Burgundy. Various spellings have been used (Chardenet, Chardonnet, Chardenay, etc.) and the modern spelling, Chardonnay, did not become common until the twentieth century.

Chardonnay was confused with PINOT BLANC until the end of the nineteenth century due to their morphological similarities. It was often called Chardonnet Pinot Blanc or Pinot Blanc Chardonnet in France; and in Germany it was called Clevner or Clävner, or even Ruländer, the current name for PINOT GRIS in western Germany.

DNA parentage analysis has unexpectedly shown that Chardonnay is a natural PINOT × GOUAIS BLANC cross (Bowers *et al.* 1999), just like its siblings GAMAY NOIR, MELON, ALIGOTÉ, etc. (see PINOT pedigree diagram). This parentage explains the confusion with Aligoté and PINOT BLANC, and also the synonyms Gamay Blanc and Melon d'Arbois in the Jura. Chardonnay unsurprisingly belongs to the ampelographic group called Noirien (see p XXVII; Bisson 2009).

Chardonnay Rose is a colour mutation found in vineyards around the village of Chardonnay. **Chardonnay Musqué** is a mutation that has grapey Muscat-like aromas and was first observed in Pouilly and Arbois (Galet 1990). Certain clones of Chardonnay, notably 77 and 809, have similar grapey aromas.

Chardonnay has been used as a parent of several different varieties: in France, it was crossed with BAROQUE to breed LILIORILA, with a Seibel hybrid to produce RAVAT BLANC, and with PALOMINO FINO to breed CHASAN; in the US, it was bred with SEYVAL BLANC to produce CHARDONEL; in Switzerland, it was crossed with CHASSELAS to breed CHARMONT and DORAL; in Serbia, it was crossed with KÖVIDINKA to produce SILA; in Ukraine with Plavaï (no longer cultivated) to produce SUKHOLIMANSKY BELY.

OTHER HYPOTHESES

According to some sources, the variety called Obaideh in Lebanon might be an ancestral form, or even an ancestor, of Chardonnay. This can be refuted because Chardonnay's parents have never been cultivated in Lebanon. Obaideh is more likely to be a local clone of Chardonnay but this has not yet been tested by DNA analysis.

VITICULTURAL CHARACTERISTICS

Early budding (so prey to spring frost) and ripening, productive. Easy to grow. Best suited to soils dominated by limestone or calcareous clay that are not too dry. Susceptible to powdery mildew and grapevine yellows (eg flavescence dorée and bois noir). It may also suffer from coulure and millerandage. Relatively thin skins make it susceptible to botrytis bunch rot in rainy harvests. Burgundy has twenty-eight authorized clones (also known as Dijon clones): 75, 78, 121, 124, 125 and 277 are particularly productive, whereas 76, 95 and 96 tend to produce higher-quality wine. The so-called Mendoza clone is known for its propensity to millerandage ('hen and chicken' uneven fruit set), though some producers consider this to be a benefit since it reduces yields.

WHERE IT'S GROWN AND WHAT ITS WINE TASTES LIKE

Chardonnay, loved or reviled, is arguably the most versatile white wine grape. Without a dominant flavour of its own, it can take on a wide array of aromas depending on where it is grown and, particularly, how it is made. Although the grapes generally reach relatively high sugar levels, the wines can taste lean, even tart, almost like a SAUVIGNON BLANC in its refreshing effect, in a cool climate such as that of Chablis. But whatever the style, through various swings of fashion, Chardonnay has shown itself a particularly malleable partner in such winemaking operations as malolactic fermentation, lees stirring, barrel fermentation and all manner of oak-related practices. It runs the gamut from subtle and savoury to rich and spicy still, dry wines, as well as providing unparalleled base wine for top-quality fizz and, occasionally, in the Mâconnais and Burgen-land for example, submitting to botrytis to produce interesting sweet wines.

As the wine world expanded dramatically in the late twentieth century, and the fashion then was for white wine, plantings of Chardonnay increased enormously. Chardonnay became the dominant white wine grape in both North and South America and in Australia, and its popularity with consumers was such that Chardonnay was virtually synonymous with white wine. But for every action there is a reaction and, by the end of the first decade of the twenty-first century, Chardonnay was considered distinctly old hat by many consumers, having been replaced by Pinot Grigio (PINOT GRIS) in the mass market and with the reputation of Chardonnay in its most glamorous form as fine white burgundy somewhat tarnished by the high incidence of premature oxidation.

France, for example, saw total plantings rise from just 7,325 ha (18,100 acres) in 1958, mainly in Burgundy and Champagne, to 44,593 ha (110,191 acres) in 2009. Chardonnay is now France's second most planted white variety, although Ugni Blanc (TREBBIANO TOSCANO), planted chiefly for distillation into cognac and armagnac, is planted on twice as much land as Chardonnay. Today Chardonnay's French strongholds are Burgundy, Languedoc-Roussillon and Champagne, in that order in terms of area.

Chardonnay is the white grape of Burgundy, even if some ALIGOTÉ and a little bit of PINOT BLANC is also grown there. The total in greater Burgundy was 14,753 ha (36,455 acres; 4,048 ha/10,003 acres in 1958), of which only 2,120 ha (5,239 acres) were on the narrow strip of Côte d'Or vineyards, 6,929 ha (17,122 acres) in the Côte Chalonnaise and Mâconnais in the Saône-et-Loire, and 5,648 ha (13,957 acres) in Chablis country in the Yonne *département*.

Chardonnay shows its versatility in Chablis, where it produces its most mouth-wateringly appetizing, longest-lived wines, especially when grown on the local kimmeridgean clay-limestone soils. The best wines are steely and precise, with notably high acidity, and can develop interest in bottle for decades.

Just as top-quality red bordeaux inspired CABERNET SAUVIGNON's spread around the world, so top-quality white burgundy powered Chardonnay's extraordinary popularity – and Chardonnay, unlike Cabernet Sauvignon, is not difficult to ripen, so it can thrive in a much wider range of climates. There are few more nuanced, longer-lived, full-bodied dry white wines than a great wine from the tiny Montrachet vineyard and its grand cru neighbours, even if overpriced, unambitious examples are not difficult to find. Ditto for the grand cru Corton-Charlemagne. Excellent, tense, similarly savoury white burgundies can be found carrying less grand but classic appellations such as Puligny-Montrachet, Chassagne-Montrachet and Meursault; and the likes of

Saint-Aubin and Pernand-Vergelesses, which once struggled to ripen Chardonnay fully, are also making some very fine white wine in the warmer summers the Côte d'Or has been experiencing. The southern half of the Côte d'Or, the Côte de Beaune, is Chardonnay's most famous heartland and the vine is found only exceptionally in the Côte de Nuits to the north.

The Côte d'Or may hold the crown for Chardonnay quality (under constant siege by ambitious producers elsewhere) but far more is grown to the south in the Côte Chalonnaise, where white wines can be good value though rarely have the tension and density of the best from the Côte d'Or, and especially in the Mâconnais, where the potential is so obvious that it is systematically being invaded by top Côte d'Or producers. Here wines can have an extra dimension of ripeness and are certainly much more precocious than their northern counterparts. Flavours in these bumptious, friendly wines can vary from melons through to ripe apples, although an increasing proportion very obviously reflect the varied soils of such appellations as Pouilly-Fuissé, Viré-Clessé and Saint-Véran. And Chardonnay is also responsible for the generally rather lighter wine sold as Beaujolais Blanc.

Chardonnay is significant to the small Jura wine region, where the wines could be described as high-altitude white burgundies.

Champagne has even more Chardonnay planted than the Mâconnais and Côte Chalonnaise combined – 9,781 ha (24,169 acres), mostly in the Marne *département* (dramatically more than the 2,664 ha/6,583 acres growing in 1958) rather than in the Aube to the south. Chardonnay is virtually the only white wine variety grown, and an increase in demand for Blanc de Blancs (all pale-skinned grapes) champagne together with the general expansion of the Champagne *vignoble* explain the dramatic increase. All-Chardonnay traditional-method sparkling wine, whether champagne or not, tends to exhibit a certain lemony quality, developing into something more biscuity or brioche-like, especially if it is given extended bottle ageing on the lees.

But the real surprise is the extent to which Chardonnay is planted in Languedoc-Roussillon – to a total of 12,156 ha (30,038 acres), making Chardonnay the most widely planted white wine variety there. With the exception of the Chardonnay grown around Limoux for sparkling wine, the great majority of this Chardonnay finds its way into relatively inexpensive varietals with varying degrees of varietal expression. Yields are often high and the resulting wines more dry white than ineluctably Chardonnay.

There are also less significant plantings in cognac and armagnac country, where it produces everyday wines, the Ardèche, where it makes full-bodied southern answers to Mâcon Blanc, and the Val de Loire, where it is grown for the sparkling Crémant de Loire as well as for some lively, if fairly lightweight, still wines. It is allowed up to 20% in Anjou Blanc and largely replaced Tressallier (SACY) in Saint-Pourçain in the late twentieth century.

For over a century **Italy** grew a lot of Pinot Bianco (PINOT BLANC) and a bit of Chardonnay and wasn't too sure which was which but in the late twentieth century Italians fell prey to Chardonnay mania, and learnt the difference, so that by 2000 there were 11,772 ha (29,089 acres) of it. Ambitious producers all over the country, particularly in Piemonte and in central Italy, felt they had to produce small volumes of expensively oaked Chardonnay – and the early-ripening variety has been useful for those exposures in the Langhe hills too cool to ripen NEBBIOLO or indeed any red wine grape. Gaja's Gaia e Rey is arguably Italy's most famous Chardonnay, although Querciabella's Batàr grown in Chianti Classico country and Antinori's Umbrian Cervaro, into which some GRECHETTO DI ORVIETO is blended, are other very superior wines made in the image of white burgundy.

Italy's greatest concentrations of Chardonnay plantings are in Lombardia (much of it for sparkling wine such as Franciacorta), Trentino (also largely for fizz) and the Alto Adige, where much of the Chardonnay seems to lack the conviction of other white wine varietals. Exceptions include Tiefenbrunner's top bottling and Elena Walch's Cardellino. Other varieties also seem to suit Friuli better, although Jermann made waves with his de luxe 'Dreams' bottling of Chardonnay at one stage. Much of southern Italy is a bit hot for the variety but that has not stopped large tracts of Puglia being planted to produce varietal Chardonnay – an Italian answer to the generally rather anodyne Languedoc version. The variety has reached greater heights in Sicilia, notably chez

Planeta, whose world-class, if not exactly quintessentially Sicilian, example inspired others to follow suit to such an extent that 4% of the island's vineyards, 4,960 ha (12,256 acres), were planted with Chardonnay in 2008.

Much of **Spain**, where there were 5,423 ha (13,401 acres) in 2008, is too hot to imbue such an early-ripening variety as Chardonnay with much flavour and interest, although Torres' Milmanda from the hills of Conca de Barberà and Chivite's Colección 125 bottling from their cooler vineyards in Navarra have had some success with varietal Chardonnay, as have Enate of Somontano.

Chardonnay is a very small player in **Portugal**, where there were just 566 ha (1,399 acres) in 2010.

Luxembourg, which in a cooler era depended on the much earlier-ripening AUXERROIS, had 16 ha (40 acres) of Chardonnay in 2008. In **Belgium**, a copy of fine white burgundy has been produced in the walled Clos d'Opleeuw in Haspengouw. Thanks to recent plantings for sparkling wine, the area of Chardonnay vines in the **UK** had risen to 225 ha (556 acres) by 2009.

The extent to which Chardonnay is planted in the land of RIESLING, **Germany**, may surprise many. By 2008 there were already 1,171 ha (2,894 acres), mainly in the Pfalz and Rheinhessen and to a lesser extent in Baden. It was perhaps a natural development in those who reckoned they had already mastered PINOT BLANC and PINOT GRIS – and certainly German consumers are far from immune to the charms of Chardonnay. Recommended producers of Germany's decidedly full-bodied Chardonnays include Knipser and Rebholz in the Pfalz, Dr Heger and Bernhard Huber in Baden, and Fürst in Franken.

Austria has more limited plantings and those mainly in the Steiermark (Styria) in the south (268 ha/662 acres in 2007), where it is generally known as Morillon. Neumeister, Tinnacher, Polz, Sabathi, Sattlerhof and Tement all have a stab at it, although their Sauvignon Blancs are arguably more successful. Kracher of Burgenland has made much-admired Trockenbeerenauslesen from both pure Chardonnay and Chardonnay blended with the crisper Welschriesling (GRAŠEVINA).

Chardonnay is also planted in **Switzerland**, where there were 321 ha (793 acres) in 2009. Recommended producers of a refined style of Chardonnay include Davaz in Graubünden, Domaine des Balisiers in Genève and Raymond Paccot at Domaine La Colombe in Vaud.

In Eastern Europe, Chardonnay is most important in the **Republic of Moldova**, where 5,134 ha (12,686 acres) were logged in 2009, but little of note is exported. **Bulgaria** also has considerable plantings – 4,624 ha (11,426 acres) in 2009 – but has had a tendency to overdo the yields and, often, oak chips. Over the border in **Romania**, plantings seem much more limited, just 704 ha (1,740 acres) in 2008, mostly around Constanța on the Black Sea coast.

Hungary had 2,862 ha (7,072 acres) of Chardonnay in 2008, spread throughout the wine regions but with the most substantial areas in Mátra, Etyek-Buda and Nezmély in the central north and around Lake Balaton in the west. The best examples are barrel fermented and oak aged from the likes of Hilltop (Neszmély), Ottó Légli (Balatonboglár), József Bock and Malatinszky (Villány-Siklós), Mátyás Szőke (Mátra) and Tibor Gál (Eger).

Slovenia, with 1,215 ha (3,002 acres) in 2009, has some very fine Chardonnay made with great care in the image of white burgundy, typically in Goriška Brda near the Italian border by Marjan Simčič, Edi Simčič and Klinec, although some is also grown in Štajerska Slovenija (Slovenian Styria) in the north east.

In **Croatia**, where there were 910 ha (2,249 acres) in 2009, Chardonnay is often blended with MALVAZIJA ISTARSKA, especially in Istra (Istria), as in Ivica Matošević's Grimalda, and the variety is presumably grown in **Serbia** too. There were 400 ha (988 acres) in 2009 in **Montenegro**, where Chardonnay was the most important non-indigenous variety. The **Czech Republic** had 756 ha (1,868 acres) in 2008, mainly in Morava in the south east.

Ukraine has significant plantings of Chardonnay, 2,984 ha (7,374 acres) in 2009, while **Russia** had just 300 ha (741 acres), most of it at Fanagoria in Krasnodar Krai on the Black Sea. **Georgia** is more interested in RKATSITELI and other local specialities and logged just 40 ha (99 acres) of Chardonnay in 2009.

White wine in general is a minority interest in the warmth of the eastern Mediterranean, although the likes of Castel and Golan Heights make some quite impressive 'international-style' Chardonnay from some cooler vineyards in **Israel**, which is estimated to have about 200 ha (494 acres) of Chardonnay. **Greece** had 140 ha (346 acres) in 2007, almost all of it in the north (eastern Makedonía and Thráki/Thrace). **Turkey** had 183 ha (452 acres), **Cyprus** 130 ha (321 acres) and **Malta** about 100 ha (247 acres) in 2010.

Chardonnay's hold on white-wine selections in **California** has been vice-like. The Golden State's plantings of the golden grape reached a peak of 94,164 acres (38,107 ha) in 2003. Thanks to greater interest in red wines, the total had fallen slightly by 2008 to 91,522 acres (37,038 ha) but rose again to 95,271 acres (38,555 ha) in 2010 so that Chardonnay is still by a sizeable margin the most widely planted white wine grape variety in the state. Thanks to its versatility, it can be grown virtually everywhere vines are grown but it ripens so early in the hottest parts such as in the Central Valley that it has no time to develop much flavour and often has to rely instead on residual sugar to make it sell. California's Chardonnays have in general tasted sweeter than most – often because of relatively high alcohol – but as cooler territories in Sonoma Coast and the Central Coast are colonized by the vine, the proportion of nervy, mineral Chardonnays produced is increasing. Although much of Napa Valley proper is too hot to produce exciting Chardonnay, Carneros has considerable acreage of the variety, which was often planted with sparkling-wine production in mind. Russian River Valley is another hotspot for Chardonnay quality, sometimes too hot nowadays. Monterey has produced more than its fair share of fine, sometimes grassy Chardonnays, and the variety spearheaded the extraordinary expansion of the vine in the Central Coast. Virtually every California wine producer makes Chardonnay, often a wide array of them. Those who have enjoyed real acclaim for theirs include Au Bon Climat, Chalone, DuMol, Dutton Goldfield, Kistler, Kongsgaard, Marcassin, Ramey, Ridge, Sandhi and Stony Hill.

Wente of Livermore played a notable part in the history of California Chardonnay, producing the first varietal example in 1936, and being responsible for the so-called Wente clone, or field selection of the variety, that was widely planted all along the Pacific coast and is well suited to warm climates. Robert Young of Alexander Valley later propagated and then sold widely his own successful clone based on a Wente selection.

Oregon's climate proved too cool for this and other selections of Chardonnay imported from California and for years Oregon's Chardonnay was wildly outshone by its PINOT GRIS. But this century, plantings of Chardonnay clones imported from Burgundy known generically as 'Dijon clones' have increased dramatically, as has the quality of the state's Chardonnay, which can display appetizing stone-fruit quality. Dijon clones 95 and 96 have smallish berries that can impart useful intensity of flavour even at relatively low ripeness levels. The state had 1,008 acres (408 ha) of Chardonnay in 2008, with those who have already proved superior Chardonnay producers including Bergström, Chehalem, Domaine Drouhin, Eyrie, Ponzi and Rex Hill.

In **Washington** State, Chardonnay is still the most planted white wine grape, with a total of 7,654 acres (3,098 ha) in the ground in 2011 – but RIESLING (6,320 acres/2,558 ha), supposedly the state's white-wine speciality, has been catching up. There has been much less experimentation with clonal diversity than in Oregon but the likes of Abeja, Arbor Crest and the dominant Château Ste Michelle have produced superior examples, helped by the crisp acidity that characterizes the state's wines.

Chardonnay is grown in virtually every American state with vineyards of note. **Idaho** has a long tradition of making particularly open examples. **Texas** had 360 acres (146 ha) in 2008 and **Virginia** 474 acres (192 ha) in 2010, making it the state's most planted variety by quite a margin. Chardonnay is even more widely planted in **New York** State (981 acres/397 ha in 2006), where it makes rather richer wines in Long Island than in the cooler climes of the Finger Lakes, where RIESLING is better suited to surviving the coldest winters. The style of American Chardonnay – big, bold, oaked and vaguely sweet – can be remarkably similar no matter where it was grown.

Across the border in **Canada**, Chardonnay, with about 2,180 acres (882 ha) in 2009, is the single most planted and most admired variety in Ontario. Indeed, Canada has long seized the global high ground for its Chardonnay with relatively subtle, sophisticated offerings from the likes of Clos Jordanne, Closson Chase, Norman Hardie and Southbrook vindicating this. Prince

Edward County's newish plantings are especially promising. There are minor plantings in Nova Scotia and Québec and there were 866 acres (350 ha) of the variety in British Columbia in the west in 2008 but PINOT GRIS was more widely planted, PINOT BLANC arguably more successful.

Chardonnay is also grown to a limited extent in **Mexico**, mainly in Baja California and Coahulia de Zaragoza, but the climate tends to be a bit hot for this early-ripening variety.

In **Argentina** in 2008 there were 6,342 ha (15,671 acres), mostly in Mendoza, with lesser plantings in San Juan, La Rioja and Neuquén, making Chardonnay the second most widely planted white wine variety after the country's speciality TORRONTÉS (if you exclude the much more basic PEDRO GIMÉNEZ). Argentine Chardonnays can show surprising finesse, which superior practitioners such as Catena put down to altitude but the Californian Paul Hobbs of Viña Cobos suggests is down to the particularly stony soils.

In **Chile** Chardonnay's total of 8,549 ha (21,125 acres) in 2008 made it the most widely planted white variety, although SAUVIGNON BLANC was almost as widely planted. The most extensive plantings are in cool Casablanca (1,845 ha/4,559 acres) but also Maule (1,597 ha/3,946 acres), Curicó (1,440 ha/3,558 acres) and Colchagua (1,288 ha/3,183 acres). It is planted in all the main wine-growing valleys to some extent, even as far south as Malleco, where Viña Aquitania's Sol de Sol has shown the potential for Chardonnay at Chilean viticulture's southern limit. The quality of Chilean Chardonnay improved considerably in the early years of this century with oak increasingly mastered. Some of the most successful have been Matetic's in San Antonio, Amayna in Leyda, Maycas del Limarí in Limarí, Haras de Pirque in Maipo and Errázuriz Wild Ferment in Aconcagua.

Brazil's 642 ha (1,586 acres) of Chardonnay in 2007 were grown mainly for sparkling wine. **Uruguay** had 135 ha (334 acres) in 2008 and **Bolivia** 21 ha (52 acres) in 2009. It was also grown to a limited extent in **Peru**.

Australia and Chardonnay have been inextricably linked ever since the 1980s, when predictable, often heavily oak-chipped, Chardonnay was Australia's chief weapon in conquering the world. By 2008 there were 31,564 ha (77,996 acres) in the ground and Chardonnay was the country's most planted white wine grape by far – although the Chardonnay explosion did not begin until 1971, when Tyrrell's launched their Vat 47 varietal called, with remarkable prescience, Pinot Chardonnay. Today the biggest plantings are in the hot, inland volume-producing regions of Riverland, Murray Darling and Riverina. Perhaps the single quickest national transformation of a wine style was the refinement of Australian Chardonnay in the mid 2000s, when it became clear that 'oaky' had become a pejorative term for the world's wine drinkers. Australian Chardonnay today is crisp, occasionally even lean and angular, but some of the finest, and certainly most technically dependable, Chardonnays in the world now come from the state of Victoria from the likes of Curly Flat and Bindi in the Macedon Ranges; Giaconda in Beechworth; Giant Steps, De Bortoli, PHI in the Yarra Valley; growers in Henty who contribute to Penfolds' Yattarna; and many small estates in the Mornington Peninsula. But subtle, sophisticated, ageworthy Chardonnays are also made in Western Australia (Cullen, Leeuwin Estate, Moss Wood), the Adelaide Hills (Grosset, Petaluma, Shaw and Smith), Tasmania (Derwent Estate, Freycinet) and cooler spots in New South Wales such as Orange.

New Zealand had 3,769 ha (9,313 acres) of Chardonnay planted in 2008, making it the second most planted white but way behind SAUVIGNON BLANC. With its naturally high acidities, the country should produce fine Chardonnay but Sauvignon has been so very much more fashionable that Chardonnay has languished. Growers in its New Zealand base, the east coast of the North Island, especially Gisborne, have been struggling to find buyers for their estimable grapes. Millton are one of the few to have established a reputation for their own bottlings. Most of New Zealand's best Chardonnays are characteristically fresh, and many of them distinctly burgundian. Those (too few) producers who seem to take Chardonnay seriously also include, in the North Island, Kumeu River north of Auckland, Dry River and Ata Rangi in Martinborough, Craggy Range and Trinity Hill in Hawke's Bay. In the South Island, Neudorf of Nelson, Sacred Hill and St Clair of Marlborough, Mountford and Bell Hill in Canterbury, and Felton Road in Central Otago keep the Chardonnay fire burning.

South Africa, like New Zealand, has several other white-wine distractions, notably CHENIN BLANC, its most planted variety of all, as well as COLOMBARD and SAUVIGNON BLANC, the latter recently overtaking Chardonnay in terms of total vineyard area. In 2008 in Cape winelands there were 8,255 ha (20,399 acres) of Chardonnay, about 15% of the white-wine vineyard area. Cooling Antarctic influence keeps much Cape Chardonnay refreshing and in some cases remarkably long lived, even if it is not hugely valued in South Africa itself. It is planted in most wine regions, though the biggest concentrations are in Robertson, Paarl, Stellenbosch and Breedekloof, in that order. Particularly refined versions come from cooler areas such as Walker Bay (Ataraxia, Bouchard Finlayson, Hamilton Russell) and Elgin (Oak Valley and Paul Cluver). Other successes, usually a little richer, include Chamonix and Boschendal in Franschhoek, and Jordan, Rustenberg and Sterhuis in Stellenbosch.

China is the land of red wine and so only 498 ha (1,231 acres) of Chardonnay were recorded in 2009, while **Japan** has a much longer history of growing Chardonnay, and the big producers such as Suntory and Château Mercian make really quite creditable copies of white burgundy from some of the country's 602 ha (1,488 acres) of the variety, mainly in the Nagano and Yamagata prefectures. A little is also grown in **India** – and virtually everywhere else in the world that claims a wine industry.

MELON À QUEUE ROUGE

Melon à Queue Rouge is not related to MELON but is a mutation of Chardonnay that has a red stem (*queue* meaning 'tail' in French) and is found in the Jura (notably in Arbois). Although it is gradually gaining in popularity, there are just a few producers who continue to make varietal versions, including Lucien and Vincent Aviet at Caveau de Bacchus, Domaine de la Pinte and Jacques Puffeney.

CHARENTSI

Cold-hardy Armenian hybrid.

BERRY COLOUR: ● ● ● ● ●

PRINCIPAL SYNONYMS: Charentzi

ORIGINS AND PARENTAGE

Charentsi is a complex hybrid obtained in 1961 by S A Pogosyan at the Armenian viticultural research centre in Merdzavan, just west of the capital, Yerevan, by crossing Seyanets C 1262 with KARMRAHYUT, in which Seyanets C 1262 is a hybrid between a vine of *Vitis amurensis* Ruprecht and CSABA GYÖNGYE. It is therefore a sibling of NERKARAT.

VITICULTURAL CHARACTERISTICS

Productive, cold-hardy to –28 °C (–18.4 °F).

WHERE IT'S GROWN

Kazumoff, based in western **Armenia**, blend Charentsi with its parent KARMRAHYUT to make a sweet fortified wine.

CHARMONT

Minor Swiss cross making slightly more aromatic wines than its parents.

BERRY COLOUR: ● ● ● ● ●

ORIGINS AND PARENTAGE

Charmont is a CHASSELAS × CHARDONNAY cross obtained in 1965 by Jean-Louis Simon at the Caudoz viticultural research centre in Pully, just outside Lausanne, Switzerland, today part of Agroscope Changins-Wädenswil. The purpose of breeding Charmont (and its sibling DORAL) was to create a more sugar-rich and aromatic version of Chasselas by crossing it with varieties such as Chardonnay, RIESLING or GEWÜRZTRAMINER, although only the crosses with Chardonnay turned out to be of sufficient interest. Charmont, initially labelled 1-33, was selected from several seedlings by André Jaquinet at Caudoz (Vouillamoz 2009c).

VITICULTURAL CHARACTERISTICS

Relatively vigorous, consistently productive (more so than CHASSELAS). Early to mid budding, early ripening. Compact bunches of smallish berries. Susceptible to botrytis bunch rot.

WHERE IT'S GROWN AND WHAT ITS WINE TASTES LIKE

Charmont produces typically soft wines that are slightly more aromatic than CHASSELAS but otherwise quite similar, lacking the elegance of its more illustrious parent. There were just 10 ha (25 acres) in **Switzerland** in 2009, mainly in the cantons of Vaud and Genève. Varietal wines, typically soft and gently aromatic, are made by Birchmeier Rebgut in the canton of Aargau, Domaine Fernand Cina in the Valais, Domaine Gaillard & Fils in Vaud and Le Petit Château in Fribourg.

CHASAN

Low-acid French cross grown mainly in the Languedoc.

BERRY COLOUR: ● ● ● ● ●

PRINCIPAL SYNONYMS: INRA 1527-78

ORIGINS AND PARENTAGE

A PALOMINO FINO × PINOT cross obtained by Paul Truel in 1958 at the INRA (Institut National de la Recherche Agronomique) research centre in Montpellier, southern France, and named after its progenitors (Palomino Fino is known as Listán in France).

VITICULTURAL CHARACTERISTICS

Early budding and ripening. Fertile and productive with very big berries so needs to be pruned short using cordon de Royat training. Resistant to botrytis bunch rot but susceptible to phomopsis and powdery mildew.

WHERE IT'S GROWN AND WHAT ITS WINE TASTES LIKE

Recommended in every French winemaking *département* except the Rhône yet there were just 862 ha (2,130 acres) in **France** in 2008, about half of that area in the Aude in the Languedoc. The wines bear some resemblance to CHARDONNAY, with which it is often blended, and may be lightly aromatic but they are prone to oxidation in the cellar and acidity tends to be uncomfortably low unless it is planted in cooler sites. Producers of varietal wines include Domaine Hautes Terres de Comberousse and Domaine de Mairan in the Languedoc. Domaine de Genouillac in Gaillac, south-west France, used to produce a varietal wine but now include it in their Burgale Blanc blend.

CHASSELAS

French Switzerland's characteristic variety producing soft, occasionally distinguished but often pretty ordinary, whites. Widely grown for juice and the table.

BERRY COLOUR: ● ● ● ● ●

PRINCIPAL SYNONYMS: Amber Chasselas (US), Bar-sur-Aube (France), Bassiraube (Germany), Blanchette⧖ (Vaud in Switzerland), Bois Rouge⧖ (Vaud), Bon Blanc (Savoie in France), Chasselas Bianco (Ticino in Switzerland as well as Italy), Chasselas Blanc (Switzerland), Chasselas Cioutat⧖ (Switzerland and France), Chasselas Croquant (Switzerland and France, especially in Alsace), Chasselas de Bar-Sur-Aube (France), Chasselas de Fontainebleau⧖ (France), Chasselas de Moissac (France), Chasselas de Montauban (France), Chasselas de Thomery (France), Chasselas Dorada (Spain), Chasselas Dorato (Ticino and Italy), Chasselas Doré⧖ (Switzerland, France, US), Chasselas Fendant⧖ (Vaud), Chasselas Giclet⧖ (Vaud), Chasselas Jaune Cire⧖ (Vaud), Chasselas Musqué⧖ (Vaud and France), Chasselas Piros (Hungary), Chasselas Plant Droit⧖ (Vaud), Chasselas Queen Victoria, Chasselat (Switzerland), Corinto (Chile), Dorin (Vaud), Elba Toro⧖ (Castilla y León in Spain), Féher Chasselas (Hungary), Fendant⧖ (Valais and formerly in Vaud, Switzerland), Fendant Blanc (formerly in Vaud), Fendant Roux (formerly in Vaud), Fendant Vert (formerly in Vaud), Franceset or Franceseta (Spain), Frauentraube (Germany and Tyrol), Gelber Gutedel (Germany), Gutedel⧖ or Gutedel Weiss⧖ (Austria, Germany and Switzerland), Junker (Austria and Germany), Krachgutedel (Germany), Lausannois (formerly in Burgundy), Mornen or Mornen Blanc (Vallée du Rhône in France), Moster (Austria), Perlan (Genève in Switzerland), Pinzutella (Corse in France), Plemenka or Plemenka Bela (Croatia), Queen Victoria White, Rdeča Žlahtina⧖ (Croatia), Rougeasse⧖ (Vaud), Royal Muscadine, Schönedel (Germany), Shasla Belaya (Krim/Crimea), Silberling (Germany), Süssling (Germany), Temprana Agosteña⧖ (Cigales in Spain), Temprana Media⧖ (Cigales in Spain), Temprana Tardía⧖ (Cigales in Spain), Tempranillo de Nav⧖ (Rueda in Spain), Temprano⧖ or Temprano Blanco (Spain), Viviser (Germany), Wälsche (Switzerland), Weisser Gutedel (Germany and Switzerland), Weisser Krachgutedel (Germany and Switzerland)

VARIETIES COMMONLY MISTAKEN FOR CHASSELAS: ALBILLO MAYOR⧖ (Spain), BARBAROSSA⧖ (Italy), Chasselas de Courtiller⧖, Chasselas de Pondichery, Fayoumi⧖ (Egypt), Madeleine Royale⧖, MARZEMINA BIANCA⧖ (Italy), PALOMINO FINO (called Golden Chasselas in California), PRIKNADI⧖ (Greece), Vrbnicka Žlahtina⧖ (Croatia), ŽLAHTINA⧖ or Žlahtina Bijela (Croatia)

ORIGINS AND PARENTAGE

Chasselas is a very old variety known under numerous synonyms across the world. Its earliest mention supposedly appeared in 1539 in the *Kreutterbuch* by the German botanist Hieronymus Bock (Aeberhard 2005): 'Dessgleichen gross und kleyn Fränkisch/eins anderen namen/Edel oder Lautterdrauben' ('The same big and small Fränkisch/also named/Edel or Lautterdrauben'). The name Edeldrauben (*edel* means 'noble' and *drauben* stands for *Trauben,* meaning 'grapes') is probably the precursor of Gutedel ('good noble') that occurs later in Württemberg in the Basler naturalist Johann Bauhin's (1650) *Historia Plantarum Universalis*, while the name Lautterdrauben most likely corresponds to the local synonyms Luter, Wyssluter and Gutluter that were still in use for Chasselas around 1850 in Aargau in northern Switzerland (Aeberhard 2005). Bauhin was also the first to cite the common Swiss synonym Fendant, meaning 'splitting', in reference to the berries of this type of Chasselas that split instead of bursting when pressed between the thumb and the index finger.

In Switzerland, the name Fendant was widely used from the eighteenth century in the canton of Vaud, both for the variety and for the wine it produced, and several subtypes of the variety were distinguished (Fendant Vert, Fendant Roux, etc.). At the beginning of the twentieth century, producers started using the name Chasselas and naming the wines after their village of origin. Introduced to the neighbouring canton of Valais around 1850, Chasselas under the name Fendant became so common and successful that the Valaisans have now protected the name and claimed it as their own.

The official ampelographic name Chasselas first appeared in 1654 in *Les délices de la campagne* by Nicolas de Bonnefons (Rézeau 1997): 'Muscats, Muscadets, Genetins, Chasselats & autres qui ont le Grain plus charnu, sont bons à faire les Vins doux & Bourus' ('Muscats, Muscadets, Genetins, Chasselats and others that have fleshier berries are good to make sweet wines and cloudy, part-fermented wines straight from the barrel'). This name undoubtedly derives from the village of Chasselas near Mâcon in Saône-et-Loire in Burgundy, whence the first cuttings were probably distributed in France.

A recent genetic study comparing Chasselas' DNA with that of more than 500 varieties from eighteen countries in Europe and the Near East (Vouillamoz and Arnold 2009) has ruled out any possibility of a Near East origin for Chasselas. Combining historical data with genetic clustering, the study points to the region around Lac Léman (Lake Geneva) as the most likely origin of Chasselas. This is consistent with Vavilov's (1926) theory that varieties usually come from the place where the greatest diversity of forms is observed. In addition, a parent–offspring relationship was found between Chasselas and MORNEN NOIR from the Vallée du Rhône (Vouillamoz and Arnold 2009), which explains why Chasselas is called Mornen Blanc near Mâcon, and why Mornen Noir was called Chasselas Noir by the French ampelographer Victor Pulliat.

DNA profiling has also shown that Chasselas Rose Royal, Chasselas Rouge and Chasselas Violet (or Königsgutedel in Austria and Germany) all have the same DNA profile as the white Chasselas and are therefore colour mutations. It has also shown that Chasselas Musqué is an aromatic mutation of Chasselas (Vouillamoz and Arnold 2009).

Chasselas was used to breed many table grapes and a few wine grape varieties: ARNSBURGER, CHARMONT, DORAL, HUXELREBE, MUSCAT FLEUR D'ORANGER, MUSCAT OTTONEL and NOBLING.

OTHER HYPOTHESES

Like many old grape varieties, Chasselas has been the subject of much speculation as to its origin (Vouillamoz and Arnold 2009). Direct domestication from Swiss wild grapes has been suggested (Hoffmann 1982) and then rejected by genetics (Perret 1997). An Egyptian origin was supported by French ampelographer Adrien Berget (1932), who worked in a French college in Cairo, where he was shown supposed Chasselas vines from the Fayoum oasis, 80 km south of Cairo. He concluded that Chasselas originated from Egypt, where it was called Fayoumi. DNA profiling has rejected this synonymy as well as the Egyptian origin of Chasselas (Vouillamoz and Arnold 2009; Myles *et al.* 2011). Galet (2000) recounts the suggestion that it was introduced to Fontainebleau near Paris in 1523 by the French ambassador to Constantinople (today's Istanbul), as witness the huge and famous Treille du Roy that still survives. Others support the hypothesis that Jehan del Rival dict Prince, a vintner from Cahors, planted the Treille du Roy in 1531 at the request of the French king, François I (Dion 1959). However, historical records indicate that the Treille du Roy was planted around 1750, well after the death of François I, whose role in planting the vine is more fiction than fact (Galet 1990). The village of Chasselas near Mâcon in Burgundy, France, is often proposed as the birthplace of the variety (for example, in Goethe 1887) but it is more likely that the variety came from Switzerland and then spread throughout France via the village of Chasselas (Galet 1990).

Chasselas de Courtiller, mistaken for Madeleine Royale (see MÜLLER-THURGAU) by Regner

(1996), is a distinct variety with a unique DNA profile, whereas Golden Chasselas from California is identical to PALOMINO FINO (although some vines going by this name might also be the true Chasselas, called Chasselas Doré in California, *doré* meaning 'golden'). In Greece, Chasselas might also be cultivated under the name PRIKNADI, unless the sample has been mislabelled in the grape collection (see Priknadi).

VITICULTURAL CHARACTERISTICS

Mid budding, early ripening. Moderate to high vigour, very high but occasionally irregular yields since it is particularly prone to coulure and millerandage. Also susceptible to magnesium deficiency, chlorosis, esca, eutypa dieback, desiccation of the stems and especially to phomopsis. Largish, thin-skinned berries prone to premature shrivelling.

WHERE IT'S GROWN AND WHAT ITS WINE TASTES LIKE

Chasselas is widely planted, and also widely disparaged – except in its probable homeland of Switzerland, where it is said to be particularly expressive of its place of origin, the wine changing markedly according to soil and climate – but this may be largely due to the inherent lack of aroma and flavour in the wines.

Plantings of Chasselas in **France** – where it is commonly grown for the table and for juice – are in decline but still extensive (2,442 ha/6,034 acres in 2009). In Alsace it is in sharp decline and tends to disappear into Edelzwicker blends, and while most of the varietal Chasselas labelled Pouilly-sur-Loire are distinctly inferior to the Pouilly-Fumé produced in the same region, there are a few exceptions such as Serge Dagueneau et Filles' bottling from centenarian vines. In Haute-Savoie, however, on the south side of Lac Léman, it is the main variety in the Vin de Savoie crus of Crépy, Marignan, Marin and Ripaille, where it has a long history. The best examples are from Château de Ripaille and Delalex in Marin. A little is also found in **Italy** and **Spain**.

Chasselas has been known in **Germany** since the sixteenth century, and known as Gutedel since the seventeenth. Most of the 1,136 ha (2,807 acres) in 2008 were in the Markgräflerland in southern Baden. Here the best wines, particularly those from old vines on calcareous soils, have flavours of almonds and hay and a distinct minerality. Ziereisen take the trouble to make several single-vineyard bottlings, the best being Steingrüble; other recommended producers include Lämmlin-Schindler and Dörflinger. There are a few outposts beyond Baden, such as the Frenchman Frédéric Fourré's in Sachsen.

Chasselas is the symbol of Swiss white wine, being the most widely planted white wine variety in **Switzerland**, with a total of 4,013 ha (9,916 acres) in 2009, although the overall area is in decline. More than half of it is planted in the canton of Vaud, where Chasselas represented 62% of the production area in 2009 (2,346 ha/5,797 acres), distributed in six main regions: Vully, Bonvillars, Côtes-de-l'Orbe, La Côte, Lavaux and Chablais, each one with its own terroir. Some of the villages' or appellations' names have acquired a local reputation almost as great as that of the crus in neighbouring Burgundy. Notable examples are (from west to east) Mont-sur-Rolle, Féchy, Morges, Villette, Cully, Epesses, Calamin, Dézaley, Saint-Saphorin, Chardonne, Yvorne and Aigle. Recommended producers include La Colombe, Henri & Vincent Chollet, Henri Cruchon, Louis Bovard and Pierre Monachon. The second most important canton for Chasselas is the Valais (1,051 ha/2,597 acres in 2009), where the mosaic of terroirs combined with the warm, dry climate result in more powerful, aromatic, 'sunny' Chasselas, called here Fendant. The best villages for Fendant are (from west to east) Martigny, Fully, Saillon, Chamoson, Ardon, Vétroz, Sion, Saint-Léonard and Sierre, and recommended producers include Cave des Bernunes, Simon Maye & Fils, Rouvinez and Domaine des Muses. Then comes the canton of Neuchâtel, with Caves du Château d'Auvernier being the best-known producer. In January, just a few months after harvest, they bottle a traditional and eagerly awaited *non filtré* ('unfiltered') Chasselas. Genève also produces interesting Chasselas in Satigny, Peissy and Dardagny, and a much smaller amount is produced in German-speaking eastern Switzerland under the name Gutedel.

The variety is also known as Gutedel in **Austria**, where it was once an important ingredient in the Viennese speciality Gemischter Satz, a multi-variety white blend often sourced from interplanted vineyards. However, it is now more commonly used as a table grape; there are probably fewer than 50 ha (124 acres) planted for wine and it is rarely found as a varietal.

Chasselas is widely planted in **Romania** but mostly for the table, also quite widely in **Hungary** (1,804 ha/4,458 acres in 2008), and to a lesser extent in **Serbia** and **Russia**.

In **California**, where it is known as Chasselas Doré, that grown in the Pagani Vineyard in Sonoma is bottled as a varietal. Quails' Gate produce a Chasselas in British Columbia, western **Canada**, and there were 404 ha (998 acres) in **Chile** in 2008.

A little is also grown in **New Zealand**.

CHATUS

Ancient, rare tannic Ardèche variety also found in Piemonte.

BERRY COLOUR: ● ● ● ● ●

PRINCIPAL SYNONYMS: Bolgnino or Bourgnin (Pinerolese in north-west Italy), Brachet (Canavese in north-west Italy), Brunetta (Val di Susa in north-west Italy), Chanu (Isère in south-eastern France), Chatelos or Chatelus (Ardèche and Lozère in south-central France), Chatos, Châtut, Corbeil or Corbel or Corbelle (Drôme in south-central France), Corbès or Corbesse (Isère), Mouraud or Mouret or Mourre (Isère), Nebbiolo di Dronero (Monregalese, Bassa Val Maira, Colline Saluzzesi in north-west Italy), Nebbiolo Pairolè, Neiret or Neiret Pinerolese or Neretto (Pinerolese), Ouron or Houron (Saint-Péray in southern Rhône), Persagne-Gamay (Rhône), Scarlattin (Val di Susa)

VARIETIES COMMONLY MISTAKEN FOR CHATUS: NEBBIOLO (when Chatus appears under the name Nebbiolo di Dronero in Piemonte, north-west Italy), NERET DI SAINT-VINCENT (Valle d'Aosta)

ORIGINS AND PARENTAGE

Chatus is an old variety mentioned by Olivier de Serres (1600), the 'father' of agronomy, as one of the best varieties of the Vivarais (today's Ardèche). Before the arrival of phylloxera, Chatus was widespread in France, from the Alps to the Massif Central, mainly in the Ardèche, where it was the main variety in the southern part of the region, and in the Drôme, Isère and Savoie. Its name might derive from Chatus, a lieu-dit in Verclause in the Drôme.

DNA profiling has established that Neiret (or Bourgnin), planted from the Valle Maira to the Val d'Ossola along the Alpine foothills of Piemonte, north-west Italy, is identical to Chatus (Schneider *et al.* 2001). DNA parentage analysis strongly suggests a parent–offspring relationship with Pougnet, an old variety from the Ardèche that is no longer cultivated, and shows that Sérénèze de Voreppe, also no longer cultivated, is a likely natural progeny of Chatus and GOUAIS BLANC (Vouillamoz), which tends to favour south-eastern France as the centre of origin for Chatus, most likely in the Ardèche, where it used to be cultivated with Pougnet. Indeed, Chatus is a member of the of Sérine ampelographic group (see p XXVII), which does not include any Piemontese varieties (Bisson 2009).

Thanks to DNA profiling, Chatus was recently revealed as the true parent in eight crosses made by the vine breeder Dalmasso in the 1930s that had previously been said to have NEBBIOLO in their pedigree: ALBAROSSA, CORNAREA, NEBBIERA, PASSAU, SAN MARTINO, SAN MICHELE, SOPERGA and VALENTINO NERO (Torello Marinoni, Raimondi, Ruffa *et al.* 2009).

VITICULTURAL CHARACTERISTICS

Well adapted to acidic, siliceous soils. Mid to late ripening, resistant to autumn rains. Small berries. Poor basal-bud fertility.

WHERE IT'S GROWN AND WHAT ITS WINE TASTES LIKE

In **France**, Chatus has almost disappeared, except in the Ardèche, where it is officially recommended and where the Cave Coopérative Vinicole de Rosières and other local producers have recently been working towards its revival. It has been replanted on terraces in the Cévennes, between the villages of Rosières, Payzac and Largentière. Total plantings in 2006 were 57 ha (141 acres), down from 150 ha (371 acres) in 1958. The small berries produce wines that are deeply coloured, fresh, notably tannic and suited to oak maturation. It is sometimes blended with SYRAH.

In Piemonte, north-west **Italy**, it is still cultivated under various synonyms, mainly in the Valle Maira, Saluzzo and Pinerolo regions, where it is usually blended with AVANÀ, BARBERA, PERSAN (under the name Becuét), NERETTA CUNEESE or PLASSA. At Poderi del Palas, Ambrogio Chiotti makes a rare Nebbiolo di Dronero varietal wine called Desmentia, aged one year in barrel.

CHELOIS

French-American hybrid found in Canada and the north-eastern US; wine is rather rustic on its own.

BERRY COLOUR: ● ● ● ● ●

PRINCIPAL SYNONYMS: Seibel 10878

ORIGINS AND PARENTAGE

A hybrid made in Aubenas in the Ardèche between Seibel 5163 and Seibel 5593 obtained by Albert Seibel (see PRIOR for the full pedigree of Seibel 5163), where:

- Seibel 5593 is a Seibel 880 × Seibel 4202 cross (see PRIOR for the full pedigree of Seibel 880)
- Seibel 4202 is a Bayard (or Couderc 28-112) × Afus Ali (a Lebanese table grape also known as Dattier de Beyrouth) cross (see PRIOR for the full pedigree of Bayard).

Chelois was used to breed VINCENT.

VITICULTURAL CHARACTERISTICS

Early ripening, vigorous and productive with a tendency to overcrop unless the clusters are thinned. Susceptible to winter freeze, and to berry splitting followed by botrytis bunch rot if it is wet at harvest, but otherwise good disease resistance. Compact bunches of small blue-black berries.

WHERE IT'S GROWN AND WHAT ITS WINE TASTES LIKE

Wines are medium-bodied and can be rather earthy or rustic so that the variety is better blended with other hybrids such as CHANCELLOR or CHAMBOURCIN or made as a rosé. Introduced into **Canada** in 1946 and to the north-eastern US in 1948 and trialled around Lake Eyrie in **Pennsylvania**. There are now fewer than 40 acres (16 ha) in **New York** State, where Bear Pond produce a good example. Its fortunes tend to wax and wane with the variable demand for red wines.

CHELVA

Spanish variety more popular for the table than for wine.

BERRY COLOUR: ● ● ● ● ●

PRINCIPAL SYNONYMS: Chelva de Cebreros, Chelva de Guareña, Eva, Forastera Blanca, Gabriela (Pajarete and Arcos), Guarena, Mantúo (Arribes del Duero), Mantúo de Pilas, Montúa, Montúo de Villanueva, Montúo Gordo, Uva Rey or Uva del Rey, Uva de Puerto Real, Villanueva

VARIETIES COMMONLY MISTAKEN FOR CHELVA: BEBA (Andalucía), DONA BRANCA (Portugal and Spain), DORADILLA

ORIGINS AND PARENTAGE

Chelva probably originates from the province of Badajoz in Extremadura, south-west Spain, where it used to be widespread. DNA profiling has confirmed that it is the same variety as Mantúo or Montúa from various regions in continental Spain (Fernández-González, Mena *et al.* 2007).

Although Chelva is sometimes called Mantúo Vigiriego or Vigiriego Blanco in Granada, it is not related to VIJARIEGO, which is found both in Granada and on the Islas Canarias, the name spelt in many different ways depending on producer and location. And it should not be confused with DORADILLA or DONA BRANCA, which share the synonym Forastera Blanca.

VITICULTURAL CHARACTERISTICS

Big bunches of big berries. Mid ripening.

WHERE IT'S GROWN AND WHAT ITS WINE TASTES LIKE

Chelva is widely grown in Extremadura, **Spain**, where it is authorized, among many other varieties, in the Ribera del Guadiana DO. It is also grown, to a much lesser extent, further south in Andalucía, where it is authorized in regional wines such Vino de la Tierra de Sierra de Alcaraz but not in any of the autonomous community's DOs. There were 7,490 ha (18,508 acres) in Spain in 2008, the vast majority in Extremadura (6,495 ha/16,049 acres), the rest in Castilla-La Mancha (845 ha/2,088 acres) and Castilla y León (150 ha/371 acres). Unusually, Chelva is used both for the table and for wine but most of these hectares are for wine grapes, producing rather neutral wines that generally disappear in blends.

Under the names Mantúa or Mantúo Vigiriego or – more confusing still – Vigiriego Blanco, it is also planted in Granada, southern Spain, where is it one of the main white varieties. Producers of varietal wines here include Cuatro Vientos, Cortijo Fuentezuelas, Fuente Victoria (with 5% MACABEO), Los Garcia de Verdevique and Los Martos.

CHENANSON

Montpellier cross planted to a limited extent around the Mediterranean.

BERRY COLOUR: ● ● ● ● ●

PRINCIPAL SYNONYMS: Chenançon

ORIGINS AND PARENTAGE

A GARNACHA × JURANÇON NOIR cross bred at INRA (Institut National de la Recherche

Agronomique) in Montpellier, southern France, its parentage recently confirmed by DNA analysis (Di Vecchi Staraz, This *et al.* 2007).

VITICULTURAL CHARACTERISTICS

Early budding and mid ripening. Very fertile and needs to be pruned short. At home in warm, dry sites. Big bunches and small berries.

WHERE IT'S GROWN AND WHAT ITS WINE TASTES LIKE

Recommended throughout the French Mediterrean *départements* and in the Vaucluse with total plantings of 539 ha (1,332 acres) in 2008, although it seems to be favoured in blends rather than as a solo player. Its popularity in **France** most likely stems from the fact that it is more productive and more deeply coloured than its GARNACHA parent, producing supple, fruity and generous wines. Mas de l'Oncle in the Pic Saint-Loup region produce a varietal wine vinified using carbonic maceration.

In a rare outpost in north-east **Spain**, Carlos Esteva at Can Rafols dels Caus in Penedès planted Chenanson in the 1980s and blends it with CALADOC and PORTAN in his Petit Caus red.

CHENEL

Light-skinned South African cross still faltering at the starting blocks.

BERRY COLOUR:

ORIGINS AND PARENTAGE

Chenel is a CHENIN BLANC × TREBBIANO TOSCANO (under the name Ugni Blanc) cross bred between 1958 and 1964 by Christiaan Johannes Orffer at the Department of Viticulture and Oenology of the University of Stellenbosch, South Africa. It was released in 1974.

Like PINOTAGE, Chenel was obtained by means of open-air hybridization in vineyards and not under controlled greenhouse conditions, which is why Orffer requested shortly before his death in 2008 that the parentage of some of his crosses be verified by DNA analysis.

VITICULTURAL CHARACTERISTICS

Early ripening and good resistance to botrytis bunch rot.

WHERE IT'S GROWN AND WHAT ITS WINE TASTES LIKE

There were 100 ha (247 acres) planted in **South Africa** in 2009, about half of that area in Robertson. Landzicht in the Northern Cape make a varietal wine but there are few signs of the variety catching on.

CHENIN BLANC

Loire, and South African, speciality making crisp, sometimes long-lived, wines with varying degrees of sweetness that deserve more recognition.

BERRY COLOUR: ● ● ● ● ●

PRINCIPAL SYNONYMS: Agudelo⚮ or Agudillo (Spain), Anjou, Blanc d'Aunis, Capbreton Blanc (Landes), Franc Blanc (Aveyron), Gros Chenin (Maine-et-Loire and Indre-et-Loire), Gros Pineau (Touraine), Pineau d'Anjou (Mayenne), Pineau de la Loire (Indre-et-Loire), Plant d'Anjou (Indre-et-Loire), Ronchalin, Rouchelein or Rouchelin (Gironde and Périgord), Steen (South Africa)

VARIETIES COMMONLY MISTAKEN FOR CHENIN BLANC: ALBILLO REAL⚮ (Australia), VERDELHO⚮ (Madeira and Açores in Portugal)

ORIGINS AND PARENTAGE

The first mention of this variety, which comes from Anjou in the Val de Loire, France, most likely appeared under its old synonym Plant d'Anjou, as suggested by historical documents from the famous Château de Chenonceau and reported by Casimir Chevalier (1864):

On 3 January 1496, Thomas Bohier (often erroneously spelt Gohier) bought lands around Chenonceaux, where he planted 'neuf arpens de vignes, de plants, qu'il fit venir à grand frais d'Orléans, d'Arbois, de Beaune et de l'Anjou' ('nine arpents of vineyards, with grapevines imported at great expense from Orléans, Arbois, Beaune and Anjou'). The variety from Anjou was white and was almost certainly Chenin Blanc (Bouchard 1901a).

On 18 May 1523, a vineyard on the Chenonceau estate was described as 'un autre cloz de vigne blanche appelé le Plant d'Anjou contenant quatre arpens ou environ' ('another vineyard of a white variety called Plant d'Anjou containing four arpents or so'). Again, the Plant d'Anjou was in all probability Chenin Blanc (Bouchard 1901a).

Between 1520 and 1535, Denis Briçonnet, the Abbot of Corméry in Touraine and brother-in-law of Thomas Bohier, planted various varieties from all over France in Montchenin (or Mont-Chenin), a monastery near Corméry. Among these varieties, the one called Plant d'Anjou turned out to be the best suited to this climate.

This variety was most likely named Chenin after the monastery of Montchenin, from where it was then propagated across the Touraine region, and whence it returned to its birthplace in Anjou under its new name, Chenin.

This is consistent with the earliest written mention of the name Chenin Blanc in 1534 in chapter 25 of the first book of *Gargantua*, written by the medieval Touraine poet François Rabelais: 'Et, auecques groz raisins chenins, estuuarent les iambes de Forgiez mignonnement, si bien quil feut tantoust guary' ('And with large chenin grapes they carefully covered Forgiez's legs, so that he soon recovered'). In the same book, Rabelais praised the 'vin pineau. O le gentil vin blanc! et, par mon ame, ce nest que vin de taffetas' ('pineau wine. Oh the nice white wine! and for my soul, it is nothing else than taffeta'), which is also undoubtedly Chenin Blanc.

In Touraine, Chenin Blanc was also called Gros Pineau, as opposed to MENU PINEAU (the latter known historically as Arbois Blanc in the Val de Loire). Both are distinct from PINEAU D'AUNIS, another variety of the Loire that is also confusingly called Chenin Noir. This might be the reason Comte Odart (1845) considered Chenin Blanc to be a white mutation of Pineau d'Aunis, which is today strongly rejected by DNA profiling.

A parent–offspring relationship with SAVAGNIN has been discovered by DNA parentage analysis (Regner 1999), and a probabilistic analysis of DNA data has shown that TROUSSEAU, SAUVIGNON BLANC and Chenin Blanc are most likely siblings – the offspring of Savagnin and an unknown parent (Myles *et al.* 2011; Vouillamoz), which makes Chenin Blanc an uncle/aunt of CABERNET SAUVIGNON (see the Cabernet Sauvignon and PINOT pedigree diagrams)! This is consistent with Chenin Blanc being a member of the Messile ampelographic group (see p XXVII; Bisson 2009). In addition, Chenin Blanc and GOUAIS BLANC have been shown to be the natural parents of BALZAC BLANC, COLOMBARD and MESLIER SAINT-FRANÇOIS (Bowers *et al.* 2000).

Cipriani *et al.* (2010) have recently suggested through DNA parentage analysis that Chenin Blanc is a possible natural cross between SAUVIGNON BLANC and SAVAGNIN (under the name Traminer Rot) but this is not consistent with Myles *et al.* (2011), and it can be rejected by a thorough parentage analysis using sixty DNA markers (Vouillamoz).

Unexpectedly, the DNA profile of the Spanish variety Agudelo, said to be native to the Betanzos area in Galicia and also cultivated in Alella and Penedès, matched that of Chenin Blanc (Martín *et al.* 2006).

Chenin Blanc was recently used to breed CHENEL in South Africa.

OTHER HYPOTHESES

The first mention of Chenin Blanc supposedly appeared as early as 845 in two royal grants in which Charles II le Chauve, King of France, made the following gifts to the abbey of Saint-Maur de Glanfeuil in Anjou: on 14 August, the land of Soulangé with its vineyards; on 10 October, the land of Bessé with its vineyards. Soulangé and Bessé are likely to be the names of the previous owners rather than place names but there is absolutely no evidence that these vineyards were planted to Chenin Blanc.

Two alternative and unconvincing etymologies have been proposed. First that the name Chenin could derive from the French *chien* ('dog'), its grapes often being eaten by dogs. This led some authors to suggest that Chenin Blanc was domesticated from local wild grapevines. The second is that the variety's name is derived from Chenonceau, the name of the château where Chenin Blanc was supposedly planted by Thomas Bohier at the end of the fifteenth century.

VITICULTURAL CHARACTERISTICS

Vigorous and fertile, depending on the soil type. Early budding, thus at risk from spring frosts, and mid ripening. Very susceptible to botrytis bunch rot, powdery mildew and diseases of the wood. The bunches are medium-sized to big but the berries are relatively small.

WHERE IT'S GROWN AND WHAT ITS WINE TASTES LIKE

Chenin Blanc's influence in **France** is declining. Its total area of 9,828 ha (24,286 acres) in 2008 (down from 16,594 ha/41,005 acres in 1958) represented less than 1.2% of the total French vineyard. Chenin Blanc is most widespread in the *départements* of Maine-et-Loire, in the regions around Angers (5,044 ha/12,464 acres), Indre-et-Loire, around Tours (3,186 ha/7,873 acres), and Loir-et-Cher upriver towards Blois (296 ha/731 acres). Further upriver is SAUVIGNON BLANC country, whereas MELON dominates the Muscadet vineyards at the mouth of the Loire.

Most of the Chenin Blanc is grown in the middle of the Val de Loire, where it produces one of the widest arrays of wine styles of any single variety. Chenin is by far the most important base of the Loire's many sparkling wines, Crémant de Loire and sparkling Saumur and Vouvray in particular, characterized by a tight bead, fairly light body and the slightest edge of honey. The best sparkling Vouvrays can have a long and nuanced life. This is also a feature of arguably Chenin's most glorious incarnations, the late-picked sweet wines of Vouvray, Montlouis, Bonnezeaux and Quarts de Chaume in which high sugar levels, usually concentrated by botrytis, are counterweighted by notably high acid levels that can keep the wines tangy and developing positively in bottle for decades. Common or garden Vouvray, Saumur and Anjou can be little more than a medium-dry to medium-sweet white wine with marked acidity and, sometimes, sulphur, but sulphur levels have been falling along with yields in the Loire so that more and more wines have real Chenin flavours of honey, straw and apples. There was a time when Savennières was virtually the only dry Chenin of real interest in the Loire but this has been changing and now most appellations are producing fine dry Chenins, too, often leaning, sometimes too heavily, on oak and lees stirring. Few of them have quite the cordite minerality of a top Savennières though.

Producers with a long history of high quality include Foreau and Huet in Vouvray; François Chidaine, and Jacky Blot of Taille aux Loups in Montlouis; Philippe Delesvaux in Saint-Aubin-de-Luigné; and Baumard, Closel and Nicolas Joly in Savennières.

Just about the only other place where Chenin Blanc grows in France is Limoux in the hills of the western Languedoc, where it's a required ingredient (20–40%) in Crémant de Limoux along with CHARDONNAY, MAUZAC BLANC and PINOT NOIR, and is also part of the blend in still wines, with Chardonnay and Mauzac Blanc. It also contributes to Côtes de Duras and may play a role in Vins d'Entraygues et du Fel and Vins d'Estaing, again usually with Mauzac Blanc. About 60 ha (148 acres) of Chenin Blanc were grown on Corse in 2008.

In **Spain** there were about 100 ha (247 acres) in 2008, almost exclusively in Catalunya. The variety is declining in **Israel**, where it is minor and less than half as important as it was in 1996.

One part of the world where Chenin Blanc has been cultivated to a considerable extent is **California**, where there were still 7,223 acres (2,923 ha) in 2010, but this total has fallen by well over a half in the last decade and there have been almost no new plantings in recent years. Chenin Blanc has very few proponents in the US and is widely regarded as a blending variety valued only for its acidity. The majority of vineyards are in the hot Central Valley in Fresno, Madera and Kern counties where Chenin Blanc high yields – four or five times the usual yield for a fine Loire Chenin – and low flavour quotients are the norm. But superior, if not exactly world-shattering, varietal examples have emerged from the Clarksburg AVA, just south of Sacramento, where Dry Creek Vineyard and Ehrhardt source theirs. Others include Lucas and Lewellen (Santa Barbara County), Foxen's old-vine Chenin from the Ernesto Wickenden Vineyard (Santa Maria) and Casa Nuestra (Napa Valley). Chappellet for long produced a lively example from vines planted in the 1960s on Pritchard Hill high above the Napa Valley. Those vines were ripped out in the mid 2000s but some are said to have been replaced.

Washington State has just over 200 acres (81 ha) of Chenin, and benefits from the support of the dominant producer, Château Ste Michelle, which produces both late-harvest and icewine versions. Kiona also make sweet wines from it. About 300 acres (121 ha) are grown in **Texas** but it is not regarded with much enthusiasm. Paumanok grow some successfully on **New York** State's Long Island and there are various small plantings in several other states but it does have an image problem – not least because the great Chenins of the Loire are, for the moment, a well-kept secret. A small amount is grown in Baja California, Aguascalientes and Coahulia de Zaragoza in **Mexico**, where presumably its high acidity is useful.

There are very limited plantings in **Chile** (76 ha/188 acres), **Brazil** (30 ha/74 acres) and **Uruguay** (7 ha/17 acres) but, also in 2008, **Argentina** had 2,908 ha (7,186 acres) of Chenin Blanc planted, the vast majority in Mendoza and blended with other, more valuable, varieties such as CHARDONNAY.

Nor do Australians have much respect for Chenin Blanc. In 2008 there were just 642 ha (1,586 acres) in **Australia**, mostly in the irrigated inland Riverland and in Western Australia, both the hot Swan District north of Perth and, to a lesser extent, in cooler Margaret River. For long it was an ingredient in the popular branded wine once known as Houghton's White Burgundy but nowadays it is generally used as a cheap but fresh blending partner for CHARDONNAY, SÉMILLON and SAUVIGNON BLANC.

In **New Zealand** there is virtually only one proponent of Chenin Blanc, the Millton Vineyard of Gisborne, who make the consistently superior Te Arai from it. In 2008 the country had only 50 ha (124 acres), half as much as in 2004.

Chenin Blanc is grown by vine growers in the tropical vineyards of **India** and **Thailand**, presumably once again treasured for its high acidity.

It is **South Africa** that has the biggest area of Chenin Blanc of all, 18,852 ha (46,584 acres) in 2008, making it the most widely planted variety in the country with 33% of white wine plantings and 18.6% of all plantings. It is cultivated in every wine-growing region, and is often treated as a workhorse grape, with the biggest areas in Paarl (3,326 ha/8,219 acres), Malmesbury (3,317 ha/8,196 acres) and Olifants River (2,521 ha/6,230 acres). Chenin's naturally high acidity is particularly useful in the warm Cape climate, and its relative neutrality when grown at high yields, as much of it is, means that it has been used for just about everything here, from stretching blends of more fashionable varieties to base wines for distillation. But there is a nucleus of ambitious winemakers who recognize that South Africa's old Chenin Blanc bushvines are one of the wine industry's most valuable resources and have been making truly fine, concentrated, rich but usually dry, ageworthy whites from them. Excellent examples include those of Adi Badenhorst (Swartland), De Morgenzon (Stellenbosch), De Trafford (Stellenbosch), Ken Forrester (Stellenbosch), Mullineux (Swartland, including their sumptuous sweet straw wine), Rudera (Stellenbosch) and Eben Sadie (Swartland, in blends). It is worth noting that it was only in the mid 1960s that South Africans realized that the predominant variety they called Steen, which had apparently been grown on the Cape for centuries, was in fact Chenin Blanc.

CHICHAUD

Ardèche variety that can still, just, be found in southern France.

BERRY COLOUR: ● ● ● ● ●

PRINCIPAL SYNONYMS: Brunet (Privas), Tsintsào (Aubenas)

VARIETIES COMMONLY MISTAKEN FOR CHICHAUD: BRUN FOURCA, CINSAUT

ORIGINS AND PARENTAGE

Chichaud is an old and almost extinct variety from the Ardèche in southern France, once grown in Aubenas, Vals, Joyeuse, Argentière and Privas. Chichaud belongs to the Sérine ampelographic group (see p XXVII; Bisson 2009). Its etymology is unknown but the name could derive from the family name Chichaud.

OTHER HYPOTHESES

Galet (1990) suggests Chichaud might be a clone of BRUN FOURCA, while others believe it is closely related or identical to CINSAUT, possibly because they share the common dialect synonym Tsintsào. However, ampelographic and genetic data have confirmed that they are all distinct varieties.

VITICULTURAL CHARACTERISTICS

Late budding and early to mid ripening. Likes dry, stony, well-exposed slopes. Susceptible to downy mildew and botrytis bunch rot.

WHERE IT'S GROWN

This variety lost out in the post-phylloxera replanting in **France** and has more or less disappeared. However, Mas de La Salle in the Cévennes have recently planted a few experimental ares of Chichaud, BOURRISQUOU and Vituaigne and will increase the area if the results of vinification are encouraging.

CHIDIRIOTIKO

Greek variety revived by one producer on the island of Lésvos (Lesbos).

BERRY COLOUR: ● ● ● ● ●

PRINCIPAL SYNONYMS: Kalloniatiko

ORIGINS AND PARENTAGE

Chidiriotiko is said to be an indigenous variety from the village of Chidira in the western part of the Greek island of Lésvos in the Aegean, where it was almost completely destroyed by phylloxera in the 1950s. Convinced that this local variety was responsible for what was once one of the most famous and expensive wines in ancient Greece, Chidira-born Dimitris Lambrou replanted a vineyard from the few surviving vines on the island in 1985. It was claimed that the DNA profile of Chidiriotiko was unique but the analysis remained unpublished and our request to check it was turned down. It is therefore not possible to confirm that Chidiriotiko is a unique variety and, according to the viticulturist Haroula Spinthiropoulou, it may even be a blend of varieties.

VITICULTURAL CHARACTERISTICS

Very productive, mid to late ripening. Large bunches of big, thin-skinned berries. Good disease resistance on the volcanic soils around Chidira but not elsewhere on the island.

WHERE IT'S GROWN AND WHAT ITS WINE TASTES LIKE

Planted only on the island of Lésvos, **Greece**, where Methymneos produce both red wines and a white (blanc de noirs) that are low in acidity. The red is light in colour with floral and mineral notes and marked tannins; the white is more citrusy. The total area planted had expanded to around 60 ha (148 acres) by 2010.

CHINURI

High-acid Georgian variety used for both still and sparkling wines.

BERRY COLOUR: ● ● ● ● ●

PRINCIPAL SYNONYMS: Chinabuli, Kaspura, Kaspuri Tetri, Okroula, Tchinouri

ORIGINS AND PARENTAGE

Chinuri is indigenous to the Kartli region of southern central Georgia. Its name derives from *chinebuli*, meaning 'excellent' in Georgian (Chkhartishvili and Betsiashvili 2004).

VITICULTURAL CHARACTERISTICS

Vigorous and productive. Late budding, late ripening. Resistant to downy mildew but susceptible to powdery.

WHERE IT'S GROWN AND WHAT ITS WINE TASTES LIKE

Chinuri is cultivated mainly in the Mukhrani, Ateni and Kaspi subregions of Kartli, southern central **Georgia**, where it is generally the dominant variety in the fresh-fruited and crisp sparkling wines of the Atenuri appellation, which may also include GORULI MTSVANE and possibly

ALIGOTÉ in the blend. Although the must tends to have moderate sugar levels and high acidity, it shows good potential for the production of still as well as sparkling wines. Bagrationi make a typical sparkling version while Iago Wine, Ateni Sioni Monastery and Pheasant's Tears all make wines vinified in the traditional clay vessels known as qvevri. There were 859 ha (2,123 acres) of Chinuri in Georgia in 2004.

CHISAGO

Winter-hardy hybrid producing soft, fruity wines for its breeder in Minnesota.

BERRY COLOUR: ● ● ● ● ●

ORIGINS AND PARENTAGE

A ST CROIX × SWENSON RED hybrid created by Kevin and Kyle Peterson in Chisago City north east of Minneapolis, Minnesota, US. It was patented in 2008 and is therefore currently a proprietary variety made only at Peterson's Winehaven winery.

VITICULTURAL CHARACTERISTICS

Vigorous, winter-hardy to –40 °F (–40 °C) and with good all-round disease resistance.

WHERE IT'S GROWN AND WHAT ITS WINE TASTES LIKE

Chisago wines are fruity, deeply coloured and have soft tannins with flavours of cherry, raspberry and blackberry. Winehaven in **Minnesota** make both dry and semi-sweet versions as well as a red icewine and are keen to keep it as their exclusive signature variety.

CHKHAVERI

Minor but versatile pink-skinned Georgian variety.

BERRY COLOUR: ● ● ● ● ●

PRINCIPAL SYNONYMS: Tchkhaveri, Vanis Chkhaveri

ORIGINS AND PARENTAGE

Chkhaveri is indigenous to the regions of Guria and Adjara in western Georgia and to Apkhazeti, north-western Georgia.

OTHER HYPOTHESES

According to Gouriel in Galet (2000), Chkhaveri and OJALESHI used to play a significant role in the wines from Colchis that were praised by Homer but there is no evidence for this.

VITICULTURAL CHARACTERISTICS

Mid budding, very late ripening. Medium-sized bunches of thin-skinned berries that can reach high sugar levels and still retain a very high level of acidity.

WHERE IT'S GROWN AND WHAT ITS WINE TASTES LIKE

This versatile variety, used to produce dry, fresh, naturally sweet and sparkling white and pink wines with red-fruit flavours, is grown mainly in the regions of Guria (particularly in the Bakhvi, Askana and Sachamiaseri subregions) and Adjara in south-western **Georgia**, and to a lesser extent further north in Apkhazeti (Akhalisopeli, Likhni and Pshapi). There were 20 ha (49 acres) of Chkhaveri in Georgia in 2004.

CHONDROMAVRO

Very rare Greek variety planted in Makedonía.

BERRY COLOUR: ● ● ● ● ●

PRINCIPAL SYNONYMS: Chondromavrouda, Chondromavroudi, Chondromavroudo, Khondromavroud

VARIETIES COMMONLY MISTAKEN FOR CHONDOMAVRO: SEFKA

ORIGINS AND PARENTAGE

Chondromavro is local to the region of Kozáni in western Makedonía in the northern part of mainland Greece.

OTHER HYPOTHESES

It is sometimes considered to be identical to SEFKA from Bulgaria because in Thráki (Thrace) Sefka is also known as Chondromavro but this has not yet been tested by DNA profiling and the two varieties appear to be ampelographically distinct.

VITICULTURAL CHARACTERISTICS

Late ripening, compact clusters of thin-skinned berries, therefore susceptible to botrytis bunch rot.

WHERE IT'S GROWN AND WHAT ITS WINE TASTES LIKE

Chondromavro is grown only in Siatista, south east of Kozáni, **Greece**, where it is found in just a few of the region's many old interplanted vineyards alongside XINOMAVRO and MOSCHOMAVRO. Diamantis have replanted the variety at 800–920 m above sea level and include it in their Liasto blend, a sweet red wine made from sun-dried grapes, as well as in a rosé and a blanc de noirs. The high sugar level in the berries is very important for Liasto, which generally has an aroma of dried figs. Acidity is generally low and tannins soft.

CHOUCHILLON

Light-skinned variety recently rescued in central France.

BERRY COLOUR: ● ● ● ● ●

PRINCIPAL SYNONYMS: Faux Viognier

VARIETIES COMMONLY MISTAKEN FOR CHOUCHILLON: VIOGNIER

ORIGINS AND PARENTAGE

Chouchillon was first mentioned by Rougier in 1899 in the vineyard of Chagnon, between Saint-Étienne and Lyon in central France (Galet 1990).

OTHER HYPOTHESES

Galet (1990) suggests that Chouchillon might be a white colour variation of BRUN FOURCA but this has been disproved by DNA profiling (Bowers *et al.* 1999).

VITICULTURAL CHARACTERISTICS

Mid to late ripening.

WHERE IT'S GROWN

Only a few Chouchillon vines remain, scattered among old VIOGNIER vineyards in Condrieu, northern Rhône, **France**, which might explain its confusion with Viognier and its synonym Faux Viognier. These few vines were recently propagated by the Association pour la Restauration et le Développement du Vignoble des Coteaux du Gier and an experimental 100% Chouchillon vineyard is planned for Genillac in the Coteaux du Gier area, between Lyon and Saint-Étienne. Chouchillon is not yet included in the French official register of varieties.

In **North Carolina**, Davesté Vineyards is notable for its cultivation of uncommon varieties such as COT, Chouchillon and RKATSITELI; the last two are often included in their Trillium blend.

CIANORIE

Old and very minor dark-skinned variety from Friuli, north-east Italy.

BERRY COLOUR: ● ● ● ● ●

PRINCIPAL SYNONYMS: Canore, Canorie, Chianorie, Cianoria, Cjanorie, Rossarie (Manzinello in the province of Udine in Friuli), Vinosa (Manzano in the province of Udine)

ORIGINS AND PARENTAGE

Known in the province of Pordenone in Friuli as early as the seventeenth century, Cianorie is said to originate from Gemona del Friuli in the province of Udine. Its name comes from the local dialect word *cjane*, Italian *canna*, supposedly related to the expression *bere a canna*, meaning 'to drink straight from the bottle', hinting at the lightness of the wine.

VITICULTURAL CHARACTERISTICS

Vigorous and often trained on pergolas. Resistant to fungal diseases. Late ripening.

WHERE IT'S GROWN AND WHAT ITS WINE TASTES LIKE

Cianorie was formerly cultivated all over Friuli in north-east **Italy**, in particular in the provinces of Pordenone and Udine, mainly in the communes of Gemona del Friuli, Castelnuovo del Friuli and Pinzano al Tagliamento. Together with other old local varieties CIVIDIN, FORGIARIN, PICULIT NERI, SCIAGLÌN and UCELÙT, it was saved from extinction by Emilio Bulfon, with the help of the ampelographers Antonio Calò and Ruggero Forti. Emilio Bulfon is one of the few producers to craft varietal wines from these grapes. Cianorie is light-bodied with aromas of black fruits, cloves and flowers and moderate acidity. Together with Cividin, Cianorie was included in the Italian national register of varieties in 2006.

CIENNA

Australian cross with tannins so aggressive it is unsuited to dry wines.

BERRY COLOUR: ● ● ● ● ●

ORIGINS AND PARENTAGE

Like its sister TYRIAN, this is a cross of the Spanish variety SUMOLL with CABERNET SAUVIGNON, created by A J Antcliff at the Merbein station in Victoria, Australia, in 1972, with the aim of creating a good-quality variety suited to hot, dry Australian conditions. Released in 2000.

VITICULTURAL CHARACTERISTICS

Good yields, needing a warm climate to ripen well but tannins may still be excessive. Resistant to mould thanks to its loose bunches.

WHERE IT'S GROWN AND WHAT ITS WINE TASTES LIKE

As with TARRANGO, Brown Brothers, who were involved in the early trials with CSIRO (Commonwealth, Scientific and Industrial Research Organisation) in **Australia** dating back to 1995, are the main proponents of Cienna. Early experiments showed that it needed a warm climate so all their recent plantings have been in the Swan Hill/Robinvale district of Victoria but even there the tannins of this deeply coloured, minty-flavoured variety can be unpleasantly tough. Yalumba did trials in Wrattonbully in South Australia but the dry wines they produced were astringent and excessively marked by methoxypyrazines in this relatively cool climate so the vines were pulled up. To get round the aggressive tannins, Brown Brothers produce a low-alcohol, lightly fizzy red for drinking young and chilled. It is now planted in three Brown Brothers vineyards as well as in those of three contract growers.

CIGÜENTE

See SÍRIA.

CILIEGIOLO

Cherry-flavoured Italian red with untapped potential, especially in Toscana. Parent of Sangiovese.

BERRY COLOUR: ● ● ● ● ●

PRINCIPAL SYNONYMS: Albana Nera☒ (Emilia-Romagna), Brunellone☒ (Grosseto), Canaiolo Romano (Pistoia), Ciliegino, Ciliegiolo di Spagna, Mazzèse, Riminese Nero☒ (Pisa), Sangiovese Polveroso☒ (Firenze)

VARIETIES COMMONLY MISTAKEN FOR CILIEGIOLO: AGLIANICO☒, AGLIANICONE☒, ALEATICO☒, DOUX D'HENRY☒ (Piemonte), SANGIOVESE☒ (Emilia-Romagna)

ORIGINS AND PARENTAGE

Probably the earliest mention of Ciliegiolo is in Soderini's (1600) treatise on the varieties cultivated around Firenze in Toscana, Italy, in which a Ciriegiuolo Dolce is described as a grape with long clusters, large berries and a sweet (*dolce*) and fragrant flavour, best grown on warm soils. This description reasonably matches the modern Ciliegiolo. The name Ciliegiolo (Italian for 'small cherry') comes from the cherry aroma of the berries.

According to tradition, Ciliegiolo was introduced from Spain around 1870 by pilgrims coming back from Santiago de Compostela, which would justify its synonym Ciliegiolo di Spagna. There does not seem to be any relationship between Ciliegiolo and modern Spanish grapes, however (Vouillamoz).

Ciliegiolo has often been mistaken for AGLIANICO, which might explain why Crespan *et al.* (2002) erroneously identified Ciliegiolo with AGLIANICONE by DNA profiling. In fact a Ciliegiolo plant had been mislabelled as Aglianicone in their collection, whereas the true Aglianicone has a quite distinct DNA profile (Manna Crespan, personal communication).

In a study carried out in 2006 at the University of Bologna, three accessions named Albana Nera in Emilia-Romagna (in the Imola, Riolo Terme and Ravenna regions) turned out to be identical to Ciliegiolo, although the latter is not related to the white-berried ALBANA. More recently, several unexpected synonyms of Ciliegiolo were found by DNA profiling of vines found in Toscana (Di Vecchi Staraz, This *et al.* 2007): Brunellone, Mazzèse, Riminese Nero and Sangiovese Polveroso. This last and misleading synonym might explain why Ciliegiolo and SANGIOVESE are often wrongly considered to be identical in Emilia-Romagna.

By DNA parentage analysis, Crespan *et al.* (2002) discovered a parent–offspring relationship between Ciliegiolo and SANGIOVESE. Vouillamoz, Imazio *et al.* (2004) later showed that Sangiovese is a natural progeny of Ciliegiolo and CALABRESE DI MONTENUOVO (see Sangiovese for details).

Recent DNA parentage studies have suggested that Ciliegiolo may be a natural cross between SANGIOVESE and Muscat Rouge de Madère, aka Moscato Violetto (Di Vecchi Staraz, This *et al.* 2007; Cipriani *et al.* 2010), the latter a possible natural MAMMOLO × MUSCAT BLANC À PETITS GRAINS cross, but this does not fit with the parentage of Sangiovese (Vouillamoz, Monaco *et al.* 2007), and since the proposed parentage showed some DNA discrepancies, it would need further investigation.

OTHER HYPOTHESES

As for many old Tuscan varieties, it has been argued that Ciliegiolo was domesticated from local wild grapes by the Etruscans.

VITICULTURAL CHARACTERISTICS

Early to mid ripening. Susceptible to botrytis bunch rot, especially in fertile soils, moderately susceptible to downy and powdery mildews and sour rot.

WHERE IT'S GROWN AND WHAT ITS WINE TASTES LIKE

Mostly cultivated in Toscana, **Italy**, where it is often used to soften SANGIOVESE's edges in Chianti wines. Since the early 1990s, as the variety has been reappraised and not found wanting, several 100% Ciliegiolo wines have appeared on the market, especially in the Maremma, where Ciliegiolo is often made as a nouveau-style wine called Novello Toscano using carbonic maceration. Ciliegiolo is also cultivated in many other regions of Italy (Valle d'Aosta, Piemonte, Liguria, the Veneto, Emilia-Romagna, Marche, Umbria, Lazio, Abruzzo, Molise, Campania, Puglia, Basilicata, Sicilia). The wines have an intense red colour, aromas of cherries, strawberries and spices; they may be juicy or full-bodied and show a certain resemblance to wines made from CINSAUT. They occasionally lack acidity, which is why they complement Sangiovese so well.

The Italian census (2000) recorded 3,078 ha (7,606 acres) but only about 10% of that total is used in DOC or DOCG wines. In Toscana, these include: Parrina, Colline Lucchesi, Montecarlo, Val di Cornia, Morellino di Scansano, Chianti and Chianti Classico. The Petraviva and Val di Cornia DOCs also allow for a varietal Ciliegiolo, as does Umbria's Rosso Orvietano DOC and Abruzzo's Controguerra DOC.

In Liguria, Ciliegiolo is authorized in Golfo del Tigulio (varietal or blended, still, sparkling or novello), Colli di Luni and Val Polcevera DOCs.

Since the 1990s the Maremma producer Sassotondo has contributed greatly to the revival of Ciliegiolo, producing a straight Ciliegiolo plus the single-vineyard San Lorenzo made from forty-year-old vines. Other recommended producers include Antonio Camillo and Rascioni & Cecconello.

CINSAUT

Underrated Mediterranean-loving variety making characterful rosés and flirtatious reds.

BERRY COLOUR:

PRINCIPAL SYNONYMS: Black Malvoisie (California), Black Prince (Australia), Blue Imperial (Australia), Cinqsaut or Cinq-saou (Languedoc), Cinsault (Languedoc as well as Algeria, Morocco, Tunisia), Grecaù or Grecu Masculinu (Sicilia in Italy), Hermitage (South Africa), Marroquin or Marrouquin, Ottavianello (Puglia in Italy), Picardan Noir (Var in France), Piquepoul d'Uzès, Prunelat or Prunellas (Gironde), Samsó (Catalunya), Sinsó (Spain), Sinsón or Sinseur (Riviera di Ponente in Italy), Uva Spina (Riviera di Ponente)

VARIETIES COMMONLY MISTAKEN FOR CINSAUT: BRUN ARGENTÉ, CHICHAUD, MAZUELO (under the name Carignan), OEILLADE NOIRE (Languedoc and Algeria), PIQUEPOUL NOIR, PLANT DROIT (Vaucluse), PRUNELARD (Gironde, where Cinsaut is called Prunelat)

ORIGINS AND PARENTAGE

In France, the earliest mention of Cinsaut was allegedly made under its old synonym Marroquin by Olivier de Serres (1600) or as Marrouquin by Magnol (1676) in his *Botanicum Monspeliense*: 'sequitur deinde marrouquin, antiquorum duracina, acini duri' ('then comes marrouquin, the duracina from the ancients, with hard berry skin'). In 1829 it was mentioned under the name Sinsâou by Bouschet in the Hérault in southern France: 'le sinsâou, le bois ressemble à l'œillade' ('sinsâou, the wood looks like œillade'). Its modern spelling, Cinsaut, did not appear until 1888 (Rézeau 1997) and the origin of the name is unknown.

There are two main reasons why the geographic origin of Cinsaut can be confidently assigned to southern France, possibly Languedoc-Roussillon:

- It is a member of the Piquepoul ampelographic group (see p xxvii; Bisson 2009), all from the Vaucluse in Provence or from the Languedoc-Roussillon.
- DNA parentage analysis based on twenty-three DNA markers suggests a possible parent–offspring relationship with RIVAIRENC (to be confirmed with more DNA markers) from the Languedoc-Roussillon (Vouillamoz).

In Italy, Cinsaut was already cultivated in the seventeenth century. Indeed, the DNA profile of Grecaù (Carimi *et al.* 2009), an old variety from the Etna area on the Italian island of Sicilia that was already cited by Cupani in 1696, unexpectedly turned out to be identical to Cinsaut (Vouillamoz). It has also been known in Puglia since at least the nineteenth century under the name Ottavianello (Giavedoni and Gily 2005). In Spain, the DNA profile of the variety Sinsó analysed by Martín *et al.* (2003) proved not surprisingly to be identical to Cinsaut (Vouillamoz).

Cinsaut was used to breed the successful South African variety PINOTAGE, named after its parents PINOT NOIR and Hermitage, as Cinsaut was once inaccurately but widely known in the Cape. South Africa was also home to two colour mutations of Cinsaut: one with grey berries and one with white berries but these seem to have disappeared and have been deleted from the official variety designations.

Cinsaut should not be confused with BOUCHALÈS even though they share the misleading synonym Picardan.

VITICULTURAL CHARACTERISTICS

Generally rather delicate but copes well with heat and drought. Not especially vigorous and the wood ages quickly; susceptible to diseases of the wood such as esca and eutypa dieback. Fertile, productive and very resistant to drought and resistant to browning of the leaves (*brunissure*). May be subject to chlorosis in calcareous soils. Also susceptible to mites and grape moths. Late budding, mid ripening. Big bunches and big berries.

WHERE IT'S GROWN AND WHAT ITS WINE TASTES LIKE

Like Grenache (GARNACHA), Cinsaut is ideally suited to the warm, dry soils of southern **France**, although its natural tendency to high yields needs to be curbed for it to show its finer side: reds that are soft, fruity and aromatic in youth and surprisingly long-lived. It is also ideally suited to the production of fresh, perfumed and fruity rosé wines. As a red, it is more often found in blends than alone, often with the tougher Carignan (MAZUELO), although Domaine des Terres Falmet and Domaine d'Aupilhac both produce excellent examples of 100% Cinsaut in the Languedoc.

When it is cultivated as a table grape in southern France, Cinsaut is often confusingly called Oeillade Noire but the true OEILLADE NOIRE is a distinct variety.

Total plantings in France in 2009 were 20,800 ha (51,398 acres), making it the ninth most planted red variety. It is widely planted in Languedoc-Roussillon and Provence-Côte d'Azur but the biggest areas are in the Hérault, Var, Gard and Aude *départements*. It is one of the permitted varieties in Châteauneuf-du-Pape but its contribution is generally limited to around 5%.

It was considerably more popular in 1970s France when the vineyard area devoted to Cinsaut tripled (there were more than 50,000 ha/123,553 acres in 1979) thanks in large part to official encouragement to plant it in place of ARAMON NOIR and ALICANTE HENRI BOUSCHET. In the 1950s and 1960s, large volumes of Cinsaut wine were imported from Algeria, then constitutionally part of France, to beef up some of the more weedy burgundies of the day.

In the south of **Italy**, Cinsaut is cultivated under the name Ottavianello around Brindisi in Puglia, notably in the tiny Ostuni DOC, where it appears as a generally light, unremarkable varietal wine (minimum 85% with other local varieties). Total plantings in Italy in 2000, according to the agricultural census, were 288 ha (712 acres).

In **Morocco** it is still the most widely planted variety, and in the **Lebanon** it is probably the third most planted wine grape after CABERNET SAUVIGNON and SYRAH, having long played an important role in Château Musar there.

Limited plantings are also found in parts of Eastern Europe such as **Turkey**, where there were 444 ha (1,097 acres) in 2010

There were 115 acres (47 ha) recorded in **California** in 2008, treasured by the likes of Bonny Doon for light reds and rosés. It is also planted in **Washington** State.

Official vineyard statistics in **Australia** do not mention Cinsaut but there are around 60 ha (148 acres) all told. Spinifex's Papillon from the Barossa Valley includes a good whack of Cinsaut, as does Foggo's McLaren Vale blend and it can be found in several other southern-Rhône-style cuvées. Foggo also produce a 100% Cinsaut rosé from vines planted in 1958. It is used by producers such as Chambers Rosewood and Morris in Rutherglen in their fortified wines.

In **South Africa**, Cinsaut was the country's most planted red wine variety and still occupies about one tenth of its vineyard area in France (2,241 ha/5,538 acres in 2008), mostly in Paarl, Breedekloof and Malmesbury. It arrived from France in the mid nineteenth century. Not until 1993 did it lose the top red slot to CABERNET SAUVIGNON. Once little more than a productive workhorse, today it provides producers such as Adi Badenhorst in the Swartland with lower-yielding old vines for increasingly fine-tuned wines although, perhaps because of its ubiquity, it is not highly regarded by most Cape growers.

CIRFANDLI

See ZIERFANDLER.

CITRONNY MAGARACHA

Citrus-flavoured Ukrainian cross used mostly in blends.

BERRY COLOUR: ● ● ● ● ●

ORIGINS AND PARENTAGE

Citronny Magaracha was made in 1978 by P Y Golodrigi at the Magarach research centre in Yalta, southern Ukraine, by crossing MADELEINE ANGEVINE with (Magarach 124-66-26 × Novoukrainsky Ranny), where:

- Magarach 124-66-26 is a RKATSITELI × Magarach 2-57-72 cross (see PERVENETS MAGARACHA for Magarach 2-57-72 pedigree)
- Novoukrainsky Ranny is a Dzhura Usyum × CSABA GYÖNGYE cross
- Dzhura Usyum is a cross between two table grape varieties from Uzbekistan, Nimrang and Taifi Rozovy.

VITICULTURAL CHARACTERISTICS

Potentially productive, mid budding, early to mid ripening. Medium-sized, thin-skinned berries. Winter-hardy to –25 °C (–13 °F) and good resistance to fungal diseases.

WHERE IT'S GROWN AND WHAT ITS WINE TASTES LIKE

Citronny Magaracha is planted in **Ukraine** (approximately 110 ha/272 acres in 2009) and in **Russia**'s Krasnodar wine region near the Black Sea (307 ha/759 acres in in 2010). As its name suggests, the wines – variously dry and sweet or sparkling though the sweet wines are the most successful – have a lemony citrus character. They are also Muscat-like with peach flavours and fresh acidity. Like many producers, Fanagoria in Krasnodar use this variety in blends. Ukrainian producers include Magarach.

CIURLESE

Virtually extinct white from around Rimini in northern Italy.

BERRY COLOUR: ● ● ● ● ●

PRINCIPAL SYNONYMS: Ciurlès

VITICULTURAL CHARACTERISTICS

Robust, productive variety, resistant to botrytis bunch rot.

WHERE IT'S GROWN

A speciality of the province of Rimini, eastern **Italy**, Ciurlese used to be cultivated mainly in Valconca, or the Valle del Conca, near Rimini. Today only a few vines survive in the area of San Giovanni in Marignano (near Rimini), eg in the vineyards of the Rocche Malatestiane co-operative in Rimini.

CIVIDIN

Ancient, rare white from Friuli, north-east Italy.

BERRY COLOUR: ● ● ● ● ●

PRINCIPAL SYNONYMS: Cividin Bianco, Cividino

ORIGINS AND PARENTAGE

Cividin is said to come from the Valli del Natisone, east of Udine in north-east Italy, and to take its name from the nearby municipality of Cividale. It was mentioned in Friuli as early as the seventeenth century.

VITICULTURAL CHARACTERISTICS

Late ripening and resistant to sour rot but susceptible to powdery mildew.

WHERE IT'S GROWN AND WHAT ITS WINE TASTES LIKE

Cividin historically spread from the Cividale region to the hillsides of Maniago, Meduno, Navarons and Vito d'Asio in Friuli, north-east **Italy**. Today it is mainly cultivated in the province of Pordenone and in Vipava, **Slovenia** (Vipacco in Italian). Together with other old local varieties CIANORIE, FORGIARIN, PICULIT NERI, SCIAGLÌN and UCELÙT, it was saved from extinction by Emilio Bulfon, with the help of the ampelographers Antonio Calò and Ruggero Forti. Emilio Bulfon is one of the few producers to craft varietal wines from these grapes, though the vines are interplanted with other varieties. The wines have spicy aromas and are light-bodied with good acidity. Together with Cianorie, Cividin was included in the Italian national register of varieties in 2006.

CLAIRETTE

Once very popular and now useful, crisp southern French white.

BERRY COLOUR: ● ● ● ● ●

PRINCIPAL SYNONYMS: Blanc Laffite, Blanquette (Aude and Gard), Clairet (Krasnodar in Russia), Clairette Blanche, Clarette, Fehér Clairette, Kleret, Kleret Belyi (former Yugoslavia), Muscade (Loupiac), Oeillade Blanche (Provence and Drôme), Osianka, Ovsyanka, Petit Blanc (Aubenas), Petit Kleret, Uva Gijona, Vivsyanka

VARIETIES COMMONLY MISTAKEN FOR CLAIRETTE: PICARDAN (Provence and Drôme), PIQUEPOUL BLANC

ORIGINS AND PARENTAGE

Clairette is one of the oldest varieties of France's Midi, possibly originating in the Hérault, where it is said to have been planted as early as 1490 in the commune of Adissan but there is no historical evidence for this date. The first reliable citation of Clairette was in 1575 in Ginestas, in the Aude: 'vint deux pleines desques [= *corbeilles*] de clarette' ('twenty-two full baskets of clarette') (Rézeau 1997). Clairette was later mentioned by Olivier de Serres (1600), using the spelling *clerete*, and also a few years later by Magnol (1676): 'vulgatissima alba dicitur clarette, ex qua fit optimum vinum album' ('very widespread is the white called clarette, from which is made an excellent white wine').

The name Clairette, meaning 'light white', might come from the colour of the berries, but since there were many light-white-berried varieties at that time, it is more likely that the name refers to the white hairs on the underside of the leaves, giving them a bright (*clair*) appearance.

Clairette belongs to the Claret ampelographic group (see p XXVII). In addition, Clairette is ampelographically close to PLANT DROIT (Bisson 2009).

DNA analysis has established that the PICAPOLL BLANCO cultivated in Pla de Bages in Spain's Catalunya does not correspond to the French PIQUEPOUL but it does have a possible genetic relationship with, or may even be identical to, Clairette (Puig *et al.* 2006).

OTHER HYPOTHESES

In Italy, DNA studies suggested a close genetic relationship between Clairette, AIRÉN from Spain and Rhoditis (RODITIS) from Greece (Labra *et al.* 1999) but this is highly questionable (see INZOLIA). Labra *et al.* (2001) also suggested a genetic relationship between Clairette and ERBALUCE from Piemonte in north-west Italy, which has a similar etymology ('clear', 'bright'), but this has yet to be confirmed.

VITICULTURAL CHARACTERISTICS

Vigorous and needs to be pruned short. Suited to poor, dry, calcareous soils. Grows particularly straight; can withstand strong winds well and does not need a stake. Late ripening. Susceptible to mites and grape moths but not to powdery mildew and botrytis bunch rot.

WHERE IT'S GROWN AND WHAT ITS WINE TASTES LIKE

Clairette's stronghold and most obvious home in southern **France** is in the Drôme, especially around Die, north-east of Montélimar. In this area it is the sole variety in appellations such as Coteaux de Die and the sparkling Clairette de Die (though those specified as Méthode Ancestrale are based on MUSCAT BLANC À PETITS GRAINS) and is the majority component in Crémant de Die.

France had 2,405 ha (5,943 acres) of Clairette in 2009 but the total has declined drastically in the last fifty or so years (14,099 ha/34,839 acres in 1958). It is planted mainly in the southern Rhône and the Vaucluse – and then further south west in the Gard region of Languedoc, where it is the star of the Clairette de Bellegarde appellation, and further south east in the Var. In the Hérault, it has its own appellation in the form of Clairette de Languedoc, where, in two non-contiguous areas north and south of Clermont-l'Hérault, it may be used to produce dry, sweet or rancio versions.

It can best be described as an old-fashioned variety, producing fairly simple but crisp, sometimes mineral, wines for relatively early drinking. Aside from its more well-known fizzy guise, it is often used as a blending partner in the southern Rhône, eg in Châteauneuf-du-Pape or Côtes du Rhône, generally contributing about 30–35% of the mix for white wines and sometimes, as at Château de Beaucastel, adding some freshness to red blends. Château Rayas' white Châteauneuf comprises 50% Grenache Blanc (GARNACHA BLANCA) and 50% Clairette and Le Vieux Donjon's white generally has equal parts of Clairette and ROUSSANNE. Many other southern Rhône appellations such as Lirac, Tavel, Vacqueyras, Grignan lès Adhémar and Ventoux may include Clairette.

In the Languedoc it is frequently blended with Ugni Blanc (TREBBIANO TOSCANO) and TERRET; early picking is common to catch the fresh aromas and acidity. It is also a permitted ingredient in several Provençal appellations such as Côtes de Provence, Palette, Cassis and Bellet, and in Saint-Mont in south-west France. Domaine Richeaume in the Côtes de Provence produces a fragrant, fresh but full-bodied blend of Clairette and Rolle (VERMENTINO) and, in Bandol, Bunan's Mas de la Rouvière is a bold, appley 100% Clairette.

Its success in the seventeenth and eighteenth centuries, blended with the more acidic PIQUEPOUL in the wine known as Picardin that was exported in large volumes north from the Languedoc, may have encouraged its wider distribution to, for example, Algeria and South Africa (see below).

In **Italy**, there were 150 ha (371 acres) at the time of the last agricultural census (2000), planted mainly in Toscana and on the island of Sardegna, where it is an officially recommended variety in the south-eastern province of Cagliari and may be found to a limited extent in Nuragus di Cagliari (which must include at least 85% NURAGUS).

Although statistics are not available for the **Lebanon**, producers such as Massaya and Ch Kefraya include Clairette in their white blends.

Under the name Clairet, this variety is still planted to a limited extent in Krasnodar on **Russia**'s Black Sea coast. It was apparently particularly popular with the Kuban Cossacks in the nineteenth century and first half of the twentieth.

The few growers in **Australia** who have planted Clairette (once known there as Blanquette) have been very successful: Honeytree in the Hunter Valley increased their plantings from 0.4 to 2 ha (1 to 5 acres) by top-grafting their Traminer (SAVAGNIN BLANC) vines to try to keep pace with demand for their early-drinking version. JB Wines also do well with the variety in the Barossa Valley.

In 2008, there were still 328 ha (811 acres) in **South Africa**, quite widely dispersed but with more planted in Worcester than anywhere else. Eben Sadie's multi-variety blend Palladius includes a dash of Clairette, as does Tulbagh Mountain Vineyard's white blend and Mullineux's Swartland white.

A colour mutation called **Clairette Rose** was already known in 1834–5 in the Bouches-du-Rhône (Rézeau 1997) and is still cultivated in the Gard, mainly in Lirac. Total French plantings of this pink version were 260 ha (642 acres) in 2009.

CLAVERIE

Virtually extinct, disease-prone white from south-west France.

BERRY COLOUR:

PRINCIPAL SYNONYMS: Bouguieu (Tursan), Chalosse Blanche, Chaloussenc (Vic-Bilh), Clabarien, Clabérieu, Claverie Blanc, Claverie Verte, Galia Zuria (Pays Basque)

VARIETIES COMMONLY MISTAKEN FOR CLAVERIE: Claverie Coulard (Bellocq in the Pyrénées-Atlantiques)

ORIGINS AND PARENTAGE

Claverie is a variety from the Landes, in the far south west of France, where it was first mentioned in 1827: 'le claverie produit de très-bon vin, sur-tout le vin de Cap-Breton, dans le département des Landes' ('Claverie produces very good wines, especially the wine from Cap-Breton in the Landes *département*'). Its name could derive from the Béarnais dialect word *claberia*, meaning 'enclosure, fence', and could correspond to the names of several lieux-dits in the Landes or in the Pyrénées-Atlantiques, more specifically to Lasclaveries north of Pau (Lavignac 2001). Claverie belongs to the Gras ampelographic group (see p XXVII). In addition, Claverie is ampelographically close to the Cot ampelographic group (Bisson 2009). While Claverie Coulard is a distinct variety that is not commercially cultivated, a black mutation of Claverie (called Claverie Noire) has been observed (Galet 2000).

VITICULTURAL CHARACTERISTICS

Vigorous and productive. Late budding and ripening. Highly susceptible to powdery mildew, botrytis bunch rot and anthracnose.

WHERE IT'S GROWN AND WHAT ITS WINE TASTES LIKE

Claverie used to be cultivated mainly in Chalosse in the Landes, south-west **France**, but also in Vic-Bilh and Jurançon in the Pyrénées-Atlantiques. Although it is included in the official French register of varieties, and recommended in the Gers and the Landes, it is almost extinct, mainly because of its susceptibility to fungal diseases. It is planted in the Saint-Mont conservatoire and recently yielded a wine that was relatively neutral, thin and eye-wateringly acidic (pH 2.78), with a particularly high level of malic acid. The conservatory is probably the best place for it.

CLINTON

Pre-Civil War natural American hybrid of historical interest only.

BERRY COLOUR: ● ● ● ● ●

PRINCIPAL SYNONYMS: Bacchus, Clinton Rose, Klinton, Plant des Carmes, Plant Pouzin, Vorthington, Worthington, Zephirin

ORIGINS AND PARENTAGE

Clinton is a spontaneous *Vitis riparia* × *Vitis labrusca* hybrid that was first planted in 1821 by Hugh White, then a student of Hamilton College, in the grounds of Professor Noyes' house on College Hill in the town of Clinton, east of Syracuse, New York State. Two years earlier, White had planted hundreds of grape seeds in his father's garden in Whitesboro, and selected the one that looked most promising, although its origin remains unknown. Noyes named it after the town of Clinton, where the original vine was still thriving eighty years later on a great elm tree (Bailey 1906).

VITICULTURAL CHARACTERISTICS

Hardy, with good general disease resistance and immune to phylloxera. Small bunches and berries.

WHERE IT'S GROWN

This old American variety produces deeply coloured wines that have good acidity but are blighted by their foxy flavour. There are apparently some pockets of Clinton vines in the Cévennes in **France**, north-east **Italy**, **Switzerland** and in **Brazil**, though their fruit is not authorized for the production of quality wine in the EU. Because of its disease resistance, it was quite popular in central southern France (Galet 2000) until it was prohibited – new plantings were banned in the mid 1930s and then the variety was outlawed completely in the 1950s – along with other *Vitis labrusca* descendants such as ISABELLA and CONCORD. Schiavo in the Veneto use Clinton as a component in their grappa and the Association Fruits Oubliés in the Languedoc make a varietal wine for 'home consumption'.

COCOCCIOLA

Fruity, herbal white from Abruzzo and Puglia.

BERRY COLOUR: ● ● ● ● ●

PRINCIPAL SYNONYMS: Cacciola, Cocacciara

ORIGINS AND PARENTAGE

The origins of this variety and of its strange name are unclear and its earliest mention seems to be in Viala and Vermorel's *Ampélographie* (1901–10) under the synonym Cacciola.

VITICULTURAL CHARACTERISTICS

Mid to late ripening. High yields.

WHERE IT'S GROWN AND WHAT ITS WINE TASTES LIKE

The majority of vineyards are found in Abruzzo, southern **Italy** (mainly in the province of Chieti, in the communes of Vacri, Ari and Rocca San Giovanni), where Cococciola is often blended with TREBBIANO TOSCANO and other varieties, but it is also present in northern Puglia. Authorized in all IGTs in Abruzzo and in IGT Daunia in Puglia and also as a minor component in Trebbiano d'Abruzzo DOC, Cococciola has only fairly recently been made as a varietal wine. These tend to be fruity and herbal with marked acidity and are sometimes aged in big barrels. Current producers include Cantine Ciampoli in Ortona, Cantina Frentana in Rocca San Giovanni, Masseria Coste di Brenta in Lanciano, Silene in Chieti and Valle Martello in Villamagna. There were 893 ha (2,207 acres) recorded in the Italian census of 2000.

CODA DI CAVALLO BIANCA

Barely cultivated and generally blended white variety from Campania, southern Italy.

BERRY COLOUR: ● ● ● ● ●

PRINCIPAL SYNONYMS: Cavalla🧬, Codacavallo, ?Latina

ORIGINS AND PARENTAGE

Named after its long clusters shaped like a horse's tail. The relationship between Coda di Cavallo from the island of Ischia and Cavalla from the Napoli region has been the subject of debate but a recent DNA study has shown them to be one and the same (Migliaccio *et al.* 2008).

VITICULTURAL CHARACTERISTICS

Early ripening, not much resistance to botrytis bunch rot.

WHERE IT'S GROWN

Used to be widespread on the slopes of Vesuvio in southern **Italy**, now scarcely cultivated on the islands of Ischia and Procida and in Campi Flegrei (west of Napoli, Campania), always ungrafted. Never vinified alone, always blended.

CODA DI PECORA

Obscure white from Campania in southern Italy.

BERRY COLOUR: ● ● ● ● ●

VARIETIES COMMONLY MISTAKEN FOR CODA DI PECORA: CODA DI VOLPE BIANCA

ORIGINS AND PARENTAGE

Coda di Pecora has long been considered identical to CODA DI VOLPE BIANCA in Campania, southern Italy. Careful ampelographic observations and recent DNA profiling have shown that they are two distinct and unrelated varieties, and that Coda di Pecora might be closely related to SAN PIETRO, an obscure Campanian variety sampled in Caserta, north of Napoli (Costantini *et al.* 2005). Coda di Pecora means 'goat's tail' and probably refers to the shape of the clusters.

WHERE IT'S GROWN AND WHAT ITS WINE TASTES LIKE

This relatively neutral grape grown in Campania, southern **Italy**, is cultivated mainly in the province of Caserta, between Monte Maggiore and Roccamofina in the communes of Conca della Campania, Galluccio, Mignago Monte Lungo and Piccilli. Always blended with other local varieties, and the confusion with CODA DI VOLPE BIANCA – even given as a synonym in the official Italian census of 2000 – makes it difficult to track down.

CODA DI VOLPE BIANCA

Ancient, full-bodied white from southern Italy coming into its own as a varietal wine.

BERRY COLOUR: ● ● ● ● ●

PRINCIPAL SYNONYMS: Durante, Falerno, Guarnaccia Bianca (Calabria)

VARIETIES COMMONLY MISTAKEN FOR CODA DI VOLPE BIANCA: CAPRETTONE, CODA DI PECORA, ?CODIVARTA, PALLAGRELLO BIANCO, TREBBIANO TOSCANO

ORIGINS AND PARENTAGE

The earliest mention of Coda di Volpe Bianca from Campania dates back to Giovan Battista della Porta (1592). The synonym Falerno probably came about because Coda di Volpe Bianca is thought by some to have produced the famous Falerno wine in Roman times.

Coda di Volpe Bianca, meaning 'white foxtail', has sometimes been considered identical to

CAPRETTONE, CODA DI PECORA, PALLAGRELLO BIANCO or TREBBIANO TOSCANO. Careful ampelographic observations and recent DNA profiling have shown that they are all distinct and unrelated cultivars, and that Coda di Volpe Bianca might be closely related to Nerella and Uva Rosa, two obscure and not commercially cultivated Campanian varieties sampled in the region of Napoli (Costantini *et al.* 2005). The epithet Bianca is useful to distinguish it from Coda di Volpe Nera, a synonym for PALLAGRELLO NERO.

VITICULTURAL CHARACTERISTICS

Moderate yields, mid to late ripening (mid to end October), high sugar levels and medium acidity.

WHERE IT'S GROWN AND WHAT ITS WINE TASTES LIKE

Widespread in but exclusive to Campania in southern **Italy**, mainly in the provinces of Benevento and Avellino. Until recently Coda di Volpe Bianca was considered a minor grape and was generally used in blends, but over the last few years varietal wines have demonstrated real quality, eg in Sannio, Irpinia and Taburno DOCs. Coda di Volpe is recommended in the provinces of Avellino, Benevento, Caserta and Napoli; it is authorized in the province of Salerno. The wines are pale to medium gold, full-bodied, quite soft, with aromas ranging from delicate stone fruit to exotic spice, sometimes with a mineral or salty flavour. A total of 1,027 ha (2,538 acres) were recorded in the 2000 Italian agricultural census.

Producers of varietal wines include D'Antiche Terre, Astroni, Cantine del Taburno, La Casa dell'Orco, Cavalier Pepe, Crogliano, Ocone, Vadiaperti and Vini Antico Palazzo.

CÓDEGA

See SÍRIA.

CÔDEGA DE LARINHO

Very fruity but soft variety from north-east Portugal.

BERRY COLOUR: ● ● ● ● ●

PRINCIPAL SYNONYMS: Côdega do Larinho

VARIETIES COMMONLY MISTAKEN FOR CÔDEGA DE LARINHO: SÍRIA (Douro)

ORIGINS AND PARENTAGE

This variety probably comes from the Douro and Trás-os-Montes regions in north-eastern Portugal (Rolando Faustino, personal communication). DNA profiling has shown that Côdega de Larinho and SÍRIA, also called Códega in the Douro, are distinct varieties (Vouillamoz).

VITICULTURAL CHARACTERISTICS

Relatively high vigour and productivity. Large, compact bunches of medium-sized berries. Mid ripening. Susceptible to downy mildew and best suited to slopes with warm soils.

WHERE IT'S GROWN AND WHAT ITS WINE TASTES LIKE

Côdega de Larinho, planted mainly in the Douro and Trás-os-Montes regions of north-eastern **Portugal**, produces wines with intense aromas of tropical fruit, sometimes also floral, but they can lack freshness so are often blended with higher-acid varieties such as RABIGATO, Arinto no

Douro (DORINTO) and Gouveio (GODELLO). There were 582 ha (1,438 acres) in Portugal in 2010. Recommended producers include the Adega Cooperativa de Valpaços in Trás-os-Montes and VDS Vinhos do Douro Superior (Castello d'Alba) in the Douro.

CODIVARTA

Soft northern Corsican.

BERRY COLOUR: ● ● ● ● ●

PRINCIPAL SYNONYMS: Codivarte Blanc, Codivertola Blanc, Cudiverta

VARIETIES COMMONLY MISTAKEN FOR CODIVARTA: ?CODA DI VOLPE BIANCA

ORIGINS AND PARENTAGE

Codivarta is a Corsican variety apparently known in Tomino at the northern tip of the French island of Corse since 1800 but its first reliable mention dates back to 1896 in Bastia, a little to the south (Rézeau 1997). The name comes from the Corsican dialect words *coda* ('tail') , probably in reference to the shape of the bunch, and *verta* ('green'). According to Galet (2000), it could be identical to the Italian CODA DI VOLPE BIANCA, but this has not yet been confirmed by DNA profiling.

VITICULTURAL CHARACTERISTICS

Late ripening and not susceptible to botrytis bunch rot.

WHERE IT'S GROWN AND WHAT ITS WINE TASTES LIKE

Codivarta is included in the official register of varieties in **France** but recommended only for the island of Corse, where it may be included in Vins de Pays de l'Île de Beauté and in the white wines of the appellation Vin de Corse and Vin de Corse-Something (eg Vin de Corse-Coteaux du Cap Corse), though only as a supplementary variety. Clos Nicrosi in Rogliano at the northernmost tip of the island produce a blend of VERMENTINO, Ugni Blanc (TREBBIANO TOSCANO) and Codivarta. Wines tend to be relatively low in acidity.

COLOBEL

Almost extinct French hybrid that can make very dark but astringent reds.

BERRY COLOUR: ● ● ● ● ●

PRINCIPAL SYNONYMS: Seibel 8357

ORIGINS AND PARENTAGE

A complex hybrid obtained at the beginning of the twentieth century by Albert Seibel in Aubenas in the Ardèche, southern France, by crossing Seibel 6150 × PLANTET, where:

- Seibel 6150 is a Seibel 405 × Flot Rouge hybrid (see PRIOR for the pedigree of Seibel 405)

– Flot Rouge is a Munson × ARAMON NOIR hybrid (see PRIOR for the pedigree of Munson).

The name Colobel was most likely created by putting together the beginning of the French word *colorant* ('dye') with the end of Seibel (Rézeau 1997).

VITICULTURAL CHARACTERISTICS

Very productive and late ripening. Susceptible to anthracnose.

WHERE IT'S GROWN AND WHAT ITS WINE TASTES LIKE

Although it is authorized in many wine regions in south and central **France**, there were only about 10 ha (25 acres) left in 2008 (down dramatically from 1,269 ha/3,136 acres in 1958), but it is also grown to a very limited extent in **New York** State even though it is only slightly cold-hardy. It tends to be bitter and astringent but very deeply coloured and therefore useful only as a good colouring agent in a blend.

COLOMBANA NERA

Minor, central Italian red that generally disappears into blends.

BERRY COLOUR: ● ● ● ● ●

VARIETIES COMMONLY MISTAKEN FOR COLOMBANA NERA: Basgnano, Bazano, Bersegano, Besgano di S. Colombano, Besgano Nero

ORIGINS AND PARENTAGE

Colombana Nera is a very old variety from the Toscana and Emilia-Romagna regions, possibly cited as early as 1600 by Soderini under the name San Colombane. Indeed, the name Colombana Nera could derive from the San Colombano monastery in Bobbio (in the province of Piacenza in Emilia-Romagna). It has some morphological similarities with Besgano Nero (not commercially cultivated), but it is a distinct variety. Recent DNA studies suggested a possible parent–offspring relationship with MAMMOLO (Di Vecchi Staraz, This *et al.* 2007), which has other Tuscan offspring (BIANCONE DI PORTOFERRAIO, CALORIA and POLLERA NERA), supporting a Tuscan origin for Colombana Nera.

VITICULTURAL CHARACTERISTICS

Vigorous, late ripening.

WHERE IT'S GROWN

Historically cultivated in the north of Toscana, **Italy**, mainly in the province of Massa-Carrara, where it is recommended, and in the provinces of Pisa, Lucca, Livorno and Pistoia, where it is authorized. It is usually blended with other varieties, with SANGIOVESE and other local varieties in Colli di Luni DOC, for example. It has no connection with San Colombano al Lambro DOC in Lombardia. 172 ha (425 acres) were recorded in the Italian census of 2000.

COLOMBARD

Productive workhorse variety now more valued for easy-drinking blended whites than for brandy.

BERRY COLOUR: ● ● ● ● ●

PRINCIPAL SYNONYMS: Colombar (Gironde, South Africa), Colombier (Gironde), French Colombard (California), Queue Tendre (Charente), Tourterelle

VARIETIES COMMONLY MISTAKEN FOR COLOMBARD: SÉMILLON

ORIGINS AND PARENTAGE

Colombard is a variety from the Charente in western central France, with a historical centre of cultivation in Borderies, a famous brandy-producing region once in Dutch hands and now a specific appellation for cognac. The earliest mention of Colombard appeared between 1706 and 1716 in La Rochelle under the old spelling Collombar and Coulombard (Rézeau 1997). Its name probably derives from *colombe* ('dove'), as corroborated by its synonyms (*colombier* = dovecote, *tourterelle* = turtledove), referring to the colour of the berries rather than to their ability to attract doves. Colombard has sometimes been confused with SÉMILLON (Mas and Pulliat 1874–5).

DNA parentage analysis has established that Colombard is the progeny of a natural CHENIN BLANC × GOUAIS BLANC cross, just like its siblings BALZAC BLANC and MESLIER SAINT-FRANÇOIS (Bowers *et al.* 2000). Indeed, Colombard is morphologically similar to Balzac Blanc (Bisson 2009). It was used to breed COLOMINO.

VITICULTURAL CHARACTERISTICS

Fertile, vigorous and productive. Can be pruned short or long but the wood is hard and so pruning is difficult. Mid ripening. Susceptible to powdery mildew on the leaves and botrytis bunch rot on the ripe fruit.

WHERE IT'S GROWN AND WHAT ITS WINE TASTES LIKE

France had 7,790 ha (19,250 acres) in 2009, making it the seventh most planted white variety in France. Although it is recommended in much of the south-west quadrant of France, south from the Charente and the Dordogne and across to parts of the Languedoc, the biggest concentrations are in the Charente-Maritime, Landes, Charente and Lot-et-Garonne *départements*. Although it was not considered as worthy as Ugni Blanc (TREBBIANO TOSCANO) and FOLLE BLANCHE because of its higher alcohol and lower acidity, Colombard was traditionally – and still is – a component in the dry base wines distilled into cognac and armagnac. It tumbled from popularity in the 1970s, losing more than half of its vineyard area between 1968 (11,892 ha/29,386 acres) and 1979 (5,829 ha/14,404 acres), but after a low point in the 1980s the vineyard area is again increasing gradually, most likely because of the growing popularity of the fresh, inexpensive, sometimes lightly floral whites now being produced as Vins de Pays des Côtes de Gascogne and Vins de Pays Charentais. It is also a permitted ingredient in certain Bordeaux appellations such as Bourg or Bordeaux *tout court*.

Colombard's decline in France in the 1970s contrasts with its rise to fame in **California** as French Colombard, particularly in the hot Central Valley where reliably high yields – anywhere between 8 and 13 tons/acre (140–220 hl/ha) – gave vast quantities of blended, often rather sweet, inexpensive wine. Increasing use of stainless steel and cool fermentation techniques resulted in fresh if rather neutral wines. Here the hot, dry climate significantly reduced the risks of fungal disease faced by French growers and high yields tended to retain fresh acidity. It was seeing the popularity of such wines in the US that inspired the Plaimont co-op in south-west France to make dry white wines from it in the Côtes de Gascogne.

It was first introduced to California in the 1850s and was originally and appropriately known as West's White Prolific, after George West, a grower in San Joaquin County. The annual rate of planting in California slowed to a standstill towards the end of the 1980s as CHARDONNAY became more and more popular but then picked up slightly in the mid 2000s. The 25,965 acres (10,508 ha) planted at the end of the first decade were not much more than half what they had been in 2000 and it is hard to imagine that at one time Colombard was the state's most planted white variety – more than 90,000 acres (36,420 ha) in the late 1980s, representing almost 40% of white wine varieties. In the US, Colombard's popularity was not restricted to the Golden State: it is also planted in **Texas**, **Virginia**, **Alabama** and **Arizona**.

In **South Africa** and **Australia**, using the same modern fermentation methods that were so successful in California, it has served a similar purpose, providing usefully crisp blending material, generally with CHENIN BLANC or CHARDONNAY. In 2008 South Africa had 11,877 ha (29,349 acres) of what is known locally as Colombar, also used in the production of brandy; Australia's Colombard plantings totalled 2,669 ha (6,595 acres), mostly in the hot, irrigated inland regions of Riverland, Riverina and Murray Darling. However, there are some honorable exceptions to cask-wine oblivion such as Primo Estate's La Biondina from McLaren Vale.

Also found in **Israel**.

COLOMBAUD

Ancient but virtually extinct white related to the Swiss variety Humagne.

BERRY COLOUR: ● ● ● ● ●

PRINCIPAL SYNONYMS: Aubié (Provence), Colombaud du Var

ORIGINS AND PARENTAGE

Colombaud is an old Provençal variety that was mentioned in 1623 in Vaison (today's Vaison-la-Romaine) near Avignon in France's southern Rhône as a 'treille de raisins colombaux' ('a vine of colombaux grapes') (Rézeau 1997). Several subsequent mentions in this area were recorded under various forms of the name (Colombeau, Colonbau) and the modern spelling Colombaud appeared in 1844 (Rézeau 1997). Unlike COLOMBARD, to which it is not related, there is no suggestion that its name derives from the French *colombe* ('dove') and its etymology is unknown.

Colombaud is said to resemble Mourvèdre (MONASTRELL; Reich 1902b) but it belongs to the Claret ampelographic group (see p XXVII; Bisson 2009). In addition, it is ampelographically close to BOUTEILLAN NOIR, which might explain why Colombaud is often erroneously considered to be a white mutation of Bouteillan Noir (Bisson 2009).

DNA profiling has surprisingly revealed a parent–offspring relationship between Colombaud and HUMAGNE, an old variety from the Valais in Switzerland (Vouillamoz and Moriondo 2011). However, this is only half a surprise since Humagne has recently been discovered in south-east France under the name Miousat (Bordenave *et al.* 2007).

VITICULTURAL CHARACTERISTICS

Very vigorous, and fertile if pruned short. Needs poor soils, very warm conditions. Generally susceptible to fungal diseases but especially to botrytis bunch rot.

WHERE IT'S GROWN

It is virtually extinct in **France** but is still authorized in the Var.

COLOMINO

This light-skinned PALOMINO FINO × COLOMBARD cross, bred in 1978 at the University of Stellenbosch, **South Africa**, barely registers in vineyard statistics: there were 5 ha (12 acres) in South Africa in 2009, down from 8 ha (20 acres) in 2002.

COLORAILLO

Central Spanish variety with confusing homonyms.

BERRY COLOUR: ● ● ● ● ●

ORIGINS AND PARENTAGE

Coloraillo is a variety from the province of Castilla-La Mancha in central Spain but DNA profiling has shown that at least four distinct varieties have been found under the name Coloraillo in this region (Fernández-González, Mena *et al.* 2007):

- the true, red-berried Coloraillo, also analysed by Martín *et al.* (2003)
- light-skinned ALARIJE, known as Coloraillo Gordo or Coloraillo Pequeño in Campillo de Altobuey in Cuenca
- two distinct and unidentified varieties.

WHERE IT'S GROWN

Official statistics in **Spain** report 272 ha (672 acres) of Coloraillo in 2008 but this figure could easily include other varieties that go under the same name. It is an authorized variety in regional wines such as Viña de la Tierra de Castilla and Viña de la Tierra de Sierra de Alacaraz, both in Castilla-La Mancha.

COLORINO DEL VALDARNO

Rare, dark-skinned Tuscan variety often confused with others called Colorino Something and especially useful for adding colour to blends.

BERRY COLOUR: ● ● ● ● ●

PRINCIPAL SYNONYMS: Colore, Colorino

VARIETIES COMMONLY MISTAKEN FOR COLORINO DEL VALDARNO: ABRUSCO⊗, CANAIOLO NERO⊗, Colorino Americano⊗, Colorino di Lucca⊗, Colorino di Pisa⊗

ORIGINS AND PARENTAGE

The name Colorino comes from the dark-blue colour of the berries and has been given to varieties such as ABRUSCO, ANCELLOTTA and CANAIOLO NERO, all with distinct DNA profiles (Vignani *et al.* 2008).

Four types of Colorino are usually distinguished in its homeland of Toscana, central Italy: Colorino Americano, Colorino del Valdarno, Colorino di Lucca and Colorino di Pisa. Ampelographic and genetic analyses have established that the four types correspond to four distinct varieties (see pp XIV–XVI for a definition of variety), with Colorino Americano being the most genetically distant (Sensi *et al.* 1996; Vignani *et al.* 2008). Colorino del Valdarno is the most widespread. In addition, a parent–offspring relationship between Colorino di Lucca and Colorino di Pisa cannot be ruled out (Vouillamoz).

As a consequence, it is not impossible that so-called Colorino vineyards are in fact made of a mix of several distinct varieties (Colorino Americano, Colorino del Valdarno, Colorino di Lucca and Colorino di Pisa).

OTHER HYPOTHESES

Said to derive from the domestication of wild local grapes several centuries ago but no evidence has yet been found to support this.

VITICULTURAL CHARACTERISTICS

Mid ripening, susceptible to powdery mildew. Pale-fleshed.

WHERE IT'S GROWN AND WHAT ITS WINE TASTES LIKE

Traditionally used in Toscana, **Italy**, to add colour to SANGIOVESE-dominated blends such as Chianti or Vino Nobile di Montepulciano (around 5–10%), it became particularly popular for this purpose in the 1980s and 1990s as a local alternative to adding CABERNET SAUVIGNON. Recommended in the whole of Toscana except in the provinces of Massa-Carrara and Livorno. It is also cultivated in Umbria, Marche, Lazio and Liguria. Some rare varietal IGT wines are made in Toscana, eg by Fattoria di Bagnolo in Montemurlo, Tenuta il Corno in San Casciano Val di Pesa, Fattoria del Cerro in Acquaviva and La Spinetta in Terricciola. They are deeply coloured with firm tannins and tend to have either red- or black-fruit flavours.

There were 436 ha (1,077 acres) recorded in the Italian agricultural census in 2000 – though there is no knowing which Colorino the statistics refer to.

COMPLETER

Rare Swiss variety making very distinctive, ageworthy whites.

BERRY COLOUR: ● ● ● ● ●

PRINCIPAL SYNONYMS: Lindauer (formerly in Schaffhausen, Thurgau and Zürich), Malanstraube (Graubünden), Zürirebe (Zürichsee)

VARIETIES COMMONLY MISTAKEN FOR COMPLETER: LAFNETSCHA (Valais), RÄUSCHLING (northern Switzerland)

ORIGINS AND PARENTAGE

Completer is a very old variety from the Malans region in Graubünden, eastern Switzerland, where it was first mentioned in 1321 in a document from the Chapter of Chur cathedral (Jenny 1938): 'quoad vinea vini completorii sisa in Malans' ('regarding the wines of the vineyard of completer based in Malans'). Its name comes from *completorium*, the evening office during which the Benedictine monks were traditionally allowed to drink a glass of wine in silence. Completer was confused with LAFNETSCHA from the Haut-Valais until DNA parentage analysis revealed that the latter is in fact a natural cross between Completer and HUMAGNE (Vouillamoz, Maigre and Meredith 2004) – surprising since Completer had never been recorded in the Valais. A thorough search uncovered a few old vines of Completer in the region of Visp, Haut-Valais, under the name Grosse Lafnetscha, providing evidence that Completer has been unwittingly cultivated in the Valais (Vouillamoz, Frei and Arnold 2008). Furthermore, DNA parentage analysis has also shown that two distinct natural crosses between Completer and Ticino's BONDOLA gave birth to the siblings BONDOLETTA in Ticino and HITZKIRCHER in the canton of Luzern (Vouillamoz and Moriondo 2011; see pedigree diagram opposite).

The parents of Completer are unknown and have most likely disappeared but historical data suggest that Completer was introduced from Italy by Benedictine monks from Kloster Pfäfers (Pfäfers Abbey in Sankt Gallen, north-eastern Switzerland), who used to own vineyards in Malans and in Italy (Aeberhard 2005). (This is consistent with Completer having close genetic links with the Italian varieties MARZEMINO and Pignolo Spano (no longer cultivated); Vouillamoz and Moriondo 2011.) If so, it is likely that they introduced Completer independently to the Haut-Valais, where traditional vineyards, including some old Completer vines, are still trained on Italian-style pergolas, and to Graubünden.

OTHER HYPOTHESES

Some authors claim that Completer was mentioned in Malans as early as 926 (Bellasi *et al.* 1993) but this is disproved by Jenny's (1938) thorough historical study.

VITICULTURAL CHARACTERISTICS

Mid budding, mid to late ripening. Medium to high, regular yields. Best suited to light soils in wind-freshened sites.

WHERE IT'S GROWN AND WHAT ITS WINE TASTES LIKE

Completer wines typically have complex aromas of quince, ripe apple, plums and honey, becoming more honeyed as they age gracefully, plus a powerful but supple structure, very high natural acidity and a tendency towards an (attractive) oxidative character. Alcohol levels are usually high but this may be due to the way the local Föhn wind dries and concentrates the grapes as much as to any varietal characteristic. There were just 3 ha (7 acres) of Completer in **Switzerland** in 2009, mainly in Graubünden in the south east, where a dozen producers make varietal wines. Schloss Reichenau and Completer-Kellerei make traditional, often slightly oxidative versions, whereas Donatsch, Peter & Rosi Hermann, and Thomas Studach produce a more modern style using small, new oak barrels. In Ticino, MERLOT stars Werner Stucky and Daniel Huber use some Completer in their white blends; and in the canton of Zürich, Hermann Schwarzenbach makes a varietal wine from the single tiny Completer vineyard in the area.

COMPLETER PEDIGREE DIAGRAM

Transalpine natural crossings between COMPLETER (Graubünden) and BONDOLA (Ticino) and between Completer and HUMAGNE (Valais) gave birth in Switzerland to varieties of strictly local interest, with Humagne making the French connection. The connections with unknown varieties (?) show that HIMBERTSCHA and COLOMBAUD could also theoretically be parent and offspring, respectively, of Humagne (see p x).

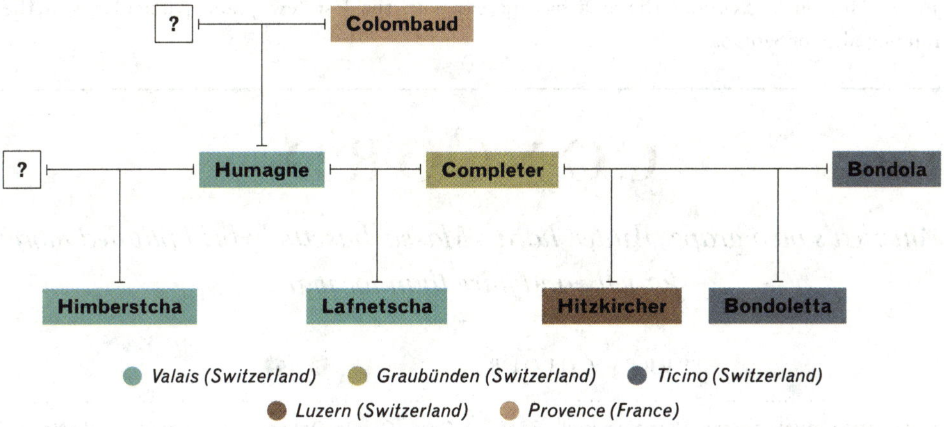

COMPLEXA

Portuguese cross making light, workaday wines on the island of Madeira.

BERRY COLOUR: ● ● ● ● ●

ORIGINS AND PARENTAGE

This (CASTELÃO × Tintinha) × MUSCAT OF HAMBURG cross was bred by José Leão Ferreira de Almeida at the Estação Agronómica Nacional in Oeiras near Lisboa, Portugal, in the 1960s and introduced to Madeira in the 1970s as an alternative to Tinta Negra (NEGRAMOLL). The true identity of Tintinha is uncertain since it is both a synonym for PETIT BOUSCHET and the name of a distinct variety whose DNA profile in Almadanim *et al.* (2004) is unique.

VITICULTURAL CHARACTERISTICS

Moderate vigour, late ripening, medium to high yields. Medium-sized bunches with a small number of red-fleshed berries and a low level of anthocyanins in the skins. Adaptable to the range of environments on the island but susceptible to phomopsis.

WHERE IT'S GROWN AND WHAT ITS WINE TASTES LIKE

Table wines are moderate in alcohol, fresh, quite light in colour, even though Complexa berries have red flesh, and not suited to ageing. The wines are in general a little paler than varietal Tinta Negra (NEGRAMOLL) and slightly higher in alcohol but they are less astringent. It is also used to make fortified wines but is nearly always better in a blend than solo. According to Ricardo Diogo of Barbeito, Complexa wines have higher acidity than Tinta Negra and are more elegant, so he prefers the results for his Madeira wines. There were 33 ha (82 acres) on Madeira, **Portugal**, in 2010, mainly on the south side of the island around Calheta and on the north side around São Jorge (Mayson 2003), and the area has increased in the last few years, particularly in the municipality of Santana.

CONCORD

America's own grape. Winter-hardy, Massachusetts hybrid planted more for jelly and juice than for wine.

BERRY COLOUR: ● ● ● ● ●

PRINCIPAL SYNONYMS: Bergerac, Bull's Seedling, Corin, Cornin, Dalmadin, Feherhatu, Fekete Noah, Furmin Noir, Gorin, Gurin, Kek Olasz, Konkordi, Konkordia, Nyarfalevelue, Nyarlevelue, Olasz Kek

ORIGINS AND PARENTAGE

The historical origin of this variety was made clear by Liberty Hyde Bailey (1906) and Wilfrid Wheeler (1908): Ephraim Wales Bull, an inhabitant of the town of Concord in Massachusetts, US, related that in 1840 some boys brought up from the river some wild grapes and scattered them around his house. A seedling appeared and Bull obtained the first bunch of grapes from the vine in 1843. He then planted seeds from this bunch and the resulting plant first fruited in 1849. The fruit had such merit that all other seedlings were destroyed. The variety was named Concord and was first officially recorded in 1853 by the Massachusetts Horticultural Society: 'E W Bull exhibited his new seedling grape, which, under the name of Concord, is now so generally cultivated throughout the country.'

From a genetic point of view, Concord is a grandchild of an unknown grapevine that was growing wild along the Concord River in 1840, most likely deriving from *Vitis labrusca* that still grows abundantly in the area. However, like many other North American hybrid varieties (such as CATAWBA or NORTON), Concord is probably a cross between *Vitis labrusca* and an unidentified *Vitis vinifera* subsp. *vinifera* variety (Stover *et al.* 2009).

Concord Seedless is a sterile mutation of Concord. White Concord is not a colour mutation of Concord but a synonym for NIAGARA.

Concord was used to breed MOORE'S DIAMOND and NIAGARA.

VITICULTURAL CHARACTERISTICS

Winter-hardy, vigorous and productive. Mid to late ripening. Highly susceptible to black rot, phomopsis and eutypa dieback, less so to downy mildew, botrytis bunch rot, anthracnose and crown gall. Medium-sized bunches of large berries. In warmer climates, prone to uneven ripening. Sensitive to sulphur application in the vineyard and the berries are also prone to cracking.

WHERE IT'S GROWN AND WHAT ITS WINE TASTES LIKE

Concord is the most widely planted variety in the eastern United States, especially in **New York** State, where there were 20,217 acres (8,181 ha) in 2006 and it is grown on a greater variety of soils and under more different climatic conditions than any other American vine. However, the juice has the powerful 'foxy' flavour typical of *Vitis labrusca* and the great majority of the fruit is harvested for juice and jelly: as little as 5–10% of New York's total is used for wine. **Pennsylvania** had 1,063 acres (430 ha) planted for wine in 2008 and it is also found in **Ohio** (665 acres/269 ha in 2008), **Missouri** (140 acres/57 ha in 2009), **Michigan** (60 acres/24 ha in 2003), **Illinois** (21 acres/8.5 ha in 2007), **Iowa** (21 acres/8.5 ha in 2006), **Indiana** (20 acres/8 ha in 2010) and **Nebraska** (1 acre/0.4 ha in 2007), for example. In the Pacific Northwest, it is widely planted in **Washington** state (25,068 acres/10,145 ha in 2011).

Wines are medium-bodied and dark bluish purple in colour and made in a range of styles, some decidedly sweet and some usefully kosher, but they typically have the foxy flavour found in the juice, limiting the variety's popularity beyond the eastern US, where its winter hardiness and high yields may have been more appealing than its flavour. Any American product sold as 'grape'-flavoured is likely to exhibit Concord's powerful aroma, not unlike a blend of damp fur and wild strawberries. Stone Hill in Missouri and Oliver Winery in Indiana make good examples, both in a sweeter style.

Concord is also authorized, along with three other American hybrids – Ripatella (thought to be a *Vitis riparia* × Taylor (see NOAH) hybrid), ELVIRA and DELAWARE – in Uhudler, an idiosyncratic table-wine speciality of the Südburgenland, **Austria**, of which around 160,000 litres are produced annually.

There are also significant plantings in **Brazil** – 2,483 ha (6,136 acres) in the state of Rio Grande do Sul in 2007 – and lesser amounts in **Peru** and **Japan**, where it is used for wine and juice.

CORBEAU

See DOUCE NOIRE.

CORDENOSSA

Recently rescued and still rare Friulian red.

BERRY COLOUR: ● ● ● ● ●

PRINCIPAL SYNONYMS: Cordenos

ORIGINS AND PARENTAGE

Cordenossa is an old variety from Friuli-Venezia Giulia in north-east Italy, probably originating from Cordenons near Pordenone, hence the name.

VITICULTURAL CHARACTERISTICS

Vigorous, productive, mid to late ripening. Small, loose bunches of thin-skinned berries. Very resistant to powdery mildew but susceptible to downy mildew.

WHERE IT'S GROWN AND WHAT ITS WINE TASTES LIKE

Cordenossa used to be grown in the area around Cordenons, Zoppola and San Vito al Tagliamento in Friuli-Venezia Giulia, **Italy**. It had almost disappeared when Emilio Bulfon, Antonio Calò and Ruggero Forti rescued it, along with other old local varieties such as CIANORIE, FORGIARIN,

PICULIT NERI, SCIAGLÌN and UCELÙT. Emilio Bulfon makes a rare varietal wine, which is said to have aromas of violets, fruits of the forest and a hint of tobacco and to taste of blackcurrants and plums.

CORNALIN

Ancient red from the Valle d'Aosta in northern Italy, widespread in the Valais in Switzerland as Humagne Rouge.

BERRY COLOUR: ● ● ● ● ●

PRINCIPAL SYNONYMS: Broblanc (Valle d'Aosta), Cargnola (Valle d'Aosta), Cornalin d'Aoste, Cornalino, Cornallin or Corniola (Valle d'Aosta), Humagne Rouge⚭ (Valais in Switzerland)

VARIETIES COMMONLY MISTAKEN FOR CORNALIN: PETIT ROUGE⚭ (Valle d'Aosta), ROUGE DU PAYS⚭ (Valais in Switzerland).

ORIGINS AND PARENTAGE

Cornalin is an old variety from the Valle d'Aosta, north-eastern Italy, on the border with Switzerland, where it was widely cultivated in the nineteenth century. Its name might derive from *corniolo*, the Italian name for the cornelian cherry (*Cornus mas*). Two types have been described by Gatta (1838), having slightly distinct morphology and distribution: Cornalin and Corniola. It is not clear whether they represented clonal variations or distinct varieties but a link with the southern Italian white Corniola di Milazzo can be ruled out.

In the Valais in southern Switzerland, the variety is called Humagne Rouge, although it is not related to the white-berried HUMAGNE. Humagne Rouge was long thought to be identical to the PETIT ROUGE of the Valle d'Aosta but morphological and DNA evidence has shown that Humagne Rouge is in fact identical to Cornalin (Moriondo 1999; Vouillamoz *et al.* 2003). In addition, DNA paternity testing has demonstrated that Cornalin is a natural progeny of ROUGE DU PAYS, which also happens to be called Cornalin in the Valais. Since Rouge du Pays is a natural offspring of Petit Rouge and MAYOLET, Cornalin is their grandchild (Vouillamoz *et al.* 2003; see PRIÉ for the pedigree diagram).

OTHER HYPOTHESES

Some authors have suggested that Cornalin and other Valle d'Aosta varieties were introduced from Burgundy by the Baron de Champorcher in around 1750. However, absolutely no link with Burgundy has been detected for any Valle d'Aosta variety (Vouillamoz and Moriondo 2011).

VITICULTURAL CHARACTERISTICS

Medium to high vigour, late ripening. Generally good resistance to fungal diseases.

WHERE IT'S GROWN AND WHAT ITS WINE TASTES LIKE

The variety is cultivated on a little over 1 ha (2.5 acres) in the Valle d'Aosta, **Italy**, according to the Italian agricultural census of 2000, whereas it covers ten times that area in the Valais, **Switzerland**, under the name Humagne Rouge. Varietal Cornalin qualifies as Valle d'Aosta DOC. Producers include Maison Anselmet and Grosjean Frères. Valaisan producers of varietal Humagne Rouge include Cave La Romaine, Defayes & Crettenand and Maurice Zufferey. The wines are well structured with somewhat rustic tannins and aromas of smoke, bark, pepper and fresh black fruits, pairing well with game dishes.

CORNALIN DU VALAIS

See ROUGE DU PAYS.

CORNAREA

Obscure dark-skinned Italian cross whose true parentage has only recently been revealed.

BERRY COLOUR:

PRINCIPAL SYNONYMS: Incrocio Dalmasso IV/28

ORIGINS AND PARENTAGE

The Italian vine breeder Giovanni Dalmasso claimed that Cornarea was a NEBBIOLO × BARBERA cross created in 1936 at the Conegliano research centre in the Veneto, northern Italy, with the objective of combining the quality of Nebbiolo with the resistance and naturally high yields of Barbera. However, DNA profiling recently showed that the Nebbiolo used by Dalmasso was actually Nebbiolo di Dronero, which is not Nebbiolo but a synonym for CHATUS (Torello Marinoni, Raimondi, Ruffa *et al.* 2009). Cornarea is therefore a Chatus × Barbera cross, just like its siblings ALBAROSSA, NEBBIERA, SAN MICHELE and SOPERGA. It entered the official Italian register of varieties in 1977.

Cipriani *et al.* (2010) have recently suggested through DNA parentage analysis that Cornarea was indeed a NEBBIOLO × BARBERA cross but this parentage showed some DNA discrepancies and is not consistent with Torello Marinoni, Raimondi, Ruffa *et al.* (2009), possibly because the Nebbiolo used by Cipriani *et al.* (2010) might not be true-to-type.

VITICULTURAL CHARACTERISTICS

Mid ripening.

WHERE IT'S GROWN

In **Italy** just 25 ha (62 acres) were reported in the agricultural census of 2000. It is planted mainly in Piemonte and generally used to enhance colour in blends.

CORNIFESTO

Minor Douro variety used for both port and unfortified wines.

BERRY COLOUR:

PRINCIPAL SYNONYMS: Cornifeito, Cornifesta (Dão), Cornifesto no Dao, Cornifesto Tinto, Cornifresco, Tinta Bastardeira (Regua)

ORIGINS AND PARENTAGE

Recent DNA parentage analysis has shown that Cornifesto is a natural cross between CAYETANA BLANCA, widespread throughout the south and centre of the Iberian peninsula, and

ALFROCHEIRO from the south or centre of Portugal (Zinelabidine *et al.* 2012), which possibly occurred in the Douro region in north-east Portugal, its main area of cultivation. Cornifesto is therefore a sibling of CASTELÃO, MALVASIA PRETA, JUAN GARCÍA and CAMARATE (see Cayetana Blanca for the pedigree diagram). Its name literally means 'curved (*festo*) horn (*corni*)', referring to the shape of the bunches (Galet 2000).

Santana *et al.* (2010) have collected a variety called Cornifesto in Arribes del Duero just over the border in Spain but it is not yet known whether it is identical to its Portuguese counterpart.

VITICULTURAL CHARACTERISTICS

Small, compact bunches of smallish, thin-skinned berries. Susceptible to powdery mildew and moderately so to powdery mildew and botrytis bunch rot. Regular yields, mid ripening.

WHERE IT'S GROWN AND WHAT ITS WINE TASTES LIKE

The variety is grown mainly in the Douro, where it is authorized for both port and dry table wines. Varietal wines tend to have red-fruit and light floral aromas. There were 135 ha (334 acres) in **Portugal** in 2010.

Campbells of Rutherglen in Victoria, **Australia**, have a small patch of Cornifesto in a vineyard block dating back to the 1960s, generally used in a port-style blend, and a very small area of this variety remains in **South Africa**, used for the same purpose.

COROT NOIR

Minor hybrid bred in upstate New York for the eastern US.

BERRY COLOUR: ● ● ● ● ●

PRINCIPAL SYNONYMS: NY 70.0809.10

ORIGINS AND PARENTAGE

A complex hybrid obtained in 1970 by Bruce Reisch and Thomas Henick-Kling at New York State Agricultural Experiment Station, Cornell University, Geneva, by crossing Seyve-Villard 18-307 with STEUBEN (Reisch *et al.* 2006a), where Seyve-Villard 18-307 is a Chancellor × Subéreux hybrid (see PRIOR for the complete pedigrees of Chancellor and Subéreux). It was selected in 1978 as NY 70.0809.10 and finally named and released in 2006 together with NOIRET and VALVIN MUSCAT (Reisch *et al.* 2006c). Corot Noir is therefore a complex *Vitis riparia*, *Vitis labrusca*, *Vitis vinifera*, *Vitis lincecumii* and *Vitis rupestris* hybrid.

VITICULTURAL CHARACTERISTICS

Vigorous, late budding, mid to late ripening. Moderately winter-hardy and moderately resistant to fungal diseases though powdery mildew may be a problem. Trials so far suggest grafting is unnecessary.

WHERE IT'S GROWN AND WHAT ITS WINE TASTES LIKE

Corot Noir is grown mainly in **New York** State, where there were 59 acres (24 ha) in 2006, and to a lesser extent in **Pennsylvania** (4 acres/1.6 ha in 2008), **Ohio** and **Illinois**. Wines are deep red with aromas of cherries and berries and big, rounded tannins. It's often used in blends but producers of varietal wines include Deere Run, Fulkerson and Otter Creek in New York and Ridge View in Illinois.

CORREDERA

Very minor cross currently being assessed in Jerez, though mainly for brandy.

BERRY COLOUR: ● ● ● ● ●

ORIGINS AND PARENTAGE

A PALOMINO FINO × CARDINAL cross obtained at the Andalucía government research centre in southern Spain.

VITICULTURAL CHARACTERISTICS

Early budding and mid ripening. Fertile and productive (more than 4 kg per vine). Berries can achieve high sugar levels with moderate acidity.

WHERE IT'S GROWN

Corredera, along with several other varieties, has been the subject of official research in Jerez, **Spain**, into the most suitable varieties for distillation.

CORTESE

Generally rather bland northern-Italian white that shows its best varietal face in Gavi.

BERRY COLOUR: ● ● ● ● ●

PRINCIPAL SYNONYMS: Bianca Fernanda (Veneto), Corteis, Courteis, Courtesia

ORIGINS AND PARENTAGE

Cortese was first mentioned in an inventory of the cellars at the castle of Casale Monferrato in the province of Alessandria, in south-eastern Piemonte, in 1614 (Nada Patrone 1991). It was mentioned again in 1658 in Montaldeo (also in Alessandria), where vineyards were planted with 'cortese', some 'fermentino' (ie VERMENTINO) and 'nebioli dolci' ('sweet NEBBIOLOs'; Nada Patrone 1991). In the eighteenth century, Cortese was found mainly in the province of Alessandria as well as in the province of Novara (Demaria and Leardi 1875). No parentage was found for Cortese in northern Italy (Schneider *et al.* 2003).

WHERE IT'S GROWN AND WHAT ITS WINE TASTES LIKE

Cultivated in the provinces of Asti and Alessandria in south-eastern Piemonte, north-west **Italy**, but known particularly as the star of Gavi or Cortese di Gavi DOCG. It appears as a varietal in several DOCs in Piemonte such as Colli Tortonesi and Monferrato – though ripeness and quality are rarely on a par with Gavi's – and is also cultivated in Oltrepò Pavese in Lombardia and in Garda in the Veneto. According to the Italian agricultural census of 2000, total plantings were 3,135 ha (7,747 acres). Gavi was originally produced to supply the fish restaurants of Genova and the Ligurian coast and the wines are generally rather neutral but rescued by their refreshing acidity. At their best, they can be gently aromatic with restrained mineral and citrus flavours.

CORVA

Old and virtually extinct dark-skinned variety surviving in the middle of the city of Brescia.

BERRY COLOUR: ● ● ● ● ●

VARIETIES COMMONLY MISTAKEN FOR CORVA: CORVINA VERONESE, LAMBRUSCA DI ALESSANDRIA (Pinerolese)

ORIGINS AND PARENTAGE

Old variety from the province of Brescia in Lombardia, northern Italy, where it was once widely cultivated. Because of their similar names, Corva has been confused with CORVINA VERONESE and LAMBRUSCA DI ALESSANDRIA, known as Corva in Pinerolese, but it has a distinct and unique DNA profile (Vouillamoz).

VITICULTURAL CHARACTERISTICS

Mid to late ripening.

WHERE IT'S GROWN AND WHAT ITS WINE TASTES LIKE

Right in the middle of the city of Brescia in central Lombardia, **Italy**, at the foot of the hill up to the castle, is the historic Vigneto della Pusterla, owned by the Pusterla winery. This 4 ha (10 acre) vineyard is trained in a local pergola style and produces not only a varietal Invernenga but also 2,000 bottles per year of Pusterla Rosso IGT Ronchi di Brescia, a blend of old, local and sometimes rare varieties, including Corva as well as MARZEMINO, GROPPELLO GENTILE, MAIOLINA, SCHIAVA GROSSA and BARBERA.

CORVINA VERONESE

Bright, fresh, cherry-scented red encountered most often in Valpolicella and Bardolino.

BERRY COLOUR: ● ● ● ● ●

PRINCIPAL SYNONYMS: Corvina Comune, Corvina Gentile, Corvina Grossa, Corvina Nostrana, Corvina Reale, Cruina

VARIETIES COMMONLY MISTAKEN FOR CORVINA VERONESE: Corbina, CORVA, CORVINONE

ORIGINS AND PARENTAGE

The name Corvina might not simply derive from *corvo* ('crow') in reference to the colour of the berries but rather from the local dialect *cruina* (Italian *crua* means 'unripe') in reference to its late maturation. Corvina Veronese could be identical to the Acinatico grape already mentioned in the Middle Ages for the production of Recioto.

The name Corvini appeared in 1627 among the wines of Verona, and a Corvina was found in 1755 (Calò and Costacurta 2004), but those names could represent distinct varieties. The full name Corvina Veronese first appeared in Pollini (1818), in which it is described as widespread in Val Pullicella (Valpolicella in the province of Verona), where it is considered an autochthonous variety.

Corvina Veronese has long been thought to be identical to Corbina, an ancient and almost-forgotten variety of Friuli-Venezia Giulia, rescued and included in the national register of varieties in 2007. However, morphological and molecular studies showed that Corbina and Corvina Veronese are distinct and unrelated varieties (Cancellier *et al.* 2007). Similarly, CORVINONE was generally considered to be a clonal mutation of Corvina Veronese but ampelographic, isoenzymatic and DNA profiling have shown that they are distinct varieties (Cancellier and Angelini 1993), although they might be genetically related.

The genetic group of Corvina Veronese includes DINDARELLA, GARGANEGA, OSELETA and RONDINELLA (Vantini *et al.* 2003). Parentage analysis at sixty DNA markers provided evidence that Corvina Veronese has parent–offspring relationships with at least two northern Italian cultivars: REFOSCO DAL PEDUNCOLO ROSSO and Rondinella (Grando *et al.* 2006). Since Refosco dal Peduncolo Rosso is a progeny of MARZEMINO (Grando *et al.* 2006), Corvina Veronese must be a progeny of Refosco dal Peduncolo Rosso and a parent of Rondinella (see the Refosco dal Peduncolo Rosso pedigree diagram). As a consequence, Corvina Veronese is also a grandchild of Marzemino.

VITICULTURAL CHARACTERISTICS

Late budding and mid to late ripening. Good resistance to winter cold. Vigorous but often fails to bear fruit on the first few buds. Very sensitive to downy mildew, drought stress and sunburn on the berries. Particularly suited to drying (as in the production of Amarone and Recioto).

WHERE IT'S GROWN AND WHAT ITS WINE TASTES LIKE

Corvina Veronese is blended in the Veneto, north-east Italy, with its progeny RONDINELLA and with MOLINARA to produce Valpolicella, including Amarone and Recioto styles (ranging from 40% to 80% of the blend), and Bardolino (33–65%) but it is generally considered to be the superior component in such blends. Garda DOC allows for a 100% Corvina Veronese wine. A few producers such as Allegrini make varietal wines (eg the single-vineyard La Poja) that demonstrate the variety's ability to produce serious, barrel-aged reds. Varietal wines or those in which Corvina Veronese dominates are marked by their bright sour-cherry flavour, fresh acidity and occasionally a note of bitter almonds. They are light-bodied only if the yields are high, and tannins tend to be light to moderate. There were 4,867 ha (12,027 acres) in Italy in 2000.

Freeman Vineyards in the Hilltops region of New South Wales have pioneered this variety in **Australia**. They have 4 ha (10 acres) planted and they blend it with RONDINELLA to make an Amarone-style wine.

In high-altitude vineyards in Tupungato, Mendoza, **Argentina**, the Veneto producer Masi make a similar style of wine from Corvina Veronese and COT (Malbec).

CORVINONE

A dark-skinned variety with fresh acidity that contributes to Veneto blends.

BERRY COLOUR: ● ● ● ● ●

PRINCIPAL SYNONYMS: Corvinon, Cruinon

VARIETIES COMMONLY MISTAKEN FOR CORVINONE: Corbina✗, CORVINA VERONESE✗

ORIGINS AND PARENTAGE

Corvinone has generally been considered to be a clonal mutation of CORVINA VERONESE but ampelographic, isoenzymatic and DNA profiling have shown that they are distinct varieties (Cancellier and Angelini 1993), although they might be genetically related.

VITICULTURAL CHARACTERISTICS

Late ripening and vigorous. Loose bunches and large berries make it particularly suited to drying. Susceptible to downy mildew.

WHERE IT'S GROWN AND WHAT ITS WINE TASTES LIKE

In Valpolicella, in **Italy**'s Veneto, Corvinone may be substituted for up to 50% of the 40–80% of CORVINA VERONESE allowed in the blends for straight Valpolicella DOC as well as for Amarone and Recioto DOCGs. It may also form part of the blend in Bardolino DOC and Bardolino Superiore DOCG. It is not generally made as a varietal wine. There were 95 ha (235 acres) recorded in the Italian agricultural census in 2000.

COT

Makes dark, flavoursome, well-structured reds more celebrated as Argentine Malbec than in its home Cahors.

BERRY COLOUR: ● ● ● ● ●

PRINCIPAL SYNONYMS: Agreste (Lorraine), Auxerrois (Quercy), Bouyssales (Tarn-et-Garonne), Cagors (Republic of Moldova), Cahors (Gironde and Loir-et-Cher), Coq Rouge (Loir-et-Cher), Cor or Cors (Indre-et-Loire), Cos (Vienne and Indre-et-Loire), Côt (Vienne and Touraine), Cots, Estrangey or Étranger (Gironde), Lutkens (Gironde), Malbec or Malbeck (Gironde and South America), Malbech (Italy), Mancin (Gironde), Nuar de Presac (Republic of Moldova), Pied de Perdrix (Tarn-et-Garonne), Pressac or Noir de Pressac (Saint-Émillion), Prunelat (Gironde), Quercy (Charente)

VARIETIES COMMONLY MISTAKEN FOR COT: ABOURIOU (called Malbec Argenté in the Médoc), BOUCHALÈS, CANARI NOIR (Ariège), MENOIR (Hungary), NÉGRETTE (Gers), PRUNELARD (Gironde, where Cot is called Prunelat)

ORIGINS AND PARENTAGE

Cot is an old variety from what used to be known as the province of Quercy in France, centred on Cahors and corresponding to the modern *département* of Lot and part of the Tarn-et-Garonne. Yet it was first mentioned under the name Noir de Pressac by Fontemoing in 1761 with respect to his Château Trochau (today's Château Haut-Tropchaud) in Pomerol (Garde 1946), and then further north in 1783–4 in La Châtre (due east of Poitiers) and in Issoudun, near Bourges, under the name Cor: 'Cor [. . .] Baye grosse, ovale, cassante sous la dent, d'un noir bien foncé, grappe très longue d'une grosseur moyenne, très lâche, donnant beaucoup de vin d'une bonne qualité' ('Cor [. . .] Large berry, oval, easily crunched between the teeth, dark black, very long, loose, medium-sized bunch, giving a lot of good-quality wine') (Rézeau 1997). However, it is unlikely that the Indre *département*, where Cot is also called Cau, Cor, Cors or Cos, is its birthplace, because these names are all contractions of Cahors in Quercy, the most likely cradle of this variety.

From Quercy, Cot was introduced to the Gironde in the eighteenth century although prior to that 'the black wines of Cahors' were sometimes used to add colour and body to the more delicate products of the Bordeaux region. According to Auguste Petit-Lafitte (1868), it was first of all known as Estrangey or Étranger ('stranger') in the Médoc and the Graves, suggesting it had been introduced from another wine region. Later the variety was known there as Lutkens (often misspelt Luckens) after a physician from Bordeaux who died in 1782 and who planted it in Camblanes just south of the city (today's Camblanes et Meynac). At the same time, Cot was also called Malbeck after the person who cultivated it in Sainte-Eulalie d'Ambarès in what is now the Premières Côtes de Bordeaux (today's Sainte-Eulalie) and who propagated it throughout the Médoc, its first citation appearing in 1783–4 in Pauillac: 'Malbeck ou Cahors. Il se multiplie beaucoup dans les nouvelles complantations des terres fortes' ('Malbeck or Cahors. It is widely propagated in the new plantings in robust soils'; Rézeau 1997). In Saint-Émilion, Cot was called Pressac or Noir de Pressac from at least 1784, after the person who introduced this variety to that area.

Cot has numerous additional names all over France and around the world, reflecting the significant clonal diversity within the same variety and its wide and ancient distribution.

Cot was often confused with BOUCHALÈS, perhaps because both belong to the Cot ampelographic group (see p XXVII; Bisson 2009).

Recent DNA parentage analysis has revealed that Cot is a natural progeny of PRUNELARD, an old variety from the Tarn, and Magdeleine Noire des Charentes, the mother of MERLOT (Boursiquot *et al.* 2009; see CABERNET SAUVIGNON and PINOT pedigree diagrams). This makes Cot a half-sibling of Merlot, its birthplace somewhere between its parents' homes (the Tarn and the Charente), which is consistent with its originating from Quercy. By a natural cross with FOLLE BLANCHE, Cot is also a parent of JURANÇON NOIR.

Cot was used to breed CALADOC.

Although the Italian variety MALBO GENTILE has a name similar to Malbec, they are unrelated.

VITICULTURAL CHARACTERISTICS

Vigorous and sensitive to coulure, though this has been minimized by clonal selection. Young vines are at risk from spring frosts and the secondary buds are not very fertile. When ripe, the berries sometimes drop. May be susceptible to phomopsis and leafhoppers. Mid ripening.

WHERE IT'S GROWN AND WHAT ITS WINE TASTES LIKE

Although Cot (sometimes Côt) is a native of south-west **France**, its centre of gravity moved to the southern hemisphere in the late twentieth century thanks to the success of Argentine Malbec (formerly Malbeck) and the fact that the Malbec vines lost to the 1956 frosts in Bordeaux were not generally replaced. Today French plantings are mainly in the south-western *département* of the Lot, where it must constitute at least 70% of any wine labelled Cahors (and even here there is now an increasing tendency to call the variety Malbec in recognition of Argentine Malbec's popularity in the US). It plays a more minor role in other south-western appellations such as Bergerac, Buzet, Côtes de Duras, Fronton, Côtes du Marmandais, Pécharmant and Côtes du Brulhois. It is also theoretically allowed into Cabardès and Côtes de la Malepère but is rarely found this far from Atlantic influence. If picked before full maturity, it tastes herbaceous, and tannins may be very astringent, leading to a certain hard bitterness. Château du Cèdre and Domaine Cosse Maisonneuve produce some particularly fine examples in Cahors with the drama of fully ripe Malbec given a sort of French polish. Total plantings in France in 2009 were 6,155 ha (15,209 acres).

Only 934 ha (2,308 acres) remain in the Gironde (compared with nearly 5,000 ha/12,350 acres at the end of the 1960s) although it is still permitted in all major red Bordeaux appellations. It is most persistent in Bourg, Blaye and the Entre-Deux-Mers but even there the vines are being replaced by less demanding varieties such as MERLOT.

Further north in the middle Loire it is even harder to get Cot to ripen fully than in Bordeaux, although it does tend to ripen about ten days earlier than CABERNET FRANC if yields are controlled. Growers have found that it is much better as a varietal than blended with GAMAY NOIR, CABERNET SAUVIGNON or Cabernet Franc. It is permitted in appellations such as Anjou, Coteaux du Loir, Touraine of various sorts and sparkling Saumur but it is being gradually replaced by Cabernets Sauvignon and Franc. Even so, there were still 560 ha (1,384 acres) planted in the region in 2009. At Domaine de la Charmoise, Henry Marionnet's Vinifera Cot, produced from 1 ha (2.5 acres) of ungrafted vines, produces one of the silkiest examples of this variety in the region. Other good producers include Jean-François Mérieau and Le Cellier du Beaujardin.

According to the 2000 agricultural census, **Italy** had 267 ha (660 acres) of what is known as

Malbech, mainly in the north east. Paladin's Gli Aceri Malbech from the Veneto and Spinelli's Malbec from Abruzzo are rare examples of varietal wines although they are permitted in the Lison-Pramaggiore DOC in the Veneto. In the same region, it may also play a minor part in Montello–Colli Ascolani and much further south, in Puglia, it may be included in the blend for Rosso Barletta and Rosso di Cerignola.

Malbec is also found in **Turkey** (21 ha/52 acres in 2010), the **Republic of Moldova** (39 ha/96 acres in 2009) and **Israel**.

Much of **California**'s 1,469 acres (594 ha) is used to add colour and acidity to Meritage blends, and to soften the tannins of other Meritage components such as CABERNET SAUVIGNON, and it generally plays a bit part in Opus One. However, varietal wines are produced by, for example, Chateau St Jean (Sonoma), Ferrari Carano (Dry Creek Valley), Geyser Peak (Alexander Valley) and Rutherford Hill (Napa Valley). Producers such as Ironstone use it to make a varietal rosé. **Washington** State produces one of the widest ranges of varietal Malbecs in the US but the success of Argentine Malbec there has encouraged plantings in many other states.

Canada's British Columbia has 72 acres (29 ha) and plantings appear to be on the rise, and the variety is bottled as a varietal in **Mexico**.

Malbec was introduced to **Argentina** by the French agricultural engineer Michel Pouget in 1868. Today it is the most widely planted good-quality variety in the country with 26,845 ha (66,335 acres), way ahead of Bonarda (DOUCE NOIRE; 18,759 ha/46,354 acres) and CABERNET SAUVIGNON (17,746 ha/43,851 acres), found in just about every wine-growing region but with the vast majority of plantings in Mendoza (22,885 ha/56,550 acres). Only CEREZA, used mainly for basic-quality rosé or grape concentrate, has a greater area under vine. Bunches are smaller and tighter and berries smaller than on the Cot vines of south-west France; wines are deeply coloured, intensely flavoured with black fruits and more than a hint of game, sometimes with a note of violets, and range in style from soft and fruity to more structured, sophisticated and ageworthy. Vineyards in Lujan de Cuyo in the upper Mendoza Valley produce some of the very best wines at altitudes between 800 and 1,100 m (2,640 and 3,630 ft) although Colomé way up in Salta has shown that Malbec can be made in a very fine, restrained style at even higher altitudes. Top producers include Achaval Ferrer, Alta Vista, Susanna Balbo's Dominio del Plata, Catena Zapata, Colomé, Paul Hobbs' Cobos, Cuvelier Los Andes, Fabre Montmayou, Norton, O Fournier, Pulenta, Salentein, Terrazas de los Andes, Pascual Toso, Trapiche and, in a more traditional style, Weinert.

Chile is a minor Malbec grower with 1,027 ha (2,538 acres) in 2008, of which the majority – 385 ha (951 acres) – is in Colchagua, the home of Chile's Malbec specialist Viu Manent. On this side of the Andes it is often blended with CABERNET SAUVIGNON. Other producers of varietal wines include Montes and Valdivieso. Malbec is also grown in **Bolivia** and **Peru** – hardly surprising in view of its importance in Argentina.

Australia's 363 ha (897 acres) are mostly in South Australia (eg Clare Valley and Langhorne Creek), but also in Western Australia (eg Margaret River and Great Southern). The variety is not as popular as it once was but there are signs of a minor comeback. Some of the more notable of more than 200 Malbec producers are Bleasdale in Langhorne Creek, Henschke in Eden Valley, Tahbilk in Victoria's Nagambie Lakes and Wendouree in Clare Valley, whose Malbec blends can be wonderfully dark and elegant.

New Zealand had just 156 ha (385 acres) in 2008 but has some serious supporters, including Fromm on the South Island and Kumeu River, Millton, CJ Pask and Villa Maria on the North Island. And Argentina's success with the variety has encouraged plantings in **South Africa**, where Diemersfontein and Signal Hill take it more seriously than most.

COUDERC NOIR

Undistinguished French hybrid with few remaining fans.

BERRY COLOUR: ● ● ● ● ●

PRINCIPAL SYNONYMS: Contassot 20, Couderc 7120, Plant Verni

ORIGINS AND PARENTAGE

A natural hybrid of Munson (or Jaeger 70, a *Vitis lincecumii* × *Vitis rupestris* cross) × *Vitis vinifera* subsp. *vinifera* (ie an unknown variety) obtained in 1886 by Contassot, a pastry cook from Aubenas who gave some of his seeds to Georges Couderc (series 71) and others to Albert Seibel (series 1-254; Galet 1988). This hybrid was in series 71 and is therefore named in honour of Couderc. See PRIOR for the complete pedigree of Munson.

VITICULTURAL CHARACTERISTICS

Early budding and late ripening. Vigorous, productive and prone to chlorosis. Compact bunches and not much juice in the berries.

WHERE IT'S GROWN AND WHAT ITS WINE TASTES LIKE

Along with other hybrids such as CHAMBOURCIN and PLANTET, Couderc Noir was widely planted in **France** in the early twentieth century. So much so that it was not until well into the 1970s, when Couderc was being determinedly uprooted, that its plantings were outnumbered by CABERNET SAUVIGNON. It is still authorized in most of southern and south-western France but there were only 214 ha (529 acres) remaining in France in 2009 compared with 26,616 ha (65,770 acres) at the end of the 1950s.

A Couderc is grown in **Missouri**. **Brazil**'s 498 ha (1,231 acres) in Rio Grande do Sul are Couderc 13, a different variety and grown mainly for the table.

Wines are deeply coloured and often off-puttingly non-*vinifera* in flavour and the least worst results are achieved in rosé.

COUNOISE

Minor but valued southern Rhône ingredient, occasional varietal.

BERRY COLOUR: ● ● ● ● ●

PRINCIPAL SYNONYMS: Coneze, Connoges, Connoise, Counèse, Counoïse, Counoïso, Counoueiso, Guénoise, Quenoise (Vaucluse), Rivier (Ardèche)

VARIETIES COMMONLY MISTAKEN FOR COUNOISE: AUBUN (Vaucluse)

ORIGINS AND PARENTAGE

Counoise is a very old variety from southern France, cultivated mainly in the southern Rhône, where it was mentioned as early as 1626 in Avignon: 'berardy, espagne, barboulenque, col tort et aultre bon plant sans aulcune counoyse' ('berardy, espagne, barboulenque, col tort and other good varieties without any counoyse'). Its name is of uncertain dialectal origin and has various old spellings. Counoise belongs to the Piquepoul ampelographic group (see p XXVII), which also includes AUBUN (Bisson 2009), which explains why Counoise is often confused with Aubun in the vineyards, although ampelography and DNA profiling show that they are distinct varieties (Galet 1990; Bowers *et al.* 1999).

OTHER HYPOTHESES

According to the Provençal poet Frédéric Mistral, Counoise could have been introduced from Spain to Châteauneuf-du-Pape by a papal vice-legate called Counesa, who would have offered the vine to Pope Urbain V at the time when popes resided in Avignon (Foëx 1901).

VITICULTURAL CHARACTERISTICS

Vigorous, best pruned short and may be bushvine trained. Mid to late ripening and well suited to hot, stony slopes. Tends to be more productive in alternate years. Susceptible to botrytis bunch rot. Big berries.

WHERE IT'S GROWN AND WHAT ITS WINE TASTES LIKE

Counoise is recommended around the southern Rhône, **France**, authorized in the *départements* further east and west, and permitted in appellations such as Châteauneuf-du-Pape, Côtes du Rhône-Villages, Gigondas, Côteaux d'Aix en Provence and Ventoux. In older vineyards it is often mixed with the lesser AUBUN. Although it is one of the less used varieties in the Châteauneuf-du-Pape blend, producers such as Château de Beaucastel and Domaine de Pegaü do include it. Wines have moderate alcohol and tannins and are generally fruity but pale, small amounts adding spice and acidity to the blend. Total French plantings in 2009 were down to 443 ha (1,095 acres). Domaine Montpertuis' La Counoise de Jeune is 90% Counoise and therefore classified as a Vin de Pays du Gard. The wine is fairly pale and light and, typically, has an aroma of white pepper. Counoise can oxidize easily and is generally better in blends than as a varietal.

Thanks to the enthusiasm and original plant material of the Perrins of Beaucastel from the southern Rhône, Counoise has been exported to **California**'s Central Coast. Tablas Creek (part owned by the Perrins) in Paso Robles use it in blends but also produce a varietal wine, which they liken to a Beaujolais cru. Here and further north in Santa Cruz, other Rhône enthusiasts such as Randall Grahm at Bonny Doon are disseminating the variety. Further north, in Livermore Valley, Wente also produce a small-lot varietal wine sold only at the cellar door.

There are also limited plantings (15 acres/6 ha) in **Washington** State: McCrea in Red Mountain and Wines of Substance in Walla Walla both bottle varietal Counoise, the latter's vines coming from Beaucastel via Tablas Creek.

COURBU BLANC

Minor variety often consorting with the Mansengs in French Basque Country.

BERRY COLOUR: ● ● ● ● ●

PRINCIPAL SYNONYMS: Courbeau, Courbi or Courbis Blanc, Courbu, Courbut Blanc, Courtoisie (Portet in the Pyrénées-Atlantiques), Vieux Pacherenc (Lembeye in the Pyrénées-Atlantiques)

VARIETIES COMMONLY MISTAKEN FOR COURBU BLANC: ARRUFIAC (Jurançon), CROUCHEN (Irouléguy in the Pays Basque and Txakoli vineyards in the Spanish País Vasco), NOAH (País Vasco), PETIT COURBU

ORIGINS AND PARENTAGE

Courbu is a name given to several distinct varieties in the Pyrénées-Atlantiques and in the Basque Country, both in the French Pays Basque and in the Spanish País Vasco. A white-berried Courbu was first mentioned in 1783–4 in Jurançon and its name probably alludes to its curved (*courbé* in French) vine shoots. Galet (1990) considers Courbu Blanc and COURBU NOIR to be colour mutations of the same variety, and PETIT COURBU to be simply a clonal variant of Courbu Blanc, but both these hypotheses have been disproved by DNA profiling (Bowers *et al.* 1999).

According to Bordenave *et al.* (2007), Courbu Blanc was introduced to the Spanish area of Hondarribia (Txakoli vineyards, País Vasco), where it was given the name Hondarribi Zuri (*zuri* is Basque for 'white'), a name also used for CROUCHEN. The use of this name for both Courbu Blanc and Crouchen is a source of confusion because DNA profiling has surprisingly shown that another reference sample of Hondarribi Zuri is in fact identical to NOAH, an American hybrid (Vouillamoz). In addition, Courbu Blanc is often confused with ARRUFIAC in Jurançon, an error revealed by ampelographic observations and DNA profiling (Bowers *et al.* 1999; Bordenave *et al.* 2007).

Courbu Blanc belongs to the Courbu ampelographic group (see p XXVII), which also includes CROUCHEN, possibly explaining why Courbu Blanc is often confused with Crouchen in the Basque Country.

OTHER HYPOTHESES

Just like many old varieties, Courbu is often described as a 'population variety', which implies that it is composed of different morphological forms, some of them being distinct varieties, but this is not the definition of variety used in this book.

VITICULTURAL CHARACTERISTICS

Prone to wind damage and best trellised and pruned long, ideally grafted onto a low-vigour rootstock. Mid ripening. Susceptible to botrytis bunch rot. Small bunches and very small berries.

WHERE IT'S GROWN AND WHAT ITS WINE TASTES LIKE

The far south-west corner of **France** is home to Courbu Blanc, which is generally a minor ingredient, with GROS MANSENG and/or PETIT MANSENG and other local varieties, in the Jurançon, Béarn and Irouléguy appellations, although it may in theory play a more prominent role in Pacherenc du Vic-Bilh, again with the Mansengs. Clos Lacabe's Compère Lutin is one of the few varietal wines available. Clos Lapeyre, who have a few Courbu vines from the 1940s, have recently planted both Courbu Blanc and CAMARALET DE LASSEUBE. Wines are reasonably alcoholic with moderate acidity and are made in sweet or dry styles. There were 43 ha (106 acres) in France in 2008.

It is likely that there are some plantings of Courbu Blanc, mistakenly known as Hondarribi Zuri, in the País Vasco in **Spain**.

COURBU NOIR

Virtually extinct speciality of Béarn, south-west France, unrelated to Courbu Blanc.

BERRY COLOUR:

PRINCIPAL SYNONYMS: Courbu Rouge, Courbut, ?Dolceolo, Noir du Pays (Pays Basque)

VARIETIES COMMONLY MISTAKEN FOR COURBU NOIR: HONDARRIBI BELTZA, MANSENG NOIR

ORIGINS AND PARENTAGE

Courbu, a name referring to the curved (*courbé* in French) shape of the vine shoots, has been used for several distinct varieties in the Pyrénées, but contrary to common belief (Galet 1990), Courbu Noir and COURBU BLANC are not colour variants. The first mention of a black-berried Courbu appeared in 1880 in the Hautes-Pyrénées (Rézeau 1997). Interestingly, Courbu Blanc belongs to the Courbu ampelographic group (see p XXVII) while Courbu Noir belongs to the Mansien ampelographic group, named after Manseng and composed of CAMARALET DE LASSEUBE, CAMARAOU NOIR, Courbu Noir, GROS MANSENG, Mancin and PETIT MANSENG. Courbu Noir and MANSENG NOIR, the latter paradoxically a member of the Courbu group, are both called Courbu Rouge and were collectively named 'Noirs du Pays' (Herriko Beltza in Basque dialect, meaning 'blacks of the area'). In the Txakoli vineyards in Hondarribia in the País Vasco close to the French border, the local variety HONDARRIBI BELTZA is often confused either with CABERNET FRANC or with Courbu Noir.

VITICULTURAL CHARACTERISTICS

Late ripening, not very productive and needs to be pruned long. Susceptible to powdery mildew. Small bunches and berries.

WHERE IT'S GROWN AND WHAT ITS WINE TASTES LIKE

Although Courbu Noir is in the official French register of varieties, there was little more than a hectare (2.5 acres) remaining in 2008 in the south-west corner of **France**, where it is a permitted ingredient in the red and rosé versions of the Béarn appellation – though its partners, CABERNET FRANC and CABERNET SAUVIGNON plus TANNAT, will always dominate. Wines tend to be light in colour and alcohol but relatively tannic.

CRÂMPOȘIE SELECȚIONATĂ

High-quality, high-acid Romanian variety producing zesty, mineral-laden whites.

BERRY COLOUR: ● ● ● ● ●

PRINCIPAL SYNONYMS: Crîmpoșie Selecționată

ORIGINS AND PARENTAGE

Crâmpoșie Selecționată is a seedling of Crâmpoșie obtained in 1972 by Emilian Popescu, Marin Neagu and Petre Banita at the viticultural research centre in Drăgășani, south-west Romania, with the aim of overcoming the problem of uneven berry size caused by the functionally female flowers in Crâmpoșie, an old variety that probably originated in the Drăgășani region in Oltenia, south-west Romania, and has now been supplanted by its more reliable progeny. The selection was obtained by open fertilization of Crâmpoșie so the other parent is unknown.

VITICULTURAL CHARACTERISTICS

Relatively vigorous, productive, late ripening. Big bunches of thick-skinned berries. Resistant to rain, heat and fungal diseases.

WHERE IT'S GROWN AND WHAT ITS WINE TASTES LIKE

Crâmpoșie Selecționată, generally referred to as Crâmpoșie even though the full name must appear on the label, is grown mainly in the Drăgășani region and further east in Panciu and in Constanța near the Black Sea coast of **Romania**. The best wines are refined, zesty, high in acidity and have a strong minerality, made even more refreshing by a light bitterness on the finish. It is sometimes blended with other varieties such as SAUVIGNONASSE to increase acidity. Recommended producers include Casa Isarescu and Prince Ştirbey, the latter also experimenting with a sparkling version. There were 473 ha (1,169 acres) in Romania in 2008. It is also popular for the table.

CRIOLLA

In South America, Criollas or Criollos is the generic name for a group of distinct grape varieties that are considered to be descended from European parents, either born *in situ* or introduced as cuttings or seeds by the Spanish and Portuguese conquistadors (Martínez *et al.* 2003). *Criollo* is Spanish for 'Creole'.

They were supposedly first introduced in 1551 to the Valle de la Concepción in Peru by Francisco de Caravantes and Hernando de Montenegro from Spain, possibly from the Islas Canarias, a regular stop-over en route to South America (Rodríguez and Matus 2002). Criollas were then introduced to northern Chile, and then from La Serena to Argentina in 1556, in today's provinces of Mendoza, San Juan and La Rioja (Martínez, Cavagnaro *et al.* 2006).

Criollas include CEREZA, CRIOLLA GRANDE, Criolla Chica (LISTÁN PRIETO), PEDRO GIMÉNEZ, Moscatel Rosado, Moscatel Amarillo, TORRONTÉS MENDOCINO, TORRONTÉS RIOJANO and TORRONTÉS SANJUANINO. Through DNA profiling, Criolla Chica has been identified as Listán Prieto, an old Spanish variety (see Listán Prieto for details), but no other Criollas have been identified as European varieties, probably because they were born in South America. Indeed, DNA parentage analysis has shown that Cereza, Moscatel Amarillo, Torrontés Riojano and Torrontés Sanjuanino are all natural crosses between MUSCAT OF ALEXANDRIA and Listán Prieto, while Torrontés Mendocino is a natural cross between Muscat of Alexandria and an unidentified variety (Agüero *et al.* 2003; Myles *et al.* 2011). These varieties can thus be considered to be indigenous because both Muscat of Alexandria and Listán Prieto have been cultivated in Argentina and Chile for centuries under the names Moscatel (or Uva do Italia) and Criolla Chica, respectively (Lacoste *et al.* 2010), so they had plenty of time and opportunity to cross.

The parentages described above mean that CEREZA, Moscatel Amarillo, TORRONTÉS RIOJANO and TORRONTÉS SANJUANINO are all siblings, and that the South American TORRONTÉS MENDOCINO and the Italian GRILLO, MOSCATELLO SELVATICO, MALVASIA DEL LAZIO plus MUSCAT OF HAMBURG are their half-siblings as well as grandchildren of Axina de Tres Bias and MUSCAT BLANC À PETITS GRAINS (see that entry for the pedigree diagram). In addition, the Spanish PERRUNO is either a half-sibling or a grandparent (depending on which way the parent–offspring relationship goes) of Cereza, Moscatel Amarillo, Torrontés Riojano and Torrontés Sanjuanino.

Criollas have been used routinely for table grapes, juice, and raisins as well as wine and pisco. Together, CEREZA and CRIOLLA GRANDE account for more than 50% of the surface planted with Criollas in Argentina (Martínez, Cavagnaro *et al.* 2006).

VITICULTURAL CHARACTERISTICS

According to Martínez, Cavagnaro *et al.* (2006), Criollas are more tolerant of drought and salt than European varieties.

CRIOLLA CHICA

See LISTÁN PRIETO.

CRIOLLA GRANDE

Declining but still very widely planted Argentine pink-skinned variety making humdrum wine.

BERRY COLOUR:

PRINCIPAL SYNONYMS: Criolla Grande Sanjuanina

ORIGINS AND PARENTAGE

Criolla Grande is indigenous to Argentina and is a member of the Criollas (see CRIOLLA entry for details). According to Martínez, Cavagnaro *et al.* (2003, 2006), Criolla Grande could be related to CEREZA and TEMPRANILLO.

VITICULTURAL CHARACTERISTICS

Mid to late ripening. Big, loose bunches of big berries.

WHERE IT'S GROWN AND WHAT ITS WINE TASTES LIKE

Although it is in decline in terms of total area planted for wine, Criolla Grande is still the third most planted variety in **Argentina** after CEREZA and Malbec (COT), with a total of 20,892 ha (51,625 acres) in 2009, the vast majority in Mendoza province. It is used mainly for inexpensive and generally mediocre pink and deeply coloured white wine for early domestic consumption, often sold in litre bottles or cardboard cartons, as well as for the table.

CRLJENAK KAŠTELANSKI

See TRIBIDRAG.

CROATINA

A fruity dark-skinned variety grown in many different areas of northern Italy and used both solo and in blends.

BERRY COLOUR:

PRINCIPAL SYNONYMS: Bonarda (Oltrepò Pavese, Roero), Bonarda di Rovescala, Crovattina, Nebbiolo di Gattinara, Neretto, Spanna di Ghemme, Spanna-Nebbiolo (Novara, Vercelli), Uga del Zio, Uva Vermiglia

VARIETIES COMMONLY MISTAKEN FOR CROATINA: BONARDA PIEMONTESE, HRVATICA, NEBBIOLO

ORIGINS AND PARENTAGE

Croatina is believed to originate from Rovescala in the Valle del Versa in the Oltrepò Pavese, south of the River Po in northern Italy, where it was first mentioned in the late nineteenth century. In the Oltrepò Pavese it is called Bonarda, but this name can be misleading because it is used for at least six distinct varieties (see BONARDA PIEMONTESE). Although Croatina and the old Croatian variety HRVATICA both mean 'Croatian girl', DNA profiling rejected any possible link between them (Maletić *et al.* 1999). Croatina was used to breed ERVI.

VITICULTURAL CHARACTERISTICS

Particularly resistant to powdery mildew. Mid to late ripening. Doesn't like summer drought.

WHERE IT'S GROWN AND WHAT ITS WINE TASTES LIKE

It is grown in **Italy**, mainly in the Oltrepò Pavese in Lombardia, in Piemonte in the provinces of Novara, Cuneo, Alessandria (eg in Colli Tortonesi DOC) and Vercelli, and also in Emilia-Romagna, in the Veneto and on Sardegna. Oltrepò Pavese Bonarda DOC is 100% Croatina. Several DOCs such as Colline Novaresi and Coste della Sesia permit varietal Croatina wines. In the hills above Tortona, Walter Massa makes an excellent varietal example, and at Le Piane in Boca, Christophe Künzli carefully tends his ancient Croatina vines in a mixed planting trained on the traditional maggiorina trellis. Wines are well coloured, fruity with a certain bite and best drunk young. In 2000, there were 3,280 ha (8,105 acres) in Italy.

CROUCHEN

Widely travelled neutral French variety from the western Pyrenees that for long benefited from naming confusion.

BERRY COLOUR: ● ● ● ● ●

PRINCIPAL SYNONYMS: Cape Riesling (South Africa), Clare Riesling (Merbein in Australia), Cougnet (Portet), Cruchen Blanc, Cruchenta (Pays Basque), Messanges Blanc (Landes), Paarl Riesling (South Africa), Sable Blanc (Landes), Trouchet Blanc (Monein in the Pyrénées-Atlantiques)

VARIETIES COMMONLY MISTAKEN FOR CROUCHEN: COURBU BLANC (Txakoli and Irouléguy), NOAH (País Vasco in Spain), RIESLING (South Africa), SÉMILLON (Australia)

ORIGINS AND PARENTAGE

Crouchen is from the Pyrénées-Atlantiques in south-western France, where it was first mentioned in 1783–4 (Rézeau 1997): 'Crouchen blanc [. . .], le goût est très doux et bon a manger' (Crouchen blanc [. . .] the taste is very sweet and good to eat'). Its name supposedly means 'crunchy'.

It has been confused with COURBU BLANC in the Txakoli vineyards in País Vasco on the north coast of Spain, where both are mistakenly called Hondarribi Zuri, and in Irouléguy in south-west France. (See Courbu Blanc for the confusion with NOAH.)

Crouchen belongs to the Courbu ampelographic group (see p XXVII; Bisson 2009). In addition, Crouchen is ampelographically close to the Carmenet group (see CABERNET FRANC). In spite of having very similar names, Crouchen and Cruixen are distinct varieties, the latter found only in vine collections. Crouchen was used to breed NOUVELLE.

VITICULTURAL CHARACTERISTICS

Mid ripening and suited to well-exposed gravelly or sandy soils. Sensitive to powdery mildew.

WHERE IT'S GROWN AND WHAT ITS WINE TASTES LIKE

Crouchen has had its time, and its time was prolonged mainly because the variety was so long, and so conveniently, confused with more noble varieties – with RIESLING in South Africa and with SÉMILLON (once known as Clare Riesling) in Australia. It produces neutral wines that share only one thing with Riesling – the ability to age. It used to be cultivated as both a table and a wine grape in its home territory, the sandy vineyards in Messanges and Capbreton north of Bayonne on the Atlantic coast of **France**, but today only isolated vines remain there.

It is likely that there are some plantings of Crouchen in the Spanish País Vasco on the coast in north-west **Spain**, where it is mistakenly known as Hondarribi Zuri.

Crouchen was introduced around 1850 to **Australia** under the name Sales Blanc (presumably a misspelling of Sables, referring to the sandy vineyards where the cuttings originated) and was first cultivated by the Jesuits of Sevenhill Monastery in Clare Valley (Galet 1990). Since its identity was unknown, it was named Clare Riesling (Riesling was often used by Australians to refer to SÉMILLON) until 1976, when the French ampelographer Truel recognized it as Crouchen. Today in the Clare Valley it has been largely replaced by true RIESLING. Of the 117 ha (289 acres) that remain in Australia, almost all are in the warm inland regions of Swan Hill, Riverland and Murray Darling, where their produce presumably disappears into anonymous white blends. One exception is Brown Brothers in Victoria, who make a medium-sweet blend of Crouchen with Riesling. There were just five growers of the variety in 2010.

South Africa has more Crouchen planted than any other country. It was probably introduced to the Cape in 1656, initially named Groenblaarsteen ('Green Leaf Steen') and then renamed Cape Riesling, Paarl Riesling and South African Riesling (Beuthner 2009). True RIESLING, which became known as Weisser Riesling or Rhine Riesling in South Africa, did not reach the Cape until the late nineteenth century. In the 1950s, Christiaan Johannes Orffer realized that there were two different varieties named Riesling in South Africa but only one was the true Riesling. Although he identified the other, more widely planted, variety as Crouchen Blanc, it continued to be sold under the name Riesling. However, a law passed in 2009, with effect from the 2010 harvest, stipulates that Crouchen must be labelled as such and Riesling no longer needs to be qualified as Weisser or Rhine. (This was already in effect for exports but not domestically.) There were 895 ha (2,212 acres) in 2009, mostly in Breedekloof and Paarl. According to Gary Jordan, of the eponymous wine estate in Stellenbosch, good Crouchen is an oxymoron but Theuniskraal produce a serviceable varietal example, until recently labelled Riesling, of course.

CROVASSA

Very minor red variety in the Valle d'Aosta.

BERRY COLOUR: ● ● ● ● ●

PRINCIPAL SYNONYMS: Croassa

ORIGINS AND PARENTAGE

Crovassa has been cultivated in the Valle d'Aosta in north-east Italy for at least two centuries but has never been an important variety. It has no genetic link with other Valle d'Aosta varieties but shows some DNA similarities with NEBBIOLO, suggesting it could originate in Piemonte (Vouillamoz and Moriondo 2011).

VITICULTURAL CHARACTERISTICS

Vigorous and late ripening.

WHERE IT'S GROWN

Crovassa is currently planted only in the Valle d'Aosta, in the far north west of **Italy**, mainly around the towns of Issogne and Donnas, where it is usually blended with other local varieties. Just 15 ha (37 acres) were recorded in Italy in 2000.

CRUJIDERA

See MARUFO.

CRYSTAL

As yet unidentified variety producing labrusca-*like wines in China's Yunnan province.*

BERRY COLOUR: ● ● ● ● ●

ORIGINS AND PARENTAGE

According to an unverified online source (www.cnwinenews.com), Crystal was apparently imported into China in 1965 by Dongfeng Farm in Mile county, Yunnan province, southern central China, along with ten other grape varieties, including ROSE HONEY and FRENCH WILD. Since DNA profiling has not yet been done, it is not known whether Crystal is related to European varieties or is identical to the *Vitis labrusca* variety called Crystal that was selected by S D Case in Canton Center, Connecticut, US, which could explain the funky aromas of the wine. According to Sun (2004), it is a *Vitis vinifera* × *Vitis labrusca* hybrid more suitable for eating and processing than for wine.

OTHER HYPOTHESES

Crystal was allegedly introduced from Europe to the Shangri-La region in Yunnan province by missionaries at the beginning of the nineteenth century, along with ROSE HONEY and FRENCH WILD but there is no documentary evidence to confirm this.

VITICULTURAL CHARACTERISTICS

Early ripening and generally easy to cultivate, with good resistance to cold and diseases.

WHERE IT'S GROWN AND WHAT ITS WINE TASTES LIKE

Found mainly in Yunnan province, **China**, where it is the most widely planted variety and has the longest history. Yunnanhong (Yunnan Red) Winery is the biggest producer. Wines have the highly distinctive sweet, foxy smell and taste of those made from *labrusca* varieties as well as rose-like aromas.

CSABA GYÖNGYE

Hungarian table grape occasionally turned into light, grapey whites.

BERRY COLOUR: ● ● ● ● ●

PRINCIPAL SYNONYMS: Cabski Biser (Slovenia), Csabagyöngye (Hungary), Julski Muskat (Slovenia),

Pearl of Csaba, Perl do Saba (Bulgaria), Perla Czabanska, Perla di Csaba, Perle de Csaba (France), Perle von Csaba (Austria, Germany), Rindunicaz Strugurilor (Romania), Vengerskii Muskatnii Rannüj, Zemcug Saba (Russia)

ORIGINS AND PARENTAGE

Csaba Gyöngye, literally Pearl of Csaba, is a cross obtained by the private breeder János Mathiász (1838–1921) around 1890 in Hungary. According to Kozma *et al.* (2003), Mathiász presented the seeds to Adolf Stark (1834–1910), another – more famous – grape breeder, who grew it in his garden in Békéscsaba in southern Hungary, hence the name. Csaba Gyöngye was long assumed to be a Bronnertraube × MUSCAT OTTONEL cross (Galet 2000) but DNA profiling has shown that Csaba Gyöngye is a progeny of MADELEINE ANGEVINE and, putatively, MUSCAT FLEUR D'ORANGER (Kozma *et al.* 2003; Galbács *et al.* 2009).

Csaba Gyöngye was used by Hungarian breeders to create many table grapes such as Szőlőskertek Királynője (known in Germany as Königin der Weingärten), as well as a few wine-grape varieties such as IRSAI OLIVÉR and ZALAGYÖNGYE.

OTHER HYPOTHESES

According to Kozma *et al.* (2003), some authors have suggested that Csaba Gyöngye was serendipitously discovered in 1904 growing in a garden in Békéscsaba.

VITICULTURAL CHARACTERISTICS

Early budding, very early ripening. Low yields. Good winter hardiness but thin-skinned berries prone to cracking.

WHERE IT'S GROWN AND WHAT ITS WINE TASTES LIKE

Although Csaba Gyöngye was bred for the table and is quite widely grown in Romania, Hungary and Bulgaria for that purpose, its thin skins make it difficult to transport without damage to the berries. It is occasionally used in **Hungary** to make light, grapey, aromatic wines, such as the semi-sweet, early-bottled example produced by Varga in Badacsony in western Hungary, but it is difficult to say what proportion of the 92 ha (227 acres) cultivated are destined for wine.

CSERSZEGI FŰSZERES

Popular Hungarian cross making aromatic everyday whites.

BERRY COLOUR:

ORIGINS AND PARENTAGE

An IRSAI OLIVÉR × SAVAGNIN ROSE cross bred in 1960 by Károly Bakonyi at the Pannon University of Agriculture in Hungary. It was named after Cserszegtomaj, a village in the county of Zala in western Hungary, and *fűszeres*, meaning 'spicy', and was officially released in 1982.

VITICULTURAL CHARACTERISTICS

Early budding, mid ripening, productive. Small berries with some reddish berries in the bunch. Winter-hardy and resistant to botrytis bunch rot.

WHERE IT'S GROWN AND WHAT ITS WINE TASTES LIKE

Cserszegi Fűszeres, quite widely dispersed in **Hungary**, produces aromatic, fruity and spicy wines, both dry and off dry although the best are dry, with lively acidity and hints of spring flowers or a grapey Muscat aroma, combining the characteristics of its parents. These are wines for everyday drinking rather than for the cellar. Recommended producers include Font, Frittmann, Gál and Benő Kökény in Kunság in southern central Hungary, Hilltop in Neszmély in the north west, Benedek in Mátra in the north and Jásdi just north of Lake Balaton in the west. There were 3,057 ha (7,554 acres) in 2008, with well over two thirds of that area in Kunság. Although it is very popular in Hungary, its unpronounceable name has not helped sales on export markets.

CSÓKASZŐLŐ

Old Hungarian variety in the throes of rediscovery.

BERRY COLOUR:

PRINCIPAL SYNONYMS: Čavčica, Čavkaas, Csóka Szőlő or Csóka, Fekete Magyarka, Kleinhungar Blauer (Austria), Magyarka Neagra (Romania), Rácfekete, Vadfekete

VARIETIES COMMONLY MISTAKEN FOR CSÓKASZŐLŐ: Cigányszőlő

ORIGINS AND PARENTAGE

Csókaszőlő, named after *csóka*, meaning 'jackdaw', and *szőlő*, meaning 'grape', in reference to its almost black berries, is an old, almost forgotten Hungarian variety that used to be widespread in the Carpathian Basin.

VITICULTURAL CHARACTERISTICS

Mid budding and ripening, low yields. Small bunches of small, waxy berries. Resistant to botrytis bunch rot but susceptible to downy mildew, less so to powdery mildew.

WHERE IT'S GROWN AND WHAT ITS WINE TASTES LIKE

Neglected and close to extinction, this old variety is currently being re-evaluated and replanted by a handful of growers in **Hungary** who are making wines that are well coloured, spicy and vividly fruity and have bright acidity, fairly high alcohol and marked but velvety tannins. They fall somewhere between PINOT NOIR and Kékfrankos (BLAUFRÄNKISCH) in style, showing a certain elegance, although producers are still feeling their way towards a typical style. József Szentesi, just south west of Budapest, was the first to replant: his 2004 was probably the first Csóka wine for 150 years. Other recommended producers include Bussay in Zala in the far west of Hungary (although his production of this variety is tiny and not readily available), Kaló Imre in Eger in the north, and Vylyan and Attila Gere in Villány in the south west. Csóka has now also been registered for the Villány appellation and is no longer relegated to table-wine status.

CYGNE BLANC

Literally 'White Swan', this is a white-berried chance seedling of CABERNET SAUVIGNON incidentally spotted in 1989 by Sally Mann, wife of the winemaker Dorham Mann, in her garden in Baskerville, Swan Valley, Western **Australia**. The garden was close to a vineyard planted mainly to Cabernet Sauvignon, which indicates that the seedling is most likely the result of selfing (see p 1182), just like the myriad pips produced each year on planet earth. The berry colour may have surprised those who were not aware that Cabernet Sauvignon has a white-berried parent, SAUVIGNON BLANC. Unlike Shalistin, a colour mutation of Cabernet Sauvignon also found in Australia (see Cabernet Sauvignon), Cygne Blanc is a distinct variety that Dorham and Doris

Elsie Mann officially registered in 1997. Planted in Mount Benson on the Limestone Coast in Southern Australia early this century, it has so far produced whites without any very strong characteristics other than the vague leafiness of Cabernet Sauvignon.

D

DAFNI
DAKAPO
DALKAUER
DAMASCHINO
DE CHAUNAC
DEBINA
DEBINE E BARDHË
DEBINE E ZEZË
DEBIT
DECKROT
DELAWARE
DELISLE
DEVÍN
DIAGALVES
DIMRIT
DIMYAT
DINDARELLA
DIOLINOIR
DIŠEĆA RANINA
DNESTROVSKY ROZOVY
DOBRIČIĆ
DOINA
DOLCETTO
DOLCIAME
DOMINA
DONA BRANCA
DONZELINHO BRANCO
DONZELINHO TINTO
DORADILLA
DORAL
DORINTO
DORNFELDER
DORONA DI VENEZIA
DOUCE NOIRE
DOUX D'HENRY
DRAGON BLUE
DRNEKUŠA
DRUPEGGIO
DUNAJ
DUNKELFELDER
DURANIJA
DURAS
DURELLA
DUREZA
DURIF
DUTCHESS
DZVELSHAVI OBCHURI

ᴥ this symbol on the following pages indicates that the synonyms, and mistaken identities, are confirmed by DNA evidence

DAFNI

Distinctively laurel-scented variety from Kríti (Crete) recently rescued from oblivion.

BERRY COLOUR: ● ● ● ● ●

PRINCIPAL SYNONYMS: Dafnia, Daphni (Irákleio/Heraklion), Daphnia

VARIETIES COMMONLY MISTAKEN FOR DAFNI: Daphnato or Daphnata

ORIGINS AND PARENTAGE

Dafni is an old variety from the Greek island of Kríti. The name derives from *daphne*, meaning 'laurel', referring to the distinctive aroma of the wine.

VITICULTURAL CHARACTERISTICS

Productive and very late ripening. Loose bunches of thick-skinned grapes and therefore very disease-resistant.

WHERE IT'S GROWN AND WHAT ITS WINE TASTES LIKE

Dafni is grown exclusively and to a very limited extent on the island of Kríti in **Greece**, where it has been rescued and rehabilitated by Lyrarakis, who make a distinctive laurel-scented varietal wine which is moderate in alcohol (usually around 12% or 12.5%) with flavours of rosemary, tropical fruit and ginger spice. The aroma seems to be mainly in the thick skins and intensive pruning is necessary to keep the yields down. Dafni had virtually disappeared at the end of the 1980s but there are now around 15 ha (37 acres) on the island, mostly in the centre, and a few producers are currently experimenting with this variety elsewhere in Greece.

DAKAPO

Red-fleshed German cross adds colour to blends.

BERRY COLOUR: ● ● ● ● ●

PRINCIPAL SYNONYMS: Geisenheim 7225-8

ORIGINS AND PARENTAGE

A DECKROT × BLAUER PORTUGIESER cross bred by Helmut Becker at Geisenheim in the Rheingau, Germany, in 1972.

VITICULTURAL CHARACTERISTICS

Vigorous, consistently productive. Mid budding, early ripening. A teinturier variety, ie one with red-fleshed berries. Good resistance to frost and fungal diseases but prone to berry splitting towards harvest.

WHERE IT'S GROWN AND WHAT ITS WINE TASTES LIKE

The name Dakapo, based on the Italian *da capo*, meaning 'extra' or 'encore', hints at the purpose of this intensely coloured wine: to darken blends. This tannic but otherwise pretty neutral variety was included on the official German register of varieties in 1999 and there are now 59 ha (146 acres) planted in **Germany**, with the biggest plantings in Baden, Rheinhessen and the Pfalz. Haxel in Cochem in the Mosel region produce a sweet version and Rothweiler and the Staatsweingut in Hessische Bergstrasse produce dry versions.

Switzerland had 12 ha (30 acres) in 2009, mainly in the canton of Schaffhausen.

DALKAUER

Rare German offspring of Riesling and A N Other.

BERRY COLOUR: ● ● ● ● ●

PRINCIPAL SYNONYMS: Beutelrebe

VARIETIES COMMONLY MISTAKEN FOR DALKAUER: RIESLING

ORIGINS AND PARENTAGE

This variety was allegedly imported from the Caucasus to Germany after 1945 by the owner of Weingut Görz in Bad Kreuznach in Rheinland-Pfalz, whence it spread to the Nahe as Beutelrebe (literally 'bag grape'; Ambrosi *et al.* 1997; Hillebrand *et al.* 2003). It was registered in 1982 by Georg Dalkowski, a private vine breeder in Bad Kreuznach, hence its name. Some think it could be a RIESLING × GRÜNER VELTLINER cross, others claim it is a cross between Riesling and an unknown variety, and others think it is a mutation of Riesling. DNA profiling could easily prove or disprove these hypotheses.

VITICULTURAL CHARACTERISTICS

Late budding and late ripening. Winter frost-resistant.

WHERE IT'S GROWN

There are very limited plantings in Rheinhessen and the Nahe in **Germany**.

DAMASCHINO

Ancient but neutral white included in blends around the Mediterranean and in Portugal.

BERRY COLOUR: ● ● ● ● ●

PRINCIPAL SYNONYMS: Alicante Branco (Portugal), Beldi🧬 (Tunisia), Damaschena (Sicilia), Farana 🧬 or Faranah or Farranah (Algeria), Mayorquin (France), Planta Fina🧬 (Spain), Planta Fina de Pedralba (Spain), Planta Pedralba (Spain)

VARIETIES COMMONLY MISTAKEN FOR DAMASCHINO: TREBBIANO TOSCANO🧬 (Australia), VERDEJO🧬

ORIGINS AND PARENTAGE

Damaschino was first documented on Sicilia in the mid nineteenth century but it is said to have been introduced at the time of Arab rule (ninth and tenth centuries). Although its name might suggest a link with the city of Damascus, there is no historical or genetic evidence so far that the variety originates from Syria; its origin is still unknown but it probably comes from somewhere on the Mediterranean coast.

The belief that Planta Fina, the Spanish name for Damaschino, and VERDEJO are the same (Ibañez *et al.* 2003) seems to be a mistake, because their DNA profiles are distinct.

VITICULTURAL CHARACTERISTICS

Fertile and productive. Biggish bunches of medium-sized to large, thin-skinned berries. Mid budding and ripening. Resistant to drought and well suited to light, well-drained soils in a maritime climate. Susceptible to downy mildew.

WHERE IT'S GROWN AND WHAT ITS WINE TASTES LIKE

Damaschino is planted all over the Mediterranean under various names and used both as a table grape and as a wine grape. On the island of Sicilia, **Italy**, it used to be widely planted but has gradually been replaced by more-resistant varieties such as CATARRATTO BIANCO and more recently by GRILLO. At present, Damaschino is cultivated mainly in the Marsala area, where it is authorized in the DOC of the same name, blended with local varieties such as Grillo, Catarratto Bianco and INZOLIA. Elsewhere it may be blended with GARGANEGA (under the name Grecanico Dorato). The Delia Nivolelli DOC, established in 1998, allows for a varietal Damaschino. The Italian agricultural census recorded 384 ha (949 acres) in 2000 but there were 444 ha (1,097 acres) on Sicilia in 2008.

Mayorquin is authorized for the Bellet AC in south-east **France** but there was less than 1 ha (2.5 acres) planted in 2006.

In **Spain**, it is cultivated in Valencia under the names Planta Fina or Planta Fina de Pedralba and authorized in the Valencia DO. In 2008 there was a total of 227 ha (561 acres) in Spain. Here varietal wines are light, not especially aromatic, with an appley flavour. In **Portugal**, under the name Alicante Branco, there were 1,167 ha (2,884 acres) in 2010 and recommended producers include Barbeito on Madeira and Vinolive in Beira Interior.

Plantings in **Algeria**, under the names Farana, Faranah or Farranah, have decreased since independence, but Farana is still used to make dry whites in the mountainous regions of Algeria, eg by Atlasian Cellars in the Coteaux de Tlemcen appellation. In **Tunisia**, Beldi is among the most widespread table and wine grape varieties.

Under the name Farana, a small amount of this variety was produced in **Australia** but not made as a varietal wine and the production has dwindled to virtually nothing since 2005.

DE CHAUNAC

French hybrid now at home in Canada and New York State.

BERRY COLOUR: ● ● ● ● ●

PRINCIPAL SYNONYMS: Cameo, Seibel 9549

ORIGINS AND PARENTAGE

A hybrid between Seibel 5163 and Seibel 793 obtained by Albert Seibel (see PRIOR for the full pedigree of Seibel 5163), where:

- Seibel 793 is a Seibel 41 × *Vitis rupestris* 'Gigantesque' Jaeger hybrid
- Seibel 41 is a Munson × *Vitis vinifera* subsp. *vinifera* hybrid (ie an unknown variety) (see PRIOR for the full pedigree of Munson).

De Chaunac was the name of the French Canadian chemist and later director of research at Brights Wines in Niagara Falls, Ontario, between 1933 and 1961. After the Second World War, he imported and experimented with a range of hybrids, including De Chaunac.

VITICULTURAL CHARACTERISTICS

Early ripening, productive, disease-resistant and relatively winter-hardy.

WHERE IT'S GROWN AND WHAT ITS WINE TASTES LIKE

Although De Chaunac was bred in France, its current home is quite definitely in Canada and North America. It doesn't even feature in official French vineyard statistics but there were over 100 acres (40 ha) in Ontario, **Canada**, in 2006 and about the same in **New York** State, plus minor plantings in Québec and the American Northeast and Midwest. At one time it was the most widely planted hybrid in Canada. Buckingham Valley in **Pennsylvania** make a good example.

DEBEJAN

See GEGIĆ.

DEBINA

Highly localized, high-acid Greek variety making light, fresh and digestible wines.

BERRY COLOUR: ● ● ● ● ●

PRINCIPAL SYNONYMS: Debina Metsovou, Debina Palea, Dempina, Ntempina, Zítsa

VARIETIES COMMONLY MISTAKEN FOR DEBINA: Debina Kala (Skrapar in Albania), Debina Teki (Skrapar)

ORIGINS AND PARENTAGE

Debina is probably indigenous to the Zítsa area in Ioánnina in the Ípeiros (Epirus) region, northwest Greece, where it has traditionally been cultivated. According to Manessis (2000), Debina was praised by Lord Byron and his companion Hobhouse in the nineteenth century.

Contrary to local belief, Debina (under the synonym Dempina) is not identical to Debina Kala, nor to Debina Teki, two varieties from Skrapar in south-eastern Albania that are no longer commercially cultivated, although they may well be genetically closely related to Debina (Ladoukakis *et al.* 2005).

VITICULTURAL CHARACTERISTICS

Big, compact bunches. Mid to late budding, late ripening. Vigorous, moderately high yields. Prone to water stress and susceptible to botrytis bunch rot.

WHERE IT'S GROWN AND WHAT ITS WINE TASTES LIKE

Debina is the sole authorized variety for the Zítsa appellation in Ípeiros so it is not surprising that it is the most widely planted variety there (459 ha/1,134 acres in 2008). In this remote part of north-west **Greece**, at altitudes of 600–700 m above sea level and with a climate influenced by the Ionian Sea, the wines are generally light in every way, not particularly aromatic and high in acidity, which is probably why many are made in off-dry styles or used as the base for sparkling wines. The fruit is also at risk from unpredictable September weather so that it can be difficult to ripen the grapes fully (Lazarakis 2005). However, when yields are restricted, the wines may be fresh, elegant and thirst-quenching, in some instances having an aroma of apples and a refreshing, lightly sour aftertaste. Recommended producers include Glinavos, the Ioánnina Co-operative of Zítsa, and Katogi & Strofilia. It is also found in Attalandi near Lamía, south and east of Zítsa.

DEBINE E BARDHË

Declining Albanian of mysterious origin.

BERRY COLOUR: (first of five swatches marked)

PRINCIPAL SYNONYMS: Debin

ORIGINS AND PARENTAGE

Debine e Bardhë, literally 'white Debine', possibly originates from the area around the towns of Përmet and Këlcyrë in south-east Albania. However, the ampelographic identity of this variety is still obscure because two reference vines called Debina Teki and Debina Kala (berry colour not given), sampled in Skrapar in south-eastern Albania, turned out to have distinct DNA profiles in Ladoukakis *et al.* (2005) so it is not known which is the true-to-type reference: Debina Teki is genetically close to the Greek DEBINA, which would explain the name correspondence, while Debina Kala is genetically close to Bora, an Albanian variety no longer cultivated.

DNA profiling is needed to verify whether Debine e Bardhë is a colour mutation of DEBINE E ZEZË, and whether Debine e Bardhë is identical to Debina Teki or Debina Kala.

WHERE IT'S GROWN

The variety is grown mainly in south-east **Albania** but is losing ground to international varieties.

DEBINE E ZEZË

Dark-skinned south-eastern Albanian also known as Debin or Debine e Zi. See DEBINE E BARDHË.

DEBIT

Vine producing alcoholic wine throughout Dalmacija (Dalmatia).

BERRY COLOUR: ● ● ● ● ●

PRINCIPAL SYNONYMS: Belan (Republic of Macedonia), Čarapar (Croatia), Debit Bijeli (Croatia), Puljižanac

ORIGINS AND PARENTAGE

Debit is thought to be indigenous to Croatia. The name Debit means 'debt', alluding to its high yields, which supposedly allow growers to pay their debts, like Pagadebiti in Italy, although they are genetically distinct (Vouillamoz).

OTHER HYPOTHESES

Debit is said by some to have been introduced to Dalmacija from Puglia in Italy, as illustrated by its synonym Puljižanac (literally 'from Puglia'), but this is likely to be a confusion with Pagadebiti, an incorrect synonym. Debit is thought by others to originate in Turkey but this is not consistent with genetic analysis, which groups Debit together with other Croatian varieties such as GEGIĆ, ŠKRLET and BOGDANUŠA (Maletić *et al.* 1999).

VITICULTURAL CHARACTERISTICS

Late ripening, productive. Susceptible to downy mildew.

WHERE IT'S GROWN AND WHAT ITS WINE TASTES LIKE

Debit is widespread in the whole of Dalmacija, **Croatia**, and authorized in all the main wine regions there: Sjeverna Dalmacija (Northern Dalmatia), Dalmatinska Zagora (Dalmatian Hinterland), Srednja i Južna Dalmacija (Central and Southern Dalmatia). Varietal wines tend to be powerful and structured but remain fresh, with flavours of green fruits; the variety can add concentration and depth to blends. It represented 1.7% of the Croatian vineyard area in 2009, about 550 ha (1,359 acres). Recommended producers include Bibich and Vinoplod.

DECKROT

Red-fleshed German cross bred for colour.

BERRY COLOUR: ● ● ● ● ●

PRINCIPAL SYNONYMS: Freiburg 71-119-39

ORIGINS AND PARENTAGE

A PINOT GRIS (Ruländer) × TEINTURIER (Färbertraube) cross created in 1938 in Freiburg by Johannes Zimmermann. It was used to breed DAKAPO.

VITICULTURAL CHARACTERISTICS

Red-fleshed berry, good general disease resistance, moderate yields, high levels of sugar and acidity in the grapes.

WHERE IT'S GROWN

This very deeply coloured, red-fleshed variety (*Deckrot* is German for *teinturier*) entered the official variety register in **Germany** in 1991 and is used mainly to darken red blends. There are 21 ha (52 acres) planted in Germany, mainly in Baden in the south, where the Winzergenossenschaft Achkarren produce a varietal wine.

There are 5 ares in **Switzerland** – just 5% of a hectare.

DELAWARE

Historic variety of obscure origin once popular in the Midwest and in the eastern US states.

BERRY COLOUR: ● ● ● ● ●

PRINCIPAL SYNONYMS: Delavar, Heath Grape, Ladies Choice, Powell, Ruff Heath

ORIGINS AND PARENTAGE

The historical origin of this variety is still obscure but it is thought to come from Frenchtown in Delaware County, Ohio, US. In 1792, a French-speaking Swiss aristocrat, Paul Henri Mallet-Prevost, fled first from France and then from Switzerland, across Europe, wanted by the French for having hidden his countrymen during the French Revolution.

In 1794 he arrived in New York city and later in the same year bought from Thomas Lowrey a piece of land containing several mills in Alexandria on the Delaware River, between Milford and Stockton, at Kingwood township, Hunterdon County, New Jersey. The settlement that developed around the mills became known as Frenchtown, even though Mallet-Prevost was actually Swiss.

When he first came to the US, Mallet-Prevost brought some European vines, allegedly from Burgundy, and made some experimental crossings with native American grapes. The most interesting of them was supposedly obtained from a crossing of a native vine, possibly *Vitis labrusca*, *Vitis aestivalis* or CATAWBA, with a European variety, but this has not yet been analysed by DNA profiling.

Mallet-Prevost gave this vine to several of his Frenchtown neighbours, including a blacksmith and wheelwright named Benjamin Heath. When the latter moved to the township of Concord in Delaware County in 1837, he took the vine with him and cultivated it on his farm on Freshwater Road. His neighbour Abraham Thompson, editor of the *Delaware Gazette* and horticulturist, was so impressed by Heath's grapes that he sent a sample for study to the Massachusetts Horticultural Society in 1855, and it is said that the Society named it 'Grape from Delaware city, Ohio'. Thompson, along with pioneers such as Campbell from Courtland County in New York, who had moved to Delaware, and Frederick P Vergon from Columbus (Delaware County) later distributed the grape around the country, where it became highly popular under the simple name Delaware.

OTHER HYPOTHESES

The variety has been variously considered to be an Italian grape or red Traminer (Bailey 1906; see SAVAGNIN). It has also been called 'Lisbon wine grape' or 'red Riesling'. Some have even suggested that Delaware was sent from France to Joseph (brother of Napoleon) Bonaparte's garden in Bordentown, New Jersey, in the nineteenth century but it shows little sign of *Vitis vinifera* characteristics. For this reason, others think that Delaware is a seedling of a native grape, without any genetic input from *Vitis vinifera*. However, a partial *vinifera* parentage was recently supported by ampelographic observations at Michigan University, not least because its roots are highly susceptible to phylloxera.

VITICULTURAL CHARACTERISTICS

Relatively hardy but not as hardy as CONCORD. Early ripening. Small bunches of small, thin-skinned pink berries that have less pronounced foxy flavours than most North American varieties and hybrids and are attractive to birds. Low yields and susceptible to fungal diseases, especially downy mildew and phomopsis but also to black rot, powdery mildew and botrytis bunch rot. Susceptibility to phylloxera means it should ideally be grafted.

WHERE IT'S GROWN AND WHAT ITS WINE TASTES LIKE

Delaware is grown mainly in the Northeast and Midwest of the United States. The biggest area is found in **New York** State (265 acres/107 ha in 2006) but it is also planted in states such as **Michigan, Ohio** and **Pennsylvania**. Used for sparkling wine as well as for still wine in a wide range of sweetness levels, from dry to icewine, Delaware is losing ground to other, more disease-resistant and higher-yielding, American hybrids. The wines are generally white or slightly pink and simple but fruity.

Delaware is often part of a blend but varietal wines in various styles are produced by Barrington Cellars, Keuka Lake, Mayers Lake Ontario, Thousand Islands and Wagner in New York State; Nassau Valley in **Delaware**; St Croix in **Minnesota**; and Klingshirn in Ohio (an icewine).

Delaware is also authorized, along with three other American hybrids, for southern **Austria**'s Uhudler (see CONCORD).

It is quite widely planted in **Japan**, although the vineyards dedicated to wine grapes amounted to not much more than 50 ha (124 acres) in 2009, divided between Kagawa and Hokkaido. **Korea** also had around 80 ha (198 acres) in 1997.

DELISLE

Very minor, cold-hardy North American hybrid with marked petrol-like aromas.

BERRY COLOUR: ● ● ● ● ●

PRINCIPAL SYNONYMS: ES 7-5-41

ORIGINS AND PARENTAGE

An Elmer Swenson 2-2-22 × ESPIRIT hybrid made by Elmer Swenson in Osceola, Wisconsin, US, in which Elmer Swenson 2-2-22 is an Elmer Swenson 283 × Elmer Swenson 193 hybrid (see SABREVOIS for the pedigrees). Delisle is therefore an extremely complex *Vitis riparia*, *Vitis labrusca*, *Vitis vinifera*, *Vitis aestivalis*, *Vitis lincecumii*, *Vitis rupestris*, *Vitis cinerea* and *Vitis berlandieri* hybrid.

VITICULTURAL CHARACTERISTICS

Medium to small bunches and berries. Early ripening. Cold-hardy.

WHERE IT'S GROWN AND WHAT ITS WINE TASTES LIKE

Delisle is planted to a very limited extent in **Canada**, particularly in Ontario and Québec. Domaines et Vins Gélinas in Québec is one of the few producers of varietal wines. Wines are said to have citrus or tropical flavours and a strong petrol-like character.

DEVÍN

Slovakian cross producing full-bodied, spicy whites in Slovakia and the Czech Republic.

BERRY COLOUR: ● ● ● ● ●

PRINCIPAL SYNONYMS: Děvín (Czech Republic), TCVCB 15/4

ORIGINS AND PARENTAGE

Devín is a GEWÜRZTRAMINER × ROTER VELTLINER cross (with Slovakian names Tramín Červený × Veltlínske Červené) obtained in 1958 by Dorota Pospíšilová at the VUVV research centre for oenology and viticulture in Bratislava, Slovakia. It was initially grown in Moravian vineyards (Czech Republic) under the name Ryvola before being named after a ruined castle near Bratislava. Authorized in Slovakia in 1997 and included in the Czech register of varieties in 1998.

VITICULTURAL CHARACTERISTICS

High and consistent yields, vigorous, late budding, mid to late ripening. Largish, thick-skinned berries. Resistant to spring and winter frosts. Susceptible to botrytis bunch rot but in dry autumn conditions can be used to produce raisins.

WHERE IT'S GROWN AND WHAT ITS WINE TASTES LIKE

Devín is planted throughout **Slovakia** (144 ha/356 acres in 2009) and is growing in popularity. It gives its best results in the Južnoslovenská and Malokarpatská wine regions in the south west. It produces full-bodied, low-acid, lightly aromatic – spicy, fruity and floral – wines such as those made by Mrva & Stanko (a sweet version) and Karpatská Perla (dry). Sugar levels can be very high and it is well suited to highly aromatic botrytized wines. However, if picked too late, it gives flabby wines, while high yields and poor ripening lead to a bitter aftertaste. There were 19 ha (47 acres) in Morava in the south east of the **Czech Republic** in 2009.

DIAGALVES

Southern Portuguese variety generally used in blends or eaten fresh.

BERRY COLOUR: ● ● ● ● ●

PRINCIPAL SYNONYMS: Carnal, Dependura, Diego Alves, Diogalves, Fernan Fer, Formosa, Formosa Dourada, Formosa Portalegre, Pendura, Pendura Amarela, Villanueva (Spain)

VARIETIES COMMONLY MISTAKEN FOR DIAGALVES: PALOMINO FINO ☒ (Spain)

ORIGINS AND PARENTAGE

Not much is known about the origin of Diagalves, a Portuguese variety that should not be confused with the Spanish PALOMINO FINO, which is sometimes also called Diagalves but has a distinct DNA profile in Veloso *et al.* (2010). It was used to breed RIO GRANDE and SEARA NOVA.

VITICULTURAL CHARACTERISTICS

Early to mid budding, late ripening and benefits from hot, dry summers. High but irregular yields. Large, thick-skinned berries. Susceptible to coulure, millerandage, botrytis bunch rot, esca and eutypa dieback.

WHERE IT'S GROWN AND WHAT ITS WINE TASTES LIKE

Diagalves is grown throughout **Portugal** both for the table and for wine and is authorized in regional wines such as IGP Alentejano and IGP Tejo. Typically wines have a slight grassy aroma, are low in most things – acidity, alcohol and structure – and should be drunk young. In the Alentejo, Enoforum's Real Forte includes Diagalves in the blend. There were 1,102 ha (2,723 acres) in Portugal in 2010.

DIEGO

See VIJARIEGO.

DIMRIT

Poor-quality, dark-skinned Turkish variety little used for wine so far.

BERRY COLOUR: ● ● ● ● ●

PRINCIPAL SYNONYMS: Dimlit, Dimrit Kara, Dirmit Kara

ORIGINS AND PARENTAGE

Dimrit is a variety from central Turkey, probably originating from the Nevşehir, Kırşehir, Kayseri and Niğde regions of Kapadokya (Cappadocia) in central Turkey. Akdimrit, Burdur Dimriti and Dimrit are often considered to be members of a so-called Dimrit group but DNA profiling has recently shown that twenty-two accessions with a name based on Dimrit corresponded to twenty-one distinct and mostly unrelated varieties (Şelli *et al.* 2007).

VITICULTURAL CHARACTERISTICS

Early ripening. Susceptible to botrytis bunch rot due to thin skins. Berries can attain fairly high sugar levels.

WHERE IT'S GROWN AND WHAT ITS WINE TASTES LIKE

Widely planted in Kapadokya, **Turkey**, but only a small proportion is used to make wine, which is generally poor quality and hidden in blends. The variety may have greater potential, however. There were 729 ha (1,801 acres) in Turkey in 2010.

DIMYAT

Productive, everyday Bulgarian white.

BERRY COLOUR: ● ● ● ● ●

PRINCIPAL SYNONYMS: Ahorntraube, Bekaszőlő (Hungary), Dertonia (Turkey), Dimiat (Romania, France), Galan (Russia), Grobweisser (Austria), Misket Slivenski (Bulgaria), Roşioară (Republic of Moldova), Semendria (Romania), Smederevka☒ (Republic of Macedonia, Croatia, Slovenia, Serbia),

Smederevka Bianca (Italy), Szemendriai Fehér (Hungary), Töröklugas (Hungary), Yapalaki (Greece), Zoumiatiko⚮ or Zumiatico (Greece), Zumyat

VARIETIES COMMONLY MISTAKEN FOR DIMYAT: Parmak⚮ (Bulgaria)

ORIGINS AND PARENTAGE

Dimyat is a very old Balkan variety, possibly originating in Bulgaria, where it shows the greatest biodiversity, with at least three main types: Dimyat Cherven with red-violet berries, Dimyat Cheren with black berries and Dimyat Edar, an oversized mutation with large leaves and berries and twice as many chromosomes (Katerov 2004). Its Bulgarian origin is supported by recent genetic analyses which place Dimyat (under the name Smederevka) among 'nonauthentic Macedonian grapevines' (Štajner, Angelova *et al.* 2009) and in the middle of seventy-four other varieties from Bulgaria (Hvarleva *et al.* 2004). In addition, DNA parentage analysis has suggested a likely parent–offspring relationship with the prolific GOUAIS BLANC (Boursiquot *et al.* 2004; see PINOT pedigree diagram).

Dimyat was used to breed CABERNET SEVERNY, MISKET VARNENSKI, NEOPLANTA and SIRMIUM.

OTHER HYPOTHESES

The variety was allegedly introduced from Egypt to Thráki (Thrace) by the Crusaders and named after the Egyptian city of Damiette, Dimyat in Arabic, on the Nile Delta. However, this is doubtful because Egypt is outside the natural distribution range of wild grapevines and cannot be home to any ancient indigenous variety (Katerov 2004).

VITICULTURAL CHARACTERISTICS

Very fertile, vigorous and productive, sometimes leading to poor ripening. Medium-sized to large thin-skinned berries. Late ripening. Sensitive to cold winter temperatures and susceptible to downy and powdery mildews but shows some resistance to botrytis bunch rot. Vulnerable to winter frosts. Its synonym Zumiatico derives from medieval Greek and alludes to 'sugar-rich fruit' (Katerov 2004).

WHERE IT'S GROWN AND WHAT ITS WINE TASTES LIKE

Dimyat produces crisp, everyday whites, at various levels of sweetness, often with a delicate vanilla perfume, although it is also made into sparkling wine, used as a table grape and distilled. It is widely planted across **Bulgaria**, with a total of 6,057 ha (14,967 acres) in 2009. Although the majority is in the eastern provinces of Varna and Burgas that adjoin the Black Sea coast, the better wines come not only from Varna and Pomorije on the coast but also from Preslav and Shumen in the north east and Chirpan in the centre of the country. Sparkling wines are made in Lyaskovets, Varna and Chirpan.

Under the name Smederevka, it is the leading white wine variety in the **Republic of Macedonia**, where it produces unexceptional everyday wines. Also found in Smederevo on the Dunav (Danube) in **Serbia**.

DINDARELLA

Rare, dark-skinned Veneto variety well suited to dried-grape wine styles.

BERRY COLOUR: ● ● ● ● ●

PRINCIPAL SYNONYMS: Dindarella Rizza, Pelada, Pelara⚮

ORIGINS AND PARENTAGE

In the nineteenth century, Dindarella and Pelara were considered to be two distinct varieties from the Veneto in northern Italy because Pelara often has far fewer berries. However, morphological studies in the twentieth century and recent DNA profiling have shown that they are one and the same variety (Vantini *et al.* 2003).

VITICULTURAL CHARACTERISTICS

Mid ripening but ripening is rapid, with a high capacity for both sugar synthesis and loss of acid. The thick skins and openness of the bunches make it resistant to botrytis bunch rot and therefore ideal for raisining and thus for use in dried-grape wines such as Amarone.

WHERE IT'S GROWN AND WHAT ITS WINE TASTES LIKE

Formerly widespread in Valpolicella, northern **Italy**, Dindarella is now scarcely cultivated but a small percentage is occasionally included in Valpolicella blends, especially those for Amarone and Recioto DOCGs. Tedeschi always include about 1% Dindarella in their Monte Olmi Amarone. It is authorized in Garda Orientale, Valdadige and Valpolicella DOCs. The variety produces wines that are light ruby in colour and lightly structured but notably aromatic: floral, fruity and spicy. Cantine Aldegheri in Sant'Abrogio Valpolicella produce a varietal Dindarella, sold as an IGT Rosso Veronese, for which the grapes are dried for several months after harvest and then undergo a very long, slow fermentation. There were just 9 ha (22 acres) in Italy in 2000. Brigaldara produce a Dindarella rosato.

DINKA ALBA

See KÖVIDINKA.

DIOLINOIR

Swiss cross that produces deeply coloured wines with some quality potential.

BERRY COLOUR: ● ● ● ● ●

ORIGINS AND PARENTAGE

Diolinoir is a Robin Noir × PINOT NOIR cross obtained in 1970 by André Jaquinet at the Caudoz viticultural research centre in Pully, near Lausanne, Switzerland, today part of Agroscope Changins-Wädenswil. Robin Noir is an old and abandoned variety from the Drôme in France that was found by chance in the 1920s in a grape collection in Diolly near Sion in the Valais, Switzerland. At first nicknamed Rouge de Diolly, it was identified as Robin Noir in 1995 (Vouillamoz 2009c). Diolinoir, a contraction of Diolly and Noir, was initially named 4-42 when the seedling was selected by its breeder.

VITICULTURAL CHARACTERISTICS

Medium to high vigour, reliably fertile, inconsistently productive. Early to mid ripening. Small berries. Sometimes prone to coulure and millerandage and sensitive to winter frosts. Good resistance to botrytis bunch rot.

WHERE IT'S GROWN AND WHAT ITS WINE TASTES LIKE

Initially bred as a blending component to enhance the colour of some Swiss wines, Diolinoir has in recent years been used to make deeply coloured, barrel-aged varietal wines with firm but rounded tannins and dark-fruit flavours such as blackberry. Producers include the Cave des Tilleuls and Provins (Domaine Evêché), both in the Valais. Clos de Crétaz in Vaud even make a late-harvest sweet version. Diolinoir was planted on 112 ha (277 acres) in **Switzerland** in 2009, with almost 80% (88 ha/217 acres) in the Valais.

DIŠEĆA RANINA

Rare and very localized Croatian variety from south of Zagreb.

BERRY COLOUR: ● ● ● ● ●

PRINCIPAL SYNONYMS: Dišeća Ranina Bijela, Petrinjska Ranina or Petrinjska Bijela

ORIGINS AND PARENTAGE

Dišeća Ranina is a rare variety from the Moslavina and Pokuplje wine regions, just south of Zagreb, where it has been grown for centuries. This variety is very similar to BOUVIER, an unknown natural cross that has a Muscat aroma and is also called Ranina or Ranina Bela in Slovenia. Ampelographic observations and DNA profiling are needed to determine whether or not they are identical.

VITICULTURAL CHARACTERISTICS

Late ripening. Functionally female flowers and therefore often planted with ŠKRLET for pollination.

WHERE IT'S GROWN

Dišeća Ranina, which produces wines with a slight Muscat aroma, is found and officially authorized in the Moslavina and Pokuplje regions in **Croatia**, south of Zagreb. Producers of varietal wines include Florijanović in Potok and Boris Mesarić in Kutina.

DNESTROVSKY ROZOVY

A (Nimrang × *Vitis amurensis* Ruprecht) × Mathiasz Janos red-berried hybrid created in **Ukraine** in which Nimrang is a table grape from Uzbekistan and Mathiasz Janos is a MUSCAT OF ALEXANDRIA × Chasselas Rouge (a red-berried mutation of CHASSELAS) cross. Vityazevo Winery in the Krasnodar region, **Russia**, for example, blend it with the Hungarian variety ZALAGYÖNGYE (under the name Zhemchug Zala).

DOBRIČIĆ

Rare Croatian variety high in sugar, rich in colour and used mainly in red blends on the island of Šolta.

BERRY COLOUR: ● ● ● ● ●

PRINCIPAL SYNONYMS: Čihovac, Crljenak Slatinski, Dobričić Crni, Dobrovoljac, Krucalin, Okručanac, Sholtanats, Slatinjac, Slatinski

ORIGINS AND PARENTAGE

Dobričić is an old variety from southern Dalmacija (Dalmatia) in Croatia, possibly originating from the island of Šolta, where it is still cultivated to a very limited extent though not as widely as it once was. Maletić *et al.* (2004) have proved through DNA parentage analysis that Dobričić and TRIBIDRAG (also known as Primitivo or Zinfandel) are the parents of PLAVAC MALI. In addition, a possible parent–offspring relationship between BABIĆ and Dobričić has recently been discovered (Zdunić, Pejić *et al.* 2008; see Tribidrag for the full pedigree).

VITICULTURAL CHARACTERISTICS

Late budding, mid to late ripening. Susceptible to downy mildew. Moderate but regular yields. Berries reach high sugar levels and are rich in anthocyanins.

WHERE IT'S GROWN AND WHAT ITS WINE TASTES LIKE

Dobričić is cultivated almost exclusively on the island of Šolta, south of Split and west of the island of Brač in **Croatia**, but since 2004 it has been authorized throughout Srednja i Južna Dalmacija (Central and Southern Dalmatia). Varietal wines, which tend to be very rich in colour and extract and low in acidity, are generally used to add colour in blends.

DOINA

Also spelt Dojna, this is a Coarnă Neagră × (a mixture of pollens of Seyanets 35 + Varousset) black-berried hybrid bred by D D Verderevskii in 1968 at the Chisinau research centre in the **Republic of Moldova** in which Coarnă Neagră is a table grape from Moldova, Seyanets 35 is an unknown cross (*seyanets* means 'seedling' in Russian) and Varousset (or Seyve-Villard 23-657) is a Seibel 4668 × SUBÉREUX hybrid (see MUSCAT SWENSON for the complete Varousset pedigree). It is used in blends by Vityazevo Winery in the Krasnodar region of **Russia**, for example.

DOLCETTO

Deeply coloured, low-acid, fragrant Piemontese red.

BERRY COLOUR: ● ● ● ● ●

PRINCIPAL SYNONYMS: Dolcetto Nero, Nibièu or Nibiò (Oltrepò Pavese in Lombardia and around Tortona in Piemonte), Ormeasco (Ormea in Piemonte and Pieve di Teco in Liguria)

VARIETIES COMMONLY MISTAKEN FOR DOLCETTO: Charbono☒ (Piemonte), DOUCE NOIRE☒ (Savoie in France), NEBBIOLO☒ (province of Alessandria in Piemonte)

ORIGINS AND PARENTAGE

The earliest mention of Dolcetto appears in 1593 in Dogliani near Cuneo (Comba and Dal Verme 1990). The name Dolcetto, meaning 'little sweet', comes from the low natural acidity of the berries but Dolcetto wines are almost always dry.

Galet's (2000) contention that Dolcetto is identical to the DOUCE NOIRE of Savoie, known as Charbono in California but distinct from the true (but no longer commercially cultivated) Charbono of Piemonte, has been disproved by DNA profiling at the University of California at Davis. Some authors considered Dolcetto to be morphologically similar to Zinfandel (TRIBIDRAG) but recent DNA studies contradict this hypothesis (Calò, Costacurta *et al.* 2008).

Dolcetto was crossed with CHATUS to produce PASSAU, SAN MARTINO and VALENTINO NERO.

A white-berried Dolcetto Bianco does exist but it has, paradoxically, nothing to do with the black-berried Dolcetto (Schneider and Mannini 2006).

VITICULTURAL CHARACTERISTICS

Early ripening (up to four weeks before NEBBIOLO), making it suitable for the cooler, higher sites in Piemonte where BARBERA and Nebbiolo fail to ripen fully. In Alba, it is thought to prefer the lighter white-marl soils on the right bank of the River Tanaro. It is susceptible to fungal diseases and in some locations tends to drop its bunches before harvest (Botta, Vallania *et al.* 1995).

WHERE IT'S GROWN AND WHAT ITS WINE TASTES LIKE

Formerly widely cultivated in Piemonte, north-west **Italy**, Dolcetto has been gradually losing ground to BARBERA since the phylloxera plague but is still widespread in the provinces of Cuneo (Langhe and Ormea regions), Alessandria (Acqui and Ovada) and Asti (Colli del Monferrato). It is authorized as a blending partner for Barbera in several Piemonte DOCs but as a varietal in many more, including Acqui, Alba, Asti, Diano d'Alba, Langhe and Monferrato, and has reached the heights of DOCG in Dogliani. Alba, Ovada and Dogliani are quantitatively the most significant DOCs for varietal Dolcetto but the finest quality is generally said to come from Alba.

Marcarini in La Morra (Barolo country) make a fine, unusually full-bodied wine, Boschi di Berri, from ungrafted centenarian vines – the oldest Dolcetto vineyard in Italy – that survived phylloxera thanks to the sandy soils. Other good producers include G D Vajra and Enrico e Marziano Abbona.

In Liguria, Ormeasco is a particular clone of Dolcetto historically cultivated in Pornassio, where the DOC allows both a dry and a sweet passito version, and in Pieve di Teco in the province of Imperia, although it takes its name from the village of Ormea in Piemonte. Dolcetto is also grown in various other parts of Italy, including the Oltrepò Pavese, as well as further south, in Umbria.

Varietal Dolcetto wines are generally deeply coloured, soft, round and fruity; the best are fragrant with flavours of liquorice and almonds. Although some well-made and more highly structured wines – from Alba or Ovada, for example – may last five years or more, most are best drunk within two or three years of the harvest. The variety's unusual combination of notably low acidity and high tannins creates a challenge for winemakers and softer extraction and shorter fermentation times generally produce the best results.

Total plantings in Italy in 2000 amounted to 7,450 ha (18,409 acres). Mas de Daumas Gassac in southern **France** have 0.1 ha (0.25 acres)!

Dolcetto is scattered in tiny pockets around the United States, particularly in **California** (eg Acorn, Bonny Doon, Cartlidge & Brown, Duxoup, Jacuzzi, Palmina, Kent Rasmussen) and to a lesser extent in **Washington** State (Woodward Canyon) and **Oregon** (eg Ponzi, Stag Hill). It is also grown in **Texas** on a very limited scale.

A variety known as Dolcetto is grown on an extremely limited scale in **Argentina** but it is not known if this is the true Dolcetto or not.

In **Australia**, more than twenty producers have planted Dolcetto, in regions as diverse as the

Hunter Valley, the Barossa Valley and the Adelaide Hills. However, total plantings are not significant enough to feature in official statistics. In **New Zealand**, Heron's Flight in Matakana is one of the few who produce a varietal Dolcetto.

DOLCIAME

Virtually extinct Umbrian white.

BERRY COLOUR: ● ● ● ● ●

PRINCIPAL SYNONYMS: Malfiore, Parlano, Uva delle Vecchie

VITICULTURAL CHARACTERISTICS

Vigorous, mid to late ripening.

WHERE IT'S GROWN

Almost extinct variety, scarcely cultivated in Umbertide and Città di Castello in the province of Perugia, **Italy**, and used in blends. Just 7 ha (17 acres) were recorded in the Italian agricultural census in 2000.

DOMINA

Modern red German cross planted mainly in Franken.

BERRY COLOUR: ● ● ● ● ●

PRINCIPAL SYNONYMS: Geilweilerhof 4-25-7

ORIGINS AND PARENTAGE

A BLAUER PORTUGIESER × PINOT NOIR (Spätburgunder) cross created by Peter Morio and Bernhard Husfeld in 1927 at the Geilweilerhof research centre in the Pfalz in Germany. Domina, the Latin word for 'lady' or 'mistress', was used to breed CALANDRO.

VITICULTURAL CHARACTERISTICS

A productive variety with good disease resistance, particularly tolerant of virus diseases, and with higher sugar levels than BLAUER PORTUGIESER.

WHERE IT'S GROWN AND WHAT ITS WINE TASTES LIKE

Domina was added to the official register of varieties in **Germany** in 1974 and there are now 404 ha (998 acres) planted, 347 ha (857 acres) of those in Franken. In good years, wines are full-bodied, deeply coloured with firm tannins and blackberry or cherry flavours. They are generally fresh but lack the finesse of their PINOT NOIR parent, though they are generally well suited to barrel ageing. Recommended producers include Brennfleck and Roth in Franken.

There are also minuscule plantings in **Switzerland** (8 ares/1,000 square yards).

DOÑA BLANCA

See SÍRIA.

DONA BRANCA

Low-acid Portuguese variety whose name is used for several others.

BERRY COLOUR: ● ● ● ● ●

VARIETIES COMMONLY MISTAKEN FOR DONA BRANCA: CHELVA (Spain), DORADILLA (Spain), MALVASIA FINA, SÍRIA, TAMAREZ

ORIGINS AND PARENTAGE

Dona Branca, meaning 'white lady', is a confusing name given to distinct varieties in Portugal, but the Dona Branca that is cultivated in the Beiras, Douro and Trás-os-Montes regions has a distinct and unique morphology (Rolando Faustino, personal communication) and – based on the sample at the United States Department of Agriculture's National Clonal Germplasm Repository in Davis, California – DNA profile (Vouillamoz).

VITICULTURAL CHARACTERISTICS

Susceptible to powdery mildew and botrytis bunch rot. Compact bunches of medium-sized, thick-skinned berries. Mid to late ripening.

WHERE IT'S GROWN AND WHAT ITS WINE TASTES LIKE

Dona Branca, grown mainly in the northern half of **Portugal**, in Beiras, the Douro and Trás-os-Montes, produces soft, fruity wines without any great distinction. Total plantings in Portugual in 2010 were 293 ha (724 acres). Recommended producers include Quinta da Comenda in Beiras.

DONZELINHO BRANCO

Waning vine adding perfume to blends in north-eastern Portugal.

BERRY COLOUR: ● ● ● ● ●

VARIETIES COMMONLY MISTAKEN FOR DONZELINHO BRANCO: RABIGATO

ORIGINS AND PARENTAGE

Donzelinho Branco is an old Portuguese variety that was mentioned in the Douro as early as 1531 by Rui Fernandes (1531–2) under the spelling Donzelyhno but it is not known if he was referring to the white- or the black-berried variety. Donzelinho Branco is not a colour mutation of DONZELINHO TINTO, both showing distinct DNA profiles in Veloso *et al.* (2010), and although Donzelinho Branco shares the synonym Terrantez with FOLGASÃO and is also sometimes confused with RABIGATO, they are all genetically distinct (Ferreira Monteiro *et al.* 2000).

VITICULTURAL CHARACTERISTICS

Early budding, mid ripening. Small, fairly compact bunches of thick-skinned berries. Susceptible to frost. Poor fertility necessitates long pruning.

WHERE IT'S GROWN AND WHAT ITS WINE TASTES LIKE

Donzelinho is found mainly in old interplanted vineyards in north-eastern **Portugal** and is therefore rarely made as a varietal wine even though it is authorized for port, Douro and Trás-os-

Montes appellation wines as well as in the Duriense and Trasmontano IGPs. Wines are light-bodied and lavender-scented. The field blends used to make Niepoort's Tiara and Redoma Reserva, for example, include a small amount of Donzelinho Branco, but that's mainly because they happen to be in the vineyard. There were just 42 ha (104 acres) in Portugal in 2010 (presumably of vineyards devoted entirely to the variety), much less than 10–15 years ago.

DONZELINHO TINTO

Old, rare Douro variety playing a very small, inglorious part in blended wines, fortified and unfortified.

BERRY COLOUR: ● ● ● ● ●

PRINCIPAL SYNONYMS: Donzelinho do Castello, Donzelynho, Tinta do Minho

VARIETIES COMMONLY MISTAKEN FOR DONZELINHO TINTO: TROUSSEAU

ORIGINS AND PARENTAGE

Donzelinho Tinto is an old Portuguese variety that was mentioned in the Douro as early as 1531 by Rui Fernandes (1531–2) under the spelling Donzelyhno but, since the berry colour was not given, this could refer to DONZELINHO BRANCO, which is a distinct variety.

VITICULTURAL CHARACTERISTICS

Small, compact bunches of small, thin-skinned berries.

WHERE IT'S GROWN AND WHAT ITS WINE TASTES LIKE

Donzelinho Tinto, like the unrelated but equally old and rare DONZELINHO BRANCO, is found mainly in field blends in north-eastern **Portugal**, where it is authorized for appellation wines such as Douro and Trás-os-Montes and regional wines such as Duriense and Trasmontano but is used mainly as a minor ingredient in port. The Real Companhia Velha recently planted the variety separately but were disappointed with the results and pulled the vines out after a few years. There were 87 ha (215 acres) in Portugal in 2010.

DORADILLA

Very minor variety from Málaga in Andalucía frequently confused with other varieties.

BERRY COLOUR: ● ● ● ● ●

PRINCIPAL SYNONYMS: Forastera Blanca, Plateado

VARIETIES COMMONLY MISTAKEN FOR DORADILLA: BEBA, CAYETANA BLANCA (also misleadingly called Plateado in Spain and Doradillo in Australia), CHELVA, DONA BRANCA

ORIGINS AND PARENTAGE

Doradilla is a variety from Málaga that is distinct from Doradillo, the name under which CAYETANA BLANCA (Jaén Blanco or Pardina in Spain) was introduced from Spain to Australia in the nineteenth century (Galet 2000). Indeed, recent DNA parentage analysis has found a parent–offspring relationship between Doradilla (under the synonym Plateado) and CAYETANA BLANCA, a variety that is widespread throughout the south and centre of the Iberian peninsula (Zinelabidine *et al.* 2012; see Cayetana Blanca for the pedigree diagram). Moreover, Doradilla most likely has a parent–offspring relationship with ALBILLO MAYOR from Ribera del Duero in north-central Spain (Santana *et al.* 2010). In addition, it should not be confused with CHELVA, BEBA or DONA BRANCA, which are also sometimes called Forastera Blanca though they are not commonly confused in the vineyard.

WHERE IT'S GROWN

Doradilla is permitted in both the Málaga and Sierras de Málaga DOs in **Spain**'s far south but is planted only to an extremely limited extent in the north of that area.

DORAL

Minor Swiss cross producing slightly less neutral wines than its Chasselas parent.

BERRY COLOUR: ● ● ● ● ●

ORIGINS AND PARENTAGE

Doral is a CHASSELAS × CHARDONNAY cross obtained in 1965 by Jean-Louis Simon and selected from among several seedlings (selection number 1-2-1) by André Jaquinet at the Caudoz viticultural research centre in Pully, near Lausanne, Switzerland, today part of Agroscope Changins-Wädenswil. Like its sibling CHARMONT, Doral was created to produce a more aromatic version of CHASSELAS.

VITICULTURAL CHARACTERISTICS

Low to moderate fertility, early to mid budding, early ripening. Medium-sized, very compact bunches. Some susceptibility to millerandage, botrytis bunch rot and magnesium deficiency.

WHERE IT'S GROWN AND WHAT ITS WINE TASTES LIKE

Doral wines can be quite full-bodied, thanks to the high sugar content in the berries, show some similarity with CHARDONNAY and tend to be fresher than wines made from its sibling CHARMONT, with aromas of citrus and apricot. The variety was planted on 27 ha (67 acres) in **Switzerland** in 2009, predominantly in Vaud (75%), where Domaine de la Fornelette and Christian Dugon make good varietal wines. Other examples are found in Ticino (Davide Cadenazzi) and in the Valais (Cave du Paradou and Cave des Amis). It seems to have potential for the production of sparkling and late-harvest wines.

DORINTO

Recently renamed Portuguese white.

BERRY COLOUR: ● ● ● ● ●

PRINCIPAL SYNONYMS: Arinto Branco, Arinto do Douro, Arinto do Interior, Arinto no Douro☒, Arinto de Trás-os-Montes

VARIETIES COMMONLY MISTAKEN FOR DORINTO: ARINTO DE BUCELAS☓

ORIGINS AND PARENTAGE

There are several varieties with names or synonyms referring to Arinto (see MALVASIA FINA and BICAL, for example) but the DNA profile of Arinto no Douro (Lopes *et al.* 1999; Veloso *et al.* 2010), recently renamed Dorinto, from Portugal's Dão region, is distinct from all of them although it has often been confused with ARINTO DE BUCELAS (Lopes *et al.* 1999). The new name Dorinto, a contraction of Douro and Arinto, is an attempt to reduce the confusion.

VITICULTURAL CHARACTERISTICS

Mid budding, early ripening. Moderately susceptible to downy and powdery mildews.

WHERE IT'S GROWN AND WHAT ITS WINE TASTES LIKE

Dorinto wines are typically citrus-scented and fruity with a long aftertaste. The variety is authorized in the Trasmontano and Duriense IGPs in northern **Portugal**, where the ability of its large leaves to resist the heat is much appreciated. Most of the 20 ha (49 acres) remaining in 2010 are found in old vineyards in Baixo Corgo, western Douro.

DORNFELDER

The most successful of the modern red German crosses, making dark, velvety wines.

BERRY COLOUR: ● ● ● ● ●

PRINCIPAL SYNONYMS: Weinsberg S 341

ORIGINS AND PARENTAGE

A HELFENSTEINER × HEROLDREBE cross bred in 1956 by August Herold at the Weinsberg research centre in Baden-Württemberg, southern Germany. Its extended pedigree is therefore: (PINOT NOIR PRÉCOCE × SCHIAVA GROSSA) × (BLAUER PORTUGIESER × BLAUFRÄNKISCH). Named after Immanuel August Ludwig Dornfeld (1796–1869), one of the founders of the Weinsberg school of viticulture, it was used to breed ACOLON, CABERNET DORIO, CABERNET DORSA and MONARCH.

VITICULTURAL CHARACTERISTICS

Vigorous, consistently productive (up to 120 hl/ha), early to mid budding and ripening. Good resistance to botrytis bunch rot thanks to thick skins, which also contribute to its prized deep colour.

WHERE IT'S GROWN AND WHAT ITS WINE TASTES LIKE

Dornfelder's popularity increased steadily after its official authorization in **Germany** in 1980 and then surged ahead in the first five or six years of the twenty-first century, during which time the planted area doubled. The German total is now fairly stable at 8,101 ha (20,018 acres), dominated by plantings of over 3,000 ha (7,400 acres) in both Rheinhessen and the Pfalz. It produces dark, rich, velvety wines with good acidity and occasionally an attractively floral note. However, yields must be controlled to get good fruit concentration if the wine is to be aged in barrel. In a country where red wines are often pale, the variety is prized for its deep colour. Recommended producers include Deutzerhof in the Ahr and Gutzler, Knipser and Lingenfelder in the Pfalz.

Switzerland had 21 ha (52 acres) in 2009, with producers such as Fredi Clerc and Brigitte

Bamert in Nuolen (Schwyz), Gottlieb Welti in Küsnacht (Zürich) and Landoldt Weine (Zürich) making varietal wines.

The **Czech Republic** has a little. In 2007 there were 16 ha (40 acres) in **England**, where it was first planted in the late 1980s and it is sometimes made as a rosé. Camel Valley, for example, produce a sparkling version. It can also produce lightly spicy reds but on the whole is less successful than REGENT. Some is planted in Santa Rita Hills in **California**'s Central Coast and also in **Pennsylvania**. If you look hard enough, you may find a few rows of Dornfelder in **Brazil**'s Rio Grande do Sul and **Japan** had around 5 ha (12 acres) in Hokkaido in 2009.

DORONA DI VENEZIA

Recently rescued but not especially ancient Venetian white.

BERRY COLOUR: ● ● ● ● ●

PRINCIPAL SYNONYMS: Dorona, Dorona Veneziana, Uva d'Oro

VARIETIES COMMONLY MISTAKEN FOR DORONA DI VENEZIA: GARGANEGA ☒

ORIGINS AND PARENTAGE

Dorona di Venezia is local to the Venetian lagoon, where it has almost exclusively been cultivated, mainly as a table grape. It was thought to be identical to GARGANEGA but DNA analysis has both disproved this and explained the confusion (Di Vecchi Staraz, This *et al.* 2007): Dorona di Venezia was shown to be a natural Garganega × Bermestia Bianca cross, where Bermestia Bianca is a table grape from Emilia-Romagna (see Garganega for the complete pedigree). The name Dorona derives from the Italian *d'oro*, meaning 'gold', referring to the golden colour of the berries at harvest. Although there is documentary evidence of vines planted in the Republic of Venice at the time of its settlement and a reference to a *vinea*, or vineyard, on the Venetian island of Mazzorbo in the mid fifteenth century, the first reference to Dorona was in Eden (1903).

OTHER HYPOTHESES

Dorona di Venezia is said to have been known since the fifteenth century but this is probably the result of the frequent confusion with the very old GARGANEGA.

VITICULTURAL CHARACTERISTICS

Hardy, relatively resistant to fungal diseases, including botrytis bunch rot. Mid ripening but well suited to late harvesting or drying.

WHERE IT'S GROWN AND WHAT ITS WINE TASTES LIKE

In **Italy**, just a few vines of Dorona di Venezia remained on the Venetian island of Sant'Erasmo before Gianluca Bisol, of the eponymous Prosecco producer, restored the ancient estate now known as Venissa on the neighbouring island of Mazzorbo and replanted the variety in the warm, saline environment of the small estate vineyard. It was traditionally grown pergola-style both for the table and for wine but the Bisol vines are vertically trained and a single-vineyard Dorona wine is produced. It is hard to speak of a typical style when there is just one producer but Bisol's example has a strong minerality and notes of dried peach and apricot as well as iodine and a deliberate rusticity, intensified by fermentation on the skins without added sulphites.

DOUCE NOIRE

Savoie variety with several aliases causing considerable confusion, most widely planted in Argentina.

BERRY COLOUR:

PRINCIPAL SYNONYMS: Bathiolin (Albertville), Bonarda (Argentina), Charbonneau (Jura), Charbono (California in US), Corbeau or Corbeau Noir (Ain, Isère and Jura), Mauvais Noir, Plant de Montmélian, Plant de Turin (Jura), Plant Noir (Haute-Savoie), Turca (Trentino in Italy), Turin (Jura)

VARIETIES COMMONLY MISTAKEN FOR DOUCE NOIRE: BONARDA PIEMONTESE, Charbono (Piemonte), DOLCETTO (Piemonte)

ORIGINS AND PARENTAGE

Until the end of the nineteenth century, Douce Noire was one of the most widespread red varieties of Savoie in eastern France, notably in the region of Montmélian and Arbin south east of Chambéry. Outside of Savoie, it was called Corbeau (meaning 'crow') because of the deeply coloured wine it produced. The earliest mention of this variety was found in a letter dated 24 November 1803 (24 Brumaire XII in the Republican calendar) and sent by the mayor of Saint-Pierre-d'Albigny to the prefect of Savoie, in which Douce Noire was cited as the main variety in his commune (Durand 1901b).

Douce Noire was often said to originate from Piemonte in north-west Italy, hence its synonym Plant de Turin or simply Turin. Furthermore, for most of the twentieth century, Douce Noire (meaning 'sweet black' (see DOLCETTO)) was erroneously identified with Dolcetto Nero (meaning 'small sweet black') from Piemonte, until careful ampelographic observations and DNA profiling definitively rejected this hypothesis (Schneider and Mannini 2006; Vouillamoz, Frei and Arnold 2008). Consequently, the birthplace of Douce Noire is most likely to be Savoie, where the variety was historically found and where it has so many synonyms.

Recent DNA profiling has also shown that the variety cultivated in California under the name Charbono is distinct from the true (but no longer commercially cultivated) Charbono from Piemonte, and identical to Douce Noire (Martínez *et al.* 2008), which was probably introduced to California under the synonym Charbonneau once used in the Jura, eastern France.

In Argentina, DNA profiling has established that the variety known there as Bonarda, the second most common red variety after Malbec (COT), is identical to neither the Italian BONARDA PIEMONTESE nor any of the five other distinct varieties that are misleadingly called Bonarda in Italy because it is in fact identical to Douce Noire (Martínez *et al.* 2008). In addition, comparison of the DNA profiles of Turca (Grando 2000; Crespan, Giannetto *et al.* 2008), cultivated in the Veneto in north-eastern Italy since the early twentieth century, and Douce Noire has unexpectedly shown them to be one and the same variety (Vouillamoz).

Galet (2000) mentions a variety called Douce Noire Grise, which is no longer cultivated but is unrelated to Douce Noire.

VITICULTURAL CHARACTERISTICS

Late ripening.

WHERE IT'S GROWN AND WHAT ITS WINE TASTES LIKE

The very limited plantings remaining in **France** (about 2 ha/5 acres in 2007), where it is officially known as Corbeau, are in Savoie, where producers use the most ancient name of all, Douce Noire. Domaine Ganevat in Rotalier in the Jura have replanted a small area of Corbeau and include it in an early-drinking red blend. Most include it in blends (it was traditionally used to soften wines made from PERSAN) but Jean-Paul Finas has made a varietal Vin de Pays d'Allobrogie.

In **California**, under the name Charbono, there were 88 acres (36 ha) in 2008 and there are almost no new plantings. It is thought to have been imported by Italian immigrants in the 1800s but it came into the country as BARBERA and was planted and bottled as such by wineries including Inglenook and Parducci until the 1930s. Charbono was at one stage a speciality of the historic Napa Valley winery Inglenook, whose first vintage labelled Charbono was 1941 (Savoie 2003). About half the California Charbono vines are in Napa, particularly in the hotter parts of Calistoga, with the rest in Mendocino, Monterey and Madera. The best vineyards are Frediani, Heitz, Andriana (Summers), Tofanelli, Meyer and Cooke; the biggest producers are Summers and Pacific Star and others include Chameleon, Coturri, Duxoup, Robert Foley, Thomas Michael, Shypoke and Turley.

But these totals are swamped by the 18,759 ha (46,354 acres) planted under the name Bonarda in **Argentina**, the vast majority in Mendoza and San Juan, where it generally produces good-value fruity quaffing wines, although if allowed to ripen fully, there is more quality potential than is generally realized in the bottle, and alcohol tends to remain within bounds. Argentine producers of varietal Bonarda include Anubis, Chakana, Cristóbal, Augusto Pulenta, R J Vinedos and Zuccardi.

DOUX D'HENRY

Rare, light-bodied red exclusive to the Pinerolese region of Piemonte.

BERRY COLOUR: ● ● ● ● ●

PRINCIPAL SYNONYMS: Doux d'Enry, Gros d'Henry

VARIETIES COMMONLY MISTAKEN FOR DOUX D'HENRY: CILIEGIOLO

ORIGINS AND PARENTAGE

This curiously named variety is typical of the Pinerolese region and the Val Chisone in Piemonte, north-west Italy, where it is sometimes confused with CILIEGIOLO (Schneider and Mannini 2006). It is said to have been named in honour of Henri IV of France, who allegedly praised this wine when he met the Duke of Savoy, Charles Emanuel I, in the early seventeenth century. The epithet *doux* (meaning 'sweet') probably refers to the residual sugar that producers used to leave in Doux d'Henry wines.

OTHER HYPOTHESES

A French origin has often been suggested but so far there is no evidence of a match or a link with French varieties.

VITICULTURAL CHARACTERISTICS

Vigorous but has functionally female flowers and therefore needs the presence of other varieties in the neighbourhood for fertilization. Resistant to frost, but prone to millerandage and to botrytis bunch rot and sour rot if the bunches are particularly compact. Early to mid ripening. It is also grown as a table grape.

WHERE IT'S GROWN AND WHAT ITS WINE TASTES LIKE

Cultivated exclusively in the Pinerolese area (mainly in San Secondo di Pinerolo and in Prarostino) and in the Val Chisone and the Val Germanasca in Piemonte, north-west **Italy**, Doux d'Henry is authorized in the Pinerolese Rosso DOC and in the varietal Pinerolese Doux d'Henry DOC. The wines are light and fresh with floral aromas. Producers of the varietal version include Bruno Daniela, Cantina Dora Renato and Il Tralcio. Just 28 ha (69 acres) were recorded in the Italian agricultural census in 2000.

DRAGON BLUE

Strongly flavoured Canadian hybrid best hidden in a blend.

BERRY COLOUR: ● ● ● ● ●

ORIGINS AND PARENTAGE

A hybrid obtained in Québec, Canada, by the private vine breeder Mario Cliche.

VITICULTURAL CHARACTERISTICS

Very productive.

WHERE IT'S GROWN AND WHAT ITS WINE TASTES LIKE

In eastern **Canada**, Dragon Blue produces wines with the markedly sweet candy flavour typical of American hybrids. This is probably why Les Murmures use it as part of a fortified blend with FRONTENAC and ST CROIX and it is not found as a varietal.

DRNEKUŠA

Rare vine from the Croatian island of Hvar.

BERRY COLOUR: ● ● ● ● ●

PRINCIPAL SYNONYMS: Darnekuša, Darnekuša Mala, Darnekuša Vela, Dernakuša

ORIGINS AND PARENTAGE

Drnekuša is from the island of Hvar off Croatia's Dalmatian coast and south of Split.

VITICULTURAL CHARACTERISTICS

Late budding and late ripening. Moderate to high, stable yields. Big loose bunches of thin-skinned berries. Susceptible to powdery and downy mildews.

WHERE IT'S GROWN AND WHAT ITS WINE TASTES LIKE

Drnekuša is grown almost exclusively on the island of Hvar in **Croatia** although it is authorized for the whole of Srednja i Južna Dalmacija (Central and Southern Dalmatia). Plančić is one of the few producers to make a varietal wine, this one from fruit grown at 580 m (1,900 ft) above the Adriatic.

DRUPEGGIO

Neutral, central Italian white also known as Canaiolo Bianco.

BERRY COLOUR: ● ● ● ● ●

PRINCIPAL SYNONYMS: Bottaio Bianco (Toscana), Cacinello or Cacciumo (Campobasso), Canaiolo

Bianco☒ (Toscana), Canajola, Canina or Uva dei Cani (Ascoli Piceno), Drupeccio (Orvieto), Lupeccio, Trupeccio (Orvieto), Volpicchio (Toscana)

VARIETIES COMMONLY MISTAKEN FOR DRUPEGGIO: VERMENTINO☒, VERNACCIA DI SAN GIMIGNANO☒, Zuccaccio☒

ORIGINS AND PARENTAGE

Known as Drupeggio in the Orvieto region of Umbria, this variety has also been known in Toscana as Canaiolo Bianco since at least 1817. However, a recent DNA profiling survey by Storchi *et al.* (2011) has established that the name Canaiolo Bianco was mistakenly given to at least six distinct varieties in Toscana, not only to Drupeggio but also to VERNACCIA DI SAN GIMIGNANO and to Zuccaccio, a very old local variety that is no longer cultivated. Since the official clone currently available in nurseries – Canaiolo Bianco ARSIAL-CRA 402 – is in fact Drupeggio, and since Canaiolo Bianco is a source of confusion, we have chosen to follow Storchi *et al.* (2011) by using Drupeggio as the prime name for this variety.

VITICULTURAL CHARACTERISTICS

Mid to late ripening.

WHERE IT'S GROWN AND WHAT ITS WINE TASTES LIKE

Drupeggio is grown mainly in Orvieto in Umbria, central **Italy**, where it is generally blended with other local white varieties. It is also planted in other parts of Umbria and in Lazio. Under the name Canaiolo Bianco, it is also grown in Toscana in the provinces of Firenze, Grosseto and Pistoia, where it is usually blended with TREBBIANO TOSCANO and MALVASIA BIANCA LUNGA (Malvasia Toscana), although a maximum of 10% is permitted in the red and rosé Barco Reale di Carmignano DOC. Occasionally included in Vin Santo, though the grapes are difficult to raisin. A full 674 ha (1,665 acres) were recorded in the 2000 Italian census under the name Canaiolo Bianco. However, given the possible confusion of Canaiolo Bianco with other varieties such as VERNACCIA DI SAN GIMIGNANO, it is difficult to know exactly which variety contributes to a given DOC.

DUNAJ

Minor but potentially high-quality Slovak cross responsible for soft, full-bodied reds.

BERRY COLOUR: ● ● ● ● ●

PRINCIPAL SYNONYMS: MBOP × SV 6/10

ORIGINS AND PARENTAGE

Dunaj is a (Muscat Bouschet × BLAUER PORTUGIESER) × SANKT LAURENT cross (with Slovakian names: (Muškát Bouchet × Oporto) × Svätovavrinecké) obtained in 1958 by Dorota Pospíšilová at the VUVV research centre for oenology and viticulture in Bratislava, Slovakia, in which Muscat Bouschet is a Muscat Rouge de Madère × PETIT BOUSCHET cross and Muscat Rouge de Madère is a likely natural progeny of MUSCAT BLANC À PETITS GRAINS × MAMMOLO from Toscana (Di Vecchi Staraz, This *et al.* 2007) that is no longer commercially cultivated. It was christened with the Slovakian and Czech name for the Danube and officially registered in 1997.

VITICULTURAL CHARACTERISTICS

Productive, late budding, mid to late ripening. Resistant to frost and to botrytis bunch rot. Susceptible to powdery mildew and sometimes to poor fruit set.

WHERE IT'S GROWN AND WHAT ITS WINE TASTES LIKE

Wines are typically deeply coloured and full-bodied with soft tannins and flavours of ripe cherries and chocolate. It is adaptable to a range of sites in **Slovakia** but it does need a good supply of water. There were 62 ha (153 acres) in Slovakia in 2011. Producers include Joseph Képeš, Kmet'o, Masaryk, Chateau Modra and Mrva & Stanko.

DUNKELFELDER

Dark-skinned German cross usefully adding colour and body to blends.

BERRY COLOUR: ● ● ● ● ●

PRINCIPAL SYNONYMS: Fröhlich V. 4.4, Fröhlich V 4, Purpur

ORIGINS AND PARENTAGE

Unknown cross made by Gustav Adolf Fröhlich (1847–1912), a private vine breeder and wine producer in Edenkoben in the Pfalz, Germany. It was transplanted to several grape collections under the name Fröhlich V. 4.4 in the early twentieth century, notably at the Alzey research centre. In 1948, several vines were introduced to the vine improvement nursery at Rhodt, south of Neustadt in the Pfalz, by Fritz Uhl, whence the selections 276a and 276b were further passed on to the Neustadt and Geisenheim research centres. Ampelographical study was carried out in Geisenheim, where Helmut Becker (1927–90) named this variety Dunkelfelder in the 1970s, with reference to its unknown origin and deep colour (*Felder* means 'fields' and *dunkel* means 'dark').

OTHER HYPOTHESES

Dunkelfelder is often considered to be a BLAUER PORTUGIESER × TEINTURIER cross but DNA profiling has shown this to be false (Vouillamoz).

VITICULTURAL CHARACTERISTICS

Red-fleshed, small-berried, early ripening and not very good in a drought. Early budding makes it vulnerable to late spring frosts and early ripening attracts the wasps. Also susceptible to shrivelling and powdery mildew.

WHERE IT'S GROWN AND WHAT ITS WINE TASTES LIKE

Germany has 352 ha (870 acres), including 182 ha (450 acres) in the Pfalz, 67 ha (124 acres) in Rheinhessen and 50 ha in Baden, but the total is currently pretty stable, perhaps slightly in decline. Wines are very deeply coloured, full-bodied but rather neutral in flavour, making them useful for adding colour to blends. Minges in the Pfalz produce a good varietal example and Aloisiushof in the same region produce a slightly sweet version.

Dunkelfelder was also planted on 24 ha (59 acres) in **Switzerland** in 2009, and to a limited extent in **England** and British Columbia in **Canada**.

DURANIJA

Almost extinct Croatian variety from the Istrian peninsula.

BERRY COLOUR: ● ● ● ● ●

PRINCIPAL SYNONYMS: Brajdenica (Croatia and the Slovenian part of Istra/Istria), Duranija Bijela, Duronija (Buzet in Istra)

VITICULTURAL CHARACTERISTICS

Mid to late ripening. Very susceptible to powdery mildew and botrytis bunch rot. Big berries (bigger than MALVAZIJA ISTARSKA) that attain relatively low sugar levels.

WHERE IT'S GROWN AND WHAT ITS WINE TASTES LIKE

Duranija is a very rare Istrian variety from **Croatia**, on the verge of extinction, and is also grown to a limited extent near Koper in **Slovenia** under the name Brajdenica. It is sometimes blended with MALVAZIJA ISTARSKA in central Istra, mainly around the town of Pazin. Producers include Antun Ivaninić and the beekeeper Boris Hrvatin.

DURAS

Structured and peppery variety playing a minor role in the Gaillac appellation.

BERRY COLOUR: ● ● ● ● ●

PRINCIPAL SYNONYMS: Durade, Duras Rouge (Tarn), Durazé (Ariège)

ORIGINS AND PARENTAGE

Duras is an old variety from the Midi-Pyrénées, where it has long been cultivated in the Ariège and the Tarn *départements*. Its first mention in the region – as Duras Noir – was in 1842 in Gaillac (Rézeau 1997). According to Tallavignes (1902), Duras had already been referred to in a vineyard lease held by Maître de Pachino, notary in Lavaur (Tarn), dated 8 November 1484, but there is no reliable historical evidence for this document.

The name Duras derives from the French *dur*, meaning 'hard', referring to the hard wood of the vine.

According to DNA parentage analysis, Duras is a parent of TRESSOT (Bowers *et al.* 2000; see PINOT pedigree diagram).

OTHER HYPOTHESES

Some authors have suggested that Duras could derive from the Duracina mentioned by Cato and Columella but they could at most share the same etymology.

VITICULTURAL CHARACTERISTICS

Fertile and best pruned short. Early budding, mid ripening. Susceptible to phomopsis, powdery mildew, mites and eutypa dieback.

WHERE IT'S GROWN AND WHAT ITS WINE TASTES LIKE

France had 923 ha (2,281 acres) in 2009, almost exclusively in the Tarn *département*. It is one of the obligatory main varieties in the Gaillac appellation, along with FER and SYRAH, and a secondary variety in Côtes de Millau and Vins d'Estaing, both red and rosé. Wines are deeply coloured, well structured with highish alcohol and a peppery taste. Since the Gaillac appellation does not allow a varietal Duras, Patrice Lescarret at Domaine de Causse-Marines has to label his version, Rasdu, as a vin de France.

DURELLA

Makes crisp, light-bodied Veneto whites.

BERRY COLOUR: ● ● ● ● ●

PRINCIPAL SYNONYMS: Cagnina (Vicenza), Caina, Durella Gentile (La Spezia in Liguria and Massa-Carrara in Toscana), Rabiosa (Treviso in the Veneto)

VARIETIES COMMONLY MISTAKEN FOR DURELLA: NOSIOLA ☒ (Trentino)

ORIGINS AND PARENTAGE

Durella comes from the Monti Lessini area in the provinces of Verona and Vicenza in the Veneto region of north-east Italy. Its earliest mention is found in Acerbi (1825), as a grape cultivated in the province of Vicenza and in the Oltrepò Pavese. Its name may refer to the thickness of the berries' skin. Durella has often been confused with NOSIOLA from Trentino but ampelographic (Giavedoni and Gily 2005) and DNA studies (Vouillamoz) have shown that they are distinct varieties. Recent DNA parentage analysis has suggested that Durella is a possible parent of BIANCHETTA TREVIGIANA from Treviso in the Veneto (Cipriani *et al.* 2010).

OTHER HYPOTHESES

It has been suggested that the grape named Duracina by Roman authors and the Uva Durasena variety found in 1292 in the Colli Berici could be identical to the modern Durella (Giavedoni and Gily 2005) but there is no firm evidence.

VITICULTURAL CHARACTERISTICS

Vigorous and late ripening.

WHERE IT'S GROWN AND WHAT ITS WINE TASTES LIKE

Durella appears as a varietal wine in the Monti Lessini (or Lessini) DOC in the hills between Verona and Vicenza in north-east **Italy**. Marcato in Roncà produce a range of sparkling wines from 85% Durella blended with PINOT NOIR and/or CHARDONNAY as well as an excellent sweet passito version which is 100% Durella. Other producers include Casa Cecchin (still and dry), Fongaro (sparkling), Colli Vicentini, Cantina di Montecchia di Crosara, Cantina di Monteforte d'Alpone, Cantina di Gambellara and Cantina de Vallegora. The variety is also found in Lombardia and Toscana. Wines are light-bodied with high acidity and therefore well suited for sparkling-wine production. There were 723 ha (1,787 acres) in Italy in 2000.

DUREZA

Virtually extinct Ardèche variety most famous as a parent of Syrah.

BERRY COLOUR: ● ● ● ● ●

PRINCIPAL SYNONYMS: Duré, Duret, Durezza (Annonay, south of Lyon), Petit Duret (Drôme, Isère), Serène, Serine

VARIETIES COMMONLY MISTAKEN FOR DUREZA: DURIF, PELOURSIN (sometimes called Duresa, Dureza or Durezi), SYRAH

ORIGINS AND PARENTAGE

Dureza is an almost extinct variety from the northern Ardèche in France's greater Vallée du Rhône, whence it spread east and north to the northern Drôme and to the Isère. Although PELOURSIN is sometimes called Duresa, Dureza or Durezi, the two varieties are distinct, even if both belong to the Pelorsien ampelographic group (see p XXVII; Bisson 2009). Dureza's greatest claim to fame, revealed by DNA parentage analysis, is its offspring SYRAH (Bowers *et al.* 2000; see the Syrah entry for the pedigree diagram), a relationship which explains why they share the synonyms Serène and Serine, and why Dureza and Syrah used to be cultivated together in the northern Ardèche (Rougier 1905).

VITICULTURAL CHARACTERISTICS

Mid to late ripening, vigorous, rustic and productive.

WHERE IT'S GROWN

Dureza is virtually extinct in **France** but the first varietal version is eagerly awaited from the Saint-Joseph producer Pascal Jamet. It qualifies as Collines Rhodaniennes.

DURIF

Originally French but particularly famous, as Petite Sirah, in California.

BERRY COLOUR: ● ● ● ● ●

PRINCIPAL SYNONYMS: Dure, Duret, Gros Noir (Savoie), Petite Sirah or Petite Syrah (Cerdon in Ain, France, and also in Australia, California and Israel), Pinot de l'Hermitage, Pinot de Romans (Romans in the Drôme), Plant Durif, Serine des Mauves (Ampuis), Sirane Fourchue

VARIETIES COMMONLY MISTAKEN FOR DURIF: BÉCLAN, BÉQUIGNOL NOIR (Lot-et-Garonne), DUREZA, PELOURSIN, PINOT, SYRAH

ORIGINS AND PARENTAGE

This variety was discovered in the 1860s in the canton of Tullins in the Isère, eastern France, in the experimental vineyard of François Durif (often incorrectly spelt Duriff), a French botanist and grape breeder. It was first mentioned in 1868 by the ampelographer Victor Pulliat under the name Plant du Rif as a variety propagated in Tullins by said Durif (Di Rovasenda 1877; Goethe 1878; Rézeau 1997). Durif was praised for its resistance to downy mildew, a fungal disease that had been a serious threat to French vineyards since 1878. In 1884 Charles McIver of Linda Vista Winery near San Jose in California inadvertently imported it under the name Petite Sirah, a name he is said to have used for marketing purposes, but the name could also be the result of a mislabelling and a misspelling because Petite Syrah is the name commonly used in the Isère and the Ardèche for the true SYRAH, a variety that had been introduced to California in 1876.

Through DNA profiling, Meredith *et al.* (1999) of California established that the vast majority of the vines called Petite Sirah in California vineyards were identical to Durif, and although a few of them from old vineyards were identical to PELOURSIN, Petite Sirah and Durif can be considered synonymous. While Durif apparently knew that Peloursin was the mother of his variety, he did not know the identity of the father. The same California researchers have also established by DNA parentage analysis that Durif is most likely a Peloursin × SYRAH natural cross that occurred by chance in Durif's nursery. This explains why Durif has often been confused with both its parents, and also why Durif belongs to the Pelorsien ampelographic group (see p XXVII; Bisson 2009).

OTHER HYPOTHESES

According to Conte di Rovasenda (1877), this variety (under the name Plant du Rif) was also widespread in the region of Saluzzo in Piemonte, north-west Italy, where Durif has never been recorded since.

VITICULTURAL CHARACTERISTICS

Late ripening, the berries susceptible to sunburn, shrivelling and winter frost. Also susceptible to diseases of the wood, botrytis bunch rot and black rot.

WHERE IT'S GROWN AND WHAT ITS WINE TASTES LIKE

Durif has virtually disappeared from its homeland in **France** but thrives in **California.** In 2008 there were 6,584 acres (2,664 ha) misleadingly known as Petite Sirah, occasionally Petite Syrah, a considerable increase on the 1999 total of 2,282 acres (923 ha).

The variety's fans in California are active proselytes, as their PS I Love You website reveals. The largest areas are in the San Joaquin Valley, San Luis Obispo, Napa and Sonoma. Wines are deeply coloured and full-bodied and have good ageing potential, making them useful blending components as well as varietal wines. The most vocal and active supporter of the variety is Foppiano in Healdsburg, which produces both a straight Petite Sirah and a reserve bottling. Old-vine Petite Sirah is the distinguishing mark of the giant Concannon of Livermore Valley. Others who take this variety seriously include Guenoc, Ridge, Lava Cap and J C Cellars.

Petite Sirah has travelled from California to a wide range of other American states, particularly **Washington**.

It also spread south to **Mexico**, where L A Cetto is a keen proponent, **Brazil** (12 ha/30 acres in Rio Grande do Sul in 2007) and **Chile** (31 ha/77 acres in 2008). Perhaps partly because of its reputation in the US, and its suitability for hot, dry climates, it is also quite popular in **Israel** (721 tons harvested in 2007).

The plantings in **Australia** (433 ha/1,070 acres) and **South Africa** (14 ha/35 acres in 2008) appear on stage under their original French name. In Australia, the majority of plantings are in Riverina and in Rutherglen, Victoria, where the climate is ideally suited to this late-ripening variety. Here varietals can be intense, structured, inky wines with masses of dark fruit and well suited to oak ageing. Although the variety is officially known as Durif, Rutherglen Estates produce a Durif and a more ambitious 'Petit Sirah' (sic) as well as blending it with Shiraz (SYRAH). De Bortoli make several varietal wines from Riverina fruit.

DURIZE

See ROUGE DE FULLY.

DUTCHESS

Disappearing, low-yielding, sensitive North American hybrid also found in Brazil.

BERRY COLOUR: ● ● ● ● ●

PRINCIPAL SYNONYMS: Duchess

ORIGINS AND PARENTAGE

Named after its county of origin, Dutchess is a hybrid created when a seedling of NIAGARA (under the synonym White Concord) was pollinated by mixed pollen from DELAWARE and Walter vines by a commercial nurseryman, Andrew Jackson Caywood, in 1868 in Poughkeepsie, NY (Hedrick 1919). Since Walter is itself a DELAWARE × Diana hybrid (Diana being a CATAWBA seedling), Dutchess' pedigree therefore comprises *Vitis labrusca*, *Vitis aestivalis* and *Vitis vinifera*.

VITICULTURAL CHARACTERISTICS

Not particularly hardy and susceptible to many grape diseases, especially black rot but also downy and powdery mildews, phomopsis and botrytis bunch rot, so it needs to be treated more like a *Vitis vinifera* vine in terms of disease control and site selection. Late ripening.

WHERE IT'S GROWN AND WHAT ITS WINE TASTES LIKE

Planted to a very limited extent in **New York** State (only 27 acres/11 ha remained in 1996), where it has largely been replaced by hybrids better suited to the north-eastern United States, and, much further from home, in **Brazil**.

Quai du Vin in Elgin County, Ontario, Canada, like others, have pulled up their Dutchess vines to unsatisfactory yields and quality but they describe the wines as having a range of citrus flavours, much like VIDAL or SEYVAL BLANC, and with less foxy character than ELVIRA or CONCORD, plus a surprising ability to age in bottle for four or five years.

DZVELSHAVI OBCHURI

Mediocre Georgian variety producing intensely coloured reds.

BERRY COLOUR: ● ● ● ● ●

PRINCIPAL SYNONYMS: Zelscavi

VARIETIES COMMONLY MISTAKEN FOR DZVELSHAVI OBCHURI: Dzvelshavi Sachkheris

ORIGINS AND PARENTAGE

Dzvelshavi Obchuri is an old variety from the Imereti region in central western Georgia. Its name combines *dzveli*, meaning 'old', and *shavi*, meaning 'black' (Chkhartishvili and Betsiashvili 2004).

VITICULTURAL CHARACTERISTICS

Vigorous, mid ripening. Generally susceptible to fungal diseases.

WHERE IT'S GROWN AND WHAT ITS WINE TASTES LIKE

Dzvelshavi is grown mainly in the Imereti and Racha regions of **Georgia**, where in the past it was occasionally blended with SAPERAVI for everyday wines. On its own, it produces wines that are very deeply coloured and fresh and have red-fruit flavours but the average quality is low, making it generally unpopular.

E

ᛝ this symbol on the following pages indicates that the synonyms, and mistaken identities, are confirmed by DNA evidence

EARLY MUSCAT

California cross bred for the table but also producing aromatic wines in Oregon.

BERRY COLOUR:

PRINCIPAL SYNONYMS: California K 4-19

ORIGINS AND PARENTAGE

A cross of MUSCAT OF HAMBURG × Szőlőskertek Királynője (a Hungarian table grape also known as Königin der Weingärten) obtained by Harold P Olmo in 1943 at the University of California at Davis, US, and released in 1958 (Walker 2000). This parentage has been confirmed by DNA profiling (Myles *et al.* 2011).

VITICULTURAL CHARACTERISTICS

Early ripening. Large bunches, vigorous but with thin canes.

WHERE IT'S GROWN AND WHAT ITS WINE TASTES LIKE

Early Muscat was bred as a table grape in California but has had some commercial success as a wine grape in **Oregon**'s generally cooler climate. In the Willamette Valley, for example, Silvan Ridge produce a medium-sweet varietal with flavours of peach, orange and grape; Sokol Blosser's off-dry bottling is floral and citrus-flavoured; Bridgeview make a lightly sweet, semi-sparkling version; and Ribbon Ridge in Yamhill County produce a sweet, fortified style.

EDELWEISS

Minor, usefully disease-resistant and winter-hardy; not too foxy American hybrid.

BERRY COLOUR:

PRINCIPAL SYNONYMS: Elmer Swenson, ES 40

ORIGINS AND PARENTAGE

A Minnesota 78 × Ontario hybrid obtained by Elmer Swenson (1913–2004), a grape breeder from Osceola, Wisconsin, US (see BRIANNA for the pedigree of Minnesota 78 and see CAYUGA WHITE for the pedigree of Ontario). This complex *Vitis labrusca*, *Vitis vinifera* and *Vitis riparia* hybrid was selected in 1955 and released by Elmer Swenson and the University of Minnesota in 1978 together with SWENSON RED (Swenson *et al.* 1980). It was used to breed ESPIRIT and SWENSON WHITE.

VITICULTURAL CHARACTERISTICS

Early budding and therefore at risk from spring frosts. Vigorous and productive. Large loose bunches. The foxy flavours increase with grape maturity so it is best to pick early for wine production. Susceptible to wind damage. Good general disease resistance and only moderately susceptible to anthracnose, botrytis bunch rot and powdery mildew. Cold-hardy to –15 °F (–26 °C) (Smiley 2008).

WHERE IT'S GROWN AND WHAT ITS WINE TASTES LIKE

Small areas of Edelweiss are found in the Midwest, notably 53 acres (21 ha) in **Iowa** in 2006 and around 10 acres (4 ha) in **Nebraska** in 2007, and also in **Pennsylvania**. It may be used both for wine and for the table. Producers of varietal wines include Canon River in **Minnesota**; Cuthills and Mac's Creek in Nebraska; and Snus Hill, Tassel Ridge and Two Saints in Iowa. Wines are pleasant if the fruit is not picked too late and the wine is made semi-sweet (Swenson *et al.* 1980).

EDERENA

Bordeaux cross being trialled in California.

BERRY COLOUR: ● ● ● ● ●

PRINCIPAL SYNONYMS: Édéréna

ORIGINS AND PARENTAGE

A MERLOT × ABOURIOU cross obtained in 1952 by Pierre Marcel Durquéty at INRA (Institut National de la Recherche Agronomique) in Bordeaux, south-west France. Ederena means 'the most beautiful', from the Basque word *ederrena* (*eder* means 'beautiful'; Rézeau 1997).

VITICULTURAL CHARACTERISTICS

Fertile, mid ripening.

WHERE IT'S GROWN AND WHAT ITS WINE TASTES LIKE

Ederena is not in the official French catalogue of varieties and does not appear to be planted to any significant extent in **France** (less than 1 ha/2.5 acres in 2008). In **California** it is available from Sunridge Nurseries in Bakersfield, who have sold it to Gallo for experimental plantings, and it is one of the many varieties being trialled by the University of California at Davis in the San Joaquin Valley. Closer to its home, La Toveyre, an adventurous Swiss producer in Chardonne in Vaud, crafts an obscure blend of Malbec (COT), CALADOC, MARSELAN, ARINARNOA, EGIODOLA, Ederena and CARMINOIR called, less obscurely, Numéro 2. Ederena produces wines that are light and perfumed, sometimes slightly herbaceous.

EGIODOLA

Recent productive Bordeaux cross starting to show some potential at home and abroad. Quite tannic.

BERRY COLOUR: ● ● ● ● ●

PRINCIPAL SYNONYMS: Égiodola

ORIGINS AND PARENTAGE

Egiodola was presented as a FER × ABOURIOU cross obtained in 1954 by Pierre Marcel Durquéty at INRA (Institut National de la Recherche Agronomique) in Bordeaux, south-west France, but recent DNA parentage analysis at INRA in Montpellier corrected it to an Abouriou ×

NEGRAMOLL (under the name Tinta da Madeira) cross. The name Egiodola is a contraction of two Basque words, *egiazko odola*, meaning 'true blood' (Rézeau 1997).

VITICULTURAL CHARACTERISTICS

Early budding and ripening, fertile, sometimes vigorous, productive. Good general disease resistance.

WHERE IT'S GROWN AND WHAT ITS WINE TASTES LIKE

Wines tend to be deeply coloured, generous, tannic and low in acidity, although Domaine de la Chevrue in the Muscadet region in **France** describe their wine as light with raspberry aromas. The variety is reported to be relatively high in catechins, phenolic compounds that are said to help prevent coronary heart disease (Teissedre and Landrault 2000).

Egiodola has been recommended in just about every wine-growing *département* in France since 1983. Total plantings in 2009 were 271 ha (670 acres), down from 342 ha (845 acres) in 2006, and a few producers are making varietal wines, eg Domaine Millet in Gascogne, who produce a rosé version. In the south west, it is more often blended with FER and/or CABERNET FRANC, eg in Coteaux de Chalosse or Côtes de Gascogne. In Châteauneuf-du-Pape, Domaine Roger Perrin blend it with MERLOT, Grenache (GARNACHA), Cinsault (CINSAUT), CABERNET SAUVIGNON, etc.

In Jumilla, south-eastern **Spain**, Casa de la Ermita have planted some Egiodola vines in their experimental vineyard.

In **Switzerland**, La Toveyre in Chardonne in Vaud craft an obscure blend of Malbec (COT), CALADOC, MARSELAN, ARINARNOA, Egiodola, EDERENA and CARMINOIR, called simply Numéro 2.

Brazil had 56 ha (138 acres) in the state of Rio Grande do Sul in 2007, where producers such as Cave de Pedra and Pizzato in the Vale dos Vinhedos wine region, near the city of Bento Gonçalves, make varietal wines.

EHRENFELSER

Frost-resistant German cross bred as an alternative to Riesling.

BERRY COLOUR: ● ● ● ● ●

PRINCIPAL SYNONYMS: Geisenheim 9-93

ORIGINS AND PARENTAGE

Ehrenfelser is said to be a RIESLING × SILVANER cross bred at Geisenheim in 1929 by Heinrich Birk. However, DNA profiling has shown that Silvaner is not one of the parents (Grando and Frisinghelli 1998) so one side of the family remains unknown. Ehrenfelser was used to breed PRINZIPAL.

VITICULTURAL CHARACTERISTICS

Good resistance to disease and frost. Yields about the same as RIESLING but with slightly higher sugar levels and lower acidity.

WHERE IT'S GROWN AND WHAT ITS WINE TASTES LIKE

Ehrenfelser, named after the ruined Schloss Ehrenfels near Rüdesheim in the Rheingau, was bred with the aim of creating a RIESLING-like grape that would ripen in a wider range of sites but it was only partially successful: it does ripen better and more productively but the acidity is much lower than in Riesling and the wines therefore develop quickly in bottle even though they are attractively fruity at the outset. It also lacks Riesling's aromatic complexity.

Total **German** plantings stand at 91 ha (225 acres), scattered across the German wine regions, with the biggest areas in Rheinhessen and the Pfalz, but the overall area is in decline, probably because the more recently developed KERNER is a better alternative to RIESLING since it has higher acidity and is more versatile with regard to site. Varietal wines are made by Hermann Jourdan & Söhne, Lohmühle and Rothweiler, all in Hessische Bergstrasse.

A few plantings exist in the US, mainly in **California** and in **New York** State.

Ehrenfelser was first imported into British Columbia, western **Canada**, in 1968, along with RIESLING, and together they constituted the first major planting of *vinifera* in the southern Okanagan Valley. Today there are 74 acres (30 ha); Cedar Creek's version is off dry, clean and refreshing.

One small producer in **Australia**, Palmara in Tasmania, has had Ehrenfelser planted since 1984 but usually blends it with SÉMILLON and SAUVIGNON BLANC to improve body and acidity.

EKIGAÏNA

Recent Bordeaux cross yet to make inroads in the vineyard.

BERRY COLOUR: ● ● ● ● ●

PRINCIPAL SYNONYMS: Ékigaïna

ORIGINS AND PARENTAGE

A TANNAT × CABERNET SAUVIGNON cross obtained at INRA (Institut National de la Recherche Agronomique) in Bordeaux, south-west France, in 1955, most probably by Pierre Marcel Durquéty. The name derives from two Basque words, *ek(h)i* ('sun') and *gaïn* ('top'), meaning 'quintessence of the sun' (Rézeau 1997).

VITICULTURAL CHARACTERISTICS

Mid ripening. Needs to be pruned long. Good general level of disease resistance.

WHERE IT'S GROWN AND WHAT ITS WINE TASTES LIKE

Although it is officially recommended for several *départements* in the south and south west of **France**, it certainly doesn't seem to have caught on yet; in 2008 there were fewer than 3 ha (7 acres) planted. The wines are generally deeply coloured – not surprising given the pedigree – warm and relatively low in acidity. Ekigaïna could be a useful blending partner with the highly structured TANNAT, eg in vins de pays such as Terres Landais.

There are minuscule plantings in **Argentina**, thanks to experiments by Zuccardi in Mendoza.

EKIM KARA

Late-ripening Ukrainian variety making fortified red dessert wines.

BERRY COLOUR:

PRINCIPAL SYNONYMS: Echim Kara

VARIETIES COMMONLY MISTAKEN FOR EKIM KARA: KEFESSIYA

ORIGINS AND PARENTAGE

Ekim Kara, literally 'doctor's black', is a variety from Solnechnaya Dolina in eastern Krym (Crimea), in the far south of Ukraine. It was allegedly brought here at the end of the nineteenth century by an unknown Russian doctor.

Ekim Kara is sometimes confused with KEFESSIYA, occasionally known as Doktorsky Chernyi, which also means 'doctor's black'.

VITICULTURAL CHARACTERISTICS

Late ripening. Medium-sized to large bunches of relatively large, thick-skinned berries. Susceptible to downy and powdery mildews.

WHERE IT'S GROWN AND WHAT ITS WINE TASTES LIKE

Ekim Kara is grown in **Ukraine** (27 ha/67 acres in 2009) and is often blended with CEVAT KARA in rich, sweet fortified wines such as Massandra's and Solnechnaya Dolina Winery's popular Black Doctor labels.

ELBLING

Ancient, once ubiquitous, now unfashionable German variety used mainly as a base wine for fizz.

BERRY COLOUR:

PRINCIPAL SYNONYMS: Aelbinen (old name in Baden-Württemberg), Albich (old name in the Pfalz), Burger (Alsace and Lorraine in France), Elbling Weiss, Elsässer (Biel in Switzerland), Grobriesling, Grossriesling, Haussard (Neuchâtel in Switzerland), Raisin Blanc des Allemands (Franche-Comté in France), Rheinelbe (Alsace), ?Weisser Silvaner, Ysèle (Neuchâtel in Switzerland).

VARIETIES COMMONLY MISTAKEN FOR ELBLING: GOUAIS BLANC, PEDRO XIMÉNEZ, SILVANER

ORIGINS AND PARENTAGE

As suggested by its numerous synonyms, Elbling is one of the most ancient German grape varieties. Its name is thought to derive either from the Latin *alba* ('white'), which is also the etymology of the River Elbe, or from the Latin *helvus* ('yellow'). Just like RIESLING, Elbling was most likely born in the Rheintal (Rhine Valley), as suggested by the synonym Rheinelbe used in France.

Nevertheless, the earliest mention of Elbling is thought to be under another name, near Zürich in Switzerland at the end of the thirteenth century: a *Hofrodel* (a type of register) of the Engelberg Abbey records the request of a visiting local landowner for meat and good *Elseser* wine, because he didn't like the local Zürich wine (*Landwein*; Bluntschli 1838; Grimm 1840; Aeberhard 2005).

Elseser, meaning Alsatian, is an attested old synonym for Elbling in Switzerland that can be found in many later sources such as Daniel Rhagor's (1639) Wein-Gärten, in which the variety is described in Oberhofen am Thunersee (in the German-speaking canton of Bern) under the names Elber, Elbelen or Elsisser. In addition, DNA profiling recently confirmed that the variety still known as Elsässer near the Bielersee is identical to Elbling (Vouillamoz, Frei and Arnold *et al.* 2008), suggesting that it most likely came to Switzerland from Germany via Alsace, where it was formerly cultivated.

In Germany, the earliest mention of Elbling appeared under the old name Aelbinen on 8 August 1483. The Bebenhausen monastery allotted vineyards in Kriegsberg near Stuttgart to local citizens, stipulating that only Frennsch, Traminer and Aelbinen should be grown (Krämer 2006). Numerous additional mentions of this variety occurred later in Germany, eg in Hieronymus Bock's (1539) famous *Kreutterbuch*, where it appears under the name Albich in Laudau in the Pfalz.

Regner *et al.* (1998) established by DNA profiling that Elbling has a parent–offspring relationship with GOUAIS BLANC, one of the most ancient and prolific wine grapes of Western Europe. Since Gouais Blanc also has parent–offspring relationships with at least eighty additional grape varieties, including CHARDONNAY, GAMAY NOIR, FURMINT, RÄUSCHLING and RIESLING, they all turn out to be half-siblings, grandparents or grandchildren of Elbling; see PINOT for the complete pedigree. The close genetic link with Riesling probably explains why Elbling is sometimes called Grobriesling or Grossriesling. Elbling has also been shown to be in some way genetically close to Pinot (Imazio *et al.* 2002).

The variety named Blauer Elbling or Elbling Blau, once cultivated in Germany and in Switzerland, is morphologically and genetically distinct from the white Elbling, while the variety called **Roter Elbling**, Elbling Rot or Elbling Rose is a simple colour mutation of Elbling (Galet 2000; Vouillamoz, Frei and Arnold 2008).

OTHER HYPOTHESES

Elbling is said to have been brought by the Romans to the Rheintal via Gaul in the fourth century AD because it is often identified with the ambiguous *Vitis albuelis* mentioned by Pliny the Elder and Columella.

Based on DNA data, Regner *et al.* (1998) have speculated that Elbling, just like RIESLING, could be a cross of GOUAIS BLANC with a parent of SAVAGNIN (or Traminer), this ancestor being long dead and possibly related to wild grapevines (*Vitis silvestris* L.) from the Rheingau. However, this hypothesis has not been confirmed by subsequent research.

VITICULTURAL CHARACTERISTICS

Prolific early ripener, susceptible to powdery and downy mildews, botrytis bunch rot and desiccation of the stems and at risk from grape moths.

WHERE IT'S GROWN AND WHAT ITS WINE TASTES LIKE

Elbling used to be planted in north-east **France** but has now effectively disappeared and it is not authorized in any appellation contrôlée wines, although it is still authorized in the Moselle and Bas-Rhin and Haut-Rhin *départements*.

At one time Elbling was effectively the only variety planted in **Luxembourg** but it is in decline here too: there are currently 116 ha (287 acres) with virtually no new plantings since the end of the 1980s. However, at Château Pauqué, Abi Duhr is experimenting with a still varietal version.

Elbling apparently constituted three quarters of all vines in **Germany** in the first half of the nineteenth century (Maul 2006) but today there are only 578 ha (1,428 acres) of this ancient variety, planted almost entirely in the upper reaches of the Mosel-Saar-Ruwer region, where it is difficult to ripen RIESLING. Even there it is declining, although Steinmetz produce a good varietal example. The wines are light and fresh with high acidity and low alcohol and are therefore often used to make sparkling wine.

In **Switzerland**, Elbling was the major variety near the Biel, Spiez and Thun lakes until the end of the nineteenth century, and it used to be widespread across German-speaking Switzerland, where plantings have now shrunk to just 65 ares (1.6 acres). Producers of varietal wines include Schmidheiny in the canton of Sankt Gallen and Reinhard and Bettina Bachmann at Weinbau zum Stäckerösseler in the canton of Aargau, who also make a 100% Roter Elbling in good vintages.

The pink-skinned mutation Roter Elbling is rarely found in Germany, though Frieden-Berg in the Mosel produce a varietal wine. Switzerland has the grand total of 10 ares, or one tenth of a hectare (1,200 square yards).

ELVIRA

Very productive, winter-hardy North American hybrid producing mediocre wine best hidden in blends.

BERRY COLOUR: ● ● ● ● ●

VARIETIES COMMONLY MISTAKEN FOR ELVIRA: NOAH

ORIGINS AND PARENTAGE

Elvira is a seedling of Taylor (see NOAH) obtained in 1862 by Jacob Rommel of Morrison, Missouri, that was introduced onto the market by Bush & Son & Meissner of Bushberg, Missouri, in 1874–5. The other parent is said to be Martha, a white seedling of CONCORD, making Elvira a complex *Vitis labrusca*, *Vitis riparia* and *Vitis vinifera* hybrid.

VITICULTURAL CHARACTERISTICS

Mid ripening (the same time as CONCORD), very productive and very hardy. Small, compact bunches of thin-skinned berries prone to cracking and therefore often harvested early when the acidity is high and before botrytis bunch rot can take hold. Resistant to downy mildew.

WHERE IT'S GROWN AND WHAT ITS WINE TASTES LIKE

Wine quality is at best mediocre and most of the fruit is used for blended bulk wines. The variety's main advantage is its winter hardiness. There were 587 acres (238 ha) in **New York** State in 2006 and minor plantings elsewhere, eg 26 acres (11 ha) in **Pennsylvania** in 2006. Producers of varietal wines, often on the sweet side, include Montgomery Underground Winery in Pennsylvania, Table Mountain Vineyards in **Wyoming** and Prairie Vine in **Nebraska**.

In **Canada**, Ontario's acreage, already small, is in decline, down from 40 acres (16 ha) in 2002 to around 10 acres (4 ha) in 2006.

Elvira is also authorized, along with three other American hybrids, for southern **Austria**'s Uhudler (see CONCORD for details).

EMERALD RIESLING

California cross that has travelled widely but not gloriously and turns out to be no Riesling.

BERRY COLOUR: ● ● ● ● ●

PRINCIPAL SYNONYMS: California 1139E29, Emerald Rizling

ORIGINS AND PARENTAGE

Emerald Riesling was originally said to be a Muscadelle (CA) × RIESLING cross obtained by Harold P Olmo in 1935 at the University of California at Davis and released in 1948 (Walker 2000), but DNA parentage analysis by Lacombe *et al.* (2013) has shown that this is a Malvasia Aromatica di Parma x GARNACHA cross. The confusion most likely comes from the fact that Malvasia Aromatica di Parma was initially called Muscadelle du Bordelais Faux (distinct from the true MUSCADELLE) at the INRA grape collection in Montpellier.

VITICULTURAL CHARACTERISTICS

Productive.

WHERE IT'S GROWN AND WHAT ITS WINE TASTES LIKE

Since its initial success in the 1960s and 1970s, when it was valued for adding freshness and aroma to white blends in **California**'s hot San Joaquin Valley, it has declined in acreage although a few better-quality varietal wines are produced in slightly cooler locations such as Monterey. California's 235 acres (95 ha) in 2008 were mainly in Kern and Fresno counties.

It is also found in **South Africa** (139 ha/343 acres in 2008) and in **Israel** (150 ha/371 acres in 2009), where 2,041 tons were harvested in 2007, down from a peak of 3,508 tons in 2004, mainly for an off-dry style. Producers include Barkan, Binyamina, Carmel and Tishbi. A few growers in **Australia** have experimented with this variety but it doesn't seem to have caught on.

EMIR

Light, fresh variety common in central Turkey.

BERRY COLOUR: ● ● ● ● ●

ORIGINS AND PARENTAGE

Emir comes from Kapadokya (Cappadocia) in central Turkey. The name can mean several things, eg 'ruler' or 'command'.

VITICULTURAL CHARACTERISTICS

Best suited to continental climates, likes heights (900–1,200 m) and poor sandy/volcanic soils. Late ripening in this type of environment. Productive but very susceptible to botrytis bunch rot.

WHERE IT'S GROWN AND WHAT ITS WINE TASTES LIKE

Emir is the main white variety in the provinces of Nevşehir and Niğde in Kapadokya and, along with NARINCE, is one of the most important white wine grapes in Turkey, even though there were only 92 ha (227 acres) planted in 2010. Wines are typically light in colour, with a greenish cast, delicate, fresh, light- to medium-bodied, sometimes mineral-scented and best drunk young. They are used in the production of both still and sparkling wines, don't take well to oak or malolactic fermentation and are quite often blended with Narince or SULTANIYE. Producers include Kavaklidere, Kocabağ and Turasan.

ENANTIO

Fairly recent official name for a Trentino variety once called Lambrusco a Foglia Frastagliata.

BERRY COLOUR: ● ● ● ● ●

PRINCIPAL SYNONYMS: Foja Tonda, Lambrusco a Foglia Frastagliata, Lambrusco Nostrano

VARIETIES COMMONLY MISTAKEN FOR ENANTIO: CASETTA, FOGLIA TONDA

ORIGINS AND PARENTAGE

Formerly called Lambrusco a Foglia Frastagliata (meaning 'toothed leaf'), although it is distinct from other Lambruscos of Italy, this variety was officially renamed Enantio in the mid 1980s by Mario Fregoni, in reference to a grape name used by Pliny the Elder. The renaming was felt necessary in order to distinguish it from Lambrusco a Foglia Tonda (meaning 'round leaf'), another autochthonous variety from Vallagarina now called CASETTA. Indeed, those two varieties differ considerably in morphology (Calò *et al.* 2006), and comparison of their DNA profiles rules out any parent–offspring relationship between Casetta and Enantio (Vouillamoz). Genetic studies suggested a close relationship between Enantio and other Trentino varieties such as MARZEMINO, TEROLDEGO, LAGREIN and GROPPELLO GENTILE (Grando and Frisinghelli 1998; Grando 2000). In addition, a more recent DNA parentage analysis has shown with great confidence that Enantio is either a parent or a progeny of NEGRARA TRENTINA (Grando *et al.* 2006), which anchors Enantio's roots in the Trentino region. On the other hand, Enantio does not show any relationship with traditional varieties of the province of Verona, where it was probably introduced relatively recently (Vantini *et al.* 2003), most likely from Trentino.

OTHER HYPOTHESES

According to Scienza *et al.* (1990), Enantio or Lambrusco a Foglia Frastagliata is morphologically and genetically similar to local wild grapes of Trentino and could be the result of on-site domestication. This hypothesis is usually proposed for all Lambruscos in Italy, but it has not been convincingly proven yet. DNA profiling erroneously established the identity between Enantio (under the name Lambrusco a Foglia Frastagliata) and CASETTA (or Lambrusco Casetta), also cultivated in Trentino (Grando and Frisinghelli 1998; Grando 2000), simply because the Lambrusco a Foglia Frastagliata sample was mislabelled in the grape collection (Stella Grando, personal communication).

VITICULTURAL CHARACTERISTICS

Late ripening. Disease and frost-resistant. High in colour and acidity.

WHERE IT'S GROWN AND WHAT ITS WINE TASTES LIKE

Traditionally cultivated in lower Vallagarina (between the Veneto and Trentino in northern **Italy**), Enantio is today the typical grape of Terra dei Forti between Monte Baldo and the Lessinia mountains. Letrari and Masso Roveri make creditable varietals in Terra dei Forti.

It is registered as a monovarietal wine in Valdadige-Terra dei Forti Enantio DOC and in blends in Valdadige-Terra dei Forti Rosso o Casteller DOC. Wines have a medium and balanced structure and must be drunk young.

The Italian agricultural census of 2000 recorded 1,184 ha (2,926 acres) of Enantio, then called Lambrusco a Foglia Frastagliata.

ENCRUZADO

High-quality, well-structured Portuguese variety with the potential to produce ageworthy wines.

BERRY COLOUR: ● ● ● ● ●

PRINCIPAL SYNONYMS: Salgueirinho

ORIGINS AND PARENTAGE

This rare variety probably comes from the Dão region in central northern Portugal since it is almost exclusively grown here.

VITICULTURAL CHARACTERISTICS

Small bunches of medium-sized berries. Vigorous, mid budding, early to mid ripening.

WHERE IT'S GROWN AND WHAT ITS WINE TASTES LIKE

Encruzado has great potential – possibly more than any other white variety in Dão – but is a challenge in the winery since it is very prone to oxidation and the loss of its delicate aromas, which can be a complex mix of capsicum, roses and violets, flinty minerality and lemon (Rolando Faustino, personal communication). As these full-bodied wines age, the wines develop aromas of hazelnut and pine resin. Encruzado also marries well with oak in barrel fermentation and lees ageing, giving a well-structured wine a little like white burgundy. The best examples age well in bottle – up to thirty years in some cases – although most of today's wines are intended for early consumption. There were just 295 ha (729 acres) in **Portugal** in 2010 and the variety is authorized in the Dão DOC and in various regional wines. Growing recognition for the variety's quality potential seems likely to lead to increased plantings further afield. Recommended producers include Casa de Mouraz, Julia Kemper, Filipa Pato, Quinta dos Carvalhais, Quinta das Marias and Quinta dos Roques.

ENFARINÉ NOIR

Ancient and virtually extinct variety that survives in the Jura.

BERRY COLOUR: ● ● ● ● ●

PRINCIPAL SYNONYMS: Enfariné du Jura, Gaillard (Val de Saône, Burgundy), Gouais Noir (Aisne, Aube, Marne, Meuse, Seine-et-Marne), Grison (Dôle)

ORIGINS AND PARENTAGE

The first mention of Enfariné appeared in an order of the parliament of Besançon dated 3 February 1731 (Bousson de Mairet 1856): 'Ordre du Parlement d'arracher toutes les vignes plantées depuis l'an 1702, et d'extirper dans les vieilles vignes tous les plants dénommés: *grapenoux*, *foirard noir et blanc*, *maldoux*, *gamé noir et blanc*, *valet noir*, *enfariné*, *mesy*, *plant qui tache*, *bregin*, *pinot ou plant des haies*, et *barclan blanc.* Il est aussitôt procédé à l'exécution de l'arrêt; près du tiers du vignoble est arraché et rendu à la culture des céréales; l'opération n'est achevée qu'au mois de septembre de l'année suivante' ('Order of Parliament to pull out all the vineyards planted since 1702 and to eradicate from old vineyards the old vines of all the plants known as: … *enfariné* … The execution of the decree was immediately carried out; almost a third of the vineyard is pulled out and dedicated to the cultivation of cereals; the process is not completed until September of the next year'). The name Enfariné Noir is based on the French word *farine* ('flour') in reference to the bloom covering the berries. It has also long been known as Gouais Noir, although it is not a colour mutation of GOUAIS BLANC. Similarly, the pink-berried mutation Enfariné Gris is also called Gouais Gris. According to Guicherd, Enfariné could originate in the Aube and is similar to GUEUCHE NOIR (Galet 1990).

VITICULTURAL CHARACTERISTICS

Very vigorous and productive, late ripening. Big bunches of big, thick-skinned berries. Good all-round disease resistance.

WHERE IT'S GROWN AND WHAT ITS WINE TASTES LIKE

This ancient variety is almost extinct – less than 1 ha (2.5 acres) in **France** in 2008 – but there are a few vines in the Jura, mostly interplanted with POULSARD. However, the variety, which is usefully high in natural acidity, is not forgotten: Domaine Jean Bourdy planted some in their conservation vineyard in 2006; Domaine Ganevat in Rotalier have deliberately replanted a small area and include about 20% in an early-drinking red blend (and occasionally in a sparkling wine); and, at Domaine des Cavarodes, Étienne Thiebaud uses some in his red Vin de Pays de Franche-Comté blend.

EONA

South Dakotan hybrid just about surviving in Québec.

BERRY COLOUR: ● ● ● ● ●

PRINCIPAL SYNONYMS: Éona

ORIGINS AND PARENTAGE

A Lady Washington × Beta hybrid obtained in the early twentieth century by the Danish-born Niels Ebbesen Hansen (1866–1950) at the South Dakota Agricultural Experiment Station in Brookings. Lady Washington is a hybrid of CONCORD and Allen's Hybrid, which is an ISABELLA × CHASSELAS hybrid, and Beta is a hybrid of Carver and CONCORD, where Carver is a selected *Vitis riparia* vine. Eona, a complex *Vitis riparia*, *Vitis labrusca* and *Vitis vinifera* hybrid introduced commercially in 1925 (Plocher 1993), has a native American name, like many of Hansen's creations.

VITICULTURAL CHARACTERISTICS

Early ripening, hardy.

WHERE IT'S GROWN AND WHAT ITS WINE TASTES LIKE

Eona, like Hensen's other new varieties, never really prospered, probably as a result of being released at the peak of Prohibition (Plocher 1993). It is planted to a limited extent in Québec, **Canada**, where Isle de Bacchus use it in several blends and Domaine Félibre make a dry varietal wine. Sainte Pétronille have just a few vines and use the fruit in a blend. Wines are said to have flavours of flowers and tropical fruit. It is also used for the table and for juice and is planted by home winemakers.

ERBALUCE

Minor white Piemontese variety, occasionally making distinctive sweet wines.

BERRY COLOUR: ● ● ● ● ●

PRINCIPAL SYNONYMS: Albaluce, Bian Roustì, Bianchera, Greco Bianco di Novara (province of Novara), Uva Rustìa

VARIETIES COMMONLY MISTAKEN FOR ERBALUCE: TREBBIANO TOSCANO

ORIGINS AND PARENTAGE

The earliest mention of Erbaluce appears in Croce (1606) under the name Elbalus, possibly deriving from *alba luce* (meaning 'dawn light') because the berries gleam when they are ripe. Erbaluce is thought to originate in the Canavese region in the province of Torino, in the foothills of the Alps. Genetic analysis suggests a close relationship with CASCAROLO BIANCO from Piemonte (Vouillamoz). Although the synonyms Greco and Greco Bianco are used in Ghemme in the province of Novara, probably in reference to its ability to make sweet wines (ie similar to historic Greek wine styles), Erbaluce has no relationship with the Campanian variety GRECO.

OTHER HYPOTHESES

According to DNA studies, Erbaluce could be morphologically and genetically close to CLAIRETTE, which has a similar etymology ('clear, bright') (Labra *et al.* 2001). This relationship has yet to be confirmed.

VITICULTURAL CHARACTERISTICS

Vigorous and often trained on *topia*, a particular type of pergola, in the Canavese region. The berries have thick skins and they turn amber when ripe (hence the Bian Roustì and Uva Rustìa synonyms – *arrostita* means 'roasted'). Early to mid ripening. Poor resistance to powdery mildew.

WHERE IT'S GROWN AND WHAT ITS WINE TASTES LIKE

Erbaluce is cultivated in the provinces of Torino (in the Canavese area and in Serra d'Ivrea and Lago di Viverone), Biella and Novara (where it is called Greco) but it is a speciality of Caluso, north east of the city of Torino, **Italy**. The variety is also authorized in the following DOCs: Canavese, Colline Novaresi, Coste della Sesia and Erbaluce di Caluso (or just Caluso). In the last it is used to make dry, sparkling and sweet passito wines. In 2000 there were 342 ha (845 acres) in Italy, according to the agricultural census.

Although the dry wines used to be rather light-bodied and acidic, there are now some impressive, vibrant examples that challenge the local Arneis and Gavi from négociant houses such as Orsolani and Bava and from Paolo de Marchi's Proprietà Sperino in Lessona. The passito version, which must be aged in barrels for at least four years from the harvest, is sweet, golden and relatively rare. Some producers allow botrytis to develop towards the end of the drying period, which lasts until March following the harvest, to add further complexity. Cieck's Alladium combines aromas of saffron, almonds, acacia honey and fresh figs, its sweetness counterbalanced by light bitterness and acidity.

ERBAMAT

Virtually extinct local white from Brescia, northern Italy.

BERRY COLOUR: ● ● ● ● ●

PRINCIPAL SYNONYMS: Albamatto, Erbamatto, Verdealbara

ORIGINS AND PARENTAGE

Erbamat was already mentioned in the sixteenth century as one of the best white varieties in Valtènesi, a region in the Lago di Garda area in northern Italy. The name Erbamat probably derives from the green colour of the berries (*erba* means 'herb'). DNA profiling unexpectedly showed that Erbamat is identical to Verdealbara (Vouillamoz), an ancient and almost extinct variety of Avio in the province of Trento. The name Verdealbara also comes from the green colour of the berries at dawn (*verde* is Italian for 'green' and *alba* means 'dawn').

VITICULTURAL CHARACTERISTICS

Late ripening.

WHERE IT'S GROWN AND WHAT ITS WINE TASTES LIKE

In Lombardia, northern **Italy**, Erbamat was almost extinct but a few adventurous producers in the Lago di Garda region have persisted with it. Comincioli's vino da tavola Perlì, for example, is a blend of 40% Erbamat with 60% Trebbiano Valtenesi (VERDICCHIO BIANCO). In addition, the Franciacorta *consorzio* (wine growers' association) has recently introduced experimental plantings. Varietal wines are lightly aromatic.

As Verdealbara in Trentino, Erbamat was included in the national register of varieties, together with LAGARINO BIANCO, in 2007. Historically it was cultivated in Val Lagarina, north of Verona, but also in Vallarsa and in the Valle di Terragnolo. Today, Verdealbara is still found in old vineyards near Avio, closer to the Lago di Garda, although there don't appear to be any commercial examples of a varietal Verdealbara wine.

ERMITAGE

See MARSANNE.

ERVI

Minor and relatively recent dark-skinned cross from Emilia-Romagna.

BERRY COLOUR: ● ● ● ● ●

PRINCIPAL SYNONYMS: Barbera × Bonarda 108, Incrocio Fregoni 108

ORIGINS AND PARENTAGE

Ervi was obtained in 1970 by Mario Fregoni at the Università Cattolica di Piacenza by crossing BARBERA and CROATINA, the latter under its misleading synonym Bonarda, as used in Oltrepò

Pavese and Roero. The aim was to avoid the weaknesses of the parents: poor ripening, high acidity and poor bud fertility.

VITICULTURAL CHARACTERISTICS

Moderately vigorous with good bud fertility. Mid ripening. The deeply coloured berries have high sugar levels and moderate acidity.

WHERE IT'S GROWN AND WHAT ITS WINE TASTES LIKE

Ervi tends to be blended with BARBERA to add colour and increase alcoholic strength. The 2000 agricultural census in **Italy** recorded just 6 ha (15 acres).

ESCURSAC

Recently rescued Mallorcan making promising reds.

BERRY COLOUR: ● ● ● ● ●

PRINCIPAL SYNONYMS: Corçac, Cursach, Escorçac, Escursag, Excursach

ORIGINS AND PARENTAGE

Not much is known about Escursac, a variety indigenous to Mallorca in the Islas Baleares, Spain, where it was already mentioned in 1871 (Favà i Agud 2001) and then almost eradicated when phylloxera hit the island in 1891. It was recently rescued from extinction by Bodegas Galmés i Ribot, who isolated cuttings from Colònia de Sant Pere d'Artà to replant a vineyard in 2006.

VITICULTURAL CHARACTERISTICS

Low vigour, late budding and ripening. Medium-sized bunches of small, thin-skinned berries. Relatively resistant to downy and powdery mildews but susceptible to botrytis bunch rot.

WHERE IT'S GROWN AND WHAT ITS WINE TASTES LIKE

Thanks to Galmés i Ribot, there is now about half a hectare (1.2 acres) of this variety in Santa Margalida in north-east Mallorca and they produce a varietal wine, although by late 2011 Escursac had not yet been included in the official register of varieties in **Spain**. It seems well suited to the island and produces an elegant, aromatic wine that has red-fruit flavours, moderate alcohol, soft tannins and noticeable persistence.

ESGANA CÃO

See SERCIAL.

ESPADEIRO

Minho vine producing rosé.

BERRY COLOUR: ● ● ● ● ●

PRINCIPAL SYNONYMS: Espadeiro Tinto, Padeiro Tinto (Vinho Verde)

VARIETIES COMMONLY MISTAKEN FOR ESPADEIRO: CAMARAOU NOIR ☓ (known as Espadeiro in Galicia in Spain), PADEIRO ☓ (Minho), TRINCADEIRA ☓ (Setúbal)

ORIGINS AND PARENTAGE

Espadeiro is from the Minho region in north-western Portugal. It should not be confused with TRINCADEIRA, known as Espadeiro in Setúbal, nor with PADEIRO, sometimes called Espadeiro in Minho.

Espadeiro is also regarded as an old indigenous variety in Galicia but its DNA profile in Martín *et al.* (2003; 2006), confirmed by the sample at the Misión Biológica de Galicia (Maria-Carmen Martínez, personal communication), is different from the Portuguese Espadeiro in Almadanim *et al.* (2007), and the Espadeiro in Galicia could well be a mistake because it is most probably identical to CAMARAOU NOIR from south-west France (Vouillamoz).

VITICULTURAL CHARACTERISTICS

Vigorous, productive, late budding and ripening. Tolerates wind and humidity but needs plenty of heat to ripen fully. Well suited to granitic soils.

WHERE IT'S GROWN AND WHAT ITS WINE TASTES LIKE

Espadeiro is grown in **Portugal**'s Vinho Verde region for the production of crisp rosés, such as the one made by Quinta de Gomariz, or to contribute to a red blend. There were 295 ha (729 acres) in Portugal in 2010. It was traditionally grown up trees in an attempt to moderate its vigour.

ESPIRIT

Very minor but cold-hardy American hybrid producing mild, fruity wines.

BERRY COLOUR: ● ● ● ● ●

PRINCIPAL SYNONYMS: ES 422, Esprit

ORIGINS AND PARENTAGE

A VILLARD BLANC × EDELWEISS hybrid bred by Elmer Swenson in Osceola, Wisconsin, US. This extremely complex *Vitis riparia*, *Vitis labrusca*, *Vitis vinifera*, *Vitis aestivalis*, *Vitis lincecumii*, *Vitis rupestris*, *Vitis cinerea* and *Vitis berlandieri* hybrid was named and released in 1984, and patented to Swenson Smith Vines in 1986 (Smiley 2008). It was used to breed DELISLE.

VITICULTURAL CHARACTERISTICS

Large bunches of large berries. Moderately vigorous and productive. Mid budding and ripening. Highly susceptible to powdery mildew, moderately so to botrytis bunch rot and downy mildew. Hardy (–15 °F/–26 °C to –20 °F/–29 °C).

WHERE IT'S GROWN AND WHAT ITS WINE TASTES LIKE

Espirit is grown to a very limited extent in the American Midwest. Wines are said to be free of the foxy character typical of *Vitis riparia* (Smiley 2008) and to have a mild fruity flavour. Two Saints Winery in **Iowa** produce a varietal wine. It has also been recommended for **Colorado**.

ESQUITXAGOS

See MERSEGUERA.

ETRAIRE DE L'ADUÏ

Isère speciality still cultivated to a limited extent in Savoie.

BERRY COLOUR: ● ● ● ● ●

PRINCIPAL SYNONYMS: Beccu de l'Aduï (Isère), Betu (Isère), Etraire de la Dot, Etraire de la Dû or Due, Etraire de la Duï, Etraire de la Duy, Grosse Etraire

VARIETIES COMMONLY MISTAKEN FOR ETRAIRE DE L'ADUÏ: PERSAN

ORIGINS AND PARENTAGE

According to Rougier (1902b), this variety was found, in the form of an enormous and productive vine, at the beginning of the nineteenth century by the inhabitants of Saint-Ismier in the Vallée de Grésivaudan north east of Grenoble while they were fetching water from a spring near the Mas de l'Aduï. Its olive-shaped berries looked very much like those of PERSAN, and since Persan was called Etraire in Isère, probably based on the local Dauphinois dialect word *eitreïri*, meaning 'grape producing a lot of must' (Rézeau 1997), this vine was named Etraire de l'Aduï. Initially propagated in Saint-Ismier-en-Grésivaudan by a certain Monsieur Fourrier, Etraire de l'Aduï was subsequently propagated throughout Isère, where – under several different spellings – it became one of the most important varieties. It also spread to the Ain, Savoie and Ardèche *départements*. Its first written mention appeared in 1842 under the name Etraire de la Due (Rézeau 1997).

Rougier (1902b) suggested that Etraire de l'Aduï could be a natural seedling of PERSAN, which has now been confirmed by DNA parentage analysis that strongly suggests a parent–offspring relationship between Etraire de l'Aduï and Persan (Vouillamoz). Indeed, both varieties belong to the Sérine ampelographic group (see p XXVII; Bisson 2009).

VITICULTURAL CHARACTERISTICS

Very vigorous and may be pruned short. Rather susceptible to winter frosts. Mid ripening and well suited to deep, hillside clay-limestone soils. When ripe, the berries sometimes drop.

WHERE IT'S GROWN AND WHAT ITS WINE TASTES LIKE

Etraire de l'Aduï used to be widespread in Isère, between the Rhône and Savoie, eastern **France**, in particular after the phylloxera crisis in 1880 because it was erroneously thought to be resistant to this pest. Although it is still a recommended variety in Isère, and has been one of the main varieties permitted in red and rosé versions of Vin de Pays des Coteaux du Grésivaudan, in 2008 there were just 6 ha (15 acres) compared with 406 ha (1,003 ha) in 1958. Only one clone has been approved but a dozen more were planted in Savoie in 1999. Domaine Grisard, also in Savoie, planted Etraire de l'Aduï in their nursery in 2004 along with other local varieties such as CHATUS, PERSAN and VERDESSE. Wines may be deeply coloured, full-flavoured and tannic, sometimes even astringent. Rare varietal wines are produced by the Cave Coopérative de Bernin and by Domaine Michel Magne, who make a rosé version.

EVA

See BEBA.

EVMOLPIA

Recent Bulgarian cross valued for its productivity and early ripening as well as for its easy-drinking reds.

BERRY COLOUR: ● ● ● ● ●

PRINCIPAL SYNONYMS: Evmolpiya, Thracian Mavrud

ORIGINS AND PARENTAGE

Evmolpia is a MAVRUD × MERLOT cross recently made by Dimitar Babrikov and C Kukunov at the Vasil Kolarov Institute of the University of Plovdiv, Bulgaria, and officially registered in 1991.

VITICULTURAL CHARACTERISTICS

Productive, early budding, early to mid ripening with high sugar accumulation in the berries. Large bunches of medium-sized, thin- but tough-skinned berries. Comparatively resistant to botrytis bunch rot.

WHERE IT'S GROWN AND WHAT ITS WINE TASTES LIKE

There have been many new plantings of this increasingly (economically) important Bulgarian variety in the last few years, with a total of 92 ha (227 acres) in **Bulgaria** in 2009. The variety is also patented in Australia and the US. Wines have fresh acidity, medium to full body, well-balanced tannins and moderate alcohol and may be drunk young but are not particularly aromatic.

EYHOLZER ROTE

Ancient and extremely rare variety vinified by just one southern Swiss producer.

BERRY COLOUR: ● ● ● ● ●

PRINCIPAL SYNONYMS: Eyholzer Roter, Gross Roth, Grossroter

VARIETIES COMMONLY MISTAKEN FOR EYHOLZER ROTE: AVANÀ ☒ (under the name Hibou Noir)

ORIGINS AND PARENTAGE

Eyholzer Rote is a rare variety from the area of Visp in the Haut-Valais, Switzerland, named after the village of Eyholz, between Visp and Brig. Although DNA analysis failed to identify its parents, Eyholzer Rote could have some genetic relationship with VIEN DE NUS from the neighbouring Valle d'Aosta and with FREISA from Piemonte, suggesting it may have been introduced long ago from northern Italy, which would be consistent with the use of pergola training in the Haut-Valais (Vouillamoz and Moriondo 2011). One ancient vine, 150–250 years old, has been found near Visp in the village of Stalden, and an even older one in the middle of the town of Sion, some 50 km from Visp (Vouillamoz and Moriondo 2011).

VITICULTURAL CHARACTERISTICS

Early budding, late ripening. Big bunches of big berries.

WHERE IT'S GROWN AND WHAT ITS WINE TASTES LIKE

Just 0.25 ha (0.6 acres) of Eyholzer Rote existed around Visp in the Haut-Valais, **Switzerland**, in 2009. Chanton, the only commercial producer of this variety, make a dry version, with high acidity and strong aromas of strawberries, as well as an extremely rare Eiswein that takes 6–7 months to ferment in barrique, with notes of crème brûlée, dried cherries and very high acidity.

EZERFÜRTŰ

Neutral Hungarian white cross generally hidden in blends.

BERRY COLOUR: ● ● ● ● ●

PRINCIPAL SYNONYMS: Kecskemét 5, Miklóstelepi 5

ORIGINS AND PARENTAGE

Ezerfürtű is a HÁRSLEVELŰ × SAVAGNIN ROSE (under the name Piros Tramini) cross obtained in 1950 by András Kurucz and István Kwaysser at the Kecskemét research centre in Hungary.

VITICULTURAL CHARACTERISTICS

Vigorous, productive, early ripening. Compact bunches of small berries. Susceptible to botrytis bunch rot, sensitive to drought and prone to frost damage. Hopes for a winter-hardy and rot-resistant cross were disappointed.

WHERE IT'S GROWN AND WHAT ITS WINE TASTES LIKE

Ezerfürtű was particularly popular in large-scale vineyards in **Hungary** in the 1970s and 1980s and can still be found on the north shore of Lake Balaton and on the Alföld (Great Plain). It produces neutral, medium-bodied wines with a slight vegetal aroma and is generally blended, with Olaszrizling (GRAŠEVINA), for example, and rarely bottled on its own. Total plantings in 2008 in Hungary were 478 ha (1,181 acres).

EZERJÓ

Once widely planted Hungarian variety making high-acid whites from dry to sweet.

BERRY COLOUR: ● ● ● ● ●

PRINCIPAL SYNONYMS: Budai Fehér, Fehér Bakator, Kolmreifler, Korponai, Tausendgute (Steiermark/Styria in Austria), Tausent Güte (Steiermark), Trummertraube (Steiermark), Zátoki

ORIGINS AND PARENTAGE

Ezerjó, meaning 'a thousand boons', as attested by its German synonym Tausendgute, is an old Hungarian variety from the region of Mór between Budapest (hence the synonym Budai Fehér, ie 'white of Buda') and Lake Balaton, where wine has been made since the Middle Ages. It used to be one of the most common varieties in Hungary, well known throughout the Carpathian Basin. Ezerjó has often been used to breed new varieties, including GENEROSA (HUNGARY), ZENGŐ, ZENIT and ZEUSZ.

VITICULTURAL CHARACTERISTICS

High yielding and early ripening. Big bunches of big, thick-skinned berries that tend to redden in the sun when fully ripe. Susceptible to botrytis bunch rot and sensitive to low winter temperatures.

WHERE IT'S GROWN AND WHAT ITS WINE TASTES LIKE

Ezerjó was one of the most common varieties in **Hungary** in the first half of the twentieth century (Rohály *et al.* 2003) and, although the area is in decline, there were still 1,655 ha (4,090 acres) in 2008. Today it is grown mainly in Kunság on the Alföld (Great Plain) in the central south of Hungary but also in Mór and Neszmély in the north west. Varietal wines are high in acidity, potentially also in alcohol, but are otherwise typically neutral. The most refreshing and lively dry wines come from the deeper subsoils of the small wine region of Mór, which are richer in clay, marl and limestone. In the best locations, berries left on the vine tend to shrivel and develop botrytis, allowing producers such as Bozóky to make high-quality sweet wines known as Mór Aszú. Other recommended producers of varietal wines include Frittmann in Kunság and Maurus and Miklóscsabi in Mór.

F

FABERREBE
FALANGHINA BENEVENTANA
FALANGHINA FLEGREA
FENILE
FER
FERNÃO PIRES
FERTILIA
FETEASCĂ ALBĂ
FETEASCĂ NEAGRĂ
FETEASCĂ REGALĂ
FIANO
FIOLETOVY RANNY
FLAVIS
FLORA
FLORENTAL
FOÇA KARASI
FOGARINA
FOGLIA TONDA
FOGONEU
FOKIANO
FOLGASÃO
FOLIGNAN
FOLLE BLANCHE
FONTE CAL
FORASTERA
FORCALLAT TINTA
FORGIARIN
FORSELLINA
FORTANA
FRANCAVIDDA
FRÂNCUŞĂ
FRAPPATO
FRAUELER
FREDONIA
FREISA
FREISAMER
FRENCH WILD
FRONTENAC
FRÜHROTER VELTLINER
FUBIANO
FUELLA NERA
FUMIN
FURMINT

ᛝ this symbol on the following pages indicates that the synonyms, and mistaken identities, are confirmed by DNA evidence

FABERREBE

High-yielding, lightly aromatic German cross losing ground.

BERRY COLOUR:

PRINCIPAL SYNONYMS: Alzey 10375, Faber

ORIGINS AND PARENTAGE

Faberrebe is said to be either a PINOT BLANC × MÜLLER-THURGAU or a SILVANER × PINOT BLANC cross bred by Georg Scheu in 1929, an uncertainty which could easily be resolved by DNA analysis. Its original name was Faber, the Latin for German 'Schmied', or 'Smith' in English, in honour of Karl Schmitt in Landau in the Pfalz, who carried out the field trials. It was later officially renamed Faberrebe (*Rebe* is German for 'vine').

VITICULTURAL CHARACTERISTICS

Early ripening and high yielding but very susceptible to powdery mildew.

WHERE IT'S GROWN AND WHAT ITS WINE TASTES LIKE

Germany has 587 ha (1,450 acres), principally in Rheinhessen and to a much lesser extent in the Pfalz and the Nahe, but the area is fast declining, down from 1,657 ha (4,095 acres) in 1998 as part of German vintners' general disaffection with twentieth-century crosses. Wines have a greenish tinge and tend to be fruity with light Muscat-like aromas. Acidity levels are a little higher than in MÜLLER-THURGAU (a common blending partner) and high ripeness levels are possible. Hilgert and Metzler in Rheinhessen make varietal wines at various sweetness levels.

There are a few hectares in **England**.

FALANGHINA BENEVENTANA

The recuperated, versatile 'other' Falanghina from Benevento in inland Campania.

BERRY COLOUR:

VARIETIES COMMONLY MISTAKEN FOR FALANGHINA BENEVENTANA: FALANGHINA FLEGREA

ORIGINS AND PARENTAGE

Until recently, there was thought to be just one Falanghina variety in Campania in southern Italy, ie the same variety in Campi Flegrei west of the city of Napoli and in the province of Benevento, north east of Napoli. Thanks to DNA profiling, Costantini *et al.* (2005) showed that FALANGHINA FLEGREA, the most widespread white cultivar in the province of Napoli, and Falanghina Beneventana, more limited and recently rediscovered in the province of Benevento, were two distinct varieties. Genetic comparison also suggested that Falanghina Beneventana is closely related to BIANCOLELLA, while Falanghina Flegrea is closely related to SUPPEZZA (Costantini *et al.* 2005).

Falanghina Beneventana may originate in Bonea in the province of Benevento. It was rediscovered in the province by Leonardo Mustilli in the 1970s.

VITICULTURAL CHARACTERISTICS

Mid to late ripening.

WHERE IT'S GROWN AND WHAT ITS WINE TASTES LIKE

Falanghina is listed in many DOCs in Campania, southern **Italy**, some blended, some varietal. The regulations do not make the distinction between Falanghina Beneventana and FALANGHINA FLEGREA but it seems likely that the former is planted and used in those north and east of Napoli, such as Taburno, Guardia Sanframondi (or Guardiolo), Sannio, Sant'Agata dei' Goti and Solopaca. There is also an IGT called Falanghina Beneventana. The variety is well suited for sparkling and passito wines as well as still, dry wines, which are usually unoaked and generally less aromatic than Falanghina Flegrea.

The official Italian census of 2000 lists just 'Falanghina' so it is impossible to tell how much of the 1,721 ha (4,253 acres) is Falanghina Beneventana.

FALANGHINA FLEGREA

Campania's signature white wine grape.

BERRY COLOUR:

PRINCIPAL SYNONYMS: Falanghina Pigna Piccola, Falernina (Caserto), Uva Falerna (Caserto)

VARIETIES COMMONLY MISTAKEN FOR FALANGHINA FLEGREA: FALANGHINA BENEVENTANA

ORIGINS AND PARENTAGE

First mentioned in 1666 in a poem (Canto VII) in Neapolitan dialect by Giulio Cesare Cortese, Falanghina is one of the oldest and most widespread varieties of Campania. The name Falanghina could derive from the Latin *falangae*, the stakes supporting the vines. Until recently, the Falanghina of Campi Flegrei and that of Benevento in Campania were thought to be one and the same, although ampelographers had noticed some morphological differences. Thanks to DNA profiling, Costantini *et al.* (2005) discovered that Falanghina Flegrea, the most widespread white cultivar in the province of Napoli, and FALANGHINA BENEVENTANA, more limited and recently rediscovered in the province of Benevento, are two distinct varieties. Genetic comparison also suggested that Falanghina Flegrea is closely related to SUPPEZZA, while Falanghina Beneventana is closely related to BIANCOLELLA.

OTHER HYPOTHESES

Like many other Campanian varieties, Falanghina Flegrea is said to have been imported from Greece around the seventh century BC. This hypothesis is far from proven, since no genetic link with modern Greek varieties has been found (Vouillamoz). Similarly, some authors have suggested that, based on its synonym Uva Falerna, Falanghina was used to produce the famous Falerno wine in Roman times. This is doubtful because Falerno was the name of a wine-growing region where several grape varieties were probably cultivated. It is therefore impossible to identify Falerno with Falanghina or any other modern variety.

VITICULTURAL CHARACTERISTICS

Mid to late ripening.

WHERE IT'S GROWN AND WHAT ITS WINE TASTES LIKE

The leafily scented Falanghina from Campi Flegrei, west of Napoli in Campania, southern **Italy**, usefully retains its acidity. After the devastating effect of phylloxera at the beginning of the twentieth century, cultivation of Falanghina Flegrea decreased dramatically until Francesco Avallone, local historian and founder of Villa Matilde, recuperated old vines in the Massico region in the 1970s. His Falerno del Massico is 100% Falanghina Flegrea and has flavours of stone fruit, considerable weight in the mouth, and a fine apricot-kernel aftertaste and fresh acidity. The Caracci version, which includes a very small percentage fermented in oak, is a bigger wine, a little less aromatic, but still elegant.

Falanghina Flegrea is recommended throughout Campania and Molise and in the province of Foggia in Puglia. It is authorized as a varietal wine in DOCs such as Falerno del Massico, Galluccio and Campi Flegrei and may contribute to the blend in such DOCs as Campi Flegrei, Capri, Penisola Sorrentina, Lacryma Christi del Vesuvio and Costa d'Amalfi.

The official Italian census of 2000 lists just 'Falanghina' so it is impossible to tell how much of the 1,721 ha (4,253 acres) is Falanghina Flegrea.

FAVORITA

See VERMENTINO.

FENILE

Minor and highly localized white from the Almafi coast in southern Italy.

BERRY COLOUR: ● ● ● ● ●

ORIGINS AND PARENTAGE

Apparently from the Amalfi coast in Campania, southern Italy, Fenile is not mentioned in any historical book about Campanian viticulture. Local growers suggest that the name refers to the straw (*fieno*) colour of the berries (Manzo and Monaco 2001).

DNA profiling has shown that Fenile has a unique DNA profile and is genetically close to Pellecchiona and CACAMOSCA, both traditional Campanian varieties, the former not cultivated commercially (Costantini *et al.* 2005).

VITICULTURAL CHARACTERISTICS

Trained on pergolas. Low yields, small clusters, early ripening. Susceptible to botrytis bunch rot. High sugar levels in the berries.

WHERE IT'S GROWN AND WHAT ITS WINE TASTES LIKE

Old and often ungrafted Fenile vines are found interplanted with other old varieties in the vineyards of Furore, Positana and Amalfi in Campania in southern **Italy**. Fenile is allowed as a supplementary grape in the Furore subzone of Costa d'Amalfi DOC. It has been tested for a few years and it may in future be included in the national register of varieties. The wines have moderate acidity and flavours of apricots, candied and exotic fruits and honey, sometimes aromas of broom. Marisa Cuomo's Furore Bianco Fiorduva, Costa d'Amalfi DOC, is a dry, barrel-fermented blend of roughly equal proportions of late-harvested Fenile, RIPOLO and GINESTRA.

FER

The untamed, tannic variety of Marcillac, south-west France.

BERRY COLOUR: ● ● ● ● ●

PRINCIPAL SYNONYMS: Braucol (Gaillac), Caillaba (Pyrénées-Atlantiques), Camaralet Noir, Camirouch (Ardèche), Estronc (Lot), Fer Servadou, Gragnelut⊗ (Friuli-Venezia Giulia in northern Italy), Hère (Gironde), Mansois or Samençois (Marcillac), Mourac, Pinenc (Madiran), Servadou (Lot-et-Garonne)

VARIETIES COMMONLY MISTAKEN FOR FER: ARROUYA⊗ (Jurançon), BÉQUIGNOL NOIR⊗ (Lot-et-Garonne), CASTETS⊗ (Tarn), PETIT VERDOT⊗ (Gironde)

ORIGINS AND PARENTAGE

The earliest references to this variety appeared in 1783–4 under no fewer than five distinct synonyms (Rézeau 1997):

- as Fer in Montauban, north of Toulouse in the Tarn-et-Garonne: 'Fer noir ... il donne une bonne qualité au vin rouge pourvu qu'il ne domine pas' ('Fer noir ... it gives good quality to red wine as long as it does not dominate'), as well as further north and west in Bergerac in the Dordogne and in Caumont (now Caumont-sur-Garonne) in the Lot-et-Garonne
- as Servadou in Agen in the Lot-et-Garonne, between Bordeaux and Toulouse: 'Servadou gros, Servadou petit'
- as Samençois in Rodez in Aveyron, between Toulouse and Saint-Étienne: 'Le Samençois est aussi bon à manger, le vin qui en provient est de seconde qualité' ('Samençois is also good to eat, the wine made from it is of secondary quality')
- as Pinenc and Hère in Condom in the Gers, south west of Agen, although with rather strange spellings: 'Piek. On le nomme haire dans bien des cantons' ('Piek. It is called haire in many a canton').

The name Fer and its synonym Hère both derive from the Latin *ferus*, meaning 'wild, savage', which is consistent with the common belief that Fer was domesticated from local wild grapevines (Lavignac 2001). This etymology is more plausible than the unconvincing derivation from the French *fer* ('iron'), supposedly referring to the hardness of its wood and stem (Galet 1990). In the local dialect, *servadou* means 'well suited to conservation', which most likely refers to the grapes' keeping well during winter (Rézeau 1997).

Confusingly, Fer is described in Viala and Vermorel (1901–10) under the name Mourac, while BÉQUIGNOL NOIR is described under the name Fer.

DNA parentage analysis has established that Fer and HONDARRIBI BELTZA, a variety from the País Vasco, ie Spain's Basque Country, spontaneously crossed to give birth to Gros Cabernet, which in turn naturally crossed with CABERNET FRANC to produce CARMENÈRE (Boursiquot *et al.* 2009; see CABERNET SAUVIGNON's pedigree). Fer and Hondarribi Beltza are therefore grandparents of Carmenère, which confirms Fer's membership of the Carmenet ampelographic group (see p XXVII; Bisson 2009). Consequently, although Fer is usually said to come from the Gironde, the parentage of Gros Cabernet corroborates Lavignac's (2001) hypothesis that both its parents, Fer and Hondarribi Beltza, originate in the Spanish Basque Country.

Curiously enough, the DNA profile of five distinct plants called Gragnelut in Friuli-Venezia Giulia in northern Italy (Crespan *et al.* 2011) turned out to be identical to Fer (Vouillamoz).

VITICULTURAL CHARACTERISTICS

Mid ripening, erratic fertility and should be pruned long. Good resistance to botrytis bunch rot. Susceptible to powdery mildew and attractive to leafhoppers. Does well on poor, stony soils.

WHERE IT'S GROWN AND WHAT ITS WINE TASTES LIKE

After a dip in the 1970s and 1980s, French plantings have risen thanks to better clonal selection, particularly in the last fifteen years or so. In 2009 there were 1,610 ha (3,978 acres), mainly in the Tarn and the Gers in south-west **France**, although it is authorized or recommended across much of southern France. It is the most widely planted variety in the Aveyron *département* in southern central France, home to the Marcillac appellation, near the town of Rodez, the only one in which Fer – or Mansois as it is known locally – is the dominant or sole variety (regulations permit up to 10% CABERNET SAUVIGNON or MERLOT or PRUNELARD but most growers stick to 100% Fer). In the nearby appellations of Entraygues et le Fel and Estaing, it plays a more minor role. Beyond the Aveyron, the variety is most likely to surface in blended appellation wines such as Gaillac (with DURAS and SYRAH), Madiran, Fronton, Tursan, Saint-Mont and several others from Côte du Marmandais in the Lot-et-Garonne down to Béarn in the Pyrénées-Atlantiques. In the Languedoc it may play third fiddle to GAMAY NOIR and Syrah in the Côtes de Millau.

Wines are generally well coloured, relatively tannic and sometimes rather rustic, a little on the wild side, with some similarity to CABERNET FRANC, though if the berries are fully ripe, they give flavours of cassis and red fruits. Research by the Institut National de la Recherche Agronomique's Institut des Produits de la Vigne in Montpellier, southern France, has shown that Fer is rich in catechins and procyanidols, said to help inhibit heart disease (Strang 2009).

Domaine Laurens in Marcillac make varietal red and rosé wines as well as a fortified ratafia (grape must fortified with grape spirit). Gaillac's Domaine Genouillac have bottled a varietal Fer Servadou as a Vin de Pays du Comté Tolosan because the local appellation regulations do not permit 100% Fer. Other good producers in Marcillac include Domaine du Cros, the Cave Coopérative des Vignerons du Vallon, Le Vieux Porche, Domaine des Costes Rouges and Domaine du Mioula.

The vine is rarely found outside France but Hillsborough Vineyards in **Virginia**, US, apparently blends Fer with TANNAT and PETIT VERDOT.

FERNÃO PIRES

Most common light-skinned grape in Portugal. Adaptable, high-yielding, making aromatic whites.

BERRY COLOUR: ● ● ● ● ●

PRINCIPAL SYNONYMS: Fernão Pirão, Fernão Pires de Beco, Gaeiro or Gaieiro (Oeste), Maria Gomes (Bairrada), Molinha (Setúbal), Torrontés

VARIETIES COMMONLY MISTAKEN FOR FERNÃO PIRES: BICAL (Ançã-Cantanhede), Fernão Pires de Colares, TREBBIANO TOSCANO

ORIGINS AND PARENTAGE

Fernão Pires is an old Portuguese variety that was mentioned in writing before the end of the eighteenth century (Rolando Faustino, personal communication). Until recently, received opinion was that Fernão Pires originated in the Tejo region but the greater morphological diversity found in Bairrada, in the west of Beiras, northern Portugal, suggests the vine has been planted here even longer. This theory is supported by historical records showing significant migration from Bairrada to the Tejo region in the eighteenth century (Rolando Faustino, personal communication). Fernão Pires used to be a common surname in Portugal but it is not known after whom the variety was named.

DNA profiling has established that Fernão Pires is identical to Maria Gomes, both cultivated in Bairrada (Lopes *et al.* 1999). Fernão Pires was used to breed GENEROSA (PORTUGAL), RIO GRANDE and SEARA NOVA.

A pink-berried mutation called **Fernão Pires Rosado** has also been observed.

Fernão Pires should not be confused with TREBBIANO TOSCANO, with which it shares the synonym Fernão Pires de Beco, nor with Fernão Pires de Colares, which is a distinct but extinct variety, nor with BICAL, which is called Fernão Pires Galego in the Ançã-Cantanhede region.

VITICULTURAL CHARACTERISTICS

Early budding and ripening, productive. Basal buds are fertile so it is best pruned short. Medium-sized compact bunches of small berries. Sensitive to water stress, leading to loss of leaves and shrivelled berries. Acidity drops rapidly at the end of maturation.

WHERE IT'S GROWN AND WHAT ITS WINE TASTES LIKE

Fernão Pires, more commonly known as Maria Gomes in the north of the country, is the most widely planted white grape variety in **Portugal**, with a total of 16,800 ha (41,514 acres) in 2010. It is particularly important in Bairrada in the west and in the Tejo but also plays a significant role throughout central and southern Portugal, eg in Lisboa, in the Península de Setúbal and in the Alentejo, as well as in the Douro. It is authorized in many different DOCs, from the Douro in the north to Palmela, south of Lisboa.

Its popularity is due in large part to its high yields and easy ripening but also to its distinctive aromatic character – from oranges, orange and linden blossom, mimosa and laurel to spice and honey – whereas its detractors point to low acidity, susceptibility to oxidation, lack of ageing potential and over-intense, sometimes heavy, aromas. Much of the produce of the fertile plains used to be distilled but the arrival of temperature-controlled, stainless-steel fermentation tanks has allowed producers to make inexpensive, simple, fruity dry wines (Mayson 2003). Recommended producers include Companhia das Lezírias in the Tejo and Quinta das Bágeiras (Mário Sérgio Alves Nuno) in Bairrada.

Fernão Pires is one of the few Portuguese varieties to have escaped Portugal's borders: there were 165 ha (408 acres) in **South Africa** in 2009 and limited plantings in **California** and **Australia**.

FERRÓN

See MANSENG NOIR.

FERTILIA

Very minor red blending component in Treviso, northern Italy.

BERRY COLOUR: ● ● ● ● ●

PRINCIPAL SYNONYMS: Incrocio Cosmo, Merlot × Raboso Veronese 108

CHASSELAS

pp 229–32

CHENIN BLANC

pp 236–9

CODA DI VOLPE BIANCA

pp 254–5

CONCORD

pp 263–4

COT

pp 272–4

DUREZA

p 316

ELBLING

pp 325–7

FETEASCĂ ALBĂ

pp 347–8

FIANO

pp 350–51

FURMINT

pp 373–5

PETIT GAMAY

see GAMAY NOIR

pp 383–6

GRENACHE NOIR

see GARNACHA

pp 396–403

GOUAIS BLANC

pp 419–22

GRACIANO

pp 425 – 7

HÁRSLEVELŰ

pp 455–6

INZOLIA

pp 472–3

ORIGINS AND PARENTAGE

Obtained in 1976 at the Conegliano research centre in the Veneto in northern Italy by Italo Cosmo, who crossed MERLOT with RABOSO VERONESE.

VITICULTURAL CHARACTERISTICS

Vigorous, mid to late ripening.

WHERE IT'S GROWN AND WHAT ITS WINE TASTES LIKE

Fertilia is authorized in the province of Treviso in the Veneto region of northern **Italy**, where it is cultivated to a limited extent in Conegliano and Pieve di Soglio. The 2000 Italian census recorded just 15 ha (37 acres). It is generally blended with Bordeaux varieties.

FETEASCĂ ALBĂ

Second most planted Fetească and even more important in Romania than in the Republic of Moldova.

BERRY COLOUR: ● ● ● ● ●

PRINCIPAL SYNONYMS: Dievcie Hrozno (Slovakia), Fetiasca Belii (Russia, Ukraine), Fetişoară (Republic of Moldova, Romania), Fetyaska Alba (Ukraine), Mädchentraube (Hungary), Păsărească Albă (Republic of Moldova, Romania), Poamă Fetei Albă (Republic of Moldova), Văratic (Republic of Moldova)

VARIETIES COMMONLY MISTAKEN FOR FETEASCĂ ALBĂ: LEÁNYKA (Hungary)

ORIGINS AND PARENTAGE

Fetească Albă, meaning 'young girl's white', is an old variety that probably originates in the historic region of Moldavia (including today's Republic of Moldova and the Romanian region of Moldova), where it has traditionally been cultivated, then spreading west to Transilvania and Hungary. Fetească Albă is genetically close to other varieties from Moldavia such as GRASĂ DE COTNARI (Bodea *et al.* 2009).

According to Dejeu (2004), Fetească Albă was 'obtained from' (not clear if this means a clone or a descendant) FETEASCĂ NEAGRĂ in historic Moldavia between the third and thirteenth centuries but, although recent studies have suggested that they are morphologically and genetically close (Bodea *et al.* 2009), comparison of the DNA profiles of Fetească Albă and Fetească Neagră in Gheţea *et al.* (2010) tends to disprove any parent–offspring relationship (Vouillamoz).

The Hungarian variety LEÁNYKA has often been considered to be identical to Fetească Albă (Galet 2000; Rohály *et al.* 2003) but comparison of their DNA profiles (Galbács *et al.* 2009; Gheţea *et al.* 2010) disproves this hypothesis (Vouillamoz).

OTHER HYPOTHESES

Like FETEASCĂ NEAGRĂ, Fetească Albă has often been said to have been directly domesticated by the Dacians from local wild grapevines but there is no evidence so far.

VITICULTURAL CHARACTERISTICS

Vigorous, early budding, mid ripening. Small, compact bunches of small, thin-skinned berries. Susceptible to downy and powdery mildews, botrytis bunch rot and mites but resistant to cold winter temperatures (to –20 °C/–4 °F).

WHERE IT'S GROWN AND WHAT ITS WINE TASTES LIKE

The **Republic of Moldova**'s 4,334 ha (10,710 acres) of Fetească Albă are in the centre and south of the country, producing wines with medium acidity, floral and citrus flavours that are often used to make sparkling wine. Recommended producers include Chateau Vartely.

Fetească Albă is even more important in neighbouring **Romania**, where there were 10,529 ha (26,018 acres) in 2008, more than 10% of the total vineyard area and second only to FETEASCĂ REGALĂ. It is spread throughout most of the wine regions but the greatest concentrations are in Transilvania in the centre of the country and in Romanian Moldova in the east. Wines, both still and sparkling, are generally dry or medium dry but some excellent sweet, still versions, alone or blended with GRASĂ DE COTNARI and other local varieties, are produced in Cotnari in the far north east. Styles of wine vary considerably but dry wines are typically more full-bodied than Fetească Regală and combine citrus and light peach or apricot flavours. Vines planted in the south often produce wines lacking in acidity. Producers of varietal wines include Budureasca, Basilescu, Davino, Gîrboiu, Dominiile Săhăteni, Senator and SERVE.

The variety is said to be planted in **Hungary** but this may be due to the commonly held but incorrect belief that it and LEÁNYKA are one and the same.

Ukraine's official vineyard statistics recorded approximately 1,600 ha (3,954 acres) in 2009 but they do not distinguish between the various Fetească.

FETEASCĂ NEAGRĂ

The red-wine Fetească, a resurgent variety from the historic region of Moldavia producing good-quality wine mainly in Romania.

BERRY COLOUR: ● ● ● ● ●

PRINCIPAL SYNONYMS: Coada Rândunicii (Romania, Republic of Moldova), Fetyaska Chernaya (Ukraine), Păsărească Neagră (Republic of Moldova), Poamă Fetei Neagră (Republic of Moldova), Schwarze Mädchentraube (Romania)

ORIGINS AND PARENTAGE

Fetească Neagră, literally 'black young girl', is an old variety that probably originates in the historical region of Moldavia (including today's Republic of Moldova and the Romanian region of Moldova), where it has been traditionally cultivated, then spreading west to Transilvania and Hungary. Contrary to Roy-Chevrier's (1903a) hypothesis, Fetească Neagră is not a colour mutation of FETEASCĂ ALBĂ, as evidenced by DNA profiling (Gheţea *et al.* 2010). Fetească Neagră shows significant biodiversity, with at least four distinct types (Dejeu 2004), suggesting it is relatively old.

OTHER HYPOTHESES

Like FETEASCĂ ALBĂ, Fetească Neagră has often been said to have been domesticated from local wild grapevines by the Dacians but there is so far no evidence.

VITICULTURAL CHARACTERISTICS

Very vigorous, mid budding and mid to late ripening. Compact bunches of thick-skinned berries. Resistant to drought, cold winter temperatures (to –22 °C/–8 °F) and botrytis bunch rot but susceptible to powdery mildew.

WHERE IT'S GROWN AND WHAT ITS WINE TASTES LIKE

Although Fetească Neagră is indigenous to the region that is now divided between Romania and the **Republic of Moldova**, it was not planted during the Soviet era and more or less disappeared. However, producers such as Cricova, Equinox, Et Cetera, Purcari and Chateau Vartely began replanting the variety in the late 2000s.

In **Romania**, Fetească Neagră produces some of the country's best reds from the 1,088 ha (2,689 acres) planted mostly in Romanian Moldova in the east and in Muntenia in the south. They are typically dry and full-bodied with intense, complex aromas of spice, both red and black fruit, especially ripe plums, and tannins that become velvety as they age. The variety has some similarity to BLAUFRÄNKISCH and can easily be over-oaked. Some semi-sweet and sweet wines are also produced, particularly for the domestic market. Recommended producers include Prince Ştirbey, SERVE and Vinterra.

Ukraine's official vineyard statistics recorded approximately 1,600 ha (3,954 acres) in 2009 but they do not distinguish between the various Feteascăs.

FETEASCĂ REGALĂ

The most planted Fetească, responsible for fresh, aromatic whites in Romania.

BERRY COLOUR: ● ● ● ● ●

PRINCIPAL SYNONYMS: Dănăşană or Danesana (Transilvania in Romania), Dunesdörfer Königsast (Transilvania), Galbenă de Ardeal (Romania), Königliche Mädchentraube (Hungary), Pesecká Leánka (Slovakia).

VARIETIES COMMONLY MISTAKEN FOR FETEASCĂ REGALĂ: KIRÁLYLEÁNYKA☒ (Hungary)

ORIGINS AND PARENTAGE

Fetească Regală, literally 'royal young girl', was first observed in the 1920s in the village of Daneş near Sighişoara, Transilvania, Romania, hence its synonym Dănăşană. It was initially cultivated by a nurseryman called Gaşpari from the town of Mediaş, who distributed it under the German name Dunesdörfer Königsast (Dejeu 2004), literally 'king's branch from Dunes village'. The wine obtained from this variety was then presented by Gaşpari at the National Wine and Fruit Exhibition in Bucharest in 1928 under the name Fetească Regală, which was then adopted (Galet 2000). Two morphological types have been distinguished: one with long clusters and yellowish, thin-skinned berries, the other with winged clusters and yellow-green, thick-skinned berries (Dejeu 2004).

It has long been suggested that Fetească Regală is a natural cross between FETEASCĂ ALBĂ and GRASĂ DE COTNARI (aka Kövérszőlő) obtained in the 1930s (Galet 2000; Ţârdea and Rotaru 2003) but, even if recent studies have suggested that Fetească Regală and Fetească Albă are morphologically and genetically close (Bodea *et al.* 2009), comparison of their DNA profiles (Gheţea *et al.* 2010; Galbács *et al.* 2009) tends to contradict any strict parent–offspring relationship between Fetească Regală and either of its alleged parents (Vouillamoz).

VITICULTURAL CHARACTERISTICS

Vigorous, mid to late ripening. Compact bunches of thin-skinned berries. Susceptible to botrytis bunch rot and drought but resistant to cold winter temperatures (to –20 °C/–4 °F) and hot summers. Clone 21 Bl has higher yields and a higher sugar content in the berries.

WHERE IT'S GROWN AND WHAT ITS WINE TASTES LIKE

Fetească Regală, the most widely planted variety in **Romania** (16,363 ha/40,434 acres in 2008), tends to produce aromatic wines that are dry and fresh, occasionally floral, sometimes with more exotic-fruit flavours – both still and sparkling but also used for distillation into brandy. Thanks to the tannins in the skins, it takes better to oak than does FETEASCĂ ALBĂ. Recommended producers include Recas and Prince Ştirbey.

In the **Republic of Moldova** it has been common to make no distinction between the less common Fetească Regală and the more widely planted FETEASCĂ ALBĂ. The former is not even listed in the official statistics and is almost always blended with Fetească Albă, the final wine labelled simply Fetească.

There has been some confusion in **Slovakia** because Fetească Regală (391 ha/966 acres in 2009) and Pesecká Leánka (516 ha/1,275 acres, named after the region of Pesek, where it has traditionally been grown) have been thought by some to be separate varieties even though growers are now beginning to recognize them as one and the same (no DNA evidence yet). Producers of varietal wines include Štiglic, Josef Yhnák and Josef Zalaba. Hacaj also make a sparkling version.

Ukraine's official vineyard statistics recorded approximately 1,600 ha (3,954 acres) in 2009 but they do not distinguish between the various Feteascăs.

FIANO

Rich, waxy, strongly flavoured, fashionable southern Italian.

BERRY COLOUR: ● ● ● ● ●

PRINCIPAL SYNONYMS: Fiano di Avellino

VARIETIES COMMONLY MISTAKEN FOR FIANO: Fiano Aromatico🧬 or Fiano di Puglia🧬 now called MINUTOLO (Puglia), Greco Aromatico🧬 (Puglia), SANTA SOFIA🧬 (Basilicata)

ORIGINS AND PARENTAGE

Fiano is an old variety from Campania in southern Italy whose presence near Foggia was mentioned as early as 1240 in a register of purchases by the Emperor Frederic II: 'de vino greco saumas III, de vino grecisco saumas III, de vino fiano saumas III' (Huillard-Bréholles 1859), *saumas* being an old unit of measurement. The name Fiano is said to derive from a place named Appia (today's Lapiò) near Avellino (Scienza and Boselli 2003), where the grape supposedly has its origin.

Ampelographic studies and DNA profiling established that Fiano Aromatico or Fiano di Puglia cultivated in Puglia (specifically in Ostuni in the province of Brindisi and in the Valle d'Itria) is distinct from Fiano or Fiano di Avellino cultivated in Campania (Calò *et al.* 2001). This prompted a proposal that Fiano Aromatico be renamed MINUTOLO and included as a distinct variety in the Italian national register of grape varieties.

OTHER HYPOTHESES

Some authors suggested that Fiano was identical to the *Vitis apiane* mentioned by Pliny the Elder and Columella, and that the name Fiano was a corruption of *apianis*. However, trying to establish botanical identity between grape names of the Roman period and modern grape varieties is somewhat futile because Roman authors often used the same name for several varieties.

VITICULTURAL CHARACTERISTICS

Vigorous, susceptible to downy and powdery mildews, particularly during flowering. Mid ripening.

WHERE IT'S GROWN AND WHAT ITS WINE TASTES LIKE

Fiano was widespread before the arrival of phylloxera in the early twentieth century but since then its cultivation has been drastically reduced. It was revived in the 1970s by Antonio Mastroberardino, whose winery in the Taurasi zone has been a pioneer in the cultivation of traditional and forgotten varieties, and plantings are now on the increase.

It is recommended throughout Campania and in Potenza in Basilicata in southern **Italy** and also in the provinces of Ancona, Macerata and Ascoli Piceno in the Marche in central Italy. It is responsible for Fiano di Avellino DOCG and may be a varietal or blending component in DOCs such as Irpinia and Sannio. It has also been enthusiastically planted on Sicilia for varietal wine by the likes of Planeta and Settesoli, so that there were 236 ha (583 acres) on the island by 2008.

Wines are strongly flavoured and full-bodied but less aromatic than those made from GRECO or MINUTOLO. They have an attractively waxy texture and subtle aromas and the best have the capacity to age. Feudi di San Gregorio in Campania produce a textbook example of this style. Di Majo Norante produce a varietal version in Molise. There was a national total of 783 ha (1,935 acres) in the Italian census of 2000.

Fiano has more recently found favour in **Australia** with growers looking for varieties that withstand the heat, and there are now at least ten producers who claim to have the variety planted, including Jeffrey Grosset in the Clare Valley. Coriole's rich, dense but fresh example from McLaren Vale was one of the first and has flavours of lemon and pineapple and some complexity.

FIANO AROMATICO

See MINUTOLO

FINDLING

See MÜLLER-THURGAU.

FIOLETOVY RANNY

Russian variety used for the table, grape juice and nutmeg-flavoured red wine.

BERRY COLOUR: ● ● ● ● ●

PRINCIPAL SYNONYMS: Filetovyi Ranii

ORIGINS AND PARENTAGE

Fioletovy Ranny, meaning 'early violet', is a SEVERNY × MUSCAT OF HAMBURG hybrid bred by Ya I Potapenko, I P Potapenko and E L Zakharova at the all-Russia research and development centre for viticulture and winemaking in Novocherkassk, in the province of Rostov, Russia. It was included in the official register of varieties in 1965.

VITICULTURAL CHARACTERISTICS

Productive, early ripening. Smallish, thick-skinned berries with a light Muscat aroma. Some resistance to downy mildew and botrytis bunch rot but susceptible to powdery mildew. Cold-hardy to –25 to –27 °C (–13 to –16.5 °F).

WHERE IT'S GROWN AND WHAT ITS WINE TASTES LIKE

Grown in **Russia** (50 ha/124 acres in Rostov in 2010) and across southern **Ukraine**, Fioletovy Ranny is said to produce wines with a certain nutmeg-like flavour. Tavria in Kherson include it with RKATSITELI and ALIGOTÉ in a rosé blend but it is generally found in sweet reds, as well as in juice and on the table.

FLAVIS

Minor white northern Italian cross.

BERRY COLOUR: ● ● ● ● ●

PRINCIPAL SYNONYMS: Incrocio Cosmo 76

ORIGINS AND PARENTAGE

A VERDISO × Welschriesling (GRAŠEVINA) cross obtained in 1976 by Italo Cosmo at the Conegliano research centre in the Veneto in northern Italy.

VITICULTURAL CHARACTERISTICS

Resistant to rot. Mid to late ripening.

WHERE IT'S GROWN

Authorized in the province of Treviso, **Italy**, where it may be used as a replacement for VERDISO. The Italian agricultural census of 2000 recorded just 13 ha (32 acres).

FLORA

Minor, pink-skinned California cross capable of producing aromatic, full-bodied, sometimes flabby wines.

BERRY COLOUR: ● ● ● ● ●

PRINCIPAL SYNONYMS: California H59-90

ORIGINS AND PARENTAGE

A SÉMILLON × GEWÜRZTRAMINER cross bred by Harold P Olmo in 1938 at the University of California at Davis, US, and released in 1958 (Walker 2000). This parentage has been confirmed by DNA profiling (Myles *et al.* 2011).

VITICULTURAL CHARACTERISTICS

Early ripening, brittle canes (like GEWÜRZTRAMINER). Productive but the fruit has a tendency to split in a wet vintage (like SÉMILLON).

WHERE IT'S GROWN AND WHAT ITS WINE TASTES LIKE

Of all Olmo's crossings, this is the most delicately aromatic, though it shows more SÉMILLON characteristics in warmer climates. It was bred for the hot San Joaquin Valley in **California** and, although the wine is full-bodied and retains its aromatic qualities well, it lacks acidity, particularly in warmer sites, and fares better in cooler sites such as Mendocino's Yorkville Highlands. Since it was first produced in 1972, Schramsberg's Napa Valley Crémant Demi-Sec has had Flora as its core component (currently 85%, with 15% CHARDONNAY).

Its other well-known expression, this time as the minority (20%) component, is Brown Brothers' delicate dessert wine Orange Muscat (MUSCAT FLEUR D'ORANGER) and Flora. Brown Brothers have the only 7 ha (17 acres) of the variety in **Australia**, near Swan Hill, Victoria, and say its floral lightness, even when the fruit is picked late and shrivelled, is a foil to the powerful citrus character of Orange Muscat. The distinctive variety also found limited favour in **New Zealand** in the 1970s but only 2 ha (5 acres) remained in 2009.

FLORENTAL

Minor eastern French hybrid widely recommended but apparently hardly grown.

BERRY COLOUR: ● ● ● ● ●

PRINCIPAL SYNONYMS: Burdin 7705

ORIGINS AND PARENTAGE

A hybrid obtained by father and son Joanny and Rémy Burdin around 1925 in Iguerande, north west of Lyon, in the Saône-et-Loire, central France, by crossing Seibel 8365 × GAMAY NOIR, in which:

- Seibel 8365 is a Seibel 5410 × Seibel 4643 hybrid
- Seibel 5410 is a Seibel 405 × Seibel 3004 hybrid (see PRIOR for the complete pedigree of Seibel 405)
- Seibel 3004 is a Bayard (ie Couderc 28-112) × Afus Ali (a Lebanese table grape also known as Dattier de Beyrouth) hybrid
- Bayard is an Emily × *Vitis rupestris* hybrid
- Emily is a *Vitis vinifera* subsp. *vinifera* × *Vitis labrusca* hybrid
- Seibel 4643 or Roi des Noirs is a Seibel 29 × Danugue hybrid, the latter being a French black-berried table grape
- Seibel 29 is a Munson × *Vitis vinifera* subsp. *vinifera* hybrid (see PRIOR for the complete pedigree of Munson).

VITICULTURAL CHARACTERISTICS

Early budding and ripening, low vigour, prone to botrytis bunch rot. Even though it is a hybrid, it needs to be grafted. Berries are small to medium and round. May be Guyot or bushvine trained.

WHERE IT'S GROWN AND WHAT ITS WINE TASTES LIKE

There were only 28 ha (69 acres) in **France** in 2008 even though it is officially recommended in about thirty *départements*. Galet (2000) suggests plantings are mainly in Burgundy and the Loire but no producers seem to own up to it in their vineyards. Wines tend to taste rather similar to GAMAY NOIR and have relatively high levels of colour and alcohol.

FOÇA KARASI

Almost-extinct variety from Turkey's Aegean coast.

BERRY COLOUR: ● ● ● ● ●

PRINCIPAL SYNONYMS: Foça, Foçakarası

ORIGINS AND PARENTAGE

Foça Karası, literally the 'black of Foça', is an almost extinct variety that was named after the Foça district near İzmir on Turkey's Aegean coast.

VITICULTURAL CHARACTERISTICS

Mid ripening, good general disease resistance. Vines grown in hot conditions tend to produce more aromatic wines.

WHERE IT'S GROWN

There have been recent attempts to replant and re-evaluate this variety with promising results. The Öküzgözü Winery (sharing a name with the grape variety) in İzmir, **Turkey**, makes a varietal version.

FOGARINA

Mysterious, twice-rescued northern Italian red that adds colour and acidity in blends.

BERRY COLOUR: ● ● ● ● ●

PRINCIPAL SYNONYMS: Fogarina di Gualtieri, Fugarina

VARIETIES COMMONLY MISTAKEN FOR FOGARINA: Uva Fogarina☒

ORIGINS AND PARENTAGE

Fogarina probably comes from Emilia-Romagna. According to local tradition, it was found in 1820 by a certain Carlo Simonazzi, who was fishing in the Crostolina, a small tributary of the River Crostolo near Gualtieri, about 70 km north west of Bologna. Another legend has it that Fogarina was dumped a few centuries ago at a place called Fogarin when the River Po flooded. (There is a place north of Gualtieri called Fogare.)

DNA profiling has shown that it is distinct from the non-cultivated Uva Fogarina from Bologna and that Fogarina could be closely related to LAMBRUSCO MARANI (Boccacci *et al.* 2005); it has also suggested a possible parent–offspring relationship with RABOSO PIAVE (Myles *et al.* 2011).

VITICULTURAL CHARACTERISTICS

Vigorous, highly productive, late ripening.

WHERE IT'S GROWN AND WHAT ITS WINE TASTES LIKE

Fogarina was one of the most widely planted varieties in the province of Reggio Emilia in northern

Italy at the beginning of the twentieth century, especially in the commune of Gualtieri, and also in those of Boretto and Brescello. Fogarina then spread to the provinces of Modena and Reggio Emilia in Emilia-Romagna and Mantova in Lombardia. It is even the subject of a popular Italian folk song.

Although Fogarina was included in the national register of varieties in the 1960s, by the 1970s it had almost disappeared from Reggio Emilia and, although it did still exist around Modena, Mantova and Verona, it was considered to be extinct and was taken off the register. At the end of the 1990s, the Cantina Sociale di Gualtieri rescued Fogarina with the help of local scholars and public authorities and it has now been reinserted in the register and experimental plantings are in progress.

In 2001, Fabio Simonazzi (no relation to Carlo), one of the producers who helped rescue the almost-forgotten grape, planted a vineyard of Fogarina from cuttings taken from an old, ungrafted vine in Pieve di Gualtieri. He is now one of the major producers of Fogarina, which is generally used to increase colour and acidity in blends. The variety also has firm tannins. The Gualtieri co-op produce both a passito and a semi-sparkling dry rosé wine.

FOGLIA TONDA

Minor but well-connected central Italian red generally used to bolster Sangiovese.

BERRY COLOUR: ● ● ● ● ●

VARIETIES COMMONLY MISTAKEN FOR FOGLIA TONDA: CASETTA, ENANTIO, MALBO GENTILE

ORIGINS AND PARENTAGE

Foglia Tonda was apparently found by the famous ampelographer Conte di Rovasenda in the vineyards of Barone Ricasoli's Castello di Brolio in Gaiole in Chianti in Toscana, central Italy. Its name derives from the shape of the leaf: *foglia* means 'leaf' and *tonda* means 'round'.

Ampelographic studies have revealed the distinction between Foglia Tonda and two Trentino varieties with misleadingly similar names: Foja Tonda, which is a synonym for ENANTIO, and Lambrusco a Foglia Tonda, which is a synonym for CASETTA (Storchi 2007).

Its supposed synonymy with MALBO GENTILE (Boccacci *et al.* 2005) can be rejected by DNA profiling (Vouillamoz). Recent DNA studies suggest a parent–offspring relationship between Foglia Tonda and SANGIOVESE, from which Foglia Tonda may have inherited its anthocyanin profile (Crespan, Calò *et al.* 2008). Foglia Tonda would therefore be a grandchild of CILIEGIOLO and CALABRESE DI MONTENUOVO, the parents of Sangiovese.

VITICULTURAL CHARACTERISTICS

Mid ripening with a high level of phenolics in the berries.

WHERE IT'S GROWN AND WHAT ITS WINE TASTES LIKE

Foglia Tonda was added to the national register of varieties in **Italy** in 1970 but was almost extinct in the 1980s. However, the researcher Paolo Storchi rescued it and the variety is currently authorized in the Tuscan provinces of Firenze, Siena and Pistoia, although it is mainly cultivated in the Chianti Classico zone (by Arceno, Brolio, Mannucci Droandi, for example) and in the Valdorcia area and is generally blended with SANGIOVESE. Wines are deeply coloured, powerful and ageworthy.

Varietal wines, classified as IGT Toscana or IGT Colli della Toscana Centrale, are made by a few producers such as Guido Gualandi.

FOGONEU

Minor Mallorcan, usually blended.

BERRY COLOUR: ● ● ● ● ●

PRINCIPAL SYNONYMS: Fogoneau, Fogonet, Fogonetxo, Fogoneu Frances, Fogoneu Mallorquí

ORIGINS AND PARENTAGE

Fogoneu originates from the Spanish island of Mallorca and is said to have some similarities with GAMAY NOIR, which might explain its synonym Fogoneu Frances. Recent DNA parentage analysis has suggested that Fogoneu is a natural cross between two obscure varieties from Mallorca, namely Excursach (ESCURSAC) and Mansés de Capdell, and a parent of CALLET (García-Muñoz *et al.* 2012). However, these parentages were based on too few DNA markers (twenty) and need to be confirmed.

The name Fogoneu is said to be a diminutive of *fogo*, meaning 'oven', but there's no rational explanation for that etymology.

WHERE IT'S GROWN AND WHAT ITS WINE TASTES LIKE

Fogoneu is the most important variety in the Felanitx area on the island of Mallorca, **Spain**, where it is authorized in both the Binissalem and Plà i Llevant appellations as well as in the regional wine Vino de la Tierra de Formentera on the island of the same name. On Mallorca it is often blended with its putative parent CALLET and with MANTO NEGRO, eg in Àn Negra wines, or with international varieties in Cap de Barbaria's blends. Official statistics suggest there are also 5 ha (12 acres) planted in Extremadura.

FOKIANO

Rare Greek variety almost lost in the confusion over its name and synonyms.

BERRY COLOUR: ● ● ● ● ●

PRINCIPAL SYNONYMS: Damaskino, Fokiana, Fokiano Kokkino, Fokiano Mavro, Giouroukiko ⚮ or Ghiouroukiko ⚮ (Kykládes/Cyclades), Phokiano, Phokiano Kokkineli, Phokiano Mavro

VARIETIES COMMONLY MISTAKEN FOR FOKIANO: İRI KARA ⚮ (Hadım in Turkey)

ORIGINS AND PARENTAGE

According to Lefort and Roubelakis-Angelakis (2001), several grape varieties with black, red or white berries are cultivated in Greece under the name Fokiano – in the Kykládes, the Dodekánisa (Dodecanese), the western and northern Aegean Islands, Makedonía and Thráki (Thrace). Its name is supposedly based on Phocaea or Phokaia, an ancient Ionian Greek city on the western coast of Anatolia corresponding to the modern-day Foça in Turkey.

The 'true' Fokiano discussed here is a black-berried variety that probably originates from the Kykládes, where it has been shown by DNA profiling to be identical to Giouroukiko (Lefort and Roubelakis-Angelakis 2001). In addition, comparison of the DNA profiles in the Greek Vitis Database for Skiadopoulo, a white-berried variety that is cultivated in the regions of Ilía, Aitolía-Akarnanía and Zákynthos (Nikolau and Michos 2004), and for Giouroukiko shows that they are identical (Vouillamoz). **Skiadopoulo** is therefore likely to be a colour mutation of the Fokiano described here.

OTHER HYPOTHESES

Fokiano is often said to originate in the region of İzmir, historically Smyrna, on the Turkish coast of the Aegean Sea, and it has been considered to be identical to İRI KARA (literally 'black plum') from Hadım in Turkey, which is why synonyms such as Ari Karas, Erikaras and Irikaras are usually ascribed to Fokiano (Galet 2000; Lefort and Roubelakis-Angelakis 2001; Nikolau and Michos 2004; Greek Vitis Database). However, comparison of the DNA profiles of Fokiano (Lefort and Roubelakis-Angelakis 2001) and İRI KARA (Vouillamoz *et al.* 2006) show that they are distinct varieties (Vouillamoz).

VITICULTURAL CHARACTERISTICS

Fertile and productive. Mid to late budding, early to mid ripening. Relatively resistant to diseases and drought. Compact bunches of big berries.

WHERE IT'S GROWN AND WHAT ITS WINE TASTES LIKE

Given the confusion over this variety and all its synonyms, it is difficult to be sure where the 'real' Fokiano is grown. However, official statistics show that there were just 52 ha (128 acres) in Attikí in the south of mainland **Greece** in 2008, and it is also found on various Aegean Islands. For example, Afianes on the island of Ikaría, south west of Sámos, blend Fokiano with a small amount of MANDILARIA for their red and with Begleri (ie THRAPSATHIRI) for their rosé. Varietal wines generally lack colour, have moderate to low acidity and are prone to oxidation, so that they often taste of plums, figs and raisins.

The light-skinned Skiadopoulo features along with other local varieties such as GOUSTOLIDI in the traditional white Verdea blends of both Comoutos and Callinico on the island of Zákynthos in the Ionian Sea.

FOLGASÃO

Robust Iberian variety with complex origins.

BERRY COLOUR:

PRINCIPAL SYNONYMS: Cagarrizo, Silveiriña

VARIETIES COMMONLY MISTAKEN FOR FOLGASÃO: ARINTO DE BUCELAS (Lisboa), MALVASIA FINA (Dão and Douro), TERRANTEZ (Madeira), TERRANTEZ DO PICO (Açores), TERRANTEZ DA TERCEIRA (Açores)

ORIGINS AND PARENTAGE

Folgasão is said to originate in the Dão region but comparison of the DNA profile of Folgasão in Almadanim *et al.* (2007) with other published data shows that it is probably identical to two Spanish varieties that are no longer cultivated – Silveiriña in Gago *et al.* (2009) and Cagarrizo in Santana *et al.* (2010) – thereby undermining the theory that Folgasão is Portuguese. In addition, Folgasão was until recently considered to be identical to TERRANTEZ on the island of Madeira, and with TERRANTEZ DA TERCEIRA and TERRANTEZ DO PICO in the Açores, but this has been disproved by DNA profiling (Veloso *et al.* 2010).

VITICULTURAL CHARACTERISTICS

Low yields, early budding, mid ripening. Small, compact bunches of small, thin-skinned berries. Susceptible to coulure and moderately so to powdery mildew and botrytis bunch rot.

WHERE IT'S GROWN AND WHAT ITS WINE TASTES LIKE

Folgasão wines are generally full-bodied and high in acid, with good ageing potential. In the Trás-os-Montes region in northern **Portugal**, where the vine is most commonly planted, it is usually blended, and there are some remaining vines in the Douro to the south. There were 357 ha (882 acres) in Portugal in 2010.

FOLIGNAN

Recent Bordeaux cross authorized for the production of cognac.

BERRY COLOUR: ● ● ● ● ●

PRINCIPAL SYNONYMS: INRA 8476

ORIGINS AND PARENTAGE

A TREBBIANO TOSCANO × FOLLE BLANCHE cross obtained at INRA (Institut National de la Recherche Agronomique), Bordeaux, south-west France, in 1965. Its name seems to be a contraction of the names of its parents, Folle and Ugni Blanc (the name of TREBBIANO TOSCANO in the Charente).

VITICULTURAL CHARACTERISTICS

Mid ripening, generally vigorous.

WHERE IT'S GROWN

After forty years of research, Folignan was authorized for the production of cognac in 2005 and was therefore the first INRA cross to be authorized in an appellation contrôlée, though its contribution to any blend is capped at 10%. There were 44 ha (109 acres) planted in **France** in 2008, all intended for the production of the base wine for cognac.

FOLLE BLANCHE

Productive but rot-prone and very acidic white variety.

BERRY COLOUR: ● ● ● ● ●

PRINCIPAL SYNONYMS: Chalosse Blanche (Gers, Landes), Dame Blanche (Lot-et-Garonne), Enrageat or Enragé (Gironde), Folle (Charente), Gros Plant (Pays Nantais, Vendée), Matza Zuri⌘ (País Vasco in Spain), Mune Mahatsa (País Vasco), Piquepoul (Gascogne), Piquepoul du Gers (Lot-et-Garonne)

VARIETIES COMMONLY MISTAKEN FOR FOLLE BLANCHE: KNIPPERLÉ⌘, PIQUEPOUL BLANC⌘ (Folle Blanche is called PIQUEPOUL in Gascogne but is not related in any way to the true PIQUEPOUL BLANC of southern France)

ORIGINS AND PARENTAGE

Folle Blanche was first mentioned in 1696 in Chérac in the Charente-Maritime, western France, where it has traditionally been used to make cognac and armagnac: 'lequel plan sera tout de visan blanc, savoir un tiers de folle blanche, un autre de Gros Bouillau et Gros-Blanc, l'autre tiers de Blanche-Ramée' ('which vineyard will be all visan blanc, ie one third folle blanche, another Gros Bouillau and Gros-Blanc, the other third Blanche-Ramée'; Rézeau 1997). The name Folle (*folle* is the feminine form of the French word *fou*, meaning 'mad') probably refers to the vigour and productivity of the vine, as do its synonyms Gros Plant (meaning 'big vine'), widely used in the Loire-Atlantique and first mentioned in 1732, and Enrageat (French *enragé* meaning 'mad' or 'rabid'), widely used in the south west of France, where it was first mentioned in 1736 (Rézeau 1997).

DNA parentage analysis has shown that Folle Blanche is a likely progeny of GOUAIS BLANC as well as a parent of MONBADON (aka Burger) from the Charente and of Merlot Blanc (no longer commercially cultivated) from the Gironde (Bowers *et al.* 2000; Boursiquot *et al.* 2009; see CABERNET SAUVIGNON and PINOT pedigree diagrams), two regions where Folle Blanche has traditionally been cultivated. However, Folle Blanche is a member of the Folle ampelographic group (see p XXVII) and is morphologically close to PETIT MESLIER and MESLIER SAINT-FRANÇOIS (Bisson 2009), varieties from south-west France. This is consistent with Lavignac's (2001) hypothesis that Folle Blanche originates from the Gers or the Landes in south-west France, whence it spread to the Charente and later to the Pays Nantais at the Atlantic end of the Val de Loire. This theory is supported by parent–offspring and sibling (or half-sibling) relationships recently discovered with JURANÇON BLANC from the Gers and the Pyrénées-Atlantiques, JURANÇON NOIR from the Lot-et-Garonne, as well as with MONTILS from the Charente and Meslier Saint-François (under its Charente synonym Blanc Ramé) from the Val de Loire (Myles *et al.* 2011).

Folle Blanche is not a colour mutation of JURANÇON NOIR nor of FUELLA NERA, both often called Folle Noire.

Folle Blanche was used to breed BACO BLANC, BACO NOIR and FOLIGNAN.

VITICULTURAL CHARACTERISTICS

Fertile, productive, mid ripening. Early budding makes it susceptible to spring frosts. Very susceptible to botrytis bunch rot due to the compact bunches and also prone to black rot and mites and, to a lesser extent, to downy mildew and diseases of the wood.

WHERE IT'S GROWN AND WHAT ITS WINE TASTES LIKE

Folle Blanche was once much more widely planted for the production of eau de vie, particularly in regions near the coast that were easily accessible to the Dutch merchants. Today it is recommended in several *départements* in the Loire and the south west of **France** and authorized as far south as the Pyrénées. Folle Blanche is nevertheless declining considerably: down from 15,865 ha (39,203 acres) in 1958 to a total of 1,770 ha (4,374 acres) in 2009. Most of the vines are at the western end of the Loire, where it is used to produce the varietal Gros Plant du Pays Nantais in roughly the same area as Muscadet. Wines have searingly high acidity and may be described as Muscadet's country cousin. In south-west France, Folle Blanche is more likely to find its way into the base wines for cognac or armagnac, although it was generally replaced by the more disease-resistant hybrid BACO BLANC after the region was hit by phylloxera. Tariquet, Domaine de Mirail, Godet and a few other producers make a varietal armagnac from Folle Blanche.

As Mune Mahatsa, the variety is sometimes blended with other local varieties such as Hondarribi Zuri (see COURBU BLANC) in the País Vasco, **Spain**'s Basque Country on the northern coast, although it is not recorded (under any name) in Spain's official vineyard statistics. It is also used to a limited extent in the production of Catalan brandy by the likes of Torres.

There are also minor plantings in **California**.

FONTE CAL

Minor Portuguese variety that rarely takes a leading role.

BERRY COLOUR: ● ● ● ● ●

PRINCIPAL SYNONYMS: Fonte de(a) Cal

ORIGINS AND PARENTAGE

Under the old original name Fonte de(a) Cal that was in use until the end of the nineteenth century, this variety was first mentioned in 1790 by Lacerda Lobo in the region of Pinhel in Beira Interior in north-east Portugal, quite close to the Spanish border, where it most likely originates (Rolando Faustino, personal communication).

VITICULTURAL CHARACTERISTICS

Vigorous, early budding, mid ripening. Averagely susceptible to downy and powdery mildews.

WHERE IT'S GROWN AND WHAT ITS WINE TASTES LIKE

Although Fonte Cal is rarely made as a varietal wine, the few examples that exist indicate that it produces wines that are aromatic (both fruity and floral), a little like ALVARINHO, with moderate acidity, although the grapes are often harvested late, after the acidity has dropped a little too far. Harvested earlier, the fruit produces well-structured wines with good ageing potential that become interesting once the primary aromas have been replaced by more complex tertiary ones. It is usually blended with SÍRIA and occasionally with ARINTO DE BUCELAS.

Fonte Cal expanded from Pinhel throughout the Beira Interior region in **Portugal** in the twentieth century but total Portuguese plantings in 2010 were just 134 ha (331 acres). It is authorized for regional wines in Beiras and in the Península de Setúbal, as well as in Beira Interior DOC, but varietal wines are extremely rare. Alcohol and acidity are slightly higher than in FERNÃO PIRES wines although the acidity is still relatively low.

FORASTERA

Ischian speciality.

BERRY COLOUR: ● ● ● ● ●

PRINCIPAL SYNONYMS: Forastiera, Forestiera, Frastera, Uva dell'Isola (island of Procida in the bay of Napoli)

ORIGINS AND PARENTAGE

Conte di Rovasenda (1877) was the first to mention Forastera, planted on the island of Ischia, just off Napoli in Campania, southern Italy, as well as in Grumello di Monte in the province of Bergamo, whence it has now disappeared. A year later, the ampelographer Giuseppe Froio described Forastera as a variety especially grown in the vineyards of the island of Ischia (Migliaccio *et al.* 2008). Some authors have suggested that the variety was introduced from Grumello to Ischia in the nineteenth century, as indicated by its name (*forestiero* means 'foreign').

The variety known as Forastera on the Islas Canarias but no longer commercially cultivated is not the same variety and has a distinct and unique DNA profile (Vouillamoz).

VITICULTURAL CHARACTERISTICS

Vigorous, mid to early ripening.

WHERE IT'S GROWN AND WHAT ITS WINE TASTES LIKE

Recommended throughout Campania and provisionally authorized on Sardegna, Forastera is in practice cultivated only on the islands of Ischia and Procida in southern **Italy**. Ischia DOC allows for a varietal wine (minimum 85%) as well as a blend of Forastera (45–70%) with BIANCOLELLA, which may be still or sparkling. Varietal wines are made by producers such as Casa d'Ambra and Antonio Mazzella. Wines tend to be light and fresh.

There was a total of 102 ha (252 acres) in the Italian census of 2000.

FORCALLAT BLANCA

See AIRÉN.

FORCALLAT TINTA

Late-ripening central Spanish variety used in blends.

BERRY COLOUR: ● ● ● ● ●

PRINCIPAL SYNONYMS: Alcabril di Gualadin (León), Forcalla, Forcalla Negra, Forcalla Prieta, Verdal (Orense), Verdalejo (Murcia), Verdalla (Islas Canarias)

ORIGINS AND PARENTAGE

Forcallat Tinta comes from Castilla-La Mancha. Forcallat Blanca is not a white mutation of Forcallat Tinta, since it is a distinct variety and its DNA profile (Martín *et al.* 2003) does not match any other known variety (Vouillamoz). (It is also used as a synonym for AIRÉN.) Verdal is also used as a synonym for other Spanish varieties and may therefore be misleading.

VITICULTURAL CHARACTERISTICS

Vigorous and very late ripening.

WHERE IT'S GROWN AND WHAT ITS WINE TASTES LIKE

There were 930 ha (2,298 acres) of Forcallat Tinta in **Spain** in 2008, the majority of the vines in the provinces of Murcia and Valencia. It is authorized in various regional wines such as Vino de la Tierra de Abanilla and Vino de la Tierra Campo de Cartagena in Murcia and Vino de la Tierra el Terrerazo in Valencia. Wines tend to be pale in colour and to go into blends.

FORGIARIN

Rare dark-skinned variety of Friuli, north-east Italy.

BERRY COLOUR: ● ● ● ● ●

ORIGINS AND PARENTAGE

Forgiarin takes its name from Forgaria near Pinzano al Tagliamento in Friuli.

VITICULTURAL CHARACTERISTICS

Susceptible to powdery mildew.

WHERE IT'S GROWN AND WHAT ITS WINE TASTES LIKE

Forgiarin was formerly cultivated in the provinces of Udine and Pordenone in Friuli, north-east **Italy**, especially in San Daniele, Spilimbergo and Maniago. Today, it is mainly cultivated in Pinzano al Tagliamento and Castelnovo del Friuli in Pordenone. Together with other old local varieties CIANORIE, CIVIDIN, PICULIT NERI, SCIAGLÌN and UCELÙT, it was saved from extinction by Emilio Bulfon, who is one of the very few producers crafting varietal wines from these grapes. Forgiarin wines have light tannins and aromas of herbs, red fruits, toasted almonds and vanilla. Along with Piculit Neri, Sciaglìn and Ucelùt, Forgiarin was included in the Italian national register of varieties in 1991.

FORSELLINA

Rarity occasionally blended into Valpolicella or Bardolino.

BERRY COLOUR: ● ● ● ● ●

PRINCIPAL SYNONYMS: Forcelina, Forselina, Forsella, Forsellana, Forzelina

ORIGINS AND PARENTAGE

Mentioned in the early nineteenth century in the Valle d'Illasi near Verona in northern Italy.

VITICULTURAL CHARACTERISTICS

Mid ripening.

WHERE IT'S GROWN AND WHAT ITS WINE TASTES LIKE

Historically the variety has been limited to Valpantena, Valpolicella, the Val d'Illasi, the Val Tramigna and south of Lago di Garda in the Veneto region of northern **Italy**. Its cultivation decreased drastically during the twentieth century but in the 1970s the research centre at Conegliano rescued some plants from old vineyards and in 1987 the variety was added to the Italian national register. A small percentage may be blended with CORVINA VERONESE, CORVINONE and RONDINELLA and other local varieties in DOCs such as Bardolino (maximum 10%) and Valpolicella, including Amarone and Recioto styles (maximum 15%). Wines are lightly aromatic and fruity.

FORTANA

Late-ripening, declining Italian red prized for its high yields.

BERRY COLOUR: ● ● ● ● ●

PRINCIPAL SYNONYMS: Canèna, Canina Nera⊗, Cannina, Costa d'Oro, Oliva⊗, Rapa⊗ (Ticino in Switzerland), Uva Canina, Uva Cornetta (Umbria), Uva d'Oro, Uva d'Oro di Comacchio, Uva Francese Nera, Uva Vecchia⊗

VARIETIES COMMONLY MISTAKEN FOR FORTANA: BRUGNOLA⊗ (Valtellina in Lombardia)

ORIGINS AND PARENTAGE

This variety was known in Emilia-Romagna in northern central Italy as early as 1550 as Uva d'Oro ('golden grape'). Some have suggested that this rather strange name for a black-berried variety may have something to do with the variety's high yields, which meant that it was cherished as if it were golden. The name Fortana is much more recent, chosen when the variety was included in the national register in 1970, and some argue for a related etymology since Fortana means 'very fruitful'.

DNA profiling and ampelographic observations unexpectedly proved that Fortana is identical to Canina Nera and possibly has a parent–offspring relationship with LAMBRUSCO MAESTRI, and that it could also be genetically close to MALBO GENTILE from Modena (Boccacci *et al.* 2005). Canina Nera is not related to the white-berried Canina, a synonym for DRUPEGGIO in Ascoli Piceno. Fortana was also shown to be identical to Rapa, a rare variety of the Italian-speaking Swiss canton of Ticino (Vouillamoz).

OTHER HYPOTHESES

Some suggested that the original name Uva d'Oro comes from Côte d'Or in France and presumed that Uva d'Oro/Fortana was a French grape introduced from Bourgogne. This hypothesis can be rejected for three reasons:

- Fortana has parent–offspring relationships with the typically Italian varieties LAMBRUSCO MAESTRI and UVA TOSCA
- Fortana's DNA shows no relation to that of PINOT or any major Burgundian variety (Vouillamoz)
- Fortana's DNA clustered with many Lambrusco and other central Italian varieties (ENANTIO, MALBO GENTILE, etc.) in a genetic study (Boccacci *et al.* 2005).

VITICULTURAL CHARACTERISTICS

Some susceptibility to powdery mildew, fruitful, late ripening (harvested the first two weeks of October).

WHERE IT'S GROWN AND WHAT ITS WINE TASTES LIKE

Formerly cultivated in the Oltrepò Pavese and the Veronese, Fortana is now limited to the areas of Ravenna (Lugo, Russi, Bagnacavallo, Alfonsine) and Ferrara in **Italy**. Varietal Fortana wines can be found only in the Bosco Eliceo DOC but a small percentage is allowed in the blend of the various Lambrusco DOCs further west. Fortana is also cultivated to a very limited extent in the provinces of Reggio Emilia and Parma. Under the name Canina Nera, it is on the verge of extinction and used only in blends. Fortana wines can be dry, sparkling or sweet. They have aromas of red fruits, not much alcohol, good acidity that occasionally gives a sour note and somewhat rustic tannins, softened by residual sugar in the sweet version. The Italian census of 2000 recorded 1,109 ha (2,740 acres).

FRANCAVIDDA

Minor and dwindling vine found in Brindisi, generally blended.

BERRY COLOUR: ● ● ● ● ●

PRINCIPAL SYNONYMS: Francavilla

VARIETIES COMMONLY MISTAKEN FOR FRANCAVIDDA: CATARRATTO BIANCO

ORIGINS AND PARENTAGE

Most likely originates from Francavilla Fontana in the province of Brindisi in Puglia, southern Italy, since Francavidda is the dialect version of the place name.

VITICULTURAL CHARACTERISTICS

Susceptible to fungal diseases and winter frost. Mid ripening.

WHERE IT'S GROWN AND WHAT ITS WINE TASTES LIKE

Cultivation in **Italy** is restricted to the province of Brindisi (notably in Francavilla Fontana, Carovigno, San Vito dei Normanni, San Michele Salentino and Ostuni) but Francavidda is slowly disappearing. Up to 50% Francavilla is allowed in Ostuni DOC, usually blended with IMPIGNO, BIANCO D'ALESSANO and/or VERDECA. Francavidda wines tend to have a delicate perfume and only moderate alcohol.

FRÂNCUŞĂ

North-east Romanian variety producing fresh and relatively light, dry whites for early drinking.

BERRY COLOUR: ● ● ● ● ●

PRINCIPAL SYNONYMS: Frâncuşe, Frîncuşă, Mustoasă or Mustoasă de Moldova, Poamă Creaţă, Poamă Franchie, Poamă Muştei, Târţără, Vinoasă

VARIETIES COMMONLY MISTAKEN FOR FRÂNCUŞĂ: KREACA (Serbia), MUSTOASĂ DE MĂDERAT (Romania)

ORIGINS AND PARENTAGE

Frâncuşă was cultivated in the areas of Cotnari, Odobeşti, Panciu and Huşi in Romanian Moldova well before phylloxera hit the region at the turn of the twentieth century. Two different types have been distinguished morphologically: one in Odobeşti with evenly sized berries and one in Cotnari, called Târţără, with higher sugar content and berries of irregular size due to coulure (Dejeu 2004).

Recent DNA parentage analysis has suggested that Frâncuşă is one of the numerous progenies of GOUAIS BLANC, like FURMINT in Hungary (Boursiquot *et al.* 2004).

Frâncuşă should not be confused with KREACA from Serbia, which is sometimes known as Frâncuşă.

VITICULTURAL CHARACTERISTICS

Late ripening with compact bunches (Galet 2000).

WHERE IT'S GROWN AND WHAT ITS WINE TASTES LIKE

Francușă is today found only in the northern part of **Romanian** Moldova, where there were 352 ha (870 acres) in 2008. Cotnari Winery is the main proponent of varietal wines, which are typically crisp, light and direct with flavours of lemon and green apple.

FRAPPATO

Fruity, fresh and floral Sicilian often blended with Nero d'Avola.

BERRY COLOUR: ● ● ● ● ●

PRINCIPAL SYNONYMS: Frappato di Vittoria⚮ (Sicilia), Frappato Nero, Frappatu (Sicilia), Nero Capitano (Calatino region), Surra (province of Siracusa)

VARIETIES COMMONLY MISTAKEN FOR FRAPPATO: GAGLIOPPO⚮, NERELLO CAPPUCCIO⚮, NERELLO MASCALESE⚮

ORIGINS AND PARENTAGE

Frappato was first mentioned in 1760 in Vittoria in the province of Ragusa in the south of Sicilia, where it is said to have originated. The name Frappato could derive from a corruption of the adjective *fruttato*, meaning 'fruitful'. Recent DNA studies suggested a parent–offspring relationship between Frappato and SANGIOVESE (Di Vecchi Staraz, This *et al.* 2007; Crespan, Calò *et al.* 2008), making it a likely grandchild of CILIEGIOLO and CALABRESE DI MONTENUOVO. In addition, Frappato and GAGLIOPPO from Calabria are likely to be siblings (Di Vecchi Staraz, This *et al.* 2007).

OTHER HYPOTHESES

Some authors have proposed a Spanish origin for Frappato. This is inconsistent with the fact that Frappato clusters with most ancient and traditional Sicilian grape varieties in genetic studies (such as Di Vecchi Staraz, This *et al.* 2007).

VITICULTURAL CHARACTERISTICS

Mid to late ripening. Thin, delicate skins and compact clusters make Frappato prone to botrytis bunch rot.

WHERE IT'S GROWN AND WHAT ITS WINE TASTES LIKE

Frappato is cultivated mainly in the province of Ragusa in southern Sicilia, **Italy**, in particular near Vittoria, but also in the province of Trapani. It tends to be blended with other Sicilian varieties such as NERO D'AVOLA, NERELLO MASCALESE, NOCERA, and even white-berried grapes such INZOLIA and CATARRATTO BIANCO. Frappato contributes 30–50% of the blend in Cerasuolo di Vittoria DOCG and is authorized as a component in the Alcamo, Eloro, Erice and Vittoria DOCs but may also be produced as a varietal wine in the last two (or as IGT Sicilia from elsewhere on the island). The Italian census of 2000 recorded total plantings of 785 ha (1,940 acres) but there were 846 ha (2,091 acres) on Sicilia in 2008.

Producers of varietal wines include: Avide, Calò Giordano, Casaventura, Cos, Cantine Rallo and Le Terre del Gattopardo. Wines are far from heavy but are generally fresh and succulent and can taste of strawberries – often both fruity and floral.

FRAUELER

Rare, high-yielding and acidic Tyrolean white.

BERRY COLOUR: ● ● ● ● ●

PRINCIPAL SYNONYMS: Frauler, Vezzaner

ORIGINS AND PARENTAGE

Obscure and old variety of the Val Venosta (Vinschgau in German) in the Alto Adige (Südtirol) in northern Italy. Its name could derive from *Friaul* (German for Friuli), suggesting a north-eastern Italian origin. Fraueler is thought to be identical to Grüner Barthainer, an ancient Styrian variety from southern Austria that is no longer cultivated, but DNA profiling has not yet been done.

VITICULTURAL CHARACTERISTICS

Vigorous and high-yielding. Very susceptible to powdery mildew. Mid to late ripening and susceptible to botrytis bunch rot during the later stages of ripening.

WHERE IT'S GROWN AND WHAT ITS WINE TASTES LIKE

Fraueler is cultivated to a very limited extent in the Val Venosta/Vinschgau, where, only just in **Italy**, a few producers continue to make a varietal wine, eg Schnalshuberhof and Befehlhof. Unterortl's Juval Glimmer blend includes both Fraueler and another rare variety, BLATTERLE, with either RIESLING or MÜLLER-THURGAU. Varietal wines are rather neutral, light-bodied and high in acid, though they can have a certain minerality when fully ripe, which is why Fraueler is generally better as a blending partner.

FREDONIA

New York hybrid found mainly in Pennsylvania.

BERRY COLOUR: ● ● ● ● ●

PRINCIPAL SYNONYMS: Early Concord

ORIGINS AND PARENTAGE

A Champion × Lucille hybrid obtained in 1915 by Fred E Gladwin at the Department of Pomology and Viticulture at New York State Agricultural Experiment Station, Geneva, where:

- Champion is a CONCORD × Hartford hybrid, in which Hartford (or Hartford Prolific) is a likely spontaneous ISABELLA × *Vitis labrusca* hybrid found around 1846 in the garden of Paphro Steele & Son in West Hartford, Connecticut (see BRIANNA for the complete pedigree of Champion)
- Lucille (or Lucile) is a seedling of Wilmington Red
- Wilmington Red is a selected *Vitis labrusca* vine.

This *Vitis labrusca* and *Vitis vinifera* hybrid was released in 1927.

Confusingly, Fredonia is also the name of a tetraploid CONCORD (ie with artificially doubled chromosomes). Fredonia was used to breed VALIANT and VAN BUREN.

VITICULTURAL CHARACTERISTICS

Large to medium-sized compact bunches of thick-skinned berries. Cold-hardy but highly susceptible to anthracnose, downy mildew and phomopsis cane and leaf spot. Ripens earlier than CONCORD (hence the synonym Early Concord) and is therefore better suited to regions with short growing seasons.

WHERE IT'S GROWN AND WHAT ITS WINE TASTES LIKE

Fredonia is used mainly for the table and for jelly and juice but also occasionally to make mostly rosé or light-red wines, sometimes sweet, with a foxy *labrusca* flavour. **Pennsylvania** can boast the largest area – 35 acres (14 ha) in 2008. **Michigan**'s total is down from 21 acres (8 ha) in 1997 to just 7 acres (3 ha) in 2006. Varietal wines are produced by, among others, Arrowhead, Heritage Wine Cellars, Montgomery Underground Cellars, Star Hill and Winfield, all in Pennsylvania.

FREISA

Light, fragrant, unusual Nebbiolo relative, made in a wide range of styles, including frothy.

BERRY COLOUR: ● ● ● ● ●

PRINCIPAL SYNONYMS: Freisa di Chieri, Freisa Piccola, Freisetta, Fresia, Monferrina, Mounfrina, Spannina

VARIETIES COMMONLY MISTAKEN FOR FREISA: FUMIN (Valle d'Aosta), NERETTA CUNEESE (under the name Freisa Grossa or Freisa di Nizza), NERETTO NOSTRANO (Canavese)

ORIGINS AND PARENTAGE

Freisa is one of the oldest and most important varieties of Piemonte. Although there is some doubt, the earliest mention of Freisa is thought to be in a 1517 document speaking of a toll fee in Pancalieri, about 30 km south west of Torino; twice as much was paid for *fresearum* wine compared with other wines (Nada Patrone 1988). The first reliable ampelographic description of Freisa was made by Giuseppe Nuvolone Pergamo (1787–98), Count of Scandaluzza.

DNA profiling showed that Freisa and NEBBIOLO have a parent–offspring relationship, and that Freisa could have a close genetic relationship with VIOGNIER from the northern Rhône (Schneider, Boccacci *et al.* 2004). In addition, Freisa might be closely related to RÈZE, an old variety from the Valais in Switzerland (Vouillamoz, Schneider *et al.* 2007).

VITICULTURAL CHARACTERISTICS

A robust vine with high levels of anthocyanins in the berries. Mid ripening. Subject to coulure.

WHERE IT'S GROWN AND WHAT ITS WINE TASTES LIKE

Freisa is cultivated in various regions of Piemonte in north-west **Italy**, particularly in the provinces of Asti, Torino and Cuneo and in the alpine regions, including the subalpine areas. East of Piemonte, there are scattered plantings of Freisa in the province of Vicenza in the Veneto. The Italian agricultural census of 2000 records total plantings of 1,453 ha (3,590 acres).

It is permitted as a minor component in various NEBBIOLO-, BARBERA- or GRIGNOLINO-dominant DOCs and as part of the blend in a wide range of Piemonte DOCs. It also has several DOCs carrying its own name, eg Freisa d'Asti and Freisa di Chieri.

Varietal examples are generally quite light in colour and unexpectedly high in acidity and tannins but they are typified by their distinctive wild-red-fruit fragrance, especially strawberries and raspberries, which, together with a certain bitterness, is not universally appreciated. The Freisa d'Asti and Freisa di Chieri DOCs allow for a bewildering range of permutations: dry or sweet, still, lightly sparkling (frizzante), or fizzy (spumante). The most traditional style, lightly sparkling from a secondary fermentation, has some residual sugar to balance that lightly bitter note. Modern winemaking now allows for better control of fermentation and the level of residual sweetness. Freisa is also allowed as a minor component in several other DOCs such as Valle d'Aosta, Colli Tortonese and Barbera d'Asti.

Borgogno make good examples of still, dry, perfumed Freisa, and other worthwhile examples are made by Aldo Vajra and Coppo.

Freisa has been bottled by the likes of Bonny Doon and Viansa in **California.**

FREISAMER

Early-twentieth-century German cross in severe decline.

BERRY COLOUR: ● ● ● ● ●

PRINCIPAL SYNONYMS: Freiburg 25-1, Freiburger

ORIGINS AND PARENTAGE

A SILVANER × PINOT GRIS cross created in 1916 by Karl Müller in Freiburg in southern Germany, hence its original name, Freiburger. DNA profiling has confirmed this parentage (Vouillamoz). Its later name is a combination of Freiburg and Dreisam, the river running through that town.

VITICULTURAL CHARACTERISTICS

Mid to late ripening.

WHERE IT'S GROWN AND WHAT ITS WINE TASTES LIKE

In **Germany**, Freisamer is still grown to a very limited extent (4 ha/10 acres) in its home state of Baden-Württemberg but is nowhere near as popular as in its heyday in the early 1970s. A rare varietal example is produced by Dr Lawall in Rheinhessen.

About the same area is planted to this high-acid variety in German-speaking **Switzerland** and in Fribourg/Freiburg, and sweeter versions are a speciality of the Bündner Herrschaft in Graubünden. Domaine Chervet in Vully in the canton of Fribourg produce a varietal wine. Wines are crisp, rich in extract but relatively neutral and the grapes need to reach high sugar levels to produce good-quality wine.

It is also planted to a very limited extent in **England.**

FRENCH WILD

Pink-skinned French Wild is not related to the truly wild *Vitis vinifera* subsp. *silvestris* growing naturally in France and elsewhere. According to an unverified online source (www.cnwinenews.

com), it was imported in 1965 by Dongfeng Farm in Mile county in Yunnan province, southern central **China**, along with ten other grape varieties, including ROSE HONEY and CRYSTAL. Since DNA profiling has not yet been done, it is not known whether French Wild is related to European varieties. There are supposedly four clones of French Wild but only two are cultivated, and to a minor extent.

FRONTENAC

Successfully cold-hardy American hybrid thriving particularly in its home state of Minnesota.

BERRY COLOUR: ● ● ● ● ●

PRINCIPAL SYNONYMS: MN 1047

ORIGINS AND PARENTAGE

Bred in 1978 by Peter Hemstad, James Luby and Patrick Pierquet at the University of Minnesota, Frontenac is a hybrid of LANDOT NOIR and *Vitis riparia* selection #89, where *Vitis riparia* #89 (also known as MN 89) was a vine found growing wild near Jordan in Minnesota. An extremely complex *Vitis riparia*, *Vitis labrusca*, *Vitis vinifera*, *Vitis aestivalis*, *Vitis lincecumii*, *Vitis rupestris*, *Vitis cinerea* and *Vitis berlandieri* hybrid, Frontenac was selected in 1983 and first released in 1996, one of the first of a new wave of cultivar introductions.

Frontenac Gris is a colour mutation of Frontenac first observed by Peter Hemstad at the University of Minnesota in 1992 and released in 2003.

Frontenac Blanc is a more recent and more promising colour mutation first observed at Vignoble Les Blancs Coteaux near Dunham, Québec, in 2005, and later found further afield, in Minnesota, for example.

VITICULTURAL CHARACTERISTICS

Productive, vigorous, cold-hardy (even more so than MARÉCHAL FOCH). Good all-round disease resistance, especially immune from downy mildew, but highly susceptible to foliar phylloxera infestation. Loose bunches of small to medium-sized berries that can attain high levels of sugar and acidity. Moderately early budding, mid ripening.

WHERE IT'S GROWN AND WHAT ITS WINE TASTES LIKE

Thanks especially to its winter resilience and good yields, allied with better quality than most hybrids and an absence of foxy flavour, Frontenac is the mostly widely planted red wine grape in **Minnesota** (116 acres/47 ha in 2007), and is popular in the northern Midwest, particularly in **Illinois** (58 acres/23 ha in 2007), **Iowa** (47 acres/19 ha in 2006), **Nebraska** (14 acres/6 ha in 2007) and **Indiana** (10 acres/4 ha in 2010). It is also thriving in northern **Nevada**. Wines are deeply coloured with flavours of cherry and some blackcurrant or plum, even chocolate as they age, but the acidity can be very high. Frontenac is used to make a wide variety of styles, including rosé, red and port-like wines, with the last style gaining in popularity since the grapes can be harvested later with lower acidity. Successful varietal wines are made by Falconer, Fieldstone and St Croix in Minnesota; 5 Trails in Nebraska; and Snus Hill and Summerset in Iowa. Despite its popularity, it may well lose ground to newer, lower-acid varieties such as MARQUETTE for the production of dry table wines.

Frontenac is also planted in **Canada**, notably in Québec (52 acres/21 ha in 2009).

The grey-berried mutation Frontenac Gris has the same useful viticultural characteristics as the dark-skinned Frontenac, yielding wines that have aromas of peach, apricot and citrus together with refreshing acidity, although the wines are quite deeply coloured. There were 75 acres (30 ha) in Minnesota in 2007 and more limited plantings in Iowa. Fieldstone, Indian Island, Parley Lake and St Croix in Minnesota and Prairie Berry in **South Dakota** make varietal Frontenac Gris, which is also found in Québec, Canada.

Commercial plantings of Frontenac Blanc began in US states such as Minnesota, Iowa, Illinois, **North Dakota** and **Montana** in around 2009–10, earlier in Québec. Thanks to its high acidity, it is best suited to dessert and fortified, port-style wines and may well start to displace Frontenac Gris because its wines are not pink-tinged.

FRÜHBURGUNDER

See PINOT NOIR PRÉCOCE *in the* PINOT *entry.*

FRÜHROTER VELTLINER

Tentacular in its relations but Grüner Veltliner is not one of them.

BERRY COLOUR:

PRINCIPAL SYNONYMS: Früher Roter Malvasier (Germany), Korai Piros Veltelini (Hungary), Malvasier (Austria), Malvoisie Rouge d'Italie (Savoie), Velteliner Rouge Précoce (France), Veltlínské Červené Rané (Czech Republic), Veltlínske Červené Skoré (Slovakia)

ORIGINS AND PARENTAGE

DNA parentage analysis has shown this Austrian variety to be a natural cross between ROTER VELTLINER and SILVANER, just like its sibling NEUBURGER (Sefc, Steinkellner *et al.* 1998). Frühroter Veltliner is consequently a half-sibling of ROTGIPFLER and ZIERFANDLER, as well as a grandchild of SAVAGNIN (see PINOT pedigree diagram). Surprisingly, Roter Veltliner and its progeny Frühroter Veltliner are not related to GRÜNER VELTLINER. In addition, Frühroter Veltliner was crossed with Grauer Portugieser (see BLAUER PORTUGIESER) to produce JUBILÄUMSREBE, and with MÜLLER-THURGAU to produce MORAVA and PANONIA.

VITICULTURAL CHARACTERISTICS

Early budding and early ripening, as its name suggests, susceptible to downy and powdery mildews and to botrytis bunch rot. Vigorous and best pruned long. Small bunches and berries.

WHERE IT'S GROWN AND WHAT ITS WINE TASTES LIKE

In **France**, Velteliner Rouge Précoce – also confusingly known as Malvoisie Rouge d'Italie – is an authorized variety in the Savoie and Haute-Savoie *départements* and listed in the Vin de Savoie regulations but is little planted, with only 5 ha (12 acres) remaining in 2008. In **Italy**, where it is known as Frühroter Veltliner or simply Veltliner, there are a few old vineyards in the far north. Valle Isarco Veltliner is an authorized variant of the Alto Adige appellation but here it has been largely replaced by GRÜNER VELTLINER (Veltliner Verde). In **Germany**, where this red-skinned white wine variety is known as Früher Roter Malvasier, there are extremely limited plantings (4 ha/10 acres) in Rheinhessen.

Plantings in **Austria** are gradually declining and currently stand at 605 ha (1,495 acres), the vast majority of which are in Niederösterreich (Lower Austria), particularly in the Weinviertel north of Wien (Vienna) but also west along the Donau (Danube) in Donauland and Kamptal, for example. The wines it produces are relatively neutral, for early consumption, and tend to have lower acidity than those made from the more widespread and valued GRÜNER VELTLINER.

It is also sparsely grown in Sopron in north-west **Hungary** and occasionally appears as Veltlinské Červené Rané in the **Czech Republic** (251 ha/620 acres) and as Veltlinske Cervene Skore in **Slovakia** (363 ha/897 acres).

FUBIANO

Barely significant white cross from northern Italy.

BERRY COLOUR: ● ● ● ● ●

PRINCIPAL SYNONYMS: Incrocio Dalmasso II/32

ORIGINS AND PARENTAGE

A FURMINT × TREBBIANO TOSCANO cross – hence the name – obtained by the Italian vine breeder Giovanni Dalmasso in 1936 at the Conegliano research centre in the Veneto, northern Italy, in order to improve quality and reduce the risk of millerandage.

VITICULTURAL CHARACTERISTICS

Mid to early ripening.

WHERE IT'S GROWN

There were just 3 ha (7 acres) in **Italy** in 2000 according to the agricultural census.

FUELLA NERA

Minor but aromatic variety at home on the French Riviera.

BERRY COLOUR: ● ● ● ● ●

PRINCIPAL SYNONYMS: Dame Noire, Folle de Nice, Folle Noire, Fuella, Fuola, Jurançon Rouge

VARIETIES COMMONLY MISTAKEN FOR FUELLA NERA: JURANÇON NOIR (called Folle Noire in the Lot-et-Garonne), NÉGRETTE

ORIGINS AND PARENTAGE

Fuella Nera is an old variety from the French Alpes-Maritimes *département* (mainly around Grasse, Antibes, La Gaude and Bellet), where it was first mentioned under its synonym Folle Noire in 1783–4 in Cannes: 'Folles, grain de moyenne grosseur, rond et noir. Pas bon à manger, faisant de mauvais vin, mais produisant beaucoup, la grappe est ronde et serrée' ('Folles, medium-sized berry, round and black. Not good to eat, making bad wine, but very productive, the bunch is round and compact'; Rézeau 1997). However, the name Folle Noire was also given to JURANÇON NOIR, NÉGRETTE and Négret Castrais (no longer cultivated), three varieties with which Fuella Nera is still often confused. According to Rézeau (1997), the first occurrence of the local spelling Fuella appeared in Odart in 1859 (see FOLLE BLANCHE for the etymology).

Fuella Nera is a member of the Folle ampelographic group (see see p XXVII; Bisson 2009).

OTHER HYPOTHESES

According to Reich (1902a), Fuella Nera could have been introduced from the Levant by sailors, those 'great propagators of the best grape varieties'.

VITICULTURAL CHARACTERISTICS

Early budding, mid ripening. Susceptible to powdery mildew and botrytis bunch rot. Large bunches but small berries.

WHERE IT'S GROWN AND WHAT ITS WINE TASTES LIKE

In 2008 there were 19 ha (47 acres) of Fuella Nera – more often called Folle Noire – in **France**'s Alpes-Maritimes, the only French *département* in which it is authorized. The tiny Bellet appellation around Nice stipulates a minimum of 60% Fuella Nera and BRAQUET NOIR combined, along with CINSAUT and Grenache (GARNACHA). The best-known producers in Bellet are Châteaux de Bellet and de Crémat but Clos Saint-Vincent make a rare varietal Fuella Nera called Vino di Gio. Wines are generally deeply coloured and aromatic, especially when blended with Braquet Noir, with moderate acidity but still with the ability to age for several years – so rather better than the eighteenth-century assessment above.

FUMIN

Deeply coloured, ancient and rustic dark-skinned variety exclusive to Valle d'Aosta.

BERRY COLOUR: ● ● ● ● ●

VARIETIES COMMONLY MISTAKEN FOR FUMIN: FREISA ✗

ORIGINS AND PARENTAGE

Fumin is one of the oldest varieties of the Valle d'Aosta, mentioned in 1785 under its dialect name Fumen (Vouillamoz and Moriondo 2011). The name Fumin could derive from the French word *fumée*, meaning 'smoke' (the Valle d'Aosta was a French-speaking area until the nineteenth century), in reference to the waxy layer covering the ripe berries.

DNA studies have established that Fumin is a parent of the obscure Valle d'Aosta variety VUILLERMIN, and a sibling of PETIT ROUGE, the most widespread grape in the Valle d'Aosta (Vouillamoz and Moriondo 2011).

OTHER HYPOTHESES

It has been suggested that Fumin might be identical to FREISA but this hypothesis, already rejected by some ampelographers (Gatta 1838; Dalmasso and Reggio 1963), was firmly disproved by DNA profiling (Vouillamoz).

VITICULTURAL CHARACTERISTICS

Late ripening, in the last week of October.

WHERE IT'S GROWN AND WHAT ITS WINE TASTES LIKE

Very little Fumin was cultivated in **Italy** during most of the twentieth century, but since the 1990s there have been considerable new plantings and some varietal wines have been produced, mainly in the centre of the Valle d'Aosta. The well-structured, cherry-scented wines benefit from the softening effect of oak ageing but they are also used to add colour and acidity to blends. Alcohol levels are sometimes a little high. There were 76 ha (188 acres) in Italy in 2000.

The Valle d'Aosta DOC (registered in 1971) allows for a varietal Fumin. Recommended producers include Di Barrò, Les Crêtes, L'Atoueyo (Fernanda Saraillon) and Grosjean Frères.

FURMINT

The principal variety of the fiery, high-acid wines of Tokaj in Hungary.

BERRY COLOUR:

PRINCIPAL SYNONYMS: Fehér Furmint (Tokaj), Lazafürtű Furmint (Tokaj), Moslavac or Moslavac Bijeli (Croatia), Mosler (Austria), Šipon (Slovenia, northern Croatia), Som (Transilvania), Szigeti, Tokay (France), Változó Furmint (Tokaj), Zapfner or Zopfner (Germany)

VARIETIES COMMONLY MISTAKEN FOR FURMINT: ALTESSE (Savoie in France), GRASĂ DE COTNARI, KÉKNYELŰ (Hungary), POŠIP BIJELI (island of Korčula in Croatia), SAUVIGNONASSE (northern Italy), ŽILAVKA (Bosnia and Herzegovina)

ORIGINS AND PARENTAGE

Furmint originates from the Tokaj region in north-eastern Hungary, the home of the famous sweet Aszú wines (made from grapes affected by the benevolent form of botrytis bunch rot), whose existence was first documented on 15 May 1571 in the Hétszőlő vineyard in the village of Tokaj (Zelenák 2002). This wine most probably contained Furmint grapes because this variety was described as 'die eigentliche Tokayer Trockenbeertraube' ('the genuine Tokaji Aszú grape') by János Dercsényi (1796) and also because the variety name was recorded as early as 1611 in the Gyepű Valley near Erdőbénye in the Zempléni Mountains, about 20 km north of the village of Tokaj (Dienes 2001): 'A Prédikátorsághoz vagyon három szőlő. Egyik Bakfű szőlő, másik Furmint szőlő a Giöpü völgyön, harmadik Hosszú szőlő' ('There are three vineyards for the preacher's salary. One is the Bakfű vineyard, the other is the Furmint vineyard in the Gyepű Valley, the third one is the Hosszú vineyard'). According to the historian Dénes Dienes of the Sárospatak Calvinist College, who kindly provided translations and historical data, the first vineyard was named Bakfű, meaning 'Purple Betony' (*Stachys officinalis*), because this herb was grown on the land; the second vineyard, in the Gyepű Valley near Erdőbénye, was named Furmint because this variety was planted there; and the third vineyard was named Hosszú, meaning 'long', referring to its size. (In the seventeenth century the word *szőlő* meant not only 'grape' but also 'vineyard'.)

However, and despite a few dubious hypotheses (see below), the etymology of Furmint is still unknown.

Several DNA parentage analyses (see PINOT pedigree diagram) have established that:

- Furmint has a parent–offspring relationship with GOUAIS BLANC (Boursiquot *et al.* 2004). Since Gouais Blanc was mentioned much earlier and was much more widespread throughout Europe in the Middle Ages, it is more likely that Furmint is its offspring, making Furmint a half-sibling of at least eighty other progenies of Gouais Blanc such as CHARDONNAY, GAMAY NOIR, ELBLING and RIESLING.
- Furmint has a parent–offspring relationship with HÁRSLEVELŰ (Calò, Costacurta *et al.* 2008; Vouillamoz), the other key variety in Tokaji wines. Since DNA profiling is not consistent with Furmint's being a putative cross between GOUAIS BLANC and Hárslevelű, Furmint is therefore a parent of Hárslevelű.
- The same applies to the obscure PLANTSCHER in Switzerland, which is a natural progeny of Furmint (Vouillamoz, Maigre and Meredith 2004) and is a likely sibling of HÁRSLEVELŰ.

Studying the old varieties from Tokaj, Varga *et al.* (2008) have found four distinct types of Furmint with the same DNA profile: Fehér ('white'), Lazafürtű ('loose-bunched'), Piros ('pink-berried') and Változó ('variable'). This high degree of genetic diversity speaks strongly in favour of the Tokaj region as the birthplace of Furmint.

Furmint was used to breed the wine-grape varieties BUSSANELLO, FUBIANO and VEGA in Italy and ZÉTA (formerly Oremus) in Hungary.

OTHER HYPOTHESES

Legend has it that Furmint was brought to Tokaj in the first half of the twelfth century by Italian missionaries who had been invited by King Stephen II (Fabbro *et al.* 2002), or around 1250 by Italian colonists from the city of Formia (Latin *Formianum*) in Lazio (Galet 2000). However, this is highly doubtful because Furmint has never been observed in Italy and has no genetic link with any other Italian varieties. The supposed etymology that derives Furmint from *fiore dei monti* ('flower of the hills') is equally improbable. Even more dubious is the story set during the Seven Years' War (1756–63) about a soldier from Collio in Friuli, nicknamed Forment (Italian *fromento* means 'wheat') because of his reddish-blond beard, who was made Count of Formentin by Empress Maria Teresa in Tokaj and introduced Furmint from his homeland (Fabbro *et al.* 2002). This nice story can be eliminated simply because Furmint was mentioned in Hungary much earlier than the Seven Years' War. Others have suggested a Balkan origin for Furmint (Syrmia, today's Serbia), and even a Near Eastern origin has been proposed (Galet 2000), based on the mistaken belief that Furmint is the same as ALTESSE (see Altesse for details).

VITICULTURAL CHARACTERISTICS

Early budding, late ripening. Loose bunches of medium-sized, thick-skinned berries. Susceptible to botrytis bunch rot and downy and powdery mildews and prone to frost damage, but good drought tolerance.

WHERE IT'S GROWN AND WHAT ITS WINE TASTES LIKE

Furmint is an intense variety producing world-class, fiery, full-bodied, high-acid, long-lived sweet and dry wines almost exclusively in Central Europe, especially Hungary. By 2010 there were just 9 ha (22 acres) of Furmint in **Austria**, mainly in Rust, near the Neusiedlersee in the far east of the country, where producers such as Heidi Schröck use it to make fine, sweet Ausbruch wines. Although it is still a rarity here, Furmint seems to appreciate the humidity, and plantings are likely to increase. Both Schröck and Triebaumer also make even rarer dry versions.

In **Hungary** it is the leading variety in Tokaj, with Somló in the far west of Hungary the only other area in which it is significant. Total plantings amounted to 4,006 ha (9,894 acres) in 2006, more than 97% in Tokaj. The combination of high acid, susceptibility to botrytis bunch rot and high sugar levels in the berries makes Furmint ideal raw material for intensely rich, long-lived Aszú dessert wines, often blended with the more aromatic HÁRSLEVELŰ and sometimes also a lesser amount of Sárga Muskotály (MUSCAT BLANC À PETITS GRAINS) and/or the more recent ZÉTA.

Furmint's powerful dry wines, perhaps better able to reflect their exact place of origin than the sweet marvels, have received increased attention and renewed recognition in recent years. In Tokaj the best sites for this style – around Mád, Tolcsva and Rátka – have long been recognized, even before the sweet wines gained such a strong foothold. Recommended producers of both sweet and dry wines include István Balassa, Zoltán Demeter, Disznókő, Királyudvar, the Royal Tokaji Wine Company and István Szepsy in Tokaj plus Hollóvár in Somló.

There is a good deal of clonal variation and the better clones, generally those with smaller bunches that produce more concentrated wines, tend to be those planted before the Second World War: Madárkás, Holyagos and Féher Furmint (Rohály *et al.* 2003).

Just over the border from Tokaj, in **Slovakia**'s Tokajská, similar if more muted styles of sweet wines are produced from the more scattered vineyards there.

In **Slovenia** there were 694 ha (1,714 acres) of Šipon in 2009, mainly in Štajerska Slovenija

(Slovenian Styria), in Podravje in the north east. Traditionally made as a dry wine, the variety can in good years yield sweet wines but not in the Tokaj style. Producers of fine dry wines include Dveri-Pax, Jeruzalem Ormož, Kupljen, Ljutomerčan, Pullus and Verus; Čurin-Prapotnik in the south east of the country and a few other smaller producers make sweet varietal wines.

Furmint is known as both Moslavac and Šipon in **Croatia**, where there were 422 ha (1,043 acres) in 2008 and the wines are almost always dry. It is generally known as Šipon in Medjimurje in the north, close to the Slovenian border. The Moslavac synonym derives from the Moslavina region, south east of Zagreb, where a handful of producers such as Miklavžić occasionally make a varietal wine, although it is more often found in blends and frequently overcropped. Miklavžić is also experimenting with a sparkling blend of Moslavac, CHARDONNAY and PINOT BLANC.

In **South Africa**, where Furmint was imported in tandem with the other Tokaji grape, HÁRSLEVELŰ, some fine wine has been made from low-yielding vines in the Swartland, notably Signal Hill's Eszencia 2002.

Such a high-quality variety seems ripe for experimentation in other regions with a suitably long growing season.

G

☿ this symbol on the following pages indicates that the synonyms, and mistaken identities, are confirmed by DNA evidence

GOUGET NOIR
GOUSTOLIDI
GOUVEIO REAL
GR 7
GRACIANO
GRAISSE
GRAMON
GRAND NOIR
GRAPARIOL
GRASĂ DE COTNARI
GRAŠEVINA
GRECHETTO DI ORVIETO
GRECO
GRECO BIANCO
GRECO NERO
GRECO NERO DI SIBARI
GRECO NERO DI VERBICARO
GRIGNOLINO
GRILLO
GRINGET
GRISA NERA
GRK
GROLLEAU NOIR
GROPPELLO DI MOCASINA
GROPPELLO DI REVÒ
GROPPELLO GENTILE
GROS MANSENG
GROS VERDOT
GROSSA
GRUAJA
GRÜNER VELTLINER
GUARDAVALLE
GUEUCHE NOIR
GUTENBORNER

GAGLIOPPO

The characterful red wine grape of Calabria.

BERRY COLOUR:

PRINCIPAL SYNONYMS: Arvino Nero, Cirotana (Calabria), Gaglioppo di Cirò (Calabria), Galloppo (Calabria), Morellino Pizzuto (Toscana), Navarna (Calabria)

VARIETIES COMMONLY MISTAKEN FOR GAGLIOPPO: FRAPPATO, GALLIOPPO DELLE MARCHE (Marche), MAGLIOCCO CANINO (Calabria), MAGLIOCCO DOLCE (Calabria), MAIOLICA (Abruzzo)

ORIGINS AND PARENTAGE

The earliest mention of Gaglioppo as a wine appeared in 1240, when on 21 January Emperor Frederic II ordered a hundred barrels of good 'galloppo' wine of Napoli ('centum barrilia de bono vino de galloppo usque Naepolim'; Huillard-Bréholles 1859). The name Gaglioppo is said to have Greek roots, *kalós podos* meaning 'good/beautiful foot' (the 'foot' referring to the grape cluster).

DNA profiling has shown that GALLIOPPO DELLE MARCHE is not identical to Gaglioppo (Schneider, Raimondi, Grando *et al.* 2008). Morellino Pizzuto was once thought to be a clone of SANGIOVESE in Toscana, just like Morellino di Scansano (Scalabrelli *et al.* 2001), but the same authors recently discovered by DNA profiling that Morellino Pizzuto is in fact identical to Gaglioppo (Scalabrelli *et al.* 2008). This confusion is explained by recent DNA studies suggesting a parent–offspring relationship between Gaglioppo and Sangiovese (Di Vecchi Staraz, This *et al.* 2007; Crespan, Calò *et al.* 2008), which was confirmed recently by the discovery of the parentage of Gaglioppo (Cipriani *et al.* 2010): it is a possible Sangiovese × MANTONICO BIANCO natural cross, the latter being a typical variety of the Locride in Calabria, southern Italy. This makes Gaglioppo a sibling of NERELLO MASCALESE from Sicilia. As a consequence, Gaglioppo is a grandchild of CILIEGIOLO and CALABRESE DI MONTENUOVO, the parents of Sangiovese; and Gaglioppo and FRAPPATO from Sicilia are likely to be siblings (Di Vecchi Staraz, This *et al.* 2007). A possible genetic relationship with ALEATICO has also been suggested (Filippetti, Silvestroni, Thomas and Intrieri 2001).

OTHER HYPOTHESES

It is often suggested that Gaglioppo was introduced by the Greeks between the eighth and third centuries BC, when Calabria was called Magna Grecia. However, comparison of DNA profiles shows that Gaglioppo has no genetic relationship with modern Greek varieties (Vouillamoz).

VITICULTURAL CHARACTERISTICS

Good resistance to both frost and drought. Yields well and consistently but susceptible to downy and powdery mildews. It is vigorous, prefers siliceous or clay soils, ripens mid to late and can reach high sugar levels.

WHERE IT'S GROWN AND WHAT ITS WINE TASTES LIKE

Gaglioppo is the predominant red variety in Calabria in southern **Italy**, where it comprised 95–100% of Cirò DOC (until the rules were changed in 2011 to allow the addition of other, generally 'international' varieties) and is authorized as a significant part of the blend in many other Calabrian reds such as Bivongi, Donnici and Lamezia. It is also planted in the Marche, Umbria, Abruzzo and Campania and on Sicilia. The Italian agricultural census of 2000 recorded a total of 3,703 ha (9,150 acres).

The wines are characterized by a ruby colour of only moderate intensity with quite pronounced dusty tannins and, often, haunting aromas of roses. Librandi is the most famous exponent of Cirò but other good examples have been made by Francesco e Laura de Franco and Ippolito. Cerraudo have made dramatic 100% Gaglioppo IGT Val di Neto.

GAIDOURIA

Very rare variety just surviving on the Greek island of Santoríni.

BERRY COLOUR: ● ● ● ● ●

PRINCIPAL SYNONYMS: Gaidouricha, Gaidouriha, Gaydura (Turkey), Guydourina

ORIGINS AND PARENTAGE

Gaidouria comes from the Kykládes (Cyclades) in Greece. Its name derives from *gaidouri*, the Greek word for 'donkey', as the animal is apparently partial to this vine's foliage (Galet 2000). Recent DNA profiling (under the name Guydouria) has suggested a possible parent–offspring relationship with ASSYRTIKO (Myles *et al.* 2011).

VITICULTURAL CHARACTERISTICS

Early budding and ripening. Vigorous but moderate to low yielding. Medium-sized, loose bunches of largish, thick-skinned berries. Relatively resistant to diseases.

WHERE IT'S GROWN AND WHAT ITS WINE TASTES LIKE

Gaidouria is rare even on the island of Santoríni, **Greece**, but Gavalas, a producer intent on protecting and promoting local indigenous varieties, blend about 15% Gaidouria with the equally rare KATSANO. Gavalas describe it as 'a stubborn variety'. Wines are aromatic, have lower acidity than ASSYRTIKO and may have a lightly tannic texture.

In the Urla region of western **Turkey**, near the coast west of İzmir, Can Ortabaş of Urla Sarapçılık has propagated this variety from a single vine and is now expanding the cultivation to higher vineyards (1,050 m/3,445 ft). It is difficult to graft and here the berries are very small.

GALATENA

Red variety grown by just one producer in north-east Sicilia.

BERRY COLOUR: ● ● ● ● ●

ORIGINS AND PARENTAGE

This variety does not appear in any ampelographic source and further investigation is needed to establish how Galatena may be related to another registered Italian variety.

VITICULTURAL CHARACTERISTICS

Generally still grown on pergolas.

WHERE IT'S GROWN AND WHAT ITS WINE TASTES LIKE

Galatena is a local variety in Sicilia, **Italy**, grown only by Salvatore Geraci, owner of Palari near Messina. Palari include it as a minor component in their red Faro DOC wine, based on NERELLO MASCALESE (though Galatena is not actually mentioned in the Faro regulations), and in Rosso del Soprano, their IGT Sicilia red blend.

GALBENĂ DE ODOBEȘTI

Gradually declining Romanian variety producing everyday whites.

BERRY COLOUR:

PRINCIPAL SYNONYMS: Bucium de Poamă Galbenă, Galbenă di Căpătanu, Galbenă Grasă, Galbenă Uriașă, Poamă Galbenă

VARIETIES COMMONLY MISTAKEN FOR GALBENĂ DE ODOBEȘTI: Narancsszőlő (Hungary), ORANGETRAUBE (Germany)

ORIGINS AND PARENTAGE

Galbenă de Odobești is an old variety that probably originates in Căpătanu, near Odobești in eastern Romania. At least three berry-colour variations have been described, suggesting that the variety is very old: Galbenă Măruntă (pale yellow), Galbenă Verde (yellow-green) and Galbenă Aurie (golden-yellow). Galbenă de Odobești is said to be morphologically similar to other obscure or almost extinct local varieties: Alb Românesc, Berbecel, Cruciulița, Bătută Neagră, Cabasma Neagră, Cabasma Albă and ZGHIHARĂ DE HUȘI (Dejeu 2004).

VITICULTURAL CHARACTERISTICS

Mid to late budding and late ripening. Compact bunches of thin-skinned berries. Very susceptible to botrytis bunch rot, downy and powdery mildews and drought. Winter-hardy to around –18 °C (–0.4 °F).

WHERE IT'S GROWN AND WHAT ITS WINE TASTES LIKE

During the Soviet era, Galbenă de Odobești was valued in **Romania** for its productivity rather than its quality and today is still used mainly for bulk wine and/or blended, although Vincon Vrancea bottle a varietal version. Wines are light, high in acidity and designed for early drinking. Total plantings in 2008 were 407 ha (1,006 acres), almost all in the southern half of Romanian Moldova in the east of the country, down from 500 ha (1,236 acres) in 1995.

GALEGO DOURADO

Rare Portuguese clings on.

BERRY COLOUR:

PRINCIPAL SYNONYMS: Dourado, False Pedro (South Africa), Gallego, Moscato Galego Dourado, Olho de Lebre (Setúbal), Pedro Luis (South Africa), Rutherglen Pedro (Australia)

VARIETIES COMMONLY MISTAKEN FOR GALEGO DOURADO: ALVARINHO (north-eastern Portugal and Galicia in Spain), CAÑOCAZO (Australia), LOUREIRO (Vinho Verde), PEDRO XIMÉNEZ (South Africa)

ORIGINS AND PARENTAGE

Galego Dourado, from Carcavelos and Colares near Lisboa in western Portugal, has since 2004 been part of a rescue programme at the national agronomy centre in Oeiras (Brazão *et al.* 2005).

Now also virtually extinct in South Africa, it was imported as PEDRO XIMÉNEZ, later called

Pedro Luis and finally named False Pedro, which is also the name used for CAÑOCAZO in Australia (Kerridge and Antcliff 2009).

VITICULTURAL CHARACTERISTICS

Fertile, vigorous, not very productive, early budding and ripening. Susceptible to coulure, downy and powdery mildews and especially to botrytis bunch rot. Best suited to clay-limestone soils.

WHERE IT'S GROWN AND WHAT ITS WINE TASTES LIKE

Galego Dourado continues to survive on **Portugal**'s Atlantic coast around Lisboa despite severe urban encroachment on its vineyards. It is one of the principal varieties in the sweet, fortified blended wines from the Carcavelos appellation, such as that produced by Quinta dos Pesos. Wines are generally rounded, full-bodied and aromatic. There were just 6 ha (15 acres) in Portugal in 2010 even though it is also authorized in Lisboa regional wines.

GALLIOPPO DELLE MARCHE

Recently rescued and replanted central Italian vine that may be related to Sangiovese.

BERRY COLOUR: ● ● ● ● ●

PRINCIPAL SYNONYMS: Balsamina Galloppa, Balsamina Grossa (Ancona), Gaglioppa, Galloppa (Macerata), Lancianese Nero, Moretta

VARIETIES COMMONLY MISTAKEN FOR GALLIOPPO DELLE MARCHE: GAGLIOPPO

ORIGINS AND PARENTAGE

DNA profiling has shown that Gallioppo delle Marche, local to the provinces of Ascoli Piceno and Macerata in the Marche, central Italy, is not identical to GAGLIOPPO from Calabria (Schneider, Raimondi, Grando *et al.* 2008), as had long been thought. DNA profiling at the University of Bologna suggests a possible parent–offspring relationship with SANGIOVESE.

VITICULTURAL CHARACTERISTICS

Early budding, mid ripening. Medium-sized to small, thin-skinned berries.

WHERE IT'S GROWN

On the brink of extinction in the 1990s, Gallioppo delle Marche was rescued by the Agenzia Servizi Settore Agroalimentare delle Marche, central **Italy**, a local-government agency supporting the food services industry. There have been limited replantings recently by Lino Silvestri in Cossignano, the Aurora co-op in Affida, Helmut Kindermann in Monterinaldo, Daniele Longanesi in Bagnacavallo and Dante Marucci in Pontano.

GALLIZZONE

Tuscan red on the verge of extinction.

BERRY COLOUR: ● ● ● ● ●

PRINCIPAL SYNONYMS: Gallazzone

ORIGINS AND PARENTAGE

Gallizzone is a rare variety that most likely originates from the region between Castelnuovo Magra and Arcola in La Spezia (Liguria, Italy), where it was mentioned in the nineteenth century (Torello Marinoni, Raimondi, Ruffa *et al.* 2009).

VITICULTURAL CHARACTERISTICS

Early ripening.

WHERE IT'S GROWN AND WHAT ITS WINE TASTES LIKE

Gallizzone is not (yet) included in the official register of varieties in **Italy** but it is being grown by at least one producer: Alessandro Cagnasso, owner of La Fontanina in Massa-Carrara, includes Gallizzone in his red blend of SCHIAVA GROSSA, MERLOT, CILIEGIOLO and BONAMICO. It is also said to be found in the provinces of Lucca and La Spezia.

GALOTTA

Recent Swiss cross useful for adding colour and tannins to blends.

BERRY COLOUR: ● ● ● ● ●

ORIGINS AND PARENTAGE

Galotta is an ANCELLOTTA × GAMAY NOIR cross named after its parents and obtained in 1981 at the Caudoz viticultural research centre in Pully, just outside Lausanne, Switzerland, today part of Agroscope Changins-Wädenswil (Vouillamoz 2009c).

VITICULTURAL CHARACTERISTICS

Medium to high vigour and fertility. Mid budding and ripening. Small to medium-sized compact bunches of small to medium-sized berries. Good resistance to botrytis bunch rot.

WHERE IT'S GROWN AND WHAT ITS WINE TASTES LIKE

Galotta wines are deeply coloured and full-bodied with firm tannins, which is why a small proportion has been used since the 2005 harvest to increase the structure and colour of blends in French **Switzerland**. Bolle in Vaud use a higher than average percentage to make a barrel-aged blend of 60% GAMARET with 40% Galotta and may make a varietal wine. Other producers include Edmond Giroud (some vintages only) in the Valais. There were 13 ha (32 acres) in Switzerland in 2009, mainly in the cantons of Vaud, Genève and Valais.

GAMARET

Increasingly popular Swiss cross valued for its early ripening and resistance to rot.

BERRY COLOUR: ● ● ● ● ●

PRINCIPAL SYNONYMS: Pully B-13

ORIGINS AND PARENTAGE

Gamaret is a GAMAY NOIR × REICHENSTEINER cross obtained in 1970 by André Jaquinet at the Caudoz viticultural research centre in Pully, just outside Lausanne, Switzerland, today part of Agroscope Changins-Wädenswil (Vouillamoz 2009c). It was initially labelled Pully B-13 before being selected and christened by its breeder. Gamaret is therefore a sibling of GARANOIR and MARA. It was officially released in 1990 together with Garanoir.

VITICULTURAL CHARACTERISTICS

Early budding and ripening. Smallish bunches of small berries. Extraordinarily resistant to botrytis bunch rot but susceptible to flavescence dorée and to diseases of the wood such as esca.

WHERE IT'S GROWN AND WHAT ITS WINE TASTES LIKE

Gamaret provides deep purple colour, spice, power and tannins to blends, eg with PINOT NOIR or GAMAY NOIR, but it is also often made as a barrel-aged varietal wine. It is planted mostly in the cantons of Vaud, Genève and Valais (total 380 ha/939 acres in **Switzerland** in 2009) and is increasing in popularity. Producers of varietal wines include Klosterhof in the canton of Luzern, Domaine Les Perrières in the canton of Genève, Domaine Rosset in Vaud and Saint-Gothard in the Valais.

Gamaret was added to the official French catalogue of varieties in 2008 and is authorized in Beaujolais, **France**; it is also found in **Italy**.

GAMAY NOIR

Greater Burgundian refresher making wines generally but not always for early consumption.

BERRY COLOUR: ● ● ● ● ●

PRINCIPAL SYNONYMS: Beaujolais, Bourguignon Noir, Gamai, Gamai Chatillon, Gamay Beaujolais, Gamay Charmont, Gamay d'Arcenant, Gamay d'Auvergne, Gamay de la Dôle, Gamay de Liverdun, Gamay de Saint-Romain, Gamay de Sainte-Foix, Gamay de Toul, Gamay de Vaux, Gamay d'Orléans, Gamay du Gâtinais, Gamay Labronde, Gamay Noir à Jus Blanc, Gamay Ovoïde, Gamay Précoce, Gamé or Gammé, Grosse Dôle (Vaud and Valais in Switzerland), Liverdun Grand, Lyonnais (Allier), Petit Gamay, Plant Robert⌘ (Vaud)

VARIETIES COMMONLY MISTAKEN FOR GAMAY NOIR: ABOURIOU⌘ (called Gamay du Rhône or Gamay Saint-Laurent or simply Beaujolais), BLAUFRÄNKISCH⌘ (called Gamé in Bulgaria and Borgonja in Croatia), GOUGET NOIR⌘, GROLLEAU NOIR⌘ (called Gamay de Châtillon in Anjou), PINOT NOIR⌘ (called Gamay Beaujolais in California and New Zealand), VALDIGUIÉ⌘ (called Napa Gamay or Gamay 15 in California)

ORIGINS AND PARENTAGE

Gamay Noir is a very old Burgundian variety whose name first appeared in an official ban promulgated by Duc Philippe le Hardi in Dijon on 31 July 1395 (Rossignol 1854; Vermorel 1902): 'un très-mauvaiz et très-desloyaulx plant nommėz *Gaamez*, duquel mauvaiz plant vient très-grant habondance de vins ... Et lequel vin de Gaamez est de tel nature qu'il est moult nuysible a creature humaine, mesmement que plusieurs, qui au temps passé en ont usé, en ont esté infestés de griez maladies, si comme entendu avons; car le dit vin qui est yssuz du dit plant, de sa dite nature, est plein de très-grant et horrible amertume ... Pour quoi nous ... vous mandons ... sollempnellement à touz cilz qui ont les diz plans de vigne des diz Gaamez, que yceulx coppent ou fassent copper en quelque part qu'ilz soient en nostre dit païs dedens cing mois' ('a very bad and disloyal variety called Gaamez, from which come abundant quantities of wine ... And this wine of Gaamez is of such a kind that it is very harmful to human creatures, so much so that many people who had it in the past were infested by serious diseases, as we've heard; because said wine from said plant of said nature is full of significant and horrible bitterness. For this reason we solemnly command you ... all who have said vines of said Gaamez to cut them down or have them cut down, wherever they may be in our country, within five months').

After that, numerous subsequent orders were issued with the aim of banning plantings of Gamay Noir (in 1567, 1725 and in 1731), underlining the poor reputation that this variety has today only partially overcome. Numerous spellings were used (Gaamez in 1395, Gamey in 1444, Gamé in 1496 and 1650, Gamay in 1564, Gamez in 1567, Gamet in 1667, etc.) until 1896, when the spelling Gamay was adopted by the ampelographic congress in Chalon-sur-Saône (Rézeau 1997). Although this variety is usually said to have been named after the village of Gamay near Saint-Aubin in the Côte d'Or, where vineyards still exist, it would be interesting to see whether the name of the village followed the same spelling evolution as that of the grape variety: Gaamez to Gamey to Gamé to Gamay.

Gamay Noir is a member of the Noirien ampelographic group (see p XXVII), and is ampelographically close to SACY (Levadoux 1956; Bisson 2009). These ampelographic observations were confirmed in 1999 by researchers at the University of California at Davis and at INRA (Institut National de la Recherche Agronomique) in Montpellier, southern France, who found by DNA parentage analysis that Gamay Noir is a natural progeny of PINOT and GOUAIS BLANC (Bowers *et al.* 1999), just like at least twenty other sibling varieties (see the pedigree diagram in the PINOT entry). Gamay Noir was therefore born before the fourteenth century, somewhere in Burgundy, where both its parents were already cultivated in the Middle Ages.

Gamay Teinturier Fréaux, sometimes known as Gamay Fréaux, Teinturier de Couchey or Teinturier Fréaux, is said to be a mutation of Gamay Teinturier de Bouze found in 1841 in Couchey in the Côte d'Or by Antoine Fréaux, who then brought it back home to Saint-Denis-de-Vaux, west of Chalon-sur-Saône, and propagated it in the area (Galet 1990). However, bunches of the white-juiced 'regular' Gamay Noir, or Gamay Noir à Jus Blanc, have often been observed on Gamay Teinturier Fréaux vines, which indicates that it is a mutation and not an offspring of Gamay Noir (Galet 1990).

Gamay Blanc Gloriod, no longer cutlivated today, is not a colour mutation of Gamay Noir but its brother, since it is also a natural progeny of PINOT and GOUAIS BLANC (Bowers *et al.* 1999).

Gamay Noir was used to breed FLORENTAL, GALOTTA, GAMARET, GARANOIR, MARA and REGNER.

OTHER HYPOTHESES

Some authors have suggested that this variety was named after the village of Gamay near Saint-Sernin-du-Bois (Saône-et-Loire), but this is unlikely because Gamay has never been planted in this area. It has also been suggested that Gamay or its direct ancestors were introduced from Dalmacija (Dalmatia) by the legions of Probus (Vermorel 1902) but this has now been refuted by DNA parentage analysis.

VITICULTURAL CHARACTERISTICS

Early budding and ripening. Although it is not vigorous, it is fertile and yields must be restricted, particularly in fertile soils in a hot climate. Best pruned short. Susceptible to sunburn, botrytis bunch rot, phomopsis, diseases of the wood, grape moths and grapevine yellows.

WHERE IT'S GROWN AND WHAT ITS WINE TASTES LIKE

France had 30,443 ha (75,226 acres) of Gamay in 2009, making it the seventh most planted red variety and it is recommended or authorized in almost every wine-growing region with the exception of the Bordeaux area, the island of Corse and Alsace. About two thirds of total plantings are in the greater Vallée du Rhône and particularly in Beaujolais, where it is *the* red variety. In 2009, of the 18,682 ha (46,164 acres) of red wine varieties in the Rhône *département* (not to be confused with the Vallée du Rhône), only 312 ha (771 acres) were not Gamay (they were in fact SYRAH). A few other varieties are permitted in Beaujolais' red and rosé wines but only in small amounts: GAMAY TEINTURIER DE BOUZE and Gamay Teinturier de Chaudenay (no more than 10% separately or together), ALIGOTÉ, CHARDONNAY, MELON, PINOT GRIS and PINOT NOIR (only if the vines are interplanted with Gamay in the vineyard and up to a maximum of 15%). In practice, most Beaujolais is 100% Gamay, or Gamay Noir à Jus Blanc as it is sometimes called to distinguish it from the quite different, red-fleshed varieties Gamay Teinturier de Bouze and Gamay Teinturier de Chaudenay (the latter a mutation of the former).

Until the early years of the twenty-first century, recommended Gamay producers would have been entirely made up of those producers with a long tradition in Beaujolais such as Jean-Marc Burgaud, Lapierre, Lapalu, Château de Thivin, Vissoux and Jean-Paul Brun. But there is now a perceptible influx of serious, hopeful newcomers, several of them from further north, in Burgundy proper, following in the footsteps of Louis Jadot, who took over the Château des Jacques.

In the Côte d'Or *département*, Gamay is generally considered inferior to PINOT NOIR, as the vineyard area and grape prices show. In the half-century since 1958, the area planted to Pinot Noir has doubled to 6,579 ha (16,257 acres) whereas that for Gamay is little more than one fifth of what it was (193 ha/477 acres). In greater Burgundy, hectares planted to Pinot Noir outnumber those of Gamay by about five to one. The long-standing confusion between Beaujolais and Bourgogne appellations – used by some producers, it was suggested, to give the impression their wines were Pinot Noir rather than Gamay – is being resolved by changes to labelling regulations. Beaujolais wines may state the main variety on the label and the Côteaux Bourguignons appellation may be applied to any red Beaujolais.

In the past, Gamay was generally light and fruity, and those rushed through the cellar by vinification techniques such as carbonic maceration in order to reach the market in November after the harvest as Beaujolais Nouveau were often thin and tasted like bananas and bubble-gum. However, with the decline in fashion and market demand for this style and the increasing number of well-made and ambitious wines vinified more traditionally, sometimes with oak ageing, Gamay is showing more purely its fine, refreshing, sometimes peppery, red fruit – and surprising longevity in the case of some wines from the ten crus of Beaujolais, in approximately ascending order of substance: Chiroubles, Saint-Amour, Fleurie, Régnié, Brouilly, Côte de Brouilly, Juliénas, Chénas, Morgon and Moulin-à-Vent.

The vine's precocious behaviour means that it is at risk of spring frosts in the cooler vineyards of the Loire but it is easier to ripen than, say, CABERNET SAUVIGNON. It is planted across the region but especially in the Loir-et-Cher *département* upstream of Tours, where there were 1,707 ha (4,218 acres) in 2009. Although the Gamay de Touraine appellation is the variety's most self-evident expression here, providing a light and generally cheaper alternative to Beaujolais, the variety is widely planted further west, alongside SAUVIGNON BLANC, to produce the likes of Cheverny and Coteaux de Vendômois. Further upstream and around Lyon, the variety contributes to a number of less well-known appellations such as Châteaumeillant, Coteaux du Lyonnais, Coteaux du Giennois, Côtes d'Auvergne, Côtes du Forez, Côtes Roannaises and Saint-Pourçain. On the whole, Gamay has been planted for quantity rather than quality but Henry Marionnet at Domaine de la Charmoise, Domaine Jacky Marteau and Domaine Robert Serol et Fils are exceptions. To the east, Gamay is the most widely planted red variety in Savoie, with the best examples coming from the cru of Chautagne.

It is also known in **Germany** and in **England**, where Biddenden have made a competent one.

Gamay is the second most planted red in **Switzerland**, after PINOT NOIR, with which it is often blended – most famously in the wine known as Dôle in the Valais. Swiss plantings in 2009 totalled 1,514 ha (3,741 acres) and were mainly in the cantons of Vaud, Valais and Genève. Good examples are made by Gérald Besse (Martigny, Valais), Frédéric Dumoulin at L'Orpailleur (Uvrier, Valais), Domaine Les Hutins (Dardagny, Genève), especially their La Briva Vieille Vigne, and Domaine Le Grand Clos (Satigny, Genève). In Vaud, producers such as Henri & Vincent Chollet and Domaine du Daley are particularly proud to label their wine Plant Robert. The best producers of Dôle include Marie-Thérèse Chappaz (Fully), Denis Mercier (Sierre) and the Cave du Rhodan (Salgesch). The variety is also grown in the Valle d'Aosta in the far north west of **Italy**, where there were 159 ha (393 acres) in 2000.

Gamay has been confused with BLAUFRÄNKISCH in Eastern Europe, not helped by the use of the synonym Gamé for Blaufränkisch in Bulgaria. There are limited plantings of the real Gamay in **Serbia**, **Kosovo** and the **Republic of Macedonia**. **Turkey** had a total of 236 ha (583 acres) in 2010 and it is known in both **Lebanon** and **Israel**.

In **California** in the 1980s, Charles F Shaw (who was to give his name to the trail-blazing recession-beater Two Buck Chuck) tried to launch a 'Gamay' but it turned out that neither the variety known there as Napa Gamay nor that called Gamay Beaujolais was in fact Gamay Noir. Today, Edmunds St John, V Sattui, J Rochioli but few others persist with Gamay.

Oregon is a more successful exponent of true Gamay, notably with examples from Brick House and Evening Land. It also pops up in bottles from **Michigan**, **Texas** and **New York** State.

Gamay is grown and produced in **Canada** as a varietal from a wide array of producers in both British Columbia (which had 144 acres/58 ha in 2009) and Ontario (185 acres/75 ha in 2006). It is not a South American phenomenon, however.

In **Australia** there are about twenty producers in various parts of the country, with Sorrenberg of Beechworth, Victoria, having the best track record, although William Downie has had recent success in Yarra Valley. Plantings are limited. Te Mata is one of the few **New Zealand** producers to take this variety (12 ha/30 acres in 2008) seriously.

South Africa had 19 ha (47 acres) in 2009 with Klein Zalze taking it more seriously than most.

GAMAY TEINTURIER DE BOUZE

Red-fleshed variety possibly related to the more common Gamay Noir but much less widely planted.

BERRY COLOUR: ● ● ● ● ●

PRINCIPAL SYNONYMS FOR GAMAY TEINTURIER DE BOUZE: Gamay de Bouze, Moureau (Rhône), Mourot (Saône-et-Loire), Plant de Bouze (Côte d'Or), Rouge de Bouze (Côte d'Or)

FOR GAMAY TEINTURIER DE CHAUDENAY: Gamay de Chaudenay, Gamay Six Pièces, Gros Mourot, Plant Rouge de Chaudenay, Teinturier de Chaudenay

ORIGINS AND PARENTAGE

Gamay Teinturier de Bouze was the first red-juiced (*teinturier* means 'dyer') Gamay to be observed, in the village of Bouze, near Beaune, in the nineteenth century. Its origin is disputed: some suggest it is a mutation of GAMAY NOIR, while others believe it could be a natural offspring of Gamay Noir and TEINTURIER because the latter was often planted with the former to enhance the colour of the wines. DNA parentage analysis would resolve the debate. It might better be described as a 'semi-teinturier' because the juice is lighter than that of Gamay Teinturier Fréaux.

Gamay Teinturier de Bouze is a member of the Noirien ampelographic group (see p XXVII; Bisson 2009).

Gamay Teinturier de Chaudenay is a red-fleshed mutation of Gamay Teinturier de Bouze observed in 1832 in Chaudenay in the Saône-et-Loire *département* by Jean-Marie Bidault, who had introduced some vines of red-juiced Gamay from Puligny in the Côte d'Or. From 1850, Bidault commercialized this mutation across the Saône-et-Loire under the name Gros Mourot. This variety was so productive that it was called Gamay Six Pièces, *pièce* being a measure corresponding to 300 hl/ha (Galet 1990).

VITICULTURAL CHARACTERISTICS

See GAMAY NOIR. The main difference is that the juice is coloured.

WHERE IT'S GROWN AND WHAT ITS WINE TASTES LIKE

Official statistics for **France** for 2008 list the mutations separately: Gamay Teinturier de Bouze 232 ha (573 acres), Gamay Teinturier de Chaudenay 183 ha (452 acres) and Gamay Teinturier Fréaux (see GAMAY NOIR) 55 ha (136 acres). They also list 'Gamay Teinturiers', whatever that may be: 81 ha (200 acres).

Henry Marionnet at Domaine de la Charmoise makes not only a very good GAMAY NOIR but also one of the rare varietal examples of Gamay Teinturier de Bouze, under the brand name Les Cépages Oubliés, or the Forgotten Varieties. He notes that the wine has predominantly red-fruit flavours and tends to be more tannic than Gamay Noir à Jus Blanc, which is why he vinifies it using carbonic maceration.

The **Republic of Moldova** had a small amount of both Gamay Teinturier Fréaux and Gamay Teinturier de Bouze in 2009.

GAMAY TEINTURIER DE CHAUDENAY

See GAMAY TEINTURIER DE BOUZE.

GAMAY TEINTURIER FRÉAUX

See GAMAY NOIR.

GAMBA DI PERNICE

Rare, aromatic and spicy Piemontese.

BERRY COLOUR: ● ● ● ● ●

PRINCIPAL SYNONYMS: Gamba Rossa, Neretto degli Alteni🜲 (Cuneo), Pernice

ORIGINS AND PARENTAGE

From Astigiano in the province of Asti, Gamba di Pernice is an old and rare Piemontese variety, mentioned by Nuvolone Pergamo, Conte di Scandaluzza (1787–98). Its name, meaning 'partridge's leg', points to the intense red colour of the stalks before veraison. DNA profiling has established that Gamba di Pernice is cultivated under the name Neretto degli Alteni around Saluzzo in the province of Cuneo in Piemonte (Raimondi *et al.* 2006).

VITICULTURAL CHARACTERISTICS

Grown for both wine and table. Late ripening. Yields well and consistently.

WHERE IT'S GROWN AND WHAT ITS WINE TASTES LIKE

Gamba di Pernice is cultivated mainly in Calosso, Costigliole and Asti Canelli in the province of Asti in north-west **Italy**. A group of six producers – Tenuta dei Fiori, Maurizio Domanda, La Canova, Fea, Carlo Benotto and Villa Giada – are working together to promote this ancient variety and obtain recognition for it as a DOC. However, it is argued that Gamba di Pernice cannot be included in the national register of varieties because there is a risk of confusion with Occhio di Pernice, a protected name for an expensive Vin Santo in Toscana. Plantings are so limited that the variety did not feature in the Italian agricultural census of 2000. The Cantina Collina d'Oro, also make varietal Gamba di Pernice. Wines are usually spicy, peppery and fruity, with hints of tar as well as notes of green pepper. They generally benefit from a year or more in bottle.

GAMZA

See KADARKA.

GÄNSFÜSSER

Barely surviving ancient variety, probably German.

BERRY COLOUR: ● ● ● ● ●

PRINCIPAL SYNONYMS: Argan (Jura in France), Argant (Franche-Comté in France), Bockshorn (Rheinland-Pfalz in Germany), Buchser (Rheinland-Pfalz), Espagnol (Jura), Gros Margillien (Arbois in France), Margillien (Arbois)

VARIETIES COMMONLY MISTAKEN FOR GÄNSFÜSSER: BARBAROUX (Provence)

ORIGINS AND PARENTAGE

Gänsfüsser, meaning 'goose-feet', referring to the shape of the leaves, is an old variety from Rheinland-Pfalz in Germany. Its wine was mentioned as early as 1505, when citizens of Deidesheim offered *Gänsefüsserwein* to their bishop (Schumann 1983). The variety was also mentioned by the famous German botanist Hieronymus Bock (1546) in his *Kreutterbuch*: 'Genssfüssel wachsen umb die Newenstatt' ('Gänsfüsser grows around Neustadt [an der Weinstrasse, in Württemberg]'). In 1584, Pfalzgrafen (Count Palatine) Johann Kasimir ordered that no Gänsfüsser vineyard be grubbed up if no new one was to be planted. Numerous subsequent mentions in Rheinland-Pfalz and in Württemberg indicate that Gänsfüsser was widely cultivated in south-western Germany in the Middle Ages (Schumann 1983).

The French ampelographers Adrien Berget and Paul Pacottet established that Gänsfüsser and Argant, considered an old variety of the Franche-Comté in north-east France, were one and the same variety (Pacottet 1904), which was recently confirmed by DNA profiling (Vouillamoz, Frei and Arnold 2008). Pacottet rejected the hypothesis of a Franche-Comté origin for Argant, arguing that neither Argant nor any of its synonyms had been mentioned in the Franche-Comté before the eighteenth century. Indeed, the earliest mention of this variety in France is likely to have been in 1774 under the name Margillien in a list of the seven best varieties for winemaking in Arbois (Franche-Comté). However, the name Margillien has been used for several distinct varieties in the Jura, and the name Argant appeared much later.

DNA parentage analysis has revealed that CÉSAR, a variety cultivated in the Yonne (west of the Franche-Comté), is a natural progeny of Gänsfüsser × PINOT (Bowers *et al.* 2000; see Pinot pedigree diagram), which could support a north-eastern France origin. However, César is also known in Rheinland-Pfalz under the name Römer, and Berget found that Argant (Gänsfüsser) was morphologically distinct from any French variety and considered it to be of German origin.

DNA profiling has disproved the synonymy between the Provençal variety BARBAROUX, sometimes called Gänsfüssler, and Gänsfüsser (Vouillamoz).

More recent DNA profiling (under the name Argant) has suggested a possible parent–offspring relationship with LEÁNYKA (although the reference sample might be FETEASCĂ ALBĂ; Myles *et al.* 2011).

OTHER HYPOTHESES

Since Argant is sometimes called Espagnol in the Jura, local tradition has it that the vine was introduced from Spain to north-eastern France by the Romans. Adrien Berget noticed a morphological similarity between Gänsfüsser and Zimmettraube, an old variety from the Steiermark (Styria) in southern Austria that is no longer cultivated. Several nineteenth-century authors mention the cultivation of Gänsfüsser in the Südtirol and in the Steiermark but its presence has not been confirmed and it is likely that it was mistaken for Zimmettraube.

VITICULTURAL CHARACTERISTICS

Vigorous and productive, late ripening. Big bunches of small, thick-skinned berries.

WHERE IT'S GROWN AND WHAT ITS WINE TASTES LIKE

Étienne Thiebaud, who established Domaine des Cavarodes in the Jura, eastern **France**, at the age of twenty, has rescued an old vineyard in Liesle, south west of Besançon in the Doubs region, and produces a red Vin de Pays de Franche-Comté which is a blend of PINOT NOIR, TROUSSEAU, PINOT MEUNIER, GAMAY NOIR, POULSARD, GUEUCHE NOIR, Gänsfüsser (ie Argant), BLAUER PORTUGIESER, ENFARINÉ NOIR and MÉZY, all from interplanted vines aged 50–100 years. Domaine Ganevat in Rotalier have replanted a small area and include it in an early-drinking red blend.

Plantings of Gänsfüsser, like those of many other late-ripening varieties in south-western **Germany** (AFFENTHALER, GOUAIS BLANC, ORLEANS GELB, ELBLING, etc.), have declined since the Middle Ages, probably because the climate became cooler, and it had almost disappeared from Rheinland-Pfalz as well as from the Franche-Comté across the French border in the twentieth century. In the village of Hassloch in Rheinland-Pfalz, Gänsfüsser vines several centuries old are still growing on the walls of houses in the Gillergasse. A Hasslocher 'Edition Gillergasse' special cuvée was made from these vines in 2007 for the first time and sold in aid of the local museum. The same year, 1,500 vines were planted in the Leisböhl vineyard by Fritz Braun and an equal number by the DLR viticulture school in Mussbach. A varietal wine is already produced by the Staatsweingut mit Johannitergut in Neustadt-Mussbach in the Pfalz.

GANSON

Recent, dark-skinned French cross that has not been widely planted.

BERRY COLOUR: ● ● ● ● ●

ORIGINS AND PARENTAGE

A GARNACHA × JURANÇON NOIR cross obtained in 1958 by Paul Truel at INRA (Institut National de la Recherche Agronomique) in Montpellier, southern France. Its name is a partial contraction of Grenache and Jurançon, with *ç* becoming *s* (Rézeau 1997).

VITICULTURAL CHARACTERISTICS

Big berries and bunches. Mid ripening. Fertile enough to be pruned short and pruned as a bushvine. Not unduly susceptible to bortrytis bunch rot.

WHERE IT'S GROWN AND WHAT ITS WINE TASTES LIKE

Although the variety is recommended in many parts of **France**, there were fewer than 5 ha (12 acres) planted in 2008. According to the official French register of varieties, wines are generally light in colour without particularly strong characteristics.

GARA ALDARA

Also called Qara Aldara, Sev Aldara or Aldara Kara, this is a mid- to late-ripening black-berried variety from **Azerbaijan** used to make a port-like wine called Arpachay.

GARA IKENI

Azeri grape used for both dry and sweet reds.

BERRY COLOUR: ● ● ● ● ●

PRINCIPAL SYNONYMS: Gara Lkeni, Qara Lkeni, Lkeny Tchernyi

ORIGINS AND PARENTAGE

Gara Ikeni is indigenous to Azerbaijan.

VITICULTURAL CHARACTERISTICS

Mid to late ripening.

WHERE IT'S GROWN AND WHAT ITS WINE TASTES LIKE

Gara Ikeni is used to make dry, dessert and fortified wines and also for juice. Producers of the first, which typically have just 9.5–10.5% alcohol, include Ganja Sharab 2 in the Ganja region of north-western **Azerbaijan**.

GARA SARMA

Also known as Qara Sarma, this black-berried, late-ripening variety from **Azerbaijan** is used to make both dry and dessert wines.

GARANDMAK

Old, popular and multi-purpose Armenian variety.

BERRY COLOUR: ● ● ● ● ●

PRINCIPAL SYNONYMS: Alani Chagog, Alivoruk, Dik Chardji, Garan Dmak

ORIGINS AND PARENTAGE

Garandmak is an old variety from the province of Armavir in western central Armenia that is said still to be propagated by seeds (Chkhartishvili and Betsiashvili 2004), which would result in vineyards composed of Garandmak and many of its progenies. Garan Dmak means 'fatty tail'.

VITICULTURAL CHARACTERISTICS

Productive, early budding, mid to late ripening. Compact bunches. Susceptible to powdery mildew but resistant to downy. Winter-hardy to –15 °C (5 °F).

WHERE IT'S GROWN AND WHAT ITS WINE TASTES LIKE

One of **Armenia**'s most popular varieties, Garandmak is used in blended table wines and dessert wines as well as for brandy and grape juice. According to Galet (2000), it is also found in **Ukraine**.

GARANOIR

Swiss cross producing fruity but low-acid wines.

BERRY COLOUR: ● ● ● ● ●

PRINCIPAL SYNONYMS: Pully B-28

ORIGINS AND PARENTAGE

Garanoir is a GAMAY NOIR × REICHENSTEINER cross obtained in 1970 by André Jaquinet at the Caudoz viticultural research centre in Pully, just outside Lausanne, Switzerland, today part of Agroscope Changins-Wädenswil (Vouillamoz 2009c). It was initially labelled Pully B-28 before being selected and christened Gastar, then Granoir and finally Garanoir by its breeder. Garanoir is therefore a sibling of GAMARET and MARA. It was officially released in 1990 together with Gamaret.

VITICULTURAL CHARACTERISTICS

Low vigour, good fertility. Early ripening, good resistance to botrytis bunch rot. Best suited to cooler sites.

WHERE IT'S GROWN AND WHAT ITS WINE TASTES LIKE

Garanoir is fruitier and less concentrated than its sibling GAMARET and has not been as widely planted: 203 ha (502 acres) in **Switzerland** in 2009, mostly in the cantons of Vaud, Genève and Valais. Acidity can be low so it is generally blended with varieties such as GAMAY NOIR or PINOT NOIR or Gamaret, and contributes colour, fruit and spice. It is sometimes made as a varietal wine, eg by Cidis in Vaud, Cave du Paradou in the Valais and Domaine de la Vigne Blanche in the canton of Genève.

It is also found to a very limited extent in **Germany**, where Weingut Kuhnle in the Schwaben (Swabia) region, just west of Stuttgart, make a popular varietal wine.

GARGANEGA

A delicate white closely associated with Soave whose vigour and yields need strict control.

BERRY COLOUR:

PRINCIPAL SYNONYMS: Grecanico Dorato (Sicilia), Malvasía de Manresa (Catalunya in Spain)

VARIETIES COMMONLY MISTAKEN FOR GARGANEGA: DORONA DI VENEZIA (Veneto), RIBOLLA GIALLA, VITOVSKA (Italy and Slovenia)

ORIGINS AND PARENTAGE

Garganega is an old variety from the Veneto in north-east Italy, first mentioned in the thirteenth-century treatise of the Italian agronomist Pietro de Crescenzi as a grape cultivated in the Bologna and Padova regions.

DNA profiling provided surprising evidence that Garganega is identical to Grecanico Dorato from Sicilia and, even more surprising, to Malvasía de Manresa from Catalunya in Spain (Di Vecchi Staraz, This *et al.* 2007; Crespan, Calò *et al.* 2008), which is no longer cultivated. Comparison of DNA profiles also suggested some genetic similarities between Garganega and several other old varieties from the province of Verona: CORVINA VERONESE, OSELETA, RONDINELLA and DINDARELLA (Vantini *et al.* 2003), thereby confirming that Garganega has its roots in this area. Garganega is also a parent of SUSUMANIELLO through a natural crossing with Uva Sogra (Di Vecchi Staraz, This *et al.* 2007), a variety that is no longer cultivated but has the same DNA profile as Uva Sacra, a local table grape of Puglia (Zulini *et al.* 2002).

In addition, DNA parentage analysis has shown that parent–offspring relationships exist between Garganega and at least eight other varieties: ALBANA, CATARRATTO BIANCO, DORONA DI VENEZIA, MALVASIA BIANCA DI CANDIA, MARZEMINA BIANCA, MONTONICO BIANCO, MOSTOSA and TREBBIANO TOSCANO (Di Vecchi Staraz, This *et al.* 2007; Crespan, Calò *et al.* 2008). Theoretically, any of the varieties shown in the diagram overleaf as progenies of Garganega could also be a parent of Garganega, except Dorona di Venezia and SUSUMANIELLO, for which the full parentage has been established. Although those possible relationships have yet to be confirmed, they illustrate the importance of Garganega throughout the length of Italy.

Garganega was used to breed INCROCIO BIANCO FEDIT 51.

OTHER HYPOTHESES

It has been argued that this variety has a polyclonal origin (Scienza *et al.* 2008), meaning that the vines of this variety derive from more than one original seed. However, see pp XIV–XVI for the definition of variety used in this book.

Like many other old Italian grapes, Garganega is supposed by some to be of Greek origin, which could explain its name, Grecanico Dorato, on Sicilia. However, comparison of DNA profiles did not show any relationship with modern Greek varieties (Vouillamoz). It is more likely that grapes used to make sweet wines were called Greco Something in reference to the famous sweet Greek wines, not because they have any relationship with Greek varieties.

VITICULTURAL CHARACTERISTICS

Vigorous, very productive (so needs careful management) and late ripening.

WHERE IT'S GROWN AND WHAT ITS WINE TASTES LIKE

Garganega, from the Veneto region in **Italy**, is most well known as the mainstay (minimum 70%) of Soave in all its forms, including Classico and Recioto. It is also the predominant variety in Gambellara DOC and is an authorized ingredient in several others such as Bianco di Custoza, Colli Berici and Colli Euganei. Its most common blending partner is Trebbiano di Soave (VERDICCHIO BIANCO) but many producers include a small percentage of CHARDONNAY to fill it out. Varietal versions are allowed in the Veneto DOCs of Arcole, Colli Berici and Garda and it is authorized in the blend of a few DOCs further afield, in Lombardia and Trentino-Alto Adige. Further south, in Umbria, it can reach higher ripeness levels than in its more familiar territory in the north.

The best examples of Soave, generally from the Classico zone when yields have been restricted to allow the grapes to ripen fully, are characterized by lemon and almond flavours and a fine grainy texture like that of fresh pears. This delicacy combines with fresh acidity to give wines that may be steely and yet still tantalizingly spicy. The best exponents of the variety include Anselmi (who choose the greater freedom of working outside the Soave DOC regulations), Ca' Rugate, Inama, Cantina di Monteforte and Pieropan.

Under the name Grecanico Dorato, or sometimes just Grecanico, this variety has been cultivated on Sicilia for several centuries, mainly in the north west and west, around Trapani, Palermo and Agrigento. It features in the Alcamo, Contea di Sclafani, Contessa Entellina, Delia Nivolelli, Erice, Menfi, Monreale, Santa Margherita di Belice and Sciacca DOCs. There were 5,358 ha (13,240 acres) on Sicilia in 2008. COS make an excellent blend of 50% Grecanico, 50% INZOLIA, which is rich and nutty and full of flavour.

A total of 11,637 ha (28,756 acres) was recorded in the Italian agricultural census in 2000.

Further afield in **Australia**, Politini Wines in King Valley, Victoria, harvested their first Grecanico crop in 2009 and are experimenting with both still and sparkling versions, although Domain Day at Mount Crawford in the Barossa were the first to plant this variety outside Italy (first vintage 2004).

GARGANEGA PEDIGREE DIAGRAM

The old Italian variety GARGANEGA has natural progenies throughout the boot-shaped peninsula. The presence of unknown (and probably extinct) varieties (?) means that the inverse relationships are also theoretically possible (see p x).

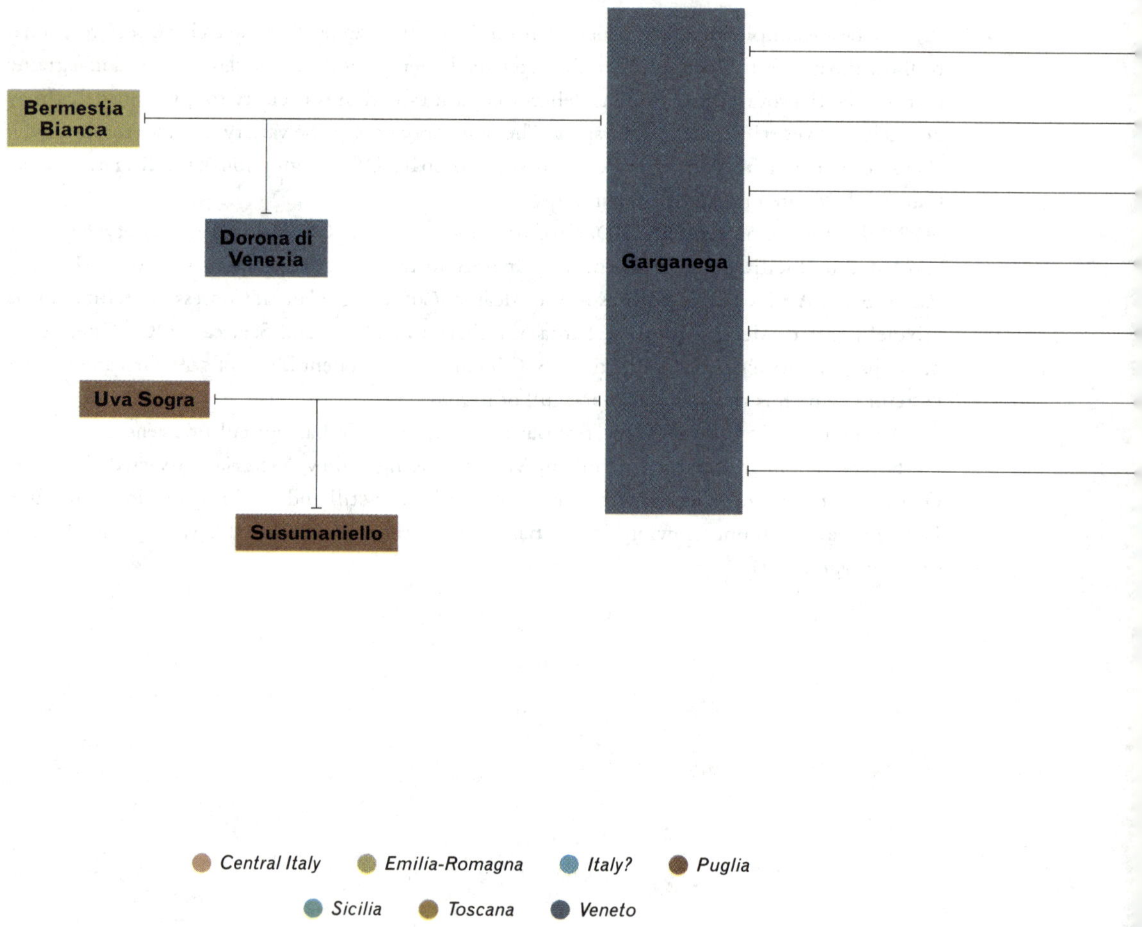

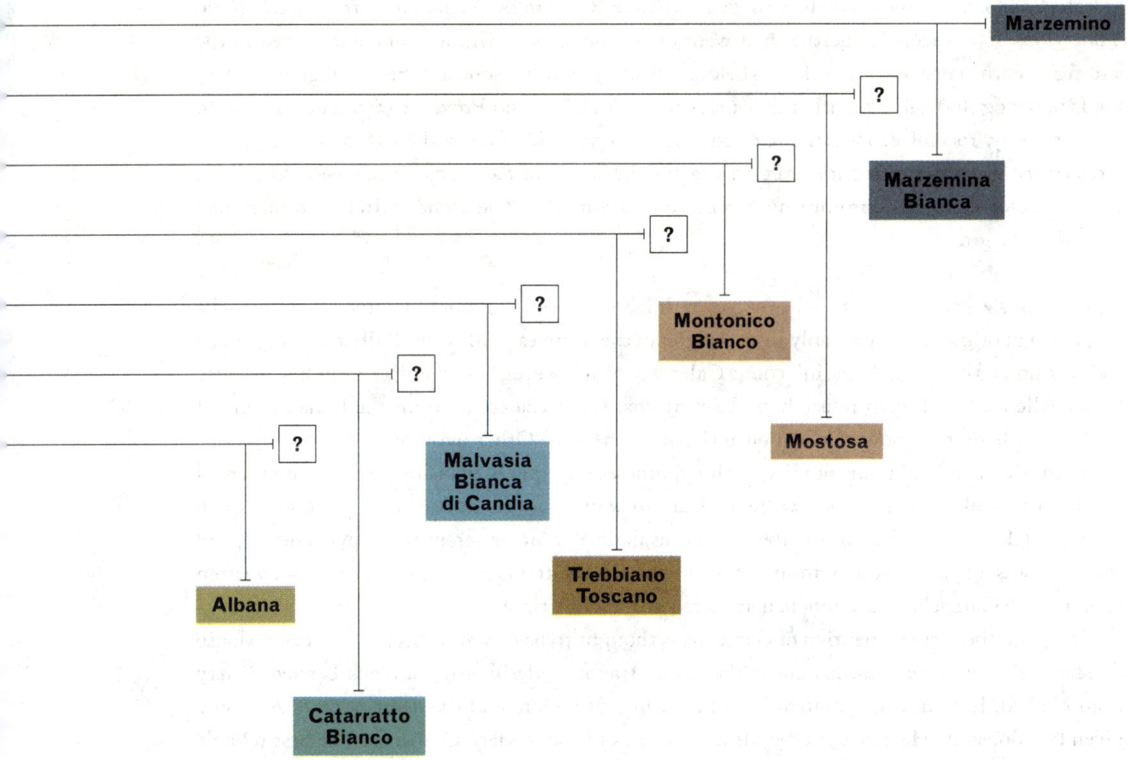
Marzemino
?
?
Marzemina Bianca
?
?
Montonico Bianco
?
Mostosa
?
?
Malvasia Bianca di Candia
Trebbiano Toscano
Albana
Catarratto Bianco

GARNACHA

The origins and parentage of the various mutations of Garnacha are dealt with together at the start of this entry, followed by subsections on specific colour mutations.

ORIGINS AND PARENTAGE

Garnacha is an old variety that has undergone several mutations affecting berry colour (ie Garnacha Blanca, Garnacha Roja) and the downiness of the underside of the leaves (ie Garnacha Peluda), as described in separate subsections below. These all show the same DNA profile (Martín *et al.* 2003; Santana *et al.* 2007; Ibañez *et al.* 2009).

Outside Spain, Garnacha has been known under various synonyms in many countries, the most important of which are Grenache in France (and, subsequently, many other countries) and, in Italy, Cannonau on the island of Sardegna and Tocai Rosso in the Veneto in northern Italy (Calò *et al.* 1990). In France, Garnacha at first went under the name of the area where it arrived in the late eighteenth century – Roussillon – before spreading to other southern French regions such as the Languedoc, and subsequently further east to the Vaucluse and Provence (Semichon 1905). In northern Italy, Tocai Rosso was not referred to until 1950 (Giavedoni and Gily 2005), suggesting a relatively recent introduction. Yet on the Italian island of Sardegna, it has been known as Cannonau for centuries, prompting recent suggestions that Garnacha is Italian rather than Spanish in origin.

SPANISH RATHER THAN SARDINIAN It has long been commonly assumed that Garnacha is Spanish in origin, most probably from Aragón in the north east (Mas and Pulliat 1874–5; Viala and Vermorel 1901–10; Mondini 1903; Galet 2000). However, its Spanish origin has recently been challenged by Italian researchers claiming that Garnacha comes from the Italian island of Sardegna, where it is known as Cannonau (Lovicu 2006) (see Other hypotheses below).

From the historical point of view, both hypotheses are equally plausible because commercial and cultural links between the two regions are so well established. Sardinian populations had already settled in the south of the Iberian peninsula in 800 BC (references in Lovicu 2006), and Sardegna was a Spanish colony from 1479 to 1720 (De Mattia *et al.* 2009). In addition, written references to Garnacha and Cannonau appeared at the same time.

In Spain, the earliest mention of Garnacha is thought to have been in 1513 in Gabriel Alonso de Herrera's *Agricultura general*, under the name Aragones, designating a black-berried variety from Madrid. Indeed, this synonym is still used in that region, and the description of Aragones given by Alonso de Herrera corresponds more or less to the modern Garnacha. The first reliable mention of Garnacha seems to be in Estevan de Corbera's book *Cataluña illustrada* (1678): 'En lo demás del campo de Tarragona Iustolines, Xerelos, Verdieles, Garnachas, y otros vinos blancos tintos' ('For the rest of the plain of Tarragona, Iustolines, Xerelos, Verdieles, Garnachas and other white and red wines').

The etymology of the variety's name is also ammunition in the Spanish–Italian debate: the name Garnacha is widely thought to be derived from the Italian VERNACCIA (see Other hypotheses below) but it may equally be derived from the name of a gown (*garnaxa* in Catalan) worn by the king of Spain's judges and advisers (Torre y Ocón 1728) because its deep colour resembled that of the official regalia. On Sardegna, the earliest mention of Cannonau appeared in 1549 in Cagliari under the name Canonat in an act of the notary Bernardino Coni dated 21 October. It was later called Cañonates in 1612 by a certain Martín Carrillo who was visiting the king, and also in 1677 by a Franciscan called Giorgio Aleo. The etymology of the name Cannonau is unknown.

Based on historical data, it is therefore impossible to be categorical about the origin of Garnacha. However, from the genetic point of view, Spain appears to be the birthplace of Garnacha for two main reasons:

- All three colour variants (black, grey and white) as well as other morphological mutations (Garnacha Peluda) have been observed in Spain but not on Sardegna.
- Recent morphological and genetic studies have provided evidence of significant clonal diversity among the Garnacha reference varieties in Spain but not among those of Cannonau on Sardegna (Cabezas *et al.* 2003; De Mattia *et al.* 2009).

According to the Russian agronomist Vavilov (1926), the area of greatest diversity is usually the centre of origin of a crop, which is why on balance we consider Spain to be the homeland of Garnacha.

GARNACHA AND ITS RELATIVES DNA profiling has demonstrated that Cannonau Bianco is not identical to the white mutation Garnacha Blanca, and that it could be a progeny of Garnacha (De Mattia *et al.* 2009). Similarly, the Italian Granaccia should not be confused with Garnacha Blanca, because it is a synonym for VERNACCIA DI ORISTANO. Recent DNA research has suggested that MENCÍA could have been 'obtained from different crosses between Garnacha and another cultivar' (Martínez, Santiago *et al.* 2006) but this was rejected by De Mattia *et al.* (2009), who concluded that Mencía is not related to Garnacha.

An important study of genetic relationships among Spanish grape varieties (Ibañez *et al.* 2003) grouped together Garnacha with several other old Spanish varieties such as ALARIJE, VERDEJO, AIRÉN, etc. This speaks in favour of a Spanish origin for Garnacha because if Garnacha had been brought in the eighteenth century from Sardegna to Spain, where it then became widespread thanks to its resistance to powdery mildew, as suggested by Lovicu (2006), then it could not be genetically related to so many old Spanish varieties. However, no parent–offspring relationship has yet been found for Garnacha.

Garnacha was used to breed ARANEL, CALADOC, CARNELIAN, CENTURIAN, CHENANSON, EMERALD RIESLING, GANSON, GRAMON, MARSELAN, MONERAC, PORTAN and SYMPHONY.

OTHER HYPOTHESES

Lovicu (2006) explicitly claims that the Sardinian origin of Garnacha/Cannonau is proven. In support of this assertion, he puts forward several arguments:

- Several experts in linguistics and Romance languages (cited in Lovicu 2006) are convinced that the Spanish name Garnacha (as well as Grenache, Garnaxa, etc.) derives from the Italian grape name VERNACCIA, whose wine was reputed all over Europe from the Middle Ages. This would explain why Garnacha is still called Bernacha in Teruel in Aragón.
- The first mention of the variety Garnacha, Lovicu suggests, appeared in 1348 in the *Regiment de preservaciò de pestilencia* in the commune of Lleida in Catalunya as a hot white wine recommended as a cure: 'axí com grech ho vernaça' ('be it either greco or vernaça').
- The first mention of the name Garnacha appeared in 1613 in Miguel de Cervantes' short story *El licenciado vidriera* in a list of Italian white wines that the protagonist enjoyed in a tavern in Genova.

Lovicu concluded that the name Garnacha derives from the name of a white grape (VERNACCIA) and that it designated a white-berried variety until the nineteenth century. However, it is much more likely that the *vernaça* cited in 1348 as well as the *guarnacha* of Cervantes corresponded to the already famous Vernaccia wine rather than to a Garnacha wine. Indeed, Lovicu himself reported that at the end of the fourteenth century, the Catalan Franciscan Eiximenis listed Vernaccia among the white wines in his *Lo Crestià* book. In addition, molecular analyses have excluded any genetic relationship between Garnacha and Vernaccia as well as any possibility of direct domestication from wild grapevines in Spain or on Sardegna (De Mattia *et al.* 2009).

The name Cannonau is sometimes thought to derive from a variety called Canonazo, said to be cultivated in Jerez and Trebujena in the province of Cádiz in Spain. However, Lovicu (2006) has shown that this etymology is based on Di Rovasenda's (1877) misspelling of CAÑOCAZO, which was described by Clemente y Rubio (1807) as a white-berried variety from Andalucía.

GARNACHA TINTA

Very widely planted, long-lived vine making strong, sweetish reds and some successful rosés.

BERRY COLOUR:

PRINCIPAL SYNONYMS FOR GARNACHA TINTA: Abundante (Portugal), Alicante or Licante (Spain), Aragones (Madrid region in Spain), Bernacha Negra, Bois Jaune (France), Cannonao or Cannonau (Sardegna in Italy), Crannaxia or Granaxia or Vrannaxia (Italy), Elegante (Corsica), Gamay del Trasimeno (Perugia in Italy), Garnacha (Ribera del Duero and Catalunya in Spain), Garnaxa (Catalunya), Gironet (Spain), Granaccia (Italy), Granache (Hérault, Gard, Aude and Pyrénées-Orientales in south-western France), Granaxa (Aragón in Spain), Grenache (France), Grenache Crni (Croatia), Grenache Noir (France), Lladoner (Catalunya), Redondal (Haute-Garonne in France), Ranaccio (Sicilia in Italy), Roussillon (Var and Bouches-du-Rhône in France), Sans Pareil (Basses-Alpes in France), Tai Rosso (Veneto), Tinto Basto (Castilla-La Mancha in Spain), Tocai Rosso (province of Vicenza in the Veneto in northern Italy), Ullet de Perdiu (Valencia), Vernaccia Nera (province of Macerata in the Marche, and also Umbria, in Italy)

VARIETIES COMMONLY MISTAKEN FOR GARNACHA TINTA: ALICANTE HENRI BOUSCHET (also called Garnacha Tintorera in Spain), MORRASTEL BOUSCHET (called Garnacho in Valladolid in Spain), PLANT DROIT (called Garnacha Francesa in Zaragoza in Spain), TEMPRANILLO, VIDADILLO DE ALMONACID (called Garnacha Basta in Huesca in Spain)

VITICULTURAL CHARACTERISTICS

Relatively early budding but late ripening so has to be grown in fairly warm climates. Vigorous (less so on sandy soils), susceptible to downy mildew, phomopsis, botrytis bunch rot, grapevine bacterial necrosis and grape moth and may suffer from millerandage. Potentially very high sugar levels in the berries, which lose colour if yields are too high. Acidity levels tend to be moderate at best. Equally well suited to lightly acid, gravelly or stony soils and to soils high in limestone. Best pruned short and trained as bushvines. Good drought resistance but may suffer from lack of magnesium. Good resistance to diseases of the wood such as esca and eutypa dieback, which may partly explain the vine's remarkable longevity.

WHERE IT'S GROWN AND WHAT ITS WINE TASTES LIKE

This is one of the world's most planted grape varieties, not least because it can thrive in hot, dry climates – so it is likely to maintain this position in a world experiencing climate change and where drought seems an increasingly common phenomenon. Considering how common the variety is, it is remarkable that the words Garnacha and Grenache are not seen more often on labels, although the variety is admittedly often blended, particularly with SYRAH (Shiraz) and Mourvèdre (MONASTRELL) – so often that 'GSM' has become an established name for such a blend.

As Grenache Noir, this is **France**'s second most planted variety after MERLOT, with a total of 94,240 ha (232,872 acres) in 2009, restricted by its heat- and sun-loving temperament to the south of the country, especially the Vaucluse *département* (27,466 ha/67,870 acres), where most of the southern Rhône wines are grown. There were also 3,675 ha (9,081 acres) in the Bouches-du-

Rhône to its immediate south, 8,928 ha (22,062 acres) in the Drôme, 2,325 ha (5,745 acres) in the Ardèche, and 8,359 ha (20,656 acres) in the Var, centre of Provence's wines.

The single wine that most obviously represents southern French Grenache is Châteauneuf-du-Pape, most of which is made substantially and sometimes exclusively of Grenache Noir, with some of the many other varieties allowed in red Châteauneuf. In these dry vineyards, berries are small, grape skins are thick and the resulting wines are deep-coloured and often tannic in youth. They also dance along the spectrum between herby and spicy, perhaps as much because of the terroir as the grape itself, but if the variety's wines have a characteristic it is high alcohol and a certain ripe sweetness. This can be seen, sometimes in rather more dilute form, in all the satellite southern Rhône wines such as Gigondas, Vacqueyras and Côtes du Rhône (Villages). There are literally hundreds of producers of fine examples of these wines, some of which can be great value. Substantially north of this area in the Drôme and the Ardèche, the wines are a little lighter and less sweet. To the south east in the heart of Provence, Grenache is more likely to be grown to produce herby dry rosé, some of the most characterful in the world, as well as some of the least distinguished. And to the west, in the Gard *département*, where 16,987 ha (41,976 acres) grew in 2009, Grenache is responsible for some of the world's most substantial rosés in Tavel and its neighbour Lirac. In the Gard it seems to take on an extra juiciness and polish, perhaps because rainfall is a little higher.

In Languedoc-Roussillon proper, total plantings of Grenache are now greater than those of Carignan (MAZUELO) but rather less than those of fashionable SYRAH. In 2009 there were 10,372 ha (25,639 acres) in the Hérault, 8,494 ha (20,989 acres) in the Aude, and 6,769 ha (16,727 acres) in the Pyrénées-Orientales, where Grenache of all three colours has long been grown to produce strong vins doux naturels from appellations such as Maury, Banyuls and Rivesaltes, often made in a rancio (ie oxidative) style by ageing in glass demi-johns or old barrels in the sun. Throughout the warmer, more Mediterranean parts of Languedoc-Roussillon, Grenache Noir is also increasingly used for table wines as an ingredient with Syrah, Carignan, Cinsault (CINSAUT) and Mourvèdre (MONASTRELL) in the likes of Faugères, Saint-Chinian, Minervois, Corbières, Fitou and Côtes Catalanes. Varietal Grenache Noir is a relative rarity here but if yields are too high and the land unpromising, Grenache can taste a little too jammy.

In **Italy**, Sardegna is Garnacha's (and Carignano's/MAZUELO's) Italian stronghold. Known here as Cannonau, it is the island's most important grape and is grown mostly in the province of Nuoro in the east and is particularly successful on the hillside areas of Barbagia, Ogliastra, Sarrabus and Romangia and also in the Nurra and Campidano areas. The 2000 agricultural census found a total of 6,288 ha (15,538 acres). Most wines are sold as varietal Cannonau di Sardegna, a DOC which may be produced anywhere on the island and accounts for about 20% of total production. It is made in dry and sweet, both fortified and not, but alcohol levels are naturally high, often more than 15%, anyway. Superior producers include Argiolas, whose Turriga blend is probably the single most exciting Italian expression of the grape, Dettori and Alberto Loi.

As Tocai Rosso it is authorized in Colli Berici DOC in the Veneto, where the census found a total of 384 ha (949 acres) in 2000.

A grand total of 75,399 ha (186,315 acres) of **Spain** were planted with Garnacha Tinta in 2008, less than 7% of the Spanish vineyard area, predominantly in the autonomous communities of Aragón in the north east (21,202 ha/52,391 acres, which represented 45% of the region's vineyard area) and Castilla-La Mancha in the centre (23,518 ha/58,114 acres; only 4.5% of vineyard area) but also in Castilla y León, Catalunya, La Rioja, Madrid, Extremadura, Navarra and Valencia, in fact pretty much everywhere except the north west and the deep south. It is now in third place among the dark-skinned varieties in Spain, after TEMPRANILLO (206,988 ha/511,478 acres) and BOBAL (85,124 ha/210,346 acres).

Overall plantings are in decline (eg 89,045 ha/220,035 acres in 2001) but it is particularly significant for rosé production in Navarra. For years Garnacha's ubiquity meant that Spaniards viewed it as necessarily inferior to TEMPRANILLO but, partly because it is so important in Priorat, which since the 1990s has managed to command some of the highest wine prices in Spain, its reputation has been reassessed. It is now much more likely to feature on Spanish wine labels than it was even ten years ago, and many export markets treasure the concentrated, sweet fruit in wines made from old Garnacha bushvines which proliferate in regions such as Empordà, Campo de Borja, Cariñena, Costers del Segre, Madrid, La Mancha, Méntrida, Penedès, Somontano, Tarragona, Terra Alta, Utiel-Requena and Valdeorras.

Even in Rioja, where the more structured TEMPRANILLO reigns supreme, it is now recognized that Garnacha, the signature grape of the lower, warmer Rioja Baja subregion, can provide the flesh to complement the bones of Tempranillo.

Older plantings of Garnacha Tinta – and average vine age of this variety is high for it belongs to another era – are often intermingled with some Garnacha Peluda, or 'downy Garnacha'.

Producers who make superior Garnacha, or Garnacha-based blends, include Celler Mas Doix, Celler Val Llach, Clos i Terrasses, Mas Alta, Mas Martinet, Alvaro Palacios and Terroir al Limit in Priorat; Torres in Catalunya; Chivite and Guelbenzu in Navarra; Can Blau and Porto del Montsant in Montsant; Baigorri, Bretón and Dinastía Vivanco for varietal versions in Rioja; and the Australian-run Alto Moncaya plus Bodegas Aragonesas in Campo de Borja.

Grenache is also found in **Croatia** (292 ha/722 acres of Grenache Crni in 2009), in **Turkey** (34 ha/84 acres in 2010), on **Cyprus** (95 ha/235 acres in 2010) and on **Malta** (24 ha/59 acres in 2010) as well as, to a strictly limited extent, in **Morocco**, **Algeria** and **Israel.**

Grenache Noir is said to have been introduced to **California** by Charles Lefranc, a Santa Clara grower, in the late 1850s and its versatility made it a popular choice during the planting boom towards the end of the nineteenth century. After Prohibition, plantings increased in the Central Valley, mainly for dessert and rosé wines, and some Grenache-based commercial pink wines have more recently been marketed as an alternative to White Zinfandel. Today, however, it is being taken increasingly seriously, particularly by those with treasured old vines, as witness Keplinger's various bottlings.

The state total in 2010 was 6,170 acres (2,497 ha), mostly in the Sierra Foothills, especially El Dorado and Madera counties. Favia's Rompecabezas from the Sierra Foothills is a good example of a GSM blend while superior varietal or Grenache-dominant wines are produced by Ridge in Dry Creek Valley, Sine Qua Non in Santa Barbara and Tablas Creek in Paso Robles.

Grenache, one of the first varieties to be planted in the state, has recently blossomed in **Washington** as a cult for Rhône wines has developed, and total plantings had crept up to 261 acres (106 ha) by 2011. McCrae Cellars (who also make a varietal Grenache Blanc), Cayuse and K Vintners are some of the leading exponents.

It is also grown to a limited extent in southern **Oregon**, **Arizona** and **Texas**. And in **Mexico**, some drought-proof Grenache is grown mainly in Baja California. It is not of much importance in South America although it is known in **Argentina** (22 ha/54 acres in 2008, mainly for rosé), **Chile** (1 ha/2.5 acres in 2008) and **Peru**.

Grenache has known at least two reversals of fortune in **Australia** in modern times. Until the 1960s it was the country's most important variety. It was then overtaken by Shiraz (SYRAH), which then made way for CABERNET SAUVIGNON for a French-influenced while. A global re-evaluation of Rhône varieties has led to a fashion for Grenache varietals and, especially, GSM blends, with considerable tracts of old-vine Grenache having unfortunately already been pulled out of the inland regions and some new plantings in more admired, but necessarily warm, regions, mainly in South Australia. In 2008 plantings totalled 2,011 ha (4,969 acres), predominantly in the Barossa Valley, McLaren Vale, Riverland and Langhorne Creek, with lesser plantings in the Clare Valley, Swan District, Adelaide Hills and Plains, Eden Valley and Heathcote.

Recommended producers of varietal Grenache include Schwarz Wine Company, Torbreck, Turkey Flat and Yalumba in the Barossa Valley, d'Arenberg in McLaren Vale, Kilikanoon in the Clare Valley, Longhop on the Adelaide Plains and Jasper Hill in Heathcote. However, as in the southern Rhône, some of the finest expressions of Grenache are in classic blends, usually with Shiraz (SYRAH) or with Shiraz and Mourvèdre (MONASTRELL), such as those produced by Glaetzer, Henschke, Charlie Melton, Steve Pannell, Rockford, Teusner and Spinifex.

As in southern France, Grenache is also an important ingredient in some of Australia's finest fortified wines such as the venerable 'Tawny' wines produced by Grant Burge, Penfolds and Seppeltsfield.

Grenache is not a particularly significant variety in **South Africa**, where 144 ha (356 acres) were found in 2009 quite widely dispersed around the wine regions. Varietal wines are produced by Citrusdal, Neil Ellis, Sequillo and Tierhoek, among others, but Grenache Noir is more often found in blends, eg with Shiraz (SYRAH), on the Australian model, or with a whole range of other varieties, including Carignan (MAZUELO), Mourvèdre (MONASTRELL) and sometimes even CABERNET SAUVIGNON.

GARNACHA BLANCA

Full-bodied whites, pale version of the reds.

BERRY COLOUR: ● ● ● ● ●

PRINCIPAL SYNONYMS FOR GARNACHA BLANCA: Alicant Belyi, Alicante Blanca, Belan (Croatia and Republic of Macedonia), Belon, Bernacha Blanca, Fehér Grenache (Hungary), Garnacha, Garnacho Blanco, Garnatxa, Garnatxa Blanca, Gkrenas Mplan, Grenache Blanc, Grenash Beli, Grenash Belyi, Grenash Bjal, Lladanor Blanca, Rool Grenache, Silla Blanc, Sillina Lanc

VITICULTURAL CHARACTERISTICS

Similar to Garnacha Tinta but mid budding and ripening and less sensitive to coulure.

WHERE IT'S GROWN AND WHAT ITS WINE TASTES LIKE

The big, sometimes fat, wines are rich in extract but can be prone to premature oxidation. Aromas are likely to be of very ripe green fruits (greengages?) with some floral notes. Alcohol levels tend to be high so blending in crisper varieties can be helpful.

It is recommended, like Grenache Noir, in much of southern **France**, including the southern Rhône. There were 4,976 ha (12,296 acres) in 2009, spread right around the Mediterranean basin and up into the southern Rhône, with the biggest area in Roussillon, where it has long been grown for vins doux naturels. Its importance has declined so that it is now the ninth most planted white wine variety in France, just ahead of VIOGNIER. But the variety's qualities are increasingly recognized in a Rhône-besotted era, even though it is more often encountered in a blend than as a varietal wine. If yields are checked and it is vinified carefully, often in oak, it can produce rich, substantial wine. In Roussillon, Clos des Fées produce an impressively fragrant and densely textured old-vine Grenache Blanc (including 10% Grenache Gris) and Gauby's high-toned Coume Gineste is a 50/50 blend of Blanc and Gris. Diffonty's white Châteauneuf blends a significant proportion of Grenache Blanc with ROUSSANNE while Château Rayas combine 50% Grenache Blanc with 50% CLAIRETTE in their dense and savoury rendition of white Châteauneuf.

Grenache Blanc is still a significant ingredient in various vins doux naturels such as Rasteau from the southern Rhône and Maury, Rivesaltes and Banyuls from Roussillon, particularly in the white version of Maury, where it is blended with Grenache Gris, Macabeu (MACABEO) and Tourbat (TORBATO).

The 2008 census found 2,100 ha (5,189 acres) of Garnacha Blanca in **Spain**, mostly in Catalunya and Aragón in the north east, and especially in the provinces of Tarragona, Zaragoza and Teruel, where it produces full-bodied wines with a high alcohol content. It is classified as one of the principal varieties in DOs such as Alella, Costers del Segre, Priorat, Rioja, Tarragona and Terra Alta, although it is the least planted of Rioja's authorized white varieties.

In Priorat, Viñedos de Ithaca produce a sweet varietal version of Garnacha Blanca, while Gran Clos' Vinya Llisarda is a blend based on Garnacha Blanca.

It is found in **Croatia** and, under the name Belan, in the **Republic of Macedonia**, where it tends to be planted only in the warmer vineyards because of its sensitivity to cold winter temperatures. It is also grown to a limited extent in **California** (266 acres/108 ha in 2010) with Santa Ynez Valley being a hotspot. There is also a little in the Sierra Foothills. The number of exuberantly varietal California Grenache Blancs has been increasing in line with CHARDONNAY fatigue.

South Africa had 46 ha (114 acres) of the variety in 2008. The Foundry and Signal Hill take it seriously.

GARNACHA ROJA (GRIS)

Perfumed and full-bodied.

BERRY COLOUR:

PRINCIPAL SYNONYMS FOR GARNACHA ROJA: Garnacha Gris, Garnacha Rioja, Garnacho Rojo, Garnatxa Gris (Catalunya), Garnatxa Roja, Grenache Gris, Grenache Rouge, Grey Grenache, Lledoner Roig (Empordà), Piros Grenache, Rosco dos Pinheiro, Szuerke Grenache

VITICULTURAL CHARACTERISTICS

Similar to Garnacha Tinta, and a little more productive than GARNACHA BLANCA.

WHERE IT'S GROWN AND WHAT ITS WINE TASTES LIKE

Grenache Gris, almost a combination of Grenache Blanc and PINOT GRIS, is authorized in exactly the same *départements* of southern **France** as Grenache Blanc, with total plantings of 1,699 ha (4,198 acres) of Grenache Gris in 2009, mostly in Roussillon and some in the Languedoc and the southern Rhône. More often than not it contributes to white blends, particularly with Grenache Blanc or Macabeu (MACABEO) but also with any number of different varieties from the Châteauneuf-du-Pape recipe. In Roussillon, Matassa blend 70% Grenache Gris with 30% Macabeu to create a racy, complex white while Le Roc des Anges' Iglesia Vella is a perfumed and polished varietal Grenache Gris. Domaine Jones sometimes add a tiny percentage of Muscat to the majority Grenache Gris to create a rich and expressive slightly off-dry version.

Spain grew just 71 ha (175 acres) of this grey- or pink-skinned Garnacha in 2008, mainly in Navarra and Castilla-La Mancha. Producers such as Espelt in Catalunya blend Garnatxa Gris with Garnatxa Blanca as well as Garnatxa Negra; Acústic in Montsant make a white blend that includes Garnacha Blanca and Garnacha Gris; Castell de Biart in Empordà make a rancio-style blend of Garnatxa Roja and Gris. Rare varietal wines made from Garnacha Roja include the Pau i Roses co-operative's sweet, fortified version from Empordà, Magí Baiget's dessert wine (Montsant) and Vinyes dels Aspres' dried-grape dessert wine made in a rancio style in Empordà.

It is also grown to a limited extent in **California**, mainly in the Central Coast and McDowell Valley, and in **South Africa** (3 ha/7 acres in 2008).

GARNACHA PELUDA

Downy-leaved variant of Garnacha Tinta.

BERRY COLOUR: ● ● ● ● ●

PRINCIPAL SYNONYMS FOR GARNACHA PELUDA: Garnatxa Pelud, Garnatxa Peluda, Grenache d'Afrique, Grenache Noir Tomenteux, Grenache Poilu, Grenache Velu, Lladoner Pelut, Lledoner Pelut

VITICULTURAL CHARACTERISTICS

Similar to Grenache Noir but with furry down on the underside of the leaves. Smaller-berried with thicker skins and more acidity. Slightly later budding and said to be more vigorous, less sensitive to coulure (it therefore crops more consistently) and downy mildew and more sensitive to mites. Less prone to millerandage than Garnacha Tinta.

WHERE IT'S GROWN AND WHAT ITS WINE TASTES LIKE

The grapes reach phenolic ripeness at a lower sugar levels than Garnacha Tinta so the wines are generally lower in alcohol (no more than 13.5%), and this hairy-leaved version was not widely replanted after phylloxera when vins doux naturels were all the rage. The 'fur' (*pelut*, which gives it its French name) evolved to protect the vine from transpiration in the heat, conserve moisture and reduce stress, just like the corresponding fuzz on rosemary or other Mediterranean plants. Another adaptation is that it never breaks in high winds, so it will survive in even the most exposed sites. Its wine tends to have a higher acid level and, as it ages, spicy and savoury characters develop more quickly than with Grenache Noir.

In 2008 there were 433 ha (1,070 acres) of Lledoner Pelut in **France.** It is treated as a separate variety from Grenache Noir in the French official register of varieties but it tends to be recommended alongside it and is authorized in such southern French appellations as Corbières, Languedoc, Côtes du Roussillon and Côtes du Roussillon-Villages, Faugères, Minervois and Saint-Chinian. In the Languedoc, Domaine La Colombette produce a varietal version while the likes of Les Clos Perdus and Domaine du Comte del Roc make significant use of it in blends.

Spain has much more, 903 ha (2,231 acres) in total in 2008, in Castilla-La Mancha, Aragón, Catalunya and, to a lesser extent, in Castilla y León. It is often blended with CABERNET SAUVIGNON or TEMPRANILLO in oak-aged wines. Lighter, younger wines are prone to oxidation and should be drunk young. Varietal versions are rare but Viñedos de Ithaca in Priorat produce an exception and Capafons-Ossó in Montsant include it their Mas de Massos blend.

GARNACHA TINTORERA

See ALICANTE HENRI BOUSCHET.

GARRIDO FINO

Very minor, acidic variety from Huelva in southern Spain.

BERRY COLOUR: ● ● ● ● ●

PRINCIPAL SYNONYMS: Charrido Fino, Garrido, Garrido Fino de Villanueva, Garrio Fino, Palomino Garrio

VARIETIES COMMONLY MISTAKEN FOR GARRIDO FINO: Garrido Macho

ORIGINS AND PARENTAGE

Garrido Fino is an old Andalusian variety from the province of Huelva. It should not be confused with the almost-extinct Garrido Macho as the two varieties have distinct DNA profiles (Martín *et al.* 2003).

VITICULTURAL CHARACTERISTICS

Big bunches of big berries. Fertile. Mid budding and mid to late ripening, susceptible to powdery and downy mildews and botrytis bunch rot.

WHERE IT'S GROWN AND WHAT ITS WINE TASTES LIKE

Garrido Fino is found in the Condado de Huelva DO in south-western Andalucía in **Spain**'s deep south, where its fresh acidity, smooth texture and moderate alcohol make it a valued but minor blending component, particularly in sweeter styles and with the variety ZALEMA. So little is now planted that it does not feature in official Spanish statistics but Bodegas Góngora (founded 1682) make both dry and off-dry varietal wines.

GASCON

Rare Touraine vine.

BERRY COLOUR: ● ● ● ● ●

PRINCIPAL SYNONYMS: Franc Noir de l'Yonne (Yonne, Aube), Franc Noir du Gâtinais (Loiret), Noirien, Plant de Moret (Yonne)

VARIETIES COMMONLY MISTAKEN FOR GASCON: TRESSOT (shares the synonym Noirien with Gascon)

ORIGINS AND PARENTAGE

In spite of its name, Gascon is unlikely to come from Gascogne, that part of south-west France which corresponds to the modern Aquitaine and Midi-Pyrénées regions, but from around Orléans in the Loiret *département* in northern central France, where this variety was first mentioned by Olivier Jullien (1816): 'les plants que l'on cultive plus généralement dans le Loiret sont, en rouge, ... le gascon, le gammé ...' ('the varieties that are more generally cultivated in the Loiret are, in red, ... gascon, gammé ...'). Alternatively, it may come from the Yonne *département*, to the east of the Loiret, where it used to be widely cultivated before the phylloxera crisis and where it got its synonym Franc Noir de l'Yonne (Galet 1990).

Gascon is a member of the Cot ampelographic group (see p XXVII; Bisson 2009).

OTHER HYPOTHESES

According to Viala and Vermorel (1901–10), Gascon was formerly known as Roche Noire or Rochelle Noire and was already cited in 1667 by Jean Merlet, but this is far from proven.

VITICULTURAL CHARACTERISTICS

Small to medium-sized compact bunches of small berries. Susceptible to powdery mildew and botrytis bunch rot. Mid to late ripening.

WHERE IT'S GROWN AND WHAT ITS WINE TASTES LIKE

Gascon produces deeply coloured red wines with lively acidity. In **France** a few limited plantings of Gascon can still be found in Touraine. Pascal Simonutti, in Mesland, south west of Blois, makes a funky vin de table varietal Gascon. Claude Courtois in Soings-en-Sologne to the west has a plot of young Gascon vines and describes the wine he makes from them in some vintages as 'the most beautiful of the domaine'.

GEGIĆ

Speciality of the Croatian island of Pag.

BERRY COLOUR: ● ● ● ● ●

PRINCIPAL SYNONYMS: Debejan or Debljan (islands of Krk and Rab), Gegić Bijeli, Paška (island of Pag), Paškinja (Zadar and Pag), Žutina

ORIGINS AND PARENTAGE

Gegić most likely originates from the islands of the Kvarner Gulf (between Istra (Istria) and the Croatian mainland), including the island of Pag, where it has long been grown to produce the locally famous Paška Žutica wine, from which the local synonyms Paška or Paškinja derive.

VITICULTURAL CHARACTERISTICS

Late budding, late ripening. Yields may be irregular due to coulure in the difficult climatic conditions of Pag. Relatively disease-resistant. Medium-sized, thick-skinned berries.

WHERE IT'S GROWN AND WHAT ITS WINE TASTES LIKE

Gegić is officially authorized in **Croatia** in the Hrvatsko Primorje (Croatian Littoral) region but its irregular yields have led to a decline in the area planted. Producers include Boškinac, Cissa and Vina Otaka, all on the island of Pag. Anton Katunar on the island of Krk blends Debejan with SYRAH and SUŠĆAN (known locally as Sansigot).

GEISENHEIM 318-57

Minor German hybrid surviving but not especially valued in Canada.

BERRY COLOUR: ● ● ● ● ●

ORIGINS AND PARENTAGE

A complex hybrid created in 1957 at the Geisenheim research centre in Germany by crossing RIESLING with CHANCELLOR.

VITICULTURAL CHARACTERISTICS

Reliable yields. Early to mid ripening. Generally good resistance to powdery mildew but susceptible to downy mildew thanks to its CHANCELLOR parentage.

WHERE IT'S GROWN AND WHAT ITS WINE TASTES LIKE

The results of the four trial plantings of this variety in **Germany** in the 1970s and 1980s were not satisfactory so the vines were pulled up and the remaining fifty-two vines are in Geisenheim's collection.

Although it has not succeeded in its homeland, generally because better alternatives were subsequently created, Geisenheim 318-57 can be found in **Canada**: approximately 205 acres (83 ha) in Ontario and 20 acres (8 ha) in Québec, as well as in Nova Scotia. It is not authorized for VQA wines, however. Varietal wines are generally rather neutral with a slight hybrid flavour, sometimes flowery and lightly spiced, although Geisenheim 318-57 is often part of a blend. Producers include Artisans du Terroir, Clos Napierois (an oak-aged version) and Le Royé Saint-Pierre in Québec.

It is also grown in **New York** State's Finger Lakes.

ĠELLEWŻA

Second most planted Maltese variety but being edged out by international invaders.

BERRY COLOUR: ● ● ● ● ●

PRINCIPAL SYNONYMS: Gelleuza

ORIGINS AND PARENTAGE

A recent study has shown that Ġellewża has a unique DNA profile (Giannetto *et al.* 2010) and is therefore likely to be indigenous to Malta or Gozo. The name Ġellewża is said to derive from an Arabic word and is Maltese for 'hazelnut', perhaps a reference to the shape and small size of the berries.

Ġellewża was said to be a seedling of the Italian MAMMOLO (Borg 1922) but this is contradicted by DNA analysis (Vouillamoz).

VITICULTURAL CHARACTERISTICS

Mid to late ripening, susceptible to fungal diseases, especially botrytis bunch rot. Best pruned long.

WHERE IT'S GROWN AND WHAT ITS WINE TASTES LIKE

Ġellewża, the most popular indigenous red variety on **Malta** and Gozo, produces light, red-fruited reds and easy-drinking rosé wines with modest alcohol levels. Rosés tend to be particularly successful since they avoid Ġellewża's sometimes astringent tannins.

Estimates as to the area planted range from 40 ha (99 acres) to 100 ha (247 acres), since not all vineyards are registered, and they are mainly unirrigated bushvines. The area has been in decline, Ġellewża being an unfortunate victim of the general preference for international varieties, especially since the EU-funded plantings began after Malta's accession in 2004. However, increasing interest in indigenous varieties may soon halt that trend. Delicata and Marsovin both make a medium-dry sparkling rosé but several producers blend Ġellewża with varieties such as Shiraz (SYRAH).

GENEROSA (HUNGARY)

An EZERJÓ × SAVAGNIN ROSE (under the name Piros Tramini) pink-berried cross obtained in 1951 by Bíró Károly in **Hungary**, tested since 1976 by Edit Hajdú at the Kecskemét viticulture research centre, and finally christened Generosa in 2004. Szőke Winery produce a varietal example. It is not related to the Portuguese GENEROSA.

GENEROSA (PORTUGAL)

Rare Portuguese cross used in blends on the Açores.

BERRY COLOUR: ● ● ● ● ●

ORIGINS AND PARENTAGE

A FERNÃO PIRES × Sultana Moscata cross obtained in Portugal, in which Sultana Moscata (or Pirovano 75) is a MUSCAT OF ALEXANDRIA × SULTANIYE cross.

WHERE IT'S GROWN

Official vineyard statistics record less than 1 ha (2.5 acres) of Generosa in **Portugal** but it is planted on the island of Pico in the Açores and included in a white blend by the local co-operative. There are also a few scattered vines on Madeira, mainly in Santana in the north, whose produce goes into Madeira not table wine.

GENOUILLET

Recently rescued central French vine.

BERRY COLOUR: ● ● ● ● ●

PRINCIPAL SYNONYMS: Genouilleret (Issoudun in Indre), Genouillet Noir (Aube), Moret Noir (Bourges)

VARIETIES COMMONLY MISTAKEN FOR GENOUILLET: MONDEUSE NOIRE

ORIGINS AND PARENTAGE

Genouillet used to be cultivated in the Indre (for example, in Issoudun, La Châtre and Châteaumeillant) and the Cher *départements* in central France until the phylloxera crisis. Its name probably derives from Genouilly, a commune near Bourges in the Cher.

Genouillet is a member of the Noirien ampelographic group, to which PINOT belongs (see p XXVII; Bisson 2009). However, this ampelographic classification has been challenged by DNA parentage analysis, which has revealed that Genouillet is a natural TRESSOT × GOUAIS BLANC cross (Bowers *et al.* 2000), neither of the parents being a member of the Noirien group. Since Tressot is a variety from the Yonne, a *département* bordering the Cher, and since Gouais Blanc was cultivated almost everywhere in central and north-eastern France, Genouillet was therefore most likely born somewhere in either the Cher or the Yonne *département*.

OTHER HYPOTHESES

Some think that Genouillet could be the Biturica cited by Columella.

VITICULTURAL CHARACTERISTICS

Early budding, late ripening. Susceptible to powdery and downy mildews and botrytis bunch rot.

WHERE IT'S GROWN AND WHAT ITS WINE TASTES LIKE

Genouillet was almost extinct in **France** when it was rescued in 1990 in the region of Issoudun, south west of Bourges in the Indre *département*. Some 150 vines were planted in 2005 in an experimental vineyard at the Domaine de Villalin in the Quincy appellation in collaboration with the Union pour la Préservation et la Valorisation des Ressources Génétiques du Berry. The aim is to obtain official authorization to plant, and make commercial wine from, Genouillet in the near future.

GENOVESE

See SCIMISCIÀ.

GEWÜRZTRAMINER

See SAVAGNIN.

GF-GA 48-12

Disease-resistant German hybrid that has, despite its (lack of) name, found some favour in Switzerland.

BERRY COLOUR: ● ● ● ● ●

ORIGINS AND PARENTAGE

A hybrid obtained by Gerhardt Alleweldt at the Geilweilerhof research centre at Siebeldingen in the Pfalz in Germany (part of the Julius Kühn-Institut) from undisclosed parents and so far without an official name. Gf stands for Geilweilerhof and Ga for the breeder.

VITICULTURAL CHARACTERISTICS

Good general resistance to fungal diseases.

WHERE IT'S GROWN AND WHAT ITS WINE TASTES LIKE

Gf-Ga 48-12, cultivated mainly in **Switzerland**, produces wines that are aromatic and show a good balance between fruit and acidity. According to the sales pitch of the Julius-Kühn Institut, the wines taste rather like a blend of SCHEUREBE and GEWÜRZTRAMINER and are superior to wines made from PHOENIX, created by the same breeder. Hans-Peter Baumann at Diroso makes a varietal wine and Lenz use it in two of their white blends.

GIBI

Ancient Spanish table grape still planted in Argentina with significant offspring.

BERRY COLOUR: ● ● ● ● ●

PRINCIPAL SYNONYMS: Alzibib, Aparia, Augibi (Gard in France), Jubi Blanc, Maccabeu à Gros Grains (Pyrénées-Orientales in France), Panse Blanche (Pyrénées-Orientales), Passerille Blanche (Hérault in France), Tercia Blanc

ORIGINS AND PARENTAGE

Gibi is an old table grape from Spain that was supposedly introduced by the Arabs under the name Alzibib, deriving from the Arabic word *zabīb*, meaning 'raisin', just like Zibbibo, the name for MUSCAT OF ALEXANDRIA on Sicilia. DNA parentage analysis has shown that Gibi is the likely mother of PEDRO XIMÉNEZ (Vargas *et al.* 2007), an old white grape variety from Andalucía, and that Gibi and PLANTA NOVA (under the name Tortozón) are the parents of ALARIJE (under the name Subirat Parent; Lacombe *et al.* 2007), an old variety from Extremadura, where it was known as long ago as 1513. These relationships with ancient Spanish varieties suggest that Gibi originates in south-western Spain, or at least has been cultivated there since the Middle Ages.

VITICULTURAL CHARACTERISTICS

Female-only flowers.

WHERE IT'S GROWN AND WHAT ITS WINE TASTES LIKE

Gibi was once grown in **Spain** and southern **France**, where it is still occasionally found as a table grape (Galet 2000), but in 2009 it was grown on 1,080 ha (2,669 acres) in **Argentina**. Here it is used both for wine (generally for the domestic market) and for grape juice.

GINESTRA

Minor but interesting and refreshing Campanian.

BERRY COLOUR: ● ● ● ● ●

PRINCIPAL SYNONYMS: Biancatenera⚥ (Scala), Biancazita⚥ (Furore, Tramonti, Corbara and Positano), Ginestro, Nocella

ORIGINS AND PARENTAGE

Ginestra was already mentioned by Acerbi (1825) as one of the main varieties of the Napoli area in Campania in southern Italy. Its name comes from the broom-like flavour in the grapes (Italian *ginestra*). The synonymy between Ginestra and Biancazita, a variety grown in the same area that is also called Biancatenera, had been suspected by Froio (1875) and was recently confirmed by DNA profiling (Costantini *et al.* 2005). The same DNA study also established a possible genetic relationship between Ginestra and two other Campanian varieties: CAPRETTONE and Piedirosso Napoletano, a non-cultivated variety that should not be confused with PIEDIROSSO.

VITICULTURAL CHARACTERISTICS

Ripens mid September. Its susceptibility to botrytis bunch rot has limited its spread. Berries have high sugar levels and notably high acidity.

WHERE IT'S GROWN AND WHAT ITS WINE TASTES LIKE

Ginestra is cultivated exclusively in the province of Salerno along the Amalfi coast in southern **Italy**, in particular in the municipalities of Scala, Ravello, Amalfi, Minori, Maiori, Furore, Tramonti, Corbara and Positano. Ginestra wines have crisp acidity and can taste similar to FALANGHINA FLEGREA. With some ageing they tend to develop the kerosene-like aromas sometimes found in older RIESLING. Marisa Cuomo's Costa d'Amalfi DOC Furore Bianco Fiorduva is a blend of FENILE, Ginestra and RIPOLO.

GIOUROUKIKO

See FOKIANO.

GIRGENTINA

The most planted Maltese variety, producing soft, light whites but under threat from Chardonnay.

BERRY COLOUR: ● ● ● ● ●

PRINCIPAL SYNONYMS: Ghirghentina, Insolja Tal-Girgenti

ORIGINS AND PARENTAGE

Girgentina probably comes from the village of Girgenti in south-west Malta. Recent DNA profiling of several Maltese varieties has shown it to be a distinct variety (Giannetto *et al.* 2010).

OTHER HYPOTHESES

A connection to the historical city of Agrigento (Grigenti in Sicilian) has been suggested but no relationship has so far been found with Sicilian varieties (Giannetto *et al.* 2010).

VITICULTURAL CHARACTERISTICS

Productive, late ripening, susceptible to fungal diseases, especially botrytis bunch rot. Best pruned long.

WHERE IT'S GROWN AND WHAT ITS WINE TASTES LIKE

Wines are typically light, low in alcohol, often light on acidity and prone to oxidation but more modern examples are clean, fresh and delicate. Estimates of the Girgentina vineyard area on the islands of **Malta** and Gozo in 2010 range from 90 ha (222 acres) to more than 200 ha (500 acres), mainly in the north of Malta and typically trained as unirrigated bushvines. The overall area has been in decline thanks to competition from international varieties. Both Delicata and Marsovin make varietal wines and the former make a dry, sparkling version. The variety is also used to enliven international varieties such as CHARDONNAY.

GIRÒ

Cherry-flavoured Sardinian red used to make dry, sweet and fortified wines.

BERRY COLOUR: ● ● ● ● ●

PRINCIPAL SYNONYMS: Girò Comune, Girò Rosso di Spagna, Girone di Spagna, Zirone

ORIGINS AND PARENTAGE

Girò was most probably introduced to the Italian island of Sardegna during Spanish rule (1323–1720), together with MAZUELO (Bovale Grande), GRACIANO (Cagnulari) and PASCALE, as suggested by its synonym Girò Rosso di Spagna. Comparison of the DNA profiles of Girò from Sardegna (De Mattia *et al.* 2007) and GIRÓ BLANC from Mallorca (Ibañez *et al.* 2003) shows that they are not colour mutations but, as suggested by Galet (2000), distinct varieties. Recent DNA parentage analysis has suggested that Girò may be a natural parent of ALBARANZEULI BIANCO from Sardegna (Cipriani *et al.* 2010).

VITICULTURAL CHARACTERISTICS

Mid to late ripening. Best suited to hot, dry climates.

WHERE IT'S GROWN AND WHAT ITS WINE TASTES LIKE

Widespread in Sardegna, **Italy**, until phylloxera struck at the end of the nineteenth century, Giròis now cultivated in the provinces of Cagliari and Oristano. Girò di Cagliari DOC allows dry and sweet wines and both may be fortified. In addition, the grapes may be dried – on or off the vine – to increase the sugar concentration. Sweet wines such as by Meloni's Donna Jolanda have cherry aromas and a firm tannic structure but may sometimes be lacking in acidity.

There were 552 ha (1,364 acres) of Giròplanted in Italy in 2000.

GIRÓ BLANC

Recently rescued, high-quality, aromatic, full-bodied Mallorcan.

BERRY COLOUR: ● ● ● ● ●

PRINCIPAL SYNONYMS: Giròò Roz

ORIGINS AND PARENTAGE

Giró Blanc is indigenous to Mallorca in the Islas Baleares, Spain, where it was recently rescued from virtual extinction by Toni Gelabert, thanks to the 200 vines that Juaquin Monserrat, a local grower with faith in the variety, had propagated from the few plants that remained on the island. Mutations have given rise to considerable variation in berry colour, hence the synonym Giró Roz.

Giró Blanc is not a colour mutation of GARNACHA, called Giró or Gironet on the island. Similarly, comparison of the DNA profiles of GIRÒ from Sardegna (De Mattia *et al.* 2007) and Giró Blanc from Mallorca (Ibañez *et al.* 2003) shows that they are distinct, as suggested by Galet (2000).

VITICULTURAL CHARACTERISTICS

Early budding, late ripening, moderate vigour. Largish bunches of small, thick-skinned berries. Relatively resistant to downy and powdery mildews but slightly susceptible to botrytis bunch rot.

WHERE IT'S GROWN AND WHAT ITS WINE TASTES LIKE

The new generation of aromatic wines made on the island of Mallorca, **Spain**, are dry and some are barrel aged. Flavours range from citrus, pineapple and mango and through to pâtisserie, with medium acidity, quite high alcohol and noticeable power for a white wine. There are now 6 ha (15 acres) of Giró Blanc planted on Mallorca. Toni Gelabert was the first to produce a commercial wine, in 1995, from his vines near Manacor in the west of the island, followed by Can Majoral in Algaida, closer to Palma. Wines had to be produced on an experimental basis until the variety's official authorization in 2010 after a ten-year campaign.

GLAVINUŠA

Very rare Croatian variety sometimes useful for the Dalmatian sweet wine Prošek.

BERRY COLOUR: ● ● ● ● ●

PRINCIPAL SYNONYMS: Carnjenak, Glavanjuša, Glavinka, Okatac or Okatac Crni

ORIGINS AND PARENTAGE

Glavinuša comes from the Split region in Dalmacija (Dalmatia), the southern part of coastal Croatia.

VITICULTURAL CHARACTERISTICS

Late maturing, the berries often high in sugar but low in acid. Good general disease resistance.

WHERE IT'S GROWN AND WHAT ITS WINE TASTES LIKE

Glavinuša was at one time considered to be one of the finest varieties in central Dalmacija, **Croatia**. Today it is authorized in the Dalmatinska Zagora (Dalmatian Hinterland) and Srednja i Južna Dalmacija (Central and Southern Dalmatia) wine regions but cultivated only sparsely around Omiš, Sinj and Kaštela near Split, and to a lesser extent on the islands of Šolta and Vis (Zdunić 2005). Varietal wines are rarely made but the variety can be a useful component in Prošek, the sweet Dalmatian wine made from dried grapes. Along with several other autochthonous varieties, Glavinuša is part of a revitalization project involving clonal selection and replanting. (Although it is still officially known as Okatac Crni, as a result of its current re-evaluation, it is now more commonly known as Glavinuša.)

GLERA

See PROSECCO.

GODELLO

Very high-quality variety undergoing a revival, under several names, in north-western Iberia.

BERRY COLOUR: ● ● ● ● ●

PRINCIPAL SYNONYMS: Agodello, Agodenho (Portugal), Agudanho (Portugal), Agudelha or Agudelho (Portugal), Agudello, Agudelo, Agudenho (Portugal), Berdello, Godelho, Godella, Godenho, Gouveio (Douro in Portugal), Ojo de Gallo, Prieto Picudo Blanco (Tierra de León), Trincadente, Verdelho or Verdelho do Dão (Dão in Portugal)

VARIETIES COMMONLY MISTAKEN FOR GODELLO: Gouveio Estimado, GOUVEIO REAL, VERDEJO, VERDELHO or Verdelho da Madeira (Madeira)

ORIGINS AND PARENTAGE

Godello most likely originates from the banks of the Río Sil in Galicia, north-west Spain. Its earliest mention was in the Douro in 1531 by Rui Fernandes (1531–2) in the 'Descripção do terreno em roda da cidade de Lamego duas leguas' under the synonyms Agudelho and Trincadente. Comparison of the DNA profiles of Godello (Ibañez *et al.* 2003; Martín *et al.* 2003), Gouveio and Gouveio Roxo from the Douro (Lopes *et al.* 2006) shows that they are identical (Martín *et al.* 2006; Vouillamoz). **Gouveio Roxo** is therefore a colour mutation of Gouveio, and in consequence the same variety as Godello.

The vine called Verdelho in Dão has been shown to be identical to Gouveio (Lopes *et al.* 2006) and therefore also to Godello while the true VERDELHO planted on Madeira (also planted in Australia), which was also thought to be identical to Gouveio (Galet 2000) and Godello, has a distinct DNA profile (Vouillamoz).

However, it should be noted that the reference variety named GOUVEIO REAL in Lopes *et al.* (2006) has a unique DNA profile and it is not known to what extent the Douro vineyards are composed of Godello (under the name Gouveio) or of Gouveio Real. Another DNA study has shown that the variety named Prieto Picudo Blanco in the Tierra de León DO (Yuste *et al.* 2006; Santana *et al.* 2007) is not a colour mutation of PRIETO PICUDO profiled in other studies (Ibañez *et al.* 2003; Martín *et al.* 2003) but it is identical to Godello, which illustrates the wide distribution of this variety in the Iberian peninsula. DNA parentage analysis has also recently suggested that Godello is a sibling of VERDEJO from the Valladolid region (Santana *et al.* 2010).

According to Myles *et al.* (2011), the reference sample of Gouveio in the United States Department of Agriculture's National Clonal Germplasm Repository, which most probably corresponds to Godello, could be identical to HÁRSLEVELŰ from Hungary but this can be rejected by comparison with other published DNA profiles for both GOUVEIO REAL and Godello (Vouillamoz).

VITICULTURAL CHARACTERISTICS

Well suited to dry sites. Small, compact bunches of small berries with medium-thick skins and succulent flesh. Fertile and more productive than its Galician compatriot Albariño (see ALVARINHO). Early budding and ripening, susceptible to powdery mildew and botrytis bunch rot, less so to downy mildew. Berries can attain high sugar levels and medium to high acidity.

WHERE IT'S GROWN AND WHAT ITS WINE TASTES LIKE

Godello was on the verge of extinction in the 1970s, down to little more than several hundred vines, but thanks in large part to Horacio Fernández and Luis Hidalgos' Godello Revival (REstructuring of the VIneyards of VALdeorras) project, Godello continues to gain in popularity and reputation in its homeland of Galicia, north-west Spain, where the vast majority of **Spain**'s 1,153 ha (2,849 acres) were found in 2008 (up from 880 ha/2,175 acres in 2004), notably in the Valdeorras DO, where distinctive, high-quality varietal wines are increasingly common. This resurgence is all the more impressive given that the first varietal Godello wine was released only in the mid 1980s by Godeval, founded by Fernández. It is also authorized in other Galician DOs such as Ribeiro, Ribeira Sacra and Monterrei and in the nearby Bierzo DO in Castilla y León, although here it may well be part of a blend with varieties such as Treixadura (TRAJADURA; Coto de Gomoriz in Ribeiro is a good example). Fine, mineral-rich wines are produced by Godeval, Rafael Palacios, Quinta do Buble, Telmo Rodriguez, La Tapada and Valdesil. The best wines are intense and concentrated with sleek minerality, well suited to barrel fermentation, and are often long-lived due to the highly reductive nature of the variety.

In **Portugal** in 2010 there were 970 ha (2,397 acres) of Gouveio, mainly in the Douro in the north, but also in Dão, a little further south, sometimes under the misleading name Verdelho. Wines are generally complex and quite full-bodied, in the Douro typically part of a rewarding blend with other local varieties such as VIOSINHO, RABIGATO, Códega (SÍRIA), CÔDEGA DE LARINHO and MALVASIA FINA, although it is also sometimes used in a white port blend. Recommended Douro producers of dry table wines that have Gouveio in the mix include Quinta do Crasto, Quinta de la Rosa, Quinta do Vallado and Wine and Soul. Caves Transmontas make still and sparkling varietal wines.

There were 487 ha (1,203 acres) of the red mutation Gouveio Roxo in Portugal in 2009.

GOLDBURGER

Generally unexciting pale Austrian cross used mainly for sweet wines.

BERRY COLOUR: ● ● ● ● ●

PRINCIPAL SYNONYMS: Klosterneuburg 16-8, Orangeriesling

ORIGINS AND PARENTAGE

A GRAŠEVINA (under the name Welschriesling) × ORANGETRAUBE cross obtained by Fritz Zweigelt in 1922 at the Klosterneuburg research centre in Austria.

VITICULTURAL CHARACTERISTICS

Mid ripening, productive, susceptible to powdery mildew.

WHERE IT'S GROWN AND WHAT ITS WINE TASTES LIKE

Austria has 246 ha (608 acres), the majority planted in Neusiedlersee in Burgenland but also scattered across Niederösterreich (Lower Austria) and the Steiermark (Styria). Wines made from these golden-skinned grapes are relatively neutral-tasting but full-bodied and high in extract with moderate acidity. They are used mainly for sweet wines.

GOLDEN MUSCAT

Minor American hybrid producing grapey dessert wines.

BERRY COLOUR: ● ● ● ● ●

PRINCIPAL SYNONYMS: New York 10303

ORIGINS AND PARENTAGE

This is a complex *Vitis labrusca* and *Vitis vinifera* hybrid obtained in 1915 at Cornell University in New York State by crossing MUSCAT OF HAMBURG with MOORE'S DIAMOND (see BRIANNA for Moore's Diamond's pedigree), released in 1927.

VITICULTURAL CHARACTERISTICS

Hardy and productive. Large bunches of large berries. Late ripening and hard to get to full maturity in much of New York State. Susceptible to cracking and botrytis bunch rot. More suitable for home than commercial use.

WHERE IT'S GROWN AND WHAT ITS WINE TASTES LIKE

Golden Muscat is found to a limited extent in the eastern US, in states such as **Indiana** and **Illinois** and even in the more southerly state of **Tennessee**. Wines tend to be rich, grapey and usually sweet and may have a foxy (*labrusca*) flavour. Producers of varietal wines include Carousel in Indiana, Simon Creek in **Wisconsin**, and August Hill and Inheritance Valley in Illinois.

GOLDMUSKATELLER

See MOSCATO GIALLO.

GOLDRIESLING

Alsace cross producing light, white quaffing wine, now a speciality of Sachsen in eastern Germany.

BERRY COLOUR: ● ● ● ● ●

PRINCIPAL SYNONYMS: Franzosentraube, Gelbriesling, Goldmuskat, Riesling Doré, Risling Khativ, Risling Zolotistyi

ORIGINS AND PARENTAGE

Goldriesling was obtained in 1893 by Christian Oberlin in the Institut Viticole Oberlin in Colmar in Alsace, north-eastern France, by crossing RIESLING with an as yet unidentified variety. Indeed, Goldriesling was said by its breeder to be a cross between Riesling and Muscat Précoce de Saumur (or Courtiller Musqué Précoce), a French table grape, but this has been disproved by DNA profiling (Regner *et al.* 2001).

Goldriesling was used to breed LÉON MILLOT, LUCIE KUHLMANN and MARÉCHAL FOCH.

In Austria, a GRAŠEVINA (ie Welschriesling) × MÜLLER-THURGAU cross obtained by David Schantl was also and erroneously named Goldriesling but this variety has not yet been cultivated.

VITICULTURAL CHARACTERISTICS

Productive. Early budding, therefore at risk from spring frosts, and early ripening. Not especially cold-hardy. Susceptible to downy mildew and botrytis bunch rot. Small bunches and berries.

WHERE IT'S GROWN AND WHAT ITS WINE TASTES LIKE

Goldriesling barely survives in **France** (less than 10 ares, or one tenth of a hectare/1,200 square yards, in 2006) but is a little more widely planted in eastern **Germany**, exclusively in the state of Sachsen, where there were 17 ha (42 acres) in 2008. Varietal wines are made by producers such as Karl Friedrich Aust, Joachim Lehmann, Steffen Loose, Schloss Proschwitz, Schloss Wackerbarth, Weinhaus Schuh, Jan Ulrich and by the Winzergenossenschaft Meissen. Wines are generally light, easy-drinking, dry or off-dry, gently aromatic with a range of citrus flavours, sometimes slightly herbaceous, sometimes with a hint of Muscat.

GOLUBOK

Red-fleshed Ukrainian hybrid also found in Russia.

BERRY COLOUR: ● ● ● ● ●

PRINCIPAL SYNONYMS: Golubuk

ORIGINS AND PARENTAGE

Golubok, meaning 'pigeon', is a hybrid obtained in 1958 by P K Ayvazyan, E N Dokuchaeva, A P Ablyazova, M I Tulaeva, L F Meleshko and A K Samborskaya at the Tairov research centre in Odessa, Ukraine, by fertilizing the Russian hybrid SEVERNY with a mixture of pollen from several varieties, 40 Let Oktyabrya, Odessky Ranny and 1-17-54 (ALICANTE HENRI BOUSCHET × CABERNET SAUVIGNON), all three being non-cultivated local crosses. Officially registered in 1981, there are now three clones in common use.

VITICULTURAL CHARACTERISTICS

Productive, early to mid ripening. Red-fleshed berries with thin but strong skins. Winter-hardy and resists frost to –25 °C (–13 °F). Not particularly susceptible to powdery mildew or botrytis bunch rot.

WHERE IT'S GROWN AND WHAT ITS WINE TASTES LIKE

Golubok is planted in the Odessa and Kherson regions of **Ukraine** and used here and in Russia for dry, sweet and fortified wines, which are typically deeply coloured, often spicy, rich in tannins and showing flavours that range from cassis and red berries to dried fruits. In Krasnodar, southern **Russia**, where there were 37 ha (91 acres) in 2010, Château Le Grand Vostock blend it with several other varieties, including KRASNOSTOP ZOLOTOVSKY and CABERNET SAUVIGNON in their Terres du Sud dry red. In Rostov, closer to the Ukraine border, Vedernikov Winery blend it with TSIMLYANSKY CHERNY. The variety is also found in the **Czech Republic**.

Golubok was among several varieties trialled at Tabor Home in **Iowa**, on their own roots, but all succumbed to the cold winters or crown gall within five years.

GORDO

See MUSCAT OF ALEXANDRIA.

GORGOLLASA

Recently resurrected, high-quality, red-fruited Mallorcan.

BERRY COLOUR: ● ● ● ● ●

PRINCIPAL SYNONYMS: Gargollasa, Gargollosa, Gorgollosa

ORIGINS AND PARENTAGE

Gorgollasa is indigenous to Mallorca in the Islas Baleares, Spain, where it was already mentioned in 1839 in Pollença and was the main variety of the Raiguer area. Recent DNA parentage analysis has suggested that Gorgollasa is a natural cross between MONASTRELL from Valencia on Spain's east coast and Hebén, an almost extinct Andalusian variety (García-Muñoz *et al.* 2012). This would make it a half-sibling of MANDÓN from the Bierzo region in the province of León in north-west Spain and MACABEO from Catalunya, but these parentages were based on too few DNA markers (twenty) and need to be confirmed.

VITICULTURAL CHARACTERISTICS

Late budding and ripening, low vigour. Medium-sized bunches of small, thin-skinned berries. Relatively resistant to downy and powdery mildews but susceptible to botrytis bunch rot.

WHERE IT'S GROWN AND WHAT ITS WINE TASTES LIKE

There are now 4 ha (10 acres) of Gorgollasa planted on the island of Mallorca, **Spain**, in Consell, Algaida and Manacor, and the area is increasing. It was virtually extinct until Can Ribas rescued it by planting 200 vines in 1998, propagated from the four remaining vines found near Celler Can Amer, in Inca. They planted an additional 1,800 vines in 2000 but the variety was not officially authorized until 2011, even though it is a traditional Mallorcan variety and despite strong demand for its official accreditation.

Gorgollasa is well suited to barrel ageing, producing wines that are dry and elegant but with a tendency to oxidation, which is why larger (eg 500-litre) barrels are preferred. Producers include Can Majoral, Can Ribas and Toni Gelabert. Wines typically have strawberry and violet flavours, fairly low acidity but an unexpected freshness nonetheless, moderate alcohol and soft tannins. According to Araceli Servera Ribas, the variety is closer to PINOT NOIR than to other Mallorcan varieties such as MANTO NEGRO or CALLET.

GORULI MTSVANE

Fresh, fruity, light-skinned Georgian variety used to make both still and sparkling wines.

BERRY COLOUR: ● ● ● ● ●

PRINCIPAL SYNONYMS: Goruli Mcvané, Kvishkhuri, Mtsvane, Mtsvané, Suramula, Tetrpotola

VARIETIES COMMONLY MISTAKEN FOR GORULI MTSVANE: MTSVANE KAKHURI ☒

ORIGINS AND PARENTAGE

Goruli Mtsvane, meaning 'green from Gori', referring to the colour of the berries, is an old indigenous Georgian variety from the Kartli region, named after the town of Gori close to the Greater Caucasus mountains. It has also been suggested the name means 'green from the hills'. Goruli Mtsvane should not be confused with the more heavily planted MTSVANE KAKHURI from the Kakheti region in south-eastern Georgia; the latter is commonly referred to simply as Mtsvane but they have distinct DNA fingerprints (Vouillamoz *et al.* 2006).

OTHER HYPOTHESES

Along with KHIKHVI, RKATSITELI and MTSVANE KAKHURI, Goruli Mtsvane is said to have been mentioned as early as the fifth century AD but there is no evidence for this.

VITICULTURAL CHARACTERISTICS

Medium-sized to large bunches of thick-skinned berries. Resistant to downy mildew but susceptible to powdery. Late budding and ripening.

WHERE IT'S GROWN AND WHAT ITS WINE TASTES LIKE

Planted mainly in Kartli and Imereti in **Georgia**, Goruli Mtsvane produces high-quality wines with fresh acidity and is also widely used as a base for sparkling wines. It has the potential to make very good fortified wines but, with such limited plantings, they are rarely found today. Chateau Mukhrani make a good varietal example that has attractive flavours of honey and wild flowers. There were 221 ha (546 acres) in Georgia in 2004.

GOUAIS BLANC

Extremely important progenitor (with Pinot) even if almost extinct.

BERRY COLOUR: ● ● ● ● ●

PRINCIPAL SYNONYMS: Belina or Belina Drobna (Croatia), Best's no 4 (Australia), Blanció (Val Maira in Piemonte in Italy), Bouilleaud (Charente and south-west France), Enfariné Blanc (Jura in France), Foirard (Jura), Gôt, Gouais Jaune (Valais in Switzerland), Gouget Blanc (Allier and Cher in France), Grobe (Austria), Gueuche Blanc (Jura), Gwäss (Haut-Valais in Switzerland), Hajnos (Hungary), Heunisch Weiss or Heinsch or Heinisch (Germany and Austria), Krapinska Belina (Croatia), Liseiret (Val di Susa in Piemonte), Plant de Séchex (Marin in France), Président (France), Preveiral (Val di Susa), Provereau Blanc (Isère in France), Weisser Heunisch (Germany), Wippacher (Croatia)

VARIETIES COMMONLY MISTAKEN FOR GOUAIS BLANC: ELBLING (Rheintal in Germany), RANFOL (Slovenia)

ORIGINS AND PARENTAGE

Gouais Blanc is one of Western Europe's most ancient and prolific grape varieties. It was extremely widespread in the Middle Ages under many different names that were mainly variations on Gouais (Gau, Gauche, Geuche, Goe, Goet, Gohet, Goi, Goin, Goix, Got, Gouay, Gouche, Gouest, Gueuche, etc.) and Heunisch (Heinisch, Heinsch, Heinschen, Hennische, Hensch, Heunsch, Hinschen, Hintsch, Huensch, Huntsch, Hyntsch, etc.). The earliest plausible, albeit derogatory, mention of a white-berried Heunisch appeared in Hieronymus Bock's (1539) *Kreutterbuch*: 'die grossen feiste Hynische Drauben, welche umb ihrer schnelle würckung willen, von etlichen scheiss Drauben genandt werden' ('the big large Hynsch grapes, which are known for their rapid growth, are called shit grapes by some'). The earliest reliable mention of a white-berried Gouais appeared in 1540 in the Haut-Valais, Switzerland, under the Germanized name Gewess, today Gwäss (Vouillamoz and Moriondo 2011): 'blantschier und gewess, die 9 Gross gelten' ('blantschier [ie PLANTSCHER] and gewess, which are worth 9 gros'). Although it was most likely present in France much earlier than in Switzerland, the first reliable mention of a white-berried Gouais appeared in Charles Estienne and Jean Liébault (1564): 'Sur tous complants, nul n'est moins tendre à la gelée que le Gouëst' ('among all varieties, none is less sensitive to frost than Gouëst').

FRANCO-GERMAN ORIGIN Gouais Blanc is often said to originate in Eastern Europe (see Other hypotheses) but for several reasons it seems more likely that Gouais Blanc comes from the area encompassing central north-eastern France and south-western Germany:

- Vavilov's (1926) theory says that the place of origin of a variety usually corresponds to the place where the most important biodiversity is observed, and various forms were reported in that area: Gouais Long, Gouais Rond, Gouais Jaune, Gouais Saugé, a pink-berried mutation mentioned by Guicherd in the Aube (Galet 1990), and another one mentioned by Boursiquot *et al.* (2004) in south-west Germany under the name Dreifarbiger Heunisch.

- The French ampelographer Adrien Berget (1903) had also suggested northern France as the origin of Gouais Blanc and similarly named varieties: 'Nul de nos cépages n'a été signalé de plus longue date et ne semble plus vraiment français d'origine ... Les Gouais paraissent, en effet, indigènes des provinces de l'Île-de-France et de la Champagne où ... ils composaient, au moyen âge, la majeure partie des vignobles les plus communs au Nord-Est de la France' ('None of our grape varieties has been mentioned over a longer period and seems more genuinely French ... Indeed, the Gouais seem to be indigenous to the provinces of Île-de-France and Champagne where ... they formed the major part of the ordinary vineyards in north-eastern France in the Middle Ages').
- In the Middle Ages, not only were PINOT, SAVAGNIN and Gouais Blanc the dominant varieties in north-eastern France but they also gave birth to many offspring there (see below).
- The name Gouais is very likely to derive from the name of one or several French villages (see below).

At that time, in Germany as well as in France, the noble PINOT and SAVAGNIN (the Fränkisch group, see below) were assigned the best vineyards, while the feeble Gouais Blanc and ELBLING (the Heunisch group) were given the worst expositions, so that orders to pull out Gouais Blanc were given as early as 1598 and the variety had almost disappeared from France by the twentieth century (Berget 1903).

PROGENIES DNA parentage analyses have revealed that Gouais Blanc is the parent of at least eighty-one distinct grape varieties in Western Europe (see the pedigree diagram in PINOT), justifying its nickname 'Casanova of the grapes':

- Bowers *et al.* (1999) discovered that Gouais Blanc and PINOT are the natural parents of sixteen grape varieties from north-eastern France, including CHARDONNAY and GAMAY NOIR (see Pinot for the pedigree), which came as a huge surprise. These crosses probably occurred in different places and at different times.
- Bowers *et al.* (2000) found that AUBIN BLANC and PETIT MESLIER from northern France are natural crosses between Gouais Blanc and SAVAGNIN, while BALZAC BLANC, COLOMBARD and MESLIER SAINT-FRANÇOIS are natural progenies of Gouais Blanc and CHENIN BLANC.
- Five additional offspring of the Gouais Blanc × PINOT cross were found later, along with fifty-five possible offspring of Gouais Blanc with unknown parents, the most famous being FURMINT, BLAUFRÄNKISCH, ELBLING and RIESLING (Regner *et al.* 1998; Boursiquot *et al.* 2004; Vouillamoz).

According to Bisson (2009), Gouais Blanc is the reference variety for the Gouais ampelographic group (see p XXVII).

Gouais Blanc was recently used to breed MENNAS.

ETYMOLOGY Berget (1903) has suggested two distinct origins for the word Gouais, possibly deriving from:

- the name of a village, such as Gouaix in the Seine-et-Marne, Gouais-les-Saint-Bris in the Yonne, Gouex in the Vienne or Goix in the Nièvre
- the adjective *gou* used as a derogatory term in central and eastern dialects in France, corresponding to the poor quality of Gouais wines.

Berget favours the second hypothesis but the first seems more likely because the ancient names of these villages corresponded to old names of Gouais (for example, Gouais-les-Saint-Bris was called Goys in 1408, Goez in *c.*1550 and Goix in 1782), and also because old variety names were often based on the characteristics of the vine or the grape but rarely on those of the wine.

Ampelography and DNA profiling have established that Gouais Noir (or Heunisch Rot or Heunisch Schwarz) is not a colour mutation of Gouais Blanc (Boursiquot *et al.* 2004; Frei *et al.* 2006).

OTHER HYPOTHESES

Most authors consider that the earliest mention of Gouais Blanc appeared in 1283 under the spelling Goet in a legal tome, *Les coutumes de Beauvoisis* by the lawyer Philippe de Beaumanoir (1246/7–96) in the Beauvais region, north of Paris (Beugnot 1842; see PINOT for the complete citation): 'le vin de gros noirs ou de goet chascun mui. VI. sols de rente' ('the wine of gros noirs or goet 6 sols each for income'). However, this refers to a black-berried variety, which makes it difficult to identify because a black mutation of Gouais Blanc has never been observed and because the name Gouais Noir was used for several black-berried varieties, including ENFARINÉ NOIR, GUEUCHE NOIR and GAMAY NOIR, all closely related to yet genetically distinct from Gouais Blanc. Similarly, the mention of Gouais in the anonymous poem 'La disputoison du vin et de l'iaue', written *c.* 1300, and the 1394 mention by Eustache Deschamps ('le bon plant ne fait que changier. Gouais devient le Morillon', ie 'the good grapevine keeps changing. Gouais becomes Pinot') also relate to black-berried varieties.

In medieval Germany, grapevines were separated into two general groups: Fränkisch (*Vinum franconicum*) for the noble varieties giving good wines (eg CHASSELAS, PINOT, SAVAGNIN) and Heunisch (*Vinum hunicum*) for the poor varieties giving mediocre wines (eg Gouais Blanc, ELBLING, etc.; Bassermann-Jordan 1923). It is therefore often impossible to identify which variety is referred to under the name Heunisch (and its variant forms) in old German documents. Since the name Heunisch possibly derives from the Huns, a group of nomads from Asia, it has often been speculated that Gouais Blanc was introduced by the Huns from Hungary, where it was called Hajnos (Aeberhard 2005; Maul 2005). This would be consistent with Gouais Blanc being a parent of the Hungarian grape FURMINT but it would not explain why Gouais Blanc has now disappeared from Hungary, where it had only one progeny, and why the vast majority of Gouais' progenies are in central northern France and south-western Germany. Similarly, Gouais Blanc is said to originate from Dalmacija (Dalmatia) and to have been given to the Gauls by the third-century Roman emperor Probus (Goethe 1887; Bowers *et al.* 1999), but Gouais Blanc has never been mentioned in Croatia and has no progeny there.

In Slovenia, Štajerska Belina, meaning 'white from Styria', is often mistakenly considered to be a synonym of Gouais Blanc, whereas it is identical to RANFOL (Edi Maletić, personal communication).

VITICULTURAL CHARACTERISTICS

Vigorous, very productive and fertile. Mid budding and ripening. Big bunches of big, thin-skinned berries. Susceptible to botrytis bunch rot but resistant to winter frosts.

WHERE IT'S GROWN AND WHAT ITS WINE TASTES LIKE

Gouais Blanc's significance to today's wine drinker lies solely in its famous offspring and it is now very hard to find this variety in the vineyard. It has been banned several times in **France** and only one small vineyard remains, in Marin in the Haute-Savoie, but the wine is not commercially bottled.

In **Italy**, Preveiral is still grown in some of the alpine valleys in northern Piemonte such as Val di Susa, Val Bormida (where it is known as Liseiret) and Pinerolo. Enzo Berger makes a small but, he hopes, increasing number of bottles of varietal wine from some old Liseiret vines in Val Germanasca and the Conte Reyneri di Lagnasco makes a varietal wine in Saluzzo, south-western Piemonte.

Georg Breuer in Rüdesheim in the Rheingau, **Germany**, planted a small parcel of Heunisch Weiss and the first wine was made in 2007.

In the Haut-Valais, **Switzerland**, Gouais Blanc has been continuously cultivated since the late Middle Ages under the name Gwäss and it is made today by a handful of producers. Josef-Marie Chanton's very good example is made from vines carefully tended to control yields and has aromas of green apple, pear blossom and a marked yet attractive citrus acidity. A handful of other Haut-Valais producers also make a varietal Gwäss: Daniela Kramberger, Lengen Weine and Weinkeller zum Leysche. Chambers Rosewood in Rutherglen, **Australia**, occasionally produce a varietal wine from their 0.6 ha (1.5 acres) of centenarian vines.

GOUGET NOIR

Rare and disappearing central French vine.

BERRY COLOUR: ● ● ● ● ●

PRINCIPAL SYNONYMS: Gouge Noir (Allier, Cher), Gouget, Goujet, Lyonnais (Indre), Nérou, Neyran (Allier), Neyrou (Puy-de-Dôme)

VARIETIES COMMONLY MISTAKEN FOR GOUGET NOIR: GAMAY NOIR, PINOT NOIR

ORIGINS AND PARENTAGE

Gouget Noir is a rare variety of the Allier and the Cher *départements* in central France, where it was first documented in 1843 (Rézeau 1997): 'le Nérou ou Gouget est celui de tous les cépages dont le raisin est le plus sucré, le plus agréable à manger' ('Nérou or Gouget is the sweetest, the most pleasant to eat among grape varieties'). It could simply have been named after its propagator, Gouget being a common local family name. Gouget Noir is a member of the Noirien ampelographic group (see p XXVII), along with AUXERROIS, CHARDONNAY, GAMAY NOIR, GAMAY TEINTURIER DE BOUZE, GENOUILLET, MELON, PINOT and Troyen, which explains why it is sometimes erroneously considered to be a clone of Gamay Noir or Pinot. According to the official French register of varieties, DNA analysis indicates that Gouget Noir is probably a descendant of GOUAIS BLANC (see Pinot pedigree diagram).

VITICULTURAL CHARACTERISTICS

Early budding and ripening. Vigorous and generally pruned long. Fruits well even after spring frosts but very susceptible to botrytis bunch rot. Small bunches and berries

WHERE IT'S GROWN AND WHAT ITS WINE TASTES LIKE

Gouget Noir used to be the main variety in Montluçon, Domérat and Huriel in the Allier *département* of central **France**, where it was largely replaced by GAMAY NOIR after the phylloxera crisis towards the end of the nineteenth century. In the mid nineteenth century, when the Allier had a much more significant vineyard area (nearly 17,000 ha/42,008 acres), Gouget Noir accounted for more than half the total plantings in the region (Kelley 2010). Official French statistics for 2008 record just 10 ha (25 acres), compared with 739 ha (1,826 acres) fifty years ago. Gilles Desgranges at the Ferme des Barchauds near Huriel seems to be the only producer persevering with this variety.

GOUSTOLIDI

Traditional, rustic Greek variety.

BERRY COLOUR: ● ● ● ● ●

PRINCIPAL SYNONYMS: Augoustelidi (Aitolía, Achaïa), Augoustelli (Arkadía), Avgoustolidi, Bostilidas (Kefaloniá, Itháki), Goustoulidi, Goystolidi, Rompola, Vostilidi or Vostilida or Voustolidi (Kefaloniá)

VARIETIES COMMONLY MISTAKEN FOR GOUSTOLIDI: ROBOLA

ORIGINS AND PARENTAGE

Goustolidi is a Greek variety whose exact origin is unknown but whose name comes from the Greek *avgoustolidi*, meaning 'August's jewel' in reference to its early ripening (Boutaris 2000). It is genetically very close but not identical to ROBOLA (Biniari and Stavrakakis 2007), even though they share the synonym Rompola, and it has a possible parent–offspring relationship with the dark-skinned THEIAKO MAVRO (Boutaris 2000). Similarly and even more confusingly, it also shares many synonyms in the Vitis International Variety Catalogue with the dark-skinned AVGOUSTIATIS, of which it is not a colour mutation. On Kefaloniá, Goustolidi is known as Vostilidi.

VITICULTURAL CHARACTERISTICS

Vigorous and productive. Large bunches of medium-sized to large, thick-skinned berries. Late budding and mid to late ripening. Good resistance to drought and botrytis bunch rot but susceptible to powdery and downy mildews.

WHERE IT'S GROWN AND WHAT ITS WINE TASTES LIKE

Wines from this variety, whether they be called Goustolidi or Vostilidi, are generally rustic, robust and fairly alcoholic – as Marianna Kosmetatos of Gentilini says, 'traditionally part of the daily diet ... to fortify the farmers in their heavy manual labour'.

Goustolidi is grown mainly on the island of Zákynthos and on Kefaloniá, **Greece**, where it is known as Vostilidi, although it is also found to a limited extent in the Pelopónnisos (Peloponnese) and the south west of the mainland. On Zákynthos it contributes to the island's Verdea wine – a traditional dry white blend, now less oxidized and lower in alcohol than it used to be. Producers such as Comoutos and Callinico make a Verdea, each with a slightly different combination of varieties but generally including Goustolidi, Skiadopoulo (see FOKIANO) and Pavlos (see MALVASIA BIANCA LUNGA), among others.

On Kefaloniá, there are around 30 ha (74 acres) of Vostilidi, much of it in old mixed vineyards. The must is rich in sugar, moderate in acidity (depending on the altitude). It was traditionally made and drunk locally but more recently there have been attempts to make it in a more modern style for a wider market, although it oxidizes easily, which is probably why producers such as Divino, Domaine Foivos and Melissinos blend it with other local varieties, including ROBOLA and ZAKYNTHINO. Sclavus is one of the few producers to make a varietal wine: orange in colour, quite tannic, with marked flavours of apricots and considerable length. Experiments with oak maturation have shown good results at Melissinos.

GOUVEIO

See GODELLO.

GOUVEIO REAL

Minor light-skinned variety used for still and sparkling wines in northern Portugal.

BERRY COLOUR:

VARIETIES COMMONLY MISTAKEN FOR GOUVEIO REAL: GODELLO, Gouveio Estimado

ORIGINS AND PARENTAGE

Gouveio Real most probably comes from the Douro region in northern Portugal, where it is frequently confused with Gouveio, the local name for GODELLO, but recent DNA profiling has shown that the two varieties are quite distinct (Lopes *et al.* 2006).

VITICULTURAL CHARACTERISTICS

Vigorous, early budding, mid ripening. Small, compact bunches of small berries. Susceptible to powdery mildew and botrytis bunch rot, less so to downy mildew.

WHERE IT'S GROWN AND WHAT ITS WINE TASTES LIKE

Gouveio Real wines are relatively full-bodied but remain lively, typically with floral and fruity aromas and freshness in the mouth. There were 955 ha (2,360 acres) of Gouveio Real in the northern half of **Portugal** in 2009, mainly in the Douro. It is used mainly for sparkling wines and for white blends. Murganheira in the Vale do Varosa, for example, produce a sparkling varietal wine as well as a blend with MALVASIA FINA and CERCEAL BRANCO. Not all wine regions make the distinction between the different Gouveios in their regulations, though Gouveio Real is specifically authorized both for port wine and for unfortified table wines in the Douro.

GR 7

Minor but sturdy, winter-hardy complex American hybrid used mostly in blends.

BERRY COLOUR: ● ● ● ● ●

PRINCIPAL SYNONYMS: Geneva Red 7, NY 34791

ORIGINS AND PARENTAGE

A Buffalo × BACO NOIR hybrid made in 1947 at New York State's Cornell University, in which:

- Buffalo is a black-berried Herbert × Watkins hybrid obtained in 1921
- Herbert is a Carter × SCHIAVA GROSSA cross obtained in 1865
- Carter (also called Mammoth Globe) is an offspring of ISABELLA (*Vitis vinifera* × *Vitis labrusca*) and another unknown parent selected by Charles Carter in Virginia
- Watkins is a Mills × Ontario cross obtained in 1911 (see CAYUGA WHITE for the pedigree of Ontario)
- Mills is a MUSCAT OF HAMBURG × Creveling cross
- Creveling is possibly a *Vitis vinifera* × *Vitis labrusca* hybrid.

GR 7 is therefore a complex *Vitis labrusca* and *Vitis vinifera* hybrid.

Fruit was first observed in 1953 and vines were propagated that year and tested under the number NY 34791 (later renamed GR 7 or Geneva Red 7). After further testing it was eventually released in 2003 (Reisch *et al.* 2003).

VITICULTURAL CHARACTERISTICS

Early budding (therefore sometimes at risk from spring frosts), early to mid ripening. Productive, vigorous, winter-hardy and more resistant to ringspot virus than BACO NOIR, DE CHAUNAC and CHELOIS. Well adapted to mechanized production.

WHERE IT'S GROWN AND WHAT ITS WINE TASTES LIKE

GR 7 is valued mainly for its contribution to blends; Glenora in **New York** State's Finger Lakes region, for example, combine it with BACO NOIR in their Bobsled Red. In warmer years, wines show cherry and red-fruit aromas, as well as some foxy *labrusca* notes, but tend to be herbaceous in cooler years. They have better tannin structure but lower acidity than Baco Noir but higher acidity than, for example, DORNFELDER (Reisch *et al.* 2003). There were 64 acres (26 ha) in New York State in 2006 and minor plantings in states such as **Iowa** and **Pennsylvania**.

GRACIANO

Low-yielding, richly coloured, perfumed variety that retains acidity and is regaining favour in Rioja.

BERRY COLOUR:

PRINCIPAL SYNONYMS: Bovale, Bovaleddu, Cagliunari (Alghero), Cagniulari, Cagnulari, Caldareddhu or Calda Reio (Gallura on Sardegna), Courouillade (Languedoc), Graciana (Argentina), Graciano de Haro (Rioja), Minustellu (Corse), Monastrell Menudo or Monastrell Verdadero (Spain), Morastell or Morrastel (Languedoc), Moristell, Tinta Miúda, Tintilla de Rota (Jerez), Xeres (Australia), and many more

VARIETIES COMMONLY MISTAKEN FOR GRACIANO: BOBAL, BOVALE SARDO (Sardegna), MAZUELO, MONASTRELL, MORISTEL, MORRASTEL BOUSCHET, MURISTELLU, NÉGRETTE (Camargue in southern France), PARRALETA, PASCALE

ORIGINS AND PARENTAGE

Graciano has a huge number of synonyms in Spain and all around the Mediterranean, suggesting it is an old and widespread variety. Spanish researchers were astonished when they found by DNA profiling that Graciano, PARRALETA and Tintilla de Rota are one and the same variety (Borrego *et al.* 2001) because such a hypothesis had never previously been advanced. However, the synonymy with Parraleta was recently disproved by more thorough studies showing that the two varieties are distinct yet closely related (Montaner *et al.* 2004; Buhner-Zaharieva *et al.* 2010). In addition, DNA profiling has provided evidence that Graciano is identical to Bovale Sardo and Cagnulari from Sardegna (Nieddu *et al.* 2007). Indeed, several scholars since the nineteenth century have considered Cagnulari to be a clone of Bovale Sardo, and its identity with Graciano had already been suggested by the French ampelographer Truel (Galet 2000). In addition, the synonymy between Bovale Sardo, Cagnulari and Graciano is consistent with their similar ampelographic descriptions (Galet 2000) and with their clustering together in genetic analyses (Nieddu *et al.* 2007).

On the other hand, some authors suggested that Bovale Sardo was simply a clone of Bovale Grande, which was categorically refuted by ampelographic and molecular studies, not least because Bovale Grande is identical to Carignano (MAZUELO) on Sardegna (Reale *et al.* 2006; Nieddu *et al.* 2007). Moreover, a tentative parentage analysis based on very limited DNA data suggests that Bovale Sardo could be a parent or a progeny of Bovale Grande (Reale *et al.* 2006). In other words, Graciano and Mazuelo could have a parent–offspring relationship, but this can be rejected by the analysis of their respective DNA profiles (Vouillamoz). Nevertheless, it is clear that the cross that gave birth to Graciano or to Mazuelo originally occurred in Spain.

Recent DNA kinship analysis has suggested that Graciano and MONASTRELL could be siblings (Santana *et al.* 2010), which might explain the existence of the synonyms Muristeddu or Muristellu for Bovale Sardo on Sardegna and of the synonyms Monastrell Menudo and Monastrell Verdadero for Graciano in Spain. However, comparison of DNA profiles from independent sources disproves any close relationship between Graciano and Monastrell (Vouillamoz). This erroneous hypothesis most probably comes from the confusion between Monastrell and Moristell: Monastrell is identical to Mourvèdre, while Moristell is identical to Graciano, and consequently with Bovale Sardo and Cagnulari. DNA parentage analysis has also suggested that Graciano is a parent of MANDÓN from the Bierzo region in the province of León in north-west Spain (García-Muñoz *et al.* 2012).

After it was enfeoffed to the Aragonese, the kingdom of Sardegna remained Spanish from 1323 to 1720. In all likelihood, this was the period during which Graciano was introduced from Spain to the island, where it took the name Bovale Sardo, possibly together with Carignan (MAZUELO), which took the name Bovale di Spagna (or Bovale Grande).

Graciano was used to breed MORRASTEL BOUSCHET.

OTHER HYPOTHESES

Some authors have suggested that this variety was introduced from France to Sardegna in the nineteenth century, initially and erroneously labelled as Mourvèdre, which took the name Cagnulari.

VITICULTURAL CHARACTERISTICS

Mid to late budding, late ripening. Vigorous, drought-resistant but difficult to grow due to low fertility and yields, susceptibility to downy mildew and rot. Its wine is valued for its acidity and perfume, however. Best suited to clay and limestone soils, cooler sites and short pruning.

WHERE IT'S GROWN AND WHAT ITS WINE TASTES LIKE

The fragrant, fresh, potentially powerful and occasionally spicy Graciano was far more widely planted in **Spain** before phylloxera struck at the end of the nineteenth century and has been in decline since then. However, increased appreciation of its fresh and aromatic contribution to Rioja blends, and the small but growing number of varietal wines, is highlighted by total plantings in 2008 of 1,478 ha (3,650 acres). Contino's single-vineyard Graciano, Valdemar's Inspiración Valdemar Graciano and Dinastia Vivanco's Colección Vivanco Graciano are notable examples of varietal Graciano produced in the Rioja region. As Tintilla de Rota, it is grown to a very limited extent in Jerez and is listed as an authorized variety for Vino de la Tierra de Cádiz and de Córdoba.

Similarly, in the nineteenth century under the name Morrastel it was relatively widespread in Languedoc-Roussillon, **France**, where it has now become a rarity.

It is more fashionable in Sardegna, **Italy**, under the names Bovale Sardo and Cagnulari. As Bovale Sardo, it is used in blends with other local varieties to provide colour, acidity and body (eg in the Mandrolisai DOC in Nuoro and Oristano and in the Campidano di Terralba DOC in Cagliari and Oristano). The excellent Korem made by Argiolas is a blend of mainly Bovale Sardo with Carignano and Cannonau – in other words, a blend of the Spanish varieties Graciano, MAZUELO and GARNACHA. As Cagnulari, a particular clone of Bovale Sardo that was rescued from extinction a few years ago by Giocanni Cherchi from Usini, it is mainly grown in the Sassari area. Cherchi was the first to produce a 100% Cagnulari, a full-bodied and well-structured wine with intense aromas of crushed red fruits, fresh mint leaves, pepper, ink, dry but rounded tannins and an attractive aftertaste of black cherries. Not every consumer of Sardinian wines is aware that a blend of Bovale Sardo and Cagnulari (eg in an Isola dei Nuraghi IGT wine) is a blend of two clones of the same variety, namely Graciano from Rioja!

In **Portugal**, Tinta Miúda, meaning 'small red one', was traditionally grown around Lisboa but can now also be found further north and in the Alentejo to the east. Total plantings in 2010 amounted to 461 ha (1,139 acres). When the summer remains dry, maturation is long, producing juice high in sugar and phenolics. In sites with cooler soils and exposed to the sea, it is particularly susceptible to rot. The best examples have good ageing potential thanks to the level of acidity and tannins. Recommended producers include Herdade dos Grous in the Alentejo and Quinta do Sanguinhal in Lisboa.

In **California**, there are some limited plantings and an increasing number of varietals bottled, mainly from Central Valley or Sierra Foothills fruit. There were 31 ha (90 acres) of Graciana in **Argentina** in 2007.

There are also pockets of Graciano vines, sometimes bottled as varietals, around **Australia.** Word of the variety's qualities is clearly spreading.

GRAISSE

Minor south-west French variety used mostly for armagnac.

BERRY COLOUR: ● ● ● ● ●

PRINCIPAL SYNONYMS: Blanquette Grise (Gers), Chalosse (Landes), Gras or Gras Blanc (Béarn), Grèce Blanche (Gers), Plant de Graisse, Plant de Grèce, Président (Jurançon, Armagnac), Ramassou Blanc (Landes), Tizourine Bou-Afrara (Algeria)

ORIGINS AND PARENTAGE

Graisse was first mentioned in 1783–4 in the Gers under its misspelt synonym Grèce Blanche (Rézeau 1997): 'Grèce Blanche, ce raisin est blanc sale, le grain gros, la grappe longue et serrée' ('Grèce Blanche, this grape is dirty white, the berry is big, the bunch is long and tight'). It was named after the French word *graisse*, meaning 'fat', alluding to the oily viscosity of the must. This becomes *grechu* in the Gascon dialect, later wrongly interpreted as Grèce (French for 'Greece') or Plant de Grèce.

Graisse is a member of the Gras ampelographic group (see p XXVII; Bisson 2009). DNA parentage analysis has suggested that Graisse is one of the numerous natural progenies of GOUAIS BLANC (Boursiquot *et al.* 2004; see PINOT pedigree diagram), which explains why it is sometimes called Président, just like Gouais Blanc.

VITICULTURAL CHARACTERISTICS

Productive, late ripening, tending to produce many shoots and suckers and low sugar levels in the berries. Susceptible to botrytis bunch rot.

WHERE IT'S GROWN AND WHAT ITS WINE TASTES LIKE

Wines tend to be neutral and acidic and better suited to armagnac production than to wine. According to Lavignac (2001), it is rarely planted today, because of its susceptibility to botrytis bunch rot, and is now found only in the Condom region of the Gers in south-west **France**. There were 17 ha (42 acres) remaining in 2008. Château de Neguebouc in Haut-Armagnac blend 20% Graisse with 80% Ugni Blanc (TREBBIANO TOSCANO) in the base wine for their Les Très Rares bottling.

GRAMON

Montpellier cross that seems to have found few proponents.

BERRY COLOUR: ● ● ● ● ●

ORIGINS AND PARENTAGE

A GARNACHA × ARAMON NOIR cross obtained in 1960 by Paul Truel at the Institut National de la Recherche Agronomique in Montpellier, southern France. Its name is a contraction of Grenache and Aramon (Rézeau 1997).

VITICULTURAL CHARACTERISTICS

Late ripening and sufficiently fertile to be pruned short and trained as a bushvine. Occasionally susceptible to millerandage and mites but shows good resistance to botrytis bunch rot.

WHERE IT'S GROWN AND WHAT ITS WINE TASTES LIKE

It is recommended in much of south-eastern France and on the island of Corse but official statistics for 2008 record fewer than 4 ha (10 acres). Wines are usually full-bodied with moderate tannins and colour.

GRAND NOIR

Pale red-fleshed nineteenth-century French cross with potential. Still found around the world.

BERRY COLOUR: ● ● ● ● ●

PRINCIPAL SYNONYMS: Gran Negro (Valdeorras in Spain), Grand Bouschet (Alentejo in Portugal), Grand Noir de la Calmette, Gros Noir (Pyrénées-Atlantiques), Sumo Tinto (Portugal)

VARIETIES COMMONLY MISTAKEN FOR GRAND NOIR: ALICANTE HENRI BOUSCHET⚭, MORRASTEL BOUSCHET⚭

ORIGINS AND PARENTAGE

A PETIT BOUSCHET × ARAMON NOIR cross obtained in 1855 by Henri Bouschet at the Domaine de la Calmette in Mauguio, near Montpellier in the Hérault *département*, in southern France. It is often mistaken for its half-sibling MORRASTEL BOUSCHET (Viala and Vermorel 1901–10), as in the grape collection at the University of California at Davis, for example (Vouillamoz).

VITICULTURAL CHARACTERISTICS

Red-fleshed though paler than teinturier varieties such as ALICANTE HENRI BOUSCHET. High yields. Tends to shrivel on low fertility soils. Susceptible to powdery mildew.

WHERE IT'S GROWN AND WHAT ITS WINE TASTES LIKE

Grand Noir is gradually disappearing from the Languedoc and the Cognac region (there was less than 1 ha/2.5 acres in **France** in 2008) but **Spain** had 885 ha (2,187 acres) in 2008, virtually all in Galicia in the green north west. It is treasured for its colour in DOs such as Valdeorras and Ribeira Sacra, where it has been replacing Garnacha Tintorera (ALICANTE HENRI BOUSCHET) because it seems much better adapted to the humid climate.

It is also still found in southern **Portugal**, mainly in older vineyards, where it may be interplanted with ALICANTE HENRI BOUSCHET. There were 347 ha (857 acres) in 2009 and the variety is authorized in the Portalegre subregion of the Alentejo appellation along with Aragonez (TEMPRANILLO), TRINCADEIRA and Alicante Henri Bouschet, although there are no new plantings and newer vineyards tend to favour Alicante Henri Bouschet.

A small block, planted in the early 1900s, has been found in Windsor in the Russian River Valley in **California**, where Grand Noir was commonly planted at one point, but they were pulled out over the years in favour of other, more fashionable varieties. It is reported to exhibit more interesting – often peppery – flavours than ALICANTE HENRI BOUSCHET, with which it is often compared.

GRAPARIOL

Veneto variety recently rescued from extinction.

BERRY COLOUR: ● ● ● ● ●

PRINCIPAL SYNONYMS: Grappariol, Rabosa Bianca, Rabosina Bianca, Rabosino Grappariol

ORIGINS AND PARENTAGE

Grapariol is a rare grape from the province of Treviso in the Veneto, northern Italy, that was believed to be a white mutation of RABOSO PIAVE, but recent DNA profiling has shown no direct relationship either with Raboso Piave or with RABOSO VERONESE (Crespan *et al.* 2009).

WHERE IT'S GROWN AND WHAT ITS WINE TASTES LIKE

Grapariol was rescued by the research centre at Conegliano in the Veneto, **Italy**. Thanks to the efforts of researchers and adventurous producers, Grapariol entered the national register of grape varieties in 2007 and a few hectares are now being cultivated. It produces wines with aromas of white flowers, white peaches and celery. Producers of varietal wines include Barbaran Vigne e Vini and Le Rive.

GRASĂ DE COTNARI

Rare Romanian variety once famous for its sweet Cotnari wines, occasionally found in those of Tokaj in Hungary.

BERRY COLOUR: ● ● ● ● ●

PRINCIPAL SYNONYMS: Bajor, ?Fehér Kövérszőlő, ?Fejérszőlő (Hungary), Gras, Grasă (Romania), Grasă de Cotnar (Romania), Grasă Mare (Romania), Grasă Mică (Romania), Grasi (Romania), Grassa (Romania), ?Kövérszőlő (Hungary), Poamă Grasă (Romania), Resertraube

VARIETIES COMMONLY MISTAKEN FOR GRASĂ DE COTNARI: FURMINT (Hungary)

ORIGINS AND PARENTAGE

Although ampelographers (Galet 2000; Dejeu 2004) seem to agree that the variety known in Romania as Grasă, meaning 'fat', or Grasă de Cotnari, and the variety known in Hungary as Kövérszőlő, meaning 'fat grape', are one and the same, this has not yet been confirmed by DNA profiling. In Hungary, this variety has been known since the nineteenth century but it could be much older because, according to Balassa in Varga (2008), Kövérszőlő is identical to Fejérszőlő, an old and unidentified variety that was frequently mentioned in old variety lists in Tokaj. In Romania, it has long been cultivated in the region of Cotnari, hence Grasă de Cotnari, and at least three morphological types have been described (Dejeu 2004): Grasă Galbenă (yellow berries), Grasă Verde (green berries) and Grasă Crocantă (crunchy berries). This biodiversity points to a Romanian origin for Grasă, supported by a recent study showing that it is genetically close to other varieties of the historical Moldavia region (including today's Republic of Moldova and the Romanian region of Moldova) such as FETEASCĂ ALBĂ and FETEASCĂ REGALĂ (Bodea *et al.* 2009).

Grasă de Cotnari is sometimes said to be identical to FURMINT (Roy-Chevrier 1903b; Galet 2000) but the two varieties might simply be morphologically similar (Dejeu 2004).

OTHER HYPOTHESES

In Romania, legend has it that Stephen the Great (1433–1504), Prince of Moldavia, was so impressed when he tasted the wine in Alba Iulia (Transilvania) while visiting Matthias Corvinus (1443–90), King of Hungary and Croatia, that he was given cuttings of what would later be called Grasă de Cotnari (Dejeu 2004). However, a Transilvanian origin is doubtful because Grasă de Cotnari has traditionally never been cultivated in this region.

According to another legend (Roy-Chevrier 1903b), this variety was introduced from Hungary at the end of the fifteenth century by a German named Gutnar, who gave his name to the village of Kutnar, which eventually became Kotnar and then Cotnari, where the original vines planted by Gutnar were supposedly still growing in the Valea Ungurului ('Valley of the Hungarians').

VITICULTURAL CHARACTERISTICS

Early budding and mid to late ripening. Low yields of fat, rounded, sugar-rich berries that are a welcoming host to noble rot. Moderately resistant to cold winter temperatures and drought but susceptible to downy mildew. Best suited to warm, volcanic soils and slopes.

WHERE IT'S GROWN AND WHAT ITS WINE TASTES LIKE

There were 415 ha (1,025 acres) of Grasă de Cotnari in **Romania** in 2008, predominantly in the hills of Romanian Moldova in the far north east of the country, where it is used to make the sweet botrytized Cotnari wines that once rivalled Tokaji in the courts of northern Europe. It may be made as a varietal wine but its 'fat', full-bodied character is better balanced and fresher when blended (minimum 30%) with its traditional partners MUSCAT BLANC À PETITS (sometimes confusingly called Tămâioasă Românească), FRÂNCUȘĂ and FETEASCĂ ALBĂ. Semi-sweet and dry styles are also found and Cotnari wines are generally less exposed to oxygen and oak than Tokaji is, producing wines with a greenish tinge even after some years in bottle.

In **Hungary**, Kövérszőlő had almost disappeared after phylloxera until it experienced a limited revival in the late 1990s, although there were only 37 ha (91 acres) planted in 2008, almost exclusively in Tokaj. Its high sugar levels and susceptibility to noble rot make it a useful if minor ingredient in the sweet botrytized wines of the Tokaj appellation, in which it is the fifth authorized variety. Varietal wines are full-bodied, soft, slightly aromatic and not suited to ageing, which is why producers such as Tokaj Nobilis use only a small percentage in some of their sweet FURMINT-dominated wines. Rare varietal examples are made by Dorogi, Hétszőlő (mostly Aszú wines but also late-harvest styles in some vintages) and Tokaj Nobilis.

GRAŠEVINA

Unfairly maligned variety widely planted in Central Europe that has in its time suffered in comparison to Riesling.

BERRY COLOUR:

PRINCIPAL SYNONYMS: Borba (Ribera del Guadiana in Spain), Graševina Bijela (Croatia), Italian Riesling (Romania), Laški Rizling (Slovenia, Croatia and Vojvodina in Serbia), Olasz Rizling or Olaszrizling (Hungary), Riesling Italico (northern Italy, Croatia), Riesling Italian (Romania), Rismi (province of Treviso in Italy), Rismi or Risli (Treviso), Rizling Vlašský (Slovakia), Ryzlink Vlašský (Czech Republic), Taljanska Graševina (Croatia), Wälschriesling (Germany), Welschriesling (Austria, Canada, Germany and Switzerland)

VARIETIES COMMONLY MISTAKEN FOR GRAŠEVINA: ALIGOTÉ, GRECO, PETIT MESLIER, PIGNOLETTO, RIESLING

ORIGINS AND PARENTAGE

The origin of this old and widely disseminated variety has long been the subject of speculation (see Other hypotheses below) and it goes under many different synonyms, mainly but not exclusively in Central and Eastern Europe. The name Graševina (used in Croatia) was chosen as prime name in this book because:

- It is the most widespread white grape variety in Croatia.
- The names Laški Rizling, Olasz Rizling, Riesling Italico, Ryzling Vlašský and Welschriesling confusingly suggest a link to the white RIESLING from Germany even though they are genetically and organoleptically quite distinct.
- The name Welschriesling (*welsch* meaning 'foreigner') suggests that the variety was introduced from abroad to German-speaking countries such as Austria, where it is widely grown.
- Graševina was not introduced to northern Italy until the nineteenth century, where it became known as Riesling Italico (Calò *et al.* 2006).

For all the above reasons, Croatia or the Danube basin appear to be the most likely places of origin of Graševina.

Most surprisingly, the DNA profile of the Spanish variety Borba cultivated in Ribera del Guadiana in south-west Spain that was published by Martín *et al.* (2003) matches Graševina (Vouillamoz).

Graševina was used (under different synonyms) to breed several cultivated varieties, including BUSSANELLO, FLAVIS, GOLDBURGER, ITALICA and ŞARBĂ.

OTHER HYPOTHESES

The ampelographer Lambert von Babo (1843–4) suggested that Graševina (under the name Wälschriesling) was introduced to Heidelberg in Germany from the Champagne region, where he tentatively identified it with ALIGOTÉ or PETIT MESLIER. However, Graševina is not related to those varieties and has never been cultivated in France. The names Welsch, Laški and Vlašský preceded or followed by Riesling in different languages all supposedly mean 'from Wallachia' (a historical region in Romania), suggesting a Romanian origin. However, this is doubtful because the Romanians themselves call it Italian Riesling, and their Hungarian neighbours simply translated it (*olasz* means 'Italian'). For some authors, the name Riesling Italico suggests an Italian origin for this variety, and a similarity with GRECO (aka Greco di Tufo) was proposed because both supposedly resemble the Roman *Vitis aminea gemella* (Giavedoni and Gily 2005). DNA profiling, however, clearly rejects any genetic link between Graševina and Greco.

VITICULTURAL CHARACTERISTICS

Productive, late budding and ripening but able to retain acidity, especially if yields are controlled. Suited to dry climates and warmer soils but relatively cold-hardy. Susceptible to powdery mildew, less so to downy mildew and botrytis bunch rot.

WHERE IT'S GROWN AND WHAT ITS WINE TASTES LIKE

The country having given its name to several of the variety's synonyms, it is not surprising that there is a considerable area of this variety planted in **Italy**: 2,030 ha (5,016 acres) in 2000, according to the agricultural census, predominantly in Lombardia in the north west. Recommended producers include Cà del Gè, Cantina di Casteggio, Martilde, Pieropan of Soave and Mazzolino of Oltrepò Pavese, although in general Italian examples are light, crisp and dry but not long on character.

Spain had just 8 ha (20 acres) of Borba in 2008, mostly in the province of Cáceres in Extremadura in the south west but also to a limited extent to the east in Castilla-La Mancha.

Austria recorded 3,597 ha (8,888 acres) of Welschriesling in 2009, representing 7.8% of the total vineyard area and almost double the area devoted to RIESLING. The finest wines, sometimes blended with CHARDONNAY or other, locally grown varieties, are from late-harvested or botrytized grapes on the shores of the Neusiedlersee in Burgenland, where nearly half of Austria's Welschriesling is planted – although they lack the longevity of the finest Riesling TBAs. Schröck & Kracher's Greiner Welschriesling is a very good late-harvest example. The other main expression of this variety is less dramatic: light, dry or off-dry wines for easy, early drinking. It is also used for Austrian Sekt. Recommended producers include Feiler-Artinger, Kracher and Velich.

Under various different names, it is a widely planted workhorse white in much of Eastern Europe, with some impressive wines, potentially both powerful and refined, where the variety is taken seriously and the variety's productivity is curbed by the site and/or by work in the vineyard. When yields are high, the wines tend to be harsh or neutral, sometimes mildly floral, needing some residual sugar to make them more palatable. There were 2,482 ha (6,133 acres) of Laški Rizling in **Slovenia** in 2009, the great majority of vines in Štajerska Slovenija (Slovenian Styria) in the north east. Recommended producers include Dveri-Pax, Jamšek 1887, Kupljen, Marof, P&F Jeruzalem Ormož, Pullus and Valdhuber for dry wines and RA-VinO and KZ Metlika for sweet wines.

Recommended in every wine-growing region of Kontinentalna Hrvatska (Continental Croatia) but not in those of Primorska Hrvatska (Coastal Croatia), Graševina is the most widely planted variety in **Croatia**, with nearly a quarter of the vineyard area, around 8,405 ha (20,769 acres) in 2009. There is a very wide range of styles in terms of both sweetness and winemaking, from dry or off-dry and simple to rich and sweet, sometimes botrytized, via lightly oaked and elegant. The best wines generally come from Slavonija in the east, notably from the Kutjevo, Slavonski Brod and Djakovo subregions. Graševina is less aromatic than RIESLING but this is less of a disadvantage in intensely sweet wines (like Austria's Trockenbeerenauslesen). Recommended producers include Adžić (particularly their lightly oaked version), Belje, Ivan Enjingi, Vlado Krauthaker, Kutjevo co-operative and Mihalj. The variety is also found in neighbouring Vojvodina, **Serbia**, and in **Albania**.

Ryzling Vlašský comprised more than 16% (3,132 ha/7,739 acres) of **Slovakia**'s vineyard area in 2009, second only to GRÜNER VELTLINER among white wine varieties. The **Czech Republic** reported 1,247 ha (3,081 acres) of Ryzlink Vlašský in 2009 – almost exactly the same area as that dedicated to Ryzlink Rýnský (RIESLING). Recommended producers for dry styles include Michlovský (Rakvice), Mikrosvín (Mikulov) and Zámecké Vinarske Bzenec (Bzenec), as well as Galant, Silinek and Volarik (all in Mikulov), Reisten (Pavlov) and Sebesta (Brezi). For sweet wines, Marcincak (Mikulov), Oldrich Drapal Sr and Perina (Klentice).

Known in **Hungary** as Olasz Rizling, it is the most widely planted white variety (4,909 ha/12,130 acres in 2008), grown throughout the country, in gardens and extensive vineyards, but most important in the regions around Lake Balaton (Rohály *et al.* 2003). There are some very good examples of dry wines but also sweet, late-harvest versions and the best tend to come from volcanic sites such as Balaton-Felvidék and Somló north of the lake. Recommended producers include Bussay, Figula, Jásdi, Ottó Légli, Scheller, Szeremley around Lake Balaton; Lajor Takács in Somló; and Lajos Gál and Vilmos Thummerer in Eger. Wines sometimes have a bitter-almond aroma.

Romania's Riesling Italian vineyards, concentrated in Transilvania, Moldova, Muntenia and Oltenia, totalled a substantial 7,061 ha (17,448 acres) in 2008. **Bulgaria**'s Italian Riesling is grown mostly in the Varna and Shumen districts in the north east.

There are limited plantings in both **Canada** and **China**.

GRAUBURGUNDER

See PINOT GRIS *in the* PINOT *entry.*

GRAUER PORTUGIESER

See BLAUER PORTUGIESER.

GRAUER VELTLINER

See GRÜNER VELTLINER.

GRECANICO DORATO

See GARGANEGA.

GRECHETTO DI ORVIETO

Characterful and versatile Umbrian.

BERRY COLOUR: ● ● ● ● ●

PRINCIPAL SYNONYMS: Grechetto, Grechetto Bianco, Grechetto Spoletino

VARIETIES COMMONLY MISTAKEN FOR GRECHETTO DI ORVIETO: MACERATINO (Marche), PIGNOLETTO (also known as Grechetto di Todi)

ORIGINS AND PARENTAGE

The names Greco and Grechetto have been given to several distinct southern Italian varieties used to produce sweet wines, with more than a nod to the traditional sweet-wine styles of ancient Greece (Costacurta *et al.* 2004), but this is not necessarily an indication that these varieties are Greek in origin.

Ampelographic and DNA studies have established that Grechetto di Orvieto and Grechetto di Todi, a synonym for PIGNOLETTO, are two distinct varieties (Filippetti *et al.* 1999). In addition, parentage studies suggested a parent–offspring relationship between them (Costacurta *et al.* 2004), which might explain why they were previously considered to be clonal variations of the same variety.

VITICULTURAL CHARACTERISTICS

Thick skins and therefore good resistance to downy mildew. Mid to late ripening.

WHERE IT'S GROWN AND WHAT ITS WINE TASTES LIKE

Speciality of Umbria, central **Italy**. Orvieto DOC, Umbria's best-known white wine, is a blend of TREBBIANO TOSCANO and Grechetto (minimum 60% combined) plus up to 40% other local varieties, and 15–40% is allowed in Torgiano DOC, another Umbrian Trebbiano-dominated blend. Grechetto is authorized or required in many Umbrian DOC wines, including several varietal versions such as Assisi, Colli Martani, Colli del Trasimeno and Colli Perugini in the province of Perugia and Colli Amerini in the province of Terni. However, since only the Colli Martani regulations distinguish between Grechetto di Orvieto and Grechetto di Todi, it is often not possible to know which variety is intended.

Grechetto is also authorized in various DOCs in the Marche, Lazio and Toscana, though again it is not clear which variety is intended; in the last it may also be a useful component in Vin Santo thanks to its resistance to downy mildew. It is occasionally included in the blend for Umbrian sparkling wines, with a minimum of 50% required in Colli Altotiberini. Antinori's top white Cervaro uses Grechetto as a support to CHARDONNAY, an indication of the growing recognition of its quality potential. Lungarotti's varietal Grechetto (IGT Umbria) is a good, straightforward example of this interesting variety, with citrus and white fruit flavours as well as a slightly herbal, almond character and a delicately creamy texture.

There were 1,204 ha (2,975 acres) planted in Italy in 2000, according to the agricultural census. However, the census, like the DOC regulations, does not distinguish between Grechetto di Orvieto and Grechetto di Todi.

GRECHETTO DI TODI

See PIGNOLETTO.

GRECO

Potentially fine, aromatic but notably firm Campanian.

BERRY COLOUR: ● ● ● ● ●

PRINCIPAL SYNONYMS: Asprinio (Caserta), Greco del Vesuvio (Napoli), Greco della Torre, Greco di Napoli, Greco di Tufo (Avellino, Benevento), ?Grieco di Castelvenere (Costa d'Amalfi south of Napoli)

VARIETIES COMMONLY MISTAKEN FOR GRECO: GRAŠEVINA, GRECO BIANCO, GUARDAVALLE (under the name Greco Bianco di Cirò), MACERATINO (Marche), MALVASIA DI LIPARI (under the names Greco Bianco di Gerace or Greco di Bianco), ROSSETTO, TREBBIANO GIALLO (Lazio)

ORIGINS AND PARENTAGE

The name Greco has been applied to many varieties used to make sweet wines like those of Ancient Greece (Costacurta *et al.* 2004) but this is not necessarily an indication that these varieties are Greek in origin. Greco is distinct from GRECO BIANCO in Calabria and also from several other varieties known as Greco Something (Costacurta *et al.* 2004). DNA profiling recently demonstrated that Greco, or Greco di Tufo, is unexpectedly identical to Asprinio, already mentioned in the seventeenth century, and is genetically close to ALEATICO (Costantini *et al.* 2005).

OTHER HYPOTHESES

It has been suggested that Greco, or Greco di Tufo, was introduced by the Greeks in the eighth or seventh century BC or that it is a descendant of the Aminea Gemina Minor described by classical Roman authors as a variety from the Vesuvio area.

VITICULTURAL CHARACTERISTICS

Late ripening. Susceptible to downy and powdery mildew.

WHERE IT'S GROWN AND WHAT ITS WINE TASTES LIKE

Mostly cultivated and recommended in Campania in southern **Italy**, especially in the provinces of Avellino, Benevento, Napoli and Salerno but also beyond Campania to the east in Puglia (provinces of Bari, Foggia and Taranto), and north of Campania in the province of Campobasso in Molise, in Lazio (provinces of Latina and Viterbo), in Toscana (Grosseto, Lucca and Massa-Carrara) and as far north as La Spezia in Liguria. In the Capri DOC, up to 50% Greco is permitted.

The best-known and most widely admired varietal Greco is Greco di Tufo DOCG but varietal Greco is also permitted in various Campanian DOCs such as Irpinia, Sannio and Sant'Agata de' Goti. In addition, it is authorized in several blended DOCs in Campania and beyond, eg in Gravina DOC, which straddles Puglia and Basilicata and where the higher altitude enhances the aromatic character of Greco. The best wines are full-bodied with some minerality and only delicately aromatic with a note of apricot kernels, or more herbal and racy, as in Mastroberadino's examples. Other notable producers of Greco di Tufo include Feudi di San Gregorio, Pietracupa, Terredora, Vadiaperti and Villa Raiano.

According to the Italian census of 2000, total plantings of Greco were 941 ha (2,325 acres). It also records another 414 ha (1,023 acres) of Asprinio because the discovery that the varieties are identical is so recent.

GRECO BIANCO

Calabrian used to make dessert wines.

BERRY COLOUR: ● ● ● ● ●

PRINCIPAL SYNONYMS: Greco Bianco di Cosenza, Greco Bianco di Rogliano, Pecorello Bianco

VARIETIES COMMONLY MISTAKEN FOR GRECO BIANCO: GRECO, GUARDAVALLE (also known as Greco Bianco di Cirò), MACERATINO (Marche), MALVASIA DI LIPARI (also known as Greco Bianco di Gerace or Greco di Gerace), ROSSETTO

ORIGINS AND PARENTAGE

The names Greco and Grechetto have been given to several distinct southern Italian varieties used to produce sweet wines, probably only with reference to the traditional sweet-wine styles of ancient Greece (Costacurta *et al.* 2004) rather than because they are Greek in origin. Greco Bianco is distinct from GRECO in Campania and from Greco di Bianco, named after a Calabrian coastal village which was once part of the village of Gerace (hence the synonym), which has been shown to be identical to MALVASIA DI LIPARI (Crespan, Cabello *et al.* 2006). It is also distinct from Greco Bianco di Cirò, which is identical to GUARDAVALLE (Costacurta *et al.* 2004). Pecorello Bianco of Calabria was also recently found to be identical to Greco Bianco (Schneider, Raimondi, Grando *et al.* 2008).

OTHER HYPOTHESES

As with GRECO, it has been suggested that Greco Bianco was introduced by the Greeks in the eighth or seventh century BC.

WHERE IT'S GROWN AND WHAT ITS WINE TASTES LIKE

Since several distinct varieties are named Greco Bianco in Calabria, southern **Italy**, sometimes with an additional epithet (eg di Cirò or di Consenza), it is impossible to have a clear picture of the distribution of the Greco Bianco dicussed in this entry. However, the Greco Bianco in the area around Cosenza and Rogliano definitely corresponds to the variety described here. Under the name Pecorello Bianco, it is included in the Savuto and Donnici DOCs in Calabria. Total plantings of 686 ha (1,695 acres) were recorded in Italy in 2000.

GRECO BIANCO DI GERACE

See MALVASIA DI LIPARI.

GRECO BIANCO DI NOVARA

See ERBALUCE.

GRECO NERO

A group of different Calabrians confusingly sharing one name.

BERRY COLOUR: ● ● ● ● ●

PRINCIPAL SYNONYMS: Collective name for different varieties, so synonyms are misleading.

ORIGINS AND PARENTAGE

The name Greco Nero has been used in Calabria, southern Italy, for several distinct varieties differing in their morphology and distribution. Recent ampelographic and DNA studies detected at least five different black-berried varieties named Greco Nero in Calabria (Schneider, Raimondi, Grando *et al.* 2008):

- Greco Nero found in the Cirò area, identical to a vine known as Nerello found near Lamezia Terme
- Greco found in Scilla, identical to CASTIGLIONE
- Greco Nero found in the Lametino, identical to MAGLIOCCO DOLCE
- Greco Nero found in Verbicaro and Scalea
- Greco Nero found in Sibari in north-eastern Calabria.

The last two varieties have distinct morphologies and DNA profiles, and must be considered as two distinct varieties: GRECO NERO DI VERBICARO and GRECO NERO DI SIBARI.

OTHER HYPOTHESES

Supposedly introduced by the Greeks, but there is so far no evidence to support this.

VITICULTURAL CHARACTERISTICS

Few generalizations can be applied to five different varieties, and confusion has been such that hard facts are thin on the ground.

WHERE IT'S GROWN AND WHAT ITS WINE TASTES LIKE

Greco Nero is recommended in the whole of Calabria, southern **Italy**, but it is usually impossible to know which of the five or more varieties mentioned above goes into any one wine.

The Italian agricultural census of 2000 records 3,041 ha (7,514 acres) of 'Greco Nero' but makes no further distinction.

GRECO NERO DI SIBARI

One of the many Calabrian Greco Neros.

BERRY COLOUR: ● ● ● ● ●

VARIETIES COMMONLY MISTAKEN FOR GRECO NERO DI SIBARI: GRECO NERO DI VERBICARO ⌛

ORIGINS AND PARENTAGE

Greco Nero di Sibari was identified around Sibari on the Ionian coast in the province of Cosenza, Calabria. It is distinct from GRECO NERO DI VERBICARO (Schneider, Raimondi, De Santis and Cavallo 2008).

WHERE IT'S GROWN AND WHAT ITS WINE TASTES LIKE

Greco Nero di Sibari is cultivated only in the north east of the province of Cosenza in Calabria, southern **Italy**.

GRECO NERO DI VERBICARO

Another distinct Calabrian Greco Nero.

BERRY COLOUR: ● ● ● ● ●

VARIETIES COMMONLY MISTAKEN FOR GRECO NERO DI VERBICARO: GRECO NERO DI SIBARI ⌛

ORIGINS AND PARENTAGE

Greco Nero di Verbicaro was identified in the area around Verbicaro in the province of Cosenza, Calabria. It is distinct from GRECO NERO DI SIBARI (Schneider, Raimondi, De Santis and Cavallo 2008).

WHERE IT'S GROWN AND WHAT ITS WINE TASTES LIKE

Cultivation of Greco Nero di Verbicaro is limited to the north west of the province of Cosenza, in southern **Italy**, mainly in Verbicaro and Scalea (under the misleading name Mantonico Nero Italico).

GREGU BIANCO

See BARIADORGIA.

GRENACHE

See GARNACHA.

GRENACHE BLANC

See GARNACHA.

GRENACHE GRIS

See GARNACHA.

GRENACHE NOIR

See GARNACHA.

GRIGNOLINO

Waning, distinctively pale, light, local (ancient) Monferrato speciality.

BERRY COLOUR: ● ● ● ● ●

PRINCIPAL SYNONYMS: Barbesino, Barbisino, Berbesino

ORIGINS AND PARENTAGE

A document dated 7 November 1249 records the fact that the church of Sant'Evasio (in Casale Monferrato in Piemonte) lent a piece of land to a certain Guglielmo Crova on condition that it be planted 'bonis vitibus berbexinis' ('with good vines of berbexinis'; Nada Patrone 1988). Although some authors have identified Berbexinis with BARBERA, most ampelographers agree that Berbexinis was in fact Grignolino, because Berbesino is still used as a synonym for Grignolino in the Monferrato area. This document is therefore most likely to be the earliest mention of Grignolino.

Even if there is any doubt that Grignolino and Berbexinis are one and the same, then Grignolino still has medieval credentials, mentioned in Almese in the province of Torino in 1337–8 in a document from the abbey of San Giusto di Susa, which refers to the production of 'sextarios XII vini albi tam muscatelli quam grignolerii' (Comba and Dal Verme 1990).

The name Grignolino could derive from the Asti dialect word *grignòle*, meaning 'pips', because the berries of Grignolino usually have more pips (3–4 times more) than many other varieties.

VITICULTURAL CHARACTERISTICS

Mid to late ripening but sometimes ripens unevenly. Susceptible to powdery mildew and especially to botrytis bunch rot and sour rot

WHERE IT'S GROWN AND WHAT ITS WINE TASTES LIKE

Once cultivated throughout Piemonte in north-west **Italy**, Grignolino is today grown mainly in the Monferrato hills that straddle the provinces of Asti and Alessandria, and to a much more limited extent in the province of Cuneo. The most important DOCs are Grignolino del Monferrato Casalese, Grignolino d'Asti and Piemonte Grignolino. Up to 10% FREISA is allowed in the first two and the best wines tend to be found in the first. It is also authorized in small or moderate amounts in other Piemonte DOCs such as Barbera del Monferrato and Colli Tortonesi.

Varietal wines are generally pale red and light in alcohol (11–12%) with a herbal/floral alpine scent and tangy acidity. They are intended for early drinking but they can be quite tannic thanks to the high ratio of seeds. Recommended producers include: Bricco Mondalino, Marchesi Incisa della Rocchetta, Agostina Pavia, Carlo Querello, Rovero, San Sebastiano, Cantine Sant'Agata and Luigi Spertino.

There were 1,336 ha (3,301 acres) in Italy in 2000 according to the agricultural census.

In **California**, Heitz of the Napa Valley persist with their Grignolino plantings, making a rosé and sometimes a port-style wine from them.

GRILLO

Increasingly popular high-quality, full-bodied western Sicilian white.

BERRY COLOUR:

PRINCIPAL SYNONYMS: Ariddu, Riddu, Rossese Bianco (Riomaggiore in Liguria)

VARIETIES COMMONLY MISTAKEN FOR GRILLO: ROSSESE BIANCO DI SAN BIAGIO

ORIGINS AND PARENTAGE

Grillo is a classic Sicilian variety, although its earliest mention is in 1873 (Bica 2007). This is probably because the variety is not that old: DNA studies have shown that Grillo is the offspring of a natural cross between CATARRATTO BIANCO and MUSCAT OF ALEXANDRIA, the latter known as Zibibbo on Sicilia (Di Vecchi Staraz, This *et al.* 2007). See the MUSCAT entry for the complete pedigree. Recent DNA profiling has surprisingly revealed that the Rossese Bianco grown in Riomaggiore in Liguria, along the north-western Italian coast (not to be confused with ROSSESE BIANCO from Piemonte), is identical to Grillo (Torello Marinoni, Raimondi, Ruffa *et al.* 2009).

OTHER HYPOTHESES

Some suggest that Grillo was introduced from Puglia towards the end of the nineteenth century (Giavedoni and Gily 2005) but its parentage clearly locates its birthplace on Sicilia.

VITICULTURAL CHARACTERISTICS

Vigorous, productive and resistant to winter cold. Mid ripening. Susceptible to powdery mildew.

WHERE IT'S GROWN AND WHAT ITS WINE TASTES LIKE

After phylloxera reached Sicilia in **Italy** in the early 1880s, Grillo was widely planted to replace its parent CATARRATTO BIANCO but thereafter the planted area declined in favour of more productive varieties. Today Grillo is cultivated mainly in the province of Trapani in the north west of the island, where it is one of the main components of Sicilia's famous fortified wine Marsala DOC, though the higher-yielding Catarratto Bianco and INZOLIA and are now more often favoured. It is also cultivated in the provinces of Agrigento, Palermo, Messina, Caltanissetta and Siracusa.

Varietal Grillo is catered for in DOCs such as Alcamo, Contea di Sclafani, Delia Nivolelli, Erice, Monreale and Salaparuta or it may well be classified as IGT Sicilia. The variety also features in blended DOCs such as Contessa Entellina, Mamertino di Milazzo and Menfi. The wines are full-bodied and slightly herbaceous or floral but not generally as aromatic as INZOLIA. Recommended producers include Calatrasi, Cusumano, Marco De Bartoli, Fondo Antico, Foraci, Rallo and Spadafora.

There were 1,808 hectares (4,468 acres) in Italy in 2000 but in Sicilia alone there were 4,118 ha (10,176 acres) by 2008.

GRINGET

Rare, lightly floral and high-quality variety exclusive to Savoie.

BERRY COLOUR: ● ● ● ● ●

PRINCIPAL SYNONYMS: Gringet Gras, Gros Gringet, Petit Gringet

VARIETIES COMMONLY MISTAKEN FOR GRINGET: ROUSSANNE, SAVAGNIN BLANC

ORIGINS AND PARENTAGE

Gringet is an old variety of the Vallée de l'Arve in the Haute-Savoie in eastern France, where it was first mentioned in a manuscript written by Pierre-Joseph Decret in 1766 (Durand 1905). In his *Topographie de tous les vignobles connus*, André Jullien (1832) wrote about the apparently curious properties of Gringet from Ayze (now a Savoie cru): 'on fait, près de cet endroit, un vin blanc nommé *gringet*, du nom du plant qui le produit. Ce vin a la singulière propriété de ne pas enivrer tant qu'on ne quitte pas la table; mais, aussitôt que l'on prend l'air, on perd l'usage de ses jambes, et l'on est forcé de s'asseoir' ('we make near this place a white wine named *gringet*, after the name of the variety that produces it. This wine has the unique property of not causing inebriation so long as one does not leave the table; but as soon as one takes the fresh air, one loses the use of one's legs, and one is forced to sit down').

The exact origin of the name Gringet is unknown, but the variety might have inherited the family name of the grower who propagated it since Gringet was a common local patronym (Rézeau 1997).

Gringet has long been considered a clone of SAVAGNIN (Durand 1905; Galet 1990, 2000) but DNA profiling of a sample from Le Feu vineyard at Domaine Belluard in Ayze has established irrefutably that they are distinct varieties and that Gringet does not match any other known variety (Vouillamoz). In addition, DNA parentage analysis shows that Gringet is not genetically related to Savagnin and strongly suggests that MOLETTE, another rare Savoie variety, is a natural Gringet × GOUAIS BLANC cross (Vouillamoz).

OTHER HYPOTHESES

According to local legend, Gringet was supposedly introduced by bishops from Cyprus in the thirteenth century.

VITICULTURAL CHARACTERISTICS

Mid ripening. Best vertically trained and pruned short or to single Guyot. Well suited to scree-covered slopes. Susceptible to the grape erineum mite, to powdery mildew and especially to downy mildew. Big bunches and small berries.

WHERE IT'S GROWN AND WHAT ITS WINE TASTES LIKE

Gringet used to be cultivated in several places in the Vallée de l'Arve in Haute-Savoie (Ayze, Bonneville, Contamine, Côte d'Hyot, Faucigny, Marignier, Thiez), **France**, and, since the wines tended to be light and lively, were often used for sparkling wine. Today it is almost exclusively found in the commune of Ayze, where Domaine Belluard is currently the only producer of commercial varietal wines. French official statistics record 26 ha (64 acres) in 2008 and, according to Belluard, there are 22 ha (54 acres) in Ayze. Belluard makes two still versions, Les Alpes and Le Feu, the latter a full-bodied single-vineyard wine with complex aromas of dried apple peel, wisteria and yellow plums and a rounded texture but with significant acidity, both elegant and persistent. In addition he makes two sparkling wines: an Ayse (sic) Méthode Traditionelle and the Ayse (sic) Mont-Blanc Brut Zéro.

GRISA NERA

Old and rare Piemontese.

BERRY COLOUR: ● ● ● ● ●

PRINCIPAL SYNONYMS: Grisa, Grisa di Cumiana (Pinerolese)

VARIETIES COMMONLY MISTAKEN FOR GRISA NERA: Grisa (Alto Canavese in the province of Torino)

ORIGINS AND PARENTAGE

Grisa Nera is an old variety of the Pinerolo area south west of Torino in Piemonte, where it is still cultivated, often interplanted with other local varieties. DNA parentage analysis has clearly established a parent–offspring relationship between Grisa Nera and AVARENGO, another old Piemontese variety, thus anchoring the origin of both varieties in this area (Anna Schneider and José Vouillamoz, unpublished data). It should not be confused with Grisa, found here and there in the Alto Canavese (Schneider and Mannini 2006).

VITICULTURAL CHARACTERISTICS

Mid to late ripening. Vigorous but with low fertility and subject to coulure. Low susceptibility to fungal diseases. Suitable for both table and wine grapes.

WHERE IT'S GROWN AND WHAT ITS WINE TASTES LIKE

Grisa Nera is found throughout the foothills and alpine vine-growing regions in the province of Torino in Piemonte, north-west **Italy**, especially west of Pinerolo and in the lower part of the Val di Susa, and to a lesser extent in the province of Cuneo. It is usually blended with other local varieties such as NERETTA CUNEESE, PLASSA or BARBERA.

GRK

Rare, high-quality and highly localized island white from Croatia.

BERRY COLOUR: ● ● ● ● ●

PRINCIPAL SYNONYMS: Gark, Grk Bijeli, Grk Korčulanski, Grk Mali, Grk Veli, Lumbarajski Grk

VARIETIES COMMONLY MISTAKEN FOR GRK: PRČ (Hvar)

ORIGINS AND PARENTAGE

Grk is an old variety from the island of Korčula whose name most likely derives from *gark*, meaning 'bitter', referring to the taste of the must or the wine. DNA parentage analysis has shown that Grk, Crljenak Crni (no longer cultivated), PLAVINA and VRANAC all have a parent–offspring relationship with TRIBIDRAG (Maletić *et al.* 2004), also called Primitivo and Zinfandel, and are therefore grandparents or half-siblings of PLAVAC MALI (see Tribidrag for the full pedigree).

OTHER HYPOTHESES

Since Grk literally means 'Greek', some writers assume it was brought to Croatia by the ancient Greeks in the fourth century BC (Žunec 2009) but this hypothesis can be rejected because of the direct parentage with TRIBIDRAG and the genetic relationship with several other Croatian varieties.

Although it is often listed as a synonym of PRČ, comparison of the DNA profiles of Prč (Benjak *et al.* 2005) and Grk (Pejić *et al.* 2000) clearly establishes that they are distinct varieties.

VITICULTURAL CHARACTERISTICS

Mid to late ripening, erratic yields due to deformed stamens (so that the flowers are effectively female), not especially sensitive to fungal diseases. Suited to warm, well-exposed, sunny sites. The berries can reach high levels of sugar concentration.

WHERE IT'S GROWN AND WHAT ITS WINE TASTES LIKE

Grk is found in **Croatia**, mostly around Lumbarda on the island of Korčula within the Srednja i Južna Dalmacija (Central and Southern Dalmatia) wine region, although it is also planted further south on the islands of Mljet and Šipan and in the coastal area near Dubrovnik. Wines are full-bodied, generous and well structured but retain well-balanced acidity – unusual in this area – and have an attractive bitterness on the finish. Grk has the potential to make high-quality dry and sweet wines. Recommended producers include Bartul Batistić, Ivica-Fidulić Batistić, Vinarija Bire, Stipan Cebalo, Branimir Cebalo and Milina.

GROLLEAU GRIS

See GROLLEAU NOIR.

GROLLEAU NOIR

Widely planted though little recognized dark-skinned Loire variety that makes a lot of rosé and some supple reds.

BERRY COLOUR:

PRINCIPAL SYNONYMS: Bourdalès (Madiran), Gamay de Châtillon (Savennières), Gamay Groslot (Maine-et-Loire), Grolleau de Cinq-Mars, Grolleau de Tourraine, Grolleau des Mahé, Grolleau de Tours, Groslot or Gros-Lot, Groslot de Vallères (Indre-et-Loire, Maine-et-Loire), Groslot Noir, Moinard, Pineau de Saumur

VARIETIES COMMONLY MISTAKEN FOR GROLLEAU NOIR: GAMAY NOIR

ORIGINS AND PARENTAGE

According to Bouchard (1901b), this variety was planted in 1810 by a certain M Lothion, a farmer at the Grande Gaudrière in the commune of Mazières-de-Touraine, near Cinq-Mars-la-Pile, north of the Loire and west of Tours, in the Indre-et-Loire, France, hence its synonym Grolleau de Cinq-Mars. After a few years, this vine became large enough to cover a pear tree and cuttings were dispersed around the neighbourhood and then quickly reached the left bank of the Loire (eg the communes of Vallères and Louerre). The name Grolleau supposedly derives from the old French *grolle* meaning 'black crow', referring to the colour of the berries.

DNA parentage analysis has suggested that Grolleau Noir is one of the numerous natural progenies of GOUAIS BLANC (Boursiquot *et al.* 2004; see PINOT pedigree diagram). Grolleau Gris (pink-berried) and Grolleau Blanc are colour mutations of Grolleau Noir.

OTHER HYPOTHESES

Its synonym Gros-Lot, literally 'jackpot', was the starting point of a doubtful etymology related to the fortune that this productive grape made for some vine growers. However, this is not consistent with the original spelling and can be ruled out (Rézeau 1997).

VITICULTURAL CHARACTERISTICS

Early budding, mid ripening. Fertile and needs to be pruned short with the yields kept low. Susceptible to red leaf (also known as *flavescence* or *brunissure*), stem rot and phomopsis. Grolleau Gris is susceptible to botrytis bunch rot.

WHERE IT'S GROWN AND WHAT ITS WINE TASTES LIKE

Grolleau is the third most cultivated dark-skinned variety in the Loire after CABERNET FRANC and GAMAY NOIR and is mainly used to produce off-dry rosé wines in Anjou and drier styles in Touraine. Yields must be controlled if the aim is to produce the relatively pale, light-bodied, supple, expressive, red-fruited red wines such as those made by Stéphane Bernadeau, Domaine Chahut & Prodiges, Domaine Cousin-Leduc and Domaine de Château-Gaillard. Domaine Cousin-Leduc also produce a rare varietal Grolleau Gris – a rosé.

There were 2,350 ha (5,807 acres) of Grolleau Noir in **France**'s greater Loire region in 2009, mostly west of Tours but especially in the Maine-et-Loire. This marks a serious decline for Grolleau in the last fifty or so years, down from 11,409 ha (28,192 acres) in 1958, though, thanks largely to its high yields, it is still responsible for one sixth of the Val de Loire's total production and the decline seems to have slowed in the last few years. A maximum of 25% is allowed in Coteaux du Loir rosé but Rosé d'Anjou is usually made predominantly from Grolleau Noir, often blended with GAMAY NOIR. It is also a permitted variety in Saumur Brut and Rosé de Loire. The latter may also include Grolleau Gris.

Although the area of Grolleau Gris has halved from 908 ha (2,244 acres) in 2000 down to 454 ha (1,122 acres) in 2009, it is still way up on the 19 ha (47 acres) recorded in 1958. Domaine James et Nicolas Paget in Azay-le-Rideau make a good traditional-method sparkling Touraine rosé.

GROPPELLO DI MOCASINA

Minor northern Italian whose wine is generally blended.

BERRY COLOUR: ● ● ● ● ●

PRINCIPAL SYNONYMS: Groppello di San Stefano☒, Groppello Moliner

VARIETIES COMMONLY MISTAKEN FOR GROPPELLO DI MOCASINA: Groppello dei Berici (Veneto), GROPPELLO DI REVÒ☒ (Trentino), GROPPELLO GENTILE☒ (Lombardia), PIGNOLA VALTELLINESE☒ (Lombardia and the Veneto)

ORIGINS AND PARENTAGE

The name Groppello was given to several distinct varieties in northern Italy (see etymology in GROPPELLO GENTILE).

Recent ampelographic studies at the Conegliano research centre have established that

Groppello di Mocasina, from Lombardia, is distinct from Groppello dei Berici (not commercially cultivated), although the results of DNA profiling have not yet been published. Genetic studies have shown that Groppello di Mocasina, named after a village west of Lago di Garda, is distinct from GROPPELLO DI REVÒ, GROPPELLO GENTILE and PIGNOLA VALTELLINESE (called Groppello di Breganze in the Veneto), and that it is identical to Groppello di San Stefano (Costantini *et al.* 2001; Cancellier *et al.* 2009). The DNA profiles of Groppello di Mocasina and Groppello Gentile suggest that they are closely related.

VITICULTURAL CHARACTERISTICS

Mid ripening. Susceptible to botrytis bunch rot.

WHERE IT'S GROWN

It is grown mainly in the Lago di Garda area of Lombardia, northern **Italy**, and is often blended with GROPPELLO GENTILE, SANGIOVESE, MARZEMINO or BARBERA. In particular, it is blended with Groppello Gentile in the Riviera del Garda Bresciano DOC.

GROPPELLO DI REVÒ

Very minor northern Italian that was far more popular in the eighteenth and nineteenth centuries.

BERRY COLOUR: ● ● ● ● ●

PRINCIPAL SYNONYMS: Gropel, Gropel Nones, Groppello Anaune

VARIETIES COMMONLY MISTAKEN FOR GROPPELLO DI REVÒ: Groppello dei Berici (Veneto), GROPPELLO DI MOCASINA (Lombardia), GROPPELLO GENTILE (Lombardia), PIGNOLA VALTELLINESE (Lombardia and the Veneto).

ORIGINS AND PARENTAGE

The name Groppello was given to several distinct varieties in northern Italy (see etymology in GROPPELLO GENTILE).

Recent ampelographic studies at the Conegliano research centre have established that Groppello di Revò is distinct from Groppello dei Berici (not commercially cultivated) found in the Veneto, although DNA profiling has not yet been published. Genetic studies have shown that Groppello di Revò is distinct from GROPPELLO DI MOCASINA, GROPPELLO GENTILE and PIGNOLA VALTELLINESE (called Groppello di Breganze in the Veneto; Costantini *et al.* 2001; Cancellier *et al.* 2009). More recently, DNA parentage tests unexpectedly showed that Groppello di Revò is a natural progeny of RÈZE, an old variety from the Valais in Switzerland (Vouillamoz, Schneider *et al.* 2007). Since Rèze is also a parent of the white Trentino variety NOSIOLA, Groppello di Revò and Nosiola are half-siblings, which might explain why Nosiola is sometimes called Groppello Bianco and why all three are said to be descendants of the Roman grape Raetica.

VITICULTURAL CHARACTERISTICS

Mid to late ripening. Good resistance to winter frost.

WHERE IT'S GROWN

Groppello di Revò is cultivated exclusively in the Val di Non in Trentino in northern **Italy**, in the communes of Cagnò, Cloz, Revò and Romallo, where it was the most important variety in the eighteenth and nineteenth centuries. In the early twentieth century, its cultivation declined dramatically but, since the end of the twentieth century, interest in this autochthonous variety has been revived by local growers and academics. Groppello di Revò has recently been authorized for use in IGT Vigneti delle Dolomiti.

Varietal wines are produced by Cantina Rotaliana and by Augusto Zadra (El Zeremia).

GROPPELLO DI SAN STEFANO

See GROPPELLO DI MOCASINA.

GROPPELLO GENTILE

Ancient, usually blended, northern Italian red.

BERRY COLOUR:

PRINCIPAL SYNONYMS: Groppella, Groppello Comune, Groppellone

VARIETIES COMMONLY MISTAKEN FOR GROPPELLO GENTILE: Groppello dei Berici (Veneto), GROPPELLO DI MOCASINA (Lombardia), GROPPELLO DI REVÒ (Trentino), PIGNOLA VALTELLINESE (Lombardia and the Veneto), RABOSO PIAVE (Veneto)

ORIGINS AND PARENTAGE

The name Groppello has been given to several distinct varieties in northern Italy. The etymology probably comes from *grop* or *gropo*, meaning 'knot' in the local Veneto dialect, perhaps because the grape cluster looks like a knot.

Groppello Gentile is the oldest and most widespread of the varieties named Groppello. It was mentioned in the sixteenth century in the area of Brescia and Bergamo as a fine grape whose wine was exported to Germany (Bacci 1596). Recent ampelographic studies at the Conegliano research centre have shown that Groppello Gentile is distinct from Groppello dei Berici (not commercially cultivated), although DNA profiling has not been published yet. In addition, genetic studies have established that Groppello Gentile is distinct from GROPPELLO DI REVÒ, GROPPELLO DI MOCASINA and PIGNOLA VALTELLINESE (called Groppello di Breganze in the Veneto), and that it is identical to Groppellone (Costantini *et al.* 2001; Cancellier *et al.* 2009). The DNA profiles of Groppello Gentile and Groppello di Mocasina suggest that they are closely related.

VITICULTURAL CHARACTERISTICS

Mid ripening. Good resistance to winter frost but susceptible to powdery mildew and thrips.

WHERE IT'S GROWN

Mainly cultivated in Lombardia (provinces of Brescia and Bergamo) in northern **Italy**. Often blended with GROPPELLO DI MOCASINA, SANGIOVESE, MARZEMINO or BARBERA. In particular, it is blended with Groppello di Mocasina in the Rivera del Garda Bresciano DOC.

GROS MANSENG

Full-flavoured, thick-skinned white speciality of the far south west of France.

BERRY COLOUR: ● ● ● ● ●

PRINCIPAL SYNONYMS: Gros Mansenc, Izkiriot Haundi or Iskiriota Zuri Handia (País Vascos in Spain), Manseng Gros Blanc

ORIGINS AND PARENTAGE

The earliest mention of a Manseng was recorded in 1562 in Jurançon in a text in Occitan: 'vinhe mansengue'. However, it is not known whether this was referring to Gros Manseng or PETIT MANSENG, and the distinction between the two varieties was not recorded until 1783–4: 'Manseng gros, blanc ... couleur dorée, bon gout quoy que piquant. Manseng petit, blanc ... le gout plus piquant que tout autres' ('Gros Manseng, white ... golden colour, good taste although spicy. Petit Manseng, white ... the taste is spicier than any others'). The name Manseng could derive from *mans*, meaning 'mansion'.

According to Galet (1990), Gros Manseng is morphologically close to PETIT MANSENG, and Bisson (2009) groups them in the Mansien ampelographic group (see p XXVII). DNA parentage analysis and reconstruction of the most likely pedigree (see PINOT pedigree diagram) strongly suggests that Gros Manseng is a progeny of Petit Manseng and a grandchild of SAVAGNIN (Vouillamoz; Myles *et al.* 2011).

VITICULTURAL CHARACTERISTICS

Early budding, late ripening. Vigorous, reasonably fertile and generally though not necessarily pruned long. Susceptible to powdery mildew though shows good resistance to botrytis bunch rot thanks to thick skins, which also means it is well suited to late harvesting for sweet wines. Very small berries which can reach high sugar levels while retaining a high level of acidity.

WHERE IT'S GROWN AND WHAT ITS WINE TASTES LIKE

Gros Manseng has tended to be the main grape variety for the dry whites in Jurançon in the south west of **France** and in 2009 there was almost three times as much Gros Manseng planted in France as the smaller-berried PETIT MANSENG (2,919 ha/7,213 acres v 1,019 ha/2,518 acres). However, Gros Manseng is starting to give way in the vineyard to Petit Manseng thanks to the increasing popularity of the sweet Jurançon wines in which Petit generally plays a more significant role (Strang 2009), although there are exceptions such as Château Joly's Épiphanie, a vendanges tardives (sic) harvested as late as January, which is 100% Gros Manseng, and Alain Brumont's sweet Vin de Pays des Côtes de Gascogne. The majority of the vines are planted in the Gers.

Gros Manseng is permitted as a principal variety in other appellations of the south west such as Pacherenc du Vic-Bilh, Béarn and the rare Irouléguy whites but is also recommended further afield in *départements* such as the Landes and the Aude. It is sometimes used for the *mistelle* Floc de Gascogne. The yields are more generous though the wines may be less elegant and rich than those produced from PETIT MANSENG but they are not without power. Flavours of citrus fruit, apricot and light spice, occasionally floral notes, are intensified in late-harvest wines. Those thick skins have to be handled carefully in the winery to avoid astringent or bitter notes.

Particularly good examples of varietal wines are produced by Domaine Cauhapé: the dry Chant des Vignes and Sève d'Automne and the sweet Ballet d'Octobre. Other recommended producers, mostly for blends of Gros and PETIT MANSENG and other local varieties, include Domaine Bellegarde, Clos Lapeyre and Producteurs Plaimont. Blending with SAUVIGNON BLANC is not uncommon, especially in Vin de Pays des Côtes de Gascogne.

Over the border in northern **Spain**, Gros Manseng is authorized for two of the three appellations in the País Vasco (Basque Country): Arabako Txakolina/Chacolí de Álava and Bizkaiko Txakolina/Chacolí de Viscaia (a principal variety in the former and a complementary one in the latter). However, total plantings in Spain in 2008 amounted to no more than 11 ha (27 acres).

Gros Manseng is also planted to a very limited extent in **Uruguay** (three vineyards totalling 0.7 ha/1.7 acres in 2009), where Familia Deicas produce a noble sweet wine from a blend of Gros and PETIT MANSENG. Plantings in **Brazil**'s Rio Grande do Sul are even more limited at 0.29 ha (0.7 acres) in 2007. **Australia** has very limited plantings, originally thought to be Petit Manseng.

GROS PLANT

See FOLLE BLANCHE.

GROS VERDOT

Very minor dark-skinned variety inferior to Petit Verdot.

BERRY COLOUR: ● ● ● ● ●

PRINCIPAL SYNONYMS: Hère (Dordogne), Plant des Palus (Gironde), Verdot Colon (Gironde)

VARIETIES COMMONLY MISTAKEN FOR GROS VERDOT: CABERNET PFEFFER, PETIT VERDOT

ORIGINS AND PARENTAGE

Gros Verdot and PETIT VERDOT are usually said to come from the Gironde in south-west France (see Petit Verdot for details and etymology) but recent ampelographic and genetic research has placed them in an ampelographic group together with Ardonnet and Petit Verdot Faux that most likely comes from the Pyrénées-Atlantiques, further south, possibly the result of the domestication of local wild grapevines (Bordenave *et al.* 2007).

VITICULTURAL CHARACTERISTICS

Productive and late ripening. Smaller berries than those of PETIT VERDOT.

WHERE IT'S GROWN AND WHAT ITS WINE TASTES LIKE

In **France**, Gros Verdot has been banned from plantings in the Gironde since 1946 (Lafforgue 1947) and is not included in the official French register of varieties, even though it was an important variety in the Queyries vineyards of nineteenth-century Bordeaux (Bordenave *et al.* 2007), what is now the Bastide area of the city. It lacks the interest and concentration of its namesake PETIT VERDOT and has almost disappeared.

Gros Verdot is one of the lesser varieties permitted in **California**'s Meritage. There is considerable confusion between Gros Verdot and CABERNET PFEFFER in California. DNA tests at Kenneth Volk Vineyards in San Benito have shown that what was thought to be the latter was in fact the former, even though their fresh, cherry-fruited and peppery wine is still labelled Cabernet Pfeffer.

Although it is sometimes difficult to be confident as to which variety is referred to in official statistics in South America, as elsewhere, there are apparently 7 ha (17 acres) of 'Verdot' (said to be Gros Verdot) in **Chile** compared with the 257 ha (635 acres) given over to PETIT VERDOT. **Argentina**'s 455 ha (1,124 acres) of 'Verdot', on the other hand, are said to be all Petit Verdot.

GROSSA

Alentejo variety used to improve red blends.

BERRY COLOUR: ● ● ● ● ●

PRINCIPAL SYNONYMS: Tinta Grossa

VARIETIES COMMONLY MISTAKEN FOR GROSSA: MARUFO, TINTA BARROCA

ORIGINS AND PARENTAGE

Grossa, literally 'big', probably comes from the Alentejo in southern Portugal. The synonym Tinta Grossa is also given to TINTA BARROCA in the Douro, and to MARUFO, but their DNA profiles in Almadanim *et al.* (2004, 2007) are distinct.

VITICULTURAL CHARACTERISTICS

Vigorous, mid budding, late ripening. Small bunches of small, thick-skinned berries. Best suited to poor, well-drained soils.

WHERE IT'S GROWN AND WHAT ITS WINE TASTES LIKE

Grossa is authorized in various DOC and regional wines in the Lisboa, Alentejo and Tejo regions of **Portugal**, although it is found almost exclusively in the Vidigueira subregion of the Alentejo. Wines are deeply coloured, fresh and rounded with exotic aromas of sour cherries, vanilla sweetness and blackberry fruits, and are generally used to add complexity, longevity and elegance to a blend. Paulo Loreano, for example, uses it in this way but also produces a rare varietal wine. There were 182 ha (450 acres) in Portugal in 2010.

GRUAJA

Rescued from extinction in northern Italy in the early 1990s but still very rare.

BERRY COLOUR: ● ● ● ● ●

PRINCIPAL SYNONYMS: Crovaja, Cruaia, Cruara, Cruvaio, Gruaia, Gruaio, Gruajo

VARIETIES COMMONLY MISTAKEN FOR GRUAJA: RABOSO VERONESE

ORIGINS AND PARENTAGE

Gruaja is an old variety of the Vicenza and Breganze areas of the Veneto in northern Italy, mentioned as long ago as the eighteenth century. A recent DNA study suggested a close relationship between Gruaja and NEGRARA VERONESE, a rare variety from Breganze (Salmaso *et al.* 2008).

VITICULTURAL CHARACTERISTICS

Mid to late ripening. Needs a warm year and restricted yields to ripen fully.

WHERE IT'S GROWN AND WHAT ITS WINE TASTES LIKE

Gruaja was almost extinct until it was rescued in 1992 by Severina Cancellier and fellow researchers at the Conegliano research centre in the Veneto, northern **Italy**, together with three producers (Avogadro, Miotti and Vitacchio) who still had some old Gruaja vines. There are now even some new plantings. The wines have aromas of maraschino cherries and forest fruits, are well structured and finish slightly bitter. Firmino Miotti in Breganze produce a varietal version.

GRÜNER VELTLINER

Fashionable, versatile, top-quality Austrian white wine variety.

BERRY COLOUR:

PRINCIPAL SYNONYMS: Grauer Veltliner⧖ (Austria), Grün Muskateller (Austria), Veltliner (Alto Adige), Veltliner Grau⧖ (Austria), Veltliner Grün (Austria), Veltlinske Zelené (Slovakia), Veltlinské Zelené (Czech Republic), Weissgipfler (Austria), Zeleni Veltinec (Slovenia), Zöld Veltlini (Hungary)

ORIGINS AND PARENTAGE

The earliest mention of Grüner Veltliner dates back to the eighteenth century. Despite its name, Grüner Veltliner has no link with Valtellina. Recent DNA profiling unexpectedly uncovered the parentage of Grüner Veltliner: it is a natural cross between Traminer (SAVAGNIN) and, very likely, an obscure vine found in St Georgen am Leithagebirge near Eisenstadt (Regner 2007). This old vine has a unique DNA profile and is currently the only known survivor of this variety. It used to be called Grün Muskateller in St Georgen, but in order to avoid confusion with the heterogeneous group referred to as MUSCAT, it was simply named 'St Georgener'. The vine, attacked by vandals in February 2011 and now a protected natural monument, is currently being propagated and the first wine is expected to be made in a few years.

Going further back up the family tree, Grüner Veltliner is either a grandchild or a half-sibling of PINOT, because SAVAGNIN and Pinot have a parent–offspring relationship.

Given its parentage, Grüner Veltliner does not belong to the 'Veltliner' family, which includes ROTER VELTLINER, FRÜHROTER VELTLINER, NEUBURGER and ROTGIPFLER (Sefc, Regner *et al.* 1998), even if it is a half-sibling of Rotgipfler (they have SAVAGNIN as common parent).

DNA profiling has shown that, contrary to common belief, **Grauer Veltliner** (or Veltliner Grau) is not a distinct variety but just a colour mutation of Grüner Veltiner, with greyish berries (Regner *et al.* 1996).

VITICULTURAL CHARACTERISTICS

Very fruitful, with small greenish-yellow berries. Susceptible to downy and powdery mildews and to grape rust mite. Mid ripening but too late for much of northern Europe.

WHERE IT'S GROWN AND WHAT ITS WINE TASTES LIKE

Plantings of Grüner Veltliner in **France** are so far strictly experimental but **Italy** has one small area where the variety grows. Virtually any vine called Veltliner high up in the Valle Isarco of the Alto Adige in the far north of the country is Grüner Veltliner, which has all but completely replaced the unrelated FRÜHROTER VELTLINER there.

In **Germany**, the historical Hansenwein of Plochingen (on the outskirts of Stuttgart) is made with Grüner Veltliner, and not, as is sometimes suggested, with the true Hans or Hansen Rot, an old variety that is no longer in cultivation.

It is **Austria** that is inextricably associated with Grüner Veltliner (sometimes referred to simply as Grüner or GV and more recently marketed as gru-vee, particularly in the US). It is the country's most widely planted variety, occupying 17,034 ha (42,092 acres), almost one third of the total vineyard area. It is most at home, and produces many of its best wines, along the Donau (Danube), where the vines lap up the heat funnelled west along the valley from the Pannonian Plain, but there are also substantial plantings throughout the northern half of Austria's wine-growing region, notably in the Weinviertel north of the river (8,529 ha/21,076 acres) but also in Burgenland, particularly in Neusiedlersee (1,272 ha/3,143 acres) and Neusiedlersee-Hügelland (882 ha/2,179 acres), the two main wine regions around the lake of the same name.

When yields are high (up to 100 hl/ha or 5.7 tonnes/acre are possible), the wines are crisp, fresh and inoffensive, as in large swathes of the Weinviertel. But in the hands of the best growers and on the best sites in the Wachau, Kamptal, Kremstal, Weinviertel and Wagram, they can be stunningly concentrated and ageworthy: typically dry and full-bodied with a peppery perfume and firm, minerally citrus and sometimes spicy flavours, becoming more burgundian with age. Wines produced on the plain are often more peachy and fruit-dominated. It's a versatile variety, capable of producing anything from base wine for Sekt via light, medium-bodied peppery wines for everyday refreshment through to concentrated and powerful expressions of the steep, river-hugging sites. Some sweet Auslese and Trockenbeerenauslese bottlings can also be found.

Grüner Veltliner is currently recognized in five Austrian DACs, the classification used to indicate quality and typicality for a combination of grape and region: Kremstal, Kamptal, Leithaberg, Traisental and Weinviertel.

Best producers include Bründlmayer, Schloss Gobelsburg, Hirsch, Jurschitsch and Loimer in Kamptal; Felsner, Malat, Sepp Moser, Nigl, Salomon, Dr Unger and Winzer Krems in Kremstal; Alzinger, Domäne Wachau, Donabaum, Hirtzberger, Jamek, Knoll, F X Pichler, Rudi Pichler, Prager and Schmelz in Wachau; Graf Hardegg and Pfaffl in Weinviertel; and Fritsch and Ott in Wagram. Mainhard Forstreiter's Tabor Grüner Veltliner is made from 150-year-old vines in Krems-Hollenburg just east of Krems, probably the oldest vineyard in Austria.

Grüner Veltliner has a long history across the border in the **Czech Republic**, where there are now 1,713 ha (4,233 acres) of Veltlinské Zelené. The variety has even greater significance in **Slovakia**, where Veltlinske Zelené is the most widely planted white variety, with 3,805 ha (9,402 acres) and nearly 20% of the total vineyard area. In **Hungary**, there are 1,439 ha (3,556 acres) of Zöld Veltlini, scattered across the country but with the greatest concentrations in Kunság and Tolna in the south, Dél-Balaton in the west, and Mátra further north and east. It is also found in **Bulgaria**.

This internationally fashionable variety is beginning to be planted in the US with **Oregon** and then **California** leading the way, so far on a modest scale. Grüner Veltliner is also found to a very limited extent in **Washington** State and in British Columbia, western **Canada**.

In **Australia** the first varietal Grüner Veltliner was Lark Hill of Canberra's 2009. Hahndorf Hill in the Adelaide Hills bottled a 2010. There are some new plantings in **New Zealand**, where Coopers Creek released the first commercial bottling, from Gisborne fruit, in 2008. It has also been planted at Mount Edward and Quartz Reef in Central Otago and by Yealands and a few growers in Marlborough.

Thanks to its presence on restaurant wine lists around the world, it seems likely to continue to make an increasing impact in the world's vineyards.

GUARDAVALLE

Makes a minor, nutty white in Italy's deep south.

BERRY COLOUR:

PRINCIPAL SYNONYMS: Greco Bianco di Cirò, Greco del Cirotano, Montonico di Rogliano (Rogliano), Uva da Passito (Savelli), Uva Greca (Bivongi), Vardavalli

VARIETIES COMMONLY MISTAKEN FOR GUARDAVALLE: GRECO, GRECO BIANCO, MANTONICO BIANCO

ORIGINS AND PARENTAGE

Guardavalle probably comes from and is named after the eponymous town on the Ionian coast, between Locri and Soverato in the province of Catanzaro, Calabria. DNA profiling has established that Greco Bianco di Cirò, which is distinct from GRECO and GRECO BIANCO, is in fact identical to Guardavalle (Costacurta *et al.* 2004). Recent ampelographic studies have shown that local Calabrians such as Montonico di Rogliano, Uva da Passito and Uva Greca are also identical to Guardavalle (Schneider, Raimondi, Grando *et al.* 2008).

VITICULTURAL CHARACTERISTICS

Late ripening.

WHERE IT'S GROWN AND WHAT ITS WINE TASTES LIKE

Under the name Guardavalle, it is recommended in the province of Reggio Calabria on **Italy**'s 'toe' but is planted only to a limited extent, mainly around the communes of Bianco, Locride and Bivongi, whereas under the name Greco Bianco di Cirò it is one of the most widespread white wine varieties of Calabria. Guardavalle is authorized in the blend for Bivongi DOC. Wines have a nutty character and may be slightly astringent. Santa Venere's Vescovado is a varietal wine made from organically grown grapes and sold as IGT Calabria.

There were 177 ha (437 acres) of Guardavalle in Italy in 2000, according to the agricultural census, but this is unlikely to include the area planted to Greco Bianco di Cirò, which the census fails to isolate.

GUARNACCIA BIANCA

See CODA DI VOLPE BIANCA.

GUEUCHE NOIR

Virtually extinct, late-ripening variety from eastern France, related to Gouais Blanc.

BERRY COLOUR:

PRINCIPAL SYNONYMS: Foirard or Foirard Noir (Poligny), Plant d'Arlay (Salins), Plant de Treffort (Ain)

ORIGINS AND PARENTAGE

Gueuche Noir is an almost extinct variety from the Franche-Comté in eastern France, where it has been cultivated since the eighteenth century. It is a member of the Gouais ampelographic group (see p XXVII; Bisson 2009), which explains why GOUAIS BLANC is sometimes called Gueuche Blanc, and why Gueuche Noir is a likely offspring of Gouais Blanc (Boursiquot *et al.* 2004; see PINOT pedigree diagram). In addition, Guicherd considers Gueuche Noire to be similar to ENFARINÉ NOIR from the Aube (Galet 1990).

VITICULTURAL CHARACTERISTICS

Fertile, mid to late ripening. Compact bunches of small, thin-skinned berries, susceptible to botrytis bunch rot and other fungal diseases.

WHERE IT'S GROWN AND WHAT ITS WINE TASTES LIKE

Étienne Thiebaud, who established Domaine des Cavarodes in the Jura, eastern **France**, at the age of twenty, has rescued an old vineyard in Liesle, south west of Besançon in the Doubs region, and produces a red Vin de Pays de Franche-Comté which is a blend of PINOT NOIR, TROUSSEAU, PINOT MEUNIER, GAMAY NOIR, POULSARD, Gueuche Noir, Argant (GÄNSFÜSSER), BLAUER PORTUGIESER, ENFARINÉ NOIR and MÉZY, all from interplanted vines aged 50–100 years. Domaine Ganevat in Rotalier in the Jura have replanted a small area of Gueuche Noir (and also what they refer to as Gueuche Blanc) and include it in an early-drinking red blend.

According to Galet (2000), wines made solely from Gueuche Noir are generally rather hard and acidic because it is too cool in the Franche-Comté for the variety to ripen fully. Gueuche Noir is not yet included in the official French register of varieties.

GUTEDEL

See CHASSELAS.

GUTENBORNER

Minor German cross with very limited plantings in the UK.

BERRY COLOUR: ● ● ● ● ●

PRINCIPAL SYNONYMS: Geisenheim 17-52

ORIGINS AND PARENTAGE

A MÜLLER-THURGAU × Bicane (under the name Chasselas Napoleon) cross bred by Heinrich Birk at the Geisenheim research centre in 1928, where Bicane is a French wine and table grape that is no longer cultivated.

VITICULTURAL CHARACTERISTICS

At risk from frost and yields irregularly. Late ripening and not notably disease-resistant though more so than MÜLLER-THURGAU, hence its appeal in the UK in the 1970s and 1980s.

WHERE IT'S GROWN AND WHAT ITS WINE TASTES LIKE

Thanks to its ability to ripen in cool climates, this variety has had more success in sheltered sites in **England** than in Germany, where it is not authorized for commercial planting. However, plantings are still no more than a few hectares. Wines tend to be rather neutral, sometimes mildly grapey.

H

HAMASHARA
HÁRSLEVELŰ
HASANDEDE
HELFENSTEINER
HELIOS
HERBEMONT
HEROLDREBE
HETERA
HIBERNAL
HIMBERTSCHA
HITZKIRCHER
HÖLDER
HONDARRIBI BELTZA
HOROZKARASI
HRON
HRVATICA
HUMAGNE
HUXELREBE

ᛝ this symbol on the following pages indicates that the synonyms, and mistaken identities, are confirmed by DNA evidence

HAMASHARA

Early-ripening Azeri red wine grape.

BERRY COLOUR: ● ● ● ● ●

PRINCIPAL SYNONYMS: Aronova Boroda (meaning 'Aaron's beard'), Gamashara

ORIGINS AND PARENTAGE

Hamashara, the historical name of the village of Hasıllı, is a variety local to the Jalilabad region in south-east Azerbaijan. It is said to be the result of natural pollination of wild vines whose offspring were cultivated from seeds by the local people. If this were the case, then Hamashara would not be a single grape variety but a group of several distinct and closely related varieties.

VITICULTURAL CHARACTERISTICS

Early ripening. Large bunches of large, thick-skinned berries.

WHERE IT'S GROWN AND WHAT ITS WINE TASTES LIKE

Jalilabad Winery in **Azerbaijan** make dry, dessert and fortified wines from this variety, whose produce is naturally low in alcohol and high in acidity.

HANEPOOT

See MUSCAT OF ALEXANDRIA.

HÁRSLEVELŰ

Aromatic, descriptively named Hungarian variety used for both sweet and dry whites.

BERRY COLOUR: ● ● ● ● ●

PRINCIPAL SYNONYMS: Budai Féher, Feuille de Tilleul (France), Harslevleue, Harzevelu, Lindenblättrige (Germany, Austria), Lipolist (Croatia), Lipovina (Czech Republic, Slovakia)

VARIETIES COMMONLY MISTAKEN FOR HÁRSLEVELŰ: EZERJÓ

ORIGINS AND PARENTAGE

Hárslevelű, meaning 'linden leaf', in reference to the wine's aroma, is a variety from Hungary, first mentioned there in 1744. Using thirteen DNA markers, Calò, Costacurta *et al.* (2008) had posited a parent–offspring relationship with FURMINT, which was confirmed with a total of sixty markers (Vouillamoz), making Hárslevelű an offspring of Furmint (see that entry) and a sibling of PLANTSCHER, an old Swiss variety of Hungarian origin, confirming the variety's Hungarian roots.

Hárslevelű should not be confused with EZERJÓ even though they share the synonym Budai Fehér ('white of Buda').

Hárslevelű was used to breed EZERFÜRTŰ, KABAR and ZEFÍR.

OTHER HYPOTHESES

Like FURMINT, Hárslevelű is often said, without evidence, to have been introduced from Italy.

VITICULTURAL CHARACTERISTICS

Productive, mid budding, late ripening. Big, loose bunches of small to medium-sized, thin-skinned berries. Prone to frost damage, sensitive to drought and susceptible to powdery mildew. Best suited to warmer volcanic soils. Marginally less prone to noble rot than FURMINT.

WHERE IT'S GROWN AND WHAT ITS WINE TASTES LIKE

Wines are aromatic and delicately spicy, adding perfume to the more fiery FURMINT in sweet Tokaji wines. The wines are softer than Furmint and mature more quickly. This century has seen an increasing number of very successful varietal late-harvest and Aszú varietal wines (Rohály *et al.* 2003). Dry and off-dry varietal wines are made at all quality levels but the best are viscous and full-bodied and have a distinctive flavour of linden blossom and linden honey.

In **Austria**'s Burgenland, Josef Umathum has planted a small parcel of Lindenblättrige, planning to expand it, and makes a dry varietal wine but also sees the potential for botrytized sweet wines. The variety was officially authorized in Austria in 2011.

Hárslevelű is grown throughout **Hungary** but particularly in Tokaj in the far north east (about 65% of the total), Eger (notably in the Debrő district) slightly further west and Somló in the far west, north of Lake Balaton, where wines tend to be more mineral and less aromatic. Wines from Villány in the south are softer and more perfumed. Recommended producers include Győrgyovács and Hollóvár in Somló, St Andrea in Eger, Berés and Zoltán Demeter in Tokaj and Bock and Plgár in Villány. There were 1,612 ha (3,983 acres) in Hungary in 2008.

Hárslevelű is also grown over the border in **Slovakia**; and in **Romania** (20 ha/49 acres in 2008), where it is used mainly for dry wines.

In **South Africa**, there were 75 ha (185 acres) in 2008 scattered across the wine regions but with the biggest area in Robertson. Lammershoek in the Swartland make a dry varietal wine and have high hopes for this variety, picked early to retain acidity, from their 1 ha (2.5 acres) of 42-year-old bushvines. Elsewhere the fruit tends to be swallowed up in blends.

HASANDEDE

Humdrum Turkish variety found east of the capital, Ankara.

BERRY COLOUR: ● ● ● ● ●

PRINCIPAL SYNONYMS: Ahmet Bey, Aşeri☒ (Nevşehir in Kapadokya/Cappadocia), Hasan Dede, Hasandede Beyazi

VARIETIES COMMONLY MISTAKEN FOR HASANDEDE: ?KALECIK KARASI, Sungurlu☒

ORIGINS AND PARENTAGE

Hasandede, literally 'grandfather Hasan', probably originates from the Kırıkkale province east of Ankara in central Turkey. DNA profiling has shown that Aşeri from Nevşehir in Kapadokya is identical to Hasandede, and DNA parentage analysis has suggested a possible parent–offspring relationship between Hasandede and Sungurlu, another variety from the Kırıkkale region that is not commercially cultivated (Vouillamoz *et al.* 2006). Confusingly, Sungurlu is sometimes used as a synonym for Hasandede.

OTHER HYPOTHESES

Sungurlu is listed as a synonym for Hasandede in the Vitis International Variety Catalogue but

DNA profiling rejects this hypothesis (Vouillamoz *et al.* 2006). Nevertheless, KALECIK KARASI is also listed among the synonyms, and so far no DNA profiling has been done to verify this hypothesis.

VITICULTURAL CHARACTERISTICS

Thin-skinned, medium-sized berries. Mid ripening, good general disease resistance.

WHERE IT'S GROWN AND WHAT ITS WINE TASTES LIKE

Grown in the Kırıkkale province east of Ankara in central **Turkey**, mostly for the table and to produce *pekmez*, concentrated grape juice, though it is also used to make very basic-quality white wine.

HEIDA

See SAVAGNIN BLANC *in the* SAVAGNIN *entry.*

HELFENSTEINER

German cross eclipsed by its offspring Dornfelder.

BERRY COLOUR: ● ● ● ● ●

PRINCIPAL SYNONYMS: Blauer Weinsberger, Weinsberg S 5332

ORIGINS AND PARENTAGE

A PINOT NOIR PRÉCOCE × SCHIAVA GROSSA cross created in 1931 by August Herold. This variety was named after the ruined castle of Helfenstein close to Geislingen an der Steige, south east of Stuttgart. Helfensteiner is now more famous as a parent of the highly successful DORNFELDER than in its own right.

VITICULTURAL CHARACTERISTICS

Mid to late ripening (but usefully earlier than its parent SCHIAVA GROSSA) with good resistance to downy mildew and botrytis bunch rot but erratic yields. Its susceptibility to coulure has played a major part in its decline since the 1970s.

WHERE IT'S GROWN

Whereas DORNFELDER has been widely planted within (and to a lesser extent without) Germany, there are just 19 ha (47 acres) of Helfensteiner, exclusively in southern **Germany**. In Württemberg, Dieterich, Jaillet and Wangler make varietal wines, the first producing a Weissherbst (pink) version. Kern (Baden) also bottles a varietal Helfensteiner.

HELIOS

Little-known, very complex, disease-resistant German hybrid.

BERRY COLOUR: ● ● ● ● ●

PRINCIPAL SYNONYMS: Freiburg 242-73

ORIGINS AND PARENTAGE

A complex hybrid obtained in 1973 by Norbert Becker at the Freiburg research centre in Baden-Württemberg, southern Germany, by crossing MERZLING × Fr 986-60, where:

- MERZLING is itself a complex hybrid (see PRIOR for the full pedigree)
- Fr 986-60 is a Seyve-Villard 12-481 × MÜLLER-THURGAU hybrid
- Seyve-Villard 12-481 is a Seibel 6468 × Seibel 6746 hybrid
- Seibel 6468 is a Seibel 4614 × Seibel 3011 hybrid, and the parents of Seibel 6746 were not recorded
- Seibel 4614 is a Seibel 752 × (Vivarais × *Vitis berlandieri*) hybrid (see PRIOR for the complete Vivarais pedigree)
- Seibel 752 is a Panse Précoce (Sicilien) × Clairette Dorée Ganzin hybrid
- Panse Précoce is a Bicane × PASCAL BLANC cross, the former being a non-cultivated French wine and table grape variety
- Clairette Dorée Ganzin is a Ganzin 60 × BOURBOULENC hybrid
- Ganzin 60 is an ARAMON NOIR × *Vitis rupestris* Ganzin hybrid
- Seibel 3011 is a Bayard × Afus Ali (ie Dattier de Beyrouth) hybrid, where Bayard is an Emily × *Vitis rupestris* hybrid and Emily is a *Vitis vinifera* subsp. *vinifera* × *Vitis labrusca* hybrid
- Afus Ali is a table grape variety from Lebanon.

Helios was named after the Greek God of the sun. Helios is also the name of an unrelated table grape grown in Slovakia.

VITICULTURAL CHARACTERISTICS

Mid ripening. Good resistance to botrytis bunch rot, downy and powdery mildews.

WHERE IT'S GROWN AND WHAT ITS WINE TASTES LIKE

Helios was authorized for commercial production in **Germany** in 2005 and plantings are still insufficient for it to feature in official German vineyard statistics, although varietal wines are made by, for example, Dr Benz, Feuerstein and Isele in Baden in the south. They can be rather nondescript and are sometimes improved by a few grams of residual sugar. Acidity levels are similar to MÜLLER-THURGAU although must weights can be higher.

A few experimental vines (0.23 ha/0.57 acres) are also planted in **Switzerland**. Also found in **Belgium** at the Domaine Viticole du Chenoy.

HERBEMONT

Almost extinct American hybrid with a complex and disputed parentage.

BERRY COLOUR: ● ● ● ● ●

PRINCIPAL SYNONYMS: Black Herbemont, Bottsi, Brown French, Dunn, Herbemon, Herbemont's Madeira, Hunt, Kay's Seedling, Madeira, Mcknee, Neal Grape, Neil Grape, Warren, Warrenton, White Herbemont

ORIGINS AND PARENTAGE

The exact origin of this variety is obscure and disputed. Thomas Volney Munson (1843–1903), a horticulturist and grape breeder from Texas, considered Herbemont and JACQUEZ (under the synonym Lenoir) to be members of *Vitis bourquiniana*, a doubtful botanical group of varieties that are closely related to the native *Vitis aestivalis* (Munson 1909). Bailey (1906) described these *bourquiniana* varieties as 'ameliorated forms of the native summer-grape *Vitis aestivalis*', and thought that some of them 'may be hybrids of *Vitis aestivalis* and the European wine-grape', the last hypothesis being supported by Galet (1988). It is likely that the European varieties came from Madeira.

Nicholas Herbemont (1771–1839), a French-born grower from Columbia, South Carolina, collected one such *bourquiniana* variety in 1798 in the plantation of Judge John Huger in Columbia, South Carolina, and called it Madeira. Herbemont was so active in its promotion and propagation that it was later known as Herbemont's Madeira, and then simply Herbemont. According to Thomas McCall's record, this variety was brought to Georgia before the American Revolution by Henry Hunt in St Paul's parish north of Augusta, where it was known as the Hunt grape. In the early nineteenth century it spread to Warren County, Georgia, where it was then simply known as the Warren grape (Bonner 2009). These separate mentions clearly indicate that the variety had been cultivated and propagated long before Herbemont found it in Columbia.

OTHER HYPOTHESES

According to Munson, Herbemont and JACQUEZ were introduced from southern France around 1750 as 'Brown French' (Herbemont) and 'Blue French' (Jacquez) by the Bourquin family of Savannah, Georgia, hence the name *Vitis bourquiniana,* in honour of Gougie Bourquin. However, their characteristics and disease resistance differ widely from pure *Vitis vinifera* and they were considered to be hybrids of *Vitis aestivalis*, *Vitis cinerea* and most likely some unidentified *Vitis vinifera* subsp. *vinifera* (Munson 1909). If Herbemont and Jacquez are not pure *Vitis vinifera*, then some *Vitis aestivalis* and *Vitis cinerea* must previously have been introduced from America to Europe, where they were hybridized spontaneously or deliberately with some European *vinifera* to generate Herbemont and Jacquez, which were imported to the New World in the mid eighteenth century. This hypothesis can be rejected because the hybridization would predate the earliest deliberate cross made with grapes (around 1750) and because there was no reason to make interspecific hybrids in Europe at that time, simply because powdery mildew, downy mildew and phylloxera had not yet been introduced.

WHERE IT'S GROWN AND WHAT ITS WINE TASTES LIKE

In the humid climate of southern **Brazil** the variety found favour because of its disease resistance and there were still 846 ha (2,091 acres) in Rio Grande do Sul in 2007. Salton, for example, produce a rosé that blends Herbemont with Isabel (ISABELLA) and SEYVAL BLANC. There are also still a few vines in the Cévennes, even though the variety is not authorized in **France**, and some of the fruit contributes to the Jacquez-dominated Cuvée des Vignes d'Antan made by the Association Mémoire de la Vigne, which cannot legally be sold in the EU since almost all hybrids are banned there.

HEROLDREBE

Unexceptional, productive but inconveniently late-ripening German cross.

BERRY COLOUR: ● ● ● ● ●

PRINCIPAL SYNONYMS: Weinsberg S 130

ORIGINS AND PARENTAGE

A BLAUER PORTUGIESER × BLAUFRÄNKISCH cross bred by and named after August Herold at the Weinsberg research centre in Baden-Württemberg in southern Germany in 1929. Since Blaufränkisch is a progeny of GOUAIS BLANC, Heroldrebe is its grandchild. Heroldrebe was used to breed DORNFELDER. This parentage has been confirmed by DNA profiling (Myles *et al.* 2011).

VITICULTURAL CHARACTERISTICS

Yields are high at around 140 hl/ha (8 tons/acre) and regular but it ripens so late that it can be grown only in warmer regions of Germany. Good resistance to downy mildew and botrytis bunch rot.

WHERE IT'S GROWN AND WHAT ITS WINE TASTES LIKE

There are 155 ha (383 acres) in **Germany**, mainly in the Pfalz, where it is used predominantly for rosé, but it's also planted in Württemberg and Rheinhessen. The wines are reasonably light but can have noticeable tannins and overall wine quality is lower than for its parent BLAUER PORTUGIESER. It is more famous as a parent of DORNFELDER than in its own right. Varietal wines are made by Anselmann in the Pfalz and Bernecker further south in Baden.

HETERA

Recently authorized Slovakian cross making promising sweet wines.

BERRY COLOUR: ● ● ● ● ●

PRINCIPAL SYNONYMS: HTCVCB 4/13

ORIGINS AND PARENTAGE

Hetera is a GEWÜRZTRAMINER × ROTER VELTLINER cross (with Slovakian names: Tramín Červený × Veltlínske Červené) obtained in 1965 by Dorota Pospíšilová at the VUVV research centre for oenology and viticulture in Bratislava, Slovakia, and authorized in 2011.

VITICULTURAL CHARACTERISTICS

Vigorous, regularly productive, late budding and ripening. Largish bunches of small, flavourful berries. Susceptible to botrytis bunch rot and powdery mildew.

WHERE IT'S GROWN AND WHAT ITS WINE TASTES LIKE

Although total plantings in **Slovakia** are still small (5 ha/12 acres in 2011), Hetera seems best suited to the Južnoslovenská and Malokarpatská wine regions in the south west and to botrytized sweet wines. Žitavské Vinice make a varietal wine in this style.

HIBERNAL

German hybrid planted more widely in the Czech Republic than at home.

BERRY COLOUR: ● ● ● ● ●

PRINCIPAL SYNONYMS: Geisenheim 322-58

ORIGINS AND PARENTAGE

Complex self-crossed hybrid obtained in 1944 by Helmut Becker at the Geisenheim research centre in the Rheingau, Germany, by crossing CHANCELLOR × RIESLING with itself (see PRIOR for the complete pedigree of Chancellor).

VITICULTURAL CHARACTERISTICS

Good winter-frost resistance and reasonably good resistance to powdery and downy mildews. Sugar level and yield are generally higher than for RIESLING but acidity is lower. Ripens in the first half of October.

WHERE IT'S GROWN AND WHAT ITS WINE TASTES LIKE

Very little Hibernal is planted in **Germany** (1 ha/2.5 acres) but there are 17 ha (42 acres) – and rising – in the **Czech Republic**, where it was registered in 2004, mainly in Morava in the south east. Knab in Baden, southern Germany, produce a rare varietal wine. It is also found to a very limited extent in **Canada**.

Wines are fresh and fruity, in some years rather like SCHEUREBE, and the fruit purity and high levels of extract of the best, fully ripe examples reveal its RIESLING parent. When unripe, it can have a slightly foxy note but this is not usually pronounced. Owing to its complex family history, it is not 100% *Vitis vinifera* but it is registered as such in the German official list of varieties and is therefore authorized for making quality wine because the wine generally tastes like a *vinifera* product. This is the case for several other German hybrids such as RONDO.

HIBOU NOIR

See AVANÀ.

HIMBERTSCHA

Swiss variety recently brought back from the brink of extinction.

BERRY COLOUR: ● ● ● ● ●

VARIETIES COMMONLY MISTAKEN FOR HIMBERTSCHA: LAFNETSCHA

ORIGINS AND PARENTAGE

Himbertscha is a rare and old variety originating from the region of Visp in the Haut-Valais, Switzerland. It was rescued from extinction by Josef-Marie Chanton, who in the early 1970s discovered a few vines in an old vineyard that was itself rescued in 2010 by the Association VinEsch. The name Himbertscha has nothing to do with the German *Himbeer* (strawberry) and most likely comes from *im Bercla*, a dialect version of the Italian *in pergola*, which is consistent with the common practice of training the variety in this way. DNA parentage analysis has established that Himbertscha is a progeny of the old Valais variety HUMAGNE, and therefore a half-sibling of LAFNETSCHA (Vouillamoz, Maigre and Meredith 2004). See the pedigree diagram in the COMPLETER entry.

VITICULTURAL CHARACTERISTICS

Mid budding and ripening. Susceptible to botrytis bunch rot.

WHERE IT'S GROWN AND WHAT ITS WINE TASTES LIKE

With the exception of a few vines in the original VinEsch vineyard in Vispertal, the sole surviving Himbertscha vineyard in the world, just 0.17 ha (0.42 acres) in Varen, is owned by the man who rescued the variety, Josef-Marie Chanton in the Haut-Valais, **Switzerland**. His varietal wine is floral and citrusy with crisp acidity.

HITZKIRCHER

Swiss rarity found only in the village after which it was named.

BERRY COLOUR: ● ● ● ● ●

PRINCIPAL SYNONYMS: Grosse Blaue Mörsch, Hitzkirchener, Hitzkirchler

VARIETIES COMMONLY MISTAKEN FOR HITZKIRCHER: BONDOLA

ORIGINS AND PARENTAGE

Hitzkircher is an extremely rare variety from the village of Hitzkirch in the canton of Luzern in central Switzerland. It has often been confused with Briegler, which was recently shown to be identical to BONDOLA from Ticino (Frei *et al.* 2006; Vouillamoz, Frei and Arnold 2008). This makes more sense when we know from DNA parentage analysis that both Hitzkircher and BONDOLETTA from Ticino are distinct natural crosses between COMPLETER from Graubünden and Bondola from Ticino (Vouillamoz and Moriondo 2011).

WHERE IT'S GROWN AND WHAT ITS WINE TASTES LIKE

The only varietal Hitzkircher wine is made by Rebbaugesellschaft Hitzkirch in the village in **Switzerland** of the same name.

HÖLDER

Very minor, distinctly ordinary German cross.

BERRY COLOUR: ● ● ● ● ●

PRINCIPAL SYNONYMS: Hoelder, Weinsberg S 397

ORIGINS AND PARENTAGE

A RIESLING × PINOT GRIS (Ruländer) cross obtained in 1955 by August Herold. It was apparently named after the Lauffen am Neckar poet Friedrich Hölderlin.

VITICULTURAL CHARACTERISTICS

Mid to late ripening, regular yields with good sugar levels and moderate acidity. Good resistance to downy mildew and botrytis bunch rot, less so to powdery mildew.

WHERE IT'S GROWN AND WHAT ITS WINE TASTES LIKE

Germany has just 6 ha (15 acres) of this variety, authorized in 1987, scattered across its more southerly wine regions. Wines are unexceptional – reasonably fresh and fruity and lightly aromatic.

HONDARRIBI BELTZA

Rare speciality of northern Spain's Basque Country related to, and sometimes reminiscent of, Cabernet Franc.

BERRY COLOUR: ● ● ● ● ●

PRINCIPAL SYNONYMS: Chacolí (País Vasco), Cruchen Nègre (Béarn), Hondarrabi Beltza (País Vasco), Kurixketu Beltza (Guipúzcoa and Biscaye in Spain), Ondarrabi Beltza (País Vasco), Txakoli (País Vasco), Verde Matza, Xerratu Beltza (Guipúzcoa and Biscaye)

VARIETIES COMMONLY MISTAKEN FOR HONDARRIBI BELTZA: CABERNET FRANC (Txakoli vineyards in País Vasco), COURBU NOIR (Béarn in France)

ORIGINS AND PARENTAGE

Hondarribi Beltza was named after the town of Hondarribia in the province of Guipúzcoa in northern Spain, close to the border with France, and the colour of its berries (*beltza* means 'black' in Basque). It is also called Txakoli or Chacolí, after the name of the vineyards where it is cultivated. It is often interplanted with CABERNET FRANC, with which it is frequently and reasonably confused: DNA parentage analysis has revealed a parent–offspring relationship between Hondarribi Beltza and Cabernet Franc (Boursiquot *et al.* 2009). In addition, Hondarribi Beltza and FER spontaneously crossed to give birth to Gros Cabernet (Boursiquot *et al.* 2009; see CABERNET SAUVIGNON's pedigree), an almost extinct variety from the Gironde in south-west France. Hondarribi Beltza is therefore a member of the Carmenet ampelographic group (see p XXVII; Bisson 2009).

Hondarribi Beltza is not a colour mutation of Hondarribi Zuri (*zuri* means 'white'), a name which is, confusingly, used for three varieties planted in the País Vasco: COURBU BLANC, CROUCHEN and the hybrid NOAH. Official Spanish statistics reported 412 ha (1,018 acres) of Hondarribi Zuri in 2008, almost exclusively in the País Vasco, where it is the main white variety in the three principal Basque DOs (Bizkaiko Txakolina, Getariako Txakolina and Arabako Txakolina), but it is difficult to know which of the three varieties are referred to in the statistics and in the varietal wines produced (see below).

VITICULTURAL CHARACTERISTICS

Vigorous. Small to medium-sized, compact bunches.

WHERE IT'S GROWN AND WHAT ITS WINE TASTES LIKE

There were just 7 ha (17 acres) in 2008, almost exclusively in the País Vasco (Basque Country) in northern **Spain**, where it is the principal variety in the red and rosé wines of the Bizkaiko Txakolina (Chacolí de Bizcaia) DO and in the reds of the Getariako Txakolina (Chacolí de Guetaria) DO. Doniene Gorrondona make a varietal wine thanks to cuttings from ancient pre-phylloxera vines. Other producers of varietal wines include Zabala, Abio and Talai Berri, while Txomín Etxaniz use it in their Txakoli white blend. Some rosé wines are a blend of the two Hondarribis. Wines are typically high in both alcohol and acidity, potentially also in colour and tannins, and are generally aromatic with some herbaceous notes, sometimes reminiscent of CABERNET FRANC.

HONDARRIBI ZURI

See COURBU BLANC, CROUCHEN *and* NOAH.

HONIGLER

See MÉZES FEHÉR.

HOROZKARASI

A stay-at-home variety used more for the table than for wine.

BERRY COLOUR: ● ● ● ● ●

PRINCIPAL SYNONYMS: Horoz Karası

VARIETIES COMMONLY MISTAKEN FOR HOROZKARASI: ADAKARASI⊗ (Avşa), ÇALKARASI⊗ (Denizli), KALECIK KARASI (Ankara), PAPAZKARASI (Avşa)

ORIGINS AND PARENTAGE

Horozkarası, literally 'black rooster', most likely originates from Kilis province near Gaziantep in south-eastern Turkey, near the border with Syria.

OTHER HYPOTHESES

See ADAKARASI.

VITICULTURAL CHARACTERISTICS

Mid ripening, large berries, productive. Susceptible to botrytis bunch rot and is best suited to a hot climate.

WHERE IT'S GROWN AND WHAT ITS WINE TASTES LIKE

Horozkarası has not travelled far. Today it is grown almost exclusively around Gaziantep, near the Syrian border in the far south of **Turkey**. Wines are typically deeply coloured, and high in tannin and alcohol, but it is more widely used for the table than for wine and is not generally regarded as a good-quality wine grape. Bırıcık make a varietal wine.

HRON

Recent but promising Slovakian cross producing full-bodied, ageworthy reds.

BERRY COLOUR: ● ● ● ● ●

PRINCIPAL SYNONYMS: CAAB 3/22

ORIGINS AND PARENTAGE

Hron, named after a tributary of the Dunaj (Danube), is a CASTETS × ABOURIOU cross obtained in 1976 by Dorota Pospíšilová at the VUVV research centre for oenology and viticulture in Bratislava, Slovakia. It is a sibling of NITRANKA, RIMAVA and VÁH and was authorized in 2011.

VITICULTURAL CHARACTERISTICS

Late budding, mid to late ripening. Susceptible to hard winter frosts and moderately so to powdery mildew but resistant to spring frosts and botrytis bunch rot. Best suited to deep, warm soils.

WHERE IT'S GROWN AND WHAT ITS WINE TASTES LIKE

Hron is not yet widely planted (5 ha/12 acres in 2011) but shows promise, particularly in the Južnoslovenská and Malokarpatská wine regions in south-west **Slovakia**, for high-quality, deeply coloured, full-bodied, ageworthy reds with some similarity to CABERNET SAUVIGNON. Producers of varietal wines include Chateau Modra and Karol Braniš.

HRVATICA

Recently rescued Croatian variety making pale reds.

BERRY COLOUR: ● ● ● ● ●

PRINCIPAL SYNONYMS: Hrvatica Crna, Jarbola🧬 (Matulji), Karbonera, Markolina, Negrara, Negrona

VARIETIES COMMONLY MISTAKEN FOR HRVATICA: CROATINA🧬

ORIGINS AND PARENTAGE

Hrvatica, meaning 'Croatian girl', is an indigenous variety from the area of Kaštelir in Istra (Istria) in north-west Croatia. Having exactly the same etymology, the Italian variety CROATINA has long been considered identical to Hrvatica, but DNA profiling not only rejected this identity (Maletić *et al.* 1999) but also later showed that Jarbola, a black-berried variety cultivated exclusively in Matulji, north west of Rijeka and close to the Adriatic coast, is identical to Hrvatica.

The white-berried Jarbola is not a colour mutation of the black-berried Jarbola and represents a distinct variety (Sladonja *et al.* 2007). New plantings were initiated in 2004 and it has been suggested that it be called Jarbola Bijela (*bijela* meaning 'white'; Sladonja *et al.* 2007).

VITICULTURAL CHARACTERISTICS

Mid budding, early to mid ripening, high yields. Very susceptible to downy mildew.

WHERE IT'S GROWN AND WHAT ITS WINE TASTES LIKE

Recommended in all the Istrian wine regions in north-west **Croatia** but is currently planted only in Bujštine in the far north-western corner of Istra. A project to preserve and rehabilitate Jarbola in the Matulji area was begun in 2003. The grapes are not deeply coloured and are therefore well suited to the production of rosé.

HUMAGNE

Ancient, tangy Swiss variety recently revived in the Valais.

BERRY COLOUR: ● ● ● ● ●

PRINCIPAL SYNONYMS: Humagne Blanc☒, Humagne Blanche☒, Miousat☒ or Miousap or Mioussat (Pyrénées-Atlantiques in France)

ORIGINS AND PARENTAGE

Humagne is a very old variety from the Valais, southern Switzerland, where it was first mentioned in 1313 as 'humagny' (see also RÈZE) on a parchment known as the 'Registre d'Anniviers' (Ammann-Doubliez 2007; Vouillamoz 2009a). Recent ampelographic observations and DNA profiling have surprisingly shown that Miousat, still scattered in the vineyards of Jurançon and Monein in the Pyrénées-Atlantiques in south-western France, is identical to Humagne (Bordenave *et al.* 2007). This makes more sense in the light of DNA parentage analysis revealing a parent–offspring relationship between Humagne and COLOMBAUD (see COMPLETER for the pedigree diagram), an almost extinct variety from Provence in southern France (Vouillamoz and Moriondo 2011). It is therefore quite plausible that Humagne was introduced from southern France to the Valais via the Vallée du Rhône before the thirteenth century, which would also explain its possible Greek etymology based on *hylomaneus*, meaning 'vigorous', the Phocaeans having probably introduced viticulture to southern France (Aebischer 1937; Vouillamoz and Moriondo 2011).

In the Haut-Valais, a natural cross between Humagne and COMPLETER gave birth to LAFNETSCHA, while another natural cross with an unknown and probably extinct variety produced HIMBERTSCHA (Vouillamoz, Maigre and Meredith 2004).

Humagne, often called Humagne Blanc, is not related to Humagne Rouge, the Valais name for CORNALIN from the Valle d'Aosta.

OTHER HYPOTHESES

The name Humagne is often said to derive from the Latin *Vinum humanum*, allegedly mentioned in old documents and pointing to a Roman origin for this variety. However, no documentary evidence has been found.

VITICULTURAL CHARACTERISTICS

Late ripening, extremely vigorous, average but irregular yields. Susceptible to botrytis bunch rot and magnesium deficiency. Smallish berries.

WHERE IT'S GROWN AND WHAT ITS WINE TASTES LIKE

Almost extinct by the mid twentieth century, Humagne was revived at the end of the 1980s and by 2009 its plantings in **Switzerland** totalled 30 ha (74 acres), exclusively in the Valais. Humagne wines are typically dry and fresh with subtle hints of lime-blossom, moderate alcohol and an elegant texture, developing a resin-like aroma as they age, although they can be very ordinary if yields are not kept low. Recommended producers include Defayes-Crettenand, Mabillard-Fuchs and Montzuettes.

Local legend has it that Humagne wines contain more iron than any other, which is why it is often called *vin des accouchées* ('wine for new mothers') and has traditionally been offered to women who have just given birth. However, chemical analysis has shown this to be a myth and historical documents suggest that it was the medicinal herbs added to hot Humagne (or any other wine) that were responsible for its restorative properties (Pont 2005).

HUMAGNE ROUGE

See CORNALIN.

HUXELREBE

German cross scores highly for yields, ripeness, aroma and acidity but not for subtlety.

BERRY COLOUR: ● ● ● ● ●

PRINCIPAL SYNONYMS: Alzey S 3962

ORIGINS AND PARENTAGE

A CHASSELAS (aka Weisser Gutedel) × Muscat Précoce de Saumur (aka Courtiller Musqué) cross bred in 1927 by Georg Scheu at Alzey, although it is named after the nurseryman Fritz Huxel, who was the first to propagate this variety widely in the 1950s. Muscat Précoce de Saumur is a selfling of PINOT NOIR PRÉCOCE obtained by Auguste Courtiller in 1842.

VITICULTURAL CHARACTERISTICS

Very high yielding and manages to produce very high sugar levels – Auslese and above – even in an average year if yields are controlled. This is sometimes the result of its tendency towards millerandage. Good resistance to downy and powdery mildews. Thin skins make it susceptible to botrytis bunch rot but also allow it to benefit from noble rot in the making of sweet wines. Ripens early.

WHERE IT'S GROWN AND WHAT ITS WINE TASTES LIKE

In **Germany** there are 635 ha (1,569 acres), predominantly in Rheinhessen (421 ha/1,040 acres) and in the Pfalz (189 ha/467 acres). Wines are often very sweet and tend to be pungently aromatic and full-flavoured, if not desperately subtle, with flavours and aromas of apricot, passion fruit, honey, spice and musk. Only the best examples from producers such as Gysler and Seehof have ageing potential thanks to the variety's marked acidity.

Huxelrebe was introduced to **England** in 1972 but the current total of 25 ha (62 acres) is about half what it was twenty years ago. In such a cool climate, Huxelrebe's reliable ripeness helps to balance the high acidity. It is frequently added to blends such as Chapel Down's Flint Dry, for example, but also used to make dessert wines.

I

IMPIGNO
INCROCIO BIANCO FEDIT 51
INCROCIO BRUNI 54
INCROCIO MANZONI 2.15
INCROCIO TERZI 1
INVERNENGA
INZOLIA
İRI KARA
IRSAI OLIVÉR
ISABELLA
ITALIA
ITALICA
IVES

ɤ this symbol on the following pages indicates that the synonyms, and mistaken identities, are confirmed by DNA evidence

IMPIGNO

Important only in the Ostuni DOC.

BERRY COLOUR:

ORIGINS AND PARENTAGE

The name Impigno, meaning 'lazybones', is derived from the nickname of a grower from Ostuni in the province of Brindisi in Puglia, southern Italy, who took the variety to the Martina Franca region in the province of Taranto in the nineteenth century.

Cipriani *et al.* (2010) have recently suggested through DNA parentage analysis that Impigno may be a natural cross between BOMBINO BIANCO and QUAGLIANO. However, this parentage showed some DNA discrepancies and needs further investigation.

VITICULTURAL CHARACTERISTICS

Mid ripening (mid September). Good resistance to frost and powdery mildew.

WHERE IT'S GROWN AND WHAT ITS WINE TASTES LIKE

Impigno is cultivated in the province of Brindisi, Puglia, **Italy**, in the communes of Ostuni, Carovigno, Brindisi, San Vito dei Normanni and San Michele Salentino and to a lesser extent in Latiano and Ceglie Messapica. It is usually blended with other local grapes such as VERDECA, BIANCO D'ALESSANO or FRANCAVIDDA and it comprises 50–85% of wines labelled Ostuni DOC. Plantings in Martina Franca in the province of Taranto are few and far between. The wines are generally fresh and simple. There were 62 ha (153 acres) in Italy in 2000.

INCROCIO BIANCO FEDIT 51

Rare Veneto 1950s white cross good for passito wines.

BERRY COLOUR:

PRINCIPAL SYNONYMS: Fedit 51

ORIGINS AND PARENTAGE

Note that *incrocio* is Italian for 'crossing', which is why there are several unrelated varieties called Incrocio Something (with Something usually being the breeder's name).

This is a GARGANEGA × MALVASIA BIANCA LUNGA (under the name Malvasia del Chianti) cross obtained by the Federazione Italiana dei Consorzi Agrari (Italian Federation of Agricultural Consortiums) in Padova in 1951.

VITICULTURAL CHARACTERISTICS

Mid to late ripening. Robust berries are resistant to botrytis bunch rot and well suited to drying to make passito-style wines.

WHERE IT'S GROWN

Authorized in the provinces of Padova and Vicenza in the Veneto, northern **Italy**. Only 13 ha (32 acres) were recorded in Italy in 2000.

INCROCIO BRUNI 54

Minor but high-yielding central Italian white cross that generally disappears into blends.

BERRY COLOUR: white

PRINCIPAL SYNONYMS: Dorico, Sauvignon × Verdicchio

ORIGINS AND PARENTAGE

Obtained in 1936 by Bruno Bruni at the ministry of agriculture in the Marche region by crossing SAUVIGNON BLANC and VERDICCHIO BIANCO.

VITICULTURAL CHARACTERISTICS

A vigorous and abundant early ripener. Good resistance to botrytis bunch rot and therefore suitable for late-harvested or passito-style wines.

WHERE IT'S GROWN AND WHAT ITS WINE TASTES LIKE

It is cultivated in the provinces of Ancona and Macerata in the Marche in central **Italy** and, to a lesser extent, in the province of Perugia in Umbria. However, total plantings in Italy in 2000, according to the agricultural census, were just 13 ha (32 acres). It is invariably blended with other varieties such as PECORINO, TREBBIANO TOSCANO, VERDICCHIO BIANCO, etc. but La Montata have produced a varietal wine classified as IGT Marche that combines the aromatic character of SAUVIGNON BLANC with the structure of Verdicchio Bianco. The wines tend to be aromatic, dry and slightly bitter.

Incrocio Bruni 54 may comprise up to 30% of the blended Colli Maceratesi DOC just south of Ancona.

INCROCIO MANZONI 2.15

Very minor black-skinned cross confined to the Veneto in northern Italy.

BERRY COLOUR: black

PRINCIPAL SYNONYMS: Manzoni Nero, Prosecco × Cabernet Sauvignon 2-15

ORIGINS AND PARENTAGE

Incrocio Manzoni 2.15 is a cross obtained by Luigi Manzoni at the Conegliano research centre in the Veneto between 1924 and 1930. Manzoni's initial intention was to produce a white variety by crossing PROSECCO and SAUVIGNON BLANC but he used the pollen of CABERNET SAUVIGNON instead of Sauvignon Blanc and ended up with a black-berried variety.

VITICULTURAL CHARACTERISTICS

Vigorous, productive and robust. Good resistance to winter cold. Late ripening.

WHERE IT'S GROWN AND WHAT ITS WINE TASTES LIKE

Cultivation of Incrocio Manzoni 2.15 is restricted to the province of Treviso in the Veneto, **Italy**, mainly in the vineyards of Conegliano and Montello. Up to 10% may be included in the Colli di Conegliano Rosso DOC, blended with CABERNET FRANC, CABERNET SAUVIGNON, MERLOT or MARZEMINO. The Italian agricultural census of 2000 recorded total plantings of 169 ha (418 acres). Varietal wines are rare but Collalto produce one, classified as IGT Colli Trevigiani, and Perlage produce a traditional-method sparkling wine. The wines tend to be slightly herbaceous and low in tannins, with aromas of red and black fruits.

INCROCIO TERZI 1

Minor Lombardia crossing that disappears into blends.

BERRY COLOUR: ● ● ● ● ●

PRINCIPAL SYNONYMS: Barbera × Cabernet Franc 1

ORIGINS AND PARENTAGE

Obtained by Riccardo Terzi, a wine grower in the province of Bergamo in Lombardia, north-west Italy, by crossing BARBERA with CABERNET FRANC.

VITICULTURAL CHARACTERISTICS

Vigorous with generous and reliable yields. Mid to late ripening.

WHERE IT'S GROWN AND WHAT ITS WINE TASTES LIKE

Incrocio Terzi 1 is cultivated only in Lombardia, north-west **Italy**, where it is particularly recommended in the provinces of Bergamo and Brescia. Authorized as part of the blend in the following DOCs, it does not appear to be produced as a varietal wine: Cellatica (10–15%) and Capriano del Colle (maximum 15%). It is also planted to a limited extent in Franciacorta and on the shores of Lago di Iseo and Lago di Garda. The wines are deeply coloured with highish alcohol.

There were 69 ha (171 acres) in Italy in 2000.

INVERNENGA

Old and rare Brescia white, ideal for drying off the vine.

BERRY COLOUR: ● ● ● ● ●

PRINCIPAL SYNONYMS: Invernesca, Ua'mbrunesca

ORIGINS AND PARENTAGE

Invernenga was first described in 1826 as one of the most widely cultivated varieties around Brescia in Lombardia (Giavedoni and Gily 2005). Its name probably comes from *inverno* (Italian for 'winter') because Invernenga grapes used to be dried during winter, and later rehydrated by putting the bunches in warm water, for consumption as table grapes.

VITICULTURAL CHARACTERISTICS

Mid to late ripening.

WHERE IT'S GROWN AND WHAT ITS WINE TASTES LIKE

Right in the middle of the city of Brescia in central Lombardia, **Italy**, at the foot of the hill that leads up to the castle, is the historic Vigneto della Pusterla vineyard, owned by the Pusterla Winery. This 4 ha (10 acres) vineyard is trained in a local pergola style and produces the Pusterla Bianco IGT Ronchi di Brescia, the only existing example of varietal Invernenga. This unique 'city wine', supported by Slow Food, is an elegant white with aromas of almonds. Outside Pusterla, there are just a few pockets of old, generally interplanted Invernenga vines.

INZOLIA

Nutty Sicilian white.

BERRY COLOUR: ● ● ● ● ●

PRINCIPAL SYNONYMS: Ansolica (Toscana), Ansonica (Toscana), Ansora (Toscana), Anzonica (Toscana), Insolia (Sicilia), Insolia di Palermo (Sicilia), Insora (Sicilia), 'Nzolia (Sicilia), Zolia Bianca (Sicilia)

VARIETIES COMMONLY MISTAKEN FOR INZOLIA: Inzolia Imperiale (Sicilia), Regina (Sicilia), Valenci Bianco (Sicilia)

ORIGINS AND PARENTAGE

Under a number of names, this variety has been cultivated since the sixteenth century in Toscana and on Sicilia. Inzolia is usually said to have its roots on Sicilia, where it was first cited by Cupani (1696) and whence it spread to Sardegna and then to the islands of Elba and Giglio in Toscana (Giavedoni and Gily 2005). This island origin has been confirmed by a recent genetic study relating Inzolia to such other Sicilian varieties as GRILLO, FRAPPATO and NERELLO MASCALESE, as well as the closely related but distinct (no longer cultivated) Inzolia Imperiale and Inzolia Nera and the more widespread MOSCATO GIALLO (known as Muscatedda on Sicilia) (Carimi *et al.* 2009). This refutes the hypothesis of a Greek origin for Inzolia supported by Labra *et al.* (1999), who have suggested that it (under the name Ansonica) could be related to Greek varieties SIDERITIS and RODITIS (without specifying which one), as well as the supposed genetic relationship with AIRÉN from Spain or with CLAIRETTE from France. This is probably a consequence of the low sampling number of Mediterranean (twenty-five) and Greek (forty) varieties analysed by Labra *et al.* (1999) and of the lack of other Tuscan or Sicilian varieties in their study.

OTHER HYPOTHESES

According to some scholars, Inzolia/Ansonica could be identical to the Irziola mentioned by Pliny the Elder but there's no convincing evidence to support this hypothesis. Some think Inzolia was introduced in several separate events, first to Elba from Spain, then in 1561 to the island of Giglio by the Minoan Greeks, but again no historical data can confirm this.

Contrary to common belief, the black-berried Inzolia Nera is not a colour mutation of Inzolia, as evidenced by DNA profiling (Carimi *et al.* 2009). Since four distinct types of Inzolia have been described in Sicilia, it has been argued that this variety has a polyclonal origin (Scienza *et al.* 2008) but the comparison of those four DNA profiles with other studies provided evidence that the most widespread one is the true Inzolia, while the other three are identical to other known varieties such as Regina or Valenci Bianco (Vouillamoz). (See pp XIV–XVI for more on polyclonal origin and what constitutes a variety.)

VITICULTURAL CHARACTERISTICS

Early flowering and veraison, early to mid ripening. Large, juicy berries. Very susceptible to downy mildew but relatively drought-tolerant.

WHERE IT'S GROWN AND WHAT ITS WINE TASTES LIKE

Inzolia is one the most important and ancient white grapes of Sicilia, **Italy**, long cultivated in the province of Palermo, as well as in a few spots in the provinces of Trapani, Agrigento, Caltanisetta and Girgenti. It has been a valued ingredient in Marsala but is now more often blended with CATARRATTO BIANCO and GRILLO as part of numerous Sicilian DOCs (Alcamo, Contea di Sclafani, Contessa Entellina, Delia Nivolelli, Erice, Mamertino di Milazzo, Marsala, Menfi, Monreale, Riesi, Salaparuta, Sambuca di Sicilia, Santa Margherita di Belice, Sciacca, Vittoria) and IGTs.

Inzolia has also been cultivated for a long time in Toscana, where it is called Ansonica, in particular on the Tuscan coast (Val di Cornia, Monte Argentario and Valle dell'Albegna) and in the Tuscan archipelago (the islands of Elba and Giglio). A few Inzolia vines have apparently also been found on Sardegna and in Campania and Lazio. Total Italian plantings in 2000 were 9,500 ha (23,500 acres) although the area is declining. Sicilia recorded 7,795 ha (19,262 acres) in 2008.

The best wines have a nutty character but the grapes must be picked early enough to retain sufficient acidity.

İRI KARA

Very minor, mediocre, pink-skinned Turkish variety.

BERRY COLOUR: ● ● ● ● ●

PRINCIPAL SYNONYMS: Arikaras, Focea, Fodja

VARIETIES COMMONLY MISTAKEN FOR İRI KARA: FOKIANO ☓ (Greece), Sergi Karası (Gaziantep province), YEDIVEREN (Şanlıurfa province)

ORIGINS AND PARENTAGE

İri Kara, literally meaning 'big black', is a variety from the region of Hadım in the province of Konya, Central Anatolia. According to Vouillamoz *et al.* (2006), İri Kara could be genetically closely related to Ekşi Kara, a variety from the province of Karaman, south east of Konya, that is no longer cultivated.

VITICULTURAL CHARACTERISTICS

Small to medium-sized pink berries with thick skins. Mid to late ripening, productive and with generally good disease resistance. Acidity and sugar levels in the berries tend to be low. Best suited to hot, dry conditions.

WHERE IT'S GROWN AND WHAT ITS WINE TASTES LIKE

İri Kara is grown on the Baklana plateau between Baklan-Bekilli in the province of Denizli and Karahallı in the province of Uşak, south-west **Turkey**. It is not ideally suited as a table grape and is therefore also made into low-quality table wine.

IRSAI OLIVÉR

Hungarian table grape promoted to making soft, aromatic white wines.

BERRY COLOUR: ● ● ● ● ●

PRINCIPAL SYNONYMS: Irsay Oliver, Irsay Oliver Muskotaly, Muscat Oliver (Croatia, Germany, German-speaking Switzerland), Oliver Irsay

ORIGINS AND PARENTAGE

Irsai Olivér is a Pozsonyi Fehér × CSABA GYÖNGYE cross obtained in 1930 by Pál Kocsis in Kecskemét, central Hungary (Galbács *et al.* 2009), in which Pozsonyi Fehér (Pozsony is the Hungarian name for Bratislava and Fehér means 'white') is a white wine variety from Hungary that is no longer cultivated but is often confused with KÖVIDINKA because they share the synonym Dinka Fehér. It was officially registered in 1975.

Irsai Olivér was used to breed AGNI and CSERSZEGI FŰSZERES.

VITICULTURAL CHARACTERISTICS

Vigorous, early budding and ripening, needs hard pruning. Medium-sized bunches of small, firm-skinned berries. Sensitive to low winter temperatures, susceptible to powdery mildew and mites, resistant to botrytis bunch rot. Enjoyed by wasps and birds.

WHERE IT'S GROWN AND WHAT ITS WINE TASTES LIKE

Until the mid 1980s, Irsai Olivér was grown exclusively for the table but by 2008 there were also 997 ha (2,464 acres) planted to make soft, aromatic, grapey wines in many parts of **Hungary**, although the best results come from cooler areas in the north and west such as Sopron, Neszmély, Etyek-Buda, Mátra and around Lake Balaton (Rohály *et al.* 2003). The wines are best drunk young and are often improved by blending with a more structured variety. Recommended producers include Hilltop (Neszmély), Nyakas (Etyek-Buda), Font (Kunság), Mátyás and Zoltán Szőke (Mátra).

Irsai Olivér is quite widely planted in **Slovakia** (460 ha/1,137 acres in 2009) and also found in Morava in the south east of the **Czech Republic** (80 ha/198 acres in 2009), where it is often blended with Ryzlink Vlašský (GRAŠEVINA) and used in the production of sparkling wines.

Muscat Oliver is also found in **Switzerland** (fewer than 5 ha/12 acres). Producers include Schmidheiny and Lindenamann in the German-speaking canton of Argau and Rosenau in the canton of Luzern.

Irsai Oliver is also found in Krasnodar Krai in **Russia** (24 ha/59 acres).

ISABELLA

Once widely planted, old American variety now found mostly in Brazil and India.

BERRY COLOUR: ● ● ● ● ●

PRINCIPAL SYNONYMS: Americano (Ticino in Switzerland, Madeira), Ananas, Bangalore Blue (India), Bellina, Black Cape, Bromostaphylo, Captraube, Constantia, Dorchester, Fragola, Framboisier, Fraula (Corse), Fraulaghju, Frutilla (Uruguay), Gibb's Grape, Gros Framboisé (Switzerland), Isabel (Brazil) , Isabella Nera (Italy), Isabelle (France), Isabellinha, Izabella (Hungary), Kepshuna, Kerkyraios, Kokulu

Uezuem, Koreos, Lidia, Loipe Makedonia, Moschostaphylo, New Hanover, Nostrano (Ticino in Switzerland), Odessa (Krym/Crimea), Paign's Isabella, Raisin du Cap (France), Raisin Fraise (France), Raisin Framboise (France), Sainte Helene, Saluda, Schuykill, Strawberry Grape, Tjortjidica, Tzampela, Tzortzidika Chakidike, Tzortzines, Utkopro, Uva Cimice (Italy), Uva Fragola (Italy), Vernet, Woodward, Zampela

ORIGINS AND PARENTAGE

This is one of the oldest native American varieties, but its exact origin has been the subject of much speculation. It is most probably descended from the seedling of a wild native, *Vitis labrusca*, sometimes known as 'fox grapes' because of their so-called foxy flavour, found in a garden near Dorchester, South Carolina. According to Munson (1909), it is not a pure *Vitis labrusca* but a cross between *Vitis labrusca* and an unidentified *Vitis vinifera* subsp. *vinifera* variety. It was taken north under an unknown name as early as 1816 and was introduced to New York State by Colonel George Gibbs, an amateur grower from Brooklyn, Long Island. Gibbs then passed it on to William Robert Prince of the Linnaean Botanic Garden in Flushing, Long Island, who named it Isabella (*Vitis isabellae*) in 1822 in honour of the Colonel's wife, Mrs Isabella Gibbs (Prince 1830; Bailey 1906; Pinney 1989; Mabberley 1999).

Isabella was used to breed LYDIA.

OTHER HYPOTHESES

In the nineteenth century, Isabella was thought to be of Spanish origin: Bernard Laspeyre of Wilmington, North Carolina, claimed to have discovered this variety in the garden of another Frenchman, at Charleston, South Carolina, who had apparently brought it from Spain (Bailey 1906).

VITICULTURAL CHARACTERISTICS

Moderately hardy, late ripening, and yet able to withstand tropical and semi-tropical conditions. High yields.

WHERE IT'S GROWN AND WHAT ITS WINE TASTES LIKE

Isabella and CATAWBA were the two dominant varieties in North American viticulture for the first half of the nineteenth century but there have been few recent plantings and only a few hectares remain, eg in **New York** State (22 acres/9 ha in 2009), where it is used for juice and sparkling wine but has largely been replaced by CONCORD.

It was once found in many different regions of the world, particularly in the southern hemisphere and also in Eastern Europe and the former Soviet Union, but the two main countries in which Isabella is still widely planted are **Brazil**, where, as Isabel, it is the most widely planted variety of all (10,691 ha/26,418 acres in 2007), and India. In Brazil, Isabel is used for juice and jelly but also for very large volumes of the sweetish, inexpensive red that is still enormously popular on the domestic market. In **India**, is it known as Bangalore Blue and the majority of the 4,500 ha (11,120 acres) recorded in 2000 were used for juice rather than wine. Wines are pale, more like a rosé, and have the tell-tale sweet, foxy notes of *labrusca*.

Isabella was at one time also grown in **France** but new plantings of this and other American hybrids such as JACQUEZ were proscribed in the mid 1930s and existing vineyards compulsorily grubbed up (with compensation) in the 1950s. One objection was that they produced higher levels of methanol in the wines due to greater amounts of skin pectins. It is still quite widely planted in the **Republic of Moldova** (10,802 ha/26,692 acres in 2009) and in **Ukraine** (2,126 ha/5,253 acres in 2009).

ITALIA

Important, extremely large-berried table grape only very rarely used to make wine.

BERRY COLOUR: ● ● ● ● ●

PRINCIPAL SYNONYMS: Moscatel Italia (Portugal), Moscatel Italiano (Spain), Muscat d'Italie (Tunisia, Morocco)

ORIGINS AND PARENTAGE

A Bicane × MUSCAT OF HAMBURG cross obtained in 1911 by Alberto Pirovano at the arboriculture research centre in Roma, where Bicane is a French wine and table grape that is no longer cultivated.

VITICULTURAL CHARACTERISTICS

Extremely vigorous, very large berries, late ripening and susceptible to downy mildew and botrytis bunch rot.

WHERE IT'S GROWN

Widely grown around the world as a table grape. At one time it was used to make wine in **Australia** but this is now very rare and the area under this variety is no longer even recorded for wine-growing. It is authorized for wine in **France** but the area is limited (152 ha/376 acres in 2006) and declining.

ITALICA

A 1950s white Italian cross of interest only in the province of Treviso.

BERRY COLOUR: ● ● ● ● ●

PRINCIPAL SYNONYMS: (V × RI) 103, Incrocio Cosmo 103, Verdiso × Riesling Italico 103

ORIGINS AND PARENTAGE

A VERDISO × GRAŠEVINA (Riesling Italico) cross obtained in the 1950s by Italo Cosmo at the Conegliano research centre in the Veneto.

VITICULTURAL CHARACTERISTICS

Early to mid ripening.

WHERE IT'S GROWN

Cultivated on the hills near Conegliano and Pieve di Soligo in the province of Treviso in the Veneto, northern **Italy**, the only province in which the variety is authorized. There were 184 ha (455 acres) of Italica in Italy in 2000.

IVES

American hybrid that needs fresh air.

BERRY COLOUR:

PRINCIPAL SYNONYMS: Black Ives, ?Bordô (Brazil), Ives Madeira, Ives' Madeira Seedling, Ives Seedling, Kittredge

ORIGINS AND PARENTAGE

The origin of Ives has long been debated. Most accounts suggest it is a chance seedling, the result of open pollination of Hartford (or Hartford Prolific), that was planted in 1840 by Henry Ives in his garden near Cincinnati, Ohio. However, Hartford is also an accidental seedling that was most probably a spontaneous ISABELLA × *Vitis labrusca* hybrid that sprang up around 1846 in the garden of Paphro Steele & Son in West Hartford, Connecticut, thereby postdating Ives' birth by six years and contradicting this theory. Others have suggested that Ives is a *Vitis labrusca* × *Vitis aestivalis* hybrid.

OTHER HYPOTHESES

Ives himself claimed that the variety was grown from a seed of a *vinifera* vine ambiguously called Malaga or Madeira but this theory was firmly rejected by ampelographers in the nineteenth century.

VITICULTURAL CHARACTERISTICS

Mid ripening (earlier than CONCORD), very prone to ozone damage, leading to loss of vigour and productivity. Best grafted onto a vigorous rootstock. Doesn't cope well with water stress as the root system is not very strong. Very sensitive to sulphur-based sprays.

WHERE IT'S GROWN AND WHAT IT TASTES LIKE

Wines are deeply coloured, making them useful in blends, and marked by the foxy sweetness of *labrusca* varieties, similar to those made from CONCORD but lighter in colour. The grapes are also used for juice. After Prohibition, the variety was a common ingredient in sweet port-style wines but plantings have declined considerably in recent years due to the vine's sensitivity to air pollution. Johnson Estate close to Lake Eyrie in **New York** State have nearly 7 acres (3 ha) and have seen improved quality since the power stations downwind in Michigan and Ohio cleaned up their emissions. Varietal wines, often made in a blush, sweet or semi-sweet style, are also produced by Americana and Bully Hill. Total plantings in New York amounted to 50 acres (20 ha) in 1996. In **Arkansas**, there were around 15 acres (6 ha) in 2011, almost all of those owned and vinified by Post Familie Vineyards, where some of the vines are over fifty years old. Ives can also be found to a very limited extent in **Pennsylvania**.

J

ȣthis symbol on the following pages indicates that the synonyms, and mistaken identities, are confirmed by DNA evidence

JACQUÈRE

Savoie speciality producing fresh, light, mountain whites.

BERRY COLOUR: ● ● ● ● ●

PRINCIPAL SYNONYMS: Cugnette (Isère), Jacquerre, Jaquère, Martin-Cot (La Rochette in Savoie), Molette de Montmélian, Plant des Abîmes or Plant des Abymes de Myans (Chambéry), Raisin des Abîmes, Redin, Robinet (Conflans in Savoie), Roussette (Montmélian)

VARIETIES COMMONLY MISTAKEN FOR JACQUÈRE: ALTESSE (Montmélian), MOLETTE (Montmélian), ROUSSETTE D'AYZE (Savoie)

ORIGINS AND PARENTAGE

Jacquère is the most important white wine variety of Savoie in eastern France. According to Rougier (1903), Jacquère was first mentioned by Fleury Lacoste under the name Redin in 1855. However, this mention is doubtful because a few years later the same Fleury Lacoste (1861) says that Redin is also called Sarvagnin, which undoubtedly corresponds to SAVAGNIN in Savoie (Galet 1990). Therefore, the first occurrence of the name Jacquère is more likely to be the one found in Pierre Tochon's 1868 description of the vineyard in the hilly area known as Abîmes de Myans (Rézeau 1997): 'le cépage qui recouvre ces blocs énormes de calcaire ... s'appelle Jacquère, martin-cot et raisin des Abymes' ('the variety that covers these huge blocks of limestone ... is called Jacquère, martin-cot and raisin des Abymes'). According to Rougier (1903), this variety was supposedly named after the person who introduced it to Savoie in the thirteenth century, but although both Jacquère and Jacquerre do exist as family names in France, they are not common in this area. Jacquère is a member of the Pelorsien ampelographic group (see p XXVII).

DNA parentage analysis suggested that Jacquère is one of the numerous natural progenies of GOUAIS BLANC (Boursiquot *et al.* 2004; see PINOT pedigree diagram).

VITICULTURAL CHARACTERISTICS

Very productive, hence its synonym Robinet (meaning 'tap'). Early budding, mid to late ripening. Best pruned short. Well suited to clay-limestone and rocky soils. Susceptible to black rot and botrytis bunch rot but more resistant to downy and powdery mildews.

WHERE IT'S GROWN AND WHAT ITS WINE TASTES LIKE

On its home territory in Savoie, eastern **France**, Jacquère produces light, fresh, lightly scented, mostly dry, alpine wines for drinking young, sometimes with a light spritz. Jacquère is authorized for the Vin de Savoie and Bugey appellations but is also officially recommended in the Ain and Isère *départements* to the west. The most flavourful wines tend to come from the crus of Apremont and Abymes, where Jacquère must constitute at least 80% of the vineyard for whites, as it does in Jongieux. It has also been planted further south west, in Condrieu in the northern Rhône, although it is not entitled to that appellation. Total plantings in France in 2009, predominantly in Savoie, were a fairly stable 1,027 ha (2,538 acres).

Recommended producers of varietal wines include Didier et Denis Berthollier (Chignin), Domaine Frédéric Giachino (Apremont), Domaine Grisard (Fréterive), Domaine de l'Idylle (Cruet) and André et Michel Quenard (Chignin). Domaine Jean Masson et Fils in Apremont make a range of single-vineyard wines with good ageing potential.

It is also planted to a very limited extent in **Portugal** (3 ha/7 acres in 2009).

JACQUEZ

Once-popular American hybrid with good resistance to Pierce's disease and now surviving mainly in Brazil and Texas.

BERRY COLOUR: ● ● ● ● ●

PRINCIPAL SYNONYMS: Black Spanish (Texas in US), Cigar Box (US), French Grape (US), Jacquet (Cévennes in France), Lenoir (US), Longworth's Ohio (US), Troya (Australia)

VARIETIES COMMONLY MISTAKEN FOR JACQUEZ: LISTÁN PRIETO (US)

ORIGINS AND PARENTAGE

The origin of this variety is obscure and disputed. It is said to be a member of *Vitis bourquiniana*, a botanically dubious group of *Vitis aestivalis* × *Vitis vinifera* hybrids (see HERBEMONT for details).

Jacquez possibly originates from around the town of New Bordeaux in McCormick County, close to the Savannah River, which divides South Carolina from Georgia. Native *Vitis aestivalis* vines were growing here and French Huguenots, most of whom had settled in the region between 1730 and 1780, had introduced European *Vitis vinifera* varieties. The Huguenots were experienced vine growers and probably selected this chance seedling because it was more resistant to diseases than their European varieties. It was then apparently taken to Natchez in Mississippi, by a Spaniard called Jacques, hence the name Jacquez. It is often known as Lenoir in the United States, after the county of that name in South Carolina.

OTHER HYPOTHESES

Some sources suggest that Jacquez was brought from France under the name Blue French by the Bourquin family around 1750, but this hypothesis in not convincing (see HERBEMONT). Others have said that it was introduced from the island of Madeira, where it is still grown today, by an unknown sailor who transported it in a cigar box that was handed to Nicholas Longworth – shoemaker, millionaire banker and horticulturist sometimes referred to as the 'father of American grape culture' – in Cincinnati, Ohio, hence two of its synonyms, Cigar Box and Longworth's Ohio. It is more likely that Jacquez was taken to Madeira by settlers who did not stay in the US but returned to their homeland. It is very unlikely that it originates from Madeira, because this presupposes that the *Vitis aestivalis* × *Vitis vinifera* hybridization that generated Jacquez took place on Madeira, which is unlikely for chronological reasons (see HERBEMONT).

Jacquez and LISTÁN PRIETO (known as Mission in the United States) share the synonym Black Spanish but should not be confused.

VITICULTURAL CHARACTERISTICS

Large to medium-sized loose bunches of very dark, red-fleshed berries. Mid ripening, moderately vigorous and productive. Tolerates Pierce's disease but is susceptible to black rot, downy mildew and summer bunch rot. Best in deep sandy soils.

WHERE IT'S GROWN AND WHAT ITS WINE TASTES LIKE

Jacquez was at one time popular in southern **France**, used mainly in blends from the end of the nineteenth to the middle of the twentieth century (it was even authorized for Châteauneuf-du-Pape until 1935), but new plantings of this and other hybrids such as ISABELLA were proscribed in the mid 1930s and existing vineyards compulsorily grubbed up (with compensation) in the 1950s. However, some centenarian vines still exist in the Cévennes, known here as Jacquet, and their fruit constitutes the main ingredient in the Cuvée des Vignes d'Antan produced by the

Association Mémoire de la Vigne, which cannot legally be sold within the EU because of its hybrid component. They also produce a varietal Cuvée Le Jacquet.

Once widely planted in **Texas** for the purposes of communion wine, the red-fleshed Jacquez – under the name Lenoir or Black Spanish – is still grown there to a limited extent: 130 acres (53 ha) in 2008. The bearing acreage has more than doubled since 2005 thanks to the variety's marked resistance to Pierce's disease, which is a serious problem in the heat and humidity of the state. It is used for the production of both Texas port-style wines and table wines with the former more typical today, showing the spicy black fruit characteristics typical of the variety. Producers of varietal table wines include Chisholm Trail, Dry Comal Creek and Inwood.

The variety's surest stronghold is **Brazil**, where there were 1,397 ha (3,452 acres) in 2007 in Rio Grande do Sul. It is used mainly for juice, jelly and the inexpensive, slightly sweet red wines popular on the domestic market.

The variety is sometimes used as a rootstock but its resistance to phylloxera is only moderate so that it is restricted to sandy soils or to those areas where phylloxera is so far unknown.

JAEN or JAEN DU DÃO

See MENCÍA.

JAÉN BLANCO

See CAYETANA BLANCA.

JAÉN TINTO

See MENCÍA.

JAMPAL

Minor Portuguese vine recognized for its quality but not widely planted.

BERRY COLOUR:

PRINCIPAL SYNONYMS: Cercial (Pinhel in Dão), Jampaulo, João Paolo, Pinheira Branca

VARIETIES COMMONLY MISTAKEN FOR JAMPAL: CERCEAL BRANCO (Dão), SÍRIA

ORIGINS AND PARENTAGE

Jampal probably originates in Portugal's Colares region north west of Lisboa, whence it spread north to the rest of the Lisboa region and Beiras and east to the Tejo. Recent DNA profiling has suggested a possible sibling relationship with the widely planted black-berried CASTELÃO from the Douro (Myles *et al.* 2011).

VITICULTURAL CHARACTERISTICS

Average but irregular yields. Mid budding, mid to late ripening. Susceptible to botrytis bunch rot, coulure and powdery mildew.

WHERE IT'S GROWN AND WHAT ITS WINE TASTES LIKE

Jampal, authorized in a whole host of regional and DOC appellations from Setúbal and Lisboa in the west up to the Douro in the north east of **Portugal**, produces full-bodied, perfumed wines but has not been widely planted despite its potential quality. Biomanz produce a varietal wine within the Lisboa region. There were 106 ha (262 acres) in Portugal in 2010.

JOHANNITER

Recent German hybrid with very good disease resistance.

BERRY COLOUR: ● ● ● ● ●

PRINCIPAL SYNONYMS: Freiburg 177-68

ORIGINS AND PARENTAGE

Complex self-crossed hybrid bred in 1968 by Johannes Zimmermann, former director of the Freiburg research centre in Baden-Württemberg, southern Germany, by crossing RIESLING × Freiburg 589-54 with itself, where:

- Freiburg 589-54 is a Seyve-Villard 12-481 × Freiburg 153-39 hybrid (see HELIOS for the complete Seyve-Villard 12-481 pedigree)
- Freiburg 153-39 is a PINOT GRIS × CHASSELAS cross.

It was named in honour of Johannes Zimmermann, former director of the Freiburg research centre.

VITICULTURAL CHARACTERISTICS

High yields that necessitate bunch thinning. Generally good disease resistance, particularly to powdery mildew. Mid to late ripening.

WHERE IT'S GROWN AND WHAT ITS WINE TASTES LIKE

Authorized in **Germany** in 2001, where there are now 65 ha (161 acres), mainly in Baden and Rheinhessen. Wines are fresh and lightly aromatic in the citrus/grapefruit spectrum, sometimes a little bitter on the finish, although this can be overcome with a few grams of residual sugar. Producers of varietal wines include Clemens Lang, Schlossgut Hohenbeilstein and Zähringer in Baden, Anselmann in the Pfalz, Weingut Im Hagenbüchle in Baden-Württemberg and Zang in Franken.

It is found to a very limited extent (about 10 ha/25 acres) in **Switzerland**, where Hasenheide, for example, produce a varietal wine south east of Zürich. It is also grown in **Belgium** by Domaine Viticole du Chenoy and was authorized in **Italy** in 2011.

JUAN GARCÍA

Intriguing, perfumed speciality of north-western Spain.

BERRY COLOUR: ● ● ● ● ●

PRINCIPAL SYNONYMS: Gorda (Coimbra in central Portugal), Malvasía Negra, Mouratón (Galicia), Negrera (Bierzo), Negrón de Aldán (Bierzo), Nepada, Tinta Gorda (Dão and Douro in Portugal), Villarino

VARIETIES COMMONLY MISTAKEN FOR JUAN GARCÍA: ARAMON NOIR (Languedoc-Roussillon in France), MENCÍA (Bierzo), TEMPRANILLO

ORIGINS AND PARENTAGE

Juan García is said to come from the Los Arribes del Duero national park in western Spain along the River Duero that forms the border with Portugal. The origin of its name is unknown. DNA parentage analysis has recently revealed that Juan García (under the synonym Mouratón) is a natural cross between CAYETANA BLANCA, widespread throughout the south and centre of the Iberian peninsula, and ALFROCHEIRO from the south or centre of Portugal (Zinelabidine *et al.* 2012). Juan García is therefore a sibling of CORNIFESTO, MALVASIA PRETA, CASTELÃO and CAMARATE (see Cayetana Blanca for the pedigree diagram).

Although both Juan García and PETIT BOUSCHET are sometimes called Negrón de Aldán (González-Andrés *et al.* 2007), they are not related. Juan García and MALVASIA NERA DI BRINDISI share the synonym Malvasía Negra but should not be confused.

VITICULTURAL CHARACTERISTICS

Highish yields, susceptible to botrytis bunch rot and powdery mildew. Early ripening (especially when planted near the river) and grapes often shrivel when ripe, giving higher sugar levels in the berries.

WHERE IT'S GROWN AND WHAT ITS WINE TASTES LIKE

Juan García is a speciality of the Fermoselle area in the province of Zamora, western **Spain**, west of Toro and just upstream from Portugal's Douro Valley, where it is sometimes blended with the Portuguese variety RUFETE. It is one of the principal dark-skinned varieties of the Arribes DO. Originally planted on the rocky slopes down to the Duero, it has since spread to flatter sites and now comprises more than half of the region's vineyard area with a total of 1,260 ha (3,114 acres) in 2008, mostly in Castilla y León but also further north in Galicia. Producers of varietal wines include Almajora, Bodega Ribera de Pelazas, La Setera and Terrazgo. Old vines planted on the best sites generally produce wines that are deeply coloured and highly perfumed with herbal and red-fruit flavours and moderate alcohol.

In northern **Portugal**, it is know as Gorda (Tinta Gorda until 2000) and is grown predominantly in the Trás-os-Montes and Beiras regions (Rolando Faustino, personal communication), where it produces early-drinking wines that are relatively low in alcohol and have simple red-fruit aromas. There were 77 ha (190 acres) in Portugal in 2010.

JUAN IBÁÑEZ

See MORISTEL.

JUBILÄUMSREBE

Austrian cross best suited to very sweet whites.

BERRY COLOUR: ● ● ● ● ●

ORIGINS AND PARENTAGE

This variety was obtained in 1922 by Fritz Zweigelt at the Klosterneuburg research centre in Austria and was initially named Klosterneuburg 24-125. It got its modern name in 1960 when it was presented as part of the 100th anniversary celebrations (*Jubiläum* in German) of the institute of Klosterneuburg. Fritz Zweigelt declared Jubiläumsrebe to be a BLAUER PORTUGIESER × BLAUFRÄNKISCH cross.

This crossing of two dark-berried cultivars giving birth to a white-berried cultivar was a little bit surprising, but nobody really questioned what Zweigelt had said until 1998, when Austrian researchers discovered through DNA paternity tests that Jubiläumsrebe is in fact a FRÜHROTER VELTLINER × Grauer Portugieser cross (Sefc, Steinkellner *et al.* 1998). Although it is impossible to distinguish BLAUER PORTUGIESER from Grauer or Grüner Portugieser by DNA typing (they are the result of clonal mutations affecting berry colour), researchers found clues that the clone used by Zweigelt was very likely Grauer Portugieser. Indeed, the original vineyard where Zweigelt did the crossings was destroyed during the Second World War, and the remaining plants might have been misidentified. In the 1920s, Zweigelt reported several Blauer Portugieser × BLAUFRÄNKISCH crosses, and one single cross between Grauer Portugieser and Frühroter Veltliner: this has to be Jubiläumsrebe.

In support of this parentage, crosses between Jubiläumsrebe and RIESLING (white berries) yield progenies with white as well as grey and red berries, which supports its descent from Grauer Portugieser (grey berries) and FRÜHROTER VELTLINER (pink berries). As a consequence, Jubiläumsrebe is a grandchild of ROTER VELTLINER and SILVANER, the parents of Frühroter Veltliner.

VITICULTURAL CHARACTERISTICS

Mid to late ripening, very good botrytis tolerance and able to reach high sugar levels through dehydration.

WHERE IT'S GROWN AND WHAT ITS WINE TASTES LIKE

There are just 26 ha (64 acres) planted in **Austria**, mainly in the Neusiedlersee region in the far east of the country, where it is particularly suited to sweet Stroh- and Schilfweine (made from grapes dried on straw or reeds) thanks to the high sugar levels it can obtain even without the influence of botrytis. It is generally used as part of a blend since it is not particularly aromatic and has low acidity but M & M Kögl produce a varietal TBA.

JUHFARK

Minor Hungarian variety that needs ageing and aeration to swap asperity for elegance.

BERRY COLOUR: ● ● ● ● ●

PRINCIPAL SYNONYMS: Lämmerschwanz

VARIETIES COMMONLY MISTAKEN FOR JUHFARK: Csomorika☒

ORIGINS AND PARENTAGE

Juhfark, meaning 'sheep's tail' and referring to the long cylindrical shape of the bunches, is an old variety from Hungary whose ampelographic identity is still unclear because Juhfark reference samples in Varga *et al.* (2008), Galbács *et al.* (2009) and Jahnke *et al.* (2009) have yielded three distinct DNA profiles.

It should not be confused with Csomorika, a very old but no longer cultivated mediocre Hungarian variety from the Baranya region in southern Hungary (Galbács 2009) which in the past was mistakenly planted instead of Juhfark by growers in Somló (Rohály *et al.* 2003).

VITICULTURAL CHARACTERISTICS

Early budding, mid ripening, relatively high yields. Big bunches of small berries. Sensitive to low winter temperatures and susceptible to downy mildew and botrytis bunch rot.

WHERE IT'S GROWN AND WHAT ITS WINE TASTES LIKE

Once grown quite widely in **Hungary**, this distinctive variety is now found almost exclusively in Somló, north of Lake Balaton in the west of the country. It's perhaps the least modern or internationally appealing of Hungarian varieties, producing relatively neutral, high-acid, sometimes rustic but ageworthy wines that need time in barrel and plenty of aeration to become more elegant and admirable. Recommended producers include Imre Györgykovács, Inhauser, Kreinbacher, Meinklang (based over the border in Austria), Somlói Apátsági Pince and Tornai. There were 156 ha (385 acres) of Juhfark in Hungary in 2008.

JURANÇON BLANC

Inferior and increasingly neglected variety traditionally used as a base for spirits.

BERRY COLOUR:

PRINCIPAL SYNONYMS: Braquet (Landes), Dame Blanc (Lot-et-Garonne), Plant de Dame, Plant Debout (Sables d'Olonne), Quillat or Quillard (Gers), Secal (Tarn)

ORIGINS AND PARENTAGE

Despite its name, Jurançon Blanc has no connection with the Jurançon appellation, although it was supposedly named after the commune of Jurançon in the Pyrénées-Atlantiques, south-west France, where it allegedly originated. However, the birthplace of Jurançon Blanc is more likely to be in the Tarn-et-Garonne, north east of the Pyrénées-Atlantiques, where it was first mentioned in 1839 under the name Jurançon (in Moissac) and under the name Plant de Dame or Quillat (in Auvillar; Rézeau 1997).

Jurançon Blanc is not a colour mutation of JURANÇON NOIR, even though both are members of the Folle ampelographic group (see p XXVII; Bisson 2009). DNA parentage analysis at INRA (Institut National de la Recherche Agronomique) in Montpellier has shown that Jurançon Blanc is most likely a progeny of Folle Blanche and the obscure white-berried French variety Pruéras (no longer cultivated).

VITICULTURAL CHARACTERISTICS

Mid ripening. May be pruned short. Small bunches and berries. Very susceptible to downy mildew and botrytis bunch rot.

WHERE IT'S GROWN AND WHAT ITS WINE TASTES LIKE

The wines tend to be low in alcohol, acidic and generally poor quality, which, along with its susceptibility to fungal diseases, probably accounts for the variety's decline in **France** in the last fifty years – from 5,755 ha (14,221 acres) in 1958 down to 12 ha (30 acres) in 2006 – and for its traditional destination: armagnac or cognac.

JURANÇON NOIR

South-west French variety producing light wines but rapidly losing ground.

BERRY COLOUR: ● ● ● ● ●

PRINCIPAL SYNONYMS: Chalosse Noire (Lavilledieu in Tarn-et-Garonne), Dame Noire or Plant de Dame (Lot), Enrageat Noir (Gironde), Fola Belcha (País Vasco in Spain), Folle Noire or Folle Rouge (Lot-et-Garonne), Gouni (Hautes-Alpes), Jurançon Rouge (Dordogne), Luxuriant (Aveyron), Piquepout Rouge (Landes), Quillat or Quillard (Gers), Vidiella (Uruguay)

VARIETIES COMMONLY MISTAKEN FOR JURANÇON NOIR: CALITOR NOIR, FUELLA NERA (called Folle Noire in Bellet)

ORIGINS AND PARENTAGE

Although this variety is not related to the Jurançon appellation, it was supposedly named after the commune of Jurançon in the Pyrénées-Atlantiques, in the far south west of France, where it allegedly originated (Rézeau 1997). However, Galet (1990) locates its birthplace in the area of Agen, between Toulouse and Bordeaux in the Lot-et-Garonne, which is consistent with its first mention, by Comte Odart, in 1837: 'le nom de Jurançon noir qui servait d'étiquette à un paquet de sarments qui m'a été envoyé du département de Lot-et-Garonne' ('the name Jurançon noir that was used as a label for a bunch of vinestocks that was sent to me from the Lot-et-Garonne *département*'; Rézeau 1997). Although they share the synonym Folle Noire, Jurançon Noir and FUELLA NERA are distinct varieties that belong to the Folle ampelographic group (see p XXVII; Bisson 2009). DNA parentage analysis at INRA (Institut National de la Recherche Agronomique) in Montpellier has recently shown that Jurançon Noir is a natural FOLLE BLANCHE × COT cross (Viollet and Boursiquot 2009; see CABERNET SAUVIGNON and PINOT pedigree diagrams).

Jurançon Noir was used to breed CHENANSON, GANSON and SEGALIN.

VITICULTURAL CHARACTERISTICS

Mid ripening, fertile and productive and needs to be pruned short (bushvine). Susceptible to botrytis bunch rot but not to diseases of the wood.

WHERE IT'S GROWN AND WHAT ITS WINE TASTES LIKE

Wines are light with very moderate levels of alcohol and colour, best suited to rosé or lighter styles of red. Plantings in **France** have diminished dramatically in the last fifty years, down from 12,325 ha (30,456 acres) in 1958 to 708 ha (1,750 acres) in 2009, mostly in the Tarn and the Gers. It is authorized in appellations such as Vins d'Estaing and Vins d'Entraygues et du Fel.

Jurançon Noir was introduced to **Uruguay** in the 1870s under the name Folle Noire by Francisco Vidiella and it was known as Vidiella until the mid 1980s. However, most of the old vineyards have gone and in 2007 less than a hectare (2.5 acres) remained of what is now known as Jurançon Noir.

JUWEL

Modern German cross with patchy disease resistance.

BERRY COLOUR: ● ● ● ● ●

PRINCIPAL SYNONYMS: Jewel, Weinsberg S 378

ORIGINS AND PARENTAGE

A KERNER × SILVANER cross bred in 1951 by August Herold at Weinsberg. Expanding the pedigree, Juwel is a grandchild of SCHIAVA GROSSA and RIESLING (parents of Kerner) as well as of SAVAGNIN and ÖSTERREICHISCH WEISS (parents of Silvaner). The name Juwel is a contraction of three parts: JU in honour of the vine breeder Josef Schäffer, nicknamed Jupp; WE is the official abbreviation of the Weinsberg research centre; and L for Lauffen am Neckar, the location of the research centre.

VITICULTURAL CHARACTERISTICS

Mid ripening. High must weights and moderate acidity. Resistant to downy mildew but very susceptible to powdery mildew.

WHERE IT'S GROWN

Germany has 25 ha (62 acres), largely in Rheinhessen, but the area is gradually declining. Producers include Görgen in the Mosel and Umbrich in Württemberg.

K

KABAR
KACHICHI
KADARKA
KAKHET
KAKOTRYGIS
KALECIK KARASI
KALINA
KANGUN
KANZLER
KAPISTONI TETRI
KAPITAN JANI KARA
KAPSELSKY
KARA IZYUM ASHKHABADSKY
KARALAHNA
KARASAKIZ
KARÁT
KARMRAHYUT
KARNACHALADES
KATSAKOULIAS
KATSANO
KAY GRAY
KEFESSIYA
KÉKNYELŰ
KERATSUDA
KERNER
KHIKHVI
KHINDOGNI
KIRÁLYLEÁNYKA
KISI
KLARNICA
KNIPPERLÉ
KÖHNÜ
KOK PANDAS
KOKUR BELY
KOLINDRINO
KOLORKO
KORINTHIAKI
KORIOSTAFYLO
KÖSETEVEK
KOSHU
KOTSIFALI
KOUTSOUMPELI
KÖVIDINKA
KRAKHUNA
KRALJEVINA
KRASNOSTOP ZOLOTOVSKY
KRASSATO
KREACA
KRKOŠIJA
KRONA
KRSTAČ
KUJUNDŽUŠA
KUMSHATSKY CHERNY
KUNLEÁNY
KUPUSAR
KYDONITSA

ᛝ this symbol on the following pages indicates that the synonyms, and mistaken identities, are confirmed by DNA evidence

KABAR

Recent Hungarian cross designed for Tokaj.

BERRY COLOUR: ● ● ● ● ●

PRINCIPAL SYNONYMS: Tarcal 10

ORIGINS AND PARENTAGE

Kabar is a BOUVIER × HÁRSLEVELŰ cross obtained in 1967 by László Brezovcsik, Gáborné Szakolczay, Ferenc Marcinkó, László Baracskai and Sándorné Éles at the Tarcali research centre in the Tokaj region.

VITICULTURAL CHARACTERISTICS

Vigorous, early ripening, lowish yields. Loose clusters of small, thick-skinned berries that can reach high sugar levels. Winter-hardy, moderately susceptible to fungal diseases but also prone to the noble form of botrytis.

WHERE IT'S GROWN AND WHAT ITS WINE TASTES LIKE

Kabar, authorized in the Tokaj region of **Hungary** since 2006, typically produces high-acid wines. Chateau Dereszla make a rare, dry varietal wine which they describe as relatively light with the taste of linden-blossom honey. Other producers such as Grof Degenfeld and Tokaj Renaissance include it as a minor component in the blend for their sweet Tokaji wines. It represents about 1% of plantings in the region and seems unlikely to increase.

KACHICHI

Minor variety producing deep red wines in north-western Georgia.

BERRY COLOUR: ● ● ● ● ●

PRINCIPAL SYNONYMS: Abkhazouri, Kagigi, Katchitchi, Katcitci

ORIGINS AND PARENTAGE

Kachichi, indigenous to Apkhazeti, north-western Georgia, was mentioned as early as the nineteenth century (Chkhartishvili and Betsiashvili 2004).

VITICULTURAL CHARACTERISTICS

Mid budding, very late ripening, susceptible to powdery mildew.

WHERE IT'S GROWN

Kachichi is not widely planted (25 ha/62 acres in **Georgia** in 2004) but produces deeply coloured, dry red wines in Samegrelo in the north west of Georgia and in neighbouring Apkhazeti.

KADARKA

Variety of uncertain origin producing gentle, fresh reds in Eastern Europe.

BERRY COLOUR:

PRINCIPAL SYNONYMS: Branicevka (Croatia), Cadarcă or Cadarcă Neagră (Romania), Cadarcă de Miniş (Romania), Fekete Budai (Hungary), Gamza (Bulgaria), Gamza de Varna (Bulgaria), Gomza (Croatia), Gumza, Gymza (Bulgaria), Kadarka Kék, Kallmet (Albania), Lugojană (Romania), Skadarka (Croatia, Serbia), Törökszőlő (Hungary), Varnenska Gimza (Bulgaria)

ORIGINS AND PARENTAGE

Kadarka is said to have been introduced to Hungary from the Balkans, either by Serbians, which fits with Kadarka being the Slavic name for Scutari, a lake (also known as Skadar) between Montenegro and Albania, and with the fact that the variety was once cultivated under the synonym Skadarka in Croatia and Serbia (Levadoux 1956; Galet 2000; Rohály *et al.* 2003), or by Turks, hence its synonym Törökszőlő ('Turkish grape'). It is conceivable that it made its way to Hungary via Bulgaria, where it is still widely planted under the synonym Gamza, and where it is considered to be indigenous. It is also said to be indigenous to the region of Miniş near Arad in western Romania, where the first sweet Aszú-style red wine was made from shrivelled Cadarcă grapes in 1744 (Dejeu 2004). The exact origin of Kadarka remains unknown but it lies somewhere within the Balkan-Pannonian area.

Kadarka was used to breed BÍBORKADARKA, PROBUS and RUBINTOS.

OTHER HYPOTHESES

Kadarka is sometimes said to originate from Asia Minor (Dejeu 2004) but there is no evidence to support this hypothesis.

VITICULTURAL CHARACTERISTICS

Productive, late budding and ripening. Compact bunches of medium-sized berries. Good drought resistance but sensitive to low winter temperatures and susceptible to botrytis bunch rot. Best suited to loess soils and bushvine growth.

WHERE IT'S GROWN AND WHAT ITS WINE TASTES LIKE

Statistically Kadarka has been in decline in **Hungary** in recent decades but it is still popular with consumers for its medium body, gentle tannins, fresh acidity, lightly spicy aromas and overall elegance, with some similarity to PINOT NOIR (Rohály *et al.* 2003). Nor is it without champions among producers, even though it is sometimes difficult to ripen fully. Wines are not typically deep in colour and Kadarka is also very good for rosé. Cultivated in most of Hungary's wine regions, its heartland is in the south, in Szekszárd and Villány, but it is quite widely planted in Eger, where it is used in the red blend called Egri Bikavér or Bull's Blood of Eger. However, BLAUFRÄNKISCH, known as Kékfrankos in Hungary, is easier to grow and vinify and has therefore been gaining ground at the expense of Kadarka. Recommended produers include Dúzsi, Eszterbauer, Heimann, Takler, Ferenc Vesztergombi and Péter Vida in Szekszárd and St Andrea in Eger. Total plantings of Kadarka in Hungary in 2008 were down to 666 ha (1,646 acres) but plantings are likely to increase, particularly in Szekszárd, as progress is made in clonal research to find healthier plant material. Some older vines such as those in Oszkár Maurer's centenarian vineyard over the border in **Serbia** show better disease resistance and produce more richly flavoured wines.

Under the name Gamza, the variety is widely planted in **Bulgaria** (3,169 ha/7,831 acres in

2009), particularly in Vidin, Pleven and Suhindol in the north, and it is also found in the **Republic of Macedonia**. As Kallmet, this variety is one of the more important red varieties in **Albania**, planted mainly around Lake Skadar, which straddles the border with Montenegro in the north, and to a lesser extent in the centre of the country. Arbëri, for example, produce a varietal wine. **Romania**'s 47 ha (116 acres; 2008) are in the west of the country, both in the southerly region of Oltenia and further north – closer to Hungary – in Miniş Măderat (eg Wine Princess).

KAKHET

Possibly Georgian grape best known for the sweet Kagor reds in Armenia.

BERRY COLOUR: ● ● ● ● ●

PRINCIPAL SYNONYMS: Cakhete, Kachet, Kakheti

ORIGINS AND PARENTAGE

Kakhet is a traditional variety of the provinces of Ararat and Armavir in western central Armenia but, as its name suggests, may have been introduced to Armenia long ago from the Kakheti region of Georgia. Further, Kakhet appears to be genetically close to Georgian varieties, in particular KHIKHVI and KISI, and not to Armenian varieties (Vouillamoz *et al.* 2006).

VITICULTURAL CHARACTERISTICS

Late ripening. Medium-sized to large compact berries.

WHERE IT'S GROWN AND WHAT ITS WINE TASTES LIKE

Ijevan Wine Factory make a dry Kakhet rosé and a semi-sweet red but it is used by many other producers in **Armenia** such as Brest and Vedi Alco to make the traditional red dessert wine known as Kagor as well as dry reds.

KAKOTRYGIS

Very minor variety found on the Greek island of Kérkyra (Corfu).

BERRY COLOUR: ● ● ● ● ●

PRINCIPAL SYNONYMS: Kako Tryghi, Kakotriguis, Kakotriki (Ilía), Kakotryghis

ORIGINS AND PARENTAGE

Kakotrygis is an old variety from the Ionian Islands, possibly from Kérkyra, where a red-berried mutation (Kokkino Kakotrygis) is also grown (Nikolau and Michos 2004). Its name refers to the fact that it is difficult to harvest because the stems are *kakotryghos*, ie hard to cut (Manessis 2000).

VITICULTURAL CHARACTERISTICS

Large, compact bunches of small berries. Vigorous, fertile, productive, late budding and early ripening. Susceptible to downy mildew.

WHERE IT'S GROWN AND WHAT ITS WINE TASTES LIKE

Kakotrygis is grown mainly on the island of Kérkyra, **Greece**, where it is the dominant variety in the island's dwindling vineyards, producing wines that are typically delicate, citrus-flavoured, with relatively high alcohol, medium acidity and good length, often blended with other varieties. Livadiotis, one of the few producers determined to craft serious wines from Kérkyra's indigenous vines, make a varietal wine from ungrafted plants in the sandy southern tip of the island. Halikouna regional wine must be 100% Kakotrygis whereas Kérkyra regional wines must include at least 60% with other local varieties.

KALECIK KARASI

Recuperated Turkish variety making very fresh, fruity reds for relatively early drinking.

BERRY COLOUR: ● ● ● ● ●

VARIETIES COMMONLY MISTAKEN FOR KALECIK KARASI: ADAKARASI (Avşa), ÇALKARASI (Denizli), ?HASANDEDE, HOROZKARASI (Kilis), PAPAZKARASI (Avşa), ?Sungurlu

ORIGINS AND PARENTAGE

Kalecik Karası, the 'black of Kalecik' or 'black from small castle', logically originates from the Kalecik district near Ankara in central Turkey, most likely from the vineyards along the River Kızılırmak.

OTHER HYPOTHESES

Kalecik Karası is said to have been grown in Turkey since the Hittite Empire (1650–1200 BC) but there is no evidence for this. Surprisingly, Kalecik Karası is listed among the synonyms of HASANDEDE in the Vitis International Variety Catalogue but DNA profiling has not yet been done to verify this ampelographically doubtful hypothesis.

See also ADAKARASI.

VITICULTURAL CHARACTERISTICS

Compact clusters of medium-sized berries with thick skins. Mid ripening and productive. Susceptible to botrytis bunch rot.

WHERE IT'S GROWN AND WHAT ITS WINE TASTES LIKE

Kalecik Karası's home is in the Kalecik district, north east of Ankara, the capital of **Turkey**, and it is the most important red wine variety in Central Anatolia. It was virtually wiped out by phylloxera in the 1960s but the combined efforts of Y Sabit Ağ aoğ lu at the University of Ankara and Kavaklidere in the 1970s and 1980s led to the selection and propagation of three clones. Kavaklidere marketed their first varietal wine as recently as 1989. The vines are planted at around 600–900 m above sea level, generally on the steep slopes above the River Kızılırmak, which means that the nights are cold but the winters relatively mild. Plantings have now reached around 3,500 ha (8,649 acres) and the trend towards higher yields and planting on the valley floor have had a negative effect on quality in the last ten years. The variety is also planted further south and west on the Denizli plateau and to a lesser extent in Kapadokya (Cappadocia), Trakya (Thrace), Manisa in the Aegean region and Konya (Central Anatolia). Total plantings in 2010 were 729 ha (1,801 acres).

Varietal wines can have a candied aroma but the best examples have aromas of sour cherries and raspberries, are medium-bodied, soft and fresh. The wines are best unoaked and drunk young.

They have been likened to PINOT NOIR but are probably closer to GAMAY NOIR in style. Producers of this refreshingly uncomplicated wine include Pamukkale, Kavaklidere, Kocabağ and Turasan.

KALINA

Very minor recent red-skinned Swiss hybrid originally bred for the table.

BERRY COLOUR: ● ● ● ● ●

ORIGINS AND PARENTAGE

Kalina is a hybrid obtained in the 1970s by Anton Meier at Rebschule Meier in Würenlingen in the canton of Aargau, northern central Switzerland, from undisclosed parents.

VITICULTURAL CHARACTERISTICS

Early to mid ripening, vigorous, generally resistant to frost and fungal diseases but susceptible to downy mildew. Large, loose bunches of medium-sized, reddish, thin-skinned berries.

WHERE IT'S GROWN AND WHAT ITS WINE TASTES LIKE

Although it is described as a white table grape, Kalina has red-violet berries and is sometimes used to make white wine. Less than half a hectare (1 acre) grew in **Switzerland** in 2009 but ck-Weine make a varietal wine and Schödlerweine blend it with JOHANNITER.

KALLMET

See KADARKA.

KANGUN

Armenian complex hybrid used to produce brandy and sweet, fortified whites.

BERRY COLOUR: ● ● ● ● ●

PRINCIPAL SYNONYMS: 2-17-22, Cangoune, Kangoon, Kangoun

ORIGINS AND PARENTAGE

Kangun is a complex hybrid obtained in 1979 by P K Aivazyan at the Armenian viticultural research centre in Merdzavan, just west of Armenia's capital, Yerevan, by crossing SUKHOLIMANSKY BELY with RKATSITELI.

WHERE IT'S GROWN

Kangun is used to make brandy and sweet fortified wines, mainly in **Armenia**'s Ararat Valley.

KANZLER

Low-yielding and declining German cross.

BERRY COLOUR: ● ● ● ● ●

PRINCIPAL SYNONYMS: Alzey S 3983

ORIGINS AND PARENTAGE

A MÜLLER-THURGAU × SILVANER cross bred in 1927 by Georg Scheu at the Oppenheim research centre and included in the official list of varieties in 1987. The name Kanzler (meaning 'chancellor') is said to honour the first three post-war German chancellors, and at the same time the grape's qualities: it is refined like Konrad Adenauer, full and round like Ludwig Erhard and elegant like Kurt Georg Kiesinger!

VITICULTURAL CHARACTERISTICS

The variety's high vigour produces a dense canopy, which makes it susceptible to both downy and powdery mildews, although it has good resistance to botrytis bunch rot. It starts to ripen early but can be picked very late with commensurately high must weights.

WHERE IT'S GROWN AND WHAT ITS WINE TASTES LIKE

Wines are full-bodied and potentially long-lived, needing a year in bottle to develop their aromas. There are 34 ha (84 acres) planted in **Germany**, clustered in Rheinhessen and the Pfalz, but it does not yield well and plantings are declining. Producers of sweet versions include Fleischmann in the Pfalz and Schales in Rheinhessen.

KAPISTONI TETRI

Ancient Georgian variety responsible for still and sparkling wines.

BERRY COLOUR: ● ● ● ● ●

PRINCIPAL SYNONYMS: Capistoni Tetri, Kabistoni Tetri, Kapistona, Kapistoni, Zekroula Kapistoni

ORIGINS AND PARENTAGE

Kapistoni Tetri (*tetri* means 'white') is indigenous to Imereti in western central Georgia and is reputedly one of the oldest Georgian varieties (Chkhartishvili and Betsiashvili 2004). According to Galet (2000), the other white Georgian varieties Kapistoni Imeretinsky from Imereti and Kapistoni Rgvali from Racha and Kutaisi are distinct from Kapistoni Tetri. This is consistent with DNA profiling that shows Kapistoni Imeretinsky to be identical to Dondglabi (Vouillamoz *et al.* 2006).

The black-berried ALEKSANDROULI is also known as Kabistoni but it is not a dark-skinned version of Kapistoni Tetri, as shown by morphological and genetic studies (Chkhartishvili and Betsiashvili 2004; Maghradze *et al.* 2009).

VITICULTURAL CHARACTERISTICS

Mid ripening. Small, compact bunches of medium-sized berries. Good resistance to downy mildew but susceptible to powdery.

WHERE IT'S GROWN AND WHAT ITS WINE TASTES LIKE

Planted in the western part of **Georgia**, particularly in Imereti, this ancient but relatively neutral variety is used to make both still and sparkling wines.

KAPITAN JANI KARA

Minor Ukrainian component in sweet red blends.

BERRY COLOUR: ● ● ● ● ●

PRINCIPAL SYNONYMS: Adzhi Ibram Kara, Agii Ibram, Capitan Kara, Chaban Khalil Kara, Kapitan Yani Kara, Ridzhaga

ORIGINS AND PARENTAGE

Kapitan Jani Kara, literally 'Captain Jani's black', comes from Solnechnaya Dolina in eastern Krym (Crimea), in the far south of Ukraine.

VITICULTURAL CHARACTERISTICS

Productive, mid budding and ripening. Large, compact bunches of fairly large, thick-skinned berries. Susceptible to powdery mildew.

WHERE IT'S GROWN AND WHAT ITS WINE TASTES LIKE

Wines are typically full-bodied with rounded tannins and are generally blended with other local **Ukraine** varieties such as EKIM KARA to make dessert wines. Kapitan Jani Kara plays a bit part in Massandra's Black Doctor, for example.

KAPSELSKY

Minor Ukrainian white.

BERRY COLOUR: ● ● ● ● ●

PRINCIPAL SYNONYMS: Kapsel Skii, Kapselski, Kapselskii, Matvienkovsky, SD-62

ORIGINS AND PARENTAGE

Kapselsky comes from Krym (Crimea) in southern Ukraine and is morphologically similar to SOLNECHNODOLINSKY, another local variety.

VITICULTURAL CHARACTERISTICS

Productive but yields may be variable. Compact bunches of thick-skinned berries. Relatively resistant to fungal diseases.

WHERE IT'S GROWN AND WHAT ITS WINE TASTES LIKE

Sun Valley (Solnechnaya Dolina) include Kapselsky in the blend of a sweet wine from Krym in **Ukraine** that comprises up to thirty different varieties, both indigenous and international.

KARA ERIK

See ÖKÜZGÖZÜ.

KARA IZYUM ASHKHABADSKY

Also known as Karaisumor Tara Uzum Ashkhabadski, this black-berried, late-ripening variety comes from **Turkmenistan**, where it is one of the main varieties planted.

KARALAHNA

Found mainly on the Turkish island of Bozcaada in the Aegean, makes tannic but fresh reds.

BERRY COLOUR: ● ● ● ● ●

PRINCIPAL SYNONYMS: Kara Lahna, Lahna Kara

ORIGINS AND PARENTAGE

Karalahna, literally 'black pumpkin', probably originates from Çanakkale on the southern coast of the Dardanelles or from the island of Bozcaada in north-western Turkey.

VITICULTURAL CHARACTERISTICS

Large berries that do not attain very high sugar levels. Mid to late ripening. Generally quite susceptible to diseases. Well suited to clay-loam soils and dry conditions.

WHERE IT'S GROWN AND WHAT ITS WINE TASTES LIKE

Used mainly to make brandy until the early twentieth century, Karalahna makes wines that are typically deeply coloured, medium to full-bodied and tannic with high acidity and flavours of black and red berries in **Turkey.** They need time in bottle and marry well with oak – and care is needed not to over-extract the tannins. The main producers of varietal wines – Corvus, Talay and Yunatçılar – are on the Aegean island of Bozcaada, in the far west of Turkey, south of Çanakkale, but the variety is also used to add colour in blends. Further east and north in Trakya (Thrace), Seyit Karagözoğlu is producing a Karalahna-dominant blend, with MERLOT, from vineyards near the Sea of Marmara. The Karalahna seems more fragrant here. There were just 43 ha (106 acres) of Karalahna in Turkey in 2010 but it shows promise when handled gently.

KARASAKIZ

Turkish variety making soft, light reds and being overtaken by the likes of Cabernet and Merlot.

BERRY COLOUR: ● ● ● ● ●

PRINCIPAL SYNONYMS: Kara Sakız, Karakız, Karassakýz, Kuntra (Turkey and Greece), Makbule, Mavrupalya, Sakız Kara

ORIGINS AND PARENTAGE

Karasakız, literally meaning 'black chewing gum', also called Kuntra in Greece as well as in parts of Turkey, is a variety from the Dardanelles, the strait dividing Europe and Asia near Istanbul. It probably originates from the island of Bozcaada in the Aegean, at the south-western end of the strait.

VITICULTURAL CHARACTERISTICS

Large berries that tend to have high acidity but do not reach particularly high sugar levels. Late ripening, good general disease resistance and well suited to clay-loam soils and hot conditions.

WHERE IT'S GROWN AND WHAT ITS WINE TASTES LIKE

Found mostly on the island of Bozcaada (Ténedos in Greek) in the Aegean Sea, Karasakız is also planted in Çanakkale, Lapseki and Bayramiç on mainland **Turkey** to the east. It was once much more widely planted because it was the main ingredient in the Turkish brandy made in large volumes by the former state monopoly Tekel. Although their brandy plant in Çanakkale was closed down in 2007, Mey (who bought Tekel) still use Karasakız in the production of the anise-flavoured spirit raki, in Tekirdag. It is losing ground to international varieties such as CABERNET SAUVIGNON, with which it is sometimes blended. Producers of varietal wines include Çamlıbağ, Corvus and Talay; the last also make sweet and pink versions (the former confusingly called Kuntra). Wines tend to be light- to medium-bodied with red-fruit and candy aromas, low acidity and soft tannins; they don't really stand up to oak. There were just 43 ha (106 acres) in Turkey in 2010.

KARÁT

Hungarian cross well suited to dry conditions but making rather lacklustre wines.

BERRY COLOUR: ● ● ● ● ●

ORIGINS AND PARENTAGE

A KÖVIDINKA × PINOT GRIS cross obtained in 1950 by András Kurucz and István Kwaysser at the Kecskemét research centre near Budapest, Hungary.

VITICULTURAL CHARACTERISTICS

Mid to late ripening. Drought-resistant and moderately resistant to botrytis bunch rot.

WHERE IT'S GROWN AND WHAT ITS WINE TASTES LIKE

Karát produces wines that are relatively high in alcohol and fresh but lack concentration. The variety's chief attribute is its suitability for dry sites (Gabriella Mészáros, personal communication). Total plantings in **Hungary** in 2008 amounted to 61 ha (151 acres), mainly in the Kunság, Etyek-Buda and Badacsony wine regions. Öreghegy Szőlőbirtok make a varietal wine.

KARMRAHYUT

Dark-skinned, red-fleshed Armenian hybrid.

BERRY COLOUR: ● ● ● ● ●

PRINCIPAL SYNONYMS: Karmrahiut, Karmraiute

ORIGINS AND PARENTAGE

Karmrahyut, meaning 'red juicy', is a complex hybrid obtained in 1950 by S A Pogosyan at the Armenian viticultural research centre at Merdzavan, west of the capital, Yerevan, by crossing Adisi with NO 15-7-1, in which Adisi is an Armenian variety that is no longer cultivated and NO 15-7-1 is a hybrid between a vine of *Vitis amurensis* Ruprecht and Sladky Cherny (meaning 'sweet black'), the latter being a selfling of PINOT NOIR. Karmrahyut was used to breed CHARENTSI and NERKARAT.

VITICULTURAL CHARACTERISTICS

Compact bunches of medium-sized, red-fleshed berries.

WHERE IT'S GROWN AND WHAT ITS WINE TASTES LIKE

Karmrahyut is a red-fleshed variety planted mainly in Armavir in western central **Armenia**. Kazumoff, for example, make a deeply coloured varietal wine and MAP blend it with RKATSITELI and MEGRABUIR to produce a semi-sweet dessert wine.

KARNACHALADES

Very rare variety from north-eastern Greece.

BERRY COLOUR: ● ● ● ● ●

PRINCIPAL SYNONYMS: Karnachalas

ORIGINS AND PARENTAGE

Obscure variety from the Evros region in north-eastern Thráki (Thrace), north-east Greece.

VITICULTURAL CHARACTERISTICS

Vigorous and fertile. Late budding and ripening. Smallish, compact bunches of big, spotted berries.

WHERE IT'S GROWN

Bellas in the city of Soufli in north-east Thráki, **Greece**, produce both a varietal wine and a blend of Karnachalades with CABERNET SAUVIGNON and MERLOT.

KATSAKOULIAS

Rare Greek variety almost exclusive to the island of Zákynthos.

BERRY COLOUR: ● ● ● ● ●

PRINCIPAL SYNONYMS: Gyftokoritho (Zákynthos), Kaltsakouli, Kartsakouli, Katsacoul, Katsakouli, Katsakoulia

ORIGINS AND PARENTAGE

Katsakoulias is a rare variety from the Greek island of Zákynthos in the Ionian Sea, off the western coast of the Pelopónnisos (Peloponnese). A white-berried Katsakoulias has been observed in Chalkída on the island of Évvoia (Euboea) and in Mesolóngi in Aitolía-Akarnanía (Galet 2000).

VITICULTURAL CHARACTERISTICS

Vigorous, productive and sensitive to drought. Mid budding and mid to late ripening. Large bunches of thick-skinned berries that achieve relatively high acidity.

WHERE IT'S GROWN AND WHAT ITS WINE TASTES LIKE

Katsakoulias is found mainly on the island of Zákynthos, **Greece**, where it is used in blends. Comoutos, for example, make both a red and a rosé based on traditional local varieties that include AVGOUSTIATIS and SKYLOPNICHTIS. Solomos and Callinico also include it in their red blends. The variety is found to a very limited extent in the Pelopónnisos, eg in Aitolía-Akarnanía and Ilía.

KATSANO

Rare variety just surviving on Santoríni.

BERRY COLOUR: ● ● ● ● ●

PRINCIPAL SYNONYMS: Katsamon

ORIGINS AND PARENTAGE

Katsano is probably indigenous to the Greek island of Santoríni. It is genetically close to PLATANI from the Kykládes (Cyclades; Biniari and Stavrakakis 2007).

VITICULTURAL CHARACTERISTICS

Low yields, small berries. Mid budding, early ripening and relatively resistant to diseases.

WHERE IT'S GROWN AND WHAT ITS WINE TASTES LIKE

Katsano is found on the Kykládes and particularly on Santoríni, **Greece**. Gavalas is possibly the only producer making a commercially available example, which is a field blend of approximately 85% Katsano and 15% GAIDOURIA. He describes the wine as having 'sweet aromas of flowers, honey and lemon blossom'. The grassy character and lightness of the wine may be more to do with the early harvest than the nature of the grape itself since acidity is relatively low if the grapes are left to ripen fully.

KAY GRAY

Minor and not terribly successful American hybrid.

BERRY COLOUR:

PRINCIPAL SYNONYMS: Elmer Swenson 1-63, ES 1-63

ORIGINS AND PARENTAGE

An Elmer Swenson 217 seedling obtained by Elmer Swenson (1913–2004), a grape breeder from Osceola, Wisconsin, US, in which the pollen parent (ie the 'father') is probably Onaka, which is an old South Dakota hybrid (see BRIANNA for the complete pedigree of Kay Gray). Kay Gray is therefore a complex *Vitis riparia*, *Vitis labrusca* and *Vitis vinifera* hybrid. It was released in 1981 and named after a family friend.

Kay Gray was used to breed BRIANNA and LOUISE SWENSON.

VITICULTURAL CHARACTERISTICS

Small, compact bunches of medium-sized to large berries. Very vigorous. Cold-hardy and only slight susceptibility to fungal diseases. Early to mid ripening and best harvested before full ripeness.

WHERE IT'S GROWN AND WHAT ITS WINE TASTES LIKE

There are minor plantings of Kay Gray in the US Midwest (3 acres/1.2 ha in **Minnesota** in 2007) and Northeast, and also in Québec in **Canada**. Wines are relatively neutral though they can have some floral notes. Unpleasant aromas may develop during winemaking if the grapes are picked fully ripe (Smiley 2008). Tom Plocher of the Minnesota Grape Growers Association suggests it is best blended with a small percentage of a better-quality *vinifera* wine such as SAUVIGNON BLANC. Varietal wines are produced by Bob Cedergren, La Vitacée and Vignobles Les Vents d'Ange in Québec; Wilbur Thomas and Bob Williams in Minnesota; and Amazing Grace in **New York** State.

KEFESSIYA

Late-ripening variety from southern Ukraine that needs help to reproduce.

BERRY COLOUR:

PRINCIPAL SYNONYMS: Cefecia, Cefesia, Doktorsky Chernyi, Kefe Izyum, Kefesia, Kefesiya, Kethessia

VARIETIES COMMONLY MISTAKEN FOR KEFESSIYA: EKIM KARA

ORIGINS AND PARENTAGE

Kefessiya was probably named after Feodosiya (or Theodosia), a city on the Black Sea coast in Krym (Crimea), southern Ukraine, where the variety most likely originates. It is sometimes confused with EKIM KARA, meaning 'doctor's black', because of its similar synonym Doktorsky Chernyi.

VITICULTURAL CHARACTERISTICS

Functionally female-only flowers. Productive if planted with other varieties such as SAPERAVI for pollination. Intensely coloured, thick-skinned berries. Late ripening. Susceptible to downy and powdery mildews and attractive to grape moths, but good drought tolerance.

WHERE IT'S GROWN AND WHAT ITS WINE TASTES LIKE

Grown in **Ukraine** mainly around the city of Feodosiya and in the Sudak region to the south west. Massandra and Sun Valley (Solnechnaya Dolina) both include Kefessiya in their Black Doctor dessert wines.

KÉKFRANKOS

See BLAUFRÄNKISCH.

KÉKMEDOC

See MENOIR.

KÉKNYELŰ

Rare, potentially high-quality western Hungarian variety producing fragrant, high-acid whites.

BERRY COLOUR: ● ● ● ● ●

PRINCIPAL SYNONYMS: Blaustängler

VARIETIES COMMONLY MISTAKEN FOR KÉKNYELŰ: Balafánt (Hungary), FURMINT, PICOLIT (northern Italy)

ORIGINS AND PARENTAGE

Kéknyelű, meaning 'blue stalk' and referring to the slightly blue tint of the petiole, is an old variety probably originating from the Badacsony region in western Hungary. The alleged synonymy between Kéknyelű and the Italian PICOLIT mentioned in the Vitis International Variety Catalogue has been disproved by DNA profiling (Jahnke *et al.* 2007).

VITICULTURAL CHARACTERISTICS

Prone to poor fertilization and therefore often planted alongside BUDAI ZÖLD to improve yields. Mid budding, late ripening. Medium-sized loose bunches of small, thick-skinned berries. Winter-hardy, relatively resistant to botrytis bunch rot but susceptible to downy mildew.

WHERE IT'S GROWN AND WHAT ITS WINE TASTES LIKE

Total plantings of Kéknyelű in **Hungary** in 2009 amounted to just 41 ha (101 acres), although this was twice as much as at the turn of the century. It is grown mainly in Badacsony on the north side of Lake Balaton in the west but to a much more limited extent than in the past, thanks mainly to its poor and unreliable yields. Kéknyelű at its best produces full-bodied, high-acid but fresh and highly fragrant wines which benefit from some time in bottle. Planting alternate rows with a variety such as BUDAI ZÖLD has encouraged more consistent pollination. Recommended producers include Laposa, Endre Szászi and Szeremley.

KERATSUDA

Minor Bulgarian variety making simple whites in the Struma Valley.

BERRY COLOUR: ● ● ● ● ●

PRINCIPAL SYNONYMS: Breza (Bulgaria, France), Breznik (Bulgaria), Govedina (Makedonía in Greece), Keratsouda (Greece), Keratuda (Bulgaria), Mirizlivka (Bulgaria), Tsarevitsa (Bulgaria)

ORIGINS AND PARENTAGE

Keratsuda, meaning 'girl' in Greek, is an indigenous Bulgarian variety that was probably born on the banks of the Struma, or Strymónas, which runs through Bulgaria and Greece, where it has traditionally been cultivated (Katerov 2004).

VITICULTURAL CHARACTERISTICS

Fertile and productive. Very compact bunches of medium-sized, thick-skinned berries. Late ripening. Sensitive to low winter temperatures but resistant to drought. Susceptible to downy and powdery mildews but moderately resistant to botrytis bunch rot.

WHERE IT'S GROWN AND WHAT ITS WINE TASTES LIKE

Keratsuda is used to make simple, low-acid wines for early domestic consumption, although it is also made into a semi-sweet Jerez-style fortified wine. There were just 8 ha (20 acres) in **Bulgaria** in 2008, almost all in the province of Blagoevgrad in the far south west.

KERNER

Qualitatively the most successful modern German cross, versatile, producing Riesling-like wines.

BERRY COLOUR: ● ● ● ● ●

PRINCIPAL SYNONYMS: Weinsberg S 2530

ORIGINS AND PARENTAGE

A SCHIAVA GROSSA × RIESLING cross bred by August Herold at the Weinsberg research centre in Baden-Württemberg, southern Germany, in 1929 and named after the nineteenth-century Württemberg medical doctor and writer of drinking songs Justinus Kerner, who used to recommend that his patients drink a glass of wine as the best natural medicine. Parentage has been confirmed by DNA profiling (Grando and Frisinghelli 1998). Kerner and ROTBERGER are siblings and Kerner was used to breed JUWEL.

VITICULTURAL CHARACTERISTICS

Less fussy about site and higher yielding than RIESLING. Susceptible to powdery mildew. It buds late, giving it good frost resistance, and ripens mid to late.

WHERE IT'S GROWN AND WHAT ITS WINE TASTES LIKE

Kerner is currently **Germany**'s fifth most planted white variety, only just less popular than Grauburgunder (PINOT GRIS). Its 3,712 ha (9,173 acres), planted in many of Germany's wine-growing regions but predominantly in Rheinhessen and the Pfalz, are particularly impressive given that the variety was not authorized until 1969, even though the total area is now a good deal less than the 5,000 ha (12,355 acres) of 2003. However, it is still the most widely planted of the new German crosses.

The wines are in many ways quite similar to RIESLING, making them suitable for varietal wines and not just blends, although the acidity is lower and the texture slightly coarser, and they can have a lightly grapey, leafy or candied note. Its popularity with growers is mainly due to its reliable yields and ability to produce higher levels of sugar and acidity than MÜLLER-THURGAU in a wide range of sites, although its vigour needs curbing to produce high-quality wines. Some of the best examples are grown on the clay-limestone soils of Franken. Recommended producers include Karl Haidle and Staatsweingut Weinsberg in Württemberg; Divino Nordheim Winzergenossenschaft, Reiss and Rainer Sauer in Franken; Didinger in the Mittelrhein; Kees-Kieren in the Mosel; and Zimmerling in Sachsen.

In the Alto Adige in northern **Italy**, Manfred Nössing regularly wins Gambero Rosso's top Tre Bicchieri award for his Kerner although total plantings in Italy amount to no more than 25 ha (62 acres). **Switzerland** had 18 ha (44 acres) in 2008. **England** has just 9 ha (22 acres), down from over 20 ha (49 acres) in 1990, and its less vigorous mutation, known as Kernling, is heading in the same direction. British Columbia in **Canada** has 33 acres (13 ha) and there are very limited plantings in both **California** and **South Africa**. In the extreme conditions of **Japan**'s northernmost island, Hokkaido, Kerner is producing some surprisingly good wines (Sapporo make more than one example); there was a total of 357 ha (882 acres) in Japan in 2009.

KHIKHVI

Rare Georgian variety used for both dry and sweet wines.

BERRY COLOUR: ● ● ● ● ●

PRINCIPAL SYNONYMS: Chichvi, Djananura, Janaani, Jananura, Khichvi

ORIGINS AND PARENTAGE

Khikhvi is indigenous to the Kakheti region in south-eastern Georgia. The synonym Janaani is a village in Kakheti today called Ojio.

OTHER HYPOTHESES

Along with RKATSITELI, MTSVANE KAKHURI and GORULI MTSVANE, Khikhvi is said to have been mentioned as early as the fifth century AD (Chkhartishvili and Betsiashvili 2004) but there is no evidence for this.

VITICULTURAL CHARACTERISTICS

Medium-sized to small bunches that can achieve quite high sugar levels while retaining acidity. Late budding, mid ripening, conservative yields. Susceptible to powdery mildew.

WHERE IT'S GROWN AND WHAT ITS WINE TASTES LIKE

Khikhvi, found mainly in the Kakheti region in south-eastern **Georgia**, particularly in the villages of Kardenakhi, Gurjaani, Tibaani and Ojio, makes high-quality dry and sweet wines that are typically full-bodied and soft in acidity and have aromas of box (*Buxus*) but plantings are few and far between, with a total of around 20 ha (49 acres). Khikhvi is also well suited to vinification in the traditional Georgian qvevri (clay vessels), showing more mineral and floral tones. Recommended producers include Kindzmarauli Marani, Vinoterra and Winiveria. Galet (2000) suggests it is also found in the **Republic of Moldova** and **Ukraine**.

KHINDOGNI

Versatile grape from the southern Caucasus used to make just about every style of red wine.

BERRY COLOUR: ● ● ● ● ●

PRINCIPAL SYNONYMS: Chindogni, Khendorni, Khindogny, Khndogni, Scireni (Russia), Sveni (Russia)

ORIGINS AND PARENTAGE

Khindogni, meaning 'laughing' or 'joyful', is indigenous to the breakaway region of Nagorno-Karabakh in south-west Azerbaijan, where it has traditionally been cultivated.

VITICULTURAL CHARACTERISTICS

Mid to late ripening. Large, compact bunches. Fairly resistant to summer drought, sensitive to cold winter temperatures, relatively resistant to fungal diseases but susceptible to parasites.

WHERE IT'S GROWN AND WHAT ITS WINE TASTES LIKE

Khindogni is found mainly in Nagorno-Karabakh, the disputed enclave within south-west **Azerbaijan**, but also, according to Galet (2000), in **Armenia**, **Uzbekistan**, **Iran** and **Russia**. It produces full-bodied dry reds with high alcohol and is often harvested late to produce sweet, fortified styles. Stepanakert Brandy Factory in Nagorno-Karabakh produces both dry and semi-sweet varietal examples, and sparkling and rosé wines can also be found. Other producers of dry wines include Absheron Sharab, Qabala Sharab and Gyoy-Gyol.

KIRÁLYLEÁNYKA

Hungarian variety producing light, aromatic whites.

BERRY COLOUR: ● ● ● ● ●

PRINCIPAL SYNONYMS: Dánosi Leányka

VARIETIES COMMONLY MISTAKEN FOR KIRÁLYLEÁNYKA: FETEASCĂ REGALĂ (Romania), Királyszőlő (Hungary, but no longer cultivated)

ORIGINS AND PARENTAGE

Királyleányka, meaning 'princess', is said by some to be identical to FETEASCĂ REGALĂ from Romania (Galet 2000) and by others to be a natural cross between Kövérszőlő (GRASĂ DE COTNARI) and LEÁNYKA originating from Transilvania, now in central Romania, that was introduced to Hungary in the 1970s (Rohály *et al.* 2003). However, comparison of their DNA profiles shows that Királyleányka and Feteascǎ Regalǎ from Romania are distinct varieties (Vouillamoz), and the proposed parentage has been rejected by DNA profiling (Galbács *et al.* 2009), thus challenging the possible Romanian origin of Királyleányka. Leányka is possibly one of the parents of Királyleányka.

VITICULTURAL CHARACTERISTICS

Mid budding and ripening, medium to high yields. Small, compact bunches of small berries. Susceptible to botrytis bunch rot and downy mildew.

WHERE IT'S GROWN AND WHAT ITS WINE TASTES LIKE

This widely grown variety (942 ha/2,328 acres in **Hungary** in 2008) produces wines that are fresh and lightly perfumed, often grapey, making it particularly popular in the last ten years as a typical summer or party wine. Its fresh acidity makes it a good blending partner for other, softer aromatic varieties and it seems to do best in Mátra and Eger. Recommended producers include Bolyki (Eger), Bujdosó and Ottó Légli (Balatonboglár), Etkeki Kúria (Etyek), Tamás Szecskő and Mátyás and Zoltán Szőke (Mátra).

KISHMISH

See SULTANIYE.

KISI

Floral-scented Georgian variety made in both modern and traditional styles.

BERRY COLOUR: ● ● ● ● ●

PRINCIPAL SYNONYMS: Kissi, Maghranuli

ORIGINS AND PARENTAGE

Kisi is indigenous to Kakheti in south-eastern Georgia. According to the Vitis International Variety Catalogue, Kisi is a cross between Mtsvane (it's not clear whether this refers to MTSVANE KAKHURI or GORULI MTSVANE) and RKATSITELI but this has been disproved by DNA analysis (Vouillamoz *et al.* 2006).

VITICULTURAL CHARACTERISTICS

Late budding, mid ripening. Small to medium-sized loose bunches of thin-skinned berries. Very sensitive to climate and difficult to grow.

WHERE IT'S GROWN AND WHAT ITS WINE TASTES LIKE

Kisi wines have intense floral aromas as well as flavours of dried pear and apples that give the impression of sweetness even when the wine is dry. There are two main styles: modern European and traditional Georgian, the latter fermented in qvevri, traditional clay vessels, showing more apricot, mint and orange characteristics. There were 50 ha (124 acres) of Kisi in Kakheti, south-eastern **Georgia**, in 2008, notably in the villages of Maghraani and Argokhi, and it is often part of a local blend. Recommended producers of varietal wines include Kindzmarauli Marani, Vashadze and Winiveria in a European style and Alavardi Monastery Cellar, Vinoterra and Pheasant's Tears in the more traditional Georgian style. Many producers blend Kisi with other local varieties such as RKATSITELI. Telavi Wine Cellar blend Kisi with MTSVANE KAKHURI.

KLARNICA

Minor but characterful Slovenian variety making sweet-smelling whites.

BERRY COLOUR: ● ● ● ● ●

PRINCIPAL SYNONYMS: Klarna Mieja, Klarnca, Klarnitza, Mejina

ORIGINS AND PARENTAGE

Klarnica is an old variety most likely originating from Vipavska Dolina in the Primorska region in south-west Slovenia.

VITICULTURAL CHARACTERISTICS

Very compact bunches of medium-sized berries.

WHERE IT'S GROWN AND WHAT ITS WINE TASTES LIKE

There were just 3 ha (7 acres) of Klarnica in **Slovenia** in 2009, all in the Vipavska Dolina wine region in Primorska in the south west. Producers include Mansus and Sveti Martin and the wines are typically full-bodied and dry or semi-sweet with surprisingly sweet, floral or fruity (canned strawberries?) aromas. It is also used to make sparkling wine.

KLEVENER DE HEILIGENSTEIN

See SAVAGNIN ROSE *in the* SAVAGNIN *entry.*

KNIPPERLÉ

Light-skinned, rot-prone Alsace variety on the way out.

BERRY COLOUR: ● ● ● ● ●

PRINCIPAL SYNONYMS: Ettlinger (Colmar), Gelber Ortlieber (Württemberg in Germany), Kleiner Räuschling (Ribeauvillé and Baden), Kniperlé, Ortlieber (Württemberg), Petit Mielleux, Petit Räuschling (Ribeauvillé and Baden), Reichenweiherer (Württemberg), Strassburger (Landau in Germany), Türckheimer (Ribeauvillé and Baden), Wesser Ortlieber

VARIETIES COMMONLY MISTAKEN FOR KNIPPERLÉ: FOLLE BLANCHE, PINOT BLANC, RÄUSCHLING

ORIGINS AND PARENTAGE

Knipperlé most probably comes from Türckheim in Alsace, north-east France. It was first mentioned in 1780 in nearby Riquewihr as 'Knipperling ... improprie kleine Rüschling' ('Knipperling ... incorrectly Kleine Rüschling'), although the modern spelling Knipperlé, meaning 'small size' in the Alsatian dialect, did not appear until 1848 (Rézeau 1997). According to Jean-Louis Stoltz (1852) from Sélestat, between Colmar and Strasbourg, this variety was initially cultivated under the name Petit-Mielleux in the area around Türckheim, where the most widespread variety was ELBLING. Between 1760 and 1770, the early ripener Elbling produced extremely low yields due to the colder climate, while the late-ripening and more resistant

Knipperlé produced riper and more abundant fruit. In the space of a few years, Knipperlé spread to the rest of Alsace, mainly near Colmar (Katzenthal, Kaysersberg, Ingersheim, Niedermorschweyer, Mittelweyer, Siegolsheim and Türckheim) and Ribeauvillé (Beblenheim, Hunawihr and Riquewihr).

From 1780, the Riquewihr vine grower Jean-Michel Ortlieb (1730–1807) actively advertised this variety for its good yields, disease resistance and precocity, and exported large quantities of Knipperlé cuttings to neighbouring Baden-Württemberg. In 1789, Ortlieb even claimed rather pompously: 'daignerait-on pardonner à ma vanité, si j'osais former le désir, que ce *petit Rœuschlinger* fût nommé de préférence, le cru d'Ortlieb' ('would my vanity be forgiven if I dared to formulate the desire that this *petit Rœuschlinger* be preferably named the cru of Ortlieb'; Rézeau 1997). Since that time, the variety has been known in Württemberg under the name Ortlieber or Reichenweiherer (meaning 'from Riquewihr'), and then under numerous additional synonyms (see above).

The German ampelographer Lambert von Babo (1843–4) suggested a genetic relationship between Knipperlé (under the name Ortlieber) and PINOT BLANC. More than a century later, his hypothesis was confirmed by DNA parentage analysis, which established that Knipperlé is one of the twenty-one natural progenies of PINOT × GOUAIS BLANC (Bowers *et al.* 1999; see Pinot pedigree diagram). This spontaneous cross most likely occurred in the neighbourhood of Türckheim, some time before the eighteenth century. DNA profiling has also disproved the supposed identity of Knipperlé with the German variety RÄUSCHLING, which is in fact a half-sibling, as well as with FOLLE BLANCHE and Pinot Blanc.

Knipperlé was used to breed TRIOMPHE.

VITICULTURAL CHARACTERISTICS

Well adapted to winter cold but susceptible to chlorosis and botrytis bunch rot and attractive to grape moths. Small bunches and berries, mid ripening.

WHERE IT'S GROWN AND WHAT ITS WINE TASTES LIKE

After a peak in the late eighteenth century, the area planted to Knipperlé in what is now **France** has decreased continuously – not helped by the fact that it is not permitted in any Alsace appellation wine – and the variety has virtually disappeared from Alsace and Baden-Württemberg in **Germany**. However, Alois Raubal has 1,000 Gelber Ortlieber vines on his estate in Gumpoldskirchen, south-west of Wien (Vienna) in **Austria**, and produces a varietal wine. Knipperlé tends to produce soft, pleasant, easy-drinking wines with highish alcohol.

KÖHNÜ

Possibly named after a person, the black-berried Köhnü probably originates from the Arapgir district in the province of Malatya in the south west of eastern Anatolia, **Turkey**, and is found mainly in the province of Elazığ, just north and east of Malatya. DNA profiling has shown that Köhnü is distinct from KÖSETEVEK (Gökhan Söylemezoğlu, personal communication), with which it is often blended.

KOK PANDAS

Vine making full-bodied whites in southern Ukraine.

BERRY COLOUR: ● ● ● ● ●

PRINCIPAL SYNONYMS: Coc Pandas, Kok Pandasse, Pandas Kok, Tken Izyum

ORIGINS AND PARENTAGE

Kok Pandas is an old variety from Solnechnaya Dolina in the Sudak district of Krym (Crimea), southern Ukraine.

VITICULTURAL CHARACTERISTICS

Vigorous, late budding, mid ripening. Relatively compact bunches of medium-sized thick-skinned berries. Average resistance to fungal diseases.

WHERE IT'S GROWN AND WHAT ITS WINE TASTES LIKE

Kok Pandas, planted mainly in the Sudak region in Krym, southern **Ukraine**, produces full-bodied dry wines and is used in sweet and/or fortified blends such as Sun Valley's eponymous dessert wine comprising up to thirty different varieties and in their fortified RKATSITELI and Kok Pandas blend.

KOKUR BELY

Multi-purpose Ukrainian variety with possible Greek origins.

BERRY COLOUR: ● ● ● ● ●

PRINCIPAL SYNONYMS: Belji Dolgi, Dolgi, Kokour Blanc, Kokuri Belji

ORIGINS AND PARENTAGE

Kokur Bely is an old variety of unknown origin that was supposedly cultivated as early as the fourteenth and fifteenth centuries in Krym (Crimea), southern Ukraine, where it used to be the most important white grape variety (Chkhartishvili and Betsiashvili 2004). It is said to have been introduced from the Greek island of Kérkyra (Corfu; Galet 2000).

VITICULTURAL CHARACTERISTICS

Very vigorous, mid budding and mid to late ripening. Compact bunches of largish, thin-skinned berries. Susceptible to fungal diseases, especially downy mildew.

WHERE IT'S GROWN AND WHAT ITS WINE TASTES LIKE

Kokur Bely is a versatile grape, used in **Ukraine** for the table, grape juice and a range of wine styles, including dry and sweet still wines, fizz and fortified wines. It is quite widely planted around Alushta and Feodosiya in the far south of Krym, along the Black Sea coast, with total plantings of around 918 ha (2,268 acres) in Ukraine in 2009. Massandra Winery, for example, make a sweet varietal wine, and Sun Valley's Kokur is dry with spice, floral and honey aromas.

It is also found in Rostov, south-western **Russia**, and in **Kyrgyzstan**.

KOLINDRINO

Rare Greek variety named after its place of origin.

BERRY COLOUR: ● ● ● ● ●

PRINCIPAL SYNONYMS: Kolindros

ORIGINS AND PARENTAGE

Obscure grape variety grown in Makedonía, most likely named after the village of Kolindros. There has so far been no DNA analysis to determine its identity or uniqueness.

WHERE IT'S GROWN AND WHAT ITS WINE TASTES LIKE

Pieria Eratini's Gymnos Vasilias ('Naked King') Syrah from Kolindros, south west of Thessaloníki in Makedonía, **Greece**, includes a small percentage of this variety, which Pavlos Argyropoulos first planted in 1999 in the hilly Kolindros region, encouraged by the area's past reputation for good-quality wine. According to Argyropoulos, Kolindrino contributes firm tannins and an attractively bitter aftertaste to the blend.

KOLORKO

Very rare, high-quality Turkish variety from southern Trakya (Thrace).

BERRY COLOUR: ● ● ● ● ●

ORIGINS AND PARENTAGE

Kolorko is an almost extinct variety from the region between Uçmakdere and Şarköy in southern Trakya, along the northern coast of the Sea of Marmara, Turkey. Since Kolorko does not appear in any official list, it is not yet known whether it is a unique variety or if it is a local name for another registered variety. This could be easily clarified by DNA profiling.

VITICULTURAL CHARACTERISTICS

Low yields, late ripening, thick skins. High level of catechins in the skins, so pressing must be gentle.

WHERE IT'S GROWN AND WHAT ITS WINE TASTES LIKE

Kolorko has virtually disappeared from vineyards in **Turkey** and is exclusive to a few villages in southern Trakya on the north coast of the Sea of Marmara, south west of Tekirdağ: Güzelköy, Kirazlı, Iğdebağlari and Şarköy. A varietal wine is made by Pasaeli, who are attempting to restore the variety's fortunes by collecting vines from the few growers who still have it in their vineyards and grafting them onto the rootstocks planted in their own vineyard. The wine, of which 276 bottles were made in 2009, has firm acidity, some citrus and wild herb flavours, notable minerality and great depth. Melen used to make a Madeira-style wine in the 1950s and then, until the mid 1990s, a dry wine.

KORINTHIAKI

The currant grape.

BERRY COLOUR: ● ● ● ● ●

PRINCIPAL SYNONYMS: Alga Passera (Italy), Black Corinth (California), Corinthe Noir (France), Corinto Negro (Spain, Uruguay), Corinto Nero (Italy), Crni Korint (Croatia), Currant Grape (Australia, US), Korinthiaki Stafida, Korinthiaki Stafis, Korinthusi Kék (Hungary), Lianorogi, Mavri Stafis, Passerilla (Italy), Passula di Corinto (Italy), Raisin de Corinthe (France), Stafida, Stafidambelo, Zante Currant (Australia, US)

VARIETIES COMMONLY MISTAKEN FOR KORINTHIAKI: TERMARINA ROSSA

ORIGINS AND PARENTAGE

Korinthiaki, literally 'Corinthian' in Greek, is a seedless grape that allegedly comes from Kórinthos (Corinth) in north-east Pelopónnisos (Peloponnese). However, its true origin may not be in Greece at all because a morphological and genetic study has distinguished it from most other Mediterranean wine grape varieties (Aradhya *et al.* 2003). DNA profiling has shown that neither Corinto Bianco from Italy nor Corinthe Blanc from Greece are colour mutations of Black Corinth (Vargas *et al.* 2007) but more recent profiling (under the name Black Corinth) has suggested a possible parent–offspring relationship with LIATIKO (Myles *et al.* 2011).

OTHER HYPOTHESES

Korinthiaki (under the name Corinthe Noir) is said to be a seedless mutation of LIATIKO (Krimbas in Galet 2000) but this has been contradicted by DNA profiling (Vouillamoz).

VITICULTURAL CHARACTERISTICS

Long bunches of very small, sweet, thin-skinned seedless berries.

WHERE IT'S GROWN AND WHAT ITS WINE TASTES LIKE

In the nineteenth century, the northern Pelopónnisos, **Greece**, grew rich from Korinthiaki raisins, particularly thanks to exports to the UK. Today Korinthiaki is still used for drying but the fruit is also blended (up to a maximum of 49%) with MAVRODAFNI to make the sweet, fortified red wine traditional in this area and recognized under the Mavrodafni of Pátra appellation. The inclusion of sweet-fruited Korinthiaki often compensates for the low sugar levels of Mavrodafni grown in Pátra but does little for the quality of the final wine (Lazarakis 2005). It was still the second most planted variety on the Ionian Islands in 2008, with a total of 478 ha (1,181 acres), and there were 357 ha (882 acres) in western Greece.

The vast majority, if not all, of **California**'s 1,682 acres (681 ha) of Black Corinth are for the production of raisins. When a shortage of wine grapes has coincided with a surplus of Black Corinth, in the 1970s and 1990s, for example, the fruit has occasionally been used for blending or colour.

KORIOSTAFYLO

Very rare Greek variety that seems to be exclusive to the island of Ikaría.

BERRY COLOUR: ● ● ● ● ●

PRINCIPAL SYNONYMS: Koriostaphylo

ORIGINS AND PARENTAGE

Obscure local variety on the island of Ikaría in the Aegean Sea, just west of Sámos. Its identity has not yet been revealed by DNA analysis.

WHERE IT'S GROWN

Koriostafylo appears to be grown exclusively on Ikaría, **Greece**, where Afianes blend it with other local varieties such as FOKIANO.

KÖSETEVEK

Black-berried variety from the province of Elazığ, **Turkey**. DNA profiling has shown that Kösetevek is distinct from KÖHNÜ (Gökhan Söylemezoğlu, personal communication), with which it is often blended.

KOSHU

Japan's delicate, light-bodied signature variety that is starting to gain international recognition.

BERRY COLOUR:

ORIGINS AND PARENTAGE

Koshu is considered to be indigenous to Japan, where it has allegedly been known since the seventeenth century. Legend has it that Koshu was discovered growing wild at Katsunuma in Yamanashi prefecture, central Honshu, in 1186 as a seedling of RYUGAN or Huotianhong, both supposedly introduced earlier from China by Buddhist monks (Yokotsuka *et al.* 1984; Galet 2000). However, this hypothesis has been rejected by Goto-Yamamoto *et al.* (2015), who used DNA analysis based on a new kind of marker (single nucleotide polymorphism, or SNP) to show that Koshu is 70% *Vitis vinifera*, the European grapevine species, and that the remaining 30% of its ancestry is most likely derived from wild East Asian *Vitis* species, probably *Vitis davidii* or a closely related species.

VITICULTURAL CHARACTERISTICS

Vigorous, mid to late budding, late ripening. Medium-sized bunches of large, thick-skinned berries (generally if grown for the table). Susceptible to downy and powdery mildews but usefully resistant to botrytis bunch rot in Japan's humid climate.

WHERE IT'S GROWN AND WHAT ITS WINE TASTES LIKE

Koshu wines are typically light, delicate and fresh, with a certain zen-like purity about them, and are generally best drunk young. Recent quality improvements have included a move to viticultural practices suited to growing grapes for wine rather than for the table. Vines have traditionally been planted on pergola-style trellising owing to the variety's high vigour but different approaches to canopy management may well yield better results (Kobayashi, Fujita *et al.* 2009).

Koshu is widely planted in central **Japan**, and about 90% of Japan's 2008 total of 175 ha (432 acres) for wine production is in Yamanashi prefecture, although Koshu is also found in Yamagata, Osaka, Tottori and Shimane prefectures. Recommended producers include Grace, Katsunuma-

Jozo, Marufuji and Mercian. Katsunuma-Jozo make one of the more impressive barrel-fermented versions while others are venturing into sparkling wines and experimenting with winemaking techniques designed to give more flavour and texture such as leaving the wine on the yeast lees, cryoextraction and skin contact.

KOTSIFALI

Oft-maligned variety widely planted on Kríti (Crete), producing soft, pale, alcoholic but potentially aromatic and distinctive reds.

BERRY COLOUR: ● ● ● ● ●

PRINCIPAL SYNONYMS: Kotrifali, Kotsiphali, Kotzifali

ORIGINS AND PARENTAGE

This variety most likely originates from Kríti or the Kykládes (Cyclades; Nikolau and Michos 2004). In the Greek Vitis Database, the DNA profile of Korfiatiko (or Korfiatis), a non-cultivated dark-skinned variety from the Ionian Islands, is, surprisingly, identical to Kotsifali, although they are ampelographically quite distinct (Vouillamoz).

VITICULTURAL CHARACTERISTICS

Vigorous, productive and good all-round disease resistance although some susceptibility to downy mildew and botrytis bunch rot. Medium-sized compact bunches of small, thin-skinned berries. Early to mid ripening.

WHERE IT'S GROWN AND WHAT ITS WINE TASTES LIKE

Wines tend to be pale, fairly high in alcohol but soft and generous so that they generally benefit from the addition of the more deeply coloured, firmer structured and less alcoholic MANDILARIA, a blend that is stipulated in the Peza and Achárnes appellations on the variety's home territory on the island of Kríti, **Greece**. More recently producers have been experimenting with Kotsifali/SYRAH blends, but it is easy to see why the variety is not treated with great respect, especially when it is older vines that seem to show the variety's potential. On the other hand, a proportion of Kotsifali can add a distinctive perfume of strawberries, spices and herbs to a red blend. The colour is not only hard to extract from the berries' thin skins but it is also unstable and wines often look quite developed even when young.

Kotsifali is rarely found outside Kríti, where it is the second most planted variety (1,264 ha/3,123 acres in 2008) after LIATIKO, with the greatest concentration of plantings on the eastern side of the island in Irákleio (Heraklion). Recommended producers of varietals and blends include Boutaris, Lyrarakis, Mediterra and Tamiolakis. Lyrarakis make a sour-cherry-flavoured varietal wine that manages to retain sufficient acidity to be long and fresh – but quite demanding.

KOUTSOUMPELI

Rare Greek pink-skinned variety.

BERRY COLOUR: ● ● ● ● ●

PRINCIPAL SYNONYMS: Koutsoubeli, Koutsoumbeli or Koutsoumbeli Kokkino, Koutsoumpeli Kokkino

ORIGINS AND PARENTAGE

Koutsoumpeli is from the Greek island of Zákynthos. According to the viticulturist Haroula Spinthiropoulou, there is also a white-berried clone which is more common than the pinky red one.

VITICULTURAL CHARACTERISTICS

Productive, early budding and late ripening, resistant to drought and downy mildew. Large bunches of medium-sized berries.

WHERE IT'S GROWN

Koutsoumpeli is found in the regions of Ilía, north-west Pelopónnisos (Peloponnese), **Greece**, in Aitolía-Akarnanía (south-west mainland) and especially on the island of Zákynthos, where Logothetis make a varietal rosé from this pink-skinned variety.

KÖVÉRSZŐLŐ

See GRASĂ DE COTNARI.

KÖVIDINKA

Common pink-skinned Hungarian variety making commonplace whites.

BERRY COLOUR: ● ● ● ● ●

PRINCIPAL SYNONYMS: Dinka Alba (Croatia, Hungary), Dinka Fehér (Hungary), Kevedinka (Serbia), Kevidinka (Croatia), Kövidinka Rose, Rosentraube (Germany), Ružica (Macedonia), Ruzsitza (Hungary), Steinschiller (Germany)

VARIETIES COMMONLY MISTAKEN FOR KÖVIDINKA: Pozsonyi Fehér, SAVAGNIN ROSE ☓

ORIGINS AND PARENTAGE

Kövidinka is a Hungarian variety that was allegedly cultivated in the Middle Ages. It is said to originate in the Carpathian Basin, possibly from Slavonija in Croatia (Kocsis *et al.* 2005).

According to Constantinescu in Galet (2000), Kövidinka may have come from Germany and been introduced to the historical Central European Banat region by German colonists. However, the DNA profile of Kövidinka in Galbács *et al.* (2009) does not correspond to any modern German variety (Vouillamoz).

Kövidinka was used to breed KARÁT and SILA.

VITICULTURAL CHARACTERISTICS

Fertile, late budding and ripening. Small bunches of small berries. Winter-hardy and resistant to botrytis bunch rot and drought.

WHERE IT'S GROWN AND WHAT ITS WINE TASTES LIKE

Kövidinka is widely planted in Central and Eastern Europe but particularly in central southern **Hungary** in the Kunság and Csongrád regions of the Alföld (Great Plain), where it makes light, soft and typically fairly neutral whites for everyday drinking. Total plantings in Hungary in 2008 were 858 ha (2,350 acres) and recommended producers include Font in Kunság.

KRAKHUNA

Minor but well-established Georgian variety producing fresh, fruity whites.

BERRY COLOUR:

PRINCIPAL SYNONYMS: Chkovra🞩, Krakhuna Shavi

ORIGINS AND PARENTAGE

Krakhuna is indigenous to the Imereti region in western central Georgia and was already known in the eighteenth century.

VITICULTURAL CHARACTERISTICS

Medium-sized compact bunches of thin-skinned berries. Moderate to high yields, mid budding, late ripening. Good sugar accumulation in the berries. Susceptible to powdery mildew but resistant to downy.

WHERE IT'S GROWN AND WHAT ITS WINE TASTES LIKE

Krakhuna wines typically have flavours of ripe pear as well as more savoury and herbal notes and fresh acidity, with the best examples ageing well in bottle. There were just 35 ha (86 acres) planted in 2004, mostly in the villages of Dimi and Sviri in the Imereti region of western central **Georgia**, where Khareba make a good varietal example. The variety also shows good potential for high-quality fortified wine.

KRALJEVINA

Old Croatian variety producing simple, refreshing everyday whites.

BERRY COLOUR:

PRINCIPAL SYNONYMS: Imbrina, Kraljevina Crvena, Moravina

ORIGINS AND PARENTAGE

Despite the linguistic contraindication in the synonym Kraljevina Crvena (*crvena* means 'red' in Serbo-Croatian and *kraljevina* means 'kingdom'), Kraljevina is a white-berried variety, possibly from Sveti Ivan Zelina near Zagreb, Croatia. With exposure to the sun, the berries tend to take on a slightly red tint at harvest, hence the synonym. Kraljevina shows a significant diversity of types with white, green, reddish and speckled berries (Edi Maletić, personal communication), suggesting that this is an old variety.

Kraljevina is also known as Portugieser Rot, but it is not related to BLAUER PORTUGIESER, known in Croatia as Portugizac Crni, nor to any Portuguese variety (Vouillamoz).

VITICULTURAL CHARACTERISTICS

Late budding and ripening. Big, thin-skinned berries. Vigorous and consistently productive. Susceptible to botrytis bunch rot if yields are too high and bunches are crowded.

WHERE IT'S GROWN AND WHAT ITS WINE TASTES LIKE

Wines tend to be moderate to low in alcohol, non-aromatic, high in acidity, sometimes sour, but are popular for refreshing, simple, everyday drinking. Kraljevina is grown only in northern **Croatia**, around Zagreb, and is recommended in the Prigorje-Bilogora subregion. The wines may be labelled Kraljevina Zelina after the Sveti Ivan Zelina region. It is also found in **Slovenia**, where it is one of the many varieties permitted in the traditional, light-red blend known as Cviček but rarely appears as a varietal wine.

KRASNOSTOP ZOLOTOVSKY

Russian variety making full-bodied dry reds as well as sweet fizz.

BERRY COLOUR: ● ● ● ● ●

PRINCIPAL SYNONYMS: Krasnostop, Krasnostop Anapsky

VARIETIES COMMONLY MISTAKEN FOR KRASNOSTOP ZOLOTOVSKY: PLECHISTIK

ORIGINS AND PARENTAGE

Krasnostop Zolotovsky is a variety from the Rostov wine region in south-western Russia, near the border with Ukraine, where it was first mentioned in 1814. It was traditionally grown in the Krasnodar and Anapa regions, hence its name and synonyms. Krasnostop Anapsky is not a distinct variety but a clone of Krasnostop Zolotovsky. The synonym Cherny Vinny is also given to PLECHISTIK, but the two varieties are distinct (Chkhartishvili and Betsiashvili 2004).

OTHER HYPOTHESES

Like other varieties along the River Don, Krasnostop is said to have been introduced from Dagestan in the Middle Ages (Vladimir Tsapelik, personal communication).

VITICULTURAL CHARACTERISTICS

Mid budding, mid ripening. Winter-hardy. Small bunches of small to medium-sized thin-skinned berries. Susceptible to downy mildew in some regions.

WHERE IT'S GROWN AND WHAT ITS WINE TASTES LIKE

Wines are typically rich in tannins, alcohol and acidity, with flavours of dark plums, redcurrant and green pepper. The variety is found in both the Rostov (66 ha/163 acres in 2010) and Krasnodar wine regions in **Russia**. In the former, Tsimlyansky Winery include it in the blend for their sweet sparkling red wine and, in the latter, Chateau Le Grand Vostock blend it with varieties such as

SAPERAVI, CABERNET SAUVIGNON and GOLUBOK for dry reds. Vedernikov Winery make a dry varietal wine and also a blend with Cabernet Sauvignon on the banks of the River Don.

KRASSATO

Greek variety integral to the Rapsáni appellation.

BERRY COLOUR: ● ● ● ● ●

PRINCIPAL SYNONYMS: Krasata, Krasato

ORIGINS AND PARENTAGE

Krassato is considered to be indigenous to the Rapsáni region in the foothills of Mount Ólympos (Óros Ólimbos in Greek) in the province of Larísa in Thessalía in the east of central mainland Greece (Manessis 2000; Nikolau and Michos 2004).

VITICULTURAL CHARACTERISTICS

Vigorous, fertile and productive. Late budding and ripening. Medium-sized bunches of medium-sized berries with thickish skins. Susceptible to powdery mildew.

WHERE IT'S GROWN AND WHAT ITS WINE TASTES LIKE

Krassato is blended with equal parts of XINOMAVRO and STAVROTO within the Rapsáni appellation in **Greece** – usually a field blend – to make a rich, well-structured, spicy, sometimes earthy wine that is often aged in oak. Krassato wines are rich in alcohol but have medium acidity and tannins and do not generally age well, hence the requirement to blend with Xinomavro in particular. Krassato is also found further to the south in Magnisía and occasionally to the north in Makedonía. Recommended producers of Rapsáni include Dougos and Tsantali.

KRATOŠIJA

See TRIBIDRAG.

KREACA

Rare Balkan pale-skinned vine now found mainly in Serbia.

BERRY COLOUR: ● ● ● ● ●

PRINCIPAL SYNONYMS: Banat Riesling, Bánáti Rizling (Hungary), Banatski Rizling (former Yugoslavia), Creaţă or Creaţă de Banat (Romania), Franchie (Romania), Kreáca, Kreatza (Serbia, Montenegro), Kriaca, Riesling Banatsky (Slovakia), Zakkelweiss (Germany)

VARIETIES COMMONLY MISTAKEN FOR KREACA: FRÂNCUŞĂ, RIESLING ꕤ (Serbia)

ORIGINS AND PARENTAGE

Like SLANKAMENKA, Kreaca is a very old Balkan variety, possibly originating in the Banat, a historical region spanning parts of today's Romania, Serbia and Hungary and once part of

Hungary. The variety shows significant morphological diversity within each country (Avramov and del Zan 2004), another indication of its age.

OTHER HYPOTHESES

Some authors have suggested that Kreaca was introduced to Tomnatic in the Banat by French colonists, hence the synonym Franchie (Dejeu 2004) but its DNA profile is not similar to that of any French variety (Nataša Štajner, personal communication).

VITICULTURAL CHARACTERISTICS

Productive, mid budding and mid to late ripening. Highly susceptible to downy and powdery mildews, less so to botrytis bunch rot.

WHERE IT'S GROWN AND WHAT ITS WINE TASTES LIKE

Serbia's Banat Rizling appellation allows a blend of GRAŠEVINA (Italian Riesling), Smederevka (DIMYAT), Župljanka (CHASSELAS) and Kreaca. In **Hungary**, Bánáti Rizling has almost disappeared – there were 3 ha (7 acres) in 2008 and the vineyards of the leading producer, Oszkár Maurer, are over the border in Serbia. His Édes Erdély is a blend of MÉZES FEHÉR and Bánáti Rizling. Wines are light on aroma but may be rich, sometimes heavy, in the mouth. The variety is also found in **Romania**.

KRKOŠIJA

Minor blending partner for Žilavka in Bosnia and Herzegovina.

BERRY COLOUR: ● ● ● ● ●

PRINCIPAL SYNONYMS: Krkochia, Krkoshia, Kyrkochia

ORIGINS AND PARENTAGE

Krkošija possibly originates in the Mostar region in the south of Bosnia and Herzegovina.

WHERE IT'S GROWN AND WHAT ITS WINE TASTES LIKE

Krkošija usually plays a bit part alongside **Bosnia and Herzegovina**'s principal indigenous white variety, ŽILAVKA – up to 15% is permitted in the appellation Žilavka Mostar. Producers include Čitluk Winery and Hepok.

KRONA

Minor Ukrainian variety used in sweet red wines.

BERRY COLOUR: ● ● ● ● ●

ORIGINS AND PARENTAGE

Not much is known about Krona, meaning 'crown', an obscure variety from Solnechnaya Dolina in the Sudak district of Krym (Crimea), southern Ukraine (Solnechnaya Dolina).

WHERE IT'S GROWN

Krona is one of the many varieties in **Ukraine**'s traditional sweet Black Doctor blend produced by Massandra and Sun Valley (Solnechnaya Dolina).

KRSTAČ

Old Montenegrin variety producing light, summery whites.

BERRY COLOUR: ● ● ● ● ●

PRINCIPAL SYNONYMS: Bijeli Krstač, Krata Bijela, Krstač Bijela, Loza Bijela, Vinogradarska Bijela

ORIGINS AND PARENTAGE

Krstač is an ancient variety from Montenegro, possibly from the Podgorica region in the south. *Krst* means 'cross' in Montenegrin, a reference to the shape of the bunch. There is also a black-berried and non-cultivated variety, Krstač Crna, but so far DNA profiling has not been done to verify if one is a colour mutation of the other, or if Krstač Crna is identical to TRIBIDRAG (Zinfandel), called Kratošija in Montenegro with Krstač Crna a common and confusing synonym.

VITICULTURAL CHARACTERISTICS

Vigorous, mid budding, late ripening. Large to medium-sized bunches of thin-skinned berries. Susceptible to powdery and downy mildews and especially to botrytis bunch rot. Very site-specific.

WHERE IT'S GROWN AND WHAT ITS WINE TASTES LIKE

Wines are light-bodied and fresh, with flavours of green apples, pear and elderflower. Plantaže are currently the only producers of a varietal wine, from vines planted in Podgorica, **Montenegro**.

KUJUNDŽUŠA

Rare Dalmatian variety.

BERRY COLOUR: ● ● ● ● ●

PRINCIPAL SYNONYMS: Kojundžuša, Kujundžuša Bijela, Tvrdac, Žutac, Žutka

VARIETIES COMMONLY MISTAKEN FOR KUJUNDŽUŠA: ŠKRLET ☒

ORIGINS AND PARENTAGE

Kujundžuša probably comes from the Imotski region of Croatia, close to the Bosnia and Herzegovina border, south east of Split.

OTHER HYPOTHESES

Although Kujundžuša and ŠKRLET are listed as the same variety in the Vitis International Variety Catalogue, they are distinct.

VITICULTURAL CHARACTERISTICS

Late budding, mid to late ripening. Consistently abundant yields and not particularly susceptible to diseases. Medium-sized loose bunches.

WHERE IT'S GROWN AND WHAT ITS WINE TASTES LIKE

Kujundžuša is authorized in the Dalmatinska Zagora (Dalmatian Hinterland) wine region and is the main white variety in the area around the town of Imotski, south east of Split in southern **Croatia**. It represented 2.2% of the total Croatian vineyard area in 2009, ie around 715 ha (1,767 acres). Varietal wines are produced by, for example, Imota (Imotski), Jerković-Promet (Runović) and Grabovac (Donji Proložac). Wines tend to be relatively light-bodied.

KUMSHATSKY CHERNY

Old, dark-skinned Russian variety from the Tsimlyansky district in the Rostov wine region in south-western **Russia**, near the border with Ukraine, named after a local town and used mainly in sweet red sparkling wines. There's also a white-berried version (Kumshatsky Bely), which is important in the Rostov wine region.

KUNLEÁNY

Recent but already widely planted, high-yielding Hungarian hybrid.

BERRY COLOUR: ● ● ● ● ●

ORIGINS AND PARENTAGE

A hybrid obtained in 1960 by István Tamássy and István Koleda at Corvinus University in Budapest, Hungary, by crossing a *Vitis amurensis* × *Vitis vinifera* vine (previously bred from two unidentified vines of these species) with Afus Ali (a Lebanese table grape also called Dattier de Beyrouth).

VITICULTURAL CHARACTERISTICS

High yielding, mid to late ripening, big berries. Not particularly susceptible to botrytis bunch rot.

WHERE IT'S GROWN AND WHAT ITS WINE TASTES LIKE

Kunleány was widely propagated in the 1970s and by 1997 there were 1,526 ha (3,771 acres) planted in **Hungary**, especially for sparkling wine, although this area had come down to 1,320 ha (3,262 acres) by 2008, mainly in Kunság, Hajós-Baja and Csongrád in central southern Hungary. Varietal wines are relatively rare and high in acidity and may have notes of citrus or apple, but are produced by Molnár és Társa Családi Borászat and Öreghegy Szőlőbirtok.

KUNTRA

See KARASAKIZ.

KUPUSAR

Very rare, tannic variety from coastal Croatia.

BERRY COLOUR:

PRINCIPAL SYNONYMS: Plavac Kupusar

ORIGINS AND PARENTAGE

A rare variety from the Kaštela region, north of Split, Croatia, also found on the nearby islands of Hvar and Brač. Kupusar is often called Plavac Kupusar although it is unrelated to the more famous PLAVAC MALI. The name Kupusar points to the very large leaves of this variety: *kupus* means 'cabbage' in Croatian.

VITICULTURAL CHARACTERISTICS

Medium to high vigour and productive. Very large leaves. Mid to late ripening. High tolerance of powdery and downy mildews and botrytis bunch rot but susceptible to some virus diseases (eg leafroll type 3). High tannin content in the skins.

WHERE IT'S GROWN AND WHAT ITS WINE TASTES LIKE

Kupusar is not currently made as a varietal wine. It is generally found in mixed vineyards in **Croatia** and is often blended with PLAVAC MALI. Tolić Tomis in Trogir, just west of Split, includes it in a blend with BABIĆ (under the name Rogoznička), BABICA, TRIBIDRAG (under the name Crljenak) and GLAVINUŠA.

KYDONITSA

Promising but still very minor variety producing aromatic, finely textured wines near Monemvasia.

BERRY COLOUR:

PRINCIPAL SYNONYMS: Kidonitsa

ORIGINS AND PARENTAGE

An almost forgotten variety from the southern Pelopónnisos (Peloponnese). Its name is the diminutive of quince (*kydoni*).

VITICULTURAL CHARACTERISTICS

Mid to late ripening, alternate years of high and low yields, and not always easy to ripen fully. In some years there is a second budding that bears fruit. Good all-round disease resistance. Smallish berries.

WHERE IT'S GROWN AND WHAT ITS WINE TASTES LIKE

Kydonitsa is grown in Lakonía, **Greece**, and particularly in the eastern finger of the southern Pelopónnisos near Monemvasia, where it was found in old vineyards and replanted by pioneering producers such as Yiannis Vatistas and George Tsibidis (Monemvasia Winery). Both make varietal wines, which range from strongly quince-like, as the name might predict, to more mineral- and citrus-flavoured, with a chalky texture, both powerful and fresh. Even wines from young vines seem to have remarkable concentration and suggest a promising future. Although local growers have been encouraged by government subsidies to plant Kydonitsa, without any marketing support it is still little known and hard to promote. Antonopoulos have also planted this variety at the opposite, northern, extreme of the Pelopónnisos.

L

L'ACADIE BLANC
LA CRESCENT
LA CROSSE
LACRIMA DI MORRO D'ALBA
LADO
LAFNETSCHA
LAGARINO BIANCO
LAGREIN
LAIRÉN
LALVARI
LAMBRUSCA DI ALESSANDRIA
LAMBRUSCA VITTONA
LAMBRUSCHETTO
LAMBRUSCO BARGHI
LAMBRUSCO DI FIORANO
LAMBRUSCO DI SORBARA
LAMBRUSCO GRASPAROSSA
LAMBRUSCO MAESTRI
LAMBRUSCO MARANI
LAMBRUSCO MONTERICCO
LAMBRUSCO OLIVA
LAMBRUSCO SALAMINO
LAMBRUSCO VIADANESE
LANDAL
LANDOT NOIR
LANZESA
LAPA KARA
LASINA
LAUROT
LAUZET
LEÁNYKA
LEN DE L'EL
LÉON MILLOT
LIATIKO
LILIORILA
LIMNIO
LIMNIONA
LISTÁN DE HUELVA
LISTÁN NEGRO
LISTÁN PRIETO
LJUTUN
LONGYAN
LOUISE SWENSON
LOUREIRO
LUCIE KUHLMANN
LUGLIENGA
LUMASSINA
LYDIA

ℵthis symbol on the following pages indicates that the synonyms, and mistaken identities, are confirmed by DNA evidence

L'ACADIE BLANC

1950s Canadian very complex hybrid performing well in the chilly climes of Nova Scotia and Québec.

BERRY COLOUR:

PRINCIPAL SYNONYMS: Acadie, L'Acadie, V 53261

ORIGINS AND PARENTAGE

A CASCADE × Seyve-Villard 14-287 hybrid made in 1953 by Ollie A Bradt at the Vineland research centre (now part of Guelph University) in Niagara, Ontario, Canada, in which:

- Seyve-Villard 14-287 is a Seibel 6746 × Muscat du Moulin hybrid
- Seibel 6746 is a hybrid made by Albert Seibel in Aubenas in the Ardèche, France, with unknown parents
- Muscat du Moulin is a Couderc 603 × PEDRO XIMÉNEZ hybrid
- Couderc 603 is a BOURRISQUOU × *Vitis rupestris* hybrid.

Under the code V 53261, this variety was sent for testing at the Kentville research centre in Nova Scotia, where it was named after the region of Canada and north-eastern North America that was once a French colony. The pedigree was revealed by Helen Fisher at the Department of Plant Agriculture at the University of Guelph.

L'Acadie is therefore an extremely complex *Vitis riparia*, *Vitis labrusca*, *Vitis vinifera*, *Vitis aestivalis*, *Vitis lincecumii*, *Vitis rupestris*, *Vitis cinerea* and *Vitis berlandieri* hybrid.

VITICULTURAL CHARACTERISTICS

Vigorous, productive, early to mid ripening. Cold-hardy to –7.6 °F (–22 °C), perhaps even to –13 °F (–25 °C) (ie hardier than SEYVAL BLANC). Loose bunches help prevent botrytis bunch rot.

WHERE IT'S GROWN AND WHAT ITS WINE TASTES LIKE

L'Acadie Blanc is grown mostly in Nova Scotia, eastern **Canada**, with just a few plantings in Québec and some very recent ones in northern central Ontario. Its wine is generally blended with other varieties such as VANDAL-CLICHE, SEYVAL BLANC and VIDAL, eg by Domaine St-Jacques for a dry, still wine and Vignoble Carone for a frizzante style. Varietal wines are more full-bodied than Vandal-Cliché and may have honey and floral aromas. The variety may suffer from low acidity in the warmer parts of Ontario, hence a slow increase in plantings in cooler areas of the province, but it shows promise in Québec thanks to its good disease resistance and overall balance and quality potential.

LA CRESCENT

Promising and increasingly popular cold-hardy American hybrid.

BERRY COLOUR:

PRINCIPAL SYNONYMS: MN 1166

ORIGINS AND PARENTAGE

A complex hybrid of ST PEPIN and ES 6-8-25 made in 1988 by Peter Hemstad and James Luby at the University of Minnesota's Horticultural Research Center in Excelsior, in which ES 6-8-25 is a *Vitis riparia* × MUSCAT OF HAMBURG hybrid selected by Elmer Swenson, a grape breeder from Osceola, Wisconsin. La Crescent was selected in 1992, tested under the name MN 1166 and released in 2002. This interspecific hybrid includes 45% *Vitis vinifera*, 28% *Vitis riparia*, and less than 10% each of *Vitis rupestris*, *Vitis labrusca* and *Vitis aestivalis*.

VITICULTURAL CHARACTERISTICS

Smallish berries that tend not to split even in wet years but some may drop before harvest. Vigorous and slightly erratic yields. Early budding and early to mid ripening. Moderately susceptible to black rot and powdery mildew, possibly also to anthracnose. Very hardy (–20 °F/–29 °C to –36 °F/–38 °C; Smiley 2008). Berries have relatively high levels of both sugar and acidity.

WHERE IT'S GROWN AND WHAT ITS WINE TASTES LIKE

The variety's high acidity makes it suitable for sweet wines and the intense aromas contribute usefully to an otherwise neutral variety or blend. Wines are said to have aromas of apricot, peach and citrus and to lack any foxy flavours, provoking increasing interest in this variety among the more recent cold-hardy American hybrids in the Midwest and Northeast. There were around 160 acres/65 ha of La Crescent in **Minnesota** in 2007 and small areas in **Iowa** (19 acres/8 ha in 2006), **Nebraska** (2 acres/0.8 ha in 2007), **Michigan** and even in Ontario, **Canada**. Producers of varietal wines include Falconer, Fieldstone, Parley Lake and St Croix Vineyards in Minnesota; and Lincoln Peak in **Vermont**.

LA CROSSE

Complex American hybrid starting to achieve some success in the US Midwest.

BERRY COLOUR: ● ● ● ● ●

PRINCIPAL SYNONYMS: ES 294, LaCrosse

ORIGINS AND PARENTAGE

La Crosse is an Elmer Swenson 114 × SEYVAL BLANC hybrid obtained around 1970 by Elmer Swenson (1913–2004), a grape breeder from Osceola, Wisconsin, US (see ST PEPIN for the pedigree of Elmer Swenson 114). A very complex *Vitis riparia*, *Vitis labrusca*, *Vitis vinifera*, *Vitis lincecumii* and *Vitis rupestris* hybrid, La Crosse was released in 1983, presumably named after the city of that name in western Wisconsin. It is a sibling of ST PEPIN.

VITICULTURAL CHARACTERISTICS

Small to medium-sized compact clusters of thin-skinned berries. Vigorous, mid budding and mid ripening. Highly susceptible to black rot and botrytis bunch rot. Relatively hardy (–15 °F/–26 °C to –20 °F/–29 °C; Smiley 2008).

WHERE IT'S GROWN AND WHAT ITS WINE TASTES LIKE

Aromas of La Crosse wines range from pear and apricot to citrus and floral, making them good as varietals but also as a blending component that adds body to lighter varieties. The most successful styles are dry and oak aged or fruity and semi-sweet. The variety is planted mainly in the Midwest, eg in **Iowa** (33 acres/13 ha in 2006), **Nebraska** (around 28/11 ha acres in 2007), **Minnesota** (16 acres/6 ha in 2007), **Illinois** (10 acres/4 ha in 2007), but also to a very limited extent in **Pennsylvania** and even **Nevada**. Producers of varietal wines include Breezy Hills, Prairie Crossing and Two Saints in Iowa; Cuthills and Glacial Till in Nebraska; Long Creek in Illinois. Tahoe Ridge in Nevada even produce a La Crosse Ice Wine.

LACRIMA DI MORRO D'ALBA

Recovered, very local Marche speciality used for both dry and sweet early-maturing reds

BERRY COLOUR: ● ● ● ● ●

PRINCIPAL SYNONYMS: Lacrima

ORIGINS AND PARENTAGE

The name Lacrima (English 'teardrop') was probably given to this variety because when the berries are fully ripe, they exude small drops of juice. However, several distinct varieties in various regions of central and southern Italy (Campania, Marche, Toscana, Umbria, Puglia, etc.) were given names based on Lacrima. In the Marche, Lacrima is a local variety of the Morro d'Alba region mentioned as early as the eighteenth century. DNA studies suggest a close relationship between Lacrima di Morro d'Alba and ALEATICO (Filippetti, Silvestroni, Thomas and Intrieri 2001).

The table grape Lacrima di Maria is not related to Lacrima.

VITICULTURAL CHARACTERISTICS

Vigorous and susceptible to botrytis bunch rot.

WHERE IT'S GROWN AND WHAT ITS WINE TASTES LIKE

In the twentieth century, Lacrima di Morro d'Alba was close to extinction but was rescued in 1985 when it was finally included in the national register of varieties in **Italy** for the Lacrima di Morro d'Alba DOC, in which it must comprise at least 85% (potentially blended with SANGIOVESE, MONTEPULCIANO or even the white VERDICCHIO BIANCO). Its total area jumped from 1 ha (2.5 acres) in 1985 to 102 ha (252 acres) in 2000 and it is now cultivated in the following villages in the province of Ancona in the Marche: Morro d'Alba, Monte San Vito, San Marcello, Belvedere Ostrense, Ostra and Senigallia. The wines are scented with wild strawberries, sometimes darker red fruit, may be dry or sweet (passito) and tend to mature quickly. Tannins are usually fine and the acidity refreshing. Recommended producers include: Colonnara, Conti di Buscareto, Piergiovanni Giusti, Luciano Landi, Stefano Mancinelli, Marotti Campi.

As well as being the dominant variety in Lacrima di Morro d'Alba, the variety is authorized as part of the blend in the dry red Colli Maceratesi DOC.

The Italian agricultural census of 2000 recorded 722 ha (1,784 acres) of Lacrima but this is misleading because it includes varieties other than Lacrima di Morro d'Alba.

This variety is quite separate from the Lacryma Christi del Vesuvio DOC in Campania, which is made from PIEDIROSSO (minimum 80%) sometimes blended with SCIASCINOSO and/or AGLIANICO.

LADO

Minor, high-acid Galician variety.

BERRY COLOUR: ● ● ● ● ●

ORIGINS AND PARENTAGE

Lado comes from Galicia in north-west Spain and, according to ampelographic observations and DNA profiling, it could be related to TRAJADURA (under the name Treixadura) and VIOSINHO (Pinto-Carnide *et al.* 2003; Martín *et al.* 2006; Gago *et al.* 2009).

VITICULTURAL CHARACTERISTICS

Mid to late budding, mid ripening. Small, compact bunches of medium-sized berries. Highly susceptible to botrytis bunch rot and powdery mildew, less so to downy mildew.

WHERE IT'S GROWN AND WHAT ITS WINE TASTES LIKE

Plantings of Lado are extremely limited (just 1 ha/2.5 acres recorded in **Spain** in 2008) but is currently the subject of research and re-evaluation. It was added to the official register of varieties in 2007 and has experimental status in the Ribeiro DO in Galicia in the far north west, where it is a component of blends in the Val do Arnoia. It is also permitted in neighbouring Valdeorras. Wines are high in acidity, can reach respectable levels of alcohol but are low in dry extract and are therefore very useful for adding freshness to a blend but less suitable as varietal wines.

LAFNETSCHA

Old but quantitively unimportant Swiss variety producing well-structured, ageworthy whites.

BERRY COLOUR: ● ● ● ● ●

PRINCIPAL SYNONYMS: Laffnetscha, Lafnätscha, Lavenetsch

VARIETIES COMMONLY MISTAKEN FOR LAFNETSCHA: COMPLETER ☒ (Graubünden, Haut-Valais)

ORIGINS AND PARENTAGE

Lafnetscha is indigenous to the Haut-Valais in southern Switzerland, where it was mentioned as early as 1627 under the name Lachneschen (Vouillamoz and Moriondo 2011). Its name is the dialect form of the German *laff nicht schon*, meaning 'don't drink too early', probably referring to the wine's high natural acidity, which benefits from ageing.

DNA parentage analysis has shown that Lafnetscha is a natural cross between HUMAGNE from the Valais in the south and COMPLETER from Graubünden in the south east (Vouillamoz, Maigre and Meredith 2004; see Completer for the pedigree). Old Completer vines were found near Visp in the Haut-Valais after this parentage had been discovered (Vouillamoz, Frei and Arnold 2008). Lafnetscha is therefore a half-sibling of BONDOLETTA, HIMBERTSCHA and HITZKIRCHER and a likely grandchild of COLOMBAUD from Provence.

VITICULTURAL CHARACTERISTICS

Mid budding and ripening. Medium-sized, thick-skinned berries.

WHERE IT'S GROWN AND WHAT ITS WINE TASTES LIKE

The best example of Lafnetscha, made by Chanton, is dry with notes of green apple, pear and elderflower, a dense structure, high acidity and a lingering floral finish. The only two other producers are Gregor Kuonen and Papillon and there were fewer than 2 ha (5 acres) in the Haut-Valais, **Switzerland**, in 2009. Lafnetscha's structure gives it good ageing potential.

LAGARINO BIANCO

Old but only recently registered crisp Trentino white wine grape.

BERRY COLOUR: ● ● ● ● ●

PRINCIPAL SYNONYMS: Chegarèl (Trentino dialect), Sghittarella (Trentino dialect)

ORIGINS AND PARENTAGE

Lagarino Bianco most probably comes from the Val Lagarino in the province of Trento in northern Italy. DNA profiling disproves any direct parent–offspring relationship between Lagarino Bianco and LAGREIN (Vouillamoz).

OTHER HYPOTHESES

It has been suggested that Lagarino Bianco corresponds to the 'weiss Lagrein' mentioned in the Middle Ages in the Alto Adige (see LAGREIN).

VITICULTURAL CHARACTERISTICS

Early ripening. Good resistance to winter cold and to fungal diseases. Vigorous growth and high yields.

WHERE IT'S GROWN AND WHAT ITS WINE TASTES LIKE

Lagarino Bianco has historically been cultivated mainly in the Val di Cembra, north east of Trento, **Italy**, and to a lesser extent in Valsugana near Pergine. Thanks to studies carried out at the research centre in San Michele all'Adige, Lagarino Bianco was added to the Italian national register of varieties, together with Verdealbara (a local synonym for ERBAMAT), in 2007. Varietal Lagarino Bianco wines, such as Alfio Nicolodi's crisp, lemony Cimbrus, are few and far between.

LAGORTHI

See VERDECA.

LAGREIN

Very well-connected, distinctively fruity but sometimes rustic feature of the Alto Adige.

BERRY COLOUR: ● ● ● ● ●

PRINCIPAL SYNONYMS: Lagrain

ORIGINS AND PARENTAGE

The earliest mention of Lagrein refers, surprisingly, to a white wine: in a document dated 1318 in Gries near Bolzano, people were told to give some white Lagrein ('weiss Lagrein') to the poor or to the church. In 1370, Emperor Karl IV pronounced 'Lagreiner' to be the best wine of the area, but it is not known if it was white or red. Several references to white Lagrein appear later at different locations: Tramin (1379), Eppan (1497), Schenna (1507), Kaltern and Terlan (1532), etc. (Scartezzini 2005). The true identity of this white Lagrein has not been discovered: some have suggested it could be SAVAGNIN BLANC (called Traminer in this area), others have suggested it could be the LAGARINO BIANCO cultivated in Cembra, Faver and Grumes (Trentino), or even that the name 'weiss Lagrein' was simply used for a wine blend (Barbara Raifer, personal communication).

Even though the designation 'weiss Lagrein' presumes the existence of a dark-berried version, the earliest mention of a red Lagrein, 'rott Lagrein', dates back to 1526 (Scartezzini 2005). It is most likely to come from the Val Lagarina in Trentino in northern Italy, from which its name probably derives.

Recent DNA parentage analysis and reconstruction of a complex pedigree (Vouillamoz and Grando 2006) have established that Lagrein is:

- a natural cross between TEROLDEGO and an unknown variety (although see below in Other hypotheses)
- a sibling of MARZEMINO (see the pedigree diagram for REFOSCO DAL PEDUNCOLO ROSSO)
- a nephew/niece of DUREZA, the parent of SYRAH
- a grandchild of PINOT
- a cousin of SYRAH.

This disproves the supposed Greek or Albanian origin of Lagrein, which can be considered to be an ancient and autochthonous variety of the Alto Adige.

OTHER HYPOTHESES

There has been speculation that Lagrein was domesticated from local wild grapes but the recently discovered parentage powerfully disproves this hypothesis. Others have suggested a Greek origin, based on the similarity between the name Lagrein and the Greek word *lagaros*, meaning 'clear' or 'bright', or on the existence of Lagarinos, a wine made in the ancient city of Lagaria on the Ionian coast in the district then known as Lucania (now Basilicata) in southern Italy. From there, the grape is said to have moved northwards to the Alto Adige. This too is rejected by DNA discoveries. More recently, Cipriani *et al.* (2010) have proposed that Lagrein is a natural cross between TEROLDEGO and SCHIAVA GENTILE from the Alto Adige. However, this parentage is not verified at two out of the thirty-nine DNA markers analysed, a discrepancy level that has also been observed in non parent–offspring pairs by Vouillamoz *et al.* (2006), and the proposed parentage is not consistent with the pedigree in Vouillamoz *et al.* (2006) based on sixty DNA markers, because MARZEMINO, a sibling of Lagrein, could not be a progeny of Schiava Gentile (see the PINOT pedigree).

VITICULTURAL CHARACTERISTICS

Quite vigorous and may be subject to poor fruit set in some seasons. Needs a warm site.

WHERE IT'S GROWN AND WHAT ITS WINE TASTES LIKE

Lagrein is cultivated mainly in the Alto Adige and in Trentino in the far north of **Italy**. It is often blended with SCHIAVA GROSSA but varietal wines are permitted in the Alto Adige and Trentino DOCs. Rosé wines are labelled Rosato or Kretzer, red wines Lagrein or Lagrein Dunkel or Lagrein Scuro. The reds are deeply coloured, have intense berry-fruit flavours and, although the tannin levels are not necessarily high, they can be slightly rustic. Some producers have shortened maceration times and used oak barrels to make the wines more rounded. They quite often seem to have a slight taste of iodine. They generally benefit from a year or two in bottle and are sweeter, more Germanic and less savoury than many Italian reds. Recommended producers include Barone de Cles Maso Scari, Kellerei Gries (who use some American oak), Hofstätter, Alois Lageder, Manincor, Muri-Gries and Tiefenbrunner.

Total plantings of Lagrein in the Italian agricultural census in 2000 were 477 ha (1,179 acres). The Alto Adige recorded 412 ha (1,018 acres) in 2008, making it the region's second most widely planted dark-skinned variety, some way behind SCHIAVA GROSSA.

A number of **California** producers bottle the Lagrein grown in Paso Robles, notably Imagery and Tobin James. The state's total was 77 acres (31 ha) in 2008.

In **Australia** there are currently fifteen or so producers who make a Lagrein, generally grown in cooler climates than that of its homeland, but this does not seem to result in herbaceous flavours. Producers include Cobaw Ridge at Keyneton in Victoria, Domaine Day at Mt Crawford, South Australia, and Bassham Wines in the Riverland. Tannin levels are generally lower than in the Alto Adige.

LAIRÉN

Very minor Andalusian variety often mistaken for the much more common Airén.

BERRY COLOUR: ● ● ● ● ●

PRINCIPAL SYNONYMS: Laeren del Rey (Arcos, Espera, Paxarete), Layrenes (Tarifa, Ronda), Malvar✗, Mantuo Laéren or Mantuo Layrenes (Sanlúcar de Barrameda, Jerez de la Frontera, Trebujena, Arcos, Espera, Paxarete), Temprana Agosteña✗ (Cigales)

VARIETIES COMMONLY MISTAKEN FOR LAIRÉN: AIRÉN✗ (La Mancha)

ORIGINS AND PARENTAGE

Lairén is a very old variety from Andalucía in southern Spain that was mentioned by the Spanish agronomist Gabriel Alonso de Herrera (1470–1539) in 1513 under the name Layrenes in his description of the varieties of Castilla, Extremadura and Andalucía (Alonso de Herrera 1790). The name Lairén possibly derives from the Mozarabic name of a place in the region of Sevilla in Andalucía (Favà i Agud 2001). Lairén and AIRÉN from Castilla-La Mancha have often been confused but DNA profiling has shown that they are unquestionably distinct (Ibañez *et al.* 2003; Santana *et al.* 2010).

According to Schneider *et al.* (2010), Lairén could have a parent–offspring relationship with Legiruela, a variety from Valladolid that is identical to PRIÉ, which has been cultivated in the Valle d'Aosta for centuries (see the complete pedigree of Prié).

VITICULTURAL CHARACTERISTICS

Large bunches of thin-skinned, medium-sized berries. Early budding, mid to late ripening. Fertile, productive and very vigorous. Highly resistant to drought, adapts to most soils but prefers those with more clay. Best pruned long. Very resistant to most pests and diseases but sometimes at risk from spring frosts.

WHERE IT'S GROWN AND WHAT ITS WINE TASTES LIKE

It is hard to find information on Lairén because it is so often confused with AIRÉN, although the former is more likely to be found in Andalucía, southern **Spain**, whereas the latter is much more likely to be in La Mancha in the centre of the country. Lairén is permitted in the Montilla-

Moriles, Málaga and Sierras de Málaga DOs and may be blended with other local varieties such as PEDRO XIMÉNEZ and Baladi (aka Pardina or CAYETANA BLANCA) to make early-drinking dry whites, but it does not feature in the official Spanish statistics.

However, under the Spanish synonym Malvar, information is easier to come by. It is planted to a limited extent across central Spain: a total of 266 ha (657 acres) in 2008, around three-quarters of that in the Madrid area, where it is one of the principal varieties in the Vinos de Madrid DO.

Wines are greenish in colour, high in alcohol, with a light aroma of bitter almonds and medium to high acidity and may be bottled as varietal wines or blended with AIRÉN, for example, contributing acidity and aroma. Malvar is also used in the production of late-harvest sweet wines and may be fermented or aged in oak. Vinos Jeromín produce two varietal examples: Puerta de Alcalá Blanco and the barrel-fermented Puerta del Sol Blanco. Laguna make a good late-harvest example. Other producers of varietal wines include Jesús Díaz, Bodegas Orusco and Bodegas Peral.

LALVARI

Versatile Armenian white wine grape.

BERRY COLOUR:

PRINCIPAL SYNONYMS: Burra, Dana Bouroun, Dana Burnu, Dana Burun, Glglan

ORIGINS AND PARENTAGE

Lalvari possibly comes from the Tavush region in north-east Armenia.

VITICULTURAL CHARACTERISTICS

Mid to late budding and ripening. Medium-sized to large, compact bunches of big, thick-skinned berries. Susceptible to downy and powdery mildews and to botrytis bunch rot.

WHERE IT'S GROWN AND WHAT ITS WINE TASTES LIKE

Lalvari is planted mainly in the province of Tavush, north-eastern **Armenia**, where it is used to produce a range of styles, from light and sparkling to dry, medium dry and oaked. Unlike most other Armenian varieties, it is rarely used to make brandy. Wines tend to be fresh, relatively neutral, with moderate alcohol. Producers include 365 Wines, Brest and Ijevan Wine Factory.

LAMBRUSCA or LAMBRUSCO

BERRY COLOUR:

ORIGINS AND PARENTAGE

Many distinct Italian varieties are called Lambrusca or Lambrusco or Lambrusco Something, especially in Piemonte and in Emilia-Romagna. As the name *lambrusco* means 'wild grape' in

Italian, it has often been supposed that all Lambruscos were domesticated from local wild vines, which seems to be confirmed by recent genetic evidence (Schneider, Torello Marinoni *et al.* 2009).

LAMBRUSCA DI ALESSANDRIA

Adaptable but minor dark-skinned variety grown in parts of Piemonte, north-west Italy, and generally blended.

BERRY COLOUR:

PRINCIPAL SYNONYMS: Anrè (Acqui), Caruét (Monferrato Casalese), Croetto (Alessandria), Crova (Pinerolo), Crovet (Asti), Crovìn (Roero), Moretto (Alessandria), Lambrusco di Alessandria, Nebbiolo de Baja (Mexico), Neretto di Alessandria (Val Chisone), Stupèt (Asti)

VARIETIES COMMONLY MISTAKEN FOR LAMBRUSCA DI ALESSANDRIA: CÉSAR (northern France), CORVA

ORIGINS AND PARENTAGE

Lambrusca di Alessandria most likely – and most logically – comes from the province of Alessandria in Piemonte. It was first mentioned by Nuvolone Pergamo, Conte di Scandaluzza (1787–98), under the name Crovet, cultivated in the Casale hills and in the province of Asti. Some years later, Acerbi (1825) describes five distinct 'Lambrusche', two of which are named 'Crouet' and are similar to Lambrusca di Alessandria (Torello Marinoni *et al.* 2006). The first occurrence of the name Lambrusca di Alessandria was in 1879 in *Ampelografia italiana*.

DNA profiling has suggested likely parent–offspring relationships between Lambrusca di Alessandria and two autochthonous varieties of the province of Alessandria in Piemonte: Neretto di Marengo (not commercially cultivated) and MALVASIA DI CASORZO (Torello Marinoni *et al.* 2006).

OTHER HYPOTHESES

It has sometimes been suggested that Lambrusca di Alessandria is identical to the French variety CÉSAR, but this can be disproved by DNA analysis (Vouillamoz).

VITICULTURAL CHARACTERISTICS

A robust vine, adaptable to a range of sites and high yielding. Buds late so generally safe from spring frosts. Ripens late; at maturity, the berries tend to fall off, hence the synonyms Caruét, Crova, Crovìn (*crua* is Italian for 'unripe'). Good resistance to downy mildew and botrytis bunch rot.

WHERE IT'S GROWN AND WHAT ITS WINE TASTES LIKE

Lambrusca di Alessandria has historically been cultivated in the province of Alessandria in Piemonte, **Italy**, but also in the provinces of Asti and, to a lesser extent, Cuneo. It was later introduced in the province of Torino (in the regions of Pinerolo, Val Chisone, Val Germanasca, Ivrea and Val di Susa) after phylloxera (Schneider and Mannini 2006) and it is authorized throughout Piemonte, where it is usually blended with DOLCETTO, BARBERA, FREISA or BONARDA PIEMONTESE. It is specifically mentioned as a possible blending component in Colli Tortonesi DOC (red and rosé). There were 900 ha (2,224 acres) of Lambrusca di Alessandria in Italy in 2000, according to the agricultural census. It is not available as a varietal.

LAMBRUSCA VITTONA

Very minor Piemontese variety.

BERRY COLOUR: ● ● ● ● ●

PRINCIPAL SYNONYMS: Vittona

ORIGINS AND PARENTAGE

Lambrusca Vittona is a typical variety of the Pinerolo area in the province of Torino, north-west Italy, possibly first mentioned as Lambrusca in Pinerolo in 1834, although there is some doubt as to whether it was Lambrusca Vittona (Torello Marinoni *et al.* 2006).

DNA parentage analysis has suggested a parent–offspring relationship between Lambrusca Vittona and Berla Grossa, a local and almost extinct variety of the Pinerolo region.

VITICULTURAL CHARACTERISTICS

Mid ripening. High yielding but grapes are small. Good resistance to sour rot and botrytis bunch rot.

WHERE IT'S GROWN AND WHAT ITS WINE TASTES LIKE

Historically well distributed in Cumìana, Bricherasio, San Secondo di Pinerolo and Prarostino in the province of Torino, north-west **Italy**, Lambrusca Vittona is today also cultivated in Barge, the Val Chisone and the Val Germanasca, where it is occasionally used in a blend with AVANÀ, AVARENGO and Neretto (a name for several distinct varieties), plus up to 35% of other local red varieties, to produce a wine known as Ramìe (Schneider and Mannini 2006). Lambrusca Vittona is not listed in the Italian agricultural census of 2000 because it is not officially authorized for the region. Its inclusion in Ramìe is a classic example of the creative use of synonyms to get round the regulations. There are just five producers of Ramìe. Lambrusca Vittona is thus generally declared as LAMBRUSCA DI ALESSANDRIA, another very rare local (but authorized) variety, and CHATUS is declared as Neretto!

LAMBRUSCHETTO

Old and very localized variety arousing current interest in Piemonte.

BERRY COLOUR: ● ● ● ● ●

PRINCIPAL SYNONYMS: Crovino, Lambruschetta, Malaga (Tortona region)

ORIGINS AND PARENTAGE

Lambruschetto is a typical variety of south-east Piemonte, in particular around Castelnuovo Bormida (Schneider and Mannini 2006). Its first mention may be as early as 1831, when Italian ampelographer Gallesio described this variety under the synonym Crovino, while the first use of the name Lambruschetto is found in Demaria and Leardi (1875). These authors suggested that Lambruschetta was identical to Lambruschetta di Sorbara (ie LAMBRUSCO DI SORBARA) but this has been disproved by DNA analysis (Torello Marinoni *et al.* 2006). A possible parent–offspring relationship has been established between Lambruschetto and TIMORASSO, an old white wine variety from south-east Piemonte (Torello Marinoni *et al.* 2006).

VITICULTURAL CHARACTERISTICS

Mid to late ripening. Resistant to botrytis bunch rot but prone to millerandage.

WHERE IT'S GROWN AND WHAT ITS WINE TASTES LIKE

Formerly cultivated in the provinces of Asti and Alessandria, north-west **Italy**, Lambruschetto is today cultivated almost exclusively in Castelnuovo Bormida in the province of Alessandria, where there have recently been some new plantings.

LAMBRUSCO A FOGLIA FRASTAGLIATA

See ENANTIO.

LAMBRUSCO BARGHI

Very minor, recently recognized, dark-skinned blending component in Reggio.

BERRY COLOUR: ● ● ● ● ●

PRINCIPAL SYNONYMS: Lambrusco Bardi, Lambrusco Corbelli, Lambrusco di Rivalta

VARIETIES COMMONLY MISTAKEN FOR LAMBRUSCO BARGHI: MARZEMINO

ORIGINS AND PARENTAGE

Lambrusco Barghi probably comes from the province of Reggio Emilia, where it is also called Lambrusco Corbelli or Lambrusco di Rivalta because until the 1960s it was widely cultivated in the Castelnuovo di Sotto and Rivalta properties of the Conte Corbelli.

OTHER HYPOTHESES

A Tuscan origin has been suggested. Some have considered Lambrusco Barghi to be identical, or closely related, to MARZEMINO but DNA profiling has disproved this theory (Boccacci *et al.* 2005).

VITICULTURAL CHARACTERISTICS

High yielding and mid ripening.

WHERE IT'S GROWN AND WHAT ITS WINE TASTES LIKE

Lambrusco Barghi entered **Italy**'s national register of varieties as recently as 2009 as a component in the Reggiano Lambrusco DOC. The berries attain high sugar levels and the wines are deeply coloured, thanks to a high level of anthocyanins. Wine is especially aromatic.

LAMBRUSCO DI FIORANO

Obscure, highly localized Emilia-Romagna variety unrelated to other Lambruscos.

BERRY COLOUR: ● ● ● ● ●

PRINCIPAL SYNONYMS: Lambruscone

ORIGINS AND PARENTAGE

Lambrusco di Fiorano takes its name from the commune of Fiorano in the province of Modena. Recent studies have shown that Lambrusco di Fiorano is morphologically and genetically distinct from other Lambruscos of the Emilia-Romagna region (Boccacci *et al.* 2005).

VITICULTURAL CHARACTERISTICS

High yielding. Good sugar/acid balance in the must.

WHERE IT'S GROWN

Still extensively planted on the higher slopes in the commune of Fiorano, Emilia-Romagna, **Italy**.

LAMBRUSCO DI SORBARA

Perhaps the most respected of the varieties responsible for Lambrusco, Emilia-Romagna's potentially refreshing, lightly frothy, food-friendly red.

BERRY COLOUR: ● ● ● ● ●

PRINCIPAL SYNONYMS: Lambruschetta di Sorbara, Lambrusco Sorbarese

ORIGINS AND PARENTAGE

Lambrusco di Sorbara is a typical variety of the area between the Secchia and Panaro rivers in the province of Modena in Emilia-Romagna. It is named after the small village of Sorbara in the commune of Bomporto.

VITICULTURAL CHARACTERISTICS

Functionally female flowers, so the vines need a pollinator, a role often played by LAMBRUSCO SALAMINO. Mid to late ripening.

WHERE IT'S GROWN AND WHAT ITS WINE TASTES LIKE

Lambrusco di Sorbara is cultivated mainly in the province of Modena, northern central **Italy**, in particular in Sorbara, Comporto, San Prospero and Nonantola. It is also cultivated in the province of Reggio Emilia, near San Martino, in Rio, Rubiera and Correggio. As well as being the dominant variety in the Lambrusco di Sorbara DOC (usually blended with 40% LAMBRUSCO SALAMINO), the variety is an authorized component in DOCs such as Reggiano Lambrusco, Reggiano Lambrusco Salamino and Lambrusco Mantovano (in Lombardia). There were 1,460 ha (3,608 acres) of Lambrusco di Sorbara in Italy in 2000.

The wines, which are usually dry or off dry and should be drunk young, are deep cherry red, lightly sparkling and have bright, zesty red fruit flavours and high acidity that makes them ideal with the rich pork-based dishes of Emilia-Romagna. Smaller producers making high-quality Lambrusco di Sorbara that contradict the widespread image of Lambrusco as a sweet industrialized fizz include Cavicchioli, Chiarli 1860 and Villa di Corlo.

LAMBRUSCO GRASPAROSSA

Makes fuller, more tannic Lambrusco than most, both dry and medium sweet.

BERRY COLOUR: ● ● ● ● ●

PRINCIPAL SYNONYMS: Lambrusco di Castelvetro, Scorzamara🧬 (Coviolo)

ORIGINS AND PARENTAGE

Traditionally cultivated in the commune of Castelvetro in Emilia-Romagna, where Lambrusco Grasparossa was first mentioned in 1867 (Giavedoni and Gily 2005). DNA profiling has established that Scorzamara from Coviolo, a rare black variety first recorded in the eighteenth century, is identical to Lambrusco Grasparossa (Boccacci *et al.* 2005).

VITICULTURAL CHARACTERISTICS

Late ripening.

WHERE IT'S GROWN AND WHAT ITS WINE TASTES LIKE

Lambrusco Grasparossa is widespread throughout the province of Reggio Emilia in the region of Emilia-Romagna, **Italy**, and in particular in the communes of Coviolo Scandiano, Casalgrande and Quattro Castella, where it is made as varietal wine in the Colli di Scandiano e di Canossa DOC, either still or sparkling. Under its synonym Scorzamara, it is cultivated to a very limited extent in Coviolo. In the province of Modena, it is grown mainly in the Castelvetro region, where it is the dominant variety (minimum 85%) of the sparkling Lambrusco Grasparossa di Castelvetro DOC. There are also scattered plantings in the provinces of Parma, Mantova and Bologna. The variety is authorized as a minor blending component, usually with other varieties called Lambrusco Something, in Reggiano DOC (various styles) and also in Lombardia's Lambrusco Mantovano DOC (rosé and red sparkling wines). There were 1,897 ha (4,688 acres) of Lambrusco Grasparossa in Italy in 2000.

The wines, which are deeply coloured, fruity and more tannic than most Lambrusco varieties, with good acidity, may be dry or semi-sweet. Lambrusco Grasparossa tends to be fuller-bodied and have higher alcohol than, for example, Lambrusca di Sorbara. Good examples of varietal wines are produced by Corte Manzini, Gavioli Antica Cantina (semi-sweet, blended with 10% MALBO GENTILE), Rinaldini Rinaldo's Azienda Agricola Moro and Villa di Corlo.

LAMBRUSCO MAESTRI

One of the lesser Lambruscos, from Parma, and making rather rustic wines.

BERRY COLOUR: ● ● ● ● ●

PRINCIPAL SYNONYMS: Grappello Maestri, Lambrusco di Spagna

ORIGINS AND PARENTAGE

Already mentioned at the end of the nineteenth century, Lambrusco Maestri probably comes from the small village of Villa Maestri in the commune of San Pancrazio in the province of Parma.

DNA parentage analysis suggested a possible parent–offspring relationship between Lambrusco Maestri and FORTANA (Boccacci *et al.* 2005).

VITICULTURAL CHARACTERISTICS

Vigorous and high-yielding.

WHERE IT'S GROWN AND WHAT ITS WINE TASTES LIKE

Lambrusco Maestri is widespread throughout the province of Reggio Emilia in the region of Emilia-Romagna, **Italy**, in particular in the communes of Montecchio, Boretto and Gualtieri, and it is also planted to a limited extent in the provinces of Parma and Modena. The variety is said to be cultivated in Puglia (Zulini *et al.* 2002), and authorized in the blend for the Orta Nova DOC, although it is not known for sure if it is really the same variety. It can be made as a varietal wine in the Colli di Parma DOC and may be part of the blend in Lambrusco Mantovano DOC and in various types of Reggiano DOC. Dall'Asta's Mefistofele is a deeply coloured, rustic, strawberry-flavoured and tannic sparkling red made from 100% Lambrusco Maestri and classified as Lambrusco dell'Emilia IGT. According to the agricultural census, there were 1,460 ha (3,608 acres) of Lambrusco Maestri in Italy in 2000 – almost exactly the same area as for the more well known and generally more highly regarded LAMBRUSCO DI SORBARA.

According to official vineyard statistics, there were 98 ha (242 acres) of this variety planted in **Argentina** in 2007, in Mendoza and San Juan. Tittarelli, for example, produce a varietal version.

LAMBRUSCO MARANI

Second most widely planted of the Lambrusco vines, making wines with no shortage of colour.

BERRY COLOUR: ● ● ● ● ●

ORIGINS AND PARENTAGE

The earliest mention of Lambrusco Marani was in the mid nineteenth century in Reggio Emilia. DNA profiling has shown that Lambrusco Marani could be closely related to FOGARINA (Boccacci *et al.* 2005).

VITICULTURAL CHARACTERISTICS

Vigorous and productive, susceptible to winter cold, and adapted to damp soils. Mid to late ripening.

WHERE IT'S GROWN AND WHAT ITS WINE TASTES LIKE

Lambrusco Marani is cultivated in the province of Reggio Emilia, Emilia-Romagna, **Italy**, and in particular in Rolo, Rio Saliceto, Fabbrico, Campagnola and Novellara. It is also cultivated to a lesser extent in the provinces of Mantova, Parma and Bologna. It produces lightly sparkling reds – dry and not so dry – that are deeply coloured, fruity, fresh and often tannic. The variety is the mainstay of Reggiano Lambrusco red and rosé and may comprise up to 100% of Reggiano Bianco Spumante. It is authorized as part of the blend in both still and sparkling Reggiano Rosso DOC, generally with ANCELLOTTA, and in Lombardia's Lambrusco Mantovano DOC, as well as in the sparkling reds of Colli di Scandiano e di Canossa blended with LAMBRUSCO MONTERICCO, for example. There was a total of 2,318 ha (5,728 acres) of Lambrusco Marani in Italy in 2000.

LAMBRUSCO MONTERICCO

Minor Lambrusco variety.

BERRY COLOUR: ● ● ● ● ●

ORIGINS AND PARENTAGE

Lambrusco Montericco, first mentioned in the late nineteenth century, takes its name from the small village of Montericco. DNA parentage analysis suggested a possible parent–offspring relationship between Lambrusco Montericco and UVA TOSCA (Boccacci *et al.* 2005).

VITICULTURAL CHARACTERISTICS

High yields, late ripening. Susceptible to botrytis bunch rot, powdery and downy mildews.

WHERE IT'S GROWN AND WHAT ITS WINE TASTES LIKE

Initially cultivated between Broletto and Montericco, Lambrusco Montericco spread to the whole province of Reggio Emilia in Emilia-Romagna, **Italy**. It produces rosé and red wines that are fresh and light-bodied, suitable for still or sparkling, dry or sweet. It is the main variety in the Lambrusco Montericco variant of Colli di Scandiano e di Canossa DOC, usually blended with up to 15% of other varieties called Lambrusco Something. It is also authorized in the blend of white Reggiano Spumante and red or rosé Reggiano Lambrusco DOCs. This variety is not listed in the Italian agricultural census of 2000.

LAMBRUSCO OLIVA

Relatively minor, occasionally bitter Lambrusco variety.

BERRY COLOUR: ● ● ● ● ●

PRINCIPAL SYNONYMS: Lambrusco Mazzone (Reggio Emilia), Olivone (Mantova)

ORIGINS AND PARENTAGE

Lambrusco Oliva, first mentioned in 1886 (Gozzi *et al.* 2002), takes its name from the typically olive-like shape of its berries.

VITICULTURAL CHARACTERISTICS

Productive, late ripening.

WHERE IT'S GROWN AND WHAT ITS WINE TASTES LIKE

Lambrusco Oliva was added to the national register of grape varieties in **Italy** in December 2000 and is now officially accepted for cultivation in the provinces of Modena and Reggio Emilia in Emilia-Romagna. Still thriving in old vineyards in these areas, it is generally blended with LAMBRUSCO MARANI and LAMBRUSCO SALAMINO to produce deeply coloured wines, often sweet or sparkling, that may be bitter if over-extracted. It is authorized as a minor part of the blend in Lambrusco and Rosso versions of Reggiano DOC.

LAMBRUSCO SALAMINO

The most widely planted of the Lambrusco varieties, making the most substantial wines.

BERRY COLOUR: ● ● ● ● ●

PRINCIPAL SYNONYMS: Lambrusco Galassi (red-leaved type, see below), Lambrusco di Santa Croce

ORIGINS AND PARENTAGE

Lambrusco Salamino takes its name from the cylindrical shape of its bunches, which resembles a small salami. It most likely originates from the commune of Santa Croce di Carpi in the province of Modena in Emilia-Romagna.

VITICULTURAL CHARACTERISTICS

The variety shows significant clonal diversity, with up to five different morphological types distinguished by ampelographers: tender, red-leaved, green-leaved, red-stalked, green-stalked (Giavedoni and Gily 2005). Susceptible to esca vine wood disease. Late ripening.

WHERE IT'S GROWN AND WHAT ITS WINE TASTES LIKE

Lambrusco Salamino is cultivated in the provinces of Modena and Reggio Emilia (notably in Correggio, Fabbrico, San Martino in Rio and Rio Saliceto) in Emilia-Romagna, **Italy**, and also planted to a very limited extent in the provinces of Mantova, Bologna and Ferrara. It is the most widely planted Lambrusco variety, with 4,273 ha (10,559 acres) recorded in Italy in 2000. It is also planted to a limited extent on the island of Sardegna.

In addition to being the main constituent in Lambrusco Salamino di Santa Croce DOC (minimum 90% blended with up to 10% ANCELLOTTA) and in the Lambrusco Salamino version of the Reggiano DOC (minimum 85%), Lambrusco Salamino plays an important part in Lambrusco di Sorbara DOC and several other DOCs such as Lambrusco Mantovano, Colli di Scandiano e di Canossa and Reggiano. The wines may be dry or sweet, intensely coloured with fruity aromas and, in the best examples, their naturally high level of tannin is offset by fresh acidity. The wine is quite high in alcohol if fermented dry. This is generally the most full-bodied and aromatic of all the Lambruscos.

Cavicchioli & Figli's Tre Medaglie is a good off-dry example of Lambrusco Salamino di Santa Croce and Chiarli make dry, off-dry and sweet versions within the same DOC.

LAMBRUSCO VIADANESE

Minor but characterful Lambrusco variety.

BERRY COLOUR:

PRINCIPAL SYNONYMS: Grappello Ruperti, Mantovano, Montecchio, Viadanese

ORIGINS AND PARENTAGE

Lambrusco Viadanese takes its name from the commune of Viadana in the province of Mantova in Emilia-Romagna, where it is also called Grappello Ruperti, after the agronomist Ugo Ruperti, who considered it to be one of the province's best varieties and encouraged its planting.

VITICULTURAL CHARACTERISTICS

Vigorous, consistently productive and late ripening.

WHERE IT'S GROWN AND WHAT ITS WINE TASTES LIKE

Lambrusco Viadanese is cultivated mainly in the provinces of Mantova and Cremona in Lombardia, **Italy**. The wines are ruby coloured, tannic and robust and have intense aromas of cherries and violets. It is authorized on its own or in a blend in the Lambrusco Mantovano DOC. It is also permitted in the blend of Rosso and Lambrusco versions of Reggiano DOC.

There were just 294 ha (726 acres) of Lambrusco Viadanese recorded in the Italian agricultural census in 2000 and it is less significant and much less widely planted than, say, LAMBRUSCO SALAMINO or LAMBRUSCO DI SORBARA but its wines can be much more interesting than the average Lambrusco.

LANDAL

Disease-resistant, waning French hybrid.

BERRY COLOUR:

PRINCIPAL SYNONYMS: Landot 244

ORIGINS AND PARENTAGE

A complex hybrid obtained between 1929 and 1942 by Pierre Landot in Conzieu in the Ain, eastern France, by crossing Seibel 5455 (also known as PLANTET) × Seibel 8216, in which Seibel 8216 is a Seibel 5410 × Seibel 5001 hybrid (see FLORENTAL for the pedigree of Seibel 5410), Seibel 5001 is a Seibel 2510 × Seibel 867 hybrid, the latter being a Vivarais × NOAH hybrid (see PRIOR for the pedigrees of Seibel 2510 and Vivarais). Landal was named after its breeder.

VITICULTURAL CHARACTERISTICS

Early budding and ripening. Vigorous and fertile and needs to be grafted even though it has American vines in its pedigree. Reasonably cold-hardy. Small bunches and berries.

WHERE IT'S GROWN AND WHAT ITS WINE TASTES LIKE

There were 49 ha (121 acres) in **France** in 2008, mainly in the Loire and the Vendée, but it is definitely a dying breed. It is authorized in many French wine-growing regions but recommended in none, being known more for its cold hardiness and good level of disease resistance (Basler 2003) than for the quality of the wine it produces. It is also planted to a minuscule extent in the canton of Genève in **Switzerland** (0.06 ha/0.15 acres in 2009). Naylor Wines in **Pennsylvania** make a light Landal/CHAMBOURCIN blend. It's also in **Canada**, both in Ontario, where it was first planted in the 1980s, and Québec, where Carone use Landal in two of their red blends.

LANDOT NOIR

Minor French hybrid, son of Landal, also found in a few pockets in the US and Canada.

BERRY COLOUR: ● ● ● ● ●

PRINCIPAL SYNONYMS: Landot 4511

ORIGINS AND PARENTAGE

A complex hybrid obtained between 1929 and 1942 by Pierre Landot in Conzieu in the Ain, eastern France, by crossing LANDAL × VILLARD BLANC. It was used to breed FRONTENAC.

VITICULTURAL CHARACTERISTICS

Reasonably cold-hardy and early to mid ripening.

WHERE IT'S GROWN AND WHAT ITS WINE TASTES LIKE

Although this is a French hybrid, it is not included in the official French register of varieties, but it seems to have found a little home in North America. In **Canada**, where it was introduced in the mid 1950s, Carone in Québec include Landot Noir (as well as LANDAL and CABERNET SEVERNY) in their FRONTENAC blend. In the north-eastern US, where it arrived a few years later, Jewell Towne in **New Hampshire** make a varietal wine, as do Newport Vineyards in **Rhode Island** and Farinacci in **Ohio**. Wines are generally soft and lightly spicy. It is also planted to a very limited extent in **Illinois**.

LANZESA

Very rare white wine speciality of Ravenna in northern Italy.

BERRY COLOUR: ● ● ● ● ●

PRINCIPAL SYNONYMS: Lanzés, Lanzesca

ORIGINS AND PARENTAGE

Lanzesa's origin and etymology are unknown.

VITICULTURAL CHARACTERISTICS

Vigorous and susceptible to powdery mildew. Fairly resistant to botrytis bunch rot.

WHERE IT'S GROWN AND WHAT ITS WINE TASTES LIKE

This variety was once used for both table and wine grapes. Today it is cultivated for wine on fewer than 5 ha (12 acres) in the communes of Brisighella, Faenza and Castel Bolognese in the province of Ravenna in Emilia-Romagna, **Italy**. The wines are dry, floral and fruity with a touch of bitterness. Tenuta Uccellina's Alma Luna is one example.

LAPA KARA

Ukrainian variety that generally plays a very minor part in traditional sweet red blends.

BERRY COLOUR:

ORIGINS AND PARENTAGE

Lapa Kara, literally 'black hand', is a variety from Solnechnaya Dolina in the Sudak district of Krym (Crimea), southern Ukraine.

VITICULTURAL CHARACTERISTICS

Mid to late budding, late ripening. Compact bunches of very thick-skinned berries.

WHERE IT'S GROWN AND WHAT ITS WINE TASTES LIKE

Lapa Kara's violet-coloured, lightly aromatic and full-bodied wines play a minor part in the sweet red Black Doctor bottlings of Sun Valley (Solnechnaya Dolina) and Massandra from Krym in the very south of **Ukraine**.

LASINA

Very rare vine found only in Dalmacija (Dalmatia) and generally used in blends.

BERRY COLOUR:

PRINCIPAL SYNONYMS: Krapljenica, Kutlarica, Lasin, Lasina Crna, Vlasina

ORIGINS AND PARENTAGE

Lasina possibly comes from the regions of Promina and Stankovaca in northern Dalmacija.

VITICULTURAL CHARACTERISTICS

Late ripening, loose clusters of berries that can reach high sugar levels. Not particularly susceptible to fungal diseases.

WHERE IT'S GROWN AND WHAT ITS WINE TASTES LIKE

A rare variety still cultivated in some parts of the Sjeverna Dalmacija (Northern Dalmatia) wine region in **Croatia**, the only one in which it is officially authorized. It is found mainly around Promina, between Split and Zadar, but is little known because it tends to be used in blends, such as the one produced by Bibich (Lasina plus PLAVINA and BABIĆ). Producers who grow this variety include Ivica Džapo and Marko Duvančić, both in Oklaj. The latter makes perhaps the only varietal wine available.

LAŠKI RIZLING

See GRAŠEVINA.

LAUROT

Recently bred and privately owned disease-resistant Czech dark-skinned hybrid.

BERRY COLOUR: ● ● ● ● ●

PRINCIPAL SYNONYMS: MI 5-106

ORIGINS AND PARENTAGE

Laurot is a Merlan × Fratava hybrid recently bred for disease resistance by Miloš Michlovský, Vilém Kraus, Lubomír Glos, Vlastimil Peřina and František Mádl, in which:

- Merlan is a MERLOT × Seibel 13666 hybrid
- Seibel 13666 is a PLANTET × Seibel 6468 hybrid (see HELIOS for the pedigree of Seibel 6468)
- Fratava is a BLAUFRÄNKISCH × SANKT LAURENT cross.

Pollen was obtained from the Chisinau research centre in the **Republic of Moldova** and the crossing and selection were done in Lednice, Breclav and Perná research centres in the Czech Republic. The variety rights are owned by Vinselekt Michlovský, a breeding company created in 1995 by Miloš Michlovský in Rakvice, southern Morava, in the south east of the Czech Republic. It was included in the official Czech register of varieties in 2004.

VITICULTURAL CHARACTERISTICS

Mid to late budding and mid ripening. Moderate yields and good resistance to downy mildew and botrytis bunch rot, moderate resistance to powdery mildew

WHERE IT'S GROWN AND WHAT ITS WINE TASTES LIKE

Wines are deeply coloured, with fruit flavours ranging from wild cherries to plums and blackcurrants; acidity is moderate and tannins are light. Vinselekt Michlovský make both juice and wine from their organic Laurot vineyards in the **Czech Republic**.

LAUZET

Almost extinct variety in Jurançon, south-west France.

BERRY COLOUR: ● ● ● ● ●

PRINCIPAL SYNONYMS: Laouset, Lauzet Blanc

VARIETIES COMMONLY MISTAKEN FOR LAUZET: PETIT COURBU ☒ (Jurançon)

ORIGINS AND PARENTAGE

Lauzet is an endangered variety from the Béarn region in south-west France, often confused with PETIT COURBU in Jurançon. Its first mention appeared only recently in Viala and Vermorel (1901–10) under the local synonym Laouset. The origin of the name Lauzet is unknown and, curiously enough, Lauzet does not have a Basque name, although for Bisson (2009) it clearly belongs to the Courbu ampelographic group (see p XXVII). According to Lavignac (2001), it is morphologically close to BAROQUE from the Landes and to TANNAT from Madiran.

OTHER HYPOTHESES

The ampelographer Louis Levadoux suggested that Lauzet could be related to TANNAT but this is doubted by Pierre Galet (1990).

VITICULTURAL CHARACTERISTICS

Vigorous and usually pruned long. Small bunches and very small berries. Late ripening.

WHERE IT'S GROWN AND WHAT ITS WINE TASTES LIKE

Lauzet is a minor, now virtually extinct, variety in the Jurançon area of **France**, a permitted 'accessory' variety in the Jurançon and Béarn appellations and cultivated mainly in Monein, Lasseube, Gan and Aubertin. Domaine Cauhapé still have 1 ha (2.5 acres) planted and Domaine Nigri, for example, generally include 10% of Lauzet in their dry white Jurançon, along with 80% GROS MANSENG and 10% CAMARALET DE LASSEUBE. It has the potential to produce lightly spicy wines that have quite high alcohol yet still retain good acidity. However, given the potentially high sugar levels in both Mansengs, Camaralet de Lasseube is of greater value as a blending component, particularly in dry Jurançon. The total area recorded in France for 2006 was less than 2 ha (5 acres).

LEÁNYKA

Widely planted Hungarian white, producing soft, unexciting, frequently blended whites.

BERRY COLOUR: ● ● ● ● ●

VARIETIES COMMONLY MISTAKEN FOR LEÁNYKA: FETEASCĂ ALBĂ (Romania), Leányszőlő (Hungary)

ORIGINS AND PARENTAGE

Leányka, meaning 'maiden', is an old Hungarian variety that has long been considered identical to FETEASCĂ ALBĂ (Galet 2000; Rohály *et al.* 2003) but this has been disproved by DNA profiling (Regner 1999; Galbács *et al.* 2009). Leányka is possibly one of the parents of KIRÁLYLEÁNYKA (Galbács *et al.* 2009). It was used to breed ZEFÍR.

VITICULTURAL CHARACTERISTICS

Mid budding, early ripening, high yields. Small to medium-sized compact bunches of small berries with thin, chewy skins. Winter-hardy and resistant to drought but susceptible to downy and powdery mildews as well as botrytis bunch rot.

WHERE IT'S GROWN AND WHAT ITS WINE TASTES LIKE

Leányka is rather less popular than it once was so that total plantings are down but there were still nevertheless 880 ha (2,175 acres) in **Hungary** in 2008, mostly in Eger and in the neighbouring wine regions of Mátra and Bükk in north-eastern Hungary, with minor plantings further west in Mór and Neszmély and further south in Kunság. When yields are restricted, wines are aromatic – often floral – full-bodied but generally rather soft. Vines planted on calcareous soils yield fresher and therefore better-balanced styles. Many producers use it in blends sold in bulk but some see greater potential and are experimenting with barrel fermentation (Rohály *et al.* 2003). Recommended producers include Tibor Gál, Imre Kaló, Orsolya and Tasmás Pók in Eger, and Maurus in Mór.

LEGIRUELA

See PRIÉ.

LEMBERGER

See BLAUFRÄNKISCH.

LEN DE L'EL

Highly localized and well suited to the production of sweet wines in its native Gaillac.

BERRY COLOUR: ● ● ● ● ●

PRINCIPAL SYNONYMS: Cavaillès (Castres), Cavalier, Len-de-Lel, Len de l'Elh (Toulouse), Lenc dé l'El, Loin de l'Oeil

ORIGINS AND PARENTAGE

Len de l'El is a highly localized variety from Gaillac in the Tarn, south-central France, where it was first mentioned in 1842 (Rézeau 1997). Its rather odd name comes from the Occitan *luènh de l'uèlh* (*loin de l'oeil* in standard French), literally 'far from the eye', the eye here referring to the vine bud. The bunches have long stems so that they are far from where the bud was.

OTHER HYPOTHESES

According to Cousteaux and Plageoles (2001), Len de l'El was possibly domesticated from wild grapevines growing in the forest of Grésigne, between Albi and Montauban in the Tarn, but there is no evidence for this hypothesis.

VITICULTURAL CHARACTERISTICS

Vigorous, early budding and productive. Susceptible to botrytis bunch rot and to mites. Early to mid ripening (before MAUZAC BLANC). If yields are not restricted, leaves may brown in the summer due to a lack of potassium.

WHERE IT'S GROWN AND WHAT ITS WINE TASTES LIKE

Before 1870, Len de l'El was planted on 30% of the Gaillac vineyards of **France** (Lavignac 2001) but it gave way to the higher-yielding Mauzac (see MAUZAC BLANC) after the ravages of phylloxera. Nevertheless, there were still 640 ha (1,581 acres) in the Tarn in 2009 (down from 730 ha/1,804 acres in 2000) and it is today being given a little more attention than in the recent past thanks to its ability to retain acidity better than Mauzac, thus making it better suited to sweeter wines (Strang 2009). It is one of the principal varieties in the Gaillac appellation and may be picked early for fresh, dry whites or left on the vine for sweet versions, offering flavours that range from floral to citrus.

For dry white Gaillac, Len de l'El is usually part of a blend with other local varieties such as MAUZAC BLANC, Mauzac Rose (see MAUZAC NOIR) and MUSCADELLE, sometimes also with SAUVIGNON BLANC. Sweet Gaillac wines, however, are more often 100% Len de l'El. Producers of this style include Robert and Bernard Plageoles at Domaine des Très Cantou, Domaine Gineste, Domaine Mayragues and Domaine Rotier.

LÉON MILLOT

French hybrid producing deeply coloured reds. Popular in regions with short growing seasons.

BERRY COLOUR: ● ● ● ● ●

PRINCIPAL SYNONYMS: Frühe Schwarze (Ahr in Germany), Kuhlmann 194-2, Millot

ORIGINS AND PARENTAGE

A complex hybrid obtained in 1911 by Eugène Kuhlmann in Colmar, Alsace, by crossing Millardet et Grasset 101-14 OP with GOLDRIESLING, where Millardet et Grasset 101-14 OP is a selfling of a *Vitis riparia* × *Vitis rupestris* cross generally used as a rootstock. The variety was named in honour of Léon Millot, president of the Société Vosgienne de Viticulture, and commercialized in 1921.

The same crossing produced Léon Millot's siblings, LUCIE KUHLMANN and MARÉCHAL FOCH, and Léon Millot was used to breed MILLOT-FOCH.

VITICULTURAL CHARACTERISTICS

Early budding and very early ripening, thus a short ripening season. Vigorous and needs to be pruned long. Good overall disease resistance.

WHERE IT'S GROWN AND WHAT ITS WINE TASTES LIKE

Léon Millot's popularity in parts of **France** such as Alsace was mainly due to its ability to add colour to pale reds, though it has received greater attention in the last thirty years from those looking for disease-resistant varieties that require less fungicide (Basler 2003). Since it cannot be used in any appellation wine, the vineyard area is in decline but there were still 85 ha (210 acres) in 2008. It is officially authorized in many French wine regions but recommended in none and has had greater success outside the hexagon, notably in cooler climates with shorter growing seasons where many reds are reluctant to ripen such as eastern **Switzerland** (9 ha/22 acres in 2009), where varietals are produced by Cultiva and Schmidheiny. It is now planted in **Denmark** (4 ha/10 acres in 2010), where it is a popular blending component, and in **Sweden**.

Its early ripening is also appreciated in US Midwest states including **Illinois** (15 acres/6 ha projected for 2009), **Iowa** (14 acres/6 ha in 2006) and **Nebraska** (precisely 528 vines in 2006, apparently). It is also planted to a limited extent in **New York** State and as far north as Nova Scotia in **Canada**. Varietal wines tend to be deeply coloured and are produced by, among others, Keuka Lake in New York's Finger Lakes region, China Bend in **Washington** State and Jewell Towne in **New Hampshire**.

LIATIKO

Idiosyncratic and underrated dark-skinned Cretan variety making aromatic dry and sweet wines.

BERRY COLOUR: ● ● ● ● ●

PRINCIPAL SYNONYMS: Aleatiko, Liatico, Liatis (Zákynthos), Stafili Tu Louliou

VARIETIES COMMONLY MISTAKEN FOR LIATIKO: ALEATICO (Italy), Mavrodiates, Mavroliatis (Serifos in the Kykládes/Cyclades), Mayrodiates (the last three no longer cultivated)

ORIGINS AND PARENTAGE

Liatiko is mostly found on the island of Kríti (Crete), where it possibly originates. Its name is short for *Iouliatiko*, meaning 'of July' (Boutaris 2000), in reference to its early ripening. There is also a smaller-berried clone that local growers rate more highly.

Recent DNA profiling has suggested a possible parent–offspring relationship with KORINTHIAKI (under the name Black Corinth; Myles *et al.* 2011).

OTHER HYPOTHESES

It has been suggested that KORINTHIAKI (under the name Corinthe Noir) is a mutation of Liatiko (Krimbas in Galet 2000) but this can be rejected by DNA profiling (Vouillamoz).

Although Aleatiko is a synonym for Liatiko, the variety is not related to the Italian ALEATICO, as their distinct DNA profiles show (Vouillamoz). Two alternative and less likely etymologies have been proposed for Liatiko: it could be short for *iliatiko*, meaning 'of the sun', or derive from Eleatiko on the island of Sámos, where it is said to originate (Boutaris 2000).

Liatiko is believed to have been one of the few red components of the historical blended wine known as Malvasia but there are many myths and uncertainties about this wine.

VITICULTURAL CHARACTERISTICS

Vigorous, fertile and very productive. Susceptible to downy mildew, sour rot and viruses and sensitive to drought. Early to mid budding and early ripening. Medium-sized compact bunches of medium-sized to large berries.

WHERE IT'S GROWN AND WHAT ITS WINE TASTES LIKE

Liatiko is a thoroughly Cretan variety, used for both dry and sweet wines, and the most widely planted variety on this **Greek** island (2,104 ha/5,199 acres in 2008). Wines are not deeply coloured and tend quickly to lose what colour they have during maturation, which is why producers in Sitía in the east of the island opted to change their appellation regulations so that 20% of the more deeply coloured MANDILARIA must be added to the paler Liatiko. However, the wines have distinct – often floral and delicately spicy – aromas, fresh to moderate acidity and soft tannins and although they show their age, they can age gracefully to produce pale, complex wines not unlike mature burgundy with a hint of mature NEBBIOLO in their perfume.

Yannis Economou, the champion of such styles, releases his initially more tannic wines (his old Liatiko vines, grown ungrafted on the 600-metre Ziros plateau, have smaller berries) around ten years after the harvest, once they are silky and refined. Less idiosyncratic but still relatively traditional dry Liatiko is made by producers such as Boutari, Douloufakis, Silva (Irini Daskalaki) and the Sitía co-operative. Producers more focused on the international market and working outside these appellations may include varieties such as SYRAH or CABERNET SAUVIGNON in the blend.

The Achárnes appellation in the centre of the island permits dry wines only, whereas the

Dafnés appellation (also in the centre) and Sitía in the east allow both dry and sweet wines. The latter, made from raisined grapes, are soft but fresh and the best demonstrate Liatiko charm and fragrance. All the producers mentioned above also make good examples of sweet Liatiko.

LILIORILA

Rare light-skinned Bordeaux cross with potential.

BERRY COLOUR:

ORIGINS AND PARENTAGE

A BAROQUE × CHARDONNAY cross obtained in 1956 by Pierre Marcel Durquéty at INRA (Institut National de la Recherche Agronomique) in Bordeaux, south-west France. It was supposedly named after the Basque *lili hori*, meaning 'yellow flower', with the suffix *-la* probably added by analogy with EGIODOLA (Rézeau 1997).

VITICULTURAL CHARACTERISTICS

Early ripening, generally low yielding. Small bunches and berries.

WHERE IT'S GROWN AND WHAT ITS WINE TASTES LIKE

Liliorila has the potential to produce powerful and aromatic wines with relatively low acidity. Château Plaisance's sweet Cuvée Maëlle, a Vin de Pays du Comté Tolosan from Vacquiers just north of Toulouse, **France**, is a blend of 30% Liliorila with 30% CHENIN BLANC and 40% SÉMILLON. France recorded fewer than 4 ha (10 acres) in 2008.

LIMNIO

Ancient Greek variety that has successfully travelled from the island of Límnos to the mainland and is often a minor partner to Cabernets Sauvignon and Franc.

BERRY COLOUR:

PRINCIPAL SYNONYMS: Kalabaki (Límnos), Kalambaki (Límnos), Kalampaki (Límnos), Lembiotiko, Lemnia, Lemnio

VARIETIES COMMONLY MISTAKEN FOR LIMNIO: LIMNIONA ☒

ORIGINS AND PARENTAGE

Limnio probably originates from the Greek island of Límnos in the northern Aegean, where it is also known by its Turkish name, Kalambaki (Galet 2000; Manessis 2000).

LIMNIONA from Thessalía is not identical to Limnio, the two varieties showing distinct DNA profiles in the Greek Vitis Database.

OTHER HYPOTHESES

Limnio is said to be one of the most ancient varieties still cultivated in **Greece**, allegedly mentioned by Aristotle, Pollux and Hesiod as Lemni (*Λημνι*), used to make the ancient Maronias wine (Mangafa and Kotsakis 1996; Galet 2000; Kourakou-Dragona 2001; Manessis 2000). However, there's no credible botanical evidence that the varieties referred to by the ancient Greeks are the same as modern ones.

VITICULTURAL CHARACTERISTICS

Generous yields, generally late ripening though berries tend to ripen at varying times. Small to medium-sized compact bunches of small berries. Resistant to water stress, susceptible to downy mildew and botrytis bunch rot.

WHERE IT'S GROWN AND WHAT ITS WINE TASTES LIKE

As expected, Limnio is grown on the island of Límnos in the northern Aegean and was the fifth most planted variety in that region in 2008 (but still only 72 ha/178 acres in total). However, it is now also planted in much of northern **Greece** and further south in Thessalía; it seems to produce superior wines in the cooler climes of Makedonía and Thráki (Thrace). Wines typically have fairly high alcohol, moderate to fresh acidity, a distinctive aroma of fresh herbs and unexceptional tannins. The variety has not been granted appellation status on Límnos but on the mainland the Côtes de Meliton appellation in Halkidikí requires 70% Limnio and 30% CABERNET SAUVIGNON and/or CABERNET FRANC; elsewhere in various regional wines the proportion of Limnio may be small compared with those of its French blending partners. Gerovassiliou's Avaton, a regional wine of Epanomí, is a blend of Limnio, Mavroudi (see MAVROUDA) and MAVROTRAGANO, while Tsantali and Domaine Carras blend Limnio with Cabernet Sauvignon (Mount Áthos and Côtes de Meliton respectively).

LIMNIONA

Recently resurrected and promising Greek variety from Thessalía.

BERRY COLOUR: ● ● ● ● ●

PRINCIPAL SYNONYMS: Lemniona

VARIETIES COMMONLY MISTAKEN FOR LIMNIONA: LIMNIO ☒

ORIGINS AND PARENTAGE

Limniona comes from Tyrnavos in the province of Larísa in Thessalía, in the east of central mainland Greece. Although Limniona is often confused with the more widespread LIMNIO, they are completely separate varieties, as their DNA profiles in the Greek Vitis Database show.

VITICULTURAL CHARACTERISTICS

Big, thick-skinned berries. Long growing season, late ripening.

WHERE IT'S GROWN AND WHAT ITS WINE TASTES LIKE

Limniona was recently rescued from virtual extinction by Christos Zafirakis in Tyrnavos, who makes both a rare varietal wine, fermented in large oak casks, and a 50/50 blend with SYRAH. In the past Limniona tended to be found only in field blends with XINOMAVRO. Zafirakis' varietal Limniona, from vines densely planted in 2005 at 100 metres above sea level, is well coloured and spicy, medium-bodied, fresh, fragrant and silky, with good ageing potential. According to Zafirakis, the variety thrives when the days are hot and the nights are cool. In 2009 there were around 10 ha (25 acres) in **Greece** but plantings are increasing. Other producers who have planted Limniona include Domaine Karanika in Amýnteo, western Makedonía, and George Tsibidis at Monemvasia Winery in the southern Pelopónnisos (Peloponnese).

LISEIRET

See GOUAIS BLANC.

LISTÁN

See PALOMINO FINO.

LISTÁN DE HUELVA

Minor variety used mostly for fino-like wines in western Andalucía.

BERRY COLOUR: ● ● ● ● ●

PRINCIPAL SYNONYMS: Listán (Andalucía in Spain), Listán Blanca (Andalucía), Malvasia Rasteiro (Portugal), Manteúdo or Manteúdo Branco or Mantheudo (Alentejo and Algarve in Portugal), Manteúdo do Algarve (Algarve), Vale Grosso (Alentejo, Algarve)

VARIETIES COMMONLY MISTAKEN FOR LISTÁN DE HUELVA: PALOMINO FINO (under the name Listán or Listán Blanco)

ORIGINS AND PARENTAGE

As its name suggests, Listán de Huelva comes from the province of Huelva in Andalucía in the south-west corner of Spain. According to Hidalgo in Galet (2000), it is close to PALOMINO FINO, which is also called Listán, but their DNA profiles published in Martín *et al.* (2003) are not consistent with a parent–offspring relationship (Vouillamoz). In addition, a comparison of the DNA profiles of Listán de Huelva (Ibañez *et al.* 2003) and Manteúdo Branco from the Alentejo and the Algarve in Portugal (Almadanim *et al.* 2004) unexpectedly shows that they are the same variety (Vouillamoz). Manteúdo Branco is genetically distinct from other Alentejo varieties (Almadanim *et al.* 2004), which suggests it may have been introduced fairly recently, possibly from Huelva in Spain (Vouillamoz).

VITICULTURAL CHARACTERISTICS

Vigorous and fertile with thin-skinned berries. Resistant to drought and best suited to soils with low limestone content. May be pruned short or long. Susceptible to powdery mildew, less so to botrytis bunch rot.

WHERE IT'S GROWN AND WHAT ITS WINE TASTES LIKE

Listán de Huelva is grown to a limited extent across southern **Spain** but predominantly in the province of Huelva in the west, where there were more than 200 ha (494 acres) in the early twenty-first century. It is authorized as one of many principal varieties in the Condado de Huelva DO, where it is used mainly in the blended, fortified wines classified as Palido (fino-like) and Viejo (oloroso-like), along with varieties such as PALOMINO FINO, GARRIDO FINO (plus ZALEMA in the Viejo). Bodegas Privilegio del Condado also make a varietal Palido version.

In **Portugal** there were 692 ha (1,710 acres) of Manteúdo Branco in the Alentejo and the Algarve in 2010, used mainly in blends with other local varieties such as DIAGALVES. Wines are generally neutral, low in acidity and high in alcohol.

LISTÁN NEGRO

The red wine grape of Spain's Islas Canarias. Potentially aromatic wines.

BERRY COLOUR: ● ● ● ● ●

PRINCIPAL SYNONYMS: Almuñeco (La Palma), Listán Morado (Islas Canarias), Negra Commún (Lanzarote)

VARIETIES COMMONLY MISTAKEN FOR LISTÁN NEGRO: LISTÁN PRIETO (Islas Canarias), NEGRAMOLL

ORIGINS AND PARENTAGE

Listán Negro is indigenous to the Spanish Islas Canarias, off the coast of north-west Africa, where it is often confused with LISTÁN PRIETO (called Moscatel Negro on the islands) and with NEGRAMOLL, but a recent study has shown that Listán Negro has a distinct DNA profile that does not match any other known variety (Vouillamoz; Jorge Zerolo, personal communication).

DNA profiling confirms that Listán Negro is not a dark-skinned mutation of PALOMINO FINO, which is called Listán Blanco in Andalucía in southern Spain.

VITICULTURAL CHARACTERISTICS

Vigorous and productive.

WHERE IT'S GROWN AND WHAT ITS WINE TASTES LIKE

Listán Negro is the main red grape variety on the Islas Canarias, **Spain**, particularly in northern Tenerife, with 4,698 ha (11,609 acres) planted in 2008. On the grey volcanic soils, individual vines are planted in man-made, partially walled craters or hollows to capture moisture and break the prevailing winds from the Sahara.

Carbonic maceration is used to produce distinctively aromatic wines but others are vinified traditionally, some aged in oak. Producers of varietal wines include Los Bermejos, El Grifo, Bodegas Insulares Tenerife, El Lomo, Monje, El Penitente, Tajinaste, Valleoro and Viñátigo.

LISTÁN PRIETO

Historically important, well-travelled, dark-skinned Spanish variety now more common in the Americas but generally in retreat and, with some recent exceptions, producing inferior wines.

BERRY COLOUR: ● ● ● ● ●

PRINCIPAL SYNONYMS: Criolla Chica (Argentina), El Paso (California and Mexico), Hariri (Morocco), Listrão (Madeira), Misión (Mexico), Mission or Mission's Grape (California), Moscatel Negro (Islas Canarias), Negra Antigua (Chile), Negra Corriente (Peru), Negra Peruana (Peru), País (Chile), Palomina Negra (Castilla-La Mancha), Rosa del Perú (Peru), Uva Chica Negra (Chile), Uva Negra or Uva Negra Vino (Chile and Argentina), Uva Tinta (Chile), Viña Blanca (Chile), Viña Negra (Chile)

VARIETIES COMMONLY MISTAKEN FOR LISTÁN PRIETO: JACQUEZ, LISTÁN NEGRO, NEGRAMOLL (sometimes mistakenly called Listán Prieto)

ORIGINS AND PARENTAGE

Listán Prieto is an old variety from Castilla-La Mancha that was probably first described in 1513 by Alonso de Herrera (1790) under the name Palomina Negra. The etymology of the name Listán, also used as a synonym for PALOMINO FINO, is unknown, and the Spanish word *prieto* simply means 'dark' or 'black' (Galet 2000). Phylloxera almost eradicated Listán Prieto from Spain and the variety would have been lost in oblivion if DNA profiling hadn't recently discovered that Listán Prieto had been taken in the sixteenth and seventeenth centuries from Spain to at least four geographically distinct regions where it was given different names:

- Around 1540, it was taken to Mexico by Spanish Franciscan priests who founded several missions, and the grape was simply named Misión. This almost certainly represented the first *Vitis vinifera* variety to be introduced to the Americas from the Old World. By 1620 it had moved northwards to Baja California.
- In the mid sixteenth century, it was introduced to Chile and to Cuyo in western central Argentina by the Spanish conquistadores and was simply called Uva Negra ('black grape'). Later, in the mid nineteenth century, it was named País ('country') in Chile and Criolla Chica ('little Creole') in Argentina (Tapia *et al.* 2007; Lacoste *et al.* 2010), countries where it remained the most important variety until it was overtaken by the recently introduced French varieties such as COT (Malbec), CABERNET SAUVIGNON, CABERNET FRANC, MERLOT, etc.
- In the mid sixteenth century, it made its way to the Islas Canarias, where it was misleadingly called Moscatel Negro, although it is not related to any form of Muscat.
- In 1629, Misión was planted in the Rio Grande Valley in southern New Mexico by Fray Antonio de Arteaga, a Capuchin monk, and Fray García de San Francisco y Zúñiga, a Franciscan priest, who had founded a mission at Senecú, near the modern-day town of San Antonio, and another one at Pilabó, near what is now Socorro, due south of Albuquerque. For the needs of their religious offices, the brothers planted some vines that had been smuggled out of Spain, in spite of the law forbidding the export of Spanish grapevines that was designed to protect the export of Spanish wines. This makes Mission the first *Vitis vinifera* grape to have been cultivated in the United States.

In California, Father Junípero Serra is credited with having founded the first mission in Alta California, in San Diego in 1769, and with having brought the Mission vine from Mexico, although there is documentary evidence to suggest the Mission vines were not planted until ten years later (Pinney 1989). It remained California's main variety until the introduction of other European varieties in the second half of the nineteenth century.

Despite the old synonym Palomina Negra, Listán Prieto is not the black version of PALOMINO FINO, although their DNA profiles do not exclude the possibility of a close genetic relationship (Vouillamoz).

DNA parentage analysis (see MUSCAT for the pedigree diagram) has established that CEREZA, Moscatel Amarillo, TORRONTÉS RIOJANO and TORRONTÉS SANJUANINO from Argentina are progenies of Listán Prieto × MUSCAT OF ALEXANDRIA (Agüero *et al.* 2003), that QUEBRANTA from Peru is a progeny of Listán Prieto × NEGRAMOLL, that Listán Prieto (under the synonyms Moscatel Negro or Mission) is likely to have a parent–offspring relationship with PERRUNO from Extremadura in western Spain (Ibañez *et al.* 2003; This *et al.* 2006), and that a natural cross between Listán Prieto and CAYETANA BLANCA from the Iberian peninsula gave birth to Jaén Tinto (Zinelabidine *et al.* 2012), an old and almost extinct variety from southern Spain that should not be confused with MENCÍA (also called Jaén Tinto in Bierzo in Castilla y León; see Cayetana Blanca for the pedigree diagram).

OTHER HYPOTHESES

The Mission grape variety of California was long believed to be a local seedling descended from a European variety that was introduced in the sixteenth century (Galet 2000) but DNA profiling has shown that it is identical to Listán Prieto (Tapia *et al.* 2007). It has also been thought to have a link with MONICA NERA from Sardegna but this is based on a mistake, with Monica having been used incorrectly as a synonym for CRIOLLA, and this can be rejected by DNA analysis (Vouillamoz).

VITICULTURAL CHARACTERISTICS

Vigorous, productive and usefully tolerant of drought. Susceptible to Pierce's disease and to crown gall.

WHERE IT'S GROWN AND WHAT ITS WINE TASTES LIKE

Listán Prieto has disappeared from its homeland in Castilla-La Mancha, **Spain**, but it is still grown on the Islas Canarias: there were 29 ha (72 acres) planted there in 2008 and the variety is authorized in several DOs such as Albona, Gran Canaria and La Palma, although only in the Tacoronte-Acentejo denomination is it referred to as Listán Prieto; in all the others it is listed as Moscatel Negro.

California still had 656 acres (265 ha) of Mission in 2009, although there have been almost no plantings in recent years. The vast majority of these vines are in the Central Valley, mainly in Fresno County, and are used mostly for sherry- or port-style dessert wines.

In **Chile** in 2008, País was still the second most planted variety after CABERNET SAUVIGNON, with 14,995 ha (37,053 acres) and nearly 13% of the vineyard area. Grown mainly in Maule and in Bío Bío in the south, it is mostly used for cheap domestic rosé. However, in the last few years, Frenchman Louis-Antoine Luyt of Clos Ouvert (with the support of the late Marcel Lapierre of Beaujolais) has been on a mission to re-evaluate the variety and coax the best from the very old vineyards in Maule, using carbonic maceration to tame the potentially astringent tannins and highlight the fruit. These wines are light in colour, thirst-quenchingly fresh, with tart red-fruit flavours and a texture that becomes silky with age, not unlike some old-vine South African CINSAUT.

In **Argentina**, Criolla Chica is less common than either CRIOLLA GRANDE or CEREZA and its wine is a little paler though better quality. In 2008 there were 538 ha (1,329 acres), mostly in Mendoza and San Juan but also minor plantings in La Rioja and Salta, where it is generally used to make rosé.

This ancient variety is thought to be planted in **Peru**, too.

LIVORNESE BIANCA

See ROLLO.

LJUTUN

Rare, tart Croatian variety exclusive to the area just north west of Split on the Dalmatian coast.

BERRY COLOUR: ● ● ● ● ●

PRINCIPAL SYNONYMS: Ljutac, Plavac, Plavac Bedalovac

ORIGINS AND PARENTAGE

Ljutun is an indigenous variety from the Kaštela region north west of Split. The name comes from the Croatian word *ljut* meaning 'acid', which is appropriate given the high acidity in the berries.

VITICULTURAL CHARACTERISTICS

Late ripening (about the same time as PLAVAC MALI) and retains good acidity even when picked late. Good general disease resistance.

WHERE IT'S GROWN AND WHAT ITS WINE TASTES LIKE

Ljutun is found exclusively around the town of Kaštela (Zdunić 2005) in **Croatia** and is often blended with other varieties such as PLAVAC MALI, BABIĆ and BABICA to make *opolo*, a local rosé-style wine. Producers include Svetin Biliškov, Dalibor Bučan, Marin Remetin, Branko Glumac and Neven and Vinko Vujina.

LLEDONER PELUT

See GARNACHA.

LONGYAN

Old and widely planted red-skinned Chinese variety used for the table and for white wine.

BERRY COLOUR: ● ● ● ● ●

PRINCIPAL SYNONYMS: Czhi-Pu-Tao, Hun-Juan-Sin, Long Yan, Lungyen

VARIETIES COMMONLY MISTAKEN FOR LONGYAN: ?RYUGAN (Japan)

ORIGINS AND PARENTAGE

Longyan, named after a city and meaning 'dragon's eyes', is an old Chinese variety of unknown origin. Although Longyan is also called Ryugan in China, it is not yet known whether it is the same as the Japanese variety RYUGAN.

VITICULTURAL CHARACTERISTICS

Vigorous, lowish yields, late ripening. Big, compact bunches of medium-sized to large berries.

WHERE IT'S GROWN AND WHAT ITS WINE TASTES LIKE

Although Longyan is grown mainly for the table, it also widely used to make wine (and some surprisingly good brandy) in **China** by producers such as China Great Wall Wine and is successful as a base for sparkling wines. Wines made from this red-skinned variety are typically bright, light green-yellow in colour, soft, with refreshing fruity aromas and full-bodied. It is widely planted in the Huaizhuo Basin (Hebei) as well as in Pingdu (Shandong), Qinhuangdao (Hebei), Qingxu (Shanxi) Province and Yulin (Shaanxi). Japanese oenologists produced Longyan wines in Shacheng back in the 1940s and sweet wines were made there in the 1960s, which is partly why the relationship between Longyan and RYUGAN is unclear.

LOUISE SWENSON

Complex American hybrid with a significant degree of cold hardiness.

BERRY COLOUR:

PRINCIPAL SYNONYMS: ES 4-8-33

ORIGINS AND PARENTAGE

An Elmer Swenson 2-3-17 × KAY GRAY hybrid bred in 1980 by Elmer Swenson in Osceola, Wisconsin, in which:

- Elmer Swenson 2-3-17 is an Elmer Swenson 283 × Elmer Swenson 193 hybrid
- Elmer Swenson 283 is an Elmer Swenson 114 × SEYVAL BLANC hybrid (see ST PEPIN for the pedigree of Elmer Swenson 114)
- Elmer Swenson 193 is a Minnesota 78 × Seneca hybrid (see BRIANNA for the complete pedigree of Minnesota 78)
- Seneca is a LUGLIENGA × Ontario hybrid (see CAYUGA WHITE for the pedigree of Ontario).

This complex *Vitis labrusca*, *Vitis vinifera*, *Vitis lincecumii*, *Vitis riparia* and *Vitis rupestris* hybrid was selected in 1984 and released in 2001, named after Elmer Swenson's wife (Smiley 2008).

VITICULTURAL CHARACTERISTICS

Early to mid ripening. Slightly susceptible to drought but generally good disease resistance. Winter-hardy to around –28 °F (–33 °C).

WHERE IT'S GROWN AND WHAT ITS WINE TASTES LIKE

Wines are typically light-bodied with some floral, pear and honey aromas and only moderate acidity (Smiley 2008). Plantings are limited but the variety is found in **Vermont**, **Wisconsin** and **Minnesota** in the US and in Québec in **Canada**. Producers of varietal wines include East Shore in Vermont and Parallel 44 in Wisconsin. Coteau St-Paul in Québec include it a blend with ADALMIINA, PRAIRIE STAR and SWENSON WHITE.

LOUREIRA

See LOUREIRO.

LOUREIRO

High-quality, aromatic Vinho Verde variety.

BERRY COLOUR:

PRINCIPAL SYNONYMS: Branco Redondos, Loureira (Galicia in Spain), Loureiro Blanco, Marqués or Marquez (Galicia in Spain)

VARIETIES COMMONLY MISTAKEN FOR LOUREIRO: ARINTO DE BUCELAS, GALEGO DOURADO (Monção and Viana do Castelo)

ORIGINS AND PARENTAGE

Loureiro comes from the Vale do Lima (known as Limia where the river rises in Spain) in the northern coastal part of the Minho region of north-west Portugal. *Loureiro* is Portuguese for 'laurel', referring to the aroma of its berries, which is similar to that of laurel flowers and leaves. Loureiro shows medium to high genetic variability, which suggests it is a fairly old variety (Rolando Faustino, personal communication) and C B de Lacerda Lobo refers to the existence of the variety at the end of the eighteenth century in Melgaço and Vila Nova de Cerveira (Böhm 2005). Comparison of DNA profiles indicates a possible close relationship with ALVARINHO (under the name Albariño; Ibañez *et al.* 2003) and a recent DNA parentage analysis has suggested a possible parent–offspring relationship with AMARAL (Castro *et al.* 2011; Díaz-Losada *et al.* 2011).

The almost extinct Loureiro Tinto, still found in Spain, was thought by Galet (2000) to be a colour mutation of Loureiro, or vice versa. It has also in the past been used as a synonym for MENCÍA but there is no genetic evidence for either of these connections.

VITICULTURAL CHARACTERISTICS

Relatively productive thanks to heavy, compact bunches; fertile. Mid budding but early ripening. Susceptible to downy and powdery mildews, phomopsis, botrytis bunch rot and to mites.

WHERE IT'S GROWN AND WHAT ITS WINE TASTES LIKE

In the 1960s and 1970s, Loureiro spread from the Ponte de Lima area to the rest of the Minho region of northern **Portugal**, contributing to the relatively recent revival of the Vinho Verde DOC. Varietal wines are fresh and have distinctive aromas not only of laurel leaves and flowers but also of linden, orange and acacia blossom, plus flavours of orange and peach and occasionally green apple. This intensity is often tamed by blending with ARINTO DE BUCELAS and/or TRAJADURA. It is also authorized in the Tejo, Beiras and Península de Setúbal regions of Portugal and trials have been carried out in Dão, where the variety reaches higher alcohol levels of around 13%. Total Portuguese plantings in 2010 were 2,428 ha (6,000 acres). Recommended producers of varietal wines include Aphros, Quinta do Ameal, Quinta de Curvos, Quinta de Gomariz and Quinta dos Termos.

Further north, in **Spain**'s north-western extremity Galicia, there were 584 ha (1,443 acres) of Loureiro in 2008 and the variety is authorized for the Rías Baixas, Ribeira Sacra and Ribeiro DOs, where it is usually part of a blend with varieties such as Albariño (ie ALVARINHO), Treixadura (ie TRAJADURA) and GODELLO. Valmiñor's Davila L·100, made in the O Rosal subregion of southern Galicia, is a very good example – perhaps the only one – of a 100% Loureiro wine from Spain.

There were also 13 ha (32 acres) of the very rare Loureiro Tinto in Galicia, 30 km north of most of the Loureiro vineyards, and Forja del Salnés in Rías Baixas make a varietal wine.

A grape called Loureiro is grown in **California**'s Central Coast.

LUCIE KUHLMANN

Minor French hybrid grown mainly in Canada.

BERRY COLOUR: ● ● ● ● ●

PRINCIPAL SYNONYMS: Kuhlmann 149-3, Lucy-Kuhlman

ORIGINS AND PARENTAGE

A complex hybrid made in 1911 by Eugène Kuhlmann at the Institut Oberlin in Colmar, Alsace, eastern France, by crossing Millardet et Grasset 101-14 OP with GOLDRIESLING, where

Millardet et Grasset 101-14 OP is a selfling of a *Vitis riparia* × *Vitis rupestris* cross. This variety was named in honour of the vine breeder's wife and was commercialized in 1921.

The same crossing produced the siblings of Lucie Kuhlmann (the vine): LÉON MILLOT and MARÉCHAL FOCH.

VITICULTURAL CHARACTERISTICS

Vigorous, very early ripening, suited to cooler climates with short growing seasons. Semi winter-hardy.

WHERE IT'S GROWN AND WHAT ITS WINE TASTES LIKE

Like its more popular siblings LÉON MILLOT and MARÉCHAL FOCH, this French hybrid is more widely planted in cooler climates such as **Canada**'s Québec, Nova Scotia and Ontario than it is in **France**, though plantings are very limited even there. Like Léon Millot, the variety produces deeply coloured wines but they are a little more firmly structured and are likely to taste herbaceous if not fully ripe. Varietal wines are produced by, for example, Blomidon Estate and Gaspereau Vineyards in Nova Scotia, and Québec's Domaine des Côtes d'Ardoises is one of several to include Lucie Kuhlmann in their red blends.

LUGLIENGA

Historic, widespread but mainly table grape.

BERRY COLOUR: ● ● ● ● ●

PRINCIPAL SYNONYMS: Agostenga, Bona in Ca (Trentino), Guštana⚥ (Slovenia), Jouanenc (France), Lignan Blanc⚥ (France), Lignenga (Piemonte), Lugliatica (Piemonte), Luglienga Blanca, Luigese⚥ (Liguria), Seidentraube (Germany and Switzerland), Uva di Sant'Anna (Piemonte)

VARIETIES COMMONLY MISTAKEN FOR LUGLIENGA: PRIÉ⚥ (Valle d'Aosta)

ORIGINS AND PARENTAGE

Luglienga is a very old table and wine grape, mentioned in Piemonte, Italy, as long ago as 1329 under the old name Luglienchis (Favà i Agud 2001). Its numerous synonyms bear witness to the fact that it has been widely distributed all over Europe for several centuries. Luglienga is often grown on pergolas in front of houses, used mainly for table grapes and sometimes to produce homemade wine.

The variety's name comes from *Luglio* (Italian for 'July'), in reference to its early ripening, and this etymology finds an echo in an older French synonym, Jouanenc, from *juin* meaning 'June'. Ampelographic studies had established that Luglienga was identical to France's Lignan Blanc and to Seidentraube from Germany and German-speaking Switzerland (Galet 2000) and this has been confirmed by DNA profiling (Vouillamoz, Frei and Arnold 2008). Parentage analysis has shown that Luglienga is a parent of PRIÉ, with which it was often confused, in the Valle d'Aosta (Vouillamoz 2005). Luglienga was used as a parent in the breeding of REGNER.

VITICULTURAL CHARACTERISTICS

Vigorous, resistant to winter frost, susceptible to botrytis bunch rot, early ripening.

WHERE IT'S GROWN

As a table grape, Luglienga is grown all over Europe, from Spain to Hungary. It does not seem to be used to produce varietal wines but is still occasionally vinified.

LUMASSINA

Rare, light, crisp white exclusive to Liguria in north-west Italy.

BERRY COLOUR:

PRINCIPAL SYNONYMS: Acerbina (Noli), Buzzetto (Finale Ligure, Quiliano), Garella (Savona), Mataòsso, Mataòssu (Noli, Varigotti), Uga Matta (Spotorno)

ORIGINS AND PARENTAGE

Lumassina probably comes from the Genova region of Liguria in north-west Italy, where it is said to have been cultivated for centuries. Its name could derive from *lumasse*, local dialect for *lumache*, meaning 'snails', because it was traditionally paired with snail dishes. Its synonyms Acerbina (*acerbo* is Italian for 'acidic') and Buzzetto (*buzzo* means 'acidic' in the local dialect) both refer to the tradition of harvesting the grapes slightly before full maturity.

VITICULTURAL CHARACTERISTICS

Robust and productive vine. Ripens late (end of September to the beginning of October) and retains a high level of acidity.

WHERE IT'S GROWN AND WHAT ITS WINE TASTES LIKE

Lumassina has historically been cultivated on the steep slopes of Liguria, **Italy**, overlooking the sea and blended with BOSCO and VERMENTINO to produce the local Nostralino wine (a wine still celebrated by an annual festival). On the brink of extinction a few decades ago, the variety is today grown almost exclusively in the province of Savona, in particular in the communes of Finale Ligure, Quiliano, Noli, Spotorno and Varigotti. Today Lumassina is often blended with other local varieties to produce dry or sparkling wines but is also made as a varietal wine, eg by Cantine Ravera, Cascina delle Terre Rosse (L'Acerbina) and Isetta Macella (Buzzetto IGT Colline Savonese). The wines tend to have aromas of flowers, herbs and green apples, fresh acidity and low alcohol of around 10–11%.

LYDIA

Lydia is a dark-berried hybrid, sometimes spelt Lidia, created in the US by open pollination of ISABELLA and apparently brought to **Azerbaijan** at the end of the nineteenth century, then taken in the early twentieth century to **Ukraine** (270 ha/667 acres in 2009) and **Russia**. It is generally used to produce sweet pink or light-red wines, sometimes blended with its parent Isabella. Producers include Kuban-Vino in Russia's Krasnodar Krai and Frantsuzky Bulvar in Odessa in the Ukraine. Also found in the **Republic of Moldova**.

M

ꙫ this symbol on the following pages indicates that the synonyms, and mistaken identities, are confirmed by DNA evidence

MENNAS
MENOIR
MENU PINEAU
MÉRILLE
MERLOT
MERSEGUERA
MERZIFON KARASI
MERZLING
MESLIER SAINT-FRANÇOIS
MÉZES FEHÉR
MÉZY
MILGRANET
MÍLIA
MILLOT-FOCH
MINELLA BIANCA
MINUTOLO
MISKET CHERVEN
MISKET VARNENSKI
MISKET VRACHANSKI
MLADINKA
MOLETTE
MOLINARA
MOLLARD
MONARCH
MONASTRELL
MONBADON
MONDEUSE BLANCHE
MONDEUSE NOIRE
MONEMVASSIA
MONERAC
MONICA NERA
MONSTRUOSA
MONTEPULCIANO
MONTILS
MONTONICO BIANCO
MONTÙ
MONVEDRO
MOORE'S DIAMOND
MORADELLA
MORAVA
MORAVIA AGRIA
MORETO DO ALENTEJO
MORIO-MUSKAT
MORISTEL
MORNEN NOIR
MORONE
MORRASTEL BOUSCHET
MOSCATELLO SELVATICO
MOSCATO DI SCANZO
MOSCATO DI TERRACINA
MOSCATO GIALLO
MOSCATO ROSA DEL TRENTINO
MOSCHOFILERO
MOSCHOMAVRO
MOSTOSA
MOUYSSAGUÈS
MSKHALI
MTSVANE KAKHURI
MUJURETULI
MÜLLER-THURGAU
MUSCADELLE
MUSCARDIN
MUSCAT BAILEY A
MUSCAT BLANC À PETITS GRAINS
MUSCAT BLEU
MUSCAT FLEUR D'ORANGER
MUSCAT ODESSKY
MUSCAT OF ALEXANDRIA
MUSCAT OF HAMBURG
MUSCAT OTTONEL
MUSCAT SWENSON
MUŠKÁT MORAVSKÝ
MUSTOASĂ DE MĂDERAT

MACABEO

Often underestimated, widely grown variety that can make ageworthy whites as Viura in Rioja and Maccabeu in Roussillon.

BERRY COLOUR:

PRINCIPAL SYNONYMS: Charas Blanc (California), Lardot (Drôme in France), Macabeu (Catalunya), Maccabéo (France), Maccabeu (Roussillon in France), Viura (Rioja and Rueda in Spain), Vuera (Rioja and Rueda)

VARIETIES COMMONLY MISTAKEN FOR MACABEO: ALCAÑÓN (Somontano), XARELLO (Catalunya)

ORIGINS AND PARENTAGE

Macabeo most likely originates from the Vilafranca del Penedès region in Catalunya, where it was first mentioned in the early seventeenth century by Fray Miquel Agustí (1617): 'Lo vi Moscat, y lo vi Macabeu se han de cullir molt madurs los Rahims' ('Grapes from the Muscat vine and from the Macabeu vine must be harvested very ripe'). It was then introduced to Roussillon in southern France, where it was first mentioned in 1816 in Salses-le-Château, north of Perpignan (Gournay 1790): 'c'est à Salses, près Perpignan, que se fait l'excellent vin blanc, connu sous le nom de Macabeu, qui est d'une délicatesse & d'un goût exquis' ('it is in Salses near Perpignan that the excellent white wine known under the name Macabeu, which has an exquisite delicacy and taste, is made'). According to Favà i Agud (2001), the Catalan name Macabeu could derive from the French *macab(r)é* used to designate a certain type of cleric.

Contrary to popular belief, ALCAÑÓN from Somontano is not identical to Macabeo: the DNA profiles of Alcañón (Martín *et al.* 2003; Fernández-González, Mena *et al.* 2007) and Macabeo (Ibañez *et al.* 2003; Martín *et al.* 2003) are quite distinct (Vouillamoz).

According to DNA profiling, Macabeo could be related to XARELLO (Ibañez *et al.* 2003), with which it is often blended and sometimes even confused. Recent DNA parentage analysis has also suggested that Macabeo is a natural cross between Hebén, an almost extinct Andalusian variety, and a mysterious variety called Brustiano Faux (ie False Brustiano, distinct from BRUSTIANO BIANCO on Sardegna and Corse), which would make it a half-sibling of MANDÓN from the Bierzo region in the province of León in north-west Spain and GORGOLLASA from Mallorca (García-Muñoz *et al.* 2012). However, these parentages were based on too few DNA markers (twenty) and need to be confirmed.

OTHER HYPOTHESES

According to Comte Odart (1845), Macabeo could come from the Middle East.

VITICULTURAL CHARACTERISTICS

Big, compact bunches of medium-sized, thick-skinned berries. Late budding and late ripening. Productive, shoots sometimes broken off by the wind. Cool, damp sites are unsuitable. Very susceptible to botrytis bunch rot and vine bacterial necrosis, less so to downy mildew. Responds well to low yields and relatively early picking.

WHERE IT'S GROWN AND WHAT ITS WINE TASTES LIKE

In **France**, the variety is much less widely planted than in Spain but seems to have earned greater respect, particularly in Roussillon, where at least 85% of the 2,628 ha (6,494 acres) of Maccabéo/ Maccabeu in France in 2009 was planted (significantly down on the 5,200 ha/12,849 acres reported in 2000). In and around the cauldron of wine experimentation that is the Roussillon's

Vallée de l'Agly, the variety frequently turns up in generally barrel-aged blends with the likes of Grenache Blanc (see GARNACHA) and Carignan Blanc (see MAZUELO), while very successful varietal examples have emerged from Domaine Gayda (even if they have been labelled without much geographical precision).

Macabeo/Viura is northern **Spain**'s most planted white variety, although it is widely grown throughout the country, except in the far south and the north west. In 2008, there were 34,401 ha (85,007 acres) in Spain, predominantly in Catalunya (13,718 ha/33,898 acres), Castilla-La Mancha (5,311 ha/13,124 acres) and Aragón (4,987 ha/12,323 acres), although it is also quite widely planted in Rioja, Valencia and Extremadura. Its prevalance in Catalunya is due mainly to the role it plays in Cava, where it is generally blended with PARELLADA and XARELLO. In Rioja it more or less displaced Malvasía Riojana (ALARIJE) and GARNACHA BLANCA, which were much more widely planted before phylloxera hit the region in 1901, but has generally not been planted recently, so average vine age is high. Although it has often suffered from lack of respect, some of the most distinguished traditional oak-aged white wines in Rioja, such as López de Heredia's Viña Tondonia and Marqués de Murrieta's Capellanía, are based on or are exclusively Macabeu.

Varietal wines are typically lightly floral and relatively aromatic when young and unoaked but acidity tends to be low and they can soon lose their aroma or develop a certain bitter-almond character. Despite this, low-cropped Macabeo responds well to barrel fermentation and oak ageing – both in the more traditional oxidative styles and in the rather more modern versions of producers such as Finca Allende, whose wines can drink well at about five years. In both Catalunya and Rioja it seems to be shown to best advantage in a blend, though frequently with Macabeo as the dominant partner. Other recommended producers in Rioja include Beronia, Palacios Remondo, Remírez de Ganuza, Benjamin Romeo and Tobía, as well as Cercavins further east in the Costers del Segre DO.

MACERATINO

Rare white found in the Marche on Italy's Adriatic coast.

BERRY COLOUR: ● ● ● ● ●

PRINCIPAL SYNONYMS: Bianchetta Montecchiese, Greco delle Marche, Greco Maceratino, Maceratese, Matelicano, Montecchiese, Ribona (Macerata in the Marche), Uva Stretta, Verdicchio Marino, Verdicchio Tirolese

VARIETIES COMMONLY MISTAKEN FOR MACERATINO: GRECHETTO DI ORVIETO, GRECO, GRECO BIANCO, TREBBIANO TOSCANO, VERDICCHIO BIANCO

ORIGINS AND PARENTAGE

This variety originates from the province of Macerata in the Marche, from which its name derives. DNA profiling has shown that, contrary to local belief, Maceratino is distinct from GRECHETTO DI ORVIETO, GRECO, GRECO BIANCO and TREBBIANO TOSCANO and suggests a close relationship with VERDICCHIO BIANCO (Filippetti, Silvestroni, Thomas and Intrieri 2001).

OTHER HYPOTHESES

Maceratino is often considered to be a member of the 'Greco family' but Greco is a general term used for several distinct varieties which do not constitute a family.

VITICULTURAL CHARACTERISTICS

Vigorous, mid to late ripening. Low resistance to winter frost, downy mildew and powdery mildew.

WHERE IT'S GROWN AND WHAT ITS WINE TASTES LIKE

Maceratino is planted almost exclusively in the province of Macerata and in the commune of Loreto in the province of Ancona, **Italy**. Colli Maceratesi DOC (still, sparkling or passito styles are all possible) stipulates a minimum of 70% Maceratino but also allows for a varietal version if the variety, under the synonym Ribona, is mentioned on the label. Varietal wines have good acidity and aromas of flowers, almonds and sometimes aniseed. Producers include Accattoli and Boccadigabbia, whose Mont'Anello is an example of Colli Maceratesi Ribona, Villa Forano and Azzoni Avogadro Carradori.

MADELEINE ANGEVINE

Minor, fresh, citrusy French variety found mainly in Canada and very cool parts of Europe, sometimes grown for table grapes.

BERRY COLOUR: ● ● ● ● ●

PRINCIPAL SYNONYMS: Äugstler Weiss (Basel in Switzerland), Maddalena Angevina (Italy), Madlen Anževin (former Yugoslavia)

VARIETIES COMMONLY MISTAKEN FOR MADELEINE ANGEVINE: Madeleine Angevine Oberlin, MADELEINE × ANGEVINE 7672

ORIGINS AND PARENTAGE

Madeleine Angevine is a seedling obtained in Angers in 1857 by Robert and Moreau, the successors to the famous vine breeder Jean-Pierre Vibert, and released in 1863 by their successor, who gave himself the nickname Moreau-Robert. This early-ripening variety was named after Sainte-Madeleine's Day (22 July), the period when the grapes were often ready for harvest, as well as after the city of Angers in the Val de Loire.

Madeleine Angevine was originally said to be a Madeleine Royale × PRÉCOCE DE MALINGRE cross but recent DNA parentage analysis has shown that it is in fact a Circé × Madeleine Royale (see MÜLLER-THURGAU) cross (Vargas *et al.* 2009), where Circé is a Schiras × CHASSELAS cross obtained earlier by Vibert, and where Schiras is a French table grape obtained by Dr Houbdine.

Madeleine Angevine was used to breed several wine or table grape varieties, including SIEGERREBE, CITRONNY MAGARACHA, CSABA GYÖNGYE and MADELEINE × ANGEVINE 7672.

OTHER HYPOTHESES

According to DNA parentage analysis at INRA (Institut National de la Recherche Agronomique) in Monpellier, southern France, Madeleine Angevine is a cross between Madeleine Royale and Blanc d'Ambre, a seedling obtained by Robert in 1853 with the same parents as Circé.

VITICULTURAL CHARACTERISTICS

Early budding and very early ripening. Female-only flowers and therefore particularly susceptible to coulure and millerandage.

WHERE IT'S GROWN AND WHAT ITS WINE TASTES LIKE

Plantings in **France** in 2006 were a mere 7 ha (17 acres). It is also planted to a limited extent in

Washington State's cool, damp Puget Sound AVA, where San Juan Vineyards produce a varietal wine which they describe as very aromatic and citrusy. Other Puget Sound producers of this fresh varietal include Perennial Vintners and Lopez Island Vineyards. **Canada**'s British Columbia recorded 13 acres (5 ha) in 2007 and varietal wines are made in the Okanagan Valley by, among others, Larch Hills. It is apparently grown for table grapes in **Kyrgyzstan**.

There are limited plantings in **Denmark** (less than 1 ha/2.5 acres in 2010), eg at Domain Aalsgaard just north of Helsingør, the first accredited vineyard in the country, as well as in Jutland and on the islands of Zealand and Fyn, and in the very south of **Sweden**.

MADELEINE × ANGEVINE 7672

White German variety with a tangled and confused parentage, quite popular in England.

BERRY COLOUR: ● ● ● ● ●

VARIETIES COMMONLY MISTAKEN FOR MADELEINE × ANGEVINE 7672: MADELEINE ANGEVINE

ORIGINS AND PARENTAGE

Madeleine × Angevine 7672 is a freely pollinated seedling of MADELEINE ANGEVINE whose other parent is unknown (self-pollination can be ruled out because Madeleine Angevine has female-only flowers). It appeared some time before 1930, possibly in the Alzey research centre in Rheinhessen, Germany.

VITICULTURAL CHARACTERISTICS

Early ripening, reliable yields and low acidity.

WHERE IT'S GROWN AND WHAT ITS WINE TASTES LIKE

The wine's low acidity is valued far from the equator. Introduced to **England** in the late 1950s, where, according to viticulturist Stephen Skelton, it is affectionately known as Mad Angie, it was thought to have promise but its reputation and spread have been impeded by confusion with the table grape MADELEINE ANGEVINE. Nevertheless, there were 48 ha (119 acres) in 2009. The wines have a light Muscat flavour, sometimes closer to citrus, and Sharpham produce good examples, both oaked and unoaked. It is also found in **Denmark**'s diminutive vineyards (less than 1 ha/2.5 acres in 2010) and in **Sweden**.

MADRASA

Widely planted Azeri variety producing full-bodied and well-structured wines.

BERRY COLOUR: ● ● ● ● ●

PRINCIPAL SYNONYMS: Kara Shirei or Qara Shira, Matrasa, Matrassa, Medrese, Sevi Shirai

ORIGINS AND PARENTAGE

Madrasa probably originates from the village of Madrasa in the Shamakhi region in north-eastern Azerbaijan, where it has been cultivated since the nineteenth century or earlier (Chkhartishvili and Betsiashvili 2004).

VITICULTURAL CHARACTERISTICS

Mid ripening. Compact bunches of thick-skinned berries. Good resistance to drought, the main fungal diseases and winter frosts.

WHERE IT'S GROWN AND WHAT ITS WINE TASTES LIKE

Wines are typically deeply coloured, high in alcohol and tannins and full-bodied and used – often blended – in the production of dry, sweet and fortified wines in **Azerbaijan**. Producers include Absheron Sharab, Ganja Sharab 2, Ismailli Wine and Shemakha Farm.

MAGLIOCCO CANINO

Makes tannic wines mainly for blends in Calabria.

BERRY COLOUR: ● ● ● ● ●

PRINCIPAL SYNONYMS: Magliocco Ovale (Lamezia Terme), Maglioccolone (Lamezia Terme)

VARIETIES COMMONLY MISTAKEN FOR MAGLIOCCO CANINO: CASTIGLIONE, GAGLIOPPO, MAGLIOCCO DOLCE, NERELLO MASCALESE, NOCERA (Scilla)

ORIGINS AND PARENTAGE

A variety named Magliocca was mentioned in Calabria by Marafioti as early as the end of the 15th century. However, the name Magliocco has been used for several distinct varieties in southern Italy, often referred to as the 'Magliocchi family'. Recent ampelographic and DNA studies have shown that only two distinct 'Magliocchi' exist: Magliocco Canino and MAGLIOCCO DOLCE (Schneider, Raimondi *et al.* 2009). The other grapes named 'Magliocco' correspond to other varieties, mainly GAGLIOPPO. As a consequence, there's no such thing as a 'Magliocchi family'.

In addition, DNA parentage analysis suggests that Magliocco Canino could have a parent–offspring relationship with CALABRESE DI MONTENUOVO, which gave birth to SANGIOVESE through a natural cross with CILIEGIOLO (Vouillamoz, Monaco *et al.* 2007). As a consequence, Magliocco Canino is either a grandparent or a half-sibling of Sangiovese, thus further supporting the theory that part of Sangiovese's roots are in Calabria.

OTHER HYPOTHESES

As for many southern Italian varieties, an ancient Greek origin has often been suggested for the various Magvarioccos, the name meaning 'tender knot' in Greek, supposedly in reference to the shape of the bunches. However, there is no evidence of a genetic link between Magliocco Canino and modern Greek grape varieties (Vouillamoz).

VITICULTURAL CHARACTERISTICS

Late ripening (first ten days of October). Maglioccolone, one particular clone of this variety, has an earlier maturation period. Sensitive to drought.

WHERE IT'S GROWN AND WHAT ITS WINE TASTES LIKE

Magliocco Canino is cultivated on the Tirrenian coast of Calabria, **Italy**, mainly in the provinces of Catanzaro and Cosenza. Savuto is the only DOC to officially authorize Magliocco Canino, in a blend with other local varieties such as GAGLIOPPO and NERELLO CAPPUCCIO. The wine is generally tannic and contributes structure and ageing potential to a blend. Malaspina's Patros Pietro (IGT Calabria) is a blend of 80% Magliocco Canino and 20% CABERNET SAUVIGNON aged in acacia wood; Tenuta Terre Nobili's Cariglio (IGT Valle dei Crati) is a blend of 80% Magliocco Canino with 20% MAGLIOCCO DOLCE.

Magliocco Canino is also said to be planted in the Marche and in Sicilia, but closer ampelographic study is probably needed.

The Italian agricultural census (2000) records 616 ha (1,522 acres) of Magliocco Canino but since the census also describes MAGLIOCCO DOLCE as a synonym of Magliocco Canino, the figure cannot be taken at face value. Magliocco Dolce seems to be more widespread in Calabria than Magliocco Canino.

MAGLIOCCO DOLCE

Widely planted Calabrian and the more common of the two Maglioccos.

BERRY COLOUR: ● ● ● ● ●

PRINCIPAL SYNONYMS: Arvino (Cosenza, Crotone), Catanzarese, Gaddrica (Longobardi), Greco Nero (Lamezia Terme), Guarnaccia Nera (Verbicaro), Lacrima Cristi Nera (Cosenza, Reggio Calabria), Magliocco Tondo, Maglioccuni (Bivongi), Mangiaguerra (Capo Vaticano), Marcigliana, Marsigliana Nera (Lamezia Terme), Merigallo (Terravecchia), Nera di Scilla (Scalea), Petroniere

VARIETIES COMMONLY MISTAKEN FOR MAGLIOCCO DOLCE: CASTIGLIONE (Locride), GAGLIOPPO, MAGLIOCCO CANINO, NERELLO MASCALESE (Sicilia), NOCERA (the Nocera sample at the University of California at Davis is in fact Magliocco Dolce; Vouillamoz)

ORIGINS AND PARENTAGE

See MAGLIOCCO CANINO for more detail on the supposed 'Magliocchi family'.

The name Arvino, one of Magliocco Dolce's oldest synonyms, most likely derives from Lago Arvo in La Sila, a plateau that overlaps the Calabrian provinces of Crotone, Cosenza and Catanzaro and corresponding to the probable cradle of Magliocco Dolce. Ampelography and DNA profiling have shown that Marsigliana Nera, often considered to be a distinct variety, is in fact identical to Magliocco Dolce (Schneider, Raimondi, De Santis and Cavallo 2008).

OTHER HYPOTHESES

See MAGLIOCCO CANINO. There is no evidence of a genetic link between Magliocco Dolce and modern Greek grape varieties (Vouillamoz).

VITICULTURAL CHARACTERISTICS

There are many different clonal variations of this variety (as with SAVAGNIN, for instance). High tannin content in the berries.

WHERE IT'S GROWN

Under numerous synonyms, Magliocco Dolce is one of the most widespread red wine grapes in Calabria, **Italy**, cultivated mainly in the provinces of Crotone, Catanzaro and Cosenza. Librandi's Magno Megonio is a good example of a varietal wine.

The Italian agricultural census (2000) records 616 ha (1,522 acres) of MAGLIOCCO CANINO but includes Magliocco Dolce in this total so the actual figure for Magliocco Dolce is uncertain. However, Magliocco Dolce seems to be more widespread in Calabria than Magliocco Canino.

MAGNOLIA

Minor Muscadine variety that ripens unevenly.

BERRY COLOUR:

PRINCIPAL SYNONYMS: North Carolina 60-60

ORIGINS AND PARENTAGE

A ((Hope × Thomas) × SCUPPERNONG) × (Topsail × Tarheel) Muscadine hybrid (mainly *Vitis rotundifolia* with a bit of *Vitis labrusca* and *Vitis vinifera*) selected in 1954 at the North Carolina Agriculture Experiment Station, in which Hope is a selected vine of *Vitis rotundifolia* (see CARLOS for the pedigrees of Thomas, Topsail and Tarheel). It was released in 1961.

VITICULTURAL CHARACTERISTICS

Vigorous and productive but uneven ripening and susceptible to botrytis bunch rot. Small bronze-coloured berries. Relatively cold-hardy.

WHERE IT'S GROWN AND WHAT ITS WINE TASTES LIKE

There were 20 acres (8 ha) of Magnolia in **North Carolina** in 2006 but overall plantings are very limited. Producers of varietal wines include Cypress Bend, Duplin and Uwharrie in North Carolina and Piney Woods in **Texas**. Others such as Colony Cellars in Texas and Wills Creek in **Alabama** use it in blends. Most descriptions of the wine refer to apples and pears.

MAGYARFRANKOS

Dark-skinned escapee from experimental plantings during Hungary's Communist era.

BERRY COLOUR:

PRINCIPAL SYNONYMS: Magyar Frankos

ORIGINS AND PARENTAGE

Magyarfrankos, meaning 'Hungarian-Franconian', is a Muscat Bouschet × BLAUFRÄNKISCH (or Kékfrankos) cross obtained in 1953 by Pál Kozma and József Tusnádi at the Szent István University in Hungary, in which Muscat Bouschet is a black-berried Muscat Rouge de Madère × PETIT BOUSCHET cross that is no longer cultivated (see SANGIOVESE for the origin of Muscat Rouge de Madère). It was officially registered in 1974 together with BÍBORKADARKA.

VITICULTURAL CHARACTERISTICS

Late ripening, high yields. Big bunches of small berries. Winter-hardy but susceptible to botrytis bunch rot.

WHERE IT'S GROWN AND WHAT ITS WINE TASTES LIKE

Bálint Losonci and a few other growers in the village of Gyöngyöspata in Mátra, central northern **Hungary**, cultivate the last 2–3 ha (5–7 acres) of Magyarfrankos in the country, the remains of an experiment during the 1970s, rejected because it did not meet the two critical requirements of Communist-era viticulture: high yields and ease of cultivation. Losonci makes the only varietal example currently on the market, as well as a deeply coloured Kékfrankos/Magyarfrankos rosé blend.

MAIOLICA

Minor dark-skinned Abruzzo variety, usually blended with Montepulciano.

BERRY COLOUR: ● ● ● ● ●

PRINCIPAL SYNONYMS: Gaglioppa, Ortonese

VARIETIES COMMONLY MISTAKEN FOR MAIOLICA: GAGLIOPPO

ORIGINS AND PARENTAGE

Maiolica is often confused with GAGLIOPPO because the bunches are similarly shaped.

VITICULTURAL CHARACTERISTICS

Productive, susceptible to winter cold, very late ripening.

WHERE IT'S GROWN AND WHAT ITS WINE TASTES LIKE

Formerly widespread throughout Abruzzo, **Italy**, Maiolica is today restricted to the provinces of Pescara and Chieti, with the exception of a few plantings in the province of Macerata (Marche). There were 76 ha (188 acres) planted in Italy in 2000. It is usually blended with MONTEPULCIANO, providing colour, spicy aromas and soft tannins but it is also authorized as a varietal wine in IGT Colline Pescaresi.

MAIOLINA

Rare, productive, northern Italian red rescued from extinction in the 1980s.

BERRY COLOUR: ● ● ● ● ●

PRINCIPAL SYNONYMS: Majolina

ORIGINS AND PARENTAGE

Maiolina DNA is unique and fairly similar to that of other old Lombardian varieties (Geuna *et al.* 1997).

OTHER HYPOTHESES

Some authors have suggested a similarity between Maiolina and TEMPRANILLO but this has been disproved by DNA profiling (Vouillamoz).

VITICULTURAL CHARACTERISTICS

Very productive. Berries are generally low in sugar but intensely red.

WHERE IT'S GROWN AND WHAT ITS WINE TASTES LIKE

Right in the middle of the city of Brescia in central Lombardia, **Italy**, at the foot of the hill up to the castle, there's a unique and historic vineyard called Vigneto della Pusterla, owned by the Pusterla Winery. This 4 ha (10 acres) vineyard is trained in a local pergola style and produces not only a varietal INVERNENGA but also 2,000 bottles per year of Pusterla Rosso IGT Ronchi di Brescia, a blend of old, local and sometimes rare varieties, including Maiolina as well as MARZEMINO, GROPPELLO GENTILE, SCHIAVA GROSSA, BARBERA and CORVA.

In addition to its inclusion in this unique 'city wine' supported by Slow Food, Maiolina has also long been cultivated in Ome in the Franciacorta area of Lombardia, where the appropriately named Majolini brothers rescued the variety from extinction in the early 1980s. Although the variety's name predates them, it could indeed derive from this common local family name. A few years ago, the brothers produced 360 magnums of red-fruited, well-structured but smooth varietal Maiolina to support their successful application to the ministry of agriculture to officially register Maiolina as an indigenous variety.

MAJARCĂ ALBĂ

See SLANKAMENKA.

MALAGA BLANC

Malaga Blanc is the main variety in **Thailand**, said to have been imported as a table grape (now used for the table and for wine) from southern France in 1685, a gift from Louis XIV to King Narai the Great of Siam. It could be identical to the Spanish variety Teneron, also called Panse de Provence, which would make sense because Malaga Blanc is grown together with COLOMBAUD from Provence and CHENIN BLANC from the Loire, eg at Siam Winery. Along with the other 'local' variety, Pokdum (BLACK QUEEN), Malaga Blanc is cultivated in spectacular floating vineyards in the Chao Praya Delta in Thailand, its thick skins making it resistant to Thailand's heavy rains. It is also used for winemaking in **Vietnam**.

Many distinct varieties, including Dabouki in Armenia and MUSCAT OF ALEXANDRIA in Cyprus, have also been given the vague name Malaga Something.

MALAGOUSIA

Aromatic, high-quality Greek recently rescued from oblivion.

BERRY COLOUR: ● ● ● ● ● (first underlined)

PRINCIPAL SYNONYMS: Malagouzia, Malagoyzia, Malaouzia, Melaouzia

ORIGINS AND PARENTAGE

Malagousia probably comes from around Náfpaktos in Aitolía-Akarnanía on the southern coast of mainland Greece (Manessis 2000; Nikolau and Michos 2004).

VITICULTURAL CHARACTERISTICS

Vigorous and productive, early budding and ripening. Large compact bunches of large berries, often fairly thin-skinned. Highly susceptible to botrytis bunch rot and downy mildew but resistant to drought.

WHERE IT'S GROWN AND WHAT ITS WINE TASTES LIKE

Malagousia produces full-bodied, aromatic wines, their perfume ranging from basil, lime, orange and exotic fruits to rose petals. Although acidity is generally low, the best are zesty and elegant but poor examples can be unattractively heavy. Some wines have a finely grained texture thanks to the phenolics in the skins, increased by skin contact. There is some debate as to whether there is more than one clone since some of the vines seem to produce smaller berries and have higher acidity, others have bigger berries and are more aromatic, but this could also be due to pruning or site. Unlike certain aromatic varieties such as SAUVIGNON BLANC, the aromas do not seem to be lost – burnt off – in hotter sites. Although the acidity is higher in cooler regions, the wines can be thinner and less impressive.

It is thanks to Evangelos Gerovassiliou's work when he was at Domaine Carras in the 1980s and 1990s that Malagousia was recognized for its great potential, the variety having been found in western **Greece** by Professor Logothetis of the Agricultural University of Thessaloníki and planted in an experimental plot that he rented at Carras (Lazarakis 2005). However, Roxani Matsa close to the city of Athens has also been particularly influential in the continuing process of selection and distribution from her vineyards to many other growers around the country. At his own estate in Epanomí, Gerovassiliou continues the work of selection and makes both varietal wines (dry and barrel-fermented late harvest) and blends Malagousia with ASSYRTIKO to add perfume to the latter higher-acid and more mineral variety.

Malagousia is now being planted in many different regions as far south as Attikí and the Pelopónnisos (Peloponnese), generally on a small, experimental scale at first. Most of the more significant areas currently are in Kentrikí Makedonía (Central Macedonia). Recommended producers of varietal wines include Antonopoulos (Pátra), Argatia (Náoussa), Boutari/Matsa (Pallini), Gerovassiliou (Epanomí), Papagiannakos (Attikí), Claudia Papagianni (Halkidikí) and Vatistas (Monemvasia).

MALBEC

See COT.

MALBO GENTILE

Minor Italian red exclusive to Modena.

BERRY COLOUR: ● ● ● ● ●

PRINCIPAL SYNONYMS: Amabile di Genova, Tubino, Turbino

VARIETIES COMMONLY MISTAKEN FOR MALBO GENTILE: FOGLIA TONDA

ORIGINS AND PARENTAGE

Despite its synonym Amabile di Genova, Malbo Gentile is not morphologically or genetically related to the varieties of the Ligurian Riviera. Recent DNA studies have shown that Malbo Gentile could be closely related to FORTANA from Emilia-Romagna (Boccacci *et al.* 2005).

OTHER HYPOTHESES

A local tradition has it that Malbo Gentile was introduced from California at the end of the nineteenth century under the name Tubino by a returning Ligurian emigrant but there is no documentary support for this. Although the name Malbo is similar to the French Malbec (COT), the two varieties have nothing in common. Boccacci *et al.* (2005) suggested Malbo Gentile and FOGLIA TONDA were synonyms but this can be refuted by DNA profiling (Vouillamoz).

VITICULTURAL CHARACTERISTICS

Well coloured, vigorous, productive, mid ripening. Berries have a high sugar content and are susceptible to powdery and downy mildews but resistant to botrytis bunch rot.

WHERE IT'S GROWN AND WHAT ITS WINE TASTES LIKE

Malbo Gentile is cultivated on hillsides in the provinces of Modena and Reggio Emilia, and to a much lesser extent in south-eastern Emilia-Romagna, **Italy**, where it is often blended with the various Lambruscos to increase the sugar content of the juice, eg in the Lambrusco Grasparossa di Castelvetro or Reggiano DOCs. The Colli di Scandiano e di Canossa DOC allows Malbo Gentile to be used in various blends or as a varietal wine and the latter is often produced in a precocious novella style. Sweet versions, such as Villa Bonluigi's passito-style La Pinona, and/or sparkling wines, such as Casali Viticoltori's Campo delle More, are also possible though less common.

The Italian census of 2000 recorded 110 ha (270 acres) of Malbo Gentile.

MALIGIA

Rare vine makes alcoholic white in Emilia-Romagna.

BERRY COLOUR: ● ● ● ● ●

PRINCIPAL SYNONYMS: Malese, Malige, Maligia Omalise, Malis (Dozza), Malisa, Malise, Malisia, Malixa, Malixe, Malixia

ORIGINS AND PARENTAGE

The earliest mention of this variety supposedly dates back to the Middle Ages, when in *c.* 1300 Pietro de Crescenzi wrote about a Malixia in the province of Bologna, but some morphological differences make this mention doubtful. Tanara's reference to a Maligia in Dozza (1644) is more reliable since the variety is still scattered in old vineyards in this area.

OTHER HYPOTHESES

Maligia is often said to belong to the 'Malvasia family' but ampelography and genetics provide strong counter-evidence for this 'family' concept (see MALVASIA).

VITICULTURAL CHARACTERISTICS

Vigorous and fertile, with poor resistance to fungal diseases. Late ripening.

WHERE IT'S GROWN AND WHAT ITS WINE TASTES LIKE

Maligia used to be widespread in the Imola region in the province of Bologna in Emilia-Romagna, **Italy**, extending westwards to Modena, in the province of the same name, and eastwards to Forlì in the province of Forlì-Cesena. Today, Maligia is cultivated on fewer than 3 ha (7 acres) in the communes of Imola, Castel San Pietro Terme, Dozza and Castel Bolognese in the province of

Bologna and Faenza in the province of Ravenna. Wines are lightly aromatic with high alcohol and are generally used to enrich other local varieties.

MALVAR

See LAIRÉN.

MALVASIA

Malvasia, and its several synonyms, is a name commonly applied to a wide array of generally unrelated grape varieties, both as prime name and in a host of synonyms.

PRINCIPAL SYNONYMS: These are basically translations of Malvasia, rather than synonyms, since the name Malvasia is used for so many different varieties, as explained below: Malmsey (English), Malvagia (Spanish), Malvasier (German), Malvasijie (Croatian), Malvelzevec (Slovenian), Malvoisie (French)

ORIGINS AND PARENTAGE

Malvasia is a generic name given to a wide range of distinct white-, pink-, grey- or black-skinned varieties which share an ability to produce sweet wines high in alcohol. It supposedly derives from the Greek island port of Monemvasia off the east coast of the Pelopónnisos (Peloponnese), from which the various Malvasia grapes are supposed to have gone out into the world. Monemvasia was corrupted to Malfasia and then to Malvasia in Italian and Portuguese, Malvagia in Spanish, Malvasier in German, Malmsey in English, Malvasijie in Croatian and Malvoisie in French (Galet 2000).

The port of Monemvasia used to be the trading hub for the wines produced in the region and on the neighbouring islands of the southern Aegean. The earliest mention of the sweet wines of Monemvasia dates back to 1214 in Éfesos (Ephesus, now in Turkey), and in 1278 the Venetians were already importing *vinum de Malvasias*. The identities of the grape varieties making up this sweet wine are unknown, although the various suggestions include some permutation of the following: AÏDANI, ASSYRTIKO, ATHIRI, KYDONITSA, LIATIKO, MONEMVASSIA, THRAPSATHIRI and VILANA.

However, the Greek origin of the Malvasias is not supported by recent DNA studies showing that the varieties named Malvasia Something are genetically very distinct and do not usually have a common origin (Fanizza *et al.* 2003; Lacombe *et al.* 2007). It therefore makes no sense to speak of a 'Malvasia family'.

MALVASIA BIANCA DI BASILICATA

The Malvasia exclusive to Basilicata.

BERRY COLOUR: ● ● ● ● ●

ORIGINS AND PARENTAGE

Malvasia Bianca di Basilicata is morphologically very distinct from MALVASIA BIANCA LUNGA though may have some similarities with MALVASIA BIANCA DI CANDIA.

VITICULTURAL CHARACTERISTICS

Productive, mid ripening.

WHERE IT'S GROWN AND WHAT ITS WINE TASTES LIKE

Long cultivated in Basilicata, **Italy**, Malvasia Bianca di Basilicata can add aromatic interest and acidity to other varieties such as Moscato Bianco (MUSCAT BLANC À PETITS GRAINS) and used to be added to AGLIANICO. Malvasia Bianca di Basilicata may be used to produce dry, sparkling or sweet wines and is recommended in the provinces of Potenza and Matera. It constitutes the majority of the blend in white and sparkling versions of Matera DOC such as Dragone's Brut and has recently been officially registered for the Grottino di Roccanova DOC. Cervino Donata Maria's bottling of the latter is a dry blend of Malvasia Bianca di Basilicata, TREBBIANO TOSCANO and CHARDONNAY.

Total plantings in Italy in 2000 amounted to a not insignificant 959 ha (2,370 acres).

MALVASIA BIANCA DI CANDIA

Most planted Malvasia, making neutral white. In decline but still quite widely planted and blended in central Italy.

BERRY COLOUR: ● ● ● ● ●

PRINCIPAL SYNONYMS: Malvasia Bianca (Calabria), Malvasia Candia, Malvasia di Candia, Malvasia Rossa

VARIETIES COMMONLY MISTAKEN FOR MALVASIA BIANCA DI CANDIA: MALVASIA BIANCA DI PIEMONTE, MALVASIA DI CANDIA AROMATICA ⧖, MALVASIA DI LIPARI ⧖ (aka Malvasia Càndida)

ORIGINS AND PARENTAGE

Candia is an old name for the island of Kríti (Crete), where this variety supposedly comes from. However, so far there's no genetic evidence to support this idea. Recent DNA profiling has shown that MALVASIA DI CANDIA AROMATICA is not a mutation of Malvasia Bianca di Candia but a distinct variety, genetically close to MALVASIA DI CASORZO (Lacombe *et al.* 2007). More recently, DNA parentage analysis has explained these relationships by showing that Malvasia Bianca di Candia is a likely progeny of GARGANEGA (see that entry for the pedigree diagram), the other parent being unknown (Crespan, Calò *et al.* 2008). As a consequence, Malvasia Bianca di Candia is either a half-sibling or a grandparent of Garganega's other progenies: ALBANA, CATARRATTO BIANCO, DORONA DI VENEZIA, MARZEMINA BIANCA, MONTONICO BIANCO, MOSTOSA (aka Empibotte Bianco), SUSUMANIELLO and TREBBIANO TOSCANO.

VITICULTURAL CHARACTERISTICS

The surprising synonym Malvasia Rossa is linked not to the colour of the berries but to the colour of the apical buds. Mid to late ripening (second half of September).

WHERE IT'S GROWN AND WHAT ITS WINE TASTES LIKE

Malvasia Bianca di Candia is the most widely planted Malvasia in **Italy**. It is cultivated mainly in Lazio and to a lesser extent in Emilia-Romagna, Liguria, Toscana, Campania, Umbria and on Sardegna. It is usually blended with other varieties, particularly various types of TREBBIANO, to create what might be described as a typical central Italian blend, eg in DOCs such as Colli Lanuvini, Colli Martani, Castelli Romani, Colli Piacentini, Frascati and so on. It is prone to oxidation, thereby becoming deeper in colour and losing freshness, and the wines are often rather neutral-tasting and high in alcohol.

Total plantings in Italy in 2000 amounted to 8,788 ha (21,716 acres) – down significantly from around 50,000 ha (123,553 acres) in the early 1990s.

MALVASIA BIANCA DI PIEMONTE

California's Malvasia. Aromatic, tough-skinned, originally Italian.

BERRY COLOUR: ● ● ● ● ●

PRINCIPAL SYNONYMS: Caccarella, Greco (Roero in Piemonte), Malvasia Bianca (California), Malvasia Greca (province of Alessandria in Piemonte), Moscatella, Moscato Greco (province of Asti), Mosella (province of Alessandria)

VARIETIES COMMONLY MISTAKEN FOR MALVASIA BIANCA DI PIEMONTE: MALVASIA BIANCA DI CANDIA, MALVASIA BIANCA LUNGA, MALVASIA DEL LAZIO, MALVASIA DI LIPARI

ORIGINS AND PARENTAGE

This member of the heterogeneous group of varieties known as Malvasia Something may have been mentioned near Torino in 1606 by agronomist and polymath Croce, under the name Malvasia. It was widely cultivated in Piemonte, north-west Italy, until it was replaced by MUSCAT BLANC À PETITS GRAINS (known there as Moscato Bianco) after powdery mildew devastated the vineyards at the end of the nineteenth century (Schneider, Mannizi and Cravero 2004).

VITICULTURAL CHARACTERISTICS

Very vigorous, early to mid budding, mid ripening. Medium-sized to large, compact bunches of large, tough-skinned, Muscat-scented berries that are well suited to late harvesting. More susceptible to powdery mildew than MUSCAT BLANC À PETITS GRAINS but less so to botrytis bunch rot.

WHERE IT'S GROWN AND WHAT ITS WINE TASTES LIKE

A few old vines of Malvasia Bianca di Piemonte can still be found in scattered vineyards in the provinces of Alessandria, Asti, Cuneo and Torino in Piemonte, **Italy**.

Today it is much more important in **California**, where it was introduced by immigrants from Piemonte under the misleading name of Malvasia Bianca (see the MALVASIA entry). Wines here are typically dry to off dry with floral, herbal, citrus, sometimes more tropical, aromas and flavours and a certain viscosity or oiliness on the palate. Producers of varietal wines include Ballentine in Napa, Birichino in Monterey, Palmina in Santa Ynez Valley, Kenneth Volk and Wild Horse in San Bernabe. There were 1,384 acres (560 ha) in California in 2009, mainly in Merced, Madera and San Joaquin counties in the Central Valley but also in cooler Sonoma.

MALVASIA BIANCA LUNGA

Widely distributed neutral Tuscan variety used in blends for both dry wines and Vin Santo.

BERRY COLOUR: ● ● ● ● ●

PRINCIPAL SYNONYMS: Krizol (Croatia), Malvasia Bianca, Malvasia Bianca Siciliana, Malvasia Bianca Toscana, Malvasia del Chianti (Toscana), Malvasia di Arezzo, Malvasia di Brolio, Malvasia di San Nicandro, Malvasia Lunga, Malvasia Toscana, Malvasia Verace, Maraškin (Croatia), Maraština or Mareština (Croatia), Menuetta (Slovenia), Pavlos (Greece), Prosecco Nostrano (Conegliano in the Veneto), Racina du Monacu Bianca, Rukatac (the island of Korčula in Croatia), Tundulillu Bianco (Calabria), Višana (Croatia)

VARIETIES COMMONLY MISTAKEN FOR MALVASIA BIANCA LUNGA: MALVASIA BIANCA DI PIEMONTE, MALVASÍA DE LANZAROTE, PROSECCO, PROSECCO LUNGO (Conegliano)

ORIGINS AND PARENTAGE

This variety has for several centuries been found in the Chianti region of Toscana, under the entirely logical name of Malvasia del Chianti. Recent DNA profiling has shown that it is identical to Maraština in Croatia and Pavlos in Greece (Šimon *et al.* 2007), and that it is genetically close to MALVAZIJA ISTARSKA and MALVASIA DI LIPARI (aka Malvasia di Sardegna; Lacombe *et al.* 2007). In addition, DNA parentage analysis has shown that MALVASIA NERA DI BRINDISI is a natural cross between Malvasia Bianca Lunga and NEGROAMARO (Crespan *et al.* 2008). Parentage analysis has also revealed that VITOVSKA, a white grape of the Karst region (north-east Italy and western Slovenia), is the progeny of Prosecco Tondo (see PROSECCO), also from the Karst region, and Malvasia Bianca Lunga (Crespan *et al.* 2007), suggesting that the latter was also cultivated in north-east Italy at the time Vitovska was born, possibly under its local synonym Prosecco Nostrano di Conegliano. It was used to breed INCROCIO BIANCO FEDIT 51.

VITICULTURAL CHARACTERISTICS

Mid to late ripening, highly susceptible to downy mildew but also to powdery mildew, botrytis bunch rot and sour rot.

WHERE IT'S GROWN AND WHAT ITS WINE TASTES LIKE

Malvasia Bianca Lunga, under the name Malvasia del Chianti, is cultivated in almost every province of Toscana in central **Italy** and is still sometimes used in Chianti DOCG wines. It was traditionally blended with SANGIOVESE and CILIEGIOLO in the famous Chianti recipe established by Barone Bettino Ricasoli in the nineteenth century but significantly modified in the 20th. (Since 2006, white varieties have not been permitted in Chianti Classico DOCG.) It is authorized in several Tuscan DOCs, particularly for Vin Santo. More recently, Malvasia Bianca Lunga has also been cultivated in the Veneto, in Puglia in the provinces of Brindisi and Bari (and must constitute at least 50% of the blend in Leverano DOC), in Umbria, Lazio and the Marche. In some DOCs, it is restricted to a maximum of 10%, eg in Biferno (Molise) and Lizzano (Puglia). Wines tend to be rather neutral but full-bodied thanks to their relatively high alcohol. Total plantings in Italy in 2000 were 4,741 ha (11,715 acres).

As Maraština (and sometimes as Rukatac), this variety is widely but not intensively cultivated along the coast in **Croatia** and on several islands, including Korčula and Lastovo, often interplanted as a secondary variety. It is officially authorized for the regions of Hrvatsko Primorje (Croatian Littoral), Sjeverna Dalmacija (Northern Dalmatia), Dalmatinska Zagora (Dalmatian Hinterland) and Srednja i Južna Dalmacija (Central and Southern Dalmatia). There were just over 500 ha (1,236 acres) in 2009, ie just over 1.5% of the total vineyard area. Wines are golden yellow, full-bodied and sometimes lacking in acidity. Producers of varietal wines include Mato Antunović, Marko Duvančić (Oklaj) and Toreta (Smokvica).

MALVASIA BRANCA DE SÃO JORGE

Portuguese variety of unknown parentage but high quality used in the production of Madeira.

BERRY COLOUR: ● ● ● ● ●

PRINCIPAL SYNONYMS: Malvasia Branca de S Jorge

ORIGINS AND PARENTAGE

This variety was bred from unknown parents by José Leão Ferreira de Almeida at the Estação Agronómica Nacional research centre near Lisboa in Portugal. It was introduced to the island of Madeira in the 1970s and was cultivated mainly in the parish of São Jorge in the district of Santana on the north coast of the island, hence the name.

VITICULTURAL CHARACTERISTICS

Productive and highly susceptible to botrytis bunch rot.

WHERE IT'S GROWN AND WHAT ITS WINE TASTES LIKE

The variety was found to be well suited to the production of **Portugal**'s Madeira because the grapes have high levels of both sugar and acidity and it expanded rapidly on the island in the 1990s. The planted area is now stable at around 35 ha (86 acres) and this is now the main variety used to make Malmsey, or Malvasia, the sweetest of the island's fortified wines. Ricardo Diogo of Barbeito is very enthusiastic about the quality of this variety and is experimenting with drier styles, although such wines cannot be labelled Malvasia.

MALVASIA CÂNDIDA

See MALVASIA DI LIPARI.

MALVASÍA CASTELLANA

See SÍRIA.

MALVASIA DE COLARES

The principal, if demanding, variety of an unusual, slowly shrinking wine region on the sandy coastline north west of Lisboa.

BERRY COLOUR: ● ● ● ● ●

ORIGINS AND PARENTAGE

As the name suggests, Malvasia de Colares is exclusive to the Colares wine region north west of Lisboa in Portugal, and it is genetically distinct from any other known Malvasia (Rolando Faustino, personal communication; Vouillamoz). Officially known in Portugal simply but less than helpfully as Malvasia.

VITICULTURAL CHARACTERISTICS

High yields, prone to sunburn.

WHERE IT'S GROWN AND WHAT ITS WINE TASTES LIKE

Wines within the Colares DOC, on the Atlantic coast north west of Lisboa, **Portugal**, must include at least 80% of this variety. They are typically both fruity and floral but also have a mineral edge, sometimes with tropical or herbal notes. The vines are generally ungrafted in the sandy vineyards, rooted in the clay subsoils and then trailed along the sand; the fruit needs to be covered with straw towards harvest time to protect the berries from sunburn – all in all a huge amount of effort and a major reason for this wine region's slow demise. There were just 3 ha (7 acres) planted in 2010. Quinta das Vinhas de Areia and Monte Cascas make a varietal wine.

MALVASÍA DE LANZAROTE

Recently identified variety found on the Islas Canarias, easily confused with other Malvasía Somethings.

BERRY COLOUR: ● ● ● ● ●

PRINCIPAL SYNONYMS: Málaga (Tacoronte-Acentejo on Tenerife), Malvasía (Lanzarote, Gran Canaria and Valle de Güímar on the Islas Canarias), Malvasía Portuguesa (La Palma on the Islas Canarias), Malvasía Volcánica (Islas Canarias), Peregil or Perejil (Islas Canarias), Sebastián García (Islas Canarias)

VARIETIES COMMONLY MISTAKEN FOR MALVASÍA DE LANZAROTE: MALVASIA BIANCA LUNGA (Italy), MALVASIA DI LIPARI (or Malvasía de Sitges on the Islas Canarias)

ORIGINS AND PARENTAGE

This variety is found on the Islas Canarias under the ambiguous name of Malvasía but since it has a unique DNA fingerprint (Rodríguez-Torres *et al.* 2009), it was christened Malvasía de Lanzarote. DNA parentage analysis by Rodríguez-Torres *et al.* (2009) has suggested that Malvasía de Lanzarote could be a natural and local cross between MALVASIA DI LIPARI, cultivated on the Islas Canarias under the names Malvasía de Sitges and Malvasía de la Palma, and the local MARMAJUELO.

VITICULTURAL CHARACTERISTICS

Mid ripening. Small but relatively loose bunches. Better resistance to viruses, pests and diseases than MALVASIA DI LIPARI (Rodríguez-Torres *et al.* 2009).

WHERE IT'S GROWN AND WHAT ITS WINE TASTES LIKE

Malvasía de Lanzarote has not yet been included in the register of varieties in **Spain** and does not therefore appear in the official vineyard statistics but it has been found in several of the appellations on the Islas Canarias such as Lanzarote, La Palma and Valle de Güímar. Wines show typical Malvasia characteristics of light aromas, full body and full flavour. La Geria on Lanzarote make both a full-bodied sweet, tropical-fruited, rather wonderfully named Malvasía Volcánica and a dry version.

MALVASÍA DE SITGES

See MALVASIA DI LIPARI.

MALVASIA DEL LAZIO

Muscat of Hamburg sibling, grapey wines.

BERRY COLOUR: ● ● ● ● ●

PRINCIPAL SYNONYMS: Malvasia col Puntino, Malvasia Gentile, Malvasia Puntinata

VARIETIES COMMONLY MISTAKEN FOR MALVASIA DEL LAZIO: MALVASIA BIANCA DI PIEMONTE

ORIGINS AND PARENTAGE

Recent DNA profiling has shown that Malvasia del Lazio is a natural cross between MUSCAT OF ALEXANDRIA and SCHIAVA GROSSA (Lacombe *et al.* 2007), which explains the light Muscat aroma of its berries. The crossing most likely took place in Lazio, where both parents must have been cultivated in the past. Since Malvasia del Lazio and MUSCAT OF HAMBURG share the same parents, they are full siblings (see MUSCAT and Schiava Grossa pedigree diagrams).

The synonym of Malvasia Puntinata comes from the typical single dots (*punti* in Italian) on the berries.

OTHER HYPOTHESES

Supposedly imported from Greece by Genoese merchants.

VITICULTURAL CHARACTERISTICS

Susceptible to fungal diseases, especially powdery and downy mildews. Mid to late ripening. The berries have a light Muscat flavour.

WHERE IT'S GROWN AND WHAT ITS WINE TASTES LIKE

Malvasia del Lazio is cultivated predominantly in Lazio, **Italy**, mainly in the Castelli Romani area south of the capital, and usually blended with other local grapes in various DOCs such as Cerveteri, Colli Albani, Colli Etruschi Viterbesi, Colli Lanuvini and Frascati. Marino DOC also allows for a varietal Malvasia del Lazio. It is also cultivated to a limited extent in Abruzzo.

Varietal wines, mostly classified as IGT Lazio, generally have aromas of flowers and exotic or white fruits and are generally best drunk young. Examples include Pallavicini's Stillato, now made from grapes dried off the vine, Casale Pilozzo's Passione, Fontana Candida's Terre dei Grifi, Cantine San Marco's Solo Malvasia and Cantine Colli di Catone's Colle Gaio and Villa Ari Fiore di Malvasia. Its susceptibility to botrytis also makes it a useful – and sometimes the main – component in nobly rotten sweet wines such as those made by Castel de Paolis.

There were 2,626 ha (6,489 acres) planted in Italy in 2000.

MALVASIA DI CANDIA AROMATICA

Aromatic, widely distributed Malvasia found in Italy and Greece.

BERRY COLOUR: ● ● ● ● ●

VARIETIES COMMONLY MISTAKEN FOR MALVASIA DI CANDIA AROMATICA: MALVASIA BIANCA DI CANDIA

ORIGINS AND PARENTAGE

Recent DNA profiling has shown that Malvasia di Candia Aromatica is not a mutation of MALVASIA BIANCA DI CANDIA but a distinct variety, and genetically close to MALVASIA DI CASORZO (Lacombe *et al.* 2007).

Malvasia Rosa is a colour mutation of Malvasia di Candia Aromatica found in 1967 by Mario Fregoni in Val Nure in the province of Piacenza in Emilia-Romagna.

OTHER HYPOTHESES

Candia is an old name for the island of Kríti (Crete), where this variety supposedly comes from, and is grown today. However, so far there is no genetic evidence to support this theory.

VITICULTURAL CHARACTERISTICS

Mid to late ripening. Berries have light Muscat-like aromas and are susceptible to downy mildew.

WHERE IT'S GROWN AND WHAT ITS WINE TASTES LIKE

Malvasia di Candia Aromatica is authorized in the provinces of Piacenza, Parma and Reggio Emilia in Emilia-Romagna in northern **Italy** and features as part of the blend or as the dominant variety in the Colli Piacentini, Colli di Parma and Colli di Scandiano e di Canossa DOCs. It is also planted in Oltrepò Pavese in Lombardia, bottled as a varietal under the DOC of that name, and further south in Lazio, and in Campania in the Solopaca DOC (though some sources say this last is in fact MALVASIA BIANCA DI CANDIA). Like so many Malvasia Somethings, it appears in various guises: dry, sparkling, passito-style sweet wines and Vin Santo. La Stoppa and Pizzarotti produce a sweet sparkling version from 100% Malvasia di Candia Aromatica and La Stoppa's Vigna del Volta, an ageworthy, slightly sherry-like passito wine, is predominantly made from this variety, with up to 20% Moscato Bianco (MUSCAT BLANC À PETITS GRAINS). Barattieri's

Vin Santo di Albarola, Val di Nure, is a varietal wine made from sun-dried grapes and aged in barrels for about ten years. Wines are generally floral and weighty, sometimes oily if the grapes are picked late.

In 2000, 1,756 ha (4,339 acres) of Malvasia di Candia Aromatica were recorded in Italy.

The variety is also planted in various parts of **Greece**: from the island of Kríti in the southern Aegean, where Douloufakis, for example, make rose-scented dry and sweet versions, to Makedonía in the north of the mainland.

MALVASIA DI CASORZO

Rare red-berried variety used to make sweet and sparkling wines in Piemonte.

BERRY COLOUR: ● ● ● ● ●

VARIETIES COMMONLY MISTAKEN FOR MALVASIA DI CASORZO: MALVASIA DI SCHIERANO (Piemonte), MALVASIA NERA LUNGA (Piemonte)

ORIGINS AND PARENTAGE

The earliest mention of a red-berried Malvasia in Piemonte is to be found in De Maria and Leardi (1875) but it is not known whether it was a reference to MALVASIA NERA LUNGA, MALVASIA DI SCHIERANO or Malvasia di Casorzo.

Recent DNA profiling revealed a likely parent–offspring relationship between Malvasia di Casorzo and LAMBRUSCA DI ALESSANDRIA (Torello Marinoni *et al.* 2006) and suggested that Malvasia di Casorzo could be genetically close to MALVASIA DI CANDIA AROMATICA (Lacombe *et al.* 2007).

VITICULTURAL CHARACTERISTICS

Mid to late ripening.

WHERE IT'S GROWN AND WHAT ITS WINE TASTES LIKE

Malvasia di Casorzo is cultivated in Piemonte, north-west **Italy**, in the provinces of Asti (notably in Casorzo d'Asti and Grazzano Badoglio) and Alessandria (in Altavilla Monferrato, Olivola, Ottiglio and Vignale Monferrato), and probably also in Emilia-Romagna. The Casorzo or Malvasia di Casorzo DOC requires a minimum of 90% of this variety, which may be blended with FREISA, GRIGNOLINO, BARBERA or other local varieties. The red or rosé wines are generally sweet and sparkling, with low aclohol. Bricco Mondalino's Molignano, made from 100% Malvasia di Casorzo, has aromas of cinnamon, black fruits and citrus peel. Other producers include Fracchia and La Sera. Tenuta Montemagno's Nectar and Accornero's Brigantino are examples of the even rarer passito style.

Total plantings in Italy in 2000 amounted to 107 ha (264 acres).

MALVASIA DI LIPARI

Very widespread variety used for a wide array of generally sweet, often dried-grape, wines.

BERRY COLOUR:

PRINCIPAL SYNONYMS: Greco Bianco di Gerace or Greco di Gerace (Calabria in Italy), Malmsey (Madeira), Malvagia (Sardegna in Italy), Malvasia Cândida (Madeira), Malvasía de la Palma (Islas Canarias in Spain), Malvasía de Sitges (Catalunya in Spain), Malvasía de Tenerife (Islas Canarias), Malvasia delle Lipari (Sicilia in Italy), Malvasia di Cagliari (Sardegna), Malvasia di Sardegna (Sardegna), Malvasija Dubrovačka (Croatia)

VARIETIES COMMONLY MISTAKEN FOR MALVASIA DI LIPARI: GRECO (Campania), GRECO BIANCO (Calabria), MALVASIA BIANCA DI CANDIA, MALVASIA BIANCA DI PIEMONTE, MALVASÍA DE LANZAROTE

ORIGINS AND PARENTAGE

Recent DNA profiling has shown that the Italian varieties Malvasia di Lipari, Malvasia di Sardegna and Greco Bianco di Gerace plus Malvasía de Sitges (Spain), Malvasia Cândida (Madeira) and Malvasija Dubrovačka (Croatia), previously believed to be distinct cultivars, are in fact one and the same variety (Crespan, Cabello *et al.* 2006; Lopes *et al.* 2006), and that Malvasia di Lipari (under the name Malvasia di Sardegna) is genetically close to MALVAZIJA ISTARSKA, MALVASIA BIANCA LUNGA (also known as Malvasia del Chianti) and MALVASIA NERA DI BRINDISI (Lacombe *et al.* 2007). This widespread distribution strongly suggests that Malvasia di Lipari is a very old variety, but its origin and itinerary remain obscure (Crespan, Cabello *et al.* 2006).

Malvasía Rosada is a colour mutation found on the Islas Canarias, Spain; **Malvasia Cândida Roxa** is similarly a mutation with thin reddish skin grown on the island of Madeira and also found in Dão, Portugal, though the total area is less than 2 ha (5 acres).

OTHER HYPOTHESES

The synonym Greco Bianco di Gerace and the supposed introduction of the variety to Sicilia by the Greeks, using this name, in the sixth century BC suggest a Greek origin but genetic studies do not support this hypothesis (Crespan, Cabello *et al.* 2006).

VITICULTURAL CHARACTERISTICS

Best pruned long. Susceptible to powdery mildew and at risk from spring frosts. Early budding, late ripening.

WHERE IT'S GROWN AND WHAT ITS WINE TASTES LIKE

This variety is cultivated under various synonyms all round the western Mediterranean, and on the Islas Canarias and Madeira in the Atlantic:

- Malvasia di Lipari, typical of Lipari's archipelago just north of Sicilia, **Italy**, where producers such as the Barone Villagrande and Tasca d'Almerita make the famous orange-flavoured Malvasia delle Lipari passito wines. The Italian agricultural census recorded total plantings of 133 ha (329 acres) in 2000. In Sicilia in 2008, there were 122 ha (301 acres).
- Greco Bianco di Gerace, cultivated mainly in the villages of Bianco and Casignara in Calabria, where it also produces the less famous but locally acclaimed Greco di Bianco DOC passito wine.
- Malvasia di Sardegna, used to produce dry and sweet wines, sometimes fortified (and,

if so, aged for at least two years), in the DOCs of Malvasia di Cagliari and Malvasia di Bosa. Producers include Battista Colombu, Fratelli Porcu and Silattari.

- Malvasia Cândida was introduced to Madeira, **Portugal**, in the 15th century and has been used to produce Malmsey or Malvasia, the sweetest of the island's wines, although today very little remains (4 ha/10 acres in 2010) and MALVASIA BRANCA DE SÃO JORGE is the dominant variety for such wines. It is also found to a lesser extent in regions of the Portuguese mainland such as Beiras, Trás-os-Montes, the Tejo and Lisboa (just 1 ha/2.5 acres in 2010).
- Malvasía de Sitges survives on 2.5 ha (6 acres) in Catalunya, north-east **Spain**. It is bottled as an aromatic, fresh, sweet passito wine within the Penedès DO by the Hospital Sant Joan Baptista and also found very occasionally in the Catalunya DO. It may also be used to add perfume and freshness to more neutral dry whites.
- Malvasía de Tenerife or Malvasía Rosada, used on the Islas Canarias in DOs such as Abona and Valle de Güímar.
- Malvasija Dubrovačka, recognized in **Croatia** since 1385 (Archive of the Republic of Dubrovnik) but today rarely found except in the Konavle wine region around Dubrovnik, is the subject of current clonal research at the Faculty of Agriculture in Zagreb and plantings are increasing from a very small base. Wines tend to be powerful with moderate acidity and only lightly aromatic. The variety's loose bunches make it suitable for drying to make sweet wines. Recommended producers include Karaman and Crvik.

MALVASIA DI SARDEGNA

See MALVASIA DI LIPARI.

MALVASIA DI SCHIERANO

Minor, dark-skinned Malvasia from Piemonte that produces lightly aromatic sweet reds.

BERRY COLOUR: ● ● ● ● ●

PRINCIPAL SYNONYMS: Malvasia a Grappolo Corto, Malvasia di Castelnuovo Don Bosco

VARIETIES COMMONLY MISTAKEN FOR MALVASIA DI SCHIERANO: MALVASIA DI CASORZO⊗, MALVASIA NERA LUNGA⊗, Malvasia Rosa⊗ (colour mutation of MALVASIA DI CANDIA AROMATICA, Piacenza)

ORIGINS AND PARENTAGE

The earliest mention of a red-berried Malvasia in Piemonte is to be found in De Maria and Leardi (1875), but it is not known whether it was a reference to MALVASIA NERA LUNGA, Malvasia di Schierano or MALVASIA DI CASORZO. Recent DNA profiling has shown that Malvasia di Schierano is genetically close to MALVASIA DEL LAZIO and MUSCAT OF ALEXANDRIA (Lacombe *et al.* 2007).

VITICULTURAL CHARACTERISTICS

Early to mid ripening (late September). Quite good resistance to powdery mildew and botrytis bunch rot but low tolerance of downy mildew and prone to leafhoppers. Good frost resistance.

WHERE IT'S GROWN AND WHAT ITS WINE TASTES LIKE

Malvasia di Schierano is cultivated in the province of Torino in Piemonte, north-west **Italy** (in Schierano, Castelnuovo Don Bosco and Pino Torinese), where it must comprise at least 85% of the still or sparkling red Malvasia di Castelnuovo Don Bosco DOC (with up to 15% FREISA) and at least 85% of a wine labelled Collina Torinese Malvasia. Producers of varietal wines include Cascina Gilli and Cantine Bava.

There were 183 ha (452 acres) of this variety planted in Italy in 2000.

MALVASIA FINA

High-quality Portuguese variety that goes by many names, notably Boal on Madeira, and produces many styles of wine.

BERRY COLOUR:

PRINCIPAL SYNONYMS: Arinto do Dão (Dão), Assario or Assario Branco (Dão), Boal (Madeira), Boal Branco (Algarve), Boal Cachudo (Algarve), Boal da Graciosa (Açores), Boal da Madeira (Madeira), Cachudo (Tejo), Galego (Douro), Gual (Islas Canarias in Spain), Terrantez do Pico (Açores), Torrontés (Galicia in Spain)

VARIETIES COMMONLY MISTAKEN FOR MALVASIA FINA: ARINTO DE BUCELAS (Lisboa), Boal Ratinho, DONA BRANCA (Douro), FOLGASÃO (Dão and Douro), RABIGATO (Douro), SÉMILLON (Douro), SÍRIA (Guarda), VITAL

ORIGINS AND PARENTAGE

Malvasia Fina is a very old variety from either the Douro or the Dão region in Portugal. The particularly high degree of genetic diversity found in the Douro suggests this may be the variety's birthplace but neither Dão nor Lisboa can be ruled out (Rolando Faustino, personal communication). Malvasia Fina is known under various names throughout the country. The synonym Boal used on the island of Madeira is especially confusing since it is also used for several other white wine varieties such as ARINTO DE BUCELAS, Boal Ratinho (no longer cultivated), DONA BRANCA, RABIGATO, SÉMILLON and SÍRIA. The official name adopted in Portugal in 2000 for this variety was Malvasia Fina. The synonym Boal Branco is equally confusing because it may also refer to Sémillon in the Douro and to another obscure but distinct variety, now scarcely cultivated (Rolando Faustino, personal communication).

DNA parentage analysis has shown that Boal Ratinho from the Carcavelos region just west of Lisboa is an offspring of Malvasia Fina (under the name Boal da Madeira) and SÍRIA (under the name Códega; Lopes *et al.* 1999).

VITICULTURAL CHARACTERISTICS

Late budding and therefore not at risk from spring frosts and yet fairly early ripening. Quite vigorous, generally easy to grow, only moderately susceptible to downy mildew and botrytis bunch rot, more so to powdery mildew. Sometimes affected by coulure, and severe water stress may interfere with ripening.

WHERE IT'S GROWN AND WHAT ITS WINE TASTES LIKE

Malvasia Fina is the basis of Dão whites in **Portugal** and produces varietal wines that are often relatively high in alcohol with average levels of acidity. The best are fruity and perfumed, elegant and ageworthy, developing fine aromatic complexity as they age in bottle. They respond well to barrel fermentation and ageing. Fruit from cooler regions produces wines that are particularly elegant and early-harvested grapes make a good base for sparkling wines, such as those from

Varosa, Lamego and Encostas da Nave in the far north of Beiras. Total plantings in Portugal in 2010 were 2,217 ha (5,478 acres) and Malvasia Fina is authorized in DOCs such as Porto, Douro, Távora-Varosa, Dão, Beira Interior and Lagos and in many regional wines, under various synonyms, as far from its birthplace as the Algarve and the Açores. Quinta das Maias make an unfortified varietal wine in Dão, Murganheira a sparkling version in the Beiras subregion Távora-Varosa.

On the island of Madeira, Boal is grown predominantly on the southern coast, between the municipalities of Ribeira Brava and Calheta. The name Boal (known as Malvasia Fina only in official documents) was once used to identify a style of wine that was generally sweeter than Verdelho and less sweet than Malmsey. However, today a Madeira labelled Boal must comprise 85% of this variety and there were just 20 ha (49 acres) left on the island in 2010, though that area has remained stable over the last decade. The several producers of varietal Boal Madeiras include Barbeito, Blandy's and Henriques & Henriques.

MALVASIA NERA DI BASILICATA

The less common of the two dark-skinned Malvasias of southern Italy.

BERRY COLOUR: ● ● ● ● ●

PRINCIPAL SYNONYMS: Malvasia Nera

ORIGINS AND PARENTAGE

Malvasia Nera di Basilicata is not a colour mutation of MALVASIA BIANCA DI BASILICATA. It has morphological similarities with MALVASIA NERA DI BRINDISI, and some authors have suggested they might be one and the same variety but this has not yet been established by DNA testing. On the other hand, DNA profiling has established that Malvasia Nera di Brindisi and Malvasia Nera di Lecce are identical (Crespan, Colleta *et al.* 2008). Malvasia Nera di Basilicata is said to have been introduced to Basilicata from Puglia. In Toscana and in Basilicata, a few old vines of the Spanish variety TEMPRANILLO have been found under the misleading names Malvasia Nera and Malvasia Nera di Lecce (Storchi *et al.* 2009).

VITICULTURAL CHARACTERISTICS

Mid ripening. Slightly aromatic. Good resistance to downy and powdery mildews and botrytis bunch rot.

WHERE IT'S GROWN AND WHAT ITS WINE TASTES LIKE

Malvasia Nera di Basilicata is widespread in the provinces of Potenza and Matera in Basilicata in southern **Italy**, where it is sometimes blended with other local grapes, including AGLIANICO, to increase aroma, alcohol and acidity in IGT Basilicata or Grottino di Roccanova DOC. It is also authorized in the rosé version of Terre dell'Alta Val d'Agri DOC. The Italian agricultural census of 2000 distinguishes between Malvasia Nera di Basilicata, MALVASIA NERA DI BRINDISI and Malvasia Nera di Lecce and records total plantings of 848 ha (2,095 acres) for the first, with di Brindisi being much more widely planted (2,238 ha/5,530 acres).

MALVASIA NERA DI BRINDISI

The most common of the two dark-skinned Malvasias of southern Italy, commonly blended with Negroamaro, one of its parents.

BERRY COLOUR: ● ● ● ● ●

PRINCIPAL SYNONYMS: Malvasia Negra, Malvasia Nera (Alberese and Grosseto in Toscana), Malvasia Nera di Bari, Malvasia Nera di Bitonto, Malvasia Nera di Lecce☒, Malvasia Nera di Trani, Malvasia Niura

ORIGINS AND PARENTAGE

Until recently, Malvasia Nera di Brindisi and Malvasia Nera di Lecce were thought to be two distinct varieties from the Salento region in Puglia, southern Italy, but DNA profiling has recently shown that they are identical and that this variety is a natural cross between MALVASIA BIANCA LUNGA and NEGROAMARO (Crespan, Colleta *et al.* 2008). Since both parents have been cultivated in Puglia for centuries, the cross most likely happened there. Whether Malvasia Nera di Brindisi and MALVASIA NERA DI BASILICATA are the same variety has not yet been tested by DNA profiling and therefore cannot be ruled out.

In Toscana and in Basilicata, a few old vines of the Spanish variety TEMPRANILLO have been found under the misleading names Malvasia Nera (MALVASIA NERA DI BASILICATA) or Malvasia Nera di Lecce (a synonym for Malvasia Nera di Brindisi; Storchi *et al.* 2009).

VITICULTURAL CHARACTERISTICS

Mid to late ripening. Slightly aromatic.

WHERE IT'S GROWN AND WHAT ITS WINE TASTES LIKE

Malvasia Nera di Brindisi is cultivated mainly in Puglia (notably in the provinces of Lecce, Taranto and Brindisi), southern **Italy**, but also in Toscana and Calabria. It is often used to produce rosé wines in a blend with its parent NEGROAMARO and with SUSUMANIELLO. Conti Zecca produce both red and rosé versions. In most of the DOCs in which it is authorized, such as Lizzano, Nardò, Salice Salentino and Squinzano, it is the junior partner (15–30% maximum) in blends with Negroamaro. Producers of varietal wines within IGT Puglia or IGT Salento include Duca Carlo Guarini, Feudi di San Marzano, Bortrugno (both dry and passito styles) and Lomazzi & Sarli. Varietal wines' fruit can taste sweet and rather dusty.

The Italian agricultural census of 2000 distinguishes between Malvasia Nera di Brindisi and di Lecce because the DNA results (see above) were not available at that time. It records total plantings of 2,238 ha (5,530 acres) for the first and 669 ha (1,653 acres) for the second.

MALVASIA NERA DI LECCE

See MALVASIA NERA DI BRINDISI.

MALVASIA NERA LUNGA

Rare, dark-skinned Malvasia exclusive to the province of Asti.

BERRY COLOUR:

PRINCIPAL SYNONYMS: Moscatella (Alessandria)

VARIETIES COMMONLY MISTAKEN FOR MALVASIA NERA LUNGA: MALVASIA DI CASORZO (Piemonte), MALVASIA DI SCHIERANO (Piemonte), Malvasia Rosa (colour mutation of MALVASIA DI CANDIA AROMATICA in Piacenza)

ORIGINS AND PARENTAGE

In Piemonte, Malvasia Nera Lunga is often confused with MALVASIA DI SCHIERANO but it has bigger and longer clusters, and it is also very distinct from MALVASIA DI CASORZO (Schneider and Mannini 2006). Although it is also called Malvasia Rosa, it should not be confused with the bud mutation of MALVASIA DI CANDIA AROMATICA discovered by Mario Fregoni in 1967 in the province of Piacenza.

VITICULTURAL CHARACTERISTICS

Productive, with big, elongated bunches. Good resistance to mites and leafhoppers.

WHERE IT'S GROWN AND WHAT ITS WINE TASTES LIKE

Malvasia Nera Lunga is cultivated almost exclusively in the area of Castelnuovo Don Bosco and Albugnano in the province of Asti in Piemonte, north-west **Italy**, together with MALVASIA DI SCHIERANO, often interplanted. The area of Malvasia Nera Lunga was not recorded in the Italian agricultural census in 2000 because it had not been entered in the official register of varieties. However, it has been now been added to the register and, since 2008, both varieties have been authorized in the Collina Torinese and Malvasia di Castelnuovo Don Bosco DOCs.

MALVASIA PRETA

Douro variety used for port and table wines.

BERRY COLOUR:

PRINCIPAL SYNONYMS: Moreto, Mureto, Pinheira Roxa

VARIETIES COMMONLY MISTAKEN FOR MALVASIA PRETA: MORETO DO ALENTEJO, Moreto do Dão

ORIGINS AND PARENTAGE

Recent DNA parentage analysis has shown that Malvasia Preta, literally 'black Malvasia', is a natural cross between the Iberian variety CAYETANA BLANCA and the Portuguese ALFROCHEIRO (Zinelabidine *et al.* 2012), which possibly occurred in the Douro in northern Portugal, its main area of cultivation. Malvasia Preta is therefore a sibling of CORNIFESTO, CAMARATE, JUAN GARCÍA and CASTELÃO (see Cayetana Blanca for the pedigree diagram). The link between Malvasia Preta and the other Malvasia Something varieties is unknown. A large-berried mutation known as Pinheira Roxa has been observed but any link with JAMPAL, sometimes called Pinheira Branca (Lopes *et al.* 2006), is uncertain.

According to Truel in Galet (2000), Malvasia Preta could be a synonym of Moreto do Dão (no longer cultivated), but this has been disproved by DNA profiling (Veloso *et al.* 2010).

VITICULTURAL CHARACTERISTICS

Medium-sized, compact bunches of small berries. Medium to high yields, late ripening. Susceptible to powdery mildew and botrytis bunch rot, less so to downy mildew.

WHERE IT'S GROWN AND WHAT ITS WINE TASTES LIKE

Malvasia Preta is grown mostly in the Douro, often in old mixed vineyards, where it is authorized for both port and unfortified wines; it is also found to a lesser extent in Beiras and is authorized for regional wines and for the Távora-Varosa DOC. Varietal wines show aromas of red fruits (often raspberry) and a light floral note of violets. They are well structured and have the potential to age in bottle. There were 1,144 ha (2,827 acres) in **Portugal** in 2010.

MALVASIA REI

See PALOMINO FINO.

MALVASIJA DUBROVAČKA

See MALVASIA DI LIPARI.

MALVAZIJA ISTARSKA

The emphatic white wine grape of Croatia's Istra (Istria) peninsula.

BERRY COLOUR: (first of five)

PRINCIPAL SYNONYMS: Istrska Malvazija (Primorje in Slovenia), Malvasia del Carso (north-east Italy), Malvasia Friulana (north-east Italy), Malvasia Istriana (north-east Italy), Malvazija or Malvazija Istarska Bijela (Istra), Polijšakica Drnovk (Slovenia)

ORIGINS AND PARENTAGE

As the Croatian name Malvazija Istarska and the Italian name Malvasia Istriana suggest, this variety is likely to originate in Istra, the peninsula in north-west Croatia that used to belong to Italy. Although it was not mentioned in Istra until 1891, Malvazija Istarska has certainly been cultivated there for centuries. It is ampelographically quite distinct from other Malvasias such as MALVASIA DEL LAZIO and MALVASIA BIANCA DI CANDIA (Giavedoni and Gily 2005) but recent DNA profiling has shown that Malvazija Istarska is genetically close to MALVASIA DI LIPARI (aka Malvasia di Sardegna), MALVASIA BIANCA LUNGA and MALVASIA NERA DI BRINDISI (Lacombe *et al.* 2007).

Malvazija Istarska was used to breed VEGA.

OTHER HYPOTHESES

Like all Malvasia Something varieties, Malvazija Istarska is often said to originate in Greece, possibly having been introduced by the Venetians during the Venetian Republic's rule over Istra, but this can be rejected by recent genetic comparisons showing that Malvazija Istarska has no relationship to twenty-eight Greek varieties that were supposedly close to some Malvasias (Pejić *et al.* 2005).

VITICULTURAL CHARACTERISTICS

Easy to grow, vigorous, reliably good yields and not notably susceptible to any diseases except powdery mildew. It is adaptable in terms of site but produces better quality on lighter soils. Mid to late ripening. Berries can achieve high sugar levels but acidity is generally moderate.

WHERE IT'S GROWN AND WHAT ITS WINE TASTES LIKE

Although Malvazija Istarska has spread down the **Croatian** coast to northern Dalmacija (Dalmatia) and is still found in some old vineyards in Kontinentalna Hrvatska (Continental Croatia) (eg in Kalnik and Križevci), the variety's homeland is Istra, where it is the dominant variety; just about every grower bottles one or several examples. After GRAŠEVINA, by a very long way, it is Croatia's second most planted variety, with its 3,410 ha (8,426 acres) in 2009 comprising 10.5% of the total vineyard area. It is officially authorized in the Istra and Hrvatsko Primorje (Croatian Littoral) wine regions.

Varietal wines (and they usually are) vary enormously in style, from full and powerful, sometimes oaked and a little spicy, to more taut, zesty and refreshing examples, most of them having a certain lightly honeyed tang about them. They often have a slightly green cast and seem generally fresher than many other varieties with Malvasia in their name, although the alcohol can reach intrusively high levels. Recommended producers include Bienvenuti, Giorgio Clai (Clai Bijele Zemlje), Moreno Coronica, Kabola (Marino Markežić), Gianfranco Koslović, Ivica Matošević, Roxanich and Bruno Trapan. Wines that have been made using long skin maceration such as Clai's and Roxanich's show that Malvazija Istarska is well suited to this winemaking technique, adding further depth and complexity yet without bitterness or excess of tannin.

There were 466 ha (1,152 acres) in **Italy** in 2000, mostly in the north east, according to the agricultural census. This seems like a variety that could be of real interest elsewhere.

MALVERINA

Recent, late-ripening, disease-resistant Czech hybrid producing aromatic whites.

BERRY COLOUR: ● ● ● ● ●

PRINCIPAL SYNONYMS: BV-19-143

ORIGINS AND PARENTAGE

Malverina is a Rakisch × Merlan hybrid recently bred for disease resistance by Miloš Michlovský, Vilém Kraus, Lubomír Glos, Vlastimil Peřina and František Mádl in which Rakisch is a VILLARD BLANC × FRÜHROTER VELTLINER hybrid (see LAUROT for Merlan). Pollen was obtained from the Chisinau research centre in the Republic of Moldova and the crossing and selection were carried out in the Lednice, Břeclav and Perná research centres in the Czech Republic. The variety rights are owned by Vinselekt Michlovský, a breeding company created in 1995 by Miloš Michlovský in Rakvice, southern Morava, in the south east of the Czech Republic. It was included in the official Czech register of varieties in 2001.

VITICULTURAL CHARACTERISTICS

Vigorous, consistently productive, with good resistance to fungal diseases. Early budding, late ripening. Fairly large bunches of small, light-skinned berries with a pinkish hue. Well suited to fertile clay soils.

WHERE IT'S GROWN AND WHAT ITS WINE TASTES LIKE

Wines are aromatic, spicy and relatively full-bodied with fairly high acidity and typically develop an aroma of cinnamon, then bread, as they age in bottle. Sweet, later-harvest wines are possible when yields are reduced. Recommended producers include Michlovský and Rovenius. There were just 5 ha (12 acres) planted in Morava in the south east of the **Czech Republic** in 2009.

MALVOISIE

See MALVASIA.

MAMMOLO

Minor but historically significant perfumed central Italian Sangiovese blending partner, more important on the French island of Corse, as Sciaccarello.

BERRY COLOUR: ● ● ● ● ●

PRINCIPAL SYNONYMS: Broumest (Aveyron in France), Malvasia Montanaccio (Sartène on Corse), Mammolo Toscano, Mammolone di Lucca, Montanaccia (Corse), Muntanaccia (Sartène), Schiorello (Corse), Sciaccarello (Corse), Sciaccarellu (Corse)

VARIETIES COMMONLY MISTAKEN FOR MAMMOLO: BARBERA (under the name Barbera Rotonda in Piemonte), PERRICONE

ORIGINS AND PARENTAGE

Mammolo is a very old Tuscan variety: Soderini (1600) referred to both the red and white versions: the white one either corresponded to another variety or has since disappeared, the red being considered a good wine grape. The long history of Mammolo is confirmed by the significant morphological variability of its numerous clonal variations, all found and described in Toscana: Mammolo Nero Primaticcio, Mammolo Piccolo, Mammolo Grosso, Mammolo Asciuto, Mammolo Sgrigliolante, Mammola Tonda, etc. The name supposedly derives from *Viola mammola*, the Italian name for *Viola odorata* or sweet violet, a reflection of the wine's perfume.

DNA profiling has established that two varieties grown on the French island of Corse, Sciaccarellu and Malvasia Montanaccio, are identical to Mammolo (Di Vecchi Staraz, This *et al.* 2007). It is likely that Mammolo was brought from Toscana to Corse, rather than vice versa, probably during the rule of the Republic of Pisa (1077–1284) or the Republic of Genova (1284–1768), not least because parentage analysis has revealed a possible parent–offspring relationship between Mammolo and several Tuscan varieties: BIANCONE DI PORTOFERRAIO, CALORIA, COLOMBANA NERA and POLLERA NERA (Di Vecchi Staraz *et al.* 2007). These relationships illustrate the significant genetic influence of Mammolo on the Tuscan grape heritage. In addition, DNA profiling suggests a possible parent–offspring relationship with ROLLO from Liguria (Vouillamoz).

VITICULTURAL CHARACTERISTICS

Mid to late ripening and may be vigorous. Some clonal variations (see above) are particularly susceptible to botrytis bunch rot. As Sciaccarello it thrives in the dry soils of the French island of Corse.

WHERE IT'S GROWN AND WHAT ITS WINE TASTES LIKE

Once widespread in Toscana, Mammolo began to decrease in the 1960s, and today it is grown to a very limited extent mainly in the regions of Montepulciano, Siena, Arezzo and Lucca. There were just 147 ha (363 acres) in **Italy** in 2000 and it appears to be on the slippery slope to extinction.

It is authorized in the Carmignano, Chianti, Colli dell'Etruria Centrale, Monteregio di Massa Marittima, Morellino di Scansano, Parrina, Pomino, Rosso di Montepulciano and Vino Nobile di Montepulciano DOCs and is usually blended with SANGIOVESE. Wines are medium-bodied, with moderate alcohol, and become more perfumed with age. (Cennatoio's Mammolo di Toscana IGT is, confusingly, a 100% MERLOT.)

On the island of Corse, **France**, Sciaccarello (often Sciaccarellu and sometimes Sciacarello), meaning 'crunchy', is one of the main varieties in all red and rosé appellations except Patrimonio and is the dominant variety in the south of the island, particularly in the Ajaccio and Sartène regions. It is also frequently sold as an IGP L'Île de Beauté. Plantings totalled 783 ha (1,935 acres) in 2008, climbing back to the level of the late 1970s after a dip in the late 1980s. Varietal wines tend to be rather pale in colour but combine a highish level of alcohol with fresh acidity and aromas of red fruit and spices, which explains why they are often made as rosés. Notable producers include Clos Canarelli and Comte Peraldi. Some producers, such as Domaine Saparale, add a bit of VERMENTINO to their rosé, others, such as Clos Ornasca, add Nielluccio (SANGIOVESE) or Grenache (GARNACHA) in their reds.

In **Australia**, in the Barossa Valley in the state of South Australia, Thorn-Clarke co-ferment about 5% of Mammolo with NEBBIOLO to add fragrance and colour to the blend, and in the far east of Gippsland, in Victoria, Norinbee Selection Vineyards blend Mammolo with Nebbiolo and MERLOT for similar reasons.

MANDILARIA

Greek island variety producing intensely coloured, tannic wines that tend to lack body.

BERRY COLOUR: ● ● ● ● ●

PRINCIPAL SYNONYMS: Amorghiano or Amorgiano (Ródos/Rhodes), Dombrena Mavri, Doubraina Mavri or Doumpraina Mavri☒ (Attikí), Kontoura (Sámos), Koudouro or Kountoura or Koundoura or Kountoura Mavri (Achaïa, Attikí, Chíos, Évvoia/Euboea, Ikaría, Halkidikí, Lésvos/Lesbos, Magnisía, Samothráki, Thessaloníki), Mandelaria, Mandilari, Mantilari or Mantilaria (Kríti/Crete), Montoyra (Sámos)

ORIGINS AND PARENTAGE

Mandilaria originates from the islands of the eastern Aegean (Nikolau and Michos 2004) and its name may derive from the Greek *mandili*, meaning 'handkerchief' (Boutaris 2000), although no one seems to know why.

There is some confusion among the reference vines in grape collections: for example, there are two distinct references for Vilana in the Greek Vitis Database: one has a unique DNA profile, most likely the true VILANA, the other is identical to a vine labelled Mandilaria (but not true to type) in the same database and to Mantilaria in Sefc *et al.* (2000), itself distinct from the true Mandilaria (Greek Vitis Database; Benjak *et al.* 2005).

Although SAVATIANO is sometimes called Kountoura Aspri (*aspro* means 'white') and

Mandilaria is also called Kountoura Mavri (*mavro* means 'black'), Savatiano is not a colour mutation of Mandilaria, as evidenced by their distinct DNA profiles in the Greek Vitis Database.

VITICULTURAL CHARACTERISTICS

Vigorous and productive, resistant to drought but susceptible to downy mildew, sour rot and botrytis bunch rot. Late ripening. Wines are rarely more than 12% potential alcohol except in the best sites.

WHERE IT'S GROWN AND WHAT ITS WINE TASTES LIKE

Mandilaria produces wines that are very deeply coloured and tannic but generally lack body, which is why they are often blended with the softer but more alcoholic KOTSIFALI, eg in the Peza and Achárnes appellations on the island of Kríti (Crete), **Greece**. However, Lyrarakis have shown that, on the right site, a varietal Mandilaria can achieve balance while remaining imposingly tannic. On the island of Ródos, where the variety has a longer growing season and berries can achieve higher sugar levels, varietal Mandilaria qualifies for the Ródos appellation. There also seems to be a superior, smaller-berried clone known as Diminitiko growing on the higher slopes of Mount Attáviros, close to the village of Ágios Isídorus. So deeply coloured are Mandilaria wines that on the island of Páros the appellation of the same name allows the blending of white MONEMVASSIA with 20% Mandilaria for the production of red wine (Lazarakis 2005).

The variety is also planted on Santoríni and Pýlos as well as in the southern Pelopónnisos (Peloponnese) and, to a limited extent, further north on the mainland, although its produce tends to be lean and dry if not blended, eg with MAVROTRAGANO on Santoríni. Overall it is the second most planted variety on the islands in the southern Aegean (682 ha/1,685 acres in 2008).

Recommended producers include Boutari, Cair, Emery (using the synonym Amorgiano) and Mediterra. Sigalas' Apiliotis is a rare and outstanding example of an intensely sweet but tannic red made from sun-dried Mandilaria.

MANDÓN

Virtually extinct variety from Bierzo, north-west Spain.

BERRY COLOUR: ● ● ● ● ●

PRINCIPAL SYNONYMS: Galmeta, Galmete, Mandó, Valenciana Tinta (Arribes del Duero)

ORIGINS AND PARENTAGE

Mandón is an endangered variety from the Bierzo region in the province of León in north-west Spain that was recently suggested by DNA parentage analysis to be a natural cross between GRACIANO (from Rioja and Navarra) and Hebén, an almost extinct Andalusian variety, which would make it a half-sibling of GORGOLLASA from Mallorca and MACABEO from Catalunya (García-Muñoz *et al.* 2012). While the parent–offspring relationship with Graciano is backed up by Santana *et al.* (2010), the complete parentage was based on too few DNA markers (twenty) and needs to be confirmed.

WHERE IT'S GROWN

Official 2008 statistics recorded just 1 ha (2.5 acres) of Mandón in 2008, in Castilla-La Mancha, **Spain**, but there have recently been new plantings in Arribes del Duero in an attempt to halt its disappearance. Celler del Roure in Valencia use the variety in their Maduresa blend.

MANDRÈGUE

Ancient and recently rescued Ariège variety.

BERRY COLOUR: ● ● ● ● ●

PRINCIPAL SYNONYMS: Manrègue, Néral

VITICULTURAL CHARACTERISTICS

Mid to late ripening.

WHERE IT'S GROWN AND WHAT ITS WINE TASTES LIKE

Isolated vines of this almost forgotten variety were discovered in Marmande, between Bordeaux and Toulouse, **France**, under the synonym Néral (Galet 1990). Recently, Les Vignerons Ariégeois in Montégut-Plantaurel (just west of Pamiers) came to the rescue by planting 30–40 vines of this and other endangered local varieties, including BERDOMENEL, CAMARALET DE LASSEUBE, CANARI NOIR and Tortozon (PLANTA NOVA; Carbonneau 2005). They aim to multiply these and to produce a commercial wine if their experiments yield encouraging results.

MANSENG NOIR

Rare but characterful red wine grape from south-west France, also found in north-west Spain.

BERRY COLOUR: ● ● ● ● ●

PRINCIPAL SYNONYMS: Caíño do Freixo (Galicia in Spain), Ferrón or Ferrol (Galicia), Mansenc Noir, Noir du Pays (Pyrénées)

VARIETIES COMMONLY MISTAKEN FOR MANSENG NOIR: ARROUYA (Jurançon region), COURBU NOIR (Pyrénées-Atlantiques)

ORIGINS AND PARENTAGE

Manseng Noir used to be widespread in the Basque Country, where it was first mentioned in 1783–4 (Rézeau 1997): 'Manseng noir … il est bon à manger et fait le vin bon, il est monté en couleur' ('Manseng noir … it is good to eat and makes good wine, it is highly coloured'). In the Pyrénées-Atlantiques, Manseng Noir and COURBU NOIR were sometimes confused and both were called Noirs du Pays (Herriko Beltza in Basque dialect), in particular in Irouléguy. Manseng Noir is a member of the Courbu ampelographic group (see p XXVII), so it is not surprising that a possible parent–offspring relationship between Manseng Noir and TANNAT (of the same group) was recently suggested by the analysis of twenty-two DNA markers (Vouillamoz; to be confirmed with more markers).

Unexpectedly, the DNA profile of Ferrón published in Ibañez *et al.* (2003) and in Martín *et al.* (2003) turned out to match the profile of Manseng Noir (Vouillamoz). This is surprising because Ferrón is considered to be indigenous in the region of Orense in Galicia, north-western Spain, where it is also known as Caíño do Freixo (Santiago, Boso, Martín *et al.* 2005). This could be explained by the similarities between the climate of Galicia and that of the Basque region, but the historical presence of Manseng Noir in Galicia under different names does not challenge its

Basque origin, not least because it is a member of the Courbu ampelographic group, typical of south-west France.

Manseng Noir is morphologically and genetically distinct from GROS MANSENG and PETIT MANSENG.

VITICULTURAL CHARACTERISTICS

Vigorous but needs to be pruned long because the basal buds are not very fertile. Mid to late ripening. Good resistance to downy mildew but not to botrytis bunch rot or powdery mildew. Bunches are compact.

WHERE IT'S GROWN AND WHAT ITS WINE TASTES LIKE

Wines tend to be very deeply coloured, astringent and fairly acidic but with good ageing potential and thus make good blending components. According to Lavignac (2001), plantings were on the increase but in 2008 there were just 3 ha (7 acres) in the Pyrénées-Atlantiques of **France**, where it is a permitted minor ingredient in the Béarn appellation. It was originally interplanted with COURBU NOIR in the Irouléguy region. The variety has been included in the Saint-Mont conservation vineyard and Antonin Nicollier at Domaine de Pouypardin in nearby Condom, between Bordeaux and Toulouse in the Gers, planted 10 ares, or one tenth of a hectare (1,200 square yards) in hot, clay-limestone soils in 2005.

Ferrón is listed as one of the principal varieties permitted in Galicia's Ribeiro appellation in **Spain** and producers such as O'Ventosela, Viña Mein and Docampo Diéguez Colleteiro include it in their red blends, generally with MENCÍA and other local varieties. Spain recorded a total of 4 ha (10 acres) in 2008.

MANTEÚDO

See LISTÁN DE HUELVA.

MANTO NEGRO

Principal dark-skinned variety on the Spanish island of Mallorca, producing pale, fruity, high-alcohol wines.

BERRY COLOUR: ● ● ● ● ●

PRINCIPAL SYNONYMS: Cabelis, Mantonegro, Mantuo Negro

ORIGINS AND PARENTAGE

Manto Negro most likely originates from Mallorca in the Islas Baleares. Recent DNA parentage analysis has suggested that Manto Negro is a natural cross between two other, obscure and no longer cultivated, varieties from Mallorca, namely Sabaté and Callet Cas Concos, the latter being proposed as a natural cross between BEBA from Andalucía and GIRÒ from Sardegna (García-Muñoz *et al.* 2012). This would make Manto Negro a grandchild of Beba and Girò and a half-sibling of CALLET, also from Mallorca. However, these parentages were based on too few DNA markers (twenty) and need to be confirmed.

Although it is also known as Mantúo Negro, it is not a black mutation of CHELVA, called Mantúo in Extremadura.

VITICULTURAL CHARACTERISTICS

Mid budding, mid to late ripening, very productive in rich soils. Relatively resistant to downy and powdery mildews but susceptible to botrytis bunch rot. Relatively large, thick-skinned berries.

WHERE IT'S GROWN AND WHAT ITS WINE TASTES LIKE

All 320 ha (791 acres) in 2010 of Manto Negro are to be found on the island of Mallorca, **Spain**, where it is the most widely planted variety with just over 20% of the total vineyard area. It is the main variety in the Binissalem-Mallorca DO (at least 30% of any blend) and is authorized for the Plà i Llevant DO, also on Mallorca. Although it is widely planted on the island, it is much more successful and common in the more westerly Binissalem region than in the east and does particularly well on stony soils. Wines are generally soft, light in colour – some berries are closer to pink than black – and high in alcohol and therefore often blended with a small amount of the more structured and lower-alcohol CALLET. Young vines tend to produce red-fruit flavours in the wine but older plants with lower yields result in greater concentration, dark-fruit flavours and more marked aromas – wines that respond well to maturation in large oak barrels but do not age particularly well because the variety is prone to oxidation. Recommended producers include Binigrau, José Luis Ferrer, Macià Batle, Nadal and Ribas.

MANTONICO BIANCO

Minor light-skinned Calabrian.

BERRY COLOUR:

PRINCIPAL SYNONYMS: Mantonico Pizzutella, Mantonacu Viru della Locride, Mantonico Vero

VARIETIES COMMONLY MISTAKEN FOR MANTONICO BIANCO: GUARDAVALLE (Cosenza), MONTONICO BIANCO

ORIGINS AND PARENTAGE

Mantonico Bianco is typical of the Locride area, around Locri, in the province of Reggio Calabria in southern Italy, mentioned perhaps as early in 1601 by Marafioti (Schneider, Raimondi, De Santis and Cavallo 2008). It should not be confused with MONTONICO BIANCO, cultivated in central and southern Italy, or with GUARDAVALLE, which is called Mantonico in Cosenza. Recent morphological and DNA studies have established that Mantonico Pizzutella, also cultivated in Calabria, is identical to Mantonico Bianco (Zappia *et al.* 2007). Recent DNA parentage analysis has suggested that Mantonico Bianco is a possible parent of GAGLIOPPO from Calabria and of Nerello Calabrese (aka NERELLO MASCALESE) from Sicilia (Cipriani *et al.* 2010).

Mantonico Bianco is not a white mutation of any of the four distinct grapes named Mantonico Nero (Schneider, Raimondi, De Santis and Cavallo 2008).

VITICULTURAL CHARACTERISTICS

Very late ripening. Good resistance to fungal diseases.

WHERE IT'S GROWN AND WHAT ITS WINE TASTES LIKE

Mantonico Bianco is cultivated along the Ionian coast of Calabria, **Italy**, mainly in the region of Bianco and Casignana. It is used to produce both dry and passito-style sweet wines and varietal examples include Librandi's elegant, minerally and oak-aged Efeso (IGT Val di Neto) and their lightly sweet, exotic-fruit-flavoured Le Passule. Statti's floral and herbal Nosside Passito (IGT Calabria) is predominantly Mantonico Bianco but they also make a dry varietal wine.

Mantonico Bianco is mentioned in the regulations for DOCs such as Bivongi and Donnici but, unsurprisingly, it is not clear whether this refers to Mantonico Bianco or MONTONICO BIANCO.

MANZONI BIANCO

Potentially high-quality northern Italian white cross which has inherited the positive qualities of its parents.

BERRY COLOUR: ● ● ● ● ●

PRINCIPAL SYNONYMS: I M 6.0.13, Incrocio Manzoni 6.0.13, Manzoni

ORIGINS AND PARENTAGE

A RIESLING × PINOT BLANC cross obtained by Luigi Manzoni at the Conegliano research centre in the Veneto in northern Italy between 1930 and 1935. Some recent studies suggested that Manzoni could have used CHARDONNAY instead of Pinot Blanc (Giavedoni and Gily 2005) but DNA profiling has disproved this hypothesis (Vouillamoz).

VITICULTURAL CHARACTERISTICS

Early to mid ripening, adaptable to a range of different soils and climates, good resistance to powdery mildew, esca and sour rot.

WHERE IT'S GROWN AND WHAT ITS WINE TASTES LIKE

Manzoni Bianco is the most successful of Manzoni's crosses and for DOC wines is grown mainly in the Veneto, **Italy**, in particular in the province of Treviso, where it is made as varietal wine in Vincenza DOC and included in the blend for Colli di Conegliano Bianco DOC (minimum 30%). Since the end of the 1960s it has also been planted in Friuli and Trentino and listed as a possible component in Trentino Bianco DOC. Today it is cultivated throughout Italy, especially in the south (Calabria, Puglia, Molise). Total plantings of 38,668 ha (95,550 acres) were recorded in the Italian agricultural census in 2000. The wines are usually dry but in the last few years some producers have started making sparkling wines (spumanti and frizzanti) from late-harvest grapes. Giuseppe Fanti makes a concentrated, full-bodied, ripe, orange-peel-scented off-dry varietal wine classified as IGT Vigneti delle Dolomiti. Other examples of varietals include Ca' Di Rajo, Italo Cescon (Treviso), Paladin (Annone Veneto) and Colvendra (Treviso).

Carlos Esteva of Can Rafols dels Caus has imported the variety to Penedès in north-east **Spain** and makes an impressive oak-aged varietal Manzoni Bianco called El Rocallis. It is rich, smooth and highly aromatic with notes of jasmine, verbena, lime and honey.

MANZONI MOSCATO

One of the least planted of Manzoni's crosses, this dark-skinned version is used to make sweet, sparkling rosé in Trentino.

BERRY COLOUR: ● ● ● ● ●

PRINCIPAL SYNONYMS: Incrocio Manzoni 13.0.25

ORIGINS AND PARENTAGE

This is a RABOSO VERONESE × MUSCAT OF HAMBURG cross obtained by Luigi Manzoni at the Conegliano research centre in Italy's Veneto between 1930 and 1935, in order to combine the robustness of the former with the softness, vigour and productivity of the latter.

VITICULTURAL CHARACTERISTICS

Vigorous, productive and late ripening. Good resistance to powdery mildew.

WHERE IT'S GROWN AND WHAT ITS WINE TASTES LIKE

Manzoni Moscato is cultivated to a very limited extent in the province of Treviso (particularly in Tezze, Fontanelle, Mareno di Piave and Cessalto) in the Veneto, **Italy**. Casa Roma and Villa Almè, for example, both make a light, fresh, off-dry varietal sparkling rosé; Tenuta San Giorgi's Bizzarro version is sweeter and more deeply coloured. The variety is also blended with the pink-berried MANZONI ROSA to produce Manzoni Liquoroso.

MANZONI ROSA

Very rare Veneto pink-berried cross.

BERRY COLOUR: ● ● ● ● ●

PRINCIPAL SYNONYMS: Incrocio Manzoni 1-50

ORIGINS AND PARENTAGE

A TREBBIANO TOSCANO × GEWÜRZTRAMINER (Traminer Aromatico) cross created by Luigi Manzoni at the Conegliano research centre in Italy's Veneto between 1924 and 1930 with the aim of combining the robustness and productivity of Trebbiano Toscano with the finesse of Gewürztraminer. From the four crossings obtained, Manzoni selected the one with pink and aromatic berries.

VITICULTURAL CHARACTERISTICS

Early to mid ripening; a robust and fertile vine with consistent yields; good resistance to fungal diseases.

WHERE IT'S GROWN

Manzoni Rosa was planted in the 1950s and 1960s around Conegliano and between Treviso and Venezia in the Veneto, northern **Italy**, but today there are only a few vineyards left, although Casa Roma does make a varietal wine. At the oenology school at Conegliano it is blended with MANZONI MOSCATO to produce an experimental wine known as Manzoni Liquoroso.

MARA

Recent, promising, early-ripening Swiss cross.

BERRY COLOUR: ● ● ● ● ●

PRINCIPAL SYNONYMS: C41, C41 (Gamay × Reichensteiner)

ORIGINS AND PARENTAGE

Mara is a GAMAY NOIR × REICHENSTEINER cross obtained in 1970 by André Jaquinet at the Caudoz viticultural research centre in Pully, just outside Lausanne, Switzerland, today part of Agroscope Changins-Wädenswil. It was initially labelled C41 before being selected and christened Mara on its release in 2009, nineteen years after its siblings GAMARET and GARANOIR.

VITICULTURAL CHARACTERISTICS

Fertile and consistently productive. Early budding and ripening. Resistant to botrytis bunch rot. Adaptable but not suited to sites that are too hot or dry. Loose bunches of smallish berries.

WHERE IT'S GROWN AND WHAT ITS WINE TASTES LIKE

Mara wines are generally well coloured and full-bodied with flavours of dark fruit and spice, the tannins slightly softer and more rounded than those of GAMARET. Recommended producers, all in the canton of Vaud, **Switzerland**, include Domaines Henri & Vincent Chollet, Christian Dugon and Pierre Monachon.

MARAŠTINA

See MALVASIA BIANCA LUNGA.

MARATHEFTIKO

Rare, high-quality dark-skinned Cyprus variety that is starting to gain in popularity despite difficulties in the vineyard.

BERRY COLOUR: ● ● ● ● ●

PRINCIPAL SYNONYMS: Aloupostaphylo, Bambakada (Pitsilia), Bambakina (Pitsilia), Maratheftico, Marathefticon, Marathophiko, Mavrospourtiko, Pambakada (Pitsilia), Pambakina (Pitsilia), Pampakia (Pitsilia), Vambakadha, Vambakina, Vamvakada

ORIGINS AND PARENTAGE

Maratheftiko is indigenous to the island of Cyprus, where it was first mentioned by the French ampelographer Pierre Mouillefert (1893) under the name Marathophiko as a grape with large berries, easy to crush and very susceptible to powdery mildew. The name Maratheftiko probably derives from Marathasa, the name of a valley in the Troodos mountains on Cyprus. According to DNA profiling, Maratheftiko is genetically close to Aspro X (Hvarleva, Hadjinicoli *et al.* 2005), an obscure variety from Cyprus that is distinct from the more common Aspro, which is a clone of XYNISTERI.

VITICULTURAL CHARACTERISTICS

Vigorous and prone to poor fruit set (flowers are anatomically hermaphrodite but physiologically female only), often leading to loose bunches with thick-skinned berries of various sizes. Better set on later lateral shoots but leading to varying levels of ripeness and high yields in some years, making picking particularly difficult. Early budding, late ripening. Good general disease resistance but susceptible to powdery mildew.

WHERE IT'S GROWN AND WHAT ITS WINE TASTES LIKE

Maretheftiko vines have traditionally been planted on **Cyprus** in mixed vineyards to increase the colour and body of the lower-quality but far more productive MAVRO, which has always been more popular with growers paid by weight. The greatest concentration is found in the mountainous region of Páfos and also in Pitsilia, where it is commonly known as Vamvakada. KEO also has a large area of older vines in Malia and their Heritage is one of the oldest examples of a varietal Maretheftiko. Increasing recognition of Maretheftiko's quality potential has encouraged single-vineyard plantings but there were only 178 ha (440 acres) in 2010, just 1.8% of the vineyard area, and many of these vines are still quite young. Further work is needed to improve clonal selection and resolve the issues of poor pollination and inconsistent yields (by alternate-row plantings with SPOURTIKO, for example) but already the best wines, often made from the oldest, ungrafted, interplanted vines at higher altitudes, are deeply coloured, fresh, rich in dark berry and cherry fruit, sometimes slightly herbal or scented with violets. They have rounded tannins when fully ripe, an affinity with oak and the potential to age gracefully – although producers are still experimenting and a 'typical' Maretheftiko does not yet exist. Recommended producers include Aes Ampelis, Ezousa, KEO, Tsiakkas (labelled Vamvakada), Vardalis, Vasa, Vlassides and Zambartas. Aes Ampelis and Ezousa also make good rosés.

MARCHIONE

Very rare, recently rescued Puglian.

BERRY COLOUR: ● ● ● ● ●

ORIGINS AND PARENTAGE

Several eighteenth- and nineteenth-century documents mention the cultivation of Marchione in the Valle d'Itria in Puglia, its most likely place of origin, together with other local varieties such as VERDECA and BIANCO D'ALESSANO.

Cipriani *et al.* (2010) have recently suggested through DNA parentage analysis that Marchione might be a Ragusano × MONTONICO BIANCO natural cross but the true identity of Ragusano is unknown here because the name is given to at least three other varieties (MOSCATELLO SELVATICO, GRECO/Asprinio, and MALVASIA DI LIPARI/Malvasia di Sardegna).

VITICULTURAL CHARACTERISTICS

Susceptible to fungal diseases.

WHERE IT'S GROWN AND WHAT ITS WINE TASTES LIKE

Formerly widespread, Marchione has been largely abandoned and today thrives only in Ostuni (in the Valle d'Itria, province of Brindisi), **Italy**, thanks to the Santoro family, who rescued it and now cultivate just 150 vines. They blend it with VERDECA and BIANCO D'ALESSANO to produce a fresh and aromatic white wine, and they have had promising results with experimental sparkling wines. The Santoros are planning to expand the cultivation of Marchione and other obscure local varieties.

MARÉCHAL FOCH

French hybrid that has been very successful in the colder climes of the US and Canada.

BERRY COLOUR:

PRINCIPAL SYNONYMS: Foch, Kuhlmann 188-2, Marechal Foch, Marshal Fosh

ORIGINS AND PARENTAGE

A complex hybrid obtained in 1911 by Eugène Kuhlmann in Colmar by crossing Millardet et Grasset 101 OP with GOLDRIESLING, where Millardet et Grasset 101 OP is an open pollination of Millardet et Grasset 101.

The same crossing produced Maréchal Foch's siblings, LÉON MILLOT and LUCIE KUHLMANN, and Maréchal Foch was used to breed MILLOT-FOCH.

OTHER HYPOTHESES

Some have argued that Maréchal Foch might in fact be an Oberlin Noir (or Oberlin 595) × PINOT NOIR hybrid, while others have suggested it is an Oberlin Noir × KNIPPERLÉ hybrid, or even an Oberlin Noir × GOLDRIESLING hybrid. The DNA data available to date is not sufficient to confirm or reject any of these hypotheses (Vouillamoz).

VITICULTURAL CHARACTERISTICS

Vigorous and needs to be pruned long. Early ripening and good winter hardiness (to about −25 °F/−32 °C). Small berries and bunches that attract the birds.

WHERE IT'S GROWN AND WHAT ITS WINE TASTES LIKE

Maréchal Foch is well suited to cold climates and is widely planted in Canada and the eastern and Midwestern US. Wines may be deeply coloured and quite tannic, sometimes on the herbaceous side or slightly smoky, but they avoid the foxy character of some hybrids that have American species in their pedigree. For lighter styles, carbonic maceration is often used but the variety can also be aged in oak for a more sturdy result.

European plantings are relatively small: 13 ha (32 acres) in **France** (2008), though it was once quite widely planted in the Loire, and 12 ha (30 acres) in **Switzerland** (2009), mainly in the eastern, German-speaking part of the country.

The variety was introduced to **Canada** in the late 1940s and there are now 171 acres (69 ha) in Ontario (2006), 122 acres (49 ha) in British Columbia (2008) and small areas in Nova Scotia and Québec, where it is the most widely planted red variety. **Iowa** (93 acres/38 ha in 2006) and **Illinois** (80 acres/32 ha in 2007) have the most significant plantings in the US Midwest but its most popular and successful second home is **New York** State, where there were 144 acres (58 ha) in 2006. There are also significant plantings in **Oregon.** Producers of varietal wines include Kittling Ridge (Niagara Escarpment, Ontario), Prejean (Finger Lakes, New York), Wollersheim (**Wisconsin**) and 5 Trails (**Nebraska**).

MARIA GOMES

See FERNÃO PIRES.

MARMAJUELO

Rare variety from Spain's Islas Canarias producing crisp, aromatic and potentially high-quality whites.

BERRY COLOUR:

PRINCIPAL SYNONYMS: Bermejuela (Islas Canarias), Marmajuela (Islas Canarias), Vermejuelo (Islas Canarias)

ORIGINS AND PARENTAGE

Marmajuelo most probably originates on the Islas Canarias in Spain, where DNA parentage analysis has suggested that the local MALVASÍA DE LANZAROTE is a natural cross between MALVASIA DI LIPARI, called Malvasía de Sitges and Malvasía de la Palma on the islands, and Marmajuelo (Rodríguez-Torres *et al.* 2009).

VITICULTURAL CHARACTERISTICS

Compact bunches of berries that retain high acidity. Highly susceptible to coulure.

WHERE IT'S GROWN AND WHAT ITS WINE TASTES LIKE

Marmajuelo is permitted in almost all the Islas Canarias DOs but is one of the principal varieties in El Hierro, La Palma and Ycoden-Daute-Isora under the name Bermejuela and in Tacoronte-Acentejo as Marmajuelo. Wines are aromatic and full-bodied, have a very good level of acidity and are frequently used in blends because there is so little planted, although there is an increasing interest in producing varietal wines, such as the one by Viñátigo. There were 24 ha (59 acres) on the Islas Canarias, **Spain**, in 2008.

MARQUETTE

Promising, increasingly popular, cold-hardy complex American hybrid.

BERRY COLOUR:

PRINCIPAL SYNONYMS: MN 1211

ORIGINS AND PARENTAGE

A complex hybrid of MN 1094 and Ravat 262 created in 1989 by Peter Hemstad and James Luby at the University of Minnesota's Horticultural Research Center in Excelsior, US, in which:

- MN 1094 is a hybrid between MN 1019 and MN 1016
- MN 1019 is a Riparia 64 × CARMINE hybrid, in which Riparia 64 is a *Vitis riparia* vine selected at Geisenheim for use as a rootstock
- MN 1016 is a Mandan × LANDOT NOIR hybrid

- Mandan is a Wilder × *Vitis riparia* hybrid
- Wilder is a Carter × SCHIAVA GROSSA (under the name Black Hamburg) hybrid
- Carter (also called Mammoth Globe) is an offspring of ISABELLA and an unknown parent selected by Charles Carter in Virginia
- Ravat 262 (or Ravat Noir) is a Seibel 8365 × PINOT NOIR hybrid (see FLORENTAL for the pedigree of Seibel 8365).

This grandchild of PINOT NOIR (and cousin of FRONTENAC) is therefore an extremely complex *Vitis riparia*, *Vitis labrusca*, *Vitis vinifera*, *Vitis aestivalis*, *Vitis lincecumii*, *Vitis rupestris*, *Vitis cinerea* and *Vitis berlandieri* hybrid. It was selected in 1994 and tested as MN 1211 before it was named in 2005 after Père Marquette, a seventeenth-century Jesuit missionary and explorer, and released in 2006.

VITICULTURAL CHARACTERISTICS

Moderately vigorous with small to medium-sized clusters of smallish berries. Early budding and therefore at risk from spring frosts but very hardy (–20 °F/–29 °C to –30 °F/–34 °C). Moderately resistant to foliar phylloxera but good resistance to downy mildew, powdery mildew and black rot. Early to mid ripening.

WHERE IT'S GROWN AND WHAT ITS WINE TASTES LIKE

Marquette is one of the most promising recent cold-hardy American hybrids in the Midwest and the Northeast as well as in Québec, **Canada**, thanks to its performance in the vineyards and high potential wine quality. Alcohol can be relatively high but varietal wines have cherry and blackcurrant flavours as well as more complex notes of black pepper and spice. They have relatively high acidity, which needs to be softened by malolactic fermentation, and noticeable tannins. Early trials with barrel ageing have shown positive results. Plantings are on the increase. For example, there were 214 acres (87 ha) in **Minnesota** in 2007, 10 acres (4 ha) in **Indiana** in 2010, 6 acres (2.4 ha) in **Iowa** in 2006 and 1 acre (0.4 ha) in **Pennsylvania** in 2008. Producers of varietal wines include Hinterland, Indian Island, Lincoln Peak, Parley Lake and St Croix in Minnesota.

MARSANNE

Flavourful and potentially very high-quality light-skinned Rhône variety with notable outposts in Australia and the US.

BERRY COLOUR: ● ● ● ● ●

PRINCIPAL SYNONYMS: Avilleran (Isère), Ermitage⚭ or Hermitage (Valais in Switzerland), Grosse Roussette (Saint-Péray in the Rhône), Marsanne Blanche, Roussette de Saint-Péray (Ardèche)

VARIETIES COMMONLY MISTAKEN FOR MARSANNE: ROUSSANNE⚭

ORIGINS AND PARENTAGE

Marsanne is a traditional variety from the Vallée du Rhône, where it was first mentioned in 1781 in a description of the white wine from Hermitage (Rézeau 1997): 'ce vin blanc se fait avec la Roussane, & plusieurs personnes y mêlent de la Marsanne, autre raisin blanc qui rend le vin un peu liquoreux' ('this white wine is made from Roussane, and several people add in some Marsanne, another white-berried variety that makes the wine a little sweet'). The variety was most likely named after the commune of Marsanne near Montélimar in the Drôme, its most probable birthplace.

Marsanne is a member of the Sérine ampelographic group (see p XXVII; Bisson 2009). DNA parentage analysis strongly suggests a parent–offspring relationship between Marsanne and ROUSSANNE (Vouillamoz), which might explain why they are often confused, as in Cipriani *et al.* (2010).

VITICULTURAL CHARACTERISTICS

Late budding and mid ripening, vigorous, fertile and productive. Best pruned short and suited to poor, stony soils. Susceptible to powdery mildew, mites and botrytis bunch rot. Large bunches but small berries.

WHERE IT'S GROWN AND WHAT ITS WINE TASTES LIKE

Marsanne is often blended with ROUSSANNE, and occasionally VIOGNIER, especially in the Rhône, **France**, although it has effectively taken over from its blending partner Roussanne in northern Rhône appellations such as Saint-Péray, Saint-Joseph, Crozes-Hermitage and even Hermitage. It is one of the six principal varieties permitted in white Côtes du Rhône – although, unlike its north Rhône blending partner Roussanne, Marsanne was unknown in Châteauneuf-du-Pape in the 1930s, when its rules were drawn up, so is not permitted in Châteauneuf. Alone it produces wines that are generally deep-coloured, full-bodied, occasionally positively fat, and scented with only moderate acidity and flavours ranging from honeysuckle to rich almond paste via pear. Chapoutier's varietal Chante Alouette, Ermitage de l'Orée, Ermitage le Méal and the sweet Ermitage Vin de Paille demonstrate the potential quality of the variety and that the best examples can age well. Further south and west, too, it is gaining popularity in the Aude, where there were 239 ha (591 acres) in 2009 (compared with 270 ha/667 acres in the Drôme and 226 ha/558 acres in the Ardèche). Varietal Marsannes from the Languedoc are by no means uncommon, as well as blends of it with various Rhône and local grapes.

In **Italy**, where there were 50 ha (124 acres) planted in 2000, it is specifically mentioned in the regulations for only one wine: Vin Santo Vigoleno, a locally famous blended passito wine within the Colli Piacentini DOC in Emilia-Romagna.

There are very few growers of Marsanne in **Portugal** but Esporão in the Alentejo have 11 ha (27 acres) and use it in their Private Selection white blend.

Switzerland's 47 ha (117 acres), almost exclusively in the Valais, are used for both dry and sweet wines though the latter are the most prized. Marie-Thérèse Chappaz' Ermitage Grain Noble is one of the locally most sought-after sweet wines.

California's Rhône Rangers are the main proponents of the 93 acres (38 ha) planted in California, mostly in Monterey, San Luis Obispo, Santa Barbara and Sonoma. Recommended producers include Qupé and Beckmen in Santa Ynez Valley and Cline in Carneros. Marsanne is not restricted to the Sunshine State: Airfield Estates in **Washington** State's Yakima Valley, for example, have planted Marsanne to use as a blending partner. There are scatterings of the variety all over the US.

In **Australia**, Marsanne is planted in pockets in most wine-growing states but is most closely associated with one producer in Nagambie Lakes, Victoria: Tahbilk, who have been growing Marsanne since the 1860s and still have vines from the end of the 1920s, have the largest Marsanne vineyard area in the world. Here too is consistent evidence of the variety's ability to age. Total Australian plantings in 2008 amounted to over 200 ha (494 acres).

MARSELAN

One of the most successful relatively recent French crosses.

BERRY COLOUR: ● ● ● ● ●

PRINCIPAL SYNONYMS: INRA 1810-68

ORIGINS AND PARENTAGE

A CABERNET SAUVIGNON × GARNACHA cross obtained by Paul Truel in 1961 at INRA (Institut National de la Recherche Agronomique) and ENSAM (École Nationale Supérieure d'Arts et Métiers) in Montpellier, southern France, in order to get a productive large-berried variety. However, Marselan turned out to have small berries and was initially neglected because it did not offer the high yields prized then. With the increasing demand for varieties that combined good disease resistance with high quality, Marselan was subsequently unearthed and included in the official register of varieties in 1990. Its name refers to Marseillan, the commune on the Mediterranean coast between Montpellier and Béziers housing the grape collection of INRA Montpellier's Domaine de Vassal.

VITICULTURAL CHARACTERISTICS

Mid to late ripening. May be pruned to short cordons. Good resistance to powdery mildew, mites and especially to botrytis bunch rot as well as to coulure. Big bunches but small berries.

WHERE IT'S GROWN AND WHAT ITS WINE TASTES LIKE

Marselan wines are generally aromatic, deeply coloured, well structured but with supple tannins and the potential to age well. **France** already had 1,356 ha (3,351 acres) planted by 2006 and 2,375 ha (5,869 acres) in 2009, mainly in the Languedoc and southern Rhône. Successful varietal examples have been made by Paul Mas, Mas de Rey, Domaine de la Camarette, Château Camplazens and Domaine de Couron, while Domaine de la Mordorée have produced a Marselan/MERLOT blend.

Another early adopter was Carlos Esteva at Can Rafols dels Caus, whose first Marselan vines were planted in the mid 1990s in Penedès, north-east **Spain**. Also in Catalunya, there are experimental plantings in the Terra Alta appellation.

In **California**, Marselan has been sold by Sunrise Nurseries. In **Argentina** there were 79 ha (195 acres) in 2008 – Zuccardi seem to be enthusiastic early adopters, having planted their vines in 1998. **Brazil** had 24 ha (59 acres) in 2007; producers include Cave Antiga and Serra Gaúcha.

Even in **China**, Marselan has found friends: the Sino-French Demonstration Vineyard, close to the Great Wall in the Hualai county of Hebei province, has for several years been producing a varietal Marselan, as has Grace Vineyard in Shanxi province.

MARUFO

Iberian variety that goes by many names and whose light wine is used mainly in blends.

BERRY COLOUR:

PRINCIPAL SYNONYMS: Brujidera or Brujidero (Toledo in Castilla-La Mancha in central Spain as well as Arribes del Duero in Castilla y León in western Spain), Brujigero, Crujidera or Crujideiro (Cuenca and Sacedón near Guadalajara, as well as in Alicante near Valencia in Spain), Marufa (Pinhel region in Beira Interior Norte in Portugal), Moravia Dulce (Cuenca), Mourisco (northern Portugal), Mourisco du Douro (Douro in Portugal), Mourisco Preto (Portugal and Australia), Mourisco Tinto (Douro), Rucial (Casasimarro in Cuenca), Trujidera (Portugal), Vigorosa (Sierra de Francia in Spain)

VARIETIES COMMONLY MISTAKEN FOR MARUFO: GROSSA or Tinta Grossa (Portugal), MORAVIA AGRIA

ISABELLA

pp 474–5

KADARKA

pp 490–91

KOKOUR BLANC

see KOKUR BELY

p 508

MISSION

see LISTÁN PRIETO
pp 550–52

MACABEO

pp 560–61

BOAL

see MALVASIA FINA

pp 582–3

MARSANNE

pp 600–601

MAVRUD

pp 614–15

MAZUELO

pp 616–19

MERLOT

pp 629–34

MOURVÈDRE

see MONASTRELL

pp 646–8

MONDEUSE BLANCHE

pp 649–50

MONDEUSE NOIRE

pp 650–51

MTSVANE KAKHURI

p 674

MUSCAT BLANC À PETITS GRAINS

pp 683–7

MUSCAT OF ALEXANDRIA

pp 689–92

ORIGINS AND PARENTAGE

Marufo is a very old and widespread variety from the Iberian peninsula, where it is known under a variety of names, as DNA profiling has revealed:

- Marufo is identical to Brujidera in Arribes del Duero in Spain and Mourisco Tinto in the neighbouring Douro in Portugal (Castro *et al.* 2011).
- Comparison of the DNA profiles of Marufo (Almadanim *et al.* 2004; Castro *et al.* 2011) and Moravia Dulce (Fernández-González, Mena *et al.* 2007) shows that they are one and the same variety (Vouillamoz), which explains why Brujidera is a common synonym for both. However, the same authors have spotted a genetically distinct but so far unidentified variety in Albacete that is mistakenly called Moravia Dulce (Fernández-González, Martínez and Mena 2007).
- Moravia Dulce is also called Crujidera and Rucial (Fernández-González, Mena *et al.* 2007).

It is not possible to say whether Spain or Portugal is the birthplace of this variety but recent DNA parentage analysis has shown that TOURIGA FRANCA from the Douro is an offspring of TOURIGA NACIONAL, also from the Douro, and Marufo, which means that Marufo has long been present in the Douro, but this is not sufficient to locate its origin. Since the name Moravia Dulce is provisional in the official register of varieties in Spain and is used for at least two distinct varieties, one in Cuenca and one in Albacete (Fernández-González, Martínez and Mena 2007), it seems less ambiguous to use Marufo as the prime name.

According to DNA profiling (Fernández-González, Martínez and Mena 2007 and Fernández-González, Mena *et al.* 2007), Colgadera is a synonym for Moravia Dulce in Campillo de Altobuey in Cuenca. Although the authors do not specify Colgadera's berry colour, this synonymy is surprising because Colgadera is a pale-berried variety, and no colour mutation of Moravia Dulce has ever been reported.

VITICULTURAL CHARACTERISTICS

Vigorous and moderately productive, late budding and ripening. Large bunches of large to medium-sized berries with thick skins. Susceptible to powdery mildew and coulure due to the functionally female flowers.

WHERE IT'S GROWN AND WHAT ITS WINE TASTES LIKE

Official statistics for 2008 record 2,202 ha (5,441 acres) of Moravia Dulce in **Spain**, mostly in Castilla-La Mancha and particularly in the provinces of Albacete and Cuenca in central and south-east Spain, where it is one of the permitted varieties in appellations such as La Mancha (an enormous region covering 190,000 ha/469,500 acres) and Manchuela and is generally used in blends, although it is also used sometimes for the table.

In **Portugal**, under the name Marufo, the variety makes wines that are light in colour and have a slight aroma of red fruit and a light structure that makes them unsuitable for ageing and better suited to rosé than to red wines. There were 3,631 ha (8,972 acres) in 2010, mostly in the Douro and Pinhel regions, in the north east of the country, where it is also used as a table grape. Recommended producers include the Figueira de Castelo Rodrigo co-op and Quinta dos Termos.

MARUGGIO

Ancient, aromatic, nearly extinct Puglian.

BERRY COLOUR: ● ● ● ● ●

ORIGINS AND PARENTAGE

Maruggio is an old variety from the Puglian coastal provinces of Taranto and Bari, named after a village near Taranto.

VITICULTURAL CHARACTERISTICS

Good yields and a good level of acidity in the berries.

WHERE IT'S GROWN

It is grown in the provinces of Taranto and Bari in Puglia in southern **Italy** and has recently been evaluated for potential development at the University of Bari.

MARZEMINA BIANCA

Rare Veneto variety usually found as an ingredient in sweet blended wines.

BERRY COLOUR: ● ● ● ● ●

PRINCIPAL SYNONYMS: Berzemina di Breganze, Sciampagna

VARIETIES COMMONLY MISTAKEN FOR MARZEMINA BIANCA: CHASSELAS ☒

ORIGINS AND PARENTAGE

Marzemina Bianca has long been cultivated in the Breganze region and was first mentioned by Giacomo Agostinetti (1679). Recent DNA studies have revealed a likely parent–offspring relationship beween Marzemina Bianca and the black-berried MARZEMINO (Salmaso *et al.* 2008) as well as between Marzemina Bianca and GARGANEGA (Crespan, Calò *et al.* 2008; see Garganega for more details). Indeed, the DNA profiles available are consistent with Marzemina Bianca being a natural cross between Marzemino and Garganega (Vouillamoz), to be confirmed by further analyses. In addition, parentage analysis has shown that RABOSO VERONESE is a Marzemina Bianca × RABOSO PIAVE cross (Salmaso *et al.* 2008; see the REFOSCO DAL PEDUNCOLO ROSSO pedigree diagram).

OTHER HYPOTHESES

Some authors have suggested that Marzemina Bianca came from Burgundy, spreading from there to Germany, Italy and Switzerland.

VITICULTURAL CHARACTERISTICS

Mid ripening.

WHERE IT'S GROWN AND WHAT ITS WINE TASTES LIKE

Marzemina Bianca is cultivated to a limited extent in the hills of the provinces of Treviso, Padova and Vicenza in northern **Italy**. It lost ground to more productive varieties when vineyards were replanted in the wake of phylloxera. A small amount may be blended with Glera (PROSECCO), VERDISO and Boschera (see VERDICCHIO BIANCO) in the Torchiato di Fregona (dry or sweet, passito-style) version of the Colli di Conegliano DOC, and with VESPAIOLA in the late-harvest Torcolato version of Breganze DOC. Dell'Antonio and Francesco Tomasi make good examples of the former and Maculan is one of the best-known producers of the latter. There were 83 ha (205 acres) in Italy in 2000 according to the agricultural census.

Casa Roma's Vinegia is a rare example of a light, dry, still white from 100% Marzemina Bianca, classified as a vino da tavola. They describe it as having aromas of green apple and hay and a faintly bitter aftertaste. Firmina Miotti's Sampagna is a dry, lightly sparkling version.

MARZEMINO

Makes fruity, perfumed reds often sweet or sparkling or both.

BERRY COLOUR: ● ● ● ● ●

PRINCIPAL SYNONYMS: Balsamina, Barzemin, Berzamino, Marzemina Cenerenta, Marzemina Nera, Marzemino Comune, Marzemino Gentile

VARIETIES COMMONLY MISTAKEN FOR MARZEMINO: LAMBRUSCO BARGHI, MERZIFON KARASI, PAVANA

ORIGINS AND PARENTAGE

Marzemino is a very old northern Italian variety, mentioned in Lando (1553) as recommended for the Veneto. It is supposed to have spread from there to Trentino, Friuli, Lombardia and Emilia-Romagna (Giavedoni and Gily 2005). However, its wine was first mentioned in relation to a famous banquet held on 6 June 1409 in Cividale (Friuli) in honour of Pope Gregory XII, at which the following wines were served: Ribolla from Rosazzo, Verduzzo from Faedis, Ramandolo from Torlano, Refosco from Albana and Marzemino from Gradiscutta (Peterlunger *et al.* 2004), a small village near Varmo in the province of Udine.

Pedigree reconstruction based on DNA profiling (see REFOSCO DAL PEDUNCOLO ROSSO pedigree diagram) has revealed that Marzemino and LAGREIN from the Alto Adige are both progenies of the same parents: TEROLDEGO from Trentino and another, unknown, variety, which means that Marzemino and Lagrein are siblings (Vouillamoz and Grando 2006). Additional genetic investigations have shown that Marzemino is a likely parent of MARZEMINA BIANCA (Salmaso *et al.* 2008) and Refosco dal Peduncolo Rosso from Friuli and a likely grandchild of PINOT (Grando *et al.* 2006), thus securely anchoring Marzemino's origins in northern Italy (see the PINOT pedigree diagram).

There are two main clonal variations: Marzemino Gentile and Marzemino Comune.

OTHER HYPOTHESES

Some authors have proposed that the variety originated from Marzemin, a village in Slovenia (Galet 2000), while others have suggested that it comes from Merzifon in Asia Minor on the grounds that Marzemino is identical to MERZIFON KARASI and thus of Turkish origin. However, DNA analysis has shown that Marzemino and Merzifon Karası are unrelated (Serena Imazio, personal communication). In addition, the hypothetical genetic link with the Greek variety VERTZAMI (Labra *et al.* 2003) can be rejected (see Vertzami).

VITICULTURAL CHARACTERISTICS

Mid to late ripening and susceptible to fungal diseases, especially botrytis bunch rot and powdery mildew. Vigorous, usually trained on pergolas and can be overly productive and therefore is easily overcropped.

WHERE IT'S GROWN AND WHAT ITS WINE TASTES LIKE

Today Marzemino is cultivated mainly in Trentino, in particular in the Val Lagarina between Isera and Volano, northern **Italy**, where it is produced as a still wine within the Trentino DOC. Varietal wines are also authorized in DOCs such as Breganze, Garda and Merlara, the last in sparkling form only. Marzemino is also cultivated in Lombardia, in particular near the Lago di Garda in several DOCs, including: Botticino (minimum 20%), Capriano del Colle (minimum 35%), Cellatica (20–30%) and Riviera del Garda Bresciano (5–30%). In the Colli di Scandiano e di Canossa DOC in Reggio Emilia in Emilia-Romagna, it may be a varietal wine or part of the blend (minimum 50%), either still or sparkling. Total plantings in Italy in 2000 were 268 ha (662 acres).

Mozart apparently tasted a wine made from Marzemino when he was in the Val Lagarina for his first concert in Italy in 1769 and in Act 2, scene 17, of *Don Giovanni*, the hero extols the virtues of the wine: 'Versa il vino! Eccellente Marzemino!'

Wines may be still or lightly sparkling, sometimes off dry or even sweet, passito-style, but they are generally fragrant and fresh with aromas of red fruits, occasionally violets, and a zesty sour-cherry aftertaste and not much tannin. Successful producers include the Cantine Sociale di Norni, Eugenio Rosi and Vivallis.

In **Australia**, Michelini, in the cool Alpine Valleys area of north-east Victoria, have been producing a varietal Marzemino for more than ten years and Chrismont, in the neighbouring region of King Valley, produce both a still, dry version and an off-dry sparkling one.

In **New Zealand**, Church Road have produced a spicy varietal Marzemino.

MÁTRAI MUSKOTÁLY

Recently registered, grapey Hungarian cross.

BERRY COLOUR: ● ● ● ● ●

PRINCIPAL SYNONYMS: Mátrai Muskatály

ORIGINS AND PARENTAGE

An ARANY SÁRFEHÉR × MUSCAT OTTONEL cross obtained in 1952 by Pál Kozma at the Szent István University in Hungary. It was officially registered in 1982.

VITICULTURAL CHARACTERISTICS

Productive, vigorous, mid to late budding and ripening. Big bunches of thin-skinned berries. Moderately susceptible to fungal diseases.

WHERE IT'S GROWN AND WHAT ITS WINE TASTES LIKE

There were just 25 ha (62 acres) in **Hungary** in 2008, mostly in Kunság in the centre and south and Etyek-Buda in the north. Bárdos make a varietal wine. Wines have an aromatic Muscat-like character with flavours of ripe fruit and fresh acidity.

MATURANA BLANCA

Rare and recently rescued Rioja variety with real potential.

BERRY COLOUR: ● ● ● ● ●

PRINCIPAL SYNONYMS: Maturano, Ribadavia

ORIGINS AND PARENTAGE

Maturana Blanca is found exclusively in Rioja but there is no reliable historical documentation on this variety (Martínez de Toda and Sancha González 2000). The name Maturana most likely derives from the Spanish word *madurando*, meaning 'ripening'. Morphological and DNA studies have shown that Maturana Blanca is identical to a few vines recovered under the name Ribadavia near the villages of Navarrete and Sotés in La Rioja, west of Logroño (Cervera *et al.* 1998; Martínez de Toda and Sancha González 2000; Ulanovsky *et al.* 2002).

Ampelographic and DNA evidence has clearly shown that Maturana Blanca is not a colour mutation of the black Maturana Tinta (Martínez de Toda and Sancha González 2000), which is a synonym for TROUSSEAU.

OTHER HYPOTHESES

Maturana Blanca is considered to be the first grape variety documented in La Rioja, in 1622, cited in the small town of Nájera, west of Logroño, under the name Ribadavia (Palacios Sánchez 1991): 'Practicada la cala en Nájera, se detectaron 60.256 cántaras de vino, de las cuales 9.340 eran de vino blanco, 760 de ribadavia y el resto de tinto' ('Examining the hold in Nájera, there were found 60,256 *cántaras* [an old measure equivalent to 16.13 litres] of wine, of which 9,340 were white wine, 760 were Ribadavia and the rest red'). However, Ribadavia is a town in the province of Orense in Galicia in north-western Spain whose wines were already renowned and exported abroad in the sixteenth century (Huetz de Lemps 2009). It is therefore more likely that this 1622 mention as well as other instances of Ribadavia in old documents correspond to the wine from this area, not to Maturana Blanca. In addition, no vine of Maturana Blanca has ever been observed in Ribadavia, about 600 km west of La Rioja. It is therefore plausible that Maturana Blanca was locally called Ribadavia to benefit from the historical fame of the wine from that area.

VITICULTURAL CHARACTERISTICS

Small clusters of small berries. Very fertile, early budding and ripening. Susceptible to botrytis bunch rot.

WHERE IT'S GROWN AND WHAT ITS WINE TASTES LIKE

According to official statistics, there were just 3 ha (7 acres) of Maturana Blanca in **Spain** in 2008, all in the province of La Rioja in the centre of northern Spain. Viña Ijalba rescued this variety in 1988, helped by viticulturists Juan Carlos Sancha González and Fernando Martínez de Toda from the Universidad de La Rioja and the Rioja regulatory council. They planted the first vineyard in 1995 and are currently the sole producers of a varietal wine (first vintage 2000). Maturana Blanca, along with SAUVIGNON BLANC and several other white varieties, has recently been officially authorized by the Rioja regulatory council. Wines may have quite high alcohol but are well balanced by fresh acidity, citrus or tropical fruit flavours and light herbal notes. Plantings are expected to increase.

MATURANA TINTA

See TROUSSEAU.

MAUZAC BLANC

Characterful, apple-skin-flavoured Gaillac and Limoux variety used in dry, sweet and sparkling wines.

BERRY COLOUR:

PRINCIPAL SYNONYMS: Blanc Laffite (Entre-Deux-Mers), Gaillac or Plant de Gaillac (Gers), Gamet Blanc (Aveyron), Mausat (Languedoc)

VARIETIES COMMONLY MISTAKEN FOR MAUZAC BLANC: TORBATO (Sardegna in Italy)

ORIGINS AND PARENTAGE

Mauzac Blanc is a variety specific to the Gaillac area of south-west France. The earliest mention is said to have appeared in the *Livre de raison* by Antiquamareta (Lavignac 2001) in 1525, and later in 1564 (in an archive document cited by Cayla in Rézeau 1997): 'vigne muscade et mausague'. However, the true identity and the berry colour of this *mausague* is unknown, and the first reliable mention of Mauzac dates back to 1736 under the name Mausat (Rézeau 1997): 'En Languedoc. Mausat: il y en a de blanc et de noir, il produit un grain rond et croquant' ('In Languedoc. Mausat: there is a white and a black one, it produces a round and crunchy berry'). The name Mauzac (Mausac in Occitan) could derive from two distinct villages: Mauzac, south west of Toulouse (Haute-Garonne), or Meauzac to the north of Toulouse, near Moissac (Tarn-et-Garonne). According to the IFV (Institut Français de la Vigne et du Vin) website, Mauzac Blanc has a possible parent–offspring relationship with Négret Castrais, a local variety that is no longer cultivated.

Mauzac Blanc is not a white mutation of MAUZAC NOIR, but **Mauzac Rose, Mauzac Roux, Mauzac Vert** and **Mauzac Jaune** are all colour mutations of Mauzac Blanc.

VITICULTURAL CHARACTERISTICS

Compact bunches, late ripening and best pruned short. Well suited to limestone and clay-limestone soils. Susceptible to mites, grape moths, phomopsis and eutypa dieback but not to powdery and downy mildews.

WHERE IT'S GROWN AND WHAT ITS WINE TASTES LIKE

Mauzac Blanc (often referred to simply as Mauzac) is used to produce dry, sweet and sparkling wines exclusively in Limoux and Gaillac in the western Languedoc and south-west **France** respectively. Its distinctive dried-apple-skin flavour is lost if the fruit is picked early to retain freshness, as is so often the case today. Mauzac Blanc is sometimes rather rustic but all the more tasty for that.

In still Gaillac, Mauzac Blanc (including the colour mutation Mauzac Rose) is one of the three principal varieties, along with LEN DE L'EL and MUSCADELLE, whereas the unusual, semi-cloudy, medium-sweet, lightly sparkling wines entitled to the term Méthode Ancestrale (bottled before the end of the first fermentation) must be 100% Mauzac (Blanc or Rose). The similar Limoux Méthode Ancestrale must be 100% Mauzac Blanc while the much more common Blanquette de Limoux has to be at least 90% Mauzac. In Limoux still wines and Crémant de Limoux, however, CHARDONNAY is gaining ground and Mauzac plays a less significant role.

Despite the efforts in the late 1980s of champions of the variety such as Robert Plageoles and an increase in vineyard area in the late 1980s and early 1990s, total French plantings have slowly declined over the last fifty years, from 8,511 ha (21,031 acres) in 1958 to 1,991 ha (4,920 acres) in 2009 (although Mauzac is still more widely planted than either MARSANNE or ROUSSANNE). Top of the list of recommended producers come Domaine Robert et Bernard Plageoles but there are many others, including Domaine de Causse Marines and Château Rives-Blanques.

As well as being a permitted variety in still Gaillac, Mauzac Rose is an authorized part of the blend for armagnac.

Sparkling-wine producer Schramsberg of the Napa Valley have planted a little in **California**.

MAUZAC NOIR

Rare variety from south-west France unrelated to its white-wine namesake.

BERRY COLOUR: ● ● ● ● ●

PRINCIPAL SYNONYMS: Mauzac Rouge

VARIETIES COMMONLY MISTAKEN FOR MAUZAC NOIR: Négret Castrais

ORIGINS AND PARENTAGE

Mauzac Noir, which is not the black version of MAUZAC BLANC (see that entry for history and etymology), is a variety from the Tarn in south-western France. Mauzac Noir is genetically close to FER and it is often confused with Négret Castrais, which is no longer cultivated and, according to the IFV (Institut Français de la Vigne et du Vin) website, has a possible parent–offspring relationship with Mauzac Blanc.

VITICULTURAL CHARACTERISTICS

Fertile, vigorous but not very productive. Mid ripening.

WHERE IT'S GROWN AND WHAT ITS WINE TASTES LIKE

Wines tend to be light in colour and fruity. Plantings in Gaillac are extremely limited but Bernard and Robert Plageoles of Domaine des Très Cantou/Domaine Plageoles, who have worked hard to revive this old variety, make a rare varietal example.

MAUZAC ROSE

See MAUZAC BLANC.

MAVRO

The dominant Cyprus variety, makes undistinguished wine.

BERRY COLOUR: ● ● ● ● ●

PRINCIPAL SYNONYMS: Cipro Nero ⚥ (Italy), Cyperntraube Blaue (Germany), Cypro Nero (Italy), Korithi Mavro ⚥ (mainland Greece), Kritiko Mavro ⚥ (Kríti/Crete in Greece), Kypreico Mavro, Kypreiko Mavro, Kypriotiko (Irákleio/Heraklion on Kríti), Mavro Kyproy, Skuro Mavro, Staphili-Mavro

VARIETIES COMMONLY MISTAKEN FOR MAVRO: MAVROUDI ARACHOVIS ⚥ (Greece), MAVRUD ⚥ (Bulgaria)

ORIGINS AND PARENTAGE

Mavro, meaning 'black', was first mentioned under the name Cipro Nero ('black of Cyprus') on the island of Cyprus by the Italian ampelographer Giuseppe di Rovasenda (1877) and then by French ampelographer Mouillefert (1893) under the name Mavro or Staphili-Mavro.

It's difficult to identify Mavro's place of birth: Lefort and Roubelakis-Angelakis (2001) have shown by DNA profiling that Korithi Mavro (black Corinthian) from central Greece and the Pelopónnisos (Peloponnese) and Kritiko Mavro (black Cretan) from Kríti and the Kykládes (Cyclades) are clones of the same variety, and comparison of their DNA profile with Mavro from Cyprus (Hvarleva, Hadjinicoli *et al.* 2005) reveals that they are identical. It is therefore not known if this variety, allegedly referred to in ancient Greek texts (Lefort and Roubelakis-Angelakis 2001), originates from Cyprus, mainland Greece or Kríti.

Mavro should not be confused with other varieties in Greece and Bulgaria that have Mavro as a synonym.

VITICULTURAL CHARACTERISTICS

Productive. Mid budding, mid to late ripening. Large berries. Highly susceptible to powdery mildew and botrytis bunch rot.

WHERE IT'S GROWN AND WHAT ITS WINE TASTES LIKE

Mavro is by far the most widely planted grape variety on **Cyprus**, the 4,550 ha (11,243 acres) recorded in 2010 comprising 46% of the island's vineyard, although this is considerably down on the 80% it used to represent, with the vineyard area declining overall. Its large berries make it suitable for the table as well as for wine – perhaps more so since varietal wines are typically light in colour and flavour, being simple, low in acidity and best drunk young. The tannins rarely ripen fully. The (slightly) better wines come from vines planted at higher altitudes and in the less fertile soils of Pitsilia, Laona and Afames in the west of the island, where yields are naturally less generous.

In addition to red wines, Mavro is used to make Commandaria, the traditional sweet and almost always fortified wine of the island made from sun-dried grapes, either on its own or blended with the light-skinned XYNISTERI, although the latter is generally thought to give better results. There were 289 ha (714 acres) in the Commandaria region in 2010.

MAVRO KALAVRITINO

Very minor Greek variety from the northern Pelopónnisos (Peloponnese).

BERRY COLOUR: ● ● ● ● ●

PRINCIPAL SYNONYMS: Kalavritino Mavro, Mavro Kalavrytiko, ?Psilomavro Kalavryton[DNA]

ORIGINS AND PARENTAGE

A rare variety named after the village of Kalavryta in the Achaïa region of the Pelopónnisos, Greece. Comparison of DNA profiles in the Greek Vitis database shows that it is possibly identical to Psilomavro Kalavryton, also grown in Achaïa.

WHERE IT'S GROWN AND WHAT ITS WINE TASTES LIKE

Found only in Achaïa in the northern Pelopónnisos, **Greece**, Mavro Kalavritino is blended with CABERNET SAUVIGNON by Domaine Tetramythos, who also produce varietal wine tasting of cherries, spices, violets and leather. Domaine Mega Spileo blend it with MAVRODAFNI.

MAVRO MESSENIKOLA

Variety exclusive to the Greek region of Thessalía that benefits greatly from its partnership with Syrah.

BERRY COLOUR: ● ● ● ● ●

PRINCIPAL SYNONYMS: Mavro de Messenicolas, Mavro Mesenikola, Messenikola Mavro

ORIGINS AND PARENTAGE

This variety comes in all probability from the province of Kardítsa in Thessalía, central eastern mainland Greece. DNA profiling has not yet been done to see if it matches any other known Greek variety.

OTHER HYPOTHESES

According to Lazarakis (2005), it has been suggested that this variety was brought from France during the Ottoman occupation by a certain Monsieur Nicolas, corrupted to Messenikola, but there is no evidence to support this. Lazarakis believes it is more likely to be related to XINOMAVRO.

VITICULTURAL CHARACTERISTICS

Fertile and vigorous. Late budding and ripening. Large, compact bunches of medium-sized berries. Relatively susceptible to downy mildew and botrytis bunch rot.

WHERE IT'S GROWN AND WHAT ITS WINE TASTES LIKE

The Messenikola appellation (created in 1994) in the lowlands around the town of Kardítsa, at 250–600 m above sea level, in **Greece**, is, unsurprisingly, home to this variety. The regulations stipulate 70% Mavro Messenikola plus 30% Carignan (see MAZUELO) and SYRAH, though an equal blend of Messenikola with Syrah seems to produce far superior results (Lazarakis 2005).

Varietal wines are typically light in every way, often more like a rosé than a red, hence the obligation to blend. Karamitrou produce both a varietal wine as well as a blend with Syrah.

MAVRODAFNI

Deeply coloured, tannic yet aromatic Greek variety famous for producing ageworthy, sweet, fortified wines.

BERRY COLOUR: ● ● ● ● ●

PRINCIPAL SYNONYMS: Ahmar Mechtras☒, Fraoula Kokkini☒, Mavrodafnitsa (Arkadía), Mavrodaphne, Mavrodaphni☒, Mavrodrami (Kérkyra/Corfu), Thiniatiko

ORIGINS AND PARENTAGE

Mavrodafni, meaning 'black laurel', is thought to come from the Ionian islands of Lefkáda and Kefaloniá (Manessis 2000) or from the north-western part of the Pelopónnisos (Peloponnese; Nikolau and Michos 2004).

The DNA profile of the reference sample named Mavrodaphni at the United States Department of Agriculture's National Clonal Germplasm Repository in Davis, California, is distinct from the Mavrodafni in Sefc *et al.* (2000).

The variety known as Thiniatiko, mainly found around Thinéa in the north-western part of Kefaloniá, is believed to be a synonym for, or a clone of, Mavrodafni but this has not yet been confirmed by DNA profiling (Haroula Spinthiropoulou, Nico Manessis, personal communications). DNA profiling with only eight markers has suggested a possible genetic relationship between Mavrodafni and GOUSTOLIDI (Boutaris 2000) but this needs to be confirmed with many more markers.

VITICULTURAL CHARACTERISTICS

Productive, early ripening, susceptible to millerandage, dry conditions and downy mildew. Loose bunches of thick-skinned berries.

WHERE IT'S GROWN AND WHAT ITS WINE TASTES LIKE

Mavrodafni is probably the most widely recognized Greek variety around the world, mainly because of the sweet, fortified wines that boast its name: Mavrodafni of Pátra and Mavrodafni of Kefaloniá. The former appellation represents about 7% of total quality-wine production in Greece.

The variety's high levels of colour and tannin also make it very useful for adding colour and structure to unfortified, dry red blends, particularly in poor vintages, but on its own, it can be rather one-dimensional. Found mainly in Pátra and Ilía in the northern Pelopónnisos as well as on the islands of Kefaloniá and Lefkáda, it is the third most planted variety in western Greece (497 ha/1,228 acres in 2008). There are said to be two main clones: the smaller-berried Tsigelo on Kefaloniá and the Regnio clone, with tighter bunches, in the Achaïa (Lazarakis 2005).

While the Mavrodafni of Pátra appellation allows the addition of up to 49% KORINTHIAKI, Mavrodafni of Kefaloniá must be a varietal wine. Aside from this potential difference, styles range widely depending on the winemaking, the timing of fortification, use of oak (generally large and old although there have been experiments with small and new), whether the barrels are topped up, etc., but the best are silky, sweet, complex and sometimes tawny in colour thanks to long ageing. Traditional styles are probably more akin to Grenache (GARNACHA)-based vins doux naturels from southern France than to port, although both styles exist. The very best have remarkable longevity.

From Pátra, Parparoussis' Taos is a good example of a dry varietal wine, aromatic, full of lifted cherry fruit and softened by two years in oak. Antonopoulos' version spends less time in oak and is equally fragrant and supple. Both make good examples of sweet, fortified Mavrodafni, aged in barrel anywhere between four and six years, as do Gentilini and Sklavos on Kefaloniá. Achaïa Clauss, named after Gustav Clauss, widely regarded as the pioneer of such fortified wines in the mid nineteenth century, have been through turbulent times since the 1980s and have not produced such consistent quality in recent years but they have an impressive library of wines back to the 1880s.

MAVROTRAGANO

Distinctive variety from Santoríni starting to create a buzz.

BERRY COLOUR: ● ● ● ● ●

ORIGINS AND PARENTAGE

This variety is indigenous to the volcanic Greek island of Santoríni in the Kykládes (Cyclades). Its name means 'black and crisp'.

It should be noted that the Greek Vitis Database lists two different reference samples of Mavrotragano, both from Santoríni but with distinct morphology and DNA profiles. It is not known which one is the correct Mavrotragano.

VITICULTURAL CHARACTERISTICS

Low yields, good general disease resistance. Small berries.

WHERE IT'S GROWN AND WHAT ITS WINE TASTES LIKE

According to Santoríni winemaker Paris Sigalas, Mavrotragano was once used to add aroma to Vinsanto on Santoríni, **Greece**, probably because it can reach high sugar levels, but for a time was found only in field blends and had virtually disappeared. It currently represents around 2% of the island's vineyard and two of Santoríni's top producers, Sigalas and Hatzidakis, now make dry, unfortified varietal Mavrotragano wines. Sigalas had been making sweet wines with the variety since the early 1980s but it was not until the late 1990s that both Sigalas and Hatzidakis independently saw the potential of the variety for dry wine, and now it's becoming something of a local trend – so much so that it has been recognized by the Slow Food movement as worthy of protection and is probably one of the most expensive varieties in Greece.

The variety is hugely characterful, full of wild-berry or cherry fruit, dark spice, a certain earthiness, with fresh acidity but also potentially fierce tannins that need to ripen fully and be tamed during winemaking and maturation, otherwise it can produce wines that are tough and rustic. However, with careful harvesting and handling and some bottle age, it can give distinctive, well-balanced, velvety reds. There are some experimental plantings on the island of Tínos (near Mýkonos, ie much closer to the mainland), and Evangelos Gerovassiliou, never one to opt for run-of-the-mill varieties, has planted Mavrotragano in Epanomí and adds a little to spice up the LIMNIO and Mavroudi (see MAVROUDA) in his top red blend, Avaton.

MAVROUDA

Mavrouda, also Mavroudi or Mavroudia, is a generic name meaning 'blackish' that was given to several distinct varieties all over Greece (for example, AGIORGITIKO is also known as Mavroudi Nemeas), while the Bulgarian MAVRUD is also called Mavroudi. All have very dark skins and it is often impossible to know which variety is being referred to.

MAVROUDI ARACHOVIS

Rare variety from central Greece.

BERRY COLOUR: ● ● ● ● ●

PRINCIPAL SYNONYMS: Arachovis, Arachovitiko Mavro, Arahovitikos, Mavro Arachovitiko, Mavroudi of Arachova

VARIETIES COMMONLY MISTAKEN FOR MAVROUDI ARACHOVIS: MAVRO (Cyprus), MAVRUD or Mavroudi Boulgarias (Bulgaria)

ORIGINS AND PARENTAGE

The plethora of Mavrouda or Mavroudi Something grape varieties is confusing but Mavroudi Arachovis ('Black from Aráchova') most logically comes from the town of Aráchova in the prefecture of Voiotía in central Greece and DNA profiling has shown that it is quite distinct from the MAVRUD of Bulgaria and the MAVRO of Cyprus (Hvarleva *et al.* 2004; Hvarleva, Hadjinicoli *et al.* 2005).

VITICULTURAL CHARACTERISTICS

Vigorous and productive. Compact bunches of medium-sized berries.

WHERE IT'S GROWN AND WHAT ITS WINE TASTES LIKE

Mavroudi Arachovis is the dominant variety around Aráchova in the north east of the Fokída region, central **Greece**, where Parnassos Vineyards produce a blend of 80% Arachovis and 20% MERLOT. Lazarakis (2005) describes the variety as 'modern and promising', producing wines that are full-bodied and deeply coloured.

MAVRUD

Indigenous Bulgarian variety producing sturdy reds that can improve with age.

BERRY COLOUR: ● ● ● ● ●

PRINCIPAL SYNONYMS: Kachivela (Bulgaria), Mavro (Greece), Mavroud (France), Mavroudi or Mavroudi Boulgarias (Greece), Tsiganka (Bulgaria)

VARIETIES COMMONLY MISTAKEN FOR MAVRUD: Karvouniaris☒ (Greece), MAVRO☒ (Cyprus), MAVROUDI ARACHOVIS☒ (central Greece), Mavrud Varnenski☒ (Bulgaria)

ORIGINS AND PARENTAGE

Mavrud, from the Greek *mavro*, meaning 'black', is one Bulgaria's two best quality indigenous red wine varieties and one of its oldest. It probably comes from Asenovgrad in the province of Plovdiv, central southern Bulgaria, where it has traditionally been cultivated (Katerov 2004). This place of origin is supported by the significant morphological diversity seen there, with at least five variants (Katerov 2004): Mavrud s Edri Zarna (large-berried Mavrud), Mavrud s Drebni Zarna (small-berried), Mavrud sas Spleskani Zarna (flat-berried), Mavrud Siv (grey-berried) and Mavrud Izresliv (with a tendency to coulure). Mavrud was used to breed BUKET and EVMOLPIA.

VITICULTURAL CHARACTERISTICS

Fertile and productive. Medium-sized, relatively compact bunches of small to medium-sized berries. Very late ripening after a long vegetative period. Sensitive to cold winter temperatures and to drought. Susceptible to downy and powdery mildews but shows good resistance to botrytis bunch rot.

WHERE IT'S GROWN AND WHAT ITS WINE TASTES LIKE

There were 1,709 ha (4,223 acres) of Mavrud in **Bulgaria** in 2009, less than of Melnik (SHIROKA MELNISHKA) and much less than of PAMID. Mavrud produces tannic and deeply coloured wines with fresh acidity that respond well to oak maturation although they are generally less ageworthy than wines made from Melnik. It is grown mainly in the centre and south of Bulgaria, concentrated in the province of Plovdiv. The majority of wines are dry, often blended with the country's widely planted international varieties such as MERLOT or CABERNET SAUVIGNON but also with the local RUBIN. However, Assenovgrad Winery produce a semi-sweet Mavrud called Manastirsko Shushukane, meaning 'monastery whisper', and Santa Sarah, a boutique producer near Pomorie on the southern part of the Black Sea coast, somehow manage to make a Mavrud icewine. Recommended producers of dry red varietal wines include Assenovgrad Winery, Malkata Zvezda, Rumelia, Terra Tangra and Todoroff but ambitious new wineries are being established all the time.

MAYOLET

Extremely rare Aosta variety gradually recovering from near-extinction.

BERRY COLOUR: ● ● ● ● ●

PRINCIPAL SYNONYMS: Maïolet, Majolet

ORIGINS AND PARENTAGE

Mayolet is an indigenous variety of the Valle d'Aosta, where it has been known since 1787, mainly between Saint-Vincent and Avise. It was traditionally cultivated in the same vineyards as PETIT ROUGE and VIEN DE NUS and typically blended to produce Torrette, a full-bodied late-harvest wine famous since the nineteenth century (Gatta 1838). DNA parentage analysis has established that Mayolet and Petit Rouge are the parents of ROUGE DU PAYS, cultivated across the border in Switzerland's Valais, and that Mayolet has a parent–offspring relationship with PRIÉ, another Aostan variety (Vouillamoz and Moriondo 2011; see Prié for the pedigree diagram).

VITICULTURAL CHARACTERISTICS

Vigorous, early to mid ripening. Highly susceptible to botrytis bunch rot, less so to powdery mildew.

WHERE IT'S GROWN AND WHAT ITS WINE TASTES LIKE

Mayolet was rescued from extinction in the twentieth century and has seen new plantings in its native Valle d'Aosta in north-west **Italy** since the 1990s (Moriondo 1999). Today, a handful of varietal wines are produced in the Valle d'Aosta DOC, eg by the Cave des Onzes Communes and the Cantina di Barrò. Di Barrò's version has fresh, tangy acidity, sour-cherry flavours and quite soft tannins. Today Torrette is an official category within the Valle d'Aosta DOC which must include at least 70% PETIT ROUGE, so the Mayolet component is optional and limited. Producers include Cantina di Barrò, Feudo di San Maurizio and Franco Noussan. There were barely 4 ha (10 acres) planted in Italy in 2000.

MAYORQUIN

See DAMASCHINO.

MAZUELO

Black variety from north-east Spain high in tannins and acidity that can make fine wine from old vines but has in its time, as Carignan, blighted Languedoc-Roussillon.

BERRY COLOUR:

PRINCIPAL SYNONYMS: Bovale di Spagna, Bovale Grande (Sardegna), Bovale Mannu, Carignan Noir, Carignane (Languedoc-Roussillon), Carignano (Sardegna), Cariñano (Aragón in north-east Spain), Cariñena (Aragón), Crujillón (Aragón), Mazuela (Rioja in north-east Spain), Mollard (Rioja), Samsó (Catalunya in north-east Spain) and many more

VARIETIES COMMONLY MISTAKEN FOR MAZUELO: BOBAL, CINSAUT, GRACIANO (under the names Bovale Sardo and Cagnulari on Sardegna), MOLLARD (Hautes-Alpes in France), NIEDDERA, PARRALETA, PASCALE (aka Nieddu Mannu on Sardinia), TINTILIA DEL MOLISE

ORIGINS AND PARENTAGE

Mazuelo, probably from Aragón in north-east Spain, has dozens of synonyms in Spain and in various European countries, suggesting it is a very old variety that was dispersed a long time ago. The name Mazuelo derives from Mazuelo de Muñó, a village in the province of Burgos in Castilla y León, and the well-known synonym Cariñena (which became Carignan in French, Carignano in Italian and Carignane in American English) most likely derives from the town of Cariñena near Zaragoza in Aragón. The name Cariñena used to be prominent in Catalunya, but since it is also used for a Denominación de Origen, in the province of Zaragoza, the Catalan authorities now call it Samsó in order to avoid confusion. However, as Spanish journalist Victor de la Serna points out, Samsó is also a synonym for CINSAUT, which creates further confusion. Mazuelo is therefore not only the most unambiguous name for the variety but also the one used in official Spanish statistics. Mazuelo is increasingly seen on labels.

DNA profiling has confirmed that Cariñena and Mazuelo – or Mazuela in Rioja – are identical (Martín *et al.* 2003), and, more recently and surprisingly, that Carignan is also identical to Bovale di Spagna on the island of Sardegna (Nieddu *et al.* 2007). Comparison of DNA profiles from different sources has also established that Bovale di Spagna, often called Bovale Grande, is distinct from BOBAL, despite what the name might suggest; from NIEDDERA (Nieddu *et al.* 2007), contrary to opinions frequently expressed; and from CINSAUT, even though the synonym Samsó in Catalunya might suggest otherwise (Vouillamoz). Some ampelographers also considered Bovale di Spagna to be identical to Bovale Sardo or Cagnulari but it is now clear that this is impossible, since these last two are both synonyms for GRACIANO (Nieddu *et al.* 2007). Most authors also speculated that TINTILIA DEL MOLISE from Molise and Lazio in central Italy, which is different from varieties known as Tintilia in other parts of Italy (Campania, Sardegna) and in Spain, could be the same variety as Bovale Grande but this was recently disproved by DNA profiling (Reale *et al.* 2006).

As the name reveals, the Spanish origin of Bovale di Spagna had long been suspected. Given the recent discovery that it is identical to Mazuelo, it seems most likely that Bovale di Spagna was introduced to Sardegna during Spanish rule of the island (1323–1720), possibly together with Bovale Sardo (aka GRACIANO). Although the number of DNA markers was too low to be conclusive, Reale *et al.* (2006) also suggested that Bovale di Spagna and Bovale Sardo could have a parent–offspring relationship, making Mazuelo either a parent or a progeny of GRACIANO, a theory that can be rejected by the analysis of their respective DNA profiles (Vouillamoz). In any case, the spontaneous cross that bred Mazuelo or the one that produced Graciano took place in Spain, not on Sardegna. Mazuelo was used to breed ARGAMAN and RUBY CABERNET.

OTHER HYPOTHESES

Some Italian authors believe that Carignano was introduced to Sardegna by the Phoenicians around the ninth century BC and then spread throughout Sulcis in the south west of the island during the Roman period but there is no evidence to support this theory.

VITICULTURAL CHARACTERISTICS

Vigorous and highly productive (up to 200 hl/ha when given free rein), late budding and very late ripening, so that it needs the heat of a Mediterranean climate to thrive. It is highly prone to powdery mildew, less so to botrytis bunch rot and downy mildew; also prone to attack from grape moths. Resistant to phomopsis. Bunches cling firmly to the vine, making it less suitable for mechanical harvesting.

WHERE IT'S GROWN AND WHAT ITS WINE TASTES LIKE

Mazuelo/Carignan is an extreme variety in every way: high yielding and high in acid, colour and generally tough tannins, occasionally bitter. Even the best examples, the product of old bushvines or softened by carbonic maceration and blended with finer varieties, can provoke mixed responses in wine drinkers. It is not a subservient partner for oak.

Total plantings in **France** in 2009 were nearly nine times higher than in Spain at 53,155 ha (131,349 acres), with nearly 80% of the total area in Languedoc-Roussillon, especially in the two major Languedoc *départements* of the Aude (16,113 ha/39,816 acres) and the Hérault (16,251 ha/40,157 acres). However, thanks to a decline in popularity, a reduction in the percentage allowed in certain appellations and financial incentives from the EU to replace it with 'improving varieties' such as SYRAH, Grenache (GARNACHA) and Mourvèdre (MONASTRELL), the area has been significantly reduced since the late 1990s, when these two *départements* had more than 60,000 ha (148,000 acres) between them. There are substantial plantings all over south-eastern France, particularly in the Gard, Pyrénées-Orientales, Var and Vaucluse *départements*. It rarely travels further north because it ripens so late. Its susceptibility to rot and mildew (both downy and especially powdery) makes it unsuitable for damper maritime climates such as that of south-west France, including Bordeaux. As elsewhere, the better wines depend on ancient, hand-harvested bushvine stock or the softening influence of carbonic maceration. Top producers include Sylvain Fadat at Domaine de l'Aupilhac in Montpeyroux, Château Lastours and Maxine Magnon in Corbières, Jean-Marie Rimbert in Saint-Chinian and Domaine Bertrand Bergé in Fitou. Roc des Anges and Clos du Gravillas have been major French champions of it.

In **Italy**, more than 97% of the 1,748 ha (4,320 acres) of Carignano vines and the additional 583 ha (1,440 acres) of its synonym Bovale Grande are planted on the island of Sardegna, notably in the south west for Carignano del Sulcis DOC, though it is also authorized in Lazio's Cerveteri DOC and (as Bovale Grande) in Sardegna's Campidano di Terralba and Mandrolisai DOCs. The Santadi co-op's Terre Brune and Rocca Rubine are excellent examples, almost as good as and certainly more velvety than the best Mazuelo-dominated Priorats. Few consumers of Sardinian wines are aware that a blend of Bovale Grande (or Bovale di Spagna) and Carignano is a blend of two clones of the same variety, namely the Mazuelo from Aragón.

Mazuelo is now the official Spanish name for what many have until recently called Cariñena. In spite of its Spanish origins, Mazuelo is less widely planted in **Spain** than it is, as Carignan, in France. It is authorized in the Rioja DOCa but, although it can add useful acidity, its contribution to the blend is much less frequent and significant than in the past. A very few producers such as Marqués de Murrieta produce a varietal Mazuelo. The Rioja region had 1,193 ha (2,950 acres) planted in 2008, compared with 2,634 ha (6,510 acres) in Catalunya (where it is sometimes known as Samsó). The other main regions for this variety are Aragón (926 ha/2,290 acres), where it is not the dominant variety in the Cariñena DOC, Castilla-La Mancha (697 ha/1,720 acres) and Navarra (515 ha/1,270 acres). Total plantings in Spain in 2008 were 6,130 ha (15,150 acres). Some of the finest varietal wines such as Cims de Porrera's are from centenarian bushvines on Priorat's schistous slopes, where this variety dominates the northern part of the region, around Porrera and Poboldea. El Trío Infernal 2/3, made by Laurent Combier, Jean Michel-Gérin and Peter Fischer from seventy-year-old vines, is another excellent example.

Under the name Carignan, Mazuelo is found in **Croatia** (210 ha/519 acres in 2009). **Cyprus** had 366 ha (904 acres) in 2010, **Turkey** 134 ha (331 acres) in 2010 and **Malta** 10 ha (25 acres) in 2010. It has also been grown in **Tunisia** and **Morocco**.

Carignan was widely planted in **Israel**, from the late nineteenth century, but it has become much less common in the last thirty or so years, down to 800 ha (1,977 acres) in total in 2009, making it the third most planted red after CABERNET SAUVIGNON and MERLOT. However, in recent years, some of the best producers have kept a much tighter control on yields from their old-vine Carignan and are producing some very worthy examples.

As Carignane (sic) in **California** (3,393 acres/1,373 ha in 2010), once again it is old bushvines, prized by producers such as Ridge, Cline and Lioco, that produce the best wines. High yields in the Central Valley do it no favours. Even though the total acreage is in gradual decline, there are some signs of revival, notably nearer the coast, in Mendocino, Sonoma and Contra Costa counties. Alvin Tollini in Mendocino's Redwood Valley AVA is one example of a grower with genuine respect for his old bushvine Carignane. There is also a little in **Washington** State (where its produce has been blended with Grenache/GARNACHA). **Mexico**'s limited Carignane plantings are mostly in Aguascalientes, Sonora and Zacatecas.

Probably the most promising Carignan vines in the New World are the ancient, unirrigated bushvines in Maule in southern **Chile**, whose Carignan total was 675 ha (1,668 acres) in 2008. They have their own promotional organization Vignadores de Carignan, and can produce some particularly profound wines. In 2009, **Uruguay** had 486 ha (1,201 acres) and **Argentina** just 30 ha (74 acres). A tiny amount is grown in **Australia** in the state of South Australia but so far it has been blended.

South Africa boasts around 80 ha (200 acres), with Charles Back's Fairview Pegleg bottling from the shale slopes of the Paarderberg in the Swartland being the most successful of relatively few New World examples outside South America.

CARIÑENA BLANCA

The white mutation of this variety is known as Carignan Blanc in southern **France**, where there were 411 ha (1,016 acres) in 2009, down from 1,652 ha (4,082 acres) half a century ago, but increasingly often acknowledged as an ingredient in full-bodied whites. Like the dark-skinned version, it is late budding and late ripening and very susceptible to powdery mildew but it keeps its acidity well, and alcohol levels are usually kept in check. As Carignan Blanc in France, it produces light, citrus flavours and is generally best in a blend, as demonstrated by Clos Marie and Mas Jullien in the Languedoc, and Clot de l'Oum (where the Carignan Blanc is part of a field blend) and Roc des Anges in Roussillon.

Although Mazuelo Blanco would be the logical Spanish name for this white mutation, it is still generally called Cariñena Blanca in **Spain**, or Carinyena Blanca in Empordà, where most of the very few vines are found, and Samsó Blanco in Catalunya. There were just 3 ha (7 acres) in Spain in 2008. The Espolla co-operative in Empordà in the north-eastern corner of Spain blends Carinyena Blanca with MACABEO in its Espolla Bianca and 10 Sentits make a varietal version which is undistinguished on the nose but more luscious on the palate, and finishes refreshingly sour and dry.

CARIGNAN GRIS

Carignan Gris is also simply a colour mutation of the dark-skinned Mazuelo, as has been demonstrated by DNA profiling for Mazuelo Blanco (under the name Carignan Blanc). However, plantings of the pink-berried version are very limited, with less than 1 ha (2.5 acres) recorded in **France** in 2008.

MAZZESE

Very minor variety used as a minor component in blends.

BERRY COLOUR: ● ● ● ● ●

PRINCIPAL SYNONYMS: Massase, Mazzese di Parlascio, Orzese, Rinaldesca, Uva Mazzese, Vajano (Grosseto, Pisa)

ORIGINS AND PARENTAGE

Mazzese is an old Tuscan variety, mentioned in the manuscripts of the botanist Pier Antonio Micheli (1679–1737). It is thought to originate from either the province of Grosseto or the province of Pisa, where it is still cultivated.

VITICULTURAL CHARACTERISTICS

Early ripening and vigorous.

WHERE IT'S GROWN

Mazzese is cultivated exclusively in Toscana, mainly in the province of Grosseto and to a lesser extent in the province of Pisa. There were just 98 ha (242 acres) in **Italy** in 2000. Usually blended with other local varieties, it may be included as a minor component in Tuscan DOCs such as Montescudaio, as well as in IGT Toscana. For example, Poggio Gagliardo's Pulena is a blend of SANGIOVESE, MALVASIA NERA DI BRINDISI, COLORINO DEL VALDARNO, Mazzese, FOGLIA TONDA, CILIEGIOLO and VERMENTINO.

MÈCLE DE BOURGOIN

Neglected eastern French variety now receiving some attention.

BERRY COLOUR: ● ● ● ● ●

PRINCIPAL SYNONYMS: Mescle de Bourgoin

VARIETIES COMMONLY MISTAKEN FOR MÈCLE DE BOURGOIN: POULSARD ⚭ (called Mècle in the Ain)

ORIGINS AND PARENTAGE

Mècle de Bourgoin, named after the commune of Bourgoin-Jallieu, between Lyon and Grenoble in eastern France, has often been confused with Mècle or Mescle in Bugey, between Lyon and Genève, which is identical to POULSARD (Galet 2000). According to Rézeau (1997), the name Mècle derives from the local dialect *meiklle*, meaning 'mixing', either because the berry colour is somewhere between light- and dark-skinned or because of the frequent occurrence of pink berries in the bunch.

VITICULTURAL CHARACTERISTICS

Mid to late ripening. Quite large, compact bunches of large to medium-sized berries.

WHERE IT'S GROWN AND WHAT ITS WINE TASTES LIKE

Galet (1990) reported 1 ha (2.5 acres) of Mècle de Bourgoin in Saint-Savin in the Isère, **France**, but the variety has today almost disappeared from the vineyards. However, with the support of the Centre d'Ampélographie Alpine in Montmélian (Savoie), Nicolas Gonin in Saint-Chef (Isère) has started a project to reintroduce long-forgotten local varieties, including this one. Having found pre-Second World War plantings of Mècle de Bourgoin in Isère in 2009, he successfully petitioned for its official classification and plans to plant in 2012. According to Galet (2000), the wines are rather hard when young.

MEDNA

Rare Dalmatian variety, often with a honey aroma.

BERRY COLOUR: ● ● ● ● ●

PRINCIPAL SYNONYMS: Buboj (Dubrovnik), Bumba (Dubrovnik), Medna Bijala (Metkovic), Medva (Makarska), Rizavac, Zložder (Korčula)

ORIGINS AND PARENTAGE

Medna is known under several synonyms in Dalmacija (Dalmatia), mainly in the Vrgorac, Pelješac and Dubrovnik areas, its most likely region of origin (Pezo *et al.* 2006). *Medna* is Croatian for 'honeyed', referring to the characteristic honey aroma of the wines.

VITICULTURAL CHARACTERISTICS

Productive and reasonably resistant to most fungal diseases.

WHERE IT'S GROWN AND WHAT ITS WINE TASTES LIKE

Medna is officially authorized in the Dalmatinska Zagora (Dalmatian Hinterland) wine region in **Croatia** and is found mostly around the town of Vrgorac, between Split and Dubrovnik. Wines are generally light, with marked acidity and a characteristic aroma of honey (Pezo *et al.* 2006). Producers of varietal wines include Dabera, Opačak and Katić Veljko.

MEDOC NOIR

See MENOIR.

MEGRABUIR

Productive and cold-hardy Armenian hybrid used mainly to produce pink dessert wines.

BERRY COLOUR: ● ● ● ● ●

PRINCIPAL SYNONYMS: Meghrabuyr, Megrabouir, Megrabuyr

ORIGINS AND PARENTAGE

Megrabuir, meaning 'honey-flavoured', is a complex hybrid obtained in 1960 by S A Pogosyan at the Armenian viticultural research centre in Merdzavan, just west of Armenia's capital, Yerevan, by crossing C 484 with C 128, in which C 484 is a MADELEINE ANGEVINE × Chasselas Musqué cross and C 128 is an Ichkimar × Yanvarskii Chernyi cross. Ichkimar is a Ukrainian table grape and Yanvarskii Chernyi, literally 'January Black', is a Ukrainian variety that is no longer cultivated.

VITICULTURAL CHARACTERISTICS

Very productive. Hardy to –28 °C (–18 °F).

WHERE IT'S GROWN AND WHAT ITS WINE TASTES LIKE

Recommended for dessert wines in **Armenia** and grown mainly in the westerly province of Armavir, Megrabuir can attain high sugar levels even at high yields. Wines are generally pink in colour. Brest and Kazumoff produce varietal wines and MAP blend it with RKATSITELI and KARMRAHYUT.

MELARA

Rare, used as a component in Vin Santo di Vigoleno.

BERRY COLOUR: ● ● ● ● ●

PRINCIPAL SYNONYMS: Merlara

VARIETIES MOST COMMONLY MISTAKEN FOR MELARA: SANTA MARIA

ORIGINS AND PARENTAGE

From the province of Piacenza in Emilia-Romagna, northern Italy.

VITICULTURAL CHARACTERISTICS

Mid to late ripening; the berries have a slight Muscat flavour.

WHERE IT'S GROWN AND WHAT ITS WINE TASTES LIKE

Cultivated mainly in the province of Piacenza for the production of the locally famous Vin Santo di Vigoleno (within the Colli Piacentini DOC), a passito blend of Melara, SANTA MARIA, VERNACCIA DI SAN GIMIGNANO (Bervedino), MARSANNE, SAUVIGNON BLANC, ORTRUGO and TREBBIANO ROMAGNOLO, aged for at least five years in small casks. There are only about a dozen producers of this wine, with a combined total production of *c.*2,500 bottles per year. According to the agricultural census, there were just 13 ha (32 acres) of Melara in **Italy** in 2000.

MELNIK

See SHIROKA MELNISHKA.

MELNIK 82

Result of crossing the variety commonly called Melnik, Shiroka Melnishka, with a variety from south-west France.

BERRY COLOUR:

PRINCIPAL SYNONYMS: Melnik

VARIETIES COMMONLY MISTAKEN FOR MELNIK 82: RANNA MELNISHKA LOZA⌛, SHIROKA MELNISHKA⌛

ORIGINS AND PARENTAGE

Melnik 82 is a SHIROKA MELNISHKA × VALDIGUIÉ cross obtained in 1963 by Yane Atanasov and his colleagues Z Zankov and G Karamitev at the Sandanski research centre (now Sintica Winery) in the Struma Valley in south-west Bulgaria, near the tiny but historical town of Melnik, hence the name. The breeders made a series of crossings using a mixture of pollens from the three French varieties DURIF, Jurançon (most probably JURANÇON NOIR) and Valdiguié to create an early-ripening variety with the characteristics of Shiroka Melnishka. DNA parentage analysis established that Valdiguié was the other parent of sibling varieties RANNA MELNISHKA LOZA and Melnik 82 (Hvarleva, Russanov *et al.* 2005).

VITICULTURAL CHARACTERISTICS

Late budding, mid to early ripening. Thick-skinned berries and moderately good botrytis resistance.

WHERE IT'S GROWN AND WHAT ITS WINE TASTES LIKE

There were just 28 ha (69 acres) in 2008, all in the far south west of **Bulgaria**. The variety makes wines with good body, acid and tannin levels which are suitable for ageing. This should not be confused with its much more common parent.

MELODY

Recent complex American hybrid making fresh, vinifera-*like wines.*

BERRY COLOUR:

ORIGINS AND PARENTAGE

A complex hybrid obtained in 1965 by Bruce Reisch and team at New York State's Cornell University in Geneva by crossing SEYVAL BLANC with Geneva White 5 (GW 5), the latter being a progeny of PINOT BLANC × Ontario selected at Cornell (see CAYUGA WHITE for the pedigree of Ontario). Melody is therefore a complex *Vitis labrusca, Vitis vinifera, Vitis lincecumii* and *Vitis rupestris* hybrid. The first fruit was observed in 1969 and, after a period of testing under the code NY 65.444.4, it was officially released as Melody in 1985 (Reisch *et al.* 1985).

VITICULTURAL CHARACTERISTICS

Vigorous and productive, moderately winter-hardy (similar to SEYVAL BLANC). Mid to late ripening. Moderately resistant to powdery mildew and botrytis bunch rot and the berries resistant to cracking.

WHERE IT'S GROWN AND WHAT ITS WINE TASTES LIKE

Melody produces a *vinifera*-like wine that, according to Reisch *et al.* (1985), has 'a neutral fruitiness with floral overtones and slight hints of herbaceousness' and seems well suited to the north-eastern US. Varietal wines are made by, for example, Holy-Field in **Kansas**, and by Goose Watch and Wagner in **New York** State's Finger Lakes, where there were around 14 acres (6 ha) in 2009.

MELON

Old Burgundian variety well suited to the western, Atlantic-dominated mouth of the Loire.

BERRY COLOUR:

PRINCIPAL SYNONYMS: Gamay Blanc (Aube, Beaujolais, Haute-Marne, around Lyon, Saône-et-Loire), Latran (Anjou, named after the person who brought it from the Jura), Melon de Bourgogne (US), Muscadet (Pays Nantais, Vendée), Plant de Bourgogne or Petit Bourgogne (Loire)

VARIETIES COMMONLY MISTAKEN FOR MELON: ALIGOTÉ, CHARDONNAY (in the Jura, Chardonnay is sometimes called Melon à Queue Rouge, Melon d'Arbois or Melon d'Arlay), PINOT BLANC, SAVAGNIN BLANC (California)

ORIGINS AND PARENTAGE

Melon is an old variety from Burgundy, where it was cultivated as far back as the thirteenth century, according to Durand and Pacottet (1901a), although this is not historically documented. Melon was then introduced to the Franche-Comté, which is where it was first mentioned, in 1567, in an edict issued by Philippe II, King of Spain and Comte de Bourgogne (Rézeau 1997): 'Auons interdit, prohibé & defendu … de planter, & édifier de nouueau en quelque lieu que ce soit, gamez, melons, & autres plants de semblable nature' ('We have forbidden, prohibited and not allowed … the planting and rebuilding again in any place of gamez, melons and other similar varieties'). Like GAMAY NOIR, Melon was later banned several times in Burgundy and the Franche-Comté.

In the Middle Ages, Melon was taken to the Val de Loire under the name Plant de Bourgogne, where it won its spurs under the name Muscadet, possibly mentioned among obscure and mostly unidentified varieties around 1530 in the *Cinquième Livre* by the famous medieval writer François Rabelais (see MENU PINEAU), and then again more certainly in 1635 in Gorges in the Loire-Atlantique (Rézeau 1997): 'planter en bonne plante de vigne blanche de muscadet de quatre pieds de distance' ('to plant good plants of white muscadet vines four feet apart'). During the seventeenth century, plantings of Melon significantly increased in the Loire-Atlantique to fulfil the demand of Dutch traders for productive grapes suited to distillation (Schirmer 2010).

The name Melon could come from the Latin *misculare*, meaning 'to mix', like Meslier (see MESLIER SAINT-FRANÇOIS), or from the round shape of its leaves that are similar to those of the melon plant. The name Muscadet could come from the discreet musk aroma of the wine or from the *noix de muscade* (nutmeg) that the Dutch traders added to the wines, along with other spices, prior to distillation.

DNA parentage analysis has established that Melon is a natural PINOT × GOUAIS BLANC cross (Bowers *et al.* 1999), which makes it a sibling of CHARDONNAY, GAMAY NOIR, ALIGOTÉ, etc., as well as a half-sibling of the numerous other progenies of either Gouais Blanc or Pinot (see Pinot pedigree diagram). Melon is a member of the Noirien ampelographic group (see p XXVII).

The Melon à Queue Rouge that is still scarcely cultivated in the Jura is just a type of CHARDONNAY and should not be confused with the true Melon.

OTHER HYPOTHESES

It is often said that the presence of Melon in the region of Nantes remained marginal until 1709, the year of the coldest winter France had ever experienced, which damaged or destroyed huge areas of vines across the country, especially in the north, and that Melon was chosen for replanting because of its resistance to frost. However, this story is rigorously refuted by Schirmer (2010) because it is not supported by any historical document and because Melon was already abundant in the Pays Nantais in 1615 under the name Bourgogne.

VITICULTURAL CHARACTERISTICS

Mid ripening. Low basal-bud fertility and thus better pruned long. Well suited to clay-silica soils and to relatively cool regions thanks to its ability to withstand frost. Very susceptible to botrytis bunch rot but not to powdery mildew or eutypa dieback. Small berries.

WHERE IT'S GROWN AND WHAT ITS WINE TASTES LIKE

There were 12,384 ha (30,602 acres) in **France** in 2009, making it the fourth most planted pale-skinned variety in the country, just ahead of SÉMILLON but a long way behind SAUVIGNON BLANC. Plantings are almost exclusively at the western end of the Val de Loire, especially in the Loire-Atlantique and Maine-et-Loire *départements*, in the Pays Nantais surrounding the town of Nantes. The greatest concentration is in the Muscadet Sèvre-et-Maine appellation, south and west of Nantes, but Melon is the sole permitted variety in all the Muscadet appellations. The great success of Muscadet from the 1970s led to indiscriminate planting but now the sites are being more carefully selected and the total area is starting to decline. However, there are 3 ha (7 acres) and some recent plantings in Bourgogne-Vézelay.

Wines are archetypically crisp, marine and relatively neutral in flavour, sometimes with a touch of citrus. Lees ageing (*sur lie*) can imbue a little more flavour and the wines can make an ideal partner to the fish and seafood of the region. Recommended producers in Muscadet Sèvre-et-Maine include Guy Bossard's Domaine de l'Écu, Christophe Gadais at Domaine de la Tourmaline and Pierre Lieubeau at Domaine de la Fruitière; and in Muscadet Côtes de Grandlieu, Jérôme Choblet at Domaine des Herbauges under labels such as Fief Guérin, Château de la Pierre, Clos de la Fine, and Eric Chevalier at Domaine de l'Aujardière.

There is limited but marked enthusiasm for the variety in the US – and even a dedicated website (http://melondebourgogne.com) set up by Perennial Vintners on Bainbridge Island near Seattle, **Washington** State. The variety was brought to the US by Napa Valley winemaker Georges de Latour but was initially labelled PINOT BLANC, and there seems to be some confusion even today. Before true-to-type selections were introduced in the 1980s, most Pinot Blanc vineyards in **California** were actually planted with Melon. Today, most US producers of wines labelled Melon de Bourgogne are in **Oregon** with De Ponte Cellars a notable producer.

MENCÍA

Increasingly appreciated aromatic variety found in both Spain and Portugal.

BERRY COLOUR:

PRINCIPAL SYNONYMS: Jaen or Jaen du Dão (Dão in Portugal), Loureiro Tinto (Portugal), Mencía Pajaral (Bierzo in Spain)

VARIETIES COMMONLY MISTAKEN FOR MENCÍA: JUAN GARCÍA (Castilla-La Mancha)

ORIGINS AND PARENTAGE

Mencía most probably originates from Salamanca, in Bierzo in the province of León, north-west Spain, although it was not mentioned in this area until the late nineteenth century, after the arrival of phylloxera (Martínez, Santiago *et al.* 2006).

Recent DNA profiling has revealed that Jaen du Dão, also known simply as Jaen, cultivated in Dão in central Portugal and considered to be native to that region, is in fact identical to Mencía (Martín *et al.* 2006), and that Jaén Tinto, an old and almost extinct variety from southern Spain, should not be confused with Mencía (also misleadingly called Jaén Tinto in Bierzo in Castillo y León; Zinelabidine *et al.* 2012). It is plausible that Mencía was brought to Dão by pilgrims on their way back from Santiago de Compostela. The fact that Jaen vines in Dão are notably homogeneous and that Mencía in Spain shows a greater level of genetic variability supports the Spanish origin of this variety.

OTHER HYPOTHESES

Some have suggested that Mencía is related to CABERNET FRANC and GRACIANO (under the name Tintilla de Rota) but any close genetic link has been ruled out by morphological and DNA studies (Ibañez *et al.* 2003; Martínez, Santiago *et al.* 2006). Similarly, recent DNA research has suggested that Mencía could have been 'obtained from different crosses between Garnacha and another cultivar' (Martínez, Santiago *et al.* 2006) but this was refuted by De Mattia *et al.* (2009), who concluded that Mencía is not related to GARNACHA.

VITICULTURAL CHARACTERISTICS

Small bunches of medium-sized berries in Spain; medium to large berries in Portugal. Early budding, mid ripening. It is said to have good production potential in Portugal but to be not very productive in Spain. It is prone to wind damage and susceptible to powdery and downy mildews and botrytis bunch rot. Best pruned short.

WHERE IT'S GROWN AND WHAT ITS WINE TASTES LIKE

There were 9,055 ha (22,375 acres) of Mencía in **Spain** in 2008, mostly in the north of Castilla y León and in Galicia but also to a limited extent in Asturias and Catalunya. It is particularly important and increasingly respected in the Bierzo, Ribeira Sacra, Monterrei and Valdeorras

DOs and authorized in Rías Baixas and Ribeiro. Although it has traditionally been known to produce aromatic and fruity varietal reds as well as attractive rosé wines, the rediscovery of old vineyards with naturally low yields on deep schist soils, notably in Bierzo, has led to the more recent appearance of dense, concentrated examples. The alcohol level can be quite high and the acidity drops quickly at the end of ripening so it is not always easy to get the right balance and retain freshness. Recommended producers include Alma de Tinto, Descendientes de J Palacios, Docampo, Estefania, Mengoba, Peique, Raúl Pérez, Pittacum, J Rebolledo, Dominio de Tares and Castro Ventosa.

Portugal had 2,578 ha (6,370 acres) of Jaen in 2010, predominantly in Dão. The best wines come from the south-eastern part of the region (Seia and Gouveia): intensely coloured, very soft due to the low level of acidity but with an intense and delicate aroma of raspberries and blackberries. They vary from slightly rustic to elegant and are generally for early drinking, especially if they have been made using carbonic maceration. In order to produce wines with a firmer structure and better ageing capacity, Jaen is sometimes blended with other varieties such as TOURIGA NACIONAL. Recommended producers include Quinta das Maias, Quinta da Pellada and Quinta dos Roques.

MENNAS

Aromatic Swiss variety grown by a single producer.

BERRY COLOUR: ● ● ● ● ● (first underlined)

ORIGINS AND PARENTAGE

Mennas is a GEWÜRZTRAMINER × GOUAIS BLANC cross obtained in 1986 by Hans-Peter Baumann at Chanton Kellerei in Visp, Haut-Valais, Switzerland.

VITICULTURAL CHARACTERISTICS

Mid budding, mid to late ripening. Small bunches of medium-sized, thin-skinned berries.

WHERE IT'S GROWN AND WHAT ITS WINE TASTES LIKE

Mennas is found only at Diroso Kellerei, Valais, **Switzerland**, who grow just 0.2 ha (0.5 acres) and make a late-harvest wine with aromas of lychees, roses and marzipan from it. The variety is said to be like its aromatic parent but with higher acidity.

MENOIR

Minor, recently renamed Hungarian variety making reds with Muscat aromas.

BERRY COLOUR: ● ● ● ● ● (last underlined)

PRINCIPAL SYNONYMS: Kékmedoc, Medoc Noir, Menoire

VARIETIES COMMONLY MISTAKEN FOR MENOIR: COT ☒ (Hungary), MERLOT ☒ (Hungary), MORNEN NOIR ☒ (Hungary)

ORIGINS AND PARENTAGE

Partly because of its old name Medoc Noir, Menoir has long been thought to be a French variety brought to Hungary by grape breeder Mathiasz János (1838–1921) but recent DNA analysis strongly challenges this origin.

According to Galet (1990), the French Rhône variety MORNEN NOIR is cultivated in Hungary under the name Médoc Noir, which is French for Kékmedoc, a synonym for Menoir. The name Médoc Noir has also been used elsewhere for COT and MERLOT but the DNA profile of Kékmedoc published by Jahnke *et al.* (2009) is distinct from all others, proving that Mornen Noir and Menoir are different varieties, if their reference sample was correct.

VITICULTURAL CHARACTERISTICS

Early budding and ripening, low yields. Small bunches of small berries. Moderately hardy to drought and cold winter temperatures but susceptible to fungal diseases.

WHERE IT'S GROWN AND WHAT ITS WINE TASTES LIKE

Menoir produces early-drinking wines in **Hungary** that are typically medium-bodied with fine tannins and the smell of Muscat. The change of name from Kékmedoc, occasionally also Médoc Noir, meaning 'black Médoc', to Menoir came as a result of Hungary's accession to the EU in 2004 and the latter's hypersensitive attitude towards misleading nomenclature. Menoir is found mainly in the Eger wine region, where Lajos Gál makes a good varietal example, but total Hungarian plantings in 2008 were just 46 ha (114 acres).

MENU PINEAU

Well-connected and once important Blois speciality with a handful of advocates.

BERRY COLOUR: ● ● ● ● ●

PRINCIPAL SYNONYMS: Arbois Blanc (Loir-et-Cher), Herbois (Loir-et-Cher), Menu, Menu Pinot (Touraine), Orbois (Loir-et-Cher), Petit Pineau (Loir-et-Cher), Verdet

VARIETIES COMMONLY MISTAKEN FOR MENU PINEAU: PETIT MESLIER ✗ (Loir-et-Cher), RIESLING ✗

ORIGINS AND PARENTAGE

Menu Pineau is an old variety from the Val de Loire named after its similarity to Gros Pineau (CHENIN BLANC; *menu* means 'small'). Historically it was known as Arbois Blanc in the Loir-et-Cher, although it has nothing to do with the village called Arbois in the Franche-Comté, where it has never been cultivated. Instead the name comes from the Loir-et-Cher name Herbois, in dialect *Orboé* or *Orboué*, which became Orbois and then Arbois. To make things even more confusing, it is known as Orbois in the French official register of varieties.

It was supposedly mentioned for the first time around 1530 in the *Cinquième Livre* by the famous medieval writer François Rabelais, among other obscure and mostly unidentified varieties: 'un grand vignoble faict de toutes especes de vignes, comme Phalerne, Malvoisie, Muscadet, Taige, Beaune, Mireveaux, Orléans, Picardent, Arbois, Coussi, Anjou, Grave, Corsicque, Verron, Nerac et autres' ('a large vineyard composed of many grape varieties such as … Arbois …'). A more specific mention of this variety was found in Bauhin (1650): 'Spectandæ pulchritudinis sunt uvæ, quas Montbelgardi Arbois vocant, à loco in Burgundia, vbi vocantur à quibusdam teste d'oiselet' ('Considered fine are the grapes named Arbois in Montbéliard, a place in Burgundy where some people call it baby bird's head').

Menu Pineau is a member of the Messile ampelographic group (see p XXVII), named after Meslier (Bisson 2009). According to Galet (2000), **Arbois Rose** is a colour mutation.

DNA parentage analysis has discovered a parent–offspring relationship with GOUAIS BLANC (Boursiquot *et al.* 2004), which means that Menu Pineau is a half-sibling of at least eighty Western European varieties (see PINOT pedigree diagram).

OTHER HYPOTHESES

Some authors have erroneously identified Menu Pineau with RIESLING.

VITICULTURAL CHARACTERISTICS

Short vegetative cycle. Mid ripening. Compact bunches of small berries but not particularly susceptible to botrytis bunch rot. Suitable for regions with cool winters.

WHERE IT'S GROWN AND WHAT ITS WINE TASTES LIKE

Menu Pineau used to be widely cultivated in Touraine, northern **France**, where it has today been almost entirely replaced by SAUVIGNON BLANC. In 2006 there were 270 ha (667 acres) planted in all of France, compared with 1,455 ha (3,595 acres) in 1968. It owes its more recent renaissance to the determination of Thierry Puzelat of Clos du Tue Boeuf, one of the only producers, along with Claude Courtois, to make wines in which Menu Pineau plays a leading role. Puzelat's Pétillant Naturel is a blend of Menu Pineau and CHENIN BLANC, while his La Tesnière has an even higher proportion of the former. Puzelat describes it as 'a little more rustic but less acid than Chenin Blanc'. It certainly has both honey and tang. Courtois' Alkimya is a rare varietal wine and Domaine des Capriades produce a 100% Orbois. Because the variety tends to have relatively low acidity it is generally blended, eg as a minor component in the Sauvignon Blanc-dominated Cheverny, Valençay and Touraine appellations. When taken seriously, it can show mineral, spicy and apple flavours.

Note that the synonym Arbois Blanc has nothing to do with the white or red wines of the Arbois appellation in the Jura.

MERENZAO

See TROUSSEAU.

MÉRILLE

Virtually extinct variety found occasionally in Fronton blends in south-west France.

BERRY COLOUR: ● ● ● ● ●

PRINCIPAL SYNONYMS: Bordelais or Bordelais Noir (Fronton in south-west France), Bourdalès or Bouchalès (Pamiers in Ariège), Grand Vesparo or Vesparo à Queue Verte (Gers; also used for other varieties such as COT), Périgord (Dordogne)

VARIETIES COMMONLY MISTAKEN FOR MÉRILLE: BOUCHALÈS (Pamiers in Ariège), MILGRANET (under the name Petite Mérille)

ORIGINS AND PARENTAGE

Mérille was first mentioned in 1783–4 in the Lot-et-Garonne and in the Gers (Rézeau 1997): 'Mérille, raisin noir, petits grains ronds, grappe longue et large, les grains serrés, de bon goût'

('Mérille, black grape, small round berries, long, wide cluster, closely packed berries, good taste'). Its name is of unknown origin.

The supposed parent–offspring relationship with VALDIGUIÉ (Galet 1990) has not yet been confirmed but DNA parentage analysis has uncovered a likely parent–offspring relationship with PRUNELARD (Boursiquot *et al.* 2009), which is consistent with Mérille's being a member of the Cot ampelographic group (see p XXVII).

VITICULTURAL CHARACTERISTICS

Mid ripening, regular and generous production, high yields especially if pruned long. Susceptible to mites, grape moth and leafhoppers. Big berries and very big bunches.

WHERE IT'S GROWN AND WHAT ITS WINE TASTES LIKE

Wines are light and rather ordinary. Mérille is sometimes found in the blend of a wine from the Fronton appellation, where it is an 'accessory' variety. It is also permitted in the Vin de Pays des Côtes du Tarn. There were 55 ha (136 acres) in **France** in 2008, down from 2,422 ha (5,985 acres) around fifty years ago.

MERLOT

Fleshy, widely planted, relatively early ripening member of the Bordeaux family.

BERRY COLOUR: ● ● ● ● ●

PRINCIPAL SYNONYMS: Bigney, Crabutet, Langon, Médoc Noir, Merlau, Merlot Noir, Picard, Sémillon Rouge, Vitraille (all except Merlot Noir are historical)

VARIETIES COMMONLY MISTAKEN FOR MERLOT: CARMENÈRE (Chile), MENOIR (Hungary)

ORIGINS AND PARENTAGE

The earliest mention of Merlot appeared in 1783–4 in Libourne in the Gironde (Rézeau 1997): 'Merlau ... fait un vin noir et excellent, et il abonde dans un bon terrein' ('Merlau ... makes a black and excellent wine, and is productive in a good soil'). Its modern spelling appeared in 1824 in a treatise about Médoc wines, which also provides the most convincing hypothesis for the origin of the name (Rézeau 1997): 'on a donné le nom de Merlot à cette variété de vigne, parce que le merle (oiseau) aime beaucoup ce raisin' ('the name Merlot was given to this variety because the blackbird likes this grape very much'). Indeed, this bird is called *merlau* in Occitan. Merlot also used to be called *lou seme doù flube* ('the seedling from the river') because it was thought to have been born on the islands in the River Garonne.

In the late 1990s, DNA parentage analysis revealed a parent–offspring relationship between Merlot and CABERNET FRANC but the other parent remained unknown (Regner 1999). Ten years later, the other parent of Merlot was discovered by Boursiquot *et al.* (2009). It was an unknown cultivar, first sampled in 1996 in Saint-Suliac near Saint-Malo in northern Brittany, where vines were cultivated at the end of the Middle Ages (Droguet 1992). The cuttings were taken from an abandoned grapevine growing on the slope of a hill called Mont Garrot. A few years later, four more plants of this variety were discovered in the Charente, planted in front of houses in four distinct villages: Mainxe, Figers, Tanzac and Saint-Savinien. These vines were locally called Raisin de la Madeleine or Madeleina, referring to the precocity of the grapes (often ripe at Sainte-Madeleine, 22 July). The DNA profile of this variety did not match any other known variety from the vast collection at Vassal in Montpellier, so it had to be christened. In order to avoid confusion with other Madeleine Somethings, Boursiquot and colleagues named this cultivar 'Magdeleine Noire des Charentes'. Through DNA testing, they have also established that Cabernet Franc is the father and Magdeleine Noire des Charentes the mother of Merlot.

Merlot belongs to the Carmenet ampelographic group (see p XXVII) originating from the Gironde or south-western France or Spain's País Vasco (Bordenave *et al.* 2007; Bisson 2009). In addition, ABOURIOU, CABERNET SAUVIGNON, CARMENÈRE and COT, having at least one parent in common, are all half-siblings (Abouriou could also be a grandparent) of Merlot (see the pedigree diagram in the Cabernet Sauvignon entry).

While **Merlot Gris** is a colour mutation of Merlot, Merlot Blanc is a distinct variety found in 1891 and born from a natural cross between Merlot and FOLLE BLANCHE (Boursiquot *et al.* 2009). Merlot Blanc has today almost disappeared.

Merlot has been used to breed numerous varieties around the world, including the following half-siblings, some of which are cultivated for commercial wines: CARMINE (US), EDERENA (France), EVMOLPIA (Bulgaria), FERTILIA (Italy), Mamaia (Romania), NIGRA (Italy), PRODEST (Italy) and REBO (Italy). There is also a small, isolated vineyard near Arezzo, central Italy, where Carnasciale grow a variety called Caberlot, thought to be a crossing of Merlot with an unidentified parent (Paolo Storchi, personal communication).

VITICULTURAL CHARACTERISTICS

Early budding, mid ripening. Moderately, sometimes very, vigorous with a tendency to produce many shoots and suckers. Fertile and best if short pruned. Prone to coulure in some climates. Well suited to clay-limestone soils. At risk of winter and spring frosts and drought. Very susceptible to downy mildew and leafhoppers. More susceptible to botrytis bunch rot than the Cabernets, with which it is so frequently grown.

According to Boursiquot *et al.* (2009), Merlot's positive characteristics and the reason for its success may be due to its inheritance: high-quality phenolic compounds (tannins, anthocyanins) from CABERNET FRANC and precocity and fertility from Magdeleine Noire des Charentes.

WHERE IT'S GROWN AND WHAT ITS WINE TASTES LIKE

It would be quicker and easier to say where Merlot is not grown than to list the many regions and countries in which it has been planted. Because it is easier to ripen, it is grown even more widely than CABERNET SAUVIGNON, and the sweetness and freshness of its wine make it particularly easy to appreciate. Like its relatives Cabernet Sauvignon and CABERNET FRANC, it can exhibit a certain herbaceousness if not fully ripe, though this is much less common with Merlot than with either Cabernet. In general, a very considerable proportion of all the Merlot grown worldwide is blended with either or both Cabernets to provide fruit and flesh to counterbalance their more tannic structure.

It is one of the world's most planted wine grapes and the single most planted in **France**, where the total was 115,746 ha (286,015 acres) in 2009. It is quite widely dispersed but by far the greatest concentrations are in Aquitaine, the greater Bordeaux region and Languedoc-Roussillon. Merlot was long regarded as secondary in Bordeaux until the mid nineteenth century, when its resistance to the scourge of powdery mildew, especially compared with CABERNET SAUVIGNON, made it very much more popular. The ease with which it could be ripened even in relatively cool years also contributed to its appeal so that today Merlot far exceeds Cabernet Sauvignon in the Bordeaux *département*, Gironde (69,053 v 26,790 ha/170,634 v 66,200 acres) – and also in the Bergerac *département* to the east, Dordogne (4,303 v 1,582 ha/10,633 v 3,909 acres).

But the French total has started to decline slightly as Bordeaux's wine producers' respect for CABERNET SAUVIGNON and the majesty of its wines is revived. Wine made from Merlot generally has more obvious fruit and less aggressive tannins so that it can be drunk younger than a Cabernet. Typically it has a certain sweet plummy fruit which can vary between facile and the extraordinary velvety richness of a great Pomerol such as Petrus. See CABERNET FRANC for reasons why producers here generally like to spread their risk by planting all three varieties, although the plateau of Pomerol has shown a particular affinity with Merlot so that most Pomerols are dominated by the variety. Cabernet Sauvignon can be difficult to ripen on the (generally less temperate) right bank of the Gironde and red-wine vineyards in any of the cooler areas in the greater Bordeaux region are likely to be dominated by the Merlot vine. In the traditional Bordeaux blend, particularly on the left bank in the Médoc and Graves, Merlot provides the flesh on the bones of the Cabernets. It generally achieves sugar levels up to a full 1% of alcohol more than Cabernet.

Merlot is also widely planted in the Languedoc, where total plantings are 29,914 ha (73,919 acres), a good 50% more than the CABERNET SAUVIGNON total, with more than 11,000 ha (27,000 acres) in both the Aude and the Hérault, although the Gard is not far behind with more than 7,000 ha (17,000 acres). Most Languedoc Merlot is reasonably anodyne stuff, sweetish red with a vaguely leafy perfume, for yields tend to be relatively high here.

Italy's 2000 census found 25,614 ha (63,294 acres) of Merlot, most in the northern half of the country and two thirds of it grown for more basic IGT rather than DOC(G) wines. Italian Merlot is, typically, a mass-market, sweet-and-sour blend assembled from over-productive vines grown on the plains of the Veneto. Its other quantitative stronghold, Friuli, can boast rather more convincing examples such as Borgo del Tiglio's Rosso della Centa made in the image of a Pomerol and Graf de la Tour Merlot named after the founder of the Villa Russiz estate and children's institution, who is credited with importing top-quality French vine cuttings to Collio. Merlots from Trentino-Alto Adige tend to lie somewhere between the Friuli and Veneto styles and quality level, and are being helped by generally rising temperatures and concomitant ripeness levels.

All over Italy, small quantities of Merlot are used like seasoning in ambitious blends – and, if gossip is to be believed, in supposedly all-Italian DOCs to give them 'international appeal'. Because selection of suitable plant material is a much more developed art in France than in Italy, Merlot (and Cabernet) vines sourced from French nurseries tend to be very much more reliable and consistent than their Italian counterparts, which makes them popular with growers. But the variety's Italian stronghold is Toscana, especially but by no means exclusively the coastal Maremma. Here too it is only rarely a solo performer, as in such full-throttle, almost Californian, examples as Ornellaia's Masseto and wines such as Castello di Ama's L'Apparita, Le Macchiole's Messorio and Tua Rita's Redigaffi made in Masseto's image. Merlot is also important in a host of Bordeaux blends made throughout central Italy, especially in the Maremma and Chianti Classico, and is also occasionally blended to good effect with the local SANGIOVESE as in Castello di Fonterutoli's Siepi. Merlot's achievements in southern Italy have been less glorious although, in Campania, Riccardo Cotarella's Montevetrano Bordeaux blend has won international adulation, as has Planeta's Sicilian varietal Merlot made in distinctly international style.

Merlot is much less important to **Spain** than it is to generally cooler, damper France and Italy but the fashion for international varietals in Spain drove the country's total plantings from 8,700 ha (21,498 acres) in 2004 to 13,325 ha (32,927 acres) in 2008, with the biggest areas in Catalunya (3,360 ha/8,303 acres), Navarra (2,450 ha/6,054 acres) and Aragón (2,218 ha/5,481 acres) in the north east and Castilla-La Mancha (2,894 ha/7,151 acres) in the centre of Spain. Spanish Merlot is used mainly in blends, often to add flesh to TEMPRANILLO as well as CABERNET SAUVIGNON, and is generally most successful in Catalunya – in Penedès and cooler areas such as Conca de Barberà. Tomàs Cusiné in Costers del Segre, for example, has fashioned a Bordeaux blend with Merlot the major ingredient, labelled Geol, with considerable success, but some of the best examples of varietal Merlot have been Caus Lubis from Can Ràfols dels Caus and the old vintages of Magaña Merlot from Navarra, where they now focus on blends. As ever, **Portugal** has been more resistant to international fashion and had only 556 ha (1,374 acres) of Merlot in 2010, mainly in the Tejo.

Global warming has encouraged the planting of Merlot in **Germany**, mostly in the warmer Pfalz and Rheinhessen regions: 450 ha (1,112 acres) by 2008.

In **Austria**, CABERNET FRANC is very much more important than Merlot and there were only 112 ha (277 acres) of the latter in 2008, half in Niederösterreich (Lower Austria) and the rest mostly in Burgenland.

The warm southern, Italianate Ticino canton of **Switzerland** has long been a Merlot stronghold: 1,028 ha (2,540 acres) in 2009, the vast majority in Ticino in the south, towards the Italian border. Wines can vary from the most basic to fine subalpine answers to Pomerol from the likes of Guido Brivio (Platinum), Daniel Huber (Montagna Magica), Cantina Kopp von der Crone Visini (Balin), Werner Stucky (Tracce di Sassi) and Luigi Zanini (Castello Luigi).

Hungary values CABERNET FRANC much more than Merlot but had 1,791 ha (4,426 acres) in 2008, mainly in Szekszárd and Villány in the south, and Kunság in the south and centre, plus Dél-Balaton in the central west and Eger in the north east. Attila Gere's Solus is one of the better examples.

In 2009, **Bulgaria** had 15,202 ha (37,565 acres) devoted to Merlot, only slightly less than total plantings of CABERNET SAUVIGNON, but no definitive varietal style. Much of the best Bulgarian Merlot is used in Bordeaux blends. Merlot is the dominant variety in the ambitious, plump and spicy, Bordeaux-influenced Enira blend, for example.

Slovenia's 1,019 ha (2,518 acres) of Merlot in 2009 were predominantly in the west, in Vipavska Dolina, where it is the most planted variety, and Goriška Brda, where it comes second. There's also quite a bit in Slovenska Istra (Slovenian Istria) in the far south west. The variety constitutes 15% of the vineyards of Primorska with the likes of Movia, Edi Simčič, Marjan Simčič and Sutor arguably producing rather more serious varietal examples than their counterparts over the border in Italian Friuli.

The **Czech Republic**'s 87 ha (215 acres) in 2009 were almost all in Morava in the south east. Merlot is one of **Serbia**'s principal international varieties while **Croatia** grew 1,105 ha (2,731 acres) in 2009, and it is also a feature of **Montenegro**.

Merlot is by far the leading imported red wine grape in **Romania**, where there were 10,782 ha (26,643 acres) in 2008, mostly in the south east – eg in Dealu Mare in Muntenia and closer to the Black Sea in Dobrogea but also further west in Drăgăşani. The variety is sometimes blended with FETEASCĂ REGALĂ, as in Davino's superior bottling, although, being so widely planted, it is also available in some pretty ordinary offerings. Over the border in very similar conditions, the **Republic of Moldova** grew 8,123 ha (20,072 acres) of Merlot in 2009.

Ukraine recorded 2,820 ha (6,968 acres) of Merlot in 2009, while in **Russia** the Krasnodar Krai region had 1,588 ha (3,924 acres), ie Merlot constituted 5–6% of the region's vines.

In **Greece**, Merlot was in the top six varieties in eastern Makedonía and Thráki (Thrace; 243 ha/600 acres) and there were 86 ha (213 acres) in western Makedonía and 74 ha (183 acres) in central Greece but it needs a relatively cool site to retain any freshness.

Merlot is the second most planted variety in **Israel** after CABERNET SAUVIGNON – with about 1,000 ha (2,470 acres). Superior, concentrated, California taste-alike versions are made by the likes of Dalton, Margalit and Recanati.

Turkey had just 429 ha (1,060 acres) of Merlot in 2010 and the variety is also found on **Malta** and **Cyprus**.

Merlot has been hugely important in the United States, particularly in **California**, where 46,762 acres (18,924 ha) of vineyard was planted with the variety in 2010, declining slightly since the peak of over 51,000 acres (20,640 ha) in 2004. The 1990s saw a craze for mellow Merlot, so much fleshier than cussed CABERNET SAUVIGNON, although the pro-PINOT NOIR movie *Sideways* rubbed the shine off Merlot's reputation. Today it is most readily, and perhaps unfairly, associated with mass-market, sweet reds fed by considerable plantings in the Central Valley and on the Monterey valley floor. The more cosseted fruit from vineyards in Napa and Sonoma tends to play a supporting role in expensive wines labelled Cabernet Sauvignon and those deliberately sold as Meritage blends. Those who take wines labelled Merlot rather more seriously than most include Duckhorn, Pahlmeyer, Paloma, Pride Mountain, Shafer and Truchard, who have proved Carneros' kinship with the variety. Matanzas Creek made a similar point in what is now the Bennett Valley AVA in Sonoma in the late twentieth century. The combination of the opulence of Merlot and California sunshine can be overwhelming; the trick is to retain a certain freshness.

The natural crispness of **Washington** wines complements Merlot's fleshy fruit and only CABERNET SAUVIGNON is more planted here. By 2011 the state's plantings of Merlot totalled

8,235 acres (3,334 ha) with much of it used in blends but also in particularly winning varietals which rarely suffer the excess ripeness that can mar those in some hotter climates. Producers of superior varietal Washington Merlot, often vineyard designated, include Bookwalter, Canoe Ridge, DeLille, L'Ecole No 41, Leonetti, Pedestal, Quilceda Creek and Andrew Will.

Merlot is relatively minor in **Oregon**, where vineyards totalled a steady 508 acres (206 ha) in 2008. This easily ripened variety is planted in most American states where vines are grown, with notable concentrations in **Texas**, where there were 290 acres (117 ha) in 2008, making it the second most planted red wine grape after CABERNET SAUVIGNON. In **Virginia**, Merlot was the most planted red wine variety in 2010 with a total of 338 acres (136 ha), although much of it goes into Meritage blends. The state with the third biggest Merlot acreage after California and Washington is **New York**, with 902 acres (365 ha) back in 2006, most of it thriving in Long Island's maritime conditions.

In **Canada**, Merlot is British Columbia's most planted variety, with 1,585 acres (641 ha) planted by 2008, representing one red wine vine in three. It dominates in such successful blends as Osoyoos Larose and Mission Hill's Oculus, and Cedar Creek's Platinum Reserve shows that varietal Merlot can perform here too. In Ontario, where there were 1,230 acres (498 ha) of Merlot in 2008, it has so far only rarely achieved the intensity of BC Merlot or the precision of the best Ontario Pinot Noir.

Merlot is also grown in **Mexico**, particularly in Baja California and Coahulia de Zaragoza, but can be a bit heavy.

Merlot is extremely important in **Chile**, where it is the third most planted variety after CABERNET SAUVIGNON and País (LISTÁN PRIETO), with 13,280 ha (32,816 acres) in 2009, predominantly in the valleys that cross the Central Valley: Colchagua (3,359 ha/8,300 acres), Maule (3,019 ha/7,460 acres), Curicó (2,911 ha/7,193 acres), Cachapoal (2,007 ha/4,959 acres) and Maipo (1,168 ha/2,886 acres). See CARMENÈRE for details of how for years these two Bordeaux varieties were confused in Chilean vineyards. Today, most true Merlot vines are relatively youthful but the quality of Merlot offerings from the likes of Casa Lapostolle, Errázuriz, Viña Leyda and Montes is still extremely impressive. Successful Chilean Merlot can have real lusciousness.

The variety is much less important to the wine industry in **Argentina**, where the obvious fruit of the dominant Malbec (COT) and to a lesser extent Bonarda (DOUCE NOIRE) need Merlot's flesh much less than, say, the CABERNET SAUVIGNON of Chile. There were 7,142 ha (17,648 acres) in 2008, mostly in Mendoza and to a lesser extent in San Juan and other regions. Here Merlot is often blended with Malbec, as in successful offerings from Finca La Anita, Chacra and NQN.

Uruguay had 853 ha (2,108 acres) of Merlot in 2009, about 10% of the total vineyard area and just ahead of CABERNET SAUVIGNON, with this relatively soft variety an important, and steadily increasing, blending component with tough TANNAT. Bouza and Pizzorno make some of the better examples.

In **Brazil**, there were 1,089 ha (2,691 acres) in Rio Grande do Sul in 2007 with Miolo showing notable confidence in vinifying Merlot – perhaps not surprisingly since Pomerol-based Michel Rolland has consulted here. In 2009 there were 30 ha (74 acres) of Merlot in **Bolivia** and it was also found in **Peru**.

Although Merlot is quantitatively important in **Australia**, its total of 10,537 ha (26,037 acres) in 2008 making it the third most planted red variety after Shiraz (SYRAH) and CABERNET SAUVIGNON, Australian varietal Merlot has no clear identity or reputation. Fuelled by global Merlot mania, the variety was very widely planted in the early years of this century, with the biggest areas in the hot inland regions of Riverland, Riverina and Murray Darling but also in all the country's best-known wine regions, although Merlot fever seems to have cooled somewhat recently. Basic varietals apart, the M word is encountered on labels most often in Cabernet/Merlot blends, of which Cullen's of Margaret River in Western Australia is probably the most respected.

Merlot plantings have also tailed off in **New Zealand**, where there were 1,363 ha (3,368 acres) in 2008, constituting about 5% of the country's total vineyard area and making Merlot the second most planted red wine grape (a long way) after PINOT NOIR. The variety is very much easier to ripen this far from the equator than CABERNET SAUVIGNON and thrives in the well-drained soils of Hawke's Bay, as witness the best examples from Alluviale, Craggy Range and Trinity Hill. Ata Rangi have also had success with it in Martinborough.

Where there is CABERNET SAUVIGNON there is usually Merlot and **China** is no exception. In 2009 total area planted in north-east and western China was estimated at 3,204 ha (7,917 acres; about a sixth of all Cabernet Sauvignon plantings) but this is probably an under-estimate. It is also grown in **Japan** (816 ha/2,016 acres in 2009) and **India**.

Merlot is the third most planted red variety in **South Africa** with 6,614 ha (16,344 acres) in 2008 – almost 15% of all red vines. It is planted in most wine-growing regions, with the biggest areas in Stellenbosch (2,105 ha/5,202 acres) and Paarl (1,289 ha/3,185 acres) and wines tend to be made in a chocolatey, glossy California style. Over-achievers include Dombeya, Shannon, Thelema, Veenwouden, Vergelegen and Yonder Hill.

MERSEGUERA

Generally uninspiring variety planted in eastern central Spain.

BERRY COLOUR: ● ● ● ● ●

PRINCIPAL SYNONYMS: Exquitsagos or Esquitxagos, Verdosilla

VARIETIES COMMONLY MISTAKEN FOR MERSEGUERA: PALOMINO FINO ⚤ (called Merseguera in Algeria)

ORIGINS AND PARENTAGE

Merseguera most probably comes from Valencía in eastern central Spain and is identical to Exquitsagos, cultivated further north, in San Mateo in the province of Castellón and in the Bajo Ebro region in the province of Tarragona. Merseguera is genetically close to the Spanish TORRONTÉS (Ibañez *et al.* 2003).

The name Marisancho is also used occasionally in Valencia as a synonym for Merseguera, which is confusing as it is also a synonym for PARDILLO.

VITICULTURAL CHARACTERISTICS

Mid budding and mid to late ripening. Vigorous and fertile. Good resistance to drought and well adapted to poor soils. Susceptible to powdery mildew.

WHERE IT'S GROWN AND WHAT ITS WINE TASTES LIKE

Varietal wines are generally lacklustre with moderate alcohol levels, lightly herbaceous aromas and notes of bitter almonds. For more body and aroma, Merseguera is sometimes blended with MALVASIA and other varieties. The variety was traditionally used to make rancio-style and sweet wines. The vast majority of **Spain**'s 3,921 ha (9,689 acres) in 2008 were in Valencia, although there are lesser plantings in Murcia and Castilla-La Mancha. It is one of the principal varieties in the Valencia and Utiel-Requena DOs and permitted in DO Alicante. Varietal or Merseguera-dominant wines are produced by, for example, Castillo de Chiva and Polo Monleón. Mustiguillo make an unusually complex varietal wine.

MERWAH

See SÉMILLON.

MERZIFON KARASI

Recently revived variety from just south of Turkey's Black Sea coast.

BERRY COLOUR: ● ● ● ● ●

VARIETIES COMMONLY MISTAKEN FOR MERZIFON KARASI: MARZEMINO ☓ (Italy)

ORIGINS AND PARENTAGE

Merzifon Karası, literally the 'black of Merzifon', logically originates from the Merzifon district in the province of Amasya, in the centre of the Black Sea region of Turkey.

OTHER HYPOTHESES

A recent Italian–Turkish collaboration suggested that Merzifon Karası was identical to the Italian variety MARZEMINO, but DNA profiling has shown that they are unrelated (Serena Imazio, personal communication).

VITICULTURAL CHARACTERISTICS

Mid to late ripening, good general disease resistance. Berries can attain high sugar and acid levels. Well suited to clay-loam soil and dry conditions.

WHERE IT'S GROWN

The mistaken belief that MARZEMINO and Merzifon Karası were one and the same, and the excited press coverage referring to Mozart (see MARZEMINO), triggered a revival of interest in Merzifon Karası vines in **Turkey**. In recent years 26,000 cuttings were distributed to local growers.

MERZLING

Very minor, disease-resistant German hybrid.

BERRY COLOUR: ● ● ● ● ●

PRINCIPAL SYNONYMS: Freiburg 993-60

ORIGINS AND PARENTAGE

Complex hybrid bred in 1960 by Johannes Zimmerman at the Freiburg research centre in Baden-Württemberg, southern Germany, by crossing Seyve-Villard 5276 (see SEYVAL BLANC) with Freiburg 379-52, where Freiburg 379-52 is a RIESLING × PINOT GRIS cross (see PRIOR for the complete pedigree of Merzling). It was named after Merzhausen, the area of the city of Freiburg where the research institute had several trial vineyards. Merzling was used to breed BRONNER, HELIOS and SOLARIS, which are more widely planted than their parent.

VITICULTURAL CHARACTERISTICS

High yielding and late ripening.

WHERE IT'S GROWN AND WHAT ITS WINE TASTES LIKE

There are just 8 ha (20 acres) planted in **Germany**. Thomas Müller in Franken produces a dry varietal wine, as do Siebenhaller and Köpfer in Baden. It is also found in **Belgium** at the Domaine Viticole du Chenoy.

MESLIER NOIR

See MÉZY.

MESLIER SAINT-FRANÇOIS

Almost extinct, historic French variety.

BERRY COLOUR: ● ● ● ● ●

PRINCIPAL SYNONYMS: Blanc Ramé (Charente), Gaillac (Saint-Mont), Gros Meslier (Charente), Meslier (Gers), Purgarie or Pelegarie (Gironde)

VARIETIES COMMONLY MISTAKEN FOR MESLIER SAINT-FRANÇOIS: PETIT MESLIER ☒

ORIGINS AND PARENTAGE

The earliest mention of a grape named Meslier dates back to 1512 in the Comptes de l'Abbaye de Saint-Germain-des-Prés (Rézeau 1997): 'le vi^{e} jour de mars, pour huit cens de marguottez de meslier' ('the 6th day of March, for eight rents of meslier cuttings'). However, three different types of Meslier were described in 1564 by Charles Estienne and Jean Liébault (Rézeau 1997): 'l'vn se nomme le meslier commun, qui est de grand rapport: l'autre s'appelle le gros meslier, qui a le bois & le fruit aussi plus gros: & l'autre s'appelle le franc meslier, qui rapporte le meilleur fruict de tous, & le grain plus séparé' ('one is called meslier commun, which has high yields: the other is called gros meslier, which has bigger wood and fruit: and the other is called franc meslier, which provides the best fruit of all, and has looser berries'). Meslier Commun corresponds to the modern PETIT MESLIER (Galet 1990) and Gros Meslier is an old synonym for Meslier Saint-François (Odart 1845) but the modern identity of Franc Meslier remains unknown.

After several subsequent mentions under various spellings (Melié, Meliez, Mélier, Mellier), the name Meslier de Saint-François first appeared in Odart (1859): 'le Gros Meslier, ou Meslier de Saint-François, produit en abondance du vin passable' ('Gros Meslier, or Meslier de Saint-François, produces abundant quantities of average wine'). The name *meslier* most likely derives from the old French *mesler*, meaning 'to mix' (Latin *misculare*), owing to the ability of these varieties to be blended with others, an etymology shared with MELON.

According to Galet (1990), Meslier Saint-François originated in the area around Notre-Dame de Château-Landon and then spread out to the Val de Loire, the Charente and the south west. This is geographically consistent with DNA profiling showing that Meslier Saint-François is a natural GOUAIS BLANC × CHENIN BLANC cross (see the PINOT pedigree), just like its siblings BALZAC BLANC and COLOMBARD from the Charente (Bowers *et al.* 2000). Meslier Saint-François is a member of the Messile ampelographic group (see p XXVII; Bisson 2009).

Meslier Rose is a colour mutation of Meslier Saint-François.

OTHER HYPOTHESES

The word *meslier* is also used for the common medlar (*néflier* in French), from which the grape variety is said to have inherited its name. Another etymology relates this name to the Latin *mel* ('honey') but the berries do not have a sweet taste at maturity. According to Mouillefert (1902c), it is unlikely that the variety or the name originated in the commune of Meslier in the Eure *département*, where the variety has rarely been cultivated. Lavignac (2001) also suggests that the etymology based on *mesler*, 'to mix', mentioned above may be to do with the mix of flavours in the grape.

VITICULTURAL CHARACTERISTICS

Very early budding and thus at risk from spring frosts. Fertile, mid ripening. Susceptible to botrytis bunch rot and sour rot.

WHERE IT'S GROWN AND WHAT ITS WINE TASTES LIKE

Meslier Saint-François is a nearly extinct variety that once grew in Champagne, Basse-Bourgogne, the Franche-Comté and the Val de Loire in **France** but today it is found mainly in the Gers and used in the production of armagnac and Pineau des Charentes, a more fitting use given the light, flat, low-alcohol character of the wines. However, Charles Jumert in Villiers-sur-Loir, north of Tours in the Coteaux du Vendômois appellation, took cuttings in 1990 from what were probably some of the last Meslier Saint-François vines in the region, about to be grubbed up, and has for the last few years been producing a varietal wine, classified as a vin de table (Kelley 2009b). In eastern Touraine, Lionel Gosseaume's Climat no 2 is a blend of old-vine Meslier with SAUVIGNON BLANC.

MÉZES FEHÉR

Ancient but almost extinct Hungarian variety making rich, aromatic whites.

BERRY COLOUR: ● ● ● ● ●

PRINCIPAL SYNONYMS: Alföldi Fehér, Aranyka or Aranyka Sarga, Bieli Medenac (Croatia), Budai Fehér, Goldtraube (Germany), Honigler or Honigler Weisser (Germany), Honigler Bianco (Italy), Honigtraube (Germany), Margit, Mézédes, Mézes, Sarféjer (Austria), Sárga Margit, Sárga Szőlő, Zsige

ORIGINS AND PARENTAGE

Mézes Fehér, meaning 'white honey', is a very old Hungarian variety once widespread across Europe, mainly for table grapes. The name, and its German synonym Honigler Weisser, points to the variety's high sugar content. It is said to originate from the Balkans, but there is so far no evidence of this.

OTHER HYPOTHESES

A family relationship with ROTER VELTLINER has been suggested but the DNA profile of Mézes Fehér published in Galbács *et al.* (2009) is not consistent with a parent–offspring relationship.

VITICULTURAL CHARACTERISTICS

Late budding, early ripening, high yields. Big bunches of small berries. Drought-tolerant but susceptible to low winter temperatures, botrytis bunch rot, downy and powdery mildews.

WHERE IT'S GROWN AND WHAT ITS WINE TASTES LIKE

Mézes Fehér is virtually extinct in **Hungary** but Oszkár Maurer makes a rich, rather unusual yet fresh and fragrant varietal wine (from vineyards over the border in **Serbia**) and Fetzer's Svábbor is a field blend including Mézes Fehér.

MÉZY

Franche-Comté variety known only as a minor ingredient in a blend.

BERRY COLOUR: ● ● ● ● ●

PRINCIPAL SYNONYMS: Mési, Petit Mesi

ORIGINS AND PARENTAGE

Mézy is a minor variety from the Franche-Comté in eastern France. Although it is also called Meslier Noir, it is not the black version of MESLIER SAINT-FRANÇOIS, and it is distinct from the true (and no longer cultivated) Meslier Noir. DNA parentage analysis has shown that Mézy is one of the twenty-one crosses between PINOT and GOUAIS BLANC that have occurred in north-east France (Bowers *et al.* 1999; Boursiquot *et al.* 2004).

VITICULTURAL CHARACTERISTICS

Productive, mid to late ripening. Compact bunches of small, thin-skinned berries.

WHERE IT'S GROWN AND WHAT ITS WINE TASTES LIKE

Even in its region of origin, Mézy has never had great significance, but Étienne Thiebaud, who founded Domaine des Cavarodes in the Jura, eastern **France**, at the age of twenty, has rescued an old vineyard in Liesle, south west of Besançon in the Doubs region, and produces a red Vin de Pays de Franche-Comté, a blend of PINOT NOIR, TROUSSEAU, PINOT MEUNIER, GAMAY NOIR, POULSARD, GUEUCHE NOIR, Argant (ie GÄNSFÜSSER), BLAUER PORTUGIESER, ENFARINÉ NOIR and Mézy.

MILGRANET

Recently snatched from virtual extinction by researchers in southern France.

BERRY COLOUR: ● ● ● ● ●

PRINCIPAL SYNONYMS: Périgord Noir (Averyon), Petite Mérille (Lavilledieu)

VARIETIES COMMONLY MISTAKEN FOR MILGRANET: MÉRILLE⌘, MONDEUSE NOIRE⌘

ORIGINS AND PARENTAGE

Milgranet is a traditional variety of the Tarn, Haute-Garonne and Tarn-et-Garonne *départements* in central southern France. The name Milgranet may be derived from *mille graines* ('thousand seeds') but Lavignac (2001) suggests it comes from *milgrana*, the Occitan word for pomegranate, because the grape bunches have small, tight berries. DNA parentage analysis has established that Milgranet is a likely progeny of GOUAIS BLANC (Boursiquot *et al.* 2004).

VITICULTURAL CHARACTERISTICS

Fertile and not suited to very dry soils. Small berries and some susceptibility to powdery mildew.

WHERE IT'S GROWN AND WHAT ITS WINE TASTES LIKE

Milgranet is a complementary variety in the Vins de Lavilledieu appellation in **France** but there were just 2 ha (5 acres) recorded in France in 2008. However, local research continues: cuttings were taken from some thirty different clones in 2003 and the first official clone was agreed in 2007. If yields are not restricted, the wines can be green and astringent.

MÍLIA

Red-skinned Slovakian cross making soft, aromatic whites.

BERRY COLOUR:

ORIGINS AND PARENTAGE

Mília is a MÜLLER-THURGAU × GEWÜRZTRAMINER cross obtained in 1973 at the VSVV viticulture and oenology research centre in Šenkvice, Slovakia. It is therefore a sibling of the Czech variety PÁLAVA. It was included in the official register of varieties in 2002.

VITICULTURAL CHARACTERISTICS

Productive, mid budding and ripening. Medium-sized to large bunches of small, thin-skinned berries. Best suited to well-exposed sites with deep soils. Susceptible to drought and poor fruit set but resistant to frosts and botrytis bunch rot.

WHERE IT'S GROWN AND WHAT ITS WINE TASTES LIKE

Plantings so far are extremely limited though a few producers such as Masaryk are making commercially available wines. These are typically lightly aromatic, showing a marked similarity to MÜLLER-THURGAU, with plenty of extract and soft acidity. Sweet, sometimes botrytized, wines are particularly successful in Južnoslovenská and Malokarpatská in south-west **Slovakia**.

MILLOT-FOCH

Rare, disease-resistant Swiss hybrid producing deeply coloured, fruity wines.

BERRY COLOUR:

PRINCIPAL SYNONYMS: VB 85-1

ORIGINS AND PARENTAGE

As suggested by its name, Millot-Foch is a LÉON MILLOT × MARÉCHAL FOCH hybrid obtained by Valentin Blattner in Soyhières in the Swiss Jura.

VITICULTURAL CHARACTERISTICS

Good general resistance to frost and diseases.

WHERE IT'S GROWN AND WHAT ITS WINE TASTES LIKE

Only a few producers, including the breeder, cultivate this variety on just 2 ha (5 acres) in **Switzerland**. A deeply coloured varietal wine is made by Creux-de-la-Terre in the canton of Jura and Risch in the canton of Zug. They tend to have aromas of wild berries, cherries and hints of pepper plus silky tannins.

MINELLA BIANCA

Minor Sicilian white rarely made as a varietal.

BERRY COLOUR: ● ● ● ● ●

PRINCIPAL SYNONYMS: Eppula, Minedda Bianca, Minnella

ORIGINS AND PARENTAGE

Minella Bianca was described in 1760 by Sestini (1991) in the Etna region of the Italian island of Sicilia, where it is still cultivated. Its name could derive from the local word *minna*, meaning 'breast', referring to the elongated shape of the berries. It is considered to be indigenous to Sicilia.

VITICULTURAL CHARACTERISTICS

Ripens mid September, earlier than CARRICANTE and most Sicilian whites.

WHERE IT'S GROWN AND WHAT ITS WINE TASTES LIKE

Minella Bianca is cultivated exclusively in the province of Catania, in the Etna foothills in north-east Sicilia, **Italy**, and to a lesser extent in the province of Enna in the centre of the island. It tends to be interplanted with NERELLO MASCALESE, CARRICANTE and CATARRATTO BIANCO and is authorized in the Etna Bianco and Etna Bianco Superiore DOCs. Benanti is one of the very few producers to make a varietal wine, which is aromatic and dry and may have aromas of aniseed. The Italian agricultural census recorded 86 ha (213 acres) in Italy in 2000 but it was not listed in more recent Sicilia statistics.

MINUTOLO

Opulently grapey Puglian once known as Fiano Aromatico.

BERRY COLOUR: ● ● ● ● ●

PRINCIPAL SYNONYMS: Fiano Aromatico, Fiano della Valle d'Itria, Fiano di Puglia✗, Fiano di Salento, Fiano Minutolo, Fiore Mendillo, Greco Aromatico, Minutola, Moscatellina

VARIETIES COMMONLY MISTAKEN FOR MINUTOLO: FIANO✗

ORIGINS AND PARENTAGE

Until recently, this grape variety from the Valle d'Itria in Puglia, southern Italy, was known as Fiano Aromatico or Greco Aromatico, then Fiano Minutolo. It is now officially known as Minutolo, in order to distinguish it from Campania's FIANO (aka Fiano di Avellino), which has a distinct DNA profile (Calò *et al.* 2001). Despite its emphatically grapey flavour, Minutolo is genetically distinct from varieties with Muscat in their name (Fanizza *et al.* 2003).

VITICULTURAL CHARACTERISTICS

Mid budding and ripening. Medium-sized bunches of thick-skinned berries. Bunches are more compact than those of FIANO and berries are rounder.

WHERE IT'S GROWN AND WHAT ITS WINE TASTES LIKE

Minutolo, a speciality of Puglia, southern **Italy**, was rescued from the brink of extinction in 2000 by oenologist Lino Carparelli and colleagues, who discovered old vines of what was then called Fiano Minutolo, a local synonym for the previous name Fiano Aromatico, in some old vineyards in the Valle d'Itria in Puglia. Additional vines were later found and the variety was evaluated and subsequently included in the Italian national register of grape varieties under the name Minutolo. It is its Muscat-flavoured grapes that distinguish this so clearly from FIANO (ie Fiano di Avellino).

Minutolo is recommended throughout Puglia and is used mainly in blends in Locorotondo and Martina/Martina Franca DOCs. Torrevento produce one such blend using grapes from Carparelli's vineyard. Varietal versions such as I Pastini's Rampone and Feudi di San Marzano's Magia are proliferating. The wines, if carefully vinified so as to retain freshness, tend to be intensely grapey, somewhat reminiscent of MUSCAT OF ALEXANDRIA, with floral, citrus, or white-fruit and herbal flavours.

MIOUSAT

See HUMAGNE.

MISKET CHERVEN

Old, pink-skinned Bulgarian variety used to make dry, scented whites.

BERRY COLOUR: ● ● ● ● ●

PRINCIPAL SYNONYMS: Karlovski Misket, Kimionka, Misket Siv, Misket Starozagorski, Misket Sungurlarski, Sinja Temenuga, Songurlarski Misket, Yuzhnobalgarski Cherven Misket

VARIETIES COMMONLY MISTAKEN FOR MISKET CHERVEN: MISKET VARNENSKI, MISKET VRACHANSKI

ORIGINS AND PARENTAGE

Misket Cherven, literally Red Muscat, is an old Bulgarian variety that possibly originates in the Sungurlare Valley in central eastern Bulgaria or around Karlovo or Brezovo further west, where it has traditionally been cultivated. The colour mutations **Misket Rozov** (pink berries) and **Misket Byal** (white berries) also exist. Despite its name, Misket Cherven is not related to any other Muscat Something or to the other Miskets in Bulgaria.

VITICULTURAL CHARACTERISTICS

Relatively compact bunches of small tough-skinned berries. Fertile, productive, late ripening. Good resistance to low winter temperatures. Susceptible to downy and powdery mildews but fairly resistant to botrytis bunch rot.

WHERE IT'S GROWN AND WHAT ITS WINE TASTES LIKE

Misket Cherven produces simple, only lightly perfumed dry whites and is widely planted in **Bulgaria**, mostly in the provinces of Plovdiv, Burgas, Haskovo and Sliven in the south of the country, with a total area of 6,303 ha (15,456 acres) in 2009, making it the most widely planted indigenous white wine variety in Bulgaria. Varietal appellation wines are made in regions such as Sungurlare and Karlovo, although Misket Cherven is naturally low in acidity and is therefore often blended with crisper local varieties such as DIMYAT.

MISKET VARNENSKI

Minor Bulgarian cross from Varna.

BERRY COLOUR: ● ● ● ● ●

PRINCIPAL SYNONYMS: Muscat de Varna, Muscat Varnenski

VARIETIES COMMONLY MISTAKEN FOR MISKET VARNENSKI: MISKET CHERVEN, MISKET VRACHANSKI

ORIGINS AND PARENTAGE

Misket Varnenski, literally 'Muscat from Varna', is a DIMYAT × RIESLING cross obtained in 1971 by K Stoev at the Pleven viticulture and oenology institute in northern Bulgaria. It is not related to any other varieties named Muscat Something.

VITICULTURAL CHARACTERISTICS

Early budding and late ripening with good resistance to both botrytis bunch rot and cold winters. Productive. Thick-skinned berries.

WHERE IT'S GROWN AND WHAT ITS WINE TASTES LIKE

Not surprisingly, the majority of **Bulgaria**'s 259 ha (640 acres) of Misket Varnenski is planted in the province of Varna in eastern Bulgaria, although it is also found to a much more limited extent in other provinces such as Ruse in the north. Wines are not particularly memorable and show little RIESLING character, which is perhaps why producers such as the Varna Winery blend Misket Varnenski with its more noble parent.

MISKET VRACHANSKI

Minor Bulgarian cross planted mostly in the far east and west. The most aromatic of the Bulgarian Miskets.

BERRY COLOUR: ● ● ● ● ●

PRINCIPAL SYNONYMS: Mirizlivka, Misket Vratchanski, Tvarda Vrazhda Misket, Wrat Chanskii Musket, Wratchanski Misket

VARIETIES COMMONLY MISTAKEN FOR MISKET VRACHANSKI: MISKET CHERVEN, MISKET VARNENSKI

ORIGINS AND PARENTAGE

Misket Vrachanski is a PUKHLIAKOVKSY (under the name Coarnă Alba) × MUSCAT BLANC À PETITS GRAINS cross made in Bulgaria.

VITICULTURAL CHARACTERISTICS

Late budding and ripening. Medium-sized bunches of thick-skinned grapes with good botrytis bunch rot resistance.

WHERE IT'S GROWN AND WHAT ITS WINE TASTES LIKE

There were just 152 ha (376 acres) of this lesser Misket planted in **Bulgaria** in 2008, mostly in the province of Montana in the north west and in Varna in the east. Wines are generally dry or off dry and have relatively powerful Misket aromas of flowers, herbs and honey. Chateau Euxinograde, housed in the former royal palace on the Black Sea coast north of Varna, make a varietal wine and the Ruse Wine House in the far north of the country blend Misket Vrachanski with Traminer (SAVAGNIN).

MISSION

See LISTÁN PRIETO.

MLADINKA

Very rare variety from the Croatian coast generally used in blends.

BERRY COLOUR: ● ● ● ● ●

PRINCIPAL SYNONYMS: Mladenka

VARIETIES COMMONLY MISTAKEN FOR MLADINKA: BOGDANUŠA

ORIGINS AND PARENTAGE

A rare variety that has survived in the Kaštela and Trogir regions north of Split on the Croatian coast. Contrary to local opinion, recent ampelographic research and DNA profiling have shown that Mladinka is not identical to BOGDANUŠA (Goran Zdunić, personal communication).

VITICULTURAL CHARACTERISTICS

Late budding, mid to late ripening. Thin skins, susceptible to powdery and downy mildews. Medium vigour, good yields.

WHERE IT'S GROWN AND WHAT ITS WINE TASTES LIKE

Mladinka is found mainly in the Kaštela and Trogir regions very close to Split on the coast of **Croatia**. It is also grown in the Omiš region slightly further south. It is generally used in blends with other local varieties, sometimes interplanted in the vineyard with VLAŠKA, for example, to add acidity.

MOLETTE

Minor and rather neutral vine used mainly for sparkling wines in Savoie, eastern France.

BERRY COLOUR: ● ● ● ● ●

PRINCIPAL SYNONYMS: Molette Blanche, Molette de Seyssel

VARIETIES COMMONLY MISTAKEN FOR MOLETTE: CACABOUÉ, JACQUÈRE, MONDEUSE BLANCHE, Roussette Basse (Seyssel)

ORIGINS AND PARENTAGE

Molette is a local variety from Savoie, eastern France, where it was first mentioned by the ampelographers Alphonse Mas and Victor Pulliat in 1876–7 in Seyssel (Rézeau 1997): 'la Molette paraît être un cépage special et particulier au vignoble de Seyssel, où nous l'avons vu pour la première fois' ('Molette seems to be a particular variety of the Seyssel vineyards, where we have seen it for the first time'). DNA parentage analysis has shown that Molette is most likely a progeny of GOUAIS BLANC (Boursiquot *et al.* 2004; see PINOT pedigree diagram), and it might be a GRINGET × Gouais Blanc cross (Vouillamoz). The name Molette could derive from the Savoie dialect word *molèta*, meaning 'soft', because of its tender berries.

OTHER HYPOTHESES

Roussette Basse, formerly cultivated in the area of Seyssel and Frangy, is often considered to be a mutation of Molette, but this is rejected by DNA profiling (Vouillamoz).

VITICULTURAL CHARACTERISTICS

Vigorous, mid ripening. Well suited to both clay-limestone and sandy soils. Susceptible to downy mildew and botrytis bunch rot. Small berries.

WHERE IT'S GROWN AND WHAT ITS WINE TASTES LIKE

Molette is grown almost exclusively in the Seyssel area of Savoie, eastern **France**, and produces wines that are generally pale and rather ordinary with relatively high levels of alcohol and acidity, and greatly improved by being blended with ALTESSE. It is officially recommended in the Ain and Haute-Savoie *départements*, authorized in Savoie and may be included in the sparkling wines of the Seyssel and Bugey appellations. Domaine de Vens-le-Haut make one of the very rare examples of still varietal wines which are labelled Allobrogie. The area given over to the variety continues to decline, to 28 ha (69 acres) in 2008.

MOLINARA

Now plays distinctly third fiddle to Corvina Veronese and Rondinella in wines such as Valpolicella.

BERRY COLOUR: ● ● ● ● ●

PRINCIPAL SYNONYMS: Brepon, Mulinara, Uva del Mulino, Uva Salà

ORIGINS AND PARENTAGE

Molinara has been mentioned in the Verona area of the Veneto, northern Italy, since the early nineteenth century. Its name derives from the Italian word *mulino*, meaning 'mill', because the bloom on the surface of the berries looks like flour.

VITICULTURAL CHARACTERISTICS

Mid to late ripening. Good resistance to fungal diseases. Quite vigorous.

WHERE IT'S GROWN AND WHAT ITS WINE TASTES LIKE

Molinara is cultivated mainly in the province of Verona, **Italy**, in particular around the Lago di Garda. Its wines are generally pale in colour, high in acidity and prone to oxidation. They therefore tend to play only a minor role in blends such as Valpolicella and Bardolino, where the DOC rules permit a maximum percentage of 10% but in practice it is usually a good deal less than that. Molinara used to play a more important role in Amarone and Recioto but the regulations have changed in recognition of the fact that CORVINA VERONESE, CORVINONE and RONDINELLA give better results. However, old vines of some of the older clones, such as those in the care of Cecilia Trucchi of Villa Bellini, can produce low yields of very high-quality fruit. Masi also have superior pre-phylloxera clones at their Serego Alighieri estate. The fruit from the newer clonal selections is not as good. Carlo Boscaini produces a varietal Molinara, classified as IGT Veronese and looking very much like a rosé. There were 1,348 ha (3,331 acres) in 2000 but the variety is losing ground to Corvina Veronese, Corvinone, Rondinella and international varieties.

MOLL

See PRENSAL.

MOLLAR

See NEGRAMOLL.

MOLLARD

Very minor variety making fresh, light reds in south-east France.

BERRY COLOUR: ● ● ● ● ●

PRINCIPAL SYNONYMS: Molard, Petit Mollard, Tallardier

VARIETIES COMMONLY MISTAKEN FOR MOLLARD: MAZUELO (known as Mollard in Rioja in Spain), NEGRAMOLL (called Mollar in the province of Huelva in south-west Spain)

ORIGINS AND PARENTAGE

Mollard is a traditional variety of the Hautes-Alpes in south-east France. Its earliest mention appeared in Comte Odart (1854), but the text implies much earlier cultivation in the Hautes-Alpes (Rézeau 1997): 'le cépage le plus anciennement cultivé et plus commun encore dans ces vignes, c'est le gros mollar noir' ('the variety that has been cultivated the longest and is still the most common in these vineyards is gros mollar noir'). DNA parentage analysis has shown that Mollard is most likely a progeny of GOUAIS BLANC (Boursiquot *et al.* 2004). The name Mollard could derive from the French *mol*, meaning 'soft', because of the texture of the berries.

OTHER HYPOTHESES

According to the French ampelographer Victor Pulliat, Mollard was introduced from Andalucía in 1780 by the Marquis de Brassier de Jocas, a landowner near Tallard, just south of Gap, where Mollard used to be cultivated (Galet 1990). This unlikely hypothesis probably relies on the erroneous identity that has been suggested with the variety Mollar cultivated in the province of Huelva in Spain (Tacussel in Viala and Vermorel 1901–10), which is a synonym for NEGRAMOLL.

VITICULTURAL CHARACTERISTICS

Late budding, mid ripening. Best pruned short and may be grown as a bushvine. Susceptible to both powdery and downy mildews.

WHERE IT'S GROWN AND WHAT ITS WINE TASTES LIKE

Varietal wines tend to be fresh and light-bodied but well coloured. The main proponent in the

Hautes-Alpes, the only *département* in **France** in which this variety is recommended, is Marc Allemand at Domaine Allemand. He has been making varietal Mollard since 1984 and currently has two bottlings, one from old vines. He also works closely with Yann de Agostini at Domaine du Petit Août, created in 2009 in the Vallée de la Durance and focusing on local varieties such as Mollard. There were 23 ha (57 acres) in France in 2008.

Mollard has recently been identified in an old vineyard in Sonoma, **California**.

MONARCH

Dark, powerful recent German hybrid giving fruity and full-bodied wines.

BERRY COLOUR:

PRINCIPAL SYNONYMS: Freiburg 487-88

ORIGINS AND PARENTAGE

Complex hybrid obtained in 1988 by Norbert Becker at the Freiburg research centre in Baden-Württemberg, southern Germany, by crossing SOLARIS with DORNFELDER. The name Monarch was intended to confer nobility.

VITICULTURAL CHARACTERISTICS

Very resistant to fungal diseases.

WHERE IT'S GROWN AND WHAT ITS WINE TASTES LIKE

There are limited plantings in **Germany** and just over 1 ha (2.5 acres) in **Switzerland**. Weingut Im Hagenbüchle in Weinstadt-Schnait in Württemberg make a varietal wine and Staatsweingut Freiburg & Blankenhorstberg use it in a blend. It was authorized in **Italy** in 2011.

MONASTRELL

High-quality, heat-loving dark-skinned variety most valued for its heady, structured contribution to blends.

BERRY COLOUR:

PRINCIPAL SYNONYMS: Alcayata (Spain), Balzac Noir (Charente and Vienne in France), Catalan (Bouches-du-Rhône in France), Espagnen (Ardèche in France), Espar (Hérault in France), Garrut (Valencia and Tarragona in Spain), Gayata (Murcia in Spain), Mataro or Mataró (Catalunya in Spain, Pyrénées-Orientales in France, Australia, Cyprus, United States), Mourvede (Provence), Mourvedon (Provence), Mourvèdre (Provence), Negria (Greece), Ros (Sagunto in Spain)

VARIETIES COMMONLY MISTAKEN FOR MONASTRELL: BOBAL, GRACIANO (called Morastell or Morrastel in the Languedoc and Monastrell Menudo or Monastrell Verdadero in Spain), MORISTEL (Somontano in Spain), NÉGRETTE (Camargue in southern France)

ORIGINS AND PARENTAGE

The name Monastrell was mentioned by Francesc Eiximenis in Empordà, Catalunya, as early as

1381–6 (Favà i Agud 2001) and is also said to have been mentioned together with BOBAL among the most important grape varieties of the region of Valencia in 1460 in Jaume Roig's *Espill o llibre de les dones*. It derives from the Latin *monasteriellu*, a diminutive of *monasteriu*, meaning 'monastery', suggesting that the variety was first cultivated and propagated by monks (Favà i Agud 2001). Monastrell most likely originates from Sagunto in the Camp de Morvedre region in Valencia, eastern Spain, where it was the main variety before the arrival of phylloxera in the early twentieth century. Morvedre (Spanish Murviedro) was until 1868 the Catalan name for the town of Sagunto, an important wine port north of Valencia from the fifteenth century onwards. Outside Spain, Monastrell is often known as Mataró (or Mataro), a name also derived from a town on the Mediterranean coast, between Barcelona and Valencia. (See Bobal for the confusion caused by the synonym Beni Carlo.)

In France, Monastrell was probably introduced in the sixteenth century from Morvedre (now Sagunto) to Provence, hence the name Mourvèdre, and from Mataró to Roussillon, where it kept the name Mataro, which is now also used in Australia and California (Ganzin 1901a; Lavignac 2001). However, the earliest written mention of Mourvèdre did not appear until the end of the eighteenth century (Rézeau 1997).

Recent DNA kinship analysis has suggested that Monastrell and GRACIANO could be siblings (Santana *et al.* 2010) but any parent–offspring relationship between those two can be rejected (Vouillamoz).

OTHER HYPOTHESES

It is often supposed that Monastrell, like many other varieties, was introduced to Spain by the Phoenicians around 500 BC but there is strictly no scientific evidence for such speculation. Some scholars have suggested a doubtful genetic relationship between Monastrell and VERMENTINO NERO from Toscana but this has not yet been tested by DNA analysis.

VITICULTURAL CHARACTERISTICS

Small to medium-sized compact bunches of small to medium-sized, sweet, thick-skinned berries with a distinctive bloom. Late budding and very late ripening. A demanding and difficult variety that needs high temperatures, especially at the end of the ripening period, and a sufficient supply of magnesium and potassium to ripen. Best suited to short pruning – cordon or bushvine. Highly susceptible to drought and therefore well suited to deep calcareous soils that provide a limited but regular water supply. Low yields. Susceptible to mites, leafhoppers, esca and sour rot but good resistance to botrytis bunch rot and phomopsis.

WHERE IT'S GROWN AND WHAT ITS WINE TASTES LIKE

In Europe, the variety is typically grown within 80 km (50 miles) of the Mediterranean, where winters are mild and summers are generally long and hot. Its wines are particularly prone to reduction so great care has to be taken to expose them to oxygen during élevage. Monastrell is typically high in alcohol and tannins and may well have an intense aroma of blackberries.

France has the second biggest plantings total with 9,363 ha (23,136 acres) of Mourvèdre in 2009, predominantly along the Mediterranean coast in Provence and Languedoc-Roussillon, where it is authorized in many different appellations. This is remarkable given that fifty years ago there were just 517 ha (1,278 acres) but the variety enjoyed a certain vogue in the late twentieth century, was widely regarded as a useful and ameliorating ingredient in blends throughout southern France, and much better-yielding clones became available. The biggest plantings are in the Var (2,198 ha/5,431 acres), which is the home of the Bandol appellation, the Aude (1,712 ha/4,230 acres) and the Vaucluse (1,494 ha/3,692 acres) but there are also substantial plantings in the Pyrénées-Orientales (974 ha/2,407 acres) and the Gard (898 ha/2,219 acres). Provence's most famous red wine, Bandol, is arguably Mourvèdre's finest expression. The variety must contribute 50–95% of the blend (the rest comprising any or all of Grenache/GARNACHA, CINSAUT, SYRAH and Carignan/MAZUELO). Recommended producers include Domaines Bunan, La Suffrène, Tempier and Château Pibarnon.

It also plays a very significant structural role in southern Rhône blends – Château de Beaucastel, for example, are renowned for the high proportion of Mourvèdre (usually around 30%) in their red Châteauneuf-du-Pape and for their Hommage à Jacques Perrin made mainly from old Mourvèdre vines. Other varietal wines are occasionally found in the Languedoc, from producers such as Borie de Maurel and Condamine l'Évêque, but such examples are rare.

Spain has by far the biggest area devoted to this variety, most usually called Monastrell in Spain nowadays: 63,244 ha (156,279 acres) in 2008, making it the fourth most planted red wine variety in the country and the fifth most planted overall, behind AIRÉN, TEMPRANILLO, BOBAL and GARNACHA. Most of the vines are in the centre and south-east, in the autonomous communities of Murcia (33,710 ha/83,299 acres), Castilla-La Mancha (19,641 ha/48,534 acres) and Valencia (9,495 ha/23,463 acres). It is the principal black grape variety in DOs such as Alicante, Almansa, Jumilla, Valencia, and Yecla on or near the Mediterranean coast. Varietal versions are common in Spain and recommended producers include Casa Castillo, Hijos de Juan Gil, Luzón and Silvano García in Jumilla, Castaño in Yecla, Molino y Lagares de Bulla in Bullas, Enrique Mendoza and Laderas de Pinoso in Alicante and Rafael Cambra in Valencia.

It is also found to quite an extent on **Cyprus** (149 ha/368 acres in 2010) and to a very limited extent in **Turkey** (7 ha/17 acres in 2010).

In **California**, it was long known as Mataro, and indeed it was not until the 1990s that many growers realized that the Mataro associated with rustic old plantings was identical to the modish Mourvèdre of southern France. Total plantings in the state were 939 acres (380 ha) in 2010 and it thrives particularly in the deep, sandy, infertile, largely dry-farmed vineyards of Contra Costa County, as well as in Madera and San Luis Obispo counties. Recommended producers include Bonny Doon, Cass, Cline, Qupé, Rosenblum and Tablas Creek.

Mourvèdre plantings along with all things Rhôneish have been increasing in **Washington** State and had reached 165 acres (67 ha) by 2011. McCrae, Chateau Ste Michelle and Syncline make particularly respectable varietal examples. Plantings in **Oregon** go unrecorded officially but Torii Moor and a few others include Mourvèdre in their Rhône-style blends. Varietal Mourvèdre is also produced in **Arizona**, **Texas** and even **Virginia**.

In **Australia**, too, the variety's stock went up considerably when it turned out that Mataro could be labelled Mourvèdre and that it could be blended with the country's generally even older stocks of Shiraz (SYRAH) and Grenache (GARNACHA) to make the popular and often well-balanced blends known colloquially as GSMs. In 2008, Australia grew a total of 784 ha (1,937 acres), mainly in hotter regions such as the Riverland, Barossa and Riverina, although there are scattered plantings elsewhere. Its most common expression is in fashionable Rhône blends with Shiraz and/or Grenache such as those produced by D'Arenberg, Henschke, Charlie Melton, Rockford, Spinifex, Teusner and Torbreck. Hewitson is one of the few to produce a varietal version. Although Mourvèdre is now a sought-after ingredient in such table wines, it was once, and still is to a more limited extent, an important component in fortified wines such as Grant Burge's 30 Year Old Tawny and Seppeltsfield's Para 100 Year Old Tawny.

Under the official name Mataro – though often called Mourvèdre on labels – there were 359 ha (887 acres) in **South Africa** in 2008. Producers of varietal wines include Beaumont and Signal Hill but it is much more common for Mourvèdre to be part of a blend, most often with SYRAH (Shiraz), as in those from De Morgenzon (DMZ), Lammershoek (Roulette), Newton-Johnson, Sadie Family (Columella), Sequillo and Tulbagh Mountain Vineyard.

MONBADON

Productive, dull western French variety that once thrived in California's Central Valley.

BERRY COLOUR: ● ● ● ● ●

PRINCIPAL SYNONYMS: Burger (California), Frontignan des Charentes (Entre-Deux-Mers)

ORIGINS AND PARENTAGE

Monbadon, from the Charente in western France, could well be named after the village of Monbadon in the Gironde, or from a family name common in western France. DNA parentage analysis has shown that Monbadon is probably a FOLLE BLANCHE × TREBBIANO TOSCANO cross (see CABERNET SAUVIGNON and PINOT pedigree diagrams), the latter being known in France under the name Ugni Blanc (Bowers *et al.* 2000).

VITICULTURAL CHARACTERISTICS

Productive, mid to late ripening. Likes hot, dry climates and suited to spur pruning. Large bunches and susceptible to botrytis bunch rot.

WHERE IT'S GROWN AND WHAT ITS WINE TASTES LIKE

In **France**, Monbadon is not on the official register of varieties and has virtually disappeared.

Under the name Burger, this variety is much more widely planted in **California** than in its home in the Charente but the area of California vineyard given over to this workhorse, high-acid variety is much smaller than when in the mid twentieth century it produced more wine than any other variety in the state. There were 1,247 acres (505 ha) in 2008, cultivated mostly in the Central Valley for the production of jug wine. There have been few new plantings in the twenty-first century.

There's a rare outpost of Monbadon in **Australia**: Pfeiffer Wines in Rutherglen, Victoria, use a blend of PALOMINO FINO and Monbadon in their Apera (sherry-like) wines. The vines responsible were in an experimental planting at Charles Sturt University that has now been grubbed up.

MONDEUSE BLANCHE

Savoie variety more famous as Syrah's parent than for its wines.

BERRY COLOUR: ● ● ● ● ●

PRINCIPAL SYNONYMS: Dongine (Ain), Jongin (Savoie), Savouette

VARIETIES COMMONLY MISTAKEN FOR MONDEUSE BLANCHE: MOLETTE, ROUSSETTE D'AYZE

ORIGINS AND PARENTAGE

Mondeuse Blanche is an old variety from Savoie and Isère in eastern France. It was first mentioned in 1843 in the catalogue of the grape collection of Bouchereau Frères at Château Carbonnieux (Rézeau 1997), and then in 1868 by the ampelographer Pierre Tochon: 'la mondeuse blanche est moins féconde que la rouge' ('mondeuse blanche is less fertile than the red one').

It was long considered a white mutation of MONDEUSE NOIRE (see that entry for the etymology) but DNA profiling has disproven this hypothesis and parentage analysis has shown that Mondeuse Blanche spontaneously crossed with DUREZA to give birth to SYRAH (Bowers *et al.* 2000). More recently, Vouillamoz (2008) has shown that Mondeuse Blanche has parent–offspring relationships with two varieties: Mondeuse Noire, which was already considered by ampelographers (eg Comte de Villeneuve 1901) to be very close to Mondeuse Blanche, and, more unexpectedly, VIOGNIER, the Condrieu variety (see Syrah for the complete pedigree).

These DNA findings disprove Tochon's (1869) assertion: 'la mondeuse blanche est un plant isolé en Savoie qui, ici pas plus qu'ailleurs, n'a formé de famille' ('mondeuse blanche is an isolated variety from Savoie that has not, here not more than elsewhere, generated a family'). Despite its strong genetic links with the Sérine ampelographic group (Syrah, MONDEUSE NOIRE, VIOGNIER, etc.), Mondeuse Blanche has been placed in the Pelorsien ampelographic group (see p XXVII; Bisson 2009).

VITICULTURAL CHARACTERISTICS

Late ripening. Shows little susceptibility to botrytis bunch rot.

WHERE IT'S GROWN AND WHAT ITS WINE TASTES LIKE

There were just 5 ha (12 acres) remaining in Savoie in 2008, mostly in Bugey, where it is authorized as a secondary variety in the Bugey and Vin de Savoie appellations. The wines it produces are not very exciting: generally soft with relatively high alcohol. However, Domaine Grisard, Le Caveau Bugiste and Domaine Les Tartères all make varietal wines.

MONDEUSE NOIRE

Aromatic Savoie variety regaining recognition for its well-structured reds.

BERRY COLOUR: ● ● ● ● ●

PRINCIPAL SYNONYMS: Gros Rouge (Vaud and Genève in Switzerland), Grosse Syrah, Maldoux (Jura), Persagne (Ain), Petite Persaigne (Rhône), Plant Maldoux (Jura), Savoyan (Isère)

VARIETIES COMMONLY MISTAKEN FOR MONDEUSE NOIRE: GENOUILLET (Indre), MILGRANET (Midi-Pyrénées), PERSAN, REFOSCO DAL PEDUNCOLO ROSSO (California), SYRAH

ORIGINS AND PARENTAGE

Mondeuse Noire is an old variety from the Dauphiné, a former province of eastern France that roughly comprises the Isère, Drôme and Hautes-Alpes *départements*, where its first mention dates back no further than 1845, in the Haut de la Vallée de l'Isère (Rézeau 1997). However, its earliest mention most likely appeared in the Jura under the synonym Maldoux in an order of the Parliament of Besançon dated 3 February 1731 (see ENFARINÉ NOIR for the citation). While Maldoux undoubtedly comes from the French *mal doux*, meaning 'badly sweet', referring to the bitter taste of the berries, two equally valid hypotheses have been proposed for the etymology of the name Mondeuse (Rézeau 1997): it could derive from the Franco-Provençal dialect *moda* or *moduse* that was used to describe a grape producing a lot of must or it could come from the verb *monder* or *émonder*, meaning 'to prune', because Mondeuse Noire's leaves often fall before harvest.

In California, Mondeuse Noire has long been confused with REFOSCO DAL PEDUNCOLO ROSSO but DNA profiling has shown them to be distinct varieties (see the Refosco dal Peduncolo Rosso entry). DNA parentage analysis has shown that Mondeuse Noire is either a progeny or a parent of MONDEUSE BLANCHE (see SYRAH for the complete pedigree), which means Mondeuse Noire is either a half-sibling or a grandparent of Syrah (Vouillamoz 2008). This explains why Mondeuse Noire is often called Grosse Syrah, and also why Mondeuse Noire belongs to the Sérine ampelographic group (see p xxvii; Bisson 2009). In addition, Mondeuse Noire is ampelographically close to DOUCE NOIRE (Bisson 2009).

Mondeuse Grise is a mutation of Mondeuse Noire described by ampelographer Victor Pulliat at the end of the nineteenth century. It was thought to have disappeared but was reintroduced to the Institut National de la Recherche Agronomique's vine collection at Domaine de Vassal, Montpellier, southern France, by Pierre Galet in 1950.

OTHER HYPOTHESES

According to Pierre Tochon (1887), Mondeuse Noire could be identical to the Allobrogica, a grape cultivated in the land of Allobroges that was mentioned in the first century by Celsus, Columella and Pliny the Elder. This unlikely hypothesis has also been suggested for PINOT and SYRAH (see Syrah for more details).

VITICULTURAL CHARACTERISTICS

Mid ripening. Vigorous and best pruned short. Susceptible to drought, chlorosis, mites, downy and powdery mildews. Does well on stony clay-limestone slopes.

WHERE IT'S GROWN AND WHAT ITS WINE TASTES LIKE

Mondeuse Noire produces wines that are aromatic, tannic and deeply coloured, with good ageing potential and, in the best sites, an Italianate bitter-cherry bite. It is one of the principal varieties authorized in the Vin de Savoie and Bugey appellations of Savoie, eastern **France**. Recommended producers include Michel Grisard at the Domaine du Prieuré Saint-Christophe and Domaine Louis Magnin. Following a sharp decline in the 1970s, France had a total of 300 ha (741 acres) in 2009 and the area is now very slowly increasing in recognition of the potential quality of these wines – and their distinctiveness in the company of most Savoie reds.

In the nineteenth century, Mondeuse Noire, under the name Gros Rouge, was the most planted red variety on the shore of Lac Léman (Lake Geneva) in **Switzerland** and even though there were fewer than 4 ha (10 acres) in the Vaud and Genève cantons in 2009, there are still a few very good producers there, such as Henri & Vincent Chollet at Domaine Mermetus (especially their Le Vin de Bacouni) and Jean-François Cossy at Domaine des Rueyres, both in Vaud, and Damien et Luc Mermoud in Genève.

It is hard to get accurate planting statistics for **California** because, although the Foundation Plant Services at the University of California at Davis corrected the name of their selection from Refosco to Mondeuse in 2005, the official California grape acreage report of 2008 still refers to the varieties as synonymous. Carole Meredith at Lagier Meredith in Napa Valley (who made many exciting discoveries about grapevine DNA when she was a professor at the University of California at Davis) is probably the only producer who is growing the real thing (apart from a few vines in old, mixed vineyards). She reports that her Mondeuse vines planted alongside SYRAH in 2007 are less vigorous and more productive, and produce wine that is darker and spicier than Syrah.

Some old vines at Carlisle Winery in California's Russian River Valley that were thought to be Mondeuse turned out to be Calzin, a Zinfandel (aka TRIBIDRAG) × Mondeuse Noire cross obtained by Harold Olmo in 1937 (although at the time it was said to be a Zinfandel × Refosco cross). Calzin, bred to add tannin to Central Valley wines but in practice overly productive and brutally tannic, has now been almost entirely grubbed up, as has its white mutation Helena.

In the state of Victoria in **Australia**, both Buller (Rutherglen) and Brown Brothers (Milawa) include Mondeuse Noire in blends but the overall planted area is too small to make it into the official statistics. According to Brown Brothers' viticulturist Roland Wahlquist, Mondeuse has been planted at Milawa since 1907. They have 1.2 ha (3 acres) of late-ripening vines and co-ferment it with Shiraz (SYRAH) and CABERNET SAUVIGNON.

Mondeuse Grise, also known as Persagne Grise, has been propagated at the Grisard vine nursery in Savoie.

MONEMVASSIA

Powerful, historic variety widely planted on Páros but receiving renewed attention in the southern Pelopónnisos (Peloponnese).

BERRY COLOUR:

PRINCIPAL SYNONYMS: Artemissi (Santoríni), Klossaria (Dodekánisa/Dodecanese), Monemvasia, Monemvassitiko (Lakonía), Monovassia (Évvoia/Euboea and Náxos)

VARIETIES COMMONLY MISTAKEN FOR MONEMVASSIA: MALVASIA

ORIGINS AND PARENTAGE

This variety probably comes from the area close to Monemvasia, an island and fortified medieval port town on an imposing rock in the southern Pelopónnisos. The name of the port means 'single entrance', from the Greek *moni emvassis* (Boutaris 2000) as there is just one gate into the island-town, linked to the easternmost leg of the Pelopónnisos by a bridge, having become an island as the result of an earthquake in AD 375.

Monemvassia is believed to be one of the varieties that went into the once-famous blended sweet wine known as Malvasia that was favoured by the Venetians and widely acclaimed in the Middle Ages (Manessis 2000). The first written mention of this wine dates back to 1214 in Éfesos (Ephesus) as Monovasia or Monemvasias (Arranz *et al.* 2008). However, there is certainly no consensus and so far no firm evidence as to the make-up of the blend.

Although the name Malvasia is said to be a Venetian transliteration of Monemvassia (Boutaris 2000), the DNA profile of the variety Monemvassia (Greek Vitis Database) is not identical, nor even related, to any of the MALVASIA varieties (Vouillamoz).

VITICULTURAL CHARACTERISTICS

Vigorous, relatively high yields, early ripening. Small bunches of small and irregular-sized berries. Resistant to most diseases and drought.

WHERE IT'S GROWN AND WHAT ITS WINE TASTES LIKE

Monemvassia is found mainly on the Kykládes (Cyclades) islands as well as elsewhere in the Aegean and on the large island of Évvoia that lies parallel to southern mainland **Greece** (Lazarakis 2005). Although the most significant plantings are on the island of Páros, it is in the southern Pelopónnisos near the port that gave the variety its name that it is particularly thriving currently. George Tsibidis of Monemvasia Winery (one S) in particular aims to re-create the historical sweet blend Malvasia. Tsibidis brought his Monemvassia vines from Páros and both he and Yiannis Vatistas have new plantings of the variety in Lakonía.

The confusion over nomenclature does not stop there, since the regional wine of Monemvasia in Lakonía allows eight different varieties, including Monemvassia. On Páros, in the middle of the Kykládes, this variety plays an exclusive role in the appellation of the same name, and even the red version allows two thirds Monemvassia to one third MANDILARIA, the single exception to the rule that red and white varieties should not be blended for rosé or red wines (Lazarakis 2005).

Varietal Monemvassia tends to be relatively alcoholic, full-bodied and low in acidity, needing cooler sites to produce fresher, better-balanced wines. They can be spicy and lightly aromatic and show plenty of fruit depth on the palate. Producers of dry varietal wines include Moraitis in north-eastern Páros and Tsibidis in Lakonía. For his sweet blend, Tsibidis picks the grapes early to retain acidity but he also makes a powerful, spicy, slightly smoky dry version.

MONERAC

Fairly recent Montpellier cross yet to win favour.

BERRY COLOUR: ● ● ● ● ●

PRINCIPAL SYNONYMS: Monérac

ORIGINS AND PARENTAGE

A GARNACHA × ARAMON NOIR cross obtained by Paul Truel in 1960 at INRA (Institut National de la Recherche Agronomique) in Montpellier, southern France. Monerac is a partial anagram of its progenitors' names.

VITICULTURAL CHARACTERISTICS

Late ripening, vigorous, needs to be pruned short, either trellised or as a bushvine. Loose bunches mean that it is not particularly susceptible to botrytis bunch rot.

WHERE IT'S GROWN

Although this variety is included in the official register of varieties, and recommended in much of southern **France**, there do not as yet appear to be any commercial bottlings and there were just 3 ha (7 acres) recorded in 2008.

MONICA NERA

Generally undistinguished dark-skinned variety widely planted on the Italian island of Sardegna and made in a variety of styles.

BERRY COLOUR: ● ● ● ● ●

PRINCIPAL SYNONYMS: Manzesu⌘, Monaca, Monica di Sardegna, Niedda de Ispagna, Niedda Mora, Nieddera Manna, Pascale Sardu (Sassari)

VARIETIES COMMONLY MISTAKEN FOR MONICA NERA: NIEDDERA⌘, PASCALE⌘

ORIGINS AND PARENTAGE

Like many other Sardinian varieties, Monica Nera is said to come from Spain. During Aragón's rule over the island (1323–1720), the Spanish could have introduced this variety under the name Morillo. However, no evidence of a historical or genetic link with Spain has yet been found. Others suggest that its cultivation dates back to the eleventh century, when the Camaldolese monks (*monaci*) began to plant vines and produce wine.

OTHER HYPOTHESES

Monica Nera was thought to be related to LISTÁN PRIETO, the variety called Mission in California, País in Chile and Criolla Chica in Argentina, merely because Monica is a synonym for Criolla, but DNA analysis refutes this (Vouillamoz).

VITICULTURAL CHARACTERISTICS

Fruitful, mid ripening. Good resistance to spring frost and powdery mildew, less resistant to downy mildew.

WHERE IT'S GROWN AND WHAT ITS WINE TASTES LIKE

Monica Nera is cultivated all over Sardegna, **Italy**, for varietal wines in Monica di Sardegna (dry and still or sparkling) and Monica di Cagliari (dry, sweet or fortified) DOCs, or used in blends with Bovale Sardo and GARNACHA (Cannonau) in the Mandrolisai DOC and to a lesser extent in Campidano di Terralba DOC. The Italian agricultural census recorded total plantings of 2,909 ha (7,199 acres) in 2000. Unless yields are kept low, the wines are generally undistinguished and for early drinking. Recommended producers include Argiolas, Ferruccio Deiana, Alberto Loi, Santadi and Sardus Pater.

MONSTRUOSA

Extremely rare, recuperated Galician capable of producing light, fresh, aromatic whites.

BERRY COLOUR: ● ● ● ● ●

ORIGINS AND PARENTAGE

Monstruosa, named for the size of the vine not the wine, originates in the Monterrei region in Galicia, north-west Spain.

VITICULTURAL CHARACTERISTICS

Vigorous, mid budding, late ripening. Big bunches of big, thick-skinned berries. Susceptible to powdery mildew.

WHERE IT'S GROWN AND WHAT ITS WINE TASTES LIKE

Rescued from virtual extinction by José Luis Mateo of Quinta da Muradella, Monterrei, **Spain**. He planted 789 Monstruosa vines in 2002 and currently includes it in a blend with SÍRIA (known locally as Doña Blanca), TRAJADURA (Treixadura) and VERDELHO (Verdello). Varietal wines display aromas of wild and alpine flowers, high acidity and moderate alcohol.

MONTEPULCIANO

Productive, deeply coloured, firmly structured and widely planted.

BERRY COLOUR: ● ● ● ● ●

PRINCIPAL SYNONYMS: Africano, Angolano, Cordisco, Montepulciano Spargolo, Morellone (Toscana), Sangiovese Cordisco, Uva Abruzzese, Violone

VARIETIES COMMONLY MISTAKEN FOR MONTEPULCIANO: PUGNITELLO, SANGIOVESE (under the name Montepulciano Primaticcio)

ORIGINS AND PARENTAGE

Although this variety's name refers to the Montepulciano region in the province of Siena in Toscana, its origin is more likely to be in Abruzzo in central Italy, possibly in the area of Torre de' Passeri. Because SANGIOVESE is widely grown in the Montepulciano area of Toscana, Montepulciano has often been considered a synonym for Sangiovese but this is contradicted by ampelographic observations and DNA profiling (Vouillamoz).

VITICULTURAL CHARACTERISTICS

High yielding, late ripening. Good resistance to botrytis bunch rot and downy mildew.

WHERE IT'S GROWN AND WHAT ITS WINE TASTES LIKE

Montepulciano is a workhorse grape planted across central **Italy**, often alongside SANGIOVESE (which is the variety responsible for Vino Nobile di Montepulciano DOCG, which has nothing to do with the Montepulciano grape). Montepulciano's most well-known and often best-value manifestation is as the main or sole variety in Montepulciano d'Abruzzo DOC. The creation of the more specific DOCG Montepulciano d'Abruzzo Colline Teramane in the hills around Teramo in the northern part of Abruzzo in 2002 indicates an aspiration to higher quality but not necessarily its achievement. It is also the principal ingredient in Marche DOCs such as Rosso Conero (minimum 85%), Rosso Piceno (35–70%) and Offida (minimum 50%), and the variety is specified in varying percentages for various DOCs in Molise and further south in Puglia. It is also planted further north in Umbria and Toscana but it ripens better in the warmer and sunnier conditions further south. There were 29,828 ha (73,707 acres) in Italy in 2000, according to the agricultural census.

The best wines are deeply coloured with ripe, robust tannins, making them an ideal blending component with softer wines. Recommended producers of Montepulciano d'Abruzzo include Cataldi Madonna, Contucci Ponno, Dino Illuminati, Lepore, Masciarelli, Villa Medoro, Emidio Pepe, Bruno Nicodemi, Valentini, Valle Reale and Zaccagnini.

A wide variety of **California** producers, from Paso Robles to Lodi, make a Montepulciano and there were 74 acres (30 ha) in 2008.

It is also planted to a very limited extent in **Australia** (eg by Amadio in the Adelaide Hills, who supply grapes to Montepulciano devotees Matt Gant at First Drop and Ben Riggs), in **New Zealand**, notably from Hans Herzog, and on **Malta** (6 ha/15 acres).

MONTILS

Charentais variety used in the production of cognac.

BERRY COLOUR: ● ● ● ● ●

PRINCIPAL SYNONYMS: Chalosse

ORIGINS AND PARENTAGE

Montils is a variety from the Charente in western France that was first mentioned in Viala and Vermorel (1901–10). Its name comes from the village of Montils in the Charente-Maritime. DNA parentage analysis has shown that Montils is most likely a progeny of GOUAIS BLANC (Boursiquot *et al.* 2004). It is member of the Folle ampelographic group (see p XXVII; Bisson 2009).

VITICULTURAL CHARACTERISTICS

Fairly early budding and therefore at risk from spring frosts. Mid ripening. Berries are susceptible to botrytis bunch rot when ripe.

WHERE IT'S GROWN AND WHAT ITS WINE TASTES LIKE

Montils is used in blended base wines for cognac and for local *mistelles* such as Pineau des Charentes in western **France**. After a dip in the 1970s and 1980s, the area planted to this variety in France was back up to 165 ha (408 acres). It does not seem to be made as a varietal table wine, although Lavignac (2001) notes that it was still found among old vines in Bordeaux's Entre-Deux-Mers vineyards in 1960. Maurice O'Shea planted Montils at Mount Pleasant in **Australia**'s Hunter Valley in 1921 and cuttings from the original 0.5 ha will be used for future replanting.

MONTONICO BIANCO

Minor southern Italian white used mainly in blends.

BERRY COLOUR: ● ● ● ● ●

PRINCIPAL SYNONYMS: Chiapparù (Marche), Ciapparone (Abruzzo), Greco Bianco del Pollino⁂ (Calabria), Greco del Pollino⁂, Mantonico Bianco Italico⁂ (Calabria), Montonico⁂ (Puglia), Pagadebit⁂ (Puglia), Racciapaluta (Abruzzo), Uva della Scala⁂ (Puglia), Uva Regno (Marche)

VARIETIES COMMONLY MISTAKEN FOR MONTONICO BIANCO: MANTONICO BIANCO⁂ (Calabria)

ORIGINS AND PARENTAGE

The names Mantonico or Montonico are often applied to various varieties in southern Italy. Recent ampelographic studies and DNA profiling have shown that Montonico Bianco is identical to two Puglian reference samples (Pagadebit and Uva della Scala) and two Calabrian ones (Greco Bianco del Pollino and Mantonico Bianco Italico) but it is – surprisingly – distinct from Calabria's MANTONICO BIANCO (Schneider, Raimondi, De Santis and Cavallo 2008; Schneider, Raimondi *et al.* 2009). In addition, Montonico Bianco is not a white mutation of Mantonico Nero (aka CASTIGLIONE) and is distinct from Montonico di Rogliano (aka GUARDAVALLE; Schneider, Raimondi *et al.* 2009).

Recent DNA parentage analysis has shown that Montonico Bianco (under the name Greco del Pollino) has a parent–offspring relationship with GARGANEGA, one of the most widespread and ancient white varieties of Italy (Crespan, Calò *et al.* 2008). Since Garganega also has parent–offspring relationships with eight other varieties (ALBANA, CATARRATTO BIANCO, DORONA DI VENEZIA, MALVASIA BIANCA DI CANDIA, MARZEMINA BIANCA, MOSTOSA, SUSUMANIELLO and TREBBIANO TOSCANO), Montonico Bianco is either their half-sibling or their grandparent (see GARGANEGA for the pedigree diagram).

VITICULTURAL CHARACTERISTICS

Vigorous, good resistance to winter frost, late ripening.

WHERE IT'S GROWN AND WHAT ITS WINE TASTES LIKE

Until the 1960s, Montonico Bianco was planted mainly on the Adriatic coast in central **Italy**: in Abruzzo, notably around Chieti, Sulmona and Teramo, and in the Marche, in particular around Macerata and Fermano. Since then plantings have decreased in central Italy and today the variety is more or less restricted to Teramo. Montonico Bianco is also found, under various synonyms (see above), in Puglia (provinces of Foggia and Bari) and Calabria (around Locri). It is authorized as part of the blend in Calabrian DOCs such as Bivongi, Donnici and Pollino and it not usually made as a varietal wine, although La Quercia in the province of Teramo do make one, classified as IGT Colli Aprutini. At one time it was also grown for table grapes. There were 809 ha (1,999 acres) in Italy in 2000.

MONTÙ

Very late ripening Bolognese variety.

BERRY COLOUR: ● ● ● ● ●

PRINCIPAL SYNONYMS: Bianchetto Faentino (south-east Emilia-Romagna), Bianchino (south-east Emilia-Romagna), Montuni (Bologna), Montuno (Bologna)

ORIGINS AND PARENTAGE

Montù has long been cultivated in the province of Bologna in Emilia-Romagna under various names. The name could derive from the local dialect *molz'u'* (Italian *molta uva*), meaning 'many grapes', a reference to its productivity. DNA comparison has suggested a genetic relationship between Montù and BLAUFRÄNKISCH, called Franconia in Italy (Filippetti, Silvestroni, Thomas and Intrieri 2001).

OTHER HYPOTHESES

Some authors have suggested a similarity between Montù and MONTONICO BIANCO but this can be disproved by DNA profiling (Vouillamoz). Others have proposed a Spanish origin, based on the similarity between the name Montù and Mantùa or Montua Castigliano, an unidentified variety.

VITICULTURAL CHARACTERISTICS

Good resistance to downy and powdery mildews and, to a slightly lesser extent, to botrytis bunch rot. Vigorous, productive and very late ripening.

WHERE IT'S GROWN AND WHAT ITS WINE TASTES LIKE

Montù is cultivated mainly in the provinces of Bologna and Ravenna in Emilia-Romagna in northern **Italy**. It is the dominant variety (minimum 85%) in the dry, sweet or sparkling wines of the Montuni version of Reno DOC. Total plantings in Italy in 2000 amounted to 1,160 ha (2,866 acres). The wines tend to have good acidity.

MONVEDRO

Minor, late-budding Portuguese variety causing confusion by its synonyms.

BERRY COLOUR: ● ● ● ● ●

PRINCIPAL SYNONYMS: Bastardo, Monvedro Dão

VARIETIES COMMONLY MISTAKEN FOR MONVEDRO: PARRALETA ☒ (under the name Monvedro do Algarve)

ORIGINS AND PARENTAGE

In Portugal, the name Monvedro is a variation on Bonvedro, which is commonly used as a synonym for PARRALETA in the Algarve. This explains why Truel in Galet (2000) suggested that Monvedro could be identical to Carcajolo Noir (ie Parraleta) on Corse and Caricagiola Nera on

Sardegna, both being local synonyms for Parraleta. In Dão, Monvedro is the name of a distinct variety with a unique DNA profile in Veloso *et al.* (2010). It is therefore distinct from TROUSSEAU, even though they share the synonym Bastardo.

VITICULTURAL CHARACTERISTICS

Late budding. Small, dense bunches of thick-skinned berries.

WHERE IT'S GROWN

Monvedro is cultivated in Beiras, where it is authorized for regional but not for appellation wines. There were 8 ha (20 acres) in **Portugal** in 2010.

MOORE'S DIAMOND

Minor American hybrid similar to, but less successful than, Concord and Niagara.

BERRY COLOUR: ● ● ● ● ●

PRINCIPAL SYNONYMS: Diamond

ORIGINS AND PARENTAGE

A CONCORD × Iona hybrid obtained around 1870 by Jacob Moore, private breeder in Brighton, New York State, in which Iona is a cross between Diana (or possibly CATAWBA, also known as Diana in the US) and an unknown partner obtained in 1855 by C W Grant on Iona Island in New York State. It is therefore a *Vitis labrusca* and *Vitis vinifera* hybrid. It was used to breed GOLDEN MUSCAT.

VITICULTURAL CHARACTERISTICS

Berries susceptible to cracking in wet weather. Winter-hardy to about –20 °F (–28.9 °C).

WHERE IT'S GROWN AND WHAT ITS WINE TASTES LIKE

New York State had 93 acres (38 ha) of Moore's Diamond in 2006, mostly in the west of the state and used for sparkling blends and dry varietal wines (unusually for an American hybrid) as well as for the table. Although the plant is similar to CONCORD, the fruit is more like that of NIAGARA, but it is less successful than either. This may be partly due to the fact that the berry skins tend to crack in wet weather. It is also found to a very limited extent in **New Hampshire**, **Pennsylvania**, **Virginia**, **Kentucky** and the Midwest. Producers include Plymouth Winery in Seneca Lake, Fulkerson in Finger Lakes, Starr Hill in Pennsylvania, Wight-Meyer in Kentucky and Candia in New Hampshire. Wines are quite similar to those made from Niagara but are fresher and less foxy-tasting.

MORADELLA

Prized by growers but not recognized by officialdom.

BERRY COLOUR: ● ● ● ● ●

PRINCIPAL SYNONYMS: Croà, Moranzana, Vermiglio

ORIGINS AND PARENTAGE

Originally from around Pavia in Lombardia. A distinction is often made between three clonal variations of Moradella, with slightly different morphologies: Moradella Comune (or di San Colombano), Moradella Vermiglio (or dal Peduncolo Rosso) and Moradella Croà.

VITICULTURAL CHARACTERISTICS

Highly susceptible to powdery mildew.

WHERE IT'S GROWN AND WHAT ITS WINE TASTES LIKE

Moradella is widespread in the Oltrepò Pavese and more generally in the province of Pavia in Lombardia in northern **Italy.** It was once the most important variety in that area but its popularity has declined dramatically in favour of BARBERA, partly because of its susceptibility to powdery mildew and partly because it has never been included in the Italian national register of varieties, and is not therefore mentioned in any of the DOC regulations, though the former may explain the latter. It is now found only in old vineyards but is prized by growers and used incognito with Barbera and CROATINA in such blends as Buttafuoco and Sangue di Giuda, both within the Oltrepò Pavese DOC. Fortesi have made an experimental 100% Moradella, classified as a table wine. Cantina Storica di Montù Beccaria, Il Santo and Fattoria Mondo Antico are also making valiant efforts to preserve the variety. The wines tend to be full-bodied and spicy with fairly high alcohol.

MORAVA

Promising, aromatic white cross from Serbia making Sauvignon-like wines.

BERRY COLOUR: ● ● ● ● ●

ORIGINS AND PARENTAGE

Morava is a FRÜHROTER VELTLINER × MÜLLER-THURGAU cross obtained by P Cindrić, Nada Korać and V Kovač at the viticulture research centre in Sremski Karlovci, part of Novi Sad University in Vojvodina, Serbia. Authorized in 2003.

VITICULTURAL CHARACTERISTICS

Late ripening, loose clusters. Resistant to downy mildew and botrytis bunch rot and relatively winter-hardy.

WHERE IT'S GROWN AND WHAT ITS WINE TASTES LIKE

Grown mainly in the Valjevo region in western **Serbia**, it produces wines with some similarity to SAUVIGNON BLANC. It is one of the more successful recent Serbian crosses and plantings are increasing. Milijan Jelić and Tamuz, for example, both make fresh, aromatic varietal wines.

MORAVIA AGRIA

Hitherto neglected Spanish variety contributing to at least one distinctively fresh, tannic red in the heat of Manchuela.

BERRY COLOUR: ● ● ● ● ●

PRINCIPAL SYNONYMS: Maravia Agria, Moravia, Moravio

VARIETIES COMMONLY MISTAKEN FOR MORAVIA AGRIA: BOBAL, MARUFO

ORIGINS AND PARENTAGE

Moravia Agria is an obscure variety from eastern Castilla-La Mancha, central Spain. The name refers to its high acidity, *agria* meaning 'sour', to distinguish it from Moravia Dulce (*dulce* meaning 'sweet'), the name given to MARUFO in Castilla-La Mancha.

VITICULTURAL CHARACTERISTICS

Fertile, late budding and ripening and long growing season. Large bunches of small berries. Susceptible to fungal diseases.

WHERE IT'S GROWN AND WHAT ITS WINE TASTES LIKE

Moravia Agria, found only in **Spain**, has in the past tended to languish anonymously in blends thanks to its pale colour and poor performance in the vineyard. However, it can contribute freshness and structure as well as attractive aromas ranging from red fruits through floral to wild herbs and grasses. In Manchuela, Juan Antonio Ponce has begun to co-ferment GARNACHA with Moravia Agria to produce Buena Pinta, a structured, earthy, garrigue-scented red.

MORAVIA DULCE

See MARUFO.

MORAVIAN MUSCAT

See MUŠKÁT MORAVSKÝ.

MORETO DO ALENTEJO

Popular but generally light and lacklustre red in southern Portugal more commonly but ambiguously known as Moreto.

BERRY COLOUR: ● ● ● ● ●

PRINCIPAL SYNONYMS: Morito

VARIETIES COMMONLY MISTAKEN FOR MORETO DO ALENTEJO: BAGA (Dão), BLAUER PORTUGIESER (Austria), CAMARATE (Bairrada and Tejo), CASTELÃO (Algarve), MALVASIA PRETA (Douro), Moreto do Dão (Dão)

ORIGINS AND PARENTAGE

As suggested by the name, this variety most likely originates in the Alentejo in south-eastern Portugal (Almadanim *et al.* 2007), where it is usually called simply Moreto. However, this name is given to several distinct varieties, and Regner *et al.* (1999) have shown that Moreto from Dão and Moreto from the Alentejo have distinct DNA profiles, the latter corresponding to Moreto in Almadanim *et al.* (2007) and in Veloso *et al.* (2010), hence our use of the longer names Moreto do Alentejo and Moreto do Dão (the latter no longer cultivated) to distinguish the two varieties. In addition, BAGA, CAMARATE and MALVASIA PRETA are often called Moreto or Moreto Something but their DNA profiles in Veloso *et al.* (2010) are all distinct from those of Moreto do Alentejo and Moreto do Dão in Regner *et al.* (1999).

The Austrian variety BLAUER PORTUGIESER was once thought to be identical to Moreto but even if it is impossible to rule out the possibility that they have a common ancestor (Regner *et al.* 1999), their DNA profiles are completely distinct.

VITICULTURAL CHARACTERISTICS

Productive, early budding, late ripening. Best suited to hot, sunny climates and deep, well-drained soils. Good resistance to pests and fungal diseases.

WHERE IT'S GROWN AND WHAT ITS WINE TASTES LIKE

Moreto wines are generally light in colour and body with light, red-fruit aromas and best drunk young. The variety is popular in the Alentejo, usually as part of a blend with local and/or international varieties, since the wines can be sold before others are ready. Carmim (the Reguengos de Monsaraz co-operative), Esporão and Roquevale all produce Alentejo blends that include Moreto do Alentejo. There were 1,152 ha (2,847 acres) in **Portugal** in 2010.

MORIO-MUSKAT

Intensely grapey but rather flabby German cross that has had its day.

BERRY COLOUR:

PRINCIPAL SYNONYMS: Geilweilerhof I-28-30

ORIGINS AND PARENTAGE

Morio-Muskat was initially introduced as a SILVANER × Weissburgunder (PINOT BLANC) cross bred by Peter Morio at Geilweilerhof in 1928 yet in the Vitis International Variety Catalogue it is shown as a Silvaner × MUSCAT BLANC À PETITS GRAINS cross, which would better explain its name and its light Muscat aroma. However, analysis of the DNA profile published in Lamboy and Alpha (1998) contradicts both suggested parentages (Vouillamoz), so that Morio-Muskat's true parentage remains to be elucidated.

VITICULTURAL CHARACTERISTICS

Mid to late ripening, very susceptible to botrytis bunch rot and powdery mildew and slightly less so to downy mildew. Yields can be high but sugar levels are lower than for MÜLLER-THURGAU.

WHERE IT'S GROWN AND WHAT ITS WINE TASTES LIKE

Authorized in **Germany** in 1956, Morio-Muskat was particularly popular in the mid 1970s as providing a way of adding a dash of Muscat-like perfume to an otherwise neutral cheap German blend, frequently labelled Liebfraumilch on export markets. Today only 502 ha (1,240 acres) remain – much less than its peak of around 3,000 ha (7,413 acres) – and almost all of those are in Rheinhessen and the Pfalz.

When the grapes are fully ripe, the wines have an overpowering, almost sickly grapiness, but the variety requires a good site (which would probably be better used by a more noble variety) and sugar levels are never very high, although acidity is generally fresh. Producers include Huff in Rheinhessen, Baldau in Franken, Franz Braun in the Pfalz and Wehweck in Baden.

It is also planted to a very limited extent in **Austria** and there are around 26 ha (64 acres) in **South Africa**. **Japan** grows about 1.5 ha (4 acres) of Morio-Muskat.

MORISTEL

Makes light but distinctive reds in Somontano, north-east Spain.

BERRY COLOUR:

PRINCIPAL SYNONYMS: Arco (Valencia), Concejón, Juan Ibáñez, Moristell (Huesca in Spain)

VARIETIES COMMONLY MISTAKEN FOR MORISTEL: GRACIANO (also called Morrastel), MONASTRELL (also called Morastrell in Zaragoza), TEMPRANILLO (Moristel is known as Tempranillo Temprano in Logroño)

ORIGINS AND PARENTAGE

Moristel most likely originates from Aragón in north-east Spain, where it is sometimes erroneously considered to be a variation of MONASTRELL, or a synonym for Morrastel, itself a synonym for GRACIANO; these synonymies can be rejected by DNA profiling (Vouillamoz).

VITICULTURAL CHARACTERISTICS

Compact bunches of small to medium-sized, thin-skinned berries. Mid to late budding. Late ripening. Low yielding. Reasonably resistant to drought and best suited to fertile or clay soils. Good resistance to pests and diseases.

WHERE IT'S GROWN AND WHAT ITS WINE TASTES LIKE

Varietal wines are generally quite light but with a distinctive perfume and flavour of loganberries and thus often blended with the fuller-bodied TEMPRANILLO, PARRALETA or even CABERNET SAUVIGNON. The variety is prone to oxidation and wines are best unoaked and drunk young. The variety is also well suited to rosé production. Total plantings in 2008 amounted to 154 ha (381 acres) in **Spain**, found mainly in the provinces of Huesca and Zaragoza as one of the principal varieties in the Somontano DO, which is otherwise dominated by international interlopers. Producers include Pirineos and Viñas del Vero.

MORNEN NOIR

Chasselas relative that might just be rescued from extinction.

BERRY COLOUR: ● ● ● ● ●

PRINCIPAL SYNONYMS: Chasselas Noir, Montruchon (Savoie), Mornant (Rhône), Mornerain Noir (Loire)

VARIETIES COMMONLY MISTAKEN FOR MORNEN NOIR: MENOIR (Hungary)

ORIGINS AND PARENTAGE

Mornen Noir has long been cultivated in the Loire and in the Rhône but its first mention was not until 1872 by Pulliat (Rougier 1902a). Mornen Noir probably originates from the area between the Rhône and the Loire rivers and its name most likely derives from the canton of Mornant in the Rhône-Alpes region, where it used to be widely cultivated together with CHASSELAS (Rougier 1902a). Indeed, DNA profiling has confirmed that an obscure variety named Montruchon in the grape collection at Domaine Grisard in Savoie is identical to Mornen Noir (Vouillamoz) and DNA parentage analysis strongly suggests that Mornen Noir has a parent–offspring relationship with Chasselas (Vouillamoz and Arnold 2009). This explains why Chasselas is called Mornen Blanc near Mâcon and why Mornen Noir is often called Chasselas Noir and has sometimes been wrongly thought to be a black mutation of Chasselas.

OTHER HYPOTHESES

According to Galet (1990), Mornen Noir is cultivated in Hungary under the name Médoc Noir (or Kékmedoc; see MENOIR). However, the name Médoc Noir is also used for MERLOT and for COT and the DNA profile of Kékmedoc published by Jahnke *et al.* (2009) is distinct from that of Mornen Noir.

VITICULTURAL CHARACTERISTICS

Early to mid ripening.

WHERE IT'S GROWN

Mornen Noir is not on the register of varieties in **France**, is not therefore included in the official vineyard statistics and it does seem to be disappearing. However, a small parcel recently discovered in the commune of Chavanay, very close to the Rhône between Lyon and Valence, has given local growers in the Coteaux du Gier the hope that it can be recovered and eventually enter the register. In 2008 they also planted an experimental vineyard with both Mornen Noir and CHOUCHILLON. Les Déplaude de Tartaras make a varietal Mornen Noir.

MORONE

Very minor Tuscan occasionally used to add colour to blends.

BERRY COLOUR: ● ● ● ● ●

PRINCIPAL SYNONYMS: Morone Farinaccio, Mostaiola del Lapi, Uva Moro, Uva Morone Nera

VARIETIES COMMONLY MISTAKEN FOR MORONE: PINOT MEUNIER (known as Morone Farinaccia in Italy)

ORIGINS AND PARENTAGE

Morone is a Tuscan variety that has been recorded since the early nineteenth century.

VITICULTURAL CHARACTERISTICS

Mid ripening. Regular and highish yields. Susceptible to botrytis bunch rot.

WHERE IT'S GROWN

Morone is cultivated to a very limited extent in the province of Massa-Carrara in northern Toscana, **Italy**, in particular in the communes of Pontremoli and Aulla. Some plantings also exist in Val d'Arno in the province of Firenze but the total area in 2000 amounted to just 39 ha (96 acres). It is usually added to blends to improve their colour rather than made as a varietal wine.

MORRASTEL

See GRACIANO.

MORRASTEL BOUSCHET

Nineteenth-century southern French cross now making tiny amounts of deep red wine.

BERRY COLOUR: ● ● ● ● ●

PRINCIPAL SYNONYMS: Garnacho (Valladolid), Morrastel Bouschet à Gros Grains

VARIETIES COMMONLY MISTAKEN FOR MORRASTEL BOUSCHET: GARNACHA, GRACIANO (called Morrastel in France), GRAND NOIR, Morrastel à Sarments Érigés (obtained in 1885 from the same parents, it was also called Morrastel Bouschet but was later abandoned)

ORIGINS AND PARENTAGE

A GRACIANO × PETIT BOUSCHET cross obtained in 1855 by Henri Bouschet at the Domaine de la Calmette in Mauguio in the Hérault, southern France. It was named after Morrastel, a synonym for Graciano in south-eastern France, and after its breeder. DNA profiling has shown that the variety called Garnacho in Valladolid is identical to Morrastel Bouschet (Ibañez *et al.* 2003; Martín *et al.* 2003). In addition, Morrastel Bouschet is often confused with GRAND NOIR (Viala and Vermorel 1901–10), eg in the grape collection at the University of California at Davis (Vouillamoz).

VITICULTURAL CHARACTERISTICS

Very fertile, early budding and ripening. Big bunches of big, thick-skinned berries. Coloured juice (ie a *teinturier* variety). Very susceptible to downy mildew.

WHERE IT'S GROWN AND WHAT ITS WINE TASTES LIKE

Although there were only 7 ha (17 acres) remaining in **France** in 2008, this variety was once widely planted in the Languedoc and effectively usurped the better quality GRACIANO (Morrastel). It is not included in the official French register of varieties and so cannot be used in an appellation wine, which probably contributed to its uprooting in the 1980s. It produces deeply coloured wines and was appreciated in the past mainly for its ability to add colour to blends.

MOSCATEL

See MUSCAT.

MOSCATEL DE SETÚBAL

See MUSCAT OF ALEXANDRIA.

MOSCATEL GRAÚDO

See MUSCAT OF ALEXANDRIA.

MOSCATELLO SELVATICO

Rare, aromatic white for sweet wines usually from southern Italy.

BERRY COLOUR: ● ● ● ● ●

PRINCIPAL SYNONYMS: Moscato di Barletta

ORIGINS AND PARENTAGE

Moscatello Selvatico is a Muscat-flavoured variety from Puglia in southern Italy. Comparison of the DNA profiles of Moscatello Selvatico (Crespan and Milani 2001) and Moscato di Barletta (Zulini *et al.* 2002) shows that they are identical. DNA parentage analysis has revealed that Moscatello Selvatico is possibly a natural BOMBINO BIANCO × MUSCAT OF ALEXANDRIA cross (Crespan and Milani 2001; Cipriani *et al.* 2010). As a consequence, Moscatello Selvatico is a grandchild of MUSCAT BLANC À PETITS GRAINS (Moscato Bianco) and Axina de Tres Bias (see the MUSCAT pedigree diagram).

VITICULTURAL CHARACTERISTICS

Mid ripening.

WHERE IT'S GROWN AND WHAT ITS WINE TASTES LIKE

Moscatello Selvatico is cultivated mainly on the Adriatic coast of Puglia in southern **Italy**, between Barletta and Monopoli, in particular in the region of Andria and Trani, where it is blended with MUSCAT BLANC À PETITS GRAINS (Moscato Bianco) in the Moscato di Trani DOC. Moscatello Selvatico is also a minor component in the Moscadello di Montalcino DOC near Siena (though it was the dominant variety before phylloxera), producing still and fizzy sweet wines, sometimes late harvested. There were 118 ha (292 acres) in Italy in 2000. The wines have the characteristic grapey aroma and flavour of many Muscat varieties. Azienda Vitivinicola Speziale in Crispiano make a rare, dry, varietal passito version called I Ualani and classified IGT Tarantino.

MOSCATO BIANCO

See MUSCAT BLANC À PETITS GRAINS.

MOSCATO DI SCANZO

Rare, aromatic red exclusive to Bergamo used to make sweet passito wines and basking in the creation of its new DOCG.

BERRY COLOUR: ● ● ● ● ●

PRINCIPAL SYNONYMS: Moscatino di Scanzo

VARIETIES COMMONLY MISTAKEN FOR MOSCATO DI SCANZO: ALEATICO

ORIGINS AND PARENTAGE

Moscato di Scanzo most probably comes from Scanzorosciate in the province of Bergamo in Lombardia in northern Italy, where it is thought to have been planted since the mid fourteenth century, although the first reliable written reference to it was not until 1789. DNA parentage analysis has revealed a parent–offspring relationship between Moscato di Scanzo and MUSCAT BLANC À PETITS GRAINS (Moscato Bianco; Crespan and Milani 2001), making it a half-sibling or a grandchild of at least five other varieties that have parent–offspring relationships with Muscat Blanc à Petits Grains: ALEATICO (Moscatello Nero), MOSCATO GIALLO, MOSCATO ROSA DEL TRENTINO, MUSCAT OF ALEXANDRIA and the non-cultivated Muscat Rouge de Madère (see MUSCAT pedigree diagram).

OTHER HYPOTHESES

Some scholars suggest that the Romans introduced this grape to Scanzorosciate.

VITICULTURAL CHARACTERISTICS

Mid to late ripening. Irregular yields.

WHERE IT'S GROWN AND WHAT ITS WINE TASTES LIKE

This rare grape is cultivated almost exclusively in a very limited area around Scanzorosciate, north east of Milano, in the province of Bergamo, **Italy**, although it is also found just to the south in the hills between Seriate and Val Seriana. There were just 76 ha (188 acres) in Italy in 2000 but the creation of the Moscato di Scanzo DOCG in 2009 signifies a renewed interest in and respect for the variety. Wines within this DOCG and the Moscato Passito version of Valcalepio DOC are full-bodied, aromatic, sweet red passito wines made from 100% late-harvested Moscato di Scanzo grapes, dried off the vine for at least twenty-one days. Recommended producers include the Cantina Sociale Bergamasca, Castello di Grumello, Monzio Compagnoni and Tenuta degli Angeli.

MOSCATO DI TERRACINA

Recently rescued aromatic white from central Italy making a wide range of styles.

BERRY COLOUR:

PRINCIPAL SYNONYMS: Moscato di Maccarese

VARIETIES COMMONLY MISTAKEN FOR MOSCATO DI TERRACINA: MUSCAT OF ALEXANDRIA⁑

ORIGINS AND PARENTAGE

Moscato di Terracina is local to the province of Latina, between Napoli and Roma. It is morphologically similar to MUSCAT OF ALEXANDRIA but DNA profiling shows that they are distinct (Stella Grando and José Vouillamoz, unpublished data).

VITICULTURAL CHARACTERISTICS

Early ripening. Susceptible to fungal diseases.

WHERE IT'S GROWN AND WHAT ITS WINE TASTES LIKE

Widely cultivated in and around Terracina in the province of Latina in Lazio, central **Italy**, in the early twentieth century, Moscato di Terracina was on the verge of extinction some fifty years later until it was rescued a few decades ago. Previously labelled IGT Lazio, it was promoted to Moscato di Terracina DOC, or just Terracina, in 2007. The regulations require a minimum of 85% of this variety for the dry, amabile (sweet) and passito wines and 100% for the sparkling version (sweet or dry). The Italian agricultural census recorded total plantings of 292 ha (722 acres) in 2000.

Today, only very few producers make varietal Moscato di Terracina wines, notably Sant'Andrea, Terra delle Ginestre and Villa Gianna. The wines tend to be aromatic, with aromas of dried roses, and to taste of tropical fruits and apricots, though Terre Ginestra's two wines, Stellaria and Promessa, are both oaked.

MOSCATO GIALLO

Golden-berried Muscat variety of northern Italy making mainly passito-style wines.

BERRY COLOUR:

PRINCIPAL SYNONYMS: Fior d'Arancio (Padova), Goldenmuskateller or Goldmuskateller (Bolzano), Moscatel (Trentino), Moscato dalla Siria, Moscato Sirio, Muscat du Pays⁑ (Valais in Switzerland), Muscat Vert⁑ (Valais), Muscatedda⁑ (Sicilia)

VARIETIES COMMONLY MISTAKEN FOR MOSCATO GIALLO: MUSCAT BLANC À PETITS GRAINS⁑ (Moscato Bianco on Sicilia), MUSCAT FLEUR D'ORANGER⁑ (Colli Euganei)

ORIGINS AND PARENTAGE

DNA parentage analysis has revealed that Moscato Giallo has a parent–offspring relationship with MUSCAT BLANC À PETITS GRAINS (Moscato Bianco; Crespan and Milani 2001) and is therefore a half-sibling or a grandchild of at least five other varieties that have parent–offspring relationships with Muscat Blanc à Petits Grains (see MUSCAT pedigree diagram): ALEATICO (Moscatello Nero), MOSCATO DI SCANZO, MOSCATO ROSA DEL TRENTINO, MUSCAT OF ALEXANDRIA and the no longer cultivated Muscat Rouge de Madère. The origins of Moscato Giallo are more likely to be in northern Italy than in the Middle East.

OTHER HYPOTHESES

Supposedly introduced from Syria, but this suggestion has been disproved by the recently discovered parent–offspring relationship with MUSCAT BLANC À PETITS GRAINS.

VITICULTURAL CHARACTERISTICS

Vigorous, early to mid ripening. Largish, loose bunches of thick-skinned berries. Resistant to botrytis bunch rot, moderately so to downy and powdery mildews. Susceptible to phomopsis and chlorosis. Best suited to calcareous slopes.

WHERE IT'S GROWN AND WHAT ITS WINE TASTES LIKE

In **Italy**, Moscato Giallo is mainly cultivated in Vallagarina (Lagertal in German) in the province of Trento and in Bolzano (under the name Goldmuskateller) in northern Italy, where varietal wines are allowed in the Trentino and Alto Adige DOCs. It is also used in the sparkling and passito Fior d'Arancio wines within the Colli Euganei DOC, in the sweet and sparkling Moscato wines of the Corti Benedettine del Padovano DOC, both in the Veneto, and further east in the Friuli-Isonzo DOC. There were 360 ha (890 acres) in Italy in 2000.

Varietal wines are usually golden in colour, sweetly aromatic, and grapey or musky, with only moderate acidity. Manincor produce an intensely aromatic dry wine, Nalles Magré and Produttori Merano produce archetypal passito styles.

In **Switzerland**, Muscat vineyards in the Valais have recently been proven to be a field mix of MUSCAT BLANC À PETITS GRAINS and Moscato Giallo, locally named Muscat du Pays (Spring *et al.* 2008).

MOSCATO NERO

See MUSCAT OF HAMBURG.

MOSCATO ROSA DEL TRENTINO

A minor, red-skinned northern Italian that produces rose-scented sweet wines.

BERRY COLOUR: ● ● ● ● ●

PRINCIPAL SYNONYMS: Muškat Ruža Porečki⚥ (Croatia), Rosenmuskateller (Alto Adige, Germany, Austria)

VARIETIES COMMONLY MISTAKEN FOR MOSCATO ROSA DEL TRENTINO: Moscato Rosa di Breganze

ORIGINS AND PARENTAGE

The variety is first mentioned at the end of the nineteenth century and its name probably derives from the wine's aroma of roses rather than from the colour of the berries, which are pinkish red rather than light pink.

A thorough study of various Muscat accessions in vine collections around the world has shown that three distinct varieties were called Moscato Rosa, one in Trentino, one in Breganze (in the province of Vicenza), and another in a collection in Belgrade. The accessions from Trentino and Breganze were distinct and did not match any other Muscat, while the Moscato Rosa from Belgrade was a colour mutation of MUSCAT BLANC À PETITS GRAINS (Crespan and Milani 2001). Since the accession from Trentino is the only one with significant plantings, we have used the name Moscato Rosa del Trentino here. The same authors carried out a DNA parentage analysis and revealed that Moscato Rosa del Trentino has a parent–offspring relationship with Muscat Blanc à Petits Grains (Moscato Bianco), and is therefore a half-sibling or a grandchild of at least five other varieties that have parent–offspring relationships with Muscat Blanc à Petits Grains (see the MUSCAT pedigree diagram): ALEATICO (Moscatello Nero), MOSCATO DI SCANZO, MOSCATO GIALLO, MUSCAT OF ALEXANDRIA and Muscat Rouge de Madère (the last no longer cultivated).

Surprisingly, Moscato Rosa del Trentino turned out to be identical to Muškat Ruža Porečki from Croatia (Maletić *et al.* 1999). This could support the old hypothesis that in the nineteenth century Moscato Rosa del Trentino was introduced to Caldaro in the Alto Adige via Dalmacija (Dalmatia).

OTHER HYPOTHESES

Some authors suggest that Moscato Rosa del Trentino is identical to Apiana of Roman times but this is highly doubtful.

VITICULTURAL CHARACTERISTICS

Mid to late ripening. Low and erratic production due to poor fruit set since the flowers are female and need to be pollinated by another variety. Very sensitive to botrytis bunch rot due to thin skins though the bunches tend to be very loose.

WHERE IT'S GROWN AND WHAT ITS WINE TASTES LIKE

Moscato Rosa del Trentino is cultivated mainly in the Alto Adige in northern **Italy** as Rosenmuskateller, where it must comprise at least 85% of Alto Adige Moscato Rosa/Rosenmuskateller DOC, and to a lesser extent in Trentino along the Adige River and in the Valle dei Laghi, where it comprises at least 80% of Trentino Moscato Rosa DOC. In both these regions, it produces mainly late-harvest and/or passito wines, even fortified in the latter. Moscato Rosa del Trentino is also found in Friuli, in particular as a varietal wine within the Friuli-Isonzo DOC and in various IGTs. Rare plants are found in Piemonte in the provinces of Alessandria and Tortona. Total plantings in Italy in 2000 were 99 ha (245 acres) but this probably includes both the Trentino and the Breganze varieties.

These generally sweet, smooth red wines are marked by aromas of roses, sometimes spice. Recommended Italian producers include Riccardo Battistotti, Cantina Produttori Termeno, Franz Haas, Kettmeir, Alois Lageder (labelled Rosenmuskateller), Letrari, Maso Martis and Waldgries.

There are some small experimental vineyards in **Germany**, planted eg by Fuchs in Flörsheim-Dalsheim in Rheinhessen and Wissing in the Pfalz, and also in Württemberg. A few growers in **Austria** are experimenting with the variety, including Kloster am Spitz in Purbach in Burgenland and Gerhard Kracher in Neusiedlersee, who uses the variety to make a TBA-style wine.

As Muškat Ruža Porečki (meaning 'Rose Muscat of Poreč') it is planted to a very limited extent in Istra (Istria) in **Croatia**, where it was first known as Muškat Ruža Omiški ('Rose Muscat of Omiš').

MOSCHATO SPINAS

See MUSCAT BLANC À PETITS GRAINS.

MOSCHOFILERO

Aromatic, high-acid Greek variety that has seen a surge in popularity and plantings in the last twenty years.

BERRY COLOUR: ● ● ● ● ●

PRINCIPAL SYNONYMS: Fileri, Fileri Mantineias, Moschophilero

ORIGINS AND PARENTAGE

Moschofilero probably comes from the central western Pelopónnisos (Peloponnese) in Greece, possibly from Mantíneia in the prefecture of Arkadía, where it was allegedly mentioned as early as 1601 (Nikolau and Michos 2004).

Moschofilero, like Fileri or Fileri Mantineias, is generally used to refer to a group of clones, some quite distinctive, that are also referred to by more specific names but all have the same DNA profile (Lefort and Roubelakis-Angelakis 2001). These include Asprofilero (white), Xanthofilero (yellow-skinned), Mavrofilero (dark-skinned) and Fileri Kokkino (red-skinned). The first two are very minor – less than 1% of the Mantíneia vineyards – and the vast majority of vines are the pink/grey-skinned version (with extreme variations in skin colour, leaf shape, aroma intensity, berry size, etc.) that most people refer to simply as Moschofilero. According to Stefanos Koundouras, assistant professor of viticulture at the Aristotle University of Thessaloníki, DNA analysis and clonal selection could make a great difference to the overall quality and typicality of the wines.

Some writers use the name Fileri to refer to a table grape grown in Attikí, Ilía and Zágkli (Messinía) in the Pelopónnisos. However, this may simply be a clone that has adapted to more fertile soils to produce higher yields and bigger berries (Stefanos Koundouras, personal communication).

VITICULTURAL CHARACTERISTICS

Various according to clone but generally vigorous and relatively high yielding. Late ripening, especially at higher altitudes. Asprofilero and Xanthofilero mature later than Mavrofilero and attain higher acidity and lower sugar levels in the berries.

WHERE IT'S GROWN AND WHAT ITS WINE TASTES LIKE

This delicate, aromatic variety enjoyed a renaissance in the early 1990s, in the Pelopónnisos, **Greece**, generally but in particular on the high plateau of Mantíneia (600–700 m above sea level) in the centre of the area, where the eponymous appellation requires at least 85% Moschofilero. Many growers will have several clones interplanted in the same vineyard though there is a move towards planting them separately in new vineyards. Wines are generally light-bodied, fresh and relatively low in alcohol – often around 11% – and in some years it is a struggle to ripen this variety fully given the altitude of the best sites and the risk of poor weather at harvest.

However, some producers aim for a richer, more concentrated alternative style as well. Yiannis Tselepos, for example, makes a light, fresh, grapey and floral version from a mixed-clone vineyard as well as a more serious blanc de gris made from the darker-skinned Mavrofilero that has undergone lees stirring and is part barrel fermented. Apostolos Spiropoulos has a similar pair of wines in contrasting styles, Astala showing the influence of skin contact and some new oak in its increased weight and creamier texture, though his are both made from the Mavrofilero clone.

Moschofilero is also used in the production of sparkling wines, to make rosé and also sweet wines. It is found on many of the Ionian Islands and is useful for adding aroma and freshness to varieties such as SAVATIANO or RODITIS. It was the fourth most planted variety in western Greece in 2008, with a total of 486 ha (1,201 acres).

Recommended producers include Anton-opoulos, Skouras, Spiropoulos and Tselepos.

MOSCHOMAVRO

Minor Greek variety with good potential for rosés.

BERRY COLOUR: ● ● ● ● ●

PRINCIPAL SYNONYMS: Moschato Mavro, Moschogaltso, Xinogaltso

VARIETIES COMMONLY MISTAKEN FOR MOSCHOMAVRO: MUSCAT OF HAMBURG

ORIGINS AND PARENTAGE

Moschomavro, meaning 'Black Muscat', possibly comes from Makedonía in northern Greece. Genetic analysis (under the name Moschato Mavro) has shown that it is distinct from MUSCAT OF HAMBURG and that it is not a colour mutation of Moschato Aspro, the Greek name for MUSCAT BLANC À PETITS GRAINS (Stavrakakis and Biniari 1998).

VITICULTURAL CHARACTERISTICS

Vigorous and productive. Late budding and ripening. Susceptible to botrytis bunch rot and sour rot, relatively drought-tolerant. Medium-sized to large compact bunches.

WHERE IT'S GROWN AND WHAT ITS WINE TASTES LIKE

Moschomavro is found here and there in western and central Makedonía, **Greece**. Wines are generally light in colour, with medium to high alcohol, good acidity and aromatic potential. Voyatzis in Velvendos, western Makedonía, for example, blends Moschomavro with XINOMAVRO to make a spicy/floral dry rosé and Diamantis in Siatista, also in western Makedonía, uses some Moschomavro in his CABERNET SAUVIGNON/MERLOT blend. Tsantali, on the other hand, make a varietal wine in eastern Makedonía that they describe as full of red fruit and aromas of cherry jam. Both Nakos and Diamantis use Moschomavro to make their sweet Liastos wines. Viticulturist Haroula Spinthiropoulou, who grows this variety on her Argatia estate, believes it has a bright future as a rosé and also includes a small amount in her unoaked Xinomavro-based red blend.

MOSLAVAC

See FURMINT.

MOSTOSA

Minor white variety found on the Adriatic coast, occasionally made as a varietal wine under its synonym Pagadebit.

BERRY COLOUR: ● ● ● ● ●

PRINCIPAL SYNONYMS: Empibotte Bianco, Pagadebit or Pagadebito

VARIETIES COMMONLY MISTAKEN FOR MOSTOSA: BOMBINO BIANCO

ORIGINS AND PARENTAGE

Mostosa is from the Adriatic coast of central Italy. Its name comes from *mosto* (Italian for 'must'), a reference to the large amount of juice produced, which is confirmed by the synonyms Empibotte ('barrel filling') and Pagadebito ('debt paying'), both also used misleadingly for other productive varieties.

Recent DNA parentage analysis has shown that Mostosa (under the name Empibotte) has a parent–offspring relationship with GARGANEGA, one of Italy's most widespread and ancient white varieties (Crespan, Calò *et al.* 2008). Since Garganega also has parent–offspring relationships with eight other varieties (ALBANA, CATARRATTO BIANCO, DORONA DI VENEZIA, MALVASIA BIANCA DI CANDIA, MARZEMINA BIANCA, MONTONICO BIANCO, SUSUMANIELLO and TREBBIANO TOSCANO), Mostosa is either their half-sibling or their grandparent (see Garganega for the pedigree diagram).

Cipriani *et al.* (2010) have recently suggested through DNA parentage analysis that Mostosa could be a Ragusano × Uva Femmina natural cross but this is contradicted by Crespan, Calò *et al.* (2008) and is questionable because Ragusano is the name of at least three other varieties (MOSCATELLO SELVATICO, GRECO and MALVASIA DI LIPARI), and Uva Femmina, meaning 'female grape', is a vague name.

VITICULTURAL CHARACTERISTICS

Vigorous, mid to late ripening. Slight susceptibility to powdery mildew.

WHERE IT'S GROWN AND WHAT ITS WINE TASTES LIKE

Mostosa is cultivated in central **Italy**, mainly on the Adriatic coast (south-east Emilia-Romagna, the Marche, Abruzzo and Puglia), and to a lesser extent in Liguria and Lazio. Total plantings in Italy in 2000 were 104 ha (257 acres). It may comprise 30–50% of the blend in Colli di Rimini, along with varieties such as TREBBIANO ROMAGNOLO and BIANCAME, and contribute in a much more minor way to Colli di Rimini Rebola.

MOURISCO BRANCO

See CAYETANA BLANCA.

MOURISCO TINTO

See MARUFO.

MOURVÈDRE

See MONASTRELL.

MOUYSSAGUÈS

Virtually extinct variety once producing astringent reds in southern central France.

BERRY COLOUR: ● ● ● ● ●

PRINCIPAL SYNONYMS: Négret, Plant du Pauvre (Lot)

ORIGINS AND PARENTAGE

Mouyssaguès was mentioned as early as 1783–4 in Rodez in the Aveyron in central southern France (Rézeau 1997): 'Le Manval lanut, ou Moissagués, c'est un plan mou qui produit en quantité mais de mauvaise qualité' ('Manval lanut, or Moissagués, is a dull variety that produces large quantities of poor-quality grapes'). In the local dialect, its name literally means 'from Moissac', a village in the Tarn-et-Garonne where the variety most likely originated.

VITICULTURAL CHARACTERISTICS

Very vigorous, early ripening.

WHERE IT'S GROWN AND WHAT ITS WINE TASTES LIKE

The variety used to be cultivated in the harsh uplands of the Aveyron in southern central **France** but less than 1 ha (2.5 acres) remained in 2008. It is in the official French register of varieties, recommended in the Cantal and Aveyron *départements* and mentioned in the appellation regulations as a possible component in Vins d'Entraygues et du Fel and Vins d'Estaing but that doesn't seem to have preserved it in the vineyard, perhaps because it does not seem to graft well, according to Galet (2000), and produces astringent wines.

MSKHALI

Armenia's most planted variety, used mainly for brandy.

BERRY COLOUR: ● ● ● ● ●

PRINCIPAL SYNONYMS: Ararati ⊗, At Uzyum, Mashali, Messchaly, Mishali, Msali, Mschali, Musrali, Spitak Khagog

ORIGINS AND PARENTAGE

Mskhali probably comes from the province of Ararat in Armenia.

VITICULTURAL CHARACTERISTICS

Early to mid budding and ripening. Medium-sized to large, compact bunches of medium-sized to large, thin-skinned berries. Very susceptible to fungal diseases and (in Ararat at least) vines need to be earthed up in winter.

WHERE IT'S GROWN AND WHAT ITS WINE TASTES LIKE

Mskhali was the dominant wine-grape variety in **Armenia** at the start of the twenty-first century, planted mainly in the Ararat region and used mostly for the production of brandy, though it also produces dry whites with moderate alcohol and fruity dessert wines. Both Vedi Alco and MAP produce dry varietal wines.

MTSVANE KAKHURI

The more widespread of the Mtsvanes makes high-quality Georgian whites in both traditional and European styles.

BERRY COLOUR: ● ● ● ● ●

PRINCIPAL SYNONYMS: Dedali Mtsvane, Mamali Mtsvane, Mcknara, Mtsvane, Mtsvane Kachuri, Mtsvani, Mtzvané

VARIETIES COMMONLY MISTAKEN FOR MTSVANE KAKHURI: GORULI MTSVANE

ORIGINS AND PARENTAGE

Mtsvane Kakhuri, literally 'green from Kakheti', is a very old indigenous variety from the Kakheti region in south-eastern Georgia. Although it is often referred to simply as Mtsvane, it should not be confused with the genetically distinct GORULI MTSVANE from the Kartli region in southern central Georgia (Vouillamoz *et al.* 2006).

OTHER HYPOTHESES

Like many other Georgian varieties, Mtsvane Kakhuri is said to have been mentioned as early as the fifth century AD (Chkhartishvili and Betsiashvili 2004).

VITICULTURAL CHARACTERISTICS

Moderately vigorous and productive. Late budding, mid ripening. Capable of achieving high sugar levels while retaining fresh acidity. Susceptible to powdery mildew.

WHERE IT'S GROWN AND WHAT ITS WINE TASTES LIKE

Mtsvane Kakhuri is used, sometimes blended with up to 15% RKATSITELI, to produce dry white wines within the Manavi appellation in **Georgia** on mainly calcareous soils. These typically have fresh citrus aromas, some tropical fruit notes, relatively high alcohol and crisp acidity. The variety is found mainly in the villages of Manavi and Tsinandali within the Kakheti region of south-eastern Georgia but also in **Ukraine**, **Russia**, the **Republic of Moldova** and **Armenia**. There were 240 ha (593 acres) in Georgia in 2004. Recommended producers of European-style wines include Badagoni, Telavi Wine Cellar and Teliani Valley; those who make varietal wines using the traditional clay qvevri include Pheasant's Tears and Vinoterra.

MUJURETULI

Georgian variety used to produce dry and semi-sweet wines.

BERRY COLOUR: ● ● ● ● ●

PRINCIPAL SYNONYMS: Keduretuli

ORIGINS AND PARENTAGE

Mujuretuli comes from the Racha-Lechkhumi region.

VITICULTURAL CHARACTERISTICS

Medium-sized berries that can attain high sugar levels. Mid budding and ripening (but harvested late for semi-sweet wines) and low yields.

WHERE IT'S GROWN

Although Mujuretuli is often blended with ALEKSANDROULI (with which it is frequently interplanted) in the semi-sweet red wine Khvanchkara, it also has potential for decent dry reds. Planted mainly in Racha-Lechkhumi in northern **Georgia**, particularly in the villages of Khvanchkara, Chorjo, Sadmeli and Tola, totalling just 64 ha (158 acres) in 2004. Recommended producers include Bugeuil and Rachuli Wine.

MÜLLER-THURGAU

Prolific German invader of the world's vineyards producing soft, semi-aromatic whites in over-abundance.

BERRY COLOUR: ● ● ● ● ●

PRINCIPAL SYNONYMS: ?Findling, Riesling-Silvaner or Riesling × Silvaner (New Zealand, Switzerland, forbidden in the EU), Rivaner (Austria, Luxembourg), Rizlingszilváni (Hungary), Rizvanec (Slovenia)

ORIGINS AND PARENTAGE

A RIESLING × Madeleine Royale cross developed in 1882 by Swiss vine breeder Hermann Müller (born in the canton of Thurgau, hence the name), who was working at the Geisenheim research centre in the Rheingau, Germany, at the time. Müller-Thurgau was used to breed several new crosses that are now in cultivation: ARNSBURGER, BACCHUS, FABERREBE, GUTENBORNER, KANZLER, MÍLIA, MORAVA, OPTIMA, ORTEGA, PÁLAVA, PANONIA, PERLE, REICHENSTEINER and WÜRZER.

This variety was originally recorded as a RIESLING × SILVANER cross by its breeder, Hermann Müller, which explains the erroneous synonyms Riesling × Silvaner and Rivaner still in use today. It was later named Müller-Thurgau by a German grape expert in honour of the breeder. A first DNA study claimed that Müller-Thurgau was in fact a Riesling × CHASSELAS cross (Regner 1996), and a second one corrected the parentage to Riesling × Chasselas de Courtiller, a table grape of minor importance bred in the nineteenth century by Courtiller in Saumur in France's Val de Loire (Sefc *et al.* 1997). A third and final DNA study established that Müller-Thurgau is in reality Riesling × Madeleine Royale, the latter a nineteenth-century cross with unrecorded parents that is no longer cultivated (Dettweiler *et al.* 2000). DNA parentage analysis recently extended the pedigree by showing that Madeleine Royale is a PINOT × SCHIAVA GROSSA cross (Vouillamoz and Arnold 2010). As a consequence, Pinot and Schiava Grossa are the grandparents of Müller-Thurgau, as is GOUAIS BLANC, which is a parent of Riesling (see the Schiava Grossa pedigree diagram).

Findling is said to be a progeny of Müller-Thurgau but this has not yet been confirmed by DNA profiling. Its name means 'foundling' but it can also be used to refer to a seedling.

VITICULTURAL CHARACTERISTICS

Abundant yields but susceptible to downy and powdery mildews and to botrytis bunch rot thanks to thin skins. Ripens early (even before SILVANER) and is versatile in terms of site although the vine wood is quite soft and can easily be damaged by hard winters. It is also affected by the fungal disease Roter Brenner that can cause severe crop losses.

WHERE IT'S GROWN AND WHAT ITS WINE TASTES LIKE

Müller-Thurgau has been the most successful German cross, quantitatively though not qualitatively, and in the decades after it was officially authorized in **Germany**, in 1956, it spread far and wide around the world's wine regions. Growers in cooler climates loved its prolific yields and the fact that it ripened almost anywhere.

It is authorized in **France** but plantings are negligible. It is in the Alto Adige in northern **Italy** that the variety produces its most impressive results, far more than in Germany, particularly from old vines on steep, high stony sites such as Tiefenbrunner's Feldmarschall. The minerality, refinement and complexity of these wines demonstrate what can be achieved on prime sites and at lower yields. Recommended producers include Cantina San Michele Appiano, Manfred Nössing, Josef Briegl, the Caldaro and Santa Maddalena co-operatives and Abbazia di Novacella. The variety is also planted in Trentino, where Pojer e Sandri virtually built their reputation on their unusually firm varietal version. It has not been a success in Friuli but can be found as far south as Emilia-Romagna. Total Italian plantings in 2000 were no more than 279 ha (689 acres). Surprisingly for a variety designed for cool climates, there are even experimental plantings in the Costers del Segre region in **Spain**.

As Rivaner, it is still the most widely planted variety in **Luxembourg**, with a current total of 367 ha (907 acres) although there have been virtually no new plantings in the last ten years. It produces some of the lightest wine in Europe.

The variety has had a chequered history in **England**, from its enthusiastic introduction in the 1950s as an early-ripening high yielder with low acidity, via a peak of 184 ha (455 acres) in 1990, when it was the most widely planted vine, down to its 2009 total of 70 ha (173 acres). Its percentage of the vineyard area diminishes with the rush to plant CHARDONNAY and PINOT NOIR for sparkling wines and the certainty that its vigour and susceptibility to disease make it a poor choice for a cool, damp climate. The best examples are fruity and soft, usually off dry; the worst are flabby and/or overly herbaceous.

In **Germany** it was so popular that by the early 1970s Müller-Thurgau had overtaken RIESLING as Germany's most widely planted variety, underpinning what were then Germany's prodigious exports of bland, light white wine dismissed by some at the time as little more than sugarwater. Fortunately, the focus of German vintners has now shifted definitively back towards quality and Müller-Thurgau plantings were down to (a still considerable) 13,721 ha (33,905 acres) by 2008, compared with Riesling's 22,434 ha (55,436 acres). The fortunes of Liebfraumilch have mirrored those of the grape variety that supplied its dominant ingredient.

It is planted throughout the German wine regions but with most vines in Rheinhessen, the Pfalz, Baden and Franken. Even in the Mosel-Saar-Ruwer area, generally much better suited to RIESLING, there are still 1,263 ha (3,121 acres). Müller's aim was to breed a less pernickety version of Riesling but it falls sadly short of the latter's potential greatness, especially when volume is the grower's goal. The wines tend to be lightly aromatic, even floral, with peachy or grapey, Muscat-like aromas (often assisted by a dollop of a more aromatic variety such as MORIO-

MUSKAT), but most lack zip and can be unappetizingly flabby, particularly when sweetened with a little Süssreserve.

Austria's 3,010 ha (7,438 acres), two thirds of which are in Niederösterreich (Lower Austria), are declining overall but are still significant, particularly in the Weinviertel, north of the Donau (Danube), where it produces inexpensive, early-drinking whites of limited interest. It is also used for much sweeter wines such as Winzerhof Gmeiner's TBA and Eiswein from Wagram.

Müller-Thurgau is **Switzerland**'s second most planted white variety, at 487 ha (1,203 acres), after CHASSELAS, planted mainly in the German-speaking cantons in the north and east.

Further east, **Hungary** has 2,296 ha (5,674 acres) of Rizlingszilváni, grown just about everywhere from Mátra and Eger in the central north to Kunság in the south, with significant plantings also in the wine regions around Lake Balaton. As in Germany, the wines are generally soft and rarely exciting. Plantings in the **Czech Republic** are substantial and total 1,778 ha (4,394 acres), predominantly in the southern regions of Morava, where wines tend to be dry and crisper than average for the variety. **Slovenia**'s 160 ha (395 acres) of Rizvanec are almost exclusively in Štajerska Slovenija (Slovenian Styria) region in the north east. The **Republic of Moldova** had 173 ha (427 acres) in 2009, **Russia** 106 ha (262 acres). The wines are rarely exciting though are less likely to be heavily burdened by sugar than in Germany.

Müller-Thurgau's spread has not been limited to Europe. In North America, California seems to have resisted its dubious charms but **Oregon** has 91 acres (37 ha), mostly in the Willamette Valley. Sokol Blosser were the first to plant in the early 1970s but they now use it as part of a blend. Chateau Benoit, on the other hand, proudly continue with a varietal version. It can be found to a very limited extent in **Washington** State.

On advice from Helmut Becker of Geisenheim in Germany, it was welcomed with open arms in cool-climate **New Zealand** as an improvement on the hybrids that dominated there in the 1950s and 1960s. In the mid 1980s, it was New Zealand's most widely planted variety although the total area has been in decline for the last fifteen years or so and currently stands at just 79 ha (195 acres). The best examples show the freshness that typifies New Zealand whites, though styles are frequently off dry, as in Germany. They seem to last better in bottle than their German counterparts.

Müller-Thurgau is also found in **Japan** (171 ha/423 acres in 2009, mainly in Hokkaido).

Germany has 29 ha (72 acres) of the Findling that is thought to be a mutation of Müller-Thurgau. There's less than 1 ha (2.5 acres) of Findling in **Switzerland**, with two producers in the Genève region. Findling is also planted to a very limited extent in **England**.

MUSCADELLE

Significant in Monbazillac, less so in sweet white bordeaux and very important in north-east Victoria.

BERRY COLOUR: ● ● ● ● ●

PRINCIPAL SYNONYMS: Bouillenc (Lot), Guinlhan Musqué (Gers, Tarn), Muscat Fou (Dordogne), Sauvignon Vert🧬 (California), Tokay (Australia)

VARIETIES COMMONLY MISTAKEN FOR MUSCADELLE: Muscadelle du Bordelais🧬 (California), MUSCAT BLANC À PETITS GRAINS🧬 (called Muscadel in South Africa), SAUVIGNONASSE🧬

ORIGINS AND PARENTAGE

Muscadelle is an old variety from the Gironde and the Dordogne in south-western France, where it was first mentioned in 1736 in Bordeaux and in Cadillac (Rézeau 1997). Contrary to the opinion of Michel Cazeaux-Cazalet in Viala and Vermorel (1901–10) and although it was named for its supposed Muscat flavour, Muscadelle is not related to any Muscat variety. Indeed, DNA parentage analysis has established a parent–offspring relationship with GOUAIS BLANC (Boursiquot *et al.* 2004; see PINOT pedigree diagram), which explains why Muscadelle is a member of the Gouais ampelographic group (see p XXVII; Bisson 2009).

VITICULTURAL CHARACTERISTICS

Vigorous, early to mid ripening. Susceptible to powdery mildew, wasps and grape moths and very susceptible to botrytis bunch rot, and therefore best planted on a well-exposed site.

WHERE IT'S GROWN AND WHAT ITS WINE TASTES LIKE

Planted mainly in Bordeaux and Bergerac in south-west **France** (885 ha/2,187 acres in the Gironde in 2009), there are also 438 ha (1,082 acres) in the Dordogne and 234 ha (578 acres) further south and east in the Tarn. It plays a supporting role in many an appellation – Bordeaux, Bergerac, Sauternes, Entre-Deux-Mers, etc. for dry wines and Sauternes, Barsac, Cadillac, Loupiac, etc. for sweet ones – generally contributing scent and youthful fruitiness to blends of SAUVIGNON BLANC and SÉMILLON. Muscadelle is more highly regarded in Bergerac than in Bordeaux, especially in the sweet wines of Monbazillac. Overall, French plantings have been falling consistently, down from 6,257 ha (15,461 acres) in 1958. There are very few varietal wines but Château des Chapelains in Sainte-Foy Bordeaux and Château de Salettes make rare examples.

In **California**, producers of varietal Sauvignon Vert include Michelini (Napa). The variety may also be included in white Meritage blends, though usually in very small amounts.

The most successful and distinctive outpost of Muscadelle is **Australia**, where it contributes to very ordinary dry whites in the Riverland, for example, and yet produces stunning sweet, slowly oak-matured, fortified wines in north-east Victoria, as it has since the 1850s. These wines were once known as Liqueur Tokay (because the Muscadelle variety was originally thought to have been brought to Australia from Hungary) but were renamed Topaque as part of an Australia–EU agreement, and to appease the Hungarians. Some of the finest exponents are Morris, Seppeltsfield, McWilliams, Chambers, Pfeiffer, Stanton & Killeen, and Campbells. Total plantings in Australia were just 155 ha (383 acres) in 2008. In the Barossa Valley, Peter Lehmann co-ferment a small percentage of Muscadelle in Mudflat Shiraz, a practice they began in the 1960s.

MUSCARDIN

Used in very small amounts to season southern Rhône reds.

BERRY COLOUR: ● ● ● ● ●

ORIGINS AND PARENTAGE

Muscardin is a now-rare variety that most likely originates from the area of Châteauneuf-du-Pape in the Vaucluse *département* in Provence in the south of France, where it was first mentioned in 1895 (Rézeau 1997): 'la plus grande partie du vignoble de Châteauneuf est plantée avec d'anciens cépages méridionaux: Grenache, Piquepoul, Tinto, Terret noir, gris et blanc, Counoise, Muscardin, Vaccarese, Clairette, Picardan' ('the major part of the vineyard of Châteauneuf is planted to old southern grape varieties: ... Muscardin ... '). According to Galet (1990), who was the first to give an ampelographic description of Muscardin, this variety could be related to MONDEUSE NOIRE, but a parent–offspring relationship is ruled out by DNA analysis (Vouillamoz). However,

Muscardin is a member of the Sérine ampelographic group (see p XVII and it is also ampelographically close to AUBUN (Bisson 2009).

VITICULTURAL CHARACTERISTICS

Mid ripening, sugar accumulation in the berries is generally low to moderate.

WHERE IT'S GROWN AND WHAT ITS WINE TASTES LIKE

Although it is one of the many varieties permitted in Châteauneuf-du-Pape, Gigondas, Côtes du Rhône and other southern Rhône appellations, it is definitely a very minor player. After increasing to 25 ha (62 acres) in 1979, the total area given over to this variety in **France** was down to 18 ha (44 acres) in 2008. Estates that continue to grow Muscardin and use it in their red blends include Clos des Papes, Château de Beaucastel, Font de Michelle, Roger Sabon, Robert Usseglio, Chante-Perdrix and others – so they can't have that many vines each. Varietal wines are light in colour, relatively high in acidity and sometimes lightly floral, making this variety a useful minor ingredient in wines that are generally rich and full-bodied.

MUSCAT

PRINCIPAL SYNONYMS: These are basically translations of Muscat, rather than synonyms, since the name Muscat is used for so many different varieties, as explained below: Meski (Tunisia), Misket (Bulgaria), Moscatel (Spain and Portugal), Moscato or Moscatello (Italy), Moschato or Moschoudia (Greece), Muskat (Germany), Muskateller (Germany), Muskatoly or Muskotály (Hungary)

ORIGINS AND PARENTAGE

The name Muscat, or Muscat Something, in various languages, is given to over 200 distinct and often unrelated white, pinkish or black-berried varieties grown in many different wine regions. These varieties are used as table and/or wine grapes and all have in common a more or less distinctive grapey Muscat flavour. However, this flavour can also appear by mutation in other varieties (eg Chasselas Musqué (see CHASSELAS), GEWÜRZTRAMINER, etc.) so it is not necessarily a distinguishing characteristic for varieties with Muscat in their name.

The name Muscat most likely derives from the musky aroma produced by a gland of the male musk deer from southern Asia that was used in the fifth century AD to make a rare fragrance. The Persian word *muchk* became *moskos* in Greek, *muscus* in Latin, *musc* in French and musk in English.

The earliest mention of the name Muscat dates back to 1230–40 in the *Liber de proprietatibus rerum* by Bartholomaeus Anglicus, translated into French in 1372 by Jean Corbichon as *Le livre des propriétés des choses*: 'vin extrait de raisins muscats' ('wine extracted from muscat grapes').

Recent genetic research (Crespan and Milani 2001; Cipriani *et al.* 2010; see pedigree diagram overleaf) has shown that:

- MUSCAT OF ALEXANDRIA is probably and quite unexpectedly a natural cross between MUSCAT BLANC À PETITS GRAINS, a white variety grown all around the Mediterranean, and Axina de Tres Bias, an old black-berried table grape grown on Sardegna, Malta and the Greek islands.
- MUSCAT OF ALEXANDRIA and MUSCAT BLANC À PETITS GRAINS are the progenitors of at least fourteen grape varieties that are currently cultivated to make wine, of which nine are Italian and five are South American.

MUSCAT PEDIGREE DIAGRAM

While many varieties with a Muscat aroma are not genetically related, this pedigree of natural crosses headed by MUSCAT BLANC À PETITS GRAINS and MUSCAT OF ALEXANDRIA reveals those that actually constitute a true 'Muscat family'. In the case of unknown (and probably extinct) varieties (?), the inverse relationships are also theoretically possible (see p x).

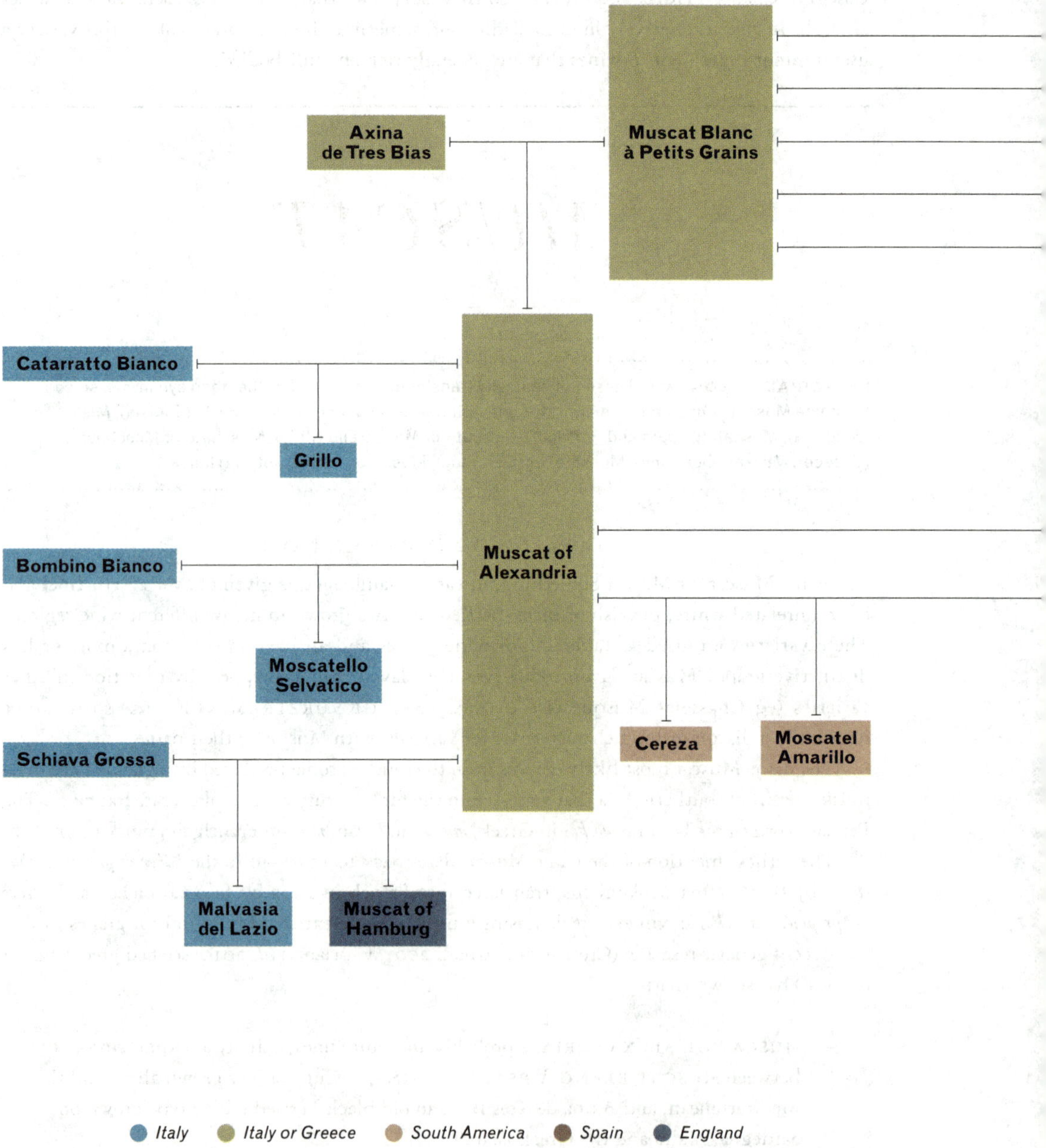

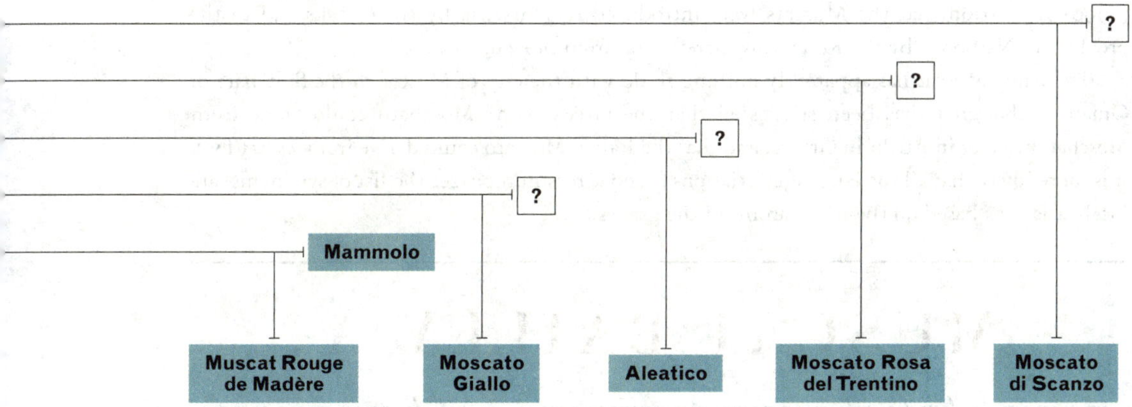
?
?
?
?
Mammolo
Muscat Rouge de Madère
Moscato Giallo
Aleatico
Moscato Rosa del Trentino
Moscato di Scanzo

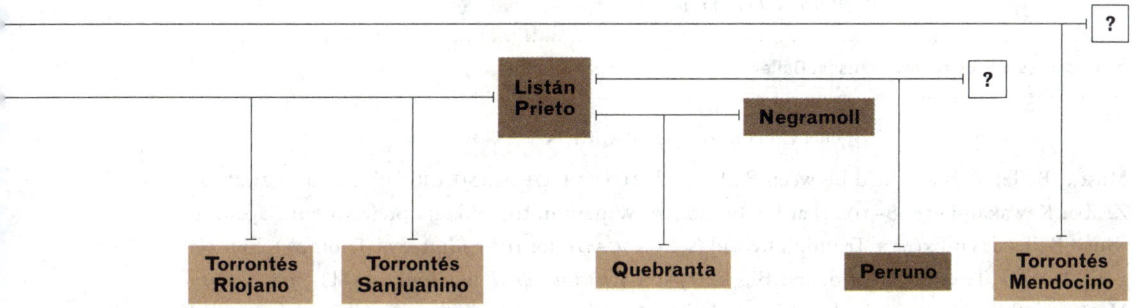
?
Listán Prieto
?
Negramoll
Torrontés Riojano
Torrontés Sanjuanino
Quebranta
Perruno
Torrontés Mendocino

MUSCAT OF ALEXANDRIA, MUSCAT BLANC À PETITS GRAINS and their fourteen respective progenies clearly constitute a close family within the Muscat group as a whole even though that group is otherwise largely heterogeneous, making it somehow misleading to speak about the 'Muscat family' for all Muscat-flavoured varieties.

OTHER HYPOTHESES

It is often said that Muscat was already known to the Persians and the Egyptians in Early Antiquity (*c.* 3000–1000 BC), purely on the basis of frescoes representing grapevines (Bronner 2003). Similarly, it is often believed that Muscat corresponds to the varieties known as Anathelicon Moschaton by the Greeks and to those described by Pliny the Elder, Columella and other writers of the Roman period as Apianae, so named because their high sugar level attracted bees (*apis* in Latin; Galet 1990). However, there is no botanical evidence for these identifications. Galet's (1990) suggestion that the Muscats were introduced to Marseille by the Greeks and to the province of Narbonne by the Romans is therefore far from proven.

The name Muscat has apparently nothing to do with the city of Muscat in the Sultanate of Oman. Although it has been suggested that the Greek name Moschato could come from Moschaton, a city in Attikí in Greece, and that the Italian Moscato could derive from *mosca* ('fly'), it is more likely that all the language variations found across Europe (see the list of synonyms) are local variations based on the musk aroma of the grapes.

MUSCAT BAILEY A

Japanese hybrid with many an American gene, good disease resistance but producing candy-flavoured wines.

BERRY COLOUR: ● ● ● ● ●

PRINCIPAL SYNONYMS: Muscat Bailey

ORIGINS AND PARENTAGE

Muscat Bailey A is a hybrid between Bailey and MUSCAT OF HAMBURG obtained in 1927 by Zenbei Kawakami (1868–1944) at his Iwanohara Winery in the Niigata prefecture in Japan, in which Bailey is an Extra × Triumph hybrid (see CASCADE for the pedigree of Triumph), Extra is a Big Berry × Triumph hybrid, and Big Berry is a *Vitis lincecumii* vine found by Thomas Volney Munson (1843–1903) in the US. Muscat Bailey A is therefore a *Vitis vinifera*, *Vitis labrusca* and *Vitis lincecumii* hybrid that was developed to obtain a variety adapted to the Japanese climate.

VITICULTURAL CHARACTERISTICS

Mid budding and ripening. Big bunches of big, thick-skinned berries (smaller if grown for the table). Resistant to powdery and downy mildews and botrytis bunch rot.

WHERE IT'S GROWN AND WHAT ITS WINE TASTES LIKE

Wines tend to be sweet-fruited, candy-like reds, with moderate acidity and soft tannins. Oak-aged wines from fully ripe grapes seem to achieve better balance. About one third of **Japan**'s 2008 total of 135 ha (334 acres) is grown in Yamanashi prefecture, central Honshu, and it is also found in Yamagata, Osaka, Shimane and other prefectures. Recommended producers include Alps Wine and Iwanohara Wine.

MUSCAT BLANC À PETITS GRAINS

The classic Muscat, the small-berried one that is widely dispersed around the world.

BERRY COLOUR:

NOTE: We have used the French name, most commonly used around the world, rather than an Italian or a Greek name – eg Moscato Bianco or Moschato Samou – because the origin is not yet completely certain. (In all other entries, we aim to use a prime name that belongs to the country of origin unless the variety is no longer grown there.)

PRINCIPAL SYNONYMS: Bela Dinka (Hungary), Beli Muskat (Hungary), Beyaz Misket (Turkey), Bornova Misketi, Brown Muscat (Australia), Franczier Veros Muscatel (Hungary), Frontignac (Australia, South Africa), Gelber Muskateller (Germany), Moscatel Branco (Portugal), Moscatel Castellano (Spain), Moscatel Commun (Spain), Moscatel de Grano Menudo (Spain), Moscatel de Grano Pequeno (Spain), Moscatel do Douro (Portugal), Moscatel Fino (Spain), Moscatel Galego Branco (Portugal), Moscatel Morisco (Spain), Moscatello Bianco (Italy), Moscatello Bianco di Basilicata (Italy), Moscatello di Saracena (Saracena in the province of Cosenza in Calabria, southern Italy), Moscatello di Taggia (Italy), Moscato Bianco (Italy), Moscato di Chambave (Valle d'Aosta in Italy), Moscato d'Asti (Italy), Moscato dei Colli Euganei (Italy), Moscato di Momiano (Italy), Moscato di Montalcino (Italy), Moscato di Tempio (Italy), Moscato di Trani (Puglia in Italy), Moscato Reale (Italy), Moschato Aspro (Greece), Moschato Kerkyras (Greece), Moschato Lefko (Kefaloniá in Greece), Moschato Mazas (Greece), Moschato Samou (Sámos in Greece), Moschato Spinas (Greece), Moschato Trani (Greek island of Ródos/Rhodes), Moschoudi (Pelopónnisos/Peloponnese in Greece), Moschoudi Proïmo (Pelopónnisos), Moscovitza, Muscadel (South Africa), Muscat à Petits Grains Blanc (France), Muscat Blanc (France, US), Muscat Canelli (US, Israel), Muscat d'Alsace (Alsace in France), Muscat de Die (Drôme in France), Muscat de Frontignan or Muscat Frontignan (south of France, South Africa, US, Chile), Muscat de Lunel (south of France), Muscat du Valais (Switzerland), Muscat Sámos (Greece), Muskadel (South Africa), Muskateller, Muskuti (Greece), Myskett (Turkey), Piros Muskotály (Hungary), Rumeni Muškat (Slovenia), Sárga Muskotály (Hungary), Tămâioasă Alba (Romania), Tamjanika or Tamnjanika (Serbia), Tamyanka (Bulgaria, Serbia, Russia), Temjanika (Republic of Macedonia), Weisse Muskattraube (Germany), Weisser Muskateller (Germany), White Frontignan (UK)

VARIETIES COMMONLY MISTAKEN FOR MUSCAT BLANC À PETITS GRAINS: MOSCATO GIALLO (Switzerland), MUSCADELLE, TĂMÂIOASĂ ROMÂNEASCĂ (Romania), TORRONTÉS RIOJANO (Mendoza in Argentina)

ORIGINS AND PARENTAGE

Muscat Blanc à Petits Grains is a very old variety that has been known for centuries under a vast number of synonyms throughout the Mediterranean belt and may come from either Italy or Greece, for the following reasons.

The earliest mention of this variety most likely appeared in Italy in 1304 in Pietro de Crescenzi's *Ruralium commodorum libri XII* under the Latin name Muscatellus, later translated into Italian as Moscadella. It referred to a table grape grown around Bologna. Countless subsequent mentions, under local synonyms, have been recorded since the Middle Ages, illustrating the expansive distribution of this variety, for example: in 1394 in France, 'vin muscat de Claira' ('Muscat wine from Claira'), Claira being the name of a place near Rivesaltes where Muscat Blanc à Petits Grains is still widely cultivated (Dion 1959); in 1513 in Spain under the name Muscatel; in 1534 in Württemberg, Germany, under the name Muscateller; in 1536 in Wallis, Switzerland, under the name Muscatelli; and so on.

The general consensus is that Greece is the homeland of Muscat Blanc à Petits Grains. The Greeks are said to have brought the variety to Italy, and from Italy the Romans would have introduced it to the region of Narbonne on France's Mediterranean coast, where it became famous under the names Muscat de Frontignan and Muscat de Lunel. What's more, DNA profiling strongly suggests that the Greek varieties Moschato Aspro, Moschato Kerkyras, Moschato Mazas, Moschato Spinas and Moschoudi, long cultivated in Greece, are all synonyms of Muscat Blanc à Petits Grains (Hvarleva *et al.* 2004; Lefort *et al.* 2000; Stavrakakis and Biniari 1998; Vouillamoz).

In a wide-ranging genetic study of 222 grape varieties from all over the world, Muscat Blanc à Petits Grains (under the name Moschato Samou) was grouped together with several Greek varieties such as MALAGOUSIA, MOSCHOMAVRO (under the name Moschato Mavro) and VOLITSA MAVRI, and with some Italian varieties such as Malvasia Bianca (whose true identity is unknown; see MALVASIA) and ALEATICO (Aradhya *et al.* 2003).

Recent DNA parentage analysis has surprisingly revealed that MUSCAT OF ALEXANDRIA is probably a natural cross between Muscat Blanc à Petits Grains and Axina de Tres Bias, an old black-berried table-grape variety on Sardegna (Cipriani *et al.* 2010), also found in Greece under the name Eftakoilo or Heftakilo (Vouillamoz; see Muscat of Alexandria), which means that the cross could have occurred either in Italy or in Greece. This parentage places Muscat Blanc à Petits Grains at the top of the Muscat pedigree. Indeed, DNA parentage analysis has shown that Muscat Blanc à Petits Grains has parent–offspring relationships with at least six other cultivated varieties of the Muscat group (see the pedigree diagram in the MUSCAT entry): ALEATICO (Toscana in Italy), MOSCATO GIALLO (northern Italy), MOSCATO ROSA DEL TRENTINO (Trentino in Italy), MOSCATO DI SCANZO (Lombardia in Italy), Muscat of Alexandria (southern Italy or Greece?) and Muscat Rouge de Madère (Italy?), the latter being a natural progeny of Muscat Blanc à Petits Grains and MAMMOLO (Crespan and Milani 2001; Di Vecchi Staraz, This *et al.* 2007) and no longer commercially cultivated. Most remarkably, with the possible exception of Muscat of Alexandria, all of them originate from Italy. This could speak in favour of an Italian rather than a Greek origin for Muscat Blanc à Petits Grains. Indeed, it has been known in Italy for several centuries under the name Moscato Bianco. Without further evidence, it is not possible to determine whether Greece or Italy is the true homeland of Muscat Blanc à Petits Grains.

Pink, red and black colour mutations of Muscat Blanc à Petits Grains also exist around the world (Crespan and Milani 2001), under various names, eg Muscat à Petits Grains Roses, Muscat à Petits Grains Rouges, Moscato Rosa, Muscat d'Alsace Rouge, Muscat Violet de Madère, Moscato Rosso, Moscatel Rosada, Moscatel Galego Roxo or Moscatel Roxo, and Brown Muscat.

Muscat Blanc à Petits Grains was used to breed MISKET VRACHANSKI and ZAGREI.

OTHER HYPOTHESES

According to the *Neustadter Chronik*, Emperor Friedrich I was served Muskateller wine in Neustadt an der Haardt (today's Neustadt an der Weinstrasse), south west of Mannheim in Rheinland-Pfalz in Germany, in 1152. However, historian Friedrich von Bassermann-Jordan (1923) raises serious doubts about this early date, and no historical document has been found to support it. According to Gervais (1902), the city of Frontignan exported its Muscat wines under the reign of Charlemagne, but there's no documentary evidence of this.

VITICULTURAL CHARACTERISTICS

Early budding, mid ripening, best pruned short. Susceptible to powdery mildew, botrytis bunch rot and hymenoptera, and very susceptible to mites. Smaller berries than other Muscats.

WHERE IT'S GROWN AND WHAT ITS WINE TASTES LIKE

Muscat Blanc à Petits Grains is the oldest and most distinguished of the Muscat Somethings but it is unfortunately also quite difficult to grow. Well-made wines, dry, sweet or sweet and fortified, have a distinctly grapey aroma which may be also delicately (sometimes pungently) floral and spicy.

Muscat Blanc à Petits Grains is by far and away the most widely planted Muscat in **France**, with 7,620 ha (18,829 acres) in 2009 compared with 2,610 ha (6,449 acres) of Muscat d'Alexandrie (MUSCAT OF ALEXANDRIA). Unlike other Muscats, this smaller-berried version has seen consistently increased plantings in France, especially in Roussillon, and total vineyard area is currently more than three times what it was fifty years ago. In 1998 in Roussillon, plantings of

Muscat d'Alexandrie were higher than for Muscat Blanc à Petits Grains. The increase in plantings of the latter is partly the result of the growing popularity for light, dry Muscat and the ability of Muscat Blanc à Petits Grains to withstand dry vintages, but also thanks to the decline in production of traditional sweet, fortified vins doux naturels and the now-rescinded rule that Muscat de Rivesaltes vineyards had to be equally divided between the two Muscats. While dry Muscat is difficult to make well, a little drop of this aromatic liquid can be a very successful addition to more neutral whites. The produce of old Muscat vines in Roussillon has been used to good effect as an ingredient in complex blended dry whites.

This finer Muscat is the third most planted white in Languedoc-Roussillon (5,737 ha/14,176 acres), the biggest plantings being in Roussillon in the Pyrénées-Orientales (3,043 ha/7,519 acres) and in the Hérault (2,234 ha/5,520 acres) in the eastern Languedoc but there are also significant plantings in the Drôme, where it is the main, sometimes the sole, component in the lightly sweet and sparkling Clairette de Die. It's also the most widely planted Muscat on the French island of Corse (230 ha/568 acres), where it is the sole variety permitted for the vin doux naturel Muscat du Cap Corse, although it can also be included in dry white wines based on VERMENTINO. Sweet fortified wines in a similar style are made in many southern French appellations in Roussillon (Muscat de Rivesaltes) and in the Languedoc and southern Rhône (Muscats of Beaumes-de-Venise, Frontignan, Lunel, Mireval and Saint-Jean-de-Minervois). Most must be made exclusively from Muscat Blanc à Petits Grains, although Beaumes-de-Venise may also include the red mutation Muscat à Petits Grains Rouges.

Italy recorded an impressive total of 13,280 ha (32,816 acres) in 2000, plus another 99 ha (245 acres) of the colour mutation Moscato Rosa. Much of this goes into the sweet and lightly sparkling Asti and the superior Moscato d'Asti as well as other north-west Italian fizz, so it's not surprising that it's the most widely planted white wine variety in Piemonte with 10,328 ha (25,521 acres) and nearly 22% of the vineyard area in 2000. Further south, Moscato wines tend to appear in the richer Mediterranean style, often made from dried grapes and labelled *passito*. Further north, in the Valle d'Aosta, the variety is known as Moscato di Chambave, named after the town very close to the Swiss border, and is responsible mainly for sweet passito-style wines. In Montalcino, Moscadello has its own DOC for dry, sparkling and sweet, late-harvest wines.

In **Spain**, most wines labelled Moscatel are made from the generally inferior and less subtle MUSCAT OF ALEXANDRIA but Moscatel de Grano Menudo was still planted on 713 ha (1,762 acres) in 2008, mostly in Navarra and Catalunya in the north east. Chivite have had great success with their botrytized version from southern Navarra.

In **Portugal**, there were 693 ha (1,712 acres) of Moscatel Galego Branco in 2010. Plantings in the Douro, most concentrated on the plateau around Alijó and Favaios, east of Vila Real, are important for Moscatel de Favaios, rather like white port, and as a component of white port itself. According to Mayson (2003), Moscatel is still grown on the island of Madeira, though the grapes are today usually destined for the table rather than the vat. **Moscatel Galego Roxo** (or just Moscatel Roxo), which has purple-pink berries and is most probably a colour mutation of Moscatel Galego Branco, is found in the Douro and on the Península de Setúbal (total plantings 62 ha/153 acres in 2010). Recommended producers of sweet, fortified varietal wines in Setúbal include Bacalhôa Vinhos de Portugal, José Maria da Fonseca and Horácio Simões.

Gelber Muskateller, as it is known in **Germany**, has a long history there, purportedly as far back as the twelfth century, but today has virtually disappeared, displaced in large part in the 1970s by the less noble but also less troublesome and higher-yielding and intensely aromatic MORIO-MUSKAT (no relation and now also in decline). The 174 ha (430 acres) that remained in 2008 are found mostly in Baden and the Pfalz, where Müller-Catoir made some fine, pungent dry wines until the turn of the century. The variety seems to earn greater respect in **Austria**, where its existence is documented back to 1400. Total plantings have risen from 143 ha (353 acres) in 1999 to 422 ha (1,043 acres) in 2007. The generally dry, light, aromatic wines are particularly successful in the Südsteiermark (South Styria), although Muskateller is also planted here and there in the regions along the Donau (Danube) known collectively as Niederösterreich (Lower Austria), and it contributes to some fine sweet wines in Burgenland.

Although **Switzerland** has only a small area of Muscat Blanc à Petits Grains (49 ha/121 acres in 2009), it can boast a varietal wine made from the red mutation: Cave Labuthe's fresh and aromatic L'Extase, from the Valais.

Slovenia's 403 ha (996 acres) of Rumeni Muškat are planted predominantly in Štajerska Slovenija (Slovenian Styria), near the border with Austria's Südsteiermark, where they tend to produce aromatic, soft and deeply coloured wines that need drinking relatively early. They can be thin in more difficult years. The variety is popular in Slovenia for both the table and the fermentation vat.

Although several synonyms are used in **Hungary**, the prime name here is Sárga Muskotály, meaning 'yellow Muscat', since it is distinguished by the deep yellow colour of its skins. Said to be the most ancient extant variety in the Carpathian Basin (Roháły *et al.* 2003), it was cultivated on 647 ha (1,599 acres) of Hungarian vineyard in 2008, mostly in Tokaj, where it tends to play a small part in the region's illustrious sweet wines, alongside FURMINT and HÁRSLEVELŰ. It is occasionally given a soliloquy, dry or sweet, by the likes of János Árvay, Zoltán Demeter and Királyudvar. Dry wines can be fresh and intensely aromatic but tend to age quickly.

In other parts of Eastern Europe, MUSCAT OTTONEL holds sway but there are limited plantings in **Romania**, where it is known as Tămâioasă Alba, in **Serbia** as Tamjanka, and in **Bulgaria** and **Russia** under the name Tamyanka. The quality and longevity of wines labelled Massandra White Muscat and Pink Muscat produced early last century in Krym (Crimea), **Ukraine**, suggest that they must have been made from Muscat Blanc à Petits Grains; in 2009, Ukraine had 674 ha (1,665 acres). There were 133 ha (329 acres) of Bornova Misketi in **Turkey** in 2010.

Greece's strict appellation controls on its most highly classified sweet Moschato wines (sometimes also fortified) from the islands and from the northern Pelopónnisos insist on 100% Muscat Blanc à Petits Grains, under the various synonyms listed above, for the Muscats of Kefaloniá, Límnos, Pátra, Rio of Pátras, Ródos and, probably the most famous, Muscat of Sámos. Here the vines are grown up to 800 m (2,600 ft) above sea level, often on steep hillside terraces, and the co-operative has a monopoly. A delicate dry version as well as the sweet one is made on the island of Límnos. Parparoussis make a very good example of the sweet Muscat Rio of Pátras and Antonopoulos' Amydalies is a delicate dry Muscat (classified as a regional wine of Achaḯa) that is particularly impressive given the young age of the vines. Boutaris' Moschato Spinas from Kríti (Crete) is a complex, silky, full-bodied and highly aromatic sweet wine made from sun-dried grapes that spends a few months in old oak.

It is also found in **Israel** and on **Malta**.

MUSCAT OF ALEXANDRIA is very much more important than this finer Muscat in **California**, where most of Muscat Blanc's 1,908 acres (772 ha) are in the central San Joaquin Valley (2010). Once variously called Muscat Frontignan and Muscat Canelli, it is now officially called Muscat Blanc and distinguished from Orange Muscat (MUSCAT FLEUR D'ORANGER). There were 177 acres (72 ha) of Muscat Canelli in **Washington** State in 2011.

There are minor plantings in **Argentina**, **Chile** and **Uruguay**, although Chile has more of the pink-berried mutation **Moscatel Rosada**.

Liqueur Muscat is a magnificent speciality of **Australia**, made especially around Rutherglen in north-east Victoria, from the mutation known as **Brown Muscat**, or Muscat à Petits Grains Rouges. There were 237 ha (586 acres) of this darker-berried version in Australia in 2008 and 263 ha (650 acres) of the pale-skinned Muscat Blanc à Petits Grains, which is generally used for lighter sweet wines. Liqueur Muscat is an extremely sweet, dark, fortified wine long aged in old wood that becomes increasingly complex with age. Some of the finest exponents are Chambers, Morris, Seppeltsfield, and Stanton & Killeen. Muscat is also found in **New Zealand** but the official statistics are less than helpful, reporting 135 ha (334 acres) in 2008 for 'Muscat varieties', which does not necessarily mean this one.

South Africa's Muscat de Frontignan, also known as Frontignac, and as Muscadel or Muskadel on the domestic market to avoid confusion with MUSCADELLE, was planted on 668 ha (1,651 acres) in 2008. It was responsible for a good proportion of the famous sweet wines of Constantia in the eighteenth and nineteenth centuries and may well have been planted in the Cape as early as 1659. Its most famous guise today is Klein Constantia's sweet Vin de Constance, an attempt to re-create those historic wines, made from 100% Muscat Blanc à Petits Grains shrivelled on the vine in the Constantia region. Both pale- and dark-skinned versions are found and used mainly for dessert and fortified wines, the latter especially when the vines are planted in hotter inland regions such as Robertson. Boplaas of Klein Karoo is one of the most successful producers of sweet, strong Muscat both pale and dark.

MUSCAT BLEU

Minor Swiss hybrid valued particularly by organic growers for its Muscat-like aroma and disease resistance but more commonly grown for the table.

BERRY COLOUR: ● ● ● ● ●

PRINCIPAL SYNONYMS: Garnier 83/2, Muscat Bleu Garnier

ORIGINS AND PARENTAGE

Muscat Bleu is a complex Garnier 15-6 × Perle Noire hybrid obtained in the 1930s (Dupraz and Spring 2010) by producer and grape breeder Charles Garnier in Peissy in the canton of Genève, Switzerland, in which:

- Garnier 15-6 is a Villard Noir × MÜLLER-THURGAU hybrid
- Villard Noir (or Seyve-Villard 18-315) is a CHANCELLOR × Subéreux hybrid (see PRIOR for the complete pedigrees)
- Perle Noire (or Seyve-Villard 20-347) is a Teneron × Seyve-Villard 12-358 hybrid, where Teneron is a table grape from Spain and Seyve-Villard 12-358 is a Seibel 6468 × Subéreux hybrid (see HELIOS for the pedigree of Seibel 6468).

VITICULTURAL CHARACTERISTICS

Mid budding, early ripening. Large to medium-sized loose bunches of large, Muscat-flavoured berries. Generally resistant to downy mildew and other fungal diseases but very attractive to wasps and some susceptibility to coulure and millerandage.

WHERE IT'S GROWN AND WHAT ITS WINE TASTES LIKE

Muscat Bleu is grown mainly by amateur growers in **Switzerland** for the table but there are also a few producers of commercial wines, including Biolenz, Diroso and Bruno Martin, with a total of less than 3 ha (7 acres) in 2009. Wines are generally soft and grapey. It's also found in **Belgium**.

MUSCAT FLEUR D'ORANGER

Highly aromatic, orange-flavoured variety of unknown geographical origin but definitely related to Muscat Blanc à Petits Grains.

BERRY COLOUR: ● ● ● ● ●

PRINCIPAL SYNONYMS: Chasselas Fleur d'Orange, Cranford Muscat, Madarski Muskat, Mirisavka, Moscatel Gordo Peludo, Moscato Fior d'Arancio, Muscat Croquant, Muscat de Jésus, Muscat d'Espagne, Muscat of Hungary, Muscat Queen Victoria, Muscat Regnier, Muscat Vengerskii, Orange Muscat, Orange Muskat, Primavis Muscat, Raisin Vanille, Vanilia Muskotaly, Weisse Vanillentraube

ORIGINS AND PARENTAGE

The parentage of this old variety was unknown until DNA parentage analysis revealed that Muscat Fleur d'Oranger is derived from a cross between CHASSELAS and MUSCAT BLANC À PETITS GRAINS (Schneider, Torello Marinoni and Crespan 2008). It is not known where the crossing took place. Muscat Fleur d'Oranger was used to breed CSABA GYÖNGYE.

OTHER HYPOTHESES

Some have suggested that Muscat Fleur d'Oranger was a simple mutation of MUSCAT BLANC À PETITS GRAINS originating from the city of Orange in the south of France, while others considered it to be identical to Chasselas Musqué (see CHASSELAS; Bronner 2003). Further hypotheses have suggested that it is not related to Muscat Blanc à Petits Grains at all or that it originated in Syria.

VITICULTURAL CHARACTERISTICS

Early to mid ripening and productive. Looser clusters than MUSCAT BLANC À PETITS GRAINS make it slightly less susceptible to botrytis bunch rot but the berries are prone to splitting in wet weather. Moderately susceptible to powdery and downy mildews.

WHERE IT'S GROWN AND WHAT ITS WINE TASTES LIKE

In **Italy**, Moscato Fior d'Arancio, authorized for dry, passito and sparkling wines in the Colli Euganei in the Veneto, has a more floral, orange-blossom character than the rich orange flavours that this variety yields in California.

California had 232 acres (94 ha) in 2009, mostly in the Central Valley but also in Amador County. The fact that it ripens about two weeks later than MUSCAT BLANC À PETITS GRAINS means that it is less well suited to cooler sites. Andrew Quady makes two Orange Muscat varietal wines: Essencia, a lightly fortified 100% Orange Muscat aged briefly in oak with strong orange and apricot flavours, and the much lighter Electra, with peach flavours and low alcohol. According to Quady, Orange Muscat has an unusually high level of phenolics, sometimes giving a tea-like aftertaste and a darker colour if the juice is left on the skins for any length of time. Other producers of varietal wines include Swanson in Napa and Barnard Griffin in **Washington** State. It is also planted to a limited extent in British Columbia, **Canada**. Cerelia is just one of several producers.

In **Australia** in 2010 there were seven producers of this variety though the actual area under vine is not recorded. The most influential proponent is Brown Brothers, who have 27 ha (67 acres) of Orange Muscat and whose popular, delicately aromatic sweet wine is a blend of 80% Orange Muscat with 20% FLORA. Other Victorian producers of sweet wines predominantly or wholly made from Orange Flora include Goorambath and Monichino; Amulet and St Leonards produce dry versions.

MUSCAT ODESSKY

A white-berried hybrid obtained by N Y Borisovskiy, Y N Dokuchaeva and L M Pismennaya at the Tairov research centre in Odessa, **Ukraine**, by crossing Siny Ranny (a Bulgarian MADELEINE ANGEVINE × MUSCAT OF HAMBURG cross) with Pierrelle (a French Panse de Provence × VILLARD BLANC hybrid). It was registered in 1993 and three clones have been selected. This winter-hardy, early- to mid-ripening variety is used for grape juice and to make sparkling and sweet Muscat-flavoured wines, including some fortified versions.

MUSCAT OF ALEXANDRIA

Light-skinned Mediterranean variety generally considered inferior to Muscat Blanc à Petits Grains, used mostly to produce very sweet wines.

BERRY COLOUR: ● ● ● ● ●

PRINCIPAL SYNONYMS: Acherfields Early Muscat (UK), Albillo di Toro (Tunisia), Aleksandrijski Muskat (Russia), Alexandriai Muskotály (Hungary), Angliko (Greece), Apostoliatiko (Greece), Argelino (Tunisia), Daroczy Musko (Hungary), Gordo (Spain, Australia), Gordo Zibibo (Sicilia), Hanepoot (South Africa), Iskendiriye Misketi (Turkey), Lexia (Australia), Malaga (Cyprus), Moscatel (South America), Moscatel Bianco (Argentina, Spain, Uruguay), Moscatel Blanco (Andalucía in Spain), Moscatel de Alejandría (Spain), Moscatel de Chipiona (Spain), Moscatel de Grano Gardo (Spain), Moscatel de Málaga (Spain, Madeira), Moscatel de Setúbal (Portugal, Spain), Moscatel Gordo (Spain), Moscatel Graúdo (Portugal), Moscatel Romano (Italy, Spain), Moscatellone (southern Italy), Moscato d'Alessandria (Italy), Moscato di Pantelleria (Italy), Moscato Francese (Italy), Moschato Alexandrias (Greece), Moschato Limnou (Límnos in Greece), Muscat à Gros Grains (France), Muscat Bowood (UK), Muscat d'Alexandrie (France), Muscat de Berkain (Algeria), Muscat de Fandouk (Algeria), Muscat de Raf-Raf (Tunisia), Muscat de Rome (France), Muscat El Adda (Algeria), Muscat Gordo Blanco (Australia), Muscat Grec (France), Muscat Romain (France), Probolinggo Putih (Hatten, Bali), Salamanca (Spain), Seralamanna (Italy), Tămâioasă de Alexandria (Romania), White Hanepoot (South Africa), Zibbibo (Sicilia), Zibibo (Sicilia), Zibibbo (Sicilia)

VARIETIES COMMONLY MISTAKEN FOR MUSCAT OF ALEXANDRIA: MOSCATO DI TERRACINA

ORIGINS AND PARENTAGE

Muscat of Alexandria is a very old variety (as witness all those synonyms) that has for centuries been cultivated in several Mediterranean countries, both for the table and for wine. Despite its name, it is unlikely that Muscat of Alexandria originated in the city of Alexandria in Egypt, as many have assumed (Galet 2000; Nikolau and Michos 2004). There is far more historical and genetic evidence to suggest that the variety comes either from what was known as Magna Graecia (southern Italy, Sardegna, Sicilia) or from Greece:

- The name Muscat of Alexandria first appeared in Paris in 1713 (Molon 1906) but was already known in Sicilia in the sixteenth century under the name Zibibo, as documented by Pietro Andrea Mattioli (1563).
- The name Zibibo or Zibibbo possibly derives from the Arabic *zabīb*, meaning 'raisin',

which might suggest that this name was given during the rule of Arabic-speaking Carthaginians over the north-western part of Sicilia (sixth to third centuries BC).

- In Greece, Muscat of Alexandria is still cultivated under the name Moschato Limnou (ie 'of Límnos'), a clone of Muscat of Alexandria according to Stavrakakis and Biniari (1998).
- In a genetic study of 222 grape varieties from all over the world, Muscat of Alexandria was grouped together with several Greek varieties such as MALAGOUSIA, MOSCHOMAVRO (under the name Moschato Mavro) and VOLITSA MAVRI, and with some Italian varieties such as Malvasia Bianca (whose true identity is unknown) and ALEATICO (Aradhya *et al.* 2003).
- In southern Italy, two old, natural crosses have involved Muscat of Alexandria (see the pedigree diagram in the MUSCAT entry): GRILLO from Sicilia is a progeny of CATARRATTO BIANCO and Muscat of Alexandria (Di Vecchi Staraz, This *et al.* 2007) and MOSCATELLO SELVATICO from Puglia is a progeny of BOMBINO BIANCO and Muscat of Alexandria (Cipriani *et al.* 2010). Two other varieties from further north in Italy are the result of natural crosses involving Muscat of Alexandria and SCHIAVA GROSSA: MALVASIA DEL LAZIO and MUSCAT OF HAMBURG (Crespan 2003; Lacombe *et al.* 2007).
- Recent DNA parentage analysis has revealed that Muscat of Alexandria is probably a natural cross between MUSCAT BLANC À PETITS GRAINS and Axina de Tres Bias (Cipriani *et al.* 2010). Like Muscat of Alexandria, Muscat Blanc à Petits Grains has for centuries been cultivated all around the Mediterranean, including Italy, where it has long been known as Moscato Bianco. Axina de Tres Bias is an old black-berried table variety that is still grown on the Italian island of Sardegna under this dialect name meaning 'three times grape' (also Trifera or Tre Volte l'Anno, in reference to its very high yields), as well as on the Maltese archipelago (Giannetto *et al.* 2010) and in the Greek islands (Logothetis in Cipriani *et al.* 2010). Comparison of the DNA profiles of Axina de Tres Bias (Giannetto *et al.* 2010) with an unnamed sample from Calabria (Stella Grando, personal communication) and with Eftakoilo (Sefc *et al.* 2000), which is considered to be an old, black-berried Greek table variety (Eftakoilo or Heftakilo means 'seven kilograms'), shows that they are identical (Vouillamoz).

Muscat of Alexandria has also been cultivated for centuries under the name Moscatel in South America (Lacoste *et al.* 2010), where DNA parentage analysis has shown that the Argentine varieties CEREZA, TORRONTÉS RIOJANO, TORRONTÉS SANJUANINO and Moscatel Amarillo are all natural crosses between LISTÁN PRIETO (called Criolla Chica in South America) and Muscat of Alexandria, and that TORRONTÉS MENDOCINO is an offspring of Muscat of Alexandria and another unknown variety (Agüero *et al.* 2003; Myles *et al.* 2011).

A pink-berried mutation called **Flame Muscat** in California and **Red Hanepoot** in South Africa and a black-berried mutation called **Black Muscat of Alexandria** in the UK have been known since the nineteenth century or earlier. The latter has sometimes been confused with MUSCAT OF HAMBURG, a black-berried offspring of Muscat of Alexandria.

Muscat of Alexandria has been used to breed many table grapes and quite few wine grapes, including SYMPHONY.

OTHER HYPOTHESES

Like MUSCAT BLANC À PETITS GRAINS, Muscat of Alexandria has often been considered to be one of the famous *Vitis apianae* of the ancient Roman authors but there is no botanical evidence for such connections. According to Myles *et al.* (2011), Muscat of Alexandria is a progeny of Black Morocco and Muscat Blanc à Petits Grains (under the name Muscat Galego). This is inconsistent with the parentage discovered by Cipriani *et al.* (2010), unless Black Morocco were identical to Axina de Tres Bias, which is not impossible because this name has been used as a synonym for many different varieties, including Alphonse Lavallée, Augster Blau, Coarnă Rosie, Geisdutte Blau, Damaszener Blau, Maroc Gros, Marocain Noir (Crespan 2003).

According to Bronner (2003), the name Zibibbo comes from a certain 'Cape Zibbib' in South Africa. Although Muscat of Alexandria has been cultivated there since the early eighteenth century under the name Hanepoot, a South African etymology or origin can be rejected not only because Cape Zibibb does not exist (there's a Cape Zebib near Metline in Tunisia) but mainly because the name Zibibbo was mentioned much earlier in Italy.

VITICULTURAL CHARACTERISTICS

Mid budding, late ripening. Likes the heat. Big bunches of big berries. Well adapted to heat and drought and often bushvine trained, achieving high sugar levels in the right conditions. Susceptible to powdery mildew, botrytis bunch rot and insects; also to zinc deficiency.

WHERE IT'S GROWN AND WHAT ITS WINE TASTES LIKE

Muscat of Alexandria tends to make less subtle, stickier sweet wines than the small-berried Muscat and to display a certain marmalade or orange-flower aroma. The main terpenic compounds responsible for typical Muscat aromas in several Muscat (and other aromatic) varieties are linalol (characterized as lily of the valley) and geraniol (rose). Although other terpenes such as nerol (also rose) are significant, they have lower perception thresholds and are therefore less influential. Ribéreau-Gayon *et al.* (1975) analysed these compounds in various Muscat varieties and found that the average level of linalol was similar for both MUSCAT BLANC À PETITS GRAINS and Muscat of Alexandria but that the latter had a higher level of geraniol. The difference may contribute to the finer aromatic qualities of Muscat Blanc à Petits Grains.

France has a relatively modest total of Muscat d'Alexandrie, also known as Muscat Romain, 2,610 ha (6,449 acres) in 2009, about one third of the area planted to the generally superior MUSCAT BLANC À PETITS GRAINS, mostly in the Languedoc and Roussillon and especially in the Pyrénées-Orientales. Here it is used to a limited extent for dry table wines but predominantly for sweet, grapey and alcoholic vins doux naturels such as Muscat de Rivesaltes, which until 2008 had to comprise equal amounts of these two Muscats. This presumably explains the dominance of Muscat d'Alexandrie in Roussillon as recently as 1998.

In **Italy**, there's about twenty times as much Moscato d'Alessandria planted for the table as there is for wine. Most of the national total of 1,592 ha (3,934 acres) in 2000 was planted in the south and islands. The wine grapes are more often called by their Sicilian name, Zibibbo. Zibibbo, grown on the volcanic island of Pantelleria, closer to Tunisia than to mainland Italy, can produce luscious, sweet wines in three styles, from late-harvested grapes, dried berries or fortified. The best are golden or orange in colour, very sweet and intense, and can taste of ripe apricots. Recommended producers for sweet Passito di Pantelleria include Marco de Bartoli, Donnafugata and Pellegrino. On Sicilia, Marco de Bartoli also make an oak-aged sweet wine, Terzavia a barrel-fermented dry white.

In 2008, **Spain** had 9,894 ha (24,449 acres) of Moscatel de Alejandría, also known simply as Moscatel, and official statistics listed 40 ha (99 acres) of Moscatel de Málaga separately, even though they are the same variety. Some of this may be destined for the table but the rest goes into sweet, simple and sometimes rather sickly dessert wines such as Moscatel de Chipiona as well as fortified versions. Producers such as Telmo Rodriguez and Jorge Ordoñez, however, have shown that refined, delicate wines, both sweet and dry, can be made from the Moscatel grapes grown around the Andalusian city of Málaga, which once used to produce extremely dark, sticky wines given extended oak ageing – not that unlike the Liqueur Muscats of Australia.

In **Portugal**, the variety is known as Moscatel Graúdo or Moscatel de Setúbal, and there were 949 ha (2,345 acres) planted in 2010. Moscatel de Setúbal is also the name of the DOC south east of Lisboa famous for its fortified Moscatel, rendered extra grapey by macerating Moscatel skins in the finished wine for several months and becoming deep golden, spiced and raisiny after several years, occasionally up to twenty or more, in large old oak casks. Recommended producers in Setúbal include Bacalhôa Vinhos de Portugal, José Maria da Fonseca, Horácio Simões and Venâncio da Costa Lima. Unfortified versions are also produced in the region and elsewhere in Portugal. On the Portuguese island of Madeira it is known as Moscatel Graúdo or Moscatel de

Málaga, and there was little more than 1 ha (2.5 acres) in 2009, although one producer planted this variety in 2010.

In **Greece**, Muscat d'Alexandrie or Moschato Alexandrias is found mainly in the northern Aegean Islands, where there were 220 ha (544 acres) in 2008, a tiny area compared with that planted to MUSCAT BLANC (À PETITS GRAINS) but significant on the island of Límnos, whence it gets its synonym Moschato Limnou and where it is the grape of the Muscat of Límnos appellation for sweet wines. In **Turkey**, **Israel** and **Tunisia**, however, plantings of this Muscat far exceed those of MUSCAT BLANC À PETITS GRAINS, although much of the fruit is for fresh consumption.

California had 3,391 acres (1,372 ha) planted in 2010, almost all in the Central Valley and especially in Fresno. First planted in the mid 1800s, it was the main variety for the raisin business until the 1920s when seedless varieties became more popular. Today it is grown for the table, home winemaking and for dessert and sparkling wines, or is used to add floral aromas to a blend.

In **Australia**, where it is known as Muscat Gordo Blanco and occasionally Lexia, a peculiarly Australian contraction of Alexandria, there were 2,363 ha (5,839 acres) in 2008. Its importance is restricted to the hot inland irrigated vineyards, where it may be made into wine or dried as raisins, depending on the prevailing demands of the market. In periods of shortage of white wine it has been used to bulk out commercial blends.

In South America, the majority of **Chile**'s 5,000 or so hectares (12,355 acres) of 'Moscato', most of which is probably Moscatel de Alejandria, is distilled into the national spirit, pisco, but the area is so substantial that it is the country's third most planted pale-skinned grape, after the various varieties described as Sauvignon Something and CHARDONNAY. It is also found in **Argentina** (4,064 ha/10,042 acres in 2009), to a lesser extent in **Peru** (361 ha/892 acres of 'Moscatel', which includes Muscat of Alexandria, in 2010), mostly for pisco, as well as in **Uruguay**, **Colombia** and **Equador**.

In **South Africa**, there are currently more uprootings than plantings of Muscat of Alexandria, known there by its Afrikaans name, Hanepoot, and used chiefly for fortified wines and table grapes. It was planted on about 13% of the entire Cape vineyard area in the late 1970s, and there were still 2,346 ha (5,797 acres) of this variety in 2008, quite widely spread but with the greatest concentration in the hot inland Olifants River and Breedekloof regions. Varietal table wines are few and far between and generally lacklustre but the likes of Caltizdorp and Constantia Uitsig make superior fortified versions.

MUSCAT OF HAMBURG

Genetic conundrum associated most readily with table grapes grown intermittently in England and widely elsewhere.

BERRY COLOUR:

PRINCIPAL SYNONYMS: Black Muscat (Australia, Cyprus, England, US), Black Muscat of Alexandria (England), Hamburg Musqué (France), Hamburgii Muskotály (Hungary), Malvasia Nera (Liguria in Italy), Moscatel de Hamburgo (Brazil, Spain, Uruguay), Moscatel Prato (Portugal), Moscato d'Amburgo (Italy), Moscato Nero (Italy), Moscato Nero d'Acqui (Italy), Moschato Amvourgou (Greece), Moschato Tyrnavou (Greece), Muscat Albertdient's (Belgium), Muscat de Hamburg (France), Muscat Hamburg (US), Muscat-Hamburgh (England), Muscat Hamburg Crni (Croatia), Muskat-Trollinger (Germany), Oeillade Musquée (Gard in France), Snow's Muscat Hamburgh (England), Tămîioasă Hamburg (Romania), Tămîioasă Neagră (Romania), Venn's Seedling (England), Venn's Seedling Black Muscat (England), Zibibbo Nero (southern Italy)

VARIETIES COMMONLY MISTAKEN FOR MUSCAT OF HAMBURG: ALEATICO, MOSCHOMAVRO

ORIGINS AND PARENTAGE

The origin of Muscat of Hamburg is unclear. According to a portrait published in the *London*

Florist (Smith 1858), 'this fine grape is a seedling, raised at Wrest Park, Bedfordshire, by Mr Seward Snow, who ... informs us that this grape originated by fertilizing flowers of the Black Hamburgh vine with those of the White Muscat of Alexandria'. In England, Black Hamburgh or Black Hamburg was the name given to SCHIAVA GROSSA, and White Muscat corresponded to MUSCAT OF ALEXANDRIA. The grape was initially named Snow's Muscat Hamburgh, combining the parents' and breeder's names, at the meeting of the British Pomological Society: 'Mr Snow having expressed a wish that the Society should name his new Grape, Mr Hogg proposed that it should be called Snow's Muscat Hamburgh, which was approved of' (Anonymous 1856).

The parentage of Muscat of Hamburg was also surmised by twentieth-century ampelographers such as Pirovano and Levadoux and was recently confirmed by DNA analysis (Crespan 2003): it is a natural cross between MUSCAT OF ALEXANDRIA and SCHIAVA GROSSA. This makes it a sibling of MALVASIA DEL LAZIO in central Italy, which tends to suggest that both Malvasia del Lazio and Muscat of Hamburg are genetically Italian, even if the latter was bred in England. Muscat of Hamburg is therefore a grandchild of MUSCAT BLANC À PETITS GRAINS and Axina de Tres Bias (see the pedigree diagram in the MUSCAT entry) and also a half-sibling of the other progenies of Muscat of Alexandria (MOSCATELLO SELVATICO and GRILLO in Italy; CEREZA, Moscatel Amarillo, TORRONTÉS MENDOCINO, TORRONTÉS RIOJANO and TORRONTÉS SANJUANINO in Argentina) and of Schiava Grossa (BUKETTRAUBE, KERNER, Madeleine Royale, Othello, ROTBERGER and UVA TOSCA – see the pedigree diagram in the Schiava Grossa entry). Muscat of Hamburg was used to breed BEICHUN, CANADA MUSCAT, COMPLEXA, EARLY MUSCAT, FIOLETOVY RANNY, GOLDEN MUSCAT, ITALIA, MANZONI MOSCATO, MUSCAT BAILEY A and NEW YORK MUSCAT.

OTHER HYPOTHESES

According to Salomon and Salomon (1902a), Muscat of Hamburg is an old table-grape variety that was cultivated in England under the name Black Muscat of Alexandria and disappeared until the 1850s, when it was rediscovered by Seward Snow, gardener to Earl De Grey. Although this is confirmed in Snow's obituary – 'Mr Snow reintroduced, under the name of Muscat Hamburgh, the fine Black Muscat of Alexandria Grape, which had almost grown out of cultivation' (Hogg *et al.* 1868) – it is not consistent with the fact that Black Muscat of Alexandria was mentioned as early as Forsyth (1802), some time before Snow accurately reported the parentage of Muscat of Hamburg (see above). It is much more likely that Black Muscat of Alexandria was the name given to a black mutation of MUSCAT OF ALEXANDRIA, as in Kenrick (1844), who says 'It resembles the White, except in regard to color', and that the same name was later given also to Muscat of Hamburg. Indeed, both were listed separately in Hogg's (1860) *The fruit manual* and they were not treated as synonyms until a few years later (Wills 1867).

According to Galet (2000), the origin of Muscat of Hamburg is unknown and it could have been created in Hamburg for glasshouse cultivation, and later introduced to England. However, this is doubtful because the word Hamburg appears in this variety's name only because it is a progeny of Black Hamburg, an old synonym of SCHIAVA GROSSA.

VITICULTURAL CHARACTERISTICS

Mid ripening though non-human animals may get there first.

WHERE IT'S GROWN AND WHAT ITS WINE TASTES LIKE

Muscat of Hamburg is far more commonly planted for the table than for wine thanks to its plump, shiny berries and its ability to withstand long journeys. It is particularly widely planted for this purpose in **France**, **Eastern Europe**, **Greece** and **Australia** and was very popular in greenhouses in Victorian England. However, it is also used to produce light, grapey wines in Eastern Europe and dessert wines elsewhere. In **California**, for example, where there were 251 acres (102 ha) of Muscat Hamburg (or Black Muscat) in 2009, it is grown almost exclusively for wine. Quady's Elysium is a dessert version with a light rose aroma that develops only when the grapes are very ripe. It is also used to make sweet wines in several European wine regions, including Thessalía in central mainland Greece, on **Cyprus** and in central **Italy**.

MUSCAT OLIVER

See IRSAI OLIVÉR.

MUSCAT OTTONEL

Soft, grapey French variety used for both dry and sweet wines, particularly in Alsace and Eastern Europe.

BERRY COLOUR: ● ● ● ● ●

PRINCIPAL SYNONYMS: Muscadel Ottonel (South Africa), Muskat Otonel (Bulgaria), Muskat Ottonel (Germany, Austria, Slovenia), Muskotály or Ottonel Muskotály (Hungary), Tămîioasă Ottonel (Romania)

ORIGINS AND PARENTAGE

Muscat Ottonel is a seedling obtained in Angers in the Val de Loire, France, by the vine breeder Jean-Pierre Vibert in 1839, and later released in 1852 by his chief gardener, Robert. It was named in honour of a certain H Ottonel. Vibert did not keep track of the parents, and the seedling was thought to be a CHASSELAS × Muscat de Saumur cross (Galet 2000) until DNA profiling at INRA (Institut National de la Recherche Agronomique) in Montpellier, southern France, corrected this to a Chasselas × Muscat d'Eisenstadt cross. It was used to breed MÁTRAI MUSKOTÁLY and MUŠKÁT MORAVSKÝ.

VITICULTURAL CHARACTERISTICS

Low vigour, early ripening. Well suited to clay-limestone soils. Susceptible to coulure, downy mildew and botrytis bunch rot.

WHERE IT'S GROWN AND WHAT ITS WINE TASTES LIKE

The berries are able to accumulate high sugar levels and produce aromatic wines, both dry and sweet, although acidity levels are generally low unless the grapes are picked very early, in which case they are much less aromatic. Ottonel is the palest of the Muscats, in terms of colour as well as character, but the fact that it ripens earlier than the more aromatic MUSCAT BLANC À PETITS GRAINS and MUSCAT OF ALEXANDRIA gives it appeal in cooler growing regions. It is the most important of the Muscat Something varieties in **France**, mainly in Alsace for exuberantly grapey wines labelled Muscat d'Alsace that finish dry or off dry. Total French plantings in 2008 were 165 ha (408 acres).

In **Switzerland** there are just 4 ha (10 acres), mainly for light, dry wines, but **Austria** had 472 ha (1,166 acres) of Muskat Ottonel in 2007 (up from 418 ha/1,033 acres in 1999), mostly in Neusiedlersee and Neusiedlersee-Hügelland for sweet wines, though there are 110 ha (272 acres) distributed in rather small pockets in Niederösterreich (Lower Austria), the regions on either side of the Donau (Danube).

In **Hungary**, the 1,232 ha (3,044 acres) of Ottonel Muskotály in 2008 were grown throughout the country except in Somló and Sopron in the north west, but mostly in Mátra and Eger, north east of Budapest, and Kunság further south. It is popular as a blending partner and for sweet wines but Szeremley in Badacsony and Mátyás Szőke in Mátra make fine dry or off-dry varietal wines.

Romania's Muscat Ottonel (2,809 ha/6,914 acres in 2008) produces some very good sweet wines and the biggest vineyard areas are in the counties of Dobregea by the Black Sea, Vrancea in eastern Muntenia and in Transilvania in the centre of the country. Nearly twice as much Muskat Otonel (sic) was planted in **Bulgaria** – 5,573 ha (13,771 acres) in 2009 – and much of its produce is labelled Misket.

There are minor plantings in Krasnodar in **Russia** (34 ha/84 acres in 2009), the **Czech Republic** (59 ha/146 acres in 2008) and more significant ones in the former Soviet Republics. The **Republic of Moldova** had a significant 1,620 ha (4,003 acres) in 2009.

Further afield, the variety is also found in **South Africa**, **New York** State and British Columbia in **Canada**.

MUSCAT SWENSON

Muscat-like hardy American hybrid.

BERRY COLOUR: ● ● ● ● ●

PRINCIPAL SYNONYMS: ES 8-2-43, Muscat de Swenson

ORIGINS AND PARENTAGE

A *Vitis riparia* × Varousset hybrid bred by Elmer Swenson in Osceola, Wisconsin, US, in which:

- Varousset (or Seyve-Villard 23-657) is a Seibel 4668 × Subéreux hybrid (see PRIOR for the pedigree of Subéreux)
- Seibel 4668 is a hybrid of unknown parents, obtained by Albert Seibel in Aubenas in the Ardèche, south-eastern France.

VITICULTURAL CHARACTERISTICS

Productive and ripens fairly late. Cold-hardy (–30 °F/–34 °C) and good general disease resistance though some anthracnose and berry splitting has been observed. Susceptible to foliar phylloxera.

WHERE IT'S GROWN AND WHAT ITS WINE TASTES LIKE

Planting is beginning in Québec, **Canada**, where Clos St-Ignace, for example, make a sweet, fortified varietal wine, and in **New England**. Wines have a grapey, Muscat-like character when harvested fully ripe but may be more neutral if harvested early.

MUSKADEL

See MUSCAT BLANC À PETITS GRAINS.

MUŠKÁT MORAVSKÝ

The Czech Republic's very own Muscat. Widely planted.

BERRY COLOUR: ● ● ● ● ●

PRINCIPAL SYNONYMS: Mopr, Moravian Muscat, Moravsky Muskat

ORIGINS AND PARENTAGE

Muškát Moravský is a MUSCAT OTTONEL × Prachttraube cross obtained by V Křivánek at the research centre in Polešovice in Morava in the south east of the Czech Republic, in which Prachttraube is a Madeleine Royale × Boskokwi cross, and Boskokwi or Boskokvi a variety, no longer in cultivation, from the Greek island of Zákynthos. (See MÜLLER-THURGAU for details of Madeleine Royale.) It was bred in order to have a locally adapted Muscat-like variety and initially included in the offical Czech register of varieties in 1987 under the name Mopr, a combination of the initial letters of its parents, before being renamed Muškát Moravský in 1993.

VITICULTURAL CHARACTERISTICS

Vigorous, early ripening and best suited to clay soils. Prone to birds, wasps and botrytis bunch rot.

WHERE IT'S GROWN AND WHAT ITS WINE TASTES LIKE

Muškát Moravský is the most widely planted of the recently bred Czech varieties, with 358 ha (885 acres) in the **Czech Republic**, predominantly in Morava in the south east, in 2009. For dry, floral and Muscat-flavoured wines, grapes must be picked early enough for sufficient acidity. Kabinett and late-harvest styles can also be found. Recommended producers include Vladimír Tetur.

MUSKATELLER

See MUSCAT BLANC À PETITS GRAINS.

MUSKOTÁLY

See MUSCAT OTTONEL.

MUSTOASĂ DE MĂDERAT

Productive, high-acid variety from western Romania making light, fresh, locally popular whites.

BERRY COLOUR: ● ● ● ● ●

PRINCIPAL SYNONYMS: Lampor? (Hungary), Mustafer, Mustosfehér

VARIETIES COMMONLY MISTAKEN FOR MUSTOASĂ DE MĂDERAT: FRÂNCUŞĂ (Romania)

ORIGINS AND PARENTAGE

The exact origin of Mustoasă de Măderat is unknown but its name suggests that it was selected a few centuries ago in the vineyards of Măderat in the Miniş area in western Romania (Dejeu 2004). Its relationship to FRÂNCUŞĂ, also called Mustoasă de Moldova, is unknown but their ampelographic descriptions are quite similar (Dejeu 2004).

VITICULTURAL CHARACTERISTICS

Vigorous, productive, mid budding, late ripening. Compact bunches of brown-speckled berries. Good resistance to botrytis bunch rot, drought and the European red mite, moderate resistance to downy and powdery mildews. Winter-hardy to –18 °C/–0.4 °F.

WHERE IT'S GROWN AND WHAT ITS WINE TASTES LIKE

The warm Miniş Măderat wine region in western **Romania**, close to the border with Hungary, is generally better suited to the red wine varieties that make up 80% of the vineyard but Mustoasă de Măderat's high acidity keeps it fresh. Wines produced tend to be simple, but good-quality aromatic whites can be made if yields are controlled. Wine Princess make a typically light, citrus-flavoured, delicately aromatic, even a touch floral varietal. The grape is also suited to the production of sparkling wines and sometimes used for brandy. There were 486 ha (1,201 acres) in Romania in 2008, virtually all in the far west.

N

NARINCE
NASCETTA
NASCO
NEBBIERA
NEBBIOLO
NEBBIOLO ROSÉ
NEGOSKA
NEGRAMOLL
NEGRARA TRENTINA
NEGRARA VERONESE
NÉGRET DE BANHARS
NÉGRETTE
NEGRETTO
NEGROAMARO
NEGRU DE DRĂGĂŞANI
NEHELESCHOL
NEOPLANTA
NER D'ALA
NERELLO CAPPUCCIO
NERELLO MASCALESE
NERET DI SAINT-VINCENT
NERETTA CUNEESE
NERETTO DI BAIRO
NERETTO DURO
NERETTO GENTILE
NERETTO NOSTRANO
NERKARAT
NERKENI
NERO BUONO DI CORI
NERO D'AVOLA
NERO DI TROIA
NERONET
NEUBURGER
NEW YORK MUSCAT
NIAGARA
NIEDDERA
NIGRA
NIGRIKIOTIKO
NINČUŠA
NITRANKA
NOAH
NOBLE
NOBLING
NOCERA
NOIR FLEURIEN
NOIRET
NORIA
NORTON
NOSIOLA
NOTARDOMENICO
NOUVELLE
NOVAC
NURAGUS

ℵ this symbol on the following pages indicates that the synonyms, and mistaken identities, are confirmed by DNA evidence

NARINCE

Turkey's best-known white wine variety, also grown for the table.

BERRY COLOUR:

PRINCIPAL SYNONYMS: Güzül Üzüm, Kazova, Narance, Nerince

ORIGINS AND PARENTAGE

Narince, literally meaning 'delicate' or 'fragile', comes from the Tokat region in northern central Turkey, south of the Black Sea coast.

VITICULTURAL CHARACTERISTICS

Large bunches of large, thin-skinned yellowish-green berries, late ripening. Susceptible to botrytis bunch rot. Well suited to permeable soils and dry conditions.

WHERE IT'S GROWN AND WHAT ITS WINE TASTES LIKE

Narince is one of the most important white wine varieties in **Turkey** although it is also widely planted for the table. Grown mainly on sandy and gravelly soils along the Yeşilırmak River in the province of Tokat, at altitudes of 600–700 m above sea level thanks to the Pontic mountains that separate northern Turkey from the Black Sea coast, it is also found in the neighbouring provinces of Çorum and Amasya to the west and further south in Sivas. There were 814 ha (2,100 acres) planted in Turkey in 2010, making it the most widely planted white wine grape.

Wines – both dry and off dry and often oaked – are typically fresh and fruity but can be quite full-bodied, with flavours ranging from orange and grapefruit through floral to apple and lime. They are probably the only Turkish whites with the capacity to age and they blend well with lighter and more piercing wines made from EMIR, as well as with international varieties such as CHARDONNAY or SÉMILLON. Producers include Doluca, Kavaklidere, Kocabağ and Turasan.

NASCETTA

Renascent semi-aromatic Piemontese made increasingly as a varietal wine.

BERRY COLOUR:

PRINCIPAL SYNONYMS: Anascetta, Nas-cëtta (in Novello in the province of Cuneo)

VARIETIES COMMONLY MISTAKEN FOR NASCETTA: NASCO, VERMENTINO

ORIGINS AND PARENTAGE

Nascetta is a rare variety from the Langhe area of Piemonte in north-west Italy and was mentioned in the mid nineteenth century as a grape suited to making quality wine. It used to be widespread between Alba and Mondovì, usually blended with VERMENTINO (Favorita) and Moscato Bianco (MUSCAT BLANC À PETITS GRAINS).

OTHER HYPOTHESES

Some writers have suggested a genetic link with VERMENTINO or with NASCO from Sardegna.

VITICULTURAL CHARACTERISTICS

Early to mid ripening.

WHERE IT'S GROWN AND WHAT ITS WINE TASTES LIKE

Nascetta was almost extinct when Elvio Cogno made the first micro-production of Nascetta in 1994. Today it is planted around the commune of Novello in the province of Cuneo, in Piemonte, **Italy**, and to a lesser extent in the neighbouring commune of Monforte d'Alba and the more distant Trezzo Tinella (Dellavalle *et al.* 2005). In the past, this semi-aromatic variety was generally blended but varietal wines are now authorized within the Langhe DOC and made by producers such as Le Strette (labelled Anas-Cëtta), La Tribuleira and Rivetto as well as by Elvio Cogno, whose example of Anas-Cëtta has aromas of acacia flowers, exotic fruits, honey, sage, rosemary and citrus. Wines are well structured with fresh acidity and amenable to barrel ageing and often have the flavour of acacia honey.

NASCO

Possibly ancient light-berried variety making soft white, mostly dessert, wines on Sardegna.

BERRY COLOUR: ● ● ● ● ●

PRINCIPAL SYNONYMS: Nascu, Nusco

VARIETIES COMMONLY MISTAKEN FOR NASCO: NASCETTA⌘ (Piemonte)

ORIGINS AND PARENTAGE

The name Nasco could derive from the characteristic aroma of musk (*nuscu* in Sardinian dialect) found in overripe berries or aged wines of this variety. As with many Sardinian varieties, it has been suggested that Nasco was introduced from Spain, but as yet there is no genetic evidence to back this up.

VITICULTURAL CHARACTERISTICS

Low yields and little resistance to parasites. Mid ripening.

WHERE IT'S GROWN AND WHAT ITS WINE TASTES LIKE

Nasco is cultivated mainly in the provinces of Cagliari in the south of Sardegna, **Italy**, and Oristano in the west, and to a lesser extent around Alghero in the province of Sassari and in Mandrolisai in the province of Nuoro in the centre of the island. The Italian agricultural census recorded 171 ha (423 acres) in 2000.

Varietal wines that qualify as Nasco di Cagliari DOC can come from almost the whole of the southern part of the island and may be dry, sweet or fortified, and the fortified wines may be dry or sweet. The sweet wines have aromas of honey and dried fruits, apricot and spice and the sweetness is well balanced by the acidity. Argiolas' Angialis, classified as IGT Isola dei Nuraghi, made predominantly from Nasco, with a little added Malvasia di Cagliari (MALVASIA DI LIPARI), is among the best dessert wines in Italy. Other producers include Picciau and Meigamma.

NEBBIERA

Very minor northern Italian dark-skinned cross whose true parentage has only recently been revealed.

BERRY COLOUR: ● ● ● ● ●

PRINCIPAL SYNONYMS: Incrocio Dalmasso XV/29

ORIGINS AND PARENTAGE

Nebbiera is a cross created by Italian vine breeder Giovanni Dalmasso in 1938 at the Conegliano research centre in the Veneto. He himself claimed that it was a NEBBIOLO × BARBERA cross and it entered the Italian register of varieties in 1977. The objective was to develop a new variety that combined the quality of Nebbiolo and the resistance and the naturally high yields of Barbera. However, DNA profiling has recently shown that the Nebbiolo used by Dalmasso was in fact Nebbiolo di Dronero, which is not Nebbiolo but a synonym for CHATUS (Torello Marinoni, Raimondi, Mannini and Rolle 2009). Nebbiera is therefore a Chatus × Barbera cross, just like its siblings ALBAROSSA, CORNAREA, SAN MICHELE and SOPERGA.

VITICULTURAL CHARACTERISTICS

Mid to late ripening.

WHERE IT'S GROWN

There are a few plantings in Piemonte in north-west **Italy** but the total area recorded in the Italian agricultural census of 2000 was just 22 ha (54 acres).

NEBBIOLO

Piemonte's, arguably Italy's, most revered wine grape; very old variety capable of producing perfumed, expressive, ageworthy wines of great beauty.

BERRY COLOUR: ● ● ● ● ●

PRINCIPAL SYNONYMS: Chiavennasca⌘ (Valtellina), Picotendro or Picoutener (Valle d'Aosta, Ivrea), Picotèner (Carema), Prunent or Prünent (Val d'Ossola, near Lago Maggiore), Spanna⌘ (Novara and Vercelli areas)

VARIETIES COMMONLY MISTAKEN FOR NEBBIOLO: CHATUS⌘ (called Nebbiolo di Dronero in the province of Cuneo), CROATINA⌘ (called Nebbiolo di Gattinara or Spanna-Nebbiolo in the provinces of Novara and Vercelli), DOLCETTO⌘ (called Nebbiolo in the province of Tortona and in the Oltrepò Pavese), NEBBIOLO ROSÉ⌘ (in the Alba area of Cuneo province)

ORIGINS AND PARENTAGE

While the vast majority of sources give 1268 as the earliest mention of Nebbiolo, the original reference to *nibiol* is in fact in a document in the *Archivio di Stato di Torino*, kindly provided by its director, Marco Carassi, who identified it as *Camerale Piemonte, Articolo 65, paragrafo 1, Conti della castellania di Rivoli, mazzo 1, rotolo n. 1*, dated 1266. Details of this document are given by the historian Cibrario (1833): in Rivoli, Conto d'Umberto de Balma, castellan of Rivoli, obtained 626 *sextarius* (*c.* 338 litres) of wine, including 306 from the grape named Nibiol ('vini receptis de exitu vincarum de nibiol hoc anno'; Archivo di Stato di Torino; Marco Carassi, personal communication).

Several subsequent mentions in the 13th and 14th centuries confirm Nebbiolo as one of the oldest and most widespread varieties of Piemonte: 1292 in Alba (*filagnos de vitibus neblorii*), 1295 in Camerano Casasco (*nebiolo*), 1303 in Canale (*vino nebiolio*) and in Moncalieri, 1304 in Asti (*nebiolus*), 1311 in Chieri, 1324-5 in Moretta (*nubliolio*), 1327–8 in Almese (*vinum nibiolii*), 1328–9 in Bricherasio (*vites de nebiolio*) and 1380 in Pinerolo (Comba and Dal Verme 1990).

The most important synonyms of Nebbiolo also appear early in the literature: Prünent (Val d'Ossola) in 1309, Chiavennasca (Valtellina) in 1595 (Bongiolatti 1996), Spanna (around Novara) in 1466 and Picotendro (Valle d'Aosta) in the nineteenth century.

The name Nebbiolo derives from *nebbia*, Italian for 'fog', probably referring to the thick, natural bloom covering the ripe berries, as if they were covered in a layer of fog. This etymology is more convincing than the alternative suggestion that the name refers to the fog that covers the Piemonte hills when these late-ripening grapes are harvested. Grape names are more likely to refer to the characteristics of the plant than to the local climate. The same etymology applies to the synonym Prünent, more probably derived from *pruina* ('bloom') than from *prugna* ('plum') although some authors suggest that it's a reference to the tradition of cultivating grapes together with plum trees. Chiavennasca derives from the city of Chiavenna in the province of Sondrio in Lombardia.

Centuries of clonal propagation have resulted in several distinct clonal selections of Nebbiolo. Four main types are usually distinguished:

- Nebbiolo Lampia, the most widespread
- Nebbiolo Michet, a virused form of Lampia (morphology and production affected by grapevine fanleaf virus or GFLV)
- NEBBIOLO ROSÉ, morphologically rather different
- Nebbiolo Bolla, once widespread and now decreasing in importance.

DNA profiling has shown that the types Lampia, Michet and Bolla have identical DNA profiles and belong to the same grape variety but, surprisingly, that NEBBIOLO ROSÉ has a different profile (Schneider, Boccacci *et al.* 2004). Nebbiolo Rosé must be considered as a separate variety and no longer as a clone of Nebbiolo.

DNA researchers have tracked down Nebbiolo's origin, but since it is a very old variety, its parents are likely to be extinct. However, DNA parentage analysis has revealed at least eight parent–offspring relationships (see pedigree diagram overleaf) between Nebbiolo and local varieties from Piemonte and Valtellina (Schneider, Boccacci *et al.* 2004). Northwards from southern Piemonte to Valtellina:

- BUBBIERASCO (Saluzzo region, province of Cuneo) is a natural cross between Nebbiolo and Bianchetta di Saluzzo, another old and no longer cultivated white variety from the same area (Schneider *et al.* 2005–6), which confirms that Nebbiolo was formerly widely cultivated in Cuneo.
- FREISA is widespread across Piemonte, where it is sometimes called Spannina, confirming its offspring relationship to Nebbiolo, which is sometimes called Spanna. DNA profiling has shown that Freisa could have a close genetic relationship with VIOGNIER from the northern Rhône in France, so that Nebbiolo and Viognier are likely to be cousins (Schneider, Boccacci *et al.* 2004).
- NERETTO DI BAIRO, also called Neretto di San Giorgio, and VESPOLINA are minor varieties of northern Piemonte and the Oltrepò Pavese, where Nebbiolo has been known since the Middle Ages.
- NEBBIOLO ROSÉ is traditionally called Chiavennaschino in Valtellina, where it may have originated. The parent–offspring relationship with Nebbiolo explains why Nebbiolo Rosé has generally been considered to be a clone of Nebbiolo.

- Negrera, ROSSOLA NERA and BRUGNOLA (the last often confused with FORTANA) are old and minor varieties from Valtellina, confirming the historical presence of Nebbiolo in this area, where it is called Chiavennasca. (Negrera is also, confusingly, a synonym for JUAN GARCÍA.)

With the exception of BUBBIERASCO, the missing parents of these varieties are likely to have disappeared. It is therefore possible that Nebbiolo is the offspring of one of these varieties, although Nebbiolo's long history makes it more likely to be the parent.

Based on these recent DNA findings, Nebbiolo could come from either Piemonte or Valtellina since it had four progenies in both regions.

Nebbiolo was used to breed RUBIN.

OTHER HYPOTHESES

Nebbiolo has been tentatively identified with Spionia, a grape described by Pliny the Elder but this hypothesis can be rejected because Spionia was cultivated in the province of Ferrara in Emilia-Romagna – as well as because there are no grounds for suggesting botanical identity between varieties known in Roman times and modern varieties that are similar in name.

VITICULTURAL CHARACTERISTICS

Early budding, very late ripening, vigorous and very fussy about soils, preferring calcareous marls such as those to the north and south of Alba on the right bank of the Tanaro in Piemonte.

WHERE IT'S GROWN AND WHAT ITS WINE TASTES LIKE

Nebbiolo is intensively cultivated in southern central and north Piemonte (Langhe, Roero, Asti, Carema, Biella, Novara and around Vercelli) in north-west **Italy** although it constituted less than 8% of the Piemonte vineyard in 2001, according to official local statistics. Salaris (2005–6) notes 338 ha (835 acres) in northern Piemonte (provinces of Novara, Vercelli, Biella, Torino and Asti) in 2004 and 3,375 ha (8,340 acres) in the province of Cuneo. It is also grown in the lower part of the Valle d'Aosta (just 26 ha/64 acres in 2004), around Donnas, and in Valtellina and Franciacorta, both in Lombardia (908 ha/2,244 acres). It is even grown in the region of Gallura on the island of Sardegna (52 ha/128 acres in 2004). The Italian agricultural census recorded 4,886 ha (12,074 acres) in 2000, about three quarters of that total in Piemonte.

Because it ripens so late, and because of the quality (and market value) of its wines, Nebbiolo is given the best hillside sites, usually facing south or south west. It's fussy when it comes to soils, too, and produces the finest wines when planted on calcareous marls to the north and south of Alba on the right bank of the Tanaro in the DOCG zones of Barbaresco and Barolo respectively. Nebbiolo is rivalled only by PINOT NOIR in its ability to express the subtleties of different terroirs, which is why these two zones have been so assiduously divided into named crus, but in very general terms unadulterated Nebbiolo tends to be light in colour, turning orange with bottle age rather faster than most other red wine varieties, to be high in both acid and (especially) tannin, and to exhibit a haunting array of aromas which might include tar, cordite, leaf mould, dried cherries, liquorice, violets and roses.

Top producers include Elio Altare, Enzo Boglietti, Cappellano, Ceretto, Michele Chiarlo, Cigliuti, Aldo Conterno, Giacomo Conterno, Gaja, Bruno Giacosa, Elio Grasso, Marcarini, Marchesi di Grésy, Bartolo Mascarello, Giuseppe Mascarello, Massolino, Armando Parusso, Pio Cesare, Produttori del Barbaresco, Fratelli Revello, Giuseppe Rinaldi, Bruno Rocca, Luciano Sandrone, Paolo Scavino, Vajra, Vietti and Roberto Voerzio.

In these DOCGs, it comprises 100% of the wine, whereas in Gattinara, Ghemme and Roero (see Matteo Correggia's wines, for example), the minimum requirement varies between 75% and 95%. The DOCG regulations refer to the subvarieties Michet, Lampia and Rosé but one wonders if these should be amended now that it has been shown that NEBBIOLO ROSÉ is a different variety (see above). Varietal (or at least 85% Nebbiolo) wines are also permitted in the Piemonte DOCs of Albugnano, Canavese, Carema, Colline Novaresi, Coste della Sesia, Langhe and Nebbiolo d'Alba and there's a host of other DOCs in Piemonte such as Bramaterra where Nebbiolo is the

NEBBIOLO PEDIGREE DIAGRAM

NEBBIOLO is so old (first mentioned in 1266) that its parents are probably extinct, and these eight natural crosses illustrate its significant impact on the grape germplasms of its traditional growing regions in Italy, Piemonte and Valtellina. However, when unknown (and probably extinct) varieties (?) are involved, the inverse relationships are also theoretically possible (see p x).

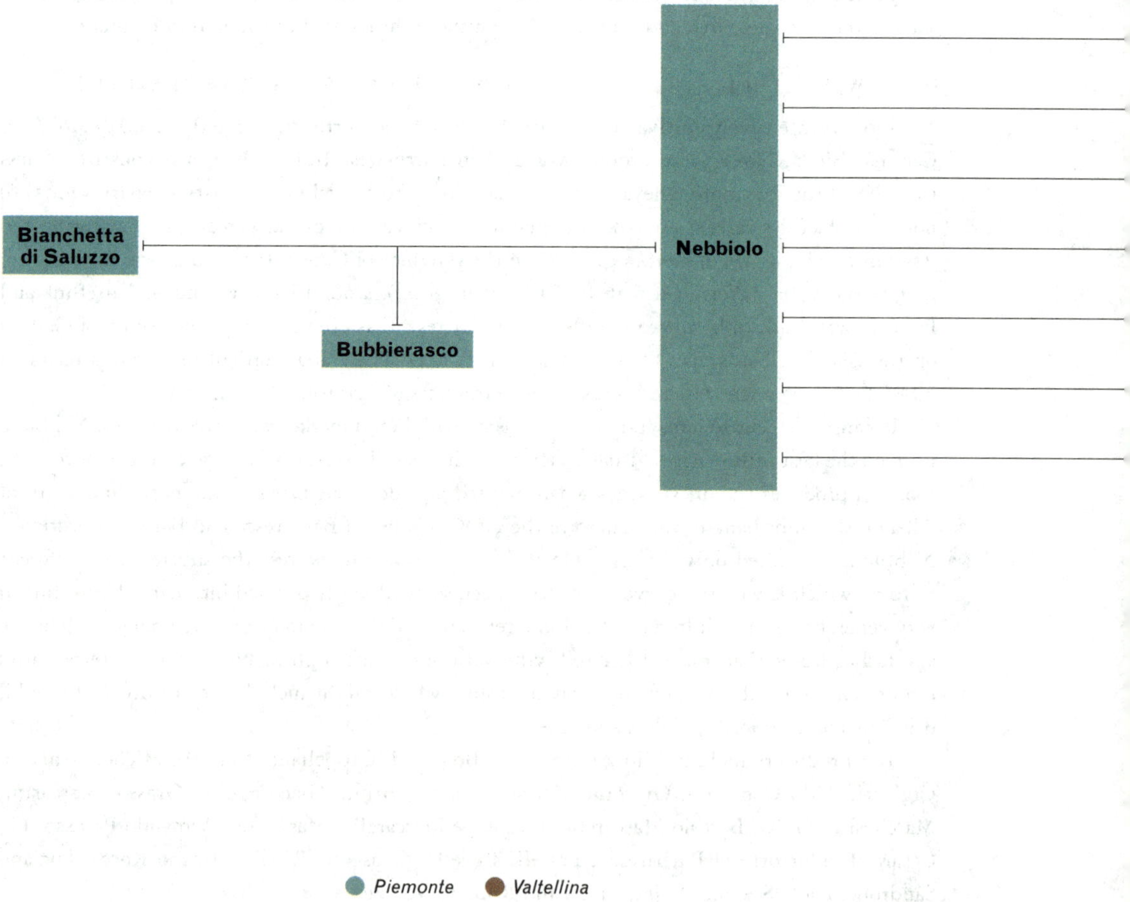

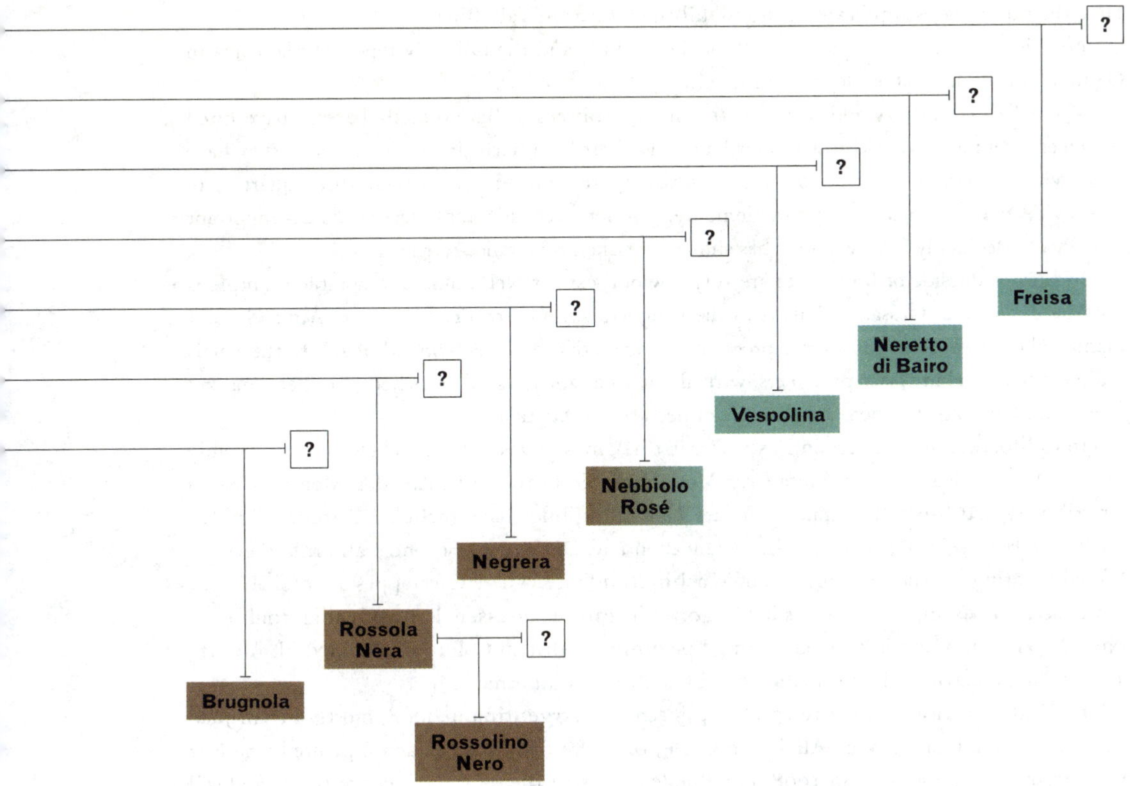
?
?
?
?
?
?
?
Freisa
Neretto di Bairo
Vespolina
Nebbiolo Rosé
Negrera
Rossola Nera
?
Brugnola
Rossolino Nero

most important or at least a significant component. However, producers insist upon local synonyms such as Spanna and use them on labels. Blending partners include VESPOLINA, BONARDA PIEMONTESE and CROATINA, depending on local regulations.

Spanna's home territory is in the north of Piemonte, in the hills above Novara and Vercelli, and wines produced there on glacial soils tend to lack the intensity of a Barolo but not its perfume. Historically more Nebbiolo was planted here than anywhere else but the area declined during Italy's industrial revolution in the 1950s. Nevertheless, good wines are once again being produced in varying soil types in the hills on the left and right banks of the Sesia River: in the province of Novara, the most important DOCs are Boca, Sizzano, Fara plus Ghemme DOCG; and in the Vercelli hills, Lessona, Bramaterra plus Gattinara DOCG. Superior producers include Antoniolo and both Valloni and Vallana.

In the far north of Lombardia, Nebbiolo – known as Chiavennasca – produces a medium-bodied wine but struggles in all but the warmest vintages to ripen sufficiently to produce the fruit flavours that will balance the variety's marked tannins and acidity. Here it is the dominant variety in Valtellina Superiore DOCG and Valtellina DOC as well as in the distinctive passito-style Sforzato di Valtellina. Beyond this narrow northerly valley, in the rest of Lombardia Nebbiolo features only as part of the Franciacorta blend and then to a maximum of just 10%. Top producers in Valtellina include Ar Pe Pe and Nino Negri.

Northern Italy's only other outpost of Nebbiolo, in the guise of Picotendro or Picoutener, is the Valle d'Aosta, in Donnas (just over the regional border from Carema) and Arnad-Montjovet. Here the variety faces similar obstacles to full ripening as in Valtellina.

Nebbiolo is pretty rare in the south of the country and the officially reported plantings in Gallura on the island of Sardegna are more likely DOLCETTO.

The Siren-like beauty and power of the finest examples of Barolo or Barbaresco have lured producers around the world to try their hand and their land with this variety, just as they have done with PINOT NOIR but so far on a much more limited scale: about three quarters of worldwide Nebbiolo plantings are in Piemonte. It is not a very adaptable variety, its all-important fragrance is decidedly fragile, and it has shown a marked reluctance to travel.

In Europe outside of Italy there are very few outposts of Nebbiolo. One notable example is the Mas de Daumas Gassac at Aniane in the Languedoc, southern **France**, where Aimé Guibert planted about twenty-five rows, the product of which goes into their top red blend. In the Valais region of **Switzerland**, two producers, with about 1 ha (2.5 acres) of vines between them, market a varietal Nebbiolo and there are a couple of hectares in **Austria**.

In **California**, there are around 150 acres (61 ha), most of those planted before 2000, notably in such AVAs as Paso Robles, Santa Cruz Mountains, Sierra Foothills, Amador, Monterey, Santa Ynez Valley and Santa Barbara. Passionate Nebbiolo-philes have included Palmina, Randall Grahm at Bonny Doon, Ahlgren, Jim Clendenen and Domenico. Beyond California there are limted plantings in the drier regions of **Washington** State, where Cavatappi's are probably the most successful examples, as well as in **Oregon**, **Virginia**, **Tennessee**, **Pennsylvania**, **Idaho** and **New Mexico**. In **Canada** there are limited plantings in British Columbia. **Mexico**'s L A Cetto also produce a particularly full-bodied Nebbiolo in Baja California.

In South America, there were 176 ha (435 acres) in **Argentina** in 2008, mostly in San Juan but also in Mendoza (eg Viña Alicia). In **Chile**, there are just 9 ha (22 acres) planted (the first vines went into the ground in 1998) but Botalcura is still the only producer to bottle and sell a Nebbiolo.

In 2007, **South Africa** recorded 18 ha (44 acres) of Nebbiolo.

There's a small but growing wave of interest in **Australia**, where it was first planted in the Hunter Valley in the early 1980s and then more successfully by Pizzini in the King Valley in Victoria in the late 1980s, although it was not until 1990 that the first wine was produced by Brown Brothers, also in the King Valley. These two have been followed by Garry Crittenden and by isolated producers in the Adelaide Hills, Langhorne Creek, Mudgee and even as far north as Queensland. It's seems that regions with long cool growing seasons and mature vineyards are most likely to succeed. In 2008 total plantings had reached 106 ha (262 acres) and, according to the

Australia and New Zealand Wine Industry Directory 2009, Nebbiolo has so far seduced eighty-four producers.

There are also very limited plantings in **New Zealand**.

NEBBIOLO ROSÉ

Piemontese similar to but genetically distinct from Nebbiolo.

BERRY COLOUR: ● ● ● ● ●

PRINCIPAL SYNONYMS: Chiavennasca Piccola (Valtellina), Chiavennaschino⊗ (Valtellina), Nebiol Matiné (Alba)

VARIETIES COMMONLY MISTAKEN FOR NEBBIOLO ROSÉ: NEBBIOLO⊗

ORIGINS AND PARENTAGE

Until recently, Nebbiolo Rosé was considered a particular clonal variation of NEBBIOLO for the region of Alba, just like Nebbiolo Lampia, Nebbiolo Michet or Nebbiolo Bolla. However, DNA profiling has established that Nebbiolo Rosé and Nebbiolo have distinct DNA profiles and must be considered distinct varieties (Schneider, Boccacci *et al.* 2004). Additional investigations uncovered a few vines of Nebbiolo Rosé in old vineyards in Roero, in the Valle d'Aosta and in Valtellina, mainly in the province of Sondrio, where Nebbiolo Rosé is called Chiavennaschino and Nebbiolo is called Chiavennasca (Schneider 2005–6).

In addition, DNA parentage analysis has revealed a parent–offspring relationship between Nebbiolo Rosé and NEBBIOLO (Schneider, Boccacci *et al.* 2004; see pedigree diagram above). Although the direction of the parentage cannot be ascertained, it is likely that Nebbiolo is a parent of Nebbiolo Rosé, the other parent being unknown and probably extinct.

VITICULTURAL CHARACTERISTICS

Ripens earlier than NEBBIOLO. Susceptible to powdery mildew and botrytis bunch rot.

WHERE IT'S GROWN AND WHAT ITS WINE TASTES LIKE

Since Nebbiolo Rosé has generally been considered to be a clone of NEBBIOLO, they have been propagated together under the name Nebbiolo so that Nebbiolo Rosé is found almost everywhere in **Italy** where you find Nebbiolo: around Alba, in Roero, the Valle d'Aosta and Valtellina. Once quite widespread in these areas, it is now grown mainly in Sassella near Sondrio in Valtellina.

Nebbiolo Rosé does not yet have its own entry in the national register of varieties and the area under vine is not recorded separately. Very few varietal wines are produced but Elvio Cogno's Barolo Vigna Elena is made exclusively from Nebbiolo Rosé, though they describe it as a 'subvariety of Nebbiolo'. Wines tend to have higher alcohol, lower acidity and less colour and structure than NEBBIOLO but attractive aromas.

NEGOSKA

Northern Greek variety making soft, full-bodied wines that are often blended with Xinomavro.

BERRY COLOUR: ● ● ● ● ●

PRINCIPAL SYNONYMS: Goumenissas Mavro, Mavro Goumenissas, Neghotska, Negkoska, Negoska Popolka, Negotska, Popolka Naoussis

ORIGINS AND PARENTAGE

Negoska probably originates from Náoussa in west Makedonía, Greece, as suggested by its name, which may derive from the Slavic word *Negush* meaning Náoussa. Negoska was believed to be a close relative of XINOMAVRO, with which it is often blended, but this is not supported by recent DNA analysis (Biniari and Stavrakakis 2007).

VITICULTURAL CHARACTERISTICS

Vigorous, productive, late ripening. Good general disease resistance but best on deep, well-drained calcareous soils.

WHERE IT'S GROWN AND WHAT ITS WINE TASTES LIKE

Negoska is grown in various parts of Makedonía, **Greece**, but is a required component – a minimum of 20% – only in the Gouménissa appellation in central Makedonía, where it is blended with XINOMAVRO. Negoska's deep colour, fairly high alcohol, and relatively soft tannins and acidity make it a good partner for the higher-acid and more highly structured Xinomavro. Recommended producers in the region include Boutari and Tatsis, the latter blending Negoska with either Xinomavro or LIMNIO and also making a varietal rosé. Dalamara make a red Negoska but only in the best vintages.

NEGRA MOLE

See NEGRAMOLL.

NEGRAMOLL

The dark-skinned grape of Iberia's Atlantic islands.

BERRY COLOUR: ● ● ● ● ●

PRINCIPAL SYNONYMS: Molar (Tejo, Oeste and Setúbal in Portugal), Mollar (Andalucía, Argentina, Bolivia, Chile and Peru), Mollar Cano (Andalucía), Mollar de América (Argentina, Chile and Peru), Mollar de Cádiz (Andalucía), Mollar de Granada (Andalucía), Mollar de Huelva (Andalucía), Mollar ICA (Mollar Valley in Peru), Mulata (Lanzarote in Islas Canarias), Negra Criolla (Bolivia), Negra Mole (Algarve, Islas Canarias) or Negramolle (Islas Canarias), Rabo de Ovelha Tinto, Saborinho (Açores in Portugal), Tinta de Madeira (Madeira), Tinta Negra or Tinta Negra Mole (Madeira)

VARIETIES COMMONLY MISTAKEN FOR NEGRAMOLL: LISTÁN NEGRO, LISTÁN PRIETO (Negramoll is sometimes mistakenly called Listán Prieto or Listán Negro in Castilla-La Mancha in central Spain), MOLLARD, PRETO MARTINHO

ORIGINS AND PARENTAGE

Negramoll was thought to be a native variety on the Spanish Islas Canarias and on the Portuguese island of Madeira (under the name Tinta Negra Mole), where it was already widespread in the nineteenth century. However, recent DNA profiling has shown that Negramoll is identical to Mollar (Martín *et al.* 2006), an old variety from Andalucía in south-western Spain, possibly from the province of Cádiz, which was first mentioned in 1787 (Favà i Agud 2001): 'me ha aportat una carbassa de vi mollar' ('has brought me a calabash of mollar wine'). It is therefore likely that this variety was introduced from Andalucía to the Islas Canarias and to Madeira in or before the eighteenth century. Here it was given the name Negramoll or (Tinta) Negra Mole (officially Tinta Negra since 2000), based on the old Andalusian synonym Mollar Negro.

The name Mollar derives from the Latin *mollis*, meaning 'soft' and referring to the berries, according to the Spanish ampelographer Simón de Rojas Clemente y Rubio (1807), who described three distinct forms of this variety:

- Mollar Negro, the base type cultivated in various places in the province of Cádiz, also known as Mollar Sevillano in the province of Málaga
- Mollar Negro Bravío, with very thin shoots
- Mollar Cano, a multicoloured mutation that can bear black, red, pink and white berries on the same vine.

A pink-berried mutation, **Negramoll Rosada**, is reported on Tenerife. Such wide diversity indicates that this variety almost certainly originates in Andalucía, where it is likely to have been cultivated long before its first mention in 1787. Indeed, Mollar was referred to in the early seventeenth century in Chile, where it had been introduced from Spain (Lacoste *et al.* 2010): the Cabildo (a Spanish colonial governor) of Santiago requested at a hearing dated 6 June 1614 that the merchants 'puedan vender vino *mollar* y *moscatel* y *albillo* ... dando tres cuartillos por un real' ('be allowed to sell wine of *mollar* and *moscatel* and *albillo* ... giving three quarts for one *real*'). There are many other early references to Mollar in South America, confirming its long history of cultivation in Argentina, Bolivia, Chile and Peru (Lacoste *et al.* 2010). The identity between the South American Mollar and the original variety from Andalucía has recently been proved by DNA profiling (Martínez, Cavagnaro *et al.* 2006; Tapia *et al.* 2007). Since the old name Mollar is hardly used today in Andalucía, the more widespread Negramoll is used here as the prime name.

Negramoll is a parent of QUEBRANTA (see MUSCAT pedigree diagram) and it was used (under the name Tinta de Madeira) to breed EGIODOLA.

OTHER HYPOTHESES

Despite the similarity of the names Mollar and MOLLARD, the latter a variety from the Hautes-Alpes in France, they have nothing in common.

VITICULTURAL CHARACTERISTICS

Mid budding and mid ripening. Big, loose bunches of thick-skinned berries that tend to taste herbaceous. Sensitive to abrupt temperature changes.

WHERE IT'S GROWN AND WHAT ITS WINE TASTES LIKE

In **Spain**, Negramoll is the main red variety in the Tacoronte-Acentejo DO on the island of Tenerife and is authorized or permitted in every other DO on the Islas Canarias, although quality-conscious growers seem to be turning their attention more towards varieties such as Vijariego Negro (SUMOLL) or Baboso Negro (ALFROCHEIRO) and the majority of varietal wines or Negramoll-dominated blends are found in La Palma DO on the island of the same name. There were 1,243 ha (3,072 acres) in Spain in 2008, all on the Islas Canarias. Wines tend to be light, soft and aromatic (though the berries can sometimes accumulate high sugar levels) and are best drunk young. Producers of varietal wines and Negramoll-dominated blends include Carballo, El Hoyo and Teneguía on La Palma and El Mocanero in the Tacoronte-Acentejo DO. In most other instances, the wines are part of a local blend.

Under the name Mollar (Mollar Cano in the official Spanish register of commercial varieties), there were 486 ha (1,201 acres) in Spain in 2008, the great majority in Extremadura in the south-west. Wines can be high in alcohol and are mostly low in acidity.

In **Portugal,** Saborinho is authorized for IGP Açores and the Pico co-operative include it in their Terras de Lava blend.

Under the name Tinta Negra Mole, the variety dominates another Atlantic island's vineyards, those of Madeira. Although according to official statistics there were just 277 ha (684 acres) in 2009, cultivated in Funchal and Câmara de Lobos in the south of the island and in São Vicente in the north, and the variety is rarely acknowledged in wine names, it was still by far the most widely planted on the island and is responsible for 80–85% of all wine produced on Madeira. (The second most planted is VERDELHO, with 47 ha/116 acres.) This dominance was particularly marked after the island was hit by phylloxera in the late nineteenth century, when growers preferred to replant the more robust, dark-skinned Tinta Negra Mole rather than the traditional, light-skinned varieties, SERCIAL, VERDELHO, Boal (MALVASIA FINA), Malvasia Cândida (MALVASIA DI LIPARI), TERRANTEZ and Moscatel (MUSCAT BLANC À PETITS GRAINS). Despite official financial encouragement to replant these, winemakers such as Ricardo Diogo Freitas of Barbeito are learning to love Tinta Negra Mole and to turn it into ambitious and high-quality Madeira. Producers of varietal Negra Mole Madeira include Barbeito and Justino Henriques.

Some in Portugal (Rolando Faustino, personal communication) suggest that Negra Mole in the Alentejo is a different variety altogether and the official statistics for 2010 list it separately (610 ha/1,507 acres).

In South America, Mollar is still found in **Peru** but mainly distilled to produce pisco.

NEGRARA TRENTINA

Trentino variety that lost significant ground post-phylloxera in the early twentieth century.

BERRY COLOUR:

PRINCIPAL SYNONYMS: Doleara (Trentino), Doveana (Trentino), Dovenzana (Trentino), Edeleschwarze (Alto Adige), Keltertraube (Alto Adige), Zoveana (Trentino)

VARIETIES COMMONLY MISTAKEN FOR NEGRARA TRENTINA: NEGRARA VERONESE, TEROLDEGO

ORIGINS AND PARENTAGE

The name Negrara, from *niger* (Latin for 'black'), refers to the dark colour of the berries but it's a name that has been applied to numerous distinct varieties (known collectively as *Negrare*), mainly NEGRARA VERONESE in the Veneto and Negrara Trentina in Trentino-Alto Adige. Negrara Trentina was first mentioned by Pollini (1818). DNA parentage analysis has shown convincingly that ENANTIO is either a parent or a progeny of Negrara Trentina (Grando *et al.* 2006).

VITICULTURAL CHARACTERISTICS

Very sensitive to powdery mildew and botrytis bunch rot. Mid to late ripening.

WHERE IT'S GROWN AND WHAT ITS WINE TASTES LIKE

Negrara Trentina was widely cultivated in Trentino-Alto Adige, **Italy**, before the phylloxera invasion, after which it was replaced by other varieties when it was found that grafting onto American rootstocks, the remedy for phylloxera, severely diminished the vine's vigour. Today Negrara Trentina plantings are much smaller than those of NEGRARA VERONESE but it is still found, trained on traditional pergolas, throughout Trentino, mainly in the Valle del Sacra, in the Valdadige (Etschtaler) and in the Valle dei Laghi. Few varietal wines are made but Pravis' Negrara, classified as IGT Vigneti delle Dolomiti, is based on Negrara Trentina blended with other local varieties. Otherwise it may be used as a minor blending component (maximum 15% but usually less) in DOCs such as Valpolicella and Garda Colli Mantovani.

NEGRARA VERONESE

Minor Veneto blending component, useful for colour.

BERRY COLOUR: ● ● ● ● ●

PRINCIPAL SYNONYMS: Terodola

VARIETIES COMMONLY MISTAKEN FOR NEGRARA VERONESE: NEGRARA TRENTINA, RABOSO VERONESE, TEROLDEGO

ORIGINS AND PARENTAGE

Negrara Veronese (see NEGRARA TRENTINA for the etymology) was first mentioned by Pollini (1824). DNA analysis suggested a close relationship between Negrara Veronese and GRUAJA, a recently rescued variety from the Vicenza and Breganze area (Salmaso *et al.* 2008).

VITICULTURAL CHARACTERISTICS

Mid to late ripening. Fertile.

WHERE IT'S GROWN AND WHAT ITS WINE TASTES LIKE

Negrara Veronese is cultivated in the province of Verona and to a lesser extent in the provinces of Vicenza and Padova in the Veneto in northern **Italy**. It is authorized as a minor blending component in the Bardolino, Breganze Rosso, Valpolicella (including Amarone and Recioto) and Valdadige DOCs and rarely made as a varietal wine. The variety was once much more widely planted and used to increase colour but is now typically restricted to small parcels of older vineyards because its quality and high yields (therefore poor ripening) have not encouraged new plantings. Tedeschi, for example, have less than 1% of this variety in their older vineyards.

NÉGRET DE BANHARS

Old and virtually extinct Aveyron vine.

BERRY COLOUR: ● ● ● ● ●

ORIGINS AND PARENTAGE

Négret de Banhars is an old variety from Entraygues and Estaing in the Aveyron, Midi-Pyrénées, southern France. The name Négret derives from *negre*, meaning 'black' in Occitan, and Banhars is a village near Truyère, close to Entraygues (Lavignac 2001).

VITICULTURAL CHARACTERISTICS

Vigorous, with fragile canes prone to wind damage. Very compact bunches of small, thin-skinned berries. Susceptible to botrytis bunch rot and phomopsis. Mid ripening.

WHERE IT'S GROWN AND WHAT ITS WINE TASTES LIKE

Although Négret de Banhars is almost extinct, there are still about 2,500 vines in the Aveyron, **France**, of which Nicolas Carmarans in Campouriez has more than half. He includes around 30% in his Mauvais Temps red blend along with FER, CABERNET SAUVIGNON and CABERNET FRANC. Wines are typically deeply coloured but low in alcohol and acidity, with what Lavignac (2001) describes as the same wild taste as Fer.

NÉGRETTE

Supple, fruity and distinctive variety that has thrived for centuries just north of Toulouse.

BERRY COLOUR: ● ● ● ● ●

PRINCIPAL SYNONYMS: Cap de More (Tarn-et-Garonne), Dégoûtant (Charente), Morelet, Morillon, Mourrelet (Tarn-et-Garonne), Négralet, Négret, Négret de Gaillac, Négret du Tarn, Négrette de Fronton, Noirien, Pinot St George (US), Vesparo Noir (Gers)

VARIETIES COMMONLY MISTAKEN FOR NÉGRETTE: COT (also called Vesparo in the Gers), FUELLA NERA (Négrette is called Folle Noire on Île de Ré and Île d'Oléron), GRACIANO (called Morrastel in southern France), MONASTRELL, Mourvaison, Négret Castrais (in Castres in the Tarn), Négrette de Nice, TRESSOT (Yonne)

ORIGINS AND PARENTAGE

The name Négret or Négrette, from the local dialect *negre* meaning 'black', is confusing because it has been given to a large number of distinct varieties such as COT, GRACIANO (aka Morrastel in the Languedoc), Mourvaison, MOUYSSAGUÈS, etc., making the task of interpreting historical documents very difficult. The Négrette variety described here, from the Toulouse area in southern France, is found in the Haute-Garonne and the Tarn-et-Garonne (Lavignac 2001) and is a member of the Cot ampelographic group (see p XXVII; Bisson 2009). According to Viala and Vermorel (1901–10), this Négrette is said to have been cultivated in Gaillac since time immemorial but its origin is unknown, and no genetic study has yet been carried out to shed light on its pedigree.

OTHER HYPOTHESES

According to legend, Négrette was introduced from the Near East by the Knights Templar. Lavignac (2001) suggests that Négrette could descend from a cross between a local wild grapevine and a Mediterranean variety, but this has not been proven by any genetic study.

VITICULTURAL CHARACTERISTICS

Late budding, early to mid ripening. Vigorous and reasonably fertile but susceptibility to coulure and millerandage can lead to erratic yields. Well suited to *boulbènes* (acidic sandy-clay soils) as well as to gravelly or sandy soils. Small, compact bunches of thin-skinned smallish berries on a short, resistant stem. Very susceptible to botrytis bunch rot, powdery mildew, leafhoppers and mites.

WHERE IT'S GROWN AND WHAT ITS WINE TASTES LIKE

Négrette is the principal variety in the vineyards north of Toulouse in south-western **France**, especially around the towns of Fronton and Villaudric, where it is the principal variety in the Fronton appellation (it must comprise 50–70% of a grower's vineyard) and must constitute at least 30% of the blend in Lavilledieu to the west. In contrast to the other main dark-skinned variety of the region, TANNAT, Négrette is also deeply coloured but is supple and perfumed, with only moderate acidity, and best drunk young, with the fruit to the fore, occasionally showing an animal or floral (violets?) note. These characteristics have also led many growers to blend it with more structured varieties such as SYRAH, COT, CABERNET SAUVIGNON or CABERNET FRANC, although the percentages are strictly limited by the appellation rules. According to Strang (2009), Fronton wines are sometimes described locally as 'the Beaujolais of Toulouse'. Total French plantings in 2009 were 1,227 ha (3,032 acres) and the area seems to be declining slightly. The Cave Coopérative de Fronton make good-value examples of Négrette blends and one varietal. Other recommended producers include Châteaux Bellevue La Forêt (who have 59 ha/196 acres of Négrette), Baudaire, Cahuzac, Domaine de Callory, Châteaux Caze, Joliet, Plaisance and le Roc.

Négrette is also, somewhat surprisingly, planted in **California**, where it was known as Pinot St George until 1997. DeRose make a varietal from old Cienega Valley vines. Kenneth Volk's from San Benito vines has plenty of sweet dark fruit but also a hint of sage.

NEGRETTO

Ancient but now minor Emilia-Romagna variety for the table or early-drinking wines.

BERRY COLOUR: ● ● ● ● ●

PRINCIPAL SYNONYMS: Negretta, Negrettino, Negrettino Bolognese, Negrettino Erioli

ORIGINS AND PARENTAGE

The name Negretto has historically been applied to several distinct black-berried varieties. The one found in south-east Emilia-Romagna was supposedly mentioned by Pietro de Crescenzi in 1303. Negretto was among the most highly praised local varieties in the nineteenth century, when it was used both as a table grape and for wine.

VITICULTURAL CHARACTERISTICS

Robust and resistant to powdery mildew but susceptible to botrytis bunch rot. Vigorous and productive. Mid to late ripening.

WHERE IT'S GROWN AND WHAT ITS WINE TASTES LIKE

After the phylloxera invasion, Negretto was widely planted because of its robustness and good resistance to powdery mildew. Today the vineyard area has decreased significantly and Negretto is cultivated mainly in the province of Bologna and in the hills near Ravenna, and to a lesser extent around the cities of Forlì and Imola in the Emilia-Romagna region of **Italy**. The wines are deeply coloured, have lightish tannins and are not generally suited to ageing. They are usually blended with ALBANA, ALIONZA, BARBERA or SANGIOVESE. There were 303 ha (749 acres) in Italy in 2000.

NEGROAMARO

Makes sweet-tasting, early-drinking reds and some good rosés on the heel of Italy.

BERRY COLOUR: ● ● ● ● ●

PRINCIPAL SYNONYMS: Abruzzese (Bari and Taranto in Puglia), Albese (Campi Salentino and Guagnano in Puglia), Jonico (Galatina in Puglia), Lacrima (Latiano, Montemasola and Squinzano in Puglia), Negro Amaro (Puglia), Nigroamaro (Puglia), Purcinara (Barletta in Puglia), Uva Olivella (Pozzuoli in Campania)

ORIGINS AND PARENTAGE

Negroamaro is one of the most important varieties in Puglia in southern Italy, where it was mentioned as early as the nineteenth century. Its numerous synonyms suggest that Negroamaro is an old and historically widespread variety. Its etymology and origin are disputed: some authors suggest a simple etymology from *negro* ('black') and *amaro* ('bitter'), while others suppose a Greek etymology and origin from *mavro* ('black'), despite the redundancy of 'black black', referring to the historical links between Puglia and Greece. Since Negroamaro's DNA shows no relation to modern Greek varieties, the 'black-bitter hypothesis' seems more logical. In addition, a variety named Negro Dolce ('black sweet') was documented in Salento in the nineteenth century, probably to distinguish it from the bitter *amaro* one.

Recent DNA profiling has suggested a possible parent–offspring relationship with SANGIOVESE (Myles *et al.* 2011) but this can be rejected by comparison with other published DNA profiles (Vouillamoz).

VITICULTURAL CHARACTERISTICS

Vigorous, high yielding, mid to late ripening, though the Negroamaro Precoce clone ripens much earlier, about twenty days before the regular Negroamaro. Good resistance to downy and powdery mildews. Dark, thick-skinned berries.

WHERE IT'S GROWN AND WHAT ITS WINE TASTES LIKE

Negroamaro is cultivated mainly in the provinces of Lecce and Brindisi in the eastern part of the Salento peninsula in Puglia in southern **Italy**, and to a lesser extent in the Puglian provinces of Bari and Taranto. It is the dominant variety in a number of DOCs, including Alezio, Brindisi, Copertino, Leverano, Salice Salentino and Squinzano, and a minor component, blended with MALVASIA NERA DI BRINDISI, for example, in several others. Cantele's Amativo (IGT Salento) is a particularly successful example of a Primitivo (TRIBIDRAG) and Negroamaro blend, Primitivo supplying the structure. There were 16,760 ha (41,415 acres) in Italy in 2000, a considerable reduction from the 31,000 ha (76,603 acres) in 1990 – thanks to the EU vine-pull scheme.

The deeply coloured wines tend to taste sweet and velvety but can age quite rapidly. The best wines come from bushvines planted near enough to the sea to be cooled at night. Notable producers include Cantele, Due Palme, Castello Monaci, Taurino, Tormaresca and Vallone. The variety is also responsible for some good, emphatically fruity rosés and some dramatic dried-grape sweet wines such as Vallone's Gratticaia.

In **Australia**, Italian specialist Parish Hill in the Adelaide Hills produced their first vintage of Negro Amaro in 2005 and are continuing to extend their plantings of this variety, which seems to retain its acidity well even in the heat – which is probably why it is so happy in southern Italy.

NEGRU DE DRĂGĂȘANI

Recent Romanian cross with the potential to produce soft, fruity reds.

BERRY COLOUR: ● ● ● ● ●

ORIGINS AND PARENTAGE

Negru de Drăgășani, meaning 'black from Drăgășani', was originally said to be a Negru Vîrtos × SAPERAVI cross obtained in 1993 by Marculescu Mircea and Vladasel Mircea at the SCPVV research centre in Drăgășani, south-western Romania, in which Negru Vîrtos ('vigorous black') is an old and non-cultivated Romanian variety. This would make Negru de Drăgășani a sibling of NOVAC. However, Saperavi's DNA profile is not consistent with this parentage. In a recent genetic study (Bodea *et al.* 2009), Negru de Drăgășani appeared to be closely related to Negru Vîrtos, Novac and Bătută Neagră, an old and non-cultivated Romanian variety that could be Negru de Drăgășani's other parent, but recent results from the viticulture research centre in Drăgășani suggest instead that the other parent could be BĂBEASCĂ NEAGRĂ (Ciprian Neascu, personal communication).

VITICULTURAL CHARACTERISTICS

Vigorous, productive, mid ripening. Loose bunches of thick-skinned berries. Not particularly susceptible to fungal diseases.

WHERE IT'S GROWN AND WHAT ITS WINE TASTES LIKE

The area planted to this variety is still very small (6 ha/15 acres in **Romania** in 2008) but it appears to be easier to grow and have more potential than NOVAC. Wines are characterized by sweet dark berry flavours, spice and soft tannins. Recommended producers include Prinz Ştirbey, as well as Casa Isarescu and Vinarte.

NEHELESCHOL

Old Middle Eastern variety with gigantic bunches that are said to resemble the heavy cluster brought back from Canaan by Moses' spies, as recounted in the Bible (Numbers 13:23). Although it is generally considered to be a table grape, Neheleschol is apparently used in **Romania** for simple table wines and it is included in the experimental plantings at Mas de Daumas Gassac in Languedoc-Roussillon, **France**.

NEOPLANTA

Recent, aromatic Serbian cross found mainly in Vojvodina.

BERRY COLOUR: ● ● ● ● ●

ORIGINS AND PARENTAGE

Neoplanta is a DIMYAT (under the name Smederevka) × SAVAGNIN ROSE (under the name Traminer Rot) cross obtained by Dragoslav Milisavljević at the viticulture research centre in

Sremski Karlovci, part of Novi Sad University (Neoplanta is Latin for Novi Sad, hence the name), in Vojvodina, Serbia. It was released in 1970 (Cindrić *et al.* 2000).

VITICULTURAL CHARACTERISTICS

Vigorous, high but variable yields, early ripening. Susceptible to frost and fungal diseases.

WHERE IT'S GROWN AND WHAT ITS WINE TASTES LIKE

Wines tend to have an intense Muscat-like aroma, high alcohol but low acidity. **Serbian** producers include Jovan Kuzmanović and Vinaria Kurjak, both in the Fruška Gora region south of Novi Sad.

NER D'ALA

Virtually extinct vine now known to come from Piemonte.

BERRY COLOUR: ● ● ● ● ●

PRINCIPAL SYNONYMS: Barau (Chiaverano), Durás (Quincinetto), Fiori (Pinerol area), Gros Vien (Arnad, Issogne and Montjovet), Neirét dal Picul Rus (Carema), Provinè or Pruinè (Castagneto Po), Uva di Biella⧖ (Biella area), Verdés, Vernassa⧖ (around Ivrea, Valle d'Aosta)

VARIETIES COMMONLY MISTAKEN FOR NER D'ALA: NERET DI SAINT-VINCENT⧖ (Valle d'Aosta)

ORIGINS AND PARENTAGE

Ner d'Ala has long been considered to be indigenous to the Valle d'Aosta, where it was first mentioned by Gatta (1838) but recent ampelographic and DNA studies have shown that it is identical to Vernassa from the Canavese region in the province of Torino in Piemonte and to Uva di Biella around Biella (Moriondo 1999; Labra *et al.* 2002; Schneider and Mannini 2006). Its name could derive from Val d'Ala (Piemonte) but today this valley has no vineyards, or it could derive from *ala* (meaning 'wing') in reference to the winged shape of the bunch. DNA parentage analysis suggests a close genetic link between Ner d'Ala and AVARENGO, confirming the Piemontese origin of Ner d'Ala (Vouillamoz and Moriondo 2011).

VITICULTURAL CHARACTERISTICS

Late ripening.

WHERE IT'S GROWN AND WHAT ITS WINE TASTES LIKE

Although it is virtually extinct in the Valle d'Aosta, Ner d'Ala does thrive in a few old vineyards in the lower part of the valley between the communes of Arnad and Montjovet, as well as in Châtillon, Saint-Denis, Quart and Saint-Christophe. Its main cultivation area is now in Piemonte, **Italy**, in the Alto Canavese (Carema, Parella, Serra and Ivrea), and to a lesser extent in the province of Biella and in Castagneto Po in the province of Torino. There were just 9 ha (22 acres) in Italy in 2000, according to the agricultural census, and the grapes are used in blends. As Vernassa, it may be found in blends with NEBBIOLO, eg in the Canavese Rosso DOC.

NERELLO CAPPUCCIO

The lesser, softer of Sicily's Nerellos, frequently planted with other, as yet unknown, varieties.

BERRY COLOUR: ● ● ● ● ●

PRINCIPAL SYNONYMS: Mantiddatu Niuru, Nerello Mantellato, Nirello Cappucio, Niureddu, Niureddu Ammatiddatu, Niureddu Capucciu

VARIETIES COMMONLY MISTAKEN FOR NERELLO CAPPUCCIO: FRAPPATO, NERELLO MASCALESE, PERRICONE

ORIGINS AND PARENTAGE

Nerello Cappuccio is typical of the Etna region on the Italian island of Sicilia and was already mentioned in the province of Catania in the eighteenth century. While Nerello obviously comes from *nero* ('black'), the name *cappuccio* ('hood') probably derives from the bloom that covers the bunches like a coat (*mantello*), hence the synonym Nerello Mantellato. A recent DNA survey of thirty-four vines of Nerello Cappuccio sampled in commercial vineyards in Sicilia showed that they consist of a mix of true Nerello Cappuccio with five other distinct and unidentified varieties (Branzanti *et al.* 2008). Nerello Cappuccio has therefore been described as a 'polyphyletic variety'. However, such terminology is not appropriate in modern ampelography (see pp XIV–XVI) and it must be concluded that Nerello Cappuccio vineyards might contain several distinct varieties.

VITICULTURAL CHARACTERISTICS

Vigorous, mid ripening. Compact bunches of medium-sized berries.

WHERE IT'S GROWN AND WHAT ITS WINE TASTES LIKE

Nerello Cappuccio is grown in both Calabria, particularly in the provinces of Reggio Calabria and Catanzaro, and on Sicilia, **Italy**, notably in the north-easterly provinces of Catania and Messina, where it is often mixed with the more widely planted and better-quality NERELLO MASCALESE (and others) – in the vineyard and/or in the vat. Wines are typically moderate in colour, black-fruited, softer and less precise than Nerello Mascalese ('the Merlot to Mascalese's Cabernet', as one Etna producer puts it). Not generally suited to a solo career. Benanti and Fessina are unusual in making a varietal wine on Sicilia. Nerello Cappuccio is authorized in several DOs, including Etna, Faro, Lamezia, Savuto and Scavigna, this last a blend with GAGLIOPPO. There were 1,559 ha (3,852 acres) in Italy in 2000.

NERELLO MASCALESE

Important, noble Sicilian of particular significance around Etna. The firmer, longer-lived Nerello.

BERRY COLOUR: ● ● ● ● ●

PRINCIPAL SYNONYMS: Mascalese Nera, Mascalisi, Nerello Calabrese, Nerello Carbunaru, Nerello di Mascali, Nerello Nostrale, Nerello Paesano, Niureddu, Niureddu Mascalese

VARIETIES COMMONLY MISTAKEN FOR NERELLO MASCALESE: FRAPPATO, MAGLIOCCO CANINO, MAGLIOCCO DOLCE, NERELLO CAPPUCCIO, PERRICONE

ORIGINS AND PARENTAGE

Nerello Mascalese, first mentioned in the eighteenth century, is indigenous to the Etna region of north-east Sicilia. While the name Nerello obviously comes from *nero*, meaning 'black', the name Mascalese derives from Mascali, a commune east of Etna and close to the coast (Zappalà 2005), its likely centre of origin.

The parentage of Nerello Mascalese (under the name Nerello Calabrese) was recently revealed by DNA analysis (Cipriani *et al.* 2010): it is a possible natural cross between SANGIOVESE and MANTONICO BIANCO. This makes Nerello Mascalese a sibling of GAGLIOPPO from Calabria and it explains why Sangiovese is sometimes called Nerello in Sicilia or Negrello in Calabria. A recent DNA survey of 111 samples of Nerello Mascalese from commercial vineyards on Sicilia showed that they consist of a mix of true Nerello Mascalese and five other distinct but undetermined varieties (Brancadoro 2006). More recently, genetic comparison has suggested that Nerello Mascalese could be related to CARRICANTE (Maitti *et al.* 2009).

OTHER HYPOTHESES

It has been argued that Nerello Mascalese has a polyclonal origin (Scienza *et al.* 2008) but this concept does not fit with the view of modern ampelography put forward in this book (see pp XIV–XVI). A more likely explanation is that Nerello Mascalese vineyards might contain several distinct and undetermined varieties.

VITICULTURAL CHARACTERISTICS

Ripens late, sometimes even into early November for vines planted on the higher slopes of Mount Etna – up to 1,200 m. Sensitive to powdery mildew and botrytis bunch rot. Yields are good but may be erratic.

WHERE IT'S GROWN AND WHAT ITS WINE TASTES LIKE

Nerello Mascalese is cultivated mainly on Sicilia, **Italy**, in particular in the Etna area in the province of Catania. It is the dominant variety in Etna Rosso DOC, although it may be blended with NERELLO CAPPUCCIO and other local varieties. It plays a significant but lesser role in DOCs such as Faro, Marsala and Contea di Sclafani, with the last allowing varietal versions. The variety is also cultivated in Calabria, mainly in Lamezia and Sant'Anna di Isola Capo Rizzuto, together with Nerello Cappuccio. There were 4,387 ha (10,840 acres) in Italy in 2000, according to the official census, and Sicilia recorded 3,985 ha (9,847 acres) in 2008.

The generally light-coloured and alcoholic wines have fresh acidity, red-fruit flavours, gentle tannins and an obvious nobility, making it possible to drink them young yet also capable of ageing gracefully. Its ability to communicate terroir is especially prized by the swelling ranks of Etna producers. Benanti and Tenuta Terre Nerre make notable varietal wines; other top producers include Calabretta, Cornelissen, Graci, Passopisciaro, Girolamo Russo and Terre di Trente. Some producers prefer to blend with the richer and fruitier NERO D'AVOLA but Etna purists can be scathing about the simplicity of the latter.

NERET DI SAINT-VINCENT

Obscure, recently identified, burly Aostan.

BERRY COLOUR: ● ● ● ● ●

PRINCIPAL SYNONYMS: Neiret, Neret, Neretto, Neyret (only the last is used uniquely for Neret di Saint-Vincent)

VARIETIES COMMONLY MISTAKEN FOR NERET DI SAINT-VINCENT: CHATUS ☒ (also known as Neret Pinerolese), NER D'ALA, NERETTO DI BAIRO, NERETTO DURO, NERETTO GENTILE, NERETTO NOSTRANO

ORIGINS AND PARENTAGE

The name Neret (or Neretto, Neiret, Neyret) has been given to several distinct varieties in Piemonte and in the Valle d'Aosta. A DNA survey of ten vines called Neret in the Valle d'Aosta showed that three of them corresponded to the three distinct varieties CHATUS (called Neret Pinerolese in Piemonte), PETIT ROUGE and BONARDA PIEMONTESE, while seven turned out to have the same and unique DNA profile (Moriondo *et al.* 2008). This variety matched the description of Neret given by Gatta (1838), and it was renamed Neret di Saint-Vincent to avoid any confusion. DNA parentage analysis has revealed that Neret di Saint-Vincent has a parent–offspring relationship with ROUGE DU PAYS, confirming that it is an old variety of the Valle d'Aosta (Vouillamoz and Moriondo 2011; see PRIÉ for the pedigree diagram).

WHERE IT'S GROWN AND WHAT ITS WINE TASTES LIKE

Neret di Saint-Vincent is scattered in old vineyards between the communes of Arnad and Montjovet in the Valle d'Aosta in northern **Italy**. It is usually blended with other local grapes to deepen colour and increase alcohol.

NERETTA CUNEESE

Minor Piemontese that disappears into blends.

BERRY COLOUR:

PRINCIPAL SYNONYMS: Freisa di Nizza or Freisa Grossa (Pinerolo, Val di Susa, Asti and Tortona regions), Freisa Mora (Asti area), Neretta, Neretto (Monregalo, Cuneo and Colli Saluzzesi regions), Neretto di Cavaglià (Ivrea)

VARIETIES COMMONLY MISTAKEN FOR NERETTA CUNEESE: BONARDA PIEMONTESE, FREISA

ORIGINS AND PARENTAGE

Neretta Cuneese is local to the provinces of Cuneo and Torino, where it was referred to at the end of the eighteenth century by Nuvolone Pergamo (1787–98), Conte di Scandaluzza. Although some of its synonyms suggest a link with FREISA, no close genetic relationship has been proved (Vouillamoz).

VITICULTURAL CHARACTERISTICS

Mid to late ripening. Productive. Often affected by grapevine fanleaf virus (GFLV).

WHERE IT'S GROWN

Neretta Cuneese is found throughout Piemonte, **Italy**, mainly in the Cuneo, Saluzzo, Pinerolo and Val di Susa areas but it is specified in only one DOC, Valsusa, and is usually blended with other local grapes such GRISA NERA, PLASSA, BARBERA or NEBBIOLO. Total plantings of 407 ha (1,006 acres) were recorded in Italy in 2000.

NERETTO DI BAIRO

Minor Piemontese used in blends.

BERRY COLOUR:

PRINCIPAL SYNONYMS: Nerét (Cuceglio, Valperga and Bairo), Nerét Gros, Neretto, Neretto di San Giorgio

VARIETIES COMMONLY MISTAKEN FOR NERETTO DI BAIRO: NERET DI SAINT-VINCENT, NERETTO DURO, NERETTO GENTILE, NERETTO NOSTRANO

ORIGINS AND PARENTAGE

Neretto di Bairo takes its name from the commune of Bairo in the province of Torino, where it was mentioned in the early nineteenth century. DNA parentage analysis has revealed a parent–offspring relationship between Neretto di Bairo and NEBBIOLO (Schneider, Boccacci *et al.* 2004), the former being the likely progeny of the latter through a natural cross with a now extinct variety.

OTHER HYPOTHESES

Another parent–offspring relationship has been suggested between Neretto di Bairo and MAMMOLO from Toscana (Di Vecchi Staraz, This *et al.* 2007). However, this is doubtful because Neretto di Bairo has never existed in Toscana and Mammolo has never been mentioned in Piemonte.

VITICULTURAL CHARACTERISTICS

Vigorous, well suited to pergola training.

WHERE IT'S GROWN

Neretto di Bairo is grown mainly in the Canavese region of Piemonte in northern **Italy**, particularly in Bairo, San Giorgio and Valperga. It is authorized as part of the blend – with other local varieties such as BARBERA, BONARDA PIEMONTESE and NEBBIOLO – in the red and rosé versions of the Pinerolese and Canavese DOCs (and with AVARENGO in Pinerolese Ramiè) but is rarely made as a varietal. However, as the various Neretti are not always distinguished in the DOC regulations, it is hard to be certain which one is planted or authorized. The Italian agricultural census of 2000 recorded just 59 ha (146 acres).

NERETTO DURO

Rare Piemontese with a disproportionate number of synonyms.

BERRY COLOUR:

PRINCIPAL SYNONYMS: Balò or Balau (Asti, Canavese and Pinerolo regions), Barbera Rotonda (Albiano in the province ofTorino and Roppolo in the province of Biella), Bonarda or Bonarda 'd Macoun (Cuceglio and San Giorgio in the province of Torino), Dolcetto di Boca (Novara region), Durasa (Novara region), Freisone (Tortona region), Peilavert (in Dorzano, Salussola and Cavaglià in the province of Biella), Uva 'd Galvan (Pinerolo area)

VARIETIES COMMONLY MISTAKEN FOR NERETTO DURO: BARBERA, BONARDA PIEMONTESE, NERET DI SAINT-VINCENT, NERETTO DI BAIRO, NERETTO GENTILE, NERETTO NOSTRANO, PELAVERGA, RASTAJOLA

ORIGINS AND PARENTAGE

Neretto Duro, literally 'hard little black', is an autochthonous red variety scattered throughout Piemonte in north-west Italy. Several synonyms of Neretto Duro misleadingly suggest a link to other varieties such as BARBERA, BONARDA PIEMONTESE and DOLCETTO. It is likely that Neretto Duro is identical to Durasa, authorized under this name in the province of Novara, but this has not yet been confirmed (Schneider and Mannini 2006).

VITICULTURAL CHARACTERISTICS

Good resistance to rot thanks to its thick skins. Its robustness, fertility, vigour and high yields make it suitable for table as well as wine grapes. The berries have a high level of anthocyanins and ripen early.

WHERE IT'S GROWN AND WHAT ITS WINE TASTES LIKE

It is cultivated mainly in the Canavese region in the province of Torino, and to a lesser extent in the whole alpine region of Piemonte, **Italy**, around Asti, Cuneo, Chieri, Pinerolo and Alessandria. The wines are both fruity and spicy, with low alcohol and low acidity, and are usually blended with other local varieties. Confusingly, Ettore Germano produce a wine named Balau, a synonym widely used for Neretto Duro, although it is a blend of DOLCETTO, BARBERA and MERLOT.

NERETTO GENTILE

Fragile, minor Piemontese.

BERRY COLOUR: ● ● ● ● ●

PRINCIPAL SYNONYMS: Nerét Cit, Neretìn (Strambinello), Neretto (Gattinara), Neretto di Cavaglià (Dorzano), Vermiglia (Vercelli)

VARIETIES COMMONLY MISTAKEN FOR NERETTO GENTILE: NERET DI SAINT-VINCENT🧬, NERETTO DI BAIRO🧬, NERETTO DURO🧬, NERETTO NOSTRANO

VITICULTURAL CHARACTERISTICS

Difficult to propagate. Vigorous and suited to pergola training. Mid ripening.

WHERE IT'S GROWN AND WHAT ITS WINE TASTES LIKE

Neretto Gentile is cultivated mainly in the Canavese area in the province of Torino in Piemonte, north-west **Italy**, and to a lesser extent in Gattinara, and is generally blended with other local varieties. The wines are usually pale and low in alcohol.

NERETTO NOSTRANO

Minor Piemontese also known in the south of the Valle d'Aosta.

BERRY COLOUR: ● ● ● ● ●

PRINCIPAL SYNONYMS: Freisa Blu (Carema), Nebbiulìn (Strambinello), Nerét 'd Rean (Castellamonte), Nerét dal Busc Bianc, Nerét di Romano, Nerét Gentil (Dorzano), Neretto della Valchiusella (Parella)

VARIETIES COMMONLY MISTAKEN FOR NERETTO NOSTRANO: FREISA, NERET DI SAINT-VINCENT, NERETTO DI BAIRO, NERETTO DURO, NERETTO GENTILE

WHERE IT'S GROWN

Neretto Nostrano is cultivated mainly in the Canavese area in the province of Torino in Piemonte, north-west **Italy** and in La Serra Morenica d'Ivrea, a glacial hill between the town of Ivrea and the Valle d'Aosta. There are also scattered plantings in Carema and in the lower part of the Valle d'Aosta.

NERKARAT

Cold-hardy Armenian hybrid used mainly for sweet reds.

BERRY COLOUR: ● ● ● ● **●**

PRINCIPAL SYNONYMS: Nerkarata

ORIGINS AND PARENTAGE

Nerkarat, meaning 'rich in colour', is a complex hybrid obtained in 1960 by S A Pogosyan at the Armenian viticultural research centre at Merdzavan, just west of Armenia's capital Yerevan, by crossing Seyanets C 1262 with KARMRAHYUT, in which Seyanets C 1262 is a hybrid between a vine of *Vitis amurensis* Ruprecht and CSABA GYÖNGYE. It is therefore a sibling of CHARENTSI.

VITICULTURAL CHARACTERISTICS

Productive, cold-hardy to –28 °C (–18.4 °F).

WHERE IT'S GROWN AND WHAT ITS WINE TASTES LIKE

Thanks to its relatively high sugar levels, Nerkarat is well suited to the production of dessert wines in **Armenia**. Ijevan Wine Factory, for example, in the north-eastern province of Tavush, produce both a dry and a semi-sweet varietal wine.

NERKENI

Armenian variety grown mainly in Ararat for sweetish reds.

BERRY COLOUR: ● ● ● ● **●**

ORIGINS AND PARENTAGE

Nerkeni is a complex hybrid obtained by S A Pogosyan and S S Khachatryan at the Armenian viticultural research centre at Merdzavan, just west of Armenia's capital Yerevan, by crossing SAPERAVI with a previously obtained cross, (GARANDMAK × Richter) × KAKHET, in which Richter might be one of the numerous hybrids obtained by French breeder Franz Richter.

WHERE IT'S GROWN AND WHAT ITS WINE TASTES LIKE

Several producers in Ararat, one of **Armenia**'s main wine-growing regions, including MAP and Kimley, produce red varietal wines at varying levels of sweetness. Others blend it with SAPERAVI and local varieties.

NERO BUONO DI CORI

Minor central Italian generally blended with Montepulciano or Cesanese.

BERRY COLOUR:

PRINCIPAL SYNONYMS: Nero Buono, Nero di Cori

VITICULTURAL CHARACTERISTICS

Regular, high yields. Mid to late ripening.

WHERE IT'S GROWN AND WHAT ITS WINE TASTES LIKE

Nero Buono di Cori, literally 'the good black of Cori', is cultivated almost exclusively in the commune of Cori in the province of Latina in Lazio, central **Italy**, and in the neighbouring region of Castelli Romani. 20–40% may be included in the blend of Cori DOC and it is authorized as the dominant variety or part of the blend in the various styles of red and rosé Castelli Romani DOC. Varietal wines are few and far between but include Marco Carpineti's rosé Os Rosae and Poggio Le Volpi's Baccarossa (both IGT Lazio). The agricultural census recorded 123 ha (304 acres) in Italy in 2000.

NERO D'AVOLA

Sicilia's most widely planted red wine variety valued for its colour, full body and ageing potential.

BERRY COLOUR:

PRINCIPAL SYNONYMS: Calabrese, Calabrese d'Avola, Calabrese di Vittoria, Calabrese Dolce, Niureddu Calavrisi, Sirkuzan (Malta)

VARIETIES COMMONLY MISTAKEN FOR NERO D'AVOLA: CALABRESE DI MONTENUOVO, SANGIOVESE

ORIGINS AND PARENTAGE

Nero d'Avola, named after and most likely originating from the city of Avola in the province of Siracusa on the Italian island of Sicilia, was first described by Sicilian botanist Francesco Cupani (1696) under the name Calavrisi, the dialect word for Calabrese ('from Calabria'), which was the name given to this variety in the eighteenth and nineteenth centuries and could point to a possible Calabrian origin. However, some have suggested that Calavrisi could derive from Calaulisi, which became Caia-Avola, meaning 'grape from Avola' (see Carimi *et al.* 2011). Although this variety is listed in the official Italian register of varieties under the name Calabrese, we have opted for Nero d'Avola because Calabrese is used for other varieties such as SANGIOVESE or CANAIOLO NERO and because Nero d'Avola is the name most often encountered on wine labels. A recent DNA survey of Nero d'Avola showed a high level of genetic diversity in Sicilia, with several distinct clones intermixed among other distinct and undetermined varieties (possibly seedlings?) in the vineyards (Carimi *et al.* 2011).

VITICULTURAL CHARACTERISTICS

Vigorous, susceptible to powdery mildew. Mid ripening but likes the heat.

WHERE IT'S GROWN AND WHAT ITS WINE TASTES LIKE

Nero d'Avola is Sicilia's most planted red variety and widely grown in the provinces of Siracusa (Avola, Noto, Pachino, etc.), Ragusa, Caltanisetta, Agrigento and Catania – all over the island, in fact. To succeed it needs a good site and warmth, which is why it is often planted close to the ground. Total plantings in **Italy** in 2000 amounted to 11,410 ha (28,195 acres) but have risen considerably since then, with 19,304 ha (47,701 acres) in Sicilia in 2008.

Until just a few decades ago, Nero d'Avola was used mainly to provide colour to other wines and was often exported to Toscana, Piemonte or even to France, where it strengthened the much lighter wines of the Languedoc plains. By the 1990s it was enjoying the spotlight of fashion, however. Although it is still often blended with FRAPPATO, and sometimes with other local varieties such as NERELLO MASCALESE and PERRICONE, Nero d'Avola is now respected in its own right and is being made successfully as a varietal wine, generally deeply coloured, full-bodied and with good ageing potential. At its best, Nero d'Avola produces wines that have a wild plum and sweet chocolate character, high levels of tannins, and decent acidity. At its worst, it can be a rather facile red with aromas of candy. An increasing number of producers, eg Abbazia Santa Anastasia, Baglio di Pianetto and Cusumano, are blending Nero d'Avola with international varieties such as CABERNET SAUVIGNON and MERLOT.

Not surprisingly given its significance on the island, it is the dominant variety in many Sicilian DOCs, including Contea di Sclafani, Erice, Riesi, and Cerasuolo DOCG. It may also be made as a varietal wine in DOCs such as Eloro, Erice, Menfi and Vittoria. Recommended producers include Calatrasi, Cos, Cusumano, Donnafugata, Feudo Montoni, Gulfi, Planeta, Rudini, Scurati, Tasca d'Almerita and Vero.

Nero d'Avola is also grown to a lesser extent in the province of Reggio Calabria, where it is authorized in the blend for Bivongi DOC.

Small outposts of this heat-loving variety can be found in **Australia**, where it has been planted by the Simians at Politini Wines in the King Valley, Victoria; by Jeffrey Grosset in the Clare Valley; and by at least four producers in McLaren Vale (Coriole, Kay Brothers, Pertaringa and Sabella). It has also found favour in one or two hotspots in **California**, eg in Tracy Hills AVA in the Central Valley, as sold by Jacuzzi. It is also found in **Turkey** and on **Malta**.

NERO DI TROIA

High-quality, flavourful, firm northern Puglian that has declined considerably in the last 40 years.

BERRY COLOUR:

PRINCIPAL SYNONYMS: Sommarrello, Sumarello, Summariello✗ (Benevento in Campania), Tranese, Troiano, Uva della Marina, Uva di Barletta, Uva di Canosa, Uva di Troia

ORIGINS AND PARENTAGE

Nero di Troia has nothing to do with the legendary city of Troy in ancient Greece, but most likely originates from the village of Troia in the province of Foggia in Puglia, southern Italy, which is coincidentally and neatly said to have been founded by Diomedes, who destroyed Troy!

Until the second decade of this century it was most commonly known as Uva di Troia but those growing it, perhaps inspired by the recent fashion for Sicilian NERO D'AVOLA, decided that renaming it Nero di Troia, already a local name for it, might make it more attractive to consumers.

OTHER HYPOTHESES

Nero di Troia is said to have been introduced from Asia Minor to Puglia by the Greeks in ancient times, or alternatively from the city of Cruja in Albania, whose name would have been 'Italianized' as Troia.

VITICULTURAL CHARACTERISTICS

Late ripening. Poor resistance to downy mildew.

WHERE IT'S GROWN AND WHAT ITS WINE TASTES LIKE

Nero di Troia is cultivated almost exclusively in northern Puglia in southern **Italy**, mainly in the provinces of Bari and Barletta-Andria-Trani, as well as in the area around Lucera and Cerignola in the province of Foggia (including Troia). It plays a significant and often dominant role in DOCs such as Cacc'è Mmitte di Lucera, Castel del Monte (particularly), Rosso di Barletta, Rosso Canosa, Rosso di Cerignola and Orta Nova. It has also been found in Benevento in Campania under the name Summariello, the identity of the varieties having been revealed by DNA profiling (Antonella Monaco and José Vouillamoz, unpublished data). The Italian agricultural census of 2000 recorded 1,782 ha (4,403 acres) but the area has declined sharply since 1970, when there were over 9,000 ha (22,240 acres).

Because the wines tend to be relatively high in tannin and to keep their freshness only in youth, most Nero di Troia is blended with other varieties but some varietal Nero di Troia wines can be impressive. Alberto Longo's Le Cruste is among the best, with aromas of camomile, verveine, liquorice, juniper, a dense and velvety texture thanks to very fine tannins, and a dry persistent finish. Rivera's Puer Apuliae is another benchmark varietal. Zagaria's Vigna Grande is full of complex flavours of sour cherries, dried herbs and olives and is remarkably fresh this far south, with slightly rustic tannins. Other superior examples include Santa Lucia's Vigna del Melograno (Castel del Monte) and Tarantino's Petrigama (IGT Puglia).

NERONET

Red-fleshed, early-ripening Czech cross producing deeply coloured wines with some vestiges of its Cabernet ancestry.

BERRY COLOUR: ● ● ● ● ●

ORIGINS AND PARENTAGE

Neronet is a (SANKT LAURENT × BLAUER PORTUGIESER) × Alibernet (ie ODESSKY CHERNY) cross obtained in 1965 by Vilém Kraus at the faculty of horticulture in Lednice, southern Morava, in the south east of the Czech Republic. Neronet was included in the official Czech register of varieties in 1991.

VITICULTURAL CHARACTERISTICS

Vigorous, consistently productive and early ripening. Loose, medium-sized bunches of small, red-fleshed berries. Good resistance to botrytis bunch rot, reasonably resistant to winter frosts.

WHERE IT'S GROWN AND WHAT ITS WINE TASTES LIKE

Wines are typically deeply coloured, as you would expect from a red-fleshed variety, medium-bodied, with aromas of cassis and cherry and may be quite tannic, though lighter, softer styles are also produced and it is also used to darken blends. Recommended producers include Kapličky, Kubík and Žůrek and there were 31 ha (77 acres) in the **Czech Republic** in 2009, almost all in

Morava in the south of the country. It is also found in **Slovakia**, where Sanvin and Michal Sadloň make varietal wines.

NEUBURGER

Nutty, full-bodied but soft Austrian white.

BERRY COLOUR: ● ● ● ● ●

PRINCIPAL SYNONYMS: Brubler, Neuburské (Czech Republic and Slovakia)

ORIGINS AND PARENTAGE

Neuburger was long been thought to be a cross between PINOT BLANC and SILVANER, but DNA profiling proved it to be a natural cross between ROTER VELTLINER and Silvaner (Sefc *et al.* 1997) that probably took place in Austria. Neuburger is therefore a sibling of FRÜHROTER VELTLINER, because they have the same parents, a half-sibling of ROTGIPFLER and ZIERFANDLER, because they have Roter Veltliner as a common parent, and a grandchild of SAVAGNIN (see Silvaner, and see the PINOT pedigree diagram).

Tradition has it that in the 1860s, a bundle of vines was found on the bank of the Donau (Danube) at Oberarnsdorf in the Wachau by the winemakers Christoph Ferstl and Franz Marchendl, who produced the first wine in 1872. It was later planted on the other side of the river in Spitz, in a place named Burg, from where it supposedly got its name. Neuburger was used to breed the Czech variety AURELIUS.

VITICULTURAL CHARACTERISTICS

Mid ripening, susceptible to downy mildew, compact bunches therefore susceptible to botrytis bunch rot.

WHERE IT'S GROWN AND WHAT ITS WINE TASTES LIKE

Austria has 990 ha (2,446 acres), a gradually declining total, with two-thirds spread across most regions of Niederösterreich (Lower Austria) but particularly in the Thermenregion, and the rest in Neusiedlersee and Neusiedlersee-Hügelland. Wines are generally golden, full-bodied but gentle, nutty, sometimes with a touch of spice. The best examples develop well over a few years in bottle, becoming more velvety and even more nutty. There are 360 ha (890 acres) in Morava in the south east of the **Czech Republic** and it is also grown in **Slovakia**, where it is known as Neuburské. It is also found in **Romania** (46 ha/114 acres), mainly in Transilvania.

NEW YORK MUSCAT

American hybrid with pronounced Muscat flavours grown mainly in Canada.

BERRY COLOUR: ● ● ● ● ●

PRINCIPAL SYNONYMS: New York 12997 , NY 12997, NY Muscat

ORIGINS AND PARENTAGE

A MUSCAT OF HAMBURG × Ontario hybrid made in 1926 by Richard Wellington at Cornell University, Geneva, US (see CAYUGA WHITE for the pedigree of Ontario) and released in 1961. New York Muscat is therefore a *Vitis labrusca* and *Vitis vinifera* hybrid.

VITICULTURAL CHARACTERISTICS

Medium-sized to large loose clusters of berries that have a heavy lilac-coloured bloom. Moderately vigorous and winter-hardy (no colder than Geneva, New York) but yields may be inconsistent. Early to mid ripening (Slate, Watson and Einset 1962).

WHERE IT'S GROWN AND WHAT ITS WINE TASTES LIKE

According to grape breeder Lon Rombough, New York Muscat has the best Muscat flavour of any hybrid. It is planted mainly in Nova Scotia and Québec in **Canada**, where it makes notable sweet wines, including icewines, and is also valued as a table grape thanks to its grapey flavours and only moderate acidity. Producers of varietal wines include Domaine de Grand Pré and Jost in Nova Scotia.

NIAGARA

The foxiest American hybrid of them all.

BERRY COLOUR: ● ● ● ● ●

PRINCIPAL SYNONYMS: Niagara White, White Concord

ORIGINS AND PARENTAGE

A CONCORD × Cassady hybrid bred in 1866 by Claudius L Hoag and B Wheaton Clark, private growers in Lockport, Niagara County, New York, in which Cassady is a selected white seedling of *Vitis labrusca*. Named after its county of origin, Niagara was first sold commercially in 1882 by the Niagara Grape Company founded in 1879 by Hoag, Clark and colleagues.

The company had vineyards in New York and in Florida and had a singular operating policy: the vines were not sold but were placed with growers who would hand over to the company the proceeds from half the net sales of the grapes. This novel and exclusive method of grape distribution proved very successful but by 1915 the market for the Niagara variety was saturated and the company folded.

VITICULTURAL CHARACTERISTICS

Medium to large bunches of thin-skinned berries. Vigorous and productive. Highly susceptible to black rot, downy mildew and phomopsis and prone to crown gall on heavy, waterlogged soil. Mid to late ripening and moderately hardy. One of the few varieties susceptible to copper injury.

WHERE IT'S GROWN AND WHAT ITS WINE TASTES LIKE

This *labrusca* variety has noticeably foxy and floral flavours, perhaps the most foxy of all the American hybrids, especially straight after picking if not fully ripe, and is best blended with a more neutral variety. Wines have lower acidity than most American hybrids. Niagara is used for juice and the table as well as for wine, which accounts for the high acreage in **New York** (3,468 acres/1,403 ha in 2006) and **Michigan** (3,520 acres/1,424 ha in 2006 but only 60 acres for wine). It is found to a lesser extent in states such as **Pennsylvania**, **Ohio**, **Indiana**, **Illinois** and **Iowa**. Producers of varietal wines in the US include Honeywood in **Oregon**, Lakewood and Schulze in New York State, and Kirkwood in **West Virginia**.

Niagara is also planted in eastern **Canada**, often for sweet wines. Ontario had about 320 acres (129 ha) in 2006, only about half of what it was in 2004. It is very significant in that other home of hybrids, **Brazil**, where 2,839 ha (7,015 acres) were grown in 2007 in Rio Grande do Sul. It is also found in **Japan** but only a small proportion is planted for wine (128 ha/316 acres in 2009), mainly in Nagano and the most northerly Hokkaido prefectures.

NIEDDERA

Recently revived and promising Sardinian.

BERRY COLOUR: ● ● ● ● ●

PRINCIPAL SYNONYMS: Nieddaera, Nireddie

VARIETIES COMMONLY MISTAKEN FOR NIEDDERA: MAZUELO (under the synonym Bovale Grande), MONICA NERA, PASCALE (also known as Nieddu Mannu), PERRICONE

ORIGINS AND PARENTAGE

Nieddera is an old and typical variety of the western coast of Sardegna. Nieddu *et al.* (2007) suggested it was closely related to Nieddu Mannu. However, a comparison of DNA profiles shows that Nieddu Mannu is identical to PASCALE (Vouillamoz), leading to the conclusion that Nieddera is probably closely related to Pascale. Other writers have suggested a Spanish origin but Nieddera's DNA does not match any known Spanish variety (Vouillamoz).

OTHER HYPOTHESES

It is often said that Nieddera was introduced by the Phoenicians or by the Punics in the eighth or seventh century BC. Others suggest, as for most varieties on Sardegna, that it was brought from Spain during Aragonese or Spanish rule (1323–1720). However, there's no historic or genetic evidence to support these hypotheses.

VITICULTURAL CHARACTERISTICS

Mid ripening.

WHERE IT'S GROWN AND WHAT ITS WINE TASTES LIKE

Nieddera is authorized in the provinces of Cagliari, Nuoro and Oristano on the island of Sardegna, **Italy**, but it is cultivated almost exclusively in the Valle del Tirso in the province of Oristano, in particular in the communes of Cabras, Solanas and Nuraxinieddu near the west coast. On the brink of extinction just a few decades ago, it was rescued by Contini, which today is one of the very few producers of Nieddera Rosso (IGT Valle del Tirso), a blend of 90% Nieddera with other local red varieties. Contini's wine, matured in a mix of French oak and chestnut casks, is typically full-bodied but well balanced with aromas of liquorice and cherries and a long finish. The Cantina Sociale della Vernaccia Oristano also make Nieddera Rosso. There were 60 ha (148 acres) planted in Italy in 2000.

NIEDDU MANNU

See PASCALE.

NIELLUCCIO

See SANGIOVESE.

NIGRA

Very minor, little-planted, dark-skinned Veneto cross.

BERRY COLOUR: ● ● ● ● ●

PRINCIPAL SYNONYMS: Incrocio Cosmo 96

ORIGINS AND PARENTAGE

Nigra is a MERLOT × BARBERA cross obtained in the 1960s by Italo Cosmo at the Conegliano research centre in the Veneto, northern Italy, with the aim of increasing the acidity of Merlot. It is a sibling of PRODEST.

WHERE IT'S GROWN

There are just a few hectares planted in the province of Treviso and fewer than 10 ha (25 acres) were recorded in **Italy** in 2000.

NIGRIKIOTIKO

Rare variety well suited to the production of dried-grape wines in Greek Makedonía.

BERRY COLOUR: ● ● ● ● ●

ORIGINS AND PARENTAGE

Nigrikiotiko is an obscure variety local to the region of Siatista in Kozáni in western Makedonía, Greece. No DNA profiling has yet been done to see whether it matches any other Greek variety.

VITICULTURAL CHARACTERISTICS

Very good all-round disease resistance, very late ripening (early October, after CABERNET SAUVIGNON). Small bunches of thick-skinned berries.

WHERE IT'S GROWN AND WHAT ITS WINE TASTES LIKE

Nigrikiotiko is grown only in Siatista, south east of Kozáni, **Greece**, where it is found along with XINOMAVRO and MOSCHOMAVRO in just a few of the region's many old interplanted vineyards. Diamantis have replanted the variety at 800–920 m above sea level, finding it a useful component in their Liasto blend, a sweet red wine made from sun-dried grapes, because the grapes are high in sugar and well suited to the open-air drying process. They also use it in a rosé and a blanc de noirs. Acidity is generally low and tannins unremarkable.

NINČUŠA

Now virtually extinct, rather ordinary variety from the region of Split, on Croatia's Adriatic coast.

BERRY COLOUR: ● ● ● ● ●

PRINCIPAL SYNONYMS: Lincuša, Mlinčevac, Vincuša

ORIGINS AND PARENTAGE

The origin of Ninčuša is unknown, and it is assumed that it was found as a wild grapevine by someone called Ninčević, and later spread throughout the Split area.

VITICULTURAL CHARACTERISTICS

Matures earlier than PLAVAC MALI and has soft, thin-skinned juicy berries that tend to drop if there is heavy rain at harvest (Zdunić 2005).

WHERE IT'S GROWN

Used both for wine and for the table, Ninčuša was one of the most widespread varieties in the Split area on **Croatia**'s Adriatic coast a hundred years ago, mainly around the towns of Omiš, Sinj and Kaštela. Today it has virtually disappeared even though it is authorized in the Dalmatinska Zagora (Dalmatian Hinterland) and Srednja i Južna Dalmacija (Central and Southern Dalmatia) wine regions.

NITRANKA

Recently authorized and increasingly popular Slovakian cross.

BERRY COLOUR: ● ● ● ● ●

PRINCIPAL SYNONYMS: CAAB 3/8

ORIGINS AND PARENTAGE

Nitranka, previously known as Nitra, is a CASTETS × ABOURIOU cross obtained in 1976 by Dorota Pospíšilová at the VSSVVM research centre for oenology and viticulture in Modra, Slovakia. It is a sibling of HRON, RIMAVA and VÁH and likely to be registered in 2012. The name was changed to avoid confusion on wine labels between the variety and its designation of origin since Nitra is a town and wine region in western Slovakia (Dorota Pospíšilová, personal communication).

VITICULTURAL CHARACTERISTICS

Consistently productive, early budding, mid to late ripening. Short bunches of very thick-skinned berries. Resistant to drought and botrytis bunch rot but susceptible to powdery and downy mildews and hard winter frosts.

WHERE IT'S GROWN AND WHAT ITS WINE TASTES LIKE

Nitranka is the most widely planted of the new **Slovakian** red wine crosses (around 5 ha/12 acres in 2011) and produces wines that are deeply coloured and relatively full-bodied but best drunk young. Producers include Igor Blaho, Mihálek Juraj, Chateau Modra and Otto Rubeš.

NOAH

American hybrid once popular in Europe, notably Spanish Basque Country; but now almost extinct.

BERRY COLOUR:

PRINCIPAL SYNONYMS: Belo Otelo, Charvat, Flaga Alba, Fraga, Noa, Noé, Noka, Nova, Tatar Rizling

VARIETIES COMMONLY MISTAKEN FOR NOAH: COURBU BLANC (País Vasco in Spain), CROUCHEN (País Vasco), ELVIRA

ORIGINS AND PARENTAGE

A seedling of Taylor planted in 1869 by Otto Wasserzieher in Nauvoo, Illinois, where Taylor is a spontaneous *Vitis riparia* × *Vitis labrusca* hybrid found in the middle of the nineteenth century by Judge Taylor of Jericho in Henry County, Kentucky. Named after the first vintner in the Bible, Noah was disseminated by Bush & Son & Meissner of Bushberg, Missouri, after it first fruited in 1873.

DNA profiling has revealed that the samples of Hondarribi Zuri from the Spanish País Vasco (Basque Country) analysed by Ibañez *et al.* (2003) and Martín *et al.* (2003) are identical to Noah (Vouillamoz). This explains why Martín *et al.* described Hondarribi Zuri as a possible hybrid variety with the foxy flavour that is specific to *Vitis labrusca* and with the cuneiform leaves that are typical of *Vitis riparia*, and why Hondarribi Zuri was completely distinct from all Spanish grape varieties in the genetic tree reconstructed by Ibañez *et al.* (2003).

Noah was used to breed BACO BLANC.

OTHER HYPOTHESES

Some think that Taylor is a natural seedling of Hartford Prolific, a likely spontaneous ISABELLA × *Vitis labrusca* hybrid found around 1846 in the garden of Paphro Steele & Son in West Hartford, Connecticut.

VITICULTURAL CHARACTERISTICS

Late budding, mid ripening. Vigorous and resistant to fungal diseases and black rot. Susceptible to chlorosis and phylloxera.

WHERE IT'S GROWN AND WHAT ITS WINE TASTES LIKE

Like so many other hybrids, Noah was planted in **France** in the wake of phylloxera, mainly but not exclusively in the Loire. It was also popular in south-west France at the end of the nineteenth century because of its resistance to black rot, which was rampant at that time (Galet 2000). However, also like other American hybrids such as ISABELLA and JACQUEZ, it was banned in the 1930s because of the higher than average level of methanol in the wine (due to greater amounts of skin pectins in some all-American hybrids). **Italy** still had 17 ha (42 acres) in 2000, according to the agricultural census. The juice has the typical foxy flavour of *labrusca* varieties, as evident from the Hondarribi Zuri still grown to a limited extent in the Basque Country, northern **Spain**. There are still limited plantings in Europe, on the island of Madeira, **Portugal**, for example, but it is not authorized. It is also planted in the **Republic of Moldova** (71 ha/175 acres in 2009).

Although the variety does not appear to have survived in its state of birth, Renault Winery and Tomasello, both in **New Jersey**, produce varietal examples.

NOBLE

Productive Muscadine variety widely grown in North Carolina for juice and wine.

BERRY COLOUR: ● ● ● ● ●

PRINCIPAL SYNONYMS: North Carolina 20-119

ORIGINS AND PARENTAGE

A Thomas × Tarheel hybrid (mainly *Vitis rotundifolia* with some *Vitis labrusca* and *Vitis vinifera*) selected at the North Carolina Agriculture Experiment Station and released in 1971 (see CARLOS for the pedigrees of Thomas and Tarheel).

VITICULTURAL CHARACTERISTICS

Highly productive, vigorous, small-berried and relatively cold-hardy. Early to mid ripening.

WHERE IT'S GROWN AND WHAT ITS WINE TASTES LIKE

Noble is the primary red Muscadine variety used in juice and wine production in **North Carolina**, with 75 acres (30 ha) planted in the state in 2006 and lesser plantings in some more southerly states. Producers of varietal wines include Bannerman, Benjamin, Uwharrie in North Carolina; Rosa Fiorelli in **Florida**; and Wills Creek in **Alabama**. Wines are generally full-bodied, soft and often sweet.

NOBLING

Demanding German cross making neutral wines.

BERRY COLOUR: ● ● ● ● ●

PRINCIPAL SYNONYMS: Freiburg 128-40

ORIGINS AND PARENTAGE

SILVANER × CHASSELAS cross bred in 1940 by Johannes Zimmerman (authorized 1971). Its name was clearly meant to evoke nobility.

VITICULTURAL CHARACTERISTICS

Vigorous, mid ripening and susceptible to botrytis bunch rot. Needs a good – not too dry – site. Reliable yields unless hit by frost.

WHERE IT'S GROWN AND WHAT ITS WINE TASTES LIKE

Germany's 63 ha (156 acres) are in Baden, particularly in the Markgräflerland. Wines are light- to medium-bodied and rather neutral, similar to those made from its parent Gutedel (CHASSELAS), although on a good site the berries can achieve respectable levels of sugar and acidity and produce wines with a bit more aroma and refinement. Producers include Bretz, Britzingen, Frick and Kaufmann, all in Baden.

In **Switzerland**, where a varietal wine is made by Räblus near the Bielersee in the west of the country, there was less than 1 ha (2.5 acres) in 2008.

NOCERA

Minor blending ingredient in Calabria and Sicilia.

BERRY COLOUR: ● ● ● ● ●

PRINCIPAL SYNONYMS: Nocera di Catania, Nocera Mantonico (Scilla in the province of Reggio Calabria), Nocera Nera di Milazzo, Nucera, Nucera Niura

VARIETIES COMMONLY MISTAKEN FOR NOCERA: MAGLIOCCO CANINO, MAGLIOCCO DOLCE

ORIGINS AND PARENTAGE

Nocera is an old variety, supposedly from the province of Messina on Sicilia and introduced to Calabria as early as 1726, although it is not known if it was the true Nocera or MAGLIOCCO DOLCE, given the frequent confusion between the two (Schneider, Raimondi, De Santis and Cavallo 2008).

OTHER HYPOTHESES

Nocera is sometimes identified with Mamertinum and Zancle, two wines mentioned by Roman authors.

VITICULTURAL CHARACTERISTICS

Mid ripening. Susceptible to millerandage but good resistance to powdery mildew.

WHERE IT'S GROWN AND WHAT ITS WINE TASTES LIKE

On Sicilia, **Italy**, it is recommended in the provinces of Siracusa and Messina, and in Calabria in all provinces except Cosenza. It used to be widely cultivated in the province of Messina in the nineteenth century, especially in the area of Milazzo, where old vines still thrive today. Its cultivation began to decrease after the Second World War, and today plantings are very limited – 30 ha (74 acres) on Sicilia and in southern Italy in 2000, according to the agricultural census.

In Sicilia, Nocera plays a minor part in the blends for Faro and Mamertino di Milazzo DOCs, blended with NERELLO MASCALESE and NERELLO CAPPUCCIO in the former and with NERO D'AVOLA in the latter. In Calabria, Nocera is cultivated mainly around the communes of Longobardi and Amantea, and may contribute to the blend in the DOCs of Bivongi (up to 30%, with GRECO NERO and/or Nero d'Avola, etc.) and Sant'Anna di Isola Capo Rizzuto (40–60% with GAGLIOPPO, Nerello Mascalese, etc.).

NOIR FLEURIEN

Ancient and very rare Auvergne vine.

BERRY COLOUR: ● ● ● ● ●

PRINCIPAL SYNONYMS: Damas Rouge⊗, Fleurien Noir, Mire-Fleurien, Mirefleurien, Noir-Fleurien

ORIGINS AND PARENTAGE

Noir Fleurien, an ancient variety from the Auvergne in central France, was first mentioned and christened in 1895 by Professor Girard-Col of Clermont-Ferrand, who had found it in the village of Mirefleurs in the Puy-de-Dôme, hence the synonym Mirefleurien (Rézeau 1997). Girard-Col propagated this variety for its resistance to downy mildew and chlorosis. It also used to be known as Bordelais and Gros Rouge, names that were given to various other varieties in the area as well. DNA profiling shows that the reference vine labelled Damas Rouge in the United States Department of Agriculture's National Clonal Germplasm Repository (NCGR) in Davis, California, is in fact identical to Noir Fleurien (Vouillamoz). In addition, DNA parentage analysis has shown that Noir Fleurien is one of the numerous progenies of GOUAIS BLANC (see the PINOT pedigree diagram; Boursiquot *et al.* 2004).

VITICULTURAL CHARACTERISTICS

Fertile, vigorous, early budding, mid ripening. Medium-sized compact bunches of small berries. Resistant to downy mildew and chlorosis and well suited to dry, calcareous soils.

WHERE IT'S GROWN AND WHAT ITS WINE TASTES LIKE

Noir Fleurien is now virtually extinct (fewer than 3 ha/7 acres remaining in **France**) but still found in its birthplace of Mirefleurs in the Puy-de-Dôme, where it is grown mostly for domestic consumption. However, a few kilometres to the north in Saint-Georges-sur-Allier, Jean Maupertuis blends Mirefleurien with Gamay d'Auvergne (GAMAY NOIR) in his Pierres Noires red blend. Traditionally wines were light in colour, alcohol and acidity but results can be much more interesting when yields are curbed (Kelley 2009a).

NOIRET

Promising new American hybrid producing peppery, well-structured dry reds.

BERRY COLOUR: ● ● ● ● ●

PRINCIPAL SYNONYMS: NY 73.0136.17

ORIGINS AND PARENTAGE

A complex hybrid obtained in 1973 by Bruce Reisch, R S Luce, B Bordelon and Thomas Henick-Kling at Cornell University, Geneva, New York State, by crossing New York 65.0467.08 with STEUBEN, where:

- New York 65.0467.08 is a CHANCELLOR × New York 33277 hybrid

- New York 33277 is a New York 10589 × Seibel 6339 hybrid
- New York 10589 is a Ripley × Golden Chasselas hybrid, the latter being PALOMINO FINO (but see also CHASSELAS)
- Ripley is a Winchell × MOORE'S DIAMOND hybrid (see CAYUGA WHITE for the pedigree of Winchell)
- Seibel 6339 is a Seibel 867 × Seibel 2524 hybrid
- Seibel 867 is a Vivarais × NOAH hybrid (see PRIOR for the complete Vivarais pedigree)
- Seibel 2524 is an Alicante Ganzin × DUTCHESS hybrid (see ROYALTY for the pedigree of Alicante Ganzin).

Noiret is therefore an extremely complex *Vitis riparia*, *Vitis labrusca*, *Vitis vinifera*, *Vitis aestivalis*, *Vitis lincecumii*, *Vitis rupestris* and *Vitis cinerea* hybrid.

The original seedling vine was germinated in 1974 and tested for wine characteristics in 1980 under the number NY 73.0136.17. Noiret was named and released in 2006 together with COROT NOIR and VALVIN MUSCAT (Reisch *et al.* 2006a, 2006c).

VITICULTURAL CHARACTERISTICS

Large, loose clusters of largish berries. Late budding, mid to late ripening. Moderately susceptible to downy mildew and moderately cold-hardy (–5 °F/–21 °C to –15 °F/–26 °C). May be best grafted.

WHERE IT'S GROWN AND WHAT ITS WINE TASTES LIKE

According to Reisch *et al.* (2006b), the wines are deeply coloured with aromas of green and black pepper, as well as raspberry, blackberry and mint. They tend to have fine tannins and to be free of the aromas typical of many hybrids. There were 39 acres (16 ha) of Noiret in **New York** State in 2006, 12 acres (5 ha) in **Pennsylvania** in 2008, as well as scattered plantings in the Northeast and Midwest. Producers of varietal wines include Two Saints in **Iowa**; Arbor Hill, Deer Run and Fulkerson in Finger Lakes, New York; Smoky Hill in **Kansas**; Peaceful Bend in **Missouri**; and Candia in **New Hampshire**.

NORIA

Minor but promising, aromatic Slovakian cross that needs a good site to ripen fully.

BERRY COLOUR:

ORIGINS AND PARENTAGE

Noria is a RIESLING × SÉMILLON cross obtained at the VSVV viticulture and oenology research centre in Veľký Krtíš, Slovakia. It was named after the grape grower Mr Nori and included in the official register of varieties in 2002.

VITICULTURAL CHARACTERISTICS

High yields, mid budding, late ripening. Small bunches. Good frost resistance, making it suitable for more marginal regions. Susceptible to botrytis bunch rot.

WHERE IT'S GROWN AND WHAT ITS WINE TASTES LIKE

Wines at varying sweetness levels are fresh and citrus-scented, with SÉMILLON-like aromas, and show their RIESLING parentage in fine acidity and high quality when fully ripe. Producers include Chateau Marco, Matyšák and Mrva & Stanko. There were about 2 ha (5 acres) in **Slovakia** in 2011.

NORTON

Old, very successful American hybrid making rich, vinifera-like reds in a wide range of US states.

BERRY COLOUR: ● ● ● ● ●

PRINCIPAL SYNONYMS: Arkansas, Cynthiana (especially in Arkansas), Norton Virginia, Norton's Seedling, Norton's Virginia Seedling, Red River, Virginia, Vitis Nortoni

ORIGINS AND PARENTAGE

This variety was named after Daniel Norborne Norton (1794–1842), a physician and horticulturist near Richmond, Virginia, who discovered it around 1820 and promoted it as a wine grape. According to Norton, it was a hybrid of Bland, a *Vitis labrusca* × *Vitis vinifera* hybrid, and PINOT MEUNIER, under the translated name of Miller's Burgundy, which were growing next to each other in his garden (Prince 1830). However, a recent DNA study has shown that Norton is a hybrid variety whose ancestry includes *Vitis aestivalis* and *Vitis vinifera*, rejecting Pinot Meunier as the *vinifera* parent but suggesting it might be a seedling of ENFARINÉ NOIR (Stover *et al.* 2009).

OTHER HYPOTHESES

Norton was also thought to be a selected vine or a seedling of a native *Vitis aestivalis* vine found growing wild by F A Lemosy on Cedar Island in James River, near Richmond, Virginia, in 1835 and then handed to Norton (Bailey 1906). However, the chronology conflicts with the first written mention of Norton, by William Robert Prince (1830).

VITICULTURAL CHARACTERISTICS

Small to medium-sized compact bunches of small, tough-skinned berries with a blue bloom. Vigorous and best on well-drained sandy or gravel soils. May be prone to magnesium deficiency. Moderately susceptible to downy mildew and attracts the birds but is tolerant of many fungal diseases. Moderately cold-hardy (–10 °F/–23 °C to –15 °F/–26 °C). Difficult to root from hardwood cuttings.

WHERE IT'S GROWN AND WHAT ITS WINE TASTES LIKE

Norton, widely grown in the American Northeast and Midwest, is said by many to be the only American vine variety capable of making premium-quality wine, lacking as it does any hint of foxiness, although it does occasionally have a grapey flavour that may come close to it. In 1873, a Norton wine from Missouri was described in a competition in Wien (Vienna) as the 'best red wine of all nations'. Europeans had high hopes for the vine after phylloxera but it was soon found to dislike the calcareous soils of many of France's best vineyards. The fact that Norton is thought to have twice as much of the antioxidant resveratrol as CABERNET SAUVIGNON has helped to maintain its popularity in the US in the context of the red wine and coronary health debate, as has its good all-round resistance to fungal diseases.

It is also the oldest native American vine still widely cultivated for wine, even if it is no longer as significant in its home state as in **Missouri**, for example, which had 307 acres (124 ha) in 2009; **Illinois**, 60 acres (24 ha) in 2007; and **Iowa**, 10 acres (4 ha) in 2006. Limited plantings are also found in **Kansas**, **Nebraska**, **New Jersey** and further south in **Kentucky**, **Tennessee**, **Texas**,

Arkansas, **Georgia**, **Louisiana** and **Florida**. **Virginia** is regaining pride in Norton and now produces about thirty varietal versions, notably those of Chrysalis and Horton. Wines tend to be ruby rather than crimson and to have rich, ageworthy flavours of spicy red fruits, soft tannins and good freshness. Other recommended producers include Augusta Winery, Crown Valley and Stone Hill in Missouri and Pontchartain in Louisiana.

NOSIOLA

Trentino-Alto Adige variety long appreciated for Vino Santo but now making waves for its characterful dry wine.

BERRY COLOUR: ● ● ● ● ●

PRINCIPAL SYNONYMS: Groppello Bianco, Nosellara, Nosiola Gentile, Nusiola, Spargelen (Alto Adige)

VARIETIES COMMONLY MISTAKEN FOR NOSIOLA: DURELLA ☒ (Veneto)

ORIGINS AND PARENTAGE

Together with GROPPELLO DI REVÒ, Nosiola is traditionally said to be a descendant of Raetica, the most widespread white wine grape in northern Italy in Roman times. DNA parentage analysis has established that both Groppello di Revò and Nosiola have a parent–offspring relationship with RÈZE, an old variety of the Valais in the Swiss Alps, whose name most likely derives from Raetica (Vouillamoz, Schneider and Grando 2007). Rèze was first mentioned in 1313, whereas both Nosiola and Groppello di Revò appear in the nineteenth century. It is therefore logical to consider Rèze as a parent of Nosiola and Groppello di Revò, which explains why Nosiola is sometimes called Groppello Bianco. Although it is now found only in Trentino in northern Italy, it was once also grown in the Alto Adige under the name Spargelen, so either region could be its place of origin.

The name Nosiola most likely derives from the Italian *nocciola*, meaning 'hazelnut', probably referring to the toasted-hazelnut flavour of the wine, or, more improbably, since the berries do not go brown, to the colour of the mature berries. An alternative and much less convincing etymology is based on the local dialect *ociolet*, meaning 'little eye', possibly referring to an unidentified eighteenth-century variety called Uva dall'Occhio Bianca (meaning 'white eye's grape'), which supposedly gave *ciaret* and then *nosiolet* by 'phonetic contamination' (Giavedoni and Gily 2005).

OTHER HYPOTHESES

Based on historical and genetic data, it is possible to suggest that Raetica might be a distant ancestor of those three varieties but it is impossible to verify this.

VITICULTURAL CHARACTERISTICS

Mid to late ripening. Its early budding makes it susceptible to spring frosts but it is also prone to powdery mildew and sour rot in humid vintages. Desiccation of the stems is another common problem.

WHERE IT'S GROWN AND WHAT ITS WINE TASTES LIKE

In Trentino, **Italy**, Nosiola is cultivated in the Val di Cembra, Vallagarina, Merano and most significantly in the Valle dei Laghi, between Lago di Garda and Trento, where Nosiola is used – usually as the sole variety – to produce the rare Vino Santo version of Trentino DOC, a slow-fermented and barrel-aged dessert wine known since the seventeenth century and made from grapes that are concentrated, losing 60–80% of their original weight, by drying on racks and by the effects of noble rot. Such wines are not usually put on the market until they are 7–10 years old.

Today, Nosiola grapes for this Vino Santo are cultivated on about 10 ha (25 acres), mainly in Calavino, Cavedine, Lasino, Padergnone and Vezzano by only a handful of determined producers: Gino Pedrotti, Pisoni, Francesco Poli, Giovanni Poli, Pravis, Cantina Toblino and Cantina Viticoltori Cavit. The total annual production is just 30,000 half-bottles. With a century of uninterrupted production, Pisoni make an outstanding Vino Santo Trentino with complex aromas of apricot, quince, lime, pineapple and orange peel – a dense mouthful superbly balanced by excellent acidity.

In recent years there has been increasing interest in growing Nosiola to produce fragrant, zesty dry wines, often with a slight aroma of hazelnuts, within the Trentino DOC, mainly in the hills around the village of Pressano. Cesconi are producing particularly good examples of this style, as are Pojer & Sandri. Nosiola is also a potential ingredient in the rare Sorni Bianco style within the same DOC and in Valdadige/Etschtaler DOC. Giuseppe Fanti make a varietal wine, classified as IGT Vigneti delle Dolomiti, that is perfumed and citrusy but also has richer, sweeter fruit flavours of peach and apricot.

Other producers of respectable dry varietal wines include Battistotti, Spagnoli, Toblino, Pravis, Concilio, Endrizzi and Zeni.

Total plantings in Italy in 2000 were 193 ha (477 acres).

NOTARDOMENICO

Minor variety of unknown origin found in Puglia today.

BERRY COLOUR: ● ● ● ● ●

PRINCIPAL SYNONYMS: Notar Domenico

ORIGINS AND PARENTAGE

Notardomenico's earliest mention appears in Di Rovasenda (1877). The orgin of the variety is unknown but it may have been named after the fourteenth-century Puglian writer Notar Domenico.

VITICULTURAL CHARACTERISTICS

Excessively vigorous on fertile soils. Early to mid ripening.

WHERE IT'S GROWN

Notardomenico is a rare variety found in Puglia in southern **Italy**, traditionally cultivated in the Brindisi area, mainly for wine but also occasionally as a table grape. Today only a few vines survive in the more common Ottavianello (CINSAUT) vineyards. Notardomenico is cultivated mainly between Cisternino, Ostuni and San Vito dei Normanni north west of Brindisi, and is authorized as a minor component (maximum 15%) in the Ottavianello version of Ostuni DOC. The Italian agricultural census recorded just 15 ha (37 acres) in 2000.

NOUVELLE

Recent South African cross that can add grassiness to a white blend.

BERRY COLOUR: ● ● ● ● ●

ORIGINS AND PARENTAGE

A Cape Riesling (CROUCHEN) × Ugni Blanc (TREBBIANO TOSCANO) cross bred between 1958 and 1964 by Christiaan Johannes Orffer at the Department of Viticulture and Oenology of the University of Stellenbosch, South Africa.

Like PINOTAGE, Nouvelle was obtained by means of open-air hybridization in the vineyard and not under controlled greenhouse conditions. There have been suggestions that one of the parents could have been SÉMILLON but this was rejected by DNA analysis at ENTAV (Établissement National Technique pour l'Amélioration de la Viticulture) in France in 2007 and the originally stated parentage was confirmed.

VITICULTURAL CHARACTERISTICS

Vigorous, productive, early ripening. Big bunches of small, very thick-skinned (almost leathery) berries. Very resistant to rot and most fungal diseases apart from powdery mildew.

WHERE IT'S GROWN AND WHAT ITS WINE TASTES LIKE

Plantings of Nouvelle are small but increasing: an additional 25 ha (62 acres) were planted in 2007/8, bringing **South Africa**'s total to 373 ha (922 acres) in 2009. When picked fully ripe, it produces wines that are waxy and have a flavour a bit like bubblegum. However, when picked early, at around 10–11% potential alcohol, the result is a wine with an intense grassy, green-pepper character that can add freshness to riper, more tropical-style SAUVIGNON BLANC. One successful example of such a blend is Hermanuspietersfontein's Die Bartho from the cool Hermanus Bay area: mainly Sauvignon Blanc plus oaked SÉMILLON and less than 10% Nouvelle that has had about 48 hours of cold maceration. Bartho Eksteen of Hermanuspietersfontein sees a future for the variety but only as a minor component in a blend. Boland did at one time make a varietal wine, from vineyards just north of Stellenbosch, but currently use it in blends.

NOVAC

Recent Romanian cross that is difficult to handle but can produce fresh, lively wines.

BERRY COLOUR: ● ● ● ● ●

ORIGINS AND PARENTAGE

Novac is a recent Negru Vîrtos × SAPERAVI cross obtained in 1987 by Marculescu Mircea at the SCPVV research centre in Drăgăşani, south-western Romania, in which Negru Vîrtos ('vigorous black') is an old Romanian variety no longer cultivated. Novac may be closely related to NEGRU DE DRĂGĂŞANI (see that entry).

VITICULTURAL CHARACTERISTICS

Vigorous, productive, mid ripening, generally difficult to grow because it is sensitive to rain and susceptible to fungal diseases, and has a short harvest window. Compact bunches of thin-skinned berries.

WHERE IT'S GROWN AND WHAT ITS WINE TASTES LIKE

Novac is not only difficult in the vineyard but also in the winery, where the dark-fruit aromas may be lost, which is probably why Prinz Ştirbey is currently the only producer of a varietal wine. More often than not it is blended with international varieties to give it more body. On its own, it can be fresh, juicy and distinctly nervy thanks to its high acidity. Tannins are generally pretty soft unless bolstered by oak but it ages surprisingly well over a few years. There were 42 ha (104 acres) in **Romania** in 2008, in Oltenia in the south west and Romanian Moldova in the north east.

NURAGUS

Ancient, fertile and generally unremarkable Sardinian.

BERRY COLOUR: ● ● ● ● ●

PRINCIPAL SYNONYMS: Abbondosa, Axina 'e Pòberus, Axina Scacciadèppidus, Preni Tineddus

VARIETIES COMMONLY MISTAKEN FOR NURAGUS: Nuragus Arrubiu, Nuragus Moscadeddu, Nuragus Moscatello, Nuragus Rosso Rompizzolla

ORIGINS AND PARENTAGE

Nuragus is one of the oldest varieties of Sardegna. Its name most likely derives from the eponymous village near Cagliari, although it is often said to derive from *nuraghe*, the name of the typical truncated conical towers dry-built from volcanic stones in prehistoric times. Some of its local synonyms reflect its high yields: Axina 'e Pòberus means 'grapes of the poor' and Axina Scacciadèppidus means 'debt-paying grapes'.

Recent DNA profiling (De Mattia *et al.* 2007) has shown that two other distinct and non-cultivated varieties share the name Nuragus: Nuragus Arrubiu and Nuragus Moscadeddu (also called Nuragus Rosso Rompizzolla or Nuragus Moscatello).

OTHER HYPOTHESES

Some authors have suggested that Nuragus was introduced to Sardegna by the Phoenicians in the twelth century BC, because the name Nuragus could derive from the Phoenician *nur*, meaning 'fire'.

VITICULTURAL CHARACTERISTICS

Good resistance to spring frosts as well as to downy mildew. More susceptible to powdery mildew. Vigorous, productive and late ripening.

WHERE IT'S GROWN AND WHAT ITS WINE TASTES LIKE

Nuragus is the second most widely cultivated variety on Sardegna, **Italy**, with total plantings of 3,272 ha (8,085 acres) in 2000, according to the agricultural census. It is found mainly in the provinces of Cagliari and Oristano in the south and west of the island. If yields of this fertile and generally unremarkable variety are not controlled, wines tend to be bland and unbalanced. Its main expression is as the varietal wine Nuragus di Cagliari DOC – dry or sweet, still or sparkling – but it may also be blended with other local varieties such as SEMIDANO, which has the potential to produce higher-quality wines but is less generous in its crop. The best examples, such as Argiolas's S'elegas and Cantine di Dolianova's Perlas, display fresh fruit flavours, often have a lightly bitter finish and are best drunk young.

It is also used to produce vermouth and dry sparkling wines.

O

ODESSKY CHERNY
OEILLADE NOIRE
OFTHALMO
OHRIDSKO CRNO
OJALESHI
ÖKÜZGÖZÜ
ONCHETTE
ONDENC
OPSIMO EDESSIS
OPTIMA
ORANGETRAUBE
ORANIENSTEINER
ORION
ORLEANS GELB
ORTEGA
ORTRUGO
OSELETA
OSTEINER
ÖSTERREICHISCH WEISS
OTSKHANURI SAPERE
OVIDIOPOLSKY

this symbol on the following pages indicates that the synonyms, and mistaken identities, are confirmed by DNA evidence

OBAIDEH

See CHARDONNAY.

ODESSKY CHERNY

Red-fleshed, versatile Ukrainian cross that reveals its Cabernet parentage.

BERRY COLOUR:

PRINCIPAL SYNONYMS: Alibernet, Oděskij Čornyj, Odessa Black, Odesskii Chernyi, Semenac 1-17-4

ORIGINS AND PARENTAGE

Odessky Cherny, meaning 'black of Odessa', is an ALICANTE HENRI BOUSCHET × CABERNET SAUVIGNON cross obtained in 1948 by M P Tsebriy, P K Ayvazyan, A N Kostyuk, E N Dokuchaeva, M I Tulaeva and A P Ablyazova at the Tairov research centre in Odessa, southern Ukraine, and officially registered in 1972. Four clones are currently cultivated. While the name Odessky Cherny is widespread in its homeland, the variety is known in Slovakia under the name Alibernet (officially registered there in 1975), a contraction of the parents' names. It was used to breed NERONET.

VITICULTURAL CHARACTERISTICS

Fertile, productive, late ripening. Loose bunches of firm-skinned, red-fleshed berries. Relatively resistant to botrytis bunch rot and downy mildew, and good winter hardiness but only moderate resistance to frost.

WHERE IT'S GROWN AND WHAT ITS WINE TASTES LIKE

Odessky Cherny, widely planted in **Ukraine** (2,426 ha/5,995 acres in 2009) and recommended in the Odessa, Nikolaev and Kherson regions and in Krym (Crimea), is used to make fresh, deeply coloured, good-quality dry, semi-sweet, sweet and fortified reds with dark-fruit flavours, sometimes a note of tomato, and moderate tannins. Sun Valley (Solnechnaya Dolina) use it in both sweet and fortified blends while Kolonist's dry version combines black-fruit flavours with a herbaceous, minty note.

Although plantings are much more limited than in Ukraine, Alibernet is popular in **Slovakia** for deepening the colour of red blends, for its Cabernet-like structure and dark-berry fruit flavours that marry well with oak as well as for late-harvested varietal wines such as Hlohovec Food Farm's example under the Golguz label.

OEILLADE BLANCHE

See PICARDAN.

OEILLADE NOIRE

Disappearing, dark-skinned, southern French variety more popular as a table grape.

BERRY COLOUR: ● ● ● ● ●

PRINCIPAL SYNONYMS: Aragnan Noir (Vaucluse), Ouillade, Ouillard, Ouliade, Passerille Noire (Saint-Péray)

VARIETIES COMMONLY MISTAKEN FOR OEILLADE NOIRE: CINSAUT

ORIGINS AND PARENTAGE

Oeillade Noire is an old variety from Provence, southern France, where it used to be cultivated in the Var, Vaucluse, Gard, Hérault and the Vallée du Rhône. According to Rézeau (1997), its first mention supposedly dates back to 1544 in the 'Chant de vendanges' ('harvest song') by Bonaventure des Périers as 'œillades, cépage de la vallée du Rhône' ('œillades, variety of the vallée du Rhône'), but it is not known whether this 'œillades' corresponded to Oeillade Blanche (a synonym for PICARDAN) or Oeillade Noire, which are distinct cultivars. The first reliable mention appeared in Pierre Magnol (1676): 'Plures aliæ [vites] sunt nigræ quibus vinum rubrum conficitur, ut tarret, piquepoule, efoirou, ouliade, aliáque plurima' ('Many other black grape varieties are used to make red wines, such as tarret, piquepoule, efoirou, ouliade, and many others').

The name Oeillade most likely derives from the French verb *ouiller*, meaning 'to top up', the very productive Oeillade Noire being used to refill the vats or barrels (Lavignac 2001). It is much less likely to be derived from the French *œil*, meaning 'eye' or 'bud', for which there's no logical explanation.

Oeillade Noire is not the black version of Oeillade Blanche, a synonym of PICARDAN, and is not identical to CINSAUT, often sold under the name Oeillade Noire, especially for table grapes in the south of France.

VITICULTURAL CHARACTERISTICS

Big berries and bunches. Mid ripening. Best pruned short. Susceptible to leafhoppers and botrytis bunch rot, and sometimes to coulure and millerandage.

WHERE IT'S GROWN AND WHAT ITS WINE TASTES LIKE

Traditionally grown around Saint-Chinian in the Languedoc, **France**, and valued for its ability to ripen, albeit at modest alcohol levels, on higher sites that did not suit CINSAUT, Oeillade virtually disappeared in the twentieth century. However, Thierry Navarre of Domaine Navarre at Roquebrun, north west of Béziers in the Languedoc, has recently started bottling a varietal wine – fruity and for early drinking, lightly chilled. It is not included in the official French register of varieties and so cannot be used in any appellation wine.

OFTHALMO

Rare Cypriot variety whose soft, light reds are generally hidden in blends.

BERRY COLOUR: ● ● ● ● ●

PRINCIPAL SYNONYMS: Oftalmo, Ophtalmo, Ophtalmon, Optalmo, Pephtalmo, Pophtalmo

ORIGINS AND PARENTAGE

Ofthalmo is indigenous to the island of Cyprus, where it was first mentioned by the French ampelographer Pierre Mouillefert (1893) under the name Pophtalmo, meaning 'bull's eye'.

According to DNA profiling, Ofthalmo is genetically relatively close to SPOURTIKO, also from Cyprus (Hvarleva, Hadjinicoli *et al.* 2005).

VITICULTURAL CHARACTERISTICS

Mid budding, mid to late ripening. Vigorous and productive, with uneven ripening. Medium-sized to large bunches.

WHERE IT'S GROWN AND WHAT ITS WINE TASTES LIKE

Ofthalmo is scattered across the wine areas in the western half of **Cyprus** but the better wines come from Pitsilia and the villages in the Lemesós (Limassol) region (Constantinou 2006). Wines are aromatic but light in colour and body, with low acidity, which is probably why they are generally blended with other local varieties such as MAVRO and MARATHEFTIKO (eg by Chrysorroyiatissa, Kolios and Zenon) or with international varieties such as CABERNET SAUVIGNON and Grenache (GARNACHA) (eg by Kamanterena, Nelion and Vasilikon). Total plantings in 2010 were just 126 ha (311 acres) and the quality of the wines suggests this variety may well slowly disappear.

OHRIDSKO CRNO

Macedonian variety that was a casualty of phylloxera.

BERRY COLOUR: ● ● ● ● ●

PRINCIPAL SYNONYMS: Prespanka

ORIGINS AND PARENTAGE

Ohridsko Crno, literally 'black from Ohrid', is probably indigenous to the area around Lake Ohrid in the far south west of the Republic of Macedonia on the border with Albania.

VITICULTURAL CHARACTERISTICS

Productive, mid to late ripening. Relatively compact, small to medium-sized bunches of large berries. Best on averagely fertile and moist soils. Some resistance to downy and powdery mildews and botrytis bunch rot.

WHERE IT'S GROWN AND WHAT ITS WINE TASTES LIKE

Before the advent of phylloxera, Ohridsko Crno was widely planted in the Ohrid region of the **Republic of Macedonia** but now survives only in pockets in Ohrid, Struga and Resen. Wines are generally light in colour, fresh and fruity with moderate alcohol and often blended with PROKUPAC and VRANAC.

OJALESHI

Georgian variety suspected of Turkish origins.

BERRY COLOUR: ● ● ● ● ●

PRINCIPAL SYNONYMS: Chonouri, Odjaleshi, Odzhaleshi, Sconuri, Svanuri

VARIETIES COMMONLY MISTAKEN FOR OJALESHI: Orbeluri Odjaleshi

ORIGINS AND PARENTAGE

Ojaleshi, meaning 'growing on trees' in the Megrelian dialect, is thought to be indigenous to Samegrelo, near Georgia's Black Sea coast, but a recent DNA study has shown it to be genetically close to several Turkish varieties (Vouillamoz *et al.* 2006), suggesting it may have been introduced from a region now part of neighbouring Turkey but once part of Georgia.

OTHER HYPOTHESES

According to Chkhartishvili and Betsiashvili (2004), Ojaleshi was mentioned by ancient Greek historian Herodotus.

VITICULTURAL CHARACTERISTICS

Small bunches of medium-sized berries. Early budding, very late ripening.

WHERE IT'S GROWN AND WHAT ITS WINE TASTES LIKE

Ojaleshi is well suited to the subtropical climate of Samegrelo, north-western **Georgia**, particularly in the villages of Salkhino, Tamakoni and Abedati, where it is used to make sweet ruby-red wines that have plenty of red-fruit flavours and sometimes a peppery or spicy note. It seems to have good potential for dry wines, too. It is also planted further east in Orbeli in the Racha-Lechkhumi region, where it develops rose aromas. As its name suggests (see above), it was traditionally grown up trees, and its survival was mainly thanks to a Frenchman, Achille Murat, who started trellising it in the mid nineteenth century (Smithsonian Institution 2009). Recommended producers include Khareba, Tbilvino and Telavi Wine Cellar. There were around 130 ha (321 acres) in Georgia in 2004.

OKATAC CRNI

See GLAVINUŠA.

ÖKÜZGÖZÜ

Turkish variety with an increasing reputation for its juicy, fresh, aromatic reds.

BERRY COLOUR: ● ● ● ● ●

PRINCIPAL SYNONYMS: ?Kara Erik

ORIGINS AND PARENTAGE

Öküzgözü, literally 'bull's eye' because of its large, black berries, most likely originates from the province of Elâzığ in eastern Anatolia. Together with BOĞAZKERE, from the same area, it counts among the most important Turkish reds. DNA profiling has shown that the variety called Kara Erik in the neighbouring provinces of Sivas, Erzincan and Malatya is identical to Öküzgözü (Vouillamoz *et al.* 2006) but recent ampelographic observations have cast doubts on the true-to-typeness of the Kara Erik sample (Yılmaz Boz, personal communication).

VITICULTURAL CHARACTERISTICS

Large, compact bunches of large black fleshy berries. Mid to late ripening. Good general disease resistance. Berries can attain fairly high sugar levels. Well suited to clay-loam soils and dry conditions.

WHERE IT'S GROWN AND WHAT ITS WINE TASTES LIKE

Grown both for the table and for wine, Öküzgözü is a widely planted wine grape in **Turkey**, with 1,656 ha (4,092 acres) in 2010, mostly in the provinces of Elâzığ and Malatâya in eastern Anatolia and, to a lesser extent, in Tunceli (north of Elâzığ), Trakya (Thrace, north of the Sea of Marmara), Kapadokya (Cappadocia), Manisa (western Turkey) and along the Aegean coast. The vineyards in eastern Anatolia, planted mostly on sandy-clay soils over limestone although there is also red clay and decomposed granite, are at 850–1,100 m above sea level but here the harsh continental climate is softened by the Euphrates and its dams.

Varietal wines are generally light in colour, medium-bodied and juicy, with bright, aromatic fruit flavours, fresh acidity, moderate alcohol and relatively soft tannins so that they are often blended with the tougher and less acidic BOĞAZKERE. Öküzgözü has been growing in stature since the 1980s and there are now some very good varietal wines, following the lead of pioneer Kavaklidere. Others include Diren, Doluca, Kayra and Melen. On its own, the variety is not particularly ageworthy and is best with little or very subtle oak influence.

OLASZ RIZLING

See GRAŠEVINA.

OLIVELLA NERA

See SCIASCINOSO.

ONCHETTE

Recently rescued Isère variety.

BERRY COLOUR: ● ● ● ● ●

PRINCIPAL SYNONYMS: Ouchette (Isère)

ORIGINS AND PARENTAGE

Onchette is a rare variety from Isère that is close to extinction. DNA parentage analysis has shown that Onchette is one of the numerous natural progenies of GOUAIS BLANC (Boursiquot *et al.* 2004; see PINOT pedigree diagram).

VITICULTURAL CHARACTERISTICS

Mid to late ripening.

WHERE IT'S GROWN

Onchette used to be cultivated mainly in Lavars and in Roissard in the Vallée du Drac in the Isère, central eastern **France**, although it has now almost disappeared. The association Vignes et Vignerons du Trièves, founded in 2008 with the aim of saving the old vineyards of the Trièves region from oblivion, planted 1.2 ha (3 acres) of Onchette in 2009 near the village of Prébois in the Isère.

ONDENC

Neglected variety persists in south-west France and in isolated spots in Australia.

BERRY COLOUR: ● ● ● ● ●

PRINCIPAL SYNONYMS: Blanquette, Irvine's White (Great Western in Australia), Oundenc or Oundenq (Gaillac), Piquepout de Moissac (Gers), Plant de Gaillac (Tarn)

VARIETIES COMMONLY MISTAKEN FOR ONDENC: PIQUEPOUL BLANC

ORIGINS AND PARENTAGE

Ondenc is an ancient variety from the Tarn region, where it was first mentioned in 1783 (Rézeau 1997): 'Ondene blanc … c'est un des meilleurs raisins que je connoisse pour manger et faire de bon vin' ('Ondene blanc … it is one of the best grapes I know to eat and make good wine'). The name Ondenc, meaning 'undulating' in the local dialect, could derive from the abundant and clear juice, or from the undulation of the leaves in the strong, local Autan wind, but this is doubtful – growers' descriptions are generally descriptive rather than poetic.

Ondenc is a member of the Folle ampelographic group (see p XXVII; Bisson 2009).

Surprisingly, DNA parentage analysis suggests a possible parent–offspring relationship between Ondenc and ROUSSETTE D'AYZE from Savoie (Vouillamoz).

VITICULTURAL CHARACTERISTICS

Early budding and thus at risk from spring frosts. Vigorous, fertile and early ripening.

Susceptible to botrytis bunch rot and sour rot and, to a lesser extent, to both downy and powdery mildews.

WHERE IT'S GROWN AND WHAT ITS WINE TASTES LIKE

Ondenc was preferred over and above MAUZAC BLANC after the phylloxera crisis because it was thought to be less susceptible to the louse responsible but its popularity waned due to poor yields and susceptibility to botrytis bunch rot. It is still officially recommended in much of the south-western quarter of **France** and permitted as a minor variety in several appellations, including Bordeaux, Bergerac, Gaillac, Montravel and Vins de Lavilledieu. However, just 10 ha (25 acres) remained in France in 2008, down from 1,589 ha (3,927 acres) in 1958.

The variety produces wines that are not very aromatic but in the right conditions may be late harvested for sweet wines. It has been saved from oblivion by a small group of growers, notably Robert and Bernard Plageoles, who, at their Domaine de Très-Cantous, produce both dry and sweet varietal wines within the Gaillac appellation.

Ondenc was taken to **Australia** at the end of the nineteenth century and known as Irvine's White at Great Western and confusingly as Sercial in South Australia. At Seppelt in Victoria it contributed to their Hans Irvine sparkling wine until it was pulled up a few years ago. In the Barossa Valley, Langmeil have produced a traditional-method sparkling wine from 100% Ondenc which they describe as having aromas of honey, beeswax and apples. They plan to replace it.

OPSIMO EDESSIS

Inferior northern Greek variety also grown for the table.

BERRY COLOUR: ● ● ● ● ●

PRINCIPAL SYNONYMS: Agriostaphylo, Chimoniatiko, Dopio, Foustani, Karatsova Naousis⚮, Karatzovitiko, Ntopia, Opsimo, Opsimos Edessis, Opsimos Lefko⚮, Pandiri, Paschalino, Raisin de Foustani, Raisin de Karatzova, Staphyli Edessis, Staphyli Karatzovas, Valandovski Drenak (Republic of Macedonia), Zimsko Belo⚮ or Belo Zimsko (Republic of Macedonia)

VARIETIES COMMONLY MISTAKEN FOR OPSIMO EDESSIS: Amasya (Turkish table grape)

ORIGINS AND PARENTAGE

Opsimo Edessis, literally 'late ripening from Édessa', probably comes from Kentrikí Makedonía (Central Macedonia) in northern Greece, where it is also called Dopio or Ntopio, meaning 'from the country' (Galet 2000). Comparison of the DNA profiles of Opsimo Edessis, Opsimos Lefko and Karatsova Naousis in the Greek Vitis Database suggests that they are identical (Vouillamoz), this last being also known as Zimsko Belo in the Republic of Macedonia (Štajner, Angelova *et al.* 2009). Both the synonyms Zismko and Chimoniatiko mean 'winter', possibly referring to the fact that the grapes will keep for up to five or six months for the table.

VITICULTURAL CHARACTERISTICS

Vigorous and productive. Mid budding but very late ripening. Susceptible to powdery mildew (especially if pergola trained) and botrytis bunch rot.

WHERE IT'S GROWN AND WHAT ITS WINE TASTES LIKE

Opsimo Edessis, originally a table grape but used to make wine when the market for table grapes dried up, seems to be confined to Kentrikí Makedonía in northern **Greece**, especially around Édessa, north west of Thessaloníki. However, the wines are generally poor quality – low in alcohol and acidity – and are typically used in the production of retsina. There were 228 ha (563 acres) in 2008.

OPTIMA

German cross designed to reach stupendous sugar levels even on poor sites.

BERRY COLOUR: ● ● ● ● ●

PRINCIPAL SYNONYMS: Optima 113

ORIGINS AND PARENTAGE

A (SILVANER × RIESLING) × MÜLLER-THURGAU cross bred by Peter Morio and Bernhard Husfeld in 1933 at the Geilweilerhof research centre, authorized in Germany in 1971. Parentage has been confirmed by DNA profiling (Grando and Frisinghelli 1998). Optima and BACCHUS are siblings and, having RIESLING twice in their pedigree, are genetically very similar to it. Optima was used to breed ORION.

VITICULTURAL CHARACTERISTICS

Vigorous and very early ripening – up to ten days before MÜLLER-THURGAU – and copes with the poorest of soils, though yields are not particularly high and it is not particularly winter-hardy. Good resistance to downy mildew but susceptible to botrytis bunch rot and a victim of wasps due to its precocity.

WHERE IT'S GROWN AND WHAT ITS WINE TASTES LIKE

Germany's 64 ha (158 acres) are planted mainly in Rheinhessen, Mosel-Saar-Ruwer and the Pfalz, a marked decline from 184 ha (455 acres) in 2001. There were more than 400 ha (988 acres) in 1990 but it has mainly been used to boost sugar levels in a blend, hence its current decline. Even though, or perhaps because, it can reach a high level of ripeness, the wines tend to be flabby and undistinguished or just plain sweet. Producers of varietal wines include Butzelt in Franken and Karlheinz Schneider in the Nahe.

It is also planted to a limited extent in **England** (2 ha/5 acres), where it is sometimes made into a late-harvest-style sweet wine, and also in British Columbia, **Canada**, where there are 14 acres (6 ha).

ORANGE MUSCAT

See MUSCAT FLEUR D'ORANGER.

ORANGETRAUBE

Variety with a much-discussed history but now strictly limited geography.

BERRY COLOUR: ● ● ● ● ●

PRINCIPAL SYNONYMS: Gelbe Orangetraube (Germany), Narancsszőlő (Hungary), Orangentraube

VARIETIES COMMONLY MISTAKEN FOR ORANGETRAUBE: GALBENĂ DE ODOBEŞTI (Romania)

ORIGINS AND PARENTAGE

Johann Philipp Bronner (1792–1864) was a German pharmacist and ampelographer from Neckargemünd in Baden-Württemberg in southern Germany. Since the 1840s, he had carefully studied the wild grapevines growing along the Rhein between Mannheim and Rastatt (Bronner 1857). Bronner found thousands of red-berried vines, and only three white-berried ones, the most interesting being the one found near Speyer (south of Mannheim in Rheinland-Pfalz) that had yellow, round and sweet berries, with a strong aroma of orange-blossom oil. Bronner had established his own classification system of varieties and gave it the Latin name *Zaehringia nobilis*, in honour of the old and influential German ruling family Zähringen, and he gave it the common name Orangetraube for its particular aroma. The first official use of this name was made by Friedrich Jakob Dochnahl (1848), who described Orangetraube as a new and valuable table grape.

Orangetraube has always been considered to be a wild grapevine, thus belonging to the *Vitis vinifera* subsp. *silvestris*. However, this is highly improbable because wild grapevines are usually dioecious, having separate male and female plants, and they never have light-skinned berries. In addition, several cultivated varieties were found to be morphologically similar to Orangetraube:

- KNIPPERLÉ (called Ortlieber in Germany), according to the famous ampelographer Hermann Goethe (1878), who promulgated Orangetraube in Austria
- PINOT (called Blauburgunder), according to August Wilhelm Freiherr von Babo and Edmund Mach (1923)
- RÄUSCHLING, according to Ambrosi *et al.* (1997).

Since the DNA profile of Orangetraube published by Sefc *et al.* (2000) does not match any other known cultivated variety or wild grapevine (Vouillamoz), and since it resembles several cultivated varieties, we suggest that Orangetraube originated from a grape seed that was randomly eaten by a bird in a neighbouring vineyard and ejected in the forest along the banks of the Rhein. A basic DNA parentage analysis based on a mere ten DNA markers suggests that the parents of Orangetraube could be PINOT, possibly PINOT GRIS because of the colour and aroma of the berries, and CHASSELAS (Vouillamoz), although at least thirty or forty markers would be necessary to confirm this hypothesis. This parentage would be absolutely consistent with historical data. Indeed, Chasselas (aka Gutedel) has been cultivated in Baden-Württemberg since the sixteenth century (Krämer 2006), and Pinot Gris is said to have been first introduced to Germany by a grape collector in Speyer (Babo 1843–4), exactly where Orangetraube was found, before being spread by a merchant called Ruland, hence its synonym Ruländer in Baden-Württemberg. *La boucle est bouclée!*

Orangetraube was crossed with Welschriesling (GRAŠEVINA) to breed GOLDBURGER.

OTHER HYPOTHESES

The variety GALBENĂ DE ODOBEŞTI cultivated in Romania is often said to be identical to

Orangetraube, but it has non-aromatic berries and its DNA profile in Hârţa *et al.* (2011) is distinct and unique (Vouillamoz).

WHERE IT'S GROWN AND WHAT ITS WINE TASTES LIKE

In **Austria**, on the south side of the capital, Wien (Vienna), Richard Zahel is one of the few producers of a varietal wine, though the variety is also planted to a very limited extent in Burgenland, further south, and around Klosterneuburg north west of Wien. Zahel describes his wine as having intense stone-fruit flavours and racy acidity.

ORANIENSTEINER

German cross that seems to have a found a niche in Canada.

BERRY COLOUR: ● ● ● ● ●

PRINCIPAL SYNONYMS: Geisenheim 11-34, Hochkroner

ORIGINS AND PARENTAGE

A RIESLING × SILVANER cross bred in 1985 at the Geisenheim research centre, probably named after Schloss Oranienstein in Rheinland-Pfalz. Oraniensteiner, OSTEINER and RIESLANER are siblings.

WHERE IT'S GROWN AND WHAT ITS WINE TASTES LIKE

Bred in Germany though not yet authorized there for quality wine production, Oraniensteiner is planted to a limited extent in the Okanagan Valley in British Columbia, **Canada**, where Sonoran Estate, for example, produce a varietal icewine and Joie include it in their off-dry Edelzwicker-style Noble Blend. It's also authorized in Ontario. In **Hungary** there are a few vines near Lake Balaton. Wines have good acidity and strong citrus notes.

ORION

Little-planted cool-climate German hybrid similar to Müller-Thurgau.

BERRY COLOUR: ● ● ● ● ●

PRINCIPAL SYNONYMS: Geilweilerhof GA-58-30

ORIGINS AND PARENTAGE

Orion is a complex OPTIMA × VILLARD BLANC hybrid bred by Gerhardt Alleweldt in 1964 at the Geilweilerhof research centre. However, Orion is categorized as *vinifera* in the UK because of its *vinifera*-like wines. It was named after the giant hunter in Greek mythology, killed by Artemis and now the name of a constellation.

VITICULTURAL CHARACTERISTICS

Cold-hardy, with quite good resistance to downy mildew but susceptible to botrytis bunch rot. Yields are high but lower than those of MÜLLER-THURGAU and the sugar levels tend to be higher.

WHERE IT'S GROWN AND WHAT ITS WINE TASTES LIKE

Orion was authorized in 1994 but there are still only a few hectares in **Germany**, mostly in Rheinhessen, where Dr Lawall produce a varietal wine, and there are a few lonely vines in **Switzerland** (one tenth of a hectare/a quarter of an acre). It has been slightly more popular in **England**, where there were 9 ha (22 acres) in 2007. **Denmark** had just 2 ha (5 acres) in 2010. There are also limited plantings in **Sweden**. Wines are fresh, sometimes slightly floral but relatively soft. According to Basler (2003), it is in decline because its resistance to disease has been disappointing.

ORLEANS GELB

Ancient but now very minor, high-acid German vine replaced by Riesling. The oldest vines in Germany may be Orleans Gelb.

BERRY COLOUR: ● ● ● ● ●

PRINCIPAL SYNONYMS: Gelber Orleans, Hart-Heunscht, Hartheinisch, Harthengst or Hart-Hängst, Harthinsch, Orleaner, Orléans Jaune, Orleanser, Orleanstraube, Weisser Orleans

ORIGINS AND PARENTAGE

Despite being named after the city of Orléans in central northern France, Orleans Gelb (*gelb* means 'yellow' in German) comes from the Rheingau region, where it was widely cultivated until the nineteenth century, especially in the Rüdesheimer Berg vineyard, near Rüdesheim, its historical base. Its first mention most likely appeared under its old synonym Harthinsch in Hieronymus Bock's (1539) *Kreutterbuch*: 'Harthinsch umb Türckheim und Wachenheim' (Türckheim here is not the village in Alsace but an old spelling of Dürkheim, a village close to Wachenheim in the Pfalz.)

The name Harthinsch, also spelt Hartheinisch in *Neuw Kreutterbuch* (1588) by Jacobus Theodorus (also known as Tabernaemontanus), is undoubtedly a contraction of *hart Heunisch* meaning 'hard Gouais Blanc'. This name refers to the thickness of the berry skins and to a similarity with GOUAIS BLANC, still called Heunisch in Germany. According to Aeberhard (2005), the name was changed to Hart-Hängst in 1697, Harthengst in 1751, Hartheinst in 1781 and then back to Hart-Heunscht in Bassermann-Jordan (1923). The name Orleans can be confusing because it has also been given to different grape varieties in the past (mainly PINOT), and the first use of this name for the variety under discussion here was in Lambert von Babo (1843–4), who described three types within the variety (Gelber Orleans, Grüner Orleans and Später Weisser Orleans), which are most likely mutations of berry colour and ripening period (*grün* meaning 'green', *weiss* meaning 'white' and *spät* meaning 'late').

OTHER HYPOTHESES

Lambert von Babo (1843–4) relates an unfounded legend saying that Orleans Gelb was brought from the region of Orléans in the Loiret (central northern France) to the Rüdesheimer Berg in the Rheingau by the emperor Charlemagne (742–814), who owned a vineyard in Rüdesheim. From there, it would have spread to the whole of the Rheintal (Rhine Valley). Similarly, it is sometimes said that monks from the Cistercian monastery in Eberbach could have brought it from Orléans in the twelfth century. However, there are no historical documents to support these hypotheses, and Orleans Gelb has never been observed in the region of Orléans.

Orleans Gelb is often said to be an offspring of GOUAIS BLANC, but this hypothesis can be rejected by DNA profiling (Vouillamoz).

VITICULTURAL CHARACTERISTICS

Vigorous and productive with very big, thick-skinned berries.

WHERE IT'S GROWN AND WHAT ITS WINE TASTES LIKE

Orleans Gelb was once widely cultivated in some of the best vineyards of the Rheingau, **Germany** (Assmannshausen, Nierstein, Rüdesheimer Berg, Scharlachberg bei Bingen and Schloss Johannisberg), where it was planted together with RIESLING, GOUAIS BLANC (Heunisch), ELBLING, SILVANER, PINOT GRIS (Ruländer), SCHIAVA GROSSA (Trollinger) and SAVAGNIN BLANC (Traminer). By the end of the nineteenth century, Orleans Gelb had been largely replaced by Riesling, which produced better-quality, aromatic wine that commanded higher prices, and there is no record of the variety being vinified after 1921 in Rüdesheim.

The variety was then assumed to be extinct until Helmut Becker of the Geisenheim research centre rediscovered old feral vines of Orleans Gelb in the 1980s on the terraces of Rüdesheimer Berg. The first plantings were made in the early 1990s in collaboration with Weingut Knipser in Laumersheim (*c.*700 vines), and in 1995 in collaboration with Weingut Georg Breuer in Rüdesheim (*c.*600 vines), who released the first commercial – oak-aged – wine in 2002. Knipser describes the wine as rather neutral with marked acidity but this is buffered by the relatively full body. Wines can be quite forceful.

Five very old vines of Orleans Gelb have been discovered growing almost wild by grape researcher Andreas Jung in Disibodenberg near Odernheim am Glan in the Nahe. Based on local historical documents, the planting date of these vines has been estimated at between 1108 and 1559, when the Disibodenberg monastery was still active, and it has been claimed that these grapevines may be the oldest in Germany.

ORTEGA

Early-ripening Gerrman cross that can achieve impressive sugar levels if not acidity.

BERRY COLOUR: ● ● ● ● ●

PRINCIPAL SYNONYMS: Würzburg 48-21-4

ORIGINS AND PARENTAGE

A MÜLLER-THURGAU × SIEGERREBE cross bred by Hans Breider in 1948 at Würzburg, Franken, named after the Spanish philosopher José Ortega y Gasset. This parentage has been confirmed by DNA analysis (Myles *et al.* 2011).

VITICULTURAL CHARACTERISTICS

Early budding and thus at risk from spring frosts. Very early ripening, with good resistance to both downy and powdery mildews but susceptible to botrytis bunch rot and coulure.

WHERE IT'S GROWN AND WHAT ITS WINE TASTES LIKE

Ortega can achieve high grape-sugar levels in most years, making it popular in **Germany** (634 ha/1,567 acres), though less so than in the late 1980s. Half of the German plantings are in Rheinhessen and the rest mostly in the Pfalz. Its tendency towards coulure makes it less appropriate than OPTIMA for the Mosel-Saar-Ruwer region, where there are still a few plantings. It produces soft, full-flavoured, occasionally spicy wines, including sweet, peachy examples from Auslese level to very sweet. Recommended producers in Germany include Scherner-Kleinhanss in Rheinhessen and Frey in the Pfalz.

Its reliable ripeness has led to its adoption in **England** (27 ha/67 acres), where Biddenden Vineyards have been particularly successful with their offerings and Chapel Down produce a barrel-aged version. It is also grown in **Switzerland**, **Sweden** and **Denmark** (1.5 ha/4 acres in 2010).

It is also found in British Columbia in **Canada** (51 acres/21 ha), where it is valued for its precociousness and its sugar content.

ORTRUGO

Emilia-Romagnan often blended with the most aromatic Malvasia.

BERRY COLOUR:

PRINCIPAL SYNONYMS: Altruga, Altrugo, Artrugo

ORIGINS AND PARENTAGE

The earliest mention of this variety in the province of Piacenza in Emilia-Romagna is to be found under its synonyms Artrugo or Altrugo in an ampelographic bulletin dated 1881, whereas the name Ortrugo does not appear until the early part of the twentieth century, in Toni (1927). The name could derive from *altrughe* or *artrugo*, meaning 'the other grape' (*altra uva*) in local dialect.

VITICULTURAL CHARACTERISTICS

Vigorous and productive. Late ripening. Susceptible to the European grapevine moth (*Lobesia botrana*).

WHERE IT'S GROWN AND WHAT ITS WINE TASTES LIKE

Ortrugo almost disappeared in the mid twentieth century, largely replaced by MALVASIA DI CANDIA AROMATICA, before being rescued in the 1960s and 1970s, thanks mainly to the perseverance of Luigi Mossi in the Val Tidone south west of the city of Piacenza, **Italy**. Today, Ortrugo is one of the most widely cultivated white varieties in the province of Piacenza, almost exclusively in the Colli Piacentini DOC (Val Tidone, Val Luretta, Val Trebbia, Val Nure, Val Chero, Val d'Arda and Val Stirone), as a sparkling, lightly sparkling or still varietal or as part of a blend (often with Malvasia di Candia Aromatica or with TREBBIANO ROMAGNOLO) and also in Vin Santo and Vin Santo di Vigoleno within the same DOC. It is also planted in the Oltrepò Pavese, close to the border with Emilia-Romagna. The Italian census recorded 526 ha (1,300 acres) in 2000.

Good examples are produced by Mossi (still and sparkling), Barattieri, Pusteria (includes 15% TREBBIANO ROMAGNOLO). Sparkling wines tend to be light and fresh whereas the still wines are generally more full-bodied and alcoholic.

OSELETA

Revived and rather fashionable in Valpolicella country.

BERRY COLOUR:

PRINCIPAL SYNONYMS: Oselina

ORIGINS AND PARENTAGE

The name Oseleta probably derives from its synonym Oselina, based on Italian *uccelli*, meaning 'birds', a name given to several distinct varieties whose berries are often eaten by birds.

OTHER HYPOTHESES

Oseleta is said to have been domesticated from wild grapes in the Verona area, but there is as yet no evidence to support this.

VITICULTURAL CHARACTERISTICS

Mid ripening. Low and variable yields. Resistant to botrytis bunch rot and therefore may be harvested late.

WHERE IT'S GROWN AND WHAT ITS WINE TASTES LIKE

Oseleta was almost extinct before it was recovered in the 1970s in the small village of Pigozzo, north east of Verona in the Veneto, north-east **Italy**. It is used by some producers in small amounts to add colour and tannins to the blend for Valpolicella, Recioto and Amarone, and Tedeschi, for example, have included the variety in their most recent plantings. Since 2002 Oseleta has been officially authorized and recommended in the Valpolicella area. Zymè and Masi produce two of the very rare examples of varietal wines, both classified as IGT Rosso del Veronese. Masi's Osar, made from vines planted in 1985 and 1990, spends twenty-four months in small new oak barrels but still has a complex aromatic profile from herbs and cinnamon through hedgerow fruit to floral and then leather notes, and pretty firm tannins even after several years in bottle. Zymè, who planted Oseleta in 1998, estimate a total of 10–20 ha (25–50 acres) in the Valpolicella region. La Cappuccina include 10% Oseleta in the CARMENÈRE-based Campo Buri and 50% in their sweet passito wine called Carmenos (blended with Carmenère, hence Carmenos).

OSTEINER

Very minor German cross with a successful outpost in New Zealand.

BERRY COLOUR: ● ● ● ● ●

PRINCIPAL SYNONYMS: Geisenheim 9-97

ORIGINS AND PARENTAGE

A RIESLING × SILVANER cross created by Heinrich Birk in 1929 at the Geisenheim research centre in the Rheingau, Germany, and authorized in 1984. This parentage has been confirmed by DNA profiling (Myles *et al.* 2011). It was named after Graf Friedrich Karl von Ostein, who constructed several buildings and gardens in Geisenheim. ORANIENSTEINER, Osteiner and RIESLANER are siblings.

VITICULTURAL CHARACTERISTICS

Mid to late ripening. Very good resistance to downy mildew and quite resistant to powdery mildew and botrytis bunch rot. Regular yields with good sugar levels. Thick skins give assertive straw colour to the wine.

WHERE IT'S GROWN AND WHAT ITS WINE TASTES LIKE

There is about as much Osteiner in **New Zealand** as there is in **Germany** – about 1 ha (2.5 acres) in each. Rolfe Mills of Rippon Vineyards in Central Otago imported the variety in the mid 1990s. Rippon's Hotere white, a 100% Osteiner, has aromas of currants and blossom plus a hint of lemon and is made in a fresh but off-dry style. More generally, the wines have good acidity and body.

ÖSTERREICHISCH WEISS

Rare light-skinned Viennese variety with a well-known offspring.

BERRY COLOUR: ● ● ● ● ●

PRINCIPAL SYNONYMS: Kahlenberger Weisse

VARIETIES COMMONLY MISTAKEN FOR ÖSTERREICHISCH WEISS: BRAUNER VELTLINER, SILVANER (called Österreichisch in Franken)

ORIGINS AND PARENTAGE

Österreichisch Weiss (literally 'Austrian White') is an ancient variety cultivated mainly in the region of Kahlenberg near Wien (Vienna), hence its synonym Kahlenberger Weisse. DNA profiling has shown that Österreichisch Weiss spontaneously crossed with SAVAGNIN to give birth to SILVANER (Sefc, Steinkellner *et al.* 1998), and that Österreichisch Weiss is one of the numerous progenies of GOUAIS BLANC, the other parent being unknown (Regner *et al.* 1998).

VITICULTURAL CHARACTERISTICS

Good general disease resistance.

WHERE IT'S GROWN

Österreichisch Weiss was quite widely planted in the vineyards of Kahlenberg on the north side of Wien, capital of **Austria**, where a popular blend with 10% Traminer (SAVAGNIN) was known as Grinzinger.

Until the 1950s, it was used for the production of sparkling wine, but its importance has greatly diminished today (Sefc, Steinkellner *et al.* 1998). Alois Raubal (Bioweinbau-Raubal) has recently planted 300 vines of Österreichisch Weiss in Gumpoldskirchen south of Wien and produces a varietal wine.

OTSKHANURI SAPERE

Minor Georgian variety responsible for deeply coloured, ageworthy reds.

BERRY COLOUR: ● ● ● ● ●

PRINCIPAL SYNONYMS: Argvetuli Sapere

ORIGINS AND PARENTAGE

Otskhanuri Sapere is indigenous to Imereti, western central Georgia, possibly named after the Otskha River. The Georgian word *sapere* means 'dye', in reference to the deep black colour of the berries.

VITICULTURAL CHARACTERISTICS

Mid budding, late ripening. Good general disease resistance.

WHERE IT'S GROWN AND WHAT ITS WINE TASTES LIKE

Grown in the western, subtropical part of **Georgia**, mostly in the villages of Rodinauli, Sviri, Dimi and Obcha in Imereti, Otskhanuri Sapere produces deeply coloured, well-structured and ageworthy red wines with distinctive varietal flavours of forest fruits, cherries and wild grasses. Khareba and Imeretian Wines produce good varietal examples.

OTTAVIANELLO

See CINSAUT.

OTTONESE

See BOMBINO BIANCO.

OVIDIOPOLSKY

Also called Nachodka or Nakhodka, Ovidiopolsky is a white-berried hybrid obtained in 1959 by P K Aivazian, Y N Dokuchaeva, M I Tulaeva, L F Meleshko, A P Abliazova and A K Samborskaya at the Tairov research centre in Odessa, Ukraine, by crossing SEVERNY with Odessky Ustoichivy, a BĂBEASCĂ NEAGRĂ × Rupestris du Lot (a variety of *Vitis rupestris* used as a rootstock) hybrid. It was registered in 1986. This vigorous, winter-hardy, mid- to late-ripening variety produces fruity, sometimes floral, dry wines in the Odessa region of **Ukraine**. It was used to breed ZAGREI.

P

PADEIRO
PÁLAVA
PALLAGRELLO BIANCO
PALLAGRELLO NERO
PALOMINO FINO
PAMID
PAMPANUTO
PANONIA
PAOLINA
PAPAZKARASI
PARDILLO
PARELLADA
PARKENT
PARRALETA
PASCAL BLANC
PASCALE
PASSAU
PASSERINA
PAVANA
PECORINO
PEDRAL
PEDRO GIMÉNEZ
PEDRO XIMÉNEZ
PELAVERGA
PELAVERGA PICCOLO
PELOURSIN
PEPELLA
PERERA
PERLE
PERRICONE
PERRUNO
PERSAN
PERVENETS MAGARACHA
PERVOMAISKY
PETIT BOUSCHET
PETIT COURBU
PETIT MANSENG
PETIT MESLIER
PETIT ROUGE
PETIT VERDOT
PETITE AMIE
PETITE PEARL
PETRA
PETROKORITHO
PETROULIANOS
PHOENIX
PICAPOLL BLANCO
PICARDAN
PICCOLA NERA
PICOLIT
PICULIT NERI
PIEDIROSSO
PIGNOLA VALTELLINESE
PIGNOLETTO
PIGNOLO
PINEAU D'AUNIS

ᛝ this symbol on the following pages indicates that the synonyms, and mistaken identities, are confirmed by DNA evidence

PINELLA
PINOT
PINOTAGE
PINOTIN
PIONNIER
PIQUEPOUL
PLANT DROIT
PLANTA NOVA
PLANTET
PLANTSCHER
PLASSA
PLATANI
PLAVAC MALI
PLAVEC ŽUTI
PLAVINA
PLECHISTIK
PLYTO
PODAROK MAGARACHA
POLLERA NERA
PORTAN
POŠIP BIJELI
POTAMISSI
POULSARD
PRAIRIE STAR
PRČ
PRÉCOCE DE MALINGRE
PRENSAL
PRETO MARTINHO
PRIÉ
PRIETO PICUDO
PRIKNADI
PRIMETTA
PRINZIPAL
PRIOR
PROBUS
PRODEST
PROKUPAC
PROSECCO
PROSECCO LUNGO
PRUNELARD
PRUNESTA
PUGNITELLO
PUKHLIAKOVSKY
PULES

PADEIRO

Productive but good-quality Portuguese variety makes rosé and light-red wines.

BERRY COLOUR: ● ● ● ● ●

PRINCIPAL SYNONYMS: Padeiro de Basto, Tinto Matias

VARIETIES COMMONLY MISTAKEN FOR PADEIRO: ESPADEIRO (Vinho Verde), TINTO CÃO (Douro and Dão)

ORIGINS AND PARENTAGE

Padeiro, or Padeiro de Basto, literally 'baker from Basto', comes from the Minho region in northwest Portugal. It should not be confused with ESPADEIRO, also called Padeiro de Basto or Padeiro Tinto in the same region, nor with TINTO CÃO, another synonym for Padeiro de Basto, all three having distinct DNA profiles in Veloso *et al.* (2010).

VITICULTURAL CHARACTERISTICS

Productive, late budding, early to mid ripening. Susceptible to water stress and downy mildew, not particularly susceptible to botrytis bunch rot.

WHERE IT'S GROWN AND WHAT ITS WINE TASTES LIKE

Padeiro juice tends to be high in sugar and low in acidity but the variety is popular for rosé and light red-fruited reds in the Minho region of **Portugal**, where Vinho Verde is produced and whose climate ensures fresher-style wines. Recommended producers include Quinta de Gomariz, who also make a varietal ESPADEIRO rosé.

PAÏEN

See SAVAGNIN BLANC *in the* SAVAGNIN *entry.*

PAÍS

See LISTÁN PRIETO.

PÁLAVA

Pink-skinned Czech cross producing full-bodied yet fresh, spicy whites.

BERRY COLOUR: ● ● ● ● ●

PRINCIPAL SYNONYMS: Nema, Veverka

ORIGINS AND PARENTAGE

Pálava is a SAVAGNIN ROSE × MÜLLER-THURGAU cross initially bred in 1953 by Josef Veverka at the ŠSV research centre in Velké Pavlovice in southern Morava, in the Czech Republic, and the seedlings then moved to Perná, as for AURELIUS. It was named after the UNESCO-registered Pálava Protected Landscape Area (abbreviated CHKO Pálava in Czech) in southern Morava, and included in the official Czech register of varieties in 1977. Pálava is a sibling of the Slovak variety MÍLIA.

VITICULTURAL CHARACTERISTICS

Thick-skinned berries. Mid to late ripening, moderate resistance to frost but susceptible to fungal diseases. Needs damp, fertile soils and good sites.

WHERE IT'S GROWN AND WHAT ITS WINE TASTES LIKE

Pálava wines show many of the characteristics of its SAVAGNIN ROSE parent – full body, spiciness, some rose petal notes – but are generally slightly higher in acidity and smoother. There were 195 ha (482 acres) in the **Czech Republic** in 2009, almost all in Morava in the south east. Recommended producers include Karpatská Perla, Vajbar Bronislav and Znovín Znojmo. The variety is also found in **Slovakia**, where Château Topoľčianky and Vino Levice make varietal wines.

PALLAGRELLO BIANCO

Old, rare Campanian, not unlike Viognier.

BERRY COLOUR: ● ● ● ● ●

PRINCIPAL SYNONYMS: Pallagrella Bianca, Pallagrello di Avellino, Pallarella, Piedimonte Bianco

VARIETIES COMMONLY MISTAKEN FOR PALLAGRELLO BIANCO: CODA DI VOLPE BIANCA, PIGNOLETTO (also known as Grechetto di Todi)

ORIGINS AND PARENTAGE

The name Pallagrello probably derives from *pagliarello*, the lattice of straw on which the grapes were traditionally dried. The earliest mention of this variety is most likely at the end of the eighteenth century under the name Pallarelli, including both white and black versions. However, Pallagrello Bianco is not a colour mutation of PALLAGRELLO NERO. Both varieties were included in the Vigna del Ventaglio, a vineyard established in 1775 by the architect Luigi Vanvitelli for King Ferdinand I of the Two Sicilies on the slopes of Monte San Leucio, near the royal palace of Caserta in Campania, southern Italy. The vineyard has now disappeared, but an 1826 document described it as a semicircle divided into ten segments so that it looked like a hand-held fan (a *ventaglio*). Each segment was planted with a different variety, among which were Piedimonte Bianco and Piedimonte Rosso, two old synonyms for Pallagrello Bianco and Pallagrello Nero. Indeed, both varieties are thought to originate in the area between Piedimonte Matese and Alife in the province of Caserta.

Pallagrello Bianco has often been confused with CODA DI VOLPE BIANCA since the bunches of both are shaped like a foxtail (*coda di volpe*) but DNA profiling has shown that they are distinct varieties, and that Pallagrello di Avellino and Pallagrello di Caserta are also distinct (Costantini *et al.* 2005). It was later established that Pallagrello di Caserta is in fact identical to PIGNOLETTO, also called Grechetto di Todi (Vouillamoz).

VITICULTURAL CHARACTERISTICS

Low yields, mid to late ripening. Good resistance to botrytis bunch rot.

WHERE IT'S GROWN AND WHAT ITS WINE TASTES LIKE

As with many other indigenous Campanian varieties, cultivation of Pallagrello was drastically reduced during the twentieth century. Both Pallagrello Bianco and PALLAGRELLO NERO were considered to be extinct when they were rediscovered in the 1990s by Peppe Mancini, a local lawyer and passionate winegrower.

Pallagrello Bianco is now cultivated to a very limted extent in the province of Caserta, in Campania, **Italy**, mainly in the communes of Caiazzo, Castel Campagnano and Castel di Sasso. Rare examples of varietal wines are made by Terre del Principe, who produce both an unoaked and a barrel-fermented version, and Castello Ducale. The wines tend to be high in alcohol and have only moderate acidity, with aromas of peaches and apricots, rather like VIOGNIER. No statistics are available because Pallagrello Bianco was treated as a synonym of CODA DI VOLPE BIANCA in the 2000 census.

PALLAGRELLO NERO

Rare dark-skinned variety from Campania,
unrelated to Pallagrello Bianco.

BERRY COLOUR: ● ● ● ● ●

PRINCIPAL SYNONYMS: Coda di Volpe Nera, Pallagrella Nera, Piedilungo, Piedimonte Rosso

ORIGINS AND PARENTAGE

The name Pallagrello probably derives from *pagliarello*, the lattice of straw on which the grapes were traditionally dried. The earliest mention of this variety is most likely at the end of the eighteenth century under the name Pallarelli, including both white and black versions. Although it is also named Coda di Volpe Nera, Pallagrello Nero is a colour mutation of neither CODA DI VOLPE BIANCA nor PALLAGRELLO BIANCO. Together with Pallagrello Bianco, Pallagrello Nero was included in the eighteenth-century Vigna del Ventaglio – see Pallagrello Bianco for more details.

DNA studies suggest a close genetic relationship between Pallagrello Nero and CASAVECCHIA, another rare and old variety of the province of Caserta in Campania (Masi *et al.* 2001).

VITICULTURAL CHARACTERISTICS

Mid to late ripening. Low yields. Rich in anthocyanins. Good resistance to botrytis bunch rot.

WHERE IT'S GROWN AND WHAT ITS WINE TASTES LIKE

As with many other indigenous Campanian varieties in **Italy**, cultivation of Pallagrello Nero was drastically reduced during the twentieth century. Both PALLAGRELLO BIANCO and Pallagrello Nero were considered to be extinct when they were rediscovered in the 1990s by Peppe Mancini, a local lawyer and passionate winegrower.

Pallagrello Nero is now cultivated in the province of Caserta, mainly in the communes of Alife, Alvignano, Caiazzo and Castel Campagnano. It is also cultivated to a lesser extent in the provinces of Avellino and Campobasso. Rare examples of varietal wines, generally classified as IGT Campania, are made by Alois, Castello Ducale and Terre del Principe. Such wines are generally cherry-fruited and sometimes peppery, with smooth but noticeable tannins that allow the wine to age some years in bottle.

PALOMINO FINO

The sherry grape.

BERRY COLOUR:

PRINCIPAL SYNONYMS: Albán, Albar, Albillo de Lucena, Bianco (Piemonte in north-western Italy), Fransdruif (South Africa), Golden Chasselas (California), Jerez, Listan or Listan de Jerez (France), Listán Blanco (Andalucía, Islas Canarias), Listán Comun, Listão or Listrão (Madeira), Madera, Malvasia Rei (Portugal), Manzanilla de Sanlucar, Ojo de Liebre, Palomina Blanca (Jerez de la Frontera), Palomino (Requena, Conila and Tarifa), Palomino Macho, Palomino Pelusón, Temprana, Tempranilla (Rota and Granada), Xeres

VARIETIES COMMONLY MISTAKEN FOR PALOMINO FINO: ALBILLO DE ALBACETE (Castilla-La Mancha), DIAGALVES (Portugal), LISTÁN DE HUELVA, MERSEGUERA (Palomino Fino is called Merseguera in Algeria), SÉMILLON (called Palomino Fino in Fréjus in France)

ORIGINS AND PARENTAGE

Palomino Fino is an old Andalucian variety from southern Spain known under various synonyms, Listán being the most widespread. According to Permartín in García de Luján (1996), Palomino was named after Fernán Yáñez Palomino, a knight of King Alfonso X of Castilla (1221–84), and was introduced to the Islas Canarias in the late fifteenth century. Two morphological variants of Palomino Fino have been described: Palomino Pelusón with hairy leaves and Palomino Macho with poor fruit set.

Palomino Fino is not the white-berried version of Palomina Negra, an old synonym for LISTÁN PRIETO, already described in 1513 by Alonso de Herrera (1790) under the name Palomina Negra, but their DNA profiles do not exclude a close genetic relationship (Vouillamoz).

In California, Golden Chasselas usually refers to Palomino Fino although some vines under that name have turned out to be the true CHASSELAS.

Palomino Fino was used to breed many varieties but of these only CHASAN, COLOMINO and CORREDERA seem to be cultivated today.

VITICULTURAL CHARACTERISTICS

Mid budding and mid to late ripening. Medium to large bunches of medium to large berries with relatively thin skins. Rapid drop in acidity at maturity. Consistently productive and resistant to rot. Well suited to hot, dry and sunny sites. Fairly susceptible to downy mildew and anthracnose but less so to powdery mildew.

WHERE IT'S GROWN AND WHAT ITS WINE TASTES LIKE

In **France** in 2009 there were 172 ha (425 acres) of what is known there as Listan or, logically, Listan de Jerez, mainly in Gascogne and the western Languedoc – a significant reduction from the 1,252 ha (3,094 acres) fifty years ago. It was once much more widely planted for distillation into armagnac but has now been largely replaced by other varieties such as Ugni Blanc (TREBBIANO TOSCANO), COLOMBARD and BACO BLANC.

Palomino Fino was once grown exclusively around Sanlúcar de Barrameda in the Jerez region north of Cádiz in south-western **Spain** and gradually became the variety of choice for sherry production. Although it is only one of three varieties authorized for such wines, it is by far the most widely planted, with 13,926 ha (34,412 acres) in 2008, almost three quarters of that total in Andalucía and the rest mostly, and surprisingly given the cooler, wetter climate, in Galicia, in north-west Spain. After phylloxera, Palomino Fino's productivity and ready availability in nurseries were regarded as sufficient attributes for widespread planting. Although it is giving way to more aromatic and more characterful local varieties in the north west, it is still officially a principal white variety in such DOs as Rueda and Bierzo in Castilla y León. In Andalucía, it is most significant in Jerez, where it thrives on the white chalky-looking albariza soils, but also contributes to the fortified wines of Condada de Huelva further north and west towards Portugal.

To this total vineyard area recorded for Palomino Fino, one should also add the 9,483 ha (23,433 acres) of what is listed separately in the official statistics as Listán Blanco, the most important and successful white variety on the Islas Canarias. Here it is one of the principal white varieties in all the islands' DOs. Producers of varietal wines include Bilma, Monje, Soagranorte and Valleoro.

In addition, the official Spanish register of commercial varieties and the official vineyard statistics both distinguish between Palomino and Palomino Fino. Although there are some ampelographic differences that have led to this division (mainly the shape or openness of the flower), the varieties are genetically identical – based on a comparison of the DNA profiles published by Ibañez *et al.* (2003), Martín *et al.* (2003) and González-Andrés *et al.* (2007) – and the synonyms are not used consistently to distinguish them. In many regions, Palomino is used for both, although Palomino Fino is more often used in Andalucía, its place of origin. The relative quality of the wines produced seems to depend more on the growing environment than on the particular clone planted. 3,358 ha (8,297 acres) of Palomino *tout court* were recorded in 2008.

Varietal wines are typically neutral and low in alcohol (both because the berries have low levels of fermentable sugar and because they tend to be picked early) and low in acidity and ideally suited to the production of fortified wines in a range of styles such as those found in the Jerez region, from fino through to amontillado. Although the variety's tendency to oxidize is not a drawback in these wines, it does mean that unfortified wines are generally dull and flabby and in need of added acidity (although the Navazos-Niepoort dry table wine aged under flor is an interesting exception).

In **Portugal** there were 3,468 ha (8,570 acres) of Malvasia Rei in 2010, mostly in northern Portugal and particularly in the Douro, where it produces large amounts of typically bland wine, although the aromas are increased by cool fermentation. It is used in both table and fortified wines and is steadily losing ground to better-quality varieties.

The variety was a popular import to **Cyprus** because of its high yields and suitability for the production of what was once known as 'Cyprus sherry' but the decline in the market for such wines led to widespread uprooting of the variety, leaving just 56 ha (138 acres) in 2009, used for both fortified and table wines.

In **California**, there were 340 acres (138 ha) in 2008, mostly blended away in Madera and Fresno counties in the the San Joaquin Valley, although a few producers such as Andrew Quady also make sherry-style wines. **Argentina** recorded 168 ha (415 acres) in 2008.

Used in **Australia** mainly for fortified, sherry-style wines, Palomino had declined to just 50 ha (124 acres) in 2008. There are also minor plantings in **New Zealand** (14 ha/35 acres in 2008), where it used to be far more popular.

At one time Palomino – as Fransdruif in Afrikaans or White French in English – was widely planted in **South Africa**, mainly for distillation, but just 381 ha (941 acres) remained in 2008.

PAMID

Old, widely planted but uninspiring Bulgarian red wine grape.

BERRY COLOUR: ● ● ● ● ●

PRINCIPAL SYNONYMS: Grechesky Rosovy (Russia), Manaluki (Albania), Pamidi (Greece), Pamitis (Greece), Piros Szlanka (Hungary), Plovdina (Republic of Macedonia), Roşioară (Romania), Saratchoubouk (Turkey), Slankamenka Crvena (Serbia)

ORIGINS AND PARENTAGE

Some ampelographers say that Pamid originates in Bulgaria, others in Thráki (Thrace), Greece, but all agree that it is an old variety from the Balkan peninsula, where it has long been cultivated. Pamid shows significant biodiversity, with at least seven cultivated forms (Katerov 2004): Pamid Rozov (pink), Pamid Siv (grey), Pamid Cheren (black), Pamid Byal (white), Pamid Edar (big), Pamid Izresliv (prone to coulure) and Pamid s Makhrovati Tsvetove (with flowers like sugar beet).

The white-berried DIMYAT is also known as Pamid but the varieties are quite distinct.

VITICULTURAL CHARACTERISTICS

Fertile, mid ripening, productive. Small, thin-skinned berries. Susceptible to downy and powdery mildews but shows some resistance to botrytis bunch rot. At risk from mites and grape moths.

WHERE IT'S GROWN AND WHAT ITS WINE TASTES LIKE

Pamid is **Bulgaria**'s most widely planted variety (15,842/39,146 acres in 2009, ie 15% of the total vineyard area) but also its least interesting, as hinted by the fact that it is also a table grape. Wines are typically light in colour, thin, lacking acidity and best drunk very young, though they do have a slight fruit sweetness to explain their popularity on the domestic market. The best wines come from hillside vineyards and some rosé is also made.

It is popular in **Romania**, where it is known as Roşioar and found predominantly in Oltenia in the south west. Total Romanian plantings in 2008 were 3,231 ha (7,984 acres). Pamid is found under various names (see synonyms above) in **Serbia**, the **Republic of Macedonia**, **Albania**, **Turkey**, **Greece** and **Hungary**. In Macedonia, the area declined considerably after the uprootings required by phylloxera but it is still found in most vineyard areas.

PAMPANUTO

Minor and largely unexciting Puglian.

BERRY COLOUR: ● ● ● ● ●

PRINCIPAL SYNONYMS: Pampanino, Pampanuta, Rizzulo

ORIGINS AND PARENTAGE

The origins of this variety and of its strange name are unknown.

VITICULTURAL CHARACTERISTICS

Vigorous, mid to late ripening.

WHERE IT'S GROWN AND WHAT ITS WINE TASTES LIKE

Long cultivated in Puglia in southern Italy, mainly in the province of Bari (Corato, Ruvo di Puglia and Minervino), Pampanuto was rescued from extinction in the 1970s and by 2000 there were 291 ha (719 acres) planted in **Italy**. Today, Pampanuto is often blended with BOMBINO BIANCO or TREBBIANO GIALLO, although the Bianco version of the Castel del Monte DOC may comprise 100% Pampanuto. Examples of such varietal wines include Crifo's Bianco Dry (a vino da tavola), Agresti's Reges Apuliae (IGT Murgia) and Marmo Maria's Castel del Monte DOC. They tend to be dry and light and can lack acidity.

PANONIA

Panonia is a FRÜHROTER VELTLINER × MÜLLER-THURGAU white-berried cross obtained by Petar Cindić at the Sremski Karlovci viticulture research centre (part of Novi Sad University), Vojvodina, **Serbia**, and released in 2003. It is a sibling of MORAVA and should not be confused with the Hungarian table grape Pannonia Kincse.

PANSA BLANCA

See XARELLO.

PAOLINA

Paolina is a very rare local white variety of the Val di Cembra in Trentino, northern **Italy**. It was traditionally cultivated in the region of Pergine near the Lago di Canzolino, east of Trento, and in Dro near the Lago di Cavedino in the Valle dei Laghi, west of Trento. Until recently, Paolina was considered to be extinct, but it has been rediscovered by amateur ampelographer Gianpaolo Girardi near the Lago di Canzolino. Pelz and Piffer in Cembra are currently the only producers of a varietal wine, classified as IGT Vigneti delle Dolomiti.

Paolina does not appear in any official catalogue or any official modern grape book. It is not yet known if Paolina is a unique variety or if it is a synonym for a more well-known grape, but DNA profiling could easily clear this up.

PAPAZKARASI

Turkish variety found mainly in Trakya (Thrace) with a very local purpose.

BERRY COLOUR: ● ● ● ● ●

PRINCIPAL SYNONYMS: Papaskara, Papaz Karası

VARIETIES COMMONLY MISTAKEN FOR PAPAZKARASI: ADAKARASI (Avşa), ÇALKARASI (Denizli), HOROZKARASI (Kilis), KALECIK KARASI (Ankara)

ORIGINS AND PARENTAGE

Papazkarası, literally 'black of the priest', was possibly named after the numerous Byzantine monasteries on the Turkish island of Avşa in the Marmara Sea, where it was supposedly born.

OTHER HYPOTHESES

See ADAKARASI.

VITICULTURAL CHARACTERISTICS

Mid ripening, good general disease resistance. Grows well in clay-loam soils and dry conditions but berries generally attain only moderate levels of acidity and sugar.

WHERE IT'S GROWN AND WHAT ITS WINE TASTES LIKE

Papazkarası is grown in those parts of Trakya, north of the Sea of Marmara, that have a continental climate. The grapes are used to make dark, powerful wines but also to produce the traditional local drink known as Hardaliye, in which the must goes through malolactic fermentation and is mixed with mustard seeds, sour cherry leaves and benzoic acid (Gülcü 2010). Melen produce an oak-aged example. There were 211 ha (521 acres) in Turkey in 2010.

PARDILLO

Local La Mancha variety producing well-structured, non-aromatic wines.

BERRY COLOUR: ● ● ● ● ●

PRINCIPAL SYNONYMS: Blanca Pequeña, Marisancho (Albacete and Cuenca), Parda, Pardilla, Pardillo de Madrid, Parill Blanco, Temprana Media (Cigales)

VARIETIES COMMONLY MISTAKEN FOR PARDILLO: ALBILLO REAL

ORIGINS AND PARENTAGE

Pardillo (until 2009 officially known as Pardilla) is probably indigenous to La Mancha in central Spain, where it has long been cultivated under its old synonyms Parill Blanco and Marisancho (Clemente y Rubio 1807). The latter name is also used occasionally for MERSEGUERA in Valencia.

Confusingly, Pardilla is a synonym for both Pardina (see CAYETANA BLANCA) and Pardillo.

VITICULTURAL CHARACTERISTICS

Small to medium-sized bunches of relatively thin-skinned, medium-sized berries. Late budding and very late ripening. Fertile, drought-resistant and well suited to low-fertility soils. Susceptible to botrytis bunch rot.

WHERE IT'S GROWN AND WHAT ITS WINE TASTES LIKE

Varietal wines are typically high in alcohol and well structured but not particularly aromatic. There were 3,547 ha (8,765 acres) in 2008, almost all in Castilla-La Mancha, **Spain**.

PARDINA

See CAYETANA BLANCA.

PARELLADA

Aromatic variety with high quality potential used mainly in the production of Cava.

BERRY COLOUR: ● ● ● ● ●

PRINCIPAL SYNONYMS: Martorella, Moltonach, Montañesa, Montona (Mallorca), Montònec, Montónega, Montonench, Verda Grossa

ORIGINS AND PARENTAGE

Parellada originates in Aragón in north-east Spain, where it has been widespread since the fourteenth century under a variety of old synonyms (Favà i Agud 2001). In the fifteenth century, Valencian winemaker, physician and writer Jaume Roig is credited with having made a white Montònec that was much appreciated by King Juan II. It may be genetically close to ALBILLO REAL (Ibañez *et al.* 2003).

VITICULTURAL CHARACTERISTICS

Big bunches of medium-sized berries. Early budding but very late ripening. Very productive and well suited to the higher zones of Penedès in Catalunya. Susceptible to powdery mildew

WHERE IT'S GROWN AND WHAT ITS WINE TASTES LIKE

Spain grew 8,786 ha (21,711 acres) of Parellada in 2008 with virtually all of it, 8,600 ha (21,251 acres), in Catalunya, where it is mainly used in the production of Cava, particularly in the provinces of Tarragona, Barcelona and Lérida. Because it is one of the three principal white varieties of the Cava DO (along with MACABEO and XARELLO), which may be made in 160 municipalities in seven different regions (even though 85% comes from the area around Sant Sadurní d'Anoia in Penedès), it is also scattered across most of central Spain. It is one of the main varieties in the Catalan DOs Conca de Barberà, Costers del Segre, Penedès and Tarragona, and is authorized on the island of Mallorca for both still and sparkling wines but is only scantily planted there.

Parellada produces aromatic wines that have moderate alcohol levels and delicate, fruity acidity, generally for drinking young. Examples of varietal wines, which are few and far between, are produced by Celler Carles Andreu in Conca de Barberà and Torres, who makes a dry (Viña Sol) and a sweet version (San Valentin). Gramona's Moustillant Brut sparkling wine is also 100% Parellada.

Parellada may be advancing north. It was added to the official register of varieties in **France** in 2011.

PARKENT

Black-berried, mid-ripening variety that, as its name suggests, probably originates from the Parkent area, near Tashkent in north-east **Uzbekistan**, where it is still planted and is used both for the table and for wine.

PARRALETA

Distinctive Somontano variety producing dark, aromatic wines in several countries and under many different names.

BERRY COLOUR: ● ● ● ● ●

PRINCIPAL SYNONYMS: Bonifaccencu or Bonifacienco (Sardegna in Italy), Bonvedro or Bomvedro (Bucelas in Portugal, Australia), Carcaghjolu Neru (Corse), Carcajolo Nero or Carcajolo Noir (Corse), Carenisca⚮ (Sardegna), Caricagiola (Gallura in Sardegna), Cua Tendra (Lérida in Spain), Espagnin Noir (France), False Carignan (Australia), Lambrusco de Alentejo (Portugal), Monvedro⚮ (Bucelas in Portugal) or Monvedro do Algarve (Algarve in Portugal), Monvedro de Sines (Portugal), Olho Branco (Dão in Portugal), Parrel (Somontano in Spain), Pau Ferro (Algarve), Perrel, Preto Foz (Dão), Preto João Mendes (Portugal), Salceño Negro⚮ (Somontano), Tinta Caiada⚮ (Alentejo in Portugal), Tinta Grossa (Alenquer in Portugal), Tinta Lameira (Douro), Tintorro (Alenquer), Torres de Algarve (Portugal)

VARIETIES COMMONLY MISTAKEN FOR PARRALETA: GRACIANO⚮, MAZUELO⚮, MONVEDRO⚮ (Dão)

ORIGINS AND PARENTAGE

This widely dispersed variety is known under various names in several countries around the Mediterranean. The Spanish name Parraleta has been chosen here as the prime name because it's the name used in its likely place of origin, the Somontano region in the province of Huesca, north-east Spain (Casanova Gascón 2008), as suggested by Vacarel's 1765 reference to it under the old synonym Parrel (Montaner *et al.* 2004). Vacarel also referred to **Ribote** in the province of Huelva, an old pink-berried variety that has been shown by DNA profiling to be a colour mutation of Parraleta (Montaner *et al.* 2004).

Comparison of the DNA profiles of Tinta Caiada (Almadanim *et al.* 2004), cultivated in the Alentejo in Portugal, and Carenisca (Zecca *et al.* 2010), grown on the Italian island of Sardegna, and Salceño Negro (Ibañez *et al.* 2003) from Somontano with that of Parraleta (Martín *et al.* 2003) suggests that they are one and the same. Tinta Caiada is also identical to Carcajolo Nero on the French island of Corse, this local name being based on the dialect word *carcaghjolu*, meaning 'yielding much'. It was then introduced as Caricagiola to the Italian island of Sardegna, probably from Bonifacio, as suggested by the synonym Bonifaccencu or Bonifacienco. These synonyms illustrate the vast distribution of this old variety across the Mediterranean basin, even though some of the synonymies are based on ampelographic similarity and have not yet been confirmed by DNA analysis.

Although Parraleta/Tinta Caiada and MONASTRELL are morphologically and genetically quite distinct (Martín *et al.* 2003), there has been some linguistic confusion between them because they both share the synonym Monvedro (or Bonvedro) in Portugal, which is also the prime name for another distinct variety in Dão (see MONVEDRO). Similarly, Parraleta/Tinta Caiada is known as Bastardo in Reguengos de Monsaraz in the Alentejo and possibly as Bastardão in Bairrada but is quite distinct from any other variety of that name (see TROUSSEAU).

OTHER HYPOTHESES

Although it is known as Carcajolo Nero on the French island of Corse, Parraleta is not a colour mutation of BARIADORGIA, which is also called Carcajolo Blanc. It has been speculated that Parraleta (under the Sardinian name Caricagiola) is closely related to VERMENTINO NERO from Toscana and Liguria, but this has not yet been tested by DNA analysis.

VITICULTURAL CHARACTERISTICS

Vigorous and relatively early ripening. Susceptible to botrytis bunch rot.

WHERE IT'S GROWN AND WHAT ITS WINE TASTES LIKE

On the island of Corse, **France**, there are very limited plantings of Carcaghjolu Neru (or Carcajolo Nero), mainly in Figari, Porto-Vecchio and Sartène in the south and, to an even lesser extent, on the east coast.

In **Spain** Parraleta vines, found only in Aragón in the north east, produce wines that tend to be deeply coloured, light, fresh and fragrant, often slightly floral. The variety is provisionally included in the official Spanish register of commercial varieties and is authorized in the Somontano DO but its plantings in that region have diminished considerably in the last thirty-five years: in 1975 it represented 22% of the vineyard area in the province of Huesca but by 2006 it was down to a meagre 0.2% (Casanova Gascón 2008) and total plantings in Aragón in 2008 were just 75 ha (185 acres). However, new plantings are underway as companies such as Viñas del Vero make an effort to conserve this distinctive cultivar. Varietal wines are produced by, for example, Ballabriga and Pirineos, both in Somontano.

In **Portugal**, under various names but officially known as Tinta Caiada, there were 299 ha (739 acres) in 2010. The variety has had a chequered history: it was long thought to produce rather mediocre and anonymous wines until in recent years producers in the Alentejo, perhaps the only region where it reaches perfect maturity, started to take it more seriously and produce much better wines. This wave of enthusiasm and the increased plantings have started to slow but producers such as Ervideira, Maria de Lourdes de Noronha Lopes and Monte Seis Reis make deeply coloured, fresh yet rounded reds, mostly for early drinking. Current work on clonal selection should help to maintain this improvement in quality.

Bonvedro/Bomvedro does not appear in **Australia**'s official statistics but in the past it was occasionally mistakenly known as Carignane (see MAZUELO).

PASCAL BLANC

Historic variety barely surviving in Provence.

BERRY COLOUR: ● ● ● ● ●

PRINCIPAL SYNONYMS: Pascal, Pascaou, Pascau

ORIGINS AND PARENTAGE

Pascal Blanc is a local Provençal variety, typically in the Var and in the Bouches-du-Rhône in southern France, where it was first mentioned by Gadriel in 1715 under its dialect name in the region of Aix-en-Provence (Rézeau 1997): 'Pascau blanc, cette espèce est très-commune' ('Pascau blanc, this species is very common'). The first use of the modern spelling appeared in 1783–4 in Toulon: 'Pascal Blanc, la grappe est longue et grosse, les grains gros, ronds, le gout en est fade' ('Pascal Blanc, the bunch is long and large, the berries are big, the taste is dull').

Its name comes from the Provençal word *pascau* that most likely derives from the Latin *paschalis*, meaning 'of Easter', but the link with the religious feast is not clear.

Pascal Blanc belongs to the Claret ampelographic group (see p XXVII; Bisson 2009).

VITICULTURAL CHARACTERISTICS

Fertile, very productive and late ripening. Best pruned short and may be trained as a bushvine. Prefers very poor, dry, well-exposed soils. Very susceptible to powdery mildew and botrytis bunch rot. Very big bunches and big berries.

WHERE IT'S GROWN

Recommended in the Vaucluse, **France**, and authorized in several other *départements* of the southern Rhône and the Languedoc, Pascal Blanc is a permitted minor variety in the tiny Cassis appellation on the Mediterranean coast south east of Marseille. However, it is virtually extinct, with less than a hectare (2.5 acres) planted in 2008. Domaine Saint-Louis Jayne continue to use it in their white Cassis blend, Laurent Jayne having defiantly replanted it in 1996.

PASCALE

Dark-skinned Sardinian rarely bottled on its own.

BERRY COLOUR: ● ● ● ● ●

PRINCIPAL SYNONYMS: Barberone, Falso Gregu, Giacomino (Gallura), Morescono (Corse), Muresconu (Corse), Nera Tomentosa, Nieddu Mannu (Pattada), Nieddu Pedra Serra, Pascale di Cagliari, Pascale Nero, Pasquale, Picciolo Rosso, Primidivu Nieddu

VARIETIES COMMONLY MISTAKEN FOR PASCALE: GRACIANO (under the names Bovale Sardo or Cagnulari), MAZUELO, MONICA NERA, NIEDDERA

ORIGINS AND PARENTAGE

Pascale was first mentioned in 1780 in the province of Sassari in the north of the Italian island of Sardegna. Like many other Sardinian varieties, Pascale is said to have been introduced during Spanish rule over the island (1323–1720).

Recent DNA profiling has revealed several synonyms of Pascale: Falso Gregu, Nieddu Mannu, Nieddu Pedra Serra, Nera Tomentosa, Picciolo Rosso, Primidivu Nieddu (De Mattia *et al.* 2007).

OTHER HYPOTHESES

Pascale is also thought to have been introduced from Toscana, because it shares with the Pisan variety BONAMICO the synonym Giacomino. This hypothesis would be supported by the commercial links that existed historically between Toscana and the province of Sassari.

VITICULTURAL CHARACTERISTICS

Highly resistant to downy mildew but not to powdery mildew. Mid ripening, abundant yields.

WHERE IT'S GROWN AND WHAT ITS WINE TASTES LIKE

Pascale is cultivated in the provinces of Sassari and Nuoro in the north of the island of Sardegna, **Italy**, and often blended with GARNACHA TINTA (known locally as Cannonau) and Bovale Sardo. It is authorized as part of the blend in the DOCs Campidano di Terralba and Carignano del Sulcis. Dettori's Ottomarzo is a very rare example of a varietal wine, classified as IGT Romangia Rosso: full-bodied with aromas of blackberries and black cherry. Official Italian statistics for 2000 record a total of 1,273 ha (3,146 acres).

PASSAU

Very minor Veneto cross whose correct parentage has only recently been identified.

BERRY COLOUR: ● ● ● ● ●

PRINCIPAL SYNONYMS: Incrocio Dalmasso 17/25

ORIGINS AND PARENTAGE

The Italian vine breeder Giovanni Dalmasso introduced Passau in 1936 at the Conegliano research centre as a cross between NEBBIOLO and DOLCETTO and it was included in the official register of Italian varieties in 1977. However, DNA profiling has recently shown that the Nebbiolo used by Dalmasso was in fact Nebbiolo di Dronero, which is not Nebbiolo at all but a synonym for CHATUS (Torello Marinoni, Raimondi, Mannini and Rolle 2009). Therefore, Passau must be considered to be a Chatus × Dolcetto cross, just like its siblings SAN MARTINO and VALENTINO NERO.

VITICULTURAL CHARACTERISTICS

Late ripening.

WHERE IT'S GROWN

Passau is cultivated in Piemonte in north-west **Italy**. The Italian census of 2000 recorded just 13 ha (32 acres).

PASSERINA

Old variety from Italy's Adriatic coast made in a variety of styles.

BERRY COLOUR: ● ● ● ● ●

PRINCIPAL SYNONYMS: Cacciadebiti, Caccione, Camplese, Pagadebito, Scacciadebito, Trebbiano di Teramo (Abruzzo), Uva d'Oro, Uva Fermana

VARIETIES COMMONLY MISTAKEN FOR PASSERINA: BIANCAME, BOMBINO BIANCO, TREBBIANO TOSCANO

ORIGINS AND PARENTAGE

Passerina is an old variety thought to come from the Adriatic coast of Italy. It has numerous, often erroneous, synonyms, many referring to its high yields: Cacciadebiti and Scacciadebito both mean 'chasing the debts', Pagadebito means 'paying the debts' and Uva d'Oro, 'golden grape'. A relationship with TREBBIANO TOSCANO has long been suspected but not yet proven by DNA testing.

VITICULTURAL CHARACTERISTICS

A robust, vigorous, high-yielding variety. Late ripening.

WHERE IT'S GROWN AND WHAT ITS WINE TASTES LIKE

Passerina is grown throughout central **Italy** and is widespread in the province of Ascoli Piceno in the Marche, where it is blended with PECORINO and TREBBIANO TOSCANO in Falerio dei Colli Ascolani DOC, and it is made as a varietal wine in Offida. In the Controguerra DOC, in the province of Teramo in Abruzzo, it is authorized on its own or as part of a blend. 767 ha (1,895 acres) were recorded in Italy in 2000. Successful producers of Offida Passerina include La Caniette (Vino Santo), Cocci Grifoni (passito and sparkling), San Giovanni and Villa Pigna. Still, dry varietal wines such as those produced by Casale della Loria and Giovanni Terenzi tend to be fresh, taste of apples and pears and have a spicy or slightly bitter aftertaste.

PAVANA

Much diminished, now highly localized, northern Italian.

BERRY COLOUR: ● ● ● ● ●

PRINCIPAL SYNONYMS: Nera Gentile di Fonzaso (Belluno in Veneto), Nostrana Nera (Puos d'Alpago and Quero, province of Belluno), Pelosetta, Ussulara, Vicentina, Visentina (the last four all in Valsugana, Vallarsa, Vallagarina in Trentino)

VARIETIES COMMONLY MISTAKEN FOR PAVANA: MARZEMINO

ORIGINS AND PARENTAGE

Pavana was first mentioned in Valsugana and near Trento in Trentino-Alto Adige by Acerbi (1825). Its name probably derives from *padovana*, suggesting that the variety came from the region of Padova in the Veneto.

Pavana is not the black version of the no longer cultivated Pavana Bianca, which is often confused with BIANCHETTA TREVIGIANA even though they have distinct DNA profiles (Crespan, Giannetto *et al.* 2008). Pavana is said to be genetically related to SCHIAVA LOMBARDA and Turca (Scienza and Failla 1996), two other northern Italian varieties. However, Turca is identical to DOUCE NOIRE so these relationships seem unlikely.

OTHER HYPOTHESES

Pavana is said to share with SCHIAVA LOMBARDA and Turca the same eastern origin (Scienza and Failla 1996), which is not supported by genetic studies.

VITICULTURAL CHARACTERISTICS

Mid to late ripening. Good resistance to powdery mildew.

WHERE IT'S GROWN

Pavana used to be cultivated in several provinces of the Veneto and Trentino, **Italy**, in the nineteenth century, but its acreage drastically decreased during the twentieth century. In Trentino, Pavana is today cultivated only in the province of Trento, in particular between Levico and Borgo Valsugana. In the Veneto, Pavana is today cultivated only in the province of Belluno, in particular between Fonzaso and Arsiè. Pavana della Valsugana by Francesco Poli is a rare example of a varietal.

PAVLOS

See MALVASIA BIANCA LUNGA.

PECORINO

Marche speciality enjoying a certain modishness. Mineral or herbal dry whites.

BERRY COLOUR: ● ● ● ● ●

PRINCIPAL SYNONYMS: Arquitano, Norcino, Pecorina Arquatanella, Pecorino di Osimo, Promotico, Uva delle Pecore, Uva Pecorina, Vissanello

ORIGINS AND PARENTAGE

Pecorino (from *pecora*, meaning 'sheep') is an old variety probably originating from the Marche, in particular in the area of Monti Sibillini south east of Perugia, where it was widespread in the nineteenth century. In Calabria, the variety named Pecorino is identical to GRECO BIANCO, also known as Pecorello Bianco.

OTHER HYPOTHESES

Pecorino is said to have been domesticated from local wild grapes.

VITICULTURAL CHARACTERISTICS

Early ripening. Good resistance to powdery and downy mildews.

WHERE IT'S GROWN AND WHAT ITS WINE TASTES LIKE

Formerly widely cultivated in the Marche, eastern **Italy**, its distribution was reduced during the twentieth century to the area of Arquata del Tronto midway between Spoleto and the coast. In the 1980s, producer Guido Cocci Grifoni introduced this variety to Offida and Ripatransone in the province of Ascoli Piceno, slightly further north east, where most plantings are today.

A varietal wine is authorized as part of the Offida DOCG. It tends to be firm, dry and minerally and producers include Aurora, Le Caniette, Cantine di Castignano, Ciù Ciù, Cocci Grifoni, Costadoro, Capecci-San Savino and San Giovanni. Pasetti's Pecorino is fresh, herbal and lemony. It is also cultivated in Abruzzo in the provinces of Teramo and Pescara and, to a lesser extent, in Umbria, Toscana, Lazio and Liguria. Umani Ronchi's Pecorino, an IGT Terre di Chieti from Abruzzo, is fresh, delicate and zesty on the nose and crisp and much more mineral on the palate. It may be included as part of a blend – often but not exclusively with TREBBIANO TOSCANO – in the Colli Maceratesi and Falerio dei Colli Ascolani DOCs and in the sparkling version of Controguerra DOC.

There were 87 ha (215 acres) in Italy in 2000 but it is doubtless much more important today.

PEDERNÃ

See ARINTO DE BUCELAS.

PEDRAL

Rare Minho variety, now more common (just) over the border in Spain and found in blends.

BERRY COLOUR: ● ● ● ● ●

PRINCIPAL SYNONYMS: Alvarinho Tinto (Monção), Cainho dos Milagres or Cainho Espanhol (Monção), Dozal, Padral (Monção), Pardal (Castelo de Paiva), Pedrol (Galicia in Spain), Pégudo (Ponte de Lima), Perna de Perdiz (Ponte de Lima), Verdejo Colorado (Arribes del Duero in Spain)

VARIETIES COMMONLY MISTAKEN FOR PEDRAL: BAGA, Periquita (CASTELÃO), RUFETE

ORIGINS AND PARENTAGE

Pedral comes from the Monção area in the Minho, Vinho Verde country, north-west Portugal. The DNA profiles of Verdejo Colorado (Martín *et al.* 2006), which is not a colour mutation of VERDEJO, and of Pedral (Gago *et al.* 2009) turned out to be identical (Vouillamoz). This Portuguese origin probably explains why Martín *et al.* (2006) found Verdejo Colorado to be genetically distant from the other Spanish varieties they analysed.

VITICULTURAL CHARACTERISTICS

Relatively late ripening.

WHERE IT'S GROWN

In spite of Pedral's Portuguese origins, current plantings in Galicia in north-west **Spain** (79 ha/195 acres in 2008), although small-scale, far exceed what remains in **Portugal** (5 ha/12 acres in 2009). The variety is provisionally included in the official Spanish list of commercial varieties and authorized in Rías Baixas.

PEDRO GIMÉNEZ

Quantitatively if not qualitatively important Argentine white unrelated to Spain's Pedro Ximénez.

BERRY COLOUR: ● ● ● ● ●

PRINCIPAL SYNONYMS: Pedro Jiménez

VARIETIES COMMONLY MISTAKEN FOR PEDRO GIMÉNEZ: PEDRO XIMÉNEZ (Spain)

ORIGINS AND PARENTAGE

Pedro Giménez is indigenous to Argentina and is one of the Criollas (see CRIOLLA entry for details). Martínez *et al.* (2003) suggested that Pedro Giménez is ampelographically distinct from the Spanish variety PEDRO XIMÉNEZ, as confirmed by DNA profiling (Vouillamoz).

WHERE IT'S GROWN AND WHAT ITS WINE TASTES LIKE

Pedro Giménez is in decline in **Argentina** but is still one of the mainstays, along with CEREZA and CRIOLLA GRANDE, of the country's production of cheap everyday wine sold in litre bottles and cartons. Fapes in the Valle de Uco make a varietal example. There were 13,476 ha (33,230 acres) in 2009, only slightly less than the much more reputable Bonarda (DOUCE NOIRE), predominantly in Mendoza province and to a lesser extent in San Juan. It is also planted to a much more limited extent in **Chile.**

PEDRO LUIS

See GALEGO DOURADO.

PEDRO XIMÉNEZ

Andalucía's variety used mainly for some of the darkest, stickiest wines made anywhere.

BERRY COLOUR: ● ● ● ● ●

PRINCIPAL SYNONYMS: Corinto Bianco🧬, Don Bueno🧬, Pedro Jimenez (Andalucía in Spain), Pedro Ximenes (Andalucía), Perrum🧬 (Alentejo in Portugal), PX (Andalucía), Verdello🧬 (Islas Canarias), Ximenes (Andalucía)

VARIETIES COMMONLY MISTAKEN FOR PEDRO XIMÉNEZ: ELBLING🧬, GALEGO DOURADO🧬 (South Africa), PEDRO GIMÉNEZ🧬 (Argentina), RIESLING🧬

ORIGINS AND PARENTAGE

Pedro Ximénez most likely originates from Andalucía, southern Spain, where it has been cultivated since the seventeenth century or even earlier: it was already famous in Málaga when Vicente Espinel (1618) spoke about 'una bota de Pedro Ximenez de Malaga' ('a barrel of Pedro Ximenez from Málaga'). Janini and Roy-Chevrier (1905) suggest the variety's name comes from a place called Jiménez in Sanlúcar de Barrameda, but it seems more likely that it was named after a famous local vintner since Ximenès/Ximenez is a common surname in Andalucía.

DNA parentage analysis has suggested a possible parent–offspring relationship between Pedro Ximénez and GIBI (Vargas *et al.* 2007), an old table grape of probable Arabic origin that was formerly cultivated in the south of France and in Spain, where it also gave birth to ALARIJE in Extremadura (Lacombe *et al.* 2007), which is therefore a half-sibling of Pedro Ximénez. In addition, DNA profiling has shown that the widely cultivated table grape Corinto Bianco is a seedless mutation of Pedro Ximénez (Vargas *et al.* 2007).

OTHER HYPOTHESES

In 1661, Dr Sachs speculated that Pedro Ximénez was introduced from the Islas Canarias or from Madeira to the Rhein and the Mosel, whence a soldier named Pedro Ximen or a cardinal named Ximenès brought it back to Málaga in Spain (Galet 2000; Vargas *et al.* 2007). This legend was propagated by many German authors (von Babo, Stoltz, Sprenger, etc.) who even suggested that Pedro Ximénez and ELBLING or RIESLING were one and the same (Janini and Roy-Chevrier 1905). This identity has been disproved by DNA analysis and the hypothesis as a whole appears highly doubtful since Pedro Ximénez has never been mentioned in south-western Germany and it would not ripen there.

VITICULTURAL CHARACTERISTICS

Large bunches of thin-skinned berries, often far from uniform in size with lots of smaller berries. Mid budding and ripening, vigorous and regularly productive. Highly susceptible to botrytis bunch rot and downy mildew, less so to powdery mildew. Susceptible to esca, eutypa dieback and termites.

WHERE IT'S GROWN AND WHAT ITS WINE TASTES LIKE

The great majority of Spain's 9,583 ha (23,680 acres) in 2008 was in the Montilla-Moriles region in Andalucía, southern **Spain**. Here the variety is used to produce fortified wines similar to sherry and also unfortified wines in various styles but generally fairly high in alcohol (though not as high as sherry) and low in acidity. The sugar-rich grapes are traditionally dried in the sun to concentrate these sugars in order to produce intensely sweet fortified wines akin to essence of alcoholic raisins, often known as PX, and to sweeten certain styles of fortified Montilla. Recommended producers include Alvear, La Aurora, Delgado, Equipo Navazos (both Jerez and Montilla), Gracia Hermanos, Moreno and Pérez Barcero.

Neighbouring Jerez uses Pedro Ximénez for similar purposes but grows far less because the producers here and in Málaga may legally buy PX wines from Montilla for blending, saving them the trouble of cultivating this rather disease-prone variety themselves. Most of the top Jerez producers make a varietal PX. Recommended producers from Málaga include Gomara and Málaga Virgen. It is one of the principal varieties in the Jerez, Málaga, Montilla-Moriles and Valencia DOs and is also planted in Extremadura (639 ha/1,579 acres), Castilla-La Mancha (117 ha/289 acres), Catalunya (12 ha/30 acres) and the Islas Canarias (less than 1 ha/2.5 acres).

As Perrum in **Portugal**, it is grown mainly in the Alentejo, especially in Évora, and to a lesser extent in the Algarve, where it is considered a late ripener and generally contributes to a blend, such as those produced by Carmim, Fundacäo Eugénio de Almeida (Cartuxa's EA) and the Borba co-operative. There were 350 ha (865 acres) in Portugal in 2010.

The wine made from the more than 3,000 ha (7,413 acres) of Pedro Jimenez (sic) grown in **Chile** is mostly distilled to make pisco but the statistics for wine grapes specifically recorded 33 ha (82 acres) of this variety in 2008. Falernia in the Elqui Valley make a refreshing and fruity dry, unfortified version, although it is not known whether the vines are this variety or Argentina's distinct PEDRO GIMÉNEZ.

Pedro Ximénez, often called simply Pedro, was once fairly widely planted in **Australia**, for unfortified blends and for very fine botrytized wines such as the rich, deep golden McWilliam's Pedro Sauterne (sic) made in irrigated vineyards near Griffith in New South Wales. There are still a few producers from a range of wine-growing regions who bottle this wine in a number of styles: rich and fortified at Turkey Flat in the Barossa Valley and at Gralyn in Western Australia, for example; dry, unfortified and bottle aged at Campbell's in Rutherglen, Victoria.

PELAVERGA

The more commonly planted, big-berried Pelaverga, often blended with Nebbiolo and/or Barbera in Piemonte.

BERRY COLOUR: ● ● ● ● ●

PRINCIPAL SYNONYMS: Cari (Chieri and Baldissero), Pelaverga di Pagno (Val Bronda), Pelaverga Grosso, Uva Coussa, Uva delle Zucche (Chieri and Baldissero)

VARIETIES COMMONLY MISTAKEN FOR PELAVERGA: NERETTO DURO (called Peilavert in Salussola and Cavaglià), PELAVERGA PICCOLO

ORIGINS AND PARENTAGE

It is believed that the first mention of Pelaverga appeared in 1511 when Pope Julius II praised a wine from Pagno and Chastellar (now spelt Castellar), in the province of Cuneo, when he was offered a few barrels by Margherita di Foix, Marchesa di Saluzzo (Comba and Dal Verme 1990), although the name Pelaverga is not specifically mentioned in the source document. Pelaverga is distinct from PELAVERGA PICCOLO.

OTHER HYPOTHESES

Pelaverga is said to have been introduced to the commune of Pagno in the Val Bronda about 50 km south west of Torino, in Piemonte, north-west Italy, in the eighth century by the monks of San Colombano di Bobbio, who came to build a convent, which is today the church of San Pietro and San Colombano.

VITICULTURAL CHARACTERISTICS

Late ripening, susceptible to botrytis bunch rot.

WHERE IT'S GROWN AND WHAT ITS WINE TASTES LIKE

Formerly widespread all over Piemonte in north-west **Italy**, grown for the table as well as for wine, Pelaverga is today cultivated for wine production mainly in the Val Bronda, south west of Torino in Piemonte, northern Italy, as well as in Chieri and Baldissero Torinese in the province of Torino, where it is called Cari. 121 ha (299 acres) were recorded in Italy in 2000.

A varietal wine (labelled Pelaverga or Cari) is authorized as part of the Collina Torinese DOC and Pelaverga may be a varietal or blended with NEBBIOLO and/or BARBERA in the Colline Saluzzesi DOC. The wines are light red, sometimes slightly fizzy and peppery, perfumed, and often taste of strawberries. Producers include Casetta, Maero Emidio and Produttori Pelaverga.

PELAVERGA PICCOLO

Fashionable if relatively minor Piemontese making light, fruity, spicy varietal reds.

BERRY COLOUR: ● ● ● ● ●

PRINCIPAL SYNONYMS: Pelaverga di Verduno

VARIETIES COMMONLY MISTAKEN FOR PELAVERGA PICCOLO: PELAVERGA ⚮

ORIGINS AND PARENTAGE

Pelaverga Piccolo is distinct from PELAVERGA (or Pelaverga Grosso), although this was not officially established until the early 1990s (Schneider and Mannini 2006). It is said to have been brought eastwards across the Piemontese province of Cuneo from Saluzzo to Verduno by the priest Sebastiano Valfrè in the eighteenth century.

VITICULTURAL CHARACTERISTICS

Vigorous, mid to late ripening.

WHERE IT'S GROWN AND WHAT ITS WINE TASTES LIKE

Pelaverga Piccolo was almost forgotten until a few producers began to craft interesting varietal wines in the 1970s. Today it is grown to a limited extent around the commune of Verduno, near Alba in Piemonte, north-west **Italy**, where it is the main variety (minimum 85%) of Verduno Pelaverga, or plain Verduno, DOC, producing light, fruity, spicy reds. It is also blended with NEBBIOLO and BARBERA. There were 26 ha (64 acres) in Italy in 2000.

Varietal wines tend to be light in colour, smooth and fruity, occasionally spicy. GB Burlotto's tastes of sweet red fruit but is also a little peppery and nicely fresh. Successful producers include Fratelli Alessandria, Ascheri, Bel Colle, Michele Reverdito and Cantina Terre del Barolo.

PELOURSIN

From Isère, now mostly of historical and genetic significance.

BERRY COLOUR: ● ● ● ● ●

PRINCIPAL SYNONYMS: Corsin (Savoie), Fumette or Feunette (Haute-Savoie), Gros Béclan (Jura), Pelossard, Péloursin (Isère), Pourret or Pourrot (Arbois in the Franche-Comté)

VARIETIES COMMONLY MISTAKEN FOR PELOURSIN: BÉCLAN (Jura), DUREZA (Peloursin is sometimes called Duresa, Dureza or Durezi), DURIF (especially in California), POULSARD (Peloursin is called Pelossard in the Isère)

ORIGINS AND PARENTAGE

Peloursin is an old variety from the Vallée du Grésivaudan near Grenoble in the Isère in eastern France that was later introduced to the Savoie and the Franche-Comté, where it took on many different names. Its prime name supposedly derives from Pelossier, the local name for the blackthorn (*prunellier*). Peloursin was used to breed DURIF through a crossing with SYRAH, and Joubertin (no longer cultivated) through a crossing with PERSAN, and it gave its name to the Pelorsien ampelographic group (see p XXVII; Bisson 2009). In addition, Peloursin is morphologically close to Bia Blanc (Bisson 1999), no longer cultivated.

VITICULTURAL CHARACTERISTICS

Vigorous, mid ripening. Susceptible to black rot and botrytis bunch rot.

WHERE IT'S GROWN AND WHAT ITS WINE TASTES LIKE

The variety seems to have more or less disappeared, apart from a few vines interplanted in old vineyards, and has been overtaken by its offspring DURIF. It has, however, been planted at the experimental domaine of the Tarn *département* at Peyrole in south-west **France** for research purposes. Domaine Ganevat in Rotalier in the Jura have replanted a small area of Gros Béclan and include it in an early-drinking red blend.

Peloursin has recently been identified in an old vineyard in Sonoma, **California**.

PENSAL BLANCA

See PRENSAL.

PEPELLA

White-berried variety from Campania, central **Italy**, grown exclusively in the Costa d'Amalfi south of Napoli, particularly in the Tramonti and Ravello subzones. Its name refers to the variety's susceptibility to millerandage, resulting in bunches with many smaller, seedless berries – *pepe* means 'peppercorn'. With low sugar and acidity, Pepella is used in blends with FALANGHINA FLEGREA and/or BIANCOLELLA in the Costa d'Amalfi DOC, eg by Tenuta San Francesco and in IGT Colli di Salerno by Azienda Agricola Reale.

PERERA

Old variety of the Veneto used to perfume passito and sparkling wines in Prosecco country.

BERRY COLOUR: ● ● ● ● ●

PRINCIPAL SYNONYMS: Perera Gialla, Pevarise

VARIETIES COMMONLY MISTAKEN FOR PERERA: PROSECCO, VERDICCHIO BIANCO (sometimes known as Pevarise and Peverella in Trentino)

ORIGINS AND PARENTAGE

Perera has traditionally been cultivated in Valdobbiadene in the province of Treviso in the Veneto. Its name derives from *pera*, meaning 'pear', because the shape of the bunch looks like an inverted pear. Some producers suggest the name might also be inspired by the pear-like flavour of the grape pulp.

VITICULTURAL CHARACTERISTICS

Vigorous, late ripening, susceptible to flavescence dorée as well as to downy and powdery mildews. Yields are often irregular due to problems at flowering.

WHERE IT'S GROWN AND WHAT ITS WINE TASTES LIKE

Once widespread in the area between Conegliano and Valdobbiadene in the province of Treviso in northern **Italy**, the heartland of Prosecco, Perera almost disappeared in the 1970s after the arrival of the vine disease flavescence dorée, to which it is very sensitive. Perera is today cultivated mainly in Valdobbiadene, in particular in the region of Cartizze. It may be included as a minor component – with Glera (see PROSECCO), VERDISO and Boschera (see VERDICCHIO BIANCO) – in the dry or sweet, passito-style Torchiato di Fregona version of the Colli di Conegliano DOC and a small amount (generally 5–10%, often together with Verdiso and BIANCHETTA TREVIGIANA) may be added to the main variety Glera in Prosecco Conegliano-Valdobbiadene DOCG sparkling wines. It is said to add perfume and contribute a pear-like flavour. There were 25 ha (62 acres) in Italy in 2000.

PERIQUITA

See CASTELÃO.

PERLE

Pink-berried German cross with a hint of its rose-like parental perfume but short on freshness.

BERRY COLOUR:

PRINCIPAL SYNONYMS: Perle von Alzey

ORIGINS AND PARENTAGE

A Roter Traminer (see SAVAGNIN ROSE) × MÜLLER-THURGAU cross bred by Georg Scheu in 1927 at the Alzey research centre in Rheinhessen, Germany, though it was developed at Würzburg in the 1950s and authorized in 1961. This parentage has been confirmed by DNA profiling (Myles *et al.* 2011).

VITICULTURAL CHARACTERISTICS

Late budding but its compact bunches of small berries make it susceptible to botrytis bunch rot. Early to mid ripening with moderate sugar levels and fair yields.

WHERE IT'S GROWN AND WHAT ITS WINE TASTES LIKE

In **Germany**, more than half of its 37 ha (91 acres) can be found in Franken, where its late budding is good insurance against spring frosts. Wines are light, lightly floral but can lack acidity, and the variety is generally in decline. Producers who persevere with it include Kistner and Herbert Schuler in Franken. Perle is authorized in **England** but plantings are too insignificant to appear in the official vineyard statistics.

PERRICONE

Dark-skinned, relatively tannic Sicilian, often blended with Nero d'Avola.

BERRY COLOUR:

PRINCIPAL SYNONYMS: Catarratto Rouge, Niuru, Perricone Nero, Pignateddu, Pignatello, Pirricuni, Tuccarinu

VARIETIES COMMONLY MISTAKEN FOR PERRICONE: BARBERA, MAMMOLO, NERELLO CAPPUCCIO, NERELLO MASCALESE, NIEDDERA, SANGIOVESE

ORIGINS AND PARENTAGE

Perricone, from the Italian island of Sicilia, has been confused with many other varieties, as listed above. Recent DNA profiling has provided evidence that they are not identical but that there is a possible parent–offspring relationship with SANGIOVESE (Di Vecchi Staraz, This *et al.* 2007).

VITICULTURAL CHARACTERISTICS

Vigorous, mid ripening.

WHERE IT'S GROWN AND WHAT ITS WINE TASTES LIKE

Perricone is cultivated mainly in the north west of the Italian island of Sicilia, in the provinces of

Palermo and Trapani, as well as in those of Agrigento on the south-west coast and Messina in the far north east. There were 340 ha (840 acres) planted on the island in 2008, according to official statistics. Figures for the whole of **Italy** in the 2000 census were 624 ha (1,542 acres).

Perricone may be made as a varietal wine or included as part of the blend – usually with NERO D'AVOLA – in DOCs such as Contea di Sclafani, Delia Nivolelli, Eloro (here called Pignatello), Erice and Monreale, and as a minor part of the blend in the Alcamo DOC.

Varietal wines vary in style but Caruso & Minini's Sachìa Perricone, classified as IGT Sicilia, has firm tannins and slightly bitter flavours of cherries and dark chocolate. Tamburello's Pietragavina Perricone, Monreale DOC, is also notably tannic and persistent. Other examples of varietal wines include Castellucci Miano di Valledolmo's Maravita (using grapes that have been dried for a month prior to vinification), Feotto dello Jato's Vigna Curria, softened by two years in oak, and Romano's Perricone IGT Sicilia.

PERRUM

See PEDRO XIMÉNEZ.

PERRUNO

Old Extremaduran variety with a lone supporter in Andalucía.

BERRY COLOUR:

PRINCIPAL SYNONYMS: Casta de Montúo, Firmissima, Getibi, Granadina, Jetibi, Morata, Perruna, Perruno Común, Perruno de Arcos, Perruno de la Sierra, Perruno Duro, Perruno Fino, Perruno Tierno

ORIGINS AND PARENTAGE

Perruno, from Extremadura, western Spain, was mentioned in 1792 by García de Lena, who said: 'ni el sol los dulcifica' ('even the sun cannot mellow it') and that 'el nombre está diciendo su ninguna bondad' ('the name speaks about its lack of mildness'), because *perruno* means 'dog' or 'dog-like' in Spanish. Clemente y Rubio (1807) later described three distinct types of this variety in Andalucía, the last two probably colour mutations: Perruno Duro with hard white berries, Perruno Común or Perruno Tierno with reddish berries, Perruno Negro with black berries. At that time, two centuries ago, it covered half of the vineyard area of Jerez de la Frontera.

DNA parentage analysis has suggested a parent–offspring relationship between Perruno and LISTÁN PRIETO (under the synonyms Moscatel Negro or Mission; Ibañez *et al.* 2003; This *et al.* 2006), which makes Perruno a likely half-sibling or a grandparent of CEREZA, Moscatel Amarillo, TORRONTÉS RIOJANO and TORRONTÉS SANJUANINO cultivated in Argentina (see MUSCAT pedigree diagram).

VITICULTURAL CHARACTERISTICS

Early budding with good general resistance to pests and diseases.

WHERE IT'S GROWN AND WHAT ITS WINE TASTES LIKE

Perruno has been provisionally included in the Spanish register of commercial varieties and is authorized in Andalucía, south-west **Spain**, as well as in the Ribera del Guadiana DO in Extremadura. There were 741 ha (1,831 acres) in Extremadura in 2008, mostly in the Ribera Alta subregion around Castuera, where it tends to disappear in bulk wines, and just 3 ha (7 acres) in Andalucía, where Lagar Ambrosio in the Sierra de Cádiz region produce a varietal wine and are working hard to resurrect Perruno there. Victor Palacios from the University of Cádiz suggests that despite the very negative picture painted by García de Lena, it could have interesting results as a wine grape even if it is unsuitable for the table. It is said to be low in both sugar and acidity.

PERSAN

Rare but high-quality Savoie variety on a voyage of rediscovery.

BERRY COLOUR: ● ● ● ● ●

PRINCIPAL SYNONYMS: Beccu (Savoie), Bécuet or Becuét (Val di Susa in Piemonte, Italy), Bécuette (Savoie), Berla'd Crava Cita or Berlo Citto (Val Chisone in Piemonte), Etraire (Isère), Etris (Isère), Princens or Prinsens (Savoie), Serine or Siranne (Isère)

VARIETIES COMMONLY MISTAKEN FOR PERSAN: ETRAIRE DE L'ADUÏ (Persan is called Etraire in the Isère), MONDEUSE NOIRE (called Persagne in the Ain and Petite Persaigne in the Rhône), PINOT NOIR, SYRAH (in Isère, Persan is called Serine, which is also a synonym for Syrah)

ORIGINS AND PARENTAGE

Persan most likely originates from Princens, a hamlet above Saint-Jean-de-Maurienne whose vineyards were reputed to give the best wines of Savoie. Indeed, the name Princens probably derives from the dialect *prin*, meaning 'prime', and *cens*, used to refer to a fee collected by landowners from their vassals, suggesting that Princens was the best land. Princens later evolved in popular usage to Persan, a name that was first mentioned by Albin Gras in 1846 in the Isère (Rézeau 1997): 'Etraire (grosse) R. ou Etrière (rive droite de l'Isère, etc.) synon. Persan (rive gauche de l'Isère, etc.)' ('Etraire (capital) R. or Etrière (right bank of the Isère, etc.) synon. Persan (left bank of the Isère, etc.').

DNA profiling has established that Persan has long been cultivated in the Valle di Susa and around Pinerolo in Piemonte in north-western Italy under the name Bécuet (Schneider *et al.* 2001). Persan has often been confused with other varieties such as ETRAIRE DE L'ADUÏ, MONDEUSE NOIRE, SYRAH or even PINOT NOIR; Guyot (1876) thought that Persan was a local mutation of this last. Rougier (1902b) suggested that Etraire de l'Aduï could be a natural seedling of Persan, which is now confirmed by DNA parentage analysis that strongly suggests a parent–offspring relationship between Etraire de l'Aduï and Persan (Vouillamoz). Indeed, both varieties belong to the Sérine ampelographic group (see p XXVII; Bisson 2009).

Persan was crossed with PELOURSIN to breed Joubertin (no longer cultivated).

OTHER HYPOTHESES

Based on a simplistic and erroneous etymology, Persan has often been thought to originate in Persia. Similarly, the name Princens was said to derive from *principis* ('prince') and *census* (an old fee) and it was suggested that Persan was brought from Cyprus by the Prince of Savoie.

VITICULTURAL CHARACTERISTICS

Vigorous, usually pruned long and trellised. Early budding, mid ripening. Performs well on calcareous stony soils. Susceptible to downy and powdery mildews. Small bunches and berries.

WHERE IT'S GROWN AND WHAT ITS WINE TASTES LIKE

In **France**, Persan used to be the dominant variety in Princens, a highly regarded vineyard near Saint-Jean-de-Maurienne, east of Grenoble, that was mentioned way back in the seventeenth century (Gros 1930). Before phylloxera hit the region at the end of the nineteenth century, Persan was widespread throughout Savoie and in the Isère. Today, even though it is recommended in the Savoie and Isère *départements*, is a permitted variety in the Vin de Savoie appellation and also appears under the local Vin de Pays d'Allobrogie, there are just 9 ha (22 acres). Nevertheless, the variety is regaining popularity and, with the help of Michel Grisard in Fréterive, one of the rare producers of varietal Persan, a local association is planning to replant Persan in the Princens vineyard. Wines tend to be tannic with good acidity and well suited to ageing. Domaine de Méjane and Domaine Saint-Germain also produce varietal wines.

In **Italy**, under the name Bécuet, Persan is usually blended with other local varieties such as AVANÀ to produce Ramiè, a historical alpine wine, but Azienda Agricola Martina in Giaglione and Azienda Sibille in Gravere both make unblended wines.

In **Switzerland**, Genève producer Jean-Pierre Pellegrin has recently planted 1,500 vines of Persan at his Domaine Grand'Cour and the first varietal wine is expected to be released in 2015 or 2016.

PERVENETS MAGARACHA

Recent Ukrainian cross notably resistant to downy mildew and used for dry and sweet wines.

BERRY COLOUR: ● ● ● ● ●

PRINCIPAL SYNONYMS: Pervenec Magaraca, Pervenyec Magaracsa

ORIGINS AND PARENTAGE

Pervenets Magaracha, meaning 'firstborn of Magarach', is a hybrid obtained in 1966 by P Y Golodriga, V T Usatov, L P Troshin, Y A Malchikov, I A Suyatinov, V A Dranovskiy and P N Nedov at the Magarach research centre in Yalta, southern Ukraine, by crossing RKATSITELI with Magarach 2-57-72, in which the latter is a MTSVANE KAKHURI × Sochinsky Cherny cross, and Sochinsky Cherny is a non-cultivated Ukrainian variety. Pervenets Magaracha is a sibling of PODAROK MAGARACHA. It was included in Ukraine's official register of varieties in 1992.

VITICULTURAL CHARACTERISTICS

Productive, mid to late ripening. Cold-hardy to –25 °C (–13 °F) and resistant to downy mildew.

WHERE IT'S GROWN AND WHAT ITS WINE TASTES LIKE

Pervenets Magaracha is used in **Ukraine** (643 ha/1,589 acres in 2009) to make amber-coloured dry and sweet wines with fresh acidity. It is also quite widely planted in **Russia** – in Temriuk and Anapa in Krasnodar in the south west (1,915 ha/4,732 acres in 2010) as well as in Stavropol and Rostov-on-Don (323 ha/798 acres in 2010) – and has also been reported in **Azerbaijan** and the **Republic of Moldova**.

PERVOMAISKY

Uzbek cross used to make deeply coloured dessert wines.

BERRY COLOUR: ● ● ● ● ●

ORIGINS AND PARENTAGE

Pervomaisky is a TEINTURIER × ALEATICO cross obtained in 1935 by A N Negrul and M S Zhuravel at the Central Asian branch of the VIR (now known as UzRIPI or the Uzbek Research Institute of the Plant Industry) in Uzbekistan and released in 1953. Pervomaisky means 'first of May' or 'May Day'.

VITICULTURAL CHARACTERISTICS

Vigorous, productive, mid to late ripening. Red-fleshed berries with thin but tough skins. Susceptible to downy and powdery mildews.

WHERE IT'S GROWN AND WHAT ITS WINE TASTES LIKE

Pervomaisky, grown in **Uzbekistan** and to a limited extent in **Ukraine** and the **Republic of Moldova**, is used to make sweet, dark red wines and also for the table.

PESECKÁ LEÁNKA

See FETEASCĂ REGALĂ.

PETIT BOUSCHET

Disappearing red-juiced nineteenth-century southern French cross.

BERRY COLOUR: ● ● ● ● ●

PRINCIPAL SYNONYMS: Aramon-Teinturier, Bouschet Petit, Negrón de Aldán☒ (Bierzo in Spain), Tintinha (Portugal)

ORIGINS AND PARENTAGE

An ARAMON NOIR × TEINTURIER cross obtained in 1824 by Louis Bouschet at the Domaine de la Calmette in Mauguio in the Hérault in the south of France. Although both Petit Bouschet and JUAN GARCÍA are sometimes called Negrón de Aldán (González-Andrés *et al.* 2007), they are not related. Recent DNA profiling has suggested a possible parent–offspring relationship with TÉOULIER NOIR (Myles *et al.* 2011) but this can be rejected by comparison with the DNA profile of the true Téoulier Noir (Vouillamoz).

Petit Bouschet is a sibling of Gros Bouschet (no longer cultivated) and was used to breed ALICANTE HENRI BOUSCHET, GRAND NOIR and MORRASTEL BOUSCHET.

VITICULTURAL CHARACTERISTICS

Early budding, fertile and vigorous, best pruned short. Tends to produce many suckers. Risk from erinose mite in some years. Early to mid ripening.

WHERE IT'S GROWN AND WHAT ITS WINE TASTES LIKE

Petit Bouschet, like all vines with TEINTURIER ancestry, produces deeply coloured wines but they are otherwise unremarkable. Popular in **France** in the second half of the nineteenth century, notably to add colour to blends, it was overtaken by the greater popularity of its offspring ALICANTE HENRI BOUSCHET, and has now largely disappeared from French vineyards, though there are minor plantings in **Portugal** (11 ha/27 acres in 2009) and **North Africa**.

PETIT COURBU

Minor but high-quality, aromatic Jurançon speciality.

BERRY COLOUR:

PRINCIPAL SYNONYMS: Courbis, Courbu Petit, Petit Courbu Blanc, Vieux Pacherenc, Xuri Zerratu (País Vasco in Spain)

VARIETIES COMMONLY MISTAKEN FOR PETIT COURBU: COURBU BLANC, LAUZET (Gan and Lasseube, near Pau)

ORIGINS AND PARENTAGE

Petit Courbu is local to Jurançon in France's far south west, considered by Galet (1990) to be a clonal variation of COURBU BLANC, another Jurançon variety with which Petit Courbu shares the synonyms Courbis and Vieux Pacherenc, while others confuse it with LAUZET. However, ampelographic observations and DNA profiling have clearly disproved these hypotheses (Bowers *et al.* 1999; Bordenave *et al.* 2007).

VITICULTURAL CHARACTERISTICS

Not very productive, late ripening. Susceptible to botrytis bunch rot. Small bunches and very small berries.

WHERE IT'S GROWN AND WHAT ITS WINE TASTES LIKE

Petit Courbu is grown only in the far south west of **France** and is generally used in blends, although dry varietal wines have the potential for great finesse and have been described as tasting like a cross between VIOGNIER and RIESLING. Rare examples of this aromatic wine (lemon, floral, sometimes honey) are produced by Alain Brumont and Domaine Laougué, both within the Pacherenc du Vic-Bilh appellation, where it is considered one of the principal varieties. It may be included in blends in the Béarn, Irouléguy, Jurançon and Pacherenc appellations, though it rarely contributes more than 20–30%. The regulations for Côtes de Saint-Mont stipulate a minimum of 30%. Total plantings in France in 2008 were 68 ha (168 acres), a considerable increase on the 10 ha (25 acres) in 1958.

PETIT MANSENG

High-quality variety making superb sweet wines in south-western France.

BERRY COLOUR: ● ● ● ● ●

PRINCIPAL SYNONYMS: IzkiriotTtipi or Iskiriota ZuriTipia or Ichiriota ZuriaTipia (País Vasco in Spain), Manseng Petit Blanc, Mansengou (Béarn)

ORIGINS AND PARENTAGE

The earliest mention of a Manseng was recorded in 1562 in Jurançon in a text in Occitan: 'vinhe mansengue'. However, it is not known whether this name referred to GROS MANSENG or to Petit Manseng, and the two varieties were not distinguished until 1783–4: 'Manseng gros, blanc ..., couleur dorée, bon gout quoy que piquant. Manseng petit, blanc..., le gout plus piquant que tout autres' ('Manseng gros, white ... golden colour, good taste although spicy. Manseng petit, white ... the taste is spicier than any others'). The name Manseng could derive from *mans*, meaning 'mansion'.

According to Galet (1990), Petit Manseng is morphologically close to GROS MANSENG, and Bisson (2009) groups them in the Mansien ampelographic group (see p XXVII). DNA parentage analysis and reconstruction of the most likely pedigree (see the PINOT pedigree) strongly suggests that Petit Manseng is a progeny of SAVAGNIN and a parent of Gros Manseng (Vouillamoz; Myles *et al.* 2011).

VITICULTURAL CHARACTERISTICS

Early budding, mid to late ripening. Vigorous and needs to be pruned long. Very resistant to botrytis bunch rot thanks to small, loose bunches and very small, thick-skinned berries.

WHERE IT'S GROWN AND WHAT ITS WINE TASTES LIKE

Thanks to its remarkable ability to concentrate sugars in the berry and still retain very high acidity, Petit Manseng is ideally suited to high-quality, aromatic sweet wines made from grapes shrivelled on the vine and picked as late as November and even early December.

There were 1,019 ha (2,518 acres) in France in 2009, predominantly in south-west **France**, where it is a principal variety in the Jurançon and Pacherenc du Vic-Bilh appellations. There have been some recent plantings in the Languedoc and the total planted area is continuing to rise. Recommended producers include Domaine Cauhapé (both dry and sweet versions), Domaine Bru-Baché and Clos Uroulat. A little is included in the blend of white Mas de Daumas Gassac in the Hérault, eastern Languedoc.

Over the border in northern **Spain**, Petit Manseng is authorized for two of the three appellations in the País Vasco: Arabako Txakolina/Chacolí de Álava and Bizkaiko Txakolina/Chacolí de Viscaia (a principal variety in the former and a complementary one in the latter). However, total plantings in Spain in 2008 amounted to no more than 4 ha (10 acres). There could be potential here.

The first varietally labelled Petit Manseng wine in the US was produced by Chrysalis Vineyards in **Virginia** (63 acres/25 ha in 2010) but others such as Horton Vineyards, Delfosse, Prince Michel, Sugarleaf Vineyards and White Hall Vineyards have followed suit, some making dry wines. The variety's naturally high acidity suits Virginia's hot summers. It is also planted to a very limited extent even further south in **South Carolina**, and **Georgia**, where Tiger Mountain Vineyards in Tiger, Georgia, make both a dry and a late-harvest version from grapes very much at home at 2,000 ft on the flinty, decayed granite of the southern Appalachians.

Basque immigrants took the variety to **Uruguay** in 1870, along with the now much more widespread TANNAT, but the total was fewer than 2 ha (5 acres) in 2010, produced by Familia Deicas, who blend it with GROS MANSENG and other aromatic varieties in their Botrytis Noble.

There are also very minor plantings in **Argentina** and **Japan**.

In **Australia**, what was thought to be Petit Manseng has now been identified by DNA testing as Gros Manseng. Churton in Marlborough, **New Zealand**, planted 0.5 ha (1.24 acres) of Petit Manseng on a hillside vineyard in 2007, the country's first commercial planting of this variety.

PETIT MESLIER

Extremely rare, light-berried potential champagne ingredient with significant relatives and a lone outpost in Australia.

BERRY COLOUR: ● ● ● ● ●

PRINCIPAL SYNONYMS: Melié Blanc (Vallée de la Vanne), Meslier Doré (Aube), Meslier Jaune, Meslier Petit

VARIETIES COMMONLY MISTAKEN FOR PETIT MESLIER: GRAŠEVINA, MENU PINEAU (Loire-et-Cher), MESLIER SAINT-FRANÇOIS

ORIGINS AND PARENTAGE

The term Meslier (variously spelt) has been used indiscriminately for Petit Meslier or for MESLIER SAINT-FRANÇOIS (see the latter for history and etymology) and its first reliable mention appeared in Sancerre in 1783–4 (Rézeau 1997): 'Petit meslier blanc. Feuille delicate, blanche par dessous, petit raisin excellent, grain rare' ('Petit meslier blanc. Delicate leaf, white underside, small and excellent grape, few berries'). According to Galet (1990), Petit Meslier originates from north-eastern France, between Champagne and the Franche-Comté, which is consistent with DNA profiling showing that Petit Meslier is a natural GOUAIS BLANC × SAVAGNIN cross (see PINOT pedigree diagram), just like its siblings AUBIN BLANC from Lorraine (Bowers *et al.* 2000) and RÄUSCHLING from the Rheintal (Rhine Valley) in Germany (Vouillamoz). In addition, DNA parentage analysis has strongly suggested that AMIGNE from the Valais, Switzerland, is a likely grandchild of Petit Meslier (Vouillamoz and Moriondo 2011). Petit Meslier is a member of the Messile ampelographic group (see p XXVII; Bisson 2009).

VITICULTURAL CHARACTERISTICS

Early budding and therefore at risk from spring frosts, mid ripening. Susceptible to millerandage and coulure and very susceptible to botrytis bunch rot. Best pruned long. Small bunches and berries.

WHERE IT'S GROWN AND WHAT ITS WINE TASTES LIKE

Once valued in the Champagne region from **France** for its ability to produce wines able to retain acidity in warm vintages, particularly in the warmer areas south of the River Aube, it fell out of favour because of its erratic and often low yields and because the timing of the harvest is particularly time-critical – the grapes are not easy to ripen and when they are ripe, they must be harvested promptly. The very high acidity and low pH mean that it is also more difficult to vinify and needs longer on the lees to attain balance.

There were just 4 ha (10 acres) remaining in France in 2008, although it is still permitted in the Champagne and still Coteaux Champenois appellations. There is clearly a growing interest in this variety. Duval-Leroy, for example, produced their first 100% Petit Meslier vintage champagne in 1998, from a vineyard in Venteuil in the Vallée de la Marne, and will continue to do so in years when the quality is high enough. Wines are generally quite floral and sometimes show aromas of rhubarb and white nettle. Tarlant have recently replanted some Petit Meslier, as well as PINOT BLANC and ARBANE, and have produced a new cuvée with 75% Petit Meslier. René Geoffroy too has included some Petit Meslier in a field blend, and a few champagne houses such as Moutard and L Aubry Fils include a small amount in particular cuvées.

More unexpected is the rare varietal sparkling wine produced by Irvine Wines in the Eden Valley, South Australia. They planted their 0.6 ha (1.5 acres) of vines in the mid 1980s and ten years later decided to start producing a varietal wine. Along with another local grower, who supplies them, they are pioneers of the variety in **Australia**. The wine has fine acidity and intense flavours of green apple.

PETIT ROUGE

Old, dark-skinned Aostan producing spicy, redcurrant-flavoured wines.

BERRY COLOUR: ● ● ● ● ●

PRINCIPAL SYNONYMS: Oriou, Picciourouzo

VARIETIES COMMONLY MISTAKEN FOR PETIT ROUGE: Humagne Rouge✕ (as CORNALIN is known in the Valais, Switzerland)

ORIGINS AND PARENTAGE

Petit Rouge, French for 'little red', is one of the oldest and most widespread varieties of the Valle d'Aosta in the Italian Alps. It was described by Gatta (1838) as the main type of the Orious, a group of closely related varieties which includes, as the most recent DNA studies have shown, Petit Rouge, CORNALIN and VIEN DE NUS in the Valle d'Aosta, as well as ROUGE DE FULLY and ROUGE DU PAYS in the Valais (Vouillamoz and Moriondo 2011). DNA parentage analysis has provided evidence that Petit Rouge is a parent of Rouge du Pays through a natural cross with MAYOLET (Vouillamoz *et al.* 2003), a sibling of FUMIN and a parent of Vien de Nus (Vouillamoz and Moriondo 2011), illustrating the key role of Petit Rouge for Valle d'Aosta viticulture.

Several clonal variations of Petit Rouge are distinguished under the group name Oriou: Oriou Lombard (the most widespread), Oriou Voirard, Oriou Picciou, etc.

OTHER HYPOTHESES

According to local tradition, Petit Rouge and other Valle d'Aosta grapes could have been introduced from Burgundy some time after the region became part of the kingdom of Burgundy in the fifth century. This hypothesis has been strongly rejected by morphological and genetic studies. Some authors have suggested that Petit Rouge is a natural cross between a grape variety introduced by the Romans and a local wild grape. However, wild grapes have never been found in the Valle d'Aosta.

VITICULTURAL CHARACTERISTICS

Late ripening.

WHERE IT'S GROWN AND WHAT ITS WINE TASTES LIKE

Petit Rouge is the most important grape variety of the Valle d'Aosta in Italy's far north west, cultivated mainly between Saint-Vincent and Arvier, in particular in Aymavilles, Saint-Pierre and Villeneuve, where it contributes at least 70% of the blend, with MAYOLET and VIEN DE

NUS, in the Torrette DOC, a historical wine already famous at the beginning of the nineteenth century, at least 85% in Enfer d'Arvier DOC and contributes to a lesser extent to the Nus Rouge and Chambave Rouge DOCs. There were 119 ha (294 acres) in Italy in 2000.

Varietal wines tend to have aromas of redcurrants and a spicy, fruity taste. Examples include Petit Rouge DOC from Anselmet, Di Barrò, Didier Gerbelle, Vinirari; Enfer d'Arvier from Albino Thomain; and Torrette DOC by Château Feuillet.

PETIT VERDOT

High-quality, late-ripening Bordeaux variety adding colour and spice to blends there but making an increasing number of solo appearances elsewhere.

BERRY COLOUR: ● ● ● ● ●

PRINCIPAL SYNONYMS: Lambrusquet, Verdau, Verdot, Verdot Petit

VARIETIES COMMONLY MISTAKEN FOR PETIT VERDOT: FER, GROS VERDOT

ORIGINS AND PARENTAGE

Petit Verdot and GROS VERDOT are usually considered to come from the Gironde in south-west France, where they were first mentioned together in Bordeaux in 1736 (Rézeau 1997): 'grand verdot, petit verdot, bons'. Although they partially share a name, derived from *vert*, meaning 'green' (in the sense of unripe and/or acidic), and although they used to be planted together, Petit Verdot and Gros Verdot are two completely different varieties. The ampelographer Bisson (2009) classifies Petit Verdot in the Carmenet ampelographic group (see p XXVII), together with Ardonnet, ARROUYA, CABERNET FRANC, CABERNET SAUVIGNON, FER, MERLOT, Merlot Blanc and Pardotte, most of them still cultivated in Bordeaux along with Petit Verdot, but recent ampelographic and genetic research has placed Petit Verdot in a distinct ampelographic group along with Ardonnet, Gros Verdot and Petit Verdot Faux (Bordenave *et al.* 2007). This as yet unnamed group most likely originates from the Pyrénées-Atlantiques, where these varieties could have been domesticated from wild grapevines, because their morphology is very similar to the *lambrusques* (Levadoux 1956), the French name for *Vitis vinifera* subsp. *silvestris* that gave Petit Verdot its synonym Lambrusquet.

DNA parentage analysis has shown that Petit Verdot and DURAS spontaneously crossed to give birth to TRESSOT (Bowers *et al.* 2000; see PINOT pedigree diagram).

VITICULTURAL CHARACTERISTICS

Early budding but late ripening – even later than CABERNET SAUVIGNON. Fertile and fairly productive. Well suited to gravel soils. Sensitive to drought. Small berries with thick skins and thus good resistance to botrytis bunch rot. Cool weather conditions can sometimes lead to a proportion of small, seedless green berries.

WHERE IT'S GROWN AND WHAT ITS WINE TASTES LIKE

When fully ripe – not always easy to achieve in Bordeaux – Petit Verdot produces wines that are powerful, rich, deeply coloured, tannic and ageworthy, often spicy and with good levels of alcohol and acidity in the best sites. Today it generally plays a small part in the Bordeaux blend but in the nineteenth century, according to Bordenave *et al.* (2007), Petit Verdot was the dominant variety in the Queyries vineyards on the left bank of the Gironde, what is now the Bastide area of the city. Even though the planted area has declined since then, the variety's qualities and contribution are highly prized in the best vintages – except that ironically the best vintages are often those when that contribution is least needed to enrich the Cabernet and MERLOT. Wines which benefit from a relatively high proportion of Petit Verdot in some years include Châteaux La Lagune, Léoville Poyferré, Margaux, Palmer and Pichon Lalande.

Petit Verdot is recommended in much of southern and south-western **France**. Of the 862 ha (2,130 acres) recorded in 2009, the majority (526 ha/1,300 acres) was in the Bordeaux region. Plantings declined in the second half of the twentieth century, down to just 338 ha (835 acres) in 1988, mostly because of its inconveniently late ripening, but they are increasing once again, perhaps encouraged by global warming. Varietal wines are very rare in Bordeaux (though Moutte Blanc is an exception) but are increasingly produced elsewhere, including the Languedoc. Domaines Preignes and de Ravanes are examples.

Italy recorded 63 ha (156 acres) in the 2000 agricultural census, predominantly in the Maremma or in Toscana more generally for seasoning Bordeaux blends chez the likes of Ornellaia, Castello del Terriccio and Cavali Tenuta degli Dei. Casale del Giglio in Lazio make a rare varietal example.

The variety was introduced to **Spain** in the 1990s by Marqués de Griñon at Dominio de Valdepusa near Toledo and the country's plantings had reached 1,042 ha (2,575 acres) by 2008. It ripens well here and has sufficient flesh to provide a satisfactory varietal in the hands of Abadía Retuerta (Castilla y León); Casa de la Ermita and Luzón (Jumilla); Friedrich Schatz, La Sangre de Ronda and Vetas (Sierras de Málaga); Arrayán (Méntrida); Finca Antigua (La Mancha); Laudum Nature, Enrique Mendoza and Vinalopó (Alicante); and Pago del Vicario (Castilla).

Portugal reported 139 ha (343 acres) in 2010, used mainly in ambitious blends such as Campolarga's Calda Bordalesa (with ALICANTE HENRI BOUSCHET) or Monte da Ravasqueira's Premium (with SYRAH and MERLOT). However, varietal versions are made by Azamor, Júlio Bastos, Lima Mayer and Monte da Ravasqueira, all in the Alentejo. This late-ripening variety does well in this hot southern region and produces fresh, lively wines with intense colour and notable fragrance.

Turkey (20 ha/49 acres in 2010) and **Israel** (50 ha/124 acres in 2009) produce a few varietal examples.

In **California**, plantings have been increasing to reach 4,396 acres (1,779 ha) by 2008, mostly in Napa, San Luis Obispo and San Joaquin counties, generally for including in Meritage (Bordeaux) blends, but the number of varietal versions has been increasing significantly, with L'Aventure's from Paso Robles being particularly well received. **Washington** State had 131 acres (53 ha) at the last count in 2006, also for Bordeaux blends in the main, but varietals are also emerging at quite a rate here, where there is rarely a problem achieving full ripeness. There are limited plantings in other American states, particularly those making Bordeaux blends such as **Virginia** (167 acres/68 ha by 2010) and a few plantings on Long Island in **New York** State. Petit Verdot is also finding friends in the Okanagan Valley in **Canada**'s British Columbia.

Argentina recorded 455 ha (1,124 acres) of Petit Verdot in 2008 (though it is generally referred to as just Verdot). Finca Anita and Landelia Mendoza produce good varietal examples. **Chile** recorded 257 ha (635 acres) in 2008, though the distinction between Petit Verdot and what the statistics refer to simply as 'Verdot' may not always be accurate. Von Siebenthal's bottling is a worthy representative of the real McCoy. **Uruguay** had 20 ha (49 acres) in 2009, on a steady upward trajectory over the last ten years or so. The variety plays a significant role at Viña Tacama in the Ica Valley in **Peru**, where it is generally blended with TANNAT and/or Malbec (COT).

Petit Verdot has become a popular varietal in **Australia**, particularly as an alternative to Shiraz (SYRAH) in warmer regions such as Riverland, Riverina and Murray Darling. Because it ripens two weeks later, it is likely to finish ripening in cooler conditions. In 2008 there were 1,354 ha (3,346 acres), mostly in these inland regions but also in McLaren Vale, Padthaway, Langhorne Creek, Margaret River, Barossa, Mudgee, Swan Hill, Clare Valley and Coonawarra. Here it is far more likely than in Bordeaux to be produced as a varietal wine and these are generally very deeply coloured, dark-fruited (more plum than cassis), rich, full-bodied and smooth, sometimes with a floral note akin to violets. Recommended producers include Capel

Vale, De Bortoli, Gemtree and Trentham Estate. It is occasionally an ingredient in **New Zealand**'s Bordeaux blends.

The variety looks set for a very positive future in **South Africa**, although here too it is more often encountered in Bordeaux blends than as a varietal. Total vineyard area has grown from little more than 20 ha (49 acres) in 1998 to 634 ha (1,567 acres) ten years later, mostly in Paarl and Stellenbosch, but it shows promise in other wine growing regions except perhaps in the Swaartland, where it may be too dry. Recommended from among the handful of varietal wines are Zorgvliet's powerful version, Asara's and a charming example from the KWV Mentors range.

PETITE AMIE

Recent and promisingly aromatic American hybrid planted in Nebraska.

BERRY COLOUR:

PRINCIPAL SYNONYMS: DM 8313.1

ORIGINS AND PARENTAGE

An Elmer Swenson 2-11-4 × DM P2-54 hybrid obtained in 1983 by David MacGregor in South Haven, Minnesota, in which:

- Elmer Swenson 2-11-4 is an Elmer Swenson 5-14 × SWENSON RED hybrid (see BRIANNA for the pedigree of Elmer Swenson 5-14)
- DM P2-54 is a Suelter × MORIO-MUSKAT hybrid
- Suelter is a *Vitis riparia* × CONCORD hybrid.

Petite Amie it therefore an extremely complex *Vitis riparia*, *Vitis labrusca*, *Vitis vinifera*, *Vitis aestivalis*, *Vitis lincecumii*, *Vitis rupestris*, *Vitis cinerea* and *Vitis berlandieri* hybrid.

The plant was selected in 1987 and tested as DM 8313.1 before being named in 2004 by grower Ed Swanson of Cuthills in Pierce, Nebraska, with the permission of David MacGregor.

VITICULTURAL CHARACTERISTICS

Small to medium-sized bunches of small, thin-skinned berries. Low to moderate vigour. Mid budding and early to mid ripening but also amenable to late harvesting. Very cold-hardy (–25 °F/–32 °C).

WHERE IT'S GROWN AND WHAT ITS WINE TASTES LIKE

Wines typically have a fine grapey flavour similar to MUSCAT BLANC À PETITS GRAINS as well as an aroma of roses and tropical flavours. There was less than an acre (0.4 ha) planted in **Nebraska** in 2007 and Cuthills is currently the only producer there. Indian Island in **Minnesota** also produce a varietal wine.

PETITE ARVINE

See ARVINE.

PETITE PEARL

Promising new cold-hardy American hybrid.

BERRY COLOUR: ● ● ● ● ●

PRINCIPAL SYNONYMS: TP 2-1-24

ORIGINS AND PARENTAGE

An MN 1094 × Elmer Swenson 4-7-26 hybrid crossed in 1996 by Tom Plocher, private grape breeder in Hugo, Minnesota, US, in which:

- MN 1094 (MN stands for Minnesota) is a hybrid between MN 1019 and MN 1016 (see MARQUETTE for the pedigrees)
- Elmer Swenson 4-7-26 is an Elmer Swenson 2-12-27 × ST CROIX hybrid
- Elmer Swenson 2-12-27 is an Elmer Swenson 5-14 × SWENSON RED hybrid (see BRIANNA for the pedigree of Elmer Swenson 5-14).

Petite Pearl is therefore an extremely complex *Vitis riparia*, *Vitis labrusca*, *Vitis vinifera*, *Vitis aestivalis*, *Vitis lincecumii*, *Vitis rupestris*, *Vitis cinerea* and *Vitis berlandieri* hybrid. The plant was tested as TP 2-1-24 and was initially named Black Pearl by Plocher, who had to change the name to Petite Pearl to avoid a trademark dispute. Petite Pearl was finally released by Plocher in 2009.

VITICULTURAL CHARACTERISTICS

Moderately vigorous, extremely winter-hardy (–31 °F/–35 °C with no injury), small, dense clusters but not prone to botrytis bunch rot and with good all-round disease resistance. Late budding and mid ripening.

WHERE IT'S GROWN AND WHAT ITS WINE TASTES LIKE

This new variety shows great promise among new winter-hardy hybrids and produces big, dense red wine with some nice soft tannins, not too high acidity and complex aromas mainly of herbs and spices. Plocher suggests it has some similarity with Malbec (COT) or Bonarda (DOUCE NOIRE). Commercial wines are expected from 2013 in the US Midwest but it is likely to be popular with growers in cold-climate regions such as **New York** State, and Ontario and Québec in **Canada**.

PETITE SIRAH

See DURIF.

PETRA

Serbian hybrid producing full-bodied, aromatic whites.

BERRY COLOUR: ● ● ● ● ●

PRINCIPAL SYNONYMS: Petka

ORIGINS AND PARENTAGE

A Kunbarat × PINOT NOIR hybrid obtained in 1977 by P Cindrić and V Kovač at the viticulture research centre in Sremski Karlovci, part of Novi Sad University in Vojvodina, Serbia, in which Kunbarat is a (*Vitis amurensis* × *Vitis vinifera*) × ITALIA hybrid. Petra was used to breed BAČKA and RUBINKA.

VITICULTURAL CHARACTERISTICS

Vigorous. Small, compact bunches of small berries. Susceptible to powdery mildew. More winter-hardy than RIESLING but early budding makes it susceptible to spring frosts.

WHERE IT'S GROWN AND WHAT ITS WINE TASTES LIKE

Grown mainly in the Fruška Gora region, **Serbia**, Petra produces wines that are high in alcohol with Muscat-like aromas and are well suited to the production of dessert wines. Producers of varietal wines include Daniel Celovški, Stanislav Đierčan, Milijan Jelić and Dušan Stefanović.

PETROKORITHO

Greek island variety of which the white mutation is the stronger survivor.

BERRY COLOUR: ● ● ● ● ●

PRINCIPAL SYNONYMS: Petrocoritho, Pietro Corinto

ORIGINS AND PARENTAGE

The dark-skinned Petrokoritho Mavro comes from the Greek island of Kérkyra (Corfu) but it is the light-skinned mutation (sometimes referred to as Petrokoritho Lefko) that survives.

VITICULTURAL CHARACTERISTICS

Late ripening. Large bunches of big berries. Relatively disease-resistant.

WHERE IT'S GROWN AND WHAT ITS WINE TASTES LIKE

Once found mainly on the island of Kérkyra but also in the Pelopónnisos (Peloponnese), **Greece**, and on Évvoia (Euboea; Krimbas in Galet 2000), the original black-berried Petrokoritho Mavro is close to extinction. However, Livadiotis make a rare varietal version from the white-berried mutation in the south west of Kérkyra that has lower alcohol than KAKOTRYGIS but higher acidity.

PETROULIANOS

Rare but promising vine from the southern tip of Greece's Pelopónnisos (Peloponnese).

BERRY COLOUR: ● ● ● ● ●

PRINCIPAL SYNONYMS: Petrolianos

ORIGINS AND PARENTAGE

Rare variety local to the southern Pelopónnisos in Greece.

VITICULTURAL CHARACTERISTICS

Vigorous, mid ripening, good resistance to drought and to downy mildew but suseptible to powdery mildew. Generally difficult to cultivate.

WHERE IT'S GROWN AND WHAT ITS WINE TASTES LIKE

Petroulianos is found mainly in Lakonía in the southern Pelopónnisos, **Greece**, but also on the island of Kýthira (Cythera) a little further south. Yiannis Vatistas was one of the first producers to select Petroulianos vines from a field blend and plant the variety in a single vineyard close to the sea. He makes a lightly oaked varietal wine that is relatively full-bodied but still quite taut and has an intense, pure lemon character, a peachy richness and a clean minerality with relatively soft acidity. He also includes it in a sweet blend with ASSYRTIKO and MALVASIA.

PEVERELLA

See VERDICCHIO BIANCO.

PHOENIX

Disease-resistant German hybrid producing fresh and lightly aromatic wines in the UK.

BERRY COLOUR:

PRINCIPAL SYNONYMS: Geilweilerhof GA-49-22

ORIGINS AND PARENTAGE

A complex hybrid obtained by crossing BACCHUS × VILLARD BLANC authorized in Germany in 1992. It was named after the mythological bird that rose from the ashes.

VITICULTURAL CHARACTERISTICS

Vigorous, mid to early ripening, good resistance to downy mildew, quite good resistance to powdery, but susceptible to botrytis bunch rot. Yields are similar to MÜLLER-THURGAU and higher than RIESLING, sugar and acid levels are lower than Riesling.

WHERE IT'S GROWN AND WHAT ITS WINE TASTES LIKE

Although **Germany** has a greater area of Phoenix vineyards than the UK (48 ha/119 acres v 16 ha/40 acres), it is declining there owing to its susceptibility to botrytis bunch rot. German plantings are mainly in Rheinhessen and the Nahe.

It has been relatively more successful and significant in **England**, where it can produce attractively herbaceous and elderflower-scented wines, similar to BACCHUS, tasting very much like wines made from *vinifera* vines, which is why the variety has been registered as *vinifera* and authorized for the production of quality wines. Three Choirs produce a good example. It is also planted to a very limited extent in **Switzerland**, **Denmark** and **Sweden**.

PICAPOLL BLANCO

Minor Catalan that may be the same as France's Clairette Blanche.

BERRY COLOUR: ● ● ● ● ●

PRINCIPAL SYNONYMS: Blanca Extra, ?Clairette Blanche, Picapoll, Picapolla Blanca

VARIETIES COMMONLY MISTAKEN FOR PICAPOLL BLANCO: PIQUEPOUL BLANC (France)

ORIGINS AND PARENTAGE

Picapoll is an indigenous variety from the area that is now designated as the Pla de Bages DO, north west of Barcelona in Catalunya, north-eastern Spain, where it was already mentioned in the mid sixteenth century and where it was the only white variety at the end of the nineteenth (Puig *et al.* 2006). DNA profiling has shown that Picapoll Blanco is not a white mutation of Picapoll Negro, which is a synonym for PIQUEPOUL NOIR, and is not identical to PIQUEPOUL BLANC from the Vaucluse in southern France (Puig *et al.* 2006). Furthermore, the DNA profiles of Picapoll Blanco and CLAIRETTE from southern France are almost identical (Puig *et al.* 2006) but it is not yet clear whether Picapoll Blanco is simply a clonal variation of Clairette or a distinct yet very closely related variety (Vouillamoz).

VITICULTURAL CHARACTERISTICS

Late ripening and thick skinned.

WHERE IT'S GROWN AND WHAT ITS WINE TASTES LIKE

There were 29 ha (72 acres) of Picapoll Blanco in **Spain** in 2008, mostly in Catalunya, where it is authorized in the Pla de Bages, Empordà, Priorat and Alella DOs. Wines tend to have moderate acidity with grapefruit and herbal flavours, floral in cooler years. Producers of varietal wines include Abadal and Mas de Sant Iscle, both in Bagés, west of Tarragona.

PICARDAN

Occasionally adds acidity to Châteauneuf.

BERRY COLOUR: ● ● ● ● ●

PRINCIPAL SYNONYMS: Aragnan, Aragnan Blanc, Araignan Blanc (Var), Gallet or Gallet Blanc (Costières de Nîmes), Grosse Clairette (Vallée de la Durance), Milhaud Blanc (Tarn), Papadoux (Ardèche), Oeillade Blanche (Provence), Picardan Blanc (Châteauneuf-du-Pape)

VARIETIES COMMONLY MISTAKEN FOR PICARDAN: CLAIRETTE (called Oeillade Blanche in Provence and in the Drôme)

ORIGINS AND PARENTAGE

Picardan is an old variety from Provence in southern France, where it used to be cultivated in the Var, Vaucluse, Gard, Hérault and in the Vallée du Rhône under various names, especially Oeillade Blanche and Araignan Blanc. The names Picardan and Oeillade Blanche were first mentioned in 1544 in the 'Chant de vendanges' by Bonaventure des Périers (Rézeau 1997): 'Les vins des blancs sperollans/Des rouvergnas, des picquardans/Des belles grappes muscades/Pellefèdes et oeillades' ('The wines from the white sperollans/The rouvergnas, the picquardans/The beautiful bunches of muscades/Pellefèdes and oeillades'). However, the name Picardan was also used for BOURBOULENC, and 'oeillades' here could also refer to CLAIRETTE. Therefore, the first indisputable mention of this variety appeared much later, in 1715, under its synonym Araignan Blanc (Rézeau 1997): '*Vitis acini oblong, subviridi, dulci & molli.* Aragnan. Cette espèce est très-commune' ('Vine with oblong, greenish, sweet and soft berries. Aragnan. This species is very common'). The name Picardan supposedly derives from the French terms *piquer* ('to sting') and *ardent* ('burning'), in reference to its acidity, and the name Araignan probably comes from the French *araignée* ('spider'), in reference to the spiderweb-like hairs on the underside of the leaves.

VITICULTURAL CHARACTERISTICS

Mid ripening. Best pruned short and vertically trained. Sensitive to lack of magnesium and well suited to hot, dry, low-fertility sites. Big bunches and berries.

WHERE IT'S GROWN AND WHAT ITS WINE TASTES LIKE

Wines tend to be moderate in both alcohol and acidity and the variety's most well-known home is in Châteuneuf-du-Pape, southern **France**, where it is one of the eighteen varieties (plus colour mutations) authorized for red and white appellation wines, albeit a very minor one. Château de Beaucastel, for example, still include very small amounts of Picardan in their red Châteauneuf (from a 15-are parcel and from old interplanted vineyards). Although there was less than a hectare (1.2 acres) planted in France in 2008, Clos de Centeilles in the Minervois have 35% Picardan (which they refer to as Araignan Blanc) in their C de Centeilles Blanc, a Vin de Pays des Côtes du Brian from the Hérault.

PICCOLA NERA

Makes very pale reds in the hinterland of Trieste.

BERRY COLOUR: ● ● ● ● ●

PRINCIPAL SYNONYMS: Mala Cerna (Slovenia), Nera Tenera

VARIETIES COMMONLY MISTAKEN FOR PICCOLA NERA: SCHIAVA GROSSA ☓

ORIGINS AND PARENTAGE

This trivially named variety (literally 'little black') originates from the border region of Italy and Slovenia, in the province of Trieste.

VITICULTURAL CHARACTERISTICS

Vigorous, late ripening.

WHERE IT'S GROWN AND WHAT ITS WINE TASTES LIKE

Piccola Nera is cultivated in the province of Trieste, **Italy**, close to the Slovenian border. It is said also to be found in the province of Trento in Trentino to the west but it is sometimes confused with SCHIAVA GROSSA. It produces a light rosé-red wine for early drinking. A small amount may be blended with TERRANO in the Carso DOC. It is also authorized as a varietal wine under the IGT Venezia classification. The Italian agricultural census of 2000 recorded just 20 ha (49 acres).

It is also found occasionally in **Slovenia.**

PICOLIT

Ancient Friulian producing small quantities of highly regarded sweet wines.

BERRY COLOUR:

PRINCIPAL SYNONYMS: Pikolit (Slovenia)

VARIETIES COMMONLY MISTAKEN FOR PICOLIT: Balafánt (Hungary), KÉKNYELŰ (Hungary)

ORIGINS AND PARENTAGE

The earliest mention of Picolit appears in 1682 in a list of wines for the wedding celebration of Alvise Contarini, the Doge of Venice. Picolit wines became famous throughout Europe in the eighteenth century when Conte Fabio Asquini began exporting to the courts of France, Holland, Austria, England and Russia the few thousand bottles produced from his vineyards at Fagagna near Udine. It was a marketing triumph, with the wines bottled in hand-made Murano glass and sold at a far higher price than any other wine from Friuli at that time. The Pope liked it, too, apparently. Asquini also left detailed notes on how the wine was made although no one picked up the baton until the twentieth century (see below). This fame prompted some to identify the Hungarian varieties Balafánt, today almost extinct, and KÉKNYELŰ with Picolit, but this can be rejected by comparison of their DNA profiles (Jahnke *et al.* 2007; Varga *et al.* 2008; Galbács *et al.* 2009; Crespan *et al.* 2011).

The name Picolit could derive from the reduced size of the clusters (*piccolo* is Italian for 'small'), or from the local dialect *pecol*, meaning 'top of a hill', the best place for late-harvest Picolit.

Picolit is not a colour mutation of PICULIT NERI.

OTHER HYPOTHESES

Like many old varieties, Picolit is said to have been cultivated since Roman times but there's no evidence of any botanical identity between the grapes used by the Romans and modern varieties.

VITICULTURAL CHARACTERISTICS

Vigorous but low yielding because Picolit flowers often produce sterile pollen, making cultivation difficult and producing loose bunches with very few berries. It needs to be planted together with other varieties such as VERDUZZO FRIULANO to improve berry set. Although it ripens at the end of September or the beginning of October, it is usually harvested much later.

WHERE IT'S GROWN AND WHAT ITS WINE TASTES LIKE

Formerly cultivated in the provinces of Udine, Treviso (Conegliano), Bassano and Vicenza in Friuli, **Italy**, Picolit vineyards were severely reduced by phylloxera and it was mainly thanks to Giuseppe Perusini and his family of the Rocca Bernarda in Rosazzo that Picolit survived the first half of the twentieth century, helped by their work on identifying hardier clones, rising to become something of a cult wine in the late 1960s and early 1970s, though plantings are still very limited. Today the variety is grown mainly in the provinces of Udine and Gorizia. There were just 97 ha (240 acres) in Italy in 2000.

It is used almost exclusively to make sweet late-harvest wines within the recently promoted Colli Orientali del Friuli Picolit DOCG. Wines from the Rosazzo and Cialla subzones are recognized as the finest but there is little consistency with regard to style. The grapes are either dried on mats or in racks after the harvest or simply left longer on the vine, harvested with or without the influence of botrytis. More recently, some producers have experimented with ageing in small oak barrels.

Many wines are less impressive than one might expect considering the prices they achieve but recommended producers include Ronchi di Cialla, Dorigo, Livio Felluga and Moschioni. The best are delicately sweet with floral aromas and the flavour of apricots and peaches, sometimes a little honeyed. Wines aged for more than two years may be labelled Riserva. Picolit may also play a major or minor role in the dry Colli Orientali del Friuli Bianco or the sweet Colli Orientali del Friuli Dolce. Le Due Terre produce a rare dry varietal Picolit called Implicito.

Pikolit is planted in **Slovenia** under two morphologically distinct types called Pikolit Italia and Pikolit Vienna, although DNA profiling has shown that both are identical to Picolit from Friuli (Štajner *et al.* 2008) and growers are unlikely to distinguish between them. Total plantings in 2009 were just 5 ha (12 acres), in Goriška Brda and Vipavska Dolina. Producers include Ščurek, who make a dry, alcoholic version, and Vinska Klet Goriška Brda and Čarga, who make sweet or medium-sweet versions.

Pizzini produce some in the King Valley, Victoria, **Australia**.

PICPOUL

See PIQUEPOUL.

PICULIT NERI

Old, rare, small-berried red wine variety unrelated to Picolit.

BERRY COLOUR: ● ● ● ● ●

PRINCIPAL SYNONYMS: Picolit Neri, Picolit Nero

ORIGINS AND PARENTAGE

Historically cultivated in Friuli. Its name simply means 'small black', in reference to its tiny dark berries. Despite the similarity in their names, Piculit Neri is not a black mutation of the white-berried PICOLIT used to make expensive sweet wines in Friuli.

VITICULTURAL CHARACTERISTICS

Highly susceptible to botrytis bunch rot. Vigorous and mid to late ripening (early October).

WHERE IT'S GROWN AND WHAT ITS WINE TASTES LIKE

Piculit Neri is cultivated in particular in the province of Pordenone in Friuli-Venezia Giulia in north-east **Italy**, mainly in Pinzano al Tagliamento and Castelnovo del Friuli. Together with other old local varieties, CIANORIE, CIVIDIN, FORGIARIN, SCIAGLÌN and UCELÙT, it was saved from extinction by Emilio Bulfon, with the help of ampelographers Antonio Calò and Ruggero Forti. Emilio Bulfon is one of the few producers to craft varietal wines from these grapes. Others include Florutis and Ronco Cliona. Piculit Neri has firm tannins and aromas of herbs, red fruits and vanilla. Together with Forgiarin, Sciaglìn and Ucelùt, Piculit Neri was included in the Italian national register of varieties in 1991.

PIEDIROSSO

Fresh, aromatic Campanian that makes an ideal blending partner for Aglianico but can also shine as a varietal.

BERRY COLOUR: ● ● ● ● ●

PRINCIPAL SYNONYMS: Palombina, Palumbina, Palumbo, Palummina, Per'e Palummo, Streppa Verde (island of Ischia)

VARIETIES COMMONLY MISTAKEN FOR PIEDIROSSO: Piedirosso Avellinese⌛, Piedirosso Beneventano⌛ or Piedirosso Napoletano

ORIGINS AND PARENTAGE

Piedirosso is a very old Campanian variety, probably originating from the area of Vesuvio, which is still the variety's heartland. It was supposedly first mentioned in the sixteenth century under its old synonyms Palomina Nera and Uva Palombina but the first reference to the name Piedirosso appears much later, in Carlucci (1905) in the provinces of Avellino and Benevento. However, Carlucci was not convinced that Palomina Nera and Uva Palombina were botanically identical to Piedirosso because the older references lacked good descriptions. Nevertheless, their names have a common origin: Piedirosso, meaning 'red feet', comes from the red colour of the stalk and the stem at harvest time, which look like the claws of a pigeon (Latin *Columba palumbus*). Interestingly, the clonal variation known as Streppa Verde ('green stalk') on the island of Ischia does not have this red character. DNA profiling has shown that Piedirosso is distinct from Piedirosso Avellinese (province of Avellino) and from Piedirosso Beneventano (province of Benevento), also known as Piedirosso Napoletano in the province of Napoli, neither commercially cultivated, and genetically close to CAPRETTONE (Costantini *et al.* 2005).

OTHER HYPOTHESES

Piedirosso is sometimes identified with Columbina, mentioned by Pliny the Elder.

VITICULTURAL CHARACTERISTICS

Vigorous, mid ripening, likes the heat and volcanic soils. Relatively resistant to powdery mildew but susceptible to downy mildew.

WHERE IT'S GROWN AND WHAT ITS WINE TASTES LIKE

Piedirosso, intensively cultivated in the area around Napoli in southern **Italy**, is the second most widely grown red in Campania after AGLIANICO, with which it is usually blended to add perfume and freshness in DOCs such as Cilento, Falerno del Massico and Sant'Agata de' Goti. It plays a more dominant role or may be a varietal in DOCs such as Capri, Ischia, Taburno, Sannio and Vesuvio. In some instances, eg Costa d'Amalfi, Penisola Sorrentina and Vesuvio, it may also be blended with the local SCIASCINOSO. Piedirosso is also authorized in the province of Bari in Puglia. Plantings halved in the 1980s but have been fairly constant since then, with 1,013 ha (2,503 acres) recorded in the Italian census in 2000.

Varietal wines tend to be relatively soft despite their fresh acidity, and aromatic, with flavours of red fruits such as plums and cherries – almost Gamay-like in some examples. They may also have herbal and spicy notes, as in Mastroberardino's Lacryma Christi del Vesuvio Rosso. Other producers of varietal wines include Annunziata, Mustilli and Cantina del Taburno. Many of the varietal wines are classified as IGT, eg IGT Campania or IGT Pompeiano.

PIGATO

See VERMENTINO.

PIGNATELLO

See PERRICONE.

PIGNOLA VALTELLINESE

Minor red wine grape grown in Lombardia and Piemonte, and in the Veneto as the recently rescued Groppello di Breganze.

BERRY COLOUR:

PRINCIPAL SYNONYMS: Groppello di Breganze (Vicenza in the Veneto), Pignola (Valtellina in Lombardia), Pignola Spanna (Vercelli in Piemonte), Pignolo Spano (Novarra in Piemonte)

VARIETIES COMMONLY MISTAKEN FOR PIGNOLA VALTELLINESE: Groppello dei Berici (Veneto), GROPPELLO DI MOCASINA (Lombardia), GROPPELLO DI REVÒ (Trentino), GROPPELLO GENTILE (Lombardia), PIGNOLO (Friuli)

ORIGINS AND PARENTAGE

The name Pignola, derived from *pigna*, meaning 'pine cone' and referring to the shape of the bunch, has been given to several varieties in central Piemonte, Valtellina, the Oltrepò Pavese and Friuli. DNA profiling has shown that Pignola from Valtellina is identical to Pignolo Spano cultivated in Piemonte (Schneider *et al.* 2001). PIGNOLO from Friuli, on the other hand, is a completely distinct variety (Fossati *et al.* 2001; Salmaso *et al.* 2008), whereas the true identity of Pignolo from the Oltrepò Pavese is still uncertain. Since DNA parentage analysis established that the Pignola grown in Valtellina has a parent–offspring relationship with ROSSOLINO NERO (Vouillamoz), a variety local to Valtellina, this Pignola is likely to originate in Valtellina and would be better named Pignola Valtellinese to avoid any confusion. In addition, DNA studies have suggested that Pignola Valtellinese could also be a cousin of NEBBIOLO (Schneider, Boccacci *et al.* 2004), called Chiavennasca in Valtellina.

A recent ampelographic and DNA profiling study has unexpectedly shown that Pignola Valtellinese is identical to Groppello di Breganze in the province of Vicenza in the Veneto, northern Italy, where it was already mentioned in the Breganze area in the eighteenth century (Cancellier *et al.* 2009). According to this study, this variety is distinct from GROPPELLO GENTILE, GROPPELLO DI MOCASINA and GROPPELLO DI REVÒ, in spite of some morphological similarities with the last.

VITICULTURAL CHARACTERISTICS

Mid ripening. Susceptible to powdery mildew and botrytis bunch rot because of its compact bunches, but good resistance to downy mildew and cold.

WHERE IT'S GROWN AND WHAT ITS WINE TASTES LIKE

Under the name Pignola Valtellinese, this variety is cultivated in the province of Sondrio in Lombardia; under the name Pignolo Spano, it is grown in Piemonte in the provinces of Biella, Novara and Vercelli, often blended with NEBBIOLO (misleadingly called Spanna in the same areas). There were 77 ha (190 acres) in **Italy** in 2000. Triacca's Pinea, classified IGT Terrazze

Retiche di Sondrio, is a rare varietal wine. In general, acidity is quite high but tannins are moderate, and certainly softer than those of Nebbiolo. Under the name Groppello di Breganze in the province of Vicenza in the Veneto, it was rescued from extinction by Severina Cancellier, an ampelographer based at the Conegliano research centre, and has recently been vinified by a few producers in Breganze, resulting in fruity, tannic wines. Cantina Beato Bartolomeo's passito wine Ezzelino is a blend of Groppello di Breganze and CABERNET SAUVIGNON, while Firmino Miotti crafts a 100% Groppello di Breganze.

PIGNOLETTO

Light, refreshing variety quite widely planted in the Bologna hills and in Umbria.

BERRY COLOUR: ● ● ● ● ●

PRINCIPAL SYNONYMS: Grechetto di Todi (Umbria), Grechetto Gentile, Occhietto (Marche), Pallagrello di Caserta (Campania), Pignoletto Bolognese, Pignolino, Pulcinculo (Marche), Rébola (Emilia-Romagna), Ribolla Riminese (Emilia-Romagna), Strozzavolpe (Marche)

VARIETIES COMMONLY MISTAKEN FOR PIGNOLETTO: GRAŠEVINA, GRECHETTO DI ORVIETO, PALLAGRELLO BIANCO, PINOT BLANC, RIBOLLA GIALLA

ORIGINS AND PARENTAGE

Pignoletto is an ancient variety from the hills in the province of Bologna in Emilia-Romagna, where it was supposedly first mentioned in 1654 under the name *uve pignole*. Like PIGNOLA VALTELLINESE, the name Pignoletto most likely derives from the shape of the bunch, which resembles a pine cone (*pigna*).

Ampelographic and DNA studies have established that Pignoletto is identical to Grechetto di Todi in Umbria and Ribolla Riminese or Rébola in Emilia-Romagna but distinct from GRECHETTO DI ORVIETO in Umbria (Filippetti *et al.* 1999). Parentage studies have suggested a parent–offspring relationship between Pignoletto (under the name Grechetto di Todi) and Grechetto di Orvieto (Costacurta *et al.* 2004), which might explain why they were previously considered as clonal variations of the same variety, and DNA analysis has suggested a close genetic relationship between SPERGOLA and Pignoletto (Filippetti, Silvestroni, Thomas and Intrieri 2001). More surprisingly, a comparison of the DNA profiles of Pignoletto (Filippetti and Intrieri 1999) and Pallagrello di Caserta (Costantini *et al.* 2005) has revealed that they are identical. In addition, Pignoletto has often been confused with Pinot Bianco (PINOT BLANC) and Riesling Italico (GRAŠEVINA) by local growers.

OTHER HYPOTHESES

Some have suggested a link between Pignoletto and Pino Lieto (meaning 'happy pine'), a wine mentioned by Pliny the Elder, but there's no botanical evidence to confirm this hypothesis.

VITICULTURAL CHARACTERISTICS

Vigorous and productive. Mid to late ripening.

WHERE IT'S GROWN AND WHAT ITS WINE TASTES LIKE

Grechetto is authorized in several varietal DOC wines in the province of Perugia in Umbria, central **Italy** (Assisi, Colli Martani, Colli Perugini and Colli del Trasimeno), as well as in the Marche and in Lazio (Colli Etruschi Viterbesi). However, since some DOC regulations simply say Grechetto, it is often impossible to be sure whether they refer to GRECHETTO DI ORVIETO or Grechetto di Todi although, logically, Grechetto found in the Orvieto area is likely to be Grechetto di Orvieto.

As Pignoletto, it is cultivated mainly in the hills south west of Bologna and produced as a varietal wine in various styles in many DOCs, particularly in the subzones of Colli Bolognesi but also in Colli di Imola and Reno. Pignoletto is the most important white variety in the Colli Bolognesi and the Italian vineyard census of 2000 recorded 6,790 ha (16,778 acres).

Varietal wines are lively and crisp and well suited to the production of sparkling wine. Recommended producers include Chiarli 1860, Il Monticino, Verigilio Sandone, Tizzano and Vallona in Emilia-Romagna and Todini and Tudernum in Umbria.

PIGNOLO

Rare and recently rescued vine showing very good potential in Friuli.

BERRY COLOUR: ● ● ● ● ●

VARIETIES COMMONLY MISTAKEN FOR PIGNOLO: AGLIANICO, PIGNOLA VALTELLINESE

ORIGINS AND PARENTAGE

Like Pignola, Pignolo, literally 'fussy', takes its name from the Italian word *pigna*, meaning 'pine cone', because of the shape of its bunches. It is most likely to come from the hills of Buttrio and Rosazzo in Friuli, where it is said to have been known since the seventeenth century. DNA profiling has shown that PIGNOLA VALTELLINESE and Pignolo from Friuli are completely distinct (Fossati *et al.* 2001; Salmaso *et al.* 2008).

VITICULTURAL CHARACTERISTICS

Mid October. Low yields. Particularly susceptible to powdery mildew.

WHERE IT'S GROWN AND WHAT ITS WINE TASTES LIKE

Like SCHIOPPETTINO and TAZZELENGHE, Pignolo had almost disappeared in the early twentieth century after Friuli was hit by phylloxera. It was rescued in the 1970s from a vineyard in the abbey of Rosazzo, where an entire row of old and ungrafted Pignolo vines had miraculously survived. Pignolo is used in blends as well as for varietal wines in Colli Orientali del Friuli DOC (in particular in Prepotto, Albana, Rosazzo, Buttrio and Premariacco). In 2000 there were just 20 ha (49 acres) in **Italy** but the total has surely increased.

Varietal wines have fresh acidity and firm but silky tannins, with fruit flavours that range from small red berries to ripe plums, developing notes of liquorice over time. They are well suited to oak ageing. Girolamo Dorigo was the first to plant Pignolo back in 1973 using cuttings from those original centenarian vines. Other producers of this rare wine include Castello di Buttrio, Giorgio Colutta, Dario e Luciano Ermacora, Adriano Gigante, Moschioni, Petrucco, Paolo Rodaro, Ronco delle Betulle and Le Vigne di Zamò.

PINEAU D'AUNIS

Ancient, light and underrated Loire variety; often used for rosé but capable of interesting reds.

BERRY COLOUR: ● ● ● ● ●

PRINCIPAL SYNONYMS: Aunis (Loir-et-Cher), Chenin Noir (Loire and California), Gros Pineau, Gros-Véronais (Loire), Plant d'Aunis (Maine-et-Loire and Loir-et-Cher), Plant de Mayet (Sarthe)

VARIETIES COMMONLY MISTAKEN FOR PINEAU D'AUNIS: PINOT NOIR

ORIGINS AND PARENTAGE

The name Pineau appeared as early as 1183 in the Val de Loire, but it has been given to at least four distinct varieties: CHENIN BLANC, MENU PINEAU, PINOT NOIR and Pineau d'Aunis. The first reliable mention of Pineau d'Aunis, a variety from the Val de Loire (mainly Sarthe, Loiret, Vienne, Maine-et-Loire, Indre and Touraine), was in Loches (Touraine) by André Jullien (1816): 'Dans l'arrondissement de Loches, on cultive le tendrier, l'auberon, le fromenteau, le bordelais, l'aunis, le viret, le salais, le fié, le côte-rôtie, le confort et la franche noire' ('In the Loches *arrondissement*, we cultivate ...' etc.).

Pineau d'Aunis was named after the French *pin*, because the bunches resemble pine cones, as do those of PINOT, and after Aunis, a former province of the Charente-Maritime *département*.

Although Pineau d'Aunis was called Chenin Noir by Odart (1845), and although CHENIN BLANC is still often called Pineau de la Loire or Gros Pineau, Chenin Blanc is not a white mutation of Pineau d'Aunis, as DNA profiling has verified. Pineau d'Aunis is not related to PINOT and belongs to the Messile ampelographic group (see p XXVII; Bisson 2009).

VITICULTURAL CHARACTERISTICS

Mid ripening. Very susceptible to browning of the leaves and sunburn. Yields may be irregular. Susceptible to chlorosis, and compact, small-berried bunches make it susceptible to botrytis bunch rot. Quality falls steeply when yields are not controlled.

WHERE IT'S GROWN AND WHAT ITS WINE TASTES LIKE

Wines are generally pale but lively and peppery, often produced as a rosé or used as base wines for fizz. **France** had 435 ha (1,075 acres) in 2009, barely changed in the previous decade but considerably down on the 1,741 ha (4,302 acres) of 1958. Plantings are predominantly around the city of Tours and in the Loir-et-Cher in the Val de Loire. Pineau d'Aunis features in various appellations: as a minor component in rosé and sparkling Touraine and in the reds, fizz and sweetish rosés of Anjou; red and rosé Coteaux du Loir require at least 65%, and in the Coteaux du Vendômois, vin gris must be 100% and reds at least 50% Pineau d'Aunis. Domaine de Montrieux's light red in this last appellation is lifted and peppery but has quite a firm tannic grip. Other recommended producers of varietal wines include Thierry Puzelat (red) and Clos Roche Blanche (rosé), both from Touraine, and Domaine Les Maisons Rouges in the Coteaux du Loir.

PINELA

See PINELLA.

PINELLA

Minor variety from the hills south of Padova, also grown in neighbouring Slovenia.

BERRY COLOUR: ● ● ● ● ●

PRINCIPAL SYNONYMS: Mattozza (Padova), Pinela or Pinjela (Slovenia), Pinola (Friuli)

ORIGINS AND PARENTAGE

Pinella probably originates from the province of Gorizia in Friuli, north-east Italy, where it was apparently first mentioned on 8 August 1324 among the grapes used in Udine, Friuli, listed in Pietro di Maniago's *Catalogo delle varietà delle viti del Regno Veneto* (Fabbro *et al.* 2002).

Under the Slovenian spelling Pinela, it is considered to be an old Slovenian variety, most likely originating from Vipavska Dolina. DNA profiling has not yet been done to confirm whether it is the same variety as Friulian Pinella.

VITICULTURAL CHARACTERISTICS

Mid ripening. Highly susceptible to botrytis bunch rot and sour rot because of the berries' thin skins, and sensitive to sun and wind. Yields well and consistently.

WHERE IT'S GROWN AND WHAT ITS WINE TASTES LIKE

Pinella has now virtually disappeared from its supposed cradle in Gorizia in Friuli, Italy, and it is found particularly in the Colli Euganei, the hills just south of the town of Padova. Here it is used to produce still or sparkling wines, as a varietal or blended (maximum 20%) with varieties such as GARGANEGA or PROSECCO, in the DOC of the same name, or as a minor part of the blend in Bagnoli di Sopra (maximum 10%) and Merlara DOCs. Some varietal wines, eg the Cantine dei Colli Tramonte's sparkling Pinello di Pinella, are classified as IGT Veneto. There were 72 ha (178 acres) planted in **Italy** in 2000.

There are also limited plantings over the border in **Slovenia** (50 ha/124 acres in 2009) under the name Pinela, occasionally Pinjela, especially in the Vipavska Dolina region in Primorska in the south west. Recommended producers include Guerila, and Vipava 1894.

PINOT

The history, origins and etymology of Pinot are discussed first, then the most important clonal variations are described separately, **in chronological order of appearance in the literature**: Pinot Noir, Pinot Meunier, Pinot Gris, Pinot Blanc, Pinot Teinturier and Pinot Noir Précoce.

CLONAL DIVERSITY

Pinot has been propagated by vegetative reproduction for several centuries so it has had plenty of time to mutate, which explains the enormous clonal diversity observed, with more than 1,000 registered clones. Given this high level of diversity, seen in berry colour, yield, organoleptic qualities, leaf shape, cluster size, etc., it is often claimed that Pinot has a particularly high mutation rate (Bernard 1995) but to date there is no scientific evidence that Pinot has a higher mutation rate than any other variety. The real reason for this considerable clonal diversity is simply the length of time the variety is thought to have existed – around 2,000 years.

Until recently, Pinot Noir, Pinot Gris, Pinot Blanc, Pinot Meunier, Pinot Noir Précoce and Pinot Teinturier, to cite the most common mutations, were considered to be distinct varieties belonging to a so-called 'Pinot family'. However, black, grey and white berries can sometimes be observed on the same Pinot vine, or even individual striped berries, and DNA profiling has now revealed that all the Pinot types mentioned above have the same genetic fingerprint when analysed at the standard set of eight DNA markers (Regner *et al.* 2000b). As a consequence, it is misleading to talk about a 'Pinot family' because all these forms are mutations that appeared within a single initial variety, Pinot. Nevertheless, clonal differences do have a genetic background, and analysis of 100 DNA markers has shown that a few genetic variations can be observed among clones of Pinot, as with CHARDONNAY (Riaz *et al.* 2002).

In the 1960s, in response to the predations of viruses in the vineyards of Burgundy, the French instituted a concerted attempt at selecting and certifying disease-free clones of Pinot Noir (and CHARDONNAY). Some of the early selections, including the large-berried so-called Pinots Droits, were chosen for quantity rather than quality, with disastrous results, but since then hundreds of superior selections have been made, numbered and imported all over the Pinot-growing world. Those currently most in demand include 113, 114, 115, 667, 777 and 828. In the US they tend to be known as 'Dijon clones'; in Australia as 'Bernard clones' after Raymond Bernard, who did much to promulgate them. Although they have usually been selected for specific attributes, in general their small berries tend to produce more complex, but more reserved, wine than the Pinot Noir produced from the clones that were originally planted outside Europe, such as the widely planted clone called Pommard (UCD 4 and 5) distributed by the University of California at Davis, the so-called Wädenswil clone (UCD 1) certified virus-free in the Swiss town of the same name, and Mariafeld (UCD 23). There are also clones planted outside Europe named after those who first propagated them (Wente in California) or who smuggled them through plant quarantines (Abel in New Zealand). Many growers today, especially in Burgundy, prefer to grow a 'mass selection' of different plants rather than a single clone.

ORIGINS AND PARENTAGE

Many legends have been told about the origins of Pinot (see below) but the true origin is still unknown. According to the French ampelographer Louis Levadoux (1956), the morphology of Pinot is very similar to that of the wild grapevine (*Vitis vinifera* subsp. *silvestris*), from which it could have been domesticated. Natural populations of *Vitis vinifera* subsp. *silvestris* are indeed still growing in the Île-de-France region of northern France (Arnold *et al.* 1998), where Pinot was first mentioned under the name Morillon in the thirteenth century (see below). However, no genetic link between Pinot and local wild grapevines has so far been found.

Joint research between the University of California at Davis and INRA (Institut National de la Recherche Agronomique) in Montpellier, southern France, has established that Pinot and GOUAIS BLANC have given birth to at least twenty-one grape varieties through spontaneous crosses that occurred at different times and in different places in north-east France (Bowers *et al.* 1999; Boursiquot *et al.* 2004), some of them no longer cultivated: ALIGOTÉ, AUBIN VERT, AUXERROIS, BACHET NOIR, BEAUNOIR, CHARDONNAY, Dameron, GAMAY NOIR, Franc Noir de la Haute-Saône, François Noir Femelle, Gamay Blanc Gloriod, Gros Bec, KNIPPERLÉ, MELON, MÉZY, Peurion, Romaine, ROMORANTIN, Roublot, Rubi and SACY (see the pedigree diagram below).

Most of these progenies belong, with Pinot, to the Noirien ampelographic group (see p XXVII), named after one of Pinot's most ancient synonyms (see below; Bisson 2009). Since Pinot Gris and Pinot Blanc are colour mutations of Pinot Noir and show the same DNA profile, the berry colour of the parent of these offspring cannot be determined by DNA parentage analysis. However, it can be assumed that Pinot Noir is the parent of all black-berried progenies (BACHET NOIR, BEAUNOIR, Dameron, GAMAY NOIR, Franc Noir de la Haute-Saône, François Noir

PINOT PEDIGREE DIAGRAM

No fewer than 156 western European grape varieties make up this huge pedigree of natural crosses, with SAVAGNIN, GOUAIS BLANC and PINOT playing the major roles. Given the number of varieties that are currently unknown (?), this pedigree is just one of the possible interpretations of all direct parent–offspring relationships that have been discovered by DNA parentage analyses (Pinot and TEROLDEGO, which have a grandparent–grandchild relationship, are included to illustrate the unexpected link between Pinot and SYRAH). It strongly challenges the commonly assumed independent origins of western European varieties and argues in favour of a small core set of founder varieties that have given birth through natural crosses to the significant biodiversity we know today. For the sake of clarity, TREBBIANO TOSCANO and FOLLE BLANCHE appear twice in the diagram.

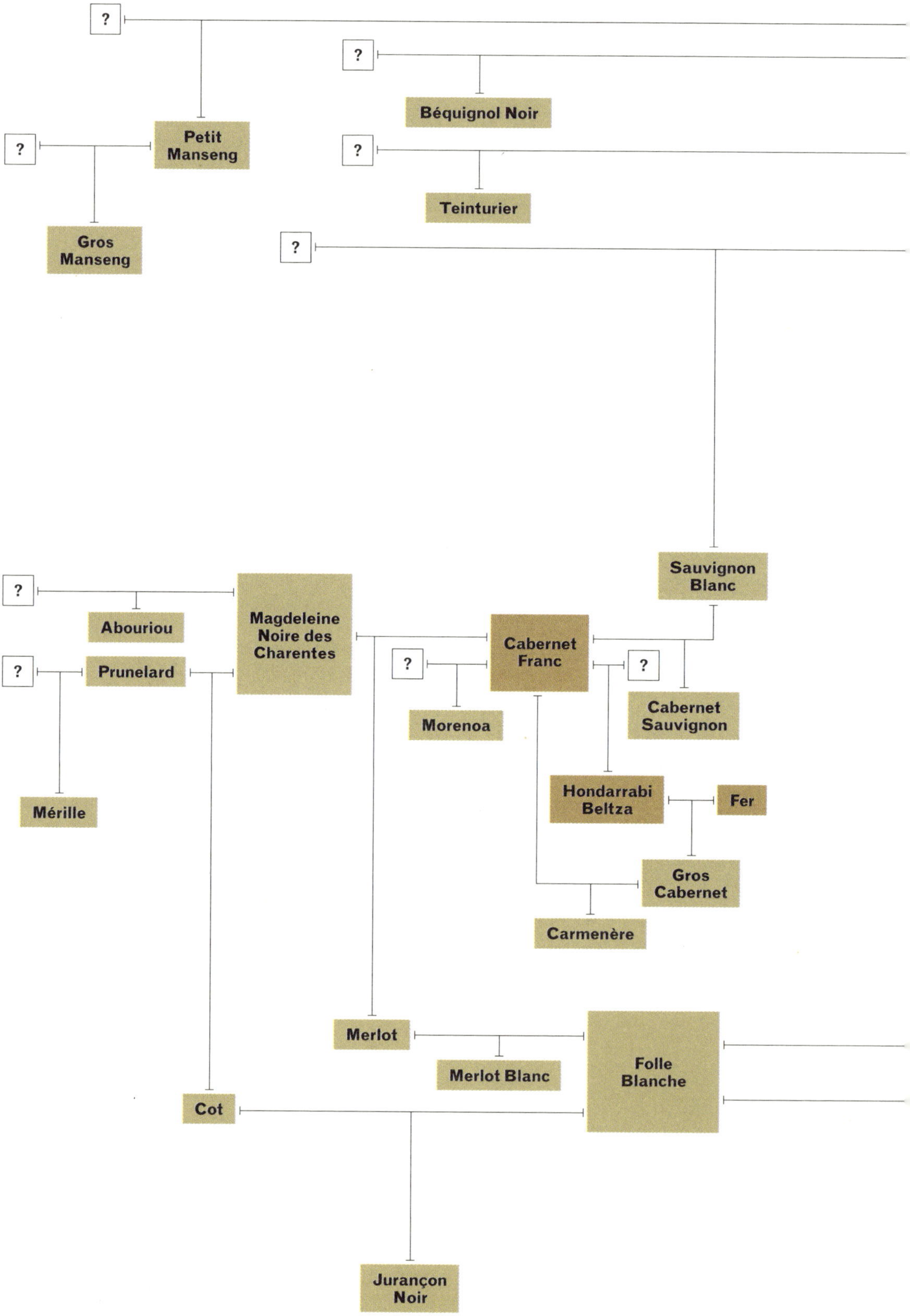

?
?
Béquignol Noir
?
Petit Manseng
?
Teinturier
Gros Manseng
?
Sauvignon Blanc
?
Abouriou
Magdeleine Noire des Charentes
Cabernet Franc
?
Prunelard
?
?
Morenoa
Cabernet Sauvignon
Mérille
Hondarrabi Beltza
Fer
Gros Cabernet
Carmenère
Merlot
Merlot Blanc
Folle Blanche
Cot
Jurançon Noir

Femelle, Gros Bec, MÉZY, Romaine, Rubi, SACY), even if the inheritance of berry colour in the grapevine is governed by a complex mechanism of transposable DNA sequences (Kobayashi *et al.* 2004; This *et al.* 2007).

Extending the pedigree further, Vouillamoz and Grando (2006) have shown that Pinot is a likely grandparent of TEROLDEGO, MARZEMINO and LAGREIN from northern Italy and of DUREZA from the Ardèche in France (see SYRAH pedigree). Since Dureza is a parent of Syrah, Pinot is therefore a likely great-grandparent of Syrah, which challenges the common assumption that they have distinct origins. Most interestingly, a study carried out at the Klosterneuburg research centre has discovered a parent–offspring relationship between two of the oldest grape varieties from Western Europe: Pinot and SAVAGNIN (or Traminer; Regner *et al.* 2000b). This explains why Pinot has been called Savagnin or Salvagnin Noir in the Jura and in western Switzerland, and why both varieties share the name Clevner (with various spellings) in Baden-Württemberg in Germany. Since the other parent in the parentage is unknown, it is not possible to tell whether Pinot is a parent or a progeny of Savagnin, and Pinot may therefore be either a grandparent or a half-sibling of all the varieties that also have a spontaneous parent–offspring relationship with Savagnin (only the cultivated ones are listed below): AUBIN BLANC, CHENIN BLANC, PETIT MESLIER, ROTGIPFLER, SAUVIGNON BLANC and SILVANER. Given the equal validity of the two possibilities in such parentages, there are several ways of reconstructing the most complete pedigree for Pinot. In the one illustrated here, Pinot is considered to be a parent of Savagnin, so that it most surprisingly turns out to be a grandparent of Rotgipfler, Chenin Blanc and Sauvignon Blanc, as well as a great-grandparent of BALZAC BLANC, COLOMBARD, MESLIER SAINT-FRANÇOIS and, most unexpectedly, of CABERNET SAUVIGNON, once again challenging the assumption that these varieties have distinct origins (Vouillamoz).

Pinot (including all clonal variations) was used to breed the following varieties still in cultivation: Bouquet (BUKET), CARMINOIR, DECKROT, DIOLINOIR, DOMINA, FABERREBE, FREISAMER, HELFENSTEINER, HÖLDER, KARÁT, MANZONI BIANCO, PETRA, PINOTAGE, PRÉCOCE DE MALINGRE, SCHÖNBURGER, VIGNOLES and ŽUPLJANKA.

Austrian researcher Ferdinand Regner and colleagues (2000b) have claimed that Pinot is a progeny of SAVAGNIN (Traminer) and Schwarzriesling. This parentage can be rejected because Schwarzriesling, which is not related to RIESLING, is simply a mutation of Pinot Noir (see Pinot Meunier below). The complete parentage of Pinot is still unknown.

ORIGIN OF THE NAME PINOT There are two equally valid hypotheses regarding the etymology of Pinot:

- It is usually assumed that the name Pinot, like Pignola (see PIGNOLA VALTELLINESE) and PIGNOLO in Italy, derives from *pin*, meaning 'pine' (*pigna* in Italian), because the clusters of this variety resemble a pine cone.
- Dion (1959) suggested an alternative explanation: Pinos or Pignols could be the name of a place in France whence came cuttings of this variety used to plant a vineyard for a king or lord. A village named Pignols does exist in the Puy-de-Dôme in the Auvergne, where Pinot has been cultivated since the Middle Ages.

In Viala and Vermorel (1901–10), the French ampelographer Eugène Durand (1901a) tells how, at a congress in Chalon-sur-Saône in 1896, he, Adrien Berget and other scholars voted in favour of the spelling Pinot (traditional in Burgundy) instead of the Pineau spelling that has since then been used for MENU PINEAU and PINEAU D'AUNIS in the Val de Loire.

PINOT NOIR

Finicky Burgundian vine produces wildly variable, relatively delicate, potentially haunting essences of place.

BERRY COLOUR:

PRINCIPAL SYNONYMS: Auvernat or Auvernas (Orléanais in northern France, Baden-Württemberg in Germany), Berligou or Berligout (Loire), Black Burgundy (US), Blauburgunder (Germany, Switzerland and Austria), Blauer Arbst (Baden-Württemberg), Blauer Spätburgunder (Germany, Switzerland and Austria), Bourguignon (Auvergne in France), Burgunder (Germany, Switzerland and Austria), Cerna (Republic of Moldova), Clevner or Klävner (Alsace in France, Baden-Württemberg in Germany, Steiermark/Styria in Austria, near Zürich in Switzerland), Cortaillod (Neuchâtel in Switzerland), Kék Burgundi (Hungary), Kisburgundi (Hungary), Klebroth (Baden-Württemberg), Moréote (Baden-Württemberg), Morillon or Morillon Noir or Mourillon (Burgundy in France), Noirien or Noirin (Burgundy), Orléanais (Touraine in France), Pineau Noir (Burgundy), Pino Fran and Pino Ceren (Republic of Moldova), Pinot Cernii (Russia), Pinot Liébault (Burgundy), Pinot Nero (northern Italy), Plant Doré (Champagne in France), Rulandské Modré (Slovakia), Savagnin Noir or Salvagnin Noir (Jura in France, Neuchâtel and Vaud in Switzerland), Servagnin (Vaud), Spätburgunder (Germany and Austria), Vert Doré (Champagne)

VARIETIES COMMONLY MISTAKEN FOR PINOT NOIR: BÉCLAN, BLAUBURGER, BLAUFRÄNKISCH (under the name Burgund Mare in Romania), BRUN FOURCA, GAMAY NOIR (both Pinot Noir and Gamay Noir were called Dôle in the Valais in Switzerland, and some Pinot Noir was once called Gamay Beaujolais in California), GOUGET NOIR, PERSAN, PINEAU D'AUNIS, SANKT LAURENT (mistakenly believed to be a mutation of Spätburgunder/Pinot Noir in Austria), TRESSOT (Yonne), TROUSSEAU

HISTORICAL REFERENCES

REFERENCES TO THE OLD SYNONYMS MORILLON, NOIRIEN AND AUVERNAT Before the names Pinot or Pineau came into use, this variety was known mainly by three old synonyms: Morillon, Noirien and Auvernat (with various spellings). These early synonyms usually referred to a black-berried variety, unless a colour name was added (eg Morillon Gris), and their identity with Pinot Noir has been documented by several authors since the nineteenth century (Odart 1854; Chevalier 1873; Durand 1901a; Galet 1990). The earliest mention appeared in 1283 under the name Moreillon (later spelt Morillon, Mourillon or Maurillon) in a legal tome, *Les coutumes de Beauvaisis*, by jurist Philippe de Beaumanoir (1246/7–96) in the region centred on Beauvais, north of Paris (Beugnot 1842): 'Droit pris de vins de rentes selonc le coustume, doit estre prisiés en trois manieres de vins, chest à savoir: vin fourmentel, vin de moreillons et vin de gros noirs. Le vin fourmentel, a la mesure de Clermont, doit estre prisiés chascun mui. XII. sols de rente, et le vin de moreillons chascun mui. IX. sols de rente chascun an, et le vin de gros noirs ou de goet chascun mui. VI. sols de rente' ('The right to wines as income, according to the custom, must be priced in three different types of wine, that is to say: fourmentel wine, moreillons wine and gros noirs wine. Wine of fourmentel, using the Clermont [Clermont-Ferrand] measure, must be paid 12 sols each for income, and the wine of moreillons 9 sols each for income every year, and the wine of gros noirs or goet 6 sols each for income'). According to Dion (1959), the name Moreillon or Morillon was given to the round-berried version of Pinot Noir in the Île-de-France and Picardie regions in northern France, where it later appeared in 1403–4 near Amiens, when the monks of Corbie were asked by the Hôtel-Dieu for a supply of 'moreillons' to renew their vineyard.

The etymology of Morillon has never been clear. Many authors have suggested that Morillon could derive from the Moors (*Maures* in French) from North Africa, whence the variety was supposedly imported, or alternatively because the dark-skinned berries resembled the Moors' skin colour. Comte Odart (1854), on the other hand, thought the name Morillon came from *mour* or *mouret*, the local name for 'blackberry', because of the shape and colour of the berries. This etymology is supported by Krämer (2006), who suggests the name derives from the Latin *maurus*, meaning 'dark brown'. However, a more plausible hypothesis is that Morillon comes from La Morée, a river in the Île-de-France region where the variety was historically widely planted.

The name Noirien (from *noir*, meaning 'black') was used contemporaneously with Morillon in the mid thirteenth century, according to a document cited by Joigneaux (1865): 'Le franc noirien était connu en Bourgogne sous le règne de nos premiers ducs; il en est parlé dans un acte qui date du milieu du treizième siècle, et par lequel il était dû à Hugues IV, fils de Eudes III et d'Alix de Vergy, une rente de quatre muids de vin de noirien Pomard' ('Franc noirien was known in Burgundy in the reign of our earliest dukes; it was mentioned in an act dating back to the mid thirteenth century, according to which Hugues IV, son of Eudes III and Alix de Vergy, was owed an income of four muids [a barrel containing 1,200 litres] of noirien wine from Pomard'). Noirien, Noyrien or Noirin was the name of this variety in the Côte d'Or, as shown by several subsequent references, confirming its identity with the modern Pinot Noir, eg describing a vineyard planted 'de noyriens et de saulvoigniens' in 1383 in Chassagne, now Chassagne-Montrachet (Rézeau 1997).

Slightly after Morillon and Noirien, the variety was mentioned under the name Auvernas in an Act passed in Beaugency in the Loiret *département* on 22 July 1302 (Rézeau 1997): 'Item une piece de terre tenent aux vignes feu Foillet … laquelle piece de terre il promistrent a edifier et planter en bons auvernas' ('Item, one piece of land bordering the vineyards feu Foillet … which piece of land they promised to build up and plant with good auvernas'). Auvernas (also Auvernat) undoubtedly means that the variety was introduced from the Auvergne.

EARLIEST MENTION OF THE NAME PINOT The first mention under the modern spelling Pinot appeared in 1375 when Duc Philippe le Hardi anticipated his forthcoming diplomatic trip to Bruges in Belgium by sending there '6 queues et 1 poinçon de vin de pinot vermeil' ('6 queues and 1 poinçon of ruby Pinot wine [queue and poinçon are measures of volume; total *c.*25 hl]'), Pinot undoubtedly being Pinot Noir here (Dion 1959; Rézeau 1997). This mention predates by almost twenty years the commonly assumed oldest mention of Pinot that appeared under the spelling *pinoz* (plural form of *pinot*) in 1394 in Saint-Briz en Aucerrois (today's Saint-Bris-le-Vineux) in the Yonne *département* (Rézeau 1997): 'ycellui suppliant dist a yceulx vendengeurs, entre lesquels estoit ledit Jehannin, que ilz meissent les pinoz a part sans y mettre autres raisins, mais ce non obstant ledit Jehannin mettoit des treceaux et autres raisins avec les pinoz, dont ledit suppliant fut moult courrauciez, pour doubte que ses vins n'en vaulsissent pis' ('this supplicant says to those harvesters, among whom was the aforesaid Jehannin, that they should keep pinoz separate without any other grapes, but nevertheless the aforesaid Jehannin did put treceaux and other grapes together with pinoz, which much irritated the aforesaid supplicant, fearing that his wines were worth less'). Although the colour of the grapes in these vineyards is not specified, the plural form *pinoz* might suggest the presence of different types of Pinot, possibly of different colours.

At the same place and in the same year of 1394, a letter of dismissal issued by Charles VI reported that a fifteen-year-old boy, who had been hired for the harvest in Saint-Bris-le-Vineux and who had disobeyed the order given to the harvesters to set aside the Pinot Noir grapes to prevent them from being mixed with other kinds of grapes, was hit by the owner of the vineyard so brutally that he died (Dion 1959). Again in the same year, an entry in *Les ordonnances du Louvre* (a collection of French royal ordinances) showed the high value placed on wines made from Pinoz. Many other documents later testified that Pinot Noir was already considered to be a variety of utmost quality in the Middle Ages. From the fifteenth century on, the name Pinot (under numerous spellings such as Pineau, Pignotz, Pinoz, Pynos, Pynotz, etc.) gradually replaced the older ones, yet without supplanting them completely.

EARLY MENTIONS OF PINOT NOIR OUTSIDE FRANCE The names of grape varieties did not appear generally in written documents until the mid to late Middle Ages. However, it is very likely that Pinot had been propagated throughout Western Europe well before its name was documented. In Germany, although earlier dates are often suggested (884, 1318 and 1330, see below in Other hypotheses), the first reliable mention of Pinot Noir was found in 1470 in Hattenheim in the Rheingau under a variant spelling of its German synonym Klebroth: 'Item 1 firtel Clebroit wyngart' ('Item 1 quarter of Clebroit vineyard'; Staab 1971). In Switzerland, Pinot Noir was first mentioned in the towns of Cortaillod in 1766 and in Auvernier in 1775 under the local synonym Salvagnin (Aeberhard 2005) still in use today in the Swiss cantons of Neuchâtel and Vaud. After the eighteenth century, Pinot Noir was then mentioned in many other countries (Austria, Italy, Hungary, etc.).

OTHER HYPOTHESES

The origin of Pinot has been the subject of much speculation, as in the case of many other very old and highly regarded grape varieties. Although the grape colour is rarely mentioned clearly, all the following suppositions are likely to refer to Pinot Noir.

EGYPTIAN ORIGIN Russian ampelographer A M Negrul (1938), who was the first to classify *Vitis vinifera* L. varieties into three large groups (see p XXV), suggested that Pinot Noir originated from the Nile Valley in Egypt, then spread to Greece and later from the Greeks to the Romans, who brought it to France around the fourth century. However, there is absolutely no historical, morphological or genetic evidence supporting this insubstantial hypothesis.

ALLOBROGICA Pinot Noir is often considered to be identical to Allobrogica, a grapevine cultivated in the land of the Allobroges and mentioned in the first century AD by Celsus, Columella and Pliny the Elder (see SYRAH). However, if Allobrogica really was Pinot Noir or one of its ancestors, then Pinot Noir would still be present in the area extending from Lac Léman (Lake Geneva) to Grenoble, and from the Haute-Savoie to Vienne (south of Lyon), but it isn't. If Allobrogica could be identified as an ancestor of a modern grape variety, then it would be Syrah rather than Pinot Noir.

DUBIOUS MENTION IN 884 NEAR THE BODENSEE (LAKE CONSTANCE) Emperor Charles III (Charles the Fat) is supposed to have introduced Pinot Noir (Spätburgunder) from Burgundy to his imperial palace in Bodman near the Bodensee (between Switzerland, Austria and southern Germany) in 884 in his *Königsweingarten* ('King's vineyard'). However, the names of the varieties planted were never mentioned, and the legend has long been propagated without evidence (Aeberhard 2005; Krämer 2006).

DUBIOUS MENTION IN 1183 IN THE CHER (FRANCE) The place name Plantes de Pinaud was mentioned in 1183 in Mennetou-sur-Cher but it is far from proven that it actually referred to Pinot Noir (Rézeau 1997). In addition, although Pinot was sometimes spelt Pineau in Burgundy, the spelling Pineau in the Vallée de la Loire most often refers to the white MENU PINEAU (or the dark-skinned PINEAU D'AUNIS).

CLEVNER FROM VALTELLINA IN ITALY Clevner is an old synonym for Pinot Noir (and also Pinot Blanc) in Alsace in north-eastern France as well as in Baden-Württemberg in Germany. This name supposedly derives from Kleven or Kläven, the German name for Chiavenna, a city near Sondrio in Valtellina in Lombardia, northern Italy, and some authors have suggested that Clevner (aka Pinot Noir) was brought from Valtellina to Germany and France. However, no trace of Pinot Noir has ever been found in Valtellina, and there was possibly a mix-up between the name of the wines made in Chiavenna in the Middle Ages (Klevener or Kleffener) and the name of the grape variety in Germany (Krämer 2006).

DUBIOUS MENTIONS IN 1318 AND 1330 IN BADEN-WÜRTTEMBERG According to Hoffmann (1982), Pinot Noir was first mentioned in Germany under its synonym

Klebroth in the archives of the monastery of Salem in 1318, and then in 1330 in Affental (Ortenau), both in Baden-Württemberg. However, the original Latin text speaks about Clauener in Salem in 1318 and apparently also in 1330, which corresponds to Clevner or Klävner. The berry colour is not mentioned in these sources, and since Clevner is now the official name for SAVAGNIN in Ortenau (Hillebrand *et al.* 2003), it is extremely doubtful that these references are to Pinot Noir rather than to Savagnin.

SAVAIGNINS NOIRS IN 1386 IN THE JURA Pinot Noir is thought to have been first mentioned in the Jura in eastern France in 1386, together with POULSARD in Chilley, today called Chilly-le-Vignoble, under the name Savaignins Noirs (see Poulsard for details), because Pinot Noir is still called Servagnin, Sarvagnin, Savagnin or Salvagnin Noir in the Jura and in Switzerland (Krämer 2006). However, the identification with Pinot Noir cannot be guaranteed because Savaignins could also correspond to SAVAGNIN ROSE, the red-berried mutation of Savagnin.

VITICULTURAL CHARACTERISTICS

Early budding, therefore susceptible to spring frosts and coulure, and early ripening. Likes temperate climates and does well on calcareous-clay soils. In hot climates, it ripens too quickly and the relatively thin-skinned berries tend to shrivel and are subject to sunburn. Best when fertility and yields are restricted, especially for the higher-yielding, larger-berried clones. Tends to produce lots of small bunches. Delicate and susceptible to downy and powdery mildews (more so the former), botrytis bunch rot and virus diseases, notably fanleaf and leafroll, and at risk from leafhoppers.

WHERE IT'S GROWN AND WHAT ITS WINE TASTES LIKE

Pinot Noir occupies a very special place in the firmament of well-travelled grape varieties. Such is the allure of top-quality examples, produced mainly in Burgundy, where the variety's ability to express minute differences in terroir is most brilliantly evident, and the frustration involved in trying to reproduce them in other climates, that making fine Pinot Noir is seen by wine producers the world over as the ultimate test. Its early ripening means that only cooler regions can provide a long enough growing season to produce interesting wines but many climates are too cool or too wet for healthy crop levels.

Outside Burgundy, regions that have earned a reputation for producing successful still Pinot Noir tend to be either continental (Jura, parts of Germany, Switzerland and Canada), low latitude (New Zealand, Tasmania, Patagonia and Bío Bío), high altitude (Alto Adige and some emerging new territories for the vine), or cooled by maritime influence (Oregon, Sonoma, Carneros, Monterey, California's Central Coast, emerging Chilean coastal regions, and parts of Victoria and Tasmania in Australia).

There may be geographical limitations on its spread but the early twenty-first century saw extraordinary commercial demand for Pinot Noir, especially in the US, as a direct result of *Sideways*, a successful Hollywood film about its unique qualities. The result was a rash of wine-based movies and particularly enthusiastic planting of the variety in most suitable spots and some unsuitable in Sonoma, Carneros and all points south from Monterey, and a scramble for varietal Pinot Noir, not all of it authentic, on the international bulk market. In parallel with what became known as 'the *Sideways* effect', warmer summers and more fastidiousness in vineyard and cellar saw the variety, as Spätburgunder, blossom in Germany, where it is hugely popular and where total plantings are surpassed only by those in France and California. Oregon and New Zealand have also staked their red-wine future on this cool-climate variety.

In contrast to Cabernet's trademark tannins, most Pinot Noir is relatively soft, fruity and easy to like. Great grand cru red burgundy can be tough in its youth and need a decade or two for its ineffable complexities to unfurl but, in general, charm is one of Pinot Noir's great assets. This can translate into a cheap commercial example that is simply sweet. At much higher prices, some practitioners have evolved a particularly distinctive rich, deep-coloured, powerful (up to 17% alcohol) California style of Pinot. Most other Pinot producers aim to capture some of its elusive delicacy, however, and see a relatively light cherry hue as a badge of honour. Oregonian Pinot Noir is typically more Burgundian than California's, albeit often betraying the easy fruit of the Pommard clone that underpinned early plantings here, while New Zealand's exuberantly fruity Pinot Noir can be as easy to appreciate as its SAUVIGNON BLANC, with overtones of dried herbs from Central Otago, occasionally beetroot from Marlborough.

Pinot Noir the wine can also be precocious, with only the very best demanding long ageing. In youth it can taste of cherries, raspberries and a wide range of other fruits even if in Burgundy it can be more obviously autumnal and savoury. As it ages, it can take on even more obviously mulch-like notes, hints of truffles and other fungi, but always with an underlying sweetness and, sometimes, a uniquely filigree appeal. There is more to Pinot than power.

But there is also more to Pinot than still wine. In Champagne, Pinot Noir (along with Pinot Meunier and CHARDONNAY) is a key ingredient, its just-ripe berries pressed lightly to yield pale-pink juice without aggressive tannins whose light charge of pigments is precipitated naturally. The worldwide increase in sparkling wines made in champagne's image has seen the variety invade territories such as Lombardia in northern Italy, Carneros in northern California, Catalunya in northern Spain, Limoux in southern France and even England.

France grew 29,576 ha (72,595 acres) of Pinot Noir in 2009 so that, despite its acclaim, Pinot Noir was only the seventh most planted red wine vine in France, rather less than the total for GAMAY NOIR. However, the area has expanded remarkably in the last fifty years, from just 8,535 ha (21,090 acres) in 1958. Much of the growth has been in Champagne, where its 12,900 ha (31,877 acres) represented 39% of all plantings in 2010 (as opposed to 34% five years earlier). It is virtually the only variety grown in the southern Aube *département* of the Champagne region and makes generally rather earthy, rustic Blanc de Noirs champagnes there.

There is more Pinot Noir grown in Champagne than in its presumed birthplace in greater Burgundy, where 10,691 ha (26,418 acres) grow today, only 6,579 ha (16,257 acres) of that in the Côte d'Or heartland. Here it reaches the peak of its performance as transmitter of terroir, the medium of expression for perceptibly different though sometimes neighbouring small vineyards. The Côte d'Or deserves a book to itself and has inspired many excellent works more detailed than we can be here. There are 3,200 ha (7,907 acres) of Pinot Noir in the Saône-et-Loire, where some first-class vignerons fashion lively and sometimes delicate southern answers to the best of the Côte d'Or but where most wines are just slightly rustic. Some Pinot Noir makes very light, sometimes tart, red in the greater Chablis area while Pinot Noir is the grape of red and pink Sancerre, Menetou Salon and other Loire appellations. Pinot Noir in marginal areas such as Sancerre and Alsace has been becoming darker in colour and deeper in flavour as general temperatures and winemaker skills have increased. The variety's 1,521 ha (3,758 acres) in Alsace in 2009 represented almost 10% of the total vineyard area.

With POULSARD, Pinot Noir is the principal red wine grape of the Jura and is allowed in many far-flung French appellations, often in conjunction with other varieties. It is found as far north as the Côtes de Toul and Moselle and as far west as Saumur for its fizzy pink version but only as far south as Châtillon en Diois (maximum 25%, mainly with GAMAY NOIR), where altitude helps to mitigate the latitude.

Italy had 3,314 ha (8,189 acres) of Pinot Nero, according to the agricultural census of 2000. It tends to be most successful in the Alto Adige, where recommended producers include Hofstatter, Haas, Gottardi and Lageder. Respectable examples are also made by Le Due Terre and Bressan in Friuli and Le Fracce in the Oltrepò Pavese, where there has been marked progress with making still wines from this variety in recent years even though a very high proportion of it is used for sparkling wine, notably Franciacorta.

Much of **Spain** is too hot for this delicate, early-ripening variety. The 968 ha (2,392 acres) planted in 2008 were used predominantly for Cava in Catalunya, with the most successful still varietal examples coming from high-altitude vineyards such as Escoda-Sanahuja's in Conca de Barberà. **Portugal** is no great enthusiast for international varieties of any sort and had just 244 ha (603 acres) of the red burgundy grape in 2010, mainly in the north. Perhaps unexpectedly, the tiny wine industry in **England** can draw on a greater total area of Pinot Noir than Portugal: 250

ha (618 acres) in 2009, making it the UK's most planted vine variety, just ahead of CHARDONNAY, thanks to the rush to make English sparkling wine in the image of champagne.

Germany is currently bursting with pride in its hugely improved Spätburgunder or Blauer Spätburgunder, of which there was a total of 11,800 ha (29,158 acres) in 2008. Thanks largely to climate change, but also to lower yields and real will and skill, most German Pinot Noir is now made from fully ripe, healthy grapes with good colour – and often a tad more sweetness than is customary in France. Baden is the centre of Pinot activity, with 5,855 ha (14,468 acres) of Spätburgunder planted. Well over 1,000 ha (2,471 acres) of the variety are grown in each of the Pfalz, Rheinhessen and Württemberg, however. The overall area increased significantly in the 1980s and 1990s but has now levelled off. Recommended producers include Rudolf Fürst in Franken; Hessische Staatsweingüter Kloster Eberbach and August Kesseler in the Rheingau; Friedrich Becker and Klaus Peter Keller in Rheinhessen; J J Adenauer, Deutzerhof, Meyer-Näkl and Jean Stodden in the Ahr; Bercher, Dr Heger, Bernhard Huber and Martin Wassmer in Baden; and Bernhart, Knipser and Philipp Kuhn in the Pfalz.

Pinot Noir, sometimes known as Blauburgunder, is a comparatively minor red wine grape in **Austria**, where there were just 649 ha (1,604 acres) in 2009, up from 409 ha (1,011 acres) in 1999 but much less popular than its descendants BLAUFRÄNKISCH and ZWEIGELT, and even less than the local SANKT LAURENT. The most substantial plantings are in Niederösterreich (Lower Austria) and Burgenland and recommended producers include Paul Achs, Gernot and Heike Heinrich, Pöckl, Claus Preisinger and Schloss Halbturn in Neusiedlersee; Prieler in Neusiedlersee-Hügelland; Meinklang and Pittnauer in Burgenland; Schloss Gobelsburg in Kamptal; and Gerhard Markowitsch in Carnuntum.

Pinot Noir, often known as Blauburgunder in eastern **Switzerland**, notched up a 2009 national total of 4,401 ha (10,878 acres), making it the most widely planted red wine variety by far, representing more than half the non-hybrid reds in the country. It is grown throughout the country but with the greatest area in the Valais in the French-speaking south west. The tradition was to blend it with GAMAY NOIR but varietal wines are on the increase. The next biggest areas are in Zürich and Graubünden in the German-speaking north and east. Recommended varietal wines are made by the likes of Martha & Daniel Gantenbein, Thomas Marugg and Markus Stäger in Graubünden and Domaine Henri Cruchon in Vaud.

Pinot Noir, and vines mistaken for Pinot Noir, are widely spread throughout Eastern Europe, where its name is usually some variant on the local word for Burgundian. The **Republic of Moldova** has most, 6,521 ha (16,114 acres) in 2009, of the variety known variously as Pino Fran, Pino Ceren and Cerna. Neighbouring **Romania** had 779 ha (1,925 acres) in 2008, more than half of it on the Black Sea. In **Hungary**, sometimes known as Kék Burgundi or Kisburgundi, it was planted on 963 ha (2,380 acres) of vineyard in 2008, with the biggest area on the hills of Balatonboglár, although Eger looks more promising to judge from the handiwork of producers such as Vilmos Thummerer and the late Tibor Gál. It has a longer tradition in Villány in the warmer south, but the vital acidity is sometimes lacking. Recommended producers here include Malatinszky, Tiffan and Vylyan. In **Croatia**, there were 290 ha (717 acres) of Pinot Crni in 2009. In Krasnodar Krai, the main wine-growing region in southern **Russia**, where it is also known as Pinot Franc, there were 533 ha (1,317 acres) in 2009. **Turkey** had 10 ha (25 acres) in 2010 and **Israel** a tiny amount. **Ukraine** had 767 ha (1,895 acres) in 2009. Pinot Noir is grown to a very limited extent in **Georgia** (171 ha/423 acres in 2004).

The *Sideways* effect boosted **California** plantings of Pinot Noir enormously in the early years of the twenty-first century so that by 2010 the state had a total of 37,290 acres (15,091 ha), especially in Sonoma (11,013 acres/4,457 ha) and Monterey (8,569 acres/3,468 ha). In the mid 2000s, there were about 24,000 acres (9,600 ha), notably in Sonoma – especially the dynamic but excessively extensive Sonoma Coast AVA and Russian River Valley, where demand for Pinot Noir has put pressure even on ancient Zinfandel (TRIBIDRAG) plantings – but the most fog-cooled pockets of the Central Coast have seen considerable Pinot plantings too.

A certain amount of California Pinot Noir is used as a sparkling wine base but there has been such huge demand for it as a still red varietal that this is now its principal incarnation. Styles vary from lighter, nervy, cherry-fruited wines from the likes of Saintsbury of Carneros, Rhys of Santa Cruz Mountains, Littorai and Kutch of various Sonoma locations, and Sandhi of Sta. Rita Hills through the firm, long-lived burgundy-inspired wines of Au Bon Climat in Santa Maria, the rather richer Pinots of Santa Lucia Highlands of Monterey to some really full-throttle, late-picked renditions from Sonoma such as those of Marcassin and Kosta Browne.

The entire **Oregon** wine industry is predicated on Pinot Noir, of which 11,201 acres (4,533 ha) were planted in 2008, up from 3,689 acres (1,493 ha) in 1998, mainly in the Willamette Valley immediately south of Portland. The state staked its wine future on Pinot Noir when David Lett planted the Wädenswil clone back in 1965, having moved north from California in search of cooler climes. The grey skies (and, often, early autumns) of the Willamette Valley suited him, and Pinot, perfectly. Today Dijon clones are all the rage although even some of the wines made from clones sourced in California can have impressive structure and longevity. Domaine Drouhin, owned by the eponymous Beaune négociant, has been a pioneer of sensitively made Pinot Noir but there are dozens of other fine practitioners – so many that, as in the Médoc, for example, it is almost invidious to name names. The south of the state can be a little hot for this variety.

Summers are too hot for Pinot Noir in eastern **Washington**, where most of the state's vines grow, but it is the most popular variety in the damp Puget Sound and has performed encouragingly in the new Lake Chelan AVA. The state total was 307 acres (127 ha) in 2011.

New York State's tally of Pinot Noir vines was 335 acres (136 ha) in 2009. There has been some tentative experimentation with Pinot Noir in the Finger Lakes but sparkling versions so far seem the best bet.

There has been some success with the variety in **Idaho** and in **Michigan** (135 acres/55 ha in 2006, up from 39 acres/16 ha in 1997), and Gruet grow it on the hot, dry mesas of **New Mexico** for their sparkling wine. **Virginia**'s total had reached 37 acres (15 ha) by 2011, including some interesting new plantings at altitude at Ankida Ridge, and there were 70 acres (28 ha) in 2008 in **Texas**.

Both Ontario (1,025 acres/415 ha in 2006) and British Columbia (793 acres/321 ha in 2008) in **Canada** can produce some very creditable examples of Pinot Noir indeed, with Le Clos Jordanne and Blue Mountain being skilled respective practitioners, but there have been no shortage of competitors.

The dazzling sunshine and low latitudes of most wine regions in **Argentina** are inimical to a variety as fragile as Pinot Noir but some intriguingly fine, if relatively rich and earthy, Pinot has emerged from Río Negro, Patagonia – notably from Chacra, whose vines are many decades old. Others making good progress with the variety include Familie Schroeder and NQN in Neuquén and Gouguenheim in Tupungato. The country's total plantings were just 1,509 ha (3,729 acres) in 2008.

Chile has more obvious terrain for Pinot Noir along its fog-cooled Pacific coast, specifically the cooler regions of Casablanca and San Antonio in the north, as well as in the cooler inland vineyards of Bío Bío in the south. Recommended producers include Casa Marin and Matetic in San Antonio; Viña Leyda in the Leyda Valley; Cono Sur (Chile's Pinot pioneer) and Ventisquero in the Casablanca Valley; and the various operations of the Viña Dos Andes group in Bío Bío, which has initiated joint ventures with such luminaries of Burgundy as Pascal Marchand and Nicolas Potel. Chile grew a total 1,382 ha (3,415 acres) of Pinot Noir in 2008 and is able to produce some of the world's best-value examples.

There were just 42 ha (104 acres) of Pinot in **Uruguay** in 2009 and 190 ha (470 acres) in **Brazil** in 2007. A little is also found in **Peru**.

As with SAUVIGNON BLANC, **Australia** is very conscious of New Zealand's aptitude for Pinot Noir and grows only about the same amount as its much cooler, trans-Tasman rival: 4,490 ha (12,118 acres) in 2008. Australia's Pinot Noir is if anything more varied than New Zealand's, even if limited to the coolest spots, many of them in the state of Victoria with the Yarra Valley (702 ha/1,735 acres) offering the greatest concentration of the variety, although only its cooler reaches are really suitable. Victorian Pinot superstars include Coldstream Hills, De Bortoli, Gembrook Hill, TarraWarra and Yering Station in the Yarra Valley; Bannockburn in Geelong;

Curly Flat and Domaine Epis in the Macedon Ranges; Bass Phillip in Gippsland; and a host of producers in the ocean-cooled Mornington Peninsula who are making very creditable Antipodean answers to red burgundy. In South Australia, the variety is most commonly planted in the cool Adelaide Hills (404 ha/998 acres) although it is starting to show form in the cool Fleurieu Peninsula. Australia's second biggest concentration of Pinot Noir is in Tasmania (658 ha/1,626 acres), where much of the Pinot Noir is grown for wine made sparkling on the mainland, notably Hardys' various fizzes. Stoney Rise, Sinapius, Freycinet (known as Wineglass Bay in Europe) and Stefano Lubiana have all produced some very pretty still Pinot Noir.

Along with SAUVIGNON BLANC, Pinot Noir is one of **New Zealand**'s proudest accomplishments. Plantings were estimated to be 5,000 ha (12,355 acres) in 2011, up from 2,624 ha (6,484 acres) in 2003 and still increasing, albeit at a slower rate than before. Only about 10% is destined for sparkling wine – the rest is for exuberantly fruity still red wine that is slowly gaining complexity as the vines age. The greatest area is in Marlborough (2,000 ha/4,942 acres), where the likes of Clos Henri, Dog Point, Fromm, Isabel Estate, Saint Clair and Seresin offer welcome individuality. Central Otago is almost entirely devoted to its cheerfully fruity still Pinot Noir, much of its 1,202 ha (2,970 acres) also having been planted during the recent boom. Felton Road, Mount Difficulty, Peregrine, Quartz Reef and Valli are just some of many reliable names here. Waitaki in North Otago has the limestone that Central lacks, but its young vines have also suffered severe frost damage. Some isolated outcrops of limestone in Canterbury such as those planted by Bell Hill and Pyramid Valley also show promise, as does Mountford. Some of New Zealand's oldest Pinot Noir vines are in Martinborough in the south east of the North Island, where producers such as Ata Rangi, Dry River and Martinborough Vineyards have already established admirable track records.

Less than 1% of **South Africa**'s total vineyard area in 2008, 727 ha (1,795 acres), was given over to Pinot Noir, most successfully in the Antarctic-cooled region of Walker Bay, near Hermanus. Recommended producers there include pioneer Hamilton Russell, Bouchard Finlayson and Newton Johnson, although there has long been frustration in Cape winelands that their Pinot Noir has yet to make much of a splash abroad.

Pinot Noir is also found in **China** (36 ha/89 acres in 2009), **Japan** (63 ha/156 acres in 2009, mainly in Nagano but results in Hokkaido are promising) and **India**. But while winemakers all over the world are fascinated by the challenge of producing this delicate variety, in most locations it tends to be too finicky to shine.

PINOT LIÉBAULT

This is simply a selection of Pinot Noir that was selected by A Liébault *c.* 1810 from Les Charmes vineyard in Gevrey in Burgundy. It is slightly more productive than others.

PINOT MEUNIER

Fruity blending ingredient for top-quality sparkling wine.

BERRY COLOUR: ● ● ● ● ●

PRINCIPAL SYNONYMS: Auvernat Blanc (Loiret in northern France), Gris Meunier (Loiret), Meunier (Loiret and Champagne), Morillon Taconné (Marne), Müllerrebe (Germany), Samtrot⚭ (Heilbronn in Baden-Württemberg, Germany), Schwarzriesling⚭ (Baden-Württemberg)

ORIGINS

The black-berried Pinot Meunier is characterized by a layer of white hairs on the underside of the leaves that look like a dusting of flour, hence the name *meunier*, meaning 'miller' in French. This mutation was first mentioned by Merlet in 1690 (Rézeau 1997): 'le Morillon taconné, ou Meunier, à cause qu'il a ses feüilles blanches et farineuses' ('Morillon taconné, or Meunier, because it has white and floury leaves').

Samtrot is a mutation of Pinot Meunier observed by Schneider in Heilbronn, north of Stuttgart in southern Germany.

VITICULTURAL CHARACTERISTICS

Early budding and ripening, though it generally buds later and ripens earlier than Pinot Noir and is therefore less prone to winter frosts and coulure and more reliably productive. Favours rich clay soils but also well suited to calcareous soil, showing little susceptibility to chlorosis. Susceptible to grape moth and botrytis bunch rot.

WHERE IT'S GROWN AND WHAT ITS WINE TASTES LIKE

Pinot Meunier makes more obviously fruity, earlier maturing wine than does Pinot Noir – although varietal versions are relatively rare, for its wine is most commonly blended with Pinot and CHARDONNAY base wines for sparkling wine.

France grew 11,088 ha (27,387 acres) of Meunier in 2009, making it the tenth most planted red wine variety (and the second most widely planted in the Champagne region after Pinot Noir), even though the word Meunier is hardly ever seen on a French label.

It was once widely planted in the far north of France but is now chiefly found, and valued, in Champagne, particularly in the cooler north-facing vineyards of the damp and frost-prone vineyards of the Vallée de la Marne. Not only does it fare better than Pinot Noir on such sites but it also tends to have higher acidity yet can achieve comparable sugar levels, making it a good blending component with Pinot Noir and CHARDONNAY. It is said to contribute youthful fruitiness but not longevity and few growers prize it enough to make a varietal Meunier-dominated wine. Notable exceptions include Françoise Bedel, Egly-Ouriet, Georges Laval, Michel Loriot, Jérôme Prévost and, notably, Krug, which long favoured the inclusion of Meunier in their highly regarded blends.

Outside Champagne, plantings are few and far between in France although it is recommended in much of the Loire and eastern France and permitted in such appellations as Côte de Toul, Moselle and Touraine, and is the principal variety in the rosé wines of Touraine Noble-Joué and the reds and rosés of Orléans.

England had 65 ha (161 acres) in 2009, an indication of the increasing interest in sparkling wines based on Pinot Noir, Meunier and CHARDONNAY there.

Pinot Meunier is known as Müllerrebe, and also rather more misleadingly as Schwarzriesling, in **Germany**, where total plantings were 2,361 ha (5,834 acres) in 2008, mostly in Württemberg (1,738 ha/4,295 acres) but also Baden (266 ha/657 acres), the Pfalz (161 ha/398 acres), Franken (90 ha/222 acres) and Rheinhessen (81 ha/200 acres). It is hard to know how much Samtrot is planted because the official German statistics include this mutation within the area for Pinot Noir (Blauer Spätburgunder). However, good examples are produced by Aldinger, Wachstetter and Weingärtner Bad Canstatt, all in Württemberg. Thomas Seeger in Baden is one of the few to produce a worthwhile wine labelled Schwarzriesling. Most others are rather light and insubstantial.

California grew just 163 acres (66 ha) in 2010 (mostly in Sonoma and to a lesser extent in Napa) with still varietal versions typically produced by sparkling-wine operations such as Domaine Chandon. There are also several producers in **Oregon**'s Willamette Valley, including Eyrie, and 8th Generation make a version in British Columbia, **Canada**.

Pinot Meunier turns up in a surprising number of countries around the world, most commonly where producers in cooler sites are following the champagne model to make sparkling wine. **Australia** had 106 ha (262 acres) in 2008. More surprising is the variety's track record for varietal red wine with some ageing ability in Australia, where it was known at one time as Miller's Burgundy, produced most successfully from old vines by Best's Great Western in the Grampians in Victoria. Other producers of varietal Pinot Meunier in Australia include a number of others in Victoria and Barringwood Park in Tasmania.

There were just 19 ha (47 acres) in **New Zealand** in 2009, used mainly for sparkling wine.

PINOT GRIS

Full-bodied and aromatic at its best but much more usually encountered enjoying international fame if not glory as anodyne Pinot Grigio.

BERRY COLOUR: ● ● ● ● ●

PRINCIPAL SYNONYMS: Auvernat Gris (Orléanais in northern France), Beurot, Burgunder Roter (Germany), Friset (Franche-Comté), Fromenteau or Fromenteau Gris (Champagne), Grauburgunder or Grauer Burgunder (Germany and Austria), Grauer Clevner (Germany), Griset (France and Switzerland), Kleiner Traminer (Rheingau in Germany), Malvoisie (Savoie and Val de Loire in France, Valle d'Aosta in Italy and Valais in Switzerland), Pinot Beurot, Pinot Grigio (Italy), Pirosburgundi (Hungary), Râjik (Republic of Moldova), Ruländer (Baden-Württemberg in Germany, Austria), Rulandské Šedé (Czech Republic), Rulandské Sivé (Slovakia), Sivi Pinot (Slovenia), Speyeren (Germany), Szürkebarát (Hungary), Tokay (Alsace until 2007)

VARIETIES COMMONLY MISTAKEN FOR PINOT GRIS: SAVAGNIN ROSE (Pinot Gris is called Kleiner Traminer in the Rheingau)

ORIGINS

Pinot Gris is a colour mutation of Pinot Noir that occurred at different times and in different places in Burgundy (notably in Chassagne-Montrachet in the Côte d'Or), but also in Rheinland-Pfalz and in Baden-Württemberg in south-west Germany (Durand 1901a). According to Dion (1959), the mention in 1283 of a 'vin fourmentel' in the region centred on Beauvais, north of Paris (see above), corresponded to Pinot Gris, which was called Fourmenteau or Fromenteau in the Middle Ages, but this interpretation is highly questionable because this old synonym was also given to SAVAGNIN. The earliest reliable mention of Pinot Gris was found in Rheinland-Pfalz in 1711, when it was discovered growing wild in a garden in Speyer by Johann Seger Ruland, after whom the variety was then named Ruländer in this area. In France, this colour mutation was mentioned one year later in the region of Orléans by Boullay (1712) under the old name Auvernat Gris (Rézeau 1997). The synonym Beurot was first mentioned in 1770 under the variant Bureau (Rézeau 1997) and the modern name Pinot Gris did not appear until 1783–4, in Flavigny in the Côte d'Or in Burgundy: 'Pinot Gris, autrement l'auvergnat gris d'Orléans et vulgairement beurot' ('Pinot Gris, otherwise grey auvergnat from Orléans, and commonly beurot').

From 1984, after the Hungarian government lodged a complaint with what was then the European Economic Community, Alsace producers agreed to stop using just the name Tokay on their labels and to use instead the full name Tokay Pinot Gris. In exchange, the Hungarians agreed to drop the name Médoc from their labels. Since 2007, the name Tokay has been banned from Alsace wine labels and only Pinot Gris may be mentioned.

The name Malvoisie, still used for Pinot Gris in some parts of France, Italy and Switzerland, appeared as early as the Middle Ages. However, this name is confusing because it has been given to many distinct varieties such as VERMENTINO, SAVAGNIN, MACABEO, BOURBOULENC, etc. (Rézeau 1997), and Malvoisie was not identified as Pinot Gris until 1800 in France (Rézeau 1997) and 1869 in Switzerland (Vouillamoz 2009d). Pinot Gris or Pinot Grigio is said to have been introduced to Italy at the beginning of the nineteenth century by a producer who cultivated it in the provinces of Alessandria and Cuneo in Piemonte (Giavedoni and Gily 2005) but its presence in the Valle d'Aosta under the name Malvoisie was already attested by Gatta (1838) and its cultivation there might well predate Pinot Gris' introduction to Piemonte.

OTHER HYPOTHESES

Legend has it that in 1375 the emperor Charles IV brought Pinot Gris from France to Hungary, where Cistercian monks planted it on the Badacsony hills near Lake Balaton. This early introduction supposedly explains why Pinot Gris in Hungary is called Szürkebarát, meaning 'grey monk'. In 1568, the theory continues, Pinot Gris was brought back from Hungary to Kientzheim in the French region of Alsace by Lazarus von Schwendi (Graff-Höfgen 2007), a general who took possession of the Hungarian city of Tokay in the reign of Charles-Quint and who owned a castle in Kientzheim, north west of Colmar. This also supposedly explains why Pinot Gris is called Tokay in Alsace, a synonym first cited in 1750 in a manuscript at Domaine Weinbach in Alsace (Krämer 2006). However, there is no historical evidence to document these hypotheses, and it is more likely that Pinot Gris, often used to make sweet wines in Alsace, was intentionally given the name Tokay to benefit from the fame already achieved throughout Europe by the Tokaji wines from Hungary, which were and still are mainly made from FURMINT and HÁRSLEVELŰ, two local varieties that are unrelated to Pinot Gris.

VITICULTURAL CHARACTERISTICS

Berry colour is very variable and much darker than that of most white wine grapes, from pinky purple to almost as dark as Pinot Noir in hotter sites. Early budding and ripening. Relatively vigorous but not very productive. Some susceptibility to botrytis bunch rot and downy mildew. Bunches and berries are small or very small. Potential for high sugar levels in the berries along with low to moderate acidity.

WHERE IT'S GROWN AND WHAT ITS WINE TASTES LIKE

When Pinot Gris' yields are not too high and it is allowed to reach full ripeness it can produce deep-coloured, fairly rich wines with relatively low acidity – occasionally running to flab – and a heady perfume, but in the mass-market arena, the name Pinot Grigio seems enough to guarantee sales of even a tart, neutral, almost colourless and flavourless white wine. In very general terms, producers tend to choose the name Pinot Grigio for less expensive, less characterful versions destined for export markets, but there are many exceptions to this rule and Friuli in particular in north-east Italy uses the term for all of its varietal wine, no matter how concentrated and exciting.

France grew 2,617 ha (6,467 acres) of Pinot Gris in 2009, mostly in Alsace (2,356 ha/5,822 acres), where it is the third most planted variety, after RIESLING (3,382 ha/8,357 acres) and GEWÜRZTRAMINER (2,928 ha/7,235 acres), and has been steadily increasing over the last ten years so that it now represents 15% of the vineyard area. Alsace Pinot Gris can be as luscious as a ripe peach or apricot, with a hint of smoke, developing biscuity, buttery flavours with age. Vendange Tardive versions can stand up to savoury foods. Particularly good examples have been made by JB Adam, Kuentz-Bas, Trimbach, Weinbach and Zind-Humbrecht. It is recommended in much of wine growing France except the south and south west.

In Burgundy, where it is also known as Pinot Beurot, it is still authorized as an 'accessory' variety (along with Pinot Blanc and CHARDONNAY) in many red-wine appellations, from basic Bourgogne Rouge (theoretically up to a maximum 10%) to Gevrey-Chambertin and Bonnes Mares, although it is very rarely found in practice.

Very limited plantings remain in the Loire, where it is known as Malvoisie or Pinot Beurot and can produce perfumed, substantial wines in a wide range of different sweetness levels. The name Malvoisie occasionally appears on the label of a wine from the Coteaux d'Ancenis appellation. Pierre Guindon's version, like many, is slightly sweet. Malvoisie must contribute at least 30% to a pale-pink wine labelled Touraine Noble-Joué, along with Pinot Meunier and Pinot Noir. Domaine Rousseau Frères in Esvres make a good one. The Frères Couillaud in the Nantais region are unusual in producing a varietal Vin de Pays du Val de Loire labelled Pinot Gris. The synonym Malvoisie also crops up in Savoie.

Italy's Pinot Grigio has been the somewhat unfathomable success story of the early twenty-first century. The area given over to the variety, mostly in the north east, almost doubled between 1990 and 2000 (6,668 ha/16,477 acres at the last agricultural census) and has continued to soar as export markets such as the UK and US have flourished thanks to the generally fresh but anodyne style of these inexpensive wines. More serious, characterful and weightier wines may still be found, eg in Friuli made by Vie di Romans and in the Alto Adige by Alois Lageder and Franz Haas. In Lombardia, Pinot Grigio is often an important component in local fizz.

Most Veneto versions are rather neutral and, at their best, inoffensive, the high yields diluting any true Pinot Grigio character – but at least they are Italian and are Pinot Grigio. Many observers are bemused by the sheer volume of wine labelled Pinot Grigio exported from Italy.

Luxembourg is relatively keen on Pinot Gris with a total of 181 ha (447 acres) planted in 2008, more than Pinot Blanc and just less than AUXERROIS. Pinot Gris' weight and relatively low acidity are useful this far north.

In **Germany**, the variety known as Grauburgunder, Grauer Burgunder or (occasionally) Ruländer is the country's fourth most planted white wine grape, with 4,481 ha (11,073 acres) in 2008, concentrated in Baden, Rheinhessen and the Pfalz. Recommended producers include Bercher, Dr Heger, Franz Keller and especially Salwey in Baden. Ruländer is generally used for sweeter styles although traditionalists Reinhold & Cornelia Schneider in Baden make a very good dry version. It needs a good site with deep, heavy soils to maximize the impressive level of extract of which it is capable.

In **Switzerland**, where the majority of the vines – known as either Malvoisie or Pinot Gris – are in the Valais, total plantings were 216 ha (534 acres) in 2009. Marie-Thérèse Chappaz's Malvoisie from Fully in the Valais is one of the country's most sought-after sweet wines. Yvon Cheseaux from Saillon in the Valais is another successful producer of sweet Malvoisie.

In **Austria**, the variety may be called Grauer Burgunder or Pinot Gris and in 2009 there were just 222 ha (549 acres), down from 293 ha (724 acres) in 1999. The majority of those vines are in Burgenland, where it is used both for dry varietal wines and as a component in sweet blends.

In **Hungary** it enjoys the memorable name Szürkebarát and in 2008 there were 1,522 ha (3,761 acres), mostly in Mátra in the north, although it is more often associated with Badacsony north of Lake Balaton in the west. The best results are produced on warmer, volcanic soils. Wines tend to be rich and sometimes oily with high alcohol, heady perfume and some residual sugar, often aged in large oak. Some of the finest examples come from Dr Bussay in Zala, south west of Lake Balaton and near the Croatian border. Other recommended producers include Mihály Figura, Szakálvin and Szent Orbán.

In **Slovenia**, it is known as Sivi Pinot and has a real Slovenian identity. Plantings in 2009 totalled 495 ha (1,223 acres), about half in Štajerska Slovenija (Slovenian Styria) in the north east and half in Primorska in the west, mainly in Goriška Brda, just over the border from Italy's Friuli. Good examples are produced by Mansus, Prinčič, Edi Simčič and Marjan Simčič in the west and Dveri Pax in the north east.

Called Rulandské Šedé in the **Czech Republic**, it was grown on a sizeable 741 ha (1,831 acres) in 2009, mostly in Morava in the south east of the country. In **Romania**, it is more usually known as Pinot Gris and was grown on 969 ha (2,394 acres) in 2008, mostly in the south west towards the Black Sea. The **Republic of Moldova** grew a considerable 2,042 ha (5,046 acres) of Pinot Gris in 2009, **Ukraine** 638 ha (1,577 acres) in 2009. Plantings in Krasnodar Krai in **Russia** were a much more modest 78 ha (193 acres) in 2009.

The fashion for Pinot Grigio has certainly not passed **California** by. There were already 12,907 acres (5,223 ha) of Pinot Gris planted by 2010, a total that had been steadily increasing even though it lost the race against SAUVIGNON BLANC to be the third most planted variety after CHARDONNAY and COLOMBARD. It is widely produced throughout the state, generally in a rich, Alsace style. Bottlings labelled Pinot Grigio tend to be lighter and, usually, more commercial. Particularly toothsome examples are made from Anderson Valley fruit by Breggo, Copain and Navarro, while Babcock's from Sta. Rita Hills and Robert Sinskey's Carneros version have also garnered acclaim.

But Pinot Gris has been regarded as the signature white wine of **Oregon** almost since David Lett of The Eyrie Vineyard first trialled cuttings from Alsace – not least because it has been only recently that suitable clones of CHARDONNAY have been planted there. The state total of 2,736 acres (1,107 ha) in 2008 was double the area planted at the end of the last century, making it by far the most widely planted light-skinned variety in the state (with more than double the area given over to Chardonnay). Virtually all of the many good Pinot Noir producers also make a competent, lightly scented Pinot Gris that can taste rather like an aromatic Chardonnay.

Pinot Gris is also popular in **Washington** State, where its 2011 total of 1,576 acres (638 ha) make it the third most planted white wine grape, expanding steadily, but a long way behind CHARDONNAY and RIESLING.

It is produced in scores and grown in dozens of other American states, particularly **New York, Michigan** and **Virginia** (70 acres/28 ha in 2010).

Pinot Gris is an important variety in **Canada**'s British Columbia, where it was the most widely planted white wine variety in 2008 with 928 acres (376 ha), just ahead of CHARDONNAY and planted on more than 10% of the total vineyard area even though its produce seems less distinctive than that of Pinot Blanc.

Pinot Gris was introduced to one of Mendoza's higher vineyards in **Argentina** when the Lurton brothers of Bordeaux planted Alsace clones 52 and 53 in 2002 and it is now firmly established there. All that sunshine results in a fairly rich wine, not unlike a California CHARDONNAY. Grape skins are often burnished by the sun but the wines can deliver great value. The country grew a total of 349 ha (862 acres) in 2008. A very small amount is grown in the far south of **Chile.**

Australia grew a total of 2,836 ha (7,004 acres) of Pinot Gris, sometimes labelled as Pinot Grigio, in 2008. The fact that this includes 758 ha (1,873 acres) that are not yet bearing fruit, and that the total in 2004 was just 300 ha (741 acres), indicates the popularity of this variety with growers. T'Gallant in Mornington Peninsula pioneered varietal versions, under both names, and they have been much followed. Care is needed to prevent the wines from flab in most warmer areas although Holly's Garden and Seppelt have risen to the challenge in Victoria, and Tim Adams makes a useful version in the Clare Valley.

In **New Zealand,** Pinot Gris has been the most successful of all the aromatic or semi-aromatic light-skinned varieties after runaway leader SAUVIGNON BLANC. The area planted to this variety has doubled in the last four years, to reach 1,501 ha (3,707 acres) in 2009, 5% of the country's total vineyard area and projected to increase further over the next few years. Although there has been a good deal of planting in Marlborough on the South Island, the warmer and more northerly Hawke's Bay is also proving to be a successful region for this variety. The producer's choice of Pinot Gris v Pinot Grigio is a strong indication of the style of wine – rich and weighty v crisp and bland, though many have been held back by a little too much residual sugar. The best achieve a balanced and aromatic richness without the need for excess sweetness. Recommended producers include Dry River in Martinborough; Brennan, Mount Difficulty and Richardson in Central Otago; Neudorf in Nelson; and Saint Clair and Seresin in Marlborough.

A small amount is grown in **South Africa**; a tiny amount in **Japan**.

PINOT BLANC

Possibly underrated producer of friendly, medium-bodied, confident whites. Easy to ripen.

BERRY COLOUR: ● ● ● ● ●

PRINCIPAL SYNONYMS: Auvernat Blanc (Orléanais in northern France), Beli Pinot (Slovenia), Burgunder Veisser (Republic of Moldova), Chardonnet Pinot Blanc or Pinot Blanc Chardonnet (Burgundy), Clevner or Klävner (Alsace), Fehér Burgundi (Hungary), Pino Belîi (Republic of Moldova), Pinot Bianco, Pinot Bijeli (Croatia), Pinot Blanc Vrai, Pinot Branco (Portugal), Rulandské Biele (Slovakia), Rulandské Bílé (Czech Republic), Weissburgunder§, Weisser Burgunder§ or Burgunder Weiss§ (Germany, Austria)

VARIETIES COMMONLY MISTAKEN FOR PINOT BLANC: AUXERROIS§, CHARDONNAY§ (Burgundy), KNIPPERLÉ§, MELON§, PIGNOLETTO§, SAVAGNIN BLANC§ (called Clevner or Klävner in Alsace in France and in Baden-Württemberg in Germany)

ORIGINS

Until the end of the nineteenth century, Pinot Blanc was frequently confused with CHARDONNAY, which was mistakenly called Chardonnet Pinot Blanc or Pinot Blanc Chardonnet. French ampelographer Victor Pulliat was the first to distinguish the two varieties, in 1868 (Rézeau 1997): 'le Pineau blanc vrai est un accident fixé d'un Pineau noir devenu blanc: il n'en diffère que par la couleur de sa grappe qui est d'un blanc jaunâtre' ('the true Pineau blanc is an irreversible mutation of a Pineau noir that became white; it differs only by the colour of its cluster, which is yellowy white'). The distinction between Pinot Blanc and Chardonnay was officially accepted at the Lyon viticultural exhibition in 1872 (Chandon de Briailles in Durand 1901a). Therefore, every mention of Pinot Blanc earlier than 1868 is ambiguous and unreliable because of the confusion that existed with Chardonnay. Needless to say, the confusion went on several decades later in both France and Switzerland.

According to Eugène Durand (1901a), Pinot Blanc is a colour mutation of Pinot Gris that was observed by Pulliat in Chassagne-Montrachet in 1895 and by Durand himself in Nuits-Saint-Georges in 1896, which means the mutation had already occurred at least twice. The existence of additional and possibly earlier mutations is quite plausible because black, grey and white berries can sometimes be observed on the same Pinot Noir vine, and some vines even have stripey berries.

VITICULTURAL CHARACTERISTICS

Early budding and early ripening. Fairly vigorous and more regularly productive than Pinot Gris or Pinot Noir. Likes deep, warm soils. Good cold resistance but rather delicate in relation to fungal diseases. Small berries.

WHERE IT'S GROWN AND WHAT ITS WINE TASTES LIKE

Many wine lovers' introduction to this variety is via Alsace, where it is regarded as a workhorse variety and is much less respected than Pinot Gris. Admirers of freshness in wine can find some Pinot Gris overbearing whereas, in the right place and hands, Pinot Blanc can offer a lip-smacking halfway house between refreshment and opulence. The wines are moderately structured and have moderate acidity and may be used for sparkling wine.

The area of **France** planted with Pinot Blanc increased steadily until it reached 1,565 ha (3,867 acres) at the end of the 1980s but is now declining slowly, reaching 1,292 ha (3,193 acres) in 2009, surprisingly far down the league table of white varieties, perhaps because so little is planted outside its home territory of Alsace. Even there, it is less significant in terms of area planted than RIESLING, SILVANER and AUXERROIS, with which it is traditionally blended, although only Pinot Blanc is mentioned on the label and the official Alsace statistics report the two varieties together to give a total of 3,331 ha (8,231 acres) in 2009, more or less steady over the last ten years. It may also be included in Crémant d'Alsace. Very little is now grown in Burgundy, although it is still technically eligible for inclusion in Bourgogne Blanc.

Pinot Blanc makes some of the most successful wine in the French Moselle region, just as it does over the border in **Luxembourg**, where there were 144 ha (356 acres) in 2008 with very few new plantings this century. AUXERROIS (184 ha/455 acres) is more highly regarded this far north because of its usefully lower acidity.

A total of 5,126 ha (12,667 acres) of Pinot Bianco were recorded in the last, 2000, agricultural census in **Italy**, the vast majority of vines in the north east, where it is used to make a popular dry white, along the lines of a lighter-style unoaked CHARDONNAY. However, good examples can have delicate floral and green-apple aromas and are equally delicate on the palate. Recommended producers include Franco Toros and Ermacora in Friuli and Cantina Terlano and Franz Haas in the Alto Adige.

The variety is probably most respected in **Germany**, where there were 3,731 ha (9,220 acres) in 2008. Known as Weissburgunder or Weisser Burgunder, it is the country's fifth most planted white wine variety, albeit some way behind the Grauburgunder (Pinot Gris) of which it is so often a stablemate. By far the biggest area is in Baden (1,165 ha/2,879 acres), followed by the Pfalz (862 ha/2,130 acres) and Rheinhessen (831 ha/2,053 acres). Producers such as Bercher, Dr Heger, Franz Keller and Martin Wassmer in Baden, Philipp Kuhn, Rebholz and Dr Wehrheim in the Pfalz and Dreissigacker in Rheinhessen make some of the most winning examples of the varietal anywhere – although there has been a tendency to overripen and overoak it – perhaps in an attempt to provide something utterly unlike RIESLING.

As Weissburgunder, the variety makes most of its richest wines in **Austria** in Burgenland, where it succumbs easily to noble rot and is traditionally blended with the rather crisper Welschriesling (GRAŠEVINA) to produce luscious, if fairly early-maturing Trockenbeerenauslesen. Prieler and Willi Opitz also make dry versions with admirable precision. As a dry white varietal, Weissburgunder is associated with an almond-like scent, medium to high alcohol, and an ability to age. It is grown throughout the country with a total of 1,995 ha (4,930 acres) planted in 2010, especially in Niederösterreich (Lower Austria) and the Steiermark (Styria).

Switzerland grows just 105 ha (259 acres) of Pinot Blanc, mostly in the Valais, Vaud and Graubünden but Weingut Bad Osterfingen in Schaffhausen make a notable exception.

In **Hungary** there were 200 ha (494 acres) as Pinot Blanc (also Fehér Burgundi) in 2008. Although it is not that widely planted, it was already officially registered in the 1930s. The most significant area is in Tolna in the central south. In **Slovenia**, it is known as Beli Pinot and 2009 total plantings were 540 ha (1,334 acres), the majority (315 ha/778 acres) being in Podravje in the north east. Vipava 1894 make an excellent rendition. **Croatia** grew 357 ha (882 acres) of Pinot Bijeli in 2008. In the **Czech Republic**, there were 812 ha (2,006 acres) of Rulandské Bílé in 2009, mostly in Morava in the south east, while in **Slovakia**, where it is also known as Rulandské Bílé, there were 1,050 ha (2,595 acres) in 2005 (5.4% of vineyard area). The **Republic of Moldova** grew 350 ha (865 acres) of Pinot Blanc in 2009. The 2009 total in Krasnodar Krai in **Russia** was 605 ha (1,495) and **Ukraine** had 338 ha (835 acres) in 2009.

The variety is not nearly as popular as Pinot Gris in **California** but is widely planted with various bottlings from the likes of the Bien Nacido vineyard in the Santa Maria Valley and some Anderson Valley examples proving there may be a future for Pinot Blanc as a still wine. Historically much of it was used as base wine for California fizz, a practice pioneered by old hand Schramsberg. In 2010 there were just 536 acres (217 ha) with the area increasing slowly, the biggest areas in Santa Barbara, Sonoma and Monterey, in that order. According to Pinder and Meredith (2003), many older Pinot Blanc vines in California are actually MELON, including those pioneered by Chalone.

There are also some very competent **Oregon** Pinot Blancs but it is the Okanagan Valley in **Canada**'s British Columbia that seems to be the variety's perfect spot. There were only 369 acres (149 ha) planted in 2009, making it BC's fifth most planted white variety, but when varieties were assessed in the late 1970s and early 1980s, Pinot Blanc was the most popular because it was as winter-hardy as the hybrids that had been planted before and it produced consistently good wines from year to year in both the north and south of the Okanagan Valley (Philip 2007). Presumably it is only Pinot Blanc's lack of international glamour that is acting as a brake, because vivacious varietals from the likes of Lake Breeze, Nk'Mip and Wild Goose suggest a particularly bright future here.

There are tiny amounts planted in South America and **Japan**.

PINOT TEINTURIER

Red-fleshed Pinot mutation.

BERRY COLOUR: ● ● ● ● ●

PRINCIPAL SYNONYMS: Farbclevner, Pinot Fin Teinturier

VARIETIES COMMONLY MISTAKEN FOR PINOT TEINTURIER: TEINTURIER

ORIGINS

Pinot Teinturier is a mutation of Pinot Noir with red-juiced berries, somewhere between GAMAY TEINTURIER DE BOUZE and TEINTURIER in terms of colour intensity (Galet 1990). It was first selected at the end of the nineteenth century by Jean Guicherd in a vineyard in Nuits-Saint-Georges in the Côte d'Or in Burgundy, and the same mutation was then also observed in Morey and in Vosne-Romanée (Durand 1901a).

VITICULTURAL CHARACTERISTICS

Similar to Pinot Noir in the vineyard but the grape juice is red.

PINOT NOIR PRÉCOCE

Early-ripening mutation of Pinot Noir that is currently popular in Germany as Frühburgunder.

BERRY COLOUR: ● ● ● ● ●

PRINCIPAL SYNONYMS: Augustclevner (Germany), Augusttraube, Blauer Frühburgunder (Germany, Switzerland, Austria), Frühburgunder (Germany and Austria), Ischia or Vigne d'Ischia, Jacobstraube (Alsace), Jakobstraube (Germany), Juliustraube, Madeleine Noire, Möhrchen (Germany), Morillon Noir Hâtif, Pinot Hâtif de Rilly, Pinot Madeleine, Pinot Précoce, Pinot Précoce Noir, Raisin de la Madeleine

ORIGINS

According to Eugène Durand (1901a), nothing is known about the origins of this mutation that ripens approximately two weeks before Pinot Noir (*précoce* means 'early'). However, it has numerous old synonyms that were mentioned in Merlet (1690): 'le Morillon hâtif, ou de la Madeleine' and in Liger (1762): 'le raisin précoce, raisin de la madeleine, ou morillon hâtif, parce qu'il est hâtif & mûr souvent dès la Madeleine, est un raisin noir, plus curieux que bon, à cause qu'il a la peau dure' ('the early grape, the madeleine grape, or early morillon, because it is early and often ripe at the Madeleine, is a black grape, more curious than good, because it has hard skin'). The earliest mention of the modern name Pinot Noir Précoce appeared in Di Rovasenda

(1877), and the modern German name Frühburgunder was first mentioned in Braun (1824): 'Unter den weißen auch schimmerte dicht von Beeren der dunkle Frühburgunder' ('among the white [grapes], the dark Frühburgunder shimmered with berries').

VITICULTURAL CHARACTERISTICS

Small bunches and berries (smaller than Pinot Noir). Ripens fourteen days earlier than Pinot Noir but susceptible to coulure and has small yields. The thickish skins provide good protection against botrytis bunch rot.

WHERE IT'S GROWN AND WHAT ITS WINE TASTES LIKE

Wines are well coloured, soft, ample and full-flavoured with flavours of cherries and blackberries and sometimes a smoky note. It is well suited to restrained barrel ageing but generally not as long-lived as Pinot Noir of comparable quality.

Plantings in **England** are so far limited but there are some promising examples such as that from Welcombe Hills near Stratford-upon-Avon.

Frühburgunder was very popular in the early nineteenth century in **Germany**, partly thanks to its ability to ripen even in poor vintages and also for its very early ripening – sometimes as precocious as late July or early August. This explains several of its synonyms: Juliustraube, meaning 'July grape', Augustclevner, Jakobstraube (the grape that ripened by St Jacob's day, ie 25 July) and Pinot Madeleine (ripening around St Mary Magdelene's day, ie 22 July).

Although it had declined to around 15 ha (37 acres) by the 1960s, it is now seeing something of a revival, thanks largely to a clonal selection programme carried out at the Geisenheim research centre in the 1970s, although it was Julius Wasem & Söhne in Rheinhessen that initially came to the variety's rescue when it was deleted from the official German register of varieties in the 1960s. Slow Food's championing of Frühburgunder can only have encouraged its survival.

From just 44 ha (109 acres) in 1994, the total area in Germany increased slowly but steadily to 251 ha (620 acres) in 2007. Although there are more hectares of Frühburgunder in Rheinhessen (83 ha/205 acres) and the Pfalz (62 ha/153 acres), the hotbed is clearly in the Ahr, where there are 35 ha (86 acres), getting on for 10% of the valley's vineyard area, and more than half of the region's producers are making wine from this variety. It is also the only region where members of the VDP, Germany's top growers' association, are allowed to produce Frühburgunder labelled Grosses Gewächs, the highest classification for dry wines. Franken had only 14 ha (35 acres) in 2008 but the variety seems to do particularly well on the red sandstone soils around Miltenberg and, historically, Bürgstadt. Among the best producers are Burggarten, Deutzerhof, Kreutzberg, Pieter Kriechel and Meyer-Näkel in the Ahr; Rudolf Fürst in Franken; Philipp Kuhn and Ludi Neiss in the Pfalz; and Villa Bäder, Wendelsheim and Julius Wasem & Söhne in Rheinhessen.

PINOT BIANCO

See PINOT.

PINOT BLANC

See PINOT.

PINOT GRIGIO

See PINOT GRIS *in the* PINOT *entry.*

PINOT GRIS

See PINOT.

PINOT MEUNIER

See PINOT.

PINOT NERO

See PINOT NOIR *in the* PINOT *entry.*

PINOT NOIR

See PINOT.

PINOT NOIR PRÉCOCE

See PINOT.

PINOT TEINTURIER

See PINOT.

PINOTAGE

South Africa's very own cross, both loved and despised.

BERRY COLOUR: ● ● ● ● ●

PRINCIPAL SYNONYMS: Perold's Hermitage × Pinot

ORIGINS AND PARENTAGE

Pinotage was bred in 1925 by Abraham Izak Perold (1880–1941), the first professor of viticulture at the University of Stellenbosch, South Africa. In the garden of his official residence at the Welgevallen experimental farm belonging to the University of Stellenbosch, Perold pollinated a PINOT NOIR plant with CINSAUT, known at the time in South Africa as Hermitage. He then planted the four resulting seeds in the same garden and forgot them when he left Stellenbosch University to work for the KWV winery in Paarl in 1927.

The seedlings were rescued by Charlie Niehaus, who transported them to CJ Theron's nursery at Elsenburg Agricultural College, where they were simply labelled Perold's Hermitage × Pinot. He then grafted material from the Pinotage seedlings onto rootstocks at Welgevallen and later showed them to Perold, who urged him to propagate them. They are said to have come up with the name Pinotage – a contraction of those of its parents – while in the Welgevallen vineyard.

The best of the four plants was then selected to become the mother material of all Pinotage vines. The first commercial planting of Pinotage was made in 1943 at Myrtle Grove farm near Sir Lowry's Pass (Pinotage Association 2010) although the name was not seen on a bottle until 1961, on Lanzerac's 1959 vintage.

Like several other South African crosses, Pinotage was obtained by means of open-air hybridization in vineyards and not under controlled greenhouse conditions. Nevertheless, the proposed parentage was confirmed by DNA analysis at ENTAV (Établissement National Technique pour l'Amélioration de la Viticulture) in France in 2007.

VITICULTURAL CHARACTERISTICS

Hardy, moderately vigorous and productive, early to mid budding and early to mid ripening. Moderately susceptible to powdery and downy mildews and to botrytis bunch rot. Adaptable in terms of climate and able to reach high sugar levels in the small, thick-skinned berries. Best suited to hillsides with moderate to good water-holding capacity.

WHERE IT'S GROWN AND WHAT ITS WINE TASTES LIKE

Depending on your point of view, Pinotage is either **South Africa**'s signature red variety or its worst vinous ambassador. Its reputation has not been helped by the effects of virus diseases in the vineyard and fermentations that were sometimes heavy handed or too hot. However, with healthier planting material, continued refinements in winemaking and the efforts of the Pinotage Association, founded in the mid 1990s, the best wines have improved enormously, ranging in style from light and red-fruited (pointing to the vine's parentage) to the rich, dark, firm and spicy but fresh produce of old vines in the hands of expert proponents such as Beyers Truter at Beyerskloof in Stellenbosch, who also uses it to make port-style wines, rosé, fizz and red and even 'white' blends (with CHENIN BLANC). Other recommended producers of varietal wines include L'Avenir, Chamonix, Kaapzicht and Kanonkop. Grangehurst and Kaapzicht, among others, have produced successful Pinotage-dominated blends.

There were 6,088 ha (15,044 acres) in 2009, that's 6% of the South African vineyard area, and grape prices have fluctuated considerably according to fashion. The variety is planted in many different regions but the biggest areas are in Malmesbury, Stellenbosch and Paarl.

According to the Pinotage Association (2010), unpleasant spray-paint aromas (due to isoamyl acetate) may develop if the vines are subject to water stress or high temperatures at harvest time, aromas that are increased by high fermentation temperatures. Others have reported an aroma of burnt rubber in many Pinotage wines. This may be linked to the continued effects of viral diseases in the vineyard but the causes are unclear and research is underway at the University of Stellenbosch to try to identify the compounds that are responsible.

Pinotage is also grown in **California** (53 acres/21 ha in 2009 mostly in the San Joaquin Valley but also in Napa Valley and Sonoma Coast), **Brazil** (112 ha/277 acres in 2007 in Rio Grande do Sul), **New Zealand** (74 ha/183 acres in 2008 in Gisborne, Hawke's Bay and Marlborough) and to a more limted extent in **Australia**, **Oregon**, **Washington** State, British Columbia in **Canada**, and in **Zimbabwe** and **Israel**.

PINOTIN

Recent, obscure Swiss hybrid.

BERRY COLOUR: ● ● ● ● ●

PRINCIPAL SYNONYMS: VB 91-26-19

ORIGINS AND PARENTAGE

The vast majority of sources define Pinotin as a cross between PINOT NOIR and an undisclosed disease-resistant variety (known in German as a *Resistenzpartner*) bred by Valentin Blattner in Switzerland in 1991, although the name was not registered until 2002. However, Blattner recently revealed that Pinotin is in fact a complex hybrid: CABERNET SAUVIGNON × (SILVANER × (RIESLING × *Vitis vinifera*) × (JS 12417 × CHANCELLOR)), in which JS 12417 is an unknown hybrid obtained by Joannès Seyve (see PRIOR for the pedigree of Chancellor). Nonetheless, the PINOT-like flavour of the wine might be due to the proximity of a row of Pinot Noir vines next to the crossed vines, and without DNA analysis it is not possible to rule out Pinot Noir as a parent.

VITICULTURAL CHARACTERISTICS

Long, thin, loose bunches contribute to good resistance to downy and powdery mildews and botrytis bunch rot. Early budding, mid ripening, good resistance to late frost but more susceptible in the spring.

WHERE IT'S GROWN AND WHAT ITS WINE TASTES LIKE

There are very minor plantings in **Switzerland** and in Germany, where Thüringer Weingut Bad Sulza in Saale-Unstrut (central **Germany**), for example, produce a varietal wine that's ruby-coloured and soft with aromas of black cherries.

PIONNIER

Minor but complex, cold-hardy American hybrid used in blends in Québec.

BERRY COLOUR: ● ● ● ● ●

PRINCIPAL SYNONYMS: ES 4-7-25

ORIGINS AND PARENTAGE

An Elmer Swenson 2-12-27 × ST CROIX hybrid obtained by Elmer Swenson in Osceola, Wisconsin, US, in which Elmer Swenson 2-12-27 is an Elmer Swenson 5-14 × SWENSON RED hybrid (see BRIANNA for the pedigree of Elmer Swenson 5-14). Pionnier is therefore an extremely complex *Vitis riparia*, *Vitis labrusca*, *Vitis vinifera*, *Vitis aestivalis*, *Vitis lincecumii*, *Vitis rupestris*, *Vitis cinerea* and *Vitis berlandieri* hybrid.

WHERE IT'S GROWN

Pionnier is planted in Québec, **Canada**, where producers such as Bouche-Art, Clos du Roc Noir and Domaine Mont Vézeau blend it with other cold-hardy varieties such as FRONTENAC and SABREVOIS.

PIQUEPOUL

The history, origins and etymology of Piquepoul are discussed first, then Piquepoul Noir and Piquepoul Blanc are described separately.

ORIGINS AND PARENTAGE

Piquepoul, from the Vaucluse in southern France, comes in three different berry colours – Noir, Gris and Blanc – the first and the last still widely planted but the second on the verge of extinction and all having identical DNA profiles (Puig *et al.* 2006). However, the name Piquepoul has been given to several distinct and unrelated white- or black-berried grape varieties under various spellings for all three colour mutations (Piccapoule, Picpoul, Picpoule, Picquepoul, Piquepout, etc.; Rézeau 1997). The earliest mention of the black-berried version was in 1384 in a lease for a piece of land near Toulouse: 'De vitibus Picapoll nigri aut de bono Canas' ('vineyards of Piquepoul noir or of good Canas'; Rézeau 1997). It was later mentioned several times in Valréas (Piquepole in 1507, Piccapoule in 1548) and other villages in the Vaucluse and the Aude (Piquepoulle in Courthézon in 1590, Piquepouhl in Carcassonne in 1619, Picapoule in Sérignan-du-Comtat in 1651, etc.), without reference to the berry colour (Rézeau 1997). The earliest mention of the white version appeared in 1667 (Rézeau 1997): 'le Piquant Paul est vn Raisin blanc, fort doux, dit Bec d'Oyseau, dont le grain est gros, très-long & en pointe des deux costez' ('Piquant Paul is a white grape, very sweet, called Bird's Beak, whose berries are large, very long and pointed on both sides').

According to Lavignac (2001), the name Piquepoul might derive from the Oc dialect *picpol* or *picapol*, possibly designating a place with a peak (pre-Latin root *pikk*) and a rock or cliff (pre-Latin root *pal*) where the variety was planted.

DNA profiling has recently shown that PICAPOLL BLANCO, a white variety from Pla de Bages in Catalunya, north-east Spain, already mentioned in the sixteenth century and long considered identical to the French Piquepoul Blanc, is in fact a distinct variety that is much more closely related to CLAIRETTE, possibly a clone (Puig *et al.* 2006). However, Picapoll Negro has been shown to be identical to Piqupoul Noir (Galet 2000). Piquepoul belongs to the Piquepoul ampelographic group (see p XXVII; Bisson 2009).

PIQUEPOUL NOIR

BERRY COLOUR:

PRINCIPAL SYNONYMS: Picapoll Negro (Catalunya), Picapoll Tinta

VARIETIES COMMONLY MISTAKEN FOR PIQUEPOUL NOIR: CALITOR NOIR (called Piquepoul de Fronton), CINSAUT (called Piquepoul d'Uzès)

VITICULTURAL CHARACTERISTICS

Mid to late ripening, relatively productive, generally in alternate years. Likes dry, low-fertility soils. Susceptible to botrytis bunch rot and, to a lesser extent, to powdery mildew. Best pruned short.

WHERE IT'S GROWN AND WHAT ITS WINE TASTES LIKE

Piquepoul Noir produces pale, alcoholic but aromatic and nervy wines well suited to rosé production and is much less widely planted than the white version – there were just 73 ha (180 acres) in southern **France** in 2008. Although it is recommended in most French wine-growing *départements*, and is a permitted variety in Châteauneuf-du-Pape and Côtes du Rhône, it is generally a very minor component in blends, unlike its more successful white mutation. A rare example of a Piquepoul Noir-dominant red wine is Minervois producer Clos Centeille's C de Centeilles, which is 90% Piquepoul Noir and 10% RIVAIRENC; Domaine La Grangette just south of Pézenas bottle a varietal rosé.

In **Spain**, there were just 3 ha (7 acres) of Picapoll Negro officially recorded in 2008 in Catalunya, where it is generally included in blends, as in those of Álvaro Palacios and Celler Francisco Castillo's from Priorat. It is authorized in DOs such as Montsant and Priorat and has experimental status in Pla de Bages.

PIQUEPOUL BLANC

BERRY COLOUR: ● ● ● ● ●

PRINCIPAL SYNONYMS: Avillo (Catalunya), Languedocien, Picpoul de Pinet or Piquepoul de Pinet (Hérault in the Languedoc)

VARIETIES COMMONLY MISTAKEN FOR PIQUEPOUL BLANC: BACO BLANC (called Piquepoul du Gers in Armagnac), CLAIRETTE (possibly called Picapoll or Picapoll Blanco in Pla de Bages in Catalunya), FOLLE BLANCHE (called Picpoul in the Gers), ONDENC (called Piquepout de Moissac in the Gers)

VITICULTURAL CHARACTERISTICS

Mid to late ripening. Fertile, productive and best pruned short. Suited to calcareous-clay or sandy soils in hot Mediterranean climates. Susceptible to botrytis bunch rot.

WHERE IT'S GROWN AND WHAT ITS WINE TASTES LIKE

PIQUEPOUL's high acidity makes it an ideal blending partner to freshen up many a Languedoc or Provençal white since it is recommended across much of southern France, and may be included in Châteauneuf-du-Pape's complex cocktail, but its most famous expression is as a varietal wine in the eponymous appellation Picpoul de Pinet, grown between Pézenas and Sète on the coast. Wines are usually lemon-scented, nervy and dry, occasionally on the floral side. There were 1,455 ha (3,595 acres) in **France** in 2009, a considerable increase on 1998's 650 ha (1,606 acres), the overwhelming majority of the vines being in the Languedoc. Production is dominated by the Cave de l'Ormarine in Pinet and the three other co-ops in the area, but smaller producers such as Domaine de Creyssels, Domaine Félines Jourdan and Domaine Gaujal de Saint Bon make slightly more intensely flavoured, in some instances lees-aged, examples.

Rhône specialists Tablas Creek in western Paso Robles were pioneers of the variety in **California**, though their Picpoul Blanc has more tropical fruit flavours than its counterpart in southern France. A range of varietal examples is now made.

PIQUEPOUL BLANC

See PIQUEPOUL.

PIQUEPOUL NOIR

See PIQUEPOUL.

PLANT DROIT

Minor and diminishing southern Rhône variety producing rather pale reds.

BERRY COLOUR: ● ● ● ● ●

PRINCIPAL SYNONYMS: Cinsaut Droit (Vaucluse), Espanenc, Garnacha Francesa (Zaragoza in Spain), Plant Dressé

VARIETIES COMMONLY MISTAKEN FOR PLANT DROIT: CINSAUT, GARNACHA TINTA

ORIGINS AND PARENTAGE

Plant Droit comes from the Vaucluse in the southern Vallée du Rhône, where it was first described by Henri Marès in 1890 (Rézeau 1997). Its name was deliberately changed in the Vaucluse to Cinsaut Droit in an attempt to include it discreetly in the Côtes du Rhône appellation (Galet 1990). A comparison of the DNA profile of Garnacha Francesa in Martín *et al.* (2003) with Plant Droit shows that they are identical (Vouillamoz). According to Bisson (2009), Plant Droit is ampelographically close to CLAIRETTE.

VITICULTURAL CHARACTERISTICS

Early budding and mid ripening. Upright growth (*droit* is French for 'upright') makes it well suited to bushvine training. Relatively productive and fertile. Survives well in very dry conditions but is very susceptible to botrytis bunch rot.

WHERE IT'S GROWN AND WHAT ITS WINE TASTES LIKE

Wines are generally pale and low in alcohol. There were 21 ha (52 acres) in **France** in 2008, down from 337 ha (833 acres) in 1968, the downward trend being most marked in the 1990s. It is pretty much confined to the Châteauneuf-du-Pape region in the southern Rhône, although it is not one of the many permitted varieties in that appellation.

PLANTA FINA
PLANTA FINA DE PEDRALBA

See DAMASCHINO.

PLANTA NOVA

Actually very old but generally undistinguished variety grown mainly in Valencia today.

BERRY COLOUR: ● ● ● ● ●

PRINCIPAL SYNONYMS: Coma, Malvasía☒ (Yecla), Tardana, Tortozon (Ariège in France), Tortozón (province of Toledo in Castilla-La Mancha)

ORIGINS AND PARENTAGE

Although the name means 'new plant', it is an old Andalucian variety from southern Spain that was first mentioned in 1513 by Gabriel Alonso de Herrera (1790) under the synonym Tortozón. DNA parentage analysis has shown that Planta Nova (under the synonym Tortozón) and GIBI are the parents of ALARIJE (under the synonym Subirat Parent; Lacombe *et al.* 2007), an old variety from Extremadura that was also cited in 1513 by Gabriel Alonso de Herrera. The synonym Tardana reflects its late-ripening habit (*tardar* means 'to take a long time' in Spanish).

VITICULTURAL CHARACTERISTICS

Thick-skinned berries. Late ripening.

WHERE IT'S GROWN AND WHAT ITS WINE TASTES LIKE

In **France**, Les Vignerons Ariégeois in Montégut-Plantaurel (just west of Pamiers) have planted 30–40 vines of Tortozon (no accent in French) and other endangered varieties (Carbonneau 2005). They aim to multiply these and to produce a commercial wine if their experiments yield encouraging results, although it seems unlikely in this case.

Authorized in neighbouring DOs Utiel-Requena and Valencia in **Spain**, Planta Nova was grown on a total of 1,385 ha (3,422 acres) in 2008, the vast majority in the autonomous community of Valencia plus very limited plantings in Castilla-La Mancha and Extremadura. Torroja's Sybarus Tardana Único, one of the very few better-quality varietal wines, is fresh and well structured with flavours of peach and lemon.

PLANTET

Hybrid once popular in the Loire, now producing slightly odd, raspberry-flavoured wines.

BERRY COLOUR: ● ● ● ● ●

PRINCIPAL SYNONYMS: Seibel 5455

ORIGINS AND PARENTAGE

A complex hybrid obtained at the beginning of the twentieth century by Albert Seibel in Aubenas in the Ardèche, south-eastern France. Its genetic origin is still not clear because Seibel initially described it as a Seibel 867 × Seibel 2524 cross but then corrected his own notes to '[Seibel] 4461 × *Berlandieri*' (Galet 1988). It was used to breed COLOBEL and LANDAL.

The origin of the name is unknown but it may be related to *plantier*, designating a young vine (Rézeau 1997).

VITICULTURAL CHARACTERISTICS

Early budding and ripening. Fertile and regularly productive. Best trellised as it is very prone to wind damage. Good general resistance to disease and may be planted ungrafted in areas not subject to phylloxera. Compact bunches and small berries, which sometimes drop when ripe. Good resistance to winter and spring frosts, although not in colder climes such as New York State.

WHERE IT'S GROWN AND WHAT ITS WINE TASTES LIKE

Plantet was one of the most popular hybrids in the Loire, **France**, but has gone from total plantings in France of 27,900 ha (68,942 acres) in 1958 to just 1,105 ha (2,731 acres) in 2009, even though it is authorized in almost every wine-growing region. At one time it was significant in western **Switzerland** but it has all but disappeared today. The grapes tend to have a marked raspberry flavour and the wines may taste odd to those not used to varieties with a part-American pedigree. Domaine Mondon-Demeure, specialists in obscure and forgotten varieties, produce both a rosé and a medium-dry version (classified as a vin de France) north west of Saint-Étienne in the Rhône.

PLANTSCHER

Very rare Swiss or possibly Hungarian variety just surviving in the Haut-Valais.

BERRY COLOUR: ● ● ● ● ●

PRINCIPAL SYNONYMS: Bordeaux Blanc☒ (Valais), Bourgogne Blanc (Valais), Gros Bourgogne☒ (Valais and Vaud)

ORIGINS AND PARENTAGE

The name Plantscher supposedly derives from Blanchier, from *blanc* ('white'), a name that has been used to designate several distinct varieties in Vaud and the Valais in southern Switzerland. DNA parentage analysis has unexpectedly revealed that Plantscher is a natural progeny of FURMINT (Vouillamoz, Maigre *et al.* 2004) and a sibling of HÁRSLEVELŰ (Vouillamoz and Moriondo 2011), making it a grandchild of GOUAIS BLANC (see PINOT pedigree diagram) and suggesting that Hungary might be the origin of Plantscher.

VITICULTURAL CHARACTERISTICS

Vigorous, consistently productive, mid budding and ripening. Big bunches of big berries. Susceptible to botrytis bunch rot.

WHERE IT'S GROWN AND WHAT ITS WINE TASTES LIKE

Despite the synonyms linking this variety to two French regions, Plantscher is found only in **Switzerland**, where it used to be scattered in the vineyards of Vaud and the Valais. In addition to the few remaining isolated vines, only one Plantscher vineyard existed in 2011: 0.75 ha (2 acres) cultivated by Josef-Marie Chanton in the Haut-Valais. Chanton's softly textured wine has aromas of wild flowers such as chamomile, a hint of honeysuckle and moderate acidity, although the variety has also been described as rather neutral (Dupraz and Spring 2010).

PLASSA

Thick-skinned red making tannic wines south west of Torino.

BERRY COLOUR:

PRINCIPAL SYNONYMS: Scarlattino

ORIGINS AND PARENTAGE

Plassa probably originates from the Pinerolo area in the province of Torino, Italy, more precisely in the region around Bibiana, Bricherasio, Fenile and San Secondo. Plassa is the local dialect for the Italian *pellaccia*, meaning 'hard skin', in reference to the thickness of the berries' skin. Its synonym Scarlattino refers to the red colour of the stalks.

VITICULTURAL CHARACTERISTICS

Vigorous, mid to late ripening. Thick skins. Susceptible to millerandage and grape moths but good resistance to fungal diseases.

WHERE IT'S GROWN AND WHAT ITS WINE TASTES LIKE

Thanks to its thick berry skin, Plassa used to be stored after the harvest for consumption as table grapes during the winter, but today it is used exclusively for wine, in blends with other local varieties such as BARBERA, NERETTA CUNEESE and CHATUS. It is cultivated mainly in the area of Pinerolo, in particular in the communes of Pinerolo, Bibiana, Bricherasio, Campiglione, Cumiana and Frossasco. There are also scattered plantings of Plassa in the Val di Susa. There were just 44 ha (109 acres) in **Italy** in 2000. The wines are tannic, thanks to those thick skins, and generally benefit from the softening effects of bottle ageing.

PLATANI

Minor blending component on the Greek island of Santoríni.

BERRY COLOUR:

PRINCIPAL SYNONYMS: Platania (Santoríni), Platanos

ORIGINS AND PARENTAGE

Platani, from the Greek Kykládes (Cyclades) islands and from Kríti (Crete), is genetically closely related to KATSANO, also cultivated in the Kykládes (Biniari and Stavrakakis 2007). Recent DNA profiling has suggested a possible parent–offspring relationship with ASSYRTIKO (Myles *et al.* 2011).

VITICULTURAL CHARACTERISTICS

Very productive, vigorous and susceptible to botrytis bunch rot. Mid budding and early to mid ripening. Large, loose clusters of thick-skinned berries.

WHERE IT'S GROWN AND WHAT ITS WINE TASTES LIKE

Platani is hidden almost entirely in blends on the island of Santoríni, **Greece**, and found in old interplanted vineyards, probably to about the same extent as ATHIRI. Both are among the earliest picked varieties on the island and are therefore frequent companions in the fermenting vat, although Platani is less often mentioned and not as highly regarded. Wine producer Paris Sigalas once tried vinifying it separately and produced a wine he describes as 'fresh and light with aromas of lemon and orange'.

PLAVAC MALI

Ancient and leading Croatian variety making powerful reds, genetically close but not identical to Tribidrag (Zinfandel).

BERRY COLOUR:

PRINCIPAL SYNONYMS: Crljenak (Split), Kasteljanak (Split), Pagadebit Crni or Pagadebit Mali (Korčula), Plavac Mali Crni, Plavac Veliki, Plavec Mal (Republic of Macedonia), Zelenak (Dubrovnik)

VARIETIES COMMONLY MISTAKEN FOR PLAVAC MALI: BABIĆ (Šibenik), PLAVINA (Dalmatian coast), TRIBIDRAG (Dalmatian coast)

ORIGINS AND PARENTAGE

Plavac Mali from the Dalmatian coast in south-western Croatia is Croatia's economically most important dark-skinned grape variety, first mentioned by Trümmer (1841) as a typical Dalmatian grape. However, the variety must be much older because recent ampelographic and genetic investigations have shown relatively significant clonal diversity within Plavac Mali (Zdunić *et al.* 2009). The name is descriptive of the berries: in Croatian *plavo* means 'blue' and *mali* means 'small'.

Plavac Mali has long been confused with TRIBIDRAG (aka Zinfandel) but recent DNA profiling has shown that they are distinct varieties (Pejić *et al.* 2000) and DNA parentage analysis has established that Plavac Mali is in fact a cross between Tribidrag and DOBRIČIĆ (Maletic *et al.* 2004), two very old Dalmatian varieties. Plavac Mali is therefore a half-sibling of PLAVINA, which is a Tribidrag × VERDECA cross, and either a grandparent or a half-sibling of GRK, Crljenak Crni and VRANAC, which all have a parent–offspring relationship with Tribidrag (Maletić *et al.* 2004; see Tribidrag for the pedigree diagram). Plavac Mali has also been confused with BABIĆ but DNA parentage analysis has shown that BABIĆ is a grandparent or a half-sibling of Plavac Mali since it has a parent–offspring relationship with Dobričić (Zdunić, Pejić *et al.* 2008). In addition, BABICA, planted only in the Kaštela Valley, has been found to be a progeny of Plavac Mali (Zdunić, Pejić *et al.* 2008).

VITICULTURAL CHARACTERISTICS

Late budding and ripening and therefore best suited to sunny terraces on south-facing coastal slopes such as those of the Pelješac peninsula. Small blue berries with thick skins that can reach very high sugar levels. Not generally susceptible to fungal diseases.

WHERE IT'S GROWN AND WHAT ITS WINE TASTES LIKE

Plavac Mali is the most widely planted red wine variety in **Croatia** even though it constitutes only 9% of the total vineyard area, roughly 2,925 ha (7,228 acres) in 2009. It is authorized in the Dalmatinska Zagora (Croatian Hinterland) and Srednja i Južna Dalmacija (Central and Southern Dalmatia) wine regions and was the first variety to have its own appellation: Dingač, created in 1961. Wines in the Postup and Dingač appellations on the Pelješac peninsula just north of Dubrovnik must be made exclusively from Plavac Mali and may be dry or off dry – the residual sugar helping to disguise aggressive tannins in some wines. The vines that make the best, most intense, wines are grown on steep, sandy terraces sloping down to the Adriatic, usually as bushvines to shield the fruit from the hot sun and protect the vine from the strong winds. Vines grown on more fertile sites that produce higher yields are likely to produce lower-quality fruit.

The combination of location and variety produces deeply coloured, powerful, alcoholic wines that are notably – sometimes aggressively – tannic with flavours of dark cherry, plum and blackberry, well suited to bottle ageing despite the generally low acidity. Regulations for Dingač stipulate a minimum of 13% alcohol but some go as high as 17%, unfortified. Vines higher up the slopes are fresher and more elegant. Other important appellations for Plavac Mali are Ivan Dolac and Sveta Nedilja on the island of Hvar. The islands of Brač, Lastovo and Vis are also capable of producing high-quality Plavac Mali. Overripe, dried grapes are sometimes used to make the dessert wine Prošek.

Recommended producers include Hrvoje Baković, Bura-Mokalo, Ivo Duboković, Kiridžija, Korta Katarina, Miličić, Saints Hills, Skaramuča and Andro Tomić.

The variety is also found – spelt Plavec Mal – in the Veles and Tikveš areas of the **Republic of Macedonia**.

PLAVEC ŽUTI

Minor variety producing light, tart whites in Croatia and Slovenia.

BERRY COLOUR:

PRINCIPAL SYNONYMS: Debeli Klešec (Slovenia), Plavac Žuti (Croatia), Plavec Rumeni (Croatia), Rumeni Plavec (Slovenia)

VARIETIES COMMONLY MISTAKEN FOR PLAVEC ŽUTI: Plavaï☒ (Republic of Moldova)

ORIGINS AND PARENTAGE

Plavec Žuti is an old variety from northern Croatia and southern Slovenia, once used for table grapes but now also to make wine. *Žuti* and *rumeni* mean 'yellow' in Croatian and Slovenian respectively.

OTHER HYPOTHESES

DNA profiling shows that Plavec Žuti is not identical to Plavaï (no longer cultivated) from the Republic of Moldova, as the Vitis International Variety Catalogue suggests (Vouillamoz).

VITICULTURAL CHARACTERISTICS

Sugar levels in the medium to large berries tend to be low to moderate. Good resistance to downy mildew.

WHERE IT'S GROWN AND WHAT ITS WINE TASTES LIKE

In **Croatia**, Plavec Žuti is authorized in three regions around Zagreb in the north of Kontinentalna Hrvatska (Continental Croatia): Plešivica, Prigorje-Bilogora and Pokuplje. The variety has been showing more promise as a wine grape since producers started to restrict yields. Those who make varietal wines, which are generally rather light at around 11–12% alcohol, include Šember and Vučinić.

In **Slovenia** the variety goes by the name of Rumeni Plavec and is planted in the Bizeljsko-Sremič and Dolenjska subregions of Posavje and to a lesser extent further north in Štajerska Slovenija (Slovenian Styria), with a total of 105 ha (259 acres) in 2009. Vino Graben produce a varietal wine but it is more often part of a blend because of its very high acidity. For the same reason, producers such as Istenič may use it in a blend for sparkling wines.

PLAVINA

Dark-skinned offspring of Tribidrag (Zinfandel) producing light reds the length of the Dalmatian coast.

BERRY COLOUR: ● ● ● ● ●

PRINCIPAL SYNONYMS: Brajda, Brajdica (Rab), Plavina Crna, Plavka or Plajka (Kotor and Korčula)

VARIETIES COMMONLY MISTAKEN FOR PLAVINA: PLAVAC MALI

ORIGINS AND PARENTAGE

Plavina is often confused with PLAVAC MALI although DNA parentage analysis has established that Plavina is a cross between VERDECA from Puglia and TRIBIDRAG (ie Primitivo or Zinfandel) from Dalmacija (Dalmatia; Lacombe *et al.* 2007). This trans-Adriatic link is further illustrated by the similarity between some Dalmatian and Puglian dialects, as well as by the numerous villages in southern Italy where there is a significant Dalmatian population. However, it is not yet known whether Plavina was born in Puglia or in Dalmacija, because the crossing that gave it birth could have taken place in either.

Southern Italy, most likely Puglia, is a possibility because VERDECA and TRIBIDRAG have traditionally been cultivated there, but Plavina seems to have disappeared from Puglia, unless it is still cultivated under another name. Dalmacija, where Plavina is today widely cultivated, is also a possible birthplace for the variety but Verdeca has never been mentioned in Dalmacija and is not grown there now, unless it has been cultivated under another name.

However, since the Dalmatian varieties GRK, Crljenak Crni (no longer cultivated), PLAVAC MALI and VRANAC also have a parent–offspring relationship with TRIBIDRAG (Maletic *et al.* 2004) and are either grandparents or half-siblings of Plavina (see Tribidrag for the pedigree diagram), it is more likely that Plavina is of Dalmatian origin. Since VERDECA has recently been identified in Greece under the name Lagorthi (Vouillamoz), it might be interesting to search for Verdeca in Dalmacija in order to confirm this hypothesis.

VITICULTURAL CHARACTERISTICS

Consistently high yields. Susceptible to downy mildew, and to botrytis bunch rot if yields are high and clusters compact. Buds late and not therefore at risk from spring frosts. Mid to late ripening. Best results come from poor, light soils and controlled yields.

WHERE IT'S GROWN AND WHAT ITS WINE TASTES LIKE

Plavina is grown throughout most of Primorska Hrvatska (Coastal Croatia), from the islands in the Kvarner Gulf down to Crna Gora (Montenegro), often planted with PLAVAC MALI. It is the most important red variety in the Sjeverna Dalmacija (Northern Dalmatia) wine region and represented 3.35% of **Croatia**'s total vineyard area in 2009, around 1,090 ha (2,693 acres). Officially authorized in all the coastal wine regions apart from Istra (Istria), Plavina makes wines that are generally light to medium-bodied, lightly coloured, with soft tannins, making them good for early drinking or to complement more powerful varieties in a blend. Recommended producers include Duvančić.

PLECHISTIK

Viticulturally difficult Russian variety used in sweet red sparkling wine.

BERRY COLOUR: ● ● ● ● ●

PRINCIPAL SYNONYMS: Bogata Kist, Goryun, Khreshchatinskii, Letun, Osipnijak, Petoun, Plechistik Tsimlyansky, Plecistik, Rogataia Kisty

VARIETIES COMMONLY MISTAKEN FOR PLECHISTIK: KRASNOSTOP ZOLOTOVSKY, TSIMLYANSKY CHERNY

ORIGINS AND PARENTAGE

Plechistik is an old variety from the Rostov wine region, along the River Don in south-western Russia, near the border with Ukraine. Under the shared synonym Cherny Vinny, it has been confused with KRASNOSTOP ZOLOTOVSKY. It is also sometimes called Plechistik Tsimlyansky, but Plechistik and TSIMLYANSKY CHERNY are distinct varieties (Chkhartishvili and Betsiashvili 2004).

OTHER HYPOTHESES

Like other varieties along the River Don, Plechistik is said to have been introduced from Dagestan in the Middle Ages (Vladimir Tsapelik, personal communication).

VITICULTURAL CHARACTERISTICS

Vigorous, mid ripening. Functionally female flowers lead to coulure. Thin, fragile skins. Susceptible to drought, cold winter temperatures and fungal diseases, especially downy mildew. Needs to be earthed up in winter.

WHERE IT'S GROWN AND WHAT ITS WINE TASTES LIKE

The **Russian** variety Plechistik needs to be grown next to another variety in order to achieve pollination – eg TSIMLYANSKY CHERNY, with which it used to be confused in the vineyard and with which it is often blended in the sweet but tannic sparkling reds from Tsimlyansky in the Rostov wine region. Tsimlyansky Winery make a benchmark version of this traditional wine. Plechistik is never produced as a varietal wine but it contributes flavours of prunes and sweet and sour cherries. It is also planted to a limited extent further south – also with Tsimlyansky Cherny – in Russia's Krasnodar region.

PLYTO

Rare, ancient, herbal-scented Cretan variety with great promise.

BERRY COLOUR: ● ● ● ● ●

PRINCIPAL SYNONYMS: Plito, Ploto (Kýthira/Cythera), Pluto (Kérkyra/Corfu), Plytó

ORIGINS AND PARENTAGE

Plyto probably comes from the eastern half of the Greek island of Kríti (Crete).

VITICULTURAL CHARACTERISTICS

Vigorous, productive, mid ripening. Susceptible to downy mildew and botrytis bunch rot, less so to powdery mildew. Large bunches of medium-sized to large berries.

WHERE IT'S GROWN AND WHAT ITS WINE TASTES LIKE

Very old Plyto vines can still be found in mixed vineyards in eastern Kríti, **Greece**, but the variety was close to extinction before Lyararkis saw its potential and planted 1.5 ha (4 acres) at around 400 m above sea level. Their varietal Plyto, a regional wine of Irákleio (Heraklion), combines a herbal, slightly grassy character with a delicate, chalky, pear-grained texture and juicy citrus flavours, at the modest alcohol level of 12.5–13%. Other producers of varietal wines include Michalakis, also in Irákleio; Tamiolakis make a VIDIANO/Plyto blend. Total plantings on the island were around 5–7 ha (12–17 acres) in 2011.

PODAROK MAGARACHA

Recent, versatile Ukrainian hybrid notably resistant to downy mildew.

BERRY COLOUR: ● ● ● ● ●

ORIGINS AND PARENTAGE

Podarok Magaracha, meaning 'gift of Magarach', is a hybrid obtained in 1966 by P Y Golodriga, V T Usatov, L P Troshin and Y A Malchikov at the Magarach research centre in Krym (Crimea), southern Ukraine, by crossing RKATSITELI with Magarach 2-57-72, in which the latter is a MTSVANE KAKHURI × Sochinsky Cherny cross, and Sochinsky Cherny is a non-cultivated Ukrainian variety. Podarok Magaracha is a sibling of PERVENETS MAGARACHA. It was included in Ukraine's official register of varieties in 1987.

VITICULTURAL CHARACTERISTICS

Vigorous, early to mid ripening. Thin-skinned berries turn pink when ripe. Cold-hardy to –25 °C (–13 °F) and resistant to downy mildew.

WHERE IT'S GROWN AND WHAT ITS WINE TASTES LIKE

Podarok Magaracha is not particularly aromatic and can be used for everything from juice to dry whites to fortified wines and brandy. It is grown in its homeland of **Ukraine** (212 ha/524 acres in 2009) and also in **Russia** (130 ha/321 acres in Krasnodar in 2010). It has also been reported in the **Republic of Moldova**.

POLLERA NERA

Old dark-skinned variety grown in Liguria and north-west Toscana.

BERRY COLOUR: ● ● ● ● ●

PRINCIPAL SYNONYMS: Corlaga, Pollora Nera

ORIGINS AND PARENTAGE

Pollera Nera was first described by Acerbi (1825) as a grape cultivated in the Cinque Terre (Liguria) and in Massa-Carrara (Toscana). DNA profiling has shown that Pollera Nera is a progeny of MAMMOLO, like several other Tuscan varieties (BIANCONE DI PORTOFERRAIO, CALORIA and COLOMBANA NERA) that turn out to be half-siblings of Pollera Nera (Di Vecchi Staraz, This *et al.* 2007). Two clonal variations of Pollera Nera are usually distinguished: Pollera Corlaga and Pollera Commune.

VITICULTURAL CHARACTERISTICS

Mid ripening. Abundant yields.

WHERE IT'S GROWN AND WHAT ITS WINE TASTES LIKE

Pollera Nera is cultivated here and there in the Cinque Terre in Liguria, between Monterosso and Riomaggiore, and in the area of Carrara in north-west Toscana, **Italy**. Wines are simple and fruity and are usually blended with other local varieties, as in the Colli di Luni DOC, which tends to be predominantly SANGIOVESE. There were 73 ha (180 acres) planted in Italy in 2000.

PONTAC

See TEINTURIER.

PORTAN

Montpellier cross.

BERRY COLOUR: ● ● ● ● ●

ORIGINS AND PARENTAGE

A GARNACHA × BLAUER PORTUGIESER cross obtained by Paul Truel in 1958 at INRA (Institut National de la Recherche Agronomique) in Montpellier, southern France. Its name is a neologism based on the first four letters of Portugais (Blauer Portugieser is called Portugais Bleu in French) and the final *-an*, a possible analogy with Chasan (Rézeau 1997).

VITICULTURAL CHARACTERISTICS

Very early budding and early ripening and therefore not generally susceptible to botrytis bunch rot. Very vigorous and best pruned long. Suited to low-fertility soils. Susceptible to powdery mildew and phomopsis.

WHERE IT'S GROWN AND WHAT ITS WINE TASTES LIKE

This precocious variety makes soft, well-coloured wines for early drinking. There were 311 ha (758 acres) in **France** in 2009, mainly in the Languedoc, down slightly in the last few years even though it is recommended in every French wine-growing region except Alsace. It is usually found blended with other varieties such as Carignan (MAZUELO), Grenache (GARNACHA), MERLOT, CABERNET FRANC, etc. in Pays d'Oc wines. Les Domaines Auriol, however, make a varietal wine in the western part of the Aude, as do the Cave La Malepère, further west near Carcassonne.

In **Spain**, Carlos Esteva at Can Rafols dels Caus in Penedès planted Portan in the 1980s and blends it with other Montpellier crosses, CALADOC and CHENANSON, in his Petit Caus red.

POŠIP BIJELI

High-quality vine from the Croatian island of Korčula making full-bodied whites.

BERRY COLOUR: ● ● ● ● ●

PRINCIPAL SYNONYMS: Pošip, Pošip Veliki (Dubrovnik), Pošipak, Vgorski Bijeli

VARIETIES COMMONLY MISTAKEN FOR POŠIP BIJELI: FURMINT

ORIGINS AND PARENTAGE

Pošip Bijeli probably originates from the Croatian island of Korčula off the Dalmatian coast because DNA parentage analysis has established that it is a cross between Zlatarica Blatska Bijeli and BRATKOVINA BIJELA (Piljac *et al.* 2002), all three varieties having been grown on Korčula for a long time. (Zlatarica Blatska Bijeli is no longer cultivated despite its reputation for quality, probably because of irregular yields.) The black-berried Pošip Crni from the same island is a distinct variety (Maletić *et al.* 1999) but is not commercially cultivated.

OTHER HYPOTHESES

Pošip Bijeli is often confused with FURMINT, but this identity is rejected by ampelography (Galet 2000) and DNA profiling (Maletić *et al.* 1999).

VITICULTURAL CHARACTERISTICS

Good, consistent yields. Early budding and early ripening, often reaching high sugar levels in the berries. Sensitive to strong winds, sunburn and, thanks to its thin berry skins, desiccates easily in drought conditions. Susceptible to downy and powdery mildews.

WHERE IT'S GROWN AND WHAT ITS WINE TASTES LIKE

Pošip Bijeli, more commonly known simply as Pošip, is qualitatively one of the most important white varieties of central Dalmacija (Dalmatia) in Primorska Hrvatska (Coastal Croatia), comprising 1.4% of the Croatian vineyard area in 2009, around 450 ha (1,112 acres). The main plantings are on the island of Korčula, particularly in Čara and Smokvica, and to a lesser extent on neighbouring islands and, now, on the coast itself. The wines, officially recognized for their high quality by the creation of the Korčula appellation in 1967, are full-bodied, rich and yet fresh, with gentle but ripe fruit aromas, depending on location. The variety is commonly used for dry white wines, generally best in southern Dalmacija. It is occasionally blended with its parent BRATKOVINA BIJELA and sometimes oaked, but the grapes are also dried for the production of the traditional sweet wine Prošek. Recommended producers include Grgić, Korta Katarina, Krajančić and Kunjas.

POTAMISSI

Minor Greek island white hidden mainly in blends.

BERRY COLOUR:

PRINCIPAL SYNONYMS: Aspropotamisio (Mýkonos and Tínos), Aspropotamissi, Potamisi, Potamisiès (Santoríni), Potamisio, Potamisio Lefko, Potamissio Lefko

ORIGINS AND PARENTAGE

Potamissi, literally 'from the river', is a variety from the Greek Kykládes (Cyclades) islands. According to Myles *et al.* (2011), the reference sample of Potamissi in the United States Department of Agriculture's National Clonal Germplasm Repository could be identical to EARLY MUSCAT but this can be rejected by comparison with the DNA profile of the true Potamissi (Vouillamoz).

The dark-skinned Potamissi Mavro is also cultivated in the same area but it is not yet known whether it's a colour mutation or a distinct variety.

VITICULTURAL CHARACTERISTICS

Vigorous, productive and susceptible to coulure, so although it is generally quite easy to grow, the berry size is very uneven, from gunshot to golfball. Mid budding and ripening. Large, loose bunches of large, thick-skinned berries.

WHERE IT'S GROWN AND WHAT ITS WINE TASTES LIKE

Potamissi is found mainly on the island of Santoríni, **Greece**, where it is used in blends, and also to a limited extent on Kykládes islands such as Náxos, Tínos, Amorgós, Mýkonos, Folégandros, Íos, Mílos and, more surprisingly, on the island of Kýthira (Cythera), off the southern tip of the Pelopónnisos (Peloponnese). It is rarely mentioned since it comprises only a small percentage of old interplanted vineyards. Potamissi generally produces rather ordinary wines, with relatively high alcohol and moderate acidity, sometimes slightly tannic, but Jerome Binda at Domaine de Kalathas on Tinos makes an exceptional example.

POULSARD

Ancient, perfumed Jura speciality.

BERRY COLOUR:

PRINCIPAL SYNONYMS: Mècle (Ain), Pelossard, Peloussard, Plant d'Arbois, Pleusart, Ploussard, Plussart, Poulsard Noir, Pulceau, Pulsard

VARIETIES COMMONLY MISTAKEN FOR POULSARD: MÈCLE DE BOURGOIN, PELOURSIN (called Pelossard in Isère), TRESSOT (Yonne)

ORIGINS AND PARENTAGE

Poulsard is an old variety from the Franche-Comté in eastern France, where it was mentioned as early as 1386 under the name Polozard in a charter written in Lons-le-Saunier and referring to a plot of land in Chilley, today called Chilly-le-Vignoble (Rézeau 1997): 'il devray planter savaignins noirs, polozard et mergeliains pour le pris et some de quatre florins de Florence d'or' ('he should plant savaignins noirs, polozard and mergeliains for the price and amount of four golden Florentine florins'). Spelt Pulsard, it later appeared in a list of good varieties established by the parliament of Besançon in 1732 (Rézeau 1997).

Like PELOURSIN, the name Poulsard could derive from the colour of the berries, which resembles that of *pelosses*, 'wild plums' in the local dialect (Galet 1990). The black-berried Poulsard, or Poulsard Noir, has undergone two colour mutations – **Poulsard Blanc**, which has white berries (13 ha/32 acres in France in 2000), and **Poulsard Rouge**, which has red berries – as well as an aromatic mutation, the **Poulsard Noir Musqué**, with black berries and a slight Muscat aroma but no longer in cultivation (Galet 2000).

VITICULTURAL CHARACTERISTICS

Very early budding and early ripening. A rather delicate vine, at risk from spring frosts, coulure and sunburn. Best pruned long and trellised. Well suited to clay and calcareous-clay soils. Highly susceptible to powdery mildew. Small bunches and large berries.

WHERE IT'S GROWN AND WHAT ITS WINE TASTES LIKE

A speciality of the Jura in eastern **France**, Poulsard flourishes under the local name Ploussard at Pupillin near Arbois in the north of the region and has the potential to produce perfumed, high-quality wines that are not particularly powerful, usually pale in colour and often blended, eg with TROUSSEAU or PINOT NOIR, or sold as rosé. It is also valued for adding colour to vin de paille. It is planted to a lesser extent in neighbouring Bugey, mainly for semi-sweet sparkling wines labelled Cerdon. There were 311 ha (768 acres) in France in 2009, an area almost unchanged over the last fifty years, and it is the second most widely planted red variety in the Jura. Poulsard is authorized as one of the main varieties in the Arbois, Macvin du Jura vin de liqueur, red Côtes du Jura and Crémant du Jura appellations and as a minor variety in those of L'Étoile and white Côtes du Jura.

Recommended producers include Philippe Bornard (Arbois-Pupillin), Domaine Emmanuel Houillon/Pierre Overnoy (Arbois-Pupillin), Domaine Puffeney (Arbois), Domaine André et Mireille Tissot (Stéphane Tissot in Arbois), Domaine de la Tournelle (Arbois).

PRAIRIE STAR

Cold-hardy American hybrid producing neutral, full whites for blending.

BERRY COLOUR: ● ● ● ● ●

PRINCIPAL SYNONYMS: ES 3-24-7

ORIGINS AND PARENTAGE

An Elmer Swenson 2-7-13 × Elmer Swenson 2-8-1 hybrid bred in 1980 by Elmer Swenson in Osceola, Wisconsin, US, in which:

- Elmer Swenson 2-7-13 is an Elmer Swenson 5-14 × SWENSON RED hybrid (see BRIANNA for the pedigree of Elmer Swenson 5-14)
- Elmer Swenson 2-8-1 (or Alpenglow) is also an Elmer Swenson 5-14 × SWENSON RED hybrid.

Prairie Star is therefore the progeny of two sibling varieties and it is an extremely complex *Vitis riparia*, *Vitis labrusca*, *Vitis vinifera*, *Vitis aestivalis*, *Vitis lincecumii*, *Vitis rupestris* and *Vitis cinerea* hybrid. It was selected in 1984 and tested as ES 3-24-7 before being named and released by Tom Plocher and Bob Parke in 2000 (Smiley 2008).

VITICULTURAL CHARACTERISTICS

Long, rather loose clusters of thick-skinned berries. Vigorous, mid budding, early ripening. Very hardy (–20 °F/–29 °C to –35 °F/–37 °C). Moderately susceptible to powdery mildew, black rot and anthracnose (Smiley 2008).

WHERE IT'S GROWN AND WHAT ITS WINE TASTES LIKE

Wines tend to be pretty neutral, sometimes with floral notes, but without any foxy characteristics and are unusually useful for adding body to lighter blends. Producers of varietal wines in **Minnesota**, where there were 40 acres (16 ha) in 2007, include Glacial Ridge, Indian Island and Millner Heritage, though it is also planted to a very limited extent elsewhere in the Midwest, eg in **Iowa** and **Nebraska**. In Québec, **Canada**, Les Petits Cailloux and Clos du Roc Noir make dry varietal wines (oaked and unoaked versions) but many others blend it with other cold-hardy varieties such as KAY GRAY, Frontenac Gris (see FRONTENAC), ST PEPIN or VANDAL-CLICHE.

PRČ

Unusually scented Croatian variety being recuperated on the island of Hvar.

BERRY COLOUR: ● ● ● ● ●

PRINCIPAL SYNONYMS: Čimavica, Parč, Prč Bijeli

VARIETIES COMMONLY MISTAKEN FOR PRČ: GRK (Makarska)

ORIGINS AND PARENTAGE

Prč is undoubtedly from the Croatian island of Hvar. *Prč* in the local slang means 'goat', supposedly referring to the smell of the berries (Galet 2000).

OTHER HYPOTHESES

Although Prč is often listed as a synonym for GRK, comparison of their DNA profiles (Benjak *et al.* 2005; Pejić *et al.* 2000) clearly establishes that they are distinct varieties (Vouillamoz).

VITICULTURAL CHARACTERISTICS

Vigorous, mid to late budding and ripening. Small to medium-sized bunches of thick-skinned berries. Yields well. Susceptible to powdery mildew.

WHERE IT'S GROWN AND WHAT ITS WINE TASTES LIKE

Although Prč is authorized for the whole of the Srednja i Južna Dalmacija (Central and Southern Dalmatia) wine region in **Croatia** and was much more widely planted at the end of the nineteenth century, it is now found almost exclusively on the island of Hvar, most often interplanted with other varieties. Wines have a distinctive aroma or odour which some describe as Muscat-like and others consider somewhat malodorous, which might explain the goat reference. The wines typically have low acidity but can be a useful addition to more neutral blends. Prč, along with several other autochthonous varieties, is part of a revitalization project involving clonal selection and replanting to save it from extinction. Such clonal selection is particularly important because almost all the vines recently studied were found to be virused (Zdunić, Pejić *et al.* 2008). Producers of varietal wines include Plančić (dry and sweet examples as well as a blend with the equally rare BOGDANUŠA) and Vujnović.

PRÉCOCE DE MALINGRE

Obscure Parisian cross still found in one old vineyard in the Ahr in Germany.

BERRY COLOUR: ● ● ● ● ●

PRINCIPAL SYNONYMS: Blanc Précoce de Malingre, Early Malingre, Madeleine Blanche de Malingre, Malinger (Ahr), Malingre Précoce

ORIGINS AND PARENTAGE

An cross obtained around 1840 by Malingre, a gardener in the Paris region of France, who did not record the parents. Recent DNA parentage analysis has shown that it is in fact a PINOT × Bicane cross (Vargas *et al.* 2009), the latter being no longer cultivated.

VITICULTURAL CHARACTERISTICS

Buds early and ripens very early. Not very vigorous and best pruned short. Thin-skinned. Very sensitive to rain especially during flowering but not to fungal diseases. Bunches are small; the smallish berries can achieve high sugar levels and are a favourite with wasps.

WHERE IT'S GROWN AND WHAT ITS WINE TASTES LIKE

In 2008 there was less than a hectare (2.5 acres) in **France**, where it is recommended for the Vendée only and there's no sign that anyone is still producing a varietal wine. However, Johannes Hostert at Weingut Jakob Hostert in the Ahr in **Germany** has a small vineyard (1,330 vines on 0.16 ha/0.4 acres) of ninety-year-old ungrafted Malinger vines and produces dry, medium and sweet versions. A dry example was non-aromatic and tasted like a muted RIESLING without such fresh acidity. It is planted in many other countries as a table grape.

PRENSAL

Robust Mallorcan speciality.

BERRY COLOUR: ● ● ● ● ●

PRINCIPAL SYNONYMS: Moll (Binissalem), Pensal Blanca, Premsal, Premsal Blanca, Prensal Blanc

VARIETIES COMMONLY MISTAKEN FOR PRENSAL: Afus Ali, XARELLO

ORIGINS AND PARENTAGE

Prensal is an indigenous variety on the Spanish Islas Baleares.

VITICULTURAL CHARACTERISTICS

Productive, mid ripening. Thick-skinned berries, which are big if the soils are fertile. Susceptible to powdery mildew.

WHERE IT'S GROWN AND WHAT ITS WINE TASTES LIKE

Premsal – as it is generally spelt on Mallorca – lacks acidity and is often blended with other varieties such as Moscatel (MUSCAT), CHARDONNAY or VIOGNIER to make it fresher, and acidification is common. Although it is the principal white variety in the Binissalem-Mallorca

DO, Spain, the minimum required is just 50%. There were about 65 ha (161 acres) planted on the island in 2008, used mainly for easy-drinking whites. It can produce a range of styles, from tropical-fruited and soft to more structured with an almond or woody note, all for early consumption, although some improve after a few months in bottle. It is also one of the principal varieties in the other DO on the Islas Baleares, Plà y Llevant, and is frequently bottled as a Vino de la Tierra Mallorca. Àn Negra's Quíbia, a blend of 50% Premsal with 50% dark-skinned CALLET (made as a white) is a dry, finely textured and minerally wine with pear and greengage flavours and a sour, refreshing aftertaste. Toni Gelabert blends Premsal with Moscatel (MUSCAT BLANC À PETITS GRAINS) to produce a perfumed, delicate and surprisingly refreshing example. It is not an easy variety to handle in the winery because it is prone to reduction when aged on the lees but also to oxidation.

PRESSAC

See COT.

PRETO MARTINHO

Portuguese variety often confused with Trincadeira.

BERRY COLOUR: ● ● ● ● ●

PRINCIPAL SYNONYMS: Amostrinha (Arruda dos Vinhos and Bucelas), Preto Martinho do Oeste

VARIETIES COMMONLY MISTAKEN FOR PRETO MARTINHO: NEGRAMOLL, TRINCADEIRA

ORIGINS AND PARENTAGE

Preto Martinho is from north-east Portugal. It has long been confused with TRINCADEIRA, and even with NEGRAMOLL (Brazão *et al.* 2005), but their DNA profiles are distinct in Veloso *et al.* (2010).

VITICULTURAL CHARACTERISTICS

Mid budding, early ripening. Compact bunches of medium-sized berries.

WHERE IT'S GROWN AND WHAT ITS WINE TASTES LIKE

Wines are typically moderate in alcohol and blended with other varieties such as CASTELÃO but Casa Santos Lima produce a soft, oaked varietal example. There was a total of 269 ha (665 acres) in **Portugal** in 2010, in the Lisboa region as well as in the north east.

PRIÉ

Hugely significant Aostan variety with links to Spain and Switzerland.

BERRY COLOUR: ● ● ● ● ●

PRINCIPAL SYNONYMS: Bernarde (Valais in Switzerland), Blanc de Morgex, Blanc du Valdigne, Legiruela (province of Ávila in Castilla y León, Spain), Plant de la Salle, Prié Blanc

VARIETIES COMMONLY MISTAKEN FOR PRIÉ: LUGLIENGA

ORIGINS AND PARENTAGE

Prié is one of the oldest varieties of the Valle d'Aosta, where it was mentioned as early as 1691 in Saint-Pierre near Aosta. It is also called Blanc du Valdigne, after the old Latin name of the upper Valle d'Aosta, at the foot of Mont Blanc, where this variety probably comes from. At one time there were two clonal variations of Prié (Gatta 1838): one in Valdigne (mainly in Morgex and La Salle) and one in Valpelline at the foot of the Grand Combin but the one from Valpelline has disappeared. Prié has also been cultivated to a very limited extent since the early nineteenth century in the Valais in Switzerland under the name Bernarde, referring to the likely route taken by those who imported the variety from the Valle d'Aosta via the Grand-Saint-Bernard pass.

Quite unexpectedly, Schneider *et al.* (2010) discovered by DNA profiling that Prié has also long been cultivated in the province of Ávila in Spain under the local synonym Legiruela and they have found parent–offspring relationships with ALBILLO REAL from the neighbouring Valladolid region and with LUGLIENGA from Piemonte in north-west Italy, while Santana *et al.* (2010) have also suggested a parent–offspring relationship with LAIRÉN from Andalucía in southern Spain. The discovery of additional parent–offspring relationships with MAYOLET and PRIMETTA from the Valle d'Aosta (Vouillamoz 2005; Vouillamoz and Moriondo 2011) results in a complex pedigree that connects northern Italy and northern Spain. Because Luglienga was mentioned in the literature much before Prié, it has been placed at the head of the genealogy shown in the diagram but since the other parents are unknown, Lairén and Albillo Real could be switched with Luglienga, and even Prié could be placed as parent of all the other five varieties (Luglienga, Albillo Real, Lairén, Primetta and Mayolet).

Morphological similarities between Prié and LUGLIENGA had already been noticed, as well as with PRIMETTA, which is sometimes (quite reasonably, given its pedigree) called Prié Rouge. However, the parent–offspring relationship with MAYOLET was totally unexpected and makes Prié a grandparent of ROUGE DU PAYS, born in Italy's Valle d'Aosta and now grown exclusively in the Valais in Switzerland, and a great-grandparent of CORNALIN, ROUSSIN and NERET DE SAINT-VINCENT (see Rouge du Pays). This illustrates the significant role played by Prié in the biodiversity of Valle d'Aosta varieties. In addition, the links with the Spanish varieties ALBILLO REAL and LAIRÉN call into question the Italian origin of Prié, and scholars have not been able to determine whether Prié was brought from Italy to Spain or vice versa (Schneider *et al.* 2010).

OTHER HYPOTHESES

Prié is sometimes called Agostenga, derived either from Italian *agosto*, meaning 'August' and alluding to the vine's precocity, or from *Augusta Praetoria*, the Latin name for the city of Aosta, but this misleading name has been applied to several distinct varieties and should not be used. It has also been suggested that Prié was introduced from the Valais (Switzerland), but this is very unlikely (Vouillamoz and Moriondo 2011).

VITICULTURAL CHARACTERISTICS

Good resistance to frost and to a cold climate. Very early ripening. Susceptible to the vine moth (*Eupoecilia ambiguella*) and to rot.

WHERE IT'S GROWN AND WHAT ITS WINE TASTES LIKE

Prié is cultivated almost exclusively in the Valle d'Aosta, north-west **Italy**, in the communes of Morgex and La Salle, between 900 and 1200 m above sea level, which puts these vineyards among the highest in continental Europe. Interestingly, phylloxera does not seem to survive in such cold climates, and all Prié vines are still ungrafted. Prié is traditionally trained on a low pergola in order to benefit from the heat accumulated in the soil. The agricultural census of 2000 recorded 39 ha (96 acres) in Italy.

The Blanc de Morgex et de La Salle version of the Valle d'Aosta DOC is a varietal Prié wine, usually dry, with aromas of mountain hay and white flowers, a fresh acidity and a persistent finish. Sparkling and dessert wines are also made. Examples include Ermes Pavese's Cuvée Nathan (dry,

PRIÉ PEDIGREE DIAGRAM

Today isolated in the Valle d'Aosta (Italy), ancient PRIÉ has natural ramifications as far away as Spain and Switzerland. Given the numerous connections via unknown (and probably extinct) varieties (?), its origin has been debated because inverse relationships are also theoretically possible (see p x).

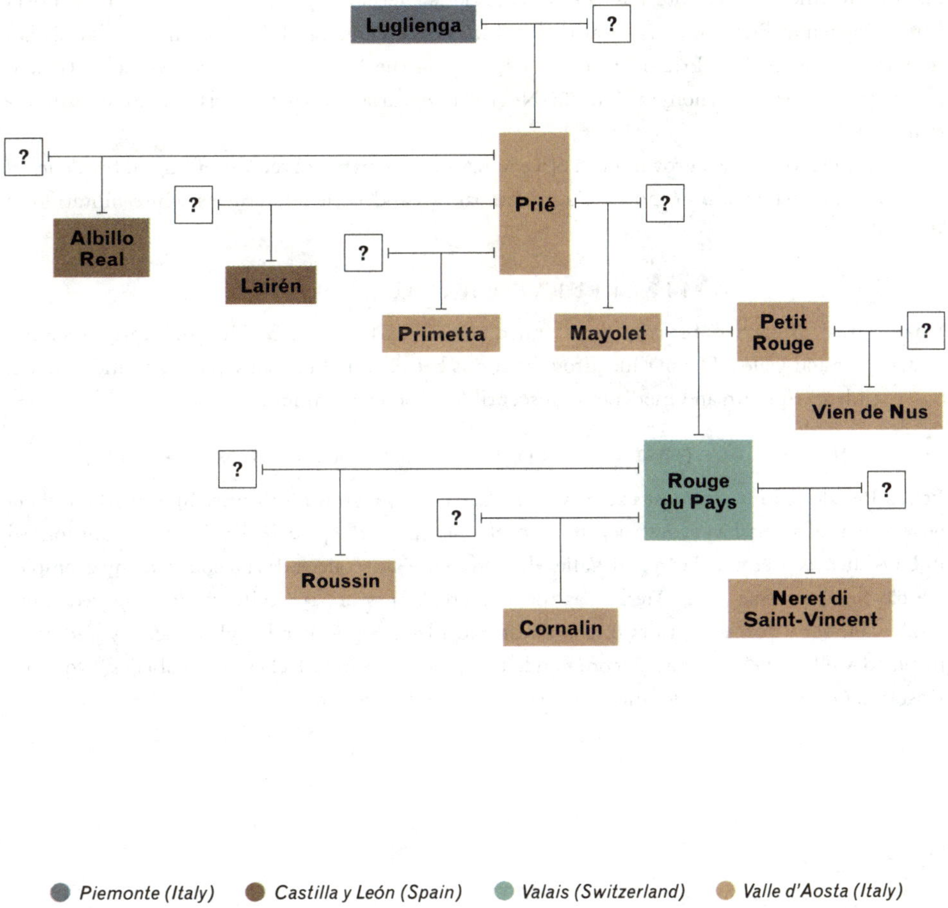

oaked) and Cuvée Ninive (sweet, grapes harvested in December), Maison Albert Vevey's dry wine and the Cave du Vin Blanc de Morgex et de La Salle's traditional-method sparkler.

A few vines (0.02 ha/0.05 acres) still exist in the Valais in **Switzerland** under the name Bernarde.

In **Spain**, it is scarcely cultivated, under the local name Legiruela, in San Esteban del Valle in the Provincia de Ávila (Schneider *et al.* 2010).

PRIETO PICUDO

Crisp, perfumed speciality of León, north-west Spain.

BERRY COLOUR: ● ● ● ● ●

PRINCIPAL SYNONYMS: Prieto Picudo Oval ☒ (Tierra de León), Prieto Picudo Tinto ☒ (Tierra de León)

ORIGINS AND PARENTAGE

Prieto Picudo is thought to be indigenous to the area south of León in north-west Spain that qualifies for the Valdevimbre-Los Oteros Viño de la Tierra designation. A mutation with oval berries known as Prieto Picudo Oval is found in the same region. DNA parentage analysis has recently suggested that Prieto Picudo is a sibling of the Portuguese variety ALFROCHEIRO, cultivated under the synonym Albarín Negro in Asturias north towards the coast (Santana *et al.* 2010).

Prieto Picudo Blanco grown south of León has been shown by DNA profiling to be identical to GODELLO (Santana *et al.* 2007; Gago *et al.* 2009) and is therefore not a white mutation of Prieto Picudo.

VITICULTURAL CHARACTERISTICS

Small compact bunches of small, thin-skinned berries. Early to mid budding and early ripening. Low vigour and yields. Doesn't like drought and is better suited to cool soils and a mild climate. Susceptible to sunburn and moderately susceptible to powdery mildew.

WHERE IT'S GROWN AND WHAT ITS WINE TASTES LIKE

Prieto Picudo is grown almost exclusively in Castilla y León in north-west **Spain**, where there were a total of 5,187 ha (12,817 acres) in 2008. It is one of the principal red varieties authorized in DOs such as Tierra de León and Valles de Benavente and Valtiendas although many examples are classified as Vino de la Tierra Castilla y León. Wines are typically cherry-coloured, high in alcohol, very aromatic (fruits of the forest and black pepper) and high in acidity and seem to blend well with MENCÍA. Recommended producers include Dehesa de Rubiales, Dominio Dostares, Gordonzello, Pardevalles, Tampesta and Vega Carriegos.

PRIKNADI

Rare, and puzzling, variety from eastern Makedonía.

BERRY COLOUR: ● ● ● ● ●

PRINCIPAL SYNONYMS: Prekiadi, Prekna, Preknadi, Preknari, Preknari Lefko, Prekniariko, Prekno

ORIGINS AND PARENTAGE

Priknadi is a rare variety local to Thessalía and Makedonía.

OTHER HYPOTHESES

The DNA profile of the reference sample of Priknadi in the Greek Vitis Database matches the Swiss variety CHASSELAS, whereas the ampelographic description in the same database differs completely from Chasselas (Vouillamoz).

VITICULTURAL CHARACTERISTICS

Large bunches of large, thick-skinned berries. Vigorous, fertile and productive, mid to late ripening. Susceptible to botrytis bunch rot but relatively resistant to drought. Best in calcareous soils.

WHERE IT'S GROWN AND WHAT ITS WINE TASTES LIKE

Priknadi wines are typically lightly aromatic, high in alcohol but low in acidity and need careful handling in the winery to avoid oxidation. The only producers currently making a varietal wine, both in Náoussa in western Makedonía, **Greece**, are Chrisohoou, who have 1.5 ha (4 acres) planted, and Diamantakos, who planted 1.6 ha in 2006. Dalamara nearby make a blend of MALAGOUSIA, Prekniariko and RODITIS.

PRIMETTA

Recuperated Aostan pink wine speciality.

BERRY COLOUR: ● ● ● ● ●

PRINCIPAL SYNONYMS: Neblou, Premetta, Prëmetta, Prié Rouge, Prometta

VARIETIES COMMONLY MISTAKEN FOR PRIMETTA: BONDA ⊗

ORIGINS AND PARENTAGE

Primetta is an old variety of the Valle d'Aosta, sometimes called Prié Rouge because it was considered to be a seedling of PRIÉ. DNA evidence has recently shown this to be true: Primetta is a progeny of Prié and is therefore a grandchild, grandparent or half-sibling of MAYOLET, LUGLIENGA, ALBILLO REAL and LAIRÉN (see Prié pedigree diagram; Vouillamoz and Moriondo 2011). Its name comes from Italian *prima*, meaning 'first', in reference to its early ripening.

VITICULTURAL CHARACTERISTICS

Early ripening. Resistant to cold climates, downy mildew and botrytis bunch rot.

WHERE IT'S GROWN AND WHAT ITS WINE TASTES LIKE

Primetta is grown only in the Valle d'Aosta, north-west **Italy**, mainly in Saint-Pierre, Saint-Denis, Aymavilles and Quart, and produces a rosé-coloured wine. In the past it was only ever blended in small amounts with other local varieties such as PETIT ROUGE in Enfer d'Arvier and Chambave or with Petit Rouge and VIEN DE NUS in Nus (all within the Valle d'Aosta DOC). In the late twentieth century it was on the verge of extinction but was rescued by Costantino Charrère, who was the first in modern times to plant a small vineyard with this old variety and the first to produce a varietal wine. A few rare varietal rosés have appeared quite recently, also within the Valle d'Aosta DOC, such as Grosjean Frères' Prëmetta, which includes 10% CORNALIN, smells of roses and strawberries and is fresh and quite firmly structured, and Les Crêtes Premetta, a light-pink traditional-method sparkling version. Under the name Prié Rouge, the Italian agricultural census recorded just 18 ha (44 acres) in 2000.

PRIMITIVO

See TRIBIDRAG.

PRINZIPAL

Recent disease-resistant German hybrid with marked Riesling influence.

BERRY COLOUR: ● ● ● ● ●

PRINCIPAL SYNONYMS: Geisenheim 7116-26

ORIGINS AND PARENTAGE

A complex hybrid obtained by Helmut Becker in 1971 at the Geisenheim research centre by crossing Geisenheim 323-58 × EHRENFELSER, where Geisenheim 323-58 is a CHANCELLOR × RIESLING hybrid (see PRIOR for the complete pedigree of Chancellor).

VITICULTURAL CHARACTERISTICS

Winter-hardy and resistant to botrytis bunch rot. Reasonable resistance to downy and powdery mildews but susceptible to the fungal disease Roter Brenner. Needs deep soils with a good water supply. Sugar levels are slightly higher than in RIESLING, acidity slightly lower. Mid to late ripening.

WHERE IT'S GROWN AND WHAT ITS WINE TASTES LIKE

Authorized in **Germany** in 1999 though there are still only a few hectares planted. Varietal wines are produced by, for example, Wolfgang Hermes in the Nahe and Hans J Becker in Rheinhessen. They show an encouraging likeness to RIESLING and are often made in off-dry or sweet styles.

PRIOR PEDIGREE DIAGRAM

The complex pedigree of PRIOR, a hybrid made in Baden-Württemberg (Germany), involves deliberate and natural crosses between seventy distinct parents encompassing the species *Vitis aestivalis*, *Vitis amurensis*, *Vitis berlandieri*, *Vitis cinerea*, *Vitis labrusca*, *Vitis lincecumii*, *Vitis riparia*, *Vitis rupestris* and *Vitis vinifera*. In this pedigree, some varieties appear more than once simply to make the illustration easier to read, and *Vitis* is abbreviated to *V.* in some instances to save space.

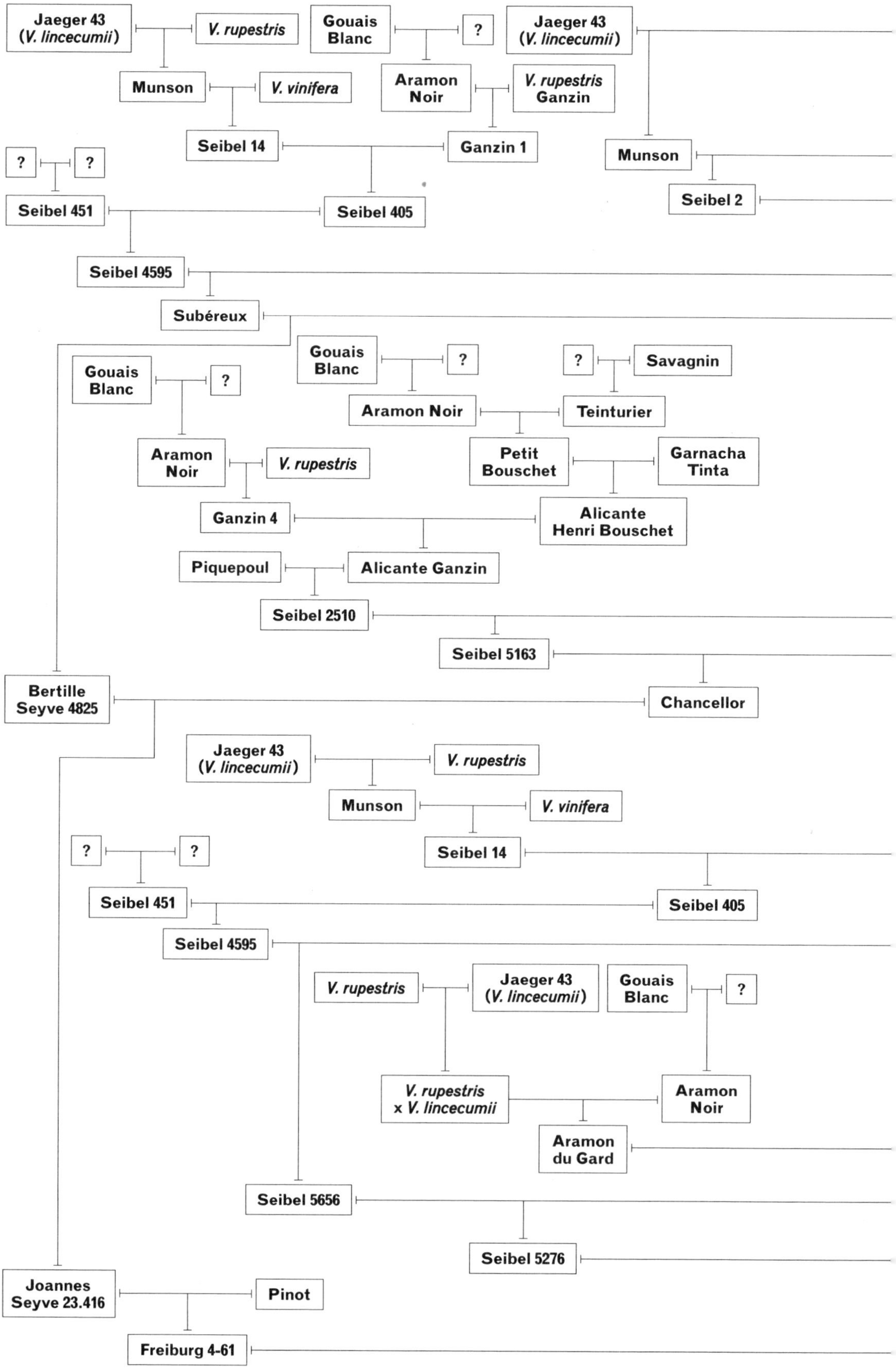
Jaeger 43 (V. lincecumii)
V. rupestris
Gouais Blanc
?
Jaeger 43 (V. lincecumii)
Munson
V. vinifera
Aramon Noir
V. rupestris Ganzin
Seibel 14
Ganzin 1
Munson
?
?
Seibel 2
Seibel 451
Seibel 405
Seibel 4595
Subéreux
Gouais Blanc
?
?
Savagnin
Gouais Blanc
?
Aramon Noir
Teinturier
Petit Bouschet
Garnacha Tinta
Aramon Noir
V. rupestris
Ganzin 4
Alicante Henri Bouschet
Piquepoul
Alicante Ganzin
Seibel 2510
Seibel 5163
Bertille Seyve 4825
Chancellor
Jaeger 43 (V. lincecumii)
V. rupestris
Munson
V. vinifera
?
?
Seibel 14
Seibel 451
Seibel 405
Seibel 4595
V. rupestris
Jaeger 43 (V. lincecumii)
Gouais Blanc
?
V. rupestris x V. lincecumii
Aramon Noir
Aramon du Gard
Seibel 5656
Seibel 5276
Joannes Seyve 23.416
Pinot
Freiburg 4-61

PRIOR

German hybrid bred for disease resistance.

BERRY COLOUR: ● ● ● ● ●

PRINCIPAL SYNONYMS: Freiburg 484-87 R

ORIGINS AND PARENTAGE

Complex hybrid obtained in 1987 by Norbert Becker at the Freiburg research centre in Baden-Württemberg, southern Germany, by crossing Freiburg 4-61 × Freiburg 236-75. Its name derives from the Latin for 'former' or 'elder', presumably intending to suggest superior quality, though it might also be a reference to its relatively early ripening. Its complete pedigree is shown in the diagram because it reveals that of many other varieties as well.

VITICULTURAL CHARACTERISTICS

Good resistance to downy mildew but susceptible to botrytis bunch rot and powdery mildew. Early to mid ripening (7–10 days later than PINOT NOIR).

WHERE IT'S GROWN AND WHAT ITS WINE TASTES LIKE

In **Germany**, the Winzergenossneschaft Achkarren in Baden produce a varietal wine, as do Arnoldweine in Arlesheim, just south of Basel, northern Switzerland. After several years of trials in the experimental vineyard of Agroscop Changins-Wädenswil at Pully in **Switzerland**, Prior is now cultivated to a very limited extent in Switzerland (0.33 ha/0.8 acres). It shows more promise than CABERNET CORTIS, CABERNET CARBON and CABERNET CAROL. Authorized in **Italy** in 2011.

PROBUS

A productive, late-ripening KADARKA × CABERNET SAUVIGNON black-berried cross obtained by D Milisavljević, S Lazić and V Kovač at the viticulture research centre in Sremski Karlovci, part of Novi Sad University in Vojvodina, **Serbia**. Released in 1983 (Cindrić *et al.* 2000), it was named after the Roman emperor Probus, who promoted vineyard planting in Fruška Gora and elsewhere along the Danube in the third century AD. Producers include Milanovic, Prekogačic and Živanovic.

PROCANICO

See TREBBIANO TOSCANO.

PRODEST

Conegliano cross; attempt at fresher Merlot.

BERRY COLOUR: ● ● ● ● ●

PRINCIPAL SYNONYMS: Incrocio Cosmo 109

ORIGINS AND PARENTAGE

Prodest is a MERLOT × BARBERA cross obtained in the 1960s by Italo Cosmo at the Conegliano research centre in the Veneto, northern Italy, with the aim of increasing the acidity of Merlot. It is a sibling of NIGRA.

VITICULTURAL CHARACTERISTICS

Offers some resistance to the main fungal diseases. Vigorous, mid ripening.

WHERE IT'S GROWN AND WHAT ITS WINE TASTES LIKE

A few hectares are cultivated around Conegliano and Pieve di Soligo in the province of Treviso in the Veneto, **Italy**, where it has been proposed as a replacement for MERLOT. The Italian agricultural census of 2000 records just 14 ha (35 acres). Wines tend to be herbaceous and lightly fruity, full-bodied, tannic and have good ageing capacity but its name is unlikely to catch on in English-speaking markets.

PROKUPAC

Balkan variety making early-drinking reds and deeply coloured rosé.

BERRY COLOUR: ● ● ● ● ●

PRINCIPAL SYNONYMS: Kameničarka, Kamenilarka, Majski Čornii (Russia and Republic of Moldova), Nikodimka, Niševka (Serbia and Republic of Macedonia), Prokupec (Republic of Macedonia), Prokupka, Rskavac (Serbia), Skopsko Crno (Republic of Macedonia), Zarčin (Bulgaria)

ORIGINS AND PARENTAGE

Prokupac is a very old Balkan variety, possibly originating in Serbia, where it has long been the dominant red variety, supposedly since the Middle Ages, and where its significant morphological diversity (Avramov and del Zan 2004) confirms its ancientness. It was used to breed ŽUPLJANKA.

OTHER HYPOTHESES

According to Biron in Galet (2000), it could be identical to PAPAZKARASI from Turkey but this theory is doubtful and not yet tested by DNA profiling.

VITICULTURAL CHARACTERISTICS

Vigorous, late ripening. Thick-skinned berries. Very susceptible to downy mildew but relatively resistant to botrytis bunch rot. Cold-hardy (–14 °C/6.8 °F to –18 °C/–0.4 °F).

WHERE IT'S GROWN AND WHAT ITS WINE TASTES LIKE

Widely planted in **Serbia**, particularly in the Zapadna Morava and Juzna Morava wine regions in the centre and south of the country, and in **Kosovo**, Prokupac produces early-drinking wines that are typically a deep ruby colour, high in alcohol, with aromas of red fruit. It is often blended with international varieties and frequently produced as a dark rosé. Producers include Ivanović and Toplički Winery. It is also found in the **Republic of Macedonia**, where it is grown to a limited extent in cooler sites, as well as in **Bulgaria** and **Russia**.

PROSECCO

The dominant, rather neutral grape for Prosecco sparkling wine, probably Istrian. Misleadingly renamed Glera for commercially protective reasons.

BERRY COLOUR:

PRINCIPAL SYNONYMS: Briška Glera (Slovenia), Glera (occasionally in Trieste), Prosecco Tondo (Friuli), Serpina (Colli Euganei), Serprina (Colli Euganei), Serprino (Colli Euganei), Števerjana (Slovenia), Teran Bijeli (Croatia)

VARIETIES COMMONLY MISTAKEN FOR PROSECCO: MALVASIA BIANCA LUNGA (sometimes called Prosecco Nostrano in Conegliano), PERERA, PROSECCO LUNGO

ORIGINS AND PARENTAGE

Since the late eighteenth century, several morphologically distinct varieties have been called Prosecco in the area of Conegliano in the province of Treviso in northern Italy, possibly taking the name of the village of Prosecco in the province of Trieste. In the mid nineteenth century, Balbi Valier selected and cultivated a particular clonal variation named Prosecco Tondo after the shape of the berries (*tondo* is Italian for 'round'). In the 1980s, two other clonal variations were selected for further study by the research centre at Conegliano: PROSECCO LUNGO, with oval berries (*lungo* means 'long') and Prosecco Nostrano (*nostro* means 'ours').

Morphological and DNA analyses proved that all three supposed clonal variations are in fact distinct varieties: Prosecco Tondo is now simply called Prosecco, PROSECCO LUNGO keeps its name and Prosecco Nostrano was shown to be identical to MALVASIA BIANCA LUNGA (Costacurta *et al.* 2003; Crespan *et al.* 2003). A strong genetic relationship was observed between Prosecco and Prosecco Lungo, although a parent–offspring relationship can be ruled out (Crespan *et al.* 2009). Interestingly, DNA parentage analysis showed that VITOVSKA (under the name Vitouska), a white grape of the Karst region on the Italian–Slovenian border, is a natural progeny of Prosecco and Malvasia Bianca Lunga (Crespan *et al.* 2007). In terms of the clonal variation names explained above, that makes Vitovska a Prosecco Tondo × Prosecco Nostrano cross!

In addition, comparison of DNA profiles shows that Prosecco is identical to Teran Bijeli (*bijeli* is Croatian for 'white'), previously considered to be a rare variety native to the Istra (Istria) region of Croatia (Maletić *et al.* 1999), and also identical to Briška Glera and Števerjana, previously thought to be two distinct varieties cultivated in Slovenia (Štajner *et al.* 2008).

Prosecco was crossed with CABERNET SAUVIGNON to obtain INCROCIO MANZONI 2.15.

As part of the promotion of Prosecco di Conegliano-Valdobbiadene to DOCG status and the enlargement of the Prosecco DOC zone in 2009, the Prosecco Consorzio set in motion an official name change so that this principal grape variety is known as Glera, its supposed Friulian synonym, and Prosecco is reserved for the designation of origin, effectively preventing producers from other regions or countries taking advantage of the name Prosecco to designate any old sparkling wine. The enlargement of the Prosecco DOC area includes the village of the same name near Trieste, the variety's alleged birthplace. This amendment is both confusing and misleading: Glera is a generic name applied to several distinct varieties in the province of Trieste, and recent studies have shown that Glera in fact usually refers to PROSECCO LUNGO and much less frequently to Prosecco (Tondo) and other local varieties from the Karst region such as VITOVSKA, or the non-cultivated Aghedone and Mocula.

By reconstructing the genetic relationships among Croatian grapes, Maletić *et al.* (1999) showed that Prosecco (under the name Teran Bijeli) occupies a key role in relation to all other Croatian grapes, and appears to be related to ŽILAVKA from Bosnia-Herzegovina. This genetic study supports an Istrian origin for Prosecco. As a consequence, it is likely that the village of Prosecco in the province of Trieste was only a stopover of this variety on its way from Istra to Friuli, and that its original name was Teran Bijeli.

OTHER HYPOTHESES

Prosecco has often been erroneously identified with the Pucinum mentioned by Pliny the Elder as a variety cultivated on rocks in Istra. Prosecco is also considered by some to be a polyclonal variety but this concept is not accepted in our approach in this book (see pp XIV–XVI).

VITICULTURAL CHARACTERISTICS

Late ripening. Susceptible to powdery and downy mildews, summer drought, millerandage and flavescence dorée.

WHERE IT'S GROWN AND WHAT ITS WINE TASTES LIKE

Prosecco is one of the most widespread varieties of the Veneto in northern **Italy**, cultivated mainly in the province of Treviso between Valdobbiadene and Conegliano to produce popular sparkling wines in the generally superior zone of the hills around Conegliano that qualifies as Prosecco di Conegliano-Valdobbiadene DOCG, in the more obscure Colli Asolani Prosecco DOCG and in the recently expanded flatter area that is now Prosecco DOC. The latter encompasses no fewer than nine provinces (Belluno, Gorizia, Padova, Pordenone, Treviso, Trieste, Udine, Venezia and Vicenza), absorbing practically all of the IGTs under which Prosecco production outside the classic hills has been marketed up to now. While the province of Treviso boasts more than 10,000 ha (24,711 acres) devoted to the variety, Padova, Vicenza and Belluno together have no more than 600 ha (1,483 acres) planted with Prosecco, and Friuli a mere 125 ha (309 acres).

Total annual production of this generally inexpensive fizz is around 1,300,000 hl and total plantings of the variety in Italy in 2000 were already 8,144 ha (20,124 acres), according to the agricultural census, which does not, however, distinguish between Prosecco and PROSECCO LUNGO. More recently, the total vineyard area has been estimated at 11,000 ha (27,182 acres). The subzone of Cartizze, even cooler than the rest of the zone, comprises 106 ha (262 acres) of vineyard, and one million bottles of Prosecco which is held to be superior to the normal DOC or DOCG production.

In the right hands, Prosecco (Tondo) can be lightly floral, compared with the slightly more spicy PROSECCO LUNGO. The variety is rather neutral, especially if yields are not controlled. The finished wines are light and frothing, sometimes lemony, sometimes slightly floral, but their neutrality and defects are too often masked by over-generous additions of sugar.

Prosecco wines exist in still, but mainly fizzy and sparkling, versions. Recommended producers of superior wines include Adami, Bisol, Bortolin, Bortolomiol, Case Bianche, Col Vetoraz, Conte Collalto, Nino Franco, Andreola Orsola and Sorelle Bronca. Still wines may be produced under the Prosecco designations described above but also in the Colli di Conegliano DOC, where it may be included with VERDISO and Boschera (aka VERDICCHIO BIANCO) in the dry or sweet, passito-style Torchiato di Fregona, and both still and sparkling wines are made in the Colli Euganei DOC (Prosecco here known as Serprino), both on its own and blended with varieties such as GARGANEGA.

A few hectares are planted over the border in Goriška Brda in **Slovenia**, and there are similarly limited plantings in **Argentina**. Brown Brothers in **Australia** have already produced several vintages of 'Prosecco' but are no longer allowed to export it without changing the name.

PROSECCO LUNGO

The spicier, less common Prosecco grape.

BERRY COLOUR: ● ● ● ● ●

PRINCIPAL SYNONYMS: Glera (Trieste), Ribolla Spizade, Tocai Nostrano (Breganze in the province of Vicenza)

VARIETIES COMMONLY MISTAKEN FOR PROSECCO LUNGO: Bela Glera (Slovenia), MALVASIA BIANCA LUNGA (sometimes called Prosecco Nostrano in Conegliano), PROSECCO, RIBOLLA GIALLA

ORIGINS AND PARENTAGE

In the province of Treviso in north-eastern Italy, Prosecco Lungo, Prosecco Tondo and Prosecco Nostrano used to be considered simple clonal variations of the same variety but with differently shaped berries (see PROSECCO). However, recent DNA profiling has shown that Prosecco Lungo is distinct from Prosecco Tondo, now simply called Prosecco, and distinct from Prosecco Nostrano, which turned out to be identical to MALVASIA BIANCA LUNGA (Costacurta *et al.* 2003; Crespan *et al.* 2003).

Prosecco Lungo had almost disappeared when it was selected and cultivated by the research centre at Conegliano in the Veneto, northern Italy, at the end of the 1980s. It was long thought to be an old and virtually extinct variety from the province of Treviso but recent DNA investigation showed unexpectedly that the majority of grapes named Glera in the Karst area on the Italy–Slovenia border are in fact identical to Prosecco Lungo, whereas only a few of them corresponded to PROSECCO and other local Italian varieties (Crespan *et al.* 2009). In addition, Prosecco Lungo was found scattered in vineyards from the Veneto to Friuli-Venezia Giulia under various synonyms: Glera, Tocai Nostrano, Ribolla Spizade (Crespan *et al.* 2003; Crespan *et al.* 2009). The same authors observed a strong genetic relationship between Prosecco and Prosecco Lungo, although a parent–offspring relationship can be ruled out. The non-cultivated Bela Glera (*bela* means 'white' in Slovenian), mistakenly identified as CHASSELAS by Galet (2000), was shown by DNA analysis to be distinct from Glera (Štajner *et al.* 2008), and therefore from Prosecco Lungo and Prosecco, and identical to Pienel (Crespan *et al.* 2011), once cultivated near Udine.

VITICULTURAL CHARACTERISTICS

Mid ripening but slightly earlier than PROSECCO and less vigorous.

WHERE IT'S GROWN AND WHAT ITS WINE TASTES LIKE

Prosecco Lungo is often confused with PROSECCO (Tondo), though wines made from the former tend to be slightly spicier than the latter's wines, which can be more floral, and a small amount of Lungo may be blended with Prosecco in the fermentation vat in the production of Prosecco sparkling wines since the varieties ripen around the same time. Costadila, for example, include some Prosecco Lungo in their single-vineyard San Lorenzo wine.

Its distribution in **Italy** is very similar to that of PROSECCO, although Prosecco Lungo is more sparsely cultivated. Exact figures are hard to come by because the Italian register of varieties, and therefore also the agricultural census, do not distinguish between Prosecco and Prosecco Lungo. Under the synonym Tocai Nostrano, this variety is grown around Breganze in the province of Vicenza in north-east Italy and may be included with VESPAIOLA and other minor local varieties in the sweet, passito-style Torcolato wine within the Breganze DOC. In the Colli di Conegliano DOC, Prosecco Lungo is one of the varieties that may be used to supplement the minimum 85% of Prosecco. Most unusually, Varaschin produce a varietal, still Prosecco Lungo using grapes from San Pietro di Barbozza.

PRUGNOLO GENTILE

See SANGIOVESE.

PRUNELARD

Old, rare, recently rescued, south-west French variety; parent of Cot (Malbec).

BERRY COLOUR: ● ● ● ● ●

PRINCIPAL SYNONYMS: Prunelart

VARIETIES COMMONLY MISTAKEN FOR PRUNELARD: BÉQUIGNOL NOIR (between the Dordogne and the Charente), BOBAL (called Prunelar in the Ariège), CINSAUT (Dordogne and Haute-Garonne), COT (Corrèze)

ORIGINS AND PARENTAGE

Prunelard is an old variety from the Haute-Garonne and the Tarn that was mentioned for the first time in Hardy (1842), and a dozen years later by Victor Rendu (1854) in his *Ampélographie française*: 'les principaux cépages du vignoble de Gaillac sont le Duras, le Taloche, le Muscat et le Prunelard, associés au Mauzac' ('the main grape varieties of the Gaillac vineyards are Duras, Taloche, Muscat and Prunelard, together with Mauzac'). The name Prunelard comes from *prunèl*, the Languedoc dialect word for 'plum', because of the shape and colour of the berries (Rézeau 1997).

Prunelard should not be confused with Prunelat, a synonym of CINSAUT and/or COT in the Gironde, that was mentioned as far back as the sixteenth century (Rézeau 1997), even though they all belong to the Cot ampelographic group (see p XXVII; Bisson 2009). Recent DNA parentage analysis has shown that Cot is a Magdeleine Noire des Charentes × Prunelard cross (see CABERNET SAUVIGNON for the complete pedigree), and that Prunelard has a parent–offspring relationship with MÉRILLE (Boursiquot *et al.* 2009).

VITICULTURAL CHARACTERISTICS

Mid ripening. Quite vigorous but not particularly productive. Very susceptible to mites in the spring but little susceptibility to botrytis bunch rot. Small, compact bunches.

WHERE IT'S GROWN AND WHAT ITS WINE TASTES LIKE

Prunelard was once widely planted from Lavilledieu to Gaillac in south-west **France** and has been recently rediscovered and nursed back to life thanks to Robert and Bernard Plageoles and a few other local backers of the region's historic indigenous varieties. Although Prunelard is recommended only in the Tarn and is not permitted in any appellation wine, the total vineyard area has increased from just over 2 ha (5 acres) in 1998 to 13 ha (32 acres) in 2008. Producers of varietal wines include Domaine Plageoles and Domaine Carcenac; Domaine de la Ramaie produce a blend of 75% Prunelard with 25% Braucol (FER). Wines are deeply coloured, full-bodied yet balanced and well suited to ageing, often with robust alcohol and flavours of ripe plum and spice.

PRUNESTA

Minor Calabrian used mainly in blends.

BERRY COLOUR: ● ● ● ● ●

PRINCIPAL SYNONYMS: Prunesta Nera

ORIGINS AND PARENTAGE

The name Prunesta might derive from the Italian *bruma*, meaning 'mist', referring to the bloom covering the berries, similar to the etymology suggested for NEBBIOLO. Galet (2000) proposed that Bermestia Bianca could be the white mutation of Prunesta Nera but comparison of their DNA profiles proves them to be distinct varieties (Vouillamoz).

Confusingly, a table grape grown in Puglia is also called Prunesta (or Prugesta Pugliese) even though it is a different variety.

OTHER HYPOTHESES

Prunesta is sometimes thought to be the variety Capnia mentioned by Pliny the Elder.

VITICULTURAL CHARACTERISTICS

Mid ripening.

WHERE IT'S GROWN AND WHAT ITS WINE TASTES LIKE

Prunesta is cultivated on the Tyrrenian coast of Calabria, **Italy**, in particular in the province of Reggio Calabria. Although it is authorized as a varietal wine in IGT Val di Neto, it is usually blended with other local varieties, as in Enopolis Costa Viola's Armacìa Rosso (IGT Costa Viola) and Tramontano's Scilla. There were 96 ha (237 acres) planted to this variety in Italy in 2000.

PUGNITELLO

Renascent dark-skinned Tuscan with good quality potential.

BERRY COLOUR: ● ● ● ● ●

VARIETIES COMMONLY MISTAKEN FOR PUGNITELLO: MONTEPULCIANO ✗

ORIGINS AND PARENTAGE

Pugnitello is a rare variety of unknown origin rescued in 1981 by researchers at the University of Firenze from old vines that were thriving in Poggi del Sasso between Grosseto and Montalcino in Toscana, central Italy. The name means 'small fist', from the Italian *pugno*, meaning 'fist', referring to the shape of the bunches. Pugnitello was at first thought to be a clonal variation of MONTEPULCIANO but DNA profiling has disproved this theory (Vignani *et al.* 2008).

VITICULTURAL CHARACTERISTICS

Vigorous but not very productive because of the small bunches, which produce musts rich in sugar, anthocyanins and tannins.

WHERE IT'S GROWN AND WHAT ITS WINE TASTES LIKE

In the mid 1980s, the rescuers from the University of Firenze, **Italy**, collaborated with the producer San Felice based in Castelnuovo Berardenga in Siena. The variety was planted in the latter's experimental vineyard and the results were so encouraging that 1,000 SANGIOVESE vines were top-grafted with Pugnitello. The first three barrels were made in 1993 and the San Felice plantings have grown to 10 ha (25 acres) in Chianti, 1 ha (2.5 acres) in Montalcino and 3 ha (7 acres) in the Maremma. The variety was eventually included in the Italian official register of varieties in 2002 and San Felice's first commercial bottling was made in 2003, classified as IGT Toscana. Since then, other producers have recognized the potential of this variety, including Fattoria Santa Vittoria, Poggio al Gello and Mannucci Droandi. The wines are deeply coloured with rich fruit flavours, firm yet smooth tannins and fresh acidity and benefit from barrel ageing. San Felice's bottling bears some resemblance to a Rioja Riserva but with fresher acidity and firmer tannins.

PUKHLIAKOVSKY

Russian variety with erratic yields producing light wines.

BERRY COLOUR: ● ● ● ● ●

PRINCIPAL SYNONYMS: Coarnă Alba (Romania), Kecskecsecsu (Hungary), Korna Belaja, Majorka Belaja, Puchljakovski (Austria)

ORIGINS AND PARENTAGE

Pukhliakovsky probably originates from the Rostov wine region in south-western Russia, near the border with Ukraine, and is probably named after the village of Pukhliakovskaya. The black-berried table grape known as Pukhliakovsky Cherny is not a colour mutation but a distinct variety.

OTHER HYPOTHESES

Like other varieties along the River Don, Pukhliakovsky is said to have been introduced from Dagestan in the Middle Ages (Vladimir Tsapelik, personal communication).

VITICULTURAL CHARACTERISTICS

Vigorous, high but irregular yields due to functionally female flowers – needs to be planted alongside a hermaphrodite variety to achieve pollination. Early budding, late ripening and needs a warm site. Vines must be earthed up from early November to protect them from cold winter temperatures. Good resistance to downy mildew

WHERE IT'S GROWN AND WHAT ITS WINE TASTES LIKE

Pukhliakovsky is grown mainly in the Rostov wine region along the River Don in **Russia**, especially in the village of Pukhliakovskaya. Wines are relatively light and low in alcohol. The grapes are often grown for the table.

MOLAR

see NEGRAMOLL

pp 708–10

NOAH

pp 731 – 2

PEDRO XIMÉNEZ

pp 776–7

PERSAN

pp 783–4

PETIT MANSENG

pp 787–8

ORIOU

see PETIT ROUGE

pp 789–90

PIEDIROSSO

p 800

PINOT NOIR

see PINOT

pp 805 – 24

RÄUSCHLING

pp 873–4

RÈZE

pp 884–5

RIESLING

pp 888–93

RKATSITELI

pp 897 – 8

ROUSSANNE

pp 920–21

SAN GIOVETO

see SANGIOVESE

pp 942–6

SAPERAVI

pp 949–50

SAUVIGNON BLANC

pp 952–8

PULES

Minor Albanian white.

BERRY COLOUR: ● ● ● ● ●

PRINCIPAL SYNONYMS: Pulës, Puls

ORIGINS AND PARENTAGE

Pules most likely originates in the Berat region in south-central Albania but its ampelographic identity is still obscure because two reference vines called Pules i Bylyshit, sampled in Tepelene in southern Albania, turned out to have distinct and unique DNA profiles in Ladoukakis *et al.* (2005) and it is not known which of the two is the true-to-type Pules.

WHERE IT'S GROWN

Found only in its region of origin in south-central **Albania** and traditionally used to make raki and trained up trees, Pules has recently been replanted and used to make wine. Çobo, for example, make a varietal version.

PX

See PEDRO XIMÉNEZ.

Q

QUAGLIANO
QUEBRANTA

⚭this symbol on the following pages indicates that the synonyms, and mistaken identities, are confirmed by DNA evidence

QUAGLIANO

Highly localized Piemontese making light, sweet, perfumed reds.

BERRY COLOUR: ● ● ● ● ●

ORIGINS AND PARENTAGE

Quagliano is an old variety from the region of Saluzzo in Piemonte, north-west Italy, first mentioned in 1721 in the commune of Busca, about 60 km south west of Torino, and then in 1749 in Costigliole Saluzzo, a few kilometres north of Busca. Recent DNA parentage analysis has suggested, very unexpectedly and questionably, that Quagliano is a possible parent of IMPIGNO from Puglia (Cipriani *et al.* 2010).

VITICULTURAL CHARACTERISTICS

Vigorous, late ripening and susceptible to botrytis bunch rot.

WHERE IT'S GROWN AND WHAT ITS WINE TASTES LIKE

Quagliano is grown only in the province of Cuneo in Piemonte, north-west **Italy**, mainly in its region of origin between Busca and Costigliole, but also in the communes of Pagno, Piasco, Manta, Verzulo, Brondello, Castellar and Saluzzo. Quagliano's local fame is illustrated by the eco-museum at the Poderi del Palas di Chiotti Diego and by the local Sagra dell'Uva Quagliano festival, which has taken place in Costigliole Saluzzo every September since 1928. However, there were fewer than 10 ha (25 acres) of these vines in Italy in 2000, according to the agricultural census.

Interesting varietal dessert wines, in the style of Brachetto (see BRACHETTO DEL PIEMONTE), with aromas of violets and roses and a sweet taste of strawberries, are produced by the likes of Tomatis Dario and Giordanino Teresio. Poderi del Palas di Chiotti Diego make a low-alcohol (7–8%) sparkling version. All are within the Colline Saluzzesi DOC.

QUEBRANTA

Also called Quebranta Peruana, this is a traditional red-skinned variety from **Peru**, where it is grown mainly to produce pisco, the national grape spirit, but also for rosé or red table wines at varying sweetness levels for domestic consumption. Recent DNA parentage analysis (This *et al.* 2006) has shown that Quebranta is a natural cross between LISTÁN PRIETO (under the name Mission) and NEGRAMOLL (under the name Negra Mole), which makes it a half-sibling of the other South American varieties CEREZA, TORRONTÉS SANJUANINO and TORRONTÉS RIOJANO (see MUSCAT pedigree diagram).

R

RABIGATO
RABO DE OVELHA
RABOSO PIAVE
RABOSO VERONESE
RAC 3209
RADISSON
RAFFIAT DE MONCADE
RAMISCO
RANAC BIJELI
RANFOL
RANNA MELNISHKA LOZA
RASPIROSSO
RASTAJOLA
RÁTHAY
RÄUSCHLING
RAVAT BLANC
RAYON D'OR
REBERGER
REBO
RECANTINA
REFOSCO DAL PEDUNCOLO ROSSO
REFOSCO DI FAEDIS
REGENT
REGNER
REICHENSTEINER
RÉSELLE
RETAGLIADO BIANCO
RÈZE
RIBOLLA GIALLA
RIESEL
RIESLANER
RIESLING
RIESUS
RIMAVA
RIO GRANDE
RIÒN
RIPOLO
RITINO
RIVAIRENC
RKATSITELI
ROBOLA
RODITIS
ROESLER
ROKANIARIS
ROLLO
ROMÉ
ROMEIKO
ROMERO DE HÍJAR
ROMORANTIN
RONDINELLA

ℵthis symbol on the following pages indicates that the synonyms, and mistaken identities, are confirmed by DNA evidence

RONDO
ROSÉ DU VAR
ROSE HONEY
ROSETTA
ROSETTE
ROSSARA TRENTINA
ROSSESE BIANCO
ROSSESE BIANCO DI MONFORTE
ROSSESE BIANCO DI SAN BIAGIO
ROSSESE DI CAMPOCHIESA
ROSSETTO
ROSSIGNOLA
ROSSOLA NERA
ROSSOLINO NERO
ROTBERGER
ROTER VELTLINER
ROTGIPFLER
ROUGE DE FULLY
ROUGE DU PAYS
ROUSSANNE
ROUSSETTE D'AYZE
ROUSSIN
ROVELLO BIANCO
ROYAL DE ALLOZA
ROYALTY
RUBIN
RUBIN GOLODRIGI
RUBIN TAIROVSKY
RUBINET
RUBINKA
RUBINOVY MAGARACHA
RUBINTOS
RUBIRED
RUBY CABERNET
RUCHÈ
RUDAVA
RUEN
RUFETE
RUGGINE
RUZZESE
RYUGAN

RABIGATO

Potentially high-quality, high-acid Douro variety.

BERRY COLOUR:

PRINCIPAL SYNONYMS: Baldsena, Carrega Besta, Estreito, Muscatel Bravo, Não Há, Puesta en Cruz (Arribes del Duero in Spain), Rabigato Respigueiro, Rabo de Asno, Rabo de Carneiro (Minho), Rabo de Gato (Minho), Rodrigo Affonso

VARIETIES COMMONLY MISTAKEN FOR RABIGATO: DONZELINHO BRANCO, MALVASIA FINA, RABO DE OVELHA

ORIGINS AND PARENTAGE

Rabigato comes from the Douro in northern Portugal, where it is grown almost exclusively (Rolando Faustino, personal communication). It was first mentioned as Rabigato Respigueiro in Trás-os-Montes in 1711 by Vicêncio Alarte in his 'Agricultura das vinhas' (Galet 2000). The name Rabigato is a contraction of one of its synonyms, Rabo de Gato, meaning 'cat's tail' (Galet 2000), referring to its long clusters. Other synonyms make similar references: Rabo de Asno means 'donkey's tail' and Rabo de Carneiro means 'sheep's tail'. Recent DNA studies have suggested a possible genetic link between Rabigato and TINTA FRANCISCA (Castro *et al.* 2011).

Rabigato from the Douro should not be confused with RABO DE OVELHA, which is also known as Rabigato in the Minho in north-western Portugal. Although the names are confusing, they are quite easy to distinguish in the vineyard because their leaves are so different.

VITICULTURAL CHARACTERISTICS

Early budding and early to mid ripening. Thin-skinned berries. Susceptible to downy and powdery mildews and botrytis bunch rot.

WHERE IT'S GROWN AND WHAT ITS WINE TASTES LIKE

Rabigato is widely grown in the Douro, particularly in the Douro Superior in the east of the region, but is rarely made as a varietal wine. Such wines tend to be fresh, vibrant, fairly high in alcohol and very high in acidity with aromas of lemon and orange blossom as well as having vegetal or mineral notes. When carefully made, the wines have impressive length and are worth ageing in bottle. There were 2,542 ha (6,281 acres) in **Portugal** in 2010. Given the variety's high acidity, it is often blended with lower-acid varieties such as FERNÃO PIRES, SÍRIA, Gouveio (GODELLO) and VIOSINHO. Varietal wines are, unusually, made by Dona Berta, CARM, Muxagat and Quinta de S José. It is also used in the production of port.

RABO DE OVELHA

Undemonstrative Portuguese variety making whites for early drinking.

BERRY COLOUR:

PRINCIPAL SYNONYMS: Fernan Piriz (Spain), Rabo de Ovella

VARIETIES COMMONLY MISTAKEN FOR RABO DE OVELHA: RABIGATO (Douro)

ORIGINS AND PARENTAGE

Rabo de Ovelha, meaning 'ewe's tail' and referring to the shape of the bunch, probably originates in the Alentejo region in southern Portugal, where the greatest genetic diversity has been observed (Lopes *et al.* 2006). It is often confused with RABIGATO from the Douro but their DNA profiles in Almadanim *et al.* (2007) are distinct, and it is not a colour mutation of Rabo de Ovelha Tinto, which is a synonym for NEGRAMOLL (Martín *et al.* 2006; Veloso *et al.* 2010).

DNA parentage analysis has shown that Rabo de Ovelha has a likely parent–offspring relationship with CAYETANA BLANCA from the Iberian peninsula (Zinelabidine *et al.* 2012; see Cayetana Blanca for the pedigree diagram).

VITICULTURAL CHARACTERISTICS

Vigorous, productive, mid budding, mid to late ripening. Large, compact bunches of thick-skinned berries. Susceptible to water stress, powdery mildew and botrytis bunch rot. Resistant to coulure.

WHERE IT'S GROWN AND WHAT ITS WINE TASTES LIKE

Widely planted in the Alentejo in southern **Portugal** and also found further north in Beiras, Rabo de Ovelha produces lightly aromatic wines without great intensity, usually blended with other varieties and best drunk young. J Portugal Ramos' Loios is a blend of Rabo de Ovelha with Roupeiro (SÍRIA), and Carmim (the Reguengos de Monsaraz co-operative) include it in several white blends. There were 957 ha (2,365 acres) in Portugal in 2010.

RABOSO PIAVE

Often tough Veneto variety needing careful handling and long ageing to produce high-quality wine. Parent of Raboso Veronese.

BERRY COLOUR: ● ● ● ● ●

PRINCIPAL SYNONYMS: Friulara di Bagnoli (Padova), Friularo (Padova), Raboso del Piave, Raboso Nostrano

VARIETIES COMMONLY MISTAKEN FOR RABOSO PIAVE: GROPPELLO GENTILE, RABOSO VERONESE

ORIGINS AND PARENTAGE

Raboso Piave and RABOSO VERONESE are very similar old varieties from the Veneto in northern Italy, where they have often been confused. However, morphological and DNA profiling has now clearly established that they are distinct varieties (Crespan, Cancellier *et al.* 2006). Recent DNA profiling has suggested a possible parent–offspring relationship with FOGARINA (Myles *et al.* 2011).

Raboso Piave most likely originates from the plains of the River Piave in the Veneto north of Venezia. The name Raboso could derive from the local dialect *rabioso*, used to describe unripe fruit, possibly in reference to the variety's astringency. Alternatively, Raboso could have been named after a tributary of the River Piave, although no Raboso vines have ever been found there. In the province of Padova, Raboso Piave is called Friulara or Friularo, probably because this grape was introduced from Friuli in the seventeenth or eighteenth century by the De Vidiman family, who owned a villa in Bagnoli. However, this does not necessarily signify a Friulian origin for Raboso Piave because at that time the River Piave marked the Friuli boundary.

DNA profiling showed unexpectedly that Raboso Friularo is not a synonym of Raboso Piave but of RABOSO VERONESE (Salmaso *et al.* 2008). Raboso Piave, already mentioned in 1679, is much older than Raboso Veronese, which was not mentioned until the nineteenth century (Calò, Francini *et al.* 2008). This is consistent with the results of DNA parentage analysis, which show that Raboso Veronese is a Raboso Piave × MARZEMINA BIANCA cross (Crespan, Cancellier *et al.* 2006; see REFOSCO DAL PEDUNCOLO ROSSO and PINOT pedigree diagrams).

GRAPARIOL, also called Rabosina Bianca, is a rare variety from the province of Treviso that was often believed to be a white mutation of Raboso Piave but recent DNA profiling has showed there is no direct relationship between Grapariol and either Raboso Piave or RABOSO VERONESE (Crespan *et al.* 2009). Finally, it should also be noted that MANZONI MOSCATO is a Raboso Piave × MUSCAT OF HAMBURG cross.

OTHER HYPOTHESES

Raboso Piave is supposed to have been domesticated from local wild vines and was supposedly mentioned in the thirteenth century under the name Vinum Plavense. There is no evidence for these hypotheses. It has also been suggested that the name derives from the Italian word *rabbioso*, meaning 'angry', presumably a reference to consumer reactions at tasting a wine with such high acidity and rough tannins.

VITICULTURAL CHARACTERISTICS

Robust and vigorous vine. Late ripening (end of October). High tannin content. Resistant to downy mildew, botrytis bunch rot, sour rot and esca but susceptible to powdery mildew.

WHERE IT'S GROWN AND WHAT ITS WINE TASTES LIKE

Raboso Piave has a reputation for astringent tannins and high acidity, though proponents claim that this reputation is partly because the grapes were in the past not fully ripe at harvest and that the grape can be tamed in the winery and by extended bottle ageing to produce high-quality wines. It is planted right across the Veneto, northern **Italy**, often interplanted with the less common RABOSO VERONESE. Of all the Raboso plantings today, about 90% are Raboso Piave and 10% Raboso Veronese. About 80% of the Raboso Piave vineyards are in the province of Treviso, the rest in the provinces of Padova, in particular in the neighbourhood of Bagnoli di Sopra, where it is known under the name Friularo, and Venezia, between the Piave and Livenza rivers.

In the 1950s and 1960s, Raboso Piave plantings decreased in favour of softer and easier-to-sell international varieties such as MERLOT and CABERNET FRANC but the Confraternita del Raboso Piave (Brotherhood of Raboso Piave), established in 1996 at the instigation of Cecchetto (which has the biggest vineyard holdings of this variety) and other local producers, is making strenous efforts to safeguard and promote the variety. There were 1,330 ha (3,287 acres) planted in Italy in 2000, according to the agricultural census, but the total had shrunk to around 1,100 ha (2,718 acres) by 2008.

Today Raboso Piave is used to make not only deeply coloured, firmly tannic reds but also rosé and sparkling wines, both dry and sweet. Varietal wines are authorized in DOCs such as Bagnoli di Sopra, Corti Benedettine del Padovano and Riviera del Brenta (labelled Friularo in the first). Although the tannins can be aggressive, emphasized by the naturally high acidity (useful as a minor component in blends), improved winemaking and maturation are showing this variety's real potential. For example, Cecchetto's full-bodied and ageworthy Amarone-style Gelsaia has complex aromas and flavours of ripe black fruits, liqueur de cassis, almonds and fresh mint, and although the tannins are firm, they are ripe and in balance with the fruit and the freshness on the finish. Villa Sandi make an interesting, lightly sweet, late-harvest passito-style wine. Other recommended producers include Bonotto delle Tezze, Casa Roma and Italo Cescon.

RABOSO VERONESE

Makes tannic Veneto red, often used for sparkling wines, and often confused with its parent, the more common Raboso Piave.

BERRY COLOUR: ● ● ● ● ●

PRINCIPAL SYNONYMS: Rabosa, Raboso di Verona, Raboso Friularo

VARIETIES COMMONLY MISTAKEN FOR RABOSO VERONESE: GRUAJA, NEGRARA VERONESE, RABOSO PIAVE

ORIGINS AND PARENTAGE

Raboso Veronese and RABOSO PIAVE are very similar old varieties from the Veneto, where they have often been confused. However, morphological and DNA profiling has now clearly established that they are distinct varieties (Crespan, Cancellier *et al.* 2006). Raboso Veronese was first mentioned in the province of Treviso in the nineteenth century whereas Raboso Piave was already mentioned in 1679 (Calò, Francini *et al.* 2008).

As Raboso Veronese is – surprisingly – not cultivated in the province of Verona today, its name was long thought to derive from a man named Veronesi who was responsible for the spread of the variety. However, DNA parentage analysis has revealed that Raboso Veronese is a RABOSO PIAVE × MARZEMINA BIANCA cross that probably occurred at the Azienda dei Conti Papadopoli in Cologna Veneta (Crespan, Cancellier *et al.* 2006), a village in the province of Verona (see REFOSCO DAL PEDUNCOLO ROSSO pedigree diagram). This recent discovery points again to Verona as the birthplace of Raboso Veronese, suggesting its name is fully deserved geographically.

Since MARZEMINA BIANCA is likely to be a natural cross between MARZEMINO and GARGANEGA (Salmaso *et al.* 2008; Vouillamoz), two important and historical northern Italian varieties, Raboso Veronese is their probable grandchild. Remarkably, DNA profiling showed that Raboso Friularo is a synonym for Raboso Veronese, and Friularo a synonym for RABOSO PIAVE (Salmaso *et al.* 2008). It should also be noted that FERTILIA is a MERLOT × Raboso Veronese cross.

VITICULTURAL CHARACTERISTICS

Robust, vigorous and late ripening; productive and consistent yields. Susceptible to powdery mildew.

WHERE IT'S GROWN AND WHAT ITS WINE TASTES LIKE

Raboso Veronese is cultivated mainly in the south east of the Veneto, **Italy**, in the provinces of Treviso, Venezia, Vicenza and Rovigo, and to a lesser extent in the provinces of Ferrara and Ravenna in Emilia-Romagna and in Friuli. Raboso Veronese is often interplanted with RABOSO PIAVE. The Italian agricultural census of 2000 recorded total plantings of 340 ha (840 acres).

Raboso Veronese is authorized in the sparkling, rosé and passito versions of the Bagnoli di Sopra DOC and in Colli Euganei DOC; a varietal wine, labelled simply Raboso, may be made within the Vicenza DOC. Sparkling rosé wines are not uncommon, presumably because of the high acidity and sweet fruit flavours. The wines tend to be less tannic than those made from RABOSO PIAVE although the two varieties are often blended since they may well be interplanted and are still widely confused, despite the DNA evidence of their separate identities.

In 2008 there were also limited plantings of Raboso Veronese of around 50 ha (124 acres) in **Argentina**, divided between the Mendoza and San Juan wine regions.

RAC 3209

Rare Swiss cross producing deeply coloured wines.

BERRY COLOUR: ● ● ● ● ●

ORIGINS AND PARENTAGE

RAC 3209 was bred from undisclosed parents at the Caudoz viticultural research centre in Pully, just outside Lausanne, Switzerland, today part of Agroscope Changins-Wädenswil, formerly called RAC (Recherche Agronomique Changins), hence the name.

VITICULTURAL CHARACTERISTICS

Early to mid ripening. Good general disease resistance but susceptible to downy mildew.

WHERE IT'S GROWN

Weingut zum Frohhof in the canton of Zürich, **Switzerland**, have a tiny parcel of RAC 3209. They value its deep colour and include it in blends.

RACHULI TETRA

See TSULUKIDZIS TETRA.

RADISSON

Another dark-skinned American hybrid that has found its way to Québec, although its exact identity is unclear.

BERRY COLOUR: ● ● ● ● ●

PRINCIPAL SYNONYMS: ES 5-17

ORIGINS AND PARENTAGE

Radisson is said to be a seedling of an Elmer Swenson 593 vine obtained by Elmer Swenson in Osceola, Wisconsin, US, in which Elmer Swenson 593 is an Elmer Swenson 80 × VILLARD BLANC hybrid (see BRIANNA for their pedigrees). However, what is cultivated in Canada under the name ES 5-17 or Radisson might not be the original ES 5-17 with female-only flowers but a distinct and unknown variety with complete male and female flowers (Hart nd).

WHERE IT'S GROWN

Like several other Swenson hybrids, this seems to be more widely planted in Québec, **Canada**, than in the American Midwest. Several producers, including Domaine de la Source à Marguerite, use it in blends with varieties such as MARÉCHAL FOCH, FRONTENAC and SABREVOIS.

RAFFIAT DE MONCADE

Rare south-west French vine often confused with Arrufiac.

BERRY COLOUR: ● ● ● ● ●

PRINCIPAL SYNONYMS: Arréfiat (Béarn), Arrufiat, Raffiat (Béarn), Refiat (Jurançon), Ruffiac

VARIETIES COMMONLY MISTAKEN FOR RAFFIAT DE MONCADE: ARRUFIAC (called Raffiat in Béarn and the Landes)

ORIGINS AND PARENTAGE

Raffiat de Moncade is an old variety from Béarn in the Pyrénées-Atlantiques and from the Landes (mainly Orthez, Salies and Lagor) in south-west France, where it shares several synonyms with ARRUFIAC, with which it has often been confused. The first reliable reference to Raffiat de Moncade was probably the one made by Jean-Alexandre Cavoleau in 1827 (Rézeau 1997): 'dans les Basses-Pyrénées, les vins blancs de Jurançon sont produits par le refiat, le menseing gros et petit, le claverie, l'aulhan, le courtoisie' ('in the Basse-Pyrénées, white wines from Jurançon are produced by refiat . . . ' etc.). Raffiat de Moncade is an offspring of GOUAIS BLANC (Boursiquot *et al.* 2004; see PINOT pedigree diagram).

Like ARROUYA and ARRUFIAC, Raffiat could derive from *arro(u)y*, meaning 'red' in the Béarnese dialect, while *rufe* means 'rough'. The name Moncade is a reference to an old and noble family from Béarn whose name was also given to the 'Tour Moncade', built in 1242 by Gaston VII de Moncade in Orthez, the historical location of the Raffiat de Moncade vineyard.

Raffiat de Moncade was used to breed ARRILOBA.

VITICULTURAL CHARACTERISTICS

Early budding, mid ripening. Relatively productive with loose bunches of very small berries. Shows little susceptibility to powdery mildew.

WHERE IT'S GROWN AND WHAT ITS WINE TASTES LIKE

Even though Raffiat de Moncade is recommended in the Gers, Landes and Pyrénées-Atlantiques *départements* and is one of the principal varieties officially authorized for the Béarn appellation, total plantings in **France** in 2008 were just 9 ha (22 acres), down from 86 ha (213 acres) in 1979. Wines tend to be rather neutral and alcoholic. Pascal Lapeyre of Domaine Guilhemas produces one of the very rare examples of varietal Raffiat, which he describes as rounded with aromas of pear, apple and flowers.

RAISIN BLANC

See SERVANT.

RAJINSKI RIESLING

See RIESLING.

RAMANDOLO

See VERDUZZO FRIULANO.

RAMISCO

Demanding and endangered Portuguese variety that produces wines aggressive in youth but elegant after considerable time in bottle.

BERRY COLOUR:

PRINCIPAL SYNONYMS: Ramisco nos Açores

ORIGINS AND PARENTAGE

Ramisco originates in Portugal's Colares region, north east of Lisboa. According to recent DNA studies, TRINCADEIRA from the Alentejo and SERCIAL from Lisboa could be genetically close to Ramisco (Almadanim *et al.* 2007).

VITICULTURAL CHARACTERISTICS

Mostly ungrafted vines planted in sand over clay. Late budding and ripening. Small, thick-skinned berries.

WHERE IT'S GROWN AND WHAT ITS WINE TASTES LIKE

Ramisco is *the* variety of the Colares DO, the threatened, Atlantic-whipped sandy vineyards on the coast north west of Lisboa, **Portugal**. They are challenged not only by the city's encroachment but also because the variety is difficult to grow since it has to be planted in trenches several metres deep to anchor the plant in the clay soils below the sand, yields are low and the wines are generally approachable only after several years in bottle thanks to high acidity and powerful tannins. The sand does, however, mean that the vines can be planted on their own roots since phylloxera cannot survive here. It is hardly surprising that there were just 23 ha (57 acres) of Ramisco remaining in 2010 but it is a shame since these rarefied wines, once aged, can become elegant and aromatic – a complex mix of flavours that include fresh meat, mushrooms, wet earth and cedar. Alcohol is generally low at around 11 or 11.5% (Rolando Faustino, personal communication). Recommended producers include Monte Cascas, Quinta das Vinhas de Areia and Paulo da Silva.

RANAC BIJELI

Almost extinct, intensely flavoured Croatian island variety.

BERRY COLOUR:

PRINCIPAL SYNONYMS: Ranac Silbijanski Bijeli, Silbijanac Bijeli

ORIGINS AND PARENTAGE

Ranac Bijeli, literally 'early white', is an almost extinct variety from the islands of Pag and Silba (hence its synonyms), north of Zadar, in Croatia.

VITICULTURAL CHARACTERISTICS

Early ripening and high sugar level in the berries, therefore at risk from wasps and children.

WHERE IT'S GROWN AND WHAT ITS WINE TASTES LIKE

A few vines of Ranac Bijeli can still be found in old vineyards on the islands of Pag and Silba in the northern Adriatic. It is currently part of a revitalization project involving clonal selection and replanting of autochthonous **Croatian** varieties. Although it has quality potential, its main advantage of full flavour and early ripening also make it unpopular with growers as they lose so much of the crop to sweet-toothed wasps.

RANFOL

Variety with Slovenian origins making humdrum whites there and in Croatia.

BERRY COLOUR: ● ● ● ● ●

PRINCIPAL SYNONYMS: Belina Pleterje (Slovenia), Plavis, Praskava Belina, Ranfol Beli, Ranfol Bijeli (Croatia), Sremska Lipovina, Štajerska Belina, Urbanka, Vrbanka

VARIETIES COMMONLY MISTAKEN FOR RANFOL: GOUAIS BLANC

ORIGINS AND PARENTAGE

This variety probably originates in Štajerska Slovenija (Slovenian Styria) or possibly from the north of neighbouring Croatia. It should not be confused with GOUAIS BLANC, of which it is a possible offspring (Štajner *et al.* 2008). The common synonym Štajerska Belina simply means 'Styria white'.

VITICULTURAL CHARACTERISTICS

Productive. Large, compact bunches.

WHERE IT'S GROWN AND WHAT ITS WINE TASTES LIKE

Ranfol produces light and relatively neutral whites. There were 210 ha (519 acres) in **Croatia** in 2009, mainly in the Slavonija, Moslavina, Prigorje-Biogora and Pokupjle wine regions in Kontinentalna Hrvatska (Continental Croatia). Plantings in **Slovenia** are smaller still (36 ha/89 acres in 2009), divided fairly evenly between the wine regions of Podravje and Posavje in the east of the country. Recommended producers there include Ptujska Klet.

RANNA MELNISHKA LOZA

Bulgarian cross with good potential but lacking recognition.

BERRY COLOUR: ● ● ● ● ●

PRINCIPAL SYNONYMS: Early Melnik, Melnik 55, Melnishka Ranna, Ranna Melnishka

VARIETIES COMMONLY MISTAKEN FOR RANNA MELNISHKA LOZA: MELNIK 82, SHIROKA MELNISHKA

ORIGINS AND PARENTAGE

Ranna Melnishka Loza, literally 'early Melnik grape', is a SHIROKA MELNISHKA × VALDIGUIÉ cross obtained in 1963 by Yane Atanasov and his colleagues Z Zankov and G Karamitev at the Sandanski research centre (now Sintica Winery) in south-west Bulgaria. The breeders made a series of crossings using a mixture of pollens from the three French varieties DURIF, JURANÇON NOIR and Valdiguié to create an early-ripening variety with the characteristics of Shiroka Melnishka. The seedling that was initially named Melnik 55 was later officially christened Ranna Melnishka Loza. DNA parentage analysis has finally established that Valdiguié was the other parent of sibling varieties Ranna Melnishka Loza and MELNIK 82 (Hvarleva, Russanov *et al.* 2005). It was not commercially vinified until the late 1990s.

VITICULTURAL CHARACTERISTICS

Mid ripening, fertile and productive. Susceptible to downy and powdery mildews but virtually resistant to botrytis bunch rot. Relatively compact bunches of thick-skinned berries.

WHERE IT'S GROWN AND WHAT ITS WINE TASTES LIKE

Although there were 546 ha (1,349 acres) of Ranna Melnishka Loza in **Bulgaria** in 2009, almost all in the Struma Valley in the far south west of the country, the variety struggles for recognition because it is often thought to be the same as SHIROKA MELNISHKA and both are often referred to simply as Melnik. Logodaj produce both red and rosé varietal wines, Sintica an oak-aged red. The wines are said to have sweet and sour cherry flavours and a note of spice, with rounded tannins, respectable levels of alcohol and fresh acidity.

RARĂ NEAGRĂ

See BĂBEASCĂ NEAGRĂ.

RASPIROSSO

Also known as Raspi Rosso or Raspo Rosso, a dark-skinned variety from Toscano, **Italy**, used to add colour to SANGIOVESE-based wines.

RASTAJOLA

Minor variety from northern Piemonte encountered in blends.

BERRY COLOUR: ● ● ● ● ●

PRINCIPAL SYNONYMS: Durera (Novara), Restajola

VARIETIES COMMONLY MISTAKEN FOR RASTAJOLA: Erbaluce Nero (Canavese), NERETTO DURO ⌘ (or Durasa in Novara)

ORIGINS AND PARENTAGE

Rastajola probably comes from northern Piemonte, where it was once widespread. It is often confused with NERETTO DURO, or with Erbaluce Nero in the Canavese in the province of Torino

(Schneider and Mannini 2006), a non-cultivated variety that is not a colour mutation of ERBALUCE.

VITICULTURAL CHARACTERISTICS

Mid ripening. Robust vine, susceptible to millerandage and downy mildew. Vigorous but with slightly erratic yields.

WHERE IT'S GROWN

Rastajola is cultivated here and there in the communes of Sizzano, Ghemme, Fara and Romagnano Sesia in the province of Novara in north-east Piemonte, north-west **Italy**. It is nearly always blended with other local varieties such as VESPOLINA and NEBBIOLO.

RÁTHAY

Rare, disease-resistant Austrian hybrid making wines for early drinking.

BERRY COLOUR: ● ● ● ● ●

PRINCIPAL SYNONYMS: Klosterneuburg 1355-3-33, Ráthay Noir

ORIGINS AND PARENTAGE

A Klosterneuburg 1189-9-77 × BLAUBURGER hybrid obtained in 1970 by Gertrude Mayer at the Klosterneuburg research centre in Austria (see ROESLER for the pedigree of Klosterneuburg 1189-9-77), making it a half-sibling of Roesler. Ráthay was named after Emerich Ráthay (1845–1900), second director of Klosterneuburg.

VITICULTURAL CHARACTERISTICS

Good resistance to downy mildew.

WHERE IT'S GROWN AND WHAT ITS WINE TASTES LIKE

Ráthay is planted mainly in Burgenland (total area *c.* 10 ha/25 acres) and produces soft, early-drinking reds. Varietal wines are made by Trabauer (Niederösterreich/Lower Austria), Taurot (Burgenland) and Edenhof (Weinviertel), for example.

RÄUSCHLING

Historic, rare German variety experiencing a small revival in Switzerland.

BERRY COLOUR: ● ● ● ● ●

PRINCIPAL SYNONYMS: Dretsch or Drutsch or Drutscht (Rheinland-Pfalz), Grosser Räuschling (north-eastern France), Klöpfer (Rheinland-Pfalz), Offenburger (Rheinland-Pfalz), Räuschling Weiss, Reuschling, Ruchelin (France), Weisser Räuschling, Zürirebe (northern Switzerland), Züriwiss (northern Switzerland)

VARIETIES COMMONLY MISTAKEN FOR RÄUSCHLING: COMPLETER (Zürich), KNIPPERLÉ (Alsace), RIESLING (Rheinland-Pfalz and Württemberg)

ORIGINS AND PARENTAGE

Räuschling is a very old variety from the Rheintal (Rhine Valley) in Germany that was grown in the Middle Ages in what is now Rheinland-Pfalz and Württemberg in southern Germany, Alsace in north-eastern France, as well as northern and western Switzerland. According to Aeberhard (2005), the earliest mention of this variety appeared under its old synonym Drutsch(t) in Hieronymus Bock's (1546) *Kreutterbuch* in the region of Landau in Rheinland-Pfalz. A clearer reference to the name Raüschling is found in Franken in a request from Graf Philip Ernest von Hohenlohe-Langenburg, dated 12 December 1614, to replace GOUAIS BLANC (Hünnisch) by a selection of better varieties including Reuschling and ELBLING (Dornfeld 1868). The modern spelling was first mentioned by Michael Sorg in 1759 in Schaffhausen, along with its synonym Zürirebe, meaning 'grape from Zürich' (Aeberhard 2005). The name Räuschling could derive from *rauschen*, referring to the sound of the wind going through the dense foliage of this variety, or alternatively from *Russling*, *Rus* meaning 'dark wood'.

A preliminary DNA parentage analysis tentatively suggesting that Räuschling could be a natural GOUAIS BLANC × SAVAGNIN cross (Maul 2006), just like its siblings PETIT MESLIER and AUBIN BLANC in northern France (see PINOT pedigree diagram), has been confirmed by the analysis of fifty DNA markers (Vouillamoz). This parentage explains why Räuschling was sometimes erroneously called Grosser Traminer (Traminer is the name used for Savagnin in German-speaking regions), and why it was called Grosser Räuschling in north-eastern France to distinguish it from Kleiner Räuschling, a synonym for KNIPPERLÉ in Alsace, which is also a progeny of Gouais Blanc.

A red-berried **Räuschling Rot**, not to be confused with Blauer Räuschling or Gelbhölzer, yet another offspring of Gouais Blanc (Erika Maul, personal communication), was often mentioned in eighteenth- and nineteenth-century literature but was thought to be extinct until it was found in a Swiss grape collection, showing the same DNA profile as Räuschling, making it a colour mutation (Vouillamoz, Frei and Arnold 2008).

Although Räuschling shares the synonym Zürirebe with COMPLETER, the two varieties are quite distinct.

VITICULTURAL CHARACTERISTICS

Mid budding, early to mid ripening. Productive if pruned long but subject to coulure and millerandage. Compact bunches of small berries with a strong tendency to split near harvest and contract botrytis bunch rot.

WHERE IT'S GROWN AND WHAT ITS WINE TASTES LIKE

Räuschling has now virtually disappeared from Alsace and Germany and is found only in German-speaking **Switzerland** (23 ha/57 acres in 2009), especially in the canton of Zürich in the north east. It produces light wines with citrus notes and typically high acidity. One of the best and most ageworthy Räuschlings is made by Hermann Schwarzenbach by the Zürichsee (Lake Zürich). Other notable producers in Switzerland include Erich Meier in Zürich and Weinstamm in Schaffhausen.

RAVAT BLANC

Mediocre light-skinned Mâconnais hybrid on its way out.

BERRY COLOUR:

PRINCIPAL SYNONYMS: Rava 6 or Ravat 6, Ravat Chardonnay 6, Ravat Chardonnay B

ORIGINS AND PARENTAGE

A complex hybrid obtained in the 1930s by J F Ravat in Marcigny, eastern central France, by crossing Seibel 5474 (or Seibel 8724 according to some authors) × CHARDONNAY, where Seibel 5474 is a Seibel 405 × Seibel 867 cross, and Seibel 867 is a Vivarais × NOAH cross (see PRIOR for the complete Seibel 405 and Vivarais pedigrees).

VITICULTURAL CHARACTERISTICS

Early budding, mid ripening. Not very fertile and best pruned long. Good resistance to downy mildew but very susceptible to powdery mildew, botrytis bunch rot and anthracnose. Must be grafted. Bunches and berries are small and the ripe berries are sometimes slightly pink.

WHERE IT'S GROWN AND WHAT ITS WINE TASTES LIKE

Wines are generally rather mediocre and susceptible to oxidation. Even though this variety is included in the official register of varieties in **France**, total plantings are shrinking, from 600 ha (1,483 acres) in 1958 to just 7 ha (17 acres) in 2008. Vin et Pic, specialists in obscure and forgotten varieties, produce a varietal version (classified as a vin de France) north west of Saint-Étienne in the Rhône, as do Les Déplaude de Tartaras a little further east.

There are still scattered plantings in the Finger Lakes of **New York** State (Bully Hill make a very sweet version), and elsewhere in the north-eastern US, but not enough to make it into the official statistics.

RAYON D'OR

Minor but relatively successful French hybrid once popular in the Loire. Begat Seyval Blanc and Vidal.

BERRY COLOUR: ● ● ● ● ●

PRINCIPAL SYNONYMS: Feherek Kiralya, Roi des Blancs, Seibel 4986, Zlatni Luc, Zokoy Loloutch

ORIGINS AND PARENTAGE

A complex hybrid obtained at the beginning of the twentieth century by Albert Seibel in Aubenas, west of Montélimar in south-eastern France, by crossing Seibel 405 × Aramon du Gard (see PRIOR for the complete pedigrees of Seibel 405 and Aramon du Gard).

Rayon d'Or was used to breed SEYVAL BLANC, SEYVAL NOIR and VIDAL, the first and the last having been more successful than their parent, especially in colder climates.

VITICULTURAL CHARACTERISTICS

Late budding and early ripening. Good resistance to powdery mildew, black rot and downy mildew.

WHERE IT'S GROWN

There were just 6 ha (15 acres) of Rayon d'Or in **France** in 2008 although it was once authorized and widely grown in the Loire (6,965 ha/17,211 acres in 1958) when vineyards were replanted in the wake of phylloxera. It has also found its way across the Atlantic and is found to a very limited extent in the American states as far apart as **Missouri** and **New Jersey**, for example, but has not achieved the success of its offspring VIDAL.

REBERGER

Recent, disease-resistant German hybrid.

BERRY COLOUR: ● ● ● ● ●

PRINCIPAL SYNONYMS: Geilweilerhof 86-2-60

ORIGINS AND PARENTAGE

A REGENT × BLAUFRÄNKISCH hybrid obtained in 1986 by Rudolf Eibach and Reinhard Töpfer at the Geilweilerhof research centre, released in 2004 but not yet authorized for quality wine production in Germany.

VITICULTURAL CHARACTERISTICS

Yields are lower than for REGENT but sugar levels are generally higher. Moderate resistance to downy and powdery mildews.

WHERE IT'S GROWN AND WHAT ITS WINE TASTES LIKE

Wines can be powerful and have well-integrated tannins, similar to BLAUFRÄNKISCH, but more deeply coloured. Among the few producers of varietal Reberger wines are Højbakke Vineyard in **Denmark** and, in **Germany**, Eberle in the Pfalz and Dr Lawall in Rheinhessen. The Geilweilerhof research centre also sell their bottling of this variety.

REBO

Rather promising red cross made in Trentino.

BERRY COLOUR: ● ● ● ● ●

PRINCIPAL SYNONYMS: Incrocio Rigotti 107-3

ORIGINS AND PARENTAGE

Rebo is number 107-3 in a series of crossings obtained in 1948 by Rebo Rigotti at the San Michele all'Adige research centre between Verona and Bolzano in Trentino, northern Italy, and selected for its disease resistance and good production. Rigotti described this cross as MERLOT × MARZEMINO but recent DNA profiling has revealed that Rebo is a Merlot × TEROLDEGO cross (Malossini *et al.* 2000).

VITICULTURAL CHARACTERISTICS

Vigorous, mid to late ripening. Resistant to fungal diseases.

WHERE IT'S GROWN AND WHAT ITS WINE TASTES LIKE

Rebo is cultivated in Trentino, northern **Italy**, mainly in San Michele all'Adige, Volano and Calavino near Trento, as well as in Cavedine and Padergnone (Rigotti's birthplace) in the Valle dei Laghi west of Trento. Although the variety is also planted in other areas, varietal wines are made only within the Trentino DOC, by producers such as Dorigati, the Istituto Agrario di San Michele all'Adige and Pisoni. They can be rich, full-bodied and well suited to barrel maturation. There were 39 ha (96 acres) in Italy in 2000, according to the agricultural census.

It is also found in **Brazil**, where Aurora are experimenting with Rebo in Serra Gaúcha.

RECANTINA

An old black-berried variety from the province of Treviso in the Veneto, northern **Italy**, where it was considered among the best as early as the seventeenth century under the name Recardina or Recandina. It was on the brink of extinction until recently, when it was rescued by viticulturists at the Conegliano research centre in the Veneto, who found at least three distinct local varieties under this equivocal name in the province of Treviso (Cancellier *et al.* 2007): Recantina a Pecolo Scuro (dark stalk), Recantina a Pecolo Rosso (red stalk) and Recantina Forner (named after the winery where it was found). It is not known which one is the true old Recantina, maybe all three of them, but for now only the Recantina Forner has been registered in the official Italian register of varieties. One of the rare varietal Recantinas is made by Serafini & Vidotto, a very smooth and confident wine with luscious blackberry fruit flavours.

REFOSCO

BERRY COLOUR: ● ● ● ● ●

PRINCIPAL SYNONYMS: Refoschin, Refošk, Rifòsc, Rifosco

ORIGINS AND PARENTAGE

The earliest mention of Refosco appeared in a document referring to a famous banquet held in Cividale on 6 June 1409 in honour of Pope Gregory XII at which the following wines were served: Ribolla from Rosazzo, Verduzzo from Faedis, Ramandolo from Torlano, Refosco from Albana and Marzemino from Gradiscutta (Peterlunger *et al.* 2004).

Refosco is later mentioned several times in Friuli (Calò 2005) but its identity is uncertain because Refosco or Refošk is a name used in northern Italy and Slovenia for several distinct varieties often referred to as the Refoschi family. Using ampelography and DNA profiling, Costacurta *et al.* (2005) distinguished six varieties among the Refoschi:

- REFOSCO DAL PEDUNCOLO ROSSO, the most widespread variety
- REFOSCO DI FAEDIS, also called Refoscone, Refosco Nostrano, Refosco di Ronchis (Cipriani *et al.* 1994)
- Refosco d'Istria, identical to Refošk in Slovenia (Kozjak *et al.* 2003) and with TERRANO or Teran in Istra (Istria), Croatia (Maletić *et al.* 1999)
- Refosco di Guarnieri, identical to TREVISANA NERA (Crespan, Giannetto *et al.* 2008)
- Refosco del Botton, identical to TAZZELENGHE (Costacurta *et al.* 2005)
- Refosco di Rauscedo, or Refosco Gentile, not officially registered in Italy and almost extinct.

In addition, both Refosco del Botton and Refosco di Rauscedo showed a likely parent–offspring relationship with REFOSCO DI FAEDIS (Costacurta *et al.* 2005).

OTHER HYPOTHESES

Like PROSECCO, Refosco has often been erroneously identified with the Pucinum mentioned by Pliny the Elder as a variety cultivated on rocks in Istra, and Refosco is also considered a polyclonal variety, a concept that does not fit with the understanding of variety followed in this book (see pp XIV–XVI).

Some authors suggest that Refosco is identical to a variety called Racimulus Fuscus by the Romans and from which its name supposedly derives.

REFOSCO DAL PEDUNCOLO ROSSO

Most widely planted Refosco with complex family history making tart but potentially interesting reds in Friuli and Slovenia.

BERRY COLOUR: ● ● ● ● ●

PRINCIPAL SYNONYMS: Rifòsc dal Pecòl Ròss

VARIETIES COMMONLY MISTAKEN FOR REFOSCO DAL PEDUNCOLO ROSSO: BONARDA PIEMONTESE, MONDEUSE NOIRE (California), REFOSCO DI FAEDIS, TERRANO

ORIGINS AND PARENTAGE

Refosco dal Peduncolo Rosso (meaning 'red pedicel', or 'red stem') is the best known of the Refoschi, a group of closely related varieties from the Karst plateau on the Italian–Slovenian border. It is not known whether the Refosco mentioned in this region from 1409 onwards was identical to Refosco dal Peduncolo Rosso or to another member of the Refoschi. The name Refosco dal Peduncolo Rosso appeared for the first time around 1870, when ampelographers started to distinguish this variety from the other Refoschi.

The widespread confusion between Refosco dal Peduncolo Rosso and MONDEUSE NOIRE probably started life some years ago at the Foundation Plant Services at the University of California at Davis (Nelson-Kluk 2005). A vine named Refosco was collected some time before 1966 by Austin Goheen from the Jackson experimental vineyard established in Amador County in the 1880s by the University of California at Berkeley's Eugene Hilgard. In the 1990s, European ampelographers identified this variety as Mondeuse Noire, as did DNA profiling some time later, so that Refosco and Mondeuse Noire were said to be synonyms. However, at the beginning of the twenty-first century, DNA profiling revealed that this American Refosco matched neither the original Italian Refosco dal Peduncolo Rosso nor any other variety of the Refosco group: this was simply a misnamed Mondeuse Noire.

Refosco dal Peduncolo Rosso is also often confused with TERRANO or Teran from Istra (Istria) but a comparison of their DNA profiles disproves this hypothesis (Vouillamoz).

DNA parentage studies have provided evidence that Refosco dal Peduncolo Rosso is a progeny of MARZEMINO (Grando *et al.* 2006). Since Marzemino and LAGREIN are both progenies of TEROLDEGO, Refosco dal Peduncolo Rosso turns out to be a grandchild of Teroldego and a nephew/niece of Lagrein, as shown in the diagram. In addition, Refosco dal Peduncolo Rosso is a parent of CORVINA VERONESE and a grandparent of RONDINELLA. The historical geographic

REFOSCO DAL PEDUNCOLO ROSSO PEDIGREE DIAGRAM

This northern Italian web of natural crosses encompasses most of the varieties used for Valpolicella as well as the traditional TEROLDEGO, LAGREIN and MARZEMINO. Inverse relationships are also theoretically possible when unknown (and probably extinct) varieties (?) are involved (see p x).

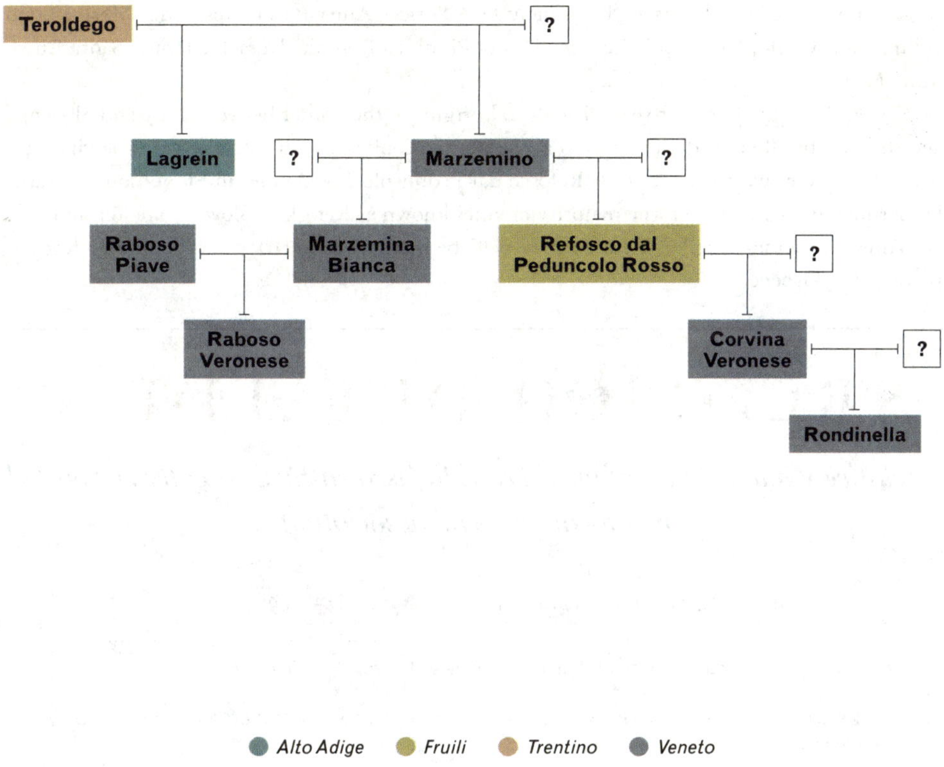

distribution of the varieties involved in the pedigree of Refosco dal Peduncolo Rosso strongly anchors its roots in north-eastern Italy.

VITICULTURAL CHARACTERISTICS

Late ripening. Susceptible to downy mildew and phomopsis, less so to powdery mildew; resistant to autumn rains, botrytis bunch rot and esca. Smaller berries so generally more phenolics than REFOSCO DI FAEDIS.

WHERE IT'S GROWN AND WHAT ITS WINE TASTES LIKE

Refosco dal Peduncolo Rosso is widely cultivated in Friuli-Venezia Giulia and in the Veneto in northern **Italy**, both on hillsides and on flatter land. Varietal wines are authorized in many DOCs in both the Veneto (eg Lison-Pramaggiore, Riviera del Brenta) and in Friuli (eg Colli Orientali del Friuli, Friuli Aquileia, Friuli Isonzo). The wines are deeply coloured with flavours of plums and sometimes a hint of almonds; they are usually medium-bodied but the acidity can be quite high because the grapes ripen notoriously late. Interest in Refosco increased in the 1980s, leading to greater care in the vineyard and winery and a general improvement in quality, although experiments with small-barrel ageing have not been universally successful. The most promising zone for Refosco is the Colli Orientali and the Koper district in Slovenia. Others include Grave del Friuli, Lison-Pramaggiore (outside Friuli), Latisana and Aquileia. There were 720 ha (1,779 acres) in Italy in 2000, according to the agricultural census.

Recommended producers in Italy include Alberice, Anna Berra, Bosco del Merlo, Livio Felluga, Iole Grillo, Moschioni, Paladin, Tenuta Pinni, La Roncaia, Russolo, Tenuta Santa Anna, Santa Margherita, Vignai da Duline.

Given the political rather than viticultural origins of the border between Italy and Slovenia, and the fact that Refosco dal Peduncolo Rosso may be confused with TERRANO in the vineyard, it is quite likely that there are some Refosco dal Peduncolo Rosso vines in **Slovenian** vineyards bordering Gorizia. However, the majority of vines known as Refošk in Slovenia are Terrano.

American wines labelled Refosco may well be MONDEUSE NOIRE. The odd Refosco is produced in **Greece**.

REFOSCO DI FAEDIS

A relatively minor but perfumed Friuli Refosco enthusiastically supported by a local growers' association.

BERRY COLOUR: ● ● ● ● ●

PRINCIPAL SYNONYMS: Refosco di Ronchis, Refosco Nostrano, Refoscone

VARIETIES COMMONLY MISTAKEN FOR REFOSCO DI FAEDIS: REFOSCO DAL PEDUNCOLO ROSSO, TERRANO

ORIGINS AND PARENTAGE

Refosco di Faedis is thought to originate in the area between Faedis, Cividale and Torreano to the west of the city of Udine in Friuli. DNA profiling has shown that Refosco di Faedis is distinct from REFOSCO DAL PEDUNCOLO ROSSO and identical to Refoscone, Refosco Nostrano and Refosco di Ronchis (Cipriani *et al.* 1994; Costacurta *et al.* 2005). DNA analysis has also provided evidence of likely parent–offspring relationships with both TAZZELENGHE (or Refosco del Botton) and Refosco di Rauscedo (Costacurta *et al.* 2005), the latter identical to Refosco Gentile in Crespan *et al.* (2011) but no longer cultivated.

VITICULTURAL CHARACTERISTICS

Vigorous, late ripening.

WHERE IT'S GROWN AND WHAT ITS WINE TASTES LIKE

Refosco di Faedis used to be widespread across Friuli, north-east **Italy**, especially in its homeland in the province of Udine, between Faedis and Torreano, but today it has virtually disappeared. In an effort to keep the variety alive, a group of enthusiastic growers in Faedis have formed the Associazione Volontaria fra viticoltori del Refosco di Faedis to promote this local variety. In general, it tends to be slightly more acidic and tannic than REFOSCO DAL PEDUNCOLO ROSSO. Varietal wines may be produced within the Colli Orientali del Friuli DOC.

Listed as Refosco Nostrano, the variety notched up a total of 263 ha (650 acres) in the 2000 Italian agricultural census. Forchir and Vigna delle Beccacce produce a varietal wine, the former labelling it Refoscone.

REFOSCO NOSTRANO

See REFOSCO DI FAEDIS.

REGENT

Highly successful and increasingly popular German hybrid.

BERRY COLOUR: ● ● ● ● ●

PRINCIPAL SYNONYMS: Geilweilerhof 67-198-3

ORIGINS AND PARENTAGE

A complex hybrid obtained in 1967 by Gerhardt Alleweldt at the Geilweilerhof research centre in the Pfalz, Germany, by crossing Diana × CHAMBOURCIN, in which Diana is a SILVANER × MÜLLER-THURGAU cross that should not be confused with the CATAWBA seedling that is also known as Diana in the US.

It was named after Le Régent, the famous diamond (140.5 carats) found by a slave in the Golkonda mine in India. Philippe II, Duke of Orléans, used the gem to decorate the crown of Louis XV for his coronation in 1722, Napoleon later embellished his sword with it and since 1887 it has sparkled in the French Royal Treasury at the Louvre.

Regent was used to breed CALANDRO and REBERGER and was authorized for quality wine production in Germany in 1996 despite its non-*vinifera* genes.

VITICULTURAL CHARACTERISTICS

Early ripening and winter-hardy. Very good resistance to downy and powdery mildews and quite resistant to botrytis bunch rot.

WHERE IT'S GROWN AND WHAT ITS WINE TASTES LIKE

Of all the recently bred, disease-resistant German varieties, Regent is the most widely planted and rapidly expanding in the German-speaking wine world. There are 2,161 ha (5,340 acres) in **Germany**, with most vines in Rheinhessen, the Pfalz and Baden. Fippinger-Wick in the Pfalz and Jürgen Ellwanger in Württemberg make good examples.

Regent has already been widely adopted outside Germany: 40 ha (99 acres) in north-east **Switzerland**, particularly in the cantons of Zürich and Schaffhausen; 15 ha (37 acres) in **England**, where it shows real promise though is currently less widely planted than RONDO; and modest plantings in **Sweden** and **Belgium**.

Its success is not due solely to its disease resistance but also to the quality of the wines it produces. The grapes ripen to higher sugar levels than Spätburgunder (PINOT NOIR) and the wines are full-bodied and approachable, suitable both for early drinking and for barrel ageing thanks to present but velvety tannins. Fruit flavours are in the cherry to redcurrant spectrum.

REGNER

Declining, light-skinned German cross with little to recommend it.

BERRY COLOUR: ● ● ● ● ●

PRINCIPAL SYNONYMS: Alzey 10378

ORIGINS AND PARENTAGE

A crossing of the white table grape LUGLIENGA (aka Seidentraube) with GAMAY NOIR (in fact the Gamay Précoce mutation), created in 1929 by Georg Scheu at Alzey. It was named in honour of Anne-Marie Regner (1911–99), who collaborated with Scheu.

VITICULTURAL CHARACTERISTICS

Buds early, ripens early and yields consistently.

WHERE IT'S GROWN AND WHAT ITS WINE TASTES LIKE

Wines are low in acidity with a slight Muscat note, reminiscent of MÜLLER-THURGAU. The variety can attain high sugar levels but at the expense of acidity and it is not surprising that it is in decline, with a current total in **Germany** of only 46 ha (114 acres), down from 124 ha (306 acres) in 2001. It has produced slightly better results in **England**, where there are just over 5 ha (12 acres), though it is probably better as a blending component than as a varietal wine.

REICHENSTEINER

Cool-climate, productive, relatively soft German cross of pan-European descent.

BERRY COLOUR: ● ● ● ● ●

PRINCIPAL SYNONYMS: Geisenheim 18-92

ORIGINS AND PARENTAGE

A MÜLLER-THURGAU × (MADELEINE ANGEVINE × Weisser Calabreser) cross bred by Heinrich Birk at the Geisenheim research centre in the Rheingau, Germany, in 1939 and authorized in 1978. (Weisser Calabreser is not a colour mutation of NERO D'AVOLA, which is often called Calabrese.) It was named after Reichenstein castle at Trechtingshausen, perched dramatically above the Rhein in Rheinland-Pfalz. Reichensteiner was used to breed the Swiss varieties GAMARET, GARANOIR and MARA.

VITICULTURAL CHARACTERISTICS

Very high yielding, early to mid ripening. Good resistance to downy mildew and botrytis bunch rot thanks to its loose bunches. Sugar levels are generally high but acidity moderate except in very cool climates, where it is at risk from winter frost.

WHERE IT'S GROWN AND WHAT ITS WINE TASTES LIKE

Germany had 106 ha (262 acres) in 2008 with more than half in Rheinhessen but also 26 ha (64 acres) in the Mosel region. Producers include Paternushof in Rheinhessen and Rainer Heil in the Mosel.

In **England** it was the third most planted white wine variety after CHARDONNAY and BACCHUS with total plantings of 95 ha (235 acres) by 2009. Wines are pretty neutral, sometimes lightly floral, and best as part of a blend. In the UK, it is often blended with SEYVAL BLANC, MADELEINE × ANGEVINE 7672, HUXELREBE, etc.; Chapel Down and Camel Valley include it in their blended sparkling wines. There are also limited plantings in **New Zealand** (72 ha/178 acres) and tiny amounts in **British Columbia**, **Japan** and **Switzerland**.

RÉSELLE

Very minor disease-resistant Swiss hybrid.

BERRY COLOUR: ● ● ● ● ●

PRINCIPAL SYNONYMS: VB 86-3

ORIGINS AND PARENTAGE

Réselle is a BACCHUS × SEYVAL BLANC hybrid obtained by Valentin Blattner in Soyhières in the Swiss Jura. Réselle and BIRSTALER MUSKAT are therefore siblings. The variety was named after the hamlet of La Réselle close to Soyhières.

VITICULTURAL CHARACTERISTICS

Good resistance to both downy and powdery mildews.

WHERE IT'S GROWN AND WHAT ITS WINE TASTES LIKE

The variety's breeder Valentin Blattner is almost the only producer of this variety (less than a hectare/2.5 acres planted in **Switzerland** in 2009), giving light wines with aromas of grapefruit, lemon and blackcurrant.

RETAGLIADO BIANCO

Rare Sardinian with almost more synonyms than vines.

BERRY COLOUR: ● ● ● ● ●

PRINCIPAL SYNONYMS: Arba Luxi, Arretallau, Arrotelas, Bianca Lucida, Co 'e Erbei, Coa de Brebèi, Erbaluxi, Mara Bianca, Rechiliau, Redaglàdu, Retagladu, Retellau, Ritelau

VARIETIES COMMONLY MISTAKEN FOR RETAGLIADO BIANCO: BRUSTIANO BIANCO

ORIGINS AND PARENTAGE

Retagliado Bianco was first cited in 1877 on Sardegna, where it has numerous synonyms depending on where it's grown.

VITICULTURAL CHARACTERISTICS

Vigorous, ripens early October.

WHERE IT'S GROWN

Today Retagliado Bianco is cultivated on a very small scale in the provinces of Sassari and Gallura, especially in old vineyards on the island of Sardegna. There were 28 ha (69 acres) in **Italy** in 2000. It is authorized as a varietal wine within the IGT Colli del Limbara but is usually blended with VERMENTINO and other local varieties.

RÈZE

Rare, very old, well-connected Swiss variety worthy of greater attention.

BERRY COLOUR:

PRINCIPAL SYNONYMS: Blanc de Maurienne (Savoie in France), Resi (Haut-Valais), Rèze Jaune (Valais), Rèze Verte (Valais)

ORIGINS AND PARENTAGE

Rèze is one of the oldest varieties in the Alps. Its earliest mention appeared in the Valais, southern Switzerland, on a Latin parchment dated 1313 known as the 'Registre d'Anniviers' (Ammann-Doubliez 2007; Vouillamoz 2009a): 'tribus generibus racemorum bene et sufficienter maturorum, scilicet de neyrun, de humagny et de regy' ('three kinds of grapes, good and ripe enough, that is to say of neyrun, of humagny and of regy'). The name Regy undoubtedly corresponds to Rèze, because several subsequent documents mentioned it under the name Regis, suggesting that it possibly derived from the family name Regis that was widespread in the region in the Middle Ages (Vouillamoz and Moriondo 2011).

Until recently, Rèze was known only in the Valais, Switzerland, but José Vouillamoz and the brothers Jean-Pierre, Philippe and Michel Grisard from Fréterive in France's Savoie discovered old Rèze vines in the Vallée de la Maurienne that were identified by DNA profiling in 2009. Local people simply called it Blanc de Maurienne, suggesting that Rèze had long been grown in the Maurienne. A similar finding was made by Gaël Delorme in the French Jura (Thierry Lacombe, personal communication), which indicates that the distribution of Rèze was once much wider than it is now. This has been confirmed by DNA parentage analyses showing that Rèze has parent–offspring relationships with at least five geographically distant varieties (Vouillamoz, Schneider and Grando 2007; Vouillamoz 2009b): in Switzerland, the now virtually extinct Diolle and Grosse Arvine in the Valais; in Italy, CASCAROLO BIANCO in Piemonte as well as GROPPELLO DI REVÒ and NOSIOLA in Trentino. In addition, Rèze might also be closely related to FREISA from Italy's Piemonte (Vouillamoz, Schneider and Grando 2007).

A rare red-berried colour mutation was recently found in the Valais (Vouillamoz, Frei and Arnold 2008).

OTHER HYPOTHESES

According to most linguists, the name Rèze derives from the Latin Raetica, the most widespread white grape variety in northern Italy in Roman times (Aebischer 1937; André 1953), and Rèze is often considered to be identical to Raetica. Since it is not botanically possible to make a firm connection between Latin grape names and modern varieties, it is more likely that Rèze is a distant relative of Raetica, just like NOSIOLA and GROPPELLO DI REVÒ (Vouillamoz and Moriondo 2011).

VITICULTURAL CHARACTERISTICS

Moderate to low vigour, average but sometimes irregular yields. Early to mid budding, mid ripening. Susceptible to botrytis bunch rot.

WHERE IT'S GROWN AND WHAT ITS WINE TASTES LIKE

Before the widespread replacement of Rèze by CHASSELAS and SILVANER in the replantings that followed the phylloxera crisis at the end of the nineteenth century, it was one of the most widespread varieties in the Valais, **Switzerland**, where it was traditionally used to make Vin du Glacier, a dry unfortified wine made in a solera system in which the casks are refilled only once a year, producing a maderized style.

Today there are little more than 2 ha (5 acres) of Rèze grown in the Valais, where Josef-Marie Chanton crafts varietal wines with typical aromas of gooseberries, green apple, crisp acidity and a light structure. Other notable producers include Cave Les Sentes and Vin du Mur.

RIBOLLA GIALLA

Ancient white from the Italy–Slovenia border.

BERRY COLOUR:

PRINCIPAL SYNONYMS: Rabola, Rabiola, Rebolla, Rebula (Goriška Brda in Slovenia), Ribolla di Rosazzo, Ribuèle (Friuli)

VARIETIES COMMONLY MISTAKEN FOR RIBOLLA GIALLA: GARGANEGA, Jarbola Bijela (Matulji in Croatia; see HRVATICA), PIGNOLETTO (under the name Rébola or Ribolla Riminese in Rimini), PROSECCO LUNGO (under the name Ribolla Spizade), Ribolla Verde, ROBOLA

ORIGINS AND PARENTAGE

Ribolla Gialla is a very old variety from Friuli-Venezia Giulia. The earliest mention of its wine dates from 1296 under the name Rabola, an old and attested synonym for Ribolla: on 20 March, Pope Boniface VIII settled a dispute between the Bishop of Trieste and the monastery of San Giorgio Maggiore in Venezia about the selling of a wine called Rabola (Di Manzano 1860). Three years later (3 February 1299), the same wine appears in the documents of the notary Ermanno di Gemona among the gifts made by the city of Udine to Girardo da Camino: one pig, one cow, two *conzi* of Rabiola (a *conz* was an old measure equal to 85 litres/22.5 US gallons) and some bread (Di Manzano 1860). Numerous subsequent references demonstrate that Ribolla Gialla was already widespread in Friuli and Istra (Istria) in the fourteenth century.

Ribolla Gialla is distinct from the almost extinct Ribolla Verde and is not a colour mutation of Ribolla Nera (ie SCHIOPPETTINO). The variety named Rébola or Ribolla Riminese in the province of Rimini is identical to PIGNOLETTO (Filippetti *et al.* 1999), and Ribolla Spizade is a synonym for PROSECCO LUNGO. A possible parent–offspring relationship with GOUAIS BLANC, based on DNA evidence, is currently under investigation (Serena Imazio and José Vouillamoz, unpublished data).

A recent DNA survey of Rebula and similar varieties in Slovenia confirmed that it is the same variety as Ribolla Gialla but also that it shows a significant level of genetic diversity in Goriška Brda (Slovenia), Collio (Italy) and Istra (Croatia), with several distinct clones interplanted with other distinct and undetermined varieties (possibly seedlings?) in the vineyard (Rusjan *et al.* 2010).

OTHER HYPOTHESES

Like PROSECCO and REFOSCO, Ribolla Gialla has often been erroneously identified with the Pucinum mentioned by Pliny the Elder as a variety grown on rocks in Istra.

VITICULTURAL CHARACTERISTICS

Mid ripening. Susceptible to millerandage and rot. Low fertility.

WHERE IT'S GROWN AND WHAT ITS WINE TASTES LIKE

Ribolla Gialla is cultivated in Friuli, north-east **Italy**, hard by the border with Slovenia – a border that has seen so many changes that it is hardly surprising that the variety is also grown, as Rebula, in Slovenia. In Italy in 2000 the agricultural census recorded 284 ha (701 acres), planted mainly on the hills between Tarcento and Istra and around Gorizia. Varietal wines are produced in the Colli Orientali del Friuli and Collio DOCs, with the villages of Cialla, Rosazzo, Capriva, Cividale, Cormons, Gradisca d'Isonzo, Manzano and Oslavia producing particularly fine examples – the first two officially recognized as subzones within the Colli Orientali del Friuli DOC. Oslavia is regarded as the home of the variety, which was traditionally fermented on the skins. Today a few days' skin contact is by no means unusual.

Traditionally, varietal wines have been light-bodied and high in acidity, sometimes slightly floral. More recently, ambitious and innovative producers have been making more concentrated and characterful versions with a deep yellow colour and rich yellow-fruit, sometimes nutty and mineral, flavours, although experiments with barrel fermentation and/or ageing have not been universally successful. Recommended producers include Josko Gravner, whose distinctive Anfora is made in clay amphorae, Miani, Primosic, Matjaz Tercic and Le Vigne di Zamò.

In **Slovenia** the majority of Rebula vines are planted in Goriška Brda north of the Italian city of Gorizia (446 ha/1,102 acres), where it is the most widely planted variety of either colour, and in Vipavska Dolina east of Gorizia (295 ha/729 acres), where it is the most widely planted white. There are also very limited plantings in Kras (Carso in Italian) and Slovenska Istra (Slovenian Istria). Recommended producers include Batič, Kabaj, Edi Simčič, Marjan Simčič and Ščurek.

A little is also planted in the Napa Valley, **California**.

RIBOLLA NERA

See SCHIOPPETTINO.

RIESEL

Minor Swiss hybrid bred mainly for its disease resistance that has found a home in the Netherlands.

BERRY COLOUR: (white/green indicated)

PRINCIPAL SYNONYMS: VB 11-11-89-12

ORIGINS AND PARENTAGE

Riesel is a hybrid obtained by Valentin Blattner in Soyhières in the Swiss Jura by crossing CABERNET SAUVIGNON with an undisclosed disease-resistant partner (*Resistenzpartner* in German).

VITICULTURAL CHARACTERISTICS

Vigorous, early budding, mid ripening. Relatively good resistance to frost, downy and powdery mildews but the compact bunches are prone to botrytis bunch rot.

WHERE IT'S GROWN AND WHAT ITS WINE TASTES LIKE

In the **Netherlands**, Colonjes in Groesbeek produce a medium-dry, lemony varietal wine and both Colonjes and Wijnboerderij De Gravin in Roswinkel make a blend of Riesel and SOLARIS. It is not unlike RIESLING, with floral and white-fruit aromas, but with less acidity and complexity.

RIESLANER

Rare, undervalued but demanding Riesling-like German cross making stunning botrytis-influenced sweet wines.

BERRY COLOUR: ● ● ● ● ●

PRINCIPAL SYNONYMS: Mainriesling, Würzburg N 1-11-17

ORIGINS AND PARENTAGE

A SILVANER × RIESLING cross bred by August Ziegler in 1921 in Würzburg. This parentage has been confirmed by DNA profiling (Grando and Frisinghelli 1998). ORANIENSTEINER, OSTEINER and Rieslaner are siblings. Rieslaner was used to breed ALBALONGA.

VITICULTURAL CHARACTERISTICS

Like RIESLING, it ripens late while retaining high acidity. It has a tendency to shrivel so it is better on unexposed sites, and in general it requires a good site to produce high quality. Yields are generally low and it has a tendency to millerandage if the weather is not perfect at flowering. If all that weren't enough, the stems are not very stable so there can be additional crop losses if grapes are left to ripen fully.

WHERE IT'S GROWN AND WHAT ITS WINE TASTES LIKE

Germany has 87 ha (215 acres), mostly in Franken and the Pfalz, but Dreissigacker and Keller both make outstanding Rieslaner TBA in Rheinhessen, where just a few hectares remain. In these wines, as in Fürst Löwenstein's and Rudolf Fürst's Auslesen from Franken, the botrytis influence concentrates the ripe grapefruit, apricot and bitter orange flavours of the variety, balanced by fine acidity. Müller Catoir in the Pfalz have shown the potential too. It is a shame that Rieslaner is becoming increasingly rare because if it reaches full ripeness, it can produce remarkable, full-bodied yet fresh, ageworthy wines. Dry versions can taste alcoholic and unbalanced if the grapes are picked fully ripe and overly acidic if not.

RIESLING

One of the world's greatest white wine grapes, capable of making particularly geographically expressive and long-lived wines at all sweetness levels.

BERRY COLOUR: ● ● ● ● ●

PRINCIPAL SYNONYMS: Beyaz Riesling (Turkey), Johannisberg (Valais in Switzerland until the 1920s before this name was given to Silvaner; California), Kleinriesling, Klingelberger (Ortenau in Baden), Lipka (Czech Republic), Petit Rhin (Valais), Raisin du Rhin (Alsace in France), Rajinski Riesling (former Yugoslavia), Rajnai Rizling (Hungary), Renski Riesling (Slovenia), Renski Rizling (Slovenia), Rheinriesling (Austria), Rhine Riesling, Riesling Edler (Germany), Riesling Gelb (Germany), Riesling Renano (Italy), Riesling Rhénan (Alsace), Riesling Weisser, Rislinoc (Republic of Moldova), Risling (Bulgaria), Rizling Rajnski (Croatia), Ryzlink Rýnský (Czech Republic, Slovakia), Starovetski (Czech Republic, Slovakia), Weisser Riesling, White Riesling (US)

VARIETIES COMMONLY MISTAKEN FOR RIESLING: CROUCHEN (South Africa), GRAŠEVINA (Italy, under the name Riesling Italico), MENU PINEAU (Loire), PEDRO XIMÉNEZ, RÄUSCHLING (Germany), SAUVIGNONASSE (Chile and Argentina)

ORIGINS AND PARENTAGE

Riesling is one of the most ancient German grape varieties, as suggested by its impressive number of synonyms. According to Bertsch in Levadoux (1956), the name Riesling could derive from *reissen* (old German *rîzan*), originally meaning 'to split', then 'to make an incision', 'to carve' or 'to engrave' and later 'to tear' or 'to write'. No explanation is provided for this rather strange etymology, but perhaps the earliest meaning, 'to split', could be a reference to the way the berries split between the fingers, similar to the use of the synonym Fendant for CHASSELAS. Alternatively, the name might possibly be related to the grape's tendency to millerandage (in German *verrieseln*, though it is stretching it a bit to say millerandage is like 'making an incision' in the flowers so that they do not form berries) or to the German word *Rissling* (also derived from *reissen*), meaning 'cutting'.

Riesling probably originated in the Rheingau, on the northern bank of the Rhein in Germany, where its earliest mention is said to be in a document dated 3 March (February according to some accounts) 1435. Klaus Kleinfisch, cellarmaster at Schloss Katzenelnbogen in Rüsselsheim near Frankfurt-am-Main in the state of Hessen, reports on the expenditures and revenues of Graf Johann IV von Katzenelnbogen (Staab 1991): 'item XXII ß umb setzreben riesslingen in die wingarten' ('22 Solidi [gold coins] for riesslingen vine cuttings for the vineyard'). Some sources have *seczreben* but the writing is hard to decipher and *setzreben* ('vine cuttings') makes more sense.

Numerous references to Riesling are to be found in other documents in Germany but the similarity with ancient mentions of RÄUSCHLING (eg *ruesseling* in Kintzheim, Alsace, in 1348 and in Trier in 1464) can be a source of confusion. The first mention of the modern Riesling spelling appeared in the Latin edition of Hieronymus Bock's (1552) *Kreutterbuch*: 'Ad Mosellam, Rhenum et in agro Wormatiensi vites procreantur Riesling appellate' ('Riesling grows in the Mosel/Rhein and in Worms'), though some sources say it appeared in the 1546 edition of Bock's book.

Regner *et al.* (1998) established by DNA profiling that Riesling has a parent–offspring relationship with GOUAIS BLANC, one of the most ancient and prolific wine grapes of Western Europe. Since Gouais Blanc also has parent–offspring relationships with at least eighty other grape varieties (see the pedigree diagram in the PINOT entry), including CHARDONNAY, GAMAY NOIR, ELBLING and FURMINT, they all turn out to be half-siblings, grandparents or grandchildren of Riesling.

Riesling has no genetic relationship with Riesling Italico or Welschriesling (both synonyms of GRAŠEVINA) and many other grape varieties have been erroneously named after Riesling: in Australia, Clare Riesling (= CROUCHEN); in South Africa, Cape Riesling (= CROUCHEN); in California, Gray Riesling (= TROUSSEAU) and so on.

Not least because Germans were prolific vine breeders, Riesling has been used in numerous crosses, many of which are cultivated today: AURELIUS, DALKAUER (possibly), EHRENFELSER, GEISENHEIM 318-57, GOLDRIESLING, HÖLDER, MANZONI BIANCO, KERNER, MISKET VARNENSKI, MÜLLER-THURGAU, NORIA, ORANIENSTEINER, OSTEINER, RIESLANER, RIESUS, ROTBERGER, SCHEUREBE and VERITAS.

The red-berried mutation **Roter Riesling** is discussed at the end of this entry.

OTHER HYPOTHESES

As in the case of many other varieties, there has been speculation that Riesling is identical to the *Vitis aminea* mentioned by various Roman authors. Some sources suggest the variety is named after the Ritzling, a river in the Wachau region of Austria, but this is probably a coincidence because Riesling was not cultivated there in the Middle Ages. The name Riesling was also said to derive from *Russling* (*Rus* meaning 'dark wood') but this etymology is true only for RÄUSCHLING.

On the basis of DNA data, Regner *et al.* (1998) speculated that Riesling, like ELBLING, could be a cross of GOUAIS BLANC with a close relative of SAVAGNIN (or Traminer), this ancestor being long dead and possibly related to wild grapevines (*Vitis vinifera* subsp. *silvestris*) in the Rheingau. However, this hypothetical parent has never been found and no genetic link between Riesling and wild grapevines could subsequently be established.

VITICULTURAL CHARACTERISTICS

Very cold-hardy, thanks to the hardness of its wood, though it needs to be planted in the best sites to ripen fully and yield economically. In the right sites, it can retain high quality even at yields of around 70 hl/ha. It buds late and therefore tends to avoid spring frosts. Mid to late ripening, depending on the style of wine being produced. Resistant to downy mildew and only slightly susceptible to powdery mildew and botrytis bunch rot.

WHERE IT'S GROWN AND WHAT ITS WINE TASTES LIKE

At long last there seems to be some real international enthusiasm for Riesling, not least in the US, where the two biggest wine companies have introduced branded varietals, imported from Germany. This is not a variety that shows much of its mettle in inexpensive versions but a hand-crafted Riesling is generally relatively low in alcohol but high in both acidity and, often, extract. There is rarely any shortage of flavour and aroma in a Riesling – in fact the bouquet in mature examples can be so strong as to deter some wine drinkers. Because Riesling is so good at expressing the terroir where it is grown, it is difficult to generalize about its aromas and general character. They can be floral, fruity, mineral and/or spicy. One generalization can be made: Riesling makes some of the longest-lived wines in the world, wines that last about as long as a red bordeaux of equivalent quality.

The quintessentially Germanic grape Riesling is not officially allowed in the appellations of **France**, except in Alsace, which was of course German once. This region between the Vosges Mountains and the Rhine had 3,382 ha (8,357 acres) of Riesling planted in 2009, a considerable increase on the 787 ha (1,945 acres) planted in 1958, and a total that makes it the most planted variety in the region now that it has so widely displaced Sylvaner (SILVANER). But there are still only three authorized Riesling clones in Alsace – 49, 1089 and 1091 – compared with more than sixty in Germany. Alsace Riesling has traditionally been chaptalized and vinified dry so has typically been fuller-bodied than German Riesling. Only Vendange Tardive and Sélection de Grains Nobles Rieslings are sweet. Riesling is a fine medium here for communicating the variation between various limestone, granite and marl sites, and their altitudes and expositions. Recommended producers include Marcel Deiss, Hugel, Josmeyer, Kreydenweiss, Meyer-Fonné, René Muré,

André Ostertag, Rolly-Gassmann, Domaines Schlumberger, Trimbach, Weinbach and Zind-Humbrecht.

Riesling Italico is much more important in **Italy** than Riesling and the national total was only 624 ha (1,542 acres) in 2000, mostly in the cooler reaches of the Alto Adige and in Friuli. Kuenhof, Cantina San Michele Appiano, Schwanburg and Tiefenbrunner produce some of the best examples, often a little grassy, in the Alto Adige; Jermann, Puiatti and Villa Russiz in Friuli. Aldo Vajra makes a distinctive 100% Riesling in Piemonte: Pietracine sold as Langhe Bianco.

Spain grew only 97 ha (240 acres) of Riesling in 2009, principally in Catalunya in the north east for wines such as Torres' Waltraud from Penedès, as well as in Castilla-La Mancha.

Riesling is much more important in little **Luxembourg**, just over the border from the upper Moseltal (Mosel Valley) in Germany after all. The country's 159 ha (393 acres) represent more than 12% of the total vineyard area. Most sophisticated exponents are Mathis Bastian, Charles Decker, Gales, Alice Hartmann and Ch Pauqué. Klaus Peter Keller of Rheinhessen in Germany is involved with a small experimental planting of Riesling in southern **Norway.**

Germany is epicentre of Riesling lore, and the country with by far the most Riesling in the ground: 22,434 ha (55,436 acres), making it the country's most planted grape variety. (There was an ignoble period in the history of German wine in the late twentieth century when plantings of the much-easier-to-ripen MÜLLER-THURGAU overtook those of the noble Riesling.) Riesling is grown in every German wine region with almost equal totals in the Pfalz (5,458 ha/13,487 acres) and Mosel-Saar-Ruwer (5,390 ha/13,319 acres) as well as 3,769 ha (9,313 acres) in Rheinhessen, 2,464 ha (6,089 acres) in the Rheingau and 2,083 ha (5,147 acres) in Württemberg. It arguably reaches its apogee, or certainly makes its most distinctive wines, in Germany's coolest wine region, the Mosel-Saar-Ruwer, where the relatively late-ripening Riesling can be ripened only on the most propitiously sited and exposed slopes, angled so as to catch maximum sunshine, some of it reradiated by slaty soils, some reflected from the surface of the river. Wines produced here are typically only about 8% alcohol with extremely high acidity (which needs time in bottle to soften) yet chock full of extract and character and, usually, with perceptible if delicate fruity sweetness.

There was a time when virtually all German wine had some sweetness, sometimes to disguise shortcomings in the wine itself, but climate change and revived concentration on quality rather than quantity have meant that Germany's Riesling, especially in southern wine regions, can now be made into perfectly respectable dry, or *trocken*, wines, with alcohol levels closer to 13% and no help from sugar added to the fermentation vat. Dry wines, particularly popular in Germany itself, have a quite different structure from fruitier ones, but in general Rheingau Rieslings are relatively firm with a lightly honeyed character. Those from the Nahe are nervier, with a certain grapefruit quality in some instances. Rieslings from the extensive Rheinhessen region vary enormously from big, luscious ferments ripened overlooking the Rhein around Nierstein to the tense, vibrant, sometimes nutty vineyard expressions now being produced by a nucleus of younger, ambitious producers in the south of the region. Pfalz Rieslings tend to be almost fat, certainly honeyed and quite full-bodied, while those of Württemberg are light and refreshing, mainly for local quaffing. Franken is SILVANER country and Baden specializes in the various PINOTS but both can produce their own very respectable renditions of this, the quintessential German grape.

Although it is so close to Germany, **Switzerland** hardly grows Riesling at all – just 12 ha (30 acres), nearly half of which is in the Valais. Amphitryon by Domaine du Mont d'Or in Sion was the first Valais Riesling and is still today among the best examples.

Austria, on the other hand, grows 1,874 ha (4,631 acres) – although Riesling routinely plays second fiddle to Austria's own GRÜNER VELTLINER. The Weinviertel has the most hectares planted to Riesling, 552 ha (1,364 acres), but this represents less than 4% of the region's vineyards (compared with almost 50% given over to Grüner Veltliner), whereas the Wachau's 198 ha (489 acres) represent 14% of the region's vines, Kamptal's 298 ha (736 acres) is 7.5% and Kremstal's 214 ha (529 acres) is nearly 9.5%. Burgenland's Riesling is found mainly in Neusiedlersee and is made into impressively concentrated sweet wines. Austrian Riesling is rather more full-bodied than most German Riesling, and often somewhat spicier and less austere than the Rieslings of

Alsace. Most of the top Riesling exponents are based on south-facing slopes on the north bank of the Donau (Danube), the likes of Bründlmayer, Hirtzberger, Knoll, Loimer, Nigl, Nikolaihof, F X Pichler, Prager, Schloss Gobelsburg, Stadt Krems, Undhof Salomon and Domäne Wachau, but in general the quality level is consistently high.

Hungary grows 1,283 ha (3,170 acres) of Riesling, mainly in the southern central regions of Kunság, Csongrád and Tolna, but there are some fine specimens from near Lake Balaton. The top producers are Bussay (Zala), Ottó Légli (Balatonboglár), József Szentesi (Pázmánd) and Villa Tolnay (Badacsony).

There are 1,270 ha (3,138 acres) of Ryzlink Rýnský in the **Czech Republic** with plantings concentrated in the southern wine regions of Morava in the south east of the country, although Velké Žernoseky is also known for its Riesling. In **Slovakia**, Pezinok and especially Modra within the Malokarpatská region on the border with Austria have a good reputation for Riesling. There was a total of 998 ha (2,466 acres) of Riesling in 2009. Egon Müller of the Saar makes the particularly precise Kastiel Belá Riesling near Stúrovo a little further south and east on the border with Hungary.

Renski Riesling ranks as the tenth most planted variety in **Slovenia**. It is grown mainly in Štajerska Slovenija (Slovenian Styria; 591 ha/1,460 acres) and Prekmurje (52 ha/128 acres) in the north east. Dveri-Pax make a creditable example.

Riesling was the sixth most planted variety in **Croatia** in 2009, under the name Rizling Rajnski, with just over 3% of the vineyard area, ie 1,072 ha (2,649 acres).

The **Republic of Moldova** had 1,343 ha (3,319 acres) of the variety, also known as Rislinoc, in 2009 while **Bulgaria**'s 2009 total of Risling was 1,170 ha (2,891 acres). Official statistics showed 882 ha (2,179 acres) of Riesling planted in 2009 in Krasnodar Krai, **Russia**'s biggest winemaking region, but it is not clear which Riesling this is. Fanagoria do have 60 ha (148 acres) of Rhine Riesling and **Ukraine** had 2,702 ha (6,677 acres) of 'Rhine Riesling' in 2009. Real Riesling is definitely grown in **Israel** to a very limited extent.

The same problem does not apply to **California**, where the variety once known as Johannisberg or White Riesling is clearly identified with the Riesling under discussion here. There were 3,831 acres (1,550 ha) of it in California vineyards in 2010, although 300 acres (121 ha) of that were not yet bearing – a sign of American commercial interest in the variety. Much California Riesling can be rather soft and dilute but Chateau Montelena, Smith Madrone and Stony Hill in Napa Valley; Dashe Cellars, Esterlina and Navarro in Mendocino County; and Ventana Vineyards in Monterey County have all produced delicious and relatively long-lived exceptions to this rule.

Cooler, cloudier **Oregon** has long had a solid tradition of producing respectable Riesling, lacking only an enthusiastic market for it. The state total of 777 acres (314 ha) provides fruit for the most interested and competent producers such as A to Z, Amity, Argyle, Brooks, Chehalem and Elk Cove.

But on the west coast it is **Washington** State that is busy styling itself as a Riesling specialist. Thanks to a long-standing co-operation (evinced by the hugely successful Eroica Riesling) between the dominant wine company there, Château Ste Michelle, and Erni Loosen of the Mosel, interest in the variety has grown enormously and has been reflected in a similar increase in plantings of Riesling in the state, from 1,900 acres (769 ha) in 1999 to 6,320 acres (2,558 ha) by 2011. CHARDONNAY is still more widely grown in Washington but the state joins the Rheingau and Australia as organizer of a triennial international Riesling event and there are dozens of producers of fine, bright-fruited, extremely crisp Riesling at all sweetness levels.

Château Grand Traverse of **Michigan** is well known for its Riesling.

On the east coast, the Finger Lakes region of **New York** State, where 683 acres (276 ha) of Riesling were grown in 2008, has a long tradition of producing seriously fine Riesling, at all sweetness levels, even if it has been somewhat ignored during the era of CHARDONNAY mania. Anthony Road, Glenora, Heron Hill and Hermann J Wiemer have all produced truly world-class Rieslings here. Not surprisingly, Riesling can also shine across the border in Ontario, **Canada**, where it is a highly significant variety and the likes of Norman Hardie and Cave Spring (related by marriage to St Urbans-Hof of the Mosel) have made extremely promising bone-dry Rieslings.

Riesling icewines can also shine. The finest (dry) Rieslings of British Columbia (367 acres/149 ha) have so far come from Tantalus.

Riesling is much more at home in the cool north east of North America than in South America. **Argentina** had only 112 ha (277 acres) planted in 2008, mainly around Mendoza. A handful of determined young winemakers are experimenting with Riesling in **Chile** so that there were 33 ha (82 acres) in Bío Bío (eg Cono Sur) in the cool far south in 2008 and Cousiño Macul have some planted near Santiago. In **Uruguay** there were just 17 ha (42 acres) in 2009.

Historically Riesling was hugely important to **Australia.** Silesian immigrants brought it to the Barossa Valley in South Australia in the early nineteenth century and it quickly showed an aptitude for the rather cooler Clare and Eden Valleys nearby. Before the advent of CHARDONNAY, it was the country's favourite white wine grape and it is still the fifth most planted with a national total of 4,401 ha (10,875 acres) in 2008. This may be dwarfed by Chardonnay's 31,564 ha (77,996 acres) but Australia can claim to have one of the world's most distinctive styles of Riesling – produced in a very much drier, warmer climate than that of any German wine region. Pioneered by the likes of Jeffrey Grosset of Clare Valley (with 862 ha/2,130 acres, home to the greatest concentration of Australian Rieslings), current offerings are some of the driest, steeliest and ambitiously tightly furled in the world, although there are signs of very slight mellowing in this respect. Australia has long made some top-quality sweet Rieslings, too. The most significant and successful regions for Riesling today are the Clare, Eden and Barossa Valleys and Great Southern, though there are also significant plantings in the Adelaide Hills and Padthaway – and Tasmania shows some potential. Crawford River, Frankland Estate, Grosset, Henschke, Leeuwin, Petaluma, Piper's Brook, Plantagenet, Tamar Ridge and Yalumba are just some of the most accomplished Riesling producers.

New Zealand is just getting to grips with varieties it calls 'aromatics', of which Riesling is undoubtedly one. The country grows a total of 917 ha (2,666 acres) of Riesling, up from 493 ha (1,218 acres) in 2001 and projected to rise further. Top producers include Dry River, Forrest Estate, Framingham (including some very fine sweet versions), Palliser, Pegasus Bay, Richardson and Seresin.

The cause of Riesling has not been helped in **South Africa** by the persistence of the term Cape Riesling, and occasionally just Riesling, for the lowly CROUCHEN. Only recently was the term Riesling reserved for the real thing and the requirement that it be called Weisser Riesling abandoned. Total plantings of 'real' Riesling were just 215 ha (531 acres) in 2008 and they are widely scattered, with the biggest areas in Stellenbosch and Robertson. Best wines tend to come from the coolest climes such as Constantia (Buitenverwachting and Klein Constantia), Elgin (Paul Cluver) and Swartberg at 1,000 m altitude in the far east of the Cape winelands (Howard Booysen).

Riesling is also grown in **China** (378 ha/934 acres in 2009), in the provinces of Shanxi, Gansu and Xinjiang, and there were 7.6 ha (19 acres) in 2009 in **Japan.**

ROTER RIESLING

Roter Riesling, also known as Riesling Rot, is a rare colour mutation of Riesling with pinkish rather than red berries (Sefc, Regner *et al.* 1998). Since Roter Riesling often produces a mixture of white and pinkish berries, and since white Riesling never does, it has been suggested that Roter Riesling is in fact the original form, the *Urform*, of Riesling. However, recent genetic research has shown that mutations of the colour gene in grapes (called *VvmybA1*) are often the result of an insertion of retrotransposons, which are small pieces of DNA able to move around the genome by themselves, thus creating variability (This *et al.* 2007). In the case of Roter Riesling, genetic investigation has established that the mix of berry colours in the same bunch is most likely due to complex mutations in the bud (Stenkamp 2009). As a consequence, Roter Riesling cannot be considered a forerunner of Riesling, in the same way that Elbling Rot, Chasselas Rouge, Sauvignon Rouge, etc. cannot be considered forerunners of their white counterparts.

There are tiny parcels in **Germany** and Austria. Ulrich Allendorf in Oestrich-Winkel in the

Rheingau is helping to keep Roter Riesling alive thanks to his 2006 plantings, undertaken after he had tasted a wine made at the Geisenheim research centre. Corvers Cauter in the Rheingau and Weingut Holzmann in the Weinviertel in **Austria** also produce Roter Riesling. The wines are similar to Riesling in their crisp acidity and aromas but slightly more powerful, rounded and full-bodied. The vine has the additional advantage of being less susceptible to botrytis bunch rot than white-berried Riesling.

RIESLING ITALICO

See GRAŠEVINA.

RIESUS

Recent and relatively robust Ukrainian hybrid with Riesling-like aromas.

PRINCIPAL SYNONYMS: Risus, Risys, Rysus

ORIGINS AND PARENTAGE

Riesus is a hybrid obtained by P Golodriga, M Kostik and V Yurchenko at the Ampelos private breeding company in Yalta, southern Ukraine, by crossing RIESLING with Roucaneuf, the latter being a Seibel 6468 × Subéreux hybrid (for the complete pedigrees, see HELIOS for Seibel 6468 and PRIOR for Subéreux). Riesus is a contraction of Riesling and *ustoichivy*, meaning 'resistant' with respect to diseases.

VITICULTURAL CHARACTERISTICS

Productive, mid budding, mid to late ripening. Dense but elastic berry skins. Good resistance to fungal diseases and cold-hardy to –26 °C (–14.8 °F).

WHERE IT'S GROWN AND WHAT ITS WINE TASTES LIKE

Riesus presumably grows in its homeland of **Ukraine** but is also planted in the Rostov and Krasnodar wine regions of **Russia**, where it is used mainly in blends. Its produce is also distilled. Wines typically have RIESLING-like aromas.

RIMAVA

Recently authorized, promising Slovakian cross.

BERRY COLOUR: ● ● ● ● ●

PRINCIPAL SYNONYMS: CAAB 3/12

ORIGINS AND PARENTAGE

Rimava, named after a river in south-eastern Slovakia, is a CASTETS × ABOURIOU cross obtained in 1976 by Dorota Pospíšilová at the VUVV research centre for oenology and viticulture in Bratislava, Slovakia. It is a sibling of HRON, NITRANKA and VÁH and was officially authorized in 2011.

VITICULTURAL CHARACTERISTICS

Productive, vigorous (best suited to poor soils), mid to late budding and ripening. Susceptible to hard winter frosts and drought but resistant to spring frosts and botrytis bunch rot. Berries are small with an intense Cabernet flavour. Good frost resistance.

WHERE IT'S GROWN AND WHAT ITS WINE TASTES LIKE

Rimava is one of the best Cabernet-like varieties in **Slovakia**, producing deeply coloured, full-bodied wines with a distinctive Cabernet flavour and good ageing potential (Dorota Pospíšilová, personal communication). Although there were only around 3 ha (7 acres) in 2011, it shows greatest promise on sloping sites in the best Slovakian wine regions of Južnoslovenská and Malokarpatská, in the south west of the country. Igor Blaho currently includes it in a red blend.

RIMINÈSE

See ALBANA.

RIO GRANDE

Very minor Portuguese cross planted mainly on Pico in the Açores.

BERRY COLOUR: ● ● ● ● ●

ORIGINS AND PARENTAGE

A DIAGALVES × FERNÃO PIRES cross obtained in Portugal. It is a sibling of SEARA NOVA.

WHERE IT'S GROWN

Rio Grande is found mainly on the Portuguese island of Pico, where the local co-operative use it in a light, easy-drinking blend with varieties such as FERNÃO PIRES, GENEROSA (PORTUGAL) and SEARA NOVA. There are also a few scattered vines on Madeira, mainly in Santana in the north, but the official total vineyard area in **Portugal** in 2010 was less than 1 ha (2.5 acres).

RIÓN

Rarity recently rescued from an old vineyard near Barcelona.

BERRY COLOUR: ● ● ● ● ●

ORIGINS AND PARENTAGE

An obscure, local variety recovered in Penedès, north-east Spain, by Albet i Noya.

VITICULTURAL CHARACTERISTICS

Mid budding and ripening. Good general disease resistance.

WHERE IT'S GROWN AND WHAT ITS WINE TASTES LIKE

Rión was part of an experimental project begun in 1999 by Josep Maria Albet of Albet i Noya, south west of Barcelona, **Spain**. An old vine was discovered in an abandoned vineyard and the variety is now being commercially cultivated on a small scale. The wine is very pale, restrained and minerally, not unlike Sancerre, with RIESLING-like freshness and a citrus aftertaste.

RIPOLO

Makes ageworthy whites on the Amalfi coast in southern Italy.

BERRY COLOUR: ● ● ● ● ●

PRINCIPAL SYNONYMS: Ripala, Ripoli, Uva Ripola

ORIGINS AND PARENTAGE

Ripolo used to be cultivated on Monti Lattari, especially in Gragnano, Lettere and Castellamare di Stabia in the province of Napoli in southern Italy. Ripolo is not mentioned in historical documents until the mid nineteenth century. It is genetically close to AGLIANICONE, FALANGHINA FLEGREA and PALLAGRELLO NERO (Costantini *et al.* 2005).

VITICULTURAL CHARACTERISTICS

Mid ripening.

WHERE IT'S GROWN AND WHAT ITS WINE TASTES LIKE

Ripolo is cultivated almost exclusively on the Amalfi coast in the province of Salerno, south of Napoli, **Italy**, in particular in the communes of Amalfi, Furore and Positano, where it is generally trained (often ungrafted) on pergolas. Ripolo may be included in the Costa d'Amalfi DOC, unofficially at the moment. The grapes have high sugar levels and moderate acidity. Varietal wines have scents of exotic fruits, apricot and honey, and with age they tend to develop kerosene aromas similar to those of some mature RIESLING wines. Marisa Cuomo's Furore Bianco Fiorduva, Costa d'Amalfi DOC, is a dry, barrel-fermented blend of roughly equal proportions of late-harvested Ripolo, FENILE and GINESTRA.

RITINO

Very minor Greek variety used in local blends.

BERRY COLOUR: ● ● ● ● ●

PRINCIPAL SYNONYMS: Aretino, Reteno, Retina, Retino, Ripno

ORIGINS AND PARENTAGE

Ritino probably comes from the Greek Aegean Islands.

WHERE IT'S GROWN AND WHAT ITS WINE TASTES LIKE

Ritino is found to a very limited extent on the **Greek** islands of Évvoia (Euboea), Ikaría, Sámos and the Sporádes in the Aegean Sea and in Magnesía. The Sámos Co-operative include it in a rosé blend with FOKIANO and the Ikarian Winery include it in a red blend with other local varieties.

RIVAIRENC

Ancient, once widespread, nearly extinct southern French vine. Better known as Aspiran Noir.

BERRY COLOUR: ● ● ● ● ●

PRINCIPAL SYNONYMS: Aspiran, Aspiran Noir, Épiran (Gard and Hérault), Esperan, Espiran (Gard and Hérault), Peyral, Peyrar, Piran (Gard and Hérault), Ribeyrenc (Aude), Riverain, Riveirenc Noir, Riveyrenc (Aude), Riveyrene (Aude), Spiran (Gard and Hérault), Verdai, Verdal

ORIGINS AND PARENTAGE

Under its old synonym Esperan, this variety may well have been mentioned for the first time some years before 1544 in the 'Chant de vendanges' ('harvest song') by Bonaventure des Périers (Rézeau 1997): 'La Gabbie/Jà rougie/Du sang des bruns esperans' ('The wine press/Already reddened/By the blood of the brown esperans'). It was later mentioned under the name Espiran in 1676 by Magnol in his *Botanicum Monspeliense*: 'Inter nigras [vites] vulgò comendatur espiran, quae tum ad esum, tum ad vinum optima est' ('Among the widespread black [varieties] is recommended espiran, which is excellent as food as well as wine'). Several additional citations were found in the eighteenth and nineteenth centuries under the variety's historical name Aspiran Noir in the region between Nîmes and Montpellier, southern France, which can be considered the birthplace of this variety, not least because several colour mutations were found in this area: Aspiran Blanc (white), Aspiran Gris (grey), Aspiran Rouge Foncé (red) and Aspiran Verdal (pink), although these are no longer found.

The etymology of Aspiran is unknown but it is unlikely that this variety was named after the village of Aspiran in the Hérault in southern France, because it has never been cultivated there. Recently, it was included in the official French register of varieties as Rivairenc (along with Rivairenc Gris and Rivairenc Blanc), surprisingly using the name local to the Aude *département* rather than the name Aspiran Noir, formerly widespread in the Gard and the Hérault. The name Rivairenc probably derives from the Occitan *ribairenc* meaning 'riparian', *riverain* in French, which is one of its synonyms, suggesting that this variety was found on river banks.

DNA parentage analysis based on twenty-three DNA markers suggests a possible parent–offspring relationship between Rivairenc and CINSAUT (Vouillamoz, to be confirmed with more DNA markers), which is consistent with Rivairenc belonging to the Piquepoul ampelographic group (see p XXVII; Bisson 2009). According to Myles *et al.* (2011), the reference sample of Rivairenc (under the name Aspiran) in the United States Department of Agriculture's National Clonal Germplasm Repository could be identical to the American table grape Queen but this can be rejected by comparison with other DNA profiles (Vouillamoz).

Aspiran Noir was used to breed ASPIRAN BOUSCHET.

OTHER HYPOTHESES

Some authors believe that Rivairenc is identical to the *vitis narbonnensis spiralis* mentioned by Pliny the Elder.

VITICULTURAL CHARACTERISTICS

Mid to late ripening and low yielding. Susceptible to downy and powdery mildews, botrytis bunch rot and winter frosts. Big, thin-skinned berries and big, compact bunches. Traditionally pruned short.

WHERE IT'S GROWN AND WHAT ITS WINE TASTES LIKE

Widely planted in the Hérault and the Gard in southern **France** under the name Aspiran Noir before phylloxera struck the region in the late nineteenth century, it is now very hard to find even though it and the grey and white colour mutations are included in the official French register of varieties. Some vines found their way to **Algeria**. Wines used to be pretty light in every way but they generally went into blends. Thierry Navarre in Saint-Chinian may be the sole producer of a varietal wine but Clos des Centeilles include 10% Riveirenc (sic) Noir with 90% PIQUEPOUL NOIR in their Minervois C de Centeilles red, and their white C de Centeilles (Côtes du Brian) includes both Riveirenc Gris and Blanc with other local varieties. Rivairenc may constitute up to 10% of the blend in the red and rosé versions of Minervois though it is now extremely rare for it so to do.

RKATSITELI

Extremely useful, versatile, widely planted, originally Georgian variety.

BERRY COLOUR:

PRINCIPAL SYNONYMS: Baiyu (China), Corolioc (Republic of Moldova), Dedali Rcatiteli (Republic of Moldova), Mamali Rcatiteli (Republic of Moldova), Rkatiteli (Republic of Moldova), Rkatziteli, Topolioc (Republic of Moldova)

ORIGINS AND PARENTAGE

Rkatsiteli is one of the most ancient grape varieties in Georgia. Its name reflects the appearance of the vine, combining *rka*, meaning 'vine shoot', and *tsiteli*, meaning 'red'. Recent DNA studies have shown that it is genetically very close to local wild grapevines (Ekhvaia *et al.* 2010), suggesting a possible direct domestication that allegedly took place centuries ago in the region of the Lekhura gorge. However, very few DNA markers were used and a much more detailed study is needed to confirm this hypothesis.

Rkatsiteli was used to breed KANGUN, PERVENETS MAGARACHA and PODAROK MAGARACHA.

OTHER HYPOTHESES

Some writers claim that clay vessels dating back to 3000 BC and containing Rkatsiteli seeds have been found in Georgia, others suggest it was the first vine planted by Noah after the flood, but there is no botanical, archaeological, historical nor genetic evidence for these assertions. Like many Georgian varieties, Rkatsiteli is said to have been mentioned as early as the fifth century AD (Chkhartishvili and Betsiashvili 2004) but there is no evidence for this.

VITICULTURAL CHARACTERISTICS

Medium-sized to large bunches of small berries. Grapes can reach high sugar levels while still retaining high acidity. Late budding and ripening, productive, good resistance to phylloxera and cold winter temperatures.

WHERE IT'S GROWN AND WHAT ITS WINE TASTES LIKE

Rkatsiteli is by far the most widely planted variety of either colour in **Georgia** (19,503 ha/48,193 acres in 2004, considerably less than before Gorbachev's vine-pull scheme), particularly throughout Kakheti in the south east of the country, with some of the best wines coming from the villages of Kardenakhi and Tibaani. It is used not only for decent table wines in appellations such as Tsinandali, Gurjaani and Vazisubani (each one allowing up to 15% MTSVANE KAKHURI in the blend) but also for fortified wines in hotter parts of Kakheti such as Kardenakhi. Recommended producers of more traditional style wines, vinified on the skins in clay qvevri and often amber in colour with apple-skin flavours, include Alaverdi Monastery, Pheasant's Tears and Vinoterra. European styles, which tend to be lightly floral or with stone-fruit, apple or quince flavours, crisp acidity and lots of fruit, are made by the likes of Badagoni, Chateau Mukhrani, Schuchman, Tbilvino, Telavi Wine Cellar and Teliani Valley.

Rkatsiteli is very popular in most of the ex-Soviet states, being valued for its productivity, cold hardiness and ability to retain acidity – and the fact that the grapes can withstand transport well. It is the most planted variety in **Ukraine** with the sizeable total area of 11,552 ha (28,546 acres) in 2009 and the **Republic of Moldova** recorded a total of 11,508 ha (28,437 acres) in 2009. It is also planted in **Russia** but to a more limited extent (569 ha/1,406 acres in Rosotov in 2010 and 133 ha/329 acres in Krasnodar Krai in 2010) and elsewhere in Eastern Europe. During the Soviet era, Rkatsiteli was the most widely planted grape in **Armenia** but plantings had fallen to around 7% of the vineyard area by the end of the twentieth century. Yet it is still the most widely planted white variety in **Bulgaria** (12,631 ha/31,212 acres in 2009) and is also found in **Romania** (354 ha/875 acres in 2008).

Konstantin Frank introduced the variety to **New York** State's Finger Lakes while Horton did the same in **Virginia** and there are tiny plantings in other American states.

Under the name Baiyu it was introduced to **China** in 1956 and has adapted well there. Today it is widely grown in northern China, as well as in Shandong, Jiangsu and the former Yellow River region of Anhui province. Chinese winemakers use it for sweet, dry and sparkling wines as well as for distillation.

ROBOLA

High-quality Greek island white producing elegant, sometimes powerful, dry lemony wines.

BERRY COLOUR: ● ● ● ● ●

PRINCIPAL SYNONYMS: Asporombola, Asprorobola, Asprorompola, Robbola, Robola Aspri, Robola Aspro, Robola Kerini, Robolla, Rombola, Rombola Aspri, Rompola, Rompola Kerine

VARIETIES COMMONLY MISTAKEN FOR ROBOLA: GOUSTOLIDI, RIBOLLA GIALLA (Friuli)

ORIGINS AND PARENTAGE

The name Robola has been used to refer to a wine made of black, red or white grapes and also to several distinct varieties, especially the light-skinned GOUSTOLIDI and the dark-skinned THEIAKO MAVRO (Boutaris 2000). This account describes the white-berried Robola that comes from either the Ionian Islands off Greece's western coast or from the Pelopónnisos (Peloponnese).

Robola was for many years presumed to be related to Friuli's RIBOLLA GIALLA. It was argued that the vine had been transported from Italy to Kefaloniá by the Venetians in the thirteenth century (Manessis 2000; Nikolau and Michos 2004). However, recent DNA profiling has shown that Robola (under the names Robola Aspri or Robolla) is not related to Ribolla Gialla (Sladonja *et al.* 2007; Štajner *et al.* 2008; Vouillamoz) but it is genetically very close to Greece's GOUSTOLIDI and THEIAKO MAVRO (Biniari and Stavrakakis 2007).

The pink-berried **Rombola Kokkino** and the black-berried **Mavro Rombola** are considered to be colour mutations of Robola (Nikolau and Michos 2004).

VITICULTURAL CHARACTERISTICS

Vigorous, high yielding and best suited to poor soils on well-drained slopes, often as bushvines. Susceptible to powdery mildew, botrytis bunch rot and virus diseases. Smallish, open bunches of small berries.

WHERE IT'S GROWN AND WHAT ITS WINE TASTES LIKE

Robola, grown mainly on the Ionian Islands, **Greece**, particularly on Kefaloniá, Zákynthos, Kérkyra (Corfu) and Lefkáda, is the most highly regarded white in the region and the third most planted variety (276 ha/682 acres in 2008). It produces wines that are fresh, lightly aromatic and citrus-flavoured – lime, lemon or grapefruit – though the variety is prone to oxidation and needs careful handling during vinification. The timing of the harvest is critical to retain acidity and also to maintaining Robola's depth of flavour. The best wines, typically from vines at altitudes above 300 m (984 ft), can be quite powerful and have a fine chalky texture and a marked persistence. Such wines are most often encountered within the Robola of Kefaloniá appellation, which is for certain communes in the south of the island and vineyards above 50 m, but the variety is now being planted further afield, eg in Arkadía in the central Pelopónnisos. Recommended producers include Domaine Foivos, Divino, Gentilini, who have vines planted in the late 1950s after the 1956 earthquake, the Robola Co-operative and Vassilakis.

RODITIS

Very widely planted Greek variety with many clonal variations and generally low ambitions.

BERRY COLOUR:

PRINCIPAL SYNONYMS: Alepou Roditis, Arilogos Roditis, Arsenikos Roditis, Kanellato, Kritsanisti, Lisitsines, Litsitsines (Mégara), Rhodites, Rhoditi, Rhoditis, Rodea Stafyli, Rodites, Roditi, Roditis Alepou, Roditis Kanellatos (Lefkáda), Roditis Rosé, Rodomoussi (Santoríni), Rogdites, Rogditis (Ionian Islands), Roidites, Roiditis, Roïdo, Roigditis, Sourviotes or Sourviotis (Kastoriá), Thilikos Roditis

ORIGINS AND PARENTAGE

The name Roditis, probably derived from the island of Ródos (Rhodes; Lefort and Roubelakis-Angelakis 2001) rather than from the colour of the berries (*rodon* means 'rose'; Manessis 2000), is given to at least three distinct varieties cultivated throughout Greece, each with a distinct berry colour and DNA profile: Roditis (pink), discussed here, Roditis Lefkos (white and no longer cultivated), also called Kolokythas Lefkos, and Roditis Kokkinos (red), which is a synonym for TOURKOPOULA (Lefort and Roubelakis-Angelakis 2001; Greek Vitis Database).

OTHER HYPOTHESES

According to Nikolau and Michos (2004), this variety was mentioned as early as the first century in the *Greek anthology* by Philippus of Thessalonica under the name Rodea Stafyli, and then in the *Geoponica* as the Roditis wine by lawyer Cassianus Bassos. In the ninth century, the Patriarch of Constantinople Photios cited Rodonia as a 'species of grapevine with almost pink berries'. However, these names might simply refer to grapes or wines from Ródos, or simply to rosé wines, and there is no botanical evidence that they corresponded to the modern Roditis.

Alepou is considered by Lefort and Roubelakis-Angelakis (2001) to be a synonym for Roditis Lefkos (*lefkos* means 'white') whereas local growers and other ampelographers use the name Roditis Alepou for the darker clone of the pink-berried Roditis.

VITICULTURAL CHARACTERISTICS

Vigour and productivity depend very much on clone and site (notably altitude). Resistant to drought but susceptible to powdery mildew.

WHERE IT'S GROWN AND WHAT ITS WINE TASTES LIKE

There is considerable confusion over this old variety, deepened by the fact that many Roditis vineyards in **Greece** are interplanted with what are said to be several different clones. For example, most growers believe that Roditis Alepou (*alepou* meaning 'fox' and referring to the colour of the berries), once far more common in the northern Pelopónnisos (Peloponnese) than it is now, is a superior clone of Roditis, with lower yields of more deeply coloured, more aromatic, smaller berries. It also ripens up to two weeks earlier than other clones. However, without DNA analysis, there is as yet no evidence to prove that it is not a distinct variety. According to viticulturist Haroula Spinthiropoulou, Roditis is very susceptible to viruses, especially closteroviruses, which alter the colour of the berries, and many of the clones are the result of adaptation to different climatic conditions.

Most wines labelled Roditis are therefore made from a field blend. The best examples are full-bodied with melon and apple aromas, a sour-citrus freshness and moderate acidity. They are not designed for ageing. The most common, pink-berried clone is both vigorous and productive, especially in more fertile soils and on lower sites, resulting in wines that are neutral and lack concentration or are used in the production of retsina.

Producers in Pátra in the northern Pelopónnisos, notably Antonopoulos, Oenoforos and Parparoussis, seem to be paying most attention to the Roditis Alepou clone, although Skouras further south is also investing in this clone, preferring its aromatic spiciness and its greater longevity. Two other clones considered worthy of greater investment are the Kanellatos ('cinammon') clone and Roditis from Mégara in Attikí (Lazarakis 2005). Oenoforos have been particularly active in trying to identify and select the various clones.

Roditis is planted in many areas of mainland Greece and the Pelopónnisos, from Thráki (Thrace) in the north east to Lakonía in the south (Lazarakis 2005). Varietal wines qualify for the Pátra appellation, where the variety is particularly significant; the Anhialos appellation in Thessalía requires a 50/50 blend of Roditis and SAVATIANO; and the Côtes de Meliton appellation in Halkidikí permits a blend of 35% Roditis with 50% ATHIRI and 15% ASSYRTIKO.

Also important in the make-up of many regional wines, Roditis was the second most planted white variety in Greece in 2008 (after SAVATIANO), with a total of 9,743 ha (24,075 acres), including 5,383 ha (13,302 acres) in western Greece and 1,407 ha (3,477 acres) in the Pelopónnisos. When yields are limited and sites are cooler, wines show much better concentration and weight at around 13.5% alcohol. Recommended producers include Babatzim (Kentrikí Makedonía/Central

Macedonia), Katogi Averoff (Pelopónnisos), Kir-Yianni (Náoussa) and Mercouri (western Pelopónnisos).

ROESLER

Relatively recent, well-coloured Austrian hybrid.

BERRY COLOUR: ● ● ● ● ●

ORIGINS AND PARENTAGE

A ZWEIGELT × Klosterneuburg 1189-9-77 hybrid obtained in 1970 by Gertrude Mayer at the Klosterneuburg research centre in Austria, in which:

- Klosterneuburg 1189-9-77 is a Seyve-Villard 18-402 × BLAUFRÄNKISCH hybrid
- Seyve-Villard 18-402 is a Seibel 7162 × Seyve-Villard 12-308 hybrid
- Seibel 7162 is a PLANTET × Seibel 5163 hybrid (see PRIOR for the complete pedigree of Seibel 5163)
- Seyve-Villard 12-308 is a Seibel 6468 × Subéreux hybrid (see BRIANNA for the complete pedigree of Seibel 6468 and PRIOR for the complete pedigree of Subéreux).

Klosterneuburg 1189-9-77 was also involved in the breeding of Roesler's half-sibling RÁTHAY. Roesler was named after Leonhard Roesler (1839–1910), former director of Klosterneuburg.

VITICULTURAL CHARACTERISTICS

Mid ripening, good overall disease resistance.

WHERE IT'S GROWN AND WHAT ITS WINE TASTES LIKE

Roesler produces deeply coloured reds with moderate tannins, well suited to barrel ageing. It is planted in **Austria** on about 137 ha (339 acres) and producers such as Vogl in the Weinviertel or Bründy in Wagram blend Roesler with RÁTHAY. Tschida produce a sweet TBA wine in Apetlon on the west side of the Neusiedlersee. Weingut zum Frohhof, the lone producer in **Switzerland**, also use Roesler in blends.

ROKANIARIS

Rarity from the eastern Pelopónnisos (Peloponnese).

BERRY COLOUR: ● ● ● ● ●

ORIGINS AND PARENTAGE

Rokaniaris is local to the Kranídi area in the southern part of Argolída in the eastern Pelopónnisos, Greece. No DNA profiling has yet been done to confirm that it is a distinct variety.

VITICULTURAL CHARACTERISTICS

Vigorous, early budding and mid ripening. Relatively resistant to disease and drought.

WHERE IT'S GROWN AND WHAT ITS WINE TASTES LIKE

Kontovraki in Argolída in the north-eastern Pelopónnisos, **Greece**, make a varietal wine described as having fresh acidity, complex aromas and tasting of peaches and similar stone fruits.

ROLLE

See VERMENTINO.

ROLLO

Rare, old Ligurian (possibly Tuscan?) with a confusing array of synonyms.

BERRY COLOUR: ● ● ● ● ●

PRINCIPAL SYNONYMS: Bruciapagliaio (or Bujapajà, Brusapajà in dialect; Cinque Terre in Liguria), Capello (Toscana), Livornese Bianca (Massa-Carrara), Occhiana (Toscana), Pagadebiti (Corse), Rollo Genovese

VARIETIES COMMONLY MISTAKEN FOR ROLLO: ?BIANCONE DI PORTOFERRAIO (island of Elba in Toscana), BIANCU GENTILE (Corse), VERMENTINO (north-west Italy, Alpes-Maritimes and Var in France)

ORIGINS AND PARENTAGE

Rollo is an old variety of Liguria in north-west Italy that was often confused with VERMENTINO, also called Rolle or Rollé in the Alpes-Maritimes and the Var in southern France. DNA profiling has recently shown that Rollo is distinct from Vermentino and that it is identical to:

- Bruciapagliaio (Torello Marinoni, Raimondi, Mannini and Rolle 2009), an old and rare Ligurian
- Livornese Bianca (Di Vecchi Staraz, This *et al.* 2007; Torello Marinoni, Raimondi, Mannini and Rolle 2009), a very minor blend component used occasionally in Toscana, central Italy
- Occhiana (Di Vecchi Staraz, This *et al.* 2007), found in the region of Gavorrano in the province of Grossetto in Toscana
- Pagadebiti from Corse (Di Vecchi Staraz, This *et al.* 2007; Torello Marinoni, Raimondi, Mannini and Rolle 2009) – this identity was already known to ampelographers.

It should be noted that Di Vecchi Staraz, This *et al.* (2007) consider Biancone from Corse to be a synonym for Rollo but its DNA profile is distinct from that shown for Pagadebiti in the same study and that given for Rollo in Torello Marinoni, Raimondi, Mannini and Rolle (2009). A more detailed DNA profiling study of Biancone and BIANCONE DI PORTOFERRAIO from the Tuscan island of Elba is therefore necessary to determine whether they are identical.

In addition, comparison of the DNA profiles of Rollo and MAMMOLO from Toscana (also found on the French island of Corse under the name Sciaccarello) suggests a possible parent–offspring relationship (Vouillamoz). This hypothesis should be further investigated to verify a possible Tuscan origin for Rollo.

VITICULTURAL CHARACTERISTICS

Mid ripening, productive but inconstant. Susceptible to botrytis bunch rot.

WHERE IT'S GROWN AND WHAT ITS WINE TASTES LIKE

Rollo is a tentacular variety cultivated under several names in various places, although the exact identity of the varieties is often uncertain.

ROLLO Within Liguria, along the coast of north-west **Italy**, Rollo is cultivated mainly in the province of Genova and in the Riviera Ligure di Ponente and virtually absent from the Riviera di Levante and the province of Savona. It is usually blended with varieties such as ALBAROLA, BOSCO and TREBBIANO TOSCANO. It may be part of the blend with VERMENTINO and Albarola (Bianchetta Genovese), among others, in the Val Polcevera DOC, as in Bruzzone's Coronata della Val Polcevera (Coronata is a subzone of the DOC). The Italian agricultural census recorded a total of 114 ha (277 acres) in 2000.

BRUCIAPAGLIAIO Bruciapagliaio is cultivated in the Cinque Terre, mainly in Val Polcevera in the province of Genova, **Italy**, and in Val Graveglia in the Riviera di Levante, though it does not appear by name in the census. Bruciapagliaio may be a minor part of the blend in the famous and expensive sweet passito-style Cinque Terre Sciacchetrà DOC, which is predominantly BOSCO, ALBAROLA and VERMENTINO.

LIVORNESE BIANCA Even though the name Livornese Bianca suggests a link to the port of Livorno in the far west of Toscana, **Italy**, the variety is cultivated mainly in the province of Massa-Carrara, further north towards La Spezia. The agricultural census of 2000 recorded just 11 ha (27 acres) planted, in particular in Versilia in the province of Lucca and in the historical region of Lunigiana, which straddles the provinces of La Spezia and Massa-Carrara. Livornese Bianca may make a minor contribution to the blend in the Colli di Luni DOC and in the Toscana and Val di Magra IGTs. Wines made from this variety tend to be low in alcohol.

PAGADEBITI Pagadebiti is a confusing name applied to several distinct varieties. Its literal meaning is 'paying the debts'. On the island of Corse, **France**, Pagadebiti usually corresponds to Rollo, but also sometimes to BIANCONE DI PORTOFERRAIO. It was once much more widespread but today it is cultivated on just a few hectares near Porto Vecchio, Figari and Sartène.

ROMÉ

Rare local Málaga speciality, possibly better at rosés than reds.

BERRY COLOUR: ● ● ● ● ●

PRINCIPAL SYNONYMS: Romé de Motril, Romé Negro (Motril in Spain)

ORIGINS AND PARENTAGE

Romé, considered to be indigenous to Málaga in south-west Spain, was mentioned by Clemente y Rubio (1807) in Motril in the province of Granada in Andalucía.

OTHER HYPOTHESES

Clemente y Rubio (1807) cites his predecessor Jean-Baptiste François Rozier, who argued that Romé could be identical to the French variety TEINTURIER: both have traditionally been used to deepen the colour of red wines (although see below) and their descriptions are fairly similar. However, unlike Teinturier, the berries of Romé do not have coloured flesh.

VITICULTURAL CHARACTERISTICS

Mid to late ripening and generally disease-resistant (though this may be due more to the climate in Málaga than to the variety).

WHERE IT'S GROWN AND WHAT ITS WINE TASTES LIKE

Romé is grown almost exclusively in the Axarquía region east of Málaga. Bodegas Bentomiz, who include the variety in their Ariyanas red blend, report that it contributes red fruit, and also some floral aromas and vanilla, and a creaminess in the mouth. They are also planning to experiment with a varietal rosé since the wine lacks colour and body and is relatively light on its own. Many locals use it for home wine production.

ROMEIKO

Greek island variety that has been important on Kríti (Crete).

BERRY COLOUR: ● ● ● ● ●

PRINCIPAL SYNONYMS: Loïssima (Kríti), Romeico, Romeiko Mavro, Tsardana (Chaniá and Réthymno in western Kríti)

VARIETIES COMMONLY MISTAKEN FOR ROMEIKO: Romeiko Machaira

ORIGINS AND PARENTAGE

It is not known whether Romeiko comes from the Aegean Islands or from Kríti, where it has traditionally been cultivated under the name Tsardana (Lefort and Roubelakis-Angelakis 2001). It should not be confused with Romeiko Machaira (no longer cultivated), which has larger berries and a distinct DNA profile (Lefort and Roubelakis-Angelakis 2001). The name Romeiko possibly derives from the Turkish word *romios* or *romioi* used to designate the Greeks (Boutaris 2000; Lefort and Roubelakis-Angelakis 2001).

VITICULTURAL CHARACTERISTICS

Productive and slightly susceptible to water stress.

WHERE IT'S GROWN AND WHAT ITS WINE TASTES LIKE

Romeiko is grownly mainly in western Kríti, **Greece**, in the Chaniá region, where it once comprised more than 80% of the vineyard area, and to a lesser extent in Réthymno. There were still almost 1,000 ha (2,470 acres) on the island in 2008. In the local Kissamos regional wine, for example, it is authorized for white wines (up to 70% of the blend and an unusual phenolic style, according to Lazarakis 2005), rosé (up to 60%) and reds (up to 60%), often blended with international rather than local varieties. Wines are generally high in alcohol, lack freshness and the colour quickly browns but this may be partly due to the overly high yields and lack of clonal selection. According to Lazarakis (2005), in the past Romeiko was used to make wines known as *marouvas*, mostly for home consumption: very ripe grapes finished fermenting in large barrels, where they were aged for four years without topping up, resulting in wines that were fierce but complex and drunk at the end of a meal. Today the variety seems out of kilter with modern wine styles.

Although Romeiko and Tsardana have been shown to be synonyms, the official Greek list of varieties for winemaking includes them as separate varieties.

ROMERO DE HÍJAR

Potentially interesting but very rare Aragonese speciality.

BERRY COLOUR:

PRINCIPAL SYNONYMS: Romero

ORIGINS AND PARENTAGE

Romero de Híjar is an obscure autochthonous variety named after the area of Híjar in the province of Teruel in Aragón, northern Spain, for which little information is available.

WHERE IT'S GROWN AND WHAT ITS WINE TASTES LIKE

Apparently only one producer, Fandos y Barriuso in Bajo Aragón, **Spain**, makes this commercially, producing about 1,250 bottles a year. They age it in barrel and describe it thus: 'It has an intense bouquet of black fruits, spices, brambles and damp soil. Fresh and flavourful in the mouth, somewhat fleshy with hints of liquorice and good acidity'. Sounds promising.

ROMORANTIN

Minor and difficult to ripen Burgundian variety now exclusive to Cour-Cheverny in the Loire.

BERRY COLOUR:

PRINCIPAL SYNONYMS: Bury, Dameret Blanc, Damery (named after a village in the Marne in the Champagne region), Dannery (possibly an alternative name for the same village), Dannezy, Maclon, Petit Dannezy

ORIGINS AND PARENTAGE

Romorantin is today cultivated exclusively in the Val de Loire but, like at least twenty other varieties, it is a PINOT × GOUAIS BLANC cross born in north-eastern France (Bowers *et al.* 1999; Boursiquot *et al.* 2004; see PINOT pedigree diagram). In fact, more accurate DNA parentage analyses have suggested that the father is not exactly PINOT NOIR but its colour mutation PINOT TEINTURIER (Bowers *et al.* 2000; Franks *et al.* 2002).

The earliest mention of Romorantin appeared in 1868 (Rézeau 1997): 'on cultive, dans les arrondissements de Blois et de Romorantin ... en cépages blancs, le romorantin ou gros pineau blanc de la Loire, le chenin, l'orbois, arbois ou meslier du Gâtinais' ('we cultivate in the *arrondissements* of Blois and Romorantin ... as white grape varieties romorantin...' etc.). It was named after the village of Romorantin (part of today's Romorantin-Lanthenay) in the Loir-et-Cher, where François I of France owned a residence and had supposedly ordered 80,000 plants of this vine from Burgundy in 1519, but there is no historical evidence for this introduction of Romorantin to the region. According to Mouillefert (1903), local tradition in the Loir-et-Cher has it that Romorantin was introduced around 1830 by an unknown vine grower in Villefranche-sur-Cher, whence it spread to Romorantin and then to the whole of the Loir-et-Cher. In any case, the Burgundian Romorantin, like MELON in Muscadet, won its spurs in the Loire. In spite of its parentage, Romorantin is considered by Bisson (2009) to be a member of the Cot ampelographic group (see p XXVII). Since it is also ampelographically close to Melon (Bisson 2009), classified among the Noirien ampelographic group based on PINOT, Romorantin's ampelographic position might change in the light of further DNA discoveries.

VITICULTURAL CHARACTERISTICS

Early budding, mid ripening. Reasonably fertile and productive, may be pruned long or short. Small berries.

WHERE IT'S GROWN AND WHAT ITS WINE TASTES LIKE

Romorantin has its own appellation, Cour-Cheverny, just west of Blois in the eastern Val de Loire, **France**, created in 1993. Once more widely cultivated, it was planted on a grand total of 73 ha (180 acres) in 2008, pretty much exclusive to this area. The wines range from dry and rather sharp to rich, quite broad-flavoured and yet still intensely fresh. It is hard to get the grapes fully ripe in some years and this is its major disadvantage since the acidity is then way too high. Examples from Thierry Puzelat and from Henri Marionnet at Domaine de la Charmoise suggest that wines from old, ungrafted vines seem to have much more character and depth. Other producers of varietal wines include Hervé Villemade at Domaine du Moulin, Michel Gendrier at Domaine des Huards, Domaine François Cazin and Franz Saumon. Cazin's Cuvée Renaissance is a late-harvested, botrytized, off-dry version.

RONDINELLA

Lesser blending partner in reds such as Valpolicella and Bardolino.

BERRY COLOUR: ● ● ● ● ●

ORIGINS AND PARENTAGE

The earliest mention of Rondinella dates back to 1882 and this relatively short history may explain why it has no common synonyms. DNA parentage analysis has provided evidence that Rondinella is an offspring of CORVINA VERONESE and another, unknown, possibly extinct, variety (Grando *et al.* 2006). This makes it a grandchild of REFOSCO DAL PEDUNCOLO ROSSO (see that entry for the pedigree diagram). Its name could derive from the Italian *rondini*, meaning 'swallows', in reference to the berry colour, which is similar to that of swallows' feathers.

VITICULTURAL CHARACTERISTICS

Very high yields, mid to late ripening. Good resistance to most fungal diseases but susceptible to esca.

WHERE IT'S GROWN AND WHAT ITS WINE TASTES LIKE

Rondinella is cultivated mainly in the province of Verona in northern **Italy**, in particular in Bardolino and Valpolicella, east of the Lago di Garda, where its prolific yields have in the past endeared it to producers of Valpolicella of all styles. Its other endearing characteristic is its resistance to fungal diseases, which is an advantage in the drying process used in the production of Amarone and Recioto. However, it is not as highly regarded or as flavourful as its main blending partner, CORVINA VERONESE, which is favoured by quality-minded producers. The DOC regulations for Bardolino and Valpolicella stipulate a maximum of 40% and 30% respectively, though it is often even less than this. Just over the border in Lombardia, the variety is a permitted component in Garda Colli Mantovani. The Italian agricultural census of 2000 recorded total plantings of 2,874 ha (7,101 acres). Rondinella produces fruity, cherry-flavoured wines but does not have the character or the stuffing to make a truly satisfactory varietal wine, which is why it is always blended with other local varieties such as Corvina Veronese, CORVINONE or, to an increasingly lesser extent, MOLINARA.

Freeman Vineyards in the Hilltops region of New South Wales are the sole growers of this variety in **Australia**. One hectare (2.5 acres) of vines produces wine that is blended with CORVINA VERONESE to make an Amarone-style wine. It grows prolifically and has to be crop-thinned by as much as 80%.

RONDO

German red-fleshed hybrid with a complicated genealogy well suited to cool climes makes light, fruity reds.

BERRY COLOUR:

PRINCIPAL SYNONYMS: Geisenheim 6494-5

ORIGINS AND PARENTAGE

A hybrid obtained by Helmut Becker in 1964 at the Geisenheim research centre in the Rheingau, Germany, by crossing Zarya Severa × SANKT LAURENT, where Zarya Severa is a Seyanets Malengra × *Vitis amurensis* hybrid, and Seyanets Malengra a seedling of PRÉCOCE DE MALINGRE. Rondo was used to breed ALLEGRO.

VITICULTURAL CHARACTERISTICS

Vigorous and early to very early ripening. Good resistance to downy mildew but more susceptible to powdery mildew. The grapes' red flesh helpfully increases the colour of the wines.

WHERE IT'S GROWN AND WHAT ITS WINE TASTES LIKE

Rondo was authorized in **Germany** in 1999 but plantings are now more numerous in **England**, where there are 39 ha (96 acres). Although Rondo represents less than 4% of the vineyard area, it is still the third most planted red wine variety in England. Yearlstone and Wroxeter, for example, both make varietal wines, as do Schönhals in Baden, southern Germany.

It is also found in other cool extremes of the wine world such as **Belgium**, **Poland** and the **Netherlands**. Rondo is the most widely planted variety in **Denmark** (20 ha/49 acres in 2010) and one of the most successful varieties in **Sweden**. Although it was bred to include certain *Vitis amurensis* genes to give it better resistance to cold winters, the light red wines it produces are sufficiently similar to the produce of *vinifera* vines for it to be officially registered as a *vinifera* variety so that it can be used in the production of quality wines. Experience in England shows that it responds well to a certain amount of oak, preferably American, although its wine may not be as good as REGENT's.

ROSÉ DU VAR

Dull and understandably declining pink-skinned Provençal variety.

BERRY COLOUR:

PRINCIPAL SYNONYMS: Barbaroux (Arcs), Grec Rose (Var), Roussanne du Var (Var), Rousseli (Var)

ORIGINS AND PARENTAGE

As its name suggests, Rosé du Var comes from the Var *département* of Provence in southern France but, despite its synonym Roussanne du Var, has nothing to do with the ROUSSANNE from the northern Rhône. DNA parentage analysis has shown that Rosé du Var is one of the numerous progenies of GOUAIS BLANC (Boursiquot *et al.* 2004; see PINOT pedigree diagram).

VITICULTURAL CHARACTERISTICS

Early budding, late ripening. Fertile and productive, best pruned short. Very susceptible to botrytis bunch rot. Big bunches and berries.

WHERE IT'S GROWN AND WHAT ITS WINE TASTES LIKE

Wines are vinified white or very pale pink, generally dull and low in alcohol, and sometimes included in rosé blends. This lack of potential quality and its exclusion from the Côtes de Provence appellation in southern **France** in 1986 probably account for the variety's continuing decline, down from 1,557 ha (3,847 acres) in 1968 to 65 ha (161 acres) in 2008. Domaine de Sauvecanne in the Var is one of the few producers to make a varietal Rosé du Var, a rosé classified as Vin de Pays des Maures.

ROSE HONEY

As yet unidentified minor variety found in China's Yunnan province.

BERRY COLOUR: ● ● ● ● ●

PRINCIPAL SYNONYMS: Honey, Red

ORIGINS AND PARENTAGE

According to an unverified online source (www.cnwinenews.com), Rose Honey was imported into China in 1965 by Dongfeng Farm in Mile county, Yunnan province, southern central China, along with ten other grape varieties, including Yunnan's two other curiosities, FRENCH WILD and CRYSTAL. Since DNA profiling has not yet been done, it is not known whether Rose Honey is related to European varieties. According to Sun (2004), it is a *Vitis vinifera* × *Vitis labrusca* hybrid more suitable for eating and processing than for wine.

OTHER HYPOTHESES

Rose Honey was allegedly introduced from Europe to the Shangri-La region of Yunnan by missionaries at the beginning of the nineteenth century, along with CRYSTAL and FRENCH WILD, but there is no documentary evidence to support this.

VITICULTURAL CHARACTERISTICS

Vigorous, fertile, mid ripening. Resistant to drought and disease and survives well in poor soils.

WHERE IT'S GROWN AND WHAT ITS WINE TASTES LIKE

Along with FRENCH WILD and CRYSTAL, Rose Honey is widely planted in Yunnan, southern **China**, and is one of the main red varieties grown by the Yunnanhong (Yunnan Red) Wine Company. Wines tend to be ruby-coloured and full-bodied and have the sweet foxy flavours typical of wines made from *labrusca* vines or their hybrids. According to Chinese statistics, Honey or Rose Honey is also grown in Sichuan (60 ha/148 acres), immediately north of Yunnan, and in Jilin province (112 ha/277 acres) in the north east.

ROSETTA

Red-skinned German hybrid used mainly for the table.

BERRY COLOUR: ● ● ● ● ●

PRINCIPAL SYNONYMS: Freiburg 991-60

ORIGINS AND PARENTAGE

A VILLARD BLANC × SIEGERREBE hybrid created in 1960 at the Freiburg research centre in Baden-Württemberg, southern Germany, by Johannes Zimmermann.

VITICULTURAL CHARACTERISTICS

Vigorous, mid to late ripening.

WHERE IT'S GROWN

Although Rosetta is used mainly as a table grape, Dale Valley in **Iowa** make a varietal wine.

ROSETTE

Ardèche hybrid still planted in New York State.

BERRY COLOUR: ● ● ● ● ●

PRINCIPAL SYNONYMS: Francuz Seibel, Seibel 1000

ORIGINS AND PARENTAGE

A complex hybrid obtained at the beginning of the twentieth century by Albert Seibel in Aubenas, southern France, by crossing *Vitis rupestris* × Munson (see PRIOR for the complete pedigree of Munson).

VITICULTURAL CHARACTERISTICS

Vigorous and generally very fertile but susceptible to coulure. Hardy and moderately productive. Early ripening.

WHERE IT'S GROWN AND WHAT ITS WINE TASTES LIKE

Widely planted in **France** prior to 1960 but eliminated from French vineyards by 1983 (Galet 2000), it was one of the first French hybrids to be grown in **New York** State, now surviving in just a few pockets (total 10 acres/4 ha) in the Finger Lakes region, where it has been largely ousted by *vinifera* varieties but is still used by producers such as Glenora and Fulkerson in light reds and rosé blends. Aracadian Estate make a varietal blush wine.

It is unrelated to the sweet-wine Rosette appellation just north of Bergerac in south-west France.

ROSSARA TRENTINA

Trentino interrelated rarity.

BERRY COLOUR:

PRINCIPAL SYNONYMS: Geschlafene, Ross Ciàr, Rossar, Rossera, Varenzasca (Piemonte)

VARIETIES COMMONLY MISTAKEN FOR ROSSARA TRENTINA: Rossara (Reggio Emilia), ROSSOLA NERA (Valtellina)

ORIGINS AND PARENTAGE

As its name suggests, Rossara Trentina most probably originates from Trentino in Italy's far north. The name Rossara refers to the clear red colour of its berries and its must. DNA profiling has established that Rossara Trentina is distinct from the variety named Rossara (or Rosséra, Rossana), today almost extinct in Reggio Emilia (Boccacci *et al.* 2005), but identical to Varenzasca in Piemonte (Anna Schneider and José Vouillamoz, unpublished data). It has a possible genetic relationship with SGAVETTA from Reggio Emilia (Boccacci *et al.* 2005) and DNA parentage studies have also revealed that Rossara Trentina is a likely progeny of GOUAIS BLANC (Vouillamoz; see PINOT pedigree diagram).

VITICULTURAL CHARACTERISTICS

Vigorous, productive and late ripening. Resistant to downy mildew but susceptible to powdery mildew, botrytis bunch rot and flavescence dorée.

WHERE IT'S GROWN AND WHAT ITS WINE TASTES LIKE

Rossara Trentina is grown mainly on the Rotaliano plain north of the city of Trento in the far north of **Italy**, where it has traditionally been planted alongside TEROLDEGO and NEGRARA TRENTINA, with which it is often blended. The Italian agricultural census recorded 34 ha (84 acres) in 2000.

One of the very few varietal wines is produced by Zeni and classfied as an IGT Vigneti delle Dolomiti. They describe it as fresh and perfumed with apple-like acidity.

ROSSESE

ORIGINS AND PARENTAGE

A variety named Rossese Bianco or Roxeise produced excellent wines in the fifteenth century, according to Bacci (1596). However, it is impossible to determine the identity of this variety because DNA profiling has recently shown that the name Rossese, probably derived from the intense colour of the berries, has been confusingly used for at least seven distinct varieties – some dark-skinned, some pale-skinned – in Liguria and other parts of northern Italy (Torello Marinoni, Raimondi, Mannini and Rolle 2009).

Dark skinned:

– ROSSESE DI CAMPOCHIESA, in the area of Campochiesa near Albenga in province of Savona in Liguria

- Rossese di Dolceacqua, in the area of Dolceacqua north of Ventimiglia in Liguria, close to the French border, now proven by DNA profiling to be identical to TIBOUREN, an old variety of Provence in the south of France. Although its exact origin remains unknown, the use of the name Tibouren helps to avoid any confusion with the other Rossese varieties.

White or pink skinned:

- Rossese Bianco, in Arcola in the province of La Spezia in Liguria (see RUZZESE, its local name)
- Rossese Bianco, in Riomaggiore in the province of La Spezia in Liguria, now known to be identical to GRILLO from Sicilia
- ROSSESE BIANCO DI SAN BIAGIO, in San Biagio della Cima and Soldano in the province of Savona in Liguria
- ROSSESE BIANCO, in Roddino and Sinio in the province of Cuneo in Piemonte (included in the Italian national register of varieties under Rossese Bianco)
- ROSSESE BIANCO DI MONFORTE, in Monforte d'Alba in the province of Cuneo in Piemonte.

It should be noted that Rossese from Campochiesa, Rossese Bianco from Arcola (see RUZZESE), Rossese Bianco from San Biagio and Rossese Bianco from Monforte d'Alba all have unique DNA profiles although none of them is yet included in the national register of varieties.

ROSSESE BIANCO

Rare Piemontese making light, fresh whites.

BERRY COLOUR:

ORIGINS AND PARENTAGE

The name Rossese is a source of confusion (see ROSSESE) but Rossese Bianco, which is not a white mutation of Rossese di Dolceacqua (also known as TIBOUREN), is a rare variety still thriving in Roddino and Sinio in the Langhe region of Piemonte.

WHERE IT'S GROWN AND WHAT ITS WINE TASTES LIKE

Rossese Bianco is cultivated mainly in Roddino and Sinio in Piemonte, north-west **Italy**, where it gives light and fresh varietal wines such as Cantine Gemme's Verdiana, Langhe DOC.

ROSSESE BIANCO DI MONFORTE

Extremely rare, perfumed white wine variety exclusive to the Langhe in Piemonte.

BERRY COLOUR: ● ● ● ● ●

ORIGINS AND PARENTAGE

The name Rossese is a source of confusion (see ROSSESE) but Rossese Bianco di Monforte is a rare variety from Monforte d'Alba in the Langhe region of Piemonte.

WHERE IT'S GROWN AND WHAT ITS WINE TASTES LIKE

As suggested by its name, Rossese Bianco di Monforte is cultivated exclusively in Monforte d'Alba in Piemonte, north-west **Italy**. It was rescued from extinction in 1980 by Giovanni Manzone from old vines that had survived in a family vineyard. (There is as yet no DNA evidence to confirm the identity of these old vines but their location fits with the known distribution of Rossese Bianco di Monforte.) Giovanni Manzone's perfumed, fresh Rosserto, Langhe DOC, is probably the only example of a varietal wine (total production 2,500 bottles per year).

ROSSESE BIANCO DI SAN BIAGIO

Extremely rare Ligurian white authorized only for table wines.

BERRY COLOUR: ● ● ● ● ●

VARIETIES COMMONLY MISTAKEN FOR ROSSESE BIANCO DI SAN BIAGIO: GRILLO (known as Rossese Bianco in Riomaggiore in Liguria)

ORIGINS AND PARENTAGE

The name Rossese is a source of confusion (see ROSSESE) but Rossese Bianco di San Biagio is a rare variety still thriving in San Biagio della Cima and in Soldano in the province of Savona in Liguria, north-west Italy.

VITICULTURAL CHARACTERISTICS

Grapes tend to have high acidity levels.

WHERE IT'S GROWN

Rossese Bianco di San Biagio is not included in the national register of grape varieties in **Italy** and can therefore be used only for table wines. Tenuta Anfosso's Rossese Bianco Vino da Tavola is a rare example of varietal wine made from vines more than a hundred years old. (There is as yet no DNA evidence to confirm the identity of these old vines but their location fits with the known distribution of Rossese Bianco di San Biagio.)

ROSSESE DI CAMPOCHIESA

Rare Ligurian producing fruity reds for early drinking.

BERRY COLOUR: ● ● ● ● ●

VARIETIES COMMONLY MISTAKEN FOR ROSSESE DI CAMPOCHIESA: TIBOUREN (also known as Rossese di Dolceacqua, Rossese Nericcio or Rossese di Ventimiglia in Liguria)

ORIGINS AND PARENTAGE

Rossese di Campochiesa was long thought to be identical to Rossese di Dolceacqua, but recent DNA profiling has shown that they are distinct varieties and that Rossese di Dolceacqua is in fact identical to TIBOUREN (Torello Marinoni, Raimondi, Mannini and Rolle 2009).

WHERE IT'S GROWN AND WHAT ITS WINE TASTES LIKE

As its name suggests, Rossese di Campochiesa is cultivated mainly around Campochiesa on the Ligurian coast of **Italy**, between Genova and Nice, where it produces Rossese Riviera Ligure di Ponente DOC wines, characterized by aromas of strawberries and raspberries and intended for early drinking. However, given the frequent confusion between Rossese di Campochiesa and Rossese di Dolceacqua (ie TIBOUREN), the Rossese Riviera Ligure di Ponente DOC wines produced further south in the province of Imperia are more likely to be made from Rossese di Dolceacqua.

Lupi and Rosella Saguato are among the few producers of Rossese Riviera Ligure di Ponente DOC wines truly made from 100% Rossese di Campochiesa.

ROSSESE DI DOLCEACQUA

See TIBOUREN.

ROSSETTO

Rare Lazio vine only recently showing its potential for quality.

BERRY COLOUR: ● ● ● ● ●

PRINCIPAL SYNONYMS: Greco di Velletri, Greco Giallo, Greco Verde, Roscetto

VARIETIES COMMONLY MISTAKEN FOR ROSSETTO: GRECO (Campania), GRECO BIANCO (Calabria), TREBBIANO GIALLO

ORIGINS AND PARENTAGE

Although this variety's berries are essentially white, it is named Rossetto because they turn pinky red (Latin *russum*) when fully ripe. A similarity between Rossetto and TREBBIANO GIALLO has been suggested (Giavedoni and Gily 2005). Recent ampelographic and DNA studies have shown that although Rossetto is also known locally as Greco, Greco Giallo or Greco di Velletri, its DNA profile is quite distinct from those of the Campanian GRECO and the Calabrian GRECO BIANCO (Muganu *et al.* 2009).

VITICULTURAL CHARACTERISTICS

Resistant to fungal diseases, suitable for late harvest.

WHERE IT'S GROWN AND WHAT ITS WINE TASTES LIKE

Historically cultivated between Lazio and Umbria in **Italy**, in particular in the regions of Castelli Romani and between Orvieto and the Lago di Bolsena. It was close to extinction in the 1960s but had never been very popular with farmers because of its low yields. It continued to be used in very small amounts (up to 15%) blended with TREBBIANO TOSCANO and MALVASIA BIANCA LUNGA in Est! Est!! Est!!! di Montefiascone. To avoid the pinkish colour, growers have traditionally picked the grapes prematurely, thus sacrificing the depth of complex citrus fruit flavours that it can produce. A varietal Rossetto is authorized within the Colli Etruschi Viterbesi DOC in Lazio.

Oenologists Ricardo and Renzo Cottarella have been influential in promoting this variety and showing its real potential from their vineyards in Montefiascone in Lazio but they found they had to flash freeze the grapes to release the yellow pigment from the skins into the must. Their oaked Falesco Ferentano is a rich but very well-balanced varietal wine, reminiscent of a top-notch Saint-Aubin.

ROSSIGNOLA

Minor and rather tart variety used to a limited extent in Veneto blends.

BERRY COLOUR: ● ● ● ● ●

PRINCIPAL SYNONYMS: Rossetta, Rossetta di Montagna, Rossola

VARIETIES COMMONLY MISTAKEN FOR ROSSIGNOLA: Rosetta di Montagna (Lago di Garda), ROSSOLA NERA (Valtellina)

ORIGINS AND PARENTAGE

The earliest mention of Rossignola as an autochthonous variety of the province of Verona was in 1818 (Pollini 1824). Although Rossignola has synonyms such as Rossetta, Rossetta di Montagna or Rossola, it should not be confused with Rosetta di Montagna, an old variety once grown near the Lago di Garda but no longer cultivated commercially, nor with ROSSOLA NERA from Valtellina.

VITICULTURAL CHARACTERISTICS

Vigorous, high yielding and late ripening. Susceptible to downy and powdery mildews, botrytis bunch rot, sour rot and esca.

WHERE IT'S GROWN AND WHAT ITS WINE TASTES LIKE

Rossignola's grapes and therefore wine tend to be high in acidity. It is grown in Valpolicella as well as in the Lago di Garda area, where a small amount may be blended with other local varieties in the Bardolino, Breganze Rosso, Garda Orientale, Valdadige and Valpolicella DOCs. It is also cultivated to a much more limited extent in the province of Vicenza. Total plantings in **Italy** in 2000 were 341 ha (843 acres).

ROSSOLA NERA

Minor, high-acid Valtellina vine.

BERRY COLOUR:

PRINCIPAL SYNONYMS: Rossera, Rossola, Rossolo

VARIETIES COMMONLY MISTAKEN FOR ROSSOLA NERA: ROSSARA TRENTINA, ROSSIGNOLA, ROSSOLINO NERO

ORIGINS AND PARENTAGE

Rossola Nera was already mentioned in Valtellina in the early seventeenth century in a Latin text under the name Rossoladure (Bongiolatti 1996). The variety's name derives from the pink-red (Italian *rosso*) colour of the berry at maturity. ROSSOLINO NERO is usually taken to be the most widespread clonal variation of Rossola Nera (Bongiolatti 1996; Fossati *et al.* 2001) but DNA profiling has established that they are distinct varieties, linked by a parent–offspring relationship (Vouillamoz). In addition, Rossola Nera has been shown to have a parent–offspring relationship with NEBBIOLO (Schneider, Boccacci *et al.* 2004). Chronologically, it makes sense to consider Rossola Nera to be an offspring of Nebbiolo and a parent of Rossolino Nero.

VITICULTURAL CHARACTERISTICS

Late ripening. Vigorous, susceptible to powdery mildew and botrytis bunch rot, good resistance to spring frosts.

WHERE IT'S GROWN AND WHAT ITS WINE TASTES LIKE

Rossola Nera is cultivated all over Valtellina in the far north of **Italy** close to the Swiss border, where it is usually blended (5–25%) with other local varieties, mainly NEBBIOLO, known here as Chiavennasca. Total plantings in Italy in 2000 were 115 ha (284 acres). Wines tend to have marked acidity.

ROSSOLINO NERO

Minor variety in Valtellina in Italy's cool far north, probably progeny of Rossola Nera.

BERRY COLOUR:

PRINCIPAL SYNONYMS: Rossolino Rosa

VARIETIES COMMONLY MISTAKEN FOR ROSSOLINO NERO: ROSSOLA NERA

ORIGINS AND PARENTAGE

Rossolino Nero is usually considered to be the most widespread clonal variation of ROSSOLA NERA (Bongiolatti 1996; Fossati *et al.* 2001) but DNA profiling has established that they are distinct varieties, linked by a parent–offspring relationship, with Rossolino Nero most probably the offspring (Vouillamoz). A pink-berried form called Rossolino Rosa can be found throughout Valtellina, in particular in Teglio.

VITICULTURAL CHARACTERISTICS

Vigorous, mid ripening. Sensitive to powdery mildew and botrytis bunch rot, good resistance to downy mildew.

WHERE IT'S GROWN

Rossolino Nero and ROSSOLA NERA are interplanted and generally confused with one another in vineyards throughout Valtellina in the far north of **Italy**, close to the Swiss border.

ROTBERGER

German cross that takes after its parent Schiava Grossa (aka Trollinger).

BERRY COLOUR: ● ● ● ● ●

PRINCIPAL SYNONYMS: Geisenheim 3-37, Redberger

ORIGINS AND PARENTAGE

A SCHIAVA GROSSA × RIESLING cross bred by Heinrich Birk in 1928 at the Geisenheim research institute in the Rheingau, Germany. Parentage has been confirmed by DNA profiling (Grando and Frisinghelli 1998). Having the same parents, Rotberger and KERNER are full siblings.

VITICULTURAL CHARACTERISTICS

Vigorous, high yielding, late ripening and good resistance to downy mildew and botrytis bunch rot but more susceptible to powdery mildew. Needs deep soils and doesn't like drought.

WHERE IT'S GROWN AND WHAT ITS WINE TASTES LIKE

There are just 16 ha (40 acres) in **Germany**, nearly half of them in the Rheingau. Notable German producers include Schlossgut Schmidt in Rheinhessen and Birkert in Württemberg.

Pojer e Sandri in the Alto Adige, **Italy**, produce a fresh and zesty varietal Rotberger called Vin dei Molini, although total Italian plantings were too few for the variety to be included in the 2000 agricultural census.

There are 3 ha (7 acres) in British Columbia, **Canada**, where Gray Monk make a varietal rosé from it in the Okanagan Valley. Wines can be reasonably full-bodied yet fresh and rosé versions are common.

ROTER ELBLING

See ELBLING.

ROTER RIESLING

See RIESLING.

ROTER VELTLINER

Ancient Austrian variety that makes powerful white wine, whatever its name suggests.

BERRY COLOUR:

PRINCIPAL SYNONYMS: Piros Veltelini (Hungary), Rote Fleischtraube, ?Roter Veltliner Baldig (Slovakia), Rotmuskateller, Rotreifler (Austria), Ryvola Cervena (Czech Republic), Veltlínske Červené (Slovakia)

ORIGINS AND PARENTAGE

The parentage of Roter Veltliner has not yet been uncovered but DNA profiling provided evidence that Roter Veltliner spontaneously crossed with SILVANER to produce FRÜHROTER VELTLINER and NEUBURGER, with SAVAGNIN to produce ROTGIPFLER (Sefc, Steinkellner *et al.* 1998), and possibly with a close relative of Savagnin to produce ZIERFANDLER. Roter Veltliner and its progeny Frühroter Veltliner are, surprisingly, not related to GRÜNER VELTLINER (see PINOT pedigree diagram). Roter Veltliner was used to breed DEVÍN and HETERA.

VITICULTURAL CHARACTERISTICS

Late ripening and subject to frost in winter and spring. Needs good sites and deep loess soils. Very compact bunches and therefore prone to botrytis bunch rot. Irregular but high yields.

WHERE IT'S GROWN AND WHAT ITS WINE TASTES LIKE

Austria has 259 ha (640 acres) of Roter Veltliner scattered across Niederösterreich (Lower Austria) but with most plantings in the Weinviertel and in Wagram. It used to be much more widely planted and can produce powerful, full-bodied wines even at relatively high yields; peppery and spicy when young, some tasting quite similar to (the unrelated) GRÜNER VELTLINER. The best wines develop considerable complexity and some almond flavours after two or three years in bottle. Top producers include Stefan Bauer, Josef Fritz, Grill, Franz Leth and Söllner.

It is also planted in **Slovakia** (363 ha/897 acres), to a very limited extent in **Hungary** (just over 3 ha/7 acres officially recorded in 2008) and in the **Czech Republic**, where it doesn't even make it into the official statistics.

ROTGIPFLER

Thermenregion speciality makes powerful whites usually blended with Zierfandler.

BERRY COLOUR:

PRINCIPAL SYNONYMS: Reifler (Austria), Slatzki Zelenac (Slovenia), Zelenac Slatki (Croatia)

ORIGINS AND PARENTAGE

DNA parentage tests revealed that Rotgipfler, most probably from Austria's Thermenregion, is a natural cross between SAVAGNIN and ROTER VELTLINER (Sefc, Steinkellner *et al.* 1998), and is therefore a half-sibling of FRÜHROTER VELTLINER, NEUBURGER and ZIERFANDLER via

Roter Veltliner, and a half-sibling of GRÜNER VELTLINER and SILVANER via Savagnin. Taking the pedigree further, Rotgipfler is either a grandchild or a half-sibling of PINOT, because Savagnin and Pinot have a parent–offspring relationship (see PINOT pedigree diagram).

VITICULTURAL CHARACTERISTICS

Berries are very small with a strongly green cast and bunches are compact so it is prone to the spread of botrytis bunch rot. Mid to late ripening and susceptible to powdery and especially downy mildew.

WHERE IT'S GROWN AND WHAT ITS WINE TASTES LIKE

Although Rotgipfler is not widely planted, with a total of 117 ha (289 acres), almost entirely in its homeland of the Thermenregion, south of Wien (Vienna) in **Austria**, local producers in Gumpoldskirchen are fiercely proud of this local speciality, which finds its most powerful, spicy and ageworthy expression in partnership with ZIERFANDLER, the other local speciality. It retains its acidity well, even when very ripe, and produces generous, mouth-filling wines high in extract and alcohol, often with flavours of peach and apricot, occasionally heavy, becoming more expressive with bottle age. Wines produced in an off-dry style can lack freshness. The variety is well suited to restrained use of oak. Recommended producers include K Alphart, Leo Aumann, Biegler, Johanneshof Rheinisch, Stadlmann and Zierer.

It is also found to a limited extent, under local synonyms, in **Slovenia** and **Croatia** (*zeleni* means 'green', *slatki*, 'sweet'). Krauthaker produces a good example of Zelenac in the latter.

ROUGE DE FULLY

Swiss, or possibly Italian, variety making rustic reds.

BERRY COLOUR: ● ● ● ● ●

PRINCIPAL SYNONYMS: Durize

ORIGINS AND PARENTAGE

Local to and traditional in the region around Fully in the Valais, Switzerland, where it was the most planted variety until the end of the nineteenth century (Vouillamoz 2009). Today it is often referred to under its dialect name, Durize, in reference to its hard (French *dur*) skin.

DNA parentage analysis has not revealed its parents but has suggested a close genetic relationship with ROUSSIN from the neighbouring Valle d'Aosta in Italy, where Rouge de Fully most likely originates (Vouillamoz and Moriondo 2011).

VITICULTURAL CHARACTERISTICS

Early to mid budding, late ripening. Low vigour, average but sometimes irregular yields. Susceptible to powdery mildew and particularly sensitive to magnesium deficiency but resistant to botrytis bunch rot. Compact bunches of smallish berries.

WHERE IT'S GROWN AND WHAT ITS WINE TASTES LIKE

Rouge de Fully is cultivated exclusively in the villages of Fully and Saillon in the Valais, **Switzerland**, on less than 1 ha (2.5 acres) by fewer than a dozen producers, including Henri Valloton and Samuel Roduit. The wines are lightly smoky, sometimes spicy, and have aromas of strawberries and blackberries, with rustic tannins and fresh acidity.

ROUGE DU PAYS

Variety exclusive to southern Switzerland also known as Cornalin but distinct from the Italian Cornalin in the Valle d'Aosta.

BERRY COLOUR:

PRINCIPAL SYNONYMS: Cornalin or Cornalin du Valais, Landroter (Haut-Valais), Rouge du Valais

VARIETIES COMMONLY MISTAKEN FOR ROUGE DU PAYS: CORNALIN (Valle d'Aosta)

ORIGINS AND PARENTAGE

Rouge du Pays (meaning 'homeland red') is possibly one of the most ancient red wine grape varieties of the Valais, southern Switzerland, where it has been cultivated for centuries. From the mid nineteenth century, it was supplanted by interlopers such as PINOT NOIR and GAMAY NOIR, and by the mid twentieth century it was almost extinct. Rouge du Pays was rescued in the 1970s by taking cuttings from the few surviving vines and was unfortunately renamed Cornalin (Nicollier 1972) – unfortunately because this name is already taken by another variety in the Valle d'Aosta, which is incidentally called Humagne Rouge in the Valais (see CORNALIN).

As it happens, it has long been suspected that Rouge du Pays originated from the Valle d'Aosta (Berget 1904), which was confirmed by DNA parentage analysis showing that Rouge du Pays is a natural cross between PETIT ROUGE and MAYOLET (Vouillamoz *et al.* 2003), both traditional varieties from the Valle d'Aosta, making it a grandchild of PRIÉ (see Prié for the pedigree diagram). Rouge du Pays is also the parent of at least three varieties from the Valle d'Aosta: ROUSSIN, NERET DI SAINT-VINCENT and CORNALIN. This last parentage fully justifies the choice of Rouge du Pays instead of Cornalin as prime name for this variety in the Valais. In addition, a close relationship with VIEN DE NUS has been suggested (Vouillamoz and Moriondo 2011).

Rouge du Pays was probably introduced from its homeland in the Valle d'Aosta, where it has today disappeared, to the neighbouring Valais several centuries ago, possibly via the Grand-Saint-Bernard pass that followed the Roman road Via Francigena.

OTHER HYPOTHESES

Rouge du Pays is often said to have been mentioned under the name Neyrun in the 'Registre d'Anniviers', an old parchment dated 1313, but there is no evidence for this synonymy (Vouillamoz 2009a; Vouillamoz and Moriondo 2011).

VITICULTURAL CHARACTERISTICS

Extremely vigorous, alternate low and high yields. Very early budding, mid to late ripening. Generally susceptible to diseases but especially to botrytis bunch rot and powdery mildew. Prone to magnesium deficiency, sunburn, desiccation of the stems and premature shrivelling of the berries. Needs severe pruning and best suited to poor, dry, well-drained soils.

WHERE IT'S GROWN AND WHAT ITS WINE TASTES LIKE

Rouge du Pays is cultivated almost exclusively in the Valais in southern **Switzerland**, where its vineyards totalled 116 ha (287 acres) in 2009. It's a difficult variety to grow, but in good hands it gives wines that rank among the most praised in the Valais, with aromas of black cherries, violets and raspberries, dense but silky tannins and a sapid, slightly bitter-acidic finish. Recommended producers include Arte Vinum, Domaine Cornulus, Anne-Catherine & Denis Mercier and Maurice Zufferey.

ROUPEIRO

See SÍRIA.

ROUSSANNE

Scented, high-quality Rhône white often blended with Marsanne. Not easy to grow.

BERRY COLOUR: ● ● ● ● ●

PRINCIPAL SYNONYMS: Barbin (Vallée du Gélon in Savoie and also on the left bank of the Isère), Bergeron (Chignin in Savoie), Fromental or Fromenteau (Isère), Martin Cot (Rochette and Vallée du Gélon in Savoie and also on the left bank of the Isère), Petite Roussette (Saint-Péray and Hermitage), Roussanne Blanc

VARIETIES COMMONLY MISTAKEN FOR ROUSSANNE: ALTESSE (Roussanne is wrongly called Roussette in Bugey, and Roussette is a widespread synonym for Altesse), GRINGET (Ayze), MARSANNE, ROUSSETTE D'AYZE (Savoie)

ORIGINS AND PARENTAGE

Roussanne is a traditional variety of the northern Rhône, where it was first mentioned in 1781 in a text about the white wines from the Hermitage region (see MARSANNE). The name Roussanne probably refers to the russet (*roux*) colour of the mature berries. Roussanne is a member of the Sérine ampelographic group (see p XXVII; Bisson 2009). DNA parentage analysis strongly suggests a parent–offspring relationship between Roussanne and MARSANNE (Vouillamoz). This might explain why Marsanne and Roussanne are often confused.

VITICULTURAL CHARACTERISTICS

Mid ripening. Does well on well-exposed, poor, stony, calcareous-clay soils although it has poor wind resistance. Highly susceptible to powdery mildew, botrytis bunch rot, mites and thrips.

WHERE IT'S GROWN AND WHAT ITS WINE TASTES LIKE

In **France**, Roussanne's most famous and revered renditions are in partnership with MARSANNE in the northern Rhône still white wines of Hermitage, Crozes-Hermitage and Saint-Joseph and in still and sparkling Saint-Péray. The regulations do not specify the exact proportions but there is much more Marsanne planted than Roussanne now because, although the latter produces more elegant wine, the former is easier to grow, more productive and less susceptible to fungal diseases. However, Roussanne is not restricted to the northern Rhône since, unlike Marsanne, it is also a permitted ingredient in white Châteauneuf-du-Pape and Côtes du Rhône and is recommended in many a southern and south-eastern *département* as well as in the Loire, and in Savoie (96 ha/237 acres in 2009), where it is the sole component of Chignin-Bergeron.

Roussanne is more aromatic than MARSANNE, often with a refreshing perfume akin to herbal tea (verbena?) reminiscent of spring blossom, and tends to have higher acidity, hence its superior ageing capacity.

The variety's fashionability is revealed by the rapid increase in total French plantings, which doubled in the last ten years to reach 1,352 ha (3,341 acres) in 2009, compared with just 71 ha (175 acres) in the 1960s. The variety's tendency to ripen late is less of a problem in Languedoc and Roussillon than in the northern Rhône although it can suffer in very dry conditions. In Languedoc-Roussillon it is often blended with BOURBOULENC, Grenache Blanc (GARNACHA BLANCA), VERMENTINO, MARSANNE and so on but it has also proved to be a good partner for

CHARDONNAY. A particularly fine example of varietal Roussanne is produced by Château de Beaucastel in Châteauneuf-du-Pape, who have demonstrated that the fruit of their old vines has great affinity with oak and the capacity to age. Grand Veneur is another example of this style.

Other recommended French producers of varietal wines or Roussanne-dominated blends include Château Pesquié (Ventoux), Domaine Sainte-Rose (Côtes de Thongue), Domaine de la Solitude (Châteauneuf) and Paul Jaboulet Aîné (Rhône).

In Savoie, Michel Quenard is one of the best exponents of the variety, here called Bergeron, in Chignin, a cru south east of Chambéry. Relative to Savoie's generally delicate whites, these wines are powerful, rich, heady and usually just dry. Other recommended Savoie producers include all the other Quénards in Chignin: André and Michel, Jean-Pierre and Jean-François, Raymond, and Pascal and Annick. Roussanne in Savoie should not be confused with Roussette, once the name used for the ALTESSE variety that constitutes Roussette de Savoie.

In **Italy** it is grown to a very limited extent in Liguria and Toscana, where it is a permitted ingredient in Italy's Montecarlo Bianco (no more than 10% allowed) but the area was too small to be included in the 2000 agricultural census. Lamoresca on the island of Sicilia sometimes include Roussanne in their VERMENTINO-dominated white.

There are very few growers of Roussanne in **Portugal** but Esporão in the Alentejo have 6 ha (15 acres) and use it in their Private Selection white blend. Ted Manousakis produces it on the island of Kríti (Crete), **Greece.**

The variety has recently been gaining ground rapidly in **California**, where there were 348 acres (141 ha) in 2009. It is particularly favoured by the state's Rhône Rangers in the Central Coast, notably Qupé in Santa Ynez Valley and Alban in Edna Valley. It is often blended with MARSANNE, while at Bonny Doon in the Santa Cruz Mountains, Randall Grahm blends Roussanne and Grenache Blanc (GARNACHA BLANCA) in his Cigare Volant white. Further north and inland, Renaissance have been particularly successful with varietal wines in North Yuba. There are also limited plantings in **Washington** State, where Syncline and DeLille, among others, make varietal wines and Airfield Estate use it in blends. In **Oregon**, Andrew Rich, Agate and Troon all make varietal examples quite successfully. In the Okanagan Valley in western **Canada**, it is usually blended with VIOGNIER.

As in the Rhône, Roussanne is more often blended with MARSANNE in **Australia**, sometimes also with VIOGNIER thrown in. In 2010 there were over forty producers of this variety, though the total area under vine was too small to be recorded. Yeringberg in Victoria's Yarra Valley have been particularly successful with their ageworthy varietal wine. Other producers in Victoria include Giaconda in Beechworth, Tahbilk (more famous for their Marsanne) in the Nagambie Lakes and Brown Brothers in the King Valley; in South Australia, d'Arenberg in McLaren Vale and Torbreck in the Barossa; in Western Australia, McHenry Hohnen and Jerusalem Hollow, among others.

South Africa had 28 ha (69 acres) of Roussanne in 2009, the majority in Paarl and to a lesser extent in Stellenbosch, where Rustenberg make a particularly powerful varietal wine. Producers such as Adi Badenhorst and Eben Sadie in the Swartland are using a small amount of Roussanne in their complex white blends.

ROUSSANNE DU VAR

See ROSÉ DU VAR.

ROUSSETTE D'AYZE

Virtually extinct and rather inferior Savoie variety used occasionally in sparkling wines.

BERRY COLOUR:

PRINCIPAL SYNONYMS: Bonne Roussette d'Ayze (Bonneville), Grosse Roussette (Bonneville), Riusse or Riussette

VARIETIES COMMONLY MISTAKEN FOR ROUSSETTE D'AYZE: ALTESSE (also called Roussette or Roussette Haute), Blanchet (Pomaretto in Piemonte in Italy), JACQUÈRE (called Roussette in Montmélian), Maclon (Roussette d'Ayze is supposedly called Maclon in Saint-Chef in Isère), MONDEUSE BLANCHE, ROUSSANNE (wrongly known as Roussette in Bugey)

ORIGINS AND PARENTAGE

Roussette d'Ayze is a variety particular to the Ayze region near Bonneville in the Haute-Savoie and should not be confused with other varieties that include Roussette among their synonyms. It is a member of the Sérine ampelographic group (see p XXVII; Bisson 2009). Surprisingly, parentage analysis of twenty-three DNA markers suggests a possible parent–offspring relationship between Roussette d'Ayze and ONDENC from the Vallée du Tarn in southern France (Vouillamoz, to be confirmed with more markers).

According to Galet (2000), Roussette d'Ayze is simply a clone of MONDEUSE BLANCHE but this can be disproved by DNA profiling (Vouillamoz).

VITICULTURAL CHARACTERISTICS

Early ripening, susceptible to botrytis bunch rot and downy mildew.

WHERE IT'S GROWN AND WHAT ITS WINE TASTES LIKE

Wines are not especially aromatic and are generally used to make sparkling wines. There was less than 1 ha (2.5 acres) in **France** in 2009. It is authorized as a complementary variety for still and sparkling wines in the Ayze cru within the Vin de Savoie appellation, along with its blending partners GRINGET and ALTESSE, but the quality of the selection has been a major factor in its decline – though even in 1988 there were only 20 ha (49 acres).

ROUSSIN

Almost extinct Aostan.

BERRY COLOUR:

PRINCIPAL SYNONYMS: Gros Roussin, Picciou Roussin, Rossé

VARIETIES COMMONLY MISTAKEN FOR ROUSSIN: Roussin de Morgex

ORIGINS AND PARENTAGE

Roussin was first mentioned in the Valle d'Aosta, north-west Italy, by Gatta (1838) as a rare vine from the area between Arnad and Verrayes, in the lower part of the valley, due north of Torino. It most likely takes its name from the red colour (Italian *rosso*) of the berries at maturity. DNA profiling has shown that Roussin is not related to Roussin de Morgex, a variety from the upper valley that is no longer cultivated, and parentage analysis has established that Roussin is an offspring of ROUGE DU PAYS (Vouillamoz 2005), an old Valle d'Aosta variety now grown only in the Valais in Switzerland (see PRIÉ for the pedigree diagram). Roussin is therefore a grandchild of the parents of Rouge du Pays, namely PETIT ROUGE and MAYOLET, and a half-sibling of the other progenies of Rouge du Pays, namely CORNALIN and NERET DI SAINT-VINCENT, as well as a half-sibling of VUILLERMIN through another genetic line (Vouillamoz and Moriondo 2011).

Labra *et al.* (2002) suggested a genetic link with AMIGNE but this has been disproved by a more detailed DNA investigation (Vouillamoz and Moriondo 2011).

VITICULTURAL CHARACTERISTICS

Late ripening (harvested end of October) and requiring low-altitude, well-exposed sites.

WHERE IT'S GROWN

Saved from extinction in the 1990s by Giulio Moriondo, ampelographer at the Institut Agricole Régional in Aosta, **Italy**, Roussin is today an endangered variety. There is just one Roussin vineyard, near Arnad, plus a few old vines in the neighbouring vineyards.

ROVELLO BIANCO

Rovello Bianco, or Roviello, is an endangered white-berried variety from Campania, southern **Italy**, that was first described in 1875. It is also known locally as Greco Muscio, or simply Grecomusc' by its aficionados, but it is distinct from all other Greco Somethings, as evidenced by DNA profiling (Francesca *et al.* 2009). A few hundred old and often ungrafted vines of this variety had survived in the areas of Taurasi, Mirabella Eclano and Bonito in the province of Avellino. With the help of the Universities of Napoli and Palermo, Contrade di Taurasi was the first winery to replant this forgotten variety in 2003 and today produces a Grecomusc' showing delicate aromas of ash and green citrus leaves along with refreshing acidity.

ROYAL DE ALLOZA

Recuperated and very limited Aragonese speciality.

BERRY COLOUR: ● ● ● ● ●

PRINCIPAL SYNONYMS: Alloza, Derechero de Muniesa

ORIGINS AND PARENTAGE

As the name suggests, Royal de Alloza is an autochthonous variety from the village of Alloza in the province of Teruel, north-east Spain.

WHERE IT'S GROWN AND WHAT ITS WINE TASTES LIKE

Royal de Alloza is grown exclusively in the province of Teruel in Aragón, northern **Spain**. Plantings are so limited that it is not mentioned in the official vineyard statistics and Fandos y Barriuso in Bajo Aragón, who rescued the variety (and ROMERO DE HÍJAR) from local vineyards, may be the only ones to make a varietal wine from this variety. They describe it as 'deep cherry red with copper edges, a potent bouquet, toasted, of black fruit confit and Mediterranean herbs. Potent in the mouth yet fresh.'

ROYALTY

Red-fleshed American hybrid on the wane in California.

BERRY COLOUR: ● ● ● ● ●

PRINCIPAL SYNONYMS: Calif 526, California S 26, Royalti, Royalty 1390

ORIGINS AND PARENTAGE

A complex *Vitis vinifera* and *Vitis rupestris* hybrid obtained by Harold P Olmo in 1938 at the University of California at Davis by crossing Alicante Ganzin with TROUSSEAU (Walker 2000), where Alicante Ganzin is a Ganzin 4 × ALICANTE HENRI BOUSCHET hybrid and Ganzin 4 an ARAMON NOIR × *Vitis rupestris* hybrid (see PRIOR pedigree diagram). RUBIRED and Royalty were released together in 1958, having been bred mainly for their deep and stable colour (Olmo and Koyama 1962).

VITICULTURAL CHARACTERISTICS

Mid ripening, relatively resistant to powdery mildew. Thick-skinned, red-fleshed. More demanding as to soil type but less vigorous and lower yielding than its stablemate RUBIRED.

WHERE IT'S GROWN AND WHAT ITS WINE TASTES LIKE

Because it is more difficult to grow, Royalty has been far less successful than RUBIRED. There were just 240 acres (97 ha) in **California** in 2009, mostly in the San Joaquin Valley. It is used in blends to add colour but also to produce food colouring. Preston Wines in **Washington** State use it to make a port-style wine.

RUBIN

Intriguing, successful but not widely planted Bulgarian cross used for both dry and sweet reds.

BERRY COLOUR: ● ● ● ● ●

PRINCIPAL SYNONYMS: Rubin Bolgarskii

VARIETIES COMMONLY MISTAKEN FOR RUBIN: Melnishki Rubin 🧬

ORIGINS AND PARENTAGE

Rubin is a NEBBIOLO × SYRAH cross obtained in 1944 at the Pleven viticulture and oenology institute in northern Bulgaria. It was officially released in 1961.

VITICULTURAL CHARACTERISTICS

Vigorous, fertile and fairly productive. Relatively compact bunches of small berries that have thin but tough skins and accumulate high sugar levels. Mid to early ripening.

WHERE IT'S GROWN AND WHAT ITS WINE TASTES LIKE

In its homeland, **Bulgaria**, Rubin produces full-bodied, soft, sweet, spicy reds with only moderate alcohol and fruit flavours ranging from blackberries and plums to figs, very ripe mulberries and raisins, especially when harvested late for sweet wines. Their deep colour is due to the level of anthocyanins, which is higher than in CABERNET SAUVIGNON. It marries well with oak but there is little evidence as to its ageworthiness. Logodaj in the Struma Valley in the south-western corner of Bulgaria produce a good varietal example, as do Brestovitsa. There were 165 ha (408 acres) in Bulgaria in 2008. Rubin is also found in **Romania** and **Slovenia**.

RUBIN GOLODRIGI

Recent and usefully disease-resistant Ukrainian hybrid producing deeply coloured reds.

BERRY COLOUR: ● ● ● ● ●

PRINCIPAL SYNONYMS: Magarach 15-74-29, Ruby of Golodryga

ORIGINS AND PARENTAGE

Rubin Golodrigi is a hybrid obtained in 1974 by P Golodriga, M Kostik and V Yurchenko at the Ampelos vine-breeding company in Yalta, southern Ukraine, by crossing RUBINOVY MAGARACHA with Magarach 6-68-27, the latter being an unspecified cross from the Magarach research centre in Krym (Crimea), southern Ukraine. The parentage was confirmed by recent DNA analysis (Goryslavets *et al.* 2010).

VITICULTURAL CHARACTERISTICS

Mid ripening. Good all-round disease resistance and winter-hardy to –26 °C (–14.8 °F).

WHERE IT'S GROWN AND WHAT ITS WINE TASTES LIKE

Wines produced in **Ukraine** are deeply coloured and appear in dry, sweet and fortified guises. The variety is also found in Krasnodar, south-west **Russia** (82 ha/203 acres in 2010).

RUBIN TAIROVSKY

A black-berried hybrid obtained in 1964 by Y N Dokuchaeva, M I Tulaeva, G P Ovchinnikov, A P Abliazova and L I Tarakhtiy at the Tairov research centre in Odessa, **Ukraine**, by crossing Odessky Ustoichivy (see OVIDIOPOLSKY for details) with Varousset, a French Seibel 4668 × Subéreux hybrid (see MUSCAT SWENSON for details). It was registered in 1990 and two clones are available. This consistently productive, winter-hardy, late-ripening variety is used to make dry and semi-sweet red wines that have soft, sweet tannins and flavours that include tomato and elderberry.

RUBINET

Very minor red-fleshed Czech cross producing usefully dark, rich wines.

BERRY COLOUR: ● ● ● ● ●

PRINCIPAL SYNONYMS: R – 27, Tintet

ORIGINS AND PARENTAGE

Rubinet is a (Revolta × Alibernet) × ANDRÉ cross recently obtained by Vilém Kraus at the faculty of horticulture in Lednice in southern Morava, in the south east of the Czech Republic, in which Revolta is a (PRÉCOCE DE MALINGRE × CHASSELAS) × (CSABA GYÖNGYE × Corinthe Rouge) cross bred in the Czech Republic, Corinthe Rouge, or Korinthusi Piros, being a Greek table grape. (For Alibernet, see ODESSKY CHERNY.) Rubinet was initially named Tintet before being officially renamed and included in the Czech register of varieties in 2004.

VITICULTURAL CHARACTERISTICS

Medium-sized bunches of smallish, red-fleshed berries. Early budding, mid ripening. Generally good resistance to fungal diseases.

WHERE IT'S GROWN AND WHAT ITS WINE TASTES LIKE

Thanks to its deep colour, richness, black-fruit flavour and soft tannins, Rubinet is a useful blending partner, although there were only 2 ha (5 acres) recorded in 2009, mostly in Morava, in the south east of the **Czech Republic**. Hradil and Císařské Sklepy Horák both make varietal examples.

RUBINKA

A PETRA × BIANCA cross obtained by P Cindrić, N Korać, V Cindrić and V Kovač at the Sremski Karlovci viticulture research centre (part of Novi Sad University), Vojvodina, **Serbia**, and released in 2002. This pinkish-berried cross is mistakenly confused with Srpski Rubin in the Vitis International Variety Catalogue. It shows good resistance to fungal diseases and frost.

RUBINOVY MAGARACHA

Robust but awkwardly late-ripening Ukrainian cross.

BERRY COLOUR: ● ● ● ● ●

PRINCIPAL SYNONYMS: Crossing 56, Magarach 56, Magarach Ruby, Ruby Magaracha

ORIGINS AND PARENTAGE

Rubinovy Magaracha is a CABERNET SAUVIGNON × SAPERAVI cross obtained by N Paponov, V Zotov, P Tsarev and P Y Golodriga in 1928 at the Magarach research centre in Krym (Crimea), southern Ukraine, and included in the official register of varieties in 1969. The parentage was confirmed by recent DNA analysis (Goryslavets *et al.* 2010). Rubinovy Magaracha was used to breed ANTEY MAGARACHSKY and RUBIN GOLODRIGI.

VITICULTURAL CHARACTERISTICS

Productive and late ripening. Good all-round disease resistance, relatively winter-hardy and drought-tolerant. Medium-sized loose bunches of thick-skinned berries.

WHERE IT'S GROWN AND WHAT ITS WINE TASTES LIKE

Grown in **Ukraine**, Rubinovy Magaracha berries can attain high sugar levels and produce dark-fruited wines at various sweetness levels as well as being used for juice and the table. Magarach, for example, include it in various semi-sweet and dessert blends. It is also found in **Kazakhstan** and **Uzbekistan**.

RUBINTOS

Rare Hungarian cross with potential for well-structured, deeply coloured reds.

BERRY COLOUR: ● ● ● ● ●

ORIGINS AND PARENTAGE

A KADARKA × BLAUFRÄNKISCH cross obtained in 1951 by Pál Kozma and József Tusnádi at the Szent István University, Hungary. It was officially registered in 1980.

VITICULTURAL CHARACTERISTICS

High yielding, late ripening. Big bunches of small berries. Moderately susceptible to low winter temperatures, some resistance to botrytis bunch rot.

WHERE IT'S GROWN AND WHAT ITS WINE TASTES LIKE

Rubintos wines are well coloured, spicy and rich in tannins but the variety is little planted: 22 ha (54 acres) in 2008, almost all in Mátra in northern **Hungary**. Tamás Szecskő's Peres-Dülő is a rare, single-vineyard example.

RUBIRED

Winemakers' secret weapon. Productive, red-fleshed American hybrid valued for juice and for adding colour.

BERRY COLOUR: ● ● ● ● ●

PRINCIPAL SYNONYMS: California S 8, Tintoria

ORIGINS AND PARENTAGE

A complex hybrid obtained by Harold P Olmo in 1938 at the University of California at Davis by crossing Alicante Ganzin (see ROYALTY for the pedigree) with TINTO CÃO. Rubired and

ROYALTY were released together in 1958, having been bred mainly for their intense and stable colour (Olmo and Koyama 1962).

VITICULTURAL CHARACTERISTICS

Medium-sized to large compact clusters of small berries with deep purple juice. Vigorous and very productive, mid ripening. Moderately resistant to powdery mildew and very low susceptibility to botrytis bunch rot, allowing late harvesting. Very susceptible to eutypa dieback, and young vines susceptible to collar rot. Vine stress may cause yields to be erratic. Danger of berry shrivel in overripe fruit (Christensen *et al.* 2003).

WHERE IT'S GROWN AND WHAT ITS WINE TASTES LIKE

Rubired took over from SALVADOR as the primary variety grown in **California** for red juice concentrate, to add colour to red-wine blends and for other food processing because the juice colour was more intense. The wine's main asset is its deep red colour but it has little character or body (Christensen *et al.* 2003). Mega Purple is the major brand of colouring grape juice concentrate dependent on Rubired. There were 11,776 acres (4,766 ha) in California in 2009, about a third as much as total plantings of PINOT NOIR, almost all in the San Joaquin Valley, particularly in Fresno, Kern and Madera counties.

RUBY CABERNET

Productive California cross with sweet-fruit flavours, generally used to bolster blends.

BERRY COLOUR: ● ● ● ● ●

PRINCIPAL SYNONYMS: California 234 F 2

ORIGINS AND PARENTAGE

A MAZUELO (under the name Carignane) × CABERNET SAUVIGNON cross bred by Harold P Olmo in 1936 at the University of California at Davis, US, and introduced to the market in 1948 (Walker 2000). It is Olmo's oldest crossing of wine varieties (Christensen *et al.* 2003). His aim was to combine the characteristics of Cabernet with the productivity and heat tolerance of Mazuelo. Ruby Cabernet was used to breed CABERINTA.

VITICULTURAL CHARACTERISTICS

Medium-sized to large berries with a noticeable CABERNET SAUVIGNON flavour. Vigorous if planted on deep, fine, sandy-loam or clay-loam soils. Very productive so that yields must be controlled. Late budding and mid to late harvesting (late if yields are too high). Moderately susceptible to powdery mildew and very resistant to botrytis bunch rot.

WHERE IT'S GROWN AND WHAT ITS WINE TASTES LIKE

Wines tend to be deeply coloured, slightly rustic, with sweet red-fruit flavours if yields are controlled, occasionally showing the darker fruit more typical of its parents. Although varietal wines exist, it is more often used in blends or to stretch CABERNET SAUVIGNON. **California** had 5,993 acres (2,425 ha) in 2009, down from the height of its popularity in the 1960s, planted mainly in the San Joaquin Valley. Ruby Cabernet is also planted in **North Carolina**, **Wisconsin** and **Virginia**, and it apparently shows potential in **Oklahoma**. In **Texas**, particularly at higher altitudes in the Texas High Plains AVA, new plantings of Ruby Cabernet are showing greater finesse thanks to the cooler temperatures. Val Verde produce a good example. Although some of the older, less well-sited vineyards are being pulled up, there were still about 100 acres (40 ha) in Texas in 2010.

It is also planted to a very limited extent in **Brazil** (40 ha/100 acres in 2009), **Uruguay** (2 ha/5 acres in 2009) and **Israel**.

Australia had 1,141 ha (2,819 acres) in 2008, mainly in the hot inland regions of Riverland, Riverina and the Murray Darling but also scattered to a much lesser extent in cooler areas. Most is used for blending.

Elsewhere, **South Africa** had almost as much as California – 2,381 ha (5,884 acres) in 2008 – but it is restricted to hot, irrigated areas and used for bulk wine. Ladismith's varietal has won some acclaim.

RUCHÈ

Rare, perfumed red wine grape with its own Piemonte DOC.

BERRY COLOUR: ● ● ● ● ●

PRINCIPAL SYNONYMS: Moscatellina, Roché, Romitagi, Rouchet

ORIGINS AND PARENTAGE

Ruchè most probably originates from the area of Castagnole Monferrato in the province of Asti in Piemonte, north-west Italy. Its name is said to derive from the local dialect word *roncet*, describing a virus infection to which Ruchè had proved to be resistant (unlike BARBERA and GRIGNOLINO). However, it seems very odd to name a variety after a disease that did not affect it.

OTHER HYPOTHESES

Some have suggested the variety was introduced from Burgundy in the eighteenth century but this theory is not supported by any historical, ampelographic or genetic studies.

VITICULTURAL CHARACTERISTICS

Early to mid ripening (and therefore attractive to wasps). Susceptible to powdery mildew. Good resistance to certain viral diseases affecting leaf formation.

WHERE IT'S GROWN AND WHAT ITS WINE TASTES LIKE

Ruchè is cultivated mainly in its birthplace in the province of Asti in Piemonte, north-west **Italy**, in the communes of Castagnole Monferrato, Grana, Montemagno, Portacomaro, Refrancore, Scurzolengo and Viarigi. Varietal wines are produced under the Ruchè di Castagnole Monferrato DOCG (with up to 10% BARBERA and/or BRACHETTO DEL PIEMONTE). There are limited plantings of Ruchè in the province of Alessandria to the east, under the names Moscatellina or Romitagi. There were 46 ha (114 acres) planted in Italy in 2000.

Varietal wines tend to be headily scented, often with aromas of roses. They can be spicy and the tannins so marked that the wines can sometimes leave a bitter aftertaste. Acidity is generally low. Recommended producers include Antica Casa Vinicola Scarpa, Bersano, Dacapo, Montalbera and Cantine Sant'Agata. It can also produce good dessert wines.

RUDAVA

Recently authorized Slovakian cross with good potential for everyday reds.

BERRY COLOUR: ● ● ● ● ●

PRINCIPAL SYNONYMS: CATAP 6/28

ORIGINS AND PARENTAGE

Rudava, named after a river in western Slovakia, is a CASTETS × I-35-9 cross obtained in 1967 by Dorota Pospíšilová at the VUVV research centre for oenology and viticulture in Bratislava, Slovakia, and authorized in 2011. I-35-9 is an obscure (TEINTURIER × ALEATICO) × PUKHLIAKOVSKY cross obtained in Russia (Dorota Pospíšilová, personal communication). Rudava is a sibling of TORYSA.

VITICULTURAL CHARACTERISTICS

Vigorous and consistently productive. Shortish bunches of thick-skinned berries. Resistant to frost and botrytis bunch rot but moderately susceptible to powdery mildew.

WHERE IT'S GROWN AND WHAT ITS WINE TASTES LIKE

Wines are dark and full-bodied, with a CABERNET SAUVIGNON-like flavour and ageworthy in the best years but also suited to early drinking. In 2011 there were just 5 ha (12 acres) of this unpretentious variety that can be grown in most of **Slovakia**'s wine regions. Karol Braniš in the west of the country is one of the few producers to make a varietal example.

RUEN

Almost extinct Bulgarian cross starting to win support for its full-bodied, ageworthy reds.

BERRY COLOUR: ● ● ● ● ●

ORIGINS AND PARENTAGE

Ruen is a SHIROKA MELNISHKA × CABERNET SAUVIGNON cross obtained in 1951 at the Pleven viticulture and oenology institute in northern Bulgaria and officially released in 1964. The parentage has been confirmed by DNA analysis (Hvarleva *et al.* 2004).

VITICULTURAL CHARACTERISTICS

Early budding and late ripening. Big bunches of rather big, thick-skinned berries. Comparatively resistant to botrytis bunch rot.

WHERE IT'S GROWN AND WHAT ITS WINE TASTES LIKE

Although there are currently just a few small plots of Ruen in south-west **Bulgaria** (a total of 7 ha/17 acres in 2008), the variety having been neglected during the Communist era, it shows promise, according to Philip Harmandjiev of Damianitza Winery, producing deeply coloured, full-bodied, velvety wines with distinctive, dark-red fruit flavours and good ageing potential.

Damianitza don't yet have enough fruit to make a varietal wine but they use Ruen in their admired ReDark blend. Logodaj make a varietal wine.

RUFETE

Late-ripening Portuguese variety that can occasionally yield fine, ageworthy, light reds.

BERRY COLOUR:

PRINCIPAL SYNONYMS: Penamacor (Castelo Branco), Preto Rifete (Pinhel), Rifete (Pinhel), Rosette (Pinhel), Rufeta (Pinhel), Tinta Pinheira (Pinhel)

VARIETIES COMMONLY MISTAKEN FOR RUFETE: PEDRAL (Vinho Verde), TINTA CARVALHA (Douro)

ORIGINS AND PARENTAGE

Rufete, also commonly called Tinta Pinheira, most likely originates from the district of Castelo Branco in central eastern Portugal, where it was already in cultivation by the end of the nineteenth century (Cunha *et al.* 2009). According to DNA analysis, Rufete could be genetically very close to TOURIGA NACIONAL from the neighbouring Dão region (Almadanim *et al.* 2007). Rufete is also very close to Puesto Mayor from Rueda in Spain, itself very close to PRIETO PICUDO from Tierra de León in Spain (Santana *et al.* 2010). These geographically distant relatives suggest that Rufete is a very old variety that spread out from central Portugal to Spain a long time ago.

Rufete Blanca is not a colour mutation of Rufete but a misleading synonym for VERDEJO SERRANO (Vouillamoz).

VITICULTURAL CHARACTERISTICS

Demanding, late-ripening variety that is susceptible to powdery mildew and botrytis bunch rot.

WHERE IT'S GROWN AND WHAT ITS WINE TASTES LIKE

Rufete is planted mainly in the regions of Pinhel, Figueira de Castelo Rodrigo and Cova da Beira in eastern Beiras as well as further north in Dão and the Douro, with a total in **Portugal** of 2,741 ha (6,773 acres) in 2010. Like BAGA, it performs well only in very specific sites where it can reach full maturity and the majority of wines are light in colour, low in alcohol and have high acidity that highlights the astringency of the tannins (Rolando Faustino, personal communication). However, in some years in the Pinhel region, given the best, ie less productive, clones, the grapes ripen fully (around the end of October) and are capable of producing very good wines that are rich in colour and body, have flavours of red fruits and the potential to age well in bottle, not unlike Saumur CABERNET FRANC. Plantings have expanded in Dão, where producers have started to make successful rosé and sparkling wines (Rolando Faustino, personal communication). Recommended producers include the Adega Cooperativa da Covilhão, Quinta dos Barreiros and Quinta dos Termos, both in Beira Interior, as well as Quinta das Carvalhas in the Douro.

RUGGINE

Recuperated Emilia-Romagnan.

BERRY COLOUR: ● ● ● ● ●

PRINCIPAL SYNONYMS: Rugginosa, Ruzninteina

ORIGINS AND PARENTAGE

Ruggine is a historical variety from the province of Modena in Emilia-Romagna, where it was once as widespread as TREBBIANO MODENESE or ALBANA. The name Ruggine derives from the rusty (Italian *arrugginito*) colour of the berries at maturity.

VITICULTURAL CHARACTERISTICS

Early ripening (end of August).

WHERE IT'S GROWN AND WHAT ITS WINE TASTES LIKE

Ruggine was almost extinct when it was rescued in the early 1990s from old, local vineyards by Italo Pedroni, who planted one hectare (2.5 acres) in Rubbiara, just west of the city of Modena, northern central **Italy**. The resulting wine is fizzy and fruity. Closer to the coast, in Faenza, Leone Conti also makes a varietal wine called An Ghin Gò.

RUKATAC

See MALVASIA BIANCA LUNGA.

RULÄNDER

See PINOT GRIS in the PINOT entry.

RUZZESE

Rare Ligurian rescued from extinction but so far authorized only for table wines.

BERRY COLOUR: ● ● ● ● ●

PRINCIPAL SYNONYMS: Razzese, Rossese Bianco (Arcola in Liguria)

ORIGINS AND PARENTAGE

Ruzzese, also known as Rossese Bianco (although this name is a source of confusion – see ROSSESE), was mentioned in Liguria by Acerbi (1825) under the spelling Razzese as one of the most important varieties of the Cinque Terre region. Its name could come from the intense pink colour of the berries at maturity. Ruzzese is currently under consideration for entry in the Italian national register of varieties. Recent DNA profiling has shown that it should not be confused with another, morphologically distinct variety called ROSSESE BIANCO, of which a few old vines still thrive in Roddino and Sinio in Piemonte (Torello Marinoni, Raimondi, Mannini and Rolle 2009). Since this latter is already included in the national register, it is clearer to stick to the name Ruzzese for the Ligurian variety found in La Spezia and keep Rossese Bianco for the Piemontese variety.

VITICULTURAL CHARACTERISTICS

Vigorous, mid ripening.

WHERE IT'S GROWN AND WHAT ITS WINE TASTES LIKE

Since Ruzzese is not yet in **Italy**'s national register of grape varieties, it can be used only for table wines. Once widespread along the Ligurian coast, it was virtually extinct by the late twentieth century but was recently rescued by Nino Picedi Benettini, who planted cuttings from old vines he had rediscovered in Baccano di Arcola, west of La Spezia. Today Ruzzese is cultivated mainly in the province of Imperia, in the far west of Liguria, usually blended with VERMENTINO and ALBAROLA. Rare examples of varietal wines are made by Conte Picedi Benettini and Maccario Dringenberg.

RYUGAN

Minor, grey-skinned Japanese variety of unknown origin.

BERRY COLOUR: ● **●** ● ● ●

VARIETIES COMMONLY MISTAKEN FOR RYUGAN: ?LONGYAN (China), Lungyen (China)

ORIGINS AND PARENTAGE

Ryugan is a Japanese variety whose exact origin is unknown. It is said to have been introduced from China in the late nineteenth century (Yokotsuka *et al.* 1984) but its description suggests that it might be distinct from the Ryugan that is also found in China under the names LONGYAN or Lungyen.

VITICULTURAL CHARACTERISTICS

Late budding and ripening. Large bunches of large berries. Susceptible to botrytis bunch rot.

WHERE IT'S GROWN AND WHAT ITS WINE TASTES LIKE

Ryugan wines generally taste of Japanese pear and have well-balanced acidity. There were just 24 ha (59 acres) in **Japan** in 2008, predominantly in Nagano prefecture. Recommended producers include Goichi Wine, Manns Wine and Shinano Wine.

S

SABATO
SABREVOIS
SACY
SAGRANTINO
SAINT-MACAIRE
SALVADOR
SAN GIUSEPPE NERO
SAN LUNARDO
SAN MARTINO
SAN MICHELE
SAN PIETRO
SANGIOVESE
SANKT LAURENT
SANT'ANTONIO
SANTA MARIA
SANTA SOFIA
SAPERAVI
SAPERAVI SEVERNY
SAPHIRA
ŞARBĂ
SARY PANDAS
SAUVIGNON BLANC
SAUVIGNONASSE
SAVAGNIN
SAVATIANO
SCHEUREBE
SCHIAVA GENTILE
SCHIAVA GRIGIA
SCHIAVA GROSSA
SCHIAVA LOMBARDA
SCHIOPPETTINO
SCHÖNBURGER
SCIAGLÌN
SCIASCINOSO
SCIMISCIÀ
SCUPPERNONG
SEARA NOVA
SEFKA
SEGALIN
SEIBEL 10868
SEMIDANO
SÉMILLON
SERCIAL
SERCIALINHO
SERINA E ZEZË
SERODIO
SERVANT
SEVERNY
SEYVAL BLANC
SEYVAL NOIR
SGAVETTA
SHAVKAPITO
SHESH I BARDHË
SHESH I ZI
SHEVKA
SHIROKA MELNISHKA
SHIRVANSHAHY
SIBI ABBAS
SIBIRKOVY
SIDALAN
SIDERITIS
SIEGERREBE
SILA
SILVANER
SIRAMÉ
SÍRIA
SIRMIUM
SKLAVA
SKOPELITIKO
ŠKRLET
SKYLOPNICHTIS
SLANKAMENKA
SOLARIS
SOLDAIA
SOLNECHNODOLINSKY
SOPERGA
SORBIGNO
SOYAKI
SPERGOLA
SPOURTIKO
ST CROIX
ST PEPIN
ST VINCENT
STANUŠINA CRNA
STAVROTO
STEUBEN
STORGOZIA
SUKHOLIMANSKY BELY
SULTANIYE
SUMOLL
SUMOLL BLANC
SUPPEZZA
SUŠĆAN
SUSUMANIELLO
SWENSON RED
SWENSON WHITE
SYKIOTIS
SYMPHONY
SYRAH
SYRIKI
SYSAK

☿this symbol on the following pages indicates that the synonyms, and mistaken identities, are confirmed by DNA evidence

SABATO

Minor Campanian vine associated with the Sorrento coast.

BERRY COLOUR: ● ● ● ● ●

PRINCIPAL SYNONYMS: ?Gigante

ORIGINS AND PARENTAGE

Sabato, Italian for 'Saturday', is an old grape variety from the area around the city of Sorrento in the bay of Napoli, southern Italy, or possibly from further north in the Vesuvio area. It may be identical to Gigante, a variety from the Val Fortore in the province of Benevento in northern inland Campania, but there is as yet no ampelographic or DNA evidence to confirm this.

VITICULTURAL CHARACTERISTICS

Mid to late ripening (early October).

WHERE IT'S GROWN

Sabato is still cultivated to a very limited extent on the Sorrento coast, south of Napoli in Campania, southern **Italy**, in particular in Vico Equense on the north of the peninsula where old ungrafted vines are still propagated, and in nearby Castellamare, Lettere and Gragnano. Sabato is being studied as part of a preservation programme. It is a secondary variety in the Penisola Sorrentina DOC.

SABORINHO

See NEGRAMOLL.

SABREVOIS

Insubstantial US hybrid prospering in Québec.

BERRY COLOUR: ● ● ● ● ●

PRINCIPAL SYNONYMS: ES 2-1-9

ORIGINS AND PARENTAGE

An Elmer Swenson 283 × Elmer Swenson 193 hybrid bred by Elmer Swenson in Osceola, Wisconsin, US, in which:

- Elmer Swenson 283 is an Elmer Swenson 114 × SEYVAL BLANC hybrid (see ST PEPIN for the pedigree of Elmer Swenson 114 and PRIOR for that of Seyval Blanc)
- Elmer Swenson 193 is a Minnesota 78 × Seneca hybrid (see BRIANNA for the pedigree of Minnesota 78)
- Seneca is a LUGLIENGA × Ontario hybrid (see CAYUGA WHITE for the pedigree of Ontario).

This complex *Vitis riparia*, *Vitis labrusca*, *Vitis vinifera*, *Vitis lincecumii* and *Vitis rupestris* hybrid has been grown in Minnesota for a long time as ES 2-1-9. It was named in August 2001 by Gilles Benoît after the village of Sabrevois on the Richelieu River, south east of Montréal, the location of Benoît's winery, and is grown in Québec under that name (Smiley 2008).

Sabrevois and ST CROIX have the same parents and are therefore siblings.

VITICULTURAL CHARACTERISTICS

Fairly loose, small to medium-sized bunches of small to medium-sized berries. Vigorous but not overly productive. Cold-hardy (–31 °F/–35 °C) with good overall disease resistance although phylloxera can be a problem (Smiley 2008).

WHERE IT'S GROWN AND WHAT ITS WINE TASTES LIKE

Wines may be deeply coloured and pleasantly fruity but tend to lack body and tannin and are thus better as a blending component, though they can make an aromatic dry rosé. Sabrevois seems to have found greater favour in Québec, **Canada** (16 acres/6 ha and the seventh most planted variety in the province), than in its homeland in the American Midwest, though there were 26 acres (11 ha) in **Minnesota** in 2007. Varietal wines are made by Gilles Benoît (see above) at Vignoble des Pins in Québec and at Oak Ridge in **Wisconsin** among others.

SACY

Light, nether Burgundian, losing ground.

BERRY COLOUR: ● ● ● ● ●

PRINCIPAL SYNONYMS: Aligoté Vert, Plant d'Essert, Plant de Sacy, Tressallier (Saint-Pourçain)

VARIETIES COMMONLY MISTAKEN FOR SACY: ALIGOTÉ, CHARDONNAY

ORIGINS AND PARENTAGE

Sacy is a variety from north-eastern France that is cultivated mainly in the area around Auxerre, between Dijon and Paris, and named after a village south of Reims in champagne country. Its first mention appeared under the synonym Plant de Sacy in 1782, when the mayor of Vermenton requested 'l'interdiction de cultiver le Plant d'Essert ou de Sacy, lequel ne donne qu'un vin plat et de mauvaise calité' ('the ban on cultivating the Plant d'Essert or de Sacy, which gives no more than a dull, poor-quality wine'; Galet 1990; Rézeau 1997). Its main synonym, Tressallier, possibly derived from *trans Alligerim* ('beyond Allier') because it used to be planted on the left bank of the Allier River, was first mentioned in Jullien (1816). These relatively recent dates could give a clue to when the variety was born. DNA parentage analysis has established that Sacy is one of the numerous progenies of PINOT and GOUAIS BLANC, this cross most likely occurring in Burgundy or at least in north-eastern France (Bowers *et al.* 1999; Boursiquot *et al.* 2004).

Solely on the basis of morphology, Sacy was included in the Folle ampelographic group (see p XXVII; Bisson 2009), and not, surprisingly, in the group of either of its parents. However, this classification might change in the light of DNA parentage analysis.

OTHER HYPOTHESES

According to Galet (1990), the first mention of this variety appeared in Bauhin (1650) under the descriptive name 'Uvae albae "enfarins" nuncupatae', ie a grape with leaves covered by a flour-like tomentum (a layer of matted woolly down: French *enfariné* means 'covered in flour'). However, many distinct varieties have been named *enfariné* in north-eastern France, and there is no clear evidence that this is the same variety as the modern Sacy. Galet also tells of a legend that Sacy could have been brought from Italy in the thirteenth century and propagated by the monks of the Abbaye de Reigny in Vermenton but this can be rejected by the recently discovered parentage.

VITICULTURAL CHARACTERISTICS

Early budding, mid ripening. Vigorous and fertile, generally pruned long but short also possible. Small bunches and berries.

WHERE IT'S GROWN AND WHAT ITS WINE TASTES LIKE

Once widely grown in the Yonne *département* of **France**, it is now considered a speciality of Saint-Pourçain in central France, where it is known as Tressallier. It is authorized as a complementary variety in the Saint-Pourçain appellation (20–40% of the blend) alongside the principal variety CHARDONNAY and possibly also SAUVIGNON BLANC (maximum 10%). It produces light, high-acid wines, sometimes tasting of pears, that are low in alcohol and well suited to the production of sparkling wine (it is a permitted variety in Crémant de Bourgogne). The vineyard area is in steady decline: just 12 ha (30 acres) in 2008 compared with 172 ha (425 acres) just ten years before and 655 ha (1,619 acres) in 1958. Domaine Nebout is one of the rare producers in Saint-Pourçain to make a varietal wine.

SAGRANTINO

Particularly tannic Umbrian speciality of Montefalco.

BERRY COLOUR: ● ● ● ● ●

ORIGINS AND PARENTAGE

Sagrantino was described in the late nineteenth century as an old variety from the Montefalco area in the province of Perugia in Umbria, central Italy, its likely origin.

OTHER HYPOTHESES

Sagrantino is said to have been introduced from Greece by Byzantine monks in the Middle Ages or from Asia Minor by Franciscans, and its name is thought to derive from the Italian *sacro* meaning 'sacred', or *sagrestia* meaning 'sacristy' from its use in a sacred context. It is also believed by some that the Itriola mentioned by Pliny the Elder could correspond to Sagrantino.

VITICULTURAL CHARACTERISTICS

Late ripening. Resistant to winter and spring frost. Susceptible to downy mildew.

WHERE IT'S GROWN AND WHAT ITS WINE TASTES LIKE

Sagrantino's home territory is Montefalco, between Perugia and Spoleto in Umbria, central **Italy**. At one time it was used to produce sweet dried-grape wines but today the wines are more usually dry, lively, sometimes tannic, with flavours of cherries and mulberries. When the tannins are well managed, the wines are firm but smooth and well suited to oak ageing.

The variety had become almost extinct in the 1960s but was rescued by Marco Caprai and other local growers who believed in the qualities and potential of this variety. After much campaigning, Montefalco Sagrantino attained DOC status in 1979 and was promoted to DOCG in 1992. Caprai has continued to produce internationally acclaimed wines, along with the likes of Còlpetrone, Perticaia and Lungarotti, but the wine quality is not universally high. The Collepiano clone is said to have higher levels of sugar and acidity than others.

Sagrantino is today cultivated in Montefalco and the surrounding communes of Bevagna, Gualdo Cattaneo, Castel Ritaldi and Giano. Although it is generally made as a varietal wine, a small amount (maximum 15%) may be blended with SANGIOVESE in the Montefalco DOC. Some sweet, passito-style wines such as Tabarrini's Colle Grimaldesco are still made and are permitted within the DOCG. There were 361 ha (892 acres) planted in Italy in 2000.

The best examples and especially those of Arnaldo Caprai have encouraged plantings in Toscana and Sicilia.

Outside Italy, the vine is grown by a handful of growers around **California** and there are limited plantings in **Australia**. Early adopters included Andrew Peace in Victoria, Heathvale in South Australia and Preston Peak in Queensland.

SAINT-LAURENT

See SANKT LAURENT.

SAINT-MACAIRE

Obscure Bordeaux variety found to a very limited extent in California and Australia.

BERRY COLOUR: ● ● ● ● ●

PRINCIPAL SYNONYMS: Bouton Blanc (Bordeaux), Moustère (Gironde), Moustouzère (Premières Côtes de Bordeaux)

ORIGINS AND PARENTAGE

Little is known about the history of this obscure Bordeaux variety named after a village near Langon in the Gironde, south-west France, where it used to be cultivated and where it probably originated. First mentioned as Saint-Macaire in the 'Livre de raison d'Antoine Feuilhade', an account book written between 1763 and 1777 by the mayor of Libourne (Garde 1946), and commonly grown in the Gironde in the nineteenth century.

VITICULTURAL CHARACTERISTICS

Early budding, mid ripening and vigorous. Susceptible to powdery mildew.

WHERE IT'S GROWN AND WHAT ITS WINE TASTES LIKE

Saint-Macaire has more or less disappeared from its Bordeaux homeland: official statistics for 2008 in **France** report less than 1 ha (2.5 acres). There are also tiny plantings in **California**, where it is one of the minor varieties permitted in a bordeaux-style Meritage blend. O'Shaughnessy in Howell Mountain, for example, include 2% Saint-Macaire in theirs. In **Australia**, Bill Calabria at Westend Estate in Griffith, Victoria, claims to be the only producer of this variety in the country, growing just 2 ha (5 acres). Varietal wines tend to be deeply coloured and relatively acidic.

SALVADOR

Red-fleshed French hybrid now largely superseded by Rubired in California's San Joaquin Valley.

BERRY COLOUR: ● ● ● ● ●

PRINCIPAL SYNONYMS: Pate Noir, Salvadore, Seibel 128

ORIGINS AND PARENTAGE

A complex hybrid obtained at the beginning of the twentieth century by Albert Seibel in Aubenas in south-eastern France by crossing Munson with *Vitis vinifera* subsp. *vinifera* (ie an unknown cultivated variety). See PRIOR for the pedigree of Munson.

VITICULTURAL CHARACTERISTICS

Small, compact bunches susceptible to botrytis bunch rot and grape thrips. Productive, resistant to powdery mildew but prone to potassium and manganese deficiency when grown on own roots or on low-fertility soils.

WHERE IT'S GROWN AND WHAT ITS WINE TASTES LIKE

There were 176 acres (71 ha) of this red-fleshed variety in **California** in 2008, mostly in Kern County at the southern end of the San Joaquin Valley. It was once more widely planted and was the main variety used in Fresno in the 1960s and 1970s for the production of red grape-juice concentrate though it was also occasionally used to add colour to wine blends. It has been largely replaced by RUBIRED, which serves the same function but does not have the same herbaceous flavours and harsh tannins and has redder – less purple – juice. Salvador is also planted in Aguascalientes and Chihuahua in **Mexico**.

SÄMLING 88

See SCHEUREBE.

SAMSÓ

See CINSAUT *and* MAZUELO.

SAMTROT

See PINOT MEUNIER *in the* PINOT *entry.*

SAN GIUSEPPE NERO

Little is known about this dark-skinned central Italian vine. It may be identical to ABBUOTO, but there is no evidence yet. There were 387 ha (956 acres) in Italy in 2000, mainly in central and southern **Italy**. It is an authorized variety in Lazio.

SAN LUNARDO

Rare Ischian white wine variety.

BERRY COLOUR:

PRINCIPAL SYNONYMS: Bianca (Ventotene), Don Lunardo (Ischia), Lunardo (Ischia)

ORIGINS AND PARENTAGE

Although the earliest reference to San Lunardo on the island of Ischia in the bay of Napoli, southern Italy, was in 1962 (Migliaccio *et al.* 2008), the existence of old vines indicates that it has been there much longer. Its name is said to derive either from Don Lunardo, the name of the priest who owned the farmhouse where it was found, or from San Leonardo, as a tribute to the patron saint of Panza, a village near Forio on Ischia.

Cipriani *et al.* (2010) have recently suggested through DNA parentage analysis that San Lunardo is a possible Albana Bordini × BIANCOLELLA natural cross, the former being an obscure black-berried variety of unknown origin, the latter a white-berried variety indigenous to Ischia.

VITICULTURAL CHARACTERISTICS

Abundant and regular yields. Good resistance to botrytis bunch rot.

WHERE IT'S GROWN AND WHAT ITS WINE TASTES LIKE

San Lunardo is authorized throughout the province of Napoli in Campania, southern **Italy**, but it is found only on the island of Ischia, in the bay of Napoli, where a few old vines thrive in old vineyards, and on the island of Ventotene, where it is simply called Bianca. A small amount – up to 15% – may be blended with FORASTERA and BIANCOLELLA in Ischia DOC. La Vigna dell'Encadde are one of the few producers of a varietal wine.

SAN MARTINO

Very minor cross made at Conegliano in the Veneto but apparently grown only in Piemonte.

BERRY COLOUR:

PRINCIPAL SYNONYMS: Incrocio Dalmasso 7/21

ORIGINS AND PARENTAGE

The Italian vine breeder Giovanni Dalmasso announced that San Martino was a NEBBIOLO ×

DOLCETTO cross obtained in 1936 at the Conegliano research centre and it was officially entered in the national register of varieties in 1977. However, DNA profiling has recently shown that the Nebbiolo used by Dalmasso was in fact Nebbiolo di Dronero, which is not Nebbiolo but a synonym for CHATUS (Torello Marinoni, Raimondi, Mannini and Rolle 2009). Therefore, San Martino must be considered to be a Chatus × Dolcetto cross, just like its siblings PASSAU and VALENTINO NERO.

WHERE IT'S GROWN

San Martino is cultivated in Piemonte, north-west **Italy**. The Italian agricultural census of 2000 recorded total plantings of 45 ha (111 acres).

SAN MICHELE

Obscure Conegliano cross whose true parentage has only recently been revealed.

BERRY COLOUR: ● ● ● ● ●

PRINCIPAL SYNONYMS: Incrocio Dalmasso 15/34

ORIGINS AND PARENTAGE

The Italian vine breeder Giovanni Dalmasso claimed that San Michele was a NEBBIOLO × BARBERA cross obtained in 1938 at the Conegliano research centre and it was included in the national register of varieties in 1977. The objective was to get a new variety that combined the quality of Nebbiolo with the resistance and high yields of Barbera. However, DNA profiling has recently shown that the Nebbiolo used by Dalmasso was in fact Nebbiolo di Dronero, which is not Nebbiolo but a synonym for CHATUS (Torello Marinoni, Raimondi, Mannini and Rolle 2009). San Michele is therefore a Chatus × Barbera cross, just like its siblings ALBAROSSA, CORNAREA, NEBBIERA and SOPERGA.

VITICULTURAL CHARACTERISTICS

Late ripening. Slightly susceptible to botrytis bunch rot.

WHERE IT'S GROWN

There were 128 ha (316 acres) in **Italy** in 2000, mainly in Piemonte.

SAN PIETRO

Rare Campanian that may be related to Muscat Blanc à Petits Grains.

BERRY COLOUR: ● ● ● ● ●

ORIGINS AND PARENTAGE

San Pietro is a rarity from the province of Caserta in Campania, southern Italy, where it was first mentioned at the beginning of the twentieth century. DNA analysis suggests that San Pietro is genetically close to CODA DI PECORA (Costantini *et al.* 2005). Cipriani *et al.* (2010) have recently suggested through DNA parentage analysis that San Pietro is a possible natural cross between

MUSCAT BLANC À PETITS GRAINS (under the name Moscato Bianco) and Bicane, the latter a French wine and table grape that is no longer cultivated.

VITICULTURAL CHARACTERISTICS

Mid to late ripening.

WHERE IT'S GROWN

San Pietro is cultivated mainly in the area between Monte Maggiore and Roccamonfina in the province of Caserta in Campania, southern **Italy**, where it is usually blended with other local varieties.

SANGIOVESE

Hugely important and wildly variable central Italian variety with many a name and surprising origins.

BERRY COLOUR: ● ● ● ● ●

PRINCIPAL SYNONYMS: Brunelletto (Grosseto in Toscana), Brunello or Brunello di Montalcino (Montalcino in Toscana), Cacchiano (Toscana), Calabrese (Toscana), Chiantino (Toscana), Corinto Nero (Isola di Lipari in Calabria), Guarnacciola (Benevento in Campania), Liliano (Toscana), Morellino or Morellino di Scansano (Scanzano in Toscana), Negrello (Calabria), Nerello (Sicilia), Nerello Campotu (Calabria), Niella (Corse), Nielluccio (Corse), Primaticcio, Prugnolo Dolce (Toscana), Prugnolo Gentile (Montepulciano in Toscana), Puttanella (Calabria), San Gioveto, San Zoveto (Toscana), Sangiogheto (Toscana), Sangiovese Grosso (Toscana), Sangiovese Piccolo (Toscana), Sangioveto (Toscana), Tabernello, Toustain (Algeria), Tuccanese (Puglia), Vigna del Conte (Calabria), Vigna Maggio (Toscana)

VARIETIES COMMONLY MISTAKEN FOR SANGIOVESE: CILIEGIOLO (Emilia-Romagna), MONTEPULCIANO, Morellino del Casentino (Toscana, not commercially cultivated), Morellino del Valdarno (Toscana, not commercially cultivated), NERO D'AVOLA, PERRICONE, Sanvicetro (Toscana, not commercially cultivated), UVA TOSCA (Emilia-Romagna)

ORIGINS AND PARENTAGE

EARLIEST MENTION Sangiovese was first mentioned in Giovan Vettorio Soderini's (1600) treatise on viticulture in Toscana under the name Sangiogheto, its oldest attested synonym: 'il Sangiogheto, aspro a mangiare, ma sugoso e pienissimo di vino' ('the Sangiogheto, bitter to eat, but juicy and very vinous'). Legend has it that Sangiovese derives from *sanguis Jovis* ('Jupiter's blood'), the name arbitrarily given to it by monks in Santarcangelo di Romagna, at the foot of Monte Giove near Rimini in Emilia-Romagna, after they had been asked by a visitor for the name of the wine they used to simply call *vino*.

SANGIOVESE IS HALF-TUSCAN, HALF-CALABRIAN The Tuscan origin of Sangiovese had long been taken for granted until Vouillamoz *et al.* (2004) revealed that DNA parentage analysis had shown the true origin of Sangiovese: it is a natural CILIEGIOLO × CALABRESE DI MONTENUOVO cross. Ciliegiolo's role did not come as a surprise because this old Tuscan variety, also known since Soderini (1600), had traditionally been blended with Sangiovese in Chianti and a possible family relationship between the two had already been suspected (Crespan *et al.* 2002).

The big surprise was CALABRESE DI MONTENUOVO. Just a few vines of this obscure variety were found near a winery on the hill of Montenuovo next to the Lago d'Averno in the Campi Flegrei west of Napoli during a survey of old local varieties (Costantini *et al.* 2005). Montenuovo is a relatively young hill, formed in one night after a volcanic eruption in 1538. The winery was established in the mid nineteenth century by the Calabrian family Strigari, who introduced several varieties from their homeland, as evidenced by the presence of MAGLIOCCO DOLCE and

NERELLO MASCALESE in the vineyards around the Lago d'Averno (Vouillamoz, Monaco *et al.* 2008). Calabrese di Montenuovo, which is unrelated to the Sicilian NERO D'AVOLA – often called Calabrese – was later found in an additional vineyard in Campi Flegrei called Soccavo (Vouillamoz, Monaco *et al.* 2008). However, its true identity remained unknown. Vouillamoz, Monaco *et al.* (2007) then analysed the DNA profiles of 194 additional old varieties from Calabria, Campania, Toscana, Basilicata and Puglia. None matched Calabrese di Montenuovo but a possible parent–offspring relationship was detected between Calabrese di Montenuovo and MAGLIOCCO CANINO, one of the most widespread red varieties in Calabria. This strongly suggests a Calabrian origin for Calabrese di Montenuovo.

In addition, DNA profiling has revealed the historical cultivation of Sangiovese in southern Italy under various local names: Corinto Nero (a seedless mutation grown in Scalea in Calabria and on the Isola di Lipari just north of Sicilia, where it is used in small amounts in Malvasia delle Lipari wines), Nerello (Savelli in Calabria), Nerello Campotu (Motta San Giovanni in Calabria), Puttanella (Mandatoriccio in Calabria), Tuccanese (Puglia), Vigna del Conte (Calabria; Zulini *et al.* 2002; Grando *et al.* 2008).

Interestingly, DNA parentage analysis has also established that Sangiovese has some more or less obscure possible progenies in Toscana – FOGLIA TONDA, Morellino del Casentino, Morellino del Valdarno (Crespan, Calò *et al.* 2008) – and several ancient and well-known progenies in southern Italy: FRAPPATO (under the name Frappato di Vittoria; Sicilia), GAGLIOPPO (Calabria), Mantonicone (Calabria), NERELLO MASCALESE (Sicilia), PERRICONE (Sicilia; Di Vecchi Staraz, This *et al.* 2007; Crespan, Calò *et al.* 2008; Cipriani *et al.* 2010). The parent–offspring relationship between Sangiovese and Gaglioppo explains why Morellino Pizzuto, supposedly a clone of Sangiovese in Toscana, was found to be identical to Gaglioppo (Scalabrelli *et al.* 2008), and the parent–offspring relationship with Nerello Mascalese explains why Sangiovese is sometimes called Nerello on Sicilia or Negrello in Calabria (Grando *et al.* 2008; Schneider, Raimondi *et al.* 2009). Another seedless mutation has also been found in Campania (Schneider, Raimondi *et al.* 2009). In addition, a possible genetic relationship with ALEATICO has been suggested (Filippetti, Silvestroni, Thomas and Intrieri 2001).

As a consequence, both CALABRESE DI MONTENUOVO and possibly Sangiovese are most likely of southern Italian origin, whence Sangiovese later spread to Toscana, the French island of Corse and elsewhere. This might explain why Vannuccini (1902) mentioned a San Gioveto called Calabrese near Arezzo in the south of Toscana.

CLONAL DIVERSITY Like many old and widespread varieties, Sangiovese is said to be of polyclonal origin (a concept that does not fit with the definition of a grape variety used in this book – see pp XIX–XVI), which means that several distinct varieties are confusingly cultivated under the name Sangiovese (Vignani *et al.* 2002; Filippetti *et al.* 2005; Di Vecchi Staraz, This *et al.* 2007). The significant biodiversity within Sangiovese is illustrated by its various synonyms (Filippetti *et al.* 2005; Di Vecchi Staraz, This *et al.* 2007), listed above.

Conventional descriptions of Sangiovese, based on the pioneering work of ampelographer G Molon in 1906, divide the variety into two groups (sometimes erroneously called families): the Sangiovese Grosso (including Brunello, Prugnolo Gentile and the Sangiovese di Lamole of Greve in Chianti) and the Sangiovese Piccolo of other zones of Toscana. This led to an implicit supposition that superior quality was found in the Grosso family. However, from the more recent evidence outlined above, such a simple distinction is no longer possible and qualitative judgements based on the size of either the berries or the bunches are not justified. Significant efforts have been made in central Italy to identify and propagate superior clones rather than the former practice of selecting those with the highest yields irrespective of fruit or wine quality.

OTHER HYPOTHESES

Some authors believe that Sangiovese was cultivated by the Etruscans and was domesticated from wild grapes. However, no genetic link has been found between Sangiovese and wild grapes of Toscana (Di Vecchi Staraz *et al.* 2004).

Di Vecchi Staraz, This *et al.* (2007) suggest, on the basis of thirty-eight DNA markers, that CILIEGIOLO is a cross between Sangiovese and the (no longer cultivated) Muscat Rouge de Madère, and that the latter is in fact a possible cross between MAMMOLO and MUSCAT BLANC À PETITS GRAINS. However, this is disputed for several reasons:

- Muscat Rouge de Madère is thought to be of Portuguese origin and used to be grown only on Madeira, in the Douro and Lisboa as well as in South Africa (Salomon and Salomon 1902b).
- Muscat Rouge de Madère has no history of cultivation in Italy.
- The usually dominant Muscat flavour is absent in CILIEGIOLO.
- CILIEGIOLO's proposed offspring–parent relationship with Sangiovese is contradicted by the theory that Sangiovese is a Ciliegiolo × CALABRESE DI MONTENUOVO cross, as discovered by Vouillamoz, Imazio *et al.* (2004) and Vouillamoz, Monaco *et al.* (2007) and strongly supported by fifty DNA markers.

A recent study based on fifty-seven DNA markers (Bergamini *et al.* 2012) has suggested yet another parentage for Sangiovese: it is said to be a natural cross between CILIEGIOLO and Negrodolce, the latter apparently an old Puglian variety. The authors claim to provide strong evidence for a southern Italian origin for Sangiovese. However, the DNA profile of their obscure Negrodolce (meaning 'sweet black') is identical to Morellino del Valdarno in Crespan, Calò *et al.* (2008), a Tuscan variety that is no longer cultivated and for which a parent–offspring relationship with Sangiovese had already been found.

VITICULTURAL CHARACTERISTICS

Vigorous and highly susceptible to botrytis bunch rot due to its thin skins. Slow and late ripening. Resistant to drought and a relatively high yielder.

WHERE IT'S GROWN AND WHAT ITS WINE TASTES LIKE

On the island of Corse in **France**, as Nielluccio, Sangiovese is the most highly regarded and widely planted red wine vine, with 1,319 ha (3,259 acres) in 2008. Nielluccio is often thought of as an indigenous variety, even though it is genetically identical to Sangiovese, and was probably brought to the island by the Genoese, who ruled the island until the eighteenth century. Often blended with Sciaccarello (MAMMOLO), it is particularly important in the Patrimonio region in the north, where it thrives on clay-limestone soils to produce tannic but ageworthy wines, though it plays a significant role in all Corsican red and rosé appellations.

The most important nursery in Italy, Rauscedo, has recently begun selling Sangiovese cuttings to producers on the French mainland, specifically the Languedoc, where they are intrigued by its drought resistance and reliable yields.

But the kingdom of Sangiovese is **Italy**, where it is the most widely planted variety by some margin – 69,790 ha (172,455 acres) in 2000, according to the agricultural census, down from more than 100,000 ha (247,000 acres) in 1990 – and responsible for some of Italy's most internationally famous wines such as the Chiantis and Brunello di Montalcino. It is recommended or authorized for more than half of Italy's wine regions – all but the most northerly and the most southerly – and is particularly important in Toscana, Emilia-Romagna, the Marche and Umbria. However, the quality of Sangiovese wines is extremely variable, influenced particularly by the clones planted and vineyard location and yield. The variety adapts well to a wide variety of soils, although the presence of limestone seems to exalt the elegant and forceful aromas that are perhaps the most attractive quality of the grape, and vines planted too high may fail to ripen fully. Many of the clones planted in the 1960s to 1980s, and some over-ambitiously located vineyard sites, resulted in wines lacking in body and colour but with considerable acid and astringency. With Chianti Classico vineyards regularly over 450 m and even up to 600 m altitude, exposition and a willingness to harvest late can be essential. Considerable and considered research in the late 1980s and 1990s into different clones of Sangiovese, particularly by the Chianti Classico Consorzio, showed, much to Tuscan surprise, that some of those producing the finest wine such as R24 and T19 originated in Emilia-Romagna, however thin, tart and uninspiring the quality of basic Sangiovese di Romagna DOC. Fully ripe Sangiovese can have alluring aromas of plums and dried underbrush. Less ripe, or less carefully made, it can exhibit farmyard aromas.

In its many guises (see above for synonyms and clones used in specific regions), Sangiovese is the principal vine variety for red wine in Toscana, sometimes unadorned, as in Brunello di Montalcino, often the dominant variety in a blend, eg in Chianti Classico and other Chiantis, Vino Nobile di Montepulciano and many of the Supertuscans, those ambitious bottlings usually classified as IGT and often using international varieties such as CABERNET SAUVIGNON, MERLOT or SYRAH to give added lustre. Since the late twentieth century, Sangiovese has depended on other varieties, traditionally CANAIOLO NERO and MAMMOLO but increasingly the well-known interlopers, to make up for its deficiencies – quite legally within Chianti's oft-changed regulations, more covertly for other wines such as Brunello di Montalcino.

But an increasing number of producers who really care about local expression, or who have learned that Toscana's distinction lies in carefully ripened Sangiovese, eschew non-Tuscan grapes and there has been a re-evaluation of Chianti Classico and particularly Riserva versions, and of top-quality all-Sangiovese wines in general, at the expense of those Supertuscans that depend on international varieties. Fine varietal Sangioveses from the Chianti zones include Fontodi's Flaccianello, Montevertine's pioneering Le Pergole Torte, Poggio Scalette's Il Carbonaione, Isole e Olena's Cepparello, La Porta de Vertine's Chianti Classico Riserva, San Giusto a Rentennano's Percarlo and Selvapiana's Riserva Bucerchiale. In theory, all Brunello di Montalcino is made from 100% Sangiovese, called here Brunello. Arguably the single most convincingly 100% Brunello example is Soldera's at Case Basse. Of Vino Nobile di Montepulciano, of which Sangiovese can constitute between 70 and 100%, Avignonesi is the leader.

All but the most brazenly 'international' reds of Emilia-Romagna are based on Sangiovese, even if the most basic wines tend to be thin and tart. Castelluccio's Ronco delle Ginestre, Drei Donà's Pruno and Zerbina's Pietramora represent the other end of the quality spectrum. In Umbria, Lungarotti's Torgiano Riserva Vigna Monticchio has upheld the glory of Sangiovese, albeit seasoned with other central Italian varieties from this vineyard, for decades. In the Marche further north, Sangiovese is the foundation for Rosso Piceno DOC (30–50%) and an important component in Rosso Conero DOC and Conero DOCG.

Sangiovese is also now found in **Switzerland**, **Malta**, **Turkey** and **Israel**. In **Greece**, Nico Lazaridi have planted Sangiovese in the north-east region of Dráma and blend 55% Sangiovese with CABERNET SAUVIGNON to produce their Perpetuus oak-aged red.

In **California**, Sangiovese benefited from a perceived fashion for 'Cal-Ital' varietals in the early years of this century but by 2010 there were only 1,950 acres (789 ha), a significant decline since a peak of around 3,000 acres (1,214 ha) in 2003. There are some significant plantings in the Napa Valley, Sonoma, San Luis Obispo, Santa Barbara and the Sierra Foothills. Pioneers of the variety included Robert Pepi, Seghesio and Atlas Peak, in which the Tuscan firm of Antinori originally had an important stake, plus Villa Ragazzi and Noceto on a smaller scale. Other successful producers of Sangiovese in California include Benessere, Dalla Valle, Shafer and Stolpman.

In **Washington** State, the generally bright fruit quality seems to suit Sangiovese rather well, filling in some of its often spare frame, but plantings had fallen to 185 acres (75 ha) by 2011. Successful producers include Leonetti and Novelty Hill.

In **Canada**, there are experimental plantings of 5–10 acres (2–4 ha) in Ontario. Pillitteri Estates even make a small amount of icewine from their one-acre (0.4 ha) vineyard in Niagara-on-the-Lake. There's about the same area of Sangiovese in British Columbia with Sal D'Angelo's a notable example.

There were 2,319 ha (5,730 acres) of Sangiovese in **Argentina** in 2008, mostly in Mendoza but with limited plantings in San Juan and La Rioja. Benegas of Mendoza have made a speciality of Sangiovese. **Chile** had only 124 ha (306 acres) in 2008. Errázuriz produce a single-vineyard version in the Aconcagua Valley. There were only 25 ha (62 acres) in 2007 in **Brazil**, where Casa de Amaro make a varietal.

Sangiovese was a late arrival in **Australia** with only one clone available until recently, the H6V9 imported from the University of California at Davis in the 1960s, though there are now ten to choose from. The first commercial plantings were in the 1980s, the first source block in Penfold's Kalimna vineyard. Early results were often pale, hollow and dry, partly because the vines were allowed to follow their natural inclination to overcrop. In 2008 there were 517 ha (1,278 acres). The pioneers were Coriole and Dromana; others succeeding with the variety today include Brown Brothers, De Bortoli, The Little Wine Company, Mount Langi Ghiran, Pizzini and Scaffidi. Today's wines are a vast improvement on the early bottlings but consistency and popular appeal both seem elusive.

New Zealand was home to just 6 ha (15 acres) of Sangiovese in 2008 with no signs of more enthusiasm. Heron's Flight in Matakana north of Auckland is one of the few who produce a varietal wine; their first release was 1998.

From **South Africa**'s 63 ha (156 acres), as counted in 2008 – mostly in Darling and Stellenbosch – Chris Mullineux and Terra Da Capo make varietals; blends include Nederburg's Ingenuity Red and Bouchard Finlayson's Hannibal.

SANKT LAURENT

Middle European variety of uncertain origin producing velvety reds.

BERRY COLOUR: ● ● ● ● ●

PRINCIPAL SYNONYMS: Laurenzitraube (Austria), Lorenztraube (Austria), Saint Laurent (Germany), Saint-Laurent (France), Saint-Lorentz (Austria), Schwarzer (Germany), Svatovavřinecké (Czech Republic), Svätovavrinecké (Slovakia), Vavřinecké (Czech Republic)

VARIETIES COMMONLY MISTAKEN FOR SANKT LAURENT: PINOT NOIR⊗, Pinot Saint-Laurent⊗

ORIGINS AND PARENTAGE

Sankt Laurent most likely originates from Austria, where it has been cultivated since the nineteenth century at the very least. Its name has nothing to do with any of the numerous villages named Saint-Laurent in France but probably derives from its old Austrian synonym Laurenzitraube, which in turn refers to St Laurentius, patron saint of chefs, whose saint's day falls on 10 August, when the variety begins to ripen. According to Regner (2000b; 2007), Sankt Laurent could be a progeny of an as yet unknown variety that is genetically close to PINOT. However, this hypothesis is doubtful because the DNA profile of Sankt Laurent is genetically distant from Pinot, and Regner possibly confuses Sankt Laurent with Pinot Saint-Laurent, two quite distinct varieties. Contrary to common belief, Pinot Saint-Laurent (which is not in commercial production) is not a synonym of Sankt Laurent, nor does it have anything to do with Pinot (Vouillamoz).

Sankt Laurent was used to breed several varieties, some of which are cultivated for commercial wine production: ANDRÉ, DUNAJ, RONDO and ZWEIGELT.

OTHER HYPOTHESES

Ampelographers have suggested controversial hypotheses about the origins of Sankt Laurent. On the one hand, Oberlin suggests that it could originate from Alsace, where it was first noticed under the name Schwarzer (*schwarz* means 'black') by the ampelographer Bronner, who introduced it to Baden-Württemberg. On the other hand, Pulliat hypothesizes that it was introduced to the botanical garden in Dijon by Weber, who had received cuttings from a grape collector from Saint-Laurent du Var, hence the name.

VITICULTURAL CHARACTERISTICS

Compact bunches of small berries. Mid ripening, good resistance to winter frosts but susceptible to blossom drop and coulure, thanks to early flowering, botrytis bunch rot and downy mildew. It needs deep soils or irrigation and is at risk from late frosts. Yields are erratic and September rains can cause grapes to split and develop sour rot.

WHERE IT'S GROWN AND WHAT ITS WINE TASTES LIKE

Sankt Laurent produces wines in several Central European countries that are generally aromatic, deeply coloured and velvety but well structured with fine tannins and flavours of sour cherry, sometimes likened to a powerful version of PINOT NOIR. The best wines are well suited to subtle oaking.

Germany's 669 ha (1,653 acres) of unhyphenated Saint Laurent were, in 2008, overwhelmingly in the Pfalz and Rheinhessen. The variety had almost disappeared in the 1960s but is now enjoying something of a comeback thanks especially to the efforts of viticultural adviser Fritz Klein, who propagated the variety from the last few remaining vines, and to the perseverance of the Pfalz wine grower Karlheinz Kleinmann, who campaigned for its inclusion in the official register of varieties. Recommended producers include Knipser, Philip Kuhn and Bernhart in the Pfalz; Gutzler and Wagner-Stempel in Rheinhessen; and Fischer in Baden.

Plantings in the variety's presumed homeland of **Austria** have almost doubled in the last nine years in keeping with the general expansion of red-wine production, reaching 795 ha (1,964 acres) in 2008. The majority of the plantings are south and east of Vienna in Burgenland's Neusiedlersee and in the Thermenregion and west of the capital in the Weinviertel. Despite the viticultural disadvantages described above, its potential quality and the fact that it can be grown on a wider range of sites than PINOT NOIR, and ripens earlier, explains its increasing popularity. Recommended producers include Gernot and Heike Heinrich, Leo Hillinger, Pimpel, Schloss Halbturn, Steinklammer and Zantho.

But the variety is even more widely planted north and east of Austria. Under the name Svatovavřinecké (it lost its sanctity during the Communist era and was known simply as Vavřinecké), it is the most widely cultivated dark-skinned variety in the **Czech Republic**, planted mainly in Morava in the south east, with a total of 1,482 (3,662 acres) in 2008. Recommended producers include Patria in Kobylí, Michlovský in Rakvice, Stapleton-Springer and Jedlička & Novak in Bořetice and ZD in Němčičky.

Further east, in **Slovakia**, where only the accents change on Svätovavrinecké, there were 1,408 ha (3,479 acres) in 2009, 7% of the vineyard area, making it the second most planted red variety after BLAUFRÄNKISCH. Most wines are varietal but it is occasionally blended with BLAUER PORTUGIESER (under the name Modrý Portugal).

SANT'ANTONIO

Very minor Campanian variety.

BERRY COLOUR: ● ● ● ● ●

ORIGINS AND PARENTAGE

Sant'Antonio is a local variety from the province of Caserta in Campania, southern Italy, where it was mentioned at the beginning of the twentieth century. DNA analysis has suggested that Sant'Antonio could be genetically close to Sanginella Salerno (Costantini *et al.* 2005), another local variety but no longer in cultivation.

VITICULTURAL CHARACTERISTICS

Mid to late ripening.

WHERE IT'S GROWN

Sant'Antonio is cultivated in a few vineyards in the province of Caserta in Campania, southern **Italy**, in particular in the commune of Galluccio, about 60 km north west of Napoli, where it is blended with other local varieties.

SANTA MARIA

Emilia-Romagnan speciality that plays a minor part in a rare Vin Santo wine.

BERRY COLOUR: ● ● ● ● ●

VARIETIES COMMONLY MISTAKEN FOR SANTA MARIA: MELARA

ORIGINS AND PARENTAGE

Santa Maria is a rare variety from the province of Piacenza in Emilia-Romagna, where it has been known since 1768. Cipriani *et al.* (2010) have recently suggested through DNA parentage analysis that Santa Maria may be a VERNACCIA DI ORISTANO × Pensicato natural cross, the latter being an obscure black-berried variety of unknown origin. This parentage, which showed some DNA discrepancies, is surprising because Vernaccia di Oristano has never been cultivated outside Sardegna.

VITICULTURAL CHARACTERISTICS

Early to mid ripening.

WHERE IT'S GROWN AND WHAT ITS WINE TASTES LIKE

Santa Maria is cultivated today exclusively in the hills of the province of Piacenza, in particular in the Val d'Adra, west of Parma, where it may be blended with Bervedino (VERNACCIA DI SAN GIMIGNANO), MARSANNE, SAUVIGNON BLANC, ORTRUGO, TREBBIANO ROMAGNOLO and MELARA to produce the rare Vin Santo di Vigoleno within the Colli Piacentini DOC. Producers of this sweet wine include: Giuseppe Ballarini, Rina Illica Magnani, Paolo Loschi, Alberto Lusignani, Giuseppe Moschini, Enzo Perini, Franco Sesenna, Giuseppe Sesenna, Cristina Sozzi, Edoardo Visconti and Massimo Visconti. There were 16 ha (40 acres) in **Italy** in 2000.

SANTA SOFIA

Minor southern Italian for long confused with Fiano.

BERRY COLOUR: ● ● ● ● ●

VARIETIES COMMONLY MISTAKEN FOR SANTA SOFIA: FIANO ☒

ORIGINS AND PARENTAGE

Santa Sofia used to be widespread between Basilicata and Cilento in the province of Salerno in

Campania, southern Italy, where it was considered to be the same as FIANO. However, this has been contradicted by DNA profiling, which shows that Santa Sofia is genetically close to Rosso Antico (Costantini *et al.* 2005), another local variety but no longer in cultivation.

VITICULTURAL CHARACTERISTICS

Usually harvested slightly overripe, in October. Low yields.

WHERE IT'S GROWN

Santa Sofia was formerly grown both for wine and for the table. Its cultivation drastically decreased after the devastating effects of phylloxera in the early twentieth century. A few old and often ungrafted vineyards of Santa Sofia can still be found today in Cilento, south of the city of Salerno, **Italy**, where the harvest is blended with FIANO, TREBBIANO TOSCANO and other local varieties in the Cilento DOC. The Italian agricultural census of 2000 does not give the vineyard area for this variety because it treats it as a synonym for Fiano.

SAPERAVI

Quintessential Georgian variety producing deeply coloured, firm, initially tart but ageworthy reds.

BERRY COLOUR: ● ● ● ● ●

PRINCIPAL SYNONYMS: Atenuri Saperavi, Meskhuri Saperavi, Obchuri Saperavi

ORIGINS AND PARENTAGE

Saperavi, meaning 'dye' in Georgian, referring to its deep black skin and pinkish juice, is a very old variety from south-western Georgia (Zemo Kartli), historically known as Meskheti and now straddling the Georgian-Turkish border. From there it spread to the Kartli region and then more widely to Kakheti, where it found its second home. It was widespread in the seventeenth century in western Georgia as well as in Kakheti (Chkhartishvili and Betsiashvili 2004).

Several distinct clones have been identified (Saperavi Budeshurisebri, Saperavi Grdzelmarcvala, Saperavi Mskhvilmarcvala, Saperavi Pachkha, etc.), illustrating the ancient cultivation of this variety.

Saperavi was used to breed AKHTANAK, BASTARDO MAGARACHSKY, NERKENI, NOVAC, RUBINOVY MAGARACHA, SAPERAVI SEVERNY and TIGRANI.

VITICULTURAL CHARACTERISTICS

Mid budding, late ripening, relatively productive. Winter-hardy as well as drought-tolerant. Unusually for a pale-fleshed variety, berries produce pinkish juice.

WHERE IT'S GROWN AND WHAT ITS WINE TASTES LIKE

The most widely planted dark-skinned variety in **Georgia** (3,692 ha/9,123 acres in 2004), Saperavi is planted throughout the Kakheti wine region in the south east of the country but produces its finest wines in the villages of Mukuzani, Khashmi, Anaga, Kardenakhi, Shilda, Napareuli, Kondoli and Tsinandali and is solely responsible for highly regarded appellation wines such as Mukuzani, Napareuli, Akhasheni and Kindzmarauli. Varietal wines are notably high in colour, acidity and tannin, full-bodied, rich in dark fruit as well as more savoury flavours and generally repay time in bottle. Saperavi also adds colour and acidity to blends and makes impressive semi-sweet and port-style fortified wines, particularly in the Kindzmarauli and Akhasheni appellations. In cooler, higher sites, the grapes fail to ripen fully and wines may be overly acidic and sappy. Recommended producers include Badagoni, Chelti, Eniseli Wines, Jakeli Wines, Kakhuri, Nika, Orovela, Maisuradze Wines, Chateau Mukhrani, Tbilvino, Telavi

Wine Cellar, Wine Man and Teliani Valley. Pheasant's Tears and Vinoterra make good examples of traditional styles, vinified in qvevri (traditional fermentation vessels made of clay).

Saperavi is widely planted throughout the wine regions of the former Soviet Union, including **Ukraine** (1,514 ha/3,741 acres in 2009), the **Republic of Moldova** (716 ha/1,769 acres in 2009), **Russia** (614 ha/1,515 acres in Krasnodar and 102 ha/252 acres in Rostov in 2010), **Azerbaijan**, **Armenia** and also in Eastern European countries such as **Bulgaria** (30 ha/74 acres in 2009, mostly in Plovdiv).

The exceptional qualities of Saperavi are so self-evident that it has even found a foothold in **Australia**, where Symphonia and Gapsted in Victoria, Domain Day and Patritti in Barossa and Hugh Hamilton in McLaren Vale all produce varietal examples.

SAPERAVI SEVERNY

Cold-hardy Russian hybrid with potentially astringent tannins.

BERRY COLOUR:

ORIGINS AND PARENTAGE

Saperavi Severny, literally 'northern Saperavi', is a hybrid bred in 1947 by Ya I Potapenko, I P Potapenko and E L Bezruchenko at the all-Russia research and development centre for viticulture and winemaking in Novocherkassk, in the province of Rostov, Russia, by crossing SEVERNY with SAPERAVI. It was released in 1965.

VITICULTURAL CHARACTERISTICS

Consistently productive, early to mid ripening. Small to medium-sized, thick-skinned berries. Some susceptibility to downy mildew and botrytis bunch rot. Winter-hardy but early buds may be affected by spring frosts.

WHERE IT'S GROWN AND WHAT ITS WINE TASTES LIKE

Although this variety has the advantage of cold hardiness so that in the cold winters of the Rostov region it does not need earthing up, thanks to its *Vitis amurensis* genes, its wines may be slightly herbaceous and the tannins need time in bottle to soften. It does not appear to be planted outside **Russia** (325 ha/803 acres in Rostov in 2010), and Fanagoria in the Krasnodar region have now replaced all their Saperavi Severny vines with those of its more noble parent, the Georgian variety SAPERAVI.

SAPHIRA

New German hybrid with moderate disease resistance making rather neutral whites.

BERRY COLOUR:

PRINCIPAL SYNONYMS: Geisenheim 7815-1

ORIGINS AND PARENTAGE

A complex hybrid obtained in 1978 by Helmut Becker at the Geisenheim research centre in

Germany by crossing ARNSBURGER × Seyve-Villard 1-72, where Seyve-Villard 1-72 is a RAYON D'OR × Bertille Seyve 450 hybrid and Bertille Seyve 450 is a NOAH × Vivarais hybrid (see PRIOR for the complete Vivarais pedigree). Authorized in 2004.

VITICULTURAL CHARACTERISTICS

Vigorous, mid to late ripening, susceptible to powdery mildew but moderately resistant to botrytis bunch rot and powdery mildew. Grapes have relatively high sugar levels and marked acidity.

WHERE IT'S GROWN AND WHAT ITS WINE TASTES LIKE

There are experimental plantings in the canton of Zürich in **Switzerland** and Martin Buser produces a varietal wine in the Swiss Jura, close to the French border. There are some vines in **Liechtenstein** and very limited commercial production in **Germany** (eg Weingut Schönhals in Rheinhessen). The wines are similar to PINOT BLANC but acidity can be very high if the grapes are not fully ripe.

ŞARBĂ

Recent Romanian cross capable of producing mouth-watering, aromatic whites.

BERRY COLOUR: ● ● ● ● ●

ORIGINS AND PARENTAGE

Şarbă, meaning 'Serbian' in Romanian, is a cross obtained by G Popescu, M Oşlobeanu, I Poenaru and M Bădiţescu at the viticulture research centre in Odobeşti, eastern Romania, by open pollination of GRAŠEVINA (under the name Italian Riesling), thus the other parent, the pollen donor, is unknown. Şarbă is genetically fairly close to FETEASCĂ NEAGRĂ and to FURMINT (Bodea *et al.* 2009). It was officially registered in 1972.

VITICULTURAL CHARACTERISTICS

Vigorous, highly productive, late ripening. Compact bunches of thick-skinned berries. Resistant to cold and drought, moderately so to powdery mildew and botrytis bunch rot.

WHERE IT'S GROWN AND WHAT ITS WINE TASTES LIKE

Şarbă produces aromatic, crisp, dry whites with flavours of orange blossom and clementine skin. It is found mainly in central eastern **Romania** and varietal wines are made by Gîrboiu and Senator.

SÁRGA MUSKOTÁLY

See MUSCAT BLANC À PETITS GRAINS.

SARY PANDAS

Old Ukrainian variety planted mainly in the south of the country for dessert wines.

BERRY COLOUR:

PRINCIPAL SYNONYMS: Sarah Pandas

ORIGINS AND PARENTAGE

Sary Pandas is an old variety from Solnechnaya Dolina in the Sudak district of Krym (Crimea), southern Ukraine. It was first described in 1803 by Palas (Chkhartishvili and Betsiashvili 2004).

VITICULTURAL CHARACTERISTICS

Late budding and ripening. Compact bunches of medium-sized, thin-skinned berries. Functionally female-only flowers so needs another variety for pollination. Good resistance to winter frosts.

WHERE IT'S GROWN AND WHAT ITS WINE TASTES LIKE

Found mainly in the Sudak district of Krym, southern **Ukraine**, Sary Pandas has the advantage of attaining high sugar levels in the berries while retaining fresh acidity as long as the vines are in warm, sunny sites. It is generally made into amber-coloured, fragrant, spicy, full-bodied dessert wines, often in a blend, as in that produced by Sun Valley (Solnechnaya Dolina), and is rarely encountered as a dry wine. It is also found in **Russia**.

SAUVIGNON BLANC

Parent of Cabernet Sauvignon producing increasingly appreciated aromatic, zesty, crisp whites.

BERRY COLOUR:

PRINCIPAL SYNONYMS: Blanc Fumé or Blanc Fumet (Sancerre and Pouilly in the Loire), Fié or Fiers (for Sauvignon Gris/Rose in the Val de Loire and Vallée de la Vienne), Fumé (Nièvre), Fumé Blanc (California), Muskat-Silvaner (Austria and Germany), Muškatni Silvanec (Slovenia), Sauternes (Indre and Cher), Sauvignon Fumé (Loire), Sauvignon Musqué or Sauvignon Blanc Musqué (US), Savagnou (Béarn), Sotern Mărunt (Republic of Moldova), Surin (Loir-et-Cher), Verdo Belîi (Republic of Moldova)

VARIETIES COMMONLY MISTAKEN FOR SAUVIGNON BLANC: AHUMAT (Jurançon), SAUVIGNONASSE (often erroneously called Sauvignon in Chile, and Sauvignon de la Corrèze or Sauvignon Vert in the centre of France), SAVAGNIN BLANC, SILVANER (Sauvignon Blanc was once called Würzel Silvaner in Austria), SPERGOLA (Reggio Emilia in Italy)

ORIGINS AND PARENTAGE

Contrary to common belief, Sauvignon Blanc is unlikely to come from the Bordeaux area, where it was first mentioned between 1710 and 1720 in Margaux: 'le blanc dont le raisin est pour la plupart du Sauvignon' ('the white [wine] in which the grape is for the most part Sauvignon'). Instead it is more likely to have originated from the Val de Loire in France, where its earliest mention appeared under the old synonym Fiers in 1534 in chapter 25 of François Rabelais' *Gargantua*: 'Car notez que cest viande celeste manger a desieuner raisin auec fouace fraische;

mesmement des pineaulx, des fiers, des muscadeaulx, de la bicane et des foyrars pour ceulx qui sont constipez du ventre' ('note that it is heavenly meat to eat grapes with fresh *fouace* [a local pastry] for lunch; also pineaulx, fiers, muscadeaulx, bicane and foyrars for those who have a constipated stomach').

Under the name Sauvignon Fumé or Blanc Fumé, it was first mentioned in 1783–4 in Sancerre and Pouilly in the Cher *département* in central France, where it has won its spurs over the last two centuries. DNA parentage analysis also supports a Val de Loire origin for Sauvignon Blanc (see CABERNET SAUVIGNON for the complete pedigree) for the reasons outlined below.

Sauvignon Blanc has a parent–offspring relationship with SAVAGNIN (Regner 1999), which explains why the two varieties have sometimes been confused. Since Savagnin is mentioned much earlier in the literature, it would logically be the parent and Sauvignon Blanc the progeny, the other parent being unknown. This makes Sauvignon Blanc a half-sibling of Savagnin's other progenies (AUBIN BLANC, BÉQUIGNOL NOIR, GRÜNER VELTLINER, PETIT MANSENG, PETIT MESLIER, RÄUSCHLING, ROTGIPFLER, SILVANER, TEINTURIER, VERDELHO) and a grandchild of PINOT (if we consider Savagnin to be an offspring of Pinot; see that entry). Since Savagnin most likely originates from north-eastern France and has never been mentioned in western France, Sauvignon Blanc was probably born somewhere in between.

Statistical analyses of the DNA profiles suggest that Sauvignon Blanc is a sibling of CHENIN BLANC (Vouillamoz), the most important variety in the Val de Loire, as well as a sibling of TROUSSEAU (analysed under the name Merenzao in Santana *et al.* 2010), an old variety from the Jura in eastern France. As a consequence, the genetic and historical links strongly support a Loire origin for Sauvignon Blanc as well.

In addition, Sauvignon Blanc belongs to the Messile ampelographic group (see p XXVII), together with BÉQUIGNOL NOIR, CHENIN BLANC, MENU PINEAU, MESLIER SAINT-FRANÇOIS, PETIT MESLIER and PINEAU D'AUNIS, all varieties that are typical of northern France, which strengthens the case for a Loire origin for Sauvignon Blanc. From there, Sauvignon Blanc spread to the Gironde, where it spontaneously crossed with CABERNET FRANC to produce CABERNET SAUVIGNON (Bowers and Meredith 1997).

This Loire origin is also supported by the etymology of the name Sauvignon, which probably comes from the French *sauvage*, meaning 'wild', because the shape of Sauvignon Blanc's leaves is similar to those of the wild grapevine. This origin and etymology are supported by the old synonyms Fié or Fiers (in the Loire and the Vienne valleys), which come from the Latin *ferus*, meaning 'wild' (Rézeau 1997). It should be noted that the wild vine *Vitis vinifera* subsp. *silvestris* is still present in the Loire, whereas it has never been observed in the Gironde (Arnold *et al.* 1998).

Recent DNA studies have suggested that Sauvignon Blanc and SÉMILLON are genetically very close (Jahnke *et al.* 2009) but a parent–offspring relationship can be rejected (Vouillamoz).

At least two colour mutations of Sauvignon Blanc have been observed: **Sauvignon Gris**, also called Fié, Fiers, Surin Gris or Sauvignon Rose, and **Sauvignon Rouge**, both grown to a very limited extent in the Val de Loire.

Sauvignon Blanc was used to breed ARRILOBA, INCROCIO BRUNI 54 and SIRMIUM.

OTHER HYPOTHESES

According to Lavignac (2001), the name Sauvignon could possibly come from a Gallo-Roman place name such as Sauvagnon in the Béarn, itself derived from a personal name, eg Salvinius, Silvinius or Salvanius.

Austrian grape geneticist Ferdinand Regner (1999) claimed that DNA parentage analysis provided evidence that Sauvignon Blanc is a CHENIN BLANC × SAVAGNIN cross, but this parentage can be rejected by a more detailed analysis of sixty DNA markers showing eighteen discrepancies (Vouillamoz).

VITICULTURAL CHARACTERISTICS

Early to mid ripening. Very vigorous and needs to be planted on low-vigour rootstocks in soils that are not too fertile, otherwise the canopy can be almost uncontrollable. Small, compact bunches of small berries are highly susceptible to botrytis bunch rot, as well as to diseases of the wood, and to powdery, but not to downy, mildew. Cordon-trained vines are susceptible to eutypa dieback.

WHERE IT'S GROWN AND WHAT ITS WINE TASTES LIKE

Sauvignon Blanc has spread rampantly around the world, just as its foliage is inclined to do if not held in check. It has been the most obvious beneficiary of a certain boredom with both CHARDONNAY and obvious oak. Sauvignon the wine is typically high in acidity with a range of aromas generally associated with things green – grass, leaves, nettles and gooseberries – provided the grapes are picked early enough. 'Cat's pee on a gooseberry bush' became a popular description of the aroma of some early-picked Sauvignon in the late twentieth century. The wine produced by overripe grapes, on the other hand, can be relatively dull.

The origin of Sauvignon Blanc's particularly distinctive aromas, and the way they may be influenced in the vineyard and the winery, have provoked a considerable amount of research over many years, particularly since the mid 1990s. Studies in the last twenty or so years focused first on the influence of flavour compounds called methoxypyrazines on the aroma and flavour of Sauvignon Blanc, and the impact of climate, sunlight and harvesting time on the intensity of the herbaceous flavours for which they are responsible.

More recently, there has been further research into the key role played by volatile thiols, compounds formed after alcoholic fermentation by the interaction between yeast and the sulphur in the lees due to certain aroma precursors barely detectable in the fruit prior to fermentation and influenced by yeast strain, enzymes and fermentation conditions (Dubourdieu 2009). Thiols typically contribute wide-ranging aromas from boxwood and cat's pee via grapefruit and passion fruit to flinty, smoky notes.

Sauvignon Blanc is the third most planted white wine variety in **France**, after Ugni Blanc (TREBBIANO TOSCANO) and CHARDONNAY, with total plantings in 2009 of 26,839 ha (66,321 acres). The French region with the greatest area planted is, rather surprisingly, Languedoc-Roussillon with 7,357 ha (18,180 acres), making Sauvignon Blanc the second most important white wine variety there after Chardonnay's 12,156 ha (30,038 acres). Southern France can be a little warm for Sauvignon Blanc and many of the wines produced, typically labelled Pays d'Oc, are simply vaguely tart white wines. Slightly zestier Sauvignon is grown in the Gers (armagnac country), where there are 1,801 ha (4,450 acres), and in the nearby Tarn (460 ha/1,137 acres). Simple Vins de Pays are also made in cognac country, where there are about 520 ha (1,285 acres).

Sauvignon Blanc makes much more serious wines in the Aquitaine region comprising Bordeaux and Bergerac, where Sauvignon Blanc's plantings total 7,117 ha (17,586 acres), making it the second most planted white wine variety after SÉMILLON's 11,348 ha (28,042 acres). The great majority, 5,142 ha (12,706 acres), are in the Gironde, the Bordeaux *département*, where it is often blended with the fatter, richer Sémillon. Sauvignon typically adds zest to sweet white bordeaux in which it is a minority partner and has become increasingly important in the serious dry white bordeaux, notably in Pessac-Léognan, where it is generally blended with Sémillon and aged in oak. All-Sauvignon dry white bordeaux has grown in popularity, however, from inexpensive wines designed to be drunk young to such rarities as Pavillon Blanc de Château Margaux. The 1,171 ha (2,894 acres) of Sauvignon Blanc grown in the Dordogne are typically blended with Sémillon, occasionally MUSCADELLE, to make Bergerac Blanc.

But it is in the Loire that Sauvignon Blanc finds its purest expression in France, notably in Sancerre in the Cher *département* (2,961 ha/7,317 acres in 2009) and Pouilly-Fumé in the Nièvre *département* (1,317 ha/3,254 acres), where the variety is presented unblended and unoaked but, in the best examples, as expressions of very local terroir. Basic wines may simply be tart, bone dry and with vaguely refreshing aromas of green fruits but the finest can demand a year or more in bottle, and then time in the glass, to unfurl sufficiently to display the sometimes austere but usually intriguing and refreshing character imbued by the particular permutations and dispositions of flint, limestone and clay that underlie each vineyard. Menetou-Salon, Reuilly, Quincy and Coteaux du Giennois are more rural versions of Sancerre but the other great concentration of Loire Sauvignon is in Touraine, where easy, gooseberry-flavoured, well-priced expressions of Sauvignon Blanc fruit can be found. More than 3,000 ha (7,413 acres) of Sauvignon Blanc are grown upstream of Tours, used mainly for wines of this sort.

Even Burgundy has a Sauvignon Blanc outpost in the light, racy, bone-dry wines of St Bris (146 ha/361 acres).

Sauvignon is not an especially important variety in **Italy**. The 2000 agricultural census identified only 3,393 ha (8,384 acres), overwhelmingly in the north east. In Piemonte, Gaja manages to display some genuine Sauvignon Blanc flavour in his Alteni di Brassica and there are isolated other examples in north-western Italy but most Italian Sauvignon Blanc is grown in the Alto Adige and Friuli. Flavours can be a little muted but there are thrilling exceptions such as the Cantino Terlano's Quarz and Colterenzio's Lafoa in the Alto Adige and Vie de Romans' bottling from Isonzo. Much of southern Italy is too hot for the preservation of Sauvignon aroma but producers as grand as Ornellaia in Bolgheri and Biondi-Santi in Montalcino have persisted with it.

Much of **Spain** is also too hot for Sauvignon. In 2008 the national total was just 2,515 ha (6,215 acres), with the biggest areas in Castilla-La Mancha (1,700 ha/4,201 acres) for often humdrum inexpensive varietals, and Castilla-León (473 ha/1,169 acres), where Sauvignon Blanc has become important, like VERDEJO, in fashionable Rueda. Good examples of Sauvignon-based Rueda are produced by Marqués de Riscal, Vinos Sanz (notably Finca La Colina) and Sitios de Bodega (Palacio de Menade). Modernizers in Catalunya have planted nearly 500 ha (1,236 acres) there, too, with Torres' Fransola vineyard in Penedès producing a distinctive, barrel-fermented example.

In 2010 only 26 ha (64 acres) of Sauvignon were grown in **Portugal**. When not used in varietal wines, it is often blended with ARINTO DE BUCELAS, which gives the wine more acidity but most of the varietal character is lost. Quinta do Cidrô (belonging to Real Companhia Velha) produce a fine, high-altitude Sauvignon on the southern edge of the Douro and it's also grown in the Tejo.

RIESLING and the like provide the great majority of **Germany**'s aromatic, unoaked white wine but a little Sauvignon Blanc, often called Muskat-Silvaner, is grown in such southern regions as Württemberg, Franken and, especially, the Pfalz. The variety is much more important, and notably successful, in **Austria**, where the total area planted more than doubled in the first decade of this century. There were 753 ha (1,861 acres) in 2007, more than half in the Steiermark (Styria) in the south and the rest scattered throughout Niederösterreich (Lower Austria) but especially in the Weinviertel. Recommended producers of wines that typically combine the refreshment of Sancerre with the creaminess of a rich white burgundy in the Südsteiermark (South Styria) include Erich and Walter Polz, Erwin Sabathi, Sattlerhof, Manfred Tement and Ewald Zweytick. R & A Pfaffl make some of the finest Sauvignon in the Weinviertel.

Switzerland's 133 ha (329 acres) of Sauvignon Blanc in 2009 were scattered across the country. Recommended producers include Peter and Rosi Hermann in Graubünden, Domaines des Balisiers and de la Planta in Genève and Stefan Gysel Saxer in Schaffhausen.

Sauvignon Blanc is relatively important in the **Czech Republic**, where 869 ha (2,147 acres) are grown, all in Morava in the south east. Recommended producers include Jiří Barabáš (Hodonice), František Mádl (Velké Bílovice), Vinselekt Michlovsky (Rakvice) and Trpělka & Oulehla (Nové Bránice).

Sauvignon Blanc, sometimes known as Muškatni Silvanec (literally, Muscat Silvaner) in **Slovenia**, grows both around Trieste in Primorska in the south west and especially in Štajerska Slovenija (Slovenian Styria) in the north-eastern Podravje region, across the border from Austria's Südsteiermark, where the variety thrives in dry, piercingly mineral unoaked styles and fuller, richer oaked ones. There are just 390 ha (964 acres) of Sauvignon Blanc in **Croatia**, mainly in Kontinentalna Hrvatska (Continental Croatia) since this is where the international varieties congregate.

Romania's 3,243 ha (8,014 acres) in 2008 were quite widely dispersed but with particular concentrations in Constanța near the Black Sea in the south east, Valcea in the centre and Vrancea further east.

The variety is widely planted in the **Republic of Moldova** (8,151 ha/20,142 acres in 2009), in Krasnodar Krai in southern **Russia** (900 ha/2,223 acres in 2009) and to a lesser extent in **Ukraine** (432 ha/1,067 acres in 2009) – although all of these locations are a little warm for really stimulating varietal expression.

Although the national total area devoted to Sauvignon in **Greece** is not that high, the variety is regarded as a usefully crisp, aromatic blending ingredient in dry whites based on indigenous varieties. It is also grown on **Malta** and **Cyprus** (35 ha/86 acres). There were 158 ha (390 acres) in **Turkey** in 2010 with Sevilen making a particularly successful Fumé Blanc grown at 900 m altitude on the Güney plateau inland from İzmir. Sauvignon Blanc was **Israel**'s most widely planted white wine grape in 2009, with around 250 ha (618 acres).

California grew 15,407 acres (6,235 ha) of Sauvignon Blanc in 2010, a marked fashion for the variety seeming to have peaked. The biggest areas are in Sonoma (2,512 acres/1,017 ha) and Napa (2,496 acres/1,010 ha) counties, closely followed by San Joaquin in the Central Valley (1,996 acres/808 ha). Much of the state's wine country is too hot to produce Sauvignon Blanc with grassy aromas and a California Sauvignon Blanc is notably richer and softer than most of the world's Sauvignon, often lightly oaked and with melon-like aromas in warmer parts of the North Coast, but there are certainly fog-cooled parts of the Central Coast, particularly Santa Ynez Valley and Happy Canyon, well able to make distinctive wines with real tension.

It was David Stare of Dry Creek Vineyard in Sonoma's Dry Creek Valley who first coined the term Fumé Blanc for Sauvignon Blanc. It was adopted in the mid 1970s by Robert Mondavi and has been used far and wide ever since, without much rigour but usually for marketing reasons. Particularly successful California Sauvignons have been made by Araujo, Benzinger Family, Brander Vineyard, Dry Creek Vineyard, Merry Edwards, Robert Mondavi, Sbragia Family, Star Lane and Westerly Vineyards.

The golden-berried clone known as Sauvignon Musqué, Sauvignon Blanc (Musqué) or clone 27 in California has been shown by DNA analysis to be identical to Sauvignon Blanc (as the French grape expert Pierre Galet had previously adjudged on ampelographic grounds) but the flavours tend to be richer, rounder, sometimes more floral or freshly tropical and less feline, and it is easier to get the grapes to ripen fully compared with those of the more common Wente clone. However, the wines do not generally have a Muscat-like aroma, as the name might suggest. Many top California Sauvignon Blancs include a percentage of this clone and its success in California has encouraged viticulturist Richard Smart to take it to Tasmania.

In **California** (and sometimes elsewhere) there has been a growing trend to encourage botrytis infections of Sauvignon Blanc in order to produce sweet, Sauternes-like wines which can be charmingly luscious in youth.

There were just 79 acres (32 ha) of Sauvignon Blanc in **Oregon** in 2008. **Washington** State's 2011 total was 1,173 acres (475 ha), making it the state's fourth most popular white variety, having been overtaken by PINOT GRIS. It is also waning in **Texas**, where only 40 acres (16 ha) remained in 2008. **New York** State produces some Sauvignon Blanc and **Virginia** had 45 acres (18 ha) in the ground in 2010. It is also found in **Mexico**, although varietal aroma must be difficult to retain here. It is not an especially significant variety in **Canada** although a certain amount goes into some convincing Bordeaux blends with SÉMILLON in British Columbia.

Sauvignon Blanc has had an interesting history in **Chile** (see SAUVIGNONASSE). Plantings of true Sauvignon Blanc continue to increase so that, according to official records, they had reached 7,922 ha (19,576 acres) by 2008, predominantly in Curicó (3,819 ha/9,437 acres) and Maule (1,744 ha/4,310 acres) but with increasing plantings in the cooler northern regions such as Casablanca (1,085 ha/2,681 acres). However, although all these newer plantings in the north are likely to be true Sauvignon Blanc, it is highly likely that more than half the so-called Sauvignon Blanc vines in Curicó and Maule, especially the older vines, are in fact SAUVIGNONASSE, because Sauvignon Blanc is a legal synonym for Sauvignonasse. Sauvignon does well in Chile and quality is improving all the time. Superior examples from regions north of Santiago, where the variety seems to perform best, include Amayna, Montes, O Fournier and Viña Leyda made from fruit grown in the Leyda Valley; and Casa Marin and Matetic in the San Antonio Valley.

Sauvignon Blanc is still a relatively minor variety in **Argentina**, where a total of 2,090 ha (5,165 acres) was planted by 2008. Most Argentine Sauvignon is rather flabby or insipid but Chakana, O Fournier, Pulenta Estate, Finca Sophenia and especially Tapiz provide exceptions to the rule. **Uruguay** had just 113 ha (279 acres) of Sauvignon and **Bolivia** 35 ha (86 acres) in 2009.

Australia's total plantings of Sauvignon Blanc, on the other hand, were 6,405 ha (15,827 acres) in 2008, making it the third most planted white variety after CHARDONNAY and SÉMILLON. It is only in very recent years that the variety has become popular there, perhaps partly driven by the success of New Zealand Sauvignon (see below); there were only about 20 ha (49 acres) in the mid 1970s. The crush doubled between 2003 and 2004 and it seems the Australians can't get enough 'Savvy', even with the tide of Kiwi Sauvignon Blanc that continues to sweep over the Tasman. Early examples in the 1980s and 1990s were sometimes a bit oily, until the variety took to cooler regions such as the Adelaide Hills, Coonawarra, Western Australia and Tasmania. Blends with Semillon are particularly successful in Western Australia, eg from producers such as Cape Mentelle, Larry Cherubino, Cullen and Stella Bella. Excellent examples are also made in the Yarra Valley, from PHI, for example; in the Adelaide Hills, in the hands of Shaw and Smith among others; and at Domaine A on Tasmania.

But it is **New Zealand** that claims the Sauvignon Blanc crown. In fact patriotic producers like to suggest that New Zealand is the home of the variety, despite its very short history there. First planted in the 1970s, Sauvignon Blanc alone was responsible for the extraordinary vineyard expansion of the late twentieth and early twenty-first centuries in New Zealand, especially Marlborough in the north of the South Island, where thousands of hectares of farmland have now been planted to this profitable variety. High yields, no expensive oak barrels, little ageing to speak of and a market primed by the extraordinary global success of Cloudy Bay, now a sister company of Moët & Chandon champagne, combined to make Sauvignon Blanc look a most attractive cash crop. Needless to say, a glut has followed and there have been virtually no new plantings since 2009, but the variety's 2011 total was officially estimated at 18,000 ha (44,479 acres), of which 15,700 ha (38,796 acres) are in Marlborough. This puts it way ahead of any other variety in the country, its nearest competitor being PINOT NOIR with around 5,000 ha (12,355 acres). The pungent, almost aggressively fruity but always razor-sharp style of Marlborough Sauvignon has been most obviously successful and the characters of specific subregions are now being identified. Although the variety is by no means restricted to the north of the South Island, the dominant style is unoaked, fruit-driven, and into the glass as quickly as possible. Recent years have seen increasing subtlety and the occasionally successful oaked version, however. Recommended producers include Babich, Dog Point, Isabel Estate, Jackson Estate, Brancott, St Clair, Seresin and Spy Valley in Marlborough; Palliser in Martinborough; and Neudorf in Nelson.

South African Sauvignon Blanc is a little more muted than that of New Zealand but can be attractively balanced, with reasonably demure fruit and Antarctic influences keeping the wines refreshing. In 2008 a total of 9,155 ha (22,622 acres) made it the country's third most planted white variety after CHENIN BLANC and COLOMBARD – more important than CHARDONNAY – with plantings widely spread around the regions but with the greatest hectarage in Stellenbosch. Recommended producers include Kaapzicht, Mulderbosch, Quoin Rock, Vergelegen and Waterkloof in Stellenbosch; Paul Cluver and Iona in Elgin; Chamonix in Franschhoek; The Berrio and Strandveld in Elim; Springfield in Robertson; Klein Constantia and Steenberg in Constantia; Cape Point Vineyards in Cape Point; and Fryer's Cove in Bamboes Bay.

Sauvignon Blanc is also found in **India** and **Japan**.

SAUVIGNON GRIS/ROSE and SAUVIGNON ROUGE

Of the colour mutations Sauvignon Gris/Rose and Sauvignon Rouge, the former is the more widely planted and seems to be increasingly fashionable. The total in **France** was 463 ha (1,144 acres) in 2009, and seems to be on the increase. In the Loire, it is known as Fié and it is not officially permitted in some areas, such as Sancerre. However, the fruit for Henri Bourgeois' Sancerre Jadis is harvested from old vineyards that include a good proportion of Fié. Other recommended producers include Domaine de l'Aujardière in Muscadet country, Paul Buisse at Montrichard, Clos des Ronceveaux in Touraine and Fred Brochet at Ampelidae in the southern Loire. But it is also increasingly found in dry white bordeaux, generally blended as an ingredient that adds rather heady perfume and a bit of weight to Sauvignon Blanc.

There are also more and more examples further afield, especially in **Chile** – for instance chez Cousiño-Macul, Viña Leyda and Casa Marin. It is also planted in **Argentina**, **Uruguay** and **New Zealand**.

SAUVIGNON GRIS

See SAUVIGNON BLANC.

SAUVIGNON ROSE

See SAUVIGNON BLANC.

SAUVIGNON ROUGE

See SAUVIGNON BLANC.

SAUVIGNON VERT

See SAUVIGNONASSE.

SAUVIGNONASSE

Lightly aromatic variety now called Friulano in Italy and for long confused with Sauvignon Blanc in Chile.

BERRY COLOUR: ● ● ● ● ●

PRINCIPAL SYNONYMS: Friulano (Friuli in northern Italy), Jakot (Slovenia), Occhio di Gatto⌘ (Reggio Emilia in north-central Italy), Sauvignon de la Corrèze (Corrèze in central France), Sauvignon Gros Grain (Corrèze), Sauvignon Vert (central France and Chile), Tai Bianco (Friuli), Tocai⌘ or Tocai Friulano⌘ (Friuli), Tocai Italico or Tocai Italico Friulano (Lombardia and the Veneto in northern Italy), Trebbianello (Custoza), Tuchì (Brescia), Zeleni Sauvignon (Slovenia)

VARIETIES COMMONLY MISTAKEN FOR SAUVIGNONASSE: FURMINT⌘ (northern Italy), MUSCADELLE⌘ (mistakenly called Sauvignon Vert in California), RIESLING⌘ (Chile and Argentina), SAUVIGNON BLANC⌘ (France, Chile, Argentina and Russia)

ORIGINS AND PARENTAGE

Sauvignonasse is an old variety from the Gironde region of south-west France, where it has never been widely cultivated. Sauvignonasse is not, as its name might suggest, a progeny of SAUVIGNON BLANC, as DNA parentage analysis has shown (Vouillamoz).

This variety was introduced to Friuli, north-east Italy, in the early nineteenth century and given the name Tokai (Acerbi 1825; Di Rovasenda 1877), the modern name Tocai Friulano appearing in the 1930s (Giavedoni and Gily 2005). As with PINOT GRIS in Alsace, it was deliberately called Tokai/Tocai in order to take advantage of the name of the internationally famed Tokaji wines made mainly from FURMINT in north-eastern Hungary. And it worked: the identity of this variety has been long debated, and until the 1990s most Italian growers and ampelographers considered Tocai Friulano to be unique, possibly originating from Hungary. However, ampelographic studies (Calò and Costacurta 1992) as well as DNA profiling (Crespan *et al.* 2003) have clearly established that Tocai Friulano and Sauvignonasse are one and the same.

Ampelography and DNA profiling show that the variety known as Sauvignon Vert in California is in fact MUSCADELLE (Galet 2000; Vouillamoz).

VITICULTURAL CHARACTERISTICS

Fertile, vigorous and productive. Early budding and early to mid ripening. Susceptible to botrytis bunch rot, sour rot, esca, downy and powdery mildews.

WHERE IT'S GROWN AND WHAT ITS WINE TASTES LIKE

Sauvignonasse was often cultivated with SAUVIGNON BLANC and SÉMILLON in the Gironde in south-west **France** but the area is now drastically reduced, to the extent that Sauvignonasse can no longer be found in the official register of varieties.

Under the name Tocai Friulano and its various synonyms, this variety has thrived in north-east **Italy** and continues to do so, even under its new name, Friulano. It is used in varietal and blended wines in many different appellations, mostly in Friuli (eg Collio, Colli Orientali del Friuli, Friuli Grave, Friuli Izonzo) but also in Lombardia (eg Garda Colli Mantovani) and the Veneto (eg Colli Berici, Colli Euganei, Lison-Pramaggiore). In 2000, when the last Italian agricultural census was taken, growers in Friuli were still nonchalantly calling their vines Tocai Friulano and a total of 4,698 ha (11,609 acres) was recorded, the vast majority in the north east. Under pressure from the Tokaj wine region in Hungary since 1993, the Council of the European Union agreed to ban the use of the name Tocai Friulano from Italian wine labels, with effect from 2008, and, after much debate and not a little dissent, the Italians adopted the name Friulano for this variety.

Recommended producers include Canus, La Castellada, Colle Duga, Livio Felluga, Adriano Gigante, Edi Keber, Doro Princic, Dario Raccaro, Ronco del Gelso, Schiopetto, Franco Toros, Venica & Venica and Volpe Pasini. Wines often have both herbal and almond flavours, sometimes more on the grassy side, and are neither as crisp nor as aromatic as SAUVIGNON BLANC but may well be fuller-bodied.

Similar naming issues exist over the border in **Slovenia**, where there were 29 ha (195 acres) of Zeleni Sauvignon in 2009, mainly in the Primorska region close to Friuli, although there are also significant plantings in Štajerska Slovenija (Slovenian Styria) in the north east. In Slovenia there has been official agreement to use either Zeleni Sauvignon, Sauvignonasse or Jakot, but some producers are unhappy with these and prefer to give their wines a fantasy name such as Pikotno or Gredič instead, without referring to the grape variety. Recommended producers include Blažič, Kabaj, Marjan Simčič and Kocijančič Zanut. It is also found in **Ukraine** (2,691 ha/6,650 acres in 2009) and to a limited extent in Krasnodar Krai in **Russia** (51 ha/126 acres in 2010).

It is highly likely that some of the older plantings of so-called SAUVIGNON BLANC in **Chile**'s Curicó and Maule valleys are in fact Sauvignonasse but exact figures are hard to come by as the latter may legally still be labelled as the former. However, many of the older Sauvignonasse vines are now being grafted over to Sauvignon Blanc and all the new plantings certainly are. Official statistics for Sauvignon Vert, as Sauvignonasse is officially known in Chile, reported 234 ha (578 acres) in 2008. In **Argentina**, there were 659 ha (1,628 acres) in 2008, almost all in the Mendoza region. Finca La Anita make a particularly good and surprisingly ageworthy example using fruit from the Alto Agrelo subregion.

SAVAGNIN

CLONAL DIVERSITY

Savagnin, as it is called in the Jura and Franche-Comté in north-east France, is a very old variety that has developed several different forms over centuries of vegetative reproduction, like other very old varieties such as PINOT or GOUAIS BLANC. Today Savagnin shows a considerable level of clonal diversity, seen in berry colour, aroma, leaf shape, cluster size and so on. Several of these forms have often been mistakenly considered to be distinct varieties, eg Gewürztraminer in Alsace and Germany; Heida or Païen in Switzerland; Traminer or Traminer Weisser in Germany; Traminer Aromatico in Trentino-Alto Adige in Italy. Yet DNA profiling has shown that they all have the same genetic fingerprint, with the exception of some minor clonal genetic differences, and that they all correspond to the same variety (Regner *et al.* 2000a; Imazio *et al.* 2002). It is consequently misleading to talk about a 'Savagnin family' or a 'Traminer family' because this suggests family ties such as parent–offspring, uncle–nephew, cousins.

For that reason, the most significant clonal mutations of this variety will be discussed as subsections within this Savagnin entry: the white-yellow-berried Savagnin Blanc, the pink-berried Savagnin Rose and the pink-berried, aromatic Gewürztraminer, **in chronological order of appearance in the literature**. Even though the historical references to Traminer are earlier than those to Savagnin, the latter is used here as the prime name because it is less misleading in terms of the variety's origin.

ORIGINS AND PARENTAGE

Historical and genetic data suggest that Savagnin was born in the area encompassing north-east France (Franche-Comté, Champagne-Ardenne, Lorraine and Alsace) and south-west Germany (Rheinland-Pfalz and Baden-Württemberg), either through a natural cross between PINOT and an unknown variety, or through a natural cross between two even more ancient, undetermined and most likely extinct varieties, or from direct domestication of wild grapevines, as suggested by Bronner (1857) and later by Regner *et al.* (2000b). This last hypothesis is valid only if Savagnin is the parent of Pinot and not vice versa (see below). It would be consistent with the theory that the name Savagnin derives from the French *sauvage*, meaning 'wild', and with the morphological similarity between the leaves of Savagnin and those of wild grapevines growing in the Rheintal (Rhine Valley), but no evidence of a genetic link between Savagnin and wild grapevines from France or Germany has yet been found. Savagnin has been mentioned under various names (see below) since the Middle Ages.

DNA parentage analyses have shown the following natural genetic relationships, ie relationships that are the result of natural rather than deliberate crosses (see PINOT for the pedigree diagram):

- In northern France, Savagnin and GOUAIS BLANC gave birth to AUBIN BLANC in the Lorraine and to PETIT MESLIER in Champagne (Bowers *et al.* 2000). Savagnin also has parent–offspring relationships with PINOT (Regner *et al.* 2000b) and with TEINTURIER (Vouillamoz) but it is not possible to tell which is the parent and which the progeny (see Pinot). In addition, it has crossed with a single, unknown and probably extinct variety to give birth to SAUVIGNON BLANC, TROUSSEAU and CHENIN BLANC, which are therefore siblings (Vouillamoz; Myles *et al.* 2011). Consequently, Savagnin is also a grandparent of CABERNET SAUVIGNON, which is a progeny of Sauvignon Blanc, and a grandparent of BALZAC BLANC, COLOMBARD and MESLIER SAINT-FRANÇOIS, which are progenies of Chenin Blanc.

- In southern France, Savagnin is a likely parent of PETIT MANSENG and a grandparent of GROS MANSENG, both in the Pyrénées-Atlantiques, and has a likely parent–offspring relationship with BÉQUIGNOL NOIR from the Gironde (Vouillamoz; Myles *et al.* 2011).
- In Austria, Savagnin and ÖSTERREICHISCH WEISS gave birth to SILVANER, while Savagnin and ROTER VELTLINER are the parents of ROTGIPFLER (Sefc, Steinkellner *et al.* 1998). Consequently, Savagnin is a grandparent of NEUBURGER and FRÜHROTER VELTLINER, which are progenies of Silvaner. It was also recently shown that Savagnin is a parent of GRÜNER VELTLINER through a cross with St-Georgener (Regner 2007).
- In Portugal, Savagnin has a likely and quite unexpected parent–offspring relationship with VERDELHO, traditionally cultivated on Madeira (Myles *et al.* 2011).

Savagnin was used to breed several commercially cultivated varieties, including CSERSZEGI FŰSZERES, DEVÍN, EZERFÜRTŰ, FLORA, GENEROSA (HUNGARY), HETERA, MANZONI ROSA, MENNAS, MÍLIA, NEOPLANTA, PÁLAVA, PERLE, SIEGERREBE, TAMINGA, TRAMINETTE and WÜRZER.

OTHER HYPOTHESES

Several dubious hypotheses have been proposed for the origin of Savagnin.

EGYPT The morphology of grape seeds found in archaeological sites in Egypt resembled that of Savagnin's seeds, which led several ampelographers to suggest an eastern origin for this variety (Hoffmann 1982). However, identification of grape varieties from excavated grape seeds is extremely difficult, and connections made recently were not very conclusive (Terral *et al.* 2010).

VITIS AMINEA, VITIS NOMENTANA* AND *VITIS EUGENIA Like many old varieties, Savagnin has been identified with so-called varieties such as *Vitis aminea*, *Vitis nomentana* or *Vitis eugenia* mentioned by Pliny the Elder and other Roman authors (Hoffmann 1982) but there is absolutely no botanical evidence that they are related or the same.

TRAMINER IS NOT FROM TRAMIN Savagnin, called Traminer in German, is usually said to originate in the village of Tramin/Termeno in the Südtirol/Trentino-Alto Adige in northern Italy, whence it spread to Switzerland and Germany before reaching Central Europe and France, in particular the Franche-Comté (Galet 1990). According to German wine historian Christine Krämer (2006), Hieronymus Bock (1539) was probably the one who sparked off this legend – still alive today – when he wrote in his *Kreutterbuch*: 'darnach Traminner drauben/deren wachsen vil inn der Etsch/als zu Tramyn und Elsaß' ('then Traminner grapes/that grow widely in the Etsch/as well as in Tramin and Alsace'). However, Krämer (2006) has gone through numerous old documents and advances several convincing arguments rejecting the Südtirol/Trentino-Alto Adige as the origin of Savagnin:

- In the nineteenth century, ampelographer Johann Philipp Bronner (1857) visited the Sütirol and was surprised not to find a single Savagnin vine.
- Neither Savagnin nor any of its synonyms was ever mentioned in Italian ampelography books before the nineteenth century, when plantings of Savagnin began in the Südtirol and Austria.

- There is no local synonym for Savagnin in the Südtirol, which there probably would have been if the variety had been cultivated there for centuries.
- The variety is called Traminer in the village of Tramin/Termeno in the Südtirol, which makes this an unlikely origin because no one in Tramin would have simply named it Traminer if it had originated there (otherwise they would have had several grapes all called Traminer).
- Since the Middle Ages, the name Traminer has been used in the Südtirol to describe a wine that was made from several distinct varieties such as Muscateller and Weisser Lagrein (see LAGREIN), and not to designate a single variety.

SAVAGNIN BLANC

Full-bodied, firm, ageworthy wines, sometimes called Traminer.

PRINCIPAL SYNONYMS: Brynšt (Czech Republic), Edler Weiss or Weissedler (Alsace), Formentin (Hungary), Fromenteau or Fourmenteau or Fourmentans (mainly in the Franche-Comté), Formentin (Hungary), Fränkisch (Rheinland-Pfalz in Germany), Frennschen or Weiss Frennschen (Baden-Württemberg in Germany), Gentil Blanc (Franche-Comté), Heida⌘ (Haut-Valais in Switzerland), Malvoisie (Tirol), Naturé or Naturel (Arbois in the Jura), Païen⌘ (Bas-Valais in Switzerland), Pryně (Czech Republic), Sauvagnin (Jura), Savagnin Jaune, Savagnin Vert, Tramín Bílý (Czech Republic), Traminec (Slovenia), Traminer⌘ (Germany), Weisser Traminer⌘ (Germany)

VARIETIES COMMONLY MISTAKEN FOR SAVAGNIN BLANC: ALBARÍN BLANCO⌘ (Asturias), ALVARINHO⌘ (as Albariño in Rías Baixas in Galicia, north-west Spain, as well as in Australia), CHARDONNAY⌘ (possible synonym for Clevner in Alsace), GRINGET⌘ (mainly in Ayze in the Haute-Savoie), MELON⌘ (called Savagnin Jaune in Lons-le-Saunier in the Jura), PINOT BLANC⌘ (still called Clevner or Klävner in Alsace and in Germany's Baden-Württemberg), SAUVIGNON BLANC⌘

HISTORICAL REFERENCES

SAVAGNIN BLANC IN FRANCE According to Rézeau (1997), the earliest reliable mention of Savagnin Blanc, spelt Sauvagnin, appeared in 1732, together with other varieties, many of them very difficult to identify, in a decree of the Besançon parliament: 'ne laisser subsister en plants noirs ou blancs que le pulsard noir, le gros & le petit noirin dit pinot, le baclans, le troussot, la malvoisie, le pulsar gris, le grappenoux, le muscat noir, le petit margillin, le pulsar blanc, le sauvagnin, le blanc court ou brun, le milleran, le luzannois ou valet blanc, & le chasselard, qui sont les seuls bons plants' ('do retain among black or white grape varieties only pulsard noir, … le sauvagnin … which are the only good varieties'). However, the grape variety Savagnin Blanc must be much older and almost certainly existed in north-eastern France under old and confusing synonyms a long time before this citation (see Other hypotheses below).

TRAMINER IN GERMANY Krämer (2006) believes that Savagnin originates from south-west Germany, where it was given the name Traminer, ie 'from Tramin', the name of the wine that was at that time among the favourites in Württemberg in southern Germany. The first reliable mention of Traminer as a grape variety appeared on 4 August 1483 at the monastery of Bebenhausen near Stuttgart (Krämer 2006): 'underm weg itel gut gesund Frennsch und Traminer stoeck und ob dem weg die zweiteil Frennsch und Traminer stoeck und das driteil Aelbinen' ('below the path good and healthy Frennsch and Traminer vines and above the path two thirds Frennsch and Traminer vines and the third part Aelbinen'). Since Frennsch and Weiss Frennschen are old synonyms for Savagnin Blanc in Baden-Württemberg, Krämer suggests that in this quotation Traminer refers to Savagnin Rose. More than a century later, Swiss botanist Johannes Bauhin (1650) mentions Muscateller and Traminner (sic) as the most widely planted grape varieties in the Rheintal (Rhine Valley).

OTHER HYPOTHESES

PRE-SAVAGNIN BLANC SYNONYMS Before it became known as Savagnin Blanc, this variety was mainly known under several old synonyms but all were sources of confusion with other grape varieties and it is often impossible to determine which varieties were designated in old documents:

- Fromenteau (or Fourmenteau or Fourmentans) in the Jura in northern France, also a synonym for PINOT GRIS in Champagne-Ardenne, Lorraine and the Île-de-France and for ROUSSANNE in Isère (the name refers to the berry colour, which is similar to wheat, *froment* in French)
- Gentil Blanc or Blanc Gentil in Alsace, also a synonym for CHARDONNAY and PINOT BLANC (Galet 1990)
- Plant d'Arbois, also a synonym for Poulsard Blanc (see POULSARD), while Arbois was also a name used for MENU PINEAU
- Klevner in Alsace, also a synonym for PINOT BLANC
- Edeltraube (literally 'noble grape'), but Edelwein was also a synonym of CHASSELAS.

TRAMINER IN THE SÜDTIROL/TRENTINO-ALTO ADIGE The first occurrence of the name Traminer for a wine was 'vini de Traminne' ('wines from Tramin/Termeno') in a document about wine prices issued on 17 September 1242 in Bozen/Bolzano in the Südtirol/Trentino-Alto Adige in northern Italy (Krämer 2006). Wines from 'Tramin' were mentioned in many subsequent documents, eg 'wein guets Traminner' ('good wine from Tramin') in the Tirol in 1362, 'Tramin' at the Council of Constance in 1414, 'Traminer' in Strasbourg in 1450, 'Tramynner' also in Strasbourg in 1514, 'Traminer' in Basel in 1532, 'Traminer Wein' in Tirol in 1558 (Sprandel 1998; Krämer 2006). However, there was often confusion between the wines from the village of Tramin/Termeno in the Südtirol and the grape variety that was called Traminer in Germany – a deliberate attempt to piggy-back on the fame the wines from Tramin had at that time. Indeed, the variety Savagnin did not exist in the Südtirol before the nineteenth century, and Traminer wines were made from several distinct varieties (MUSCAT BLANC À PETITS GRAINS and Weisser Lagrein among the oldest), but certainly not from Savagnin Blanc (Krämer 2006).

CONFUSING SAVAGNIN AND SAVAGNIN-LIKE MENTIONS In the late Middle Ages, mentions of Savagnin or similar names did not necessarily correspond to the modern variety Savagnin Blanc. In 1383, a vineyard in Chassagne (today's Chassagne-Montrachet) in the Côte d'Or was planted with 'noyriens' and with 'saulvoigniens' (Rézeau 1997). While *noyriens* almost undoubtedly corresponds to PINOT NOIR, *saulvoigniens* is more likely here to correspond to PINOT BLANC or CHARDONNAY rather than to Savagnin Blanc, as is probably the case for all such mentions in the Côte d'Or, eg the Savignien Blanc mentioned in 1512 in Meursault.

CLEVNER, OLD SYNONYM MENTIONED IN 1318 AND 1330 IN BADEN-WÜRTTEMBERG? Latin texts refer to a variety called Clauener, both in the archives of the monastery of Salem, about ten miles from Konstanz in southern Germany (1318), and in Affental, south west of Baden-Baden (1330). Clauener undoubtedly corresponds to Clevner or Klävner, but this is an old synonym for both PINOT BLANC and Savagnin Blanc, Clevner now being the official name for Savagnin Blanc in the Ortenau district of Baden-Württemberg (Hillebrand *et al.* 2003). In addition, Clevner is the Latin name for the city of Chiavenna in Valtellina in Lombardia, and this name could also designate wines coming from this area, not necessarily made from Savagnin or PINOT.

VITICULTURAL CHARACTERISTICS

Early budding and ripening. Generally pruned long. Small, sometimes very small, bunches and berries. Thick skins give it good resistance to fungal diseases, especially to botrytis bunch rot. Berries can achieve high sugar concentration while retaining good levels of acidity.

Savagnin Blanc (or Traminer) has the potential to produce high-quality varietal wines that are powerful, heady, well structured and ageworthy. In **France**, where there were 481 ha (1,189 acres) of Savagnin Blanc in 2009, it is particularly valued in the Jura for the production of vin jaune, a wine aged under flor and with some similarity to dry sherry although it is not fortified. The Château Chalon appellation is exclusively for vin jaune. Full-bodied dry white wines are also produced here from Savagnin Blanc, now in two main styles: the traditional, golden, nutty, oxidative version and the more modern, non-oxidative one, occasionally blended with CHARDONNAY. Recommended Jura producers of Savagnin-based wines include Daniel Dugois, Domaine Lornet, Jacques Puffeney, Domaine de la Renardière, André and Mireille Tissot and La Tournelle in Arbois, and Château d'Arlay, Domaine Baud and Les Chais du Vieux Bourg in the Côtes du Jura. Berthet-Bondet, Jean Bourdy and Jean Macle are particularly renowned for Château Chalon.

Austria recorded 321 ha (793 acres) of Traminer in 2009 but these vines are likely to be Roter Traminer (see Savagnin Rose below) and Gewürztraminer. In Krems and the Südoststeiermark (South-East Styria), the rare yellow-berried mutation Gelber Traminer (or Traminer Gelb) has been preserved and producers such as Knoll and Salomon Undhof make extremely floral, high-acid, dry and sweet wines from it.

In **Switzerland**, Heida or Païen, as Savagnin Blanc is known, is a traditional speciality of the high-altitude vineyards (1,100 m/3,600 ft) of Visperterminen in the Valais, which produce full-bodied, lightly aromatic, flavourful and gently spicy whites. St Jodern's Heida Veritas (Visperterminen) from ungrafted vines and Chanton's Heida (Visp) are good examples. Other recommended producers include Gérald et Patricia Besse (Martigny-Combe), Didier Joris (Chamoson) and Provins Valais (Sion).

Traminer is also grown in the **Czech Republic**, where it is known as Pryně or Brynšt, although its official name is Tramín Bílý.

In **Slovenia**, it is known as Traminec, of which there were 215 ha (531 acres) in 2009, far more than for Gewürztraminer (see below). The great majority of it is in Štajerska Slovenija (Slovenian Styria) in Podravje, in the north east of the country.

Bulgaria had 1,072 ha (2,649 acres) of Traminer in 2009 and although the official statistics do not distinguish between Gewürztraminer and Traminer, the more recent plantings in the north west are likely to be Gewürztraminer.

It is also found in Krasnodar Krai in **Russia**.

There are minor plantings of Savagnin Blanc in **Canada**, where Château des Charmes, in the Niagara-on-the-Lake appellation in Ontario, produce an icewine and occasionally a dry white from this variety.

Australia also has limited but successful plantings of Savagnin Blanc, although this success, which has been encouraging an increasing number of growers to plant the variety, was partly based on the mistaken belief that what they had was the variety known in Spanish as Albariño (ALVARINHO). The mix-up was not finally confirmed until 2009, when it was shown that the source material propagated at CSIRO, Australia's plant research institute, was derived from mis-labelled plant material imported from Spain via France. Now those who were selling wines labelled Albariño have to choose whether to change their labels to Savagnin (Blanc) or Traminer or start again and plant Albariño. However, the use of the term Traminer could also be confusing since many Australians use Traminer to refer to Gewürztraminer.

SAVAGNIN ROSE

The non-aromatic version of the much more famous Gewürztraminer.

PRINCIPAL SYNONYMS: Clevener or Clevner (Baden-Württemberg in Germany), Fromenteau Rouge or

Fromenté Rose (Doubs), Heidarot⊗ (Oberwallis in Switzerland), Klevener de Heiligenstein (Alsace), Piros Tramini (Hungary), Prinç (Bulgaria), Roter Traminer (Germany), Rotfrenschen (Württemberg), Savagnin Rose Non Musqué, Savagnin Rouge, Traminac Crveni (Croatia), Traminer Rot or Traminer Rother or simply Traminer (Germany)

VARIETIES COMMONLY MISTAKEN FOR SAVAGNIN ROSE: KÖVIDINKA⊗, PINOT GRIS⊗ (called Kleiner Traminer in the Rheingau, Fromenteau in Champagne and Lorraine)

HISTORICAL REFERENCES

In France, the name Savagnin Rose did not appear until 1868 when it was mentioned by Pulliat (Rézeau 1997): 'Savagnin rose. Diffère du precedent [Savagnin vert] par la couleur rose de son grain' ('Savagnin rose. Differs from the previous [Savagnin vert] by the pink colour of the berry'). In Germany, this colour mutation probably existed long before this date, as suggested by the 1483 mention of Frennsch (white version) and Traminer (red version) of the same grape (see above), but, according to Galet (1990), in Germany the name Traminer could refer to Savagnin Blanc as well as to Savagnin Rose, and Traminer Rot or Rother Traminer could refer to Savagnin Rose as well as to Gewürztraminer, making it hard to be sure about historical references to this pink-berried mutation. As a consequence, the most plausible first appearance of Savagnin Rose in the literature could well be Bauhin's (1650) mention of *Rotfrenschen* in Württemberg.

OTHER HYPOTHESES

In 1386, Savaignins Noirs were mentioned in Chilley-le-Vignoble in the Jura, together with POULSARD, but this could correspond to PINOT NOIR as well as to Savagnin Rose or Rouge.

VITICULTURAL CHARACTERISTICS

Early budding and ripening and therefore sometimes at risk from spring frosts. Vigorous, not very productive and sensitive to coulure. Best pruned long. Does well in clay-limestone soils but at risk from chlorosis. Also prone to desiccation of the stems. Distinguished from Gewürztraminer mainly by its more muted aromas.

WHERE IT'S GROWN AND WHAT ITS WINE TASTES LIKE

Although Savagnin Rose has pink berries, it produces wines that are closer in style to Savagnin Blanc than to the more aromatic pink-berried Gewürztraminer. There were 44 ha (109 acres) of Savagnin Rose in **France** in 2008, almost all in the Heiligenstein area of Alsace (Bourgheim, Gertwiller, Heiligenstein and Obernai) under the name Klevener de Heiligenstein, where it qualifies as a varietally named wine within the Alsace appellation.

In **Germany**, Roter Traminer is used in the official statistics to refer to both the pink-berried Savagnin Rose and the pink-berried and aromatic Gewürztraminer (see below); total plantings in 2008 were 835 ha (2,063 acres), the majority in the Pfalz, Rheinhessen and Baden. The Traminer plantings in **Austria**'s Burgenland (192 ha/474 acres in 2008) and Niederösterreich (Lower Austria; 118 ha/292 acres) are likely to be Roter Traminer. In the Neusiedlersee region of Burgenland, it is often used in sweet-wine blends.

In **Croatia**, there were around 470 ha (1,161 acres) of Traminac Crveni in 2009 but this figure could include both Savagnin Rose and Gewürztraminer. **Ukraine** had around 1,000 ha (2,470 acres) of Savagnin Rose in 2009, but again they could be either or both.

Old vines are also likely to be found in other Central European countries interplanted with Traminer or Gewürztraminer.

GEWÜRZTRAMINER

One of the most headily and distinctively aromatic varietals of all with strong lychee flavours and high alcohol levels.

PRINCIPAL SYNONYMS: Dišeči Traminec (Slovenia), Gentil Aromatique or Gentil Rose Aromatique (Jura), Rother Muskattraminer (Germany), Rusa (Romania), Savagnin Rose Aromatique (Switzerland), Tramin Červený (Czech Republic), Traminac Crveni (Croatia), Traminer Aromatico☒ (Alto Adige in Italy), Traminer Musqué, Traminer Rose (Republic of Moldova, Ukraine), Traminer Rot or Traminer Rother (Germany), Traminer Roz (Romania), Tramini or Fűszeres Tramini or Piros Tramini (Hungary)

ORIGINS AND PARENTAGE

Gewürztraminer is thought to be an aromatic mutation of Savagnin Rose (Galet 2000).

HISTORICAL REFERENCES

The first occurrence of the name Gewürztraminer (*Gewürz* is the German for 'spice') appeared in Germany in Johann Metzger (1827): 'Rother Traminer. Rother Rießling bei Oppenheim, Traminer kleiner, Gewürztraminer im Rheingau, aber äußerst selten und nur an einigen Orten gekannt' ('Rother Traminer. Rother Riessling near Oppenheim, Traminer kleiner, Gewürztraminer in Rheingau, but rare outside this area and only known in a few places'). This mention strongly suggests that the mutation occurred in this area, which is backed up by the first mention in France in 1886 in Oberlin, Alsace (cited in Rézeau 1997): 'le Gewürztraminer, rouge … du Palatinat' ('Gewürztraminer, red … from the Pfalz').

VITICULTURAL CHARACTERISTICS

Similar to Savagnin Rose (see above). Particularly prone to virus diseases, although the viticultural station at Colmar has developed several virus-free clones (eg 47, 48 and 643).

WHERE IT'S GROWN AND WHAT ITS WINE TASTES LIKE

Gewürztraminer, often written Gewurztraminer, has become by far the most planted variant of Traminer (Savagnin Blanc). The grapes are certainly notable at harvest for their variegated but incontrovertibly pink colour, which is translated into very deep golden wines, sometimes with a slight coppery tinge. Winemakers unfamiliar with the variety have been known to be panicked into extracting colour and flavour. Gewürztraminers also attain higher alcohol levels than most white wines, with over 14% being by no means uncommon, and acidities can correspondingly be precariously low. Because of this the variety is not suited to very warm climates. Malolactic fermentation is almost invariably suppressed for Gewürztraminer and steps must be taken to avoid oxidation.

If all goes well, the result is deep-golden, full-bodied wines with a substantial spine and concentrated heady aromas of lychees and rose petals whose acidity level will preserve them while those aromas unfurl. In a lesser year or too hot a climate, the result is respectively either an early-picked, neutral wine or an oppressively oily, flabby one that can easily taste bitter to boot. Really intense, ageworthy examples made from low-yielding or late-picked vines can develop flavours reminiscent of bacon fat, while some of the least intense examples may be given ersatz aroma via a small proportion of MUSCAT in the blend. The distinctive aroma of Gewürztraminer is often referred to as 'spicy' by wine tasters simply because *Gewürz* means 'spice', not because it is reminiscent of any particular spice.

France had 3,083 ha (7,618 acres) of Gewürztraminer (spelt without the accent) in 2009. The vast majority of these vines (2,928 ha/7,235 acres) are in Alsace, where the variety comprises 19%

of the vineyard area. The variety slowly gained ground in the first decade of this century and the total area is double what it was fifty years ago. The variety is also authorized in the Moselle appellation. Wines are very distinctively perfumed, often with aromas of rose petal and lychees as well as exotic fruits, whether they are made in a dry or almost-dry style or whether the grapes are picked late for the sweet Vendange Tardive classification or even later and botrytized for the ultra-sweet Sélection de Grains Nobles. Recommended producers include Paul Blanck, Léon Beyer, Marcel Deiss, Willy Gisselbricht, Hugel, Josmeyer, Gustave Lorentz, André Ostertag, Schlumberger, Trimbach, Domaine Weinbach and Zind-Humbrecht.

In **Italy**, it is known as Traminer Aromatico, of which there were 560 ha (1,384 acres) in 2000, almost exclusively in the Alto Adige with the best wines generally coming from the west of the Val d'Adige both north and south of the village of Tramin. Recommended producers include Kuenhof, Alois Lageder, Manincor, Cantina Terlano, Tiefenbrunner, Elena Walch and, appropriately, Cantina Termeno/Tramin, which produces specific crus of this heady variety.

Spain had 247 ha (610 acres) in 2009, mostly in Somontano (Aragón) and in Catalunya. Enate in Somontano and Torres and Gramona in Penedès make recognizable varietal versions.

Luxembourg had 19 ha (47 acres) of Gewürztraminer in 2008.

In **Germany**, Roter Traminer is used in the official statistics to refer to both Savagnin Rose and Gewürztraminer, although it is likely that most of the 835 ha (2,063 acres) recorded in 2008, notably in the Pfalz, Baden and Rheinhessen, south of where it is most likely to suffer frost damage, refer to the latter. Recommended producers include Bercher and Martin Wassmer in Baden and Bürgerspital in Franken. Some others tend to make rather oily, heavy examples. At Rhodt in the Pfalz is a Traminer vineyard said to be nearly 400 years old.

The majority of the 81 ha (200 acres) officially described as 'Traminer' in **Austria**'s Steiermark (Styria) is likely to be Gewürztraminer. **Switzerland** had 50 ha (124 acres) of Gewürztraminer in 2010, mainly in the Trois Lacs region and Vaud in the west of the country, in the Valais to the south, and in German-speaking Switzerland to the north and east. In Vully in the canton of Fribourg, Cru de l'Hôpital make a good varietal wine confusingly labelled Traminer.

In **Hungary**, it is known as Tramini, of which there were 720 ha (1,779 acres) in 2008, grown in most wine-growing regions except Tokaj, with the greatest concentrations in Mátra, Kunság and Mór. It does well on clay or loess soils but has better acidity when grown on volcanic rock, eg in Somló in the north west and Balatonfelvidék on the north-western side of Lake Balaton. Recommended producers include Láslό Bussay (Balatonmelléke), Eurobor (Tolna) and Szőlőskert (Mátra).

In the **Czech Republic**, it is known as Tramin Červený and there were 600 ha (1,483 acres) in 2009, almost all in Morava in the south east.

Slovakia had 363 ha (897 ha) of Tramín Červený in 2009, 1.9% of the vineyard area.

In **Slovenia**, it is known as Dišeči Traminec and there were just 12 ha (30 acres) in 2008, mostly in Štajerska Slovenija (Slovenian Styria) in Podravje, in the north east of the country.

Romania had 218 ha (539 acres) in 2008, the majority in Transilvania in the centre. It is also found across the border in the **Republic of Moldova** (2,731 ha/6,748 acres in 2009), and there were 961 ha (2,375 acres) in **Ukraine** in 2009.

In many of these Eastern European cases, yields are too high to conserve much of the variety's trademark aroma.

Plantings are not insignificant in **California**, where there were 1,735 acres (702 ha) in 2010. Once grown in many parts of the state, it is now predominantly in cool locations in coastal counties such as Monterey, Mendocino, Santa Barbara and Sonoma. Recommended producers include Navarro in Anderson Valley; Fetzer, Handly, Husch and Lazy Creek in Mendocino; Alderbrook, DeLoach, Mill Creek and Ravenswood in Sonoma; and Babcock, Bedford, Claiborne and Churchill, Firestone, Fogerty, Thompson and Ventana in the Central Coast.

Gewürztraminer played a significant part in the early years of the **Oregon** wine industry and the likes of Brooks and Erath still make fine examples. There were 217 acres (88 ha) in 2008.

In **Washington** State, the variety's acreage has been growing steadily, along with that of

RIESLING. There were 775 acres (314 ha) of the variety by 2011 and reliable producers include Chateau Ste Michelle, Snowqualmie and Tsillan Cellars of Lake Chelan.

The variety is quite well suited to the cool climes of the Finger Lakes in **New York** State, which had 143 acres (58 ha) in 2006. Some of the best-known varietals are from Dr Frank, Keuka Spring, Red Newt, Standing Stone and Herman Wiemer. **Michigan** had 45 acres (18 ha) in 2006 and there were 15 acres (6 ha) in **Virginia** in 2008, but most American states appear to produce this hugely recognizable varietal to some extent.

In **Canada**, there are only very limited plantings in Ontario but there were 644 acres (261 ha) in British Columbia in 2008. Superior producers include Blasted Church, Cedar Creek, Gray Monk, Inniskillin, Kettle Valley, Poplar Grove, Quails' Gate, Red Rooster, See Ya Later Ranch, Sumac Ridge, Thornhaven and Wild Goose. Mission Hill is one of several successful producers of Gewürztraminer icewine.

The variety is a minor one in **Chile**, where there were only 182 ha (450 acres) in 2008, but producers such as Aresti in Curicó, Carmen in Colchagua Valley, Casa Marin in San Antonio Valley, Cono Sur as far south as the Bío Bío Valley and Viña Casablanca in Casablanca Valley are all trying hard to produce particularly fresh, zesty examples of it.

Argentina had only 20 ha (49 acres) in 2008 while there were 48 ha (119 acres) in **Brazil**'s Rio Grand do Sul in 2007.

Australia has a long history with the variety. As a varietal or blended with RIESLING, 'Traminer', as Gewürztraminer has often been called in Australia, was a popular, scented, off-dry wine throughout the 1960s and 1970s. In 2008 there were still 840 ha (2,076 acres) of the variety, with the biggest areas in the inland Riverina and Riverland, often used to add aromatic spice to more neutral blends, plus limited plantings in the Hunter Valley, Mudgee, Padthaway, Clare Valley and elsewhere. The best examples are from the Clare Valley, Great Southern and Tasmania. Traditionally, official statistics referred to the variety as Traminer, but now that cuttings thought to be Albariño (ALVARINHO) have been shown to be Savagnin Blanc, perhaps more care will be taken with the terminology for Gewurz. Reliable producers of Australian Gewürztraminer include Pirie in Tasmania and Knappstein in the Clare Valley.

There were 316 ha (781 acres) of Gewürztraminer in **New Zealand** in 2008 and the total is expected to remain stable, although this 'aromatic' is a notable speciality here, partly thanks to reliably crisp acidity. Most of it grows in Gisborne and Hawke's Bay on the North Island and Marlborough on the South Island. Vinoptima in Gisborne is a producer devoted to the variety and makes an excellent late-harvest version modelled on a fine Alsace Vendange Tardive. Other recommended producers include Dry River, Framingham, Hunter's, Lawson's Dry Hills, Saint Clair, Seresin, Spy Valley and Te Whare Ra.

It is something of a minority interest in **South Africa**, where it can take on a rather simplistic, talc-like character, but Paul Cluver and Woolworths have earned respect for theirs. There were 122 ha (301 acres) in 2008, mainly in Robertson, then Stellenbosch.

Gewürztraminer is also found in **Japan** and **Israel.**

SAVAGNIN BLANC

See SAVAGNIN.

SAVAGNIN ROSE

See SAVAGNIN.

SAVATIANO

Greece's most widely planted variety, used to make retsina but capable of a good deal more.

BERRY COLOUR: ● ● ● ● ●

PRINCIPAL SYNONYMS: Dobraina Aspri (Attikí), Doubraina Aspri (Attikí), Doumpraina Lefki☒ (Attikí), Kountoura Aspri (Attikí and Évvoia/*Euboea*), Perahoritiko (Attikí and Kefaloniá), Sakeiko, Savathiano, Savvatiano, Stamatiano (Voiotía), Tsoumprena (Attikí)

VARIETIES COMMONLY MISTAKEN FOR SAVATIANO: XYNISTERI☒

ORIGINS AND PARENTAGE

The exact origin of Savatiano is unknown, but it may come from Attikí, where it has traditionally been cultivated to produce retsina and has many synonyms. The name may derive from the Greek *savvato*, meaning 'Saturday' (Boutaris 2000).

Although MANDILARIA is sometimes called Kountoura Mavri (*mavro* means 'black'), it is not the black version of Kountoura Aspri, which is a synonym for Savatiano (Vouillamoz).

VITICULTURAL CHARACTERISTICS

Big, compact bunches of medium-sized berries. Resistant to drought and to downy and powdery mildews. Very late budding and ripening, highly productive.

WHERE IT'S GROWN AND WHAT ITS WINE TASTES LIKE

Savatiano is the most widely planted grape variety in **Greece**, with an estimated total production of more than 6 million hectolitres (158.5 million US gallons) in 2010. Its heartlands are in Attikí, central Greece, and the large island of Évvoia but it is also found in Makedonía, the Kykládes (Cyclades) islands, the Pelopónnisos (Peloponnese) and, to a lesser extent, on Kefaloniá. Its popularity is due in large part to its role in the production of retsina, sometimes blended with a higher-acid variety such as ASSYRTIKO, and large-volume, inexpensive table wines. The latter tend to be bland, high in alcohol and pretty dull. However, older vines, controlled yields and cooler sites can produce broad but well-balanced wines that go well with food. Recommended producers such as Papagiannakos in Attikí and Roxani Matsa just outside Athens show what can be achieved with this underrated variety.

SCHEUREBE

Underrated and declining early-twentieth-century German cross capable of producing intense, refreshing, dry and sweet wines.

BERRY COLOUR: ● ● ● ● ●

PRINCIPAL SYNONYMS: Alzey S 88, Dr Wagnerrebe, Sämling 88 (Austria), Scheu 88

ORIGINS AND PARENTAGE

Scheurebe was bred by and named after Georg Scheu in 1916 at the Alzey research centre in Rheinhessen. DNA profiling has shown that Scheurebe is not a SILVANER × RIESLING cross, as commonly accepted, but a cross between Riesling and an unknown variety (Grando and Frisinghelli 1998).

VITICULTURAL CHARACTERISTICS

Late ripening, susceptible to powdery mildew. Not as cold-hardy as RIESLING.

WHERE IT'S GROWN AND WHAT ITS WINE TASTES LIKE

Scheurebe was developed specifically for the sandy soils of Rheinhessen, which is where the variety is most widely planted in **Germany**, followed by the Pfalz and, at quite a distance, by Franken and the Nahe. Germany's total plantings stand at 1,672 ha (4,132 acres) but they have been in persistent decline over the last few decades: as recently as 2003 there were 2,200 ha (5,436 acres).

This decline is disappointing given the high potential quality of dry and sweet Scheurebe wines. When fully ripe, they retain a refreshing, racy acidity, are filled with rich, intense flavours of blackcurrant and/or grapefruit and demonstrate an impressive ability to age. Like RIESLING, it is also able to translate the characteristics of the vineyard site into nuanced flavours in the glass. Producers such as Müller-Catoir, Pfeffingen and Lingenfelder in the Pfalz; Wirsching in Franken; and Wittman in Rheinhessen produce remarkable dry versions. Keller in Rheinhessen make an intense, grapefruit-laden botrytized Scheurebe; its sweet brilliance is matched by the best Austrian interpretations of the variety (also known as Sämling 88) in Neusiedlersee, by Kracher, for example.

A similar decline, from a smaller base, has also been seen in **Austria**, where there were 511 ha (1,263 acres) in 2007, the majority in Neusiedlersee close to the Hungarian border, to a more limited extent in the Steiermark (Styria) in the south of the country and in the Weinviertel north of Wien (Vienna). Just 5 ha (12 acres) of Scheurebe were grown in **Switzerland** in 2008.

Scheurebe is also found on 14 acres (6 ha) of vineyard in British Columbia, **Canada**; **New Zealand** recorded just 1 ha (2.5 acres) in 2009; and it has also been planted in Western **Australia** at Wills Domain.

SCHIAVA

ORIGINS AND PARENTAGE

The Italian name Schiava refers to a group of distinct types or varieties, either black- or white-berried, all vigorous and productive, grown in alpine regions under different names: Schiava in Lombardia, Vernatsch in the Südtirol's German (from a Tyrolean dialect word that derives from Latin *vernaculus*, meaning 'native' or 'indigenous'), Rossola in the Veneto and Friuli, Rossara in Trentino, Geschlafene in the Alto Adige (deriving from Sclaf, an old local Trentino name for Schiava that's based on the Latin *sclavus*, meaning 'slave'), Trollinger (derived from *Tyrolinger*, meaning 'from Tyrol') in Tyrol (Italy and Austria), Urban in Württemberg (Germany).

DNA analysis has shown that the varieties within the Schiava group are genetically very dissimilar (Fossati *et al.* 2001), and that at least four distinct varieties are called Schiava: SCHIAVA GENTILE, SCHIAVA GRIGIA, SCHIAVA GROSSA and SCHIAVA LOMBARDA (Grando 2000). It therefore seems more likely that the name Schiava refers to a common training practice (pergola) that deprives the vine of its natural vigour, making it a 'slave' (Italian *schiavo*), rather than a supposed common origin from the historical region of Slavonia (now part of eastern Croatia).

OTHER HYPOTHESES

According to Scienza and Failla (1996), the Schiava group originates from the Caucasus, in particular from Georgia and Armenia, whence it crossed Pannonia (the ancient Roman province covering parts of modern Hungary, Austria, Croatia and Slovenia) before reaching northern Italy. This hypothesis is doubtful because none of the members of the Schiava group shows any DNA similarity with modern varieties from Georgia or Armenia (Vouillamoz *et al.* 2006).

SCHIAVA GENTILE

Small-berried vine making soft, fruity reds in the Alto Adige.

BERRY COLOUR:

PRINCIPAL SYNONYMS: Edelvernatsch, Rothervernatsch

VARIETIES COMMONLY MISTAKEN FOR SCHIAVA GENTILE: SCHIAVA GRIGIA, SCHIAVA GROSSA, SCHIAVA LOMBARDA

ORIGINS AND PARENTAGE

Schiava Gentile probably originates from the Alto Adige in the far north of Italy and is a member of the heterogeneous Schiava group. DNA profiling has shown that SCHIAVA GRIGIA, SCHIAVA GROSSA and SCHIAVA LOMBARDA are all distinct varieties (Grando 2000).

OTHER HYPOTHESES

Schiava Gentile is supposedly close to several MUSCAT varieties (Scienza and Failla 1996), which has not been verified yet, and recent DNA parentage analysis has proposed that Schiava Gentile naturally crossed with TEROLDEGO to produce LAGREIN in the Alto Adige, which is not consistent with other studies (see Lagrein).

VITICULTURAL CHARACTERISTICS

Vigorous, mid ripening. Slightly sensitive to powdery mildew and sour rot. Relatively small berries.

WHERE IT'S GROWN AND WHAT ITS WINE TASTES LIKE

Schiava Gentile is included in numerous DOCs in Trentino-Alto Adige in **Italy**'s far north, including a varietal wine in the Alto Adige/Südtirol DOC, though the regulations for some DOCs do not specify which of the Schiavas they have in mind, or allow any and all. In Lombardia, it may be part of the blend in Botticino and Cellatica DOCs. Varietal wines, labelled Schiava Gentile or Edelvernatsch, are perfumed and fruity, with hints of almonds and violets, and generally soft tannins, very similar to SCHIAVA GRIGIA and SCHIAVA GROSSA. There were 1,181 ha (2,918 acres) in Italy in 2000, according to the agricultural census. Producers of varietal wines, where Gentile or one if its synonyms is specified on the label, include Cantina Aldeno, Franz Haas, Neidermayer and Pravis. Some producers emphasize the lightness of the variety and make a rosato version.

SCHIAVA GRIGIA

Particularly blueish-berried member of the Alto Adige's Schiava group of varieties.

BERRY COLOUR: ● ● ● ● ●

PRINCIPAL SYNONYMS: Cenerina, Grauervernatsch, Grauvernatsch

VARIETIES COMMONLY MISTAKEN FOR SCHIAVA GRIGIA: SCHIAVA GENTILE, SCHIAVA GROSSA, SCHIAVA LOMBARDA

ORIGINS AND PARENTAGE

Schiava Grigia probably originates from the Alto Adige in northern Italy and is a member of the heterogeneous Schiava group. As the berries are violet-blue in colour, it is likely that the name Schiava Grigia (*grigia* meaning 'grey') derives from the bloom covering them. DNA profiling has proved that Schiava Grigia is not identical to SCHIAVA GENTILE, SCHIAVA GROSSA or SCHIAVA LOMBARDA but that it is the same as the obscure Cenerina in Trentino (Grando 2000).

VITICULTURAL CHARACTERISTICS

Mid ripening.

WHERE IT'S GROWN AND WHAT ITS WINE TASTES LIKE

Schiava Grigia appears in various DOCs in Trentino-Alto Adige and in the blends for Botticino and Cellatica DOCs in Lombardia in northern **Italy**, sometimes specified precisely and sometimes more generally as Schiava *tout court*. Varietal wines are authorized in the Alto Adige/Südtirol DOC, labelled Schiava Grigia or Grauvernatsch. They are generally perfumed and fruity, with hints of almonds and violets and light tannins, just like SCHIAVA GENTILE and SCHIAVA GROSSA. Producers of varietal Schiava Grigia include Griesbauer, Cantina San Michele Appiano, Cantina Terlan and Wilhelm Walch.

SCHIAVA GROSSA

The most common Schiava, from the Tirol. Well connected and more widely planted, as Trollinger, in Germany than in Italy. Also a table grape.

BERRY COLOUR: ● ● ● ● ●

PRINCIPAL SYNONYMS: Black Hamburg or Black Hamburgh (England), Blauer Trollinger (Württemberg in Germany), Bresciana (Lombardia), Bressana (Lombardia), Frankenthaler (Württemberg), Grossvernatsch (South Tyrol), Kleinvernatsch (Südtirol), Meraner Kurtraube (Südtirol), Probolinggo Biru (Hatten, Bali), Schiavone, Trollinger (Württemberg), Tschaggele (Südtirol), Uva Meranese (Südtirol)

VARIETIES COMMONLY MISTAKEN FOR SCHIAVA GROSSA: PICCOLA NERA (Trieste), SCHIAVA GENTILE, SCHIAVA GRIGIA, SCHIAVA LOMBARDA

ORIGINS AND PARENTAGE

Schiava Grossa probably originates from the Alto Adige in the far north of Italy and is a member

SCHIAVA GROSSA PEDIGREE DIAGRAM

SCHIAVA GROSSA from the Alto Adige (Südtirol) used to be cultivated all over western Europe and has been involved in both deliberate and natural crosses. Here we show all the natural ones and the deliberate ones recently untangled by DNA analyses (MÜLLER-THURGAU and MUSCAT OF HAMBURG).

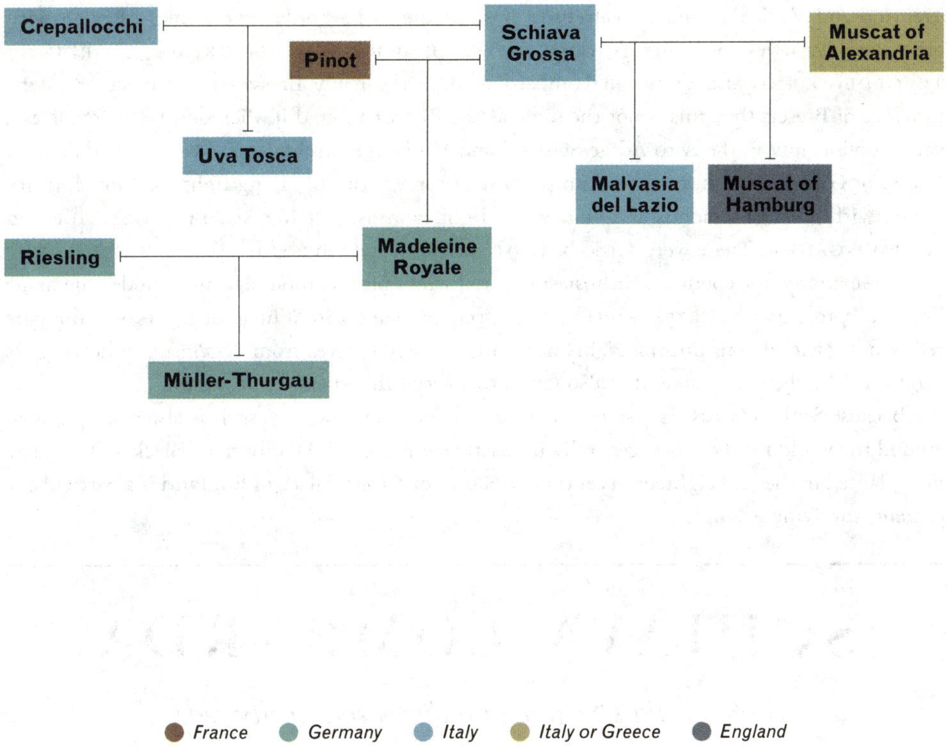

of the heterogeneous Schiava group. As its name suggests, it has larger bunches and berries than the other Schiava varieties. DNA profiling has shown that SCHIAVA GENTILE, SCHIAVA GRIGIA and SCHIAVA LOMBARDA are distinct varieties (Grando 2000). Schiava Grossa was deliberately crossed with RIESLING to produce KERNER and ROTBERGER, with SILVANER to produce BUKETTRAUBE and with PINOT NOIR PRÉCOCE to produce HELFENSTEINER. In addition, recent DNA parentage analysis has revealed that both MUSCAT OF HAMBURG and MALVASIA DEL LAZIO are natural crossings of Schiava Grossa × MUSCAT OF ALEXANDRIA (Crespan 2003; Lacombe *et al.* 2007), which makes them siblings (see pedigree diagram above). Recent pedigree reconstruction has also shown that PINOT and Schiava Grossa are the parents of Madeleine Royale (no longer cultivated), and consequently the grandparents of MÜLLER-THURGAU (Vouillamoz and Arnold 2010). Recent DNA parentage analysis has suggested that Schiava Grossa is a possible parent of UVA TOSCA from Emilia-Romagna and, more questionably, of LAGREIN from the Alto Adige (Cipriani *et al.* 2010).

VITICULTURAL CHARACTERISTICS

Mid ripening (although Trollinger is described as late ripening in Germany). Susceptible to rot and to downy and powdery mildews. Highly resistant to winter frost.

WHERE IT'S GROWN AND WHAT ITS WINE TASTES LIKE

Schiava Grossa is the most widespread variety among the Schiava Somethings. In **Italy**, it is included in various DOCs in Trentino-Alto Adige such as Alto Adige, Caldaro, Casteller, Trentino and Valdadige and in Valtellina close to the Swiss border, sometimes blended with other local varieties including SCHIAVA GENTILE and/or SCHIAVA GRIGIA, LAGREIN or TEROLDEGO. It is also grown in Lombardia under the name Bressana, in particular in the province of Brescia (but this is not the same as the Bressana found in Piemonte). It is made as a varietal wine only in the Alto Adige/Südtirol and Valdadige/Etschtaler DOCs, labelled Schiava Grossa or Grossvernatsch, but often simply Schiava or Vernatsch, giving light, perfumed, fruity wines, with hints of almonds and violets and light tannins, just like SCHIAVA GENTILE and SCHIAVA GRIGIA. There were 1,180 ha (2,916 acres) in Italy in 2000.

In **Germany**, the country's enthusiasm for red wines pushed total plantings, under the name Trollinger, to 2,491 ha (6,155 acres) by 2008, predominantly in Württemberg, producing pale red wines for local consumption. This name most likely derives from *Tirolinger*, indicating its origin. Under the same name, it is also found to a very limited extent in **Japan**.

Because Schiava Grossa is easy to cultivate and does well under glass, it is also widely grown around the world for the table, generally under the names Black Hamburg or Black Tripoli. The vine planted in the mid eighteenth century at Hampton Court Palace in **England** is a particularly ancient, surviving example.

SCHIAVA LOMBARDA

Lombardia's Schiava produces soft, fruity red.

BERRY COLOUR: ● ● ● ● ●

PRINCIPAL SYNONYMS: Schiava Nera

VARIETIES COMMONLY MISTAKEN FOR SCHIAVA LOMBARDA: SCHIAVA GENTILE, SCHIAVA GRIGIA, SCHIAVA GROSSA

ORIGINS AND PARENTAGE

As its name suggests, Schiava Lombarda probably originates from Lombardia and is a member of the heterogeneous Schiava group. DNA profiling has disproved any identity with SCHIAVA

GENTILE, SCHIAVA GRIGIA or SCHIAVA GROSSA (Grando 2000), which are genetically quite distinct.

VITICULTURAL CHARACTERISTICS

Mid to late ripening.

WHERE IT'S GROWN AND WHAT ITS WINE TASTES LIKE

Schiava Lombarda is cultivated principally in Lombardia in northern **Italy**, in the provinces of Como, Brescia and Bergamo, but also in Trentino-Alto Adige, usually blended with other varieties. The Cantina Bergamasca makes a varietal rosé, classified as IGT Bergamasca.

SCHIOPPETTINO

Perfumed and fruity Friulian rescued in the 1970s and now thriving.

BERRY COLOUR: ● ● ● ● ●

PRINCIPAL SYNONYMS: Pocalza, Poçalza, Pokalca or Pokalza (Slovenia), Ribolla Nera

ORIGINS AND PARENTAGE

The earliest mention of Schioppettino was probably under its synonym Ribolla Nera in Di Rovasenda (1877) as a grape variety from Udine in Friuli. Morphological observations and DNA profiling have confirmed that Schioppettino is not a black variant of RIBOLLA GIALLA (Crespan *et al.* 2011). Its name is said to derive from *scoppiettio*, meaning 'crackle', either because Schioppettino wines used to be slightly fizzy or because its thick-skinned berries 'crackle' in the mouth. Originally Schioppettino was the name of the wine made from Ribolla Nera but it is now used as the name of the variety itself.

OTHER HYPOTHESES

It is sometimes said that Schioppettino was cited for the first time in 1282 during the Rieppi-Caucig wedding in Prepotto. The truth is that the wedding took place in 1910, and on this occasion a 1282 document saying that there were already some vineyards in the Albana-Prepotto area was shown. However, this document did not contain any grape variety name (Fabbro *et al.* 2002). The *Catalogo delle varietà delle viti del Regno Veneto* written in 1823 by Pietro di Maniago is also said to mention the variety, possibly under the local dialect name Scopp or Sciopp, but in reality there is no trace of Schioppettino, Sciopp or Ribolla Nera in this list.

VITICULTURAL CHARACTERISTICS

Mid to late ripening. Susceptible to millerandage if spring is cold and wet. Also susceptible to downy mildew and to desiccation of the particularly fragile stems.

WHERE IT'S GROWN AND WHAT ITS WINE TASTES LIKE

Like PIGNOLO and TAZZELENGHE, Schioppettino had almost disappeared in the early twentieth century after Friuli in north-east **Italy** was hit by phylloxera. The variety was rehabilitated in the 1970s thanks to the efforts of the mayor of Prepotto and Paolo and Dina Rapuzzi from Cialla, within the commune of Prepotto very close to the Slovenian border. In 1976 the Rapuzzis won the first Nonino Risit d'Aur award, set up by the Nonino family to encourage those in Friuli who persevere with endangered indigenous varieties.

Paolo Rapuzzi had just established his winery when he heard several local senior citizens talking about a forgotten variety called Pocalza or Schioppettino. He began his search for this variety and found some vines in the mayor's garden. He then scoured many old vineyards in Friuli and Slovenia to collect more vine material. In 1972, he had a hundred vines clandestinely grafted by the Rauscedo nursery – clandestinely because the variety was not at that time authorized and fines were imposed on those who planted unauthorized varieties – and planted the first dedicated Schioppettino vineyard in his crusade to save it from oblivion. He was then followed by other growers in the commune of Prepotto, in Cialla and Albana, which is now Schioppettino's most important area of cultivation. The mayor of Prepotto continued to campaign against the outlawing of Schioppettino and it was eventually included in the list of authorized varieties for the province of Udine in 1981. By 2000 there were 96 ha (237 acres) planted in Italy. It is also grown to a very limited extent in the rest of Friuli but has pretty much disappeared over the border in Slovenia, where it is known as Pokalca. It was awarded DOC status in 1987.

Schioppettino has become a major source of fine red wine in Friuli, producing excellent and sometimes expensive varietal wines within the Colli Orientali del Friuli and the Friuli Isonzo DOCs (although an equal amount of wine is produced as IGT Venezia Giulia). In the former, it may be blended with REFOSCO DAL PEDUNCOLO ROSSO. Wines are generally deeply coloured, perfumed and peppery, with flavours of wild fruits, firm but fine tannins and fresh acidity, both distinctive and elegant. A Schioppettino festival is held annually in Prepotto in the first week of May.

Recommended producers include Bressan, Girolamo Dorigo, Petrussa and Ronchi di Cialla (the Rapuzzi family winery).

SCHÖNBURGER

A pink-skinned German cross making white wines with light Muscat aromas, mainly in England.

BERRY COLOUR: ● ● ● ● ●

PRINCIPAL SYNONYMS: Geisenheim 15-114, Schönberger, Schönburger Rose

ORIGINS AND PARENTAGE

A PINOT NOIR × Pirovano 1 cross made by Heinrich Birk at the Geisenheim research centre in the Rheingau, Germany, in 1939 and authorized in 1980. Pirovano 1 is a Chasselas Rose × MUSCAT OF HAMBURG cross, where Chasselas Rose is a colour mutation of CHASSELAS. This parentage has been confirmed by DNA profiling (Myles *et al.* 2011).

VITICULTURAL CHARACTERISTICS

Mid to early ripening, reasonable resistance to downy and powdery mildews. Winter-hardy but prefers warmer, wind-protected sites. Yield levels similar to RIESLING and more reliable than MÜLLER-THURGAU.

WHERE IT'S GROWN AND WHAT ITS WINE TASTES LIKE

Named after the town of Schönburg in the Mittelrhein, this variety is planted today on 20 ha (49 acres) of vineyard in **Germany**, predominantly in Rheinhessen. Schillinghof in the Nahe make one of the rare German varietal examples.

It is more widely planted in **England** (45 ha/111 acres in 2009) although the total English area has declined from 75 ha (185 acres) in 2003. In the cool English climate, where lack of acidity is not generally a problem, the wines are fresh and have a delicate Muscat aroma, occasionally reminiscent of GEWÜRZTRAMINER in their perfume. They generally benefit from a touch of residual sweetness but can lack concentration. Danebury (Hampshire) and Denbies (Surrey) make varietal wines; others such as Carr Taylor (Sussex) use it in blends.

Schönburger is also planted to a limited extent in British Columbia, **Canada** – 28 acres (11 ha) – and there are minuscule plantings (less than a hectare/2.5 acres) in **South Africa**.

SCIACCARELLO

See MAMMOLO.

SCIAGLÌN

Ancient and rare Friulian.

BERRY COLOUR: ● ● ● ● ●

PRINCIPAL SYNONYMS: Scjaglìn

ORIGINS AND PARENTAGE

Sciaglìn is said to have been mentioned near Spilimbergo in Friuli-Venezia Giulia, north-east Italy, as early as the fifteenth century. Its name could derive from Italian *s'ciale* (meaning 'terrace') or from *schiavolino* ('wine from Slavia').

OTHER HYPOTHESES

Possibly related to the term *schiadina* used in ancient texts to describe grapes giving delicate and sweet wines.

WHERE IT'S GROWN AND WHAT ITS WINE TASTES LIKE

Sciaglìn was formerly cultivated in Vito d'Asio and Fagagna del Friuli, north-east **Italy**. Today it is grown mainly between Maniago and Pinzano al Tagliamento in the province of Pordenone. Together with other old local varieties CIANORIE, CIVIDIN, FORGIARIN, PICULIT NERI and UCELÙT, it was saved from extinction by Emilio Bulfon, with the help of ampelographers Antonio Calò and Ruggero Forti. Emilio Bulfon is one of the few producers to craft varietal wines from these grapes. Florutis is another. Sciaglìn wines are quite juicy with aromas of apple and citrus. Together with Forgiarin, Piculit Neri and Ucelùt, Sciaglìn was included in the Italian national register of varieties in 1991.

SCIASCINOSO

Unusual Campanian making light reds.

BERRY COLOUR: ● ● ● ● ●

PRINCIPAL SYNONYMS: Crovina, Livella di Battipaglia, Livella Nera, Olivella Nera, Olivella Nostrana, Sanginoso, Uva Olivella, Uva Olivella Verace, Vulivella

VARIETIES COMMONLY MISTAKEN FOR SCIASCINOSO: Livella d'Ischia, Livella di Mirabella

ORIGINS AND PARENTAGE

It has long been debated whether Sciascinoso and Olivella were identical or not. The question is not straightforward because the name Olivella, or Livella in the dialect of Napoli, has been given to several distinct Campanian varieties with olive-shaped berries. Recent DNA profiling has shown that at least three distinct varieties share this common name: Livella d'Ischia, Livella di Mirabella and Livella di Battipaglia. While the first two are not commercially cultivated, Livella di Battipaglia, also called Olivella Nera, turned out to be identical to Sciascinoso (Costantini *et al.* 2005).

OTHER HYPOTHESES

It has been argued that its synonym Olivella Nera could correspond to the *Vitis oleaginea* described by Pliny the Elder and Columella.

VITICULTURAL CHARACTERISTICS

Late ripening. Sensitive to downy mildew and botrytis bunch rot.

WHERE IT'S GROWN AND WHAT ITS WINE TASTES LIKE

Sciascinoso is cultivated all over Campania, **Italy**, mainly in the provinces of Avellino, Benevento and Salerno, but it is also found in the province of Caserta as well as further north in the province of Frosinone in Lazio. Varietal wines are authorized in the Irpinia and Sannio DOCs and it may be part of the blend, often with PIEDIROSSO or AGLIANICO, in several others such as Costa d'Amalfi, Penisola Sorrentina and Vesuvio. Total Italian plantings in 2000 were 169 ha (418 acres).

Examples of varietal wines are made by Dedicato a Marianna and Tenuta Vitagliano (both still and sparkling). They tend to be light- to medium-bodied and not suited to ageing, which is another reason Sciascinoso is blended with more robust varieties such as Aglianico.

SCIMISCIÀ

Old and very minor Genoese variety also found on the French island of Corse.

BERRY COLOUR: ● ● ● ● ●

PRINCIPAL SYNONYMS: Çimixâ (Liguria), Frate Pelato (CinqueTerre in Liguria), Genovese or Genovèse (Corse in France), Raisin Génois (Corse), Rossala Bianca, Simixà (Liguria)

VARIETIES COMMONLY MISTAKEN FOR SCIMISCIÀ: ALBAROLA (known as Bianchetta Genovese in Liguria), BOSCO (Liguria)

ORIGINS AND PARENTAGE

Scimiscià is an old variety from the province of Genova in Liguria, near the coast in north-west Italy. Its name is a dialect form of the Italian word *cimice*, meaning 'bug', referring to the tiny stains on the berry skin that resemble small bugs.

DNA profiling has recently established that the variety cultivated on the French island of Corse since the early nineteenth century under the name Genovese, often confused with ALBAROLA (aka Bianchetta Genovese) and BOSCO from Liguria, is in fact identical to Scimiscià (Torello Marinoni, Raimondi, Ruffa *et al.* 2009), which is consistent with its synonym Genovese, meaning 'from Genova', and is a reminder of its introduction from Liguria. In addition, DNA profiling has shown that the variety Frate Pelato grown in the Cinque Terre in Liguria is also identical to Scimiscià (Torello Marinoni, Raimondi, Ruffa *et al.* 2009).

VITICULTURAL CHARACTERISTICS

Vigorous, mid to late ripening, moderate regular yields. Resistant to powdery mildew.

WHERE IT'S GROWN AND WHAT ITS WINE TASTES LIKE

Scimiscià vines are found only in the province of Genova in north-west **Italy**, more specifically in the Fontanabuona and Cicchero valleys. In the 1990s, Scimiscià was close to extinction when Marco Bacigalupo, an ex-pastry chef from Cassottana di Cicogna in the Val Fontanabuona, wanted to save the 500 old vines that some elderly locals had told him were the ancient and famous Scimiscià. With the support of the Comunità Montana della Fontanabuona, the identity of these vines was subsequently confirmed and Scimiscià was included in the national register of varieties in 2003.

Daniele Parma of La Ricolla and the Cantina Çimixâ both make 100% Scimiscià wines. It was not included in the Italian national register of varieties until 2003 and was not therefore included in the most recent agricultural census. Dry wines have aromas that range from herbs to apples and pears and marked acidity on the palate. Passito-style wines generally show aromas of dried fruits.

Under the name Frate Pelato (meaning 'bald friar') it is still sporadically found in old vineyards in Liguria's Cinque Terre and it has been recently replanted by Samuele Bonanini at Azienda Agricola Possa, whose dry white 5 Terre blend includes a small amount of the variety.

Under the names Genovese and Rossala Bianca, it is found on the island of Corse in **France**, eg in Domaine Abbatucci's Cuvée Collection Il Cavaliere Diplomate d'Empire, blended with BIANCU GENTILE, BRUSTIANO BIANCO and VERMENTINO.

SCUPPERNONG

Bronze-berried native American variety producing distinctive sweet wines said to have health benefits.

BERRY COLOUR: ● ● ● ● ●

PRINCIPAL SYNONYMS: American Muscadine, Big White Grape, Bull, Bullace, Bullet Grape, Green Muscadine, Green Scuppernong, Hickman's Grape, Pedee, Roanoke, White Muscadine, White Scuppernong, Yellow Muscadine

ORIGINS AND PARENTAGE

Scuppernong is probably the source of one of the first American wines as well as one of the rare cultivated representatives of the American *Vitis rotundifolia*, a dark-purple-berried native species of the southern United States that is often referred to as the Muscadine grape. It is thought to be a greeny bronze-berried mutation of a wild *Vitis rotundifolia* vine (Gohdes 1982) that grew on the Outer Banks of North Carolina, a chain of barrier islands in the Atlantic off the coast of North Carolina and Virginia. However, when and where the first vines were cultivated remains unknown.

Until the nineteenth century, this variety was known simply as the Big White Grape or the Roanoke Grape, after Roanoke Island, often considered to be its birthplace. In 1811, Calvin Jones and Tom Henderson, then editors of a journal called the *Southern Planter*, christened the grape Scuppernong, meaning 'place where magnolias grow' in the Algonquian language, in honour of Colonel James Blount, who found numerous vines growing along the banks of the Scuppernong

River that runs from Washington County to the Albemarle Sound, east of Roanoke Island (Gohdes 1982).

Scuppernong's popularity rapidly increased and its cultivation spread beyond North Carolina, helped particularly by Thomas Jefferson, who in 1817 praised the wine as one that would be 'distinguished on the best tables of Europe, for its fine aroma, and chrystalline transparence'. However, Jefferson's patriotic comment must be read with caution, because at that time Scuppernong was closer to fortified grape juice than to wine (Pinney 1989).

By the mid nineteenth century, the vine had been propagated across the country for table consumption and winemaking and North Carolina had become the foremost grape producer of the United States, mainly thanks to Scuppernong. In the early 1900s, Paul Garrett began selling his Scuppernong wine in North Carolina as Virginia Dare, named after the first child born to English settlers on Roanoke Island, and the wine became so popular that Garrett was the most successful eastern winegrower before Prohibition (Pinney 1989).

In 2001 the Scuppernong was named North Carolina's official state fruit.

OTHER HYPOTHESES

The name Scuppernong is sometimes used for all *Vitis rotundifolia* varieties but it should be restricted to the clonally propagated selection described above. Legend has it that the Florentine explorer Giovanni da Verrazzano (1485–1528), in the service of the French king François I, first mentioned this variety in 1524 after he had seen 'many vines growing naturally' during his exploration of the Cape Fear River valley (today in North Carolina). However, he did not mention the colour of the berries, and he may simply have seen wild *Vitis rotundifolia*.

Other sources credit the first British colonists Philip Amadas and Arthur Barlowe with the discovery of Scuppernong in 1584 on Roanoke Island during their American expedition to settle the Lost Colony on behalf of Sir Walter Raleigh. They wrote that the vine 'covered every shrub and climbed the tops of high cedars. In all the world, a similar abundance was not to be found' (Gohdes 1982). Again, they do not specify the berry colour and there is no evidence that the grapes were cultivated. Indeed, one year later, Governor Ralph Lane, describing Roanoke Island to Raleigh, stated that they had 'discovered the main to be the goodliest soil under the cope of heaven, so abounding with sweet trees that bring rich and pleasant, grapes of such greatness, yet wild, as France, Spain, nor Italy hath no greater'.

Some believe that every Scuppernong vine descends from a single 'mother plant' that is today over 400 years old, reputedly the oldest grapevine in America. Nicknamed 'The Mother Vine', this old Scuppernong was first mentioned in the 1720s when Joseph Baum secured a large land grant on Roanoke Island, including the famous 'Mother Vineyard'. Amazingly, this vine is still alive at the Manteo property on Roanoke Island, where it now covers half an acre (32 feet wide and 120 feet long). However, there's no evidence that this vine was the source of all the Scuppernong cuttings, and it is not known whether it was planted by native Americans or by the first colonists, or if it has simply developed naturally.

VITICULTURAL CHARACTERISTICS

Bronze berries low in sugar and high in acid.

WHERE IT'S GROWN AND WHAT ITS WINE TASTES LIKE

Scuppernong is still planted in the south-eastern US but it has been largely superseded by Muscadines developed in more recent breeding programmes that produce better-quality wine. However, it still has some marketing appeal thanks to its history and its name and the purported health benefits of Muscadines. According to Georgia's College of Agricultural and Environmental Sciences (2010), 'A few vineyards are using Scuppernong as a bronze juice cultivar, but mostly just for the novelty of the name, rather than the superiority of the juice. We do not recommend planting of Scuppernong in new vineyards.' However, the better Scuppernong wines are certainly distinctive and very different from wine made from *vinifera* varieties: dark gold, well structured and sweet. Several recent studies have suggested that Scuppernong and other Muscadines have a high level of antioxidants that are said to reduce the risk factors associated with degenerative diseases.

Producers of varietal wines include Benjamin Vineyards, Cypress Bend, Duplin, Hinnant Family and WoodMill in **North Carolina**; Perdido in **Alabama**; and Butterducks in **Georgia**.

SEARA NOVA

Cross quite widely planted in western Portugal and on the Açores.

BERRY COLOUR: white

PRINCIPAL SYNONYMS: H-8-51-29

ORIGINS AND PARENTAGE

A DIAGALVES × FERNÃO PIRES cross obtained in 1951 by José Leão Ferreira de Almeida at the national agronomy centre in Oeiras, east of Lisboa, Portugal. It is a sibling of RIO GRANDE.

VITICULTURAL CHARACTERISTICS

Productive, late budding, early ripening. Thick-skinned berries. Highly susceptible to diseases of the wood such as esca and eutypa dieback but more resistant to fungal diseases such as botrytis bunch rot.

WHERE IT'S GROWN AND WHAT ITS WINE TASTES LIKE

Found mainly in the Lisboa and the Tejo regions of **Portugal** but also authorized on the island of Pico in the Açores, where the local co-op include it in their Terras de Lava blend. Seara Nova adapts well to Atlantic-influenced climates to produce wines that may have quite high alcohol and average acidity but are otherwise fairly light in aroma and flavour and are therefore destined for blends. There were apparently around 700 ha (1,730 acres) in Portugal in 2010 but this may be an overestimate.

SEFKA

Undistinguished variety of Balkan origin found only in Greece.

BERRY COLOUR: black

PRINCIPAL SYNONYMS: Mavrouti, Nichevka, Sefka Nichevka, Sefko

VARIETIES COMMONLY MISTAKEN FOR SEFKA: CHONDROMAVRO

ORIGINS AND PARENTAGE

Sefka is said to originate in Bulgaria or from the Balkans (Greek Vitis Database; Nikolau and Michos 2004). However, today it is known only in Greece and its DNA profile (Greek Vitis Database) does not match that of any other variety (Vouillamoz).

In Thráki (Thrace), Sefka is also known as Chondromavro but it is ampelographically distinct from the true CHONDROMAVRO from Kozáni in western Makedonía.

VITICULTURAL CHARACTERISTICS

Vigorous and productive, mid budding and ripening. Big bunches of big berries.

WHERE IT'S GROWN AND WHAT ITS WINE TASTES LIKE

Sefka is grown in Makedonía in northern mainland **Greece** and further south in Thessalía, where it is used mainly in blends with local varieties such as KARNACHALADES and MOSCHOMAVRO, sometimes even with white varieties. Wines are typically pale with moderate alcohol and low acidity.

SEGALIN

Minor Montpellier cross useful for stiffening blends in Cahors country.

BERRY COLOUR: ● ● ● ● ●

PRINCIPAL SYNONYMS: Ségalin

ORIGINS AND PARENTAGE

A JURANÇON NOIR × BLAUER PORTUGIESER cross obtained by Paul Truel in 1957 at INRA (Institut National de la Recherche Agronomique) in Montpellier, southern France, included in the national register of varieties in 1976. It was named after the *Ségalis*, inhabitants of Ségala, a region between the Aveyron and the Tarn (Rézeau 1997).

VITICULTURAL CHARACTERISTICS

Early budding and mid ripening. Low vigour when planted in warmer climates but better suited to more moderate temperatures and pruned long and trellised. Susceptible to desiccation of the stems and magnesium deficiency, especially when grafted onto SO4 rootstocks. Very small berries.

WHERE IT'S GROWN AND WHAT ITS WINE TASTES LIKE

Wines are deeply coloured, well structured and tannic. There were 67 ha (166 acres) in **France** in 2008, marginally down on 2006. It is recommended, as so many of these recent INRA crossings, in much of southern France, from the Dordogne in the west to the Bouches-du-Rhône in the south east. Domaine de Matèle produce a Segalin/Malbec (COT) blend just north of Cahors. In the Lot, where Segalin comprises around 10% of the vineyard, along with GAMAY NOIR and MERLOT, the variety is used to increase body.

SEIBEL 10868

Ardèche hybrid found only (rarely) in Michigan today.

BERRY COLOUR: ● ● ● ● ●

PRINCIPAL SYNONYMS: Soleil Blanc

ORIGINS AND PARENTAGE

A complex hybrid obtained at the beginning of the twentieth century by Albert Seibel in Aubenas in the Ardèche in southern France, by crossing Seibel 5163 × Seibel 5593 (see PRIOR for the complete pedigree of Seibel 5163 and see CHELOIS for the complete pedigree of Seibel 5593).

VITICULTURAL CHARACTERISTICS

Vigorous, mid ripening, not very cold-hardy. Medium-sized clusters of thick-skinned bronze-coloured berries, with purple hues when fully ripe. Susceptible to downy and powdery mildews and to botrytis bunch rot. Juice has a high pH.

WHERE IT'S GROWN

Although Seibel 10868 was bred in France, it is not authorized there and does not appear to be planted. Boskydel in **Michigan** have just a few hundred vines left which they include in their Soleil Blanc blend.

SEMIDANO

Minor white variety exclusive to the Italian island of Sardegna.

BERRY COLOUR: ● ● ● ● ●

PRINCIPAL SYNONYMS: Arvusiniagu, Laconarzu, Migiu, Semidamu, Semidanu

ORIGINS AND PARENTAGE

Semidano was first mentioned by Manca dell'Arca (1780) under the name Semidamu or Laconarzu. Semidano is said to have spread across the Italian island of Sardegna, together with NURAGUS, having been introduced possibly much earlier through the ports of Karalis (today's Cagliari) and Nora.

VITICULTURAL CHARACTERISTICS

Mid to late ripening. Poor resistance to fungal diseases.

WHERE IT'S GROWN AND WHAT ITS WINE TASTES LIKE

After the ravages of phylloxera at the end of the nineteenth century, Semidano was supplanted by the more productive and disease-resistant NURAGUS and slowly disappeared from the island of Sardegna, **Italy**. It is cultivated today mainly in the Campidano area in the province of Cagliari, usually blended with Nuragus, and occasionally made as a varietal wine. The Sardegna Semidano DOC, sometimes with the subzone Mogoro on the label, allows still, sparkling and passito versions. The main producer is the local co-operative, the Cantina Sociale Il Nuraghe. There were 50 ha (124 acres) planted in Italy in 2000.

SÉMILLON

The grape variety of sweet white bordeaux and a natural blending partner of Sauvignon Blanc in dry whites too.

BERRY COLOUR: ● ● ● ● ●

PRINCIPAL SYNONYMS: Barnawartha Pinot (Australia), Blanc Doux (France), Chevrier (France),

Greengrape (South Africa), Hunter River Riesling (Australia), ?Merwah (Lebanon), Saint-Émilion (Gironde), Semilion, Semillon, Sémillon Blanc (Libourne), Sémillon Muscat (Sauternes)

VARIETIES COMMONLY MISTAKEN FOR SÉMILLON: COLOMBARD, CROUCHEN (Australia), MALVASIA FINA (Douro in Portugal), PALOMINO FINO (Fréjus in southern France), SERCIAL (Sémillon has sometimes been called Sercial in Australia), SPERGOLA, TREBBIANO TOSCANO (Ugni Blanc, the Charentais name for Trebbiano Toscano, is also known as Saint-Émilion)

ORIGINS AND PARENTAGE

Until the eighteenth century, Sémillon was cultivated only in the Sauternes region of Bordeaux, south-west France, where it was first cited in 1736 as 'semilion ou St Émilion'. Although Sauternes is commonly considered to be its birthplace, the name Sémillon probably derives from *semeljun*, the local pronunciation of Saint-Émilion (Rézeau 1997), where the variety does exist but has never been cultivated to any significant extent.

Recent DNA studies have suggested that Sémillon and SAUVIGNON BLANC are genetically very close (Jahnke *et al.* 2009) but a parent–offspring relationship can be rejected (Vouillamoz).

MERLOT and TROUSSEAU have sometimes been called Sémillon Rouge (Levadoux 1956; Galet 1990) but they have strictly no relationship with Sémillon. A colour mutation called **Red Sémillon** has been observed in South Africa (see below).

Sémillon was used to breed FLORA and NORIA.

OTHER HYPOTHESES

According to Lavignac (2001), Sémillon comes from the forests or the islands of the Gironde, where it appeared around the sixteenth century. This origin is partially consistent with an alternative etymology that has been suggested for Sémillon – that it derives from the Occitan word *sem*, meaning 'seed' or 'grape variety', with the suffix *-ilho*, possibly indicating that the variety was obtained from seeds rather than being derived from wild grapevines (Levadoux 1948; Rézeau 1997). However, this hypothesis is uncertain because every grape variety comes from a seed.

It has been suggested that the variety known as Merwah in Lebanon might be an ancestral form or even an ancestor of Sémillon but this can be refuted because Sémillon is genetically very close to SAUVIGNON BLANC, which has never been historically cultivated in Lebanon. Merwah might simply be a local clone of Sémillon but this has yet to be tested by DNA profiling.

VITICULTURAL CHARACTERISTICS

Mid ripening. Moderate vigour and may be pruned long or short. Yields vary considerably according to the fertility of the vineyard. Well adapted to gravel or calcareous-clay soils. Susceptible to botrytis bunch rot, black rot (on young leaves), mites and leafhoppers but not to downy mildew or eutypa dieback. The big berries are yellow to golden when ripe, in contrast to the vine's typically bright green leaves.

WHERE IT'S GROWN AND WHAT ITS WINE TASTES LIKE

Sémillon is not a fashionable variety, which is a little strange because at less than maximum ripeness it can taste quite herbaceous and remarkably like its probable relative SAUVIGNON BLANC, which is currently extremely popular. Fully ripe Sémillon is fuller-bodied and fatter than Sauvignon Blanc, with waxy, lanolin, sometimes lemony aromas and, often, a broad, more viscous impact on the palate rather than Sauvignon's sharper attack and high acidity.

France grows more Sémillon than any other country. Its 11,693 ha (28,894 acres) in 2009 made it the country's fifth most planted white wine variety, focused almost entirely on south-west France. There is considerably more Sémillon than SAUVIGNON BLANC planted in Bordeaux, where there were 7,384 ha (18,246 acres) in 2009. Here it is planted particularly on the left bank of the Garonne, in Sauternes and surrounding areas producing sweet white wines, for which its thin skins make it usefully prone to noble rot, and also the oaked – sometimes smoky, sometimes almost oily – dry whites whose apogee are the best of Pessac-Léognan such as Châteaux La Mission Haut-Brion and Haut-Brion. In both cases, the variety is generally blended with Sauvignon

Blanc, with four to one being the traditional Sémillon:Sauvignon ratio in sweet white wines but with the proportion of Sauvignon Blanc having increased in recent years in the dry wines. A little Sémillon is also grown in the Entre-Deux-Mers for simpler, fresher dry whites.

Thanks to its (declining) importance throughout Bergerac, Sémillon is no longer the most planted variety in the Dordogne *département*, its 3,726 ha (9,207 acres) in 2009 outnumbered by MERLOT's 4,303 ha (10,633 acres), but it is still a very important ingredient in wines such as Monbazillac (with SAUVIGNON BLANC and MUSCADELLE) and provides heft to much Bergerac Blanc. There were also 236 ha (583 acres) to the south east in neighbouring Lot-et-Garonne, where it is less important than Sauvignon Blanc and COLOMBARD. Sémillon is technically allowed in most appellations of Provence – there were 187 ha (462 acres) in the Var *département* in 2009 – but it has made little impact on the vineyards of the Midi, where acidity is at a premium.

The variety is hardly known in **Italy**, although Castel de Paolis in Lazio somehow make their own riposte to Sauternes, a botrytized Sémillon with the classic 20% of SAUVIGNON BLANC to help the acid levels. In **Spain**, it is virtually unknown and in **Portugal** is grown by one Douro producer for a late-harvest white. **Hungary** grew 55 ha (136 acres) of Sémillon in 2008, while there is also a little in Krasnodar Krai in **Russia**. It is grown by a handful of producers in **Greece** and **Cyprus** (36 ha/89 acres in 2010) – and **Turkey** had 547 ha (1,352 acres) of Sémillon in 2010, mostly in the Tekirdağ region north of the Marmara Sea. In **Israel**, it still has a limited, declining presence, having been taken from Bordeaux to help establish the wine industry in the late nineteenth century. Nowhere outside Sauternes, unfortunately, does there seem to be a groundswell of enthusiasm for this noble variety.

If you are unfashionable, **California** is hardly the place to be, but there was still a total of 890 acres (360 ha) of Semillon (no accent in the US) in 2010, dotted around the state and mainly in Napa, Sonoma, Yuba and San Joaquin counties. It is generally used in the same way as in Bordeaux: blended with SAUVIGNON BLANC, mostly for dry but also sweet and occasionally botrytized whites. It is also used to bulk out such blends as Mondavi's Woodbridge Chardonnay. Kalin make arguably the most distinctive California varietal Semillon, deliberately bottle aged, from vines planted in the Wente vineyard in the 1880s near Livermore, based on cuttings imported from Château d'Yquem by Charles Wetmore. But this is not California's only source of very old-vine Semillon. The Monte Rosso vineyard in Sonoma, also famous for its old Zinfandel (TRIBIDRAG) vines, has Semillon planted in 1886 whose concentrated fruit is bought by the likes of Carlisle and Luc Morlet, who blend it, and Morgan, who produce a 100% Semillon sweet wine from Monte Rosso fruit.

Washington State has a long history of Semillon production and has in its time produced some fine and distinctive dry wine, especially that made by L'Ecole 41, but the total area devoted to the variety has declined sharply in recent years, from 700 acres (283 ha) in 1993 to 222 acres (90 ha) in 2011. DeLille's Chaleur Estate white Bordeaux blend has long been exceptional, and a few sweet Semillons are also made. Over the border in **Canada**, British Columbia had 72 acres (29 ha) in 2008 and the likes of Sumac Ridge and Black Hills make very creditable Semillon/SAUVIGNON BLANC blends.

Sémillon was once quite widely planted in **Chile**. As recently as 2008 there were still about 1,727 ha (4,268 acres) planted and one must assume that most of its produce bulks out wines labelled Sauvignon – so rare are varietal Semillons (usually without the accent in South America) in Chile, although Viu Manent and Valdivieso have made rather delicious, and extremely well-priced, late-harvest sweet ones.

Argentina grew 973 ha (2,404 acres) of Semillon in 2008, mostly in Mendoza but also to a limited extent in San Juan and Río Negro. Finca La Anita make a rare, serious dry varietal version in Mendoza. **Uruguay** grows just 129 ha (319 acres) of Semillon.

Sémillon's real sphere of influence outside France, however, is **Australia**, where it has a noble history of being responsible for some of the most distinctive dry white wines in the world. In 2008 Australia had a total of 6,715 ha (16,593 acres) of Semillon (without the accent), making it Australia's second most planted white, albeit a long way behind CHARDONNAY's 31,564 ha

(77,996 acres). Semillon is grown throughout the country but is most famous as a varietal wine in the Hunter Valley (618 ha/1,527 acres), where the unoaked, low-alcohol (typically picked at around 10% potential alcohol) and high-acid wines are wonderfully transformed into toasty complexity by ageing over many years in bottle. Blended with SAUVIGNON BLANC in Western Australia, especially Margaret River, and in the Adelaide Hills, it can produce elegant and sophisticated wines, both oaked and unoaked, similar to but generally more refreshing than white bordeaux – although the Semillon tends to be picked early, when it has a marked flavour of grapefruit and sometimes tastes very similar to its blending partner, Sauvignon Blanc, to avoid any oily fatness. Varietal wines from the Barossa are sometimes powerful and heavily oaked but a more restrained style is now emerging. Recommended producers of varietal wines include Brokenwood, McWilliams and Tyrrells in the Hunter Valley; Mount Horrocks in the Clare Valley; Moss Wood and Vasse Felix in Margaret River; and Peter Lehmann in the Barossa Valley. Good examples of Semillon/Sauvignon blends are made by, for example, Brookland Valley, Cape Mentelle, Larry Cherubino, Cullen, Goundrey and Lenton Brae in Western Australia.

The variety is not nearly as important in **New Zealand**, where the total area planted to Semillon had declined to 201 ha (497 acres) by 2009, shared mainly among Gisborne, Marlborough and Hawke's Bay, and virtually none is exported.

Semillon (again without the accent) was once extremely important in **South Africa**, comprising over 90% of Cape vines in the early nineteenth century. It became so popular and dominant that it was called simply 'greengrape', *groendruif* in Afrikaans, because of its bright green foliage, and then later *wijndruif* or 'winegrape'. It was not identified as Bordeaux's Sémillon until the end of the nineteenth century – and when a red mutation appeared (see below), it was known logically as 'red greengrape'.

Semillon's importance in South Africa declined throughout the twentieth century and by 2008 only 1,153 ha (2,849 acres) remained, the equivalent of just over 2% of the total white vineyard area. The largest plantings are in Stellenbosch, Swartland and Franschhoek, which has some particularly ancient bushvines at, for example, Landau du Val. But even if plantings are waning, the variety's reputation may be waxing. Many of the best examples are oaked and some of South Africa's top producers such as Eben Sadie value its contribution to blends. Recommended producers of varietal wines include Boekenhoutskloof, Cederberg, Fairview, Steenberg and Stellenzicht. As in Semillon's Bordeaux homeland, many of the finest wines are Semillon/SAUVIGNON BLANC blends such as Tokara's and Vergelegen's top white, which usually has more Semillon than Sauvignon. A few oaked dessert wines are also made from botrytized grapes.

SÉMILLON ROSE

The red mutation Sémillon Rose, also known as Red Semillon or Sémillon Gris, seems to be unique to South Africa, where it was at one time widely interplanted with the light-skinned original and may even have become the more common of the two. It is not known when it first appeared but it had become common by the 1820s. Like the white version, the red- or pink-berried vine declined throughout the twentieth century and today is found only in a few old vineyards such as the few hectares of interplanted, ungrafted bushvines dating from the early 1900s discovered in the Swartland by viticulturist Rosa Kruger and vinified by Eben Sadie (James 2009).

SERCIAL

Exceptionally acid Portuguese variety most famous for its role in the driest of Madeiras.

BERRY COLOUR: ● ● ● ● ●

PRINCIPAL SYNONYMS: Esgana, Esgana Cão, Esganoso (Minho)

VARIETIES COMMONLY MISTAKEN FOR SERCIAL: ARINTO DE BUCELAS, CERCEAL BRANCO (Bairrada and Douro), SÉMILLON

ORIGINS AND PARENTAGE

Sercial probably originates from Bucelas near Lisboa, Portugal, where it was traditionally grown under the name Esgana Cão, although it has also been suggested that it might possibly come from the Douro. It was introduced from the mainland to the island of Madeira, where it was given the name Sercial. However, this variety should not be confused with CERCEAL BRANCO (also known as Cercial) from Dão, nor with SERCIALINHO. The name Esgana Cão, meaning 'dog strangler', hints at the fierce acidity of the wines.

According to recent DNA studies, Sercial could be genetically close to RAMISCO from the Colares region, to TRINCADEIRA from the Alentejo (Almadanim *et al.* 2007) and to ALVARELHÃO from the Douro or Dão (Lopes *et al.* 1999).

OTHER HYPOTHESES

Sercial is is thought by some to have its origins on the banks of the Rhein, but there is no evidence for this.

VITICULTURAL CHARACTERISTICS

Produces big bunches of medium-sized, thin-skinned berries which are prone to rot. Late ripening. Resistant to powdery and downy mildews.

WHERE IT'S GROWN AND WHAT ITS WINE TASTES LIKE

Although Esgana Cão is an official synonym for Sercial, the latter name is probably more familiar because of the variety's importance for the driest style of Madeira. However, there were fewer than 20 ha (49 acres) on the island in 2010, mostly high up on the southern coast and on the northern coast near the sea. There are around 70 ha (173 acres) on mainland **Portugal**, mostly in Bucelas just north of Lisboa and further north in the Minho region and as far north as the Douro, where it sometimes plays a minor – acidifying – role in white port and dry white table wines. As the mainland synonym suggests, wines are austere and high in acidity, on Madeira rarely achieving more than 11% alcohol before fortification, but they produce intensely aromatic, vibrant, bone-dry, nervy, extremely long-lived wonders. In Bucelas, Esgana Cão plays a minor role in blends with ARINTO DE BUCELAS and is rarely found as a varietal wine, although Portocarro make a varietal Sercial in the Setúbal region south of Lisbon.

About 0.5 ha (1.24 acres) is grown in the vineyards of Mas de Daumas Gassac in the eastern Languedoc, **France**, and adds aroma and acidity to either the dry white or, in some vintages, to Muscat to produce the late-harvest sweet Vin de Laurence.

SERCIALINHO

Rare, high-acid Portuguese cross adding freshness and aroma.

BERRY COLOUR:

PRINCIPAL SYNONYMS: Cercealinho, Sercealinho

ORIGINS AND PARENTAGE

A Portuguese cross obtained in the 1950s by Leão Ferreira de Almeida at the Alcobaça research centre in Portugal. According to the Vitis International Variety Catalogue, this is a VITAL × ALVARINHO cross but this is inconsistent with their DNA profiles in Veloso *et al.* (2010). Luis Pato of the Bairrada region, continuing in his father's footsteps as the lone promoter of this variety, believes it to be a SERCIAL × Alvarinho cross (Luis Pato, personal communication).

VITICULTURAL CHARACTERISTICS

Mid budding, late ripening. Small, thick-skinned berries. Moderately resistant to botrytis bunch rot.

WHERE IT'S GROWN AND WHAT ITS WINE TASTES LIKE

Sercialinho – all 9 ha (22 acres) of it in **Portugal** in 2010 – is found mainly close to its birthplace, in Bairrada. Its most successful proponent is Luis Pato, who likens it to RIESLING and says it adds acidity and a pine-resin/petrol character to the BICAL and CERCEAL BRANCO in his Vinhas Velhas blend. He experimented with a varietal Sercialinho in 2009 but concluded that its wine is not complete enough on its own. The vines were originally planted by his father to add freshness to sparkling wine but Luis Pato has now planted more and also makes a sweet wine from the same fruit using cryoextraction. On its own, it is aromatic with green-apple, pear and light honey flavours and a creamy texture. It can retain very high acidity even at 13.5% alcohol, making it particularly interesting in a blend as the climate gets warmer.

SERINA E ZEZË

Early-maturing Albanian red wine grape.

BERRY COLOUR:

VARIETIES COMMONLY MISTAKEN FOR SERINA E ZEZË: SHESH I ZI (Albania), SYRAH☒ (Albania)

ORIGINS AND PARENTAGE

Serina e Zezë, literally 'black Serina', probably comes from the Korçë area in south-east Albania. Given the similarity between the names Serina e Zezë and Sérine, an old synonym for SYRAH in the Côte Rôtie, some Albanian authors have suggested that Serina e Zezë and Syrah are synonyms but this has been contradicted by DNA profiling, which also showed that there are at least four distinct cultivars in Albania called Serina Something: Serina e Zezë, Serina e Zezë Prisha, Serina Landare and Serina e Drenovës (Ladoukakis *et al.* 2005). Comparison of the DNA profiles of Serina e Zezë and Heftakilo, an old black-berried Greek table grape, shows that they could be closely related, if not identical. DNA profiling has not yet been done to verify whether the non-cultivated white-berried Serina e Bardhë is a colour mutation of Serina e Zezë.

VITICULTURAL CHARACTERISTICS

Early maturing.

WHERE IT'S GROWN

Found mainly in the cooler growing regions in south-eastern **Albania**, although, like so many other indigenous vines here, it is losing ground to international ones.

SERODIO

Serodio is a rare and almost extinct black-berried variety from the Monterrei region in Galicia, north-west **Spain**, which was saved from virtual extinction by José Luis Mateo of Quinta da Muradella, who includes a small percentage of Serodio in A Trabe, his ZAMARRICA-dominant blend.

SERVANT

Minor variety in both France and South Africa, used mainly for the table.

BERRY COLOUR:

PRINCIPAL SYNONYMS: Nonay (Toulon), Raisin Blanc (South Africa), Servan

VARIETIES COMMONLY MISTAKEN FOR SERVANT: Gros Vert

ORIGINS AND PARENTAGE

In the Hérault in France's Languedoc, the name Servant was given to the old table grape Gros Vert until the nineteenth century, and was then used for the modern Servant, which is supposed to have been introduced from Greece around 1864 by Amiral Nonay in the region of Toulon, where it is cultivated under the name Nonay (Galet 1990; Rézeau 1997). The name Servant derives from the Languedoc dialect *serva*, meaning 'may be kept/stored', because Gros Vert used to be stored in lofts to remain fresh until Easter (Rézeau 1997). According to Galet (2000), its original name could be Moldavsky Bely in Bulgarian, but this hypothesis is not certain because this name is also given to at least five other varieties.

VITICULTURAL CHARACTERISTICS

Very late ripening. Suited to white calcareous soils that are not too dry. Sometimes prone to millerandage and to the production of small bunches. Good resistance to botrytis bunch rot.

WHERE IT'S GROWN AND WHAT ITS WINE TASTES LIKE

Both in **France** (143 ha/353 acres in 2008) and in **South Africa** (95 ha/235 acres in 2008), this variety is grown mainly as a table grape, but Domaine Jordy's Cert-Vent, a Vin de Pays de l'Hérault, is a Servant-dominated blend. Frédéric Jordy describes the wine as rustic, dry, dense, with a note of flint thanks to the site, and a good companion to shellfish.

SEVERNY

Russian variety valued mainly for its cold hardiness.

BERRY COLOUR: ● ● ● ● ●

PRINCIPAL SYNONYMS: Severnyi Muskat

ORIGINS AND PARENTAGE

Severny, literally 'northern', is a hybrid obtained in 1936 by Ya I Potapenko and E Zakharova at the all-Russia research and development centre for viticulture and winemaking in Novocherkassk, in the province of Rostov, Russia, by breeding Seyanets Malengra with a *Vitis amurensis* Ruprecht vine. Seyanets Malengra, literally 'seedling of Malingre', was indeed obtained by open pollination of PRÉCOCE DE MALINGRE (ie the pollen donor is unknown). Severny was used to breed FIOLETOVY RANNY, GOLUBOK, OVIDIOPOLSKY and SAPERAVI SEVERNY.

VITICULTURAL CHARACTERISTICS

Vigorous, very early ripening, cold-hardy. Functionally female flowers so needs another variety to achieve pollination. Loose bunches of thick-skinned berries.

WHERE IT'S GROWN

Thanks to the north-east Asian *amurensis* genes, the vine is very winter-hardy, which is why it does not need earthing up in winter in the Rostov wine region, unlike most other varieties, but it does not seem to have won favour beyond the borders of **Russia**.

SEYVAL BLANC

One of the more successful pale-skinned French hybrids, popular in marginal climates, especially England.

BERRY COLOUR: ● ● ● ● ●

PRINCIPAL SYNONYMS: Seyval, Seyve-Villard, Seyve-Villard 5276

ORIGINS AND PARENTAGE

A complex hybrid obtained by Bertille Seyve and Victor Villard in Saint-Vallier in the Isère in eastern France by crossing Seibel 5656 × RAYON D'OR (see the pedigree diagram in PRIOR). The same parents were used to breed the sibling variety SEYVAL NOIR. The name Seyval is a contraction of Seyve and Vallier (Rézeau 1997). Seyval Blanc was used to breed BIRSTALER MUSKAT, CAYUGA WHITE, CHARDONEL, LA CROSSE, MELODY, RÉSELLE and ST PEPIN.

VITICULTURAL CHARACTERISTICS

Early to mid ripening, vigorous and fertile, with consistently high yields and good all-round disease resistance, although it is susceptible to botrytis bunch rot later in the season. Should be grafted. Berries have a downy skin.

WHERE IT'S GROWN AND WHAT ITS WINE TASTES LIKE

Seyval is authorized in many French wine regions but, perhaps tellingly, recommended in none and not allowed in the production of quality wine (in the regulatory sense of the word) due to its hybrid ancestry. There were 111 ha (274 acres) in **France** in 2008, down from 159 ha (393 acres) in 1998 and 1,309 ha (3,235 acres) in 1958.

However, it is both more successful and more useful in cooler climates due to its productivity and relatively early ripening. Seyval and MÜLLER-THURGAU were the standard varieties for most of the early **UK** vineyards in the 1950s, for example, and it was at one time, before the rush to plant the traditional champagne varieties for fizz, the most widely planted variety in England, although the area had dropped slightly, down to 90 ha (222 acres) in 2009, around 6.5% of the total vineyard area. The sugar level in the berries is not generally high, perhaps because it is usually found in cooler climates, and the wine can be relatively neutral and high in acidity, but it is a good base for sparkling wine and the better examples take well to barrel fermentation and lees ageing, though ageing in bottle is not generally recommended. Most importantly, it does not have the foxy flavour of some hybrids. For strange bureaucratic reasons, in the UK its produce does qualify as a Quality Sparkling Wine. Recommended producers who use Seyval Blanc in their sparkling wines include Camel Valley and Breaky Bottom.

Other cooler regions that have welcomed Seyval Blanc include **Canada**, New York State and the eastern United States. Ontario had approximately 210 acres (85 ha) in 2006 and there are limited plantings in Nova Scotia and Québec. **New York** State had a sizeable 373 acres (151 ha) in 2006, and there were 75 acres (30 ha) in **Michigan** (2006), 48 acres (19 ha) in **Virginia** (2008), 30 acres (12 ha) in **Indiana** (2009) and 16 acres (6 ha) further west in **Iowa** (2006). St James Winery in **Missouri** make a good example.

SEYVAL NOIR

Isère hybrid that is less successful than its white sibling but still found in Québec.

BERRY COLOUR: ● ● ● ● ●

PRINCIPAL SYNONYMS: Seyve-Villard 5247

ORIGINS AND PARENTAGE

A complex hybrid obtained by Bertille Seyve and Victor Villard in Saint-Vallier in the Isère in eastern France by crossing Seibel 5656 × RAYON D'OR (see the pedigree diagram in PRIOR). The same parents were used to breed the sibling variety SEYVAL BLANC. The name Seyval is a contraction of Seyve and Vallier (Rézeau 1997).

VITICULTURAL CHARACTERISTICS

Susceptible to powdery mildew. Moderate cold resistance.

WHERE IT'S GROWN AND WHAT ITS WINE TASTES LIKE

Domaine Ganevat in Rotalier in the Jura, **France**, have replanted a small area of Seyval Noir and include it in an early-drinking red blend. Unlike SEYVAL BLANC, Seyval Noir is not in the official French register of varieties. It is still found to a limited extent in Québec, **Canada**, but even there it is in decline. La Mission and Domaine Les Brome, for example, include it in their rosé blends and Les Pervenches have it in their red blend. L'Orpailleur make vinegar with it!

SGAVETTA

Emilia-Romagna rarity, generally playing only a small, colouring part in blends.

BERRY COLOUR: ● ● ● ● ●

PRINCIPAL SYNONYMS: Sganetta, Sgavetta a Graspo Rosso

ORIGINS AND PARENTAGE

Sgavetta was first mentioned by Di Rovasenda (1877), who considered the name to be derived from Schiava or Schiavetta in Emilia-Romagna. Sgavetta used to be highly praised for its potential to make good wine in the provinces of Modena and Reggio Emilia. DNA analysis has suggested a genetic relationship with ROSSARA TRENTINA (Boccacci *et al.* 2005).

VITICULTURAL CHARACTERISTICS

Mid to late ripening. Good resistance to fungal diseases.

WHERE IT'S GROWN AND WHAT ITS WINE TASTES LIKE

Sgavetta is cultivated in the provinces of Modena and Reggio Emilia in Emilia-Romagna in northern **Italy**, where it is usually blended with other local grape varieties to add colour. It may be included as a minor component in both the MALBO GENTILE and the MARZEMINO versions of the Colli di Scandiano e di Canossa DOC. Varietal wines are rare but there have been some successful sweet and sparkling versions. There were 68 ha (168 acres) in Italy in 2000.

SHAVKAPITO

Rediscovered and promising dark-skinned Georgian variety.

BERRY COLOUR: ● ● ● ● ●

PRINCIPAL SYNONYMS: Chavkapito

ORIGINS AND PARENTAGE

Shavkapito is indigenous to the Kartli region of southern central Georgia.

VITICULTURAL CHARACTERISTICS

Mid budding and ripening and doesn't like damp conditions. Medium-sized, loose bunches of small, thick-skinned berries. Low yields.

WHERE IT'S GROWN AND WHAT ITS WINE TASTES LIKE

The recently rediscovered Shavkapito, planted to a very limited extent in the villages of Metekhi, Khidistavi, Mukhrani and Okami in the Kartli region of **Georgia**, north west of Tbilisi, produces red-fruited wines with a smoky flavour, even without the influence of oak. Apart from a few home-grown vines, there are about 10 ha (25 acres) now in production: Pheasant's Tears make a qvevri-fermented version, Chateau Mukhrani a more European, oaked style.

SHESH I BARDHË

Albania's most planted, late-ripening, light-skinned indigenous variety.

BERRY COLOUR: ● ● ● ● ●

PRINCIPAL SYNONYMS: Pucalla

ORIGINS AND PARENTAGE

Shesh i Bardhë, literally 'white Shesh', is an Albanian variety named after the village of Shesh just west of Tirana, from which it probably originates. DNA profiling has shown that Pucalla grown in the Berat region in south-central Albania is identical to Shesh i Bardhë (Ladoukakis *et al.* 2005) but it is not yet known whether Shesh i Bardhë is a colour mutation of SHESH I ZI.

VITICULTURAL CHARACTERISTICS

Late ripening.

WHERE IT'S GROWN AND WHAT ITS WINE TASTES LIKE

Shesh i Bardhë and SHESH I ZI are the two main indigenous varieties planted in **Albania**, mostly around Tirana, together constituting around 30% of the vineyard area at the beginning of the twenty-first century. They tend not to be planted in the cold eastern and north-eastern parts of the country (Maul *et al.* 2008). Arbëri and Çobo, among others, produce a varietal Shesh i Bardhë.

SHESH I ZI

Albania's most planted indigenous dark-skinned variety.

BERRY COLOUR: ● ● ● ● ●

PRINCIPAL SYNONYMS: Seshi Zi, Shesh

VARIETIES COMMONLY MISTAKEN FOR SHESH I ZI: SERINA E ZEZË (Albania), SYRAH

ORIGINS AND PARENTAGE

Shesh i Zi, literally 'black Shesh', is an Albanian variety named after the village of Shesh just west of Tirana, from which it probably originates. DNA profiling has not yet been done to verify whether SHESH I BARDHË is a colour mutation of Shesh i Zi.

A close genetic relationship with SYRAH has been suggested (see that entry) but this can be rejected by DNA profiling.

VITICULTURAL CHARACTERISTICS

Late ripening.

WHERE IT'S GROWN

See SHESH I BARDHË for details of where in **Albania** these two varieties are planted.

SHEVKA

Bulgarian making bland, low-alcohol, low-acid reds.

BERRY COLOUR: ● ● ● ● ●

PRINCIPAL SYNONYMS: Chevka

ORIGINS AND PARENTAGE

Shevka is considered to be indigenous to the Sliven region in central eastern Bulgaria, where it has been widely cultivated. Lack of morphological diversity (Katerov 2004) suggests it is a relatively young variety.

VITICULTURAL CHARACTERISTICS

Large to medium-sized bunches of big, thick-skinned berries. Productive, late budding and ripening. Susceptible to downy and powdery mildews but more or less resistant to botrytis bunch rot.

WHERE IT'S GROWN AND WHAT ITS WINE TASTES LIKE

Vini Sliven, founded in 1920 and currently one of the largest wineries in **Bulgaria**, was originally called Shevka, perhaps an indication of the importance of the variety at that time, although in 2008 there were just 156 ha (385 acres) of it left, split between the provinces of Sliven in central eastern Bulgaria and Burgas further east. Vini Sliven relaunched their Shevka a few years ago but the fruit is not generally rich enough in sugar, acidity or tannins to make a serious red wine. Silvenska Perla make a semi-sweet version and it is also grown for the table.

SHIRAZ

See SYRAH.

SHIROKA MELNISHKA

Very old, late-ripening, top-quality variety associated with Melnik in the far south west of Bulgaria producing powerful, ageworthy reds.

BERRY COLOUR: ● ● ● ● ●

PRINCIPAL SYNONYMS: Chiroka Melnichka, Melnik, Shiroka Melnishka Loza, Siroka Melniska

VARIETIES COMMONLY MISTAKEN FOR SHIROKA MELNISHKA: MELNIK 82, RANNA MELNISHKA LOZA

ORIGINS AND PARENTAGE

Shiroka Melnishka, often called Shiroka Melnishka Loza, literally 'broad-leaved vine of Melnik', is a very old Bulgarian variety named after the town of Melnik in south-western Bulgaria, at the foot of the Pirin Mountains and close to the Greek border.

Shiroka Melnishka was used to breed MELNIK 82, RANNA MELNISHKA LOZA and RUEN.

OTHER HYPOTHESES

It has been suggested that Shiroka Melnishka was brought from Asia by the Thracians in 2000 BC but there is no evidence for this.

VITICULTURAL CHARACTERISTICS

Very fertile and productive. Large to medium-sized bunches of small, thick-skinned berries. Late budding and late to very late ripening. Sensitive to cold winter temperatures, susceptible to downy and powdery mildews, reasonably resistant to botrytis bunch rot.

WHERE IT'S GROWN AND WHAT ITS WINE TASTES LIKE

More generally referred to simply as Melnik, this variety produces impressively dense dry red wines with high levels of extract, tannin and acidity, well suited to oak maturation and more ageworthy than reds made from the other local red wine varieties of **Bulgaria**. It is extensively and almost exclusively planted in the Struma Valley in the far south west of the country, where it is ideally suited to this the warmest wine region: 2,602 ha (6,430 acres) in 2009. Recommended producers include Damianitza and Logodaj. Semi-sweet and sweet, fortified wines are also made.

SHIRVANSHAHY

Powerful, Azeri variety used to make fortified, semi-sweet as well as dry reds.

BERRY COLOUR: ● ● ● ● ●

PRINCIPAL SYNONYMS: Shirvan Schahi, Shirvanshakhi or Shirvan Shakhi

ORIGINS AND PARENTAGE

Shirvanshahy is local to the Kurdamir region in central Azerbaijan.

VITICULTURAL CHARACTERISTICS

Mid budding and ripening. Loose bunches of medium-sized berries. Best suited to sunny, well-exposed sites.

WHERE IT'S GROWN AND WHAT ITS WINE TASTES LIKE

Shirvanshahy is grown in **Azerbaijan** to make a fortified semi-sweet wine called Kurdamir as well as deeply coloured, full-bodied, ageworthy dry reds. Producers include Kurdamir Wines.

SIBI ABBAS

Ancient **Azerbaijan** variety with very dense bunches used to make crisp, moderate-alcohol dry whites. Beilagen Winery also make a fortified version.

SIBIRKOVY

Very minor Russian variety making aromatic whites on the banks of the Don.

BERRY COLOUR: ● ● ● ● ●

PRINCIPAL SYNONYMS: Efremovsky, Sibirek, Sibir'kovy

ORIGINS AND PARENTAGE

Sibirkovy originates from the Rostov wine region along the River Don in south-western Russia, close to the border with Ukraine, not in Siberia, as its name might suggest.

OTHER HYPOTHESES

Like other varieties along the River Don, Sibirkovy is said to have been introduced from Dagestan in the Middle Ages (Vladimir Tsapelik, personal communication).

VITICULTURAL CHARACTERISTICS

Early to mid budding, mid ripening. Thin-skinned small to medium-sized berries. Susceptible to downy mildew, drought and cold winter temperatures.

WHERE IT'S GROWN AND WHAT ITS WINE TASTES LIKE

Sibirkovy is found to a limited extent (less than 15 ha/37 acres in 2011) in the Vedernikov wine region on the banks of the Don near Rostov-on-Don in **Russia**, where the Vedernikov Winery produce a soft, aromatic wine with flavours of wild grasses, thyme, green apples, citrus and white peaches.

SIDALAN

Rare Turkish variety grown mainly on the Aegean island of Bozcaada.

BERRY COLOUR: ● ● ● ● ●

PRINCIPAL SYNONYMS: Sidalak

ORIGINS AND PARENTAGE

Sidalan probably originates from the island of Bozcaada off Turkey's western coast.

WHERE IT'S GROWN

Grown mainly on the Turkish Island of Bozcaada in the Aegean, western **Turkey**, Sidalan is also found in the Ezine and Bayramiç regions on the mainland to the east. It is rarely made as a varietal wine, with the exception of Gülerada, but Talay, for example, blend Sidalan from Ezine with VASILAKI from Bozcaada in their Assos white blend and Sidalan from Bayramiç with SÉMILLON from Lapseki in their Ténedos white.

SIDERITIS

Pink-skinned Greek variety grown mainly for the table but now showing potential for crisp, peppery whites.

BERRY COLOUR:

PRINCIPAL SYNONYMS: Akaki, Chimoniatiko, Siderites, Sideritis Scopelan (on Skópelas), Sideritis Scopelou, Sidiritis

VARIETIES COMMONLY MISTAKEN FOR SIDERITIS: Sideritis ✕ (Cyprus)

ORIGINS AND PARENTAGE

Sideritis is said to be indigenous to central Greece, where it was mentioned in the nineteenth century (Nikolau and Michos 2004). In Cyprus there is also a variety called Sideritis but it has a completely distinct DNA profile (Hvarleva, Hadjinicoli *et al.* 2005) and is no longer cultivated. The name comes from the Greek word *sidero,* meaning 'iron', referring to the toughness of the skins.

VITICULTURAL CHARACTERISTICS

Vigorous and productive. Large bunches of large berries that range in colour from pink to a deeper purplish red. Very late budding and ripening. Very susceptible to disease, particularly to botrytis bunch rot and powdery mildew but good resistance to drought.

WHERE IT'S GROWN AND WHAT ITS WINE TASTES LIKE

Once grown mainly for the table in the north-western Pelopónnisos (Peloponnese), **Greece**, it has more recently been made into fresh, steely, peppery wines in the northern Pelopónnisos, particularly in Pátra, where Parparoussis is leading the way with about 25% of the region's production and some of the best varietal examples, as well as an excellent eau de vie. It is also found in the north of mainland Greece, Thessalía and Attikí.

SIEGERREBE

Overpoweringly Muscat-scented, low-acid German cross.

BERRY COLOUR:

PRINCIPAL SYNONYMS: Alzey 7957, Scheu 7957, Sieger

ORIGINS AND PARENTAGE

A MADELEINE ANGEVINE × SAVAGNIN ROSE cross bred by Georg Scheu at Alzey in 1929, confirmed by DNA profiling (Vouillamoz). Siegerrebe was used to breed ORTEGA and ROSETTA.

OTHER HYPOTHESES

Heinz Scheu, the son of the famous vine breeder, later claimed that Siegerrebe was a self-pollination of MADELEINE ANGEVINE but this can be disproved by DNA profiling (Vouillamoz).

VITICULTURAL CHARACTERISTICS

Buds early, therefore at risk from spring frost, and ripens early. The grapes have very high sugar and very low acidity levels. Its early ripening makes it a victim to wasps.

WHERE IT'S GROWN AND WHAT ITS WINE TASTES LIKE

Germany's total of 103 ha (255 acres), mainly in Rheinhessen and the Pfalz, continues to decline, thankfully. Its potential for very high natural sugar levels (the name translates as 'champion vine') makes it suitable for sweet wines, but its low acidity does not. Its heady, sometimes pungent Muscat aroma can be overwhelming and even a small percentage in a blend tends to overpower all other components. Knobloch in Rheinhessen, Schmitt-Peitz in the Nahe and Rabenhof in Baden nevertheless persist with varietal wines.

England has 10 ha (25 acres) and one notably successful producer of trophy-winning Siegerrebe is Three Choirs in Gloucestershire. It is also planted in **Denmark**, at Domain Aalsgaard, for example, and in **Switzerland**.

It is also grown in another particularly cool, wet wine region, the San Juan Islands of **Washington** State, where Lopez Island Vineyards produce a varietal version. British Columbia in **Canada** has 38 acres (15 ha) and producers include Domaine de Chaberton, Gray Monk, Larch Hills and Recline Ridge.

Palmara in Tasmania, **Australia**, make a varietal Siegerrebe. It has to be harvested early to retain enough acidity but, in small quantities, can add body and flavour to other whites.

SILA

Recent Serbian cross producing light, crisp whites.

BERRY COLOUR: ● ● ● ● ●

ORIGINS AND PARENTAGE

Sila, meaning 'power' in Serbian, is a KÖVIDINKA × CHARDONNAY cross obtained by S Lazić, V Kovač and P Cindrić at the viticulture research centre in Sremski Karlovci, part of Novi Sad University, in Vojvodina, Serbia. It was released in 1988 (Cindrić *et al.* 2000).

VITICULTURAL CHARACTERISTICS

Late ripening, productive. Loose clusters of small, thick-skinned berries confer good resistance to botrytis bunch rot. At risk from frost.

WHERE IT'S GROWN AND WHAT ITS WINE TASTES LIKE

Wines tend to be light-bodied and high in acidity. Producers in **Serbia** include Apatović, Milanović and Petrović. Sila is currently planted in the Fruška Gora vineyards in the west of Vojvodina.

SILVANER

Austrian variety making relatively neutral wines with impressively racy firmness and staying power in Germany and Alsace.

BERRY COLOUR:

PRINCIPAL SYNONYMS: Bálint, Gros Rhin (Switzerland), Grüner Silvaner (Germany), Grüner Zierfandler or Zierfandl (Austria), Johannisberg (Switzerland), Österreicher (Germany), Österreichisch (Franken in Germany), Roter Silvaner, Salfin (Republic of Moldova), Silvain Vert (France), Silvanac Zeleni (Croatia), Silvánske Zelené (Slovakia), Sylvaner (Austria, Switzerland), Sylvaner Verde (Italy), Sylvanske Zelené (Czech Republic), Zeleni Silvanec (Slovenia), Zöld Szilváni (Hungary)

VARIETIES COMMONLY MISTAKEN FOR SILVANER: ELBLING (sometimes called Weisser Silvaner), ÖSTERREICHISCH WEISS, SAUVIGNON BLANC (formerly called Würzel Silvaner)

ORIGINS AND PARENTAGE

DNA profiling has unexpectedly shown that Silvaner is a natural SAVAGNIN × ÖSTERREICHISCH WEISS cross (Sefc, Steinkellner *et al.* 1998). Silvaner is first mentioned in 1665 under the name Östareiche Rebe (Sylvaner), when it was introduced by abbot Alberich Degen at the Cistercian abbey of Ebrach in southern Germany. This means that the cross took place spontaneously more than 500 years ago and the name suggests this happened in Austria. Indeed, Savagnin was mentioned in Austria in 1349 under its well-known synonym Traminer, and Österreichisch Weiss is an ancient variety cultivated mainly in Austria in the region around Wien (Vienna). Since Österreichisch Weiss was restricted mainly to the eastern part of Austria, it is most likely that Silvaner was born in this area. This is consistent with the historical synonym Österreichisch still in use in Franken today, and this also explains why the DNA profile of Bálint (Galbács *et al.* 2009), considered indigenous to the nearby region of Syrmia (today straddling Serbia and Croatia), turns out to be identical to Silvaner (Vouillamoz).

The Silvaner family tree can be expanded further (see PINOT pedigree diagram). Pinot and SAVAGNIN have a parent–offspring relationship (Regner *et al.* 2000b), and ÖSTERREICHISCH WEISS is a progeny of GOUAIS BLANC (Regner *et al.* 1998). Therefore, Silvaner is either a grandchild or a half-sibling of Pinot, and a grandchild of Gouais Blanc. Silvaner also had offspring by two crosses with ROTER VELTLINER: FRÜHROTER VELTLINER and NEUBURGER. Silvaner was also extensively used by grape breeders and is a parent of several varieties, some of them used for commercial wines: ALBALONGA, BUKETTRAUBE, FREISAMER, JUWEL, KANZLER, ?MORIO-MUSKAT, NOBLING, ORANIENSTEINER, OSTEINER and RIESLANER.

OTHER HYPOTHESES

Several earlier hypotheses concerning the origin of Silvaner can now be rejected. Based on the fact that *silva* is Latin for 'wood', some suggested it was selected from a wild vine on the banks of the Donau (Danube). In Transcaucasia, others conjectured that it was introduced from Transilvania, from the Balkans or from Kurdistan.

VITICULTURAL CHARACTERISTICS

Mid ripening with regular yields although it buds relatively early and is therefore susceptible to spring frosts. Smallish green-yellow berries that are more aromatic than the wine they produce. Also prone to chlorosis and highly susceptible to powdery and downy mildews and botrytis bunch rot.

WHERE IT'S GROWN AND WHAT ITS WINE TASTES LIKE

Although Silvaner's origins are in **Austria**, the 44 ha (109 acres) planted there, mostly in Niederösterreich (Lower Austria), are far outstripped by those in Germany and France.

In **France** plantings of the vine called here Sylvaner rather than Silvaner are declining and are exclusively in Alsace, where there's a total of 1,446 ha (3,573 acres), about 10% of the total *vignoble*. Agathe Bursin, Rolly Gassmann, Josmeyer, René Muré, André Ostertag and Domaine Weinbach are among the finest exponents here. The variety can retain acidity rather more proficiently than the PINOT BLANC and AUXERROIS that have ousted Sylvaner because they are easier to grow. Sylvaner achieves particular distinction on the slopes of the Bas-Rhin, especially around Mittelbergheim – so much so that in 2005 the Zotzenberg vineyard was recognized as a Grand Cru, exceptionally, for its Sylvaner.

Italy has just 113 ha (279 acres) of what it calls Sylvaner Verde and it's a speciality of the Valle Isarco (Eisacktaler in German) in the Alto Adige in the far north of the country. Kuenhof is a particularly fine exponent of this nervy combination of the variety and relatively high altitude.

Germany has by far the greatest total area planted, with 5,236 (12,938 acres), including 1,276 ha (3,153 acres) in Franken, 2,467 ha (6,096 acres) in Rheinhessen and 844 ha (2,086 acres) in the Pfalz, but the area planted to Silvaner has declined considerably in the last 100 years: at the beginning of the twentieth century it was Germany's most important white wine variety; now it is only the third most planted after RIESLING and MÜLLER-THURGAU.

There may be more Silvaner vines in Rheinhessen than in Franken but the latter is where the variety finds its greatest expression and where its 350th anniversary was celebrated in 2009. The finest wines, grown on clay-limestone soils, are not particularly aromatic and tend to have only moderate acidity but they are dry, mineral, sometimes earthy, firm and full-bodied. The Würzburger Stein vineyard is one of the best sites, with a long-established reputation. Silvaner is still revered at the Castell estate, where it was introduced from Austria, and can make lovely wines on the Steigerwald slopes near Iphofen. Best producers include Horst Sauer, Hans Wirsching, Fürst Löwenstein, Juliusspital and Schloss Sommerhausen and virtually all the wines are sold in the distinctive flat flask known as the Bocksbeutel.

In Rheinhessen, the variety has a noble history and producers such as Wittmann and Keller demonstrate the 'greener', leafier side of this variety, producing wines that are slightly more aromatic than classic Franken wines, still with the dry weight and power, and ideally suited to accompany asparagus. Although Silvaner is typically made in a dry style, witness the Silvaner RS classification introduced for this style of wine in the mid 1980s, there are some excellent and ageworthy examples of sweeter wines from Auslese to TBA and everything in between.

Although **Switzerland** has just 246 ha (608 acres), mostly in the Valais region, its name here has been the cause of a certain amount of confusion and anguish: from the nineteenth century until 1928, Johannisberg was a synonym for RIESLING in the Valais. From 1928 up to now, Johannisberg has been a synonym for Silvaner in the Valais (some local agronomists were angry with this switch at that time). This synonym is now much more widespread than Gros Rhin, which was also an old Swiss (mainly Valais and Vaud) synonym for Silvaner. The variety ripens after the popular CHASSELAS but can produce fine wine from favoured sites.

It is also planted in the southern regions of Morava in the south east of the **Czech Republic** (104 ha/257 acres), where it is known as Sylvanske Zelené, in **Slovakia** as Silvánske Zelené, in **Slovenia** as Zeleni Silvanec (105 ha/259 acres in 2009), in **Croatia** as Silvanac Zeleni (195 ha/482 acres in 2009), in **Ukraine** (*c.*70 ha/173 acres in 2009) and in the **Republic of Moldova** (98 ha/242 acres) in 2009.

Beyond Europe, producers seem to have paid little attention to this variety, perhaps because it is easily underestimated given its rather neutral aromatic profile. However, Flood Family Vineyards and Rancho Sisquoc grow some in Santa Barbara, **California**, there are limited plantings in British Columbia in **Canada** – and **New Zealand** recorded 4 ha (10 acres) in 2008.

ROTER SILVANER and BLAUER SILVANER

Roter Silvaner and Blauer Silvaner cannot be genetically distinguished from Silvaner and are simple colour mutations. Roter Silvaner is also called Cirfandler, Österreicher Roth, Piros Cirfandli, Piros Szilvani, Rifai Piros, Roter Zierfahndler, Schönfeiler Roth, Zierfahndler Rot. Blauer Silvaner is restricted to the Württemberg area of Germany. It is also called Blauer Österreicher, Blauer Reifler, Blauer Schönfeilner, Blauer Zierfahndler, Bodenseeburgunder, Bodenseetraube, Schwarzer Österreicher, Schwarzer Silvaner and Sylvaner Rouge.

SIMIXÀ

See SCIMISCIÀ.

SIRAMÉ

Very minor Swiss hybrid that adapts well to marginal climates.

BERRY COLOUR: ● ● ● ● ●

PRINCIPAL SYNONYMS: Salomé

ORIGINS AND PARENTAGE

Siramé is a hybrid obtained in the 1970s by Anton Meier at Rebschule Meier in Würenlingen in the canton of Aargau, northern central Switzerland, from undisclosed parents, one of them probably a Seyve-Villard hybrid.

VITICULTURAL CHARACTERISTICS

Early ripening and good all-round disease resistance. Loose bunches of smallish berries.

WHERE IT'S GROWN

Developed as a table and wine grape, Siramé covered just 2 ares (0.05 acres) in **Switzerland** in 2009. Davinum in Zürich are perhaps the only producer to use this variety, blended with the French hybrid LÉON MILLOT. Also found in **Denmark** and southern **Sweden**.

SÍRIA

Aromatic, light-skinned variety widely planted in Portugal under various names, with as many names but fewer vines in Spain.

BERRY COLOUR: ● ● ● ● ●

PRINCIPAL SYNONYMS: Alva (Cova da Beira), Alvadurão (Dão), Alvaro de Soire (Portugal), Alvaro de Sousa (Douro), Blanca Extra (Asturias in Spain), Cigüente (Extremadura in Spain), Coda (Beira Interior), Códega (Douro), Colhão de Gallo (Portugal), Crato Branco (Algarve), Doña Blanca (Orense in Spain),

Graciosa or Gracioso (Portugal), Malvasía (Toro in Spain), Malvasía Blanca (Galicia in Spain), Malvasia Branca (Portugal), Malvasía Castellana (Spain), Malvasía Grosso, Moza Fresca (Orense in Spain), Posto Branco (Douro), Roupeiro (Alentejo), Sabro (Algarve), Valenciana (Valdeorras in Spain), Verdegudillo (Cigales in Spain)

VARIETIES COMMONLY MISTAKEN FOR SÍRIA: CÔDEGA DE LARINHO, JAMPAL, MALVASIA FINA, TAMAREZ

ORIGINS AND PARENTAGE

Síria is a very old variety that is known in the north west of the Iberian peninsula under various synonyms, mainly Códega, Doña Blanca, Malvasia Castellana and Roupeiro (Arranz *et al.* 2010). The earliest mention of this variety was found in Spain under the name Cigüente in Alonso de Herrera's 1513 description of the varieties of Castilla, Extremadura and Andalucía (Alonso de Herrera 1790). It was described as a variety that is similar to the Alvillas, the generic name of the Albillos (see ALBILLO REAL). In Portugal, it was first mentioned near Lamego in the district of Viseu in the Douro region by Rui Fernandes (1531–2) under the synonym Alvaro de Sousa. Its wide distribution makes it difficult to locate the variety's birthplace, but recent genetic investigations have suggested that it could come from around Pinhel in the district of Guarda in north-eastern Portugal, where it has shown the highest level of diversity (Almadanim *et al.* 2007) and where it is known as Síria, which is why this is its official name (along with the synonym Roupeiro).

With DNA parentage analysis, Zinelabidine *et al.* (2012) have recently found a parent–offspring relationship between Síria and CAYETANA BLANCA, a variety that is widespread throughout the south and centre of the Iberian peninsula (see Cayetana Blanca for the pedigree diagram); Lopes *et al.* (1999) found that the white and no longer cultivated Boal Ratinho from the Carcavelos region near Lisboa originates from a natural cross between MALVASIA FINA (under the name Boal da Madeira) and Síria.

Dona Branca, sometimes known as Dona Branca do Dão, is also used as a synonym for Síria, possibly because it is so similar to Doña Blanca, one of the Spanish synonyms for Síria. However, there is a distinct variety in Portugal called DONA BRANCA, grown mainly in the Beiras, Douro and Trás-os-Montes regions.

VITICULTURAL CHARACTERISTICS

Quite vigorous and initially high yielding, though the yield reduces with age and is very dependent on the weather. Mid budding, late ripening. Susceptible to downy and powdery mildews and botrytis bunch rot. Best suited to warm soils and slopes.

WHERE IT'S GROWN AND WHAT ITS WINE TASTES LIKE

The variety is controversial among producers: proponents point to the intensity of its aromas – floral notes such as linden, acacia, orange and laurel blossom as well as orange, peach and melon fruit – but its opponents highlight its tendency to oxidize, which causes the premature loss of such aromas, and its lack of ageing capacity (Rolando Faustino, personal communication). Alcohol and acidity tend to be about average. However, work on clonal selection aims to increase the variety's productivity and allow harvesting before the acidity drops but without a reduction in sugar levels. In the meantime, producers vinifying more carefully and sometimes blending with other local varieties are able to achieve these ends.

Under its various names, but officially Síria or Roupeiro, this is the fifth most planted variety in **Portugal** (and the second most planted white after FERNÃO PIRES), with 13,778 ha (34,046 acres) in 2010, from the Douro in the north, where it is known as Códega, to the Algarve in the south, where it is known as Crato Branco. Roupeiro has long been a recommended variety in the Alentejo: records from 1889 show that it was grown in thirty-two of the thirty-seven subregions, with even more synonyms then than now. In the area around Pinhel and Figueira de Castelo Rodrigo in Beira Interior in the north east, Síria is the most important white variety and the basis of many of the appellation's wines – Quinta do Cardo (part of Companhia das Quintas) make a good example there.

There were 656 ha (1,621 acres) of Doña Blanca in **Spain** in 2008, mainly in Galicia (especially Monterrei and Bierzo but also Valdeorras, where it is known as Valenciana) and Castilla y León, and 148 ha (366 acres) of Cigüente in Extremadura. Quinta da Muradella's Gorvia Blanco from Monterrei is a good example of varietal Doña Blanca: full, dense and peachy with a fabulously satin-smooth texture. Viñas del Bierzo's varietal wines tend to be more neutral.

SIRMIUM

Sirmium, sometimes known as Sirmimum, is a vigorous and productive SAUVIGNON BLANC × DIMYAT (under the name Smederevka) white-berried cross obtained by Dragoslav Milisavljević at the viticulture research centre in Sremski Karlovci, part of Novi Sad University in Vojvodina, **Serbia**. Released in 1970 (Cindrić *et al.* 2000), it is very susceptible to frost and botrytis bunch rot and is therefore not widely planted.

SKIADOPOULO

See FOKIANO.

SKLAVA

Greek variety producing lightweight wines used in blends.

BERRY COLOUR: ● ● ● ● ●

PRINCIPAL SYNONYMS: Sklaba, Sklabes, Sklabos, Sklaves (Argolída), Sklavos

VARIETIES COMMONLY MISTAKEN FOR SKLAVA: ALIONZA (Emilia-Romagna in Italy)

ORIGINS AND PARENTAGE

Sklava originates from Argolída in the eastern Pelopónnisos (Peloponnese), where it has often been mistakenly considered to be related to the Italian SCHIAVA GENTILE (Nikolau and Michos 2004); both Sklava and Schiava mean 'slave' and Schiava is sometimes used as a synonym for Sklava.

VITICULTURAL CHARACTERISTICS

Vigorous and productive. Very late budding, mid ripening. Big bunches of small berries.

WHERE IT'S GROWN AND WHAT ITS WINE TASTES LIKE

Sklava, traditionally grown in Argolída, **Greece**, for blending, produces wines that are low in alcohol but can have relatively fresh acidity. Kontovraki are unusual in making a varietal example.

SKOPELITIKO

Speciality of the island of Kérkyra (Corfu).

BERRY COLOUR: ● ● ● ● ●

PRINCIPAL SYNONYMS: Scopelitico, Skopelitis

ORIGINS AND PARENTAGE

As the name might suggest, Skopelitiko is said to come from the Aegean island of Skópelos in the Sporádes, off the east coast of mainland Greece and brought by monks to the island of Kérkyra.

VITICULTURAL CHARACTERISTICS

Late ripening, good resistance to fungal diseases. Small bunches of small to medium-sized, thick-skinned berries rich in anthocyanins and norisoprenoid aromatic compounds.

WHERE IT'S GROWN AND WHAT ITS WINE TASTES LIKE

Today Skopelitiko is considered a speciality of Kérkyra, **Greece**, grown mainly around Ano Garouna in the centre of the island and to a lesser extent in the south. Wines can be rather rustic. Livadiotis are unusual in making a deeply coloured varietal wine, which has moderate acidity, 12–13% alcohol and fine tannins.

ŠKRLET

Delicate, fresh Croatian variety enjoying a new lease of life.

BERRY COLOUR: ● ● ● ● ●

PRINCIPAL SYNONYMS: Ovnek Slatki, Ovnek Žuti, Škrlet Tusti, Škrtec

VARIETIES COMMONLY MISTAKEN FOR ŠKRLET: KUJUNDŽUŠA

ORIGINS AND PARENTAGE

Škrlet originates from the Moslavina and Pokuplje wine regions, south east of Zagreb, in the centre of Kontinentalna Hrvatska (Continental Croatia). The name Škrlet supposedly derives from the German word *Scharlach*, meaning 'scarlet fever', referring to the red speckles on the surface of the berry skin.

OTHER HYPOTHESES

Although Škrlet and KUJUNDŽUŠA are listed as identical varieties in the Vitis International Variety Catalogue, they come from different regions and are distinct.

VITICULTURAL CHARACTERISTICS

Vigorous, late budding, mid to late ripening. Not particularly susceptible to diseases. Moderate to high, stable yields.

WHERE IT'S GROWN AND WHAT ITS WINE TASTES LIKE

Škrlet is officially authorized in the Moslavina, Prigorje-Bilogora and Pokuplje wine regions east of Zagreb in northern **Croatia** but had become virtually extinct until efforts to recuperate it began in the late 1990s, resulting in a current increase in popularity and demand. Varietal wines, which are typically light and fresh with delicate aromas of spring flowers or hay, are made by producers such as Ivan Belajec, Marijan Brlić, Juren, Kezele, Vladimir Kos, Košutić, Marko Miklaužić and Vladimir Mikša in Moslavina as well as Anto Marinčić and Trdenić in Pokuplje.

SKYLOPNICHTIS

Rare Greek variety used in traditional local blends.

BERRY COLOUR: ● ● ● ● ●

PRINCIPAL SYNONYMS: Kasteliotiko, Mavros Arkadias, Skylopnichtis Kokkino, Skylopnichtra (Argolída), Skylopnihti

ORIGINS AND PARENTAGE

Skylopnichtis, meaning 'smothers the dog', is a Greek variety that may come from Zákynthos or possibly from the Argolída and Arkadía regions in the Pelopónnisos (Peloponnese). The white-berried Skylopnichtis Lefkos (no longer commercially cultivated but found occasionally on Zákynthos, Lefkáda and in Aitolía-Akarnanía on the south-western mainland) is said not to be a colour mutation (Galet 2000) but this is uncertain.

VITICULTURAL CHARACTERISTICS

Fertile, vigorous and productive. Mid budding, mid to late ripening. Large, compact, thin-skinned bunches. Susceptible to downy mildew and botrytis bunch rot.

WHERE IT'S GROWN AND WHAT ITS WINE TASTES LIKE

Skylopnichtis is found to a very limited extent on the island of Zákynthos, **Greece**, where it is used in blends. Comoutos, for example, blend it with local varieties such as AVGOUSTIATIS and KATSAKOULIAS.

SLANKAMENKA

Ancient Balkan variety producing generally uninspiring whites.

BERRY COLOUR: ● ● ● ● ●

PRINCIPAL SYNONYMS: Mađaruša (Croatia), Maghiarca (Romania), Magyarica or Magyarkă (Hungary), Majarcă Albă (Romania), Slankamenka Béla, Szlanka Fehér (Hungary)

ORIGINS AND PARENTAGE

Like KREACA, Slankamenka is a very old Balkan variety, possibly originating from the Banat, the historical region that is today part of Romania, Serbia and Hungary. Found under various different names, it shows significant morphological diversity – and therefore its age – in all three countries

(Avramov and del Zan 2004). The synonym Maghiarca means 'Hungarian' but Slankamen is the name of a village in the Srem district of Vojvodina.

The black-berried **Slankamenka Rosie**, also called Majarcă Rosie in Romania, is probably a colour mutation, while the pink-berried Slankamenka Crvena is a synonym for PAMID.

VITICULTURAL CHARACTERISTICS

Vigorous, productive, late ripening. Medium-sized, thick-skinned berries. Well suited to warm, sandy, fertile soils. Very susceptible to downy mildew but resistant to botrytis bunch rot.

WHERE IT'S GROWN AND WHAT ITS WINE TASTES LIKE

In **Serbia** Slankamenka is planted exclusively in Vojvodina in the north. Wines are typically low in alcohol, acid and flavour and are often part of a blend. Distillation is common. Producers of varietal wines include Antonijević Winery and Dejan Dejanov. There were 55 ha (136 acres) in **Romania** in 2008 under the name Majarcă Albă and the variety is also found in **Hungary** and eastern **Croatia**.

There were 27 ha (67 acres) of Majarcă Rosie in **Romania** in 2008, and an accidental planting at Amato Vino in Margaret River, Western **Australia**.

SMEDEREVKA

See DIMYAT.

SOLARIS

Modern, disease-resistant and early-ripening German hybrid that produces tooth-rottingly high sugar levels.

BERRY COLOUR: ● ● ● ● ●

PRINCIPAL SYNONYMS: Freiburg 240-75

ORIGINS AND PARENTAGE

A MERZLING × Geisenheim 6493 hybrid bred by Norbert Becker in Freiburg in Baden-Württemberg, southern Germany, in 1975, where Geisenheim 6493 is a SEVERNY × MUSCAT OTTONEL hybrid. Solaris has been authorized in Germany since 2004. The name Solaris refers to the sun, as a symbol of power and early ripening. Solaris was used to breed CABERNET CAROL, CABERNET CORTIS and MONARCH.

VITICULTURAL CHARACTERISTICS

Recommended by PiWi International (a group of viticulturists promoting new crosses that are bred or noted for their resistance to fungal diseases, ie *pilzwiderstandsfähige* varieties) for its disease resistance, especially to downy mildew. Can reach very high sugar levels very early in the season, and is therefore a magnet for wasps. Vigorous.

WHERE IT'S GROWN AND WHAT ITS WINE TASTES LIKE

Solaris has become one of the more planted varieties in the new wine regions in the far north of vinous Europe. There were 4 ha (10 acres) in **Denmark** in 2010, at Domaine Aalsgaard, for example, and it is one of the most successful varieties in **Sweden**. It is also planted to a limited extent in **Norway, Belgium, Holland** and **England**.

There are 59 ha (146 acres) in **Germany**, mainly in Baden in the south but there are also a few hectares in Rheinhessen. Early trials have produced wines with aromas of pineapple and hazelnut and pretty good acidity but which are more neutral though powerful (sometimes alcoholic) on the palate. Schindler and Engelhof in Baden and Jan Ulrich in Sachsen are already producing varietal wines.

It is grown on 11 ha (27 acres) in **Switzerland** and producers include Cultiva (Solaris Amabile from the canton of Thurgau), Chalmberger (Aargau), Räblus (Solaris Dessertwein, from the canton of Bern) and Forster (Aargau). Solaris was authorized in **Italy** in 2011.

SOLDAIA

Old, late-ripening Ukrainian variety found occasionally in blends.

BERRY COLOUR: ● ● ● ● ●

PRINCIPAL SYNONYMS: SD-56, Soldaiya

ORIGINS AND PARENTAGE

Soldaia is an old and obscure variety from Krym (Crimea), southern Ukraine. Soldaia was the Venetian name for the fortress at Sudak.

VITICULTURAL CHARACTERISTICS

Late ripening. Compact bunches of thick-skinned berries.

WHERE IT'S GROWN AND WHAT ITS WINE TASTES LIKE

Grown for the table and the vat in **Ukraine**, Soldaia is typically a minor component in blended dessert or fortified wines such as Arkhadersse by Sun Valley (Solnechnaya Dolina).

SOLNECHNODOLINSKY

Productive Ukrainian pale-skinned variety.

BERRY COLOUR: ● ● ● ● ●

PRINCIPAL SYNONYMS: DM-63, SD-2

ORIGINS AND PARENTAGE

Selected by P M Gramotenko, V V Pestretsov, N S Skripkin, A M Romanenko, M N Matvienko, L P Troshin, V F Karzov and V V Karzova in 1969 at the Magarach research centre in Krym (Crimea), southern Ukraine, because it was morphologically similar to SARY PANDAS. It was named after the Solnechnaya Dolina region and included in the official register of varieties in 1995 but its pedigree is still unknown.

VITICULTURAL CHARACTERISTICS

Late ripening, vigorous and productive. Large bunches of thick-skinned berries.

WHERE IT'S GROWN AND WHAT ITS WINE TASTES LIKE

Solnechnodolinsky's generous yields and high sugar levels make it a useful ingredient in the production of semi-sweet and dessert wines in **Ukraine** such as Sun Valley's eponymous bottling and in the sweet, fortified Arkhaderesse (Sun Valley is the direct translation of Solnechnaya Dolina).

SOPERGA

Rare Conegliano cross whose true parentage has only recently been revealed.

BERRY COLOUR: ● ● ● ● ●

PRINCIPAL SYNONYMS: Incrocio Dalmasso IV/31, Superga

ORIGINS AND PARENTAGE

Italian vine breeder Giovanni Dalmasso announced that Soperga was a NEBBIOLO × BARBERA cross created in 1936 at the Conegliano research centre in the Veneto, north-east Italy, with the aim of combining the quality of Nebbiolo and the resistance and the naturally high yields of Barbera. It was included in the official register of varieties in 1977. However, DNA profiling has recently shown that the Nebbiolo used by Dalmasso was in fact Nebbiolo di Dronero, which is not Nebbiolo but a synonym for CHATUS (Torello Marinoni, Raimondi, Mannini and Rolle 2009). Soperga is therefore a Chatus × Barbera cross, just like its siblings ALBAROSSA, CORNAREA, NEBBIERA and SAN MICHELE.

VITICULTURAL CHARACTERISTICS

Mid to late ripening. Medium resistance to fungal diseases.

WHERE IT'S GROWN

Used for wine as well as for the table in Piemonte, north-west **Italy**. There were 32 ha (79 acres) in Italy in 2000, although it is now difficult to find evidence of any commercial production.

SORBIGNO

A minor Ischian speciality.

BERRY COLOUR: ● ● ● ● ●

PRINCIPAL SYNONYMS: Sorbino, Sorvegna, Sorvigno

ORIGINS AND PARENTAGE

Sorbigno is an obscure and ancient variety from the Italian island of Ischia, where it was mentioned as long ago as 1588 (Monaco 2006). However, Sorbigno does not appear in the local atlas of varieties and its identity is still unknown (Migliaccio *et al.* 2008).

WHERE IT'S GROWN

Sorbigno is exclusive to the island of Ischia, southern **Italy**, where La Vigna dell'Encadde is probably the only producer of a varietal wine.

SOUSÃO

See VINHÃO.

SOYAKI

Also known as Soïaki, this white-berried Uzbek variety is grown in the Tashkent region in the north east and further south in Samarkand in **Uzbekistan**. It is used both for the table and for wine; the Samarkand Winery, for example, use it in blends.

SPANNA

See NEBBIOLO.

SPÄTBURGUNDER

See PINOT NOIR *in the* PINOT *entry.*

SPERGOLA

Tart Emilia-Romagnan thought until very recently to be Sauvignon Blanc.

BERRY COLOUR:

PRINCIPAL SYNONYMS: Spargoletta Bianca, Spergolà, Spergolina, Spergolinàis

VARIETIES COMMONLY MISTAKEN FOR SPERGOLA: SAUVIGNON BLANC, SÉMILLON

ORIGINS AND PARENTAGE

Spergola was first mentioned by Dalla Fossa in 1811 as one of the best grape varieties of Reggio Emilia in northern Italy (Giavedoni and Gily 2005). It was often confused with SAUVIGNON BLANC – on ampelographic rather than organoleptic grounds – until further analysis and DNA profiling proved that they were distinct varieties (Filippetti, Silvestroni, Ramazzotti and Intrieri 2001). In addition, DNA analysis suggested a close genetic relationship between Spergola and PIGNOLETTO (Filippetti, Silvestroni, Thomas and Intrieri 2001).

OTHER HYPOTHESES

Spergola was supposedly mentioned in the sixteenth or seventeenth century under the names Pomoria or Pellegrina, but there's no evidence that they are all the same variety.

VITICULTURAL CHARACTERISTICS

Mid ripening.

WHERE IT'S GROWN AND WHAT ITS WINE TASTES LIKE

Spergola was a major grape variety in Reggio Emilia in the nineteenth century, where it is now cultivated on approximately 200 ha (494 acres). Spergola is no longer listed as a synonym for SAUVIGNON BLANC in the official register of varieties in **Italy** and is included separately, authorized in Emilia-Romagna, in particular as the main variety (minimum 85%) in white Colli di Scandiano e di Canossa DOC. Spergola can produce dry, sparkling or sweet varietal wines. Tenuta di Aljano's dry version is appropriately named: La Vigna Ritrovata means 'the rediscovered vine'; the alcohol is quite high but it has very good acidity and roundness on the palate. The high level of acidity also makes it suitable for the production of sparkling wine. Ca' de Noci's Querciole is a good example of the latter but they also make a still varietal wine called Reserva dei Fratelli and use Spergola in blends with MOSCATO GIALLO and what the producer calls 'Malvasia Aromatica' (see MALVASIA).

SPOURTIKO

Rare Cypriot variety producing light, zesty whites.

BERRY COLOUR: ● ● ● ● ●

PRINCIPAL SYNONYMS: Spourtico

ORIGINS AND PARENTAGE

Spourtiko, meaning 'bursting', referring to the explosive habit of the berries, is indigenous to the district of Páfos on the island of Cyprus. According to DNA profiling, it is relatively close to the dark-skinned OFTHALMO, also from Cyprus (Hvarleva, Hadjinicoli *et al.* 2005).

VITICULTURAL CHARACTERISTICS

Early budding, mid ripening. Thin-skinned berries that easily crack but relatively good disease resistance.

WHERE IT'S GROWN AND WHAT ITS WINE TASTES LIKE

Wines typically have fresh flavours of grapefruit and lemon, sometimes lemongrass, low alcohol (rarely above 11%) and can be slightly grassy. There were just 10 ha (25 acres) on the island of **Cyprus** in 2010, a significant increase from less than 2 ha (5 acres) in 2008, but many of the Spourtiko vines, found mostly in Páfos and to a lesser extent around Lemesós (Limassol), are planted in alternate rows with MARATHEFTIKO to help improve the pollination – and thus the yields – of the latter rather than for the quality of the wine. Fikardos is one of the few producers to make a varietal wine.

ST CROIX

Very cold-hardy, complex American hybrid gaining ground in Québec and the Midwest.

BERRY COLOUR: ● ● ● ● ●

PRINCIPAL SYNONYMS: ES 2-3-21, Sainte-Croix, Ste-Croix

ORIGINS AND PARENTAGE

An Elmer Swenson 283 × Elmer Swenson 193 hybrid bred in 1974 by Elmer Swenson in Osceola, Wisconsin, US (see SABREVOIS for the pedigrees). The vine first fruited in 1977 and the variety was introduced in 1981 and patented to Elmer Swenson in 1982. This complex *Vitis riparia*, *Vitis labrusca*, *Vitis vinifera*, *Vitis lincecumii* and *Vitis rupestris* hybrid has the same parents as its sibling Sabrevois. St Croix was used to breed CHISAGO and PIONNIER.

VITICULTURAL CHARACTERISTICS

Compact medium-sized bunches of thin-skinned grapes that are prone to leak. Mid budding, early to mid ripening. Very hardy (below –20 °F/–29 °C, possibly as low as –35 °F/–37 °C) although the roots are slightly less so and need snow cover in really cold winters (Smiley 2008). Moderately susceptible to botrytis bunch rot, downy and powdery mildews.

WHERE IT'S GROWN AND WHAT ITS WINE TASTES LIKE

Wines are typically light- to medium-bodied with moderate acidity, fairly low alcohol, light tannins and are generally best in a blend. They are free from foxy *labrusca* flavours and can show heavy flavours of tobacco, tar or smoke if overripe or overly macerated (Hart nd). St Croix (often Ste Croix in **Canada**) is the most widely planted variety in the province of Québec (36 acres/15 ha in 2009) and is also found in the American Midwest (eg 29 acres/12 ha in **Iowa** in 2006, 20 acres/8 ha in **Illinois** in 2007, 18 acres/7 ha in **Minnesota** in 2007) and to a lesser extent further east in **Pennsylvania** (6 acres/2.4 ha in 2008). Varietal wines are produced by Tassel Ridge in **Indiana**; Fieldstone in Minnesota; La Roche des Brises, Le Royer St-Pierre and Vignoble Ste-Pétronille in Québec.

ST PEPIN

Fruity, low-acid American hybrid that often has a hint of foxiness.

BERRY COLOUR: ● ● ● ● ●

PRINCIPAL SYNONYMS: Elmer Swenson 282, ES 282, Saint-Pépin

ORIGINS AND PARENTAGE

A complex hybrid of Elmer Swenson 114 and SEYVAL BLANC made around 1950 by Elmer Swenson in Osceola, Wisconsin, US, in which:

- Elmer Swenson 114 is a Minnesota 78 × ROSETTE hybrid
- Minnesota 78 is a Beta × Witt hybrid (see the BRIANNA pedigree diagram for Minnesota 78)

- Beta is a Carver × CONCORD hybrid, where Carver is a selected *Vitis riparia* vine
- Witt is a seedling of CONCORD, itself descended from *Vitis labrusca*.

This complex *Vitis riparia*, *Vitis labrusca*, *Vitis vinifera*, *Vitis lincecumii* and *Vitis rupestris* hybrid was introduced in 1983 and plant patent was assigned to Swenson Smith Vines in 1986. It is a sibling of LA CROSSE and was used to breed LA CRESCENT.

VITICULTURAL CHARACTERISTICS

Vigorous with a relatively open canopy and productive (but has pistillate flowers and therefore needs to be planted nearer another vine for cross-pollination). Highly susceptible to powdery mildew and moderately so to botrytis bunch rot and downy mildew. Early to mid ripening. Cold-hardy (–15 °F/–26 °C to –20 °F/–29 °C) but probably less so than LA CROSSE (Smiley 2008).

WHERE IT'S GROWN AND WHAT ITS WINE TASTES LIKE

Varietal wines, which are fruity but often have a hint of foxy *labrusca* flavour and generally low acidity, are made by Cannon River Winery in Minnesota and Parallel 44 in **Wisconsin**. In **Canada**, where there were just a few acres in 2009, it is often blended with other cold-hardy hybrids such as PRAIRIE STAR and VANDAL-CLICHE, although Domaine les Brome in Québec produce a varietal St Pepin. Plantings in the US Midwest are limited. For example, there were 17 acres (7 ha) in **Minnesota** in 2007, 15 acres (6 ha) in **Iowa** in 2006, 10 acres (4 ha) in **Illinois** in 2007 and 2 acres (less than 1 ha) in **Nebraska** in 2007. It is occasionally used to make icewine.

ST VINCENT

Recent cherry-flavoured American variety, only possibly a hybrid, from Missouri.

BERRY COLOUR: ● ● ● ● ●

ORIGINS AND PARENTAGE

This variety was discovered in 1973 by Scott Toedesbusch, then manager of Mount Pleasant Vineyards owned by Lucian Dressel in Augusta, Missouri, US. Cuttings were sent to Philip Wagner of Maryland, who propagated and sold them. It was first called Stromboli, but given its similarity to PINOT NOIR, it was finally named after the patron saint of the Côte d'Or in Burgundy (Wagner 1988). The exact pedigree of St Vincent is unknown. It is thought by some to be a chance seedling that was a hybrid of Pinot Noir and CHAMBOURCIN vines growing together in Dressel's vineyard. Since Chambourcin has a considerable amount of *Vitis riparia* in its ancestry, the pigments in wine made from St Vincent should in theory contain diglucosides, a type of sugar that is typical for *Vitis riparia*, *Vitis rupestris* and *Vitis labrusca* but totally absent in *Vitis vinifera* (Ribéreau-Gayon 1958). However, no traces of diglucosides have been found in St Vincent (Wagner 1988), suggesting that it is not a hybrid but rather a 100% *Vitis vinifera*. DNA parentage analysis has not yet been done to determine the true parentage of St Vincent.

It was officially introduced onto the market in 2006.

VITICULTURAL CHARACTERISTICS

Loose clusters of medium-sized to large berries. Moderately vigorous and moderately susceptible to black rot, botrytis bunch rot, and downy and powdery mildews. Late ripening and moderately hardy (–10 °F/–23 °C to –15 °F/–26 °C; Smiley 2008).

WHERE IT'S GROWN AND WHAT ITS WINE TASTES LIKE

Wines are bright red with quite complex flavours of cherry, sometimes nutty and a little smoky and generally have high acidity. Producers of varietal wines include Three Squirrels, Two Saints and Whispering Oaks in **Missouri**; and Domaine Berrien in **Michigan**. The largest area of this variety is in its home state of Missouri (34 acres/14 ha in 2009) but it is also found in other states such as **Iowa** (18 acres/7 ha in 2006) and **Pennsylvania** (just 1 acre/0.4 ha in 2008).

ŠTAJERSKA BELINA

See RANFOL.

STANUŠINA CRNA

Macedonian variety in dire need of vineyard selection to improve quality.

BERRY COLOUR: ● ● ● ● ●

PRINCIPAL SYNONYMS: Gradesh, Stanušhina Crna, Strnušina, Tikvesko

ORIGINS AND PARENTAGE

Stanušina Crna, originally called Strnušina, possibly originates in Tikveš, the Republic of Macedonia's main vine-growing region in the centre of the country, hence its synonym Tikvesko. A green-berried colour mutation (Stanušina Zelena, ie 'green Stanušina'), which lacks sugar and acidity, as well as a type of Stanušina more prone to millerandage and therefore poor yields (Stanušina Rehuljava, ie 'straggly Stanušina') have been observed in the vineyards (Bozinovic *et al.* 2003; Štajner, Angelova *et al.* 2009).

VITICULTURAL CHARACTERISTICS

Productive, late ripening. Medium-sized to large bunches of largish berries. Well suited to poor, dry soils and limestone. Moderate resistance to fungal diseases.

WHERE IT'S GROWN AND WHAT ITS WINE TASTES LIKE

Like OHRIDSKO CRNO, Stanušina Crna was the dominant variety in its region of origin, now in the **Republic of Macedonia**, before the advent of phylloxera but since then plays a more minor role in Tikveš, with small pockets in other parts of the country. Alcohol is usually moderate but the wines often lack acidity, body and texture and are generally blended with PROKUPAC and/or Kratošija (TRIBIDRAG). Popova Kula, swimming against the incoming tide of international varieties, make varietal examples, both red and rosé. Selection in the vineyard is needed to weed out the inferior-quality mutations described above.

STAVROTO

Greek variety required in and virtually exclusive to the Rapsáni appellation.

BERRY COLOUR: ● ● ● ● ●

PRINCIPAL SYNONYMS: Ampelakiotiko, Ampelakiotiko Mavro, Stavromavro

ORIGINS AND PARENTAGE

Local variety of the Rapsáni region in the north east of Thessalía, eastern central Greece.

VITICULTURAL CHARACTERISTICS

Large to medium-sized compact bunches of small berries. Very early budding, mid to late ripening, vigorous.

WHERE IT'S GROWN AND WHAT ITS WINE TASTES LIKE

Stavroto is found mainly in Thessalía, **Greece**, thanks to its role as one of the three varieties required in the Rapsáni appellation, along with KRASSATO and XINOMAVRO (usually a field blend). Colour, alcohol and acidity levels are lower than for Xinomavro (Manessis 2000). It's also found to the north in Kozáni and to the south in Magnisía. Recommended producers of Rapsáni, a rich, well-structured, spicy wine, include Dougos and Tsantali.

STEEN

See CHENIN BLANC.

STEUBEN

Minor but successful American hybrid used for the table and for wine.

BERRY COLOUR: ● ● ● ● ●

PRINCIPAL SYNONYMS: New York 12696

ORIGINS AND PARENTAGE

A Wayne × Sheridan hybrid obtained in 1925 by Richard Wellington and George D Oberle in the Department of Pomology and Viticulture at Cornell University's New York State Agricultural Experiment Station, Geneva, US, in which:

- Wayne is a Mills × Ontario hybrid (see GR 7 for the pedigrees of Mills and CAYUGA WHITE for that of Ontario)
- Sheridan is a Herbert × Worden hybrid (see GR 7 for the pedigree of Herbert)
- Worden is a selected seedling of CONCORD.

This *Vitis labrusca* and *Vitis vinifera* hybrid was selected in 1931, then named and introduced in 1947. It was used to breed COROT NOIR and NOIRET.

VITICULTURAL CHARACTERISTICS

Vigorous and productive; yields must be controlled. Large compact bunches of black berries covered in a heavy lilac bloom. Mid to late ripening. The juice is sweet and spicy but not aromatic. Moderately susceptible to black rot but otherwise generally good disease resistance. Cold-hardy (–15 °F/–26 °C to –20 °F/–29 °C).

WHERE IT'S GROWN AND WHAT ITS WINE TASTES LIKE

Steuben is thinly but widely spread in the US Midwest and the Northeast regions: 45 acres (18 ha) in **Pennsylvania** in 2008, 37 acres (15 ha) in **New York** State in 2006, 25 acres (10 ha) and growing in popularity in **Indiana** and 3 acres (1.2 ha) in **Illinois** in 2007, for example. Steuben is used to make aromatic white, blush or rosé wines at various sweetness levels and is highly regarded as a table grape thanks to the shape of the bunch and the flavour of the grapes. It is also grown for juice. Varietal wines are made by, for example, Marjim Manor, Mayers Lake Ontario and Niagara Landing in New York State; Starr Hill in Pennsylvania; Buck Creek in Indiana; Jewell Town in **New Hampshire**; Tassel Ridge in **Iowa**.

STORGOZIA

Recent, usefully disease-resistant Bulgarian hybrid.

BERRY COLOUR: ● ● ● ● ●

PRINCIPAL SYNONYMS: Storgoziya

ORIGINS AND PARENTAGE

Storgozia is a BUKET × VILLARD BLANC hybrid created in 1976 by I Ivanov, V Valtchev and G Petkov at the Pleven viticulture and oenology institute in northern Bulgaria and given the old name of the ancient city of Pleven.

VITICULTURAL CHARACTERISTICS

Good resistance to downy and powdery mildews and to botrytis bunch rot, as well as cold winter temperatures. Productive. Thin-skinned grapes.

WHERE IT'S GROWN AND WHAT ITS WINE TASTES LIKE

There were 738 ha (1,824 acres) in **Bulgaria** in 2008, almost all in the province of Pleven in the north. Yantra near the town of Ruse, close to the Romanian border, and Chateau de Val, near the city of Vidin in the far north east, include Storgozia in their red blends. Wines have good levels of acid and alcohol and can be aged.

SUBIRAT PARENT

See ALARIJE.

SUKHOLIMANSKY BELY

Versatile Ukrainian cross producing fresh and lightly floral whites.

BERRY COLOUR: ● ● ● ● ●

ORIGINS AND PARENTAGE

Sukholimansky Bely is a CHARDONNAY × Plavaï cross obtained by Y S Komarova, M P Tsebriy, A N Kostyuk, P K Ayvazyan and E N Dokuchaeva in 1949 at the Tairov research centre in Odessa, southern Ukraine, in which Plavaï is a once-popular but now almost extinct Moldovan variety. Included in the official register of varieties in 1969.

VITICULTURAL CHARACTERISTICS

Consistently productive, vigorous, mid ripening. Thin but durable berry skins. Winter-hardy to –19 °C (–2 °F).

WHERE IT'S GROWN AND WHAT ITS WINE TASTES LIKE

Sukholimansky Bely is a versatile, fresh and lightly floral variety used in dry and medium-dry blends, often with ALIGOTÉ, as well as for fizz and the table. There were a considerable 1,477 ha (3,650 acres) in **Ukraine** in 2009, found mainly in the Odessa, Nikolaev and Kherson regions in the south of the country. Kolonist make a soft but aromatic varietal wine and Odessavinprom produce a semi-sweet blend with Aligoté. It is also found in Krasnodar, south-east **Russia** (80 ha/198 acres in 2010).

SULTANA

See SULTANIYE.

SULTANINA

See SULTANIYE.

SULTANIYE

The world's most widely planted light-skinned variety. The sultana grape is only rarely used for wine, fortunately.

BERRY COLOUR: ● ● ● ● ●

PRINCIPAL SYNONYMS: Banati (Egypt), Kishmish (Afghanistan and Armenia), Kis Mis Alb (Romania), Kişmiş (Turkey), Soultanina, Sultana (South Africa), Sultani Çekirdeksiz (Turkey), Sultanina (widespread), Sultanina Blanche, Sultanine (France and Greece), Thompson Seedless (US, South Africa and Australia)

ORIGINS AND PARENTAGE

Sultaniye, named after the Ottoman sultans to whom the precious grape was offered, has been cultivated in so many countries and under so many synonyms that the location of its birthplace has long been the subject of speculation: proposed origins have included Turkey, Greece, South Africa, Afghanistan and Iran.

Colour mutations were observed by Galet (2000) in Afghanistan: **Kishmish Sorh** (red) and **Kishmish Siah** (black). Sultaniye has been used to breed many table-grape varieties.

Sultaniye was used to breed numerous wine and table grapes, of which only TARRANGO is commercially cultivated for wine.

VITICULTURAL CHARACTERISTICS

Very vigorous but poor fertility except in ideally warm conditions. Mid ripening and highly susceptible to most fungal diseases, particularly in Western Europe. Berries can attain relatively high sugar levels.

WHERE IT'S GROWN AND WHAT ITS WINE TASTES LIKE

Sultaniye, a seedless variety, as its common synonym Thompson Seedless suggests, is the most widely planted variety in the world but the vast majority of the fruit is eaten dried or fresh. (The synonym Kishmish means 'grape for drying' in Pashtun.) In **Turkey**, Sultaniye is planted for the same purposes but is also used to make wine on the Manisa plain and around İzmir and Denizli in the Aegean region in the west of the country, where it is less prone to disease. At best the wines are light and rather neutral, sometimes made in an off-dry style. Some of the bigger Turkish producers such as Kavaklidere, Kayra and Doluca make varietal wines but also blend Sultaniye with other local varieties such as EMIR and NARINCE. Total plantings in 2010 were 2,545 ha (6,289 acres) but it is likely that much of this is for the table. It is also grown in neighbouring **Armenia**.

It is widely planted in **Cyprus**, where there were 516 ha (1,275 acres) of Soultanina in 2010, for the table and for wine. In **Greece** there were 1,095 ha (2,706 acres) in the Pelopónnisos (Peloponnese) in 2008.

California recorded a massive 198,094 acres (80,166 ha) of Thompson Seedless in 2008 but these were almost all for raisins. (In its time Thompson Seedless has been pressed into winemaking service only in times of severe shortage of light-skinned grapes in California and Australia.) It goes under the same name in **Australia**, where the variety is grown predominantly for table grapes and raisins.

SUMOLL

Minor Catalan found mainly in red blends, and also now on the Islas Canarias.

BERRY COLOUR: ● ● ● ● ●

PRINCIPAL SYNONYMS: Sumoi, Sumoll Negro, Sumoll Tinto, Sunier, Verijariego Negra⚥, Vigiriega, Vigiriega Negra, Vigiriego Negro, Vijariego Negra⚥, Vijariego Negro⚥ (Islas Canarias), Vijiriego Negro

ORIGINS AND PARENTAGE

Sumoll was first mentioned in 1797 in Catalunya, its name derived from *sumollar*, a local dialect word based on the Latin *submolliare*, meaning 'to ripen' or 'to mature' (Favà i Agud 2001). DNA profiling has unexpectedly shown that Sumoll is identical to Vijariego Negra from Andalucía (SIVVEM), first described by Clemente y Rubio (1807) as a variety that was cultivated mainly in Jerez de la Frontera and was rare in Sanlúcar de Barrameda. The variety was also introduced to the Islas Canarias under the name Vijariego Negro, possibly together with Vijariega Blanca (see VIJARIEGO), which is not a colour mutation.

It was used to breed varieties such as CIENNA and TYRIAN in Australia.

SUMOLL BLANC has a distinct morphology and DNA profile and is therefore not a colour mutation of Sumoll (Giralt *et al.* 2009).

VITICULTURAL CHARACTERISTICS

Fertile, productive, late ripening. Very large bunches.

WHERE IT'S GROWN AND WHAT ITS WINE TASTES LIKE

Spain's total vineyard area of Sumoll, under that synonym, was 258 ha (638 acres) in 2008, virtually all in Catalunya. It mostly contributes a small percentage to blends, with both local and international varieties, but the following producers make varietal wines, either red or rosé: Bohigas, Can Ramon Viticultors del Montgrós, Cellar Pardas, MonRubí. Sumoll has been gaining in popularity and is now authorized for Alella, Catalunya, Penedès, Pla de Bages and Tarragona DOs. Wines are typically red-fruited with fresh acidity.

Under the synonym Vijariego Negro, official statistics for 2008 recorded 59 ha (146 acres) on the Islas Canarias, where it is authorized in the El Hierro, La Gomera and Pla de Bages DOs. Tanajara in the El Hierro DO make a particularly good varietal example.

SUMOLL BLANC

Very rare but promising, refreshing Catalan variety that is unrelated to the dark-berried Sumoll.

BERRY COLOUR:

PRINCIPAL SYNONYMS: Sumoi Blanc, Sumoll Blanco

ORIGINS AND PARENTAGE

Sumoll Blanc is local to Tarragona, north-east Spain, but it is not a white mutation of SUMOLL, as DNA profiling has shown (Giralt *et al.* 2009).

WHERE IT'S GROWN AND WHAT ITS WINE TASTES LIKE

Sumoll Blanc is planted to a very limited extent in Tarragona and Penedès in Catalunya, north-eastern **Spain,** on just 5 ha (12 acres) of vineyard in 2008. 10 Sentits produce rare varietal oaked and unoaked versions, both of which show fine acidity, minerality and an array of fruit flavours – not unlike Chablis in character.

SUPPEZZA

Sorrento rarity that is an ingredient in early released fizz.

BERRY COLOUR:

ORIGINS AND PARENTAGE

It is said locally that Suppezza is named after the old farm in Gragnano in the Monti Lattari, west of Salerno, where it was found. DNA analysis suggests that Suppezza could be genetically close to FALANGHINA FLEGREA (Costantini *et al.* 2005).

VITICULTURAL CHARACTERISTICS

Mid ripening. Susceptible to botrytis bunch rot.

WHERE IT'S GROWN AND WHAT ITS WINE TASTES LIKE

Suppezza is one of the numerous minor varieties of Campania, **Italy**, and can still be found today in old and ungrafted pergola vineyards on the Sorrento peninsula, in particular in the communes of Gragnano, Pimonte, Lettere and Castellamare di Stabia. Suppezza is usually blended with other local varieties, mainly in the 'rosso frizzante naturale' version of Penisola Sorrentina DOC, a naturally fizzy red wine – a Neapolitan tradition – generally made from PIEDIROSSO and/or AGLIANICO and/or SCIASCINOSO with up to 40% other local varieties, including Suppezza. The leading local producer is Grotta del Sole.

SUŠĆAN

Minor Croatian variety found on just a few islands in the Adriatic but gaining in popularity.

BERRY COLOUR:

PRINCIPAL SYNONYMS: Brajda Velika Crna, Sansigot, Susac, Sušćan Crni

ORIGINS AND PARENTAGE

Sušćan comes from the island of Susak, in the Kvarner Gulf in north-west Croatia, south of the island of Cres. This is confirmed by its synonym Sansigot, derived from Sansego, the Italian name for the island.

VITICULTURAL CHARACTERISTICS

Vigorous, yields well and consistently, late budding, mid to late ripening. Good resistance to fungal diseases, particularly botrytis bunch rot.

WHERE IT'S GROWN AND WHAT ITS WINE TASTES LIKE

Officially authorized in the Hrvatsko Primorje (**Croatian** Littoral) wine region, Sušćan is the main variety on the tiny Croatian island of Susak and is also found on the nearby and larger islands of Lošinj, Cres, Unije and Krk. Wines are typically fairly high in alcohol and sold to the increasing number of tourists who visit these islands. At one time Susak was known as the 'floating vineyard' because 95% of the island was planted with vines. On Krk, Anton Katunar blends Sansigot with SYRAH and GEGIĆ. Its viticultural advantages and the deeply coloured, full-bodied wines it produces have led to a recent increase in popularity.

SUSUMANIELLO

Minor, deeply coloured Puglian.

BERRY COLOUR:

PRINCIPAL SYNONYMS: Cozzomaniello, Grismaniello, Somarello Nero☒, Susomaniello, Susumaniello Nero, Zingariello, Zuzomaniello

ORIGINS AND PARENTAGE

Susumaniello has long been known in the province of Brindisi in Puglia, southern Italy. Its name could derive from its synonym *somarello*, meaning 'donkey', a reference to the load this animal is able to bear, just like this vine variety.

Recent DNA parentage analysis has established that Susumaniello is a cross between GARGANEGA and Uva Sogra (Di Vecchi Staraz, This *et al.* 2007), the latter identical to Uva Sacra (Zulini *et al.* 2002), a table grape of Puglia (Vouillamoz). As a consequence, Susumaniello is either a grandparent or a half-sibling of eight other varieties having parent–offspring relationships with Garganega: ALBANA, CATARRATTO BIANCO, DORONA DI VENEZIA, MALVASIA BIANCA DI CANDIA, MARZEMINA BIANCA, MONTONICO BIANCO, MOSTOSA and TREBBIANO TOSCANO (see the pedigree diagram in the Garganega entry).

OTHER HYPOTHESES

Susumaniello is said to come from the Dalmatian coast, but no evidence has yet been provided. It has also been suggested that the variety is named after Sumanus, the Roman god of evening rains.

VITICULTURAL CHARACTERISTICS

Vigorous and mid ripening.

WHERE IT'S GROWN AND WHAT ITS WINE TASTES LIKE

Susumaniello is authorized in every Puglian province except Foggia in the north. It is cultivated mainly north west and south east of the port of Brindisi, where it is usually blended with other local grape varieties such as MALVASIA NERA DI BRINDISI and NEGROAMARO in the Ostuni and Brindisi DOCs. Tenute Rubino's Torre Testa (IGT Salento), Racemi's Sum from the Torre Guaceto nature reserve (a vino da tavola), Lomazzi & Sarli's Nomas and Cantine Due Palme's Serre (IGT Salento) are rare examples of 100% Susumaniello wines. They tend to be deeply coloured and have red-fruit flavours but can be quite tart and tough. There were 72 ha (178 acres) in **Italy** in 2000.

SWENSON RED

Minor red-berried American hybrid bred for the chill of the Northwest.

BERRY COLOUR: ● ● ● ● ●

PRINCIPAL SYNONYMS: Elmer Swenson 439

ORIGINS AND PARENTAGE

A Minnesota 78 × Rubilande hybrid bred in 1962 by Elmer Swenson in Osceola, Wisconsin (see BRIANNA for the complete pedigree of Swenson Red), selected in 1967 and released jointly by Elmer Swenson and the University of Minnesota in 1978. Swenson Red is therefore an extremely complex *Vitis riparia*, *Vitis labrusca*, *Vitis vinifera*, *Vitis aestivalis*, *Vitis lincecumii*, *Vitis rupestris* and *Vitis cinerea* hybrid. It was used to breed CHISAGO.

VITICULTURAL CHARACTERISTICS

Medium-sized bunches of large red berries. Highly susceptible to downy mildew, moderately susceptible to botrytis bunch rot and powdery mildew. Hardy (–15 °F/–26 °C) but less so than other varieties bred by Swenson. Mid ripening and can reach high sugar levels.

WHERE IT'S GROWN AND WHAT ITS WINE TASTES LIKE

Swenson Red is used mainly for the table but also for wine. Swenson did not recommend fermenting the juice on the skins (Swenson *et al.* 1980) and this variety is better suited for white or blush wines. It is found mainly in **Minnesota** (24 acres/10 ha in 2007) and **Iowa** (5 acres/2 ha in 2006) but there are few varietal wines – although Summerset Winery in Iowa produce an exception. The variety also seems to be grown in **Poland**.

SWENSON WHITE

Minor but very complex, cold-hardy American hybrid better suited to the table than to wine.

BERRY COLOUR: ● ● ● ● ●

PRINCIPAL SYNONYMS: ES 6-1-43

ORIGINS AND PARENTAGE

An EDELWEISS × Elmer Swenson 442 hybrid obtained by Elmer Swenson in Osceola, Wisconsin, US, in which Elmer Swenson 442 is a Minnesota 78 × Rubilande hybrid (see BRIANNA for the pedigrees of Minnesota 78 and Rubilande). Swenson White, selected in 1988 and released in 1994, is therefore an extremely complex *Vitis riparia*, *Vitis labrusca*, *Vitis vinifera*, *Vitis aestivalis*, *Vitis lincecumii*, *Vitis rupestris*, *Vitis cinerea* and *Vitis berlandieri* hybrid.

VITICULTURAL CHARACTERISTICS

Vigorous, mid budding, late ripening and suitable for late harvest. Medium to large bunches of large, thick-skinned berries. Moderately susceptible to downy and powdery mildews and to anthracnose. Very hardy (below –20 °F/–29 °C; Smiley 2008).

WHERE IT'S GROWN AND WHAT ITS WINE TASTES LIKE

According to Mark Hart of Mt Ashwabay Vineyard in Wisconsin (Hart nd), this is really a table grape and any wines made are neutral, sometimes with floral aromas, but have a slight foxy flavour especially when the grapes are picked ripe – though not as strong as its EDELWEISS parent. Its late ripening means that it is really only suited to Midwest states such as **Iowa** and **Nebraska**. It is planted to a very limited extent in the American Midwest and in **Canada**, where Domaine et Vins Gelinas in Québec make a rare varietal icewine.

SYKIOTIS

Greek variety used mainly for distillation.

BERRY COLOUR: ● ● ● ● ●

PRINCIPAL SYNONYMS: Chiotis, Kiotes, Skiotes, Skiotis, Sykioti

ORIGINS AND PARENTAGE

Sykiotis has long been known in Thessalía in central Greece but its origin is not clear (Nikolau and Michos 2004). The name derives from *sýko*, meaning 'fig'.

VITICULTURAL CHARACTERISTICS

Early budding and late ripening, vigorous and productive.

WHERE IT'S GROWN

Although this variety is mostly used for distillation, in **Greece**, into the spirit tsipouro, the Dimitra Co-operative in Néa Anhíalos, south-eastern Thessalía, also make a dry red Sykiotis varietal wine. Also found in Makedonía in the north and on the large island of Évvoia (Euboea).

SYLVANER

See SILVANER.

SYMPHONY

Rare and highly aromatic Muscat-flavoured California cross.

BERRY COLOUR:

PRINCIPAL SYNONYMS: Davis S15-58

ORIGINS AND PARENTAGE

A GARNACHA ROJA × MUSCAT OF ALEXANDRIA cross bred by Harold P Olmo in 1940 at the University of California at Davis and released in 1983 (Walker 2000).

VITICULTURAL CHARACTERISTICS

Vigorous, productive and adaptable to a range of climates. Medium-sized to large, often compact, bunches of Muscat-flavoured berries. Mid ripening. Susceptible to botrytis bunch rot.

WHERE IT'S GROWN AND WHAT ITS WINE TASTES LIKE

Symphony produces fruity, Muscat-flavoured wines from dry to sweet, both still and sparkling. After a certain rise in popularity as an off-dry table wine in the 1990s, the acreage in **California** reached over 700 acres (283 ha) in 2004 and in 2009 stood at 803 acres (325 ha), with the biggest plantings in Fresno, Merced, Yolo and San Joaquin counties in the Central Valley. Producers of varietal Symphony in California include Maple Creek in the Yorkville Highlands, who make an off-dry wine and a sweet late-harvest version, Sebastiani in Sonoma and Michael David Winery in Lodi. Ironstone have the most significant plantings of this variety in California, in Lodi and the Sierra Foothills, and they regularly sell out of this popular medium-dry wine. New plantings are likely since it is also in demand as a blending component. The Symphonys bottled by Winter Park in Colorado and Pahrump Valley in Nevada are based on grapes grown in California whereas Volcano Winery on **Hawaii** grow their own Symphony grapes.

SYRAH

Generally fashionable alternative to Cabernet Sauvignon with a complex and surprising family background.

BERRY COLOUR:

PRINCIPAL SYNONYMS: Candive (Bourgoin-Jallieu), Hermitage (Australia), Marsanne Noire (Saint-Marcellin), Petite Sirrah, Sérène (Isère), Serine, Sérine or Serinne (Côte Rôtie and Isère), Shiraz (Australia), Sira, Sirac, Sirah, Syra, Syrac

VARIETIES COMMONLY MISTAKEN FOR SYRAH: DUREZA, DURIF (called Petite Sirah in Cerdon in the Ain *département*, in Australia and in California), MONDEUSE NOIRE, PERSAN (called Sérine in Isère), SERINA E ZEZË (Albania), SHESH I ZI (Albania)

ORIGINS AND PARENTAGE

Syrah's most famous and historic home is in the vineyards of the northern Rhône, such as Hermitage and Côte Rôtie. It was first mentioned by Faujas de Saint-Fond (1781) under various spellings: 'la Sira de l'Hermitage ... produit un vin agréable, généreux, stomachique et qui gagne en vieillissant: l'on peut ... y mêler une petite quantité de raisins blancs, ainsi qu'on le pratique à Tain. La Serine et le Vionnier de Côte-Rôtie seroient très-bien aussi' ('Sira from Hermitage ... gives a nice, generous, appetizing and ageworthy wine: we can ... blend a small quantity of white grapes, as is done in Tain. Serine and Vionnier from Côte Rôtie would be suitable as well'). Sira de l'Hermitage and Serine from Côte Rôtie are simply local variations of Syrah (the variety and the name) although Saint-Fond clearly considered them to be different varieties. The name Syrah could derive, via its synonym Sérine, from the Indo-European root **ser-* indicating a long period, which gave the Latin *serus*, meaning 'late ripening' (André and Levadoux 1964).

Syrah begat DURIF through a natural crossing with PELOURSIN.

Zizak is occasionally used as a synonym for Syrah although it is also the name of a distinct Montenegrin variety (see ZIZAK).

SYRAH'S PARENTS UNCOVERED In 1998, DNA parentage analysis carried out at the University of California at Davis and INRA (Institut National de la Recherche Agronomique) in Montpellier, southern France, unexpectedly showed that Syrah is a natural cross between MONDEUSE BLANCHE (mother) and DUREZA (father; Bowers *et al.* 2000). Mondeuse Blanche is a Savoie variety that used also to be cultivated in the Ain, Isère and Haute-Savoie *départements*, and Dureza is an Ardèche variety that was also formerly cultivated, together with Syrah, in the Drôme and the Isère. The crossing that gave birth to Syrah had to take place in a vineyard where both parents were cultivated together, making it very likely that this happened in the French Rhône-Alpes region, probably in the Isère (Meredith and Boursiquot 2008).

PINOT, PROBABLE GREAT-GRANDPARENT OF SYRAH Using a type of probabilistic analysis of DNA data that had never before been applied to grape genetics, Vouillamoz and Grando (2006) at the Istituto Agrario di San Michele all'Adige in northern Italy have shown that DUREZA is a sibling of TEROLDEGO, an old variety from Trentino in northern Italy. This was the first evidence of a close genetic link between two varieties on different sides of the Alps. Since Syrah is a progeny of Dureza, it is therefore a nephew/niece of Teroldego. Vouillamoz and Grando have also detected a second-degree genetic relationship between PINOT and both Dureza and Teroldego, which means that Pinot could be their grandparent, grandchild, uncle/aunt, nephew/niece or half-sibling. Since Pinot was already known in France and in the Tyrol in the fourteenth century, its cultivation predates that of both Dureza and Teroldego and it is logical to consider Pinot as their ancestor, either their grandparent (illustrated in diagram 1 overleaf) or their uncle/aunt. Therefore, Pinot is a very likely great-grandparent of Syrah, which challenges the supposition that they have completely different origins.

MONDEUSE NOIRE AND VIOGNIER IN THE SYRAH PEDIGREE DNA parentage analysis has recently shown that MONDEUSE BLANCHE from Savoie has a parent–offspring relationship with both MONDEUSE NOIRE from Savoie/Haute-Savoie and VIOGNIER from the northern Vallée du Rhône (Vouillamoz 2008), which means that Mondeuse Noire and Viognier are either parents or offspring of Mondeuse Blanche. However, in the absence of the other parent, it is impossible to know which is the parent and which the offspring. As a consequence, three pedigrees are equally possible, as shown in diagram 2 opposite.

If MONDEUSE NOIRE is a parent of MONDEUSE BLANCHE through a natural cross-pollination with an unknown and probably extinct variety, then Mondeuse Noire must be a grandparent of Syrah and VIOGNIER (option *A* opposite), because the relationship Mondeuse

SYRAH PEDIGREE DIAGRAM 1

The parents of SYRAH were discovered by DNA parentage analysis in 1998. Close and unexpected genetic links with the Italian TEROLDEGO and the Burgundian PINOT were detected in 2006.

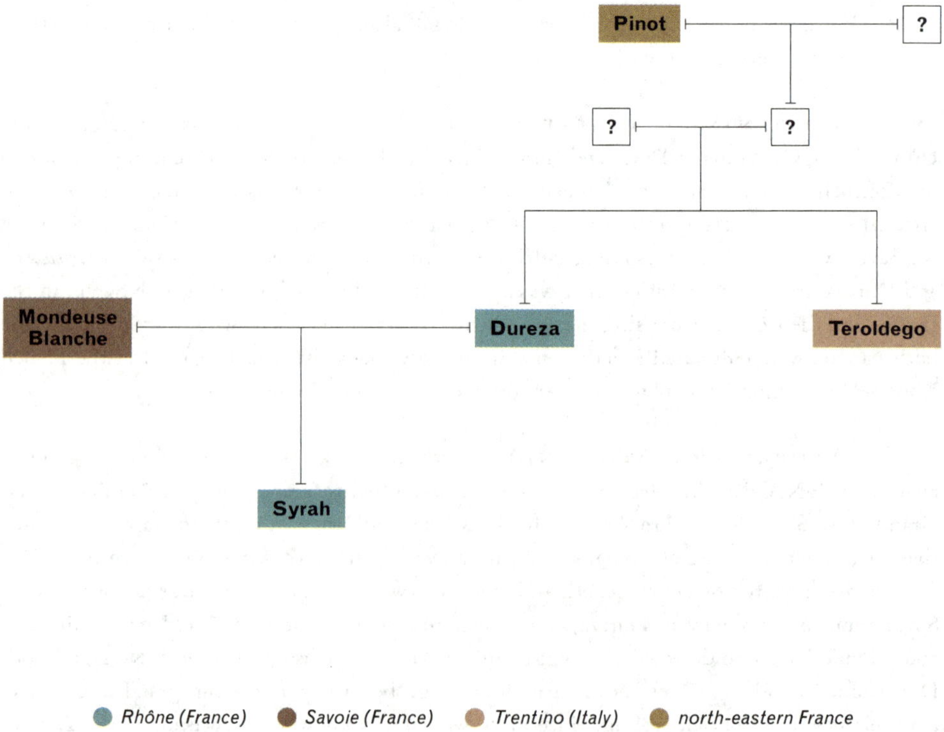

SYRAH PEDIGREE DIAGRAM 2

Parent–offspring relationships between MONDEUSE BLANCHE and MONDEUSE NOIRE and, more unexpectedly, between Mondeuse Blanche and VIOGNIER were discovered by DNA parentage analysis in 2008. Given the connections via unknown (and probably extinct) varieties (?), three genetically equivalent pedigrees could be drawn to explain these relationships.

Option A

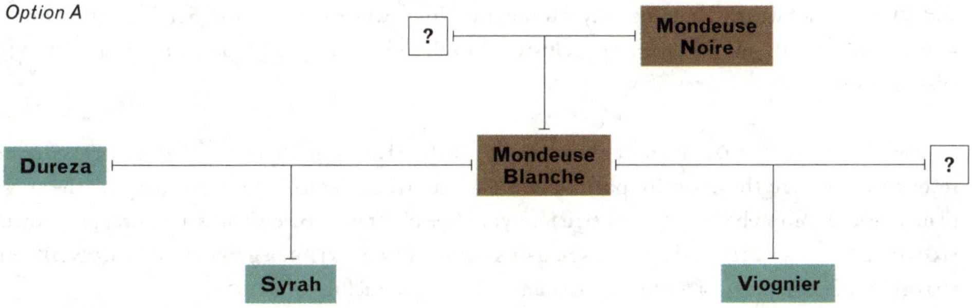

Option B

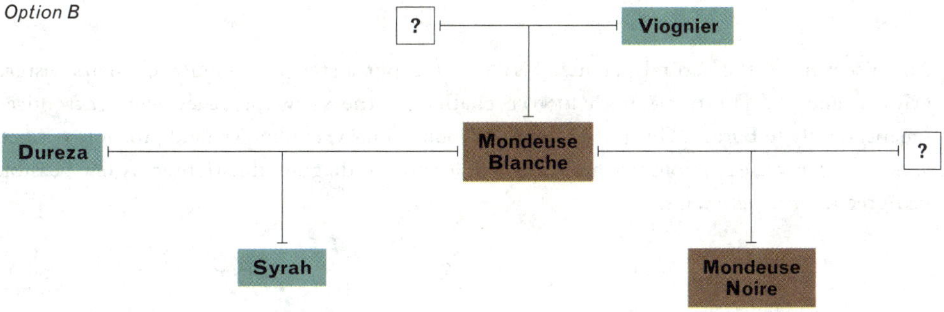

Option C

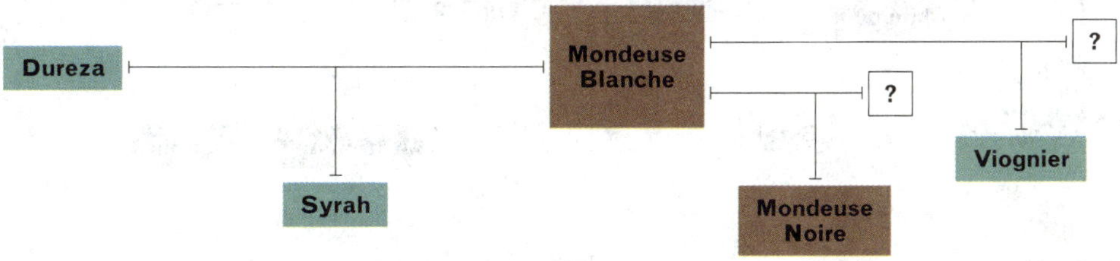

● *Rhône* ● *Savoie*

Blanche = Mondeuse Noire × Viognier is rejected by DNA analysis. In this case, Viognier is a progeny of Mondeuse Blanche and a half-sibling of Syrah.

Conversely, if MONDEUSE NOIRE is a progeny of MONDEUSE BLANCHE through a natural cross-pollination with an unknown and probably extinct variety, then Mondeuse Noire is a half-sibling of Syrah (option *B* on p 1025). In this case, VIOGNIER may either be the parent of Mondeuse Blanche, and a grandparent of Syrah and Mondeuse Noire, or Viognier could be a progeny of Mondeuse Blanche as well, and thus a half-sibling of both Mondeuse Noire and Syrah (option *C*), the other parents of Viognier and Mondeuse Noire being distinct, unknown and probably extinct.

As a consequence, Syrah is either a grandchild or a half-sibling of both MONDEUSE NOIRE and VIOGNIER, which explains why Mondeuse Noire was called Grosse Syrah in the Drôme region, and why all four varieties were clustered in the Sérine ampelographic group (see p XXVII; Bisson 2009).

SYRAH'S MOST COMPLETE FAMILY TREE Keeping in mind that other pedigree reconstructions are theoretically possible for the reasons explained above, we propose the most plausible and comprehensive family tree for Syrah (see diagram 3 below): it is a natural progeny of MONDEUSE BLANCHE and DUREZA, a half-sibling of VIOGNIER, a grandchild of MONDEUSE NOIRE, a niece/nephew of TEROLDEGO and a great-grandchild of PINOT.

SYRAH PEDIGREE DIAGRAM 3

The discovery of the natural parents of SYRAH has put a stop to rumours about its eastern origins, and the family tie with PINOT challenges the view that they were introduced independently to Europe. Given the numerous connections via unknown (and probably extinct) varieties (?), inverse relationships are also possible and this diagram illustrates only one possible pedigree reconstruction (see p X).

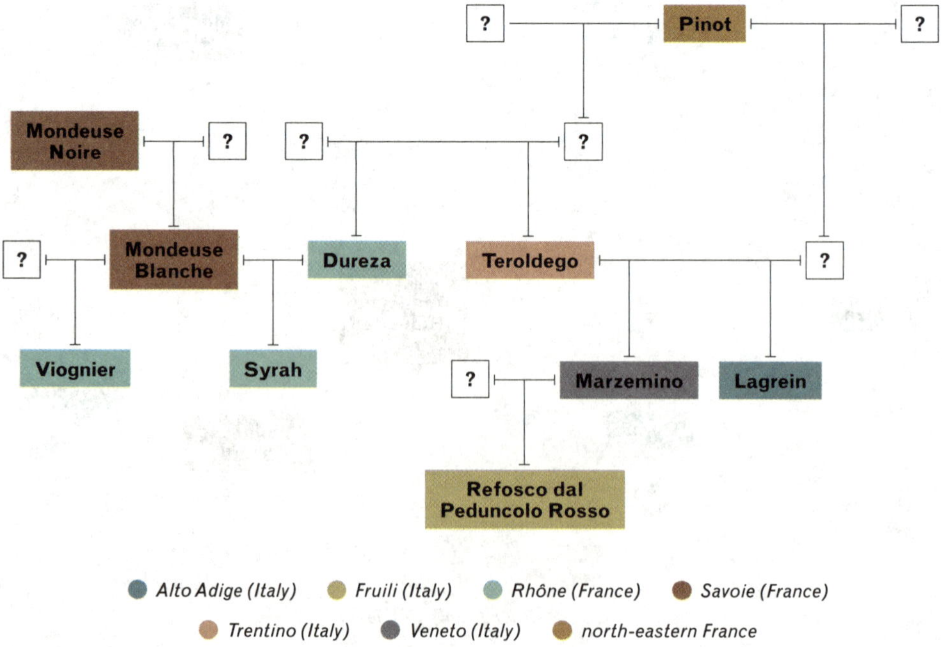

The genetic links discovered by paternity testing strongly anchor the origins of Syrah in the French Rhône-Alpes region.

OTHER HYPOTHESES

Several hypotheses have been proposed for the origin of Syrah: Roman, Syrian, Greek, Sicilian, Albanian, etc.

ALLOBROGICA Mentioned in the first century AD by Celsus, Columella and Pliny the Elder, Allobrogica was the name of a grape variety cultivated in the land of the Allobroges that stretched from Lac Léman (Lake Geneva) to Grenoble, and from the Haute-Savoie to Vienne (south of Lyon), where it was used for *picatum* (Latin for 'sticky', 'pitchy'), a resinated wine made near Vienne, the location of Côte Rôtie today. For Pliny the Elder, Allobrogica was a late-ripening, black-berried grape cultivated in cold lands. Allobrogica has been identified with PINOT by some authors, even though this variety has never been extensively cultivated in the land of the Allobroges, and by some others with Syrah or with MONDEUSE NOIRE, because their distribution in the nineteenth century perfectly matched the cultivation area of Allobrogica.

In the nineteenth century, MONDEUSE NOIRE was cultivated all over the land of the Allobroges, encompassing the entire distribution of Syrah.

However, linguist Jacques André and ampelographer Louis Levadoux suggested an alternative hypothesis: in Roman times, Allobrogica consisted of a population of 'proto-Mondeuse' that gave birth to MONDEUSE NOIRE and Syrah, both being cultivated together for a long time (André and Levadoux 1964). Their proto-Mondeuse hypothesis could also explain the etymology of Syrah or Sérine from the Latin *serus*, meaning 'late ripening' (see above), which can be said of both varieties. DNA analysis of grape pips found in archaeological sites in the land of the Allobroges could shed some light on the identity of proto-Mondeuse, which could regroup the ancestor(s) of the Sérine ampelographic group.

PLINY THE ELDER'S SYRIACA, A SYRIAN ORIGIN FOR SYRAH? For Pliny the Elder, Syriaca is a black version of the Aminea grape that grows in Syria, and his description was then used to propose that Syriaca could be an ancestor of Syrah.

SHIRAZ, THE PERSIAN LEGEND Shiraz or Chiraz was an important wine centre in ancient Persia (today's Iran). It has often been speculated that Syrah was introduced from Persia to Marseille (*Massilia*) by the Phocaeans around 600 BC, or from Persia to the Vallée du Rhône by the Crusaders between 1095 and 1291. Since Shiraz is the Australian name for Syrah, some authors even argued that the Australians had maintained the original name while the French had Frenchified it. However, the Shiraz hypothesis is doubtful, not least because the Crusades were mainly focused on the Holy Land and did not go as far as Persia.

SÝRA, THE GREEK ISLAND Sýros, also named Sýra, is a Greek island in the Kykládes (Cyclades) that has been proposed as the homeland of Syrah. However, no Syrah grape has ever been observed on this island.

SYRACUSE, THE SICILIAN TRACK Agathocles (361–289 BC), tyrant of Syracuse on the Italian island of Sicilia, is said to have initially brought Syrah to Syracuse from Egypt in 310 BC. The Roman emperor Marcus Aurelius Probus (AD 232–82), regarded as the saviour of viticulture because he abolished the edict of Domitian (AD 51–96) that required the uprooting of half the vineyards outside Italy, is thought to have brought back Syrah from Syracuse when he took control of Lyon (*Lugdunum*) in AD 281.

SERINA E ZEZË AND SHESH I ZI, TWO ALBANIAN VARIETIES SUPPOSEDLY RELATED TO SYRAH Some Albanian farmers suggest that SERINA E ZEZË, an indigenous Albanian red grape, is identical to Syrah, simply because its name is similar to Sérine, the old name for Syrah in the Côte Rôtie. However, the DNA profile of Serina e Zezë that was established by Greek-Albanian researchers Ladoukakis *et al.* (2005) shows no relation to Syrah, although it could be close or identical to Eftakoilo or Heftakilo, an old black-berried Greek table grape (Vouillamoz). Italian researchers also recently claimed to have shown by DNA analysis that Syrah is genetically close to another Albanian grape variety called SHESH I ZI, which is again rejected by DNA profiling (Vouillamoz)

In fact, thanks to DNA studies, we now know that the Syrian, Persian, Greek, Sicilian and Albanian hypotheses for the origin of Syrah are wrong.

A recent DNA study using newly developed genetic markers has suggested that Syrah and VIOGNIER are siblings (Myles *et al.* 2011) instead of half-siblings or grandparent and grandchild as above, but this would mean that Viognier is also a natural progeny of DUREZA and MONDEUSE BLANCHE, which can be rejected by DNA parentage analysis based on microsatellite markers (Vouillamoz).

Syrah was used to breed RUBIN.

VITICULTURAL CHARACTERISTICS

Vigorous. Mid ripening with a short ripening period between veraison and harvest, plus a short window for optimum harvesting. Needs careful trellising and training to protect it from the wind in spring. Very susceptible to chlorosis and unsuited to soils with a high active lime content (and should not be grafted onto 110 R rootstocks). Susceptible to mites and botrytis bunch rot, especially near harvest. Also susceptible to a disease of unknown cause described variously as Syrah/Shiraz disease/decline/disorder and found in many parts of the world, where leaves turn red, swelling and cracks appear at the graft point and vines eventually die – some clones appear more susceptible than others. Berries are small and tend to shrivel quickly once ripe.

WHERE IT'S GROWN AND WHAT ITS WINE TASTES LIKE

Until the late twentieth century, Syrah was grown chiefly in the Vallée du Rhône and, as Shiraz, in Australia. But a combination of a surge in Australian wine exports, a fashion for all things Rhôneish and a certain ennui with Bordeaux red wine grapes combined to encourage widespread planting of this variety. Its wines have a very different profile from those of the Cabernet family. Their tannins are in general much gentler and there is more obvious weight on the mid palate, especially if, as is often the case, Syrah is blended with other Rhône and southern French varieties such as Grenache (GARNACHA) and Mourvèdre (MONASTRELL). Syrah's flavours tend to be in the leather, liquorice and tar spectrum with marked black pepper or even burnt-rubber aromas in slightly underripe examples but much sweeter black-fruit flavours in Syrah picked fully ripe in warm climates. Wines made from very ripe to overripe (and therefore thoroughly shrivelled) berries can have flavours of dark chocolate and prunes, sometimes with porty overtones.

Syrah plantings have increased quite staggeringly in **France** in the last fifty years: from 1,602 ha (3,959 acres) in 1958 to 68,587 ha (169,482 acres) in 2009, making it the third most planted red wine grape overall after MERLOT then Grenache (GARNACHA). It is planted throughout south and south-eastern France (it would have difficulty ripening reliably north of Lyon) with the greatest total area in Languedoc-Roussillon (43,163 ha/106,658 acres), not least because in recent years, as the EU has offered uprooting and replanting incentives to tackle its wine surplus in southern France, Syrah has been considered the first-choice *cépage améliorateur* (improving variety) to replace the poor-quality, high-yielding varieties such as ARAMON NOIR, ALICANTE HENRI BOUSCHET and Carignan (MAZUELO) planted in the first half of the twentieth century and to add structure, complexity and longevity to blends often based on Grenache. Throughout Languedoc-Roussillon, Syrah is most often encountered in blends. In the southern Rhône, Syrah is the second most planted variety, on about a third as much vineyard as Grenache, and is generally used to add backbone and longevity to blends such as Châteauneuf-du-Pape and similar wines. Provençal vineyards to the east have a long tradition of growing Syrah and, often, blending it with CABERNET SAUVIGNON (a blend also long sanctioned in Australia).

It is in the northern Rhône that French Syrah shines as a mono-varietal wine. The magnificently concentrated, long-lived reds grown on the granite rock of Hermitage represent the most 'manly' aspect of Syrah, varying considerably according to precise soil type and aspect but usually with a notably glossy texture and luscious but bone-dry fruit. Crozes-Hermitage is a lesser, earlier-maturing version made on lower ground around the hill of Hermitage. The steep, south-east-facing terraces of Côte Rôtie represent almost the northern limit of Syrah cultivation and, not just because a small proportion of VIOGNIER is often co-fermented with the Syrah, this is stereotypically Syrah at its purest, most refreshing, lighter and more 'feminine' (although Guigal's single-vineyard trophy wines are exceptionally concentrated versions). Saint-Joseph's terrain between Côte Rôtie and Hermitage is too varied to allow generalizations.

Syrah has been planted in **Italy** since 1899, when it was introduced from Montpellier to Piemonte, but total plantings were still only 1,039 ha (2,567 acres) by 2000. Italian wine growers are generally much more besotted by Bordeaux red wine grapes but the most successful area for Syrah so far has been Cortona in southern Toscana, although several appellations on the island of Sicilia – eg Alcamo, Erice, Menfi – allow varietal Syrah wines, so it is perhaps not surprising that 5,357 ha (13,237 acres) were recorded there in 2008. Recommended producers of varietal Syrah in Italy include Gillardi in Piemonte; Isole e Olena and Fontodi in the heart of the Chianti region; La Braccesca, Ruffino and Tenimenti d'Alessandro in Cortona; and Cottanera and Planeta on Sicilia.

Wine producers in **Spain**, on the other hand, have recently been planting Syrah so enthusiastically that the national total rose from under 3,000 ha (7,413 acres) in 2004 to 16,568 ha (40,940 acres) by 2008. Most Syrah vines are in Castilla-La Mancha in the centre of the country but there are also significant areas in Aragón and Catalunya in the north east, Murcia in the south east and Extremadura in the west. Carlos Falcó, Marqués de Griñon, was one of the earliest Spanish exponents at Dominio de Valdepusa near Toledo and others who have produced Syrah of seriously notable quality, only rarely as a varietal, include Abadía Retuerta, Castell d'Encus (Thalarn), Finca Sandoval and Pago del Ama. The wines tend to be notably sweeter and plumper than French Syrahs.

Over the border in **Portugal**, Syrah is planted to no enormous extent and mainly in the Alentejo and also in the Tejo and Lisboa regions, with experimental plantings in the Douro. Recommended producers include Cortes de Cima, who pioneered it in the Alentejo, long before it was officially permitted, plus Quinta Lagoalva de Cima and Quinta da Lapa in the Tejo.

Switzerland, and especially the sunny slopes in the upper reaches of the Vallée du Rhône in the Valais, produces unexpectedly concentrated wine from mature vines. Recommended producers include Jean-René Germanier, Simon Maye & Fils, and L'Orpailleur. There was a total of just 181 ha (447 acres) in 2009.

Syrah, often labelled Shiraz in markets where Australian wine is familiar, is also found in many other European and Mediterranean wine-producing countries such as **Croatia** (110 ha/272 acres in 2009), **Romania** (16.4 ha/41 acres in 2008), **Hungary** (100 ha/247 acres in 2008), **Malta** (100 ha/247 acres in 2010), **Cyprus** (257 ha/635 acres in 2010) and **Turkey**, where there was quite a sizeable area – 1,489 ha (3,679 acres) – in 2010. **Greece** grows a limited amount of Syrah but the likes of Alpha Estate, Gerovassiliou and Spiropoulos have demonstrated that it can produce both fine varietals and the odd accomplished blend in different climates here. Syrah is also reckoned to be the second most planted red wine grape in **Lebanon**, and **Israel** had 350 ha (865 acres) in 2009. The variety likes warmth.

One would therefore expect **California** to be an enthusiastic grower of Syrah but in fact the 2010 state total of 19,283 acres (7,803 ha) was dwarfed by the more than 37,290 acres (15,091 ha) devoted to the much more fragile, supposedly cool-climate variety PINOT NOIR. Fashion rules the Golden State, but the worldwide fashion for Syrah has eluded Californians – or perhaps they just got there first. Under the noisy influence of a California movement known as the Rhône Rangers, total plantings in the decade to 2003 rose from 400 to 17,000 acres (162 to 6,880 ha), but since then there have been remarkably few new plantings. It is said that the American consumer is bemused by Syrah, doesn't really know what to expect. The advent of the *Sideways*/Pinot Noir phenomenon and a sudden rise and fall in the reputation of Australian Shiraz have combined to diminish the lustre of Syrah in California. Syrah was a major beneficiary of the new Central Coast plantings and the county with the most is San Luis Obispo with 2,770 acres (1,121 ha). Many ambitious, relatively new wine producers here are devoted to the variety and to blending it with other Rhône grapes, but often in tiny quantities. Some of the more obvious fine Syrah producers are Alban (Edna Valley), Cline (Carneros), DuMol and Pax Walker (Russian River Valley), Swanson (Napa), Lagier Meredith (Mount Veeder), Terre Rouge (Sierra Foothills), Baileyana (Edna Valley), Favia (Sierra Foothills) and Sean Thackrey's ingenious blends.

Washington State winemakers on the other hand are firmly in the grip of Rhône fever, which resulted in a dramatic increase in plantings of Syrah in recent years to a total of 3,103 acres (1,256 ha) by 2011, making it the third most planted red variety after CABERNET SAUVIGNON and MERLOT. The variety responds well to the famously bright-fruited style of the state and some of those who have already made a name for their Syrah (there are many more in the wings) are Betz Family, Cayuse, L'Ecole No 41, Hedges Family, McCrea Cellars and Charles Smith of K Vintners.

Oregon's climate is less obviously suitable for Syrah but the state had 572 acres (231 ha) planted by 2008 and the likes of Domaine Serene and newcomer Gramercy Cellars certainly show potential for the variety here. A little is also grown in **Idaho** and in **Arizona**, **Texas** and **Colorado.** Tedeschi Vineyard on the **Hawaiian** island of Maui have planted several acres of Syrah, mostly to replace the inferior CARNELIAN. **Mexico** has plantings of Syrah in Baja California and Coahulia de Zaragoza.

In **Chile**, Syrah is currently regarded with considerable hope and ambition. It was first planted there at the end of the nineteenth century but was later eradicated in the agrarian reform of the 1970s. Syrah began its second life in Chile in the mid 1990s, a trend led by Errázuriz in Aconcagua, although there is some debate about who actually planted first (Richards 2010). Thanks to the recent rush to develop this variety, plantings had already reached 3,370 ha (8,327 acres) by 2008, the vast majority in Colchagua although there have more recently been encouraging results in the more northerly and cooler San Antonio, Limarí and Elqui regions from the likes of Casa Marin, Falernia, Matetic, Maycas del Limarí and Viña Leyda. In Cachapoal, Altaïr incorporate Syrah in their impressive red blends and top-notch varietal wines are produced by the likes of Lapostolle and Santa Rita; in Colchagua, Montes, Polkura and Viu Manent have had good results.

Argentina has more Syrah planted than Chile: 12,960 ha (32,025 acres) in 2008, the majority in Mendoza, up from fewer than 1,000 ha (2,500 acres) in 1990. As in Chile, most producers call their wine Syrah rather than Shiraz and the variety is an important component in many top red blends such as Cuvelier Los Andes and Dominio del Plata as well as delivering a range of styles on its own in the hands of, for example, O Fournier. But since the Argentines already have their very own hugely successful alternative to Bordeaux grapes in Malbec (COT), there is less impetus to focus on Syrah than across the Andes. That said, San Juan has shown a particular aptitude for the north Rhône grape, which is also found in **Peru**, **Bolivia** and **Uruguay**.

Of all Syrah's many homes outside France, **Australia** is by far the most important, and has the second largest plantings after France of the variety, here called Shiraz, although when it was first taken to the Antipodes, probably from Montpellier by James Busby in 1832, it was known as Scyras. It flourished and was eagerly adopted in New South Wales and then spread quickly to other wine-growing states. It fell briefly out of favour in the late twentieth century when 'French' CABERNET SAUVIGNON seemed more exotic and glamorous but today Shiraz is the most widely planted variety in the country, with 43,977 ha (108,670 acres) in 2008 – almost 45% of the entire red wine crush and 24% of the total grape crush. Much of it is planted in the inland regions such as the Riverland, Murray Darling and Riverina, where it is heavily irrigated and makes oceans of inexpensive, sweet, lightweight red wine that have done little for the image of Australian wine in recent years.

Much more serious and certainly more concentrated wines are made in the Barossa Valley, where 5,281 ha (13,050 acres) of the variety include many of Australia's oldest vines, some of them centenarians producing extremely rich, long-lived, spicy nectars such as Penfolds Grange, Australia's most famous wine – although Henschke's Hill of Grace Shiraz, also made from some of the oldest vines in the world, grown in the Eden Valley, is a serious rival. Tar and chocolate are the predominant flavours, while the Shiraz of McLaren Vale (2,882 ha/7,122 acres) is rather more open and breezy. Langhorne Creek grows 2,097 ha (5,182 acres) of easy, spicy Shiraz but its name is relatively rarely seen on labels; Australian wine producers are inveterate blenders. It makes finer textured wines in the Clare Valley (1,614 ha/3,988 acres), cooler and tarter examples in Padthaway (1,470 ha/3,632 acres) and Coonawarra (1,241 ha/3,067 acres) in South Australia, and has its own very particular earthy style in the Hunter Valley (1,001 ha/2,474 acres) in New South Wales. There are also significant plantings in Western Australia, Queensland and Canberra District, where Clonakilla pioneered a particularly Rhône-like style of wine, introducing co-fermentation with VIOGNIER, which became a rather exaggerated trend at one stage. Virtually every Australian producer worth their salt, except those in the coolest reaches of the country, makes a worthwhile Shiraz or possibly several. Cabernet/Shiraz blends have also long been popular in Australia.

New Zealand is generally too cool for the heat-loving Syrah (there were just 278 ha/687 acres in 2008) but Hawke's Bay in the North Island has already provided some delicious exceptions to this rule, notably made by Bilancia, Craggy Range, Sacred Hill, Te Mata and Trinity Hill. The wines are based on a single clone imported from Australia and rescued from the Te Kauwhata viticultural research station by Alan Limmer of Stonecroft. Martinborough also seems to suit this variety, as Dry River, Ata Rangi and Kusuda have demonstrated. Almost all examples in New Zealand are labelled Syrah rather than Shiraz, to emphasize its cooler climate presumably.

In **South Africa**, the name Shiraz is more widely used, perhaps inspired by envy of Australia's formidable track record as a wine exporter, although two styles of the variety exist. Shiraz generally denotes the dominant full-throttle, heavily oaked style that is more Australian than French, although there are notable exceptions such as Boekenhoutskloof, The Foundry, Haskell and Lammershoek, all of them highly successful Cape renditions of a lighter, more savoury style (labelled Syrah). There were 9,907 ha (24,481 acres) of the variety in 2008, quite widely dispersed and representing nearly 10% of the total vineyard area, a huge increase since 1995, when it represented just 1%. Virus-free clones have made a big difference to the quality of the vines but there is still a major problem with what has been labelled Shiraz disease, a fatal, graft-transmissible viral disease specific to South Africa that also affects MERLOT, GAMAY NOIR and Malbec (COT) and some white varieties. Research is still underway but it is thought to be caused by several different grapevine viruses and spread by mealybugs. It is different from what is known as Syrah decline in France and Syrah/Shiraz disorder in California. These appear to be connected to specific clones and do not spread from vine to vine; the symptoms start at the graft point and the cause is as yet unknown.

As well as producing varietal wines, the variety plays a major part in many excellent blends with, for example, Grenache (GARNACHA) and Mourvèdre (MONASTRELL), particularly in the Swartland, where top producers include Eben Sadie, Mullineux and Lammershoek. In the bigger, oakier style of varietal wines some of the more successful performers include Simonsig, Haskell, Hartenberg, De Trafford and Saxenburg. Good examples from Stellenbosch include Quoin Rock and Waterkloof; cooler-climate styles are produced by, for example, Eagles' Nest (Constantia), Julien Schaal (Elgin) and Luddite (Walker Bay). Solms-Delta also make sweetish, potent wines from desiccated Syrah grapes.

China grew 195 ha (482 acres) of Syrah in 2009 and it was also known in **Thailand**.

SYRIKI

Vine variety grown spottily around Greece but rarely seen on labels.

BERRY COLOUR: ● ● ● ● ●

PRINCIPAL SYNONYMS: Cserichi, Kserikhi, Kseriki (Kýthira/Cythera), Kserukhti, Seriki, Syrike, Xerichi (Zákynthos), Xerichi Mavro, Xeruchti, Xirichi

ORIGINS AND PARENTAGE

Although Syriki is found in many parts of Greece, its exact origin is unknown.

WHERE IT'S GROWN

Syriki is grown on many of the islands of **Greece** – the Ionian Islands, the southern Kykládes (Cyclades), Kérkyra (Corfu), Kríti (Crete), Évvoia (Euboea) – as well as in the Pelopónnisos (Peloponnese) and in southern and western mainland Greece (Ladoukakis *et al.* 2005). On the island of Ikaría, between Sámos and Mýkonos, Afianes, for example, blend Syriki with other local varieties FOKIANO and KORIOSTAFYLO.

SYSAK

Also known as Sisak or Sysack, this black-berried, mid- to late-ripening variety from **Azerbaijan** is used for grape juice and to make dry and dessert wines, as well as fizz and a cognac-like spirit. Its functionally female flowers distinguish it from the ZIZAK of Montenegro.

SZÜRKEBARÁT

See PINOT GRIS in the PINOT entry.

T

TAMAREZ
TAMINGA
TAMURRO
TANNAT
TARRANGO
TATLY
TAUBERSCHWARZ
TAVKVERI
TAZZELENGHE
TECA
TEINTURIER
TEMPARIA
TEMPRANILLO
TÉOULIER NOIR
TERBASH
TERMARINA ROSSA
TEROLDEGO
TERRANO
TERRANTEZ
TERRANTEZ DA TERCEIRA
TERRANTEZ DO PICO
TERRET
THEIAKO MAVRO
THRAPSATHIRI
TIBOUREN
TIGNOLINO
TIGRANI
TIMORASSO
TINTA BARROCA
TINTA CARVALHA
TINTA CASTAÑAL
TINTA FRANCISCA
TINTILIA DEL MOLISE
TINTO CÃO
TINTO VELASCO
TINTORE DI TRAMONTI
TORBATO
TORRONTÉS MENDOCINO
TORRONTÉS RIOJANO
TORRONTÉS SANJUANINO
TORYSA
TOURIGA FÊMEA
TOURIGA FRANCA
TOURIGA NACIONAL
TOURKOPOULA
TRAJADURA
TRAMINETTE
TRBLJAN
TREBBIANO D'ABRUZZO
TREBBIANO GIALLO
TREBBIANO MODENESE
TREBBIANO ROMAGNOLO
TREBBIANO SPOLETINO
TREBBIANO TOSCANO
TREPAT
TREPAT BLANC
TRESSOT
TREVISANA NERA
TRIBIDRAG
TRINCADEIRA
TRINCADEIRA DAS PRATAS
TRIOMPHE
TRNJAK
TRONTO
TROUSSEAU
TSAOUSSI
TSIMLADAR
TSIMLYANSKY CHERNY
TSITSKA
TSOLIKOURI
TSULUKIDZIS TETRA
TUO XIAN
TURÁN
TYRIAN

☿this symbol on the following pages indicates that the synonyms, and mistaken identities, are confirmed by DNA evidence

TÁLIA

See TREBBIANO TOSCANO.

TĂMÂIOASĂ ROMÂNEASCĂ

See MUSCAT BLANC À PETITS GRAINS.

TAMAREZ

Old southern Portuguese variety authorized in many wine regions but usually distilled or blended.

BERRY COLOUR: ● ● ● ● ●

PRINCIPAL SYNONYMS: Arinto Gordo (Dão), Boal Prior, Tamares, Tamarez Branco

VARIETIES COMMONLY MISTAKEN FOR TAMAREZ: ARINTO DE BUCELAS, DONA BRANCA, SÍRIA, TRINCADEIRA DAS PRATAS (Alentejo and Setúbal)

ORIGINS AND PARENTAGE

Tamarez was first mentioned in 1712 by Vicencio Alarte in southern Portugal (Galet 2000). However, Tamarez is a rather confusing name that has been used for several distinct varieties in Portugal, so that care should be taken about the true identity of Tamarez, depending on where it's grown, especially in the Alentejo, where it is often confused with TRINCADEIRA DAS PRATAS even though they are ampelographically quite distinct.

VITICULTURAL CHARACTERISTICS

Late ripening. Compact bunches of small berries. Resistant to powdery mildew.

WHERE IT'S GROWN

Tamarez is found mainly in the Alentejo, where the Borba co-op, for example, use it in various white blends with other local varieties such as Roupeiro (SÍRIA) and ANTÃO VAZ. Although it is widely authorized for regional and DOC wines from the Algarve in the south to the Douro in the north, much of it ends up distilled. There were 602 ha (1,488 acres) in **Portugal** in 2010.

TAMINGA

Grapily aromatic Australian white cross bred for the heat.

BERRY COLOUR: ● ● ● ● ●

ORIGINS AND PARENTAGE

A Merbein 29-56 × SAVAGNIN ROSE cross, in which Merbein 29-56 is a DAMASCHINO × SULTANIYE cross, created in 1970 by CSIRO's A J Antcliff at their Merbein station in Victoria, Australia; released in 1982.

VITICULTURAL CHARACTERISTICS

High yielding, late ripening (late February to early March in the Murray Valley) and bred specifically for warm Australian conditions.

WHERE IT'S GROWN AND WHAT ITS WINE TASTES LIKE

In **Australia** by 2010 there were seven growers of this variety, planted mainly in the Murray Valley in South Australia and New South Wales, where it produces a highly aromatic, grapey wine with good acidity. Production is very small scale though Taminga can contribute aromatic interest to otherwise dull white blends. Trentham Estates make a high-quality lightly botrytized sweet version on the banks of the Murray River in NSW.

TAMURRO

Virtually extinct, thick-skinned Basilicata variety.

BERRY COLOUR: ● ● ● ● ●

PRINCIPAL SYNONYMS: Coll d'Tammurr

ORIGINS AND PARENTAGE

Tamurro was formerly cultivated in the region of Pietragalla in the province of Potenza in Basilicata, southern Italy, and was thought to be extinct until it was recently rediscovered around the town of Barile in the Vulture area between Bari and Napoli, where local growers called it Coll d'Tammurr, a name already in use in the Middle Ages when the Duca Filiberto di Savoia conquered Vulture. It is said that the duke was tired of the powerful, rich AGLIANICO wines and ordered the introduction of a more gentle variety from France. One of the first to plant this variety was a robust local grower with a neck like a bull, nicknamed Coll d'Tammurr, the local dialect for *collo di tamburo*, meaning 'drum neck', which gave the name to the variety. DNA profiling is needed to determine whether it is the same as an existing French grape variety.

VITICULTURAL CHARACTERISTICS

Late ripening. Rich in polyphenols. Thick skins increase its resistance to botrytis bunch rot but it is susceptible to powdery mildew.

WHERE IT'S GROWN AND WHAT ITS WINE TASTES LIKE

Tamurro used to be harvested overripe and blended with other local varieties to add flesh and colour. It had been effectively forgotten until it was rediscovered by Tenuta Le Querce in Barile, Vulture, **Italy**. They were the first to produce a varietal wine, under the modernized Italian name Tamurro and with an ambitious price tag.

TANNAT

Powerful, characterful, often tannic variety at home in south-west France and Uruguay but becoming global.

BERRY COLOUR:

PRINCIPAL SYNONYMS: Bordelez Beltza (País Vasco in Spain), Harriague (Uruguay), Madiran (Vallée de l'Adour), Moustrou or Moustroun (Landes), Tanat

ORIGINS AND PARENTAGE

Tannat was first mentioned in 1783–4 in Madiran in the Hautes-Pyrénées in south-west France, its most likely origin and where it is still today the dominant variety (Rézeau 1997): 'le tanat est un raisin noir ...' ('tanat is a black grape ...'). The modern spelling with double 'n' appeared in 1827 (Rézeau 1997): 'le vin de Madiran est produit par le tannat' ('the wine of Madiran is produced from tannat'). The name Tannat probably derives from the Béarn dialect *tanat*, meaning 'coloured like tan', in reference to the dark colour of the berries and also possibly to their high tannin content.

According to Lavignac (2001), Tannat could be the progeny of an undetermined variety from the Pyrénées, and it is also morphologically close to BAROQUE from the Landes and to LAUZET from the Béarn in the Pyrénées-Atlantiques, two varieties that belong, like Tannat, to the Courbu ampelographic group (see p XXVII), together with COURBU BLANC, CROUCHEN, MANSENG NOIR and PETIT COURBU (Bisson 2009). This grouping is consistent with recent parentage analysis based on twenty-two DNA markers suggesting a possible parent–offspring relationship between Tannat and Manseng Noir (Vouillamoz; to be confirmed with more markers).

Tannat was used to breed ARINARNOA and EKIGAÏNA.

VITICULTURAL CHARACTERISTICS

Mid ripening, rather vigorous, generally pruned long and trellised. Susceptible to mites and leafhoppers. Some susceptibility to botrytis bunch rot. Big bunches but small to medium-sized berries

WHERE IT'S GROWN AND WHAT ITS WINE TASTES LIKE

Tannat wines are deeply coloured, tannic with marked acidity, often powerful and ageworthy, especially from vines grown on clay soils in its Madiran heartland in south-west **France**. In the past, they were often unbroachable in youth but many of the wines made today are much softer and more supple, less rustic, although most improve with age. Wines grown on sandy soils are generally more gentle in youth but still have a surprising Peter Pan quality. It comes as no surprise that the technique of micro-oxygenation, deliberately bubbling tiny amounts of oxygen into the nascent wine to soften the tannins, was developed in the Madiran region.

There were 2,914 ha (7,201 acres) of Tannat in France in 2009, the vast majority in the Gers but also in the neighbouring regions of south-west France. It is the principal ingredient in Madiran (40–80%) and Saint-Mont (minimum 60%), its principal blending partners being CABERNET FRANC and CABERNET SAUVIGNON. It also contributes to the Irouléguy, Béarn, Tursan, Côtes du Brulhois and Cahors appellations. Recommended producers include Alain Brumont and Château d'Aydie in Madiran and Producteurs Plaimont in Saint-Mont. Tannat's popularity has increased in the twenty-first century thanks to its high level of procyanidins, reported to be the source of red wine's health benefits (Corder 2006). The Gers region of France apparently has double the national average of nonagenarian men.

Italy has a little (45 ha/111 acres in 2000), mainly on Sicilia.

It is also found to a more limited extent in **California** (238 acres/96 ha in 2008), where

Tablas Creek in Paso Robles have been one of the pioneers of this variety. Some is also grown in Lodi. Other US states such as **Virginia** (just over 20 acres/8 ha planted very recently), **Illinois** and **Georgia** are also trying their hand at Tannat. British Columbia in **Canada** grows a little.

Tannat's second home, however, is in **Uruguay**, where it used to be called Harriague after the Frenchman from the Basque region who introduced the variety to Uruguay during a period of Basque immigration. Pascal Harriague planted a Tannat vineyard 400 km north of Montevideo, on the outskirts of the city of Salto, in 1870 (González-Techera *et al.* 2004). In 2009 it was the most widely planted, and proudly trumpeted, variety in the country, with 1,784 ha (4,408 acres) in total, almost 22% of the national vineyard area. Wines produced here tend to have slightly more rounded tannins thanks to the climate. Some are blended with softer varieties such as MERLOT or PINOT NOIR but the best retain the variety's hallmark freshness and ability to age. Recommended producers include Arerunguá, Bouza, Castillo Viejo, De Lucca, Juanicó, Pisano, Pizzorno and Stagnari. Tannat has spread across the border into several regions of **Argentina** (553 ha/1,366 acres in 2008) and into **Brazil** (421 ha/1,040 acres in 2007 in Rio Grande do Sul) and Tacama grow it in **Peru**.

A tiny amount is grown in **Australia** such as by Glenguin in the Hunter Valley and there is a tiny amount in **Switzerland**, **South Africa** and **Japan**.

TARRANGO

Light, fruity Australian red variety bred for hot, inland conditions.

BERRY COLOUR:

ORIGINS AND PARENTAGE

TOURIGA NACIONAL × SULTANIYE cross created by A J Antcliff at Merbein in Victoria, Australia, in 1965; released in 1975.

VITICULTURAL CHARACTERISTICS

Slow ripening but needs heat and water to mature fully, with relatively high acidity and a low tannin level.

WHERE IT'S GROWN AND WHAT ITS WINE TASTES LIKE

Plantings are restricted to the hot inland regions of the Murray-Darling basin in **Australia** and have started to decline from a small base: down from 172 ha (425 acres) in 2006 to 102 ha (252 acres) in 2008. Ripe grapes produce a light, soft, fresh, juicy red-fruited wine that may be lightly chilled. The most persistent and successful proponents are Brown Brothers of Milawa in Victoria. Indeed varietal Tarrango has been their biggest-selling wine in the UK.

TATLY

Local to the Gazakh region in the far north west of **Azerbaijan**, Tatly is a mid- to late-ripening variety used to make dry whites with relatively low alcohol, at the Moshu Winery, for example. Also used for grape juice and to make fortified and sparkling wines as well as brandy.

TAUBERSCHWARZ

Old variety used to produce light reds in southern Germany.

BERRY COLOUR:

PRINCIPAL SYNONYMS: Blaue Hartwegstraube, Blauer Hängling, Elsässer, Frankentraube, Grobrot, Grobschwarze (Württemberg), Häusler, Süssrot

ORIGINS AND PARENTAGE

Tauberschwarz ('black of Tauber') is one of the few old, indigenous varieties of Baden-Württemberg still in cultivation today, mainly in the Tauberfranken wine-growing region. Its earliest written mention dates back to 1726 in a decree of the bishopric of Würzburg during the reign of Graf Carl-Ludwig von Hohenlohe of Weikersheim: 'Tauber schwarzen Weinbergsfexern' ('cuttings of Tauber schwarzen'). The first reference to the name Tauberschwarz appeared in 1757 in *Fränkischen Sammlungen von Anmerkungen aus der Naturlehre*, a Nürnberg newspaper.

OTHER HYPOTHESES

Based on historical documents, ampelographic descriptions and botanical paintings, Jung (2007) claims that Tauberschwarz is the same as TRIBIDRAG (aka Zinfandel) but this uncertain identification needs to be verified by DNA profiling.

VITICULTURAL CHARACTERISTICS

Early ripening. Susceptible to downy mildew but winter-hardy. Thin skins and loose bunches.

WHERE IT'S GROWN AND WHAT ITS WINE TASTES LIKE

This old, dark-skinned variety was once widely planted in Baden and Württemberg (mainly in the Kocher, Jagst and Tauber valleys) in **Germany** and had almost disappeared in the early twentieth century, remaining only in the regions of Laudenbach (Vorbachtal) and Weikersheim (Taubertal).

In the 1960s, the Weinsberg research centre in Baden-Württemberg in southern Germany selected thirty cuttings from a vineyard in Ebertsbronn (Vorbachtal), which was probably the last and only old vineyard where Tauberschwarz had survived. They carefully studied and selected the best vines and Tauberschwarz was then authorized in Germany in 1994. Today there are only 15 ha (37 acres), cultivated by a small group of producers crafting light and lightly coloured wines, with the support of the Slow Food movement. Varietal wines are made by Hubert and Renate Benz, Engelhart, and Alois and Jürgen Hofmann, all in Franken, and by Schurk and Winzergenossenschaft Beckstein in Baden.

TAVKVERI

Georgian variety producing very small amounts of light, tangy reds.

BERRY COLOUR: ● ● ● ● ●

PRINCIPAL SYNONYMS: Takweri, Tarkveri, Tavaveri, Tavkeri Kartalinsky, Tavkeri Kartlis, Tavkveri Didmartsvala

ORIGINS AND PARENTAGE

Tavkveri is indigenous to the Kakheti and Kartli regions in eastern and central Georgia.

VITICULTURAL CHARACTERISTICS

Largish compact bunches of relatively large berries. Capricious at flowering and best planted next to CHINURI for pollination because the flowers are functionally female only. Late budding and ripening.

WHERE IT'S GROWN AND WHAT ITS WINE TASTES LIKE

Tavkveri produces light and very fresh, cherry-flavoured dry reds. It is planted in **Georgia**'s Kartli region, especially in moderately warm locations such as Gori and Tskhinvali, and Pheasant's Tears produce a version fermented in traditional clay qvevri, Chateau Mukhrani a more European style rosé. It is also found in **Azerbaijan**.

TAZZELENGHE

High-acid and potentially astringent red Friulian, rescued from extinction in the 1980s.

BERRY COLOUR: ● ● ● ● ●

PRINCIPAL SYNONYMS: Refosco del Botton⚥, Tacelenghe, Tassalinghe, Taze Lunghe, Tazzalenghe Nera, Tazzalenghe Nera Friulana, Tazzalingua

ORIGINS AND PARENTAGE

Tazzelenghe, literally 'tongue cutter' in reference to its elevated acidity, was first mentioned in Friuli under its dialect name Tacelenghe in 1823 in the *Catalogo delle varietà delle viti del Regno Veneto*, together with 126 other varieties (Fabbro *et al.* 2002).

DNA profiling has shown that Refosco del Botton is identical to Tazzelenghe and suggested a parent–offspring relationship with REFOSCO DI FAEDIS (Costacurta *et al.* 2005).

VITICULTURAL CHARACTERISTICS

Vigorous and prolific. Mid to late ripening but grapes are usually harvested as late as possible to reduce the total acidity.

WHERE IT'S GROWN AND WHAT ITS WINE TASTES LIKE

Like PIGNOLO and SCHIOPPETTINO, Tazzelenghe had almost disappeared by the early twentieth century, after Friuli in north-east **Italy** was hit by phylloxera. Rescued at the end of the 1980s, the variety is today cultivated again in the province of Udine, notably in the communes of Buttrio, Manzano and Cividale east of the city of Udine. A varietal is authorized in the Colli Orientali del Friuli DOC. Wines are deeply coloured and need time in oak and/or bottle to soften the marked tannins and acidity. Producers include Gianpaolo Colutta, Conte d'Attimis-Maniago, Girolamo Dorigo, Le Due Torri, Jacùss and La Viarte. There were 75 ha (185 acres) in Italy in 2000, according to the agricultural census.

TECA

Obscure and virtually extinct pink-skinned Aragonese speciality.

BERRY COLOUR:

ORIGINS AND PARENTAGE

Teca is a particularly obscure autochthonous variety from the province of Teruel in Aragón, northern Spain.

WHERE IT'S GROWN AND WHAT ITS WINE TASTES LIKE

Fandos y Barriuso in Bajo Aragón, **Spain**, rescued this variety from old, local vineyards and are currently the only ones to vinify it. They describe it as 'a fruity and lively wine with rosé tones'.

TEINTURIER

Red-fleshed French variety still found (just) in Portugal and South Africa.

BERRY COLOUR:

PRINCIPAL SYNONYMS: Auvernat Teint (Loiret), Bourguignon Noir, Gros Noir, Moreau (Savoie), Neraut, Noir d'Orléans (Loiret), Plant d'Espagne, Pontac (South Africa), Tachard (Savoie), Teint-Vin (Meurthe), Teinturier du Cher (Loire-et-Cher), Teinturier Mâle, Tintewein (Germany)

VARIETIES COMMONLY MISTAKEN FOR TEINTURIER: PINOT TEINTURIER, TINTA FRANCISCA (Douro in Portugal)

ORIGINS AND PARENTAGE

Teinturier, literally 'dyer', is a very old red-fleshed variety possibly originating from the region of Orléans in northern France, where it was widely cultivated from the seventeenth century on. The earliest mention of this variety probably appeared under the old synonym Neraut in Charles Estienne and Jean Liébaut (1564): 'Les complans de la vigne noire sont le morillon, le samoireau, le negrier & le neraut. . . . Le neraut, nommé Bourguignon noir, a la nature du Bourguignon blanc, . . . fait force couleur de sorte que ceux qui ont pleine plante en font vin pour les teintures de draps, & le vendent fort cher' ('The black grape varieties are morillon, samoireau, negrier and neraut. . . Neraut, called Bourguignon noir, has the nature of Bourguignon blanc, . . . makes so much colour that those who have a lot of vines make wine for the dying of sheets, and sell it at a very high price'). The name Teinturier appeared somewhat later in 1667 (Rézeau 1997): 'Le Noirault, dit Plant d'Espagne ou teinturier, est vn Raisin dont le grain est fort serré, teint fort noir' ('Noirault, known as Plant d'Espagne or teinturier, is a grape whose berries are very tightly packed, deep-black colour').

Recent parentage analysis based on thirty-two DNA markers strongly suggests a possible parent–offspring relationship between Teinturier and SAVAGNIN (Vouillamoz). Consequently, Teinturier is a grandchild, a grandparent or a half-sibling of PINOT (see Pinot pedigree diagram), which explains why they share the old synonym Auvernat. Interestingly, Teinturier, Pinot and Savagnin are all said to derive from wild grapevines (Galet 1990).

Teinturier was used to breed DECKROT, PERVOMAISKY and PETIT BOUSCHET.

VITICULTURAL CHARACTERISTICS

Early budding and ripening, small bunches of small, red-fleshed berries. Low vigour.

WHERE IT'S GROWN AND WHAT ITS WINE TASTES LIKE

Teinturier has disappeared from French statistics but it is officially authorized in **Portugal**. The wines are rather harsh and its best use has always been to add colour to blends.

Teinturier is known as Pontac in **South Africa** (Galet 2000), possibly named after the de Pontac family of Bordeaux, although there are competing theories concerning Pontac's name and origin. It was widely planted in the Cape for blended and fortified wines from the seventeenth century until the arrival of phylloxera in the late nineteenth. When vineyards were replanted, it was generally replaced by higher-yielding and more fashionable varieties. Hartenberg in Stellenbosch were the last producers to make a varietal wine but their final vintage was 2000 and they have now pulled up the vines, partly because of disease but also because there was virtually no market for the wine. Allesverloren in the Swartland have 1.5 ha (4 acres) of Pontac and include the fruit in their port-style blend, as do several others in the Western Cape.

TEMPARIA

Very new, promising American hybrid bred specially for Nebraska.

BERRY COLOUR: ● ● ● ● ●

ORIGINS AND PARENTAGE

A hybrid between a native *Vitis riparia* vine and Spanish TEMPRANILLO created around 1996 by Ed Swanson of Cuthills Vineyards in Pierce, Nebraska. Swanson had worked with Elmer Swenson (1913–2004), a Wisconsin viticulturist who developed several cold-hardy varieties and was really the driving force behind much of the grape-breeding in the Midwest. Temparia was the first grape variety to be ready for release from Ed Swanson's breeding programme, which he started in 1996 in order to create varieties that thrive in Midwestern states and yield appealing, European-style wines.

VITICULTURAL CHARACTERISTICS

Early budding and therefore at risk from spring frosts. Moderately vigorous; productive if on its own roots. Swanson reports having seen only a little black rot and some powdery mildew on the fruit in bad years. Late ripening.

WHERE IT'S GROWN AND WHAT ITS WINE TASTES LIKE

Swanson bred this variety specifically for **Nebraska** and is currently the only producer of a varietal wine, at his Cuthills Winery. He plans to make it available to other growers when he has other varieties ready for release and in the meantime is enjoying his monopoly. The wines have earthy, leathery notes as well as cherry (but no foxy) flavours and supple tannins. Malic acid is often high at harvest but malolactic fermentation usually brings the wine into balance.

TEMPRANILLA

See PALOMINO FINO.

TEMPRANILLO

Spain's most famous grape, responsible for the majority of its most famous reds.

BERRY COLOUR:

PRINCIPAL SYNONYMS: Aragón, Aragones, Aragonez (Alentejo in Portugal), Arauxa (Orense), Botón de Gallo, Cencibel (Castilla-La Mancha, Madrid, Aragón, Extremadura, Murcia), Chinchillana (Extremadura), Escobera (Badajoz in Extremadura, and South America), Garnacho Foño (South America), Grenache de Logrono, Jacibiera or Jacivera (Castilla-La Mancha and South America), Negra or Negra de Mesa, Piñuela (Toledo), Santo Stefano (Pisa in Italy), Tempranilla, Tempranillo de Rioja (Extremadura), Tinta de Nava (Rueda), Tinta del País (Ribera del Duero), Tinta de Toro (Castilla-La Mancha), Tinta Madrid (León, Zamora, Ribera del Duero, Arribes), Tinta Roriz (Douro in Portugal), Tinto Aragónez (Burgos), Tinto de Madrid (Toledo, Cantabria, Salamanca, Soria, Valladolid, Madrid), Tinto del País (Castilla-La Mancha), Tinto Fino (Albacete in Castilla-La Mancha, Madrid, Ribera del Duero, Extremadura), Ull de Llebre (Catalunya), Valdepeñas (Napa Valley in California), Verdiell (Catalunya), Vid de Aranda (Burgos)

VARIETIES COMMONLY MISTAKEN FOR TEMPRANILLO: GARNACHA TINTA, JUAN GARCÍA, MORISTEL (called Tempranillo Temprano in Logroño)

ORIGINS AND PARENTAGE

The earliest reference to this variety was supposedly found under the plural *las tempraniellas* in the *Libro de Alexandre* published in the thirteenth century and referring to the region of Ribera del Duero (Casas Rigall 2007). However, *las tempraniellas* and the name Tempranillo both derive from the Spanish *temprano*, meaning 'early', and the true identity of the early-ripening varieties mentioned in the *Libro de Alexandre* is not known. The next possible earliest mention of Tempranillo is said to have appeared in 1513 under the old synonym Aragonés in Spanish agronomist Gabriel Alonso de Herrera's description of the varieties of Castilla, Extremadura and Andalucía (1790). However, the name Aragonés, referring to the medieval kingdom of Aragón (today's autonomous community of Aragón) in north-eastern Spain, has been historically used for both GARNACHA and Tempranillo, and Alonso de Herrera's description is too vague to know which he is describing.

Therefore, the first reliable mention of Tempranillo was when the Spanish ampelographer Clemente y Rubio (1807) described it as a variety that was praised in Logroño in La Rioja and in Peralta in Navarra, two adjacent regions north west of Aragón that can be considered the likely birthplace of this variety. This historical origin is consistent with genetic analyses of dozens of ancient clones of Tempranillo from all over Spain that have shown that they are all very uniform, suggesting that Tempranillo vineyards are the result of rapid multiplication and distribution from a small number of original clones (Cervera *et al.* 2002; Carcamo *et al.* 2010). These clones most likely come from the provinces of La Rioja and Navarra, where Víctor Cruz Manso de Zúñiga (1905) delineated the geographic distribution of Tempranillo in the early twentieth century.

According to Santana *et al.* (2010), Tempranillo most likely has a parent–offspring relationship with ALBILLO MAYOR from Ribera del Duero, where Tempranillo has traditionally been cultivated under the name Tinta del País, but it is not known which is the parent and which is the offspring.

In Italy, DNA profiling has surprisingly revealed that Tempranillo has traditionally been cultivated in Toscana and Basilicata under the misleading names Malvasia Nera (a synonym for MALVASIA NERA DI BASILICATA) and Malvasia Nera di Lecce (a synonym for MALVASIA NERA DI BRINDISI; Storchi *et al.* 2009). In Portugal, Tempranillo has been cultivated for centuries under the names Tinta Roriz in the Douro and Aragonez in the Alentejo, but no close relationship with Portuguese varieties could be established (Lopes *et al.* 1999; Pinto-Carnide *et al.* 2003). In South America, Tempranillo was introduced around the seventeenth century, and has been shown to be genetically close to CRIOLLA GRANDE in Argentina (Martínez *et al.* 2003).

Tempranillo was used to breed TEMPARIA.

See below for Tempranillo Blanco.

OTHER HYPOTHESES

Legend has it that Tempranillo was brought to Spain from Burgundy by Cistercian monks from the Abbaye de Cîteaux and could be related to PINOT, suggesting a link via the pilgrim route to Santiago de Compostela that crosses La Rioja and Ribera del Duero. However, this hypothesis can be rejected by DNA analysis (Vouillamoz). Clemente y Rubio (1807) included among the synonyms for Tempranillo the variety Maiolus (albeit with a question mark) cited by Italian jurist and oenophile Pietro de Crescenzi (1235–1320) and corresponding to the modern MAIOLINA from the province of Brescia in Italy but Maiolina and Tempranillo are unrelated (Vouillamoz).

VITICULTURAL CHARACTERISTICS

Medium-sized to large compact bunches of small berries rich in tannins and anthocyans. Relatively fertile. Early budding and ripening. Yields range from medium to high according to the site. Sensitive to the wind and to extreme drought. Susceptible to powdery mildew and to eutypa dieback but generally resistant to botrytis bunch rot. High yields reduce colour and fruit intensity as well as acidity.

WHERE IT'S GROWN AND WHAT ITS WINE TASTES LIKE

Although Tempranillo is essentially an Iberian variety, it can still be found in southern **France**, where there were 766 ha (1,893 acres) in 2009, about half as much as at the turn of the century, most of it in the western Languedoc and used for blending. The 2000 agricultural census in **Italy** found just 7 ha (17 acres) of Tempranillo and it was not authorized for any DOC(G) wines under that name. However, the census also shows 2,693 ha (6,655 acres) of Malvasia Nera, some or much of which may well be Tempranillo.

Spain is the kingdom of Tempranillo, a kingdom that extended to 206,988 ha (511,478 acres) in 2008, making it the country's most widely planted red wine variety and the second most planted vine after AIRÉN. It is widely distributed across the country, albeit under a host of synonyms, with the exception of Asturias in the far north, and is not as significant in Galicia in the far north west and in Andalucía in the deep south as elsewhere. In the higher, cooler subregions of Rioja, Rioja Alta and Rioja Alavesa, it can ripen a full two weeks before the GARNACHA with which it has traditionally been blended there.

Tempranillo produces wines with considerably less alcohol than GARNACHA and such other Spanish stalwarts as MONASTRELL and BOBAL, marked tannins, sometimes slightly low acidity and flavours that range over spice, leather, tobacco leaves, sometimes strawberry when it has been given extended ageing in American oak as it traditionally was in Rioja. In the distinctly less traditional wine region of Ribera del Duero, average vine age is generally much lower and producers have tended to rely on oak, often French barriques, and the concentrated bright fruit naturally produced by the cool nights at Ribera's altitude to distract from a lack of nuance in the variety known here as Tinto Fino. Modernist producers in both these leading Spanish red-wine regions have increasingly also looked to seasonings of CABERNET SAUVIGNON and occasionally MERLOT, too. Tempranillo's innate character is less obtrusive than that of these incomers from Bordeaux and it tends to submit meekly to such a blend, which has become common in Navarra and throughout Castilla-La Mancha. Perhaps the most concentrated, alcoholic examples of Tempranillo are produced in Toro.

Recommended producers of varietal wines include Artadi, Abel Mendoza, Finca Allende, Fernando Remírez de Ganuza, Roda, Señorío de San Vicente and Sierra Cantabria in Rioja; Aalto, Alión, Alonso del Yerro, Alejandro Fernández, O Fournier, Pago de los Capellanes, Pagos de Matanera, Pérez Pascuas and Pingus in Ribera del Duero; Dominio del Bendito, Maurodos, Numanthia, Pintia, Telmo Rodríguez, Teso La Monja in Toro; and, within the Vino de la Tierra de Castilla y León classification (just outside Ribera del Duero), Abadía Retuerta.

Tempranillo is one of relatively few Spanish varieties to have a significant presence in **Portugal**, which had almost 17,000 ha (42,000 acres) of Tinta Roriz in 2010, up from around 13,000 ha (32,000 acres) in 2004. It is the second most planted red variety in the Douro (both for port and for table wines), is equally significant in the Dão region, and of historical importance in the Alentejo further south, where it is known as Aragonez. Thanks to its quality potential, it has spread to other regions as well, particularly around the Tejo (Tagus) and in Lisboa.

Aragonez has long been planted in the Alentejo and is considered one of the region's most noble varieties. In this hot climate, however, great care must be taken to harvest at the right moment since any slight delay leads to a significant loss of fruit quality. Experiments carried out in vineyards at around 600 metres have demonstrated that Aragonez is at its best at those cooler altitudes and in the type of soil found there, frequently yielding better results than TOURIGA NACIONAL. Recommended producers include Ervideira and J Portugal Ramos.

In the Douro, Tinta Roriz has long been an extremely important component in the production of port, in good years adding colour and body, with elegant aromas of crushed black pepper and wild flowers and flavours of wild cherries with a hint of vegetables. The variety can bring freshness to the blends that predominate here in both port and Douro table wines. Quinta do Portal are unusual in making a varietal Tinta Roriz.

Varietal examples are much more common in Dão, where the wines can be densely coloured, full-bodied, with fine tannins and in some ways not unlike Ribera del Duero.

Tempranillo is grown to a very limited extent in both **Switzerland** and **Malta**, and in **Turkey** (7 ha/17 acres in 2010) Kavaklidere have planted Tempranillo in both Central Anatolia and Kapadokya (Cappadocia). Other producers include Melen and İdol.

California grew 957 acres (387 ha) of Tempranillo in 2010, quite widely distributed but mainly in the warmer Central Valley. Tempranillo seems to thrive where Zinfandel does and the varieties are interplanted in some older vineyards. Available clones are registered under the names Tempranillo, Valdepeñas and Tinta Roriz, the last imported from Portugal by Harold Olmo in 1984. However, it is now officially known as Tempranillo. Recommended producers include Truchard in Carneros, Curran in Santa Ynez Valley, and Jarvis and Viader in Napa Valley.

The minor but perceptible American craze for Tempranillo is demonstrated by the fact that total plantings in **Washington** State grew from zero in 2006 to 94 acres (38 ha) by 2011, with the likes of Gramercy Cellars and Cayuse leading the pack. They may have been inspired by a single grower and producer in southern Oregon, Abacela in **Oregon**'s Umqua Valley, who have more than 20 of the state total of 150 acres (8 out of 61 ha) planted and pioneered the variety in the Pacific Northwest, instigating an international Tempranillo Day. **Texas** has an increasing number of Tempranillo fans as well, more than thirty wineries drawing fruit from 75 acres (30 ha) of vineyard, encouraged by such pioneers as Inwood and Lone Oak. The variety's ability to withstand hot summers and cold winters is appreciated here. Wines tend to be pretty New World and ripe in character.

It is also planted to a very limited extent (10 acres/4 ha) in British Columbia in western **Canada**. In **Mexico**, Tempranillo is found mainly in Baja California, with Santo Tomas a notable producer.

Argentina had a very considerable total of 6,568 ha (16,230 acres) of Tempranillo in the ground in 2009, mostly in Mendoza, where it is the fifth most planted grape variety and the area continues to increase. The majority of the varietal wines are inexpensive and rather more savoury than the dominant Malbec (COT) but O Fournier's Tempranillo-dominated blends Alpha and

Beta Crux, and Zuccardi's Q bottling, show the potential for this variety in Argentina. **Chile**, on the other hand, grew only 10 ha (25 acres) of Tempranillo in 2008.

Tempranillo has been a beneficiary of **Australia**'s growing interest in 'alternative varieties'. By 2008 there were 385 ha (951 acres), very widely dispersed but with particularly sizeable plantings in the irrigated inland Riverina and Riverland, as well as the Barossa Valley, Heathcote and McLaren Vale. Recommended producers include Ross Estate in the Barossa Valley and Pokolbin Estate in the Hunter Valley. D'Arenberg blend Tempranillo with Grenache (GARNACHA) and Sousão (VINHÃO) in McLaren Vale and Yalumba have successfully blended it with Grenache and VIOGNIER in the Barossa Valley.

New Zealand had a mere 7 ha (17 acres) in 2008 but the remarkable performance of Trinity Hill's example grown on the Gimblett Gravels in Hawke's Bay may have encouraged more plantings. This is a variety currently exciting considerable interest around the world.

TEMPRANILLO BLANCO

A white mutation of Tempranillo was discovered in 1988 by grape grower Jesús Galilea Esteban in a vineyard in Rioja Baja. CIDA, the La Rioja government's research and development centre, spent four years evaluating and stabilizing the mutation and later planted 1 ha (2.5 acres) of experimental vines. The first bottling was in 2005 and the variety is now cultivated to a very limited extent (Martínez *et al.* 2007). **Tempranillo Royo**, a grey-berried mutation, has been found in Toro in Castilla y León and is being evaluated at CIDA. The names Tempranillo Gris or Tinta de Toro Gris have been proposed (Yuste *et al.* 2006).

Authorized for white Rioja in 2004, Tempranillo Blanco is identical to the dark-skinned Tempranillo apart from its slightly smaller leaves and berries. They also share the same short ripening cycle and sensitivity to pests and diseases. Yields are medium and vigour is moderate to high. Wines are generally high in alcohol and acidity and well structured, with intense flavours of citrus, tropical fruit and white flowers. Ijalba were the first to make a varietal wine in 2005 and have since been followed by Conde de Valdemar and Juan Carlos Sancha. Winemakers believe it will combine well with Viura (MACABEO).

TÉOULIER NOIR

Almost extinct Provençal variety.

BERRY COLOUR: ● ● ● ● ●

PRINCIPAL SYNONYMS: Manosquin, Plant de Manosque, Plant Dufour, Thuillier (Alpes de Haute-Provence)

ORIGINS AND PARENTAGE

Téoulier Noir is a rare variety from the Manosque area between Aix-en-Provence and Gap in Provence, southern France, where it was cited in 1715 by Garidel (Rézeau 1997): 'Le vulgaire a donné le nom de Taulier, ou de Plan de Manosquo, à cette espece qui est assez commune dans les vignes du Tholonet' ('The common language has given the name Taulier, or Plan de Manosquo, to this species that is rather common in the vineyards of Tholonet'). The modern name Téoulier Noir appeared in 1755 (Rézeau 1997). While the synonym Manosquen or Manosquin refers to its place of origin, the etymology of Téoulier is unknown. Téoulier Noir is a member of the Claret ampelographic group (see p XXVII; Bisson 2009).

VITICULTURAL CHARACTERISTICS

Mid ripening. May be trained as a bushvine and pruned short. Relatively early budding puts it at risk of spring frosts. Susceptible to powdery mildew. Big bunches and berries.

WHERE IT'S GROWN AND WHAT ITS WINE TASTES LIKE

There was less than a hectare (2.5 acres) remaining in **France** in 2009 but Château Simone, in the tiny Palette appellation, still have a few vines in a 120-year-old vineyard and these are included in their red blend. Wines are generally low in acidity and moderate in alcohol.

TERAN

See TERRANO.

TERAN BIJELI

See PROSECCO.

TERBASH

Widely planted, multi-purpose Turkmen variety.

BERRY COLOUR: ● ● ● ● ●

PRINCIPAL SYNONYMS: Irtik Iaprak, Irtuk Yaprak

ORIGINS AND PARENTAGE

Terbash, meaning 'crunchy head' in reference to the berry flesh, comes from Turkmenistan. The synonym Irtik Iaprak means 'torn leaf'.

VITICULTURAL CHARACTERISTICS

Vigorous, productive, late ripening. Medium-sized to large bunches of fairly large, thick-skinned berries. Good resistance to winter frost, drought and high summer temperatures. Susceptible to fungal diseases.

WHERE IT'S GROWN

Terbash is one of the most important and widely planted varieties in **Turkmenistan**, where it is grown both for the table and for wine – dry, dessert and fortified styles – as well as for juice and raisining. Geokdepe make varietal examples. Also found in **Russia** and **Azerbaijan**.

TERMARINA ROSSA

Unusual Emilia-Romagna seedless pink-skinned variety used in cooking as well as for wine.

BERRY COLOUR: ● ● ● ● ●

PRINCIPAL SYNONYMS: Oltremarina, Tramarina

VARIETIES COMMONLY MISTAKEN FOR TERMARINA ROSSA: KORINTHIAKI (under the name Corinto Nero)

ORIGINS AND PARENTAGE

Termarina Rossa was first cited in *Viti della Provincia di Reggio Emilia*, a manuscript written by Bertozzi in 1840. It was described as a grape variety from the province of Reggio Emilia, also called Uva Passerina, of which there was also a white-berried version. However, according to DNA and enzymatic studies, the white Termarina Bianca is not a mutation but a distinct variety. Termarina Rossa has sometimes been confused with Corinto Nero, probably because both varieties are seedless, but DNA profiling has proved they are not the same (Boccacci *et al.* 2005).

VITICULTURAL CHARACTERISTICS

Tiny seedless berries which accumulate high levels of sugar on the vine. Vigorous and productive.

WHERE IT'S GROWN AND WHAT ITS WINE TASTES LIKE

The black and white versions of Termarina were traditionally used in the province of Parma and Reggio Emilia in northern **Italy** to make jam and *saba*, a local syrup made from boiled must. Today Termarina Rossa is also used for wine, blended with other varieties. It was included in the national register of varieties only in 2007 so does not show in the most recent, 2000, agricultural census.

Varietal wines are extremely rare but Angelo Casalini, the guardian of Termarina Rossa since the 1970s, makes a distinctive, full-bodied wine with aromas of tar and ripe cherries, silky tannins, good acidity and a bitter finish.

TEROLDEGO

Well-connected, revived Trentino speciality whose acidity needs careful handling.

BERRY COLOUR: ● ● ● ● ●

PRINCIPAL SYNONYMS: Merlina (Valtellina), Teroldega, Teroldeghe, Teroldico, Teroldigo, Tiraldega, Tirodola (Lago di Garda), Tiroldegho, Tiroldigo

VARIETIES COMMONLY MISTAKEN FOR TEROLDEGO: NEGRARA TRENTINA (Trentino), NEGRARA VERONESE (Veneto)

ORIGINS AND PARENTAGE

Teroldego is a very old variety from Trentino in north-east Italy, where its wine was first mentioned on 18 January 1480 in Cognola, just east of the city of Trento, in a sales contract written in Latin:

'duarum brentarum vini teroldigi buliti, boni et sufficientis', meaning 'two *brente* [an old measure] of fermented, good and abundant teroldigo wines' (Prato 1989). There were also several other references later in historical documents from the same region. The cradle of Teroldego is therefore very likely to be the Rotaliano plain between Trento and Mezzolombardo, where it has been cultivated for centuries. Indeed, Teroldego almost certainly took its name from a place near Mezzolombardo called Alle Teroldege, where vineyards were mentioned as far back as the fifteenth century (Roncador 2006).

Pedigree reconstruction based on DNA profiling revealed that Teroldego spontaneously crossed at least twice with the same unknown and most likely extinct variety to give birth to MARZEMINO in Trentino or Lombardia and to LAGREIN in Trentino or the Alto Adige (Vouillamoz and Grando 2006). More surprisingly, DNA pedigree reconstruction revealed that DUREZA from the Ardèche, which is a parent of SYRAH, is a full-sibling of Teroldego, and that both are grandchildren of PINOT. Teroldego is therefore an uncle/aunt of Syrah, as well as a grandparent of REFOSCO DAL PEDUNCOLO ROSSO (Vouillamoz and Grando 2006; see the pedigree diagrams in the Syrah and Refosco dal Peduncolo Rosso entries).

Teroldego was used to breed REBO.

OTHER HYPOTHESES

On the basis of synonyms such as Tiroldigo, the variety's name is said to derive from Tiroler Gold, literally 'gold from Tirol', a name apparently used at the court in Wien (Vienna) to describe Trentino wines. On the other hand, some authors have suggested that Teroldego was introduced from the neighbouring plain of Verona, where a variety named Terodol'i was mentioned in the past.

VITICULTURAL CHARACTERISTICS

Mid to late ripening. Traditionally trained on a T-shaped double pergola to take advantage of the vine's high yields but in the last few decades modern vine growers have generally switched to Guyot. Slightly sensitive to downy and powdery mildews, highly susceptible to desiccation of the stems.

WHERE IT'S GROWN AND WHAT ITS WINE TASTES LIKE

Teroldego has been cultivated for centuries on the Rotaliano Plain (the Campo Rotaliano) north of Trento in northern **Italy**, in the communes of Mezzolombardo, Mezzocorona, San Michele all'Adige and Roverè della Luna. In Trentino-Alto Adige it has had its own DOC since 1971, Teroldego Rotaliano, for both red and rosé (labelled Rosato or Kretzer) made from 100% Teroldego. It may also be included in small amounts in rosé versions of DOCs such as Valdadige/Etschtaler, Trentino and Casteller. It is also planted further south, in the Veneto, and west, in Valtellina (under the name Merlina). The Italian agricultural census of 2000 recorded total plantings of 690 ha (1,705 acres) but the area is decreasing. Recommended producers include Elisabetta Foradori, who instigated a quality revolution in the mid 1980s exemplified by her straight Teroldego Rotaliano and the more concentrated and ageworthy Granato, and Marco Donati. Chianti Classico producer Poggio al Cassone has bottled a Teroldego under the name La Cattura.

Wines are deeply coloured, lively and fruity and sometimes prone to reduction. When yields are restricted and the grapes reach full maturity – which is sadly not always the case – the rich black-cherry fruit is well supported by ripe tannins, the acidity is mouth-watering rather than eye-watering, the wine marries well with oak and ages well.

In **California** a Teroldego is made by about a dozen producers all over the state, notably Central Coast producer Wolff Vineyards. Plantings totalled around 100 acres (40 ha) in 2008. Michelini in the King Valley, Victoria, are the lone proponents of this variety in **Australia**. Their first commercial vintage, 2008, was a trophy winner at the Australian Alternative Varieties Wine Show 2010. There is also some in Rio Grande do Sul in **Brazil**.

TERRANO

Significant variety making fresh, aromatic red from the Italy–Slovenia border and Croatia; often confused with the best-known Refosco.

BERRY COLOUR:

PRINCIPAL SYNONYMS: Cagnina (Friuli and Emilia-Romagna), Crodarina, Rabiosa Nera (Breganze), Refosco del Carso, Refosco d'Istria, Refosco Terrano, Refošk, Teran (Istra/Istria), Terran (Istra), Terrano del Carso, Terrano d'Istria

VARIETIES COMMONLY MISTAKEN FOR TERRANO: REFOSCO DAL PEDUNCOLO ROSSO, REFOSCO DI FAEDIS

ORIGINS AND PARENTAGE

Terrano is a very old variety from the Karst plateau (Kras in Slovenia, Carso in Italy) that includes parts of Friuli in north-eastern Italy, south-western Slovenia and Istra. The earliest mention of Terrano appeared in a document dated 13 November 1340 referring to the cultivation in Barbana (now in Slovenia) of 'Rabiole Malvasie Terrano bianco e vermiglio Moscatello e Pignolo' (Fabbro *et al.* 2002). Surprisingly, both white- (*bianco*) and red- (*vermiglio*) berried Terrano are cited, whereas only the red version has survived. Terrano wines are again referred to on 14 August 1390 when 20 *ingestariis* (an old measure of volume) of *vini terrani* were offered to Conte di Lozo, ambassador of the Emperor (Di Manzano 1860). Terrano is then mentioned in several historical documents in Friuli.

Terrano is part of the so-called Refosco group (see REFOSCO) and is often confused with REFOSCO DAL PEDUNCOLO ROSSO. However, comparison of DNA profiles disproves this hypothesis (Vouillamoz) and has shown that Terrano – Teran in Istra (Croatia) – is identical to Refosco d'Istria and to Refošk in Slovenia (Maletic *et al.* 1999; Kozjak *et al.* 2003). Comparison of DNA profiles from distinct sources has also shown that Terrano is identical to Cagnina in Friuli and in Emilia-Romagna, and with Rabiosa Nera in Breganze (Vouillamoz).

VITICULTURAL CHARACTERISTICS

Mid to late ripening. Resistant to late frosts and major fungal diseases.

WHERE IT'S GROWN AND WHAT ITS WINE TASTES LIKE

Terrano is widely planted in Friuli-Venezia Giulia and Emilia-Romagna in **Italy**, often known as Cagnina in the latter, hence the varietal DOC Cagnina di Romagna. Varietal wines are also authorized within the Carso DOC, in which Terrano may also play a minor part, as it does in Colli di Rimini. Wines are full- to medium-bodied, food-friendly, quite tannic with fresh acidity and marked aromas of forest fruits, sometimes cherries. There were 203 ha (502 acres) in Italy in 2000. Producers of varietal wines in Italy include Skerk, Castello di Rubia, Castelvecchio and Zidarich.

In **Slovenia**, plantings are mainly in the Kras region, north of Trieste, where there were 481 ha (1,189 acres) in 2008, and in Slovenska Istra (Slovenian Istria), south of Trieste, where it is the most widely planted variety, 818 ha (2,021 acres) in 2008. There are also limited plantings in Vipavska Dolina. Producers of varietal wines include Josko Rencel (Kras) and Parovel (Trieste).

As Teran, the variety is also planted in **Croatia**, on around 480 ha (1,186 acres) in 2008. It is recommended only in the Istra wine region but is very successful there, producing elegant, fresh wines with lively fruit and relatively firm tannins that respond well to barrel ageing. Recommended producers include Franc Arman, Dimitri Brečević, Kozlović and Roxanich.

Also found in the **Republic of Macedonia**.

TERRANTEZ

Historic Madeira variety that has virtually disappeared from the island.

BERRY COLOUR: ● ● ● ● ●

PRINCIPAL SYNONYMS: Terrantez da Madeira (Madeira)

VARIETIES COMMONLY MISTAKEN FOR TERRANTEZ DA MADEIRA: FOLGASÃO, TERRANTEZ DA TERCEIRA (Açores), TERRANTEZ DO PICO (Açores)

ORIGINS AND PARENTAGE

Terrantez is an old variety that used to be widespread on the Portuguese island of Madeira, as evidenced by nineteenth- and early-twentieth-century bottles still sold at auction today, but has now almost disappeared. Terrantez has long been considered to be a synonym for FOLGASÃO from the Dão region but comparison of their DNA profiles (Martín *et al.* 2006; Almadanim *et al.* 2007; Veloso *et al.* 2010) indicates that they are distinct varieties. Terrantez is also genetically distinct from TERRANTEZ DA TERCEIRA and TERRANTEZ DO PICO from the Açores, both with unique DNA profiles in Veloso *et al.* (2010), and from the variety officially known in Dão as Terrantez, which is not currently in commercial cultivation.

VITICULTURAL CHARACTERISTICS

Very low yields indeed (zero in some years), late ripening. Very good general disease resistance but compact bunches and extremely thin-skinned berries make it prone to botrytis bunch rot and berry splitting.

WHERE IT'S GROWN AND WHAT ITS WINE TASTES LIKE

Although less than 2 ha (5 acres) of Terrantez remains or has recently been replanted on Madeira, **Portugal**, it was once highly regarded and the many fine older bottles still resting in cellars show how well these sweet but initially astringent fortified wines have aged. Its demise on the island was due mainly to the lack of replanting after the destruction caused first by powdery mildew and then by phylloxera in the late nineteenth century. Any Terrantez Madeira is rare enough to be worth trying.

TERRANTEZ DA TERCEIRA

Contrary to popular belief, this is a unique white-berried variety from Terceira in the Açores, the Atlantic islands that are part of **Portugal**. DNA profiling has shown that Terrantez da Terceira is distinct from the other Açores variety TERRANTEZ DO PICO, from TERRANTEZ on Madeira, from FOLGASÃO, found mainly in Trás-os-Montes, and from Lisboa's ARINTO DE BUCELAS (Veloso *et al.* 2010).

TERRANTEZ DO PICO

Contrary to popular belief, this is a unique white-berried variety from the island of Pico in the Açores, **Portugal**. DNA profiling has shown that Terrantez do Pico is distinct from the other Açores variety TERRANTEZ DA TERCEIRA, from TERRANTEZ on Madeira, from FOLGASÃO, found mostly in Trás-os-Montes, and from MALVASIA FINA, which is sometimes mistakenly called Terrantez do Pico (Eiras Dias *et al.* 2006; Veloso *et al.* 2010). The Pico co-op use Terrantez in both dry and fortified blends.

TERRET

Old Languedoc variety found in three colours, all in quantitative decline.

BERRY COLOUR:

PRINCIPAL SYNONYMS: Bourret, Tarret

ORIGINS AND PARENTAGE

The variety Terret consists of Terret Noir, Terret Gris and Terret Blanc that are simply colour mutations. Terret, one of the oldest varieties in the Languedoc, most likely originates from the Hérault *département* in southern France, where it has long been cultivated. According to Rézeau (1997), Terret was possibly cited as early as 1619 in Carcassonne, followed by other mentions in 1639 (terren) and 1676 (tarret), but the identity of these old mentions is uncertain and the first reliable one appeared in 1736 in the Languedoc: 'Terret noir: produit beaucoup' ('Terret noir: produces a lot'). Terret belongs to the Picquepoul ampelographic group (see p XXVII; Bisson 2009).

VITICULTURAL CHARACTERISTICS

Late budding and ripening, should be pruned short but may be trained vertically or as bushvines. Productive, with large bunches of medium-sized berries. Susceptible to powdery and downy mildews. Terret Gris is also susceptible to grape moths and sunburn.

WHERE IT'S GROWN AND WHAT ITS WINE TASTES LIKE

Terret Gris used to be the most widely planted variety in the Languedoc, **France** (8,130 ha/20,090 acres in 1958, for example), and was used not only for wine but also for vermouth and eaux-de-vie but the vineyard area had declined to 104 ha (257 acres) by 2008. Now it is Terret Blanc that is the most widely planted of the three, with 1,451 ha (3,585 acres) in 2009, almost all in the Hérault. Terret Noir occupies slightly more vineyard than Gris: 189 ha (467 acres) in 2008, down from 400 ha (988 acres) in 2000.

Although there are some varietal wines – or blends with a range of other more internationally recognized varieties including CHARDONNAY, SAUVIGNON BLANC or MUSCAT BLANC À PETITS GRAINS – all three colours tend to play a minor role in appellation wines in the southern Rhône and west across the Languedoc, where they are authorized or recommended. Terret Noir may be included as a minor component in Languedoc and Provence appellations such as Fitou, Minervois, Corbières, Palette and Cassis, and also in Châteauneuf-du-Pape, Côtes du Rhône-Villages, Gigondas and Rasteau in the Rhône; Terret Blanc and Gris are limited to an equally minor role in the Cassis, Palette, Corbières, Minervois and Languedoc appellations. Clos du Gravillas produce a barrel-fermented varietal Terret Gris and Les Clos Perdus make a varietal Terret (mainly Terret Gris) with aromas of brioche and truffle from a centenarian vineyard where all three colours are interplanted. Léon Barral of Faugères make a complex, aromatic white from Terrets Blanc and Gris.

Wines made from Terret of all three hues are generally moderate in alcohol and have the useful ability to retain fresh acidity in the heat of southern France. Terret Blanc and Terret Gris are mostly rather neutral, Terret Noir tends to be light and pale.

TERRET BLANC

See TERRET.

TERRET GRIS

See TERRET.

TERRET NOIR

See TERRET.

THEIAKO MAVRO

Very minor and declining Greek variety used in blends.

BERRY COLOUR:

ORIGINS AND PARENTAGE

Theiako Mavro is an old variety from the islands of Kefaloniá and Itháki, perhaps named after the latter, or alternatively from *theio*, meaning 'sulphur'. It is possibly related to GOUSTOLIDI and ROBOLA, and a white mutation called **Theiako** or Thiako is found on the island of Lefkáda (Boutaris 2000).

VITICULTURAL CHARACTERISTICS

Mid ripening.

WHERE IT'S GROWN

Theiako Mavro is cultivated to a limited extent on the island of Kefaloniá, **Greece**, where Domaine Foivos include a small percentage of both Theiako Mavro and ARAKLINOS in a blend with MAVRODAFNI, as well as on Itháki, though the variety is in decline.

THINIATIKO

See MAVRODAFNI.

THOMPSON SEEDLESS

See SULTANIYE.

THRAPSATHIRI

Distinctive Greek island variety, travelling under two names and producing richly fruited wines.

BERRY COLOUR: ● ● ● ● ●

PRINCIPAL SYNONYMS: Begleri or Beghleri (islands of Chíos, Sámos, Ikaría, Santoríni, Lésvos/Lesbos and Kríti/Crete), Dafnato (Kríti)

VARIETIES COMMONLY MISTAKEN FOR THRAPSATHIRI: ATHIRI

ORIGINS AND PARENTAGE

Thrapsathiri probably comes from the island of Kríti. It has long been thought to be close or identical to ATHIRI (*thrapsa* means 'lots of') but this has been contradicted by DNA analysis, which has suggested instead a close relationship with VIDIANO, also from Kríti (Biniari and Stavrakakis 2007). Comparison of their respective DNA profiles in the Greek Vitis Database shows that Begleri, found on the Kykládes (Cyclades) islands and on Kríti is identical to Thrapsathiri (Vouillamoz). Therefore, the red-berried **Begleri Kokkino** (Galet 2000) may be considered to be a colour mutation of Thrapsathiri.

OTHER HYPOTHESES

Thrapsathiri was allegedly used to make Malvasia wine, the sweet blended wine famous in the Middle Ages (Galet 2000), but there is no evidence as to which varieties were included.

VITICULTURAL CHARACTERISTICS

High yielding.

WHERE IT'S GROWN AND WHAT ITS WINE TASTES LIKE

Grown mainly in the southern Aegean on **Greek** islands including the Kykládes, Kríti and Ródos (Rhodes), Thrapsathiri is very often part of a field blend. Domaine Economou, for example, have ungrafted sixty- to seventy-year-old vineyards in western Kríti and, according to wine writer Nico Manessis (personal communication), there is an increasing trend among producers in the centre of the island to blend it with the more perfumed VILANA. Lyrarakis, however, make a full-bodied and rich varietal wine, aged in acacia and French oak, that has intense tropical-fruit aromas and more acidity than is typical for the variety thanks to the altitude (500 m) and exposition of the vineyard. The Sítia co-op also make both a varietal and a blend.

There were 47 ha (116 acres) under the name Begleri scattered across the islands of the northern Aegean in 2008. Afianes are experimenting with vinifying and ageing this variety in clay amphorae.

TIBOUREN

Ancient variety making fine rosé wines in Provence and light reds over the border in Italy's Liguria.

BERRY COLOUR:

PRINCIPAL SYNONYMS: Antiboulen (Var), Rossese (Liguria), Rossese di Dolceacqua (Liguria), Rossese Nericcio (Liguria), Rossese di Ventimiglia (Liguria), Tiboulen (Var)

VARIETIES COMMONLY MISTAKEN FOR TIBOUREN: ROSSESE DI CAMPOCHIESA (Savona)

ORIGINS AND PARENTAGE

Tibouren is said to have been introduced to the region of Saint-Tropez in the Var, southern France, in the late eighteenth century by a marine captain named Antiboul, which would explain its synonym Antiboulen, later corrupted to Tibouren (Ganzin 1901b). DNA profiling recently and unexpectedly showed that Rossese di Dolceacqua, cultivated in the area of Ventimiglia in Liguria (close to the French border) and included in the Italian national register of varieties, is identical to Tibouren (Torello Marinoni, Raimondi, Mannini and Rolle 2009). This is consistent with the local tradition which says that Rossese di Dolceacqua was introduced from France to the province of Imperia in western Liguria (very close to the French border) by the soldiers of the Doria family in Dolceacqua, whence it subsequently spread along the Ligurian Riviera. It is not known whether the variety was first called Tibouren or Rossese, but given the possible confusion with other varieties named Rossese Something, it seems clearer to use Tibouren here.

OTHER HYPOTHESES

Tibouren is called Antiboulen in the Var in the south of France, which leads some authors to suggest the variety is from the Antibes area, but it has never been seen there. A link with Greece or with the Near East has been suggested but has not been confirmed by DNA analysis.

VITICULTURAL CHARACTERISTICS

Very early budding, mid ripening. Needs to be pruned short. Susceptible to millerandage and desiccation of the stems but not generally susceptible to pests and diseases, apart from possibly mites.

WHERE IT'S GROWN AND WHAT ITS WINE TASTES LIKE

In southern **France**, Tibouren's traditional growing area is the gulf of Saint-Tropez, where ungrafted centenarian vines can still be found in Saint-Tropez, Ramatuelle, Gassin, Grimaud, Cogolin, etc. (Galet 2000). Tibouren is recommended in the Var and the Alpes-Maritimes although almost all of France's 445 ha (1,100 acres) in 2009 were in the Var, slightly up on 2006's 417 ha (1,030 acres). It is generally used within the Côtes de Provence appellation to produce expressive, earthy rosé wines with the scent of garrigue, sometimes as a varietal wine, more often blended with Cinsault (CINSAUT), Grenache (GARNACHA), Mourvèdre (MONASTRELL) or SYRAH. Clos Cibonne, where the late André Roux dedicated most of his vineyards to this variety, also make full-bodied and ageworthy Côtes de Provence reds that blend Syrah, Grenache and Tibouren and in some years a rare varietal or Tibouren-dominant red cuvée.

In Liguria, **Italy**, Rossese or Rossese di Dolceacqua has long been widespread in the province of Imperia, in particular in the Valle Nervia and Vallerosia (Ventimiglia, Dolceacqua, Camporosso, Soldano, Isolabona, San Biagio and Perinaldo) very close to the Golfo di Genova and the French border. Esteemed varietal wines are produced within the Rossese di Dolceacqua and the Rossese Riviera Ligure di Ponente appellations, although the latter is more likely to be made from

ROSSESE DI CAMPOCHIESA in the area of Campochiesa and Albenga (the regulations do not specify which Rossese). The Italian agricultural census of 2000 reported 268 ha (662 acres) under the name 'Rossese' but does not distinguish between them. Recommended producers include Alessandro Anfosso, Alta Via, Giobatta Mandino Cane, Tenuta Giuncheo, Antonio Perrino, Poggio dell'Elmo, Ravera, Ramoino, Sancio, Terre Bianche. Varietal Rossese di Dolceaqua wines are typically lively, aromatic and relatively light with a distinctive and refreshing sour-cherry flavour.

TIGNOLINO

Rare and highly localized dark-skinned blending component from north-east Sicilia, **Italy**, that could simply be a clone of PERRICONE. The Palari winery in Messina include it in their red Faro DOC wines.

TIGRANI

*Armenian all-*vinifera *cross used for reds dry and sweet.*

BERRY COLOUR: ● ● ● ● ●

PRINCIPAL SYNONYMS: 1452/75

ORIGINS AND PARENTAGE

Tigrani is a SAPERAVI × ARENI cross obtained by S A Pogosyan and S S Khachatryan at the Armenian viticultural research centre at Merdzavan, just west of Armenia's capital Yerevan.

VITICULTURAL CHARACTERISTICS

Medium-sized to large, relatively compact bunches.

WHERE IT'S GROWN

MAP produce a dry varietal red in the Ararat region of **Armenia** but it is also used to produce the sweeter versions of which Armenians are so fond.

TIMORASSO

Rare, high-quality Piemontese earning renewed recognition.

BERRY COLOUR: ● ● ● ● ●

PRINCIPAL SYNONYMS: Morasso, Timoraccio, Timuassa

ORIGINS AND PARENTAGE

Timorasso is an old variety from the Colli Tortonesi region around Tortona in the province of Alessandria in Piemonte, north-west Italy. DNA parentage analysis has revealed a possible parent–offspring relationship between Timorasso and LAMBRUSCHETTO (Torello Marinoni *et al.* 2006), a variety typical of Castelnuovo Bormida, also in the province of Alessandria, thus anchoring Timorasso's roots in Piemonte.

OTHER HYPOTHESES

A Latin document from Tortona, dated 1209, mentions *vineam de gragnolato*, tentatively identified by some authors as a vineyard of Timorasso (Raimondi *et al.* 2005). A genetic relationship with NEBBIOLO has also been suggested, but this was not confirmed by recent studies (Anna Schneider and José Vouillamoz, unpublished data).

VITICULTURAL CHARACTERISTICS

Early to mid ripening. Vigorous but relatively low yielding (lower than CORTESE), small sugar-rich berries. Susceptible to botrytis bunch rot and millerandage.

WHERE IT'S GROWN AND WHAT ITS WINE TASTES LIKE

Timorasso used to be one of the most praised and widespread varieties in the province of Alessandria in Piemonte, north-west **Italy**, both for wine and for the table. When vineyards were replanted in the early twentieth century after the ravages of phylloxera, the more productive CORTESE found greater favour and plantings of Timorasso declined. However, over the last few decades, a handful of producers, notably Walter Massa, have been replanting and nurturing the variety in the Tortona region, where some of the best vineyards are on the steep slopes just east of Tortona and where the vineyard area has expanded from around 6 ha (15 acres) in 2000 to double that today. The agricultural census of 2000 shows total plantings in Italy of 20 ha (49 acres). It is authorized in the provinces of Asti and Alessandria and varietal wines are permitted within the Colli Tortonesi and Monferrato DOCs, where it may also be included in blends with other local varieties such as Cortese.

Timorasso is definitely too interesting a variety to be hidden in a blend. Even in youth, varietal wines have complex aromas of light honey and spice as well as floral, citrus and nutty characteristics and a creamy texture – they sometimes taste as if they are lightly oaked even when they are not. The acidity is fresh and well-made wines have excellent length, delicate minerality and surprising longevity. Producers of such wines include Luigi Boveri, Franco Martinetti, Walter Massa and Morgassi.

TINTA AMARELA

See TRINCADEIRA.

TINTA BARROCA

Rustic, dark-skinned, early-ripening port variety.

BERRY COLOUR:

PRINCIPAL SYNONYMS: Boca de Mina (Douro), Tinta Barocca or Tinta das Baroccas (South Africa)

VARIETIES COMMONLY MISTAKEN FOR TINTA BARROCA: GROSSA ☒ or Tinta Grossa (Douro) (Douro)

ORIGINS AND PARENTAGE

Tinta Barroca, meaning 'black baroque', comes from the Douro region of northern Portugal, where it has long been cultivated for port wines. Until the end of the nineteenth century, it was generally known as Tinta Grossa ('big black') but it should not be confused with GROSSA from the Alentejo, which is also known as Tinta Grossa. According to genetic studies, Tinta Barroca could be closely related to TOURIGA FRANCA, also from the Douro region (Almadanim *et al.* 2007).

VITICULTURAL CHARACTERISTICS

Productive, vigorous and early ripening. Susceptible to powdery and downy mildews, less so to botrytis bunch rot. Medium-sized loose bunches of medium-sized berries that are prone to heat damage and shrivel.

WHERE IT'S GROWN AND WHAT ITS WINE TASTES LIKE

Tinta Barroca is one of the five most important varieties used in the production of port in the Douro in northern **Portugal**, and the third most planted dark-skinned port variety after TOURIGA FRANCA and Tinta Roriz (TEMPRANILLO), accounting for 11.2% of vines in the Douro in 2004. Its popularity is due in large part to its abundant yields and to the fact that it ripens a good two weeks before other red varieties in the region. This precocity explains why it is planted mainly in the Baixo Corgo and Cima Corgo subregions as it does not like the heat of the Douro Superior. It is also planted in the Távora-Varosa region immediately south of the Douro for blended unfortified wines. Total plantings in 2010 were 4,455 ha (11,009 acres). Varietal wines are typically soft, full-bodied, rustic and high in alcohol with flavours of cherry and blackberry but lacking in complexity and acidity, which means Tinta Barroca is best in a blend, in both port and table wines. Its colour fades more quickly than that of Touriga Franca and TOURIGA NACIONAL. However, it can produce fresher, more elegant and aromatic wines when planted on higher, cooler sites (over 500 m above sea level). Quinta das Tecedeiras are unusual in making a varietal wine.

Tinta Barroca is one of the few port varieties that have been exported. **South Africa** recorded 261 ha (892 acres) in 2009, though more vines have been uprooted than planted in the last few years. It is used mostly for fortified wines but a few producers of port-style wines such as Allesverloren, Boplaas and DeKrans also make varietal table wines from the variety they spell Tinta Barocca, as do Lammershoek.

TINTA CAIADA

See PARRALETA.

TINTA CARVALHA

Portuguese variety that is light in every way.

BERRY COLOUR: ● ● ● ● ●

PRINCIPAL SYNONYMS: Lobão (Dão), Preto Gordo (Pinhel in Bairrada), Tinta Carvalha du Douro

VARIETIES COMMONLY MISTAKEN FOR TINTA CARVALHA: RUFETE

ORIGINS AND PARENTAGE

This variety probably comes from northern Portugal, from within the Douro or Trás-os-Montes (Rolando Faustino, personal communication).

OTHER HYPOTHESES

Although both RUFETE and Tinta Carvalha may be called Tinta Carvalha in the Douro, their DNA profiles (Martín *et al.* 2006; Almadanim *et al.* 2007; Santana *et al.* 2007; Santana *et al.* 2010) are quite distinct (Vouillamoz).

VITICULTURAL CHARACTERISTICS

Medium-sized, compact bunches of largish, thin-skinned berries. Susceptible to powdery mildew and botrytis bunch rot. Late ripening.

WHERE IT'S GROWN AND WHAT ITS WINE TASTES LIKE

Varietal wines are typically light in colour and body with low acidity, aromas of red fruits and, not surprisingly, little ageing potential. There were 1,171 ha (2,894 acres) in **Portugal** in 2010, mostly in the Douro and Trás-os-Montes regions and to a lesser extent in Beiras, the Tejo and the Alentejo, although it is authorized in most of Portugal's wine-growing regions.

TINTA CASTAÑAL

Recently resurrected Galician of unknown origin.

BERRY COLOUR: ● ● ● ● ●

PRINCIPAL SYNONYMS: Castañal☒, Rabo de Cordeiro

ORIGINS AND PARENTAGE

Tinta Castañal is an almost extinct variety from the Spanish side of the River Miño, on the border with Portugal, where it was both widely cultivated and highly praised until the early twentieth century (Santiago *et al.* 2008). Its DNA profile does not correspond to any other Spanish or Portuguese variety (Santiago *et al.* 2008; Gago *et al.* 2009) but a likely parent–offspring relationship with AMARAL (under the name Caíño Bravo) has been suggested (Díaz-Losada *et al.* 2011).

WHERE IT'S GROWN

Although this variety is not included in the official Spanish register of commercial varieties, not even provisionally, cuttings from vines more than 200 years old discovered in the areas of O Rosal, Goian and Tuy (along the Miño in the south of Pontevedra province, on the border with Portugal) were added to the collection at Misión Biológica de Galicia in 1993 and commercial vineyards were recently planted at the Viñas do Torroxal winery in the O Rosal Valley in Rías Baixas, in the north west of **Spain** (Santiago *et al.* 2008). Their Tinto Torroxal is a blend of Caíño Tinto (BORRAÇAL), Sousón (VINHÃO), Brancellao (ALVARELHÃO), Tinta Castañal and PEDRAL.

TINTA DE MADEIRA

See NEGRAMOLL.

TINTA DE TORO

See TEMPRANILLO.

TINTA FRANCISCA

Lesser but not insignificant port variety.

BERRY COLOUR: ● ● ● ● ●

PRINCIPAL SYNONYMS: Tinta de França, Tinta Francesca, Tinta Franceza

VARIETIES COMMONLY MISTAKEN FOR TINTA FRANCISCA: TEINTURIER (Douro)

ORIGINS AND PARENTAGE

Tinta Francisca is from northern Portugal, possibly originating from the Douro. According to recent DNA studies, Tinta Francisca could be genetically related to TINTO CÃO and VIOSINHO (Pinto-Carnide *et al.* 2003) as well as to RABIGATO (Castro *et al.* 2011), all from the Douro or neighbouring regions. Tinta Francisca is also cultivated under the name Teinturier Mâle (Galet 2000) but it is morphologically and genetically distinct from the real red-fleshed French variety TEINTURIER.

Tinta Francisca means 'French black' and it has been suggested that the variety was introduced from France, perhaps from Burgundy, by Robert Archibald, who planted it at his Quinta de Roriz, where it was thought to be PINOT NOIR (Galet 2000). This theory can be rejected by DNA profiling and by the genetic relationships of Tinta Francisca with TINTO CÃO, VIOSINHO and RABIGATO.

VITICULTURAL CHARACTERISTICS

Low, irregular productivity. Mid budding, late ripening. Thin skins make the berries susceptible to powdery mildew and botrytis bunch rot. Best suited to dry soils and hot, sunny sites.

WHERE IT'S GROWN AND WHAT ITS WINE TASTES LIKE

Found mostly in the Douro in north-eastern Portugal, Tinta Francisca produces aromatic wines generally high in alcohol and low in acidity but not especially concentrated. It is used mainly as a minor ingredient in the production of port, by producers such as Niepoort and Quinta das Carvalhas, often from a field blend. Quinta da Revolta make a rare varietal unfortified wine. There were 32 ha (79 acres) in **Portugal** in 2010.

There are a very few hectares in **South Africa**, where it is also included in port-style blends such as that made by Allesverloren.

TINTA MIÚDA

See GRACIANO.

TINTA NEGRA

Previously known as Tinta Negra Mole. See NEGRAMOLL.

TINTA PINHEIRA

See RUFETE.

TINTA RORIZ

See TEMPRANILLO.

TINTILIA DEL MOLISE

Minor Molise variety until recently confused with Bovale Grande (ie Mazuelo/Carignan).

BERRY COLOUR:

PRINCIPAL SYNONYMS: Tintilia (Molise)

VARIETIES COMMONLY MISTAKEN FOR TINTILIA DEL MOLISE: MAZUELO✗ (under the name Bovale Grande on Sardegna), Tintilia✗ (Campania, Sardegna), Tintilla de Rota✗ (Spain)

ORIGINS AND PARENTAGE

Tintilia is a particular variety of the Molise region in south-central Italy, where it is said to have been introduced from Spain during Bourbon rule in the middle of the eighteenth century. The variety presumably took its name from the Spanish word *tinto*, meaning 'black', although an etymology based on the Italian *tintore*, meaning 'dyer', owing to its ability to stain clothes, is also plausible.

Tintilia has long been confused with Bovale Grande (aka MAZUELO or Carignan) on Sardegna and is not listed separately in the Italian agricultural census of 2000. However, recent DNA investigations have confirmed that they are not the same and that Tintilia del Molise is also distinct from the obscure varieties known as Tintilia in Campania and Sardegna and even Tintilla de Rota from Spain (Reale *et al.* 2006). To prevent further confusion, the full name Tintilia del Molise, referring to its likely place of origin, is used here. Nonetheless, a possible Iberian origin cannot be ruled out, because Reale *et al.* (2006) found Tintilia del Molise to be relatively close to the Spanish PARRALETA.

VITICULTURAL CHARACTERISTICS

Mid to late ripening. Susceptible to botrytis bunch rot, downy and powdery mildews. Highly resistant to drought and chlorosis.

WHERE IT'S GROWN AND WHAT ITS WINE TASTES LIKE

Tintilia del Molise is cultivated mainly in Molise in south-central **Italy** and to a lesser extent in the province of Chieti in Abruzzo to the north. It is often blended with other local varieties such as MONTEPULCIANO, but a few interesting varietal wines have recently been made by, for example, Di Majo Norante, who were among the first wineries to rediscover Tintilia del Molise and make a varietal wine (as well as blending it with Montepulciano for their top cuvée, Don Luigi), Angelo d'Uva, Catabbo and Cipressi. Such wines, permitted within the Molise DOC, tend to be dense and full-bodied with aromas of ripe dark fruit – blackberries, cherries or plums – and fine tannins.

TINTO CÃO

High-quality port variety that occasionally shines in Dão.

BERRY COLOUR: ● ● ● ● ●

PRINCIPAL SYNONYMS: Castellana Negra, Farmento, Tinta Cão

VARIETIES COMMONLY MISTAKEN FOR TINTO CÃO: PADEIRO ✗ (Vinho Verde)

ORIGINS AND PARENTAGE

This is a very old variety from the Douro and Dão in northern Portugal. It has been known there since the seventeenth century (Rolando Faustino, personal communication) and some authors consider it to be one of the oldest varieties in the country. Tinto Cão means 'red dog' but the reason for the name is uncertain.

According to recent studies, Tinto Cão could be genetically related to VIOSINHO and TINTA FRANCISCA (Pinto-Carnide *et al.* 2003). Tinto Cão was used to breed the hugely successful colouring grape RUBIRED.

VITICULTURAL CHARACTERISTICS

Low yields of small, thick-skinned berries that are resistant to botrytis bunch rot. Very late ripening.

WHERE IT'S GROWN AND WHAT ITS WINE TASTES LIKE

Tinto Cão is one of the five most important port grapes planted in the Douro, **Portugal**, often described as a little more feminine than, say, TOURIGA NACIONAL, and although it is particularly valued for vintage port because of its capacity to age, it has never been widely planted – probably because of its low yields. The Visconde de Vila Maior, a viticulture expert in the Douro in the nineteenth century, noted that the variety ripens well, does not dry up or rot and, despite low production, gives rise to a very full, strong and generous wine (Vila Maior 1875). Now that yield is no longer quite such a critical attribute, plantings of this variety are on the increase and it is used in blended table wines as well as for port.

In Dão the variety is more controversial: rather like BAGA, it can produce both great and awful wines, mainly because it is not easy to ripen in all parts of the region. When the fruit reaches full maturity, however, the wines are deeply coloured, aromatic (wild flowers and red fruits), almost feminine, yet full-bodied at around 13.5% alcohol and impressively powerful, rounded and fresh on the palate and with great potential longevity, elegance and length (Rolando Faustino, personal communication). Recent plantings further south in the Beira Interior DO have shown encouraging results: wines that are austere in youth, with rather more acidity than most port grapes, but age elegantly. Recommended producers include Herdade São Miguel in the Alentejo, Luís Pato in Beiras and Quinta das Maias in Dão. There were 374 ha (924 acres) in Portugal in 2010.

Although Tinto Cão is not included in **California**'s official vineyard statistics for 2009, the variety was planted by Austin Goheen in the University of California at Davis experimental vineyards at Jackson in the 1960s, providing several selections for commercial propagation. Some producers of port-style wines use the variety in their blends and a few such as Pierce (San Antonio Valley) include it in blended unfortified reds and occasionally make a varietal. Cinquain (Paso Robles) make a sweet red from a blend of Tinto Cão plus TOURIGA FRANCA and TOURIGA NACIONAL.

Plantings in **Australia** are also too insignificant to be included in official statistics but there are a few scattered outposts. For example, Peel Estate in Western Australia use Tinto Cão in their port-style wines, as do Campbells of Rutherglen in Victoria. The latter have also had success with a dry red wine made from a blend of Portuguese varieties including Tinto Cão, all from small areas planted in a vineyard block dating back to the 1960s – so much so that they have plans to expand the area dedicated to these varieties. Mansfield Wines in Mudgee even make a varietal dry red.

TINTO DEL PAÍS

See TEMPRANILLO.

TINTO FINO

See TEMPRANILLO.

TINTO VELASCO

Generally unexciting La Mancha variety.

BERRY COLOUR: ● ● ● ● ●

PRINCIPAL SYNONYMS: Benitillo, Blasco, Frasco, Graciano de Alfaro, Granadera, Tinto de la Pámpana Blanca, Tinto Velasco Peludo

VARIETIES COMMONLY MISTAKEN FOR TINTO VELASCO: ALICANTE HENRI BOUSCHET

ORIGINS AND PARENTAGE

Tinto Velasco most likely originates from Castilla-La Mancha in central Spain. DNA profiling has confirmed Blasco, Frasco and Granadera as synonyms (Fernández-González, Mena *et al.* 2007) and has unexpectedly shown that Tinto Velasco is also identical to Benitillo and to Tinto de la Pámpana Blanca, another black-berried variety (despite its name) from La Mancha (Borrego, Cabello *et al.* 2002).

Confusingly, the name Tinto Velasco is used for both this distinct variety and also for ALICANTE HENRI BOUSCHET.

WHERE IT'S GROWN

Tinto Velasco has been provisionally included in the register of commercial varieties in **Spain**, although it also lists Tinto de la Pámpana Blanca as a separate variety, which explains why there are national 2008 vineyard totals of 2,287 ha (5,651 acres) of Tinto Velasco as well as 4,908 ha (12,128 acres) of the latter. However, since the name Tinto Velasco is also used for ALICANTE HENRI BOUSCHET, it is impossible to know which variety these statistics refer to and whether a wine labelled Tinto Velasco is made from this variety or from Alicante Henri Bouschet. Tinto de la Pámpana Blanca is authorized in DOs such as Extremadura and La Mancha. Elías López Montero at Verum in La Mancha has shown that careful selection in the vineyard can produce surprisingly good results with Tinto Velasco.

TINTORE DI TRAMONTI

Rare, deeply coloured Campanian with excellent potential for fine reds.

BERRY COLOUR: ● ● ● ● ●

PRINCIPAL SYNONYMS: Cannamelu (Ischia), Guarnaccia (Ischia), Tintora (Ischia), Tintore

ORIGINS AND PARENTAGE

Tintore means 'dyer' and this comes from the province of Avellino in Campania, southern Italy. DNA profiling has suggested genetic relationships between Tintore di Tramonti and three other minor Campanian varieties that are not significantly cultivated (Costantini *et al.* 2005): Tintiglia (distinct from TINTILIA DEL MOLISE), Livella Ischia and Mangiaguerra (distinct from MAGLIOCCO DOLCE, sometimes called Mangiaguerra in Calabria).

VITICULTURAL CHARACTERISTICS

Vigorous but poor fertility due to sensitivity to poor weather at flowering; early ripening (late September). Not susceptible to botrytis bunch rot thanks to its loose bunches and thick skins.

WHERE IT'S GROWN AND WHAT ITS WINE TASTES LIKE

As its name suggests, Tintore di Tramonti is cultivated mainly in the area around the little town of Tramonti in the Monte Lattari just north of Salerno in Campania, southern **Italy**, near the Amalfi coast, where old and ungrafted vines can still be found trained on pergolas. It is blended with PIEDIROSSO and other local varieties in the Costa d'Amalfi DOC, as well as in Ischia DOC, on the island of the same name, where it is known as Guarnaccia or Cannamelu. Varietal wines, which have to be classified as IGTs such as Campania or Colli di Salerno, are very deeply coloured, as the name suggests, rich in dark fruit but with firm acidity to give structure, sometimes peppery flavours mixed with the fruit, sometimes more savoury and earthy. They can age well thanks to the high acidity and firm tannins that become silky with time in bottle. Examples of Tintore or Tintore-dominated wines: Reale's oaked Borgo di Gete, Colli di Salerno DOC, and Alfonso Arpino's Monte di Grazia IGT Campania, a blend of 90% Tintore and 10% Piedirosso. There are less than 10 ha (25 acres) of this variety on the Amalfi coast and the total production is around 8,000 bottles a year.

TOCAI FRIULANO

See SAUVIGNONASSE.

TOCAI ROSSO

See GARNACHA.

TORBATO

Pale-berried variety found on the Italian island of Sardegna and in southern France, recovering from virtual extinction.

BERRY COLOUR:

PRINCIPAL SYNONYMS: Malvoisie du Roussillon (France), Malvoisie des Pyrénées Orientales (France), Torbat, Tourbat (France), Turbat, Uva Catalana

VARIETIES COMMONLY MISTAKEN FOR TORBATO: MAUZAC BLANC

ORIGINS AND PARENTAGE

DNA profiling recently confirmed that Torbato is identical to Malvoisie du Roussillon in southern France, also called Tourbat or Malvoisie des Pyrénées Orientales (Lacombe *et al.* 2007), and mentioned in Limoux, southern France, as early as 1804 (Galet 2000). Like many Sardinian varieties, Torbato is said to have been introduced from Spain during Catalan rule of Sardegna in the fourteenth and fifteenth centuries but there is as yet no evidence to confirm this hypothesis.

VITICULTURAL CHARACTERISTICS

Mid to late ripening. Poor resistance to powdery mildew and fanleaf virus.

WHERE IT'S GROWN AND WHAT ITS WINE TASTES LIKE

In **Italy**, Torbato is cultivated mainly in the area around Alghero west of Sassari on the north-west coast of Sardegna and also further inland in Mejlogu. It is made as varietal wine in the Alghero DOC. One of the rare examples of varietal wines is Sella & Mosca's apple and pear-flavoured Terre Bianche, which has a slightly bitter finish. There were 143 ha (353 acres) in Italy in 2000 but this is a considerable increase on the lonely 10 ha (25 acres) that remained in 1970.

In **France**, this variety had virtually disappeared at the beginning of the twentieth century because of its susceptibility to powdery mildew and fanleaf degeneration (*court-noué* in French). New and healthier plant material was reintroduced from Sardegna thanks to work done there in the 1970s and there are now around 33 ha (82 acres) of Tourbat, which is recommended in the Aude, Hérault and Pyrénées-Orientales *départements*. It is generally blended with other local varieties to produce vins doux naturels in Banyuls, Grand Roussillon, Côtes du Roussillon, Maury and Rivesaltes.

TORRONTÉS

BERRY COLOUR:

ORIGINS AND PARENTAGE

The name Torrontés is used for several distinct varieties in Spain and South America (see CRIOLLA, TORRONTÉS MENDOCINO, TORRONTÉS RIOJANO and TORRONTÉS SANJUANINO for more on South America).

In Spain, DNA profiling by Ibañez *et al.* (2003) has shown that at least four distinct varieties are known under the name Torrontés:

- in Córdoba (Montilla-Moriles in Andalucía), it has a unique DNA profile, also found under Torrontés de Montilla in Martín *et al.* (2003)
- in Navarra, it turns out to be identical to Hebén in Martín *et al.* (2003), an almost extinct Andalusian variety
- in Orense (Ribeiro in Galicia), one reference vine is identical to FERNÃO PIRES in Martín *et al.* (2003) and other sources, and the other is identical to BICAL in Sefc *et al.* (2000), the latter often confused with Fernão Pires.

In addition, the name Torrontés is also sometimes given to ALARIJE in Spain, and Borrego, De Andres *et al.* (2002) have found another Torrontés in Barbantes (Ribeiro in Galicia) that is distinct from the ones in Orense and Córdoba. Furthermore, the variety called Turruntés in Haro and Logroño (La Rioja) is identical to ALBILLO MAYOR from Valladolid, and the one called Torrontés in Galicia is identical to MALVASIA FINA (Borrego, De Andres *et al.* 2002; Gago *et al.* 2009).

In Portugal, the name Torrontés is also used for Arinto (ARINTO DE BUCELAS), DONA BRANCA, FERNÃO PIRES and MALVASIA FINA (under the names Assario and Boal Branco; Lopes *et al.* 1999; Pinto-Carnide *et al.* 2003; Vouillamoz).

Confusion reigns supreme over Torrontés in the Iberian peninsula, so much that it seems impossible for now to tell exactly which variety is cultivated under this name in the various regions where it is authorized or recommended. The Torrontés planted in Galicia is probably the most important among them. According to Galet (2000), this Torrontés is also called Monastrell Blanco and Morrastrell Blanco, but none of the Torrontés listed above is a colour mutation of either MONASTRELL or GRACIANO (also called Morrastell).

TORRONTÉS MENDOCINO

A lesser Argentine Torrontés.

BERRY COLOUR: ● ● ● ● ●

PRINCIPAL SYNONYMS: Chichera, Loca Blanca (all in Río Negro), Palet

VARIETIES COMMONLY MISTAKEN FOR TORRONTÉS MENDOCINO: Moscatel Amarillo (Chile), TORRONTÉS RIOJANO (Argentina), TORRONTÉS SANJUANINO (Argentina)

ORIGINS AND PARENTAGE

Torrontés Mendocino is indigenous to Argentina, where it was long confused with the equally obscure TORRONTÉS SANJUANINO and the more widespread TORRONTÉS RIOJANO (see the latter entry for geographic origin and etymology) until they were recently distinguished by morphological descriptions and DNA profiling (Alcalde 1989; Agüero *et al.* 2003; Lacoste *et al.* 2010). Through DNA parentage analyses, Agüero *et al.* (2003) have shown that Torrontés Mendocino is a natural cross between MUSCAT OF ALEXANDRIA (Moscatel in Argentina) and an unidentified variety, making it a member of the group of varieties known as the Criollas (see CRIOLLA entry and MUSCAT pedigree diagram).

WHERE IT'S GROWN AND WHAT ITS WINE TASTES LIKE

Torrontés Mendocino, presumably taking its name from the province of Mendoza, even though it is now a little more widely planted in the province of Rió Negro in the south, tends to lack the fragrant Muscat-like aroma of TORRONTÉS RIOJANO. There were 666 ha (1,646 acres) planted in **Argentina** in 2009.

TORRONTÉS RIOJANO

The most important of the three Argentine Torrontés Somethings, responsible for the country's highly aromatic signature white wine.

BERRY COLOUR: ● ● ● ● ●

PRINCIPAL SYNONYMS: Malvasia (San Juan)

VARIETIES COMMONLY MISTAKEN FOR TORRONTÉS RIOJANO: Moscatel Amarillo (Chile), MUSCAT BLANC À PETITS GRAINS (Mendoza), TORRONTÉS MENDOCINO (Argentina), TORRONTÉS SANJUANINO (Argentina)

ORIGINS AND PARENTAGE

Torrontés Riojano was first mentioned in the provinces of La Rioja (whence its name), San Juan and Mendoza by Damián Hudson in 1867 (Lacoste *et al.* 2010). It was often speculated that Torrontés Riojano was one of several varieties that had been introduced from Spain and that the name Torrontés was borrowed from the Spanish Torrontés (sometimes Turrontés or Turruntés), a vague name used in Spain for ALARIJE, ALBILLO MAYOR and MALVASIA FINA. However, the DNA profiles of all these varieties are quite distinct (Vouillamoz).

Through DNA parentage analyses, Agüero *et al.* (2003) have shown that Torrontés Riojano is a natural cross between MUSCAT OF ALEXANDRIA (Moscatel in Argentina) and LISTÁN PRIETO (Criolla Chica in Argentina), making it a member of the group of varieties known as the Criollas (see CRIOLLA entry and MUSCAT pedigree diagram for details).

VITICULTURAL CHARACTERISTICS

Productive, vigorous, mid budding, early ripening. Medium-sized to large bunches of relatively large berries with thickish skins. Susceptible to downy mildew and botrytis bunch rot.

WHERE IT'S GROWN AND WHAT ITS WINE TASTES LIKE

Torrontés Riojano is the most valued and widespread of the three indigenous varieties called Torrontés Something (more usually and confusingly just Torrontés) in **Argentina**, often described as the country's signature white wine variety and its mostly widely planted light-skinned variety after the uninspiring and declining PEDRO GIMÉNEZ. At its best, it produces fresh, highly aromatic wines, typically floral and Muscat-like, although poor examples can be alcoholic with a bitter aftertaste. It seems to do particularly well in the higher, cooler vineyards of Cafayate in the province of Salta in the north of the country, although there are also some very successful wines from the Valle de Uco in the Mendoza region and from Chilecito in La Rioja. Given its very distinctive flavours, it also makes a useful blending component with more neutral varieties such as PINOT GRIS. Improving the general quality of Torrontés has been a speciality of Susana Balbo of Dominio del Plata. Other recommended producers of varietal wines include Catena Zapata, Colomé, O Fournier and Mauricio Lorca. There were 8,443 ha (20,863 acres) in Argentina in 2009, mostly in Mendoza, La Rioja, San Juan and Salta but also in Catamarca and Río Negro.

Pisano have planted Torrontés Riojano in their vineyards in Progreso, **Uruguay**, and make a wine that is more powerful than most Argentine versions, less floral, with intense citrus and pear flavours.

TORRONTÉS SANJUANINO

One of the two lesser-known Torrontés Somethings in Argentina.

BERRY COLOUR: ● ● ● ● ●

PRINCIPAL SYNONYMS: Moscatel de Austria (Chile), Moscatel Romano (Mendoza in Argentina)

VARIETIES COMMONLY MISTAKEN FOR TORRONTÉS SANJUANINO: Moscatel Amarillo (Chile), TORRONTÉS MENDOCINO (Argentina), TORRONTÉS RIOJANO (Argentina)

ORIGINS AND PARENTAGE

Torrontés Sanjuanino is indigenous to Argentina, where it was long confused with the equally obscure TORRONTÉS MENDOCINO and the more widespread TORRONTÉS RIOJANO (see the latter entry for geographic origin and etymology) until they were recently distinguished by morphological descriptions and DNA profiling (Alcalde 1989; Agüero *et al.* 2003; Lacoste *et al.* 2010). Through DNA parentage analyses, Agüero *et al.* (2003) have shown that Torrontés Sanjuanino is a natural cross between MUSCAT OF ALEXANDRIA (Moscatel in Argentina) and LISTÁN PRIETO (Criolla Chica in Argentina), making it a member of the group of varieties known as the Criollas (see CRIOLLA entry and MUSCAT pedigree diagram).

VITICULTURAL CHARACTERISTICS

Medium-sized to large, compact bunches. Mid ripening. Susceptible to botrytis bunch rot.

WHERE IT'S GROWN AND WHAT ITS WINE TASTES LIKE

Not surprisingly, Torrontés Sanjuanino is found mainly in the province of San Juan in **Argentina**, although it is also planted in Mendoza and La Rioja. Wines are typically less aromatic than those made from TORRONTÉS RIOJANO. Total plantings were 2,539 ha (6,274 acres) in 2009, the majority, unsurprisingly, in San Juan. The variety known in Chile as Torrontés has recently been identified as Torrontés San Juanino.

TORTOZÓN

See PLANTA NOVA.

TORYSA

Recently authorized red-fleshed Slovakian cross.

BERRY COLOUR: ● ● ● ● ●

PRINCIPAL SYNONYMS: CATAP 9/17

ORIGINS AND PARENTAGE

Torysa, named after a river in eastern Slovakia, is a CASTETS × I-35-9 cross obtained in 1967 by Dorota Pospíšilová at the VUVV research centre for oenology and viticulture in Bratislava, Slovakia, and authorized in 2011. I-35-9 is an obscure (TEINTURIER × ALEATICO) × PUKHLIAKOVSKY cross obtained in Russia (Dorota Pospíšilová, personal communication). Torysa is a sibling of RUDAVA.

VITICULTURAL CHARACTERISTICS

Very vigorous, late budding, mid to late ripening. Susceptible to powdery mildew and affected by long dry spells but resistant to frost and botrytis bunch rot. Short, red-fleshed berries with an intense Cabernet flavour.

WHERE IT'S GROWN AND WHAT ITS WINE TASTES LIKE

Torysa is one of the few red-fleshed varieties to produce deeply coloured, ageworthy Cabernet-like wines (Dorota Pospíšilová, personal communication). Producers include Chateau Marco in south-west **Slovakia** and Igor Blaho near Trnava in the west. Less than 1 ha (2.5 acres) was grown in Slovakia in 2011.

TOURBAT

See TORBATO.

TOURIGA FÊMEA

Minor Touriga from northern Portugal.

BERRY COLOUR: ● ● ● ● ●

PRINCIPAL SYNONYMS: Tinta Coimbra (Dão), Touriga Brasileira (Douro)

VARIETIES COMMONLY MISTAKEN FOR TOURIGA FÊMEA: TOURIGA FRANCA🧬

ORIGINS AND PARENTAGE

Touriga Fêmea most probably originates from the Douro in northern Portugal. DNA parentage analysis has suggested a possible parent–offspring relationship with TOURIGA NACIONAL (Lopes *et al.* 2006), which would make it a half-sibling (or, in theory, a grandparent) of TOURIGA FRANCA, all from the same region.

VITICULTURAL CHARACTERISTICS

Mid ripening. Small, compact bunches of small, thick-skinned berries.

WHERE IT'S GROWN AND WHAT ITS WINE TASTES LIKE

Touriga Fêmea wines are deeply coloured, full-bodied yet soft with fine tannins and flavours of black fruits and scrubland. The variety is authorized for DOC Douro wines and also for various regional wines, mostly in the northern half of **Portugal** but also in the Península de Setúbal. In the Douro, Rui José Xavier Soares includes this variety, under the name Touriga Brasileira, in his Esmero red Douro blend and Quinta da Revolta make a rare varietal wine. There were 43 ha (106 acres) in Portugal in 2010, about half of that in the Douro.

TOURIGA FRANCA

High-quality, widely planted Douro variety qualitatively in the shadow of Touriga Nacional.

BERRY COLOUR: ● ● ● ● ●

PRINCIPAL SYNONYMS: Albino de Souza, Touriga Frances, Touriga Francesa, Tourigo Francês

VARIETIES COMMONLY MISTAKEN FOR TOURIGA FRANCA: TOURIGA FÊMEA

ORIGINS AND PARENTAGE

In spite of its name, Touriga Franca does not originate from France but most likely from the Douro region in northern Portugal. Indeed, recent DNA parentage analysis has shown that Touriga Franca is the offspring of TOURIGA NACIONAL and MARUFO (Castro *et al.* 2011), both also traditionally cultivated in the Douro. Since Touriga Nacional was not traditionally cultivated in Spain, where Marufo is grown under various names, the cross is likely to have occurred in Portugal (Castro *et al.* 2011). Touriga Franca shows little genetic diversity and it was not mentioned in writing until 1880, which suggests that the cross may be relatively recent (Almadanim *et al.* 2007; Rolando Faustino, personal communication). In addition, genetic studies have suggested that Touriga Franca could be closely related to TINTA BARROCA, also from the Douro region (Pinto-Carnide *et al.* 2003; Martín *et al.* 2006; Almadanim *et al.* 2007).

VITICULTURAL CHARACTERISTICS

Mid to late ripening. Generally easy to grow with good and consistent yields and not susceptible to most vine diseases. Compact bunches increase the risk of botrytis bunch rot.

WHERE IT'S GROWN AND WHAT ITS WINE TASTES LIKE

Known until 2000 as Touriga Francesa in **Portugal**, Touriga Franca is the most widely planted dark-skinned variety in the Douro region – 8,919 ha (22,039 acres) or just over 21% of the vineyard area in 2004 – and can reach almost the same quality levels as TOURIGA NACIONAL when planted on warm, south-facing slopes, but in less favourable years or sites, ripening may be less successful due to greater susceptibility to botrytis bunch rot. Varietal wines are less concentrated than Touriga Nacional but show finesse and have aromatic intensity, good depth of colour and moderate body, although for port production they are best in blends. Some 100% Touriga Franca table wines, such as the one from Quinta do Mourão, have started to appear and these show the potential of the variety for high quality and longevity: intense aromas of red fruits, forest fruits, wild flowers, violets and herbs, an excellent ruby/violet colour and firm but fine tannins.

It is also widely planted in Trás-os-Montes in the far north and is slowly being disseminated further south to other Portuguese regions such as Lisboa, the Tejo, the Alentejo and the Península de Setúbal. There are also small areas in Dão. Recommended producers of varietal wines include Ermelinda Freitas in Setúbal and Quinta da Plansel in the Alentejo. There was a total of

12,055 ha (29,789 acres) in Portugal in 2010, down from 14,000 ha (34,595 acres) in 2004, but a far greater area than TOURIGA NACIONAL's 7,268 ha (17,960 acres).

Although Tinta Barca has been used as a synonym for Touriga Franca, Tinta da Barca (occasionally just Barca) is a distinct but scarcely planted – often interplanted – variety from the Douro Superior used mainly in port blends and valued by Ramos Pinto, who have 8 ha (20 acres), and Vargellas, for example.

In **California**, although only TOURIGA NACIONAL shows up in the official statistics and Touriga *tout court* is the official synonym for that variety, growers such as Cinquain (Paso Robles), Pierce and Quinta Cruz (San Antonio Valley) and Revolution Wines (Sacramento) often include both Nacional and Franca in their port-style blends and a few, eg Morgan, make a red table-wine blend from Lodi that includes both.

Australia had 46 ha (114 acres) of 'Touriga' planted in 2008. Although this is far more likely to be TOURIGA NACIONAL, only a few producers use the full name.

TOURIGA NACIONAL

High-quality, concentrated, tannic and perfumed dark-skinned Portuguese variety increasingly cast in a starring role.

BERRY COLOUR: ● ● ● ● ●

PRINCIPAL SYNONYMS: Azal Espanhol, Carabuñeira (Ribeiro in Spain), Mortagua (Dão), Mortagua Preto (Dão), Touriga or Touriga Fina (Dão), Tourigo (Dão), Tourigo Antiguo (Dão), Tourigo do Dão (Bairrada)

VARIETIES COMMONLY MISTAKEN FOR TOURIGA NACIONAL: AMARAL ⚲ (under the name Azal Tinto)

ORIGINS AND PARENTAGE

Touriga Nacional most likely originates from the Dão region, where it was first mentioned in 1822 (Cunha *et al.* 2009) and where it shows higher morphological diversity than elsewhere (Almadanim *et al.* 2007; Rolando Faustino, personal communication). This is consistent with the suggestion that the name Touriga derives from the village of Tourigo (Galet 2000), west of Tondela in the district of Viseu in Dão.

According to recent DNA parentage analyses, Touriga Nacional and MARUFO are the parents of TOURIGA FRANCA from the Douro (Castro *et al.* 2011), and Touriga Nacional could have parent–offspring relationships with TOURIGA FÊMEA, also traditionally cultivated in the Douro and Dão regions (Pinto-Carnide *et al.* 2003; Lopes *et al.* 2006; Almadanim *et al.* 2007). In addition, Touriga Nacional could be genetically very close to RUFETE from Castelo Branco to the south east (Almadanim *et al.* 2007).

Touriga Nacional was used to breed TARRANGO.

Confusingly, the name Bical, used principally for a white-berried variety (see the BICAL entry), is also sometimes used as a synonym for Touriga Nacional in Anadia, between Porto and Coimbra. The alternative Bical Tinto is marginally less confusing.

VITICULTURAL CHARACTERISTICS

Fertile, robust and vigorous but this high vigour and typically drooping foliage increase its propensity to coulure and hamper bunch growth, so that production is generally moderate (sometimes higher in the Alentejo). Small bunches of smallish, thick-skinned berries. Mid ripening. Susceptible to powdery mildew and, to a lesser extent, to downy mildew and botrytis bunch rot.

Touriga Nacional, the most revered variety for port, has become **Portugal**'s poster child for fine dry reds too. The wines are deeply coloured, concentrated, tannic, rich in dark fruit and in some cases distinguished by their fragrance – aromas such as bergamot, rosemary, rock roses or violets. This very intensity can make such wines almost too much on their own and many believe Touriga Nacional is better in a blend. Lighter wines with lower alcohol often have a tea-like aroma. The variety also has a good level of acidity, excellent ageing potential (although the wines often close down for a few years after five years in bottle) and marries well with oak.

Until the end of the nineteenth century, Touriga Nacional accounted for almost 100% of red grape varieties in Dão and there were significant plantings in the Douro, but by the mid twentieth century the area had dropped dramatically, and was down to around 5% in Dão by 1986. This was due largely to the variety's poor productivity, exacerbated by the use of unsuitably vigorous rootstocks in the wake of phylloxera, which made it very unpopular with growers despite the very high potential quality of the fruit. In order to encourage growers to continue with Touriga Nacional, to achieve the variety's fullest potential and to be able to depend on consistent yields from year to year, there has been much investment in clonal selection since the late 1970s and attempts to find the best combinations of rootstocks and clones for higher productivity without loss of quality, in conjunction with less dense planting systems. Recent plantings in the Douro have been mainly in the Douro Superior, towards the Spanish border, and the total area in the Douro is now 1,440 ha (3,558 acres).

By 2010 there were 7,268 ha (17,960 acres) in Portugal, more than double the total area in 2004, and the improvements in planting material as well as increased global recognition for the wines have encouraged the southward migration of Touriga Nacional, which is currently planted in Bairrada, the Tejo, Lisboa, the Península de Setúbal, the Alentejo and even in the Algarve. However, wines from the hotter south tend to be less elegant and perfumed. Recommended producers of varietal wines include Quinta do Couquinho and Quinta do Noval in the Douro, Quinta dos Roques in Dão and Quinta da Sapeira in the Encostas d'Aire DOC (central western Portugal).

The recent rise in popularity of Touriga Nacional is not limited to Portugal. The variety has already spread to Priorat in **Spain**, to Australia and California.

The 220 acres (89 ha) planted in **California** in 2009, more than half in the San Joaquin Valley but also encroaching on Napa, are used in much the same way as in Australia. The TTB (Alcohol and Tobacco Tax and Trade Bureau) does not permit producers in the US to distinguish between the two Tourigas on the label even though this is legal for imported wines. Franca Morgan's Rio Tinto unfortified red blend (Santa Lucia Highlands) includes Touriga Nacional, TOURIGA FRANCA and other Portuguese varieties. Kenneth Volk's Pomar Junction is an unfortified varietal Touriga Nacional from Paso Robles (labelled Touriga *tout court*) and York Creek Vineyards (Spring Mountain, Napa) make an impressive rosé exclusively from Touriga Nacional. Ficklin Vineyards (Madera) make a fine range of port-style wines that have Touriga in the mix and Quady's Starboard (Madera) is a sweet, fortified blend of TINTO CÃO and Touriga Nacional.

There are also minor plantings in **Washington** State and **Virginia** as well as in **Argentina** and **Brazil**.

Although Australians tend to use the name Touriga to refer to both Touriga Nacional and TOURIGA FRANCA, it is likely that almost all of the scattered 46 ha (114 acres) in **Australia** are Touriga Nacional, used for both port-style and table wines. Coates and Old Mill in Langhorne Creek and Kaesler and Tscharke in the Barossa are among those who make varietal wines, the last one a fortified version. Trinity Hill in Hawke's Bay, **New Zealand**, make a fortified 100% Touriga Nacional.

Touriga Nacional has long been grown to a limited extent in the hotter, inland wine regions of **South Africa**, where it can make some quite convincing ripostes to port, generally in blends. Recommended producers include Allesverloren, Boplaas and De Krans. There was a total of 87 ha (215 acres) of Touriga Nacional in 2008, the greatest proportion in Little Karoo.

TOURKOPOULA

Rare, pinky-red-skinned Greek variety that should not be confused with Roditis.

BERRY COLOUR: ● ● ● ● ●

PRINCIPAL SYNONYMS: Rhoditis Kokkinos, Roditis Kokkinos

ORIGINS AND PARENTAGE

Tourkopoula, meaning 'young Turkish woman', comes from north-western Greece and the Pelopónnisos (Peloponnese) and has been shown by DNA profiling to be identical to Roditis Kokkinos ('red from Ródos') but distinct from the more widely planted pink-skinned RODITIS and the white-berried Roditis Lefkos (Lefort and Roubelakis-Angelakis 2001; Greek Vitis Database). The berry skins are reddish pink.

WHERE IT'S GROWN AND WHAT ITS WINE TASTES LIKE

Tourkopoula is planted to a limited extent on southern mainland **Greece**, the western Pelopónnisos and on several islands, including Zákynthos and Kérkyra (Corfu). Mercouri in the western Pelopónnisos have planted Tourkopoula and blend it with RIBOLLA GIALLA to make a wine that is light and citrus-flavoured.

TRAJADURA

High-quality variety best in blends in the north-western corner of both Spain and Portugal.

BERRY COLOUR: ● ● ● ● ●

PRINCIPAL SYNONYMS: Treixadura (Galicia in Spain)

ORIGINS AND PARENTAGE

This variety is called Trajadura in the Douro and Minho regions of northern Portugal and Treixadura in Galicia in north-west Spain. It is not genetically related to Galician varieties and is thought to have been introduced to Galicia from other vine-growing areas (Vidal *et al.* 1999), possibly from the Minho since it is genetically fairly close to ARINTO DE BUCELAS, AZAL and LOUREIRO (Lopes *et al.* 1999; Almadanim *et al.* 2007), three varieties with which Trajadura is often blended in the Vinho Verde DOC.

VITICULTURAL CHARACTERISTICS

Very vigorous and very productive. Medium-sized, compact bunches of thin-skinned grapes. Late budding and mid ripening, thus a short cycle. Highly susceptible to downy mildew and botrytis bunch rot, less so to powdery mildew.

WHERE IT'S GROWN AND WHAT ITS WINE TASTES LIKE

Trajadura is recommended throughout the Minho region in north-west **Portugal** with the exception of the Basto and Amarante subregions, although it was originally concentrated in the north west of the region, in Monção, Melgaço, Valença, Vila Nova de Cerveira and Caminha.

There was a total of 1,415 ha (3,497 acres) in 2010, down from 2,200 ha (5,436 acres) in 2004, but plantings now seem to be relatively stable.

Wines are typically golden-straw in colour, have delicate aromas of apples, pears and peaches, sometimes lemon, and are naturally low in acidity, and therefore better blended, eg with LOUREIRO and/or Padernã (ie ALVARINHO). Recommended producers include Casa de Vila Boa, Quinta da Pousada and Quinta do Regueiro.

In **Spain** there were 840 ha (2,076 acres) of Treixadura in 2008, up from 660 ha (1,631 acres) in 2004, virtually all in Galicia in the far north west. Here it is the main variety in the Ribeiro DO but is also found in Rías Baixas, where it is often blended with Albariño (ALVARINHO) and possibly Loureira (LOUREIRO; ie as over the border in Portugal). Treixadura is also authorized in the Monterrei, Valdeorras and Ribeira Sacra DOs, where it tends to be blended with other local varieties such as GODELLO. Varietal wines are relatively rare but are made by producers such as Aurea Lux, Campante (a sweet, dried-grape version, a style traditionally known as *tostado*), Cooperativa San Roque Beade, Docampo and Viño Ribeiro. Coto de Gomariz produce a fine blend of 80% Treixadura together with Godello, Loureira and Albariño.

TRAMINER

See SAVAGNIN BLANC in the SAVAGNIN entry.

TRAMINETTE

Promising, aromatic, very complex American hybrid rapidly winning favour in the Midwest and Northeast.

BERRY COLOUR: ● ● ● ● ●

ORIGINS AND PARENTAGE

A Joannes Seyve 23.416 × GEWÜRZTRAMINER hybrid bred in 1965 by Herb C Barrett at the University of Illinois, US, in which Joannes Seyve 23.416 is a Bertille Seyve 4825 × CHANCELLOR hybrid (see PRIOR for the complete pedigree of Bertille Seyve 4825). Traminette is therefore an extremely complex *Vitis riparia*, *Vitis labrusca*, *Vitis vinifera*, *Vitis aestivalis*, *Vitis lincecumii*, *Vitis rupestris* and *Vitis cinerea* hybrid. Barrett, whose intention was to produce a large-clustered table grape with Gewürztraminer flavour, sent seeds from this crossing to the New York State Agricultural Experiment Station at Cornell University in Geneva, where they were planted in 1968. Fruit was first observed in 1971 and the original vine was propagated in 1974 under the number NY65.533.13 (Reisch *et al.* 1996). Traminette was named and released for public use in 1996.

VITICULTURAL CHARACTERISTICS

Late budding, mid to late ripening and productive. Moderately winter-hardy (at Geneva, NY) and needs a long growing season. Highly susceptible to phomopsis cane and leaf spot, moderately susceptible to downy mildew and crown gall.

WHERE IT'S GROWN AND WHAT ITS WINE TASTES LIKE

Wines are typically spicy and fragrant (characteristics that are enhanced by 12–48 hours' skin contact), revealing its GEWÜRZTRAMINER parentage, and made in a dry or off-dry style. They have good body and balance and taste like a *vinifera* wine. Thanks to the encouraging combination of productivity with the potential for wine quality and reasonable cold hardiness, plantings are on the increase: 105 acres (42 ha) in **Missouri** in 2009, 80 acres (32 ha) in **Virginia** in 2008, 65 acres (26 ha) in **New York** State in 2006, 63 acres (25 ha) in **Illinois** in 2009, 55 acres (22 ha) in **Indiana** in 2009, 34 acres (14 ha) in **Pennsylvania** in 2008, 30 acres (12 ha) in **Michigan** in 2006, 20 acres (8 ha) in **Ohio** in 2008, 10 acres (4 ha) in **Nebraska** in 2007 and 5.3 acres (2 ha) in **Iowa** in 2006 – quite impressive in fifteen years.

Its popularity can be seen by the number of different states in which varietal wines are produced: eg Arbor Hill, Buck Creek, French Lick and Oliver Winery in Indiana; Deer Run and Fulkerson in New York's Finger Lakes; Adams County, Allegro, Greendance and Starr Hill in Pennsylvania; Oak Hills in **Oklahoma**; OOVVDA and Twin Oaks in Missouri; Domaine Berrien in Michigan; Meranda-Nixon and Valley Vineyards in Ohio; Gadino, Virginia Mountain Vineyards and Willowcraft Farm in Virginia; Cuthills in Nebraska; August Hill in Illinois; and Snow Farm in **Vermont**.

TRBLJAN

Variety widely planted in coastal Croatia for long thought to be a Trebbiano.

BERRY COLOUR: ● ● ● ● ●

PRINCIPAL SYNONYMS: Dobrogoština (Makarska), Grban (Split), Kuč, Pljuskavac (Imotski), Rukavina (Zadar), Šampanjol (Šibenik), Trbljan Bijeli

VARIETIES COMMONLY MISTAKEN FOR TRBLJAN: TREBBIANO TOSCANO ⊗

ORIGINS AND PARENTAGE

The exact origin of Trbljan is unknown but it is considered to be indigenous to Dalmacija (Dalmatia), southern Croatia. Contrary to common belief, recent ampelographic research and DNA profiling have shown that Trbljan is not identical to TREBBIANO TOSCANO (Goran Zdunić, personal communication) but it might be related to TREBBIANO ROMAGNOLO.

VITICULTURAL CHARACTERISTICS

Large bunches of large, thin-skinned berries, dense canopy. Very productive, mid to late ripening. Susceptible to downy and powdery mildews and botrytis bunch rot. Generally high acidity in the berries.

WHERE IT'S GROWN AND WHAT ITS WINE TASTES LIKE

Trbljan is widely planted in southern and inland Dalmacija, particularly on the island of Vis, and further north along the coast of **Croatia**. In 2009 it represented about 3% of Croatia's vineyard, with around 975 ha (2,409 acres) planted. In the region of Kaštela close to Split, the variety tends to be blended with other local varieties. Wines are typically light, crisp and rather simple, especially if yields are not controlled.

TREBBIANO

BERRY COLOUR: ● ● ● ● ●

ORIGINS AND PARENTAGE

Several Italian grape varieties which share morphological characteristics such as whitish berries, large bunches, late ripening and vigorous growth are often referred to as the 'Trebbiano family'. However, the term 'family' is not appropriate because DNA studies have shown a significant variability among these varieties and provided evidence that they do not share a common ancestor (Labra *et al.* 2001).

The earliest mention of Trebbiano appeared in 1303 in Pietro de Crescenzi's treatise on agriculture, under the name Tribiana, a grape variety cultivated in the area of Ravenna near the Adriatic coast in Emilia-Romagna, north-central Italy. However, it is not possible to tell which Trebbiano the medieval agronomist was referring to. According to the language expert Thomas Hohnerlein-Buchinger (1996), Trebbiano could derive from the medieval Frankonian term *draibio* or *draibjo*, meaning 'vigorous shoot', a common feature shared by those distinct varieties.

In this book we have included six distinct varieties with Trebbiano in their name:

- TREBBIANO D'ABRUZZO, mysterious Abruzzese
- TREBBIANO GIALLO, Lazio variety
- TREBBIANO MODENESE, used largely for vinegar
- TREBBIANO ROMAGNOLO widely planted in Emilia-Romagna
- TREBBIANO SPOLETINO, recently revived Umbrian
- TREBBIANO TOSCANO, the most widespread of all.

Other varieties of the heterogeneous Trebbiano group have been identified, although Trebbiano di Soave, Trebbiano Valtenesi and Trebbiano di Lugana are all identical to VERDICCHIO BIANCO (Calò, Costacurta *et al.* 1991; Labra *et al.* 2001; Ghidoni *et al.* 2008)

OTHER HYPOTHESES

The earliest mention of Trebbiano is often said to be found in Pliny the Elder's *Naturalis historia* under the name *vinum trebulanum*, a wine made from an ancient variety grown in 'agro Trebulanis', the area around Trebula, today's Treglia, in the province of Caserta in Campania, southern Italy, from which the variety took its name. Alternatively, some authors claim that the name Trebbiano could derive from a Tuscan town near Luni in ancient Etruria formerly named Trebulanum (Bacci 1596), while others suggest an etymology based on the Trebbia River in Liguria or on one of the many villages named Trebbo or Trebbio in Italy.

TREBBIANO D'ABRUZZO

Light-skinned variety from east-central Italy unrelated to other Trebbianos but whose identity is still unclear.

BERRY COLOUR: ● ● ● ● ●

PRINCIPAL SYNONYMS: Sbagagnina, Sbagarina, Svagadia, Trebbiano Abruzzese, Trebbiano Campolese, Trebbiano di Teramo

VARIETIES COMMONLY MISTAKEN FOR TREBBIANO D'ABRUZZO: BOMBINO BIANCO (Puglia)

ORIGINS AND PARENTAGE

As its name suggests, Trebbiano d'Abruzzo has long been known in Abruzzo, central Italy. It is generally thought to be identical to BOMBINO BIANCO in Puglia, further south. However, in 1994 they were listed as separate varieties in the Italian national register, so they should probably be thought of as distinct varieties until DNA profiling has shed some light on this ampelographic issue. A DNA study has suggested a possible genetic relationship between Trebbiano d'Abruzzo and TREBBIANO SPOLETINO (Labra *et al.* 2001). Some authors have suggested Trebbiano d'Abruzzo is a seedling of BIANCAME or TREBBIANO TOSCANO (Calò *et al.* 2006).

VITICULTURAL CHARACTERISTICS

Mid to late ripening. Sensitive to powdery mildew.

WHERE IT'S GROWN AND WHAT ITS WINE TASTES LIKE

Trebbiano d'Abruzzo is cultivated mainly in Abruzzo in east-central **Italy**, in particular in the provinces of Chieti, Teramo, Pescara and L'Aquila, and also to a more limited extent in Molise. Valentini's strongly mineral Trebbiano d'Abruzzo DOC, made from mass selection of old vines in their own vineyards, ages beautifully and is considered one of the best Italian whites but it is not typical of the DOC – and is made from Bombino Bianco. Masciarelli's Marina Cvetic Trebbiano d'Abruzzo DOC is apparently made from the variety of that name but the Trebbiano d'Abruzzo v Bombino Bianco question cannot be resolved without DNA analysis. The Italian agricultural census of 2000, which also distinguishes between these two varieties, records 418 ha (1,033 acres) of 'Trebbiano Abruzzese'.

TREBBIANO DI LUGANA

See VERDICCHIO BIANCO.

TREBBIANO DI SOAVE

See VERDICCHIO BIANCO.

TREBBIANO GIALLO

Another Trebbiano, grown and blended mainly around Roma.

BERRY COLOUR:

PRINCIPAL SYNONYMS: Greco di Velletri, Trebbiano dei Castelli (Castelli Romani), Trebbiano di Spagna (Modena), Trebbiano Giallo di Velletri

VARIETIES COMMONLY MISTAKEN FOR TREBBIANO GIALLO: GRECO (Campania), ROSSETTO (Lazio), TREBBIANO MODENESE

ORIGINS AND PARENTAGE

Trebbiano Giallo was first described by Acerbi (1825) and was later identified with Trebbiano di Spagna, a variety cultivated in the province of Modena, where it is used to produce a base wine for *aceto balsamico* or balsamic vinegar. It has often – erroneously – been suggested that there is a similarity between ROSSETTO and Trebbiano Giallo.

VITICULTURAL CHARACTERISTICS

Mid to late ripening. Good resistance to powdery and downy mildews.

WHERE IT'S GROWN AND WHAT ITS WINE TASTES LIKE

Trebbiano Giallo is cultivated mainly in the Castelli Romani and Colli Albani regions, south of Roma in Lazio, central **Italy**. It is generally blended with other varieties and authorized in many DOCs in Lazio, including Castelli Romani, Cerveteri, Cori and Frascati, as well as in Garda Colli Mantovani DOC in Lombardia. Under its synonym Trebbiano di Spagna, it is found in the province of Modena in Emilia-Romagna, exclusively for the production of *aceto balsamico*. There were 4,262 ha (10,532 acres) in Italy in 2000 for the production of wine.

TREBBIANO MODENESE

Emilia-Romagnan used more for balsamic vinegar than for wine.

BERRY COLOUR:

PRINCIPAL SYNONYMS: Terbiàn Moscatlè (Modena), Terbianella, Trebbiano Comune, Trebbiano di Collina, Trebbiano di Modena, Trebbiano Montanaro

VARIETIES COMMONLY MISTAKEN FOR TREBBIANO MODENESE: TREBBIANO GIALLO, TREBBIANO ROMAGNOLO

VITICULTURAL CHARACTERISTICS

Mid to late ripening, vigorous.

WHERE IT'S GROWN AND WHAT ITS WINE TASTES LIKE

Trebbiano Modenese has long been known on the Po plain in the provinces of Modena and Reggio Emilia in Emilia-Romagna, north-central **Italy**, mainly to produce the base wine for *aceto balsamico di Modena*. It is not listed in the regulations for any DOC but there were 667 ha (1,648 acres) for the production of wine in Italy in 2000.

TREBBIANO ROMAGNOLO

The widely planted but not especially distinguished Emilia-Romagnan Trebbiano.

BERRY COLOUR: ● ● ● ● ●

PRINCIPAL SYNONYMS: Trebbiano della Fiamma (Lugo in the province of Ravenna), Trebbiano di Romagna

VARIETIES COMMONLY MISTAKEN FOR TREBBIANO ROMAGNOLO: TREBBIANO MODENESE, VERDICCHIO BIANCO

ORIGINS AND PARENTAGE

Trebbiano Romagnolo probably originates from the province of Ravenna in Emilia-Romagna in north-central Italy, where the earliest mention of any (though not necessarily this) TREBBIANO appears in 1303 in Pietro de Crescenzi's treatise on agriculture.

VITICULTURAL CHARACTERISTICS

Mid to early ripening. Vigorous and abundant. Susceptible to leafhoppers and grape moths and various fungal diseases except powdery mildew. Berries are golden when fully ripe.

WHERE IT'S GROWN AND WHAT ITS WINE TASTES LIKE

Trebbiano Romagnolo is the most widely cultivated variety in Emilia-Romagna in north-central **Italy**, mainly in the province of Ravenna, in particular between Faenza, Lugo and Russi. It is also found in the province of Forlì and here and there in the provinces of Parma and Bologna, as well as, to a lesser extent, in Lazio and on Sardegna. The Italian agricultural census of 2000 recorded total plantings of 20,033 ha (49,503 acres).

Trebbiano Romagnolo is used in the blend of many and various dry white DOC wines, both still and sometimes sparkling, in Emilia-Romagna, including Colli Bolognesi, Colli di Rimini, Colli di Scandiano e di Canossa and Colli Piacentini, as well as an optional blending component in several Lazio DOCs such as Castelli Romani and Colli Albani. Varietal wines are relatively common, often rather bland but with fresh acidity – which is probably why they are used in the base wines for some fizz – and are authorized in Colli Romagna Centrale and the ubiquitous Trebbiano di Romagna DOCs. On the island of Sardegna, it is an alternative to TREBBIANO TOSCANO in the Arborea DOC. It is also used in the production of brandy.

Examples from Battistini, La Berta, Celli, Dalfiume, Cantina di Faenza and Zerbina are probably better than most.

TREBBIANO SPOLETINO

Recently rescued Umbrian capable of better quality, and more body, than most Trebbiano Somethings.

BERRY COLOUR: ● ● ● ● ●

PRINCIPAL SYNONYMS: Spoletino, Trebbiano di Avezzano, Trebbiano di Spoleto

VARIETIES COMMONLY MISTAKEN FOR TREBBIANO SPOLETINO: TREBBIANO TOSCANO

ORIGINS AND PARENTAGE

Trebbiano Spoletino is a local Umbrian variety first cited in 1878, which suggests it was a local natural cross that occurred some time in the early nineteenth century. A DNA study has suggested a possible genetic relationship between Trebbiano Spoletino and TREBBIANO D'ABRUZZO (Labra *et al.* 2001).

VITICULTURAL CHARACTERISTICS

Vigorous and late ripening. Resistant to winter and spring frosts. Very resistant to downy mildew, much less so to powdery mildew. Good ability to accumulate sugar in the berries without loss of acidity.

WHERE IT'S GROWN AND WHAT ITS WINE TASTES LIKE

In the early 1980s, Paolo Silvestri of the local co-operative Spoleto Ducale recognised the potential quality of Trebbiano Spoletino, which until then had been harvested mainly for domestic consumption or for bulk wine. Its renaissance was not until the early 2000s, when Giampaolo Tabarrini and Paolo Bea bottled the first modern examples using fruit from old vines, ungrafted and trained *alberate* (up trees). Both Tabarrini and Cantina Novelli have replanted the variety in Spoleto after careful selection and propagation from such old vines. It is today cultivated almost exclusively in Umbria in central Italy, mainly in the province of Perugia, in particular in the valley between Spoleto, Montefalco and Trevi. The DOC for Trebbiano Spoletino was approved in 2007. Other Umbrian DOCs such as Colli Martani and Montefalco may include a small amount of Trebbiano Spoletino but it is often TREBBIANO TOSCANO that plays the dominant role.

Varietal wines tend to be greenish yellow, full-bodied, rich and aromatic – tropical fruit, spices and mineral notes – but with good acidity, too. Tabarrini's Adarmando ages well, developing Riesling-like aromas. Antonelli San Marco and Perticaia also make varietal wines.

TREBBIANO TOSCANO

Tuscan variety also known as Ugni Blanc that probably produces more (typically tart and neutral) wine than any other vine in the world.

BERRY COLOUR: ● ● ● ● ●

PRINCIPAL SYNONYMS: Alfrocheiro Branco (Portugal), Armenian (France), Bianca di Poviglioᛝ (Reggio Emilia), Douradinha (Vinho Verde in Portugal), Morterille Blanche (France), Procanico (Umbria), Regrat (Germany), Rogoznička (Croatia), Rossola Brandica or Rossola Brandinca (Corse in France), Roussan (France), Saint-Émilion (Charente in France), St Emilion (US), Šijaka (Croatia), Táliaᛝ or Thalia (Portugal), Ugni Blancᛝ (France, Bulgaria, Croatia, Uruguay, etc.)

VARIETIES COMMONLY MISTAKEN FOR TREBBIANO TOSCANO: ALBANELLA⚥ (Marche), BIANCAME (Marche), Blanc Aube⚥ (Gironde in France, no longer cultivated), CODA DI VOLPE BIANCA⚥ (Campania), DAMASCHINO⚥ (Australia), ERBALUCE⚥ (Piemonte), FERNÃO PIRES⚥ (Portugal), MACERATINO⚥ (Marche), PASSERINA (Adriatic coast), SÉMILLON⚥, TRBLJAN⚥ (Croatia), TREBBIANO SPOLETINO (Umbria)

ORIGINS AND PARENTAGE

As its name suggests, Trebbiano Toscano has been known in Toscana, central Italy, for a long time. In his treatise on viticulture, Soderini (1600) considers Trebbiano (Toscano, in all probability) and Malvasia (MALVASIA BIANCA LUNGA, in all probability) as the most widespread varieties in the region.

Trebbiano Toscano is said to have been exported to France in the fourteenth century during the period when the papacy resided in Avignon. In France, it was given the name Ugni Blanc, first mentioned as 'uniers' in Pernes in the Vaucluse in 1514 (Rézeau 1997). Indeed, Ugni Blanc was first cultivated in Provence, then in the Languedoc and later in the Charente and the Charente-Maritime, where it found its favoured spot. The name Ugni actually comes from the Provençal dialect *uni*, meaning 'early grape', which itself possibly derives from the Greek *eugenia*, meaning 'of noble origin' (Rézeau 1997).

DNA profiling found another unsuspected synonym of Trebbiano Toscano: Bianca di Poviglio, a minor white variety of Poviglio in the province of Reggio Emilia in Emilia-Romagna (Boccacci *et al.* 2005). DNA analysis has also shown that ALIONZA could be closely related to Trebbiano Toscano (Filippetti, Silvestroni, Thomas and Intrieri 2001). More recent investigations have revealed a likely parent–offspring relationship between Trebbiano Toscano and the Soave grape GARGANEGA, another very old Italian variety (Di Vecchi Staraz, This *et al.* 2007; Crespan, Calò *et al.* 2008), which indicates Trebbiano Toscano's Italian origin. Since Garganega also has parent–offspring relationships with eight other varieties (ALBANA, CATARRATTO BIANCO, MALVASIA BIANCA DI CANDIA, DORONA DI VENEZIA, MARZEMINA BIANCA, MONTONICO BIANCO, MOSTOSA and SUSUMANIELLO), Trebbiano Toscano is either their half-sibling or their grandparent (see Garganega for the pedigree diagram).

Another parent–offspring relationship, between Trebbiano Toscano and PASSERINA, cultivated on the Adriatic coast and with which Trebbiano is sometimes confused, has long been suspected but has not yet been tested by DNA profiling.

Although Ugni Blanc is sometimes known as Clairette Ronde in the Rhône, it is not related to CLAIRETTE.

Trebbiano Toscano was used to breed CHENEL, FOLIGNAN, FUBIANO, MANZONI ROSA, NOUVELLE and VIDAL.

VITICULTURAL CHARACTERISTICS

Late budding and therefore not at risk from spring frosts in less than ideal locations. Mid to late ripening but often picked earlier to retain its usefully high acidity. Vigorous, high yielding and adaptable to a range of sites and training methods but best trellised as it is slightly sensitive to wind. Good resistance to powdery mildew and botrytis bunch rot but susceptible to downy mildew and eutypa dieback.

WHERE IT'S GROWN AND WHAT ITS WINE TASTES LIKE

Trebbiano Toscano is widely planted throughout the world and probably produces more wine than any other variety planted (Spain's AIRÉN covers far more ground but has much lower yields and is used substantially for brandy). The wines tend to be light, fresh and unremarkable, thanks partly to the vine's abundant yields. It is by far the most widespread of the heterogeneous group of varieties with Trebbiano in their name and the second most planted white in Italy after CATARRATTO BIANCO but is much less common on labels than more interesting varieties. It generally disappears into blends or in some countries into a still. Its neutral nature and high acidity make it an ideal candidate for distillation.

As Ugni Blanc, and sometimes under the name Saint-Émilion, it is the most widely planted white wine variety in **France** because of its popularity with distillers. Although the vineyard area given over to this high-acid workhorse variety has declined since its peak at the end of the 1970s, total French plantings in 2009 were still 83,892 ha (207,302 acres), with the bulk of the plantings in the Charente and the Charente-Maritime, predominantly for the production of cognac, although it also goes into many a Vin de Pays du Gers and Côtes de Gascogne.

In 2000 there were 42,456 ha (104,911 acres) of this particular, commonly planted Trebbiano in **Italy**, mainly in Toscana and Umbria, as well as in Emilia-Romagna and the Veneto. Plantings in Toscana have declined since this white wine grape began to lose favour with red-wine producers in Chianti, where it is no longer an obligatory part of the blend, and now that it is no longer sanctioned in Chianti Classico. It will surely continue to decline as wine drinkers are increasingly concerned with flavour rather than volume and price alone (there were more than 61,000 ha/150,734 acres in Italy in 1990). Attempts to promote dry whites made from the surplus have generally met with little success and much Trebbiano in Toscana has been grubbed up or hijacked for Vin Santo.

AIRÉN performs Trebbiano's functions in Spain but **Portugal** had 390 ha (964 acres) of Tália in 2010. Most of it is planted in the Tejo and the resulting wine is hardly exciting.

There were 337 ha (833 acres) in **Croatia** in 2009 while **Bulgaria** had 1,388 ha (3,430 acres).

In **Greece**, the majority of plantings are in Thessalía (72 ha/178 acres in 2007) but in Dráma, north-east Greece, Nico Lazaridi make a dry, late-harvest, oak-aged Trebbiano, blend a little in with their SAUVIGNON BLANC and are experimenting with a lightly sparkling varietal wine.

Although there were some experimental plantings in **Russia** in the late 1990s (by agronomists who had visited France and were inspired by its high yields), mainly in the Taman peninsula between the Black and Azov seas, the variety proved to be insufficiently cold-hardy and about two thirds of the *c.*150 ha (370 acres) were uprooted. The 60 or so hectares (150 acres) that remain are used, appropriately, to make base wines for brandy.

In **California**, it is called variously St Emilion, Ugni Blanc and Trebbiano and there were 197 acres (80 ha) of it in 2010, some in Paso Robles as well as the Central Valley. It is generally used as a cheap, crisp blending component, although one or two adventurous souls have managed to craft quite an interesting full-bodied varietal tasting of white currants from it.

Hester Creek bottle some in British Columbia, **Canada.**

Argentina had 2,478 ha (6,123 acres) of Ugni Blanc in 2008, mainly in Mendoza and San Juan and to a lesser extent in Río Negro in the south. Under the same name in **Uruguay** there were 719 ha (1,777 acres) in 2008, about 8% of the total vineyard area. In **Brazil**, where it is known as Trebbiano, there were 173 ha (427 acres) in Rio Grande do Sul in 2007.

Australia had 220 ha (544 acres) of Trebbiano in 2008, mainly but not exclusively in the warm, inland Riverina region where most often it is used in blends and occasionally by distillers.

Known as Ugni Blanc in **South Africa**, in 2008 it was planted on just 68 ha (168 acres) in the warmer inland regions of Robertson and, to a much lesser extent, Klein Karoo. As in France, it is used in the production of brandy as well as for wine, but this role is more often played by COLOMBARD.

It is also found in **India**.

TREBBIANO VALTENESI

See VERDICCHIO BIANCO.

TREIXADURA

See TRAJADURA.

TREPAT

Red-skinned Catalan traditionally used for still and sparkling rosés.

BERRY COLOUR: ● ● ● ● ●

PRINCIPAL SYNONYMS: Bonicaire☒, Carlina, Embolicaire, Negra Blana, Parrel, Parrel-Verdal, Trepat Negre, Traput

ORIGINS AND PARENTAGE

Trepat comes from the Conca de Barberà region in north-east Spain. DNA profiling has recently shown that Bonicaire in Murcia, south-east Spain, is identical to Trepat (SIVVEM) and that TREPAT BLANC from Priorat is not a colour mutation of Trepat (Giralt *et al.* 2009).

Parrel and Parrel-Verdal are occasional synonyms for Trepat but should not be confused with PARRALETA.

VITICULTURAL CHARACTERISTICS

Large, compact bunches of large, relatively thick-skinned berries. Robust and productive. Early budding, therefore at risk from spring frosts, and late ripening. Susceptible to phomopsis.

WHERE IT'S GROWN AND WHAT ITS WINE TASTES LIKE

Trepat, provisionally included in the Spanish register of commercial varieties, is found mainly in the Conca de Barberà and also in the Costers del Segre DOs in Catalunya, north-east **Spain**, although it is permitted in the Catalunya and Cava DOs as well. The first Festa del Trapat was held in Barberà de la Conca in 2010, suggesting there is increasing local pride in this pale-red grape. It is generally used for rosé wines (and may be included in pink Cava) but shows good potential for light reds. In 2008, total plantings had reached 953 ha (2,355 acres), virtually all in Catalunya. Varietal wines are typically fruity, lightly structured, dominated by fresh red berries (strawberries, raspberries), hay, and a certain spiciness (cinnamon).

Varietal Cava rosé wines are produced by Bohigas, Carles Andreu, Cavas del Castillo de Perelada, Portell and Agustí Torelló Mata, for example; Vilarnu's version includes 10% PINOT NOIR. Still rosés are made by producers such as the Cabanal Cooperativa Agrícola de Barberà de la Conca, the Cooperativa Agrícola de L'Espluga de Francolí, Carlania and Vinya Sanfeliu. Red wines made from 100% Trepat are harder to find, but include, for example, those made by Carles Andreu and Mas Foraster.

Under the name Bonicaire, it is authorized in such appellations as Vino de la Tierra Castelló in Castellón in Valencia, eastern Spain (just south of Catalunya), in the Valencia DO and further south in the Vinos de la Tierra Abanilla and Campo de Cartagena in Murcia, although very few vines remain. Wines tend to be light in colour and soft and have a citrus flavour, with a slightly bitter-orange aftertaste.

TREPAT BLANC

Vaguely Chenin-like, recuperated Priorat variety.

BERRY COLOUR:

ORIGINS AND PARENTAGE

Trepat Blanc was almost extinct when a few old vines were rescued in the Priorat region. DNA profiling has shown that it is not a colour mutation of TREPAT from Conca de Barberà (Giralt *et al.* 2009).

WHERE IT'S GROWN AND WHAT ITS WINE TASTES LIKE

Trepat Blanc has not yet made it into the official register of commercial varieties in **Spain**, even provisionally. However, a group of winemakers in Catalunya who have joined together behind the brand 10 Sentits have produced a varietal wine from fruit harvested in Torroja del Priorat that, with its apples and honey character, recalls CHENIN BLANC. Cartoixa de Montsalvat in Priorat make a Trepat Blanc blend and a few others in the region blend a small percentage with other local varieties.

TRESSALLIER

See SACY.

TRESSOT

Ancient but virtually extinct, struggling variety from Chablis country of intriguing parentage.

BERRY COLOUR:

PRINCIPAL SYNONYMS: Bourguignon or Bourguignon Noir (Seine-et-Marne), Morillon Noir, Noirien or Nairien (Bar-sur-Aube and Les Riceys in the Aube), Resseau (Yonne), Treceaux (Yonne), Tresseau (Yonne), Tressiot (Auxerre and Irancy), Tressot Noir, Vérot or Petit Verrot (Joigny)

VARIETIES COMMONLY MISTAKEN FOR TRESSOT: GASCON, NEGRETTE, PINOT NOIR, POULSARD, TROUSSEAU (Jura)

ORIGINS AND PARENTAGE

Tressot is a very old variety from the Yonne *département* in north-eastern France, where it was mentioned under the name Treceaux in 1394 (see PINOT for the full citation), and then in 1562 in Auxerre, the main city of the Yonne (Simpée 1902): 'Et pour ce faire, a dit Lenayn promis fornir le pland chevelée de Pynots, Terceaulx et Servigneans, bons et convenables à planter et édifier vigne' ('And to do so, it was promised to the aforesaid Lenayn to provide with rooted cuttings of pynots, terceaulx and servigneans, good and suited for planting and setting up a vineyard'). The spelling Tresseau was generally used before the twentieth century. The etymology of Tressot is uncertain, but it is usually said to mean *trois ceps* ('three vines') because it produces three times more than Pinot, with which it used to be cultivated, together with CÉSAR (Simpée 1902).

DNA parentage analysis has shown that Tressot and GOUAIS BLANC are the parents of

GENOUILLET, a local variety of the Indre and Cher *départements*, and that Tressot is a cross between DURAS and PETIT VERDOT (Bowers *et al.* 2000; see PINOT pedigree diagram. This parentage of Tressot is hard to believe from a geographical point of view: how could Duras from the Tarn or from Ariège in the Midi-Pyrénées and Petit Verdot from the Gironde or from the Pyrénées-Atlantiques in south-west France have had a progeny in the Yonne *département* in Burgundy?

The Tressot ampelographic group (see p XXVII) also includes ARBANE, BACHET NOIR and Peurion (the last no longer cultivated; Bisson 2009).

Confusion is caused by various synonyms shared by Tressot and other varieties: GASCON and NÉGRETTE share with Tressot the synonym Noirien, PINOT NOIR shares with Tressot the synonyms Noirien and Morillon Noir.

Two colour mutations of Tressot have been described: **Tressot Blanc**, which is no longer cultivated, and **Tressot Panaché** (or Tressot Bigarré or Ballon de Suisse), a spectacular chimera (a mutation affecting only some layers of tissue, in this case the berry skin) observed in Girolles in the Yonne as well as in Switzerland that results in some purple-and-green-striped berries.

VITICULTURAL CHARACTERISTICS

Mid ripening, generally pruned long. Susceptible to powdery mildew. Small berries.

WHERE IT'S GROWN AND WHAT ITS WINE TASTES LIKE

Wines are generally powerful, deeply coloured and reasonably alcoholic. There was less than a hectare (2.5 acres) in **France** in 2008 and only in the Yonne is Tressot allowed in such appellations as Bourgogne and Bourgogne-Chitry – and then only from vines already in the ground in 2009. Domaine de Cerisiers in the latter appellation is one of the few producers which still has Tressot in the vineyard and they include it in their PINOT NOIR blend.

TREVISANA NERA

Minor Veneto vine of mysterious origin.

BERRY COLOUR: ● ● ● ● ●

PRINCIPAL SYNONYMS: Borgogna (Treviso in the Veneto), Gattera (Treviso), Refosco di Guarnieri (Friuli)

ORIGINS AND PARENTAGE

Trevisana Nera is a traditional variety of the area of Feltre in the province of Belluno in the Veneto, northern Italy, where it was first mentioned as recently as the mid twentieth century. Its name suggests it may have been introduced from the neighbouring province of Treviso, but no similarity between Trevisana Nera and the traditional varieties of that area has been found so this origin is unproven.

In 2003, ampelographers of the Conegliano research centre, in the Veneto, found in Castelcucco in the province of Treviso a few vines of an obscure local variety called Gattera or Borgogna which looked very similar to Trevisana Nera (Cancellier *et al.* 2008). DNA profiling has confirmed that Gattera or Borgogna from Castelcucco as well as Refosco di Guarnieri, cultivated as a table grape in Friuli, are identical to Trevisana Nera (Crespan, Giannetto *et al.* 2008). This suggests that the cultivation area of Trevisana Nera was once much more extensive than it is now.

VITICULTURAL CHARACTERISTICS

Mid to late ripening. Vigorous and productive.

WHERE IT'S GROWN AND WHAT ITS WINE TASTES LIKE

Trevisana Nera is grown almost exclusively in the province of Belluno, between Trento and Udine in the Veneto in northern **Italy**, especially in the area of Feltre. It is usually blended with other local varieties though a varietal wine could feasibly be made as an IGT Vigneti delle Dolomiti. There were 37 ha (91 acres) planted in Italy in 2000.

TRIBIDRAG

Croatian variety making full-bodied reds, much better known as Primitivo or Zinfandel.

BERRY COLOUR: ● ● ● ● ●

PRINCIPAL SYNONYMS: Crljenak Kaštelanski⌘ (Croatia), Kratošija⌘ (Montenegro), Morellone (Puglia in Italy), Pribidrag⌘ (Croatia), Primaticcio (Puglia), Primitivo⌘ or Primitivo di Goia or Primativo (Puglia), Trebidrag⌘ (Croatia), Uva di Corato (Puglia), Zagarese (Puglia), Zinfandel⌘ (US)

VARIETIES COMMONLY MISTAKEN FOR TRIBIDRAG: BLATINA⌘ (Bosnia and Herzegovina), Crljenak Crni⌘, PLAVAC MALI⌘, VRANAC⌘

ORIGINS AND PARENTAGE

Tribidrag is the original and oldest Croatian name for this grape variety from central Dalmacija (Dalmatia) that is much better known today as Primitivo in Puglia in southern Italy and as Zinfandel in California. The respective history and peregrinations of Primitivo and Zinfandel are discussed here before an explanation of how they were found to be identical to each other and then to Tribidrag.

HISTORICAL REFERENCES TO PRIMITIVO The earliest mention of this variety in Puglia dates back to 1799 when Francesco Filippo Indellicati (1767–1831), priest of Gioia del Colle and amateur botanist and agronomist, made a note in the Gioia town records that he had found in his vineyard a particularly early ripening variety that he called Primativo, from the Latin *primativus*, meaning 'first to ripen'. Before that, this variety was apparently known as Zagarese (Molon 1906), allegedly named after the city of Zagreb in Croatia, but a similar name (Zagarolese) was also given to CASTIGLIONE in Calabria. Indellicati planted his Primativo in Liponti, a hamlet near Gioia del Colle, and with the help of local growers the variety was later spread to other areas in Puglia (Turi, Cassano delle Murge, San Michele, Noci and Castellana Grotte) around 1820. The spelling Primitivo appeared later, around 1860.

HISTORICAL REFERENCES TO ZINFANDEL The origin of California's flagship variety has been the subject of much speculation (Piljac 2005). Many an American has argued that Zinfandel is indigenous, notwithstanding the fact that *Vitis vinifera* L. is not a native species of the Americas. Many others have suggested that Zinfandel was introduced to California in the 1850s by Agoston Haraszthy, a Hungarian immigrant who planted over 300 varieties in different US states after his arrival in 1837. However, the historian Charles L Sullivan (2003) provided evidence that Zinfandel had arrived in California as a nameless vine from the Schönbrunn imperial collection in Wien (Vienna) that contained all the grape varieties grown at that time in the Austrian Empire, and that was well before Haraszthy had set foot on American soil. Indeed, Long Island grape grower George Gibbs introduced this variety to his nursery in the 1820s, and in 1829 the name Zinfardel first appeared in a catalogue of another nursery on Long Island. In the 1830s, the variety appeared in other catalogues under Zinfendal, Zinfindal or Zinfendel. The modern spelling Zinfandel first appeared in 1852 in a letter to the editor of *The Horticulturist* written by viticulturist John Fisk Allen, and in the 1860s it was agreed that this spelling should be used. Despite some doubtful hypotheses, the exact etymology of Zinfandel remains a mystery.

ZINFANDEL IDENTIFIED WITH PRIMITIVO In 1967, Austin Goheen, a plant pathologist from the United States Department of Agriculture's Agricultural Research Service (ARS), visited Bari in Puglia, southern Italy, where he happened to taste Primitivo wines that reminded him of those made from Zinfandel. He asked to see the vineyards and was so struck by the resemblance between Primitivo and Zinfandel that he had cuttings sent to the University of California at Davis to plant it next to Zinfandel. In 1972, Goheen confirmed that their morphology was very similar, and in 1975 Wade Wolfe, a PhD candidate at Davis, established that the vines showed identical isozyme patterns (isozymes are a particular type of enzymes formerly used to establish a molecular profile; modern DNA profiling based on microsatellites did not exist at that time). The news that Zinfandel and Primitivo were the same quickly spread and the 'battle over Zinfandel' immediately began (Piljac 2005). In the late 1970s, Puglian producers started selling their Primitivo in the US with the name Zinfandel on the labels, which prompted vexed California Zinfandel producers to protect the Zinfandel heritage in which they had invested effort and money. In 1985, the Bureau of Alcohol, Tobacco and Firearms (BATF) ruled that Zinfandel could not be used as a synonym for Primitivo because of lack of evidence of synonymy. However, the identity between Zinfandel and Primitivo was confirmed by DNA profiling in 1994 by Carole Meredith and her doctoral candidate John Bowers at Davis. As a consequence, in 1999 the European Union granted Italian Primitivo producers the right to use the name Zinfandel. BATF were unsuccessful in the complaint they filed in 2000 in response to this ruling. No agreement has yet been reached.

ZINFANDEL AND PRIMITIVO MISTAKEN FOR PLAVAC MALI A Dalmatian origin for Primitivo has long been suspected, as suggested by the synonym Zagarese that was already in use in the eighteenth century in Puglia. The search for the origins of Zinfandel/Primitivo in Croatia began in the mid 1970s, when Austin Goheen contacted Franco Lamberti from the University of Bari, who had recently visited Croatian vineyards together with Ana Šarić, professor at the University of Zagreb. Lamberti suggested to Goheen that Zinfandel/Primitivo might correspond to a variety grown in Dalmacija and known as Plavac Mali or Plavina (Mirošević and Meredith 2000). In 1979, Goheen had cuttings of PLAVAC MALI sent to him by Šarić, and in 1982 he was able to reject the identity with Zinfandel, helped again with isozyme analysis. Nevertheless, the word had already spread that Zinfandel and Plavac Mali were one and the same, and the rumour was fuelled by Croatians willing to sell Croatian wines in the US under the name Zinfandel, which elicited a second episode of the 'battle over Zinfandel', in which Croatian-born Californian producer Mike Grgich of Grgich Hills actively participated by supporting the identity between Plavac Mali and Zinfandel.

THE 'ZINQUEST' IN CROATIA Encouraged by Mike Grgich, Carole Meredith at the University of California at Davis in 1998 initiated a collaboration with Edi Maletić and Ivan Pejić from the University of Zagreb in order to find out the truth about Zinfandel's origins in Croatia. Their long search was later nicknamed the 'Zinquest'. Together with Jasenka Piljac, Meredith and her Croatian colleagues collected 148 samples of old PLAVAC MALI vines scattered along the Dalmatian coast. Back at the University of California at Davis, none of the DNA profiles of these samples matched that of Zinfandel (Pejić *et al.* 2000). In the following years, Maletić and Pejić continued their Zinquest in old Dalmatian vineyards and kept sending 'Zinfandel suspects' to Meredith's lab, still without any match. Finally, in December 2001, the Zinquest proved successful: a vine sampled in a vineyard owned by Ivica Radunić in Kaštel Novi, north of Split, and known locally as Crljenak Kaštelanski (literally 'the red from Kaštela') perfectly matched the DNA profile of Zinfandel!

Another nine Crljenak Kaštelanski vines were later found in Radunić's vineyard, and this variety was then found in 2002 under the name Pribidrag in the village of Svinišće near Omiš south of Split, growing in the garden of a very old lady.

TRIBIDRAG IS THE OLDEST NAME FOR THIS VARIETY Using a particular method developed for the profiling of ancient DNA, Malenica *et al.* (2011) established that the DNA profile of a ninety-year-old herbarium specimen of Tribidrag, a Croatian autochthonous variety that was mentioned as early as the fifteenth century and later in 1518, 1529 and 1546 as a reputed grape variety in the region of Split (Ambroz Tudor, Nenad Malenica, Edi Maletić and Ivan Pejić, personal communications) and has now disappeared from germplasm collections, matched Zinfandel/Primitivo/Crljenak Kaštelanski/Pribidrag. According to the rule of anteriority, whereby the oldest name takes precedence, Tribidrag is adopted here as the prime name for this variety. Most interestingly, the etymology of Tribidrag echoes that of Primitivo: according to Croatian linguist Valentin Putanec (2003), the Croatian word *tribidrag(a)* comes from the Greek *πρωι καρποζ*, meaning 'early ripening'. This perfectly corresponds to the meaning of the Latin *primativus* that gave Primativo and then Primitivo. *La boucle est bouclée!*

THE PARENTAGE OF PLAVAC MALI ANCHORS TRIBIDRAG IN CROATIA During their 'Zinquest', researchers had suspected a possible parent–offspring relationship between Tribidrag (that they obviously called Zinfandel) and PLAVAC MALI (Pejić *et al.* 2000). This relationship was confirmed four years later when DNA parentage analysis established that Plavac Mali is a natural progeny of Tribidrag and DOBRIČIĆ, another old Dalmatian variety (Maletić *et al.* 2004). In addition, Maletić *et al.* (2004) discovered additional parent–offspring relationships between Tribidrag and at least four other indigenous Dalmatian varieties (PLAVINA, GRK, VRANAC and Crljenak Crni), and Lacombe *et al.* (2007) have shown that Plavina is in fact a cross between Tribidrag and VERDECA from Puglia (see the pedigree diagram below). These close relationships with several indigenous varieties confirm the Croatian origin of Tribidrag (see PLAVINA entry for the Puglia–Croatia controversy about its birthplace).

OTHER HYPOTHESES

Using another genetic approach based on AFLP (Amplified Fragment Length Polymorphism), Fanizza *et al.* (2005) have cast doubt on the Croatian origin of Tribidrag in favour of a southern Italian one, and consider the question unresolved.

According to Maul (2006) and Jung (2007), Tribidrag was found under the name Blauer Scheuchner in old vineyards along the Baden-Württemberg wine route, north and south of Heidelberg. Jung (2007) suggests that this old variety was supposedly imported from the Hungarian town of Oedenburg in 1750 by Georg Bernhard Bilfinger for his collection of foreign grape varieties in Cannstadt near Stuttgart and then spread through Württemberg along the rivers Neckar and Rems up to the western slopes of the Rhein. Jung (2007) further claims that the following varieties also correspond to Tribidrag: in Germany, TAUBERSCHWARZ or Grobschwarze, Blaue Schaafstraube or Grosse Schwarze, and Sauerschwarz or Rosenkranz; in north-western Hungary, Sheikirn or Cirfandli; in Slovakia, Morava and Bohemia, Zierfahnler or Grobschwarze or Schertschina; in Slovenia, Blauer Palvanz. However, all these correlations were based on historical documents, ampelographic descriptions and botanical paintings and cannot be confirmed without DNA profiling.

Tribidrag was recently identified under the name Kratošija in Montenegro (Calò, Costacurta *et al.* 2008), where it has traditionally been grown and sometimes confused with its progeny VRANAC. Ampelographer Ulićević had already suspected the synonymy with Zinfandel in 1959 and considered Kratošija to be the oldest autochthonous variety in Montenegro, where it shows significant morphological diversity (Vesna Klikovac-Katanic, personal communication). However, mentions of Tribidrag largely predate those of Kratošija (early twentieth century) and a Montenegrin origin would not be consistent with the variety's pedigree involving mainly Croatian varieties, with the exception of Vranac.

TRIBIDRAG PEDIGREE DIAGRAM

The genetic influence of TRIBIDRAG (aka Primitivo or Zinfandel) on several varieties from Croatia anchors its roots on the Dalmatian coast. Where unknown (and probably extinct) varieties (?) are involved, the inverse relationships are also theoretically possible (see p x).

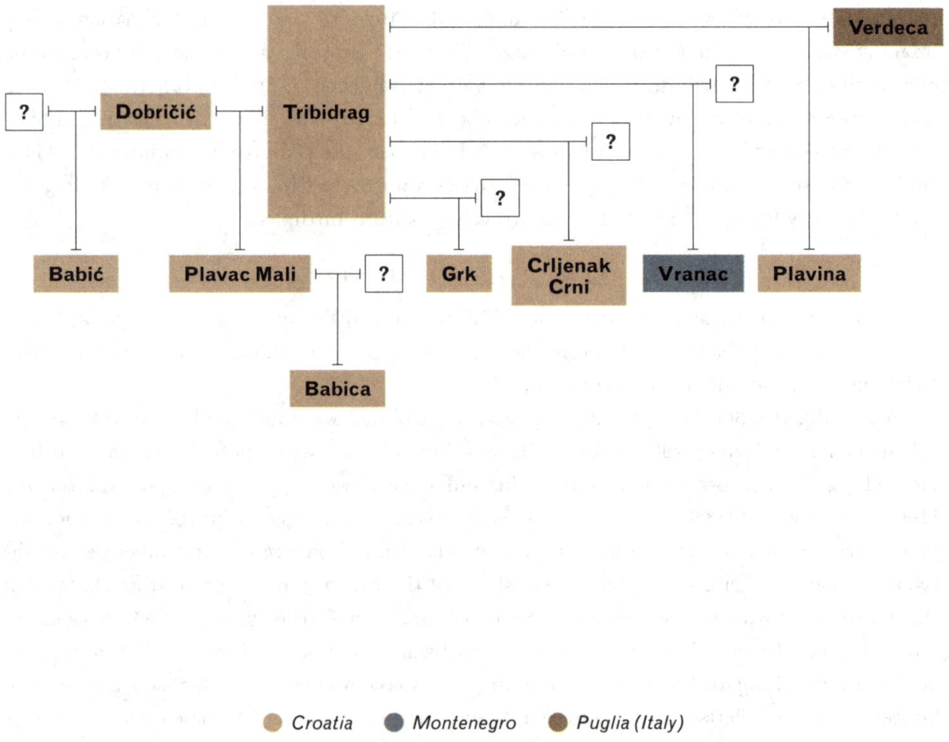

VITICULTURAL CHARACTERISTICS

Productive and therefore better suited to lower-fertility and well-drained soils. Mid to late ripening. Zinfandel bunches are medium to large and compact but there is often an inconveniently wide range of ripeness in the berries of a single bunch. Primitivo has more bunches but fewer and smaller berries and is less prone to botrytis bunch rot.

WHERE IT'S GROWN AND WHAT ITS WINE TASTES LIKE

Chiefly because of Zinfandel's importance in California, this multi-named variety has spread around the world, although, in **France**, Domaine de l'Arjolle is currently the only producer to have planted this variety and to make a varietal wine, known as Cuvée Z.

In **Italy**, Primitivo was once much more widely planted in Puglia than it is now, due mainly to the financial incentives of the EU vine-pull scheme at the end of the twentieth century: in 1990 there were around 17,000 ha (42,000 acres) compared with 7,951 ha (19,647 acres) ten years later, according to the agricultural census. However, now that it has gained popularity in its own right, rather than as an alcohol-boosting blending partner, this decline in vineyards has largely come to an end. The ovewhelming majority of Italy's Primitivo vines are in Puglia, specifically the west of the Salento peninsula on the heel of Italy, where the combination of variety and climate produces wines that are generous and rich in fruit and body. Very old bushvines on the red soils over white limestone in the Primitivo di Manduria zone produce some of the most structured and complex examples, but Gioia del Colle with its hills and higher altitude can produce rather more elegant wines. All are high in alcohol, however. Recommended producers include Cannito, Morella, Paololeo and Petrera Fatalone.

According to the ampelographer and viticulturist Edi Maletić, Crljenak Kaštelanski is the fastest growing variety in Dalmacija since the revelations about Zinfandel and Primitivo described above. From an original find of twenty-two vines, there are now more than 200,000 in **Croatia** and growing interest among producers. One of the first of the modern varietal wines was Zlatan Plenković's Crljenak, which comes from a vineyard Plenković planted near Makarska on the coast between Split and Dubrovnik with the support of Paul Draper of Ridge Vineyards in California (see below) and researchers at the University of Zagreb. Others include Bura-Mokalo.

Under the name Kratošija, Tribidrag is popular in **Montenegro**, where Zenta and Buk make varietal wines and Plantaže use it in blends. It has been neglected in the **Republic of Macedonia**, where it is found in old interplanted vineyards but shows variable and unreliable fertility. Growers are considering importing Primitivo vines from Italy to avoid the laborious process of clonal selection in these old vineyards.

As Zinfandel it is also found in **Israel**, where American influence is strong.

California is undoubtedly Tribidrag's – ie Zinfandel's – most successful outpost, previously in sheer volume but now also in terms of quality, even though a large percentage of the fruit will be harvested earlier and without much attention to detail since it is destined for oceans of off-dry 'white' (very pale pink) Zinfandel or blush wine. In 2008 there were 50,354 acres (20,377 ha) of Zinfandel, making it the state's second most planted red after CABERNET SAUVIGNON. For many years Zinfandel was California's most planted variety but was often mistreated, planted in overly hot regions and allowed to overproduce. Today there are many insect and disease problems to overcome due to the variety's tightly packed bunches, including the Willamette mite in the north of the state and the Pacific mite in the San Joaquin Valley, where there were nearly 20,000 acres (8,000 ha) in 2008, together with leafroll virus that may delay ripening and render the fruit suitable only for white Zin (Christensen *et al.* 2003).

Full-blooded red Zinfandel has been enjoying a revival of interest, not least because of the activities of ZAP, Zinfandel Advocates & Producers, the promotional organization for 'America's Heritage Wine'. The wines tend to have more obvious fruit than those made from CABERNET SAUVIGNON with a strongly spiced berry element – sometimes too simplistically so in inexpensive, overcropped, oversweet versions – reminiscent of boysenberries and often with more than a hint of dried fruit (perhaps a result of its characteristic uneven ripening). Ridge Vineyards have championed serious, ageworthy styles of Zin, typically based on some of the state's oldest vineyards, often field blends and always aged in top-quality American oak, since the 1960s. Their best bottlings can last decades. Many of the oldest Zinfandel vines, all dry farmed and typically gnarled and thick-trunked, were planted and tended by Italian immigrants, typically in Sonoma but also in Napa Valley, the Sierra Foothills and Lodi in the north of the Central Valley, which styles itself the Zinfandel Capital of the World since it grows 40% of California's total. Louis Martini's clone from the nineteenth-century Monte Rosso vineyard in Sonoma Valley enjoys special status. Other producers of currently acclaimed serious Zinfandels include Bedrock, Biale, Carlisle, Easton, Martinelli, A Rafanelli, Ravenswood (single-vineyard bottling specifically), Seghesio and Turley.

Because of Zinfandel's importance in California, and also because so many wine producers in other US states buy grapes from California, Zinfandel is bottled (even if not grown) in just about every state of the Union (even distant, chilly Maine), although only **Washington** State (89 acres/36 ha) and to an even lesser extent **Oregon** have much planted. It is also grown in British Columbia, western **Canada**, and, not surprisingly, in **Mexico**.

The variety has become quite popular in **Australia**. Once a speciality of Cape Mentelle in Western Australia, it is now a feature of South Australia with the likes of Rusden and Groom making the most of its suitability for hot, dry climates.

Glen Carlou and Zevenwacht produce slightly lighter varietal Zinfandels in **South Africa**.

TRINCADEIRA

The dark-skinned, widely planted Trincadeira goes under many aliases throughout Portugal. Potentially high-quality reds but not easy to grow.

BERRY COLOUR:

PRINCIPAL SYNONYMS: Black Alicante (Australia), Black Portugal (Australia), Crato Preto (Algarve), Mortagua or Mortagua Preto (Torres Vedras), Rosete Espalhado (Pinhel), Tinta Amarela, Tinta Amarelha, Tinta Manuola, Torneiro, Trincadeira Preta (Douro)

VARIETIES COMMONLY MISTAKEN FOR TRINCADEIRA: CASTELÃO, ESPADEIRO, PRETO MARTINHO

ORIGINS AND PARENTAGE

This variety, known under many names throughout Portugal, most likely originates from the Oeste region, north of Lisboa, where it shows the highest degree of clonal diversity (Rolando Faustino, personal communication), although it might also originate in the Alentejo region, east of Oeste. The most widespread and official Portuguese name for this variety is Trincadeira, although Tinta Amarela (meaning yellow-black in reference to the yellow colour of the mature shoots and to the black berries) is an officially authorized and still widely used synonym.

The name Trincadeira, or Trincadeira Preta (*preta* meaning black), is nevertheless a source of confusion because it is also given to other black- or white-berried varieties (CASTELÃO, TAMAREZ, TRAJADURA). The synonym Rifete used for this variety in Pinhel is equally confusing because it is also used for RUFETE in the same region. According to recent studies, Trincadeira could be genetically close to RAMISCO from the Colares region and to SERCIAL from Madeira (Almadanim *et al.* 2007). Trincadeira is not related to the light-skinned TRINCADEIRA DAS PRATAS.

VITICULTURAL CHARACTERISTICS

Requires very careful pruning to achieve balanced vigour and production. Susceptible to fungal diseases such as powdery mildew, botrytis bunch rot and, most seriously, sour rot.

WHERE IT'S GROWN AND WHAT ITS WINE TASTES LIKE

Trincadeira, though not easy to grow, can produce high-quality wines when the yields are moderate and berries are fully mature: wines that are well structured, full-bodied and full of rich, dark fruit spiced up by notes of cinnamon and clove, supported by rounded tannins, so that they can be broached in youth. However, the wine's structure is also complemented by oak ageing, resulting in wines with greater longevity. Trincadeira makes a good blending partner, eg with Aragones (TEMPRANILLO) in the Tejo or with TOURIGA NACIONAL in the Douro and in Dão. Although the variety is planted in many parts of **Portugal** and permitted in many DOCs, with a total vineyard area of 14,220 ha (35,138 acres) in 2010, it is most important in the Alentejo, Douro and Tejo (especially Bairro and Chamusca). Recommended producers in the Alentejo include Cortes de Cima, Altas Quintas and João Portugal Ramos.

TRINCADEIRA DAS PRATAS

Underrated aromatic variety from central Portugal.

BERRY COLOUR: ● ● ● ● ●

PRINCIPAL SYNONYMS: Arinto Gordo, Boal Prior

VARIETIES COMMONLY MISTAKEN FOR TRINCADEIRA DAS PRATAS: TAMAREZ (Alentejo and Setúbal)

ORIGINS AND PARENTAGE

Trincadeira das Pratas, first described by Gyrão in 1822 in his *Tratado theorico e pratico da agricultura das vinhas*, originates in the Tejo region in central Portugal (Cunha *et al.* 2009). It is still often confused with TAMAREZ in the Alentejo, although they are morphologically quite distinct. According to Galet (2000), Trincadeira das Pratas could simply be a synonym for TRINCADEIRA from the Douro but the latter is a black-berried variety, and it cannot be a colour mutation of Trincadeira because they have distinct DNA profiles in Almadanim *et al.* (2007).

VITICULTURAL CHARACTERISTICS

Very early budding and ripening. Susceptible to sunburn and botrytis bunch rot but moderately resistant to downy and powdery mildews. Best suited to deep, moist soils

WHERE IT'S GROWN AND WHAT ITS WINE TASTES LIKE

Planted mainly in the Tejo region, but also north and south of Lisboa in **Portugal**, Trincadeira das Pratas produces wines that are light in colour but well structured with aromas of fresh herbs, lemon and pear, compatible with barrel fermentation and ageing. Although many producers see little future for this variety and few are planting it, Australian Peter Bright, a consultant in southern Portugal, sees it as a variety with great promise. Recommended producers of blends that include Trincadeira das Pratas: Quinta da Aloma, Fiúza & Bright and the Cartaxo co-op. There were 250 ha (618 acres) in Portugal in 2010.

TRIOMPHE

Slightly foxy-tasting red hybrid planted mainly in the UK.

BERRY COLOUR: ● ● ● ● ●

PRINCIPAL SYNONYMS: Kuhlmann 319-1, Triomphe d'Alsace, Triumpf vom Elsass, Triumph d'Alsace

ORIGINS AND PARENTAGE

Complex French-American hybrid bred by Eugène Kuhlmann in Alsace in the early twentieth century by crossing Millardet et Grasset 101-14 OP × KNIPPERLÉ, where Millardet et Grasset 101-14 OP is a seedling of Millardet et Grasset 101-14 (OP stands for open pollination), this latter being a *Vitis riparia* × *Vitis rupestris* hybrid.

VITICULTURAL CHARACTERISTICS

Very early ripening. Vigorous, with small, red-fleshed berries and some resistance to powdery mildew.

WHERE IT'S GROWN AND WHAT ITS WINE TASTES LIKE

Little known outside of the **UK**, where there are 15 ha (37 acres). The deeply coloured wines can taste slightly foxy, revealing their part-American parentage, so it is surprising that the variety has continued to be planted when both DORNFELDER and REGENT produce better results. Also found in the nascent vineyards of **Denmark** and the **Netherlands**.

TRNJAK

Minor variety found in both Croatia and Bosnia-Herzegovina.

BERRY COLOUR: ● ● ● ● ●

PRINCIPAL SYNONYMS: Rudežuša, Trnjak Crni

ORIGINS AND PARENTAGE

Trnjak is a variety from the Croatian region of Imotski in Dalmacija (Dalmatia) south east of Split and close to the Bosnia-Herzegovina border.

VITICULTURAL CHARACTERISTICS

Mid to late budding and ripening. Compact, medium-sized bunches of thick-skinned berries. Moderate yields.

WHERE IT'S GROWN

Trnjak is authorized only in the Dalmatinska Zagora (Dalmatian Hinterland) wine region in **Croatia** and is found not only in its region of origin, Imotski, south east of Split in Primorksa Hrvatska (Coastal Croatia) but also in Makarska by the coast and in Mostar over the border in Bosnia-Herzegovina. Producers of varietal wines include Imota (Imotski) and Opačak (Makarska).

TROLLINGER

See SCHIAVA GROSSA.

TRONTO

Possibly an Aglianico relative from around Amalfi in Campania.

BERRY COLOUR:

PRINCIPAL SYNONYMS: Aglianico di Napoli, Tronta

VARIETIES COMMONLY MISTAKEN FOR TRONTO: AGLIANICO

ORIGINS AND PARENTAGE

Ampelographers had noticed many similarities between Tronto and AGLIANICO and this was confirmed by DNA profiling when Costantini *et al.* (2005) found that Tronto had a close genetic relationship with Aglianico, which would explain why it is sometimes named Aglianico di Napoli.

VITICULTURAL CHARACTERISTICS

Low sugar and low acid in the berries. Generally avoids botrytis bunch rot thanks to the relatively early harvest date.

WHERE IT'S GROWN

Tronto can still be found in some old vineyards on the Amalfi coast west of Salerno in Campania, southern **Italy**, in particular in Furore, Positano and Amalfi, where it may be included in the Costa d'Amalfi DOC. It is also occasionally found near Napoli under the name Aglianico di Napoli.

TROUSSEAU

Demanding variety making powerful, ageworthy wines and having a complex geographical distribution and many aliases, including Bastardo.

BERRY COLOUR:

PRINCIPAL SYNONYMS: Bastardhino (Portugal), Bastardo (Dão and Douro in Portugal), Bastardo (Orense in Galicia in Spain), Carnaz (Galicia), Godello Tinto (Galicia), María Ordoña (Orense), María Ordoñez (Orense and León in Spain), Maturana Tinta (La Rioja in Spain), Merenzao (Bierzo in Spain), Roibal (Paderne in La Coruña in Spain), Tinta or Tintilla (parts of Tenerife), Tinta Lisboa (Portugal), Troussé (Salins-les-Bains in the Jura), Trousseau Noir, Troussot, Verdejo (Asturias and Lugo in Spain), Verdejo Negro or Verdejo Tinto (Asturias and Rueda in Spain)

VARIETIES COMMONLY MISTAKEN FOR TROUSSEAU: ALFROCHEIRO, CABERNET PFEFFER, CASTELÃO (Portugal), DONZELINHO TINTO (Portugal), PINOT NOIR, TRESSOT (Yonne)

ORIGINS AND PARENTAGE

Trousseau is a variety from the Jura in eastern France, once widely cultivated in the regions of

Poligny and Besançon, where it was first mentioned in 1732 under the spelling Troussot. Trousseau has often been confused with TRESSOT from the Yonne *département*, undoubtedly because Trousseau was often spelt Tressau or Tresseau in the past, as in the commune of Poligny in 1774 (Rézeau 1997): 'Plants de tressaux, bon pour le vin, mediocre à manger' ('Tressaux vines, good for wine, mediocre for eating'). The name Tresseau is, rather unconvincingly, said to derive from the French word *troussé* ('trussed'), an old synonym used in Salins-les-Bains, because of the pyramid shape of the bunch (Rézeau 1997).

DNA profiling has established that Trousseau has been cultivated in Spain and Portugal for at least two centuries under different names such as Bastardo, Merenzao and Verdejo Negro (Gago *et al.* 2009), and under the name Maturana Tinta (Félix Cabello and Victor de la Serna, personal communications), but it is not known how it reached the Iberian peninsula. The most ancient and widespread name is Bastardo ('bastard'), first mentioned in 1531 by Rui Fernandes in his 'Descripção do terreno em roda da cidade de Lamego duas leguas' and described in 1790 by Rebello da Fonseca, who tentatively suggested it was the same variety as PINOT NOIR (d'Oliveira 1903).

Trousseau belongs to the Salvanien ampelographic group (see p XXVII; Bisson 2009), making it unlikely that this variety originated in Iberia. This is confirmed by DNA parentage analysis, which strongly suggests that Trousseau (under the name Merenzao), CHENIN BLANC and SAUVIGNON BLANC are siblings (Santana *et al.* 2010; Myles *et al.* 2011), and that Trousseau has a likely parent–offspring relationship with SAVAGNIN (Myles *et al.* 2011), as shown in the PINOT pedigree diagram.

The colour mutation **Trousseau Gris** was once significant in the Charente and is cultivated in California under the name Gray Riesling. Trousseau has sometimes and confusingly been called Sémillon Rouge (see SÉMILLON).

Trousseau was used to breed BASTARDO MAGARACHSKY and ROYALTY.

VITICULTURAL CHARACTERISTICS

Vigorous, early budding and ripening. Needs well-exposed sites to reach full maturity. Generally rather disease-prone, especially to botrytis bunch rot. Small, compact bunches of small to medium-sized, thick-skinned berries. High sugar potential in the berries.

WHERE IT'S GROWN AND WHAT ITS WINE TASTES LIKE

In **France**, Trousseau is one of Jura's two main dark-skinned indigenous varieties but represents only about 5% of the vineyard area because it requires more sun than other varieties and prefers warm gravel or marl soils. It is more deeply coloured and robust than the pale POULSARD, producing wines that are powerful, warm and structured, but Trousseau wines can lack colour if yields are not kept in check. Although it is one of the principal varieties authorized for red and rosé wines in the Arbois and Côtes du Jura appellations as well as for the fortified Macvin du Jura and sparkling Crémant du Jura, there were no more than 172 ha (425 acres) in France in 2009, little changed over the last few years. Recommended producers include Lucien Aviet, Daniel Dugois, Jacques Puffeney and Domaine André and Mireille Tissot in the Arbois appellation; Domaines Ganevat and Pignier in the Côtes du Jura.

In Galicia, north-west **Spain**, the variety goes under the names Merenzao, María Ordoña or Bastardo. Although it is technically one of the principal varieties authorized for reds from the Ribeira Sacra and Valdeorras DOs, along with the more well-known MENCÍA, only 12 ha (30 acres) remained in 2008. Algueira and A Tapada produce varietal wines, as do Quinta da Muradella, using the synonym Bastardo. Listed separately for Spain are the 70 ha (173 acres) of Verdejo Negro in Asturias, north east of Galicia. Here it is one of the principal varieties in the little-known Vino de la Tierra de Cangas. In the Tierra de León appellation, Gordonzillo have planted Merenzao in their trial vineyard, producing peppery-flavoured grapes.

Under the name Maturana Tinta, the variety has been provisionally included in the official Spanish list of commercial varieties as a separate entry and is authorized in the Rioja DO. Plantings are very limited indeed but Ijalba and Valdemar have both produced intense, oak-aged varietal versions that show the variety's fresh acidity and deep fruit flavours.

The variety is much more common in **Portugal**, which has 1,218 ha (3,101 acres) of

Trousseau, known here as Bastardo. Most are in the Douro, where it is one of the lesser port varieties and generally found in old mixed vineyards, in Dão and in Beiras. There are tiny plantings on the islands of Madeira and Porto Santo, mostly in experimental vineyards, although it was at one time much more widely used in the production of Madeira, sometimes even as a varietal wine. Unfortified varietal wines are typically pale in colour, high in alcohol but low in acidity, although some early-picked fruit has shown the potential for rosé.

Bastardo Tinto is a distinct and far less widely planted Portuguese variety and should not be confused with Bastardo/Trousseau.

Trousseau has recently been identified in an old vineyard in Russian River Valley, **California**, and in the same appellation it's included in Acorn Vineyards' old-vine Zinfandel bottling that includes fruit from a field blend. The variety is mentioned in the California grape acreage report, under both Bastardo and Trousseau, but no acreage is given, suggesting the area is negligible. Several North American producers include Bastardo in the blend of their port-style fortified wines, including Heitz in Napa Valley and Renwood in Sierra Nevada.

Trousseau Gris used to be planted throughout California but is now generally found only in old interplanted vineyards. However, Fanucchi Vineyards in the Russian River Valley in Sonoma, California, planted an entire Trousseau Gris block in 1981 and make a very successful varietal wine, which Peter Fanucchi describes as having a crisp, clean, flowery bouquet with flavours of fresh peaches, honeysuckle, pear, melon, delicate spice, tropical fruit ...

In **Oregon**, Abacela in the Umpqua Valley have some Bastardo planted. And a variety called Bastardo is also planted in Krym (Crimea), **Ukraine**, and in the **Republic of Moldova**.

TSAOUSSI

Greek variety at home on Kefaloniá and generally best in blends.

BERRY COLOUR: ● ● ● ● ●

VARIETIES COMMONLY MISTAKEN FOR TSAOUSSI: ÇAVUŞ (Turkey)

ORIGINS AND PARENTAGE

Tsaoussi probably originates from Kefaloniá in the Ionian Sea off the west coast of Greece. It should not be confused with the much more widely planted Turkish variety ÇAVUŞ, also known as Tsaoussi. The Greek Tsaoussi has what ampelographers call 'perfect flowers', with both male and female organs (hermaphrodite), whereas Çavuş has female flowers (Haroula Spinthiropoulou and Stefanos Koundouras, personal communications).

VITICULTURAL CHARACTERISTICS

Fairly vigorous, early to mid budding, mid ripening. Susceptible to powdery mildew. Loose bunches of large, thick-skinned berries. Prone to wasp damage close to harvest.

WHERE IT'S GROWN AND WHAT ITS WINE TASTES LIKE

The island of Kefaloniá is the main home for the variety known in **Greece** as Tsaoussi or Tsaousi. There are still around 135 ha (334 acres), although the area is diminishing as older farmers retire. It is used mainly for wine, generally blended with other local varieties such as ROBOLA, although it is good to eat. Gentilini's varietal Tsaoussi is available only to a few restaurants but their Aspro blend still shows the honey, melon and stone-fruit character of the variety, even with the added components of Robola and a small percentage of SAUVIGNON BLANC to increase the freshness. It is difficult to achieve the variety's typical fruit flavours if the fruit is picked early enough to retain acidity. Other recommended producers include Domaine Sklavos.

TSAPOURNAKO

See CABERNET FRANC.

TSARDANA

See ROMEIKO.

TSIMLADAR

This obscure, dark-berried variety from the Tsimlyansky district in the Rostov wine region, south-western **Russia**, may be a minor ingredient in the traditional sweet red fizz produced in that region and is also used in dry and fortified wines.

TSIMLYANSKY CHERNY

Convincing, if rather tannic, Russian variety used to make various styles of red.

BERRY COLOUR:

PRINCIPAL SYNONYMS: Tsimlyansky

VARIETIES COMMONLY MISTAKEN FOR TSIMLYANSKY CHERNY: PLECHISTIK

ORIGINS AND PARENTAGE

Tsimlyansky Cherny probably comes from the Rostov wine region in south-western Russia, near the border with Ukraine.

OTHER HYPOTHESES

Like other varieties along the River Don, Tsimlyansky Cherny is said to have been introduced from Dagestan in the Middle Ages (Vladimir Tsapelik, personal communication). Others believe it was introduced to the region by Cossacks returning from the Napoleonic Wars in 1812 but it seems unlikely that vines could have survived winters north of here.

VITICULTURAL CHARACTERISTICS

Mid budding and ripening. Susceptible to downy mildew.

WHERE IT'S GROWN AND WHAT ITS WINE TASTES LIKE

Tsimlyansky Cherny is grown in the Vedernikov and Tsimlyansky wine regions along the River Don north east of Rostov-on-Don in **Russia** (170 ha/420 acres in 2010 although this figure includes PLECHISTIK) and also in Krasnodar further south and west. Tsimlyansky Winery use it for sweet sparkling wines and Vedernikov Winery make a tannic dry red with flavours of dried plum, black pepper and leather. At Fanagoria winery in Krasnodar it makes slightly dusty, confident wines with good structure and flavours that locals describe as blackthorn, smoked cherry and black fruits but reminds foreign visitors of BARBERA. It is also used in dessert and fortified blends – often with Plechistik and KRASNOSTOP ZOLOTOVSKY – as well as for rosé.

TSITSKA

High-acid Georgian variety used to make still and sparkling wines.

BERRY COLOUR:

PRINCIPAL SYNONYMS: Shanti, Tsitsiko

ORIGINS AND PARENTAGE

Tsitska comes from Kolkheti, western Georgia, like TSOLIKOURI and USAKHELOURI.

VITICULTURAL CHARACTERISTICS

Vigorous and productive. Mid budding, late ripening. Relatively large, compact bunches of thin-skinned berries.

WHERE IT'S GROWN AND WHAT ITS WINE TASTES LIKE

Tsitska is found mainly in the villages of Sviri, Dimi, Persati and Obcha in the Imereti region of **Georgia**. It produces both high-quality dry table wine – on its own or blended with other varieties such as TSOLIKOURI – and, thanks to its high acidity (often 9–12 g/l at harvest), is well suited to the production of sparkling wine. Fruit flavours tend to be in the melon-apple-pear spectrum. Recommended producers of varietal wines include Imeretian Wines and Khareba. There were 2,783 ha (6,877 acres) of Tsitska in Imereti in 2004.

TSOLIKOURI

Versatile, widely planted Georgian variety.

BERRY COLOUR:

PRINCIPAL SYNONYMS: Melqos Tsolikouri, Obchuri Tsolikouri, Tsolikoouri

ORIGINS AND PARENTAGE

Tsolikouri comes from Kolkheti, western Georgia, like TSITSKA and USAKHELOURI.

VITICULTURAL CHARACTERISTICS

Medium-sized bunches of thick-skinned berries. Mid budding, late ripening.

WHERE IT'S GROWN AND WHAT ITS WINE TASTES LIKE

Tsolikouri is grown predominantly in the villages of Dimi, Baghdati, Sviri and Tvishi in **Georgia**. It is the sole variety responsible for the semi-sweet whites of the Tvishi appellation and, blended with TSITSKA (and sometimes KRAKHUNA), adds fruit and body to the dry wines of the Sviri appellation. Dry Tsolikouri wines tend to be full-bodied and to have soft acidity and flavours of quince or pear. The variety also shows good potential for sparkling wine blends despite the moderate acidity. Teliani Valley make a good varietal example. There was a total of

5,787 ha (14,300 acres) in Georgia in 2004, the vast majority in Imereti but also in Samegrelo and Racha-Lechkhumi.

TSULUKIDZIS TETRA

Minor but productive Georgian variety making both dry and sweet whites.

BERRY COLOUR: ● ● ● ● ●

PRINCIPAL SYNONYMS: Rachuli Tetra, Racthuli Tetri, Tsulukidze Tetra

ORIGINS AND PARENTAGE

Tsulukidzis Tetra comes from the Racha-Lechkhumi region of north-west Georgia.

VITICULTURAL CHARACTERISTICS

Productive, early budding and ripening.

WHERE IT'S GROWN AND WHAT ITS WINE TASTES LIKE

Tsulukidzis Tetra, used to make both semi-sweet and dry white wines, is found to a limited extent in Racha-Lechkhumi in north-west **Georgia**, particularly around the villages of Sadmeli and Chorjo. The wines are crisp, with moderate alcohol and flavours of ripe peach and pale-skinned cherry. Producers include Rachuli Wine and Bugeuli, although the former use the synonym Rachuli Tetra.

TUO XIAN

Table grape also pressed into service for wine in northern China.

BERRY COLOUR: ● ● ● ● ●

ORIGINS AND PARENTAGE

Obscure *Vitis vinifera* variety of unknown origin that has been cultivated in Tuo Xian county in northern China for more than two centuries. Also known as 'Small Carnelian'.

VITICULTURAL CHARACTERISTICS

Mid to late ripening, winter-hardy, good resistance to drought and nematodes and relatively easy to manage. Very large, long bunches.

WHERE IT'S GROWN AND WHAT ITS WINE TASTES LIKE

Tuo Xian is grown mainly as a table grape in Tuo Ke Tuo county in Nei Mongol (Inner Mongolia) in the far north of **China** but Clouds Winery in Tuo Xian county also use it to make several different styles of white wine (Plocher *et al.* 2003).

TURÁN

Dark-skinned, red-fleshed Hungarian cross valued mainly for its colour.

BERRY COLOUR: ● ● ● ● ●

PRINCIPAL SYNONYMS: Agria (North America), Bikavér 13, Eger 208

ORIGINS AND PARENTAGE

A Bikavér 8 × Gárdonyi Géza cross obtained in 1964 by József Csizmazia and László Bereznai in Hungary, in which Bikavér 8 is a TEINTURIER × KADARKA cross and Gárdonyi Géza is a MENOIR × CSABA GYÖNGYE cross.

VITICULTURAL CHARACTERISTICS

Mid ripening with big bunches of small, waxy, red-fleshed berries. Resistant to botrytis bunch rot and winter-hardy in Hungarian conditions.

WHERE IT'S GROWN AND WHAT ITS WINE TASTES LIKE

Turán produces deeply coloured wines with fine tannins, good acidity and flavours of rose, rosehip and dark forest fruits. It is sometimes blended with varieties such as KADARKA. The majority of **Hungary**'s 178 ha (420 acres) are planted in Eger, although it is also found in the neighbouring Mátra region. Recommended producers of varietal wines include Tamás Szecskő in Mátra and Csaba Demeter in Eger. Imre Kaló, also in Eger, makes the finest examples, sometimes even an Aszú-style sweet wine, but they are sold only at the cellar door.

Under the synonym Agria, this variety somehow found its way across the Atlantic to **Canada**'s British Columbia (7.5 acres/3 ha in 2008) and thence south to **Washington** State in the US, but not in sufficient quantities to make it into the official statistics. Marley Farm in BC and Glacier Peak in WA both make varietal wines.

TURCA

See DOUCE NOIRE.

TYRIAN

Full-flavoured Australian cross bred for local conditions.

BERRY COLOUR: ● ● ● ● ●

ORIGINS AND PARENTAGE

A cross of the Spanish variety SUMOLL with CABERNET SAUVIGNON, created in 1972 by A J Antcliff at the Merbein station in Victoria, Australia, with the aim of creating a good-quality variety suited to hot, dry Australian conditions. Released in 2000 along with its sister variety CIENNA.

VITICULTURAL CHARACTERISTICS

Good sugar-to-acid ratio, low pH, good colour and flavour, and reasonable yields.

WHERE IT'S GROWN AND WHAT ITS WINE TASTES LIKE

McWilliams Wines, who were involved in the initial trials of this variety, are currently its main proponent, along with one other grower in the Riverina, **Australia**. The name Tyrian – evoking its deep purple colour – was chosen for marketing purposes. McWilliams have nearly 20 ha (50 acres) planted at Hanwood near Griffith in New South Wales and 1.5 ha (4 acres) in the Hunter Valley. The wines are rich, cherry-fruited and voluptuous but with relatively firm, savoury tannins. The variety is currently used only in blends and in port-style wines.

U

UCELÙT
URLA KARASI
USAKHELOURI
UVA DELLA CASCINA
UVA LONGANESI
UVA RARA
UVA TOSCA
UVALINO

ᛝthis symbol on the following pages indicates that the synonyms, and mistaken identities, are confirmed by DNA evidence

UCELÙT

Old and rare Friulian rescued and used mainly for sweet wines.

BERRY COLOUR: ● ● ● ● ●

ORIGINS AND PARENTAGE

Historically cultivated in the hills of Valeriano, Pinzano al Tagliamento and Castelnovo del Friuli. Its name comes from *uccelline*, meaning 'birds', because they often eat its sweet berries.

OTHER HYPOTHESES

Supposedly domesticated from local wild vines.

VITICULTURAL CHARACTERISTICS

Yields abundantly and ripens late. Sensitive to botrytis bunch rot.

WHERE IT'S GROWN AND WHAT ITS WINE TASTES LIKE

Ucelùt is cultivated in particular in the province of Pordenone in Friuli-Venezia Giulia in north-east **Italy**. Together with other old local varieties CIANORIE, CIVIDIN, FORGIARIN, PICULIT NERI and SCIAGLÌN, it was saved from extinction by Emilio Bulfon, with the help of ampelographers Antonio Calò and Ruggero Forti. Sweet passito versions, which may have aromas of figs, honey and mint, are made by Bulfon, Ronco Cliona and Vicentini Orgnani, for example; Florutis make a dry example but this is unusual. Together with Forgiarin, Piculit Neri and Sciaglìn, Ucelùt was included in the Italian national register of varieties in 1991. There were 10 ha (25 acres) of Ucelùt in Italy in 2000.

UGNI BLANC

See TREBBIANO TOSCANO.

URLA KARASI

Ancient, rare and recently rescued red-skinned Turkish variety.

BERRY COLOUR: ● ● ● ● ●

ORIGINS AND PARENTAGE

Urla Karası, literally 'black from Urla', is local to the Urla peninsula in the province of İzmir in western Turkey.

VITICULTURAL CHARACTERISTICS

Late ripening. Medium-sized berries with lightly coloured flesh (Çetiner *et al.* 2009). Apparently resistant to both downy and powdery mildews (Can Ortabaş, personal communication).

WHERE IT'S GROWN AND WHAT ITS WINE TASTES LIKE

Urla Karası has recently been rescued by Can Ortabaş, owner of Urla Şarapçılık in Urla, who has planted this and other old local varieties in biodynamically farmed vineyards and is now expanding the cultivation to higher vineyards (1,050 m/3,445 ft) although the variety is resistant to grafting. Wines have flavours of blackberry and both black and sour cherry. Urla Şarapçılık combine NERO D'AVOLA with Urla Karası in an unusual but successful blend of varieties that ripen at the same time.

USAKHELOURI

Georgian variety producing semi-sweet, aromatic reds.

BERRY COLOUR: ● ● ● ● ●

PRINCIPAL SYNONYMS: Okhureshuli, Okourechouli

ORIGINS AND PARENTAGE

Usakhelouri comes from Kolkheti, western Georgia, like TSITSKA and TSOLIKOURI. The synonym Okhureshuli is taken from the name of a village in the Racha-Lechkhumi region. According to Chkhartishvili and Betsiashvili (2004), it is botanically and genetically close to the wild grapevines of the area.

VITICULTURAL CHARACTERISTICS

Early budding, late ripening. Best suited to loamy, lime-rich hillsides.

WHERE IT'S GROWN AND WHAT ITS WINE TASTES LIKE

Usakhelouri, used to produce smooth, richly flavoured, rose-scented, semi-sweet (and very occasionally dry) red wines, is grown around the villages of Zubi and Okhureshi in the Racha-Lechkhumi region of north-west **Georgia**. Recommended producers include Khareba, Paata Sharashenidze and Telavi Wine Cellar.

UVA DELLA CASCINA

Rare, perfumed, recently rescued Oltrepò Pavese variety.

BERRY COLOUR: ● ● ● ● ●

PRINCIPAL SYNONYMS: Uva Cascina

ORIGINS AND PARENTAGE

Uva della Cascina, meaning 'grape of the farm', is an ampelographic rarity of unknown origin, cultivated in the Oltrepò Pavese.

VITICULTURAL CHARACTERISTICS

Vigorous, early ripening. Low yields, probably due to the presence of viruses in the few vines that have been recovered. Berries have a slight Muscat aroma.

WHERE IT'S GROWN AND WHAT ITS WINE TASTES LIKE

Uva della Cascina was considered to be extinct but was recently rediscovered by adventurous local growers such as Fortesi in Rovescala due south of Milano in the province of Pavia, Lombardia, in **Italy**. Initially their experimental varietal wine was available for tasting only at the cellar.

Uva della Cascina could not be included in the Italian national register of varieties because of the possible confusion with the Tuscan village of Càscina. It is cultivated mainly in the Valle Staffora in Pavia and is usually blended with other local grape varieties. Cabanon's Vino del Bosco (IGT Provincia di Pavia) combines at least three ancient varieties, Uva della Cascina, UVA RARA and Ughetta (ie VESPOLINA), and Fattoria Mondo Antico's Sinodo (Oltrepò Pavese DOC) is a blend of BARBERA, CROATINA, Uva della Cascina and MORADELLA. Owner Dario Tiraboschi is planning to produce a varietal wine and describes it as perfumed and full-bodied.

UVA LONGANESI

Hardy Emilia-Romagnan named after the family that rescued it.

BERRY COLOUR:

PRINCIPAL SYNONYMS: Bursôn, Longanesi, Negretto Longanesi

ORIGINS AND PARENTAGE

This variety is said to have been cultivated incognito in the neighbourhood of Bagnacavallo, west of the city of Ravenna in Emilia-Romagna, for at least two centuries. It was almost extinct at the end of the nineteenth century and has only recently been officially named after the Longanesi family who rescued it from extinction.

In 1913, Antonio Longanesi bought a property in Bagnacavallo and found an old vine thriving in the neighbourhood, climbing on an oak tree. Given the remarkable hardiness of the plant and the resistance to fungal diseases of its thick-skinned small berries, Longanesi successfully made a house wine that was much appreciated by the locals. The vine was propagated for Antonio Longanesi and his friends, and in 1956 his son Aldo Longanesi planted the first vineyard of this variety. In 1972, one year after Aldo's death, his son Pietro tentatively named this vine Uva Longanesi in honour of his forefathers. Pietro's son Daniele, who took over the winery in 1995, obtained official recognition for Uva Longanesi in 1999 by its inclusion in the national register of varieties. Its synonym Bursôn, meaning 'big bag' in the local dialect, was Antonio Longanesi's nickname.

Recent DNA profiling at the San Michele all'Adige research centre has established that Uva Longanesi is not identical to any other known grape variety (Stella Grando and José Vouillamoz, unpublished data).

OTHER HYPOTHESES

Some local historians believe that Uva Longanesi was introduced from Spain by the Pope's soldiers, and some scholars suggest it was domesticated from local wild grapes.

VITICULTURAL CHARACTERISTICS

Mid ripening. Resistant to frost and fungal diseases.

WHERE IT'S GROWN AND WHAT ITS WINE TASTES LIKE

Remarkably, from a single vine, Uva Longanesi today covers 200 ha (494 acres) in the area of Bagnacavallo and Faenza, west and south west of Ravenna in Emilia-Romagna in north-central **Italy**, giving sometimes rich wines with slightly rustic or bitter tannins. According to local regulations, varietal wines, which qualify as IGT Ravenna, are divided into two categories: the Etichetta Nera ('black label'), signifying a more ambitious version (minimum twenty months in oak and a proportion of dried grapes), and the less oaked and less ambitious Etichetta Blu, which may be made using the technique of carbonic maceration to soften the tannins. Producers include Cantine Ercolani and the eponymous Daniele Longanesi.

UVA RARA

Often used to soften Nebbiolo in the far north of Italy.

BERRY COLOUR:

VARIETIES COMMONLY MISTAKEN FOR UVA RARA: BONARDA PIEMONTESE (Novara, Vercelli and Pavia)

ORIGINS AND PARENTAGE

Uva Rara is an old variety from Piemonte and Lombardia, northern Italy. Its name is a source of confusion: in the area of Novara, Vercelli and Pavia, south and west of the city of Milano, Uva Rara is misleadingly called Bonarda, although it is distinct from BONARDA PIEMONTESE. The name Uva Rara, literally 'rare grape', probably refers to the low number of berries per bunch, not to the rarity of the vine.

VITICULTURAL CHARACTERISTICS

Mid to late ripening. Sensitive to millerandage and powdery mildew.

WHERE IT'S GROWN AND WHAT ITS WINE TASTES LIKE

Contrary to any expectation set up by its name, Uva Rara is widespread in Lombardia (especially in the province of Pavia) and in Piemonte (in the provinces of Alessandria, Asti, Biella, Novara, Torino and Vercelli) in **Italy**. There were 608 ha (1,502 acres) in 2000 but plantings have been declining. It may be used to soften NEBBIOLO-dominated blends in northerly DOCs such as Boca, Colline Novaresi, Fara and Sizzano, Ghemme DOCG and, slightly further south, blended with mostly BARBERA and CROATINA in Oltrepò Pavese and San Colombano al Lambro DOCs. A varietal wine is authorized within the Colline Novaresi denomination.

Varietal wines tend to be perfumed and quite soft though some have a bitter aftertaste. Producers include Brigatti, the Cantina Sociale dei Colli Novarese, Fortesi and Frecciarossa.

UVA TOSCA

Minor Emilia-Romagnan variety that has moved downhill.

BERRY COLOUR:

PRINCIPAL SYNONYMS: Montanara

VARIETIES COMMONLY MISTAKEN FOR UVA TOSCA: SANGIOVESE

ORIGINS AND PARENTAGE

Uva Tosca was first described by Tanara (1644) in Emilia-Romagna, which may be its place of origin, as a grape variety giving a 'reddish wine, not very sweet, spicy, elegant and very healthy'. Despite its name, the variety does not come from Toscana because, as Cipriani *et al.* (2010) have recently suggested through DNA parentage analysis, Uva Tosca may be a Crepallocchi × SCHIAVA GROSSA natural cross (see Schiava Grossa for the pedigree diagram), the former being an obscure black-berried variety of unknown origin, the latter a variety from the Alto Adige that has spread to Lombardia. Another study has also suggested a possible parent–offspring relationship between Uva Tosca and LAMBRUSCO MONTERICCO (Boccacci *et al.* 2005), which would then have to be a progeny of Uva Tosca.

VITICULTURAL CHARACTERISTICS

Mid to late ripening. Susceptible to millerandage and powdery mildew.

WHERE IT'S GROWN

Uva Tosca used to be cultivated between 700 and 900 m above sea level in the Emilia-Romagnan stretch of the Appenines in **Italy**, where no other red variety could ripen. Today the vineyard area at this cool altitude has declined dramatically, whereas it has increased in the lower-altitude provinces of Ravenna, Modena (Zocca, Polinago) and Reggio Emilia (Ligonchio, Carpineti). However, overall plantings are down since 1990 and total plantings in Italy in 2000 were 115 ha (284 acres). Uva Tosca is the only variety grown in the Val di Secchia. It is generally used as a blending component.

UVALINO

Recently resurrected Piemontese.

BERRY COLOUR: ● ● ● ● ●

PRINCIPAL SYNONYMS: Cunaiola (Canavese), Freisone (Tortona), Lambrusca (Roero), Lambruschino (Roero)

ORIGINS AND PARENTAGE

This variety has various names even within the Piemonte region (see list of synonyms above) and it is known as Uvalino, apparently meaning 'small grape' but most likely derived from *uvario* (dialect for a blend of minor-quality wines), only in the Asti region. It was first mentioned by Gallesio in 1831 as a variety cultivated in Nizza Monferrato and Castelletto Molina south east of Asti (Torello Marinoni *et al.* 2006). DNA profiling has shown that it is most likely that Uvalino has a parent–offspring relationship with Neretto di Marengo (Torello Marinoni *et al.* 2006), an almost extinct Piemontese variety.

VITICULTURAL CHARACTERISTICS

Robust, late-ripening vine. Resistant to botrytis bunch rot and thus suited to late harvesting. High levels of the antioxidant resveratrol have been reported.

WHERE IT'S GROWN AND WHAT ITS WINE TASTES LIKE

There were at one time scattered plantings of Uvalino throughout Piemonte in north-west **Italy**, mainly in the provinces of Asti, Alessandria and Tortona, but the variety was in decline until Mariuccia Borio, owner of Cascina Castlèt in Costigliole d'Asti, between Alba and Asti, met Renato Ratti, who had planted a small vineyard of Uvalino at his Villa Pattono winery in La Morra near Barolo. Uvalino reminded Borio, like Proust's memory-stirring *madeleine*, of a precious family wine, and she started to make a determined effort in the early 1990s to save Uvalino from oblivion, in collaboration with Lorenzo Corino, head of the Institute of Experimental Viticulture in Asti. After years of historical research and vinification experiments, Uvalino was included in the official register of varieties in 2002. Today Cascina Castlèt is one of the very few producers of varietal Uvalino, which is described as having intense aromas of ripe red fruits and sweet spices plus notable tannins and good acidity.

V

ᛝ this symbol on the following pages indicates that the synonyms, and mistaken identities, are confirmed by DNA evidence

VACCARÈSE

See BRUN ARGENTÉ.

VÁH

Recently authorized Slovakian cross that needs a good site to prosper.

BERRY COLOUR: ● ● ● ● ●

PRINCIPAL SYNONYMS: CAAB 3/13

ORIGINS AND PARENTAGE

Váh, named after the longest river wholly within Slovakia, a tributary of the Dunaj (Danube), is a CASTETS × ABOURIOU cross obtained in 1976 by Dorota Pospíšilová at the VUVV research centre for oenology and viticulture in Bratislava, Slovakia, and authorized in 2011. It is a sibling of HRON, NITRANKA and RIMAVA.

VITICULTURAL CHARACTERISTICS

Vigorous, consistently productive, late budding and ripening. Fairly large bunches of small, thick-skinned berries.

WHERE IT'S GROWN AND WHAT ITS WINE TASTES LIKE

Wines are deeply coloured, full-bodied, with a marked capsicum flavour reminiscent of some CABERNET SAUVIGNONS that can improve with time in bottle. Although the variety is not yet widely planted in **Slovakia** (5 ha/12 acres in 2011), a few producers such as Karol Braniš have begun making varietal wines. Ideally it should be grown in the best sites in the Južnoslovenská and Malokarpatská wine regions in the south west of the country.

VALDIGUIÉ

High-yielding rather ordinary variety from south-west France once popular in southern France and California but now in continuing decline.

BERRY COLOUR: ● ● ● ● ●

PRINCIPAL SYNONYMS: Cot de Cheragas (Algeria), Gros Auxerrois, Valdiguier

VARIETIES COMMONLY MISTAKEN FOR VALDIGUIÉ: GAMAY NOIR ⌛ (Valdiguié is called Napa Gamay in California, Brazil and Australia, or Gamay 15 in California)

ORIGINS AND PARENTAGE

Valdiguié, from south-west France, seems to be a fairly recent variety, not mentioned until 1884 (Rézeau 1997): 'le Valdiguier, cépage encore peu connu, qui paraît mériter une attention spéciale'

('Valdiguier, still little-known grape variety that seems to deserve special attention'). Three distinct and equally uncertain hypotheses have been proposed for the origin of this strange grape name (Galet 1990):

- In Puylaroque near Montauban in the Tarn-et-Garonne, a wine grower named Valdéguier (1745–1817) could have propagated this variety that grew in his courtyard from a random seed, possibly a MÉRILLE seedling.
- Still in Puylaroque around 1845, Jean-Baptiste Valdiguié (?–1864) may have noticed this variety among others in a neighbour's vineyard and then planted it in a vineyard in Tressens.
- In Aujols in the Lot, a vine of this variety was noticed in the ruins of the Templiers monastery by a worker named Guillaume Valdiguier.

Valdiguié is a member of the Cot ampelographic group (see p XXVII; Bisson 2009). Its supposed parent–offspring relationship with MÉRILLE (Galet 1990) has not yet been confirmed by DNA analysis. It was used to breed MELNIK 82 and RANNA MELNISHKA LOZA.

VITICULTURAL CHARACTERISTICS

Mid ripening (later in the cooler districts of California's North and Central Coasts), vigorous, fertile and productive, best pruned short. Big, compact bunches of big berries and a certain tolerance of powdery mildew.

WHERE IT'S GROWN AND WHAT ITS WINE TASTES LIKE

Although Valdiguié was once widely planted in southern **France** (4,908 ha/12,128 acres in 1958), thanks in particular to its high yields and tolerance of powdery mildew, there remained just 145 ha (358 acres) in 2008, mostly in the Tarn, Languedoc and Provence. It became popular in **California** during Prohibition for the same reasons and plantings reached a high of 6,118 acres (2,476 ha) in 1977, though they have since declined dramatically to 316 acres (128 ha) in 2008. Originally misnamed Napa Gamay, it was identified as Valdigiué by Pierre Galet in 1980. Official statistics still refer to it as Gamay (Napa), as distinct from Gamay Beaujolais, incidentally, even though its approved name is now Valdiguié. Wines are generally bright and well coloured, with only moderate alcohol. They have slightly astringent tannins, countered by carbonic maceration by some California producers. It is also used to produce everyday rosés. Recommended producers include Edmunds St John and J Lohr in California and Brick House in **Oregon**. Oliver of **Indiana** also produce a varietal Valdiguié.

VALENTINO NERO

Very minor Veneto cross whose true parentage has only recently been revealed.

BERRY COLOUR: ● ● ● ● ●

PRINCIPAL SYNONYMS: Incrocio Dalmasso 16/8

ORIGINS AND PARENTAGE

The Italian vine breeder Giovanni Dalmasso announced that Valentino Nero was a NEBBIOLO × DOLCETTO cross obtained in 1936 at the Conegliano research centre, in the Veneto, northern Italy. It was included in the official register of Italian varieties in 1977. DNA profiling has recently shown that the Nebbiolo used by Dalmasso was in fact Nebbiolo di Dronero, which is not Nebbiolo but a synonym for CHATUS (Torello Marinoni, Raimondi, Mannini and Rolle 2009). This makes Valentino Nero a Chatus × Dolcetto cross, just like its siblings PASSAU and SAN MARTINO.

WHERE IT'S GROWN

Valentino Nero is grown to a very limited extent in Piemonte, north west **Italy**.

VALIANT

Very winter-hardy, foxy American hybrid best suited to juice, jelly and the table.

BERRY COLOUR: ● ● ● ● ●

PRINCIPAL SYNONYMS: South Dakota 72S15, South Dakota NF7-121, South Dakota SD7-121

ORIGINS AND PARENTAGE

A FREDONIA × South Dakota 9-39 hybrid obtained in 1967 by R M Peterson at the South Dakota Agricultural Experiment Station in Brookings, in which South Dakota 9-39 is a *Vitis riparia* seedling collected near Culbertson, Montana. This *Vitis labrusca*, *Vitis riparia* and *Vitis vinifera* hybrid was selected in 1972 and released in 1997.

VITICULTURAL CHARACTERISTICS

Vigorous and productive and may need cluster thinning. Very hardy (has been known to withstand –50 °F/–46 °C in Manitoba, Canada). Small, compact bunches of small berries with typically low acidity and high sugar levels. Early ripening and daylight-sensitive so starts dormancy early. Susceptible to downy and powdery mildews and black rot, therefore best in a dry climate.

WHERE IT'S GROWN AND WHAT ITS WINE TASTES LIKE

Although the South Dakota Agricultural Experiment Station class this as a variety for table, juice and jelly, it is being used to a limited extent for winemaking in **South Dakota**, **Minnesota** (28 acres/11 ha in 2007), **Iowa** and other parts of the Midwest, eg by Carlos Creek in Minnesota, Table Mountain in **Wyoming**, Point of View in **North Dakota**, Threefold Vine in **Michigan** and Phoenix in **Missouri**. The wines are often part of a blend or made as a light rosé. They inherit the CONCORD flavour of the FREDONIA parent but also the high sugar and deep colour typical of *Vitis riparia*.

VALVIN MUSCAT

Highly aromatic hybrid with marked Muscat flavours.

BERRY COLOUR: ● ● ● ● ●

ORIGINS AND PARENTAGE

A complex hybrid obtained in 1962 by Bruce Reisch and Thomas Henick-Kling at New York State's Cornell University by crossing Muscat du Moulin with MUSCAT OTTONEL. Muscat du Moulin, also known as Couderc 299-35, is a hybrid of Couderc 603 × PEDRO XIMÉNEZ, where

Couderc 603 is a BOURRISQUOU × *Vitis rupestris* hybrid. Valvin Muscat is therefore a *Vitis rupestris* and *Vitis vinifera* hybrid whose original seedling vine was planted in 1964 and identified as test selection NY62.0122.01 when it was propagated in 1969. This selection was made available to growers and research collaborators for testing in 1990. Valvin Muscat was not officially released until 2006 (Reisch *et al.* 2006c).

VITICULTURAL CHARACTERISTICS

Smallish compact bunches of moderately large, juicy berries. Vines tend to be quite small if not grafted. Once grafted, it is vigorous and productive. Mid ripening but it should not be harvested before the full Muscat aroma is evident. Moderately susceptible to downy and powdery mildews and black rot. Resistant to botrytis bunch rot. Moderately hardy (–5 °F/–21 °C to –15 °F/–26 °C).

WHERE IT'S GROWN AND WHAT ITS WINE TASTES LIKE

There were just 10 acres (4 ha) of Valvin Muscat in **Indiana** in 2009, and no sign yet of its arrival in the official statistics of other Midwest states. Wines are described as high quality with spicy, floral aromas and no objectionable bitterness (Reisch *et al.* 2006c). The variety's intense aromas make it suitable as a blending component as well as on its own. Producers include Hunt Country Vineyards in **New York** State and Oliver Winery's Creekbend Vineyard in Indiana.

VAN BUREN

Old, low-acid American hybrid used mostly for the table.

BERRY COLOUR: ● ● ● ● ●

PRINCIPAL SYNONYMS: Gladwin 3000

ORIGINS AND PARENTAGE

A FREDONIA × Worden hybrid obtained in the early twentieth century by Fred E Gladwin in the Department of Pomology and Viticulture at the New York State Agricultural Experiment Station, Geneva, in which Worden is a selected seedling of CONCORD, which makes Van Buren a complex *Vitis labrusca* and *Vitis vinifera* hybrid. It's not known when the cross was made but it bore its fifth crop in 1936. It was introduced to the market in 1935.

VITICULTURAL CHARACTERISTICS

Small to medium-sized compact bunches of medium-sized, thick-skinned berries with a heavy bloom. Skin tends to shatter when ripe and the flesh has a foxy taste. Vigorous and very early ripening (thus attractive to birds) with a tendency to overcrop. Highly susceptible to downy mildew. Hardy (–15 °F/–26 °C to –20 °F/–29 °C).

WHERE IT'S GROWN AND WHAT ITS WINE TASTES LIKE

Low acidity makes Van Buren more suited to table grapes than to wine production but Forks of Cheat in **West Virginia** make a sweet, fruity varietal red. The variety can also be found in Ontario, **Canada**.

VANDAL-CLICHE

Hardy Canadian complex hybrid used for a wide range of wine styles.

BERRY COLOUR: ● ● ● ● ●

PRINCIPAL SYNONYMS: Cliche, Cliche 8414, Cliche Blanc, Vandal 84-14, Vandal Blanc, Vandal Cliche

ORIGINS AND PARENTAGE

A Vandal 63 × Vandal 163 hybrid bred after 1946 by Joseph O Vandal (1907–94) at Laval University in Québec, Canada, specifically for cold-climate regions, in which:

- Vandal 63 is a Prince of Wales × *Vitis riparia* hybrid, where Prince of Wales is a *Vitis vinifera* table grape from the United Kingdom
- Vandal 163 is an AURORE × CHANCELLOR hybrid.

Vandal-Cliche was developed by Mario Cliche, professor of horticulture at the Québec Institut de Technologie Agroalimentaire, hence the name. Vandal-Cliche is an extremely complex *Vitis riparia*, *Vitis labrusca*, *Vitis vinifera*, *Vitis aestivalis*, *Vitis lincecumii*, *Vitis rupestris* and *Vitis cinerea* hybrid.

VITICULTURAL CHARACTERISTICS

Very fertile, early to mid ripening. Hardy to –31 °F (–35 °C). Thin-skinned berries tend to drop at maturity. Susceptible to downy mildew and very susceptible to anthracnose in Québec.

WHERE IT'S GROWN AND WHAT ITS WINE TASTES LIKE

Vandal-Cliche is used in eastern **Canada** to make varietal wines in a range of styles, dry to sweet as well as sparkling or semi-sparkling, but it needs to be harvested quite early to avoid an excess of foxy *labrusca* character and its thin skins and weak stems make it unsuitable for late-harvest wines or icewines. Varietal wines are generally fresh, crisp and light, often marked by flavours of lemon peel. Québec producers include Carone, Les Côtes du Gavet, Au Jardin d'Emmanuel, Les Murmures, Moulin du Petit Pré, La Rivière du Chêne and Domaine Royarnois (the commercial pioneer of this variety), among others. Plantings in Québec are unlikely to increase, because of the variety's susceptibility to disease.

VASILAKI

Low-acid but potentially high-quality white wine grape from the Turkish island of Bozcaada.

BERRY COLOUR: ● ● ● ● ●

PRINCIPAL SYNONYMS: Altıntaş

ORIGINS AND PARENTAGE

Vasilaki, often called Altıntaş (literally 'golden stone'), is a variety from the island of Bozcaada in the Aegean, south west of the port of Çanakkale in Turkey. When the island belonged to Greece (and was called Ténedos), the variety was used widely to make both dry and sweet wines.

VITICULTURAL CHARACTERISTICS

Early to mid ripening. Large loose bunches of small, thick-skinned grapes. High yielding.

WHERE IT'S GROWN AND WHAT ITS WINE TASTES LIKE

Grown mainly on the island of Bozcaada, just off **Turkey**'s western coast, Vasilaki is a low-acid variety and wines need to include some early-picked fruit to remain fresh. They have a tendency to oxidize and age quickly unless the lees are retained during vinification but they can also have interesting aromas of citrus, orange peel and sometimes clove, with a slight bitterness on the finish. Corvus produce both dry and passito varietal wines and also make a blend with ÇAVUŞ, occasionally with SAUVIGNON BLANC. Other producers on Bozcaada include Talay, Yunatçılar and Gülerada. Total plantings in Turkey in 2010 were just 4 ha (10 acres).

VB 32-7

Minor Swiss hybrid bred and valued for its disease resistance.

BERRY COLOUR: ● ● ● ● ●

ORIGINS AND PARENTAGE

VB 32-7 is a hybrid obtained by Valentin Blattner (hence VB) in Soyhières in the Swiss Jura from undisclosed parents.

VITICULTURAL CHARACTERISTICS

Good all-round resistance to diseases as well as to frost.

WHERE IT'S GROWN AND WHAT ITS WINE TASTES LIKE

In addition to the breeder himself, only a few producers made wine from VB 32-7 on just 2 ha (5 acres) in **Switzerland** in 2009. Wines show SAUVIGNON BLANC-like aromas of elderflower and passion fruit and producers include Hervins in the canton of Jura (blended sparkling wine), Hof Kasteln in Aargau (blended with SOLARIS), Geiger in Sankt Gallen, Boner-Liechti in Graubünden and Cà di Ciser in Ticino. It arguably needs another name.

VB 91-26-4

Minor and usefully disease-resistant Swiss hybrid found to a very limited extent in northern Europe.

BERRY COLOUR: ● ● ● ● ●

PRINCIPAL SYNONYMS: VB 91-26-04

ORIGINS AND PARENTAGE

VB 91-26-4 is a hybrid obtained in 1991 by Valentin Blattner (hence VB) in Soyhières in the Swiss Jura by crossing CABERNET SAUVIGNON with an undisclosed disease-resistant variety (*Resistenzpartner* in German).

SAVAGNIN JAUNE

see SAVAGNIN

pp 960–68

SCHIAVA GROSSA

pp 972–4

GENOVESE

see SCIMISCIÀ

pp 978–9

SÉMILLON

pp 983–6

SERCIAL

p 987

SILVANER

pp 999–1001

SULTANINA BLANCHE

see SULTANIYE

pp 1016–17

SYRAH

pp 1023–31

TANNAT

pp 1036–7

TEMPRANILLO

pp 1042–5

TOURIGA NACIONAL

pp 1070–71

UGNI BLANC

see TREBBIANO TOSCANO

pp 1079–81

TRINCADEIRA

pp 1090–91

TROUSSEAU

pp 1093–5

VERMENTINO

pp 1127–9

VIOGNIER

pp 1143–6

VITICULTURAL CHARACTERISTICS

Vigorous, early ripening. Resistant to winter frost, downy and powdery mildews and botrytis bunch rot. Small bunches of small, thick-skinned berries.

WHERE IT'S GROWN AND WHAT ITS WINE TASTES LIKE

In **Switzerland**, only Geiger in Sankt Gallen make wine from VB 91-26-4 on less than half a hectare (1 acre) but the variety is also found in **Belgium**, where the Domaine du Ry d'Argent use VB 91-26-4 in a blend, and in the **Netherlands**, where Wijnboerderij De Gravin make a varietal VB 91-26-4 wine, described as similar to CABERNET FRANC.

VEGA

Also known as Incrocio Dalmasso 2/26, this light-skinned variety is a FURMINT × MALVAZIJA ISTARSKA cross obtained in 1937 by vine breeder Giovanni Dalmasso at the Conegliano research centre in the Veneto, northern Italy. According to the agricultural census, there were 28 ha (69 acres) of this early-ripening in **Italy** in 2000.

VERDEA

Ancient Tuscan white, now a Lombardia speciality; used to make a range of wine styles.

BERRY COLOUR: ● ● ● ● ●

PRINCIPAL SYNONYMS: Colombana Bianca (Toscana), Colombana di Peccioli (Toscana)

VARIETIES COMMONLY MISTAKEN FOR VERDEA: VERDECA (Puglia)

ORIGINS AND PARENTAGE

Verdea is a very old variety from Toscana, where it was mentioned by Pietro de Crescenzi in 1303, and then later by Soderini (1600), who described it as a 'vitigno apprezzatissimo' ('much appreciated grape variety').

Surprisingly, DNA profiling has revealed that a black-berried variety called Sangiovese Forte is in fact identical to Verdea (Di Vecchi Staraz, This *et al.* 2007), which would mean that Sangiovese Forte is a colour mutation of Verdea. However, this is questionable because the DNA profile of Sangiovese Forte in Filippetti *et al.* (2005) is identical to that of Sangiovese.

VITICULTURAL CHARACTERISTICS

Mid ripening. Resistant to botrytis bunch rot and therefore suited to late harvesting or drying.

WHERE IT'S GROWN AND WHAT ITS WINE TASTES LIKE

Verdea is cultivated mainly in Lombardia in northern Italy, in particular in the area of San Colombano al Lambro, between the cities of Milano and Piacenza, and also further south in the province of Piacenza in Emilia-Romagna and in the province of Pisa in Toscana. Until the 1940s, Verdea was used together with Malvasia Bianca (generally MALVASIA BIANCA LUNGA), BESGANO BIANCO and Pizzutello (no longer cultivated) to produce Vino Santo in the Val

Trebbia between the cities of Genova and Piacenza and in the Val Nure near Piacenza. There were 152 ha (376 acres) in **Italy** in 2000 (though some of these may be grown for the table).

Panizzari make one of the very few dry varietal wines, fresh and delicate and classified as IGT Colline del Milanese. Nettare dei Santi produce not only a still dry wine (including 10% RIESLING) but also a sweet passito-style dessert wine and a dry fizz. Cantina Pietrasanta's Verdea also includes 15% Riesling.

VERDECA

Regarded as light, neutral and declining in Puglia but, as Lagorthi, in tantalizingly short supply in Greece.

BERRY COLOUR: ● ● ● ● ●

PRINCIPAL SYNONYMS: Alvino Verde, Lagorthi (Pelopónnisos/Peloponnese and Ionian Islands in Greece), Verdera, Verdicchio Femmina, Verdisco Bianco, Verdone, Vino Verde

VARIETIES COMMONLY MISTAKEN FOR VERDECA: ALVARINHO (Portugal), VERDEA (Lombardia, Toscana), VERDICCHIO BIANCO

ORIGINS AND PARENTAGE

Verdeca is a traditional variety in Puglia in southern Italy. DNA profiling has established that PLAVINA, widely planted in Croatia, is a cross between Verdeca and TRIBIDRAG (Lacombe *et al.* 2007), this latter being the original Croatian name for Primitivo and Zinfandel. This parentage would suggest that Verdeca could originate in Croatia, where it has never been mentioned. However, DNA profiling has shown, surprisingly, that Verdeca is identical to Lagorthi (Vouillamoz), a Greek variety from Kalávrita in the Aigiáleia mountains in the Pelopónnisos, where it is still grown to a very limited extent, as well as on the Ionian Islands. Therefore, it is also possible that Verdeca could have been brought from Greece to Puglia.

VITICULTURAL CHARACTERISTICS

Early to mid ripening (but late ripening where Lagorthi is grown in Greece, see below). Good resistance to frost, powdery and downy mildews.

WHERE IT'S GROWN AND WHAT ITS WINE TASTES LIKE

The main cultivation area of Verdeca still corresponds to its historical distribution in the provinces of Taranto and Bari in Puglia, southern **Italy**. It is the dominant variety in Locorotondo and Martinafranca DOC blends and may play a supporting role in Gravina, Ostuni and San Severo DOCs. It is also found occasionally in Basilicata, as well as in Campania's Lacryma Christi del Vesuvio DOC, and is used to produce vermouth, thanks to its relatively neutral flavour profile. There were 2,265 ha (5,597 acres) in Italy in 2000 but the vineyard area declined considerably in the last thirty years of the twentieth century (down from nearly 8,000 ha/19,768 acres in 1970) and is expected to continue to do so. Recommended varietal wines from Puglia include Leone de Castris' Messapia, I Pàstini's Faraone and Cantine Rivera's Vivace, a light, fizzy version.

Under the name Lagorthi, this variety is highly regarded but little cultivated in **Greece**. It was Konstantinos Antonopoulos, based in Pátra in the northern Pelopónnisos, who saw the potential of this variety in the late 1980s, planted high up at 600–900 m above sea level, but even now Antonopoulos don't have enough vines to make a varietal Lagorthi. Instead it is the dominant variety in their Adoli Ghis blend. Oenoforos make a varietal wine, also in Pátra; there are limited plantings on the Ionian Islands and more recent plantings in Makedonía in northern mainland

Greece. Wines from Lagorthi tend to have low to medium alcohol (particularly planted at this height), moderate to fresh acidity and a soft and slightly creamy quality. They often start off with peach aromas, becoming more mineral and citrus after six months in bottle.

VERDEJO

High-quality, aromatic, rich, nutty variety most at home in Spain's Rueda region.

BERRY COLOUR:

PRINCIPAL SYNONYMS: Albillo de Nava (Rueda), Botón de Gallo Blanco, Verdeja, Verdejo Blanco

VARIETIES COMMONLY MISTAKEN FOR VERDEJO: DAMASCHINO, GODELLO, Verdejo de Salamanca (province of Salamanca), VERDEJO SERRANO (Sierra de Francia)

ORIGINS AND PARENTAGE

Verdejo, named after the green (*verde*) colour of the berries, most likely comes from the region of Rueda, where it was supposedly introduced by the Mozarabs before Moorish rule. DNA profiling has shown that Albillo de Nava from Rueda is identical to Verdejo (Santana *et al.* 2007), although it has no relationship to ALBILLO MAYOR or ALBILLO REAL (Vouillamoz). DNA parentage analysis has recently suggested that GODELLO from Galicia in north-west Spain and Verdejo are siblings (Santana *et al.* 2010), which explains why Verdejo is sometimes – and confusingly – called Gouveio or Verdelho, both synonyms for Godello in Portugal.

OTHER HYPOTHESES

Verdejo is not a colour mutation of TROUSSEAU, which is called Verdejo Negro or Verdejo Tinto in Rueda, and in Asturias to the north, and it is not linked to PEDRAL, which is known as Verdejo Colorado in Arribes del Duero west of Rueda. The apparent discovery that DAMASCHINO (under the name Planta Fina) and Verdejo were one and the same (Ibañez *et al.* 2003) seems to have been a mistake, because their DNA profiles are quite distinct (Vouillamoz).

VITICULTURAL CHARACTERISTICS

Compact, small to medium-sized bunches of thin-skinned berries with a distinctive blue-green bloom. Early to mid budding and ripening. Low vigour and fertility. Reasonably drought-resistant and best suited to low-fertility clay soils. Best pruned long. Highly susceptible to powdery mildew.

WHERE IT'S GROWN AND WHAT ITS WINE TASTES LIKE

Verdejo, the pride and joy of the Rueda DO in Castilla y León in northern central **Spain**, is the country's fifth most planted white wine variety, with 11,352 ha (28,051 acres) in 2008. It comprises nearly 10% of the Castilla y León vineyard area but is also widely planted in Castilla-La Mancha and there's a fair amount in Extremadura. Since the 1970s, when Marqués de Riscal saw the potential for dry white wines in Rueda, once more famous for its fortified styles, Verdejo has been a rising star. Wines are very aromatic, with notes of laurel and bitter almonds, medium to high acidity, full-bodied but rich and smooth showing a hint of almonds on the finish. They become more nutty with age and respond well to barrel fermentation and oak ageing.

In Rueda, Verdejo is often blended with SAUVIGNON BLANC but seems to be easily holding its ground against the incomer (473 ha/1,169 acres in 2008) even though the DO regulations also allow a Rueda that is solely or predominantly made from Sauvignon. These regulations permit a range of styles for Verdejo-based wines: still dry table wines, sparkling wines made by the traditional method, and two different types of fortified, sherry-like wines, Pálido and Dorado.

Recommended producers of unfortified, still varietal Verdejo (or wines with at least 85% of the variety) include Belondrade et Lurton, Cerrosol, Bodegas de Crianza Castilla La Vieja, Ermita Veracruz, José Pariente, Marqués de Riscal, Naia Viña Sila, Ossian and Javier Sanz.

In Rueda, Viñedos Nieva make a Pie Franco cuvée and Bodegas Naia (Viña Sila) make Naiades, both from very old and ungrafted Verdejo vines.

Keswick produce a Verdejo in the US state of **Virginia**.

VERDEJO COLORADO

See PEDRAL.

VERDEJO NEGRO

See TROUSSEAU.

VERDEJO SERRANO

Almost extinct variety from the mountains of Salamanca province.

BERRY COLOUR: ● ● ● ● ●

PRINCIPAL SYNONYMS: Rufete Blanca

VARIETIES COMMONLY MISTAKEN FOR VERDEJO SERRANO: VERDEJO🧬, Verdejo de Salamanca🧬

ORIGINS AND PARENTAGE

Verdejo Serrano is an almost extinct variety from the mountainous Sierra de Francia region in the province of Salamanca in western central Spain near the border with Portugal. It was rediscovered in the early twenty-first century but was not described until 2006, and was officially registered in 2009 under the name Verdejo Serrano, with Rufete Blanca as synonym (Arranz *et al.* 2008). This synonym is misleading because DNA profiling shows that Verdejo Serrano is not a colour mutation of RUFETE (Vouillamoz), a Portuguese variety that is the most widespread black variety in Sierra de Francia. In addition, comparison of the DNA profiles of Verdejo Serrano, VERDEJO (Santana *et al.* 2010) and Verdejo de Salamanca (Martín *et al.* 2003), the last being an old and no longer cultivated variety from the same area, shows that they are all distinct.

WHERE IT'S GROWN

The Cooperativa de San Esteban in San Esteban de la Sierra in Sierra de Francia, **Spain**, make one of the very few varietal wines, Alma de Tiriñuelo.

VERDELHO

Portuguese variety making fine, medium-dry wines on Madeira and full-bodied table wines in Australia and the Loire.

BERRY COLOUR:

PRINCIPAL SYNONYMS: Verdelho Branco⧖ (Madeira), Verdelho da Madeira⧖ (Madeira), Verdelho dos Açores⧖ (Açores), Verdelho Pico⧖ (Pico in the Açores), Verdello no Peluda Finca Natero⧖ (Islas Canarias in Spain)

VARIETIES COMMONLY MISTAKEN FOR VERDELHO: CHENIN BLANC⧖ (Anjou in France), Doçal⧖ (Vinho Verde region), GODELLO⧖ (Galicia in north-west Spain), VERDELLO⧖ (Umbria in Italy), VERDICCHIO BIANCO⧖ (Italy)

ORIGINS AND PARENTAGE

Verdelho is said to have been widely cultivated on the island of Madeira in the seventeenth century and may have been taken there from the mainland when the island was first populated. However, the variety is rarely found on the mainland, so it may actually originate from Madeira.

DNA profiling has shown that **Verdelho Roxo** is a red-berried colour mutation of Verdelho, while Verdelho Tinto from the Minho region is a distinct variety that is no longer cultivated (Lopes *et al.* 1999; Lopes *et al.* 2006). Recent DNA parentage analysis has suggested that Verdelho could have a parent–offspring relationship with SAVAGNIN (see the pedigree of PINOT for details), a very old French variety (Myles *et al.* 2011).

Verdelho should not be confused with GODELLO, a variety from Galicia in Spain that is also cultivated in the Portuguese Dão region under the name Verdelho or Verdelho do Dão (Lopes *et al.* 2006).

OTHER HYPOTHESES

It has been suggested that Madeira's Verdelho is identical to a Sicilian variety called Verdicchio (often erroneously spelt Verdecchio, unidentified, but distinct from VERDICCHIO BIANCO), introduced to Madeira in 1420 by the Infante D Henrique, or introduced to the island of Pico in the Açores by a certain Pero Anes do Canto, according to other sources, but their DNA profiles are completely distinct (Jorge Zerolo Hernández and Vouillamoz). In Anjou in the Loire, Verdelho is often wrongly considered to be a clone of CHENIN BLANC, and legend has it that it was brought from Madeira to the Coulée de Serrant in Savennières by the famous ampelographer Comte Odart in the early eighteenth century (Bouchard 1902).

VITICULTURAL CHARACTERISTICS

Susceptible to botrytis bunch rot and coulure and to a lesser extent to downy and powdery mildews. Requires deep soils with some degree of humidity. Low yields, very early ripening. Small, compact bunches with few berries (Rolando Faustino, personal communication). In Australia it produces relatively open small bunches of thick-skinned berries and is both drought-resistant and wet-weather-tolerant (Thistlewood 2007).

WHERE IT'S GROWN AND WHAT ITS WINE TASTES LIKE

Before the arrival of phylloxera on Madeira, **Portugal**, in 1872, Verdelho constituted around two thirds of Madeira's vineyards (Mayson 2003). Although the four so-called noble varieties associated with Madeira lost out to Tinta Negra Mole (NEGRAMOLL) in the subsequent replanting, Verdelho has seen the greatest expansion and is now the most widely cultivated of the four within the DOC region – 47 ha (116 acres) on Madeira in 2010, mainly in the cooler north. The best sites are said to be around Ponta Delgada and São Vicente (Mayson 2003).

The must has moderate sugar levels, marked acidity and is used for medium-dry fortified wines that fall between Sercial and Bual on the sweetness scale. It is also used for dry table wines labelled Terras Madeirenses. In the Açores, Verdelho is the main variety in the Biscoitos, Pico and Graciosa DOCs (60 ha/148 acres in 2010) for both fortified and alcoholic but unfortified medium-dry styles. There are several producers of Madeira made from 100% Verdelho – eg Barbeito – and Paixão do Vinho make an unfortified version, as does José Maria da Fonseca in Setúbal. Verdelho is also surprisingly successful in the hot Alentejo region in southern Portugal, retaining its refreshing acidity in the hands of Peter Bright at Terras d'Alter, for example. The total area planted on mainland Portugal in 2010 was 93 ha (230 acres).

In **France** Verdelho was recently reinstated in the official register of vine varieties, from which it had been dropped in 1995. Verdelho was introduced to Savennières in the Loire in the early nineteenth century, by mistake according to some sources, and those who have persisted with the variety, making deep-flavoured wines of varying degrees of sweetness, include Baumard, Stéphane Bernaudeau and Closel.

In **California** are tiny pockets of the variety such as in the DeWitt vineyard in Amador County and in Clarksburg in the Sacramento Delta. Kenneth Volk of Paso Robles makes an impressively fresh, slightly grippy version with flavours of quince, pear and almonds. A little is planted in Mendoza, **Argentina.**

Taken from Madeira to **Australia** in the 1820s, Verdelho was enthusiastically planted and praised in the mid nineteenth century. Even though it is probably better suited to the drier heat of parts of Australia compared with the subtropical conditions of Madeira, where it is at greater risk from fungal diseases, it was once quite widespread in the Hunter Valley in New South Wales. In South Australia it has made a minor but significant come-back in the last twenty years even though there are few winemakers to sing its praises (Thistlewood 2007). In 2008 there were 1,761 ha (4,352 acres) in Australia, the biggest areas in the Hunter Valley and Riverina. Producers here, in Victoria and in the warmer parts of Western Australia such as the Swan District are making vibrant, tangy, full-bodied white wines, generally for early drinking, although some are experimenting with a fresher style from earlier-harvested fruit. Recommended producers of varietal wines include Angove (South Australia), Bimbadgen (Hunter Valley), Bremerton (Langhorne Creek), Elysium (Hunter Valley), Houghton (Swan District), Karri Grove (Western Australia) and especially David Traeger (Victoria), who deliberately makes Verdelho to be aged. Some fortified Verdelhos are also made in Australia, by Bimbadgen, for example.

Varietal Verdelho is also produced in **New Zealand** by the sister companies Esk Valley and Villa Maria.

VERDELHO DO DÃO

See GODELLO.

VERDELLO

Waning high-acid Umbrian blending ingredient.

BERRY COLOUR: ● ● ● ● ●

PRINCIPAL SYNONYMS: Verdetto

VARIETIES COMMONLY MISTAKEN FOR THE ITALIAN VERDELLO: VERDELHO (Madeira and Açores in Portugal)

ORIGINS AND PARENTAGE

Verdello comes from Umbria in central Italy. Like many other varieties, Verdello was named after the greenish colour of its berries just before maturity. It was first mentioned in 1894 in Orvieto and has no relationship with VERDELHO from Portugal. The high-acid Verdello planted on the Italian island of Sicilia is likely to be a distinct variety but this has not yet been proven.

VITICULTURAL CHARACTERISTICS

Late ripening. Very susceptible to powdery mildew and botrytis bunch rot, less so to downy mildew.

WHERE IT'S GROWN AND WHAT ITS WINE TASTES LIKE

Verdello is cultivated in various parts of Umbria, central **Italy**, where it is prized for its natural acidity but is nevertheless losing popularity. There are also minor plantings in Toscana and Sicilia, giving a total vineyard area of 678 ha (1,675 acres) in Italy in 2000 – about half the area planted to Verdello in 1970. It may play a minor role to TREBBIANO TOSCANO's lead in Colli Amerini DOC in Umbria and in Bianco di Pitigliano DOC in Toscana, and is also authorized as a minor component (up to 20–30%) in Orvieto DOC. Palazzone's Umbria Bianco IGT includes 20% Verdello in a Procanico- (ie Trebbiano Toscano-) dominated blend, Decugnano dei Barbi's traditional-method sparkling wine also contains 20% Verdello and Cerquetto's Orvieto has 30% Verdello, but the lack of varietal examples suggest it is less successful on its own.

VERDESSE

Potentially high-quality, aromatic variety being replanted in Savoie.

BERRY COLOUR: ● ● ● ● ●

PRINCIPAL SYNONYMS: Bian Ver (Piemonte in Italy), Verdasse (Voreppe in Isère), Verdèche, Verdesse Musquée

ORIGINS AND PARENTAGE

Verdesse most likely comes from the Vallée du Grésivaudan in Isère, where it was first mentioned in Sassenage in 1845 (Rézeau 1997). The name refers to the dark-green colour of the leaves and berries (*vert* is French for 'green'). Verdesse belongs to the Pelorsien ampelographic group (see p XXVII; Bisson 2009).

VITICULTURAL CHARACTERISTICS

Vigorous, late budding, mid ripening. Best pruned long and gives good results on calcareous-clay slopes. Susceptible to powdery and downy mildews. Smallish bunches of small, thick-skinned berries.

WHERE IT'S GROWN AND WHAT ITS WINE TASTES LIKE

Verdesse used to be widely cultivated in the Vallée du Grésivaudan in the Isère, south-eastern **France**, where it was the most widely planted variety at the end of the 1920s, and is still authorized for the still and sparkling whites of the Vin de Savoie appellation. The vineyard area had declined to little more than 2 ha (5 acres) by 2008 but Verdesse has recently been successfully replanted. Producers of varietal wines include the Cave Coopérative de Bernin, Michel Magne and Mas du Bruchet. According to Schneider, Mannini and Cravero (2004), it was also formerly cultivated in the Valle Chisone and the Valle di Susa in Piemonte in north-west **Italy** but it has disappeared to the extent that it is not listed in the official census statistics of 2000. Wines tend to be powerful yet lively and aromatic.

VERDICCHIO BIANCO

Famously associated with the Marche's best-known dry white, this is probably originally from the Veneto, where it is also found, as Trebbiano di Soave. Wines can age.

BERRY COLOUR: ● ● ● ● ●

PRINCIPAL SYNONYMS: Angelica (Trentino), Boschera or Boschera Bianca, Lugana (provinces of Brescia and Verona in the Lago di Garda region), Pevarise (Trentino), Peverella⌛ (Trentino and Brazil), Peverenda (Trentino), Pfefferer (Alto Adige), Pfeffertraube (Alto Adige), Pievana (Trentino), Terbiana, Trebbiano di Lonigo (Vicenza), Trebbiano di Lugana⌛ (provinces of Brescia and Verona), Trebbiano di Soave⌛ (provinces of Brescia and Verona), Trebbiano di Verona (province of Verona), Trebbiano Nostrano, Trebbiano Valtenesi⌛ (province of Brescia), Trebbiano Verde (Lazio and Umbria), Turbiana (Vicenza), Turbiano (Vicenza), Turviana (Vicenza), Verdetto (Romagna), Verdicchio Giallo (Marche), Verdicchio Marchigiano (Marche), Verdicchio Peloso (Marche), Verdicchio Verde (Marche), Verdone (Marche)

VARIETIES COMMONLY MISTAKEN FOR VERDICCHIO BIANCO: MACERATINO⌛, PERERA (sometimes called Pevarise in the Veneto), TREBBIANO ROMAGNOLO, VERDECA⌛ (Puglia), VERDELHO⌛ (Sicilia)

ORIGINS AND PARENTAGE

Verdicchio Bianco is an extremely widespread and probably very old variety from the Veneto in northern Italy or from central Italy, known under a great variety of synonyms. Among the numerous mentions of 'Trebbiano' grapes in the Marche and the Veneto since the thirteenth century, it is likely that some of them corresponded to Verdicchio. However, the earliest mention of a grape variety named Verdicchio appeared in 1569 in a document signed by Costanzo Felici (1525–85), a physician and naturalist from Piobbico (in the Marche, between Rimini and Perugia), author of a series of manuscripts (1569–72) entitled *De l'insalata e piante che in qualunque modo vengono per cibo de l'homo* ('On the salads and the plants that in whatever way are used as food for man'). It was later mentioned in 1579 in Matelica (between Perugia and Ancona) by the notary Niccolò Attucci, and then in the province of Ascoli Piceno by Bacci (1596).

Although Verdicchio has been grown in the Marche for centuries, according to Pollini (2007), it originally came from the Veneto and was most likely introduced to the Marche in the fifteenth century, when farmers moved south to repopulate the plague-devastated region. Indeed, early ampelographic and enzymatic studies have established that Trebbiano di Soave, widely planted in the provinces of Brescia (Lombardia) and Verona (Veneto), is identical to Verdicchio Bianco (Calò, Costacurta *et al.* 1991), which was subsequently confirmed by DNA profiling (Labra *et al.* 2001).

More recently, DNA studies at Milano University have proved that Verdicchio Bianco is identical to two other varieties of the heterogeneous TREBBIANO group: Trebbiano Valtenesi, a local and endangered variety from the Lago di Garda region in the province of Brescia (Lombardia), and Trebbiano di Lugana, cultivated in the provinces of Brescia and Verona (Ghidoni *et al.* 2008).

Similarly and more surprisingly, DNA profiling has shown that Peverella, an old variety traditionally cultivated in Trentino (Trento, Lavis, Salorno, Faedo and Val di Cembra) as well as in the Alto Adige under the name Pfefferer, is identical to Verdicchio Bianco (Stella Grando and José Vouillamoz, unpublished data). It had almost disappeared after the phylloxera plague at the end of the nineteenth century and was recently rescued from extinction by the joint effort of the San Michele all'Adige research centre and amateur ampelographer Gianpaolo Girardi.

In addition, DNA comparison has suggested a close relationship between Verdicchio Bianco and MACERATINO (Filippetti, Silvestroni, Thomas and Intrieri 2001). Finally, it should be noted that the pink-berried Verdicchio, for which Di Vecchi Staraz, This *et al.* (2007) have found a parent–offspring relationship with MAMMOLO, has been shown by DNA profiling to be distinct from the true Verdicchio Bianco (Vouillamoz).

Verdicchio Bianco was used to breed INCROCIO BRUNI 54.

OTHER HYPOTHESES

As with many other grape varieties, some authors have suggested that Verdicchio Bianco could be a descendant of the *uva aminea* cited in Columella's *De re rustica*.

VITICULTURAL CHARACTERISTICS

Mid to late ripening. Susceptible to powdery and downy mildews, and especially to botrytis bunch rot and sour rot.

WHERE IT'S GROWN AND WHAT ITS WINE TASTES LIKE

Under the name Verdicchio Bianco, its main area of cultivation is the Marche in eastern central **Italy**, where it is best known as a varietal wine in the DOCs of Verdicchio dei Castelli di Jesi, west of Ancona 30 km (20 miles) from the Adriatic, and the much smaller Verdicchio di Matelica, on hillier terrain further inland and close to the border with Umbria. A minimum of 50% Verdicchio is required in the less well-known Esino DOC. Further south, in the province of Ascoli Piceno, it may play a minor role in the Falerio dei Colli Ascolani DOC. The variety is recommended in Abruzzo, Umbria and Lazio and authorized in Toscana and Sardegna. There were 3,561 ha (8,799 acres) in Italy in 2000. Recommended producers of varietal wines include Fazi Battaglia and Umani Ronchi. The best wines have a marked lemony acidity and combine substance and freshness, often leaving a bitter-almond flavour on the finish. The finest examples such as Umani Ronchi's Casal di Serra and La Monacesca's Mirum can improve in bottle for at least as long as the average white burgundy. In the past, wines were often given more weight by the use of skin contact but this also tended to make them coarser and the practice has been largely abandoned. The variety's high acidity also makes it well suited to the production of sparkling wines, with a long tradition that goes back to the nineteenth century.

Further north in the Veneto, under the name Trebbiano di Soave, it adds perfume to complement GARGANEGA's steely profile in all forms of Soave, including sweet Recioto di Soave. Pieropan is one of the top producers to include the variety in his Soaves. The census records 1,802 ha (4,453 acres) of Trebbiano di Soave in 2000. Further to the west, the warmer climate produces full-bodied, perfumed wines of real interest in the guise of Trebbiano di Lugana (not listed separately in the census), mainly in the eponymous Lugana DOC south of the Lago di Garda. Zenato and Cà dei Frati make good examples.

As Peverella, the variety is still planted to a very limited extent in Trentino, eg by Francesco Poli, whose Massenza Bianco is a blend of NOSIOLA and Peverella.

Under the name Trebbiano Valtenesi (again, not mentioned in the census), the variety is rarely found but Comincioli in Puegnago sul Garda blend 60% Trebbiano Valtenesi with 40% ERBAMAT to produce their dry white Perlì IGT Garda.

Rather curiously, Peverella turns up in **Brazil**'s Rio Grande do Sul, where there were 20 ha (49 acres) in 2007. This is probably due to the arrival of numerous Trentino families in Brazil at the end of the nineteenth century. DNA profiling has recently confirmed that this Brazilian Peverella is identical to Verdicchio Bianco (Leão *et al.* 2009). Salvati & Sirena produce a varietal wine in Bento Gonçalves. It also pops up in **California**'s Sierra Foothills, where Avanguardia blend it with Friulano (SAUVIGNONASSE) and other Italian varieties in their Selvatico blend.

VERDIL

An almost extinct white-berried variety from the province of Valencia, **Spain**, that was recently rescued by Daniel Belda and replanted in Vall dels Alforins, mainly in Daniel Belda's vineyards, yielding an apple- and pineapple-scented light varietal white. Enguera freeze Verdil grapes to make a sweet version.

VERDISO

Recuperated minor variety from the Colli Euganei in the Veneto makes wines from fizz to passito.

BERRY COLOUR: ● ● ● ● ●

PRINCIPAL SYNONYMS: Pedevenda (Breganze), Peverenda (Breganze), Pexerenda (Breganze), Verdino (Conegliano and Valdobbiadene), Verdise (Colli Euganei)

ORIGINS AND PARENTAGE

Verdiso, or Verdise, is local to the Colli Euganei, south of Padova in the Veneto, north-east Italy, where it was widely planted after the terrible 1709 frost. Recent DNA profiling has confirmed that Verdiso and Pedevenda, another local variety that has long been cultivated further north around Breganze in the province of Vicenza, where it was mentioned as early as 1754 by Aureliano Acanti in a book of poetry entitled *Roccolo*, are one and the same (Crespan *et al.* 2003). Since Pedevenda is the old name for the Colli Euganei, this links the variety to its likely place of origin.

Verdiso was used to breed FLAVIS and ITALICA.

VITICULTURAL CHARACTERISTICS

Late ripening. Suitable for passito wines.

WHERE IT'S GROWN AND WHAT ITS WINE TASTES LIKE

Verdiso had almost disappeared from the Colli Euganei just south of Padova in the Veneto, **Italy**, when it was recovered in the 1960s by vine breeder and collector Giuseppe Tocchetti, who propagated and preserved it. Today Verdiso is cultivated mainly in the Refrontolo hills and between Conegliano and Valdobbiadene in the province of Treviso, where it is blended with PROSECCO and Boschera (aka VERDICCHIO BIANCO) to produce Colli di Conegliano DOC's Torchiato di Fregona sweet wines. A small percentage (maximum 10%) is also permitted in the generally superior fizz with the longest name in the Prosecco region: Conegliano Valdobbiadene Superiore di Cartizze DOCG. There were 81 ha (200 acres) in Italy in 2000, considerably less than in 1970 (1,150 ha/2,842 acres). Il Colle, Conte Collato and Gregoletto are among the few producers of varietal wines. The variety is characterized by aromas of citrus and white flowers but these generally lively wines are best drunk young.

VERDONCHO

Undistinguished La Mancha blending material.

BERRY COLOUR: ● ● ● ● ●

ORIGINS AND PARENTAGE

Verdoncho is an old variety from central Spain. Little is known about its history or origin.

WHERE IT'S GROWN

Verdoncho is scattered throughout the vineyards of La Mancha and Manchuela in central **Spain** although it does not appear to be authorized for DO wines. It is provisionally included in the official Spanish register of commercial varieties, with 2,212 ha (5,466 acres) recorded in 2008, almost all in Castilla-La Mancha. Most of its produce seems to disappear into blends.

VERDUZZO FRIULANO

Ancient, characterful, sometimes astringent, Friulian; good sweet wines.

BERRY COLOUR: ● ● ● ● ●

PRINCIPAL SYNONYMS: Ramandolo, Ramandolo Dorato, Verdùç (Slovenia), Verdùz or Verdùzz (Udine), Verduzzo di Ramandolo☒, Verduzzo Giallo (Ramandolo)

VARIETIES COMMONLY MISTAKEN FOR VERDUZZO FRIULANO: VERDUZZO TREVIGIANO☒ (Veneto)

ORIGINS AND PARENTAGE

The earliest mention of a variety named Verduzzo was in relation to a banquet held on 6 June 1409 in Cividale (Friuli) in honour of Pope Gregory XII, at which the following wines were served: Ribolla from Rosazzo, Verduzzo from Faedis, Ramandolo from Torlano, Refosco from Albana and Marzemino from Gradiscutta (Peterlunger *et al.* 2004). Since Faedis is in the province of Udine in Friuli, north-east Italy, this document most likely refers to Verduzzo Friulano. Interestingly, Ramandolo is an ancient synonym for Verduzzo Friulano, which therefore seems to have been served twice at this banquet! Comparison of DNA profiles shows that Verduzzo di Ramandolo (Cipriani *et al.* 2010) is indeed identical to Verduzzo Friulano (Vouillamoz) but distinct from VERDUZZO TREVIGIANO (Crespan *et al.* 2011).

VITICULTURAL CHARACTERISTICS

Vigorous, mid to late ripening, prefers well-exposed hillsides. Good resistance to botrytis bunch rot.

WHERE IT'S GROWN AND WHAT ITS WINE TASTES LIKE

Verduzzo Friulano is cultivated mainly in Friuli in north-east **Italy**, notably but by no means exclusively in the more easterly province of Udine (the communes of Nimis, Faedis, Tarcento and Torlano have long been associated with this variety), but also in Piave over the border in the Veneto. Total plantings in Italy, which are gradually declining, were 1,658 ha (4,097 acres) in 2000.

Varietal wines both dry and sweet, and occasionally sparkling, are produced within the DOCs of Lison-Pramaggiore, Colli Orientali del Friuli, Friuli Annia, Friuli Aquiliea, Friuli

Grave, Friuli Isonzo, Friuli Latisana and Piave. Ramandolo (named after the Ramandolo hills) has been used rather loosely in the past for a variety of sweet wines but since its establishment as a DOCG, it may refer only to sweet wines made from Verduzzo Friulano (known locally as Ramandolo) in the steeply hilly region above Nimis, at around 380 m (1,250 ft) above sea level. The most significant quantities are produced in Grave and in the Colli Orientali. The Verduzzo of the Colli Orientali is qualitatively far superior thanks to the preponderance of hillside vineyards. Some Verduzzo Friulano is planted in the Collio production zone but has to be labelled as vino da tavola.

Varietal wines are generally quite powerful, fresh, ranging from dry to fairly sweet but the sweeter wines – made from late-harvested or partially dried grapes – tend to be the more successful. They often have slightly astringent tannins and a lightly herbal or cedary aroma which works in harmony with the honeyed flavours that increase with age. In fact, the best examples are greatly improved by a few years in bottle. In the dry wines, however, that astringency may be slightly too noticeable.

Recommended producers include Bressan, Conte d'Attimis-Maniago, Adriano Gigante, Davino Meroi, Roberto Scubla, La Tunella and Valchiarò.

In **Slovenia** there were 4.15 ha (10 acres) in Gorkiška Brda in 2009. Klinec make a good example.

Pizzini make dry and sweet versions in the King Valley, Victoria, **Australia.**

VERDUZZO TREVIGIANO

The other Verduzzo, generally inferior to and often blended with Verduzzo Friulano. Waning.

BERRY COLOUR: ● ● ● ● ●

PRINCIPAL SYNONYMS: Verduc, Verduz, Verduzzo di Motta

VARIETIES COMMONLY MISTAKEN FOR VERDUZZO TREVIGIANO: VERDUZZO FRIULANO☒ (Friuli)

ORIGINS AND PARENTAGE

Despite its name (Trevigiano means 'from the province of Treviso'), the origins of Verduzzo Trevigiano are unknown. It is said to have been introduced from Sardegna in the early twentieth century, although it cannot be identified with any modern Sardinian varieties. It is distinct from VERDUZZO FRIULANO, with which it is often cultivated and confused.

VITICULTURAL CHARACTERISTICS

Vigorous, late ripening.

WHERE IT'S GROWN AND WHAT ITS WINE TASTES LIKE

Verduzzo Trevigiano is cultivated mainly in the provinces of Treviso and Venezia in the Veneto, north-east **Italy**. It is often blended with the superior VERDUZZO FRIULANO in the Piave and Lison-Pramaggiore DOCs, where wines labelled Verduzzo may be made from Verduzzo Trevignano and/or Verduzzo Friulano. Wines appear in a range of styles, from dry to medium sweet as well as sparkling, and Casa Piave are one of the few producers to make a wine from 100% Trevigiano, in both dry and late-harvest styles. Terre di Ger's Limine is a barrel-fermented, oak-aged blend of Verduzzo Trivigiano and CHARDONNAY. The agricultural census of 2000 recorded 1,734 ha (4,285 acres) in Italy, not much more than half what they were in 1970.

VERITAS

Minor Czech cross yet to make its mark.

BERRY COLOUR:

ORIGINS AND PARENTAGE

Veritas is a Roter Riesling × BOUVIER cross obtained in 1963 by Cyril Míša and Milos Zbořil at the vine research centre in Znojmo, Czech Republic, in which Roter Riesling is a colour mutation of RIESLING. It was included in the offical Czech register of varieties in 2001.

VITICULTURAL CHARACTERISTICS

Small bunches of medium-sized berries. Productive, early ripening. Moderate resistance to fungal diseases except botrytis bunch rot.

WHERE IT'S GROWN AND WHAT ITS WINE TASTES LIKE

Veritas is well suited to the more northerly areas of the Morava wine region in the south east of the **Czech Republic**, although plantings had not reached 4 ha (10 acres) in 2009. Wines are generally full-bodied with moderate acidity but otherwise not particularly remarkable. Spěvák and Znovín produce varietal wines.

VERMENTINO

Aromatic, crisp high-quality variety flourishing near the sea in northern Italy, southern France and on the islands of Corse and Sardegna.

BERRY COLOUR:

PRINCIPAL SYNONYMS: Favorita (Roero in Piemonte), Furmentin (Valle del Belbo in Piemonte), Garbesso (France), Malvasia or Malvoisie de Corse (northern Corse), Malvoisie à Gros Grains (Corse), Malvoisie du Douro (Pyrénées-Orientales in France), Pigato (Liguria), Rolle or Rollé (Alpes-Maritimes and Var in southern France), Sapaiola (Italy), Verlantin or Varlantin (Corse and southern France), Verlentin or Varlentin (Var), Vermentino di Gallura (Sardegna), Vermentinu (southern Corse)

VARIETIES COMMONLY MISTAKEN FOR VERMENTINO: BRUSTIANO BIANCO (Corse), CARICA L'ASINO (Piemonte), DRUPEGGIO (Toscana), NASCETTA (Piemonte), ROLLO (Cinque Terre in Liguria)

ORIGINS AND PARENTAGE

Vermentino is said to have been introduced to the French island of Corse in the thirteenth or fourteenth century but the earliest mention appears to have been in 1658 in Montaldeo in the province of Alessandria in Piemonte, north-west Italy, where vineyards were planted with *cortese* (CORTESE), *fermentino* (Vermentino) and *nebioli dolci* (NEBBIOLO; Nada Patrone 1991). Some scholars believe that the name Vermentino derives from *vermene*, a local word used to describe a young, thin and flexible shoot, but an etymology based on *fermento*, meaning 'ferment', probably referring to the fizzy character of the young wine, sounds more convincing, not least because it echoes the spelling of its first mention, *fermentino*.

Morphological and DNA comparisons have clearly established that Favorita (Piemonte), Pigato (Liguria) and Vermentino (Liguria, Sardegna, Toscana and Corse) are one and the same variety (Schneider and Mannini 1990; Botta, Scott *et al.* 1995). Favorita is supposed to have been brought to Piemonte as a gift from Ligurian oil merchants. It was first mentioned in 1676 in the cellar book of Conti Roero in Piemonte: three loads of 'biancho di favorie' were registered in the cellar in Vezza d'Alba. In Roero, it quickly became the 'favourite' variety for table-grape consumption, hence its name. Pigato derives from the Ligurian dialect word *pigau*, meaning 'spotted', because of the appearance of the berries at maturity. Pigato was supposedly mentioned in the Savona area in the fifteenth century, and is said to have been introduced from Toscana by merchants, this time from Firenze.

OTHER HYPOTHESES

Like many varieties traditionally grown on the Italian island of Sardegna, Vermentino is said to have been introduced from Spain to Corse and Sardegna between the fourteenth and seventeenth centuries, although this variety has never been found in Spain. Other suggested places of origin include Thessalía, in central Greece, or the Middle East, but they are not supported by any scientific evidence. A parent–offspring relationship with NASCETTA from Piemonte has been suggested, but there's no evidence yet.

VITICULTURAL CHARACTERISTICS

Mid ripening and seems to benefit from proximity to the sea. Early budding and therefore particularly sensitive to spring frosts. Susceptible to downy mildew and the European grape moth.

WHERE IT'S GROWN AND WHAT ITS WINE TASTES LIKE

This aromatic variety is widely grown, under its various synonyms, in Liguria in north-west Italy, on the Italian island of Sardegna and in Piemonte, as well as in southern France and on the French island of Corse.

In fact Vermentino/Rolle is even more widely planted in southern **France** than in its native Italy. In 2008 total plantings amounted to 3,453 ha (8,533 acres), most in the Var in Provence (1,165 ha/2,879 acres) but also on the island of Corse (814 ha/2,011 acres), where it is the most important variety in every one of the island's white wine appellations (minimum 75–80%) apart from its Muscats, and in Languedoc-Roussillon (813 ha/2,009 acres), especially in the Hérault *département*. Here Domaine du Poujol blend it with ROUSSANNE and Carignan Blanc (see MAZUELO) to produce a complex and highly drinkable wine. It is recommended in the Alpes-Maritimes, Aude, Bouches-du-Rhône, Gard, Hérault, Pyrénées-Orientales and Var, and authorized in the Alpes-de-Haute-Provence, Hautes-Alpes, Ardèche, Drôme and Vaucluse. Styles range from intensely aromatic to steely citrus.

Total plantings of Vermentino in **Italy** in 2000 were 3,000 ha (7,413 acres) and fairly stable. The census lists Favorita and Pigato separately, recording 276 ha (682 acres) and 255 ha (630 acres) respectively. Varietal wines appear in many guises but the best are fragrant, sometimes more mineral than floral, fruity or gently spicy; mostly moderate alcohol – especially when picked early enough to retain acidity – and for early drinking.

As Pigato or Vermentino it used to be the most widespread variety in Liguria in north-west Italy, where it is still traditionally cultivated in the provinces of Savona and Imperia. It is also grown in the communes of Arenzano and Cogoleto in the province of Genova, where it is known exclusively as Vermentino. It appears mainly in the following DOCs, on its own or blended with other local varieties such as ALBAROLA: Colli di Luni, Colline di Levanto, Golfo del Tigullio, Riviera Ligure di Ponente and Val Polcevera. This last DOC, incidentally, still provides for a varietal Pigato as well as a varietal Vermentino. U Munte produce a particularly good example that is relatively light but combines floral, mineral and slightly salty aromas and Laura Aschero's Pigato is seriously vibrant with intriguing scents of fresh herbs and lush vegetation and then real tension on the palate.

On Sardegna, Vermentino is the most important white variety, especially in Gallura in the north of the island, where it was awarded its own DOCG, the island's first and only, in 1996. The grapes tend to be harvested early to retain acidity but the wines are still lively and characterful.

Producers such as Capichera have successfully produced a richer, fuller style. More generally, Vermentino di Sardegna DOC can be produced anywhere on the island, as the name suggests. Alghero DOC in the north west includes a sparkling Vermentino. Other recommended producers include Argiolas, Carpante, Ferruccio Deiana, Cantina Sociale Gallura, Cantina del Giogantinu, Piero Mancini, Fratelli Pala and Pedres.

It is planted to a lesser extent in Massa-Carrara in Toscana and in Puglia, Umbria and Lazio. In Toscana, Vermentino is thriving thanks to the growing interest in indigenous varieties. It appears in the recipes for several Vin Santos and may be made as a varietal in a host of DOCs including Bolgheri, Colline Lucchese and Montecucco.

Under the name Favorita, it has traditionally been planted – formerly as both table and wine grape but today almost exclusively as a wine grape – in the regions of Alba and especially Roero in Piemonte, north-west Italy. Since the 1990s, Favorita has been slowly replaced by the more fashionable ARNEIS in Roero and by CHARDONNAY in Langhe although some fine aromatic wines are still being made by producers such as Gianni Gagliardo.

Vermentino is found on **Malta** and has recently been planted by Massaya at an altitude of 1,200 m in the Bekaa Valley in **Lebanon**.

In **California**, Tablas Creek have been producing bright, citrus-flavoured Vermentino from their acre of vines in Paso Robles. Other produces include Mahoney in Carneros, Cougar in Temecula, Indian Rock Vineyards in the Sierra Nevada, Uvaggio and Woodbridge in Lodi and, beyond the Golden State, in **Texas**, chez Raffaldini in **North Carolina** and particularly successfully at Barboursville Vineyards in **Virginia.**

Vermentino is also found in **Brazil**.

In **Australia**, there were around fifteen producers of Vermentino in 2009, most notably Yalumba, the most sustained pioneer of alternative varieties in their vine nursery in Nuriootpa, South Australia, but also Brown Brothers in Victoria, De Bortoli in the Hunter Valley, Politini in the King Valley, Spinifex in the Barossa Valley and Trentham Estate in Murray Darling. In Australia the wines tend to have riper fruit flavours – stone fruit and even more tropical aromas – but still have that refreshing acidity as long as the grapes are picked fairly early.

VERMENTINO NERO

Very minor Tuscan rescued from extinction in the 1980s.

BERRY COLOUR: ● ● ● ● ●

ORIGINS AND PARENTAGE

Vermentino Nero is local to the province of Massa-Carrara in Toscana, central Italy. Some scholars claim that it is a colour mutation of the white VERMENTINO, while others have suggested a possible genetic relationship with MONASTRELL and PARRALETA, both from Spain. However, neither hypothesis has yet been proven by DNA profiling.

VITICULTURAL CHARACTERISTICS

Mid to late ripening. Susceptible to powdery mildew and botrytis bunch rot.

WHERE IT'S GROWN AND WHAT ITS WINE TASTES LIKE

Vermentino Nero was abandoned after the Second World War and was almost extinct when it was rescued by Podere Scurtarola in Massa, the first to make a varietal wine in 1989. They were followed by the likes of Terenzuola and Cima, and there were 199 ha (492 acres) of Vermentino Nero in Italy in 2000: in Toscana, central **Italy**, in particular in the province of Massa-Carrara and, to a lesser extent, in the province of Lucca. Vermentino Nero has recently been authorized for the Candia dei Colli Apuani DOC and may be included in the Colline Lucchesi DOC. Varietal wines tend to be rounded and full-bodied with spicy flavours of both red and black fruits.

VERNACCIA

ORIGINS AND PARENTAGE

The name Vernaccia probably derives from the Latin *vernaculus*, meaning 'native' or 'indigenous', an etymology also found in Vernatsch (see SCHIAVA GROSSA) in the Alto Adige in northern Italy. That is probably why Vernaccia is used for at least six distinct varieties:

- VERNACCIA DI ORISTANO on the Italian island of Sardegna
- VERNACCIA DI SAN GIMIGNANO in Toscana
- Vernaccia Trentina (or Vernaccia Bianca), an old Trentino variety that is no longer cultivated
- Vernaccia di Pergola, identical to ALEATICO in southern Italy
- Vernaccia Nera di Valdarno, probably an offspring of SANGIOVESE with a unique DNA profile (Crespan, Giannetto *et al.* 2008), not yet commercially cultivated
- Vernaccia Nera in the Marche and in Umbria, identical to GARNACHA in Spain.

That Vernaccia Nera is identical to GARNACHA is supported by those linguists who believe that Spanish Garnacha or Garnaxa (Catalan), the French Grenache and the Sardinian Vernaccia, Crannaxia, Granaccia, Granatza, Granaxia or Vrannaxia all derive from *vernaculus*.

OTHER HYPOTHESES

For some authors, Vernaccia could derive from the Latin *ibernum*, meaning 'winter', because the grapes were consumed during this period; for others it could derive from *verrum*, meaning 'boar', since it is said the grapes were eaten by these animals. A derivation from the village name Vernazza in the Cinque Terre in Liguria has also been suggested.

VERNACCIA DI ORISTANO

Makes complex sherry-like wines on Sardegna.

BERRY COLOUR:

PRINCIPAL SYNONYMS: Aregu Biancu, Aregu Seulu, Cranaccia, Garnaccia, Granaccia, Varnaccia, Vernaccia, Vernaccia Austera, Vernaccia Orosei, Vernaccia S. Rosalia

VARIETIES COMMONLY MISTAKEN FOR VERNACCIA DI ORISTANO: VERNACCIA DI SAN GIMIGNANO

ORIGINS AND PARENTAGE

Vernaccia di Oristano, from the Italian island of Sardegna, was first mentioned in 1327 in *Breve di Villa di Chiesa*, a law book conserved in Iglesias, a town in south-western Sardegna, formerly known as Villa di Chiesa, with the following statement about the vintners: 'non debbiano nè possano tenere a mano piò di una botte de uno vino, ma sì divisi vini, cioè una di varnaccia, una di greco, una di vermiglio, et una di brusco bianco che fusse facto fuora di Sardigna, et una di vino Sardisco' ('neither should they nor could they possess more than one barrel of a type of wine, but wines are separated, that is to say one of varnaccia, one of greco, one of vermiglio, and one of brusco bianco that was made outside of Sardegna, and one of a Sardinian wine'). DNA profiling has confirmed several synonyms for Vernaccia di Oristano in Sardegna, as listed above, including many similar to 'Garnacha', but has also proved that the Spanish GARNACHA and Vernaccia di Oristano are distinct varieties (De Mattia *et al.* 2007; Zecca *et al.* 2010). Recent DNA parentage analysis has suggested that Vernaccia di Oristano is a possible parent of SANTA MARIA from Emilia-Romagna (Cipriani *et al.* 2010).

OTHER HYPOTHESES

Vernaccia di Oristano is said to have been introduced to Sardegna by the Phoenicians via the port of Tharros, which they founded around 800 BC and which is now an archaeological site in the commune of Cabras in the province of Oristano. A Spanish origin has also been proposed, probably because of its synonym Granaccia, as well as a direct domestication from wild grapevines in the Valle del Tirso, which would explain its etymology based on *vernaculus* ('native').

VITICULTURAL CHARACTERISTICS

Mid to late ripening. At risk from late frost and highly susceptible to powdery and downy mildews.

WHERE IT'S GROWN AND WHAT ITS WINE TASTES LIKE

This variety is cultivated almost exclusively in the Vernaccia di Oristano DOC in the lower part of the Valle del Tirso in the province of Oristano, on the central west coast of the Italian island of Sardegna, producing various styles, from dry and unfortified to the more complex sherry-like fortified wines both sweet and dry. Each style has strict minimum ageing requirements – ranging from twenty-nine months for the Secco (ie dry and unfortified) to fifty-three months for the Superiore Riserva (also dry and unfortified) – in a solera system using partially filled chestnut or oak barrels. There were 582 ha (1,438 acres) planted in **Italy** in 2000.

Attilio Contini make fine examples of these oxidative-style wines. Their Antico Gregori incorporates wine approximately 100 years old.

VERNACCIA DI SAN GIMIGNANO

Refreshing and generally light variety from the Tuscan town with the towers. Potential for high-quality wine.

BERRY COLOUR: ● ● ● ● ●

PRINCIPAL SYNONYMS: Bervedino (Emilia Romagna), Piccabòn (Liguria), Vernaccia

VARIETIES COMMONLY MISTAKEN FOR VERNACCIA DI SAN GIMIGNANO: DRUPEGGIO (Toscana), VERNACCIA DI ORISTANO (Sardegna)

ORIGINS AND PARENTAGE

The earliest mention of this variety dates back to 1276 in San Gimignano in the province of Siena in Toscana, central Italy: Vernaccia and Greco wines were praised in the *Ordinamenti di Gabella*, a tax document held in the archives of San Gimignano. It was later acclaimed in the sixteenth century by Sante Lancerio, the cellar keeper of Pope Paolo III, and then mentioned as a sweet Tuscan wine under the names Vernaccia and Vernacce (plural) by Soderini (1600) in his treatise on the grape varieties of Toscana. Morphological and DNA analyses have shown that Vernaccia di San Gimignano is not identical to VERNACCIA DI ORISTANO or any other similarly named variety (Vouillamoz; see VERNACCIA, above), but also that it is identical to Bervedino in Emilia-Romagna (Storchi *et al.* 2011).

OTHER HYPOTHESES

Various sources suggest that Vernaccia di San Gimignano could have been introduced from Liguria in north-west Italy, or from Spain or Greece.

VITICULTURAL CHARACTERISTICS

Mid to late ripening. Vigorous with regular and generous yields.

WHERE IT'S GROWN AND WHAT ITS WINE TASTES LIKE

Vernaccia di San Gimignano is cultivated mainly on sandstone-based soils around the tower-festooned hill town of San Gimignano in the province of Siena in Toscana, central **Italy**, where it is found as a varietal wine in the Vernaccia di San Gimignano DOCG. It may be blended with TREBBIANO TOSCANO or other local varieties in the Colli dell'Etruria Centrale DOC and in the Vin Santo version of San Gimignano DOC. Total plantings in Italy in 2000 were 784 ha (1,937 acres).

San Gimignano's claim to fame, in addition to its fortifications, is that it was awarded the first ever DOC in Italy, in 1966 (DOCG in 1993). This accolade, and Vernaccia di San Gimignano's superiority to most of the local TREBBIANO TOSCANO-based blends, helped to halt its steady decline in the first half of the twentieth century. Varietal wines are crisp and citrusy, sometimes gently floral, with a refreshing bitter-almond aftertaste and have not generally been improved by experiments with small-barrel maturation. Some producers have experimented with additions of varieties such as CHARDONNAY and SAUVIGNON BLANC but it is easy to overwhelm the inherently rather light Vernaccia di San Gimignano.

Recommended producers include Giovanni Panizzi, whose Evoè is marked by a fine astringency thanks to extended skin contact and fermentation in large wooden vats, and Fontaleoni. Others include Baroncini, La Calcinaie, Vincenzo Cesani, Il Colombaio di Santa Chiara, La Lastra, Montenidoli, La Marmoraia, Niccolai-Palagetto and Fratelli Vagnoni.

As Bervedino, it is cultivated in the Val d'Adra in the province of Piacenza in Emilia-Romagna, where it is used in the very special Vin Santo di Vigoleno (see MELARA).

VERNACCIA NERA

See GARNACHA.

VERNATSCH

See SCHIAVA GROSSA.

VERSOALN

Recent burgeoning based on a single, ancient Alto Adige giant vine.

BERRY COLOUR:

PRINCIPAL SYNONYMS: Versailler, Weiss Versoalen, Weisser Versailler

ORIGINS AND PARENTAGE

The strange name of this variety from the province of Bolzano in Trentino-Alto Adige (Südtirol), northern Italy, could simply derive from the local dialect word *verdolen*, meaning 'green', like the berries, or less straightforwardly from the dialect *versoaln*, meaning 'to secure with a rope', possibly in reference to the need to tie the numerous branches of the vine to the supporting pergolas. Another and less convincing hypothesis suggests this name refers to Versailles, whence the plant was supposedly brought to Italy in the fifteenth century by Grafen Schlandesberg, a previous owner of Castel Katzenzungen (see below). The DNA profile of Versoaln does not match any other known variety (Vouillamoz).

VITICULTURAL CHARACTERISTICS

Mid budding and ripening. Large, irregular-shaped bunches (similar to those of SCHIAVA GROSSA) of largish, thin-skinned berries. Susceptible to downy and powdery mildews and sour rot.

WHERE IT'S GROWN AND WHAT ITS WINE TASTES LIKE

This oddly named, once far more widespread, variety was until very recently represented by a single vine at Castel Katzenzungen in Tesimo in the province of Bolzano in the Südtirol in the far north of **Italy**. It is one of the world's oldest and largest vines – approximately 350 years old (according to Martin Worbes at the University of Göttingen) and covering 350 m² – and is tended by the Laimburg viticultural research centre, where 100 new vines have recently been planted. Castel Katzenzungen and researchers from Laimburg collaborate to produce about 500 numbered bottles every year, using fruit from old and young vines. The dry wine has fresh green-apple aromas, marked acidity and a touch of apricot on the finish. So far quantity has been more reliable than quality.

VERTZAMI

Promising Greek variety that produces deeply coloured, powerful, very tannic wines.

BERRY COLOUR:

PRINCIPAL SYNONYMS: Deykaditiko, Lefkada (Cyprus), Lefkaditiko, Lefkas☒ (Cyprus), Marzavi, Vartzami

VARIETIES COMMONLY MISTAKEN FOR VERTZAMI: MARZEMINO☒ (Trentino and Lombardia in Italy)

ORIGINS AND PARENTAGE

Vertzami is local to the island of Lefkáda in the Ionian Sea off the west coast of Greece, its most likely place of origin given that DNA profiling has shown it to be identical to the variety known as Lefkas on Cyprus (Hvarleva, Hadjinicoli *et al.* 2005).

There is also a white-berried Vertzami Lefko (Nikolau and Michos 2004) but it is not known if it is a colour mutation or a distinct variety.

OTHER HYPOTHESES

Several authors including Labra *et al.* (2003) suggest that Vertzami and MARZEMINO from northern Italy could have a common ancestor, which would fit with the supposed introduction of Vertzami from Italy during Venetian rule in the fourteenth century. However, the DNA profile of Vertzami in Labra *et al.* (2003) is distinct from those in the Greek Vitis Database and in the United States Department of Agriculture's National Clonal Germplasm Repository in Davis, California, disproving this relationship with Marzemino (Vouillamoz).

VITICULTURAL CHARACTERISTICS

Small compact bunches of medium-sized, thick-skinned berries. Vigorous, productive, mid budding and mid to late ripening. Good general disease resistance but susceptible to downy mildew.

WHERE IT'S GROWN AND WHAT ITS WINE TASTES LIKE

Vertzami produces deeply coloured, almost black, wines with high levels of tannin and moderate to high acidity as well as plenty of dark and forest fruits. The variety is so deeply coloured that it is also used to deepen the colour of blended reds. According to research by Kallithraka *et al.* (2006), Vertzami had by far the highest total anthocyanin content of the twenty wines tested.

Vertzami is found mainly and most densely on the island of Lefkáda, **Greece**, where it constitutes about 90% of the wine vineyards, as well as on Kérkyra (Corfu). There was a total of 269 ha (665 acres) in the Ionian region in 2008; it is also found in the western Pelopónnisos (Peloponnese) and western central Greece and although the total area is small, the variety is becoming increasingly attractive for its quality potential. Vineyards need to be carefully sited (neither too cool/high nor too dry) to ensure the tannins ripen fully, otherwise they can be just too assertive and drying. Antonopoulos have been experimenting with this variety for the last twenty years using fruit from Lefkáda but have not been able to produce a commercial wine in every vintage. They also soften the variety by blending it with CABERNET FRANC. The main producers of Vertzami on Lefkáda are the Lefkáda co-op, Lefkaditiki Gi and Vertzamo.

Vertzami is also popular on the island of **Cyprus**, where there were 96 ha (237 acres) of Lefkada in 2010, mainly in Stroumbi and Polemi in the Páfos district in the far west and in Malia and Omodos in the Lemesós (Limassol) district further east. It is often blended with international varieties such as CABERNET SAUVIGNON, CABERNET FRANC or SYRAH but it shows the same promise as in Greece. Recommended producers include Zambartas.

VESPAIOLA

Speciality of Vicenza, associated with sweet whites.

BERRY COLOUR: ● ● ● ● ●

PRINCIPAL SYNONYMS: Bresparola (province of Vicenza)

ORIGINS AND PARENTAGE

Vespaiola is considered to be an ancient variety from the province of Vicenza, but little is known about its origin and history. It was first mentioned by Acerbi (1825) in the towns of Bassano del

Grappa and Marostica. Like VESPOLINA, its name most likely comes from the attractiveness to wasps of the berries' high sugar content (*vespa* is Italian for 'wasp').

VITICULTURAL CHARACTERISTICS

Vigorous, mid ripening. Resistant to winter frost, susceptible to powdery mildew, botrytis bunch rot and sour rot, good resistance to downy mildew, cholorosis and drought.

WHERE IT'S GROWN AND WHAT ITS WINE TASTES LIKE

Vespaiola is cultivated exclusively in the province of Vicenza in the Veneto, north-east **Italy**. Within the Breganze DOC, it appears as a dry varietal wine or as the sweet passito-style Torcolato, as well as being authorized in the Friulano- (SAUVIGNONASSE-) dominated Bianco. It may be blended with GARGANEGA in the Gambellara and Vicenza DOCs. There were 112 ha (277 acres) in Italy in 2000, according to the agricultural census. Maculan's rare varietal Vespaiolo has good acidity and aromas of flowers, fruits and honey, though this may be more from the passito process since the variety itself is rather acidic and neutral, which may be why it is used by the Cantina Beato Bartolomeo da Breganze to make a sparkling version.

VESPOLINA

Low-yielding, minor Nebbiolo relative of Gattinara that is currently losing ground.

BERRY COLOUR: ● ● ● ● ●

PRINCIPAL SYNONYMS: Ughetta (Oltrepò Pavese), Uvetta di Canneto (Oltrepò Pavese)

ORIGINS AND PARENTAGE

Vespolina has been known in the Oltrepò Pavese in Lombardia and in Piemonte since the eighteenth century, especially in the area of Gattinara (Nuvolone Pergamo 1787–98). It used to be widespread in the Piemontese provinces of Novara, Vercelli and Biella, in Piacenza in Emilia-Romagna, and in Como and Pavia in Lombardia. Like VESPAIOLA, its name most likely comes from the high sugar content of the berries attracting wasps (*vespa* in Italian). DNA parentage analysis has established that Vespolina is a progeny of NEBBIOLO (Schneider, Boccacci *et al.* 2004), and that it is therefore a half-sibling of Nebbiolo's other progenies: BRUGNOLA, BUBBIERASCO, FREISA, NEBBIOLO ROSÉ, Negrera, NERETTO DI BAIRO and ROSSOLA NERA (see Nebbiolo pedigree diagram).

VITICULTURAL CHARACTERISTICS

Mid to early ripening. Susceptible to early shrivelling of the berries, rot, downy mildew, millerandage and drought.

WHERE IT'S GROWN AND WHAT ITS WINE TASTES LIKE

Vespolina's popularity and vineyard area decreased in the aftermath of phylloxera in favour of the more productive and resistant BARBERA, not least because grafting onto American rootstocks affected Vespolina's ripening. It is today cultivated in north Piemonte, **Italy**, mainly in the region of Gattinara, and in Oltrepò Pavese under the name Ughetta or Uvetta di Canneto. Rarely made as varietal wine, in Piemonte DOCs such as Boca, Bramaterra, Colline Novaresi, Coste della Sesia, Fara, Ghemme and Sizzano it is usually blended with NEBBIOLO and in the Oltrepò Pavese DOC mostly with Barbera. However, varietal wines are permitted within the Colline

Novaresi and Coste della Sesia DOCs and Vercesi del Castellazzo make a spicy, fruity, tannic varietal wine. Varietal wines are also produced by Sergio Barbaglia, Francesco Brigatti and Ioppa, among others. There were 108 ha (267 acres) in Italy in 2000, less than half of plantings in 1970.

VIDADILLO DE ALMONACID

Often simply known as Vidadillo, this is an almost extinct dark-berried variety from the region of Zaragoza in Aragón in north-eastern **Spain**. Although it is also called Garnacha Basta and Garnacha de Grano Gordo, and Ganatxa del Po in Priorat, DNA profiling (Martín *et al.* 2003; Ibañez *et al.* 2003) has shown that it is distinct from GARNACHA TINTA. Vidadillo de Almonacid is commercially cultivated under its historical name Crespiello; Bodegas y Viñedos Pablo's Pulchrum is a rare varietal example made from centenarian vines in the Cariñena area.

VIDAL

French hybrid ideally suited to the production of sweet Canadian icewine.

BERRY COLOUR: ● ● ● ● ●

PRINCIPAL SYNONYMS: Vidal Blanc, Vidal 256

ORIGINS AND PARENTAGE

A complex hybrid obtained by Jean-Louis Vidal (1880–1976) in the 1930s by crossing TREBBIANO TOSCANO × RAYON D'OR in order to produce a variety that was suitable for the production of cognac in Charente-Maritime.

VITICULTURAL CHARACTERISTICS

Moderately winter-hardy, mid ripening. Long bunches of small berries. Good resistance to downy mildew but susceptible to coulure, powdery mildew, anthracnose and botrytis bunch rot.

WHERE IT'S GROWN AND WHAT ITS WINE TASTES LIKE

Although Vidal was bred in **France**, it is not authorized and is rarely found there – yet it is, amazingly, now grown successfully in **Sweden**. It has been much more successful in Canada, particularly for icewine, and, to a lesser extent and generally for dry whites, in a wide array of American states including **Virginia** (150 acres/61 ha in 2010), the Finger Lakes region of **New York** State, **Michigan** (145 acres/59 ha in 2006), **Missouri** (118 acres/48 ha in 2009, more than 7% of the vineyard area), **Indiana** (35 acres/14 ha in 2009) and **Illinois** (32 acres/13 ha in 2009).

Canada had 1,920 acres (777 ha) planted in 2008, the vast majority in Ontario, plus minor areas in British Columbia, Nova Scotia and Québec. The variety was probably introduced to the country by Frenchman Adhémar de Chaunac, chief oenologist at T G Bright & Co, who imported both hybrid and *vinifera* varieties in the late 1940s. Originally used to make voluptuously fruity white table wines, according to Schreiner (2001) de Chaunac also left some grapes on the vine for late-harvest wine and icewine but it was not until the 1980s that further experimentation led to wider commercial production of curranty sweet wines made from grapes frozen on the vine. Vidal has the great advantage of being winter-hardy but having no foxy taste and the wines can taste very bright and pure, with the usual extreme crispness associated with icewine, for two or three years, though the wines are not generally designed to age in bottle. Recommended producers include Inniskillin (the world's largest producer of icewine), sister company Jackson-Triggs, and Peller in the Niagara Peninsula, Mission Hill in the Okanagan Valley and Vignoble de l'Orpailleur in Québec. Wines of all sweetness levels are made from Vidal but icewine has seen the greatest commercial success and achieves the highest prices.

VIDIANO

Old Greek variety showing signs of a revival on Kríti (Crete).

BERRY COLOUR: ● ● ● ● ●

PRINCIPAL SYNONYMS: Abidano, Abidiano, Abudiano, Avidiano, Bidiano

ORIGINS AND PARENTAGE

Rare variety from the Greek island of Kríti that is genetically close to THRAPSATHIRI and VILANA (Biniari and Stavrakakis 2007), both also from Kríti.

VITICULTURAL CHARACTERISTICS

Variable yields, early to mid ripening. Medium-sized to large bunches of thick-skinned berries. Moderate resistance to downy mildew but sensitive to powdery mildew and grape moths.

WHERE IT'S GROWN AND WHAT ITS WINE TASTES LIKE

Vidiano had virtually disappeared from Kríti, **Greece**, found only in old mixed vineyards, but it seems to be slowly regaining ground thanks to the efforts of growers such as Douloufakis, who has three recently planted vineyards, the original one based on mass selection from old vines in Réthymno, the later ones working with a local nursery.

It produces wines rich in fruit flavours that range from lime to apricot, sometimes quite tropical, and they retain fresh acidity even at alcohol levels of around 13%. Tamiolakis, who produce a Vidiano/PLYTO blend, may have been the first to replant this variety but they have been followed by others such as Lyrarakis, whose barrel-fermented varietal Ippodromos has hints of spice and honey as well as aromas of apricot and quince. Douloufakis always blend in some SAUVIGNON BLANC for increased aroma.

VIEN DE NUS

Minor Aostan with relatives elsewhere. Wine is always blended.

BERRY COLOUR: ● ● ● ● ●

PRINCIPAL SYNONYMS: Gros Rouge, Gros Vien, Oriou Gros

ORIGINS AND PARENTAGE

Vien de Nus literally means 'vine of Nus', named after the village in the Valle d'Aosta between the towns of Aosta and Saint-Vincent where this variety was most likely born. Its first mention and description appeared in the writings of Gatta (1838), an amateur ampelographer and professional physician from neighbouring Ivrea (Piemonte). DNA parentage analysis has established that Vien de Nus is a progeny of PETIT ROUGE from the Valle d'Aosta and another unknown and probably extinct variety, and that Vien de Nus probably has close relationships with EYHOLZER ROTE and ROUGE DU PAYS from the neighbouring Valais (Vouillamoz and Moriondo 2011).

VITICULTURAL CHARACTERISTICS

Vigorous, early to mid ripening, generous but erratic yields. Resistant to frost and humidity. Very susceptible to powdery mildew and botrytis bunch rot.

WHERE IT'S GROWN AND WHAT ITS WINE TASTES LIKE

Vien de Nus is cultivated exclusively in the Valle d'Aosta, between Donnas and Avise, in north-west **Italy**, close to the borders with Switzerland and France, especially in the centre of the valley. Never made as varietal wine, Vien de Nus is blended (minimum 40%) with PETIT ROUGE (maximum 30%) and other local red varieties in the Nus Rouge category of the Valle d'Aosta DOC. The Italian agricultural census of 2000 recorded just 25 ha (62 acres), down from 170 ha (420 acres) in 1970.

VIGNOLES

French hybrid better known in the United States, mostly for sweeter wines.

BERRY COLOUR: ● ● ● ● ●

PRINCIPAL SYNONYMS: Ravat 51

ORIGINS AND PARENTAGE

According to Galet (1979), a complex hybrid obtained in the 1930s by J F Ravat in Marcigny, eastern central France, by crossing Subéreux with PINOT NOIR (see PRIOR for the complete Subéreux pedigree). However, recent DNA analysis has raised doubts about this parentage (Bautista *et al.* 2008) and further analyses are needed.

VITICULTURAL CHARACTERISTICS

Late budding, making it less susceptible to frost damage, early ripening and good cold hardiness. Low yields of smallish bunches of very small berries, susceptible to botrytis bunch rot.

WHERE IT'S GROWN AND WHAT ITS WINE TASTES LIKE

Vignoles is not included in the official register of varieties in **France** but, like VIDAL, has found its home much further afield, this time in the United States, particularly for sweet wines in the Finger Lakes region of New York State and for fresh wines at a range of sweetness levels – often to add floral aromas and fruity flavours to blends – in the Northeast and Midwest. Introduced in the 1950s, Vignoles' biggest plantings are in **Missouri** (208 acres/84 ha in 2009 – 13% of the vineyard), **New York** (148 acres/60 ha in 2006) and **Michigan** (85 acres/34 ha in 2006), with smaller areas in **Illinois** (72 acres/29 ha in 2007), **Indiana** (only 30 acres/12 ha in 2009 but one of the state's best varieties), **Iowa** (18 acres/7 ha in 2006) and **Nebraska** (7 acres/3 ha in 2007), among others. Recommended producers include Stone Hill in Missouri (all sweetness levels), Hunt Country Vineyards, Keuka Springs, Swedish Hill and Wagner Vineyards in New York.

VIJARIEGO

Minor variety found mainly on the Islas Canarias, used for both still and sparkling wines.

BERRY COLOUR:

PRINCIPAL SYNONYMS: Bujariego, Diego (Lanzarote), Verijadiego Blanco (Islas Canarias), Vijariego Blanco, Vijiriega Blanca, Vijiriega Común, Vijiriego, Vujariego

ORIGINS AND PARENTAGE

Vijariego was first described under the name Vigiriega Común by Clemente y Rubio (1807) as a variety widespread in Andalucía. It is said to have been introduced to the Islas Canarias in the late fifteenth or early sixteenth century together with Moscatel (possibly MUSCAT OF ALEXANDRIA), TORRONTÉS, PALOMINO FINO and LISTÁN PRIETO (Tapia *et al.* 2007), in all probability from Andalucía. It is not a colour mutation of Vijariego Negro (Jorge Zerolo, personal communication; Vouillamoz).

There are numerous different spellings, depending on location and producer, for what the official Spanish regulations for authorized varieties refer to as Vijariega or Vijiriega; and although CHELVA is sometimes called Mantúo Vigiriego or Vigiriego Blanco in Granada, it is not related to the Vijariego described here.

VITICULTURAL CHARACTERISTICS

Productive, vigorous, mid to late ripening. Good general disease resistance.

WHERE IT'S GROWN AND WHAT ITS WINE TASTES LIKE

There were said to be 473 ha (1,169 acres) of Vijariego on the Islas Canarias, **Spain**, in 2008, and it is authorized in all the island's DOs, where it is used for both still and sparkling wines thanks to the generally high levels of acidity. On the island of Tenerife, producers of varietal wines include Tanajara and Viñátigo.

In Granada, the variety had virtually disappeared, partly due to its low alcohol levels, and was found only in some old interplanted vineyards in Sierra de la Contraviesa. In 1984, Manuel Valenzuela of Barranco Oscuro in the Alpujarra planted a new Vijariego vineyard (using the historical name Vigiriega) and they now produce an impressive varietal sparkling wine with no dosage and also use the variety in other white blends. A handful of growers have followed suit, both here and in La Mancha.

VILANA

The grape of Kríti (Crete) produces fresh, lightly floral whites.

BERRY COLOUR:

PRINCIPAL SYNONYMS: Velana

ORIGINS AND PARENTAGE

Vilana is a Cretan variety that is genetically close to Piperionos, which is from Achaïa but no longer cultivated (Ladoukakis *et al.* 2005), and to VIDIANO from Kríti (Biniari and Stavrakakis 2007).

VITICULTURAL CHARACTERISTICS

Vigorous, high yielding and susceptible to fungal diseases. Large, compact bunches of small berries. Late budding and ripening.

WHERE IT'S GROWN AND WHAT ITS WINE TASTES LIKE

Vilana is the most important white wine variety on the island of Kríti, **Greece**, planted mostly in the eastern half, with a total of 589 ha (1,455 acres) in 2008. The Peza appellation requires 100% Vilana, Sitía requires 70% (with THRAPSATHIRI), while the variety is authorized in all of the island's regional whites and must constitute at least 80% of Irákleio (Heraklion) regional wines. The variety's tendency to overcrop means that it often produces bland, dilute whites and the variety also has a tendency to oxidize in the winery. However, when vines are planted on dry-farmed sandy hillsides and yields are kept in check, they can produce fresh, aromatic whites with aromas of white blossom and flavours of citrus and green apple, sometimes a little spicy. Producers of creditable varietal wines include Lyrarakis, Mediterra, Michalakis and the Peza co-op. A few producers such as Douloufakis are making successful if less typical blends with a small amount of SAUVIGNON BLANC rather than with other local varieties.

VILLARD BLANC

Light-skinned French hybrid in decline but still quite widely scattered round the globe.

BERRY COLOUR: ● ● ● ● ●

PRINCIPAL SYNONYMS: Seyve-Villard 12-375, SV 12-375

ORIGINS AND PARENTAGE

A complex hybrid obtained by Bertille Seyve and Victor Villard in Saint-Vallier in the Isère, France, by crossing Seibel 6468 with Subéreux (for the complete pedigrees, see HELIOS for Seibel 6468 and PRIOR for Subéreux). It was used to breed ESPIRIT, ORION, PHOENIX, ROSETTA and STORGOZIA.

VITICULTURAL CHARACTERISTICS

Late ripening, vigorous and productive, with good resistance to downy mildew and botrytis bunch rot. Best grafted.

WHERE IT'S GROWN AND WHAT ITS WINE TASTES LIKE

There were 349 ha (862 acres) of Villard Blanc in **France** in 2008, down from 1,129 ha (2,790 ha) in 1998 and only a tiny fraction of the 21,397 ha (52,873 ha) in 1968, when it was the third most planted white wine variety in France. It is still authorized in many *départements* across southern and eastern France but recommended in none, probably because the wines are rather ordinary and can be slightly bitter, though the best examples have some citrus freshness. It is still found in some vineyards in the south of France and is also grown on pergolas for table grapes.

It is also grown in the US, generally in the cooler parts of the northern states, although in some regions, eg New York State, it may be hard to ripen fully. Plantings are too few to be included in official state statistics but producers include Goose Watch in **New York** State, Blue Sky in **Illinois**, Rushland Ridge in **Pennsylvania** and Tomasello in **New Jersey**. Further south, White Oak in **Alabama** also produce a varietal wine.

Limited plantings can still be found in **Australia**. Douglas Vale in Port Macquarie and Raleigh Vineyard, both in New South Wales, produce varietal wines. Villard Blanc has also been planted to a limited extent to resist the humid summers of KwaZulu-Natal in **South Africa**.

VINCENT

Hardy North American hybrid bred in Canada for cold climates.

BERRY COLOUR:

PRINCIPAL SYNONYMS: Vineland 49431, V 49431

ORIGINS AND PARENTAGE

A Vineland 370628 × CHELOIS hybrid obtained in 1947 by Ollie A Bradt at the Vineland research centre (now part of Guelph University) in Niagara, Ontario, Canada, in which:

– Vineland 370628 is a Lomanto × Seneca hybrid
– Lomanto is a Salado × Molinera hybrid, where Molinera is a *Vitis vinifera* table grape from Spain
– Salado is a De Grasset × Brilliant hybrid
– De Grasset is *Vitis champinii* vine selected by T V Munson
– Brilliant is a Lindley × DELAWARE hybrid
– Lindley is a Carter × CHASSELAS hybrid
– Carter (also called Mammoth Globe) is an offspring of ISABELLA and another unknown parent selected by Charles Carter in Virginia
– Seneca is a LUGLIENGA × Ontario hybrid (see CAYUGA WHITE for the pedigree of Ontario).

Vincent, selected in 1958 and released in 1967, is therefore an extremely complex *Vitis labrusca*, *Vitis vinifera*, *Vitis champinii*, *Vitis aestivalis*, *Vitis lincecumii*, *Vitis rupestris* and *Vitis cinerea* hybrid.

VITICULTURAL CHARACTERISTICS

Medium-sized to large compact clusters. Very productive and late ripening. Susceptible to powdery mildew. Hardy (–15 °F/–26 °C).

WHERE IT'S GROWN AND WHAT ITS WINE TASTES LIKE

Fulkerson in Finger Lakes, **New York** State, is one of the very few producers to make a varietal wine, which they describe as smooth, low in tannin with plum and black-cherry aromas and flavours. The juice is very dark (much darker than CABERNET SAUVIGNON), making it useful as a blending component. There were 44 acres (18 ha) in New York State in 2006. It is also used by blenders in **Canada** for its colour.

VINHÃO

High-acid, highly distinctive Portuguese variety making rustic, deeply coloured wines in north-west Iberia.

BERRY COLOUR: ● ● ● ● ●

PRINCIPAL SYNONYMS: Espadeiro Basto (Arcos de Valdeves), Espadeiro da Tinta (Valença), Espadeiro Preto (Monção), Negrão (Minho), Pinhão, Sousao (California), Sousão, Sousão de Correr (Minho), Sousão Forte (Minho), Sousón (Galicia in Spain), Souzao (California), Souzão, Tinta Nacional (Amarante), Tinta País (Galicia), Tinto de Parada (Megaço), Tinto Nacional (Amarante)

VARIETIES COMMONLY MISTAKEN FOR VINHÃO: AMARAL

ORIGINS AND PARENTAGE

According to the Visconde de Vila Maior (1875), Vinhão originates in Minho in north-west Portugal, where it is still the main variety today (Rolando Faustino, personal communication). Around 1790, it was introduced to the Douro, where it was named Sousão and was used instead of the prohibited elderflower berries to darken port wines (Rolando Faustino, personal communication). DNA profiling finally ended a long debate and confirmed that Vinhão and Sousão are one and the same (Ferreira Monteiro *et al.* 2000). A likely parent–offspring relationship with AMARAL (under the name Caíño Bravo) has recently been suggested (Díaz-Losada *et al.* 2011).

Vinhão was used to breed ARGAMAN.

VITICULTURAL CHARACTERISTICS

Mid to late budding and ripening. In Portugal, Vinhão berries are thick-skinned, bunches medium-sized; in Spain, Sousão produces small compact bunches of small to medium-sized berries. Good general resistance to pests and diseases.

WHERE IT'S GROWN AND WHAT ITS WINE TASTES LIKE

Vinhão produces what are probably the darkest wines in all of **Portugal**, even darker than those produced by truly red-fleshed varieties. There is so much colour in the skins that the colour leeches into the flesh when the berry is split, giving the impression that it is a red-fleshed variety.

It is the basis for all good reds in the Vinho Verde DOC in the north west of the country, giving well-structured, characterful wines with very high acidity and flavours of cherry and fruits of the forest that have until recently had a mainly local appeal. It is often the dominant or sole variety in some of the most expensive wines and does particularly well in the Basto and Amarante subregions in the south west of the Vinho Verde DOC. After a swing towards white varieties such as LOUREIRO that began in this region in the 1950s, there now seems to be a return to reds and to Vinhão in particular and it is now one of the most grafted varieties in local nurseries. Recommended producers of varietal wines include Aphros, Quinta da Palmirinha, Quinta da Raza, Quinta de Aguiã, Quinta de Carapeços and Quinta de Gomariz; others use it in a blend of local varieties. Quinta de Linhares, among others, turn it into a rosé.

In the Douro, Sousão (or Souzão) is found mainly in the old mixed vineyards and its primary role is still to add colour. There are Sousão vines in Quinta do Noval's ungrafted Nacional vineyard but it is the site and the age of the vines that make them special and other plantings of this variety on the Noval estate are less interesting. There are nevertheless some producers who value Sousão as a component in their port blends; Paul Symington, for example, of the eponymous group of port houses says they appreciate it for the acidity it brings in a region where freshness may be lacking. New plantings are on the increase thanks to the fashion for deeply coloured wines. Quinta do Vallado make a varietal Douro table wine. Total plantings in Portugal were 2,099 ha (5,187 acres) in 2010.

Spain recorded 573 ha (1,426 acres) of Sousón in 2008, virtually all in Galicia in the north west, where it is authorized in the Rías Baixas, Ribeiro and Valdeorras DOs. It is generally used as part of a blend but a few producers make varietal wines: Quinta da Muradella and Santa Marta in Valdeorras (the former in the south of the appellation on the Portuguese border) and Coto Redondo in Rías Baixas, just over the border from the Vinho Verde subregion of Monção.

There are also plantings of Souzao in **California** (63 acres/25 ha in 2009), mostly for inclusion in fortified wines. Belo in St Helena, California, are unusual in making a varietal port-style wine. Campbells of Rutherglen in Victoria, **Australia**, have also had success with a dry red wine made from a blend of Portuguese varieties including Sousão, all from small areas planted in a vineyard block dating back to the 1960s – so much so that they have plans to expand the area dedicated to these varieties. Others who use Sousão as part of a blend include Mazza in Geographe, Western Australia, and D'Arenberg in McLaren Vale.

South Africa had 38 ha (94 acres) of Sousão in the ground for port-style wines in 2009.

VIOGNIER

Headily aromatic variety making particularly full-bodied whites – all over the world now.

BERRY COLOUR: ● ● ● ● ●

PRINCIPAL SYNONYMS: Viogné or Vionnier, Viognier Jaune, Viognier Vert

VARIETIES COMMONLY MISTAKEN FOR VIOGNIER: ALTESSE, CHOUCHILLON, VUGAVA (Croatia)

ORIGINS AND PARENTAGE

Viognier is a variety from the northern Rhône in France, probably from the Condrieu and Ampuis wine regions, where it was first mentioned by Faujas in 1781 (Rézeau 1997): 'Deux seules especes de raisins composent ces excellent vins [côte-rôtie], la Serine & le Vionnier' ('Only two species of grapes compose the excellent wines [côte-rôtie], Serine and Vionnier'). The name Viognier supposedly derives from the French *viorne*, meaning 'viburnum', but this etymology is doubtful. Through DNA parentage analysis, a parent–offspring relationship has been discovered between Viognier and MONDEUSE BLANCHE (Vouillamoz 2008), which makes Viognier either a half-sibling or a grandparent of SYRAH (see Syrah for the complete pedigree). Viognier is therefore logically a member of the Sérine ampelographic group (see p XXVII; Bisson 2009). In addition, Viognier has a close genetic relationship with FREISA, so that Viognier is in some way related to NEBBIOLO (Schneider, Boccacci *et al.* 2004).

It has recently been suggested that Viognier is a sibling of SYRAH (Myles *et al.* 2011) rather than a half-sibling or a grandparent, as in Vouillamoz (2008), but this can be refuted because Viognier does not have a parent–offspring relationship with DUREZA, Syrah's other parent (Vouillamoz).

OTHER HYPOTHESES

Viognier was supposedly introduced from the Dalmatian coast to France by the emperor Probus, who came from Smirnium in Croatia, and Calò *et al.* (2006) suggest that Viognier is still cultivated on the island of Vis under the name VUGAVA or Bugava. However, there is no historical evidence in support of this hypothesis, and the DNA profile of Vugava Bijela that was available in the Croatian Vitis Database does not match Viognier (Vouillamoz).

VITICULTURAL CHARACTERISTICS

Early budding and therefore at risk from spring frosts, and mid ripening. Generally trellised as it is sometimes susceptible to wind, and pruned fairly long. Traditionally grown on acid soils but also adaptable to warmer regions as long as there is no water stress. Long, compact bunches with small, thick-skinned berries and reasonably good resistance to botrytis bunch rot. The juice is often high in sugar but low in acidity.

WHERE IT'S GROWN AND WHAT ITS WINE TASTES LIKE

It is extraordinary to think that this world-famous variety, now planted in virtually every wine region around the globe, including on 4,395 ha (10,869 acres) of vineyard in **France** in 2009, could be found on only 14 ha (35 acres) of vineyard as recently as fifty years ago. All of those vines were on the steep right bank of the Rhône just south of Lyon, where Viognier had long been producing headily perfumed, full-bodied, rather soft, sometimes sweet, white wines on the granite terraces of Condrieu and in Château Grillet's unusual amphitheatre just to the south. It had also long been added in small quantities (generally 5% or so, although 20% is allowed in the vineyard) to be fermented with the local SYRAH grapes in Côte Rôtie just north of Condrieu so as to help stabilize the colour and occasionally add perfume.

The variety was almost on the point of extinction in the late 1960s because the plant material then grown was highly susceptible to coulure and yields were discouragingly low. But as the particularly distinctive character of the variety – descriptors include apricots, honeysuckle, May blossom, gingerbread – became known and appreciated outside its region of origin, Viognier's fame spread so that by the 1980s curious growers from the Languedoc to California were experimenting with it and the quality of cuttings improved considerably. Condrieu today is a highly fashionable, usually dry wine that, when successful, combines a firm, rich undertow of extract with Viognier's hedonistic cocktail of aromas – even if it rarely improves with age. Château Grillet, recently acquired by the owner of Bordeaux first growth Château Latour, is in the process of being restored to its former glory when it did seem to represent a particularly lemony, mineral-scented answer to Condrieu that could improve in bottle. And many Côte Rôtie growers today still add a small proportion of Viognier to their red wine fermentation vats, from generally healthier vines than half a century ago. But the variety is also planted very much more widely in the northern and southern Rhône today. It is a permitted ingredient in the Côtes du Rhône, Côte du Rhône-Villages and Lirac appellations – although it is not allowed in Hermitage, Crozes-Hermitage, Saint-Joseph, nor Châteauneuf-du-Pape, presumably because the rules for these well-established appellations merely embraced the status quo when they were drawn up. Recommended producers in the northern Rhône include Cuilleron, Gangloff, Guigal, Niéro, Christophe Pichon, René Rostaing, Georges Vernay, François Villard and Les Vins de Vienne.

Today, however, Viognier is recommended or authorized in much of southern France. Quantitatively its greatest presence is in the Languedoc, where it is used to make aromatic vins de pays wines, too often lacking acidity and disappointing after an initial whiff of a vaguely floral and unconvincing perfume, although there are exceptions and Viognier is often an ingredient with such varieties as Grenache Blanc (GARNACHA BLANCA), sometimes CHARDONNAY, MARSANNE and/or ROUSSANNE, in some intriguing blended full-bodied dry whites, particularly in Roussillon. The variety needs to be fully ripe, and not overcropped, in order to deliver its characteristic perfume – which is usually signalled by a deep, golden colour.

Viognier has not really caught on in **Italy**, although Ascheri have an established tradition of producing it in Piemonte, as Castel de Paolis do in Lazio. It is also planted in the Maremma and on Sicilia.

In **Spain**, Viognier is even less important: just 18 ha (44 acres) in 2008, mostly in Castilla-La Mancha, often blended with GARNACHA BLANCA or other Rhône varieties. In Priorat, producers such as Celler Mas d'En Just are trying their hand at SYRAH plus a little Viognier. Quinta do Monte d'Oiro in the Lisboa region and Quinta do Lagar Novo in Alenquer are two of the few producers of varietal Viognier in **Portugal**, where there were 82 ha (203 acres) in 2010.

A little is grown in the Pfalz in **Germany**, while Graf Hardegg in the Weinviertel is **Austria**'s specialist Viognier producer. **Hungary** had 5 ha (12 acres) in 2008, with Tibor Gal producing a varietal version in Eger. Gerovassiliou is the chief proponent of the variety in **Greece**. There were also 15 ha (37 acres) in **Turkey**, 7 ha (17 acres) on **Malta** and at least 50 ha (124 acres) in **Israel** in 2010.

As in France, Viognier plantings in **California** have increased remarkably over the last few decades, up from the first 25 acres (10 ha) in 1982 to 2,993 acres (1,211 ha) in 2010, particularly in the Central Coast. In California it is moderately vigorous and generally grown on deep, alluvial soils but has to reach full maturity, generally with plenty of sun exposure, to reveal its characteristic scent. Recommended producers include Calera of Mount Harlan, who have particularly long experience of the variety, and Darioush of Napa Valley. Thomas George of Russian River Valley is a Viognier specialist.

There were only 183 acres (74 ha) of Viognier in **Oregon** in 2008, almost all planted in the last ten years, with Cristom being a notable pioneer. The fragrant, relatively full-bodied PINOT GRIS that is the state's signature white wine leaves less scope for Viognier than elsewhere. Thanks to its enthusiasm for all things Rhôneish, **Washington** has a little more of the variety (its famous RIESLING may be aromatic but is not full-bodied): 390 acres (158 ha) in the ground by 2011, a very small proportion of the state's total vineyard.

In **Virginia**, on the other hand, Viognier is regarded as the state's signature grape. The area devoted to it doubled between 2003 and 2010: there were 229 acres (93 ha) in 2010, only about half the area devoted to CHARDONNAY but much more than any other white wine grape. Horton pioneered the variety and was delighted to find how much better its thick-skinned berries withstood Virginia's humid summers than CHARDONNAY. Keswick and Veritas are other reliable producers.

The variety is far from uncommon in **Texas**, and in western **Canada** there were 165 acres (67 ha) of Viognier in British Columbia in 2008.

In South America, Viognier has been embraced slightly more enthusiastically in **Argentina**, where there were 714 ha (1,764 acres) in 2008 and successful varietal wines are produced by the likes of Chakana, Mauricio Lorca and Trapiche, than in **Chile**, where there were 263 ha (650 acres) in 2008. Notable Chilean varietal Viogniers are made by Viña Emiliana and Anakena but, as in Argentina, vines tend to be too young for very powerful varietal expression. There were 32 ha (79 acres) in **Uruguay** in 2009 and 7 ha (17 acres) in **Brazil** in 2007.

The practice of co-fermentation traditional in the northern Rhône has become fashionable in other parts of the world where SYRAH flourishes, including California and Australia. For example, Ridge's Lytton West Syrah from Dry Creek Valley contains around 6% Viognier, and in **Australia** Clonakilla (Canberra), Torbreck (Barossa Valley) and Yering Station (Yarra Valley) all include a percentage of Viognier in their top Shiraz (Syrah) wines. One South Australian winemaker has done so much for the cause of Viognier that her name has become almost synonymous with this variety that was first trialled at Heathcote Winery in central Victoria in 1978 (Rose 2006): Yalumba's Louisa Rose, who has spearheaded work on clonal selection at the company's own nursery and set up the first ever Viognier Symposium in 2002. It was Yalumba that initiated the first commercial planting of Viognier in the Eden Valley in 1980. Today more than 500 wineries produce a Viognier in Australia, whose generally warm climate develops flavours of apricot, peaches, white flowers, occasionally ginger, and luscious texture in these wines. Recommended producers in addition to Yalumba include D'Arenberg, Haselgrove, Petaluma and Temple Bruer in South Australia and Trentham Estate in New South Wales. Total plantings in Australia in 2008 amounted to 1,401 ha (3,462 acres).

New Zealand has a much smaller total area planted to Viognier, with only a very minor increase in the last eight years: from 155 ha (383 acres) in 2005 to 160 ha (395 acres) in 2013. The naturally high acidity levels in Kiwi wine counteract this variety's richness. Recommended producers include Dry River in Martinborough, Herzog in Marlborough, Millton in Gisborne and Craggy Range in Hawke's Bay. Several Hawke's Bay producers such as Trinity Hill are also following the Côte Rôtie model of co-fermenting Viognier with SYRAH.

Viognier is also increasingly fashionable in **South Africa**, where plantings had reached 859 ha (2,123 acres) by 2008, planted throughout the country's wine-growing regions. Many top producers such as Adi Badenhorst, Lammershoek, Mullineux and Eben Sadie in the Swartland,

Tulbagh Mountain Vineyard in the Coastal Region and Quoin Rock in Stellenbosch use Viognier to add perfume and texture to complex white blends, while others such as Eagle's Nest, Elgin Valley, Fairview, The Foundry and The Winery of Good Hope are producing creditable varietal wines, often oaked.

VIOLENTO

Pink-skinned variety from the Greek island of Zákynthos used mostly in blends.

BERRY COLOUR:

PRINCIPAL SYNONYMS: Violenti (Ilía)

ORIGINS AND PARENTAGE

Violento is local to the Greek island of Zákynthos in the Ionian Sea off the north-western coast of the Pelopónnisos (Peloponnese). It has traditionally been considered to be a clone of RODITIS but this has been contradicted by DNA profiling (Boutaris 2000).

VITICULTURAL CHARACTERISTICS

Vigorous, productive, mid budding, mid ripening. Particularly susceptible to downy mildew but resistant to drought. Large berries.

WHERE IT'S GROWN AND WHAT ITS WINE TASTES LIKE

Violento, grown on Zákynthos, **Greece**, produces wines that have rather high alcohol but are typically low in acidity and are usually part of a local rosé or white blend. Producers include Comoutos and Callinico.

VIOSINHO

Low-yielding but high-quality aromatic Portuguese variety.

BERRY COLOUR:

ORIGINS AND PARENTAGE

Viosinho most probably comes from the Douro region in northern Portugal. The high level of genetic variation in the vineyard suggests it is a relatively old variety, praised for its quality at least since the mid nineteenth century (Rolando Faustino, personal communication). According to recent studies, Viosinho could be genetically related to LADO, TINTA FRANCISCA and TINTO CÃO (Pinto-Carnide *et al.* 2003; Martín *et al.* 2006).

VITICULTURAL CHARACTERISTICS

Small bunches of small berries, low productivity, early ripening. Average susceptibility to downy and powdery mildews and botrytis bunch rot.

WHERE IT'S GROWN AND WHAT ITS WINE TASTES LIKE

Grown almost exclusively in the Douro and in the Planalto Mirandês subregion of Trás-os-Montes, **Portugal**, close to the Spanish border, Viosinho has very good potential for quality thanks to the balance of sugar and acidity in the berries, giving rise to fresh yet full-bodied aromatic whites with ageing potential. Some have described it as the Portuguese SAUVIGNON BLANC because of its intense aroma profile in the best vintages. However, it is not an easy variety, not just because of low yields but also because the must is prone to oxidation and consequent loss of aromatic finesse. It is used mostly in blends with other local varieties such as RABIGATO and Gouveio (GODELLO), for both white port and increasingly impressive Douro unfortified wines. José Maria da Fonseca in Azeitão have succeeded in producing a very good varietal wine from vines planted at 500 m above sea level. Other recommended producers include Caves Transmontanas and Vinilourenço. It is also authorized in the regional wines of Lisboa, the Península de Setúbal, Tejo and the Açores. There was a total of 100 ha (247 acres) in Portugal in 2010, much of it found in old interplanted vineyards, although new plantings are increasing, particularly in the Alentejo.

VITAL

Low-acid variety grown mainly in central Portugal.

BERRY COLOUR: ● ● ● ● ●

PRINCIPAL SYNONYMS: Boal Bonifacio (Oeste), Malvasia Corada☒ (Douro, Dão and Minho), Malvasia Fina do Douro

VARIETIES COMMONLY MISTAKEN FOR VITAL: MALVASIA FINA☒

ORIGINS AND PARENTAGE

Very little is known about the origins of this Portuguese variety although it is likely to come from the north or centre of the country, probably in the area north of Lisboa. DNA profiling shows that it is not identical to MALVASIA FINA (Vouillamoz), despite the confusing synonyms mentioned above.

VITICULTURAL CHARACTERISTICS

Vigorous and productive. Mid budding and mid to early ripening. Very sensitive to drought and prone to berry shrivel, leading to poor ripening in very dry conditions. Best suited to fine-grained and relatively fertile soils and a cooler summer climate (windy sites seem to work well). Moderately susceptible to powdery mildew and bunch rot.

WHERE IT'S GROWN AND WHAT ITS WINE TASTES LIKE

Wines made from Vital are generally low in acidity and pretty bland, quickly losing any youthful aroma they may have had, hence the need for blending. It is grown mainly in central **Portugal** – in the Lisboa region as well as in the Alentejo and Tejo. Sanguinhal's Quinta de São Francisco (Óbidos DOC) is a blend of Vital, FERNÃO PIRES and ARINTO DE BUCELAS. It is also planted further north in the Douro.

However, thanks to the late António Carvalho, a Portuguese winemaker passionate about indigenous varieties in his home region just north of Lisboa, Vital has now been shown to have much greater quality potential. He made his first varietal Vital from vines in the mountainous Montejunto area near the town of Cadaval, where the variety thrives thanks to the dual influences of altitude and windy proximity to the Atlantic on a northern exposure. Since Carvalho's death, his partner Marta Soares has continued to make this elegant, citrus- and mineral-scented white under the name Casal Figueira.

There were 1,080 ha (2,669 acres) in Portugal in 2010, some of its produce being distilled into brandy.

VITOVSKA

Makes rare but high-quality white in the Italy–Slovenia border region around Trieste.

BERRY COLOUR: ● ● ● ● ●

PRINCIPAL SYNONYMS: Vitouska

VARIETIES COMMONLY MISTAKEN FOR VITOVSKA: GARGANEGA, VITOVSKA GRGANJA

ORIGINS AND PARENTAGE

Vitovska is a specific variety of the Carso/Kras/Karst region which straddles north-eastern Italy and western Slovenia. Its name could derive from the local Slovenian dialect, or from the village of Vitovlje in the Vipavska Dolina region. DNA parentage analysis has recently found that Vitovska is a natural cross between MALVASIA BIANCA LUNGA and PROSECCO (Crespan *et al.* 2007). Prosecco is widespread in Karst, and although Malvasia Bianca Lunga has never been mentioned there, Karst is still the most likely region of origin for Vitovska. Contrary to the consensus in the literature, recent DNA profiling has shown that VITOVSKA GRGANJA (or Vitovska Garganija) is not identical to Vitovska, but it may be an offspring of Vitovska (Štajner *et al.* 2008).

VITICULTURAL CHARACTERISTICS

Mid ripening. A hardy variety that is resistant to winter frost and summer droughts. Large triangular bunches weighing up to 1 kg (2.2 lbs) each.

WHERE IT'S GROWN AND WHAT ITS WINE TASTES LIKE

Vitovska has long been cultivated exclusively in the Carso/Kras/Karst region, in particular in the villages of Sgónico and Duino-Aurisina. It had almost disappeared during the twentieth century and was rescued in the 1980s by winemakers such as Edi Kante and Benjamin Zidarich. Today, after MALVAZIJA ISTARSKA, Vitovska is the most widely cultivated white in the Italian province of Trieste and the neighbouring area in **Slovenia** (all one province before the Second World War). However, there is still much confusion between Vitovska and VITOVSKA GRGANJA.

A varietal Vitovska is included in the Carso DOC and made by a few determined Slovenian producers based in **Italy**. The wines tend to be fresh, dry or off dry with aromas of hay, sage, pear, quince and lemon and an attractive sour edge. They are also relatively expensive. Zidarich's version is tangy and firmly structured thanks to extended skin contact and generally hands-off winemaking. Other pioneering producers include Edi Kante, Paolo & Walter Vodopivec, who have been producing a special amphora-made version since 2005, Silvano Ferluga and Škerk.

VITOVSKA GRGANJA

Rare Slovenian white attracting increasing recognition.

BERRY COLOUR:

PRINCIPAL SYNONYMS: Grganka, Racuk, Vitovska Garganija

VARIETIES COMMONLY MISTAKEN FOR VITOVSKA GRGANJA: VITOVSKA

ORIGINS AND PARENTAGE

Recent DNA profiling has shown that, contrary to popular belief, Vitovska Grganja is not identical to VITOVSKA, of which it is a possible offspring (Štajner *et al.* 2008; Rusjan *et al.* 2010), which means that it probably also originates in the Carso/Kras/Karst region on the Italian-Slovenian border. It was first mentioned in 1844 by Matija Vertovec, a priest and teacher in Vipava (Grmek 2007). The proposed parent–offspring relationship would make Vitovska Grganja a grandchild of MALVASIA BIANCA LUNGA and PROSECCO (see VITOVSKA).

VITICULTURAL CHARACTERISTICS

Vigorous, late budding, early to mid ripening. Bunches are smaller and more elongated than those of VITOVSKA and berries are relatively thick-skinned.

WHERE IT'S GROWN AND WHAT ITS WINE TASTES LIKE

Vitovska Grganja was once far more important than it is today (Grmek 2007) but is still found in the Carso/Kras/Karst region, and in Vipavska Dolina, a little to the north in western Slovenia, and is gaining in importance once more. Given the common confusion between VITOVSKA and Vitovska Grganja, exact areas are hard to confirm although there were apparently 19 ha (47 acres) of the latter in **Slovenia** in 2009. Wines have moderate alcohol levels, quite marked acidity and peachy, even cinnamon aromas that are like a less pronounced MALVAZIJA ISTARSKA. Recommended Slovenian producers include Čotar, Furlan, Renčel and Vinakras.

VIURA

See MACABEO.

VLACHIKO

Greek variety that makes svelte, dark-fruited wines and deserves more attention.

BERRY COLOUR:

PRINCIPAL SYNONYMS: Blachiko, Blachos, Vlachico de Jannina, Vlachos (Ioánnina), Vlachs, Vlacos, Vlahico, Vlahos

ORIGINS AND PARENTAGE

Vlachiko is local to the region of Ioánnina in Ípeiros (Epirus), western mainland Greece. Its name comes from *vlacos*, meaning 'countryman' (Galet 2000).

VITICULTURAL CHARACTERISTICS

Vigorous and productive, late budding and ripening. Compact bunches of medium-sized berries.

WHERE IT'S GROWN AND WHAT ITS WINE TASTES LIKE

Vlachiko was traditionally grown in Ípeiros, **Greece**, and was often blended with DEBINA and BEKARI to make a sparkling rosé (still made by Zítsa producers such as the Ioánnina co-operative and Glinavos). However, Katogi & Strofilia planted a new Vlachiko vineyard in 2001 in the mountainous Zagóri region in Ípeiros and now make a varietal wine known as Rossiu di Munte ('red of the mountains') under the Katogi Averoff label – a wine with fresh acidity, very fine tannins and restrained, peppery, cassis fruit. Glinavos also use it in red blends with varieties such as AGIORGITIKO and Bekari.

VLAŠKA

Rare, low-acid Croatian variety found only in the area around Split.

BERRY COLOUR: ● ● ● ● ●

PRINCIPAL SYNONYMS: Maraškina Velog Zrna, Prejica, Vezuljka, Žutuja

ORIGINS AND PARENTAGE

Vlaška is Croatian for 'Wallachia' (what is now southern Romania) and this variety is said to have been introduced to the regions of Kaštela and Trogir north of Split from Romania in the twentieth century (Zdunić 2005).

VITICULTURAL CHARACTERISTICS

Early to mid ripening. Berries tend to have high sugar and low acid levels.

WHERE IT'S GROWN

Vlaška, along with several other autochthonous varieties, is part of a revitalization project involving clonal selection and replanting in **Croatia** and although it is authorized throughout the Srednja i Južna Dalmacija (Central and Southern Dalmatia) wine region, it is currently found only in Kaštela and Trogir near the coastal town of Split.

VLOSH

Dark-skinned Albanian variety in decline.

BERRY COLOUR: ● ● ● ● ●

PRINCIPAL SYNONYMS: Vloshi, Vlosk

ORIGINS AND PARENTAGE

Vlosh originates in the region around the coastal town and lagoon of Narta, south-west Albania, and may well be named after the nearby town of Vlosh, between Narta and Berat. However, it has also been suggested that the name derives from a Greek word meaning 'sweet grape'.

Lambert-Gocs (2007) suggested that Vlosh is a possible ancestor of CABERNET SAUVIGNON but this is contradicted by DNA profiling (see Cabernet Sauvignon).

VITICULTURAL CHARACTERISTICS

Late ripening.

WHERE IT'S GROWN AND WHAT ITS WINE TASTES LIKE

Vlosh is found mainly in south-west **Albania**, particularly around its birthplace, but is in decline despite the quality of the wine produced due to its lack of colour and moderate yields (Maul *et al.* 2008). Wines are typically full-bodied and astringent, often made in an oxidized, rancio style. It is also planted for the table.

VOLITSA MAVRI

Dark-skinned Greek variety unlikely to be made into wine unless healthy plant material can be found.

BERRY COLOUR: ● ● ● ● ●

ORIGINS AND PARENTAGE

Volitsa Mavri comes from Kalávrita, between Pátra and Kórinthos (Corinth) in Greece's northern Pelopónnisos (Peloponnese).

In the United States Department of Agriculture's National Clonal Germplasm Repository in Davis, California, Volitsa Mavri and the white-berried and non-cultivated Volitza Lefki (or Volitza Aspri) have distinct DNA profiles (Vouillamoz), suggesting that the latter is not a colour mutation of the former.

OTHER HYPOTHESES

Lambert-Gocs (2007) suggested that CABERNET SAUVIGNON was a descendant of Volitsa Mavri but this has been refuted by DNA profiling (see details in Cabernet Sauvignon).

WHERE IT'S GROWN

Oenoforos in Égio east of Pátra made a valiant attempt to rejuvenate this variety in the late 1990s but, according to winemaker Tasos Drosiadis, have had to give up because the plant material available was not the best clone and was severely virused. It seems that the few other plantings in the area were equally unhealthy.

VOSKEAT

Productive old Armenian variety used mostly for fortified whites.

BERRY COLOUR: ● ● ● ● ●

PRINCIPAL SYNONYMS: Hardzhi, Kanacicheni, Katviacik, Khardji or Kharji, Pishik Gezi, Pscigi, Voskeate, Voskehat, Xardji

ORIGINS AND PARENTAGE

Voskeat is a very old variety that probably originates in the Ashtarak region in western Armenia, where it shows significant morphological diversity (Chkhartishvili and Betsiashvili 2004). Traditionally called Khardji or Kharji, it was later named after the village of Voskehat, just west of the capital Yerevan.

VITICULTURAL CHARACTERISTICS

Early to mid budding, mid to late ripening. Very productive. Compact bunches of thin-skinned berries that can attain notably high sugar levels. Very susceptible to powdery and downy mildews, botrytis bunch rot and grape moths.

WHERE IT'S GROWN AND WHAT ITS WINE TASTES LIKE

Voskeat is very well suited to the production of sweet fortified wines in **Armenia**, the most famous of which is Ochakan, made in the image of Madeira (Carbonneau 1995). It is also used to make sparkling wine, usually as part of a blend, and for the table. Found mainly in the Ararat region. Ginekar, for example, produce a dry version, Ijevan Wine Factory a semi-sweet one. It is also popular in the Nakhchivan Autonomous Republic in **Azerbaijan**, where it is known as Kharji and used to make sherry- and Madeira-style wines.

VOSTILIDI

See GOUSTOLIDI.

VRANAC

Balkan variety capable of producing powerful, high-quality, ageworthy reds.

BERRY COLOUR:

PRINCIPAL SYNONYMS: Vranac Crmnichki, Vranac Crni, Vranac Prhljavac, Vranec (Republic of Macedonia)

VARIETIES COMMONLY MISTAKEN FOR VRANAC: TRIBIDRAG

ORIGINS AND PARENTAGE

Vranac, meaning 'strong black' or 'black stallion', is the most important indigenous variety of the region that includes today's Montenegro, part of Serbia and southern Croatia, where it has allegedly been cultivated since the Middle Ages and where it shows a high degree of heterogeneity (Avramov and del Zan 2004), suggesting it is an ancient variety. DNA parentage analysis has revealed a parent–offspring relationship with TRIBIDRAG (Maletić *et al.* 2004), otherwise known as Zinfandel, which makes it a half-sibling (or grandparent) of PLAVAC MALI, GRK, Crljenak Crni (no longer cultivated) and PLAVINA (see Tribidrag for the pedigree diagram). This could also explain why it has often been confused with Kratošija (Maras *et al.* 2004), the Montenegrin synonym for Tribidrag, with which it is often interplanted. Although Vranac is today the major black-berried variety in the Republic of Macedonia, genetic comparison has revealed that it is quite distinct from the other local varieties and that it was most probably introduced there in the mid twentieth century (Galet 2000; Štajner, Angelova *et al.* 2009).

VITICULTURAL CHARACTERISTICS

Vigorous, productive, early budding, mid ripening. Compact bunches of thin-skinned berries. Best suited to warm soils. Moderate resistance to downy and powdery mildews but particularly susceptible to botrytis bunch rot. Very sensitive to frosts – buds freeze at –12 °C (10 °F) to –14 °C (7 °F).

WHERE IT'S GROWN AND WHAT ITS WINE TASTES LIKE

Vranac is widely planted in **Montenegro**, the **Republic of Macedonia** (particularly in Povardarje in the south east), **Serbia** and **Kosovo**. It is also important in southern **Croatia** (390 ha/964 acres in 2008) and in **Herzegovina**. Wines are deeply coloured, full-bodied, tannic, high in alcohol and have soft acidity plus rich flavours of both black and red fruit. The best take well to oak and become more complex with age. Its wine is also used to add colour to blends. Given the heterogeneity of this variety, research is underway in Macedonia to select the best-quality clones. Producers include Milović, Plantaže and Sjekloća in Montenegro, Popova Kula, Skovin, Tikveš and Veritas in Macedonia, Stonecastle in Kosovo, Jović in Serbia and Vukoje in Herzegovina.

VUGAVA

High-quality, aromatic variety found on the Croatian island of Vis.

BERRY COLOUR: ● ● ● ● ●

PRINCIPAL SYNONYMS: Bugava, Ugava, Viškulja, Vugava Bijela

VARIETIES COMMONLY MISTAKEN FOR VUGAVA: VIOGNIER

ORIGINS AND PARENTAGE

Vugava is an old variety from the island of Vis in Dalmacija (Dalmatia), south-west Croatia.

OTHER HYPOTHESES

Vugava was allegedly introduced by the Greeks, but there is as yet no evidence for this theory. Vugava has also been considered to be identical to the French variety VIOGNIER but this can be firmly rejected because the DNA profile of Vugava Bijela in the Croatian Vitis Database does not match that of Viognier (Vouillamoz).

VITICULTURAL CHARACTERISTICS

Early ripening and thin skins make it prone to wasps, bees and sunburn. Highly susceptible to powdery mildew and phomopsis as well as to botrytis bunch rot, though less so.

WHERE IT'S GROWN AND WHAT ITS WINE TASTES LIKE

Vugava is authorized throughout the Srednja i Južna Dalmacija (Central and Southern Dalmatia) wine region in **Croatia** but is found mainly on the island of Vis, with limited plantings on neighbouring islands and on the coast. Wines are typically intense and fairly alcoholic with the flavour of ripe apricots so it is easy to see why the variety has been confused with VIOGNIER. Producers on Vis include Lipanović, the Podšpilje co-operative, Rokis and Sviličić.

VUILLERMIN

Extremely rare Aostan clinging on near the Swiss border.

BERRY COLOUR:

ORIGINS AND PARENTAGE

Vuillermin is an obscure variety of the Valle d'Aosta, first mentioned in Bich (1890) and probably named after a common local family name. DNA parentage analysis has established that Vuillermin is a natural cross between FUMIN of the Valle d'Aosta and another unknown and probably extinct variety (Vouillamoz and Moriondo 2011). Vuillermin is therefore a nephew/niece of PETIT ROUGE.

OTHER HYPOTHESES

Vuillermin could be identical to the extinct variety Spron or Eperon described by Gatta (1838), with which it shares many morphological features (Moriondo 1999).

VITICULTURAL CHARACTERISTICS

Late ripening.

WHERE IT'S GROWN

Vuillermin was almost extinct at the end of the twentieth century when it was rescued *in extremis* from old vineyards in Pontey and Châtillon by Giulio Moriondo, ampelographer at the Institut Agricole Régional in Aosta. Today, only three vineyards have been planted: in Chambave (Mario Bosc), in Aosta (Institut Agricole Régional) and in Saare (Michel Vallet at Feudo di San Maurizio). Vuillermin is rarely made as a varietal wine although Vallet produces one example in the Valle d'Aosta DOC. According to the agricultural census, there was a mere tenth of a hectare in **Italy** in 2000 but there must be more now.

W

WÜRZER

ᛉthis symbol on the following pages indicates that the synonyms, and mistaken identities, are confirmed by DNA evidence

WEISSBURGUNDER

See PINOT BLANC *in the* PINOT *entry.*

WELSCHRIESLING

See GRAŠEVINA.

WÜRZER

Declining, perfumed German cross best in small doses.

BERRY COLOUR: ● ● ● ● ●

PRINCIPAL SYNONYMS: Alzey 10487

ORIGINS AND PARENTAGE

A GEWÜRZTRAMINER × MÜLLER-THURGAU cross bred by Georg Scheu at Alzey in 1932, authorized in Germany in 1978. It was named after the spicy (*würzig*) aromas of the wine.

VITICULTURAL CHARACTERISTICS

Early to mid ripening, good resistance to downy mildew. Sensitive to winter frosts. Moderate sugar levels but acidity is low except in very cool climates.

WHERE IT'S GROWN AND WHAT ITS WINE TASTES LIKE

There are 66 ha (163 acres) in **Germany**, mostly in Rheinhessen and the Nahe, down from a peak of 121 ha (299 acres) in 1995. Wines are generally soft and too headily perfumed not to be blended with other varieties. Producers of varietal wines include Sitzius and Schillingshof in the Nahe and Schmitt's Kinder in Franken. There are also very limited plantings in **Switzerland** (0.33 ha/0.8 acres).

Just 6 ha (15 acres) of Würzer grow in **England** but in such a cool climate it ripens too late to be successful in most sites and it is declining in favour of earlier-ripening varieties such as BACCHUS.

Seifried in Nelson, **New Zealand**, have a few rows and make a varietal wine that is as spicy as its name suggests.

X

XARELLO
XINOMAVRO
XYNISTERI

⚭ this symbol on the following pages indicates that the synonyms, and mistaken identities, are confirmed by DNA evidence

XARELLO

High-quality Catalan with a strong personality making firm wines both still and sparkling.

BERRY COLOUR: ● ● ● ● ●

PRINCIPAL SYNONYMS: Cartoixà, Pansa Blanca, Pansal, Premsal Blanca (Mallorca), Xarel-lo, Xarel·lo, Xerello

VARIETIES COMMONLY MISTAKEN FOR XARELLO: CAYETANA BLANCA, MACABEO (Catalunya), PRENSAL (Islas Baleares)

ORIGINS AND PARENTAGE

Xarello was first mentioned under the name Xerello in 1785 in Sitges in Catalunya (Favà i Agud 2001), its most likely place of origin. According to this author, the name Xarello comes from the Italian *chiarello*, meaning 'claret', a term originally used to refer to wine that was light red in colour. DNA profiling suggests that Xarello might be related to MACABEO (Ibañez *et al.* 2003), with which it is often blended and sometimes even confused.

A pink-berried mutation called **Xarello Rosado**, also known as Pansa Rosada, has been observed near Barcelona (Ibañez *et al.* 2003) and is authorized in the Alella DO.

VITICULTURAL CHARACTERISTICS

Compact bunches of thick-skinned, medium-sized berries. Mid budding and ripening. Adapts to most soils. Susceptible to coulure and therefore productive only as long as weather at flowering is good. Susceptible to powdery and downy mildews. Good resistance to other diseases, notably botrytis bunch rot.

WHERE IT'S GROWN AND WHAT ITS WINE TASTES LIKE

The official variety list in **Spain** has adopted the Xarello spelling, perhaps because not many keyboards have the raised Catalan dot to hand for Xarel·lo, though the hyphenated form Xarel-lo is probably more common.

Xarello is Catalunya's highest-quality light-skinned variety for ageworthy still whites and sparkling Cavas thanks to its combination of power, density, high acidity and relatively high sugar levels. Its main role is in Cava, partnered with PARELLADA and MACABEO, throughout Catalunya – particularly in Penedès – in north-east Spain. In 2008, almost 99% of the Spain's 8,043 ha (19,875 acres) were planted in Catalunya, where it is authorized in most of the region's DOs. It is valued by Cava producers because it contributes body, structure and freshness and increases a wine's longevity. Studies carried out at Barcelona and Washington universities found that Xarello has a very high level of the anti-oxidant resveratrol, which helps to protect it against disease in the vineyard. One of Spain's finest Cavas, Recaredo's single-vineyard Turó d'en Mota, is made from 100% Xarello, and Gramona's fine Celler Batlle is dominated by Xarello.

However, it can also produce powerful still wines, sometimes as part of a blend with varieties such as GARNACHA BLANCA, MACABEO or occasionally CHARDONNAY. As a varietal wine, it can be a little austere rather than overwhelmingly fruity but it often has aromas of citrus and pear as well as a herbal note. Producers of still, varietal wines include Albet i Noya, Arç Blanc, Bolet, Can Ràfols dels Caus, Donzella, El Cep (L'Alzinar), Gramona, Mas Rabassa, Nadal, Parató, Pardal, Parés Baltà, Sabaté i Coca and Torelló.

XINOMAVRO

Top-quality, widely planted but pernickety, high-acid Greek variety.

BERRY COLOUR: ● ● ● ● ●

PRINCIPAL SYNONYMS: Csinomavro, Mavro, Mavro Naoussis, Mavro Naoustino, Mavro Xyno (western Makedonía), Naouses Mavro, Naoustiano, Naoystiro Mavro, Negroska Popolka, Niaousa, Niasoustino, Pipoliko, Pipolka, Popoliko (Véroia in Makedonía), Popolka (Gouménissa and Amýnteo in Makedonía), Xinogaltso, Xynomavro Bolgar, Xynomavro Naoussis, Xynomavro of Náoussa, Xynomavron, Zinomavro, Zynomavro

ORIGINS AND PARENTAGE

Xinomavro, meaning 'acid black' in reference to the grape's acidity, is the dominant red variety in northern Greece. At least three main clones have been distinguished in the Náoussa region, where it most likely originates from (Galet 2000; Manessis 2000; Nikolau and Michos 2004).

Some have suggested a similarity between Xinomavro and PINOT NOIR, or between Ximomavro and NEBBIOLO, but DNA analysis has shown it to be genetically distinct from both.

VITICULTURAL CHARACTERISTICS

Vigorous and very productive. Mid budding and late ripening. Susceptible to downy mildew, botrytis bunch rot and, to a lesser extent, powdery mildew. Best on light, poor sandy soils. Canopy management is critical to achieve full ripeness.

WHERE IT'S GROWN AND WHAT ITS WINE TASTES LIKE

Xinomavro was the second most planted dark-skinned variety in **Greece** in 2008 (2,389 ha/5,903 acres) after AGIORGITIKO and the most widely planted variety in central Makedonía (749 ha/1,851 acres) and in western Makedonía (1,376 ha/3,400 acres). It is best known as the sole variety allowed in the Náoussa appellation but is showing increasingly impressive results in the higher, cooler and windier vineyards further north in Amýnteo, an appellation that also requires 100% Xinomavro. In Gouménissa further east, at least 20% NEGOSKA must be blended in; and in the Rapsáni appellation to the south, near Mount Ólympos in Thessalía, Xinomavro is an equal partner with KRASSATO and STAVROTO.

Wines vary enormously in style but generally share high acidity and excellent potential for ageing. Colour stability can be a problem so that wines may be relatively pale and with time become brick-like in colour, rather like NEBBIOLO. To generalize further is difficult since there is considerable clonal variation and a wide range of winemaking influences such as use of oak during maturation. Tannins may be dry and even angular in youth but time in bottle can soften them and create complexity and elegance. In young wines the aromas are dominated by red fruits such as strawberry and plum but with age these change to more savoury aromas of tomato, olive and dried fruit. There is also a strong trend to blend Ximomavro with SYRAH or MERLOT, to fill out the palate. Recommended producers include Alpha Estate, Argatia, Boutari, Foundis, Katogi & Strofilia, Kir-Yiannis and Tsantali (particularly for Rapsáni).

Several different clones have been identified (ampelographically rather than genetically). Of these, the Velventós clone is better suited to cooler climates, ripens earlier and has softer tannins, whereas the Yiannakochori clone tends to be more powerful and tannic (Lazarakis 2005).

Xinomavro is also responsible for some increasingly good rosé wines, still or sparkling, and at various sweetness levels.

XYNISTERI

Widely planted Cypriot variety capable of making fresh, citrusy whites.

BERRY COLOUR:

PRINCIPAL SYNONYMS: Aspro Kyprou, Cipro Bianco, Hebron Blanc, Hebron White, Hibron Blanc, Koumantaria, Lefko Kyprou, Lefko Kyproy, Topiko Aspro, Xinisteri, Xynistera, Xynisteri Aspro Kyprou

VARIETIES COMMONLY MISTAKEN FOR XYNISTERI: Aspro X, SAVATIANO (Greece)

ORIGINS AND PARENTAGE

Xynisteri is indigenous to the island of Cyprus, where it was first mentioned, spelt Xinisteri, among the better quality wines of the island in 1881 in volume 17 of the Italian *Bolletino consolare pubblicato per cura del Ministero per gli Affari Esteri*, and then several years later by French ampelographer Pierre Mouillefert (1893) as the best and most widespread white variety of Cyprus.

OTHER HYPOTHESES

According to Galet (2000), Xynisteri is identical to Zitania from Israel, which might explain why it is sometimes called Hebron Blanc, but this has not been verified by DNA profiling. Although Aspro, meaning 'white', is a well-known synonym for Xynisteri (Galet 2000), an obscure reference sample called Aspro X in the collection of the ARI research centre at Zygi on Cyprus turned out to be distinct from Xynisteri (Hvarleva, Hadjinicoli *et al.* 2005).

VITICULTURAL CHARACTERISTICS

Vigorous and productive. Mid budding, mid to late ripening. Long bunches of thick-skinned berries. Susceptible to powdery mildew but resistant to botrytis bunch rot.

WHERE IT'S GROWN AND WHAT ITS WINE TASTES LIKE

Xynisteri, the most common white wine grape on **Cyprus** (2,227 ha/5,503 acres in 2010, ie nearly 23% of the total vineyard area) is used to produce both dry white wines and Commandaria, the island's sweet, usually fortified, dried-grape wine. Table wines are generally light and lemony with moderate alcohol levels. Planted at higher altitudes, the variety produces wines that are more vibrant and mineral but the variety is often refreshed and/or filled out by blending with international varieties such as SÉMILLON, CHARDONNAY or SAUVIGNON BLANC, occasionally with a light touch of oak and, in a few instances, with a dollop of MUSCAT OF ALEXANDRIA. Some examples have a slightly grippy texture, attractive in moderation. Recommended producers include Aes Ambelis, Constantinou, Ezousa, Kolios, Kyperounda, Tsiakkas, Vasilikon, Vlassides and Zambartas.

Y

YAMABUDO
YAPINCAK
YEDIVEREN

ꕤthis symbol on the following pages indicates that the synonyms, and mistaken identities, are confirmed by DNA evidence

YAMABUDO

Member of an Asian vine species producing fresh, leafy wines in Japan.

BERRY COLOUR: ● ● ● ● ●

PRINCIPAL SYNONYMS: Crimson Glory Vine, Meoru (Korea)

ORIGINS AND PARENTAGE

Yamabudo, literally 'wild grape' in Japanese, is the common name of *Vitis coignetiae*, an indigenous Asian vine species that French ampelographer Jules-Emile Planchon named in honour of M and Mme Coignet from Lyon, southern France, who brought seeds back to Victor Pulliat, another French ampelographer, after their trip to Japan in 1875.

VITICULTURAL CHARACTERISTICS

Dioecious, early budding, late ripening. Small bunches of small berries. Susceptible to downy mildew and botrytis bunch rot.

WHERE IT'S GROWN AND WHAT ITS WINE TASTES LIKE

Yamabudo produces deeply coloured reds with leafy notes and high acidity. There were 158 ha (390 acres) planted in **Japan** in 2008, mainly in Okayama, Nagano, Yamagata, Iwate and Aomori prefectures, but less than half of this area was for wine. Recommended producers include Gassan Wine, Hirizen Wine and Kuzumaki-Kogen. Also found in **Korea**.

YAPINCAK

Minor Turkish variety grown on the north coast of the Sea of Marmara.

BERRY COLOUR: ● ● ● ● ●

PRINCIPAL SYNONYMS: Erkek Yapıncak, Kınalı Yapıncak, Yapakak, Yapındjac

ORIGINS AND PARENTAGE

Yapıncak probably originates in the province of Tekirdağ on the north side of the Sea of Marmara, north-western Turkey. The synonym Kınalı Yapıncak, meaning 'henna-ed Yapıncak', makes reference to the red-flecked berries, which have given rise to the nickname Yapıncak for girls with freckles, suggesting it was once much more widely planted (Umay Çeviker, personal communication).

VITICULTURAL CHARACTERISTICS

Late budding and ripening. Largish bunches of thin-skinned, freckled berries but good general disease resistance. Well suited to clay-loam soils and dry conditions.

WHERE IT'S GROWN AND WHAT ITS WINE TASTES LIKE

It is grown exclusively in the Marmara region, around the coastal villages of Şarköy and Mürefte in **Turkey**. Wines are typically medium-bodied, low in acidity (which is why it is generally blended), with aromas of golden apple, pear and quince. Melen produce a sweet wine from this variety, whose leaves are also the preferred choice for the production of *yaprak salamura* (brined or pickled vine leaves; Gülcü 2010). Paşaeli have just produced their first vintage of dry, varietal Yapıncak, which is creamy but has respectable acidity and some minerality. It is also used for the table.

YEDIVEREN

Central Turkish vine used as much for the table as for (rather ordinary) wine.

BERRY COLOUR:

VARIETIES COMMONLY MISTAKEN FOR YEDIVEREN: IRI KARA (Central Anatolia in Turkey)

ORIGINS AND PARENTAGE

Yediveren, meaning 'produces seven times' in reference to its high yields, comes from the province of Şanlıurfa, south-eastern Anatolia. According to Karataş *et al.* (2007), Yediveren is genetically closely related to Sergi Karası from the neighbouring province of Gaziantep but no longer cultivated.

VITICULTURAL CHARACTERISTICS

Small berries. Late ripening.

WHERE IT'S GROWN AND WHAT ITS WINE TASTES LIKE

Yediveren, grown in the provinces of Kırşehir and Nevşehir in central **Turkey**, is used mostly for the table, for raisins and to produce *pekmez*, concentrated grape juice, but is also occasionally made into basic table wine.

Z

ZAČINAK
ZAGREI
ZAKYNTHINO
ZALAGYÖNGYE
ZALEMA
ZAMARRICA
ŽAMETOVKA
ZEFÍR
ZELEN
ZENGŐ
ZENIT
ZÉTA
ZEUSZ
ZGHIHARĂ DE HUŞI
ZHEMCHUZHINA OSKHI
ZIERFANDLER
ŽILAVKA
ZIZAK
ŽLAHTINA
ZLATARICA VRGORSKA
ŽUPLJANKA
ZWEIGELT

ᛝ this symbol on the following pages indicates that the synonyms, and mistaken identities, are confirmed by DNA evidence

ZAČINAK

Serbian variety contributing colour to red blends.

BERRY COLOUR: ● ● ● ● ●

PRINCIPAL SYNONYMS: Krajinsko Crno, Negotinsko Crno, Zachinak, Zazinak

ORIGINS AND PARENTAGE

Začinak is an old Balkan variety, probably originating in Serbia, where it shows significant morphological diversity (Avramov and del Zan 2004).

VITICULTURAL CHARACTERISTICS

Early budding, late to very late ripening. Small bunches of small to medium-sized berries. Some resistance to downy mildew and botrytis bunch rot. Best suited to a continental climate but hardy down to –16 °C (3 °F) to –18 °C (0 °F).

WHERE IT'S GROWN AND WHAT ITS WINE TASTES LIKE

Začinak is grown in **Serbia**, and also to a limited extent in the **Republic of Macedonia**, as a useful variety for adding colour to blends. Braća Rajković, in Župa in south-eastern Serbia, for example, blend it with the paler PROKUPAC (known locally as Rskavac) and VRANAC.

ZAGREI

A white-berried hybrid obtained by all of Y N Dokuchaeva, Y P Chebanenko, L F Meleshko, L M Pismennaya, G P Ovchinnikov, M G Bankovskaya, L I Tarakhtiy and A I Grigprishen at the Tairov research centre in Odessa, **Ukraine**, by crossing OVIDIOPOLSKY with Muscat Rozovy, a black colour mutation of MUSCAT BLANC À PETITS GRAINS. It was registered in 2006. This productive, winter-hardy, mid- to late-ripening variety is used to make dry, apple-scented, occasionally floral, citrus-flavoured whites.

ZAKYNTHINO

Ancient Greek island variety being modestly revived.

BERRY COLOUR: ● ● ● ● ●

PRINCIPAL SYNONYMS: Zachara, Zacharo, Zakinthino

ORIGINS AND PARENTAGE

As suggested by its name, Zakynthino may well come from the island of Zákynthos in the Ionian Sea, off the western coast of the Greek Pelopónnisos (Peloponnese).

VITICULTURAL CHARACTERISTICS

Good general disease resistance thanks to loose bunches of thick-skinned berries. High yielding, mid ripening.

WHERE IT'S GROWN AND WHAT ITS WINE TASTES LIKE

Producers in Kefaloniá, **Greece**, who seem to have a virtual monopoly on this variety, confirm the view that it came to their island from Zákynthos but suggest its arrival was a good 500 years ago. It has been neglected until recently, when Metaxa and Gentilini started bottling it as a varietal wine; the former also blend it with ROBOLA and make both dry and off-dry versions. Wines are lightly aromatic, showing mostly citrus and apricot characteristics but also slight floral notes. Acidity is generally moderate to fresh and there seems to be more intensity on the nose than on the palate. There are around 10 ha (25 acres) in the south west of the island and it is also found in the northern Pelopónnisos, where Antonopolous have recently planted a small experimental vineyard.

ZALAGYÖNGYE

High-yielding, widely planted Hungarian hybrid producing generally inferior wines.

BERRY COLOUR:

PRINCIPAL SYNONYMS: Ecs-24, Egri Csillagok 24, Pearl of Zala, Zala Dende, Zhemchug Zala

ORIGINS AND PARENTAGE

Zalagyöngye is an Eger 2 × CSABA GYÖNGYE hybrid obtained in 1957 by József Csizmazia and László Bereznai at the research station for viticulture and oenology in Kölyuktetö, Eger region, Hungary (see BIANCA for the pedigree of Eger 2).

VITICULTURAL CHARACTERISTICS

Vigorous, productive, early ripening. Moderately drought- and frost-tolerant but susceptible to powdery mildew.

WHERE IT'S GROWN AND WHAT ITS WINE TASTES LIKE

Zalagyöngye is widely planted on the Alföld (Great Plain) in southern central **Hungary**. Its high-acid wines are used mainly as a basis for fizz, although it is also grown for the table. There were a substantial 2,243 ha (5,543 acres) in 2008 and producers include Zoltán Farkas, József Locskai, József Nagy, Öreghegy Szőlőbirtok and József Vámosi.

ZALEMA

Southern Spanish variety that is light in every way and used to make fortified wines.

BERRY COLOUR:

PRINCIPAL SYNONYMS: Del Pipajo (Sierra de Francia), Rebazo, Salemo, Zalemo

ORIGINS AND PARENTAGE

Zalema is a rare variety from the province of Huelva in Andalucía, south-west Spain. According to DNA profiling, Zalema could be genetically close to Blanco Castellano (ALCAÑÓN) from the province of Castellón in the autonomous community of Valencia, on the opposite side of the country, where Zalema is also grown to a very limited extent (Ibañez *et al.* 2003).

Zalema is sometimes confusingly known under the synonyms Perruna (also a synonym for PERRUNO) or Torrontés de Montilla (see TORRONTÉS).

VITICULTURAL CHARACTERISTICS

Big and very compact bunches of medium-sized berries. Mid budding, late ripening. Fertile, vigorous and productive. Highly resistant to drought and adapts well to poor soils.

WHERE IT'S GROWN AND WHAT ITS WINE TASTES LIKE

Wines are light and pretty neutral with a slight note of almonds and the unfortunate tendency to oxidize. They are used mainly in the production of fortified wines in the Condado de Huelva DO in Andalucía in southern **Spain** and to a lesser extent in Montilla-Moriles. There were 5,055 ha (12,491 acres) in 2008 but it is gradually being replaced by the higher-quality PALOMINO FINO. Bodegas Iglesias make several varietal examples in Condado de Huelva.

ZAMARRICA

Virtually extinct Monterrei vine contributing aroma and freshness to one blend so far.

BERRY COLOUR: ● ● ● ● ●

ORIGINS AND PARENTAGE

Zamarrica comes from the mountain vineyards of the Monterrei region in Galicia, north-west Spain.

VITICULTURAL CHARACTERISTICS

Vigorous, low yielding, mid budding and ripening. Small bunches. Highly susceptible to powdery mildew.

WHERE IT'S GROWN AND WHAT ITS WINE TASTES LIKE

Apart from one mixed, centenarian vineyard near a place called A Trabe south of Coruña in north-west **Spain**, Zamarrica was virtually extinct until José Luis Mateo of Quinta da Muradella planted 2,300 vines in 2007. The fruit of these old vines, including Zamarrica, Bastardo (TROUSSEAU), Doña Blanca (SÍRIA) and others, is used by Raul Perez and Mateo to produce A Trabe. Zamarrica produces a wine with intense orange and balsamic, sometimes eucalyptus, aromas, high acidity, soft tannins and moderate alcohol.

ŽAMETOVKA

Ancient dark-skinned Slovenian variety traditionally goes into the tart local Cviček but also has a claim as oldest vine in the world.

BERRY COLOUR:

PRINCIPAL SYNONYMS: ?Bettlertraube (Austria), ?Blauer Hainer (Austria), ?Blauer Kölner (Austria), Kapcina, Kavčina Crna, ?Kölner Blau (Austria), Modra Kavčina, Plava Velica, Žametasta Črnina, Žametna Črnina

ORIGINS AND PARENTAGE

Žametovka is an old and obscure variety that possibly originates in the Dolenjska region in south-east Slovenia, where it used to be widely cultivated. The synonym Žametna Črnina, literally 'black velvet', could be a translation of Samtrot, the German name for PINOT MEUNIER, which is a quite distinct variety. In Austria, Žametovka is supposedly called Blauer Hainer, Blauer Kölner or Kölner Blau, or even Frankenthaler, a synonym for SCHIAVA GROSSA, but this last suggestion can be rejected by DNA profiling because the DNA profile of the old vine from Maribor (see below) in Vršič *et al.* (2011) does not match any other known local or foreign varieties (Nataša Štajner and José Vouillamoz, unpublished data).

VITICULTURAL CHARACTERISTICS

Late ripening. Largish, compact bunches. Good disease resistance but prone to spring-frost damage.

WHERE IT'S GROWN AND WHAT ITS WINE TASTES LIKE

Žametovka is planted mainly in the Dolenjska wine region in Posavje, south-eastern **Slovenia**, close to the Croatian border, where it is often blended with the white variety KRALJEVINA (and/or BLAUFRÄNKISCH) to make Cviček, a traditional, low-alcohol, pale-red, high-acid wine that is locally very popular and has its own appellation within Dolenjska. In Bela Krajina, it is part of another, more deeply coloured and powerful, blended appellation wine known as Metliška Črnina. Total plantings in Slovenia in 2009 amounted to 945 ha (2,335 acres), with only 73 ha (180 acres) further north in Podravje.

Here, however, the variety has a more striking claim to fame, in the town of Maribor, where a huge 400-year-old Žametovka vine growing in front of what is now a museum on the bank of the River Drava is said by many to be the oldest living vine in the world, notwithstanding contenders to this title in Italy and Georgia, for example. Nicknamed Stara Trta, Slovenian for 'old grapevine', its status was recognized in the 2004 *Guinness book of records*. The building that houses the museum, at 8 Vojašniška Street, appears with the vine in visual records dating back to 1657 and 1681 (Vršič *et al.* 2011). The vine yields 35–55 kg (77–121 lbs) of grapes per year, giving around a hundred 25 cl bottles of wine, a true collector's item.

ZEFÍR

Recent Hungarian cross of uncertain parentage yet to establish a track record.

BERRY COLOUR: ● ● ● ● ●

PRINCIPAL SYNONYMS: Badacsony 2, Zephyr

ORIGINS AND PARENTAGE

Zefír was initially described as a HÁRSLEVELŰ × LEÁNYKA cross created in 1951 by Ferenc Király in Pécs, Hungary, but recent DNA parentage analysis has disproved this pedigree (Gizella Jahnke, personal communication). Its true parentage is still unknown.

The Vitis International Variety Catalogue defines it as a HÁRSLEVELŰ × FETEASCĂ ALBĂ cross but this is because Feteascã Albã has often been confused with LEÁNYKA.

VITICULTURAL CHARACTERISTICS

Early ripening.

WHERE IT'S GROWN AND WHAT ITS WINE TASTES LIKE

Wines are light and mildly aromatic (floral or blackcurrant), although a 'typical' style is yet to be established. Vaszary in Pannonhalma in north-west Hungary make a varietal wine. There were 71 ha (175 acres) in **Hungary** in 2008, mostly in the Etyek-Buda and Eger wine regions in the north of the country. It is an optional component in Egri Csillag, a white blended wine devised by a group of winemakers in Eger as a pale partner to Egri Bikavér, otherwise known as Bull's Blood.

ZELEN

Old, rare Slovenian variety, of uncertain identity, producing crisp, dry whites.

BERRY COLOUR: ● ● ● ● ●

ORIGINS AND PARENTAGE

Zelen is considered to be an old and indigenous variety from Vipavska Dolina in the Primorska region in south-west Slovenia but recent DNA profiling has shown that Zelen is a heterogeneous group consisting of three or more distinct varieties, each with a unique profile (Štajner *et al.* 2008). The 'true' Zelen, if there is one, remains to be identified, possibly by comparing modern vines with old ampelographic descriptions. The name Zelen, meaning 'green' in Slovenia, referred to the emerald colour of the unfiltered must, but modern filtered wines tend to be golden yellow (Koruza and Lavrencic 2004).

OTHER HYPOTHESES

It is said to be similar to VERDUZZO FRIULANO (Koruza and Lavrencic 2004) but this has not been confirmed by genetic analysis.

VITICULTURAL CHARACTERISTICS

Mid budding, mid to late ripening. Small berries.

WHERE IT'S GROWN AND WHAT ITS WINE TASTES LIKE

Wines are light, crisp and grassy and produced exclusively in Vipavska Dolina in south-west **Slovenia**, where there were 64 ha (158 acres) in 2009. Recommended producers include Burja, Guerila and Vipava 1894.

ZELENAC SLATKI

See ROTGIPFLER.

ZENGŐ

Hungarian cross making full-bodied, high-quality whites if handled carefully.

BERRY COLOUR: ● ● ● ● ●

PRINCIPAL SYNONYMS: Badachon 8, Badacsony 8

ORIGINS AND PARENTAGE

An EZERJÓ × BOUVIER cross produced by Ferenc Király in 1951 in Pécs, south-west Hungary, making it a sibling of ZENIT and ZEUSZ. It is probably named after the highest peak in the nearby Mecsek mountains.

VITICULTURAL CHARACTERISTICS

Early budding, early to mid ripening. Medium-sized compact bunches of small, thin-skinned berries. Moderately winter-hardy and susceptible to botrytis bunch rot. Best suited to volcanic soils.

WHERE IT'S GROWN AND WHAT ITS WINE TASTES LIKE

Although Zengő is rarely made as a varietal wine, it does produce full-bodied wines that mellow with age, particularly when matured in cask (Rohály *et al.* 2003). According to Tamás Pók, a leading wine grower in Eger, **Hungary**, if Zengő is taken seriously, it can make high-quality wines with quince and pepper flavours, becoming more tropical when harvested later. However, if left on the vine, the skins have a high phenolic content, which must be carefully handled during maceration. There were 317 ha (783 acres) in 2008, mainly in Eger and Mátra but also in Bükk, Kunság and Tokaj. József Szentesi just south west of Budapest makes a rare but good varietal example.

ZENIT

Usefully early-ripening Hungarian cross with finely balanced acidity.

BERRY COLOUR: ● ● ● ● ●

ORIGINS AND PARENTAGE

An EZERJÓ × BOUVIER cross produced by Ferenc Király in 1951 in Pécs, south-west Hungary, making it a sibling of ZENGŐ and ZEUSZ. Zenit was officially registered in 1980.

VITICULTURAL CHARACTERISTICS

Early budding, early to mid ripening, medium to low yields. Medium-sized bunches of small berries. Moderately sensitive to cold winter temperatures and some resistance to botrytis bunch rot, downy and powdery mildews. Well suited to volcanic soils.

WHERE IT'S GROWN AND WHAT ITS WINE TASTES LIKE

Wines are crisp with flavours of apple and citrus but not overly aromatic; the best examples are potentially long-lived and subtle with a mineral edge. Although Zenit is widely dispersed around **Hungary**, particularly in higher sites in Mátra, Eger and around Lake Balaton, the total area in 2008 was just 560 ha (1,384 acres). Its success and popularity, particularly in the Eger region, are due in part to its early ripening, an advantage in less than ideal vintages (Varga *et al.* 1998), but also to the excellent balance between fine acidity and high sugar levels, as well as a propensity towards noble rot. Recommended producers of varietal wines include Mihály Figula near Lake Balaton, who has recently planted another 2 ha (5 acres), Vincellér Ház in Sopron and Tóth Sándor in Balatonfelvidék.

ZÉTA

Renamed daughter of Furmint used mainly in sweet Tokaji blends.

BERRY COLOUR: ● ● ● ● ●

PRINCIPAL SYNONYMS: Oremus

ORIGINS AND PARENTAGE

A FURMINT × BOUVIER cross produced by Ferenc Király in 1951 in Pécs, Hungary, making it a half-sibling of ZENGŐ, ZENIT and ZEUSZ. The variety was renamed in 1999 to avoid confusion with the Oremus vineyard in Tokaj. Zéta is a character from a novel by the well-known Hungarian novelist Géza Gárdonyi'.

VITICULTURAL CHARACTERISTICS

Mid ripening, susceptible to botrytis bunch rot and moderately sensitive to winter cold. Small, compact bunches of lightly speckled berries.

WHERE IT'S GROWN AND WHAT ITS WINE TASTES LIKE

Zéta is becoming increasingly significant in **Hungary**'s Tokaj wine region, where it is one of the four recommended ingredients in Tokaji wines – along with FURMINT, HÁRSLEVELŰ and Sárga Muskotály (MUSCAT BLANC À PETITS GRAINS) – and where all of Hungary's 114 ha

(282 acres) are planted. Known as Oremus until 1999, it ripens earlier than both Furmint and Hárslevelű, has good levels of acidity and sugar and is well suited to the production of sweet botrytized wines, hence its use in the blend of the region's Azsú specialities. Varietal wines are rare but both Balassa and Kikelet make good examples.

ZEUSZ

Hungarian cross making full-bodied but well-balanced late-harvest whites.

BERRY COLOUR:

PRINCIPAL SYNONYMS: Badacsony 10

ORIGINS AND PARENTAGE

An EZERJÓ × BOUVIER cross produced by Ferenc Király in 1951 in Pécs, south-west Hungary, making it a sibling of ZENGŐ and ZENIT. It was named in honour of the Greek god Zeus (Hungarian grape breeders seem to like Z).

VITICULTURAL CHARACTERISTICS

Late budding and ripening. Medium-sized bunches of small berries. Moderately susceptible to botrytis bunch rot and some tolerance of cold winter temperatures.

WHERE IT'S GROWN AND WHAT ITS WINE TASTES LIKE

Wines, generally made from late-harvested grapes, are full-bodied but with fine fresh acidity, tropical-fruit flavours and almost always sweet. There were just 14 ha (35 acres) in **Hungary** in 2008, almost all in the west of the country, in Badacsony and further north in Somló. Recommended producers include Szeremley, a particular champion of this variety, and Borbély, both on the north shore of Lake Balaton.

ZGHIHARĂ DE HUȘI

Old Romanian variety producing crisp, easy-drinking whites.

BERRY COLOUR:

PRINCIPAL SYNONYMS: Ghihară, Poamă Zosănească, Sghigardă Galbenă, Zghihară, Zghihară Galbenă, Zghihară Verde Bătută

ORIGINS AND PARENTAGE

As suggested by its name, Zghihară de Huși probably comes from the region of Huși in Romanian Moldova, where it was known long before phylloxera hit the country at the end of the nineteenth century and where it still shows a high level of morphological diversity (Galet 2000). It has often been considered to be a selection of GALBENĂ DE ODOBEȘTI (Dejeu 2004) but this hypothesis has not yet been tested by DNA profiling.

VITICULTURAL CHARACTERISTICS

Vigorous, early to mid budding, mid to late ripening. Medium-sized, thin-skinned berries. Susceptible to downy mildew and botrytis bunch rot. Resistant to downy mildew and drought and some resistance to cold winter temperatures.

WHERE IT'S GROWN AND WHAT ITS WINE TASTES LIKE

Wines are very fresh, relatively neutral, with light notes of apple or pear, sometimes lightly acacia-scented and designed for early drinking, although they can be quite powerful. Senator and Casa de Vinuri Huşi are two of the very few producers of varietal wines. There were 95 ha (235 acres) in 2008, all in the Huşi region in the far east of **Romania**, close to the border with the Republic of Moldova, but some of this is for home winemaking. It is also used to make brandy.

ZHEMCHUZHINA OSKHI

A (ÇAVUŞ × Mathiasz Janos) × CARDINAL white-berried cross made in **Russia**, in which Mathias Janos is a MUSCAT OF ALEXANDRIA × Chasselas Rouge (a mutation of CHASSELAS) cross. It is used by the im. Lenina co-op in Anapa west of Krasnodar, for example, to produce a varietal wine.

ZIBIBBO

See MUSCAT OF ALEXANDRIA.

ZIERFANDLER

Thermenregion speciality making powerful, aromatic and highly distinctive wines, typically in partnership with Rotgipfler.

BERRY COLOUR:

PRINCIPAL SYNONYMS: Cirfandli (Hungary), Gumpoldskirchner (Austria), Roter Zierfandler, Rubiner, Spätrot, Zierfandler Rot

ORIGINS AND PARENTAGE

DNA parentage analysis has suggested that Zierfandler could be a natural cross between ROTER VELTLINER and a relative of SAVAGNIN that took place near Gumpoldskirchen in Niederösterreich (Lower Austria; Regner 2000a; see PINOT pedigree diagram). Its supposed relationship with SILVANER, which was sometimes called Grüner Zierfandler or Zierfandl, is not confirmed by DNA analysis (Vouillamoz).

VITICULTURAL CHARACTERISTICS

Big, compact bunches of small berries with a greenish-pink cast. Susceptible to botrytis bunch rot and downy mildew. Irregular yields. Late budding, mid to late ripening.

WHERE IT'S GROWN AND WHAT ITS WINE TASTES LIKE

Wines are powerful and high in extract with flavours of orange, peach, honey and spice and a golden, sometimes lightly oily, aroma. The vast majority of **Austria**'s 102 ha (252 acres) are in the Thermenregion, particularly around the town of Gumpoldskirchen, where it is generally blended with the other local variety, ROTGIPFLER, though there are a few hectares in Wagram. Unblended, it has enough acidity to make long-lasting sweet wines – unlike its blending partner. Recommended producers include Leo Aumann, Biegler, Johanneshof Reinisch, Stadlmann and Harald Zierer.

There were 18 ha (44 acres) of Cirfandli in **Hungary** in 2008, almost all in the Pécs wine region in the south of the country, where it was first planted in the mid 1800s. Recommended producers of this full-bodied wine include Andreas Ebner, Lisicza, Matias and Radó.

ŽILAVKA

Herzegovina's most famous variety, producing full-bodied but nutty, well-balanced, high-quality whites.

BERRY COLOUR: ● ● ● ● ●

PRINCIPAL SYNONYMS: Jilavka, Mostarska, Mostarska Zilavka

VARIETIES COMMONLY MISTAKEN FOR ŽILAVKA: FURMINT

ORIGINS AND PARENTAGE

Žilavka most likely originates in the Mostar region in southern Bosnia and Herzegovina, as suggested by some of its synonyms. Its name is said to refer to the fine veins in the berry, visible through the skin in the ripe berry. DNA profiling has suggested a genetic relationship with Teran Bijeli (PROSECCO) in Istra (Maletić *et al.* 1999).

VITICULTURAL CHARACTERISTICS

Fertile, vigorous, late ripening. Large, very compact bunches; berry size varies according to clone. Good drought tolerance but susceptible to downy mildew, botrytis bunch rot and spring frosts.

WHERE IT'S GROWN AND WHAT ITS WINE TASTES LIKE

Žilavka is the main indigenous white wine variety in **Bosnia and Herzegovina** and is found predominantly in the Mostar region in the south, often blended with up to 15% KRKOŠIJA and BENA. Its wines are typically full-bodied, rich in extract with fresh acidity and a slightly nutty flavour, ageing well in bottle, although many are made in a more fruity, easy style, sometimes off dry. Most wines are unoaked but a few producers such as Hepok make both oaked and unoaked versions.

Žilavka is also found in **Croatia**, **Serbia** and in the **Republic of Macedonia**, where it gives its best results in the warmer sites. Macedonian producers include Popova Kula.

ZIMSKO BELO

See OPSIMO EDESSIS.

ZINFANDEL

See TRIBIDRAG.

ZIZAK

This old and almost forgotten black-berried **Montenegrin** variety, producing full-bodied wines, is being revived by Plantaže, who use it in blends but are planning to make a varietal wine. It should not be confused with SYSAK from Azerbaijan, which has functionally female flowers; its DNA profile is quite distinct from that of SYRAH (Cipriani *et al.* 2010), for which Zizak is sometimes used as synonym.

ŽLAHTINA

High-yielding, minor Croatian variety currently being re-evaluated and benefiting from an enthusiastic local market.

BERRY COLOUR: ● ● ● ● ●

PRINCIPAL SYNONYMS: Vrbnička Žlahtina☒, Žlahtina Bijela, Žlahtina Toljani , Žlajtina

VARIETIES COMMONLY MISTAKEN FOR ŽLAHTINA: CHASSELAS☒ (under the name Rdeča Žlahtina)

ORIGINS AND PARENTAGE

Žlahtina probably originates from the Croatian island of Krk in northern Dalmacija (Dalmatia). The name derives from the Croatian word *žlahtno*, meaning 'noble'.

The DNA profile of Rdeča Žlahtina (*rdeča* means 'pink') in Štajner *et al.* (2008) is identical to CHASSELAS (Vouillamoz), which might explain why the true Žlahtina is sometimes considered identical to Chasselas (as it is in the Vitis International Variety Catalogue).

VITICULTURAL CHARACTERISTICS

Vigorous and moderate to high yielding. Late budding, mid to late ripening. Susceptible to diseases. Large bunches of relatively large berries

WHERE IT'S GROWN AND WHAT ITS WINE TASTES LIKE

Authorized in the Hrvatsko Primorje (Croatian Littoral) and Istra (Istria) wine regions, Žlahtina is a popular variety grown mainly on the island of Krk, off the coast of northern Dalmacija in **Croatia** just south of Rijeka, specifically around the town of Vrbnik in the east of the island, hence the synonym Vrbnička Žlahtina. Along with several other autochthonous varieties, Žlahtina is part of a revitalization project involving clonal selection and replanting. Recommended producers include Anton Katunar, Šipun, Toljanić and the Vrbnik co-operative. Wines are typically dry, refreshing, medium-bodied with floral aromas and flavours of white-fleshed fruit, and the variety is occasionally used to make sparkling wine.

ZLATARICA VRGORSKA

Very rare Croatian variety.

BERRY COLOUR:

PRINCIPAL SYNONYMS: Bila loza (Vis), Dračkinja (Kotor), Plavka (Split), Zlatarica (Dubrovnik, Metković and Korčula), Zlatarica Bijela

VARIETIES COMMONLY MISTAKEN FOR ZLATARICA VRGORSKA: Zlatarica Blatska Bijeli

ORIGINS AND PARENTAGE

Zlatarica Vrgorska is a Croatian variety possibly originating from southern Dalmacija (Dalmatia) – the town of Vrgorac, whence Vrgorska, is between Split and Dubrovnik – where it has traditionally been grown. Zlatarica means 'golden'.

It should not be confused with Zlatarica Blatska Bijeli, which has a completely distinct DNA profile (Edi Maletić, personal communication) and is a parent of POŠIP BIJELI (Piljac *et al.* 2002) but is no longer cultivated.

VITICULTURAL CHARACTERISTICS

Vigorous, productive, early to mid budding and ripening. Thick-skinned berries. Susceptible to powdery mildew.

WHERE IT'S GROWN

Although Zlatarica Vrgorska is rare, it is still found in **Croatia**'s Dalmatian hinterland around Vrgorska, where it is cultivated by producers such as Vjekoslav Opačak and Ante Franič.

ŽUPLJANKA

Recent and promising Serbian light-skinned cross from dark-skinned parents.

BERRY COLOUR:

ORIGINS AND PARENTAGE

Župljanka, meaning 'parishioner', is a PROKUPAC × PINOT NOIR cross obtained by Dragoslav Milisavljević at the viticulture research centre in Sremski Karlovci, part of Novi Sad University in Vojvodina, Serbia. It was released in 1970 (Cindrić *et al.* 2000).

Župljanka is also a synonym for CHASSELAS but the two should not be confused.

VITICULTURAL CHARACTERISTICS

Regularly productive, late ripening. Good resistance to botrytis bunch rot but sensitive to cold winter temperatures and susceptible to powdery and downy mildews. Best suited to continental climates.

WHERE IT'S GROWN AND WHAT ITS WINE TASTES LIKE

Found mainly in **Serbia**'s Fruška Gora region within Vojvodina, the variety produces wines that are well balanced and refreshing. The juice is unusually high in malic acid, which can cause problems if malolactic fermentation is not prevented. Producers of varietal wines include Antonijević Winery, Dulka Winery, Stevan Novakovic and Vinu Lódi. It is sometimes used as the base wine in Bermet, Fruška Gora's traditional herb- and spice-infused dessert wine. It is also found in the **Republic of Macedonia**, where it has spread rapidly in many vineyard areas.

ZWEIGELT

Most common red wine grape in Austria producing firm, full-bodied wines if yields are controlled.

BERRY COLOUR: ● ● ● ● ●

PRINCIPAL SYNONYMS: Blauer Zweigelt (Austria), Rotburger (Austria), Zweigeltrebe⧖ (Czech Republic, Slovakia and Japan)

ORIGINS AND PARENTAGE

A BLAUFRÄNKISCH × Saint-Laurent (ie SANKT LAURENT) cross obtained in 1922 by Fritz Zweigelt at the Klosterneuburg research centre in Austria. Initially called Rotburger by Zweigelt himself, it was later named Zweigelt in honour of the breeder. Given the pedigrees of its parents, Zweigelt turns out to be a grandchild of GOUAIS BLANC and PINOT, two of the most influential varieties in the genetics of Western European grape varieties.

Zweigelt was used to breed CABERNET MORAVIA in the Czech Republic and ROESLER in Austria.

VITICULTURAL CHARACTERISTICS

Early budding, mid ripening (earlier than BLAUFRÄNKISCH). Susceptible to powdery mildew. Yields are high so careful canopy management and bunch thinning are needed to produce high-quality wine. Particularly susceptible to berry shrivel/grape wilting (*Traubenwelke*).

WHERE IT'S GROWN AND WHAT ITS WINE TASTES LIKE

Although there are experimental plantings of around 100 ha (247 acres) in **Germany** and it is planted on a much more minor scale in the **UK**, Zweigelt is most famous in **Austria**. It is the country's most widely planted dark-skinned variety: its 6,511 ha (16,089 acres) of vines are almost double the figure for BLAUFRÄNKISCH and the total has increased by nearly 50% since 2000. There are major plantings in Niederösterreich (Lower Austria), particularly in the Weinviertel north of Wien (Vienna), and in Burgenland, especially in Neusiedlersee. However, regions best known for their white wines such as Kamptal can also produce very good examples, as Schloss Gobelsburg have shown.

Wines are red-violet when young and show particularly exuberant fruit in the purple berry spectrum. As with Austria's signature white wine variety, GRÜNER VELTLINER, styles can vary enormously, largely dependent on yield, but simple, unoaked Zweigelt has its charms. If yields are kept low Zweigelt can be full-bodied and ageworthy, tasting of sour or black cherries and responding well to barrel ageing. It is also favoured in red blends for its rich fruit. Best producers include K + K Kirnbauer, Franz Leth, Claus Preisinger, Johanneshof Reinisch and Umathum.

Just over the border from Austria's red-wine heartland in Burgenland, in the Sopron wine region of **Hungary**, Ráspi makes exceptional Zweigelt but it is not a common Hungarian variety. It is also found in the **Czech Republic** (860 ha/2,125 acres mainly in the south) and **Slovakia**, where it is known as Zweigeltrebe.

Rather more unexpectedly, Zweigelt is planted on a minor scale in British Columbia, **Canada**, and in **Japan** (231 ha/571 acres, almost all in Hokkaido).

GLOSSARY

ANTHRACNOSE fungal disease of European origin also known as black spot or bird's eye rot, common in humid regions such as the eastern United States. Caused by the fungus *Elsinoe ampelina*, it can be controlled by the application of Bordeaux mixture (lime, copper sulphate and water).

BACTERIAL NECROSIS, or **BACTERIAL BLIGHT**, bacterial disease spread by rain and infected pruning tools that kills young shoots and causes loss of vigour. Its incidence is reduced by good hygiene at pruning.

BLACK ROT fungal disease caused by *Guignardia bidwelli* that attacks young shoots, leaves and berries during mild, wet weather conditions, causing heavy crop losses. Native to North America, it is now economically significant in the north-eastern US, Canada, and parts of Europe and South America. Fungicides are the most common means of control.

BLOOM the powdery white protective layer on a grape's skin, a combination of waxes and cutin, thicker and more obvious on some varieties than on others.

BOTRYTIS BUNCH ROT, in its malevolent form also known as **grey rot**, is a fungal disease caused by *Botrytis cinerea* that occurs most often in humid conditions and as grapes reach maturity, especially if berries have been damaged – by birds, for example. It results in reduced yields, off flavours and, in red wines, loss of colour. Known in its benevolent form as **noble rot**, under the right conditions and with certain grape varieties, it is also responsible for some of the greatest sweet wines in the world. Generally controlled by fungicides.

BUD MUTATION a mutation or genetic change that occurs in the bud or buds of a single cane and does not spread to the rest of the vine, resulting in one cane of a dark-skinned variety bearing light-skinned berries, for example; also known as a chimera or sport.

BUNCHSTEM NECROSIS *see* desiccation of the stems

BUSHVINE as the name suggests, a vine trained like a bush, ie free-standing, with a short trunk and without a trellis system, although they do sometimes have a central supporting stake. Also known as a gobelet-trained vine. Many of the world's oldest vines are trained in this way and it is more common in certain parts of the world and for certain varieties, eg Garnacha in southern Europe.

CANOPY the part of the vine that is above the ground.

CARBONIC MACERATION fermentation method occurring under anaerobic conditions in a closed vessel whereby a proportion of the sugar and other compounds in the uncrushed berries of whole bunches is converted to ethanol without the action of yeasts. Generally used for light, fruity wines such as Beaujolais or for tough varieties such as Mazuelo, the process is often completed by the more common, yeast-induced fermentation.

CHANCE SEEDLING a vine that grows from a seed that has fallen to the ground by accident, resulting in a **selfling** or, less commonly but more significantly if cross-pollination has occurred, a new variety.

CHAPTALIZATION adding sugar or concentrated grape must or rectified concentrated grape must to grape juice before or during fermentation to increase the final alcohol level in a wine, more common in cooler climates.

CHLOROSIS vine disorder most often caused by iron-deficient limestone-rich soils resulting in lack of chlorophyll and therefore yellow leaves, although it can also be the result of vine diseases or a lack of other nutrients.

CLONAL SELECTION involves selecting a highly performing vine for some particular characteristic (generally high yield or wine quality), taking cuttings from it and propagating them for replanting. It is the primary method of improving a vine variety and a way of reducing the incidence of virus diseases. **Mass selection**, on the other hand, takes cuttings from several plants and propagates them to retain diversity within a variety.

CLONE a single vine or a population of vines derived by taking cuttings from a single mother vine and showing the same measurable characteristics.

COULURE a French name used widely to describe a type of poor fruit set in which the very tiny berries fall to the ground soon after flowering, reducing yields. This is caused by stems shrivelling because of carbohydrate imbalance in the vine as a result of soil or climate conditions – cool, cloudy or wet weather in particular. Certain varieties are more susceptible than others.

CROSS a vine that has been bred from two other vine varieties of the same species, eg *Vitis vinifera*. Rieslaner, for example, is a cross between Silvaner and Riesling, both *vinifera* varieties. The term **natural cross** is used in this book to refer to a cross that happened spontaneously in the vineyard rather than as a result of deliberate pollination by a human hand. And a multiplication sign (×) is used to show that two varieties have been crossed, eg Nobling is a Silvaner × Chasselas cross.

CRYOEXTRACTION is the French term for the process that imitates the natural conditions required to make icewine. Grapes are held at sub-zero temperatures so that the water in the berries freezes. When the fruit is pressed, the ripest and therefore least frozen berries yield sugar-rich juice.

DESICCATION OF THE STEMS, also known as bunchstem necrosis, *dessèchement de la rafle* in French and waterberry in the US, is a physiological disorder of unknown origin that results in stems drying and shrivelling during ripening so that berries become watery, fail to ripen and shrivel on the vine before they are ripe. High vigour, high levels of nitrogen in the soil and poor weather at flowering are known to be contributory factors. Cabernet Sauvignon is particularly prone to this problem.

DNA PROFILING, also known as DNA fingerprinting or typing, is a technique that uses DNA markers to identify a vine and its parentage (if both parents are still in existence) from any part of the plant except for the embryo inside the seed, which is already the progeny of that plant. First applied to vines in 1993, simple sequence repeats (SSRs or microsatellites) are still the markers of choice, though single nucleotide polymorphism (SNP) might be favoured in the future.

DOWNY MILDEW, also known as peronospora, is a fungal disease caused by *Plasmopara viticola* that is encouraged by warm, humid summers. Introduced from the US to Europe in the late 1870s, it caused havoc when growers were still reeling from the effect of phylloxera (see below) and is now widespread throughout the wine-growing regions of the world. White cottony patches appear on the underside of the leaves and reduce photosynthesis, often causing the leaves to fall and inhibiting grape ripening, in turn producing thin, poor-quality wines. Copper sprays are a common but short-term form of protection.

ESCA a complex of fungal diseases that cause white rot and sudden death in old vines and may inhibit the establishment of new vineyards if the young plants are infected before planting. It is an increasingly serious problem around the world, not least because a cure has yet to be identified.

EUTYPA DIEBACK, also known as dead/dying arm, or *eutypiose* in French, is a fungal disease caused by *Eutypa lata* that rots the vine wood and is especially common in Mediterranean climates around the world. It attacks the vine via pruning wounds, which may be painted with fungicide, and has a greater effect on yield than on quality, as well as shortening the life of the vine.

FERAL describes a cultivated grape vine that has returned to a wild environment.

FERTILITY refers to the fruitfulness of a vine's buds and shoots, though it may also refer to a type of soil rich in nutrients that encourages vigorous growth.

FIELD BLEND a term used to describe a vineyard – usually an old one – that is populated by several different varieties, usually interplanted but occasionally separated in rows or blocks. It is now very uncommon to plant vines in this way but in the past it was a traditional practice that helped insure against wholesale loss of crop or poor ripening due to disease or climatic conditions since different varieties ripen at different times and have different susceptibilities. Such vineyards are highly prized by those who believe that such a blend results in more complex wines that are an eloquent expression of a site rather than of a grape variety.

FLOR yeasts are those that form a film on the surface of a wine after they have fermented all the available sugar in the must to alcohol, typically associated with Jerez and the production in large wooden butts of naturally pale sherry styles such as fino and manzanilla but also important in the Jura. The yeast, nourished by oxygen and alcohol, produces acetaldehyde that gives sherry its classic aroma and protects it from oxygen. Also known as film-forming yeasts.

FOXY, a term often used to describe the distinctively sweet aroma and taste – an odd mix of animal fur and wild strawberries – of wines made from American vines, particularly *Vitis labrusca*, and some American hybrids (see pp XIII–XIV for more on the vine family).

FRUIT SET occurs over a period of about a week immediately after flowering when the fertilized flowers develop into buds and the future yield of the vine is determined. On average, only about 30% of fertilized flowers develop into buds and fruit set is inhibited both by hot, dry conditions that lead to water stress and by cold, cloudy or rainy weather. *See also* coulure (above) and millerandage (below).

GRAPE MOTH, *vers de la grappe* in French because it damages the vine at the larval (*vers*) stage, is used to refer to several different types of flying insect. In France the most common ones are pyrale (*Sparganothis pilleriana*), cochylis (*Eupoecilia ambiguella*), eudemis – or European grapevine moth in the US (*Lobesia botrana*) – and eulia (*Argyrotaenia pulchellana*), the first damaging the leaves and berries, the last three breeding within the bunch and often causing botrytis bunch rot. Other types include the grape berry moth (active in the US but not in California), the orange tortrix moth and light brown apple moth (California) and the grapevine moth (Australia and New Zealand). They can be controlled by insecticides, pheromone traps or natural predators.

HYBRID a cross between two different vine species, eg *Vitis vinifera* × *Vitis labrusca*, or, more commonly, between a *Vitis vinifera* variety and another hybrid. Reberger, for example, is a cross between the hybrid Regent and the *vinifera* variety Blaufränkisch.

HYMENOPTERA an order of insects that includes wasps and bees.

INTERPLANTED *see* field blend

INTERSPECIFIC CROSS referred to in this book as a **hybrid** (see above).

INTRASPECIFIC CROSS referred to in this book simply as a **cross** (see above).

ISOENZYMATIC PROFILING is a type of genetic profiling (*see* DNA profiling) that uses isoenzymes (or isozymes) as markers.

LAYERING, *marcottage* in French, is an ancient method used to propagate vines, particularly useful for filling in gaps in the vineyard. A long cane from an existing vine is buried in the ground for it to take root, leaving the tip above the soil to form the new plant, which may later be separated from the parent.

MASS SELECTION *see* clonal selection

MILLERANDAGE, also known as 'hen and chicken', describes an abnormal fruit set (see above) resulting in different-sized berries – containing varying numbers of seeds – within the same bunch. The smallest grapes are seedless. This phenomenon is due to poor pollination, in turn the result of poor weather at flowering, lack of boron in the soil, or a virus disease known as fanleaf degeneration. The consequences vary from severe reduction in yield to what some winemakers see as an increase in quality thanks to a proportion of smaller berries. Certain varieties, or certain clones of a particular variety (eg the Mendoza clone of Chardonnay), are more susceptible than others.

MISTELLE the French name for a sweet, alcoholic drink made by mixing unfermented grape must with alcohol. Also known as *mistela* (Spanish and Portuguese) or *mistella* (Italian). The alcohol prevents the must from fermenting.

MITES, *acariens* in French, is a general term that refers to various different species of minute insects that generally feed on the surface of the leaves, thereby inhibiting photosynthesis and grape ripening, although some particularly damaging ones feed on all the green parts of the vine, causing greater destruction. Some types are controlled by sulphur sprays, others are best controlled by predators or miticides.

MORPHOLOGICAL DIVERSITY minor variation in form and structure among vines of the same variety. Greater morphological diversity in a particular location is often a good indication that a variety is old or has been planted in that area for a long time, allowing that diversity to occur naturally.

MORPHOLOGY is the study of form and structure, so **vine morphology** looks at the physical characteristics of a vine – leaves, shoot tips, bunches, etc. – to identify and describe varieties and clones.

MUTATION an accidental change in the DNA sequence of a vine that may occur at any time of cell division during plant growth, and the product of such a change. It may also be caused by external influences such as radiation, mutagenic chemicals or transposons ('jumping genes'). Most mutations appear to have no effect, some may have a minor biochemical influence and a very few result in a 'spectacular' visual change such as a modification in the skin colour of the berry, ie a colour mutation.

NATURAL CROSS *see* cross

OÏDIUM *see* powdery mildew

PASSITO is an Italian term that refers to sweet wines made from grapes dried after the harvest, so that the sugar concentration is increased, and also to the method of production of such wines. The bunches are traditionally dried on mats in the sun or hung from the rafters but sometimes the vines are simply left to raisin on the vine.

PHOMOPSIS usually refers to the more specific fungal disease **phomopsis cane and leaf spot**, *excoriose* in French and also known as dead arm. Caused by *Phomopsis viticola*, it produces spots or lesions on various parts of the vine and reduces yields. Most commonly controlled by fungicides.

PHYLLOXERA, known by various scientific names but most commonly today as *Dactulosphaira vitifoliae* (Fitch), is a small but highly destructive aphid that feeds on vine roots, causing stunted growth and eventually vine death. Native to North America, it arrived in Europe in the early 1860s and went on to devastate the vineyards of many countries. It is still a serious threat today, not only in Europe but throughout the world. The only solution is grafting: *Vitis vinifera* scions are grafted onto American rootstocks, which are more or less tolerant of the insect. Vines planted in sandy soils are less susceptible.

POWDERY MILDEW, commonly known in France as *oïdium*, is a fungal disease caused by *Erysiphe necator* (formerly *Uncinula necator*). Like downy mildew, it is native to North America but was unknowingly imported to Europe in the late 1840s and is now a worldwide problem. It affects all green parts of the vine, and the spores, spread by the wind, resemble a fine dusting of grey-white ash. The main effects are reduced fruit set and yield and delayed ripening;
infected fruit gives a mouldy, earthy taint to the wine. Controlled mainly by sulphur dust or organic fungicides.

QVEVRI (sometimes spelt kvevri) large clay vessel, lined with beeswax and buried in the ground to control the temperature, used for wine fermentation and storage. First used in Colchis in the Caucasus around 6000 BC.

RANCIO describes a style of wine, sometimes fortified, usually sweet, that has been deliberately exposed to oxygen and/or heat. Such processes result in rich and complex aromas and flavours variously described as cheesy, nutty or akin to overripe/candied fruit.

REFERENCE VARIETY example of variety planted in a vine collection, generally represented by 5–10 vines, used for analysis, identification, preservation and comparison with other vines found in the vineyard.

ROOTSTOCK the part of the vine that forms the

root system and onto which the scion is grafted. Rootstocks are generally American vine species or a hybrid of more than one and their main purpose is to protect the vine from phylloxera, although they can also modify the vine in other ways, eg by moderating growth or by inhibiting chlorosis.

ROTER BRENNER a type of **anthracnose** (see above) caused by the fungus *Pseudopezicula tracheiphila*.

SCION the section of vine grafted onto a rootstock that will determine the variety of the fruit produced by the plant.

SELFING self-pollination.

SELFLING *see* chance seedling

TEINTURIER, meaning 'dyer' in French, may be used to refer to any dark-skinned variety that is used to deepen the colour of a blend. However, we have used it in the stricter sense of a variety that produces deeply coloured juice and wine because the berries are red-fleshed.

THRIPS several species of very small winged insects that feed on vine flowers and nascent bunches causing scarring and stunted shoots in spring.

TRADITIONAL METHOD the most respected and painstaking method of producing sparkling wine, whereby the second fermentation takes place in the bottle in which the wine is sold.

VARIETY *see* pp XIV–XVI

VEGETATIVE PROPAGATION is the asexual reproduction of a plant by taking a cutting from another plant, in contrast to sexual reproduction from seed.

VIGOROUS tending naturally to produce a large amount of vegetative growth, sometimes at the cost of ripening the fruit.

VIN DOUX NATUREL, a French term meaning 'naturally sweet wine', describes wines that are sweet because the fermentation process has been stopped by the addition of spirit. Well-known examples include Muscat de Beaumes-de-Venise and port.

VIN JAUNE, French for 'yellow wine', is a speciality of the Jura made rather like sherry but without fortification.

***VITIS VINIFERA* L.**, the Latin name for the vine species used to make wine, is subdivided by many botanists into two subspecies: *Vitis vinifera* subsp. *silvestris*, the name we have used to designate an unknown wild grape variety, and *Vitis vinifera* subsp. *vinifera*, the name we use to refer to an unknown cultivated grape variety.

VITIS VINIFERA* SUBSP. *SILVESTRIS an unknown wild grape variety belonging to the species *Vitis silvestris*.

VITIS VINIFERA* SUBSP. *VINIFERA an unknown cultivated grapevine belonging to the species *Vitis vinifera*.

BIBLIOGRAPHY

Entries highlighted in red are those that have been particularly useful and referred to most often in the course of our research.

Acerbi, G, 1825, *Delle viti italiane ossia materiali per servire alla classificazione monografia e sinonimia*, Giovanni Silvestri, Milano

Aeberhard, M, 2005, *Geschichte der alten Traubensorten*, Arcadia Verlag, Solothurn

Aebischer, P, 1937, 'Les noms de trois cépages valaisans', *Vox Romanica* 2: 360–3

Agostinetti, G, 1679, *Cento, e dieci ricordi, che formano il buon fattor di villa*, Stefano Curti, Venetia

Agüero, C B, Rodríguez, J G, Martínez, L E, Dangl, G S, and Meredith, C P, 2003, 'Identity and parentage of Torrontés cultivars in Argentina', *American Journal of Enology and Viticulture* 54, 4: 318–21

Agustí, M, 1617, *Llibre dels secrets de agricultura, casa rustica y pastoril*, Estampa de Esteve Liberôs, Barcelona

Akkak, A, Boccacci, P, and Botta, R, 2007, 'Cardinal grape parentage: a case of a breeding mistake', *Genome* 50, 3: 325–8

Alcalde, A, 1989, *Cultivares vitícolas argentinos*, INTA, Mendoza

Alla, F, 2003, 'L'Albanello, vitigno dimenticato', *Il Sommelier* 3, 22

Almadanim, M C, Baleiras-Couto, M M, Pereira, H S, Carneiro, L C, Fevereiro, P, Eiras Dias, J E, Morais-Cecilio, L, Viegas, W, and Veloso, M M, 2007, 'Genetic diversity of the grapevine (*Vitis vinifera* L.) cultivars most utilized for wine production in Portugal', *Vitis* 46, 3: 116–19

Almadanim, M C, Baleiras-Couto, M M, Pereira, H S, Melo, E, Valero, E, Fevereiro, P, Eiras-Dias, J E, Morais, L, Viegas, W, and Veloso, M M, 2004, 'Os microssatélites na identificação de variedades da videira', *6º simpósio de Vitivinicultura do Alentejo, 26–8 Maio*, Évora, Portugal, pp 23–9

Alonso de Herrera, G, 1790 (first published 1513), *Agricultura general*, Don Josef de Urrutia, Madrid

Ambrosi, H, Dettweiler-Münch, E, Rühl, E H, Schmid, J, and Schumann, F, 1997, *Guide des cépages: 300 cépages et leurs vins*, Eugen Ulmer, Paris

Ammann-Doubliez, C, 2007, 'Trois vieux cépages valaisans: neyrun, humagny et regy. Édition, traduction et commentaire d'un texte de 1313', *Vallesia* 62: 221–60

Andrasovsky, J, 1933, 'La systématique de *Vitis vinifera* s-sp. *sativa* DC', *Le Monde des Plantes*, Series IV, 202: 23–30

André, J, 1953, 'Contribution au vocabulaire de la viticulture: les noms de cépages', *Revue des Études Latines* 30: 126–56

André, J, and Levadoux, L, 1964, 'La vigne et le vin des Allobroges', *Journal des Savants* 3, 3: 169–81

Anonymous, 1856, 'British Pomological Society', *The Florist, Fruitist, and Garden Miscellany* 6: 242–3

Aradhya, M K, Dangl, G S, Prins, B H, Boursiquot, J-M, Walker, M A, Meredith, C P, and Simon, C J, 2003, 'Genetic structure and differentiation in cultivated grape, *Vitis vinifera* L.', *Genetical Research* 81, 3: 179–92

Arnold, C, Gillet, F, and Gobat, J-M, 1998, 'Situation de la vigne sauvage *Vitis vinifera* ssp. *sylvestris* en Europe', *Vitis* 37: 159–70

Arranz, C, Yuste, J, Alburquerque, M V, Barajas, E, Castaño, F J, Rubio, J A, Hidalgo, E, Santana, J C, Ortiz, J M, and Martín, J P, 2008, 'Variedades de vid cultivadas en la Sierra de Francia. Importancia, identificación, sinonimias y homonimias', *Semana Vitivinícola* 3223: 1414–20

Arranz, C, Yuste, J, Ortiz, J M, Martín, J P, and Rubio, J A, 2010, 'Castas presentes no oeste de Castilla e León (Espanha) e norte de Portugal', 8° Simpósio de vitivinicultura do Alentejo, Évora, http://cvra.mikroelement.pt/media/documents/250510_1274806852.pdf

Arrigo, N, and Arnold, C, 2007, 'Naturalised *Vitis* rootstocks in Europe and consequences to native wild grapevine', *PLoS ONE* 2, 6: e521

Arroyo-García, R, Ruiz-García, L, Bolling, L, Ocete, R, López, M A, Arnold, C, Ergül, A, *et al.*, 2006, 'Multiple origins of cultivated grapevine (*Vitis vinifera* L. ssp. *sativa*) based on chloroplast DNA polymorphisms', *Molecular Ecology* 15, 12: 3707–14

Asensio Sánchez, L, 2000, *Caracterización de variedades de Vitis vinifera L. cultivadas en Extremadura, mediante estudios morfológicos, agronómicos y bioquímicos*, Universidad Politécnica de Madrid, Madrid

Avramov, L, and Del Zan, F, 2004, 'Serbia e Montenegro', in Del Zan, F, Failla, O, and Scienza, A (eds), *La vite e l'uomo dal rompicapo delle origini al salvataggio delle reliquie*, ERSA–Agenzia Regionale per lo Sviluppo Rurale, Gorizia, pp 677–705

Babo, A W F Freiherr von, and Mach, E, 1923, *Handbuch des Weinbaues und der Kellerwirtschaft*, vol 1, Parey, Berlin

Babo, L von, 1843–4, *Der Weinstock und seine Varietäten*, 2 vols, Heinrich Ludwig Brönner, Frankfurt am Main

Bacci, A, 1596, *De naturali vinorum historia*, Nich. Mutius, Roma

Bailey, L H, 1906, *Sketch of the evolution of our native fruits*, Macmillan, New York

Barnard, H, Dooley, A N, Areshian, G, Gasparyan, B, and Faull, K, 2011, 'Chemical evidence for wine production around 4000 BCE in the Late Chalcolithic Near Eastern highlands', *Journal of Archaeological Science* 38, 5: 977–84

Bartlett, J M S, and Stirling, D, 2003, 'A short history of the Polymerase Chain Reaction', *PCR Protocols* 226: 3–6

Basler, P, 2003, *'Andere' Rebsorten*, Stutz Druck, Wädenswil, Switzerland

Bassermann-Jordan, F von, 1923, *Geschichte des Weinbaus*, vol II, Frankfurter Verlags-Anstalt, Frankfurt am Main

Bauhin, J, 1650, *Historia plantarum universalis*, Ebroduni (Yverdon)

Bautista, J, Dangl, G S, Yang, J, Reisch, B, and Stover, E, 2008, 'Use of genetic markers to assess pedigrees of grape cultivars and breeding program selections', *American Journal of Enology and Viticulture* 59, 3: 248–54

Bazin, J-F, 2002, *Histoire du vin de Bourgogne*, Jean-Paul Gisserot, Plouédern

Bellasi, A, Rieder, U, and Anhorn, F, 1993, *Weine aus Graubünden*, Stutz Druck, Wädenswil

Benjak, A, Ercisli, S, Vokurka, A, Maletić, E, and Pejić, I, 2005, 'Genetic relationships among grapevine cultivars native to Croatia, Greece and Turkey', *Vitis* 44, 2: 73–8

Bergamini, C, Caputo, A R, Gasparro, M, Perniola, R, Cardone, M F, and Antonacci, D, 2012, 'Evidences for an alternative genealogy of Sangiovese', *Molecular Biotechnology*, March 2012 (e-publication ahead of print)

Berget, A, 1903, 'Gouais', in Viala, P, and Vermorel, V (eds), *Ampélographie*, vol 4, Masson, Paris, pp 94–106

Berget, A, 1904, 'Rouge du Valais', in Viala, P, and Vermorel, V (eds), *Ampélographie*, vol 5, Masson, Paris, pp 278–82

Berget, A, 1932, 'L'origine égyptienne des Chasselas', *Revue de Viticulture* LXXVI, 1,969: 181–5

Bernard, R, 1995, 'Aspects of clonal selection in Burgundy', *Proceedings of the International Symposium on Clonal Selection, 20–21 June 1995, Portland, Oregon*, American Journal of Enology and Viticulture, pp 52–9

Beugnot, Comte A-A, 1842, *Les coutumes du Beauvoisis – par Philippe de Rémi Beaumanoir*, vol 1, Jules Renouard & Cie, Paris

Beuthner, K, 2009, 'The future of Riesling in SA and what we can learn from the recent improvements in Germany (example: the Rheingau)', Cape Wine Master dissertation, Cape Wine Academy; http://www.capewineacademy.co.za

Bica, D, 2007, *Vitigni di Sicilia*, Regione Siciliana, Assessorato Regionale Agricoltura e Foreste, Palermo

Bich, L N, 1890, 'Le péronospora et la brûlure des raisins', *Bulletin du Comice Agricole* 21: 13–15

Biniari, K, and Stavrakakis, M N, 2007, 'Genetic study of 46 Greek grape cultivars by random amplified polymorphic DNA markers (RAPD-PCR)', *Proceedings of the XXXth OIV World Congress of Vine and Wine, 10–16 June 2007*, Budapest, p 7

Bisson, J, 1999, 'Essai de classement des cépages français en écogéogroupes phénotypiques', *Journal International des Sciences de la Vigne et du Vin* 33, 3: 105–10

Bisson, J, 2009, *Classification des vignes françaises*, Féret, Bordeaux

Bluntschli, J C, 1838, *Staats- und Rechtsgeschichte der Stadt und Landschaft Zürich*, vol 1: *Die Zeit des Mittelalters*, Orell Füssli, Zürich

Boccacci, P, Torello Marinoni, D, Gambino, G, Botta, R, and Schneider, A, 2005, 'Genetic characterization of endangered grape cultivars of Reggio Emilia province', *American Journal of Enology and Viticulture* 56, 4: 411–16

Bock, H, 1539, *Das Kreutterbuch*, Rihel, Straßburg

Bock, H, 1546, *Das Kreutterbuch*, second edition, Rihel, Straßburg

Bock, H, 1552, *De stirpium, maxime earum, quae in Germania nostra nascuntur, usitatis nomenclaturis, propriisque differentiis, neque non temperaturis ac facultatibus, commentarii*, Rihel, Straßburg (Argentorati)

Bodea, M, Pamfil, D, Pop, R, and Pop, I F, 2009, 'Use of random amplified polymorphic DNA (RAPD) to study genetic diversity among Romanian local vine (*Vitis vinifera* L.) cultivars', *Bulletin UASVM Horticulture* 66, 1: 17–22

Böhm, H J, 2005, *O grande livro das castas*, Chaves Ferreira, Lisboa

Böhm, H J, 2011, *Rebsortenatlas Spanien und Portugal*, Eugen Ulmer, Stuttgart

Bongiolatti, N, 1996, *Il vigneto valtellinese. Indagine sui vitigni presenti*, Fondazione Fojanini di Studi Superiori, Sondrio

Bonner, J C, 2009, *History of Georgia agriculture, 1732–1860*, University of Georgia Press, Athens, Georgia

Bordenave, L, Lacombe, T, Laucou, V, and Boursiquot, J-M, 2007, 'Étude historique, génétique et ampélographique des cépages Pyrénéo Atlantiques', *Bulletin de l'OIV* 80, 920–2: 553–86

Borg, J, 1922, *Cultivation and diseases of fruit trees in the Maltese Islands*, Government Printing Office, Malta

Borrego, J, Cabello, F, Ibañez, J, Rodriguez-Torrès, I, and Muñoz-Organero, G, 2002, 'Tinto de la Pámpana Blanca y Tinto Velasco: dos variedades de vid emparentadas', *Viticultura Enología Profesional* 82: 15–25

Borrego, J, de Andrés, M T, Gómez, J L, and Ibañez, J,

2002, 'Genetic study of Malvasia and Torrontes groups through molecular markers', *American Journal of Enology and Viticulture* 53, 2: 125–30
Borrego, J, Rodriguez, M T, Martín, J P, Chavez, J, Cabello, F, and Ibañez, J, 2001, 'Characterisation of the most important Spanish grape varieties through isozyme and microsatellite analysis', *Acta Horticulturae* 546: 371–5
Boso, S, Santiago, J L, Vilanova, M, and Martínez, M-C, 2005, 'Caractéristiques ampélographiques et agronomiques de différents clones du cultivar Albariño (*Vitis vinifera* L.)', *Bulletin de l'OIV* 78, 889–90: 143–58
Botta, R, Scott, N S, Eynard, I, and Thomas, M R, 1995, 'Evaluation of microsatellite sequence-tagged site markers for characterizing *Vitis vinifera* cultivars', *Vitis* 34, 2: 99–102
Botta, R, Vallania, R, Me, G, Luzzati, A, and Siragusa, N, 1995, 'Investigation of factors affecting early dropping in Dolcetto (*Vitis vinifera* L.)', *Acta Horticulturae* 379: 97–104
Bouchard, A, 1901a, 'Chenin Blanc', in Viala, P, and Vermorel, V (eds), *Ampélographie*, vol 2, Masson, Paris, pp 83–94
Bouchard, A, 1901b, 'Groslot Noir', in Viala, P, and Vermorel, V (eds), *Ampélographie*, vol 2, Masson, Paris, pp 118–24
Bouchard, A, 1902, 'Verdelho de Madère', in Viala, P, and Vermorel, V (eds), *Ampélographie*, vol 3, Masson, Paris, pp 88–99
Bouchereau, H-X, 1863, 'Raisins de table adoptés par le Congrès pomologique dans sa session de 1859', *Le Messager Agricole* 3: 307–11, 343–8
Boullay, J, 1712, *Manière de bien cultiver la vigne*, Fr. Borde, Orléans
Boursiquot, J-M, Lacombe, T, Bowers, J E, and Meredith, C P, 2004, 'Le Gouais, un cépage clé du patrimoine viticole européen', *Bulletin de l'OIV* 77, 875–6: 5–19
Boursiquot, J-M, Lacombe, T, Laucou, V, Julliard, S, Perrin, F-X, Lanier, N, Legrand, D, Meredith, C P, and This, P, 2009, 'Parentage of Merlot and related winegrape cultivars of southwestern France: discovery of the missing link', *Australian Journal of Grape and Wine Research* 15, 2: 144–55
Boursiquot, J-M, and This, P, 1999, 'Essai de définition du cépage', *Le Progrès Agricole et Viticole* 116, 17: 359–61
Bouschet, L, 1829, 'Observations sur quelques espèces de raisins', *Bulletin de la Société d'Agriculture du Département de l'Hérault*, 16e Année, Janvier: 15–25
Bousson de Mairet, E, 1856, *Annales historiques et chronologiques de la ville d'Arbois*, Madame Javel, Arbois
Boutard, E, 1842, 'Catalogue des espèces de vignes, cultivées dans l'arrondissement de La Rochelle et dans les pépinières de M. Boutard Aîné', *Bulletin de la Société Industrielle d'Angers et du Département de Maine et Loire* 13: 465–74
Boutaris, M, 2000, *Greek wine grape varieties. A microsatellite DNA marker analysis*. MSc thesis, University of California, Davis
Bowers, J E, Boursiquot, J-M, This, P, Chu, K, Johansson, H, and Meredith, C P, 1999, 'Historical genetics: the parentage of Chardonnay, Gamay, and other wine grapes of Northeastern France', *Science* 285: 1562–5 (the 322 cultivars analysed are listed in a supplementary table available at http://www.sciencemag.org/feature/data/1042157.dtl)
Bowers, J E, and Meredith, C P, 1997, 'The parentage of a classic wine grape, Cabernet Sauvignon', *Nature Genetics* 16, 1: 84–7
Bowers, J E, Siret, R, Meredith, C P, This, P, and Boursiquot, J-M, 2000, 'A single pair of parents proposed for a group of grapevine varieties in northeastern France', *Acta Horticulturae* 528: 129–32
Bozinovic, Z, Beleski, K, and Boskov, K, 2003, 'The influence of morphological characteristics on productive properties of cultivar Stanušina', *Acta Horticulturae* 603: 609–12
Brancadoro, L, 2006, 'Vitigni siciliani: in tre anni da 0 a 94 potenziali cloni', *L'Informatore Agrario* 13, Supplement, pp 38–40
Branzanti, E, Brancadoro, L, Raiti, G, Fichera, G, and Squadrito, M, 2008, 'Definizione della variabilità genotipica e fenotipica dei Nerelli etnei', 2° *Convegno Nazionale di Viticoltura (CONAVI)*, Marsala, Sicilia, 14–19 July 2008
Braun, G C, 1824, *Die Rheinfahrt: ein Natur- und Sittengemälde des Rheinlandes in drei Gesängen*, Joseph Stenz, Mainz
Brazão, J, Eiras Dias, J E, and Carneiro, L C, 2005, 'O encepamento da região vitivinícola de Carcavelos', *Ciência e Técnica Vitivinícola* 20, 2: 131–45
Bronner, A, 2003, *Muscats et variétés muscatées*, INRA Editions/Oenoplurimedia, Versailles/Chaintré
Bronner, J P, 1857, *Die wilden Trauben des Rheintales*, Georg Mohr, Heidelberg
Buhner-Zaharieva, T, Moussaoui, S, Lorente, M, Andreu, J, Núñez, R, Ortiz, J M, and Gogorcena, Y, 2010, 'Preservation and molecular characterization of ancient varieties in Spanish grapevine germplasm collections', *American Journal of Enology and Viticulture* 61, 4: 557–62
Cabezas, J A, Cervera, M T, Arroyo-García, R, Ibañez, J, Rodriguez-Torrès, I, Borrego, J, Cabello, F, and Martínez-Zapater, J M, 2003, 'Garnacha and Garnacha Tintorera: genetic relationships and the origin of *teinturier* varieties cultivated in Spain', *American Journal of Enology and Viticulture* 54, 4: 237–45
Calò, A, 2005, 'I *Refoschi*', in Del Zan, F (ed), *Dei Refoschi, Atti del Convegno organizzato dall'ERSA, 19 June 2004*, ERSA – Agenzia regionale per lo sviluppo rurale, Gorizia, pp 11–21
Calò, A, and Costacurta, A, 1992, 'Tocai friulano e Sauvignonasse: un unico vitigno', *Rivista di Viticoltura e di Enologia* 3: 31–40
Calò, A, and Costacurta, A, 2004, *Dei vitigni italici. Ovvero delle loro storie, caratteri e valorizzazione*, Matteo Editore, Dosson di Casier
Calò, A, Costacurta, A, Cancellier, S, and Forti, R, 1990, 'Garnacha, Grenache, Cannonao, Tocai rosso, un unico vitigno', *Vignevini* 17, 9: 45–8
Calò, A, Costacurta, A, Cancellier, S, and Forti, R, 1991, 'Verdicchio bianco, Trebbiano di Soave. Un unico vitigno', *Vignevini* 18, 11: 49–52

Calò, A, Costacurta, A, Crespan, M, Milani, N, Aggio, L, Carraro, R, Di Stefano, R, and Ummarino, I, 2001, 'La caratterizzazione dei Fiani: Fiano e Fiano aromatico', Tornata dell'Accademia Italiana della Vite e del Vino, Napoli-Atripalda-Pompei, 24–5 May

Calò, A, Costacurta, A, Maraš, V, Meneghetti, S, and Crespan, M, 2008, 'Molecular correlation of Zinfandel (Primitivo) with Austrian, Croatian, and Hungarian cultivars and Kratošija, an additional synonym', *American Journal of Enology and Viticulture* 59, 2: 205–9

Calò, A, Di Stefano, R, Costacurta, A, and Calò, G, 1991, 'Caratterizzazione di Cabernet franc e Carmenère (*Vitis* sp.) e chiarimenti sulla loro coltura in Italia', *Rivista di Viticoltura ed Enologia di Conegliano* 3: 3–25

Calò, A, Francini, F, Lauciani, P, Rorato, G, and Tomasi, D, 2008, *Il Raboso del Piave. Fascinosa realtà delle terre del Piave*, Dario de Bastiani, Vittorio Veneto

Calò, A, Scienza, A, and Costacurta, A, 2006, *Vitigni d'Italia*, 2nd edn, Edagricole Calderini, Bologna

Cancellier, S, and Angelini, U, 1993, 'Corvina veronese e Corvinone: due varietà diverse', *Vignevini* 5: 44–6

Cancellier, S, Coletti, A, Coletti, M, Soligo, S, and Michelet, E, 2007, 'Quattro "nuovi" vitigni per la viticoltura veneta', *Quaderni di Viticoltura ed Enologia dell'Università di Torino* 29: 5–25

Cancellier, S, Dalla Cia, L G, and Coletti, A, 2008, 'I vecchi vitigni ritrovati nel feltrino', in Tadiotto, A, and Lavezzo, I (eds), *Valorizzazione di aree viticole di montagna tramite scambio di know-how (Progetto Interreg IIIA Italia-Austria cod E305)*, Veneto Agricoltura, Legnaro, pp 48–80

Cancellier, S, Giannetto, S, and Crespan, M, 2009, 'Groppello di Breganze e Pignola sono lo stesso vitigno', *Rivista di Viticoltura e di Enologia* 62, 2/3: 3–9

Candolle, A de, 1883, *Origine des plantes cultivées*, Germer Baillière, Paris

Candolle, A-P de, 1820, *Catalogue des arbres fruitiers et des vignes du Jardin botanique de Genève*, J J Paschoud, Genève and Paris

Carbonneau, A, 1995, 'Vignes d'Arménie', *Progrès Agricole et Viticole* 112, 17: 359–63

Carbonneau, A, 2005, 'Vignobles d'Ariège: histoire et avenir', *Progrès Agricole et Viticole* 122, 18: 383–92

Carcamo, C, Provedo, I, and Arroyo-García, R, 2010, 'Detection of polymorphism in ancient Tempranillo clones (*Vitis vinifera* L.) using microsatellite and retrotransposon markers', *Iranian Journal of Biotechnology* 8, 1: 1–6

Carimi, F, Mercati, F, Abbate, L, and Sunseri, F, 2009, 'Microsatellite analyses for evaluation of genetic diversity among Sicilian grapevine cultivars', *Genetic Resources and Crop Evolution* 57, 5: 703–19

Carimi, F, Mercati, F, De Michele, R, Fiore, M C, Riccardi, P, and Sunseri, F, 2011, 'Intra-varietal genetic diversity of the grapevine (*Vitis vinifera* L.) cultivar Nero d'Avola as revealed by microsatellite markers', *Genetic Resources and Crop Evolution* 58, 7: 967–75

Carlucci, M, 1905, 'Piedirosso', in Viala, P, and Vermorel, V (eds), *Ampélographie*, vol 6, Masson, Paris, pp 360–5

Carreño, E, Lopez, M A, Labra, M, Rivera, D, Sancha, J, Ocete, R, and Martínez de Toda, F, 2004, 'Genetic relationship between some Spanish *Vitis vinifera* L. subsp. *sativa* cultivars and wild grapevine populations (*Vitis vinifera* L. subsp. *silvestris* (Gmelin) Hegi): a preliminary study', *Plant Genetic Resources Newsletter* 137: 42–5

Casanova Gascón, J, 2008, 'Caracterización de variedades de vid (*Vitis vinifera* L.) de la provincia de Huesca', unpublished PhD thesis, Universidad de Zaragoza, Zaragoza; http://zaguan.unizar.es/record/3013/files/TESIS-2009-040.pdf

Casas Rigall, J, 2007, *Libro de Alexandre*, Editorial Castalia, Madrid

Castro, I, Martín, J P, Ortiz, J M, and Pinto-Carnide, O, 2011, 'Varietal discrimination and genetic relationships of *Vitis vinifera* L. cultivars from two major Controlled Appellation (DOC) regions in Portugal', *Scientia Horticulturae* 127, 4: 507–14

Cazeaux-Cazalet, G, 1901, 'Castets', in Viala, P, and Vermorel, V (eds), *Ampélographie*, vol 2, Masson, Paris, pp 173–8

Çetiner, S, Çelik, S, Budak, H, Boz, Y, 2009, 'The diagnosis of historical local grape cultivars of Urla region through molecular and amphelographic methods', Istanbul (unpublished)

Cervera, M T, Cabezas, J A, Rodriguez-Torrès, I, Chavez, J, Cabello, F, and Martínez-Zapater, J M, 2002, 'Varietal diversity within grapevine accessions of cv. Tempranillo', *Vitis* 41, 1: 33–6

Cervera, M T, Cabezas, J A, Sancha, J C, Detoda, F M, and Martínez-Zapater, J M, 1998, 'Application of AFLPs to the characterization of grapevine *Vitis vinifera* L. genetic resources. A case study with accessions from Rioja (Spain)', *Theoretical and Applied Genetics* 97, 1–2: 51–9

Cervera, M T, Rodriguez, I, Cabezas, J A, Chavez, J, Martínez-Zapater, J M, and Cabello, F, 2001, 'Morphological and molecular characterization of grapevine accessions known as Albillo', *American Journal of Enology and Viticulture* 52: 127–35

Chaptal, J-A C, d'Ussieux, L, Parmentier, A A, and Rozier, F, 1801, *Traité théorique et pratique sur la culture de la vigne*, vol 2, Delalain & Fils, Paris

Chevalier, C, 1864, *Comptes des receptes et despenses faites en la Chastellenie de Chenonceau par Diane de Poitiers*, J Techener, Paris

Chevalier, F-F, 1873, *Œnologie ou discours sur le vignoble et les vins de Poligny*, Mareschal, Poligny

Chkhartishvili, N, and Betsiashvili, G, 2004, 'Mar Nero settentrionale, Transcaucasia, Asia centro-occidentale (CSI)', in Del Zan, F, Failla, O, and Scienza, A (eds), *La vite e l'uomo dal rompicapo delle origini al salvataggio delle reliquie*, 2nd edn, ERSA – Agenzia Regionale per lo Sviluppo Rurale, Gorizia, pp 181–360

Chomé Fuster, P M (co-ordinator), 2006, *Variedades de vid: registro de variedades comerciales, Ministerio de Agricultura, Pesca y Alimentación*, Madrid. Also available online: http://www.marm.es/ministerio/pags/exposiciones/vid/www/imagenes/variedades_uva_01.html

Christensen, L P, Dokoozlian, N K, Walker, M A,

Wolpert, J A, 2003, *Wine grape varieties in California*, publication 3419, University of California, Agriculture and Natural Resources, California
Cibrario, L, 1833, 'Delle finanze della monarchia di Savoia ne' secoli XIII e XIV. Delle entrate della Corona, Discorso 2°', *Memorie della Reale Academia delle Scienze di Torino*, Stampa Reale, Torino, pp 157–239
Cindrić, P, Korać, N, and Kovač, V, 2000, 'Grape breeding in the Vojvodina province', *Acta Horticulturae* 528: 499–504
Cipriani, G, Frazza, G, Peterlunger, E, and Testolin, R, 1994, 'Grapevine fingerprinting using microsatellite repeats', *Vitis* 33, 4: 211–15
Cipriani, G, Spadotto, A, Jurman, I, Di Gaspero, G, Crespan, M, Meneghetti, S, Frare, E, Vignani, R, Cresti, M, Morgante, M, Pezzotti, M, Pe, E, Policriti, A, and Testolin, R, 2010, 'The SSR-based molecular profile of 1005 grapevine (*Vitis vinifera* L.) accessions uncovers new synonymy and parentages, and reveals a large admixture amongst varieties of different geographic origin', *Theoretical and Applied Genetics* 121, 8: 1569–85
Clemente y Rubio, S d R, 1807, *Ensayos sobra las variedades de la vid comun que vegetan en Andalucia*, Imprenta de Villalpando, Madrid
Comba, R, and Dal Verme, A, 1990, 'Repertorio di vini e vitigni diffusi nel Piemonte medievale', in Comba, R (ed), *Vigne e vini nel Piemonte medievale*, L'Arciere, Cuneo, pp 334–42
Conchie, J, and Day, R, 2009, 'Barbera considered the great lady of red varieties', *Australia and New Zealand Wine Industry Journal* 24, 3: 70–1
Constantinou, Y, 2006, *The Cyprus wine guide*, Oinou Symvouleftiki, Nicosia
Corbera, E de, 1678, *Cataluña illustrada*, Antonino Gramiñani, Napoles
Corder, R, 2006, *The red wine diet*, Little, Brown, London
Costacurta, A, Calò, A, Antonacci, D, Catalano, V, Crespan, M, Carraro, R, Giust, M, Agio, L, Ostan, M, Di Stefano, R, and Borsa, D, 2004, 'La caratterizzazione di Greci e Grechetti a bacca bianca coltivati in Italia', *Rivista di Viticoltura e di Enologia* 3: 3–20
Costacurta, A, Calò, A, Carraro, R, Giust, M, Aggio, L, Borsa, D, Di Stefano, R, Del Zan, F, Fabbro, C, and Crespan, M, 2005, 'L'identificazione e la caratterizzazione dei *Refoschi*', in Del Zan, F (ed), *Dei Refoschi. Atti del Convegno organizzato dall'ERSA, 19 June 2004*, ERSA – Agenzia regionale per lo sviluppo rurale, Gorizia, pp 25–45
Costacurta, A, Calò, A, and Crespan, M, 2003, 'The varietal identification and characterisation work of the Istituto Sperimentale per la Viticoltura in the past fifteen years', in Hajdu, E, and Borbás, E (eds), *Acta Horticulturae* 603, pp 261–73
Costacurta, A, Cancellier, S, Angelini, U, Segattini, G, and Farina, C, 1980, 'Vecchi vitigni veronesi', *Rivista di Viticoltura e di Enologia* 33, 10, Supplement, p 105
Costantini, L, Monaco, A, Vouillamoz, J F, Forlani, M, and Grando, M S, 2005, 'Genetic relationships among local *Vitis vinifera* cultivars from Campania (Italy)', *Vitis* 44: 25–34
Costantini, L, Roncador, I, and Grando, M S, 2001, 'Il caso Groppello della Val di Non chiarito con le analisi del DNA', *L'Informatore Agrario* 45: 53–7
Cousteaux, F, and Plageoles, R, 2001, *Le vin de Gaillac, 2000 ans d'histoire*, Éditions Privat, Toulouse
Crespan, M, 2003, 'The parentage of Muscat of Hamburg', *Vitis* 42, 4: 193–7
Crespan, M, Cabello, F, Giannetto, S, Ibañez, J, Karoglan Kontić, J, Maletić, E, Pejić, I, Rodriguez-Torrès, I, and Antonacci, D, 2006, 'Malvasia delle Lipari, Malvasia di Sardegna, Greco di Gerace, Malvasía de Sitges and Malvasia dubrovačka – synonyms of an old and famous grape cultivar', *Vitis* 45, 2: 69–73
Crespan, M, Calò, A, Costacurta, A, Milani, N, Giust, M, Carraro, R, and Di Stefano, R, 2002, 'Ciliegiolo e Aglianicone: unico vitigno direttamente imparentato col Sangiovese', *Rivista di Viticoltura e di Enologia* 55, 2–3: 3–14
Crespan, M, Calò, A, Giannetto, S, Sparacio, A, Storchi, P, and Costacurta, A, 2008, 'Sangiovese and Garganega are two key varieties of the Italian grapevine assortment evolution', *Vitis* 47, 2: 97–104
Crespan, M, Cancellier, S, Chies, R, Giannetto, S, and Meneghetti, S, 2006, 'Individuati i genitori del Raboso veronese: una nuova ipotesi sulla sua origine', *Rivista di Viticoltura ed Enologia di Conegliano* 1: 3–12
Crespan, M, Cancellier, S, Chies, R, Giannetto, S, Meneghetti, S, and Costacurta, A, 2009, 'Molecular contribution to the knowledge of two ancient varietal populations: Rabosi and Glere', *Acta Horticulturae* 827: 217–20
Crespan, M, Cancellier, S, Costacurta, A, Giust, M, Carraro, R, Di Stefano, R, and Santangelo, S, 2003, 'Contribution to the clearing up of synonymies in some groups of Italian grapevine cultivars', *Acta Horticulturae* 603: 275–89
Crespan, M, Coletta, A, Crupi, P, Giannetto, S, and Antonacci, D, 2008, 'Malvasia nera di Brindisi/Lecce grapevine cultivar (*Vitis vinifera* L.) originated from Negroamaro and Malvasia bianca lunga', *Vitis* 47, 4: 205–12
Crespan, M, Crespan, G, Giannetto, S, Meneghetti, S, and Costacurta, A, 2007, 'Vitouska is the progeny of Prosecco tondo and Malvasia bianca lunga', *Vitis* 46, 4: 192–4
Crespan, M, Fabbro, A, Giannetto, S, Meneghetti, S, Petrussi, C, del Zan, F, and Sivilotti, P, 2011, 'Recognition and genotyping of minor germplasm of Friuli Venezia Giulia revealed high diversity', *Vitis* 50, 1: 21–8
Crespan, M, Giannetto, S, Meneghetti, S, and Cancellier, S, 2008, 'Approfondimenti sull'identità di alcuni vitigni della provincia di Belluno mediante analisi del DNA', in Tadiotto, A, and Lavezzo, I (eds), *Valorizzazione di aree viticole di montagna tramite scambio di know-how (Progetto Interreg IIIA Italia-Austria cod E305)*, Veneto Agricoltura, Legnaro, pp 81–4
Crespan, M, and Milani, N, 2001, 'The Muscats: a molecular analysis of synonyms, homonyms and genetic relationships within a large family of grapevine cultivars', *Vitis* 40, 1: 23–30
Croce, G B, 1606, *Della eccellenza e diversità dei vini, che*

nella montagna di Torino si fanno; e del modo di farli, P A Pizzamiglio, Torino
Cunha, J, Teixeira Santos, M, Fevereiro, P, and Eiras Dias, J E, 2009, 'Portuguese traditional grapevine cultivars and wild vines (*Vitis vinifera* L.) share morphological and genetic traits', *Genetic Resources and Crop Evolution* 56: 975–89
Cupani, Francesco, 1696, *Hortus catholicus*, Benzi, Napoli
Dalmasso, G, and Reggio, L, 1963, *Fumin*, Tipografia Longo & Zoppelli, Treviso
Day, R, 2009, 'Lagrein – an Italian enigma, relaxed in a new Aussie home', *Australia and New Zealand Wine Industry Journal* 24, 3: 13–14
De Maria, P P, and Leardi, C, 1875, *Ampelografia della Provincia di Alessandria*, A F Negro, Torino
De Mattia, F, Imazio, S, Grassi, F, Lovicu, G, Tardaguila, J, Failla, O, Maitti, C, Scienza, A, and Labra, M, 2007, 'Genetic characterization of Sardinia grapevine cultivars by SSR markers analysis', *Journal International des Sciences de la Vigne et du Vin* 41, 4: 175–84
De Mattia, F, Lovicu, G, Tardaguila, J, Grassi, F, Imazio, S, Scienza, A, and Labra, M, 2009, 'Genetic relationships between Sardinian and Spanish viticulture: the case of Cannonau and Garnacha', *Journal of Horticultural Science & Biotechnology* 84, 1: 65–71
Dejeu, L, 2004, 'Romania', in Del Zan, F, Failla, O, and Scienza, A (eds), *La vite e l'uomo dal rompicapo delle origini al salvataggio delle relique*, 2nd edn, ERSA – Agenzia Regionale per lo Sviluppo Rurale, Gorizia, pp 445–530
Del Zan, F, Failla, O, and Scienza, A (eds), *La vite e l'uomo dal rompicapo delle origini al salvataggio delle relique*, 2nd edn, ERSA – Agenzia Regionale per lo Sviluppo Rurale, Gorizia
Della Porta, G B, 1592, *Villae*, Book XXI, Andreae Wecheli, Claudium Marnium & Ioannem Aubrium, Napoli
Dellavalle, D, Valota, G, Cravero, M C, Pazo Alvarez, M d C, and Ubigli, M, 2005, 'Attitudini viticole ed enologiche dei vitigni Timorasso e Nascetta', *Piemonte Agricoltura* 48: 28–33
Demaria, P P, and Leardi, C, 1875, *Ampelografia della provincia di Alessandria*, A F Negro, Torino
Dercsényi, J, 1796, *Über Tokay's Weinbau, dessen Fexung und Gährung: mit geognostischen Beylagen*, A Blumauer, Wien
Dettweiler, E, Jung, A, Zyprian, E, and Topfer, R, 2000, 'Grapevine cultivar Müller-Thurgau and its true to type descent', *Vitis* 39, 2: 63–5
Dienes, D, 2001, *Református egyház-látogatási jegyzőkönyvek, 16–17 század*, Osiris Kiadó, Budapest
Dion, R, 1959, *Histoire de la vigne et du vin en France des origines au XIXe siècle*, self-published, republished in 1982 by Flammarion, Paris
Di Manzano, F, 1860, *Annali del Friuli, ossia raccolta delle cose storiche appartenenti a questa regione*, vol III, Trombetti-Murero, Udine
Di Rovasenda, G, 1877, *Saggio di una ampelografia universale*, Tipografia Subalpina di Stefano Marino, Torino
Di Vecchi Staraz, M, Boselli, M, Gerber, S, Laucou, V, Lacombe, T, This, P, and Varès, D, 2007, 'Famoz: a software for large scale parentage analysis in *Vitis vinifera* L. species', *Acta Horticulturae* 754: 79–83
Di Vecchi Staraz, M, Lacombe, T, Laucou, V, Bandinelli, R, Varès, D, Boselli, M, and This, P, 2004, 'Studio sulle relazioni genetiche tra viti selvatiche e coltivate in Toscana', *Proceedings of the II International Symposium on Sangiovese, Firenze, Italy, 17–18 November 2004*, Agenzia Regionale per lo Sviluppo e l'Innovazione in Agricoltura
Di Vecchi Staraz, M, This, P, Boursiquot, J-M, Laucou, V, Lacombe, T, Bandinelli, R, Varès, D, and Boselli, M, 2007, 'Genetic structuring and parentage analysis for evolutionary studies in grapevine: kingroup and origin of cv. Sangiovese revealed', *Journal of the American Society for Horticultural Science* 132, 4: 514–24
Díaz-Losada, E, Tato Salgado, A, Ramos-Cabrer, A M, and Pereira-Lorenzo, S, 2011, 'Determination of genetic relationships of Albariño and Loureira cultivars with the Caíño group by microsatellites', *American Journal of Enology and Viticulture* 62, 3: 371–5
Dochnahl, F J, 1848, *Die allgemeine Centralobstbaumschule, ihre Zwecke und Einrichtung*, Mauke, Jena
d'Oliveira, D, 1903, 'Bastardo', in Viala, P, and Vermorel, V (eds), *Ampélographie*, vol 4, Masson, Paris, pp 208–20
Dornfeld, J, 1868, *Die Geschichte des Weinbaues in Schwaben*, Cohen und Risch, Stuttgart
Drake, J W, Charlesworth, B, Charlesworth, D, and Crow, J F, 1998, 'Rates of spontaneous mutation', *Genetics* 148: 1667–86
Droguet, A, 1992, 'La culture de la vigne à Saint-Suliac au Moyen Âge', *Le Pays de Dinan*: 183–94
Dubé, G, and Turcotte, I, 2011, *Guide d'identification des cépages cultivés en climat froid: cépages de cuve*, Richard Grenier, Québec
Dubourdieu, D, 2009, 'Sauvignon Blanc's distinctive yet elusive aromas', *Tong* 1: 19–24
Dupraz, P, and Spring, J-L, 2010, *Cépages: principales variétés de vigne cultivées en Suisse*, Agroscope/École d'ingénieurs de Changins, Changins
Durand, E, 1901a, 'Pinots', in Viala, P, and Vermorel, V (eds), *Ampélographie*, vol 2, Masson, Paris, pp 18–42
Durand, E, 1901b, 'Douce Noire', in Viala, P, and Vermorel, V (eds), *Ampélographie*, vol 2, Masson, Paris, pp 371–5
Durand, E, 1905, 'Gringet', in Viala, P, and Vermorel, V (eds), *Ampélographie*, vol 6, Masson, Paris, pp 132–5
Durand, E, and Pacottet, P, 1901a, 'Melon', in Viala, P, and Vermorel, V (eds), *Ampélographie*, vol 2, Masson, Paris, pp 45–50
Durand, E, and Pacottet, P, 1901b, 'Aligoté', in Viala, P, and Vermorel, V (eds), *Ampélographie*, vol 2, Masson, Paris, pp 51–4
Eden, F (1903), *A garden in Venice*, Country Life, London
Einset, J, and Robinson, W B, 1972, 'Cayuga White, the first of a Finger Lakes series of wine grapes for New York', *New York's Food and Life Sciences Bulletin* 22: 1–2
Eiras Dias, J E, Paulos, V, Mestre, S, Martins, J T, and Goulart, I, 2006, 'O encepamento do arquipélago

dos Açores', *Ciência e Técnica Vitivinícola* 21, 2: 99–112
Ekhvaia, J, Blattner, F R, and Akhalkatsi, M, 2010, 'Genetic diversity and relationships between wild grapevine (*Vitis vinifera* subsp. *sylvestris*) populations and aboriginal cultivars in Georgia', poster given at the OIV's 33rd World Congress of Vine and Wine, 20–5 June 2010, Tbilisi, Georgia, http://www.oiv2010.ge/POSTER/POSR_VITICULTURE/P.I.05-No 40 P Ekhvaia et al OIV.pdf
Espinel, V, 1618, *Relaciones de la vida del escudero Marcos de Obregon*, Iuan de la Cuesta, Madrid
Estienne, C, and Liébault, J, 1564, *L'Agriculture et Maison rustique*, Jacques du Puis, Paris
Fabbro, C, Fortuna, P, Burelli, O, Bertossi, S, Molinari Pradelli, A, and Peloi, B, 2002, *Vigneto Friuli dalle origini al 1500*, Ducato dei Vini Friulani, Udine
Fanizza, G, Chaabane, R, Lamaj, F, Ricciardi, L, and Resta, P, 2003, 'AFLP analysis of genetic relationships among aromatic grapevines (*Vitis vinifera*)', *Theoretical and Applied Genetics* 107, 6: 1043–7
Fanizza, G, Lamaj, F, Resta, P, and Ricciardi, L, 2005, 'Grapevine cvs Primitivo, Zinfandel and Crljenak Kastelanski: molecular analysis by AFLP', *Vitis* 44, 3: 147–8
Faujas de Saint-Fond, B, 1781, *Histoire naturelle de la province de Dauphiné*, Veuve Giroud, Grenoble
Favà i Agud, X, 2001, *Diccionari dels noms de ceps i raïms: l'ampelonímia catalana*, Biblioteca de Dialectologia i Sociolingüística 8, Institut d'Estudis Catalans, Barcelona
Fernandes, R, 1531–2, 'Descripção do terreno em roda da cidade de Lamego duas leguas', in Corréa da Serra, J (ed), *Collecção de livros ineditos de historia portugueza*, vol V, Academia das Ciências de Lisboa, Lisboa, pp 546–613
Fernández-González, M, Martínez, J, and Mena, A, 2007, 'Morphological and molecular characterization of grapevine accessions known as Moravia/o (*Vitis vinifera* L.)', *American Journal of Enology and Viticulture* 58, 4: 544–7
Fernández-González, M, Mena, A, Izquierdo, P, and Martínez, J, 2007, 'Genetic characterization of grapevine (*Vitis vinifera* L.) cultivars from Castilla La Mancha (Spain) using microsatellite markers', *Vitis* 46, 3: 126–30
Ferreira Monteiro, F, Nunes, E, Magalhaes, R, Faria, M A, Martins, A, Bowers, J E, and Meredith, C P, 2000, 'Fingerprinting of the main *Vitis vinifera* varieties grown in the northern region of Portugal', *Acta Horticulturae* 528: 121–7
Filippetti, I, Intrieri, C, Centinari, M, Bucchetti, B, and Pastore, C, 2005, 'Molecular characterization of officially registered Sangiovese clones and of other Sangiovese-like biotypes in Tuscany, Corsica and Emilia-Romagna', *Vitis* 44, 4: 167–72
Filippetti, I, Intrieri, C, and Silvestroni, O, 1999, 'Individuazione di omonimie e di sinonimie in alcune cultivar di *Vitis vinifera* attraverso metodi ampelografici e analisi del DNA a mezzo di microsatelliti', *Revista di Frutticoltura e di Ortofloricoltura* 61, 7/8: 79–84
Filippetti, I, Ramazzotti, S, Intrieri, C, Silvestroni, O, and Thomas, M R, 2002, 'Caratterizzazione molecolare e analisi filogenetica di alcuni vitigni da vino coltivati nell'Italia Centro-Settentrionale', *Rivista di Frutticoltura e di Ortofloricultura* 64: 57–63
Filippetti, I, Silvestroni, O, Ramazzotti, S, and Intrieri, C, 2001, 'Caratterizzazione morfologica e genetica dei vitigni bianchi Spergola, Sauvignon e Sémillon', *Rivista di Frutticoltura e di Ortofloricoltura* 63, 12: 83–7
Filippetti, I, Silvestroni, O, Thomas, M R, and Intrieri, C, 2001, 'Genetic characterisation of Italian wine grape cultivars by microsatellite analysis', *Acta Horticulturae* 546: 395–9
Foëx, G, 1901, 'Altesse', in Viala, P, and Vermorel, V (eds), *Ampélographie*, vol 2, Masson, Paris, pp 110–12
Foëx, G, 1904, 'Cargajola', in Viala, P, and Vermorel, V (eds), Ampélographie, vol 5, Masson, Paris, pp 259–60
Forsyth, W, 1802, *A treatise on the culture and management of fruit-trees, in which a new method of pruning and training is fully described*, Nichols and Son, London
Fossati, T, Labra, M, Castiglione, S, Failla, O, Scienza, A, and Sala, F, 2001, 'The use of AFLP and SSR molecular markers to decipher homonyms and synonyms in grapevine cultivars: the case of the varietal group known as Schiave', *Theoretical and Applied Genetics* 102, 2–3: 200–5
Francesca, N, Monaco, A, Romano, R, Lonardo, E, De Simone, M, and Moschetti, G, 2009, 'Rovello bianco caratterizzazione di un vitigno autoctono campano', *Vignevini* 4: 106–11
Franks, T, Botta, R, and Thomas, M R, 2002, 'Chimerism in grapevines: implications for cultivar identity, ancestry and genetic improvement', *Theoretical and Applied Genetics* 104: 192–9
Freeman, B, 2000, 'Données viticoles générales sur la province viticole chinoise du Shadong', *Progrès Agricole et Viticole* 117, 3: 63–4
Frei, A, Porret, N A, Frey, J E, and Gafner, J, 2006, 'Identification and characterization of Swiss grapevine cultivars using microsatellite markers', *Mitteilungen Klosterneuburg* 56: 147–56
Frenchtowner.com: http://www.frenchtowner.com/frenchtown-nj/prevost.html
Froio, G, 1875, 'Primi studi ampelografici del Principato Citeriore e del Principato Ulteriore', *Bollettino Ampelografico del Ministero Agricoltura, Industria e Commercio* 3: 184
Gagnaire, N, 1872, 'Le cépage de vigne Carmenet-Sauvignon', *Annales de la Société d'Agriculture de la Dordogne*, vol XXXIII, p 877
Gago, P, Santiago, J L, Boso, S, Alonso-Villaverde, V, Grando, M S, and Martínez, M C, 2009, 'Biodiversity and characterization of twenty-two *Vitis vinifera* L. cultivars in the northwestern Iberian peninsula', *American Journal of Enology and Viticulture* 60, 3: 293–301
Galbács, Z, 2009, 'Szılıfajtak mikroszatellit alapu ujjlenyomatanak es pedigrejenek meghatarozasa', unpublished PhD thesis, Szent István University, Gödöllö, p 103
Galbács, Z, Molnár, S, Halász, G, Kozma, P, Hoffmann, S, Kovács, L, Veres, A, Galli, Z, Szoke,

A, Heszky, L, and Kiss, E, 2009, 'Identification of grapevine cultivars using microsatellite-based DNA barcodes', *Vitis* 48, 1: 17–24
Galet, P, 1979, *A practical ampelography: grapevine identification*, Cornell University Press, Ithaca, NY
Galet, P, 1988, *Cépages et vignobles de France*, vol I, *Les vignes américaines*, 2nd edn, Charles Déhan, Montpellier
Galet, P, 1990, *Cépages et vignobles de France*, vol II, *L'ampélographie française*, Charles Dehan, Montpellier
Galet, P, 2000, *Dictionnaire encyclopédique des cépages*, Hachette, Paris
Ganzin, V, 1901a, 'Mourvèdre', in Viala, P, and Vermorel, V (eds), *Ampélographie*, vol 2, Masson, Paris, pp 237–43
Ganzin, V, 1901b, 'Tibouren', in Viala, P, and Vermorel, V (eds), 1901–10, *Ampélographie*, vol 2, Masson, Paris, pp 179–84
García de Luján, A, 1996, *La viticultura del Jerez*, Mundi-Prensa, Madrid
García-Muñoz, S, Lacombe, T, de Andrés, M T, Gaforio, L, Muñoz-Organero, G, Laucou, V, This, P, and Cabello, F, 2012, 'Grape varieties (*Vitis vinifera* L.) from the Balearic Islands: genetic characterization and relationship with Iberian Peninsula and Mediterranean Basin', *Genetic Resources and Crop Evolution* 59, 4: 589–605
Garde, J-A, 1946, *Histoire de Pomerol*, P Gelix, Libourne
Gatta, L F, 1838, *Saggio sulle viti e sui vini della Valle d'Aosta*, reprinted 1971, Fratelli Enrico Editori, Ivrea
Georgia College of Agricultural and Environmental Sciences, 2010, http://www.caes.uga.edu/commodities/fruits/muscadines/cultivars/scuppernong/scuppernong.html
Gervais, P, 1902, 'Muscats', in Viala, P, and Vermorel, V (eds), *Ampélographie*, vol 3, Masson, Paris, pp 373–83
Geuna, F, Hartings, H, and Scienza, A, 1997, 'Discrimination between cultivars of *Vitis vinifera* based on molecular variability concerning 5′ untranslated regions of the *StSy-CHS* genes', *Theoretical and Applied Genetics* 95, 3: 375–83
Gheţea, L G, Magda Motoc, R, Popescu, C F, Barbacar, N, Iancu, D, Constantinescu, C, and Bărbării, L E, 2010, 'Genetic profiling of nine grapevine cultivars from Romania, based on SSR markers', *Romanian Biotechnological Letters* 15, 1, Supplement, pp 116–24
Ghidoni, F, Emanuelli, F, Moreira, F M, Imazio, S, Grando, M S, and Scienza, A, 2008, 'Variazioni del genotipo molecolare in Verdicchio, Trebbiano di Soave e Trebbiano di Lugana', *2° Convegno Nazionale di Viticoltura (CONAVI)*, Marsala, Sicilia, 14–19 July 2008
Giannetto, S, Caruana, R, La Notte, P, Costacurta, A, and Crespan, M, 2010, 'A survey of Maltese grapevine germplasm using SSR markers', *American Journal of Enology and Viticulture* 61, 3: 419–24
Giavedoni, F, and Gily, M, 2005, *Guida ai vitigni d'Italia. Storia e caratteristiche di 580 varietà autoctone*, Slow Food Editore, Bra
Giralt, L, Puig, A, Bertran, E, Catalina, O, and Carme, D, 2009, 'Prospección, identificación y conservación de la diversidad fitogenética de variedades de *Vitis vinifera* cultivadas en Cataluña', *Acenología* 111 (30 Nov 2009): http://www.acenologia.com/cienciaytecnologia/prospeccion_diversidad_vitis_1cien1109.htm
Goethe, H, 1878, *Handbuch der Ampelographie*, Leykam-Josefsthal, Graz
Goethe, H, 1887, *Handbuch der Ampelographie*, 2nd edn, Leykam-Josefsthal, Graz
Gohdes, C L F, 1982, *Scuppernong, North Carolina's grape and its wines*, Duke University Press, Durham, NC
González-Andrés, F, Martín, J P, Yuste, J, Rubio, J A, Arranz, C, and Ortiz, J M, 2007, 'Identification and molecular biodiversity of autochthonous grapevine cultivars in the Comarca del Bierzo, León, Spain', *Vitis* 46, 2: 71–6
González Moreno, J A, Bustillo Barroso, J A, and Casa Benítez, A, 2004, *Catálogo de clones de variedades de vid en Andalucía*, Consejería de Agricultura y Pesca, Andalucía
González-Techera, A, Jubany, S, Ponce de León, I, Boido, E, Dellacassa, E, Carrau, F M, Hinrichsen, P, and Gaggero, C, 2004, 'Molecular diversity within clones of cv. Tannat (*Vitis vinifera*)', *Vitis* 43, 4: 179–85
Goryslavets, S, Risovanna, V, Bacilieri, R, Hausman, J-F, and Heuertz, M, 2010, 'A parentage study of closely related Ukrainian wine grape varieties using microsatellite markers', *Cytology and Genetics* 44, 2: 95–102
Goto-Yamamoto, N, Sawler, J, and Myles, S, 2015, 'Genetic analysis of East Asian grape cultivars suggests hybridization with wild *Vitis*', *PLoS ONE* 10, 10: e0140841. doi:10.1371/journal.pone.0140841
Gournay, B C, 1790, *Tableau général du commerce, des marchands, négocians, armateurs, &c. de la France, de l'Europe, & des autres parties du monde*, published by the author, Paris
Gozzi, R, Fontana, M, and Schneider, A, 2002, 'Il Lambrusco oliva, un vitigno da riscoprire', *Agricoltura* 1: 56–8
Graff-Höfgen, G, 2007, 'Vom Ruländer zum Grauburgunder – Der Wein der grauen Mönche', *Schriften zur Weingeschichte* 158: 1–47
Grando, M S, 2000, 'Genotyping of local grapevine germplasm', *Acta Horticulturae* 528, 1: 183–7
Grando, M S, and Frisinghelli, C, 1998, 'Grape microsatellite markers – sizing of DNA alleles and genotype analysis of some grapevine cultivars', *Vitis* 37, 2: 79–82
Grando, M S, Moreira, F M, Vouillamoz, J F, De Santis, D, Schneider, A, and Librandi, N, 2008, 'Diversità e relazioni genetiche stabilite con l'analisi del DNA', in Fregoni, C, and Nigra, O (eds), *Il Gaglioppo e i suoi fratelli I vitigni autoctoni calabresi*, Librandi Spa, Cirò Marina
Grando, M S, Stefanini, M, Zambanini, J, and Vouillamoz, J F, 2006, 'Identità e relazioni genetiche dei vitigni autoctoni trentini', *Terra Trentina* 52, 8: 24–7
Grassi, F, Labra, M, Imazio, S, Spada, A, Sgorbati, S, Scienza, A, and Sala, F, 2003, 'Evidence of a secondary grapevine domestication centre detected by SSR analysis', *Theoretical and Applied Genetics* 107: 1315–20

Gray, R D, and Atkinson, Q D, 2003, 'Language-tree divergence times support the Anatolian theory of Indo-European origin', *Nature* 426: 435–9

Grimm, J, 1840, *Weisthümer*, vol 1, Dieterich, Göttingen

Grmek, K, 2007, 'Proučevanje tipov sorte Vitovska Grganja (*Vitis vinifera L.*) v vinorodni deželi Primorska', unpublished diploma thesis, Ljubljana

Gros, A, 1930, 'Le vignoble de Princens', *Société d'Histoire et d'Archéologie de Maurienne*, vol VII, book 2: 121–4

Gülcü, M, 2010, 'Traditional grape products of Thracian region and local production form in Turkey', poster given at the OIV's 33rd World Congress of Vine and Wine, 20–5 June 2010, Tbilisi, Georgia; http://www.oiv2010.ge/index.php?page=5&lang=1#

Guyot, J, 1876, *Étude des vignobles de France*, vol II, Imprimerie Nationale, Paris

Hardy, M, 1842, 'Notes cueillies pendant un voyage viticole dans quelques départements de la France (suite)', *L'Horticulteur Universel, Journal Général des Jardiniers et des Amateurs*, 3: 142–8

Hart, M, nd, 'Status and future of the Swenson hybrids'; http://www.mngrapes.org/wp-content/uploads/2009/03/future-of-the-swenson-hybrids-mark-hart.pdf

Hârţa, M, Pamfil, D, Pop, R, and Pop, I F, 2010, 'Identification of Romanian *Vitis vinifera* L. cultivars in must using nuclear microsatellite markers', *Bulletin UASVM Horticulture* 67, 1: 198–203

Hârţa, M, Pamfil, D, Pop, R, and VicaŞ, S, 2011, 'DNA fingerprinting used for testing some Romanian wine varieties', *Bulletin UASVM Horticulture* 68, 1: 143–8

Hedrick, U P, 1919, *Manual of American grape-growing*, Macmillan, New York

Hegi, G, 1925, 'Vitis L.', in Hegi, G (ed), *Illustrierte Flora von Mitteleuropa*, vol 5, part 1, Blackwell Wissenschafts-Verlag, Munich, pp 350–425

Hillebrand, W, Lott, H, and Pfaff, F, 2003, *Taschenbuch der Rebsorten*, Fachverlag Fraund, Mainz

Hinrichsen, P, Narváez, C, Bowers, J E, Boursiquot, J-M, Valenzuela, J, Muñoz, C, and Meredith, C P, 2001, 'Distinguishing Carmenère from similar cultivars by DNA typing', *American Journal of Enology and Viticulture* 52, 4: 396–9

Hoffmann, K M, 1982, *Der Gutedel und die Burgunder. Die Lebengeschichten alter Rebfamilien,* Gesellschaft für Geschichte des Weines, Wiesbaden, p 61

Hogg, R, 1860, *The fruit manual*, Cottage Gardener Office, London

Hogg, R, Moore, T, and Paul, W, 1868, *The Florist and Pomologist: A Pictorial Monthly Magazine of Flowers, Fruits, and General Horticulture*, 1–2: 92

Hohnerlein-Buchinger, T, 1996, *Per un sublessico vitivinicolo: la storia materiale e linguistica di alcuni nomi di viti e vini Italiani*, vol VIII, Niemeyer, Tübingen

Huetz de Lemps, A, 2009, *Les vins d'Espagne*, Presses Universitaires de Bordeaux, Bordeaux

Huillard-Bréholles, J-L-A, 1859, *Historia diplomatica Friderica Secundi: sive constitutiones, privilegia, manata instrumenta quae supersunt istitus imperatoris et filiorum ejus. Accedunt epistolae Paparum et documenta varia*, vol 5, H Plon, Paris, p 2

Hvarleva, T, Hadjinicoli, A, Atanassov, I, Atanassov, A, and Ioannou, N, 2005, 'Genotyping *Vitis vinifera* L. cultivars of Cyprus by microsatellite analysis', *Vitis* 44, 2: 93–8

Hvarleva, T, Russanov, K, and Atanassov, I, 2005, 'Microsatellite markers for characterization of grape genetic resources and identification of QTLs for important agronomical traits', *Biotechnology & Biotechnological Equipment* 19 (Special Issue, 20th Anniversary AgroBioInstitute): 116–23

Hvarleva, T, Rusanov, K, Lefort, F, Tschekov, I, Atanassov, A, and Atanassov, I, 2004, 'Genotyping of Bulgarian *Vitis vinifera* L. cultivars by microsatellite analysis', *Vitis* 43, 1: 27–34

Ibañez, J, Andres, M T, Molino, A, and Borrego, J, 2003, 'Genetic study of key Spanish grapevine varieties using microsatellite analysis', *American Journal of Enology and Viticulture* 54, 1: 22–30

Ibañez, J, Vélez, M D, De Andrés, M T, and Borrego, J, 2009, 'Molecular markers for establishing distinctness in vegetatively propagated crops: a case study in grapevine', *Theoretical and Applied Genetics* 119, 7: 1213–22

Ibañez, J, Vargas, A M, Palancar, M, Borrego, J, and De Andrés, M T, 2009, 'Genetic relationships among table-grape varieties', *American Journal of Enology and Viticulture* 60, 1: 35–42

Imazio, S, Labra, M, Grassi, F, Winfield, M, Bardini, M, and Scienza, A, 2002, 'Molecular tools for clone identification: the case of the grapevine cultivar Traminer', *Plant Breeding* 121, 6: 531

Institut Français de la Vigne et du Vin, 2007, *Catalogue des variétés et clones de vigne cultivés en France*, 2nd edn, INRA, SupAgro, Viniflhor, Entav-ITV, Montpellier

Jacquat, C, and Martinoli, D, 1999, '*Vitis vinifera* L.: wild or cultivated? Study of the grape pips found at Petra, Jordan; 150 BC – AD 40', *Vegetation History and Archaeobotany* 8, 1–2: 25–30

Jahnke, G, Korbuly, J, Májer, J, and Györffyné Molnár, J, 2007, 'Discrimination of the grapevine cultivars Picolit and Kéknyelu with molecular markers', *Scientia Horticulturae* 114, 1: 71–3

Jahnke, G, Májer, J, Lakatos, A, Györffyné Molnár, J, Deák, E, Stefanovits-Bányai, E, and Varga, P, 2009, 'Isoenzyme and microsatellite analysis of *Vitis vinifera* L. varieties from the Hungarian grape germplasm', *Scientia Horticulturae* 120, 2: 213–21

James, T, 2009, 'Red Semillon: return of the winegrape', *World of Fine Wine* 25: 70–3

James, T, forthcoming, *Wines of the new South Africa*, University of California Press, Berkeley and Los Angeles

Janini, B, and Roy-Chevrier, J, 1905, 'Pedro Ximenès', in Viala, P, and Vermorel, V (eds), *Ampélographie*, vol 6, Masson, Paris, pp 111–19

Jeffreys, A J, Wilson, V, and Thein, S L, 1985, 'Hypervariable "minisatellite" regions in human DNA', *Nature* 314: 67–73

Jenny, H, 1938, 'Der Malanser Weinbau in alten Akten', *Freien Rätiers* 303: 1–9

Johnson, H, 1989, *The story of wine*, Mitchell Beazley, London

Joigneaux, P, 1865, *Le livre de la ferme et des maisons de campagne*, vol 2, Victor Masson et Fils, Paris

Jones, G V, 2006, 'Climate and terroir: impacts of climate variability and change on wine', in

Macqueen, R W, and Meinert, L D (eds), *Fine wine and terroir – the geoscience perspective*, Geoscience Canada Reprint Series, 9, Geological Association of Canada, St John's, Newfoundland

Jullien, A, 1816, *Topographie de tous les vignobles connus*, 1st edn, Mme Huzard, Paris

Jullien, A, 1832, *Topographie de tous les vignobles connus*, 3rd edn, Mme Huzard, Paris

Jung, A, 2007, 'Zinfandel, the story continues: new insights to its ancient variety history from a German point of view', *Rivista di Viticoltura e di Enologia* 3: 37–58

Kallithraka, S, Tsoutsouras, E, Tzourou, E, and Lanaridis, P, 2006, 'Principal phenolic compounds in Greek red wines', *Food Chemistry* 99, 4: 784–93

Karataş, H, Değirmenci, D, Velasco, R, Vezzulli, S, Bodur, Ç, and Ağaoğlu, Y S, 2007, 'Microsatellite fingerprinting of homonymous grapevine (*Vitis vinifera* L.) varieties in neighboring regions of South-East Turkey', *Scientia Horticulturae* 114, 3: 164–9

Katerov, K I, 2004, 'Bulgaria', in Del Zan, F, Failla, O, and Scienza, A (eds), *La vite e l'uomo dal rompicapo delle origini al salvataggio delle reliquie*, 2nd edn, ERSA – Agenzia Regionale per lo Sviluppo Rurale, Gorizia, pp 361–444

Kelley, R, 2009a, 'Côtes d'Auvergne', www.richardkelley.co.uk

Kelley, R, 2009b, 'Jasnières, Coteaux du Loir and the Coteaux du Vendômois', www.richardkelley.co.uk

Kelley, R, 2010, 'The lost vineyards of the Allier département', www.richardkelley.co.uk

Kenrick, W, 1844, *The new American orchardist*, 7th edn, Otis, Broaders, and Company, Boston

Kerridge, G H, and Antcliff, A J, 1999, *Wine grape varieties*, CSIRO, Merbein, Victoria

Kobayashi, S, Goto-Yamamoto, N, and Hirochika, H, 2004, 'Retrotransposon-induced mutations in grape skin color', *Science* 304: 982

Kobayashi, H, Fujita, K, Suzuki, S, and Takayanagi T, 2009. 'Molecular characterization of Japanese indigenous grape cultivar Koshu (*Vitis vinifera*) leaf and berry skin during grape development', *Plant Biotechnology Reports*, 3, 3: 225–41

Kocsis, M, Járomi, L, Putnoky, P, Kozma, P, and Borhidi, A, 2005, 'Genetic diversity among twelve grape cultivars indigenous to the Carpathian Basin revealed by RAPD markers', *Vitis* 44, 2: 87–91

Koruza, B, and Lavrencic, P, 2004, 'Slovenia', in Del Zan, F, Failla, O, and Scienza, A (eds), *La vite e l'uomo dal rompicapo delle origini al salvataggio delle relique*, ERSA – Agenzia Regionale per lo Sviluppo Rurale, Gorizia, pp 750–66

Kourakou-Dragona, S, 2001, *A krater full of good cheer*, Lucy Braggiotti, Athens

Kozjak, P, Korošec-Koruza, Z, and Javornik, B, 2003, 'Characterisation of cv. Refosk (*Vitis vinifera* L.) by SSR markers', *Vitis* 42, 2: 83–6

Kozma, P, Balogh, A, Kiss, E, Galli, Z, and Koncz, T, 2003, 'Study of origin of cultivar Csaba Gyöngye (Pearl of Csaba)', in Hajdu, E, and Borbas, E (eds), *Acta Horticulturae* 603: 585–91

Krämer, C, 2006, *Rebsorten in Württemberg: Herkunft, Einführung, Verbreitung und die Qualität der Weine vom Spätmittelalter bis ins 19. Jahrhundert*, Jan Thorbecke Verlag, Ostfildern

Kroll, H, 1991, 'Südosteuropa', in Van Zeist, W, Wasylikowa, K, and Behre, K E (eds), *Progress in Old World paleoethnobotany*, Rotterdam-Brookfield, Balkema, pp 161–77

Labra, M, Failla, O, Fossati, T, Castiglione, S, Scienza, A, and Sala, F, 1999, 'Phylogenetic analysis of grapevine cv. Ansonica growing on the island of Giglio, Italy, by AFLP and SSR markers', *Vitis* 38, 4: 161–6

Labra, M, Imazio, S, Grassi, F, Rossoni, M, Citterio, S, Sgorbati, S, Scienza, A, and Failla, O, 2003, 'Molecular approach to assess the origin of cv. Marzemino', *Vitis* 42, 3: 137–40

Labra, M, Moriondo, G, Schneider, A, Grassi, F, Failla, O, Scienza, A, and Sala, F, 2002, 'Biodiversity of grapevines (*Vitis vinifera* L.) grown in the Aosta Valley', *Vitis* 41, 2: 89–92

Labra, M, Winfield, M, Ghiani, A, Grassi, F, Sala, F, Scienza, A, and Failla, O, 2001, 'Genetic studies on Trebbiano and morphologically related varieties by SSR and AFLP markers', *Vitis* 40, 4: 187–90

Lacombe, T, Boursiquot, J-M, and Audeguin, L, 2004, 'Prospection, conservation et évaluation des clones de vigne en France', *Bulletin de l'OIV* 77, 885–6: 799–810

Lacombe, T, Boursiquot, J-M, Laucou, V, Dechesne, F, Varès, D, and This, P, 2007, 'Relationships and genetic diversity within the accessions related to Malvasia held in the Domaine de Vassal grape germplasm repository', *American Journal of Enology and Viticulture* 58, 1: 124–31

Lacoste, F, 1861, *Cours élémentaire d'agriculture à l'usage des écoles primaires*, Puthod Fils, Chambéry

Lacoste, P, Yuri, J A, Aranda, M, Castro, A, Quinteros, K, Solar, M, Soto, N, Gaete, J, and Rivas, J, 2010, 'Variedades de uva en Chile y Argentina (1550–1850). Genealogía del torrontés', *Mundo Agrario – Revista de Estudio Rurale* 10: 20

Ladoukakis, E D, Lefort, F, Sotiri, P, Bacu, A, Kongjika, E, and Roubelakis-Angelakis, K A, 2005, 'Genetic characterization of Albanian grapevine cultivars by microsatellite markers', *Journal International des Sciences de la Vigne et du Vin* 39, 3: 109–19

Lafforgue, G, 1947, *Le vignoble girondin*, Louis Larmat, Paris

Laiadi, Z, Bentchikou, M M, Bravo, G, Cabello, F, and Martínez-Zapater, J M, 2009, 'Molecular identification and genetic relationships of Algerian grapevine cultivars maintained at the germplasm collection of Skikda (Algeria)', *Vitis* 48, 1: 25–32

Lambert-Gocs, M, 2007, 'Discovery: Cabernet's Ancient Greek ancestor', *Desert island wine*, Ambeli Press, Williamsburg VA

Lamboy, W F, and Alpha, C G, 1998, 'Using simple sequence repeats (SSRs) for DNA fingerprinting germplasm accessions of grape (*Vitis* L.) species', *Journal of the American Society for Horticultural Science* 123, 2: 182–8

Lando, O, 1553, *Commentario delle più notabili & mostruose cose d'Italia*, Bartholomeo Cesano, Venezia

Lavignac, G, 2001, *Cépages du Sud-Ouest: 2000 ans d'histoire*, Rouergue/INRA, Rodez

Lazarakis, K, 2005, *The wines of Greece*, Mitchell Beazley, London

Leão, P C S, Riaz, S, Graziani, R, Dangl, G, Motoike, S Y, and Walker, A, 2009, 'Characterization of a Brazilian grape germplasm collection using microsatellite markers', *American Journal of Enology and Viticulture* 60, 4: 517–24

Lefort, F, Anzidei, M, Roubelakis-Angelakis, K A, and Vendramin, G G, 2000, 'Microsatellite profiling of the Greek Muscat cultivars with nuclear and chloroplast SSRs markers', *Quaderni della Scuola di Specializzazione in Scienze Viticole ed Enologiche* 23: 57–82

Lefort, F, and Roubelakis-Angelakis, K A, 2001, 'Genetic comparison of Greek cultivars of *Vitis vinifera* L. by nuclear microsatellite profiling', *American Journal of Enology and Viticulture* 52, 2: 101–8

Levadoux, L, 1948, 'Les cépages à raisins de cuve', *Progrès Agricole et Viticole* 129: 6–14

Levadoux, L, 1956, 'Les populations sauvages et cultivées de *Vitis vinifera* L.', *Annales d'Amélioration des Plantes* 1: 59–118

Li, S H, Fan, P G, Li, S C, and Yang, M R, 2007, 'Grape cultivars obtained by Institute of Botany, the Chinese Academy of Sciences and their extension in China', *Acta Horticulturae* 754: 73–8

Li, Y, Gong, Y, and Wang, Z, 2008, 'Phylogenetic relationship analysis of the wine grape variety Shelongzhu', *Sino-Overseas Grapevine & Wine* 3 (in Chinese, with English abstract)

Liger, L, 1762, *La nouvelle maison rustique, ou économie générale de tous les biens de campagne*, vol 2, Savoye, Paris

Lobo, C B de Lacerda, 1790, 'Memoria sobre a cultura das vinhas de Portugal', *Memórias económicas da Academia Real das Sciencias de Lisboa*, vol II, Academia Real das Sciencias, Lisboa

Lopes, M S, Rodrigues dos Santos, M, Eiras Dias, J E, Mendonça, D, and Da Câmara Machado, A, 2006, 'Discrimination of Portuguese grapevines based on microsatellite markers', *Journal of Biotechnology* 127, 1: 34–44

Lopes, M S, Sefc, K M, Dias, E E, Steinkellner, H, Machado, M L D, and Machado, A D, 1999, 'The use of microsatellites for germplasm management in a Portuguese grapevine collection', *Theoretical and Applied Genetics* 99, 3–4: 733–9

Lovicu, G, 2006, 'È certa l'origine sarda del vitigno Cannonau', *L'Informatore Agrario* 49: 54–9

Luo, G, 1999, 'Red wine grape variety Cabernet Gernischt – a wrong name translated into Chinese a century ago', *Journal of Fruit Science* 16, 3: 161–4 (in Chinese, with English abstract)

Mabberley, D J, 1999, '*Vitis* × *alexanderi* Prince ex Jacques (Vitaceae), the first American hybrid grapes', *Telopea* 8, 3: 377–9

Maghradze, D, Rossoni, M, Imazio, S, Maitti, C, Failla, O, Del Zan, F, Chkhartishvili, N, and Scienza, A, 2009, 'Genetic and phenetic exploration of Georgian grapevine germplasm', *Acta Horticulturae* 827: 107–11

Magnol, P, 1676, *Botanicum Monspeliense*, Officina Francisci Carteron, Lugduni

Malenica, N, Šimon, S, Besendorfer, V, Maletić, E, Karoglan Kontić, J, and Pejić, I, 2011, 'Whole genome amplification and microsatellite genotyping of herbarium DNA revealed the identity of an ancient grapevine cultivar', *Die Naturwissenschaften* 98, 9: 763–72

Maletić, E, Pejić, I, Karoglan Kontić, J, Piljac, J, Dangl, G S, Lacombe, T, Mirošević, N, and Meredith, C P, 2004, 'Zinfandel, Dobričic and Plavac mali: the genetic relationship among three cultivars of the Dalmatian coast of Croatia', *American Journal of Enology and Viticulture* 55, 4: 174–80

Maletić, E, Sefc, K M, Steinkellner, H, Karoglan Kontić, J, and Pejić, I, 1999, 'Genetic characterization of Croatian grapevine cultivars and detection of synonymous cultivars in neighboring regions', *Vitis* 38, 2: 79–83

Malossini, U, Grando, M S, Roncador, I, and Mattivi, F, 2000, 'Parentage analysis and characterization of some Italian *Vitis vinifera* crosses', *Acta Horticulturae* 528: 139–43

Manca dell'Arca, A, 1780, *Agricoltura in Sardegna*, Orsino, Napoli

Manessis, N, 2000, *The illustrated Greek wine book*, 2nd edn, Olive Press, Corfu

Mangafa, M, and Kotsakis, K, 1996, 'A new method for the identification of wild and cultivated charred grape seeds', *Journal of Archaeological Science* 23, 3: 409–18

Manzo, M, and Monaco, A, 2001, *La risorsa genetica della vite in Campania*, Regione Campania, Napoli

Maraš, V, Pejović, L, and Milutinović, M, 2004, 'Variability in the autochthonous vine variety Kratošija', *Acta Horticulturae* 640: 237–41

Martín, J P, Borrego, J, Cabello, F, and Ortiz, J M, 2003, 'Characterization of Spanish grapevine cultivar diversity using sequence-tagged microsatellite site markers', *Genome* 46, 1: 10–18

Martín, J P, Santiago, J L, Pinto-Carnide, O, Leal, F, Martínez, M C, and Ortiz, J M, 2006, 'Determination of relationships among autochthonous grapevine varieties (*Vitis vinifera* L.) in the northwest of the Iberian Peninsula by using microsatellite markers', *Genetic Resources and Crop Evolution* 53, 6: 1255–61

Martínez, J, Vicente, T, Martínez, T, Chavarri, J B, and Garcíaescudero, E, 2007, 'Tempranillo blanco, características de una nueva variedad de vid', *Vida Rural* 244: 44–8

Martínez, L, Cavagnaro, P, Boursiquot, J-M, and Agüero, C, 2008, 'Molecular characterization of Bonarda-type grapevine (*Vitis vinifera* L.) cultivars from Argentina, Italy, and France', *American Journal of Enology and Viticulture* 59, 3: 287–91

Martínez, L, Cavagnaro, P, Masuelli, R W, and Rodríguez, J, 2003, 'Evaluation of diversity among Argentine grapevine (*Vitis vinifera* L.) varieties using morphological data and AFLP markers', *Electronic Journal of Biotechnology* 6, 3: 244–53

Martínez, L E, Cavagnaro, P, Masuelli, R W, and Zúñiga, M, 2006, 'SSR-based assessment of genetic diversity in South American *Vitis vinifera* varieties', *Plant Science* 170, 6: 1036–44

Martínez, M-C, Santiago, J L, Pérez, J-E, and Boso, S, 2006, 'The grapevine cultivar Mencía (*Vitis vinifera* L.): similarities and differences with respect to other well known international cultivars', *Journal International des Sciences de la Vigne et du Vin* 40, 3: 121–32

Martínez de Toda, F, and Sancha González, J C, 2000,

'Preservación y estudio de cinco antiguas variedades de vid en La Rioja: Maturana blanca, Maturana tinta, Monastel, Ribadavia y Turruntés: estado del conocimiento en 1999', *Investigación humanística y científica en La Rioja: homenaje a Julio Luis Fernández Sevilla y Mayela Balmaseda Aróspide*, Gobierno de La Rioja, Instituto de Estudios Riojanos, Logroño, pp 391–402

Mas, A, and Pulliat, V, 1874–5, *Le vignoble ou histoire, culture et description avec planches coloriées des vignes à raisins de table et à raisins de cuve les plus généralement connues*, vol II, G Masson, Paris

Masi, E, Vignani, R, Di Giovannantonio, A, Mancuso, S, and Boselli, M, 2001, 'Ampelographic and cultural characterisation of the Casavecchia variety', *Advances in Horticultural Science* 15, 1–4: 47–55

Mason, R, 2000, *The wines and vineyards of Portugal*, Mitchell Beazley, London

Maitti, C, Andreani, L, Geuna, F, Brancadoro, L, and Scienza, A, 2009, 'Genetic characterization of *Vitis vinifera* accessions cultivated in Sicily (Italy)', *Acta Horticulturae* 827: 177–81

Mattioli, P A, 1563, *I discorsi di M Pietro Andrea Matthioli Sanese, ne i sei libri di Pedacio Dioscoride Anazarbeo della materia medicinale*, Vincenzo Valgrisi, Venetia

Maul, E, 2005, 'Die sehr alte Rebsorte Weisser Heunisch und ihre zum Teil berühmt gewordenen Kinder, wie zB Chardonnay', *Deutsches Weinbau-Jahrbuch* 56: 129–45

Maul, E, 2006, 'Zur Herkunft alter Rebsorten', *Schweizerische Zeitschrift für Obst- und Weinbau* 6: 6–9

Maul, E, Eiras Dias, J E, Kaserer, H, Lacombe, T, Ortiz, J M, Schneider, A, Maggioni, L, and Lipman, E (compilers), 2008, 'Report of a working group on *Vitis*', first meeting 12–14 June 2003, Palicí, Serbia and Montenegro, Biodiversity International, Rome

Mayson, R, 2003, *The wines and vineyards of Portugal*, Mitchell Beazley, London

McGovern, P E, 2003, *Ancient wine: the search for the origins of viniculture*, Princeton University Press, Princeton and Oxford

McGovern, P E, Glusker, D L, Exner, L J, and Voigt, M M, 1996, 'Neolithic resinated wine', *Nature* 381: 480–1

McGovern, P E, Hartung, U, Badler, V R, Glusker, D L, and Exner, L J, 1997, 'The beginnings of winemaking and viniculture in the ancient Near East and Egypt', *Expedition* 39: 3–21

Meekers, G, 2011, *Wines of Malta: the definitive guide to the new heritage wines*, Kindle edn

Meneghetti, S, Costacurta, A, Crespan, M, Maul, E, Hack, R, and Regner, F, 2009, 'Deepening inside the homonyms of Wildbacher by means of SSR markers', *Vitis* 48, 3: 123–9

Meredith, C P, and Boursiquot, J-M, 2008, 'Origins and importance of Syrah around the world', *International Syrah Symposium*, Lyon 13–14 May 2008, Oenoplurimédia, pp 17–20

Meredith, C P, Bowers, J E, Riaz, S, Handley, V, Bandman, E B, and Dangl, G S, 1999, 'The identity and parentage of the variety known in California as Petite Sirah', *American Journal of Enology and Viticulture* 50, 3: 236–42

Merlet, J 1690, *L'abrégé des bons fruits*, Charles de Sercy, Paris

Metzger, J, 1827, *Der Rheinische Weinbau in theoretischer und praktischer Beziehung*, August Osswald, Heidelberg

Migliaccio, G, Scala, F, Monaco, A, Ferranti, P, Nasi, A, De Gennaro, T, Granato, T, Nicolella, V, Grando, M S, Vouillamoz, J F, Calice, O, and Matarese, M, 2008, *Atlante delle varietà di vite dell'isola d'Ischia*, Franco di Mauro, Sorrento, Napoli

Mirošević, N, and Meredith, C P, 2000, 'A review of research and literature related to the origin and identity of the cultivars Plavac Mali, Zinfandel and Primitivo (*Vitis vinifera* L.)', *Agriculturae Conspectus Scientificus* 65, 1: 45–9

Molon, G, 1906, *Ampelografia*, Hoepli, Milano

Monaco, A, 2006, 'Vite e Vino', in D'Ambra, A, Monaco, A, and Di Salvo, M (eds), *Storia del vino d'Ischia*, Imagaenaria, Ischia, pp 85–110

Mondini, S, 1903, *I vitigni da vino stranieri coltivati in Italia*, Barbera, Firenze

Montaner, C, Martín, J P, Casanova, J, Martí, C, Badía, D, Cabello, F, and Ortiz, J M, 2004, 'Application of microsatellite markers for the characterization of Parraleta: an autochthonous Spanish grapevine cultivar', *Scientia Horticulturae* 101, 3: 343–7

Moriondo, G, 1999, *Vini e vitigni autoctoni della Valle d'Aosta*, Institut Agricole Régional, Aosta

Moriondo, G, Sandi, R, and Vouillamoz, J F, 2008, 'Identificazione del Neret di Saint-Vincent, antico vitigno valdostano', *Millevigne Regioni* 3: 111

Mortensen, J A, 1987, 'Blanc du Bois. A Florida bunch grape for white wine', *University of Florida Agricultural Research Station Circular S-340*, 4 pp

Mosher, S, 1853, 'The Plough, the loom, and the anvil', *American Farmers' Magazine* 6, 4: 193–8

Motoc, R M, 2009, 'Caracterizarea moleculară, prin utilizarea markerilor microsatelitici, a unor soiuri de *Vitis vinifera* cultivate în România şi Republica Moldova', unpublished PhD thesis, Universitatea din Bucureşti, Bucureşti

Mouillefert, P, 1893, *Translation of a report on the vineyards of Cyprus*, Government Printing Office, Nicosia, Cyprus

Mouillefert, P, 1902a, 'Aubin', in Viala, P, and Vermorel, V (eds), *Ampélographie*, vol 3, Masson, Paris, pp 60–3

Mouillefert, P, 1902b, 'Madeleine Royale', in Viala, P, and Vermorel, V (eds), *Ampélographie*, vol 3, Masson, Paris, pp 267–71

Mouillefert, P, 1902c, 'Meslier', in Viala, P, and Vermorel, V (eds), *Ampélographie*, vol 3, Masson, Paris, pp 50–9

Mouillefert, P, 1903, 'Romorantin', in Viala, P, and Vermorel, V (eds), *Ampélographie*, vol 4, Masson, Paris, pp 328–32

Muganu, M, Dangl, G, Aradhya, M, Frediani, M, Scossa, A, and Stover, E, 2009, 'Ampelographic and DNA characterization of local grapevine accessions of the Tuscia area (Latium, Italy)', *American Journal of Enology and Viticulture* 60, 1: 110–15

Munson, T V, 1909, *Foundations of American grape culture*, Orange Judd, New York

Myles, S, Boyko, A R, Owens, C L, Brown, P J, Grassi, F, Aradhya, M K, Prins, B, Reynolds, A, Chiah, J-M, Ware, D, Bustamante, C D, and Buckler, E S, 2011, 'Genetic structure and domestication history of the grape', *Proceedings of the National Academy of Sciences*, 108, 9: 3530–5

Nada Patrone, A M, 1988, 'Bere vino in area pedemontana nel Medioevo', in *Il vino nell'economia e nella società italiana* (Quaderni della Rivista di Storia dell'Agricoltura 1), Accademia dei Georgofili, Firenze, pp 31–60

Nada Patrone, A, 1991, 'I vini in Piemonte tra Medioevo ed età moderna', in Comba, R (ed), *Vigne e vini nel Piemonte rinascimentale*, L'Arciere, Cuneo, pp 247–80

Negrul, A M, 1938, 'Evolution of cultivated forms of grapes', *Comptes Rendus (Doklady) Académie Sciences USSR* 18, 8: 585–8

Negrul, A M, 1946, 'Evropejskij i aziatskij vinograd *Vitis vinifera* L.', in Baranov, A (ed), *Ampelografija SSSR*, 1, Piščepromizdat, Moskva, pp 63ff

Nelson-Kluk, S, 2005, 'Variety identification updates at FPS', *Foundation Plant Services FPS Grape Program Newsletter* November, pp 12–13

Nicollier, J, 1972, 'Un dossier relatif aux cépages dits rouges du pays en Valais et rouges indigènes en vallée d'Aoste', *Revue Suisse de Viticulture, Arboriculture, Horticulture* 4, 4: 132–5

Nieddu, G, Nieddu, M, Cocco, G F, Erre, P, and Chessa, I, 2007, 'Morphological and genetic characterization of the Sardinian Bovale cultivars', *Acta Horticulturae* 754: 49–54

Nikolau, N, and Michos, V, 2004, 'Grecia', in Del Zan, F, Failla, O, and Scienza, A (eds), *La vite e l'uomo dal rompicapo delle origini al salvataggio delle relique*, 2nd edn, ERSA – Agenzia Regionale per lo Sviluppo Rurale, Gorizia, pp 562–634

Nuvolone Pergamo (Conte di Scandaluzza), G, 1787–98, *Sulla coltivazione delle viti e sul metodo migliore di fare e conservare i vini: istruzione*, Anno VII e VIII, Torino

Odart, Comte A-P, 1845, *Ampélographie, ou Traité des cépages les plus estimés dans tous les vignobles de quelque renom*, Bixio, Paris

Odart, Comte A-P, 1854, *Ampélographie universelle*, 3rd edn, Veuve Huzard, Paris

Odart, Comte A-P, 1859, *Ampélographie universelle*, 4th edn, Librairie Agricole, Paris

Olmo, H P, 1995, 'The origin and domestication of the *Vinifera* grape', in McGovern, P E, Fleming, S J, and Katz, S H (eds), *The origins and ancient history of wine*, Gordon and Breach, Amsterdam, pp 31–43

Olmo, H P, and Koyama, A, 1962, 'Rubired and Royalty: new grape varieties for color, concentrate, and port wine', *California Agricultural Experiment Station Bulletin* 789: 3–13

Ortoidze, T, Vashakidze, L, Bezhuashvili, M, Ramishvili, N, and Jigauri, G, 2010, 'Origin and peculiarities of rarely distributed grapevine variety Asuretuli Shavi', poster given at the OIV's 33rd World Congress of Vine and Wine, 20–5 June 2010, Tbilisi, Georgia; http://www.oiv2010.ge/POSTER/POSR_VITICULTURE/P.I.16-No 101 P Ortoidze-Poster.pdf

Pacottet, P, 1904, 'Argant', in Viala, P, and Vermorel, V (eds), *Ampélographie*, vol 5, Masson, Paris, pp 346–52

Palacios Sánchez, J M, 1991, *Historia del vino de Rioja*, La Prensa del Rioja, Logroño

Petit-Lafitte, A, 1868, *La vigne dans le Bordelais*, J Rothschild, Paris

Pejić, I, Maletić, E, and Naslov, E, 2005, 'Malvazija istarska and Malvasia dubrovacka – Croatian or Greek cultivars?', *'Whose goblet is Malmsey'. Production, proccessing and distribution of wine from the Aegean to the Adriatic Sea,* 17th Symposium of History and Art, Monemvasia, Castro, 30 May–1 June 2005, Monemvasia, Grčka

Pejić, I, Mirošević, N, Maletić, E, Piljac, J, and Meredith, C P, 2000, 'Relatedness of cultivars Plavac Mali, Zinfandel and Primitivo (*Vitis vinifera* L.)', *Agriculturae Conspectus Scientificus* 65, 1: 21–5

Perret, M, 1997, 'Polymorphisme des génotypes sauvages et cultivés de *Vitis vinifera* L. détecté à l'aide de marqueurs RAPD', *Bulletin de la Société Neuchâteloise de Sciences Naturelles* 120: 45–54

Peterlunger, E, Zurlini, L, Crespan, G, Colugnati, G, and Del Zan, F, 2004, 'Friuli Venezia Giulia', in Del Zan, F, Failla, O, and Scienza, A (eds), *La vite e l'uomo dal rompicapo delle origini al salvataggio delle relique*, ERSA – Agenzia Regionale per lo Sviluppo Rurale, Gorizia, pp 769–852

Pezo, I, Budić Leto, I, Kačić, S, Zdunić, G, and Mirošević, N, 2006, 'Medna bijela (*Vitis vinifera* L.) – ampelographic properties', *Agriculturae Conspectus Scientificus* 71, 3: 81–6

Philip, B, 2007, 'Examining Pinot Blanc as a potential signature wine in the Okanagan Valley, Canada', unpublished MW dissertation

Picot de Lapeyrouse, P-I, 1814, 'Topographie rurale du canton de Montastruc département de la Haute-Garonne', *Mémoires d'agriculture, d'économie rurale et domestique*, pp 33–128

Piljac, J, 2005, *Zinfandel: a Croatian–American wine story*, Zrinski, Čakovec

Piljac, J, Maletić, E, Karoglan Kontić, J, Dangl, G S, Pejić, I, Miroševic, N, and Meredith, C P, 2002, 'The parentage of Pošip bijeli, a major white wine cultivar of Croatia', *Vitis* 41, 2: 83–7

Pinder, R M, and Meredith, C P, 2003, 'The identity and parentage of wine grapes', in Sandler, M, and Pinder, R M (eds), *Wine: a scientific exploration*, Taylor and Francis, London, pp 260–73

Pinney, T, 1989, *A history of wine in America*, vol 1, *From the beginnings to Prohibition*, University of California Press, Berkeley, Los Angeles and London

Pinotage Association, 2010, http://www.pinotage.co.za

Pinto-Carnide, O, Martín, J P, Leal, F, Castro, I, Guedes-Pinto, H, and Ortiz, J M, 2003, 'Characterization of grapevine (*Vitis vinifera* L.) cultivars from northern Portugal using RAPD and microsatellite markers', *Vitis* 42, 1: 23–6

Pl@nt Grape: le catalogue des vignes cultivées en France, http://plantgrape.plantnet-project.org

Plocher, T, 1993, 'Observations of Niels Hansen grape varieties', *Pomona* 26, 1: 40–2

Plocher, T, Rouse, G, Hart, M, 2003, 'Discovering grapes and wine in the far north of China', http://www.northernwinework.com/images/extra/chinatrip/Chinatrip.pdf

Pollini, C, 1824, 'Osservazioni agrarie per l'anno

1818', *Memorie dell'Accademia d'agricoltura, commercio ed arti di Verona*, vol X, pp 135–41
Pollini, L, 2007, *Viaggio attraverso i vitigni autoctoni italiani*, Alsaba, Colle Val d'Elsa, Italy
Pont, S, 2005, 'La coupe et le vin de l'accouchée', in Zufferey-Périsset, A D (ed), *Quand le bois sert à boire*, Regards sur la Vigne et le Vin en Valais 2, Musée Valaisan de la Vigne et du Vin, Sierre-Salquenen, pp 23–36
Possingham, J V, 1995, 'Breeding grapes for warm climates – the Australian experience', *Acta Horticulturae* 388: 129–34
Prato, G B, 1989, 'Verso le origini del Teroldego', *La Situla*, p 5
Prince, W R, 1830, *A treatise on the vine: embracing its history from the earliest ages to the present day, with descriptions of above two hundred foreign and eighty American varieties*, T & J Swords, New York
Pszczólkowski T, P, 2004, 'La invención del cv. Carménère (*Vitis vinifera* L.) en Chile, desde la mirada de uno de sus actores', *Revista Universum* 2, 19: 150–65
Puig, A, Domingo, C, and Giralt, L, 2006, 'Caracterització genètica de la varietat Picapoll de diferents orígens mitjançant marcadors microsatèl·lits', *Dossiers Agraris ICEA: la varietat picapoll* 9: 51–9
Putanec, V, 2003, 'Etimološki Prinosi (27–35)', *Rasprave Instituta za Hrvatski Jezik i Jezikoslovlje* 29, 1: 225–58
Raimondi, S, Schneider, A, and Ferrandino, A, 2005, 'Barbassese, Barbera bianca e Timorasso vitigni bianchi minori', *L'Informatore Agrario* 7: 71–5
Raimondi, S, Torello Marinoni, D, and Schneider, A, 2006, 'Caratterizzazione ampelografica e genetica di vitigni minori del Basso Piemonte oggetto di valorizzazione: nuove proposte per i viticoltori', *Italus Hortus* 13, 2: 154–7
Reale, S, Pilla, F, and Angiolillo, A, 2006, 'Genetic analysis of the Italian *Vitis vinifera* cultivar Tintilia and related cultivars using SSR markers', *Journal of Horticultural Science & Biotechnology* 81, 6: 989–94
Regner, F, 1996, 'Müller-Thurgau = Riesling × Gutedel', *Der Deutsche Weinbau* 14: 18–19
Regner, F, 1999, 'Erkenntnisse über unsere Rebsorten mittels genetischer Analyse', *Deutsches Weinbau-Jahrbuch* 50: 83–8
Regner, F, 2000a, 'Anwendung der DNA Analytik zur genetischen Analyse von Rebsorten', *ALVA (Arbeitsgemeinschaft landwirtschaftlicher Versuchsanstalten) Jahrestagung*: 94–7
Regner, F, 2000b, 'Genetische Analyse von Rebsorten', *Deutsches Weinbau-Jahrbuch* 51: 125–32
Regner, F, 2007, 'Herkunft unserer Rebsorten: Grüner Veltliner, Blaufränkisch und St Laurent', *Der Winzer* 63, 4: 12–15
Regner, F, Eiras-Dias, J E, Stadlbauer, A, and Blahous, D, 1999, 'Blauer Portugieser, the dissemination of a grapevine', *Ciencia e Tecnica Vitivinicola* 14, 2: 37–44
Regner, F, Eisenheld, C, Kaserer, H, and Stadlbauer, A, 2001, 'Weitere Sortenanalysen bei Rebe mittels genetischer Marker', *Mitteilungen Klosterneuburg, Rebe und Wein, Obstbau und Fruchteverwertung* 51, 1: 3–14
Regner, F, Stadlbauer, A, and Eisenheld, C, 1998, 'Heunisch × Fränkisch, ein wichtiger Genpool europäischer Rebsorten (*Vitis vinifera* L. *sativa*)', *Viticultural and Enological Science* 53, 3: 114–18
Regner, F, Stadlbauer, A, Eisenheld, C, and Kaserer, H, 2000a, 'Considerations about the evolution of grapevine and the role of Traminer', *Acta Horticulturae* 528: 177–9
Regner, F, Stadlbauer, A, Eisenheld, C, and Kaserer, H, 2000b, 'Genetic relationships among Pinots and related cultivars', *American Journal of Enology and Viticulture* 51, 1: 7–14
Regner, F, Steinkellner, H, Turetschek, E, Stadlhuber, A, and Glössl, J, 1996, 'Genetische Charakterisierung von Rebsorten (*Vitis vinifera*) durch Mikrosatellite-Analyse', *Mitteilungen Klosterneuburg, Rebe und Wein, Obstbau und Fruchteverwertung* 46, 2: 52–60
Reich, L, 1902a, 'Fuëlla', in Viala, P, and Vermorel, V (eds), *Ampélographie*, vol 3, Masson, Paris, pp 141–5
Reich, L, 1902b, 'Colombaud', in Viala, P, and Vermorel, V (eds), *Ampélographie*, vol 3, Masson, Paris, pp 296–301
Reisch, B I, Luce, R S, Bordelon, B, and Henick-Kling, T, 2006a, 'Corot Noir grape', *New York Food and Life Sciences Bulletin* 159: 1–8
Reisch, B I, Luce, R S, Bordelon, B, and Henick-Kling, T, 2006b, 'Noiret™ grape', *New York's Food and Life Sciences Bulletin* 160: 1–3
Reisch, B I, Luce, R S, Bordelon, B, and Henick-Kling, T, 2006c, 'Valvin Muscat™ grape', *New York's Food and Life Sciences Bulletin* 161: 1–6
Reisch, B I, Luce, R S, Henick-Kling, T, and Pool, R M, 2003, 'GR 7 grape', *New York's Food and Life Sciences Bulletin* 157: 1–2
Reisch, B I, Pool, R M, Robinson, W B, Henick-Kling, T, Gavitt, B K, Watson, J P, Martens, M H, Luce, R S, and Barret, H C, 1996, 'Traminette grape', *New York's Food and Life Sciences Bulletin* 149: 1–2
Reisch, B I, Pool, R M, Robinson, W B, Henick-Kling, T, Watson, J P, Kimball, K H, Martens, M H, Howell, G S, Miller, D P, Edson, C E, and Morris, J R, 1990, 'Chardonel grape', *New York's Food and Life Sciences Bulletin* 132: 1–3
Reisch, B I, Pool, R M, Watson, J P, Robinson, W B, and Cottrell, T H E, 1985, 'Melody grape', *New York's Food and Life Sciences Bulletin* 112: 1–2
Rendu, V, 1854, *Ampélographie française*, Victor Masson, Paris
Rézeau, P, 1997, *Dictionnaire des noms de cépages de France*, CNRS, Paris
Rhagor, D, 1639, *Pflantz-Gart*, vol 3, *Wein-Gärten*, S Schmid, Bern
Riaz, S, Garrison, K E, Dangl, G S, Boursiquot, J-M, and Meredith, C P, 2002, 'Genetic divergence and chimerism within ancient asexually propagated winegrape cultivars', *Journal of the American Society for Horticultural Science* 127, 4: 508–14
Ribéreau-Gayon, P, 1958, 'Les anthocyanes des raisins', *Qualitas Plantarum*, 3/4: 491–9

Ribéreau-Gayon, P, Boidron, J-N, Terrier, A, 1975, 'Aroma of Muscat grape varieties', *Journal of Agricultural and Food Chemistry,* 23, 6: 1042–7

Richards, P, 2006, *The wines of Chile*, Mitchell Beazley, London

Richards, P, 2010, 'Prospects for premium Chilean Syrah – UK wine trade perspectives', unpublished MW dissertation

Rivera Nunez, D, and Walker, M J, 1989, 'A review of palaeobotanical findings of early *Vitis* in the Mediterranean and of the origins of cultivated grape-vines, with special reference to new pointers to prehistoric exploitation in the western Mediterranean', *Review of Palaeobotany and Palynology* 61, 3: 205–37

Roberts, P, 'Norton, America's true grape: whence, and whither?', http://www.missouriwinecountry.com/articles/wines/norton-true.php

Rodríguez, J G, and Matus, M S, 2002, 'Caracterización ampelográfica de Torrontés Riojano – Mendoza (Argentina)', *Revista de la Facultad de Ciencias Agrarias – Universidad Cuyo* 34, 1: 71–80

Rodríguez-Torres, I, Ibañez, J, De Andrés, M T, Rubio, C, Borrego, J, Cabello, F, Zerolo, J, and Muñoz-Organero, G, 2009, 'Synonyms and homonyms of Malvasía cultivars (*Vitis vinifera* L.) existing in Spain', *Spanish Journal of Agricultural Research* 7, 3: 563–71

Roháły, G, Mészáros, G, and Nagymarosy, A, 2003, *Terra Benedicta: the land of Hungarian wine*, Ako Publishing, Budapest

Roncador, I, 2006, 'Il Teroldego Rotaliano – vino principe del Trentino', *Teroldego: un autoctono esemplare*, Cantina Rotaliana di Mezzolombardo, 7 May 2004, Provincia Autonoma di Trento, pp 36–75

Rose, L, 2006, 'Viva Viognier!', *Australian and New Zealand Wine Industry Journal* 21, 5: 71–80

Rossignol, C, 1854, *Histoire de Beaune: depuis les temps les plus reculés jusqu'à nos jours*, Batault-Morot, Beaune

Roudié, P, 1994, *Vignobles et vignerons du Bordelais (1850–1980)*, Presses Universitaires de Bordeaux, Talence

Rouget, C, 1897, *Les vignobles du Jura et de la Franche-Comté*, Auguste Cote, Lyon

Rougier, L, 1902a, 'Mornen Noir', in Viala, P, and Vermorel, V (eds), *Ampélographie*, vol 3, Masson, Paris, pp 173–7

Rougier, L, 1902b, 'Étraire de l'Aduï', in Viala, P, and Vermorel, V (eds), *Ampélographie*, vol 3, Masson, Paris, pp 191–6

Rougier, L, 1903, 'Jacquère', in Viala, P, and Vermorel, V (eds), *Ampélographie*, vol 4, Masson, Paris, pp 122–6

Rougier, L, 1904, 'Bourrisquou', in Viala, P, and Vermorel, V (eds), *Ampélographie*, vol 5, Masson, Paris, pp 338–40

Rougier, L, 1905, 'Dureza', in Viala, P, and Vermorel, V (eds), *Ampélographie*, vol 6, Masson, Paris, pp 97–9

Roy-Chevrier, J, 1903a, 'Feteasca Neagra', in Viala, P, and Vermorel, V (eds), *Ampélographie*, vol 4, Masson, Paris, pp 132–3

Roy-Chevrier, J, 1903b, 'Grassa', in Viala, P, and Vermorel, V (eds), *Ampélographie*, vol 4, Masson, Paris, pp 134–7

Royal Horticultural Society (ed) 1826, *Catalogue of fruits cultivated in the garden of the Horticultural Society of London*, William Nicol, Cleveland Row

Rozier, F, 1823, *Nouveau cours complet d'agriculture théorique et pratique*, vol 16, Deterville, Paris

Rusjan, D, Jug, T, and Štajner, N, 2010, 'Evaluation of genetic diversity: which of the varieties can be named Rebula (*Vitis vinifera* L.)?', *Vitis* 49, 4: 189–92

Sabir, A, Tangolar, S, Buyukalaca, S, and Kafka, S, 2009, 'Ampelographic and molecular diversity among grapevine (*Vitis* spp.) cultivars', *Czech Journal of Genetics and Plant Breeding* 45, 4: 160–8

Salamini, F, Özkan, H, Brandolini, A, Shaefer-Pregl, R, and Martin, W, 2002, 'Genetics and geography of wild cereal domestication in the near east', *Nature Reviews Genetics* 3: 429–41

Salaris, C, 2005–6, 'La diffusione del vitigno Nebbiolo nel mondo', *Quaderni di Scienze Viticoltura ed Enologia della Università di Torino* 28: 85–92

Salmaso, M, Dalle Valle, R, and Lucchin, M, 2008, 'Gene pool variation and phylogenetic relationships of an indigenous northeast Italian grapevine collection revealed by nuclear and chloroplast SSRs', *Genome* 51, 10: 838–55

Salomon, E, and Salomon, R, 1902a, 'Muscat de Hamburgh', in Viala, P, and Vermorel, V (eds), *Ampélographie*, vol 3, Masson, Paris, pp 105–7

Salomon, E, and Salomon, R, 1902b, 'Muscat Rouge de Madère', in Viala, P, and Vermorel, V (eds), *Ampélographie*, vol 3, Masson, Paris, pp 319–21

Santana, J C, Heuertz, M, Arranz, C, Rubio, J A, Martínez-Zapater, J M, and Hidalgo, E, 2010, 'Genetic structure, origins, and relationships of grapevine cultivars from the Castilian Plateau of Spain', *American Journal of Enology and Viticulture* 61, 2: 214–24

Santana, J C, Hidalgo, E, Lucas, A I, Recio, P, Ortiz, J M, Martín, J P, Yuste, J, Arranz, C, and Rubio, J A, 2007, 'Identification and relationships of accessions grown in the grapevine (*Vitis vinifera* L.) Germplasm Bank of Castilla y León (Spain) and the varieties authorized in the VQPRD areas of the region by SSR-marker analysis', *Genetic Resources and Crop Evolution* 55, 4: 573–83

Santiago, J L, Boso, S, Gago, P, Alonso-Villaverde, V, and Martínez, M C, 2007 'Molecular and ampelographic characterisation of *Vitis vinifera* L. Albariño, Savagnin Blanc and Caíño Blanco shows that they are different cultivars', *Spanish Journal of Agricultural Research* 5, 3: 333–40

Santiago, J L, Boso, S, Gago, P, Alonso-Villaverde, V, and Martínez, M C, 2008, 'A contribution to the maintenance of grapevine diversity: the rescue of Tinta Castañal (*Vitis vinifera* L.), a variety on the edge of extinction', *Scientia Horticulturae* 116: 199–204

Santiago, J L, Boso, S, Martín, J P, Ortiz, J M, and Martínez, M C, 2005, 'Characterisation and identification of grapevine cultivars (*Vitis vinifera* L.) from northwestern Spain using microsatellite markers and ampelometric methods', *Vitis* 44, 2:

67–72
Santiago, J L, Boso, S, Vilanova, M, and Martínez, M C, 2005, 'Characterisation of cv. Albarín Blanco (*Vitis vinifera* L.). Synonyms, homonyms and errors of identification', *Journal International des Sciences de la Vigne et du Vin* 39, 2: 57–65
Savoie, P, 2003, 'Charbono: a grape struggles to avoid extinction', *Wine Business Monthly*, 17 May
Scalabrelli, G, D'Onofrio, C, Ferroni, G, De Lorenzis, G, Giannetti, F, and Baldi, M, 2008, 'L'identità del Morellino Pizzuto', in ARSIA (ed), *Terzo Simposio Internazionale sul Sangiovese Modelli di terroir per vini di eccellenza*, p 8
Scalabrelli, G, Vignani, R, Scali, M, Di Pietro, D, Materazzi, A, and Triolo, E, 2001, 'Il Morellino Pizzuto: un biotipo di Sangiovese?', in ARSIA (ed), *Proceedings of the International Symposium Il Sangiovese, 15–17 February*, Florence, p 107
Scartezzini, H, 2005, 'Geschichtliche Spurensuche zur Rebsorte Lagrein', *Deutsches Weinbau-Jahrbuch* 56: 146–56
Schirmer, R, 2010, *Muscadet: histoire et géographie du vignoble nantais*, Presses Universitaires de Bordeaux, Pessac
Schneider, A, 2005–6, 'Aspetti genetici nello studio dei vitigni del territorio', *Quaderni di Scienze Viticole ed Enologiche* 28: 7–16
Schneider, A, Boccacci, P, and Botta, R, 2003, 'Genetic relationships among grape cultivars from north-western Italy', *Acta Horticulturae* 603: 229–35
Schneider, A, Boccacci, P, Torello Marinoni, D, Botta, R, Akkak, A, and Vouillamoz, J F, 2004, 'Variabilità genetica e parentele inaspettate del Nebbiolo', *Convegno Internazionale sul Vitigno Nebbiolo*, Sondrio, Valtellina, Italy, 23–5 June 2004
Schneider, A, Carra, A, Akkak, A, This, P, Laucou, V, and Botta, R, 2001, 'Verifying synonymies between grape cultivars from France and Northwestern Italy using molecular markers', *Vitis* 40, 4: 197–203
Schneider, A, and Mannini, F, 1990, 'Indagine comparative su Vermentino, Pigato e Favorite in Piemonte e Liguria', *L'Informatore Agrario* 46, 8: 103–8
Schneider, A, and Mannini, F, 2006, 'Vitigni del Piemonte – varietà e cloni', *Quaderni della Regione Piemonte – Agricoltura*, Supplement 50 (332 pp)
Schneider, A, Mannini, F, and Cravero, M C, 2004, 'Nuovi vitigni per il Piemonte. 2° contributo. Esame delle attitudini colturali ed enologiche di vitigni autoctoni minori e rari', *Agricoltura* 47: 25–30
Schneider, A, Raimondi, S, De Santis, D, and Cavallo, L, 2008, 'Schede ampelografiche e analitiche', in Fregoni, C, and Nigra, O (eds), *Il Gaglioppo e i suoi fratelli I vitigni autoctoni calabresi*, Librandi Spa, Cirò Marina
Schneider, A, Raimondi, S, Grando, M S, Zappia, R, De Santis, D, Torello Marinoni, D, and Librandi, N, 2008, 'Studi per il riordino del germoplasma viticolo della Calabria', in Fregoni, C, and Nigra, O (eds), *Il Gaglioppo e i suoi fratelli I vitigni autoctoni calabresi*, Librandi Spa, Cirò Marina, pp 114–25
Schneider, A, Raimondi, S, Moreira, F M, De Santis, D, Zappia, R, Torello Marinoni, D, Librandi, N, and Grando, M S, 2009, 'Contributo all'identificazione dei principali vitigni calabresi', *Frutticoltura* 1/2: 46–55
Schneider, A, Torello Marinoni, D, Boccacci, P, and Botta, R, 2005–6, 'Relazioni genetiche del vitigno Nebbiolo', *Quaderni di Scienze Viticole ed Enologiche della Università di Torino* 28: 93–100
Schneider, A, Torello Marinoni, D, and Crespan, M, 2008, 'Genetics and ampelography trace the origin of Muscat fleur d'oranger', *American Journal of Enology and Viticulture* 59, 2: 200–4
Schneider, A, Torello Marinoni, D, de Andrés, M T, Raimondi, S, Cabello, F, Ruffa, P, Garcia-Muñoz, S, and Muñoz-Organero, G, 2010, 'Prié Blanc and Legiruela: a unique grape cultivar grown in distant European regions', *Journal International des Sciences de la Vigne et du Vin* 44, 1: 1–7
Schneider, A, Torello Marinoni, D, Raimondi, S, Boccacci, P, and Gambino, G, 2009, 'Molecular characterization of wild grape populations from north-western Italy and their genetic relationship with cultivated varieties', *Acta Horticulturae* 827: 211–16
Schreiner, J, 2001, *Icewine: the complete story*, Warwick Publishing, Toronto
Schuck, M R, Moreira, F M, Guerra, M P, Voltolini, J A, Grando, M S, and Lima da Silva, A, 2009, 'Molecular characterization of grapevine from Santa Catarina, Brazil, using microsatellite markers', *Pesquisa Agropecuária Brasileira* 44, 5: 487–95
Schumann, F, 1983, 'Der Gänsfüßer – eine alte Rebsorte der Pfalz', *Schriften zur Weingeschichte* 67: 27–38
Scienza, A, and Boselli, M, 2003, 'Vini e vitigni della Campania: tremila anni di storia', Prismi, Napoli, www.vinocampania.it/uplPagine/introduzione.pdf
Scienza, A, Brancadoro, L, Branzanti, E, Di Lorenzo, R, Di Stefano, R, Gagliano, F, Falco, V, and Ansaldi, G, 2008, 'Vitigni autoctoni siciliani, un valore aggiunto che cresce', *L'Informatore Agrario* 23, Supplement, pp 14–20
Scienza, A, and Failla, O, 1996, 'La circolazione dei vitigni in ambito Padano-Veneto ed Atesino: le fonti storico-letterarie e l'approccio biologico-molecolare', in Forni, G, and Scienza, A (eds), *2500 anni di cultura della vite nell'ambito alpino e cisalpino*, Istituto Trentino del Vino, Trento, pp 185–268
Scienza, A, Failla, O, Anzani, R, Mattivi, F, Villa, P, Gianazza, E, Tedesco, G, and Benetti, U, 1990, 'Le possibili analogie tra il Lambrusco a foglia frastagliata, alcuni vitigni coltivati e le viti selvatiche del basso Trentino', *Vignevini* 9: 25–36
Sefc, K M, Lopes, M S, Lefort, F, Botta, R, Roubelakis-Angelakis, K A, Ibañez, J, Pejić, I, Wagner, H W, Glössl, J, and Steinkellner, H, 2000, 'Microsatellite variability in grapevine cultivars from different European regions and evaluation of assignment testing to assess the geographic origin of cultivars', *Theoretical and Applied Genetics* 100, 3–4: 498–505

Sefc, K M, Pejić, I, Maletić, E, Thomas, M R, and Lefort, F, 2009, 'Microsatellite markers for grapevine: tools for cultivar identification and pedigree reconstruction', in Roubelakis-Angelakis, K A (ed), *Grapevine molecular physiology and biotechnology*, 2nd edn, Springer Science+Business Media BV, New York, pp 565–96

Sefc, K M, Regner, F, Glössl, J, and Steinkellner, H, 1998, 'Genotyping of grapevine and rootstock cultivars using microsatellite markers', *Vitis* 37, 1: 15–20

Sefc, K M, Steinkellner, H, Glössl, J, Kampfer, S, and Regner, F, 1998, 'Reconstruction of a grapevine pedigree by microsatellite analysis', *Theoretical and Applied Genetics* 97, 1–2: 227–31

Sefc, K M, Steinkellner, H, Wagner, H W, Glössl, J, and Regner, F, 1997, 'Application of microsatellite markers to parentage studies in grapevine', *Vitis* 36, 4: 179–83

Şelli, F, Bakir, M, Inan, G, Aygün, H, Boz, Y, Yaşasin, A S, Özer, C, Akman, B, Söylemezoğlu, G, Kazan, K, and Ergül, A, 2007, 'Simple sequence repeat-based assessment of genetic diversity in Dimrit and Gemre grapevine accessions from Turkey', *Vitis* 46, 4: 182–7

Semichon, L, 1905, 'Grenache', in Viala, P, and Vermorel, V (eds), *Ampélographie*, vol 6, Masson, Paris, pp 285–92

Sensi, E, Vignani, R, Rohde, W, and Biricotti, S, 1996, 'Characterization of genetic biodiversity with *Vitis vinifera* L. Sangiovese and Colorino genotypes by AFLP and ISTR DNA marker technology', *Vitis* 35, 4: 183–8

Serres, O de, 1600, *Le Theatre d'Agriculture et Mesnage des champs*, Mamet, Paris

Sestini, D, 1991, *Memorie sui vini siciliani*, Sellerio, Palermo

Silvestroni, O, Dipietro, D, Intrieri, C, Vignani, R, Filippetti, I, Delcasino, C, Scali, M, and Cresti, M, 1997, 'Detection of genetic diversity among clones of cv. Fortana (*Vitis vinifera* L.) by microsatellite DNA polymorphism analysis', *Vitis* 36, 3: 147–50

Šimon, S, Maletić, E, Karoglan Kontić, J, Crespan, M, Schneider, A, and Pejić, I, 2007, 'Cv Maraština – a new member of Malvasia group', *II Simposio Internazionale 'Malvasie del Mediterraneo', 2–6 October*, Salina, Italy

Simpée, T, 1902, 'Tressot', in Viala, P, and Vermorel, V (eds), *Ampélographie*, vol 3, Masson, Paris, pp 302–9

SIVVEM, 'El Sistema de Identificaión de Variedades de Vid Espagñolas mediante Microsatélites, Departmento de Biología Vegetal Universidad Politécnica de Madrid, Escuela Técnica Superior de Ingenieros Agrónomos'; http://www.sivvem.monbyte.com/sivvem.asp

Skelton, S, 2008, *UK vineyards guide 2008*, S P Skelton, London

Sladonja, B, Poljuha, D, Plavša, T, Peršurić, Đ, and Crespan, M, 2007, 'Autochthonous Croatian grapevine cultivar Jarbola – molecular, morphological and oenological characterization', *Vitis* 46, 2: 99–100

Slate, G, Watson, J, and Einset, J, 1962, 'Grape varieties introduced by the New York State Agricultural Experiment Station, 1928–1961', *New York State Agricultural Experiment Station Bulletin* 794: 24–5

Smiley, L A, 2008, 'A review of cold climate grape cultivars', MS project, Iowa State University, http://viticulture.hort.iastate.edu/cultivars/cultivars.html

Smith, J J, 1858, 'New grapes, etc., in England', *Horticulturist and Journal of Rural Art and Rural Taste*, New Series, 8: 167–9

Smithsonian Institution 2009, 'Dadiana dynasty', http://achp.si.edu/dadiani/ethnography.html

Soderini, G, 1600, *Trattato della coltivazione delle viti, e del frutto che se ne può cavare*, Filippo Giunti, Firenze

Soejima, A, and Wen, J, 2006, 'Phylogenetic analysis of the grape family (*Vitaceae*) based on three chloroplast markers', *American Journal of Botany* 93, 2: 278–87

Sondley, F A, 1918, *The origin of the Catawba grape: and other sketches*, Asheville, North Carolina

Song, L, Yin, K, Zhai, H, Zhao, L, and Yao, Y, 2005, 'Phylogenetic analysis of grapevine cultivar Cabernet Gernischet by RAPD', *Chinese Agricultural Science Bulletin* 7 (in Chinese, with English abstract)

Spiegel-Roy, P, Cohen, S, Baron, I, Assaf, R, and Ben-A'haron, S, 1996, 'Argaman: a new, highly colored, productive *vinifera* wine cultivar', *Horticultural Science* 31, 7: 1252–3

Sprandel, R, 1998, *Von Malvasia bis Kötzschenbroda: die Weinsorten auf den spätmittelalterlichen Märkten Deutschlands*, Franz Steiner Verlag, Stuttgart

Spring, J-L, Gugerli, P, Brugger, J-J, Pont, M, Parvex, C, and Vouillamoz, J F, 2008, 'Les Muscats en Valais', *Revue Suisse de Viticulture, Arboriculture, Horticulture* 40, 4: 257–61

Staab, J, 1971, '500 Jahre Rheingauer Klebrot = Spätburgunder', *Gesellschaft für Geschichte des Weines* 24: 1–13

Staab, J, 1991, *Der Riesling – Geschichte einer Rebsorte*, Gesellschaft für Geschichte des Weines 24, Wiesbaden

Štajner, N, Angelova, E, Božinović, Z, Petkov, M, and Javornik, B, 2009, 'Microsatellite marker analysis of Macedonian grapevines (*Vitis vinifera* L.) compared to Bulgarian and Greek cultivars', *Journal International des Sciences de la Vigne et du Vin* 43, 1: 29–34

Štajner, N, Korošec-Koruza, Z, Rusjan, D, and Javornik, B, 2008, 'Microsatellite genotyping of old Slovenian grapevine varieties (*Vitis vinifera* L.) of the Primorje (coastal) winegrowing region', *Vitis* 47, 4: 201–4

Štajner, N, Takse, J, Javornik, B, Masuelli, R W, and Martínez, L E, 2009, 'Highly variable AFLP and S-SAP markers for the identification of Malbec and Syrah clones', *Vitis* 48, 3: 145–50

Stavrakakis, M N, and Biniari, K, 1998, 'Genetic study of grape cultivars belonging to the Muscat family by random amplified polymorphic DNA markers', *Vitis* 37, 3: 119–22

Stenkamp, S, 2009, 'Entwicklung eines molekularen Markers für die Beerenfarbe bei *Vitis vinifera* L', Diplomarbeit, Institut für Weinbau und

Rebenzüchtung, Geisenheim (available on line: http://fdw.campus-geisenheim-service.de/fileadmin/fdw/2009Nov/abgeschlossene_Projekte/ARbeitskreis2/10_FDW_Abschlussbericht_Ruehl_2008_Farbmaker_Beerenfarbe.pdf)

Stoltz, J-L, 1852, *Ampélographie rhénane*, Dusacq, Paris

Storchi, P, 2007, 'Foglia tonda, ricco di colore e profumi', *Civiltà del Bere*, February, pp 76–8

Storchi, P, Armanni, A B, Crespan, M, Frare, E, De Lorenzis, G, D'Onofrio, C, and Scalabrelli, G, 2009, 'Indagine sull'identità delle Malvasie a bacca nera coltivate in Toscana', *Malvasias – III International Symposium*, La Palma, Islas Canarias, 26–30 May, pp 1–10

Storchi, P, Armanni, A B, Randellini, L, Giannetto, S, Meneghetti, S, and Crespan, M, 2011, 'Investigations on the identity of Canaiolo bianco and other white grape varieties of central Italy', *Vitis* 50, 2: 59–64

Stover, E, Aradhya, M K, Yang, J, Bautista, J, and Dangl, G S, 2009, 'Investigations into the origin of Norton grape using SSR markers', *Proceedings of the Florida State Horticultural Society* 122: 19–24

Strang, P, 2009, *South-west France: the wines and winemakers*, University of California Press, Berkeley and Los Angeles

Stummer, A, 1911, 'Zur Urgeschichte der Rebe und des Weinbaues', *Mitteilungen der Anthropologischen Gesellschaft in Wien* 41: 283–96

Sullivan, C L, 1998, *A companion to California wine: an encyclopedia of wine and winemaking from the mission period to the present*, University of California Press, Berkeley and Los Angeles

Sullivan, C L, 2003, *Zinfandel: a history of a grape and its wine*, University of California Press, Berkeley and Los Angeles

Sun, Q (ed), 2004, *Record of Chinese grapes*, China Agricultural Sciences Publishing

Swenson, E, Pierquet, P, and Stushnoff, C, 1980, 'Edelweiss and Swenson Red grapes', *Horticultural Science* 15, 1: 100

Tallavignes, C, 1902, 'Duras', in Viala, P, and Vermorel, V (eds), 1901–10, *Ampélographie*, vol 3, Masson, Paris, pp 328–33

Tanara, V, 1644, *L'economia del cittadino in villa*, vol 7, G Monti, Bologna

Tapia, A M, Cabezas, J A, Cabello, F, Lacombe, T, Martínez-Zapater, J M, Hinrichsen, P, and Cervera, M T, 2007, 'Determining the Spanish origin of representative ancient American grapevine varieties', *American Journal of Enology and Viticulture* 58, 2: 242–51

Țârdea, C, and Rotaru, L, 2003, *Ampelografie*, Ion Ionescu de la Brad, Iași, Romania

Tchernia, A, 1986, *Le vin de l'Italie romaine*, École française d'Athènes et de Rome, Roma

Teissedre, P-L, and Landrault, N, 2000, 'Wine phenolics: contribution to dietary intake and bioavailability', *Food Research International*, 33, 6: 461–7

Terral, J-F, Tabard, E, Bouby, L, Ivorra, S, Pastor, T, Figueiral, I, Picq, S, Chevance, J-B, Jung, C, Fabre, L, Tardy, C, Compan, M, Bacilieri, R, Lacombe, T, and This, P, 2010, 'Evolution and history of grapevine (*Vitis vinifera*) under domestication: new morphometric perspectives to understand seed domestication syndrome and reveal origins of ancient European cultivars', *Annals of Botany* 105, 3: 443–55

Theodorus, J, 1588, *Neuw Kreuterbuch*, Dreuter, Franckfurt-am-Mayn

This, P, Lacombe, T, Cadle-Davidson, M, and Owens, C L, 2007, 'Wine grape (*Vitis vinifera* L.) color associates with allelic variation in the domestication gene *VvmybA1*', *Theoretical and Applied Genetics* 114, 4: 723–30

This, P, Lacombe, T, and Thomas, M R, 2006, 'Historical origins and genetic diversity of wine grapes', *Trends in Genetics* 22, 9: 511–19

Thistlewood, S, 2007, 'Verdelho – the variety polarising opinion', *Australian and New Zealand Wine Industry Journal* 22, 5: 80–94

Thomas, M R, Matsumoto, S, Cain, P, and Scott, N S, 1993, 'Repetitive DNA of grapevine: classes present and sequences suitable for cultivar identification', *Theoretical and Applied Genetics* 86: 173–80

Thomas, M R, and Scott, N S, 1993, 'Microsatellite repeats in grapevine reveal DNA polymorphisms when analysed as sequence-tagged sites (STSs)', *Theoretical and Applied Genetics* 86, 8: 985–90

Tochon, P, 1869, 'Les cépages de la Savoie', *Le Messager Agricole* 9: 366–74

Tochon, P, 1887, *Monographie des vignes, des cépages et des vins des deux départements de la Savoie*, Ménard, Chambéry

Toni, G, 1927, *Viticoltura ed enologia emiliana*, L'Italia Agricola, Italy

Torello Marinoni, D, Raimondi, S, Boccacci, P, and Schneider, A, 2006, 'Lambruschi del Piemonte: aspetti storici, caratterizzazione molecolare e relazioni genetiche con vitigni autoctoni piemontesi ed emiliani', *Italus Hortus* 13, 2: 158–61

Torello Marinoni, D, Raimondi, S, Mannini, F, and Rolle, L, 2009, 'Genetic and phenolic characterization of several intraspecific crosses (*Vitis vinifera* L.) registered in the Italian national catalogue', *Acta Horticulturae* 827: 485–92

Torello Marinoni, D, Raimondi, S, Ruffa, P, Lacombe, T, and Schneider, A, 2009, 'Identification of grape cultivars from Liguria (north-western Italy)', *Vitis* 48, 4: 175–83

Torre y Ocón, F de la, 1728, *El maestro de las dos lenguas. Diccionario español y francés, francés y español*, Juan de Ariztia, Madrid

Trümmer, F, 1841, *Systematische Klassification und Beschreibung der im Herzogthum Steiermark vorkommenden Rebensorten*, Leykam'sche Erben, Grätz

Ulanovsky, S, Gogorcena, Y, Martínez de Toda, F, and Ortiz, J M, 2002, 'Use of molecular markers in detection of synonymies and homonymies in grapevines (*Vitis vinifera* L.)', *Scientia Horticulturae* 92, 3–4: 241–54

Valamoti, S M, Mangafa, M, Koukouli-Chrysanthaki, C, and Malamidou, D, 2007, 'Grape-pressings from northern Greece: the earliest wine in the Aegean?', *Antiquity* 81: 54–61

Vannuccini, V, 1902, 'San Gioveto', in Viala, P, and Vermorel, V (eds), *Ampélographie*, vol 3, Masson, Paris, pp 332–6

Vantini, F, Tacconi, G, Gastaldelli, M, Govoni, C, Tosi, E, Malacrinò, P, Bassi, R, and Cattivelli, L, 2003, 'Biodiversity of grapevines (*Vitis vinifera* L.) grown in the Province of Verona', *Vitis* 42, 1: 35–8

Varga, I, Gál, L, and Misik, S, 1998, 'Zenit, one of the most successful varieties of the Hungarian wine-grape breeding', in Botos, E P, and Hajdu, E (eds), *Acta Horticulturae* 473, pp 63–7

Varga, Z, 2008, 'Investigation of the cultivation value and the relations of origin of old grapevine cultivars in Tokaj', unpublished PhD thesis, Corvinus University, Budapest

Varga, Z, Bisztray, G, Bodor, P, and Lõrincz, A, 2008, 'Régi Tokaj-Hegyaljai fajták értékelése mikroszatellit markerekkel (Characterization of old grapevine cultivars of Tokaj by microsatellite markers)', *Szőlészet és Borászat* 40, 4: 47–53

Vargas, A M, Teresa de Andrés, M, Borrego, J, and Ibañez, J, 2009, 'Pedigrees of fifty table-grape cultivars', *American Journal of Enology and Viticulture* 60, 4: 525–32

Vargas, A M, Velez, M D, de Andrés, M T, Laucou, V, Lacombe, T, Boursiquot, J-M, Borrego, J, and Ibañez, J, 2007, 'Corinto bianco: a seedless mutant of Pedro Ximenes', *American Journal of Enology and Viticulture* 58: 540–3

Vavilov, N I, 1926, *Studies on the origin of cultivated plants*, Leningrad

Veloso, M M, Almandanim, C, Baleiras-Couto, M, Pereira, H S, Carneiro, L C, Fevereiro, P, Eiras-Dias, J E, 2010, 'Microsatellite database of grapevine (*Vitis vinifera* L.) cultivars used for wine production in Portugal', *Ciência e Técnica Vitivínicola*, 25, 2: 53–61

Vermorel, V, 1902, 'Gamay Beaujolais', in Viala, P, and Vermorel, V (eds), *Ampélographie*, vol 3, Masson, Paris, pp 5–24

Viala, P, and Vermorel, V, 1901–10, *Ampélographie*, 7 vols, Masson, Paris

Vidal, J R, Coarer, M, and Defontaine, A, 1999, 'Genetic relationships among grapevine varieties grown in different French and Spanish regions based on RAPD markers', *Euphytica* 109, 3: 161–72

Vignani, R, Bowers, J E, and Meredith, C P, 1996, 'Microsatellite DNA polymorphism analysis of clones of *Vitis vinifera* Sangiovese', *Scientia Horticulturae* 65, 2–3: 163–9

Vignani, R, Masi, E, Scali, M, Milanesi, C, Scalabrelli, G, Wang, W, Sensi, E, Paolucci, E, Percoco, G, and Cresti, M, 2008, 'A critical evaluation of SSRs analysis applied to Tuscan grape (*Vitis vinifera* L.) germplasm', *Advances in Horticultural Science* 22, 1: 33–7

Vignani, R, Scali, M, Masi, E, and Cresti, M, 2002, 'Genomic variability in *Vitis vinifera* L. Sangiovese assessed by microsatellite and non-radioactive AFLP test', *Electronic Journal of Biotechnology* 5, 1: 1–11

Vila Maior, Visconde de, 1875, *Manual de viticultura practica*, Imprensa da Universidade, Coimbra

Vilanova, M, de la Fuente, M, Fernández-González, M, and Masa, A, 2009, 'Identification of new synonymies in minority grapevine cultivars from Galicia (Spain) using microsatellite analysis', *American Journal of Enology and Viticulture* 60, 2: 236–40

Villeneuve, Comte de, 1901, 'Mondeuse Blanche', in Viala, P, and Vermorel, V (eds), *Ampélographie*, vol 2, Masson, Paris, p 284

Viollet, A, and Boursiquot, J-M, 2009, 'A la recherche des ancêtres de nos cépages', *La Vigne* 208: 34–5

Vitis International Variety Catalogue, http://www.vivc.de/index.php

Vouillamoz, J F, 2005, 'Quer durch den Rebgarten', *Marmite* 4: 20–1

Vouillamoz, J F, 2008, 'Mondeuse Noire et Viognier dans l'arbre généalogique de la Syrah', *International Syrah Symposium*, Lyon, 13–14 May 2008, Oenoplurimédia, pp 108–10

Vouillamoz, J F, 2009a, 'Premières mentions de cépages dans le Registre d'Anniviers', in Zufferey-Périsset, A-D (ed), *Histoire de la vigne et du vin en Valais: des origines à nos jours*, Musée Valaisan de la Vigne et du Vin, Sierre, pp 84–5

Vouillamoz, J F, 2009b, 'Raretés locales: Diolle, Goron de Bovernier, Grosse Arvine, Rouge de Fully', in Zufferey-Périsset, A-D (ed), *Histoire de la vigne et du vin en Valais: des origines à nos jours*, Musée Valaisan de la Vigne et du Vin, Sierre, pp 216–17

Vouillamoz, J F, 2009c, 'Les nouveaux croisements de cépages', in Zufferey-Périsset, A-D (ed), *Histoire de la vigne et du vin en Valais: des origines à nos jours*, Musée Valaisan de la Vigne et du Vin, Sierre, pp 532–3

Vouillamoz, J F, 2009d, 'Malvoisie, un nom équivoque', in Zufferey-Périsset, A-D (ed), *Histoire de la vigne et du vin en Valais: des origines à nos jours*, Musée Valaisan de la Vigne et du Vin, Sierre, p 215

Vouillamoz, J F, and Arnold, C, 2009, 'Étude historico-génétique de l'origine du Chasselas', *Revue Suisse de Viticulture, Arboriculture et Horticulture* 41, 5: 299–307

Vouillamoz, J F, and Arnold, C, 2010, 'Microsatellite pedigree reconstruction provides evidence that Müller-Thurgau is a grandson of Pinot and Schiava Grossa', *Vitis* 49, 2: 63–5

Vouillamoz, J F, Frei, A, and Arnold, C, 2008, 'Swiss Vitis Microsatellite Database: profils génétiques des vignes suisses sur Internet', *Revue Suisse de Viticulture, Arboriculture, Horticulture* 40, 3: 187–93

Vouillamoz, J F, and Grando, M S, 2006, 'Genealogy of wine grape cultivars: Pinot is related to Syrah', *Heredity* 97, 2: 102–10

Vouillamoz, J F, Grando, M S, Ergül, A, Agaoglu, S, Tevzadze, G, Meredith, C P, and McGovern, P E, 2004, 'Is Transcaucasia the cradle of viticulture? DNA might provide an answer', in Vieira, A (ed), *III Symposium of the International Association of History and Civilization of the Vine and the Wine*, Centro de Estudos de História do Atlântico, Funchal (Madeira), 5–8 October 2003, pp 277–90

Vouillamoz, J F, Imazio, S, Stefanini, M, Scienza, A, and Grando, M S, 2004, 'Relazioni genetiche del Sangiovese', *Proceedings of the II International Symposium on Sangiovese' Firenze, Italy, 17–18 November* 2004, Agenzia Regionale per lo Sviluppo e l'Innovazione in Agricoltura

Vouillamoz, J F, McGovern, P E, Ergül, A, Söylemezoğlu, G, Tevzadze, G, Meredith, C P, and Grando, M S, 2006, 'Genetic characterization and relationships of traditional grape cultivars from Transcaucasia and Anatolia', *Plant Genetic Resources* 4, 2: 144–58

Vouillamoz, J F, Maigre, D, and Meredith, C P, 2003, 'Microsatellite analysis of ancient alpine grape cultivars: pedigree reconstruction of *Vitis vinifera* L. Cornalin du Valais', *Theoretical and Applied Genetics* 107, 3: 448–54

Vouillamoz, J F, Maigre, D, and Meredith, C P, 2004, 'Identity and parentage of two alpine grape cultivars from Switzerland (*Vitis vinifera* L. Lafnetscha and Himbertscha)', *Vitis* 43, 2: 81–8

Vouillamoz, J F, Monaco, A, Costantini, L, Stefanini, M, Scienza, A, and Grando, M S, 2007, 'The parentage of Sangiovese, the most important Italian wine grape', *Vitis* 46, 1: 19–22

Vouillamoz, J F, Monaco, A, Costantini, L, Zambanini, J, Stefanini, M, Scienza, A, and Grando, M S, 2008, 'Il Sangiovese è per metà figlio del Calabrese di Montenuovo', *L'Informatore Agrario* LXIV, 5: 59–61

Vouillamoz, J, and Moriondo, G, 2011, *Origine des cépages valaisans et valdôtains: l'ADN rencontre l'Histoire*, Belvédère, Fleurier & Pontarlier

Vouillamoz, J F, Schneider, A, and Grando, M S, 2007, 'Microsatellite analysis of Alpine grape cultivars: alleged descendants of Pliny the Elder's *Raetica* are genetically related', *Genetic Resources and Crop Evolution* 54, 5: 1095–104

Vršič, S, Ivančič, A, Šušek, A, Zagradišnik, B, Valdhuber, J, and Šiško, M, 2011, 'The world's oldest living grapevine specimen and its genetic relationships', *Vitis* 50, 4: 167–71

Wagner, P, 1988, 'On the trail of St Vincent', *Wine East* 15, 6: 8–9

Walker, M A, 2000, 'UC Davis' role in improving California's grape planting materials', *Proceedings of the 50th Annual ASEV Meeting, Seattle, Washington, June 19–23, 2000*, American Journal of Enology and Viticulture, pp 209–15

Wang, L, Li, S, and Fan, P, 2010, 'The introduction, breeding and extension of wine grape in China', *10th International Conference on Grapevine Breeding and Genetics, 1–5 August*, Geneva, New York, USA

Wheeler, W, 1908, 'The Concord grape and its originator', *Transactions of the Massachusetts Horticultural Society*: 15–28

Wills, J, 1867, 'The Garston vineyard', *Journal of Horticulture, Cottage Gardener and Country Gentlemen* 12: 338–40

Yin, K, Liang, W, and Zhuge, H, 1998, 'Determination of a wine grape variety Cabernet Gernischet by ampelometric method', *Acta Horticulturae Sinica* 2 (in Chinese, with English abstract)

Yokotsuka, K, Nozaki, K, and Kushida, T, 1984, 'Comparison of phenolic compounds including anthocyanin pigments between Koshu and Ryugan grapes', *Journal of Fermentation Technology* 62, 5: 477–86

Yuste, J, Martín, J P, Rubio, C, Hidalgo, E, Recio, P, Santana, J C, Arranz, C, and Ortiz, J M, 2006, 'Identification of autochthonous grapevine varieties in the germplasm collection at the ITA of Castilla y León in Zamadueñas Station, Valladolid, Spain', *Spanish Journal of Agricultural Research* 4, 1: 31–6

Zappalà, A, 2005, 'I vitigni dell'Etna: Carricante, Catarratti, Nerello Mascalese e Nerello Capuccio', *Tecnica Agricola* 57, 3: 59–65

Zappia, R, Gullo, G, Mafrica, R, and Di Lorenzo, R, 2007, 'Mantonico vera e Mantonico pizzutella: descrizione ampelografica, analisi microsatellite e comportamento bio-agronomico', *Italus Hortus* 14, 3: 59–62

Zdunić, G, 2005, 'Ampelografska i genetička evaluacija autohtonih sorata vinove loze (*Vitis vinifera* L.) u području Kaštela', unpublished Master's thesis, Zagreb

Zdunić, G, Hančević, K, Sladonja, B, Poljuha, D, Hartl-Musinov, D, Budić-Leto, I, Bućan, L, Pezo, I, 2008, 'Ampelographic characterization and sanitary status of grapevine cultivar Prč bijeli (*Vitis vinifera* L.)', *Agriculturae Conspectus Scientificus* 73, 7: 85–8

Zdunić, G, Maletić, E, Vorkurka, A, Karoglan Kontić, J, Pezo, I, Pejić, I, and Naslov, I, 2009, 'Intravarietal variability of the cultivar Plavac Mali (*Vitis vinifera* L.)', *Acta Horticulturae* 827: 203–6

Zdunić, G, Pejić, I, Karoglan Kontić, J, Vukičević, D, Vokurka, A, Pezo, I, and Maletić, E, 2008, 'Comparison of genetic and morphological data for inferring similarity among native Dalmatian (Croatia) grapevine cultivars (*Vitis vinifera* L.)', *Journal of Food, Agriculture & Environment* 6, 2: 333–6

Zecca, G, De Mattia, F, Lovicu, G, Labra, M, Sala, F, and Grassi, F, 2010, 'Wild grapevine: silvestris, hybrids or cultivars that escaped from vineyards? Molecular evidence in Sardinia', *Plant Biology* 12, 3: 558–62

Zelenák, I, 2002, *A Hétszőlő*, privately published, Eger

Zhengping, L, 2011, *Chinese wine*, Cambridge University Press, New York

Zinelabidine, L H, Haddioui, A, Rodríguez, V, Cabello, F, Eiras-Dias, J E, Martínez Zapater, J M, and Ibánez, J, 2012, 'Identification by SNP analysis of a major role for Cayetana Blanca in the genetic network of Iberian Peninsula grapevine varieties', *American Journal of Enology and Viticulture* 63, 1: 121–6

Zohary, D, and Hopf, M, 2000, *Domestication of plants in the Old World*, Oxford University Press, Oxford and New York

Zulini, L, Russo, M, and Peterlunger, E, 2002, 'Genotyping wine and table grape cultivars from Apulia (southern Italy) using microsatellite markers', *Vitis* 41, 4: 183–7

Žunec, N (ed), 2009, *Vinogradarski and vinski atlas Hrvatske / Viticulture and wine atlas of Croatia*, Business Media Croatia, Zagreb

Zúñiga, V C M de, 1905, 'Tempranillo', in Viala, P, and Vermorel, V (eds), *Ampélographie*, vol 6, Masson, Paris, pp 242–3

INDEX

Page numbers in **bold** indicate main entries; page numbers in *italics* indicate the first occurrence of a synonym within its main entry.